DEAN'S ANALYTICAL CHEMISTRY HANDBOOK

DEAN'S ANALYTICAL CHEMISTRY HANDBOOK

Pradyot Patnaik
Director, Laboratory of Analytical Chemistry
Interstate Environmental Commission
Staten Island, New York

Second Edition

McGRAW-HILL
New York Chicago San Francisco Lisbon London Madrid
Mexico City Milan New Delhi San Juan Seoul
Singapore Sydney Toronto

The McGraw-Hill Companies

Cataloging-in-Publication Data is on file with the Library of Congress.

1 2 3 4 5 6 7 8 9 0 DOC/DOC 0 1 0 9 8 7 6 5 4

ISBN 0-07-141060-0

The sponsoring editor for this book was Kenneth P. McCombs, the editing supervisor was David E. Fogarty, and the production supervisor was Sherri Souffrance. It was set in Times Roman by International Typesetting and Composition. The art director for the cover was Margaret Webster-Shapiro.

Printed and bound by RR Donnelley.

This book was printed on acid-free paper.

CONTENTS

PREFACE

It is an especially daunting task to edit a new edition of a major reference handbook. Such a task becomes even more arduous when the original editor is no longer with us and the new editor has to retain the "heart and soul" of the text—performing the delicate balancing act of retaining the style and framework of the earlier edition that was so popular and at the same time updating the contents and incorporating new materials into the new edition with similar texture and style. The first (1995) edition of this book, edited by the late John A. Dean, has been well received by professionals around the globe, thus making it one of the most popular reference texts in analytical chemistry.

In this edition three new sections have been added: Analysis of Pesticides and Herbicides; Air Analysis; and Analysis of Trace Pollutants in the Environment. The latter replaces the section on Water Analysis in the first edition, which has been deleted in this edition. All other sections from the first edition are retained in this edition. They have been reviewed, and some have been rearranged and updated. The section on Statistics in Chemical Analysis has been expanded to include a detailed discussion of the precision and accuracy of measurements.

I hope this Handbook covering several important topics in analytical chemistry serves its purpose of providing useful information to professionals, teachers, and students in their fields of interest. All suggestions and comments from readers for any improvement of this work in future editions are cordially solicited and will be gracefully acknowledged.

My special words of thanks go to Nicole Canavari, Chirag Patnaik, and Zuhaib Kokab for their untiring efforts in the preparation of the manuscript of all the new chapters of this book in a short time. Also, I would like to thank Kenneth McCombs, Senior Editor at McGraw-Hill, for initiating this project and David Fogarty, Editing Manager, Sherri Souffrance, Production Supervisor, and the staffs of their production department for publishing this edition. Last but not the least, I would like to thank Matt Wolff from CRC Press for giving permission to reprint a few tables from my book, *Handbook of Environmental Analysis*, and Jill D. Thomas from Supelco Inc. for supplying one figure in the text.

PRADYOT PATNAIK
Burlington, New Jersey

PREFACE TO THE FIRST EDITION

The growth of modern technology has confronted the analytical chemist with a host of new and increasingly complex materials, has called on that person to provide information about constituents previously unrecognized or ignored, and has posed more stringent demands for greater sensitivity, reliability, and speed. On the other hand, developments in instrumentation and the research of colleagues in allied fields have provided the analyst with new techniques, instruments, procedures, and reagents for dealing with these problems.

This very expansion of equipment, reagents, and methodology has, however, greatly complicated the task of the chemist searching for the best way of attacking a new or unfamiliar sample. This handbook is intended to provide analytical chemists and their colleagues in related sciences with concise and convenient summaries of the fundamental data and the practical procedures that are most important and most useful among the conventional wet and instrumental methods in modern analytical chemistry. All this is presented in a convenient desk-size guide.

Without ready access to the data that describe the behaviors of the various substances present toward different techniques, it is all too easy for the special peculiarity of the one most suitable technique to escape notice. One of the hardest problems in analytical work is in choosing the right technique to solve a problem. With this handbook the reader will have a handy reference all in one place for analytical techniques. The handbook should be especially helpful in those laboratories which may not have developed a wide variety of in-house analytical methods.

Extensive application tables contain just enough information to enable a reader to reach a judgement call about the possible applicability and range or sensitivity of a method plug references that will supply more detailed directions and discussion of the method. Intercomparison of techniques within a general topical area or between one or several topical areas are provided to enable the user to reach a decision on choice of a method. Factors entering into this decision might be cost, time of analysis, sensitivity, reproducibility, and expertise required of the operator. As an aid to understanding the parameters involved in a technique, many examples involving mathematical expressions have been worked out. Illustrative worked examples and troubleshooting sections are included.

Sample preparation prior to analysis is a major analytical challenge; the book expounds the various techniques for extracting analytes from complex matrices and gives approaches to method development. The opening section on preliminary operations of analysis encompasses the important topics of sampling, mixing and sample reduction, moisture and drying, and methods for dissolving the sample. More than likely a preliminary separation will be necessary, perhaps to concentrate the analyte or to remove interference, or both. Preliminary separation methods take the reader through complex formation, masking and demasking reactions, extraction methods, ion-exchange methods, volatilization methods, and conclude with carrier coprecipitation and chemical reduction methods.

Traditional material in the realm of gravimetric and volumetric analysis constitutes the third section. Here are discussed inorganic gravimetric analysis, acid-base titrations in aqueous media and in nonaqueous media, precipitation titrations, oxidation-reduction titrations, and complexometric titrations. The remainder of the book treats the individual areas of instrumental analysis and a selected group of applications. A final section contains information of a general nature.

Although there is no specific section on methods for the analysis of technical materials, the Index is expanded to include references to specific elements or functional groups, and to types of sample matrices, that are tabulated in the many tables within the various sections of the handbook. This style eliminates unnecessary duplication of material.

In all probability, the technique for analyzing the sample for the components requested will require searching through several possible sections of the handbook for an appropriate method. Perhaps a second method will be selected to provide confirmation of the analyte concentration, a method that might be longer or more costly but one not to be used routinely. This collection of information in one book instead of having to make many referrals will assist the reader in selecting the best method(s) for a particular situation. With quick and easy access to the myriad of analytical techniques that the book provides, the reader can either choose the best one or get specific information on the method selected. Only in exceptional cases are procedures described in full. Enough information is provided to enable the user to make a judicious and rational choice from among the techniques and procedure available. Conditions and details of the final measurement are included to serve as a convenient ready reference. References given will direct the user to additional information.

Finally, as results of analyses have been accumulated, some of the statistical treatments outlined in Sec. 20 would be utilized; certainly if a calibration curve is to be prepared, the least square method for the best-fit line should be used and evaluated.

Theoretical discussions have been limited since several excellent texts devoted to instrumental methods (including one coauthored by this editor) are on the market. Likewise, the descriptions of apparatus and methodology are restricted, yet sufficiently comprehensive, to enable the user to judge intelligently between available choices.

The editor would be grateful for suggestions as to organization and scope of the handbook, and for bringing any errors to his attention.

JOHN A. DEAN

DEAN'S ANALYTICAL CHEMISTRY HANDBOOK

SECTION 1
PRELIMINARY OPERATIONS OF ANALYSIS

1.1 SAMPLING

1.1.1 Handling the Sample in the Laboratory

Each sample should be completely identified, tagged, or labeled so that no question as to its origin or source can arise. Some of the information that may be on the sample is as follows:

1. The number of the sample.
2. The notebook experiment-identification number.
3. The date and time of day the sample was received.

4. The origin of the sample and cross-reference number.

5. The (approximate) weight or volume of the sample.

6. The identifying code of the container.

7. What is to be done with the sample, what determinations are to be made, or what analysis is desired?

A computerized laboratory data management system is the solution for these problems. Information as to samples expected, tests to be performed, people and instruments to be used, calculations to be performed, and results required are entered and stored directly in such a system. The raw experimental data from all tests can be collected by the computer automatically or can be entered manually. Status reports as to the tests completed, work in progress, priority work lists, statistical trends, and so on are always available automatically on schedule and on demand.

1.1.2 Sampling Methodology

The sampling of the material that is to be analyzed is almost always a matter of importance, and not infrequently it is a more important operation than the analysis itself. The object is to get a representative sample for the determination that is to be made. This is not the place to enter into a discussion on the selection of the bulk sample from its original site, be it quarry, rock face, stockpile, production line, and so on. This problem has been outlined elsewhere.[1–5] In practice, one of the prime factors that tends to govern the bulk sampling method used is that of cost. It cannot be too strongly stressed that a determination is only as good as the sample preparation that precedes it. The gross sample of the lot being analyzed is supposed to be a miniature replica in composition and in particle-size distribution. If it does not truly represent the entire lot, all further work to reduce it to a suitable laboratory size and all laboratory procedures are a waste of time. The methods of sampling must necessarily vary considerably and are of all degrees of complexity.

No perfectly general treatment of the theory of sampling is possible. The technique of sampling varies according to the substance being analyzed and its physical characteristics. The methods of sampling commercially important materials are generally very well prescribed by various societies interested in the particular material involved, in particular, the factual material in the multivolume publications of the American Society for Testing Materials, now known simply as ASTM, its former acronym. These procedures are the result of extensive experience and exhaustive tests and are generally so definite as to leave little to individual judgment. Lacking a known method, the analyst can do pretty well by keeping in mind the general principles and the chief sources of trouble, as discussed subsequently.

If moisture in the original material is to be determined, a separate sample must usually be taken.

1.1.2.1 Basic Sampling Rules. No perfectly general treatment of the theory of sampling is possible. The technique of sampling varies according to the substance being analyzed and its physical characteristics. The methods of sampling commercially important materials are generally very well prescribed by various societies interested in the particular material involved: water and sewage by the American Public Health Association, metallurgical products, petroleum, and materials of construction by the ASTM, road building materials by the American Association of State Highway Officials, agricultural materials by the Association of Official Analytical Chemists (AOAC), and so on.

A large sample is usually obtained, which must then be reduced to a laboratory sample. The size of the sample must be adequate, depending upon what is being measured, the type of measurement being made, and the level of contaminants. Even starting with a well-gathered sample, errors can

---

[1] G. M. Brown, in *Methods in Geochemistry*, A. A. Smales and L. R. Wager, eds., Interscience, New York, 1960, p. 4.

[2] D. J. Ottley, *Min. Miner. Eng.* **2**:390 (1966).

[3] C. L. Wilson and D. W. Wilson, *Comprehensive Analytical Chemistry*, Elsevier, London, 1960; Vol. 1A, p. 36.

[4] C. A. Bicking, "Principles and Methods of Sampling," Chap. 6, in *Treatise on Analytical Chemistry*, I. M. Kolthoff and P. J. Elving, eds., Part 1, Vol. 1, 2d ed., Wiley-Interscience, New York, 1978; pp. 299–359.

[5] G. M. Brown, in *Methods in Geochemistry*, A. A. Smales and L. R. Wager, eds., Interscience, New York, 1960, p. 4.

occur in two distinct ways. First, errors in splitting the sample can result in bias with concentration of one or more of the components in either the laboratory sample or the discard material. Second, the process of attrition used in reducing particle sizes will almost certainly create contamination of the sample. By disregarding experimental errors, analytical results obtained from a sample of n items will be distributed about μ with a standard devitation

$$s = \frac{\sigma}{\sqrt{n}} \tag{1.1}$$

In general, σ and μ are not known, but s can be used as an estimate of σ, and the average of analytical results as an estimate of μ. The number of samples is made as small as compatible with the desired accuracy.

If a standard deviation of 0.5% is assigned as a goal for the sampling process, and data obtained from previous manufacturing lots indicate a value for s that is 2.0%, then the latter serves as an estimate of σ. By substituting in Eq. (1.1),

$$0.5 = \frac{2.0}{\sqrt{n}} \tag{1.2}$$

and $n = 16$, number of samples that should be selected in a random manner from the total sample submitted.

To include the effect of analytical error on the sampling problem requires the use of variances. The variance of the analysis is added to the variance of the sampling step. Assuming that the analytical method has a standard deviation of 1.0%, then

$$s^2 = \frac{\left(\sigma_s^2 + \sigma_a^2\right)}{\sqrt{n}} \tag{1.3}$$

where the numerator represents the variance of the sampling step plus the variance of the analysis. Thus

$$(0.5)^2 = \frac{[(2.0)^2 + (1.0)^2]}{n} \tag{1.4}$$

and $n = 20$, the number of samples required. The above discussion is a rather simple treatment of the problem of sampling.

1.1.2.2 Sampling Gases.[6] Instruments today are uniquely qualified or disqualified by the Environmental Protection Agency. For a large number of chemical species there are as yet no approved methods.

The size of the gross sample required for gases can be relatively small because any inhomogeneity occurs at the molecular level. Relatively small samples contain tremendous quantities of molecules. The major problem is that the sample must be representative of the entire lot. This requires the taking of samples with a "sample thief" at various locations of the lot, and then combining the various samples into one gross sample.

Gas samples are collected in tubes [250 to 1000 milliliter (mL) capacity] that have stopcocks at both ends. The tubes are either evacuated or filled with water, or a syringe bulb attachment may be used to displace the air in the bottle by the sample. For sampling by the static method, the sampling bottle is evacuated and then filled with the gas from the source being sampled, perhaps a cylinder. These steps are repeated a number of times to obtain the desired sampling accuracy. For sampling by the dynamic method, the gas is allowed to flow through the sampling container at a slow, steady rate. The container is flushed out and the gas reaches equilibrium with the walls of the sampling lines and container with respect to moisture. When equilibrium has been reached, the stopcocks on the sampling container are

[6] J. P. Lodge, Jr., ed., *Methods of Air Sampling and Analysis*, 3d ed., Lewis, Chelsea, Michigan, 1989. Manual of methods adopted by an intersociety committee.

closed—the exit end first followed by the entrance end. The sampling of flowing gases must be made by a device that will give the correct proportion of the gases in each annular increment.

Glass containers are excellent for inert gases such as oxygen, nitrogen, methane, carbon monoxide, and carbon dioxide. Stainless-steel containers and plastic bags are also suitable for the collection of inert gases. Entry into the bags is by a fitting seated in and connected to the bag to form an integral part of the bag. Reactive gases, such as hydrogen sulfide, oxides of nitrogen, and sulfur dioxide, are not recommended for direct collection and storage. However, Tedlar™ bags are especially resistant to wall losses for many reactive gases.

In most cases of atmospheric sampling, large volumes of air are passed through the sampling apparatus. Solids are removed by filters; liquids and gases are either adsorbed or reacted with liquids or solids in the sampling apparatus. A flowmeter or other device determines the total volume of air that is represented by the collected sample. A manual pump that delivers a definite volume of air with each stroke is used in some sampling devices.

1.1.2.3 Sampling Liquids. For *bottle sampling* a suitable glass bottle of about 1-L capacity, with a 1.9-centimeter (cm) opening fitted with a stopper, is suspended by clean cotton twine and weighted with a 560-gram (g) lead or steel weight. The stopper is fitted with another length of twine. At the appropriate level or position, the stopper is removed with a sharp jerk and the bottle permitted to fill completely before raising. A cap is applied to the sample bottle after the sample is withdrawn.

In *thief sampling* a thief of proprietary design is used to obtain samples from within about 1.25 cm of the tank bottom. When a projecting stem strikes the bottom, the thief opens and the sample enters at the bottom of the unit and air is expelled from the top. The valves close automatically as the thief is withdrawn. A *core thief* is lowered to the bottom with valves open to allow flushing of the interior. The valves shut as the thief hits the tank bottom.

When liquids are pumped through pipes, a number of samples can be collected at various times and combined to provide the gross sample. Care should be taken that the samples represent a constant fraction of the total amount pumped and that all portions of the pumped liquid are sampled.

Liquid solutions can be sampled relatively easily provided that the material can be mixed thoroughly by means of agitators or mixing paddles. Homogeneity should never be assumed. After adequate mixing, samples can be taken from the top and bottom and combined into one sample that is thoroughly mixed again; from this the final sample is taken for analysis.

For sampling liquids in drums, carboys, or bottles, an open-ended tube of sufficient length to reach within 3 mm of the bottom of the container and of sufficient diameter to contain from 0.5 to 1.0 L may be used. For separate samples at selected levels, insert the tube with a thumb over the top end until the desired level is reached. The top hole is covered with a thumb upon withdrawing the tube. Alternatively the sample may be pumped into a sample container.

Specially designed sampling syringes are used to sample microquantities of air-sensitive materials.

For suspended solids, zone sampling is very important. A proprietary zone sampler is advantageous. When liquids are pumped through pipes, a number of samples can be collected at various times and combined to provide the gross sample. Take care that the samples represent a constant fraction of the total amount pumped and that all portions of the pumped liquid are sampled.

1.1.2.4 Sampling Compact Solids. In sampling solids particle size introduces a variable. The *size/weight ratio b* can be used as a criterion of sample size. This ratio is expressed as

$$b = \frac{\text{weight of largest particle} \times 100}{\text{weight of sample}} \tag{1.5}$$

A value of 0.2 is suggested for b; however, for economy and accuracy in sampling, the value of b should be determined by experiment.

The task of obtaining a representative sample from nonhomogeneous solids requires that one proceeds as follows. A gross sample is taken. The gross sample must be at least 1000 pounds (lb) if the pieces are greater than 1 inch (in) (2.54 cm), and must be subdivided to 0.75 in (1.90 cm) before reduction to 500 lb (227 kg), to 0.5 in (1.27 cm) before reduction to 250 lb (113 kg), and so on, down to the 15-lb (6.8-kg) sample, which is sent to the laboratory. Mechanical sampling machines are used extensively because they are more accurate and faster than hand-sampling methods described below. One type removes part of a moving steam of material all of the time. A second type diverts all of stream of material at regular intervals.

For natural deposits or semisoft solids in barrels, cases, bags, or cake form, an auger sampler of post-hole digger is turned into the material and then pulled straight out. Core drilling is done with special equipment; the driving head should be of hardened steel and the barrel should be at least 46 cm long. Diamond drilling is the most effective way to take trivial samples of large rock masses.

For bales, boxes, and similar containers, a split-tube thief is used. The thief is a tube with a slot running the entire length of the tube and sharpened to a cutting edge. The tube is inserted into the center of the container with sufficient rotation to cut a core of the material.

For sampling from conveyors or chutes, a hand scoop is used to take a cross-sectional sample of material while in motion. A gravity-flow auger consists of a rotating slotted tube in a flowing mass. The material is carried out of the tube by a worm screw.

1.1.2.5 Sampling Metals. Metals can be sampled by drilling the piece to be sampled at regular intervals from all sides, being certain that each drill hole extends beyond the halfway point. Additional samples can be obtained by sawing through the metal and collecting the "sawdust." Surface chips alone will not be representative of the entire mass of a metallic material because of differences in the melting points of the constituents. This operation should be carried out dry whenever possible. If lubrication is necessary, wash the sample carefully with benzene and ether to remove oil and grease.

For molten metals the sample is withdrawn into a glass holder by a sample gun. When the sample cools, the glass is broken to obtain the sample. In another design the sampler is constructed of two concentric slotted brass tubes that are inserted into a molten or powdered mass. The outer tube is rotated to secure a representative solid core.

1.2 MIXING AND REDUCTION OF SAMPLE VOLUME

1.2.1 Introduction

The sample is first crushed to a reasonable size and a portion is taken by quartering or similar procedures. The selected portion is then crushed to a somewhat smaller size and again divided. The operations are repeated until a sample is obtained that is large enough for the analyses to be made but not so large as to cause needless work in its final preparation. This final portion must be crushed to a size that will minimize errors in sampling at the balance yet is fine enough for the dissolution method that is contemplated.

Every individual sample presents different problems in regard to splitting the sample and grinding or crushing the sample. If the sample is homogeneous and hard, the splitting procedure will present no problems but grinding will be difficult. If the sample is heterogeneous and soft, grinding will be easy but care will be required in splitting. When the sample is heterogeneous both in composition and hardness, the interactions between the problems of splitting and grinding can be formidable.

Splitting is normally performed before grinding in order to minimize the amount of material that has to be ground to the final size that is suitable for subsequent analysis.

1.2.2 Coning and Quartering

A good general method for mixing involves pouring the sample through a splitter repeatedly, combining the two halves each time by pouring them into a cone.

When sampling very large lots, a representative sample can be obtained by coning (Fig. 1.1) and quartering (Fig. 1.2). The first sample is formed into a cone, and the next sample is poured onto the apex of the cone. The result is then mixed and flattened, and a new cone is formed. As each successive

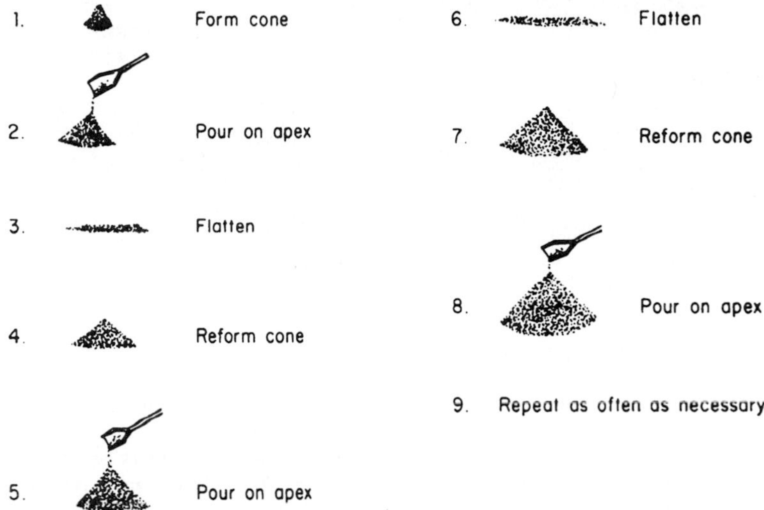

FIGURE 1.1 Coning samples. (*From Shugar and Dean,* The Chemist's Ready Reference Handbook, *McGraw-Hill, 1990.*)

sample is added to the re-formed cone, the total is mixed thoroughly and a new cone is formed prior to the addition of another sample.

After all the samples have been mixed by coning, the mass is flattened and a circular layer of material is formed. This circular layer is then quartered and the alternate quarters are discarded. This process can be repeated as often as desired until a sample size suitable for analysis is obtained.

The method is easy to apply when the sample is received as a mixture of small, equal-sized particles. Samples with a wide range of particle sizes present more difficulties, especially if the large, intermediate, and small particles have appreciably different compositions. It may be necessary to crush the whole sample before splitting to ensure accurate splitting. When a coarse-sized material is mixed with a fine powder of greatly different chemical composition, the situation demands fine grinding of a much greater quantity than is normal, even the whole bulk sample in many cases.

Errors introduced by poor splitting are statistical in nature and can be very difficult to identify except by using duplicate samples.

1.2.3 Riffles

Riffles are also used to mix and divide portions of the sample. A riffle is a series of chutes directed alternately to opposite sides. The starting material is divided into two approximately equal portions. One part may be passed repeatedly through until the sample size is obtained.

FIGURE 1.2 Quartering samples. The cone is flattened, opposite quarters are selected, and the other two quarters are discarded. (*From Shugar and Dean, 1990.*)

1.3 CRUSHING AND GRINDING

1.3.1 Introduction

In dealing with solid samples, a certain amount of crushing or grinding is sometimes required to reduce the particle size. Unfortunately, these operations tend to alter the composition of the sample and to introduce contaminants. For this reason the particle size should be reduced no more than is required for homogeneity and ready attack by reagents.

If the sample can be pulverized by impact at room temperature, the choices are the following:

1. Shatterbox for grinding 10 to 100 mL of sample
2. Mixers or mills for moderate amounts to microsamples
3. Wig-L-Bug for quantities of 1 mL or less

For brittle materials that require shearing as well as impact, use a hammer–cutter mill for grinding wool, paper, dried plants, wood, and soft rocks.

For flexible or heat-sensitive samples, such as polymers or tissues, chill in liquid nitrogen and grind in a freezer mill or use the shatterbox that is placed in a cryogenic container.

For hand grinding, use boron carbide mortars.

Many helpful hints involving sample preparation and handling are in the SPEX Handbook.[7]

1.3.2 Pulverizing and Blending

Reducing the raw sample to a fine powder is the first series of steps in sample handling. Sample reduction equipment is shown in Table 1.1, and some items are discussed further in the following sections along with containment materials, the properties of which are given in Table 1.2.

1.3.2.1 Containment Materials. The containers for pulverizing and blending must be harder than the material being processed and should not introduce a contaminant into the sample that would interfere with subsequent analyses. The following materials are available.

Agate is harder than steel and chemically inert to almost anything except hydrofluoric acid. Although moderately hard, it is rather brittle. Use is not advisable with hard materials, particularly aluminum-containing samples, or where the silica content is low and critical; otherwise agate mortars are best for silicates. Agate mortars are useful when organic and metallic contaminations are equally undesirable. Silicon is the major contaminant, accompanied by traces of aluminum, calcium, iron, magnesium, potassium, and sodium.

Alumina ceramic is ideal for extremely hard samples, especially when impurities from steel and tungsten carbide are objectionable. Aluminum is the major contaminant, accompanied by traces of calcium, magnesium, and silicon. However, because alumina ceramic is brittle, care must be taken not to feed "uncrushable" materials such as scrap metal, hardwoods, and so on into crushers or mills.

Boron carbide is very low wearing but brittle. It is probably most satisfactory for general use in mortars, although costly. Major contaminants are boron and carbide along with minor amounts of aluminum, iron, silicon, and possibly calcium. The normal processes of decomposition used in subsequent stages of the analysis usually convert the boron carbide contaminant into borate plus carbon dioxide, after which it no longer interferes with the analysis.

Plastic containers (and grinding balls) are usually methacrylate or polystyrene. Only traces of organic impurities are added to the sample.

Steel (hardened plain-carbon) is used for general-purpose grinding. Iron is the major contaminant, accompanied by traces of carbon, chromium, manganese, and silicon. Stainless steel is less subject to chemical attack, but contributes nickel and possibly sulfur as minor contaminants.

---

[7] R. H. Obenauf et al., *SPEX Handbook of Sample Preparation and Handling*, 3d ed., SPEX Industries, Edison, N. J., 1991.

TABLE 1.1 Sample Reduction Equipment

Sample composition	Hardness, mohs	Cutting mill	Jaw crusher	Cross beater mill	Rotor beater mill	Centrifugal grinder	Mortar mill	Mixer mill	Ball mills	Micro rapid mill
Basalt, carbide, carborundum, cement clinker, corundum, diabase, glass, granite, iron alloys, iron ore, quartz	Very hard and brittle, 6.5–8.5	□	■	□	□	□	■	■	■	■
Artificial fertilizers, ash, calcite, feldspar, hematite, magnetite, marble, sandstones, slags	Hard, 4.5–6.5	□	■	□	□	□	■	■	■	■
Barite, bauxite, calcite, dolomite, gneiss, kaolin, limestone, magnetite, pumice, stones	Medium hard, 2.5–4.5	□	■	■	■	■	■	■	■	■
Graphite, gypsum, hard lignite, mica, salts, talc	Soft, 1.5–2.5	□	■	■	■	■	■	■	■	■
Cardboard, cereals, feeds, fish, food, dried fruit, leather scraps, paper, plant material, textiles	Fibrous and cellulose type	■	□	□	□	■	□	■	□	□
Duroplastic and thermoplastic materials, artificial resins, rubber	Elastic	■	□	□	□	■	□	■	□	□
Maximum sample size, cm^3		<500	<300	<300	<300	10–500	10–150	<10	<10	10–300
Initial particle size, mm		4–80	<150	<20	<20	<10	<8	<6	<6	<8
Final particle size, μm		<150	<100	70	70	40	35	<35	<10	<10

■, Suitable; □, not suitable.
Source: Data supplied by Brinkmann Instruments, Inc.

TABLE 1.2 Properties of Grinding Surfaces

Material	Hardness, mohs	Knoop hardness, $kg \cdot mm^{-1}$	Density, $g \cdot cm^{-3}$
Agate	6	590	
Aluminum ceramic	9	2100	
Boron carbide	9.5	2750	2.5
Hard porcelain	8		
Silicon carbide	9.5	2480	
Stainless steel	5.7		
Tungsten carbide	8.5	1600–2050	14–15
Zirconia (97%)	8.5	1160	>5.9

Tungsten carbide containers are the most effective and versatile of all. Containers are long wearing but subject to breakage. Grinding is fast and contamination is minimal. Major contaminants are tungsten, carbon, and cobalt (a binder); minor contaminants are tantalum, titanium, and niobium. The level of contamination introduced into a hard rock or ceramic sample may well be an appreciable fraction of 1% of the total weight.

Zirconia is hard and tough, and wears slowly. Contaminants are zirconium with traces of magnesium and hafnium.

Halide-releasing compounds must be ground in agate, alumina, plastic, or tungsten carbide containers.

1.3.2.2 Ball or Jar Mill. Ball or jar mills are jars or containers that are fitted with a cover and gasket that are securely fastened to the jar. The jar is half filled with special balls, and then enough of the sample is added to cover the balls and the voids between them. The cover is fastened securely and the jar is revolved on a rotating assembly. The length of time for which the material is ground depends upon the fineness desired and the hardness of the material. After pulverization the jar is emptied onto a coarse-mesh screen to separate the balls from the ground material. For small samples, vials from 1- to 2.5-in (2.54- to 6.37-cm) long and up to 0.75 inch (1.9 cm) in diameter use methacrylate balls 0.12 to 0.38 in (0.30 to 0.97 cm) in diameter, respectively. A 1-in (2.54-cm) motion along the axis of the vial is complemented by two motions at right angles to the vial axis: a $^{3}/_{16}$-in (0.48-cm) horizontal movement and a $^{1}/_{4}$-in (0.635-cm) vertical oscillation.

1.3.2.3 Hammer–Cutter Mill. Brittle materials requiring shearing as well as impact are handled with a hammer–cutter mill. This mill can be used for grinding wool, paper, dried plants, wood, soft rocks, and similar materials. The mill utilizes high-speed revolving hammers and a serrated grinding-chamber lining to combine the shearing properties of a knife mill with the crushing action of a hammer mill. A slide at the bottom of the hopper feeds small amounts of the sample (up to 100 mL) into the grinding chamber. When the material has been pulverized sufficiently, it drops through a perforated-steel screen at the bottom of the grinding chamber and into a tightly fitted collecting tube. Particle size and rapidity of grinding are determined by interchangeable screens and a variable-speed motor control. The mill's high speed and rapid throughput allow limited medium-to-coarse grinding of flexible polymers, soft metals, and temperature-sensitive materials.

1.3.2.4 Freezer Mill. Flexible, fatty, or wet samples, such as soft polymers, fresh bone, hair, wood, and muscle tissue, are ground at liquid-nitrogen temperature in a freezer mill. The grinding vial is immersed in liquid nitrogen. An alternating magnetic field shuttles a steel impactor against the ends of the vial to pulverize the brittle material. Sample size can vary from 0.1 to 3.0 mL, with the average sample weighing 1 to 2 g.

1.3.2.5 Shatterbox. The shatterbox spins a puck and a ring inside a grinding container at 900 revolutions per minute (r/min) for rapid pulverization of sample quantities up to 100 mL. Applications include metals and cements, slags and fluxes, and fertilizers and pesticides. An auxiliary cryogenic dish extends applications for the shatterbox to liquid-nitrogen temperatures.

*1.3.2.6 **Wig-L-Bug.*** The Wig-L-Bug is an effective laboratory mill for pulverizing and blending samples in quantities of 0.1 to 1.0 mL. Sample vials are available in plastic, stainless steel, or hardened tool steel. For infrared analysis the vials contain a preweighed quantity of potassium bromide and a stainless-steel mixing ball.

*1.3.2.7 **Jaw Crusher.*** For many minerals, primary crushing with a jaw crusher is often permissible unless the sample involved has a very low iron content. A laboratory scale jaw crusher is used to prepare hard, brittle samples for further processing in laboratory mills; it rapidly reduces samples from 1 in (2.54 cm) down to about $^1/_8$ in (0.32 cm). The jaw crusher is supplied with alloy steel or alumina ceramic jaw plates and cheek plates to minimize contamination.

1.3.3 Precautions in Grinding Operations

A potential source of error arises from the difference in hardness between the sample components. The softer materials are converted to smaller particles more rapidly than the harder ones. Any sample loss in the form of dust will alter the composition.

*1.3.3.1 **Effect of Grinding on Moisture Content.*** Often the moisture content and thus the chemical composition of a solid is altered considerably during grinding and crushing. Moisture content decreases in some instances and increases in others.

Decreases in water content are sometimes observed while grinding solids containing essential water in the form of hydrates. For example, the water content of calcium sulfate dihydrate is reduced from 20% to 5% by this treatment. Undoubtedly the change results from localized heating during the grinding and crushing of the particles.

Moisture loss also occurs when samples containing occluded water are reduced in particle size. The grinding process ruptures some of the cavities and exposes the water to evaporation.

More commonly, the grinding process is accompanied by an increase in moisture content, primarily because the surface area exposed to the atmosphere increases. A corresponding increase in adsorbed water results. The magnitude of the effect is sufficient to alter appreciably the composition of a solid.

*1.3.3.2 **Abrasion of Grinding Surfaces.*** A serious error can arise during grinding and crushing due to mechanical wear and abrasion of the grinding surfaces. For this reason only the hardest materials such as hardened steel, agate, or boron carbide are employed for the grinding surface. Even with these materials, sample contamination will sometimes be encountered.

*1.3.3.3 **Alteration of Sample Composition.*** Several factors may cause appreciable alteration in the composition of the sample through the grinding step. Among these is the heat that is inevitably generated. This can cause losses of volatile components in the sample. In addition, grinding increases the surface area of the solid and thus increases its susceptibility to reactions with the atmosphere.

*1.3.3.4 **Caking.*** Caking due to moisture, heat, accumulation of static charge, and fusing of particles under pressure can be a serious problem. The following are some solutions:

1. If caking is due to moisture, as in many soils and cements, dry the sample before grinding.
2. If particles remain in suspension, as during slurry grinding, caking is unlikely. Water, alcohol, Freon, or other liquids may be added to the sample before grinding and removed afterwards. Slurry grinding is a reasonably reliable way to grind a sample to micron-sized particles.
3. Add suitable lubricants, such as sodium stearate, dry soaps, and detergents.
4. Add an antistatic agent, such as graphite (also a lubricant).

1.4 SCREENING AND BLENDING

Intermittent screening of the material increases the grinding efficiency. In this operation the ground sample is placed upon a wire or cloth sieve that passes particles of the desired size. The residual

TABLE 1.3 U.S. Standard Sieve Series*

Sieve No.	Sieve opening mm	Sieve opening inch	Sieve No.	Sieve opening mm	Sieve opening inch
	125	5.00	10	2.00	0.0787
	106	4.24	12	1.70	0.0661
	90	3.50	14	1.40	0.0555
	75	3.00	16	1.18	0.0469
	63	2.50	18	1.00	0.0394
	53	2.12	20	0.850	0.0331
	45	1.75	25	0.710	0.0278
	37.5	1.50	30	0.600	0.0234
	31.5	1.25	35	0.500	0.0197
	26.5	1.06	40	0.425	0.0165
	22.4	0.875	45	0.355	0.0139
	19.0	0.75	50	0.300	0.0117
	16.0	0.625	60	0.250	0.0098
	13.2	0.530	70	0.212	0.0083
	11.2	0.438	80	0.180	0.0070
	9.5	0.375	100	0.150	0.0059
	8.0	0.312	120	0.125	0.0049
	6.7	0.265	140	0.106	0.0041
3.5	5.60	0.223	170	0.090	0.0035
4	4.75	0.187	200	0.075	0.0029
5	4.00	0.157	230	0.063	0.0025
6	3.35	0.132	270	0.053	0.0021
7	2.80	0.111	325	0.045	0.0017
8	2.36	0.0937	400	0.038	0.0015

*Specifications are from ASTM Specification E.11-81/ISO 565. The sieve numbers are the approximate number of openings per linear inch.

particles are then reground; these steps are repeated until the entire sample passes through the screen. Screens are available in different sieve openings (Table 1.3). If there is likelihood of extreme differences in composition among the various sized particles, the sample must be prepared with special care. For the most accurate analyses, the gross sample must be sieved such that no dusting takes place and a sufficient number of fractions are obtained. Each of the fractions is then weighed, and the sample for analysis is made up of the various fractions in the same proportion that they bear to the gross sample.

Test sieves of the same diameter are made to nest one above the other. Nests of sieves can be placed on special shakers with timers and amplitude control to obtain a distribution of particle sizes. Nylon sieves contribute no metallic impurities. Each sieve consists of telescoping methacrylate cylinders over which the screen is stretched. Often the screens are 100, 200, 325, and 400 mesh that meet ASTM Specification E11-58T for size and uniformity of mesh. The sieve frames may be 70, 88, 140, or 150 mm in diameter.

1.5 MOISTURE AND DRYING

It must be determined at the start if the analysis is to be reported on the *as-received* basis or after drying to a constant weight by one of several methods described hereafter. Most analytical results for solid samples should be expressed on a dry basis, which denotes material dried at a specified temperature or corrected through a "moisture" determination made on a sample taken at the same time as the sample for analysis.

In order to cope with the variability in composition caused by the presence of moisture, the analyst may attempt to remove the water by drying prior to weighing samples for analysis. With many samples it is customary to dry the sample at 105 to 110°C. If it is difficult to obtain constant weight at such temperatures, then higher temperatures must be used. Some materials oxidize slowly when heated and require drying in a nonoxidizing atmosphere. Alternatively the water content may be determined at the time the samples are weighed out for analysis; in this way results can be corrected to a dry basis.

The presence of water in a sample represents a common problem that frequently faces the analyst. Water may exist as a contaminant from the atmosphere or from the solution in which the substance was formed, or it may be bonded as a chemical compound, a hydrate. Regardless of its origin, water plays a part in determining the composition of the sample. Unfortunately, particularly in the case of solids, water content is a variable quantity that depends upon such things as humidity, temperature, and the state of subdivision. Thus, the constitution of a sample may change significantly with environment and method of handling.

1.5.1 Forms of Water in Solids

It is convenient to distinguish among the several ways in which water can be held by a solid. The *essential water* in a substance is the water that is an integral part of the molecular or crystal structure of one of the components of the solid. It is present in that component in stoichiometric quantities. An example is $CaC_2O_4 \cdot 2H_2O$.

The second form is called *water of constitution*. Here the water is not present as such in the solid but rather is formed as a product when the solid undergoes decomposition, usually as a result of heating. This is typified by the processes

$$2KHSO_4 \longrightarrow K_2S_2O_7 + H_2O \tag{1.6}$$

$$Ca(OH)_2 \longrightarrow CaO + H_2O \tag{1.7}$$

Nonessential water is not necessary for the characterization of the chemical constitution of the sample and therefore does not occur in any sort of stoichiometric proportions. It is retained by the solid as a consequence of physical forces.

Adsorbed water is retained on the surface of solids in contact with a moist environment. The quantity is dependent upon humidity, temperature, and the specific surface area of the solid. Adsorption is a general phenomenon that is encountered in some degree with all finely divided solids. The amount of moisture adsorbed on the surface of a solid also increases with humidity. Quite generally, the amount of adsorbed water decreases as the temperature increases, and in most cases approaches zero if the solid is dried at temperatures above 112°C. Equilibrium is achieved rather rapidly, ordinarily requiring only 5 or 10 minutes (min). This often becomes apparent to a person who weighs finely divided solids that have been rendered anhydrous by drying; a continuous increase in weight is observed unless the solid is contained in a tightly stoppered vessel.

A second type of nonessential water is called *sorbed water*. This is encountered with many colloidal substances such as starch, protein, charcoal, zeolite minerals, and silica gel. The amounts of sorbed water are often large compared with adsorbed moisture, amounting in some instances to as much as 20% or more of the solid. Solids containing even this much water may appear to be perfectly dry powders. Sorbed water is held as a condensed phase in the interstices or capillaries of the colloidal solids. The quantity is greatly dependent upon temperature and humidity.

A third type of nonessential moisture is *occluded water*. Here, liquid water is entrapped in microscopic pockets spaced irregularly throughout the solid crystals. Such cavities often occur naturally in minerals and rocks.

Water may also be dispersed in a solid in the form of a solid solution. Here the water molecules are distributed homogeneously throughout the solid. Natural glasses may contain several percent moisture in this form.

The effect of grinding on moisture content is discussed in Sec. 1.3.3.1.

1.5.2 Drying Samples

Samples may be dried by heating them to 110°C or higher if the melting point of the material is higher and the material will not decompose at that temperature. This procedure will remove the moisture bound to the surface of the particles. The procedure for drying samples is as follows:

1. Fill the weighing bottle no more than half full with the sample to be dried.
2. Place a label on the beaker or loosely inside the beaker. Do not place a label on the weighing bottle as it will gradually char.
3. Place the weighing bottle in the beaker. Remove the cover from the weighing bottle and place it inside the beaker.
4. Cover the beaker with a watch glass supported on glass hooks.
5. Place the beaker with weighing bottle in a drying oven at the desired temperature for 2 hours (h).
6. Remove the beaker from the oven. Cool it somewhat before placing the weighing bottle, now covered with its cap, in a desiccator.

1.5.2.1 The Desiccator. A desiccator is a container (glass or aluminum) filled with a substance that absorbs water (a desiccant). Several desiccants and their properties are listed in Table 1.4. The ground-glass (or metal) rim is lightly greased with petroleum jelly or silicone grease. The desiccator provides a dry atmosphere for objects and substances. The desiccator's charge of desiccant must be frequently renewed to keep it effective. Surface caking signals the need to renew. Some desiccants contain a dye that changes color upon exhaustion.

Vacuum desiccators are equipped with side arms so that they may be connected to a vacuum. This type of desiccator should be used to dry crystals that are wet with organic solvents. Vacuum desiccators should not be used for substances that sublime readily.

TABLE 1.4 Drying Agents

Drying agent	Most useful for	Residual water, mg H_2O per liter of dry air (25°C)	Grams of water removed per gram of desiccant
Al_2O_3	Hydrocarbons	0.002–0.005	0.2
$Ba(ClO_4)_2$	Inert-gas streams	0.6–0.8	0.17
BaO	Basic gases, hydrocarbons, aldehydes, alcohols	0.0007–0.003	0.12
CaC_2	Ethers		0.56
$CaCl_2$	Inert organics	0.1–0.2	0.15
CaH_2	Hydrocarbons, ethers, amines, esters, higher alcohols	1×10^{-5}	0.83
CaO	Ethers, esters, alcohols, amines	0.01–0.003	0.31
$CaSO_4$	Most organic substances	0.005–0.07	0.07
KOH	Amines	0.01–0.9	
$Mg(ClO_4)_2$	Gas streams	0.0005–0.002	0.24
$MgSO_4$	Most organic compounds	1–12	0.15–0.75
Molecular sieve 4X	Molecules with effective diameter >4 Å	0.001	0.18
P_2O_5	Gas streams; not suitable for alcohols, amines, or ketones	2×10^{-5}	0.5
Silica gel	Most organic amines	0.002–0.07	0.2
H_2SO_4	Air and inert-gas streams	0.003–0.008	Indefinite

TABLE 1.5 Solutions for Maintaining Constant Humidity

Solid phase	% Humidity at specified temperatures, °C						
	10	20	25	30	40	60	80
$K_2Cr_2O_7$			98.0				
K_2SO_4	98	97	97	96	96	96	
KNO_3	95	93	92.5	91	88	82	
KCl	88	85.0	84.3	84	81.7	80.7	79.5
KBr		84	80.7		79.6	79.0	79.3
NaCl	76	75.7	75.3	74.9	74.7	74.9	76.4
$NaNO_3$			73.8	72.8	71.5	67.5	65.5
$NaNO_2$		66	65	63.0	61.5	59.3	58.9
$NaBr \cdot 2H_2O$		57.9	57.7		52.4	49.9	50.0
$Na_2Cr_2O_7 \cdot 2H_2O$	58	55	54		53.6	55.2	56.0
$Mg(NO_3)_2 \cdot 6H_2O$	57	55	52.9	52	49	43	
$K_2CO_3 \cdot 2H_2O$	47	44	42.8		42		
$MgCl_2 \cdot 6H_2O$	34	33	33.0	33	32	30	
$KF \cdot 2H_2O$				27.4	22.8	21.0	22.8
$KC_2H_3O_2 \cdot 1.5H_2O$	24	23	22.5	22	20		
$LiCl \cdot H_2O$	13	12	10.2	12	11	11	
KOH	13	9	8	7	6	5	
100% Humidity: aqueous tension (mmHg)	9.21	17.54	23.76	31.82	55.32	149.4	355.1

Source: J. A. Dean, ed., *Lange's Handbook of Chemistry*, 14th ed., McGraw-Hill, New York, 1992.

1.5.2.2 ***Humidity and Its Control.*** At times it is desirable to maintain constant humidity in an enclosed container. A saturated aqueous solution in contact with an excess of a definite solid phase at a given temperature will maintain constant humidity in an enclosed space. Table 1.5 gives a number of salts suitable for this purpose. The aqueous tension [vapor pressure, in millimeters of mercury (mmHg)] of a solution at a given temperature is found by multiplying the decimal fraction of the humidity by the aqueous tension at 100% humidity for the specific temperature. For example, the aqueous tension of a saturated solution of NaCl at 20°C is $0.757 \times 17.54 = 13.28$ mmHg. Table 1.6 gives the concentrations of solutions of H_2SO_4, NaOH, and $CaCl_2$ that give specified vapor pressures and percent humidities at 25°C. Table 1.7 gives the humidity from wet- and dry-bulb thermometer readings, and Table 1.8 gives the relative humidity from dew-point readings.

1.5.3 Drying Collected Crystals

Gravity-filtered crystals collected on a filter paper may be dried as follows:

1. Remove the filter paper from the funnel. Open up the filter paper and flatten it on a watch glass of suitable size or a shallow evaporating dish. Cover the watch glass or dish with a piece of clean, dry filter paper and allow the crystals to air-dry.
   Note: Hygroscopic substances cannot be air-dried in this way.
2. Press out excess moisture from the crystals by laying filter paper on top of the moist crystals and applying pressure with a suitable object.
3. Use a spatula to work the pasty mass on a porous plate; then allow it to dry.
4. Use a portable infrared lamp to warm the sample and increase the rate of drying. Be sure that the temperature does not exceed the melting point of the sample.
5. Use a desiccator.

TABLE 1.6 Concentrations of Solutions of H_2SO_4, NaOH, and $CaCl_2$ Giving Specified Vapor Pressures and Percent Humidities at 25°C*

Percent humidity	Aqueous tension, mmHg	H_2SO_4		NaOH		$CaCl_2$	
		Molality	Weight %	Molality	Weight %	Molality	Weight %
100	23.76	0.00	0.00	0.00	0.00	0.00	0.00
95	22.57	1.263	11.02	1.465	5.54	0.927	9.33
90	21.38	2.224	17.91	2.726	9.83	1.584	14.95
85	20.19	3.025	22.88	3.840	13.32	2.118	19.03
80	19.00	3.730	26.79	4.798	16.10	2.579	22.25
75	17.82	4.398	30.14	5.710	18.60	2.995	24.95
70	16.63	5.042	33.09	6.565	20.80	3.400	27.40
65	15.44	5.686	35.80	7.384	22.80	3.796	29.64
60	14.25	6.341	38.35	8.183	24.66	4.188	31.73
55	13.07	7.013	40.75	8.974	26.42	4.581	33.71
50	11.88	7.722	43.10	9.792	28.15	4.990	35.64
45	10.69	8.482	45.41	10.64	29.86	5.431	37.61
40	9.50	9.304	47.71	11.54	31.58	5.912	39.62
35	8.31	10.21	50.04	12.53	33.38	6.478	41.83
30	7.13	11.25	52.45	13.63	35.29	7.183	44.36
25	5.94	12.47	55.01	14.96	37.45		
20	4.75	13.94	57.76	16.67	40.00		
15	3.56	15.81	60.80	19.10	43.32		
10	2.38	18.48	64.45	23.05	47.97		
5	1.19	23.17	69.44				

* Concentrations are expressed in percentage of anhydrous solute by weight.
Source: Stokes and Robinson, *Ind. Eng. Chem.* **41**:2013 (1949).

When very small quantities of crystals are collected by centrifugation, they can be dried by subjecting them to vacuum in the centrifuge tube while gently warming the tube.

1.5.4 Drying Organic Solvents

Water can be removed from organic liquids and solutions by treating the liquids with a suitable drying agent to remove the water. The selection of drying agents must be carefully made. The drying agent selected should not react with the compound or cause the compound to undergo any reaction but should remove only the water. Table 1.4 lists drying agents.

1.5.4.1 Use of Solid Drying Agents. Solid drying agents are added to wet organic solvents in a container that can be stoppered. Add small portions of the drying agent, shaking the container thoroughly after each addition. Allow it to stand for a predetermined time. Then separate the solid hydrate from the organic solvent by decantation and filtration. Several operations may be required. Repeat if necessary.

1.5.4.2 Efficiency of Drying Operations. The efficiency of a drying operation is improved if the organic solvent is exposed repeatedly to fresh portions of the drying agent. Some dehydrating agents are very powerful and dangerous, especially if the water content of the organic solvent is high. These should be used only after the wet organic solvent has been grossly predried with a weaker agent. Drying agents will clump together, sticking to the bottom of the flask when a solution is "wet." Wet solvent solutions appear to be cloudy; dry solutions are clear. If the solution is "dry," the solid drying agent will move about and shift easily on the bottom of the flask.

TABLE 1.7 Relative Humidity from Wet- and Dry-Bulb Thermometer Readings

Dry-bulb temperature, °C	Wet-bulb depression, °C											
	0.5	1.0	1.5	2.0	2.5	3.0	3.5	4.0	4.5	5.0	5.5	6.0
	Relative humidity, %											
−10	83	67	51	35	19							
−5	88	76	64	52	41	29	18	7				
0	91	81	72	64	55	46	38	29	21	13	5	
2	91	84	76	68	60	52	44	37	29	22	14	7
4	92	85	78	71	63	57	49	43	36	29	22	16
6	93	86	79	73	66	60	54	48	41	35	29	24
8	93	87	81	75	69	63	57	51	46	40	35	29
10	94	88	82	77	71	66	60	55	50	44	39	34
12	94	89	83	78	73	68	63	58	53	48	43	39
14	95	90	85	79	75	70	65	60	56	51	47	42
16	95	90	85	81	76	71	67	63	58	54	50	46
18	95	91	86	82	77	73	69	65	61	57	53	49
20	96	91	87	83	78	74	70	66	63	59	55	51
22	96	92	87	83	80	76	72	68	64	61	57	54
24	96	92	88	84	80	77	73	69	66	62	59	56
26	96	92	88	85	81	78	74	71	67	64	61	58
28	96	93	89	85	82	78	75	72	69	65	62	59
30	96	93	89	86	83	79	76	73	70	67	64	61
35	97	94	90	87	84	81	78	75	72	69	67	64
40	97	94	91	88	85	82	80	77	74	72	69	67

Dry-bulb temperature, °C	Wet-bulb depression, °C											
	6.5	7.0	7.5	8.0	8.5	9.0	10.0	11.0	12.0	13.0	14.0	15.0
	Relative humidity, %											
4	9											
6	17	11	5									
8	24	19	14	8								
10	29	24	20	15	10	6						
12	34	29	25	21	16	12	5					
14	38	34	30	26	22	18	10					
16	42	38	34	30	26	23	15	8				
18	45	41	38	34	30	27	20	14	7			
20	48	44	41	37	34	31	24	18	12	6		
22	50	47	44	40	37	34	28	22	17	11	6	
24	53	49	46	43	40	37	31	26	20	15	10	5
26	54	51	49	46	43	40	34	29	24	19	14	10
28	56	53	51	48	45	42	37	32	27	22	18	13
30	58	55	52	50	47	44	39	35	30	25	21	17
32	60	57	54	51	49	46	41	37	32	28	24	20
34	61	58	56	53	51	48	43	39	35	30	26	23
36	62	59	57	54	52	50	45	41	37	33	29	25
38	63	61	58	56	54	51	47	43	39	35	31	27
40	64	62	59	57	54	53	48	44	40	36	33	29

Source: J. A. Dean, ed., *Lange's Handbook of Chemistry*, 14th ed., McGraw-Hill, New York, 1992.

TABLE 1.8 Relative Humidity from Dew-Point Readings

Depression of dew point, °C	Dew-point reading, °C				
	−10	0	10	20	30
	Relative humidity, %				
0.5	96	96	96	96	97
1.0	92	93	94	94	94
1.5	89	89	90	91	92
2.0	86	87	88	88	89
3.0	79	81	82	83	84
4.0	73	75	77	78	80
5.0	68	70	72	74	75
6.0	63	66	68	70	71
7.0	59	61	63	66	68
8.0	54	57	60	62	64
9.0	51	53	56	58	61
10.0	47	50	53	55	57
11.0	44	47	49	52	
12.0	41	44	47	49	
13.0	38	41	44	46	
14.0	35	38	41	44	
15.0	33	36	39	42	
16.0	31	34	37	39	
18.0	27	30	33	35	
20.0	24	26	29	32	
22.0	21	23	26		
24.0	18	21	23		
26.0	16	18	21		
28.0	14	16	19		
30.0	12	14	17		

Source: J. A. Dean, ed., *Lange's Handbook of Chemistry*, 14th ed., McGraw-Hill, New York, 1992.

Molecular sieves are an excellent drying agent for gases and liquids. In addition to absorbing water, they also absorb other small molecules. Several types have been described in Sec. 4.2.7.1. Regeneration of molecular sieves can be accomplished by heating to 250°C or applying a vacuum. In contrast to chemically acting drying agents, molecular sieves are unsuitable when the substance is to be dried in vacuo.

Calcium carbide, metallic sodium, and phosphorus(V) oxide remove moisture by chemical reaction with water. Do not use these drying agents where either the drying agent itself or the product that it forms will react with the compound or cause the compound itself to undergo reaction or rearrangement. These drying agents are useful in drying saturated hydrocarbons, aromatic hydrocarbons, and ethers. However, the compounds to be dried must not have functional groups, such as hydroxyl or carboxyl, that will react with the agents.

Some general precautions are as follows:

1. Do not dry alcohols with metallic sodium.
2. Do not dry acids with basic drying agents.
3. Do not dry amines or basic compounds with acidic drying agents.
4. Do not use calcium chloride to dry alcohols, phenols, amines, amino acids, amides, ketones, or certain aldehydes and esters.

1.5.5 Freeze-Drying

Some substances cannot be dried at atmospheric conditions because they are extremely heat-sensitive material, but they can be freeze-dried. Freeze-drying is a process whereby substances are subjected to high vacuum after they have been frozen. Under these conditions ice (water) will sublime and other volatile liquids will be removed. This leaves the nonsublimable material behind in a dried state. Use dilute solutions and spread the material on the inner surface of the container to increase the surface area.

Commercial freeze-driers are self-contained units. They may consist merely of a vacuum pump, adequate vapor traps, and a receptacle for the material in solution, or they may include refrigeration units to chill the solution plus more sophisticated instruments to designate temperature and pressure, plus heat and cold controls and vacuum-release valves. Protect the vacuum pump from water with a dry-ice trap, and insert chemical gas-washing towers to protect the pump from corrosive gases.

Freeze-drying differs from ordinary vacuum distillation in that the solution or substance to be dried is first frozen to a solid mass. It is under these conditions that the water is selectively removed by sublimation, the ice going directly to the water-vapor state. Proceed as described in the following.

1. Freeze the solution, spreading it out on the inner surface of the container to increase the surface area.
2. Apply high vacuum; the ice will sublime and leave the dried material behind.
3. Use dilute solutions in preference to concentrated solutions.
4. Protect the vacuum pump from water with a dry-ice trap, and insert chemical gas-washing towers to protect the pump from corrosive gases.

1.5.6 Hygroscopic Ion-Exchange Membrane

The Perma Pure (Perma Pure Products, Inc.) driers utilize a hygroscopic, ion-exchange membrane in a continuous drying process to remove water vapor selectively from mixed gas streams. The membrane is a proprietary extrudible desiccant in tubular form. A single desiccant tube is fabricated in a shell-and-tube configuration and sealed into an impermeable shell that has openings adjacent to the sample inlet and product outlet. If a wet gas stream flows through the tubes and a countercurrent dry gas stream purges the shell, water-vapor molecules are transferred through the walls of the tubing. The wet gas is dried, and the dry purge gas becomes wet as it carries away the water vapor.

The efficiency and capacity of a dryer at constant temperature and humidity are based on the dryer's geometry (that is, internal volume, outside surface area, and shell volume), as well as the gas flows and pressures of the wet sample and the dry purge. The reduction of water vapor in the product of a dryer may be increased by reducing the sample flow or by increasing the dryer volume (a longer tube length). Increasing the sample flow results in a higher dew point in the product. A bundle of tubes with a common header increases the volume of wet gas that can be handled.

The membrane is stable up to 160°C, as is the stainless-steel shell. Fluorocarbon and polypropylene shells are also available; their maximum use temperatures are 160 and 150°C, respectively. Table 1.9 gives the chemical resistance of a hygroscopic ion-exchange membrane. The plastic shells handle corrosive gases such as 100% chlorine, 10% HCl, and 1% sulfur dioxide.

1.5.7 Microwave Drying

Conventional microwave ovens have generally proved unsatisfactory for laboratory use because of the uneven distribution of microwave energy and the problem of excess reflected microwave energy. However, microwave dryers utilizing programmable computers are extremely versatile and easy to operate. The programmable aspect allows the microwave intensity to be varied during the drying or heating cycle.

TABLE 1.9 Chemical Resistance of a Hygroscopic Ion-Exchange Membrane

Sample	Concentration, %	Stainless steel	Polypropylene and fluorocarbons
Chlorine	100	X*	Y
HCl	10	X	Y
NO_2	0.02	Y	Y
NO_2	0.20	Y	Y
SO_2	0.50	Y	Y
SO_2	1	X	Y
EDC (liquid)	100	Y	Y
Methylene chloride	100	Y	Y

* X, Not usable; Y, usable.
Source: Data courtesy of Perma Pure Products, Inc.

Samples can be analytically dried to less than 1 mg of water in several minutes. Water selectively absorbs the microwave energy and is removed through evaporation. In some systems, moisture determination is entirely automatic. The sample is placed on the balance pan, the door closed, and the start button depressed. The initial sample weight is stored in the computer, and the microwave system is actuated for the predetermined time. The final weight of the sample is ascertained when the oven is turned off. Weight loss and percentage moisture or solid residue are displayed.

1.5.8 Critical-Point Drying

Critical-point drying offers a number of advantages over both air- and freeze-drying in the preparation of samples for examination under a transmission or a scanning electron microscope. Specimens treated by this method can be studied in the true original form without the physical deterioration and structural damage usually produced when water or other volatile materials are removed from a sample by conventional drying techniques. Preparation time is measured in minutes.

The method takes advantage of the fact that at its critical point a fluid passes imperceptibly from a liquid to a gas with no evident boundary and no associated distortional forces. Heating a liquid to its critical point in a closed system causes the density of the liquid to decrease and the density of the vapor to increase until the two phases become identical at the critical point where the liquid surface vanishes completely.

To proceed, all the water in the sample is replaced by a carefully selected transitional fluid. Then, while the sample is completely immersed in the fluid, it is heated in a sealed bomb to slightly above the critical point. The vapor is then released while holding the bomb above the critical temperature. This procedure leaves a completely dry specimen that has not been subjected either to the surface tension forces of air-drying or to the freezing and sublimation boundaries associated with freeze-drying.

A suitable fluid must have these properties:

1. Be nonreactive with the specimen
2. Have a critical temperature low enough to prevent damage to specimens that are temperature sensitive
3. Have a critical pressure low enough that conventional equipment can be used, without requiring cumbersome and bulky pressure designs
4. Be nontoxic and readily available

Carbon dioxide and nitrous oxide were used in early studies. Freons are well suited; their critical and boiling temperatures are such that the preliminary steps can be conducted in open vessels for maximum control and visibility.

TABLE 1.10 Transitional and Intermediate Fluids for Critical-Point Drying

Transitional fluid	Critical pressure, atm	Critical temperature, °C	Boiling point, °C	Intermediate liquid
CF_3CF_3 (Freon 116)	29.4	19.7	–78.2	Acetone
$CClF_3$ (Freon 13)	38.2	28.9	–81.4	Ethanol
CHF_3 (Freon 23)	47.7	25.9	–82.0	Ethanol
Carbon dioxide	72.85	31.04	Sublimes	Ethanol and pentyl acetate
Nitrous oxide	71.60	36.43	–88.47	Not required

In most critical-point drying procedures an intermediate liquid is first used to displace the moisture present in the original specimen. Ethanol and acetone are two of the more popular reagents used for this purpose. Other liquids can be used if they are fully miscible with water and with the transitional fluid. The moisture in the sample is removed by passing the specimen stepwise through a graded series of solutions starting with a 10% concentration and moving up to a moisture-free, 100% liquid. The specimen is then removed from the intermediate liquid and transferred to the transitional fluid for treatment in the bomb. Several materials that have been successfully used as intermediate liquids, together with the principal transitional fluids, are listed in Table 1.10.

1.5.9 Karl Fischer Method for Moisture Measurement

The determination of water is one of the most important and most widely practiced analyses in industry. The field of application is so large that it is the subject of a three-volume series of monographs.[8] The Karl Fischer method relies on the specificity for water of the reagent, devised by Fischer. The original reagent contained pyridine, sulfur dioxide, and iodine in an organic solvent (methanol). It reacts quantitatively with water.

$$C_5H_5N{-}I_2 + C_5H_5N{-}SO_2 + C_5H_5N + H_2O \longrightarrow 2C_5H_5N{-}HI + C_5H_5N{-}SO_3 \qquad (1.8)$$

There is a secondary reaction with the solvent (methanol):

$$C_5H_5N{-}SO_3 + CH_3OH \longrightarrow C_5H_5NH{-}O{-}SO_2{-}OCH_3 \qquad (1.9)$$

Various improvements have been suggested. The end point is usually ascertained by means of an amperometric titration using two polarizable (indicator) electrodes (see Sec. 14.5.6.1).

Iodine in the Karl Fischer reaction can be generated coulometrically with 100% efficiency. Now an absolute instrument is available, and the analysis requires no calibration or standardization (see Sec. 14.8).

An entirely automated titrimeter will determine moisture in the range from 1 $\mu g \cdot mL^{-1}$ to 100% water content. The instrument combines a burette, a sealed titration vessel, a magnetic stirrer, and a pump system for changing solvent. Liquid samples are injected through a septum; solid samples are inserted through a delivery opening in the titration head. The titration is kinetically controlled; the speed of titrant delivery is adjusted to the expected content of water. Typically 50 mg of water are titrated in less than 1 min with a precision of better than 0.3%. An optional pyrolysis system allows the extraction of moisture from solid samples.

[8] J. Mitchell and D. M. Smith, *Aquammetry*, Wiley, New York, Vol. 1, 1977; Vol. 2, 1983; Vol. 3, 1980.

1.6 THE ANALYTICAL BALANCE AND WEIGHTS

1.6.1 Introduction

If the sample is to be weighed in the air-dry condition, no special precautions are necessary, nor are any required with nonhygroscopic substances that have been dried. Slightly hygroscopic substances can be dried and weighed in weighing bottles with well-fitting covers. For those moderately hygroscopic substances that take up in a few moments most of the moisture that can be absorbed from the atmosphere, only enough for a single determination should be dried. The weighing bottle should be stoppered, cooled in a desiccator, and opened for an instant to fill it with air. Reweigh it against a similar bottle (the tare) carried through all of the operations. Pour out as much as possible of the sample without loss by dusting and without brushing the interior of the bottle. Weigh the stoppered bottle again against the tare.

Weighing is the most common and most fundamental procedure in chemical work. Today's laboratory balances incorporate the latest advancements in electronics, precision mechanics, and materials science. Gains to the chemist are unprecedented ease of use, versatility, and accuracy. A balance user should pay particular attention to the following aspects which form some of the main topics of this section.

1. Select the proper balance for a given application. Understand the technical specification of balances.
2. Understand the functions and features of the instrument and use them correctly to obtain the performance that the particular balance was designed to provide.
3. Know how to ascertain the accuracy and functionality of a balance through correct installation, care, and maintenance.
4. Use proper and efficient techniques in weighing operations.
5. Apply proper judgment in interpreting the weighing results. High accuracy may require corrections for air buoyancy.
6. Be aware that most electronic balances can be interfaced to printers, computers, and specialized application devices. It usually is more reliable and efficient to process weighing records and associated calculations electronically.

1.6.1.1 Mass and Weight. The terms *mass* and *weight* are both legitimately used to designate a quantity of matter as determined by weighing. However, the following scientific terminology should be adhered to in any technical context.

Mass: An invariant measure of the quantity of matter in an object. The basic unit of mass is the kilogram, which is embodied in a standard kept in Paris. Masses are more practically expressed in grams or milligrams.

Apparent mass: Apparent mass is the mass of an object minus the mass of the air that the object displaces. Air buoyancy corrections will be discussed in a later section. The distinction between apparent mass and absolute mass is insignificant in chemical work in all but a few special applications.

Weight: The force exerted on a body by the gravitational field of the earth, measured in units of force (the newton, abbreviation N). The weight of a body varies with geographic latitude, altitude above sea level, density of the earth at the location, and, to a very minute degree, with the lunar and solar cycles. Electronic balances must be calibrated on location against a mass standard.

Balance: A laboratory instrument used for the precise measurement of small masses.

1.6.1.2 Classification of Balances. To guide the user in selecting the correct equipment for a given application, balances are classified according to the graduation step or division of the reading

TABLE 1.11 Classification of Balances by Weighing Range

Nomenclature	Smallest division	Capacity (typical)
Ultramicroanalytical	0.1 μg	3 g
Microanalytical	0.001 mg	3 g
Semimicroanalytical	0.01 mg	30 g
Macroanalytical	0.1 mg	160 g
Precision	1 mg	160 g–60 kg

device (scale dial or digital display) and according to their weighing capacity. Balances are classified in Table 1.11. Various types of balances are discussed in the following sections.

1.6.2 General-Purpose Laboratory Balances

1.6.2.1 Top-Loading Balances. Top-loading balances are economical and easy to use for routine weighing and for educational and quality assurance applications. Because of their design, top-loading balances generally sacrifice at least one order of magnitude of readability. Models are available for many tasks; readabilities range from 0.001 to 0.1 g and capacities from 120 to 12 000 g. The latter also represent tare ranges. Balances with ranges above 5000 g are for special applications. The operating temperature is usually from 15 to 40°C. Typical specifications of single-range models are given in Table 1.12.

Many top-loading balances are dual- or polyrange balances that offer variable readability throughout their capacities for high resolution at each weight. Dual-range balances offer two levels of readability within their capacity range; polyrange balances offer a series of incremental adjustments in readability. This type of balance gives unobstructed access for tall or wide weighing loads. Electronic top-loading balances may have additional features described in Sec. 1.6.5.

1.6.2.2 Triple-Beam Balance. The *triple-beam balance* provides a modest capacity of 111 g (201 g with an auxiliary weight placed in the 100-g notch). The three tiered scales are 0 to 1 g by 0.01 g, 0 to 10 g by 1 g, and 0 to 100 g by 10 g. The 1-g scale is notchless and carries a rider; the others are notched and carry suspended weights.

Triple-beam platform balances have a sensitivity of 0.1 g and a total weighing capacity of 2610 g when used with the auxiliary weight set, or 610 g without it. Three tiered scales (front to back) are 0 to 10 g by 0.1 g, 0 to 500 g by 100 g, and 0 to 100 g by 10 g. One 500-g and two 1000-g auxiliary

TABLE 1.12 Specifications of Balances

Capacity, g	Readability, mg	Stabilization time, s	Tara range, g
40	0.01	5	0–40
60	0.1	3	0–60
160	0.1	3	0–160
400	1.0	2	0–400
800	10	3	0–800
2 200	10	3	0–2 200
5 000	100	3	0–5 000
12 000	100	5	0–12 000

weights fit in a holder on the base. The aluminum beam is magnetically damped and has a spring-loaded zero adjust.

1.6.2.3 Dial-O-Gram Balances. A dial mechanism is used to obtain the weights from 0 to 10 g in 0.1-g intervals. In use the dial is rotated to 10.0 g. After moving the 200-g poise on the rear beam to the first notch, which causes the pointer to drop, and then moving it back a notch, the same procedure is repeated with the 100-g poise. Finally, the dial knob is rotated until the pointer is centered. A vernier scale provides readings to the nearest 0.1 g.

1.6.3 Mechanical Analytical Balances

1.6.3.1 Equal-Arm Balance. The classical equal-arm balance consists of a symmetrical level balance beam, two pans suspended from its ends, and a pivotal axis (fulcrum) at its center. Ideally, the two pan suspension pivots are located in a straight line with the fulcrum and the two lever arms are of exactly equal length. A rigid, truss-shaped construction of the beam minimizes the amount of bending when the pans are loaded. The center of gravity of the beam is located just slightly below the center fulcrum, which gives the balance the properties of a physical pendulum. With a slight difference in pan loads, the balance will come to rest at an inclined position, the angle of inclination being proportional to the load differential. By reading the pointer position on a graduated angular scale, it is possible to determine fractional amounts of mass between the even step values of a standard mass set of weights.

Variations and refinements of the equal-arm balance include (1) agate or synthetic sapphire knife-edge pivots, (2) air damping or magnetic damping of beam oscillations, (3) sliding poises or riders, (4) built-in mass sets operated by dial knobs, (5) a microprojector reading of the angle of beam inclination (5) arrestment devices to disengage and protect pivots, and (6) pan brakes to stop the swing of the balance pans.

1.6.3.2 Single-Pan Substitution Balance. Substitution balances have only one hanger assembly that incorporates both the load pan and a built-in set of weights on a holding rack. The hanger assembly is balanced by a counterpoise that is rigidly connected to the other side of the beam, the weight of which equals the maximum capacity of the particular balance. The weight of an object is determined by lifting weights off the holding rack until sufficient weights have been removed to equal almost the weight of the object. In this condition the balance returns to an equilibrium position within its angular, differential weighing range. Small increments of weight between the discrete dial weight steps (usually in gram increments) are read from the projected screen image of a graduated optical reticle that is rigidly connected to the balance beam.

While single-pan substitution balances are no longer manufactured, there are a number of these products still in use. Electronic balances possess superior accuracy and operating convenience.

1.6.4 Electronic Balances[9–11]

Today two dominant types of electronic balances are in use: the hybrid and the electromagnetic force balance. The hybrid balance uses a mix of mechanical and electronically generated forces, whereas the electromagnetic force balance uses electronically generated forces entirely.

Every eletromechanical weighing system involves three basic functions:

1. The load-transfer mechanism, composed of the weighing platform or pan, levers, and guides, receives the weighing load on the pan as a randomly distributed pressure force and translates it

[9] G. W. Ewing, "Electronic Laboratory Balances," *J. Chem. Educ.* **53**:A252 (1976); **53**:A292 (1976).

[10] R. O. Leonard, "Electronic Laboratory Balances," *Anal. Chem.* **48**:879A (1976).

[11] R. M. Schoonover, "A Look at the Electronic Balance," *Anal Chem.* **54**:973A (1982).

into a measurable single force. The platform is stabilized by flexure-pivoted guides; pivots are formed by elastically flexible sections in the horizontal guide members.

2. The electromechanical force transducer, often called the load cell, converts the mechanical input force into an electrical output. A direct current generates a static force on the coil which, in turn, counterbalances the weight force from the object on the balance pan (usually aided by one or more force-reduction levers). The amount of coil current is controlled by a closed-loop servo circuit that monitors the vertical deflections of the pan support through a photoelectric sensor and adjusts the coil current as required to maintain equilibrium between weighing load and compensation force.
3. The servo system is the electronic-signal processing part of the balance. It receives the output signal of the load cell, converts it to numbers, performs computations, and displays the final weight data on the readout. In one method a continuous current is driven through the servomotor coil and in the other method the current is pulsed.

1.6.4.1 Hybrid Balance. The hybrid balance is identical to the substitution balance except that the balance beam is never allowed to swing through large angular displacements when the applied loading changes. Instead the motion is very limited and when in equilibrium the beam is always restored to a predetermined reference position by a servo-controlled electromagnetic force applied to thc beam. The most salient features that distinguish the electronic hybrid are the balance beam and the built-in weights utilized in conjunction with the servo restoring force to hold the beam at the null position.

1.6.4.2 Electromagnetic Force Balance. A magnetic force balances the entire load either by direct levitation or through a fixed-ratio lever system. The loading on the electromechanical mechanism that constitutes the balance is not constant but varies directly with applied load. With this design the sensitivity and response are largely controlled by the servo-system characteristics. The force associated with the sample being weighed is mechanically coupled to servomotor that generates the opposing magnetic force. When the two forces are in equilibrium the error detector is at the reference position and the average electric current in the servomotor coil is proportional to the resultant force that is holding the mechanism at the reference position. When the applied load changes, a slight motion occurs between the fixed and moving portions of the error-detector components, resulting in a very rapid change in current through the coil.

1.6.4.3 Special Routines for Electronic Balances. Signal processing in electronic balances usually involves special computation routines:

1. *Programmable stability control:* Variable integration permits compensation for unstable weight readings due to environmental vibration or air currents. Preprogrammed filters minimize noise due to air currents and vibration. Four settings vary integration time, update rate of display, vibration filtering, and damping.
2. *Adjustabel stability range:* Nine different settings (from 0.25 to 64 counts in the last significant digit) control the tolerance range within which the stability indicator appears (a symbol that appears when the actual sample weight is displayed within preset stability-range tolerances).
3. *Autotracking:* This routine eliminates distracting display flicker and automatically rezeros the balance during slight weight changes. Once the balance stabilizes and displays the weight, the autotracking feature takes over. Autotracking can be turned off to weigh hygroscopic samples or highly volatile substances.
4. *Serial interface:* This allows two-way communication with printers, computers, and other devices at baud rates up to 9600.
5. *Autocalibration:* When pushed, a calibration button activates a built-in test weight (up to four on some models). The balance calibrates itself to full accuracy and returns to weighing mode within seconds.

6. *Full-range taring:* Just set a container on the pan and touch the rezero button and the display automatically resets to zero.
7. *Overload protection:* Full-range taring and overload protection prevent damage to the balance if excess weight is placed on the pan.

1.6.4.4 Precautions and Procedures. With analytical balances the following guidelines should be observed:

1. Level the balance using the air-bubble float.
2. Observe the required warmup period or leave the balance constantly under power. Follow the manufacturer's recommendations.
3. Use any built-in calibration mass and microprocessor-controlled calibration cycle at the beginning of every workday.
4. Handle weighing objects with forceps. Fingerprints on glassware or the body heat from an operator's hand can influence results.
5. Always close the sliding doors on the balance when weighing, zeroing, or calibrating.
6. Accept and record the displayed result as soon as the balance indicates stability; observe the motion detector light in the display. Never attempt to average the displayed numbers mentally.
7. Check the reliability of each balance every day against certified weights (Sec. 1.6.7), as it is critical to the performance of the laboratory. Balances should be cleaned and calibrated at least twice a year, more often if the work load is unusually heavy.

1.6.5 The Weighing Station

The finer the readability of a balance, the more critical is the choice of its proper location and environment. Observe these precautions:

1. Avoid air currents. Locate the station away from doors, windows, heat and air-conditioning outlets.
2. Avoid having radiant heat sources, such as direct sunlight, ovens, and baseboard heaters, nearby.
3. Avoid areas with vibrations. Locate the station away from elevators and rotating machinery. Special vibration-free work tables may be needed. To test, the displayed weight should not change if the operator shifts his or her weight, leans on the table, or places a heavy object next to the balance.
4. Choose an area free from abnormal radio-frequency and electromagnetic interference. The balance should not be on the same line circuit with equipment that generates such interference, such as electric arcs or sparks.
5. Maintain the humidity in the range from 15% to 85%. Dry air can cause weighing errors through electrostatic charges on the weighing object and the balance windows. A room humidifier might help. Humid air causes problems because of moisture absorption by samples and container surfaces. A room dehumidifier would help.
6. Maintain an even temperature of the room and object weighed. If an object is warm relative to the balance, convection currents cause the pan to be buoyed up, and the apparent mass is less than the true mass.
7. Remember that materials to be weighed take up water or carbon dioxide from the air during the weighing process in a closed system.
8. Weigh volatile materials in a closed system.
9. Sit down when doing precise weighing. The operator should be able to plant elbows on the work table, thus allowing a steady hand in handling delicate samples.

TABLE 1.13 Tolerances for Analytical Weights

Denomination	Class M: Individual tolerance, mg	Class M: Group tolerance, mg	Class S: Individual tolerance, mg	Class S: Group tolerance, mg	Class S-1, individual tolerance, mg
100 g	0.50	None specified	0.25	None Specified	1.0
50 g	0.25		0.12		0.60
30 g	0.15		0.074		0.45
20 g	0.10		0.074	0.154	0.35
10 g	0.050		0.074		0.25
5 g	0.034	0.065	0.054	0.105	0.18
3 g	0.034		0.054		0.15
2 g	0.034		0.054		0.13
1 g	0.034		0.054		0.10
500 mg	0.0054	0.0105	0.025	0.055	0.080
300 mg	0.0054		0.025		0.070
200 mg	0.0054		0.025		0.060
100 mg	0.0054		0.025		0.050
50 mg	0.0054	0.0105	0.014	0.034	0.042
30 mg	0.0054		0.014		0.038
20 mg	0.0054		0.014		0.035
10 mg	0.0054		0.014		0.030
5 mg	0.0054		0.014		0.028
3 mg	0.0054		0.014		0.026
2 mg	0.0054		0.014		0.025
1 mg	0.0054		0.014		0.025

1.6.6 Air Buoyancy[12]

In the fields of practical chemistry and technology, there is an unspoken agreement that no correction for air buoyancy is made in the weighing results. Therefore, the apparent mass of a body of unit density determined by weighing is about 0.1% smaller than its true mass. The exact mass on a balance is indicated only when the object to be weighed has the same density as the calibration standards used and air density is the same as that at the time of calibration. For interested persons, conversions of weighings in air to those in vacuo are discussed in *Lange's Handbook of Chemistry*.[13]

Changes in apparent weight occur with changes in air density due to fluctuations in barometric pressure and humidity. In practice, this means that a 25-mL crucible is about the largest object, in point of physical size, that may be safely weighed to the nearest 0.1 mg without in some way compensating for the air-buoyancy effect. When physically large pieces of equipment are used—for example, a pycnometer or a weight burette—air-buoyancy effects can introduce a major error if atmospheric conditions change between successive weighings, particularly if the same article is weighed over an extended period of time.

1.6.7 Analytical Weights

Table 1.13 gives the individual and group tolerances established by the National Institute for Science and Technology (NIST, U.S.) for classes M, S, and S-1 weights. Individual tolerances are

[12] R. M. Schoonover and F. E. Jones, "Air Buoyancy Correction in High Accuracy Weighing on an Analytical Balance," *Anal. Chem.* **53**:900 (1981).

[13] J. A. Dean, ed., *Lange's Handbook of Chemistry*, 14th ed., McGraw-Hill, New York, 1992; pp. 2.77–2.78.

"acceptance tolerances" for new weights. Group tolerances are defined as follows: "The corrections of individual weights shall be such that no combination of weights that is intended to be used in a weighing shall differ from the sum of the nominal values by more than the amount listed under the group tolerances." For class S-1 weights, two-thirds of the weights in a set must be within one-half of the individual tolerances given.

The laboratory should check its analytical balances every day with standard certified weights, using a series of weights that bracket the expected range of weighing for which the balance will be used. If a balance is found to be out of calibration, only weighings made in the past 24 h are suspect. The balance should be taken out of service and recalibrated by a professional service engineer. Two sets of certified weights should be purchased six months apart because certified weights themselves must be recertified once a year.

1.7 METHODS FOR DISSOLVING THE SAMPLE [14–17]

1.7.1 Introduction

Relatively few natural materials or organic materials are water soluble; they generally require treatment with acids or mixtures of acids, some combustion treatment, or even a fusion with some basic or acidic flux. The procedure adopted for the solution of a material will depend on the speed and convenience of the method and reagents employed and the desirability of avoiding the introduction of substances that interfere with the subsequent determinations or are tedious to remove. Consideration must also be given to the possible loss during the solution process of the constituents to be determined. For trace analysis it is desirable to minimize the amounts of reagents that restrict the choices available for sample decomposition.

To determine the elemental composition of an organic substance requires drastic treatment of the material in order to convert the elements of interest into a form susceptible to the common analytical techniques. These treatments are usually oxidative in nature and involve conversion of the carbon and hydrogen of the organic material to carbon dioxide and water. In some instances, however heating the sample with a potent reducing agent is sufficient to rupture the covalent bonds in the compound and free the element to be determined from the carbonaceous residue.

Oxidation procedures are divided into two categories. Wet-ashing (or oxidation) uses liquid oxidizing agents. Dry-ashing involves ignition of the organic compound in air or a stream of oxygen. Oxidations can also be carried out in fused-salt media; sodium peroxide is the most common flux for this purpose.

1.7.1.1 Volatilization Losses. Continual care must be exercised that a constituent to be determined is not lost by volatilization. Volatile weak acids are lost when materials are dissolved in stronger acids. Where these weak acids are to be determined, a closed apparatus must be used. Of course, volatile materials may be collected and determined (Sec. 2.4). Volatile acids, such as boric acid, hydrofluoric acid, and the other halide acids, may be lost during the evaporation of aqueous solutions, and phosphoric acid may be lost when a sulfuric acid solution is heated to a high temperature.

Phosphorus may be lost as phosphine when a phosphide or a material containing phosphorus is dissolved in a nonoxidizing acid. A very definite loss of silicon as silicon hydride may occur by volatilization when aluminum and its alloys are dissolved in nonoxidizing acids. Mercury may be lost as the volatile element in a reducing environment.

Certain elements are lost partially or completely in wet digestion methods involving halogen compounds. These include arsenic (as $AsCl_3$ or $AsBr_3$), boron (as BCl_3), chromium (as $CrOCl_2$),

[14] Z. Sulcek and P. Povondra, *Methods of Decomposition in Inorganic Analysis*, CRC, Boca Raton, Florida (1989).

[15] R. Boch, *A Handbook of Decomposition Methods in Analytical Chemistry*, International Textbook, London, 1979.

[16] D. C. Bogen, "Decomposition and Dissolution of Samples: Inorganic," Chap. 1 in I. M. Kolthoff and P. J. Elving, eds., *Treatise on Analytical Chemistry*, Part I, Vol. 5, 2d ed., Wiley-Interscience, New York, 1978.

[17] E. C. Dunlop and C. R. Grinnard, "Decomposition and Dissolution of Samples: Organic," Chap. 2 in I. M. Kolthoff and P. J. Elving, eds., *Treatise on Analytical Chemistry*, Part I, Vol. 5, 2d ed., Wiley-Interscience, New York, 1978.

germanium (as $GeCl_4$), lead (as $PbCl_4$), mercury (as $HgCl_2$), antimony (as $SbCl_3$), silicon (as SiF_4), tin (as $SnCl_4$ or $SnBr_4$), tellurium (as $TeCl_4$ or $TeBr_4$), titanium, zinc, and zirconium.

Osmium(VIII), ruthenium(VIII), and rhenium(VII) may be lost by volatilization as the oxides from hot sulfuric acid or nitric acid solution.

Relatively few things are lost by volatilization from alkaline fusions. Mercury may be reduced to the metal and lost, arsenic may be lost if organic matter is present, and of course any gases associated with the material are expelled. Fluoride may be lost from acid fusions, carrying away with it some silicon or boron.

Not so obvious is the loss of volatile chlorides from the reduction of perchloric acid to hydrochloric acid and from the production of hydrochloric acid from poly(vinyl chloride) laboratory ware.

1.7.2 Decomposition of Inorganic Samples

The electromotive force series (Table 14.14) furnishes a guide to the solution of metals in nonoxidizing acids such as hydrochloric acid, dilute sulfuric acid, or dilute perchloric acid, since this process is simply a displacement of hydrogen by the metal. Thus all metals below hydrogen in Table 14.14 displace hydrogen and dissolve in nonoxidizing acids with the evolution of hydrogen. Some exceptions to this may be found. The action of hydrochloric acid on lead, cobalt, nickel, cadmium, and chromium is slow, and lead is insoluble in sulfuric acid owing to the formation of a surface film of lead sulfate.

Oxidizing acids must be used to dissolve the metals above hydrogen. The most common of the oxidizing acids are nitric acid, hot concentrated sulfuric acid, hot concentrated perchloric acid, or some mixture that yields free chlorine or bromine. Addition of bromine or hydrogen peroxide to mineral acids is often useful. Considerable difficulties are encountered in the dissolution of inorganic matrices such as oxides, silicates, nitrides, carbides, and borides, problems often encountered in the analysis of geological samples and ceramics.

1.7.2.1 Use of Liquid Reagents

1.7.2.1.1 Use of Hydrochloric Acid. Concentrated hydrochloric acid (about 12*M*) is an excellent solvent for many metal oxides as well as those metals that lie below hydrogen in the electro motive series. It is often a better solvent for the oxides than the oxidizing acids. Hydrochloric acid dissolves the phosphates of most of the common metals although the phosphates of niobium, tantalum, thorium, and zirconium dissolve with difficulty. Hydrochloric acid decomposes silicates containing a high proportion of strong or moderately strong bases but acidic silicates are not readily attacked. The concentrated acid dissolves the sulfides of antimony, bismuth, cadmium, indium, iron, lead, manganese, tin, and zinc; cobalt and nickel sulfides are partially dissolved. Addition of 30% hydrogen peroxide to hydrochloric acid often aids the digestion of metals due to the release of nascent chlorine.

After a period of heating in an open container, a constant-boiling 6*M* solution remains (boiling point about 112°C). The low boiling point of hydrochloric acid limits it efficiency to dissolve oxides. However, microwave technology may overcome this difficulty (Sec. 1.7.4).

1.7.2.1.2 Use of Hydrofluoric Acid. The primary use for hydrofluoric acid is the decomposition of silicate rocks and minerals in which silica is not to be determined; the silicon escapes as silicon tetrafluoride. After decomposition is complete, the excess hydrofluoric acid is driven off by evaporation with sulfuric acid to fumes or with perchloric acid to virtual dryness. Sometimes residual traces of fluoride can be complexed with boric acid.

Hydrofluoric acid dissolves niobium, tantalum, and zirconium, forming stable complexes, although the action is sometimes rather slow. Hydrofluoric acid is an excellent solvent for the oxides of these metals although the temperature to which the oxide has been heated has a notable effect. Indium and gallium dissolve very slowly.

Caution Hydrofluoric acid can cause serious damage and painful injury when brought in contact with the skin. Momentarily it acts like a "painkiller" while it penetrates the skin or works under fingernails.

1.7.2.1.3 Use of Nitric Acid. Concentrated nitric acid is an oxidizing solvent that finds wide use in attacking metals. It will dissolve most common metallic elements except aluminum, chromium, gallium, indium, and thorium, which dissolve very slowly because a protective oxide film forms. Nitric acid does not attack gold, hafnium, tantalum, zirconium, and the metals of the platinum group (other than palladium).

Many of the common alloys can be decomposed by nitric acid. However, tin, antimony, and tungsten form insoluble oxides when treated with concentrated nitric acid. This treatment is sometimes employed to separate these elements from other sample components. Nitric acid attacks the carbides and nitrides of vanadium and uranium. Nitric acid is an excellent solvent for sulfides although the sulfides of tin and antimony form insoluble acids. Mercury(II) sulfide is soluble in a mixture of nitric acid and hydrochloric acid.

Although nitric acid is a good oxidizing agent, it usually boils away before the sample is completely oxidized.

A mixture of nitric and hydrofluoric acids dissolves hafnium, niobium, tantalum, and zirconium readily. This mixture is also effective with antimony, tin, and tungsten; the carbides and nitrides of niobium, tantalum, titanium, and zirconium; and the borides of zirconium.

1.7.2.1.4 Use of Sulfuric Acid. Hot concentrated sulfuric acid is often employed as a solvent. Part of its effectiveness arises from its high boiling point (about 340°C), at which temperature decomposition and solution of substances often proceed quite rapidly. Most organic compounds are dehydrated and oxidized under these conditions. Most metals and many alloys are attached by the hot acid.

Digestions are completed often in 10 min using sulfuric acid and 50% hydrogen peroxide (4 mL + 10 mL) with the Digesdahl™ digestion apparatus (Hach Chemical Co.). Fumes are removed by connecting the fractionating column to either a water aspirator or a fume scrubber.

1.7.2.1.5 Use of Perchloric Acid. Hot concentrated perchloric acid (72%) is a potent oxidizing agent and solvent. It attacks a number of ferrous alloys and stainless steels that are intractable to the other mineral acids. In fact, it is the best solvent for stainless steel, oxidizing the chromium and vanadium to the hexavalent and pentavalent acids, respectively. In ordinary iron and steel, the phosphorus is completely oxidized with no danger of loss. Sulfur and sulfides are oxidized to sulfate. Silica is rendered insoluble, and antimony and tin are converted to insoluble oxides. Perchloric acid fails to dissolve niobium, tantalum, zirconium, and the platinum group metals.

Powdered tungsten and chromite ore are soluble in a boiling mixture of perchloric and phosphoric acids.

Cold perchloric acid and hot dilute solutions are quite safe. However, all treatment of samples with perchloric acid should be done in specially designed hoods and behind explosion shields.

1.7.2.1.6 Use of Oxidizing Mixtures. More rapid solvent action can sometimes be obtained by the use of mixtures of acids or by the addition of oxidizing agents to the mineral acids. Aqua regia, a mixture consisting of three volumes of concentrated hydrochloric acid and one of nitric acid, releases free chlorine to serve as an oxidant. Addition of bromine to mineral acids serves the same purpose.

Great care must be exercised in using a mixture of perchloric acid and nitric acid. Explosion can be avoided by starting with a solution in which the perchloric acid is well diluted with nitric acid and not allowing the mixture to become concentrated in perchloric acid until the oxidation is nearly complete. Properly carried out, oxidations with this mixture are rapid and losses of metallic ions negligible.

A suitable sample (usually 1 to 5 g) is transferred to a modified fume eradicator digestion assembly and treated with 30 mL of a 1:1 mixture of nitric and perchloric acids. Place the equipment behind a protective shield and raise the temperature gradually until fumes of perchloric acid appear. Continue heating until fumes of perchloric acid are no longer noted. Cool, wash down the sides of the beaker with distilled water, and heat again until fumes of perchloric acid are no longer noted. Continue the digestion until the sample volume is reduced to about 2 mL. Transfer to a suitable volumetric flask. The presence of chromium (or vanadium) catalyzes the decomposition and serves as an indicator since perchloric acid will oxidize chromium(III) to dichromate(VI) after the oxidation of organic matter is complete. When volatile elements are present, dissolution must be effected under reflux conditions and started with a cold hot plate. Good recovery of all elements except mercury is obtained.[18]

[18] T. T. Gorsuch, *Analyst (London)* **84**:135 (1959).

TABLE 1.14 Acid Digestion Bomb-Loading Limits

Bomb capacity, mL	Maximum inorganic sample, g	Maximum organic sample, g	Minimum and maximum volume of nitric acid to be used with an organic sample, mL
125	5.0	0.5	12–15
45	2.0	0.2	5.0–6.0
23	1.0	0.1	2.5–3.0

Another recommended procedure[19] is to weigh about 5 g of sample into an Erlenmeyer flask of suitable size so that some frothing can be tolerated. The sample is treated with 10 to 20 mL water, 5 to 10 mL nitric acid, and 0.5 mL sulfuric acid. The mixture is carefully evaporated on a hot plate to incipient fumes of sulfuric acid. If the solution is still dark or yellow, small portions of nitric acid should be added to clear the solution. Then the solution is evaporated to dryness (SO_3) fumes. If the residue is not white, the treatment with nitric acid is repeated. Finally, two or three additions of 2 to 3 mL of distilled water are evaporated to eliminate nitrosylsulfuric acid. A number of elements may be volatilized at least partially by this procedure, particularly if the sample contains chlorine; these were enumerated in Sec. 1.7.1.1.

1.7.2.2 Acid-Vapor Digestion. In a closed reaction chamber, the sample in a sample cup is held above the acid(s) in the chamber. Only the acid vapors reach the sample in its container. Trace impurities in the liquid reagent will not contaminate the sample, thus leading to considerably lower blanks. Some materials may not be fully dissolved by acid digestion at atmospheric pressure.

1.7.2.2.1 Acid Digestion Bomb. A sealed pressure vessel lined with Teflon offers an alternative method of preparing samples for analysis. The pressure vessel holds strong mineral acids or alkalies at temperatures well above normal boiling points. Often one can obtain complete digestion or dissolution of samples that would react slowly or incompletely when conducted in an open container at atmospheric pressure. Samples are dissolved without losing volatile trace elements and without adding unwanted contaminants from the container. Ores, rock samples, glass, and other inorganic samples can be dissolved rapidly by using strong mineral acids such as HF, HCl, H_2SO_4, HNO_3, and aqua regia. In all reactions the bomb must never be completely filled as there must always be adequate vapor space above the surface of the charge (sample plus digestion medium). The total volume of the charge must never exceed two-thirds of the capacity of the bomb when working with inorganic materials.

Many organic materials can be treated satisfactorily in these digestion bombs but careful attention must be given to the nature of the sample and to possible explosive reactions with the digestion media. For nitric acid digestions or organic compounds, the dry weight of organic material must not exceed the limits in Table 1.14. Note that both minimum and maximum amounts of acid are specified. If the sample contains less than the specified maximum amount of dry organic matter, the amount of nitric acid must be reduced proportionately. Fats, fatty acids, glycerol, and similar materials that form explosive compounds in an intermediate stage must not be treated with nitric acid in these bombs. When feasible, users should always try nitric acid alone and resist the temptation to add sulfuric acid as often done in wet digestions.

Caution ***Do not use perchloric acid*** in these bombs because of its unpredictable behavior when heated in a closed vessel. Also avoid any reaction that is highly exothermic or that might be expected to release large volumes of gases. Do not overheat the bomb; generally the temperature is held below 150°C. Operating temperatures and pressures up to a maximum of 250°C and 1800 pounds per square inch (psi) are permitted in bombs with a thick-walled Teflon liner with a broad, flanged seal. Safety rupture disks protect the bomb and the operator from the hazards of unexpected or dangerously high internal pressures.

[19] G. Middleton and R. E. Stuckey, *Analyst* (*London*) **53**:138 (1954).

1.7.2.3 Decomposition of Samples by Fluxes. A salt fusion is performed by mixing a sample with salts (the flux), melting the mixture with heat, cooling it, and finally, dissolving the solidified melt. Flux fusion is most often used for samples that are difficult to dissolve in acid. Quite a number of common substances—such as silicates, some of the mineral oxides, and a few of the iron alloys—are attacked slowly, if at all, by the usual liquid reagents. Recourse to more potent fused-salt media, or fluxes, is then called for. Fluxes will decompose most substances by virtue of the high temperature required for their use and the high concentration of reagent brought in contact with the sample. The basic requirement in a sample to be fused is chemically bound oxygen, as present in oxides, carbonates, and silicates. Sulfides, metals, and organics cannot be successfully fused unless they are first oxidized.

Where possible, the employment of a flux is usually avoided, for several dangers and disadvantages attend its use. In the first place, a relatively large quantity of the flux is required to decompose most substances—often 10 times the sample weight. The possibility of significant contamination of the sample by impurities in the reagent thus becomes very real.

Furthermore, the aqueous solution resulting from the fusion will have a high salt content, and this may lead to difficulties in the subsequent steps of the analysis. The high temperatures required for fusion increase the danger of loss of pertinent constituents by volatilization. Finally, the container in which the fusion is performed is almost inevitably attacked to some extent by the flux; this again can result in contamination of the sample.

In those cases where the bulk of the substance to be analyzed is soluble in a liquid reagent and only a small fraction requires decomposition with a flux, it is common practice to employ the liquid reagent first. The undecomposed residue is then isolated by filtration and fused with a relatively small quantity of a suitable flux. After cooling, the melt is dissolved and combined with the rest of the sample.

1.7.2.3.1 Sample Fusion. In order to achieve a successful and complete decomposition of a sample with a flux, the solid must ordinarily be ground to a very fine powder; this will produce a high specific surface area. The sample must then be thoroughly mixed with the flux in an appropriate ratio (usually between 1:2 and 1:20), perhaps with the addition of a nonwetting agent to prevent the flux from sticking to the crucible. This operation is often carried out in the crucible in which the fusion is to be done by careful stirring with a glass rod.

In general, the crucible used in a fusion should never be more than half-filled at the outset. The temperature is ordinarily raised slowly with a gas flame because the evolution of water and gases is a common occurrence at this point; unless care is taken there is the danger of loss by spattering. The crucible should be covered with its lid as an added precaution. The maximum temperature employed varies considerably and depends on the flux and the sample. It should be no greater than necessary in order to minimize attack on the crucible and decomposition of the flux. It is frequently difficult to decide when the heating should be discontinued. In some cases, the production of clear melt serves to indicate the completion of the decomposition. In others the condition is not obvious and the analyst must base the heating time on previous experience with the type of material being analyzed. In any event, the aqueous solution from the fusion should be examined carefully for particles of the unattacked sample.

When the fusion is judged complete, the mass is allowed to cool slowly; then just before solidification the crucible is rotated to distribute the solid around the walls of the crucible so that the thin layer can be readily detached.

The Spex-Claisse Fusion Fluxers® are automated borate fusion devices that are capable of simultaneously preparing in 5 to 10 min up to six samples either as homogeneous glass disks for x-ray fluorescence or as solutions for induction coupled plasma–atomic absorption. Each sample is heated with a borate flux in a Pt–Au crucible over a propane or butane flame (1100°C). As the flux melts, the crucible rocks back and forth or rotates. Preset fluxing programs are completely adaptable to particular applications with programmable heat level, agitation speed, and time for each step of every program. A nonwetting agent injector prevents flux sticking to crucibles. An optional oxygen injector maintains an oxidizing atmosphere in crucibles. The crucible is then emptied into a preheated mold or to a beaker of stirred dilute mineral acid. If a mold is being used, cooling fans anneal the glass disk. In solution preparation, the bead of molten flux shatters when it hits the dilute acid and is dissolved after several minutes of stirring.

Graphite crucibles are a cost-effective alternative to metal crucibles. Graphite crucibles are disposable, which eliminates the need for cleaning and the possibility of cross-sample contamination. These crucibles are chemically inert and heat-resistant, although they do oxidize slowly above 430°C; over a period of hours some erosion of the crucible can occur. Graphite is not recommended for extremely lengthy fusions or for fusion where the sample might be reduced.

Platinum crucibles may be cleaned by (a) boiling HCl, (b) sea sand with hand cleaning, and (c) blank fusion with sodium hydrogen sulfate. Chemicals to avoid are aqua regia, sodium peroxide, free elements (C, P, S, Ag, Bi, Cu, Pb, Zn, Se, and Te), ammonia, chlorine and volatile chlorides, sulfur dioxide, and gases with carbon content.

Zirconium crucibles are cleaned with boiling HCl or compatible blank fusions. HF must be absent and ceramic fusions should be avoided.

1.7.2.4 Types of Fluxes. The common fluxes used in analysis are listed in Table 1.15. Basic fluxes, employed for attack on acidic materials, include the carbonates, hydroxides, peroxides, and borates.

TABLE 1.15 The Common Fluxes

Flux	Fusion temperature, °C	Types of crucible used for fusion	Types of substances decomposed
Na_2CO_3 (mp* 851°C)	1000–1200	Pt	For silicates and silica-containing samples (clays, glasses, minerals, rocks, and slags); alumina-, beryllia-, and zirconia-containing samples; quartz; insoluble phosphates and sulfates
K_2CO_3 (mp 901°C)	1000	Pt	For niobium oxide
Na_2CO_3 plus Na_2O_2		Pt (not with Na_2O_2), Ni, Zr, Al_2O_3 ceramic	For samples needing an oxidizing agent (sulfides, ferroalloys, Mo- and W-based materials, some silicate minerals and oxides, waxes, sludge, Cr_3C_2)
NaOH or KOH (mp 320–380°C)		Au (best), Ag, Ni (<500°C)	For silicates, silicon carbide, certain minerals
Na_2O_2	600	Ni; Ag, Au, Zr	For sulfides, acid-insoluble alloys of Fe, Ni, Cr, Mo, W, and Li; Pt alloys; Cr, Sn, and Zn minerals
$K_2S_2O_7$ (mp 300°C)	Up to red heat	Pt, porcelain	Acid flux for insoluble oxides and oxide-containing samples, particularly those of aluminum, beryllium, tantalum, titanium, and zirconium
KHF_2 (mp 239°C)	900	Pt	For silicates and minerals containing niobium, tantalum, and zirconium, and for oxides that form fluoride complexes (beryllium, niobium, tantalum, and zirconium)
B_2O_3(mp 577°C)	1000–1100	Pt	For silicates, oxides and refractory minerals, particularly when alkalies are to be determined
$CaCO_3 + NH_4Cl$		Ni	For decomposing all classes of silicate minerals, principally for determination of alkali metals
$LiBO_2$ (mp 845°C)	1000–1100	Graphite, Pt	For almost anything except sulfides and metals
$Li_2B_4O_7$ (mp 920°C)	1000–1100	Graphite, Pt	Same as for $LiBo_2$

* mp denotes melting point.

TABLE 1.16 Fusion Decompositions with Borates in Platinum or Graphite Crucibles

Sample type	Sample amount, g	Fusion mixture	Elements determined	Fusion conditions
Blends of Al_2O_3, SiO_2, TiO_2, MnO, SrO, CaO, MgO	1.25	3.15 g B_2O_3 + Li_2CO_3 (1.75–1.4 g)	Na	1200°C
	2	9 g H_3BO_3 + Li_2CO_3 (6.2–2.8 g)	Ca, Fe, Mn, Si Ti, V, Zn	1 h; 1000°C
Organic polymers	1	NaF:$Na_2B_4O_7 \cdot 10H_2O$ (0.05:0.2)	Si	Heat at 750°C before fusing
Yttrium oxalate	1	Same as above (9:1)	Y	1050°C
Zr/Y oxides	1	Same as above (9:1)	Al, Fe, Si, Ti, Y	Use high heat
Zinc sulfide	1	Same as above (1:10)	Ag, Co	Treat sample with HNO_3 and fire at 750°C for 30 min to remove S before fusing
Mixtures of Ba, Ce, Mg, and Sr aluminates	1	NaF:Na_2CO_3:$Na_2B_4O_7 \cdot 10H_2O$ (1:5:10)		1050°C
Silicates	0.1	0.5 g $LiBO_2$	Ba, Sr	10 min; 950°C
	0.65	1.95 g H_3BO_3 + Li_2CO_3 (1.3–0.65 g)	Major components	1050°C
Cement manufacture, raw materials		$Li_2B_4O_7$	F	15 min; 1000°C
Slags	0.1–0.2	1 g $LiBO_2$	Al, Cu, Fe, Si, Zn	10 min; 900°C
Silica (1 part)		6–9 parts $Li_2B_4O_7$	Fe	15 min; 1200°C
CaF_2, Na_3AlF_6, apatite, dust	0.1–1	2 g $Li_2B_4O_7$	F	10 min; 1000°C
SiO_2/Al_2O_3—raw and manufactured materials	0.5	4 g H_3BO_3 + Li_2CO_3	Na, K	Fusion for few min, then 5 min with high heat

The acidic fluxes are the pyrosulfates, the acid fluorides, and boric oxide. Fluxes rich in lithium tetraborate are well suited for dissolving basic oxides, such as alumina. Lithium metaborate or a mixture of metaborate and tetraborate (Table 1.16), on the other hand, is more basic and better suited for dissolving acidic oxides such as silica or titanium dioxide, although it is capable of dissolving nearly all minerals.

The lowest melting flux capable of reacting completely with a sample is usually the optimum flux. Accordingly, mixtures of lithium tetraborate, with the metaborate or carbonate, are often selected.

If an oxidizing flux is required, sodium peroxide can be used. As an alternative, small quantities of the alkali nitrates or chlorates are mixed with sodium carbonate.

Boric oxide has one great advantage over all other fluxes in that it can be completely removed by volatilization as methyl borate by using methanol saturated with dry hydrogen chloride. It is nonvolatile and therefore can displace volatile acids in insoluble salts such as sulfates.

High-purity lithium metaborate or tetraborate is obtainable, which minimizes contamination from the flux. The entire fusion process seldom takes longer than 25 min, and is usually only 15 min. An exception is aluminum oxide minerals, which require 1 h.

1.7.3 Decomposition of Organic Compounds

Analysis of the elemental composition of an organic substance generally requires drastic treatment of the material in order to convert the elements of interest into a form susceptible to the common analytical techniques. These treatments are usually oxidative in nature, involving conversion of the carbon and hydrogen of the organic material to carbon dioxide and water. In some instances, however, heating the sample with a potent reducing agent is sufficient to rupture the covalent bonds in the compound and free the element to be determined from the carbonaceous residue.

Oxidizing procedures are sometimes divided into two categories. Wet-ashing (or oxidation) makes use of liquid oxidizing agents. Dry-ashing usually implies ignition of the organic compound in air or a stream of oxygen. In addition, oxidations can be carried out in certain fused-salt media, sodium peroxide being the most common flux for this purpose.

1.7.3.1 Oxygen Flask (Schöninger) Combustion.[20,21] A relatively straightforward method for the decomposition of many organic substances involves oxidation with gaseous oxygen in a sealed container. The reaction products are absorbed in a suitable solvent before the reaction vessel is opened. Analysis of the solution by ordinary methods follows.

A simple apparatus for carrying out such oxidation has been suggested by Schöninger. It consists of a heavy-walled flask of 300- to 1000-mL capacity fitted with a ground-gass stopper. Attached to the stopper is a platinum-gauze basket that holds from 2 to 200 mg of sample. If the substance to be analyzed is a solid, it is wrapped in a piece of low-ash filter paper cut with a tail extending above the basket. Liquid samples can be weighed in gelatin capsules that are then wrapped in a similar fashion. A tail is left on the paper and serves as an ignition point (wick).

A small volume of an absorbing solution is placed in the flask, and the air in the container is then displaced by allowing tank oxygen to flow into it for a short period. The tail of the paper is ignited, and the stopper is quickly fitted into the flask. The container is then inverted to prevent the escape of the volatile oxidation products. Ordinarily the reaction proceeds rapidly, being catalyzed by the platinum gauze surrounding the sample. During the combustion, the flask is kept behind a safety shield to avoid damage in case of an explosion. Complete combustion takes place within 20 s.

After cooling, the flask is shaken thoroughly and disassembled; then the inner surfaces are rinsed down. The analysis is then performed on the resulting solution. This procedure has been applied to the determination of halogens, sulfur, phosphorus, arsenic, boron, lanthanides, rhenium, germanium, and various metals in organic compounds. When sulfur is to be determined, any sulfur dioxide formed in the Schöninger combustion is subsequently oxidized to sulfur trioxide (actually sulfate) by treatment with hydrogen peroxide.

1.7.3.2 Peroxide Fusion. Sodium peroxide is a powerful oxidizing reagent that, in the fused state, reacts rapidly and often violently with organic matter, converting carbon to the carbonate, sulfur to sulfate, phosphorus to phosphate, chlorine to chloride, and iodine and bromine to iodate and bromate. Under suitable conditions the oxidation is complete, and analysis for the various elements may be performed upon an aqueous solution of the fused mass.

Once started (Parr method[22]), the reaction between organic matter and sodium peroxide is so vigorous that a peroxide fusion must be carried out in a sealed, heavy walled, steel bomb. Sufficient heat is evolved in the oxidation to keep the salt in the liquid state until the reaction is completed. Ordinarily the reaction is initiated by passage of current through a wire immersed in the flux or by momentarily heating the bomb with a flame.

The maximum size for a sample that is to be fused is perhaps 100 mg. The method is more suitable to semimicro quantities of about 5 mg. The recommended sample size for various bombs is shown in Table 1.17. Combustion aids and accelerators are listed in Table 1.18.

One of the main disadvantages of the peroxide-bomb method is the rather large ratio of flux to sample needed for a clean and complete oxidation. Ordinarily an approximate 200-fold excess is used. The excess peroxide is subsequently decomposed to hydroxide by heating in water. After neutralization, the solution necessarily has a high salt content.

---

[20] M. E. McNally and R. L. Grob, "Oxygen Flask Combustion Technique," *Am. Lab.*:31 (January 1981).

[21] A. M. G. MacDonald, *Analyst (London)* **86**:3 (1961).

[22] Parr Instrument Company, *Peroxide Bomb—Apparatus and Methods,* Moline, Illinois.

TABLE 1.17 Maximum Amounts of Combustible Material Recommended for Various Bombs

Bomb size, mL	Maximum total combustible matter, g	Accelerator, g	Sodium peroxide, g
22	0.5	1.0	15.0
8	0.2	0.2	4.0
2.5	0.085	0.1	1.5

Sodium peroxide is a good flux for silicates when sodium carbonate is ineffective and contamination from the crucible is permissible. It is very effective with ferrochrome, ferrosilicon, and ferroalloys that may not be attacked by acids. Chromium in chromite ore is oxidized to chromate(VI).

1.7.3.3 Oxygen–Hydrogen Burning.[23,24] An oxygen–hydrogen flame (2000°C) decomposes organic material that has been previously vaporized and swept into the flame chamber. The gaseous decomposition products are absorbed in dilute alkali (with hydrogen peroxide added to convert any sulfur to sulfate).

1.7.3.4 Low-Temperature Oxygen Plasma.[25] Reactive oxygen generated with a microwave or radiofrequency excitation source decomposes oxidizable organic compounds at a low temperature (25 to 300°C) but at a slow rate of milligrams per hour, which is a major disadvantage. It is useful for trace-metal analysis in polymers, biological specimens, and petrochemical residues.

1.7.3.5 Wet-Ashing Procedures. Wet-ashing methods are usually preferred when decomposing organic matrices for subsequent determination of trace metals. Solution in a variety of strong oxidizing agents will decompose samples. The main consideration associated with the use of these reagents is preventing volatility losses for the elements of interest. For the volatile elements, dissolution must be effected under reflux conditions. Use a 125-mL quartz flask fitted with a quartz condenser filled with quartz beads.

1.7.3.5.1 Van Slyke Method. The Van Slyke wet-ashing method[26–28] oxidizes organic samples with iodic(V) acid and chromic(VI) acid in a mixture of sulfuric and phosphoric acids. The method has had wide application in determining carbon. The evolved CO_2 is absorbed in alkali and subsequently determined.

1.7.3.5.2 Use of Sulfuric Acid. One wet-ashing procedure is the Kjeldahl method for the determination of nitrogen in organic compounds. Here concentrated sulfuric acid is the oxidizing agent.

[23] R. Wickbold, *Angew. Chem.* **66**:173 (1954).
[24] P. E. Sweetser, *Anal. Chem.* **28**:1768 (1956).
[25] J. R. Hollahan and A. T. Bell, eds., *Techniques and Applications of Plasma Chemistry*, Wiley, New York, 1974.
[26] A. Steyermark, *Quantitative Organic Microanalysis*, 2d ed., Academic, New York, 1961.
[27] D. D. Van Slyke, *Anal. Chem.* **26**:1706 (1954).
[28] E. C. Horning and M. G. Horning, *Ind. Eng. Chem. Anal. Ed.* **19**:688 (1947).

TABLE 1.18 Combustion Aids for Accelerators*

Element	Accelerator, g	Sample, g	Aid (sucrose), g
S	1.0 $KClO_4$	0.1–0.5	0.4–0.00
Cl, Br, I	1.0 KNO_3†	0.2–0.4	0.3–0.1
B	1.0 $KClO_4$	0.1–0.3	0.4–0.2
Si	1.0 $KClO_4$	0.1–0.3	0.4–0.2
Se	1.0 KNO_3	0.1–0.3	0.4–0.2
P	1.0 KNO_3	0.1–0.3	0.4–0.2

*Flame ignition, 22-mL bomb; 15 g sodium peroxide.
†KNO_3 must be used with care as it may form an explosive mixture.

This reagent is also frequently employed for decomposition of organic materials through which metallic constituents are to be determined. Commonly, nitric acid is added to the solution periodically to hasten the rate at which oxidation occurs. A number of elements may be volatilized, at least partially, by this procedure, particularly if the sample contains chlorine; these include arsenic, boron, germanium, mercury, antimony, selenium, tin, and the halogens. Organic material is decomposed with a mixture of sulfuric acid and 30% hydrogen peroxide without loss of arsenic, antimony, bismuth, germanium, gold, mercury, or silver in the absence of halogen.[29] In the presence of halogen the reaction must be carried out with a reflux condenser or liquid trap to avoid loss. Up to 100 mg of sample is placed in the flask, 5 to 10 mL of 15% fuming sulfuric acid is added, and the mixture is warmed gently. While the flask is swirled, 1 to 10 mL of 30% hydrogen peroxide is added dropwise down the sides of the flask as oxidation takes place and until the liquid becomes light yellow or clear. The heat is then increased until heavy fumes of SO_3 appear.

Transfer a suitable sample to a quartz beaker, and treat it with 5 mL of water and 10 mL nitric acid. Digest tissues overnight with the beaker covered; other material may take less digestion. Cool; add 5 mL of sulfuric acid and evaporate to fumes of sulfuric acid. Cover the beaker and add nitric acid dropwise to the hot solution to destroy any residual organic matter. Transfer the solution to a Teflon fluorinated ethylene-propylene (FEP) beaker along with any remaining siliceous material. Add 1 mL of hydrofluoric acid and evaporate the solution to fumes of sulfuric acid.

1.7.3.5.3 Use of Perchloric Acid–Nitric Acid Mixtures. The use of perchloric acid with nitric acid or nitric acid–sulfuric acid for the decompositions of organic materials has been described.[30–32] A good deal of care must be exercised in using a mixture of perchloric acid and nitric acid. Explosion can be avoided by starting with a solution in which the perchloric acid is well diluted with nitric acid and not allowing the mixture to become concentrated in perchloric acid until the oxidation is nearly complete. Properly carried out, oxidations with this mixture are rapid and losses of metallic ions negligible.

Caution This mixture should never be used to decompose pyridine and related compounds.

An alternative method involves nitric acid, perchloric acid, and sulfuric acid. The sample is heated with nitric acid in order to oxidize the more reactive matrix constituents without incurring an overly vigorous reaction. A mixture of sulfuric acid and perchloric acid is added and the digestion completed at about 200°C. Most of the nitric acid is boiled off in the second step. Sulfuric acid raises the reaction temperature so that the less reactive constituents are digested by the perchloric acid and dilutes the perchloric acid in the final solution so as to prevent the possible formation of explosive conditions.

1.7.3.6 Dry-Ashing Procedure. The simplest method for decomposing an organic sample is to heat it with a flame in an open dish or crucible until all the carbonaceous material has been oxidized by the air. A red heat is often required to complete the oxidation. Analysis of the nonvolatile components is then made after solution of the residual solid. A great deal of uncertainty always exists with respect to the recovery of supposedly nonvolatile elements when a sample is treated in this manner. Some losses probably arise from the mechanical entrainment of finely divided particular matter in the hot convection currents around the crucible. Volatile metallic compounds may be formed during the ignition. Elements that may be lost in dry-ashing at 600°C include Ag, Au, As, B, Be, Cd, Cr, Co, Cs, Cu, Fe, Ge, Hg, Ir, K, Li, Na, Ni, P, Pb, Pd, Pt, Rb, Rh, Sb, Se, Sn, Tl, V, and Zn.[33,34]

The dry-ashing procedure is the simplest of all methods for decomposing organic compounds. It is often unreliable and should not be employed unless tests have been performed that demonstrate its applicability to a given type of sample.

---

[29] D. L. Talbern and E. F. Shelberg, *Ind. Eng. Chem. Anal. Ed.* **4**:401 (1932).

[30] G. F. Smith, *The Wet Oxidation of Organic Compounds Employing Perchloric Acid*, G. Frederick Smith Chemical Co., Columbus, Ohio, 1965.

[31] A. A. Schilt, *Perchloric Acid and Perchlorates*, G. Frederick Smith Chemical Co., Columbus, Ohio, 1979.

[32] H. Diehl and G. F. Smith, *Talanta* **2**:209 (1959).

[33] R. E. Thiers, in D. Glick, Ed., *Methods of Biochemical Analysis*, Interscience, New York, 1957, Chap. 6.

[34] T. T. Gorsuch, *Analyst (London)* **84**:135 (1959).

1.7.3.6.1 Low-Temperature Dry-Ashing. This method is a departure from traditional techniques of sample decomposition. The oxidizing agent is a stream of excited high-purity oxygen that is produced in a high-frequency electromagnetic field. This method permits controlled ashing of samples in which quantitative retention of inorganic elements and retention of mineral structure is critical. Reagent contamination is eliminated.

1.7.4 Microwave Technology[35]

Microwave digestion of samples in closed containers has proven to be at least four times faster than the hot-plate method from which it was derived; in some cases it is hundreds of times faster. One of the most revolutionary aspects of microwave dissolution is the ease with which it can be automated relative to traditional flame, hot-plate, and furnace dissolution techniques.

Microwave digestion is cleaner, more reproducible, more accurate, and freer from external contamination. Extraction of impurities from the containment vessel is minimized because digestions are done in Teflon perfluoroalkoxy. Little or no acid is lost during digestion in a closed vessel so that additional portions of acid may not be required, again lowering the blank correction. Airborne particles cannot enter a sealed vessel, and cross-sample contamination caused by splattering is eliminated. Many samples can be digested directly without first being fused or ashed. This speeds analysis and reduces background contamination from the flux and potential loss of volatile elements.

1.7.4.1 Heating Mechanism. Liquids heat by two mechanisms—dipole rotation and ionic conduction. Polar molecules will tend to align their dipole moments with the microwave electric field. Because the field is changing constantly, the molecules are rotated back and forth, which causes them to collide with other nearby molecules. Ions in solution will tend to migrate in the presence of a microwave electric field. This migration causes the ions to collide with other molecules. Heat is generated when molecules or ions collide with nearby molecules or ions.

The amount of energy absorbed by the vessel contents and the power delivered are critical factors. Normally, the microwave magnetron is initially calibrated to determine these parameters and to optimize their roles during dissolution. Calibration of the microwave magnetron requires the use of 1 L of water in a Teflon or similar plastic beaker for 2 min at full power (at least 600 W); the initial and final temperatures of the water are measured.[36] For deionized water, the microwave power P in $\text{cal}\cdot\text{s}^{-1}$ is given by

$$P = 34.87\ (\Delta T) \tag{1.10}$$

The water must be stirred vigorously after removal from the microwave cavity to disperse localized superheating. The temperature must be measured to an accuracy of 0.1°C and the final temperature must be measured within 30 s after heating. To ensure linearity between absorbed power versus applied power, the microwave method requires that a calibration of the microwave magnetron be conducted at two levels, 40% and 100% power, and also requires the use of a laboratory-grade microwave oven with programmable power settings up to at least 600 W.

1.7.4.2 Low-Pressure Microwave Digestion. A pressure-control module is used in the low-pressure microwave digestion.[37] Time and power can be independently programmed in up to three stages. Power settings from 0% to 100% in 1% increments allow precise control. The turntable within the microwave oven accepts 12 digestion vessels and rotates at 6 revolutions per minute (r/min) to ensure uniform heating. The Teflon PFA vessels are much more sophisticated than the Carius tubes. The caps, which are screwed onto the canister with a torquing device, are designed to vent the container safely in case of excessive internal pressure. The vials stay sealed up to a pressure of 120 psi (830 kPa).

[35] H. M. Kingston and L. B. Jassie, *Introduction to Microwave Sample Preparation: Theory and Practice*, American Chemical Society, Washington, D. C., 1988.

[36] H. M. Kingston and L. B. Jassie, eds., *Introduction to Microwave Sample Preparation: Theory and Practice*, American Chemical Society, Washington, D. C., 1988, Chap. 6; M. E. Tatro; "EPA Approves Closed-Vessel Microwave Digestion for CLP Laboratories," *Spectroscopy* **5**[6]:17 (1990).

[37] A. C. Grillo, "Microwave Digestion Using a Closed-Vessel System," *Spectroscopy*, **5**[1]:14, 55 (1989).

TABLE 1.19A Typical Operating Parameters for Microwave Ovens

Sample	Amount	Reagent	Reagent volume, mL	Pressure, psi	Time, min
Ash, paper	0.1 g	HCl : HF (7 : 3)	10	160	5
		HNO_3	3		
Cellulose filter	. . .	$HNO_3 : H_2O$(1 : 1)	10	160	12
Cement	0.1 g	$HNO_3 : HBF_4$(1 : 1)	5	160	7
		2.5 mL + 2.5 mL			2
					(100% power)
Sediment	1 g	$HNO_3 : H_2O$ (1 : 1)	20	50	10
				160	30
Sludge	1 g	$HNO_3 : H_2O$(1 : 1)	20	30	20
				60	10
				180	15
Waste water	50 mL	HNO_3	5	160	20

TABLE 1.19B Typical Operating Parameters for Microwave Ovens (*Continued*)

Sample	Amount	Reagent	Time	Power, W
Diabase and basalt	0.1 g	HNO_3	5 min	300 W
		+ 1 mL HF	1 h	*
Oil shale	0.1 g	3 mL HNO_3 + 1 mL H_2O_2	4 × 7 min	300 W
		4 mL HF + 1 mL $HClO_4$	3 × 7 min	300 W
			2 min	600 W
			0.5 min	900 W
Hay	0.5 g	4 mL HNO_3	3–5 × 1 min	300 W
			7 min	300 W
			2 min	600 W

* Increase power gradually up to 600 W.

Above this the vessels will relieve pressure and vent fumes into a common collection vessel. Representative sample sizes, digestion acids, and times are listed in Table 1.19.

1.7.4.3 Methodology. When developing a digestion method, start with a small quantity of sample and acid apply microwave energy at varying levels for enough time to digest the sample. The container must be fabricated from microwave-transparent polymer material. The sample is considered digested when no visible solid remains and when the solution remains clear upon dilution.[38] The pressure generated in the digestion vessel is observed by a pressure monitor and compared to a value chosen by the user. If pressure in the digestion vessel exceeds the user-selected value, the pressure controller will regulate the magnetron, switching it on and off at a rate that will maintain pressure at or below the set value. This enables the use of full microwave power for a given time and leaves the system unattended to complete the digestion.

Inorganic materials including metals, water and waste water, minerals, and most soils and sediments are easily digested in acids without generating large amounts of gaseous by-products. On the other hand, samples containing a high percentage of organic material produce copious amounts of

[38] L. B. Gilman and W. G. Engelhart, "Recent Advances in Microwave Sample Preparation," *Spectroscopy* **4**[8]:14 (1989).

gaseous reaction by-products. Many organic materials require high temperatures to digest thoroughly in acid (without being dry-ashed first).

Pressure developed in a sealed vessel when microwaved can rise with dramatic suddenness. The reaction becomes exothermic above a temperature ranging from 140 to 160°C for nitric acid, the usual solvent for organic materials. When digestion is incomplete after microwaving for 15 min, the vessel is cooled, vented of gaseous by-products, and resealed. This heating, cooling, and venting cycle is repeated until a clear solution is produced that does not cloud upon dilution with water.[39]

As many as 12 samples can be digested simultaneously when using 120-mL vessels. When this size vessel is too large, particularly with biological samples when the volume of acid needed is less 2 mL, smaller size vessels (3 and 7 mL) are used. Up to three of the 3-mL or two of the 7-mL vessels are placed inside a 120-mL vessel that acts as a safety containment chamber for the sealed containers. For the smaller containers the maximum sample size and acid volume that can be used are 30 mg and 1.5 mL for the 3-mL vessels, and 70 mg and 3.5 mL for the 7-mL vessels, respectively. A double-walled vessel design permits greater heat retention and venting at 200 psi [10 343 mmHg (at 0°C); 13.61 atm; 1 378 951 Pa ($N \cdot m^2$)].

At temperatures above 250°C, virtually all polymeric materials begin to melt, flow, lose strength, or decompose. However, higher temperatures are required for digestion of some organic materials (such as polyimides) and many ceramics. For these types of samples, a ceramic turntable is inserted into the microwave oven to support open borosilicate, quartz, and vitreous carbon containers.

1.7.4.4 Dry-Ashing. The ashing block (a ceramic insert) is an accessory that absorbs microwave energy and quickly heats to a high temperature. This, in combination with the microwave energy absorbed directly by the sample, allows rapid dry-ashing of most materials in 8 to 10 min at temperatures up to 800°C.

1.7.4.5 Microwave Bomb.[40] The microwave bomb uses a microwave-transparent polymer shell with an internal Teflon (polytetrafluoroethylene, PTFE) reaction vessel. It is designed to withstand internal pressures of more than 1200 psi (81.66 atm) [actually up to 1700 psi (115.7 atm) with the lower value recommended to provide a margin between planned maximum pressure and venting pressure], and internal temperatures of more than 200°C. The internal pressure can be monitored by measuring the distance in which the pressure screw protrudes from screw cap. The internal capacity is 23 mL. Complete reaction times to less than 1 min are claimed for many samples.

Do a trial digestion with about a 0.1-g sample, assuming detection limits are adequate with this size sample. A larger sample size may be more representative of the overall sample and may yield better precision and lower detection limits. However, it will also lead to the formation of higher gas pressure during digestions. Therefore, it may cause a maximum pressure buildup before the sample is digested. The larger sample also decreases the acid-to-sample ratio. Sample rotation during microwaving will avoid potential hot-spot heating.

An obvious disadvantage is the fact that the vessel is not transparent and complete digestion is difficult to ascertain. Also the vessel is potentially more dangerous due to high pressures.

1.7.4.6 Focused Open-Vessel System.[41] A microwave digestion system without an oven consists of a magnetron to generate microwaves; a waveguide to direct and focus the microwave; and a cavity to contain the sample. The open-vessel design means that there will be no pressure buildup in the vessel during the digestion process and reagents may be added during the digestion program. Relatively large amounts, up to 10 g for solids and 50 to 100 mL for liquids (Spex 7400 and up to 20 g in the Questron Maxidigest unit), can be digested, while excess acid vapors and decomposition gases are removed with an aspirator. The action of focused microwaves causes solutions to reach high temperatures faster than with a conventional hot-plate or block-type digester. The microwaves also generate a mechanical agitation that aids in the dissolution of a sample. Times for

---

[39] W. Lautenschlaeger, "Microwave Digestion in a Closed-Vessel, High-Pressure System," *Spectroscopy* **4**[9]:16 (1989).

[40] B. D. Zehr and M. A. Fedorchak, "Microwave Acid Digestion of Inorganics Using a Bomb Vessel," *Am. Lab.* **1991**:40 (February).

[41] A. C. Grillo, "Microwave Digestion by Means of a Focused Open-Vessel System," *Spectroscopy* **4** [7]:16 (1989).

complete digestion average 10 to 30 min. Samples are run in 250-mL borosilicate glass digestion tubes. The system can be totally automated. The microwave energy is directed only at that portion of the vessel that is in the path of the focused microwaves. The neck of the vessel and the reflux unit are not within the energy flow. They remain cool and thereby ensure a refluxing action that allows the system to operate just like an ordinary everyday hot plate but with much greater energy input and superior reproducibility. Samples may be removed at any time during a digestion to inspect the progress visually.

The system retains all elements, even mercury and selenium. The reflux action allows hands-off addition of reagents between (and in some cases during) heating cycles. When there are requirements for digesting many different types of samples, when it is important to reduce the cost of labor along with its safety and insurance consequences, and when one wishes to have the digestion procedure match the precision and reproducibility of the analysis, then the focused, open-vessel concept should be considered.

1.7.4.7 Microwave Muffle Furnace. The microwave muffle furnace heats a wide variety of materials at temperatures up to 1200°C. Many materials are ashed in 5 to 10 min. Other uses include drying, fusions, loss on ignition, wax burnouts, and heat-treating processes. Results are equivalent to those obtained using conventional muffle furnaces. A temperature controller monitors temperature and switches microwave power on and off to maintain selected temperatures to within ±3°C. An exhaust system removes smoke, vapors, and gases produced inside the furnace.

A quartz fiber crucible accelerates the ashing process. The porous material rapidly cools and allows many sample types to be reweighed in 60 s or less after removal from the furnace. Crucibles fabricated of ceramic, graphite, and platinum also can be used.

1.7.5 Other Dissolution Methods

Before leaving the discussion of dissolution methods, two additional means for dissolving particular types of samples should be mentioned. Dissolution with complexing agents is treated in Table 1.20,

TABLE 1.20 Dissolution with Complexing Agents

Substance; amount	Complexing agent
AgCl; 2 g	20 mL concentrated NH_3
AgI; 0.07 g	15 mL 9*M* NH_4SCN + 3 mL of 3% $NH_2OH \cdot HCl$
AgCl, AgBr, AgI; 1 g	10 mL 4.5*M* KI
Ag halides; 10–300 mg	$K_3[Ni(CN)_4]$ (spatula tip) + 7*M* NH_3
Ag_3PO_4; 40–600 mg	0.1*M* $K_2[Ni(CN)_4]$ (excess)
$BaSO_4$; 10–120 mg	0.02*M* EDTA* in excess
$BaSO_4$; 2–6 mg	10 mL 0.01*M* EDTA + 150 mL 0.4*M* NH_3
$CaCO_3$; in 1 g fluorspar concentrate	10 mL 0.05*M* EDTA + 10 mL water
GeO_2 (hexagonal); 2 g	50 mL NaK tartrate (20% w/v)
HgS; 0.37–1.12 g	5 mL HI (57%) per g HgS
HgS (cinnabar); 0.25–1 mg	Na_2S solution (360 g Na_2S, 9 H_2O + 10 g NaOH/liter; excess)
$PbSO_4$; 30–100 mg sulfates	5 portions 5 mL 3*M* ammonium acetate (extraction from alkaline earth)
Basic lead sulfate; 1 g	30 mL ammonium acetate solution (30% w/v)
$PbCO_3$ (cerussite); 1 g	Ammoniacal ascorbic acid
$PbMoO_4$; 1 g	25 mL NaK tartrate solution (50%)
$RaSO_4$	3–5% ammoniacal EDTA solution with boiling
Se; 0.5–1 g	100 mL KCN solution (2%)
UF_3 hydrate (?)	0.2 *M* tridecylammonium fluoride in CCl_4

* EDTA denotes ethylenediaminetetraacetic acid.

TABLE 1.21 Dissolution with Cation Exchangers (H Form)

Compound; amount	Amount of exchanger	Conditions
$BaSO_4$; 0.25 g	10 g	100 mL H_2O; 80–90°C; 12 h
$CaSO_4 \cdot 2H_2O$; 0.3–0.4 g	10 g	50 mL H_2O; 25°C; 15 min
$CaHPO_4 \cdot 2H_2O$; 0.2–0.3 g	5 g	25 mL H_2O; 25°C; 15 min
$Ca_3(PO_4)_2$; 0.2 g	5 g	25 mL H_2O; 25°C; 15 min
Apatite; 0.05 g	5–10 g	35 mL H_2O; 80°C; 1–16 h
Phosphorite; 100 g	1200 mL*	400 mL H_2O; 50°C; 20 min
CaCO, dolomite; 0.2–0.3 g	50 mL	50 mL H_2O; 25°C; 10 min
$CaCO_3 + CaSO_4$; 0.2–0.3 g	50 g	300 mL H_2O; 90°C; 30 min
$MgHPO_4$; 0.2–0.3 g	5 g	25 mL H_2O; 25°C; 10–15 min
Mg borate (ascharite); 50 g	400 mL	350 mL H_2O; 70–80°C; 50–60 min
$PbSO_4$; 0.10–0.25 g	10 g	50 mL H_2O; 90–100°C; 30 min
$PbCl_2$; 0.2–0.4 g	5–10 g	50 mL H_2O; 25°C; 15 min
$RaSO_4$, $BaSO_4$; 10 mg	10 mL†	60°C; 30 min
$SrSO_4$; 0.25 g	10 g	100 mL H_2O; 80–90°C; 20 min
UO_2HPO_4; 1–2 mg	0.3 mL	0.1 mL 0.04 *M* HNO_3; 40°C; 1 mim

* Addition of NaCl.
† Cation plus anion exchanger (1:1) in H^+ and OH^- form, respectively.

TABLE 1.22 Solvents for Polymers

Substance	Solvent
Polystyrene	4-Methyl-2-hexone
Cellulose acetate	4-Methyl-2-hexone
Cellulose acetate-butyrate	4-Methyl-2-hexone
Polyacrylonitrile	Dimethylformamide
Polycarbonate	Dimethylacetamide
Poly(vinyl chloride)	Dimethylacetamide
Poly(vinyl chloride)/poly(vinyl acetate) copolymer	Cyclohexanone
Polyamides	60% formic acid
Polyethers	Methanol

and dissolution with cation exchangers in the hydrogen form are outlined in Table 1.21. In addition, solvents for polymers are summarized in Table 1.22.

1.8 FILTRATION

1.8.1 Introduction

Filtration is the process of removing material, often but not always a solid, from a substrate in which it is suspended. This process is a physical one; any chemical reaction is inadvertent and normally unwanted. Filtration is accomplished by passing the mixture to be processed through one of the many available sieves called filter media. These are of two kinds: surface filters and depth filters.

With the surface filter, filtration is essentially an exclusion process. Particles larger than the filter's pore or mesh dimensions are retained on the surface of the filter; all other matter passes through. Examples are filter papers, membranes, mesh sieves, and the like. These are frequently used when the solid is to be collected and the filtrate is to be discarded. Depth filters, however, retain particles both on their surface and throughout their thickness; they are more likely to be used in industrial processes to clarify liquids for purification.

In the laboratory, filtration is generally used to separate solid impurities from a liquid or a solution or to collect a solid substance from the liquid or solution from which it was precipitated or recrystallized. This process can be accomplished with the help of gravity alone or it can be accelerated by using vacuum techniques, to be discussed later.

Filtration efficiency depends on the correct selection of the method to be used, the various pieces of apparatus available, the utilization of the filter medium most appropriate for the particular process, and the use of the correct laboratory technique in performing the manipulations involved. Although the carrier liquid is usually relatively nonreactive, it is sometimes necessary to filter materials from high alkaline or acidic carrier liquids or to perform filtration under other highly corrosive conditions. A variety of filter media exists from which it is possible to select whichever medium best fits the particular objectives and conditions of a given process.

1.8.2 Filter Media

1.8.2.1 Paper. The general properties of filter papers from Whatman and from Schleicher and Schuell are listed in Table 1.23. There are qualitative grades, low-ash or ashless quantitative grades, hardened grades, and glass-fiber papers. The proper filter paper must be selected with regard to porosity and residue (or ash). All grades of filter paper are manufactured in a variety of sizes and in several degrees of porosity. Select the proper porosity for a given precipitate. If too coarse a paper is used, very small crystals may pass through, while use of too fine a paper will make filtration unduly slow. The main objective is to carry out the filtration as rapidly as possible while retaining the precipitate on the paper with a minimum loss.

1.8.2.1.1 Qualitative-Grade Paper. Papers of this type will leave an appreciable amount of ash upon ignition (about 0.7 to 1 mg from a 9-cm circle). Consequently, these papers are unsuitable for quantitative analysis when precipitates are ignited on the paper and weighed. Qualitative-grade papers are widely used for clarification of solutions, for filtration of precipitates that will be dissolved, and for general nonquantitative separations of precipitates from solution. Whatman grade 1 is a general purpose filter for routine laboratory applications. Grade 2 is slightly more retentive than grade 1 but with a corresponding increase in filtration time. Grade 3 is double the thickness of grade 1 with finer particle retention and possesses excellent loading capacity without clogging. The extra thickness provides increased wet strength and makes this grade suitable for using flat in Büchner funnels. Grade 4 has an extremely fast rate of filtration and has excellent retention of coarse particles and gelatinous precipitates. Grade 4 finds use as a rapid filter for routine cleanup of biological fluids or organic extracts. It is also used when high flow rates in air pollution monitoring are required, and the collection of fine particles is not critical. Grade 5 offers the maximum retention of fine particles in the qualitative range. Grade 6 offers a filtration rate that is twice as fast as grade 5, yet with almost as fine particle retention and is often specified for boiler water analysis applications. Grades 113 and 114 have extremely high loading capacity with a particle retention that is ideal for use with coarse or gelatinous precipitates. These grades are extremely strong and are the thickest fiber paper offered with a crepe surface. Grade 114 is half the thickness of grade 113 and has a smooth surface.

1.8.2.1.2 Low-Ash or Ashless Quantitative-Grade Papers. Upon ignition these papers leave little or no ash. The residue left by an 11-cm circle of low-ash paper may be as low as 0.06 mg, whereas the residue from a ashless-grade paper is 0.05 mg or less. For most quantitative work this small residue is negligible. Grade 40 is a general purpose ashless filter paper that possesses medium speed and medium retention. Grade 41 is the fastest ashless paper and is recommended for coarse particles or gelatinous precipitates. It is used in quantitative air pollution analysis as a paper tape for impregnation when gaseous compounds are determined at high flow rates. Grade 42 is used for critical gravimetric analysis that requires retention of fine particles. The retention properties of grade 43 are intermediate between grades 40 and 42, but twice as fast as grade 40. Grade 44 is the thin version of grade 42 with lower ash per circle yet is almost twice as fast in filtration rate.

TABLE 1.23 General Properties of Filter Papers and Glass Microfibers

Whatman grade	Particle retention, μm	Initial filtration, s/100 mL	Ash, %	Thickness, mm	Weight, $g \cdot m^{-2}$	Burst Wet, psi	Burst Dry, psi
			Qualitative filter papers				
1	11	40	0.06	0.18	87	0.25	14
2	8	55	0.06	0.19	97	0.29	16
3	6	90	0.06	0.39	185	0.40	28
4	20–25	12	0.06	0.21	92	0.22	10
5	2.5	250	0.06	0.20	100	0.40	25
6	3	175	0.2	0.18	100	0.25	18
			General purpose and wet strengthened				
113	30	8	NA*	0.42	125	9	29
114	25	12	NA	0.19	75	8	20
			Quantitative—ashless				
40	8	75	0.010	0.21	95	0.29	16
41	20–25	12	0.010	0.22	85	0.22	10
42	2.5	240	0.010	0.20	100	0.40	25
43	16	40	0.010	0.22	95	0.29	13
44	3	175	0.010	0.18	80	0.29	17
			Quantitative—hardened low ash				
50	2.7	250	0.025	0.12	97	6	32
52	7	55	0.025	0.18	96	7	29
54	20–25	10	0.025	0.19	90	7	19
			Quantitative—hardened ashless				
540	8	55	0.008	0.16	85	7	27
541	20–25	12	0.008	0.16	78	7	17
542	2.7	250	0.008	0.15	96	6	32
			Borosilicate glass microfiber				
934-AH	1.5	NA		0.33	64	0.54	
GF/A 1.6	13.0			0.26	53	0.29	
GF/B 1.0	5.5			0.68	143	0.47	
GF/C 1.2	10.5			0.26	53	0.29	
GF/D 2.7	16.5			0.68	120	0.47	
GF/ F 0.7	6.0			0.42	75	0.33	

1.8.2.1.3 Hardened-Grade Papers. Hardened papers are designed for use in vacuum filtrations on Büchner or three-piece filter funnels. These papers possess great wet strength and hard lintless surfaces, and will withstand wet handling and removal of precipitates by scraping. They are available in low-ash, ashless, and regular grades. Grade 54 offers fast filtration with both coarse and gelatinous precipitates. Grades 540, 541, and 542 are extremely pure and strong filter papers with a hard surface. They offer high chemical resistance to strong acids and alkalies. In other respects they parallel the properties of grades 40, 41, and 42.

1.8.2.2 Borosilicate Glass-Fiber and Quartz Filters. Glass-fiber filters are produced from borosilicate glass. Their soft texture and high water absorption provide excellent sealing in most types of filter holders. These filters give a combination of fine retention, rapid filtration, high

TABLE 1.23 **General Properties of Filter Papers and Glass Microfibers (*Continued*)**

Schleicher and Schuell grade	Precipitates retained	Initial filtration, s/10 mL	Ash, %	Thickness, mm	Weight, $g \cdot m^{-2}$	Wet strength
			Qualitative filter papers			
591-A	Med. Fine	48	0.024	0.175	83	Low
593-A	Fine	63	0.026	0.340	179	Low
597	Med. Fine	34	0.023	0.193	83	Low
598	Coarse	21	0.110	0.350	148	Low
602	Fine	125	0.039	0.165	83	Low
604	Coarse	14	0.035	0.198	74	Low
			General purpose			
405	Fine	49		0.457	220	Very high
413	Very fine	80		0.289	137	Very high
428	Fine	53		0.337	122	High
470	Gelatinous	8		0.886	303	Low
477	Coarse	8		0.228	70	High
478	Coarse	16		0.492	155	Very high
520-B	Gelatinous	8		0.401	152	High
595	Med. fine	20		0.137	63	Very high
596	Fine	52		0.147	54	High
610	Coarse	10		0.149	62	High
			Quantitative—ashless			
589-Black	Coarse	15	0.005	0.193	69	Low
589-Green	Coarse	25	0.010	0.342	146	Low
589-White	Med. fine	48	0.005	0.200	81	Low
589-Blue	Fine	147	0.005	0.180	81	Low
589-Red	Fine	151	0.007	0.139	81	Low
			Quantitative—hardened low ash			
507	Very fine	205	0.007	0.101	80	High
589-1H	Gelatinous	7	0.007	0.180	60	High
589-BH	Coarse	9	0.006	0.218	72	High
589-WH	Med. fine	44	0.005	0.195	81	High
			Glass fiber			
24	Finest	15		0.810	80	Low
25	Finest	17		0.294	73	Low
29	Finest	13		0.246	60	Low
30	Finest	16		0.241	60	Low

* NA denotes not available.

particle-loading capacity, and inertness that is not found in any cellulose-paper filter. They must always be used flat because folding can rupture the glass fiber matrix, which leads to cracks and thus poor retention.

1.8.2.2.1 Grade 934-AH. This type of filter paper is used in monitoring water pollution for suspended solids, in cell harvesting, in liquid scintillation counting, and in air-pollution monitoring.

1.8.2.2.2 Grade GF/A. This filter paper is used for high-efficiency, general-purpose laboratory filtration, for water-pollution monitoring of effluents, for filtration of water, algae, and bacteria cultures, analysis of foodstuffs, for protein filtration, and for radioimmunoassay of weak beta emitters.

It is recommended for gravimetric determination of airborne particulates, for stack sampling, and for absorption methods in air-pollution monitoring.

1.8.2.2.3 Grade GF/B. This filter paper is three times thicker than grade GF/A and has a higher wet strength and increased loading capacity. It can be used as a finely retentive membrane prefilter. It is useful where liquid clarification or solids quantification is required for heavily loaded fine-particulate suspensions.

1.8.2.2.4 Grade GF/C. This filter paper is slightly more retentive than grades GF/A and 934-AH. It is the standard filter in many parts of the world for the collection of suspended solids in potable water and natural waters and in industrial wastes. It is used in cell harvesting and in liquid scintillation counting when more retentive, but medium loading capacity, filters are required.

1.8.2.2.5 Grade GF/D. This paper is considerably faster in flow rate and overall filtration speed than are cellulose filter papers of similar particle retention. It is designed as a membrane prefilter and provides good protection for finely retentive membranes. It can be used in combination with grade GF/B to provide very efficient graded, prefilter protection for membranes.

1.8.2.2.6 Grade GF/F. Unlike membrane filters with a comparable retention value, grade GF/F has a very rapid flow rate and an extremely high loading capacity. It is very effective for filtering finely precipitated proteins, and can be used in conjunction with grade GF/D as a prefilter for the successful clarification of biochemical solutions and fluids and nucleic acids.

In addition to the filters listed in Table 1.23, Whatman EPM 2000, manufactured from pure borosilicate glass, is used in high-volume air-sampling equipment for the collection of atmospheric particulates and aerosols. Chemical analysis of trace pollutants can be done with a minimum of interference or background. Whatman quartz filters (QM-A) are available for air sampling of acidic gases, stacks, flues, and aerosols, particularly at temperatures up to 550°C. Because only low levels of alkaline earth metals are present in the filters, artifact products of sulfates and nitrates (from SO_2 and NO_2) are virtually eliminated. Grade QM-B is twice as thick as grade QM-A, which increases particle retention to 99.9985%.

1.8.2.3 Membrane Filters. There are two types of membrane filters—surface and depth. Surface membrane filters are used for final filtration or prefiltration, whereas a depth membrane filter is generally used in clarifying applications where quantitative retention is not required or as a prefilter to prolong the life of a downstream surface membrane filter.

Membrane filters are thin polymeric (plastic) structures with extraordinarily fine pores. Certain kinds of polymers can be prepared, by solvent-casting methods, into a film having a very unusual structure. One side of such a film, whose overall thickness is several thousandths of a centimeter, has an exceedingly thin layer of defect-free polymer about 0.1 to 1.0 μm thick, while the balance of the film consists of a fine-textured open-celled sponge. A thick "anisotropic" membrane, whose selective barrier properties reside solely in the this surface skin, displays nearly 1000 times higher permeability than a uniform film of the same thickness, and without loss in permselectivity. Adjustment of surface pore size from 0.3 to 10 μm is possible. Such filters are distinctive in that they remove all particulate matter or microorganisms larger than the filter pores from a gas or liquid stream that passes through them. The filtrate is ultraclean and /or sterile. Membrane filters are available in a wide variety of pore sizes in a number of different polymeric materials: hydrophilic nylon for aqueous samples, polyvinylidene fluoride (PVDF) and polysulfone for low-protein binding samples, polypropylene for low extractables with aggressive solutions, and glass microfiber for high-debris (dirty) samples (Table 1.24). For sterilization, membrane pore sizes are usually 0.2 μm. For clarification of solutions a pore size of 0.45 μm is usually used. Separation of foreign particles in sampling waters and gases utilizes a pore size of 3 to 5 μm. A membrane selection guide is given in Table 1.25.

1.8.2.4 Hollow-Fiber Membrane Cartridges. Ultrafiltration, sometimes called *reverse osmosis*, involves forcing a solution under pressure through a membrane. The usual membrane for ultrafiltration is an asymmetric cellulose diacetate or polysulfone type. It is constructed with a thin skin of polymer on a porous, spongy base. The driving force is the energy due to pressure difference. It is the solvent rather than the solute that moves through a membrane and this is against rather than with a concentration gradient. Any unwanted solutes and colloidal material unable to permeate the pores of the membrane are rejected.

TABLE 1.24 Membrane Filters

Filter pore size, μm	Maximum rigid particle to penetrate, μm	Filter pore size, μm	Maximum rigid particle to penetrate, μm
14	17	0.65	0.68
10	12	0.60	0.65
8	9.4	0.45	0.47
7	9.0	0.30	0.32
5	6.2	0.22	0.24
3	3.9	0.20	0.25
2	2.5	0.10	0.108
1.2	1.5	0.05	0.053
1.0	1.1	0.025	0.028
0.8	0.95		

TABLE 1.25 Membrane Selection Guide

Membrane type	Applications
Cellulose acetate	Very low protein retention. With a glass prefilter this membrane is ideal for filtration of tissue culture media, general biological sample filtration, clarification, and sterilization of most aqueous-based samples. Not resistant to most solvents. A glass prefilter increases filtrate yield up to 300%.
Borosilicate glass	Ideal as a prefilter when used in conjunction with a membrane such as cellulose acetate or nylon.
Nylon	For general filtration and sterilization. Has inherent hydrophilic characteristics and works well with all aqueous and most solvent-based samples. Nylon is excellent for high-performance liquid chromatography (HPLC) sample preparation. Nylon should *not* be autoclaved.
Polysulfone and polyvinylidene fluoride (PVDF)	Both membranes exhibit very low protein binding. Both can be used for general biological sample filtration and sterilization. PVDF is especially useful in HPLC sample preparation and is highly resistant to most solvents. Polysulfone has very limited resistance to solvents; it is generally used for aqueous-based biological samples. Both membranes exhibit good flow-rate characteristics.
Polytetrafluoroethylene (PTFE)	An inherently hydrophobic membrane that is ideal for filtration of gas, air, or solvents. Highly resistant to solvents, acids, and alkalies. Useful for fluid sterilization, for chromatography, and for clarification of nonaqueous solutions. Although this membrane is hydrophobic, it can be made hydrophilic by prewetting the membrane with alcohol and then flushing with deionized water.
Ultrafilter membranes	Ultrafilters (molecule weight cutoff filters) utilize cellulose triacetate or polysulfone membranes. These membranes are rated as 5 000, 10 000, 20 000, 30 000, 100 000, or 300 000 molecular weight cutoff. Ultrafilters are ideal in protein research for desalting, in sample concentration for protein electrophoresis, in deproteinization, and for buffer exchange. Ultrafilters are used for the recovery from biological samples of peptides, bacteria, yeast, virus, and certain enzymes, and for the recovery of low-molecular-weight materials from culture media, fermentation broths, or cell lysates.
Nitrocellulose	Exhibits high protein retention. Excellent for binding DNA and RNA when using this membrane alone for blotting. Ideal membrane when used as a syringe filter device for microbiological work.

Source: *Alltech Bulletin No. 190A*, 1991.

Hollow-fiber ultrafiltration membranes eliminate liquid and micromolecular species while retaining suspended particles or macromolecules. The anisotropic membrane design of the hollow-fiber membrane provides an uninterrupted polymer phase. The very thin (0.1 to 1 μm) surface skin contains the pore structure, while the integral substrate has a continuous macroporous structure of larger pores and acts only as a support for the skin. The limitations of conventional filtration are overcome through the tangential cross-flow capability. Feed material sweeps tangentially across the upstream surface of the membrane as filtration occurs. All retained particles remain on the membrane surface, and smaller particles are swept from the larger pores of the support structure, thereby eliminating depth fouling. Rejected species on the active surface of the membrane are continuously removed and returned to the feed stream. Species smaller than the membrane pore size flow through the skin and are removed from the larger pore cavities in the support structure. The membranes are configured in a "shell and tube" cartridge design. The cartridge can be backflushed easily.

Hollow-tube geometry allows pressure and flow to be easily varied to control membrane surface effects and to process shear-sensitive samples. Because the actual discrimination barrier is extremely thin, high filtration rates are easily achieved. To process dilute-to-moderately viscous solutions, 0.5- and 1.1-mm internal-diameter cartridges are used. For more viscous solutions 1.5- and 1.9-mm internal-diameter cartridges are recommended. A 20-cm-long cartridge handles most laboratory processing.

Filtrations can be modified through the use of seven different membrane pore sizes, 1 to 20 μm. Nominal molecular weight cutoffs range from 2000 to 500000 daltons, as shown in Table 1.26.

TABLE 1.26 Hollow-Fiber Ultrafiltration Cartridge Selection Guide

Fiber internal diameter, mm	Molecular-weight cutoff	Surface area, m^2	Length, cm	Recommended recirculation rate, L/min	Water flux, $L \cdot min^{-1}$
0.5	2 000	0.07	20.3	0.3–15	0.09–0.12
	5 000				0.21–0.27
	10 000				0.32–0.42
	50 000				0.53–0.69
	100 000				0.85–1.11
	2 000	0.17	45.7	2.8–8	0.20–0.26
	5 000				0.53–0.68
	10 000	0.19			0.79–1.03
	50 000				1.18–1.53
	100 000				1.92–2.50
1.1	5 000	0.03	20.3	0.6–18	0.08–0.10
	10 000				0.12–0.16
	30 000				0.12–0.16
	50 000				0.20–0.47
	100 000				0.32–0.42
	500 000				0.32–0.42
	5 000	0.09	45.7	2–16	0.26–0.34
	10 000				0.39–0.51
	30 000				0.27–0.35
	50 000				0.45–1.06
	100 000				0.71–0.92
1.5	10 000	0.03	20.3	6–29	0.12–0.16
	50 000				0.42–0.55
	10 000	0.09	47.5	5–26	0.39–0.51
	50 000				0.95–1.24
1.9	500 000	0.02	20.3	9–40	0.26–0.34
	500 000	0.07	45.7	6–30	0.84–1.09

Source: *Fine Tune*™ *Your Filtration*, Harp™ Hollow Fiber Ultrafiltration Membranes, Supelco, Bellefonte, PA, 1989.

TABLE 1.30 Pipette Capacity Tolerances

Volumetric transfer pipettes			Measuring and serological pipettes	
Capacity, mL	Tolerances,* ± mL Class A	Tolerances,* ± mL Class B	Capacity, mL	Tolerances,† ± mL Class B
0.5	0.006	0.012	0.1	0.005
1	0.006	0.012	0.2	0.008
2	0.006	0.012	0.25	0.008
3	0.01	0.02	0.5	0.01
4	0.01	0.02	0.6	0.01
5	0.01	0.02	1	0.02
10	0.02	0.04	2	0.02
15	0.03	0.06	5	0.04
20	0.03	0.06	10	0.06
25	0.03	0.06	25	0.10
50	0.05	0.10		
100	0.08	0.16		

*Accuracy tolerances for volumetric transfer pipettes are given by ASTM Standard E969 and Federal Specification NNN-P-395.

†Accuracy tolerances for measuring pipettes are given by Federal Specification NNN-P-350 and for serological pipettes by Federal Specification NNN-P-375.

After use the pipette should be thoroughly rinsed with distilled water or appropriate solvent, and dried. Store in a protective can.

1.9.3 Micropipettes

Micropipettes are available in various styles. Many are repetitive. The air-displacement design provides superior pipetting performance and eliminates carryover and contamination problems often associated with positive displacement systems (those with a plunger to displace the liquid). Some have a fixed capacity while others have multivolume capabilities that feature an adjustable, spring-loaded rotating plunger button at the top of the pipette for selection of different volumes. With digital pipettes the volume is adjusted by means of a ratchet mechanism built into the control button. Each twist of the button produces an audible click, assuring that a new volume setting is locked into place. Each of the micropipettes described use disposable polypropylene pipette tips; some feature automatic tip ejection. Typical accuracy and precision for micropipettes is shown in Table 1.31.

TABLE 1.31 Tolerances of Micropipettes (Eppendorf)

Capacity, μL	Accuracy, %	Precision, %	Capacity, μL	Accuracy, %	Precision, %
10	1.2	0.4	100	0.5	0.2
40	0.6	0.2	250	0.5	0.15
50	0.5	0.2	500	0.5	0.15
60	0.5	0.2	600	0.5	0.15
70	0.5	0.2	900	0.5	0.15
80	0.5	0.2	1000	0.5	0.15

TABLE 1.32 Burette Accuracy Tolerances

Capacity, mL	Subdivision, mL	Accuracy, ± mL	
		Class A* and precision grade	Class B and standard grade
10	0.05	0.02	0.04
25	0.10	0.03	0.06
50	0.10	0.05	0.10
100	0.20	0.10	0.20

*Class A conforms to specifications in ASTM Standard E694 for standard taper stopcocks and in ASTM Standard E287 for Teflon or poly(tetrafluoroethylene) stopcock plugs. The 10-mL size meets the requirements for ASTM Standard D664.

1.9.4 Burettes

Burettes, like measuring pipettes, deliver any volume up to their maximum capacity. Burettes of the conventional type must be manually filled; usual sizes are 10 and 50 mL. They are made from precision-bore glass tubing. Permanent graduation marks are etched into the glass and filled with fused-on enamel of some color. A blue stripe on the outside back is a distinct aid in meniscus reading. Stopcocks vary from the ground-glass barrel and glass plug (or plastic analog) to a separable polyethylene and Teflon valve assembly. One option involves an automatic self-zeroing burette with a dual stopcock plug, one to control filling and the other dispensing. Burette tolerances are given in Table 1.32.

SECTION 2
PRELIMINARY SEPARATION METHODS

2.1 COMPLEX FORMATION, MASKING, AND DEMASKING REACTIONS

Complex formation is important in two ways. It may produce a species that has more useful characteristics for a particular chemical separation method, and alternatively, the concentrations of particular species can be diminished to levels below those at which they interfere in reactions designed to separate other molecules or ions.

2.1.1 Complex Equilibria Involving Metals[1]

Practically every metal forms complex ions of some kind. Some metals form more numerous and more stable complexes than others, but in the reactions of analytical chemistry, the possibility of complex formation must always be considered. At least two sets of equilibria are involved in the process of complex formation. There is competition between solvent molecules and the ligand for the metal ion and simultaneous competition between protons and the metal ion for the ligand. Metal ions in aqueous solution possess solvent molecules in their primary solvation shell. Attraction between them is weak usually, and the number of solvent molecules immediately surrounding each metal ion is variable. However, in the transition-metal ions and higher-valent metal ions, definite complexes such as $Cu(H_2O)_4^{2+}$ and $Al(H_2O)_6^{3+}$ exist in aqueous solutions. For this reason complex formation in aqueous solutions is really a replacement process in which

[1] A. Ringbom and E. Wänninen, "Complexation Reactions," Chap. 20 in I. M. Kolthoff and P. J. Elving, eds., *Treatise on Analytical Chemistry*, Part I, Vol. 2, 2d ed., Wiley-Interscience, New York, 1979. See also K. Ueno, T. Imamura, and K. L. Cheng, *CRC Handbook of Organic Analytical Reagents*, 2d ed., CRC Press, Boca Raton, Florida, 1992.

solvent molecules in the coordination sheath surrounding a metal ion are replaced stepwise by other ligands.

Reaction rates must also be considered. The addition of aqueous ammonia to a solution of a silver salt will convert the hydrated silver ion into an ammonia complex in no more time than it takes to mix the solutions together. If acid is added, the silver ammonium complex decomposes to give back hydrated silver ions. On the other hand, the direct reaction between Fe(III) and CN^- is too slow to be useful, although the stability of $Fe(CN)_6^{3-}$, once formed, is well known. Obviously the reaction equation represents only the initial and final states and tells nothing about what goes on in between.

Taube[2] classified a large number of complexes on the basis of the rates of formation of complexes and substitution reactions. Complexes that come to equilibrium rapidly with their dissociation products are called *labile* complexes. Those that decompose very slowly and never come to equilibrium with their components are called *inert* complexes. The inert complexes are characterized by a structure in which all the inner *d* orbitals are occupied by at least one electron. These include complexes of V(II), Cr(III), Mo(III), Co(III), and some of the complexes of Fe(II) and Fe(III). The relatively few inert complexes involving outer orbital configurations are characterized by a central ion of high charge, such as SiF_6^{2-}.

2.1.1.1 Mononuclear Complexes. Only reactions leading to mononuclear compounds, ML_n, where M denotes the central metal ion and L the complexing ligand, will be considered in this brief treatment. Ringbom[3] treats the problem of mononuclear and polynuclear compounds in considerable detail.

The simplest case is represented by the reaction, for which $n = 1$,

$$M + L \rightleftarrows ML \tag{2.1}$$

For sake of simplicity here and in the following discussion, charges will be omitted as will solvated species. The stability or formation constant K_f of the reaction is

$$K_f = \beta_1 = [ML]/[M][L] \tag{2.2}$$

In older literature the inverse values, instability or dissociation constants, were used. When the complex formation occurs stepwise, each step is denoted by β's. The overall stability constant of the complex ML_n is the product of the consecutive stepwise stability constants:

$$K_f = \beta_1\beta_2 \cdots \beta_n \qquad \text{or} \qquad K_n = k_1k_2 \cdots k_n \tag{2.3}$$

Methods for the determination of stepwise and overall stability constants are described by Rossotti and Rossotti.[4]

The distribution of the various complexes with a given ligand can be obtained as a set of fractions Φ_0 to Φ_n. Each represents the ratio of the concentration of metal–ligand species to the total concentration of the central metal ion, C_m. Thus

$$\Phi_0 = [M]/C_m \tag{2.4}$$

$$\Phi_1 = [ML]/C_m \tag{2.5}$$

$$\cdots$$

$$\Phi_n = [ML_n]/C_m \tag{2.6}$$

$$\Phi_0 + \Phi_1 + \cdots + \Phi_n = 1 \tag{2.7}$$

---

[2] H. Taube, *Chem. Rev.* **50**:69 (1952).

[3] A. Ringbom, *Complexation in Analytical Chemistry*, Wiley, New York, 1963.

[4] F. J. C. Rossotti and H. Rossotti, *The Determination of Stability Constants*, McGraw-Hill, New York, 1961.

If the stepwise stability constants k_n are known, it is possible to calculate the distribution of the various complexes solely from the concentration of ligand:

$$\Phi_0 = 1/(1 + k_1[L] + k_1k_2[L]^2 + \cdots + k_n[L]^n) \tag{2.8}$$

$$\Phi_1 = k_1[L]/(1 + k_1[L] + k_1k_2[L]^2 + \cdots + k_n[L]^n) \tag{2.9}$$

$$\cdots$$

$$\Phi_n = k_n[L]^n/(1 + k_1[L] + k_1k_2[L]^2 + \cdots + k_n[L]^n) \tag{2.10}$$

In this series of expressions, each term in the denominator becomes in turn the numerator. If the ligand concentration is small, the terms with powers of $[L]$ lower than the coordination number of the metal ion are predominant, whereas if the ligand concentration is large, the terms with power of $[L]$ higher than the coordination number predominate. The overall cation concentration has relatively little effect on the fractions. Individual stepwise stability constants of metal-ion complexes are tabulated in *Lange's Handbook of Chemistry.*[5] An abbreviated version is given in Tables 2.1 and 2.2.

2.1.1.2 Conditional Constants. A ligand is often an anion or a neutral molecule with basic properties so that, at sufficiently low pH values, it becomes extensively protonated, with consequent reduction in its complex-forming ability. It is convenient to define a *conditional formation constant* K', which is really not constant but depends on the experimental conditions:

$$K' = K_{M'L'} = [ML]/[M'][L'] \tag{2.11}$$

In this expression $[M']$ denotes the concentration of all the metal in solution that has not reacted with some complexing agent. In a corresponding manner, $[L']$ represents the concentration of free ligand under the specified operating value of pH, where

$$[L'] = \alpha C_L \tag{2.12}$$

The fractions α_0, α_1, and α_2 of ligand in the forms H_2L, HL^-, and L^{2-}, respectively, are given by the expressions:

$$\alpha_0 = [H_2L]/C_L = [H^+]^2/([H^+]^2 + K_1[H^+] + K_1K_2) \tag{2.13}$$

$$\alpha_1 = [HL^-]/C_L = K[H^+]/([H^+]^2 + K_1[H^+] + K_1K_2) \tag{2.14}$$

$$\alpha_2 = [L^{2-}]/C_L = K_1K_2/([H^+]^2 + K_1[H^+] + K_1K_2) \tag{2.15}$$

The conditional constant thus gives the relationship between the quantities in which the analyst is actually interested, namely, the concentration of the product formed $[ML]$, the total concentration of uncomplexed (in terms of the principal reaction) metal $[M']$, and the total concentration of the uncomplexed reagent $[L']$.

2.1.1.3 Complexation Ability of Ligands. The most important points for the characterization of a ligand are the nature and basicity of its ligand atom.

[5] J. A. Dean, ed., *Lange's Handbook of Chemistry*, 14th ed., McGraw-Hill, New York, 1992, pp. 8.83–8.103.

TABLE 2.1 Overall Formation Constants for Metal Complexes with Organic Ligands

Metal ions	log K_f
Acetate	
Cerium(III)	3.18
Chromium(III)	4.72
Copper(II)	3.20
Iron(II)	8.3
Indium(III)	9.08
Lanthanum	2.95
Lead(II)	8.5
Mercury(II)	8.43
Titanium(III)	15.4
Yttrium	3.38
Acetylacetone	
Aluminum	15.5
Beryllium	14.5
Cadmium	6.66
Cerium(III)	12.65
Chromium(II)	11.7
Cobalt(II)	9.54
Copper(II)	16.34
Iron(II)	8.67
Iron(III)	26.7
Gallium	23.6
Hafnium	28.1
Indium	15.1
Lanthanum	11.90
Magnesium	6.27
Manganese	7.35
Nickel	13.09
Palladium	27.1
Plutonium(IV)	34.1
Rare earths	ca. 13.9
Scandium	15.2
Thorium	26.7
Uranium(IV)	29.5
Uranyl(2+)	14.19
Vanadyl(IV)(2+)	15.79
Vanadium(II)	14.7
Yttrium	13.9
Zinc	8.81
Zirconium	30.1
Aurintricarboxylic acid	
Beryllium	4.54
Copper(II)	8.81
Iron(III)	4.68
Thorium	5.04
Uranyl(VI)(2+)	4.77
Citric acid	
Aluminum	20.0
Barium	2.98
Beryllium	4.52
Calcium	4.68
Cadmium	11.3
Cerium(III)	9.65
Cobalt(II)	12.5
Copper(II)	14.2
Iron(II)	15.5
Iron(III)	25.0
Lanthanum	9.45
Lead(II)	6.50
Magnesium	3.29
Manganese(II)	3.67
Nickel(II)	14.3
Rare earths	ca. 9.8
Silver	7.1
Strontium	2.8
Uranyl(VI)(2+)	10.8
Zinc	11.4
4,5-Dihydroxybenzene-1,3-disulfonic acid (Tiron)	
Aluminum	33.5
Barium	14.6
Calcium	5.80
Cadmium	13.29
Cobalt(II)	14.4
Copper(II)	23.7
Iron(III)	46.9
Lanthanum	12.9
Lead(II)	18.28
Magnesium	6.9
Manganese(II)	8.6
Nickel(II)	14.90
Strontium	4.55
Uranyl(VI)(2+)	15.90
Vanadyl(IV)(2+)	15.88
Zinc	16.9
2,3-Dimercaptopropan-1-ol (BAL)	
Iron(II)	15.8
Iron(III)	30.6
Manganese(II)	10.43
Nickel	22.78
Zinc	23.3
Dimethylglyoxime	
Cadmium	10.7
Cobalt(II)	18.94
Copper(II)	33.44
Iron(II)	7.25
Lanthanum	12.5
Nickel	11.16
Lead	7.3
Zinc	13.9
2,2′-Dipyridyl	
Cadmium	10.47
Cobalt(II)	17.59
Chromium(II)	14.0
Copper(I)	14.2
Copper(II)	17.08
Iron(II)	17.45
Lead	3.0
Manganese(II)	11.47
Mercury(II)	19.54
Nickel(II)	16.46
Silver	7.15
Titanium(III)	25.28
Vanadium(II)	13.1
Zinc	13.63
Ethylenediamine	
Cadmium	12.09
Cobalt(II)	13.94
Cobalt(III)	48.69
Chromium(II)	9.19
Copper(I)	10.8
Copper(II)	20.00
Iron(II)	9.70
Manganese(II)	5.67
Mercury(II)	23.3
Nickel	18.33
Palladium(II)	26.90
Silver	7.70
Vanadium(II)	8.8
Zinc	14.11
Ethylenediamine-*N*,*N*,*N*′,*N*′-tetraacetic acid	
Aluminum	16.11
Americium(III)	18.18
Barium	7.78
Beryllium	9.3
Bismuth	22.8
Calcium	11.0
Cadmium	16.4
Cerium(III)	16.80
Californium(III)	19.09
Chromium(III)	23
Cobalt(II)	16.31
Cobalt(III)	36
Copper(II)	18.7
Curium(III)	18.45
Dysprosium	18.0

TABLE 2.1 Overall Formation Constants for Metal Complexes with Organic Ligands *(Continued)*

Metal ions	log K_f
Ethylenediamine-*N, N, N′, N′*-tetraacetic acid	
Erbium	18.15
Europium	17.99
Gadolinium	17.2
Gallium	20.25
Holmium	18.1
Indium	24.95
Iron(II)	14.33
Iron(III)	24.23
Lanthanum	16.34
Lead(II)	18.3
Lithium	2.79
Lutetium	19.83
Magnesium	8.64
Manganese(II)	13.8
Molybdenum(V)	6.36
Neodymium	16.6
Nickel(II)	18.56
Palladium(II)	18.5
Plutonium(III)	18.12
Plutonium(IV)	17.66
Plutonium(VI)	17.66
Praseodymium	16.55
Promethium(III)	17.45
Radium	7.4
Samarium	16.43
Scandium	23.1
Silver	7.32
Strontium	8.80
Terbium	17.6
Thullium	19.49
Tin(II)	22.1
Titanium(III)	21.7
Titanium(IV)	17.3
Uranium(IV)	17.50
Vanadium(II)	12.70
Vanadium(III)	25.9
Vanadium(IV)	18.0
Vanadium(V)	18.05
Ytterbium	18.70
Yttrium	18.32
Zinc	16.4
Zirconium	19.40
Glycine	
Beryllium	4.95
Cadmium	8.60
Cobalt(II)	10.76
Copper(II)	16.27
Dysprosium	12.2
Erbium	12.7
Gadolinium	11.9
Glycine	
Iron(II)	7.6
Iron(III)	10.3
Lanthanum	11.2
Lead	8.92
Magnesium	6.46
Manganese(II)	6.6
Mercury(II)	19.2
Nickel	15
Palladium(II)	11.5
Praseodymium	11.5
Samarium	11.7
Ytterbium	13.0
Yttrium	12.5
Zinc	9.96
8-Hydroxyquinoline	
Cadmium	13.4
Cobalt(II)	17.2
Copper(II)	23.4
Iron(II)	22.23
Iron(III)	33.9
Lanthanum	16.95
Samarium	19.50
Strontium	6.08
Thorium	38.80
Vanadium(II)	23.6
Vanadium(IV)	20.19
Yttrium	20.25
Nitrilotriacetic acid	
Alumium	>10
Barium	5.88
Calcium	11.61
Cadmium	15.2
Cerium(III)	18.67
Chromium(III)	>10
Cobalt(II)	14.5
Copper(II)	13.10
Indium	15
Iron(II)	8.84
Iron(III)	24.32
Lanthanum	17.60
Lead(II)	11.8
Magnesium	10.2
Manganese(II)	11.1
Mercury(II)	12.7
Nickel	16.0
Rare earths	ca. 21
Strontium	6.73
Thallium	3.44
Thorium	12.4
Nitrilotriacetic acid	
Zinc	13.45
Zirconium	3.6
Oxalate	
Aluminum	16.3
Barium	2.3
Beryllium	4.90
Calcium	3.0
Cadmium	5.77
Cerium(III)	11.3
Cobalt(II)	9.7
Cobalt(III)	ca. 20
Copper(II)	8.5
Erbium	10.03
Gadolinium	7.04
Iron(II)	5.22
Iron(III)	20.2
Lead	6.54
Magnesium	4.38
Manganese(II)	19.42
Mercury(II)	6.98
Neodymium	>14
Plutonium(III)	28
Plutonium(IV)	27.50
Plutonium(VI)	11.4
Silver	2.41
Thorium	24.48
Thallium(I)	2.0
Titanium(IV)	2.67
Uranium(VI)	10.57
Vanadium(II)	ca. 2.7
Vanadium(IV)	9.80
Ytterbium	>14
Yttrium	11.47
Zinc	8.15
Zirconium	21.15
1,10-Phenanthroline	
Cadmium	14.31
Cobalt(II)	19.90
Copper(II)	20.94
Iron(II)	21.3
Iron(III)	23.5
Lead	9
Manganese(II)	10.11
Mercury(II)	23.35
Nickel(II)	24.80
Silver	12.07
Vanadium(IV)	9.69
Zinc	17.55

(Continued)

TABLE 2.1 Overall Formation Constants for Metal Complexes with Organic Ligands *(Continued)*

Metal ions	log K_f
1-(2-Pyridylazo)-2-naphthol (PAN)	
Cobalt(II)	>12
Copper(II)	16
Manganese(II)	18.9
Nickel	26.0
Scandium	4.8
Titanium(III)	4.23
Zinc	23.5
Salicylic acid	
Aluminum	14.11
Beryllium	17.4
Cadmium	5.55
Chromium(II)	15.3
Cobalt(II)	11.42
Copper(II)	18.45
Iron(II)	11.25
Iron(III)	36.80
Manganese(II)	9.80
Nickel	11.75
Thorium	11.60
Titanium(IV)	6.09
Uranium(VI)	13.4
Vanadium(II)	6.3
Zinc	6.85
5-Sulfosalicylic acid	
Aluminum	28.89
Beryllium	20.81
Cadmium	29.08
Chromium(II)	12.9
Chromium(III)	9.56
Cobalt(II)	9.82
Copper(II)	16.45
Iron(II)	5.90
Iron(III)	32.12
Lanthanum	9.11
Manganese(II)	8.24
Nickel	10.24
Niobium(V)	7.7
Uranium(VI)	19.20
Zinc	10.65
Tartaric acid	
Barium	1.62
Bismuth	8.30
Calcium	9.01
Cadmium	2.8
Cobalt(II)	2.1
Copper(II)	4.78
Europium(III)	8.11
Iron(III)	7.49
Lanthanum	3.06
Lead	3.78
Magnesium	1.38
Neodymium	9.0
Zinc	8.32
Thioglycolic acid	
Cerium(III)	3.03
Cobalt(II)	12.15
Iron(II)	10.92
Lead	8.5
Manganese(II)	7.56
Mercury(II)	43.82
Nickel	13.53
Rare earths	ca. 3.1
Zinc	15.04
Thiourea	
Bismuth	11.9
Cadmium	4.6
Copper(II)	15.4
Lead	8.3
Mercury(II)	26.8
Silver	13.1
Triethanolamine	
Cobalt(II)	1.73
Copper(II)	4.30
Mercury(II)	13.08
Nickel	2.7
Silver	3.64
Zinc	2.00
3-(2′-Thenoyl)-1,1,1-trifluoroacetone (TTA)	
Barium	10.6
Copper(II)	13.0
Iron(III)	6.9
Nickel	10.0
Praseodymium	9.53
Plutonium(III)	9.53
Plutonium(IV)	8.0
Thorium	8.1
Uranium(IV)	7.2

1. Oxygen donors and fluoride are general complexing agents, combining with any metal ion with a charge more than one.

2. Acetates, citrates, tartrate, and β-diketones sequester all metals in general. Pure electrostatic phenomena predominate. The strength of the coordinating bond formed with the central metal ion increases enormously with the charge of the metal ion and decreases with its radius. Upon comparing various oxygen donors, bond stability is found to increase regularly with the basicity, as measured by proton addition, of the ligand atom.

3. Cyanide, heavy halides, sulfur donors, and to a smaller extent the nitrogen donors are more selective complexing agents than oxygen donors. These ligands do not combine with the cations of the *A* metals of the periodic table. Only the cations of *B* metals and transition-metal cations are coordinated to carbon, sulfur, nitrogen, chlorine, bromine, and iodine. The bonds of the complexes are predominately covalent. Bond strength increases with the ease by which the metal ion accepts electrons and the ease by which the ligand atom donates electrons. A decisive factor is the difference in the electronegativity of the metal ion and the donor atom. Highly polarizable

TABLE 2.2 **Overall Formation Constants for Metal Complexes with Inorganic Ligands**

Metal ions	log K_f
Ammonia	
Cadmium	7.12
Cobalt(II)	5.11
Cobalt(III)	35.2
Copper(I)	10.86
Copper(II)	12.86
Iron(II)	2.2
Manganese(II)	1.3
Mercury(II)	19.28
Nickel	8.74
Platinum(II)	35.3
Silver	7.05
Zinc	9.46
Bromide	
Bismuth(III)	9.70
Cadmium	3.70
Copper(I)	5.89
Gold(I)	12.46
Mercury(II)	21.00
Palladium(II)	13.1
Platinum(II)	20.5
Rhodium(III)	17.2
Silver(I)	8.73
Thallium(III)	31.6
Chloride	
Antimony(III)	4.72
Bismuth(III)	5.6
Cadmium	2.80
Copper(I)	5.7
Gold(II)	9.8
Indium	3.23
Mercury(II)	15.07
Palladium(II)	15.7
Platinum(II)	16.0
Silver(I)	5.04
Thallium(III)	18.00
Tin(II)	2.03
Tin(IV)	4
Cyanide	
Cadmium	18.78
Copper(I)	30.30
Gold(I)	38.3
Iron(II)	35
Iron(III)	42
Mercury(II)	41.4
Nickel	31.3
Silver(I)	21.1
Zinc	16.7
Fluoride	
Aluminum	19.84
Beryllium	12.6
Cerium(III)	3.20
Chromium(III)	10.29
Gadolinium	3.46
Gallium	5.08
Indium	9.70
Iron(III)	12.06
Manganese(II)	5.48
Plutonium(III)	6.77
Scandium	17.3
Thallium(III)	6.44
Thorium	17.97
Titanium(IV)	18.0
Uranium(VI)	11.84
Yttrium	12.14
Zirconium	21.94
Hydroxide	
Aluminum	33.03
Antimony(III)	38.3
Arsenic (as AsO^+)	21.20
Beryllium	15.2
Cadmium	8.62
Cerium(IV)	26.42
Chromium(III)	29.9
Copper(II)	18.5
Gallium	40.3
Indium	28.7
Iodine	11.24
Iron(II)	9.67
Iron(III)	29.67
Lead(II)	61.0
Manganese(II)	8.3
Nickel	11.33
Plutonium(IV)	12.39
Plutonium(VI)	20.9
Tellurium(IV)	72.0
Thallium(III)	25.37
Titanium(III)	12.71
Uranium(IV)	41.2
Uranium(VI)	32.4
Vanadium(III)	21.6
Vanadium(IV)	25.8
Vanadium(V)	58.5
Zinc	17.66
Zirconium	55.3
Iodide	
Bismuth	18.80
Cadmium	5.41
Copper(I)	8.85
Iodine	5.79
Lead(II)	4.47
Mercury(II)	29.83
Silver(I)	13.68
Thallium(III)	31.82
Pyrophosphate	
Barium	4.6
Calcium	4.6
Cadmium	5.6
Copper(II)	9.0
Lead	5.3
Magnesium	5.7
Nickel	7.8
Strontium	4.7
Yttrium	9.7
Zirconium	6.5
Sulfate	
Cerium(III)	3.40
Erbium	3.58
Gadolinium	3.66
Holmium	3.58
Indium	2.36
Iron(III)	2.98
Lanthanum	3.64
Neodymium	3.64
Nickel	2.4
Plutonium(IV)	3.66
Praseodymium	3.62
Samarium	3.66
Thorium	5.50
Uranium(IV)	5.42
Uranium(VI)	3.30
Yttrium	3.47
Ytterbium	3.58
Zirconium	7.77
Sulfite	
Copper(I)	9.2
Mercury(II)	22.06
Silver	7.35
Thiocyanate	
Bismuth	4.23
Cadmium	3.6
Chromium(III)	2.98
Cobalt(II)	3.00
Copper(I)	5.18
Gold(I)	42
Indium	4.63
Iron	3.36
Mercury(II)	17.47
Silver	9.08
Thiosulfate	
Cadmium	6.44
Copper(I)	13.84
Lead	6.35
Mercury(II)	33.24
Silver	13.46

ligands are favored, especially if the latter have suitable vacant orbitals into which some of the *d* electrons from the metal ion can be "back-bonded." This condition favors sulfur-containing ligands.

2.1.1.4 Chelates.[6] A large and important group of metal complexes contain a number of ligands that is half the usual coordination number of the metal ion involved. These organic ligands have a dual character. The ligand is usually a weak, polyfunctional organic acid. It must also possess a pair of unshared electrons on an oxygen, nitrogen, or sulfur atom that is available for coordination. Furthermore, the acidic and basic groups in the ligand must be situated so that ring formation involving the metal ion can proceed relatively free from strain. An example is 2,4-pentanedione (with its tautomeric equilibrium)

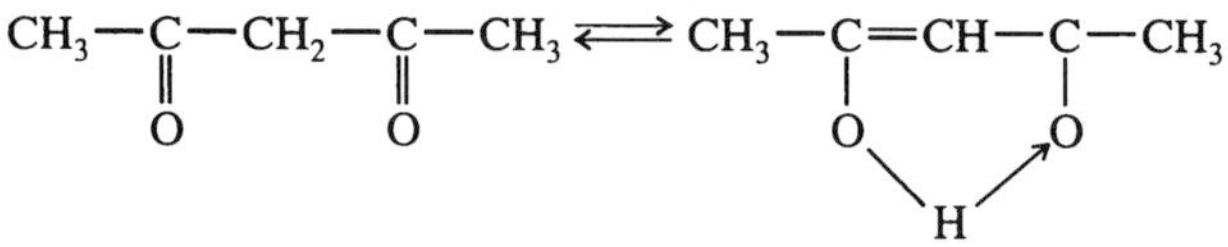

$$CH_3-\underset{\underset{O}{\|}}{C}-CH_2-\underset{\underset{O}{\|}}{C}-CH_3 \rightleftharpoons CH_3-\underset{\underset{O\diagdown_{H}\diagup O}{|\qquad\quad |}}{C=CH-C}-CH_3 \tag{2.16}$$

reacting with Cu(II) ion through replacement of the acidic proton and chelate formation

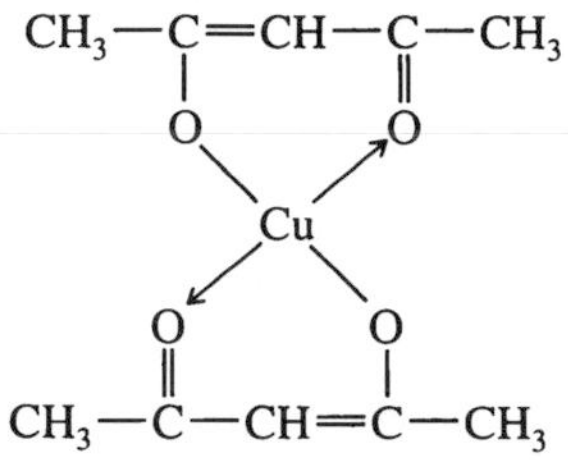

$$\begin{array}{c} CH_3-C=CH-C-CH_3 \\ |\qquad\qquad \| \\ O\qquad\quad O \\ \diagdown\ \nearrow \\ Cu \\ \swarrow\ \diagdown \\ O\qquad\quad O \\ \|\qquad\qquad | \\ CH_3-C-CH=C-CH_3 \end{array}$$

Practically all chelates have five- or six-member rings. In general, the five-member ring is more stable when the ring is entirely saturated, but six-member rings are favored when one or more double bonds are present. Multiple ring systems formed with a given metal ion markedly improve stability. When two or more donor groups are tied together to form an additional chelate ring without materially altering the donor groups, the increased stability of the chelate is due almost entirely to an increase of entropy. This is due to the increasing number of positions occupied by the chelating ligand in the metal coordination sphere and the improvement in isolation of the central metal atom from the influence of solvent hydration. Stability parallels the number of resonating structures that can be written for the chelate species. Spatial considerations, size of the metal ion, and space available between the coordination and ligand sites determine whether a given metal chelate can form.

Reagents giving uncharged complexes are precipitating and extracting agents. The pure oxygen donors among them, like the β-diketones and cupferron, are general reagents. Little selectivity is found in 8-hydroxyquinoline, which donates one oxygen and one nitrogen atom to each metal. On the other hand, the nitrogen and sulfur donors, such as dithizone and diethyldithiocarbamate, are highly selective for the *B* metals and the noble transition metals.

Reagents giving charged chelates are masking (sequestering) agents. The pure oxygen donors (oxalates, tartrates, and citrates) are nonselective. Higher complex stabilities are reached with aminopolycarboxylic acid anions, which are also general masking agents because

[6] A. E. Martell and M. Calvin, *Chemistry of the Metal Chelate Compounds*, Prentice-Hall, New York, 1952.

of the greater number of oxygens in comparison to nitrogens among their ligand atoms. Polyamines are rather selective agents, combining only with *B* metals and transition metals. Still greater selectivity is found with sulfur donors like thiourea, dithiocarbamate, and dithiophosphate.

2.1.2 Masking[7,8]

Masking, in the general sense, is the prevention of reactions whose occurrence is normally to be expected. Studies of Cheng[9] and Hulanicki[10] give the theoretical basis of application of masking agents in analytical problems. If the normal reacting molecule or ion is designated as *A*, and the masking agent as Ms, every masking reaction can be represented by the general reversible equation:

$$A + MS \rightleftarrows A \cdot Ms \tag{2.17}$$

The position of the equilibrium of the masking reaction, that is, the decreased concentration of *A* at equilibrium, determines the efficiency of masking. An excess of Ms favors the completeness of masking. Masking (and demasking) techniques are widely used in analytical chemistry because they frequently provide convenient and elegant methods by which to avoid the effects of unwanted components of a system without having to resort to physical separation. Masking techniques can increase the selectivity of many analytical methods. The best ligands to use as masking agents are those that are chemically stable and nontoxic and react rapidly to form strong, colorless complexes with the ions to be masked, but form only relatively weak complexes with other ions that are present. Tables 2.3 and 2.4 are intended as qualitative guides to the types of masking agents likely to be suitable for particular analytical problems. Although optimum working conditions can frequently be defined by calculation, it is still necessary to confirm by experiment the suitability of the calculated conditions.

Important procedures for enhancing the selective action of reagents are based on the prevention of reactions. If a selective reagent is undergoing several analogous reactions, the equilibrium positions are dependent on pH to a different extent. Adjustment of the reaction milieu to certain pH values or intervals can then prevent the occurrence of unwanted reactions. This stratagem is used especially with organic reagents. Another type of reaction impedance consists in lowering the concentration of ionic or molecular species without dilution so that certain color or precipitation reactions no longer occur. This is possible when the addition of appropriate reagents produces compounds for which the reactions are of different types. By combination with pH adjustment, masking can frequently be employed to elevate the action of less selective precipitation and color reagents to highly selective reactions or sometimes even specific reactions. Masking agents may be either inorganic or organic in nature. In the latter case, the action of certain groups plays a significant role in the same manner as with precipitation and color reagents. On the other hand, the unsuspected presence of masking agents may impair or prevent tests or methods that ordinarily can be used without hesitation.

Through deliberate selection of suitable masking agents it is possible to secure a precise, fractional masking action. For example, Cu(II) is extracted with 8-hydroxyquinoline in chloroform at a much lower pH than U(VI). If ethylenediaminetetraacetic acid (EDTA) is added, the uranium

[7] D. D. Perrin, *Masking and Demasking of Chemical Reactions*, Wiley-Interscience, New York, 1970; D. D. Perrin, "Masking and Demasking in Analytical Chemistry," Chap. 21 in I. M. Kolthoff and P. J. Elving, eds., *Treatise on Analytical Chemistry*, Part I, Vol. 2, 2d ed., Wiley-Interscience, New York, 1979.

[8] F. Feigl, *Chemistry of Specific, Selective and Sensitive Reactions*, Academic, New York, 1949.

[9] K. L. Cheng, *Anal. Chem.* **33**:783 (1961).

[10] Hulanicki, *Talanta* **9**:549 (1962).

TABLE 2.3 Masking Agents for Ions of Various Elements

Element	Masking agent
Ag	Br^-, citrate, Cl^-, CN^-, I^-, NH_3, SCN^-, $S_2O_3^{2-}$, thiourea, thioglycolic acid, diethyldithiocarbamate, thiosemicarbazide, bis(2-hydroxyethyl)dithiocarbamate
Al	Acetate, acetylacetone, BF_4^-, citrate, $C_2O_4^{2-}$, EDTA, F^-, formate, 8-hydroxyquinoline-5-sulfonic acid mannitol, 2,3-mercaptopropanol, OH^-, salicylate, sulfosalicylate, tartrate, triethanolamine, tiron
As	Citrate, 2,3-dimercaptopropanol, $NH_2OH \cdot HCl$, OH^-, S_2^{2-}, tartrate
Au	Br^-, CN^-, NH_3, SCN^-, $S_2O_3^{2-}$, thiourea
Ba	Citrate, cyclohexanediaminetetraacetic acid, *N,N*-dihydroxyethylglycine, EDTA, F^-, SO_4^{2-}, tartrate
Be	Acetylacetone, citrate, EDTA, F^-, sulfosalicylate, tartrate
Bi	Citrate, Cl^-, 2,3-dimercaptopropanol, dithizone, EDTA, I^-, OH^-, $Na_5P_3O_{10}$, SCN^-, tartrate, thiosulfate, thiourea, triethanolamine, thiourea
Ca	BF_4^-, citrate, *N,N*-dihydroxyethylglycine, EDTA, F^-, polyphosphates, tartrate
Cd	Citrate, CN^-, 2,3-dimercaptopropanol, dimercaptosuccinic acid, dithizone, EDTA, glycine, I^-, malonate, NH_3, 1,10-phenanthroline, SCN^-, $S_2O_3^{2-}$, tartrate
Ce	Citrate, *N,N*-dihydroxyethylglycine, EDTA, F^-, PO_4^{3-}, reducing agents (ascorbic acid), tartrate, tiron
Co	Citrate, CN^-, diethyldithiocarbamate, 2,3-dimercaptopropanol, dimethylglyoxime, ethylenediamine, EDTA, F^-, glycine, H_2O_2, NH_3, NO_2^-, 1,10-phenanthroline, $Na_5P_3O_{10}$, SCN^-, $S_2O_3^{2-}$, tartrate
Cr	Acetate, (reduction with) ascorbic acid + KI, citrate, *N,N*-dihydroxyethylglycine, EDTA, F^-, formate, NaOH + H_2O_2, oxidation to CrO_4^{2-}, $Na_5P_3O_{10}$, sulfosalicylate, tartrate, triethylamine, tiron
Cu	Ascorbic acid + KI, citrate, CN^-, diethyldithiocarbamate, 2,3-dimercaptopropanol, ethylenediamine, EDTA, glycine, hexacyanocobalt(III)(3–), hydrazine, I^-, NaH_2PO_2, $NH_2OH \cdot HCl$, NH_3, NO_2^-, 1,10-phenanthroline, S^{2-}, SCN^- + SO_3^{2-}, sulfosalicylate, tartrate, thioglycolic acid, thiosemicarbazide, thiocarbohydrazide, thiourea
Fe	Acetylacetone, (reduction with) ascorbic acid, $C_2O_4^{2-}$, citrate, CN^-, 2,3-dimercaptopropanol, EDTA, F^-, NH_3, $NH_2OH \cdot HCl$, OH^-, oxine, 1,10-phenanthroline, 2,2′-bipyridyl, PO_4^{3-}, $P_2O_7^{4-}$, S^{2-}, SCN^-, $SnCl_2$, $S_2O_3^{2-}$, sulfamic acid, sulfosalicylate, tartrate, thioglycolic acid, thiourea, tiron, triethanolamine, trithiocarbonate
Ga	Citrate, Cl^-, EDTA, OH^-, oxalate, sulfosalicylate, tartrate
Ge	F^-, oxalate, tartrate
Hf	See Zr
Hg	Acetone, (reduction with) ascorbic acid, citrate, Cl^-, CN^-, 2,3-dimercaptopropan-1-ol, EDTA, formate, I^-, SCN^-, SO_3^{2-}, tartrate, thiosemicarbazide, thiourea, triethanolamine
In	Cl^-, EDTA, F^-, SCN^-, tartrate, thiourea, triethanolamine
Ir	Citrate, CN^-, SCN^-, tartrate, thiourea
La	Citrate, EDTA, F^-, oxalate, tartrate, tiron
Mg	Citrate, $C_2O_4^{2-}$, cyclohexane-1,2-diaminetetraacetic acid, *N,N*-dihydroxyethylglycine, EDTA, F^-, glycol, hexametaphosphate, OH^-, $P_2O_4^{4-}$, triethanolamine
Mn	Citrate, CN^-, $C_2O_4^{2-}$, 2,3-dimercaptopropanol, EDTA, F^-, $Na_5P_3O_{10}$, oxidation to MnO_4^-, $P_2O_7^{4-}$, reduction to Mn(II) with $NH_2OH \cdot HCl$ or hydrazine, sulfosalicylate, tartrate, triethanolamine, triphosphate, tiron
Mo	Acetylacetone, ascorbic acid, citrate, $C_2O_4^{2-}$, EDTA, F^-, H_2O_2, hydrazine, mannitol, $Na_5P_3O_{10}$, $NH_2OH \cdot HCl$, oxidation to molybdate, SCN^-, tartrate, tiron, triphosphate
Nb	Citrate, $C_2O_4^{2-}$, F^-, H_2O_2, OH^-, tartrate
Nd	EDTA
NH_4^+	HCHO
Ni	Citrate, CN^-, *N,N*-dihydroxyethylglycine, dimethylglyoxime, EDTA, F^-, glycine, malonate, $Na_5P_3O_{10}$, NH_3, 1,10-phenanthroline, SCN^-, sulfosalicylate, thioglycolic acid, triethanolamine, tartrate
Np	F^-
Os	CN^-, SCN^-, thiourea
Pa	H_2O_2
Pb	Acetate, $(C_6H_5)_4AsCl$, citrate, 2,3-dimercaptopropanol, EDTA, I^-, $Na_5P_3O_{10}$, SO_4^{2-}, $S_2O_3^{2-}$, tartrate, tiron, tetraphenylarsonium chloride, triethanolamine, thioglycolic acid

TABLE 2.3 Masking Agents for Ions of Various Elements (*Continued*)

Element	Masking agent
Pd	Acetylacetone, citrate, CN^-, EDTA, I^-, NH_3, NO_2^-, SCN^-, $S_2O_3^{2-}$, tartrate, triethanolamine
Pt	Citrate, CN^-, EDTA, I^-, NH_3, NO_2^-, SCN^-, $S_2O_3^{2-}$, tartrate, urea
Pu	Reduction to Pu(IV) with sulfamic acid
Rare earths	$C_2O_4^{2-}$, citrate, EDTA, F^-, tartrate
Re	Oxidation to perrhenate
Rh	Citrate, tartrate, thiourea
Ru	CN^-, thiorurea
Sb	Citrate, 2,3-dimercaptopropanol, EDTA, I^-, OH^-, oxalate, S^{2-}, S_2^{2-}, $S_2O_3^{2-}$, tartrate, triethanolamine
Sc	Cyclohexane-1,2-diaminetetraacetic acid, F^-, tartrate
Se	Citrate, F^-, I^-, reducing agents, S^{2-}, SO_3^{2-}, tartrate
Sn	Citrate, $C_2O_3^{2-}$, 2,3-dimercaptopropanol, EDTA, F^-, I^-, OH^-, oxidation with bromine water, phosphate(3–), tartrate, triethanolamine, thioglycolic acid
Sr	Citrate, *N,N*-dihydroxyethylglycine, EDTA, F^-, SO_4^{2-}, tartrate
Ta	Citrate, F^-, H_2O_2, OH^-, oxalate, tartrate
Te	Citrate, F^-, I^-, reducing agents, S^{2-}, sulfite, tartrate
Th	Acetate, acetylacetone, citrate, EDTA, F^-, SO_4^{2-}, 4-sulfobenzenearsonic acid, sulfosalicylic acid, tartrate, triethanolamine
Ti	Ascorbic acid, citrate, F^-, gluconate, H_2O_2, mannitol, $Na_5P_3O_{10}$, OH^-, SO_4^{2-}, sulfosalicylic, acid, tartrate, triethanolamine, tiron
Tl	Citrate, Cl^-, CN^-, EDTA, HCHO, hydrazine, $NH_2OH \cdot HCl$, oxalate, tartrate, triethanolamine
U	Citrate, $(NH_4)_2CO_3$, $C_2O_4^{2-}$, EDTA, F^-, H_2O_2, hydrazine + triethanolamine, phosphate(3–), tartrate
V	(reduction with) Ascorbic acid, hydrazine, or $NH_2OH \cdot HCl$, CN^-, EDTA, F^-, H_2O_2, mannitol, oxidation to vanadate, triethanolamine, tiron
W	Citrate, F^-, H_2O_2, hydrazine, $Na_5P_3O_{10}$, $NH_2OH \cdot HCl$, oxalate, SCN^-, tartrate, tiron, triphosphate, oxidation to tungstate
Y	Cyclohexane-1,2-diaminetetraacetic acid, F^-
Zn	Citrate, CN^-, *N,N*-dihydroxyethylglycine, 2,3-dimercaptopropanol, dithizone, EDTA, F^-, glycerol, glycol, hexacyanoferrate(II)(4–), $Na_5P_3O_{10}$, NH_3, OH^-, SCN^-, tartrate, triethanolamine
Zr	Arsenazo, carbonate, citrate, $C_2O_4^{2-}$, cyclohexane-1,2-diaminetetraacetic acid, EDTA, F^-, H_2O_2, PO_4^{3-}, $P_2O_7^{4-}$, pyrogallol, quinalizarinesulfonic acid, salicylate, SO_4^{2-}, + H_2O_2, sulfosalicylate, tartrate, triethanolamine

extraction proceeds as before but the copper is not extracted until the pH is raised by more than five units.

Masking must not be identified solely with complex formation. There are numerous complex compounds in which solutions show no masking effects. On the other hand, examples can be cited in which the product of soluble principal valence compounds may lead to masking. This latter category includes the annulment of the base action of NH_2—groups in carboxylic acids by the addition of formaldehyde, the masking of the iodometric oxidation of sulfites by formaldehyde, as well as the masking of almost all reactions of molybdenum(VI), tungsten(VI), and vanadium(V) by hydrogen peroxide or fluoride ion. Sometimes the masking agent changes the valence state of the metal ion. Examples include the reduction of Fe(III) to Fe(II) with hydrazine, hydroxylamine hydrochloride, or tin(II) chloride. Hydroxylamine also reduces Ce(IV) to Ce(III), Cu(II) to Cu(I), and Hg(II) to free Hg. Ascorbic acid reduces Cu(II) to Cu(I) in the presence of chloride ion.

The reaction of the hydrogen sulfite ion in an alkaline solution with ketones and aldehydes is

$$H_2C{=}O + HSO_3^- \rightleftarrows H_2C(OH)SO_3^- \tag{2.18}$$

TABLE 2.4 **Masking Agents for Anions and Neutral Molecules**

Anion or neutral molecule	Masking agent
Boric acid	F^-, glycol, mannitol, tartrate, and other hydroxy acids
Br^-	Hg(II)
Br_2	Phenol, sulfosalicylic acid
BrO_3^-	Reduction with arsenate(III), hydrazine, sulfite, or thiosulfate
Chromate(VI)	Reduction with arsenate(III), ascorbic acid, hydrazine, hydroxylamine, sulfite, or thiosulfate
Citrate	Ca(II)
Cl^-	Hg(II), Sb(III)
Cl_2	Sulfite
ClO_3^-	Thiosulfate
ClO_4^-	Hydrazine, sulfite
CN^-	HCHO, Hg(II), transition-metal ions
EDTA	Cu(II)
F^-	Al(III), Be(II), boric acid, Fe(III), Th(IV), Ti(IV), Zr(IV)
$Fe(CN)_6^{3-}$	Arsenate(III), ascorbic acid, hydrazine, hydroxylamine, thiosulfate
Germanic acid	Glucose, glycerol, mannitol
I^-	Hg(II)
I_2	Thiosulfate
IO_3^-	Hydrazine, sulfite, thiosulfate
IO_4^-	Arsenate(III), hydrazine, molybdate(VI), sulfite, thiosulfate
MnO_4^-	Reduction with arsenate(III), ascorbic acid, azide, hydrazine, hydroxylamine, oxalic acid, sulfite, or thiosulfate
MoO_4^{2-}	Citrate, F^-, H_2O_2, oxalate, thiocyanate + Sn(II)
NO_2^-	Co(II), sulfamic acid, sulfanilic acid, urea
Oxalate	Molybdate(VI), permanganate
Phosphate	Fe(III), tartrate
S	CN^-, S^{2-}, sulfite
S^{2-}	Permanganate + sulfuric acid, sulfur
Sulfate	Cr(III) + heat
Sulfite	HCHO, Hg(II), permanganate + sulfuric acid
SO_6^{2-}	Ascorbic acid, hydroxylamine, thiosulfate
Se and its anions	Diaminobenzidine, sulfide, sulfite
Te	I^-
Tungstate	Citrate, tartrate
Vanadate	Tartrate

The carbon–oxygen double bond of the carbonyl group is opened, and the hydrogen sulfite radical is added. An increase in temperature reverses the reaction more easily for ketones than for aldehydes.

Certain organic substances have no charge at any pH but form complexes with substances that do have a charge. The sugars and polyalcohols form such complexes in the pH range between 9 and 10 with a number of anions, including borate, molybdate, and arsenite. Elegant ion-exchange methods have been devised for the sugars.

2.1.2.1 Examples of Masking. The reversible reaction involving iron(II)–iron(III) will proceed quantitatively in either direction under suitable conditions. The oxidation of iron(II) will be facilitated if the resulting iron(III) is masked by fluoride, pyrophosphate, or tartrate and thus removed from the redox system. When determining iron(II) in minerals containing fluoride, air oxidation is a problem during the dissolution process. Excess boric acid ties up the fluoride as tetrafluoroborate(1–) ion more effectively than hexafluoroferrate(III)(3–) ion.

Although iodine normally oxidizes titanium(III) to titanium(IV) with ease, the reaction is prevented when fluoride ion is present due to the formation of hexafluorotitanate(III)(3–) ion.

The precipitation of certain metals as hydrated oxides and carbonates does not occur in the presence of hydroxy acids and higher alcohols, such as citrates, glycerol, mannitol, and tartrate.

Sulfosalicylic acid is an excellent complex former for aluminum, iron(III), and titanium(IV) ions.

Beryllium ions are masked against hydroxide precipitation in ammoniacal tartrate solutions. Nevertheless, the addition of guanidine carbonate produces a quantitative precipitation of basic beryllium carbonate mixed with precipitant. Other metal ions masked by tartrate, such as aluminum, iron(III), thorium, uranium(VI), and zirconium, do not react with guanidine carbonate.

Kinetic masking utilizes differences in rates of complex formation or dissociation. For example, ice-cold solutions at pH 2 of Ni-EDTA and A1-EDTA are only slowly dissociated by added Bi(III) ion, permitting Ni to be determined in the presence of Cd, Co, Cu(II), Mn(II), Pb, and Zn.

2.1.3 Demasking

For the major part, masking reactions that occur in solutions and lead to soluble compounds are equilibrium reactions. They usually require the use of an excess of the masking agent and can be reversed again by removal of the masking agent. The freeing of previously masked ionic or molecular species has been called *demasking.* This merits consideration in regard to its use in analysis. Masking never completely removes certain ionic or molecular species, but only reduces their concentrations. The extent of this lowering determines which color or precipitation reactions can be prevented. A system masked against a certain reagent is not necessarily masked against another but more aggressive reagent. It is therefore easy to see that masked reaction systems can also function as reagents at times (e.g., Fehling's solution, Nessler's reagent).

The methods used in demasking are varied. One approach is to change drastically the hydrogen ion concentration of the solution. The conditional stability constants of most metal complexes depend greatly on pH, so that simply raising or lowering the pH is frequently sufficient for selective demasking. In most cases a strong mineral acid is added, and the ligand is removed from the coordination sphere of the complex through the formation of a slightly ionized acid as with the polyprotic (citric, tartaric, EDTA, and nitriloacetic) acids.

Another type of demasking involves formation of new complexes or other compounds that are more stable than the masked species. For example, boric acid is used to demask fluoride complexes of tin(IV) and molybdenum(VI). Formaldehyde is often used to remove the masking action of cyanide ions by converting the masking agent to a nonreacting species through the reaction

$$CN^- + HCHO \rightleftarrows OCH_2CN \tag{2.19}$$

which forms glycollic nitrile. Pertinent instances are the demasking of $Ni(CN)_4^{2-}$ ions to Ni^{2+} ions by formaldehyde and the demasking of dimethylglyoxime (dmg) from $Pd(dmg)_2^{2-}$ ions by cyanide. Selectivity is evident in that $Zn(CN)_4^{2-}$ is demasked, whereas $Cu(CN)_3^{2-}$ is not.

Destruction of the masking ligand by chemical reaction may be possible, as in the oxidation of EDTA in acid solutions by permanganate or another strong oxidizing agent. Hydrogen peroxide and Cu(II) ion destroy the tartrate complex of aluminum.

Demasking methods for a number of masking agents are enumerated in Table 2.5.

TABLE 2.5 Common Demasking Agents

Abbreviations: DPC, diphenylcarbazide; HDMG, dimethylglyoxime; PAN, 1-(2-pyridylazo)-2-naphthol; Pptn, precipitation; Dtcn, detection; Diffn, differentiation; Titrn, titration.

Complexing agent	Ion demasked	Demasking agent	Application
CN^-	Ag^+	H^+	Precipitation of Ag
	Cd^{2+}	H^+	Free Cd^{2+}
		$HCHO + OH^-$	Detection of Cd (with DPC) in presence of Cu
	Cu^+	H^+	Precipitation of Cu
	Cu^{2+}	HgO	Determination of Cu
	Fe^{2+}	Hg^{2+}	Free Fe^{2+}
	Fe^{3+}	HgO	Determination of Fe
	HDMG	Pd^{2+}	Detection of CN^- (with Ni^{2+})
	Hg^{2+}	Pd^{2+}	Detection of Pd (with DPC)
	Ni^{2+}	HCHO	Detection of Ni (with HDMG)
		H^+	Free Ni^{2+}
		HgO	Determination of Ni
		Ag^+	Detection and determination of Ni (with HDMG) in presence of Co
		Ag^+, Hg^{2+}, Pb^{2+}	Detection of Ag, Hg, Pb (with HDMG)
	Pd^{2+}	H^+	Precipitation of Pd
		HgO	Determination of Pd
	Zn^{2+}	$Cl_3CCHO \cdot H_2O$	Titration of Zn with EDTA
		H^+	Free Zn
CO_3^{2-}	Cu^{2+}	H^+	Free Cu^{2+}
$C_2O_4^{2-}$	Al^{3+}	OH^-	Precipitation of $Al(OH)_3$
Cl^- (concentrated)	Ag^+	H_2O	Precipitation of AgCl
Ethylenediamine	Ag^+	SiO_2 (amorphous)	Differentiation of crystalline and amorphous SiO_2 (with CrO_4^{2-})
EDTA	Al^{3+}	F^-	Titration of Al
	Ba^{2+}	H^+	Precipitation of $BaSO_4$ (with SO_4^{2-})
	Co^{2+}	Ca^{2+}	Detection of Co (with diethyldithiocarbamate)
	Mg^{2+}	F^-	Titration of Mg, Mn
	Th(IV)	SO_4^{2-}	Titration of Th
	Ti(IV)	Mg^{2+}	Precipitation of Ti (with NH_3)
	Zn^{2+}	CN^-	Titration of Mg, Mn, Zn
	Many ions	KMO_4^-	Free ions
F^-	Al(III)	Be(II)	Precipitation of Al (with 8-hydroxylquinoline)
		OH^-	Precipitation of $Al(OH)_3$
	Fe(III)	OH^-	Precipitation of $Fe(OH)_3$
	Hf(IV)	Al(III) or Be(II)	Detection of Hg (with xylenol orange)
	Mo(VI)	H_3BO_3	Free molybdate
	Sn(IV)	H_3BO_3	Precipitation of Sn (with H_2S)
	U(VI)	Al(III)	Detection of U (with dibenzoylmethane)
	Zr(IV)	Al(III) or Be(II)	Detection of Zr (with xylenol orange)
		Ca(II)	Detection of Ca (with alizarin S)
		OH^-	Precipitation of $Zr(OH)_4$
H_2O_2	Hf(IV), Ti(IV), or Zr	Fe(III)	Free ions
NH_3	Ag^+	Br^-	Detection of Br^-
		H^+	Detection of Ag
		I^-	Detection of I and Br
		SiO_2 (amorphous)	Differentiation of crystalline and amorphous SiO_2 (with CrO_4^{2-})
NO_2^-	Co(III)	H^+	Free Co
PO_4^{3-}	Fe(III)	OH^-	Precipitation of $FePO_4$
	UO_2^{2-}	Al(III)	Detection of U (with dibenzoylmethane)
SCN^-	Fe(III)	OH^-	Precipitation of $Fe(OH)_3$
SO_4^{2-} (conc. H_2SO_4)	Ba^{2+}	H_2O	Precipitation of $BaSO_4$
$S_2O_3^{2-}$	Ag^+	H^+	Free Ag^+
	Cu^{2+}	OH^-	Detection of Cu (with PAN)
Tartrate	Al(III)	$H_2O_2 + Cu^{2+}$	Precipitation of $Al(OH)_3$

2.2 EXTRACTION METHODS[11–13]

Most chemical reactions show poor selectivity as to the types of metal ions that take part. To improve the selectivity it is common to resort to extraction methods. Solutes have different solubilities in different solvents, and the process of selectively removing a solute from a mixture with a solvent is called *extraction*. The solute to be extracted may be in a solid or in a liquid medium, and the solvent used for the extraction process may be water, a water-miscible solvent, or a water-immiscible solvent. The selection of the solvent to be used depends upon the solute and upon the requirements of the experimental procedure. An ideal extraction method should be rapid, simple, and inexpensive to perform; should yield quantitative recovery of target analytes without loss or degradation; and should yield a sample that is immediately ready for analysis without additional concentration or class fractionation steps.

2.2.1 Solvent Extraction Systems

Extraction procedures based on the distribution of solutes among immiscible solvents are carried out for two purposes. *Exhaustive extraction* involves the quantitative removal of one solute; *selective extraction* involves the separation of two solutes.

2.2.1.1 Partition Coefficient. It is useful to consider first the behavior of a single solute between two immiscible liquids; usually these will be water and an organic liquid. The distribution equilibrium in the simplest case involves the same molecular species in each phase.

$$A_{\text{aq}} = A_{\text{org}} \tag{2.20}$$

Included in this class are neutral covalent molecules that are not solvated by either of the solvents. The partition coefficient corresponds in value to the ratio of the saturated solubilities of the solute in each phase:

$$K_d = \frac{[A]_{\text{org}}}{[A]_{\text{aq}}} \tag{2.21}$$

Ignoring any specific solute–solvent interactions, the solubility of such molecular species in organic solvents is generally at least an order of magnitude higher than that in water; that is, $K_d \geq 10$. In each homologous series, increasing the alkyl chain length increases the partition coefficient by a factor of about 4 for each new methylene group incorporated into the molecule. Branching results in a lower K_d as compared with the linear isomer. Incorporation of hydrophilic functional groups into the molecule lowers the partition coefficient as written.

2.2.1.2 Association in the Organic Phase. Dimerization in the organic phase

$$2A_{\text{org}} = [(A)_2]_{\text{org}} \tag{2.22}$$

increases the distribution ratio D,

$$D = \frac{[A]_{\text{org}} + 2[(A)_2]_{\text{org}}}{[A]_{\text{aq}}} \tag{2.23}$$

[11] T. C. Lo, H. H. I. Baird, and C. Hanson, eds., *Handbook of Solvent Extraction*, Wiley-Interscience, New York, 1983.
[12] G. H. Morrison and H. Freiser, *Solvent Extraction in Analytical Chemistry*, Wiley, New York, 1957.
[13] J. Starý, *Metal Chelate Solvent Extraction*, Pergamon, Oxford, 1965.

Incorporating the partition coefficient,

$$D = K_d(1 + 2K_2[A]_{org}) \tag{2.24}$$

where K_2 is the dimerization constant. Dimerization decreases the monomer concentration, the species that takes part directly in the phase partition, so that the overall distribution increases.

2.2.1.3 Exhaustive Extraction. Water will extract inorganic salts, salts of organic acids, strong acids and bases, and low-molecular-weight (four carbons or less) carboxylic acids, alcohols, polyhydroxy compounds, and amines from any immiscible organic solvents that contain them. The completeness of the extraction may depend upon the pH of the aqueous phase, as will be discussed in a later section.

Consider an extraction of X moles of solute A dissolved in V_w mL water, with V_o mL organic solvent. If Y moles remain in the water phase after a single extraction, the fraction remaining unextracted is

$$\frac{X}{Y} = f = \frac{V_w}{V_w + K_d V_o} \tag{2.25}$$

The fraction remaining unextracted is independent of the initial concentration. Therefore, if n successive extractions are performed with fresh portions of solvent, the fraction remaining unextracted is

$$f = \left[1 + k_d\left(\frac{V_o}{V_w}\right)\right]^{-n} \tag{2.26}$$

The amount of solute remaining in the water phase after a single extraction is dependent upon two factors: (1) the partition coefficient (in later expressions this is replaced by the distribution ratio D) and (2) the volume ratio of the phases. Exhaustive extraction will involve either repeated batch extractions or continuous extraction methods. After n extractions the fraction of solute remaining in the water phase is

$$\frac{X}{Y} = \left[\frac{V_w}{K_d(V_o + V_w)}\right]^n \tag{2.27}$$

For a given amount of organic phase, the extraction is more efficient if carried out several times with equal portions of extracting phase. However, little is gained by dividing the volume of extractant into more than four or five portions. Equation (2.27) is useful in determining whether a given extraction is practicable, with a reasonable value of V_o/V_w, or whether an extractant with a more favorable partition coefficient (or distribution ratio) should be sought.

Example 2.1 How might 99.0% of a substance with a partition coefficient of 4 be extracted into an organic phase?

Method 1 If using a single extraction, it is necessary to determine the volume ratio needed. Proceed as follows:

$$0.01 = \left[1 + 4\left(\frac{V_o}{V_w}\right)\right]^{-n}$$

$$\frac{V_o}{V_w} = 25$$

The total amount of solute would determine whether 100 mL of organic phase would be used to extract 4 mL of aqueous phase, or some other ratio such as 500 and 20 mL, respectively. In any case, the volume ratio is just feasible. Even so, too large a volume of organic solvent is required.

Method 2 How many equilibrations would be required using fresh organic phase each time and equal volumes of each phase?

$$0.01 = (1 + 4)^{-n}$$

$$\log 0.01 = n \log 0.2 \quad \text{and} \quad -2.0 = -0.7n$$

$$n = 2.8 \text{ (or 3 extractions)}$$

This second approach requires much less organic solvent, and three successive equilibrations are not particularly time consuming.

Often it is convenient to use a volume of extracting organic solvent smaller than the aqueous sample solution. Thus extraction of 50 mL of aqueous solution with 5 mL of organic solvent is easy enough. However, extraction of 250 mL of aqueous solution with 5 mL of organic solvent would require more care to prevent loss of the organic solvent on the walls of the separatory funnel and as small droplets throughout the aqueous solution. Use of a second portion of organic extracting solution overcomes these problems.

If the partition coefficient or distribution ratio is unknown, it can be obtained by equilibrating known volumes of the aqueous phase and extracting solvent, and then determining the concentration of the distributing species in both phases. This should be performed over a range of concentrations.

The requirements of a separation will set the limits on the completeness of extraction required. If removal from the aqueous phase must be 99.9%, then D must be 1000 or greater when equal volumes of the two phases are used, but D must be 10 000 or greater when the aqueous phase is 10 times the volume of the organic phase. These limits assume one equilibration only. If additional equilibration steps are employed, D could be proportionally smaller. When performed manually, the use of liquid–liquid extraction methods is limited to values of D greater than, or equal to, 1.

2.2.1.4 Selectivity of an Extraction. An extraction for separation purposes is more involved than the singular examples discussed in the preceding section. Now two or more components may be distributed between the two phases. Considering the requirements for the complete separation of one component, namely, D_1 must be 1000 or greater when the phase volumes are equal, then D_2 must be 0.001 or less if not more than 0.1% of the second component is to coextract. Do not be misled into thinking that a second equilibration will remedy an incomplete extraction when an insufficient difference exists between the two distribution ratios. Washing the extract with suitable solutions is helpful in removing undesired elements. Equilibration of an extract with fresh aqueous phase will often remove traces of coextractants whose distribution ratios are low.

Example 2.2 Assume $D_1 = 10$ and $D_2 = 0.1$, and that the phase volumes are equal. A single extraction will remove 90.9% of component 1, but also 9.1% of component 2. Now a second extraction of the residual aqueous phase will remove an additional 8.3% of component 1 (for a total of 99.2% removal) but also remove an additional 8.3% of component 2 (for a total of 17.4% removal).

What is desired—a maximum removal of component 1, or less than quantitative removal of component 1 but in as pure a state as possible?

Example 2.3 To continue with the conditions enumerated in the preceding example, but desiring pure component 1, a back-extraction should be tried. Equilibrate the organic extracting phase in step 1 with an equal volume of fresh aqueous phase. True, only 82.6% of component 1 remains in the organic phase (8.3%

is back-extracted) but only 0.7% of component 2 remains as contaminant (the remainder having been back-extracted). A second back-extraction with a second fresh aqueous phase would lower the contamination from component 2 to only 0.06% while providing a 75.0% recovery of component 1.

If the distribution ratios of two components are close in value, it becomes necessary to resort to chemical parameters, such as pH or masking agents, to improve the extraction conditions for the desired component. An alternative method would be countercurrent distribution methods in which distribution, transfer, and recombination of various fractions are performed a sufficient number of times to achieve separation.

The properties of selected solvents are given in Table 2.6.

2.2.2 Extraction of Formally Neutral Species

Ionic compounds would not be expected to extract into organic solvents from aqueous solutions. However, through addition or removal of a proton or a masking ion, an uncharged extractable species may be formed. Included in this category are the neutral metal chelate complexes.

An example is furnished by the anion of a carboxylic acid and the influence of pH upon the distribution ratio of the neutral carboxylic acid molecule between an organic solvent and water. Only the neutral molecule partitions between the contacting phases. The partition coefficient is given by

$$K_d = \frac{[\mathrm{RCOOH}]_{\mathrm{org}}}{[\mathrm{RCOOH}]_{\mathrm{aq}}} \tag{2.28}$$

In the aqueous phase, an acid–base equilibrium is involved:

$$K_{\mathrm{HA}} = \frac{[\mathrm{RCOOH}]}{[\mathrm{RCOO^-}][\mathrm{H^+}]} \tag{2.29}$$

where K_{HA}, the acid association constant, is the reciprocal of the dissociation constant K_a. The overall distribution ratio is expressed by

$$D = \frac{[\mathrm{RCOOH}]_{\mathrm{org}}}{([\mathrm{RCOOH}] + [\mathrm{RCOO^-}])_{\mathrm{aq}}} \tag{2.30}$$

Combining Eqs. (2.28), (2.29), and (2.30)

$$D = \frac{K_d}{1 + 1/K_{\mathrm{HA}}[\mathrm{H^+}]} \tag{2.31}$$

or

$$\frac{1}{D} = \frac{1}{K_d} + \frac{1}{K_d K_{\mathrm{HA}}[\mathrm{H^+}]} \tag{2.32}$$

TABLE 2.6 Properties of Selected Solvents*

The temperature in degrees Celsius at which the viscosity, dielectric constant, and dipole moment of a substance were measured is shown in this table in parentheses after the value.

Solvent	Solubility (wt. %, 20°C)		Density (20°C), $g \cdot mL^{-1}$	Boiling point, °C	Flash point, °C	Viscosity, $mN \cdot s \cdot m^{-2}$	Dielectric constant	Dipole moment (debye units)
	Solv. in H_2O	H_2O in solv.						
Acetone	miscible	miscible	0.791	56.5	−18	0.318 (20)	20.7 (25)	2.77 (22)
Acetylacetone	0.125		0.972 (25)	140.6	40	0.6 (20)	25.7 (20)	2.5 (20)
Benzene	0.18 (25)	0.06 (25)	0.878	80.1	−11	0.649 (20)	2.27 (25)	0
1-Butanol	7.8	20.0	0.810	117.7	35	2.948 (20)	17.8 (20)	1.7 (20)
Butyl acetate	1.0	1.37	0.881	126.1	37	0.734 (20)	5.0 (20)	1.86 (22)
Carbon tetrachloride	0.08 (25)	0.09 (25)	1.589 (25)	76.7	none	0.965 (20)	2.24 (20)	0
Chloroform	0.80	0.97	1.484	61.7	none	0.596 (15)	4.81 (20)	1.1 (25)
Cyclohexane	0.01	0.01	0.779	80.7	−11	0.980 (20)	2.02 (25)	0
Cyclohexanone	2.3	8.0	0.948	157	46	2.453 (15)	18.2 (20)	3.1 (20)
Dibutyl ether	0.77	0.3	0.769	142.4	25	0.602 (30)	3.06 (25)	1.19 (20)
1,2-Dichloroethane	0.87		1.253	83.5	15	0.887 (15)	10.65 (20)	1.7 (20)
2,2′-Dichloroethyl ether	1.02		1.222	178.5	55	2.41 (20)	21.2 (20)	2.61 (20)
Diethyl ether	7.4	1.26	0.713	34.6	−40	0.245 (20)	4.33 (20)	1.22 (16)
Diisobutyl ketone	0.05		0.806	165	15			
Diisopropyl ether	0.90	0.60	0.723	68.4	−12	0.379 (25)	3.88 (25)	1.26 (25)
Dipentyl acetate	6.9	1.26	0.783	186.8	57	1.19 (15)	2.77 (25)	0.98
Ethyl acetate	8.6	3.1	0.901	77.1	−3	4.26 (25)	6.11 (20)	1.84 (25)
Heptane	0.002		0.684	98.4	−4	0.416 (20)	1.92 (20)	0
Hexane	0.014		0.659	68.7	−23	0.313 (20)	1.89 (20)	0
1-Hexanol	0.70		0.819	157.5	60	3.87 (30)	13.3 (25)	1.55
Isooctane	0.0005		0.692	99.2	−7	0.502 (20)	1.94 (20)	0
Isopentyl acetate			0.876 (15)	142.0	80	0.872 (20)	4.81 (20)	1.84 (22)
Mesityl oxide	3.2	3.1	0.855	129.5	30	0.879 (25)	15.1 (20)	3.2 (25)
2-Methyl-1-butanol	2.3	9.1	0.816	128	50	5.50 (20)	14.7 (25)	1.8 (20)
Methyl isobutyl ketone	1.7 (25)	1.9 (25)	0.801	115.7	13	0.585 (20)	13.11 (20)	
4-Methyl-2-pentanol	1.6	6.4	0.808	131.7	41	4.07 (25)	13.3 (20)	
2-Methyl-1-propanol	8.5	16.4	0.802	107.9	39	2.88 (30)	17.9 (25)	2.96 (30)
Nitrobenzene	0.21	0.22	1.205 (15)	210.8	87	2.16 (15)	34.8 (25)	3.96 (25)
1-Pentanol	2.7 (22)		0.815	137.8	33	2.99 (20)	16.9 (20)	1.71 (20)
Pentyl acetate	0.181	0.9	0.875	149.2	25	0.924 (20)	4.75 (20)	1.9 (25)
Tetrahydrofuran	miscible	miscible	0.889	66	−17	0.55 (20)	7.58 (25)	1.75 (25)
Toluene	0.05	0.06	0.866	110.6	4	0.62 (15)	2.38 (20)	0.45 (20)
Tributyl phosphate	0.6		0.972	178 (22 mm)	146	3.39 (25)	7.96 (30)	3.1 (25)
m-Xylene	0.02	0.04	0.864	139.1	25	0.62 (20)	2.37 (20)	0.33 (20)

* Compiled mainly from J. A. Dean, ed., *Lange's Handbook of Chemistry*, 14th ed., McGraw-Hill, New York, 1992.

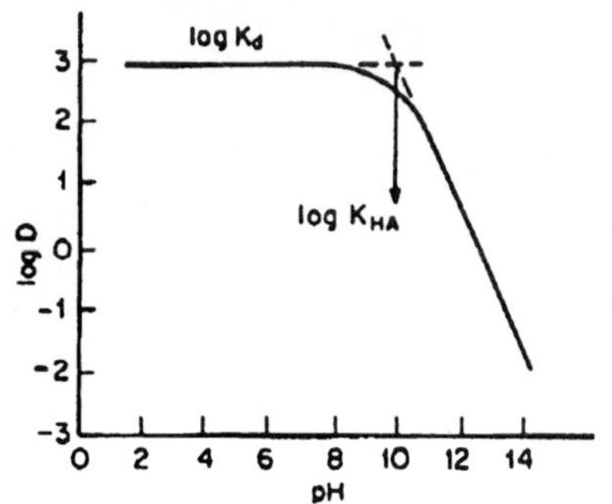

FIGURE 2.1 Log *D* vs. pH for a weak acid (RCOOH type) with $K_{HA} = 6.7 \times 10^9$ and $K_d = 720$.

In a plot of $1/D$ vs. $1/[H^+]$, the intercept gives $1/K_d$ and the slope gives $1/K_dK_{HA}$. Figure 2.1 is a logarithmic plot of Eq. (2.32). Two regions are apparent. When $1/K_{HA}[H^+]$ is much less than one, $\log D = \log K_dK_{HA} - \text{pH}$, and the plot is a line of unit slope. At low pH values the uncharged molecular species dominate in the aqueous phase and $\log D = \log K_d$. In dilute hydrochloric acid (between 5% and 10% HCl), carboxylic acids, phenols, and other weakly ionized acids will be extracted from an aqueous phase with an immiscible organic solvent, such as chloroform or carbon tetrachloride. Conversely, dilute NaOH will extract acidic solutes from an immiscible organic solvent by converting the acidic solute to the corresponding sodium salt.

Basic substances, such as organic amines, alkaloids, and cyclic nitrogen-containing ring compounds, follow just the reverse pattern from that of acidic materials. Dilute hydrochloric acid will convert basic substances into protonated species (positively charged) that will extract into the aqueous phase. Conversely, dilute NaOH will convert the protonated materials into neutral species that can be extracted from an aqueous phase into an immiscible organic phase. Control of the pH of the aqueous phase aids in separations. For example, phenols are not coverted to the corresponding salt by $NaHCO_3$, whereas carboxylic acids are converted to the corresponding salt. A NaOH extraction will convert both phenols and carboxylic acids to the corresponding sodium salts and will extract both into the aqueous phase.

The extraction of neutral organic species increases by a factor of 2 to 4 for each additional CH_2 group introduced into the molecule of a member of a homologous series. Molecules that can form hydrogen bonds with water, such as aldehydes, ethers, and ketones, will exhibit lower distribution ratios than the corresponding hydrocarbons by factors of 5 to 150. A two-solvent system can sometimes be devised involving two fairly immiscible organic solvents. The more polar substances partition into the more polar of the two phases.

In contrast to the large number of organic compounds that distribute themselves between aqueous and organic phases, the number of simple inorganic compounds that do so is quite small. The partition of iodine between water and carbon tetrachloride is well known. The high value (about 90) for the partition coefficient facilitates the extraction of small quantities of iodine from aqueous solution. Iodine in the organic phase can be removed by shaking the organic phase with an aqueous potassium iodide solution in consequence of the formation of polyiodide ions. Chlorine dioxide partitions between carbon tetrachloride and water, as does bromine and a number of simple systems involving covalent, nonpolar species (usually molecules) in both the organic and aqueous phases. Examples are OsO_4, RuO_4, $AsBr_3$, $AsCl_3$, $GeCl_4$, InI_3, $HgCl_2$, HgI_2, $GeCl_4$, $SbBr_3$, SbI_3, SnI_4, and Hg metal. These substances are all more or less volatile, but extraction methods have the advantages of speed and simplicity. The disordered structure of an inert solvent, such as CCl_4 or $CHCl_3$ (and sometimes benzene, toluene, or xylene), more easily accommodates a covalent molecule than does the ordered structure of water. Brief details for extraction of these simple molecules are given in Table 2.7.

The noble gases partition between nitrobenzene and water with the partition coefficient increasing with their atomic volume (He, 5.5; Ne, 6; Ar, 12; Kr, 19; and Xe, 28).

2.2.2.1 Oxygen-type Solvents or "Onium" Systems. Solvent molecules may participate in ionic reactions. Recognition of the hydrated hydronium ion, $H_9O_4^+$, as the cation that pairs with halometallic anions in the extraction of complex halometallic acids has clarified the role of the oxygen-containing solvent in oxonium extraction systems. This cation must be stabilized by hydrogen bonding to the organic solvent (denoted S). Thus, the coordinating ability of the solvent is important. In effectiveness, the order is roughly isobutyl methyl ketone > butyl acetate > pentyl alcohol > diethyl ether. Solvents such as benzene and carbon tetrachloride are ineffective.

TABLE 2.7 Extraction of Systems Having Simple, Nonpolar Species in Both the Organic Solvent and Aqueous Phase

Compound	Extraction conditions	Coextractants	Reference
$AsCl_3$	95% extracted when 10–12*M* HCl shaken with equal volume of benzene. Two extractions, 99.5%. Back-extracted with water.	100% $GeCl_4$; traces of Hg, Sb, Sn, Se, and Te (removed by wash with 9*M* HCl).	G. O. Brink et al., *J. Am. Chem. Soc.* **79**:1303 (1957); T. Korenaga, *Analyst* **106**:40 (1981).
$AsBr_3$	99.4% extracted by benzene from 0.03*M* HBr plus 10*M* H_2SO_4.	Ge, Hg(II), Sb(III), Se, Sn.	A. P. Grimanis and I. Hadzistelios. *Anal. Chim. Acta* **41**:15 (1968).
$AsBr_5$	Extraction with CCl_4 from 0.1*M* Br^- plus 8*M* H_2SO_4.		K. Studlar, *Coll. Czech. Chem. Commun.* **31**:1999 (1966).
AsI_3	100% extracted when 0.05*M* I^- plus 6*M* H_2SO_4 is shaken with equal volume of toluene. Scrubbing toluene with 0.05*M* I^- plus 3*M* H_2SO_4 removes 100% As and <1% Sb.	100% Sb(III), 2% Bi, Ge, 6% Hg(II), 5% Pb, Se, Sn. Insoluble iodides (such as Ag, Pb, Tl) interfere.	G. O. Brink et al., *J. Am. Chem. Soc.* **79**:1303 (1957); A. R. Byrne and D. Gorenc, *Anal. Chim. Acta* **59**:81, 91 (1972).
$GeCl_4$	Extraction coefficient is 50–500 when 8–9*M* HCl is shaken with equal volume of CCl_4 (or $CHCl_3$ or benzene). Shaking the extract with water removes the germanium.	70% $AsCl_3$ only other element. Fluoride ion decreases Ge extraction.	W. Fischer and W. Harre, *Angew. Chem.* **66**:165 (1954); W. A. Schneider and E. B. Sandell, *Mikrochim. Acta* **1954**, 263.
$HgCl_2$	Extraction is 74.1% with benzene from 0.03*M* HBr and 10*M* H_2SO_4.	As, Ge, Nb, Sb, Se, Sn.	A. P. Grimanis and I. Hadzistelios, *Anal. Chim. Acta* **41**:15 (1968).
OsO_4	Extraction coefficient is 19.1 with $CHCl_3$ from HNO_3 medium. Slight excess Fe(II) sulfate prevents coextraction of RuO_4.	RuO_4 unless Fe(II) is present.	R. D. Sauerbrunn and E. B. Sandell, *Anal. Chim. Acta* **9**:86 (1953); G. Goldstein et al., *Talanta* **7**:296 (1961).
RuO_4	Oxidation to Ru(VIII) by Ag(II) oxide or Ce(IV) ammonium sulfate in H_2SO_4 or HNO_3 medium and extraction with $CHCl_3$ or CCl_4. Partition constant is 58.4 for CCl_4. If Os:Ru ratio is <50–100, remove OsO_4 by extraction; otherwise, by distillation. Back-extract with aqueous SO_2 or 3*M* NaOH plus $NaHSO_3$.	OsO_4.	C. Surasiti and E. B. Sandell, *Anal. Chim. Acta* **22**:261 (1960); C. E. Epperson, R. R. Landolt, and W. V. Kessler, *Anal. Chem.* **48**:979 (1976); J. W. T. Meadows and G. M. Matlack, *ibid.* **34**:89 (1962).
SbI_3	100% extraction (extraction coefficient = 3000) with benzene or toluene from 0.01*M* I^- and 5*M* H_2SO_4.	Ge, 98% Hg(II), and Sn.	R. W. Ramette, *Anal. Chem.* **30**:1158 (1958).
SeI_4	95.6% extraction with benzene from 0.03*M* HBr and 10*M* H_2SO_4.	As, Ge, Hg, Nb, Sb, Sn.	A. P. Grimanis and I. Hadzistilios, *Anal. Chim. Acta* **41**:15 (1968).
SnI_4	100% extraction with benzene (or hexane) from 1*M* NaI and 4.0*M* $HClO_4$ (or 2*M* H_2SO_4).	As, Bi, In, Pb.	D. D. Gilbert and E. B. Sandell, *Microchem. J.* **4**:491 (1960); A. R. Byrne and D. Gorenc, *Anal. Chim. Acta* **59**:81 (1972).
TeI_4	Extraction coefficient is 4.0 when extracted with toluene from 0.05*M* KI and 3*M* H_2SO_4.		I. Havezov and M. Stoeppler, *Z. Anal. Chem.* **258**:189 (1972).

TABLE 2.8 Percentage Extraction of Metals as Chlorides with Oxygen-Type Solvents

From 6M HCl, *diisopropyl ether extracts essentially none of these elements: Al, Ag, Ca, Cr(III), Fe(II), Mg, Pd(II), Sn(IV), Te(IV), Zn, and Zr.*

Element	Diethyl ether 3*M* HCl	Diethyl ether 6*M* HCl	Diisopropyl ether 8*M* HCl	4-Methyl-2-pentanone (MIBK) 7*M* HCl
Antimony(III)	22	6	2	69
Antimony(V)	6	81	100 (98%, 6*M*)	94*
Arsenic(III)	7	68	67 (6*M*)	88
Arsenic(V)				3.5
Bismuth			~0	0.5
Cadmium			~0	12
Chromium(VI)			~0 (Cr^{3+})	98
Cobalt			~0 (Co^{3+})	2–3
Copper(II)	0.05	0.05	0	4
Gallium		97	99.9	>99.9
Germanium		50		97*
Gold(III)	98	95	99 (6*M*)	99*
Indium			~0	94
Iridium(IV)	0.02	5		
Iron(III)	8	99	99.9	99.996
Manganese(II)			~0	0.7
Mercury(II)	13 (0.3*M*)	0.2		5*
Molybdenum(VI)		80–90	21 (8%, 6*M*)	96
Nickel			~0	1
Niobium		30		76 (8*M*)
Platinum			~0	56
Rhenium				83
Selenium(IV)				6,* 98 (8*M*)
Selenium(VI)				4*
Tellurium(IV)	3	34		42*
Tellurium(VI)				0.6*
Thallium(III)	99	98	99	99*
Thorium				1
Tin(II)		15–30		
Tin(IV)	23	17	~0	93
Titanium(IV)			~0	1*
Uranium(VI)				22; 61 (8*M*)
Vanadium(V)			22	81
Tungsten(VI)				49†
Zinc	0.03	0.2	~0	5

* 1:1 MIBK + pentyl acetate; 7*M* HCl.

† H_3PO_4 present (1 mL concentrated acid/25 mL aqueous phase).

References

Diethyl ether: Mostly from F. Mylius, *Z. Anorg. Chem.* **70**:203 (1911); F. Mylius and C. Hüttner, *Berichte* **44**:1315 (1911).

Diisopropyl ether: R. W. Dodson, G. J. Forney, and E. H. Swift, *J. Am. Chem. Soc.* **58**:2576 (1936); F. C. Edwards and A. F. Voigt, *Anal. Chem.* **21**:1204 (1949); F. A. Pohl and W. Bonsels, *Z. Anal. Chem.* **161**:108 (1958).

4-Methyl-2-pentanone (MIBK): H. Specker, *Arch. Eisenhüttenw.* **29**:467 (1958); W. Doll and H. Specker, *Z. Anal. Chem.* **161**:354 (1958); T. Ishimore et al., *Jpn. At. Energy Res. Inst. Rep. 1106* (1966).

MIBK + pentyl acetate (1:1): A. Classen and L. Bastings, *Z. Anal. Chem.* **160**:403 (1958).

The well-known extraction of iron(III) from HCl solution is outlined with isobutyl methyl ketone as solvent. The solvent distributes between the ketone and aqueous phases subject to the "onion" association system in the aqueous phase:

$$Cl^- + H(H_2O)_4^+ + 3S_{aq} \rightleftarrows H(H_2O)(S)_3^+, Cl^- + 3H_2O \tag{2.33}$$

The tetrachloroferrate(III)(1–) ion exists in equilibrium with the hydrated iron(III) cation:

$$Fe(H_2O)_6^{3+} + 4Cl^- \rightleftarrows FeCl_4^- + 6H_2O \tag{2.34}$$

In turn the tetrachloroferrate(III)(1–) anion enters into an equilibrium with the solvent onion cation:

$$H(H_2O)(S)_3^+, Cl^- + FeCl_4^- \rightleftarrows H(H_2O)(S)_3^+,FeCl_4^- \tag{2.35}$$

The onion association complex distributes between the two phases. Extraction of iron(III) is complete in one equilibrium from 6*M* HCl when methyl isobutyl ketone is used as solvent in place of diethyl ether or other ethers. Ethers suffer from the formation of peroxides, which at best are a nuisance and at worst may cause explosions.

The iron is easily stripped from the organic phase by washing it with a 0.1*M* HCl solution. The back-extraction is expedited by adding either a reducing agent [iron(II) is not extracted from HCl solutions] or using an aqueous phase that is 1*M* phosphoric acid.

Diethyl ether or diisopropyl ether were among the early solvents of the onium type. Disadvantages were low boiling points, flammability, marked solubility in HX acids, and somewhat mediocre extracting power for metals. Generally the optimum extraction acidity will be slightly higher for diisopropyl ether than for diethyl ether, and the percentage extraction will be larger. Other oxygen-type solvents have advantages and are often substituted, particularly 4-methyl-2-pentanone (methyl isobutyl ketone or MIBK), pentyl acetate, tributyl phosphate (TBP) (viscous alone and usually dissolved in an inert diluent), and trioctylphosphine oxide (TOPO). TBP and TOPO belong to a family of powerful extraction reagents having the phosphoryl (P=O) or P→O group as the solvating group.

Tables 2.8 and 2.9 summarize the metals more readily extractable by ethers and MIBK in conjunction with HCl or SCN^- solutions. For many metals there is a rather close similarity between the extraction of bromides and chlorides, but there are some differences. Gold(III) is extracted more easily from HBr than from HCl solutions; extraction is 99% or better over the range 0.5–4*M* HBr. Whereas indium is extracted only slightly with diethyl ether from HCl solution, it is extracted >99% from 4–5*M* HBr. Thallium(III) is extracted 99% or better over the range 0.1–6*M* HBr. Thallium(I) extracts > 99% from 1–3*M* HBr. Significant amounts of these elements also extract from HBr medium: As(III), Cu(II), Fe(III), Ga, Hg, Mo(VI), Sb(III,V), Re(VII), Se(IV), Sn(II,IV), Te(IV), and Zn. Traces of Cd and Co extract.[14]

In contrast to their behavior in HCl and HBr solutions, Cd, Hg, Sb(III), and Sn(II), along with Au(III), can be extracted quantitatively by diethyl ether from 7*M* HI solution. Also quantitatively extracted are Tl(I) from 0.5–2*M* HBr and Tl(III) from 0.05–2*M* HBr. Coextractants from 7*M* HBr include As(III), 63%; Bi, 34%; In, 8%; Mo(VI), 7%; Te(IV), 6%; and Zn, 11%.

Tributyl phosphate extracts many metal chlorides from HCl solutions, including those of Mo, Nb, Sc, and Ta, most of which are not appreciably extracted by ethers or ketones (Table 2.10). Therefore, it is less selective than ethers and ketones for such metals as Fe(III) and Ga. Tributyl phosphate also is used in nitrate media for which it is a good extractant for Au(III), Lu, Np, Nb, Os, Pa, Pu, Sc, Th, U(VI), and Y (Table 2.11). Although diluted tributyl phosphate exhibits less extractive power, this loss can be easily compensated by salting the aqueous phase with nitrates, particularly $Al(NO_3)_3 \cdot 9H_2O$.

Trioctylphosphien oxide generally parallels triphenyl phosphate in its extraction of metals; it finds use for the extraction of the actinide elements (Tables 2.12 and 2.13). Metals slightly or not extracted include Al, Co, Cr(III), Cu, Fe(II), Fe(III), Mg, Mn, rare earths, and Zn.

---

[14] R. Bock, H. Kusche, and E. Bock, *Z. Anal. Chem.* **138**:167 (1953); H. Irving and F. J. C. Rossotti, *Analyst* **77**:801 (1952); L. Kosta and J. Hoste, *Mikrochim. Acta* **1956,** 790; I. Wada and R. Ishii, *Bull. Inst. Phys. Chem. Res. Tokyo* **13**:264 (1934).

TABLE 2.9 Percent Extraction of Elements as Thiocyanates with Diethyl Ether from 0.5*M* HCl Solutions

Elements slightly extracted (<0.3%): As(V), Ge, Mn, Ni, Th, and Y.

Element	1*M* SCN^-	3*M* SCN^-	5*M* SCN^-	7*M* SCN^-
Aluminum		1	9	49
Antimony(III)	hydrolyze	hydrolyze	hydrolyze	2
Arsenic(III)	0.3		0.1	0.4*
Beryllium	4	50	84	92
Bismuth	0.3			0.1
Cadmium	0.1			0.2
Chromium(III)	0.06			3
Cobalt(II)	4	58	75	75
Copper(I)		3†	0.4	
Gallium	65	91		99
Indium	52	75	68	48
Iron(III)	89	84	76	53
Mercury(II)	0.2			0.7
Molybdenum(V)	99	97		97
Palladium(II)	2			<0.1
Scandium	13	80	85‡	90
Titanium(III)	59	84	80	76
Titanium(IV)			42§	13
Tin(IV)	99	100	100	100
Uranium(VI)	45	29	14	7
Vanadium(IV)	15	9		2
Zinc	96	97	95	93

* 0.8*M* HCl also present.
† 2*M* SCN^-.
‡ 4*M* SCN^-.
§ 3*M* HCl and 5.7*M* SCN^-.
References: R. Bock, *Z. Anorg. Chem.*, **249**:146 (1942); *Z. Anal. Chem.* **133**:110 (1951).

Di-(2-ethylhexyl)-*o*-phosporic acid extracts elements of the IVb, Vb, and VIb groups of the periodic table from HCl and HNO_3 solutions (Tables 2.14 and 2.15).

2.2.3 Metal-Chelate Systems

To convert a metal ion in aqueous solution into an extractable species, the charge has to be neutralized and any waters of hydration have to be displaced. If the metal ions are effectively surrounded by hydrophobic ligands that are able to bring about charge neutralization and occupy all the positions in the coordination sphere of the metal ion, distribution will strongly favor an organic phase. Chelate systems offer one type of metal-extraction system.

The equilibria involved in a metal-chelate-extraction system can be outlined as follows. Between the organic and aqueous phases the chelating agent distributes and dissociates in the aqueous phase:

$$(\mathrm{H}L)_{\mathrm{org}} \rightleftarrows (\mathrm{H}L)_{\mathrm{aq}} \rightleftarrows \mathrm{H}^+ + L^- \tag{2.36}$$

The ligand interacts with the hydrated metal cation to form the metal chelate

$$M(\mathrm{H_2O})_m^{n+} + nL^- \rightleftarrows ML_n + m\mathrm{H_2O} \tag{2.37}$$

TABLE 2.10 Percentage Extraction of Metals into Tributyl Phosphate from HCl Solutions

Elements extracted (<1%) from 1M HCl: Al, Ba, Mg, Mn, Ni, Th, Y.

Element	1*M* HCl	4*M* HCl	6*M* HCl	9*M* HCl
Antimony		98	99.8	99.9
Arsenic(III)	15	72	95	98
Bismuth	95	67	33	9
Bromide	36			
Cadmium	61	95	95	85
Cerium	0.1	0.16	0.4	
Cobalt	<1	1.6	4.8	56
Copper(II)	1	28	44	44
Gallium	56	99.8	99.9	99.9
Germanium	3	72	98.4	98
Gold(III)	100	100	100	100
Indium	80	99.6	99.4	97.5
Iodide	99.8			
Iridium	6			
Iron(III)	96.9	99.9	99.98	99.98
Lead	16			
Mercury	98.7	98.7	98.6	89
Molybdenum	76	99.7	99.7	99.5
Niobium	72	93	99.8	99.9
Osmium	80	89	86	76
Palladium	33	72	64	44
Platinum	85	95	95	76
Protactinium(V)	0.6	85	99.8	99.8
Rhenium	99.2	99.7	98	91
Scandium	<0.1	3.8	72	99.1
Selenium(IV)	9			
Silver	93	80	36	9
Tantalum	99.8	99.3	99.7	99.97
Technetium	98.3	98.4	99.0	97
Tellurium(IV)	36	99.8	99.8	99.8
Thallium(I)	2			
Thallium(III)	99.98	99.98	99.99	99.98
Tin(II)	91		72	72
Uranium(VI)	24	96.9	98.8	98.4
Vanadium(V)	3.8	86	96	98.3
Tungsten	99.7	99.7	99.4	99.0
Zinc	86	96	91	72
Zirconium	0.01	0.09	86	99.97

References: K. Kimura, *Bull. Chem. Soc. Jpn.* **33**:1044 (1960); T. Ishimori and E. Nakamura, *Jpn. At. Energy Res. Inst. Rep. 1047* (1963); D. F. Peppard, J. P. Gray, and G. W. Mason, *Anal. Chem.* **27**:296 (1953).

which distributes between the two phases

$$(ML_n)_{aq} \rightleftarrows (HL_n)_{org} \tag{2.38}$$

where M represents a metal ion of charge $n+$ and with m waters of hydration, and HL represents a chelating agent. The chelating agent, usually a slightly dissociated, polyfunctional organic acid, distributes between the two phases. An additional acid–base dissociation system exists in the aqueous

TABLE 2.11 Percent Extraction of Elements from Nitric Acid by Tributyl Phosphate

Element	1*M* HNO_3	6*M* HNO_3	11*M* HNO_3	14*M* HNO_3
Bismuth	67	17	3	5
Gold(III)	100	99	80	67
Hafnium	20 (3*M*)	89		
Iron(III)	1	1	34	94
Lutetium			97	99
Mercury	91	28	20	24
Molybdenum	2	1	14	76
Neptunium	86	99	99	
Niobium	17	83	95	98
Osmium	97	83	76	76
Palladium	76	5	1	2
Polonium	34	2	5	14
Platinum	64	2	0.5	0.5
Plutonium(VI)	93	98	99	
Protactinium	64	97		
Rhenium	94	24	9	7
Ruthenium	7	2	1	
Scandium	24	66	97	100
Thallium(III)	24	6	1	1
Titanium	0.9	1	3	80
Thorium	80	98	100	100
Uranium(VI)	97	100		
Vanadium	3	1		
Tungsten	97	28	7	14
Yttrium	9	50	93	99
Zirconium		80		

References: B. Bernström and J. Rydberg, *Acta Chem. Scand.* **11**:1173 (1957); T. Ishimori and E. Nakamura, *Jpn. At. Energy Res. Inst. Rep. 1047* (1963).

phase. Some of the free ligand ions in the aqueous phase react with the hydrated metal ions to form the extractable chelate ML_n, which partitions between the two phases.

The expression for the distribution ratio of the metal between the two phases is

$$\frac{1}{D} = \frac{1}{(K_d)_c} + \frac{K^n_{\mathrm{HL}}(K_d)^n_r[\mathrm{H}^+]^n}{K_f(K_d)_c[\mathrm{H}L]^n_{\mathrm{org}}} \tag{2.39}$$

where $(K_d)_c$ = partition coefficient of the metal chelate
K_{HL} = partition coefficient of the chelating agent
K_f = formation constant of the metal chelate in the aqueous phase
$(K_d)_r$ = partition coefficient of the reagent

The distribution ratio is dependent on the nature of the chelating reagent and its concentration, the pH of the aqueous phase, and the solubility of the chelating reagent in the organic solvent. The extent of extraction increases with the equilibrium (excess, unused) concentration of chelating reagent in the organic phase. Starting with a low pH, the logarithm of the distribution ratio increases with a slope of *n*, eventually reaching a constant, pH-independent value determined by the partition coefficient of the metal chelate. The distribution of zinc as a function of aqueous

TABLE 2.12 Percentage Extraction of Metals from HCl Solution by a 5% Solution of Trioctylphosphine Oxide in Toluene

Element	1*M* HCl	4*M* HCl	8*M* HCl	12*M* HCl
Arsenic(III)		44	95	95
Bismuth	99.2	44	3.1	
Cadmium	80*	4		
Cobalt	60*		70	
Copper	30*	9†		
Gallium	5.9	99.9	100	100
Gold(III)	100	100	100	100
Indium	97	98.4	98.5	72
Iron(III)	67	95	94	94
Mercury(II)	93	98	93	24
Molybdenum(VI)	98	99.8	96	85
Niobium	39	98.4	99.7	99.7
Platinum	44	95	86	5.9
Protactinium	67	61	95	95
Rhenium	91	95	89	28
Scandium	4.8	95	99.8	39
Selenium	3.0	5.9	50	93
Technetium(VII)	97	99.3	98	91
Tin(IV)	99	99	99.5	97
Titanium		17	96	99
Uranium(VI)	99.9	100	100	95
Vanadium(V)	9	44	86	86
Tungsten	61	86	91	86
Zinc	98.4	99.2	80	7
Zirconium	15	99.5	99.0	99.6

* 2*M* HCl.
† 5*M* HCl.

References: T. Ishimori and E. Nakamura, *Jpn. At. Energy Res. Inst. Rep. 1047* (1963); T. Sato and M. Yamatake, *Z. Anorg. Allg. Chem.* **391**:174 (1972).

TABLE 2.13 Percent Extraction of Elements by 5% Trioctylphosphine Oxide (in Toluene) from Nitric Acid Solution

Element	1*M* HNO_3	3.5*M* HNO_3	6*M* HNO_3	11*M* HNO_3
Bismuth	83	0.3	0.01	
Gold(III)	100	99	67	24
Hafnium	95	93	91	91
Iron(III)	0.4	0.3	0.6	6
Mercury	56	24	17	28
Molybdenum	20	2	2	25
Niobium	20	39	50	72
Osmium	39	17	20	20
Protactinium	99	98	99	99
Scandium	17	24	33	44
Thallium	7	9	9	
Thorium	99	99	98	94
Tin	17	24	33	44
Titanium	2	1	6	24
Uranium(VI)	100	99	100	97
Yttrium	5	0.1	0.1	
Zirconium	91	96		98

Reference: T. Isimori and E. Nakamura, *Jpn. At. Energy Res. Inst. Rep. 1047* (1963).

TABLE 2.14 Percentage Extraction of Metals from HCl Solution with Di-(2-Ethylhexyl)phosphoric Acid (50% in Toluene)

Negligible amounts extracted from 0.01M HCl: alkali metals, As(III), Au, Pd, Pt, Ru, Se(IV), and Se(VI).

Negligible amounts extracted from 0.1M HCl: alkali metals, As(V), Cd, Co, Ge, Mg, Pt, Re, Se(IV), Se(VI), and Tc.

Negligible amounts extracted from 1M HCl: Al, alkali metals, As(III and V), Ba, Bi, Cd, Co, Cr, Ge, Ir, La, Mg, Mn, Ni, Pb, Pd, Pt, Ra, Re, Ru, Se(IV and VI), Sr, Tc, and Tl.

Element	0.01*M* HCl	0.1*M* HCl	1*M* HCl
Actinium	99.7	20	1.1
Americium	100	93	
Antimony		95	76
Bismuth		99.5	
Calcium		11	
Chromium	9	0.5	
Copper	50	1.6	0.1
Gallium		99.3	14
Gold	0	1	6
Hafnium		100	100
Indium		100	28
Iridium	11	3	0
Iron(III)	100	100	72
Lanthanum	100	91	0.2
Lead	97	24	0.1
Magnesium	6		0.01
Manganese	36	4	
Mercury		33	8
Molybdenum		97	98
Niobium			99
Neptunium		100	100
Nickel		2	
Osmium	91	89	86
Protactinium	100	100	100
Promethium	100	98	5
Scandium	100	100	100
Silver	56	9	0.7
Strontium	3		
Tantalum		6	11
Tellurium			9
Thorium	100	100	100
Thullium	100	100	96
Tin		83	50
Titanium			39
Uranium		100	100
Vanadium		17	2
Tungsten		24	17
Yttrium		100	97
Zinc	99	76	3
Zirconium	99	99	99

***Reference*:** K. Kimura, *Bull. Chem. Soc. Jpn.* **33**:1038 (1960).

TABLE 2.15 Extraction of Elements from Nitric Acid Solution with Di-(2-Ethylhexyl)phosphoric Acid

One percent or less of these elements were extracted: As(III, V), Ce(III), Co, Mn, rare earths, and Tc.

Element	Percent extracted	
	$0.5M$ HNO_3	$10M$ HNO_3
Antimony(III)	30	
Antimony(V)	2	98 ($9M$ H_2SO_4, 2 extractions)
Arsenic(III)	0.1	
Arsenic(V)	0.2	
Bismuth	98	
Cadmium	6.7	
Cerium(IV)		>99
Chromium(II)	4.5	
Chromium(VI)	47	
Cobalt	0.5	
Copper	2	
Gallium	4 (pH 3)	
Indium	98	
Iron(III)	99	
Manganese	0.6	
Molybdenum		>99 (>5*M*)
Nickel	7.5	
Niobium		~99 (>3*M*)
Ruthenium	<2	<2
Scandium		99 (1–11*M*)
Selenium	2.3	
Tellurium	3	~0
Thorium		>98
Titanium		95 (1–11*M*)
Uranium(VI)		>98
Zinc	2	
Zirconium		>99

Reference: V. G. Goryushima and E. Y. Biryukova, *J. Anal. Chem. USSR* **24**:443 (1969).

hydrogen-ion concentration for three different concentrations of chelating agent is shown in Fig. 2.2.

Collecting the constants in the right-hand portion of Eq. (2.39) into one term and ignoring the plateau, the equation when expressed in logarithmic form becomes

$$\log D = K_{ext} + n \log[HL]^n_{org} + n\,pH \tag{2.40}$$

A plot of log D versus pH ($[HL]_{org}$ being kept constant) will be a straight line of slope n and intercept log K_{ext}.

For 50% extraction, $D = 1$ and the equation becomes

$$0 = \log K_{ext} + n \log[HL]_{org} + n\,pH_{1/2} \tag{2.41}$$

where $pH_{1/2}$ is the pH value for which the extraction is 50% at any given reagent concentration. The extraction curves are of the sigmoid form, becoming increasingly steeper for $n = 1, 2, 3,$

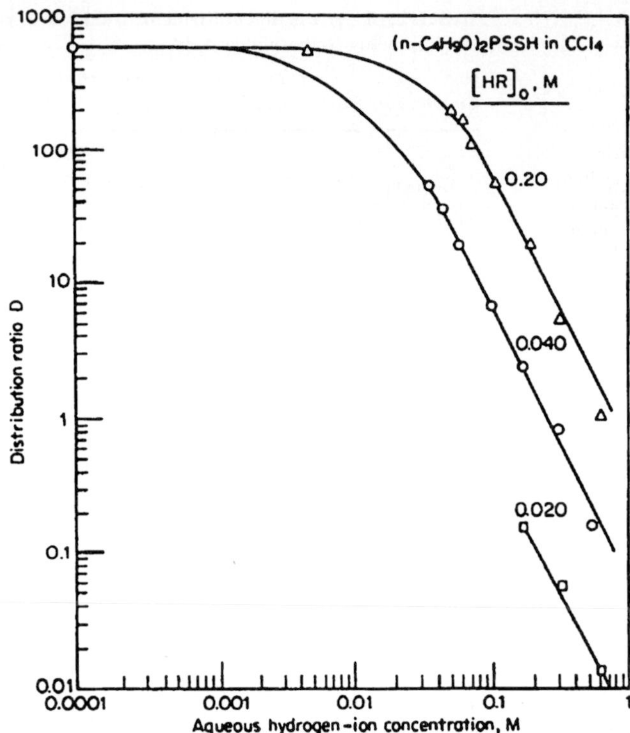

FIGURE 2.2 Distribution of zinc as a function of aqueous hydrogen-ion concentration for three stated concentrations of chelating agent in CCl_4.

respectively. Thus the pH range for extraction covering the range 1% to 99% will be 4 units when $n = 1$, 2 units when $n = 2$, and only 1.3 units when $n = 3$. To separate two metals (both $n = 2$ forming the same type chelate), the $pH_{1/2}$ separation must be 2 pH units (3 pH units for the range 0.1% to 99.9%).

It is seldom possible to change the concentration of most reagents by more than 10- or 100-fold because of their limited solubility. When possible a small improvement in the extent of extraction can be achieved. However, the hydrogen-ion concentration can readily be altered over a range of more than 10 powers of 10, and thus becomes the most effective variable for controlling selectivity in analytical procedures involving the extraction of chelate complexes.

Among the commonest chelating agents used in analytical chemistry are the anions of 8-hydroxyquinoline (Table 2.16), diphenylthiocarbazone (dithizone), acetylacetone, 2-thenoyltrifluoroacetone, nitrosophenylhydroxylamine (cupferron), nitrosonaphthols, and diethyldithiocarbamate. The properties and uses of these reagents and those of the numerous variants on their basic structure are described in Starý's book.[15]

The extractability of diethyldithiocarbamates by CCl_4 is summarized in Table 2.17.[16] Sodium diethyldithiocarbamate cannot, in general, be used for extraction of metals from mineral acid solutions because of the rapid decomposition of diethyldithiocarbamic acid in aqueous solution. By comparison, extractions with diethylammonium diethyldithiocarbamate can be made from an acidic

[15] J. Starý, *The Solvent Extraction of Metal Chelates*, Pergamon, New York, 1964.
[16] H. Bode, *Z. Anal. Chem.* **143**:182 (1954).

TABLE 2.16 Extraction of Metal 8-Hydroxyquinolates (Oxines) with $CHCl_3$

Element	pH of complete extraction	*M* HOx	Remarks
Aluminum	4.5–10	0.01	
Barium	ca. 11	1	Incomplete extraction
Beryllium	6–10	0.5	Incomplete extraction (maximum 87%)
Bismuth	2.5–11	0.1	
Cadmium	5–9	0.1	Amines aid extraction
Calcium	ca. 11	0.5	
Cerium(III)	ca. 10		Citrate solution
Chromium(III)			Boiling required to form chelate; extraction in faintly acidic solution
Cobalt(II)	4.5–10.5	0.1	
Copper(II)	2–12	0.1	
Gallium	2–12	0.01	
Hafnium	ca. 2	0.1	
Holmium	8–10	0.01	
Indium	3–11	0.01	
Iron(III)	2–10	0.01	
Lanthanum	7–10	0.1	
Lead	6–10	0.1	
Magnesium	8.5 (90%)	0.5	Hydrated Mg(Ox)2 insoluble in CHCl3 but soluble in MIBK
Manganese(II)	ca. 7 (90%)	0.01	
Manganese(III)	9		Selective extraction in presence of NH_4F
Mercury(II)	>3	0.1	Incomplete extraction; Cl^- interferes
Molybdenum(VI)	1–5.5	0.01	Slow
Nickel	5–8.5	0.1	Slow equilibrium
Niobium	4–9		HOx added to 0.3% tartrate solution and extraction made; recovery almost complete
Neptunium(V)			Partly extracted by 1-butanol at pH 10
Palladium	0	0.1	Very slow equilibrium
Plutonium(IV)			Extracted into pentyl acetate, pH < 8
Plutonium(VI)			Extracted into pentyl acetate, pH 4–8
Polonium			Partly extracted at pH 3–4
Protactinium			Incomplete extraction by 1-pentanol from basic medium
Promethium(III)			Partly extracted from basic solutions
Rhodium(III)	6–9		Chelate formed on boiling
Samarium	6–8.5	0.5	
Scandium	4.5–10	0.1	
Silver	8–9	0.1	Maximum extraction only 90%
Strontium	11.3	5%	4% Butylamine present
Tantalum			Poor extraction
Terbium	>7	0.1	Almost complete extraction
Thallium(I)			Incomplete extraction from basic solution
Thallium(III)	3.5–9	0.01	
Thorium	4–10	0.1	
Tin(II)			Quantitative extraction from weakly basic solution
Titanium(IV)	2.5–9	0.1	
Uranium(VI)	5–9	0.1	
Vanadium(V)	2–4	0.01	No extraction above pH 9
Tungsten	2.5–3.5	0.01–0.14	0.01*M* EDTA present; no extraction pH > 7
Yttrium	7–10	0.2	Almost completely extracted
Zinc	3.6	0.1	Extraction only 90%; amines aid
Zirconium	1.5–4; also 9	0.1	

TABLE 2.17 pH Ranges for Extractability of Diethyldithiocarbamates with Carbon Tetrachloride

Rhodium, ruthenium, and uranium are incompletely extracted but no pH ranges were given.

Element	Complete extraction	Partial extraction	Complete extraction, EDTA present	Complete extraction, cyanide present
Ag	5–11		5–11	None
As(III)	6.5–8			
Au		5–11	5–11	None
Bi	5–11		5–11	5–11
Cd	5–11		5–55; partial 5.5–7	7–11
Co(III)	5–11		Partial, 5–5.5	5–11
Cu(II)	5–11		5–11	9–11; partial 7–9
Fe(II)	5–9.5		Partial, 5	Partial, 7–11
Ga		5–6	Partial, 5–7	
Hg(II)	5–11		5–11	7–8, partial 8–10
Iridium		5–11		
Mn(III)	6–9	5–6, 9–10.5	None	7–8; partial 8–10
Nb		5–6	None	
Ni	5–11			None
Os		5–11	Partial, 5–11	None
Pb(II)	5–11		5–5.5; partial, 5.5–6.5	7–11
Pd(II)	5–11		5–11	None
Pt		5–11		5–11
Sb	5–9	9–11	5–9; partial, 9–11	8.5; partial 8.5–10
Se(IV)	5–6.5	6.5–8	5–5.5; partial, 5.5–8	
Sn(IV)	5–5.5	5.5–7	Partial 5–5.5	
Te(IV)	5–9	9–10	5–9; partial, 9–10	7–8.5
Tl(I,III)	5–11		5–6.5; partial, 6.5–8	7–11
V(V)	5–5.5	5.5–7	Partial 5–6	7–11
Zn	5–11		Partial 5–5.5	Partial, 7–7.5

Reference: H. Bode, *Z. Anal. Chem.* **143**:182 (1954).

aqueous solution because most of the reagent remains undecomposed in the organic solvent. Metals quantitatively extracted from 0.5*M* H_2SO_4 include antimony(III), arsenic(III), bismuth, cadmium, copper(II), indium, lead, mercury, molybdenum(VI), palladium(II), platinum(II), selenium(IV), silver, thallium(III), and tin(II). From 6*M* HCl the list narrows to antimony(III), arsenic(III), copper, mercury(II), palladium(II), platinum(II), selenium, and thallium(III). In 10*M* HCl, only palladium and platinum are extractable. In addition, such metals as cobalt(II), iron(II and III), manganese(II), nickel, and zinc are extracted from weakly acidic and basic solutions.[17]

Elements extracted as cupferrates by chloroform from 1*N* mineral acid include (elements with asterisks can be extracted quantitatively): antimony(III)*(H_2SO_4, HNO_3), copper(II), hafnium,* iron(III),* molybdenum(VI),* niobium, palladium,* tantalum, technetium, titanium,* tin(II and IV),* uranium(IV),* vanadium(V),* tungsten, and zirconium.* Those extracted from 0.1*N* mineral acid include: bismuth* (H_2SO_4, HNO_3), copper(II),* gallium,* indium, mercury(II), scandium, and thorium* (ethyl acetate). At pH 5.5 to 5.7 in the presence of EDTA and citrate, the cupferrates extracted by MIBK include aluminum, beryllium, cerium(III and IV), hafnium, niobium, rare-earth elements, tantalum, tin(II and IV), titanium, uranium(IV and VI), vanadium, and zirconium.[18]

The pH of complete extraction for some common chelate solvent extraction systems are presented in Table 2.18.

[17] H. Bode and F. Neumann, *Z. Anal. Chem.* **172**:1 (1960). Using $CHCl_3$, C. L. Luke, *Anal. Chem.* **28**:1276 (1956).
[18] K. L. Cheng, *Anal. Chem.* **30**:1941 (1958).

TABLE 2.18 **Chelate Solvent Extraction Systems for the Separation of Elements**

APDC: ammonium pyrrolidine dithiocarbamate
Cupferron: N-*nitrosophenylhydroxylamine*
Oxine: 8-hydroxyquinoline
NaDDC: sodium diethyldithiocarbamate
TTA: 2-thenoyltrifluoroacetone
AA: acetylacetone
DDDC: diethylammonium diethyldithiocarbamate
PAN: 1-(2-pyridylazo)-2-naphthol
EHD: 2-ethyl-1,3-hexanediol in chloroform
DMG: dimethylglyoxime

	pH of complete extraction						Other
Element	APDC	Cupferron	Oxine	NaDDC	TTA	AA	reagents
Ag			3–6* 8–9†	5–11†			DDDC
Al		4–6*	5–6* 4.5–10†		5–6*	4–9‡ (80%)	
As	2–6*			6.5–8†			DDDC
B							EHD
Be		4–8†	6*		6–7‡	3.5–8‡	
Bi		2–12†	2.5–11†	5–11†	2–5‡		DDDC
Cd	1–6*		5–9†	5–11†			DDDC PAN
Co	2–4*	5–12†	5–6* 4.5–10.5†	4* 5–11†	7–9*		DDDC PAN
Cr	3–9*			4–6*			
Cs							$Na(C_6H_5)_3B$
Cu	0.1–8*	2–7*	2–12†	4–6* 5–11†	3–6*	4–10‡ (80–90%)	DDDC
Fe	2–5*	0–12*	6* 2–10†	4* 5–9.5†	2–5*	2.2–8.2‡	
Ga		1.5–6†	2–12*,†		5–7‡		PAN
Hg	2–4*	2–5†	3–7†	5–11†			PAN
In		2.5–8†	4–10*,† 3–11†	4–9†	4–6*		
Mg			8.5*,† (90%)				
Mn(II)	2–4*	7*	7*,† (90%)	4*			PAN
Mn(III)				6–9†			
Mo	3–4*	0–1.5*,†	2–3* 1–5.5†				
Ni	2–4*	7*	6* 5–8.5†	4* 5–11†	6–8†		DMG
Pb	0.1–6*	3–9*,†	6* 6–10†	4* 5–11†	5–7*		
Pd				5–11†			DMG
Pt							DDDC
Sb		1.5–8†		5–9†			
Sc			4.5–10†			3.8‡	
Se	3–6*			5–6.5†			
Sn		(1.5*M* H^+)†,‡		5–5.5†		6‡ (75%)	
Te				5–9†			

(*Continued*)

TABLE 2.18 Chelate Solvent Extraction Systems for the Separation of Elements (*Continued*)

	pH of complete extraction						
Element	APDC	Cupferron	Oxine	NaDDC	TTA	AA	Other reagents
Th			4–10†			>5‡	
Ti			2.5–9†	5–9*,†			
Tl	3–10*			5–11†	4–7†		
V	1–2*	0–2*,†	2–4*,†,‡	4–6† 5–5.5†			
Zn	2–6*		3.6* (90%)	5–11†			
Rare earths			7–10†	3–5*			

* MIBK = methyl isobutyl ketone; pentyl acetate is an alternate.
† Chloroform or carbon tetrachloride.
‡ Benzene.

2.2.3.1 ***Selective Extraction.*** Addition of a masking (complexing) agent moves the distribution curve to higher pH values, the shift being greater as the amount of masking agent is increased and as the magnitude of the formation constant of the masked metal increases. For example, copper(II) is extracted with 8-hydroxyquinoline in chloroform at a much lower pH than uranium(IV); yet, if EDTA is added to the system, the uranium extraction proceeds as before, whereas the copper extraction is delayed until the pH is raised by more than 5 units.

Sometimes greater selectivity can be achieved through the use of a particular metal chelate, rather than the chelating reagent itself, as the ligand source. For example, copper(II) is selectively extracted in the presence of iron(III) and cobalt(II) using lead diethyldithiocarbamate, whereas all three metals would be coextracted with sodium diethyldithiocarbamate. The lead chelate is more stable (has a larger formation constant) than the chelates of iron and cobalt but less stable than the copper chelate.

2.2.4 Ion-Association Systems

The simplest category of ion-association systems involves large and bulky cations and anions the size and structure of which are such that they do not have a primary hydration shell. Relatively simple, stoichiometric equilibria are sufficient to describe these ion-association systems. There is dissociation of the ion pairs in the aqueous phase and association in the organic phase. An example involves the extraction of rhenium as the perrhenate(VII) anion from an aqueous phase with tetraphenylarsonium chloride contained in the organic phase as an ion pair (ion-pair formation is indicated by placing a comma between the cation and anion that are in association with each other).

$$[(C_6H_5)_4As^+,Cl^-]_{org} + ReO_4^- \rightleftarrows [(C_6H_5)_4As^+,ReO_4^-]_{org} + Cl^- \qquad (2.42)$$

Similar extractions of other large anions such as TcO_4^-, MnO_4^-, ClO_4^-, and IO_4^- are possible.[19] The general rule (to which there are some exceptions) is that singly charged bulky cations form the most easily extractable ion-association compounds with large anions of single charge (Table 2.19).

Tetraphenylphosphonium chloride behaves much the same as its arsonium analog in the extraction of anions, both qualitatively and quantitatively.

Because polyvalent anions are more easily hydrated and thus less readily extracted, separations of the foregoing anions from chromate(VI)(2−), molybdate(VI)(2−), and tungstate(VI)(2−) are

[19] S. Tribalat, *Anal. Chim. Acta* **3**:113 (1949); **4**:228 (1950).

TABLE 2.19 Percent Extraction of Tetraphenylarsonium Anions with $CHCl_3$

Element	HCl (*M*)* 2	HCl (*M*)* 4	Anion	pH 1.5†
Antimony(III)	97.5	98	F^-	<1
Bismuth	24	4	Cl^-	18
Cadmium	67	64	Br^-	82
Gallium	91	100	I^-	99.7
Gold	100	100	SCN^-	97.2
Indium	33	72	ClO_3^-	99.3
Iron(III)	83	96	BrO_3^-	48
Mercury(II)	99.5	98	IO_3^-	0.4
Molybdenum(VI)	0.4	24	ClO_4^-	99.5
Osmium	96	95	MnO_4^-	99.7
Protactinium	0.2	28	ReO_4^-	99.5
Rhenium(VII)	99.4	99	IO_4^-	2 (pH 12)
Ruthenium(III)	20	17	NO_3^-	95
Silver	99	95	NO_2^-	18 (pH 12)
Technetium	99.8	99.4	SO_4^{2-}	<0.1
Thallium(I)	11	9	CrO_4^{2-}	98.6
Thallium(III)	99.6	99.6	AsO_3^{3-}	0.6
Tin(IV)	39	61	SO_3^{2-}	4.7 (pH 8)
Vanadium(V)	7	39	$S_2O_3^{2-}$	1.6 (pH 7)
Tungsten(VI)	56	56	SeO_3^{2-}	8 (pH 8.5)
Zinc	33	33	TeO_3^{2-}	11.7 (pH 12)

* 0.05*M* Tetraphenylarsonium chloride.
† 0.02*M* Tetraphenylarsonium hydroxide.

References: Metals—K. Ueno and C. Chang, *J. Atm. Energy Soc. Jpn.* **4**:457 (1962); anions—R. Bock and G. M. Beilstein, *Z. Anal. Chem.* **192**:44 (1963).

readily achieved. Chromium(VI) is extracted from acidic solution, no doubt because of the presence of $HCrO_4^-$ or $HCr_2O_7^-$ in acidic medium, but hardly at all from basic solution in which the divalent anions predominate.

If metals can coordinate halide ions to form bulky complex anions, often they can form extractable ion pairs with suitably bulky cations. Thus the basic dye cation of rhodamine B extracts into benzene these anions: $AuBr_4^-$, $AuCl_4^-$, $FeCl_4^-$, $GaCl_4^-$, $SbCl_6^-$, $SbBr_6^-$, SbI_4^-, $TlCl_4^-$, and $TlBr_4^-$. The cationic forms of malachite green, methyl violet, and other dyestuffs and indicators can be used as extractants. Triphenylmethylarsonium cation will extract $Cu(SCN)_4^{2-}$, $Co(SCN)_4^{2-}$, and $Fe(SCN)_6^{3-}$.

Salts of high-molecular-weight (long-chain) amines in acid solution form large cations capable of forming extractable ion pairs with a variety of complex metal anions (e.g., halo, nitrato, sulfato) and oxo-anions.[20] Triisooctylamine, tri(2-ethylhexyl)amine, methyldioctylamine, and tribenzylamine are useful and versatile in the extraction of mineral, organic, and complex metal acids. The anion attached to the organic ammonium cation is exchanged for other anionic species, thus giving rise to the term *liquid ion exchangers.* Both inert and oxygenated solvents (such as MIBK) find use in these extractions. This technique provides an ideal method to

[20] Reviews: H. Green, *Talanta* **11**:1561 (1964): **20**:139 (1973). For tabulations of exchangers, see C.F. Coleman, C.A. Blake, and K. B. Brown, *ibid.* **9**:297 (1962). See also H. A. Mahlman, G. W. Leddicotte, and F. L. Moore, *Anal. Chem.* **26**:1939 (1954); W. L. Ott, H. R. MacMillan, and W. R. Hatch, *ibid.* **36**:363 (1964); F. L. Moore, *ibid.* **30**:908 (1958); J. Y. Ellenburg, G. W. Leddicotte, and F. L. Moore, *ibid.* **26**:1045 (1954).

TABLE 2.20 Percentage Extraction of Metals from HCl Solution with 0.11*M* Triisooctylamine in Xylene

Essentially none of these elements extract at all acidities: Al, Ba, Ca, Ir, Mg, Ni, rare earths, Sc, Sr, and Y.

Element	1*M* HCl	4*M* HCl	8*M* HCl	12*M* HCl
Bismuth	99.9	98.4	44	1.2
Cadmium	99.2	99.4	97.5	28
Cobalt	0.1	20	94	39
Copper	2	76	86	20
Europium		<0.01	0.03	0.05 (11*M*)
Gallium	36	100	99	86
Gold(III)	100	100	100	100
Indium	76	98.4	98.8	61
Iron(III)	91	97	98.8	96
Lead	80	67	5	0.2
Molybdenum	11	94	97	72
Niobium	0.3	2	99	86
Osmium	99.9	99.8	99.8	99.8
Platinum	99.4	99.5	98.4	56
Protactinium		91	99.9	100
Rhenium	99.7	99.4	95	28
Ruthenium	75	72	59	6
Selenium(IV)	7	14	91	91
Silver	98.4	72	9	0.4
Tantalum	0.3	2	0.6	5
Technetium	99.5	99.4	98.4	95
Tellurium(IV)	17	98	97	94
Thallium(I)	24	24	24	99.8
Thorium		0.005	0.016	0.02
Uranium(VI)	93	97	99.2	96
Vanadium	11	80	98.8	67
Tungsten	89	85	76	67
Zinc	99.3	99.6	98	61
Zirconium		0.4	89	99 (10*M*)

References: T. Ishimori et al., *Jpn. At. Energy Res. Inst. Rep. 1106* (1966); F. L. Moore, *Anal. Chem.* **30**:908 (1958).

remove mineral acids from biological preparations under mild conditions. Extractabilities of metals with triisooctylamine in xylene from HCl solutions of various concentrations are given in Table 2.20.

The removal of complex metal anions is illustrated by the extraction of zinc, as $ZnCl_4^{2-}$ from an HCl solution using a benzene solution of tribenzylamine, R_3N (R stands for the benzyl group). There are a number of interlocking equilibria. First there is the distribution of the extractant between benzene and the aqueous phase:

$$[R_3NH^+,Cl^-]_{org} \rightleftarrows [R_3NH^+,Cl^-]_{aq} \tag{2.43}$$

In the aqueous phase there is the association–dissociation equilibrium:

$$R_3NH^+,Cl^- \rightleftarrows R_3NH^+ + Cl^- \tag{2.44}$$

and the formation–dissociation of the tribenzylammonium cation:

$$H^+ + R_3N \rightleftarrows R_3NH^+ \tag{2.45}$$

The tribenzylammonium cation is involved in the tribenzylammonium tetrachlorozincate ion-association species that distributes between the aqueous phase and benzene. These two equilibria are, respectively,

$$2R_3NH^+ + ZnCl_4^{2-} \rightleftarrows (2R_3NH^+,ZnCl_4^{2-})_{aq} \tag{2.46}$$

$$(2R_3NH^+,ZnCl_4^{2-})_{aq} \rightleftarrows (2R_3NH^+,ZnCl_4^{2-})_{org} \tag{2.47}$$

One final equilibrium exists in the aqueous phase and involves the formation of the tetrachlorozincate anion from the tetraaquazincate cation:

$$Zn(H_2O)_4^{2+} + 4Cl^- \rightleftarrows ZnCl_4^{2-} + 4H_2O \tag{2.48}$$

Sufficient chloride ion must be present to form the tetrachlorozincate anion, and hydrogen ion is needed to form the tributylammonium cation.

The less hydrated anion displaces the more hydrated anion, giving the orders $I^- > Br^- > Cl^- > F^-$ and $ClO_4^- > ClO_3^- > Cl^-$. Smaller anions are displaced by larger anions of the same charge, giving $PF_6^- > BF_4^-$.

Trade names of some commercial long-chain amines are as follows:

Alamine or aliquat 336	Tertiary amine with mainly octyl and decyl alkyl groups
Amberlite LA-1	*N*-Dodecenyl(trialkylmethyl)amine
Amberlite LA-2	*N*-Dodecyl(trialkylmethyl)amine
TBA	Tribenylaime

Quaternary ammonium salts are more commonly used than secondary or tertiary amines for the extraction of metals as complex anions from chloride, nitrate, thiocyanate, and occasionally other solutions. When necessary, quaternary amines can be used for extractions from basic medium. Extractability approximately follows the trend of the stabilities of the complex anions. Although the extraction of singly charged anions is preferred, extraction is not confined to these. Extractability generally increases with the molecular weight of the reagent; the substitution of an ethyl group in the 2 position seems to shield the ion-association site from the effect of solvent (water) hydration in addition to increasing the "bulkiness" of the reagent.

2.2.5 Chelation and Ion Association

Polyvalent cations can be extracted if the cation is made large and hydrocarbonlike. This can be accomplished with certain chelate reagents that have two uncharged coordinating atoms, such as 1,10-phenanthroline, which forms cationic complex chelates with metal ions, such as iron(II), iron(III), cobalt(II), and copper(II). In conjunction with a large anion such as perchlorate, tris(1,10-phenanthroline)iron(II) perchlorate extracts moderately well into chloroform and quite well into nitrobenzene. The extraction is improved when long-chain alkyl sulfonate ions replace the perchlorate ion.

Another general method for making small, polyvalent metal cations extractable is to transform them into large negatively charged chelate complexes. An example is the formation of an anionic complex of magnesium with excess 8-hydroxyquinoline ion (Ox) in a carbonate solution,

$$Mg^{2+} + 3Ox^- \rightleftarrows MgOx_3^- \tag{2.49}$$

and extraction with the large tetrabutylammonium cation,

$$[(C_4H_9)_4N^+,Cl^-]_{org} + MgOx_3^- \rightleftarrows [(C_4H_9)_4N^+,MgOx_3^-]_{org} + Cl^- \tag{2.50}$$

2.2.6 Summary of Extraction Methods for the Elements

Table 2.21 summarizes the more significant extraction procedures for the elements. The elements are listed alphabetically by their full name. The composition of the organic and aqueous phases is given along with the significant elements from which separation is achieved. Pertinent references are included.

2.2.7 Laboratory Manipulations

The usual apparatus for batch extractions is a pear-shaped separatory funnel of the appropriate volume. The solute is extracted from one layer by shaking with a second immiscible phase in a separatory funnel until equilibrium has been attained (either from literature directions or by running a series of equilibrations at several time intervals to ascertain the time for attainment of equilibrium). After equilibration the two layers are allowed to separate completely, and the layer containing the desired component is removed.

Simple inversions of the separatory funnel, repeated perhaps 50 times during 1 to 2 min, suffice to attain equilibrium in most extractions. After each of the first several inversions, the stopcock should be momentarily vented. Both the distribution ratio and the phase volumes are influenced by temperature changes.

Droplets of aqueous phase entrained in the organic phase can be removed by centrifugation or by filtering through a plug of glass wool. Whatman phase-separating papers, available in grade 1PS, are silicone-treated papers that serve as a combined filter and separator. The organic phase passes through the silicone-treated paper. The water phase will be held back at all times in the presence of an immiscible organic solvent. These papers function best when used flat on a Büchner or Hirsch type funnel. The separator is low cost so that it can be discarded after use. The phase separator is unaffected by mineral acids to 4*M*. It will tolerate alkalies to 0.4*M*. Successful separations have been achieved at 90°C. However, surface tension is inversely proportional to temperature so that temperature limits for a given separation must be experimentally determined.

After allowing the organic phase to filter completely through the paper, wash the retained aqueous phase with a small volume of clean organic solvent. This is normally necessary when separating lighter-than-water solvents in order to clean the meniscus of the aqueous phase.

Do not allow the aqueous phase to remain in the funnel after phase separation is complete, or it will begin to seep through.

2.2.7.1 Breaking Emulsions. Emulsions are more easily produced when the densities of the two solvents are similar. It is also desirable to select solvents that have a high interfacial tension for rapid separation of phases.

Suggestions for breaking emulsions follow:

1. Emulsions caused by too small a difference in the densities of the water and organic layer can be broken by the addition of a high-density organic solvent such as carbon tetrachloride.
2. Pentane can be added to reduce the density of the organic layer, if so desired, especially when the aqueous layer has a high density because of dissolved salts.
3. Sometimes troublesome emulsions can be broken by introducing a strong electrolyte. Saturated NaCl or Na_2SO_4 salt solutions will increase the density of aqueous layers.
4. Emulsions formed with ethereal solutions can be broken with the addition of small quantities of ethanol or 2-propanol.
5. Add a few drops of silicone defoamer.

2.2.8 Continuous Liquid–Liquid Extractions

When many extractions would be needed to remove a solute with a small distribution ratio, continuous liquid–liquid extractions may be used. However, to be feasible, impurities must not be extractable and the solute must be stable at the temperature of the boiling solvent. A limited volume of organic solvent is used, its circulation usually being maintained by vaporization and condensation. The effective ratio V_{org}/V_{aq} can thus be made large. Good recoveries can be secured even when the distribution coefficient is small (Table 2.22). Two approaches are described.

TABLE 2.21 Extraction Procedures for the Elements

Element	Organic phase	Aqueous phase	Separated from	Reference
Actinium	0.25*M* TTA/C_6H_6	pH 5.5	Ra	
Aluminum	0.01*M* oxine/$CHCl_3$	pH 4.5–5.0	Alkaline earths, Be, Cr(III), Mg, lanthanoids, Mn	
	0.01*M* oxine/$CHCl_3$	pH9, CN^-	Cu, Fe, Ni, Zn	
	0.01*M* oxine/$CHCl_3$	pH 8.5–9.0, CN^- + EDTA + HSO_3^-	Ag, As, Au, B, Ba, Cd, Ca, Ce, Cs, Cr(III), Co, Cu, F, Ge, I, La, Pb, Li, Mg, Mn, Hg(I and II), Mo, Ni, P, Pt(IV), Se(IV and VI), Si, Sc, Sr, Te(IV and VI), Th, Tl, Sn(IV), W, V(V), and Zn	A. Classen, L. Bastings, and J. Visser, *Anal. Chim. Acta* **10**:373 (1954)
	0.01*M* oxine/$CHCl_3$	pH 7.5–8.5, H_2O_2 after returning Al to aqueous phase from preceding extraction	The above plus Nb, Ta, Ti, U, and V	J. L. Kassner and M. A. Ozier, *Anal. Chem.* **23**:1453 (1951)
	$CHCl_3$	1% cupferron, pH 3.5–5.0 acetate buffer	Fluoride and phosphate	H. G. Short, *Analyst* **75**:420 (1950); E. B. Sandell and P. F. Cummings, *Anal. Chem.* **21**:1356 (1949); H. Thaler and F. H. Mühlberger, *Z. Anal. Chem.* **144**:241 (1955)
	MIBK	pH 2.5–4.5, acetate buffer		H. C. Eshelman, J. A. Dean, O. Menis, and T. C. Rains, *Anal. Chem.* **31**:183 (1959)
	HTTA in MIBK	pH 5.5–6, acetate buffer		H. C. Eshelman, J. A. Dean, O. Menis, and T. C. Rains, *Anal. Chem.* **31**:183 (1959)
Americium	0.2*M* HTTA/C_6H_6	pH > 3.5		
Antimony(V)	Diisopropyl ether	6.5*M*–8.5*M* HCl	Ag, Al, Be, Bi, Ca, Cd, Co, Cr(III), Cu, Fe(II), In, Mg, Mn, Ni, Pb, Pd(II), Pt(IV), rare earths, Ru(III,IV), Sn(IV), Te(IV), Ti, U, Zn, Zr	F. C. Edwards and A. F. Voigt, *Anal. Chem.* **21**:1204 (1949); N. H. Nachtrieb and R. E. Fryxell, *J. Am. Chem. Soc.* **71**:4035 (1949); P. A. Pohl and W. Bonsels, *Z. Anal. Chem.* **161**:108 (1958)
Antimony(III)	Benzene	0.002*M*–0.06*M* I^-, 5*M* H_2SO_4	Extracted 0.1% or less: Al, Cd, Co, Cr(III), Fe(III), Ga, Ge, Hf, La, Mn(II), Mo(VI), Re(VII), Sc, Ta(V), Ti	A. P. Grimanis and I. Hadzistelios, *Anal. Chim. Acta* **41**:15 (1968)
		0.005*M*–0.1*M* Br^-, 10*M* H_2SO_4	Extracted 0.2% or less: Al, Co, Cu, Fe(III), Mn, Mo(VI), Ni, Zn	A. P. Grimanis and I. Hadzistellios, *Anal. Chim. Acta* **41**:15 (1968)

(*Continued*)

TABLE 2.21 Extraction Procedures for the Elements (*Continued*)

Element	Organic phase	Aqueous phase	Separated from	Reference
Antimony(III)	*O, O*-bis(2-ethylhexyl) hydrogen phosphorodithioate in cyclohexane	0.5*M*–7*M* HCl; 0.25*M*–9*M* H_2SO_4	Ag, As(III), Bi, Cd, Co, Fe, In, Ni, Pb, Sn(II), Zn	Yu. M. Yukhin, I. S. Levin, and N. M. Meshkova, *Zh. Anal. Khim.* **33**: 1791 (1978)
Antimony(III,V)	Ethyl acetate	1*M*–2*M* HCl, oxalate and citrate	Copper alloys and zinc; Cd, Fe, Ge, Pb, Sn, Te	C. E. White and H. J. Rose, *Anal. Chem.* **25**:351 (1953)
Antimony	$CHCl_3$	Cupferron (1:9) H_2SO_4	Arsenic	
Antimony(III)	$CHCl_3$	Cupferron, 1*N* HNO_3 or H_2SO_4	Ag, Al, As, Ba, Be, Ca, Cd, Ce(III), Co, Cr(III), Ge, Mg, Mn, Ni, Pb, Pt, rare earths, Sb(V), Se(IV), Sr, Te(IV), Tl(I), U(VI), Y, Zn	
Arsenic(III)	Benzene	10*M*–12*M* HCl	B, Bi, Cu, Hg(II), Sb(III), Se(IV), Te(IV), Sn(IV)	W. Fischer et al., *Angew. Chem.* **66**: 165 (1954); T. Korenaga, *Analyst* **106**:40 (1981); F. Puttemans, P. van den Winkel, and D. L. Massart, *Anal. Chim. Acta* **149**:123 (1983); G. O. Brink et al., *J. Am. Chem. Soc.* **79**:1303 (1957)
Arsenic(V)	CCl_4	14*N*–17*N* H_2SO_4, 0.067*M* KBr		K. Studlar, *Talanta* **8**:272 (1961)
Arsenic(III)	Toluene: also bis(2-chloroethyl) ether, $CHCl_3$, CCl_4	3*M* HI; or 0.05*M* KI + 6*M* H_2SO_4; or 0.8*M* KI + 6*M* H_2SO_4 (with CCl_4)		A. R. Bryne and D. Gorenc, *Anal. Chim. Acta* **59**:81, 91 (1972); K. Tanaka, *Jpn. Anal.* **9**:700 (1960); G. O. Brink et al., *J. Am. Chem. Soc.* **79**:1303 (1957); V. Stara and J. Starý, *Talanta* **17**:341 (1970)
	Ethylxanthate in $CHCl_3$	11*M* HCl; also after coprecipitation with $Fe(OH)_3$	Cu, Cu-Ni, Mo, Pb, and Zn concentrates, seawater, silicate rocks, and biomaterials	E. M. Donaldson, *Talanta* **24**:105 (1977); I. Sugawara, M. Tanaka, and S. Kanamori, *Bull. Chem. Soc. Jpn.* **29**:670 (1956).
	Diethylammonium diethyldithiocarbamate in $CHCl_3$	1*N*–10*N* H_2SO_4	Cd, Fe(II), Ni, Zn	P. F. Wyatt, *Analyst* **80**:368 (1955); N. Stafford, P. F. Wyatt, and F. G. Kershaw, *ibid.* **70**:232 (1945); C. L. Luke, *Anal. Chem.* **28**:1276 (1956)
Barium	Tributyl phosphate	Thiocyanate + EDTA	Lanthanoids	F. P. Gorbenko et al., *Radiokhim.* **12**:661 (1970)
	18-crown-6		Ca	Y. Takeda, S. Suzuki, and Y. Ohyagi, *Chem. Lett.* **1978**, 1377
	Dibenzo-24-crown-8		Sr	Y. Takeda, K. Oshio, and Y. Segawa,

				Chem. Lett. **1979**, 601
Beryllium	Acetylacetone in benzene	pH 4–5	Alkaline earths, phosphate	T. Y. Toribara and P. S. Chen, Jr., *Anal. Chem.* **24**:539 (1952); T. Y. Toribara and R. E. Sherman, *ibid.* **25**:1594 (1953)
	Acetylacetone	pH 7–8 plus EDTA	Al, Cu, Fe(III), Zn	I. P. Alimarin and I. M. Gibalo, *Zh. Anal. Khim.* **11**:389 (1956); C. W. Sill and C. P. Willis, *Anal. Chem.* **31**:598 (1959)
	$CHCl_3$	pH 9.7 + butyrate + EDTA	Al, Fe(III); bronze and rocks	S. Banerjee, A. K. Sundaram, and H. D. Sharma, *Anal. Chim. Acta* **10**:256 (1954); A. L. Markman and L. L. Galkina, *Anal. Abst.* **10**:5071 (1963); **12**:1628 (1965)
Bismuth	Dithizone in CCl_4	>1*M* $HClO_4$ or HNO_3	Al, alkaline earths, As, Cr(III), Fe(III), Ga, Ge, Ir, lanthanoids, Mn, Mo, P, Rh, Ru, Sb(III), Si, Ta, Ti, U, V, W, Zr	T. Bidleman, *Anal. Chim. Acta* **56**:221 (1971)
	$CHCl_3$	0.005*M* cupferron, 0.1 *N* H_2SO_4 or HNO_3	A, Al, Be, Cd, Co, Cr, Mg, Mn, Ni, U, Zn	H. Bode and G. Henrich, *Z. Anal. Chem.* **135**:98 (1952)
	Isopentyl acetate	0.005*M* KI and 3*M* $HClO_4$ or H_2SO_4 plus 3% hypophosphite		
Boron	Tetraphenylarsonium ion with $CHCl_3$	0.01*M* Tetrafluoroborate ion		L. Ducret and P. Sequin, *Anal. Chim. Acta* **17**:213 (1957)
	Methylene blue with $CHCl_3$ or 1,2-dichloroethane	0.01*M* Tetrafluoroborate ion		L. C. Pasztor, J. D. Bode, and Q. Fernando, *Anal. Chem.* **32**:277 (1960)
Cadmium	Dithizone and $CHCl_3$	1*M*–2*M* NaOH plus citrate or tartrate	Pb, Zn	Z. Marczenko, M. Mojski, and K. Kasiura, *Zh. Anal. Khim.* **22**:1805 (1967)
	Dithizone and CCl_4	pH 8.3, cyanide	Seawater	J. B. Mullin and J. P. Riley, *Nature* **174**:42 (1954)
	Diethyldithiocarbamate and CCl_4	pH 11, cyanide-tartrate	Zn	See also S. Bajo and A. Wyttenbach, *Anal. Chem.* **49**:158 (1977)
	Diethyl phosphorodithioate into CCl_4	pH 4–6	Zn	H. Bode and K. Wulff, *Z. Anal. Chem.* **219**:32 (1966)
Cerium(IV)	Diethyl ether	6*M*–8*M* HNO_3	Lanthanoids	A. W. Wylie, *J. Chem. Soc.* **1951**, 1474
	MIBK	8*M*–10*M* HNO_3	Lanthanoids	L. E. Glendenin et al., *Anal. Chem.* **27**:59 (1955)
	40% v/v tributyl phosphate in isooctane	8*M*–10*M* HNO_3	Silicate rocks; Fe, Ti	F. Culkin and J. P. Riley, *Anal. Chim. Acta* **32**:197 (1965)

(*Continued*)

TABLE 2.21 Extraction Procedures for the Elements (*Continued*)

Element	Organic phase	Aqueous phase	Separated from	Reference
Cerium(IV)	Thenoyltrifluoroacetone and xylene	0.5*M* H_2SO_4	Lanthanoids, Th, U	G. W. Smith and F. L. Moore, *Anal. Chem.* **29**:448 (1957)
Chromium(VI)	MIBK	1*M*–2*M* HCl, 5°C, 30 s shaking	V; Fe if F^- present	H. A. Bryan and J. A. Dean, *Anal. Chem.* **29**:1289 (1957); J. A. Dean and M. L. Beverly, *ibid.* **30**:977 (1958); P. D. Blundy, *Analyst* **83**:555 (1958); E. S. Pilkington and P. R. Smith, *Anal. Chim. Acta* **39**:321 (1967)
	MIBK	0.02*M* H_2O_2; pH 1.74, 0°C	Hg, Mo(VI), V(V)	R. K. Brookshier and H. Freund, *Anal. Chem.* **23**:1110 (1951); N. Ichinose et al., *Anal. Chim. Acta* **96**:391 (1978)
Chromium	Tribenzylamine in $CHCl_3$	1*M* HCl	Alkaline earths, As(V), Cd, Ce(IV), Co, Cr(III), Cu, Fe(III), Ga, lanthanoids, Mn, Mo, Ni, Sc, U(VI), V(V), Zn	G. B. Fasolo, R. Malvano, and A. Massaglia, *Anal. Chim. Acta* **29**:569 (1963); E. M. Donaldson, *Talanta* **27**:779 (1980)
Chromium(III)	Aliquat 336 in CCl_4	Thiocyanate	Co, Ni	G. J. de Jong and U. A. T. Brinkman, *J. Radioanal. Chem.* **35**:223 (1977)
Cobalt	1% 2-Nitroso-1-naphthol in $CHCl_3$, or CCl_4, or toluene	pH 4.5–8.0	Fe(III) and Ni when back extracted with strong HCl	E. Boyland, *Analyst* **71**:230 (1946); O. K. Borggaard et al., *ibid.* **107**:1479 (1982)
	0.05% dithizone in CCl_4	0.02*M* citrate and pH 8	Cr, Fe(III), Ti, V; seawater, biomaterials, rock, and soils	H. R. Marston and D. W. Dewey, *Austr. J. Exptl. Biol. Med. Sci.* **18**:343 (1940); V. D. Anand, G. S. Deshmukh, and C. M. Pandey, *Anal. Chem.* **33**:1933 (1961)
	MIBK	5*M* NaSCN	Ni	R. A. Sharp and G. Wilkinson, *J. Am. Chem. Soc.* **77**:6519 (1955)
	Diethyldithiocarbamate and CCl_4	pH 6.5	Fe, Ca, P	G. H. Ellis and J. F. Thompson, *Ind. Eng. Chem. Anal. Ed.* **17**:254 (1945)
Copper	0.05% dithizone	0.1*M*–0.2*M* HCl	As, Cd, Ga, Ge, Fe, Mg, Mn, Mo, Ni, P, Pb, Sb, Se, V, W, Zn; steel and biomaterials	K. J. Hahn, D. J. Tuma, and J. L. Sullivan, *Anal. Chem.* **40**:974 (1968)
	5% triphenylphosphite in CCl_4	0.1*M*–1*M* HCl + ascorbic acid	Highly selective; Hg, platinum metals, Zn	T. H. Handley and J. A. Dean, *Anal. Chem.* **33**:1087 (1961)

	Neocuproine		Many metal ions	P. J. Jones and E. J. Newman, *Analyst* **87**:637 (1962); C. L. Luke, *Anal. Chim. Acta* **32**:286 (1965); P. Battistoni et al., *Talanta* **27**:623 (1980)
	0.04% diethylammonium diethyldithiocarbamate in CCl_4	<7*M* HCl or <5*M* H_2SO_4		H. Bode and F. Neumann, *Z. Anal. Chem.* **172**:1 (1960); H. T. Delves, G. Shepherd, and P. Vinter, *Analyst* **96**:260 (1971)
Cyanide	Diisopropyl ether (four extractions)	Acetic acid solution	Thiocyanate	J. M. Kruse and M. G. Mellon, *Sewage Ind. Wastes* **23**:1402 (1951)
Gallium	Diisopropyl ether	5.5*M*–6*M* HCl	Ag, Al, Be, Bi, Ca, Cd, Cr(III), Co, Cu(II), Fe(II), Mg, Mn, Ni, Os, Pb, rare earths, Re, Rh, Sc, Ti, Th, U, V(IV), W, Zn, Zr	J. A. Scherrer, *J. Res. Natl. Bur. Std.* **15**:585 (1935); R. W. Dobson, G. J. Forney, and E. H. Swift, *J. Am. Chem. Soc.* **46**:668 (1926); F. C. Edwards and A. F. Voigt, *Anal. Chem.* **21**:1204 (1949); N. H. Nachtrieb and R. E. Fryxell, *J. Am. Chem. Soc.* **71**:4035 (1949); F. A. Pohl and W. Bonsels, *Z. Anal. Chem.* **161**:108 (1958)
	MIBK	6*M*–8*M* HCl	Ag, Al, alkali metals, Ba, Be, Bi, Ca, Cr(III), Mg, Mn, Ni, Pb, Pd, Th, Tl(I), rare earths, Zr	W. Doll and H. Specker, *Z. Anal. Chem.* **161**:354 (1958)
	$CHCl_3$	Cupferron, 1*M* H_2SO_4	Al, Cr, lanthanoids, Sc, U(VI), Zn	E. Gastinger, *Z. Anal. Chem.* **140**:244, 252 (1953); J. A. Scherrer, *J. Res. Natl. Bur. Stds.* **15**:585 (1935)
Germanium	CCl_4	9*M* HCl	As(V), Hg(II), Sb, Se(IV), Sn(IV)	W. Fischer and W. Harre, *Angew. Chem.* **66**:165 (1954); W. A. Schneider and E. B. Sandell, *Mikrochim. Acta* **1954**, 263
Gold	Diethyl ether	3*M*–6*M* HCl	Cu, Ni, Pt(IV), Zn	F. Mylius, *Z. Anorg. Chem.* **70**:203 (1911); F. Mylius and C. Hüttner, *Berichte* **44**:1315 (1911); G. K. Schweitzer and W. N. Bishop, *J. Am. Chem. Soc.* **75**:6330 (1953)

(Continued)

TABLE 2.21 Extraction Procedures for the Elements (*Continued*)

Element	Organic phase	Aqueous phase	Separated from	Reference
Gold	Diisopropyl ether	2.5*M*–3*M* HBr	Platinum metals (except Os)	W. A. E. McBryde and J. H. Yoe, *Anal. Chem.* **20**:1094 (1948)
	MIBK	6*M* HCl	Ir, Pd, Pt, Rh	A. Diamantatos and A. A. Verbeek, *Anal. Chim. Acta* **91**:287 (1977); **131**:53 (1981); F. W. E. Strelow et al., *Anal. Chem.* **38**:115 (1966)
Gold(I)	Methyl trialkyl(C_8-C_{10})-NCl + diisobutylketone	$Au(CN)_2^-$		T. Groenewald, *Anal. Chem.* **41**: 1012 (1969)
Indium	Diethyl ether	4*M*–6*M* HBr	Zn	I. Wada and R. Ishii, *Sci. Papers Inst. Phys. Chem. Res. (Tokyo)* **34**:787 (1938); J. E. Hudgens and L. C. Nelson, *Anal. Chem.* **24**: 1472 (1952); T. A. Collins and J. H. Kanzelmeyer, *ibid.* **33**:245 (1961)
	Diisopropyl ether	1.5*M* HI	Al, Be, Ga, Fe(II)	H. M. Irving and F. J. C. Rossotti, *Analyst* **77**:801 (1952)
Iodine (inorganic)	CCl_4, toluene	Water		H. J. Almquist and J. W. Givens, *Ind. Eng. Chem. Anal. Ed.* **5**:254 (1933); J. J. Custer and S. Natelson, *Anal. Chem.* **21**:1003 (1941)
Iodine (organic)	Butanol or isopentyl alcohol–trimethyl-pentane	Acidified aqueous solution	Inorganic iodine	J. Benotti, S. Pino, and H. Gardyna, *Clin. Chem.* **12**:491 (1966); I. Posner, *Anal. Biochem.* **23**:492 (1968)
Iridium(IV)	Tributyl phosphate	6*M* HCl	Rh	E. W. Berg and W. L. Senn, *Anal. Chim. Acta* **19**:109 (1958); R. B. Wilson and W. D. Jacobs, *Anal. Chem.* **33**:1650 (1961)
Iron(III)	MIBK or MIBK and pentyl acetate (1 : 1 or 2 : 1)	5.5*M*–7*M* HCl	Ag, Al, Ba, Be, Bi, Ca, Cr(III), alkali metals, Mg, Mn, Ni, Pb, Pd, rare earths, Th, Tl(I), Zr	H. Specker and W. Doll, *Z. Anal. Chem.* **152**:178 (1956); **161**:354 (1958); A. Claassen and L. Bastings, *ibid.* **160**:403 (1958)
	Tributyl phosphate	Thiocyanate	Al, Ni	M. Aven and H. Freiser, *Anal. Chim. Acta* **6**:412 (1952); E. Jackwerth, *Z. Anal. Chem.* **206**:335 (1964)
	Acetylacetone	pH 1.5	Many metals	R. A. Chalmers and D. M. Dick, *Anal. Chim. Acta* **32**:117 (1965)
Lanthanoids	Butyric acid in $CHCl_3$	pH 4.0–4.4 in presence of sulfosalicylic acid	Al, Fe(III), Mo, Nb, Sn, Ta, Ti, W, Zr	

Lead	0.005% dithizone in $CHCl_3$	pH 8.5–11, 0.5*M*–1*M* citrate plus 0.1*M* cyanide	Only Bi, In, Sn(II), and Tl(I) extract; Fe(III), Cu, Ca, Mg, and P cause problems	P. A. Clifford and H. J. Wichmann, *J. Assoc. Off. Agr. Chem.* **19**:134 (1936); O. B. Mathre and E. B. Sandell, *Talanta* **11**:295 (1964)
	1% Diethylammonium diethyldithiocarbamate in $CHCl_3$	1*M* HCl (also pH 3–4); add ascorbic acid to reduce Fe(III)	Al, Ca, Ce(III), Cr(III), Mg, Mn, P, Sb(V), Ti, Tl, V(V), U(VI), Zr	H. V. Hart, *Analyst* **76**:695 (1951); Analytical Methods Committee, *ibid.* **84**:127 (1959); R. K. Roschnik, *ibid.* **98**:596 (1973)
Magnesium	2-Thenoyltrifluoroacetone in tetrahydrofuran plus benzene	pH 9	Alkali metals	K. J. Hahn, D. J. Tuma, and J. L. Sullivan, *Anal. Chem.* **40**:974 (1968); see also **39**:1169 (1967)
	8-Hydroxyquinoline and tetrabutylammonium iodide in $CHCl_3$	pH > 10.8 plus tartrate and phosphate		K. Kaneko, M. Yoshida, and T. Ozawa, *Anal. Chim. Acta* **132**:165 (1981)
Manganese(III)	$CHCl_3$	Diethyldithiocarbamic acid, pH 3–6; back-extraction with acetate buffer, pH 3.3	Be, Cd, Ce, Co, Ni, Pb, Ti, Zn	B. Morsches and G. Tölg, *Z. Anal. Chem.* **250**:81 (1970)
Manganese	$CHCl_3$ (or MIBK) and 8-hydroxyquinoline	pH 9	Seawater	G. P. Klinkhammer, *Anal. Chem.* **52**:117 (1980)
Mercury	Dithizone in CCl_4	1*N* mineral acids	Cd, Co, Fe, Ni, Pb, Zn; also Ag in 0.2*M* chloride	T. Kato, S. Takei, and A. Okagami, *Jpn. Anal.* **5**:689 (1956)
	MIBK plus tributyl phosphate	1*M*–4*M* HCl solution		D. C. F. Morris and J. H. Williams, *Talanta* **9**:623 (1962)
	CCl_4 and diethyldithiocarbamic acid	pH 9.3–10, EDTA and cyanide		E. A. Hakkila and G. R. Waterbury, *Anal. Chem.* **32**:1340 (1960); J. C. Yu, J. M. Lo, and C. M. Wai, *Anal. Chim. Acta* **154**:307 (1983)
	5% Triphenylphosphine sulfide in CCl_4	0.1*M* HCl	Co, In, Fe(III), Mn, Sb(III), Se, Zn	R. B. Hitchcock J. A. Dean, and T. H. Handley, *Anal. Chem.* **35**:254 (1963)
Molybdenum	α-Benzoin oxime in $CHCl_3$	0.5*M* H^+ plus Fe(II) to reduce Cr(VI) and V(V)	Biomaterials, ores, rocks, steel, ThO_2, UO_2SO_4	G. B. Jones, *Anal. Chim. Acta* **10**:584 (1954); P. G. Jeffrey, *Analyst* **81**:104 (1956); G. Goldstein, D. L. Manning, and O. Menis, *Anal. Chem.* **30**:539 (1958); L. Wish, *ibid.* **34**:625 (1962); P. Y. Peng and E. B. Sandell, *Anal. Chim. Acta* **29**:325 (1963)
	$CHCl_3$	0.05*M* cupferron in 1*N* mineral acid plus HF	W	C. L. Luke, *Anal. Chim. Acta* **34**:302 (1966)

(*Continued*)

TABLE 2.21 Extraction Procedures for the Elements (*Continued*)

Element	Organic phase	Aqueous phase	Separated from	Reference
Molybdenum	MIBK	7*M* HCl	Coextractants: As, Cd, Cr(VI), Cu, Fe(III), Ga, In, Sb(III), Sn(IV), Te(IV), U(VI), V(V), Zn	W. Doll and H. Specker, *Z. Anal. Chem.* **161**:354 (1958); G. R. Waterbury and C. F. Metz, *Talanta* **6**:237 (1960)
Nickel	$CHCl_3$	Dimethylglyoxime, pH 4.8–12 (tartrate medium) or pH 7.2–12 (citrate medium)	Biomaterials, silicate rocks, water	E. B. Sandell and R. W. Perlich, *Ind. Eng. Chem. Anal. Ed.* **11**:309 (1939); A Claassen and L. Bastings, *Rec. Trav. Chim.* **73**:783 (1954)
	Dithizone in $CHCl_3$	pH 9.2 (tartrate–NaOH)		G. L. Hubbard and T. E. Green, *Anal. Chem.* **38**:428 (1966); B. Morsches and G. Tölg, *Z. Anal. Chem.* **250**:81 (1970)
Niobium	Methyl isobutyl ketone	3.3*M* HF plus 6*M* HCl		J. R. Werning et al., *Ind. Eng. Chem.* **46**:646 (1954)
	Pentyl acetate, isopentyl acetate, tributyl phosphate, trioctylphosphine in cyclohexane	>10*M* HCl		T. V. Ramakrishna et al., *Talanta* **16**:847 (1969); A. K. De and A. K. Sen, *ibid.* **13**:853 (1966); K. Ilsemann and R. Bock, *Z. Anal. Chem.* **274**:185 (1975)
Osmium	$CHCl_3$ or CCl_4	1*M* $NaClO_4$	Everything but Ru unless Fe(II) is added	R. D. Sauerbrunn and E. B. Sandell, *J. Am. Chem. Soc.* **75**:4170 (1953); *Anal. Chim. Acta* **9**:86 (1953); G. Goldstein, D. L. Manning, O. Menis, and J. A. Dean, *Talanta* **7**:296 (1961)
	$CHCl_3$	As $OsCl_6$ with tetraphenylarsonium chloride	Ru	W. Geilmann and R. Neeb, *Z. Anal. Chem.* **156**:420 (1957)
Palladium	Dimethylglyoxime in $CHCl_3$	0.2*M*–0.3*M* HCl or 0.5*M* H_2SO_4	Ag; Cu and Co if organic phase is washed with dilute aqueous ammonia	R. S. Young, *Analyst* **76**:49 (1951); see F. E. Beamish, *Talanta* **14**:991 (1967); S. J. Al-Bazi and A. Chow, *ibid.* **31**:815 (1984)
	Dimethylglyoxime in $CHCl_3$	1*M* HNO_3	Ag, Au, Ir, Rh, Ru, Pt	F. I. Danilova et al., *Zh. Anal. Khim.* **29**:2150 (1974)
	MIBK	As $[PdI_4]^{2-}$	Nb, Zr	E. M. Donaldson and M. Wang, *Talanta* **33**:35 (1986)
	Mesityl oxide	Salicylic acid at pH 4.5–5.0	Au, Ir, Os, Pt, Rh, Ru	A. D. Langade and V. M. Shinde, *Analyst* **107**:708 (1982)
Phosphorus	20% 1-butanol in $CHCl_3$	As molybdophosphoric acid; 1*M* HCl	Arsenate, germanate, and silicate ions	C. Wadelin and M. G. Mellon, *Anal. Chem.* **38**:1668 (1953)

Platinum	4-Ocytlaniline in diisobutyl ketone	3*M* HCl	Co, Cu, Fe, Ni, and Au (if back-extracted with 7*M* $HClO_4$)	A. A. Vasilyeva et al., *Talanta* **22**:745 (1975); C. Pohlandt, *ibid.* **26**:199 (1979); see S. J. Al-Bazi and A. Chow, *ibid.* **31**:815 (1984); F. E. Beamish, *ibid.* **14**:991 (1967)
Potassium	Benzene with crown ethers	<1*M* $HClO_4$	Li and Na	I. M. Kolthoff, *Anal. Chem.* **51**:1R (1979); M. Yoshio and H. Noguchi, *Anal. Lett.* **15**(A15):1197 (1982)
	Nitrobenzene	Dipicrylamine, pH 7.0–9.0	River water	M. Koeva and J. Hala, *J. Radioanal. Chem.* **51**:71 (1979)
	Nitrobenzene	0.01*M* Na tetraphenylborate, 0.1*M* $NaClO_4$		T. Sekine and D. Dyrssen, *Anal. Chim. Acta* **45**:433 (1969); T. Omori et al., *J. Radioanal. Chem.* **67**:299 (1981)
Rhenium(VII)	$CHCl_3$	Tetraphenylarsonium chloride, pH 7–13. Tetraphenylphosphonium chloride similar	Cu, Fe(III), Mo(VI), W(VI)	S. Tribalat, *Anal. Chim. Acta* **3**:113 (1949); K. Beyermann, *Z. Anal. Chem.* **183**:91 (1961)
	MIBK	HCl		V. Yatriajam, *Z. Anal. Chem.* **219**:128 (1966); methyl ethyl ketone: T. M. Cotton and A. A. Woolf, *Anal. Chem.* **36**:248 (1964)
Rhodium(III)	1-Butyl-3-methyl pyrazole in toluene	0.01*M* HCl	Cu, Ni, Fe(III), Ir(III,IV), Pt(IV)	V. A. Pronin et al., *Zh. Anal. Khim.* **31**:1767 (1976); see S. J. Al-Bazi and A. Chow, *Talanta* **31**:815 (1984)
Ruthenium(VIII)	$CHCl_3$ or CCl_4	As RuO_4: Ag(II) oxide, Ce(IV), or KIO_4 in HNO_3 or H_2SO_4	Highly selective	F. S. Martin, *J. Chem. Soc.* **1954**, 2564; C. Surasiti and E. B. Sandell, *Anal. Chim. Acta* **22**:261 (1960); C. E. Epperson, R. R. Landolt, and W. V. Kessler, *Anal. Chem.* **48**:979 (1976)
Scandium	Diethyl ether (three extractions)	7*M* NH_4SCN and 0.5*M* HCl; add 5 mL 2*M* HCl per 100 mL for each extraction	Ca, lanthanoids, Mg, Mn, Zr	W. Fischer and R. Bock, *Z. Anorg. Allg. Chem.* **249**:146 (1942)
	Tributyl phosphate	10*M* HCl	Al, Be, Cr(III), lanthanoids, and Ti (if H_2O_2 is added)	A. R. Eberle and M. W. Lerner, *Anal. Chem.* **27**:1551 (1955); G. F. Kirkbright, T. S. West, and C. Woodward, *Analyst* **91**:23 (1966)

(*Continued*)

TABLE 2.21 Extraction Procedures for the Elements (*Continued*)

Element	Organic phase	Aqueous phase	Separated from	Reference
Scandium	0.5*M* 2-thenoyltrifluoroacetone in benzene	pH 1.5 ± 0.2 for 10 min	Co, Cr(III), Fe(II), lanthanoids, Ni	H. Onishi and C. V. Banks, *Anal. Chim. Acta* **29**:240 (1963)
Selenium(IV)	CCl_4	Na diethyldithiocarbamate, EDTA, pH 5–6		H. Bode, *Z. Anorg. Allg. Chem.* **289**:207 (1957)
Silicon	1-Butanol	Molybdosilicic acid at pH 1.8	Molybdophosphoric acid	W. S. Clabaugh and A. Jackson, *J. Nat. Bur. Stand.* **62**:201 (1959)
Silver	0.001% dithizone in CCl_4	6*M* HNO_3		N. Suzuki, *Jpn. Anal.* **8**:283 (1959)
	Trialkyl thiophosphates			T. H. Handley and J. A. Dean, *Anal. Chem.* **32**:1878 (1960)
	Trioctylphosphine oxide in MIBK	Iodide media	Al-, Fe-, and Ni-base alloys	K. E. Burke, *Talanta* **21**:417 (1974); E. M. Donaldson and M. Wang, *ibid.* **33**:35, 233 (1986)
Strontium	1-(2-pyridylazo)-2-naphthol in CCl_4–tributyl phosphate			V. V. Sachko, V. D. Anikeeva, and F. P. Gorbenko, *Zh. Anal. Khim.* **34**:2119 (1979)
Sulfur (free)	Acetone	Hydrocarbons		J. K. Bartlett and D. A. Skoog, *Anal. Chem.* **26**:1008 (1954)
Tantalum	Diisopropyl ketone	3.9*M* HNO_3 and 0.4*M* HF	Nb	S. N. Bhattacharyya and B. Ganguly, *Solv. Extr. Ion Exch.* **2**:699 (1984); P. C. Stevenson and H. G. Hicks, *Anal. Chem.* **25**:1517 (1953)
Tantalum	Diisopropyl ketone	6*M* H_2SO_4 and 0.4*M* HF	Only Se(VI) and Te(VI) tend to follow	P. C. Stevenson and H. G. Hicks, *Anal. Chem.* **25**:1517 (1953)
Tellurium(IV)	Ethyl acetate and MIBK	5*M* HCl	Te(VI)	B. Zmbora and J. Bzenic, *Anal. Chim. Acta* **59**:315 (1972)
Thallium	Dithizone in $CHCl_3$; three extractions needed	pH 10–11 with cyanide and citrate (and ascorbic acid for Fe)	Bi, Pb, Sn accompany	G. K. Schweitzer and A. D. Norton, *Anal. Chim. Acta* **30**:119 (1964); G. K. Schweitzer and J. E. Davidson, *ibid.* **35**:467 (1966)
Thallium(III)	Diisopropyl ether	2*M*–4*M* HCl		G. F. Kirkbright, T. S. West, and C. Woodward, *Talanta* **12**:517 (1965)
	Diisopropyl ether	1*M* HBr	Cu, Pb, and Zn ores and concentrates	E. A. Jones and A. F. Lee, Natl. Inst. Metall. Rep. No. 2036 (1980)
	MIBK	2*M* HBr		J. A. Dean and J. B. Eskew, *Anal. Lett.* **4**:737 (1971)

Thorium	MIBK	1.2*M* HCl and 2.5*M* $Al(NO_3)_3$	Arsenate, borate, fluoride, phosphate, sulfate; alkaline earths, Be, Co, Mg, Mn, Mo, Ni, Pb, Ti, Zn	H. Levine and F. S. Grimaldi, *U.S. Geol. Surv. Bull.* **1006**:177 (1954)
	0.5*M* 2-Thenoyltrifluoroacetone in benzene	pH 1	Lanthanoids, U(VI), Zr	M. O. Fulda, U.S. AEC Rep. DP-165 (1956)
	0.1*M* trioctylphosphine oxide in cyclohexane	1*M* in HNO_3 and $NaNO_3$	Ce(III), lanthanoids	W. J. Ross and J. C. White, *Anal. Chem.* **31**:1847 (1959); W. C. Johnson and M. H. Campbell, *ibid.* **37**:1440 (1965); J. C. Guyon and B. Madison, *Mikrochim. Acta* **1975(I)**:133
Tin(IV)	Benzene	1*M* NaI, 2–4*M* H_2SO_4	Only As(III), Ge, Sn are coextracted	D. D. Gilbert and E. B. Sandell, *Microchem. J.* **4**:492 (1960); E. J. Newman and P. D. Jones, *Analyst* **91**:406 (1966): K. Tanaka and N. Takagi, *Anal. Chim. Acta* **48**:357 (1969); E. M. Donaldson and M. Wang, *Talanta* **33**:35 (1986)
	0.5*M* 2-thenoyltrifluoroacetone in MIBK	1*M*–2*M* HCl, 0.5*M*–1*M* H_2SO_4, and 1% H_2O_2	Fe, Sb, U, Zn	J. R. Stokeley and F. L. Moore, *Anal. Chem.* **36**:1204 (1964)
	5% cupferron in $CHCl_3$	1 : 10 H_2SO_4	As, Ge, Sb(V)	C. L. Luke, *Anal. Chem.* **28**:1276 (1956); see J. Aznarez et al., *Talanta* **33**:458 (1986)
Tin(II)	0.04% diethylammonium diethyldithiocarbamic acid in $CHCl_3$	0.5*M* H_2SO_4	Sn(IV); steel and foods	C. L. Luke, *Anal. Chim. Acta* **37**:97 (1967); M. H. Thompson and G. McClellan, *J. Assoc. Off. Agr. Chem.* **45**:979 (1962); B. Morsches and G. Tölg, *Z. Anal. Chem.* **250**:81 (1970)
Titanium	Cupferron in diisopropyl ether or $CHCl_3$	0.05*M* acid	Coextracted: Cu, Fe(III), Hf, Mo(VI), Nb, Pd, Sb(III), Sn, Ta, Tc, Ti, U(IV), V(V), W, Zr	T. C. J. Ovenston, C. A. Parker, and C. G. Hatchard, *Anal. Chim. Acta* **6**:7 (1952); C. K. Kim and W. W. Meinke, *Talanta* **10**:83 (1963)
	Cupferron in $CHCl_3$	pH 6.0 and EDTA	Mo(VI) and V(V)	J. A. Corbett, *Anal. Chim. Acta* **30**:126 (1964)
	Cupferron in $CHCl_3$	pH 8 and tartrate–EDTA	Mo and W	E. M. Donaldson, *Talanta* **16**:1505 (1969)
	$CHCl_3$	pH 2 and diethyldithiocarbamic acid	Al	R. C. Rooney, *Anal. Chim. Acta* **19**:428 (1958)
Tungsten	$CHCl_3$ or benzene	α-Benzoin oxime, alkaline solution	Silicate rocks and ores, iron and steel, and Ni	P. G. Jeffery, *Analyst* **81**:104 (1956); P. Y. Peng and E. B. Sandell, *Anal. Chim. Acta* **29**:325 (1963); V. S. Sastri, *Talanta* **29**:405 (1982)

(*Continued*)

TABLE 2.21 Extraction Procedures for the Elements (*Continued*)

Element	Organic phase	Aqueous phase	Separated from	Reference
Uranium(VI)	Ethyl acetate	9.5 g $Al(NO_3)_3 \cdot 9H_2O$ added to 5 mL sample (0.5*M* in HNO_3)	Coextracted: Th, Fe, V, Zr; interferences: Cl^-, F^-, PO_4^{3-}, SO_4^{2-}	F. S. Grimaldi and H. Levine, *U.S. Geol. Surv. Bull.* **1006**:43 (1954); R. J. Guest and J. B. Zimmerman, *Anal. Chem.* **27**:931 (1955)
	25% tributyl phosphate in toluene	5*M* HNO_3	Al, Cd, Co, Cr(III), Cu, Fe(III), Mn, Mo, Nb, Ni, Ti, V, and Zn	M. A. Mair and D. J. Savage, *U.K. At. Energy Agency Rep. ND-R-134* (1986)
	Diethyldithiocarbamate in $CHCl_3$	pH 7–8 with EDTA	Th and Fe(III)	J. S. Fritz and M. Johnson-Richard, *Anal. Chim. Acta* **20**:164 (1959)
	MIBK	7*M*–8*M* HCl plus 1*M* $MgCl_2$	Th	N. Ichinose, *Talanta* **18**:21 (1971)
Vanadium(V)	Ethyl acetate	Cupferron and 1.2*M* HCl	Al, Be, Cd, Co, Cr(III), Mn, Ni, Pb, Zn	B. Moresches and G. Tölg, *Z. Anal. Chem.* **250**:81 (1970)
	α-Benzoin oxime in $CHCl_3$	pH 2–3 (HCl); back extraction with 0.02*M* NaOH	Cu ores and steel; Cu, Fe, Pb, Zn	R. Bock and B. Jost, *Z. Anal. Chem.* **250**:358 (1970); C. L. Luke, *Anal. Chim. Acta* **37**:267 (1967)
Zinc	Dithizone in $CHCl_3$	pH 8.3, citrate buffer	Biomaterials, silicates, water	J. Cholak, D. M. Hubbard, and R. E. Burkey, *Ind. Eng. Chem. Anal. Ed.* **15**:754 (1943); D. Monnier and G. Prod'hom, *Anal. Chim. Acta* **30**:358 (1964)
	Diethyldithiocarbamate in $CHCl_3$	pH 8.5 plus citrate	Biomaterial and water	R. A. Chalmers and D. M. Dick, *Anal. Chim. Acta* **32**:117 (1965)
Zirconium	0.05*M* 2-thenoyltrifluoroacetone in xylene	*3M–6M* HCl	Al, Fe, lanthanoids, Th, U(VI)	F. L. Moore, *Anal. Chem.* **28**:997 (1956); S. F. Marsh et al., *ibid.* **33**:870 (1961)
	Dibutyl phosphate	HNO_3 solutions	Al–Mg alloys and U alloys	R. F. Rolf, *Anal. Chem.* **33**: 125, 149 (1961)
	Trioctylphosphine oxide in cyclohexane	7*M* HCl + NH_4SCN in 7*M* HNO_3	U(VI) slight extraction Mo and Ti	K. Ilsemann and R. Bock, *Z. Anal. Chem.* **274**:185 (1975)

TABLE 2.22 Recoveries by Solvent Extraction under Various Conditions

	Number of equal portions for extraction	$D = 1$	$D = 2$	$D = 3$	$D = 5$
$V_{org} = V_{aq}$	2	0.56	0.75	0.84	
	4	0.59	0.80	0.89	
	10	0.62	0.84	0.93	
$V_{org} = 2V_{aq}$	2	0.75	0.89	0.94	0.97
	4	0.80	0.94	0.974	0.993
	10	0.84	0.966	0.991	0.9990
	100	0.86	0.980	0.9971	0.99993
$V_{org} = 3V_{aq}$	100	0.948	0.997	0.9998	

2.2.8.1 ***Higher-Density Solvent Extraction.*** In this method the extracting solvent has a higher density than the immiscible solution being extracted. The condensate from the total reflux of the extracting heavier solvent is diverted through the solution to be extracted, passes through that solution, extracting the solute, and siphons back into the boiling flask (Fig. 2.3). Continuous heating vaporizes the higher-density solvent, and the process is continued as long as is necessary.

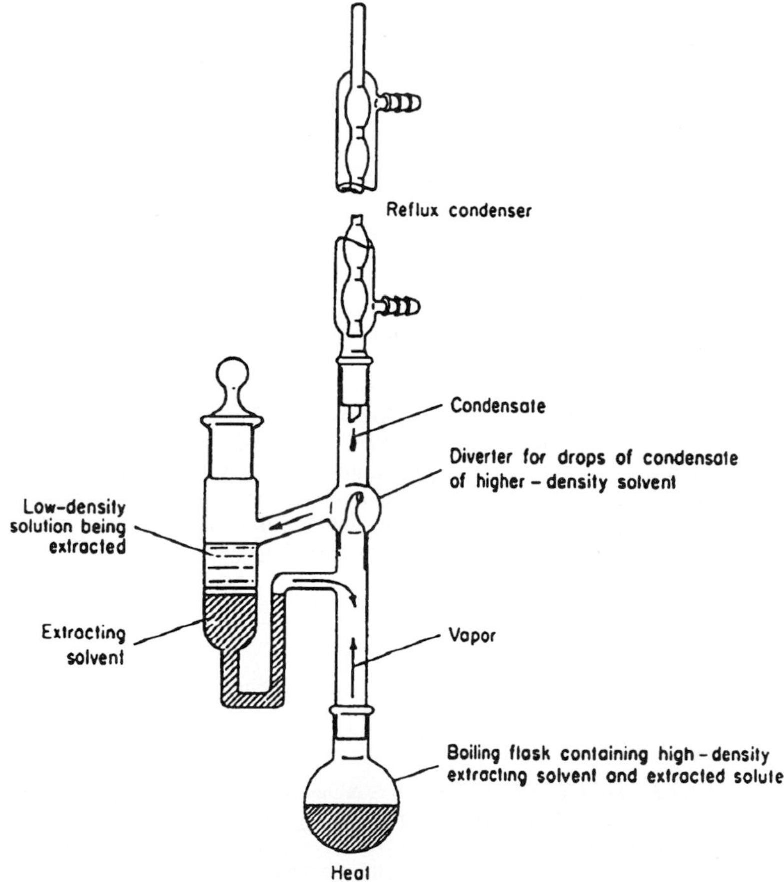

FIGURE 2.3 High-density liquid extractor.

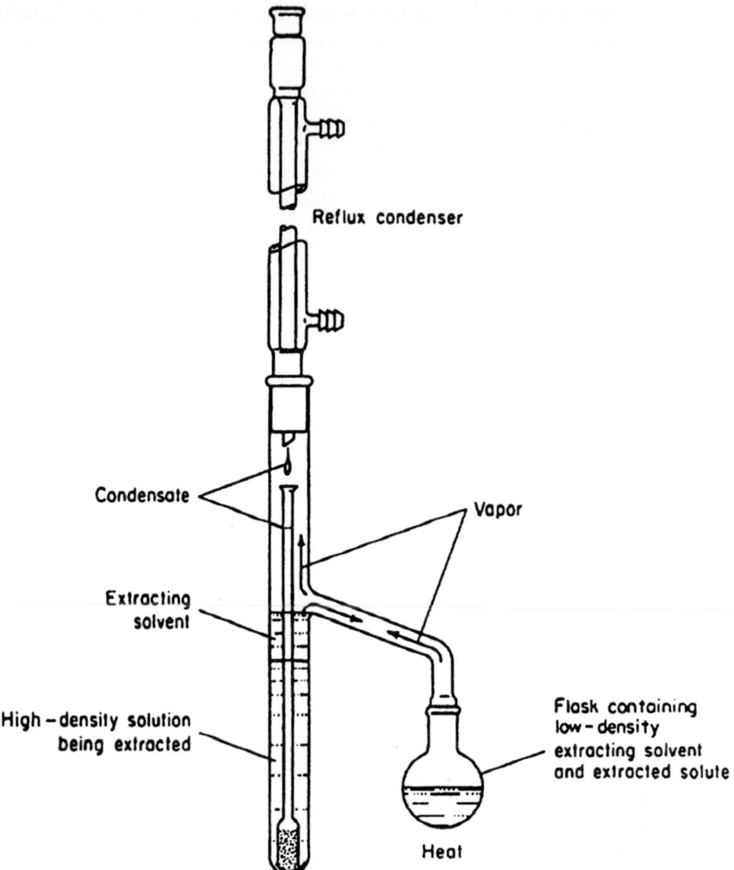

FIGURE 2.4 Low-density liquid extractor.

*2.2.8.2 **Lower-Density Solvent Extraction.*** The extracting solvent has a lower density than the immiscible solution being extracted. The condensate from the total reflux of the extracting lower-density solvent is caught in a tube (Fig. 2.4). As the tube fills, the increased pressure forces some of the lower-density solvent out through the bottom. It rises through the higher-density solvent, extracting the solute, and flows back to the boiling flask. Continuous heating vaporizes the low-density solvent, and the process is continued as long as necessary.

2.2.9 Extraction of a Solid Phase

Sample preparation is often required for cleanup and concentration. Traditionally, liquid–liquid extraction has been used for these purposes. However, this manual method using separatory funnels is inefficient, tedious, and costly.

*2.2.9.1 **Soxhlet Extraction.*** A Soxhlet extractor (Fig. 2.5) can be used to extract solutes from solids contained in a porous thimble, using any desired volatile solvent, which can be water-miscible or water-immiscible. The solvent is vaporized. When it condenses, it drops on the solid substance contained in the thimble and extracts soluble compounds. When the liquid level fills the

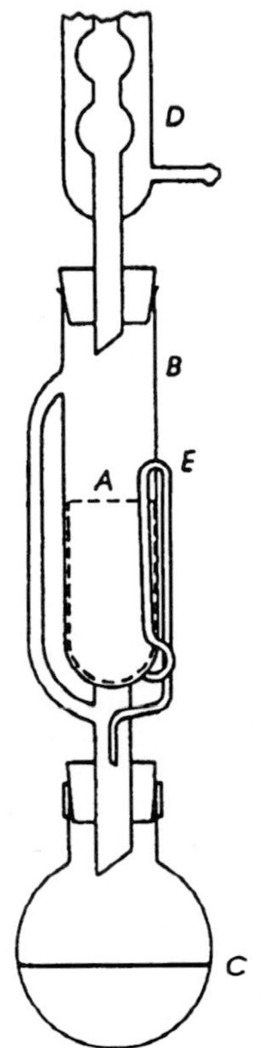

FIGURE 2.5 Soxhlet extractor.

body of the extractor, it automatically siphons into the flask. This process continues repeatedly as the solvent in the flask is vaporized and condensed.

When the extraction is complete, the extraction solvent containing the solute is poured into a beaker. The extracted components are isolated by evaporation or distillation of the solvent; the solute must be nonvolatile and thermally stable. Additional methods applicable to solvent removal include microwave drying (see Sec. 1.5.7) and critical-point drying (see Sec. 1.5.8).

Soxhlet extractors are standard equipment in laboratories that analyze fats and oils in biological samples. Separations can be achieved at low temperatures in inert atmospheres on a micro- or macroscale by a discontinuous or continuous process.

In the Soxtec® version (Fig. 2.6), extraction is faster, usually requiring one-fifth to one-tenth the time for a traditional Soxhlet extraction. In the rapid preextraction step the porous glass thimble is immersed in hot solvent, the vapor of which is condensed and returned to the thimble. After the bulk of the extractable material has been removed, the thimble is lifted from the solvent. Condensed solvent continues to drop onto the sample to leach the remaining soluble matter. Finally, the stopcock allowing the hot solvent to drip onto the sample is closed and the solvent is distilled away, leaving the soluble matter in the removable container.

2.2.9.2 Solid-Phase Extraction. Solid-phase (or liquid–solid) extraction (SPE) is becoming widely accepted as an excellent substitute that offers a number of advantages over liquid–liquid extraction. Similar to low-pressure liquid chromatography (*q.v.*), SPE uses membranes or small, disposable syringe-barrel columns or cartridges. Depending on the analytes of interest, the cartridges or membranes would contain silica coated with a compound that is either polar, moderately polar, or nonpolar (Table 2.23). The packed column is conditioned with solvent to activate the coating, and finally the column is further conditioned with the sample matrix solvent. The sample is forced through by aspiration or positive pressure from an air or nitrogen line. Then the packing is washed to elute any impurities. Finally, a solvent strong enough to displace the adsorbed analytes is forced through the packing. Alternately, the adsorbent is chosen to retain impurities and to let compounds of interest slowly pass through the column. Use about one tube volume of the sample solvent to remove any residual, desired material from the tube. Generally the flow rate should not exceed 5 mL $\cdot$ min^{-1}. Dropwise flow is best when time is not a factor. The method differs from high-performance liquid chromatography (HPLC) in that the adsorbent is used to remove a group of undesired materials from the sample, materials that might be interferences in the subsequent determinative step. SPE avoids the problems associated with the need to use immiscible solvents.

In general, a 1-mL column is used if the sample volume is no more than 1 mL; a 3-mL column if the sample volume is 1 to 250 mL and extraction speed is not critical; and a 6-mL column for fastest extraction of samples larger than 1 mL. As sample complexity increases, a column with more packing may be needed. Accurately transfer the sample to the column or reservoir above the column. Adjust the pH, salt concentration, and/or organic solvent content of the sample solution, as necessary. Add internal standards to the sample before or after transfer to the tube reservoir.

If compounds of interest are retained on the packing, unwanted and unretained materials are washed off with the same solution in which the sample was dissolved, using usually no more than a tube volume of wash solution. To remove unwanted, but weakly retained material, the packing is washed with solutions stronger than the sample matrix, but weaker than needed to remove compounds

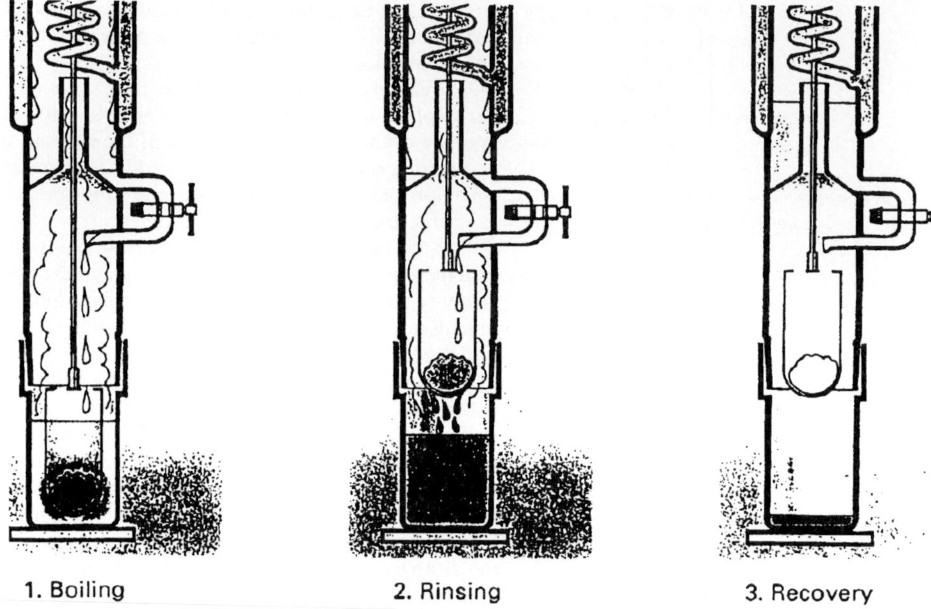

FIGURE 2.6 Extraction of solids with the Soxtec® system.

of interest. Finally, the packing is washed with 200 μL to 2 mL (depending on tube size) of a solution that removes analytes of interest. Two small aliquots generally elute analytes of interest more efficiently than one larger aliquot. Each aliquot should remain in contact with the packing for 20 s to 1 min.

TABLE 2.23 Solid-Phase Extraction Packings and Polarity Classification

Tube group	Packing type	Phase description	Polarity or classification of compound for extraction
1	C-8	Octyl bonded silica	Nonpolar to moderately polar compounds
	C-18	Octadecyl bonded silica	
	C-phenyl	Phenyl bonded silica	
	C-CN	Cyanopropyl bonded silica	
2	C-CN	Cyanopropyl bonded silica	Polar compounds
	Si	Silica gel	
	C-diol	Diol bonded silica	
	Florisil®	Magnesium silicate	
	$C\text{-}NH_2$	Aminopropyl bonded silica	
3	C-CN	Cyanopropyl bonded silica	Carbohydrates, cations
	$C\text{-}NH_2$	Aminopropyl bonded silica	Carbohydrates, weak anions, organic acids
	C-SCX	Sulfonic acid bonded silica (strong cation exchanger)	Strong cations, organic bases
	C-SAX	Quaternary amine bonded silica (strong anion exchanger)	Strong anions, organic acids
	C-WCX	Weak cation exchanger	Weak cations

SPE membranes significantly reduce extraction time and solvent use. Specifically, membrane extractions use up to 90% less solvent than liquid–liquid extractions and up to 20% less solvent than cartridges, and they cut processing time by hours. Because they are diluted with less solvent, analytes in solutions filtered through membranes are more concentrated. Lastly, sorbent particles embedded in a membrane are small and evenly distributed, thus eliminating the problem of channeling associated with columns.

2.2.10 Supercritical-Fluid Extraction

Real gases can be liquified by exceeding a certain pressure, which depends on the temperature of the gas. Above a certain temperature (the critical temperature T_c) no liquid phase is formed even when extremely high pressures are applied. A substance that is above its critical temperature and pressure is defined as a supercritical fluid. The density of gases increases continuously with pressure (at constant temperature) and at the critical point the densities of liquid and gas are identical. Table 2.24 summarizes the critical properties of gases suitable for supercritical-fluid extraction (SFE). The solvation power of gases increases with pressure (density) at constant temperature. The solvent strength of a supercritical fluid can easily be controlled by pressure and/or temperature. Consequently, the selectivity of extraction can easily be adjusted.

The solubility of solutes in supercritical fluids varies synchronously with the density. At constant temperature, in the region just above the critical pressure, the solubility of solutes increases strongly with increasing pressure (density). At higher temperature the increase in density with pressure is much less marked. At a constant temperature, extraction at lower pressures will favor less polar analytes,

TABLE 2.24 Characteristics of Selected Supercritical Fluids

Fluid	Critical temperature, °C	Critical pressure, psi	Critical density, $g \cdot cm^{-3}$	Density of liquid, $g \cdot cm^{-3}$
Acetonitrile	275	682	0.25	0.786
Ammonia	132.5	1636	0.23	
Argon	−122	707	0.53	
Benzene	289	710	0.30	0.874
Butane	152	551	0.23	0.601 (0°C)
Carbon dioxide	31	1070	0.47	
Carbon disulfide	279	1147	0.44	1.226
Chlorotrifluoromethane	106	588	0.53	
Dichlorodifluoromethane	111.8	598	0.52	
Ethane	32	706	0.20	
Ethylene	11	732	0.22	
Fluoromethane	45	853	0.30	
Heptane	267	397	0.23	0.684
Hexane	234	431	0.23	0.659
Methane	−82	667	0.16	
Methanol	240	1175	0.27	0.791
Nitrogen(I) oxide	38.5	1051	0.45	
Pentane	196.6	490	0.23	0.626
Propane	197	616	0.22	
Propene	92	658	0.23	
Sulfur hexafluoride	45.6	545	0.74	
Sulfur dioxide	158	1144	0.52	
Tetrahydrofuran	267	732	0.32	0.889
Trichloromethane	263	807	0.50	1.499
Trifluoromethane	25.9	701	0.52	

and extraction at higher pressures will favor more polar and higher-molecular-weight analytes. On the other hand, at constant pressure, the solubility decreases with increasing temperature and hence decreasing density. It is thus possible to precipitate solutes from fluid media by increasing the temperature. This allows an extraction to be optimized for a particular compound class by simply changing the pressure and, to a lesser extent, the temperature of the extraction.

The supercritical fluid exhibits combined gaslike mass transfer and liquidlike solvating characteristics. Mass-transfer limitations ultimately determine the rate at which an extraction can be performed. Because supercritical fluids have solute diffusion coefficients on the order of 2×10^{-4} to 7×10^{-4} $cm^2 \cdot s^{-1}$ (an order of magnitude 10^4 $cm^2 \cdot s^{-1}$ larger) and viscosities an order of magnitude lower (10^{-4} $N \cdot s \cdot m^2$) than liquid solvents, they have much better mass-transfer characteristics. Quantitative SFEs generally are complete in 10 to 60 min, whereas liquid-solvent extraction times can range from several hours to days. Supercritical-fluid extracts can be analyzed by any technique that is appropriate for the extracted analytes.

Class-selective extractions can be performed by extracting a single sample at different pressures. The temperature can usually be rather low and thus suitable for the extraction of thermolabile solutes, and there is no additional thermal strain for thermolabile solutes during removal of the extractor. The extraction gas protects against oxygen and the danger of oxidation is minimized. There is no formation of artifacts due to a chemical reaction with the extracting gas.

SFE is simple to perform. A pump is used to supply a known pressure of the extraction fluid to the extraction vessel, which is placed in a heater to maintain the vessel at a temperature above the critical temperature of the supercritical fluid. During the extraction, the soluble analytes are partitioned from the bulk sample matrix into the supercritical fluid, then swept through a flow restrictor into a collection device that is normally at ambient pressure. The fluids used for SFE are usually gases at ambient conditions and are vented from the collection device while the extracted analytes are retained.

Experimental variables that must be considered and optimized include the choice of supercritical fluid, pressure and temperature conditions, extraction time, sample size, the method used to collect the extracted analytes, and the equipment that is needed.

2.2.10.1 Choosing a Supercritical Fluid. The characteristics of some fluids that have been used for SFE are listed in Table 2.24. Although fluids with a range of polarities are available to optimize an extraction based on the polarity of the target analyte, the choice of fluid for analytical SFE frequently depends on practical considerations. The toxic, hazardous, explosive, and corrosive properties of the fluids have to be considered. Ammonia is too corrosive on instrument components, especially pumps. For practical extraction of aroma and food constituents only the low-molecular-weight hydrocarbons (ethylene to butane), fluorochloromethane, nitrous oxide, sulfur hexafluoride, and carbon dioxide are suitable. Supercritical CO_2 is the usual choice because of its relatively low critical temperature and pressure, low toxicity and reactivity, and high purity at low cost. The removal of carbon dioxide from the extract and its disposal present no problems. Unfortunately, supercritical CO_2 does not have sufficient solvent strength at typical working pressures (80 to 600 atm) to extract analytes quantitatively that are quite polar. In general terms, CO_2 is an excellent extraction medium in the pressure range up to 4400 psi for nonpolar species such as alkanes and terpenes. It is reasonably good for moderately polar species, including polycyclic aromatic hydrocarbons, polychlorinated biphenyls, ketones, aldehydes, esters, alcohols, organochlorine pesticides, and fats up to molecular masses of 300 to 400 amu. As a rule of thumb, compounds that can be analyzed by conventional gas chromatographic techniques can be quantitatively extracted using supercritical CO_2.

When fairly polar analytes need to be extracted, a problem arises in choice of supercritical fluid. Benzene derivatives with three phenolic hydroxyl groups are still extractable with superfluid CO_2, as are components with one carboxylic or two hydroxyl groups. More polar groups in the molecule hinder extraction. Supercritical methanol is an excellent solvent but possesses a high critical temperature and is a liquid at ambient conditions, which complicates sample concentration after extraction. Supercritical N_2O yields better efficiencies for some samples (N_2O does have a small dipole moment). Extraction of highly polar analytes has generally been done using CO_2 containing a few percent added organic modifier. Modifiers can be introduced as mixed fluids in the pumping system, with the aid of a second pump, or by simply injecting the modifier as a liquid onto the sample before beginning the extraction. Although methanol has been a widely used modifier, a variety of organic compounds have been used, ranging from alcohols, propylene carbonate, 2-methoxyethanol, and organic acids to methylene chloride and carbon disulfide. The selection of modifiers and their concentration is largely empirical. Modifiers are tested that possess different polarities and concentrations as well as determining optimal temperature and pressure conditions.

2.2.10.2 SFE Techniques and Hardware. Analytical SFE is typically performed using syringe pumps (see HPLC, Sec. 4.3.1.1.2) without the pressure or density ramp controllers required in HPLC. For samples larger than 50 g, gas compressors are useful. Extraction cells are available commercially. The temperature of the extraction cell is normally controlled by placing the cell in a chromatographic oven or a simple tube heater.

For dynamic SFE, the supercritical fluid is constantly flowing through the cell, and a flow restrictor is used to maintain pressure in the extraction vessel and allow the supercritical fluid to depressurize into the collection device. Both micrometering valves and short lengths of fused-silica tubing [~10 to 50 μm inner diameter (i.d.)] have been used. Static SFE is performed by pressurizing the cell and extracting the sample with no outflow of the supercritical fluid. After a set period of time, a valve is opened to allow the analytes to be swept into the collection device.

SFE of samples less than 10 g is most common. Quantitative collection of relatively volatile analytes is convenient with supercritical-fluid flow rates of up to at least 1 mL/min. At these rates, assuming that a typical sample has a void volume of approximately 30%, passing 10 void volumes of extraction fluid through 1-, 10-, and 100-g samples would require about 3, 30, and 300 min, respectively. Extraction cells greater than 1 mL should be packed full of the sample to avoid significant void volumes in the extraction system, which would increase extraction times.

With collection vessels, the analytes are often collected in a few milliliters of a liquid solvent. The cooling of the solvent caused by the expanding supercritical fluid prevents rapid evaporation of the collection solvent, and even relatively volatile analytes are quantitatively recovered. When large quantities of bulk matrix are extracted, direct depressurization into an empty receiving vessel has been successful, but with trace analytes, trapping in empty vessels suffers because of losses from aerosol formation.

On-line SFE has most often been coupled with capillary gas chromatography and both packed and capillary SFC (see Sec. 4.2.8).

2.3 ION-EXCHANGE METHODS (NORMAL PRESSURE, COLUMNAR)

Ion-exchange methods are based essentially on a reversible exchange of ions between an external liquid phase and an ionic solid phase. The solid phase consists of a polymeric matrix, insoluble, but permeable, which contains fixed charge groups and mobile counterions of opposite charge. These counterions can be exchanged for other ions in the external liquid phase. Enrichment of one or several of the components is obtained if selective exchange forces are operative. The method is limited to substances at least partially in ionized form.

2.3.1 Chemical Structure of Ion-Exchange Resins

An ion-exchange resin usually consists of polystyrene copolymerized with divinylbenzene to build up an inert three-dimensional, cross-linked matrix of hydrocarbon chains. Protruding from the polymer chains are the ion-exchange sites distributed statistically throughout the entire resin particle. The ionic sites are balanced by an equivalent number of mobile counterions. The type and strength of the exchanger is determined by these active groups. Ion exchangers are designated anionic or cationic, according to whether they have an affinity for negative or positive counterions. Each main group is further subdivided into strongly or weakly ionized groups. A selection of commercially available ion-exchange resins is given in Table 2.25. These are available in a range of particle sizes and with different degrees of cross-linking. Resin conversions are outlined in Table 2.26.

The cross-linking of a polystyrene resin is expressed as the proportion by weight percent of divinylbenzene in the reaction mixture; for example, "×8" for 8% cross-linking. As the percentage is increased, the ionic groups come into effectively closer proximity, resulting in increased selectivity. Intermediate cross-linking, in the range of 4% to 8%, is usually used. An increase in cross-linking decreases the diffusion rate in the resin particles; the diffusion rate is the rate-controlling step in column operations. Decreasing the particle size reduces the time required for attaining equilibrium, but at the same time decreases the flow rate until it is prohibitively slow unless pressure is applied.

In most inorganic chromatography, resins of 100 to 200 mesh size are suitable; difficult separations may require 200 to 400 mesh resins. A flow rate of 1 $mL \cdot cm^{-2} \cdot min^{-1}$ is often satisfactory. With HPLC (see Sec. 4.3), the flow rate in long columns of fine adsorbent can be increased by applying pressure.

2.3.1.1 Macroreticular Resins. Macroreticular resins are an agglomerate of randomly packed microspheres that extend through the agglomerate in a continuous nongel pore structure. The channels throughout the rigid pore structure render the bead centers accessible even in nonaqueous solvents, in which microreticular resins do not swell sufficiently. Because of their high porosity and large pore diameters, these resins can handle large organic molecules.

2.3.1.2 Microreticular Resins. Microreticular resins, by contrast, are elastic gels that, in the dry state, avidly absorb water and other polar solvents in which they are immersed. While taking up solvent, the gel structure expands until the retractile stresses of the distended polymer network balance the osmotic effect. In nonpolar solvents, little or no swelling occurs and diffusion is impaired.

2.3.1.3 Ion-Exchange Celluloses. For organic molecules of molecular weight greater than 500 and up to such macromolecules as serum proteins, nucleic acids, and enzymes, ion-exchange separations are carried out on the ion-exchange celluloses, dextrans, or polyacrylamide gels. The separation of large ionic molecules becomes possible without side effects from denaturation or irreversible absorption.

Cellulose ion exchangers are prepared by imparting anion and cation exchange properties to cellulose by treatment with various reagents. The exchange groups are bound to the fibrous surface of the cellulose matrix by ether or ester linkages through the hydroxyl groups of the anhydroglucose rings. Several types are described in Table 2.27. The major available types are diethylaminoethyl (DEAE) cellulose and polyethyleneimine (PEI) cellulose strong base anion exchangers, triethanolamine (ECTEOLA) cellulose weak base anion exchanger, phosphorylated (P) cellulose strong acid cation exchanger, and carboxymethyl (CM) cellulose weak acid cation exchanger. Exchange capacities range from 0.2 $meq \cdot g^{-1}$ for types sulfoethyl (SE) and ECTEOLA to 0.9 $meq \cdot g^{-1}$ for strongly basic types. The ether derivatives are reasonably stable and may be used under a wide range of conditions. The ester derivatives are subject to hydrolysis at pH values less than 2 and greater than 10, especially above ambient temperature.

2.3.1.4 Dextran and Polyacrylamide Exchangers. Ion exchangers made from the dextran and polyacrylamide gels are strongly hydrophilic and swell in water and salt solutions. Sephadex ion exchangers are prepared from the polysaccharide dextran cross-linked by reaction with epichlorohydrin, followed by the introduction of the desired functional groups on the fiber and pore surfaces. The separation of proteins, nucleic acids, and other large ionic molecules becomes possible without side effects from denaturation or irreversible absorption. Each monomer has three hydroxymethyl groups that impart hydrophilic character.

The gel porosity is governed by the degree of cross-linkage. Maximum uptake of a macromolecule will be facilitated by using a gel that has a pore size larger than that of the macromolecule.

TABLE 2.25 Guide to Ion-Exchange Resins

Dowex is the trade name of Dow resins; X (followed by a numeral) is percent cross-linked. Mesh size (dry) is available in the range 50–100, 100–200, 200–400, and sometimes smaller than 400.

S-DVB is the acronym for styrene-divinylbenzene.

MP is the acronym for macroporous resin. Mesh size (dry) is available in the range 20–50, 100–200, and 200–400.

Bio-Rex is the trade name for certain resins sold by Bio-Rad Laboratories.

Amberlite and Duolite are trade names of Rohm & Haas resins.

Resin type and nominal percent cross-linkage	Minimum wet capacity, $meq \cdot mL^{-1}$	Density (nominal), $g \cdot mL^{-1}$	Comments
Anion exchange resins–gel type–strongly basic–quaternary ammonium functionality			
Dowex 1-X2	0.6	0.65	Strongly basic anion exchanger with S-DVB matrix for separation of small peptides, nucleotides, and large metal complexes. Molecular weight exclusion is <2700.
Dowex 1-X4	1.0	0.70	Strongly basic anion exchanger with S-DVB matrix for separation of organic acids, nucleotides, phosphoinositides, and other anions. Molecular weight exclusion <1400.
Dowex 1-X8	1.2	0.75	Strongly basic anion exchanger with S-DVB matrix for separation of inorganic and organic anions with molecular weight exclusion <1000. 100–200 mesh is standard for analytical separations.
Dowex 2-X8	1.2	0.75	Strongly basic (but less basic than Dowex 1 type) anion exchanger with S-DVB matrix for deionization of carbohydrates and separation of sugars, sugar alcohols, and glycosides.
Amterlite IRA-400	1.4	1.11	8% cross-linkage. Used for systems essentially free of organic materials.
Amberlite IRA-402	1.3	1.07	Lower cross-linkage than IRA-400; better diffusion rate with large organic molecules.
Amberlite IRA-410	1.4	1.12	Dimethylethanolamine functionality and slightly lower basicity than IRA-400.
Amberlite IRA-458	1.2	1.08	Has an acrylic structure rather than S-DVB; hence more hydrophilic and resistant to organic fouling.
Anion exchange resins–gel type—intermediate basicity			
Bio-Rex 5	2.8	0.70	Intermediate basic anion exchanger with primary tertiary amines on a polyalkylene–amine matrix for separation of organic acids.
Anion exchange resins–gel type—weakly basic—polyamine functionality			
Dowex 4-X4	1.6	0.70	Weakly basic anion exchanger with tertiary amines on an acrylic matrix for deionization of carbohydrates. Use at pH < 7.
Amberlite IRA-68	1.6	1.06	Acrylic-DVB with unusually high capacity for large organic molecules.
Cation exchange resins–gel type—strongly acidic—sulfonic acid functionality			
Dowex 50W-X2	0.6	0.70	Strongly acidic cation exchanger with S-DVB matrix for separation of peptides, nucleotides, and cations. Molecular weight exclusion <2700.

(Continued)

TABLE 2.25 Guide to Ion-Exchange Resins (*Continued*)

Resin type and nominal percent cross-linkage	Minimum wet capacity, meq · mL^{-1}	Density (nominal), g · mL^{-1}	Comments
Cation exchange resins–gel type—strongly acidic—sulfonic acid functionality			
Dowex 50W-X4	1.1	0.80	Strongly acidic cation exchanger with S-DVB matrix for separation of amino acids, nucleosides and cations. Molecular weight exclusion is <1400.
Dowex 50W-X8	1.7	0.80	Strongly acidic cation exchanger with S-DVB matrix for separation of amino acids, metal cations, and cations. Molecular weight exclusion is <1000. 100–200 mesh is standard for analytical applications.
Dowex 50W-X12	2.1	0.85	Strongly acidic cation exchanger with S-DVB matrix used primarily for metal separations.
Dowex 50W-X16	2.4	0.85	Strongly acidic cation exchanger with S-DVB matrix and high cross-linkage.
Amberlite IR-120	1.9	1.26	8% styrene-DVB type; high physical stability.
Amberlite IR-122	2.1	1.32	10% styrene-DVB type; high physical stability and high capacity.
Weakly acidic cation exchangers—gel type—carboxylic acid functionality			
Duolite C-433	4.5	1.19	Acrylic-DVB type; very high capacity. Used for metals removal and neutralization of alkaline solutions.
Bio-Rex 70	2.4	0.70	Weakly acidic cation exchanger with carboxylate groups on a macroreticular acrylic matrix for separation and fractionation of proteins, peptides, enzymes, and amines, particularly high molecular weight solutes. Does not denature proteins as do styrene-based resins.
Selective ion-exchange resins			
Duolite GT-73	1.3	1.30	Removal of Ag, Cd, Cu, Hg, and Pb.
Amberlite IRA-743A	0.6	1.05	Boron-specific ion-exchange resin.
Amberlite IRC-718	1.0	1.14	Removal of transition metals.
Chelex® 100	0.4	0.65	Weakly acidic chelating resin with S-DVB matrix for heavy metal concentration.
Anion exchanger—macroreticular type—strongly basic—quaternary ammonium functionality			
Amberlite IRA-910	1.1	1.09	Dimethylethanolamine styrene-DVB type which offers slightly less silica removal than Amberlite IRA resin, but offers improved regeneration efficiency.
Amberlite IRA-938	0.5	1.20	Pore size distribution between 2500 and 23 000 nm; suitable for removal of high-molecular-weight organic materials.
Amberlite IRA-958	0.8		Acrylic-DVB; resistant to organic fouling.
AG MP-1	1.0	0.70	Strongly basic macroporous anion exchanger with S-DVB matrix for separation of some enzymes, radioactive anions, and other applications.

TABLE 2.25 Guide to Ion-Exchange Resins (*Continued*)

Resin type and nominal percent cross-linkage	Minimum wet capacity, $meq \cdot mL^{-1}$	Density (nominal), $g \cdot mL^{-1}$	Comments
			Cation exchange resin—macroreticular type—sulfonic acid functionality
Amberlite 200	1.7	1.26	Styrene-DVB with 20% DVB by weight; superior physical stability and greater resistance to oxidation by factor of three over comparable gel-type resin.
AG MP-50	1.5	0.80	Strongly acidic macroporous cation exchanger with S-DVB matrix for separation of radioactive cations and other applications.
			Weak cation exchanger—macroreticular type—carboxylic acid or phenolic functionality
Amberlite DP-1	2.5	1.17	Methacrylic acid–DVB; high resin capacity. Use pH > 5.
Amberlite IRC-50	3.5	1.25	Methacrylic acid–DVB. Selectivity absorbs organic gases such as antibiotics, alkaloids, peptides, and amino acids. Use pH > 5.
Duolite C-464	3.0	1.13	Polyacrylic resin with high capacity and outstanding resistance to osmotic shock.
Duolite A-7	2.2	1.12	Phenolic-type resin. High porosity and hydrophilic matrix. pH 0–6.
Duolite A-368	1.7	1.04	Styrene-DVB; pH 0–9.
Amberlite IRA-35	1.1		Acrylic-DVB; pH 0–9.
Amberlite IRA-93	1.3	1.04	Styrene-DVB; pH 0–9. Excellent resistance to oxidation and organic fouling.
			Liquid amines
Amberlite LA-1			A secondary amine containing two highly branched aliphatic chains of M.W. 351 to 393. Solubility is 15 to 20 mg/mL in water. Used at 5–40% solutions in hydrocarbons.
Amberlite LA-2			A secondary amine of M.W. 353 to 395. Insoluble in water.
			Microcrystalline exchanger
AMP-1	4.0		Microcrystalline ammonium molybdophosphate with cation exchange capacity of 1.2 meq/g. Selectively adsorbs larger alkali-metal ions from smaller alkali-metal ions, particularly cesium.
			Ion retardation resin
AG 11 A8		0.70	Ion retardation resin containing paired anion (COO^-) and cation [$(CH_3)_3N^+$ sites. Selectively retards ionic substances.

Type P-2 is suitable for the separation of low-molecular-weight materials. For solutes of molecular weight less than 10 000, the P-30 or G-25 type is best, whereas the G-50 or P-100 will have the highest available capacity for solutes of a molecular weight greater than 10 000. Exchange capacity ranges from 2 to 6 $meq \cdot g^{-1}$.

2.3.1.5 Ion-Exchange Membranes. Ion-exchange membranes are extremely flexible, strong membranes, composed of analytical grade ion-exchange resin beads (90%) permanently enmeshed

TABLE 2.26 Conversion of Ion-Exchange Resins

Resin	Conversion: from→	Reagent used	Volumes of solution per volume of resin	Flow rate, $mL \cdot min^{-1} \cdot cm^2$ of bed	Test for completeness of conversion	Rinse volume of water per volume of resin	Test for completeness of rinsing
AG 1	Cl^- to OH^-	1*M* NaOH	20		Acidify sample and add a few drops of 1% $AgNO_3$ until no white precipitate of AgCl appears.	4	$pH < 9$
AG 2	Cl^- to OH^-	1*M* NaOH	2	2	Same as AG 1	4	$pH < 9$
	Cl^- to NO_3^-	0.5*M* $NaNO_3$	5		Same as AG 1	4	
AG MP-1	OH^- to formate	1*M* formic acid	2	2	$pH < 2$	4	$pH > 4.5$
AG 4	Cl^- to free base + OH^-	0.5*M* NaOH	2	1	Same as AG 1	4	$pH < 9$
Bio-Rex 5							
AG 50	H^+ to Na^+	1*M* NaOH	2	2	$pH < 9$	4	$pH > 9$
Bio-Rex 70	H^+ to Na^+	0.5*M* NaOH	3	1	$pH > 9$		
Chelex® 100	Cu^{2+} to H^+	1*M* HCl	3	1	Test for Cu^{2+} in effluent	4	$pH < 8$
AG 11	NaCl to self-adsorbed	H_2O	20	2	Same as for AG 1		$pH < 10$
	HCl to self-adsorbed	NaOH + H_2O	20	2	$pH < 10$		

TABLE 2.27 Gel Filtration Media

Bio-Gel is a registered trade mark of Bio-Rad Laboratories.
Sephadex is a registered trade mark of Pharmacia Fine Chemicals.

Polyacrylamide gels

Product designation	Fractionation range, atomic mass units	Hydrated bed volume, $mL \cdot g^{-1}$ dry gel	Water regain, $mL \cdot g^{-1}$ dry gel
Bio-Gel® P-2	100 to 1 800*	3.5	1.6
P-4	800 to 4 000	3.5	2.6
P-6	1 000 to 6 000	7	3.2
P-10	1 500 to 20 000	9	5.1
P-30	2 500 to 40 000	11	6.2
P-60	3 000 to 60 000	14	6.8
P-100	5 000 to 100 000	15	7.5
P-200	30 000 to 200 000	25	13.5
P-300	60 000 to 400 000	30	22
Sephadex® G-10	up to 700†	2.5	1.0
G-15	up to 1 500†	3	1.5
G-25	100 to 5 000‡	5	2.5
G-50	500 to 10 000	10	5.0
G-75	3 000 to 70 000	13	7.5
G-100	4 000 to 150 000	17	10
G-150	5 000 to 400 000	24	15
G-200	5 000 to 800 000	30	20

* Determined with polysaccharides.
† Determined with polyethylene glycols.
‡ Determined with globular proteins; values considerably lower with polysaccharides.

Agarose gels

Product description	Fractionation range, atomic mass units	Estimated nucleic acid exclusion limit (base pairs)
Bio Gel A—0.5 m gel	<10 000 to 500 000	200
A—1.5 m gel	<10 000 to 1 500 000	750
A—5 m gel	10 000 to 5 000 000	2 000
A—15 m gel	40 000 to 15 000 000	7 000
A—50 m gel	100 000 to 50 000 000	20 000
A—150 m gel	1 000 000 to 150 000 000	70 000

Nonaqueous gel permeation

Product description	Molecular weight operating range	Bed volume, $mL \cdot g^{-1}$ (in benzene)
Bio-beads S-X1	600 to 14 000	9.8
Bio-beads S-X2	100 to 2 700	5.2
Bio-beads S-X3	up to 2 000	5.0
Bio-beads S-X4	up to 1 400	4.0
Bio-beads S-X8	up to 1 000	3.0
Bio-beads S-X12	up to 400	2.5

Ion-exchange celluloses and agarose gels

Product description	Exchange group	Hemoglobin capacity, $mg \cdot mL^{-1}$	Ionic capacity, $\mu mols \cdot mL^{-1}$
DEAE Bio-Gel A agarose	Weak base, diethylaminoethyl	45 ± 10	20 ± 5
CM Bio-Gel A agarose	Weak acid, carboxymethyl	45 ± 10	20 ± 5

(*Continued*)

TABLE 2.27 Gel Filtration Media (*Continued*)

Product designation	Exchange group	Exchange capacity meq · g^{-1}	Settled bed volume, mL · dry g^{-1}	Approx. pK_a
Cellex D cellulose	DEAE (diethylaminoethyl), weak base	0.7 ± 0.1	8	9.5
Cellex E cellulose	ECTEOLA (mixed amines), intermediate base	0.3 ± 0.05	3.5	7.5
Cellex CM cellulose	CM (carboxymethyl), weak acid	0.7 ± 0.1	7	4
Cellex P cellulose	P (phosphoryl), intermediate acid	0.85 ± 0.1	4.5	1.5, 6

in a poly(tetrafluoroethylene) membrane (10%). The membranes offer an alternative to column and batch methods, and can be used in many of the same applications as traditional ion-exchange resins. Three ion-exchange resin types have been incorporated into membranes: AG 1-X8, AG 50W-X8, and Chelex 100.

2.3.2 Functional Groups

Sulfonate exchangers contain the group $—SO_3^-$, which is strongly acidic and completely dissociated whether in the H form or the cation form. These exchangers are used for cation exchange.

Carboxylate exchangers contain —COOH groups, which have weak acidic properties and will only function as cation exchangers when the pH is sufficiently high (pH > 6) to permit complete dissociation of the —COOH site. Outside this range the ion exchanger can be used only at the cost of reduced capacity.

Quaternary ammonium exchangers contain $—R_4N^+$ groups, which are strongly basic and completely dissociated in the OH form and the anion form.

Tertiary amine exchangers possess $—R_3NH_2$ groups, which have exchanging properties only in an acidic medium when a proton is bound to the nitrogen atom.

Aminodiacetate exchangers have the $—N(CH_2COOH)_2$ group, which has unusually high preference for copper, iron, and other heavy-metal cations and, to a lesser extent, for alkaline-earth cations. The resin selectivity for divalent over monovalent ions is approximately 5000 to 1. The resin functions as a chelating resin at pH 4 and above. At very low pH, the resin acts as an anion exchanger. This exchanger is the column packing often used for ligand exchange.

2.3.3 Exchange Equilibrium

Retention differences among cations with an anion exchanger, or among anions with a cation exchanger, are governed by the physical properties of the solvated ions. The stationary phase will show these preferences:

1. The ion of higher charge.

2. The ion with the smaller solvated radius. Energy is needed to strip away the solvation shell surrounding ions with large hydrated radii, even though their crystallographic ionic radii may be less than the average pore opening in the resin matrix.

3. The ion that has the greater polarizability (which determines the van der Waals attraction).

To accomplish any separation of two cations (or two anions) of the same net charge, the stationary phase must show a preference for one over the other. No variation in the eluent concentration will

TABLE 2.28 Relative Selectivity of Various Counter Cations

Counterion	Relative selectivity for 50W-X8 resin	Counterion	Relative selectivity for 50W-X8 resin
H^+	1.0	Zn^{2+}	2.7
Li^+	0.86	Co^{2+}	2.8
Na^+	1.5	Cu^{2+}	2.9
NH_4^+	1.95	Cd^{2+}	2.95
K^+	2.5	Ni^{2+}	3.0
Rb^+	2.6	Ca^{2+}	3.9
Cs^+	2.7	Sr^{2+}	4.95
Cu^+	5.3	Hg^{2+}	7.2
Ag^+	7.6	Pb^{2+}	7.5
Mn^{2+}	2.35	Ba^{2+}	8.7
Mg^{2+}	2.5	Ce^{3+}	22
Fe^{2+}	2.55	La^{3+}	22

improve the separation. However, if the exchange involves ions of different net charges, the separation factor does depend on the eluent concentration. The more dilute the counterion concentration in the eluent, the more selective the exchange becomes for polyvalent ions.

In the case of an ionized resin, initially in the H form and in contact with a solution containing K^+ ions, an equilibrium exists:

$$\text{resin}, H^+ + K^+ \rightleftarrows \text{resin}, K^+ + H^+ \tag{2.51}$$

which is characterized by the selectivity coefficient, $k_{K/H}$:

$$k_{K/H} = \frac{[K^+]_r [H^+]}{[H^+]_r [K^+]} \tag{2.52}$$

where the subscript r refers to the resin phase. Table 2.28 contains selectivity coefficients for cations and Table 2.29 for anions. Relative selectivities are of limited use for the prediction of the columnar exchange behavior of a cation because they do not take account of the influence of the aqueous phase. More specific information about the behavior to be expected from a cation in a column elution experiment is given by the equilibrium distribution coefficient K_d.

The partitioning of the potassium ion between the resin and solution phases is described by the concentration distribution ratio D_c:

$$(D_c)_K = \frac{[K^+]_r}{[K^+]} \tag{2.53}$$

Combining the equations for the selectivity coefficient and for D_c,

$$(D_c)_K = k_{K/H} \frac{[H^+]_r}{[H^+]} \tag{2.54}$$

The foregoing equation reveals that essentially the concentration distribution ratio for trace concentrations of an exchanging ion is independent of the respective solution of that ion and that the uptake of

TABLE 2.29 Relative Selectivity of Various Counter Anions

Counterion	Relative selectivity for 1-X8 resin	Relative selectivity for 2-X8 resin
OH^-	1.0	1.0
Benzene sulfonate	500	75
Salicylate	450	65
Citrate	220	23
I^-	175	17
Phenate	110	27
HSO_4^-	85	15
ClO_3^-	74	12
NO_3^-	65	8
Br^-	50	6
CN^-	28	3
HSO_3^-	27	3
BrO_3^-	27	3
NO_2^-	24	3
Cl^-	22	2.3
HCO_3^-	6.0	1.2
IO_3^-	5.5	0.5
$H_2PO_4^-$	5.0	0.5
Formate	4.6	0.5
Acetate	3.2	0.5
Propanoate	2.6	0.3
F^-	1.6	0.3

each trace ion by the resin is directly proportional to its solution concentration. However, the concentration distribution ratios are inversely proportional to the solution concentration of the resin counterion.

To accomplish any separation of two cations (or two anions), one of these ions must be taken up by the resin in distinct preference to the other. This preference is expressed by the separation factor (or relative retention), $\alpha_{K/Na}$, using K^+ and Na^+ as the example.

$$\alpha_{K/Na} = \frac{(D_c)_K}{(D_c)_{Na}} = \frac{k_{K/H}}{k_{Na/H}} = K_{K/Na} \tag{2.55}$$

The more α deviates from unity for a given pair of ions, the easier it will be to separate them. If the selectivity coefficient is unfavorable for the separation of two ions of the same charge, no variation in the concentration of H^+ (the eluent) will improve the separation.

The situation is entirely different if the exchange involves ions of different net charges. Now the separation factor does depend on the eluent concentration. For example, the more dilute the counterion concentration in the eluent, the more selective the exchange becomes for the ion of higher charge.

In practice, it is more convenient to predict the behavior of an ion, for any chosen set of conditions, by employing a much simpler distribution coefficient D_g, which is defined as the concentration of a solute in the resin phase divided by its concentration in the liquid phase, or

$$D_g = \frac{\text{concentration of solute, resin phase}}{\text{concentration of solute, liquid phase}} \tag{2.56}$$

$$D_g = \frac{\text{\% solute within exchanger}}{\text{\% solute within solution}} \times \frac{\text{volume of solution}}{\text{mass of exchanger}} \tag{2.57}$$

D_g remains constant over a wide range of resin to liquid ratios. In a relatively short time, by simple equilibration of small known amounts of resin and solution followed by analysis of the phases, the distribution of solutes may be followed under many different sets of experimental conditions. Variables requiring investigation include the capacity and percentage cross-linkage of resin, the type of resin itself, the temperature, and the concentration and pH of electrolyte in the equilibrating solution.

By comparing the ratio of the distribution coefficients for a pair of ions, a separation factor (or relative retention) is obtained, Eq. (2.55), for a specific experimental condition.

Instead of using D_g, separation data may be expressed in terms of a volume distribution coefficient D_v, which is defined as the amount of solution in the exchanger per cubic centimeter of resin bed divided by the amount per cubic centimeter in the liquid phase. The relation between D_g and D_v is given by

$$D_v = D_g \rho \tag{2.58}$$

where ρ is the bed density of a column expressed in the units of mass of dry resin per cubic centimeter of column. The bed density can be determined by adding a known weight of dry resin to a graduated cylinder containing the eluting solution. After the resin has swelled to its maximum, a direct reading of the settled volume of resin is recorded.

Intelligent inspection of the relevant distribution coefficients will show whether a separation is feasible and what the most favorable eluent concentration is likely to be. In the columnar mode, an ion, even if not eluted, may move down the column a considerable distance and with the next eluent may appear in the eluate much earlier than indicated by the coefficient in the first eluent alone. A distribution coefficient value of 12 or lower is required to elute an ion completely from a column containing about 10 g of dry resin using 250 to 300 mL of eluent. A larger volume of eluent is required only when exceptionally strong tailing occurs. Ions may be eluted completely by 300 to 400 mL of eluent from a column of 10 g of dry resin at D_g values of around 20. The first traces of an element will appear in the eluate at around 300 mL when its D_g value is about 50 to 60.

Example 2.4 Shaking 50 mL of 0.001*M* cesium salt solution with 1.0 g of a strong cation exchanger in the H form (with a capacity of 3.0 meq · g^{-1}) removes the following amount of cesium. The selectivity coefficient $k_{\text{Cs/H}}$ is 2.56; thus

$$\frac{[\text{Cs}^+]_r[\text{H}^+]}{[\text{Cs}^+][\text{H}^+]_r} = 2.56$$

The maximum amount of cesium that can enter the resin is 50 mL × 0.001*M* = 0.050 equiv. The minimum value of $[\text{H}^+]_r = 3.00 - 0.05 = 2.95$ meq, and the maximum value, assuming complete exchange of cesium ion for hydrogen ion, is 0.001*M*. The minimum value of the distribution ratio is

$$(D_c)_{\text{Cs}} = \frac{[\text{Cs}^+]_r}{[\text{Cs}^+]} = \frac{2.56 \times 2.95}{0.001} = 7550$$

$$\frac{\text{Amount of Cs, resin phase}}{\text{Amount of Cs, solution phase}} = \frac{7550 \times 1.0 \text{ g}}{50 \text{ mL}} = 151$$

Thus, at equilibrium the 1.0 g of resin removed

$$\frac{100\% - x}{x} = 151$$

all but 0.66% of cesium ions from solution. If the amount of resin were increased to 2.0 g, the amount of cesium remaining in solution would decrease to 0.33%, half the former value. However, if the depleted solution were decanted and placed in contact with 1 g of fresh resin, the amount of cesium remaining in solution decreases to 0.004%. Two batch equilibrations would effectively remove the cesium from the solution.

2.3.4 Applications

Ion-exchange resins are particularly useful and convenient to handle. Columns can be made in any desired size. The diameter of the column depends on the amount of material to be treated; the length depends on the difficulty of the separations to be accomplished. See also the discussion on ion chromatography in Sec. 4.4.

2.3.4.1 Column Filtration. The column filtration technique involves continuous and discontinuous methods. In the continuous method the solution is fed continuously into the top of the ion-exchange column. The ions of the same electrical charge as the counterion of the exchanger are retained in the resin phase if their selectivity coefficients exceed that of the counterion. Oppositely charged ions, as well as uncharged species, remain in the external solution phase. Gradually the original counterion is forced out of the column by the species in the feed solution that have more affinity for the resin phase. This displacement process continues until the column capacity approaches exhaustion. In this manner, undesirable cations may be removed by a cation exchanger and, similarly, unwanted anions by an anion exchanger. Deionization and water softening fall into this category.

If interest centers around an ion ionic species present in relatively low concentration, a fixed volume of sample is introduced onto the top of the column, as before, but now a washing solution of an appropriate composition is passed through the column to remove the nonexchanged (unwanted) material from the interstitial volume of the column. After the unwanted material has been washed from the column, the adsorbed component is desorbed by an appropriate eluent and washed out of the column with a few column volumes. The species of interest is obtained in a relatively small volume of eluent that is free of all other ions present in the original sample. In this manner trace elements present in waters from springs and rivers can be concentrated in a resin column, and later eluted back in the laboratory and determined by any suitable method. It may be advantageous from the standpoint of keeping volumes small to remove the adsorbed constituent at the end by running the displacing solution through the column in the reverse direction, since the adsorbed constituent will be at the top of the column. Valuable ionic materials present at low concentration can be recovered from processing solutions; examples include citric and ascorbic acids from citrus wastes, fatty acids from soap wastes, tartaric acid from wine residues, and nicotinic acid from vitamin-product wastes. Stream pollution may be mitigated with the recovered metals partially offsetting the cost of the waste-disposal method.

In the pharmaceutical field, the antibiotic streptomycin is adsorbed from the filtered fermentation broth onto the salt form of the carboxylic cation resin and is eluted, highly purified and at high concentration, with dilute mineral acid.

Group separation of weak acids from strong acids can be accomplished on weakly basic anion exchangers. Weak acids will be retained to only a slight extent or not at all.

Carbonyl groups form addition compounds with the hydrogen sulfite ion to give $R_1R_2C(OH)(SO_3^-)$, which is strongly adsorbed by anion exchange resins. Since the addition product of ketones is less stable than that of aldehydes at elevated temperature, a group separation is possible. Ketones are desorbed by hot water and the aldehydes are eluted with a NaCl solution.

Polyhydroxy compounds (for example, sugars) and borate ions form a series of complexes with varying stabilities that dissociate as weak acids. The eluent is a borate buffer with a pH gradient that rises from pH 7 to 10. The packing is a strong anion exchanger.

Amino acids are separated on a strong cation exchanger using gradient elution with buffers of increasing pH. The most acidic amino acids emerge first.

Through prior concentration on an exchanger, the applicability of many spectrophotometric methods is extended. For example, copper in milk occurs in the range $10^{-4}\ \mu g \cdot mL^{-1}$; after concentration on a cation exchanger, the effluent will be $0.1\ \mu g \cdot mL^{-1}$, a 1000-fold concentration.

2.3.4.2 Removal of Interfering Ions. An ion exchanger may be used to retain major constituents while letting the trace constituent pass through. Although a less favorable case than retaining the trace constituent, it sometimes is useful. The amount of resin should be such that there are at least three times as many equivalents of replaceable ions in it as there are ions to be held on the column. The column length must be adequate to avoid leakage of retained ions. If the ion-exchange resin provides 1 meq of exchangeable ion per milliliter of bed, for the removal of 10 meq, about 3(10 meq ÷ 2 meq · mL^{-1}) or 15 mL of exchange resin bed would be required.

Undesirable cations in a sample may be removed by passage through a cation-exchange column in the H form or some other innocuous ionic form. In this manner cations such as calcium, iron, and aluminum, which interfere in the determination of phosphate, are easily retained on a bed of sulfonic acid resin in the H form. Iron and copper, which interfere in the iodometric determination of arsenic [as arsenate(V)(3–)], are separated.

With an anion exchanger, usually in the Cl form, interfering anions can be removed prior to the determination of a cation. Thus phosphate is removed before the volumetric determination of calcium and magnesium with EDTA.

Special chelating resins, containing iminodiacetate active groups, are useful in removing traces of heavy metals from a wide range of product streams and in separating various heavy metals. This type of resin will remove traces of iron, copper, or zinc from concentrated solutions of alkali-metal and alkaline-earth-metal salts. The heavy metals are easily eluted by mineral acids.

2.3.4.3 Stoichiometric Substitutions. An analyst may be concerned with the determination of the total concentration of cations (or anions) without any breakdown into individual cations. Passage of the sample through a column of resin in the H form (after preliminary washing of the column with distilled water) results in the retention of the cations and the release of an equivalent amount of hydrogen ions to the effluent. Rinsing the column and titration of the effluent and rinsing gives the total cation content of the sample. Any "free" acid or original alkalinity in the sample is determined by a separate titration.

The ability of ion exchangers to trade their counterions for other ions in solution makes possible the preparation either of acids from salts or a desired metal salt from the corresponding salt of another metal. In the pharmaceutical field conversions of one salt into another might involve potassium to sodium (penicillin), bromide to chloride (some vitamins), and chloride to sulfate (streptomycin). Preparation of carbonate-free hydroxide solution is accomplished by passing a known weight of dissovled sodium chloride through a strongly basic anion exchanger in the OH form. This conversion of salts to their corresponding bases (or acids) is particularly advantageous for standardizing solutions that may be prepared from pure salts, but salts that cannot be dried to a definite weight or whose composition is not known exactly.

2.3.4.4 Cation-Exchange Separation of Metals. Distribution coefficients of many metals in hydrochloric, perchloric, nitric, and sulfuric acid are given in Tables 2.30 to 2.34, in which the elements are arranged alphabetically by their symbol. The numerical values given in the foregoing tables can be used to plot the D_g–acid-concentration curves of the cations. Positions of elution peaks can be calculated using the equation[21]

$$v = D_g g \tag{2.59}$$

where v = volume of eluate (in mL) at elution peak
D_g = distribution coefficient for particular acid concentration
g = mass of dry resin (in g) in the column

The equation is valid only when the total amount of the ion is less than about 3% of the total column capacity.

Generally, distribution coefficients in nitric acid are slightly higher than those in HCl. Some cations that form relatively stable chloride complexes, such as Bi(III), Cd, Fe(III), Hg(II), In, and Pd(II), show very substantial differences. Thus, a separation of Fe(III) from the divalent heavy

[21] E. R. Tompkins and S. W. Mayer, *J. Am. Chem. Soc.* **69**:2859, 2866 (1947).

TABLE 2.30 Distribution Coefficients (D_g) of Metal Ions on AG 50W-8X Resin in HCl Solutions

Total resin capacity q = 0.4

Metal ion	0.1*M*	0.2*M*	0.5*M*	1.0*M*	4.0*M*
Al	8200	1900	318	61	3
As(III)	1.4	1.6	2.2	3.8	
Au(III)	0.5	0.1	0.4	0.8	0.2
Ba	>10^4	2930	590	127	12
Be	255	117	42	13	2
Bi(III)	*	*	<1	1	1
Ca	3200	790	151	42	5
Cd	510	84	6	1	
Ce(III)	>10^5	>10^5	2460	265	11
Co(II)	1650	460	72	21	3
Cr(III)	1130	262	73	27	3
Cs	182	99	44	19	
Cu(II)	1510	420	65	18	2
Fe(II)	1820	370	66	20	2
Fe(III)	9000	3400	225	35	2
Ga	>10^4	3040	260	43	0.4
Hg(II)	1.6	0.9	0.5	0.3	0.2
K	106	64	29	14	
La	>10^5	10^5	2480	265	11
Li	33	19	8	4	
Mg	1720	530	88	21	3
Mn(II)	2230	610	84	20	2
Na	52	28	12	6	
Ni(II)	1600	450	70	22	3
Pt(IV)				1.4	
Rb	120	72	33	15	
Sb(III)	*	*	*	*	3 (2*M*)
Se(IV)	1.1	0.6	0.8	0.6	0.7
Sn(IV)	10^4	45	6	2	
Sr	4700	1070	217	60	7
Th(IV)	>10^5	>10^5	10^5	2050	67
Ti(IV)	>10^4	297	39	12	2
U(VI)	5460	860	102	19	3
V(IV)		230	44	7	
V(V)	14	7	5	1	0.3
Y	>10^5	>10^4	1460	145	9
Zn	1850	510	64	16	2
Zr	>10^5	>10^5	10^5	7250	15

* Precipitate forms.

References: F. W. E. Strelow, *Anal. Chem.* **32**:1185 (1960); see also J. Korkisch and S. S. Ahluwalia, *Talanta* **14**:155 (1967).

metals Co, Cu, Mn, Ni, and Zn, which is unsatisfactory in HNO_3, becomes feasible in HCl. Separation of Fe(III) from Ti(IV) or Be is considerably better in HNO_3 than in HCl. K and Na, which accompany Cd and In when these are eluted with 0.5*M* HCl to separate them from Al, Cu, Fe(III), Ni, and Zn, can be separated easily from Cd and In in HNO_3 medium. A separation of Hg(I) and Hg(II) is feasible in HNO_3.

Distribution coefficients in H_2SO_4 are normally distinctly higher than those in either HCl or HNO_3. On the other hand, a number of cations show marked complex formation or ion association

TABLE 2.31 Distribution Coefficients (D_g) of Metal Ions on AG 50W-X8 Resin in Perchloric Acid Solutions

Metal ion	0.2*M*	1.0*M*	4.0*M*
Ag(I)	90	20	5.8
Al(III)	5250	106	11
Ba	2280	127	19
Be	206	14	1.9
Bi(III)	$>10^4$	243	42
Ca	636	50	7.7
Cd	423	36	6.3
Ce(III)	$>10^4$	459	53
Co(II)	378	31	4.8
Cr(III)	8410	120	11
Cu(II)	378	30	4.5
Dy(III)	$>10^4$	258	39
Fe(II)	389	32	5.2
Fe(III)	7470	119	12
Ga(III)	5870	112	11
Hg(I)	4160	147	9
Hg(II)	937	85	23
In(III)	6620	128	14
La	$>10^4$	475	58
Mg	312	24	3
Mn(II)	387	32	4.7
Mo(VI)	22	5.5	4.5
Mo(VI)*	0.7	0.4	1.3
Ni(II)	387	32	5
Pb(II)	1850	117	17
Sn(IV)	Precipitate	Precipitate	7.5
Sr	870	67	10
Th(IV)	$>10^4$	5870	686
Ti(IV)	549	19	5.7
Tl(I)	131	23	2.7
Tl(III)	1550	176	41
U(VI)	276	29	18
V(IV)	201	18	4.4
V(V)	9.8	2.2	0.8
V(V)*	9.3	3	1
W(VI)*	0.4	0.4	0.4
Y	$>10^4$	246	24
Yb(III)	$>10^4$	205	20
Zn	361	30	5
Zr	$>10^4$	$>10^4$	333

* Hydrogen peroxide present.
Reference: F. W. E. Strelow and H. Sondorp, *Talanta* **19**:1113 (1972).

in sulfuric acid. Cations such as Zr, Th(IV), U(VI), Ti(IV), Sc, and to a lesser extent Fe(III), In, and Cr(III) show the sulfate complexing effect. Uranium(VI) can be separated from Be, Mg, Co, Cu, Fe(II), Fe(III), Mn(II), Al, rare earths, Th, and other elements, and Ti(IV) can be separated from the same elements, as well as Nb(V), V(V), and Mo(VI). Scandium can readily be separated from Y, La, and the other rare-earth elements. V(IV) is most easily separated from V(V) or V(IV) from Mo(VI) and Nb(V) with H_2SO_4 as eluent.

TABLE 2.32 Distribution Coefficients (D_g) of Metal Ions on AG 50W-X8 Resin in Nitric Acid Solutions

Total resin capacity ratio q = 0.4, except for Te(IV) where q = 0.2

Metal ion	0.1*M*	0.2*M*	0.5*M*	1.0*M*	4.0*M*
Ag(I)	156	86	36	18	4
Al	>10^4	3900	392	79	5
As(III)	<0.1	<0.1	<0.1	<0.1	<0.1
Ba	5000	1560	271	68	4
Be	553	183	52	15	3
Bi(III)	893	305	79	25	3
Ca	1450	480	113	35	2
Cd	1500	392	91	33	3
Ce(III)	>10^4	>10^4	1840	246	8
Co(II)	1260	392	91	29	5
Cr(III)	5100	1620	418	112	11
Cs	148	81	35	17	3
Cu(II)	1080	356	84	27	3
Er	>10^4	>10^4	1100	182	8
Fe(III)	>10^4	4100	362	74	3
Ga	>10^4	4200	445	94	6
Gd	>10^4	>10^4	1000	167	7
Hf	>10^4	>10^4	>10^4	2400	21
Hg(I)	>10^4	7600	640	94	14
Hg(II)	4700	1090	121	17	3
In(III)	>10^4	>10^4	680	118	6
K	99	59	26	11	3
La	>10^4	>10^4	1870	267	9
Li	33	19	8	4	1
Mg	794	295	71	23	4
Mn(II)	1240	389	89	28	3
Mo(VI)	ppt	5	3	2	1
Na	54	29	13	6	1
Nb(V)	12	6	1	0.2	0.1
Ni(II)	1140	384	91	28	7
Pb(II)	>10^4	1420	183	36	4
Pd(II)	97	62	23	9	2
Rb	118	68	29	13	3
Rh(III)	78	45	19	8	1
Sc	>10^4	3300	500	116	8
Se(IV)	<0.5	<0.5	<0.5	<0.5	<0.5
Sm	>10^4	>10^4	1000	168	7
Sr	3100	775	146	39	5
Te(IV)	40	20	8	5	0.2
Th(IV)	>10^4	>10^4	>10^4	1180	25
Ti(IV)	1410	461	71	15	3
Tl(I)	173	91	41	22	3
U(VI)	659	262	69	24	7
V(IV)	495	157	36	14	2
V(V)	20	11	5	2	1
Y	>10^4	>10^4	1020	174	10
Yb	>10^4	>10^4	1150	193	9
Zn	1020	352	83	25	4
Zr	>10^4	>10^4	10^4	6500	31

References: F. W. E. Strelow, R. Rethemeyer, and C. J. C. Bothma, *Anal. Chem.* **37**:106 (1965); J. Korkisch, F. Feik, and S. S. Ahluwalia, *Talanta* **14**:1069 (1967).

TABLE 2.33 Distribution Coefficients (D_g) of Metal Ions on AG 50W-X8 Resin in H_2SO_4 Solutions

Total resin capacity q = 0.4, except q = 0.06 for Bi(III)

Metal ion	0.1*N*	0.2*N*	0.5*N*	1.0*N*	4.0*N*
Al	$>10^4$	8300	540	126	5
As(III)	<0.1	<0.1	<0.1	<0.1	<0.1
Be	840	305	79	27	3
Bi(III)	$>10^4$	$>10^4$	6800	235	6
Cd	1420	540	144	46	4
Ce(III)	$>10^4$	$>10^4$	1800	318	12
Co(II)	1170	433	126	43	5
Cr(III)	198	176	126	55	0.2
Cs	175	108	52	25	3
Cu(II)	1310	505	128	42	4
Er	$>10^4$	$>10^4$	1300	242	8
Fe(II)	1600	560	139	46	7
Fe(III)	$>10^4$	2050	255	58	2
Ga	$>10^4$	3500	618	137	5
Gd	$>10^4$	$>10^4$	1390	246	9
Hf	2690	1240	160	12	1
Hg(II)	7900	1790	321	103	12
In	$>10^4$	3190	376	87	12
K	138	86	41	19	3
La	$>10^4$	$>10^4$	1860	329	12
Li	48	28	12	6	1
Mg	1300	484	124	42	3
Mn(II)	1590	610	165	59	5
Mo(VI)	ppt	5	3	1	0.2
Na	81	48	20	9	2
Nb(V)	14	7	4	2	0.3
Ni(II)	1390	590	140	46	3
Pd(II)	109	71	33	14	3
Rb	148	91	44	21	3
Rh(III)	80	49	29	16	1
Sc	5600	1050	141	35	3
Se(IV)	<0.5	<0.5	<0.5	<0.5	<0.5
Sm	$>10^4$	$>10^4$	$>10^4$	1460	10
Te(IV)	ppt*	31	10	5	0.3
Th(IV)	$>10^4$	3900	263	52	2
Ti(IV)	395	225	46	9	0.4
TI(I)	452	236	97	50	9
TI(III)	6500	1490	205	47	5
U(VI)	596	118	29	10	2
V(IV)	1230	490	140	47	0.4
V(V)	27	15	7	3	0.4
Y	$>10^4$	$>10^4$	1380	253	9
Yb	$>10^4$	$>10^4$	$>10^4$	1330	9
Zn	1570	550	135	43	4
Zr	546	474	98	5	1

* ppt denotes precipitate.

Reference: F. W. E. Strelow, R. Rethemeyer, and C. J. C. Bothma, *Anal. Chem.* **37**:106 (1965).

TABLE 2.34 Distribution Coefficients (D_g) of Metal Ions on AG 50W-X8 Resin in 0.2N Acid Solutions

Metal ion	HCl	$HClO_4$	HNO_3	H_2SO_4
Ag(I)		90	86	
Al	2	5250	3900	8300
As(III)			<0.1	<0.1
Au(III)	0.1			
Ba	2930	2280	1560	
Be	117	206	183	305
Bi(III)	precipitate	$>10^4$	305	$>10^4$
Ca	790	636	480	
Cd	84	423	392	540
Ce(III)	10^5	$>10^4$	$>10^4$	$>10^4$
Co(II)	460	378	392	433
Cr(III)	262	8410	1620	176
Cu(II)	420	378	356	505
Dy		$>10^4$		
Fe(II)		389		
Fe(III)	3400	7470	4100	2050
Ga	3040	5870	4200	3500
Hg(I)		4160	7600	
Hg(II)	1	937	1090	1790
In		6620	$>10^4$	3190
La	10^5	$>10^4$	$>10^4$	$>10^4$
Mg	530	312	295	484
Mn(II)	610	387	389	610
Mo(VI)		22	5	5
Ni(II)	450	387	384	590
Pb(II)		1850	1420	
Pd(II)			62	71
Rh(III)			45	49
Sc			3300	1050
Sn(IV)	45	precipitate		
Sr	1070	870		
Th(IV)	$>10^5$	$>10^4$	$>10^4$	3900
Ti(IV)	297	549	461	225
Tl(I)		131	91	236
Tl(III)		1550		1490
U(VI)	860	276	262	118
V(IV)		201		
V(V)	7	10	11	15
W(VI)		0.4		
Y	$>10^4$	$>10^4$	$>10^4$	$>10^4$
Yb		$>10^4$		
Zn	510	361	361	550
Zr	10^5	$>10^4$	$>10^4$	474

References

For HCl: F. W. E. Strelow, *Anal. Chem.* **32**:1185 (1960); J. Korkisch and S. S. Ahluwalia, *Talanta* **14**:155 (1967).

For $HClO_4$: F. W. E. Strelow and H. Sondorp, *Talanta* **19**:1113 (1972).

For HNO_3: F. W. E. Strelow, R. Rethemeyer, and C. J. C. Bothma, *Anal. Chem.* **37**:106 (1965); J. Korkisch, F. Feik, and S. S. Ahluwalia, *Talanta* **14**:1069 (1967).

For H_2SO_4: F. W. E. Strelow, R. Rethemeyer, and C. J. C. Bothma, *Anal. Chem.* **37**:106 (1965).

2.3.4.5 Anion Exchange of Metal Complexes. Many metal ions may be converted to a negatively charged complex through suitable masking systems. This fact, coupled with the greater selectivity of anion exchangers, makes anion exchange a logical tool for handling certain metals. The negatively charged metal complexes are initially adsorbed by the exchanger from a high concentration of complexing agent, the eluted stepwise by lowering the concentration of the complexing agent in the eluent sufficiently to cause dissociation of the least stable of the metal complexes, and so on. If interconversion of the anion complex is fast, control of ligand concentration affords a powerful tool for control of absorbability, since ligand concentration controls the fraction of the metal present as adsorbable complex. For each metal and complexing ligand, there is a characteristic curve of log D_v versus molarity of complexing agent. Examples are given in Tables 2.35 and 2.36 for metals that form chloride and sulfate complexes. Similar studies are available for fluoride solution[22] and nitrate solutions.[23]

It is possible to devise a number of separation schemes in which a group of metals is adsorbed on a resin from a concentration of the complexing agent, and then each metal in turn is eluted by progressively lowering the complexing agent concentration (Table 2.37). Thus a cation that forms no, or only a very weak, anionic complex is readily displaced by several column volumes of the complexing agent, while its companions are retained. For separating two ions, it is advisable to choose a concentration of the complexing agent for which the separation factor is maximal, yet, at the same time, it is important that the volume distribution ratio of the eluting ion not be higher than unity. When $D_v = 1$, the peak maximum emerges within approximately two column volumes. The separation of nickel, manganese, cobalt, copper, iron(III), and zinc ions is done as follows.[24] The mixture of cations in 12*M* HCl is poured onto the column bed, which has been previously washed with 12*M* HCl. Nickel, which forms no chloro complex, elutes within several column volumes of a 12*M* HCl solution. The receiver is changed and manganese(II) is eluted with several column volumes of 6*M* HCl. This procedure is repeated using successively 4*M* HCl to elute cobalt(II), 2.5*M* HCl for copper(II), 0.5*M* HCl for iron(III), and lastly, 0.005*M* HCl for zinc.

Table 2.38 contains selected applications of ion exchange for the separation of a particular element from other metals or anions. A discussion of ion-exchange chromatographic methods for analysis is reserved for Sec. 4.5.

2.3.4.6 Ligand-Exchange Chromatography.[25] In this method a cation-exchange resin, saturated with a complex-forming metal, such as Cu(II), Fe(III), Co(II), Ni(II), or Zn(II), acts as a solid adsorbent. Thus, even though they are bound to an exchanger, these metals retain their ability to be the central atom of a coordination compound. Ligands, which may be anions or neutral molecules such as ammonia, amines, amino acids, or olefins, are removed from the liquid phase by formation of complexes with the metal attached to the resin and subsequent displacement of water or other solvents coordinated to the metal ion. Although the ordinary strongly acidic and weakly acidic cation exchangers undergo very satisfactory ligand-exchange reactions, the chelating resins that have iminodiacetate functional groups attached to a styrene matrix are ideally suited for ligand-exchange work. Strong complex formers, such as nickel, copper, or zinc, are tightly bound to the iminodiacetate exchanger. Consequently, leakage of metal ions from chelating resins by ordinary ion-exchange reactions with cationic materials in eluting solutions is held to a minimum.

Ligands with a stronger complexing tendency are more strongly retained. It is an efficient way to separate ligands from nonligands. Elution development uses a ligand in the eluent that complexes with the metal less strongly than the ligands of the mixture. Separations by ligand-exchange chromatography are outlined in Table 2.39.

---

[22] J. P. Faris, *Anal. Chem.* **32**:520 (1960).
[23] J. P. Faris and R. F. Buchman, *Anal. Chem.* **36**:1157 (1964).
[24] K. A. Kraus and G. E. Moore, *J. Am. Chem. Soc.* **75**:1460 (1953).
[25] F. Helfferich, *Nature* **189**:1001 (1961).

TABLE 2.35 Distribution Coefficients (D_v) of Metal Ions on AG 1-10X in HCl Solutions

Slight adsorption observed for Cr(III), Sc, Ti(III), Tl(I), and V(IV) in 12M HCl ($0.3 \le D_v \le 1$). No adsorption observed for Al, Ba, Be, Ca, Cs, K, La, Li, Mg, Na, Ni(II), Po, Rb, Th, and Y in 0.1–12M HCl.

Metal ions	2*M*	4*M*	6*M*	8*M*
Ag	100	10	3	2
As(III)		0.3	6	10
As(V)		2	2	2
Au(III)	6×10^5	2×10^5	79 000	30 000
Bi(III)	5 000	630	100	50
Cd	1 300	630	100	50
Co(II)		63 (10*M*)	<1	30
Cr(VI)			Strong adsorption	
Cu(II)		2.5	10	10
Fe(III)	10	250	2 000	20 000
Ga	2	1 300	10^5	63 000
Hf		63 (10*M*)	1 000 (12*M*)	<1
Hg(II)	13 000	2 500	790	250
In	12	16	12	10
Mn(II)	10	100	160	100
Mo(VI)	10	100	160	100
Nb(V)	16	10	25	5 000
Os(IV)	4 000	2 000	1 000	630
Pb(II)	16	5	1.4	1
Pd(II)	500	200	50	20
Pt(IV)	1 600	1000	400	200
Re(VII)	320	100	40	18
Rh(IV)			Strong adsorption	
Ru(IV)	630	250	100	50
Sb(III)	2 500	1 000	320	79
Sb(V)	1	250	13 000	7 9000
Se(IV)		1.3	2	13
Sn(II)	630	400	160	63
Sn(IV)	1 600	10^4	10^4	7 900
Ta(V)	8	2.5	2.5	10
Te(IV)			Strong adsorption	
Ti(IV)			10 (11*M*)	2
Tl(III)	10^5	32 000	10 000	3 200
U(IV)			1	40
U(VI)	3	100	400	630
V(V)			100 (11*M*)	1 600 (12*M*)
W(VI)		10	32	63
Zn	1 000	630	320	100
Zr			32 (10*M*)	2 000 (12*M*)

Reference: K. Kraus and F. Nelson, *Proc. Internatl. Conf. Peaceful Uses At. Energy* **7**:118 (1956).

2.4 DISTILLATION OR VAPORIZATION METHODS

Vaporization is the process of separating a mixture into its components by utilizing differences in the boiling points or partial pressures of the constituents. It is a useful method for the isolation, purification, and identification of volatile compounds. For use in analysis, the process may be simple batch

TABLE 2.36 Distribution Coefficients (D_g) of Metal Ions on AG 1-8X in H_2SO_4 Solutions

For As(V), Ga, and Yb, $D_g = 0.6$ in 0.05M H_2SO_4 and <0.5 in higher concentrations. For the alkali metals, Be, Cd, Cu, Mg, Mn, Ni, and Zn, D_g was <0.5 in 0.005–2.0M acid concentrations.

Metal ion	0.05*M*	0.1*M*	0.25*M*	0.5*M*	1.5*M*
As(III)	0.9	0.6	<0.5	<0.5	<0.5
Bi(III)	18	5	2	1	<0.5
Cr(III)	2.1	0.7	0.5	<0.5	<0.5
Cr(VI)	12 000	7 800	4 400	2 100	435
Fe(III)	16	9	4	1	0.6
Hf	4 700	701	57	12	2
In	2.4	0.8	<0.5	<0.5	<0.5
Ir(III)	388	270	218	160	92
Ir(IV)	450	310	220	180	160
Mo(VI)	533	671	484	232	14
Mo(VI)*	2 560	1 400	451	197	43
Nb(V)*	120	96	3	<0.5	<0.5
Rh(III)	13	5	1	<0.5	<0.5
Sc	22	11	5	3	1
Se(IV)	1.1	<0.5	<0.5	<0.5	<0.5
Ta(V)*	1 860	1 070	310	138	11
Th	35	21	8	4	1
Ti(IV)*	hydrolyse	0.5	<0.5	<0.5	<0.5
U(VI)	521	248	91	27	5
V(IV)	1	<0.5	<0.5	<0.5	<0.5
V(V)	6.5	3.3	2	1	<0.5
V(V)*	45	11	5	2.5	2
W(VI)*	528	457	337	222	96
Zr	1 350	704	211	47	5

* 0.3% H_2O_2 present.

Reference: F. W. E. Strelow and C. J. C. Bothma, *Anal. Chem.* **39**:595 (1967).

distillation without a fractionating column or fractional distillation involving use of a column of some sort. Inorganic applications of vaporization involve batch distillation, whereas organic applications will usually require some type of fractional distillation. Distillation is not as widely used for organic analytical purposes as was the case before chromatography was developed. Chromatography and mass spectrometry have largely supplanted distillation for organic materials.

2.4.1 Simple Batch Distillation

In simple batch distillation material is placed in a distillation flask, boiling is initiated, and the vapors are then continuously removed, condensed, and collected. The process is sometimes surprisingly simple and usually gives a very complete separation. The only parameters that need to be varied are temperature and pressure. Few reagents are required, which is an important consideration in trace analysis. Separating elements by vaporization is less convenient than separating them by solvent extraction, but it is usually clean and quantitative. Separations may often be accomplished by vaporization of the constituent being determined, as a prelude to the analytical determination of the vaporized element. Vaporization is also used to remove unwanted elements, the presence of which interferes with an analytical determination. The analyst must be continually aware that vaporization may lead to unintentional losses, as shown in Table 2.40.[26] Although no osmium is volatilized from

[26] J. I. Hoffman and G. E. F. Lundell, *J. Res. Nat. Bur. Stds.* **22**:465 (1939); C. Ballaux, R. Dams, and J. Hoste, *Anal. Chim. Acta* **47**:397 (1969).

TABLE 2.37 Metal Separations on Ion Exchangers

Metals	Ion exchanger	Eluent and eluted ions	Reference
		Anion-exchange resins	
Ni, Mn(II), Co(II), Cu(II), Fe(III), Zn	AG 1-X8	Ni–12*M* HCl; Mn–6*M* HCl; Co–4*M* HCl; Cu–2.5*M* HCl; Fe–0.5 *M* HCl; Zn–0.005*M* HCl	K. A. Kraus and G. E. Moore, *J. Am. Chem. Soc.* **75**:1460 (1953).
Ni, Co(II), Cu(II), Zn	AG 1-X8	Ni–96% MeOH, 0.2*M* HCl; Co–55% 2-propanol, 1.3*M* HCl; Cu–55% 2-propanol, 0.1*M* HCl; Zn – 0.005*M* HCl	J. S. Fritz and D. J. Pietrzyk, *Talanta* **8**:143 (1961)
Mn, Co, Ni, Fe, Mo (also Cr, Zn, Cd, Hg)	AG 1-X8	Mn, Co, Ni–0.0085*M* tartrate; Fe–tartaric acid in 0.1*M* HCl; Mo–3*M* NaOH	G. P. Morie and T. R. Sweet, *J. Chromatogr.* **16**:201 (1964)
Zn, Pb(II), Cd, Bi(III)	AG 1-X8	Zn–0.2*M* HBr + 0.5*M* HNO_3; Pb–0.05*M* HBr + 0.5*M* HNO_3; Cd–0.02*M* HBr + 2*M* HNO_3; Bi–0.05*M* EDTA + 0.1*M* NH_4NO_3	F. W. E. Strelow, *Anal. Chem.* **50**:1359 (1978)
Th, Hf, Zr, Mo	AG 1-X8	Th–0.7*N* H_2SO_4; Hf–1.25*N* H_2SO_4; Zr–2*N* H_2SO_4; Mo–2*N* NH_4NO_3, 0.5*N* NH_3	F. W. E. Strelow and C. J. C. Bothma, *Anal. Chem.* **39**:595 (1967)
Ba, Sr, Ca, Mg	AG 1-X8	Ba, Sr, Ca–0.05*M* $(NH_4)_3$Cit; Mg–0.5*M* H_3Cit*	F. N. Kraus and K. A. Kraus, *J. Am. Chem. Soc.* **77**:801 (1955)
Th, Fe, V	AG 1-X8	Adsorbed as citrate complexes; Th–8*M* HCl; Fe–MIBK, acetone, 1*M* HCl (1:8:1 v/v); V–1*M* HCl	J. Korkisch and H. Krivanec, *Anal. Chim. Acta* **83**:111 (1976)
Th, U	AG 1-X8	Th–9.6*M* HCl; U–0.1*M* HCl	S. S. Berman, L. E. McKinney, and M. E. Bednas, *Talanta* **8**:143 (1961)
Fe(III), Pu(IV)	AG 1-X8	Adsorbed in 9*M* HCl; Fe–7.2*M* HNO_3; Pu–1.2*M* HCl, 0.6% H_2O_2	N. A. Talvitie, *Anal. Chem.* **43**:1827 (1971)
Pt, Au, Os, Ir, Rh, Pd, Ru	AG 1-X8		J. Korkisch and H. Klaki, *Talanta* **15**:339 (1968)
Ag, As, Cd, Co, Cr, Cu, Fe, Hg, In, Mo, Mn, Sb, Sc, W, Zn	AG 1-X2		W. Zmijewska, H. Polkowski-Motrenki, and J. Stokowski, *J. Radio Anal. Nucl. Chem.* **84**:319 (1984)
Nb, Ta (from Mo, Ti, W, Zr)	AG 1-X8	Adsorbed in 3*M* HCl + 3.8*M* HF. Wash with one column volume HCl-HF solution. Elute Nb with 350 mL (14% NH_4Cl + 0.072*M* HF) at 125 mL/h. Elute Ta with 350 mL of same eluate adjusted to pH 6 with NH_3	S. Kallmann, H. Oberthin, and R. Liu, *Anal. Chem.* **34**:609 (1962)
Pt, Pd	AG 1-X8		C. H. Branch and D. Hutchinson, *J. Anal. At. Spectrom* **1**:433 (1986)
Bi, Cd, Fe, Cu, Mn, Ni	AG 50W-X8	Bi–60% acetone, 0.1*M* HCl; Cd–70% acetone, 0.2*M* HCl; Fe–80% acetone, 0.5*M* HCl; Cu–90% acetone, 0.5*M* HCl; Mn–92% acetone, 1*M* HCl: Ni–aqueous 3*M* HCl	J. S. Fritz and T. A. Rettig, *Anal. Chem.* **34**:1562 (1962)

TABLE 2.37 Metal Separations on Ion Exchangers (*Continued*)

Metals	Ion exchanger	Eluent and eluted ions	Reference
		Cation-exchange resins	
V, U, Sc, Y	AG 50W-X8	V–0.5*N* H_2SO_4; U–1*N* H_2SO_4; Sc–2*N* H_2SO_4; Y–4*M* HCl	F. W. E. Strelow, R. Rethemeyer, and C. J. C. Bothma, *Anal. Chem.* **37**:106 (1965)
Be, Ba, Sr	AG 50W-X8	Be, Ba–9*M* $HClO_4$; Sr–5*M* HNO_3	F. Nelson, T. Murase, and K. A. Kraus, *J. Chromatogr.* **13**:503 (1964)
K, Ti, Sc	AG 50W-X8	K–9*M* $HClO_4$; Ti–9*M* HCl; Sc–4*M* HCl, 0.1*M* HF	F. Nelson, T. Murase, and K. A. Kraus, *J. Chromatogr.* **13**:503 (1964)
Y and rare earth elements	AG 50W-X8		J. G. Crock et al., *Talanta* **33**:601 (1986)
Eu, Dy, Sm, Gd	AG 50W-X8		F. Flavelle and A. D. Westland, *Talanta* **33**:445 (1986)
As(V), As(III), methylarsonate, dimethylarsinate	AG 50W-X8 (top) and AG 1-X8 (bottom)	Sample adsorbed at pH 2.5–7. As(III) and methylarsonate–0.006*M* Cl_3CCOOH (two bands): As(V)–0.2*M* Cl_3CCOOH: dimethylarsinate–1.5*M* NH_3 followed by 0.2*M* Cl_3CCOOH	A. A. Grabinski, *Anal. Chem.* **53**:966 (1981); see also S. Tagawa and Y. Kojima, *Jpn. Anal.* **29**:216 (1980)
Ca, Sr, Ba	AG 50W-X8	Ca–0.1*M* EDTA, pH 5.25; Sr–0.1*M* EDTA, pH 6.0; Ba–0.1*M* EDTA, pH 9.0	J. J. Bouquiaux and J. H. C. Gillard, *Anal. Chim. Acta* **30**:273 (1964)
		Macroporous and special resins	
Cd, Zn, Ga, Yb, Sc	AG MP-1	Cd–0.7*M* HCl; Zn–1.5*M* HCl; Ga–2.5*M* HCl; Yb–4*M* HCl; Sc–1*M* NH_4Ac	F. W. E. Strelow, *Anal. Chem.* **56**:1053 (1984)
Cr(III), Cr(VI)	AG MP-1	Cr(III)–$HClO_4$; Cr(VI)–NH_4Cl	C. Sarzanini et al., *Annal Chim.* **73**:385 (1983)
Cr(III), Cr(VI)	Chelex 100	pH 4.0–8.0, Cr(III) retained; Cr(VI) passes through	D. E. Leyden, R. E. Channell, and C. W. Blount, *Anal. Chem.* **44**:607 (1972)
46 metals	AG MP-50		F. W. E. Strelow, *Anal. Chem.* **56**:1053 (1984)

* Cit denotes citrate.

H_2SO_4 at 200 to 220°C, it is completely volatilized at 270 to 300°C. These elements are not volatilized in any of the procedures: Ag, alkalies, Al, Ba, Be, Ca, Cd, Co, Cu, Fe, Ga, Hf, In, Ir, Mg, Nb, Ni, Pb, Pd, Pt, rare earths, Rh, Si, Ta, Th, Ti, U, W, Zn, and Zr.

The distilling flask should accommodate twice the volume of the liquid to be distilled. The thermometer bulb should be slightly below the side arm opening of the flask. The boiling point of the corresponding distillate is normally accepted as the temperature of the vapor. If the thermometer is

TABLE 2.38 Selected Applications of Ion Exchange for the Separation of a Particular Element from Other Elements or Ions

Element	Resin phase	Aqueous phase	Separated from	Reference
Aluminum	Not adsorbed from 9*M* HCl by Dowex 1-10X		Cd, Co, Cr(VI), Cu, Fe(II, III), Mn(VII), Mo(VI), Sb(III,V), Sn(II,IV), U(VI), V(V), W(VI), Zn	A. D. Horton and P. F. Thomason, *Anal. Chem.* **28**:1326 (1956)
	Dowex 50W		Phosphate	M. Seibold, *Z. Anal. Chem.* **173**:388 (1960)
Arsenic(III)	Dowex 2	8*M* HCl	Phosphate	F. Nelson and K. A. Kraus, *J. Am. Chem. Soc.* **77**:4508 (1955)
	Anion exchange resin	pH 4	Germanium	R. M. Dranitskaya and C.-C. Liu, *Zh. Anal. Khim.* **19**:769 (1964)
Arsenic(III,V)	DEAE*-cellulose	Oxalic acid	Each other	R. Kuroda et al., *Talanta* **26**:211 (1979)
Barium	Dowex 50W-X8	1*M* HCl, then 4*M* HCl	Cd, Cs, Cu, Hg(II), U(VI), and Zn	S. M. Khopkar and A. K. De, *Anal. Chim. Acta* **23**:441 (1960)
	Dowex 50W-X8	5% citric acid at pH 2.7, then 4*M* HCl	Ce(IV) and Zr	S. M. Khopkar and A. K. De, *Anal. Chim. Acta* **23**:441 (1960)
	Dowex 50W-X8	EDTA solution at pH 2.0–2.2, then 4*M* HCl	Bi, Fe(III), and Th	S. M. Khopkar and A. K. De, *Anal. Chim. Acta* **23**:441 (1960)
Beryllium	Dowex 50 in Na^+ form	EDTA solution at pH 3.5–4.0; elute Be with 3*M* HCl	Al, Fe(III), Mn, Ti, and complexes of other elements with EDTA	M. N. Nadkarni, M. S. Varde, and V. T. Athavale, *Anal. Chim. Acta* **16**:421 (1957)
	Dowex 50W-X8	HCl and HNO_3 solutions	Al, Fe(III), lanthanoids	F. W. E. Strelow, *Anal. Chem.* **33**:542 (1961)
	Dowex 50W-X8	After adsorption, elute Be with 0.2*M* $(NH_4)_2SO_4$–0.025*M* H_2SO_4	Al, Ca, Cd, Cr(III), Cu, Zn, and others	K. Kawabachi, T. Ito, and R. Kuroda, *J. Chromatogr.* **39**:61 (1969)
Bismuth	Adsorbed on Dowex 1	0.25*M* HCl; Bi eluted with 1*M* H_2SO_4	Not adsorbed: Al, Co, Fe(II,III), Ga, In, Hf, Mn, Ni, Ti(III,IV), Th, V(IV,V), U(IV,VI), and Zr. Not eluted with Bi: Ag, Au, Ir(IV), Pd, and Pt	H. A. Mottola, *Anal. Chim. Acta* **29**:261 (1963)
	DEAE* cellulose	Methanol–HCl; elute with 0.5*M* HCl, then Bi with 1*M* HCl	Cd, Hg(II), Zn adsorbed with Bi but eluted by 0.5*M* HCl	R. Kuroda and N. Yoshikuni, *Talanta* **18**:1123 (1971)
Boron	Dowex 50W-X8 in the sulfate form	Rinse with 0.05*M* H_2SO_4 to remove boron	Removal of cations from aluminum-silicon alloys, steels, effluents, rocks, and minerals	J. R. Martin and J. R. Hayes, *Anal. Chem.* **24**:182 (1952); R. C. Calkins and V. A. Stenger, *ibid.* **28**:399 (1956); R. Capelle, *Anal. Chim. Acta* **25**:59 (1961)
	Dowex 1-X8	pH 7.5–9.5; eluted with 5*M* HCl	Other cations and anions, such as phosphate, arsenate	R. A. Muzzarelli, *Anal. Chem.* **39**:365 (1967)

TABLE 2.38 Selected Applications of Ion Exchange for the Separation of a Particular Element from Other Elements or Ions (*Continued*)

Element	Resin phase	Aqueous phase	Separated from	Reference
Cadmium	Dowex 50W-X8	Adsorption from <0.2*M* HCl; Cd eluted with 0.5*M* HCl	Co, Cu, Mn, Ni, Ti, U, and Zn	F. W. E. Strelow, *Anal. Chem.* **32**:363 (1960)
	Dowex 50W-X8	After adsorption, Cd eluted with 81% dioxane–10% EtOH–9% 12*M* HCl	Al, Cr, Mn, Ni, and Pb matrices	R. R. Ruch, F. Tera, and G. H. Morrison, *Anal. Chem.* **37**:1565 (1965)
	Dowex 50	0.3*M* HI–0.075*M* H_2SO_4; CdI_4^{2-} passes	Large amounts of Zn	S. Kallmann, H. Oberthin, and R. Liu, *Anal. Chem.* **32**:58 (1960)
	Dowex 1-X8	1.2*M* HCl; Zn removed with 0.15*M* HBr; Cd eluted with 2*M* HNO_3	Natural water	J. Korkisch and D. Dimitriadis, *Talanta* **20**:1295 (1973)
	Dowex 1-X8	Elute Ag with 6.7*M* HBr and Cd with 2*M*–3*M* HNO_3	Ag	F. W. E. Strelow, W. J. Louw, and C. H. -S. W. Weinhert, *Anal. Chem.* **40**:2021 (1968)
Calcium	Dowex A-1 (Chelex 100)	Basic solution containing citrate and triethanolamine; Mg eluted with ethanolic $(NH_4)_2C_2O_4$ and Ca with 0.3*M* HCl	Al, Fe, Mn, and Ti	R. Christova and Z. Ivanova, *Z. Anal. Chem.* **253**:184 (1971)
	Dowex 50W-X8	0.1*M* HNO_3; elute Ca with 1.5*M* HNO_3	Al and lanthanoids	J. S. Fritz and B. B. Garralda, *Talanta* **10**:91 (1963)
	Dowex 50W-X8	EDTA solution of pH 3.6; elute (Ca + Mg) with 4*M* HCl	Al, Cu, Fe, Mn, and Zn	J. Chwastowska and S. Szymczak, *Chem. Anal.* **14**:1161 (1969)
	Dowex 50W-X8	0.1*M* EDTA of pH 5.0 containing 0.3*M* NH_4 acetate	Strontium	F. W. E. Strelow and C. H. S. W. Weinert, *Talanta* **17**:1 (1970)
Cerium	Dowex 1	>4*M* HCl; Ce not adsorbed	Cast iron	J. E. Roberts and M. J. Ryterband, *Anal. Chem.* **37**:1585 (1965)
Chromium(III)	Dowex 1-X8	0.1*M* HCl plus 5% H_2O_2	V(V), Mo, and W; Cr(III) not adsorbed	R. Kuroda and T. Kiriyama, *Jpn. Anal.* **19**:1285 (1970)
Chromium(VI)	Dowex 1	Cr eluted with 0.4*M* $NaClO_4$ at pH 3.75	Al alloy and steel	J. S. Fritz and J. P. Sickafoose, *Talanta* **19**:1573 (1972)
Chromium(III)	Chelex 100	pH 4.0–8.0	Chromium(VI) not retained	D. E. Leyden, R. E. Channell, and C. W. Blount, *Anal. Chem.* **44**:607 (1972)
	Anion exchange	Oxalate	Al and Fe(III) retained	A. M. Mulokozi and D. M. S. Mosha, *Talanta* **22**:239 (1975)
Cobalt	Dowex 1-X8	9*M* HCl; elution with 4*M* HCl removes Co	Cu, Fe(III), Mn, Ni, and Zn	
Copper(I)	Dowex 1-X8	0.1*M* HCl; eluted with 1*M* HNO_3	Natural (fresh) waters	J. Korkish, L. Gödl, and H. Gross, *Talanta* **22**:289 (1975)

(*Continued*)

TABLE 2.38 Selected Applications of Ion Exchange for the Separation of a Particular Element from Other Elements or Ions (*Continued*)

Element	Resin phase	Aqueous phase	Separated from	Reference
Gallium	Dowex 1	7*M* HCl then Ga eluted with 1*M* HCl	Al and Tl	K. A. Kraus, F. Nelson, and G. W. Smith, *J. Phys. Chem.* **58**:11 (1954)
	Dowex 50	0.8*M* HCl; Ga eluted with 1.5*M* HCl	Cu, Fe(II), Pb, Sb, and Zn	R. Lkement and H. Sandmann, *Z. Anal. Chem.* **145**:325 (1955)
	Dowex 1-X8	90% MeOH†–10% 4.5*M* HBr containing ascorbic acid; Ga eluted with 0.45*M* HBr	Al, alkaline earths, Co, Fe(II), U in effluent, and Bi, Cd, Cu, Pb remain on resin after Ga elution	J. Korkisch and I. Hazan, *Anal. Chem.* **37**:707 (1965)
Germanium	Dowex 1-X8	9:1 acetic acid–9*M* HCl	Adsorbs As(III,V), Fe(III), Pb, Sb(III), Sn(II,IV) but not Ge	J. Korkisch and F. Feik, *Sep. Sci.* **2**:1 (1967)
	Mixed bed with weak base and strong acid resins	pH 2.0; Ge not retained	Ag, As(III), Bi, Ca, Cr(III), Fe(III), Mg, Ni, Mo(VI), Pb, Sb(III), Sn(IV), and Zn	T. R. Cabbell, A. A. Orr, and J. R. Hayes, *Anal. Chem.* **32**:1602 (1960)
Gold	Dowex 1-X8	0.5*M* HCl then elution of Au with 0.3*M* thiourea in 0.1*M* HCl at 50°C	Cu, Fe(III), and Ni	L. L. Kocheva and R. Gecheva, *Talanta* **20**:910 (1973)
	Dowex 1	30 vol% tributyl phosphate–60 vol% methyl glycol–10 vol% 12*M* HCl; Au not adsorbed	Co, Cu, Hg(II), platinum metals, and U(VI)	W. Koch and J. Korkisch, *Mikrochim. Acta* **1973**, 117
	Dowex 50W-X8	6*M* HBr–0.0035*M* bromine; elute Au with acetylacetone	Ir, Pd, Pt, and Rh	R. Dybdznski and H. Maleszewska, *Analyst* **94**:527 (1971)
Indium	Dowex 50W-X8	30% acetone–0.2*M* HNO_3; In eluted with 0.5*M* HCl containing 30% acetone	Al, Be, Ca, Co, Cu, Fe(III), Ga, Mg, Mn, Ni, Pb, Ti, U(VI), and Zn	F. W. E. Strelow et al., *Talanta* **21**:1183 (1974)
Iridium	Dowex 50W-X8	HCl solutions of pH 1.3–1.5; Ir (and other platinum metals) are not adsorbed	Base metals	A. G. Marks and F. E. Beamish, *Anal. Chem.* **30**:1464 (1958)
	Dowex 50W-X8	Thiourea solution 0.3*M* in HCl	Rhodium	E. W. Berg and W. L. Senn, Jr., *Anal. Chem.* **27**:1255 (1955)
Iron	Dowex 50W-X8	Dilute HCl solution	Ni and Zn	I. K. Tsitovich and Zh. N. Goncharenko, *Zh. Anal. Khim.* **23**:705 (1968)
	Dowex 1-X8	Echelon elution with HCl; see under copper	Co, Cu, Mn, Ni, and Zn	K. A. Kraus and G. E. Moore, *J. Am. Chem. Soc.* **75**:1460 (1953)
Lanthanoids	Cation exchange	Nitric acid solution	Ba, Ca, Mg, Sr	F. W. E. Strelow, *Anal. Chim. Acta* **120**:249 (1980)

TABLE 2.38 Selected Applications of Ion Exchange for the Separation of a Particular Element from Other Elements or Ions (*Continued*)

Element	Resin phase	Aqueous phase	Separated from	Reference
Lanthanoids	Dowex 50W-X8	Adsorbed in 0.1*M* HCl; (1) elution with 1.75*M* HCl and (2) finally with 3*M* HCl	(1) Al, Be, Ca, Cd, Co, Cu, Fe(III), Ga, In, Mn, Ni, Ti, Tl(III), U(VI), and Zn. (2) Lanthanoids and Sc. Th retained	F. W. E. Strelow, *Anal. Chim. Acta* **34**:387 (1966); F. W. E. Strelow and P. F. S. Jackson, *Anal. Chem.* **46**:1481 (1974)
	Amberlyst XN-1002	(1) 1.5*M* HNO_3 in 85% 2-propanol and (2) 1.5*M* HNO_3 in 45–55% 2-propanol washes out lanthanoids	(1) Al, Co, Cu, Fe(III), Ga, In, Mg, Mn, Ni, V(IV), Zn. (2) Bi, Pb, Th	J. S. Fritz and R. G. Greene, *Anal. Chem.* **36**:1095 (1964)
Lead	Dowex 50W-X8; heated column	Elution with 0.6*M* HBr	Alkaline earths, Cr(III), U(VI), V(IV), and Zn	J. S. Fritz and R. G. Greene, *Anal. Chem.* **35**:811 (1963)
	Dowex 1	Adsorbed from 1*M* HCl; Pb eluted with 8*M* HCl	Bi and Fe(III)	F. Nelson and K. A. Kraus, *J. Am. Chem. Soc.* **76**:5916 (1957)
	Dowex 1	Adsorbed from 1*M* HCl; Pb eluted with 0.01*M* HCl	Alkaline earths, Cu, Fe, Sn, Tl, phosphate, and sulfate not adsorbed (foodstuffs)	E. I. Johnson and R. D. A. Polhill, *Analyst* **82**:238 (1957)
	Dowex 1-X8	0.1*M* HBr; Pb eluted with 0.3*M* HNO_3–0.025*M* HBr	Al, Ba, Be, Ca, Fe(III), lanthanoids, Mg, Mn(II), Th, Ti, U(VI), Zn, and Zr pass through; Au, Bi, Cd, Hg(II), Pd, Pt, and Tl(III) retained by resin	F. W. E. Strelow and F. von S. Toerien, *Anal. Chem.* **38**:545 (1966); F. W. E. Strelow, *ibid.* **39**:1454 (1967)
	Dowex 1-X8	90% tetrahydrofuran–10% HNO_3; Pb eluted with 80% tetrahydrofuran–20% 2.5*M* HNO_3	Bi, Ca, Fe(III), heavier lanthanoids (Sm to Lu), Mg, Th, Tl(I), and U(VI) not retained	J. Korkisch and F. Feik, *Anal. Chem.* **36**:1793 (1964); S. S. Ahluwalia and J. Korkisch, *Z. Anal. Chem.* **208**:414 (1965)
Lithium	Dowex 50W-X8	0.5*M* HNO_3 elutes alkali metals	Al, alkaline earth metals, Bi, Cd, Cu, Fe(III), Mg, Mn, Pb	F. W. E. Strelow, J. H. J. Coetzee, and C. R. Van Zyl, *Anal. Chem.* **40**:196 (1968); F. W. E. Strelow, F. von S. Toerien, and C. H. S. W. Weinert, *Anal. Chim. Acta* **50**:399 (1970); **71**:123 (1974)
Magnesium	Dowex 50-X12	Weakly acid solution and elution with 2*M* HCl	Ca	C. K. Mann and J. H. Yoe, *Anal. Chem.* **28**:202 (1956)
	Dowex 50W-X8	Al eluted with 0.5*M* oxalic acid and Mg with 2*M* HCl	Al	T. N. van der Walt and F. W. E. Strelow, *Anal. Chem.* **57**:2889 (1985)
	Dowex 50W-X8	1.5*M* HNO_3 (Mg) and 3*M* HNO_3 (Al and lanthanoids)	Al and lanthanoids	J. S. Fritz and B. B. Garralda, *Talanta* **10**:91 (1963)
Manganese	Dowex 50W-X8	1.5*M* HNO_3	Fe	F. W. E. Strelow, *Anal. Chem.* **33**:994 (1961)

(*Continued*)

TABLE 2.38 Selected Applications of Ion Exchange for the Separation of a Particular Element from Other Elements or Ions (*Continued*)

Element	Resin phase	Aqueous phase	Separated from	Reference
Manganese	Dowex 1-X8	1:4:6 water–HCl–acetone and elution with water	Ca, Mg, K, and phosphate	R. A. Webb, D. G. Hallas, and H. M. Stevens, *Analyst* **94**:794 (1969)
Mercury(II)	Dowex 50W-X8	0.5*M* HCl as eluant	See Table 2.30	C. K. Mann and C. L. Swanson, *Anal. Chem.* **33**:459 (1961); A. K. De and S. K. Majumdar, *Z. Anal. Chem.* **814**:356 (1961)
	Weak-base anion exchanger DEAE-cellulose	0.01*M* NH_4SCN–0.1*M* HCl or methanol–HCl mixtures	As(III), Cu, Fe(III), Te(IV), and many others	R. Kuroda, T. Kiriyama, and K. Ishida, *Anal. Chem. Acta* **40**:305 (1968); R. Kuroda and N. Yoshikuni, *Talanta* **18**:1123 (1971)
Molybdenum	Dowex 1-X8	Adsorbed from 0.05*M* HNO_3–0.2% H_2O_2; eluted with 2*M* HNO_3	Biomaterials	I. Matsubara, *Jpn. Anal.* **32**:T96 (1983)
	Dowex 1-X8	Elution with 0.5*M* NaCl–0.5*M* NaOH	(1) Seawater and (2) silicate rocks	(1) K. Kawabuchi and R. Kuroda, *Anal. Chim. Acta* **46**:23 (1969); (2) *ibid.* **17**:67 (1970)
Nickel	Dowex 1-X8	HCl solutions	Nonadsorption of Ni offers many possibilities	See Table 2.28
	Chelex 100		For concentrating Ni	A. J. Paulson, *Anal. Chem.* **58**:183 (1986); K. Brajter and J. Gravarek, *Analyst* **103**:632 (1978)
Niobium	Cation exchanger	In $HClO_4$–HCl solution	Hf, Sc, and Ti	F. Nelson and K. A. Kraus, *J. Chromatogr.* **178**:163 (1979)
	Dowex 50-X8	0.25*M* H_2SO_4	Al, Cr(III), Co, Cu, Fe(III), Mn(II), Ni, Ti, and Zr	J. S. Fritz and L. H. Dahmer, *Anal. Chem.* **37**:1272 (1965)
Osmium	Dowex 50-X8	HCl solution of pH 0.8–1.5	Cu, Fe, and Ni	J. C. Van Loon and F. E. Beamish, *Anal. Chem.* **36**:1771 (1964); G. H. Faye, *ibid.* **37**:296 (1965); C. Pohlandt and T. W. Steele, *Talanta* **21**:919 (1974)
Palladium	Dowex 50-X8	0.18*M* HCl solution	Cu, Fe, Ni; chloro-complexes of Ir, Pd, Pt, and Rh pass through	R. J. Brown and W. R. Biggs, *Anal. Chem.* **56**:646 (1984)
	Reviews: Separation of platinum metals from each other by cation and anion exchange			F. E. Beamish, *Talanta* **14**:991 (1967); S. J. Al-Bazi and A. Chow, *ibid.* **31**:815 (1984)
Platinum metals	Dowex 1-X8	Elution: (Rh) 2*M* HCl, (Pd) 9*M* HCl, (Pt) 2.4*M* $HClO_4$		S. S. Berman and W. A. E. McBryde, *Can. J. Chem.* **36**:835(1958)
Rhenium	Dowex 1-X8	Re eluted with 0.5*M* NH_4SCN–0.5*M* HCl	Alkali metals, alkaline earth metals, lanthanoids, Al, As(III), Cr(III), and Ge elute earlier; ions forming SCN complexes retained	H. Hamaguchi, K. Kawabuchi, and R. Kuroda, *Anal. Chem.* **36**:1634 (1964); K. Kawabuchi, H. Hamaguchi, and R. Kuroda, *J. Chromatogr.* **17**:567 (1965)

TABLE 2.38 Selected Applications of Ion Exchange for the Separation of a Particular Element from Other Elements or Ions (*Continued*)

Element	Resin phase	Aqueous phase	Separated from	Reference
Rhodium	Dowex 50W-X8	HCl solution, pH 1.0	Cu, Fe, Ni	A. G. Marks and F. E. Beamish, *Anal. Chem.* **30**:1464 (1958); B. R. Sant and F. E. Beamish, *ibid,* **33**:304 (1961)
Ruthenium	Dowex 50W-X8	HCl solution, pH 1.0	Cu, Fe, Ni	H. Zachariasen and F. E. Beamish, *Anal. Chem.* **34**:964 (1962); C. Pohlandt and T. W. Steele, *Talanta* **21**:919 (1974)
Scandium	Dowex 50W-X8	1*M* H_2SO_4 eluent (flow rates critical)	Lanthanoids and yttrium	F. W. E. Strelow and C. J. C. Bothma, *Anal. Chem.* **36**:1217 (1964)
	Dowex 50W-X8	Adsorbed from dilute mineral acid solution; Sc eluted with 0.3*M* $(NH_4)_2SO_4$ plus 0.025*M* H_2SO_4	Al, Ca, Cd, Co, Cr(III), Cu, Fe(III), Ga, In, lanthanoids, Mg, Mn, Ni, Y, and Zn	R. Kuroda, Y. Nakagomi, and K. Ishida, *J. Chromatogr.* **22**:143 (1963)
	Dowex 1-X8	Adsorbed from 0.1*M* $(NH_4)_2SO_4$–0.025*M* H_2SO_4; Sc eluted by 1*M* HCl	Al, Be, Cd, Co, Cu, Ga, Ge, lanthanoids, Mg, Mn, Ni, V(IV), Y, and Zn	H. Hamaguchi et al., *Anal. Chem.* **36**:2304 (1964)
Silver	AG MP-50	Nitric acid solution	Cd, Cu, Ni, and Zn	F. W. E. Strelow, *Talanta* **32**:953 (1985)
	Dowex 50	HCl solution, pH 1.5	Al, Cu, In, Mg, and Pb	K. Kasiura, *Chem. Anal* **29**:809 (1975)
Strontium	Dowex 50W-X8	1*M* HCl; Sr eluted 4*M* HCl	Cd, Cs, Cu, Hg(II), U(VI), and Zn	S. K. Majumdar and A. K. De, *Anal. Chim. Acta* **24**:356 (1961)
	Dowex 50W-X8	Dilute HNO_3; Sr eluted with 1.5*M* HNO_3	Al, lanthanoids, and Y	J. S. Fritz and B. B. Garralda, *Talanta* **10**:91 (1963)
	Dowex 50W-X8	Adsorption from 2% EDTA, pH 5.1; elution of alkali metals with 0.5*M* HCl, then Sr with 3*M* HCl	Alkali metals and Ca in milk ash and food ash	A. B. Strong, G. L. Rehnberg, and U. R. Moss, *Talanta* **15**:73 (1968)
Sulfur (SO_4)	Dowex 50		Removal of cations from (1) waters and (2) coal ash, precipitator dust, and boiler deposits	(1) R. M. Carlson, R. A. Rosell, and W. Vallejos, *Anal. Chem.* **39**:688 (1967); (2) H. N. S. Schafer, *ibid.* **39**:1719 (1967)
Thallium(I)	Dowex 50	EDTA solution, pH 4; Tl eluted with 2*M* HCl	Bi, Cu, Fe, Hg(II), Pb, and Zn	T. Nozaki, *J. Chem. Soc. Jpn. Pure Chem. Sect.* **77**:493 (1956)
Thallium(III)	Dowex 1	Adsorption from 0.1*M*–0.5*M* HCl; Tl eluted with H_2SO_4	Minerals, rocks, and sediments	A. D. Mathews and J. P. Riley, *Anal. Chim. Acta* **48**:25 (1969); J. P. Riley and S. A. Siddiqui, *ibid.* **181**:117(1986); G. Calderoni and T. Ferri, *Talanta* **29**:371 (1982)

(*Continued*)

TABLE 2.38 Selected Applications of Ion Exchange for the Separation of a Particular Element from Other Elements or Ions (*Continued*)

Element	Resin phase	Aqueous phase	Separated from	Reference
Thorium	Dowex 50	Adsorption from 1*M* H_2SO_4; other elements eluted with 2*M* HCl, Th with 3*M* H_2SO_4	Co, Cr(VI), Cu, Fe(III), Mo, Ni, U(VI), V(IV), and Ti	O. A. Nietzel, B. W. Wessling, and M. A. DeSesa, *Anal. Chem.* **30**:1182 (1958)
	Dowex 50W-X4	Other elements eluted with 5.5*M* HBr; Th eluted with 5*M* HNO_3	Lanthanoids, Zr, and other elements; rocks	F. W. E. Strelow and M. D. Boshoff, *Anal. Chim. Acta* **62**:351 (1972); A. H. Victor and F. W. E. Strelow, *ibid.* **138**:285 (1982)
Tin(IV)	Dowex 50	Thiourea solution	Ag, Cu, and Ga	C. H.-S. W. Weinert, F. W. E. Strelow, and R. G. Böhmer, *Talanta* **33**:481 (1986)
	Dowex 50	Tartaric acid solution	Cd, In, and Zn	F. W. E. Strelow and T. N. van der Walt, *Anal. Chem.* **54**:457 (1982)
	Dowex 1-X8 (oxalate form)	Adsorbed from HCl –oxalic acid solution; eluted with 1*M* H_2SO_4	Silicate rocks	C. Huffman, Jr., and A. J. Bartel, *U.S. Geol. Surv. Prof. Pap.* **501-D**:131 (1964)
Titanium	Dowex 50-X8	Adsorbed from <0.3*M* acid solution; Ti eluted with 0.5*M* H_2SO_4–1% H_2O_2	Al, Be, Cd, Co, Cu, Fe(III), La, Mg, Mn, Ni, Sc, Th, Th, Y, and Zn	F. W. E. Strelow, *Anal. Chem.* **35**:1279 (1963)
	Dowex 4	Other metals eluted with 0.1*M* H_2SO_4–0.15% H_2O_2, then Ti with 2*M* H_2SO_4	Al, Cr(III), Cu, Ga, lanthanoids, and Mg	R. Kuroda et al., *Anal. Chim. Acta* **62**:343 (1972)
	Dowex 1	Adsorbed from 1*M* NH_4SCN–1*M* HCl, eluted with 4*M* HCl	Biomaterials and seawater	T. Kiriyama et al., *Z. Anal. Chem.* **307**:352 (1981); **313**:328 (1982)
Tungsten	Dowex 50	0.1*M* HCl plus low concentration of oxalic acid; W not adsorbed	Bivalent metals	G. Gottschalk, *Z. Anal. Chem.* **187**:164 (1962)
	Dowex 50-X8	Elute W with 0.084*M* HF	Fe(III) and Ga	A. R. Eberle, *Anal. Chem.* **35**:669 (1963)
	Dowex 1-X8	Adsorbed from 0.07*M* $(NH_4)_2SO_4$–0.025*M* H_2SO_4; W eluted with 0.5*M* NaCl–0.5*M* NaOH	Al, Cu, Fe(III), Ni, Sc, Th, U(VI), V(IV), Zn, and Zr	T. Shimizu et al., *Jpn. Anal.* **15**:120 (1966)
Uranium(VI)	Dowex 50	U retained from 1*M* $HClO_4$ and eluted with 1*M* HCl	As(III), Co, Cr(III), Cu, Mn(II), Mo(VI), W(VI), V(V), and Zn	V. Pekarek and S. Maryska, *Collect. Czech Chem. Commum.* **33**:1612 (1968)
	Dowex 50	U eluted with 1*M* HCl	Thorium and phosphate	S. M. Khopkar and A. K. De, *Anal. Chim. Acta* **22**:153 (1960)

TABLE 2.38 Selected Applications of Ion Exchange for the Separation of a Particular Element from Other Elements or Ions (*Continued*)

Element	Resin phase	Aqueous phase	Separated from	Reference
Uranium	Dowex 50W-X8	U eluted with 0.1*M* HF	Bi, Fe(III), Pb, Zn	J. S. Fritz, B. B. Garralda, and S. K. Karraker, *Anal. Chem.* **33**:882 (1961)
	Dowex 50W-X8	U eluted with sulfosalicylic acid at pH 6–7; other metals retained	Cd, Co, La, Mg, and Ni	J. S. Fritz and T. A. Palmer, *Talanta* **9**:393 (1967)
	Dowex 1	U adsorbed from acetate medium at pH 4.3–5.2 and eluted with 0.8*M* HCl	Ca, Ce, Fe(III), and Th	F. Hecht et al., *Mikrochim. Acta* **1956**:1283
	Dowex 1	U adsorbed from ascorbic acid solution at pH 4.0–4.5, then eluted with 1*M* HCl	Al, Co, Cr, Fe, lanthanoids, Mn, Ni, and Pb	J. Korkisch, A. Farag, and F. Heckt, *Mikrochim. Acta* **1958**:415; **1959**:693
Vanadium	Dowex 1	0.1*M* HCl–0.1*M* NH_4SCN	Biomaterials and seawater	T. Kiriyama and R. Kuroda, *Anal. Chim. Acta* **62**:464 (1972); *Analyst* **107**:505 (1982); *Mikrochim. Acta* **1985:(I)** 405; *Chromatographia* **21**:12 (1986)
Vanadium(IV)	Dowex 1	12*M* HCl	Fe(III) and Ti	K. A. Kraus, F. Nelson, and G. W. Smith, *J. Phys. Chem.* **58**:11 (1954); C. Michaelis et al., *Anal. Chem.* **34**:1764 (1962)
Zinc	Dowex 1	0.1*M* HCl plus 10% NaCl	Al, alkaline earth metals, Be, Cd, Co, Cr(III), Cu, Fe(III), lanthanoids, Mn, and Zn unadsorbed; Zn eluted with 2*M* NaOH	S. Kallmann, C. G. Steele, and N. Y. Chu, *Anal. Chem.* **28**:230 (1956); J. L. Burguera, M. Burguera, and A. Townshend, *Anal. Chim. Acta* **127**:199 (1981)
	Dowex 1	1*M* HCl for adsorption and wash, then Zn eluted with 0.01*M* HCl	(1) Co, Cu, Fe(III), and Ni; (2) silicate rock, (3) sewage and sewage sludge, (4) canned juices	(1) R. M. Rush and J. H. Yoe, *Anal. Chem.* **26**:1345 (1954); (2) C. Huffman et al., *Geochim. Cosmochim. Acta* **27**:209 (1963); (3) E. V. Mills and B. L. Brown, *Analyst* **89**:551 (1964); (4) A. M. Aziz-Alrahman, *Int. J. Environ. Anal. Chem.* **19**:55 (1984)
Zirconium	Dowex 50W-X8	Adsorption from dilute HCl; other cations eluted with 2*M* HCl and Zr with 5*M* HCl	Al, Fe(III), Ti, and other elements	F. W. E. Strelow, *Anal. Chem.* **31**:1974 (1959)
	Dowex 1 or 2	Adsorbed from ascorbate solution of pH 4.0–4.5, then eluted with 1*M* HCl	Al, alkaline earth metals, Fe, lanthanoids	J. Korkisch and A. Farag, *Z. Anal. Chem.* **166**:181 (1959)

*DEAE denotes diethylaminoethyl.
† Me denotes methyl.

TABLE 2.39 Separations by Ligand-Exchange Chromatography

Compound types	Resin type, coordinating metal	Eluent	Reference
Amphetamine drugs	Carboxylic resin, Cu(II)	0.1*M* NH_3 in 33% ethanol	C. M. de Hernandez and H. F. Walton, *Anal. Chem.* **44**:890 (1972)
Purine and pyrimidine derivatives	Chelating resin, Cu(II)	1*M* NH_3	G. Goldstein, *Anal. Biochem.* **20**:477 (1967)
Terminal nucleoside assay of transfer ribonucleic acid	Chelating resin, Cu(II)	1*M* NH_3	C. A. Burtis and G. Goldstein, *Anal. Biochem.* **23**:502 (1968)
Oxypurines	Chelating resin, Cu(II)	1*M* NH_3	J. C. Wolford, J. A. Dean, and G. Goldstein, *J. Chromatogr.* **62**:148 (1971)
Oligopeptides and amino acids	Chelating resin, Cu(II). Condition column with ammoniacal solution, pH 10.3	Water (200 mL); 1.5*M* NH_3 (300 mL); 6*M* NH_3 (250 mL)	J. Boisseau and P. Jouan, *J. Chromatogr.* **54**:231 (1971)
Peptides and amino acids in urine	Chelating resin, Cu(II): borate buffer at pH 11	Borate buffer, pH 11 (50 mL) strips dipeptides and polypeptides	N. R. M. Buist and D. O'Brien *J. Chromatogr* **29**:L398 (1967)
Acid and neutral amino acids	Sulfonated resin, Zn(II); 56°C	55 m*M* sodium acetate and 0.4 m*M* zinc acetate, pH 4.08	F. Wagner and S. L. Shepherd, *Anal. Biochem.* **41**:314 (1971)
Basic amino acids	Sulfonated resin, Zn(II); 60°C; conditioned with sodium acetate buffer	Sodium acetate buffer with 1 m*M* zinc acetate	F. Wagner and S. L. Shepherd, *Anal. Biochem.* **41**:314 (1971)

not positioned correctly, the temperature reading will be inaccurate. If the entire bulb of the thermometer is placed too high, such as above the side arm leading to the condenser, the entire bulb will not be heated by the vapor of the distillate and the temperature reading will be too low. If the bulb is placed too low, too near the surface of the boiling liquid, there may be a condition of superheating, and the thermometer will show too high a temperature.

When evaporating a solution to recover the solute or when distilling off large volumes of solvent to recover the solute, do not evaporate completely to dryness. The residue may be superheated and begin to decompose.

When it becomes necessary to distill off large volumes of solvent to recover very small quantities of the solute, it is advisable to use a large distilling flask at first. (Never fill any distilling flask over one-half full.) When the volume has decreased, transfer the material to a smaller flask and continue the distillation. This minimizes losses caused by the large surface area of large flasks.

2.4.2 Inorganic Applications

The main concern in this section will be with elements other than carbon, hydrogen, nitrogen, and oxygen in organic compounds. These elements will be treated in Sec. 17 as will be the discussion of

TABLE 2.40 Approximate Percentage of Element Volatilized from 20- to 100-mg Portions at 200 to 220°C by Distillation with Various Acids

Element	HCl–$HClO_4$	HCl–H_2SO_4	HBr–H_2SO_4	HCl–H_3PO_4–$HClO_4$
As(III)	30	100	100	30
As(V)	5	5	100	5
Au	1	0.5	8.3	0.5
B	20	50	10	10
Bi	0.1	0	1	0
Cr(III)	99.7	0	0	99.8
Ge	50	90	95	10
Hg(I)	75	75	97.2	75
Hg(II)	75	75	97.2	75
Mn	0.1	0.02	0.02	0.02
Mo	3	5	1.2	0
Os	100	0	0	100
P	1	1	1	1
Re	100	90	100	80
Ru	99.5	0	0	100
Sb(III)	2	33	93.5	2
Sb(V)	2	2	98	0
Se(IV)	4	30	100	5
Se(VI)	4	20	100	5
Sn(II)	99.8	1	100	0
Sn(IV)	100	30	100	0
Te(IV)	0.5	0.1	15.7	0.1
Te(VI)	0.1	0.1	15.7	0.1
Tl(I)	1	1	1.6	1
V	0.5	0	0	0

high-temperature volatilization of certain elements from original sample matrices. The inert gases will also be ignored.

Distillation from solutions involves compounds in which covalent bonds prevail. Volatile inorganic substances form typical molecular lattices in which there exists an intimate association of a small number of atoms. This association is preserved on volatilization. Only small cohesive forces exist between individual molecules. Forces holding the individual atoms together within the molecule are much stronger. Thus, application of small amounts of energy to these systems easily disrupts the weak molecular bonds between these molecules and those of the solvent. Methods for individual elements are summarized in the following sections.

2.4.2.1 Antimony. After arsenic has been distilled, phosphoric acid is added to the distilling flask to combine with any tin (and to raise the boiling point), additional hydrazine is added, and the antimony is distilled as $SbCl_3$ at 160°C (at 140°C if bismuth is present) by dropping concentrated HCl into the distilling flask.[27,28]

Simultaneous removal of As, Sb, and Sn is accomplished by dropping a mixture of HCl–HBr (3:1) into the sample that contains a reducing agent such as hydrazine. The volatile halides are swept from the distilling flask with a stream of nitrogen or carbon dioxide.

[27] J. A. Scherrer, *J. Res. Nat. Bur. Stds.* **16**:253 (1936); **21**:95 (1938).
[28] C. J. Rodden, *J. Res. Nat. Bur. Stds.* **24**:7 (1940).

Distillation at 200 to 220°C with $HBr–HClO_4$ and CO_2 as carrier requires 10 min. Collect As and Sb in 10*M* HNO_3; dilute to 50 mL for colorimetry.

Gaseous SbH_3 may be generated with 0.2 g sodium borohydride, dispensed in pellet form, as the reducing agent added to a 1–6*M* HCl solution. A flow of inert gas transports the metallic hydride from the generation unit. The hydride-generating system finds considerable use as an adjunct to flame atomic absorption analyses. Generally, isolation of antimony as stibine is not suited for metal samples, but can be applied to igneous and sedimentary rocks, biosamples, water, and materials low in heavy metals. For references, see the arsenic section (Sec. 2.4.2.2).

2.4.2.2 Arsenic. From alloys and biological materials dissolved and in a solution 1 – 6*M* in HCl as As(III), gaseous AsH_3 may be generated with 0.2 g sodium borohydride $NaBH_4$, dispensed in pellet form, as the reducing agent added to an acid solution. A flow of inert gas transports the arsine from the generation unit. The hydride-generating system finds considerable use as an adjunct to flame atomic absorption analyses. The arsine is swept into a liquid-nitrogen trap or a sodium hydrogen carbonate solution of iodine, or absorbed in mercury(II) chloride solution. Bismuth, cadmium, copper, nickel, and silver interfere but may be removed with a cation-exchange resin.[29] The separation of As(III), As(V), methylarsonic acid, and dimethylarsinic acid by hydride generation has been described.[30]

From alloys, after adding hydrazine as reductant, arsenic can be distilled quantitatively as $AsCl_3$ from 6*M* HCl at 110 to 112°C. It may be carried over in a stream of carbon dioxide or by adding HCl from a dropping funnel into the distilling flask.[31–33] Nitric acid and other strong oxidants must be absent. The distillate is collected in ice-cold water, in solutions of hydrogen peroxide, or in nitric acid. Germanium accompanies arsenic in the distillation, and antimony may partially distill as the temperature rises above 107°C.

Arsenic is separable from antimony by distillation from H_2SO_4–HCl–HBr medium at 92 to 95°C.[34] Where a little arsenic has to be separated from much germanium, a solvent extraction method is best.

From biological materials, arsenic can be volatilized as $AsBr_5$ from boiling H_2SO_4 solution to which solid KBr is added.[35]

2.4.2.3 Bismuth. To the sample in 1*M* to 6*M* HCl, add 0.2 g $NaBH_4$. The volatile BiH_3 is swept from the flask and collected in liquid nitrogen.[36,37] Bismuth can be separated from germanium, osmium, and ruthenium by evaporation with HCl, HNO_3, and $HClO_4$ to fumes of perchloric acid. Separation from antimony, arsenic, mercury, rhenium, selenium, and tin is accomplished by evaporation with HCl and HBr several times, then finally to fumes of perchloric acid.

2.4.2.4 Boron. Except with the simplest of mixtures, it is necessary to separate boron from other constituents before it can be determined, and this is almost always done by distillation of the ester methyl borate $B(OCH_3)_3$ [boiling point (bp) 68.5°C]. As little as 0.1 $\mu g \cdot mL^{-1}$ of boron from dissolved metals and alloys can be distilled with methanol from sulfuric or hydrochloric acid with a minimum of water. Excess of methanol is added and the liquid distilled at 75 to 80°C in all-silica ware. Methanol is continually added directly through a funnel during the distillation or nitrogen is passed through the distilling flask at the rate of 50 $mL \cdot min^{-1}$. The distillate is collected in an alkaline solution of sodium hydroxide (0.1*M* to 0.5*M*) contained in a 100-mL platinum (Teflon or Vycor) dish so the hydrolysis of the ester regenerates the borate ion.[38–40] For microgram and submicrogram quantities, glycerol is added to the receiver solution to give better retention of boric

[29] R. S. Braman, L. L. Justen, and C. C. Foreback, *Anal. Chem.* **44**:2195 (1972).
[30] A. G. Howard and M. H. Arbab-Zavar, *Analyst* **106**:213 (1981).
[31] J. A. Scherrer, *J. Res. Nat. Bur. Stds.* **16**:253 (1936); **21**:95 (1938).
[32] C. J. Rodden, *J. Res. Nat. Bur. Stds.* **24**:7 (1940).
[33] W. C. Coppins and J.W. Price, *Metallurgia* **46**:52 (1952).
[34] G. Norwitz, J. Cohen, and M. E. Everett, *Anal. Chem.* **32**:1132 (1960).
[35] H. J. Magnuson and E. B. Watson, *Ind. Eng. Chem. Anal. Ed.* **16**:339 (1944).
[36] F. J. Fernandez, *At. Abs. Newsletter* **12**:93 (1973).
[37] F. J. Schmidt and J. L. Royer, *Anal. Lett.* **6**:17 (1973).
[38] C. L. Luke, *Anal. Chem.* **30**:1405 (1958).
[39] A. R. Eberle and M. W. Lerner, *Anal. Chem.* **32**:146 (1960).
[40] J. Cartwright, *Analyst* **87**:214 (1962).

acid. The deleterious effect of fluoride ion in the distillation of methyl borate is eliminated by the addition of anhydrous aluminum chloride in methanol before the distillation.[41] Vanadium, if present, distills with methyl borate, presumably as a methyl vanadate ester.[42] When the sample contains much silica, an extraction technique is preferable.

2.4.2.5 Bromine. Bromide is oxidized to free bromine by a chromic acid–sulfuric acid mixture and swept into the collecting flask by a stream of air. More than 5 mg of chloride interferes; iodine must be converted to iodate.[43]

2.4.2.6 Chromium. From solutions of alloys or ores, chromium can be removed as chromyl(VI)(2+) chloride, CrO_2Cl_2, by evaporating the sample to fumes with perchloric acid, which converts the chromium to the hexavalent state; then concentrated hydrochloric acid is dropped into the boiling solution until no more orange-red vapor is evolved and the color of the solution shows that the chromium has been removed.[44–46]

2.4.2.7 Chlorine. Hydrogen chloride can be distilled from a sulfuric acid solution at 150°C using a stream of nitrogen.

Using a Conway microdiffusion method, chlorine is liberated by the addition of permanganate, dichromate, or other oxidizing agent to the outer cell; the liberated chlorine diffuses into the inner cell and is absorbed by 0.1*M* KOH solution.[47]

2.4.2.8 Cyanide. Total cyanide is separated as hydrogen cyanide by distillation from concentrated phosphoric acid. EDTA may be added to the distillation mixture to aid in decomposing the more stable metal cyanides. The distilled hydrogen cyanide is collected in a dilute sodium acetate solution.[48]

Free cyanide is separated from thiocyanate by simple aeration in aqueous acidic solution.[49]

2.4.2.9 Fluoride. Fluoride ion is easily removed as HF by evaporation with sulfuric or perchloric acid. Separation is accomplished by distillation as H_2SiF_6. The sample is placed in a 250-mL (125-mL for small samples) distilling flask with glass beads, sulfuric or perchloric acid is added, and the mixture steam distilled at 135 to 140°C.[50] About 100 mL of distillate are collected for 10 mg of fluorine. Much gelatinous silica retards distillation. More serious interference is given by aluminum.

2.4.2.10 Germanium. $GeCl_4$ (bp 84°C) may be distilled at the rate of $1 mL \cdot min^{-1}$ from a 3*M* to 4*M* HCl solution in an atmosphere of chlorine that is maintained by a stream of chlorine or chlorine plus carbon dioxide. Collect the distillate in ice water. It is best to continue distillation until constant-boiling hydrochloric acid forms (or half the initial volume remains), or else some of the germanium may stick to the walls of the condenser.[51] Arsenic is kept in the pentavalent state, which is not volatile. Tin is kept in solution as the stable hexachlorostannate(IV)(2–) ion.[52–54] If handled carefully, $GeCl_4$ may be expelled before As.[55]

2.4.2.11 Iodine. After reduction to hydriodic acid or iodine, either form is quantitatively distilled into an alkaline scrubber. The efficacy of distillation has been tested.[56]

---

41 C. Gaestel and J. Huré, *Bull. Soc. Chim. France* **16**:830 (1949).
42 G. Weiss and P. Blum, *Bull. Soc. Chim. France* **14**:1077 (1947).
43 A. H. Neufeld, *Can J. Res.* **14B**:160 (1936).
44 C. W. Sill and H. E. Peterson, *Anal. Chem.* **24**:1175 (1952).
45 R. E. Heffelfinger, E. R. Blosser, O. E. Perkins, and W. M. Henry, *Anal. Chem.* **34**:621 (1962).
46 L. G. Bricker, S. Weinberg, and K. L. Proctor, *Ind. Eng. Chem. Anal. Ed.* **17**:661 (1945).
47 C. J. Rodden, *Analytical Chemistry of the Manhattan Project,* McGraw-Hill, New York, 1950, p. 297.
48 J. M. Kruse and M. G. Mellon, *Anal. Chem.* **25**:456 (1953).
49 L. S. Bark and H. G. Higson, *Talanta* **11**:621 (1964).
50 H. H. Willard and O. B. Winter, *Ind. Eng. Chem. Anal. Ed.* **5**:7 (1933).
51 P. G. Harris, *Anal. Chem.* **26**:737 (1954).
52 W. Geilmann and K. Brunger, *Z. Anorg. Chem.* **196**:312 (1931).
53 P. G. Harris, *Anal. Chem.* **26**:737 (1954).
54 L. M. Dennis and E. B. Johnson, *J. Am. Chem. Soc.* **45**:1380 (1923).
55 C. L. Luke and M. E. Campbell, *Anal. Chem.* **25**:1589 (1953).

2.4.2.12 ***Mercury.*** Ignition in a porcelain tube for several hours at 800°C (often with added CaO and $PbCrO_4$) in a stream of air will remove mercury from its compounds or ores. The metal is collected on gold or silver foil under water or in a liquid-nitrogen trap.[57]

After reduction to metal with tin(II) in mineral acid solution, mercury can be volatilized by aeration at room temperature. Expelled mercury is absorbed in an aqueous solution containing permanganate and sulfuric acid, or by amalgamation with gold or silver foil.[58]

After ashing organic compounds with perchloric acid (or fuming sulfuric acid plus ammonium peroxodisulfate), mercury(II) chloride can be distilled by passing hydrogen chloride gas through the distilling flask.[59]

2.4.2.13 ***Osmium.*** Osmium is easily separated from the other platinum metals as OsO_4. The sample is treated in the distilling flask with 8*M* HNO_3 or boiling 70% $HClO_4$, and a slow stream of air passed through the boiling solution carries away the OsO_4 (bp 129°C). The distillate is absorbed in a solution of thiourea in 6*M* HCl (or 6*M* HCl saturated with sulfur dioxide) to immediately reduce osmium to the nonvolatile lower valence states.[60–62] The absorption is somewhat slow, and at least two traps should be used in series. If chloride is present, the distillation is conducted from concentrated H_2SO_4 containing a small amount of HNO_3. Ruthenium will be distilled as RuO_4 from boiling $HClO_4$ but will not distill from an HNO_3 solution so long as the acid concentration is less than 40% by volume.

2.4.2.14 ***Ruthenium.*** From platinum metals, RuO_4 (bp 130°C) is volatilized from 8*M* HNO_3 plus hot $HClO_4$ or $KBrO_3$ to convert ruthenium into the octavalent oxide, which can be collected in one of several cold solutions of a suitable reducing agent—6*M* HCl that is saturated with SO_2 or hydroxylammonium chloride, or a dilute solution (3%) of hydrogen peroxide. The distillation provides a separation from most metals except osmium, which can be removed beforehand.[63]

2.4.2.15 ***Selenium.*** In soils, ores, and agricultural products, selenium can be distilled as $SeBr_4$ from a mixture of HBr and Br_2 (10 : 1), or HBr and $KBrO_3$, plus H_2SO_4.[64] Sufficient HBr and Br_2 are added to give a permanent bromine color. Add 15 mL HBr plus 0.5 mL Br_2 dropwise; heat the mixture to 210°C and distill into 10 mL HNO_3. Antimony, arsenic, germanium, and mercury accompany selenium.

2.4.2.16 ***Silicon.*** Silicon is removed where it is unwanted by evaporating with hydrofluoric and sulfuric (or perchloric) acids to copious fumes of sulfuric or perchloric acid in a platinum dish. The final residue is free from fluoride as well as silica. Where much silica has to be removed, a second or third evaporation can always be performed. Elements lost as volatile fluorides during this evaporation include As, B, Ge, Sb, Se, and also the metals Cr, Mn, and Re.[65]

When the gravimetric determination of silica is desired, but fluoride is present in the sample, excess boric acid is added and the fluoride volatilized as BF_3. Silica is recovered quantitatively.

2.4.2.17 ***Sulfur.*** When it is necessary to determine sulfur in the sulfide form, as distinct from total sulfur, the sample is treated with hydrochloric acid and the liberated hydrogen sulfide gas is absorbed in ammoniacal zinc acetate solution. Oxidative and reductive methods for pyritic sulfur involve treatment with chromium powder plus hydrochloric acid, which completely reduces pyritic sulfur without attacking organic sulfur.[66]

---

[56] F. Nesh and W. C. Peacock, *Anal. Chem.* **22**:1573 (1950).
[57] S. Asperger and D. Pavlovic, *Anal. Chem.* **28**:1761 (1956); D. H. Anderson et al., *ibid.* **43**:1511 (1971).
[58] Y. Kimura and V. L. Miller, *Anal. Chim. Acta* **27**:325 (1962).
[59] E. P. Fenimore and E. C. Wagner, *J. Am. Chem. Soc.* **53**:2468 (1934).
[60] R. Gilchrist, *J. Res. Nat. Bur. Stds.* **6**:421 (1931); R. Gilchrist and E. Wichers, *J. Am. Chem. Soc.* **57**:2565 (1935); J. J. Russell, F. E. Beamish, and J. Seath, *Ind. Eng. Chem. Anal. Ed.* **9**:475 (1937).
[61] W. J. Allen and F. E. Beamish, *Anal. Chem.* **24**:1608 (1952).
[62] A. D. Westland and F. E. Beamish, *Anal. Chem.* **26**:739 (1954).
[63] J. M. Kavanagh and F. E. Beamish, *Anal. Chem.* **32**:490 (1960); D.D. DeFord, *U.S. AEC Rep. No. NP-1104* (1948); W. J. Rogers, F. E. Beamish, and D. S. Russell, *Ind. Eng. Chem. Anal. Ed.* **12**:561 (1940).
[64] D. N. Fogg and N. T. Wilkinson, *Analyst (London)* **81**:525 (1956); K. L. Cheng, *Talanta* **9**:501 (1962).
[65] F. W. Chapman, G. G. Marvin, and S. Y. Tyree, *Anal. Chem.* **21**:700 (1949).
[66] W. Radmacher and P. Mohrhauer, *Glückauf* **89**:503 (1953).

The reduction of thiosulfate to sulfide involves first an alkaline reduction with sodium stannate(II), and second an acid reduction with aluminum. The evolved hydrogen sulfide is absorbed in ammoniacal cadmium chloride solution.[67]

Sulfates are reduced to sulfides in a mixture of hydroiodic acid, formic acid, and red phosphorus; then nitrogen is bubbled through the boiling mixture. The evolved hydrogen sulfide is passed through pyrogallol–sodium phosphate solution and then to an ammoniacal zinc acetate solution.[68]

2.4.2.18 Tellurium. Following the procedure for gaseous hydrides described under arsenic, TeH_2 can be volatilized from 1*M* to 6*M* HCl by adding 0.2 g $NaBH_4$ pellets.[69]

2.4.2.19 Tin. Tin can be distilled as $SnBr_4$ by dropping a mixture of HCl and HBr (3 : 1) into a H_2SO_4 or $HClO_4$ solution at 165°C (at 140°C if bismuth is present).[70] Antimony, arsenic, and germanium must be removed first.

A sample of silicate rocks and soils can be heated with ammonium iodide to decompose the sample and sublime tin as SnI_4.[71]

Following the procedure for volatile halides described under arsenic, $SnII_4$ may be volatilized and collected in a liquid-nitrogen trap.

2.4.3 Distillation of a Mixture of Two Liquids

Simple distillation of a mixture of two liquids will not effect a complete separation. If both are volatile, both will vaporize when the solution boils and both will appear in the condensate. The more volatile of the two liquids will vaporize and escape more rapidly and will form a larger proportion of the distillate initially. The less volatile constituent will concentrate in the liquid that remains in the distilling flask, and the temperature of the boiling liquid will rise. One approach to overcome this problem is to provide a very long column. As boiling-point differences become very small, this approach is no longer feasible. A more efficient means of obtaining higher separating capability involves use of the principle of rectification; the process by which condensed liquid and heated vapor are forced to contact one another inside the distillation flask.

In batch rectification the equipment consists of the distillation flask, a column (packed, plate, or spinning band), and a condenser. The system is first brought to steady state under total reflux, after which an overhead product is continuously withdrawn. The entire column thus operates as an enriching section. The resultant vapor-to-liquid contact enables low-boiling-point vapors to extract trapped low boilers from the condensed liquid. Conversely, vaporized high boilers are liquified as they contact similar molecules in the liquid state. High rectifying power provides high separating capability. When the difference in the volatility of the two liquids is large enough, the first distillate may be almost pure. As time proceeds, the composition of the material being distilled becomes less rich in the more volatile component, and the distillation must be stopped after a certain time to attain a desired average composition. The distillate collected between the first portion and the last portion will contain varying amounts of the two liquids. This basic still will not separate mixtures completely with boiling-point differences of less than 25°C.

2.4.4 Fractional Distillation

The separation and purification of a mixture of two or more liquids, present in appreciable amounts, into various fractions by distillation is called *fractional distillation*. Essentially it consists of the systematic redistillation of distillates (fractions of increasing purity). Fractionations can be carried out

[67] M. S. Budd and H. A. Bewick, *Anal. Chem.* **24**:1536 (1952).
[68] C. M. Johnson and H. Nishita, *Anal. Chem.* **24**:736 (1952).
[69] F. J. Fernandez, *At. Abs. Newsletter* **12**:93 (1973).
[70] K. Terada and M. Takahata, *Z. Anal. Chem.* **321**:760 (1985).
[71] B. Martinet, *Chem. Anal.* **43**:483 (1961); A. J. McDonald and R. E. Stanton, *Analyst* **87**:600 (1962).

with an ordinary distilling flask; but, with components that do not have widely separated boiling points, it is a very tedious process.

The separation of mixtures by this means is a refinement of ordinary separation by distillation. Thus a series of distillations involving partial vaporization and condensation concentrates the more volatile component in the first fraction of the distillate and leaves the less volatile component in the last fraction or in the residual liquid. The vapor leaves the surface of the liquid and passes up through the packing of the column. There it condenses on the cooler surfaces and redistills many times from the heat of the rising vapors before entering the condenser. Each minute distillation causes a greater concentration of the more volatile liquid in the rising vapor and an enrichment of the residue that drips down through the column. By means of long and efficient columns, two liquids may be completely separated.

Only brief details will be given in the following sections; for more information consult the literature.[72]

2.4.4.1 *Efficiency of Fractionating Columns.* The enrichment of finite samples can be achieved by using a distillation column so designed that excellent contact is made between vapor rising through the column and condensate falling through the column. The efficiency of such columns is usually expressed in terms of the number of theoretical plates in the column. A *theoretical plate* is that length of column required to give the same change in the composition of liquid as that brought about by one equilibrium stage in the temperature–composition diagram. The more of these plates (or stages) there are in a column, the higher the efficiency or separating power. Theoretical plate numbers are useful in predicting the number of liquid–vapor equilibrations necessary to separate a given pair of components.

The number of theoretical plates (n) can be determined from one form of the Fenske equation.

$$n - 1 = \frac{\log\left(\dfrac{X_a}{X_b}\right)\left(\dfrac{Y_b}{Y_a}\right)}{\log \alpha} \tag{2.60}$$

where X_a = percent of low boiler in head
X_b = percent of low boiler in distillation flask
Y_a = percent of high boiler in head
Y_b = percent of high boiler in distillation flask
α = ratio of vapor pressures of two components

All that is necessary is to distill a mixture of normal solvents the relative volatility α of which is known and then analyze the distillate and residue by withdrawing small samples while the column is operating under total reflux (that is, no distillate taken off). Analysis of the samples by gas chromatography will provide the data needed. An approximate value of α can be calculated from the expression

$$\log \alpha = \frac{\Delta T}{85} \tag{2.61}$$

where ΔT is the difference in boiling points of the components. This approximation is valid for materials with boiling points near 100°C. Column efficiency is often evaluated

[72] R. W. Yost, "Distillation Primer: A Survey of Distillation Systems," *Am. Laboratory* **6**:65 (1974); A Rose and E. Rose, "Distillation," Chap. 10 in I. M. Kolthoff and P. J. Elving, eds., *Treatise on Analytical Chemistry*, Part I, Vol. 5, 2nd ed., Wiley-Interscience, New York, 1978.

TABLE 2.41 Theoretical Plates Required for Separation in Terms of Boiling-Point Difference and α

30:70 v/v in flask and 99.6:0.4 v/v in distillate

ΔT, °C	α	Plates required
40	3.00	5
25	2.00	8
15	1.50	12
7	1.20	26
3.5	1.10	50
1.8	1.05	97
1.1	1.03	159
0.4	1.01	470

experimentally with a mixture of hexane and methylcyclohexane; these test mixtures are also used frequently:

1. Benzene–carbon tetrachloride, for a column of 0 to 20 plates.
2. *n*-Heptane–cyclohexane, for a column of 20 to 100 plates.
3. Chlorobenzene–ethylbenzene, useful at 1 atm up to 60 plates. May be used at reduced pressures.

The theoretical plates required to obtain 99.6 : 0.4 separation of a 30 : 70 starting mixture for various boiling-point differences is shown in Table 2.41.

Example 2.5 A mixture of 21.5% dibutyl phthalate and 78.5% dibutyl nonanedioate was distilled at a takeoff rate of 3 drops · min^{-1}. Analysis of an early takeoff gave a head concentration of 98.4% dibutyl phthalate and 1.6% of the other ester. The vapor pressure of dibutyl phthalate at 342°C is 760 mm and the vapor pressure of dibutyl nonanedioate at 342°C is 723.5 mm. Thus

$$\alpha = \frac{760}{723.5} = 1.047$$

Using the preceding information in the Fenske equation:

$$n = 1 + \frac{\log\left(\frac{98.4}{21.5}\right)\left(\frac{78.5}{1.6}\right)}{\log 1.047} = 118$$

for a total of 118 plates.

As a general rule, to obtain maximum efficiency, the reflux ratio must closely approximate the number of plates to perform the separation. The ratio of the number of drops of reflux from the condenser to number of drops of distillate allowed to pass into the receiver is called the *reflux ratio*. In the example, the reflux ratio should be at least 100 : 1 and preferably somewhat larger.

The distillation behavior of binary mixtures of organic compounds whose normal boiling points differ by 5°C or less is given in Table 2.42. The compounds are listed approximately in the order of their normal boiling points, and are given index numbers that are also employed in Tables 2.43 and 2.44.

TABLE 2.42 Distillation Behavior of Binary Mixtures of Organic Compounds

Normal b.p., °C.	Component	No.
32.0	Methyl formate	1
32.6	Isoprene	2
34.5	Propylene oxide	3
34.6	Diethyl ether	4
35.5	Ethanethiol	5
36.0	Dimethyl sulfide	6
36.1	Pentane	7
36.5	2-Chloropropane	8
38.4	Ethyl bromide	9
38.5	2-Methyl-2-butene	10
39.1	Methyl propyl ether	11
40.7	Dichloromethane	12
42.2	Methylal	13
42.2	Methyl iodide.	14
44.6	3-Chloropropene	15
46.4	1-Chloropropane	16
46.5	Carbon disulfide	17
48.5	Propylamine	18
48.7	Propionaldehyde	19
49.3	Cyclopentane	20
49.7	2,2-Dimethyl butane	21
51.0	2-Chloro-2-methyl propane	22
52.5	Acrolein	23
54.3	Ethyl formate	24
55.5	Diethylamine	25
56.5	Acetone	26
57.4	1,1-Dichloroethane	27
57.8	Methyl acetate	28
58.0	2,3-Dimethyl butane	29
59.6	Hexadiene	30
60.0	2-Bromopropane	31
60.3	2-Methyl pentane	32
61.3	Trichloromethane	33
61.7	Ethyl propyl ether	34
63.3	3-Methyl pentane	35
64.7	Methyl alcohol	36
63.5	Hexene-1	37
67.1	Isobutyl nitrite	38
67.4	1-Propanethiol	39

TABLE 2.42 **Distillation Behavior of Binary Mixtures of Organic Compounds (*Continued*)**

Normal b.p., °C.	Component	No.
67.5	Diisopropyl ether	40
68.0	2-Chlorobutane	41
68.3	Isopropyl formate	42
68.6	Isobutylamine	43
68.7	Hexane	44
68.9	Isobutyl chloride	45
71.0	1-Bromopropane	46
71.8	Methylcyclopentane	47
72.4	Iodoethane	48
76.5	Carbon tetrachloride	49
77.1	Ethyl acetate	50
77.8	1-Chlorobutane	51
78.2	Butyl nitrite	52
78.4	Ethyl alcohol	53
78.5	Acrylonitrile	54
79.2	2,2-Dimethylpentane	55
79.6	2-Butanone	56
79.8	Methyl propionate	57
80.1	Benzene	58
80.5	2,4-Dimethyl pentane	59
80.7	Cyclohexane	60
80.9	2,2,3-Trimethyl butane	61
81.3	Propyl formate	62
81.8	Acetonitrile	63
82.5	2-Propanol	64
83.47	1,2-Dichloroethane	65
82.9	*tert*-Butyl alcohol	66
84.4	Thiophene	67
84.7	Fluorobenzene	68
86.1	3,3-Dimethyl pentane	69
87.1	Trichloroethylene	70
87.9	Diethoxymethane	71
88.0	Diethyl sulfide	72
88.9	3-Methyl-2-butanone	73
89.0	Isopropyl acetate	74
89.4	Triethylamine	75
89.5	2-Iodopropane	76
89.5	Dipropyl ether	77
89.8	2,3-Dimethylpentane	78

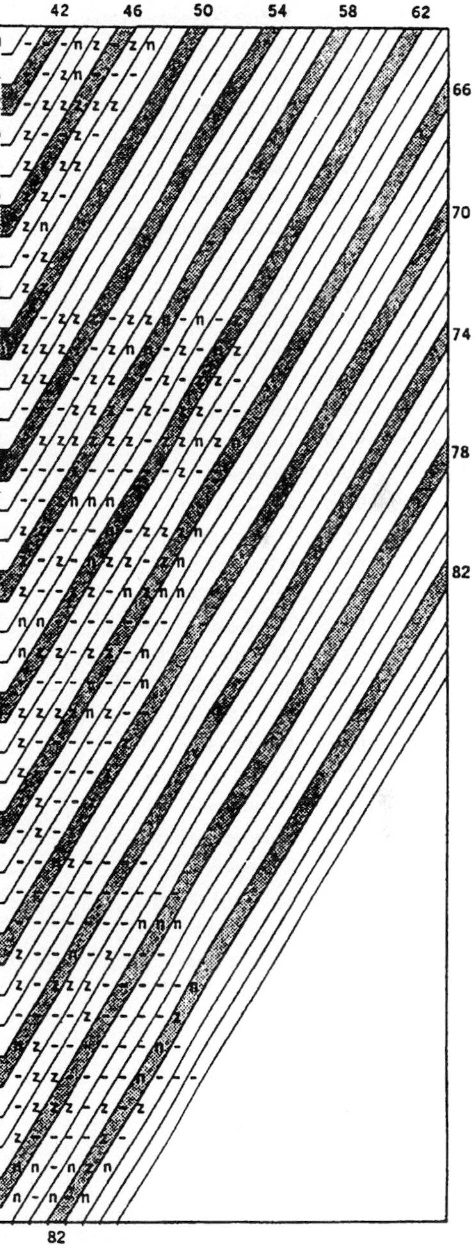

(*Continued*)

TABLE 2.42 Distillation Behavior of Binary Mixtures of Organic Compounds (*Continued*)

Normal b.p., °C.	Component	No.
90.0	2-Methylhexane	79
91.5	1-Bromo-2-methylpropane	80
91.9	3-Methylhexane	81
92.6	Methyl isobutyrate	82
93.5	3-Ethylpentane	83
95.7	*trans* - 3 - Heptene	84
95.8	*cis* -3-Heptene	85
96.8	Allyl alcohol	86
96.8	1,2-Dichloropropane	87
97.2	Isoamylnitrite	88
97.1	Propionitrile	89
97.5	1-Butanethiol	90
97.7	Chloral	91
97.8	1-Propanol	92
98.0	*trans*-2-Heptene	93
98.2	Isobutyl formate	94
98.4	Heptane	95
98.5	*cis* -2-Heptene	96
98.5	1-Chloroethyl ethyl ether	97
98.6	Dibromomethane	98
99.1	Ethyl propionate	99
99.2	2,2,4-Trimethylpentane	100
99.4	1-Chloro-3-methyl butane	101
99.5	*sec* -Butyl alcohol	102
99.5	*cis* -1,2-Dimethylcyclopentane	103
100.6	Formic acid	104
100.9	Methylcyclohexane	105
101.1	1,4-Dioxane	106
101.2	Nitromethane	107
101.6	1-Bromobutane	108
101.7	*tert*-Amyl alcohol	109
101.8	Propyl acetate	110
101.8	3-Iodopropene	111
102.0	Acetal	112
102.2	Crotonaldehyde	113
102.3	Methyl butyrate	114
102.5	1-Iodopropane	115
102.7	3-Pentanone	116
103.3	2-Pentanone	117

TABLE 2.42 Distillation Behavior of Binary Mixtures of Organic Compounds (*Continued*)

Normal b.p., °C.	Component	No.
103.9	Isobutyronitrile	118
106.0	Butyl formate	119
106.2	Pinacolone	120
106.3	2,2,3,3-Tetramethylbutane	121
106.7	1-Bromo-2-chloroethane	122
106.8	2,2-Dimethylhexane	123
108.0	Isobutyl alcohol	124
109.1	2,5-Dimethylhexane	125
109.2	Dipropylamine	126
109.4	2,4-Dimethylhexane	127
109.8	2,2,3-Trimethylpentane	128
110.1	Ethyl isobutyrate	129
110.6	Toluene	130
111.9	Trichloronitromethane	131
112.0	3,3-Dimethylhexane	132
112.9	3-Methyl-2-butanol	133
113.5	2,3,4-Trimethylpentane	134
113.9	1,1,2-Trichloroethane	135
114.0	Nitroethane	136
114.8	2,3,3-Trimethylpentane	137
115.4	Pyridine	138
115.6	2,3-Dimethylhexane	139
115.6	2-Methyl-3-ethylpentane	140
115.6	3-Pentanol	141
116.7	Methyl isovalerate	142
117.5	Butyronitrile	143
117.5	Butyl alcohol	144
117.6	2-Methylheptane	145
117.7	3,4-Dimethylhexane	146
117.7	4-Methylheptane	147
117.9	Epichlorohydrin	148
118.0	Isobutyl acetate	149
118.1	Acetic acid	150
118.3	3-Methyl-3-ethylpentane	151
118.5	3-Ethylhexane	152
118.9	3-Methylheptane	153
119.0	4-Methyl-2-pentanone	154
119.0	1-Chloro-2-propanone	155
119.3	*trans*-1,4-Dimethylcyclohexane	156

(*Continued*)

TABLE 2.42 **Distillation Behavior of Binary Mixtures of Organic Compounds** (*Continued*)

Normal b.p., °C.	Component	No.
119.5	1,1-Dimethylcyclohexane	157
119.7	2-Pentanol	158
120.0	2-Iodobutane	159
120.1	*cis*-1,3-Dimethylcyclohexane	160
120.4	1-Bromo-3-Methylbutane	161
120.4	1-Iodo-2-Methylpropane	162
120.5	Isopropyl isobutyrate	163
120.8	Tetrachloroethylene.	164
121.0	Ethyl butyrate	165
122.3	Isobutyl ether	166
122.4	Propyl propionate	167
123.3	3-Hexanone	168
123.3	Isoamyl formate	169
123.4	*trans*-1,2-Dimethylcyclohexane	170
123.5	Isobutyl nitrate	171
124.3	*cis*-1,4-Dimethylcyclohexane	172
124.4	2-Methoxy ethanol	173
124.4	*trans*-1,3-Dimethylcylohexane	174
125.0	Butyl acetate	175
125.6	Octane	176
125.8	Diethyl carbonate	177
127.0	1-Chloro-2-propanol	178
127.5	2-Hexanone	179
128.8	2-Chloroethanol	180
129.7	*cis*-1,2-Dimethylcyclohexane	181
130.0	Mesityl oxide	182
130.0	Methyl chloroacetate	183
130.4	1-Iodobutane	184
130.6	Isoamyl alcohol	185
130.7	Cyclopentanone	186
131.5	1,2-Dibromoethane	187
131.8	Ethylcyclohexane	188
131.7	Chlorobenzene	189
133.7	2-Chloro-1-propanol	190
133.9	Propyl isobutyrate	191
134.3	Ethyl isovalerate	192
135.5	2-Ethoxyethanol	193
136.2	Ethyl benzene	194
136.8	Isobutyl propionate	195

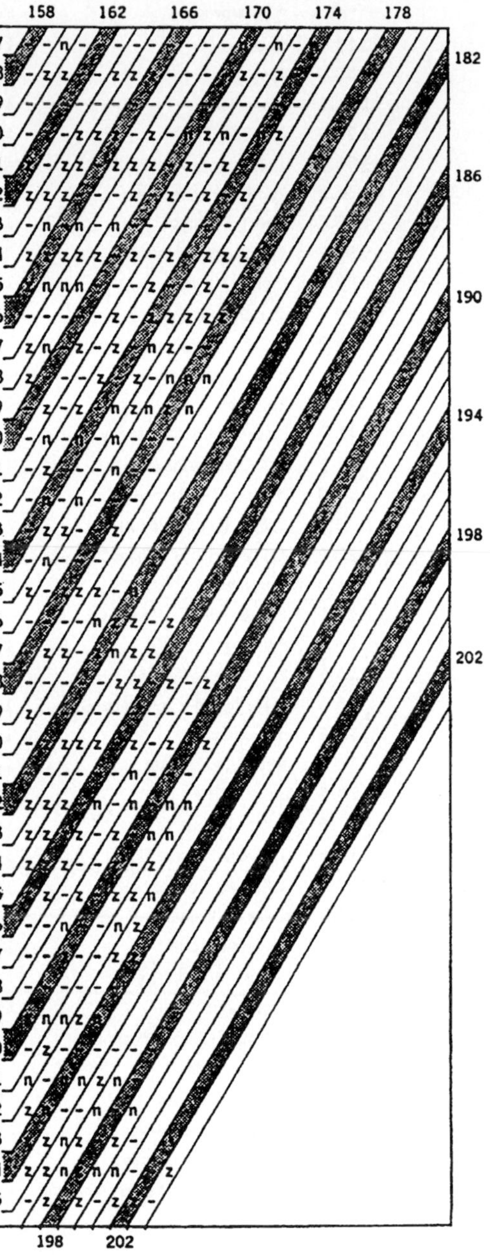

158 162 166 170 174 178 182 186 190 194 198 202 198 202

TABLE 2.42 **Distillation Behavior of Binary Mixtures of Organic Compounds (*Continued*)**

Normal b.p., °C.	Component	No.
137.8	Amyl alcohol	196
138.3	p-Xylene	197
138.6	Diallyl sulfide	198
139.1	m-Xylene	199
139.5	Diisobutylamine	200
140.8	Valeronitrile	201
140.9	Cyclopentanol	202
141.1	Propionic acid	203
141.6	1,2-Dibromopropane	204
142.1	Isoamyl acetate	205
142.1	Butyl ether	206
142.7	Propyl butyrate	207
143.7	4-Heptanone	208
144.0	3-Picoline	209
144.2	Ethyl chloroacetate	210
144.4	o-Xylene	211
144.6	2-Methoxy ethyl acetate	212
145.9	1,1,2,2-Tetrachloroethane	213
146.0	Triethyl orthoformate	214
146.8	Butyl propionate	215
147.5	Isoamyl nitrate	216
147.5	Isobutyl isobutyrate	217
148.2	1-Iodo-3-methylbutane	218
148.8	Amyl acetate	219

Source: This chart is based primarily on data given by Horsley, Azeotropic Data, *Advances Chem. Ser.* **6**, American Chemical Society, Washington, D.C., 1952.

To use the chart in Table 2.42 follow the row beginning with the lower-boiling constituent horizontally across the chart, and follow the diagonal beginning with the higher-boiling constituent upward. At the intersection of the row and the diagonal, the behavior of the mixture is described by one of the following symbols:

n	nonazeotrope	z	azeotrope
r	react	—	data not available

Where the formation of an azeotrope is indicated by the appearance of a "z," its composition may be found by consulting Table 2.44.

Table 2.44 gives the value, at a temperature that is the mean of the normal boiling points of the two compounds involved, of the ratio of the vapor pressures of each pair of pure constituents the mixture of which is described by the appearance of an "n" in Table 2.42. It can be used to determine the approximate number of theoretical plates to which a fractionating column must be equivalent in order to attain a desired separation at total reflux. The number of plates are calculated by using Eq. (2.60), where X is the mole fraction of the lower-boiling components. It is the value of the denominator of the right-hand side of this equation that is provided by Table 2.44 under the

TABLE 2.43 Azeotropic Data

A	B	B.P.	%A	A	B	B.P.	%A	A	B	B.P.	%A
1	2	22	50	25	28	53	. . .	42	48	66	62
	4	28	56		29	55	62	43	44	66	52
	5	27	30		30	55	. . .		47	68	59
	7	22	53	26	27	58	30	44	45	66	45
	8	28	60		28	55	50		46	67	50
2	3	32	40		29	46	42		47	68	75
	4	33	52		30	47	45		48	68	24
	7	34	90		31	54	42	45	46	69	95
3	7	27	57		33	65	20		47	68	63
4	5	31	60	27	28	56	. . .	46	47	69	58
	7	33	70		29	56	58	47	49	72	68
	10	34	88		30	56	77	48	49	Min.	. . .
5	7	32	50	28	29	51	50		50	70	78
	8	36	45		30	51	60	49	52	75	70
	10	33	60		31	56	68		53	65	80
6	7	33	45		33	65	23		54	66	79
	8	36	. . .	29	30	57	58		56	74	71
	10	34	45		31	58	50		57	76	75
7	8	32	48		33	55	53		62	75	60
	9	33	50	30	33	55	68	50	51	76	35
	11	35	75		34	60	95		52	76	71
	12	36	51	31	33	62	65		53	72	69
8	10	34	61		36	46	86		56	77	82
9	10	35	60	32	36	50	74		60	73	54
	12	38	80	33	36	53	87		63	75	77
10	11	36	72	34	36	55	76	51	52	77	48
	12	36	48	35	36	50	74		53	66	80
	13	35	68		39	61	66		56	77	62
11	12	45	43	36	37	49	27		57	79	48
12	13	45	41		39	58	35		60	78	64
	14	40	79		41	53	20		62	76	62
13	14	39	43		42	57	33		63	67	67
	15	41	80		44	49	27	52	56	77	70
	17	37	54		45	53	23		57	78	88
14	16	42	85	38	41	66	62		58	78	75
	17	41	60		42	65	60		60	76	63
15	17	41	50		44	65	54		62	77	65
	20	44	63		45	66	67		63	77	. . .
16	17	42	44		46	67	95	53	55	. . .	26
	19	46	. . .		47	66	68		56	76	46
	20	44	64	39	40	66	65		57	72	33
17	20	44	67		44	64	53		58	68	32
	22	43	62		47	66	64		59	. . .	29
18	20	47	52	40	45	69	. . .		60	65	30
20	22	47	50		47	68	80		62	72	41
	24	42	55	41	44	66	57		63	72	56
22	24	48	65	42	44	57	48		65	70	37
24	29	45	52		45	65	48	54	58	73	47
25	26	51	62		46	66	55		64	72	56
	27	52	55		47	61	55	55	58	76	46

TABLE 2.43 Azeotropic Data (*Continued*)

A	B	B.P.	%A
56	57	79	60
	58	78	44
	64	78	68
	65	Max.	. . .
	66	79	69
57	58	79	52
	60	75	52
	63	76	70
	64	76	62
	66	78	63
58	59	75	48
	62	78	53
	63	73	66
	66	74	63
60	62	75	52
	63	62	67
	65	74	50
	66	71	63
62	63	76	67
	64	76	64
	65	84	10
	66	78	60
	68	79	78
63	64	74	52
	70	75	29
64	65	75	57
	70	75	30
65	66	76	78
	67	83	. . .
	70	82	43
66	68	76	30
	70	76	67
67	71	83	. . .
70	71	89	54
	75	86	88
71	72	86	65
	74	88	58
	75	86	. . .
	76	86	63
72	77	89	25
	82	91	56
73	75	88	. . .
74	76	87	40
	77	86	50
75	77	89	. . .
76	77	89	65
	82	88	80
77	82	90	25
80	82	90	61
82	86	90	72

A	B	B.P.	%A
86	92	97	74
	94	93	52
	95	84	37
	98	86	20
	99	93	54
	101	88	29
	105	85	42
	107	89	57
	108	89	30
	110	95	52
87	99	Max.	. . .
	106	Max.	. . .
88	95	95	52
	98	96	. . .
	101	97	20
	105	95	79
	109	94	. . .
	111	96	. . .
89	92	90	50
	95	80	. . .
	99	94	40
	105	85	45
	109	95	55
90	92	92	59
	95	95	49
	98	95	72
	100	95	50
	103	96	48
	105	97	58
	107	93	. . .
91	94	100	60
	95	93	53
	99	101	. . .
	101	97	85
	105	94	57
	107	93	65
	108	96	. . .
	110	103	50
	114	103	45
	115	97	. . .
	116	103	23
92	94	93	40
	95	87	36
	98	90	26
	99	93	51
	100	85	41
	101	89	31
	105	86	35
	106	95	55

A	B	B.P.	%A
92	107	89	53
	110	94	40
	111	90	29
	113	97	. . .
	114	94	47
	115	90	30
94	95	90	50
	101	94	50
	102	95	60
	107	95	68
	108	95	65
	109	97	81
	111	96	62
95	97	96	52
	98	95	42
	99	93	53
	102	89	62
	104	78	44
	106	92	56
	107	80	63
	108	97	50
	109	92	74
	110	94	. . .
	111	97	52
	112	98	72
	114	95	65
	115	97	60
	116	93	65
	117	93	66
97	98	96	28
	105	97	65
98	105	96	75
99	101	98	55
	102	96	53
	105	94	53
	107	96	65
	109	98	62
	111	98	65
	113	98	75
101	102	91	71
	104	80	67
	105	98	64
	106	97	44
	107	88	52
	109	96	73
	110	98	60
	116	98	75
	118	91	65
102	105	90	41
	106	98	40

(*Continued*)

TABLE 2.43 Azeotropic Data (*Continued*)

A	B	B.P.	%A	A	B	B.P.	%A	A	B	B.P.	%A
102	107	99	54	108	119	100	75	129	130	110	...
	108	93	30		120	101	86		136	108	73
	110	96	52	109	110	99	42	130	133	106	62
	114	98	59		111	97	25		134	109	60
	116	98	58		114	102	...		136	106	75
104	105	80	46		115	97	30	131	133	106	80
	106	113	43		116	101	40		141	107	82
	107	97	45		117	101	42	133	148	109	52
	108	81	65		118	99	58	135	138	Max.	...
	111	85	35		119	101	65		150	106	70
	113	95	...	110	111	99	44	136	144	108	55
	115	82	36		112	101	68		149	112	60
	116	105	33		115	99	54		150	112	70
	117	105	32		116	101	60	138	141	117	45
105	106	94	55		117	101	65		142	115	52
	107	81	61	111	112	100	67		144	119	29
	108	99	45		114	101	65		149	114	...
	109	92	60		116	100	35		150	140	47
	110	95	...		117	101	66		154	115	60
	111	99	30		118	93	68	141	148	111	46
	112	100	60		119	100	75		154	115	35
	113	100	...	112	114	102	45	142	144	113	60
	114	97	55		115	101	40		148	115	55
	115	99	60		116	102	25		154	116	45
	116	95	60		122	108	35		158	116	80
	117	95	60	113	114	101	...		159	116	72
	118	85	60		115	100	...	143	144	113	50
106	107	101	44		116	101	...		161	110	50
	108	98	47		117	101	...		162	108	46
	109	101	80	114	115	101	44	144	148	112	43
	111	98	44		116	101	45		149	114	50
	114	101	...		117	102	50		154	114	30
	115	99	40	115	116	101	65		155	112	43
107	108	90	50		117	101	65		160	108	43
	109	93	49	119	120	106	38		161	111	31
	110	98	45		124	103	60		162	110	30
	111	89	...		130	106	70		163	115	54
	112	95	65	120	124	105	42		164	109	29
	113	99	...		126	104	...		165	116	64
	114	98	50		130	106	85		166	113	48
	115	89	42	122	124	100	...	148	149	115	50
	116	99	55	124	125	99	42		150	115	64
	117	99	56		129	105	52		158	113	60
	119	99	60		130	101	45		160	114	65
	120	101	...		131	102	32		161	111	63
108	109	98	74	125	130	107	65		162	111	47
	110	100	52		131	107	45		164	110	52
	114	99	65		133	104	32		165	116	75
	116	100	63	126	127	108	54	149	154	116	...
	117	100	63		130	108	53		155	117	70

TABLE 2.43 Azeotropic Data (*Continued*)

A	B	B.P.	%A	A	B	B.P.	%A	A	B	B.P.	%A
149	158	116	68	164	165	119	57	180	182	130	33
	159	116	70		166	119	65		183	128	85
	160	114	62		167	120	. . .		184	119	38
	161	117	72		168	118	55		185	128	75
	162	116	50		169	118	65		187	122	66
	164	115	53		171	117	70		189	120	42
150	159	111	30		173	110	75		191	128	94
	160	109	45		175	120	79	182	183	129	58
	161	109	62		176	120	92		184	128	44
	162	109	37		177	119	74		185	129	76
	164	107	39	165	173	118	68	183	184	125	42
154	160	112	53		176	118	60		185	125	60
155	156	114	. . .	166	171	121	. . .		186	130	. . .
	157	114	. . .		173	115	52		187	128	44
	158	116	68		174	120	28		189	126	60
	160	114	. . .		176	122	10	184	185	123	72
	163	117	50		177	121	35		186	129	60
	164	118	. . .		178	118	65		187	129	35
	165	117	53	167	168	122	60		190	123	70
	170	114	. . .		171	122	59		192	130	. . .
158	160	113	38		173	118	62	185	186	130	42
	161	115	26		176	118	60		187	124	31
	164	113	34	168	169	123	50		189	124	34
	165	119	47		173	119	57		191	130	53
	166	115	41		175	123	. . .		192	130	58
	171	115	52	169	171	122	46	186	193	130	73
	173	120	96		173	119	60	187	190	128	67
160	162	119	40		176	116	55		193	128	77
	164	118	. . .		178	123	70		194	131	90
	165	117	50	171	173	115	56	189	190	126	64
	166	120	72	173	175	119	48		193	127	68
	168	116	63		176	110	48	191	196	133	81
	171	114	59		179	121	56	192	193	130	58
	175	118	63	175	176	119	52	193	194	128	48
161	162	119	. . .		178	125	75		195	131	35
	164	119	52		179	125	68		197	129	50
	165	120	65		180	126	69		199	129	51
	167	120	75	176	182	121	65	194	195	136	30
	168	120	45		183	123	60		196	130	60
	169	120	76		185	120	65		198	136	89
	171	118	68	177	179	126	65		203	131	72
	173	111	20		180	126	72	195	197	137	85
162	163	119	53		182	126	6		199	134	. . .
	164	119	60		184	124	70		201	136	73
	165	119	64		185	125	73		202	136	72
	169	117	70	178	184	120	45	196	197	131	42
	171	117	60		185	127	81		201	136	58
	173	110	75		187	125	62		206	134	50
	175	120	. . .		189	122	55	197	202	132	62
163	164	119	55	179	180	129	25		203	132	. . .

(*Continued*)

TABLE 2.43 Azeotropic Data (*Continued*)

A	B	B.P.	%A	A	B	B.P.	%A	A	B	B.P.	%A
198	199	138	52	203	213	140	60	212	216	144	87
	200	135	68	205	206	141	55		218	141	65
	202	135	67		208	142	75		219	144	92
	203	135	60		210	142	60	213	214	151	61
	207	139	70		212	141	80		215	152	55
	208	138	25		213	150	32		217	151	65
199	200	137	49	206	207	142	55		219	153	40
	201	136	. . .		210	140	55	216	217	147	40
	202	139	60		211	142	78	217	218	146	42
	204	138	80		212	138	70		219	148	90
	205	136	50		213	148	70	218	219	146	60
	207	139	. . .	207	208	143	53				
	208	139	90		210	142	53				
	209	Min.	. . .		211	143	55				
	210	137	68		212	143	68				
200	208	137	68	208	210	143	53				
201	206	130	42		211	142	42				
202	206	137	39		213	143	. . .				
	210	138	50		218	143	65				
	212	139	25	210	211	140	58				
203	204	134	33		212	145	38				
	206	136	45		213	147	27				
	210	140	61		218	140	49				
	211	135	43	211	212	141	50				
	212	147	36	212	214	143	51				

Source: L. Meites, ed., *Handbook of Analytical Chemistry*, McGraw-Hill, New York, 1963, pp. 10-32–10-36.

column heading "log (P_A^0/P_B^0)." Note that the compound appearing under A is always the lower-boiling compound.

For example, suppose that a mixture of 2-methylpentane and 3-methylpentane is to be distilled. Reference to Table 2.42 gives the index number of these compounds as 32 and 35, respectively, and indicates that their mixture is nonazeotropic. By consulting the row in Table 2.44 in which these two index numbers are paired, it is found that log (P_A^0/P_B^0) is = 0.041 34. If the mixture to be distilled contains 0.2 mole fraction of component A, and if it is desired to obtain a distillate containing 0.9 mole fraction of component A, then the approximate number of plates required at total reflux is

$$n = \frac{\log\left(\frac{0.9}{0.1}\right)\left(\frac{0.8}{0.2}\right)}{0.04134} = 38 \tag{2.62}$$

Good data for vapor pressures are available, but the quantity log (P_A^0/P_B^0) can be used with accuracy only when the constituents of the test mixture follow Raoult's law closely. Hence, a distinction is to be drawn between this quantity and log α. A stepwise method of integration should be used where the values of log α vary with composition.

TABLE 2.44 **Vapor-Pressure Ratios of Binary Mixtures**

Component A	Component B	$\log\left(\frac{P_A^0}{P_B^0}\right)$	Component A	Component B	$\log\left(\frac{P_A^0}{P_B^0}\right)$	Component A	Component B	$\log\left(\frac{P_A^0}{P_B^0}\right)$
4	9	0.08276	73	82	0.04267	96	103	0.00931
	11	0.06897	74	82	0.04182		105	0.02672
7	10	0.03454	77	78	0.00370	98	108	0.03989
11	14	0.04955		79	0.00724	99	114	0.03541
16	22	0.09392		81	0.03055		116	0.04805
20	21	0.00674		83	0.05192		117	0.05604
24	26	0.03361	78	79	0.00344	100	103	0.00363
	28	0.05729		81	0.02642		105	0.02079
27	33	0.05328		83	0.04706	102	109	0.03575
	34	0.06173	79	81	0.02312	103	105	0.01728
29	34	0.05219		83	0.04388	110	114	0.00696
31	34	0.02428	81	83	0.02077		119	0.05682
32	34	0.02018		84	0.04670		120	0.05860
	35	0.04134		85	0.04741	114	119	0.04986
34	35	0.02215	83	84	0.05395		120	0.05256
35	40	0.05870		85	0.02655	116	117	0.00788
40	44	0.01722		93	0.06212		119	0.04387
	48	0.06816	84	85	0.00071		120	0.04579
41	45	0.01040		93	0.03635	117	119	0.03587
46	48	0.00258		95	0.03740		120	0.03792
49	58	0.04736		96	0.04200	119	129	0.05311
	60	0.05518		100	0.04691	120	129	0.05146
50	57	0.03870	85	93	0.03564	121	123	0.00452
	58	0.04173		95	0.03667		125	0.38892
	62	0.06030		96	0.04127		127	0.05591
53	64	0.07080		100	0.04621		128	0.04051
	66	0.07766	86	102	0.04167	123	125	0.02875
55	59	0.01698		106	0.06044		127	0.03177
	60	0.02008	88	106	0.05188		128	0.03647
	61	0.02169		108	0.05741	125	127	0.00401
56	62	0.02408		110	0.06176		128	0.00897
57	62	0.02161	92	102	0.02860		132	0.03490
	67	0.06053	93	95	0.00104		134	0.05302
58	65	0.04863		96	0.00566	127	128	0.00498
	67	0.05427		100	0.01109		132	0.03087
	68	0.06745		103	0.01488		134	0.04896
59	60	0.00310		105	0.03225	128	132	0.02562
	61	0.00493	94	99	0.01244		134	0.04355
60	61	0.00186		104	0.03192	132	134	0.01802
	69	0.06851		106	0.03910		137	0.03331
61	69	0.13134		110	0.04885		139	0.04392
62	67	0.03954		116	0.06032		140	0.04433
69	77	0.04478	95	96	0.00461	134	137	0.01540
	78	0.04729		100	0.01015		139	0.02579
	79	0.05098		103	0.01391		140	0.02988
70	74	0.02624		105	0.03132		145	0.05056
73	74	0.00144	96	100	0.00556		146	0.05116

(*Continued*)

TABLE 2.44 Vapor-Pressure Ratios of Binary Mixtures (*Continued*)

Component A	Component B	$\log\left(\frac{P_A^0}{P_B^0}\right)$	Component A	Component B	$\log\left(\frac{P_A^0}{P_B^0}\right)$	Component A	Component B	$\log\left(\frac{P_A^0}{P_B^0}\right)$
134	147	0.04790		152	0.00977		167	0.02455
	151	0.05702		153	0.01449		169	0.03632
	152	0.06111		156	0.01924	165	167	0.01803
	153	0.06587		157	0.02148		168	0.02928
137	139	0.01013		160	0.02800		169	0.03051
	140	0.01062	147	151	0.00658	167	169	0.01802
	145	0.03466		152	0.01002		175	0.04004
	146	0.03536		153	0.01478	168	179	0.05201
	147	0.03538		156	0.01956	169	175	0.02683
	151	0.04133		157	0.02181		179	0.05294
	152	0.04517		160	0.02840	170	172	0.01045
	153	0.04989	149	165	0.03916		174	0.01196
	156	0.05389		166	0.05485		176	0.02652
	157	0.05605	150	163	0.03123	172	174	0.00151
139	140	0.00052		165	0.03776		176	0.01591
	145	0.02478		167	0.05585	174	176	0.01439
	146	0.02554		169	0.06768	175	182	0.07002
	147	0.02551	151	152	0.00329	176	181	0.04776
	151	0.03168		153	0.00798	181	188	0.02360
	152	0.03544		156	0.01280	182	187	0.00871
	153	0.04020		157	0.01503		191	0.02692
	156	0.04446		160	0.02150		192	0.04600
	157	0.04666		170	0.06021	189	191	0.01272
	160	0.05333	152	153	0.00474		192	0.02152
140	145	0.02419		156	0.00971		194	0.05243
	146	0.02496		157	0.01197	191	192	0.00508
	147	0.02492		160	0.01851		194	0.02777
	151	0.03109		170	0.05772		195	0.03656
	152	0.03482	153	156	0.00506		197	0.05392
	153	0.03958		157	0.00733	192	194	0.02287
	156	0.04384		160	0.01385		195	0.03148
	157	0.04603		170	0.05317		197	0.04902
	160	0.05268	154	165	0.02554		199	0.05824
141	144	0.03107		166	0.04122	194	197	0.02520
142	149	0.01714	156	157	0.00223		199	0.03401
	163	0.04980		160	0.00840	197	199	0.00876
	165	0.05628		170	0.04718	199	206	0.03535
144	158	0.03231		172	0.05782		211	0.06145
145	146	0.00094		174	0.05939	203	209	0.02259
	147	0.00076	157	160	0.00637	204	208	0.02468
	151	0.00733		170	0.04488	205	207	0.00880
	152	0.01077		172	0.05549		211	0.02879
	153	0.01554		174	0.05706		214	0.04946
	156	0.02030	160	170	0.03871		215	0.05922
	157	0.02255		172	0.04936	206	214	0.04725
	160	0.02916		174	0.05093		215	0.05681
146	151	0.00636	163	165	0.00650	207	214	0.03845

TABLE 2.44 Vapor-Pressure Ratios of Binary Mixtures (*Continued*)

Component A	Component B	$\log\left(\frac{P_A^0}{P_B^0}\right)$	Component A	Component B	$\log\left(\frac{P_A^0}{P_B^0}\right)$	Component A	Component B	$\log\left(\frac{P_A^0}{P_B^0}\right)$
207	215	0.05041		219	0.05206	215	217	0.00870
208	215	0.03738	214	215	0.00998		219	0.02451
211	213	0.01723		217	0.01846	217	219	0.01582
	215	0.02839		219	0.03429			

Source: L. Meites, ed., *Handbook of Analytical Chemistry,* McGraw-Hill, New York, 1963, pp. 10-37–10-39.

2.4.5 Azeotropic Distillation

Azeotropic mixtures distill at constant temperature without change in composition.[73] Obviously, one cannot separate azeotropic mixtures by normal distillation methods.

Azeotropic solutions are nonideal solutions. Some display a greater vapor pressure than expected; these are said to exhibit positive deviation. Within a certain composition range such mixtures boil at temperatures higher than the boiling temperature of either component; they are maximum-boiling azeotropes.

Mixtures that have boiling temperatures much lower than the boiling temperature of either component exhibit negative deviation; when such mixtures have a particular composition range, they behave as though a third component were present. For example, pure ethanol cannot be obtained by fractional distillation of aqueous solutions that contain less than 95.57% of ethanol because this is the azeotropic composition; the boiling point of this azeotropic mixture is 0.3°C lower than that of pure ethanol.

Table 2.43 gives approximate azeotropic data on the binary mixtures denoted by the symbol "z" in Table 2.42. To use this table, locate the two components of the binary mixture in Table 2.42 and note their index numbers. Then, in Table 2.43, locate the row in which these index numbers appear in the columns headed A and B. The following two columns then give the boiling point of the azeotrope, in degrees Centigrade at 760 mm, and its composition, in percent by weight of the component whose index number appears in column A. For example, the first line of the table gives the boiling point (22°C) and composition (50 wt % methyl formate) of the azeotrope of methyl formate (index number 1) and isoprene (index number 2).

The term "Max" means that the boiling point is a maximum; "Min" means that it is a minimum. These are the only data available for the azeotropes for which these symbols are used. These data were taken from Horsley, *Azeotropic Data*, Adv. Chem. Ser. 6, American Chemical Society, Washington, D.C., 1952.

2.4.5.1 Use of an Entrainer.

In azeotropic distillation a third solvent, called an entrainer, is added to the original azeotropic mixture, which will form a constant-boiling mixture with one of the two original components of the mixture. Thus, a shift in the boiling point is often large enough to effect a further separation of close-boiling materials. The quantitative removal of one component of the original mixture is therefore dependent upon the addition of an excess of the entrainer. The second requirement is that the entrainer be easily removed by simple distillation.

Absolute ethanol can be obtained by distilling the azeotropic ethanol–water mixture with benzene. A new lower-boiling (65°C) ternary azeotrope is formed (74% benzene, 7.5% water, and 18.5% ethanol). Distillation of the ternary azeotrope accomplishes the quantitative removal of water from the system but leaves the ethanol contaminated with benzene. However, the benzene–ethanol

[73] Binary and ternary azeotropic mixtures are tabulated in J. A. Dean, ed., *Lange's Handbook of Chemistry,* 14th ed., McGraw-Hill, 1992; pp. 5.57–5.83.

azeotrope (67.6% benzene and 32.4% ethanol) boils at 68.3°C and distills over by simple fractional distillation from the anhydrous ethanol, which can be collected at 78.5°C.

Other examples are the use of acetic acid to separate ethylbenzene from vinylbenzene, and butyl acetate for the dehydration of acetic acid. In fact, the judicious selection of entrainer liquids can facilitate the separation of complex mixtures.

2.4.6 Column Designs

Distillation columns provide the contact area between streams of descending liquids and ascending vapor and, thereby, furnish an approach toward vapor–liquid equilibrium. Depending upon the type of distillation that is used, the contact may occur in discrete steps, called plates or trays, or in a continuous differential contact on the surface of the column walls or packing. The following discussion will be centered around the several basic column designs usually employed in the laboratory. In the final selection of a column a compromise must be found that combines high separating power per unit length of column and high capacity with low holdup. The holdup should be less than 10% of the amount of material to be separated, but must not be of the same order of magnitude as the material to be resolved. There is not much point in using columns with more than 40 to 50 theoretical plates for normal laboratory work because of the unusually long time required for the column to come to equilibrium and the difficulty of operating such columns efficiently.

In order for any column of laboratory size to operate properly, it is necessary that it be made approximately adiabatic. The loss of heat from the column may be rendered negligible by use of a silvered vacuum jacket or by ordinary insulation (often with electrical resistance winding with controls). The trick is to avoid excessive condensation of the rising vapor, otherwise only a small quantity of vapor would reach the upper end of the column and little or no distillate would be produced.

2.4.6.1 Packed Columns. The column can be filled with a variety of packings (Fig. 2.7) or the packings may be indented prongs that are built in as with the Vigreux or Bruun column. A Vigreux column 50 cm in length and 8 mm in diameter will have five to six plates at total reflux and will handle 10 to 50 mL of liquid. Although the number of theoretical plates is small, their versatility in terms of capacity and pressure variation makes them useful.

Glass, single-turn helices, 4 mm in diameter, are perhaps the most generally useful material for packed columns. Glass helices have high throughput capacity and low liquid holdup. Because of the low pressure drop of these columns, they are well adapted for either atmospheric or vacuum fractionations. The primary objective is the formation of an extremely tortuous vapor path to achieve good contact between the liquid descending the column and the simultaneously rising vapor. Although high efficiency can be achieved in this way, the holdup is extremely large. Higher pot temperatures are required to overcome the pressure drop of the tightly packed column; this may be destructive to many compounds. It is possible to separate materials with boiling points as close as 0.5 to 1.0°C, but only at takeoff rates of one drop every 4 h. The Podbielniak or heligrid type of packing has also found use. For a column 100 cm in length with a 17 mm diameter, the number of plates at total reflux are 30 to 50; 150 to 160 mL of liquid can be handled.

For small-diameter columns of 5 mm or less, the spiral packing is useful. One type consists of a precision-made Monel rod, 1.5 mm in diameter, which is wound with 2.2 turns per cm of the same metal. The center rod is removable for vacuum fractionation. Columns 60 cm in length exhibit 20 to 50 plates at total reflux and accommodate 20 to 200 mL of sample with only a small holdup.

2.4.6.2 Plate and Concentric Tube Columns. Plate fractionating columns have definite numbers of trays or plates and are fitted with either bubble caps or sieve perforations, or modifications of these two, which bring about intimate vapor–liquid contact. In the crossflow plate the liquid flows across the plate and from plate to plate. In the counterflow plate, the liquid flows down through the same orifices as the rising vapor. There are three designs for counterflow plates: the sieve plate, the valve plate, and the bubble-cap plate.

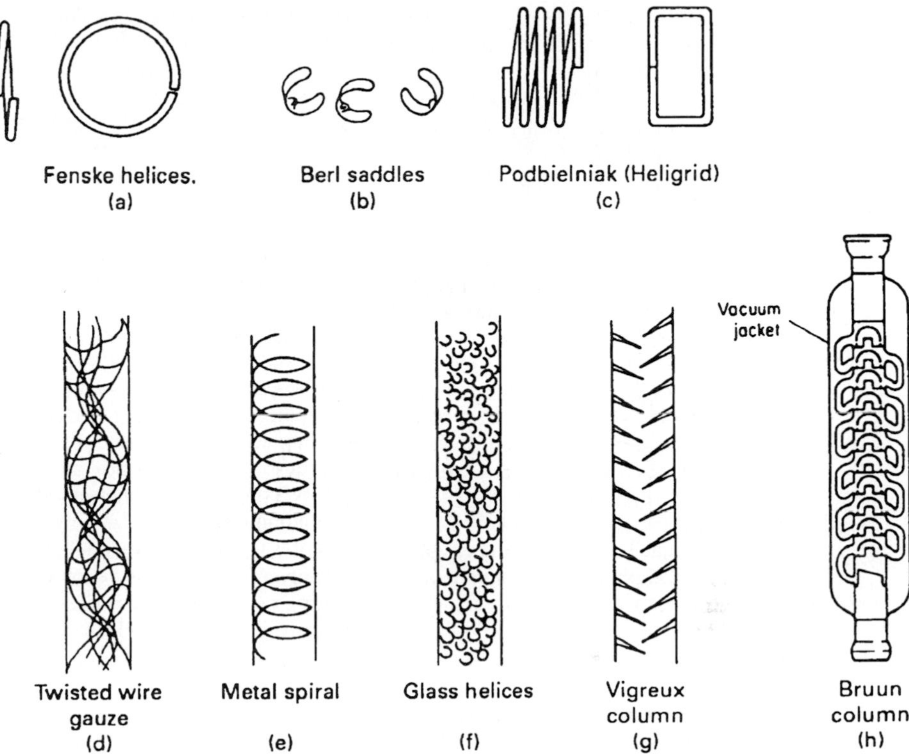

FIGURE 2.7 Packings for fractionating columns.

The sieve plate has contacting holes that measure 3.2 to 50 mm in diameter. Valve plates have valves that open or close to expose the holes for vapor passage. The latter arrangement will maintain efficient operation over a wider operating range. In the bubble-cap plate the vapor flows through a hole in the plate through a riser, reverses direction in the dome of the cap, flows downward in an annular area between the rise and the cap, and exits through slots in the cap.

A column 1 m in length containing 30 trays or plates can possess as many as 15 to 18 theoretical plates. Generally, a small 10-plate column will provide fairly rapid separation of materials with boiling-point differences of 15 to 20°C, a 25-plate column will function efficiently down to an 8 to 12°C difference, and a 45-plate column provides separation of compounds with differences of 5 to 10°C. Operating conditions with such a column are highly reproducible, but the holdup per plate is greater than that in a packed column.

In the concentric-tube column, the center of a long narrow distillation column is filled with a tube, leaving only the doughnut-shaped annular space (about 1 mm wide) for vapor to rise in. This type of column can have over 75 plates with almost no pressure drop and can provide purified material at rates of several milliliters per hour.

2.4.6.3 Spinning-Band Systems. Spinning-band systems have a motor-driven band that rotates inside the column (Fig. 2.8). This configuration combines efficiencies exceeding those of the best packed columns with low holdup, high throughput, and low pressure drop of the most basic unpacked columns. Metal mesh columns are constructed of spirally wound wire mesh, permanently fastened to a central shaft, which is connected to a motor. As the tightly fitted band is rotated, the

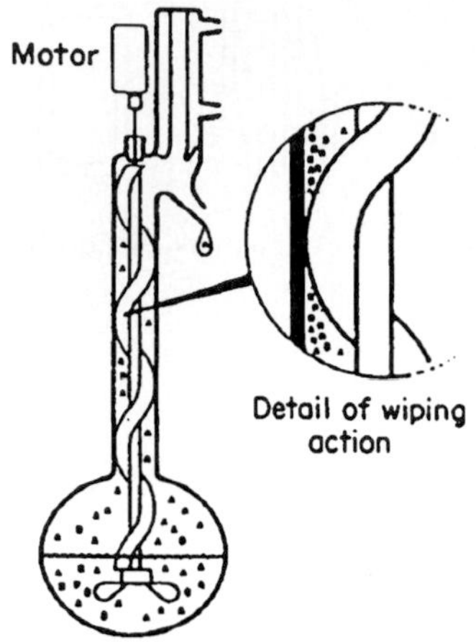

FIGURE 2.8 Teflon spinning band column.

exposed wire at the side of the band "combs" through the layer of liquid on the walls, continuously forming hundreds of vapor–liquid interfaces in each small section of the column. The band is fashioned so that it is constantly pumping downward. Since the spinning band is constantly whirling through the ascending vapors, any tendency to form vapor channels is eliminated. Mesh spinning columns can be used to obtain complete separations of materials with boiling-point differences of 50°C with relatively high throughput.

Annular Teflon spinning bands provide superior performance as compared with mesh spinning bands. The tightly fitted Teflon spiral contacts the inner glass walls of the column, wiping the descending refluxed liquid into a thin film only several molecular layers deep. This provides the maximum amount of exposure of all liquid molecules to the heated vapor, which is being hurled into it as a result of the spinning action of the band. Rotation speeds can be higher than those with metal bands since the natural "slipperiness" of the Teflon allows constant, high-speed rotation against the column walls without wear of the band or the glass wall.

Annual Teflon stills with a column length of 1 m provide a separating capability in excess of 200 theoretical plates. One example of a difficult separation is the separation of an equal percentage mixture of *m*-xylene and *p*-xylene, where the boiling-point difference is less than 1°C. With a boil-up rate of 60 drops · min^{-1} and a 100 : 1 reflux ratio, distillate can be collected at a rate of 1.7 mL · h^{-1}.

2.4.7 Total-Reflux Partial Takeoff Heads

For truly effective fractionation, use a total-reflux partial-takeoff distilling head, as shown in Fig. 2.9. With the stopcock *S* completely closed, all condensed vapors are returned to the distilling column, a total-reflux condition. With the stopcock partially opened, the number of drops of condensate falling from the condenser that returns to the fractionating column can be adjusted. The ratio of the number of drops of reflux to the number of drops of distillate allowed to pass through stopcock *S* into the receiver is called the *reflux ratio*. With an efficient column, reflux ratios as high as 100 : 1 can be used to separate effectively compounds that have very close boiling points.

2.4.8 Vacuum Distillation

Many substances cannot be distilled satisfactorily at atmospheric pressure because they are sensitive to heat and decompose before the boiling point is reached. Vacuum distillation—distillation under reduced pressure—makes it possible to distill at much lower temperatures. The boiling point of the material is affected by the pressure in the system. The lower the pressure, the lower the boiling point. For each twofold pressure reduction, the boiling point of most compounds is lowered by about 15°C. Thus, by lowering the normal 760-mm atmospheric pressure to about 3 mm, it is possible to subtract roughly 120°C from the boiling point of a mixture.

It is necessary that all glass joints, stopcocks, and hose connections be thoroughly lubricated with a vacuum-type lubricant. In addition, the system should be tested for leaks before placing the sample in the distilling flask. When the system will hold a vacuum of less than 1 mmHg, the assembly is ready to use. For many purposes, a vacuum maintained at 10 mmHg is satisfactory for laboratory fractional distillation.

Vacuum fractionation is recommended only for liquids that are affected by oxidation or are decomposed at their boiling points. One disadvantage of vacuum fractionation is that the throughput capacity of any column is considerably less than that at atmospheric pressure. Consequently, much

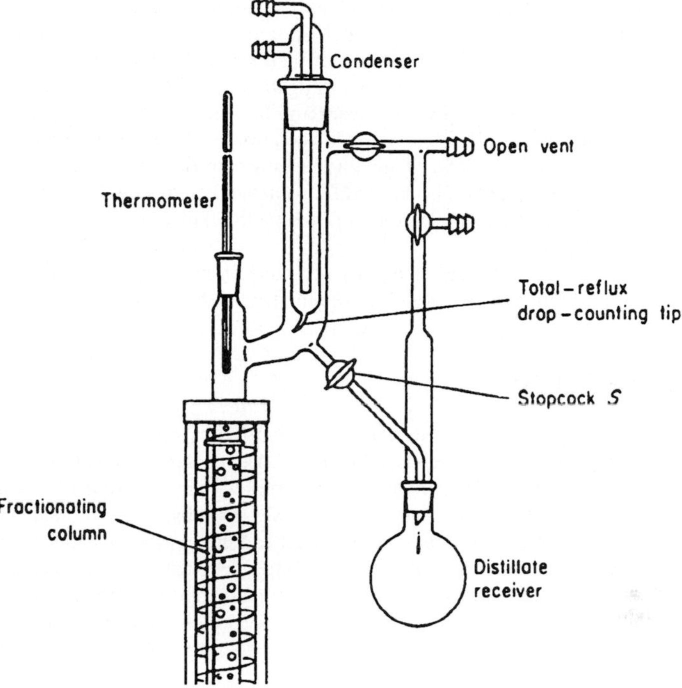

FIGURE 2.9 Total-reflux partial-takeoff still head.

more time is required to fractionate at a desired reflux ratio. Of course, more precautions are necessary to maintain a vacuum-tight system at the desired reduced pressure.

2.4.9 Steam Distillation

Steam distillation is a simple distillation where vaporization of the charge is achieved by flowing live steam directly through it. The practice has special value as a means of separating and purifying heat-sensitive organic compounds because the distillation temperature is lowered. The organic compound must be essentially immiscible with water. In these cases, both the compound and water will exert their full vapor pressure upon vaporization from the immiscible two-component liquid.

When steam is passed into a mixture of the compound and water, the compound will distill with the steam. In the distillate, this distilled compound separates from the condensed water because it is insoluble (or nearly so) in water. Both the temperature and the pressure of the distillation can be chosen at any desired value (except under a condition in which liquid water forms in the distillation flask). If the steam is superheated and remains so during its travel through the liquid, then

$$\frac{p_s}{P} = \frac{L_S}{L_T} \tag{2.63}$$

where p_s = partial pressure of steam
P = system total pressure
L_S = moles of steam issuing from liquid
L_T = total moles of vapor generated from liquid

Both L_S and L_T may be independently varied by varying the rate of steam supplied and its degree of superheat, and, in some cases, by supplying heat from an external source.

Most compounds, regardless of their normal boiling point, will distill by steam distillation at a temperature below that of pure boiling water. For example, naphthalene is a solid with a boiling point of 218°C. It will distill with steam and boiling water at a temperature less than 100°C.

Some high-boiling compounds decompose at their boiling point. Such substances can be successfully distilled at a lower temperature by steam distillation. Steam distillation can also be used to rid substances of contaminants because some water-insoluble substances are steam-volatile and others are not. Commercial applications of steam distillation include the distilling of turpentine and certain essential oils.

If a source of piped steam is not readily available, the steam can be generated in an external steam generator and then passed into the mixture to be steam distilled.

2.4.10 Molecular Distillation

Many organic substances cannot be distilled by any of the ordinary distilling methods because (1) they are extremely viscous, and any condensed vapors would plug up the distilling column, side arm, or condenser or (2) their vapors are extremely heat sensitive or susceptible to condensation reactions. For the distillation of high-molecular-weight substances (molecular weights around 1300 for hydrocarbons and around 5000 for silicones and halocarbons), molecular stills are used.

In molecular distillation the vapor path is unobstructed and the condenser is separated from the evaporator by a distance less than the mean free path (the average distance traveled by a molecule between collisions) of the evaporating molecules. Thus, molecular distillation differs from other distillations in operating procedure (mainly the pressure, or vacuum, employed).

1. Distillation is conducted at low pressures, typically 0.001 mmHg or lower. At this low pressure the boiling point of high-molecular-weight, high-boiling substances may be reduced as much as 200 to 300°C. It is the degree of thermal exposure that is so markedly less in molecular distillation.

2. The distance from the surface of the liquid being vaporized to the condenser is arranged to be less than the mean free path of the vapor at the operating pressure and temperature. For air at 25°C and 0.001 mmHg, the mean free path is 5.09 cm.

3. All condensed vapor flows to the distillate receiver with very few vapor molecules ever returning to the original material.

4. No equilibrium exists between the vapor and condensed phases. Ideal distillation conditions are attained when the rate of evaporation is equal to the rate of condensation.

A simple form of apparatus involves a cooled condensing surface that is supported a few centimeters (or even millimeters) above a thin, heated layer of sample and the whole enclosed in a highly evacuated chamber. For many organic materials the distance between condenser and evaporator surface is usually only several millimeters. The condenser is kept at liquid air or dry-ice temperatures to minimize the rebound of vapor molecules from the condenser surface.

A useful expression for the distilling rate W in grams per second per square meter of liquid surface is

$$W = \frac{0.0583PT^{1/2}}{M^{1/2}} \tag{2.64}$$

where P is the saturation vapor pressure in millimeters of mercury at the solution temperature T in kelvins, M is the molecular weight of the distilling substance, and the constant 0.0583 is the result of combining the molar gas constant and various conversion factors.

The degree of separation effected by molecular distillation is comparable to that produced by a simple batch distillation and is greatest when components differ in boiling point by 50°C or more.

Of course, when fractionation is not complete, a greater degree of separation can be attained by redistillation of the distillate.

Since no laboratory pump can achieve the reduced pressures required in one stage, two or more pumps are employed in series.

In a rotating still, materials are fed at slightly above their melting point into the still. They are distributed evenly and thinly over the heated evaporating surface. Since the fractionating power of the molecular still is restricted to the preferential evaporation of the most volatile constituents from the surface layer of liquid, it is imperative that the surface layer be continually replenished.

2.4.11 Sublimation

Some solids can go from the solid to the vapor state without passing through the liquid state. This phenomenon is called *sublimation*. The vapor can be resolidified by reducing the temperature. The process can be used for purifying solids if the impurities in the solids have a much lower vapor pressure than that of the desired compound. A sublimation point is a constant property like a melting or boiling point. Examples of materials that sublime under ordinary conditions are iodine, naphthalene, and dry ice (solid carbon dioxide).

Many solids do not develop enough vapor pressure at atmospheric pressure to sublime, but they do develop enough vapor pressure to sublime at reduced pressure. For this reason, most sublimation equipment is constructed with fittings making it adaptable for vacuum connections.

2.4.11.1 Methods of Sublimation. There are two basic approaches to the practical sublimation of substances. (1) Evacuate the system while holding the temperature below the melting point of the substance. (2) For substances that sublime at atmospheric pressure, the system is heated but the temperature is held below the melting point.

A simple laboratory setup for conducting sublimation at atmospheric pressure involves gently heating the sublimable compound in a container with a loosely fitting cover that is chilled with ice. A cold finger can be used as a condenser; the coolant can be a mixture of ice and water.

2.4.11.2 Advantages of Sublimation. The advantages of sublimation as a purification method are the following:

1. No solvent is used.
2. It is faster than crystallization. It may not be as selective if the vapor pressures of the sublimable solids are very close together.
3. More volatile impurities can be removed from the desired substance by subliming them off.
4. Substances that contain occluded molecules of solvation can sublime to form the nonsolvated product, losing the water or other solvent during the sublimation process.
5. Nonvolatile or less volatile solids can be separated from the more volatile ones.

2.5 CARRIER COPRECIPITATION AND CHEMICAL REDUCTION METHODS

2.5.1 Coprecipitation and Gathering

Coprecipitation is understood to mean the carrying down by a precipitate of substances normally soluble under the conditions employed. The carrying down of an insoluble substance by a precipitate is called *gathering*. Coprecipitation of traces with macroquantities of a precipitate is useful to isolate microcomponents by their precipitation on carriers. In Table 2.45 elements are listed the traces of which can be concentrated by means of coprecipitation on carriers. Elements with properties similar to the trace elements but not interfering in the course of further analysis are most frequently chosen as suitable carriers. The quantity of the carrier element introduced to the solution before precipitation must be large enough that the carrier element precipitates quickly in a visible quantity that is convenient for separation by filtration, or centrifugation, and washing. Experimentally, 1 to 3 mg of a carrier element is a suitable quantity for a solution of 50 to 250 mL.

For elements precipitated as hydroxides these carrier elements are often used: hydrous MnO_2, iron(III), aluminum, lanthanum, and titanium. These are good collectors for such metals as Bi, Cr(III), Ga, Ge, In, Pb, Sn, Te, and Ti in a neutral or faintly basic medium.[74] Ammonia may not be a good precipitant when ammine-forming metals are to be collected. The elements that can be isolated on carriers in the form of a hydroxide are also well isolated by organic reagents of the R—OH types, for example, 8-hydroxyquinoline and cupferron. The elements that form difficultly soluble sulfides can be precipitated with carriers not only as the sulfides, but also as complexes with organic reagents of the R—SH type, such as dithiocarbamic acid derivatives and thionalide. An advantage of organic carriers is that the carrier can be easily separated from inorganic traces by simple ignition.

For the group separation of traces, mixtures of precipitating agents can be used. 8-Hydroxyquinoline plus tannin plus thionalide precipitates traces of copper, cobalt, and molybdenum using aluminum and indium as carriers.

Quantitative precipitation of traces of metals as sulfides without the use of collectors is hardly possible. Sulfide precipitation in mineral acid solution has some degree of selectivity and may be useful for separating traces of sulfophilic elements from elements such as Al, Ca, Cr, Fe, Mg, Mn, Ti, and P. Biomaterials fall in this class, as do silicate and carbonate rocks, ferrous metals and alloys, and some nonferrous metals and alloys. Copper sulfide is a common collector for metals the sulfides of which are insoluble in dilute mineral acid solutions.[75]

Fusions may be superior to precipitations in eliminating matrix elements without serious loss of trace constituents. The insoluble material is usually obtained in a more compact form and with weaker adsorptive power.

2.5.2 Reduction to the Metal

By the application of reducing agents of different strength, certain elements may be differentially reduced to their metallic state and thus separated from other elements. Hydroquinone in an acid solution will reduce gold, mercury, platinum, and silver to the metal. Hydrazine, in strong HCl solution, reduces these same elements to the metallic state, and also palladium, selenium, and tellurium. Hypophosphorous acid, HPH_2O_2, extends the list to include antimony, bismuth, and copper. Only small amounts of the foregoing elements are advantageously determined by chemical reduction to the metallic state.

For larger quantities, electroreduction is preferable. The most widely used method is electrolysis with a mercury cathode (Sec. 14.6.3). From a 0.15*M* sulfuric acid electrolyte, without special control of the cathode potential, these metals may be quantitatively deposited into a mercury cathode: bismuth, cadmium, chromium, cobalt, copper, gallium, germanium, gold, indium, iridium, mercury, molybdenum, nickel, palladium, platinum, polonium, rhenium, rhodium, silver, technetium, tin, thallium, and zinc. In addition, quantitative reduction but incomplete deposition occurs with the following elements: arsenic, lead, osmium, selenium, and tellurium. Incomplete separation occurs for antimony, lanthanum, manganese, neodymium, and ruthenium.[76] Although controlled potential

[74] P. Strohal, K. Molnar, and I. Bacic, *Mikrochim. Acta* **1972**, 586; P. Strohal and D. Nothig-Hus, *ibid.* **1974**, 899 studies the coprecipitation of various metals with hydroxides of Be, Fe, La, Ti, and Zn at pH 3 to 11.

[75] H. Kamada, Y. Ujihira, and K. Fukada, *Radioisotopes* **14**:206 (1965); R. L. Lukas and F. Grassner, *Mikrochemie* **1930**, 203.

[76] J. A. Maxwell and R. P. Graham, *Chem. Rev.* **46**:471 (1950); J. A. Page, J. A. Maxwell, and R. P. Graham, *Analyst* **87**:245 (1962); H. O. Johnson, J. R. Weaver, and L. Lykken, *Anal. Chem.* **19**:481 (1947); T. D. Parks, H. O. Johnson, and L. Lykken, *ibid.*, **20**:148 (1948); G. E. F. Lundell and J. I. Hoffman, *Outlines of Methods of Chemical Analysis,* Wiley, New York, 1938, p. 94; W. F. Hillebrand, G. E. F. Lundell, H. A. Bright, and J. I. Hoffman, *Applied Inorganic Analysis*, Wiley, New York, 1953, p. 138.

TABLE 2.45 Preconcentration by Coprecipitation and Gathering

Element	Matrix	Collecting precipitate	Remarks	Reference
Aluminum	Silver	$La(OH)_3$		Z. Marczenko and K. Kasiura, *Chem. Anal.* **9**:87 (1964)
	Chromium	$Cr(OH)_3$		A. Kawase, *Jpn. Anal.* **11**:844 (1962)
Antimony		$MnO_2 \cdot xH_2O$	$MnO_4^- + Mn^{2+}$ in hot acid solution	D. Ogden and G. F. Reynolds, *Analyst* **89**:538, 579 (1964)
Arsenic	Seawater	Hydrous oxides of Al, Ce(IV), Fe(III), In, Ti, or Zr	Up to pH 8.5	K. Sugawara, M. Tanaka, and S. Kanamori, *Bull. Chem. Soc. Jpn.* **29**:670 (1956)
	Lead	$MnO_2 \cdot xH_2O$	0.03*M*–0.5*M* HNO_3 solution	M. Tsuiki, *J. Electrochem. Soc. Jpn.* **29**:42 (1961); C. L. Luke, *Ind. Eng. Chem. Anal. Ed.* **15**:626 (1943)
	Iron and steel	$MnO_2 \cdot xH_2O$	HNO_3 or $HClO_4$ solution	S. Maekawa and K. Kato, *Jpn. Anal.* **18**:1204 (1969)
	Copper	$MgNH_4PO_4$ or Ca phosphates		J. Meyer, Z. *Anal. Chem.* **210**:84 (1965); W. Gann, *ibid.* **221**:254 (1966)
Beryllium	Urine	$Ca_3(PO_4)_2$		C. W. Sill and C. P. Willis, *Anal. Chem.* **31**:598 (1959); T. Y. Toribara and P. S. Chen, Jr., *Anal. Chem.* **24**:539 (1952)
	Urine	Tannin and methylene blue	pH 8 in presence of EDTA	S. R. Desai and K. K. Sudhalatha, *Talanta* **14**:1436 (1959)
	Plant material	$Fe(OH)_3$	pH 11 in presence of EDTA	T. Shigematsu, M. Tabushi, and F. Isojima, *Jpn. Anal.* **11**:752 (1962)
	Urine, blood, and tissue	$AlPO_4$		P. A. Rozenberg, *Anal. Abst.* **11**:2708 (1964)
	Blood, tissue, and bone	$Fe(OH)_3$	pH 7–10	N. M. Alykov and A. I. Cherkesov, *Zh. Anal. Khim.* **31**:1104 (1976)
Bismuth	Copper and nickel	Hydrous MnO_2 · Double precipitation	Dilute (>0.1*N*) HNO_3, H_2SO_4, and $HClO_4$ solutions	Y. Yao, *Anal. Chem.* **17**:114 (1945); K. E. Burke, *ibid.* **42**:1536 (1970); B. Park, *Ind. Eng. Chem. Anal. Ed.* **6**:188 (1934)
	Copper, lead, and lead ore	Hydrous MnO_2	<0.07*M* HNO_3	S. Kallmann and F. Prestera, *Ind. Eng. Chem. Anal. Ed.* **13**:8 (1941)
	Copper	$La(OH)_3$	Ammonia precipitation at pH 9–10	W. Reichel and B. G. Bleakley, *Anal. Chem.* **46**:59 (1974)

(Continued)

TABLE 2.45 Preconcentration by Coprecipitation and Gathering (*Continued*)

Element	Matrix	Collecting precipitate	Remarks	Reference
Bismuth	Copper	Co phosphate	0.1*N* acid	V. T. Chuiko and N. I. Reva, *Zav. Lab.* **22**:1502 (1967)
Cadmium	Seawater	CuS		M. Ishibashi et al., *J. Chem. Soc. Jpn. Pure Chem. Sect.* **83**:295 (1962)
Calcium	Steel	Ce(III) fluoride		S. Maekawa, Y. Yoneyama, and E. Fujimori, *Jpn. Anal.* **11**: 981 (1962)
Cerium	Iron and steel	Ca, La, and Th collectors for cerium(III) as oxalate		R. H. Steinberg, *Appl. Spectrosc.* **7**:163 (1953)
	Plutonium(VI)	LaF_3	CeF_4 precipitating	C. J. Rodden, ed., *Analysis of Essential Nuclear Reactor Materials,* U.S. Govt. Printing Office, Washington, 1964, p. 226; G. R. Waterbury and C. F. Metz, *Talanta* **6**:237 (1960)
	Cast iron	Hydrous MnO_2	ca. 1*M* HNO_3	A. A. Amsheev, *Zav. Lab.* **34**: 789 (1968)
Cesium	Seawater	Ammonium 12-molybdo-phosphate	pH 6–7, plus Al heated to 40°C	C. Feldman and T. C. Rains, *Anal. Chem.* **36**:405 (1964)
Chromium		$Zn(OH)_2$; also hydroxides of Fe(III), Zr, Ti, Th, and Ce at pH 5.5–10.5	$Cr(OH)_3$ precipitating	D. L. Fuhrman and G. W. Latimer, Jr., *Talanta* **14**: 1199 (1967)
		Eriochrome Black T plus fuchsin	pH 5.5	G. Tölg, *Talanta* **19**:1489 (1972)
		Al or Sr phosphate; for Cr(III)	pH 4–10	I. Bacic, N. Radakovic, and P. Strohal, *Anal. Chim. Acta* **54**: 149 (1971)
Cobalt	Seawater	Hydrous MnO_2		B. R. Harvey and J. W. R. Dutton, *Anal. Chim. Acta* **67**:377 (1973)
		Al(III) and In(III)	8-Hydroxyquinoline + tannin + thionalide	W. D. Silvey and R. Brennan, *Anal. Chem.* **34**:784 (1962)
Copper		Tin(IV) hydroxide		A. Mizuike et al., *Mikrochim. Acta* **1974**:915
		Al(III) and In(III)	8-Hydroxyquinoline + tannin + thionalide	W. D. Silvey and R. Brennan, *Anal. Chem.* **34**:784 (1962)

Gallium		Iron(III) hydroxide	pH 6–8 using NaOH	R. Rafaeloff, *Radiochem. Radioanal. Lett.* **9**:373 (1969)
	Aluminum, zinc	Hydrous MnO_2	pH 1.5	V. S. Biskupsky, *Anal. Chim. Acta* **46**:149 (1969)
Germanium	Seawater	$Fe(OH)_3$		J. D. Burton, F. Culkin, and J. P. Riley, *Geochim. Cosmochim. Acta* **16**:151 (1959)
Gold	Aqueous solutions	Activated carbon or carbon black	Adsorbent	J. A. Lewis and P. A. Serin, *Analyst* **78**:385 (1953)
Iron		$Al(OH)_3$ or $La(OH)_3$	pH 6.5–8.5 with aqueous ammonia; excess ammonia keeps Cu and Ni in solution	H. Onishi and Y. Kashwagi, *Jpn. Anal.* **28**:619 (1979); J. M. Scarborough, C. D. Bingham, and P. F. DeVries, *Anal Chem.* **39**:1394 (1967)
		Al or Sr phosphate	Ph 4–10	I. Bacic, N. Radakovic, and P. Strohal, *Anal. Chim. Acta* **54**:149 (1971)
	Aluminum	$Zr(OH)_4$	pH > 10	P. N. W. Young, *Analyst* **99**: 588 (1974)
		Hydrous MnO_2	0.01–6.0*M* $HClO_4$	K. E. Burke, *Anal. Chem.* **42**: 1536 (1970)
		CdS	Slightly ammoniacal solution with ammonium sulfide	H. N. Stokes and J. R. Cain, *J. Am. Chem. Soc.* **29**:409, 443 (1907)
Lead		CuS	pH 3, citrate present	E. C. Dawson and A. Rees, *Analyst* **71**:417 (1946); C. Chow, *Analyst* **104**:154 (1979)
	Bi, In, Tl	$SrSO_4$	4*M*–8*M* HNO_3; good at high dilutions	G. Tölg, *Talanta* **19**:1489 (1972); B. C. Flann and J. C. Barlett, *J. Assoc. Off. Anal. Chem.* **51**:719 (1968)
	Nickel	Hydrous MnO_2	0.008*M*–0.1*M* HNO_3 or $HClO_4$	K. E. Burke, *Anal. Chem.* **42**: 1536 (1970)
	Cu, Ni, Zn; drinking water	Hydrous MnO_2		P. N. Vijan and R. S. Sadana, *Talanta* **27**:321 (1980)
Magnesium	Nickel	CaF_2 or $Ca_3(AsO_4)_2$		M. Cyrankowska, *Chem. Anal.* **19**:309 (1974); A. J Hegedüs and M. Bali, *Mikrochim. Acta* **1961**:721
Manganese	Aluminum	$Zr(OH)_4$	pH > 10	P. N. W. Young *Analyst* **99**:588 (1974)

(*Continued*)

TABLE 2.45 Preconcentration by Coprecipitation and Gathering (*Continued*)

Element	Matrix	Collecting precipitate	Remarks	Reference
Manganese		Al or Sr phosphate	pH 4–10	I. Bacic, N. Radakovic, and P. Strohal, *Anal. Chim. Acta* **54**:149 (1971)
Mercury		Te precipitated with $SnCl_2$ in hot solution		H. Hamaguchi, R. Kuroda, and K. Hosohara, *Jpn. Anal.* **9**: 1035 (1960)
		Cu powder in microcolumn		S. Dogan and W. Haerdi, *Anal. Chim. Acta* **76**:345 (1975); **84**:89 (1976)
	Aqueous solutions	Activated carbon	For Hg(II) and methylmercury	H. Koshima and H. Onishi, *Talanta* **27**:795 (1980)
Molybdenum	Copper	Hydrous MnO_2	0.016*M* HNO_3	B. Park, *Ind. Eng. Chem. Anal. Ed.* **6**:189 (1934)
	Seawater	Hydrous MnO_2	pH 3.8	K. M. Chan and J. P. Riley, *Anal. Chim. Acta* **36**:220 (1966); M. Tanaka, *Mikrochim. Acta* **1958**:204
		$Fe(OH)_3$	pH 4	L. F. Rader and F. S. Grimaldi, *U.S. Geol. Surv. Prof. Paper No. 391-A* (1961); Y. S. Kim and H. Zeitlin, *Anal. Chim. Acta* **46**:1 (1969)
		Al(III) and In(III)	8-Hydroxyquinoline + tannin + thionalide	W. D. Silvey and R. Brennan, *Anal. Chem.* **34**:784 (1962)
Niobium	Dilute HCl solution	W as collector; precipitation with sodium sulfite and tannin		T. Sakaki, *J. Jpn. Inst. Metals* **32**: 913 (1968)
	Ores containing Ti, W, Mo, and Cr	Hydrous MnO_2	Dilute H_2SO_4 solution	V. M. Dorosh, *Zh. Anal. Khim.* **16**:250 (1961)
	Be and BeO	Zr collector	Precipitation with cupferron	J. O. Hibbits, et al. *Talanta* **8**: 209 (1961)
	Mo and W	$Zr(OH)_4$		D. F. Wood and J. T. Jones, *Analyst* **93**:131 (1968)
	Mo, Re, V, and W	Fe_2O_3 plus MgO	To retain Nb in residue from NaOH fusion on leaching the melt with water	F. S. Grimaldi, *Anal. Chem.* **32**:119 (1960)
Palladium	Base metals	Te by reduction of a tellurite with $SnCl_2$		E. R. R. Marhenke and E. B. Sandell, *Anal. Chim. Acta* **28**:259 (1963)

		$Fe(OH)_3$	Precipitated as PdO_2	J. H. Yoe and L. G. Overholser, *J. Am. Chem. Soc.* **61**:2058 (1939)
		PbS	Hot 1% HOAc* plus H_2S	G. Lunde, *Mikrochem.* **5**:119 (1927)
Palladium and Platinum		Se(IV) or Te(IV)	HCl solution with $SnCl_2$	H. Bode, *Z. Anal. Chem.* **153**: 335 (1956); K. Beyermann, *ibid.* **200**:183 (1964); I. Palmer and G. Streichert, *Natl. Inst. Metall. Rep. No. 1273* (1971)
Rhenium		As(III)	$0.5M$ H_2SO_4, saturated with H_2S, stand 12 h	J. I. Hoffman and G. E. F. Lundell, *J. Res. Natl. Bur. Std.* **23**:497 (1939)
Scandium	Minerals and seawater	Calcium oxalate	pH 1.8; stand 12 h	T. Shigematsu et al., *J. Chem. Soc. Jpn. Pure Chem. Sect.* **84**:336 (1963)
		Calcium plus methyl oxalate	pH 3.8–5	J. G. Sen Gupta, *Anal. Chim. Acta* **138**: 295 (1982)
Silicon	Aluminum	$Zr(OH)_4$	pH > 10	P. N. W. Young, *Analyst* **99**:588 (1974)
Silver		Te from Te(IV)	$SnCl_2$ added to $1M$ HCl solution	E. B. Sandell and J. J. Neumayer, *Anal. Chem.* **23**:1863 (1951)
Strontium	Seawater	Calcium oxalate	pH 4; heat 1 h	K. Uesugi et al., *Jpn. Anal.* **13**:440 (1964)
Tantalum	B, U_3O_8, Zr, and U–Zircaloy-2 alloy	$Fe(OH)_3$		A. R. Eberle and M. W. Lerner, *Anal. Chem.* **39**:662 (1967)
Thallium		Hydrous MnO_2, $Fe(OH)_3$, $Zr(OH)_4$, and $Mg(OH)_2$	Aqueous NH_3 plus H_2O_2	W. Geilmann and K.-H. Neeb, *Z. Anal. Chem.* **165**:251 (1959); G. Tölg, *Talanta* **19**:1489 (1972)
	Urine and viscera	CdS, Ag_2S, PbS	As Tl_2S in neutral or ammoniacal medium	J. Duvivier et al., *Clin. Chim. Acta* **9**:454 (1964)
Thorium	Phosphate	$Fe(OH)_3$ or $Al(OH)_3$	pH 3.3	T. Y. Toribara and L. Koval, *Talanta* **14**:403 (1967); S. A. Reynolds, *ibid.* **10**:611 (1963)
	Most metals except lanthanoids; in water	Calcium oxalate or lanthanoids	Oxalic acid in <$0.5M$ HCl	A. E. Taylor and R. T. Dillon, *Anal. Chem.* **24**:1624 (1952); M. Cospito and L. Rigali, *Anal. Chim. Acta* **106**:385 (1979)
	Urine, U ore, Pu, silicate rocks	Lanthanum or calcium	As ThF_4 in HNO_3 solutions	R. W. Perkins and D. R. Kalkwarf, *Anal. Chem.* **28**: 1989 (1956); R. Ko and M. R. Weiler, *ibid.* **34**:85 (1962); R. G. Bryan and G. R. Waterbury, *U.S. AEC Rep. No. LA-3468* (1966)

(*Continued*)

TABLE 2.45 Preconcentration by Coprecipitation and Gathering (*Continued*)

Element	Matrix	Collecting precipitate	Remarks	Reference
Tin	Al, Mg, and Zn alloys; Cu and Sb	CuS	Thioacetamide in 1:10 H_2SO_4	S. Ambujavalli and N. Premavathi, *Anal. Chem.* **48**:2152 (1976)
	Al and Cu alloys	Hydrous MnO_2		R. Tanaka, *Jpn. Anal.* **10**:336 (1961)
	Steel	$Be(OH)_2$	EDTA present	K. Hiiro et al., *Jpn. Anal.* **18**:563 (1969)
	Iron and unalloyed steel	Hydrous MnO_2 or MoS_3		E. Kovacs and H. Guyer, *Z. Anal. Chem.* **208**:255 (1965)
	Nickel	Hydrous MnO_2	<1.2*M* HNO_3 solution	K. E. Burke, *Anal. Chem.* **42**: 1536 (1970)
		Al^{3+} using 8-hydroxyquinoline	Tannic acid and Na sulfide present	J. C. Burridge and I. J. Hewitt, *Analyst* **110**:795 (1985)
	Water	1,10-phenanthroline and tetraphenylborate		S. Dogan and W. Haerdi, *Int. J. Environ. Anal. Chem.* **8**:249 (1980)
	Lead and lead alloys	Hydrous MnO_2	0.5*M*–1.5*M* HCl	C. L. Luke, *Anal. Chem.* **28**: 1276 (1956); J. Aznarez et al., *Talanta* **33**:458 (1986)
Titanium	V, Mo, and W	Little $Fe(OH)_3$	Hot 1*M* NaOH	W. F. Hillebrand, G. E. F. Lundell, H. A. Bright, and J. I. Hoffman, *Applied Inorganic Analysis,* 2d ed., Wiley, New York, 1953, p. 579
	Aluminum	$Zr(OH)_4$	pH > 10	P. N. W. Young, *Analyst* **99**: 588 (1974)
	Alloy steels; sulfides of Cu, Pb, and Zn	$Mg(OH)_2$	EDTA present	W. F. Pickering, *Anal. Chim. Acta* **9**:324 (1953); *ibid.* **12**: 572 (1955); B. J. Shelton et al., *Natl. Inst. Metall. Rep. No. 1857* (1977)
	Ni, Co, Cr, U(VI), and P	Fe(III) or Zr(IV)	Cupferron	J. O. Hibbitts et al., *Talanta* **11**:1464, 1509 (1964)
Tungsten	Seawater and silicate rocks	Hydrous MnO_2		K. M. Chan and J. P. Riley, *Anal. Chim. Acta* **39**:103 (1967)
Uranium		Fe(III) or Al phosphate	Acetic acid–acetate buffer	K. H. Reinhardt and H. J. Miller, *Z. Anal. Chem.* **292**:359 (1978); A. P. Smith and F. S. Grimaldi, *U.S. Geol. Surv. Bull.* **1006**:125

		$BaSO_4$	As U(IV)	C. W. Sill and R. L. Williams, *Anal. Chem.* **41**:1624 (1969); C. C. Bertrand and T. A. Linn, Jr., *ibid.* **44**:383 (1972); T. Kimura and Y. Kobayashi, *J. Radioanal. Nucl. Chem. Articles* **91**:59 (1985)
Vanadium	River water and seawater	Fe(III)	Ammonia precipitation of V(V) at pH 6–8	C. P. Weisel, R. A. Duce, and J. L. Fashing, *Anal. Chem.* **56**:1050 (1984)
Zinc		Cu(II) and H_2S	Citrate medium at pH 2.5	H. A. Bright, *J. Res. Natl. Bur. Std.* **12**:383 (1934)
		Al or Sr phosphate	pH 4–10	I. Bacic, N. Radakovic, and P. Strohal, *Anal. Chim. Acta* **54**:149 (1971)
	Seawater	$Fe(OH)_3$		R. Chakravorty and R. Van Grieken, *Int. J. Environ. Anal. Chem.* **11**:67 (1982)
	Aluminum	$Zr(OH)_4$	pH > 10	P. N. W. Young, *Analyst* **99**:588 (1974)
Zirconium	Silicate rocks	Ti hydroxide		H. Degenhardt, *Z. Anal. Chem.* **153**:327 (1956)

* HOAc denotes acetic acid.

electrolysis is sometimes required for a separation, most mercury cathode electrolyses are carried out by the constant-current technique, often with the simultaneous evolution of hydrogen gas. Even in these cases, the cathode potential is kept nearly constant by the addition of oxidation–reduction buffers, such as large amounts of uranium(IV-III), titanium(IV-III), vanadium(III-II), or hydrogen ion(I-0).

Because removal by electrolysis involves a logarithmic relation between the concentration of the residual metal remaining in solution and the original concentration, using one or two portions of fresh mercury often lessens the amount remaining undeposited. Even so, microgram quantities can remain and their effect on subsequent determination methods must be considered. Also trace amounts of mercury will be introduced into solution because of its atomic solubility.

If desired, certain metals can be removed from the mercury by anodic stripping (see Sec. 14.5.5). In a suitable salt solution, the mercury is made the anode and maintained at a potential a little lower (by a few tenths of a volt) than its solution potential.

Without electrolysis, noble metals such as Ag, Au, and Pt in trace concentrations can be collected in mercury metal.[77]

[77] A. Mizuike, *Talanta* **9**:948 (1962).

SECTION 3
STATISTICS IN CHEMICAL ANALYSIS

3.1 INTRODUCTION

Each observation in any branch of scientific investigation is inaccurate to some degree. Often the accurate value for the concentration of some particular constituent in the analyte cannot be determined. However, it is reasonable to assume the accurate value exists, and it is important to estimate the limits between which this value lies. It must be understood that the statistical approach is concerned with the appraisal of experimental design and data. Statistical techniques can neither detect nor evaluate constant errors (bias); the detection and elimination of inaccuracy are analytical problems. Nevertheless, statistical techniques can assist considerably in determining whether or not inaccuracies exist and in indicating when procedural modifications have reduced them.

By proper design of experiments, guided by a statistical approach, the effects of experimental variables may be found more efficiently than by the traditional approach of holding all variables constant but one and systematically investigating each variable in turn. Trends in data may be sought to track down nonrandom sources of error.

3.1.1 Errors in Quantitative Analysis

Two broad classes of errors may be recognized. The first class, *determinate* or *systematic* errors, is composed of errors that can be assigned to definite causes, even though the cause may not have been located. Such errors are characterized by being unidirectional. The magnitude may be constant from sample to sample, proportional to sample size, or variable in a more complex way. An example is the error caused by weighing a hygroscopic sample. This error is always positive in sign; it increases with sample size but varies depending on the time required for weighing, with humidity and temperature. An example of a negative systematic error is that caused by solubility losses of a precipitate.

The second class, *indeterminate* or *random* errors, is brought about by the effects of uncontrolled variables. Truly random errors are likely to cause high as well as low results, and a small random error is much more probable than a large one. By making the observation coarse enough, random errors would cease to exist. Every observation would give the same result, but the result would be less precise than the average of a number of finer observations with random scatter.

The *precision* of a result is its reproducibility; the *accuracy* is its nearness to the truth. A systematic error causes a loss of accuracy, and it may or may not impair the precision depending upon whether the error is constant or variable. Random errors cause a lowering of reproducibility, but by making sufficient observations it is possible to overcome the scatter within limits so that the accuracy may not necessarily be affected. Statistical treatment can properly be applied only to random errors.

3.1.2 Representation of Sets of Data

Raw data are collected observations that have not been organized numerically. An *average* is a value that is typical or representative of a set of data. Several averages can be defined, the most common being the arithmetic mean (or briefly the mean), the median, the mode, and the geometric mean.

The *mean* of a set of N numbers, $x_1, x_2, x_3, \ldots, x_N$, is denoted by $\bar{x}$ and is defined as

$$\bar{x} = \frac{x_1 + x_2 + x_3 + \cdots + x_N}{N} \tag{3.1}$$

It is an estimation of the unknown true value μ of an infinite population. We can also define the *sample variance* s^2 as follows:

$$s^2 = \frac{\sum_{i=1}^{N} (x_i - \bar{x})^2}{N-1} \tag{3.2}$$

The values of $\bar{x}$ and s^2 vary from sample set to sample set. However, as N increases, they may be expected to become more and more stable. Their limiting values, for very large N, are numbers characteristic of the frequency distribution and are referred to as the *population mean* and the *population variance*, respectively.

The *median* of a set of numbers arranged in order of magnitude is the middle value or the arithmetic mean of the two middle values. The median allows inclusion of all data in a set without undue influence from outlying values; it is preferable to the mean for small sets of data.

The *mode* of a set of numbers is the value that occurs with the greatest frequency (the most common value). The mode may not exist, and even if it does exist it may not be unique. The empirical relation that exists between the mean, the mode, and the median for unimodal frequency curves that are moderately asymmetrical is

$$\text{Mean} - \text{mode} = 3(\text{mean} - \text{median}) \tag{3.3}$$

The *geometric mean* of a set of N numbers is the Nth root of the product of the numbers:

$$\sqrt[N]{x_1 x_2 x_3 \cdots x_N} \tag{3.4}$$

The *root mean square* (RMS) or quadratic mean of a set of numbers is defined by

$$\text{RMS} = \sqrt{\overline{x}^2} = \sqrt{\sum_{i=1}^{N} x_i^2/N} \tag{3.5}$$

3.2 THE NORMAL DISTRIBUTION OF MEASUREMENTS

The normal distribution of measurements (or the normal law of error) is the fundamental starting point for analysis of data. When a large number of measurements are made, the individual measurements are not all identical and equal to the accepted value μ, which is the mean of an infinite population or universe of data, but are scattered about μ, owing to random error. If the magnitude of any single measurement is the abscissa and the relative frequencies (i.e., the probability) of occurrence of different-sized measurements are the ordinate, the smooth curve drawn through the points (Fig. 3.1) is the *normal* or *gaussian distribution curve* (also the *error curve* or *probability curve*). The term *error curve* arises when one considers the distribution of errors $(x - \mu)$ about the true value.

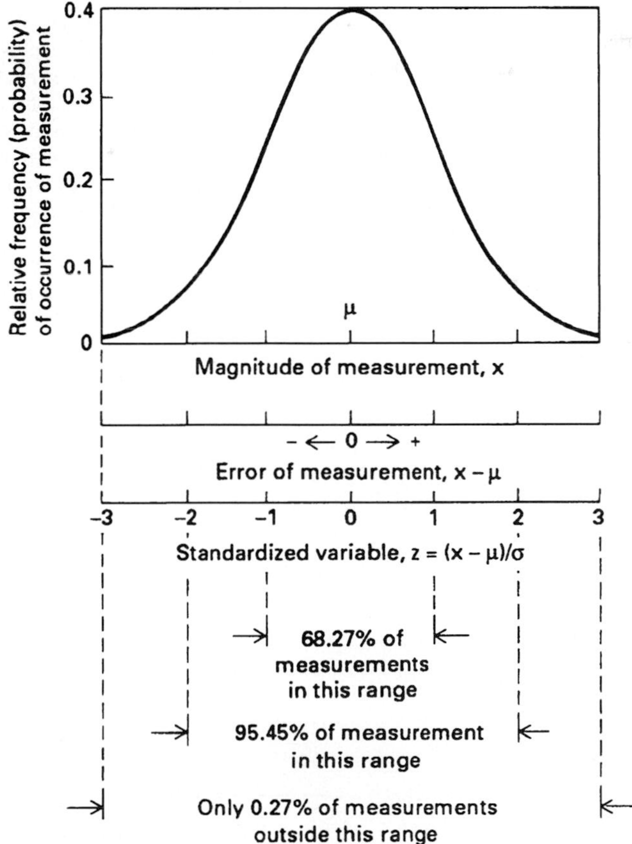

FIGURE 3.1 The normal distribution curve.

The breadth or spread of the curve indicates the precision of the measurements and is determined by and related to the standard deviation, a relationship that is expressed in the equation for the normal curve (which is continuous and infinite in extent),

$$Y = \frac{1}{\sigma\sqrt{2\pi}} \exp\left[-\frac{1}{2}\left(\frac{x-\mu}{\sigma}\right)^2\right] \tag{3.6}$$

where σ is the standard deviation of the infinite population. The population mean μ expresses the magnitude of the quantity being measured. In a sense, σ measures the width of the distribution, and thereby also expresses the scatter or dispersion of replicate analytical results. When $(x - \mu)/\sigma$ is replaced by the standardized variable z, then

$$Y = \frac{1}{\sqrt{2\pi}} e^{-(1/2)z^2} \tag{3.7}$$

The standardized variable (the z statistic) requires only the probability level to be specified. It measures the deviation from the population mean in units of standard deviation. Y is 0.399 for the most probable value, μ. In the absence of any other information, the normal distribution is assumed to apply whenever repetitive measurements are made on a sample, or a similar measurement is made on different samples.

Table 3.1*a* lists the height of an ordinate (Y) as a distance z from the mean, and Table 3.1*b* the area under the normal curve at a distance z from the mean, expressed as fractions of the total area, 1.000. Returning to Fig. 3.1, we note that 68.27% of the area of the normal distribution curve lies within 1 standard deviation of the center or mean value. Therefore 31.73% lies outside those limits and 15.86% on each side. Ninety-five percent (actually 95.43%) of the area lies within 2 standard deviations, and 99.73% lies within 3 standard deviations of the mean. Often the last two areas are stated slightly different; viz, 95% of the area lies within 1.96σ (approximately 2σ) and 99% lies within approximately 2.5σ. The mean falls at exactly the 50% point for symmetric normal distributions.

Example 3.1 The true value of a quantity is 30.00, and σ for the method of measurement is 0.30. What is the probability that a single measurement will have a deviation from the mean greater than 0.45; that is, what percentage of results will fall outside the range 30.00 ± 0.45?

$$z = \frac{x-\mu}{\sigma} = \frac{0.45}{0.30} = 1.5$$

From Table 3.1 the area under the normal curve from -1.5σ to $+1.5\sigma$ is 0.866, meaning that 86.6% of the measurements will fall within the range 30.00 ± 0.45 and 13.4% will lie outside this range. Half of these measurements, 6.8%, will be less than 29.55, and a similar percentage will exceed 30.45. In actuality the uncertainty in z is about 1 in 15; therefore, the value of z could lie between 1.4 and 1.6; the corresponding areas under the curve could lie between 84% and 89%.

Example 3.2 If the mean value of 500 determinations is 151 and $\sigma = 15$, how many results lie between 120 and 155 (actually any value between 119.5 and 155.5)?

$$z = \frac{119.5-151}{15} = -2.10 \qquad \text{Area: } 0.482$$

$$z = \frac{155.5-151}{15} = 0.30 \qquad 0.118$$

$$\text{Total area: } \overline{0.600}$$

$$500(0.600) = 300 \text{ results}$$

TABLE 3.1*a* **Ordinates (Y) of the Normal Distribution Curve at Values of z**

z	0	1	2	3	4	5	6	7	8	9
0.0	0.3989	0.3989	0.3989	0.3988	0.3986	0.3984	0.3982	0.3980	0.3977	0.3973
0.1	0.3970	0.3965	0.3961	0.3956	0.3951	0.3945	0.3939	0.3932	0.3925	0.3918
0.2	0.3910	0.3902	0.3894	0.3885	0.3876	0.3867	0.3857	0.3847	0.3836	0.3825
0.3	0.3814	0.3802	0.3790	0.3778	0.3765	0.3752	0.3739	0.3725	0.3712	0.3697
0.4	0.3683	0.3668	0.3653	0.3637	0.3621	0.3605	0.3589	0.3572	0.3555	0.3538
0.5	0.3521	0.3503	0.3485	0.3467	0.3448	0.3429	0.3410	0.3391	0.3372	0.3352
0.6	0.3332	0.3312	0.3292	0.3271	0.3251	0.3230	0.3209	0.3187	0.3166	0.3144
0.7	0.3123	0.3101	0.3079	0.3056	0.3034	0.3011	0.2989	0.2966	0.2943	0.2920
0.8	0.2897	0.2874	0.2850	0.2827	0.2803	0.2780	0.2756	0.2732	0.2709	0.2685
0.9	0.2661	0.2637	0.2613	0.2589	0.2565	0.2541	0.2516	0.2492	0.2468	0.2444
1.0	0.2420	0.2396	0.2371	0.2347	0.2323	0.2299	0.2275	0.2251	0.2227	0.2203
1.1	0.2179	0.2155	0.2131	0.2107	0.2083	0.2059	0.2036	0.2012	0.1989	0.1965
1.2	0.1942	0.1919	0.1895	0.1872	0.1849	0.1826	0.1804	0.1781	0.1758	0.1736
1.3	0.1714	0.1691	0.1669	0.1647	0.1626	0.1604	0.1582	0.1561	0.1539	0.1518
1.4	0.1497	0.1476	0.1456	0.1435	0.1415	0.1394	0.1374	0.1354	0.1334	0.1315
1.5	0.1295	0.1276	0.1257	0.1238	0.1219	0.1200	0.1182	0.1163	0.1145	0.1127
1.6	0.1109	0.1092	0.1074	0.1057	0.1040	0.1023	0.1006	0.0989	0.0973	0.0957
1.7	0.0940	0.0925	0.0909	0.0893	0.0878	0.0863	0.0848	0.0833	0.0818	0.0804
1.8	0.0790	0.0775	0.0761	0.0748	0.0734	0.0721	0.0707	0.0694	0.0681	0.0669
1.9	0.0656	0.0644	0.0632	0.0620	0.0608	0.0596	0.0584	0.0573	0.0562	0.0551
2.0	0.0540	0.0529	0.0519	0.0508	0.0498	0.0488	0.0478	0.0468	0.0459	0.0449
2.1	0.0440	0.0431	0.0422	0.0413	0.0404	0.0396	0.0387	0.0379	0.0371	0.0363
2.2	0.0355	0.0347	0.0339	0.0332	0.0325	0.0317	0.0310	0.0303	0.0297	0.0290
2.3	0.0283	0.0277	0.0270	0.0264	0.0258	0.0252	0.0246	0.0241	0.0235	0.0229
2.4	0.0224	0.0219	0.0213	0.0208	0.0203	0.0198	0.0194	0.0189	0.0184	0.0180
2.5	0.0175	0.0171	0.0167	0.0163	0.0158	0.0154	0.0151	0.0147	0.0143	0.0139
2.6	0.0136	0.0132	0.0129	0.0126	0.0122	0.0119	0.0116	0.0113	0.0110	0.0107
2.7	0.0104	0.0101	0.0099	0.0096	0.0093	0.0091	0.0088	0.0086	0.0084	0.0081
2.8	0.0079	0.0077	0.0075	0.0073	0.0071	0.0069	0.0067	0.0065	0.0063	0.0061
2.9	0.0060	0.0058	0.0056	0.0055	0.0053	0.0051	0.0050	0.0048	0.0047	0.0046
3.0	0.0044	0.0043	0.0042	0.0040	0.0039	0.0038	0.0037	0.0036	0.0035	0.0034
3.1	0.0033	0.0032	0.0031	0.0030	0.0029	0.0028	0.0027	0.0026	0.0025	0.0025
3.2	0.0024	0.0023	0.0022	0.0022	0.0021	0.0020	0.0020	0.0019	0.0018	0.0018
3.3	0.0017	0.0017	0.0016	0.0016	0.0015	0.0015	0.0014	0.0014	0.0013	0.0013
3.4	0.0012	0.0012	0.0012	0.0011	0.0011	0.0010	0.0010	0.0010	0.0009	0.0009
3.5	0.0009	0.0008	0.0008	0.0008	0.0008	0.0007	0.0007	0.0007	0.0007	0.0006
3.6	0.0006	0.0006	0.0006	0.0005	0.0005	0.0005	0.0005	0.0005	0.0005	0.0004
3.7	0.0004	0.0004	0.0004	0.0004	0.0004	0.0004	0.0003	0.0003	0.0003	0.0003
3.8	0.0003	0.0003	0.0003	0.0003	0.0003	0.0002	0.0002	0.0002	0.0002	0.0002
3.9	0.0002	0.0002	0.0002	0.0002	0.0002	0.0002	0.0002	0.0002	0.0001	0.0001

3.3 STANDARD DEVIATION AS A MEASURE OF DISPERSION

Several ways may be used to characterize the spread or dispersion in the original data. The *range* is the difference between the largest value and the smallest value in a set of observations. However, almost always the most efficient quantity for characterizing variability is the *standard deviation* (also called the *root mean square*).

TABLE 3.1*b* **Areas under the Normal Distribution Curve from 0 to *z***

z	0	1	2	3	4	5	6	7	8	9
0.0	0.0000	0.0040	0.0080	0.0120	0.0160	0.0199	0.0239	0.0279	0.0319	0.0359
0.1	0.0398	0.0438	0.0478	0.0517	0.0557	0.0596	0.0636	0.0675	0.0714	0.0754
0.2	0.0793	0.0832	0.0871	0.0910	0.0948	0.0987	0.1026	0.1064	0.1103	0.1141
0.3	0.1179	0.1217	0.1255	0.1293	0.1331	0.1368	0.1406	0.1443	0.1480	0.1517
0.4	0.1554	0.1591	0.1628	0.1664	0.1700	0.1736	0.1772	0.1808	0.1844	0.1879
0.5	0.1915	0.1950	0.1985	0.2019	0.2054	0.2088	0.2123	0.2157	0.2190	0.2224
0.6	0.2258	0.2291	0.2324	0.2357	0.2389	0.2422	0.2454	0.2486	0.2518	0.2549
0.7	0.2580	0.2612	0.2642	0.2673	0.2704	0.2734	0.2764	0.2794	0.2823	0.2852
0.8	0.2881	0.2910	0.2939	0.2967	0.2996	0.3023	0.3051	0.3078	0.3106	0.3133
0.9	0.3159	0.3186	0.3212	0.3238	0.3264	0.3289	0.3315	0.3340	0.3365	0.3389
1.0	0.3413	0.3438	0.3461	0.3485	0.3508	0.3531	0.3554	0.3577	0.3599	0.3621
1.1	0.3643	0.3665	0.3686	0.3708	0.3729	0.3749	0.3770	0.3790	0.3810	0.3830
1.2	0.3849	0.3869	0.3888	0.3907	0.3925	0.3944	0.3962	0.3980	0.3997	0.4015
1.3	0.4032	0.4049	0.4066	0.4082	0.4099	0.4115	0.4131	0.4147	0.4162	0.4177
1.4	0.4192	0.4207	0.4222	0.4236	0.4251	0.4265	0.4279	0.4292	0.4306	0.4319
1.5	0.4332	0.4345	0.4357	0.4370	0.4382	0.4394	0.4406	0.4418	0.4429	0.4441
1.6	0.4452	0.4463	0.4474	0.4484	0.4495	0.4505	0.4515	0.4525	0.4535	0.4545
1.7	0.4554	0.4564	0.4573	0.4582	0.4591	0.4599	0.4608	0.4616	0.4625	0.4633
1.8	0.4641	0.4649	0.4656	0.4664	0.4671	0.4678	0.4686	0.4693	0.4699	0.4706
1.9	0.4713	0.4719	0.4726	0.4732	0.4738	0.4744	0.4750	0.4756	0.4761	0.4767
2.0	0.4772	0.4778	0.4783	0.4788	0.4793	0.4798	0.4803	0.4808	0.4812	0.4817
2.1	0.4821	0.4826	0.4830	0.4834	0.4838	0.4842	0.4846	0.4850	0.4854	0.4857
2.2	0.4861	0.4864	0.4868	0.4871	0.4875	0.4878	0.4881	0.4884	0.4887	0.4890
2.3	0.4893	0.4896	0.4898	0.4901	0.4904	0.4906	0.4909	0.4911	0.4913	0.4916
2.4	0.4918	0.4920	0.4922	0.4925	0.4927	0.4929	0.4931	0.4932	0.4934	0.4936
2.5	0.4938	0.4940	0.4941	0.4943	0.4945	0.4946	0.4948	0.4949	0.4951	0.4952
2.6	0.4953	0.4955	0.4956	0.4957	0.4959	0.4960	0.4961	0.4962	0.4963	0.4964
2.7	0.4965	0.4966	0.4967	0.4968	0.4969	0.4970	0.4971	0.4972	0.4973	0.4974
2.8	0.4974	0.4975	0.4976	0.4977	0.4977	0.4978	0.4979	0.4979	0.4980	0.4981
2.9	0.4981	0.4982	0.4982	0.4983	0.4984	0.4984	0.4985	0.4985	0.4986	0.4986
3.0	0.4987	0.4987	0.4987	0.4988	0.4988	0.4989	0.4989	0.4989	0.4990	0.4990
3.1	0.4990	0.4991	0.4991	0.4991	0.4992	0.4992	0.4992	0.4992	0.4993	0.4993
3.2	0.4993	0.4993	0.4994	0.4994	0.4994	0.4994	0.4994	0.4995	0.4995	0.4995
3.3	0.4995	0.4995	0.4995	0.4996	0.4996	0.4996	0.4996	0.4996	0.4996	0.4997
3.4	0.4997	0.4997	0.4997	0.4997	0.4997	0.4997	0.4997	0.4997	0.4997	0.4998
3.5	0.4998	0.4998	0.4998	0.4998	0.4998	0.4998	0.4998	0.4998	0.4998	0.4998
3.6	0.4998	0.4998	0.4999	0.4999	0.4999	0.4999	0.4999	0.4999	0.4999	0.4999
3.7	0.4999	0.4999	0.4999	0.4999	0.4999	0.4999	0.4999	0.4999	0.4999	0.4999
3.8	0.4999	0.4999	0.4999	0.4999	0.4999	0.4999	0.4999	0.4999	0.4999	0.4999
3.9	0.5000	0.5000	0.5000	0.5000	0.5000	0.5000	0.5000	0.5000	0.5000	0.5000

The standard deviation is the square root of the average squared differences between the individual observations and the population mean:

$$\sigma = \sqrt{\frac{\sum_{i=1}^{N}(x_i - \mu)^2}{N}} \tag{3.8}$$

The standard deviation σ may be estimated by calculating the standard deviation s drawn from a small sample set as follows:

$$s = \sqrt{\frac{\sum_{i=1}^{N} (x_i - \bar{x})^2}{N}} \quad \text{or} \quad s = \sqrt{\frac{x_1^2 + x_2^2 + \cdots - [(x_1 + x_2 + \cdots)^2]/N}{N-1}} \tag{3.9}$$

where $x_i - \bar{x}$ represents the deviation of each number in the array from the arithmetic mean. Since two pieces of information, namely s and $\bar{x}$, have been extracted from the data, we are left with $N - 1$ *degrees of freedom* (df); that is, independent data points available for measurement of precision. If a relatively large sample of data corresponding to $N > 30$ is available, its mean can be taken as a measure of μ, and s as equal to σ.

So basic is the notion of a statistical *estimate* of a physical parameter that statisticians use Greek letters for the *parameters* and Latin letters for the estimates. For many purposes, one uses the *variance*, which for the sample is s^2 and for the entire population is σ^2. The variance s^2 of a finite sample is an unbiased estimate of σ^2, whereas the standard deviation s is not an unbiased estimate of σ.

Because the standard deviation σ for the universe is a characteristic of the measuring procedure, it is possible to get a good estimate not only from a long series of repeated analyses of the same sample, but also by taking together several short series measured with slightly different samples of the same type. When a series of observations can be logically arranged into k subgroups, the variance is calculated by summing the squares of the deviations for each subgroup, and then adding all the k sums and dividing by $N - k$ because one degree of freedom is lost in each subgroup. It is not required that the number of repeated analyses in the different groups be the same. For two groups of observations consisting of N_A and N_B members of standard deviations s_A and s_B, respectively, the variance is given by

$$s^2 = \frac{(N_A - 1)\, s_A^2 + (N_B - 1)\, s_B^2}{N_A + N_B - 2} \tag{3.10}$$

Another measure of dispersion is the *coefficient of variation*, which is merely the standard deviation expressed as a fraction of the arithmetic mean, viz., $s/\bar{x}$. It is useful mainly to show whether the relative or the absolute spread of values is constant as the values are changed.

3.4 THEORETICAL DISTRIBUTIONS AND TESTS OF SIGNIFICANCE

If the data contained only random (or chance) errors, the cumulative estimates $\bar{x}$ and s would gradually approach the limits μ and σ. The distribution of results would be normally distributed with mean μ and standard deviation σ. Were the true mean of the infinite population known, it would also have some symmetrical type of distribution centered around μ. However, it would be expected that the dispersion or spread of this dispersion about the mean would depend on the sample size.

3.4.1 Student's Distribution or *t* Test

The standard deviation of the distribution of means equals $\sigma/N^{1/2}$. Since σ is not usually known, its approximation for a finite number of measurements is overcome by the Student t test. It is a measure of error between μ and $\bar{x}$. The Student t takes into account both the possible variation of the value of $\bar{x}$ from μ on the basis of the expected variance $\sigma^2/N^{1/2}$ and the reliability of using s in place of σ. The distribution of the statistic is

$$\pm t = \frac{\bar{x} - \mu}{s / \sqrt{N}} \qquad \text{or} \qquad \mu = \bar{x} \pm \frac{ts}{\sqrt{N}} \tag{3.11}$$

The distribution of the t statistic $(\bar{x} - \mu)s$ is symmetrical about zero and is a function of the degrees of freedom. Limits assigned to the distance on either side of μ are called *confidence limits*. The percentage probability that μ lies within this interval is called the *confidence level*. The *level of significance* or *error probability* (100 – confidence level or 100 – α) is the percent probability that μ will lie outside the confidence interval and represents the changes of being incorrect in stating that μ lies within the confidence interval. Values of t are given in Table 3.2 for any desired degrees of freedom and various confidence levels.

An analytical procedure is often tested on materials of known composition. These materials may be pure substances, standard samples, or materials analyzed by some other more accurate method. Repeated determinations on a known material furnish data for both an estimate of the precision and a test for the presence of a constant error in the results. The standard deviation is found from Eq. (3.9) (with the known composition replacing μ). A calculated value for t [Eq. (3.11)] in excess of the appropriate value in Table 3.2 is interpreted as evidence of the presence of a constant error at the indicated level of significance.

Example 3.3 A new method for the analysis of iron using pure FeO was replicated with five samples giving these results (in % Fe): 76.95, 77.02, 76.90, 77.20, and 77.50. Does a systematic error exist?

From Eq. (3.1), $\bar{x}$ is 77.11; and from Eq. (3.2), s is 0.24 for 4 degrees of freedom. Because σ is not known, the Student $t_{0.975}$ (2.78 for 4 degrees of freedom) is used to calculate the confidence interval at the 95% probability level.

$$\mu = \bar{x} \pm \frac{ts}{\sqrt{N}} = 77.11 \pm \frac{(2.78)\,(0.24)}{\sqrt{5}} = 77.11 \pm 0.30$$

We used a two-tailed test. Upon rereading the problem, we realize that this was pure FeO whose iron content was 77.60%, so that μ = 77.60 and the confidence interval does not include the known value. Since the FeO was a standard, a one-tailed test should have been used since only random values would be expected to exceed 77.60%. Now the Student t value of 2.13 (for $-\,t_{0.05}$) should have been used, and now the confidence interval becomes 77.11 ± 0.23. A systematic error is presumed to exist.

The t test can be applied to differences between pairs of observations. Perhaps only a single pair can be performed at one time, or possibly one wishes to compare two methods using samples of differing analytical content. It is still necessary that the two methods possess the same inherent standard deviation. An average difference $\bar{d}$ calculated, and individual deviations from $\bar{d}$ are used to evaluate the variance of the differences.

Example 3.4 From the following data do the two methods actually give concordant results?

Sample	Method *A*	Method *B*	Difference
1	33.27	33.04	$d_1 = 0.23$
2	51.34	50.96	$d_2 = 0.38$
3	23.91	23.77	$d_3 = 0.14$
4	47.04	46.79	$d_4 = 0.25$
			$\bar{d} = 0.25$

TABLE 3.2 **Percentile Values for Student *t* Distribution**

df	$t_{0.995}$	$t_{0.99}$	$t_{0.975}$	$t_{0.95}$	$t_{0.90}$	$t_{0.80}$	$t_{0.75}$	$t_{0.70}$	$t_{0.60}$	$t_{0.55}$
1	63.66	31.82	12.71	6.31	3.08	1.376	1.000	0.727	0.325	0.158
2	9.92	6.96	4.30	2.92	1.89	1.061	0.816	0.617	0.289	0.142
3	5.84	4.54	3.18	2.35	1.64	0.978	0.765	0.584	0.277	0.137
4	4.60	3.75	2.78	2.13	1.53	0.941	0.741	0.569	0.271	0.134
5	4.03	3.36	2.57	2.02	1.48	0.920	0.727	0.559	0.267	0.132
6	3.71	3.14	2.45	1.94	1.44	0.906	0.718	0.553	0.265	0.131
7	3.50	3.00	2.36	1.90	1.42	0.896	0.711	0.549	0.263	0.130
8	3.36	2.90	2.31	1.86	1.40	0.889	0.706	0.546	0.262	0.130
9	3.25	2.82	2.26	1.83	1.38	0.883	0.703	0.543	0.261	0.129
10	3.17	2.76	2.23	1.81	1.37	0.879	0.700	0.542	0.260	0.129
11	3.11	2.72	2.20	1.80	1.36	0.876	0.697	0.540	0.260	0.129
12	3.06	2.68	2.18	1.78	1.36	0.873	0.695	0.539	0.259	0.128
13	3.01	2.65	2.16	1.77	1.35	0.870	0.694	0.538	0.259	0.128
14	2.98	2.62	2.14	1.76	1.34	0.868	0.692	0.537	0.258	0.128
15	2.95	2.60	2.13	1.75	1.34	0.866	0.691	0.536	0.258	0.128
16	2.92	2.58	2.12	1.75	1.34	0.865	0.690	0.535	0.258	0.128
17	2.90	2.57	2.11	1.74	1.33	0.863	0.689	0.534	0.257	0.128
18	2.88	2.55	2.10	1.73	1.33	0.862	0.688	0.534	0.257	0.127
19	2.86	2.54	2.09	1.73	1.33	0.861	0.688	0.533	0.257	0.127
20	2.84	2.53	2.09	1.72	1.32	0.860	0.687	0.533	0.257	0.127
21	2.83	2.52	2.08	1.72	1.32	0.859	0.686	0.532	0.257	0.127
22	2.82	2.51	2.07	1.72	1.32	0.858	0.686	0.532	0.256	0.127
23	2.81	2.50	2.07	1.71	1.32	0.858	0.685	0.532	0.256	0.127
24	2.80	2.49	2.06	1.71	1.32	0.857	0.685	0.531	0.256	0.127
25	2.79	2.48	2.06	1.71	1.32	0.856	0.684	0.531	0.256	0.127
26	2.78	2.48	2.06	1.71	1.32	0.856	0.684	0.531	0.256	0.127
27	2.77	2.47	2.05	1.70	1.31	0.855	0.684	0.531	0.256	0.127
28	2.76	2.47	2.05	1.70	1.31	0.855	0.683	0.530	0.256	0.127
29	2.76	2.46	2.04	1.70	1.31	0.854	0.683	0.530	0.256	0.127
30	2.75	2.46	2.04	1.70	1.31	0.854	0.683	0.530	0.256	0.127
40	2.70	2.42	2.02	1.68	1.30	0.851	0.681	0.529	0.255	0.126
60	2.66	2.39	2.00	1.67	1.30	0.848	0.679	0.527	0.254	0.126
120	2.62	2.36	2.98	1.66	1.29	0.845	0.677	0.526	0.254	0.126
∞	2.58	2.33	1.96	1.645	1.28	0.842	0.674	0.524	0.253	0.126

$$s_d \pm \frac{\sqrt{\Sigma (d-\bar{d})^2}}{N-1} = 0.099$$

$$t = \frac{0.25}{0.099}\sqrt{4-1} = 4.30$$

From Table 3.2, $t_{0.975} = 3.18$ (at 95% probability) and $t_{0.995} = 5.84$ (at 99% probability). The difference between the two methods is probably significant.

If the t value falls short of the formal significance level, this is not to be interpreted as proving the absence of a systematic error. Perhaps the data were insufficient in precision or in number to establish the presence of a constant error. Especially when the calculated value for t is only slightly short of the tabulated value, some additional data may suffice to build up the evidence for a constant error (or the lack thereof).

Should there be more than one known material, a weighted average of the individual differences ($\bar{x}$) should be taken. The value of s should be based on the combined estimate from the two or more materials (perhaps different primary standards for bases). Should the materials differ markedly in composition, a plot of the individual constant errors against composition should be made. If the constant errors appear to depend upon the composition, they should not be pooled in a weighted average.

The t test is also used to judge whether a given lot of material conforms to a particular specification. If both plus and minus departures from the known value are to be guarded against, a two-tailed test is involved. If departures in only one direction are undesirable, then the 10% level values for t are appropriate for the 5% level in *one* direction. Similarly, the 2% level should be used to obtain the 1% level to test the departure from the known value in one direction only; these constitute a one-tailed test. More on this subject will be in the next section.

Sometimes just one determination is available on each of several known materials similar in composition. A single determination by each of two procedures (or two analysts) on a series of material may be used to test for a relative bias between the two methods, as in Example 3.4. Of course, the average difference does not throw any light on which procedure has the larger constant error. It only supplies a test as to whether the two procedures are in disagreement.

3.4.2 Hypotheses About Means

Statistical methods are frequently used to give a "yes" or "no" answer to a particular question concerning the significance of data. When performing hypothesis tests on real data, we cannot set an absolute cutoff as to where we can expect to find no values from the population against which we are testing data, but we can set a limit beyond which we consider it very unlikely to find a member of the population. If a measurement is made that does in fact fall outside the specified range, the probability of its happening by chance alone can be rejected; something beyond the randomness of the reference population must be operating. In other words, hypothesis testing is an attempt to determine whether a given measured statistic could have come from some hypothesized population.

In attempting to reach decisions, it is useful to make assumptions or guesses about the populations involved. Such assumptions, which may or may not be true, are called *statistical hypotheses* and in general are statements about the probability distributions of the populations. A common procedure is to set up a *null hypothesis*, denoted by H_0, which states that there is no significant difference between two sets of data or that a variable exerts no significant effect. Any hypothesis that differs from a null hypothesis is called an *alternative hypothesis*, denoted by H_1.

Our answer is qualified by a confidence level (or level of significance) indicating the degree of certainty of the answer. Generally confidence levels of 95% and 99% are chosen to express the probability that the answer is correct. These are also denoted as the 0.05 and 0.01 level of significance, respectively. When the hypothesis can be rejected at the 0.05 level of significance, but not at the 0.01 level, we can say that the sample results are probably significant. If, however, the hypothesis is also rejected at the 0.01 level, the results become highly significant.

The abbreviated table on the next page, which gives critical values of z for both one-tailed and two-tailed tests at various levels of significance, will be found useful for purposes of reference. Critical values of z for other levels of significance are found by the use of Table 3.1. For a small number of samples we replace z, obtained from above or from Table 3.2, by t from Table 3.2, and we replace σ by

$$[\sqrt{N/(N-1)}]s$$

Level of significance, α	0.10	0.05	0.01	0.005	0.002
Critical values of z for one-tailed tests	−1.28 or 1.28	−1.645 or 1.645	−2.33 or 2.33	−2.58 or 2.58	−2.88 or 2.88
Critical values of z for two-tailed tests	−1.645 and 1.645	−1.96 and 1.96	−2.58 and 2.58	−2.81 and 2.81	−3.08 and 3.08

Procedures that enable us to decide whether to accept or reject hypotheses or to determine whether observed samples differ significantly from expected results are called *tests of hypotheses, tests of significance*, or *rules of decision*. For example, a set of z values outside the range −1.96 to 1.96 (at the 0.05 level of significance for a two-tailed test, constitute what is called the critical region or region of rejection of the hypothesis. The set of z results inside the range −1.96 to 1.96 could then be called the region of acceptance of the hypothesis.

Example 3.5 In the past a method gave $\mu = 0.050\%$. A recent set of 10 results gave $\bar{x} = 0.053\%$ and $s = 0.003\%$. Is everything satisfactory at a level of significance of 0.05? Of 0.01?

We wish to decide between the hypotheses

$$H_0: \quad \mu = 0.050\% \quad \text{and the method is working properly, and}$$

$$H_1: \quad \mu \neq 0.050\% \quad \text{and the method is not working properly.}$$

A *two-tailed test* is required; that is, both tails on the distribution curve are involved:

$$t = \frac{0.053 - 0.050}{0.003}\sqrt{10 - 1} = -3.00$$

Enter Table 3.2 for nine degrees of freedom under the column headed $t_{0.975}$ for the 0.05 level of significance, and the column $t_{0.995}$ for the 0.01 level of significance. At the 0.05 level, accept H_0 if t lies inside the interval $-t_{0.975}$ to $t_{0.975}$, that is, within −2.26 and 2.26; reject otherwise. Since $t = -3.00$, we reject H_0. At the 0.01 level of significance, the corresponding interval is −3.25 to 3.25, which t lies within, indicating acceptance of H_0. Because we can reject H_0 at the 0.05 level but not at the 0.01 level of significance, we can say that the sample results are probably significant and that the method is working properly.

Let us digress a moment and consider when a two-tailed test is needed, and what a one-tailed test implies. We "assume" that the measurements can be described by the curve shown in Fig. 3.1. If so, then 95% of the time a sample from the specified population will fall within the indicated range and 5% of the time it will fall outside; 2.5% of the time it is outside on the high side of the range, and 2.5% of the time it is below the low side of the range. Our assumption implies that if μ does not equal the hypothesized value, the probability of its being above the hypothesized value is equal to the probability of its being below the hypothesized value.

There will be incidences when the foregoing assumptions for a two-tailed test will not be true. Perhaps some physical situation prevents μ from ever being less than the hypothesized value; it can only be equal or greater. No results would ever fall below the low end of the confidence interval; only the upper end of the distribution is operative. Now random samples will exceed the upper bound only 2.5% of the time, not the 5% specified in two-tail testing. Thus, where the possible values are restricted, what was supposed to be a hypothesis test at the 95% confidence level is actually being performed at a 97.5% confidence level. Stated in another way, 95% of the population data lie within the interval below $\mu + 1.65\sigma$ and 5% lie above. Of course, the opposite situation might also occur and only the lower end of the distribution is operative.

Example 3.6 Six samples from a bulk chemical shipment averaged 77.50% active ingredient with $s = 1.45\%$. The manufacturer claimed 80.00%. Can his claim be supported?

A one-tailed test is required since the alternative hypothesis states that the population parameter is equal to or less than the hypothesized value.

$$t = \frac{77.50 - 80.00}{1.45}\sqrt{6-1} = 3.86$$

Since $t_{0.95} = -2.01$, and $t_{0.99} = -3.36$, the hypothesis is rejected at both the 0.05 and the 0.01 levels of significance. It is extremely unlikely that the claim is justified.

3.4.3 The Chi-Square (χ^2) Distribution

The χ^2 distribution describes the behavior of variances. Actually there is not a single χ^2 distribution but a whole set of distributions. Each distribution depends upon the number of degrees of freedom (designated variously as df, d.f., or f) in that distribution. Table 3.3 is laid out so that the horizontal axis is labeled with probability levels, while the vertical axis is listed in descending order of increasing number of degrees of freedom. The entries increase both as you read down and across the table. Although Table 3.3 does not display the values for the midrange of the distributions, at the 50% point of each distribution, the expected value of χ^2 is equal to the degrees of freedom. Estimates of the variance are uncertain when based only on a few degrees of freedom. With the 10 samples in Example 3.7, the standard deviation can vary by a large factor purely by random chance alone. Even 31 samples gives a spread of standard deviation of 2.6 at the 95% confidence level.

Understanding the χ^2 distribution allows us to calculate the expected values of random variables that are normally and independently distributed. In least-squares multiple regression, or in calibration work in general, there is a basic assumption that the error in the response variable is random and normally distributed, with a variance that follows a χ^2 distribution.

Confidence limits for an estimate of the variance may be calculated as follows. For each group of samples a standard deviation is calculated. These estimates of σ possess a distribution called the χ^2 distribution:

$$\chi^2 = \frac{s^2}{\sigma^2/\text{df}} \tag{3.12}$$

The upper and lower confidence limits for the standard deviation are obtained by dividing $(N-1)s^2$ by two entries taken from Table 3.3. The estimate of variance at the 90% confidence limits use the entries $\chi^2_{0.05}$ and $\chi^2_{0.95}$ (for 5% and 95%) with N degrees of freedom.

Example 3.7 The variance obtained for 10 samples is $(0.65)^2$. σ^2 is known to be $(0.75)^2$. How reliable is s^2 as an estimate of σ^2?

$$\frac{s^2(N-1)}{\chi^2_{0.975}} < \sigma^2 < \frac{s^2(N-1)}{\chi^2_{0.025}}$$

$$\frac{(0.65)^2(10-1)}{19.02} < \sigma^2 < \frac{(0.65)^2(10-1)}{2.70}$$

$$2.20 < \sigma^2 < 1.43$$

Thus, only one time in 40 will $9s^2/\sigma^2$ be less than 2.70 by chance alone. Similarly, only one time in 40 will $9s^2/\sigma^2$ be greater than 19.02. Consequently, it is not unlikely that s^2 is a reliable estimate of σ^2.

Stated differently:

Upper limit: $\sigma^2 = 9s^2/2.7 = 3.3s^2$

Lower limit: $\sigma^2 = 9s^2/19.02 = 0.48s^2$

Ten measurements give an estimate of σ^2 that may be as much as 3.3 times or only about one-half the true variance.

TABLE 3.3 **Percentile Values for the χ^2 Distribution**

df	$\chi^2_{0.995}$	$\chi^2_{0.99}$	$\chi^2_{0.975}$	$\chi^2_{0.95}$	$\chi^2_{0.90}$	$\chi^2_{0.75}$	$\chi^2_{0.50}$	$\chi^2_{0.25}$	$\chi^2_{0.10}$	$\chi^2_{0.05}$	$\chi^2_{0.025}$	$\chi^2_{0.01}$	$\chi^2_{0.005}$
1	7.88	6.63	5.02	3.84	2.71	1.32	0.455	0.102	0.0158	0.0039	0.0010	0.0002	0.0000
2	10.6	9.21	7.38	5.99	4.61	2.77	1.39	0.575	0.211	0.103	0.0506	0.0201	0.0100
3	12.8	11.3	9.35	7.81	6.25	4.11	2.37	1.21	0.584	0.352	0.216	0.115	0.072
4	14.9	13.3	11.1	9.49	7.78	5.39	3.36	1.92	1.06	0.711	0.484	0.297	0.207
5	16.7	15.1	12.8	11.1	9.24	6.63	4.35	2.67	1.61	1.15	0.831	0.554	0.412
6	18.5	16.8	14.4	12.6	10.6	7.84	5.35	3.45	2.20	1.64	1.24	0.872	0.676
7	20.3	18.5	16.0	14.1	12.0	9.04	6.35	4.25	2.83	2.17	1.69	1.24	0.989
8	22.0	20.1	17.5	15.5	13.4	10.2	7.34	5.07	3.49	2.73	2.18	1.65	1.34
9	23.6	21.7	19.0	16.9	14.7	11.4	8.34	5.90	4.17	3.33	2.70	2.09	1.73
10	25.2	23.2	20.5	18.3	16.0	12.5	9.34	6.74	4.87	3.94	3.25	2.56	2.16
11	26.8	24.7	21.9	19.7	17.3	13.7	10.3	7.58	5.58	4.57	3.82	3.05	2.60
12	28.3	26.2	23.3	21.0	18.5	14.8	11.3	8.44	6.30	5.23	4.40	3.57	3.07
13	29.8	27.7	24.7	22.4	19.8	16.0	12.3	9.30	7.04	5.89	5.01	4.11	3.57
14	31.3	29.1	26.1	23.7	21.1	17.1	13.3	10.2	7.79	6.57	5.63	4.66	4.07
15	32.8	30.6	27.5	25.0	22.3	18.2	14.3	11.0	8.55	7.26	6.26	5.23	4.60
16	34.3	32.0	28.8	26.3	23.5	19.4	15.3	11.9	9.31	7.96	6.91	5.81	5.14
17	35.7	33.4	30.2	27.6	24.8	20.5	16.3	12.8	10.1	8.67	7.56	6.41	5.70
18	37.2	34.8	31.5	28.9	26.0	21.6	17.3	13.7	10.9	9.39	8.23	7.01	6.26
19	38.6	36.2	32.9	30.1	27.2	22.7	18.3	14.6	11.7	10.1	8.91	7.63	6.84
20	40.0	37.6	34.2	31.4	28.4	23.8	19.3	15.5	12.4	10.9	9.59	8.26	7.43
21	41.4	38.9	35.5	32.7	29.6	24.9	20.3	16.3	13.2	11.6	10.3	8.90	8.03
22	42.8	40.3	36.8	33.9	30.8	26.0	21.3	17.2	14.0	12.3	11.0	9.54	8.64
23	44.2	41.6	38.1	35.2	32.0	27.1	22.3	18.1	14.8	13.1	11.7	10.2	9.26
24	45.6	43.0	39.4	36.4	33.2	28.2	23.3	19.0	15.7	13.8	12.4	10.9	9.89
25	46.9	44.3	40.6	37.7	34.4	29.3	24.3	19.9	16.5	14.6	13.1	11.5	10.5
26	48.3	45.6	41.9	38.9	35.6	30.4	25.3	20.8	17.3	15.4	13.8	12.2	11.2
27	49.6	47.0	43.2	40.1	36.7	31.5	26.3	21.7	18.1	16.2	14.6	12.9	11.8
28	51.0	48.3	44.5	41.3	37.9	32.6	27.3	22.7	18.9	16.9	15.3	13.6	12.5
29	52.3	49.6	45.7	42.6	39.1	33.7	28.3	23.6	19.8	17.7	16.0	14.3	13.1
30	53.7	50.9	47.0	43.8	40.3	34.8	29.3	24.5	20.6	18.5	16.8	15.0	13.8
40	66.8	63.7	59.3	55.8	51.8	45.6	39.3	33.7	29.1	26.5	24.4	22.2	20.7
50	79.5	76.2	71.4	67.5	63.2	56.3	49.3	42.9	37.7	34.8	32.4	29.7	28.0
60	92.0	88.4	83.3	79.1	74.4	67.0	59.3	52.3	46.5	43.2	40.5	37.5	35.5
70	104.2	100.4	95.0	90.5	85.5	77.6	69.3	61.7	55.3	51.7	48.8	45.4	43.3
80	116.3	112.3	106.6	101.9	96.6	88.1	79.3	71.1	64.3	60.4	57.2	53.5	51.2
90	128.3	124.1	118.1	113.1	107.6	98.6	89.3	80.6	73.3	69.1	65.6	61.8	59.2
100	140.2	135.8	129.6	124.3	118.5	109.1	99.3	90.1	82.4	77.9	74.2	70.1	67.3

3.4.4 The *F* Statistic

The *F* statistic, along with the *z*, *t*, and χ^2 statistics, constitutes the group that is thought of as fundamental statistics. Collectively they describe all the relationships that can exist between means and standard deviations. To perform an *F* test, we must first verify the randomness and independence of the errors. If $\sigma_1^2 = \sigma_2^2$, then s_1^2/s_2^2 will be distributed properly as the *F* statistic. If the calculated *F* is outside the confidence interval chosen for that statistic, then this is evidence that $\sigma_1^2 \neq \sigma_2^2$.

The *F* statistic describes the distribution of the ratios of variances of two sets of samples. It requires three table labels: the probability level and the two degrees of freedom. Since the *F* distribution requires a three-dimensional table that is effectively unknown, the *F* tables are presented as large sets of two-dimensional tables. The *F* distribution in Table 3.4 has the different numbers of

TABLE 3.4 ***F* Distribution**

Interpolation should be performed using reciprocals of the degrees of freedom.

Upper 5% points ($F_{0.95}$)

Degrees of freedom for denominator	Degrees of freedom for numerator																		
	1	2	3	4	5	6	7	8	9	10	12	15	20	24	30	40	60	120	∞
1	161	200	216	225	230	234	237	239	241	242	244	246	248	249	250	251	252	253	254
2	18.5	19.0	19.2	19.2	19.3	19.3	19.4	19.4	19.4	19.4	19.4	19.4	19.4	19.5	19.5	19.5	19.5	19.5	19.5
3	10.1	9.55	9.28	9.12	9.01	8.94	8.89	8.85	8.81	8.79	8.74	8.70	8.66	8.64	8.62	8.59	8.57	8.55	8.53
4	7.71	6.94	6.59	6.39	6.26	6.16	6.09	6.04	6.00	5.96	5.91	5.86	5.80	5.77	5.75	5.72	5.69	5.66	5.63
5	6.61	5.79	5.41	5.19	5.05	4.95	4.88	4.82	4.77	4.74	4.68	4.62	4.56	4.53	4.50	4.46	4.43	4.40	4.37
6	5.99	5.14	4.76	4.53	4.39	4.28	4.21	4.15	4.10	4.06	4.00	3.94	3.87	3.84	3.81	3.77	3.74	3.70	3.67
7	5.59	4.74	4.35	4.12	3.97	3.87	3.79	3.73	3.68	3.64	3.57	3.51	3.44	3.41	3.38	3.34	3.30	3..27	3.23
8	5.32	4.46	4.07	3.84	3.69	3.58	3.50	3.44	3.39	3.35	3.28	3.22	3.15	3.12	3.08	3.04	3.01	2.97	2.93
9	5.12	4.26	3.86	3.63	3.48	3.37	3.29	3.23	3.18	3.14	3.07	3.01	2.94	2.90	2.86	2.83	2.79	2.75	2.71
10	4.96	4.10	3.71	3.48	3.33	3.22	3.14	3.07	3.02	2.98	2.91	2.85	2.77	2.74	2.70	2.66	2.62	2.58	2.54
11	4.84	3.98	3.59	3.36	3.20	3.09	3.01	2.95	2.90	2.85	2.79	2.72	2.65	2.61	2.57	2.53	2.49	2.45	2.40
12	4.75	3.89	3.49	3.26	3.11	3.00	2.91	2.85	2.80	2.75	2.69	2.62	2.54	2.51	2.47	2.43	2.38	2.34	2.30
13	4.67	3.81	3.41	3.18	3.03	2.92	2.83	2.77	2.71	2.67	2.60	2.53	2.46	2.42	2.38	2.34	2.30	2.25	2.21
14	4.60	3.74	3.34	3.11	2.96	2.85	2.76	2.70	2.65	2.60	2.53	2.46	2.39	2.35	2.31	2.27	2.22	2.18	2.13
15	4.54	3.68	3.29	3.06	2.90	2.79	2.71	2.64	2.59	2.54	2.48	2.40	2.33	2.29	2.25	2.20	2.16	2.11	2.07
16	4.49	3.63	3.24	3.01	2.85	2.74	2.66	2.59	2.54	2.49	2.42	2.35	2.28	2.24	2.19	2.15	2.11	2.06	2.01
17	4.45	3.59	3.20	2.96	2.81	2.70	2.61	2.55	2.49	2.45	2.38	2.31	2.23	2.19	2.15	2.10	2.06	2.01	1.96
18	4.41	3.55	3.16	2.93	2.77	2.66	2.58	2.51	2.46	2.41	2.34	2.27	2.19	2.15	2.11	2.06	2.02	1.97	1.92
19	4.38	3.52	3.13	2.90	2.74	2.63	2.54	2.48	2.42	2.38	2.31	2.23	2.16	2.11	2.07	2.03	1.98	1.93	1.88
20	4.35	3.49	3.10	2.87	2.71	2.60	2.51	2.45	2.39	2.35	2.28	2.20	2.12	2.08	2.04	1.99	1.95	1.90	1.84
21	4.32	3.47	3.07	2.84	2.68	2.57	2.49	2.42	2.37	2.32	2.25	2.18	2.10	2.05	2.01	1.96	1.92	1.87	1.81
22	4.30	3.44	3.05	2.82	2.66	2.55	2.46	2.40	2.34	2.30	2.23	2.15	2.07	2.03	1.98	1.94	1.89	1.84	1.78
23	4.28	3.42	3.03	2.80	2.64	2.53	2.44	2.37	2.32	2.27	2.20	2.13	2.05	2.01	1.96	1.91	1.86	1.81	1.76
24	4.26	3.40	3.01	2.78	2.62	2.51	2.42	2.36	2.30	2.25	2.18	2.11	2.03	1.98	1.94	1.89	1.84	1.79	1.73
25	4.24	3.39	2.99	2.76	2.60	2.49	2.40	2.34	2.28	2.24	2.16	2.09	2.01	1.96	1.92	1.87	1.82	1.77	1.71
30	4.17	3.32	2.92	2.69	2.53	2.42	2.33	2.27	2.21	2.16	2.09	2.01	1.93	1.89	1.84	1.79	1.74	1.68	1.62
40	4.08	3.23	2.84	2.61	2.45	2.34	2.25	2.18	2.12	2.08	2.00	1.92	1.84	1.79	1.74	1.69	1.64	1.58	1.51
60	4.00	3.15	2.76	2.53	2.37	2.25	2.17	2.10	2.04	1.99	1.92	1.84	1.75	1.70	1.65	1.59	1.53	1.47	1.39
120	3.92	3.07	2.68	2.45	2.29	2.18	2.09	2.02	1.96	1.91	1.83	1.75	1.66	1.61	1.55	1.50	1.43	1.35	1.25
∞	3.84	3.00	2.60	2.37	2.21	2.10	2.01	1.94	1.88	1.83	1.75	1.67	1.57	1.52	1.46	1.39	1.32	1.22	1.00

Upper 1% points ($F_{0.99}$)

Degrees of freedom for denominator	Degrees of freedom for numerator																		
	1	2	3	4	5	6	7	8	9	10	12	15	20	24	30	40	60	120	∞
1	4052	5000	5403	5625	5764	5859	5928	5982	6023	6056	6106	6157	6209	6235	6261	6287	6313	6339	6366
2	98.5	99.0	99.2	99.2	99.3	99.3	99.4	99.4	99.4	99.4	99.4	99.4	99.4	99.5	99.5	99.5	99.5	99.5	99.5
3	34.1	30.8	29.5	28.7	28.2	27.9	27.7	27.5	27.3	27.2	27.1	26.9	26.7	26.6	26.5	26.4	26.3	26.2	26.1
4	21.2	18.0	16.7	16.0	15.5	15.2	15.0	14.8	14.7	14.5	14.4	14.2	14.0	13.9	13.8	13.7	13.7	13.6	13.5
5	16.3	13.3	12.1	11.4	11.0	10.7	10.5	10.3	10.2	10.1	9.89	9.72	9.55	9.47	9.38	9.29	9.20	9.11	9.02
6	13.7	10.9	9.78	9.15	8.75	8.47	8.26	8.10	7.98	7.87	7.72	7.56	7.40	7.31	7.23	7.14	7.06	6.97	6.88
7	12.2	9.55	8.45	7.85	7.46	7.19	6.99	6.84	6.72	6.62	6.47	6.31	6.16	6.07	5.99	5.91	5.82	5.74	5.65
8	11.3	8.65	7.59	7.01	6.63	6.37	6.18	6.03	5.91	5.81	5.67	5.52	5.36	5.28	5.20	5.12	5.03	4.95	4.86
9	10.6	8.02	6.99	6.42	6.06	5.80	5.61	5.47	5.35	5.26	5.11	4.96	4.81	4.73	4.65	4.57	4.48	4.40	4.31
10	10.0	7.56	6.55	5.99	5.64	5.39	5.20	5.06	4.94	4.85	4.71	4.56	4.41	4.33	4.25	4.17	4.08	4.00	3.91
11	9.65	7.21	6.22	5.67	5.32	5.07	4.89	4.74	4.63	4.54	4.40	4.25	4.10	4.02	3.94	3.86	3.78	3.69	3.60
12	9.33	6.93	5.95	5.41	5.06	4.82	4.64	4.50	4.39	4.30	4.16	4.01	3.86	3.78	3.70	3.62	3.54	3.45	3.36
13	9.07	6.70	5.74	5.21	4.86	4.62	4.44	4.30	4.19	4.10	3.96	3.82	3.66	3.59	3.51	3.43	3.34	3.25	3.17
14	8.86	6.51	5.56	5.04	4.70	4.46	4.28	4.14	4.03	3.94	3.80	3.66	3.51	3.43	3.35	3.27	3.18	3.09	3.00
15	8.68	6.36	5.42	4.89	4.56	4.32	4.14	4.00	3.89	3.80	3.67	3.52	3.37	3.29	3.21	3.13	3.05	2.96	2.87
16	8.53	6.23	5.29	4.77	4.44	4.20	4.03	3.89	3.78	3.69	3.55	3.41	3.26	3.18	3.10	3.02	2.93	2.84	2.75
17	8.40	6.11	5.19	4.67	4.34	4.10	3.93	3.79	3.68	3.59	3.46	3.31	3.16	3.08	3.00	2.92	2.83	2.75	2.65
18	8.29	6.01	5.09	4.58	4.25	4.01	3.84	3.71	3.60	3.51	3.37	3.23	3.08	3.00	2.92	2.84	2.75	2.66	2.57
19	8.19	5.93	5.01	4.50	4.17	3.94	3.77	3.63	3.52	3.43	3.30	3.15	3.00	2.92	2.84	2.76	2.67	2.58	2.49
20	8.10	5.85	4.94	4.43	4.10	3.87	3.70	3.56	3.46	3.37	3.23	3.09	2.94	2.86	2.78	2.69	2.61	2.52	2.42
21	8.02	5.78	4.87	4.37	4.04	3.81	3.64	3.51	3.40	3.31	3.17	3.03	2.88	2.80	2.72	2.64	2.55	2.46	2.36
22	7.95	5.72	4.82	4.31	3.99	3.76	3.59	3.45	3.35	3.26	3.12	2.98	2.83	2.75	2.67	2.58	2.50	2.40	2.31
23	7.88	5.66	4.76	4.26	3.94	3.71	3.54	3.41	3.30	3.21	3.07	2.93	2.78	2.70	2.62	2.54	2.45	2.35	2.26
24	7.82	5.61	4.72	4.22	3.90	3.67	3.50	3.36	3.26	3.17	3.03	2.89	2.74	2.66	2.58	2.49	2.40	2.31	2.21
25	7.77	5.57	4.68	4.18	3.86	3.63	3.46	3.32	3.22	3.13	2.99	2.85	2.70	2.62	2.53	2.45	2.36	2.27	2.17
30	7.56	5.39	4.51	4.02	3.70	3.47	3.30	3.17	3.07	2.98	2.84	2.70	2.55	2.47	2.39	2.30	2.21	2.11	2.01
40	7.31	5.18	4.31	3.83	3.51	3.29	3.12	2.99	2.89	2.80	2.66	2.52	2.37	2.29	2.20	2.11	2.02	1.92	1.80
60	7.08	4.98	4.13	3.65	3.34	3.12	2.95	2.82	2.72	2.63	2.50	2.35	2.20	2.12	2.03	1.94	1.84	1.73	1.60
120	6.85	4.79	3.95	3.48	3.17	2.96	2.79	2.66	2.56	2.47	2.34	2.19	2.03	1.95	1.86	1.76	1.66	1.53	1.38
∞	6.63	4.61	3.78	3.32	3.02	2.80	2.64	2.51	2.41	2.32	2.18	2.04	1.88	1.79	1.70	1.59	1.47	1.32	1.00

degrees of freedom for the denominator variance placed along the vertical axis, while in each table the two horizontal axes represent the numerator degrees of freedom and the probability level. Only two probability levels are given in Table 3.4: the upper 5% points ($F_{0.95}$) and the upper 1% points ($F_{0.99}$). More extensive tables of statistics will list additional probability levels, and they should be consulted when needed.

It is possible to compare the means of two relatively small sets of observations when the variances within the sets can be regarded as the same, as indicated by the F test. One can consider the distribution involving estimates of the true variance. With s_1^2 determined from a group of N_1 observations and s_2^2 from a second group of N_2 observations, the distribution of the ratio of the sample variances is given by the F statistic:

$$F = \frac{s_1^2/\sigma_1^2}{s_2^2/\sigma_2^2} \tag{3.13}$$

The larger variance is placed in the numerator. For example, the F test allows judgment regarding the existence of a significant difference in the precision between two sets of data or between two analysts. The hypothesis assumed is that both variances are indeed alike and a measure of the same σ.

The fact that each sample variance is related to its own population variance means that the sample variance being used for the calculation need not come from the same population. This is a significant departure from the assumptions inherent in the z, t, and χ^2 statistics.

Example 3.8 Suppose Analyst A made five observations and obtained a standard deviation of 0.06, where Analyst B with six observations obtained $s_B = 0.03$. The experimental variance ratio is

$$F = \frac{(0.06)^2}{(0.03)^2} = 4.00$$

From Table 3.4 with four degrees of freedom for A and five degrees of freedom for B, the value of F would exceed 5.19 five percent of the time. Therefore, the null hypothesis is valid, and comparable skills are exhibited by the two analysts.

As applied in Example 3.8, the F test was one-tailed. The F test may also be applied as a two-tailed test in which the alternative to the null hypothesis is $\sigma_1^2 \neq \sigma_2^2$. This doubles the probability that the null hypothesis is invalid and has the effect of changing the confidence level, in the above example, from 95% to 90%.

If improvement in precision is claimed for a set of measurements, the variance for the set against which comparison is being made should be placed in the numerator, regardless of magnitude. An experimental F smaller than unity indicates that the claim for improved precision cannot be supported. The technique just given for examining whether the precision varies with the two different analytical procedures also serves to compare the precision with different materials or with different operators, laboratories, or sets of equipment.

3.5 CURVE FITTING

Very often in practice a relationship is found (or known) to exist between two or more variables. It is frequently desirable to express this relationship in mathematical form by determining an equation connecting the variables.

The first step is the collection of data showing corresponding values of the variables under consideration. From a scatter diagram, a plot of Y (ordinate) versus X (abscissa), it is often possible to visualize a smooth curve approximating the data. For purposes of reference, several types of approximating curves and their equations are listed. All letters other than X and Y represent constants.

1. $Y = a_0 + a_1X$	Straight line
2. $Y = a_0 + a_1X + a_2X^2$	Parabola or quadratic curve
3. $Y = a_0 + a_1X + a_2X^2 + a_3X^3$	Cubic curve
4. $Y = a_0 + a_1X + a_2 + \cdots + a_nX^n$	nth Degree curve

As other possible equations (among many) used in practice, these may be mentioned:

5. $Y = (a_0 + a_1\mathrm{X})^{-1}$ or $1/Y = a_0 + a_1X$	Hyperbola
6. $Y = ab^X$ or $\log Y = \log a + (\log b)_X$	Exponential curve
7. $Y = aX^b$ or $\log Y = \log a + b \log X$	Geometric curve
8. $Y = ab^X + g$	Modified exponential curve
9. $Y = aX^n + g$	Modified geometric curve

When we draw a scatter plot of all X versus Y data, we see that some sort of shape can be described by the data points. From the scatter plot we can take a basic guess as to which type of curve will best describe the $X - Y$ relationship. To aid in the decision process, it is helpful to obtain scatter plots of transformed variables. For example, if a scatter plot of log Y versus X shows a linear relationship, the equation has the form of number 6 above, while if log Y versus log X shows a linear relationship the equation has the form of number 7. To facilitate this we frequently employ special graph paper for which one or both scales are calibrated logarithmically. These are referred to as *semilog* or *log–log graph paper*, respectively.

3.5.1 The Least Squares or Best-Fit Line

The simplest type of approximating curve is a straight line, the equation of which can be written as in form number 1 above. It is customary to employ the above definition when X is the independent variable and Y is the dependent variable.

To avoid individual judgment in constructing any approximating curve to fit sets of data, it is necessary to agree on a definition of a *best-fit line*. One could construct what would be considered the best-fit line through the plotted pairs of data points. For a given value of X_1, there will be a difference D_1 between the value Y_1 and the constituent value $\hat{Y}$ as determined by the calibration model. Since we are assuming that all the errors are in Y, we are seeking the best-fit line that minimizes the deviations in the Y direction between the experimental points and the calculated line. This condition will be met when the sum of squares for the differences, called residuals (or the sum of squares due to error),

$$\sum_{i=1}^{N} (Y_i = \hat{Y}_i)^2 = \sum \left(D_1^2 + D_2^2 + \cdots + D_N^2\right)$$

is the least possible value when compared to all other possible lines fitted to that data. If the sum of squares for residuals is equal to zero the calibration line is a perfect fit to the data. With a mathematical treatment known as linear regression, one can find the "best" straight line through these real world points by minimizing the residuals.

This calibration model for the best-fit line requires that the line pass through the "centroid" of the point ($\overline{X}$, $\overline{Y}$). It can be shown that

$$b = \frac{\sum_i (X_i - \overline{X})(Y_i - \overline{Y})}{\sum_i (X_i - \overline{X})^2} \tag{3.14}$$

$$a = \overline{Y} - b\overline{X} \tag{3.15}$$

The line thus calculated is known as the line of regression of Y on X, that is, the line indicating how Y varies when X is set to chosen values.

If X is the dependent variable, the definition is modified by considering horizontal instead of vertical deviations. In general these two definitions lead to different least-squares curves.

Example 3.9 The following data were recorded for the potential E of an electrode, measured against the saturated calomel electrode as a function of concentration C (moles liter^{-1}).

– log *C*	*E*, mV	– log *C*	*E*, mV
1.00	106	2.10	174
1.10	115	2.20	182
1.20	121	2.40	187
1.50	139	2.70	211
1.70	153	2.90	220
1.90	158	3.00	226

Fit the best straight line to these data; X_i represents –log C, and Y_i represents E. We will perform the calculation manually, using the following tabular layout.

X_i	Y_i	$X_i - \overline{X}$	$(X_i - \overline{X})^2$	$Y_i - \overline{Y}$	$(X_i - \overline{X})(Y_i - \overline{Y})$
1.00	106	–0.975	0.951	–60	58.5
1.10	115	–0.875	0.766	–51	44.6
1.20	121	–0.775	0.600	–45	34.9
1.50	139	–0.475	0.226	–27	12.8
1.70	153	–0.275	0.076	–13	3.6
1.90	158	–0.075	0.006	–8	0.6
2.10	174	+0.125	0.016	8	1.0
2.20	182	0.225	0.051	16	3.6
2.40	187	0.425	0.181	21	8.9
2.70	211	0.725	0.526	45	32.6
2.90	220	0.925	0.856	54	50.0
3.00	226	1.025	1.051	60	61.5
ΣX_i 23.7	ΣY_i 1992	Σ 0	Σ 5.306	Σ 0	Σ 312.6
$\overline{X}$ = 1.975	$\overline{Y}$ = 166				

Now substituting the proper terms into Eq. (3.14), the slope is

$$b = \frac{312.6}{5.306} = 58.91$$

and from Eq. (3.15), and substituting the "centroid" values of the points $(\overline{X}, Y)$, the intercept is

$$a = 166 - 58.91(1.975) = 49.64$$

The best-fit equation is therefore

$$E = 49.64 - 58.91 \log C$$

3.5.2 Errors in the Slope and Intercept of the Best-Fit Line

Upon examination of the plot of pairs of data points, the calibration line, it will be obvious that the precision involved in analyzing an unknown sample will be considerably poorer than that indicated by replicate error alone. The scatter of these original points about the calibration line is a good measure of the error to be expected in analyzing an unknown sample. And this same error is considerably larger than the replication error because it will include other sources of variability due to a variety of causes. One possible source of variability might be the presence of different amounts of an extraneous material in the various samples used to establish the calibration curve. While this variability causes scatter about the calibration curve, it will not be reflected in the replication error of any one sample if the sample is homogeneous.

The scatter of the points around the calibration line or random errors are of importance since the best-fit line will be used to estimate the concentration of test samples by interpolation. The method used to calculate the random errors in the values for the slope and intercept is now considered. We must first calculate the standard deviation $s_{Y/X}$, which is given by

$$s_{Y/X} = \sqrt{\frac{\sum_i (Y_i - \hat{Y})^2}{N-2}} \tag{3.16}$$

Equation (3.16) utilizes the *Y residuals*, $Y_i - \hat{Y}$, where $\hat{Y}_i$ are the points on the calculated best-fit line or the fitted Y_i values. The appropriate number of degrees of freedom is $N - 2$; the minus 2 arises from the fact that linear calibration lines are derived from both a slope and an intercept which leads to a loss of two degrees of freedom.

Now we can calculate the standard deviations for the slope and the intercept. These are given by

$$s_b = \frac{s_{Y/X}}{\sqrt{\sum_i (X_i - \overline{X})^2}} \tag{3.17}$$

$$s_a = s_{Y/X}\sqrt{\frac{\sum_i X_i^2}{N\sum_i (X_i - \overline{X})^2}} \tag{3.18}$$

The confidence limits for the slope are given by $b \pm t_b$, where the t value is taken at the desired confidence level and $N-2$ degrees of freedom. Similarly the confidence limits for the intercept are given by $a \pm ts_a$. The closeness of $\hat{x}$ to x_i is answered in terms of a confidence interval for x_0 that extends from an upper confidence (UCL) to a lower confidence (LCL) level. Let us choose 95% for the confidence interval. Then, remembering that this is a two-tailed test (UCL and LCL), we obtain from a table of Student's t distribution the critical value of t_c ($t_{0.975}$) and the appropriate number of degrees of freedom.

Example 3.10 For the best-fit line found in Example 3.9, express the result in terms of confidence intervals for the slope and intercept. We will choose 95% for the confidence interval.

The standard deviation $s_{Y/X}$ is given by Eq. (3.16), but first a supplementary table must be constructed for the Y residuals and other data that will be needed in subsequent equations.

$\hat{Y}$	$Y_i - \hat{Y}$	$(Y_i - \hat{Y})^2$	X_i^2
108.6	2.55	6.50	1.00
114.4	– 0.56	0.31	1.21
120.3	– 0.67	0.45	1.44
138.0	– 1.00	1.00	2.25
149.8	– 3.21	10.32	2.89
161.6	3.57	12.94	3.61
173.4	– 0.65	0.42	4.41
179.2	– 2.76	7.61	4.84
191.0	4.02	16.16	5.76
208.7	– 2.30	5.30	7.29
220.5	0.48	0.23	8.41
226.4	0.40	0.16	9.00
		Σ 61.20	Σ 52.11

Now substitute the appropriate values into Eq. (3.16) where there are 12 – 2 = 10 degrees of freedom:

$$s_{X/Y} = \sqrt{\frac{61.20}{10}} = 2.47$$

We can now calculate s_b and s_a from Eqs. (3.17) and (3.18), respectively.

$$s_b = \frac{s_{Y/X}}{\sqrt{5.31}} = 1.07$$

and

$$s_a = 2.47\sqrt{\frac{52.11}{12(5.306)}} = 2.23$$

Now, using a two-tailed value for Student's t:

$$b \pm ts_b = 58.91 \pm 2.23(1.07) = 58.91 \pm 2.39$$
$$a \pm ts_a = 49.64 \pm 2.23(2.23) = 49.64 \pm 4.97$$

The best-fit equation expressed in terms of the confidence intervals for the slope and intercept is

$$E = (49.6_4 \pm 5.0) - (58.9_1 \pm 2.43)\ \log\ C$$

To conclude the discussion about the best-fit line, the following relationship can be shown to exist among $Y, \hat{Y}$, and $\overline{Y}$

$$\sum_{i=1}^{N} (Y_i - \overline{Y})^2 = \sum_{i=1}^{N} (\hat{Y}_i - \overline{Y})^2 + \sum_{i=1}^{N} (Y_i - \hat{Y}_i)^2 \tag{3.19}$$

The term on the left-hand side is a constant and depends only on the constituent values provided by the reference laboratory and does not depend in any way upon the calibration. The two terms on the right-hand side of the equation show how this constant value is apportioned between the two quantities that are themselves summations, and are referred to as the sum of squares due to regression and the sum of squares due to error. The latter will be the smallest possible value that it can possibly be for the given data.

3.6 CONTROL CHARTS

It is often important in practice to know when a process has changed sufficiently so that steps may be taken to remedy the situation. Such problems arise in quality control where one must, often quickly, decide whether observed changes are due simply to chance fluctuations or to actual changes in the amount of a constituent in successive production lots, mistakes of employees, etc. Control charts provide a useful and simple method for dealing with such problems.

The chart consists of a central line and two pairs of limit lines or simply of a central line and one pair of control limits. By plotting a sequence of points in order, a continuous record of the quality characteristic is made available. Trends in data or sudden lack of precision can be made evident so that the causes may be sought.

The control chart is set up to answer the question of whether the data are in statistical control, that is, whether the data may be regarded as random samples from a single population of data. Because of this feature of testing for randomness, the control chart may be useful in searching out systematic sources of error in laboratory research data as well as in evaluating plant-production or control-analysis data.[1]

To set up a control chart, individual observations might be plotted in sequential order and then compared with control limits established from sufficient past experience. Limits of $\pm 1.96\sigma$ corresponding to a confidence level of 95%, might be set for control limits. The probability of a future observation falling outside these limits, based on chance, is only 1 in 20. A greater proportion of scatter might indicate a nonrandom distribution (a systematic error). It is common practice with some users of control charts to set inner control limits, or warning limits, at $\pm 1.96\sigma$ and outer control limits of $\pm 3.00\sigma$. The outer control limits correspond to a confidence level of 99.8%, or a probability of 0.002 that a point will fall outside the limits. One-half of this probability corresponds to a high result and one-half to a low result. However, other confidence limits can be used as well; the choice in each case depends on particular circumstances.

Special attention should be paid to one-sided deviation from the control limits, because systematic errors more often cause deviation in one direction than abnormally wide scatter. Two systematic errors of opposite sign would of course cause scatter, but it is unlikely that both would have entered at the same time. It is not necessary that the control chart be plotted in a time sequence. In any situation where relatively large numbers of units or small groups are to be compared, the control chart is a simple means of indicating whether any unit or group is out of line. Thus laboratories, production machines, test methods, or analysts may be put arbitrarily into a horizontal sequence.

Usually it is better to plot the means of small groups of observations on a control chart, rather than individual observations. The random scatter of averages of pairs of observations is $1/(2)^{1/2} = 0.71$ as great as that of single observations, and the likelihood of two "wild" observations in the same direction is vanishing small. The groups of two to five observations should be chosen in such a way that only chance variations operate within the group, whereas assignable causes are sought for variations between groups. If duplicate analyses are performed each day, the pairs form logical groups.

Some measure of dispersion of the subgroup data should also be plotted as a parallel control chart. The most reliable measure of scatter is the standard deviation. For small groups, the range becomes increasingly significant as a measure of scatter, and it is usually a simple matter to plot the range as a vertical line and the mean as a point on this line for each group of observations.

[1] G. Wernimont, *Ind. Eng. Chem., Anal. Ed.* **18**:587 (1946); J. A. Mitchell, *ibid.* **19**:961 (1947).

3.7 CONCEPT OF QUALITY ASSURANCE AND QUALITY CONTROL PROGRAMS

Quality assurance (QA) and quality control (QC) programs are key components of all analytical protocols in all areas of analysis, including environmental, pharmaceutical, and forensic testing, among others. These programs mandate that the laboratories follow a set of well-defined guidelines to achieve valid analytical results to a high degree of reliability and accuracy within an acceptable range. Although such programs may vary depending on the regulatory authority, certain key features of these programs are more or less the same and are briefly outlined below.

3.7.1 Quality Assurance Plans

The objective of a quality assurance plan is to obtain reliable and accurate analytical results that may be stated with a high level of confidence (statistically), so that such results are legally defensible. The key features of any plan involve essentially documentation and record keeping. The program involves documentation of sample collection for testing, the receipt of samples in the laboratory, and their transfer to the individuals who perform the analyses. The information is recorded on chain-of-custody forms stating the dates and times along with the names and signatures of individuals who carry out these tasks. Also, other pertinent information is recorded, such as any preservatives added to the sample to prevent degradation of test analytes, the temperature at which the sample is stored, the temperature to which the sample is brought prior to its analysis, the nature of the container (which may affect the stability of the sample), and its holding time prior to testing. There is also considerable overlap between certain aspects of quality assurance and quality control programs.

3.7.2 Quality Control

Unlike quality assurance plans that mostly address regulatory requirements involving comprehensive documentation, quality control programs are science-based, the components of which may be defined statistically. The two most important components of quality control are

1. Determination of precision of analysis
2. Determination of accuracy of measurement

Whereas *precision* measures the reproducibility of data from replicate analyses, the *accuracy* of a test estimates how accurate are the data, that is, how close the data would fall to probable true values or how accurate is the analytical procedure to give results that may be close to true values. Both the precision and accuracy are measured on one or more samples selected at random from a given batch of samples for analysis.

3.7.2.1 Determination of Precision in Measurements. The precision of analysis is usually determined by running duplicate or replicate testings on one of the samples in a given batch of samples. It is expressed statistically as standard deviation, relative standard deviation (RSD), coefficient of variance (CV), standard error of the mean (*M*), and relative percent difference (RPD). The definition and formulas for standard deviation are discussed above. The standard deviation in measurements, however, can vary with the concentrations of the analytes. On the other hand, RSD, which is expressed as the ratio of standard deviation to the arithmetic mean of replicate analyses and is given as a percent, does not have this problem and is a more rational way of expressing precision:

$$\mathrm{RSD} = \frac{\text{standard deviation}}{\text{arithmetic mean of replicate analysis}} \times 100\%$$

The standard error of the mean, M, is the ratio of the standard deviation and the square root of the number of measurements (n). That is,

$$M = \frac{\text{standard deviation}}{n}$$

This scale, too, will vary in the same proportion as standard deviation with the size of the analyte in the sample.

In routine testing, many repeat analyses of a sample aliquot may not be possible. Alternatively, therefore, the precision of a test may be determined from duplicate analyses of sample aliquots and expressed as RPD:

$$\text{RPD} = \frac{(a_1 - a_2) \text{ or } (a_2 - a_1)}{(a_1 + a_2)/2} \times 100\%$$

where a_1 and a_2 are the results of duplicate analyses of a sample. Since only two tests are performed on a selected sample, RPD may not be as accurate a measure of precision as RSD. Since RSD does not vary with sample size, it should be used whenever possible to estimate precision of analysis from replicate tests.

3.7.2.2 Measuring Accuracy of Analysis. The accuracy of an analysis can be determined by several procedures. One common method is to analyze a "known" sample, such as a standard solution or a QC check standard solution that may be commercially available or a laboratory-prepared standard solution made from a neat compound, and compare the test results with the true values (theoretically expected values). Such "known" samples must be subjected to all analytical steps, including sample extraction, digestion, or concentration, similar to regular samples. Alternatively, accuracy may be estimated from the recovery of a known standard solution "spiked" or added into the sample. That is, a known amount of the same substance that is to be tested is added to an aliquot of the sample, usually as a solution, prior to the analysis. The concentration of the analyte in the spiked solution of the sample is then measured. The percent spike recovery is then calculated. A correction for the bias in the analytical procedure can then be made, based on the percent spike recovery. However, in most routine analysis such bias correction is not required. Percent spike recovery then may be calculated as follows:

$$\%\ \text{recovery} = \frac{\text{measured concentration}}{\text{theoretical concentration}} \times 100\%$$

Theoretical concentration may be calculated as

$$\frac{C_u \times V_u}{V_u + V_s} + \frac{C_s \times V_s}{V_u + V_s}$$

where C_u is the measured concentration of the unknown sample and C_s is the concentration of the standard solution, while V_u and V_s are the volumes of the sample and the standard solution, respectively.

Percent spike recovery to measure the accuracy of analysis may also be determined by the U.S. Environmental Protection Agency (EPA) method often used in environmental analysis:

$$\%\ \text{recovery} = \frac{100(X_s - X_u)}{K}$$

where X_s = measured value for the spiked sample

X_u = measured value for the unspiked sample adjusted for the dilution of the spike

K = known value of spike in the sample

There are also other methods for monitoring the accuracy of measurement. Accuracy of routine tests may be monitored periodically from the control charts, discussed earlier in this section.

3.8 METHOD DETECTION LIMIT (MDL)

Method detection limit (MDL), detection limit of measurement, instrument detection limit (IDL), and practical quantitation limit (PQL) are some of the terms used in trace analysis. Among these the MDL and the IDL are more commonly used. Instrument detection limit refers to the smallest quantity or concentration of a substance that the instrument can measure. IDL depends on the type of instrument and its sensitivity, and on the physical and chemical properties of the test substance. For example, analysis of a chlorinated organic compound by gas chromatography (GC) using an electron-capture detector (ECD) offers a much higher degree of sensitivity than most other GC detectors. Similarly, IDL for most metals by furnace-atomic absorption (AA) spectrophotometer is much lower than that obtained by flame-AA technique. For most GC methods, IDL may be interpreted as the concentration of a standard solution that gives a response (peak) distinguishable from the noise background. For spectrophotometric methods it is the lowest concentration of a substance that gives a measurable response (i.e., absorbance or transmittance).

Method detection limit, on the other hand is a statistical concept, applicable only in trace analysis of certain type of substances, such as organic pollutants by GC methods. MDL measures the minimum detection limit of the method and involves all analytical steps including sample extraction, concentration, and determination by an analytical instrument; unlike IDL, it is not confined only to the detection limit of the instrument. In environmental analysis of organic pollutants, MDL is defined as the minimum concentration of a substance that can be measured and reported with 99% confidence that the analyte concentration is greater than zero and is determined from the analysis of a sample in a given matrix containing the analyte. For determination of MDL, several replicate analyses are performed at the concentration level of the IDL or at a level equivalent to two to five times the background noise level. The standard deviation of the replicate tests is found. The MDL is determined by multiplying the standard derivation by the *t*-factor (Student's *t*-test).

The term MDL is not widely used in other areas of chemical analysis. In environmental analysis, however, periodic determination of MDL (once a year or with any change in personnel, location, or instrument) is part of the QC requirement.

Bibliography

Alder, H. L., and E. B. Roessler, *Introduction to Probability and Statistics*, W. H. Freeman, San Francisco, 1972.

Bergmann, B., B. von Oepen, and P. Zinn, *Anal. Chem.*, **59**:2532 (1987).

Box, G., W. Hunter, and J. Hunter, *Statistics for Experimenters*, Wiley, New York, 1978.

Caulcutt, R., and R. Boddy, *Statistics for Analytical Chemists*, Chapman and Hall, London, 1983.

Clayton, C. A., J. W. Hines, and P. D. Elkins, *Anal. Chem.*, **59**:2506 (1987).

Dixon, W. J., and F. J. Massey, *Introduction to Statistical Analysis*, McGraw-Hill, New York, 1969.

Hirsch, R. F. "Analysis of Variance in Analytical Chemistry," *Anal. Chem.*, **49**:691A (1977).

Jaffe, A. J., and H. F. Spirer, *Misused Statistics—Straight Talk for Twisted Numbers*, Dekker, New York, 1987.

Linnig, F. J., and J. Mandel, "Which Measure of Precision?," *Anal. Chem.* **36**:25A (1964).

Mark, H., and J. Workman, *Statistics in Spectroscopy*, Academic, San Diego, Calif., 1991.
Meier, P. C., and R. E. Zund, *Statistical Methods in Analytical Chemistry*, Wiley, New York, 1993.
Miller, J. C., and J. N. Miller, *Statistics for Analytical Chemists*, Halsted, Wiley, New York, 1984.
Moore, D. S., *Statistics: Concepts and Controversies*, W. H. Freeman, New York, 1985.
Mulholland, H., and C. R. Jones, *Fundamentals of Statistics*, Plenum, New York, 1968.
Patnaik, P., *Handbook of Environmental Analysis*, CRC Press, Boca Raton, Fla., 1997.
Taylor, J. K., *Statistical Techniques for Data Analysis*, Lewis, Boca Raton, Fla. 1990.
Youden, W. J., *Statistical Methods for Chemists*, Wiley, New York, 1951.
Youden, W. J., "The Sample, the Procedure, and the Laboratory," *Anal. Chem.* **32**:23A (1960).
Youden, W. J., *Statistical Manual of the AOAC*, AOAC, 1111 North 19th St., Arlington, Va. 22209.

SECTION 4
GRAVIMETRIC AND VOLUMETRIC ANALYSIS

4.1 INORGANIC GRAVIMETRIC ANALYSIS

TABLE 4.1 Ionic Product Constant of Water

This table gives values of p*K*w *on a molal scale, where* Kw *is the ionic activity product constant of water. Values are from W. L. Marshall and E. U. Franck,* J. Phys. Chem. Ref. Data ***10****:295 (1981).*

Temp., °C	p*Kw*	Temp., °C	p*Kw*	Temp., °C	p*Kw*
0	14.938	45	13.405	95	12.345
5	14.727	50	13.275	100	12.264
10	14.528	55	13.152	125	11.911
15	14.340	60	13.034	150	11.637
18	14.233	65	12.921	175	11.431
20	14.163	70	12.814	200	11.288
25	13.995	75	12.711	225	11.207
30	13.836	80	12.613	250	11.192
35	13.685	85	12.520	275	11.251
40	13.542	90	12.431	300	11.406

TABLE 4.2 Solubility Products

The data refer to various temperatures between 18 and 25°C and were primarily compiled from values cited by Bjerrum, Schwarzenbach, and Sillen, Stability Constants of Metal Complexes, *Part II, Chemical Society, London, 1958.*

Substance	pK_{sp}	K_{sp}
Actinium		
$Ac(OH)_3$	15	1×10^{-15}
Aluminum		
$AlAsO_4$	15.8	1.6×10^{-16}
Cupferrate, AlL_3	18.64	2.3×10^{-19}
$Al(OH)_3$ amorphous	32.9	1.3×10^{-33}
$AlPO_4$	18.24	6.3×10^{-19}
8-Quinolinolate, AlL_3	29.00	1.00×10^{-29}
Al_2S_3	6.7	2×10^{-7}
Al_2Se_3	24.4	4×10^{-25}
Americium		
$Am(OH)_3$	19.57	2.7×10^{-20}
$Am(OH)_4$	56	1×10^{-56}
Ammonium		
$NH_4UO_2AsO_4$	23.77	1.7×10^{-24}
Arsenic		
$As_2S_3 + 4H_2O \rightarrow 2HAsO_2 + 3H_2S$	21.68	2.1×10^{-22}
Barium		
$Ba_3(AsO_4)_2$	50.11	8.0×10^{-15}
$Ba(BrO_3)_2$	5.50	3.2×10^{-6}
$BaCO_3$	8.29	5.1×10^{-9}
$BaCO_3 + CO_2 + H_2O \rightarrow Ba^{2+} + 2HCO_3^-$	4.35	4.5×10^{-5}
$BaCrO_4$	9.93	1.2×10^{-10}
$Ba_2[Fe(CN)_6] \cdot 6H_2O$	7.5	3.2×10^{-8}
BaF_2	5.98	1.0×10^{-6}
$BaSiF_6$	6	1×10^{-6}
$Ba(IO_3)_2 \cdot 2H_2O$	8.82	1.5×10^{-9}
$Ba(OH)_2$	2.3	5×10^{-3}
$Ba(MnO_4)_2$	9.61	2.5×10^{-10}
$BaMoO_4$	7.40	4.0×10^{-8}
$Ba(NbO_3)_2$	16.50	3.2×10^{-17}
$Ba(NO_3)_2$	2.35	4.5×10^{-3}
BaC_2O_4	6.79	1.6×10^{-7}
$BaC_2O_4 \cdot H_2O$	7.64	2.3×10^{-8}
$BaHPO_4$	6.5	3.2×10^{-7}
$Ba_3(PO_4)_2$	22.47	3.4×10^{-23}
$Ba_2P_2O_7$	10.5	3.2×10^{-11}
$BaHPO_3 \cdot 0.5H_2O$	3	1×10^{-3}
8-Quinolinolate, BaL_2	8.3	5.0×10^{-9}
$Ba(ReO_4)_2$	1.28	5.2×10^{-2}
$BaSeO_4$	7.46	3.5×10^{-8}
$BaSO_4$	9.96	1.1×10^{-10}
$BaSO_3$	6.1	8×10^{-7}
BaS_2O_3	4.79	1.6×10^{-5}
Beryllium		
$BeCO_3 \cdot 4H_2O$	3	1×10^{-3}
Beryllium		
$Be(OH)_2$ amorphous	21.8	1.6×10^{-22}
$Be(OH)_2 + OH^- \rightarrow HBeO_2^- + H_2O$	2.50	3.2×10^{-3}
$BeMoO_4$	1.5	3.2×10^{-2}
$Be(NbO_3)_2$	15.92	1.2×10^{-16}
Bismuth		
$BiAsO_4$	9.36	4.4×10^{-10}
Cupferrate	27.22	6.0×10^{-28}
$Bi(OH)_3$	30.4	4×10^{-31}
BiI_3	18.09	8.1×10^{-19}
$BiPO_4$	22.89	1.3×10^{-23}
Bi_2S_3	97	1×10^{-97}
BiOBr	6.52	3.0×10^{-7}
BiOCl	30.75	1.8×10^{-31}
BiOOH	9.4	4×10^{-10}
$BiO(NO_2)$	6.31	4.9×10^{-7}
$BiO(NO_3)$	2.55	2.82×10^{-3}
BiOSCN	6.80	1.6×10^{-7}
Cadmium		
Anthranilate, CdL_2	8.27	5.4×10^{-9}
$Cd_3(AsO_4)_2$	32.66	2.2×10^{-33}
$[Cd(NH_3)_6](BF_4)_2$	5.7	2×10^{-6}
Benzoate $\cdot$ $2H_2O$	2.7	2×10^{-3}
$Cd(BO_2)_2$	8.64	2.3×10^{-9}
$CdCO_3$	11.28	5.2×10^{-12}
$Cd(CN)_2$	8.0	1.0×10^{-8}
$Cd_2[Fe(CN)_6]$	16.49	3.2×10^{-17}
$Cd(OH)_2$ fresh	13.6	2.5×10^{-14}
$CdC_2O_4 \cdot 3H_2O$	7.04	9.1×10^{-8}
$Cd_3(PO_4)_2$	32.6	2.5×10^{-33}
Quinaldate, CdL_2	12.3	5.0×10^{-13}
CdS	26.1	8.0×10^{-27}
$CdWO_4$	5.7	2×10^{-6}
Calcium		
$Ca_3(AsO_4)_2$	18.17	6.8×10^{-19}
Acetate $\cdot$ $3H_2O$	2.4	4×10^{-3}
Benzoate $\cdot$ $3H_2O$	2.4	4×10^{-3}
$CaCO_3$	8.54	2.8×10^{-9}
$CaCO_3$ calcite	8.35	4.5×10^{-9}
$CaCO_3$ aragonite	8.22	6.0×10^{-9}
$CaCrO_4$	3.15	7.1×10^{-4}
CaF_2	8.28	5.3×10^{-9}
$Ca[SiF_6]$	3.09	8.1×10^{-4}
$Ca(OH)_2$	5.26	5.5×10^{-6}
$Ca(IO_3)_2 \cdot 6H_2O$	6.15	7.1×10^{-7}
$Ca[Mg(CO_3)_2]$ dolomite	11	1×10^{-11}
$CaMoO_4$	7.38	4.2×10^{-8}
$Ca(NbO_3)_2$	17.06	8.7×10^{-18}

(Continued)

TABLE 4.2 Solubility Products (*Continued*)

Substance	pK_{sp}	K_{sp}
Calcium		
$CaC_2O_4 \cdot H_2O$	8.4	4×10^{-9}
$CaHPO_4$	7.0	1×10^{-7}
$Ca_3(PO_4)_2$	28.70	2.0×10^{-29}
8-Quinolinolate, CaL_2	11.12	7.6×10^{-12}
$CaSeO_4$	3.09	8.1×10^{-4}
$CaSeO_3$	5.53	8.0×10^{-6}
$CaSiO_3$	7.60	2.5×10^{-8}
$CaSO_4$	5.04	9.1×10^{-6}
$CaSO_3$	7.17	6.8×10^{-8}
Tartrate dihydrate	6.11	7.7×10^{-7}
$CaWO_4$	8.06	8.7×10^{-9}
Cerium		
CeF_3	15.1	8×10^{-16}
$Ce(IO_3)_3$	9.50	3.2×10^{-10}
$Ce(IO_3)_4$	16.3	5×10^{-17}
$Ce(OH)_3$	19.8	1.6×10^{-20}
$Ce(OH)_4$	47.7	2×10^{-48}
$Ce_2(C_2O_4)_3 \cdot 9H_2O$	25.5	3.2×10^{-26}
$CePO_4$	23	1×10^{-23}
$Ce_2(SeO_3)_3$	24.43	3.7×10^{-25}
Ce_2S_3	10.22	6.0×10^{-11}
(III) Tartrate	19.0	1×10^{-19}
Cesium		
$CsBrO_3$	1.7	5×10^{-2}
$CsClO_3$	1.4	4×10^{-2}
$Cs_2[PtCl_6]$	7.5	3.2×10^{-8}
$Cs_3[Co(NO_2)_6]$	15.24	5.7×10^{-16}
$Cs[BF_4]$	4.7	5×10^{-5}
$Cs_2[PtF_6]$	5.62	2.4×10^{-6}
$Cs_2[SiF_6]$	4.90	1.3×10^{-5}
$CsClO_4$	2.4	4×10^{-3}
$CsClO_4$	2.36	4.3×10^{-3}
$CsMnO_4$	4.08	8.2×10^{-5}
$CsReO_4$	3.40	4.0×10^{-4}
Chromium(II)		
$Cr(OH)_2$	15.7	2×10^{-16}
Chromium(III)		
$CrAsO_4$	20.11	7.7×10^{-21}
CrF_3	10.18	6.6×10^{-11}
$Cr(NH_3)_6(BF_4)_3$	4.21	6.2×10^{-5}
$Cr(OH)_3$	30.2	6.3×10^{-31}
$Cr(NH_3)_6(ReO_4)_3$	11.11	7.7×10^{-12}
$CrPO_4 \cdot 4H_2O$ green	22.62	2.4×10^{-23}
violet	17.00	1.0×10^{-17}
Cobalt		
Anthranilate, CoL_2	9.68	2.1×10^{-10}
$Co_3(AsO_4)_2$	28.12	7.6×10^{-29}
$CoCO_3$	12.84	1.4×10^{-13}
$Co_2[Fe(CN)_6]$	14.74	1.8×10^{-15}
$Co(NH_3)_6[BF_4]_2$	5.4	4×10^{-6}
$Co(OH)_2$ fresh	14.8	1.6×10^{-15}
Cobalt		
$Co(OH)_3$	43.8	1.6×10^{-44}
$Co(IO_3)_2$	4.0	1.0×10^{-4}
Quinaldate, CoL_2	10.8	1.6×10^{-11}
$Co[Hg(SCN)_4]$	5.82	1.5×10^{-6}
α-CoS	20.4	4.0×10^{-21}
β-CoS	24.7	2.0×10^{-25}
8-Quinolinolate, CoL_2	24.8	1.6×10^{-25}
$CoHPO_4$	6.7	2×10^{-7}
$CO_3(PO_4)_2$	34.7	2×10^{-35}
$CoSeO_3$	6.8	1.6×10^{-7}
Copper(I)		
CuN_3	8.31	4.9×10^{-9}
$Cu[B(C_6H_5)_4]$ tetraphenylborate	8.0	1×10^{-8}
CuBr	8.28	5.3×10^{-9}
CuCl	5.92	1.2×10^{-6}
CuCN	19.49	3.2×10^{-20}
CuI	11.96	1.1×10^{-12}
CuOH	14.0	1×10^{-14}
Cu_2S	47.6	2.5×10^{-48}
CuSCN	14.32	4.8×10^{-15}
Copper(II)		
Anthranilate, CuL_2	13.22	6.0×10^{-14}
$Cu_3(AsO_4)_2$	35.12	7.6×10^{-36}
$Cu(N_3)_2$	9.2	6.3×10^{-10}
$CuCO_3$	9.86	1.4×10^{-10}
$CuCrO_4$	5.44	3.6×10^{-6}
$Cu_2[Fe(CN)_6]$	15.89	1.3×10^{-16}
$Cu(IO_3)_2$	7.13	7.4×10^{-8}
$Cu(OH)_2$	19.66	2.2×10^{-20}
CuC_2O_4	7.64	2.3×10^{-8}
$Cu_3(PO_4)_2$	36.9	1.3×10^{-37}
$Cu_2P_2O_7$	15.08	8.3×10^{-16}
Quinaldate, CuL_2	16.8	1.6×10^{-17}
8-Quinolinolate, CuL_2	29.7	2.0×10^{-30}
Rubeanate	15.12	7.67×10^{-16}
CuS	35.2	6.3×10^{-36}
$CuSeO_3$	7.68	2.1×10^{-8}
Dysprosium		
$Dy_2(CrO_4)_3 \cdot 10H_2O$	8	1×10^{-8}
$Dy(OH)_3$	21.85	1.4×10^{-22}
Erbium		
$Er(OH)_3$	23.39	4.1×10^{-24}
Europium		
$Eu(OH)_3$	23.05	8.9×10^{-24}
Gadolinium		
$Gd(HCO_3)_3$	1.7	2×10^{-2}
$Gd(OH)_3$	22.74	1.8×10^{-23}
Gallium		
$Ga_4[Fe(CN)_6]_3$	33.82	1.5×10^{-34}

TABLE 4.2 Solubility Products (*Continued*)

Substance	pK_{sp}	K_{sp}
Gallium		
$Ga(OH)_3$	35.15	7.0×10^{-36}
8-Quinolinolate, GaL_3	40.8	1.6×10^{-41}
Germanium		
GeO_2	57.0	1.0×10^{-57}
Gold(I)		
AuCl	12.7	2.0×10^{-13}
AuI	22.8	1.6×10^{-23}
Gold(III)		
$AuCl_3$	24.5	3.2×10^{-25}
$Au(OH)_3$	45.26	5.5×10^{-46}
AuI_3	46	1×10^{-46}
$Au_2(C_2O_4)_3$	10	1×10^{-10}
Hafnium		
$Hf(OH)_3$	25.4	4.0×10^{-26}
Holmium		
$Ho(OH)_3$	22.3	5.0×10^{-23}
Indium		
$In_4[Fe(CN)_6]_3$	43.72	1.9×10^{-44}
$In(OH)_3$	33.2	6.3×10^{-34}
Quinolinolate, InL_3	31.34	4.6×10^{-32}
In_2S_3	73.24	5.7×10^{-74}
$In_2(SeO_3)_3$	32.6	4.0×10^{-33}
Iron(II)		
$FeCO_3$	10.50	3.2×10^{-11}
$Fe(OH)_2$	15.1	8.0×10^{-16}
$FeC_2O_4 \cdot 2H_2O$	6.5	3.2×10^{-7}
FeS	17.2	6.3×10^{-18}
Iron(III)		
$FeAsO_4$	20.24	5.7×10^{-21}
$Fe_4[Fe(CN)_6]_3$	40.52	3.3×10^{-41}
$Fe(OH)_3$	37.4	4×10^{-38}
$FePO_4$	21.89	1.3×10^{-22}
Quinaldate, FeL_3	16.9	1.3×10^{-17}
$Fe_2(SeO_3)_3$	30.7	2.0×10^{-31}
Lanthanum		
$La(BrO_3)_3 \cdot 9H_2O$	2.5	3.2×10^{-3}
LaF_3	16.2	7×10^{-17}
$La(OH)_3$	18.7	2.0×10^{-19}
$La(IO_3)_3$	11.21	6.1×10^{-12}
$La_2(MoO_4)_3$	20.4	4×10^{-21}
$La_2(C_2O_4)_3 \cdot 9H_2O$	26.60	2.5×10^{-27}
$LaPO_4$	22.43	3.7×10^{-23}
La_2S_3	12.70	2.0×10^{-13}
$La_2(WO_4)_3 \cdot 3H_2O$	3.90	1.3×10^{-4}
Lead		
Acetate	2.75	1.8×10^{-3}
Anthranilate, PbL_2	9.81	1.6×10^{-10}
$Pb_3(AsO_4)_2$	35.39	4.0×10^{-36}
$Pb(N_3)_2$	8.59	2.5×10^{-9}
Lead		
$Pb(BO_2)_2$	10.78	1.6×10^{-11}
$PbBr_2$	4.41	4.0×10^{-5}
$Pb(BrO_3)_2$	1.70	2.0×10^{-2}
$PbCO_3$	13.13	7.4×10^{-14}
$PbCl_2$	4.79	1.6×10^{-5}
PbClF	8.62	2.4×10^{-9}
$PbCrO_4$	12.55	2.8×10^{-13}
$Pb(ClO_2)_2$	8.4	4×10^{-9}
$Pb_2[Fe(CN)_6]$	14.46	3.5×10^{-15}
PbF_2	7.57	2.7×10^{-8}
PbFI	8.07	8.5×10^{-9}
$Pb(OH)_2$	14.93	1.2×10^{-15}
PbOHBr	14.70	2.0×10^{-15}
PbOHCl	13.7	2×10^{-14}
$PbOHNO_3$	3.55	2.8×10^{-4}
PbI_2	8.15	7.1×10^{-9}
$Pb(IO_3)_2$	12.49	3.2×10^{-13}
$PbMoO_4$	13.0	1.0×10^{-13}
$Pb(NbO_3)_2$	16.62	2.4×10^{-17}
PbC_2O_4	9.32	4.8×10^{-10}
$PbHPO_4$	9.90	1.3×10^{-10}
$Pb_3(PO_4)_2$	42.10	8.0×10^{-43}
$PbHPO_3$	6.24	5.8×10^{-7}
Quinaldate, PbL_2	10.6	2.5×10^{-11}
$PbSeO_4$	6.84	1.4×10^{-7}
$PbSeO_3$	11.5	3.2×10^{-12}
$PbSO_4$	7.79	1.6×10^{-8}
PbS	27.9	8.0×10^{-28}
$Pb(SCN)_2$	4.70	2.0×10^{-5}
PbS_2O_3	6.40	4.0×10^{-7}
$PbWO_4$	6.35	4.5×10^{-7}
Lead(IV)		
$Pb(OH)_4$	65.5	3.2×10^{-66}
Lithium		
Li_2CO_3	1.60	2.5×10^{-2}
LiF	2.42	3.8×10^{-3}
Li_3PO_4	8.5	3.2×10^{-9}
$LiUO_2AsO_4$	18.82	1.5×10^{-19}
Lutetium		
$Lu(OH)_3$	23.72	1.9×10^{-24}
Magnesium		
$MgNH_4PO_4$	12.6	2.5×10^{-13}
$Mg_3(AsO_4)_2$	19.68	2.1×10^{-20}
$MgCO_3$	7.46	3.5×10^{-8}
$MgCO_3 \cdot 3H_2O$	4.67	2.1×10^{-5}
MgF_2	8.19	6.5×10^{-9}
$Mg(OH)_2$	10.74	1.8×10^{-11}
$Mg(IO_3)_2 \cdot 4H_2O$	2.5	3.2×10^{-3}
$Mg(NbO_3)_2$	16.64	2.3×10^{-17}
$Mg_3(PO_4)_2$	23–27	10^{-23} to 10^{-27}

(Continued)

TABLE 4.2 Solubility Products (*Continued*)

Substance	pK_{sp}	K_{sp}
Magnesium		
8-Quinolinolate, MgL_2	15.4	4.0×10^{-16}
$MgSeO_3$	4.89	1.3×10^{-5}
$MgSO_3$	2.5	3.2×10^{-3}
Manganese		
Anthranilate, MnL_2	6.28	5.3×10^{-7}
$Mn_3(AsO_4)_2$	28.72	1.9×10^{-29}
$MnCO_3$	10.74	1.8×10^{-11}
$Mn_2[Fe(CN)_6]$	12.10	8.0×10^{-13}
$Mn(OH)_2$	12.72	1.9×10^{-13}
$MnC_2O_4 \cdot 2H_2O$	14.96	1.1×10^{-15}
8-Quinolinolate, MnL_2	21.7	2.0×10^{-22}
$MgSeO_3$	6.9	1.3×10^{-7}
MnS amorphous	9.6	2.5×10^{-10}
crystalline	12.6	2.5×10^{-13}
Mercury(I)		
$Hg_2(N_3)_2$	9.15	7.1×10^{-10}
Hg_2Br_2	22.24	5.6×10^{-23}
Hg_2CO_3	16.05	8.9×10^{-17}
$Hg_2(CN)_2$	39.3	5×10^{-40}
Hg_2Cl_2	17.88	1.3×10^{-18}
Hg_2CrO_4	8.70	2.0×10^{-9}
$(Hg_2)_3[Fe(CN)_6]_2$	20.07	8.5×10^{-21}
$Hg_2(OH)_2$	23.7	2.0×10^{-24}
$Hg_2(IO_3)_2$	13.71	2.0×10^{-14}
Hg_2I_2	28.35	4.5×10^{-29}
$Hg_2C_2O_4$	12.7	2.0×10^{-13}
Hg_2HPO_4	12.40	4.0×10^{-13}
Quinaldate, Hg_2L_2	17.9	1.3×10^{-18}
Hg_2SeO_3	14.2	8.4×10^{-15}
Hg_2SO_4	6.13	7.4×10^{-7}
Hg_2SO_3	27.0	1.0×10^{-27}
Hg_2S	47.0	1.0×10^{-47}
$Hg_2(SCN)_2$	19.7	2.0×10^{-20}
Hg_2WO_4	16.96	1.1×10^{-17}
Mercury (II)		
$Hg(OH)_2$	25.52	3.0×10^{-26}
$Hg(IO_3)_2$	12.5	3.2×10^{-13}
1,10-Phenanthroline	24.70	2.0×10^{-25}
Quinaldate, HgL_2	16.8	1.6×10^{-17}
$HgSeO_3$	13.82	1.5×10^{-14}
HgS red	52.4	4×10^{-53}
HgS black	51.8	1.6×10^{-52}
Neodymium		
$Nd(OH)_3$	21.49	3.2×10^{-22}
Neptunium		
$NpO_2(OH)_2$	21.6	2.5×10^{-22}
Nickel		
$[Ni(NH_3)_6][ReO_4]_2$	3.29	5.1×10^{-4}
Anthranilate, NiL_2	9.09	8.1×10^{-10}
$Ni_3(AsO_4)_2$	25.51	3.1×10^{-26}
$NiCO_3$	8.18	6.6×10^{-9}
Nickel		
$Ni_2(CN)_4 \rightarrow Ni^{2+} + Ni(CN)_4^{2-}$	8.77	1.7×10^{-9}
$Ni_2[Fe(CN)_6]$	14.89	1.3×10^{-15}
$[Ni(N_2H_4)_3]SO_4$	13.15	7.1×10^{-14}
$Ni(OH)_2$ fresh	14.7	2.0×10^{-15}
$Ni(IO_3)_2$	7.85	1.4×10^{-8}
NiC_2O_4	9.4	4×10^{-10}
$Ni_3(PO_4)_2$	30.3	5×10^{-31}
$Ni_2P_2O_7$	12.77	1.7×10^{-13}
8-Quinolinolate, NiL_2	26.1	8×10^{-27}
Quinaldate, NiL_2	10.1	8×10^{-11}
$NiSeO_3$	5.0	1.0×10^{-5}
α-NiS	18.5	3.2×10^{-19}
β-NiS	24.0	1.0×10^{-24}
γ-NiS	25.7	2.0×10^{-26}
Palladium		
$Pd(OH)_2$	31.0	1.0×10^{-31}
$Pd(OH)_4$	70.2	6.3×10^{-71}
Quinaldate, PdL_2	12.9	1.3×10^{-13}
Platinum		
$PtBr_4$	40.5	3.2×10^{-41}
$Pt(OH)_2$	35	1×10^{-35}
Plutonium		
PuO_2CO_3	12.77	1.7×10^{-13}
PuF_3	15.6	2.5×10^{-16}
PuF_4	19.2	6.3×10^{-20}
$Pu(OH)_3$	19.7	2.0×10^{-20}
$Pu(OH)_4$	55	1×10^{-55}
$PuO_2(OH)$	9.3	5×10^{-10}
$PuO_2(OH)_2$	24.7	2×10^{-25}
$Pu(IO_3)_4$	12.3	5×10^{-13}
$Pu(HPO_4)_2 \cdot xH_2O$	27.7	2×10^{-28}
Polonium		
PoS	28.26	5.5×10^{-29}
Potassium		
$K_2[PdCl_6]$	5.22	6.0×10^{-6}
$K_2[PtCl_6]$	4.96	1.1×10^{-5}
$K_2[PtBr_6]$	4.2	6.3×10^{-5}
$K_2[PtF_6]$	4.54	2.9×10^{-5}
K_2SiF_6	6.06	8.7×10^{-7}
K_2ZrF_6	3.3	5×10^{-4}
KIO_4	3.08	8.3×10^{-4}
$K_2Na[Co(NO_2)_6] \cdot H_2O$	10.66	2.2×10^{-11}
$K[B(C_6H_5)_4]$	7.65	2.25×10^{-8}
KUO_2AsO_4	22.60	2.5×10^{-23}
$K_4[UO_2(CO_3)_3]$	4.2	6.3×10^{-5}
Praseodymium		
$Pr(OH)_3$	21.17	6.8×10^{-22}
Promethium		
$Pm(OH)_3$	21	1×10^{-21}

TABLE 4.2 Solubility Products (*Continued*)

Substance	pK_{sp}	K_{sp}
Radium		
$Ra(IO_3)_2$	9.06	8.7×10^{-10}
$RaSO_4$	10.37	4.2×10^{-11}
Rhodium		
$Rh(OH)_3$	23	1×10^{-23}
Rubidium		
$Rb_3[CO(NO_2)_6]$	14.83	1.5×10^{-15}
$Rb_2[PtCl_6]$	7.2	6.3×10^{-8}
$Rb_2[PtF_6]$	6.12	7.7×10^{-7}
$Rb_2[SiF_6]$	6.3	5.0×10^{-7}
$RbClO_4$	2.60	2.5×10^{-3}
$RbIO_4$	3.26	5.5×10^{-4}
Ruthenium		
$Ru(OH)_3$	36	1×10^{-36}
Samarium		
$Sm(OH)_3$	22.08	8.3×10^{-23}
Scandium		
ScF_3	17.37	4.2×10^{-18}
$Sc(OH)_3$	30.1	8.0×10^{-31}
Silver		
AgN_3	8.54	2.8×10^{-9}
Ag_3AsO_4	22.0	1.0×10^{-22}
$AgBrO_3$	4.28	5.3×10^{-5}
AgBr	12.30	5.0×10^{-13}
Ag_2CO_3	11.09	8.1×10^{-12}
$AgClO_2$	3.7	2.0×10^{-4}
AgCl	9.75	1.8×10^{-10}
Ag_2CrO_4	11.95	1.1×10^{-12}
$Ag_3[Co(NO_2)_6]$	20.07	8.5×10^{-21}
Cyanamide, Ag_2CN_2	10.14	7.2×10^{-11}
AgOCN	6.64	2.3×10^{-7}
AgCN	15.92	1.2×10^{-16}
$Ag_2Cr_2O_7$	6.70	2.0×10^{-7}
Dicyanimide, $AgN(CN)_2$	8.85	1.4×10^{-9}
$Ag_4[Fe(CN)_6]$	40.81	1.6×10^{-41}
AgOH	7.71	2.0×10^{-8}
$Ag_2N_2O_2$	18.89	1.3×10^{-19}
$AgIO_3$	7.52	3.0×10^{-8}
AgI	16.08	8.3×10^{-17}
Ag_2MoO_4	11.55	2.8×10^{-12}
$AgNO_2$	3.22	6.0×10^{-4}
$Ag_2C_2O_4$	10.46	3.4×10^{-11}
Ag_3PO_4	15.84	1.4×10^{-16}
Quinaldate, AgL	17.9	1.3×10^{-18}
$AgReO_4$	4.10	8.0×10^{-5}
Ag_2SeO_3	15.00	1.0×10^{-15}
Ag_2SeO_4	7.25	5.7×10^{-8}
AgSeCN	15.40	4.0×10^{-16}
Ag_2SO_4	4.84	1.4×10^{-5}
Ag_2SO_3	13.82	1.5×10^{-14}
Ag_2S	49.2	6.3×10^{-50}
AgSCN	12.00	1.0×10^{-12}
Silver		
$AgVO_3$	6.3	5×10^{-7}
Ag_2WO_4	11.26	5.5×10^{-12}
Sodium		
$Na[Sb(OH)_6]$	7.4	4.0×10^{-8}
Na_3AlF_6	9.39	4.0×10^{-10}
$NaK_2[Co(NO_2)_6]$	10.66	2.2×10^{-11}
$Na(NH_4)_2[Co(NO_2)_6]$	11.4	4×10^{-12}
$NaUO_2AsO_4$	21.87	1.3×10^{-22}
Strontium		
$Sr_3(AsO_4)_2$	18.09	8.1×10^{-19}
$SrCO_3$	9.96	1.1×10^{-10}
$SrCrO_4$	4.65	2.2×10^{-5}
SrF_2	8.61	2.5×10^{-9}
$Sr(IO_3)_2$	6.48	3.3×10^{-7}
$SrMoO_4$	6.7	2×10^{-7}
$Sr(NbO_3)_2$	17.38	4.2×10^{-18}
$SrC_2O_4 \cdot H_2O$	6.80	1.6×10^{-7}
$Sr_3(PO_4)_2$	27.39	4.0×10^{-28}
8-Quinolinolate, SrL_2	9.3	5×10^{-10}
$SrSeO_3$	5.74	1.8×10^{-6}
$SrSeO_4$	3.09	8.1×10^{-4}
$SrSO_3$	7.4	4×10^{-8}
$SrSO_4$	6.49	3.2×10^{-7}
$SrWO_4$	9.77	1.7×10^{-10}
Terbium		
$Tb(OH)_3$	21.70	2.0×10^{-22}
Tellurium		
$Te(OH)_4$	53.52	3.0×10^{-54}
Thallium(I)		
TlN_3	3.66	2.2×10^{-4}
TlBr	5.47	3.4×10^{-6}
$TlBrO_3$	4.07	8.5×10^{-5}
$Tl_2[PtCl_6]$	11.4	4.0×10^{-12}
TlCl	3.76	1.7×10^{-4}
Tl_2CrO_4	12.00	1.0×10^{-12}
$Tl_4[Fe(CN)_6] \cdot 2H_2O$	9.3	5×10^{-10}
$TlIO_3$	5.51	3.1×10^{-6}
TlI	7.19	6.5×10^{-8}
$Tl_2C_2O_4$	3.7	2×10^{-4}
Tl_2SeO_3	38.7	2×10^{-39}
Tl_2SeO_4	4.00	1.0×10^{-4}
Tl_2S	20.3	5.0×10^{-21}
TlSCN	3.77	1.7×10^{-4}
Thallium(II)		
$Tl(OH)_3$	45.20	6.3×10^{-46}
8-Quinolinolate, TlL_3	32.4	4.0×10^{-33}
Thorium		
$ThF_4 \cdot 4H_2O + 2H^+ \rightarrow ThF_2^{2+} + 2HF + 4H_2O$	7.23	5.9×10^{-6}

(Continued)

TABLE 4.2 Solubility Products (*Continued*)

Substance	pK_{sp}	K_{sp}	Substance	pK_{sp}	K_{sp}
Thorium			Vanadium		
$Th(OH)_4$	44.4	4.0×10^{-45}	$VO(OH)_2$	22.13	5.9×10^{-23}
$Th(C_2O_4)_2$	22	1×10^{-22}	$(VO)_3PO_4$	24.1	8×10^{-25}
$Th_3(PO_4)_4$	78.6	2.5×10^{-79}	Ytterbium		
$Th(HPO_4)_2$	20	1×10^{-20}	$Yt(OH)_3$	23.6	2.5×10^{-24}
$Th(IO_3)_4$	14.6	2.5×10^{-15}	Yttrium		
Thulium			YF_3	12.14	6.6×10^{-13}
$Tm(OH)_3$	23.48	3.3×10^{-24}	$Y(OH)_3$	22.1	8.0×10^{-23}
Tin			$Y_2(C_2O_4)_3$	28.28	5.3×10^{-29}
$Sn(OH)_2$	27.85	1.4×10^{-28}	Zinc		
$Sn(OH)_4$	56	1×10^{-56}	Anthranilate, ZnL_2	9.23	5.9×10^{-10}
SnS	25.0	1.0×10^{-25}	$Zn_3(AsO_4)_2$	27.89	1.3×10^{-28}
Titanium			$Zn(BO_2)_2 \cdot H_2O$	10.18	6.6×10^{-11}
$Ti(OH)_3$	40	1×10^{-40}	$ZnCO_3$	10.84	1.4×10^{-11}
$TiO(OH)_2$	29	1×10^{-29}	$Zn_2[Fe(CN)_6]$	15.39	4.0×10^{-16}
Uranium			$Zn(IO_3)_2$	7.7	2.0×10^{-8}
UO_2HAsO_4	10.50	3.2×10^{-11}	$Zn(OH)_2$	16.92	1.2×10^{-17}
UO_2CO_3	11.73	1.8×10^{-12}	ZnC_2O_4	7.56	2.7×10^{-8}
$(UO_2)_2[Fe(CN)_6]$	13.15	7.1×10^{-14}	$Zn_3(PO_4)_2$	32.04	9.0×10^{-33}
$UF_4 \cdot 2.5H_2O$	21.24	5.7×10^{-22}	Quinaldate, ZnL_2	13.8	1.6×10^{-14}
$UO_2(OH)_2$	21.95	1.1×10^{-22}	8-Quinolinolate, ZnL_2	24.3	5.0×10^{-25}
$UO_2(IO_3)_2 \cdot H_2O$	7.5	3.2×10^{-8}	$ZnSeO_3$	6.59	2.6×10^{-7}
$UO_2C_2O_4 \cdot 3H_2O$	3.7	2×10^{-4}	α-ZnS	23.8	1.6×10^{-24}
$(UO_2)_3(PO_4)_2$	46.7	2.0×10^{-47}	β-ZnS	21.6	2.5×10^{-22}
UO_2HPO_4	10.67	2.1×10^{-11}	$Zn[Hg(SCN)_4]$	6.66	2.2×10^{-7}
UO_2SO_3	8.59	2.6×10^{-9}	Zirconium		
$UO_2(SCN)_2$	3.4	4×10^{-4}	$ZrO(OH)_2$	48.2	6.3×10^{-49}
			$Zr_3(PO_4)_4$	132	1×10^{-132}

Source: J. A. Dean, ed., *Lange's Handbook of Chemistry*, 14th ed., McGraw-Hill, New York, 1992.

TABLE 4.3 Elements Precipitated by General Analytical Reagents

This table includes the more common reagents used in gravimetric determinations. The lists of elements precipitated are not in all cases exhaustive. The usual solvent for a precipitating agent is indicated in parentheses after its name or formula. When the symbol of an element or radical is italicized, the element may be quantitatively determined by the use of the reagent in question.

Reagent	Conditions	Substances precipitated
Ammonia, NH_3 (aqueous)	After removal of acid sulfide groups, B, and F.	*Al*, Au, *Be*, Co, *Cr*, *Cu*, *Fe*, Ga, *In*, Ir, *La*, Nb, Ni, Os, P, *Pb*, *rare earths*, *Sc*, Si, *Sn*, Ta, *Th*, *Ti*, *U*, V, *Y*, Zn, *Zr*.
Ammonium polysulfide, $(NH_4)_2S_x$ (aqueous)	After removal of acid sulfide and $(NH_4)_2S$ groups, B, and F.	Co, Mn, Ni, Si, Tl, V, W, Zn.
Anthranilic acid, $NH_2C_6H_4COOH$ (aqueous)	1% aqueous solution (pH 6); Cu separated from others at pH 2.9.	Ag, *Cd*, *Co*, *Cu*, Fe, *Hg*, *Mn*, *Ni*, *Pb*, *Zn*.
α-Benzoin oxime, $C_6H_5CHOHC(=NOH)C_6H_5$ (1–2% alcohol)	(a) Strongly acid medium. (b) Ammoniacal tartrate medium.	(a) Cr(VI), *Mo(VI)*, Nb, Pd(II), Ta(V), V(V), *W(VI)*. (b) Above list.
Benzidine, $H_2NC_6H_4C_6H_4NH_2$ (alcohol), 0.1*M* HCl		Cd, Fe(III), IO_3^-, PO_4^{3-}, SO_4^{2-}, *W(VI)*.
N-Benzoylphenylhydroxylamine, $C_6H_5CO(C_6H_5)NOH$ (aqueous)	Similar to cupferron (q.v.). Cu, Fe(III), and Al complexes can be weighed as such; Ti compound must be ignited to the oxide.	See under Cupferron.
Cinchonine, $C_{19}H_{21}N_2OH$, 6*M* HCl		Ir, Mo, Pt, *W*.
Cupferron, $C_6H_5N(NO)ONH_4$ (aqueous)	Group precipitant for several higher-charged metal ions from strongly acid solution. Precipitate ignited to metal oxide.	*Al*, *Bi*, *Cu*, *Fe*, *Ga*, La, Mo, *Nb*, Pd, rare earths, Sb, *Sn*, Ta, *Th*, *Ti*, Tl, *U*, *V*, W, *Zr*.
1,2-Cyclohexanedionedioxime	More water soluble than dimethylglyoxime; less subject to coprecipitation with metal chelate.	See under Dimethylglyoxime.
Diammonium hydrogen phosphate, $(NH_4)_2HPO_4$ (aqueous)	(a) Acid medium. (b) Ammoniacal medium containing citrate or tartrate.	(a) *Bi*, *Co*, Hf, In, Ti, *Zn*, *Zr*. (b) Au, Ba, *Be*, Ca, Hg, In, La, *Mg*, *Mn*, Pb, rare earths, Sr, Th, U, *Zr*.
Dimethylglyoxime, $[CH_3C(NOH)]_2$ (alcohol)	(a) Dilute HCl or H_2SO_4 medium. (b) Ammoniacal tartrate medium about pH 8. Weighed as such.	(a) Au, *Pd*, Se. (b) *Ni* (and Co, Fe if present in large amounts).
Hydrazine, N_2H_4 (aqueous)		Ag, Au, *Cu*, *Hg*, Ir, *Os*, Pd, Pt, *Rh*, Ru, *Se*, *Te*.
Hydrogen sulfide, H_2S	(a) 0.2–0.5*M* H^+. (b) Ammoniacal solution after removal of acid sulfide group.	(a) Ag, *As*, Au, Bi, Cd, *Cu*, *Ge*, *Hg*, In, *Ir*, *Mo*, Os, Pb, Pd, *Pt*, Re, *Rh*, Ru, Sb, Se, Sn, Te, Tl, V, W, Zn. (b) Co, Fe, Ga, In, Mn, Ni, Tl, U, V, Zn.
4-Hydroxyphenylarsonic acid, $C_6H_4(OH)AsO(OH)_2$ (aqueous)	Dilute acid solution.	Ce, *Fe*, *Sn*, *Th*, *Ti*, *Zr*.
8-Hydroxyquinoline (oxine), C_9H_6NOH (alcohol)	(a) HOAc–OAc^- buffer. (b) Ammoniacal solution.	(a) Ag, *Al*, Bi, *Cd*, *Co*, Cr, *Cu*, *Fe*, Ga, Hg, *In*, La, *Mn*, *Mo*, Nb, *Ni*, Pb, Pd, rare earths, Sb, Ta, Th, *Ti*, *U*, V, W, *Zn*, *Zr*. (b) Same as in (a) except for Ag; in addition, Ba, *Be*, *Ca*, *Mg*, Sn, Sr.

(Continued)

TABLE 4.3 Elements Precipitated by General Analytical Reagents (*Continued*)

Reagent	Conditions	Substances precipitated
2-Mercaptobenzothiazole, $C_6H_4(SCN)SH$, (acetic acid solution)	Ammoniacal solution, except for Cu, when a dilute acid solution is used.	Ag, *Au*, *Bi*, *Cd*, *Cu*, Hg, *Ir*, *Pb*, *Pt*, *Rh*, Tl.
Nitron (diphenylenedianilohydrotriazole), $C_{20}H_{16}N_4$ (5% acetic acid)	Dilute H_2SO_4 medium.	*B*, ClO_3^-, ClO_4^-, NO_3^-, ReO_4^-, *W*.
1-Nitroso-2-naphthol, $C_{10}H_6(NO)OH$ (very dilute alkali)	Selective for Co; acid solution. Precipitate ignited to Co_3O_4.	Ag, Au, B, *Co*, Cr, *Cu*, *Fe*, Mo, Pd, Ti, V, W, Zr.
Oxalic acid, $H_2C_2O_4$ (aqueous)	Dilute acid solution.	*Ag*, *Au*, Cu, *Hg*, *La*, Ni, *Pb*, *rare earths*, *Sc*, *Th*, U(IV), W, *Zr*.
Phenylarsonic acid, $C_6H_5AsO(OH)_2$ (aqueous)	Selective precipitants for quadrivalent metals in acid solution. Metals weighed as dioxides.	*Bi*, Ce(IV), Fe, *Hf*, *Mg*, *Sn*, *Ta*, *Th*, Ti, U(IV), W, *Zr*.
Phenylthiohydantoic acid, $C_6H_5N = C(NH_2)SCH_2COOH$ (aqueous or alcohol)		Bi, Cd, *Co*, Cu, Fe, Hg, Ni, Pb, Sb.
Picrolonic acid, $C_{10}H_7O_5N_4H$ (aqueous)	Neutral solution.	*Ca*, Mg, *Pb*, *Th*.
Propylarsonic acid, $C_3H_9AsO(OH)_2$ (aqueous)	Preferred for W; see Phenylarsonic acid.	
Pyridine plus thiocyanate	Dilute acid solution.	*Ag*, *Cd*, *Cu*, *Mn*, *Ni*.
Quinaldic acid, C_9H_6NCOOH (aqueous)	Dilute acid solution.	Ag, *Cd*, Co, *Cu*, Fe, Hg, Mo, Ni, Pb, Pd, Pt(II), *U*, W, *Zn*.
Salicylaldoxime, $C_7H_5(OH)NOH$ (alcohol)	Dilute acid solution.	Ag, *Bi*, Cd, Co, *Cu*, Fe, Hg, Mg, Mn, Ni, *Pb*, *Pd*, V, Zn.
Silver nitrate, $AgNO_3$ (aqueous)	(a) Dilute HNO_3 solution. (b) Acetate buffer, pH 5–7.	(a) Br^-, Cl^-, I^-, SCN^-. (b) *As(V)*, CN^-, OCN^-, IO_3^-, *Mo(VI)*, N_3^-, S^{2-}, *V(V)*.
Sodium tetraphenylborate, $NaB(C_6H_5)_4$ (aqueous)	Specific for K group of alkali metals from dilute HNO_3 or HOAc solution (pH 2), or pH 6.5 in presence of EDTA.	*Cs*, *K*, NH_4^+, *Rb*.
Tannic acid (tannin), $C_{14}H_{10}O_9$ (aqueous)	Acts as negative colloid that is a flocculent for positively charged hydrous oxide sols. Noteworthy for W in acid solution, and for Ta (from Nb in acidic oxalate medium).	*Al*, *Be*, Cr, Ga, *Ge*, Nb, Sb, *Sn*, *Ta*, *Th*, *Ti*, *U*, V, *W*, *Zr*.
Tartaric acid, $HOOC(CHOH)_2COOH$ (aqueous)		*Ca*, *K*, *Mg*, *Sc*, *Sr*, *Ta*.
Tetraphenylarsonium chloride, $(C_6H_5)_4AsCl$ (aqueous)	$(C_6H_5)_4AsTlCl_4$ and $(C_6H_5)_4AsReO_4$ weighed as such.	*Re*, *Tl*.
Thioglycolic-β-aminonaphthalide, thionalide, $C_{10}H_7NHCOCH_2SH$ (alcohol)	(a) Acid solution. (b) Carbonate medium containing tartrate. (c) Carbonate medium containing tartrate and cyanide. (d) Strongly alkaline medium containing tartrate and cyanide.	(a) Ag, As, Au, Bi, *Cu*, *Hg*, *Os*, *Pb*, Pd, *Rh*, *Ru*, Sb, Sn, Tl. (b) Au, Cd, Cu, Hg(II), Tl(I). (c) Au, Bi, Pb, Sb, Sn, Tl. (d) *Tl*.

TABLE 4.4 Summary of the Principal Methods for the Separation and Gravimetric Determinations of the Elements

In this table the elements are listed by symbol in alphabetical order. Methods accepted as "standard" by the majority of chemists are generally included. More detailed descriptions of the procedures employed should be sought in standard textbooks or in the literature referred to in review articles, particularly the biannual reviews appearing in Analytical Chemistry. *Details of "homogeneous" precipitation methods should be sought in L. Gordon, M. L. Salutsky, and H. H. Willard,* Precipitation from Homogeneous Solution, *Wiley, New York, 1959.*

Element	Conditions of Separation, precipitation, and filtration	Noteworthy separations	Principal interferences	Weighing form
Ag	(1) Add HCl to 1% HNO_3 solution at 70°C. Let stand several hours in the dark. Use filtering crucible; wash with 0.06% HNO_3. Homogeneous precipitation involves NH_3 plus β-hydroxyethyl acetate.	Most metals	Cu(I), Hg(I), Tl(I)	AgCl
	(2) Heat 0.1*N* HNO_3 solution to 80°C and add 1% thionalide in alcohol (10 mL per 10 mg metal). Filter hot through filter paper; wash with hot water. Ignite to Ag or dissolve in HNO_3 and follow Procedure 1.	Pb and Tl	See under Thionalide	Ag or AgCl
Al	(1) Add urea to boiling, slightly acid solution (pH 3.1–3.5) containing succinic acid, until pH 5–6. Filter without delay through paper; wash with 1% succinic acid. Ignite to Al_2O_3. If Cu is present, reduce with H_2NOH. If Fe is present, reduce with NH_4HSO_3 and add phenylhydrazine.	Alkali and alkaline-earth metals, Cd, Co, Cu(I), Fe(II), Mg, Mn, Ni	Bi, Cr(III), Cu, Fe(III), Hg(II), Pb, Th, Ti, Zn, Zr	Al_2O_3
	(2) To boiling 0.05% HCl containing NH_4OAc, add 10-fold excess of $(NH_4)_2HPO_4$. Add paper pulp and filter through paper; wash with hot 5% NH_4NO_3.	Ca, Fe(II), Mn	Bi, Cd, Co, Cu, Fe(III), Hg, Ni, Th, Ti, Zn, Zr	$AlPO_4$
	(3) To neutral solution at 50–60°C, add 5% 8-hydroxyquinoline in 4*M* HOAc, then NH_4OAc and boil for 1 min. Use a filtering crucible, wash with cold water.	As, B, Be, F^-, Mo, Nb, PO_4^{3-}, Ta, V(V), U; also Cd, Co, Cu, Fe, Mg, Mn, Sn, Zn in presence of NH_3, CN^-, and EDTA.	Fe(III)	$Al(C_9H_6ON)_3$
	(4) At pH 3.5–4.0 in presence of NH_4OAc, add NH_4 benzoate and heat to 80°C. Dissolve precipitate with HCl and then add NH_3 dropwise until bromophenol blue changes color. Ignite to Al_2O_3.	Ca, Co, Mg, Mn, Ni, Zn		Al_2O_3
As	(1) To As(III) in 9*M* HCl solution at 15–20°C, add H_2S or thio-acetamide and boiling. Filter through filtering crucible, wash successively with cold 8*M* HCl saturated with H_2S, alcohol, CS_2, and alcohol.	Alkali and alkaline-earth metals, Mg	Bi, Cd, Cu, Ge, Hg(II), Mo, Pb, Re, Sb, Se, Sn, Te, Pt group metals	Usually volumetrically

(Continued)

TABLE 4.4 Summary of the Principal Methods for the Separation and Gravimetric Determinations of the Elements (*Continued*)

Element	Conditions of Separation, precipitation, and filtration	Noteworthy separations	Principal interferences	Weighing form
As	(2) Distillation of $AsCl_3$ (q.v.).	Sb, Sn, all other metals	Ge	Volumetrically
	(3) To ice-cold solution of As(V), add NH_4Cl, $MgCl_2$, and NH_3 until slightly alkaline. Then add concentrated NH_3. (a) Use filtering crucible and wash with cold 1% NH_3 solution. Ignite. (b) Wash with ethanol, then ether. Dry in vacuum desiccator.		Phosphate, all heavy metals, organic matter	(a) $Mg_2As_2O_7$ (b) $MgNH_4AsO_4 \cdot 6H_2O$
Au	To a hot 2*M* HCl, add 5% hydroquinone solution. Keep hot for 2 h. Filter through paper and wash with hot water. Ignite to the metal.	Pt metals, Cu, Ni, Sn, Zn	HNO_3, alkaline earths, Pb, Se, Te	Au metal
Ba	(1) To a boiling solution, slowly add H_2SO_4. Addition of ethanol or acetone decreases solubility of $BaSO_4$. Use fine paper or filtering crucible; wash with cold water (ethanol). Homogeneous precipitation involves adding dimethyl sulfate to a boiling solution.	Most metals	Ca, Pb, Sr	$BaSO_4$
	(2) To a boiling, very dilute HOAc solution, add $(NH_4)_2CrO_4$. Cool. Use filtering crucible; wash with 0.5% NH_4OAc. Dissolve the $BaCrO_4$ in HCl and repeat. Ethanol decreases solubility. Homogeneous method involves $K_2Cr_2O_7$ and urea in a boiling solution.	Ca, Sr		$BaCrO_4$ or volumetrically
Be	To a boiling solution containing NH_4Cl and EDTA, add aqueous NH_3 but avoiding excess. Filter the $Be(OH)_2$ through paper; wash with hot 2% NH_4NO_3. Ignite to BeO. Cool over P_2O_5 desiccant and weigh immediately.	Al, Ba, Bi, Ca, Cd, Cr(III), Cu, Fe, K, Mg, Mn, Na, Pb, Sr, Zn	Al, Cr(III), Fe, Ti	BeO
Bi	(1) To a neutral, boiling solution (nitrate or perchlorate), add a little dilute HCl. Filter the BiOCl through paper; wash with hot water, then ethanol.	Most metals	Ag, As, Sb, Sn	BiOCl
	(2) To cold solution 1*N* in Na_2Co_3 and containing tartrate and cyanide, add fourfold excess of 1% thionalide in alcohol. Heat to boiling and stir until precipitate is crystalline. Wash cooled solution by decantation with water, then transfer to sintered-glass crucible and wash with 10% alcohol. Dry at 105°C.	Ag, Al, As, Cd, Cr(III), Co, Cu, Fe, Hg, Ni, Pd, Pt, Ti, V, Zn	Au, Pb, Sb(III), Sn(IV), Tl(III)	$Bi(C_{12}H_{10}NOS)_3$
	(3) Add $(NH_4)_2HPO_4$ to boiling dilute HNO_3. Use filtering crucible; wash with hot slightly acid 2% NH_4NO_3.	Ag, Al, Cd, Cu, Hg, Zn	As(V), Sb, Sn, Ti, Zr	$BiPO_4$

Br^-	Add $AgNO_3$ to boiling dilute HNO_3 solution. Use filtering crucible; wash with water containing few drops of HNO_3. Protect from light.		Cl^-, CN^-, SCN^-, I^-, S^{2-}, $S_2O_3^{2-}$	AgBr
Ca	To a boiling slightly acid solution, add $(NH_4)_2C_2O_4$ and slowly neutralize with NH_3 (or urea) to form CaC_2O_4. Alternatively, add dimethy1 oxalate to a boiling solution.	Alkali metals, Mg	All other metals	$CaCO_3$
Cd	(1) Add thioacetamide (or H_2S) to cold 3*M* H_2SO_4. Filter through paper. Reprecipitation usually necessary. Wash with slightly acid H_2SO_4 solution.	Al, alkali and alkaline-earth metals, Co, Fe, Mn, Ni, Zn	As, Au, Bi, Cu, Ge, Hg, Mo, platinum metals, Pb, Re, Sb, Se, Sn, Te	Ignite to $CdSO_4$
	(2) 3% sodium quinaldate added to weakly acid solution. Add 2*N* NaOH until pH is 4–5. Decant through filtering crucible of fine porosity; wash with cold water. Dry at 125°C.	Alkali and alkaline-earth metals	All others	$Cd(C_{10}H_6NO_2)_2$
Ce	(1) Add $H_2C_2O_4$. Filter through paper, wash with 1% $H_2C_2O_4$.	Ti, Zr	Rare earths, Th	Ignite to CeO_2
	(2) To Ce(III) and iodate, add $(NH_4)_2S_2O_8$ or $KBrO_3$ to the boiling solution. Filter Ce(IV) iodate. Ignite to CeO_2.	Rare earths	Th, Ti, Zr	CeO_2
Cl^-	Add $AgNO_3$ to boiling slightly acid HNO_3 solution. Use filtering crucible; wash with 0.5% HNO_3.		Br^-, I^-, CN^-, $Fe(CN)_6^{3-}$, $Fe(CN)_6^{4-}$, S^{2-}, SCN^-, $S_2O_3^{2-}$	AgCl
Co	(1) To a boiling, slightly acid solution, add 1-nitroso-2-naphthol (7% in HOAc). Use a filtering crucible; wash with hot 33% HOAc, then hot water. In the homogeneous method, KNO_2 and 2-naphthol are added to the boiling, slightly acid solution.	Al, As, Ba, Be, Ca, Cd, Hg, Mg, Mn, Ni, Pb, phosphate, rare earths, Sb, Sr, Zn	Ag, Bi, Cr(III), Cu, Fe, Sn(IV), Ti, U(VI), V(V), W(VI), Zr	Ignite to $CoSO_4$ or Co_3O_4
	(2) To 50 mL (≤175 mg Co) and 5 mL HOAc, heat to boiling, and add 30 mL 50% $NaNO_2$ (95°C). Let stand 1 h, then filter through fritted crucible. Wash with 100 mL 2% $NaNO_2$, then five 10-mL portions of 80% EtOH, and once with acetone. Dry at 110°C.			$K_3Co(NO_2)_6$
Cr	(1) In the homogeneous method add Pb(II) and $KBrO_3$ to a Cr(III) solution in an acetate buffer, and heat to boiling until precipitation is complete.	Ba, Bi, Hg(I,II)	NH_4	$PbCrO_4$

(Continued)

TABLE 4.4 Summary of the Principal Methods for the Separation and Gravimetric Determinations of the Elements (*Continued*)

Element	Conditions of Separation, precipitation, and filtration	Noteworthy separations	Principal interferences	Weighing form
Cr	(2) Volatilize as CrO_2Cl_2 (*q.v.*).	Most metals	Fe, Os, Ru, Sb, Sn	Volumetrically
Cu	(1) Electrodeposition as Cu (see Sec. 14.6).	Cd, Pb, Zn	Ag, As, Bi, Hg, Sb, Sn	Cu
	(2) To slightly acid HCl or H_2SO_4 solution, add NH_4HSO_3 and slight excess of NH_4SCN. Use filtering crucible; wash the CuSCN with cold 1% NH_4SCN containing a little NH_4HSO_3, then 20% ethanol.	Most metals; tartrate needed if Bi, Sb, Sn present	Ag, Hg(I), Pb, precious metals, Se, Te	CuSCN, CuO, or volumetrically
	(3) Salicylaldoxime added to solution of pH 2.6–3.1. Use filtering crucible; wash with cold water.	Most metals		$Cu(C_7H_6NO_2)_2$ or volumetrically
	(4) Add ammonia until a deep blue tetrammine copper ion formed. Bring to boiling; add tartrate and 2% alcoholic α-benzoin oxime. Use a fine porosity filtering crucible; wash with hot 1% ammonia, then hot ethanol.	Al, Cd, Co, Fe, Ni, Pb, Zn		$Cu(C_{13}H_{11}NO_2)$ or volumetrically
	(5) To boiling H_2SO_4 solution (pH 2.5) quinalidic acid is added until 25% excess. Use filtering crucible; wash with hot water.	Alkalie and alkaline-earth metals	Fe, Zn	$Cu(C_{10}H_6NO_2)_2 \cdot H_2O$
	(6) To 0.1*N* HNO_3 at 80°C, add 1% thionalide in alcohol. Filter hot through filtering crucible; wash with hot water.	Al, Cd, Cr(III), Co, Fe(II), Mn, Ni, Pb, Tl, Zn	All others	$Cu(C_{12}H_{10}NOS)_2 \cdot H_2O$
F^-	To a neutral solution, add successively 1 mL HCl, 0.3 g NaCl, $Pb(NO_3)_2$, and 5 g NaOAc. Use filtering crucible; wash with saturated PbClF solution, then a little water.			PbClF
Fe	(1) Basic formate or benzoate method involves adding urea to a boiling solution containing either formic acid or benzoic acid until pH 5. Add H_2O_2 near the end to reoxidize any Fe(II) formed. Filter through paper. Ignite precipitate.	Ba, Ca, Co, Cu, Mg, Mn, Ni, Sr, Zn	Al, As, Au, Bi, Ge, Hg, Mo, Pb, platinum metals, Re, Sb, Se, Sn, Te	Fe_2O_3
	(2) To a strongly acid solution at 0°C, add 5% aqueous solution of cupferron. Add paper pulp and filter; wash with 3.5% HCl containing 0.15% cupferron, then 9% NH_3, then H_2O. Ignite precipitate.	Al, Co, Cr, Mn, Ni, Zn, and phosphate	Sn, Ti, Zr	Fe_2O_3
Ga	(1) Extraction from HCl by diethyl ether.	Most metals	Au(III), Fe(III)	
	(2) Add NH_3 (or urea) to boiling slightly acid H_2SO_4. Add paper pulp and filter. Wash with 2% NH_4NO_3.			Ga_2O_3
Ge	Following distillation and diethyl ether extraction, then back-extraction with 40 mL water, add 20 mL 25%			$(C_{19}H_{22}ON_2)_4N_4$–$GeMo_{12}O_{40}$;

	NH_4NO_3, 20 mL 2% $(NH_4)_2MoO_4$, and 3 mL HNO_3. To the clear solution add 10 mL 0.25% cinchonine. Let stand several hours. Filter through porcelain fritted crucible, wash with 2.5% NH_4NO_3 plus 0.1*M* HNO_3. Dry at 150°C.			Ge: 0.0239
Hg	(1) Reduction to Hg metal either by adding hydrazine to an ammoniacal solution containing tartrate, or adding $SnCl_2$ to fairly strong HCl solution. Use filtering crucible; wash with acetone and then diethyl ether (coagulate Hg into single drop).	Most common metals	Ag, Au, Cu, Pt	Hg
	(2) To 0.1*N* HNO_3 at 80°C, add threefold excess 1% thionalide in alcohol. Filter hot solution through filtering crucible; wash with hot water.	Al, Cd, Cr(III), Co, Fe(II), Mn, Ni, Pb, Tl, Zn	Cl^- (< 0.1*N*), other metals	$Hg(C_{12}H_{10}NOS)_2$
I^-	Add $AgNO_3$ to ammoniacal solution, then acidify with HNO_3. Use filtering crucible; wash with 1% HNO_3, then water.		Cl^-, Br^-	AgI
In	Add NH_4NO_3 and adjust to pH 5–6 by adding aqueous NH_3. Add paper pulp; wash with slightly ammoniacal NH_4NO_3.		Al, Fe, Ti, Zr	In_2O_3
Ir	To a dilute acid solution containing $KBrO_3$, add $NaHCO_3$ until pH 6. Filter dioxide through paper; wash with 1% $(NH_4)_2SO_4$.	Pt(IV)		IrO_2
K	(1) Chloride solution evaporated until just moist with $HClO_4$. Soluble perchlorates extracted with ethyl acetate leaving $KClO_4$. Fitler through paper; wash with ethyl acetate. Dissolve in hot water and repeat precipitation. Use filtering crucible.	Al, Ca, Na, Li	Cs, Rb, NH_4^+, Tl(I), SO_4^{2-}	$KClO_4$
	(2) Chloride solution evaporated to a syrup with H_2PtCl_6. Cool, add 25 mL 80% ethanol. Use filtering crucible; wash with 80% ethanol until filtrate is colorless.	Al, alkaline earths, Fe, Li, Na, borate, phosphate, sulfate	NH_4^+, Rb, Cs, most heavy metals	K_2PtCl_6
	(3) Add HIO_4 to small volume, then adding larger volume of ethanol–ethyl acetate (1 : 1) at 0°C. Filtering crucible; wash with ethanol–ethyl acetate at 0°C.	Al, Ca (SO_4^{2-} absent), Co, Li, Mg, Ni, Na, Zn	Cr, Cs, Fe, Mn, NH_4^+, Rb	KIO_4 or volumetrically
	(4) Acidify with HNO_3 or HOAc to pH 2 and add 3% aqueous Na tetraphenylborate. Maintain constant temperature of 20°C throughout. Use filtering crucible; wash with cold water. Dry at 100°C.	Most metals	NH_4^+, Rb, Cs, Ag, Hg(II), Tl(I)	$KB(C_6H_5)_4$

(Continued)

TABLE 4.4 Summary of the Principal Methods for the Separation and Gravimetric Determinations of the Elements (*Continued*)

Element	Conditions of Separation, precipitation, and filtration	Noteworthy separations	Principal interferences	Weighing form
La (and rare earths)	Precipitation from slightly acid solution with $H_2C_2O_4$. Filter through paper; wash with 1% $H_2C_2O_4$. Homogeneous method involves adding dimethyl oxalate and boiling.	Al, Be, Fe, Nb, Ta, Ti, Zr	Th	La_2O_3, R_2O_3
Li	Extraction of LiCl from dry residue by 1-octanol.	Na		Li_2SO_4
Mg	(1) To a slightly acid solution, add $(NH_4)_2HPO_4$ and render ammoniacal. Filter and wash with cold 10% NH_4NO_3.	Na, NH_4^+, K	All other metals	Ignite $MgNH_4PO_4$ to $Mg_2P_2O_7$
	(2) To a boiling NH_4OAc solution, add 2% 8-hydroxyquinoline in 2*M* HOAc. Render ammoniacal. Use filtering crucible; wash with hot water.	Alkali metals	All other metals	Volumetrically
Mn	(1) To a boiling concentrated HNO_3 solution, add $NaBrO_3$ or $NaClO_3$. Use filtering crucible; ignite MnO_2 to Mn_3O_4.	Most metals	Cl^-, much SO_4^{2-}, organic matter	Mn_3O_4 or volumetrically
	(2) To a slightly acid solution, add $(NH_4)_2HPO_4$ and render ammoniacal. Filter and wash with cold 10% NH_4NO_3.	Alkali metals	All heavy metals	$Mn_2P_2O_7$
Mo	(1) To a boiling acetate buffered solution (pH 4), add $Pb(OAc)_2$. Filter through paper; wash with hot 2% NH_4OAc.	Co, Cu, Hg, Mg, Mn, Ni, Zn	As, Cr(VI), Fe, PO_4^{3-}, Sb, Si, Sn, SO_4^{2-}, Ti, V, W(VI)	$PbMoO_4$
	(2) Add 2–5 times the calculated amount of α-benzoinoxime (2% in ethanol) to 5% H_2SO_4 solution at 5–10°C. Add Br_2 water. Wash paper with cold, acidified 0.1% α–benzoinoxime solution. Ignite to MoO_3.	Cr(III), Fe, V(IV)	Cr(VI), Pd, Ta, V(V), W	MoO_3
N (as NO_3^-)	Add nitron acetate to hot slightly acid (H_2SO_4 or HOAc) solution, then cool to 0°C. Wash filtering crucible with cold nitron acetate solution. Dry at 110°C.	Chloride, sulfate	Br^-, I^-, SCN^-, ClO_4^-, Cr(VI), NO_2^-, W(VI)	Nitron nitrate, $C_{20}H_{10}N_4HNO_3$
Na	To 5 mL sample, add 15 mL 30% Mg (or Zn) uranyl(2+) acetate previously saturated with sodium salt. Temperature must be controlled at 20 ± 1°C for precipitation and wash solutions. Wash filtering crucible with reagent solution, then 95% ethanol saturated with the precipitate, and finally with diethyl ether.	NH_4^+, K, Rb, Cs, Ca, Sr, Ba, Mg, Al, Cr(III), Fe(III)	Li, phosphate	$NaMg(UO_2)_3(OAc)_9 \cdot 6.5H_2O$

Nb	(1) Phenylarsonic acid concentrates Nb and Ta in silicate analysis and steel. Ignited in silica crucible.			Nb_2O_3
	(2) Cupferron from dilute mineral acid solution with H_3BO_3 to complex fluoride.		Ta, Ti, Zr	Nb_2O_3
Ni	Add 1% dimethylglyoxime in ethanol to acid solution at 60–80°C and render ammoniacal (pH 8) with tartrate present. Filter and wash filtering crucible with cold water. In the homogeneous method, biacetyl and hydroxylamine are added to a hot solution at pH 7.1.	Bi (pH > 11), Fe(II), Co, Cu(I), most metals	Fe(III) and Co together, Pd, Pt	Ni dimethylglyoxime
Os	(1) Distillation of OsO_4.	All metals	Ru	
	(2) Add $NaHCO_3$ to boiling acid solution until pH 4. Hydrous oxide filtered, washed with 1% NH_4Cl, then ethanol. Ignite in H_2 atmosphere.			Os metal
P (PO_4^{3-})	(1) Add $MgCl_2$ plus NH_4Cl to neutral or slightly acid solution, then render ammoniacal. Wash paper with cool 1.5% aqueous ammonia.	Na, K	All heavy metals, SiO_2, Cl^-, SO_4^{2-}	$Mg_2P_2O_7$
	(2) To hot (80°C) HNO_3 containing 5–15% NH_4NO_3; add $(NH_4)_2MoO_4$. Wash filtering crucible with cold (1:50) HNO_3.	Al, Fe, most metals	As(V)	$(NH_4)_3[P(Mo_3O_{10})_4]$
Pb	(1) Evaporate to dense fumes of H_2SO_4; dilute to 8% H_2SO_4. Wash filtering crucible with 6% H_2SO_4 saturated with $PbSO_4$.	Many metals; As, Sn, and Sb if HF is present	Ba, Ca, Sr	$PbSO_4$
	(2) Add $(NH_4)_2MoO_4$ to boiling slightly acid HNO_3 solution. Neutralize with NH_3 and acidify with HNO_3. Wash filtering crucible with 2% NH_4NO_3.	Co, Mn, Ni, Zn	Alkaline-earth metals, Cd, Cu	$PbMoO_4$
	(3) In the homogeneous method, add Cr(III) plus $KBrO_3$ and heat to boiling in an acetate buffer.	Ba, Bi, Hg(I,II)	Ag, NH_4^+	$PbCrO_4$
	(4) Add dimethyl sulfate or sulfamic acid to a boiling system.	Many metals; As, Sb, and Sn if HF present	Ba, Ca, Sr	$PbSO_4$
	(5) Follow thionalide procedure for bismuth except wash with 50% acetone.	Ag, Al, As, Cd, Cr(III), Co, Cu, Fe(II), Ni, Ti(IV), Zn	<4% Cl^- or 1% sulfate, Au, Pb, Sb(III), Sn(IV), Tl	$Pb(C_{12}H_{10}NOS)_2$
Pd	Add dimethylglyoxime (1% in ethanol) to very hot slightly acid HCl solution; let stand 1 h. Wash with hot water. In the homogeneous precipitation method, add furfural plus hydroxylamine and boil to precipitate Pd furfuraldoximate.	Ir, Ni, Os, Rh, Ru, and most metals	Au, Pt	Pd dimethylglyoxime- or Pd furfuraldoximate

(Continued)

TABLE 4.4 Summary of the Principal Methods for the Separation and Gravimetric Determinations of the Elements (*Continued*)

Element	Conditions of Separation, precipitation, and filtration	Noteworthy separations	Principal interferences	Weighing form
Pt	Evaporate solution to syrup. Dilute to 100 mL. Add NaOAc and formic acid (or 2% HPH_2O_2). pH should be 6. Boil several hours. Filter through paper; ignite.		Other platinum groups metals and Au	Pt metal
Rb	Proceed as with potassium.		Chloride	Volumetrically
Re	(1) Volatilization of Re_2O_7.			
	(2) Add 1% solution of $(C_6H_5)_4AsCl$ to hot acid solution, 0.5*M* in NaCl. Wash filtering crucible with ice water.		ClO_4^-, I^-, NO_3^-	Tetraphenyl-arsonium per-rhenate
	(3) 1 mL 2*N* H_2SO_4 added to 50 mL sample. Heat to 80°C and add 5% nitron acetate solution. Cool in ice bath 2 h with stirring. Use fultering crucible and wash with 0.3% nitron acetate solution. Dry 2–3 h at 110°C.	Lactate, oxalate, succinate	NO_2^-, NO_3^-, Br^-, Cl^-, ClO_3^-, ClO_4^-, I^-, Cr(VI), SCN^-, W(VI)	Nitron perrhenate, $C_{20}H_{10}N_4HReO_4$
Rh	Add 2% thionalide in acetic acid in excess. Heat to boiling. Filter precipitate and wash with HOAc.			
Ru	(1) Distillation of RuO_4.	Most metals	Os	
	(2) Excess of 2% thionalide in alcohol is added to solution 0.2–0.5*N* in HCl. Solution is boiled until precipitate coagulates, filter through paper, washed with hot water. Ignited in hydrogen to Ru metal.	Follows distillation		Ru metal
S (SO_4^{2-})	(1) See under Ba.			
	(2) Benzidine hydrochloride (8 g/L) added to weakly acid medium (pH 3) at 0°C for 30 min. Filter by suction through paper; wash with 80% alcohol.	Al, Co, Cr(III), Cu, Fe, Mn, Ni, Zn	SCN^- with Ag, Cu, Zn; W(VI)	Volumetrically
Sb	(1) Distillation of $SbCl_3$ (see Sec. 2.4.2).	Most metals	Ge, As	Volumetrically
	(2) Electrodeposition of metal (see Sec. 14.6).			
	(3) Follow procedure for Bi using thionalide.	Ce(III), Cr(III), Co, Fe(II), Ti(IV)	Alkaline earths, As, Au, Bi, Cd, Hg, Pb, Th	$Sb(C_{12}H_{10}NOS)_3$
Sc	Add 100 mL 20% NH_4 tartrate, heat to 95°C, add NH_3 dropwise until excess is present. Filter, wash with 2% NH_4 tartrate.	Rare earths, Al, Fe, Th, Ti, Zr	Y	

SCN^-	(1) Add $0.1M$ $AgNO_3$ to dilute HNO_3 solution. Use filtering crucible; wash with water, then alcohol.		Halides, CN^-	$AgSCN$
	(2) Add 50 mL H_2SO_3 to cold, slight acid solution; then add $0.1M$ $CuSO_4$ plus additional H_2SO_3. Let stand 4 h. Use filtering crucible; wash with H_2O saturated with SO_2 until Cu removed, then with alcohol.			$CuSCN$
Se	(1) Distillation of $SeBr_4$ (see Sec. 2.4.2).	Most metals	As, Ge	
	(2) Precipitated as the metal with SO_2, hydrazine, hydroxylamine, or $SnCl_2$ from HCl solution. Wash in succession with cold concentrated HCl, cold water, ethanol, and diethyl ether.	Most metals	Au, Pt	Se
Si	Evaporate $HClO_4$ to fumes and boil gently 15 min. Wash with 1.5% HCl.		SnO_2, TiO_2, WO_3	SiO_2
Sn	(1) Distillation of $SnCl_4$.	All common elements	Ge, As, Sb	SnO_2
	(2) Digestion with HNO_3; dissolving sample in H_2SO_4, diluting, and boiling; or, in HCl solution, neutralizing with NH_3, acidifying slightly with HNO_3, adding a large amount of NH_4NO_3 and boiling. Wash SnO_2 with 2% NH_4NO_3 and ignite. After weighing SnO_2, add 15 times its weight of NH_4I and heat at 425–475°C until fumes cease. Weight impurities and subtract from weight of SnO_2.	Fe, Pb, Zn, Zr, all common metals	As, PO_4^{3-}, Sb, SiO_2 (unless precipitated heated with NH_4I to volatilize SnI_4), WO_3	SnO_2
Sr	(1) Slowly add 100% HNO_3 to water solution until acid concentration is 80%. Use filtering crucible; dry 2 h at 130–140°C.	Ca, Mg, and most other metals	Ba, Pb	$Sr(NO_3)_2$
	(2) To very slightly acid (HCl) solution, add 10-fold excess H_2SO_4 and dilute with equal volume of ethanol. Filter through paper or crucible; wash with 75% ethanol slightly acid with H_2SO_4, then with pure ethanol.	Na, K, NH_4^+, Al	Ca, Ba, Pb	$SrSO_4$
Ta	See under Nb.			
Te	To a boiling 9% HCl solution (50 mL), add 15 mL saturated SO_2 solution and 10 mL 15% $N_2H_4 \cdot 2HCl$ solution, and finally 25 mL more SO_2 solution. Use filtering crucible; wash with hot water, followed quickly by ethanol.	Most metals	Se, Au, Pt	Te
Th	(1) Add $H_2C_2O_4$ (not oxalates) or dimethyl oxalate to boiling $0.5M$ HCl. Filter through paper; wash with 2.5% $H_2C_2O_4$ in $0.5M$ HCl. Ignite to ThO_2.	Al, Fe, Ti, Zr	Rare earths	ThO_2

(Continued)

TABLE 4.4 Summary of the Principal Methods for the Separation and Gravimetric Determinations of the Elements (*Continued*)

Element	Conditions of Separation, precipitation, and filtration	Noteworthy separations	Principal interferences	Weighing form
Th	(2) Precipitate with KIO_3 in < 1*M* HNO_3 at 0°C in presence of tartrate and H_2O_2. Filter; wash with cold 0.5*M* HNO_3 plus 0.8% KIO_3. Complete volumetrically.	Ce(III), rare earths, phosphate	Ce(IV), Ti, Zr, U(IV)	Volumetrically
	(3) Precipitation as ThP_2O_7 by $Na_4P_2O_7$ in HCl solution.	Ce(III), Ti, rare earths, common metals	Ce(IV), Zr	ThP_2O_7
	(4) Add aqueous 1% phenylarsonic acid to slightly acid solution, then acetic acid plus ammonium acetate added to boiling solution. Dissolve precipitate in 1:1 HCl and reprecipitate. Filter hot and wash with water. Ignite.	Al, Ce(III), rare earths	Ce(IV), Hg, Sn(IV), Ti, Zr	ThO_2
Ti	(1) To hot HCl solution, add a little KI and then NH_4HSO_3 dropwise until all iron is reduced (disappearance of iodine color).	Al, Fe(III)	Bi, Sb, Sn, Zr	TiO_2
	(2) From ice-cold acid solutions, precipitate Ti with 6% aqueous cupferron (ammonium salt of phenylnitrosohydroxylamine). Filter through paper; wash with 1% HCl containing cupferron.	Al, Cr(II), Fe(II)	Fe(III)	TiO_2
	(3) Add 4% aqueous *p*-hydroxyphenylarsonic acid to a boiling dilute mineral acid solution. Filter through paper; wash with 0.5% reagent solution in 0.25*N* mineral acid. Ignite to TiO_2.	Al, Ba, Be, Ca, Ce(III), Cr(III,VI), Co, Fe(II,III), Mg, Mn(II,VII), Mo(VI), Ni, Sr, Tl(III), U(VI), V(V), Zn, phosphate	Ce(IV), Sn, and Th (unless H_2SO_4 medium and excess reagent used)	TiO_2
Tl	Adjust alkalinity to 1*N* with NaOH. Add tartrate and cyanide and tenfold excess of 5% thionalide to cold solution. Heat to boiling until precipitate is coagulated; cool and filter through filtering crucible. Wash with cold water until free of cyanide and with 30% acetone until free from thionalide [and with 10% $(NH_4)_2CO_3$ if U is present].	Ag, Al, alkali metals, As, Au, Bi, Cd, Co, Cu, Fe, Hg, Mo, Ni, Pb, Pd, Pt, Sn, Te, U, V, W, Zn	Other metals	$Tl(C_{12}H_{10}NOS)_3$
U	(1) Uranium(IV) precipitated with 6% aqueous cupferron from dilute HCl solution. Filter through paper; wash with 1% HCl containing cupferron. Dry and ignite to U_3O_8.	Al, Cr, Mn, Zn, PO_4^{3-}	Fe, Ti, V(V), Zr	U_3O_8
	(2) Precipitation with mixture of Na_2CO_3 and K_2CO_3 (10% each) added to solution at 80°C; soluble $(NH_4)_4UO_2(CO_3)_3$ remains in solution.	Al, Fe, Mn, Zn, common metals		

V	Add 6% cupferron solution to cold (1:10) H_2SO_4 solution. Filter through paper; wash with cold 1% H_2SO_4 containing a little cupferron.	U(VI), Al, Cr(III), As, phosphate	Fe(III), Ti, Zr, some common metals	V_2O_5
W	(1) Evaporate HCl–HNO_3 solution to small volume, add 15 mL HNO_3, heat to boiling, dilute to 100 mL, heat to boiling and add 5 mL of 10% cinchonine hydrochloride. Add paper pulp and filter; wash with 2.5% acidified cinchonine solution. Ignite at 750°C.	Most metals	SiO_2, Sn	WO_3
	(2) To almost neutral solution add 2% benzidine in dilute H_2SO_4. Boil solution. Add 1% HCl and cool to 25°C. Use filtering crucible; wash with 0.008% benzidine. Ignite.			WO_3
Zn	(1) Pass H_2S (or add thioacetamide and heat) into acid solution buffered to pH 2.6–2.7 with either formate, hydrogen sulfate, citrate, or chloroacetate, and in presence of gelatin or agar-agar plus acrolein. Filter through paper and wash with water.	Al, Cr, Cu, Fe, Mg, Mn, Ni, Sr, Na, Ti, and Co (with acrolein present)	Acid hydrogen sulfide group	ZnO
	(2) To a hot slightly acid solution, add 10- to 15-fold excess of $(NH_4)_2HPO_4$. Use a filtering crucible; wash with 1% $(NH_4)_2HPO_4$, then 50% ethanol.	Na, K, NH_4^+, Ni	Other metals forming insoluble phosphates	$Zn_2P_2O_7$
	(3) Add 3% sodium quinaldate to a medium containing H_2SO_3 and thiourea [to complex Ag, Cu(I), Hg(I)] at pH 3–4, using 25% excess reagent.	Ag, Al, Cu(I), Hg(I), alkali and alkaline-earth metals, Mg		$Zn(C_{10}H_6NO_2)_2 \cdot H_2O$
Zr	(1) To a 20% H_2SO_4 solution maintained at 40–50°C, add 10-fold or greater excess of $(NH_4)_2HPO_4$. Filter through paper; wash with 5% NH_4NO_3 solution. Ignite to ZrO_2. Homogeneously, the precipitate can be formed by boiling trimethyl phosphate or metaphosphoric acid solution.	Al, As(V), B, Cd, Ce(III), Co, Cr(III), Cu, Hg, Mg, Mn, Ni, Zn	Bi, Fe(III), Sn, V; Th (uless double precipitation), Ti (unless H_2O_2 present), ClO_4^-	$Zr_2P_2O_7$
	(2) Cupferron (6% aqueous) added to 10% (v/v) H_2SO_4 or HCl solution. Add paper pulp and filter; wash with cold 3.5% HCl.	Al, bivalent metals	Fe(III), Ti	ZrO_2
	(3) Add aqueous 10% propylarsonic acid to 10% HCl solution. Boil 1 min and filter hot. Wash with hot 1% HCl. Dry and ignite. Reprecipitate for Fe and Th.	Al, Be, Bi, Ce, Co, Cu, Cr, Fe(II), Ni, Mo(VI), rare earths, Th, U, V, W, Zn	Sn (unless removed by igniting with NH_4I), Ti (unless H_2O_2 present)	ZrO_2

TABLE 4.5 Heating Temperatures, Composition of Weighing Forms, and Gravimetric Factors

The minimum temperature required for heating a pure precipitate to constant weight is frequently lower than that commonly recommended in gravimetric procedures. However, the higher temperature is very often still to be preferred in order to ensure that contaminating substances are expelled. The thermal stability ranges of various precipitates as deduced from thermograms are also tabulated. Where a stronger ignition is advisable, the safe upper limit can be ascertained.

In this table, the numbering of the methods refers to the corresponding entries in Table 3.4, which outlines the techniques. Gravimetric factors are based on the 1988 International Atomic Weights. The factor Ag: 0.7526 given in the first line of the table indicates that the weight of precipitate obtained is to be multiplied by 0.7526 to calculate the corresponding weight of silver.

Element	Method number	Thermal stability range, °C	Final heating temperature, °C	Composition of weighing form	Gravimetric factors
Ag	1	70–600	130–150	AgCl	Ag: 0.7526
	2	70–600	130–150	AgCl	Ag: 0.7526
Al	1,4	>475	1200	Al_2O_3	Al: 0.5293
	2	>743	>743	$AlPO_4$	Al: 0.2212; Al_2O_3: 0.4180
	3	102–220	110	$Al(C_9H_6NO)_3$	Al: 0.0587; Al_2O_3: 0.1110
As	1	200–275	105–110	Al_2S_3	As: 0.6090; As_2O_3: 0.8041
	3a		850	$Mg_2As_2O_7$	As: 0.4827; As_2O_3: 0.6373
	3b		vacuum at 25	$MgNH_4AsO_4 \cdot 6H_2O$	As: 0.2589
Au	1	20–957	1060	Au	
Ba	1	780–1100	780	$BaSO_4$	Ba: 0.5884; BaO: 0.6570
	2	< 60	< 60	$BaCrO_4$	Ba: 0.5421; BaO: 0.6053
Be	1	> 900	1000	BeO	Be: 0.3603
Bi	1		100	BiOCl	Bi: 0.8024; Bi_2O_3: 0.8946
	2		100	$Bi(C_{12}H_{10}NOS)_3$	Bi: 0.2387
	3	379–961	800	$BiPO_4$	Bi: 0.6875; Bi_2O_3: 0.7665
Br	1	70–946	130–150	AgBr	Br: 0.4256
Ca	1	478–635	475–525	$CaCO_3$	Ca: 0.4004; CaO: 0.5601
		838–1025	950–1000	CaO	Ca: 0.7147
			Air-dried	$Ca(picrolonate)_2 \cdot 8H_2O$	Ca: 0.05642
Cd	1		>320	$CdSO_4$	Cd: 0.5392; CdO: 0.6159
	2		125	$Cd(C_{10}H_6NO_2)_2$	Cd: 0.2462
		218–420		CdS	Cd: 0.7781; CdO: 0.8888
Ce	1	>360	500–600	CeO_2	Ce: 0.8141
Cl	1	70–600	130–150	AgCl	Cl: 0.2474
Co	1	285–946	750–850	Co_3O_4	Co: 0.7342
			130	$Co(C_{10}H_6NO_2)_3 \cdot 2H_2O$	Co: 0.09639; CoO: 0.1226
			450–500	$CoSO_4$	Co: 0.3802
Cr	1		120	$PbCrO_4$	Cr: 0.1609
Cu	2		105–120	CuSCN	Cu: 0.5225; CuO: 0.6540
	3	<115	100–105	$Cu(C_7H_5NO_2)_2$	Cu: 0.1891
	4		105–115	$Cu(C_{13}H_{11}NO_2)$	Cu: 0.2201
	5		110–115	$Cu(C_{10}H_6NO_2)_2 \cdot H_2O$	Cu: 0.1494
	6		105	$Cu(C_{12}H_{10}NOS)_2 \cdot H_2O$	Cu: 0.1237
F	1	66–538	130–140	PbClF	F: 0.07261
Fe	1,2	470–946	900	Fe_2O_3	Fe: 0.6994
Ga	1,2	408–946	900	Ga_2O_3	Ga: 0.7439
Hg	2		105	$Hg(C_{12}H_{10}NOS)_2$	Hg: 0.3169
I	1	60–900	130–150	AgI	I: 0.5405
In	1	345–1200	1200	In_2O_3	In: 0.8271
Ir	1			IrO_2	Ir: 0.8573
K	1	73–653	< 653	$KClO_4$	K: 0.2822; K_2O: 0.3399
	2		< 270	K_2PtCl_6	K: 0.1609; K_2O: 0.1938
	3			KIO_4	K: 0.1700
	4		120	$KB(C_6H_5)_4$	K: 0.1091

TABLE 4.5 Heating Temperatures, Composition of Weighing Forms, and Gravimetric Factors (*Continued*)

Element	Method number	Thermal stability range, °C	Final heating temperature, °C	Composition of weighing form	Gravimetric factors
Li	1		200	Li_2SO_4	Li: 0.1263; Li_2O: 0.2718
Mg	1		1050–1100	$Mg_2P_2O_7$	Mg: 0.2184; MgO: 0.3622
	2	88–300	155–160	$Mg(C_9H_6NO)_2$	Mg: 0.07775; MgO: 0.1289
Mn	1	>946	1000	Mn_3O_4	Mn: 0.7203
	2		1000	$Mn_2P_2O_7$	Mn: 0.3871; MnO: 0.4998
Mo	1		>505	$PbMoO_4$	Mo: 0.2613; MoO_3: 0.3291
	2		500–525	MoO_3	Mo: 0.6666
N (as NO_3^-)	1	20–242	105	Nitron nitrate	N: 0.3732; NO_3: 0.1652
Na	1	360–674	125	$NaMg(UO_2)_3(C_2H_3O_2)_9 \cdot 6H_2O$	Na: 0.01527; Na_2O: 0.02058
Nb	2	650–950	900	Nb_2O_3	Nb: 0.6990
Ni	1	79–172	110–120	$Ni(C_4H_7N_2O_2)_2$	Ni: 0.2032; NiO: 0.2586
Os	2		800 (in H_2)	Os metal	
P	1		>477	$Mg_2P_2O_7$	P: 0.2783; PO_4: 0.8536
	2	160–415	110	$(NH_4)_3[P(Mo_3O_{10})_4]$	P: 0.0165; P_2O_5: 0.0378
Pb	1	271–959	500–600	$PbSO_4$	Pb: 0.6832; PbO: 0.7359
	2		600	$PbMoO_4$	Pb: 0.5643; PbO: 0.6078
	3		120	$PbCrO_4$	Pb: 0.6411
	4	271–959	600–800	$PbSO_4$	Pb: 0.6832; PbO: 0.7359
	5		105	$Pb(C_{12}H_{10}NOS)_2$	Pb: 0.3240
Pd	1	45–171	110	$Pd(C_4H_7N_2O_2)_2$	Pd: 0.3162
Rb	2	70–674	<674	Rb_2PtCl_6	Rb: 0.2954; Rb_2O: 0.3230
Re	2		130	$(C_6H_5)_4AsReO_4$	Re: 0.2939
	3		110	Nitron perrhenate	Re: 0.3306
S			>780	$BaSO_4$	S: 0.1374; SO_3: 0.3430; SO_4: 0.4116
Sb	3		100	$Sb(C_{12}H_{10}NOS)_3$	Sb: 0.1581
SCN^-	1		130	AgSCN	SCN: 0.3500
	2		110–120	CuSCN	SCN: 0.4775
Se	2		120–130	Se metal	SeO_2: 1.4052
Si	1	358–946	>358	SiO_2	Si: 0.4675
Sn	2	>834	900	SnO_2	Sn: 0.7877
Sr	1		130–140	$Sr(NO_3)_2$	Sr: 0.4140
	2	100–300	100–300	$SrSO_4$	Sr. 0.4770; SrO: 0.5641
Te	1		105	Te metal	
Th	1	610–946	700–800	ThO_2	Th: 0.8788
	3		900	ThP_2O_7	Th: 0.5863
Ti	1,2,3	350–946	900	TiO_2	Ti: 0.5992
Tl(III)	1		100	$Tl(C_{12}H_{10}NOS)$	Tl: 0.4860
U	1		1000	U_3O_8	U: 0.8480; UO_2: 0.9620
V	1	581–946	700–800	V_2O_5	V: 0.5602
W	1,2	>674	800–900	WO_3	W: 0.7930
Zn	1	>1000	950–1000	ZnO	Zn: 0.8034
	2		1000	$Zn_2P_2O_7$	Zn: 0.4292; ZnO: 0.5342
	3		125	$Zn(C_{10}H_6NO_2)_2 \cdot H_2O$	Zn: 0.1529
Zr	1		>850	ZrP_2O_7	Zr: 0.3440; ZrO_2: 0.4647
	2,3		1200	ZrO_2	Zr: 0.7403

TABLE 4.6 Metal 8-Hydroxyquinolates

Element	pH Range for complete precipitation	From weighed and drying conditions	Gravimetric conversion factor	Bromometric equivalent weight	Reference
Aluminum	4.2–9.8	$Al(C_9H_6NO)_3$; 120–140°C	Al: 0.0587	Al: 2.2485	I. M. Kolthoff and E. B. Sandell, *J. Am. Chem. Soc.* **50**:1900 (1928); R. C. Chirnside, C. F. Pritchard, and H. P. Rooksby, *Analyst* **66**:399 (1941)
Antimony	ca. 6	$Sb(C_9H_6NO)_3$; 110°C			T. I. Pirtea, *Z. Anal. Chem.* **118**:26 (1939)
Bismuth	4.8–10.5	$Bi(C_9H_6NO)_3$; 130–140°C			H. G. Haynes, *Analyst* **70**:129 (1945)
Cadmium	5.7–14.6	$Cd(C_9H_6NO)_2$; 130°C	Cd: 0.2805	Cd: 14.051	R. Berg, *Z. Anal. Chem.* **71**:321 (1927)
Calcium	9.2–>13				
Cerium(III)	9–10 (ammoniacal tartrate)	$Ce(C_9H_6NO)_3$; 110°C			R. Berg and E. Becker, *Z. Anal. Chem.* **119**:1 (1940)
Chromium(III)	ca. 10	$Cr(C_9H_6NO)_3$; 110°C	Precipitation not quite complete		E. Taylor-Austin, *Analyst* **63**:710 (1938)
Cobalt(II)	4.3–14.5	$Co(C_9H_6NO)_2$; 130°C	Co: 0.09639		R. Berg, *Z. Anal. Chem.* **76**:195 (1929)
Copper(II)	5.3–14.6	$Cu(C_9H_6NO)_2$; 105–110°C	Cu: 0.1808		R. Berg, *Z. Anal. Chem.* **170**:341 (1947)
Gallium	3.6–11	$Ga(C_9H_6NO)_3$; 110–150°C			T. Moeller and A. J. Cohen, *Anal. Chem.* **22**:686 (1950)
Hafnium	4.5–11.3				
Indium	2.5–3.0	$In(C_9H_6NO)_3$; 120–160°C	In: 0.2098	In: 9.5683	T. Moeller and A. J. Cohen, *Anal. Chem.* **22**:686 (1950)
Iron(III)	2.8–11.2	$Fe(C_9H_6NO)_3$; 130–140°C	Fe: 0.1144	Fe: 4.6539	R. Berg, *Z. Anal. Chem.* **76**:191 (1927); H. V. Moyer and W. R. Remington, *Ind. Eng. Chem. Anal. Ed.* **10**:212 (1938)
Lanthanum	6.5–>10.3				
Lead(II)	8.4–12.3				
Magnesium	9.4–12.7	$Mg(C_9H_6NO)_2$; 160°C	Mg: 0.07775	Mg: 3.0381	J. Redmond and H. Bright, *Bur. Standards J. Res.* **6**:113 (1931); **10**:823 (1933)

Manganese(II)	5.9–10	$Mn(C_9H_6NO)_2$; 160°C	Mn: 0.1600	Mn: 6.8673	R. Berg, *Z. Anal. Chem.* **76**:195 (1929)
Molybdenum(VI)	3.3–7.6	$MoO_2(C_9H_6NO)_2$; 130–140°C	Mo: 0.2305	Mo: 11.99	G. Balanescu, *Z. Anal. Chem.* **83**:470 (1931); R. Niericker and W. D. Treadwell, *Helv. Chem. Acta* **29**:1472 (1946)
Nickel(II)	4.3–14.5	$Ni(C_9H_6NO)_2$; 160°C	Ni: 0.1692		R. Berg, *Z. Anal. Chem.* **76**:195 (1929)
Palladium	3–>11.6	$Pd(C_9H_6NO)_2$; 110°C			
Scandium	6.5–8.5	$Sc(C_9H_6NO)_3 \cdot CH_6NOH$; 110– <125°C	Sc: 0.0722		L. Pokras and P. M. Bernays, *Anal. Chem.* **23**:757 (1951)
Silver	6.1–11.6				
Thallium(III)	6.5–7.0	$Tl(C_9H_6NO)_3 \cdot 3H_2O$; 100°C	Th: 0.2433		F. Feigl and L. Baumfeld, *Anal. Chim. Acta.* **3**:83 (1949)
Thorium	4.4–8.8	$Th(C_9H_6NO)_4 \cdot C_9H_6NOH$; <80°C			F. J. Frere, *J. Am. Chem. Soc.* **55**:4362 (1933); F. Hecht and W. Ehrmann, *Z. Anal. Chem.* **100**:98 (1935)
Titanium(IV)	<3.7–8.7	$TiO(C_9H_6NO)_2$ (variable)			R. Berg and M. Teitelbaum, *Z. Anal. Chem.* **81**:1 (1930); A. Claassen and J. Visser, *Rec. Trav. Chim.* **60**:715 (1941)
Tungsten	5.0–5.6	$WO_2(C_9H_6NO)_2$; 120°C			S. Halberstadt, *Z. Anal. Chem.* **92**:86 (1932); R. Niericker and W. D. Treadwell, *Helv. Chem. Acta* **29**:1472 (1946)
Uranium(VI)	5.7–9.8	$UO_2(C_9H_6NO)_2 \cdot C_9H_6NOH$; 130–140°C	U: 0.3384		A. Claassen and J. Visser, *Rec. Trav. Chim.* **65**:211 (1946)
Vanadium(V)	2.7–6	$VO(OH)(C_9H_6NO)_2 \cdot C_9H_6NOH$; 140–200°C			M. Borrel and R. Pâris, *Anal. Chim. Acta* **4**:279 (1950); R. Niericker and W. D. Treadwell, *Helv. Chem Acta* **29**:1472 (1946)
Zinc	4.6–13.4	$Zn(C_9H_6NO)_2$; 160°C	Zn: 0.1849	Zn: 8.174	H. V. Moyer and W. R. Remington, *Ind. Eng. Chem. Anal. Ed.* **10**:212 (1938)
Zirconium	4.7–12.5	Variable			R. Süe and G. Wétroff, *Bull. Soc. Chim.* [5] **2**:1002 (1935); G. Balanescu, *Z. Anal. Chem.* **101**:101 (1935)

References: J. F. Flagg, *Organic Reagents*, Interscience, New York, 1948; H. Goto, *J. Chem. Soc. Jpn.* **54**:725 (1933); *Sci. Rept. Tohoku Imp. Univ.* **26**:391 (1937); **26**:418 (1938); H. R. Fleck and A. M. Ward, *Analyst* **58**:388 (1933); **62**:378 (1937); R. Bock and F. Umland, *Angew. Chem.* **67**:420 (1955).

4.2 ACID–BASE TITRATIONS IN AQUEOUS MEDIA

4.2.1 Primary Standards

TABLE 4.7 Compositions of Constant-Boiling Hydrochloric Acid Solutions

Constant-boiling hydrochloric acid is prepared by distilling the acid of specific gravity 1.18 (approximately 38% HCl by weight) at a rate of 3 to 4 mL · min^{-1}, discarding the first 75% and the last 5% to 10% of the material. The following data are taken from C. W. Foulk and M. Hollingsworth, J. Am. Chem. Soc. ***45***:*1220 (1923) and from W. D. Bonner and A. C. Titus,* ibid. ***52***:*634 (1930). Barometric pressures are corrected to 0°C and to sea level at 45° latitude.*

According to King, J. Assoc. Offic. Agr. Chemists ***25***:*653 (1942), these data may be reproduced with satisfactory accuracy by the equation*

$$G = \frac{P + 7680}{46.8336}$$

where G *is the air weight, in grams, of constant-boiling acid required to contain 1 mole of HCl, and* P *is the corrected barometric pressure, in mmHg, at which it was collected.*

Pressure during distillation, mmHg	Percent HCl by weight (in vacuo) in distillate	Grams of distillate (weighed in air vs. brass weights) containing 1 mol HCl
600	20.638	176.55
640	20.507	177.68
680	20.413	178.50
700	20.360	178.96
730	20.293	179.55
740	20.269	179.77
760	20.221	180.19
770	20.197	180.41
780	20.173	180.62

TABLE 4.8 Densities and Compositions of Hydrochloric Acid Solutions

This table gives information on the relationship between the density of a hydrochloric acid solution and its composition by weight. The first column gives the density at 25°C, and the fourth column gives the corresponding number of moles of acid per kilogram of solution. The second and fifth columns give the differences between the successive values in the first and fourth columns, respectively, for convenience in interpolation. The third column gives the change in density for a change in temperature of 1°C.

For example, a solution has a density of 1.11192 at 23°C. At 25°C its density would be

$$1.11192 - (2)(0.00048) = 1.11096$$

It contains

$$6.3117 + \left(\frac{65}{519}\right)(0.2744) = 6.3461$$

moles of acid per kilogram. To prepare 1000 mL of a 1M solution, it would be necessary to dilute

$$\left(\frac{1000}{6.3461}\right) = 157.58$$

grams of the above solution to 1 L.

Density, g · mL^{-1} at 25°C	Δ	$-\Delta d/\Delta t$	Moles HCl per kg solution	Δ
1.09497		0.00045	5.4884	
	508			2744
1.10005		0.00046	5.7628	
	511			2744
1.10516		0.00047	6.0372	
	515			2745
1.11031		0.00048	6.3117	
	519			2744
1.11550		0.00049	6.5861	
	523			2744
1.12073		0.00050	6.8605	

TABLE 4.9 Primary Standards for Aqueous Acid–Base Titrations

Standard	Formula weight	Preparation
		Basic substances for standardizing acidic solutions
$(HOCH_3)_3CNHH_2$	121.137	Tris(hydroxymethyl)aminomethane is available commercially as a primary standard. Dry at 100–103°C (<110°C). In titrations with a strong acid the equivalence point is at about pH 4.5–5. Equivalent weight is the formula weight. J. H. Fossum, P. C. Markunas, and J. A. Riddick, *Anal. Chem.* **23**:491 (1951).
HgO	216.59	Dissolve 100 g pure $HgCl_2$ in 1 L H_2O, and add with stirring to 650 mL 1.5*M* NaOH. Filter and wash with H_2O until washings are neutral to phenolphthalein. Dry to constant weight at or below 40°C, and store in a dark bottle. To 0.4 g HgO (≡ 40 mL 0.1*N* acid) add 10–15 g KBr plus 20–25 mL H_2O. Stir, excluding CO_2, until solution is complete. Titrate with acid to pH 5–8. Equivalent weight is one-half formula weight.
$Na_2B_4O_7 \cdot 10H_2O$	381.372	Recrystallize reagent-grade salt twice from water at temperatures below 55°C. Wash the crystals with H_2O, twice with ethanol, and twice with diethyl ether. Let stand in a hygrostat over saturated $NaBr \cdot 2H_2O$ or saturated NaCl–sucrose solution. Use methyl red indicator. Equivalent weight is one-half the formula weight.
Na_2CO_3	105.989	Heat reagent-grade material for 1 h at 255–265°C. Cool in an efficient desiccator. Titrate the sample with acid to pH 4–5 (first green tint of bromocresol green), boil the solution to eliminate the carbon dioxide, cool, and again titrate to pH 4–5. Equivalent weight is one-half the formula weight.
NaCl	58.45	Accurately weigh about 6 g NaCl and dissolve in distilled water. Pass the solution through a well-rinsed cation exchange column (Dowex 50W) in the hydrogen form. The equivalent amount of HCl is washed from the column (in 10 column volumes) into a volumetric flask and made up to volume. Equivalent weight is the formula weight.
		Acidic substances for standardizing basic solutions
C_6H_5COOH	122.125	Pure benzoic acid is available from NIST (National Institute for Science and Technology). Dissolve 0.5 g in 20 mL of neutral ethanol (run a blank), excluding CO_2, add 20–50 mL, and titrate using phenolphthalein as indicator.
o-$C_6H_4(COOK)(COOH)$	204.22	Potassium hydrogen *o*-phthalate is available commercially as primary standard, also from NIST. Dry at <135°C. Dissolve in water, excluding CO_2, and titrate with phenolphthalein as indicator. For $Ba(OH)_2$ solution, perform the titration at an elevated temperature to prevent precipitation of Ba phthalate.
$KH(IO_3)_2$	389.915	Potassium hydrogen bis(iodate) is available commercially in a primary standard grade. Dry at 110°C. Dissolve a weighed amount of the salt in water, excluding CO_2, and titrate to pH 5–8. I. M. Kolthoff and L. H. van Berk, *J. Am. Chem. Soc.* **48**:2800 (1926).
NH_2SO_3H	97.09	Hydrogen amidosulfate (sulfamic acid) acts as a strong acid. Primary standard grade is available commercially. Since it does undergo slow hydrolysis, an acid end point (pH 4–6.5) should be chosen unless fresh reagent is available, then the end point can be in the range pH 4–9. W. F. Wagner, J. A. Wuellner, and C. E. Feiler, *Anal. Chem.* **24**:1491 (1952); M. J. Butler, G. F. Smith, and L. F. Audrieth, *Ind. Eng. Chem. Anal. Ed.* **10**:690 (1938).

4.2.2. Indicators

TABLE 4.10 Indicators for Aqueous Acid–Base Titrations and pH Determinations

This table lists selected indicators for aqueous acid–base titrations. The pH range of transition interval given in the third column may vary appreciably from one observer to another, and in addition it is affected by ionic strength, temperature, and illumination; consequently only approximate values can be given. They should be considered to refer to solutions having low ionic strengths and a temperature of about 25°C. In the fourth column the pK_a *of the indicator as determined spectrophotometrically is listed. In the fifth column the wavelength of maximum absorption is given first for the acidic and then for the basic form of the indicator, and the same order is followed in giving the colors in the sixth column. The method of preparation given in the last column refers to the ordinary commercially available form of the indicator, which in the case of many indicator acids is the sodium salt.*

The abbreviations used in the "color change" column are as follows:

B, blue	*O, orange*	*P, purple*
V, violet	*G, green*	*R, red*
Y, yellow	*C, colorless*	*O-Br, orange-brown*

Scientific name	Trade name	pH range	pK_a	λ_{max}, nm	Color change	Preparation
o-Cresolsulfonephthalein	Cresol red (acid range)	0.2–1.8			R–Y	0.1 g in 13.1 mL 0.02*N* NaOH, diluted to 250 mL with water; or 0.1 g in 100 mL ethanol
Tetra-, penta-, and hexamethyl-*p*-rosaniline HCl	Methyl violet	0.15–3.2			Y–V	0.25% in water
4-(Phenylamino)azobenzene-3′-sulfonic acid	Metanil yellow	1.2–2.3			R–Y	0.25% in ethanol
m-Cresolsulfonephthalein	Cresol purple (acid range)	1.2–2.8	1.51	533, …	R–Y	0.1 g in 13.1 mL 0.02*N* NaOH, diluted to 250 mL with water; or 0.1% in ethanol
1,4-Dimethyl-5-hydroxy-benzenesulfonephthalein	*p*-Xylenol blue (acid range)	1.2–2.8			R–Y	0.04% in ethanol
Thymolsulfonephthalein	Thymol blue (acid range)	1.2–2.8	1.65	544,430	R–Y	0.1 g in 10.8 mL 0.02*N* NaOH, diluted to 250 mL with water; or 0.1 g in 100 mL ethanol
Diphenylamino-*p*-benzene sodium sulfonate	Tropeolin OO	1.3–3.2	2.0	527, …	R–Y	0.1 g in 100 mL water
Ditolyldiazobis(α-naphthyl-amine-4-sulfonic acid	Benzopurpurine 4B	1.3–4.0			BV–R	0.1 g in 100 mL water
2-(*p*-Dimethylaminostyryl)-quinoline ethiodide	Quinaldine red	1.4–3.2			C–R	0.1 g in 100 mL ethanol
2,6-Dinitrophenol	2,6-Dintrophenol	2.4–4.0	3.69		C–Y	0.1 g in 5 mL ethanol diluted to 100 mL with water
2,4-Dinitrophenol	2,4-Dinitrophenol	2.5–4.3	3.90		C–Y	0.1 g in 5 mL ethanol diluted to 100 mL with water
Dimethylaminoazobenzene	Methyl yellow	2.9–4.0	3.3	508, …	R–Y	0.1 g in 100 mL ethanol
Dimethylaminoazobenzene sodium sulfonate	Methyl orange	3.1–4.4	3.40	522,464	R–O	0.1 g in 100 mL water

(Continued)

TABLE 4.10 Indicators for Aqueous Acid–Base Titrations and pH Determinations (*Continued*)

Scientific name	Trade name	pH range	pK_a	λ_{max}, nm	Color change	Preparation
Tetrabromophenolsulfone-phthalein	Bromophenol blue	3.0–4.6	3.85	436,592	Y–BV	0.1 g in 7.45 mL 0.02*N* NaOH, diluted to 250 mL with water
Diphenyldiazobis(1-naphthyl-amine)-4-sodium sulfonate	Congo Red	3.0–5.0			B–R	0.1 g in 100 mL water
4′-Ethoxy-2,4-diaminoazo-benzene	*p*-Ethoxychrysoidine	3.5–5.5			R–Y	0.1 g in 100 mL ethanol
α-Naphthylaminoazobenzene	α-Naphthyl red	3.7–5.0			R–Y	0.1 g in 100 mL ethanol
Dihydroxyanthraquinone sodium sulfonate	Na alizarinsulfonate	3.7–5.2			Y–V	0.1 g in 100 mL water
Tetrabromo-*m*-cresolsulfo-nephthalein	Bromocresol green	4.0–5.6	4.68	444,617	Y–B	0.1 g in 7.15 mL 0.02*N* NaOH, diluted to 250 mL with water
2,5-Dinitrophenol	2,5-Dinitrophenol	4.0–5.8	5.22		C–Y	0.1 g in 20 mL ethanol, then dilute to 100 mL with water
o-Carboxybenzeneazo-dimethylaniline	Methyl red	4.4–6.2	4.95	530,427	R–Y	0.1 g in 18.6 mL 0.02*N* NaOH, diluted to 250 mL with water; or 0.1 g in 100 mL ethanol
	Litmus	4.5–8.3			R–B	0.5 g in 100 mL water
Dichlorophenolsulfone-phthalein	Chlorophenol red	5.4–6.8	6.0	…,573	Y–R	0.1 g in 11.8 mL 0.02*N* NaOH, diluted to 250 mL with water; or 0.04 g in 100 mL ethanol
Dibromo-*o*-cresolsulfone-phthalein	Bromocresol purple	5.2–6.8	6.3	433,591	Y–P	0.1 g in 9.25 mL 0.02*N* NaOH, diluted to 250 mL with water; or 0.02 g in 100 mL ethanol
Dibromophenolsulfone-phthalein	Bromophenol red	5.2–6.8		…,574	Y–R	0.1 g in 9.75 mL 0.02*N* NaOH, diluted to 250 mL with water; or 0.04 g in 100 mL ethanol
p-Nitrophenol	*p*-Nitrophenol	5.3–7.6	7.15	520,405	C–Y	0.25 g in 100 mL water
Dibromothymolsulfone-phthalein	Bromothymol blue	6.2–7.6	7.1	433,617	Y–B	0.1 g in 8 mL 0.02*N* NaOH, diluted to 250 mL with water; or 0.1 g in 1:1 ethanol–water
Phenolsulfonephthalein	Phenol red	6.4–8.0	7.9	433,558	Y–R	0.1 g in 14.2 mL 0.02*N* NaOH, diluted to 250 with water
m-Nitrophenol	*m*-Nitrophenol	6.4–8.8	8.3	…,570	C–Y	0.25 g in 100 mL water or 0.1 g in 100 mL ethanol
2-Methyl-3-amino-6-dimethyl-aminophenazine	Neutral red	6.8–8.0	7.4		R–Y	0.1 g in 70 mL ethanol, diluted to 100 mL with water
Cyanine	Quinoline blue	7.0–8.0			C–V	1 g in 100 mL ethanol
o-Cresolsulfonephthalein	Cresol red (alkaline range)	7.2–8.8	8.2	434,572	Y–R	0.1 g in 13.1 mL 0.02*N* NaOH, diluted to 250 mL with water; or 0.1 g in 100 mL ethanol
α-Naphtholphthalein		7.3–8.7			P–G	0.1 g in 1 : 1 ethanol–water
m-Cresolsulfonephthalein	*m*-Cresol purple (alkaline range)	7.6–9.2	8.32	…,580	Y–P	0.1 g in 13.1 mL 0.02*N* NaOH, diluted to 250 mL with water; or 0.1 g in 100 mL ethanol
Thymolsulfonephthalein	Thymol blue (alkaline range)	8.0–9.6	8.9	430,596	Y–B	0.1 g in 10.75 mL 0.02*N* NaOH, diluted to 250 mL with water; or 0.1 g in 100 mL ethanol
3,3-Bis(*p*-hydroxyphenyl)-phthalide	Phenolphthalein	8.0–10.0	9.4	…,553	C–R	0.1 g in 100 mL ethanol

Dimethylphenolphthalein	α-Naphtholbenzein	9.0–11.0			Y–B	0.1 g in 100 mL ethanol
Thymolphthalein	Thymolphthalein	9.4–10.6	10.0	…,598	C–B	0.1 g in 100 mL ethanol
Aminonaphthodiethylamino-phenoxazine sulfate	Nile blue A	10–11			B–Pink	0.1 g in 100 mL water
3-Carboxy-4-hydroxy-4′-nitroazobenzene, Na salt	Alizarin Yellow R	10.2–12.0	11.16		Y–V	0.1 g in 100 mL water
2,4,6-Trinitrophenylmethyl-nitroamine	Nitramine	10.8–13.0			C–O-Br	0.1 g in 70% ethanol
p-Benzenesulfonic acid-azo-resorcinol	Tropeolin O	11.0–13.0			Y–O-Br	0.1 g in 100 mL water
Na triphenylrosaniline sulfonate	Poirrier blue C4B	11–13			B–P	0.2 g in 100 mL water
Na indigodisulfonate	Indigo carmine	11.6–14			B–Y	0.25 g in 50% ethanol

TABLE 4.11 Mixed Indicators for Acid–Base Titrations

Mixed indicators give sharp color changes and are especially useful in titrating to a given transition pH. The list is from I. M. Kolthoff and V. A. Stenger, Volumetric Analysis, *2d ed., Interscience, New York, 1947. The first column gives the composition of the mixture. For example, 3 parts 0.1% bromocresol green and 1 part 0.1% methyl red in ethanol denotes three volumes of an 0.1% ethanolic solution of bromocresol green mixed with one volume of an 0.1% ethanolic solution of methyl red. The second column gives the approximate pH at the midpoint of the transition interval. The final three columns give the colors in solutions considerably more acid and basic than the transition color. An asterisk appearing after a transition pH indicates an excellent indicator with a sharp color change. See Table 4.10 for preparation of indicator solutions.*

Composition of mixture	Transition pH	Acid color	Transition color	Basic color
Equal parts: 0.1% methyl yellow and 0.1% methylene blue in ethanol	3.25*	Blue-violet	—	Green
Equal parts: 0.1% methyl orange in water and 0.14% xylene cyanol FF in ethanol	3.8	Violet	Gray	Green
Equal parts: 0.1% methyl orange in water and 0.25% indigo carmine in water	4.1	Violet	Gray	Green
Equal parts: 0.02% methyl orange in water and 0.1% Na bromocresol green in water	4.3	Orange	Light green	Green
3 parts 0.1% bromocresol green in ethanol and 1 part 0.2% methyl red in ethanol	5.1*	Wine red	—	Green
Equal parts: 0.2% methyl red in ethanol and 0.1% methylene blue in ethanol	5.4	Red-violet	Gray-blue	Green
Equal parts: 0.1% Na chlorophenol red and 0.1% aniline blue in water	5.8	Green	Pale violet	Violet
Equal parts: 0.1% Na bromocresol green and 0.1% Na chlorophenol red in water	6.1	Yellow-green	Bluish	Blue-violet
Equal parts: 0.1% Na bromocresol purple and 0.1% Na bromothymol blue in water	6.7	Yellow	Violet	Blue violet
2 parts 0.1% Na bromothymol blue and 1 part azolitmin in water	6.9	Violet	—	Blue
Equal parts: 0.1% neutral red and 0.1% methylene blue in ethanol	7.0*	Violet-blue	—	Green
Equal parts: 0.1% neutral red and 0.1% bromothymol blue in ethanol	7.2	Rose	Gray-green	Green
2 parts 0.1% cyanine and 1 part 0.1% phenol red; both in 50% ethanol	7.3	Yellow	Orange (7.2)	Violet
Equal parts: 0.1% Na bromothymol blue and 0.1% Na phenol red in water	7.5*	Yellow	Pale violet	Violet
3 parts 0.1% Na thymol blue and 1 part 0.1% Na cresol red in water	8.3*	Yellow	Rose	Violet
2 parts 0.1% α-naphthalein and 1 part 0.1% cresol red in ethanol	8.3	Pale rose	Pale violet	Violet
3 parts 0.1% phenolphthalein and 1 part 0.1% α-naphthalein in ethanol	8.9	Pale rose	Pale green	Violet
2 parts 0.1% methyl green and 1 part 0.1% phenolphthalein in ethanol	8.9	Green	Pale blue	Violet
3 parts 0.1% phenolphthalein and 1 part 0.1% thymol blue in 50% ethanol	9.0*	Yellow	Green	Violet
Equal parts: 0.1% phenolphthalein and 0.1% thymolphthalein in ethanol	9.9	Colorless	Rose	Violet
2 parts 0.2% Nile blue and 1 part 0.1% phenolphthalein in ethanol	10.0*	Blue	Violet	Red
2 parts 0.1% thymolphthalein and 1 part 0.1% alizarin yellow in ethanol	10.2	Yellow	—	Violet
2 parts 0.2% Nile blue and 1 part 0.1% alizarin yellow in water	10.8	Green	—	Red-brown

TABLE 4.12 Fluorescent Indicators for Acid–Base Titrations

Indicator solutions: 1, 1% solution in ethanol; 2, 0.1% solution in ethanol; 3, 0.05% solution in 90% ethanol; 4, sodium or potassium salt in distilled water; 5, 0.2% solution in 70% ethanol; 6, distilled water.

Name	pH range	Color change acid to base	Indicator solution
Benzoflavine	–0.3 to 1.7	Yellow to green	1
3,6-Dihydroxyphthalimide	0 to 2.4	Blue to green	1
	6.0 to 8.0	Green to yellow-green	
Eosin (tetrabromofluorescein)	0 to 3.0	Nonfl.* to green	4, 1%
4-Ethoxyacridone	1.2 to 3.2	Green to blue	1
3,6-Tetramethyldiamino-xanthone	1.2 to 3.4	Green to blue	1
Esculin	1.5 to 2.0	Weak blue to strong blue	
Anthranilic acid	1.5 to 3.0	Nonfl. to light blue	2 (50% ethanol)
	4.5 to 6.0	Light blue to dark blue	
	12.5 to 14	Dark blue to nonfl.	
3-Amino-1-naphthoic acid	1.5 to 3.0	Nonfl. to green	2 (as sulfate in 50% ethanol)
	4.0 to 6.0	Green to blue	
	11.6 to 13.0	Blue to nonfl.	
1-Naphthylamino-6-sulfonamide	1.9 to 3.9	Nonfl. to green	3
(also the 1-, 7-)	9.6 to 13.0	Green to nonfl.	
2-Naphthylamino-6-sulfonamide (also	1.9 to 3.9	Nonfl. to dark blue	3
the 2-, 8-)	9.6 to 13.0	Dark blue to nonfl.	
1-Naphthylamino-5-sulfonamide	2.0 to 4.0	Nonfl. to yellow-orange	3
	9.5 to 13.0	Yellow-orange to nonfl.	
1-Naphthoic acid	2.5 to 3.5	Nonfl. to blue	4
Salicylic acid	2.5 to 4.0	Nonfl. to dark blue	4 (0.5%)
Phloxin BA extra (tetrachloro-tetrabromofluorescein)	2.5 to 4.0	Nonfl. to dark blue	2
Erythrosin B (tetraiodofluor-escein)	2.5 to 4.0	Nonfl. to light green	4 (0.2%)
2-Naphthylamine	2.8 to 4.4	Nonfl. to violet	1
Magdala red	3.0 to 4.0	Nonfl. to purple	
p-Aminophenylbenzene-sulfonamide	3.0 to 4.0	Nonfl. to light blue	3
2-Hydroxy-3-naphthoic acid	3.0 to 6.8	Blue to green	4 (0.1%)
Chromotropic acid	3.1 to 4.4	Nonfl. to light blue	4 (5%)
1-Naphthionic acid	3 to 4	Nonfl. to blue	4
	10 to 12	Blue to yellow-green	
1-Naphthylamine	3.4 to 4.8	Nonfl. to blue	1
5-Aminosalicylic acid	3.1 to 4.4	Nonfl. to light green	1 (0.2% fresh)
Quinine	3.0 to 5.0	Blue to weak violet	1 (0.1%)
	9.5 to 10.0	Weak violet to nonfl.	
o-Methoxybenzaldehyde	3.1 to 4.4	Nonfl. to green	4 (0.2%)
o-Phenylenediamine	3.1 to 4.4	Green to nonfl.	5
p-Phenylenediamine	3.1 to 4.4	Nonfl. to orange-yellow	5
Morin (2′,4′,3,5,7-penta-hydroxyflavone)	3.1 to 4.4	Nonfl. to green	6 (0.2%)
	8 to 9.8	Green to yellow-green	
Thioflavine S	3.1 to 4.4	Dark blue to light blue	6 (0.2%)

(continued)

TABLE 4.12 Fluorescent Indicators for Acid–Base Titrations (*Continued*)

Name	pH range	Color change acid to base	Indicator solution
Fluorescein	4.0 to 4.5	Pink-green to green	4 (1%)
Dichlorofluorescein	4.0 to 6.6	Blue green to green	1
β-Methylesculetin	4.0 to 6.2	Nonfl. to blue	1
	9.0 to 10.0	Blue to light green	
Quininic acid	4.0 to 5.0	Yellow to blue	6 (satd)
β-Naphthoquinoline	4.4 to 6.3	Blue to nonfl.	3
Resorufin (7-oxyhenoxazone)	4.4 to 6.4	Yellow to orange	
Acridine	5.2 to 6.6	Green to violet	2
3,6-Dihydroxyxanthone	5.4 to 7.6	Nonfl. to blue-violet	1
5,7-Dihydroxy-4-methyl-coumarin	5.5 to 5.8	Light blue to dark blue	
3,6-Dihydroxyphthalic acid dinitrile	5.8 to 8.2	Blue to green	1
1,4-Dihydroxybenzene-disulfonic acid	6 to 7	Nonfl. to light blue	4 (0.1%)
Luminol	6 to 7	Nonfl. to blue	
2-Naphthol-6-sulfonic acid	5.7 to 8.9	Nonfl. to blue	4
Quinoline	6.2 to 7.2	Blue to nonfl.	6 (satd)
1-Naphthol-5-sulfonic acid	6.5 to 7.5	Nonfl. to green	6 (satd)
Umbelliferone	6.5 to 8.0	Nonfl. to blue	
Magnesium-8-hydroxy-quinolinate	6.5 to 7.5	Nonfl. to yellow	6 (0.1% in 0.01*M* HCl)
Orcinaurine	6.5 to 8.0	Nonfl. to green	6 (0.03%)
Diazo brilliant yellow	6.5 to 7.5	Nonfl. to blue	
Coumaric acid	7.2 to 9.0	Nonfl. to green	1
β-Methylumbelliferone	>7.0	Nonfl. to blue	2 (0.3%)
Harmine	7.2 to 8.9	Blue to yellow	
2-Naphthol-6,8-disulfonic acid	7.5 to 9.1	Blue to light blue	4
Salicylaldehyde semicarbazone	7.6 to 8.0	Yellow to blue	2
1-Naphthol-2-sulfonic acid	8.0 to 9.0	Dark blue to light blue	4
Salicylaldehyde acetylhydrazone	8.3	Nonfl. to green-blue	2
Salicylaldehyde thiosemicarbazone	8.4	Nonfl. to blue-green	2
1-Naphthol-4-sulfonic acid	8.2	Dark blue to light blue	4
Naphthol AS	8.2 to 10.3	Nonfl. to yellow-green	4
2-Naphthol	8.5 to 9.5	Nonfl. to blue	2
Acridine orange	8.4 to 10.4	Nonfl. to yellow-green	1
Orcinsulfonephthalein	8.6 to 10.0	Nonfl. to yellow	
2-Naphthol-3,6-disulfonic acid	9.0 to 9.5	Dark blue to light blue	4
Ethoxyphenyl-naphthostilbazonium chloride	9 to 11	Green to nonfl.	1
o-Hydroxyphenylbenzothiazole	9.3	Nonfl. to blue-green	2
o-Hydroxyphenylbenzoxazole	9.3	Nonfl. to blue-violet	2
o-Hydroxyphenylbenzimidazole	9.9	Nonfl. to blue-violet	2
Coumarin	9.5 to 10.5	Nonfl. to light green	
6,7-Dimethoxyisoquinoline-1-carboxylic acid	9.5 to 11.0	Yellow to blue	0.1% in glycerine–ethanol–water in 2:2:18 ratio
1-Naphthylamino-4-sulfonamide	9.5 to 13.0	Dark blue to white or blue	3

* "Nonfl." denotes nonfluorescent.

Reference: G. F. Kirkbright, "Fluorescent Indicators," in E. Bishop, ed., *Indicators*, Pergamon, Oxford, 1972, Chap. 9.

4.2.3 Equilibrium Constants of Acids

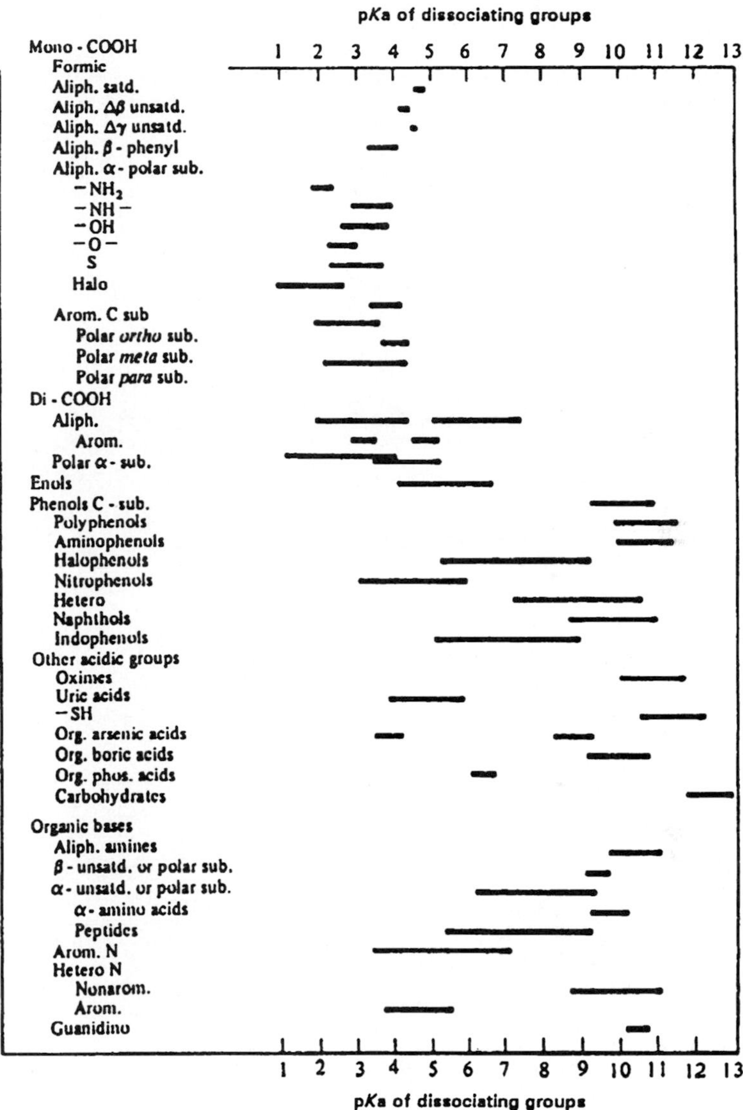

FIGURE 4.1 Range of pK_a values of dissociating groups. [After T. V. Parke and W. W. Davis, *Anal. Chem.* **26**:642 (1954). Copyright 1954 American Chemical Society.]

TABLE 4.13 Selected Equilibrium Constants in Aqueous Solutions at Various Temperatures

Abbreviations used in the table:

(+1), protonated cation — pK_{auto}, *negative logarithm (base 10) of autoprotolysis constant*
(0), neutral molecule
(−1), singly ionized anion — pK_{sp}, *negative logarithm (base 10) of solubility product*
(−2), doubly ionized anion

Substance	Temperature, °C									
	0	5	10	15	20	25	30	35	40	50
Acetic acid (0)	4.780	4.770	4.762	4.758	4.757	4.756	4.757	4.762	4.769	4.787
DL-*N*-Acetylalanine (+1)		3.699	3.699	3.703	3.708	3.715	3.725	3.733	3.745	3.774
β-Acetylamino-propionic (+1)		4.479	4.465	4.465	4.449	4.445	4.444	4.443	4.445	4.457
N-Acetylglycine (+1)		3.682	3.676	3.673	3.667	3.670	3.673	3.678	3.685	3.706
α-Alanine										
(+1)	2.42		2.39		2.35	2.34	2.33	2.33	2.33	2.33
(0)	10.59		10.29		10.01	9.87	9.74	9.62	9.49	9.26
2-Aminobenzenesulfonic acid (0), pK_2	2.633	2.591	2.556	2.521	2.448	2.459	2.431	2.404	2.380	2.338
3-Aminobenzenesulfonic acid (0), pK_2	4.075	4.002	3.932	3.865	3.799	3.738	3.679	3.622	3.567	3.464
4-Aminobenzenesulfonic acid (0), pK_2	3.521	3.457	3.398	3.338	3.283	3.227	3.176	3.126	3.079	2.989
3-Aminobenzoic acid (0)					4.90	4.79	4.75		4.68	4.60
4-Aminobenzoic acid (0)					4.95	4.85	4.90		4.95	5.10
2-Aminobutyric acid										
(+1)			2.334			2.286		2.289 (37.5°C)		2.297
(0)			10.530			9.380		9.518 (37.5°C)		9.234
4-Aminobutyric acid										
(+1)			4.057	4.046	4.038	4.031	4.027	4.025	4.027	4.032
(0)			11.026	10.867	10.706	10.556	10.409	10.269	10.114	9.874
2-Aminoethylsulfonic acid (0)			9.452	9.316	9.186	9.061	8.940	8.824	87.12	9.499
2-Amino-3-methyl-pentanoic acid										
(+0)	2.365 (1°C)		2.338 (12.5°C)			2.320		2.317 (37.5°C)		2.332
(0)	10.460 (1°C)		10.100 (12.5°C)			9.758		9.439 (37.5°C)		9.157
2-Amino-2-methyl-1,3-propanediol	9.612	9.433	9.266	9.104	8.951	8.801	8.659	8.519	8.385	8.132

2-Amino-2-methyl-propionic acid										
(+1)	2.419 (1°C)		2.380 (12.5°C)			2.357		2.351 (37.5°C)		2.356
(0)	10.960 (1°C)		10.580 (12.5°C)			10.205		9.872 (37.5°C)		9.561
2-Aminopentanoic acid										
(+1)	2.376 (1°C)		2.347			2.318			2.309	2.313
(0)	10.508 (1°C)			10.154 (12.5°C)		9.808		9.490 (37.5°C)		9.198
3-Aminopropionic acid										
(+1)	3.656	3.627		3.583		3.551		3.524	3.517	
(0)	11.000	10.830		10.526		10.235		9.963	9.842	
4-Aminopyridine (+1)	9.873	9.704	9.549	9.398	9.252	9.114	8.978	8.846	8.717	8.477
Ammonium ion (+1)	10.081	9.904	9.731	9.564	9.400	9.245	9.093	8.947	8.805	8.539
Arginine										
(+1)	1.914	1.885	1.870	1.849	1.837	1.823	1.814	1.801	1.800	1.787
(0)	9.718	9.563	9.407	9.270	9.123	8.994	8.859	8.739	8.614	8.385
Barbituric acid										
(+1)				3.969	3.980	4.02	4.00	4.008	4.017	4.032
(0)				8.493	8.435	8.372	8.302	8.227	8.147	7.974
Benzoic acid (0)		4.231	4.220	4.215	4.206	4.204	4.203	4.207	4.219	4.223
Boric acid (0)	9.508	9.439	9.380	9.327	9.280	9.236	9.197	9.161	9.132	9.080
Bromoacetic acid (0)				2.875	2.887	2.902	2.918	2.936		
3-Bromobenzoic acid (0)				3.818	3.813	3.810	3.808	3.810	3.813	
4-Bromobenzoic acid (0)				4.011	4.005	3.99	4.001	4.001	4.003	
Bromopropynoic acid (0)			1.786	1.814	1.839	1.855	1.879	1.900	1.919	
3-*tert*-Butylbenzoic acid (0)				4.266	4.231	4.199	4.170	4.143	4.119	
4-*tert*-Butybenzoic acid (0)				4.463	4.425	4.389	4.354	4.320	4.287	
2-Butynoic acid (0)			2.618	2.626	2.611	2.620	2.618	2.621	2.631	
Butyric acid (0)	4.806	4.804	4.803	4.805	4.810	4.817	4.827	4.840	4.854	4.885
DL-*N*-Carbamoylalanine (+1)		3.898	3.894	3.891	3.890	3.892	3.896	3.902	3.908	3.931
N-Carbamoylglycine (+1)		3.911	3.900	3.889	3.879	3.876	3.874	3.873	3.875	3.888
Carbon dioxide + water										
(0)	6.577	6.517	6.465	6.429	6.382	6.352	6.327	6.309	6.296	6.285
(−1)	10.627	10.558	10.499	10.431	10.377	10.329	10.290	10.250	10.220	10.172
Chloroacetic acid (0)				2.845	2.856	2.867	2.883	2.900		
3-Chlorobenzoic acid (0)				3.838	3.831	3.83	3.825	3.826	3.829	
4-Chlorobenzoic acid (0)				4.000	3.991	3.986	3.981	3.980	3.981	
Chloropropynoic acid (0)			1.766	1.796	1.820	1.845	1.864	1.879	1.893	
Citric acid										
(0)	3.220	3.200	3.176	3.160	3.142	3.128	3.116	3.109	3.099	3.095
(−1)	4.837	4.813	4.797	4.782	4.769	4.761	4.755	4.751	4.750	4.757
(−2)	6.393	6.386	6.383	6.384	6.388	6.396	6.406	6.423	6.439	6.484

(Continued)

TABLE 4.13 Selected Equilibrium Constants in Aqueous Solutions at Various Temperatures *(Continued)*

Substance	Temperature, °C									
	0	5	10	15	20	25	30	35	40	50
Cyanoacetic acid (0)		2.445	2.447	2.452	2.460	2.460	2.482	2.496	2.511	
2-Cyano-2-methyl-propionic acid (0)		2.342	2.360	2.379	2.400	2.422	2.446	2.471	2.498	
5,5-Diethylbarbituric acid (0)	8.40	8.30	8.22	8.169	8.094	8.020	7.948	7.877	7.808	7.673
Diethylmalonic acid										
(0)			2.129	2.136	2.144	2.151	2.160	2.172	2.187	
(−1)			7.400	7.401	7.408	7.417	7.428	7.441	7.457	
2,3-Dimethylbenzoic acid (0)				3.663	3.687	3.771	3.726	3.762	3.788	
2,4-Dimethylbenzoic acid (0)				4.154	4.187	4.217	4.244	4.268	4.290	
2,5-Dimethylbenzoic acid (0)				3.911	3.954	3.990	4.020	4.045	4.065	
2,6-Dimethylbenzoic acid (0)				3.234	3.304	3.362	3.409	3.445	3.472	
3,5-Dimethylbenzoic acid (0)				4.292	4.299	4.302	4.304	4.306	4.306	
N,N′-Dimethylethyleneamine-*N,N′*-diacetic acid										
(0)	6.294		6.169		6.047		5.926		5.803	
(−1)	10.446		10.268		10.068		9.882		9.684	
N,N-Dimethylglycine (0)		10.34		10.14		9.94		9.76		
3,5-Dinitrobenzoic acid (0)			2.60		2.73		2.85		2.96	3.07
2-Ethylbutyric acid (0)	4.623		4.664		4.710	4.751	4.758		4.812	4.869
5-Ethyl-5-phenyl-barbituric acid (0)				7.592	7.517	7.445	7.377	7.311	7.248	7.130
Fluoroacetic acid (0)				2.555	2.571	2.586	2.604	2.624		
Formic acid (0)	3.786	3.772	3.762	3.757	3.753	3.751	3.752	3.758	3.766	3.782
2-Furancarboxylic acid (0)						3.164	3.200	3.216	3.239	
Glucose-1-phosphate (0)		6.506	6.500	6.499	6.500	6.504	6.510	6.519	6.531	6.561
Glycerol-1-phosphoric acid (−1)		6.642	6.641	6.643	6.648	6.656	6.666	6.679	6.695	6.733
Glycerol-2-phosphoric acid										
(0)		1.223	1.245	1.271	1.301	1.335	1.372	1.413	1.457	1.554
(−1)		6.657	6.650	6.646	6.646	6.650	6.657	6.666	6.679	6.712
Glycine										
(+1)			2.397	2.380	2.36	2.351	2.34	2.33	2.327	2.32
(0)		10.34	10.193	10.044	9.91	9.780	9.65	9.53	9.412	9.19
Glycolic acid (0)	3.875		3.844 (12.5°C)			3.831		3.833 (37.5°C)		3.849
Glycylasparagine (+1)		2.968	2.958	2.952	2.943	2.942	2.942	2.944	2.947	2.959
N-Glycylglycine (+1)	3.201					3.126				3.159
			8.594 (12.5°C)			8.252		7.948 (37.5°C)		7.668

Hexanoic acid (0)	4.840		4.839		4.849		4.865		4.890	4.920
Hydrogen cyanide (0)			9.63	9.49	9.36	9.21	9.11	8.99	8.88	
Hydrogen peroxide (0)	12.23			11.86	11.75	11.65	11.55	11.45		11.21
Hydrogen sulfide										
(0)		7.33	7.24	7.13	7.05	6.97	6.90	6.82	6.79	6.69
(−1)		13.5		13.2		12.90	12.75	12.6		
4-Hydroxybenzoic acid (0)				4.596	4.586	4.582	4.577	4.576	4.578	
Hydroxylamine (0)				6.186	6.063	5.948		5.730		
2-Hydroxy-1-naphthoic acid										
(0)					3.29		3.24		3.19	3.26
(−1)					9.68		9.65		9.61	9.58
4-Hydroxyproline										
(+1)	1.900 (1°C)		1.850 (12.5°C)			1.818		1.798 (37.5°C)		1.796
(0)	10.274 (1°C)		9.958 (12.5°C)			9.662		9.394 (37.5°C)		9.138
2-Hydroxypropionic acid (0)	3.880	3.873	3.868	3.861	3.857	3.858	3.861	3.867	3.873	3.895
DL-2-Hydroxysuccinic acid										
(0)	3.537	3.520	3.494	3.482	3.472	3.458	3.452	3.446	3.444	3.445
(−1)	5.119	5.108	5.098	5.096	5.096	5.097	5.099	5.104	5.117	5.149
Hypobromous acid (0)				8.83		8.60		8.47	8.37 (45°C)	
Hypochlorous acid (0)	7.82	7.75	7.69	7.63	7.58	7.54	7.50	7.46		7.05
Imidazole (+1)	7.581	7.467	7.334	7.216	7.103	6.993	6.887	6.784	6.685	6.497
Iodoacetic acid (0)				3.143	3.158	3.175	3.193	3.213		
DL-Isoleucine										
(+1)	2.365		2.338 (12.5°C)			2.318		2.317 (37.5°C)		2.332
(0)	10.460		10.100 (12.5°C)			9.758		9.439 (37.5°C)		9.157
Isopropylmalonic acid, mononitrile (0)		2.299	2.320	2.343	2.365	2.401	2.427	2.452	2.481	
Lactic acid (0)	3.880	3.873	3.868	3.862	3.857	3.858	3.861	3.867	3.873	3.895
Lead sulfate, pK_{sp}	8.01			7.87		7.80		7.73		7.63
DL-Leucine										
(+1)	2.383 (1°C)		2.348 (12.5°C)			2.328		2.327 (37.5°C)		2.333
(0)	10.458 (1°C)		10.095 (1.5°C)			9.744		9.434 (37.5°C)		9.142
Malonic acid (−1)	5.670	5.665	5.667	5.673	5.683	5.696	5.710	5.730	5.753	5.803
Mannose (0)			12.45			12.08			11.81	
Mercury(I) chloride, pK_{sp}			18.65	18.48	18.27	17.88		16.79		
Methanol (solvent), pK_{auto}		17.12		16.84		16.71	16.65	16.53		
Methylamine (+1)	11.496		11.130		10.787	10.62	10.466		10.161	9.876

(Continued)

TABLE 4.13 Selected Equilibrium Constants in Aqueous Solutions at Various Temperatures *(Continued)*

	Temperature, °C									
Substance	0	5	10	15	20	25	30	35	40	50
Methylaminodiacetic acid										
(0)	2.138		2.142		2.146		2.150		2.154	
(−1)	10.474		10.287		10.088		9.920		9.763	
3-Methylbenzoic acid (0)				4.303	4.285	4.269	4.256	4.244	4.235	
4-Methylbenzoic acid (0)				4.390	4.376	4.362	4.349	4.336	4.322	
3-Methylbutyric acid (0)	4.726		4.742		4.767		4.794		4.831	4.871
4-Methylpentanoic acid (0)	4.827		4.827		4.837		4.853		4.879	4.908
5-Methyl-5-phenyl-barbituric acid (0)				8.104	8.057	8.011	7.966	7.922	7.879	7.797
2-Methylpropionic acid (0)	4.825		4.827		4.840	4.853	4.886		4.918	4.955
2-Methyl-2-propylamine (+1)		11.439	11.240	11.048	10.862	10.682	10.511	10.341		
Nitric acid (0)	−1.65					−1.38				−1.20
Nitrilotriacetic acid										
(0)	1.69		1.65		1.65		1.66		1.67	
(−1)	2.95		2.95		2.94		2.96		2.98	
(−2)	10.59		10.45		10.33		10.23			
4-Nitrobenzoic acid (0)				3.448	3.444	3.441	3.441	3.442	3.445	
Nitrous acid (0)				3.244	3.177	3.138		3.100		
DL-Norleucine										
(+1)	2.394		2.356 (12.5°C)			2.335		2.324 (37.5°C)		2.328
(0)	10.564		10.190 (12.5°C)			9.834		9.513 (37.5°C)		9.224
Oxalic acid (−1)	4.210	4.216	4.227	4.240	4.254	4.272	4.295	4.318	4.349	4.409
2,4-Pentanedione (0)	9.07					8.95			8.90	
Pentanoic acid (0)	4.823		4.763		4.835	4.842	4.851		4.861	4.906
Phenylalanine (0)			9.75			9.31			8.96	
Phosphoric acid (0)	2.056	2.073	2.088	2.107	2.127	2.148	2.171	2.196	2.224	2.277
(−1)	7.313	7.282	7.254	7.231	7.213	7.198	7.189	7.185	7.181	7.183
o-Phthalic acid										
(0)	2.925	2.927	2.931	2.937	2.943	2.950	2.958	2.967	2.978	3.001
(−1)	5.432	5.418	5.410	5.405	5.405	5.408	5.416	5.427	5.442	5.485
Piperidine (+1)	11.963	11.786	11.613	11.443	11.280	11.123	10.974	10.818	10.670	10.384
Proline										
(+1)	2.011		1.964 (12.5°C)			1.952		1.950 (37.5°C)		1.958
(0)	11.296		10.972 (12.5°C)			10.640		10.342 (37.5°C)		10.064

Propenoic acid (0)				4.267	4.250	4.247	4.249	4.267	4.301	
N-Propionylglycine (+1)		3.728	3.723	3.718	3.716	3.718	3.721	3.725	3.731	3.750
Propynoic acid (0)			1.791	1.829	1.867	1.887	1.940	1.932	1.963	
Pyrrolidine (+1)	12.17	11.98	11.81	11.63	11.43	11.30	11.15	10.99	10.84	11.56
Serine										
(+1)	2.296 (1°C)		2.232 (12.5°C)			2.186		2.154 (37.5°C)		2.132
(0)	9.880 (1°C)		9.542 (12.5°C)			9.208		8.904 (37.5°C)		8.628
Silver bromide, pK_{sp}		13.33		12.83	12.57	12.30	12.07	11.83	11.61	11.19
Silver chloride, pK_{sp}		10.595		10.152		9.749		9.381	9.21	8.88
Succinic acid										
(0)	4.285	4.263	4.245	4.232	4.218	4.207	4.198	4.191	4.188	4.186
(−1)	5.674	5.660	5.649	5.642	5.639	5.635	6.541	5.647	5.654	5.680
Sulfuric acid (−1)	1.778	1.812 (4.3°C)		1.894		1.987	2.05	2.095	2.17	2.246
Sulfurous acid (0)	1.63		1.74			1.89		1.98		2.12
D-Tartaric acid										
(0)	3.118	3.095	3.075	3.057	3.044	3.036	3.025	3.019	3.018	3.021
(−1)	4.426	4.407	4.391	4.381	4.372	4.366	4.365	4.367	4.372	4.391
2,3,5,6-Tetramethyl-benzoic (0)				3.310	3.367	3.415	3.453	3.483	3.505	
Threonine										
(+1)	2.200 (1°C)		2.132 (12.5°C)			2.088		2.070 (37.5°C)		2.055
(0)	9.748 (1°C)		9.420 (12.5°C)			9.100		8.812 (37.5°C)		8.548
o-Toluidine (0)				4.58	4.495	4.45	4.345	4.28	4.20	
1,2,4-Triazole										
(+1)				2.451	2.418	2.386	2.327			
(0)				10.205	10.083	9.972	9.768			
3,4,5-Trihydroxybenzoic acid (0)					4.19		4.30		4.38	4.53
Tris(2-hydroxyethyl)amine (+1)	8.290	8.173	8.067	7.963	7.861	7.762	7.666	7.570	7.477	7.299
2,4,6-Trimethylbenzoic (0)				3.325	3.391	3.448	3.498	3.541	3.577	
3-Trimethylsilylbenezene acid (0)				4.142	4.116	4.089	4.060	4.029	3.996	
4-Trimethylsilylbenzoic acid (0)				4.270	4.230	4.192	4.155	4.119	4.084	
β-Ureidopropionic acid (0)		4.514	4.505	4.497	4.490	4.487	4.486	4.486	4.488	4.500
DL-Valine										
(+1)	2.320		2.297 (12.5°C)			2.296		2.292 (37.5°C)		2.310
(0)	10.413		10.064 (12.5°C)			9.719		9.405 (37.5°C)		9.124

Source: J. A. Dean, ed., *Lange's Handbook of Chemistry*, 14th ed., McGraw-Hill, New York, 1992.

4.2.4 Titration Curves and Precision in Aqueous Acid–Base Titrations

4.2.4.1 Symbols Employed. The following symbols are used in this subsection:

C_a^0 = Initial concentration of acid titrated (mol/L).
C_a = Concentration of acid (assumed to be monobasic) used as reagent (mol/L).
C_b^0 = Initial concentration of base titrated (mol/L).
C_b = Concentration of base (assumed to be monoacidic) used as reagent (mol/L).
f = Millimoles of reagent added per millimole of substance being titrated; ratio between volume of reagent add and volume of reagent required to reach the first equivalence point of the titration. Typically, for the titration of an acid with a base, $f = V_b C_b / V_a^0 C_a^0$.
K_a = Formal dissociation constant of a monobasic weak acid (=$[H^+][A^-]/[HA]$).
K_b = Formal dissociation constant of a monoacidic weak base (=$[B^+][OH^-]/[BOH]$).
K_w = Formal ion product constant of water (=$[H^+][OH^-]$).
K_1, K_2, K_i, K_n = Successive formal dissociation constants of the basic weak acid H_nA.

$K_1 = [H^+][H_{n-1}A^{-1}]/[H_nA]$
$K_2 = [H^+]H_{n-2}A^{-2}]/[H_{n-1}A^{-1}]$
$K_i = [H^+][H_{n-i}A^{-i}]/[H_{n-i+1}A^{1-i}]$

rp = Relative precision corresponding to an uncertainty of ±0.1 pH unit in the location of an equivalence point.
ρ = Dilution ratio; ratio of initial volume of solution titrated to total volume of titration mixture at point under consideration. Typically, for the titration of an acid with a base, $\rho = V_a^0/(V_a^0+V_b)$.
V_a^0 = Initial volume of acid titrated (mL).
V_a = Volume of acid used as reagent (mL).
V_b^0 = Initial volume of base titrated (mL).
V_b = Volume of base used as reagent (mL).

In the derivation of the equations given below it is assumed that

1. No loss of any acid or base occurs (e.g., by volatilization or precipitation) at any stage of the titration.
2. The values of all formal dissociation constants remain unchanged throughout the titration (i.e., the changes in ionic strength, temperature, and other variables that affect the formal dissociation constants are assumed to be negligible).
3. Volumes are additive (see definition of ρ above).

4.2.4.2 Titrations of Strong Acids with Strong Bases. The complete equation for the titration curve is

$$[H^+]^2 + (f-1)\rho C_a^0[H^+] - K_w = 0 \qquad (4.1)$$

but this is needed only when $[H^+]$ is very nearly equal to square root of K_w, which will be the case only if C_a^0 is extremely small and/or if f is very nearly equal to 1. The following special cases are important:

1. *At the start of the titration*, $f = 0$ and $\rho = 1$; then

$$[\mathrm{H}^+] = \frac{C_a^0 + \sqrt{(C_a^0)^2 + 4K_w}}{2} \tag{4.2}$$

In any practical titration $C_a^0 \gg 2(K_w)^{-1/2}$, so that

$$[\mathrm{H}^+] = C_a^0 \tag{4.3}$$

2. *Up to the equivalence point*, the simplified equation

$$[\mathrm{H}^+] = (1 - f)\rho C_a^0 \tag{4.4}$$

which neglects the hydrogen and hydroxyl ions produced by the autoprotolysis of water, may be used unless it gives $[\mathrm{H}^+] < 10(K_w)^{1/2}$, in which case Eq. (4.1) should be used instead.

3. *At the equivalence point*, $f = 1$; then

$$[\mathrm{H}^+] = \sqrt{K_w} \tag{4.5}$$

The relative precision of the titration is given approximately by

$$\mathrm{rp} = \frac{\sqrt{K_w}}{C_a^0} = \frac{\sqrt{K_w}}{C_b} \tag{4.6}$$

if C_a^0 and C_b are approximately equal. Taking their common value as C, the inflection point occurs at $f = 1 - 16K_w/C^2$, which is so nearly equal to 1 that no difference between the inflection point and the equivalence point can be discerned in any practical titration.

4. *Beyond the equivalence point*, the equation

$$[\mathrm{H}^+] = \frac{K_w}{(f - 1)\rho C_a^0} \tag{4.7}$$

usually suffices. Like Eq. (4.4), it neglects the ions produced by the autoprotolysis of water, and the exact Eq. (4.1) should therefore the used instead if Eq. (4.7) gives $[\mathrm{H}^+] > 0.1(K_w)^{1/2}$.

4.2.4.3 Titrations of Strong Bases with Strong Acids. Use Eqs. (4.1) to (4.7), but replace $[\mathrm{H}^+]$ by $[\mathrm{OH}—]$, C_a^0 by C_b^0, and C_b by C_a.

4.2.4.4 Titrations of Weak Monobasic Acids with Strong Bases. The complete equation for the titration curve is

$$[\mathrm{H}^+]^3 + (K_a + fC_a^0\rho)[\mathrm{H}^+]^2 + [(f - 1)C_a^0\rho K_a - K_w][\mathrm{H}^+] - K_w K_a = 0 \tag{4.8}$$

The following special cases are important:

1. *At the start of the titration*, where $f = 0$ and $\rho = 1$, Eq. (4.8) becomes

$$[\mathrm{H}^+]^3 + K_a[\mathrm{H}^+]^2 - (C_a^0 K_w)[\mathrm{H}^+] - K_a K_w = 0 \tag{4.9}$$

If the acid is so weak that $C_a^0 K_w$ is comparable with or smaller than K_w, but yet concentrated enough so that $4C_a^0 \gg K_w$,

$$[\mathrm{H}^+] = \sqrt{C_a^0 K_a + K_w} \tag{4.10}$$

If, on the other hand, the acid is so strong that $4C_a^0$ is comparable with or smaller than K_w, but yet concentrated enough so that $C_a^0 K_a \gg K_w$,

$$[\mathrm{H}^+] = \frac{-K_a + \sqrt{K_a^2 + 4C_a^0 K_a}}{2} \tag{4.11}$$

Finally, for a moderately concentrated solution of an acid of intermediate strength, in which $4C_a^0 \gg K_a$ and $C_a K_a \gg K_w$,

$$[\mathrm{H}^+] = \sqrt{C_a^0 K_a} \tag{4.12}$$

If neither of the above conditions was satisfied, the exact equation (4.9) would have to be used, but this can occur only if the solution is so dilute that its titration is utterly unfeasible.

2. *From the start of the titration to the equivalence point* ($0 \le f \le 1$), Eq. (4.8) may conveniently be used in the form

$$[\mathrm{H}^+] = \frac{(1-f)\rho C_a^0 [\mathrm{H}^+] + [\mathrm{OH}^-]}{f\rho C_a^0 + [\mathrm{H}^+] - [\mathrm{OH}^-]} K_a \tag{4.13}$$

Usually it is possible to neglect one of the last two terms in both the numerator and the denominator of the fraction, thereby reducing the equation from a cubic to a quadratic. A widely, though not universally, applicable simplification is the familiar equation

$$[\mathrm{H}^+] = \frac{(1-f)}{f} K_a \tag{4.14}$$

which is obeyed in titrations of moderately concentrated solutions of acids that are neither fairly strong nor extremely weak, provided that f is not too close to either 0 or 1.

Equation (4.14) predicts that pH = pK_a and that $d(\mathrm{pH})/df$, the slope of the titration curve, is smallest (i.e., the buffer capacity of the titration mixture is greatest) at $f = 0.5$. Under ordinary analytical conditions these predictions are correct within the usual limits of experimental error if, but only if, pK_a is between about 4 and 10. Progressively increasing deviations are observed with both stronger and weaker acids, and with increasing dilution of either acid or base.

3. *At the equivalence point*, where $f = 1$, the general equation (4.8) becomes

$$[\mathrm{H}^+]^3 + (K_a + C_a^0 \rho)[\mathrm{H}^+]^2 - K_w[\mathrm{H}^+] - K_w K_a = 0 \tag{4.15}$$

If the acid is fairly strong but not too dilute, so that $4\rho C_a \gg K_w K_a$ while $\rho C_a^0 / K_a$ is comparable with or smaller than 1,

$$[\mathrm{OH}^-] = \sqrt{(\rho C_a^0 K_w / K_a) + K_w} \tag{4.16}$$

If, however, the acid is fairly weak but not too dilute, so that $\rho C_a / K_a \gg 1$ while $4\rho C_a$ is comparable with or smaller than K_w / K_a,

$$[\mathrm{OH}^-] = \frac{-K_w + \sqrt{(K_w/K_a)^2 + 4\rho C_a^0 K_w / K_a}}{2} \tag{4.17}$$

Finally, if $4\rho C_a^0 \gg K_w/K_a$ and $\rho C_a^0/K_a \gg 1$, as is usually true in titrations of acids of intermediate strength,

$$[\mathrm{OH}^-] = \sqrt{\rho C_a^0 K_w / K_a} \qquad \text{or} \qquad [\mathrm{H}^+] = \sqrt{K_w K_a / \rho C_a^0} \tag{4.18}$$

In Eqs. (4.15) and (4.18), the product ρC_a^0 is equal to the formal concentration, at the equivalence point, of the conjugate base of the acid titrated.

If C_a^0 and C_b are approximately equal, the relative precision of the titration is roughly

$$\mathrm{rp} = \sqrt{K_w / C_a^0 K_a} \tag{4.19}$$

In titrations of $0.1F$ acid with $0.1F$ base, the inflection point at which the buffer capacity of the titration mixture is smallest coincides with the equivalence point within the usual limits of experimental error if K_a is larger than about 10^{-9}. For $K_a = 10^{-10}$, the inflection point occurs about 1% before the equivalence point; for $K_a = 10^{-11}$, the inflections point at $f = 0.875$, 12.5% before the equivalence point. The difference between the inflection point and the equivalence point increases with decreasing concentration of acid and base.

4. *Beyond the equivalence point*, Eq. (4.7) can frequently be used, although it fails if the acid is very weak or if f is only very slightly greater than 1.

4.2.4.5 Titrations of Weak Monoacidic Bases with Strong Acids. Use Eqs. (4.8) to (4.19), but replace $[\mathrm{H}^+]$ by $[\mathrm{OH}^-]$, $[\mathrm{OH}^-]$ by $[\mathrm{H}^+]$, K_a by K_b, C_a^0 by C_b by C_a.

4.2.4.6 Titrations of Polybasic Weak Acids with Strong Bases. The complete explicit equation for a titration of this kind,

$$[\mathrm{H}^+]^2 + f\rho C_a^0[\mathrm{H}^+] - K_w = \frac{\rho C_a^0}{\sigma}\{K_1[\mathrm{H}^+]^{n-1} + 2K_1K_2[\mathrm{H}^+]^{n-1} + \cdots + (nK_1K_2\cdots K_n)\} \tag{4.20}$$

where

$$\rho = [\mathrm{H}^+]^n + K_1[\mathrm{H}^+]^{n-1} + K_1K_2[\mathrm{H}^+]^{n-2} + \cdots + (K_1K_2\cdots K_n) \tag{4.21}$$

is too complicated to find any practical use. For many purposes the following simple equations suffice; they involve the assumption that the acid is fairly concentrated (i.e., $C_a^0 \gg K_1$) and that each of the successive dissociation constants, K_i, is much larger than K_{i+1}.

1. *At the start of the titration,*

$$[\mathrm{H}^+] = \sqrt{C_a^0 K_1} \tag{4.22}$$

subject to the conditions outlined in connection with Eq. (4.12). If K_1 is appreciable in comparison with C_a^0, Eq. (4.13) should be used instead, taking $K_a = K_1$.

2. *Up to the first equivalence point* $(0 < f < 1)$,

$$[\mathrm{H}^+] = \frac{1-f}{f}K_1 \tag{4.23}$$

[cf. the discussion of Eq. (4.14).]

3. *At the first equivalence point, $f = 1$,* and

$$[\mathrm{H}^+] = \frac{\sqrt{K_1(\rho C_a^0 K_2 + K_w)}}{K_1 + \rho C_a^0} \tag{4.24}$$

which, if $\rho C_a^0 K_2 \gg K_w$ and $\rho C_a^0 \gg K_1$, can be further simplified to

$$[\mathrm{H}^+] = \sqrt{K_1 K_2} \tag{4.25}$$

4. *Between the first and second equivalence points* ($1 < f < 2$),

$$[\mathrm{H}^+] = \frac{2-f}{f-1} K_2 \tag{4.26}$$

[cf. Eqs. (4.23) and (4.14).]

5. *At the second equivalence point, $f = 2$,* and

$$[\mathrm{H}^+] = \frac{\sqrt{K_2(\rho C_a^0 K_3 + K_w)}}{K_2 + \rho C_a^0} \tag{4.27}$$

[cf. Eq. (4.25)]; if $\rho C_a^0 K_3 \gg K_w$ and $\rho C_a^0 \gg K_2$, this can be simplified to

$$[\mathrm{H}^+] = \sqrt{K_2 K_3} \tag{4.28}$$

[cf. Eq. (4.25)], and so on until:

6. *At the last equivalence point*, where $f = n$,

$$[\mathrm{H}^+] = \sqrt{K_w K_n / \rho C_a^0} \tag{4.29}$$

At the ith equivalence point ($i < n$), the relative precision is given approximately by

$$\mathrm{rp} = \sqrt{K_{i+1}/K_i} \tag{4.30}$$

At each equivalence point the value of i is equal to f, the number of moles of base added per mole of acid originally present. Thus, at the first equivalence point, $i = f = 1$ and

$$\mathrm{rp} = \sqrt{K_2/K_1} \tag{4.31}$$

At the last equivalence point, however, the relative precision is approximately

$$\mathrm{rp} = \sqrt{K_w / K_a C_a^0} \tag{4.32}$$

Titrations of polyacidic bases may be described by equations obtained from these by means of the substitutions described in Sec. 4.2.4.5.

4.2.5 Calculation of the Approximate pH Value of Solutions

Strong acid	$pH = -\log [\text{acid}]$
Strong base	$pH = 14.00 + \log [\text{base}]$
Weak acid	$pH = \frac{1}{2}pK_a - \frac{1}{2} \log [\text{acid}]$
Weak base	$pH = 14.00 - \frac{1}{2}pK_b + \frac{1}{2}\log [\text{base}]$
Salt formed by a weak acid and a strong base	$pH = 7.00 + \frac{1}{2}pK_a + \frac{1}{2}\log [\text{salt}]$
Acid salts of a dibasic acid	$pH = \frac{1}{2}pK_1 + \frac{1}{2}pK_2 - \frac{1}{2}\log [\text{salt}] + \frac{1}{2}\log (K_1 + [\text{salt}])$
Buffer solution consisting of a mixture of a weak acid and its salt	$pH = pK_a + \log \left(\frac{[\text{salt}] + [H_3O^+] - [OH^-]}{[\text{acid}] - [H_3O^+] + [OH^-]} \right)$

4.2.6 Calculation of Concentrations of Species Present at a Given pH

$$\alpha_0 = \frac{[H^+]^n}{[H^+]^n + K_1[H^+]^{n-1} + K_1K_2[H^+]^{n-2} + \cdots + K_1K_2\cdots K_n} = \frac{[H_nA]}{C_{\text{acid}}}$$

$$\alpha_1 = \frac{K_1[H^+]^{n-1}}{[H^+]^n + K_1[H^+]^{n-1} + K_1K_2[H^+]^{n-2} + \cdots + K_1K_2\cdots K_n} = \frac{[H_{n-1}A^-]}{C_{\text{acid}}}$$

$$\alpha_2 = \frac{K_1K_2[H^+]^{n-2}}{[H^+]^n + K_1[H^+]^{n-1} + K_1K_2[H^+]^{n-2} + \cdots + K_1K_2\cdots K_n} = \frac{[H_{n-2}A^{2-}]}{C_{\text{acid}}}$$

$$\vdots$$

$$\alpha_n = \frac{K_1K_2\cdots K_n}{[H^+]^n + K_1[H^+]^{n-1} + K_1K_2[H^+]^{n-2} + \cdots + K_1K_2\cdots K_n} = \frac{[A^{n-}]}{C_{\text{acid}}}$$

4.2.7 Volumetric Factors for Acid–Base Titrations

Volumetric factors for various substances that can be titrated with acids or alkalies are given in Table 4.14.

4.3 ACID–BASE TITRATIONS IN NONAQUEOUS MEDIA

Nonaqueous titrations may be performed with the same ease as titrations in aqueous solution.[1] Many substances that are of insufficient basic or acidic strength to give sharp end points in aqueous solution become susceptible to titration in appropriate nonaqueous solvents. The resolution of mixtures,

[1] J. Kucharsky and L. Safarik, *Titrations in Nonaqueous Solvents*, Elsevier, Amsterdam, 1965; W. Huber, *Titrations in Nonaqueous Solvents*, Academic, New York, 1967; A. H. Beckett and E. H. Tinley, *Titrations in Nonaqueous Solvents*, 3d ed., British Drug Houses, Poole, England, 1962; J. S. Fritz, *Acid-Base Titrations in Nonaqueous Solvents*, GFS Chemicals, Columbus, Ohio, 1952. Also, "Titrations in Nonaqueous Solvents" appearing (even years until after 1982) in the *Annual Fundamental Reviews* of the journal *Analytical Chemistry*.

TABLE 4.14 Volumetric (Titrimetric) Factors for Acid–Base Titrations

Acids

The following factors are the equivalent of 1 mL of *normal acid.* Where the normality of the solution being used is other than normal, multiply the factors given in the table below by the normality of the solution employed.

The equivalents of the esters are based on the results of saponification.

The indicators methyl orange and phenolphthalein are indicated by the abbreviations MO and PH, respectively.

Substance	Formula	Grams
Ammonia	NH_3	0.017 031
Ammonium	NH_4	0.018 039
Ammonium chloride	NH_4Cl	0.053 492
Ammonium hydroxide	NH_4OH	0.035 046
Ammonium oleate	$C_{17}H_{33}CO_2NH_4$	0.299 50
Ammonium oxide	$(NH_4)_2O$	0.026 038
Amyl acetate	$CH_3CO_2C_5H_{11}$	0.130 19
Barium carbonate (MO)	$BaCO_3$	0.098 67
Barium hydroxide	$Ba(OH)_2$	0.085 677
Barium oxide	BaO	0.076 67
Bornyl acetate	$CH_3CO_2C_{10}H_{17}$	0.196 29
Calcium carbonate (MO)	$CaCO_3$	0.050 04
Calcium hydroxide	$Ca(OH)_2$	0.037 047
Calcium oleate	$(C_{17}H_{33}CO_2)_2Ca$	0.301 50
Calcium oxide	CaO	0.028 04
Calcium stearate	$(C_{17}H_{35}CO_2)_2Ca$	0.303 52
Casein (N 6.38)	....	0.089 371
Ethyl acetate	$CH_3CO_2C_2H_5$	0.088 107
Glue (N 5.60)	....	0.078 445
Hydrochloric acid	HCl	0.036 461
Magnesium carbonate (MO)	$MgCO_3$	0.042 16
Magnesium oxide	MgO	0.020 16
Menthyl acetate	$CH_3CO_2C_{10}H_{19}$	0.198 31
Methyl acetate	$CH_3CO_2CH_3$	0.074 080
Nicotine	$C_{10}H_{14}N_2$	0.162 24
Nitrogen	N	0.014 007
Potassium carbonate (MO)	K_2CO_3	0.069 11
Potassium carbonate, acid (MO)	$KHCO_3$	0.100 12
Potassium nitrate	KNO_3	0.101 11
Potassium oleate	$C_{17}H_{33}CO_2K$	0.320 57
Potassium oxide	K_2O	0.047 10
Potassium stearate	$C_{17}H_{35}CO_2K$	0.322 58
Protein (N 5.70)	....	0.079 846
Protein (N 6.25)	....	0.087 550
Sodium acetate	CH_3CO_2Na	0.082 035
Sodium acetate	$CH_3CO_2Na \cdot 3H_2O$	0.136 08
Sodium borate, tetra- (MO)	$Na_2B_4O_7$	0.100 61
Sodium borate, tetra- (MO)	$Na_2B_4O_7 \cdot 10H_2O$	0.190 69
Sodium carbonate (MO)	Na_2CO_3	0.052 994
Sodium carbonate (MO)	$Na_2CO_3 \cdot H_2O$	0.062 002
Sodium carbonate (MO)	$Na_2CO_3 \cdot 10H_2O$	0.143 07
Sodium carbonate, acid (MO)	$NaHCO_3$	0.084 007
Sodium hydroxide	NaOH	0.399 97
Sodium oleate	$C_{17}H_{33}CO_2Na$	0.304 45

TABLE 4.14 **Volumetric (Titrimetric) Factors for Acid–Base Titrations *(Continued)***

Substance	Formula	Grams
	Acid	
Sodium oxalate	$Na_2C_2O_4$	0.067 000
Sodium oxide	Na_2O	0.030 990
Sodium phosphate (MO)	Na_2HPO_4	0.141 96
Sodium phosphate (MO)	$Na_2HPO_4 \cdot 12H_2O$	0.358 14
Sodium phosphate (MO)	Na_3PO_4	0.081 970
Sodium phosphate (PH)	Na_3PO_4	0.163 94
Sodium silicate	$Na_2Si_4O_9$	0.151 11
Sodium stearate	$C_{17}H_{35}CO_2Na$	0.306 47
Sodium sulfide (MO)	Na_2S	0.039 022

Alkali

The following factors are the equivalent of the milliliter of *normal alkali*. Where the normality of the solution being used is other than normal, multiply the factors given in the table below by the normality of the solution employed.

The equivalents of the esters are based on the results of saponification.

The indicators methyl orange and phenolphthalein are indicated by the abbreviations MO and PH, respectively.

Substance	Formula	Grams
Abietic acid (PH)	$HC_{20}H_{29}O_2$	0.302 46
Acetic acid (PH)	CH_3CO_2H	0.060 05
Acetic anhydride (PH)	$(CH_3CO)_2O$	0.051 045
Aluminum sulfate	$Al_2(SO_4)_3$	0.057 02
Amyl acetate	$CH_3CO_2C_5H_{11}$	0.130 19
Benzoic acid (PH)	$C_6H_5CO_2H$	0.122 12
Borate tetra- (PH)	B_4O_7	0.038 81
Boric acid (PH)	H_3BO_3	0.061 833
Boric anhydride (PH)	B_2O_3	0.034 86
Bornyl acetate	$CH_3CO_2C_{10}H_{17}$	0.196 29
Butyric acid (PH)	$C_3H_7CO_2H$	0.088 107
Calcium acetate	$(CH_3CO_2)_2Ca$	0.079 085
Calcium oleate	$(C_{17}H_{33}CO_2)_2Ca$	0.301 50
Calcium stearate	$(C_{17}H_{35}CO_2)_2Ca$	0.303 52
Carbon dioxide (PH)	CO_2	0.022 005
Chlorine	Cl	0.035 453
Citric acid (PH)	$H_3C_6H_5O_7 \cdot H_2O$	0.070 047
Ethyl acetate	$CH_3CO_2C_2H_5$	0.088 107
Formaldehyde	HCHO	0.030 026
Formic acid (PH)	HCO_2H	0.046 026
Glycerol (sap. of acetyl)	$C_3H_5(OH)_3$	0.030 698
Hydriodic acid	HI	0.127 91
Hydrobromic acid	HBr	0.080 917
Hydrochloric acid	HCl	0.036 461
Lactic acid (PH)	$HC_3H_5O_3$	0.090 079
Lead acetate	$(CH_3 \cdot CO_2)_2Pb \cdot 3H_2O$	0.189 66
Maleic acid (PH)	$(CHCO_2H)_2$	0.058 037
Malic acid (PH)	$H_2C_4H_4O_5$	0.067 045
Menthol (sap. of acetyl)	$C_{10}H_{19}OH$	0.156 27
Menthyl acetate	$CH_3CO_2C_{10}H_{19}$	0.198 31

(Continued)

TABLE 4.14 Volumetric (Titrimetric) Factors for Acid–Base Titrations *(Continued)*

Alkali		
Substance	Formula	Grams
Methyl acetate	$CH_3CO_2CH_3$	0.074 080
Nitrate	NO_3	0.062 005
Nitric acid	HNO_3	0.063 013
Nitrogen	N	0.014 007
Nitrogen pentoxide	N_2O_5	0.054 005
Oleic acid (PH)	$C_{17}H_{33}CO_2H$	0.282 47
Oxalic acid (PH)	$(CO_2H)_2$	0.045 018
Oxalic acid (PH)	$(CO_2H)_2 \cdot 2H_2O$	0.063 033
Phosphoric acid (MO)	H_3PO_4	0.097 995
Phosphoric acid (PH)	H_3PO_4	0.048 998
Potassium carbonate, acid (MO)	$KHCO_3$	0.100 12
Potassium oleate	$C_{17}H_{33}CO_2K$	0.320 56
Potassium oxalate, acid (PH)	KHC_2O_4	0.128 13
Potassium phthalate, acid (PH)	$HC_8H_4O_4K$	0.204 23
Potassium stearate	$C_{17}H_{35}CO_2K$	0.322 58
Sodium benzoate	$C_6H_5CO_2Na$	0.144 11
Sodium borate, tetra- (PH)	$Na_2B_4O_7$	0.050 305
Sodium borate, tetra- (PH)	$Na_2B_4O_7 \cdot 10H_2O$	0.095 343
Sodium carbonate, acid (MO)	$NaHCO_3$	0.084 007
Sodium oleate	$C_{17}H_{33}CO_2Na$	0.304 45
Sodium salicylate	$C_6H_5OCO_2Na$	0.160 11
Stearic acid (PH)	$C_{17}H_{35}CO_2H$	0.284 49
Succinic acid (PH)	$(CH_2CO_2H)_2$	0.059 045
Sulfate	SO_4	0.048 031
Sulfur dioxide (PH)	SO_2	0.032 031
Sulfur trioxide	SO_3	0.040 031
Sulfuric acid	H_2SO_4	0.049 039
Sulfurous acid (PH)	H_2SO_3	0.041 039
Tartaric acid (PH)	$H_2C_4H_4O_6$	0.075 044
Tartaric acid (PH)	$H_2C_4H_4O_6 \cdot H_2O$	0.084 052

Source: J. A. Dean, ed., *Lange's Handbook of Chemistry*, 14th ed., McGraw-Hill, New York, 1992.

particularly dibasic acids, may be improved in a leveling solvent that will enhance their acidic or basic properties. It is also possible to determine the individual constituents in a mixture of either bases or acids of different strengths by differentiating titrations carried out in a solvent that does not exert a leveling effect. The potentials of some reference electrodes in nonaqueous media are listed in Table 14.4.

Potentiometric titrations are necessary for highly colored solutions in which the color change of a visual indicator cannot be detected and also for substances that remain weakly basic or weakly acidic in spite of the leveling effect of the solvent employed. Visual indicators may be employed with substances that behave as sufficiently strong acids or bases in suitable nonaqueous solvents.

4.3.1 Solvents

The major considerations in the choice of a solvent are its acidity and basicity, its dielectric constant, and the physical solubility of a solute. Acidity is important because it determines to a large extent whether or not a weak acid can be titrated in the presence of a relatively high concentration of solvent molecules. Phenol, for example, cannot be titrated as an acid in aqueous solution because water is too acidic and present in too high a concentration to permit the phenolate ion to be formed stoichiometrically by titration with the hydroxide ion (the strongest base that can exist in water). In less acidic solvents, such as the lower alcohols, dimethyl formamide, and the amines, or the aprotic solvents such as acetone, acetonitrile, and 4-methyl-2-pentanone, the titration of phenol is readily achieved with the alkoxide ion or tetrabutylammonium hydroxide as titrant.

Amphiprotic solvents possess both acidic and basic properties. Their autoprotolysis produces a solvonium ion, SH_2^+ and a solvate ion, S^-:

$$SH + SH \rightleftharpoons SH_2^+ + S^- \tag{4.33}$$

The product of the ion concentrations gives the autoprotolysis constant

$$K_{auto} = [SH_2^+][S^-] \tag{4.34}$$

pK_{auto} is 14 in water, varies from 15 to 19 in alcohols, and is about 15 in glacial acetic acid. Values of pK_{auto} are listed in Table 4.15 for common solvents. Of course, the acidic and basic limits for amphiprotic solvents do differ, as illustrated in Figs. 4.2 and 4.3.

When an acid or base is dissolved in a solvent, the equilibrium ions depend upon the relative acidic and basic strengths of the solute acid (or base) and the solvonium (or solvate) ion. The position of the autoprotolysis ranges, relative to the intrinsic strength of index acids, is indicated schematically in Fig. 4.3.

4.3.1.1 Glacial Acetic Acid. Glacial acetic acid is used more frequently than any other solvent for the titration of basic substances. Commercial samples containing not more than 1% of water may be

TABLE 4.15 Properties of Common Acid–Base Solvents

Solvent	Potential span, mV	pK_{auto}	Dielectric constant, 25°C
Acetic acid	400	14.5	6.1 (20°C)
Acetic anhydride	800	14.5	20.7 (20°C)
Acetone	1600		20.7
Acetonitrile	1600	26.5	37.5 (20°C)
1-Butanol	900		17.1
1-Butylamine	500		4.88 (20°C)
Chlorobenzene	1500		5.62
Chloroform			4.81 (20°C)
N,N-Dimethylformamide	1300	18.0	36.71
Dimethylsulfoxide		17.3	46.6
Ethanol	800	19.1	24.55
Ethanolamine		5.1	37.7
Ethyl acetate	1500		6.02
Ethylenediamine (1,2-Ethanediamine)	500	15.3	14.2 (20°C)
Methanol	800	16.7	32.7
4-Methyl-2-pentanone (methyl isobutyl ketone)	1600	25.0	13.1 (20°C)
2-Propanol	900		19.92
Pyridine	1000		12.3
Water	800	14.0	78.3

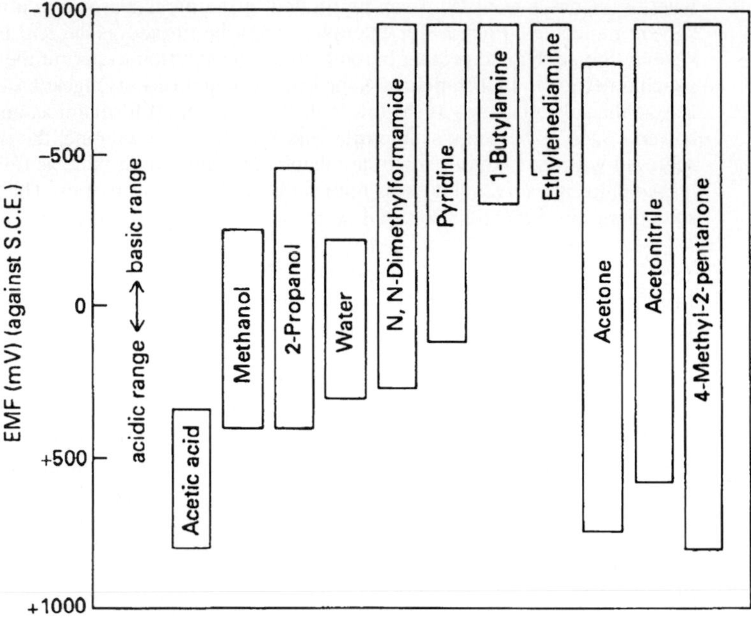

FIGURE 4.2 Approximate potential ranges in nonaqueous solvents. Basic limit is due to the titrant used: tetrabutylammonium hydroxide. *Source:* J. A. Dean, ed., *Lange's Handbook of Chemistry*, 14th ed., McGraw-Hill, New York, 1992.

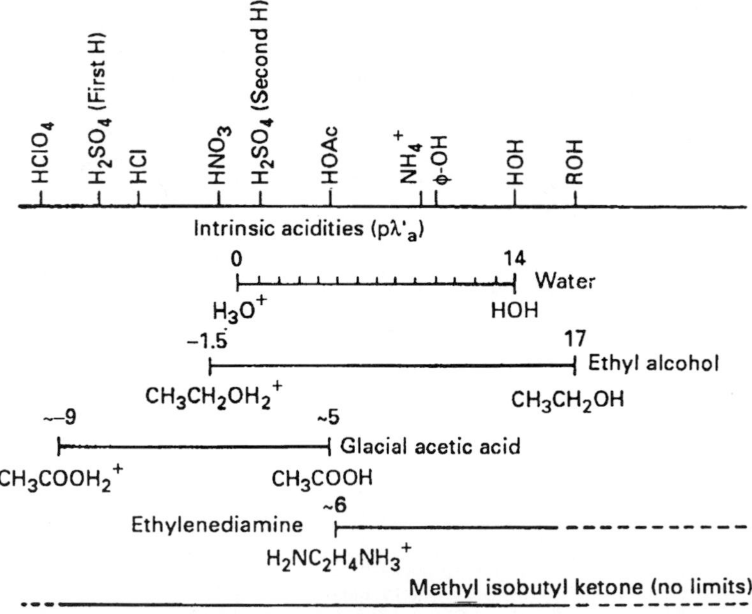

FIGURE 4.3 Schematic representation of autoprotolysis ranges of selected solvents, in relation to the intrinsic strength of certain index acids. The influence of the dielectric constant is not included. *Source:* H. H. Willard et al., *Instrumental Methods of Analysis*, 7th ed., Van Nostrand, New York, 1981.

used for more routine purposes. For work of the highest accuracy a water content between 0.01% and 0.1% is desirable. This can be achieved by performing a Karl Fischer water determination and then adding sufficient acetic anhydride to react with the surplus water and allowing the liquid to stand for 24 h to complete this reaction. Excess acetic anhydride is detrimental if primary or secondary amines are to be analyzed.

Mixtures of acetic acid with nitromethane or acetonitrile are suitable solvents for the potentiometric titration of bases.

4.3.1.2 Acetic Acid and Acetic Anhydride. Acetic acid, to which an excess of 5% acetic anhydride has been added, is a good solvent for use in the titration of tertiary amines and their salts. Good results are obtained with amides, purines, tertiary amines, and sulfoxides. Very weak bases such as urea and diphenylamine may be titrated in this mixture provided that temperature is maintained at 0°C and the titration is performed rapidly.

4.3.1.3 Acetone. Acetone, an aprotic solvent, generally needs only to be dried with fused calcium chloride, filtered, and distilled with calcium chloride. To eliminate organic impurities, add solid potassium permanganate and let stand 3 to 4 days, then filter and dry as above, and distill.

4.3.1.4 Acetonitrile. Acetonitrile, an aprotic solvent, is purified by shaking 24 h with Dowex 1 resin and filtering through Whatman 42 paper.

4.3.1.5 Benzene and Methanol. A mixture of benzene and methanol, using the minimum amount of methanol necessary to produce a clear solution, is used for the preparation of standard solutions of the alkali methoxides. Benzene may be freed from moisture by cutting metallic sodium into minute portions and letting stand in benzene with shaking for 2 days, and then distilling. Methanol can be dehydrated by fractional distillation and storing over Drierite.

4.3.1.6 Chloroform. Purify by shaking 1 L with 120 mL concentrated H_2SO_4. Remove the acid layer. Shake $CHCl_3$ layer with four portions of water followed by 20% NaOH solution. Shake again with water until the extract is neutral to bromothymol blue. Dry $CHCl_3$ with sodium sulfate, filter, and distill. Protect from sunlight during distillation and storage. Stabilize with 1% petroleum ether.

4.3.1.7* N,N-*Dimethylformamide. Many acidic substances are soluble in this solvent. When the solvent contains small amounts of acidic impurities, it must be neutralized immediately by running a blank titration. If larger amounts exist, stir the solvent for several hours with Dowex 1-X10 in the hydroxide form, then decant and filter.

4.3.1.8 1,4-Dioxane. Shake for 24 h with Dowex 1, and filter though Whatman 42 paper.

4.3.1.9 Ethylenediamine or 1-Butylamine. These are strongly basic solvents that exert a leveling effect and enhance the acid strength of substances such as phenols. The titrant is an alkali methoxide in benzene–methanol, using a visual indicator. Ethylenediamine dissolves a greater range of substances than does 1-butylamine. Ethylenediamine must be virtually free from water if it is to be used with very weak acids. Acidic impurities may be removed by passing the solvents through a column of activated aluminum oxide. Both solvents absorb carbon dioxide (a strong acid in these solvents) readily from the atmosphere and must be protected with suitable closed systems.

4.3.1.10 Glycol–Hydrocarbon (G-H) Solvents. Such solvents are made by mixing any glycol with any hydrocarbon, chlorinated hydrocarbon, or higher alcohol. A 1 : 1 mixture of ethyl glycol or propylene glycol with 2-propanol or butanol provides good solvent power for salts in general, and soaps in particular. Chloroform is often substituted for most of the hydrocarbon portion since it has good solvent properties and favors sharp color changes of visual indicators.

4.3.1.11 4-Methyl-2-Butanone (Methyl Isobutyl Ketone). This is a nonleveling (aprotic) solvent suitable for potentiometric differentiation of bases or of acids with widely differing strength. Acidic impurities can be removed by passing the solvent through a column of activated aluminum oxide.

4.3.1.12 2-Propanol. The solvent resembles ethanol in its useful range. Dehydrate by shaking 2-propanol with 10% of its weight of flake sodium hydroxide. Separate the alcohol layer and repeat with fresh sodium hydroxide. Decant and distill.

4.3.1.13 Pyridine. This solvent is used in the potentiometric titration of acidic substances, especially where the titrant is tetrabutylammonium hydroxide in benzene–methanol. Acids not readily soluble in pyridine may be dissolved in 1.5 mL of water prior to the addition of 50 mL of pyridine. In spite of its basic nature pyridine can be used in differentiating titrations. Acidic impurities can be removed by shaking pyridine with powdered potassium hydroxide for several days. Decant the pyridine layer and distill over barium oxide.

4.3.1.14 Tetrahydrofuran. Tetrahydrofuran is the solvent of choice for the titration of extremely weak acids. Small batches of solvent must be purified immediately before use by repeated distillation with lithium tetrahydridoaluminate. Oven-dried ware must be used. The solvent must be carefully protected against moisture and carbon dioxide.

4.3.2 Preparation and Standardization of Reagents

4.3.2.1 0.1M Perchloric Acid. This reagent may be prepared in any solvent with which 72% $HClO_4$ is miscible. To 950 mL of chosen solvent, add 8.5 mL of 72% $HClO_4$ and also 20 mL of acetic anhydride if a completely dry reagent is desired. Let stand 24 h. Standardize against 0.2 g of potassium hydrogen phthalate dissolved in 25 mL acetic acid with gentle warming, a reaction in which dihydrogen phthalate is the final product. The end point is ascertained either potentiometrically or with two drops of 0.2% methyl violet in glacial acetic acid or chlorobenzene. In the latter case the color changes from violet to blue at the end point. Run a blank titration. The equivalent weight of potassium hydrogen phthalate is 204.22 g.

With 1,4-dioxane solvent, dissolve the potassium hydrogen phthalate in glacial acetic acid as described above. Add two drops 1% methyl violet in glacial acetic acid and titrate with the acid until the color changes to blue-green. Perform a blank titration.

With ethylene glycol/2-propanol (G-H solvent), standardize against sodium carbonate as described for 0.1*N* HCl. Suitable indicators are methyl orange, methyl red, or thymol blue.

4.3.2.2 01M Hydrochloric Acid. Mix 500 mL of ethylene glycol with 500 mL of 2-propanol (G-H solvent). Add 90 mL of concentrated HCl and mix thoroughly. Standardize against freshly dried anhydrous sodium carbonate (0.025 g) dissolved in the smallest possible amount of acetic acid. Avoid loss during the effervescence. Evaporate to dryness in a water bath and dissolve the residue in 25 mL of G-H solvent. Titrate with 0.1*N* HCl; the end point is ascertained either potentiometrically or with methyl orange, methyl red, or thymol blue as indicator. The equivalent weight of anhydrous sodium carbonate is 106.00 g.

4.3.2.3 0.1M Tetrabutylammonium Hydroxide. Commercial reagent dissolved in methanol is available and can be diluted to the desired working concentrations. Tetrabutylammonium hydroxide, 0.1*M* in benzene–methanol (9:1) is also available. Standardize against benzoic acid (equivalent weight is 122.12 g) using thymol blue as indicator.

Preparation, where necessary, follows. Dissolve 40 g of tetrabutylammonium iodide (or bromide) in 90 mL of absolute methanol. Add 20 g of silver oxide and let stand in an ice bath for 1 h with occasional stirring. Centrifuge a few milliliters of the mixture and test the supernatant liquid for iodide. If a positive reaction is obtained, add an additional 2 g of silver oxide. When the liquid is free from iodide, filter the mixture through a fine-porosity glass filter, wash with 150 mL of dry benzene, and dilute the filtrate and washings to 1 L with benzene. Standardize potentiometrically against benzoic acid dissolved in pyridine (0.3 g/60 mL). Protect the standard hydroxide solution from carbon dioxide and moisture.

If a very pure hydroxide is desired (e.g., for work with strong acids), the reagent should be purified by passage through an ion-exchange column. Half fill a 25-mm by 400-mm chromatographic column with Dowex 1 (OH form). Pass 2*M* NaOH through the column until the effluent gives a negative test for halide; then rinse with distilled water until the effluent is neutral. Wash the column with 500 mL of methanol and then with 500 mL of benzene–methanol (10 : 1). Finally pass the reagent through the column at a rate of 15 to 20 mL per minute and collect and store the effluent when it is alkaline from universal indicator paper. Protect the effluent from carbon dioxide and moisture.

4.3.2.4 0.1M Potassium Hydroxide in Methanol. Dissolve 3 g of potassium hydroxide in 500 mL anhydrous methanol. Filter and store, protected from carbon dioxide and moisture. Standardize against benzoic acid by dissolving 0.2 g benzoic acid in 50 mL chloroform and 1 mL anhydrous methanol. Add four drops 0.5% thymol blue in anhydrous methanol and titrate until color changes to violet. Perform a blank titration. The equivalent weight of benzoic acid is 122.12 g.

Anhydrous benzene may be substituted for portions of methanol. Enough methanol must be present to produce a clear one-phase solution.

4.3.2.5 0.1M Sodium Aminoethoxide. Dissolve 2.3 g of sodium metal in ethanolamine and dilute the solution to volume with ethylenediamine. Standardize potentiometrically against benzoic acid.

4.3.2.6 0.1M Sodium Acetate in Glacial Acetic Acid. Dissolve 8.2 g of anhydrous sodium acetate in glacial acetic acid and dilute to 1 L. Equivalent weight of sodium acetate is 82.04 g.

4.3.2.7 0.1M Aniline in Glacial Acetic Acid. A volume of 9.1 mL of double-distilled aniline is dissolved in glacial acetic acid and diluted to 1 L. Standardize against benzoic acid or standard perchloric acid solution.

4.3.2.8 0.1M Lithium Tetraamidoaluminate (Lithium Aluminum Amide). Dissolve 1.0 g of lithium aluminum amide in 100 mL of tetrahydrofuran. Prepare a fresh solution every few days. All equipment must be oven-dried glassware and all solutions must be scrupulously protected from moisture and carbon dioxide. Standardize potentiometrically against dry 2-propanol (0.3 g/5 mL) in tetrahydrofuran.

4.3.3 Acidities and Basicities of Acids and Bases in Nonaqueous Solvents

Although only limited study has been devoted to the quantitative evaluation of acidity constants in solvents other than water, some data are summarized in Table 4.16. As a first approximation, the relative acidities of a group of structurally related compounds appear to be merely modified, rather than fundamentally changed.

When evaluating the titration possibilities in nonaqueous solvents, consider first what are the actual ionic or nonionized species present after the solute(s) are dissolved in the solvent selected. Any acid lying outside the acidic limit of the solvent (the SH_2^+ ion) will undergo a "leveling" process

TABLE 4.16 pK_a Values for Various Acids and Indicators in Nonaqueous Systems

Acid	Methanol	Ethanol	Other solvents
Acetic acid	9.52	10.32	11.4,* 9.75†
p-Aminobenzoic acid	10.25		
Ammonium ion	10.7		6.40‡
Anilinium ion	6.0	5.70	
Benzoic acid		10.72	10.0*
Bromocresol purple	11.3	11.5	
Bromocresol green	9.8	10.65	
Bromophenol blue	8.9	9.5	
Bromothymol blue	12.4	13.2	
di-*n*-Butylammonium ion			10.3*
o-Chloroanilinium ion	3.4		
Cyanoacetic acid		7.49	
2,5-Dichloroanilinium ion			9.48‡
Dimethylaminoazobenzene		5.2	6.32‡
N,N′-Dimethylanilinium ion		4.37	
Formic acid		9.15	
Hydrobromic acid			5.5§
Hydrochloric acid			8.55,‡ 8.9§
Methyl orange	3.8	3.4	
Methyl red (acid range)	4.1	3.55	
(alkaline range)	9.2	10.45	
Methyl yellow	3.4	3.55	
Neutral red	8.2	8.2	
o-Nitrobenzoic acid	7.6		
m-Nitrobenzoic acid	8.3		
p-Nitrobenzoic acid	8.4		
Perchloric acid			4.87‡
Phenol	14.0		
Phenol red	12.8	13.4	
Phthalic acid, pK_2	11.65		11.5,† 6.10† (pK_1)
Picric acid	3.8	3.8	8.9§
Pyridinium ion			6.1‡
Salicylic acid	8.7	7.9	
Stearic acid	10.0		
Succinic acid, pK_2	11.4		
Sulfuric acid, pK_1			7.24‡,§
Tartaric acid, pK_2	9.9		
Thymol blue (alkaline range)	14.0	15.2	
(acid range)	4.7	5.35	
Thymolbenzein (acid range)	3.5		
(alkaline range)	13.1		
p-Toluenesulfonic acid			8.44‡
p-Toluidinium ion		6.24	
Tribenzylammonium ion			5.40‡
Tropeoline 00	2.2	2.3	
Urea (protonated cation)			6.96‡
Veronal	12.6		

* Dimethylsulfoxide.
† Acetone + 10% water.
‡ Glacial acetic acid.
§ Acetonitrile.

Source: J. A. Dean, ed., *Lange's Handbook of Chemistry*, 14th ed., McGraw-Hill, New York, 1992.

to the solvonium ion. For example in ethanol, perchloric acid, the first hydrogen of sulfuric acid, and hydrochloric acid appear equal in strength because each is leveled to the $C_2H_5OH_2^+$plus ClO_4^-, HSO_4^-, or Cl^- anion. In contrast, glacial acetic acid only levels perchloric acid among the group of three acids. Nitric acid is no longer completely dissociated in ethanol and, in fact, is a weak acid in glacial acetic acid.

The magnitude of the autoprotolysis constant determines the number of acids that can be differentiated from each other. If ideally spaced apart by roughly 300 mV (or 5 pK_a units), three could be titrated successively in ethanol: an acid leveled by the solvent (perhaps HCl), a second located at an apparent pK_a of 5 (perhaps HOAc), and a third located at an apparent pK_a of 10 to 11 (perhaps phenol). In this respect, an aprotic solvent such as 4-methyl-2-pentanone would seem to be the ideal solvent. Its one drawback is its rather low dielectric constant. Indeed, these acid mixtures have been differentially titrated successfully in 4-methyl-2-pentanone using a quaternary ammonium hydroxide as titrant and glass–calomel electrodes: (1) $HClO_4$–HCl–salicylic acid–HOAc–phenol, (2) $HClO_4$ and the first and second hydrogens of sulfuric acid, and (3) $HClO_4$–HCl–HNO_3.[2]

Suitable strong acid titrants are $HClO_4$ or *p*-toluenesulfonic acid for any solvent. Suitable basic titrants would be NaOAc in glacial acetic acid, and sodium alkoxide or 2-aminoethanolate for the alcohols, ethylenediamine, and 4-methyl-2-pentanone. Quaternary ammonium hydroxides is also a strong basic titrant. The lack of any stronger basic titrant coupled with the need to exclude carbon dioxide and moisture limits the useful base range of these latter solvents (Figure 4.2).

In general, the apparent difference between the acidities of two compounds, one an uncharged acid and the other a positively charged acid (such as the solvonium cation) increases as the dielectric constant becomes smaller. The difference is about three pK_a units for methanol (dielectric constant 32.7) and about five pK_a units for ethanol (dielectric constant 24.5). Below a dielectric constant of 25, ion-pair formation becomes significant. The half-neutralization point becomes both concentration and salt dependent.

The glass–calomel electrode system is suitable where the solvent is acetonitrile, an alcohol, or a ketone provided that the sodium is absent in the titrant. It is advisable to replace the aqueous salt bridge of the calomel electrode by either a saturated solution of potassium chloride in methanol or tetraalkylammonium chloride solution. A pair of antimony electrodes forms a satisfactory combination, one dipping into the titrant and the other into the solution. Solvents with low dielectric constants exhibit such high internal resistance that it is difficult to find electrodes that function satisfactorily. When the dielectric constant is 5 or less, potentiometric methods are unsuitable. A review of potentiometric methods in nonaqueous solvents has been published.[3]

Selected titration methods in nonaqueous media are given in Table 4.17.

4.3.4 Titration Curves in Nonaqueous Acid–Base Systems

Acid–base titration curves in protolytic solvents having dielectric constants above about 25 can be closely approximated by means of the equations given for aqueous media (Sec. 4.2.4). Naturally it is essential to use the values of K_a or K_b appropriate to the solvent under consideration, and also to replace K_w wherever it occurs in those equations with the autoprotolysis constant of the solvent employed.

These equations do not apply to solvents having low dielectric constants because of the extensive ion-pair formation which then occurs. Kolthoff and Bruckenstein[4] have given a complete quantitative treatment of titrations of bases with perchloric acid in glacial acetic acid.

[2] D. B. Bruss and G. E. A. Wyld, *Anal. Chem.*, **29**:232 (1957).
[3] J. T. Stock and W. C. Purdy, *Chem. Rev.*, **57**:1159 (1957).
[4] I. M. Kolthoff and S. Bruckenstein, *J. Am. Chem. Soc.*, **79**:1 (1957).

TABLE 4.17 Selected Titration Methods in Nonaqueous Media

Substances titratable	Solvent medium	Titrant	Indicator
Acetate, aliphatic amine, and aromatic amine	Dissolve sample in smallest possible volume of acetic acid, then dilute with chloroform	Perchloric acid in 1,4-dioxane	Potentiometrically
Alcohols (organic hydroxyl groups)	(1) Tetrahydrofuran	(1) Add excess lithium tetrahydrido-aluminate, back titrate with butanol in benzene	(1) Potentiometrically
	(2) Pyridine; protect system from moisture	(2) Add 100% excess of 3,5-dinitro-benzoyl chloride in pyridine to sample; titrate with tetrabutyl-ammonium hydroxide	(2) Potentiometrically; use volume between first and second end point to evaluate percent hydroxyl
(1) Aldehydes and ketones	(1) Form 2,4-dinitrophenylhydrazones and dissolve in pyridine	(1) Tetrabutylammonium hydroxide in benzene/methanol	Potentiometrically
(2) Aldehydes, only aromatic, except formaldehyde	(2) Add excess dodecylamine in ethylene glycol/2-propanol solvent	(2) Titrate excess dodecylamine with salicylic acid in ethyleneglycol/2-propanol	
Alkaloids	(1) Glacial acetic acid; (2) benzene/chloroform; (3) ethylene glycol/2-propanol	(1) and (2) Perchloric acid in glacial acetic acid or 1,4-dioxane; (3) HCl in ethylene glycol/2-propanol	Potentiometrically
Alkoxyl groups	Pyridine	Sample in xylene, add HI, pass N_2 and reflux 45 min (methoxyl), 85 min (ethoxyl), 145 min (propoxyl and butoxyl) and 3 h for *S*-methyl	Potentiometrically or azo violet (red for excess HI end point and violet for pyridinium iodide end point)
Amides	Acetic anhydride	Perchloric acid in glacial acetic acid	Potentiometrically
Amines, primary, secondary and tertiary singly	Glacial acetic acid (no free acetic anhydride)	Acetous $HClO_4$	Oracet blue B or potentiometrically
(1) Amines, total	(1) Ethylene glycol and 2-propanol (1:1)	0.1*M* HCl in ethylene glycol and 2-propanol (1:1)	Potentiometrically
(2) Amines, tertiary	(2) Ethylene glycol/2-propanol (1:1) plus sufficient acetic anhydride to acetylate primary and secondary amines		
(3) Amines, secondary and tertiary	(3) Ethylene glycol/2-propanol (1:1) plus salicylaldehyde to form nontitratable Schiff's base with primary amine		

Amino acids	Glacial acetic acid	Acetous $HClO_4$ added in excess, mixture warmed under reflux until sample dissolved, excess $HClO_4$ titrated with acetous NaOAc	Crystal violet to full violet color
Anhydrides of carboxylic acids	(1) 1,4-Dioxane or (2) ethylene glycol/2-propanol	(1) Pyridine added, then sodium methoxide or (2) excess of aniline added and then determined by back titration with HCl in ethylene glycol/2-propanol (1:1)	Thymol blue (blue end point)
Antihistamines	Glacial acetic acid/chloroform	Perchloric acid in glacial acetic acid	Crystal violet
Barbiturates	(1) Dimethylformamide or (2) benzene/methanol	Potassium methoxide in benzene–methanol (10:1)	Thymol blue
Carboxylic acids	Dimethylformamide	Potassium methoxide in benzene–methanol (10:1)	Thymol blue
Enols and imides	Ethylenediamine or dimethylformamide (if not weakly acidic)	Potassium methoxide in benzene–methanol	Thymol blue or azo violet; *o*-nitroaniline for weakly acidic samples
Halide salts	Glacial acetic acid plus $Hg(OAc)_2$	Perchloric acid in glacial acetic acid	Purple color of Oracet blue B
Hydrazides	Acetic anhydride–glacial acetic acid	Perchloric acid in glacial acetic acid	Crystal violet
Phenols	Ethylenediamine	(1) Sodium aminoethoxide in ethylenediamine or (2) potassium methoxide in benzene–methanol (10:1)	(1) Potentiometrically or (2) *o*-nitroaniline to orange-red color
Purines	Ethylenediamine	Potassium methoxide	Azo violet
Sulfonamides—more strongly acidic	Dimethylformamide	Potassium methoxide in benzene–methanol (10:1)	Thymol blue
Sulfonamides—less acidic	Butylamine	Potassium methoxide in benzene–methanol (10:1)	Azo violet
Salts of stronger acids: perchlorates, sulfates, iodides, citrates, picrates, benzoates	Dimethylformamide (if salt is soluble or else minimum amount of water to produce a clear solution, then 15 times the volume of ethylenediamine	Potassium methoxide in benzene–methanol (10:1)	Azo violet or thymol blue (unless solute is slightly acidic)

4.4 PRECIPITATION TITRATIONS

Many precipitation reactions that are useful as separation techniques or for gravimetric analysis (Sec. 4.1) fail to meet one or both of two requirements for titrimetry.

1. For titrimetry, the reaction rate must be sufficiently rapid, particularly in the titration of dilute solutions and in the immediate vicinity of the end point. To increase the precipitation rate, it is sometimes beneficial to change solvents or to raise the temperature. By adding an excess of reagent and back-titrating, it may be possible to take advantage of a more rapid precipitation in the reverse direction. By choosing an end-point detection method that does not require equilibrium to be reached in the immediate vicinity of the end point, advantage may be taken of a faster reaction rate at points removed from the end point. Examples are amperometric titrations (Sec. 14.5.11), conductometric titrations (Sec. 14.9.2), and photometric titrations (Sec. 6.4)

2. For titrimetry, the stoichiometry must be exact. Coprecipitation by solid-solution formation, foreign ion entrapment, and adsorption are possible sources of error.

Considering these limitations, it is not surprising that the applications of titration methods based on precipitation reactions are less important than those based on acid–base, complex formation, or oxidation–reduction reactions.

4.4.1 Titration Curves and Precision in Precipitation Titrations

Almost all of the precipitation titrations listed in Table 4.20 involve a reaction of the form X + R = XR and that the solubility product of XR is $K_{sp} = [X][R]$. A reasonably well-fitting titration curve can be selected or obtained by interpolation from the curves shown in Fig. 4.4. It is customary, by analogy with pH titration curves of acids and bases, to plot the quantity pM (defined as $-\log [M^{n-1}]$) against titration volume (V_R, mL) or fraction of the stoichiometric quantity of reagent added. For metals that form reversible electrodes with their ions, the measured electrode potential is a linear function of the metal-ion concentration (strictly metal-ion activity), so the titration curve can be realized experimentally in a potentiometric titration (Sec. 14.4.3).

4.4.1.1 Titration Curves

1. *Up to the equivalence point*:

$$[X] = \frac{V_X C_X^0 - V_R C_R}{V_X + V_R} + \frac{K_{sp}}{[X]} \tag{4.35}$$

2. *At the equivalence point*:

$$[X] = [R] = \sqrt{K_{sp}} \tag{4.36}$$

3. *Beyond the equivalence point*:

$$[R] = \frac{V_R C_R - V_X C_X^0}{V_X + V_R} + \frac{K_{sp}}{[R]} \tag{4.37}$$

In using Eqs. (4.35) and (4.37), the second term on the right-hand sides can often be omitted. It is a correction for the solubility of the precipitate and is necessary only if (1) K_{sp} is rather large, (2) the solutions are quite dilute, or (3) the point being considered is very near to the equivalence point, as for example, the value of [R] or [X].

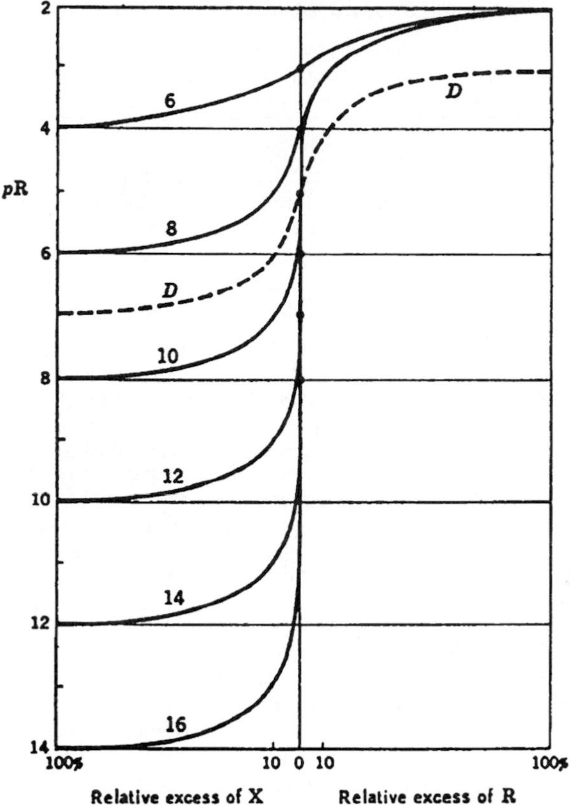

FIGURE 4.4 Titration curves for the precipitation titration X + R = XR. The numbers beside the curves give the assumed values of pK_{sp}, where K_{sp}, the solubility of XR, is given by the equation K_{sp} = [X][R]. For each numbered curve, the initial concentration of the solution being titrated is 0.01M and the concentration of the titrant is 0.1M. For curve D, the assumed value of pK_{sp} is 10, and the concentrations are 0.001M and 0.01M, respectively. *Source*: L. Meites, ed., *Handbook of Analytical Chemistry*, Mc-Graw-Hill, New York, 1963.

Figure 4.4 shows that the slope of the curve at the equivalence point, which determines the relative precision of the titration, depends upon both the value of K_{sp} and the concentrations of the solutions employed. For the titration of the 0.01M solution of X with 0.1M solutions of R, curves are given for pK_{sp} = 6, 8, 10, 12, 14, and 16. At this dilution, the titration will not be precise if K_{sp} exceeds about 10^{-8}. The undesirable effect of low concentration may be seen by comparing curves 10 and D, both of which were calculated for pK_{sp} = 10.

4.4.1.2 Relative Precision. The relative precision rp, which is the fraction of the stoichiometric quantity of reagent required to traverse the region ±0.1 pX unit on either side of the end point (i.e., the precision of the titration), may be estimated from the equation

$$\text{rp} = \pm \frac{\sqrt{K_{\text{sp}}}}{C_{\text{X}}^{0}} \tag{4.38}$$

TABLE 4.18 Standard Solutions for Precipitation Titrations

The list given below includes the substances that are most used and most useful for the standardization of solutions for precipitation titrations. Primary standard solutions are denoted by the letter (P) in the first column.

Standard	Formula weight	Preparation
$AgNO_3$ (P)	169.89	Weigh the desired amount of ACS reagent grade* $AgNO_3$, dried at 105°C for 2 h, and dissolve in double distilled water. Store in an amber container and away from light. Check against NaCl.
$BaCl_2 \cdot 2H_2O$	244.28	Dissolve clear crystals of the salt in distilled water. Standardize against K_2SO_4 or Na_2SO_4.
$Hg(NO_3)_2 \cdot H_2O$	342.62	Dissolve the reagent-grade salt in distilled water and dilute to desired volume. Standardize against NaCl.
KBr	119.01	The commercial reagent (ACS) may contain 0.2% chloride. Prepare an aqueous solution of approximately the desired concentration and standardize it against $AgNO_3$.
$K_4Fe(CN)_6 \cdot 3H_2O$	422.41	Dissolve the high-purity commercial salt in distilled water containing 0.2 g/L of Na_2CO_3. Kept in an amber container and away from direct sunlight, this solution is stable for a month or more. Standardize against zinc metal.
KSCN	97.18	Prepare aqueous solutions having the concentration desired. Standardize against $AgNO_3$ solution. Protect from direct sunlight.
K_2SO_4 (P)	174.26	Dissolve about 17.43 g, previously dried at 150°C and accurately weighed, in distilled water and dilute exactly to 1 L.
NaCl (P)	58.44	Dry at 130–150°C and weigh accurately, from a closed container, 5.844 g, dissolve in water, and dilute exactly to 1 L.
NaF (P)	41.99	Dry at 110°C and weigh and appropriate amount of ACS reagent. Dissolve in water and dilute exactly to 1 L.
Na_2SO_4 (P)	142.04	Weigh accurately 14.204 g, dried at 150°C, and dissolve in distilled water. Dilute to exactly 1 L.
$Th(NO_3)_4 \cdot 4H_2O$	552.12	Weigh the appropriate amount of crystals and dissolve in water. Standardize against NaF.

* Meets standards of purity (and impurities) set by the American Chemical Society.

where C_X^0 is the hypothetical end point concentration of RX that would exist if no precipitation had occurred. The precision may be improved by keeping C_X^0 high or by decreasing K_{sp}. The latter may be accomplished in many cases by changing the solvent by adding ethanol or some other water-miscible organic solvent; this often involves an increased danger of coprecipitation of other materials.

For a precipitate of the type R_2X, the inflection point of the titration of X with R corresponds to

$$1 - X_i = \frac{3}{4}\frac{(K_{sp})^{3/2}}{C} \tag{4.39}$$

where C is the hypothetical molar concentration of R_2X at the equivalence point, and X_i is the fraction of the stoichiometric quantity of reagent that has been added up to the inflection point. The quantity $100(X_i - 1)$ represents the theoretical percentage error.

If two substances X and Y are present that give the less soluble precipitate XR and the more soluble precipitate YR, the titration curve acquires a singular point. The direction of the titration curve changes abruptly when the precipitate YR appears as a second permanent solid phase. The titration curve for $C_X^0 = C_Y^0 = 0.01$, $K_{sp,XR} = 16$, and $K_{sp,YR} = 10$ may readily be obtained from Fig. 4.4 by using the part of curve 16 lying below pR = 8, and moving curve 10 to the right so that it starts in the (vertical) center of the graph. The condition for the start of the precipitation of YR is

$$[X] = C_Y^0 \frac{K_{sp,YR}}{K_{sp,XR}} \tag{4.40}$$

If the start of the precipitation of YR is made the end point of the titration, and if it may be assumed that YR is not coprecipitated with XR, the unavoidable error may be calculated from the solubilities of XR and YR. The relative error in the result for X is

$$C_Y^0 \left(\frac{K_{sp,XR}}{K_{sp,YR}} \right) \left(\frac{C_Y^0}{C_X^0} \right) \tag{4.41}$$

The relative error in result for Y is

$$\frac{K_{sp,XR}}{K_{sp,YR}} \tag{4.42}$$

4.4.2 Applications

Standard solutions for precipitation titrations are enumerated in Table 4.18, and indicators for precipitation titrations are listed in Table 4.19. Information on a number of commonly performed precipitation titrations is given in Table 4.20.

4.5 OXIDATION–REDUCTION TITRATIONS

4.5.1 Titration Curves and Precision in Redox Titrations

Redox titration curves are most conveniently obtained by plotting the redox potential E of the titrated solution against the relative excess of reductant or oxidant shown on the x axis. Assuming that both half-reactions are reversible, the redox potentials may be computed by means of the Nernst equation, in the following manner. It is assumed that V_A mL of a solution originally containing C_0^A mol/L of the reducing agent A is titrated with a solution containing C_B mol/L of the oxidizing agent B, that the half-reactions involved, with their formal potentials, are

$$A \longrightarrow C + n_A e; E_{C,A}^0$$

and

$$B + n_B e \longrightarrow D;\ E_{B,D}^0$$

so that the equation for the chemical reaction taking place is

$$n_B A + n_A B \longrightarrow n_B C + n_A D$$

and that V_B mL of B has been added up to the point under consideration. Then, at any point during

TABLE 4.19 Indicators for Precipitation Titrations

Indicator	Preparation and use
	Specific reagents
$NH_4Fe(SO_4)_2 \cdot 12H_2O$	Use reagent-grade (ACS)* salt, low in chloride. Dissolve 175 g in 100 mL 6*M* HNO_3 that has been gently boiled for 10 min to expel nitrogen oxides. Dilute with 500 mL water. Use 2 mL per 100 mL end-point volume.
K_2CrO_4	Use reagent-grade (ACS)* salt, low in chloride. Prepare 0.1*M* aqueous solution (19.421 g/L). Use 2.5 mL per 100 mL end-point volume.
Tetrahydroxy-1,4-benzoquinone (THQ)	Prepare fresh as required by dissolving 15 mg in 5 mL of water. Use 10 drops for each titration.
	Adsorption indicators
Bromophenol blue	Dissolve 0.1 g of the acid in 200 mL 95% ethanol.
2′,7′-Dichlorofluorescein	Dissolve 0.1 g of the acid in 100 mL 70% ethanol. Use 1 mL for 100 mL of initial solution.
Eosin, tetrabromofluorescein	See Dichlorofluorescein above.
Fluorescein	Dissolve 0.4 g of the acid in 200 mL 70% ethanol. Use 10 drops.
Potassium rhodizonate, $C_4O_4(OK)_2$	Prepare fresh as required by dissolving 15 mg in 5 mL of water. Use 10 drops for each titration.
Rhodamine 6G	Dissolve 0.1 g in 200 mL 70% ethanol.
Sodium 3-alizarinsulfonate	Prepare a 0.2% aqueous solution. Use 5 drops per 120 mL end-point volume.
Thorin	Prepare a 0.025% aqueous solution. Use 5 drops.
	Protective colloids
Dextrin	Use 5 mL of 2% aqueous solution of chloride-free dextrin per 25 mL of 0.1*M* halide solution.
Polyethylene glycol 400	Prepare a 50% (v/v) aqueous solution of the surfactant. Use 5 drops per 100 mL end-point volume.

* Meets standards as set forth in *Reagent Chemicals,* American Chemical Society, Washington, D.C.; revised periodically.

the titration,

$$[A] = V_A C_A^0 - \frac{n_B}{n_A} V_B C_B + \frac{n_B}{n_A}[B] \tag{4.43}$$

$$[C] = \frac{n_B}{n_A} V_B C_B - \frac{n_B}{n_A}[B] \tag{4.44}$$

$$[D] = V_B C_B - [B] \tag{4.45}$$

The equilibrium constant of the reaction can be obtained by setting the redox potentials of two half-reactions (see Table 4.21) equal to each other—a condition that must exist whenever chemical equilibrium is established—and solving for *K*, which yields, for the above reaction at 25°C,

$$\frac{0.05915}{n_A n_B} \log K = E^0_{B,D} - E^0_{C,A} \tag{4.46}$$

Since

$$K = \frac{[\mathrm{C}]^{n_\mathrm{B}}[\mathrm{D}]^{n_\mathrm{A}}}{[\mathrm{A}]^{n_\mathrm{B}}[\mathrm{B}]^{n_\mathrm{A}}} \tag{4.47}$$

combining Eqs. (4.43) to (4.45) and Eq. (4.46) with the numerical value of K obtained from Eq. (4.46), yields an equation containing [B] as the only unknown quantity. Solving this and combining the result with Eqs. (4.43) to (4.45) yield values for the other concentrations. The potential can then be obtained from either of the following equations:

$$E = E^0_{\mathrm{C,A}} - \frac{0.05915}{n_\mathrm{A}}\log\frac{[\mathrm{A}]}{[\mathrm{C}]} \tag{4.48}$$

$$E = E^0_{\mathrm{B,D}} - \frac{0.05915}{n_\mathrm{B}}\log\frac{[\mathrm{D}]}{[\mathrm{B}]} \tag{4.49}$$

This procedure is complex, and is required only if points very near the end point must be considered, or if K is quite small (i.e., if the formal potentials of the two half-reactions are quite close together). Usually it can be simplified greatly by considering [B], the concentration of unreacted B, to be negligibly small before the equivalence point is reached, and similarly considering [A] to be negligible after the equivalence point has been passed. These additional assumptions yield the following equations.

1. *Up to the equivalence point*:

$$E = E^0_{\mathrm{C,A}} - \frac{0.05915}{n_\mathrm{A}}\log\frac{V_\mathrm{A}C^0_\mathrm{A} - \left(\frac{n_\mathrm{B}}{n_\mathrm{A}}\right)V_\mathrm{B}C_\mathrm{B}}{\left(\frac{n_\mathrm{B}}{n_\mathrm{A}}\right)V_\mathrm{B}C_\mathrm{B}} \tag{4.50}$$

2. *Beyond the equivalence point*:

$$E + E^0_{\mathrm{B,D}} - \frac{0.05915}{n_\mathrm{B}}\log\frac{\left(\frac{n_\mathrm{A}}{n_\mathrm{B}}\right)V_\mathrm{A}C^0_\mathrm{A}}{V_\mathrm{B}C_\mathrm{B} - \left(\frac{n_\mathrm{A}}{n_\mathrm{B}}\right)V_\mathrm{A}C^0_\mathrm{A}} \tag{4.51}$$

These can be further simplified for purposes of practical calculation by introducing the quantity f defined by the equation

$$f = \left(\frac{n_\mathrm{B}}{n_\mathrm{A}}\right)\left(\frac{V_\mathrm{B}C_\mathrm{B}}{V_\mathrm{A}C^0_\mathrm{A}}\right) \tag{4.52}$$

It is the ratio between the volume V_B of reagent that has actually been added at the point under consideration and the volume of reagent required to reach the equivalence point. Combining Eqs. (4.50) and (4.51) with Eq. (4.52), the following equations are immediately obtained.

1. *Up to the equivalence point*:

$$E = E^0_{\mathrm{C,A}} - \frac{0.05915}{n_\mathrm{A}}\log\frac{1-f}{f} \tag{4.53}$$

TABLE 4.20 Titration Methods Based on Precipitation

This table contains information on a number of commonly performed precipitation titrations. Values of pK_{sp} are available in Table 4.2. Informative discussions will be found in standard textbooks and monographs, also the biannual "Reviews" appearing in Analytical Chemistry.

Substance determined	Titrant	Indicator	Procedure	Remarks	Interferences
Ag^+	KSCN or NH_4SCN	$NH_4Fe(SO_4)_2 \cdot 12H_2O$	0.3 g Ag per 200 mL 0.8*M* HNO_3. Vigorous agitation at end point.	Precise. As, Bi, Cd, Co, Fe, Mn, Ni, Pb, Sb, Zn may be present.	AgCl, Hg(II), NO_2^-, Pd, SO_4^{2-}
	KBr	Rhodamine 6G	≤0.3*M* HNO_3	≥0.005*M* Ag^+	
AsO_4^{3-}	KSCN	$NH_4Fe(SO_4)_2 \cdot 12H_2O$	Precipitate Ag_3AsO_4 with $AgNO_3$ from weakly acidic HOAc. Filter, wash with water until $AgNO_3$ removed. Dissolve in 1*M* HNO_3, dilute to 120 mL, add indicator and titrate with KSCN.	Recommended after distillation or precipitation with H_2S. Ge and small amounts of Sb and Sn do not interfere.	All ions precipitated by Ag^+ in slightly acidic HOAc medium
Br^-	$AgNO_3$ and KSCN	$NH_4Fe(SO_4)_2 \cdot 12H_2O$	Precipitate AgBr from acid solution with small excess of $AgNO_3$, add indicator, back titrate with KSCN.	Recommended. Protect from light and remove AgBr before back titration in exacting work.	Chloride, cyanide, iodide, sulfide, thiocyanate, thiosulfate
	$AgNO_3$	Eosin	pH ≥ 1	≥0.0005*M* Br^-	Other ions precipitated by Ag^+ in acid solution
	$Hg_2(NO_3)_2$	Bromothymol blue	Add a few drops indicator and titrate to lilac color.	≤0.04*M* acid	Ag, Hg(II), sulfate
Cl^-	$AgNO_3$ and KSCN	$NH_4Fe(SO_4)_2 \cdot 12H_2O$	Precipitate AgCl from acid solution with small excess $AgNO_3$, filter, add indicator, and titrate with KSCN.	Recommended. Rapid method. Filtration may be avoided by adding nitrobenzene if $Cl^- \geq 0.02M$.	Bromide, cyanide, iodide, sulfide, thiocyanate, thiosulfate
	$AgNO_3$ and NaCl	Equal turbidity method	Titration in red light until supernatant liquid gives equal turbidity with $AgNO_3$ and NaCl titrants.	Highest accuracy and precision.	Same as above
	$AgNO_3$	K_2CrO_4	Add $NaHCO_3$ (pH 5–7). Titrate until near end point, add sufficient indicator to make final concentration 0.00025*M*.	Precise. In presence of NH_4^+, pH < 7.	Arsenate, Ba, Bi, Br^-, CN^-, I^-, phosphate, Pb, sulfide, SCN^-, $S_2O_3^{2-}$, Sr, reducing agents

	$AgNO_3$	Dichlorofluorescein	pH 4; add protective colloid.	≥0.025*M* Cl^-. Avoid exposure to light.	Anions forming insoluble silver salts; CN^-, $S_2O_3^{2-}$
Dithionites	KSCN	$NH_4Fe(SO_4)_2 \cdot 12H_2O$	Add weighed solid directly to ammoniacal $AgNO_3$ solution. Filter metallic Ag; wash with dilute NH_3 plus NH_4NO_3. Dissolve Ag in HNO_3, boil to remove nitrogen oxides, and titrate with KSCN using 2 mL indicator near end point.		
F^-	$AgNO_3$ and KSCN	$NH_4Fe(SO_4)_2 \cdot 12H_2O$	Precipitate PbClF at pH 3.6–5.6, wash, dissolve in 100 mL 0.8*M* HNO_3, add small excess $AgNO_3$, filter, titrate filtrate with KSCN.	Excellent results with pure fluorides.	Al, Be, Fe(III), and large amounts of K, NH_4, Na, phosphate or sulfate.
	$Th(NO_3)_4$	Na alizarinsulfonate	pH 3.0	Use H_2SiF_6 distillate.	
Hg_2^{2+}	NaCl	Bromothymol blue	Titrate in acid solution.	≥0.05*M* Hg(I)	
I^-	$AgNO_3$ and KSCN	$NH_4Fe(SO_4)_2 \cdot 12H_2O$	Add excess $AgNO_3$ to slightly acid solution slowly with stirring; add indicator and titrate with KSCN.	Recommended. Rapid.	Br^-, Cl^-, S^{2-}, SCN^-, $S_2O_3^{2-}$
	$AgNO_3$	Dichlorofluorescein	Solution 0.1*M* HNO_3	≥0.01*M* iodide	
SO_4^{2-}	$BaCl_2$	Tetrahydroxy-1,4-benzoquinone, K rhodizonate, or Na 3-alizarinsulfonate	To 45 mL solution of pH 7–8 containing ca. 0.0025*M* sulfate, add equal volume ethanol, 5 drops indicator and titrate.	Fair precision with ≥8 mg sulfate. At pH 4, small amounts of phosphate may be present.	All ions precipitated by barium in neutral solution
Zn^{2+}	$AgNO_3$ and KSCN	$NH_4Fe(SO_4)_2 \cdot 12H_2O$	Add $Hg(SCN)_2$ to precipitate $Zn[Hg(SCN)_4]$. Filter and dissolve in 1.6*M* HNO_3. Add excess $AgNO_3$. After 30 min add indicator and titrate with KSCN.	Recommended for small and moderate amounts of Zn. As, Fe(III), Pb, Sb, and Sn(IV) may be present.	Bi, Cd, Co, Cu, Fe(II), Mn, Ni
	$K_4Fe(CN)_6$	Diphenylbenzidine	0.18*M* HCl, 0.75*M* NH_4Cl, total 200 mL, 60°C, control rate of titration near the end point.	Reliable routine method in experienced hands.	Cd, Mn, much Fe, most heavy metals, NO_3^-, oxidants

TABLE 4.21 Potentials of Selected Half-Reactions at 25°C

A summary of oxidation–reduction half-reactions arranged in order of decreasing oxidation strength and useful for selecting reagent systems.

Half-reaction	E^0, V	Half-reaction	E^0, V
$F_2(g) + 2H^+ + 2e^- = 2HF$	3.053	$I_2 + 2e^- = 2I^-$	0.536
$O_3 + H_2O + 2e^- = O_2 + 2OH^-$	1.246	$Cu^+ + e^- = Cu$	0.53
$O_3 + 2H^+ + 2e^- = O_2 + H_2O$	2.075	$4H_2SO_3 + 4H^+ + 6e^- = S_4O_6^{2-} + 6H_2O$	0.507
$Ag^{2+} + e^- = Ag^+$	1.980	$Ag_2CrO_4 + 2e^- = 2Ag + CrO_4^{2-}$	0.449
$S_2O_8^{2-} + 2e^- = 2SO_4^{2-}$	1.96	$2H_2SO_3 + 2H^+ + 4e^- = S_2O_3^{2-} + 3H_2O$	0.400
$HN_3 + 3H^+ + 2e^- = NH_4^+ + N_2$	1.96	$UO_2^+ + 4H^+ + e^- = U^{4+} + 2H_2O$	0.38
$H_2O_2 + 2H^+ + 2e^- = 2H_2O$	1.763	$Fe(CN)_6^{3-} + e^- = Fe(CN)_6^{4-}$	0.361
$Ce^{4+} + e^- = Ce^{3+}$	1.72	$Cu^{2+} + 2e^- = Cu$	0.340
$MnO_4^- + 4H^+ + 3e^- = MnO_2(c) + 2H_2O$	1.70	$VO^{2+} + 2H^+ + e^- = V^{3+} + H_2O$	0.337
$2HClO + 2H^+ + 2e^- = Cl_2 + H_2O$	1.630	$BiO^+ + 2H^+ + 3e^- = Bi + H_2O$	0.32
$2HBrO + 2H^+ + 2e^- = Br_2 + H_2O$	1.604	$UO_2^{2+} + 4H^+ + 2e^- = U^{4+} + 2H_2O$	0.27
$H_5IO_6 + H^+ + 2e^- = IO_3^- + 3H_2O$	1.603	$Hg_2Cl_2(c) + 2e^- = 2Hg + 2Cl^-$	0.2676
$NiO_2 + 4H^+ + 2e^- = Ni^{2+} + 2H_2O$	1.593	$AgCl + e^- = Ag + Cl^-$	0.2223
Bi_2O_4(bismuthate) $+ 4H^+ + 2e^- = 2BiO^+ + 2H_2O$	1.59	$SbO^+ + 2H^+ + 3e^- = Sb + H_2O$	0.212
$MnO_4^- + 8H^+ + 5e^- = Mn^{2+} + 4H_2O$	1.51	$CuCl_3^{2-} + e^- = Cu + 3Cl^-$	0.178
$2BrO_3^- + 12H^+ + 10e^- = Br_2 + 6H_2O$	1.478	$SO_4^{2-} + 4H^+ + 2e^- = H_2SO_3 + H_2O$	0.158
$PbO_2 + 4H^+ + 2e^- = Pb^{2+} + 2H_2O$	1.468	$Sn^{4+} + 2e^- = Sn^{2+}$	0.15
$Cr_2O_7^{2-} + 14H^+ + 6e^- = 2Cr^{3+} + 7H_2O$	1.36	$S + 2H^+ + 2e^- = H_2S$	0.144
$Cl_2 + 2e^- = 2Cl^-$	1.3583	$Hg_2Br_2(c) + 2e^- = 2Hg + 2Br^-$	0.1392
$2HNO_2 + 4H^+ + 4e^- = N_2O + 3H_2O$	1.297	$CuCl + e^- = Cu + Cl^-$	0.121
$N_2H_5^+ + 3H^+ + 2e^- = 2NH_4^+$	1.275	$TiO^{2+} + 2H^+ + e^- = Ti^{3+} + H_2O$	0.100
$MnO_2 + 4H^+ + 2e^- = Mn^{2+} + 2H_2O$	1.23	$S_4O_6^{2-} + 2e^- = 2S_2O_3^{2-}$	0.08
$O_2 + 4H^+ + 4e^- = 2H_2O$	1.229	$AgBr + e^- = Ag + Br^-$	0.0711
$ClO_4^- + 2H^+ + 2e^- = ClO_3^- + H_2O$	1.201	$HCOOH + 2H^+ + 2e^- = HCHO + H_2O$	0.056
$2IO_3^- + 12H^+ + 10e^- = I_2 + 3H_2O$	1.195	$CuBr + e^- = Cu + Br^-$	0.033
$N_2O_4 + 2H^+ + 2e^- = 2HNO_3$	1.07	$2H^+ + 2e^- = H_2$	0.0000
$2ICl_2^- + 2e^- = 4Cl^- + I_2$	1.07	$Hg_2I_2 + 2e^- = 2Hg + 2I^-$	−0.0405
$Br_2(lq) + 2e^- = 2Br^-$	1.065	$Pb^{2+} + 2e^- = Pb$	−0.125
$N_2O_4 + 4H^+ + 4e^- = 2NO + 2H_2O$	1.039	$Sn^{2+} + 2e^- = Sn$	−0.136
$HNO_2 + H^+ + e^- = NO + H_2O$	0.996	$AgI + e^- = Ag + I^-$	−0.1522
$NO_3^- + 4H^+ + 3e^- = NO + 2H_2O$	0.957	$N_2 + 5H^+ + 4e^- = N_2H_5^+$	−0.225
$NO_3^- + 3H^+ + 2e^- = HNO_2 + H_2O$	0.94	$V^{3+} + e^- = V^{2+}$	−0.255
$2Hg^{2+} + 2e^- = Hg_2^{2+}$	0.911	$Ni^{2+} + 2e^- = Ni$	−0.257
$Cu^{2+} + I^- + e^- = CuI$	0.861	$Co^{2+} + 2e^- = Co$	−0.277
$OsO_4(c) + 8H^+ + 8e^- = Os + 4H_2O$	0.84	$Ag(CN)_2^- + e^- = Ag + 2CN^-$	−0.31
$Ag^+ + e^- = Ag$	0.7991	$PbSO_4 + 2e^- = Pb + SO_4^{2-}$	−0.3505
$Hg_2^{2+} + 2e^- = 2Hg$	0.7960	$Cd^{2+} + 2e^- = Cd$	−0.4025
$Fe^{3+} + e^- = Fe^{2+}$	0.771	$Cr^{3+} + e^- = Cr^{2+}$	−0.424
$H_2SeO_3 + 4H^+ + 4e^- = Se + 3H_2O$	0.739	$Fe^{2+} + 2e^- = Fe$	−0.44
$HN_3 + 11H^+ + 8e^- = 2NH_4^+$	0.695	$H_3PO_3 + 2H^+ + 2e^- = HPH_2O_2 + H_2O$	−0.499
$O_2 + 2H^+ + 2e^- = H_2O_2$	0.695	$2CO_2 + 2H^+ + 2e^- = H_2C_2O_4$	−0.49
$Ag_2SO_4 + 2e^- = 2Ag + SO_4^{2-}$	0.654	$U^{4+} + e^- = U^{3+}$	−0.52
$Cu^{2+} + Br^- + e^- = CuBr(c)$	0.654	$Zn^{2+} + 2e^- = Zn$	−0.7626
$Au(SCN)_4^- + 3e^- = Au + 4SCN^-$	0.636	$Mn^{2+} + 2e^- = Mn$	−1.18
$2HgCl_2 + 2e^- = Hg_2Cl_2(c) + 2Cl^-$	0.63	$Al^{3+} + 3e^- = Al$	−1.67
$Sb_2O_5 + 6H^+ + 4e^- = 2SbO^+ + 3H_2O$	0.605	$Mg^{2+} + 2e^- = Mg$	−2.356
$H_3AsO_4 + 2H^+ + 2e^- = HAsO_2 + 2H_2O$	0.560	$Na^+ + e^- = Na$	−2.714
$TeOOH^+ + 3H^+ + 4e^- = Te + 2H_2O$	0.559	$K^+ + e^- = K$	−2.925
$Cu^{2+} + Cl^- + e^- = CuCl(c)$	0.559	$Li^+ + e^- = Li$	−3.045
$I_3^- + 2e^- = 3I^-$	0.536	$3N_2 + 2H^+ + 2e^- = 2HN_3$	−3.10

Source: J.A. Dean, ed., *Lange's Handbook of Chemistry*, 14th ed., McGraw-Hill, New York, 1992.

2. *Beyond the equivalence point*:

$$E = E^0_{\mathrm{B,D}} + \frac{0.05915}{n_\mathrm{B}} \log(f-1) \tag{4.54}$$

Since E is determined by the ratio of concentrations, the titration curves are little affected by the absolute values of the concentrations. Redox titrations are therefore well suited to the determination of substances present at low concentrations. The limit of usefulness is set by the magnitude of the end-point blank, i.e., by the volume of standard solution needed to give the end-point indication.

In general, it is advisable to construct the titration curve for an appraisal of the efficiency of a method and the indicator used. For a reaction of the type considered above, the electrode potential at the equivalence point will be given by

$$E_{\mathrm{equiv}} = \frac{n_\mathrm{A} E^0_{\mathrm{C,A}} + n_\mathrm{B} E^0_{\mathrm{B,D}}}{n_\mathrm{A} + n_\mathrm{B}} \tag{4.55}$$

As a rule of thumb, the relative precision of the titration, rp, may be roughly estimated from the equation

$$\log \mathrm{rp} = -8.5 n_\mathrm{A} n_\mathrm{B} (E^0_{\mathrm{B,D}} - E^0_{\mathrm{C,A}}) \tag{4.56}$$

or, which is the same thing,

$$\mathrm{rp} = \frac{1}{\sqrt{K}} \tag{4.57}$$

If the solution being titrated contains several substances, say A_1, A_2, and A_3, all of which are oxidized by the reagent B, satisfactory end point will generally be obtained if the above conditions are satisfied by the pairs of potentials:

$$E^0_{\mathrm{C_1A_1}} \text{ and } E^0_{\mathrm{C_2A_2}},\; E^0_{\mathrm{C_2A_2}} \text{ and } E^0_{\mathrm{C_3A_3}},\; \text{and } E^0_{\mathrm{C_3A_3}} \text{ and } E^0_{\mathrm{B,D}}$$

Titration curves are computed from equations similar to Eq. (4.53), but with f appropriately redefined for each successive stage of the titration, and finally from an equation similar to Eq. (3.54) after the last equivalence point has been passed. The equivalence-point potentials in such a titration depend upon both the formal potentials of the half-reactions involved and the ratios of the initial concentration of the various reducing (or oxidizing) agents present. The standard literature should be consulted for details. As a useful approximation if the initial concentrations are not too widely different, the potentials at the successive equivalence points can be approximated by the following equations.

First (A_1) *equivalence point*:

$$E = \frac{n_{\mathrm{A_1}} E^0_{\mathrm{C_1,A_1}} + n_{\mathrm{A_2}} E^0_{\mathrm{C_2,A_2}}}{n_{\mathrm{A_1}} + n_{\mathrm{A_2}}} \tag{4.58}$$

Second (A_2) *equivalence point*:

$$E = \frac{n_{\mathrm{A_2}} E^0_{\mathrm{C_2,A_2}} + n_{\mathrm{A_3}} E^0_{\mathrm{C_3,A_3}}}{n_{\mathrm{A_2}} + n_{\mathrm{A_3}}} \tag{4.59}$$

Last (A_i) *equivalence point*:

$$E = \frac{n_{A_i} E^0_{C_i,A_i} + n_B E^0_{B,D}}{n_{A_i} + n_B} \tag{4.60}$$

4.5.2 Indicators for Redox Titrations

A redox indicator should be selected so that its E^0 (Table 4.22) is approximately equal to the electrode potential at the equivalence point, or so that the color change will occur at an appropriate part of the titration curve. If n is the number of electrons involved in the transition from the reduced to the oxidized form of the indicator, the range in which the color change occurs is approximately given by $E^0 \pm 0.06/n$ volt (V) for a two-color indicator whose forms are equally intensely colored. Since hydrogen ions are involved in the redox equilibria of many indicators, it must be recognized that the color-change interval of such an indicator will vary with pH.

In Table 4.22, E^0 represents the redox potential at which the color change of the indicator would normally be perceived in a solution containing approximately $1M$ H^+. For a one-color indicator this is the potential at which the concentration of the colored form is just large enough to impart a visible color to the solution and depends on the total concentration of indicator added to the solution. If it is the reduced form of the indicator that is colorless, the potential at which the first visible color appears becomes less positive as the total concentration of indicator increases. For a two-color indicator, the potential at which the middle tint appears is independent of the total indicator concentration, but may differ slightly from the potentiometrically determined formal potential of the indicator in either direction, depending on which of the two forms is more intensely colored. If the reduced form is the more intense color, the middle tint will appear at a potential more positive than the potentiometrically measured formal potential, which is the potential at which the two forms are present at equal concentrations.

In addition to those indicators listed in Table 4.22, there are indicators for bromometric and iodometric titrations:

Specific reagents for titrations with bromine or bromate	
Methyl orange or methyl red	Use acid–base indicator solutions (Table 3.10). Oxidation causes bleaching of indicator to colorless.
Bordeaux acid red 17	Dissolve 2 g dye in 1 L water. The red solution is oxidized to pale yellowish green or colorless.
Naphthol blue black	Dissolve 2 g dye in 1 L water. The blue solution is oxidized to pale red.
Specific reagents for iodometric titrations	
Organic solvents such as CCl_4, $CHCl_3$	Up to 5 mL solvent is usually added per titration. Near the end point the mixture is shaken vigorously after each addition of titrant, and the appearance or disappearance of the I_2 color in the organic layer is observed.
Starch	Suspend 5 g of soluble starch in 50 mL of saturated NaCl solution, and stir slowly into 500 mL of boiling saturated NaCl solution. Cool and bottle. Free iodine produces a blue-black color.

TABLE 4.22 Selected List of Oxidation–Reduction Indicators

Name	Reduction potential (30°C) in volts at pH = 0	pH = 7	Suitable pH range	Color change upon oxidation
Bis(5-bromo-1,10-phenanthroline) ruthenium(II) dinitrate	1.41*			Red to faint blue
Tris(5-nitro-1,10-phenanthroline) iron(II) sulfate	1.25*			Red to faint blue
Iron(II)-2,2′,2″-tripyridine sulfate	1.25*			Pink to faint blue
Tris(4,7-diphenyl-1,10-phenanthroline) iron(II) disulfate	1.13 (4.6*M* H_2SO_4)* 0.87(1.0*M* H_2SO_4)*			Red to faint blue
o,m′-Diphenylaminedicarboxylic acid	1.12			Colorless to blue-violet
Setopaline	1.06 (*trans*)†			Yellow to orange
p-Nitrodiphenylamine	1.06			Colorless to violet
Tris(1,10-phenanthroline)-iron(II) sulfate	1.06 (1.00*M* H_2SO_4)* 1.00 (3.0*M* H_2SO_4)* 0.89 (6.0*M* H_2SO_4)*			Red to faint blue
Setoglaucine O	1.01 (*trans*)†			Yellow-green to yellow-red
Xylene cyanole FF	1.00 (*trans*)†			Yellow-green to pink
Erioglaucine A	1.00 (*trans*)†			Green-yellow to bluish red
Eriogreen	0.99 (*trans*)†			Green-yellow to orange
Tris(2,2′-bipyridine)-iron(II) hydrochloride	0.97*			Red to faint blue
2-Carboxydiphenylamine [*N*-phenylanthranilic acid]	0.94			Colorless to pink
Benzidine dihydrochloride	0.92			Colorless to blue
o-Toluidine	0.87			Colorless to blue
Bis(1,10-phenanthroline)-osmium(II) perchlorate	0.859 (0.1*M* H_2SO_4)			Green to pink
Diphenylamine-4-sulfonate (Na salt)	0.85			Colorless to violet
3,3′-Dimethoxybenzidine dihydrochloride [*o*-dianisidine]	0.85			Colorless to red
Ferrocyphen	0.81			Yellow to violet
4′-Ethoxy-2,4-diaminoazobenzene	0.76			Red to pale yellow
N,N-Diphenylbenzidine	0.76			Colorless to violet
Diphenylamine	0.76			Colorless to violet
N,N-Dimethyl-*p*-phenylenediamine	0.76			Colorless to red
Variamine blue B hydrochloride	0.712‡	0.310	1.5–6.3	Colorless to blue
N-Phenyl-1,2,4-benzenetriamine	0.70			Colorless to red
Bindschedler's green	0.680‡	0.224	2–9.5	
2,6-Dichloroindophenol (Na salt)	0.668‡	0.217	6.3–11.4	Colorless to blue
2,6-Dibromophenolindophenol	0.668‡	0.216	7.0–12.3	Colorless to blue
Brilliant cresyl blue [3-amino-9-dimethylamino-10-methyl-phenoxyazine chloride]	0.583	0.047	0–11	Colorless to blue
Iron(II)-tetrapyridine chloride	0.59			Red to faint blue
Thionine [Lauth's violet]	0.563‡	0.064	1–13	Colorless to violet
Starch (soluble potato, I_3^- present)	0.54			Colorless to blue
Gallocyanine (25°C)		0.021		Colorless to violet-blue

(Continued)

TABLE 4.22 Selected List of Oxidation–Reduction Indicators (*Continued*)

Name	Reduction potential (30°C) in volts at pH = 0	pH = 7	Suitable pH range	Color change upon oxidation
Methylene blue	0.532‡	0.011	1–13	Colorless to bluc
Nile blue A [aminonaphthodiethylamino-phenoxazine sulfate]	0.406‡	−0.119	1.4–12.3	Colorless to blue
Indigo-5,5′,7,7′-tetrasulfonic acid (Na salt)	0.365‡	−0.046	<9	Colorless to blue
Indigo-5,5′,7-trisulfonic acid (Na salt)	0.332‡	−0.081	<9	Colorless to blue
Indigo-5,5′-disulfonic acid (Na salt)	0.291‡	−0.125	<9	Colorless to blue
Phenosafranine	0.280‡	−0.252	1–11	Colorless to violet-blue
Indigo-5-monosulfonic acid (Na salt)	0.262‡	−0.157	<9	Colorless to blue
Safranine T	0.24‡	−0.289	1–12	Colorless to violet-blue
Bis(dimethylglyoximato)-iron(II) chloride	0.155		6–10	Red to colorless
Induline scarlet	0.047‡	−0.299	3–8.6	Colorless to red
Neutral red		−0.323	2–11	Colorless to red-violet

* Transition point is at higher potential than the tabulated formal potential because the molar absorptivity of the reduced form is very much greater than that of the oxidized form.

† *Trans* = first noticeable color transition; often 60 mV less than E^0.

‡ Values of E^0 are obtained by extrapolation from measurements in weakly acid or weakly alkaline systems.

Source: J. A. Dean, ed., *Lange's Handbook of Chemistry*, 14th ed., McGraw-Hill, New York, 1992.

4.5.3 Standard Volumetric (Titrimetric) Redox Solutions

Alkaline arsenite, 0.1*N* As(III) to As(V). Dissolve 4.9460 g of primary standard grade As_2O_3 in 40 mL of 30% NaOH solution. Dilute with 200 mL of water. Acidify the solution with 6*N* HCl to the acid color of methyl red indicator. Add to this solution 40 g of $NaHCO_3$ and dilate to 1 L.

Ceric sulfate, 0.1*N* Ce(IV) to Ce(III). Dissolve 63.26 g of cerium(IV) ammonium sulfate dihydrate in 500 mL of 2*N* sulfuric acid. Dilute the solution to 1 L and standardize against the alkaline arsenite solution as follows: Measure, accurately, 30 to 40 mL of arsenite solution into an Erlenmeyer flask and dilute to 150 mL. Add slowly, to prevent excessive frothing, 20 mL of 4*N* sulfuric acid, 2 drops of 0.01*M* osmium tetraoxide solution, and 4 drops of 1,10-phenanthroline iron(II) complex indicator. Titrate with the ceric sulfate solution to a faint blue end point. Compute the normality of the ceric solution from the normality of the arsenite solution.

Iron(II) ammonium sulfate hexahydrate, 0.1*N* Fe(II) to Fe(III). Dissolve 39.2139 g of $FeSO_4 \cdot 2(NH_4)_2SO_4 \cdot 6H_2O$ in 500 mL of 1*N* sulfuric acid and dilute to 1 L. If desired, check against standard dichromate or permanganate solution.

Iodine, 0.1*N* (0 to 1−). Dissolve 12.690 g of resublimed iodine in 25 mL of a solution containing 15 g of KI that is free from iodate. After all the solid has dissolved, dilute to 1 L. If desired, check against a standard arsenite or standard thiosulfate solution.

Potassium bromate, 0.1*N* (5 + to 1 −). Weigh out 2.7833 g of $KBrO_3$, dissolve in water, and dilute to 1 L.

Potassium dichromate, 0.1*N* Cr(VI) to Cr(III). Weigh out 4.9030 g of $K_2Cr_2O_7$ that has been dried at 120°C, dissolve in water, and dilute to 1 L.

Potassium iodate, 0.1*N* (5+ to 1−). Weigh out exactly 3.5667 g of KIO_3 (free from iodide), dried at 120°C, dissolve in water containing about 15 g of KI, and dilute to 1 L.

Potassium permanganate, 0.1*N* (7+ to 2+). Dissolve about 3.3 g in a liter of distilled water. Allow this to stand for two or three days, then siphon it carefully through clean glass tubes or filter it through a Gooch crucible into the glass container in which it is to be kept, discarding the first 25 mL and allowing the last inch (2.5 cm) of liquid to remain in the bottle. In this way any dust or reducing substance in the water is oxidized, and the MnO_2 formed is removed. Permanganate solutions should never be allowed to come into contact with rubber, filter paper, or any other organic matter and should be stored away from light. To standardize the $KMnO_4$, weigh accurately samples of about 0.3 g of primary standard grade of $Na_2C_2O_4$ into Erlenmeyer flasks, add 150 mL of distilled water and 4 mL of concentrated H_2SO_4, and heat to 70°C and maintain at this temperature throughout the titration with the permanganate solution. The end point is a faint, permanent pink color throughout the solution. The equivalent weight of $Na_2C_2O_4/2$ is 67.000 g.

Sodium thiosulfate, 0.1*N*. Weigh 24.818 g of fresh crystals of $Na_2S_2O_3 \cdot 5H_2O$ and dissolve in distilled water. Add 0.5 g of Na_2CO_3 and 0.5 mL of chloroform as preservative. Dilute to 1 L.

Equations for the principal methods for the redox determinations of the elements are given in Table 4.23. Procedures for redox titration of elements and the corresponding equivalent weights are given in Table 4.24. Volumetric factors in redox titrations are given in Table 4.25.

4.6 COMPLEXOMETRIC TITRATIONS

A complexometric titration is based on the essentially stoichiometric reaction of a complexing agent (termed *chelon*) with another species to form a complex species (termed a *chelonate*) that is only slightly dissociated and that is soluble in the titration medium. In such a titration, either the chelon or the chelonate may serve as the limiting reagent (that is, as the titrant). The end point is detected by measuring or observing some property that reflects the change, in the vicinity of the equivalence point, in the concentration of the chelon or the chelonate.

Among the complexing agents that find use as titrating agents are ethylenediamine-*N,N,N′,N′*-tetraacetic acid, or EDTA; 1,2-di(2-aminoethoxy)ethane-*N,N,N′,N′*-tetraacetic acid, or EGTA; and 1,2-diaminocyclohexanetetraacetic acid, or CDTA. Many reagents of these types have been studied, and a few have found use in specialized titrations. EDTA is by far the most important, and it is used in the vast majority of complexometric titrations.

4.6.1 Types of Chelometric Titrations

Chelometric titrations may be classified according to their manner of performance: direct titrations, back titrations, substitution titrations, or indirect methods.

4.6.1.1 Direct Titrations. The most convenient and simplest manner is the measured addition of a standard chelon solution to the sample solution (brought to the proper conditions of pH, buffer, etc.) until the metal ion is stoichiometrically chelated. Auxiliary complexing agent such as citrate, tartrate, or triethanolamine are added, if necessary, to prevent the precipitation of metal hydroxides or basic salts at the optimum pH for titration. For example, tartrate is added in the direct titration of lead. If a pH range of 9 to 10 is suitable, a buffer of ammonia and ammonium chloride is often added in relatively concentrated form, both to adjust the pH and to supply ammonia as an auxiliary complexing agent for those metal ions that form ammine complexes. A few metals, notably iron(III), bismuth, and thorium, are titrated in acid solution.

Direct titrations are commonly carried out using disodium dihydrogen ethylenediaminetetraacetate, Na_2H_2Y, which is available in pure form. Reilley and Porterfield[5] devised an electrolytic

[5] C. N. Reilley and W. W. Porterfield, *Anal. Chem.* **28**:44 (1956).

TABLE 4.23 Equations for the Principal Methods for the Redox Determinations of the Elements

The numbers in column 2 are keyed to the procedures in Table 4.24.

Al	(1)	$Al(C_9H_6NO)_3 + 3\ HCl = AlCl_3 + 3\ C_9H_7NO$ (8-hydroxyquinoline)
		$3\ C_9H_7NO + 6\ Br_2 = 3\ C_9H_5Br_2NO + 6\ HBr$
As^0	(1)	$As + 5\ Ce(IV) + 4\ H_2O = H_3AsO_4 + 5\ Ce(III) + 5\ H^+$
	(2a)	$5\ H_3AsO_3 + 2\ KMnO_4 + 6\ HCl = 5\ H_3AsO_4 + 2\ MnCl_2 + 3\ H_2O$
	(2b)	$H_3AsO_3 + 2\ Ce(SO_4)_2 + H_2O = H_3AsO_4 + Ce_2(SO_4)_3 + H_2SO_4$
	(3)	$3\ H_3AsO_3 + KBrO_3(+\ HCl) = 3\ H_3AsO_4 + KBr$
	(4)	$H_3AsO_3 + I_2 + 2\ H_2O = H_3AsO_4 + 2\ I^- + 2\ H^+$
	(5)	$H_3AsO_4 + 2$ KI (excess) $+ 2\ HCl = H_3AsO_3 + I_2 + 2\ KCl + H_2O$
		$I_2 + 2\ Na_2S_2O_3 = 2\ NaI + Na_2S_4O_6$
Ba	(1)	$2\ BaCrO_4 + 6$ KI (excess) $+ 16\ HCl = 2\ BaCl_2 + 3\ I_2 + 6\ KCl + 2\ CrCl_3 + 8\ H_2O$
		$I_2 + 2\ Na_2S_2O_3 = 2\ NaI + Na_2S_4O_6$
	(2)	$BaCrO_4 + 3\ Fe^{2+} + 8\ H^+ = Ba^{2+} + Cr^{3+} + 3\ Fe^{3+} + 4\ H_2O$
		Titrate excess Fe^{2+} with permanganate or dichromate.
Br_2	(1)	$Br_2 + 2$ KI (excess) $= 2\ KBr + I_2$
		$I_2 + 2\ Na_2S_2O_3 \rightarrow 2NaI = Na_2S_4O_6$
Br^-	(1)	$Br^- + 3\ HClO = BrO_3^- + 3\ Cl^- + 3\ H^+$
BrO_3^-	(1)	$BrO_3^- + 6\ I^-$ (excess) $+ 6\ H^+ = Br^- + 3\ I_2 + 3H_2O$
		$I_2 + 2\ Na_2S_2O_3 = 2\ NaI + Na_2S_4O_6$
CO	(1)	$5\ CO + I_2O_5 = 5\ CO_2 + I_2$ (at 125°C; adsorbed and measured colorimetrically)
$C_2O_4^{2-}$	(1)	Titrate as for CaC_2O_4.
$C_2O_6^{2-}$	(1)	Acidify and titrate as for H_2O_2.
		$C_2O_6^{2-} + 2\ H^+ = H_2O_2 + 2\ CO_2$
Ca	(1)	$5\ CaC_2O_4 + 2\ KMnO_4 + 8\ H_2SO_4 = 5\ CaSO_4 + 10\ CO_2 + K_2SO_4 + 2\ MnSO_4 + 8\ H_2O$
Cd	(1)	$Cd(anthranilate)_2 + 4\ Br_2 = 2\ NH_2C_6H_2Br_2COOH + 4\ Br^-$
Ce	(1)	Oxidize Ce(III) to Ce(IV) with $(NH_4)_2S_2O_8$
		$2\ Ce(SO_4)_2 + 2\ FeSO_4 = Ce_2(SO_4)_3 + Fe_2(SO_4)_3$
Cl_2	(1)	Same as for Br_2.
ClO^-	(1)	$ClO^- + 2\ I^- + 2\ H = Cl^- + I_2 + H_2O$
		Titrate liberated I_2 with thiosulfate.
ClO_2^-	(1)	$ClO_2^- + 4\ I^- + 4\ H^+ = Cl^- + 2\ I_2 + 2\ H_2O$
		Titrate liberated I_2 with thiosulfate.
ClO_3^-	(1)	$ClO_3^- + 6\ I^- + 6\ H_2O = Cl^- + 3\ I_2 + 3\ H_2O$
		Titrated liberated I_2 with thiosulfate.
	(2)	$ClO_3^- + 3\ H_3AsO_3$ (excess) $= Cl^- + 3\ H_3AsO_4$
		Titrate excess H_3AsO_3 with bromate.
Co	(2)	$Co(NH)_6^{2+} + Fe(CN)_6^{3-}$ (Citrate–NH_3 buffer) $= Co(NH_3)_6^{3+} + Fe(CN)_6^{4-}$
Cr	(1,2,3)	$Cr_2O_7^{2-} + 6\ Fe^{2+} + 14\ H^+ = 2\ Cr^{3+} + 6\ Fe^{3+} + 7\ H_2O$
Cu	(1)	$2\ Cu^{2+} + 2\ I^- + 2SCN^- = 2\ CuSCN + I_2$
		Titrate the liberated iodine with thiosulfate.
	(2)	$4\ CuSCN + 7\ IO_3^- + 14\ H^+ + 7\ Cl^- = 4\ Cu^{2+} + 4\ SO_4^{2-} + 7\ ICl + 4\ HCN + 5\ H_2O$
Fe(II)	(1)	$5\ Fe^{2+} + MnO_4^- + 8\ H^+ = 5\ Fe^{3+} + Mn^{2+} + 4\ H_2O$
	(2)	$Fe^{2+} + Ce(IV) = Fe^{3+} + Ce(III)$
	(3)	$6\ Fe^{2+} + Cr_2O_7^{2-} + 14\ H^+ = 6\ Fe^{3+} + 2\ Cr^{3+} + 7\ H_2O$
Fe(III)	(1)	$Fe^{3+} + 4\ SCN^- = Fe(SCN)_4^-$ and $Fe(SCN)_4^- + Ti(III) = Fe^{2+} + Ti(IV) + 4\ SCN^-$

TABLE 4.23 Equations for the Principal Methods for the Redox Determinations of the Elements (*Continued*)

Fe(III)	(2)	$2\ Fe^{3+} + Zn = 2\ Fe^{2+} + Zn^{2+}$ See under Fe(II).
	(3)	$Fe^{3+} + Ag + Cl^- = Fe^{2+} + AgCl$ See under Fe(II).
	(4)	$2\ Fe^{3+} + SnCl_2 + 4\ Cl^- = 2\ Fe^{2+} + SnCl_6^{2-}$ $2\ HgCl_2 + SnCl_2 + 2\ Cl^- = Hg_2Cl_2 + SnCl_6^{2-}$
	(5)	$2\ Fe^{3+} + 2\ I^- = Fe^{2+} + I_2$ Titrate liberated iodine with thiosulfate. $H^+(1)IO_3^- + 5\ I^- + 6\ H^+ = 3\ H_2O + 3\ I_2$ Titrate liberated iodine with neutral (no carbonate) thiosulfate.
I_2	(1)	$I_2 + 2\ S_2O_3^{2-} = 2\ I^- + S_4O_6^{2-}$
	(2)	$I_2 + H_3AsO_3 + H_2O = 2\ I^- + H_3AsO_4 + 2\ H^+$
I^-	(1)	$2\ I^- + Br_2 = I_2 + 2Br^-$
IO_3^-	(1)	$IO_3^- + 5\ I^- + 6\ H^+ = 3\ I_2 + 3\ H_2O$
IO_4^-	(1)	$IO_4^- + 7\ I^- + 8\ H^+ = 4\ I_2 + 4\ H_2O$
K	(1)	$K_2Na[Co(NO_2)_6]$ See procedure for nitrite.
Mg	(1)	$Mg(oxine)_2$ See procedure for $Al(8\text{-hydroxyquinoline})_3$.
Mn(II)	(1)	$2\ Mn^{2+} + 5\ BiO_3^- + 14\ H^+ = 2\ MnO_4^- + 5\ Bi^{3+} + 7\ H_2O$ $2\ MnO_4^- + 5\ AsO_3^{3-} + 6\ H^+ = 2\ Mn^{2+} + 5\ AsO_4^{3-} + 3\ H_2O$
	(2)	$2\ Mn^{2+} + 5\ S_2O_8^{2-} + 8\ H_2O\ (Ag^+\ catalyst) = 2\ MnO_4^- + 10\ SO_4^{2-} + 16\ H^+$ Titrate the permanganate with iron (II) as under iron.
	(3)	$2\ Mn^{2+} + 5\ IO_4^- + 3\ H_2O = 2\ MnO_4^- + 5\ IO_3^- + 6\ H^+$ Titrate excess Fe^2+ added with standard $KMnO_4$ solution.
	(4)	$MnO_4^- + 4\ Mn^{2+} + 15\ H_2P_2O_7^{2-}$ (pH range 4 – 7) $= 5\ Mn(H_2P_2O_7)_3^{3-} + 4\ H_2O$
Mn(IV)	(1a)	$MnO_2 + 2\ Fe^{2+} + 4\ H^+ = Mn^{2+} + 2\ Fe^{3+} + 2\ H_2O$
	(1b)	$MnO_2 + H_2C_2O_4 + 2\ H^+ = Mn^{2+} + 2\ CO_2 + 2\ H_2O$
Mn(VI)	(1)	$MnO_4^{2-} + 2\ H_2C_2O_4 + 4\ H^+ = Mn^{2+} + 4\ CO_2 + 4\ H_2O$
Mn(VII)	(1)	$2\ MnO_4^- + 5H_2C_2O_4\ 6\ H^+ = 2\ Mn^{2+} + 10\ CO_2 + 3\ H_2O$
Mo	(1)	$Mo(VI) + Zn = Mo(III) + Zn^{2+}$ $Mo(III) + 3\ Fe^{3+} + 4\ H_2O = MnO_4^{2-} + 3\ Fe^{2+}\ 8\ H^+$
	(2)	$Mo(VI) + Ag + Cl^- = Mo(V) + AgCl$ $Mo(V) + Ce(IV) = Mo(VI) + Ce(III)$
N_2H_4	(1)	$3\ N_2H_4 + 2\ BrO_3^- = 3\ N_2 + 2\ Br^- + 6\ H_2O$
NH_2OH	(1)	$NH_2OH + BrO_3^- = NO_3^- + Br^- + H^+ + H_2O$
HN_3	(1)	$2\ HN_3 + 2\ Ce(IV) = 3\ N_2 + 2\ Ce(III) + 2\ H^+$
NO_2^-	(1)	$5\ NO_2^- + 2MnO_4^- + 6\ H^+ = 5\ NO_3^- + 2\ Mn^{2+} + 3\ H_2O$
	(2)	$NO_2^- + 2\ Ce(IV) + H_2O = NO_3^- + 2\ Ce(III) + 2\ H^+$
Nb(V)	(1)	$Nb(V) + Zn = Nb(III) + Zn^{2+}$ $Nb(III) + 2\ Fe^{3+} = Nb(V) + 2\ Fe^{2+}$
O_2	(1)	$O_2 + 2\ Mn^{2+} + 2\ OH^- = 2\ MnO_2 + 2\ H^+$
O_3	(1)	$O_3 + 2\ I^- + H_2O = O_2 + I_2 + 2\ OH^-$
H_2O_2	(1)	$5\ H_2O_2 + 2\ MnO_4^- + 6\ H^+ = 5\ O_2 + 2\ Mn^{2+} + 8\ H_2O$
	(2)	$H_2O_2 + 2\ Ce(IV) + 2\ H^+ = 2\ Ce(III) + 2\ H_2O$
	(3)	$H_2O_2 + 2\ I^- + 2\ H^+ = I_2 + 2\ H_2O$
	(4)	$H_2O_2 + 2\ Ti(III) + 2H^+ = 2\ Ti(IV) + 2\ H_2O$

(*Continued*)

TABLE 4.23 Equations for the Principal Methods for the Redox Determinations of the Elements (*Continued*)

P	(1)	See under molybdenum; 12 mol Mo per P.
HPH_2O_2	(1)	$HPH_2O_2 + 2\ I_2 + 2\ H_2O = H_3PO_4 + 4\ I^- + 4\ H^+$
H_3PO_3	(1)	$H_3PO_3 + I_2 + H_2O = H_3PO_4 + 2\ I^- + 2\ H^+$
Pb	(1)	$2\ PbCrO_4 + 6\ I^- + 16H^+ = 2\ Pb^{2+} + 2\ Cr^{3+} + 3\ I_2 + 8\ H_2O$
S^{2-}	(1) (2)	$H_2S + I_2 = S + 2\ I^- + 2\ H^+$ $H_2S + 4\ Br_2 + 4\ H_2O = SO_4^{2-} + 8\ Br^- + 10\ H^+$
SO_2, SO_3^2	(1) (2)	$SO_2 + I_2 + 2\ H_2O = SO_4^{2-} + 2\ I^- + 4\ H_+$ $SO_2 + 4\ Br_2 + 2\ H_2O = SO_4^{2-} + 2\ Br^- + 4\ H^+$
$S_2O_3^{2-}$	(1)	$2\ S_2O_3^{2-} + I_2 = S_4O_6^{2-} + 2\ I^-$
H_2SO_5	(1)	$SO_5^{2-} + H_3AsO_3 = SO_4^{2-} + H_3AsO_4$
$S_2O_8^{2-}$	(1) (2)	$S_2O_8^{2-} + H_3AsO_3 + H_2O = 2\ SO_4^{2-} + H_3AsO_4 + 2\ H^+$ $S_2O_8^{2-} + 2\ Fe^{2+} = 2\ SO_4^{2-} + 2\ Fe^{3+}$
Sb	(1) (2) (3) (4)	$5\ Sb(III) + 2\ MnO_4^- + 16\ H^+ = 5\ Sb(V) + 2\ Mn^{2+} + 8\ H_2O$ $3\ Sb(III) + BrO_3^- + 6\ H^+ = 3\ Sb(V) + Br^- + 3\ H_2O$ $Sb(III) + I_2$ (tartrate buffer, pH > 7) $= Sb(V) + 2\ I^-$ $Sb(III) + 2\ Ce(IV) = Sb(V) + 2\ Ce(III)$
SeO_4^{2-}	(1) (2) (3)	$5\ H_2SeO_3 + 2\ MnO_4^- + 6\ H^+ = 5\ H_2SeO_4 + 2\ Mn^{2+} + 3\ H_2O$ $H_2SeO_3 + 4\ I^- + 4\ H^+ = Se + 2\ I_2 + 3\ H_2O$ $H_2SeO_3 + 4\ S_2O_3^{2-} + 4\ H^+ = SeS_4O_6^{2-} + S_4O_6^{2-} + 3\ H_2O$
SeO_4^{2-}	(1)	$SeO_4^{2-} + 2\ H^+ + 2\ Cl^- = SeO_3^{2-} + Cl_2 + H_2O$ $Cl_2 + 2\ I^- = 2\ Cl^- + I_2$
Sn(IV)	(1)	$SnCl_6^{2-} + Pb = Sn^{2+} + Pb^{2+} + 6\ Cl^-$ $Sn^{2+} + I_2 + 6\ Cl^- = SnCl_6^{2-}$
Sn(II)	(1)	$Sn(II) + 2\ Ce(IV) = Sn(IV) + 2\ Ce(III)$
Te(IV)	(1)	$3\ H_2TeO_3 + Cr_2O_7^{2-} + 8\ H^+ = 3\ H_2TeO_4 + 2\ Cr^{3+} + 4\ H_2O$
Te(VI)	(1)	$H_2TeO_4 + 2\ Cl^- + 2\ H^+ = H_2TeO_3 + Cl_2 + H_2O$
Ti	(1) (1a) (1b) (1c)	$2\ Ti(IV) + Zn = 2\ Ti(III) + Zn(II)$ $Ti(III) + Fe^{3+} = Ti(IV) + Fe^{2+}$ $Ti(III) + Fe(SCN)_4^- = Ti(IV) + Fe(II) + 4\ SCN^-$ Ti(III) + Methylene blue = Ti(IV) + leuco base
Tl	(1a) (1b)	$2\ Tl^+ + MnO_4^- + 8H^+ = 2Tl^{3+} + Mn^{2+} + 4\ H_2O$ $Tl^+ + 2\ Ce^{4+} = Tl^{3+} + 2\ Ce^{3+}$
U	(1)	$U(VI) + Zn = U(III) + U(IV) + Zn(II)$ Pair air through solution to oxidize U(III) to U(IV) $5\ U^{4+} + 2\ MnO_4^- + 2\ H_2O = 5\ UO_2^{2+} + 2\ Mn^{2+} + 4\ H^+$
V	(1) (2) (3)	Oxidize V(IV) to V(V) with permanganate. Destroy excess with sodium azide and boiling. $VO_2^+ + Fe^{2+} + 2\ H^+ = VO^{2+} + Fe^{3+} + H_2O$ Reduce V(V) with SO_2 and bubble CO_2 through boiling solution to remove excess SO_2. $5\ VO^{2+} + MnO_4^- + H_2O = 5\ VO_2^+ + Mn^{2+} + 2\ H^+$ Reduce V(V) to V(II) with Zn; catch eluate in excess Fe^{3+}. $V^{2+} + 2\ Fe^{3+} + H_2O = VO^{2+} + 2\ Fe^{2+} + 2\ H^+$ Titrate $VO^{2+} - Fe^{2+}$ mixture with permanganate to $VO_2^+ - Fe^{3+}$
Zn	(1) (3)	Dissolve precipitate of $Zn[Hg(SCN)_4]$. $2\ SCN^- + 3\ IO_3^- + 2\ H^+ + CN^- = 2\ SO_4^{2-} + 3\ ICN + H_2O$ $2\ Fe(CN)_6^{3-} + 2\ I^- + 3\ Zn^{2+} + 2\ K^{2+} = K_2Zn_3[Fe(CN)_6]_2 + I_2$ Remove I_2 as formed by standard thiosulfate.

TABLE 4.24 Procedures for Redox Titrations of Elements and Equivalent Weights

Element	Procedure	Remarks	Equivalent weights
Al	Precipitate Al with 8-hydroxyquinoline. Dissolve precipitate in 4*M* HCl, add excess KBr and standard $KBrO_3$ solution. Determine excess by adding KI and titrating liberated iodine with standard $Na_2S_2O_3$ solution.	As, Be, Ge, Pb, alkali and alkaline-earth metals, and the metalloids do not interfere.	Al/12 = 2.2485 Al_2O_3/24 = 4.2483
As^0	(1) After precipitation as arsenic metal with HPH_2O_2, add excess standard Ce(IV) and adjust solution to 1*M* H_2SO_4. After all metal has reacted, titrate excess Ce(IV) with standard arsenite solution, using 1.10-phenanthroline as indicator.	Well adapted to small amounts of arsenic.	As/5 = 14.9843
As(III)	(2) The solution should contain 1*M* HCl and 1 drop 0.002*M* KIO_3 (catalyst). Titrate with (a) standard $KMnO_4$ solution to pink color or (b) standard Ce(IV) solution with 1,10-phenanthroline.	Recommended for standardization of $KMnO_4$ solution.	As/2 = 37.4608 As_2O_3/4 = 49.460
	(3) Titrate a strong (≥10%) HCl solution kept at 80°C with standard $KBrO_3$ solution to disappearance of the pink color of the methyl orange indicator. Allow several seconds near the end point for each drop to react.	Excellent method following distillation of $AsCl_3$.	As/2 = 37.4600
	(4) Neutralize solution and maintain pH 7–8 with excess solid $NaHCO_3$ or use a boric acid–borate buffer. Titrate with a standard I_2 solution, using starch as an indicator.	Arsenic must be present or isolated as As(III). Sb(III) and other reducing agents interfere.	As/2 = 37.4608 As_2O_3/4 = 49.460
As(V)	(5) Add excess KI to HCl solution and titrate liberated I_2 with standard thiosulfate solution.	All substances able to oxidize iodide.	As/2 = 37.4608
Ba	(1) Precipitate $BaCrO_4$, filter and wash. Dissolve in HCl and add KI in excess; titrate I_2 with standard thiosulfate solution.	See Table 4.4.	Ba/3 = 45.78
	(2) Dissolve $BaCrO_4$ precipitate in HCl, add excess standard Fe^{2+} solution, and back-titrate with standard $KMnO_4$ or $K_2Cr_2O_7$.	See Table 4.4.	Ba/3 = 45.78
Br_2	Add excess KI to solution in a closed flask, adjust acidity to 0.1–2*M*, and titrate with standard $Na_2S_2O_3$ using starch.	Cl_2 and I_2 codetermined. N oxides and other oxidizing agents interfere.	Br_2/2 = 79.904

(Continued)

TABLE 4.24 Procedures for Redox Titrations of Elements and Equivalent Weights (*Continued*)

Element	Procedure	Remarks	Equivalent weights
Br^-	(1) Bromide is oxidized to bromate by hypochlorite at pH 6–6.5, the excess is removed by sodium formate, and bromate is determined iodometrically as described below.	Avoids volatilization of bromine.	Br/6 = 13.317
BrO_3^-	Add excess KI to 1*N* HCl or H_2SO_4, add few drops of 3% $(NH_4)_2MoO_4$ solution (catalyst), and titrate with standard $Na_2S_2O_3$ to disappearance of iodine color.	Standardize $Na_2S_2O_3$ against $KBrO_3$. Iodate and other oxidizing agents interfere.	$KBrO_3/6$ = 27.835
CO	Pass over I_2O_5 and estimate I_2 from color strips.	As a micromethod for determination of CO.	$^5/_2$ CO = 70.02
$C_2O_6^{2-}$	Acidify and titrate liberated H_2O_2 as under H_2O_2 (q.v.).		$K_2C_2O_6/2$ = 99.11
Ca	Dissolve CaC_2O_4 precipitate in 2*M* H_2SO_4, and titrate with standard $KMnO_4$ solution at 25–30°C.	Other insoluble oxalates interfere. Standardize $KMnO_4$ against $CaCO_3$ carried through the same procedure.	Ca/2 = 20.039 CaO/2 = 28.04
Cd	Dissolve Cd anthranilate precipitate in 4*M* HCl and titrate with $KBrO_3$–KBr until the color of indigo changes to yellow. Add KI and back-titrate iodine liberated with standard $Na_2S_2O_3$.	Co, Cu, Mn, Ni, Pb, and Zn also give insoluble anthranilates and interfere.	Cd/8 = 14.05
Ce(III)	To a 200-mL solution containing 5 mL of 16*M* HNO_3, add 1–5 g ammonium peroxodisulfate, 2–5 mL 0.015*M* $AgNO_3$ and boil 10 min. Cool, add excess $FeSO_4$, and titrate with $KMnO_4$.	Other rare earths do not interfere.	Ce/1 = 140.12 $Ce_2O_3/2$ = 164.12
Cl_2	Add excess KI solution in closed flask; titrate liberated I_2 with standard $Na_2S_2O_3$ solution in a neutral or slightly acid solution.	Br_2 and I_2 are codetermined; N oxides and other oxidizing agents interfere.	$Cl_2/2$ = 35.453
ClO^-	Add excess KI, acidify solution with HOAc and titrate liberated I_2 with standard $Na_2S_2O_3$ solution.	ClO_2^- does not react in weakly acid solution. Other oxidizing agents interfere.	HClO/2 = 26.230
ClO_2^-	Add excess KI to solution in a closed flask. Make strongly acid with HCl, and titrate with standard $Na_2S_2O_3$ solution.	Other oxidizing agents interfere.	$HClO_2/4$ = 17.115
ClO_3^-	(1) Boil with strong HCl, absorbing Cl_2 in KI solution. Titrate iodine with standard thiosulfate solution.	Other oxidizing agents interfere.	$HClO_3/6$ = 14.077

	(2) Boil with HCl and excess standard arsenite solution. Cool and back-titrate with standard $KBrO_3$ solution.	Other oxidizing agents interfere.	$HClO_3/6 = 14.077$
Co	(1) Precipitate Co anthranilate and treat as for cadmium.		$Co/8 = 7.3667$
	(2) Titration to the trivalent state in an ammoniacal citrate solution with standard ferricyanide solution. The end point must be determined potentiometrically and is somewhat sharper if an excess of ferricyanide is added and back-titrated with standard cobalt sulfate solution.	Cr, Fe, Ni, and V do not interfere. Mn is oxidized quantitatively to the trivalent state, so that titration yields sum of Co and Mn.	$Co/1 = 58.9332$
Cr	(1) Fuse insoluble compounds with Na_2O_2 at 650°C. Boil an aqueous solution of melt to destroy excess peroxide; titrate with standard Fe^{2+} solution.		$Cr/3 = 17.332$
	(2) For Cr(III) in a 300-mL solution containing 1*M* H_2SO_4 plus 0.15*M* HNO_3, add 1 mg $AgNO_3$ per mg Cr, then 20 mL 0.5*M* $(NH_4)_2S_2O_8$ and boil until excess peroxide is destroyed. $HMnO_4$ must be destroyed with (a) NaN_3 added dropwise to a boiling solution with two drops in excess and boiling 5 min or (b) a few milliliters of HCl added and liberated Cl_2 removed by boiling. Titrate with standard Fe^{2+} solution using diphenylaminesulfonic acid indicator or potentiometrically.	As, Co, Mn, Mo, Ni, and U do not interfere. Titration yields sum of Cr and V. Large amounts of tungsten make the end point uncertain unless W is taken into solution with HF and HNO_3 to form green $H_2WO_2F_4$ prior to the peroxodisulfate oxidation step.	$Cr/3 = 17.332$ $Cr_2O_3/6 = 25.337$
	(3) For Cr(VI) solution, add excess standard Fe^{2+} solution and back-titrate with (a) $KMnO_4$ if Cl^- is absent or (b) $K_2Cr_2O_7$ if Cl^- is present.	Addition of H_3PO_4 suppresses Fe^{3+} color and keeps W in solution, but large amounts of W make the end point uncertain.	$Cr/3 = 17.332$ $K_2Cr_2O_7/6 = 49.032$
Cu	(1) Boil a solution of 0.5*M* HNO_3 or 1*M* $HClO_4$ with urea to destroy nitrogen oxides. Neutralize with NH_3, buffer with NH_4HF_2, add excess KI and then standard $Na_2S_2O_3$ until the solution is light yellow. Then add KSCN and starch and continue titration until the blue color is discharged.	Applicable to oxide and sulfide ores. Sb interferes, but As and Fe do not.	$Cu/1 = 63.456$
	(2) Precipitate and wash CuSCN. In a glass-stoppered flask add 20 mL water, 30 mL 12*M* HCl, and 5 mL $CHCl_3$. Titrate with excess standard KIO_3 solution, shaking after each addition, until a definite I_2 color appears in the organic layer. Back-titrate the excess with standard thiosulfate solution until I_2 disappears.	Precipitation of CuSCN from a tartrate solution permits separation from As, Bi, Co, Fe, Ni, Sb, Sn, and others.	$Cu/7 = 9.078$ $KIO_3/4 = 53.505$

(continued)

TABLE 4.24 Procedures for Redox Titrations of Elements and Equivalent Weights (*Continued*)

Element	Procedure	Remarks	Equivalent weights
Fe(II)	(1) The solution should be 0.5–1*M* in H_2SO_4. Titrate with standard $KMnO_4$ until a pink color persists.	Most other reducing agents interfere. Chloride must be absent.	Fe/1 = 55.847 $Fe_2O_3/2$ = 79.845
	(2) Titrate a solution containing about 1*M* acid with standard Ce(IV) solution using 1,10-phenanthroline iron(II) indicator. Add more acid if phosphate is present.	Moderate amounts of alcohols and organic acids may be present. Fluoride interferes in more than trace amounts.	Fe/1 = 55.847 $Fe_2O_3/2$ = 79.845
	(3) The solution should contain 1*M* HCl or H_2SO_4 and diphenylamine sulfonate indicator. Titrate with standard $K_2Cr_2O_7$ solution. Addition of H_3PO_4 suppresses color of Fe^{3+}.	Reducing agents interfere, but V(IV) is not titrated if diphenylamine is used as indicator.	Fe/1 = 55.847 $Fe_2O_3/2$ = 79.845
Fe(III)	(1) The solution should contain 1*M* HCl or H_2SO_4 and 1 g NH_4SCN per 100 mL. Titrate with standard Ti(III) solution to disappearance of the red color of the $FeSCN^{2+}$ complex.	Cu, Mo, nitrate, Pt, Sb, Se, V, W, and organic materials interfere.	Fe/1 = 55.847 $Fe_2O_3/2$ = 79.845
	(2) Percolate an H_2SO_4 solution through a column of amalgamated zinc metal; titrate with $KMnO_4$, Ce(IV), or $K_2Cr_2O_7$ as described under Fe(II).	HNO_3 and other compounds reduced by zinc will interfere.	Fe/1 = 55.847 $Fe_2O_3/2$ = 79.845
	(3) Percolate a 1*M* HCl solution through a column of precipitated silver metal. Titrate as described under Fe(II).	Recommended for solutions containing chloride ion. Pt must be absent. Cr(III), Re(VII), and Ti(IV) are not reduced. Moderate amounts of V and nitrate are tolerated.	Fe/1 = 55.847 $Fe_2O_3/2$ = 79.845
	(4) Make solution 6*M* in HCl, heat to boiling, add $SnCl_2$ solution dropwise until color changes to pale green and then 2 drops excess. Cool, add 10 mL 0.25*M* $HgCl_2$. Pour the mixture into an H_3PO_4 plus $MnSO_4$ solution, and titrate with $KMnO_4$.	As, Au, Cu, Mo, Pt, Sb, V, and W are also reduced and interfere.	Fe/1 = 55.847 $Fe_2O_3/2$ = 79.845
H^+ (from hydrolysis)	Add a solution of KI plus KIO_3. Titrate I_2 with $Na_2S_2O_3$.		H/1 = 1.0080
I_2	(1) Titrate a solution (pH ≤ 7) with standard $Na_2S_2O_3$ until the color is pale yellow. Add KI and starch and titrate to disappearance of the blue color.	Cl_2 and Br_2 interfere; Br_2 oxidizes formic acid but I_2 does not. Oxidizing agents must be absent.	$I_2/2$ = 126.9045
	(2) Titrate solution (pH 5–9) with standard arsenite, using starch and KI for indicator.	Carbonate should not be used to neutralize or buffer system.	$I_2/2$ = 126.9045

I^-	(1) Treat a neutral or weakly acid solution with excess Br_2, remove excess reagent with formic acid. Titrate iodine with standard thiosulfate solution.	Suited to determination of small amounts of iodide in presence of chloride and bromide.	I/1 = 126.9045
IO_3^-	Add the sample to a flask containing 2*M* HCl and excess KI, and titrate the iodine with standard $Na_2S_2O_3$ solution.	Other oxidizing agents reacting with iodide and substances reacting with I_2 interfere.	$KIO_3/6 = 35.67$
IO_4^-	Reduce to iodate by addition of excess KI in a neutral buffered solution. Titrate liberated I_2 with standard arsenite solution.		$KIO_4/2 = 115.00$
K	Precipitate $K_2Na[Co(NO_2)_6]$. Dissolve in H_2SO_4 and titrate with either standard $KMnO_4$ or Ce(IV) solution.	Careful adherence to a standard method of precipitation is essential.	ca. K/5.5 but use an empirical factor
Mg	Dissolve precipitate of Mg(8-hydroxyquinoline)$_2$ in 8*M* HCl and proceed as for aluminum.	Only As, Ge, phosphate, Se, Si, Te, alkali metals, and the common anions do not interfere.	Mg/8 = 3.0381
Mn(II)	(1) Agitate a solution containing 50 mg Mn per 100 mL 4*M* HNO_3 for 1 min at 10–20°C with 1.3 g 80% $NaBiO_3$. Dilute with an equal volume of water and filter through an inorganic matte into excess $FeSO_4$ solution. Wash with 0.5*M* HNO_3, and titrate with (a) standard $KMnO_4$ solution or (b) standard arsenite solution, which avoids interference by chromate and vanadate.	Mo, Pb, Pt, Ti, U, and 250 mg W or 10 g $(NH_4)_2SO_4$ may be present. Ce(IV), Cr(VI), and V(V) are also reduced and titrated with $KMnO_4$ but not arsenite solution. Ag and Sb, and phosphate over 0.5 g affect results. Cl^-, Co, F^-, and NO_2^- interfere.	Mn/5 = 10.9876
	(2) To 100 mL of solution containing 1*M* H_2SO_4 plus 0.5*M* H_3PO_4, add 10 mL 0.05*M* $AgNO_3$ plus 10 mL 1*M* $(NH_4)_2S_2O_8$ and boil briefly. Cool to <25°C, add 75 mL cold water and sufficient chloride to precipitate the silver. Titrate with standard arsenite solution to a clear yellow end point.	Useful up to 100 mg Mn. Best standardized by controls. Sb and moderate amounts of Ce, Co, and Cr do not interfere.	Mn/5 = 10.9876
	(3) To a solution containing 0.2*M* HNO_3 plus either 2*M* H_3PO_4 or 1.5*M* H_2SO_4, add 0.3 g KIO_4, and boil 15 min. Dilute and slowly precipitate excess KIO_4 with 2–3 g $Hg(NO_3)_2$ dissolved in a little water. Filter, add excess standard Fe^{2+} solution and back-titrate with standard $KMnO_4$.	Accurate method for ≤ 10 mg Mn in bauxite, bronze, Fe ore, or steel. ≤ 0.1% Cr does not interfere.	Mn/5 = 10.9876
	(4) Titrate Mn(II) to Mn(III) with standard $KMnO_4$ solution in a medium of 0.3*M* $Na_4P_2O_7$, pH 6–7 [or pH 3–3.5 if large amounts of V(V) are present]. Use a Pt–SCE (saturated calomel electrode) indicator system.	Cl^-, Co(II), Cr(III), Fe(III), Mo(VI), Ni, NO_3^-, SO_4^-, W(VI), and U(VI) do not interfere, nor does V if titration done at pH 3–3.5.	Mn/1 = 54.9380
	(5) Dissolve precipitate of Mn anthranilate in 4*M* HCl and titrate with standard $KBrO_3$ plus KBr mixture until the color of indigo changes to yellow. Add KI and back-titrate with $Na_2S_2O_3$.	Precise separation from Ba, Mg, and Sr.	Mn/8 = 6.86725

(Continued)

TABLE 4.24 Procedures for Redox Titrations of Elements and Equivalent Weights *(Continued)*

Element	Procedure	Remarks	Equivalent weights
MnO_2	In CO_2 atmosphere, treat the solid sample with excess standard acidic Fe^{2+} or standard $Na_2C_2O_4$ solution plus 1.5*M* H_2SO_4. Back-titrate excess with standard $KMnO_4$ solution.	Cl^- must be absent. Any Fe coprecipitated in gravimetric separation will not interfere.	Mn/2 = 27.469 MnO_2/2 = 43.47
MnO_4^-	Add the sample to excess standard Fe^{2+} solution containing 1*M* H_2SO_4. Back-titrate with standard $KMnO_4$ solution.	Other oxidizing agents interfere.	Mn/5 = 10.9876 $KMnO_4$/5 = 31.606
Mo	(1) To a solution containing 0.5–1*M* H_2SO_4, add $KMnO_4$ until pink (to destroy reducing agents), and pass through a column of amalgamated Zn metal. Catch the eluate in a large excess $Fe_2(SO_4)_3$, solution, add H_3PO_4, and titrate with standard $KMnO_4$ solution.	As, Cr, Fe, nitrite, Nb, Sq, Ti, U, V, W, organic matter, and polythionic acids interfere and must be removed.	Mo/3 = 31.98
Mo	(2) To a 50-mL solution containing < 480 mg Mo, add 10 mL 12*M* HCl and 3 mL 85% H_3PO_4, and pass through an Ag reductor at 60–80°C. Wash with 150 mL 2*M* HCl and titrate with standard Ce(IV) solution.	Cr(III), perrhenate, and Ti(IV) are not reduced. V is incompletely reduced to V(III).	Mo/1 = 95.94
N_2H_4	Add excess standard $KBrO_3$ plus HCl, let stand 15 min in stoppered flask, add excess KI, and titrate with standard $Na_2S_2O_3$ solution.	Sum of N_2H_4 plus NH_2OH determined. N_2H_4 may be calculated from volume of N_2 liberated.	N_2H_4/4 = 8.01
NH_2OH	(1) Proceed as above for N_2H_4.	See above.	NH_2OH/6 = 5.505
	(2) In CO_2 atmosphere, treat 100 mg of sample with some water, add 30 mL cold saturated $Fe(NH_4)_2(SO_4)_2$ and 10 mL 3.5*M* H_2SO_4. Boil 5 min, cool, dilute to 300 mL, and titrate with standard $KMnO_4$ solution.		NH_2OH/2 = 16.515
HN_3	Shake a solution with excess standard Ce(IV) solution in a flask under inert atmosphere. Add excess KI and titrate with standard $Na_2S_2O_3$ solution.	NH_4^+ does not interfere, but NH_2OH and N_2H_4 must be absent.	HN_3/1 = 43.03
NO_2^-	(1) Mix 50 mL standard 0.02*M* $KMnO_4$ solution, 5 mL 3*M* H_2SO_4, and 25 mL of approximately 0.05*M* NO_2^-. Shake for 15 min and determine excess $KMnO_4$ with standard $Na_2C_2O_4$ solution.		$NaNO_2$/1 = 69.00
	(2) Add the sample solution below the surface of a known excess of standard Ce(IV) solution acidified with H_2SO_4 and warmed to 50°C. Add excess Fe^{2+} solution and back-titrate with standard Ce(IV) solution using erioglaucine indicator.	Method useful for K determination after precipitation as $K_2Na[Co(NO_2)_6] \cdot 6H_2O$.	$NaNO_2$/1 = 69.00

Substance	Procedure	Remarks	Equivalent weight
NO_3^-	React a solution with excess Fe^{2+} in the presence of a Mo catalyst and back-titrate excess Fe^{2+} with $K_2Cr_2O_7$ after adding H_3PO_4.		$NaNO_3/3 = 28.34$
Na	Precipitate $NaZn(UO_2)_3(OAc)_9 \cdot 6H_2O$, wash precipitate, dissolve in acid, and proceed as under uranium(VI).	Rapid routine method requiring standardization by controls.	$Na/6 = 3.8316$ $Na_2O/12 = 5.165$
Nb(V)	Reduce with Zn metal; catch reduced solution under the surface of large excess Fe^{3+} solution. Titrate with standard $KMnO_4$ solution using 1,10-phenanthroline iron(II) indicator.	Blank titration required. Ta does not interfere.	$Nb/2 = 46.453$ $Nb_2O_5/4 = 66.455$
Ni	Precipitate $Ni(anthranilate)_2$ and titrate with standard $KBrO_3$ plus KBr solution. See under Cd.	Good precision.	$Ni/8 = 7.336$
O_2	Treat the solution in a stoppered flask with KI plus NaOH plus $MnCl_2$, acidify, and titrate with standard $Na_2S_2O_3$ solution.	Method for determination of O_2 in waters. Remove NO_2^- by adding excess $KMnO_4$ and reducing excess with oxalic acid.	$O_2/4 = 7.9997$
O_3	React a gas mixture with KI solution, acidify, and titrate with standard $Na_2S_2O_3$ solution.	Br_2, Cl_2, and N oxides interfere.	$O_3/2 = 24.00$
H_2O_2	(1) Titrate a solution containing about 0.003*M* H_2O_2 plus 0.3*M* H_2SO_4 with standard $KMnO_4$ solution. (2) Titrate solution containing 0.5–3*M* HOAc, HCl, HNO_3, or H_2SO_4 with standard Ce(IV) solution using 1,10-phenanthroline indicator.	$S_2O_8^{2-}$ does not interfere. Preservatives in commercial H_2O_2 may interfere. H_2SO_5 and peroxodisulfate do not interfere when the titration is made at 0°C.	$H_2O_2/2 = 17.01$ $O_2/4 = 7.9997$ $H_2O_2 = 34.02$
	(3) Add the sample with stirring to 2 g KI in 1*M* H_2SO_4. Let stand 5 min, and titrate with standard $Na_2S_2O_3$ solution.	Preservatives like glycerol and salicylic acid do not interfere.	$H_2O_2/2 = 17.01$
	(4) Acidify the solution and titrate with standard Ti(III) solution to disappearance of the yellow color of peroxotitanic acid.	Peroxodisulfate is also reduced.	$H_2O_2/2 = 17.01$
P	Dissolve the yellow precipitate of $(NH_4)_3[P(Mo_3O_{10})_4]$ in NH_4OH and acidify strongly with H_2SO_4. Pass the solution through a column of amalgamated Zn and catch the eluate under a $Fe_2(SO_4)_3$ solution. Titrate with standard $KMnO_4$ solution.		$P/36 = 0.86038$
HPH_2O_2	Acidify the solution with dilute H_2SO_4, add excess standard I_2 solution, and let stand in a stoppered flask for 10 h. Make it alkaline with $NaHCO_3$ and titrate with standard arsenite solution.	Sum of H_3PO_3 + HPH_2O_2 is determined. Subtract I_2 used below for amount of HPH_2O_2.	$HPH_2O_2/4 = 16.499$
H_3PO_3	Add $NaHCO_3$, CO_2, and excess standard I_2 solution. Let stand in a stoppered flask 40–60 min, and titrate excess I_2 with standard arsenite solution.	See above.	$H_3PO_3/2 = 41.00$

(Continued)

TABLE 4.24 Procedures for Redox Titrations of Elements and Equivalent Weights *(Continued)*

Element	Procedure	Remarks	Equivalent weights
Pb	Isolate Pb as $PbSO_4$, dissolve it in NaOAc, and precipitate with $K_2Cr_2O_7$. Wash $PbCrO_4$, dissolve it in NaCl–HCl solution, add KI, and titrate with standard $Na_2S_2O_3$ solution.	Ag, Ba, Bi, Sb, and SiO_2 should be removed before precipitation of $PbSO_4$.	Pb/3 = 69.1 PbO/3 = 74.4
S^{2-}, H_2S	(1) Add sample to dilute HCl containing excess standard I_2 and back-titrate with standard thiosulfate solution. Collect small amounts of H_2S in ammoniacal Cd^{2+}, then proceed as before.	CdS must be protected from strong light. Thiosulfate and sulfite are also oxidized, but thiocyanate is not.	S/2 = 16.03 H_2S/2 = 17.04
	(2) To a 3*M* HCl solution in a closed flask, add excess KBr and standard $KBrO_3$ solution. Let it stand until clear, add excess KI, and titrate with standard thiosulfate solution.	Suitable for determination of sum of several substances: H_2S, SO_2, SCN^-.	H_2S/8 = 4.260 SO_2/2 = 32.03 SCN/6 = 9.681
SO_2, H_2SO_3	(1) Add gas or solution to a known volume of standard I_2 solution until it becomes colorless; or add excess I_2 and back-titrate with standard thiosulfate solution.	All reducing agents stronger than I_2.	SO_2/2 = 32.03
	(2) To an acid solution add standard $KBrO_3$–KBr solution until the methyl orange color is bleached.		SO_2/2 = 32.03
$S_2O_3^{2-}$	Titrate with standard I_2 solution using a starch indicator.	Sulfide and sulfite may be removed before titration by shaking with $CdCO_3$ and filtering.	$Na_2S_2O_3$/1 = 158.11
SO_4^{2-}	Precipitate first with $BaCl_2$ solution and then with standard dichromate solution, and determine excess with standard Fe^{2+} solution, using diphenylaminesulfonic acid as indicator.	Suitable for small amount of sulfate obtained from oxidation of any type of sulfur compound.	SO_4/3 =32.02
H_2SO_5	To a solution containing 0.25*M* H_2SO_4, add excess standard arsenite solution and KBr. Let it stand 2 min, then titrate to yellow with standard $KBrO_3$ solution, then to colorless with arsenite solution.	Thiosulfate and H_2O_2 do not react and may be determined after the titration.	H_2SO_5/2 = 57.04
$S_2O_8^{2-}$	Acidify solution with H_2SO_4 and boil 10 min with excess standard arsenite solution. Titrate excess with $KBrO_3$–KBr.	May be applied to solutions in which H_2SO_5 and H_2O_2 have been titrated.	$H_2S_2O_8$/2 = 97.07
Sb(III)	(1) Titrate solution containing 1.2–3*M* HCl plus 1.8*M* H_2SO_4 at 5–10°C with standard $KMnO_4$ solution to a pink color.	As, Fe, SO_2, V and organic materials interfere.	Sb/2 = 60.88 Sb_2O_3/4 = 72.88
	(2) Titration solution in 3*M* HCl with standard $KBrO_3$ solution, using a methyl orange or an indigo indicator to disappearance of color.		Sb/2 = 60.88
	(3) To a solution containing tartrate add $NaHCO_3$ to keep pH above 7, and titrate with standard I_2 solution with starch.	As(III) also titrated.	Sb/2 = 60.88

	(4) Titrate solution containing 2–4*M* HCl plus methyl orange indicator with standard Ce(IV) to disappearance of dye color.	As(III) is not titrated if its concentration is low and less than that of Sb.	Sb/2 = 60.88
SeO_3^{2-}	(1) To an acid solution add excess standard $KMnO_4$ solution and back-titrate with standard Fe^{2+} solution or $H_2C_2O_4$.	TeO_2^- is also oxidized.	$Na_2SeO_3/2 = 86.47$
	(2) To an acid solution add excess KI and titrate with standard thiosulfate solution.	Suitable for small amounts of Se in steel after isolation of Se.	$Na_2SeO_3/2 = 86.47$
	(3) Acidify solution with HCl, $HClO_4$, or H_2SO_4. Add small excess standard thiosulfate and back-titrate with standard I_2 solution.	HNO_3 must be absent. Good accuracy.	$Na_2SeO_3/4 = 47.23$
SeO_4^{2-}	Boil with strong HCl or HBr. Absorb liberated halogen in KI solution, and titrate I_2 with standard thiosulfate solution.	Tellurate is also titrated.	$Na_2SeO_4/2 = 94.47$
Sn(IV)	In CO_2 atmosphere, boil 300 mL of solution containing 1*M* H_2SO_4, 4*M* HCl, and ≤ 200 mg Sn with 10-g Pb granules for 40 min. Cool to 0–3°C and titrate with standard I_2 or iodate–iodide solution. Pb (or Ni) need not be removed before titration.	Br^-, Co, F^-, Fe, I^-, phosphate, sulfate, and moderate amounts of Al, As, Bi, Cu, Ge, Mn, Ni, Pb, Sb, U, and Zn do not interfere. HNO_3 must be absent.	Sn/2 = 59.35 $SnO_2/2 = 67.35$
Sn(II)	Titrate an acid solution with standard Ce(IV), using diphenylaminesulfonic acid or KI plus starch as indicator.		Sn/2 = 59.35
Te(IV)	To solution containing < 300 mg Te in 200 mL and 0.6*M* HCl, and excess standard $K_2Cr_2O_7$ solution, then excess standard Fe^{2+} solution, and back-titrate with standard $K_2Cr_2O_7$ solution.		Te/2 = 63.80
Te(VI)	See under SeO_4^{2-}.		Te/2 = 63.80
Ti(IV)	Pass a solution in dilute H_2SO_4 or HCl through a column filled with amalgamated zinc giving Ti(III). (a) Catch the eluate in excess $Fe_2(SO_4)_3$ solution, and titrate Fe^{2+} with standard $KMnO_4$ solution. (b) Titrate the eluate in CO_2 atmosphere with standard Fe^{3+} solution using KSCN as indicator. (c) In CO_2 atmosphere heat the eluate and titrate with standard methylene blue solution (reduced to colorless leuco base).	(a) As, Cr, Fe, Mo, Nb, Sb, Sn, U, V, W, nitrate, and organic substances interfere. Of the above, Fe does not interfere in methods (b) and (c).	Ti/1 = 47.88 $TiO_2/1 = 143.80$
Tl	Titrate a 60-mL solution containing 6–110 mg Tl and 1.2*M* HCl at room temperature with $KMnO_4$ to a pink color or with Ce(IV) to a yellow color (or use 1,10-phenanthroline indicator).	Up to 100 mg As(V), Bi, Cd, Cr(III), Cu(II), Fe(III), Hg(II), Pb, Sb(V), Se(IV), Sn(IV), Te(IV), or Zn do not interfere.	Tl/2 = 102.19

(Continued)

TABLE 4.24 Procedures for Redox Titrations of Elements and Equivalent Weights *(Continued)*

Element	Procedure	Remarks	Equivalent weights
U(VI)	To 1*M* H_2SO_4 containing ≤10 g/L of U add $KMnO_4$ to a permanent pink color, then pass through a column of amalgamated Zn. Pass air through the eluate 5 min to oxidize U(III) to U(IV), then titrate with standard $KMnO_4$ solution, or at 50°C with Ce(IV) using 1,10-phenanthroline indicator.	Nitrate and metals reduced by Zn must be removed.	U/2 = 119.01 $UO_2/2$ = 135.03 $U_3O_8/6$ = 140.35
V	(1) To H_2SO_4 and HF if W is present, then $KMnO_4$ at <25°C to a permanent pink color. Destroy excess $KMnO_4$ by (a) adding $NaNO_2$, then urea or sulfamic acid, and letting it stand 5 min or (b) boiling with NaN_3. Adjust pH and titrate with standard Fe^{2+} solution using diphenylaminesulfonic acid indicator.	Very satisfactory for V in steel; Cr, Fe, and W do not interfere.	V/1 = 50.94 $V_2O_5/2$ = 90.95
	(2) To boiling 0.4*M* H_2SO_4 solution add $KMnO_4$ to a permanent pink color. Destroy excess $KMnO_4$ by passing SO_2 through the solution for 5–10 min. Remove excess SO_2 by bubbling CO_2 through the boiling solution, cool to 60–80°C, and titrate with standard $KMnO_4$ solution.	Suitable for determination of large or small amounts of V. As, Fe, Pt, and Sb interfere. Cool solution before titration if Cr is present.	V/1 = 50.94
	(3) To a solution containing 1.8*M* H_2SO_4 add H_3PO_4 to suppress Fe^{3+} color, then $KMnO_4$ to a permanent pink color to oxidize organic substances, then excess Fe^{2+} solution. Stir with $(NH_4)_2S_2O_8$ to oxidize excess Fe^{2+}, and titrate with standard $KMnO_4$ solution.	Rapid method, less accurate than the preceding methods and not suited to small amounts of V. As, Co, Cr, Fe, Mn, Ni, and U do not interfere; W does.	V/1 = 50.94
	(4) Pass a dilute H_2SO_4 solution through amalgamated Zn and catch the eluate under the surface of air-free solution containing excess $Fe_2(SO_4)_3$, and titrate a Fe^{2+}–VO^{2+} mixture with standard $KMnO_4$ solution.	All metals reduced by Zn must be absent.	V/3 = 16.98 $V_2O_5/6$ = 30.32
Zn	(1) Precipitate $ZnHg(SCN)_4$, filter, and dissolve it in 45 mL 4*M* HCl in a stoppered flask. Add 7 mL $CHCl_3$ and titrate with standard KIO_3 solution with shaking until the $CHCl_3$ layer becomes colorless.	Standardization should be by controls.	Zn/24 = 2.725 ZnO/24 = 3.391
	(2) Precipitate $Zn(anthranilate)_2$ and proceed as with Cd.	Cd, Co, Cu, Mn, Ni, and Pb also give insoluble anthranilates and interfere.	Zn/8 = 8.174 ZnO/8 = 10.17
	(3) Add excess KI and use successive additions of small amounts of $K_3Fe(CN)_6$ solution, followed in each case by the titration of liberated I_2 by standard thiosulfate solution until the solution retains a faint yellow color of ferricyanide.	Cl^- and Mn in more than small amounts, interfere Al, Ca, Co, Cu, Fe (if F^- is added), Mg, and Ni do not interfere.	3Zn/2 = 98.07 but empirical value of 99.07 is recommended

TABLE 4.25 Volumetric Titrimetric Factors in Oxidation–Reduction Reactions

Iodine

The following factors are the equivalent of 1 mL of *normal iodine*. Where the normality of the solution being used is other than normal, multiply the factors given in the table below by the normality of the solution employed.

Substance	Formula	Grams
Acetone	$(CH_3)_2CO$	0.009 680 1
Ammonium chromate	$(NH_4)_2CrO_4$	0.050 690
Antimony	Sb	0.060 88
Antimony trioxide	Sb_2O_3	0.072 87
Arsenic	As	0.037 461
Arsenic pentoxide	As_2O_5	0.057 460
Arsenic trioxide	As_2O_3	0.049 460
Arsenite	AsO_3	0.061 460
Bleaching powder	$CaOCl_2$	0.063 493
Bromine	Br	0.079 909
Chlorine	Cl	0.035 453
Chromic oxide	Cr_2O_3	0.025 33
Chromium trioxide	CrO_3	0.033 331
Copper	Cu	0.063 54
Copper oxide	CuO	0.079 54
Copper sulfate	$CuSO_4$	0.159 60
Copper Sulfate	$CuSO_4 \cdot 5H_2O$	0.249 68
Ferric iron	Fe^{3+}	0.055 85
Ferric oxide	Fe_2O_3	0.079 85
Hydrogen sulfide	H_2S	0.017 040
Iodine	I	0.126 904
Lead chromate	$PbCrO_4$	0.107 73
Lead dioxide	PbO_2	0.119 59
Nitrous acid	HNO_2	0.023 507
Oxygen	O	0.007 999 7
Potassium chlorate	$KClO_3$	0.020 426
Potassium chromate	K_2CrO_4	0.064 733
Potassium dichromate	$K_2Cr_2O_7$	0.049 032
Potassium nitrite	KNO_2	0.042 554
Potassium permanganate	$KMnO_4$	0.031 608
Red lead	Pb_3O_4	0.342 78
Sodium chromate	Na_2CrO_4	0.053 991
Sodium dichromate	$Na_2Cr_2O_7$	0.043 661
Sodium dichromate	$Na_2Cr_2O_7 \cdot 2H_2O$	0.049 666
Sodium nitrite	$NaNO_2$	0.034 498
Sodium sulfide	Na_2S	0.039 022
Sodium sulfide	$Na_2S \cdot 9H_2O$	0.120 09
Sodium sulfite	Na_2SO_3	0.063 021
Sodium sulfite	$Na_2SO_3 \cdot 7H_2O$	0.126 07
Sodium thiosulfate	$Na_2S_2O_3$	0.158 11
Sulfur	S	0.016 032
Sulfur dioxide	SO_2	0.032 031
Sulfurous acid	H_2SO_3	0.041 039
Tin	Sn	0.059 345

(Continued)

TABLE 4.25 Volumetric Titrimetric Factors in Oxidation–Reduction Reactions (*Continued*)

Potassium dichromate

The following factors are the equivalent of 1 mL of *normal potassium dichromate*. Where the normality of the solution being used is other than normal, multiply the factors given in the table below by the normality of the solution employed.

Substance	Formula	Grams
Chromic oxide	Cr_2O_3	0.025 332
Chromium trioxide	CrO_3	0.033 331
Ferrous iron	Fe^{2+}	0.055 847
Ferrous oxide	FeO	0.071 846
Ferroso-ferric oxide	Fe_3O_4	0.077 180
Ferrous sulfate	$FeSO_4$	0.151 91
Ferrous sulfate	$FeSO_4 \cdot 7H_2O$	0.278 02
Glycerol	$C_3H_5(OH)_3$	0.006 578 2
Lead chromate	$PbCrO_4$	0.107 73
Zinc	Zn	0.032 685

Potassium permanganate

The following factors are the equivalent of 1 mL of *normal potassium permanganate*. Where the normality of the solution being used is other than normal, multiply the factors given in the table below by the normality of the solution employed.

Substance	Formula	Grams
Ammonium oxalate	$(NH_4)_2C_2O_4$	0.062 049
Ammonium oxalate	$(NH_4)_2C_2O_4 \cdot H_2O$	0.071 056
Ammonium peroxydisulfate	$(NH_4)_2S_2O_8$	0.114 10
Antimony	Sb	0.060 875
Barium peroxide	BaO_2	0.084 669
Barium peroxide	$BaO_2 \cdot 8H_2O$	0.156 73
Calcium carbonate	$CaCO_3$	0.050 045
Calcium oxide	CaO	0.028 04
Calcium peroxide	CaO_2	0.036 039
Calcium sulfate	$CaSO_4$	0.068 071
Calcium sulfate	$CaSO_4 \cdot 2H_2O$	0.086 086
Ferric oxide	Fe_2O_3	0.079 846
Ferroso-ferric oxide	Fe_3O_4	0.077 180
Ferrous ammonium sulfate	$Fe(NH_4)_2(SO_4)_2 \cdot 6H_2O$	0.392 14
Ferrous oxide	FeO	0.071 846
Ferrous sulfate	$FeSO_4$	0.151 91
Ferrous sulfate	$FeSO_4 \cdot 7H_2O$	0.278 02
Formic acid	HCO_2H	0.023 013
Hydrogen peroxide	H_2O_2	0.017 007
Iodine	I	0.126 904
Iron	Fe	0.055 847
Manganese	Mn	0.010 988
Manganese dioxide	MnO_2	0.043 468
Manganous oxide (Volhard)	MnO	0.035 469
Molybdenum trioxide titration from yellow precipitate after reduction	MnO_3	0.047 979
Oxalic acid	$(CO_2H)_2$	0.045 018
Oxalic acid	$(CO_2H)_2 \cdot 2H_2O$	0.063 033

TABLE 4.25 Volumetric Titrimetric Factors in Oxidation–Reduction Reactions (*Continued*)

Potassium permanganate		
Phosphorus titration from yellow precipitate after reduction	P	0.000 860 4
Phosphorus pentoxide to titration from yellow precipitate after reduction	P_2O_5	0.001 971 5
Potassium dichromate	$K_2Cr_2O_7$	0.049 032
Potassium nitrite	KNO_2	0.042 552
Potassium peroxodisulfate	$K_2S_2O_8$	0.135 16
Sodium nitrite	$NaNO_2$	0.034 498
Sodium oxalate	$Na_2C_2O_4$	0.067 000
Sodium peroxodisulfate	$Na_2S_2O_8$	0.119 05
Tin	Sn	0.059 345

Sodium thiosulfate

The following factors are the equivalent of 1 mL of *normal sodium thiosulfate*. Where the normality of the solution being used is other than normal, multiply the factors given in the table below by the normality of the solution employed.

Substance	Formula	Grams
Acetone	$(CH_3)_2CO$	0.009 680 1
Ammonium chromate	$(NH_4)_2CrO_4$	0.050 690
Antimony	Sb	0.060 88
Antimony trioxide	Sb_2O_3	0.072 87
Bleaching powder	$CaOCl_2$	0.063 493
Bromine	Br	0.079 909
Chlorine	Cl	0.035 453
Chromic oxide	Cr_2O_3	0.025 33
Chromium trioxide	CrO_3	0.033 331
Copper	Cu	0.063 54
Copper oxide	CuO	0.079 54
Copper sulfate	$CuSO_4$	0.159 60
Copper sulfate	$CuSO_4 \cdot 5H_2O$	0.249 68
Iodine	I	0.126 904
Lead chromate	$PbCrO_4$	0.107 73
Lead dioxide	PbO_2	0.119 59
Nitrous acid	HNO_2	0.023 507
Potassium chromate	K_2CrO_4	0.064 733
Potassium dichromate	$K_2Cr_2O_7$	0.049 032
Red lead	Pb_3O_4	0.342 78
Sodium chromate	Na_2CrO_4	0.053 991
Sodium dichromate	$Na_2Cr_2O_7$	0.043 661
Sodium dichromate	$Na_2Cr_2O_7 \cdot 2H_2O$	0.049 666
Sodium nitrite	$NaNO_2$	0.034 498
Sodium thiosulfate	$Na_2S_2O_3$	0.158 11
Sodium thiosulfate	$Na_2S_2O_3 \cdot 5H_2O$	0.248 18
Sulfur	S	0.016 032
Sulfur dioxide	SO_2	0.032 031
Tin	Sn	0.059 345

Source: J. A. Dean, ed., *Lange's Handbook of Chemistry*, 14th ed., McGraw-Hill, New York, 1992.

method of generation of EDTA that can be used in place of the usual titration. By means of the cathode reaction

$$HgY^{2-} + 2e^- \rightleftharpoons Hg + Y^{4-} \tag{4.61}$$

the EDTA anion Y^{4-} can be generated at a rate proportional to the current and used for the titration of metal ions such as calcium, copper(II), lead(II), and zinc.

The reaction of the chelon with the indicator must be rapid for a practical, direct titration. Where it is slow, heating of the titration medium is often expedient, or another indicator is employed. Often the sample may contain traces of metal ions that "block" the indicator. This is especially serious in titrations performed in alkaline medium. Thus, in the direct titration of magnesium and calcium, traces of copper, cobalt, nickel, and iron are frequently encountered. The metal in this case may be masked by the addition of potassium cyanide and a reducing agent, such as ascorbic acid or hydroxylamine. Cyanide complexes more strongly with these metal ions than does EDTA or the indicator. Aluminum may be masked by triethanolamine. Such masking agents should be added and time permitted for masking reactions to occur before the addition of the metal indicator.

When an indicator produces no color, or only a faint color, with the metal begin directly titrated, another metal ion may be deliberately added that does give an adequate color with the indicator. To avoid a correction in the titration result, this metal ion is usually added as its chelonate. For example, in many EDTA titrations the Cu–PAN indicator system is employed; this consists of the metal indicator 1-(2-pyridylazo)-2-naphthol (PAN) together with a small amount of copper(II)–EDTA complex.

4.6.1.2. Back Titrations. In the performance of a back titration, a known but excess quantity of EDTA or other chelon is added, the pH is then properly adjusted, and the excess of the chelon is titrated with a suitable standard metal salt solution. Back-titration procedures are especially useful when the metal ion to be determined cannot be kept in solution under the titration conditions or when the reaction of the metal ion with the chelon occurs too slowly to permit a direct titration, as in the titration of chromium(III) with EDTA. Back-titration procedures sometimes permit a metal ion to be determined by the use of a metal indicator that is blocked by that ion in a direct titration. For example, nickel, cobalt, or aluminum form such stable complexes with Eriochrome Black T that the direct titration would fail. However, if an excess of EDTA is added before the indicator, no blocking occurs in the back titration with a magnesium or zinc salt solution. These metal-ion titrants are chosen because they from EDTA complexes of relatively low stability, thereby avoiding the possible titration of EDTA bound by the sample metal ion.

Other examples would be the determination of metals in precipitates; for example, lead in lead sulfate, magnesium in magnesium ammonium phosphate, or calcium in calcium oxalate. Lastly, although thallium forms a stable EDTA complex, it does not respond to the usual metal-ion indicators. The determination becomes feasible only by the back-titration procedure.

In a back titration, a slight excess of the metal salt solution must sometimes be added to yield the color of the metal–indicator complex. If metal ions are easily hydrolyzed, the complexing agent is best added at a suitable, low pH and only when the metal is fully complexed is the pH adjusted upward to the value required for the back titration. In back titrations, solutions of the following metal ions are commonly employed: Cu(II), Mg, Mn(II), Pb(II), Th(IV), and Zn. These solutions are usually prepared in the approximate strength desired from their nitrate salts (or the solution of the metal or its oxide or carbonate in nitric acid), and a minimum amount of acid is added to repress hydrolysis of the metal ion. The solutions are then standardized against an EDTA solution (or other chelon solution) of known strength.

4.6.1.3 Substitution Titrations. Upon the introduction of a substantial or equivalent amount of the chelonate of a metal that is less stable than that of the metal being determined, a substitution occurs, and the metal ion displaced can be titrated by the chelon in the same solution. This is a direct titration with regard to its performance, but in terms of the mechanism it can be considered as a substitution titration (or replacement titration).

In principle any ion can be used if it forms a weaker EDTA complex than the metal ion being determined. Still weaker metal–EDTA complexes would not interfere. Exchange reactions are also possible with other metal complexes to permit application of the chelometric titration to nontitrable cations and anions. The exchange reagent can be added and the titration performed in the sample solution without prior removal of the excess reagent. A most important example is the exchange of silver ion with an excess of the tetracyanonickelate ion according to the equation

$$2Ag^+ + Ni(CN)_4^{2-} \rightleftharpoons 2Ag(CN)_2^- + Ni^{2+} \tag{4.62}$$

The nickel ion freed may then be determined by an EDTA titration. Note that two moles of silver are equivalent to one mole of nickel and thus to one mole of EDTA. Other examples of replacement titrations include the determination of calcium by displacement of magnesium, which uses to advantage the superior magnesium–Eriochrome Black T color reaction, and the determination of barium by displacement of zinc ions in a strongly ammoniacal solution.

4.6.1.4 Redox Titrations. Redox titrations can be carried out in the presence of excess EDTA. Here EDTA acts to change the oxidation potential by forming a more stable complex with one oxidation state than with the other. Generally the oxidized form of the metal forms a more stable complex than the reduced form, and the couple becomes a stronger reducing agent in the presence of excess EDTA. For example, the Co(III)–Co(II) couple is shifted about 1.2 V, so that Co(II) can be titrated with Ce(IV). Alternatively, Co(III) can be titrated to Co(II), with Cr(II) as a reducing agent.

Manganese(II) can be titrated directly to Mn(III) using hexacyanoferrate(III) as the oxidant. Alternatively, Mn(III), prepared by oxidation of the Mn(II)–EDTA complex with lead dioxide, can be determined by titration with standard iron(II) sulfate.

4.6.1.5 Indirect Procedures. Numerous inorganic anions that do not form complexes with a complexing agent are accessible to a chelatometric titration by indirect procedures. Frequently the anion can be precipitated as a compound containing a stoichiometric amount of a titrable cation. Another indirect approach employing replacement mechanism is the reduction of a species with the liquid amalgam of a metal that can be determined by a chelometric titration after removal of excess amalgam. For example,

$$2Ag^+ + Cd(Hg) \rightleftharpoons Cd^{2+} + 2Ag(Hg) \tag{4.63}$$

The equivalent amount of cadmium ion exchanged for the silver ion can readily be determined by EDTA titration procedures.

Other examples include the determination of sodium by titration of zinc in disodium zinc uranyl acetate, and phosphate by titration of magnesium in magnesium ammonium phosphate.

4.6.2 Metal-Ion Indicators

The most important approach to the detection of the visual end point in chelometric titrations is the use of metal indicators. These are indicators that exhibit one visual response in the presence of a metal ion through complexation with it and a different one in its absence. A metal-ion indicator, in general, is a dyestuff that can form a colored metal-ion complex at some characteristic range of pM values. Examples of the applications of metal indicators in common use are listed in Table 4.26.

During the major portion of a direct chelatometric titration, the metal indicator exists almost entirely in its metal complex form and the free metal ion is progressively chelated by the chelon titrant. At the equivalence point, the chelon removes the traces of metal from the metal–indicator complex and the indicator changes to its free form, thus signaling the end point.

Indicators often show a significant acid–base indicator activity. Indeed the color changes exhibited upon chelation usually fall within the ranges that can be elicited by pH changes. The existence of such acid–base indicator activity often limits the usefulness of a metal–indicator complex to a definite (and frequently restricted) pH range.

For a metal indicator to be useful, a proper sequence of effective stabilities must be met. On the one hand, the metal–indicator complex must be sufficiently stable to maintain itself in extremely dilute solution; otherwise the end-point color change will be spread over a broad interval of the titration, owing to the extended dissociation. On the other hand, the metal–indicator complex must be less stable than the metal chelonate; otherwise a sluggish end point, a late end point, or no end point at all will be obtained. Furthermore, the metal–indicator complex must react rapidly with the chelon. Only a limited number of the numerous chromogenic agents for metals allow this sequence and have useful indicator properties in chelometric titrations.

For the sake of brevity, only the indicator Eriochrome Black T, H_3In, will be considered in detail. Since the sulfonate group in completely ionized at all pH values, the indicator anion exhibits the following acid–base behavior:

$$\underset{\text{Red}}{H_2In^-} \underset{}{\overset{\text{pH 6.3}}{\rightleftarrows}} \underset{\text{Blue}}{HIn^{2-}} \overset{\text{pH 11.5}}{\rightleftarrows} \underset{\text{Yellow-orange}}{In^{3-}} \tag{4.64}$$

In the pH range 7 to 11, in which the dye itself exhibits a blue color, many metal ions form red complexes. These include Al, Ca, Cd, Co, Cu, Fe, Ga, Hg, In, Mg, Ni, Pb, Ti, Zn, as well as the rare earths and platinum metals. The colors are extremely sensitive; for example, $10^{-6} - 10^{-7}M$ solutions of magnesium ion give a distinct red color with Eriochrome Black T.

With calcium and magnesium ions the color-change reaction is represented by

$$\underset{\text{Blue}}{HIn^{2-}} + M^{2+} \rightleftarrows \underset{\text{Red}}{MIn^-} + H^+ \tag{4.65}$$

One of the chief difficulties in using Eriochrome Black T is that solutions of it decompose rather rapidly. Only with relatively freshly prepared solutions can the proper end-point color change be realized. Calmagite can replace Eriochrome Black T in all instances without any modification of the procedures. A solution of Calmagite is exceedingly stable, having a shelf life of years.

Nickel(II) and copper(II) can be determined by direct titration with EDTA in an ammoniacal medium if murexide is used as the metal-ion indicator, whereas EDTA cannot displace either Eriochrome Black T or Calmagite from the metal–indicator species.

4.6.3 Ethylenediaminetetraacetic Acid

A frequently employed titrant in complexation titrations is ethylenediamine-*N,N,N′,N′*-tetraacetic acid (acronym EDTA, and equation abbreviation, H_4Y). It is the example used in the following discussion.

4.6.3.1 Ionization of EDTA. The successive acid pK_a values of ethylenediaminetetraacetic acid, H_4Y, are $pK_1 = 2.0$, $pK_2 = 2.67$, $pK_3 = 6.16$, $pK_4 = 10.26$ at 20°C and an ionic strength of 0.1. The fraction of EDTA in each of its ionic forms can be calculated at any pH value by the method described in Sec. 2.1.1. The fraction α_4 present as the tetravalent anion is of particular importance in equilibrium calculations. Its magnitude at various pH values is given in Table 4.27.

TABLE 4.26 Properties and Applications of Selected Metal-Ion Indicators

Indicator	Chemical name	Dissociation constants and colors of free indicator species	Colors of metal–indicator complexes	Applications
Calmagite 0.05 g/100 mL water; stable 1 year	1-(6-Hydroxy-*m*-tolylazo)-2-naphthol-4-sulfonic acid	H_2In^- (red; pK_2 = 8.1 HIn^{2-} (blue); p K_3 = 12.4 In^{3-} (orange)	Wine red	Titrations performed with Eriochrome Black T as indicator may be carried out equally well with Calmagite
Eriochrome Black T 0.1 g/100 mL water; prepare fresh daily	1-(2-Hydroxy-1-naphthyl-azo)-6-nitro-2-naphthol-4-sulfonic acid	H_2In^- (red); pK_2 = 6.3 HIn^{2-} (blue); pK_3 = 11.5 In^{3-} (yellow-orange)	Wine red	*Direct titration:* Ba, Ca, Cd, In, Mg, Mn, Pb, Sc, Sr, Tl, Zn, and lanthanides *Back titration:* Al, Ba, Bi, Ca, Co, Cr, Fe, Ga, Hg, Mn, Ni, Pb, Pd, Sc, Tl, V *Substitution titration:* Au, Ba, Ca, Cu, Hg, Pb, Pd, Sr
Murexide Suspend 0.5 g in water; use fresh supernatant liquid each day	5-[(Hexahydro-2,4,6-trioxo-5-pyrimidinyl)imino]-2,4,6(1*H*,3*H*,5*H*)-pyrimidinetrione monoammonium salt	H_4In^- (red-violet); pK_2 = 9.2 H_3In^{2-} (violet); pK_3 = 10.9 H_2In^{3-} (blue)	Red with Ca^{2+} Yellow with Co^{2+}, Ni^{2+}, and Cu^{2+}	*Direct titration:* Ca, Co, Cu, Ni *Back titration:* Ca, Cr, Ga *Substitution titration:* Ag, Au, Pd
PAN	1-(2-Pyridylazo)-2-naphthol	HIn (orange-red); pK_1 = 12.3 In^- (pink)	Red	*Direct titration:* Cd, Cu, In, Sc, Tl, Zn *Back titration:* Cu, Fe, Ga, Ni, Pb, Sc, Sn, Zn *Substitution titration:* Al, Ca, Co, Fe, Ga, Hg, In, Mg, Mn, Ni, Pb, V, Zn
Pyrocatechol violet 0.1 g/100 mL; stable several weeks	Pyrocatecholsulfonephthalein	H_4In (red); pK_1 = 0.2 H_3In^- (yellow): pK_2 = 7.8 H_2In^{2-} (violet); pK_3 = 9.8 HIn^{3-} (red-purple); pK_4 = 11.7	Blue, except red with Th(IV)	*Direct titration:* Al, Bi, Cd, Co, Fe, Ga, Mg, Mn, Ni, Pb, Th, Zn *Back titration:* Al, Bi, Fe, Ga, In, Ni, Pd, Sn, Th, Ti
Salicylic acid	2-Hydroxybenzoic acid	H_2In; pK_1 = 2.98 HIn^-; pK_2 = 12.38	$FeSCN^{2+}$ at pH 3 is reddish-brown	Typical uses: Fe(III) titrated with EDTA to colorless iron–EDTA complex
Xylenol orange	3,3′-Bis[*N*,*N*-di(carboxy methyl)aminomethyl]-*o*-cresolsulfonephthalein	—COOH groups: pK_3 = 0.76; pK_4 = 1.15; pK_5 = 2.58; pK_6 = 3.23		Typical uses: Bi, Pb Th

TABLE 4.27 Variation of α_4 with pH

pH	$-\log \alpha_4$	pH	$-\log \alpha_4$
2.0	13.44	7.0	3.33
2.5	11.86	8.0	2.29
3.0	10.60	9.0	1.29
4.0	8.48	10.0	0.46
5.0	6.45	11.0	0.07
6.0	4.66	12.0	0.00

4.6.3.2 Formation of Metal–EDTA Complexes. The formation of metal–EDTA complexes may be represented by the equations

$$M^{n+} + H_2Y^{2-} \rightleftharpoons MY^{(n-4)+} + 2H^+ \qquad \text{pH 4 to 5} \tag{4.66}$$

$$M^{n+} + HY^{3-} \rightleftharpoons MY^{(n-4)+} + H^+ \qquad \text{pH 7 to 9} \tag{4.67}$$

in which the metals of charge n^+ displace the remaining protons from Y^{4-} ions to form the complex $MY^{(n-4)+}$.

The formation constants of the EDTA complexes with various metals, which are equilibrium constants of reactions of the type

$$M^{n+} + Y^{4-} \rightleftharpoons MY^{(n-4)+} \tag{4.68}$$

$$K_{MY} = \frac{[MY^{(n-4)+}]}{[M^{n+}][Y^{4-}]} \tag{4.69}$$

are listed in Table 4.28.

TABLE 4.28 Formation Constants of EDTA Complexes at 25°C and Ionic Strength Approaching Zero

Metal ion	$\log K_{MY}$	Metal ion	$\log K_{MY}$
Co(III)	36	V(IV)	18.0
V(III)	25.9	U(IV)	17.5
In	24.95	Ti(IV)	17.3
Fe(III)	24.23	Ce(III)	16.80
Th	23.2	Zn	16.4
Sc	23.1	Cd	16.4
Cr(III)	23	Co(II)	16.31
Bi	22.8	Al	16.13
Tl(III)	22.5	La	16.34
Sn(II)	22.1	Fe(II)	14.33
Ti(III)	21.3	Mn(II)	13.8
Hg(II)	21.80	Cr(II)	13.6
Ga	20.25	V(II)	12.7
Zr	19.40	Ca	11.0
Cu(II)	18.7	Be	9.3
Ni	18.56	Mg	8.64
Pd(II)	18.5	Sr	8.80
Pb(II)	18.3	Ba	7.78
V(V)	18.05	Ag	7.32

In practice, an auxiliary complexing agent is usually added during EDTA titrations to prevent the precipitation of heavy metals as hydroxides or basic salts. The concentration of auxiliary complexing agents is generally high compared with the metal-ion concentration, and the solution is sufficiently well buffered so that the hydrogen ions produced in Eqs. (4.66) and (4.67) do not cause an appreciable change in pH. It is, therefore, convenient to define a *conditional* formation constant K'_{MY}, which is a function of the pH and the concentration of the auxiliary complexing agent.

By definition,

$$K'_{MY} = \frac{[MY^{(n-4)+}]}{C_M C_Y} \tag{4.70}$$

where C_M = total concentration of metal not complexed with EDTA
= $[M^{n+}]/\beta$, where β is the fraction of metal ion not complexed with EDTA that is present as the aquated ion

C_Y = total concentration of EDTA not complexed with metal
= $[Y^{4-}]/\alpha_4$, where α_4, is the fraction of free EDTA present as the tetravalent anion

From Eqs. (4.69) and (4.70), we have

$$K'_{MY} = K_{MY}\alpha_4\beta \tag{4.71}$$

or

$$\log K'_{MY} = \log K_{MY} + \log \alpha_4 + \log \beta \tag{4.72}$$

where α_4 is a function of pH (Table 4.27) and where β is a function of the nature and concentration of auxiliary complexing agent.

Many EDTA titrations are carried out in ammonia–ammonium chloride buffers, which serve also to provide ammonia as an auxiliary complexing agent. The successive formation constants of metal–ammonia complexes are listed in Table 4.29. From these constants β can be calculated, using the equation

$$\frac{1}{\beta} = \sum K_i[NH_3]^i \tag{4.73}$$

where K_i is the successive overall formation constants summed from $i = 0$ through $i = n$, and $K_0 = 1$, by definition.

From Eq. (4.71), the quantity $K_{MY}\alpha_4$ is the conditional formation constant for $\beta = 1$ (absence of auxiliary complexing agents). From Eq. (4.70), the value $K'_{MY} = 10^8$ corresponds to a 99.9% conversion of metal ion to EDTA complex; that is, $[MY^{(n-4)+}]/C_M = 1000$, at a concentration C_Y of reagent equal to $10^{-5}M$, which corresponds to 0.1% relative excess of reagent in the titration of $0.01M$ metal ion. Such a curve is shown in Fig. 4.5; it represents the *minimum* pH for effective titration of various metal ions in the absence of competing complexing agents.

Example 4.1. Calculate the conditional formation constant of NiY^{2-} in a buffer containing $0.057M$ NH_3 and $0.10M$ NH_4Cl.

From Eq. (4.73) and Table 4.29, $1/\beta = 1 + 36 + 356 + 1090 + 962 + 308 + 19 = 2772$; $\beta = 3.61 \times 10^{-4}$. Taking K_a for $NH_4^+ = 5.7 \times 10^{-10}$, we calculate $[H^+] = 1.0 \times 10^{-9}$ and $\alpha_4 = 5.1 \times 10^{-2}$. From Table 4.29 and Eq. (4.71), $K'_{MY} = (10^{18.62})(5.1 \times 10^{-2})(3.6 \times 10^{-4}) = 7.8 \times 10^{13}$ or $10^{13.89}$.

4.6.4 Titration Curves

It is convenient to plot pH = –log [M] against the fraction titrated, in analogy to a pH titration curve.[6] Under the usual titration conditions, in which the concentration of metal ions is small compared with

[6] C. N. Reilley and R. W. Schmid, *Anal. Chem.* **30**:947 (1958).

TABLE 4.29 Cumulative Formation Constants of Ammine Complexes at 20°C, Ionic Strength 0.1

Cation	log K_1	log K_2	log K_3	log K_4	log K_5	log K_6
Cadmium	2.65	4.75	6.19	7.12	6.80	5.14
Cobalt(II)	2.11	3.74	4.79	5.55	5.73	5.11
Cobalt(III)	6.7	14.0	20.1	25.7	30.8	35.2
Copper(I)	5.93	10.86				
Copper(II)	4.31	7.98	11.02	13.32	12.66	
Iron(II)	1.4	2.2				
Manganese(II)	0.8	1.3				
Mercury(II)	8.8	17.5	18.5	19.28		
Nickel	2.80	5.04	6.77	7.96	8.71	8.74
Platinum(II)						35.3
Silver(I)	3.24	7.05				
Zinc	2.37	4.81	7.31	9.46		

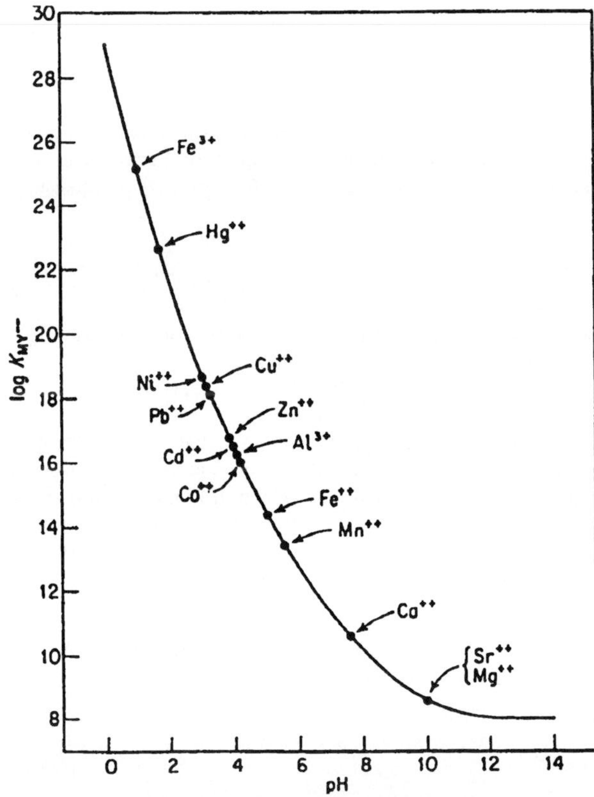

FIGURE 4.5 Minimum pH for EDTA titration of various metal ions. [With permission from C. N. Reilly and R. W. Schmid, *Anal. Chem.* **30**, 947 (1958).]

the buffer and auxiliary complexing agent concentration, α_4 and β are essentially constant during the titration, so that the conditional formation constant also remains constant. It is then a simple matter to calculate the titration curve.

Before the end point:

$$C_M = C_M^0(1-f) \tag{4.74}$$

where C_M^0 = the initial concentration of metal
f = the fraction of the stoichiometric quantity added

At the equivalent point:

$$C_M = C_Y = \sqrt{\frac{C_M^0}{K'_{MY}}} \tag{4.75}$$

Beyond the equivalence point:

$$C_Y = C_M^0(f-1) \tag{4.76}$$

From Eq. (4.70),

$$C_M = 1/(f-1)K'_{MY} \tag{4.77}$$

or

$$[M^{n+}] = 1/(f-1)\beta K'_{MY} \tag{4.78}$$

Example 4.2. Calculate the value of pNi at the following percentages of the stoichiometric amount of EDTA added in a buffer containing 0.057*M* NH_3 and 0.1*M* NH_4Cl, taking the initial concentration of nickel to be 0.001*M* and neglecting dilution: 0%, 50%, 90%, 99%, 99.9%, 100%, 100.1%,101%, and 110%.

At all points pNi = −log $[Ni^{2+}]$ = − log βC_M = 3.44 − C_M. The results are as follows: pNi = 6.44, 6.74, 7.44, 8.44, 9.44, 11.88, 14.33, 15.33, and 16.33.

To assess the feasibility of any titration with a complexing agent, it is desirable to construct the complete titration curve, as has been done in Example 4.1. Such a curve enables one to predict the accuracy of the titration and to select an appropriate method for locating the equivalence point. The calculated titration curves of 0.001*M* nickel ion at various pH values in buffers of ammonia–ammonium ion at a total buffer concentration of 0.1*M* are shown plotted in Fig. 4.6. The following features are of interest.

1. *Before the end point*, the curves at low pH values of 4 to 6 coincide because no appreciable complexation between nickel ions and ammonia occurs. At higher pH values the amount of the "lifting" of the curves is determined by the stabilities of the various nickel-ammine complexes.Beyond pH 11, essentially all the buffer is present as ammonia so no further pH effect exists.

2. *After the end point*, the placement of the curves does not depend upon β but only upon K_{MY} and α_4. Therefore, the titration curves beyond the end point for a given metal depend only on the pH and not on the nature of the auxiliary complexing agent.

For a metal not forming ammine complexes, such as calcium ion, the pM–pH curves differ from those for nickel in two ways. Before the end point, the curves are essentially independent of pH.

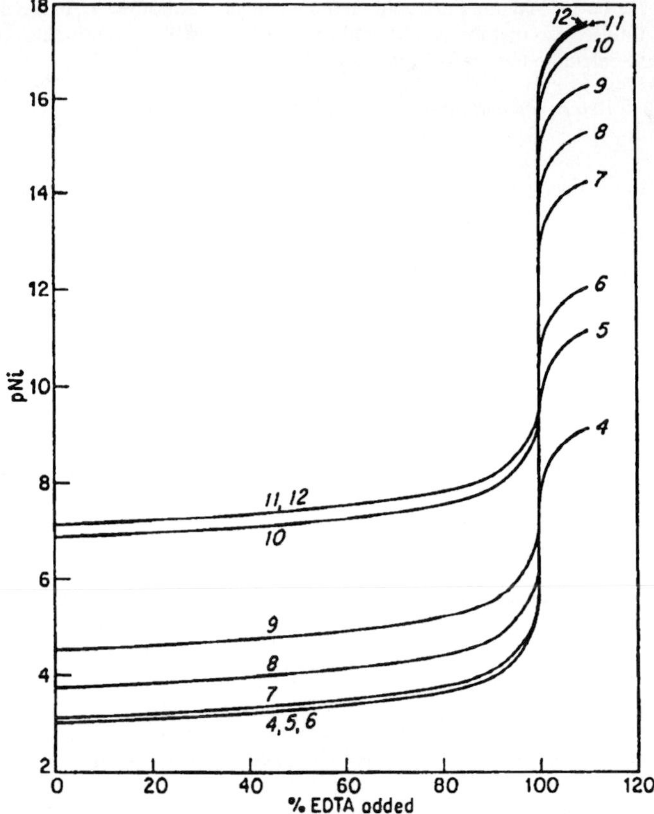

FIGURE 4.6 Titration curves of Ni(II) with EDTA. $C^0_{Ni} = 10^{-3}$.$[NH_4^+] + [NH_3] = 0.1$. The numbers refer to pH values. *Source:* H. H. Laitinen, *Chemical Analysis*, McGraw-Hill, New York, 1960.

After the end point, the jump in pM is smaller (or larger) than it is for nickel, depending on whether the formation constant of the metal is smaller (or larger) than the value for nickel.

At low pH values α_4 is so small that little or no pM break occurs unless K_{MY} is very large. At high pH, α_4 approaches unity, so it is often advantageous to perform the titrations at pH 9 to 12. On the other hand, for ions such as Fe(III), a low pH range (3 to 4) is favorable for preventing hydrolysis of the Fe(III) ion. Titrations at a low pH value are permissible only because of the great stability of the complex FeY^-.

4.6.5 Selectivity in Complexometric Titrations

EDTA forms stable complexes with many cations. Therefore, this reagent (and most chelons) frequently lack selectivity if one is interested in determining a single metal cation in a mixture or in performing a successive titration of several species in the same solution. An obvious route is to separate the desired species from the other components of the sample by one of the separation methods discussed in Sec. 2.

4.6.5.1 Control of Acidity. In the resolution of metal-ion mixtures by chelometric titrations, the problem of selectivity must be faced. Some degree of selectivity can be secured for some elements by the application of an especially favorable chelon. Furthermore, some degree of selectivity can be secured with a given chelon by taking advantage of the pH effect (*see* Fig. 4.5).

Based on their stability, the EDTA complexes of the most common metal ions may be roughly divided into three groups:

$\log K > 20$	Tri- and tetravalent cations including Bi, Fe(III), Ga, Hg(II), In, Sc, Th, U(IV), V(III), and Zr
$\log K = 15$ to 18	Divalent transition metals, rare earths, and Al
$\log K = 8$ to 11	Alkaline earths and Mg

The more stable the metal complex, the lower the pH at which it can be quantitatively formed. Elements in the first group may be titrated with EDTA at pH 1 to 3 without interference from cations of the last two groups, while cations of the second group may be titrated at pH 4 to 5 without interference from the alkaline earths. Roughly speaking, the apparent stability constant, corrected for auxiliary complex formers, should be at least 10^8 for successful titration.

4.6.5.2 Use of Masking Agents. The most fruitful approach to selectivity (other than the employment of prior separations) in chelometric titrations is the application of masking agents. Some established possibilities to masking and demasking in EDTA titrations are summarized in Table 4.30; also review the discussion in Sec. 2.1.

Probably the most extensively applied masking agent is cyanide ion. In alkaline solution, cyanide forms strong cyano complexes with the following ions and masks their action toward EDTA: Ag, Cd, Co(II), Cu(II), Fe(II), Hg(II), Ni, Pd(II), Pt(II), Tl(III), and Zn. The alkaline earths, Mn(II), Pb, and the rare earths are virtually unaffected; hence, these latter ions may be titrated with EDTA with the former ions masked by cyanide. Iron(III) is also masked by cyanide. However, as the hexacyanoferrate(III) ion oxidizes many indicators, ascorbic acid is added to form hexacyanoferrate(II) ion. Moreover, since the addition of cyanide to an acidic solution results in the formation of deadly hydrogen cyanide, the solution must first be made alkaline, with hydrous oxide formation prevented by the addition of tartrate. Zinc and cadmium may be demasked from their cyanide complexes by the action of formaldehyde.

Masking by oxidation or reduction of a metal ion to a state that does not react with EDTA is occasionally of value. For example, Fe(III) ($\log K_{MY} = 24.23$) in acidic media may be reduced to Fe(II) ($\log K_{MY} = 14.33$) by ascorbic acid; in this state iron does not interfere in the titration of some trivalent and tetravalent ions in a strong acidic medium (pH 0 to 2). Similarly, Hg(II) can be reduced to the metal. In favorable conditions, Cr(III) may be oxidized by alkaline peroxide to chromate, which does not complex with EDTA.

In resolving complex metal-ion mixtures, more than one masking or demasking process may be utilized with various aliquots of the sample solution, or applied simultaneously or stepwise with a single aliquot. In favorable cases, even four or five metals can be determined in a mixture by the application of direct and indirect masking processes. Of course, not all components of the mixture need be determined by chelometric titrations. For example, redox titrimetry may be applied to the determination of one or more of the metals present. Moreover, rapid prior separations, including ion exchange and solvent extraction, may represent the most expeditious approach.

4.6.5.3 Titration of Hydrolyzable Cations. The direct titration of readily hydrolyzable cations such as aluminum and zirconium presents difficulties, including a sluggish end point and low results, even in hot or boiling solutions. Hence, a back titration is usually preferred. The key to a successful determination lies in employing a proper method of neutralizing the strongly acidic sample solution. Strongly alkaline solutions (ammonia or NaOH) should not be used, but instead the solution should be brought to the specified pH by the addition of sodium acetate. In a back titration, the excess of

TABLE 4.30 Some Masking and Demasking Agents Used in Titration with EDTA

Masking agents	Applications
Ascorbic acid	(1) Cr(III) masked in boiling acidic solution; Ca, Mn(II), and Ni determined in concentrated ammonia by the back-titration method using a Ca solution. (2) In an ammoniacal solution Cu(II) is reduced to Cu(I); Zn determined. (3) Fe(III) reduced at pH 2 to Fe(II); Bi(III), Th, and Zr determined. (4) Hg(II) reduced at pH 2–2.5 to metal; Bi, Ga, In, Pd(II) determined by back titration with Bi.
Citrate	(1) Al in small amounts masked at pH 8.5–9.5 for Zn determination. (2) Fe(III) masked at pH 8.5 for Cd, Cu(II), and Pb determination. (3) Mo(VI) masked at pH 9 for Cu(II) determination. (4) Sn(II), Th, and Zr masked for Cd, Co, Cu(II), Ni, and Zn titrations.
Carbonate	Masks UO_2^{2-} at pH 8–8.5 in determination of zinc.
Cyanide	(1) Titrate Pb in an ammoniacal tartrate medium in the presence of Cu masked as $Cu(CN)_3^{2-}$ or Co masked as $Co(CN)_6^{4-}$. (2) Titrate Ca at pH 10 in the presence of many divalent heavy metals [Ag, Cd, Co, Cu, Fe(II), Hg, Ni, Pd, Pt, and Zn] masked as cyanide complexes. (3) Titrate In in an ammoniacal tartrate medium in the presence of Cd, Co, Cu, Hg, Ni, Pt, Pd, and Zn masked as cyanide complexes. (4) Ca, Mg, Mn, Pb, and Sr determined in the presence of Ag, Cd, Co, Cu, Hg(II), Ni, platinum metals, and Zn.
2,3-Dimercaptopropanol-1	(1) Titrate Mg in an ammoniacal medium of pH 10 in presence of Bi, Cd, Co (in small amounts), Cu, Hg, Ni (small amounts), Pb, Sb(III), Sn(IV), and Zn masked as complexes with 2,3-dimercaptopropanol-1. (2) Titrate Th at pH 2.5–3 in the presence of Bi and Pb masked as complexes with 2,3-mercaptopropanol-1.
Fluoride	(1) Titrate Zn in an ammoniacal medium in the presence of Al masked as AlF_6^{3-} or Mg and Ca masked by precipitation as MgF_2 and CaF_2, respectively. (2) Titrate Ga in acetic acid at pH 2.8 in the presence of Al masked as AlF_6^{3-}. (3) Al, Ba, Be, Ca, Fe(III) (in small amounts), Mg, Nb, rare earths, Sr, Ta, Ti(IV), and Zr masked with ammonium fluoride enables the determination of Bi, Cd, Cu, Ga, In, Mn, Ni, Pb, and Zn.
Formic acid or formaldehyde	(1) Hg reduced to metal, removing interference with Bi and Th. (2) Tl(III) reduced to Tl(I) removing interference with In.
Hydrazine	Fe(III) reduced to Fe(II) removing interference with Al and Bi.
Hydrogen peroxide	(1) Ti(IV) masked in titration of Mg and Zn. (2) W(VI) masked in determination of Cd, Cu, Fe(III), Ni, Th, and Zn.
Hydrazine	(1) Cu(II) reduced to Cu(I) removes interference with Bi and Ni. (2) Fe(III) reduced to Fe(II) for Th titration. (3) Reduce Mo(VI) to Mo(V) for determination of Mo(V).
Iodide	Hg(II) masked in titration of Cu and Zn.
2,4-Pentanedione	Masks Al, Fe(III), and Pd (by precipitation) in titration of Pb and Zn.
1,10-Phenanthroline	Masks Cd, Co, Cu(II), Mn, Ni, UO_2^{2+}, and Zn in titrations of Al, In, and Pb.
Phosphate	Masks W(VI) in determination of Cd, Cu(II), Fe(III), Ni, Mo(VI), V(V), and Zn.
Pyrophosphate	Masks Fe(III) and Cr(III) in titration of Co.
Sulfate	Masks Th and Ti(IV) in titration of Bi, Fe(III), and Zr.
Sulfosalicylic acid	Masks Al in Mn titration and (small amounts of Al) in Zn titration.

TABLE 4.30 Some Masking and Demasking Agents Used in Titration with EDTA (*Continued*)

Masking agents	Applications
Tartrate	(1) Masks Al in titration of Ca, Mn, and Zn. (2) Masks Fe(III) in titration of Ca and Mn. (3) Masks Nb(V), Ta(V), Ti(IV), and W(VI) in titration of Mo(V). (4) Masks Sb(III), Sn(IV), and Zr for titration of Zn, Bi, Cd, Co, Cu(II), Ni, and rare earths.
Thiourea	Masks Cu(II) in titrations of Fe(III), Ni, Pb, Sn, and Zn.
Thiosulfate	Masks Cu(II) in titrations of Cd, Ni, Pb, and Zn.
Triethanolamine	(1) Titrate Ni in an ammoniacal medium in the presence of Al, Fe, Ti(IV), and small amounts of Mn masked as triethanolamine complexes. (2) Titrate Mg in an ammoniacal medium of pH 10 in presence of Al masked as a triethanolamine complex. (3) Titrate Zn or Cd in an ammoniacal medium in the presence of Al masked as a triethanolamine complex. (4) Al, Ti(IV), Fe(III), and Sn(IV) masked in determination of Cd, Mg, Mn (add ascorbic acid), Pb, rare earths, and Zn in an ammoniacal medium at pH 10.

EDTA should be added at a sufficiently low pH to ensure complete complexation (with boiling), and only when this is accomplished should the pH be raised.

4.6.6 Preparation of Standard Solutions

4.6.6.1 Standard EDTA Solutions. Disodium dihydrogen ethylenediaminetetraacetate dihydrate of analytical reagent purity is available commercially. After drying at 80°C for at least 24 h, its composition agrees exactly with the dihydrate formula (molecular weight 372.25). It may be weighed directly. If an additional check on the concentration is required, it may be standardized by titration with nearly neutralized zinc chloride or zinc sulfate solution.

4.6.6.2 Standard Magnesium Solution. Dissolve 24.647 g of magnesium sulfate heptahydrate in water and dilute to 1 L for 0.1*M* solution.

4.6.6.3 Standard Manganese(II) Solution. Dissolve exactly 16.901 g ACS reagent grade manganese(II) sulfate hydrate in water and dilute to 1 L.

4.6.6.4 Standard Zinc Solution. Dissolve exactly 13.629 g of zinc chloride, ACS reagent grade, or 28.754 g of zinc sulfate heptahydrate, and dilute to 1 L for 0.1000*M* solution.

4.6.6.5 Buffer Solution, pH 10. Add 142 mL of concentrated ammonia solution (specific gravity 0.88 to 0.90) to 17.5 g of analytical reagent ammonium chloride, and dilute to 250 mL.

4.6.6.6 Water. Distilled water must be (a) redistilled in an all-Pyrex glass apparatus or (b) purified by passage through a column of cation-exchange resin in the sodium form. For storage, polyethylene bottles are most satisfactory, particularly for very dilute (0.001*M*) EDTA solutions.

4.6.6.7 Murexide Indicator. Suspend 0.5 g of powdered murexide in water, shake thoroughly, and allow the undissolved solid to settle. Use 5 to 6 drops of the supernatant liquid for each titration. Decant the old supernatant liquid daily and treat the residue with water to provide a fresh solution of the indicator.

TABLE 4.31 Representative Analyses Involving EDTA Titrations

Cation	Procedure	Remarks
Ag(I)	(1) To an ammonia buffer (pH 9–10), add an excess of $Ni(CN)_4^{2-}$. Titrate Ni(II) ion liberated with standard EDTA solution using murexide indicator (yellow to pink). Equiv. wt. Ag = 2 × 107.868 = 215.736. (2) Reduce Ag^+ with Bi amalgam and titrate liberated Bi(III) with EDTA. Equiv. wt. Ag = 107.868 ÷ 3 = 35.956.	
Al(III)	(1) Add slight excess of standard EDTA. Adjust pH to 6.5; dissolve salicylic acid indicator in the solution and back-titrate unreacted EDTA with standard Fe(III) solution to reddish-brown. Equiv. wt. Al = 26.98. (2) Add excess EDTA to HOAc solution adjusted to pH 3. Warm the solution, readjust to pH 4–6, and back-titrate with standard Cu(II) solution or with Zn using PAN indicator (yellow to pink). (3) Direct titration with EDTA in hot HOAc solution, pH 3, using Cu(II)–PAN indicator (red to yellow).	1,10-phenanthroline masks Cd, Co, Cu(II), Mn, Ni, UO_2^{2-}, and Zn. Methods applicable to alloys, clays, cryolite ores, cracking catalysts.
Ba(II)	(1) Direct titration with EDTA in an ammoniacal buffer at pH 10.5 with Eriochrome Black T indicator (red to blue). Equiv. wt. Ba = 137.33.	Sr is also titrated. Tartrate masks Al and Fe(III); cyanide masks Ag, Cd, Co, Cu, Fe(II), Hg, Ni, Pd, Pt, and Zn.
Bi(III)	(1) Adjust HNO_3 solution to pH 2.5–3.5 and titrate with EDTA using pyrocatechol violet (blue to yellow) indicator; or adjust to pH 1–3 using PAN (red to yellow) or xylenol orange (red to yellow). Equiv. wt. Bi = 208.98. (2) Add excess EDTA to borate buffer at pH 10 and back-titrate with standard Mg or Zn using Eriochrome Black T (blue to red).	Applicable to alloys, pharmaceuticals; divalent metals do not interfere, Hg(II) → metal and Fe(III) → Fe(II) by ascorbic acid at pH 2. Al, Ba, Be, Ca, Mg, Nb, rare earths, Sr, Ta, Ti(IV), and Zr masked by fluoride. Hydrazine reduces Cu(II) → (Cu(I). Sulfate masks Th and Ti(IV). Tartrate masks Sb(III), Sn(IV), and Zr.
Ca(II)	(1) Add NaOH to a neutral sample solution to obtain pH > 12. Add murexide indicator (pink or red to violet) and titrate with EDTA. Equiv. wt. Ca = 40.078; $CaCO_3$ = 100.09. (2) In an ammoniacal buffer of pH 10, add excess EDTA and back-titrate with Mg or Zn using Eriochrome Black T (blue to red) indicator.	Applicable to biological fluids, pharmaceuticals, phosphate rocks, water; Mg does not interfere. Cyanide masks Ag, Cd, CO, Cu, Fe(II), Hg, Ni, Pd, Pt, and Zn. Tartrate masks Al and Fe(III). With Ca alone, no sharp color change occurs in a direct titration.
Cd(II)	(1) In an ammoniacal buffer of pH 10, titrate with EDTA using Eriochrome Black T (red to blue) or pyrocatechol violet (green-blue to red-violet) as indicator. Equiv. wt. Cd = 112.41. (2) In an acetate buffer of pH 5, titrate with EDTA using PAN as indicator (pink to yellow).	Applicable to alloys and plating baths. Citrate masks Fe(III) at pH 8.5. Fluoride masks Al, Ba, Be, Ca, Mg, Nb, rare earths, Sr, Ta, Ti(IV), and Zr. H_2O_2 and/or phosphate masks W(VI). Tartrate masks Sb(III), Sn(IV), and Zr. Thiosulfate masks Cu(II). Triethanolamine masks Al, Fe(III), Sn (IV), and Ti(IV).
Co(II)	(1) Adjust the solution to pH 6 and add murexide indicator. Add ammonia to obtain the orange color of cobalt–murexide complex; titrate with EDTA to purple color. Equiv. wt. Co = 58.933.	Applicable to paint driers, magnet alloys, cemented carbides. Citrate masks Sn(II), Th, and Zr. Pyrophosphate masks Cr(III) and Fe(III). Tartrate masks Sb(III), Sn(IV),

	(2) Add excess EDTA and adjust solution to pH 9–10; back-titrate with standard Zn with Eriochrome Black T (blue to red-violet). (3) Add xylenol orange; if solution is red, add acid until color is yellow. Add powdered hexamethylenediamine until a deep red color appears (pH 6). Warm to 40°C and titrate with EDTA until color changes to yellow.	and Zr. Thiosulfate masks Cu(II). Triethanolamine masks Al, Fe(III), Sn(IV), and Ti(IV).
Co(III)	Add H_2O_2 to a Co(II) solution containing HNO_3; adjust to pH 2 and add excess EDTA. Back-titrate with Th(IV) using xylenol orange (yellow to red). Equiv. wt. Co = 58.933.	
Cr(III)	Boil an acidic solution (pH 4–4.5) with excess EDTA for 15 min. Adjust to pH 4–5 with ammonium acetate and back-titrate with standard Zn using xylenol orange (yellow to red) or adjust to pH 9–10 if using Eriochrome Black T (blue to red). Equiv. wt. Cr = 51.996.	
Cu(II)	Titrate with EDTA in acetate buffer at pH 4 with murexide (orange to red) indicator or use excess EDTA, heat the solution, and back-titrate with standard Cu using PAN (yellow to red-violet).	Applicable to alloys, ores, electroplating baths. Citrate masks Fe(III) and Mo(VI) at pH 8.5–9. Fluoride masks Al, Ba, Be, Ca, Fe(III) in small amounts, Mg, Nb, rare earths, Sr, Ta, Ti(IV), and Zr. H_2O_2 and phosphate masks W(VI). Iodide masks Hg(II). Tartrate masks Sb(III), Sn(IV), and Zr.
Fe(II)	Adjust to pH 5–6 with hexamethylenetetramine, add ascorbic acid, and titrate with EDTA using xylenol orange (red to yellow). Equiv. wt. Fe = 55.847.	
Fe(III)	(1) Titrate with EDTA in acetate buffer at pH 4.5 using Cu–PAN indicator (red to yellow). Equiv. wt. Fe = 55.847. (2) Adjust to pH 4, add salicylic acid and titrate with EDTA to disappearance of reddish-brown indicator color. (3) Adjust acetate buffer to pH 5, heat the solution to 70–80°C, and titrate with EDTA using Cu–PAN indicator (red to violet).	H_2O_2 masks W(VI). Thiourea masks Cu(II).
Ga(III)	Adjust the ammonia buffer to pH 9, add Eriochrome Black T and EDTA in excess. Back-titrate with standard Mn(II) or Zn solution (blue to red). Equiv. wt. Ga = 69.723.	Al, Ba, Be, Ca, Fe(III) in small amounts, Mg, Nb, rare earths, Sr, Ta, Ti(IV), and Zr masked with fluoride.
Hf(IV)	See procedure for Al(III). Equiv. wt. Hf = 178.49.	
Hg(II)	(1) To the sample solution add a known excess of standard Mg (or Zn)–EDTA solution. Neutralize the solution, add ammonia buffer, and adjust to pH 10, and titrate free Mg (or Zn) with EDTA using Eriochrome Black T (blue to red). Equiv. wt. Hg = 200.59. (2) Using HOAc, adjust to pH 3–3.5 and titrate with EDTA using Cu(II)–PAN (red-violet to yellow) indicator. (3) Add excess EDTA and adjust to pH 9 with NH_3. Back-titrate with Ni or Cu using murexide (pink to yellow) indicator.	Applicable to mercury-containing pharmaceuticals, organomercury compounds, ores
In(III)	(1) Add tartaric acid to an acidic sample solution. Neutralize, add ammoniacal buffer (pH 8–10), heat to boiling, add Eriochrome Black T (red to blue) or pyrocatechol violet (blue to yellow), and titrate with EDTA. Equiv. wt. In = 114.82.	Ascorbic acid reduces Hg(II) to metal. Cyanide in ammoniacal tartrate masks Cd, Co, Cu, Hg, Ni, Pd, Pt, and Zn. Fluoride masks Al, Ba, Be, Ca, Fe(III) in small amounts, Mg, Nb, rare earths, Sr, Ta, Ti(IV), and Zr.

(Continued)

TABLE 4.31 Representative Analyses Involving EDTA Titrations (*Continued*)

Cation	Procedure	Remarks
	(2) Proceed as above except add excess EDTA and back-titrate (without boiling) with Mg, Pb, or Zn standard solution using Eriochrome Black T.	Tl(III) interference removed by reduction to Tl(I) by formic acid or formaldehyde.
K	Precipitate $K_2Na[Co(NO_2)_6]$; determine Co (*q.v.*) in precipitate via EDTA titration. Equiv. wt. 2 K = 78.196.	
Mg	Heat an ammoniacal solution (pH 10) to 40°C and titrate with EDTA using either Eriochrome Black T (red to blue) or pyrocatechol violet (green-blue to red-violet). Equiv. wt. Mg = 24.305.	Applicable to Al alloys, soils, plant materials, water, biological fluids, pharmaceuticals. Cyanide masks Ag, Cd, Co, Cu, Hg(II), Ni, platinum metals, and Zn. 2,3-Dimercaptopropanol-1 masks Bi, Cd, Co in small amounts, Cu, Hg, Ni in small amounts, Pb, Sb(III), Sn(IV), and Zn. Triethanolamine masks Al, Fe(III), Sn(IV), and Ti(IV).
Mn(II)	(1) Add triethanolamine and ascorbic acid to an acidic sample solution. (a) Neutralize and adjust an ammoniacal buffer to pH 9–10; add Eriochrome Black T (red to blue) and titrate with EDTA. (b) Adjust to pH 7 and add murexide. Titrate with EDTA until near the end point, then add ammonia until pH 10–11, and complete titration (yellow to blue-violet). Equiv. wt. Mn = 54.938. (2) Proceed as above, heat to 50°C, add Cu–PAN indicator (violet-red to yellow), and titrate with EDTA. (3) Using an acetate buffer, adjust to pH 4.5, sand heat to 60°C. Add excess EDTA and back-titrate with standard Cu(II) solution using PAN indicator (yellow to red-violet).	Applicable to alloys, ferromanganese, metallurgical slags, silicate rocks. Al, Fe(III), Sn(IV), and Ti(IV) masked by triethanolamine. Fluoride masks Al, Ba, Be, Ca, Fe(III), Mg, Nb, rare earths, Sr, Ta, Ti(IV), and Zr. Cyanide masks Ag, Cd, Co, Cu, Hg(II), Ni, platinum metals, and Zn.
Mo(V)	Add excess EDTA to ammonia buffer (pH 10), heat to boiling, cool to 40–60°C, add Eriochrome Black T (green to red-brown) and back-titrate with standard Zn solution. Equiv. wt. Mo = 95.94.	Tartrate masks Nb(V), Ta(V), Ti(IV), and W(VI). H_2O_2 reduces Mo(VI) to Mo(V).
Ni(II)	(1) Add murexide indicator to a neutral sample solution; add NH_3 to obtain the orange color of Ni–murexide complex. Titrate with EDTA until the solution becomes colorless. Equiv. wt. Ni = 58.69. (2) To a neutral solution add an ammoniacal buffer to pH 10. Add slight excess EDTA and back-titrate excess with standard Mg or Zn using Eriochrome Black T (blue to red-violet). (3) To an acid solution, add excess standard EDTA and 10 mL of 20% triethanolamine solution. Add, with stirring, 1*M* NaOH solution until the pH is 11.6 (use a pH meter). Add Eriochrome Black T (color very pale blue) and titrate with standard Ca solution until the color becomes intense blue.	Applicable to alnico (aluminum–nickel–cobalt alloy), electroplating baths, manganese catalysts, and ferrites. Procedures are useful for $Ni(dimethylglyoxime)_2$ after dissolving in a minimum volume of hot dilute HCl. In an ammoniacal medium triethanolamine masks Al, Fe(III), Ti(IV), and small amounts of Mn. H_2O_2 masks W(VI). Hydrazine reduces Cu(II) and removes its interference. Citrate masks Sn(II), Th, and Zr. Tartrate masks Sb(III), Sn(IV), and Zr.

Pb(II)	Add several drops of xylenol orange. If the solution is red, add very dilute HNO_3 until the solution is yellow. Add powdered hexamethylenediamine until the solution is red (pH 6). Titrate with EDTA until the color changes to yellow. Equiv. wt. Pb = 207.2.	Applicable to alloys, gasoline, ores, paints, and pharmaceuticals. Citrate masks Fe(III). Cyanide masks Ag, Cd, Co, Cu, Hg(II), Ni, platinum metals, and Zn. 2,4-Pentanedione masks Al and Fe(III). 1,10-Phenanthroline masks Cd, Co, Cu, Mn, Ni, UO_2^{2+}, and Zn. Thiourea masks Cu(II). Tartrate masks Al, Fe(III), Sn(IV), and Ti(IV).
Pd(II)	(1) Add excess standard EDTA. Adjust pH to 8–10 with ammoniacal buffer, add ascorbic acid, cool to 5°C, and back-titrate unreacted EDTA with standard Mn using Eriochrome Black T (blue to red). Equiv. wt. Pd = 106.42. (2) Add excess EDTA. Adjust acetate buffer to pH 3, add xylenol orange (yellow to red), and titrate unreacted EDTA with standard Th(IV) solution.	
Rb	See under K.	
Rare earths and La	(1) Add triethanolamine and adjust to pH 7. Titrate with EDTA using Eriochrome Black T (red to blue). 1 mole EDTA ≡ 1 mole rare eath. (2) Add hexamethylenetetramine and adjust pH to 4.5–6.0. Add xylenol orange (red to yellow) and titrate with EDTA.	Triethanolamine masks Al, Fe(III), Sn(IV), and Ti(IV). Tartrate masks Sb(III), Sn(IV), and Zr.
Sb(III)	Adjust acetate buffer to pH 4, add excess EDTA, and titrate unreacted EDTA with Tl(III) using xylenol orange (yellow to pink). Equiv. wt. Sb = 121.75.	
Sc	(1) Adjust to pH 2–3 with HOAc, add Cu–PAN indicator (red to yellow) and titrate with EDTA, or use murexide (yellow to violet). Equiv. wt. Sc = 44.956. (2) In acetate buffer at pH 2.5, add excess EDTA and PAN (yellow to red-violet); titrate unreacted EDTA with standard Cu(II) solution.	
Sn(IV)	To a strongly acidic solution add a known excess of EDTA. Buffer solution at pH 5 with acetate. Heat the solution to 70–80°C, add pyrocatechol violet (yellow to blue), and titrate unreacted EDTA with standard Zn solution. Equiv. wt. Sn = 118.71.	Applicable to alloys, electroplating baths.
Sr	See under Ba. Equiv. wt. Sr = 87.62.	Cyanide masks Ag, Cd, Co, Cu, Hg(II), Ni, platinum metals, and Zn.
Th(IV)	(1) Adjust HNO_3 solution to pH 3–3.5; add pyrocathechol violet (rose to yellow). Warm to 40°C and titrate with EDTA. Equiv. wt. Th = 232.04. (2) Adjust HNO_3 solution to pH 2–3.5; add PAN (red to yellow), and titrate with EDTA; or use xylenol orange (red-violet to yellow).	Applicable to alloys, glasses, minerals, ores, reactor fuels. H_2O_2 masks W(VI). Formic acid reduces Hg to metal. Ascorbic acid reduces Fe(III) to Fe(II). 2,3-Mercaptopropanol-1 masks Bi and Pb at pH 3.
Ti(IV)	(1) Add excess EDTA and hexamethylenetetramine; adjust pH to 5–7. Add pyrocatechol violet (yellow to deep blue) and titrate unreacted EDTA with standard Cu(II) solution. Equiv. wt. Ti = 47.88. (2) Add excess EDTA, NH_4OAc, and adjust pH to 4–5.5. Titrate unreacted with standard Tl(III) solution using xylenol orange (yellow to red).	

(Continued)

TABLE 4.31 Representative Analyses Involving EDTA Titrations (*Continued*)

Cation	Procedure	Remarks
Ti(IV)	(3) Add H_2O_2 and excess EDTA; adjust pH to 4.5. Add PAN (orange to orange-red) and titrate unreacted EDTA with standard Cu(II).	
Tl(III)	Add chloroacetic acid and adjust pH to > 1.8. Heat to 80°C, add PAN (red-violet to yellow) and titrate with EDTA. Or use xylenol orange (red to yellow) in hot acetate buffer at pH 4–5. Equiv. wt. Tl = 204.38.	
U(IV)	Heat to boiling an acetate solution adjusted to pH 2–3. Add excess EDTA and back-titrated unreacted EDTA with standard Th(IV) solution using xylenol orange (yellow-green to red). Equiv. wt. U = 238.03.	
U(VI)	Add hexamethylenetetramine, adjust pH to 4.4, and add twice the solution volume of 2-propanol. Heat to 90°C and titrate with EDTA using PAN (red to yellow). Equiv. wt. U = 238.03.	
V(IV)	(1) Add ascorbic acid and adjust acetate buffer to pH > 3.5. Add Cu(II)–PAN (red-violet to yellow or green) and titrate with EDTA. Equiv. wt. V = 50.941. (2) Adjust solution to pH 6, heat, and add excess EDTA. Add PAN and titrate unreacted EDTA with standard Cu(II) solution.	
V(V)	Use 0.03*M* $HClO_4$, adjust to pH 1.8, cool, and titrate with EDTA using xylenol orange (red to yellow). Equiv. wt. V = 50.941.	Phosphate masks W(VI).
Zn	(1) Use ammoniacal buffer, pH 10. Titrate with EDTA using Eriochrome Black T (red to blue) or pyrocatechol violet (blue to red-violet), or an ammoniacal buffer (pH 8–9), and titrate with EDTA using murexide (pink to violet). Equiv. wt. Zn = 65.39. (2) Use an acetate buffer (pH 4–6) or maleate buffer (pH 6.8), and titrate with EDTA using PAN (pink or red to yellow); or add xylenol orange and adjust pH to 6 with hexamethylenediamine (intense red color). Titrate with EDTA until the color changes to yellow.	Ascorbic acid reduces Cu(II); citrate masks Al; carbonate masks UO_2^{2-}; fluoride masks Al, Ba, Be, Ca, Fe(III) (in small amounts), Mg, Nb, rare earths, Sr, Ta, Ti(IV), and Zr.
Zr	(1) (a) Heat 0.5–2*M* HCl solution to boiling or (b) heat 1*M* HNO_3 or 0.05–3*M* H_2SO_4 solution to 90°C. (a) Add Eriochrome Black T (blue-violet to red-violet) or (b) add xylenol orange (red to yellow) and titrate with EDTA. Equiv. wt. Zr = 91.224. (2) To hot HNO_3 solution, pH 2–3, add excess EDTA and titrate unreacted EDTA with standard Th(IV) solution using xylenol orange (yellow to red).	Reduces Fe(III) with ascorbic acid; masks Th and Ti(IV) with sulfate.

TABLE 4.32 Indirect Methods for Determination of Anions by Visual EDTA Titrations

Anion determined	Procedure
AsO_4^{3-}	(1) Precipitate $MgNH_4AsO_4 \cdot 6H_2O$ and determine Mg in precipitate. (2) Precipitate $ZnNH_4AsO_4$ and determine Zn in precipitate. (3) Precipitate $BiAsO_4$ and determine excess Bi(III) in filtrate.
$B(C_6H_5)_4^-$	Exchange with Hg(II)–EDTA complex and titrate liberated EDTA.
BO_3^{3-}	Precipitate barium borotartrate and determine Ba(II) in precipitate or excess Ba(II) in filtrate.
Br^-	Precipitate AgBr, exchange Ag(I) for Ni(II) in $Ni(CN)_4^{2-}$, and titrate liberated Ni(II).
BrO_3^-	Reduce to bromide with H_3AsO_3 and proceed as for bromide ion (under Halides).
CN^-	React with nickel ion [giving $Ni(CN)_4^{2-}$] and titrate excess Ni(II).
CO_3^{2-}	(1) Precipitate $CaCO_3$ and determine calcium ion in precipitate or excess calcium ion in clear supernatant solution. (2) Precipitate $SrCO_3$, and ethanol, and determine strontium ion in filtrate. (3) Precipitate $BaCO_3$ and titrate excess barium ion in presence of precipitate.
$CO_3^{2-} + HCO_3^-$	To one aliquot add Sr^{2+} and boil, converting 0.5 of HCO_3^- to CO_3^{2-} and volatilizing 0.5 as CO_2. Titrate unreacted Sr^{2+} in presence of $SrCO_3$ precipitate. To a second aliquot add NaOH, converting HCO_3^- to CO_3^{2-} , add Sr^{2+}, and titrate unreacted Sr^{2+} in presence of precipitate.
Cl^-	Precipitate AgCl, filter, exchange excess Ag(I) in filtrate with $Ni(CN)_4^{2-}$, and titrate liberated Ni(II).
ClO_3^-	Reduce with Fe(II), precipitate AgCl, and proceed as for halides.
ClO_4^-	Ignite with excess NH_4Cl, and proceed as for halides.
CrO_4^{2-}	(1) Precipitate $BaCrO_4$ and determine Ba(II) precipitate or excess Ba(II) in filtrate. (2) Reduce original sample or solution of $BaCrO_4$ precipitate to Cr(III) with ascorbic acid and determine Cr(III). (3) Precipitate $PbCrO_4$ and determine Pb(II) in precipitate.
F^-	(1) Precipitate CaF_2 and determine excess Ca(II) in filtrate. (2) Precipitate PbClF and determine Pb(II) in precipitate or excess Pb(II) in filtrate. (3) Precipitate LaF_3 and determine excess La(III) in filtrate. (4) Precipitate CeF_3 and determine excess Ce(III) in filtrate.
$Fe(CN)_6^{3-}$	Reduce with KI, discharge color with thiosulfate, and proceed as for $Fe(CN)_6^{4-}$.
$Fe(CN)_6^{4-}$	Precipitate $K_2Zn[Fe(CN)_6]$ and determine Zn(II) in precipitate of excess Zn(II) in filtrate.
I^-	(1) Precipitate PdI_2, react excess Pd(II) in filtrate with $Ni(CN)_4^{2-}$, and titrate liberated Ni(II). (2) Proceed as with Br^-.
IO_3^-	(1) Reduce with sulfite and determine I^- as above. (2) Precipitate $Pb(IO_3)_2$ from 50% ethanol or acetone and determine Pb(II) in precipitate.
MnO_4^-	Reduce to Mn(II) with hydroxylamine and titrate Mn(II) formed.
MoO_4^{2-}	(1) Precipitate $CaMoO_4$ and determine Ca(II) in precipitate. (2) Precipitate $PbMoO_4$ and determine Pb(II) in precipitate. (3) Reduce with hydroxylamine and determine Mo(III).
PO_4^{3-}	(1) Precipitate $MgNH_4PO_4 \cdot 6H_2O$ and determine Mg(II) in precipitate or excess Mg(II) in supernatant solution or filtrate. (2) Precipitate $ZnNH_4PO_4$ and determine Zn(II) in precipitate. (3) Precipitate $BiPO_4$ and determine Bi(III) in precipitate, or titrate excess Bi(III) in supernatant (1) Precipitate $MgNH_4AsO_4$ ($6H_2O$ and determine Mg in precipitate. solution or filtrate. (4) Precipitate $Th_3(PO_4)_4$ in acid solution and determine excess Th(IV) in supernatant solution.
$P_2O_7^{4-}$	Precipitate $Zn_2P_2O_7$ or $Mn_2P_2O_7$ and determine Zn(II) or Mn(II) in precipitate or excess Mn(II) in filtrate.

(Continued)

TABLE 4.32 Indirect Methods for Determination of Anions by Visual EDTA Titrations (*Continued*)

Anion determined	Procedure
ReO_4^-	Precipitate $TlReO_4$ from slightly acid or neutral solution, redissolve it in acid (add Br_2), and titrate Tl(III) in solution.
S^{2-} or HS^-	(1) Precipitate CuS and determine excess Cu(II) in filtrate or supernatant liquid. (2) Precipitate CdS by adding excess Cd(II)–EDTA complex and titrating liberated EDTA in filtrate. (3) Oxidize to sulfate and proceed as for that anion.
SO_3^{2-}	Oxidize to sulfate with Br_2 and proceed as for sulfate anion.
SO_4^{2-}	(1) Precipitate $BaSO_4$ and determine Ba(II) in precipitate or excess Ba(II) in filtrate or supernatant liquid. (2) Precipitate $PbSO_4$ from 30% ethanol and determine Pb(II) in precipitate or excess Pb(II) in filtrate. (3) Precipitate $BaSO_4$ homogeneously by adding Ba(II)–EDTA complex and slowly acidifying; titrate liberated EDTA in filtrate.
$S_2O_3^{2-}$	Oxidize to sulfate with Br_2 water and proceed as for that anion.
$S_2O_8^{2-}$	Reduce by boiling the alkali metal salts or by treating ammonium salt with Zn + HCl, and determine the sulfate produced.
SCN^-	(1) Treat sample with Cu(II) giving CuSCN precipitate and determine excess Cu(II) in filtrate. (2) Oxidize to sulfate and proceed as for that anion.
SeO_3^{2-}	Oxidize by boiling with excess $KMnO_4$ and proceed as for selenate below.
SeO_4^{2-}	Precipitate $PbSeO_4$ at pH 2–3 in the presence of 30% ethanol, and determine Pb(II) in precipitate.
VO_3^-	Reduce to VO^{2+} with hydroxylamine or ascorbic acid in an acid solution and determine VO^{2+}.
WO_4^{2-}	Precipitate $CaWO_4$ and determine Ca(II) in precipitate.

Alternatively, grind 0.1 g of murexide with 10 g of ACS reagent grade sodium chloride; use about 50 mg of the mixture for each titration.

4.6.6.8 Pyrocatechol Violet Indicator Solution. Dissolve 0.1 g of the solid dyestuff in 100 mL of water.

SECTION 5
CHROMATOGRAPHIC METHODS

5.1 CHROMATOGRAPHIC TECHNIQUES

Chromatographic techniques have probably the widest and most versatile applications in analytical chemistry. They include a number of distinct separation techniques differing from each other. In a broader terminology chromatography is a technique for separating a sample into various fractions, and then measuring or identifying the fractions in some manner. The components to be separated are distributed between two mutually immiscible phases. The heart of any chromatography is the stationary phase, which is sometimes a solid but most commonly a liquid. The stationary phase is attached to a support, a solid inert material. The sample, often in vapor form or dissolved in a solvent, is moved across or through the stationary phase. It is pushed along by a liquid or a gas—the mobile phase. As the mobile phase moves through the stationary phase, the sample components undergo a series of exchanges (partitions) many times between the two phases. What is exploited are differences in the chemical and physical properties of the components in the sample. These differences govern the rate of movement (called *migration*) of the individual components. When a sample component has emerged from the outlet of a chromatograph, it is said to have been *eluted*. Components emerge from the system ideally as gaussian-shaped peaks and in the order of increasing interaction with the stationary phase. Separation is obtained when one component is retarded sufficiently to prevent overlap with the peak of an adjacent neighbor.

5.1.1 Classification of Chromatographic Methods

The mobile phase can be a gas or a liquid, whereas the stationary phase can only be a liquid or a solid. When the stationary phase is contained in a column, the term *column chromatography* applies. The stationary phase can also occupy a plane surface, such as filter paper. This is called *planar chromatography* and includes thin-layer and paper chromatography and electrophoresis.

Column chromatography can be subdivided into *gas chromatography* (GC) and *liquid chromatography* (LC) to reflect the physical state of the mobile phase. If the sample passing through the chromatograph is in the form of a gas, the analytical technique is known as gas chromatography. Gas chromatography comprises gas–liquid chromatography (GLC) and gas–solid chromatography (GSC), names that denote the nature of the stationary phase.

Liquid-column chromatography embraces several distinct types of interaction between the liquid mobile phase and the various stationary phases. When the separation involves predominantly a simple partition between two immiscible liquid phases, one stationary and one mobile, the process is called *liquid–liquid chromatography* (LLC). In *liquid–solid* (or *adsorption*) *chromatography* (LSC) physical surface forces are mainly involved in the retentive ability of the stationary phase. Ionic or charged species are separated in *ion chromatography* (IC) by selective exchange with counterions of the stationary phase; this may be by *ion-exchange chromatography* (IEC), *ion-pair chromatography*, or *ion exclusion chromatography*. In columns filled with porous polymers, components may be separated by *exclusion chromatography* (EC) [also called *gel-permeation chromatography* (GPC)]; separation is based largely on molecular size and geometry.

5.1.2 Terms and Relationships in Chromatography

The chromatographic behavior of a solute can be described by its retention volume V_R (or the corresponding retention time t_R) and the partition ratio (or capacity ratio) k'. Distance on the recorder chart can be converted to time by multiplying it by the chart speed and then to volume by multiplying the time by the flowrate. When constant flow is assumed, retention time and volume can be used interchangeably.

5.1.2.1 Retention Time. The time required by the mobile phase to convey a solute from the point of injection onto the stationary phase, through the stationary phase, and to the detector (to the apex

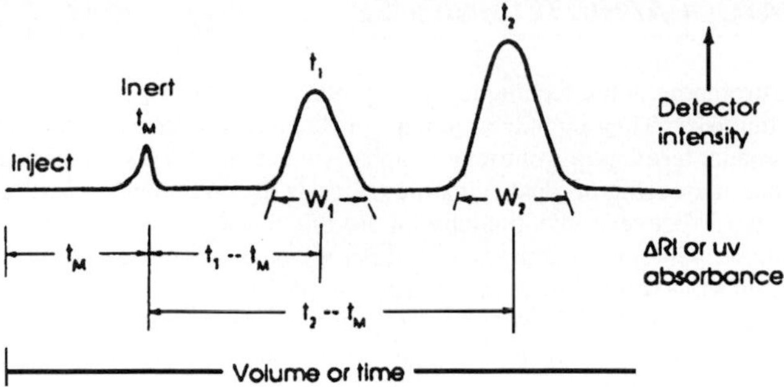

FIGURE 5.1 Chromatogram illustrating retention times and band widths *W* of a nonretained solute t_M and two retained materials 1 and 2.

of the solute peak in Fig. 5.1) is defined as the retention time. The retention volume is the retention time multiplied by the volumetric flowrate, F_c, discussed later:

$$V_R = t_R F_c \tag{5.1}$$

5.1.2.2 Nonretained Solute Retention Time. The quantity t_M or t_0 (see Fig. 5.1) is the time required for a material to pass through the system without being adsorbed or partitioned, that is, the transit time of a nonretained solute through the column. It represents the time for the average mobile phase molecule to pass through the stationary phase, that is, traverse the column or planar phase. When converted to volume, V_M, it represents the void volume (or dead space or holdup volume) of the column. The column dead volume or time is often recognized as the first disturbance in the baseline, but in LC separations there is often no baseline disturbance at t_M.

In practice, t_M is obtained by injecting a solute with $k' = 0$ (that is, all the solute is in the mobile phase, none partitions into the stationary phase). In gas chromatography, and when using a thermal conductivity detector, air can be injected to obtain t_M. For other GC detectors, the peak of a solute whose boiling point is 90°C or more below the column temperature will give an estimate of t_M; methane is often used for this purpose with a flame ionization detector.

5.1.2.3 Adjusted Retention Time. The retention time, V'_R, or volume, V'_R, that has the mobile phase time (or volume) subtracted out is of interest for theoretical work.

$$t'_R = t_R - t_M \quad \text{or} \quad V'_R = V_R - V_M \tag{5.2}$$

5.1.2.4 Volumetric Flowrate. The volumetric flowrate F_c, in terms of the column parameters, is:

$$F_c = \frac{\pi d^2}{4} \epsilon \frac{L}{t_M} \tag{5.3}$$

where d = inner diameter of column
L = column length
ϵ = total porosity of column packing

For solid packings the total porosity is 0.35 to 0.45, whereas for porous packings it is 0.70 to 0.90. In capillary columns the total porosity is unity.

From Eq. (5.3), the proper flowrate for columns of differing diameters can be approximated by assuming that the packing densities of the two columns are the same. If so, then

$$(F_c)_2 = (F_c)_1 \frac{d_2}{d_1} \tag{5.4}$$

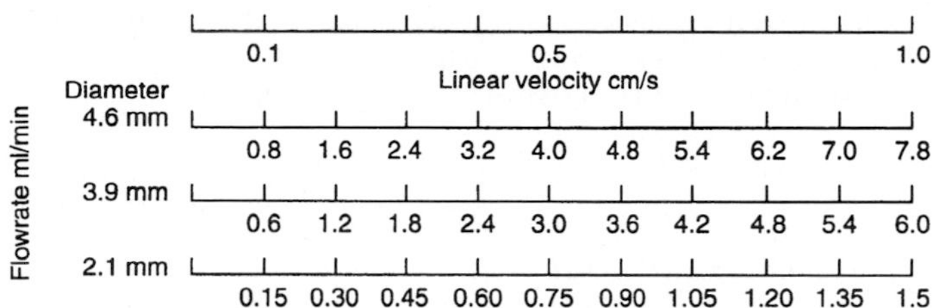

FIGURE 5.2 Comparison of linear velocity with flowrate for columns with different internal diameters.

For example, an analysis has been performed on a 4.6-mm-inside-diameter (i.d.) column at 2 $mL \cdot min^{-1}$. If the same linear velocity is desired on a 9.4-mm-i.d. preparative column, the appropriate flowrate is $2 \times (9.4/4.6)^2$ or 8.4 $mL \cdot min^{-1}$. Lengthening the column proportionally increases efficiency and analysis time but does not affect the flowrate.

5.1.2.5 Velocity of the Mobile Phase. The average linear velocity u of the mobile phase

$$u = \frac{L}{t_M} \tag{5.5}$$

is measured by the transit time of a nonretained solute through the column.

Example 5.1 The linear velocity was 43 $cm \cdot s^{-1}$ through a 15-m column. What is the value of t_M?

$$t_M = \frac{L}{u} = \frac{1500 \text{ cm}}{43 \text{ cm} \cdot \text{s}^{-1}} = 34.8 \text{ s}$$

At the same flowrate, columns of varying i.d.'s operate at different linear velocities. To compare performance objectively, it is necessary to operate the columns at the same linear velocity. Figure 5.2 gives the relationship between linear velocity and flowrate for selected diameters of packed columns (assuming that each column has the same packing density).

5.1.2.6 Partition Coefficient. The partition coefficient K is given by the ratio of the solute concentration in the stationary (liquid) phase to that in the mobile (gas or liquid) phase:

$$K = \frac{\text{concentration of solute in mobile phase}}{\text{concentration of solute in stationary phase}} \tag{5.6}$$

It is a thermodynamic quantity that depends on the temperature and on the change in the standard free energy of the solute going from the mobile to the stationary phase.

5.1.2.7 Partition Ratio. The partition ratio (or capacity factor) k' is a measure of how well the sample molecule is retained by the column during an isocratic separation. It is the additional time a solute takes to elute, as compared with an unretained solute (for which $k' = 0$), divided by the elution time of an unretained solute:

$$k' = \frac{t_R - t_M}{t_M} = \frac{t'_R}{t_M} = \frac{KV_S}{V_M} \tag{5.7}$$

where V_S and V_M are, respectively, the volume of the stationary and the mobile phases. For diagnostic purposes, k' values accurate to the nearest integer are satisfactory.

Retention times are also related to k' by the equation

$$t_R = t_M(1 + k') = \frac{L}{u}(1 + k') \tag{5.8}$$

The partition ratio k' and the solute retention time are affected by solvent composition (if a liquid), the stationary phase, and the temperature (via K) at which the separation occurs. Thick stationary films increase solute resolution by increasing the relative time each solute spends in the stationary phase. This is accomplished by increasing the stationary phase film thickness d_f of the column. Equation (4.9) illustrates the relationship between the phase ratio β of the column, the partition coefficient, and the partition ratio.

$$K = \beta k' = \frac{r}{2d_f}k' \tag{5.9}$$

The increased retention obtained on a very thick film column can be reduced by raising the column temperature. This can be of benefit to the chromatographer if an analysis requires subambient column temperatures on standard film thickness.

The solute retention time is also dependent on the mobile phase velocity and the column length. Remember that although t_M changes when the flowrate is changed, k' remains constant.

Example 5.2 If $k' = 2.25$ for the material in Example 5.1, what are the values of t_R and t'_R?

$$t_R = t_M(1 + k') = (34.8 \text{ s})(1 + 2.25) = 113 \text{ s (or 1.88 min)}$$

$$t'_R = t_R - t_M = 113 - 34.8 = 78 \text{ s (or 1.31 min)}$$

5.1.2.8 Column Selectivity or Relative Retention. The relative retention α, which is a selectivity term, is given variously by

$$\alpha = \frac{k'_2}{k'_1} = \frac{t'_{R,2}}{t'_{R,1}} \tag{5.10}$$

where solute 1 elutes before solute 2. The larger the relative retention, the better the resolution between adjacent analytes. The relative retention is dependent on (1) the nature of the stationary and mobile phases and (2) the column operating temperature. For a given column, α is only a function of the column temperature.

5.1.2.9 Column Efficiency: Plate Height and Plate Number. The most common measure of the efficiency of a chromatographic system is the plate number N. Efficiency governs how narrow the peaks will be when elution occurs. It is usually measured in terms of the number of plates N that a column can deliver for a given peak. The plate height H is given by L/N, where L, the column length, represents the distance a solute moves while undergoing one partition. Plate height is a convenient measure when making comparisons among columns of different length.

The *effective plate number* N_{eff} reflects the number of times the solute partitions between the stationary and mobile phases during its passage through the column.

$$N_{\text{eff}} = \frac{L}{H} = \left(\frac{t'_R}{\sigma}\right)^2 \tag{5.11}$$

where σ^2 is the band variance in time units.

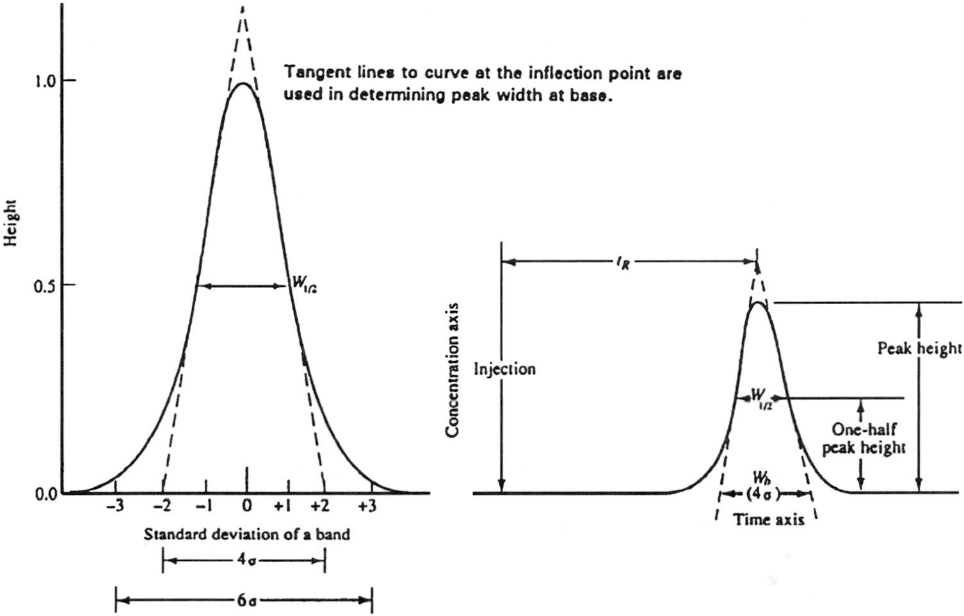

FIGURE 5.3 Profile of a solute band and evaluation of a chromatographic band for column efficiency.

Under ideal operating conditions the profile of a solute band closely approaches that given by a gaussian distribution curve (Fig. 5.3). At the baseline the span of 4σ includes 95% of the solute band; that of 6σ encompasses 99% with only 0.5% left out on each wing.

The plate number can be measured for a test chromatogram and compared with the column manufacturer's value under the same working conditions. The width at the base of the peak, W_b, is ascertained experimentally from the intersections of tangents to the inflection points with the baseline. It is equal to four standard deviations (thus $\sigma = W_b/4$) and

$$N_{\text{eff}} = 16\left(\frac{t'_R}{W_b}\right)^2 \tag{5.12}$$

Oftentimes it is preferable to measure the width at half the peak height $W_{1/2}$, particularly if the peak is not symmetrical. Then

$$n_{\text{eff}} = 5.54\left(\frac{t'_R}{W_{1/2}}\right)^2 \tag{5.13}$$

The apparent (or theoretical) plate number N_{app} differs from the foregoing treatment only in that the unadjusted retention time (or volume) is used; it is a measure of the efficiency of the entire system, not just the efficiency of the column. Typically, instrument dead-volume contribution is small relative to a standard packed column, yet as column internal diameters and lengths are reduced, instrument contribution becomes significant.

It was noted earlier that the adjusted retention time has more theoretical significance than the retention time. Consequently, the effective plate number is often a better parameter for use in comparing columns, particularly when comparing packed columns to open tubular columns.

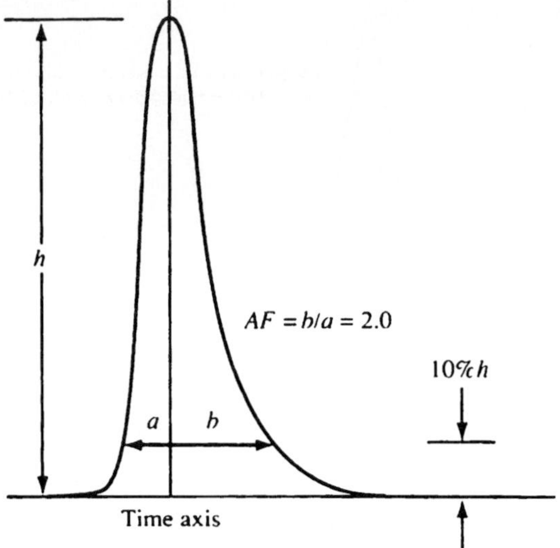

FIGURE 5.4 **Band asymmetry.**

Columns with large N values will produce narrow peaks and better resolution than columns with lower N values. When measuring N, k' should be at least 3 and preferably greater than 5. The effective plate number will increase as k' increases, and it will approach N_{app} at high k' values where t_M is no longer of significant size compared to t_R.

Example 5.3 What is the effective plate number of a solute whose $t_R = 9.55$ min when a 1.5-m column is operated at 25°C? The value of $W_b = 0.71$ min. On the same chromatogram the retention time of methane was 0.39 min.

To find t_M, use the retention time of methane (b.p. – 182.5°C). Consequently,

$$t'_R = 9.55 - 0.39 = 9.16 \text{ min}$$

$$N_{eff} = 16\left(\frac{9.16}{0.71}\right)^2 = 2660$$

and

$$H = 1500 \text{ mm}/2660 = 0.56 \text{ mm}$$

5.1.2.10 Band Asymmetry. The peak *asymmetry ratio* or *tailing factor* TF is defined as the ratio of the peak half-widths at a given peak height, usually at 10% of peak height. As shown in Fig. 5.4,

$$\text{TF} = \frac{b}{a} \tag{5.14}$$

A symmetrical peak will have a value of 1 and tailed peaks a value of greater than 1. A *fronted* peak (one with a leading edge and $a > b$) will have a value less than 1.

When the asymmetry factor lies outside the range of 0.95 to 1.15 for a peak of $k' = 2$, the apparent plate count for a column, as calculated by Eq. (5.12) or (5.13) will be too high and should be calculated by the Dorsey–Foley[1] equation:

$$N_{eff} = \frac{41.7(t'_R/W_{0.1})^2}{(b/a) + 1.25} \tag{5.15}$$

[1] J. P. Foley and J. G. Dorsey, *Anal. Chem.* **55**:730 (1983).

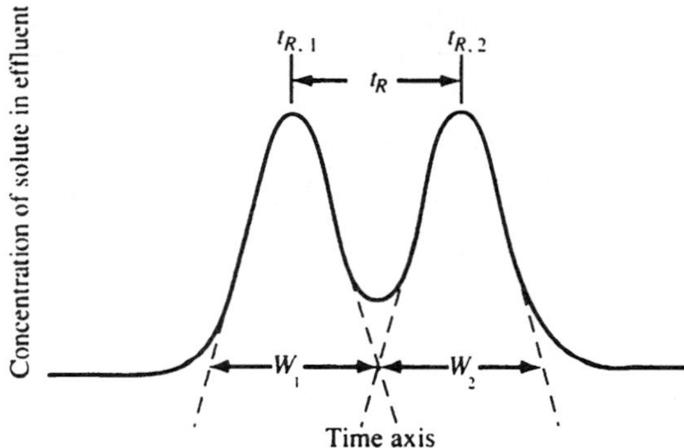

FIGURE 5.5 **Peak resolution.**

5.1.2.11 Resolution. The degree of separation or resolution, Rs, of two adjacent bands is defined as the distance between their peaks (or centers) divided by the average bandwidth, as shown in Fig. 5.5. When measured in time units

$$\mathrm{Rs} = \frac{t_{R,2} - t_{R,1}}{0.5(W_2 + W_1)} \tag{5.16}$$

For simplicity, the expression is often reduced to

$$\mathrm{Rs} = \frac{t_{R,2} - t_{R,1}}{W_2} \tag{5.17}$$

Strictly speaking, Eq. (5.16) or (5.17) is only valid when both peaks have the same height. A practical approach has been provided by Snyder[2]; his paper includes computer-drawn pairs of peaks for a variety of resolutions and peak-height ratios. Compare a given experimental chromatogram with those provided in the paper and choose the best match, assigning the value of resolution to the experimental chromatogram.

The foregoing equations define resolution in a given situation, but they do not relate resolution to the conditions of separation nor do they suggest how to improve resolution. For the latter purpose, the fundamental resolution equation is

$$\mathrm{Rs} = \frac{1}{4}\left(\frac{\alpha - 1}{\alpha}\right)\left(\frac{k'}{1 + k'}\right)\left(\frac{L}{H}\right)^{1/2} \tag{5.18}$$

The number of plates required ($N = L/H$) for a given separation of two specific (usually adjacent) solutes varies inversely with the relative retention (Table 5.1) and with the partition ratio. Increasing k' decreases the magnitude of the $(k' + 1)/k'$ (Table 5.2).

A column may have adequate selectivity but may exhibit poor efficiency (adjacent bands well resolved but not sufficiently narrow). On the other hand, bands may be narrow but poorly resolved (good efficiency but poor selectivity).

Example 5.4 What is the resolution between *trans*-2-butene ($t_R = 3.12$ min) and *cis*-2-butene ($t_R = 3.43$ min; $W_b = 0.31$ min)?

$$\mathrm{Rs} = \frac{t_{R,cis} - t_{R,trans}}{W_{b,cis}} = \frac{343\ \text{min} - 3.12\ \text{min}}{0.31\ \text{min}} = 1.00$$

[2] L. R. Snyder, *J. Chromatogr. Sci.* **10**:200 (1972).

TABLE 5.1 Values Related to Relative Retention

α	$\left(\frac{\alpha}{\alpha-1}\right)^2$	N_{req} for $Rs = 1.5$ and $k' = 2$	L_{req}, m for $H = 0.6$ mm
1.01	10 201	826 281	495
1.02	2 601	210 681	126
1.03	1 177	95 377	52
1.04	676	54 756	33
1.05	441	35 721	21
1.10	121	9 801	5.8
1.15	58	4 418	2.6
1.20	36	2 916	1.7
1.25	25	2 025	1.2
1.30	19	1 514	1.0

This is not a baseline resolution of 1.5, but because the resolution is proportional to the square root of the column length, increase the column length 2.25 times (square root of 2.25 equals 1.5).

$$\frac{(\mathrm{Rs})_2}{(\mathrm{Rs})_1} = \left(\frac{1.5}{1.0}\right)^2 = 2.25$$

Section 5.2.8 and 5.3.4 will deal with optimization of operating conditions in gas chromatography and liquid-column chromatography, respectively.

5.1.2.12 Trennzahl (Separation) Number. The *Trennzahl* (TZ) or *separation number* in English, has been used to express the separation efficiency of a column. It is the number of peaks that can be theoretically resolved between two consecutive members of a paraffin homologous series.

$$\mathrm{TZ} = \left(\frac{t_{R,b} - t_{R,a}}{W_{0.5,a} + W_{0.5,b}}\right) - 1 \tag{5.19}$$

where a and b are adjacent homologs, and $W_{0.5}$ are the bandwidths at half height. The two peaks are commonly two straight-chain alkanes differing by one methylene group. Resolution and a TZ value are related by the expression

$$\mathrm{Rs} = 1.177\ (\mathrm{TZ} + 1) \tag{5.20}$$

TABLE 5.2 Number of Theoretical Plates Required to Achieve a Given Resolution

Capacity factor, k'	Rs = 1.5, $\alpha = 1.05$	Rs = 1.5, $\alpha = 1.10$	Rs = 1.0, $\alpha = 1.05$	Rs = 1.0, $\alpha = 1.10$
0.2	571 500	156 800	254 000	69 700
0.5	142 900	39 200	63 500	17 400
1.0	63 500	17 400	28 200	7 700
2.0	35 700	9 800	15 900	4 400
5.0	22 900	6 300	10 200	2 800
10.0	19 200	5 300	8 500	2 300
20.0	17 500	4 800	7 800	2 100

The concept is useful for temperature programmed runs in which plate number or effective plate number would be meaningless. Regardless of operating conditions, a TZ number of 5 means that 5 analytes eluting adjacent to each other (Rs = 1) can be resolved between two consecutive paraffins in that region of the chromatogram.

Giddings[3] has calculated peak capacities for different chromatographic techniques.

5.1.2.13 The van Deemter Equation. Column efficiency, unlike selectivity, is a function of the average mobile phase velocity, the column i.d. or average particle diameter, the type of carrier gas (in GC), as well as the type of solute and its retention and the stationary phase film thickness. The van Deemter equation relates the plate height to the mobile phase velocity and experimental variables. There are two *C* terms—one for mass transfer in the stationary phase and the other for mass transfer in the mobile phase. Hawkes[4] has evaluated the various rate equations that have been proposed and presented a modern summary.

$$H = A + \frac{B}{u} + C_{\text{stationary}}u + C_{\text{mobile}}u \tag{5.21}$$

The *A* term is defined as

$$A = \lambda d_p \tag{5.22}$$

where d_p is the particle diameter and λ is a function of the packing uniformity and column geometry. The *A* term (a multipath term) arises from the inhomogeneity of flow velocities and path lengths around packing particles. In modern rate theories, the *A* term is coupled with the C_{mobile} term. In open tubular columns, the *A* term is zero.

The *B* term is defined as

$$B = 2\gamma D_M \tag{5.23}$$

where γ is an obstruction factor that recognizes that axial diffusion is hindered by the packed bed structure, and D_M is the solute diffusion coefficient in the mobile phase. Molecules will diffuse from the region of high concentration in the center of the zone to the region of lower concentration in proportion to D_M. The effect is a function of time; as the time increases, the zone spreads and its maximum is lowered. This results in zone broadening as the analyte proceeds through the system. In open tubular columns γ is unity; in packed columns its value is 0.6 to 0.7 for glass beads, 0.74 for Chromosorb W, 0.46 for Chromosorb P, and 0.7 to 0.9 for paper. In many chromatographic operations, the mobile phase velocity is sufficiently large that the *B* term is not of major importance.

The $C_{\text{stationary}}$ term is proportional to d_f/D_S, where d_f is the thickness of the stationary phase film and D_S is the diffusion coefficient of the solute in the stationary phase.

The C_{mobile} term is proportional to d_p/D_M, where d_p is the particle diameter of the packing material and D_M is the diffusion coefficient of the solute in the mobile phase. An inverse plot of the van Deemter equation that illustrates the column efficiency versus flow rate (linear velocity) is shown in Fig. 5.6 for a particular analyte.

5.1.2.14 Time of Analysis and Resolution. The retention time is related to the plates required for a given resolution, the partition ratio, the plate height, and the linear velocity of the mobile phase:

$$t_R = N_{\text{req}}(1+k')\left(\frac{H}{u}\right) \tag{5.24}$$

or

$$t_R = 16\,\text{Rs}^2\left(\frac{\alpha}{\alpha-1}\right)\frac{(1+k')^3}{(k')^2}\left(\frac{H}{u}\right) \tag{5.25}$$

[3] J. C. Giddings, *Anal. Chem.* **39**:1027 (1967).

[4] S. J. Hawkes, *J. Chem. Educ.* **60**:393 (1983).

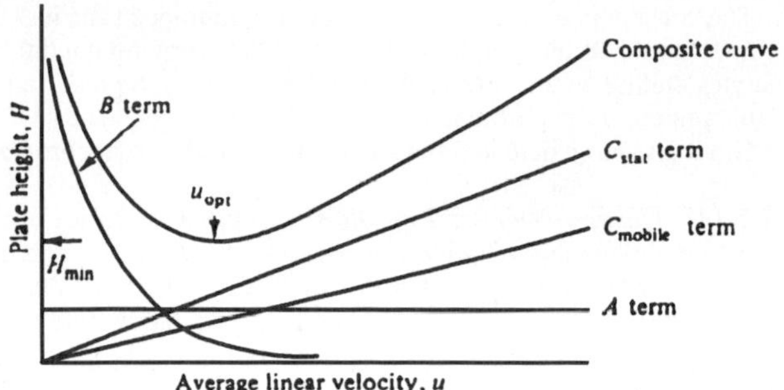

FIGURE 5.6 Typical *H/u* (van Deemter) curve for a gas-chromatographic column.

Each column has an optimum mobile phase velocity. Now t_R is a minimum when $k' = 2$, that is, when $t_R = 3t_M$. As mobile-phase velocities increase above optimum, efficiency is gradually lost. However, speed of analysis is increased (shorter retention times). Actually there is little increase in analysis time when k' lies between 1 and 10. A twofold increase in the mobile-phase velocity roughly halves the analysis time (actually it is the ratio H/u that influences the analysis time). The ratio H/u can be obtained from the experimental plate height–velocity graph. Thus, a trade-off is possible; efficiency can be sacrificed for speed when the column has excess resolving power (large gaps between relatively sharp peaks).

5.1.3 Quantitative Determinations

Chromatographic detectors that respond to the concentration of the solute yield a signal that is proportional to the solute concentration that passes through the detector. For these detectors the peak area is proportional to the mass of the component and inversely proportional to the flowrate of the mobile phase. Thus, the flowrate must be kept constant if quantitation is to be performed.

In differential detectors that respond to mass flowrate the peak area exhibits no dependency on the flowrate of the mobile phase.

5.1.3.1 Peak Area Integration

5.1.3.1.1 Peak Height. Measurements are made from the peak maximum to the baseline. Although inherently simple, peak heights are sensitive to small changes in operating conditions and sample injection. The precision is better than measuring the peak area, particularly of narrow peaks.

5.1.3.1.2 Height Times Width at Half-Height. These measurements are based on the assumed triangular shape of an ideal gaussian peak. Draw the baseline of the peak and measure the height of the peak. Position the measuring scale parallel to the baseline at half the height and measure the bandwidth at this position. The area is the product of the height times the width.

5.1.3.1.3 Disk Integrator. Good accuracy, independent of peak shape, is provided by a disk integrator. Figure 5.7 illustrates how to read the integrator trace. A full stroke of the "sawtooth" pattern (either up or down) represents 100 counts. Every horizontal division crossed by the trace has a value of 10. Values that are less than 10 are estimated. On some models the space between "blips" that project slightly above the uppermost horizontal line is equivalent to 600 counts, making it easier to count the number of full strokes.

5.1.3.2 Computing Integrator. A computing integrator automatically determines the peak area bound by the point at which the chromatographic trace leaves the baseline and the point at which it returns to the baseline. In the case of overlapping peaks, special algorithms allot areas to each component. During isothermal runs the software can automatically alter the slope sensitivity with time. This allows both sharp narrow peaks and low flat peaks to be measured with equal precision.

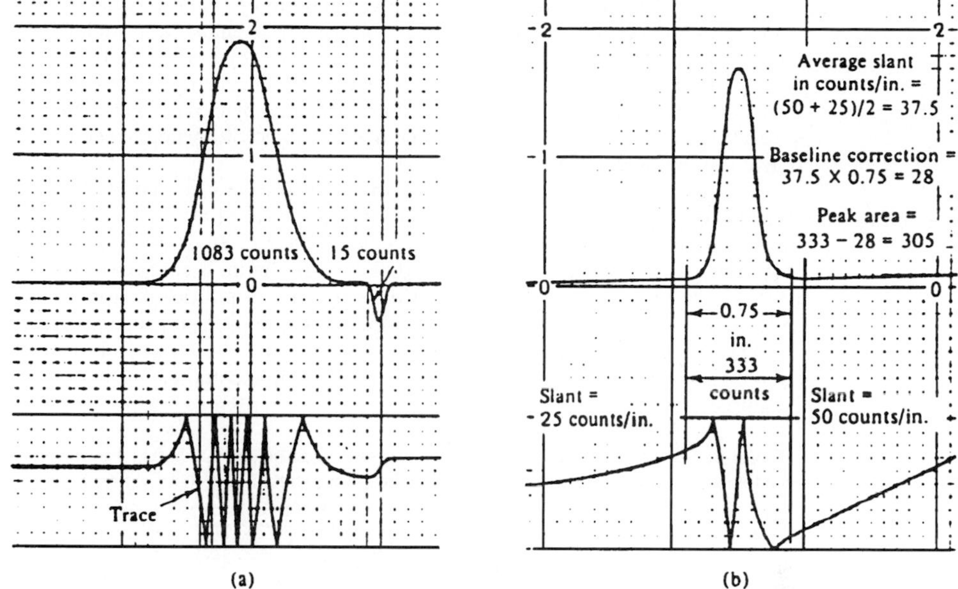

FIGURE 5.7 (*a*) **Estimation of peak areas with ball-and-disk integrator.** (*b*) **Method for handling baseline correction.**

5.1.3.3 Evaluation Methods. Once the peak height or peak areas have been measured, there are four principal evaluation methods that can be used to translate these numbers to amounts of solute.

5.1.3.3.1 Calibration by Standards. Calibration curves for each component are prepared from pure standards, using identical injection volumes and operating conditions for standards and samples. The concentration of solute is read from its calibration curve or, if the curve is linear,

$$X = K(\text{area})_x \tag{5.26}$$

where X is the concentration of solute and K is the proportionality constant (slope of the calibration curve). In this evaluation method only the area of the peaks of interest need to be measured. However, the method is very operator-dependent and requires good laboratory technique.

Relative response factors must be considered when converting area to volume and when the response of a given detector differs for each molecular type of compound.

Example 5.5 The relative response factors for *o*-xylene and toluene (relative to the value for benzene, assigned unity) were found to be 0.570 and 0.793, respectively. An unknown mixture of these three solutes gave these peak heights (in millimeters): benzene, 98; *o*-xylene, 87; and toluene, 86.

$$\text{Total adjusted response} = \frac{H_{\text{bz}}}{1.00} + \frac{H_{\text{xyl}}}{0.570} + \frac{H_{\text{tol}}}{0.793}$$

$$= \frac{98}{1.00} + \frac{87}{0.570} + \frac{86}{0.793} = 98 + 153 + 108 = 359$$

For benzene: (98/359)(100) = 27.3%
For *o*-xylene: (153/359)(100) = 42.6%
For toluene: (108/359)(100) = 30.1%

5.1.3.3.2 Area Normalization. For this method to be applicable, the entire sample must have eluted, all components must be separated, and each peak must be completely resolved. The area under each peak is measured and corrected, if necessary, by a response factor as described. All the peak areas are added together. The percentage of individual components is obtained by multiplying each individual calculated area by 100 and then dividing by the total calculated area. Results would be invalidated if a sample component were not able to be chromatographed on the column or failed to give a signal with the detector.

5.1.3.3.3 Internal Standard. In this technique a known quantity of the internal standard is chromatographed, and area versus concentration is ascertained. Then a known quantity of the internal standard is added to the "raw" sample prior to any sample pretreatment or separator operations. The peak area of the standard in the sample run is compared with the peak area when the standard is run separately. This ratio serves as a correction factor for variation in sample size, for losses in any preliminary pretreatment operations, or for incomplete elution of the sample. The material selected for the internal standard must be completely resolved from adjacent sample components, must not interfere with the sample components, and must never be present in samples.

Example 5.6 Assume that 50.0 mg of internal standard is added to 0.500 g of the sample. The resulting chromatogram shows five components with areas (in arbitrary units) as follows: $A_1 = 30$, $A_2 = 18$, $A_{\text{std}} = 75$, $A_3 = 80$, $A_4 = 45$, and the area sum equals 248. The amount of component 3 in the sample is

$$W_3 = W_{\text{std}}\left(\frac{A_3}{A_{\text{std}}}\right) = 0.0500\left(\frac{80}{75}\right) = 0.0533 \text{ g}$$

Percent component 3: Although component 3 appeared to be a major component, it represents only about 10% of the total sample.

$$\frac{0.0533 \text{ g}}{0.0500 \text{ g}} \times 100 = 10.66\%$$

A large part of the sample does not appear on the chromatogram, as would be the case if the organic mixture included some inorganic salts.

Usually it will be necessary to ascertain the ratio of response factors (such as K_{std}/K_3). When this is so,

$$\frac{K_{\text{std}}}{K_3} = \frac{W_3 A_{\text{std}}}{W_{\text{std}} A_3}$$

5.1.3.3.4 Standard Addition. If only a few samples are to be chromatographed, it is possible to employ the method of standard addition(s). The chromatogram of the unknown is recorded. Then a known amount of the analyte(s) is added, and the chromatogram is repeated using the same reagents, instrument parameters, and procedures. From the increase in the peak area (or peak height), the original concentration can be computed by interpolation. The detector response must be a linear function of analyte concentration and yield no signal (other than background) at zero concentration of the analyte. Sufficient time must elapse between addition of the standard and actual analysis to allow equilibrium of added standard with any matrix interferant.

If an instrumental reading (area or height), R_x, is obtained from a sample of unknown concentration x and a reading R_1 is obtained from the sample to which a known concentration a of analyte has been added, then x can be calculated from the relation

$$\frac{x}{x+a} = \frac{R_x}{R_1} \tag{5.27}$$

A correction for dilution must be made if the amount of standard added changes the total sample volume significantly. It is always advisable to check the result by adding at least one other standard.

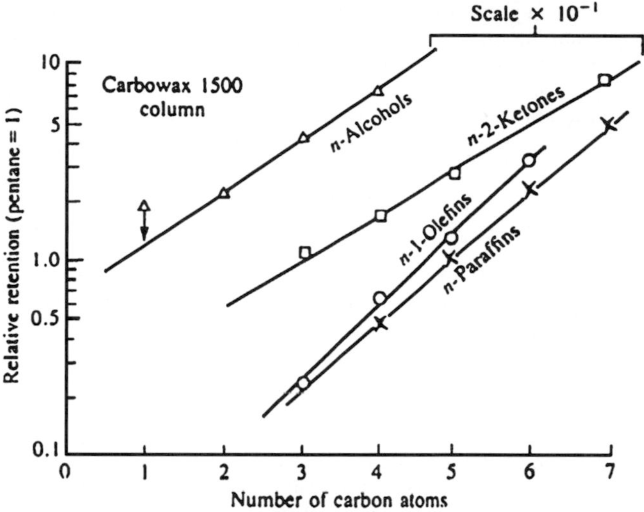

FIGURE 5.8 Plot of retention time (log scale) vs. number of carbon atoms for several homologous series of compounds.

Additions of analyte equal to twice and to one-half the amount of analyte present in the original sample are optimum statistically.

5.1.4 Sample Characterization

5.1.4.1 Use of Retention Data. The retention time under fixed operating conditions is a constant for a particular solute and can, therefore, be used to identify that solute. This is accomplished by comparing the retention times of the sample components with the retention times of pure standards.

For isocratic (HPLC) or isothermal (GC) elution, retention times usually vary in a regular and predictable fashion with repeated substitution of some group i into the sample molecule as, for example, the $—CH_2—$ group in a homologous series.

$$\log t_{R,i} = mN_i + \text{constant} \tag{5.28}$$

where m is a constant and N_i is the number of repeating groups (or the number of carbon atoms) in the homologous series (Fig. 5.8). Partition ratio k' values are actually superior to retention data because k' values are not influenced by mobile-phase flowrate or column geometry.

A suspected solute can be verified by "spiking" the sample with a added amount of the pure solute. Only the peak height should vary if the two compounds are the same. Of course, pure standards must be available if spiking is to be used.

5.1.4.2 Kovats Retention-Index System.[5] There is no single standard to which data have been ratioed, and consequently, there are no tabulations of relative retention data in the literature. Kovats suggested that a series of standards be used, and he proposed the *n*-paraffins. By definition, the retention index (RI) for these hydrocarbons are assigned a reference number that is 100 times the number of carbon atoms in the molecule. Thus, the retention indices of butane, pentane, hexane, and octane are 400, 500, 600, and 800, respectively, regardless of the column used or the operating conditions. (However, the exact conditions and column must be specified, such as liquid loading,

[5] L. S. Ettre, "The Kovats Retention Index System," *Anal. Chem.* **36**:31A (1964).

particular support used, and any pretreatment.) When they are run isothermally on a specified column, and the logarithms of their adjusted retention times (flowrate and pressure drop must be constant) are plotted versus carbon number, a straight line results, as expected from Eq. (5.27). If any other compound is run under the same conditions, its index can be read from the graph. Alternatively, the index can be calculated as follows:

$$I = 100\left[\frac{(\log t'_R)_{\text{unk}} - (\log t'_R)_x}{(\log t'_R)_{x+1} - (\log t'_R)_x}\right] + 100x \tag{5.29}$$

where x stands for the paraffin with x carbons and eluting just before the unknown, and $x + 1$ for the paraffin with $x + 1$ carbons and eluting just after the unknown.

The retention index has become the standard method for reporting GC data. By definition, the members of any homologous series should differ from each other by 100 units just as the standards do. This relationship is not always exact, but it is very popular. Other homologous series have also been used as standards in specific applications.

5.1.4.3 Chromatographic Cross Check. The ability to identify a sample component by means of retention times (or k' values) is significantly enhanced by the use of different stationary phases (in gas chromatography). The use of a polar liquid phase in one column and a nonpolar liquid phase in a second column will provide much information. If the retention times for the two stationary phases are plotted against each other, lines that radiate from the origin are obtained (one for each homologous series). If the logarithms of the retention times are plotted against each other, a corresponding series of parallel lines is obtained with points spaced linearly according to the number of repeating groups (or carbon number) for each homologous family (Fig. 5.9).

Retention indexing systems in gas chromatography, such as the Kovats indexing system and use of Rohrschneider or McReynolds constants, are also useful for qualitative analysis. These will be discussed in Sec. 5.2.4.2

5.1.4.4 Identification by Ancillary Techniques. None of the foregoing techniques based on retention times is definitive because many compounds have similar retention times. However, structural information can be independently obtained from the several spectroscopic techniques. This has led to hyphenated techniques, such as gas chromatography–mass spectroscopy and gas chromatography–infrared spectroscopy. If spectroscopic reference spectra are available, conformation of solute structure is likely.

5.1.5 Derivatization Reactions[6]

The bulk of analytical derivatization reactions used for chromatography fall into three general reaction types: *silylation, alkylation*, and *acylation*. These reactions are employed to (1) improve the thermal stability of compounds, particularly compounds that contain certain polar functional groups, (2) adjust the volatility of a compound, and (3) introduce a detector-oriented tag into a molecule.

For analysis by gas chromatography, compounds containing functional groups with active hydrogen atoms, such as —COOH, —OH, —NH, and —SH, are of primary concern because of their tendency to form intermolecular hydrogen bonds that affects the inherent volatility of compounds containing them, their tendency to interact deleteriously with column packing materials, and their thermal stability. The most common derivatization methods in HPLC are intended to enhance detectability by ultraviolet absorption, fluorescence, or electrochemistry.

[6] D. R. Knapp, *Handbook of Analytical Derivatization Reactions*, Wiley, New York, 1979; K. Blau and J. M. Halket, eds., *Handbook of Derivatives for Chromatography*, Wiley, New York, 1993.

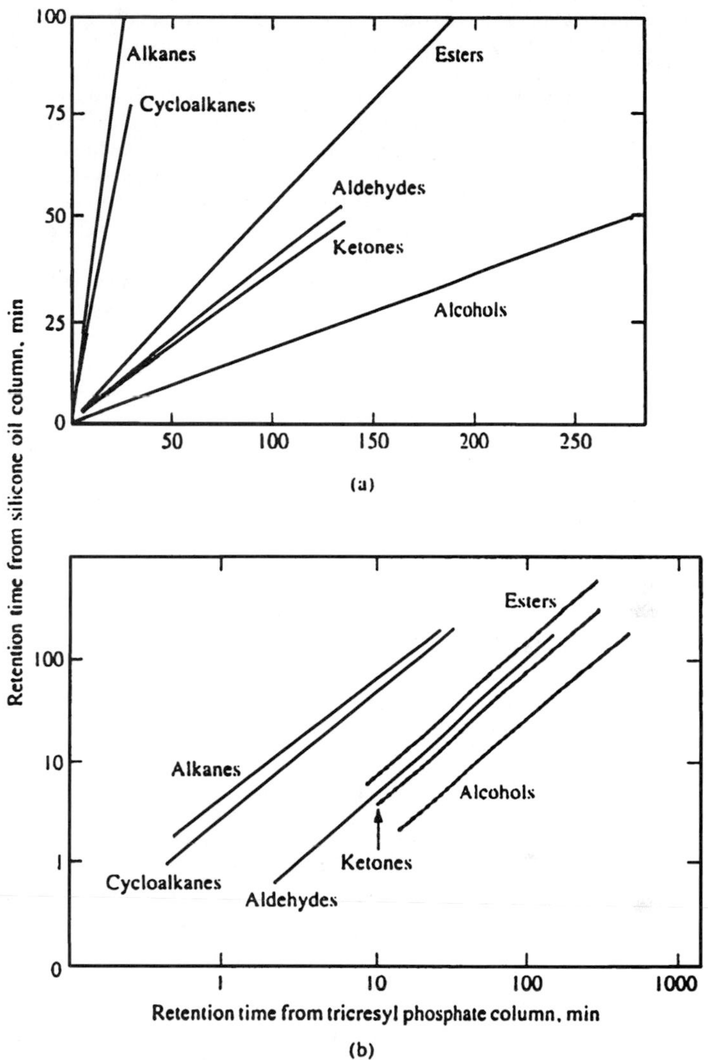

FIGURE 5.9 Two-column plots: (*a*) linear and (*b*) logarithmic.

5.1.5.1 ***Silylation.*** Silyl derivatives, which are widely used for gas-chromatographic applications, are formed by the replacement of active hydrogens from acids, alcohols, thiols, amines, amides, and enolizable ketones and aldehydes with the trimethylsilyl groups. A wide variety of reagents is available. These reagents differ in their reactivity, selectivity, side reactions, and the character of the reaction by-products from the silylation reagent.[7–9] The trifluoro group is commonly used for sensitizing

[7] A. E. Pierce, *Silylation of Organic Compounds*, Pierce Chemical, Rockford, IL, 1968. Also *Pierce Chemical Company Guide to Derivatization*, 48 pps.

[8] J. A. Perry and C. A. Feit, "Derivatization Techniques in Gas-Liquid Chromatography," in *GLC and HPLC Determination of Therapeutic Agents*, K. Tsuji and W. Morozowich, eds., Part I, Dekker, New York, 1978.

[9] C. F. Poole and A. Zlatkis, "Derivatization Techniques for the Electron-Capture Detector," *Anal. Chem.* **52**:1002A (1980).

substances to detection by electron capture. A derivatization method makes feasible the quantitative and qualitative analysis of amino acids.

5.1.5.2 Alkylation. As used in derivatization for gas chromatography, alkylation represents the replacement of an active hydrogen by an aliphatic or aliphatic-aromatic (e.g., benzyl) group. This method is used to modify compounds containing acidic hydrogen, such as carboxylic acids and phenols, which are converted into esters. Alkylation reactions can also be used to prepare ethers, thioethers, and thioesters, *N*-alkylamines, amides and sulfonamides. Although silyl derivatives of carboxylic acids are easily formed, these compounds suffer from limited stability.

5.1.5.3 Acylation. Acylation is the conversion of compounds containing active hydrogens into esters, thioesters, and amides through the action of carboxylic acid or a carboxylic acid derivative. The use of perdeuterated derivatives assists materially in interpreting the mass spectra of silylated compounds. *O*-Alkylhydroxylamines are used to prepare *O*-alkyloximes of aldehydes and ketones.

5.1.5.4 Reagents to Enhance Detectability. When the detection sensitivity to the 254-nm wavelength in ultraviolet absorption is zero or very low, detection can be enhanced by attaching to the solute a chromophore with high absorption at 254 nm. All these reagents make reductive electrochemical detection and fluorescence feasible.

Reagents	Reactants
N-Succinimidyl-4-nitrophenylacetate	Amines and amino acids
3,5-Dinitrobenzoyl chloride	Alcohols, amines, and phenols
4-Nitrobenzyloxyamine hydrochloride	Aldehydes and ketones
O-4-Nitrobenzyl-*N,N′*-diisopropylisourea	Carboxylic acids
4-Nitrobenzyl-*N*-propylamine hydrochloride	Isocyanate monomers

The formation of fluorescent derivatives allows the sensitive detection of otherwise nonfluorescent molecules and it exploits the selectivity of fluorescence. The trifluoro group was mentioned earlier. These additional reagents permit fluorescent derivatization:

Reagents	Reactants
4-Bromomethyl-7-methoxycoumarin	Carboxylic acids
7-Chloro-4-nitrobenzyl-2-oxa-1,3-diazole	Amines (primary and secondary) and thiols
1-Dimethylaminonaphthalene-5-sulfonyl chloride	Amines (primary) and phenols
1-Dimethylaminonaphthalene-5-sulfonyl hydrazine	Carbonyls

5.1.6 Preparative Chromatography

Certain procedures require that the component(s) of a sample be collected after they have passed through the detector, so that they can be isolated and studied further or collected for industrial use. The object of preparative chromatography is to separate the compound(s) of interest from a mixture in the desired purity and amount within a reasonable time and at a reasonable cost. The trade-offs here are no different than those in analytical chromatography.

In order to determine the proper conditions for preparative chromatography, use an analytical column of 2 to 4 mm i.d. Be sure that the type of packing used in the analytical column is also available in the preparative column. A match between the surface chemistry of the analytical and the preparative columns is required in order to allow a direct transition from one column to the other. In HPLC the overall cost of column packing and solvents (including the problem and cost of disposal) should be considered. Once the mode and packing are selected, the scale-up process is relatively straightforward. The procedure is as follows:

1. Determine the loading capacity of the stationary phase. This is dependent on the degree of purity required for the final product, which in turn influences the resolution needed. Gradually increase the amount of sample per injection just until resolution begins to suffer.

2. Use the throughput desired per injection to scale up the analytical column to the preparatory column by the ratio of the injection amounts. For example, if the overload conditions for a normal HPLC analytical column (4.6 mm i.d. × 25 cm) is 15 mg, and 300 mg per injection is needed for the final product, the preparative column volume should be 20 times that of the analytical column. Since the volume ratio is

$$\frac{V_{\text{prep}}}{V_{\text{anal}}} = \frac{(\pi r^2 L)_{\text{prep}}}{(\pi r^2 L)_{\text{anal}}} \tag{5.30}$$

$$20 = \frac{r^2 L}{(0.23\ \text{cm})^2 (25\ \text{cm})}$$

$$r^2 L = 26.4\ \text{cm}^3$$

Thus any combination of column radius squared times column length that equals 26.4 would be suitable. Since semipreparative columns are generally 7 mm i.d., a 53 cm length would work. The actual column dimension will be dictated by what is available commercially.

To maintain a given separation time after scale-up, use the same mobile-phase linear velocity for the analytical and the preparative columns. Thus, the mobile-phase flowrate should be increased in direct proportion to the square of the ratio of the preparative-to-analytical column diameters. For optimum resolution, the flowrate used on the preparative column should be lower than that calculated for the analytical column.

From commercially available materials, average particle diameters can start at 3 μm for fast LC applications, increase to 5 μm for conventional separations, and extend to 15 or 20 μm for preparative applications. If the separation on a small-particle column is inadequate, it is unlikely to improve on a large-particle wide-bore column. Of course, if excellent resolution can be achieved on the analytical column, then the use of a larger-particle wider-bore column is warranted.

When scaling up in preparative chromatography, sample capacity is often the factor of major importance. If load is critical, then a large-particle-sized column is probably the best choice, whereas if resolution is the most important criterion, then a small-particle-sized packing materials is the best option. Overload is a function of the surface area and, in the case of reversed phase packing, the carbon load. The majority of the surface area in a packing material is contained in the pore structure of the packing material, so changing the particle size has only a very small impact on the surface area. If resolution is less than Rs = 1.2, the column probably cannot be usefully overloaded and, therefore, sample capacity will be limited. In all cases, time spent optimizing α and k' both by varying the mobile phase and the stationary phase will pay large dividends. The system may actually be run in an overloaded or less-than-optimum condition to increase throughput. With particularly valuable samples, for which purity is of the highest concern and the best resolution is demanded, speed and capacity can be sacrificed.

Bibliography

Giddings, J. C., "Principles and Theory," in *Dynamics of Chromatography*, Part 1, Dekker, New York, 1965.

Katz, E., ed., *Quantitative Analysis Using Chromatographic Techniques*, Wiley, New York, 1987.

Miller, J. M., *Chromatography: Concepts and Contrasts*, Wiley, New York, 1988.

Poole, C. F., and S. K. Poole, *Chromatography Today*, Elsevier, Amsterdam, 1991.

5.2 GAS CHROMATOGRAPHY

Since its introduction in 1952, the technique of gas chromatography (GC) and its younger offspring capillary (open tubular) gas chromatography has grown spectacularly. Any substance, organic or inorganic, which exhibits a vapor pressure of at least 60 torr (the column temperature may be raised to 350°C) can be eluted from a GC column. The major limitation of GC is that samples, or derivatives thereof, must be volatile at the column operating temperature.

A large percentage of gas chromatographs in use today were originally supplied as packed-column instruments; some of these have been converted for operation with open tubular columns. A number of kits are available that greatly simplify retrofitting the packed-column instruments. Before conversion of an older instrument, check whether the response of the electronics is sufficiently fast.

The basic components of a gas chromatograph are as follows:

1. A supply of carrier gas with attendant pressure regulator and flow controller
2. An injection port or valve and followed possibly by a splitter
3. A separation column
4. A detector
5. A thermostatically controlled oven that is also able to be programmed for various heating rates
6. A recorder or other readout device

Each of the principal modules is composed of several electronic and pneumatic subsystems. The inlet conducts carrier gas to the column and is also heated. The detector contains excitation and amplification circuitry. The column module is made up of the gas-chromatographic column itself and the column oven. Parts most often replaced in these subsystems are injector liners, septa, columns, ferrules, detector parts, heat assemblies, and amplifier boards.

Modern gas chromatographs span a wide range of complexity, capability, and options. More than 90% of the systems available are microprocessor-controlled and have temperature-programmable column ovens. The vast majority have dual-channel detection capability. Approximately half are sold with a capillary inlet system, and many are sold with companion data-handling systems and autosamplers.

5.2.1 Carrier-Gas Considerations

The carrier gas is chosen for its inertness. Its only purpose is to transport the analyte vapors through the chromatographic system without interaction with the sample components. The gas is obtained from a high-pressure gas cylinder and should be free from oxygen and moisture. Premium grades of carrier gas are usually less expensive in the long run. Helium is the most popular in the United States but it is expensive. Hydrogen is increasing in popularity. Sometimes the choice of carrier gas is dictated by the detector.

Use a single-stage regulator when your instrument has a built-in regulator that can provide second-stage pressure reduction. Install a two-stage stainless-steel diaphragm regulator on the cylinder when there is no secondary pressure control downstream. Either type of regulator is followed by both a hydrocarbon trap and a high-capacity indicating oxygen scrubber. Gas flow is monitored by a stainless-steel mass flow controller. The pressure drop across a packed column can range from 50 to 300 kPa (7.2 to 43.5 psi), but much less for open tubular columns. When using hydrogen gas, flow controllers connected in parallel (rather than in series) should be used to direct the reduced-pressure hydrogen to each function: approximately 30 cm^3/min for a flame ionization detector, and 3 cm^3/min for an on-column injector attached to a 0.32-mm capillary.

Carrier-gas lines should be metal rather than plastic; a length of loosely coiled stainless-steel tubing suffices. Plastic and O-rings allow both permeation and leaks.

The superiority of a light carrier gas such as helium or hydrogen (as compared to nitrogen or argon) is discussed in Sec. 5.2.8.4. Traditionally, chromatographers have used helium and nitrogen as the carrier gases of choice in gas chromatography. A laboratory-size generator (Balston, Packard) produces ultrapure and ultradry (99.999%) hydrogen gas by the electrolysis of deionized or distilled water and separation of hydrogen from other electrolysis products by permeation through a palladium membrane. A replaceable desiccant cartridge reduces water in the generated hydrogen to less than 80 ppm. Gaseous hydrogen can be stored safely and conveniently using solid metal hydrides; the hydrogen is stored at the pressure of a comparable cylinder. With adequate equipment, hydrogen poses no greater hazard in your laboratory than any other compressed gas. Only 50 mL of hydrogen are stored at any time in the unit, ensuring complete safety and compliance with OSHA and NFDA regulations.

Hydrogen and helium always permit faster analysis than a denser carrier gas such as nitrogen or argon. This is so because the van Deemter curve (Fig. 5.6) for the same sample component obtained on the same column will have its minimum at higher velocities with the light carrier gases, and at the same time the slope of the ascending part of the curve will be smaller with the light carrier gas. Additional benefits with the use of hydrogen as a carrier gas include the apparent ability of hydrogen to counteract trace amounts of oxygen (as opposed to nitrogen and helium, which accumulate trace amounts of oxygen). The columns last longer because they are subjected to shorter runs of cooler temperatures.

5.2.1.1 Gas Purifiers. Moisture and oxygen must be removed from the carrier gases by using appropriate scrubbers. Even small amounts of oxygen or water can damage GC columns and detectors. Thin films of stationary phase in capillary columns are especially vulnerable to oxidation or hydrolysis. This is good practice for any phase but is essential for very polar phases. To ensure moisture-free gas (levels below 10 ppb), install a molecular sieve 5A drying tube and an indicating tube, in series, in the line. A gas purifier specifically designed to ensure maximum gas purity should be inserted in the carrier gas line ahead of the injection port. Oxygen and moisture removal should be just as efficient at concentrations up to 2000 ppm as at levels below 100 ppm and should handle gas flowrates up to 1100 $mL \cdot min^{-1}$. The solid chemical compounds should be placed in an all-metal tube with metal Swagelok fittings. The purifier tube is placed in an oven so that oxygen and water are chemically reacted with the material in the tube. Once trapped, these contaminants cannot be returned to the gas stream. It is also desirable to install an indicating tube (scrubber) downstream from the high capacity purifier. Oxygen-free carrier gas is critically important in GC for both packed and capillary columns. Problems encountered by oxygen entering the system are evident as excessive column bleed caused by oxidation of the liquid phase, short column life, loss of sensitivity in electron-capture operation, and loss of column retention from phase breakdown.

5.2.2 Sample-Introduction System

The functions of the injection port are to provide an entry for the syringe, and thus the sample, into the carrier gas stream and to provide sufficient heat to vaporize the sample. For optimum performance the sample must be deposited on the column in the narrowest band width possible. Sample introduction is most often accomplished with a microsyringe through a self-sealing silicon elastomer septum. Special septa are used for high-temperature work. Valves are more reproducible and there are no septum leaks, failures, bleeding, or septum particles in the inlet. The GC column lifetime is lengthened. Septum-base inlets are the weak link in autosampling applications. Septums simply cannot withstand the constant wear-and-tear of autosampling and often fail after relatively few injections. By contrast, the JADE valve's (JADE Systems, Austin) typical maintenance interval is once every several thousand injections, which is the needle guide's usual lifetime.

In gas chromatography, usually liquid samples are injected onto a heated block the function of which is to convert the liquid sample into the gas phase instantly (flash vaporization) without decomposition or fractionation. The flash-vaporization chamber of the injection port should be as small as possible to preserve efficiency. Sufficient volume is required, however, to accommodate the sudden vaporization and expansion of the sample after injection (1 μL of methanol produces 0.31 mL of vapor at 200°C and 30 lb · in^{-1}). Sometimes the sample is deposited directly on the top of the stationary phase (on-column injection) from where it is then evaporated by heating the column at a programmed rate. Gas samples are introduced into the carrier-gas stream by special gas syringes or by a rotary valve with sample loop. It is desirable to purge the septum with a flow of carrier gas that is vented. This prevents sample absorption onto the septum and eliminates tailing and ghost peaks caused by septum bleed.

For a capillary column, required injection volumes are on the order of 0.1 nL (nanoliter), which will require splitting the carrier-gas stream so that only a small fraction of the sample actually enters the column. Briefly, the sample is introduced into a heated injector, where it undergoes rapid vaporization and thorough mixing of the volatilized components. The carrier gas flows to the injection port liner. After going through the injection port liner, the flow is split into two highly unequal parts; a small portion to the column and the remainder vented to the atmosphere after passing through a charcoal trap (to remove sample vapors) and flowing through a back pressure regulator. By increasing or decreasing the flow through the injection port, the split ratio is increased or decreased (assuming a fixed inlet pressure.

Requirements for a sample splitter are stringent. Every component in the sample must be split in exactly the same ratio. The concentric tube design dominates. The two basic requirements for linear splitting are (1) no sample loss should be encountered during evaporation and (2) sample vapor and carrier gas must be homogeneously mixed prior to splitting.

Splitless injection is widely used for trace analysis for which maximum sensitivity is desirable. The sample is vaporized and slowly transferred onto the column. The solute bands are reconcentrated at the head of the column either through phase ratio focusing or cold trapping. Chromatographic development occurs as the column temperature increases. An important point to remember when using splitless injection in LLC is that the initial column temperature should be approximately 20°C below the boiling point of the solvent in order to realize the solvent effect, which reduces band broadening.

5.2.2.1 ***Injecting the Sample with a Syringe.*** With a syringe the sample size can be quickly selected and reproducibility adjusted (±5% variation routinely and ±1% variation with skilled users). When injecting the sample with a microliter syringe, follow these recommendations.

1. Develop a smooth rhythm that allows you to inject the sample as quickly as possible but with accuracy. The syringe should be left in the injection port for about 2 s after depressing the plunger.
2. Hold the syringe as close to the flange (in the face or unmarked area) as possible. This will prevent the heat transfer that occurs when the needle or barrel is held with the fingers. A syringe guide prevents heat transfer from the finger and the guide makes septum penetration easier.
3. Handle the plunger by the button, not the plunger shaft. This reduces the possibility of damage or contamination.
4. Use the syringe at less than maximum capacity for greatest accuracy. When samples containing components with a wide range of boiling points are injected, fractionation may occur unless the syringe is prepared with a solvent slug (separated by an air bubble) that follows the sample into the injection port.
5. Wet the interior surfaces (barrel and plunger) or the syringe with the sample by pumping the plunger before filling the syringe in order to ensure an accurate measurement.
6. Overfill the syringe in the sample bottle, withdraw from the bottle, and move the plunger to the desired calibration line, and discharge the excess sample. Do not leave any drop hanging from the syringe tip.
7. Wipe the needle clean with a lint-free tissue before injecting; use a quick motion and take care neither to wipe the sample out of the needle nor to transfer body heat from your fingers to the needle.

8. Check the syringe visually for bubbles or foreign matter in the sample.
9. Take extra care when filling syringes with a detachable needle because there is dead volume in the needle. Pressurize your sample bottle by using a gastight syringe filled with inert gas. Repeat as needed to build pressure in the bottle.
10. Use a larger bore needle when handling viscous samples.
11. Do not use dull or damaged syringe needles or overtighten the septum retainer because septum life will be shortened. Since high temperature also reduces septum life, the use of finned septum retainers that stay cooler is recommended. Since all septa start to leak eventually, it is good practice to replace the septum each day, or even more often during heavy use. Some septa may last up to 150 injections, but they should be routinely replaced well before this.

5.2.2.2 ***Care and Repair of Syringes.*** A syringe is a delicate instrument and should be treated carefully. Carefully clean the syringe before and after use. Cleaning the small bore of the needle, the glass barrel, and the closely fitting plunger is not a simple matter, but it is essential for a long syringe life. Moreover, cleaning is more effective and easier to accomplish if it is done immediately after the syringe has been used. The syringe must never be used when the needle is blocked or the barrel is dirty.

A simple cleaning procedure consists of pumping a solution of a surface-active cleaning agent through the syringe, and then rinsing both the syringe and plunger through the needle with distilled water followed by an organic solvent such as acetone or other ketone. Never touch the plunger surface with your fingers. Fingerprints, perspiration or soil from the fingers can cause the plunger to freeze in the glass barrel.

Another recommended cleaning procedure is (1) pump room-temperature chromic acid solution through the syringe with the plunger, (2) rinse both with distilled water, (3) blow the syringe dry with oil-free compressed gas, and (4) carefully wipe the plunger with lint-free tissue. For more persistent contamination, dismantle the syringe and soak the parts in a cleaning solution. A small ultrasonic bath will speed the cleaning process.

Never rapidly cool or heat assembled syringes, and never heat over 50°C as the different expansion coefficients of the metal needles and plungers and the glass barrel may cause the barrel to fracture.

5.2.2.3 ***Pyrolysis Gas Chromatography.***[10] Pyrolysis gas chromatography is a technique that has long been used in a variety of investigative fields because it produces volatile compounds from macromolecules that are themselves neither volatile nor soluble. Examples are polymers, rubbers, paint films, resins, bacterial strain differentiation, soil and rock characterization, coals, textiles, and organometallics. Volatile fragments are formed and introduced into the chromatographic column for analysis.

Let us use a polymer as an example to illustrate the use of this technique, which consists of two steps. The injection port is heated to perhaps 270°C; when the sample is injected, the volatile ingredients that are driven off provide a fingerprint of the polymer formulation. Then the pyrolysis step develops the fingerprint of the nonvolatile ingredients. Known monomers of suspected polymers can be injected with a microsyringe for identification of the peaks of an unknown pyrogram. Specific identification of the peaks appearing in pyrograms is most effectively carried out by directly coupled gas chromatography–mass spectrometry together with the retention data of the reference samples. Gas chromatography–Fourier transform infrared spectrometry also can provide effective and complementary information.

Pyrolyzers can be classified into three groups: (1) resistively heated electrical filament types, (2) high-frequency induction (Curie-point) types, and (3) furnace types.

The *filament type* uses either a metal foil or a coil as the sample holder. Heat energy is supplied to the sample holder in pulses by an electric current. This permits stepwise pyrolysis at either fixed

[10] C. J. Wolf, M. A. Grayson, and D. L. Fanter, "Pyrolysis Gas Chromatography," *Anal. Chem.* **52**:348A (1980).

or varied pyrolysis temperatures. This feature sometimes permits a discriminative analysis of volatile formulations and high polymers in a given compound without any preliminary sample treatment.

The *Curie-point type* uses the Curie points of ferromagnetic sample holders to achieve precisely controlled temperatures when the holder containing the sample is subjected to high-frequency induction heating. The Curie point is the temperature at which the material loses its magnetic property and ceases to absorb radio-frequency energy. Foils of various ferromagnetic materials enable an operator to select pyrolysis temperatures from 150 to 1040°C.

In the *furnace (continuous) pyrolyzer type*, the sample is introduced into the center of a tubular furnace held at a fixed temperature. Temperature is controlled by a proportioning controller that utilizes a thermocouple feedback loop.

5.2.2.4 Purge-and-Trap Technique. The analysis of volatile organic compounds in samples is commonly performed using the technique of purge-and-trap gas chromatography. It is the required technique for a number of the Environmental Protection Agency (EPA) methods for drinking water, source and waste water, soils, and hazardous waste. Using this technique it is possible to detect and identify various flavors, fragrances, off odors, and manufacturing by-products in a wide diversity of liquid matrix formulations, colloidal suspensions, and liquid pastes, which include perfumes, food products, blood, urine, latex paint, shampoo, and beverages.

In the purge-and-trap method, samples are contained in a gastight glass vessel (5 mL). With a sparging needle dipping nearly to the bottom of the sample vessel, the sample is purged by bubbling high-purity helium or nitrogen through the sample. If desired the sample can be heated for the removal of higher boilers, but the temperature should be kept well below the boiling point of the liquid. Volatile compounds are swept out of the sample and carried into and trapped in an adsorbent tube packed with Tenax-GC, a porous polymer based on 2,6-diphenyl-*p*-phenylene oxide, or activated carbon. Tenax has a low affinity for water. Also, a dry-gas purge is added just before the adsorption tube. Volatile organic materials can be efficiently collected from a relatively large sample, producing a concentration factor that is typically 500- to 1000-fold greater than the original. The trapped material on the adsorbent is heated (thermally or in a microwave oven) to release the sample, and then back-flushed using the GC carrier gas. This sweeps the sample directly onto the GC column for separation and detection by normal GC procedures. Trapped samples can easily be stored or shipped to another site for analysis.

5.2.2.5 Headspace Sampling.[11,12] When only the vapor above the sample is of interest and the partition coefficient allows a sufficient amount of analyte into the gaseous phase, headspace sampling can be done. Samples may be solid or liquid. A measured amount of sample, and often an internal standard, are placed in a vial, and then the septum and cap are crimped in place. The vials, contained in a carousel, are immersed in a silicone oil bath operating from ambient 15 to 150°C. A heated flexible tube that terminates in a needle samples each vial in turn. A gas-sampling valve provides a standard 1.0-mL vapor sample for transfer into the GC injection port.

Many plants and animals as well as commercial products may produce odors or volatile emissions that can be trapped on adsorbent traps and then analyzed via a thermal desorption system. The air from the samples or materials is passed through the adsorption tube packed with Tenax or some other adsorbent resin. The packed adsorption tubes must be previously cleaned and conditioned to remove any contaminants from the adsorbent. Air samples can be passed through the adsorbent tubes in two manners. One method is to use a vacuum pump, which is particularly useful for field testing but has the disadvantage of analyzing the background volatiles normally present in the air sample. The second method uses an enclosed container (such as the Wheaton Purge & Trap System) with an outlet port to which the adsorbent tube is attached and an inlet port into which clean, high-purity purge gas is pumped. A thermal desorption system completes the analysis.

---

[11] B. V. Joffe and A. G. Vitenberg, *Headspace Analysis and Related Methods in Gas Chromatography*, Wiley, New York, 1984.

[12] M. E. NcNally and R. L. Grob, "Static and Dynamic Headspace Analysis," *Am. Lab.* **17** (January):20 (1985); **17**(February):106 (1985).

5.2.2.6 ***Short-Path Thermal Desorption.*** Thermal desorption (or extraction) permits the analysis of solid samples without any prior solvent extraction or other sample preparation. Solid samples between 1 and 500 mg are placed inside a glass-lined, stainless-steel desorption tube between two glass-wool plugs. After attaching a syringe needle to the tube and then placing in the thermal desorption system (Scientific Instrument Services, Inc.), the desorption tube is purged with carrier gas to remove all traces of oxygen. The preheated heater blocks are closed around the adsorption tube to desorb samples at temperatures from 20 to 350°C and for program desorption times from 1 s to 100 min. The procedure permits the thermal extraction of volatiles and semivolatiles from the sample directly into the GC injection port (much like a syringe). The GC column, either packed or capillary, is maintained at subambient temperatures (or at a low enough temperature to retain any samples at the front of the GC column) during the desorption step. This enables the desired components to be collected in a narrow band on the front of the GC column. As an alternative to cryofocusing, one can use a thick-film capillary column or a packed column with a high loading capacity. After the desorption is complete, the needle is removed and the GC gas turned on. The components trapped on the front of the GC column are separated and eluted via a temperature program in the GC oven. This technique is useful for the analysis of a wide variety of low-moisture-content solid samples including vegetation, food products, pharmaceuticals, building materials, forensic samples, and packaging products. By selection of the desorption temperature, the number and molecular-weight distributions of components in the samples can be selected.

Adsorbent tubes with samples collected during either dynamic headspace purging or purge and trap of liquids are placed within the short-path thermal desorption unit and the analysis continued as already described. A selection of applications, given in Table 5.3, provides typical operating conditions.

5.2.3 Gas-Chromatographic Columns

Separation of the sample components takes place in packed or open tubular columns through which the carrier gas flows continuously. The separation column is placed immediately after the injection port and any attendant sample splitter. The separation column contains the stationary phase, which can be either (1) an adsorbent (GSC) or (2) a liquid distributed over the surface of small-diameter particles or the interior of capillary tubing. Columns are made of metal (stainless steel, copper, or aluminum), glass, or fused silica. Since inertness is of prime importance, glass and fused silica have become increasingly popular; stainless steel is easier to handle and coil.

GC columns can be divided into three broad categories: (1) packed columns, (2) open-tubular columns, and (3) micropacked columns. Open tubular columns are divided further into (a) wall-coated (WCOT) and (b) porous layer (PLOT); the support-coated (SCOT) column is now classified with WCOT columns. Micropacked may be either packed or micropacked capillary columns. Two factors, selectivity and efficiency, should be considered when comparing different types of columns.

5.2.3.1 ***Packed Columns.*** Packed columns for analytical work are available in diameters of 2, 3, or 4 mm. If the stationary phase is a liquid, it is held on the surface of a inert solid support. In GLC the support should not take part in the separation. Especially with low liquid loadings (10%) and nonpolar phases, the support must be thoroughly inactivated when polar compounds are to be analyzed. If the support is not inactivated at all, it is called regular or the abbreviation NAW (non-acid-washed) is used. A first treatment is acid washing (AW). Sometimes base washing (BW) is applied. A better inactivation is having the active sites that remain after acid washing react with dimethyldichlorosilane (DMCS). Surface mineral impurities of the substrate, which can serve as adsorption sites, must be removed by acid washing. Surface silanol groups (—Si—OH) tend to adsorb polar solutes and must be *end-capped*, that is, converted to silyl ethers (—Si—O—Si—) by treating the column packing with DMCS (the unused silylating agent is removed with methanol). The chromatographic behavior of non-end-capped materials most commonly noticed is the stronger retention of polar compounds. In LLC mobile-phase modifiers are added to reduce tailing but this is not possible in GC.

TABLE 5.3 Thermal Desorption Methods

Sample	Trapping details	Thermal desorption	GC operating conditions
Environmental air and dynamic headspace sampling			
Air	70 mL garage air, 100 mL building air, or 2 mL air inside an instrument pumped through Tenax at –40°C.	Block: 250°C. Purge: 2 mL/min He for 10 min.	25 m × 0.25 mm; 0.25-μm film DB-5; –40° to 280°C at 10°/min.
Foliage	Volatiles from leaves trapped on Tenax at –40°C.	Block: 200°C. Purge: 10 mL/min He for 10 min.	60 m × 0.25 mm; 0.25-μm film DB-1; –20° to 280°C 10°/min.
Shampoo	5 mL liquid purged by passing 200 mL He over surface at 20 mL/min and a dry purge at 20 mL/min. Tenax trap at –40°C.	Block: 150°C. Purge: 2 mL/min He for 10 min.	25 m × 0.25 mm; 0.25-μm film DB-5; –40° to 280°C at 10°/min.
Purge and trap			
Beverages, carbonated	25 mL sample purged with 150 mL He at 15 mL/min and a dry purge of 15 mL/min. Volatiles trapped on Tenax at –40°C.	Block: 150°C. Purge 2 mL/min He for 10 min.	25 m × 0.25 mm; 0.25-μm film DB-5; –40° to 280°C at 10°/min.
Beverages, flavored drink	25-mL sample purged with 300 mL He at 30 mL/min and a dry purge of 30 mL/min. Volatiles trapped on Tenax at –40°C.	Block: 150°C. Purge: 2 mL/min He for 10 min.	25 m × 0.25 mm; 0.25-μm film DB-5; –40° to 280°C at 10°/min.
Latex paint, VOCs* in	1.0-mL sample purged with 150 mL He at 15 mL/min and a dry purge of 15 mL/min. Volatiles trapped on Tenax at –40°C.	Block: 150°C. Purge: 2 mL/min He for 10 min.	25 m × 0.25 mm; 0.25-μm film DB-5; –40° to 280°C at 10°/min.
Olive oil	5-mL sample purged with 2000 mL of He at 20 mL/min and a dry purge of 20 mL/min. Volatiles trapped on Tenax at –40°C.	Block: 150°C. Purge: 2 mL/min He for 10 min.	25 m × 0.25 mm; 0.25-μm film DB-5; –40° to 280°C at 10°/min.
Toothpaste	Sample diluted 1 : 5 with water and 5 mL of solution purged with 120 mL He at 20 mL/min and dry purge at 20 mL/min. Tenax trap at –40°C.	Block: 150°C. Purge: 2 mL/min He for 10 min.	25 m × 0.25 mm; 0.25-μm film DB-5; –40° to 280°C at 10°/min.
Water, VOCs	50-mL sample purged with 300 mL He at 20 mL/min and dry purge at 20 mL/min. Tenax trap at –40°C.	Block: 150°C. Purge: 2 mL/min He for 10 min.	25 m × 0.25 mm; 0.25-μm film DB-5; –40° to 280°C at 10°/min.

Direct thermal desorption			
Arson material	Threads from Molotov cocktail wick. GC column trap at –10°C.	Block 150°C. Purge: 6 mL/min He for 10 min.	30 m × 0.75 mm; 1.0-μm film SPB-1; –10° to 250°C at 12°/min.
Cardboard, recycled	250-mm² sample. GC column trap at –40°C.	Block: 200°C. Purge: 2 mL/min He for 10 min.	25 m × 0.25 mm; 0.25-μm film DB-5; –40° to 280°C at 10°/min.
Corn meal	10-g sample flushed with N_2 (160°C) in sampler oven; volatiles trapped on Tenax.	Block: 220°C. Purge: 10 mL/min He for 5 min.	60 m × 0.32 mm; 0.25-μm film DB-1; split 10 : 1; –20° to 40°C at 10°/min, then 280°C at 4°/min.
Food wrap	5-mm² sample. GC column trap at –40°C.	Block: 125°C. Purge: 2 mL/min He for 10 min.	25 m × 0.25 mm; 0.25-μm film DB-5; –40° to 280°C at 10°/min.
Ink	A line 1 in long on white bond paper. GC column trap at –40°C.	Block: 150°C. Purge: 2 mL/min He for 10 min.	25 m × 0.25 mm; 0.25-μm film DB-5; –40° to 280°C at 10°/min.
Laundry detergent	1.5-mg sample. GC column trap at –40°C.	Block: 100°C. Purge: 2 mL/min He for 10 min.	25 m × 0.25 mm; 0.25-μm film DB-5; – 40° to 280°C at 10°/min.
Marijuana	3-g sample. GC column trap at –40°C.	Block: 200°C. Purge: 2 mL/min He for 10 min.	25 m × 0.25 mm; 0.25-μm film DB-5; –40° to 280°C at 10°/min.
Pepper, black	2-mg crushed sample.	Block: 150°C. Purge: 2 mL/min He for 6 min.	60 m × 0.25 mm; 0.25-μm film DB-5; –40° to 280°C at 10°/min.
Pharmaceuticals	2–25-mg crushed sample.	Block: 100°C. Purge: 2 mL/min He for 10 min.	25 m × 0.25 mm; 0.25-μm film DB-5; –40° to 280°C at 10°/min.
Soil	2-g sample flushed with N_2 (70°C) in sampler oven. Volatiles trapped on Tenax at –40°C.	Block: 220°C. Purge: 2 mL/min He for 10 min.	25 m × 0.25 mm; 0.25-μm film DB-5; –40° to280°C at 10°/min.
Tire, automobile	2-mg sample. GC column trap at –40°C.	Block: 125°C. Purge: 2 mL/min He for 10 min.	25 m × 0.25 mm; 0.25-μm film DB-5; –40° to 280°C at 10°/min.
Wood	0.5-g dry wood flushed with N_2 (70°C) in sampler oven. GC column trap at –40°C.	Block: 220°C. Purge: 10 mL/min He for 5 min.	25 m × 0.25 mm; 0.25-μm film DB-5; –40° to 280°C at 10°/min.

* VOCs denote volatile organic chemicals.

TABLE 5.4 Characteristics of GC Porapak™ Porous Polymer Packings

For each type of Packing, three particle size meshes are available: 50–80, 80–100, and 100–120.
Information in the table is courtesy of Waters Chromatography.
The HayeSep N, P, Q, R, S, and T polymers (Supelco) are interchangeable with the corresponding Porapak polymers.

Type	Polarity	Surface area, m^2/g	Density, g/cm^3	Maximum temperature, °C
P	Nonpolar	100–200	0.26	250
PS	Nonpolar	100–200	0.26	250
Q	Nonpolar to moderate	500–600	0.34	250
QS	Nonpolar to moderate	500–600	0.34	250
R	Moderate	450–600	0.32	250
S	Moderate	300–450	0.35	250
N	Polar	250–350	0.41	190
T	Highly polar	225–350	0.39	190

Adsorbents such as silica gel and molecular sieves are used for the separation of gases. The porous polymers, like the Porapaks (Table 5.4), HayeSep, GasChrom, Tenax-GC, and the Chromosorb Century Series, are used in special field applications such as the analyses of solvents, acids, and alcohols in water. The analysis temperature is often 50 to 100°C higher than that when a liquid phase is used. Carbopack and Graphpac graphitized carbons are ideal for analyzing many kinds of C_1 and C_{10} compounds, including alcohols, free acids, amines, ketones, phenols, and aliphatic hydrocarbons. Carbopack B (Graphpac-GB) and Carbopack C (Graphpac-GC) carbons have surface areas of about 100 and 12 m^2/g, respectively. Therefore, Carbopack B packings will have a larger sample capacity than Carbopack C packings. Carbopack B HT or C HT are hydrogen treated for deactivation.

When the stationary phase is a liquid, it is usually held on a porous particle. Equivalent supports are displayed in Table 5.5. A larger mesh size is desirable for long column to minimize the pressure

TABLE 5.5 Equivalent Gas-Chromatographic Column Supports

Non-acid-washed	Acid washed	Acid washed, end capped	Acid washed, end capped High performance (B = extra base washed)
		White supports	
Chromosorb W	Chromosorb W AW	Chromosorb W AW DMCS	Chromosorb W HP
Gaschrom S	Gashrom A	Gashrom Z	Gashrom Q (B)
Anakrom U	Anakrom A	Anakrom AS	Anakrom ABS (B)
		Anakrom Q	Anakrom SD
	Supelcon AW	Supelcon AW DMCS	Supelcoport
			Varaport 30
			Diatoport S
Diatomite C			Diatomite CLQ
			Diatomite CQ
Chromosorb G	Chromosorb G AW	Chromosorb G AW DMCS	Chromosorb G HP
			Chromosorb 750
		Pink supports	
Chromosorb P	Chromosorb P AW	Chromosorb P AW DMCS	
Gaschrom R	Gaschrom RA	Gaschrom RZ	
Anakrom C22	(Anakrom) C22 A(W)		
Diatomite S			
Sil-O-Cel			
Firebrick C22			

TABLE 5.6 Properties of Chromosorb Supports

Support	Packed density, g/cm^3	Surface area, m^2/g	Surface area, m^2/cm^3	Loading capacity, %	Upper pH limit
Chromosorb A	0.48	2.7	1.3	40	7.1
Chromosorb P	0.47	4.0	1.9	30	6.5
Chromosorb G	0.58	0.5	0.3	5	8.5
Chromosorb W	0.24	1.0	0.3	25	8.5
Chromosorb T	0.49	7.5		5	

drop across the column. The particles should have a large surface area. Chromosorb P (P for pink) and its equivalents have the largest surface area (4.0 m^2/g) and accept the maximum liquid phase (30%); pore size ranges from 0.4 to 2 μm. Chromosorb W (for white) has one-fourth the surface area, accepts 15% liquid phase, and has a larger pore size (8 to 9 μm). Chromosorb A, intended for preparative chromatography, has a surface area of 2.7 m^2/g and accepts a maximum 25% liquid phase. Properties of Chromosorb supports are given in Table 5.6. Glass microbeads possess a very small surface area of 0.01 m^2/g, have no pores, and accept a maximum liquid-phase load of 3%. Recommended liquid loadings for GC supports are given in Table 5.7. Relations between column inside diameter, column length, mesh size, and carrier-gas flowrate are shown in Table 5.8.

Compared to the capillary column, which is described later, the packed column has a variety of mobile-phase flow-path lengths and the discontinuous film of its stationary phase is much less uniform. Both of these differences contribute to the increased standard deviation of the packed-column peak. Poor heat-transfer properties that characterize almost all column packings increase the standard deviation of the peaks. There is a temperature gradient across any transverse section of a packed column, so identical molecules of each given solute, ranged transversely across a short segment of column length, are at any given moment exposed to different temperatures. As a consequence, identical molecules of each solute also exhibit a range of volatilities, which, in turn, leads to a larger standard deviation (wider peak) for the molecules that make up that peak.

Now to sum up the general performance characteristics of packed columns: Efficiency (plates per meter) is a function of particle size and is about equal to that of a capillary column. Selectivity is excellent due to the very wide range of phases that are available, including adsorbents (GSC). Packed columns are applicable for all gases, are excellent for preparative-scale analysis, but very limited for trace analysis. The practical length needs to be fairly short due to low permeability—a major weakness. Speed of analysis is the slowest of all column types. Compatibility with a mass spectrometer is very limited.

TABLE 5.7 Recommended Liquid Loadings for GC Supports

		Maximum loading, w/w %	
Support material	Typical loading, w/w %	Typical liquid phase	Sticky or gum phases*
Anakrom	3–10	25	20
Chromosorb G	2–6	12	7
Chromosorb P	10–30	35	20
Chromosorb T	1–2	6	3
Chromosorb W	3–10	25	20
Chromosorb 101 through 108	3–10	25	10
Gas Chrom Q	3–10	25	20
Porapak N, Q, Q-S, R, S	1–5	8	4
Porapak T	1–2	6	4
Teflon	1–2	6	3
Tenax	1–4	5	4

* Apiezon L, DEGS, OV-1, OV-275, SE-30, and SE-52 are typical sticky or gum phases.

TABLE 5.8 Relations between Inside Diameter, Column Length, Mesh Size, and Carrier-Gas Flow for Packed Columns

Inside diameter, mm	Mesh size for length up to 3 m	Mesh size for length over 3 m	Carrier-gas flow, N_2, mL/min	Carrier-gas flow, He or H_2, mL/min
2	100–120	80–100	8–15	15–30
3	100–120	80–100	15–30	30–60
4	80–100	60–80	30–60	60–100

5.2.3.2 Open Tubular (Capillary) Columns.[13] The flexible, mechanically durable, and chemically inert fused-silica capillaries offer many advantages over packed columns and are becoming the dominant type of column design. With the use of special coating techniques in combination with suspension technology, a 10- to 30-μm layer of a porous polymer can be coated on the inner wall of a fused-silica capillary column. Porous polymers are cross-linked polymers that are produced by copolymerizing styrene and divinylbenzene. The pore size and surface area can be varied by changing the amount of divinylbenzene added to the polymer. The introduction of functional groups such as acrylonitrile or vinyl pyrrolidone controls the selectivity of the polymers.

A highly efficient capillary column does not require as much selectivity toward sample components as a less efficient packed column to achieve the same resolution. Peaks are sharper. Sharper peaks provide better separation and also deliver the solutes to the detector at higher concentrations per unit time, thus enhancing sensitivity.

The major advantage of capillaries regardless of the separation is an increase in the speed of analysis. When packed-columns flowrates are used, the capillary column produces separation efficiencies that are equal to those of packed column but at roughly three times the speed. When the flowrate is optimized for the tubing diameter, the capillary column produces far superior efficiencies with analysis times approximately equal to those for packed columns. Table 5.9 provides a comparison of packed columns and two types of open-tubular columns.

Although efficiency (plates per meter) is the same for both packed and capillary columns, capillary columns are open tubes. Therefore, they are more permeable and can be made much longer before the inlet pressure requirement becomes too large. Capillary columns with more than 100 000 plates are relatively common. Because of the greater permeability of capillary columns, the van Deemter (reduced height)/(reduced velocity) (h/v) curves are also flatter, which means that the column can be operated at two to three times the optimum flowrate without losing much efficiency. Consequently, the time of analysis is shorter. Better peak shape and increased column stability are other advantages. Nothing is superior to a capillary column for resolving mixtures with many

[13] R. T. Wiedemer, S. L. McKinley, and T. W. Rendl, "Advantages of Wide-Bore Capillary Column," *Am. Lab.* **18** (January):110 (1986).

TABLE 5.9 Comparison of 1/8-in (0.316-cm) Packed, Wide-Bore, and WCOT Columns

	1/8-in Packed	Wide bore	WCOT
Inside diameter, mm	2.2	0.53	0.25
Film thickness, μm	5	1–5	0.25
Phase volume ratio (β)	15–30	130–250	250
Column length, m	1–2	15–30	15–60
Flowrate, mL/min	20	5	1
Effective plate height (H_{eff}), mm	0.5	0.6	0.3
Effective plates (N_{eff}) per meter	2000	1200	3000
Typical sample size		15 μg	50 ng

components. It has obvious advantages even in relatively simple applications that are usually solved with packed columns.

In their original and simplest form, open tubular columns contain a thin film of stationary liquid on their inside walls. Hence the name *wall-coated open-tubular* (WCOT) columns. The inner diameter of these columns is either nominally 0.25 or 0.32 mm. The inner surface the column is coated with the stationary polymer phase to a thickness of 1 to 2 μm. The coating is both surface bonded and cross-linked so it cannot be disturbed by repeated injections of a polar solvent or by prolonged heating. Thicker films permit the analysis of volatile materials without the use of subambient column temperatures. Thinner films are valuable for analyzing high-molecular-weight compounds. Columns 10 m in length closely approximate the capacity and separation of the standard analytical packed column (20 m × 2 mm i.d.) with a 3% to 5% loading. A variety of functional groups can be blended into the polysiloxane chain to provide stationary phases of different polarity or selectivity.

So-called *megabore* or *wide-bore* columns, with their wider bore (0.53 mm i.d.), thick films, and direct on-column injection, serve as a good compromise between capillary columns and packed columns. They can often be substituted directly for packed columns. If a 3% to 5% loading of a standard packed-column stationary phase currently separates your components, then the wide-bore column equivalent will separate them under the same conditions of temperature and flow. If a 10% loading is being used, then the wide-bore phase will usually perform the same separation at an oven temperature that is approximately 15 to 20°C lower.

Wide-bore columns give better resolution (because of their longer lengths) and more symmetrical peaks (because of their decreased adsorption) than packed columns. They also combine the high capacity and ease of use of packed columns with the high efficiency, rapid analyses, and greater versatility of narrow-bore columns. Sample capacity and flowrates are similar to those for packed columns, so experimental parameters need not be altered much. Virtually all commonly used solvents can be injected onto these wide-bore bonded-phase capillary columns, but water does cause column deterioration (hydrolysis of the stationary phase).

For a period of time, open tubular columns that had characteristics intermediate between those of WCOT and packed columns were popular. *Porous layer open-tubular* (PLOT) columns have diameters from 0.05 to 0.35 mm, film thicknesses from about 0.2 up to 5 to 6 μm, and lengths from 10 m or even shorter up to 100 m. The influence of column diameter on performance is shown in Table 5.10. The availability of such a wide range readily permits the selection of the optimum parameters for a given application. For high resolution, long (30 to 50 m), narrow (0.1 to 0.25 mm i.d.), thin-film (0.2 μm) columns are used to generate 250 000 plates for 0.5- to 1.5-h separations of complex materials. These columns are operated at low mobile-phase velocity near the optimum. High resolution is possible because of the fast mass transfer of sample within the thin stationary-phase films.

Other PLOT columns are coated with specially prepared gums, aluminum oxide, and molecular sieves. With aluminum oxide excellent separations of C_1 to C_{10} hydrocarbons are obtained.

TABLE 5.10 Operational Guidelines for Open Tubular GC Columns

	WCOT narrow	WCOT intermediate	Wide bore
Column inner diameter, mm	0.25	0.32	0.53
Maximum sample volume, μL	0.5	1	1
Maximum amount for one component, ng	2–50	3–75	5–100
Effective plates (N_{eff}) per meter	3000–5000	2500–4000	1500–2500
Trennzahl (separation) number per 25 m	40	35	25
Optimum flow for nitrogen, mL/min*	0.5–1	0.8–1.5	2–4
Optimum flow for helium, mL/min†	1–2	1–2.5	5–10
Optimum flow for hydrogen, mL/min‡	2–4	3–7	8–15

* Optimum velocity is 10 to 15 cm/s for each column.
† Optimum velocity is 25 cm/s for each column.
‡ Optimum velocity is 35 cm/s for each column.

With molecular sieve 5A the quantification of permanent gases is made very easy. The 13X-type molecular-sieve column easily separates aliphatic and naphthenic hydrocarbons.

Support-coated open-tubular (SCOT) columns have a porous layer of stabilizing inert support built up or deposited on the original tubing wall. This layer is coated with the liquid phase. The SCOT column finds use in combination with wide-bore columns for multidimensional separations.

5.2.4 The Stationary Phase

For any resolution to be achieved, the components of the sample must be retained by the stationary phase. The longer and more selective the retention, the better will be the resolution. In gas chromatography the inert carrier gas plays no active role in solute selectivity, although it does affect resolution. Selectivity can be varied only by changing the polarity of the stationary phase or by changing the column temperature. The higher efficiency of open-tubular columns, as compared to packed columns, has reduced the necessity for a large number of selective liquids.

5.2.4.1 Liquid Phases. In the early days of GC, the stationary liquid phase was coated on a substrate such as a diatomaceous earth. Problems arose because the liquids would pool and the surface coverage was incomplete. The former led to longer but irregular residence times in the stationary phase (leading to decreased resolution), and the latter gave rise to adsorption onto the substrate and thereby mixed partition effects. Furthermore, many of the early liquid phases suffered from low maximum temperature limits before column bleed became excessive and affected the baseline and column lifetime. When GC activities were dominated by packed columns, which are restricted to relatively low numbers of theoretical plates, separation efficiencies were very much dependent on stationary-phase selectivity. As a result a large number of stationary phases were necessary with packed columns. Over 200 stationary phases have been listed by supply houses; some are included in Table 5.11 for reference purposes when delving into the older literature.

Bonded (cross-linked or chemically bonded) liquid phases are much superior. They can be cleaned by rinsing with strong solvents and baking at high temperatures. Column life can be extended, dirtier samples can be tolerated, and sample cleanup can be reduced. Most analyses can be accomplished using columns with standard film thickness of 0.25 μm for 0.25 and 0.32 mm i.d. columns. For wide-bore (megabore) columns the standard film thicknesses are 1 or 1.5 μm depending on the phase. General recommendations on film thickness would be to use standard film columns for most applications. Use thin-film columns for high boiling solutes (petroleum waxes, glycerides, and steroids) and use thick-film columns for very volatile solutes (gases, light solvents, and purgables).

Table 5.11 provides a list of the common stationary phases, their useful temperature range, and their McReynolds constants, a polarity indicator discussed in Sec. 5.2.4.2. Also tabulated in Table 5.11 is the minimum temperature, which is represented by the melting point or glass transition temperature. Below that temperature, the stationary phase will be a solid or an extremely viscous gum. A persistent problem with liquid phases is their upper temperature limits. The maximum temperature is that above which the bleed rate will be excessive as a result of solvent vaporization. Thus, a major objective has been to find liquids with increasing boiling points.

With the present state of column technology, a few facts hold true: methylpolysiloxane (5% column loading; OV-101, SE-30, and SP-2100), the least polar of the silicones, is the first choice of stationary phase for general analytical use. Compounds are eluted primarily in order of their boiling points. Only a few stationary phases are statistically different in their chromatographic behavior. Changes in selectivity are achieved by replacing the methyl groups with more polar groups. As the percent phenyl substitution increases, the selectivity and polarity (the Σ in the right-hand column of Table 5.11) of the phase also increases. Both OV-225 (polytrifluoropropylmethylsiloxane) and SP-2300 (polyethylene glycol) are more polar and exhibit more selective retention for polar solutes. For most new analyses they should be tried second and third, respectively, after methyl silicones.

Intermediate polarity is provided by 50% poly(phenylmethylsiloxane) (3% loading; OV-17, SP 2250). Moderately polar phases often have the trifluoropropyl group. Polyalkylene glycols, such as

TABLE 5.11 McReynolds Constants for Stationary Phases in Gas Chromatography

Stationary phase	Chemical type	Similar stationary phases	Temp., °C		McReynolds constants						USP code
			Min	Max	x'	y'	z'	u'	s'	Σ	
Boiling-point separation of a broad molecular-weight range of compounds; nonpolar phases											
Squalane	2,6,10,15,19,23-Hexamethyltetracosane		20	150	0	0	0	0	0	0	
Paraffin oil					9	5	2	6	11	33	
Apiezon® L			50	300	32	22	15	32	42	143	
SPB-1	Poly(dimethylsiloxane)	SA-1, DB-1	–60	320	4	58	43	56	38	199	
SP™-2100	Poly(dimethylsiloxane)	DC-200, SE 30, UC W98, DC 200	0	350	17	57	45	67	43	229	G 9
OV-1	Methylsiloxane gum		100	350	16	55	44	65	42	227	G 2
OV-101	Methylsiloxane fluid		20	350	17	57	45	67	43	234	G 1
SPB-5	1% Vinyl, 5% phenyl methyl polysiloxane	SA-5, DB-5	–60	320	19	74	64	93	62	312	
SE-54	1% Vinyl, 5% phenyl methyl polysiloxane	PTE-5	50	300	19	74	64	93	62	312	G 36
SE-52	5% Phenyl methyl polysiloxane		50	300	32	72	65	98	67	334	G 27
OV-73	5.5% Phenyl methyl polysiloxane	SP-400	0	325	40	86	76	114	85	401	G 27
OV-3	Poly(dimethyldiphenyl-siloxane); 90%:10%		0	350	44	86	81	124	88	423	
Dexsil® 300	Carboranemethyl silicone		50	450	47	80	103	148	96	474	G 33
Dexsil® 400	Carborane–methylphenyl silicone		50	400	72	108	118	166	123	587	
OV-7	20% Phenyl methyl polysiloxane	DC 550	0	350	69	113	111	171	128	592	
SPB-20	20% Phenyl methyl polysiloxane	SPB-35, SPB-1701, DB-1301	<20	300	67	116	117	174	131	605	
Di-(2-ethylhexyl)-sebacate			–20	125	72	168	108	180	125	653	G 11
DC 550	25% Phenyl methyl polysiloxane		20	225	81	124	124	189	145	663	G 28

(Continued)

TABLE 5.11 McReynolds Constants for Stationary Phases in Gas Chromatography (*Continued*)

Stationary phase	Chemical type	Similar stationary phases	Temp., °C		McReynolds constants						USP code
			Min	Max	x'	y'	z'	u'	s'	Σ	
Unsaturated hydrocarbons and other compounds of intermediate polarity											
Diisodecyl phthalate			20	150	84	173	137	218	155	767	G 24
OV-11	35% Phenyl methyl polysiloxane		0	350	102	142	145	219	178	786	
OV-1701	Vinyl methyl polysiloxane	SPB-1701, SA-1701, DB-1701	0	250	67	170	152	228	171	789	
Poly-1 110				275	115	194	122	204	202	837	G 37
SP-2250	Poly(phenylmethyl-siloxane); 50% phenyl	OV-17, DB-17	0	375	119	158	162	243	202	884	G 3
Dexsil® 410	Carborane–methylcyano ethyl silicone		50	400	72	286	174	249	171	952	
UCON® LB-550-X	Polyalkylene glycol		20	200	118	271	158	243	206	996	
UCON LB-1880-X	Polyalkylene glycol			200	123	275	161	249	212	1020	G 18
Poly-A 103				275	115	331	144	263	214	1072	G 10
OV-22	Poly(diphenyldimethyl-siloxane); 65% : 35%		0	350	160	188	191	283	253	1075	
Di(2-ethylhexyl) phthalate				150	135	254	213	320	235	1157	G 22
OV-25	Poly(diphenyldimethyl-siloxane); 75% : 25%		0	350	178	204	208	305	280	1175	G 17
Moderately polar compounds											
DC QF-1			0	250	144	233	355	463	305	1500	
OV-210	50% Trifluoropropyl-methylpolysiloxane	SP-2401, DB-210	0	275	146	238	358	468	310	1520	G 6
OV-215	Poly(trifluoropropyl-methylsiloxane)		0	275	149	240	363	478	315	1545	
UCON-50-HB-2000	Polyalkylene glycol		0	200	202	394	253	392	341	1582	
Triton® X-100	Octylphenoxy poly-ethoxy ethanol		0	190	203	399	268	402	362	1634	

UCON 50-HB-5100	Polyglycol		0	200	214	418	278	421	375	1706	
XE-60	Poly(cyanoethylphenyl-methylsiloxane)		0	250	204	381	340	493	367	1785	G 26
OV-225	25% Cyanopropyl 25% phenyl methyl polysiloxane	DB-225, DB-23	0	265	228	369	338	492	386	1813	G 19
Ipegal CO-880	Nonylphenoxypoly-(ethyleneoxy)ethanol		100	200	259	461	311	482	426	1939	G 31
Triton® X-305	Octylphenoxy poly-ethoxy ethanol		20	250	262	467	314	488	430	1961	
Polar compounds											
Hi-EFF-3BP	Neopentylglycol succinate		50	230	272	469	366	539	474	2120	G 21
Carbowax 20M-TPA	Polythyleneglycol + terephthalic acid		60	250	321	367	368	573	520	2149	G 25
Supelcowax™ 10	Polyethyleneglycol + terephthalic acid	DB-WAX, SA-WAX	50	280	305	551	360	562	484	2262	
SP-1000	Polyethyleneglycol + terephthalic acid		60	220	304	552	359	549	498	2262	
Carbowax 20M	Polyethyleneglycol	SP-2300	25	275	322	536	368	572	510	2308	G 16
Nukol™		SP-1000, FFAP, OV-351			311	572	374	572	520	2349	
Carbowax 3350		Formerly Carbowax 4000	60	200	325	551	375	582	520	2353	G 15
OV-351	Polyethyleneglycol + nitroterephthalic acid	SP-1000	50	270	335	552	382	583	540	2392	
SP-2300	36% Cyanopropyl		25	275	316	495	446	637	530	2424	
Silar 5 CP	50% Cyanopropyl phenyl silicone	SP-2300	0	250	319	495	446	637	531	2428	G 7
FFAP			50	250	340	580	397	602	627	2546	G 35
Hi-EFF-10BP	Phenyldiethanolamine succinate		20	230	386	555	472	674	656	2744	G 21

(*Continued*)

TABLE 5.11 McReynolds Constants for Stationary Phases in Gas Chromatography (*Continued*)

Stationary phase	Chemical type	Similar stationary phases	Temp., °C		McReynolds constants						USP code
			Min	Max	x'	y'	z'	u'	s'	Σ	
Carbowax 1450		Formerly 1540	50	175	371	639	453	666	641	2770	G 14
SP-2380					402	629	520	744	623	2918	
SP-2310	55% Cyanopropyl	Silar 7 CP	25	275	440	637	605	840	670	3192	
SP-2330	68% Cyanopropyl	SP-2331, SH-60	25	275	490	725	630	913	778	3536	
Silar 9 CP	90% Cyanopropyl phenyl		50	250	489	725	631	913	778	3536	G 8
Hi-EFF-1BP	Diethyleneglycol succinate		20	200	499	751	593	840	860	3543	G 4
SP-2340	75% Cyanopropyl phenyl	OV-275, SH-80	<25	275	520	757	659	942	800	3678	
Silar 10 CP	100% Cyanopropyl silicone	SP-2340	25	275	523	757	659	942	801	3682	G 5
THEED	Amino alcohol		0	125	463	942	626	801	893	3725	
OV-275	Dicyanoallylsilicone		25	250	629	872	763	110	849	4219	
Absolute index values on squalane for reference compounds					653	590	627	652	699		

TABLE 5.12 Stationary Phases for Gas Chromatography Listed Alphabetically in Each Polarity Group

Nonpolar	Slightly polar	Moderately polar	Polar	Very polar
Apiezon L	CP SIL 8CB	CP SIL 13CB	Carbowax 1540	OV-275
CP SIL 5CB	DB-17, 1701	DB-624	Carbowax 20M	Silar 10 CP
DB-1	DC 710	DC QF-1	Carbowax 4000	SP-2330
DC 11, 200, 550	Dexsil 410	007-624	DB-WAX	SP-2340
Dexsil 300, 400	007-2	OV-210	FFAP	THEED
HP-1	OV-11, 17, 22, 25, 215, 1701	OV-215	Hi-EFF-3BP	
007-1	Poly-A 103	OV-225	Hi-EFF-10BP	
OV-1, 3, 7, 73, 105	Poly-1 110	SA-624	HP-20M	
Rtx 1	PPE-20, 21	SP-2401	Nukol	
SA-1	Rtx 5	Triton X-100	007-CW	
SE-30	SA-5	Triton X-305	OV-351	
SE-52	SP-2250	UCON-50-HB-280X	SA-WAX	
SE-54	SPB-5, 20, 1701	UCON-50-HB-2000	Silar 5 CP	
SP-2100	SPB-35	UCON-50-HB-5100	SP-1000	
SPB-1	UCON LB-550-X	VOCOL	SP-2300	
UC W98X		XE-60	SP-2380	
			Stabilwax	
			Supelcowax 10	

Suppliers:

CP SIL	Chrompack	OV	Ohio Valley Chemicals
SA	Sigma-Aldrich Chemicals	Rtx, Stabilwax	Restek
DB	Durabond, J&W Scientific	SE, XE	General Electric
DC	Dow Corning	SP, SPB, PTE, VOCOL	Supelco
HP	Hewlett Packard		
007	Quadrex	UCON	Union Carbide

the Carbowaxes, are even more polar and may be chemically bonded with polysiloxanes to increase the range of available polarities. For the separation of high-boiling-point compounds, a polar phase should be used so they can be eluted at a lower temperature. Very polar phases will have increasing amounts of cyanopropyl or cyanoallyl substituents.

One of the best ways to identify potential columns and stationary phases is to review the large number of application examples provided by column manufacturers and suppliers. Their catalogs contain many sample chromatograms obtained with a variety of columns. To help in identifying commercial phases of similar polarity, Table 5.12 is provided. Selected GC applications are also given in Tables 5.13 and 5.14. Application of GC techniques for analyzing trace pollutants in environmental samples is discussed in detail in Sec. 21.

5.2.4.2 Rohrschneider and McReynolds Constants.[14,15] To systematically categorize the multitude of liquid phases that have been suggested or that are commercially available, Rohrschneider and later McReynolds introduced a set of probes as a way of estimating the polarity of a stationary phase. From suitably chosen probes and the overall effects due to hydrogen bonding, dipole moment, acid–base properties, and molecular configuration one can get a measure of the relative magnitude of that interaction from its retention index. Squalane was chosen as the least polar and assigned constants of zero; London dispersion forces are the principal solute–solvent interaction.

To get the polarity constants for any other stationary phase, each of the five probes is run on squalene and on the liquid phase of interest at 100°C and 20% liquid loading. The difference between the retention index for each probe is obtained. The ΔI summed for all five probes is given as

$$\Delta I = ax' + by' + cz' + du' + es' \tag{5.31}$$

[14] L. Rohrschneider, *J. Chromatogr.* **22:**6 (1966).
[15] W. O. McReynolds, *J. Chromatogr. Sci.* **8:**685 (1970).

TABLE 5.13 Stationary Phase and Analyte-Type Cross Reference

See Tables 5.11 and 5.12 for phases similar to those listed.

Analyte types	Column	Analyte types	Column
Alcohols	SPB-1, SA-WAX, Carbowax 20M	Fermentation gases	SP-1000/1% H_3PO_4
Aldehydes	SPB-1, SA-1	Flavors and fragrances	SA-1, SA-WAX
Alkaloids	SP-2100, SP-2250	Glycerides	SA-5
Amines	SPB-1, SA-1	Glycols	THEED
Amino alcohols	SPB-1, SA-1	Halocarbons	SA-1, SA-5
Amphetamines	Apiezon L/KOH	Hydrocarbons	SA-1
Anticonvulsants	SA-1	Ketones	SA-1, SA-WAX
Antidepressants	SP-2250	Naphthols	SA-5
Antiepileptic drugs	SP-2510	Nitro compounds	SA-WAX
Barbiturates	SP-2250	Organic acids	SA-5, SA-1
Beta-blockers	SA-5	Polychlorobenzene	SA-5
Bile acids	SA-5	Pesticides	SA-1, SA-5
Drugs	SA-5	Phenols	SA-1, SA-5
Epoxides	SA-1	Polyaromatic hydrocarbons	SA-5
Esters	SA-WAX	Solvents, industrial	SP-1510
Fatty acids, C_1 to C_7	SA-1	Sulfur gases	XE-60/1% H_3PO_4
Fatty acids, methyl esters	SA-5, SA-WAX, SP-2100		

where *a, b, c, d,* and *e* represent the five solvents, and $x' = \Delta I$ for benzene (Table 5.11), which represents the intermolecular forces typical of aromatics and olefins, $y' = \Delta I$ for 1-butanol, which represents electron attraction that is typical of alcohols, nitriles, acids, and nitro and alkyl monochlorides, dichlorides, and trichlorides, $z' = \Delta I$ for 2-pentanone, which reflects electron repulsion that is typical of ketones, ethers, aldehydes, esters, epoxides, and dimethylamino derivatives, $u' = \Delta I$ for 1-nitropropane, which is typical of nitro and nitrile derivatives, and $s' = \Delta I$ for pyridine (or 1,4-dioxane). The last entry in the table shows the absolute retention indices for the individual reference compounds on squalane (see Sec. 5.1.4.2).

While the five values will not tell you which liquid phase is best for a given separation, any one large value can indicate a particularly strong interaction. For example, to elute an alcohol before an ether when both have nearly identical boiling points, a stationary phase with a high z' value (relative to the y' value) is needed. Stationary phases of SP-2401 or OV-210 should be tried. Just the reverse is needed if the ether is to be eluted before the alcohol, which suggests an OV-275 phase.

5.2.4.3 Separation of Enantiomers.[16] A molecule must lack symmetry for it to exhibit chirality. Separate two-mirror image molecules that cannot be superimposed are called enantiomers. Enantiomers or enantiomeric derivatives can be chromatographed on a chiral stationary phase. The separation of chiral compounds is significant from both preparative and analytical perspectives. The selective interaction of enantiomers with optically active stationary phases forms the basis for chromatographic separation of chiral compounds. Differences in the activity coefficient and the vapor pressure of the enantiomeric pair make GC separation possible.

Mono- and diamide and mono-, di-, and polypeptide stationary phases are used for separations of α-amino esters, alcohols, and other chiral compounds with hydrogen-bonding substituents. Upper temperature limits are 100 to 140°C. Higher limits are achieved by incorporating longer alkyl side chains and the larger amino acids. Chemical bonding to a poly(alkylsiloxane) backbone improves the upper temperature limits to 220°C of this class of stationary phase; a commercial packing is Chirasil-Val (Chrompack Inc.).

[16] J. V. Hinshaw, *LC-GC*, **11:**644 (1993).

TABLE 5.14 Selected Gas-Chromatography Applications

Sample	Column	Stationary phase	Operating conditions	Detector
		Clinical applications		
Alkaloids	3 ft (91.4 cm) × 2 mm, glass	100/120 Supelcoport; 3% SP-2250-DB	N_2: 40 mL/min; 230°C	FID
Amphetamines	3 ft (91.4 cm) × 2 mm, glass	80/100 Chromosorb W AW; 10% Apiezon L/2% KOH	N_2: 40 mL/min; 125°C	FID
Anesthetics	30 m × 0.53 mm	5-μm film DB-1	He: 6 mL/min; 3.8 min at 50°C, then to 125°C at 30°C/min	FID
Anticonvulsant drugs	15 m × 0.25 mm	0.15-μm film DB-1701	H_2: 37 cm/s; 1 min at 140°C, then to 260°C at 8°C/min	FID
Antidepressants	6 ft (183 cm) × 2 mm, glass	80/100 Supelcoport; 3% SP-2250	He: 40 mL/min; 250°C	NPD
Antiepileptic drugs	3 ft (91.4 cm) × 2 mm, glass	100/120 Supelcoport; 2% SP-2510-DA	He: 50 mL/min; 150° (16°C/min to 265°C), hold 4 min	FID
Antihistamines	30 m × 0.53 mm	1.5-μm film DB-1	He: 18 mL/min; 215°C to 275°C at 5°C/min	FID
Bacterial acid methyl esters	30 m × 0.25 mm	0.25-μm film DB-5	H_2: 42 cm/s; 4 min at 150°C, then to 250°C at 4°C/min	FID
Barbiturates	30 m × 0.75 mm wide bore	1.0-μm film of SPB-1	He: 10 mL/min; 150°C (8°C/min to 290°), hold 10 min	FID
Barbiturates—low molecular weight	6 ft (183 cm) × 2 mm glass	100/120 Supelcoport; 3% SP-2250-DA	N_2: 20 mL/min; 230°C	FID
Blood alcohols	6 ft (183 cm) × 2 mm glass	60/80 Carbopack B; 5% Carbowax 20M	He: 20 mL/min; 85°C	FID
Fentanyl derivatives	30 m × 0.53 mm	1.0-μm film DB-1701	H_2: 15 mL/min; 270°C	FID
Sedatives and hypnotics	30 m × 0.53 mm	1.5-μm film DB-1	He: 30 mL/min; 150°C to 250°C at 10°C/min	FID
Steroids, free	30 m × 0.25 mm	0.15-μm film DB-17	H_2: 44 cm/s; 260°C	FID
Valproic acid and ethosuximide	3 ft (91.4 cm) × 2 mm glass	80/100 Supelcoport; 10% SP-1000	He: 40 mL/min; 190°C	FID
		Environmental analyses—air		
Chlorinated hydrocarbons	6 ft (183 cm) × $^1/_8$ in stainless steel	80/100 Carbopack; 0.1% SP-1000	4 mL/min He to thermally desorb; He: 20 mL/min; 2 min at 35°C, then 16°C/min to 220°C	FID
Ethylene oxide and its derivatives	3 ft (91.4 cm) × 2 mm glass	80/100 Carbopack; 0.8% THEED	He: 20 mL/min; 70°C	FID
Hydrocarbons (thermally desorbed TO-1 compounds)	60 m × 0.75 mm capillary	1.0-μm film Supelcowax 10	10 mL/min He to thermally desorb; N_2: 40 mL/min; 4 min at 35°C, then 8°C/min to 220°C	FID

(*Continued*)

TABLE 5.14 Selected Gas-Chromatography Applications (*Continued*)

Sample	Column	Stationary phase	Operating conditions	Detector
		Clinical applications		
Hydrocarbons (thermally desorbed TO-2 and TO-3 compounds)	60 m × 0.75 mm capillary	1.0-μm film Supelcowax 10	8 mL/min He to thermally desorb; N_2: 50 mL/min; 4 min at 35°C, then 8°C/min to 160°C	FID
	8 ft (2.4 m) × 1/8 in (0.31 cm) stainless steel	60/80 Carbopack B; 1% SP-1000	He: 20 mL/min; 2 min at 35°C, then 16°C/min to 220°C	FID
Kraft mill stack gases	30 in (75 cm)[1/8 in (45 cm) packed] × 1/8 in (0.31 cm) Teflon	Supelpak-S	He: 30 mL/min; 1 min at 25°C, then 30°C/min to 210°C	FPD (S)
Organic solvents	30 m × 0.53 mm fused silica	1.0-μm film Supelcowax 10	He: 5 mL/min; 5 min at 40°C, then 5°C/min to 200°C	FID
	30 m × 0.32 mm fused silica	1.0-μm film PTE-5	He: 20 cm/s (at 135°C); 5 min at 40°C, then 4°C min to 130°C	FID
Sulfides and disulfides	1.4 m × 1/8 in (0.31 cm) Teflon	40/60 Carbopack B HT 100	N_2: 20 mL/min; 35°C	FPD (S)
Sulfides and thiols	8 ft (2.4 m) [6 ft (1.8 m) packed] × 1/8 in (0.31 cm) Teflon	Chromosil 330	N_2: 20 mL/min; 40°C (C_1 to C_3), higher molecular weight compounds at 65°C	FPD (S)
Sulfur gases (standard mixture)	30 m × 0.32 mm fused silica	4-μm film SPB-1 Sulfur	N_2: 3 min at –10°C, then 10°C/min to 300°C	Sulfur chemiluminescent
Sulfur gases: COS, H_2S, CS_2, and SO_2	8 ft (2.4 m) [6 ft (1.8 m) packed] × 1/8 in (0.31 cm) Teflon	Chromosil 310	N_2: 20 mL/min; 50°C	FPD (S)
Sulfur C_2 isomers, SF_6, H_2S, and COS	6 ft (1.8 m) × 2 mm glass	60/80 Carbopak B, 1.5% XE-60/1.0% H_3PO_4	N_2: 35 mL/min; 50°C	FPD (S)
		Environmental analysis—drinking water		
Halogenated hydrocarbons	60 m × 0.53 mm	5.0-μm film SA-1	H_2: 165 cm/s; 1 min at 50°C, then to 250°C at 10°C/min	FID
Herbicides—nitrogen containing	30 m × 0.25 mm fused silica	0.25-μm film PTE-5	He: 30 cm/s; 60°C to 300°C at 4°C/min	TSD
Organic compounds (EPA method 525)	30 m × 0.25 mm fused silica	0.25-μm film PTE-5	He: 40 cm/s; 120°C for 4 min, then to 320°C at 10°C/min	FID
Organophosphorus compounds	30 m × 0.25 mm fused silica	0.25-μm film PTE-5	He: 20 cm/s; 100°C to 300°C at 4°C/min	TSD

Environmental analysis—drinking water				
Pesticides, chlorinated (EPA method 508)	30 m × 0.25 mm fused silica	0.25-μm film PTE-5	He: 30 cm/s; 60°C to 300°C at 4°C/min	ECD
Pollutants, volatile (EPA method 505.2)	105 m × 0.53 mm VOCOL fused	3.0-μm film	He: 10 mL/min; 10 min at 35°C, then to 200°C at 4°C/min	PID & ELCD
Pollutants, volatile (EPA method 524.1)	8 ft (2.4 m) × 1/8 in (0.31 cm) stainless steel	60/80 Carbopack B, 1% SP-1000	He: 40 mL/min; 3 min at 45°C, then to 220°C at 8°C/min	FID
Pollutants, volatile (EPA method 524.2)	60 m × 0.75 mm VOCOL capillary	1.5-μm film	He: 10 mL/min; 8 min at 40°C, then to 200°C at 4°C/min; MS range: 33 to 250 m/z	MS
Polychlorinated biphenyls and halogenated pesticides	30 m × 0.32 mm fused silica	1.0-μm SPB-1	He: 25 cm/s; 180°C to 260°C at 4°C/min	ECD
Environmental analysis—wastewater				
Aromatic pollutants	30 m × 0.25 mm	0.25-μm film DB-WAX	H_2: 47 cm/s; 40°C	FID
Chlorinated hydrocarbons	15 m × 0.53 mm fused silica	1.5-μm film SPB-5	He: 10 mL/min; 50°C to 175°C at 8°C/min	ECD
Dioxins	60 m × 0.25 mm fused silica	0.2-μm film SP-2331	He: 40 cm/s; 200°C to 265°C at 8°C/min, and hold 15 min	MS or ECD
Herbicides	15 m × 0.53 mm fused silica	0.5-μm film SPB-608	He: 5 mL/min; 1 min at 60°C, then to 280°C at 16°C/min, and hold 5 min	ECD
Herbicides	30 m × 0.25 mm	0.25-μm film DB-1301	H_2: 40 cm/s; 1 min at 160°C, then to 205°C at 3°C/min	FID
Insecticides, organophosphorus	30 m × 0.25 mm	0.25-μm film DB-1301	H_2: 40 cm/s; 1 min at 140°C, then to 160°C (6°C/min) and to 180°C (10°C/min)	FID
Nitrosamines	15 m × 0.53 mm fused silica	1.5-μm film SPB-5	He: 20 mL/min; 2 min at 35°C, then to 200°C at 20°C/min	NPD
Pesticides	2 m × 4 mm glass	100/120 Supelcoport; 1.5% SP-2250/1.95% SP-2401	N_2: 60 mL/min; 200°C	ECD
Pesticides, chlorinated	Dual 30 m × 0.25 mm fused silica	0.25-μm film: (a) SPB-608 or (b) PTE-5	He: 25 cm/s at 290°C; 4 min at 150°C, then 6°C/min to 290°C, and hold 5 min	ECD
	30 m × 0.25 mm	0.25-μm film DB-17	H_2: 42 cm/s; 1 min at 110°C, then to 200°C at 20°C/min, and to 250°C at 4°C/min	FID
Phenols	15 m × 0.53 mm fused silica	1.5-μm film SPB-5	He: 15 mL/min; 2 min at 75°C, then to 180°C at 8°C/min, and hold 1 min	FID

(*Continued*)

TABLE 5.14 Selected Gas-Chromatography Applications (*Continued*)

Sample	Column	Stationary phase	Operating conditions	Detector
	2 m × 2 mm glass	100/120 Supelcoport; 1% SP-1240-DA	He: 30 mL/min; 2 min at 70°C, then to 200°C at 8°C/min, and hold 10 min	FID
Phthalates	15 m × 0.53 mm fused silica	1.5-μm film SPB-51	He: 30 mL/min; 4 min at 115°C, then to 250°C at 16°C/min, and hold 5 min	FID
Volatile pollutants; see under Environmental analyses—drinking water				
		Flavors and fragrances		
Bergamot oil	30 m × 0.25 mm	0.25-μm film Supelcowax 10	He: 20 cm/s; 2 min at 50°C, then to 230°C at 2°C/min and hold 1 min	FID
Cassia oil	40 m × 0.18 mm	0.4-μm film DB-1	H_2: 46 cm/s; 120°C (8 min), to 190°C at 6°C/min and hold 20 min	FID
Fragrance components frequently used	30 m × 0.25 mm	0.25-μm film Supelcowax 10	He: 25 cm/s; 2 min at 50°C, then to 280°C at 2°C/min and hold 20 min	FID
Geranium oil	30 m × 0.25 mm	0.25-μm film Supelcowax 10	He: 20 cm/s; 2 min at 50°C, then to 230°C at 2°C/min and hold 1 min	FID
Lemon oil	30 m × 0.25 mm	0.25-μm film SPB-5	He: 25 cm/s; 8 min at 75°C, then to 200°C at 4°C/min and hold 4 min	FID
Lime oil	30 m × 0.25 mm	0.25-μm film SPB-5	He: 25 cm/s; 8 min at 75°C, then to 200°C at 4°C/min and hold 4 min	FID
Peppermint oil	60 m × 0.25 mm	0.25-μm film Supelcowax 10	He: 25 cm/s; 8 min at 75°C, then to 200°C at 4°C/min and hold 5 min	FID
Spearmint oil	60 m × 0.25 mm	0.25-μm film Supelcowax 10	He: 25 cm/s; 4 min at 75°C, then to 200°C at 4°C/min and hold 5 min	FID
		Food and beverages		
Alcohols, C_2 to C_6	30 m × 0.25 mm fused silica	0.25-μm film SP-1701	He: 20 cm/s; 60°C	FID
Alcoholic beverages	2 m × 2 mm glass	80/120 Carbopack B AW; 5% Carbowax 20M	N_2: 20 mL/min; 70°C to 170°C at 5°C/min	FID
Alditol acetates of amino sugars, monosaccharides	6 ft (1.83 m) × 2 mm glass	100/120 Supelcoport; 3% SP-2330	N_2: 20 mL/min; 225°C	FID
Beer	30 m × 0.25 mm	0.25-μm film DB-WAX	H_2: 45 cm/s; 5 min at 35°C, then to 230°C at 6°C/min	FID
Bile acid, methyl esters	3 ft (0.91 m) × 2 mm glass	100/120 Supelcoport; 3% SP-2250	He: 40 mL/min; 275°C	FID
Fatty acids, free C_2 to C_{22}	15 m × 0.53 mm fused silica	0.5-μm film Nukol	He: 20 mL/min; 110°C to 220°C at 8°C/min	FID
Fatty acids, free C_2 to C_7	30 m × 0.25 mm fused silica	0.25-μm film Nukol	He: 20 cm/s; 185°C	FID

Fatty acids, water soluble	2 (0.6 m) × 2 mm glass	80/120 Carbopack B DA, 4% Carbowax 20M	N_2: 24 mL/min; 175°C	FID
Fatty acids, methyl esters, bacterial	30 m × 0.25 mm fused silica	0.25-μm film SPB-1	He: 20 cm/s; 4 min at 150°C, then to 250°C at 4°C/min	FID
Fatty acids, methyl esters, dibasic	30 m × 0.25 mm fused silica	0.20-μm film SP-2380	He: 20 cm/s; 170° to 200°C at 4°C/min	FID
	6 ft (1.8 m) × 1/8 in (0.31 cm) stainless steel	100/120 Chromosorb W AW, 10% SP-2340	N_2: 20 mL/min; 170° to 220°C at 4°C/min	FID
Fatty acids, methyl esters, polyunsaturated	30 m × 0.25 mm fused silica	0.25-μm film SP-2330	He: 20 cm/s (at 200°C); 200°C (injection and detector at 250°C)	FID
Fatty acids, methyl esters, tall oil	30 m × 0.25 mm fused silica	0.20-μm film SP-2380	He: 20 cm/s; 170° to 260°C at 4°C/min	FID
Fermentation products	30 m × 0.25 mm	0.25-μm film SPB-1701	He (at 220°C) 20 cm/s; 220°C	FID
		0.20-μm film SP-2380	He (at 275°C) 20 cm/s; 275°C	FID
	2 m × 2 mm	80/120 Carbopack B AW; 6.6% Carbowax 20M	N_2 carrier gas; 80°C to 200°C at 4°C/min	FID
Peppers, Jalapeno	30 m × 0.25 mm	0.25-μm film DB-5	H_2: 50 cm/s; 50°C to 300°C at 4°C/min	FID
Vermouth, dry	30 m × 0.25 mm	0.25-μm film DB-WAX	H_2: 43 cm/s; 5 min at 35°C, then to 200°C at 6°C/min	FID
Whiskey, Scotch	30 m × 0.53 mm	1.0-μm film DB-WAX	He: 3 mL/min; 3 min at 35°C, then to 170°C at 7.5°C/min	FID
Wine, alcohols, and esters	60 m × 0.32 mm	1.2-μm film SA-1	H_2: 3.3 cm/s; 60°C to 200°C at 3°C/min	FID
Wine, white	30 m × 0.25 mm	0.25-μm film DB-WAX	H_2: 48 cm/s; 5 min at 35°C, then to 230°C at 6°C/min	FID
		Petroleum and chemicals		
Acetates, C_1 to C_8	30 m × 0.53 mm	2-μm film SA-WAX	N_2: 10 mL/min; 50°C to 200°C at 10°C/min; injector 250°C; detector 275°C	FID
Acetic acid impurities	Same as for Acetates, C_1 to C_8			
Alcohols, purity analyses	30 m × 0.53 mm	5-μm film SPB-1	He: 20 cm/s; ethanol: 5 min at 70°C; 1- and 2-Propanol: 10 min at 70°C; 1-Butanol and 1-pentanol: 10 min at 110°C	FID
Alcohols, C_1 to C_5	2 m × 2 mm	80/100 Carbopack C, 2% Carbowax 1500	N_2: 20 mL/min; 135°C	FID

(Continued)

TABLE 5.14 Selected Gas-Chromatography Applications (*Continued*)

Sample	Column	Stationary phase	Operating conditions	Detector
		Petroleum and chemicals		
Alcohols, C_1 to C_6	60 m × 0.53 mm	5-μm film SA-1	H_2: 25 cm/s; 2 min at 45°C, then to 80°C (at 5°C/min) and on to 200°C at 10°C/min	FID
Alcohols, C_4 to C_8	30 m × 0.53 mm	SA-WAX	N_2: 10 mL/min; 50°C to 200°C at 10°C/min	FID
Alcohols, ketones, and ethers	60 m × 0.32 mm	1.2-μm film SA-1	H_2: 40 cm/s; 3 min at 40°C, then to 250°C at 10°C/min	FID
Alkaloids		100/120 Supelcoport, 3% SP-2250 DB		
Alkylbenzenes	30 m × 0.25 mm fused silica	0.25-μm film SPB-20	He: 20 cm/s; 170°C	FID
Amines, aliphatic	6 ft (1.8 m) × 2 mm	60/80 Carbopack B, 4% Carbowax 20M/0.8% KOH	N_2: 20 mL/min; 90°C to 150°C at 4°C/min	FID
Amines, C_2 to C_8	60 m × 0.32 mm	5-μm film SA-1	H_2: 165 cm/s; 1 min at 40°C, then to 200°C at 10°C/min	FID
Amines, heterocyclic	6 ft (1.8 m) × 2 mm	60/80 Carbopack B, 4% Carbowax 20M/0.8% KOH	N_2: 20 mL/min; 140°C	FID
Amino alcohols, C_2 to C_5	60 m × 0.53	5-μm film SA-1	H_2: 25 cm/s; 65°C to 100°C at 10°C/min	FID
Cholesteryl esters	0.5 m × 2 mm	100/120 Supelcoport, 1% Dexsil 300	N_2: 40 mL/min; 300°C to 350°C at 6°C/min; injection (325°C), detector (350°C)	FID
Cresol and dimethylphenols	30 m × 0.25 mm	Nukol	He: 20 cm/s; 200°C	FID
	6 ft (1.8 m) × 2 mm glass	80/100 Carbopack C, 1% SP-1000	N_2: 20 mL/min; 225°C	FID
Ethers	30 m × 0.53 mm	2.0 μm film SA-WAX	N_2: 47 cm/s (10 mL/min); 50°C to 200°C at 10°C/min	FID
Ethylene oxide	1 m × 2 mm	80/100 Carbopack C, 8% THEED	N_2: 20 mL/min; 115°C	FID
Freons	10 ft (3.2 m) × 1/8 in (0.31 cm) stainless steel	60/80 Carbopack B, 5% Fluocol	N_2: 30 mL/min; 30 mL/min	FID
Gases, permanent, and C_1, C_2 hydrocarbons	10 ft (3.2 m) × 1/8 in (0.31 cm) stainless steel	100/120 Carbosieve S-11	He: 30 mL/min; 7 min at 35°C, then to 225°C at 32°C/min	TCD
Gasoline	100 m × 0.25 mm fused silica	0.5-μm film Petrocol DH	He: 20 cm/s; 15 min at 35°C, then to 200°C at 2°C/min	FID
Glycols	15 m × 0.53 mm	0.5-μm film Nukol	He: 15 mL/min; 110°C to 220°C at 8°C/min	FID
Halogenated hydrocarbons (EPA method 624)	30 m × 0.53 mm	2.0-μm film SA-624	He: 3 mL/min; 2 min at 35°C, then to 80°C at 2°C/min	FID

Hydrocarbons, C_1 to C_6, and C_1 to C_4 alcohols	60 m × 0.53 mm	5-μm film SPB-1	He: 20 cm/s; 5 min at 30°C, then to 200°C at 20°C/min; injection and detector at 200°C	FID
	30 ft (9.6 m) × 1/8 in (0.31 cm) stainless steel	80/100 Chromosorb P AW, 23% SP-1700	He: 25 mL/min; 70°C	FID
Hydrocarbons, C_1 to C_3, by degree of unsaturation	5 ft (1.5 m) × 1/8 in (0.31 cm) stainless steel	60/80 Carbosieve G	N_2: 50 mL/min; 145°C to 195°C at 6°C/min and hold 5 min	FID
Hydrocarbons, C_5 to C_{13}	50 m × 0.20 mm fused silica	0.50-μm Petrocol DH 50.2 (methyl silicone)	He: 20 cm/s; 30 min at 35°C, then to 200°C at 2°C/min; detector at 225°C	FID
Hydrocarbons, aromatic, and phenols	60 m × 0.53 mm	2.0-μm film SA-1701	H_2: 70 cm/s; at 10°C/min, 50°C to 130°C, hold 2 min, then to 225°C	FID
Ketones, C_3 to C_{10}	30 m × 0.53 mm	2.0-μm film SA-WAX	N_2: 47 cm/s (10 mL/min); 50°C to 200°C at 10°C/min	FID
N-Nitrosoamines	2 m × 4 mm	80/100 Chromosorb W AW, 10% Carbowax 20M/2% KOH	He: 40 mL/min; 150°C	NPD
Nitroaromatics	30 m × 0.53 mm fused silica	1.5-μm film SPB-5	He: 10.5 mL/min; 1 min at 120°C, then to 210°C at 3°C/min	ECD
Nitro compounds	30 m × 0.53 mm	2.0-μm SA-WAX	N_2: 47 cm/s; 50°C to 200°C at 5°C/min	FID
Oak leaves extract	30 m × 0.32 mm	1.0-μm film DB-5	H_2: 45 cm/s; 4 min at 40°C, then to 320°C at 4°C/min	FID
Paraffins, high boiling	15 m × 0.53 mm	0.1-μm film SPB-1	He: 20 mL/min; 50°C to 350°C at 15°C/min	FID
Phenols	30 m × 0.25 mm	0.25-μm film PTE-5	He: 20 mL/s (at 115°C); 4 min at 50°C, then to 220°C at 8°C/min	FID
Solvents, industrial (and impurities in)	30 m × 0.32 mm	1.0-μm film SPB-1	He: 25 cm/s; 8 min at 30°C, then to 125°C at 4°C/min	FID
	10 ft (3.2 m) × 1/8 in (0.31 cm) stainless steel	Carbopack B, 3% SP-1500	N_2: 20 mL/min; 70°C to 225°C at 4°C/min	FID
Solvents, pharmaceutical	60 m × 0.53 mm	5.0-μm film SPB-1	He: 20 cm/s; 5 min at 30°C, then to 200°C at 20°C/min	FID
Sulfur compounds in diesel fuel, gasolines and sour natural gases	30 m × 0.32 mm	4-μm film SPB-1 SULFUR	−10°C for 3 min, then to 300°C at 10°C/min; dual sulfur chemiluminescence detector and FID	FID & S
Sulfur compounds	30 m × 0.53 mm	5-μm film SA-5	H_2: 3.8 mL/min; 30–110°C at 5°C/min	FPD
Triglycerides	18 in (45 cm) × 1/8 in (0.31 cm) stainless steel	100/120 Supelcoport, 1% Dexsil 300	N_2: 20 mL/min; 275°C to 350°C at 8°C/min; injection 325°C, detector 375°C	FID
Triglycerides, butter	15 m × 0.25 mm	0.1 μm film DB-5	H_2: 65 cm/s; 180°C to 230°C at 25°C/min, then to 340°C at 5°C/min	FID

The cyclodextrins are the most universally applicable group of chiral GC phases yet identified. For the most part, columns are prepared from the heptamer β-cyclodextrin, derivatized with various side chains at the multiple hydroxyl sites. Selectivity is controlled by varying the side-chain substitution and the size of the cyclodextrin ring. Specific molecular functional groups do not seem to be required. Selectivity is thought to be due primarily to formation of inclusion complexes of chiral solutes in the large molecular cavity. Thus, shape recognition appears to be the most significant chiral interaction for cyclodextrins.

5.2.4.4 Multidimensional Separations. In multidimensional gas chromatography, the components of a sample are separated by using series-connected columns of different capacity or selectivity. Two common multicolumn configurations are the packed-column and capillary-column combination and two capillary columns in series. Two independently controlled ovens may be needed, and such a configuration is available commercially. In addition to decreased analysis time, this arrangement provides an effective way of handling samples containing components that vary widely in concentration, volatility, and polarity. Used in conjunction with techniques such as heart-cutting, back-flushing, and peak switching, useful chromatographic data have been obtained for a variety of complex mixtures.

5.2.5 Detectors

After separation in the column, the sample components enter a detector. The detector should have the following characteristics:

1. High sensitivity
2. Low noise level (background level)
3. Linear response over a wide dynamic range
4. Good response for all organic component classes
5. Insensitivity to flow variations and temperature changes
6. Stability and ruggedness
7. Simplicity of operation
8. Positive compound identification

In Fig. 5.10 gas-chromatographic detectors are compared with respect to sensitivity and their linear dynamic ranges. What is desired is either a universal detector that is sufficiently sensitive or a dedicated detector that is very sensitive and specific to particular classes of molecules. Hill and McMinn discuss a large number of detectors in their monograph.[17]

5.2.5.1 Thermal Conductivity Detector. The thermal conductivity detector (TCD) is the most common universal detector used in GC. It is rugged, versatile, and relatively linear over a wide range. The thermal conductivity detector, however, is less sensitive than most other GC detectors. In operation it measures the difference in the thermal conductivity between the pure carrier gas and the carrier gas plus components in the gas stream (effluent) from the separation column. The detector uses a heated filament (often rhenium–tungsten) placed in the emerging gas stream. The amount of heat lost from the filament by conduction to the detector walls depends on the thermal conductivity of the gas. When substances are mixed with the carrier gas, its thermal conductivity goes down (except for hydrogen in helium); thus, the filament retains more heat, its temperature rises, and its electrical resistance goes up. Monitoring the resistance of the filament with a Wheatstone bridge circuit provides a means of detecting the presence of the sample components. The signals, which are fed to a chart recorder, appear as peaks on the chart, which provides a

[17] H. H. Hill and D. G. McMinn, *Detectors for Capillary Chromatography*, Wiley-Interscience, New York, 1992.

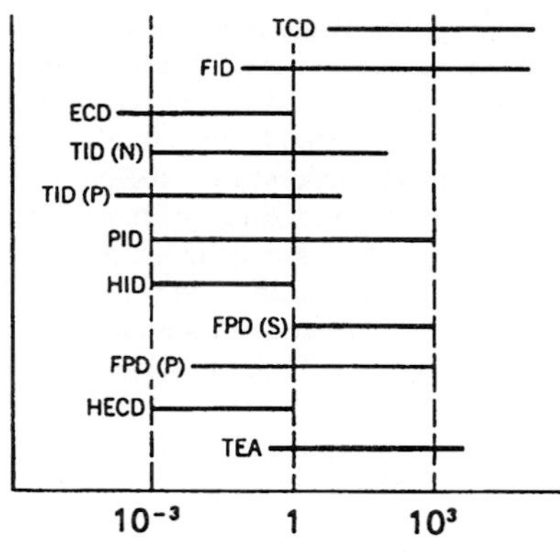

FIGURE 5.10 Relative response of GC detectors: ECD, electron capture; FID, flame ionization; FPD (P), flame photometric (phosphorus); FPD (S), flame photometric (sulfur); HECD, Hall electrolytic conductivity; HID, helium ionization; PID, photoionization; TCD, thermal conductivity; TID (N), thermal ionization (nitrogen); TID (P), thermal ionization (phosphorus).

visual representation of the process. The cavity volume of the detector should be matched with the separatory column used: 2.5 mL for detectors coupled to packed columns down to 30 μL in a detector designed for use with capillary columns.

Of all the detectors, only the thermal conductivity detector responds to anything mixed with the carrier gas. Being nondestructive, the effluent may be passed through a thermal conductivity detector and then into a second detector or a fraction collector in preparative gas chromatography.

The linearity of the detector is good at the lower concentration range but not in the high percent range. In the high percent range a multipoint calibration is the only way to ensure accurate measurements. At the low parts-per-million concentration, the trace impurities present in the carrier gas could be a limiting factor. Oxygen is the most detrimental carrier-gas impurity. Even gold- and nickel-coated tungsten–rhenium hot wires are susceptible to oxidation, which may unbalance the bridge to the point where it cannot be rezeroed. Also, oxide formation on the hot-wire surface will minimize the detector's ability to sense changes in thermal conductivity and thus decrease its sensitivity.

5.2.5.2 Flame-Ionization Detector. The flame-ionization detector (FID) is the most popular detector because of its high sensitivity [0.02 coulomb (C) per g of hydrocarbon], wide linear dynamic range, low dead volume, and responsiveness to trace levels of almost all organic compounds. This detector adds hydrogen to the column effluent and passes the mixture through a jet, in which it is mixed with entrained air and burned. The ionized gas (charged particles and electrons produced during combustion) passes through a cylindrical electrode. A voltage applied across the jet and the cylindrical electrode sets up a current in the ionized particles. An electrometer monitors this current to derive a measure of the component concentration. An ignitor coil and flame-out sensor are placed above the jet to reignite the flame should it become extinguished. The entire assembly is enclosed within a chimney so that it is unaffected by drafts and can be heated sufficiently to avoid condensation of water droplets resulting from the combustion process.

The response of the flame ionization detector is proportional to the number of $—CH_2—$ groups that enter the flame. The response to carbons attached to hydroxyl groups and amine groups is lower. There is no response to fully oxidized carbons such as carbonyl or carboxyl groups (and thio analogs) and to ether groups. Gaseous substances giving little or no response are listed in Table 5.15.

TABLE 5.15 Gaseous Substances Giving Little or No Response in the Flame-Ionization Detector

He	CS	NH_3
Ar	COS	CO
Kr	H_2S	CO_2
Ne	SO_2	H_2O
Xe	NO	$SiCl_4$
O_2	N_2O	$SiHCl_3$
N_2	NO_2	SiF_4

Source: G. F. Shugar and J. A. Dean, *The Chemist's Ready Reference Handbook*, McGraw-Hill, New York, 1990.

The detector's insensitivity to moisture and to the permanent gases is advantageous in the analysis of moist organic samples and in air-pollution studies. Modest changes in flow, pressure, or temperature have only a small effect on the response characteristics. Column operating temperatures can vary from 100 to 420°C, an obvious advantage in programmed temperature applications.

The quality of both the carrier gas and the hydrogen support gas is more critical than that of the flame support air. The presence of organic substances in any of these gases will increase the detector noise and the minimum detectable limit and, therefore, decrease the dynamic range. There are more stringent requirements on the detector when capillary columns are used because of the reduced sample size. The electrometer must be able to display current pulses in the picoampere to nanoampere range. In addition, a faster response time is required because of improved resolution and shorter retention times.

The position of the capillary column in the detector is important; the column should extend to within a few millimeters of the FID flame but not to extend into the flame.[18] The design of the jet is important so that the flow does not convolute and cause a turbulent flame. In most applications, addition of a makeup gas (hydrogen is better than helium) is recommended in order to optimize the detector and to sweep the detector volume to minimize band spreading.

5.2.5.2.1 Cleaning the Detector. Problems such as excessive noise or random spikes on the chromatogram or a general lack of detector sensitivity may indicate a dirty detector. In-place cleaning is done with a halocarbon liquid, Freon 113. Inject 5 μL once or twice a day into the chromatographic column while the flame is lit. The combustion products of the cleaner remove silica deposits from the detector electrodes.

For light cleaning, turn off the power to the detector and remove the cap or top. Use a reamer to clean the jet opening, brush the jet with a wire brush.

If more thorough cleaning is required, disassemble the detector carefully. Use a brush to remove all deposits (use stainless steel for hard deposits and brass for softer ones). Clean all parts including the insulators, especially where Teflon or ceramic insulators contact metal. If necessary, the detector parts can be immersed in a surfactant cleaning solution, preferably in a ultrasonic bath. Commercial cleaning kits are available.

5.2.5.3 Thermionic Ionization (NP) Detector.[19,20] The thermionic ionization detector (TID) responds only to compounds that contain nitrogen or phosphorus. Its fabrication is similar to that of a flame-ionization detector and, consequently, nitrogen–phosphorus detector (NPD) equipment is usually designed to mount on an existing FID-type detector base. The thermionic source has the shape of a bead or cylinder centered above the flame tip. This bead is composed of an alkali-metal compound impregnated in a glass or ceramic matrix. The body of the source is molded over an electrical heating wire. A typical operating temperature is between 600 and 800°C. A fuel-poor hydrogen flame is used to suppress the normal flame-ionization response of compounds that do not contain nitrogen or phosphorus. With a very small hydrogen flow, the detector responds to both nitrogen and phosphorus compounds. Enlarging the flame size and changing the polarity between the jet and collector limit the response to only phosphorus compounds. Located in proximity to the ionization source is an ion-collector electrode that usually is cylindrical. The thermionic source is also polarized at a voltage that causes ions formed at the source to move toward the ion collector.

Compared with the flame-ionization detector, the thermionic emission detector is about 50 times more sensitive for nitrogen and about 500 times more sensitive for phosphorus.

5.2.5.4 Electron-Capture Detector. The electron-capture detector (ECD) responds only to electrophilic species, such as nitrogenated, oxygenated, and halogenated compounds. The ECD is known as a "selective" electrode that is capable of providing extremely sensitive responses to specific sought-for substances that might be present in a sample containing a large excess of little or no responsive substances. It consists of two electrodes. On the surface of one electrode is a radioisotope

[18] K. J. Hyver, *High Resolution Gas Chromatography*, 3d ed., Hewlett-Packard Co., Avondale, PA, 1989, Chap. 4.
[19] B. Kolb and J. Bischoff, *J. Chromatogr. Sci.* **12:**625 (1974).
[20] P. L. Patterson, *J. Chromatogr.* **167:** 381 (1978); *Chromatographica,* **16:**107 (1982); *J. Chromatogr. Sci.* **24:**41 (1986).

(usually nickel-63, although tritium has been used) that emits high-energy electrons as it decays. Argon mixed with 5% to 10% methane is added to the column effluent. The high-energy electrons bombard the carrier gas (which must be nitrogen when this detector is used) to produce a plasma of positive ions, radicals, and thermal electrons. A potential difference applied between the two electrodes allows the collection of the thermal electrons. The resulting current when only carrier gas is flowing through the detector is the baseline signal. When an electron-absorbing compound is swept through the detector, there will be a decrease in the detector current, that is, a negative excursion of the current relative to the baseline as the effluent peak is traced. The potential is applied as a sequence of narrow pulses with a duration and amplitude sufficient to collect the very mobile electrons but not the heavier, slower negative ions. The ^{63}Ni sources can be safely heated up to 400°C with no loss of activity. Residual oxygen and water must be rigorously removed from the carrier gas and makeup gases.

Next to the TCD and FID, the ECD has the highest usefulness in the GC field. Unlike the FID, the ECD has neither ease of operation nor dynamic range. What it does have is detectability on the order of 1×10^{-13} g and good specificity. Trace residues of chlorinated pesticides, herbicides, and polychlorinated hydrocarbons can be measured by ECD with very high sensitivity. Steroids, biological amines, amino acids, and various drug metabolites can be converted to perfluoro derivatives, which will give a signal with this detector.

5.2.5.5 Flame Photometric Detector. In the flame photometric detector (FPD), the column effluent passes into a hydrogen-enriched, low-temperature flame contained within a shield. Both air and hydrogen are supplied as makeup gases to the carrier gas. Two flames are used to separate the region of sample decomposition from the region of emission. Flame blowout is no problem because the lower flame quickly reignites the upper flame. Phosphorus compounds emit green band emissions at 510 and 526 nm that are due to HPO species. Sulfur compounds emit a series of bands from excited diatomic sulfur (S_2^*); the most intense is centered around 394 nm, but other bands overlap the phosphorus spectrum. Phosphorus and sulfur can be detected simultaneously by attaching a photomultiplier (PM) tube and an interference filter for sulfur on one side of the flame, and a PM tube with an interference filter for phosphorus on the opposite side of the flame. The detector response to phosphorus is linear, whereas the response to sulfur depends on the square of its concentration. Carbon dioxide and organic impurities in the makeup and carrier gases must be less than 10 ppm. The quenching effect of carbon dioxide is very significant.

The FPD has found application for the determination of pesticides and pesticide residues containing sulfur and phosphorus. It has also been used to detect gaseous sulfur compounds and to monitor air for traces of nerve gases (phosphorus compounds).

5.2.5.6 Far-UV Photoionization Detectors. The photoionization detector[21] (PID) is a concentration-sensitive detector with a response that varies inversely with the flowrate of the carrier gas. A typical PID has two functional parts: an excitation source and an ionization chamber. The excitation source may be a discharge lamp excited by a direct current (1 to 2 kV), a radio frequency (75 to 125 kHz), a microwave (2450 MHz), or a laser. A discharge lamp passes ultraviolet radiation through the column effluent from one of several lamps with energies ranging from 8.3 to 11.7 eV. Photons in this energy range are energetic enough to ionize most organic species but not the permanent gases. A potential of 100 to 200 V is applied to the accelerating electrode to push the ions formed by UV ionization to the collection electrode at which the current (proportional to concentration) is measured. The most popular PID lamp is the 10.2-eV model, which has the highest photon flux and therefore the greatest sensitivity. There are certain applications in which the 9.5-eV lamp is preferable to the 10.2-eV lamp; these include aromatics in an aliphatic matrix, mercaptans in the presence of H_2S, and amines in the presence of ammonia.[22]

The discharge ionization detector (DID) uses far-ultraviolet photons to ionize and detect sample components. Helium gas is passed through a chamber in which high-voltage electrodes generate a

[21] J. N. Driscoll et al., *Am. Lab.* **10**(5):137 (1978).
[22] J. N. Driscoll, *J. Chromatogr. Sci.* **20**:91 (1982).

glow discharge and cause it to emit a high-energy emission line at 58.84 nm (having an energy of 21.2 eV). This energy passes through an aperture to a second chamber in which it ionizes all gas or vapor species present in the sample stream that have an ionization potential less than 21.2 eV (which embraces practically all compounds, including hydrogen, argon, oxygen, nitrogen, methane, carbon monoxide, nitrous oxide, ammonia, water, and carbon dioxide). A polarizing electrode directs the resulting electrons to a collector in which they are quantitated with a standard electrometer.

5.2.5.7 ***Electrolytic Conductivity Detector.***[23] In the electrolytic conductivity detector (EICD), also called the Hall conductivity detector, organic compounds in the effluent are first converted to carbon dioxide by passing the column eluate through a high-temperature reactor in which the heteroatoms of interest (halogen-, sulfur-, and nitrogen-containing compounds) are converted to small inorganic molecules. The reaction-product stream is then directed into a flow-through electrolytic conductivity cell. Changes in electrolytic conductivity are measured. Ionic material is removed from the system by water that is continuously circulated through an ion-exchange column.

The combustion products may be mixed with hydrogen gas and hydrogenated over a nickel catalyst in a quartz-tube furnace. Ammonia is formed from organic nitrogen, HCl from organic chlorides, and H_2S from sulfur compounds. If one wants to detect halogen compounds, a nickel reaction tube, hydrogen reaction gas, a reactor temperature of 850 to 1000°C, and 1-propanol are used. Under these conditions, compounds containing chlorine will be converted to HCl, methane, and water. The HCl will dissolve in 1-propanol and change its electrolytic conductivity, whereas the nonhalogen products will not dissolve in the alcohol and not change its conductivity to any significant degree.

In the detection of sulfur compounds, the compound must be converted to SO_2, which is usually accomplished in the reaction tube heated between 950 and 1000°C. Collection of SO_2 in methanol containing a small amount of water converts the SO_2 into ionic species. Although water is a satisfactory solvent for the sulfur or halogen modes, water containing 10% to 20% organic solvent (*tert*-butyl alcohol) is preferred for the nitrogen mode.

The ElCD detector finds use in analysis of pesticides, herbicides, and alkaloids and certain pharmaceuticals. When compared to other selective detectors, such as the nitrogen–phosphorus detector or the electron-capture detector, electrolytic conductivity detector chromatograms typically are much cleaner.

5.2.5.8 ***Chemiluminescence–Redox Detector.*** This detector is based on specific redox reactions coupled with chemiluminescence measurement. An attractive feature of this detector is that it responds to compounds such as ammonia, hydrogen sulfide, carbon disulfide, sulfur dioxide, hydrogen peroxide, hydrogen, carbon monoxide, sulfides, and thiols that are not sensitively detected by flame-ionization detection. Moreover, compounds that typically constitute a large portion of the matrix of many environmental and industrial samples are *not* detected, thus simplifying matrix effects and sample cleanup procedures for some applications.

5.2.6 Temperature Control

The column is either kept at a constant temperature (isothermal GC) or programmed during the run (PTGC). The temperature should be monitored, adjusted, and regulated at the injection port, in the oven surrounding the column, and at the detector. The temperature of the injection port must be sufficiently high to vaporize (virtually instantly) the sample, yet not so high that thermal decomposition or molecular rearrangements occur. The column temperature need not exceed the boiling point of the sample in order to keep the analytes in their vapor phases. Actually the column will produce better separations if the temperature is below the sample's boiling point (but above its condensation point) in order to increase the interaction with the stationary phase. The smaller the amount of the stationary phase, the lower the temperature at which the column can operate; open tubular columns are usually run at lower temperatures than packed columns.

[23] R. C. Hall, *J. Chromatogr. Sci.* **12:**152 (1974).

The temperature of the detector housing should be sufficiently high so that no condensation of the effluent occurs, yet not so high that the detector malfunctions.

5.2.6.1 ***Isothermal Operation.*** Selecting the column temperature for isothermal operation is a complex problem, and a compromise is usually the answer. A sample whose components have a wide range of boiling points cannot be satisfactorily chromatographed in a single isothermal run. A scouting run at a moderate column temperature may provide good resolution of the lower-boiling-point compounds but requires a lengthy period for the elution of high-boiling-point material. One solution is to raise the column temperature to a higher value at some point during the chromatogram so that the higher-boiling-point components will be eluted more rapidly and with narrower peaks. The better solution to this problem is to change the band migration rates during the course of separation by using temperature programming.

5.2.6.2 ***Temperature Programming.*** In temperature programming, the sample is injected into the chromatographic system when the column temperature is below that of the lowest-boiling-point component of the sample, preferably 90°C below. Then the column temperature is raised at some preselected heating rate. As a general rule, the retention time is halved for a 20 to 30°C increase in temperature. The final column temperature should be near the boiling point of the final solute but should not exceed the upper temperature limit of the stationary phase. Heating rates of 3 to 5°C · min^{-1} should be tried initially and then fine-tuned to achieve optimum separation.

Example 5.7 Find the temperature increase that typically halves the k′ value of one or several sample components. From sets (k',T) of data and realizing that, approximately,

$$\ln k' = \frac{B}{T} + C$$

where T is the temperature in degrees Kelvin and B and C are constants, determine the values of the constants. Next, estimate k' at both higher and lower temperatures. If k' = 30 at 80°C and 15 at 100°C, then a 20°C increase in temperature will halve k'. After a temperature rise of 100°C to a column temperature of 180°C, the value of k' will be approximately 1 and the solute will have been eluted. Finally, try a heating rate of 5°C · min^{-1}. Increase the heating rate if the peaks are well separated but there is excessive space between the peaks. Consult catalogs of suppliers of chromatographic columns for sample chromatograms run with various temperature programs.

5.2.7 Gas–Solid Chromatography

In gas–solid chromatography (GSC) the column-packing material may be a molecular sieve, which consists of interconnected cavities with uniform openings (Table 5.16). Only molecules smaller than

TABLE 5.16 Column Packings for Gas–Solid Chromatography

Maximum column temperature: 400°C.

Packing type	Pore diameter, nm	Typical applications
Molecular sieve 3A	0.3	Absorbs molecules such as water and ammonia
Molecular sieve 4A	0.4	Absorbs carbon dioxide, sulfur dioxide, hydrogen sulfide, ethane, ethylene, propene, and ethanol
Molecular sieve 5A	0.5	Separate straight-chain hydrocarbons (through C_{22}) from branched-chain and cyclic hydrocarbons
Molecular sieve 13X	1.0	Absorbs all molecules up to a critical diameter of 1.0 nm
Graphitized carbon B	1–3	Water is eluted quickly and is well separated from methanol, ethanol, and formaldehyde

these openings are able to enter freely the cavities, and in so doing, their passage through the column is delayed. Solutes denied entrance into the solid packing will travel through the column with the speed of the mobile gas phase. Solutes with borderline molecular size may squeeze in some openings and therefore suffer some retardation. Although not a molecular sieve, silica gel exhibits high retention for carbon dioxide and it is often used for prior removal of this gas.

Other adsorbents for GSC are porous polymers, uncoated materials that are used in GLC after suitable coatings had been applied to the surface. The characteristics of the Porapak series are given in Table 5.4. These polymers are sold under a variety of trade names: Porapak, Chromosorb Century Series, and Carbopack graphitized carbons; the list is not exhaustive.

*5.2.7.1 **Operating Techniques.*** Many gas mixtures contain some components that cannot be separated on a particular column or cannot pass through a column in a reasonable time. There are three remedies.

1. *Back-flushing*. In one scheme the flow of carrier gas is reversed after the last peak of interest has eluted from the column. The higher-molecular-weight hydrocarbons are back-flushed from the upper end the column packing and passed through the detector. In a second method the column is back-flushed before any component has eluted; after this, forward flow of carrier gas is resumed so that the lower-molecular-weight hydrocarbons can be eluted.

2. *Series-bypass method*. Certain sample components are temporarily stored in a second column while separations of the remainder are made on the first column. Then the carrier gas is switched to pass through the storage column and elute those components. For example, in a programmed time interval after sample injection, oxygen, nitrogen, methane, and carbon monoxide will have passed through the porous polymer column and will have entered the molecular sieve 5A column for temporary storage. Next, the carrier is directed only through the porous polymer column to elute hydrogen, carbon dioxide, ethylene, ethane, acetylene, and hydrogen sulfide. Finally, the molecular-sieve column is switched back into the carrier-gas stream to elute the stored components in the order listed.

3. *Concentric packed tubes*. With Alltech's CTR columns, the analysis of CO, CH_4, CO_2, O_2, and N_2 can be handled in one step. The CTR column, essentially a 1/8-in (0.32-in) o.d. column within a 1.4-in (0.63-in) column, permits the use of two different packings for sample analysis. The inner packing, a molecular sieve, gives the first three peaks: air plus CO, CH_4, and CO_2. The outer packing, a porous polymer mixture gives the last four peaks: O_2, N_2, CH_4, and CO. The composite column packing overcomes the problems of irreversible absorption of CO_2 on the molecular sieve which must be used to obtain the oxygen–nitrogen split and to separate the CO peak from the other components. The entire analysis is done under isothermal conditions at room temperature and in about 9 min using a helium flowrate of 65 $mL \cdot min^{-1}$.

5.2.8 Supercritical-Fluid Chromatography[24–27]

Supercritical-fluid chromatography (SFC) employs a mobile phase that is a highly compressed gas above its critical temperature and critical pressure points. A gas cannot be liquefied when it is above its critical temperature, irrespective of the pressure. Above the critical temperature, the substance exists as a supercritical fluid with physical properties that are intermediate between those of the gas and liquid states. Thus, in some respects SFC bridges GC and LC techniques and enables SFC to carve out a niche not adequately served by either GC or HPLC. The main advantage of SFC lies in its ability to complement both LC and GC and to overcome some of the detection

[24] P. R. Griffiths, *Anal. Chem.* **60**:593A (1988).
[25] J. C. Fjeldsted and M. L. Lee, *Anal. Chem.* **56**:619A (1984).
[26] D. W. Later et al., *Am. Lab.* **1986** (August): 108.
[27] R. C. Smith, B. W. Wright, and C. R. Yonker, *Anal. Chem.* **60**:1323A (1988).

and separation problems associated with each. See Table 2.24 for properties for selected supercritical fluids.

5.2.8.1 ***Supercritical-Fluid Solvent Properties.*** Carbon dioxide is currently the fluid of choice in many SFC applications because of its low critical temperature (31.3°C), nontoxic nature, and lack of interference with most detection methods. The critical pressure is 7.29×10^6 Pa and critical density is $0.448\ g \cdot cm^{-3}$. A supercritical fluid exhibits viscosities and densities between those of gases and liquids, and solvating powers similar to liquids. For a chromatographer, the significance of the supercritical state is that the solvent has the potential to separate thermally labile and nonvolatile compounds at low temperatures and with high efficiencies, thus complementing and extending the ranges of GC and LC analysis.

The increased diffusion coefficients of supercritical fluids compared with liquids can lead to greater speed in separations or greater resolution in complex mixture analyses. The properties of supercritical fluids can be very useful in chromatography. Low viscosities allow for efficient operation at high flow velocities, yielding separation speed and resolution comparable to those attainable with GC. The physical and chemical properties of supercritical fluids, such as solvation, diffusion, and viscosity are a function of density. Thus, by controlling the fluid's density through changes in pressure or temperature, it is possible to vary the fluid's chromatographic properties. As density is increased from the gas-phase limit, the dense gas will increasingly exhibit the solvating properties of a liquid under conditions in which transport properties are still significantly advantageous compared with liquids. The temperature and density of a particular solvent will govern the extent of solute solubility.

Unfortunately, highly polar or high-molecular-weight solutes generally have limited solubilities in carbon dioxide. Nitrous oxide and NH_3 can be used if a more polar mobile phase is required, or with compounds such as primary alkyl amines that are incompatible with supercritical CO_2. Often pentane is the fluid of choice for high-molecular-weight hydrocarbons, such as polycyclic aromatics. However, polar fluids (such as NH_3) pose problems because of complications resulting from their reactivity. As a result, fluid modifiers (entrainers) have been used to increase the solvent strength of low-polarity fluids such as CO_2. Such modified fluids include CO_2 doped with methanol, 2-propanol, methylene chloride, tetrahydrofuran, dimethylsulfoxide, and acetonitrile; trifluoromethane modified with ammonia; and pentane modified with benzene. A desirable feature that results is independent programming of the SFC composition, which is analogous to solvent programming (gradient elution) in LC, while avoiding the separation of the mixed SFC into two phases. Packed-column separation efficiency is greatly improved by the ability to vary modifier concentration. A typical modifier program might vary the methanol or 2-propanol content in the carbon dioxide carrier from 0 to 10 mol % over a period of 0 to 20 min during an isobaric chromatographic run. A rule of thumb in selecting a solvent modifier is to first try a substance that is a good solvent for the analytes.

The enhancement of selectivity or the solubilization of highly polar compounds can be accomplished by introducing secondary chemical equilibria involving the formation and use of reverse micelles, microemulsions (micelles swollen with water), and metal chelates. When using microemulsions, selection of fluid density and water-to-surfactant ratio can provide significant selectivity in separations of water-soluble dyes and biological macromolecules. The use of complexing agents in supercritical fluids provides the basis for high selectivities for metals. When the separation is based on molecular size and shape rather than on polarity, a liquid-crystal phase, such as liquid-crystalline polysiloxane, is suggested.

5.2.8.2 ***Instrumentation.*** SFC combines the instrumental complexities of GC and HPLC with some of the advantages of both. Often an existing gas chromatograph can be retrofitted to serve SFC. For applications with capillary columns and in which flame-ionization detection or an absorbance detector is appropriate, the fluid delivery system used in HPLC and related hardware can be used along with a precision oven that is equipped with an injector, capillary column, and detection system. A high-pressure pumping system delivers a fluid mobile phase under supercritical pressure. Syringe pumps can deliver a pulseless fluid flow to the capillary column under either pressure or flow control. The third component is a microprocessor-based control system that should be able to control pump pressure and oven temperature functions for pressure and/or density ramping and temperature regulation of the capillary-column oven.

The SFC instrument should have valves that permit independent control of flowrate and pressure. This is important because density controls the solubility of the sample. The higher the density of the fluid, the greater the solubility. Now SFC is no longer limited to separations of nonpolar or weakly polar samples because of lack of solubility. The valves allow the user to optimize the pressure program for solubility and then adjust the flowrate depending on the kinetics of the column. Pressure gradients are most widely used in SFC for manipulating solute retention. Rapid pressure programming with short capillaries can reduce analysis time to a few minutes.

In typical 1- to 5-mm i.d. packed columns, large sample volumes can be injected. Sample introduction poses a greater challenge for capillary columns, particularly for the 25- to 75-μm i.d. columns commonly used. Only very small sample volumes (1 to 30 nL) can be introduced to avoid serious band-broadening and column-overloading problems. This makes serious demands on detector sensitivity. The most widely used injection procedure in capillary SFC involves sample splitting after ambient temperature injection. Sample splitting is necessary to obtain a narrow injection band on the column and, sometimes, to prevent overloading the column or generation of an excessively wide solvent peak. With a "timed split" good reproducibility can be obtained and the amount of material injected into the column can easily be adjusted. Fast pneumatic or electronic control of the valve actuation permits injection of only a portion of the sample loop.

5.2.8.3 Columns. Both capillary (borrowed from GC) and packed (from HPLC) columns are used in SFC. Usage of each type depends upon the goal of the separation. Packed columns have more capacity, but capillary columns have many more plates. The lower thermal stability of many HPLC phases limits their application to low-temperature fluid systems (100°C). Generally, SFC capillary diameters must be smaller than 100 μm to maintain both reasonable analysis times and high resolution. Practical column dimensions are 50-μm i.d. and 10- to 20-m lengths. The stationary phases must be cross-linked and bonded to the column wall. Capillary columns are popular because they are compatible with almost all LC and GC detectors and are easily interfaced with a mass spectrometer.

5.2.8.4 Detectors for SFC. Currently, the most universal detectors for SFC include UV absorption and the FID, along with mass spectrometry (MS) and Fourier transform infrared (FTIR) spectrophotometry. The limitations of UV detection are related to sensitivity considerations, whereas FID is constrained to SFC fluids yielding no significant ionization during detection (such as CO_2). Flame-based detectors must be modified for fluid decompression and expansion into the flame jet. The compatibility of SFC with FID is almost entirely eliminated by the use of polar or mixed fluid systems necessary to extend SFC to separation of nonvolatile analytes.

The UV detector responds only to solutes that have UV-absorbing functional groups. The UV detector can provide compound-specific information if tuned to an absorption wavelength that is specific to the analyte of interest. In addition, with a photodiode-array UV detector, on-the-fly spectra can be collected to assist in compound identification. Interfacing the UV detector with an SFC instrument that uses 4.6-mm i.d. packed columns is straightforward. The major concern is containment of the fluid at high pressure. Capillary columns impose the same pressure requirements in addition to a stringent detector dead-volume limitation. Optical detector cell volumes must be extremely small and able to withstand high pressures. These small cell volumes (usually 50 nL) can be attained, but only at the expense of detector path length and sensitivity.

Similar requirements apply to coupled systems such as SFC-FTIR and SFC-MS. At the outlet of the column, both flow restriction and detection are required. In the case of UV detection, restriction takes place after the capillary flow cell. When GC-type detectors are used, the restriction takes place such that the carrier expands to a gas and passes into the detector in the normal fashion. Flow restrictors may be a small-bore fused-silica (5- to 10-μm i.d.) tubing, ranging from 2 to 20 cm in length, or platinum–iridium tubing. The former is useful for flow restriction in GC-type detectors whereas MS interfacing has primarily used the latter.

The off-line approach for FTIR detection offers enhanced detection limits and practical elimination of the solvent modifier, which greatly limits on-line detection. Major disadvantages include mechanical complexity, expense, and sensitivity (in some cases).

Combination of SFC and MS, based on expansion of the fluid directly into the mass spectrometer ion source, is widely practiced. Tremendous selectivity is possible. The use of high-resolution MS can further enhance this capability to resolve components of the same nominal molecular weight, to aid identification, and to eliminate background complications. SFC-MS is limited to analytes having some vapor pressure in the ion source.

The hyphenated techniques of SFC-MS and SFC-FTIR are discussed in Secs. 13.10 and 6.5.4, respectively.

5.2.8.5 Coupling SFE with SFC. Supercritical-fluid extraction (Sec. 2.2.9.3) can be coupled with capillary GC by directly depressurizing the supercritical-fluid extract inside a conventional split–splitless injection port or by inserting the extraction cell restrictor through an on-column injection port into the capillary column itself. The extracted analytes are cryogenically focused in the GC column stationary phase during the SFE step, and then analysis is performed in a normal manner.

Coupling has also been achieved by flowing the extract through an injector loop, by collecting the analytes on an accumulator trap containing a sorbent material, or by thermally trapping the analytes at the head of the SFC column. Once SFE is completed, the extracted species are swept from the accumulating device into the SFC column (packed or capillary) using the SFC mobile phase, and the analysis is carried under appropriate chromatographic conditions.

5.2.8.6 Advantages of SFC. The availability of both FID and UV absorbance detectors facilitates the determination of aromatics and saturates in diesel fuel. In addition, SFC can be used for simulated distillation and hydrocarbon class separation. Thiophosphate pesticides, which previously required derivatization before GC analysis, can now be done in SFC without derivatization.

Polymers and polymer additives are another area of application in which the high efficiency can provide oligomeric resolution of low-molecular-weight fractions. Pharmaceutical firms may find the very short analysis times offered by SFC to be advantageous.

5.2.9 The Optimization of Separations

What are the best ways to optimize separations? One way to answer this question is to consider Eq. (5.18) in Sec. 5.1.2.11. The three variables are N (column efficiency; also expressed as H/L), α (column selectivity), and k' (partition ratio for the second of two analytes being separated). Since good resolution is indicated by a large value for Rs, the three terms should be maximized. Time is a fourth variable that also needs to be considered [Eq. (5.25)]. The process involves two sets of closely interrelated parameters: column characteristics and GC operational parameters. The question of what column should be used can be answered by selecting such variables as stationary phase, film thickness, column diameter and/or particle diameter, and column length. The operational parameters include carrier-gas type, linear gas velocity, and oven temperature profile (isothermal or programmed).

Efficiency and selectivity cannot be considered as separate entities. They combine to produce the observed resolution for a pair of solutes. When resolution is insufficient, one or both must be increased. If greater selectivity is not possible, efficiency must be increased.

5.2.9.1 Increasing the Plate Number. Increasing the column length will increase the number of theoretical (apparent) plates, but also will increase the column inlet pressure and the analysis times. However, the resolution is increased only by the square root of the column length, and this is accomplished only at the expense of a linear increase in analysis time. So to double the resolution, the column length must be increased by a factor of 4. At a constant inlet pressure any increase in length will result in an increase in the retention time. However, if the plot of H versus u is fairly flat, then the column length and the carrier-gas velocity can be increased. This action will improve the resolution and help keep the analysis time reasonable. When the increase in plate height is large as the carrier gas velocity increases, then only the column length should be changed to improve the resolution. The limiting factors here are the pressure drops needed across the column or the long analysis times.

The plate number can also be increased by preparing a better column according to Eq. (5.21) and the concepts of the rate theory. The most important parameters are stationary film thickness d_f and particle diameter of the stationary phase d_p. The rate equation for GC tells us that our packed column should have thin films on the solid support. This is achieved by using supports with large surface areas and small amounts of stationary phase (low percentage of liquid phase). However, at very low phase loading, not all the support surface (or the capillary walls) will be coated. Adsorption of the solute on the bare particles or wall may occur with a deterioration in the efficiency. This limit is reached at about 0.3% by weight for a typical diatomaceous earth solid support.

The plate height is an explicit function of the diameter of the support particle. In addition, H is related (at least in packed columns) to the geometry of the packed bed, which, in turn, is support-size dependent. In general, the smaller the support particle, the better the column can be packed. Under normal conditions, a column packed with 100 to 120 mesh (149- to 125-μm) particles is more efficient than one packed with 60 to 80 mesh (250- to 177-μm) particles. As column diameters or particle sizes decrease, increased pressures are required to maintain optimum flowrates. Moreover, carrier-gas compressibility acts to further reduce gains in column efficiency as higher inlet pressures are utilized. A practical maximum inlet pressure of approximately 100 $\text{lb} \cdot \text{in}^{-2}$ (700 kPa) limits the useful capillary column inside diameter to about 100 μm at a length of about 50 m.

As a rough guide when using Chromosorb W or P (or related supports), the optimum loading is around 15%. The optimum is lower for Chromosorb G. With untreated glass beads the optimum is probably around 0.1%. With open tubular (capillary) columns the optimum film thickness is roughly between 0.25 and 1 μm. The distribution of a solute between the stationary phase and the gas phase is directly proportional to the film thickness and inversely proportional to the column diameter. To improve resolution, especially of low-molecular-weight materials, a thick film may be advisable. This will result in a longer analysis time. A good starting point is 0.33 μm.

A smaller column diameter increases the partition ratio. For increased resolution, a smaller diameter is beneficial. However, this results in longer analysis time since flows are lower. To accelerate an analysis, use a wider bore column or increase the oven temperature. Diameters of capillary columns available are 0.1, 0.25, 0.32, and 0.53 mm i.d. A 0.25-mm column is a good starting point.

Generally one should choose the shortest column capable of generating the required separation efficiency. Short columns are ideal for sample profiling, samples with few peaks, and the analysis of high-boiling-point material. Longer columns are used when a detailed analysis of complex samples is necessary, or when families of closely related isomers must be resolved.

Conventional capillary columns with inner diameters of 0.25 or 0.32 mm come in 15-, 30-, and 60-m lengths. The 30-m column is most widely used. It offers a good compromise between column efficiency and speed of analysis. The 60-m column offers the analyst high efficiency and capacity, whereas the 15-m columns are for those who have developed a method and know exactly what it takes to do the separation. Wide-bore (megabore) columns come in 15- and 30-m lengths. The shorter column is useful if the sample is fairly simple; the 30-m column is better if complex samples are being analyzed or if the operator is inexperienced, because this length has enough separation efficiency to overcome less than optimum flow and temperature parameters. Minibore columns have an inner diameter of 0.18 mm and come in 10-, 20-, and 40-m lengths. Column capacity is roughly 40 ng per component. The 10-m columns are ideal for fast analysis. The 40-m column generates over 200 000 apparent plates, which will provide excellent separation of complex mixtures; practical carrier-gas velocities are 35 cm/s (0.6 mL/min) of helium. The 20-m columns are suitable for general analysis and provide a good compromise between high-efficiency and high-speed analysis.

5.2.9.2 Maximizing the Selectivity Factor. The first term in Eq. (5.18) will be maximized if the selectivity α is maximized, as shown in Table 5.1. Generally it is desirable to select values of α within the range 1.05 to 2.0. For example, an increase in α from 1.05 to 1.10 will improve resolution by a factor of 4. When α is quite close to 1, it becomes impractical to operate the chromatographic system because the required column length and column inlet pressure become difficult or impossible to achieve.

In the separation system, the stationary and mobile phases should be chosen to show the greatest selectivity between the two analytes most difficult to separate. Select a stationary phase with a polarity similar to the sample. Use the McReynolds constants as a guide. If the sample is a mixture of various polarities, choose an intermediate polarity phase such as 5% or 50% phenylmethyl silicone. A 5% phenylmethyl silicone capillary column should handle two-thirds of all applications.

5.2.9.3 Optimizing the Partition Ratio. The retention time of a solute is a minimum when k' is 2. There is little increase in analysis time when k' lies between 1 and 10, provided k' has no effect on the other variables. Little improvement in resolution is gained as k' is increased above about 10.

Both H and k' are functions of the amount of stationary phase. Both quantities will decrease with decreasing volume of the liquid phase. Frequently u_{min} increases with decreasing volume of the liquid phase, which means shorter analysis time. From the resolution equation [Eq. (5.18)] we see that a decrease in k' is detrimental, especially if k' becomes fractional, while a decrease in plate height improves the resolution.

5.2.9.4 Carrier-Gas and Linear Velocity. The effect of carrier gas is shown in Fig. 5.2, which gives typical van Deemter plots for the common carrier gases helium, hydrogen, and nitrogen. There is a carrier-gas velocity that minimizes the plate height; it is u_{min} in a graph of H vs. u. The only parameter that is velocity dependent in the resolution expression [Eq. (5.18)] is H, hence Rs will have a maximum at u_{min}. It is good experimental practice to determine u_{min} by obtaining H at several carrier-gas velocities for some solute. Since

$$H = \frac{L}{N} \quad \text{and} \quad N = 5.54\left(\frac{t'_R}{W_{1/2}}\right)^2 \tag{5.32}$$

ascertain the plate number from the adjusted retention time and the peak width at half the maximum height. The carrier-gas velocity is measured from the expression

$$u = \frac{L}{t_M} \tag{5.33}$$

where L is the column length and t_M is the retention time of a nonretained solute. Although nitrogen is slightly more efficient and gives the smallest H_{min}, this advantage is offset by the better performance of helium and hydrogen at higher velocities for which the B term is not important. For fast analyses, helium and hydrogen and definitely preferred.

Once a suitable isocratic temperature or temperature program is found, change the carrier-gas velocity to operate around u_{min} or slightly higher if the van Deemter (H/u) plot is rather flat. Hydrogen gas is the best choice of carrier gas if analysis time is important (and adequate ventilation is available). Hydrogen is not recommended for use with a nitrogen–phosphorus detector nor while temperature programming with a flame detector since a change in hydrogen flow may affect the flame-detector response. Nitrogen is not recommended as a carrier gas for capillary chromatography, except when using ECD and NP detectors, because its use results in long analysis times compared to helium or hydrogen.

5.2.9.5 Temperature Effects. Since α, k', D_S, D_M, and u are all functions of the temperature, resolution is a strong function of temperature. With rising temperature, the diffusion coefficients increase, while k' and the retention time decrease. Most frequently, the relative volatility, or selectivity, decreases as the temperature increases. The resolution can be improved by programming the temperature during a chromatographic run.

Reduce the column temperature. Do not hesitate to work at temperatures far below the boiling points of the solutes since, once evaporated in the injector, condensation is a function of the dew point. Working at low temperature and small loadings of the stationary phase is especially effective.

Lower temperatures are conducive to longer contact periods between the solute and stationary phase, which results in increased retention and analysis times. Of course, a decrease in temperature can be coupled with a decrease in column length or in film thickness of the liquid phase. Higher temperatures lower the partition ratio and tend to reduce the effect of phase selectivity.

Several different temperature program rates may be needed through the run to tailor the profile for the particular analysis. A suitable initial trial rate is $5°C \cdot min^{-1}$ for capillary columns.

5.2.9.6 Column Conditioning. After a new column has been installed, it will need to be conditioned. Before the oven is heated, it is absolutely necessary to verify that there is a flow of carrier gas through the column and an absence of leaks. The column is conditioned by setting the oven temperature 10 to 20°C above the highest temperature that will be used or the column's isothermal maximum temperature, whichever is the lowest. After 30 to 60 min at the conditioning temperature, a steady baseline should be obtained. If the baseline is unstable or excessively high after 60 to 90 min, there probably is a contaminated section in the GC system.

5.2.9.7 Factors Influencing the Gas-Chromatographic System. The observed variance of the chromatographic peak, σ^2, is composed of the variance due to the partition process in the column (σ_c^2) and the variance due to band spreading outside the column (σ_i^2). For example, if the ratio of the two standard deviations is 0.5, the actual number of theoretical plates will be 20% lower than could be obtained if no band spreading occurred outside the column.

The bandwidth outside the column increases in several ways. This spreading includes the dilution factor caused by (1) the injector, (2) any connecting tubing, and (3) the detector volume. Some problems are as follows:

1. The sample volume is too large, which causes the original bandwidth of the sample "plug" to be excessive.

2. The sample injection is not carried out rapidly enough; thus the width of the sample vapor plug is increased.

3. The (void) volume between the point of injection and the entrance to the stationary phase is too large, which causes excessive band spreading as a result of diffusion of the sample vapor into the carrier gas.

4. The tubing connecting the column outlet and the detector must be kept to an absolute minimum because diffusion in the carrier gas results in the remixing of the separated components. Addition of makeup gas to the column effluent will increase the gas velocity and sweep the column effluent through this "dead" or void volume.

5. The volume of the detector is too large compared to the actual gas volumes corresponding to the peak width at half height. The detector volume should be roughly 0.1 the peak width.

6. All couplings should possess zero dead volume.

5.2.10 Systematic Troubleshooting

To locate a failure or a fault, work from detector to column to inlet, following the signal path back to its origin. If the problem persists, and is electronic, it is time to call the manufacturer's service technician.

Before any testing is done, a set of expected performance criteria must be established. The instrument manufacturer often will supply standard performance benchmarks along with the conditions under which they were measured. Usually a test column and a test mixture are involved. Detector sensitivity, noise, minimum detectable quantity, peak resolution, and other general measurements can be specified. These criteria should be determined by the user when the instrument is installed and periodically thereafter even though the test may not be directly related to the intended application.

Instrument manuals usually include a troubleshooting section. This should be read both as a reference and to learn how to operate the instrument properly.

Establish performance standards for the applications for which the instrument was purchased. Then check these performance standards regularly. Perform frequent and regular calibration of detector response factors along with visual assessment of the calibration chromatogram. Control samples interspersed with the application samples will enable the user to detect decreasing detector response, gradually broadening peaks, or drifting retention times.

Anticipate trouble. Keep on hand a good supply of nuts, ferrules, and fittings as well as inlet and detector spares, such as septa, inlet liners, and flame jets. Extra syringes and autosampler parts are always useful. A set of tools purchased specifically for your instrument is a good investment; store them safely in a drawer. Perform the recommended periodic maintenance procedures on schedule.

An instrument log is an essential tool for problem diagnosis. This includes everything done with the instrument and auxiliary units. Keep notes on peculiarities that occur—they may telegraph upcoming trouble.

If the problem appears to be chromatographic in origin, check the various inlet, detector, and column temperatures, pressure, and flows and compare them to their normal values. If possible, measure some basic chromatographic parameters such as unretained peak times, capacity factors, or theoretical plate numbers. Check for loose column or tubing connections. If the problem appears to be electronic in nature, carefully inspect the instrument (with the power cord unplugged). Are the cables plugged in? Are boards seated firmly in their connectors?

If the column is determined to be at fault, consider these problems and suggested remedies:

1. Loss of resolution due to band-broadening effects may have been caused by large solvent injections displacing the stationary phase. Although bonded-phase columns are relatively immune to this problem, back-flushing such columns with solvent is recommended for restoration. (Do not back-flush nonbonded columns.)

2. If solvent flushing is inappropriate or fails, remove the first two or three column turns of a capillary or the upper centimeter or so of a packed column.

3. Excessive column heating may cause the stationary film to coalesce into uneven "hills and valleys" on the inner column wall. Usually a replacement column is needed.

4. Microcontaminants may gradually deposit at the beginning of the column and interact with solutes, which could lead to adsorption and possibly to catalytic decomposition or rearrangement. Removing the beginning column section is an effective remedy. A guard column in front of the regular column in the best solution.

5. Oxygen contamination in the carrier gas can cause excessive column bleed. The remedy is to install a fresh, high-quality oxygen trap.

6. Septum bleed or residues, or high-boiling-point, strongly retentive solutes from a previous run, may raise the baseline. Try several bakeout temperature cycles with no sample injection. If large peaks from previous injections elute as broad peaks during a subsequent run, a thinner stationary phase film should be used if higher column temperatures are not possible.

Bibliography

Allenmark, S. G., *Chromatographic Enantioseparation: Methods and Applications*, 2d ed., Prentice Hall, Old Tappan, NJ, 1991.

Berezkin, V. G., *Chemical Methods in Gas Chromatography*, Elsevier, New York, 1983.

Blau, K., and J. M. Halket, eds., *Handbook of Derivatives for Chromatography,* Wiley, New York, 1993.

Cowper, C. J., and A. J. DeRose, *The Analysis of Gases by Gas Chromatography*, Pergamon, New York,1983.

Dressler, M., *Selective Gas Chromatographic Detectors*, Elsevier, New York, 1986.

Fjeldsted, J. C., and M. L. Lee, "Capillary Supercritical Fluid Chromatography," *Anal. Chem.* **56**:619A (1984).

Griffiths, P. R., "Contemporary SFC: Accomplishments and Limitations," *Anal. Chem.* **60**:593A (1988).

Grob, R. L., and M. Kaiser, *Environmental Problem Solving Using Gas and Liquid Chromatography*, Elsevier, New York, 1982.

Hill, H. H., and D. G. McMinn, eds., *Detectors for Capillary Chromatography*, Wiley-Interscience, New York, 1992.

Jennings, W., *Gas Chromatography with Glass Capillary Columns,* 2d ed., Academic, New York, 1980.

Jennings, W., *Analytical Gas Chromatography*, Academic, New York, 1987.

Knapp, D. R., *Handbook of Analytical Derivatization Reactions*, Wiley, New York, 1979.

Lee, M. L., F. J. Yang, and K. D. Bartie, *Open Tubular Column Gas Chromatography: Theory and Practice*, Wiley, New York, 1984.

Loffe, B. V., and A. G. Vitenberg, *Head-Space Analysis and Related Methods in Gas Chromatography*, Wiley, New York, 1984.

Pacakova, V., and L. Felti, eds., *Chromatographic Retention Indices: An Aid to Identification of Organic Compounds*, Harwood, Chichester, UK, 1992.

Smith, R. D., B. W. Wright, and C. R. Yonker, "Supercritical Fluid Chromatography: Current Status and Prognosis," *Anal. Chem.* **60**:1323A (1988).

Zief, M., and L. J. Crane, *Chromatographic Chiral Separations*, Dekker, New York, 1988.

5.3 HIGH-PERFORMANCE LIQUID CHROMATOGRAPHY

The range of operating modes is much greater in high-performance liquid chromatography (HPLC) because two immiscible phases, stationary and mobile, in contact with one another affect the separation. The variety of phase systems for LC allows for a wide range of selectivities, which can be adjusted to the components of interest in two ways: by varying the stationary phase or by modifying the mobile phase. Chromatographic separation in HPLC is the result of specific interactions of the sample molecules with both the stationary and mobile phases. With an interactive liquid mobile phase, another parameter is available for selectivity in addition to an active stationary phase. In practice, it is preferable to change the selectivity and the capacity by modifying the mobile-phase polarity for different samples while using the same stationary phase.

The operations of sample introduction, chromatographic separation, and detection are performed in exactly the same way in HPLC as in gas chromatography except for those modifications necessary to accommodate a liquid rather than a gas as the mobile phase. That is, the sampling is sequential, the separated components are eluted with the mobile phase, and detection is a dynamic time-dependent process. HPLC can be used whenever the sample can be dissolved in a liquid.

HPLC comprises a number of different LC methods, covering a large number of applications. LC methods can be divided into two forms. The first is retentive LC, in which separation is achieved through interaction of the solutes with the support surface or with a stationary phase bound to that surface. This form includes normal phase, reversed phase, and ion chromatography. The second is nonretentive LC, in which separation is achieved on the basis of solute size through interaction with the pore network of the packing material. This form is known as size-exclusion chromatography. Affinity chromatography uses immobilized biochemicals as the stationary phase to achieve separations via the "lock and key" binding that is prevalent in biological systems.[28] "Rules of thumb" for method selection are suggested as follows:

1. All sample components are less than 2000 daltons (atomic mass units).
   a. The sample is water insoluble and possess an aliphatic or aromatic character.
      (1) Liquid–solid (adsorption) chromatography (LSC) is suggested for class separations or for the separation of isomeric compounds.
      (2) Liquid–liquid chromatography (LLC) is suggested for the separation of homologs.

[28] R. R. Walters, "Affinity Chromatography," *Anal. Chem.* **57**:1099A (1985).

b. The sample is ionic or possesses ionizable groups; ion-chromatographic methods are likely candidates along with reversed-phase procedures.

2. Some or all of the sample components exceed 2000 daltons.

a. Exclusion (gel permeation, GPC) chromatography (EC) is suggested.

(1) If the sample is water soluble, an aqueous mobile phase is used.

(2) If water insoluble, a nonaqueous mobile phase is used.

5.3.1 Components of a Liquid Chromatograph

The essential components for HPLC instrumentation include the following:

1. A pump or pumps to force the mobile phase through the system. Suitable pressure gauges and flow meters are placed in the system.
2. Sampling valves and loops to inject the sample into the mobile phase just at the head of the separation column.
3. A separation column in which the sample components are separated into individual peaks before elution.
4. A detector and readout device to detect the presence of solutes in the mobile phase and record the resulting chromatogram. To collect, store, and analyze the chromatographic data, computers, integrators, and other data-processing equipment are being used more frequently in conjunction with the strip chart recorder.

5.3.1.1 Mobile-Phase Delivery Systems. Pumps are used to deliver the mobile phase to the column. The pump, its seals, and all connections in the chromatographic system must be constructed of materials that are chemically resistant to the mobile phase. A degassing unit is needed to remove dissolved gases from the solvent.

Less expensive LC systems require a pump that is capable of operating to at least 1500 $lb \cdot in^{-2}$ (103 atm or bar). A more desirable upper pressure limit is 6000 $lb \cdot in^{-2}$ (414 bar). The pump's holdup volume should be small. Flowrates for many analytical columns fall within 0.5 to 2 $mL \cdot min^{-1}$. For microbore columns the flowrates need only be a few microliters per minute.

5.3.1.1.1 Reciprocating Piston Pumps. The reciprocating piston pump is the most popular type because it is relatively inexpensive and permits a wide range of flowrates. A hydraulic fluid transmits the pumping action to the solvent via a flexible diaphragm, which minimizes solvent contamination and corrosion problems. Some of the characteristics of this pump are as follows: (1) Flowrates can be varied either by altering the stroke volume during each cycle of the pump or the stroke frequency. (2) Solvent delivery is continuous. (3) No restriction exists on the reservoir size or operating time. (4) Solvent changes are rapid and accurate, an advantage when doing gradient elution or solvent scouting. (5) Some type of pulse-dampening system must be used. (6) Dual-head and triple-head pumps having identical piston-chamber units operating 180° or 120° out of phase smooth out all but small pulsations from the solvent delivery.

5.3.1.1.2 Syringe-Type Pumps. Syringe-type pumps operate through positive solvent displacement and use a piston mechanically driven at a constant rate. The characteristics of this pump are as follows: (1) Solvent delivery is controlled by changing the voltage on the digital stepping motor. (2) Solvent chamber capacity is finite (250 to 500 mL) but is adequate with small-bore columns. (3) Pulseless flow is obtained. (4) High-pressure capability (207 to 483 bar or 3000 to 7000 $lb \cdot in^{-2}$) can be achieved. (5) Solvent gradients are made feasible by the tandem operation of two or more pumps.

5.3.1.1.3 Constant-Pressure Pumps. The mobile phase is driven by the pressure from a gas cylinder, which is delivered through a large piston. Characteristics of this type of pump are as follows: (1) A low-pressure gas source (1 to 10 atm) can generate high liquid pressures (up to 400 atm, 6000 $lb \cdot in^{-2}$), because of the large difference in area of the two pistons. (2) Pulseless flow is provided. (3) Large flowrates are available for preparative applications; a valving arrangement permits the rapid refill of the solvent chamber. (4) Gradient elution is inconvenient.

5.3.1.1.4 Pulse Dampers. Some detectors are sensitive to variations in flow; all detectors will benefit from the decreased noise resulting from pulseless operation. The following types of pulse dampers are available: (1) Flexible bellows or a compressible gas in the capped upright portion of a T tube. These types have large fluid volumes but require pressures in excess of 1000 lb · in^{-2} (6.89 × 10^{6} Pa, 68 atm) for effective operation. (2) A compressible fluid separated from the mobile phase by a flexible diaphragm. This system offers easy mobile-phase changeover and minimal dead volume and is effective at low system pressures. (3) Electronic pulse dampers are useful with reciprocating pumps. They provide a small, rapid forward stroke of the piston following the pump's rapid refill stroke.

5.3.1.1.5 Connecting Tubing. Polymeric tubing is limited in its use to supply lines that connect the mobile-phase reservoirs to the pump, to the outlet side of the detector, and to lines such as the injector waste. Stainless-steel tubing must be used in regions subject to high pressure; that is, between the pump outlet and the detector outlet. LC-grade tubing should be used. The tubing is clean and ready to use, and the vendor takes care to ensure concentricity (an important feature in small-diameter tubing) so that the tubing bore will align properly with the hole in connecting fittings. If tubing is cut in your laboratory, be certain to remove any burrs or filings and to clean the tubing thoroughly. Before using the tubing, flush it with dichloromethane or tetrahydrofuran and then with a detergent solution to remove grease and oil residues.

5.3.1.1.6 Sampling Valves and Loops. A calibrated sample loop (volume usually 10 or 20 μL) is filled by means of an ordinary syringe that thoroughly flushes the sample solution through the loop. Manual rotation of the valve rotor places the sample-filled loop into the flowing stream of the mobile phase. A sliding valve uses an internal sample cavity consisting of an annular groove on a sliding rod that is thrust into the flowing stream. With either injecting system, the samples are dissolved, if possible, in the mobile phase to avoid an unnecessary solvent peak in the final chromatogram.

5.3.1.2 Columns.[29,30] Separation columns are available in various lengths and diameters. To withstand the high pressures involved, columns are constructed of heavy-wall, glass-lined metal tubing or stainless-steel tubing. Connectors and end fittings must be designed with zero void volume. Column packing is retained by frits inserted into the ends of the column.

Columns with an internal diameter of 4 to 5 mm are standard in HPLC. Packings should be uniformly sized and mechanically stable. Particle diameters lie in the range from 4 to 7 μm for the most popular columns. Column lengths range from 10 to 30 cm. When analytical speed is of prime consideration, as in quality-control work, a short (3- to 6-cm) column is useful.

With narrow-bore (2-mm) columns, the mobile-phase velocity remains the same and the analysis time remains unchanged in comparison with standard columns. Narrow-bore columns offer several advantages over the standard columns: (1) The detector signal of a sample component is increased by a factor of 4 when the internal diameter of the column is decreased by about a factor of two. (2) Solvent consumption is 4 times less, which is a considerable savings in solvent purchase and waste-disposal costs. (3) Unconventional or high-purity solvents can be used because solvent consumption is lower. (4) Packing density is more homogeneous. (5) Better column permeability allows smaller particles to be packed without exceeding conventional pressures used in HPLC. (6) Frictional heat is dissipated better, so the temperature gradients across the column are smaller. (7) A number of short columns can be joined together to increase the total column length without loss of efficiency (plate count).

Extracolumn effects must also be considered. Generally, this calls for use of a detector cell with a smaller volume and for very short lengths of 1.3- to 1.8-mm i.d. tubing if short columns with 3-μm particle packings are used.

5.3.1.2.1 Guard Columns and Filters. Particulate matter from the sample or solvent(s) will plug the frit (0.5- to 2-μm pore diameter) at the column entrance or, in the absence of a frit, the upper portion of the column packing. The cause is inadequate sample cleanup or contaminated solvents.

---

[29] E. Katz, K. Ogan, and R. P. W. Scott, "LC Column Design," *J. Chromatogr.* **289:**65 (1984).

[30] B. L. Karger, M. Martin, and G. Guiochon, "Role of Column Parameters and Injection Volume on Detection Limits in Liquid Chromatography," *Anal. Chem.* **46:**1640 (1974).

The guard column acts as a chemical filter to remove strongly retained materials that might otherwise foul the analytical column and thus shorten its lifetime. Guard columns, 1 to 5 cm in length, are available in disposable cartridge designs or as pack-it-yourself kits. They contain a stationary phase similar to that in the analytical column; they fit between the injector and the analytical column. Guard columns possess a low dead volume.

The mobile phase and sample should usually be filtered through a 0.45-μm membrane filter, especially if they contain buffers, salts, or any other dissolved solid chemicals. Only HPLC-grade solvents need not be filtered.

Plugged frits can be cleaned by back-flushing or by cleaning the frit in an ultrasonic bath. Frits are easily replaced. Upper column packing that has become contaminated can be removed and replaced by fresh packing material.

5.3.1.3 Stationary Phases in HPLC. The stationary phase can be a solid, a liquid, or a bonded phase. In the latter two cases, the phase must be coated on, or bonded to, particles of a porous solid support. The solid support may be a totally porous particle, such as silica or a macroporous polymer, a superficially porous support (porous-layer beads), or a thin film covering a solid core (pellicular supports). Each type may have a polymer bonded to its surface (bonded-phase supports). Spherical particles are more popular, as are the smaller particle sizes (<15 μm; usually about 3, 5, or 7 μm). It is very important that stationary phases exhibit good long-term stability when used with highly aggressive mobile phases.

Represented by the silicas, totally porous particles have a large surface area with pores throughout the structure. These supports generate large surface-per-unit-weight values for fast mass transfer during chromatography. Particles can be packed in columns to give 800 plates per centimeter when a 5-μm particle size is used. Wide-pore silicas are the basis of many families of specialty packings. The use of large-pore microparticulate column materials have overcome, to some degree, the problems of excessive band broadening and tailing of chromatographic peaks, but poor resolving power and long analysis times present serious limitations to protein chemists. A pore size of 30 nm is popular for proteins and is suitable for the separation of most biopolymers. For biopolymers not fully resolved on the smaller-size materials, 50-, 100-, and 400-nm pore sizes are available. Above a pH of about 8, silica is sufficiently soluble in aqueous solutions that significant column degradation results with extended usage.

Macroporous polymers have large channels as well as micropores leading off the channels and surface. Beads of these polymers do not swell or shrink appreciably with changes in the ionic strength of the mobile phase or deform at high flow velocities. They are well suited to separations conducted in nonaqueous media and for gel-permeation chromatography. When functional ion-exchange groups are incorporated in the polymer structure, these packings are used in ion chromatography.

5.3.1.3.1 Bonded-Phase Supports. Bonded-phase supports were made initially from microparticulate silica (usually with a 5- or 10- μm average particle diameter) by the chemical bonding of an organic moiety to the surface through a siloxane (Si—O—Si—C) bond. After the functional group has been bonded to the silica surface, there are residual unbonded silanol groups on the silica surface that have not reacted due to steric hindrance. In these instances the residual silanol groups can interact with basic and/or polar compounds and result in tailing peaks, which make quantitation difficult. These residual silanol groups (and hydrolyzed end groups from bonded phases) are removed with a smaller silanizing reagent, such as trimethylchlorosilane, which is less sterically hindered, in a process known as *end-capping*. The bonded phase acts like an immobilized stationary liquid phase without the problems that plagued the older columns in which the stationary phase simply coated the walls of the column or the substrate. Silica-based bonded phases are not recommended for use in aqueous solutions that have pH values above about pH 7.5. Moreover, hydrolysis of the silica matrix occurs at any pH value, and bonded phases ultimately will be degraded by aqueous mobile phases, the rate being enhanced at high salt concentrations and in the presence of some ion-pairing reagents.

Unlike polysiloxanes, a polystyrene–divinylbenzene (PS-DVB) matrix is stable from pH 0 to 14. These polymers are rigid, macroporous structures with pore sizes of approximately 8 nm.

Aluminas of suitable particulate geometry and pore structure can be polymer-coated by cross-linking of a polybutadiene. Such stationary support phases have excellent separation efficiency and exhibit reversed-phase retention behavior.

Nonpolar hydrocarbonaceous moieties of low polarity are ethyl (or butyl), octyl, and octadecyl linear alkanes. The octadecyl packing is useful when maximum retention is required. By contrast, the ethyl group is useful in applications that involve very strongly retained solutes. Octyl packings are a good compromise for the separation of samples with wide-ranging polarities.

Bonded phases of medium polarity will have functional groups, such as cyanoethyl ($—CH_2CH_2CN$) groups. These packings are useful in separations involving ethers, esters, nitro compounds, double-bond isomers, and ring compounds that differ in double-bond content. Several other bonded phases have unique selectivity. The phenyl group is used for separations that rely on the interaction of aromatic compounds with the stationary phase. The aminoalkyl group provides a highly polar surface. It may function as either a Brönsted acid or base, or it may interact with solutes through hydrogen bonding. When a tertiary nitrogen is bonded into the polymer (such as Interaction ACT-2™, a vinylpyridine-DVB copolymer), the packing displays ion-exchange character at low pH values depending upon the degree of protonation of both the polymer and the analyte (determined by eluent pH and analyte pK_a). Bases will be somewhat excluded by ion-exclusion processes and acids somewhat retained, particularly at pH 2. The packing shows some weak ionic character at intermediate pH; at elevated pH values, the polymer is deprotonated and reversed-phase interactions dominate. Under these circumstances, both neutral and basic compounds are retained, and can be separated by normal partitioning mechanisms.

Among the more selective of the specialty columns are the chiral phases developed for the separation of enantiomers; for details see Sec. 5.2.4.3.

The column length should be selected so that resolution is acceptable; lengths from 50 to 300 mm are usual. An ideal diameter is 3.9 mm for an analytical column although 4.6-mm diameter columns have been conventional. Some typical bonded-phase packing characteristics are given in Table 5.17.

5.3.1.3.2 Extracolumn Effects. Most commercial LC systems are designed so that extracolumn effects are insignificant if tubing runs are kept short and if standard columns (15 cm × 4.6 mm, 5-μm particles) are used. If smaller-dimension or smaller-particle columns are used, however, extracolumn effects can be important. These extracolumn effects are the added volume contributions from the injector sample loop, the detector cell and time constant, and the tubing connecting the column to the injector and detector.

5.3.1.3.3 Void-Volume Markers. Suitable void-volume markers are few in number. The best is a D_2O-enriched mobile phase. Uracil gives the most uniform results with changes in either solute concentration or mobile-phase composition, and it is easily detected at 254 nm. The use of ionic

TABLE 5.17 Typical Bonded-Phase Packing Characteristics

Functional group	Base material	Particle shape	Particle size, μm	Pore size, nm	Surface area, $m^2 \cdot g^{-1}$	Carbon load, %
C_4	HEMA*	Spherical	10	25		
C_8	HEMA	Spherical	10	25		3.5
C_{18}	HEMA	Spherical	10	25		10
C_8	Silica	Irregular	5,10	6	450	10
C_{18}	Silica	Irregular	5,10	6	450	15
C_8	Silica	Spherical	3,10	8	200	8
C_{18}	Silica	Spherical	3,10	8	200	12
CN	Silica	Irregular	5,10	6	450	
CN	Silica	Spherical	3,10	8	200	
Phenyl	Silica	Spherical	4	6		5
Phenyl	Silica	Irregular	10	12.5		8
NH_2	Silica	Spherical	3,10	8	200	
NH_2	Silica	Irregular	10	12.5		3.5

* HEMA denotes (hydroxyethyl)methacrylate polymer.

species is ill-advised; depending upon the partial charge present on the residual silanol groups, negatively charged ionic species can be excluded from the pores of the packing material and from the surface in general. This exclusion process leads to a significant decrease in the apparent void volume of the column.

5.3.1.3.4 Radial Compression Technology. As the particle diameter is reduced, particle–particle interactions during the packing process become more pronounced, and it becomes increasingly difficult to form a highly efficient packed bed that is stable over time. Radially compressing the cartridge (column) reforms the packed, removing any voids and channels that have formed. Near the rigid wall of a conventional steel column, there is a lower resistance to flow due to the lower density of the packed bed. This "wall effect" constitutes a significant source of band broadening; it can be minimized with Waters™ radial compressed cartridges.

5.3.1.4 Detectors. Before considering individual types of detectors, general properties are discussed.

1. *Response time.* Response time should be at least 10 times less than the peak width of a solute in time units to avoid distortion of the peak area.

2. *Detection limit.* Typically the concentration of the solute peak in the detector is $^1/_5$ to $^1/_{350}$ of the initial sample concentration at injection because of the dilution factor. For precise quantitation, a tenfold greater concentration than this estimate is needed. Keep in mind that narrow-bore columns dilute small samples less than large-bore columns.

3. *Linearity and dynamic range.* For quantitative work the signal output of the detector should be linear with concentration (concentration-sensitive detector) or mass (mass-sensitive detector). A wide linear dynamic range, perhaps 5 orders of magnitude, is desirable in order to handle major and trace components in a single analysis.

Microprocessors are often used to control the operations for HPLC. The chromatographer sets a threshold detector level at a value desired to trigger the control subroutine. The threshold is the level to which the detector signal must rise from the baseline signal for the controller to recognize that a peak is entering the sensitive volume of the detector. Several tasks can be performed:

1. Ascertaining the completion of an analysis
2. Ascertaining and controlling the proper times for stop-flow analysis methods
3. Developing methods
4. Collecting individual peaks with a fraction collector
5. Monitoring of column performance

5.3.1.4.1 Ultraviolet-Visible Detectors. Optical detectors based on ultraviolet-visible absorption constitute over 70% of HPLC detection systems. It is not necessary that the wavelength selected for HPLC coincide with the wavelength where maximum absorption occurs. True, sensitivity will suffer, but any wavelength within the absorption envelope will be usable. Section 6.2 contains a number of pertinent tabular information.

Detectors are available that span the full range from single-wavelength to simultaneous multiwavelength detection capabilities. Two key advantages of this mode of detection are as follows:

1. Wavelength selectivity
2. Excellent analyte sensitivity

Restrictive operating conditions are as follows:

1. Thermostatic control to 0.01°C is required if the detector is to approach the shot (random) noise limitation of 10^{-6} absorbance units.

2. The detector cell volume should not exceed 8 μL per centimeter of optical path length for the standard separation columns, and 2 μL for narrow-bore columns. The usefulness of UV absorbance is limited to capillaries having inner diameters of 25 μm or greater. Slit widths for the UV light source should be sufficiently small that only the capillary is illuminated, thereby reducing stray light levels that could result in excessive background noise.

3. The time constant should be 0.04 s.

The mobile-phase solvent should absorb only weakly or not at all. Water, methanol, acetonitrile, and hexane all permit operation in the far-ultraviolet to at least 210 nm. A list of common HPLC solvents and their transmittance cutoffs in the ultraviolet is given in Table 5.19 in Sec. 5.3.3.

The concentration detection limit in units of mol · cm^{-1} · L^{-1} is given by

$$\frac{2\times(\text{noise})}{b\epsilon}$$

where b is the path length of the optical cell and ϵ is the molar absorptivity.

The detection limit in terms of the sample weight is given by

$$\frac{2\times(\text{noise})\times(\text{dilution factor})\times(\text{sample size, L})\times\ (\text{mol. wt.})}{b\epsilon}$$

For conventional HPLC conditions, typical detectability with commercial absorbance detectors is 1 ng of injected analyte or an injected concentration of $5 \times 10^{-7}M$.

The basic types of ultraviolet-visible detectors are listed in order of increasing complexity (and cost) along with their advantages and limitations. *A fixed-wavelength detector* uses a light source that emits maximum light intensity at one or several discrete wavelengths that are isolated by appropriate optical filters. Its *advantages* are (1) low cost, (2) a minimum of noise, usually less than 0.0001 absorbance unit, and (3) sensitivity at the nanogram level for compounds that absorb at an available fixed wavelength. Its *disadvantage* is that there is no free choice of wavelength in many situations.

A *variable-wavelength detector* is usually a wide-band-pass ultraviolet-visible spectrophotometer that is coupled to the chromatographic system. Its *advantage* is the wide selection of wavelengths from 190 to 600 nm that permits the user to choose the wavelength at which solute absorbance is maximum. *Disadvantages* are (1) increased cost and (2) increased detector noise level that is usually a factor of 5 to 10 greater than that of filter photometers.

Although the *scanning-wavelength detector* is the most costly of the three types, it has many advantages:

1. It offers a real-time spectrum from 190 to 600 nm that can be obtained in as little as 0.01 s for each solute as it elutes.

2. Spectra are available for visual presentation as three-dimensional chromatograms of time versus wavelength versus absorbance.

3. When diode arrays are used as the detector, it is possible later to extract data at other wavelengths from the memory.

4. Some instruments can be configured to monitor simultaneously a number of wavelength intervals with bandwidths ranging between 4 and 400 nm. This permits the integration of all signals between two present wavelengths.

5. With simultaneous absorbance detection at several wavelengths, the presence of co-eluting peaks is diagnosed from the shifting of retention times as a function of detection wavelength.

6. Peak purity can be determined by taking a spectrum at the front and at the rear of a peak.

If solutes do not contain a chromophore, an absorbance signal can be obtained by *indirect detection*. In this detection mode the eluant solution must contain a UV-visible chromophore. Nonabsorbing species are revealed by a decrease in absorbance when transparent analyte species substitute for the chromophoric displacing material. Sample solutes are revealed and quantified by the decrements they produce in eluant concentration. Ions useful as eluents in indirect photomatric detection are the following: (1) for anions, benzoate, Cu(EDTA), iodide, nitrate, phthalate, sulfobenzoate, and 1,3,5-benzenetricarboxylate; and (2) for cations, benzyltrimethylammonium, copper ethylenediamine, and copper(II) aqua ion. For many ions the area of the peak is independent of the solute ion and dependent only on the amount. Monitoring the effluent at two wavelengths may resolve overlapping cation and anion peaks.[31]

5.3.1.4.2 Fluorometric Detector. Fluorescence detection in HPLC provides excellent selectivity and sensitivity. If the detector will be used exclusively for a specific task and the chromatography is fairly clean, a filter fluorometer is quite appropriate. On the other hand, if the detector will be used for many different purposes, a spectrofluorometer will be best. An instrument that employs a monochromator for the excitation wavelength and a filter system for isolating the emission provides selectivity for excitation and sensitivity for emission. Fluorescence analysis is considered in Sec. 6.6.

Design criteria for flow cells involve a compromise between the cell volume and the excitation-emission collection efficiency. One commercial unit (Kratos) employs a 5-μL flow cell with a narrow depth (1.07 mm) and large surface area for excitation-emission collection. The emitted radiation is collected by a concave mirror placed around the flow cell, and the rear of the cell is reflective. This configuration captures greater than 75% of the fluorescence while minimizing inner filtering. With flow cells, scattered radiation from the excitation source is removed with cutoff filters placed before the photomultiplier tube.

The mobile-phase composition should be considered as some compounds can quench fluorescence. Avoid chlorinated hydrocarbons and salts that include heavy atoms (such as Br, I, and Cl, and heavy metals). Organic solvents such as methanol, acetonitrile, or hexane and aqueous buffers do not present a problem.

In practice, fluorescence detection is limited by the presence of background light, which includes various types of light scattering, luminescence from the flow cell walls, and emission from impurities in the solvent. All of these increase with excitation intensity to produce no net gain.

By comparison with an absorption detector, the fluorescence detector measures a signal against a blank of zero (assuming the solvent does not fluoresce). This leads to considerably less noise and thus significantly better sensitivity. Selectivity is also improved because the analyst sets the desired excitation wavelength and the desired emission wavelength. Three-dimensional (excitation, emission, and retention time) chromatograms often distinguish species that are not chromatographically resolved. All species that absorb light do not necessarily fluoresce. Two or more solutes may absorb at the same wavelength, but emit at different wavelengths (or some may not emit). A wide variety of derivative-forming reagents have been developed to extend the realm of fluorescent detection to nonfluorescing compounds (see Sec. 5.1.5.4).

5.3.1.4.3 Electrochemical Detectors.[32] Electrochemical detection depends on the voltammetric characteristics of solute molecules. Sensitive detection is possible for species exhibiting a reversible electron transfer for a particular functional group. Detectabilities are quite favorable for aromatic amines and phenols, ranging from about 10 pg to 1 ng injected material. This type of detector has found its greatest application when polar mobile phases are used. The detector offers considerable selectivity, since relatively few components in a complex mixture are likely to be electroactive; see Sec. 14.5 for pertinent tables.

The flow cell (5 μL) is a channel in a thin polyfluorocarbon gasket sandwiched between two blocks, one plastic and the other stainless steel, which serves as the auxiliary electrode. A working electrode is positioned along one side of the channel. Farther downstream a reference electrode is connected to the working region by a short length of tubing.

5.3.1.4.4 Differential Refractometers. A differential refractometer monitors the difference in refractive index between the mobile phase and the column eluent. In LC it comes closest to being a universal detector. Unless the analyte happens to have exactly the same refractive index as the mobile phase, a signal will be observed. Unfortunately, the detection sensitivity, which typically is in micrograms, is poor compared with other detectors, and the detector is extremely sensitive to temperature and flow changes. Temperature must be controlled to within 0.001°C. Solvent delivery systems must be pulse-free to avoid noise. Use of a gradient is restricted to solvent pairs that have virtually identical refractive indices, a number of which are listed in Table 5.18. These detectors find use mostly for the initial survey of samples and in exclusion (gel-permeation) chromatography.

---

[31] Z. Iskandarani and T. E. Miller, Jr., *Anal. Chem.* **57:**1591 (1985).

[32] D. A. Roston, R. E. Shoup, and P. T. Kissinger, "Liquid Chromatography/Electrochemistry: Thin-Layer Multiple Electrode Detection," *Anal. Chem.* **54:**1417A (1982).

TABLE 5.18 Solvents Having the Same Refractive Index and the Same Density at 25°C

Solvent 1	Solvent 2	Refractive index 1	Refractive index 2	Density, g/mL 1	Density, g/mL 2
Acetone	Ethanol	1.357	1.359	0.788	0.786
Ethyl formate	Methyl acetate	1.358	1.360	0.916	0.935
Ethanol	Propionitrile	1.359	1.363	0.786	0.777
2,2-Dimethylbutane	2-Methylpentane	1.366	1.369	0.644	0.649
2-Methylpentane	Hexane	1.369	1.372	0.649	0.655
Isopropyl acetate	2-Chloropropane	1.375	1.376	0.868	0.865
3-Butanone	Butyraldehyde	1.377	1.378	0.801	0.799
Butyraldehyde	Butyronitrile	1.378	1.382	0.799	0.786
Dipropyl ether	Butyl ethyl ether	1.379	1.380	0.753	0.746
Propyl acetate	Ethyl propionate	1.382	1.382	0.883	0.888
Propyl acetate	1-Chloropropane	1.382	1.386	0.883	0.890
Butyronitrile	2-Methyl-2-propanol	1.382	1.385	0.786	0.781
Ethyl propionate	1-Chloropropane	1.382	1.386	0.888	0.890
1-Propanol	2-Pentanone	1.383	1.387	0.806	0.804
Isobutyl formate	1-Chloropropane	1.383	1.386	0.881	0.890
1-Chloropropane	Butyl formate	1.386	1.387	0.890	0.888
Butyl formate	Methyl butyrate	1.387	1.391	0.888	0.875
Methyl butyrate	2-Chlorobutane	1.392	1.395	0.875	0.868
Butyl acetate	2-Chlorobutane	1.392	1.395	0.877	0.868
4-Methyl-2-pentanone	Pentanonitrile	1.394	1.395	0.797	0.795
4-Methyl-2-pentanone	1-Butanol	1.394	1.397	0.797	0.812
2-Methyl-1-propanol	Pentanonitrile	1.394	1.395	0.798	0.795
2-Methyl-1-propanol	2-Hexanone	1.394	1.395	0.798	0.810
2-Butanol	2,4-Dimethyl-3-pentanone	1.395	1.399	0.803	0.805
2-Hexanone	1-Butanol	1.395	1.397	0.810	0.812
Pentanonitrile	2,4-Dimethyl-3-pentanone	1.395	1.399	0.795	0.805
2-Chlorobutane	Isobutyl butyrate	1.395	1.399	0.868	0.860
Butyric acid	2-Methoxyethanol	1.396	1.400	0.955	0.960
1-Butanol	3-Methyl-2-pentanone	1.397	1.398	0.812	0.808
1-Chloro-2-methylpropane	Isobutyl butyrate	1.397	1.399	0.872	0.860
1-Chloro-2-methylpropane	Pentyl acetate	1.397	1.400	0.872	0.871
Methyl methacrylate	3-Methyl-2-pentanone	1.398	1.398	0.795	0.808
Triethylamine	2,2,3-Trimethylpentane	1.399	1.401	0.723	0.712
Butylamine	Dodecane	1.399	1.400	0.736	0.746
Isobutyl butyrate	1-Chlorobutane	1.399	1.401	0.860	0.875
1-Nitropropane	Propionic anhydride	1.399	1.400	0.995	1.007
Pentyl acetate	1-Chlorobutane	1.400	1.400	0.871	0.881
Pentyl acetate	Tetrahydrofuran	1.400	1.404	0.871	0.885
Dodecane	Dipropylamine	1.400	1.400	0.746	0.736
1-Chlorobutane	Tetrahydrofuran	1.401	1.404	0.871	0.885
Isopentanoic acid	2-Ethoxyethanol	1.402	1.405	0.923	0.926
Dipropylamine	Cyclopentane	1.403	1.404	0.736	0.740
2-Pentanol	4-Heptanone	1.404	1.405	0.804	0.813
3-Methyl-1-butanol	Hexanonitrile	1.404	1.405	0.805	0.801
3-Methyl-1-butanol	4-Heptanone	1.404	1.405	0.805	0.813
Hexanonitrile	4-Heptanone	1.405	1.405	0.801	0.813
Hexanonitrile	1-Pentanol	1.405	1.408	0.801	0.810
Hexanonitrile	2-Methyl-1-butanol	1.405	1.409	0.801	0.815
4-Heptanone	1-Pentanol	1.405	1.408	0.813	0.810
2-Ethoxyethanol	Pentanoic acid	1.405	1.406	0.926	0.936
2-Heptanone	1-Pentanol	1.406	1.408	0.811	0.810

(Continued)

TABLE 5.18 Solvents Having the Same Refractive Index and the Same Density at 25°C (*Continued*)

Solvent 1	Solvent 2	Refractive index		Density, g/mL	
		1	2	1	2
2-Heptanone	2-Methyl-1-butanol	1.406	1.409	0.811	0.815
2-Heptanone	Dipentyl ether	1.406	1.410	0.811	0.799
2-Pentanol	3-Isopropyl-2-pentanone	1.407	1.409	0.804	0.808
1-Pentanol	Dipentyl ether	1.408	1.410	0.810	0.799
2-Methyl-1-butanol	Dipentyl ether	1.409	1.410	0.815	0.799
Isopentyl isopentanoate	Allyl alcohol	1.410	1.411	0.853	0.847
Dipentyl ether	2-Octanone	1.410	1.414	0.799	0.814
2,4-Dimethyldioxane	3-Chloropentene	1.412	1.413	0.935	0.932
2,4-Dimethyldioxane	Hexanoic acid	1.412	1.415	0.935	0.923
Diethyl malonate	Ethyl cyanoacetate	1.412	1.415	1.051	1.056
3-Chloropentene	Octanoic acid	1.413	1.415	0.932	0.923
2-Octanone	1-Hexanol	1.414	1.416	0.814	0.814
2-Octanone	Octanonitrile	1.414	1.418	0.814	0.810
3-Octanone	3-Methyl-2-heptanone	1.414	1.416	0.830	0.818
3-Methyl-2-heptanone	1-Hexanol	1.415	1.416	0.818	0.814
3-Methyl-2-heptanone	Octanonitrile	1.415	1.418	0.818	0.810
1-Hexanol	Octanonitrile	1.416	1.418	0.814	0.810
Dibutylamine	Allylamine	1.416	1.419	0.756	0.758
Allylamine	Methylcyclohexane	1.419	1.421	0.758	0.765
Butyrolactone	1,3-Propanediol	1.434	1.438	1.051	1.049
Butyrolactone	Diethyl maleate	1.434	1.438	1.051	1.064
2-Chloromethyl-2-propanol	Diethyl maleate	1.436	1.438	1.059	1.064
N-Methylmorpholine	Dibutyl decanedioate	1.436	1.440	0.924	0.932
1,3-Propanediol	Diethyl maleate	1.438	1.438	1.049	1.064
Methyl salicylate	Diethyl sulfide	1.438	1.442	0.836	0.831
Methyl salicylate	1-Butanethiol	1.438	1.442	0.836	0.837
1-Chlorodecane	Mesityl oxide	1.441	1.442	0.862	0.850
Diethylene glycol	Formamide	1.445	1.446	1.128	1.129
Diethylene glycol	Ethylene glycol diglycidyl ether	1.445	1.447	1.128	1.134
Formamide	Ethylene glycol diglycidyl ether	1.446	1.447	1.129	1.134
2-Methylmorpholine	Cyclohexanone	1.446	1.448	0.951	0.943
2-Methylmorpholine	1-Amino-2-propanol	1.446	1.448	0.951	0.961
Dipropylene glycol monoethyl ether	Tetrahydrofurfuryl alcohol	1.446	1.450	1.043	1.050
1-Amino-2-methyl-2-pentanol	2-Butylcyclohexanone	1.449	1.453	0.904	0.901
2-Propylcyclohexanone	4-Methylcyclohexanol	1.452	1.454	0.923	0.908
Carbon tetrachloride	4,5-Dichloro-1,3-dioxolan-2-one	1.459	1.461	1.584	1.591
N-Butyldiethanolamine	Cyclohexanol	1.461	1.465	0.965	0.968
D-α-Pinene	*trans*-Decahydronaphthalene	1.464	1.468	0.855	0.867
Propylbenzene	*p*-Xylene	1.490	1.493	0.858	0.857
Propylbenzene	Toluene	1.490	1.494	0.858	0.860
Phenyl 1-hydroxyphenyl ether	1,3-Dimorpholyl-2-propanol	1.491	1.493	1.081	1.094
Phenetole	Pyridine	1.505	1.507	0.961	0.978
2-Furanmethanol	Thiophene	1.524	1.526	1.057	1.059
m-Cresol	Benzaldehyde	1.542	1.544	1.037	1.041

Source: J. A. Dean, ed., *Lange's Handbook of Chemistry*, 14th ed., McGraw-Hill, NY, 1992.

5.3.2 Gradient Elution (Solvent Programming)

5.3.2.1 Equipment. Either low- or high-pressure systems may be used in gradient elution. In the low-pressure system, two or more solvents are blended in the desired proportions by a precise valving system, which often is controlled by a microprocessor. The mixing vessel is packed with an inert fiber that creates sufficient turbulence to mix the solvents while simultaneously causing the release of any generated gas bubbles. A low-pressure system requires only one pressurization pump—a distinct advantage.

In a system in which the gradient is formed after the HPLC pump, the output from two or more high-pressure pumps is programmed into a low-volume mixing chamber before flowing into the column. One pump is required for each solvent involved.

5.3.2.2 Methodology. Gradient elution is normally used for separating samples that vary widely in polarity (the k' ratio exceeds 20), as well as for those in which the isocratic retention bunches peak together at the beginning of the chromatogram and spread them out (with broadening) at the end of the chromatogram. The gradient separation provides a more even spacing of bands and better resolution, as well as narrower bands at the end of the chromatogram. Gradient elution works by changing k' during the separation (k' programming) because the mobile-phase strength changes. Weakly retained compounds leave the column first in a weak mobile phase. Strongly retained bands leave last in a strong mobile phase.

With commercial equipment a variety of gradients are available. The linear gradient is best for scouting initially. If nothing elutes at the beginning of the run, start the gradient at an organic content that is 5% lower than that needed to elute the first peak; this will eliminate the wasted solvent at the beginning of the run. Stopping the gradient as soon as the last band elutes will eliminate the waste at the end of the run. Thus the gradient range will have been changed without adversely affecting the separation.

The starting mobile-phase composition should be sufficiently weak to give good separation of the early bands. The final percentage of strong solvent in the gradient should be adjusted so that the last band leaves the column at about the time the gradient is completed. If early bands are poorly resolved, the starting percentage of strong solvent is too large and/or the gradient shape is too convex. The remedy is to decrease the starting percentage of strong solvent. If the early bands are well resolved and sharp and there is no excessive space between bands, the gradient should be started only after they have eluted or else a convex-shape gradient should be used.

Resolution can be improved in gradient separations by optimizing k', N, and/or α. The k' value in gradient elution is given by

$$k' = \frac{(\text{constant}) \times t_G F_c}{(\Delta\% B) \times V_M} \tag{5.34}$$

where t_G = gradient retention time
F_c = flowrate
V_M = column volume of mobile phase
$\Delta\% B$ = difference between starting and final percentage of strong solvent in gradient

Here, the term denoted by "(constant)" is approximately 20 for reversed-phase HPLC. An increase in the flowrate for a gradient separation reduces the resolution, the bandwidths, and the retention times and causes k' to be higher. In order to change the flowrate without changing k', other parameters in Eq. (5.34) must be changed. One possibility is to double the flowrate while halving the gradient time. This speeds the separation without altering the selectivity. If the column length is doubled (which doubled V_M), either the flowrate or the gradient time must also be doubled (or some suitable variation in each quantity to yield a factor of 2 for their product).

Selectivity is achieved by altering α and is accomplished by changing mobile-phase composition, column type, or temperature. Experience is the main guide.

When running with high-molecular-weight samples or very shallow gradients, special problems may occur. Small changes in mobile-phase composition affect the retention times as if they were varying within a run and spurious peaks appear.

5.3.3 Liquid–Solid Chromatography[33] (LSC)

In liquid–solid (or adsorption) chromatography the solute and solvent molecules are in competition for adsorption sites on the surface of the column packing, usually silica gel, plus alumina and carbon in special situations. The adsorption sites are the slightly acidic silanol groups (Si—OH) groups that extend out from the surface of the porous particles and from the internal channels of the pore structure. These hydroxyl groups interact with polar or unsaturated solutes by hydrogen bonding or dipole interaction. The competition between the solute molecules and the solvent molecules for an active site provides the driving force and selectivity in separations.

Variations in solute retention are achieved by changes in the composition of the mobile phase. It is the solvent strength that controls the k' value of solute peaks. Table 5.19 lists the common solvents used in liquid–solid chromatography in order of increasing solvent strength (and roughly in the order of increasing polarity). This listing also ranks the adsorption strength of the various functional groups (acting singly) of solute molecules; log k' varies linearly with ϵ^0, the solvent strength parameter. Not included in Table 5.19 are sulfides (less polar than ethers), aldehydes (similar to esters and ketones), amines (similar to alcohols), and sulfones and amides (between the alcohols and acetic acid, with their polarities increasing in the order listed). Saunders[34] discusses the extension of ϵ^0 values to more complex molecules. Snyder[35] reviews the role of the solvent in LC.

In practice, the mobile phase is chosen to match the most polar functional group in the sample. If the k' values are too small (sample elutes too rapidly), a weaker (less polar) solvent is substituted. A stronger (more polar) solvent is used if the sample does not elute in a reasonable time because of high k' values. Two solvents may be blended together in various proportions to provide continuous variation in solvent strength between that of each pure solvent. An increase of 0.05 unit in the value of the solvent strength parameter usually decreases all k' values by a factor of 3 to 4. Nearly always we seek a solvent mixture that will give $k' > 1$ to obtain separation of the component of interest from other sample components of similar composition. Generally, we also seek a solvent mixture that will give $k' < 10$ in order to reduce the total separation time and minimize the sample dilution that occurs at high k'.

Binary solvent mixtures offer additional selectivity through the adjustment of the dipole, proton-acceptor, and proton-donor forces. For example, the substitution of diethyl ether or methanol for one of the other strong components in a solvent mixture can often improve selectivity by the formation of hydrogen bonds. Snyder and Saudners[36] recommend that an optimum solvent program should have a change in ϵ^0 of 0.04 unit per column volume of solvent. The advantage of a gradient run is similar to that achieved using programmed temperature in GC. The big disadvantage is the time required to reequilibrate the LC column. To recondition the column back to its original nonpolar status, about 10 column volumes of each solvent are required.

Liquid–solid chromatography is influenced more by specific functional groups and less by molecular-weight differences than liquid–liquid chromatography. This makes possible the separation of complex mixtures into classes of compounds with similar chemical functionality. For example, polynuclear aromatics are easily separated from aliphatic hydrocarbons, and the triglycerides are easily separated from a lipid extract. On the other hand, separation among members of a homologous series is usually poor.

Liquid–solid chromatography excels in the separation of positional isomers that differ in the geometrical arrangement of functional groups with respect to the adsorption sites. LSC is also popular for preparative separations.

[33] L. R. Snyder, *Principles of Adsorption Chromatography*, Dekker, New York, 1968.
[34] D. L. Saunders, *Anal. Chem.* **46**:470 (1974).
[35] L. R. Snyder, *Anal. Chem.* **46**:1385 (1974).
[36] L. R. Snyder and D. L. Saunders, *J. Chromatogr. Sci.* **7**:195 (1969).

TABLE 5.19 Solvents of Chromatographic Interest

Solvent	Solvent strength parameter $\epsilon^0(SiO_2)$	$\epsilon^0(Al_2O_3)$	Boiling point, °C	UV cutoff, nm	Viscosity, $mN \cdot s \cdot m^{-2}$ (20°C)	Refractive index (20°C)
Fluoroalkanes	−0.19	−0.25				1.25
Pentane	0.00	0.00	36	210	0.24 (15°C)	1.358
Isooctane	0.01		99	215	0.50	1.392
Hexane		0.01	69	210	0.31	1.404
Cyclohexane	0.03	0.04	81	210	0.98	1.426
Carbon disulfide	0.11	0.15	46	380	0.36	1.626
Carbon tetrachloride	0.14	0.18	77	265	0.97	1.466
1-Chlorobutane		0.26	78	220	0.43	1.402
o-Xylene	0.20	0.26	144	290	0.81	1.505
Diisopropyl ether	0.22	0.28	68	220	0.38 (25°C)	1.369
Toluene	0.22	0.29	111	286	0.59	1.497
Chlorobenzene	0.23	0.40	132		0.80	1.525
Benzene	0.25	0.32	80	280	0.65	1.501
Diethyl ether	0.29	0.38	35	218	0.25	1.353
Diethyl sulfide		0.38	92	290	0.45	1.443
Chloroform	0.31	0.40	62	245	0.57	1.443
Dichloromethane	0.32	0.42	40	235	0.44	1.425
4-Methyl-2-pentanone	0.33	0.43	118	335	0.42(15°C)	1.396
Tetrahydrofuran	0.35	0.45	66	220	0.55	1.407
1,2-Dichloroethane		0.49	84	228	0.80	1.445
2-Butanone	0.39	0.51	80	330	0.42 (15°C)	1.379
1-Nitropropane		0.53	131	380	0.80 (25°C)	1.402
Acetone	0.43	0.56	56	330	0.32	1.359
1,4-Dioxane	0.43	0.56	107	215	1.44 (15°C)	1.420
Ethyl acetate	0.45	0.58	77	255	0.45	1.372
Methyl acetate		0.60	56	260	0.48 (15°C)	1.362
1-Pentanol	0.47	0.61	138	210	4.1	1.410
Dimethylsulfoxide	0.48	0.62	189	265	2.47	1.478
Diethylamine		0.63	56	275	0.33	1.386
Nitromethane	0.49	0.64	114	380	0.67	1.394
Acetonitrile	0.50	0.65	82	190	0.37	1.344
Pyridine		0.71	115	330	0.97	1.510
2-Butoxyethanol		0.74	170	220	3.15 (25°C)	1.420
2-Propanol	0.63	0.82	82	210	2.50	1.377
Ethanol	0.68	0.88	78	210	1.20	1.361
Methanol	0.78	0.95	64	210	0.59	1.328
Ethylene glycol	0.86	1.11	197	210	21.8	1.432
Acetic acid	Large	Large	118	260	1.23	1.372

5.3.4 Liquid–Liquid Chromatography

The separations in liquid–liquid chromatography are derived from the partitioning of analytes between two liquids, one of which is held immobile on a stationary solid support. Both phases must be immiscible in each other. This requirement mandates that one phase will be polar and the other nonpolar. Separation is achieved in LLC by matching the polarities of the sample and stationary phase and by using a mobile phase that has a markedly different polarity. Selected methods in HPLC are listed in Table 5.20.

*5.3.4.1 **Normal-Phase Chromatography.*** In normal-phase chromatography a polar stationary phase is used in conjunction with a less polar mobile phase. A polar modifier, such as 2-propanol or water, is often added to the mobile phase to minimize secondary retention by interaction with the silica surface thereby deactivating the column.

In early days it was difficult to keep the stationary phase from washing off the column. Also, the column temperature had to be carefully controlled. Bonded-phase supports overcame these difficulties. The nitrile (cyano) and amine bonded phases are popular for normal-phase operation. Typical commercial packings might be cyanopropyldimethylsilyl and aminopropylsilyl, respectively. Nonpolar liquids such as the hydrocarbons and chlorocarbons are used as the mobile phase, although even alcohols and aqueous mixtures can be used. The nitrile phase produces separations somewhat similar to plain silica in LSC.

5.3.4.2 ***Reversed-Phase Chromatography.*** The distinction between reversed-phase and normal-phase chromatography is based on the nature of the stationary and mobile phases. Reversed-phase chromatography is by far the most popular HPLC method in use today. A hydrophobic (nonpolar) packing, usually with an octadecyl or octyl functional group, is used in conjunction with a polar mobile phase, often a partially or fully aqueous mobile phase. An aqueous mobile phase allows the use of secondary solute chemical equilibria (such as ionization control, ion pairing, and complexation) to control retention and selectivity. Chemically modified silica packings are often preferred for reversed-phase separations because of the low reduced plate heights that can be achieved and the possibility of simple chemical modification by the silanization reaction. By varying the organic moiety in the silanization reagent, many different stationary phase polarities can be realized: trimethylsilyl (C_1), butyldimethylsilyl (C_4), octyldimethylsilyl (C_8), octadecyldimethylsilyl (C_{18}), and diphenylmethylsilyl.

The inner surfaces of fused-silica capillaries cannot be modified by silanization for LC or supercritical-fluid chromatography applications because of the very low concentration of SiOH groups on such surfaces. Procedures that are used in capillary GC, which involve the immobilization of special still-reactive prepolymers, can be applied.

In designing laboratory methods, keep in mind these general principles:

1. The elution order often is predictable because retention time usually increases as the hydrophobic character of the solute increases. For the functional groups listed in Table 5.19, the elution order is reversed, thus the name *reversed-phase chromatography*. The nonpolar solutes are retained more strongly.

2. The eluent strength of the mobile phase follows the reverse order given in Table 5.19. The predominant mobile phase, water, is the weakest eluent. Methanol and acetonitrile are popular modifiers; they are commercially available with excellent chromatographic purity and have low viscosity. In order for the analytes to penetrate into the bonded phase, the mobile phase used must "wet" the surface of the bonded phases. The addition of an organic modifier like methanol or acetonitrile to the mobile phase will usually solve this problem.

When scouting for the optimum pairing of column packing and mobile-phase composition, try these steps:

1. If the sample components are of low to moderate polarity (that is, soluble in aliphatic hydrocarbons), use an octadecyl bonded-phase column and a methanol–water mixture as eluent.

2. For solutes of moderate polarity (soluble in methyl ethyl ketone), use an octyl bonded-phase packing in conjunction with an acetonitrile–water mobile phase.

3. High-polarity solutes (soluble in the lower alcohols) are handled best with a bonded ethyl packing and 1,4-dioxane–water mobile phase.

If the sample components elute too rapidly with a 1:1 mixture of organic and water, a lower concentration of the stronger eluent (methanol, acetonitrile, or 1,4-dioxane) should be tried. Of course, the reverse should be tried if the solutes elute too slowly. How much stronger or weaker? The 10% rule is a handy guide: k' changes two- to threefold for a 10% change in mobile-phase organic solvent. For example, if the last band elutes with a k' of approximately 20 in a 60 : 40 acetonitrile–water

TABLE 5.20 Selected Methods in HPLC

Sample	Column	Packing	Flowrate, mL/min	Mobile phase	Detector
Alkaloids	150 × 4.6 mm	C_{18}, 7 μm	2.0	76% Buffer (0.02*M* KH_2PO_4, pH 3.0), 24% acetonitrile	UV 254 nm
Alkyl benzenes	250 × 1 mm	C_{18}, 10 μm	0.030	65% Acetonitrile, 35% water	UV 254 nm
Analgesics	30 × 4.6 mm	C_{18}, 3 μm	2.0	38% MeOH, 62% water, (1% HOAc)	UV 254 nm
Anti-arrhythmics	250 × 4.6 mm	C_{18}, 7 μm	1.0	68% Water, 30% MeOH, 2% HOAc	UV 254 nm
	150 × 4.6 mm	CN	2.0	60% Acetonitrile, 15% MeOH, 25% 0.01*M* K_2HPO_4, pH 7.0	UV 215 nm
Anticonvulsants	150 × 4.6 mm	C_{18}, 5 μm	2.0	61% Buffer (0.1*M* KH_2PO_4, pH 4.4), 39% MeOH, 0.01% triethanolamine	UV 210 nm
Anticonvulsants and metabolites	150 × 4.6 mm	C_8	1.5	MeOH, 0.1*M* KH_2PO_4, triethanolamine (45:55:0.01)	UV 254 nm
Antidepressants, tricyclic	150 × 4.6 mm	CN	2.0	Acetonitrile, MeOH, 0.01*M* K_2HPO_4 pH 7.0 (H_3PO_4) (60:15:25)	UV 215 nm
Antihistamines	250 × 4.6 mm	C_8, 5 μm	1.3	A: 0.05*M* KH_2PO_4, pH 3.2; B: acetonitrile; gradient: (T,%B), (0,25), (5,25), (20,70), (25,70), (30,25)	UV 220 nm
Antihypertensive drugs	30 × 4.6 mm	C_{18}, 3 μm	2.0	MeOH, 80 m*M* KH_2PO_4 (pH 2.8) (5:95)	UV 280 nm
Aromatic hydrocarbons, polynuclear	100 × 4.6 mm	C_{18}, 3 μm	1.26	86.4% Acetonitrile, 13.6% water	UV 254 nm
	150 × 4.6 mm	C_{18}, 5μm	1.5	80:20 Acetonitrile:water 4 min; linear gradient to 100% acetonitrile 20 min	UV 254 nm
Barbiturates	150 × 4.6 mm	C_{18}, 7 μm	2.0	51% MeOH, 49% water	UV 220 nm
Carbamate and urea pesticides	150 × 4.6 mm	C_8	2.0	Acetonitrile:water (35°C) gradient, 18:82 to 65:35 in 9 min and hold 3 min	UV 240 nm
Carboxylic acids	250 × 4.6 mm	C_8, 5 μm	0.8	0.2*M* H_3PO_4	UV 210 nm
Chlorinated pesticides	250 × 4.6 mm	PAC, 5 μm	1.0	97.3% Isooctane, 2.7% ethyl acetate	UV 254 nm
Cough syrup components	150 × 4.6 mm	C_{18}, 5 μm	1.5	A: 0.05*M* KH_2PO_4, pH 3.2; B: acetonitrile; (T,%B), (0,15), (1,15), (15,50), (20,50), (25,15)	UV 254 nm
Fatty acids, *p*-bromophenacyl esters	250 × 4.6 mm	C_{18}, 5 μm	1.0	A: Acetonitrile, water, H_3PO_4 (70:30:1); 100% A for 5 min, program to 100% acetonitrile at 25 min	UV 254 nm

Hydrocarbons, aromatic	100 × 8 mm	C_{18}, 7% carbon	2.0	Acetonitrile: water (70:30)	UV 254 nm
Insecticides	250 mm × 4.6 μm	C_{18}, 10 μm	1.0	75% MeOH, 25% water	UV 254 nm
Ketones, aromatic	250 × 4.6 mm	C_{18}, 10 μm	1.0	70% MeOH, 30% water	UV 254 nm
Lubricating oil additives	250 × 4.6 mm	NH_2, 5 μm	1.6	0.21% 2-Propanol in heptane	UV 210 nm
Opium alkaloids	250 × 4.6 mm	CN, 5 μm	1.5	80% 0.015*M* NH_4OAc pH 5.8, 10% acetonitrile, 10%, 1,4-dioxane	UV 210 nm
Phenols	125 × 4 mm	C_{18}, 3 μm	1.0	(50:50) MeOH, 0.1% triethylamine acetate, pH 4.0	UV 220 nm
Phenols, priority pollutant	150 × 4.6 mm	C_8	1.5	MeOH:1% HOAc: water:1% HOAc gradient, 35:65 to 100:0 over 20 min, return to 35:65 over 5 min.	UV 280 nm
Phthalate esters	150 × 4.6 mm	C_8, 5 μm	1.5	70% Acetonitrile, 30% water	UV 254 nm
Polyacrylic acid	250 × 4.6 mm	C_{18}, 7 μm	1.0	25% Acetonitrile, 74% water, 1% H_3PO_4	UV 214 nm
Polyphenols	250 × 4.6 mm	C_{18}, 10 μm	1.0	90:10 MeOH:H_2O + 0.5% H_3PO_4	UV 254 nm
Steroids	250 × 4.6 mm	C_{18}, 5 μm	2.0	70% MeOH, 30% water	UV 254 nm
Sugars	300 × 4.6 mm	NH_2, 10 μm	3.0	85% Acetonitrile, 15% water	Refractive index
Theophylline	150 × 4.6 mm	C_{18}, 7 μm	2.0	94% 10 mM KH_2PO_4, pH 4.8, 4% acetonitrile, 2% tetrahydrofuran	UV 254 nm
Triazine pesticides	150 × 4.6 mm	C_8	1.5	Acetonitrile: 0.1*M* NH_4OAc (pH 6.0), 33:67; 30°C	UV 220 nm
Vitamins, water soluble	250 × 4.6 mm	C_{18}, 5 μm	1.0	A: 0.05*M* NaOAc (pH 5.2), B: acetonitrile; (T,%B): (0,5), (4,5), (15,50)	UV 254 nm
Xanthines	250 × 4 mm	C_{18}, 3 μm	0.7	Acetonitrile:0.4% NaOAc (15:85)	UV 254 nm

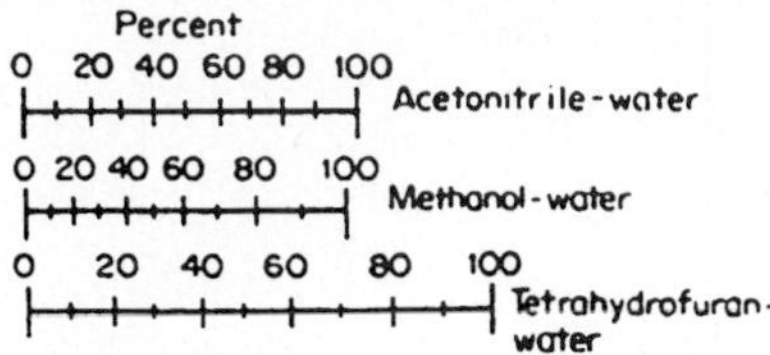

FIGURE 5.11 Nomogram for conversion from one solvent composition to another. (Percent scale is from 0 to 100% of the organic component.)

mobile phase, k' would be expected to be between about 6 and 10 for a 70:30 mobile phase. When establishing binary solvent gradients, the polarity of the eluent is continuously decreased during the chromatographic run—for example, by gradually increasing the organic solvent in methanol–water, acetonitrile–water, or 1,4-dioxane–water mixtures.

Even if k' is in the proper region, the separation still may not be adequate. In that case, try another organic solvent in the mobile phase—for example, change from acetonitrile to methanol or tetrahydrofuran. The nomogram in Fig. 5.11 can be used to convert from one solvent mixture to another. Locate the present solvent composition on the appropriate line, then draw a vertical line through the other solvent lines. The intersection of these lines indicates the equivalent other mobile-phase compositions. Thus, a 40:60 acetonitrile–water mixture is equivalent to a 50:50 methanol–water or about 25:75 tetrahydrofuran–water mixture. Changing to ternary mixtures, such as methanol–acetonitrile–water, or binary mixtures, such as acetonitrile–2-propanol or 1,4-dioxane–methanol, can often improve selectivity.

5.3.5 Ion Chromatography

The name *ion chromatography* refers to any modern and efficient method of separating and determining ions utilizing HPLC. Modern ion chromatography involves using conventional LC equipment with a variety of polymeric columns for determining the ions in question. Standard HPLC sample injectors may be used. To enhance relative sensitivity, sample loops as large as 200 μL may be used without loss of efficiency. Single-headed piston pumps work well for less demanding applications; however, a low-pulsation, dual-headed piston pump will give the best baseline and best sensitivity with most detection systems. An automatic detector is always used and the entire chromatogram is recorded. Up to 8 or 10 different anions (or cations) in a single chromatographic run takes only a few minutes. Some of its applications in environmental analysis are discussed in Sec. 21.

Ion-exchange chromatography has been used for many years to separate various organic and inorganic ions. These classical methods, described in Sec. 2.3, are often slow and required the collection and analysis of numerous fractions to delineate the chromatogram. One difficulty encountered with classical ion-exchange chromatography was the lack of suitable detectors. A second was the adaptability of the resin packings for HPLC equipment.

5.3.5.1 Column Packings Used in Ion Chromatography. Column packings for ion chromatography (IC) have charge-bearing functional groups attached to a spherical polymer matrix of uniform particle size. *Surface agglomeration* is a term used to describe the attachment of colloidal ion-exchange resin of one charge to a much larger substrate particle of the opposite charge. The capacity may be controlled in essentially three ways: by the size of the substrate particles, by the size of the colloid, and by the extent of coverage of the substrate by the colloidal resin. Present-day resins employ substrates of less than 10 μm in diameter and colloidal resins of as low as a few hundred angstroms in diameter. Ion exchangers of the surface-agglomerated type have been described as *pellicular* resins; the term *superficial* ion-exchange resin is applied for materials formed by the surface modification of a preformed polymer. Pellicular structures permit the use of slightly cross-linked resins in the pellicle layer. This extends the applicability of these resins to higher-molecular-weight ions. Users now have a wide range of columns of capacity, efficiency, and selectivity that may be matched to application demands.

Silica-based ion exchangers are less rugged than polymeric resins, and their exchange capacity tends to decrease gradually. This type of exchanger should not be used in conjunction with eluents that are very basic.

Column packings prepared by coating a resin with an ion-exchange material also work quite well in ion chromatography. Efficient column packings can also be prepared by coating porous polymeric beads with a suitable monomer such as cetylpyridinium chloride. The coating is done either in a static or dynamic mode. These ion exchangers are stable and can be prepared in a range of exchange capacities by varying the coating conditions.

Ion interaction reagents and neutral nonpolar stationary phase will be considered in Sec. 5.3.5.3.

5.3.5.1.1 Exchange Equilibrium. See the discussion in Sec. 2.3.3 on ion exchange conducted under normal atmospheric pressure and in batch and columnar modes. Cation exchangers are lightly sulfonated so that the exchange sites are in a thin zone at the outer perimeter of the beads. The exchange sites are chemically attached to benzene rings of the polymer and have little, if any, ability to move. The H^+ counterions, however, are electrostatically attracted to the negative sulfonate groups but are free to move about in the water-filled pores and channels inside the resin.

Ion exchange occurs when a cation from the outside solution enters the resin and a hydrogen ion leaves the resin. Anions from the outside solution are prevented from entering the cation-exchange resin by the "wall" of fixed sulfonate groups. The negative charges of the sulfonated groups repel the solution anions. Hydrophilic molecular species in the sample solution are not affected by this sulfonate anion barrier and can enter and leave the ion-exchange beads. Molecules that are more hydrophobic can also enter the aqueous channels of the resin, but they tend to be retained inside the resin by the van der Waals forces of the polymer, especially the benzene rings.

Anion exchangers have quaternary ammonium functional groups, usually $—CH_2N^+(CH_3)_3A^-$, where A^- is an exchangeable anion. With an anion-exchange resin, cations from the sample solution are excluded from the resin by the positive charges on the fixed quaternary ammonium groups, but anions from the outside solution can readily exchange with the mobile anions of the resin. Hydrophilic bases and other molecular species can enter and leave the aqueous channels and pores of the resin.

5.3.5.2 Detectors for Ion Chromatography. The general detectors used in HPLC are discussed in Sec. 5.3.1.4. All are applicable to ion chromatography. Conductivity detectors are generally the most versatile and useful (see Sec. 5.4.3). Compared to conductivity, spectrophotometric detectors are much less affected by small temperature changes. They can be used in a direct or indirect mode, or in conjunction with a post-column reactor. In the direct mode, a detection wavelength is chosen for which the sample ions absorb but the eluent does not. In the indirect mode an eluent is chosen that has a strong absorbance at a wavelength for which the sample ions absorb weakly or not at all. Since the concentration of ions in the column (the eluent) remains constant, the elution of a sample ion results in an equivalent decrease in the concentration of the absorbing eluent ion and in a negative chromatographic peak. A resin of low capacity and a dilute eluent make excellent sensitivity possible. This vacancy or indirect UV chromatographic method is convenient because UV detectors are commonly available with liquid chromatographs. Problems only occur if UV-absorbing materials are contained in the sample; these would need to be removed by an appropriate sample pretreatment procedure.

A post-column reactor, coupled with spectrophotometric detection of the absorbing product, is capable of great versatility. To detect metal cations, the column effluent is mixed with a buffered color-forming reagent that is also pumped into the reactor. Suggestions for suitable post-column systems will be found in Sec. 6.3.2.

Other detectors employed in modern ion chromatography include refractive index, fluorescence, electrochemical, and visible wavelength photometers. Electrochemical detectors are appropriate for oxidizable species such as sulfide and cyanide.

5.3.5.3 Ion-Pair Chromatography.[37] Ion-pair chromatography uses a neutral nonpolar stationary phase in conjunction with mobile phases containing a lipophilic electrolyte. These are the same highly efficient reversed-phase columns that are used in ordinary liquid chromatography. Typical of the stationary phases used are porous silica, the surface of which has been modified by attaching long alkyl groups (C_8 or C_{18}, for example), and macroporous styrene–DVB polymers. For cation analysis the

[37] M. T. W. Hearn, ed., *Ion-Pair Chromatography*, Dekker, New York, 1985.

typical ion interaction agent is heptane sulfonic acid, while tetrabutylammonium hydrogen sulfate (or phosphate) is commonly used for anion separations.

The eluent contains an ionic modifier, which acts as a movable ion-exchange site that can be sorbed by the column packing either alone or as an ion pair with a sample ion. An ion-pair reagent (a large organic counterion that is ionized) is added at low concentration (usually 0.005*M*) to the mobile phase. The stationary phase is a monolayer octadecyl or octyl bonded-phase packing. Two popular ion-pair reagents are triethylalkyl quaternary amines (deliberately unsymmetrical to provide a better association of the long alkyl chain with the paraffinic surface of the octadecyl stationary phase) and alkyl sulfonates. The former bind solute anions; the later bind organic cations to form an ion, pair, as follows:

$$RNH_3^+ + C_8H_{17}SO_3^- \rightleftarrows [RNH_3^+, C_8H_{17}O_3^-]^0 \tag{5.35}$$

Although the exact mechanisom has not been established, either the solute molecule forms a reversible ion-pair complex (a coulombic association species of zero charge fromed between two ions of opposite electrical charge), which partitions into the nonpolar stationary phase, or the counterion is loaded onto the packing via the alkyl moiety with its ionic group oriented at the surface and able to participate in the formation of an ion-pair complex. The sample ion is pushed down the column by a competing ion of the same charge (positive or negative) in the eluent. Ion-pair chromatography is much like ion-exchange chromatography except that the modifier must be renewed continuously by incorporation in the eluent. See also Sec. 2.2.4.

Water–methanol mixture is the common mobile phase. Any buffer required should be 0.001*M* to 0.005*M* and should have good solubility but poor ion-pair properties. Three factors that influence retention are the following.

1. *Control of pH*. Maintaining a pH around 2.0 ensures that both strong and weak bases are in their protonated form and that any weak acids present are primarily in their nonionic forms. At a pH around 7.5 both strong and weak acids are in their ionic form and weak bases are in their nonionic form.
2. *Alkyl chain length*. The longer the alkyl chain on the counterion, the greater the retention of given ions. For basic samples, choose among Na pentanesulfonate through Na octanesulfonate.
3. *Counterion concentration*. Increasing the concentration of the counterion increases retention up to the limit set by the solubility of the counterion in the mobile phase.

If one is separating nonionic and ionic compounds in the same sample, follow these steps:

1. Optimize the separation of the nonionic solutes as described in Sec. 4.3.4.2.
2. Select and add the counterion to the mobile phase.
3. Fine-tune the separation by changing the chain length of the counterion (or use mixtures of two counterions) and perhaps altering the counterion concentration.

Ion-pair chromatography overcomes difficulties with ionized or ionizable species that are very polar, multiply ionized, and/or strongly basic. The main advantage of ion-pair chromatography over regular reversed-phase LC or ion-exchange HPLC is that it facilitates the analysis of samples that contain both ions and molecular species. Unlike conventional ion exchange, ion-pair chromatography can separate nonionic and ionic compounds in the same sample.

Disadvantages are that the ionic solutions are often corrosive and result in a short column life, some of them absorb in the UV and limit the use of the UV detector, and the silica-based supports are limited to pH values below about 7.5. Mobile phases should not be left standing overnight but be replaced by water.

5.3.5.4 Anion and Cation Chromatography. In a technique from which this main subsection is named, anion and cation chromatography, or simply ion chromatography (IC), low-capacity ion-exchange resins are used to separate analyte ions. A dilute aqueous solution is used as eluent. A typical packing is 3-quaternaryaminopropyl for anion exchange and 3-propylsulfonic acid for cation exchange. Also available are a weakly basic anion-exchange packing composed of diethylaminoethyl (DEAE) groups and a weakly acid carboxymethyl (CM) cation exchanger. Ionic analytes are then detected by a conductivity meter or by indirect UV (vacancy) detection.

Anions can be classified into four groups:

1. Weakly held anions such as fluoride or monofunctional organic acids with a range of relative selectivities from 0.05 to 1
2. Mono- and divalent inorganic anions, and divalent organic acids with a range of relative selectivites from 1 to 15
3. Mono- and divalent highly polarizable inorganic anions, such as iodide, thiocyanate, and thiosulfate, and trivalent organic acids with a range of relative selectivities from 15 to 25
4. Metal oxides, such as chromate and perrhenate, with selectivities that exceed 25

Cations can similarly be classified:

1. Alkali metals and ammonium ions that possess selectivities from 5 to 10
2. Primary, secondary, and tertiary amines whose relative selectivities range from 10 to 30
3. Alkalinc-carth metals and tertiary amines that possess selectivities from 90 to 120
4. Transition metals with selectivities from 150 to 3300

5.3.5.4.1 Suppressed Anion and Cation Chromatography (Dual-Column Method). Early ion-chromatography systems detected ions eluted by strong eluents from high-capacity ion-exchange columns by measuring changes in conductivity. To achieve reasonable sensitivity, it is necessary to suppress the conductivity of the eluent prior to detection by passing the column effluent through a bed of cation exchange resin in the H^+ form. This "suppressor" column removed sodium from the effluent and replaced it with hydrogen. The result was the conversion of sodium (hydrogen carbonate, benzoate, salicylate, or phthalate) into weakly ionized carbonic acid, benzoic acid, salicylic acid, or phthalic acid (low equivalent conductance). When sample anions emerge from the column, they are converted into high-equivalent-conductance acids (HCl, HNO_3, and so on). A similar amplification occurs in cation analysis when the column effluent (typically a dilute acid solution) is passed through an anion exchanger in the hydroxide form. This converts the acid eluent to water and the sample ions (which elute as salts) into the corresponding hydroxides.

Chemical suppression does improve sensitivity, but introduces its own set of problems. It is limited to strongly dissociated sample ions, and it adds dead volume (and band spreading) to the system. The ion-exchange sites in a conventional suppressor column gradually become saturated so that the column has to be regenerated periodically. Hollow-fiber membrane suppressor units (made of a sulfonated polymer) have been developed that can be continuously regenerated. Separator-column effluent flows in one direction and a regenerating solution (dilute sulfuric acid) flows continuously around the outside of the membrane.

5.3.5.4.2 Eluents. Anion eluents have included these anions: borate, hydrogen carbonate, 4-cyanophenate, glycine, hydroxide, salicylate, silicate, and tyrosine. Cation eluents have included these cations: H^+ as HCl, HNO_3, and triethanolamine-HCl.

5.3.5.4.3 Single-Column Method. The single-column technique for anion separation uses a low-capacity ion-exchange column packed with hydroxyethyl methacrylate–based macroporous polymer with quaternary amine functionalities. The low-capacity resin allows the anions to be separated using a very dilute eluent. For the separation of anions, the resin must have an exchange capacity between about 0.005 and 0.10 meq · g^{-1}. Typical eluents are 0.10 to 0.4 mM solutions of sodium or potassium salts of phthalic acid, succinic acid, benzoic acid, *p*-hydroxybenzoic acid, or borate–gluconate. Most sample anions have a higher equivalent conductance than that of the eluent anion and can therefore be detected even when present in the low parts-per-million range. A dilute solution of any of several carboxylic acids also makes an effective eluent; in fact the detection limit is improved by a factor of almost 10.

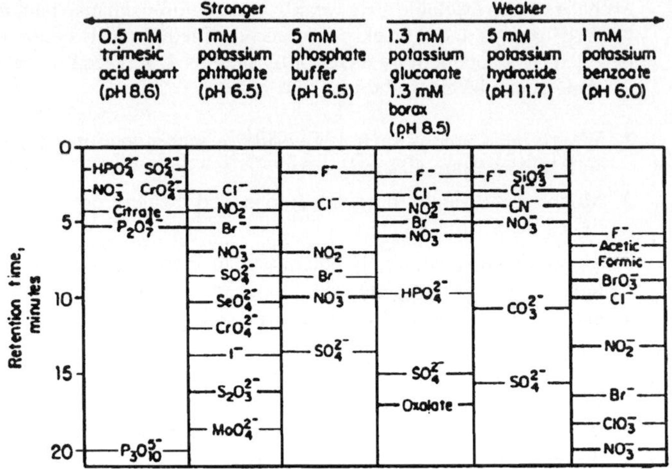

FIGURE 5.12 Anion separations using various eluants for ion chromatography. Trimesic acid is 1,3,5-benzenetricarboxylic acid.

Anions of weak acids, such as cyanide, silica, and borate, cannot be separated unless poly(styrene-divinylbenzene) packings are used. These packings make it possible to operate the column in the pH range from 2 to 11.7 (whereas with the former type of packings one is restricted to a pH from 2 to 8.5).

For the separation of cations, a cation-exchange column of low capacity is used. A dilute solution of nitric acid (or perchloric acid or sulfuric acid) is used for separation of monovalent cations, and a solution of an ethylenediammonium salt is used for separation of divalent cations. Both of these eluents are more highly conducting than the sample cations; therefore, the sample peaks are negative relative to the background (that is, decreasing conductivity).

5.3.5.4.3.1 ELUENT SELECTION AND DETECTOR. The first rule for eluent selection with conductivity detection is to choose an eluent buffer whenever possible that maximizes the difference in equivalent conductance between the sample and the buffer. The most generally useful buffer ions have either very low or very high equivalent conductance. In anion analysis, this usually means big, bulky organic acid or their salts or, in the other extreme, the hydroxide ion. In cation analysis, quaternary ammonium compounds are, in principle, suitable, but problems with adsorption onto the ion-exchange resin limit their utility. Most cation separations are carried out with hydronium or ethylenediammonium ions.

In the direct mode of UV detection, one selects an ion-exchange column, a non-UV absorbing buffer system, and a wavelength set at or near the absorption maxima of the ion of interest. UV-invisible ions cannot be detected by direct UV absorption. For anions, an anion-exchange column with a capacity range of 0.05 to 1.0 meq $\cdot$ g^{-1} can be utilized. Typical buffer systems consist of sulfuric, perchloric, or phosphoric acids or their conjugate salts. A common wavelength used is 210 nm at which the following anions have UV absorption: azide, bromate, bromide, iodate, iodide, nitrate, nitrite, thiocyanate, thiosulfate, and most organic acids.[38] For separation of weak acid species such as borate, silicate, and iodide, an eluent in the pH range of 11 can be used.

The indirect photometric method of UV detection in ion chromatography utilizes a low-capacity ion-exchange resin (to allow the use of low buffer concentrations), a high-UV-absorbing eluent, and a wavelength set within the absorption range of the eluent. For anion determination, an anion-exchange column with a capacity of 0.05 to 0.20 meq $\cdot$ g^{-1} is useful. Aromatic organic acids such as phthalic, benzoic, salicylic, 4-hydroxybenzoic, and 1,3,5-benzenetricarboxylic acid and their salts have been used as eluents. Separation possibilities are shown in Fig. 5.12. The most commonly used wavelengths are 254 and 280 nm. Very strongly retained species, such as transition-metal (divalent)

[38] H. Small and T. E. Miller, Jr., *Anal. Chem.* **54**:462 (1982).

cations, can be separated using eluents based on the divalent ethylenediammonium ion as the driving force, but which include complexing agents to reduce elution time and to control selectivity.[39] A cerium(III) mobile phase in conjunction with a moderate-capacity cation-exchange column permits the simultaneous determination of alkali and alkaline-earth metals.[40]

A poly(butadiene-maleic acid) copolymer–coated silica stationary phase (Schomburg columns) has made it possible to separate both the alkali and alkaline-earth metals in a reasonable time, especially with the use of mildly acidic complexing eluents in the unsuppressed conductometric mode.

5.3.5.4.4 Applications. Anions and cations can be separated and measured (Table 5.21) in extremely dilute samples provided a preconcentration step is included in the analytical procedure. A relatively large aqueous sample is passed through a short-bed ion-exchange "guard" or precolumn. Then a switching valve is turned so that the eluent sweeps the accumulated ions from the precolumn onto the separation column. This process permits enrichment factors from 10^3 to 10^4 for many metals. Samples can be concentrated in the field using small ion-exchange cartridges.

Using chelation concentration, a sample (either neat or acid digest) is pumped through a chelation resin precolumn that selectively retains transition metals but not the alkali or alkaline-earth metals or common anions. The transition metals concentrated in the column are then eluted and quantitated.

Ion chromatography offers potential insight into the actual coordination chemistry of transition metals. The technique has the ability to speciate oxidation states of several metals such as Fe(III) and Fe(II), Cr(VI) and Cr(III), and Sn(IV) and Sn(II). Stable metal complexes can be determined directly in their complex form. One example is the determination of cyano complexes in plating solutions. Also the ability exists to determine free versus bound metal by selective preconcentration of samples on a short cation "guard" column prior to performing the chromatography step. Free versus bound metal is often a key piece of information in environmental and biomedical studies where in many cases only the free metal can be implicated as a toxin.

The relative versatility of ion chromatography is further enhanced by its ability to determine various organic acids and amines, carbohydrates, alcohols, amines, amino acids, saccharin, sugars and sugar alcohols, and surfactants (both cationic and anionic). Table 5.22 lists selected anions and cations that may be determined by ion chromatography, usually in the single-column mode.

5.3.5.5 Ion-Exclusion Chromatography. Strictly speaking, ion-exclusion chromatography is not a form of ion chromatography because molecular species rather than ions are separated (Table 5.23). However, ion-exclusion chromatography is an excellent way to separate molecular substances from larger amounts of ionic material in the sample. Separations are carried out on high-capacity polystyrene-based exchange resins. Ionized analytes are excluded from the polymer.

A number of separation mechanisms are involved. Donnan exclusion seems to predominate in the early part of the chromatogram in which organic acids, for example, elute roughly in order of pK_a values. Later-eluting peaks are affected by hydrophobic interactions; short-chain fatty acids elute in the order of hydrocarbon chain length. Typical eluents for anion-exclusion systems include water for species such as fluoride, hydrogen carbonate, or sulfide, dilute mineral acids for organic acid analysis, and acid–acetonitrile mixtures for C_1 to C_8 fatty acids. In the latter eluent, acetonitrile serves as an organic modifier to decrease the retention time of relatively nonpolar compounds.

Ion-exclusion chromatography is superior to ion-exchange chromatography for the separation of weak acids—especially carboxylic acids from strong inorganic acids. The separation is done on a high-performance cation-exchange column in the H^+ form. Ionic material is rejected by the resin and passes through the column at the same rate as the eluent. Carboxylic acids are largely molecular rather than ionic; consequently, they partition between the occluded liquid inside the resin and the aqueous eluent. Ionization of the carboxylic acids is repressed by the high concentration of H^+ within the resin beads and by the low concentration of acid in the aqueous eluent. Weak inorganic acids, such as carbonic acid, HF, and HPO_2, can also be determined, as can salts of weak acids because they are converted to the corresponding acid by the hydrogen ions in the ion-exchange column. Weak molecular bases and their salts can be separated on an anion-exchange column in the

[39] G. J. Sevenich and J. S. Fritz, *Anal. Chem.* **55**:12 (1983).

[40] J. H. Sherman and N. D. Danielson, *Anal. Chem.*, **59**:490 (1987); (with J. W. Hazey), *J. Agr. Food Chem.*, **36**:966 (1988).

TABLE 5.21 Ions Determinable by Ion Chromatography

Inorganic anions	Organic anions	Cations
Arsenate	Acetate	Aluminum
Arsenite	Acrylate	Ammonium
Azide	Adipate	Barium
Borate	Alkene sulfates	Cadmium
Bromate	Alkyl sulfates	Calcium
Bromide	Ascorbate	Cerium
Carbonate	Citrate	Cesium
Chlorate	Ethylphosphonate	Chromium(III)
Chloride	Formate	Cobalt
Chlorite	Fumarate	Copper
Chromate	Gluconate	Diethylammonium
Cyanide	Glutarate	Dimethylammonium
Dibutylphosphate	Glycolate	Erbium
Fluoride	Glyoxylate	Ethanolammonium
Fluoroborate	Lactate	Ethylammonium
Fluorophosphate	Malate	Gadolinium
Fluorosilicate	Maleate	Guanidinium
Hydrogen carbonate	Malonate	Holmium
Hypochlorite	Methylphosphonate	Hydrazinium
Hypophosphite	Myristate	Hydroxylammonium
Iodate	Palmitate	Iron(II and III)
Iodide	Phthalate	Lanthanum
Nitrate	Propanoate	Lead
Nitrite	Propylphosphonate	Lithium
Oxalate	Pyruvate	Lutetium
Perchlorate	Saccharinate	Magnesium
Phosphate	Stearate	Manganese
Phosphite	Succinate	Methylammonium
Pyrophosphate	Sulfonates	Molybdenum
Selenate	Tartrate	Neodymium
Selenite	Chlorophenoxy acids	Nickel
Silicate		Palladium
Sulfate		Piperazinium
Sulfide		Platinum
Sulfite		Potassium
Thiocyanate		Praseodymium
Thiosulfate		Rubidium
Trimetaphosphate		Silver
Tripolyphosphate		Sodium
		Strontium
		Tin(II and IV)
		Triethanolammonium
		Trimethylammonium
		Urea
		Ytterbium
		Zinc

OH^- form. For example, the bulk of the sorbed phenol can be eluted in about one bed volume with 1.0*M* NaOH.

When the sample contains a mixture of cations as well as anions, it is necessary to pretreat the sample to avoid "scrambling" into a confusing variety of ion pairs. Pretreatment involves a simple ion exchange that converts all cations to a common cation, if anion analysis is the objective. The reverse would be done when cation analysis is the objective.

TABLE 5.22 Selected Methods in Ion Chromatography

(Analytes are eluted in the order listed)

Analytes	Column type	Eluent	Operating conditions	References
Mono-, di-, trichloroacetic acids	Sulfate-form, polystyrene	0.01*M* HCl		*Anal. Chem.* **50**:1420 (1978)
Alkali thiocyanates	Dibenzo-21-crown-7-resin	Water or MeOH		*Z. Anal. Chem.* **284**:337 (1977)
K salts of Cl, Br, I, SCN	Dibenzo-21-crown-7-resin	Water or MeOH		*Z. Anal. Chem.* **284**:337 (1977)
Na^+, NH_4^+, K^+, Rb^+, Cs^+, Mg^{2+}, Ca^{2+}	C_{18} Si columns modified with crown ethers	Water		*Anal. Chem.* **58**:2233 (1986)
	Universal anion (Alltech)	5.0 m*M* HCl	Conductivity	*J. Chromatogr.* **549**:257 (1991)
Alkyl quaternary ammonium compounds	Cation exchanger			*Anal. Chem.* **55**:393 (1983)
F^-, HCO_3^-, Cl^-, NO_2^-, Br^-, NO_3^-, HPO_4^{2-}, SO_4^{2-}	Anion exchanger	3.0 m*M* salicylate, pH 8.2	UV, 254 nm	
		1.5 m*M* phthalate, pH 8.9	UV detector, 260 nm	
Inorganic anions and carboxylic acids	Mixed-mode RP-C_{18}/anion (Alltech)	4 m*M* phthalic acid, pH 4.5	Conductivity or UV vacancy at 260 nm	*Anal. Chem.* **64**:2283 (1992)
Mono-, di-, trimethylamines	Cation/R (Wescan)	3.2 m*M* HNO_3	Conductivity	
$C_2O_4^{2-}$, maleate, malate, succinate, $HCOO^-$, OAc^-	Anion exclusion	1 m*M* H_2SO_4	Conductivity	
CO_3^{2-}, S^{2-}	Anion exclusion	Deionized water	Conductivity	
Strong and weak acids	Ion-exchange column coupled with ion exclusion	Na octanesulfonate, then octanesulfonic acid		*J. Chromatogr.* **473**:171 (1989)
Br^- in brine	Anion exchanger	0.01*M* $HClO_4$	UV detector, 210 nm	
N_3^- in tris buffer	Cation exchanger, H form	0.001*M* H_2SO_4	37°C; 210 nm	
Cl^-, sulfoacetate, sulfopropanoate, sulfosuccinate	Anion exchanger, R_4N^+ form	10 m*M* phthalic acid	Conductivity	
Citrate, H_3PO_4, Cl^-	Anion exchanger	4 m*M* phthalic acid with 8 mL benzoic acid	Conductivity	
Cl^-, NO_3^-, SO_4^{2-}, $HONH_2$, disulfonate, dithionate	Anion exchanger	4 m*M* phthalic acid, pH 3.8	Conductivity	
Li^+, Na^+, NH_4^+, K^+	Cation exchanger	2.0 m*M* picolinic acid, pH 2.0	Conductivity	
Mg^{2+}, Ca^{2+}, Sr^{2+}	Cation exchanger (Wescan)	2 m*M* phenylenediamine, pH 6 (with heart cut)	Conductivity	*Anal. Chem.* **59**:624 (1987)
Pyridine, isonicotinic acid, nicotinic acid	Anion exchanger	HNO_3, pH 2.3	Conductivity or UV (better)	

(*Continued*)

TABLE 5.22 Selected Methods in Ion Chromatography (*Continued*)

Analytes	Column type	Eluent	Operating conditions	References
Cu^{2+}, Ni, Zn, Co^{2+}, Fe^{2+}, Mn^{2+}, Cd, Pb^{2+}	Anion exchanger	Step gradient: 2.3 m*M* EDTA + 10 m*M* citric acid, then 3.5 m*M* EDTA + 10 m*M* citric acid	Post column reaction with PAR; 500 nm	
cis-Aconitic, oxaloacetic, δ-ketoglutamic, pyruvic, citric, malic, lactic, succinic, fumaric acids	Ion exclusion, ION-300 (Interaction)	0.005*M* H_2SO_4, 0.4 mL/min; 42°C	UV, 210 nm	
Cr(VI), as chromate, in water and solid waste extracts	Anion exchanger (Dionex AS7)	0.25*M* $(NH_4)_2SO_4$, 0.1*M* NH_4OH	Post column: 2 m*M* diphenylcarbonhydrazide, 10% MeOH, 0.5*M* H_2SO_4; 520 nm	*Am. Environ. Lab.* (November 1989), p. 52
Oxalic, citric, shikimic, fumaric, butyric, homoprotocatechuic, gallic, protocatechuic, gentisic, 4-hydroxybenzoic, benzoid, salicylic acids	Ion exclusion, ARH-610 (Interaction)	0.005*M* H_2SO_4, 0.6 mL/min; 45°C	UV, 210 nm	
Citric, ascorbic, sorbic, benzoic acids	Ion exclusion, ARH-610 (Interaction)	0.005*M* H_2SO_4, 0.6 mL/min; 45°C	UV, 228 nm	
Phosphate, saccharin, fluoride, glycerol, EtOH	Ion exclusion, ORH-801 (Interaction)	1.5 m*M* H_2SO_4, 0.5 mL/ min	Conductivity	
Maleic, malic, fumaric acids	Ion exclusion, ORH-801 (Interaction)	0.005*M* H_2SO_4, 0.6 mL/min; 37°C	UV, 210 nm	
Maltotriose, maltose, glucose, fructose, lactic acid, glycerol, HOAc, MeOH, EtOH	Ion exclusion, ORH-801 (Interaction)	0.0005*M* H_2SO_4, 0.6 mL/min; 65°C	Refractive index	

TABLE 5.23 Capacity Factors and Retention Times for Ion-Exclusion Packing ORH-801 (Interaction)

t_0 = 3.59 min; column: Interaction® ORH-801; eluent: 0.0025N H_2SO_4; flowrate: 0.6 mL/min; temperature: 35°C; detection: UV at 210 nm, conductivity for fluoride and phosphate; injection: 10 μL

Compound	k'	t_R, min	Compound	k'	t_R, min
Maltotriose	0.18	4.23	Succinic acid	1.27	8.16
Maltose	0.32	4.74	Lactic acid	1.34	8.41
Lactose	0.36	4.88	Fumaric acid	1.54	9.13
D-Glucuronic acid	0.36	4.87	Formic acid	1.56	9.18
Lactulose	0.42	5.10	Acetic acid	1.87	10.30
D-Galactouronic acid	0.52	5.44	Adipic acid	2.30	11.86
D-Glucose	0.62	5.83	Propionic acid	2.41	12.24
D-Tagatose	0.69	6.07	Butyric acid	3.25	15.25
D-Galactose	0.76	6.31	Sorbic acid	13.60	52.40
D-Fructose	0.80	6.46	Benzoic acid	17.90	68.00
Mannitol	0.82	6.52			
Sorbitol	0.85	6.65	Glycerol	1.46	8.82
L(+)-Arabinose	0.95	6.99	Ethylene glycol	1.96	10.63
D-Arabitol	0.98	7.10	Diethylene glycol	2.03	10.88
L-Fucose	1.05	7.35	Methanol	2.48	12.43
			Ethanol	2.88	13.92
Oxalic acid	0.08	3.89	2-Propanol	3.21	15.13
Maleic acid	0.25	4.49	1-Propanol	3.90	17.61
Citric acid	0.40	5.01			
Isocitric acid	0.43	5.13	Phosphate	0.19	4.28
L-Tartaric acid	0.48	5.31	Sulfite	0.52	5.45
Malonic acid	0.64	5.89	Fluoride	1.04	7.33
Ascorbic acid	0.66	5.95	Azide	3.12	14.80
(±)Malic acid	0.74	6.25			

For simultaneous determination of strong- and weak-acid anions, ion-exchange and ion-exclusion modes are coupled. In the unsuppressed mode, sodium octanesulfonate is the eluent for the ion-exchange mode and octanesulfonic acid for the ion-exclusion mode.

5.3.6 Optimization or Method Development in HPLC[41–44]

Diffusion in liquid mobile phase is orders of magnitude slower than when a gas in the carrier. This gives rise to slower optimized mobile phase linear velocities, lower efficiencies, and relatively poor resolving power in HPLC. Method development for an HPLC separation usually begins with retention optimization. A column type is selected bearing in mind the information in the introduction to Sec. 5.3. Any arbitrary column geometry is selected, although a standard-length (15 or 25 cm) column with an internal diameter of 4.6 mm and packed with 5- or 10-μm particles would be a good first choice. The mobile-phase composition is varied as described later under the individual HPLC methods to achieve a uniform spacing of sample bands within a range $1 < k' < 10$.

The next step is the selection of best column and packing configuration; that is, (1) length, (2) internal diameter, and (3) particle size. The separation will require a certain plate number to achieve adequate resolution between the most poorly resolved band pair. The plate number can be varied without changing retention by changing (1) the column dimension, (2) the particle size of the

[41] J. L. Glajck and J. J. Kirkland, "Optimization of Selectivity in Liquid Chromatography," *Anal. Chem.*, **55**:319A (1983).

[42] J. H. Knox, "Practical Aspects of LC Theory," *J. Chromatogr. Sci.*, **15**:352 (1977).

[43] M. Martin, G. Blu, C. Eon, and G. Guiochon, "Optimization of Column Design and Operating Parameters in High Speed Liquid Chromatography," *J. Chromatogr., Sci.*, **12**:438 (1975).

[44] M. Martin, C. Eon, and G. Guiochon, "Trends in Liquid Chromatography," *Res/Dev.*, **1975** (April): 24.

packing, or (3) the mobile-phase flowrate. Although the final choice can be made by trial and error, this is not recommended.

The general theory presented in Sec. 5.1.2 is the basis for predicting the plate number. Values of N can be related by Eq. (5.18) to the required resolution of the most poorly separated band pair in the chromatogram. Here k' refers to the average partition ratio value for the band pair.

5.3.6.1 Generation of the Required Plate Number in the Minimum Separation Time. The pressure drop ΔP across a bed packed with spherical particles of diameter d_p is

$$\Delta P = \frac{N^2 h^2 \phi \eta}{t_M} \tag{5.36}$$

Inserting the desired plate count (assumed to be $N = 10\ 000$) and $t_M = 60$ s, and using an optimum value of the reduced plate height, $h = 3$, $\phi = 1000$ (for fully porous packings; use 500 for pellicular packings), and η (viscosity of the mobile phase) = 10^{-3} N · s · m^{-1}, Eq. (5.37) gives

$$\Delta P = \frac{(10\ 000)^2 (3)^2 (1000)(10^{-3}\ \text{N}\cdot\text{s}\cdot\text{m}^{-2})}{60\ \text{s}} \tag{5.37}$$

$$= 1.5 \times 10^7\ \text{N}\cdot\text{m}^{-2} \quad (\text{or } 1500\ \text{lb}\cdot\text{in}^{-2}) \tag{5.38}$$

The required particle size is given by Eq. (5.39):

$$d_p = \frac{L}{Nh} = \frac{1}{(10\ 000)(3)} = \frac{L}{30\ 000} \tag{5.39}$$

For columns 10 and 25 cm long, the corresponding particle diameters needed are 3.3 and 8.3 μm, respectively. The combination of column lengths and particle size, plus operating pressures for different plate counts and retention times, are given in Table 5.24. Of course, in practice most laboratories must choose among available commercial columns, and they are limited to a small number of configurations.

A rapid estimate of the column plate number can be calculated from the standard equation

$$N = \frac{L}{hd_p} \tag{5.40}$$

where L = column length, cm
h = reduced plate height ($h = H/d_p$) (= 3.0 to 3.5 for real samples)
d_p = particle diameter, μm

Taking 3.3 as an average value for h, a 5-cm long column packed with 3-μm particles should give about 5000 plates. To minimize measurement errors, remember that for the measurement of the plate number the k' value should be at least 3 and preferably greater than 5 (but it should not exceed 20).

5.3.7 Exclusion (Gel-Permeation) Chromatography

Exclusion chromatography is based on the ability of controlled-porosity substrates to sort and separate sample mixtures according to the size and shape of the sample molecules. Compounds can be separated according to their molecular size by exclusion chromatography, which is based on the diffusion or permeation of solute molecules into the inner pores of the column packing. The size and shape of the molecules to be separated govern their ability to enter the pore. The smaller molecules enter the

TABLE 5.24 Typical Performances for Various Experimental Conditions*

Performances		Column parameters			
N	t_M, s	L, cm	d_p, μm	ΔP, atm (lb · in^{-2})	Peak bandwidth (4σ), μL
2 500	30	2.3	3	18.4 (270)	23
2 500	30	3.7	5	18.4 (270)	37
2 500	30	7.5	10	18.4 (270)	75
5 000	30	4.5	3	74 (1088)	41
5 000	30	7.5	5	74 (1088)	68
5 000	30	15.0	10	74 (1088)	136
10 000	30	9.0	3	300 (4410)	83
10 000	30	15.0	5	300 (4410)	136
10 000	30	30.0	10	300 (4410)	272
10 000	30	9.0	3	300 (4410)	82
10 000	60	9.0	3	150 (2200)	82
10 000	90	9.0	3	100 (1470)	82
15 000	90	2.3	3	223 (3275)	23
15 000	120	2.3	3	167 (2459)	23
11 100	30	10.0	3	360 (5420)	91
11 100	37	10.0	3	100 (4410)	91
11 100	101	10.0	3	100 (1470)	91
27 800	231	25.0	3	300 (4410)	75

* Assumed reduced parameters: $h = 3$, $v = 4.5$.

pore without hindrance and are the last to be eluted. Molecules that are too large to enter the pores are completely excluded and must travel with the solvent front. Between these two extremes, intermediate-size molecules can penetrate some passages but not others and consequently are retarded in their progress down the column and exit at intermediate times. The selection of column packings, each with its corresponding exclusion limit, enables separations to be achieved. This technique is well suited for the separation of polymer, proteins, natural resins and polymers, cellular components, viruses, steroids, and dispersed high-molecular-weight compounds. Useful detectors include the differential refractometer and the spectrophotometric detectors that operate in the ultraviolet and infrared spectral regions.

5.3.7.1 Column Packings. Column packings are either cross-linked macromolecular polymers or controlled-pore-size glasses or silicas. Semirigid materials will swell slightly; these materials are limited to a maximum pressure of 300 lb · in^{-2} (20.4 atm) because of bed compressibility. Packings prepared from methacrylate polymers can withstand pressures up to 300 lb · in^{-2} (204 atm). Hydrophilic packings are usable with aqueous system and polar organic solvents. Bead diameters are usually 5 μm.

Inorganic packings have advantages over organic packings. After calibration, columns can be used routinely and indefinitely. The bed volume remains constant at high flowrates and high pressures.

5.3.7.2 Retention Behavior. Totally excluded molecules elute in one column-void volume, and so the distribution coefficient $K = 0$. For small molecules that can enter all the pores of the packing, $K = 1$. Intermediate-size molecules elute between these two limits. An elution graph is shown in Fig. 5.13. The upper portion is the graph of the logarithm of molecular weight versus retention volume. There is a linear range of effective permeation between the limiting values that correspond to exclusion and to total permeation. The maximum elution volume is often only twice the column-void volume.

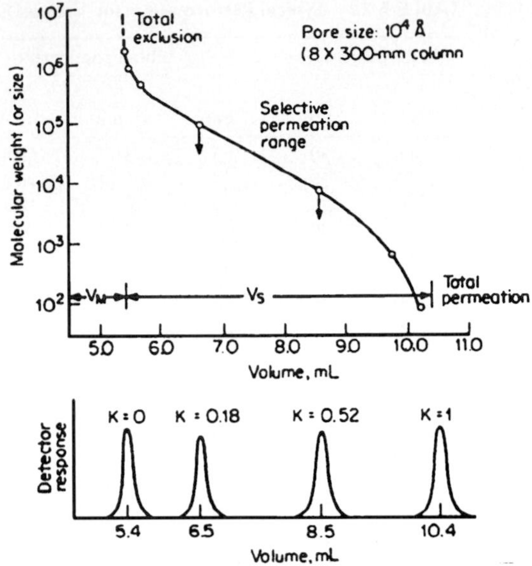

FIGURE 5.13 Retention behavior in exclusion chromatography.

The various pore sizes available in commercial packings provide selective permeation ranges that permit separation of small molecules with molecular weights less than 100 to compounds with molecular weights up to 500 million. For example, one series of Styrogel polymers has the following diameters, and the corresponding operating ranges:

Pore size, Å	Effective molecular weight range, amu
100	50 to 1500
500	100 to 10 000
10^3	500 to 30 000
10^4	5 000 to 600 000
10^5	50 000 to 4×10^6
10^6	200 000 to 1×10^7

If nothing is known regarding the probable molecular dimensions or weights of the sample components, a preliminary run can be made using a column packed with 100-nm material. Use a flow rate of 3 mL · min^{-1}. If most of the sample elutes near the exclusion limit, a column packed with larger-pore-size material should be tried. Elution of the majority of the sample halfway between the exclusion and total permeation volume suggests a column packing of 50-nm pore size. Near-total permeation indicates a smaller pore-size packing, perhaps 10 nm. If sample components elute over a wide range, a series of column cartridges, each with a specific packing, should be used.

5.3.7.3 Column Calibration. Exclusion columns are calibrated by eluting calibration standards and monitoring the elution volume. Narrow dispersed standards of polystyrene, polytetrahydrofuran, and polyisopropene are available for use in organic solvents. Samples of dextrans, polyethylene glycols, polystyrene sulfonates, and proteins are available for use in hydrophilic solvents.

5.3.8 Hydrophobic Interaction Chromatography

Hydrophobic interaction chromatography (HIC) uses a salt gradient of descending ionic strength to elute proteins without the use of organic solvents. Thus, the protein can be recovered without denaturation. HIC has become one of the more common techniques of proton HPLC.

The technique uses a packed column (4 to 10 mm i.d.) with stationary phases of low hydrophobic character. The trick of HIC is to make the hydrophobicity of the column packing strong enough for retention, but weak enough so that it does not denature the protein. Thus C_{18} and even C_8 are too hydrophobic. The supports are prepared by bonding fairly short hydrophobic ligands such as hexyl, phenyl, and butyl as well as moderately polar phases such as ethers (glycerylpropyl) on a 6.5-μm macroporous silica that has been previously bonded with a polyamide coating. The separation of proteins is usually accomplished using a descending salt gradient that preserves enzymatic activity and tertiary structure.

The gel picture of the mechanism of HIC is that the high salt concentration decreases the activity of water, which decreases the solubility of the protein. That is, the large amount of salt takes up the free water, and it is not available to solvate the protein; hence it will be squeezed out onto the surface of the column packing. When the salt concentration is lowered, the water concentration increases. This permits the solubilization and elution of the analyte.

5.3.9 Troubleshooting[45]

Most problems result from bubbles, dirt, or normal wear. Preventive maintenance is the key to eliminating them.

5.3.9.1 General Precautions. For reliable HPLC operation, take these precautions:

1. Use only clean, degassed (by helium sparging) HPLC-grade solvents. Filter all salt- or buffer-containing solvents through a 0.45-μm filter before use. Use an inlet filter on the reservoir end of the solvent inlet lines to prevent dust from entering the system.

2. Check the septum daily for leaks and change frequently.

3. Use a guard column to act as a superfilter to trap particular matter and chemical contaminants before they reach the analytical column.

4. Check the flowrate regularly at a specified pressure to detect buildup of pressure (or decrease of flow. Pressure buildup can be caused by small pieces of spetum that become deposited at the head of the column (if no guard column is used) after many injections. To correct this situation, remove a few millimeters of packing from the top of the column and repack with new material.

5. Dissolve samples in the mobile phase or in a less polar solvent than the mobile phase, if possible. This technique tends to concentrate the injection on the top of the column and yields better resolution. Many times it is possible to inject very large samples when more sensitivity is needed with no deleterious effects apparent in the separation.

6. Flush the system after each day's use with about 10 column volumes of the strong component of the mobile phase. If buffers are being used, flush with unbuffered mobile phase before changing to pure organic. This precaution prevents salt-deposit buildup, precipitation of crystals, microbial growth, and corrosion of LC hardware.

7. Replace pump seals regularly (about every 3 months).

8. Check detector (lamp life 0.5 to 1 year). Replace deuterium lamps every 6 months and mercury lamps once a year.

9. Beware of the effects of preservatives in solvents when using normal-phase chromatography. In the case of chloroform, common preservatives are ethanol and 2-methyl-2-butene at levels of

---

[45]. J. W. Dolan and L. R. Snyder, *Troubleshooting LC Systems*, Human, Clifton, NJ, 1989.

about 0.5% to 1.0% and 0.01% to 0.02%, respectively. Ethanol in amounts of 0.5% will significantly modify the retention characteristics of a chloroform–hexane mobile phase with a silica column. If possible, avoid solvents that tend to form peroxides upon storage such as ethers, chloroform, or cyclohexane.

10. Check for stabilizers (antioxidants) in solvents used for reversed-phase chromatography because the stabilizers absorb in the ultraviolet region below 320 nm, which causes problems when ultraviolet-visible detectors are used. Butylated hydroxytoluene is often added to tetrahydrofuran (THF) to prevent peroxide formation; only the unstabilized THF should be used in conjunction with UV detectors.

11. In adsorption chromatography the presence of trace levels of transition-metal ions in the adsorbent can catalyze oxidation of easily oxidized samples. Acid washing (1*M* HCl) removes transition-metal impurities from the adsorbent.

5.3.9.2 Precautions with Bonded-Phase Columns. Bonded-phase columns, as with any silica-based column, may show secondary retention by interaction with the silica surface. In reversed-phase systems, the silica is often end-capped and amine modifiers are added to the mobile phase to minimize the extent of silanol interactions. For a normal-phase system, a polar modifier, such as 2-propanol or water, is often added to the mobile phase to deactivate the column. The primary aminopropyl bonded-phase column is not inert and precludes concomitant use of strongly oxidizing samples and samples that can condense with a primary amine group such as aldehydes, ketones, and ketosteroids (including reducing carbohydrates, such as aldoses and ketoses). The condensation is accelerated by polar alcohol solvents (so use acetonitrile rather than methanol). Peroxide and hydroperoxide samples will suffer loss on this type of column and will slowly transform the amine groups into nitro groups. An amine salt column is more resistant to oxidation; conversion to the ammonium phosphate form is recommended. Also derivatizing agents such as danysl chloride should be avoided.

5.3.9.3 Extracolumn Effects. Extracolumn effects should be suspected under these laboratory observations:

1. Early bands broaden more than later ones (smaller plate numbers).
2. Coupled columns give higher plate numbers than the sum of the individual columns; also the asymmetry factor values are closer to 1.0.
3. Shorter and/or narrower columns give more band-broadening problems than normal columns.
4. Conversion from gradient to isocratic operation gives poor plate numbers and/or band shape.
5. Tailing bands are present under ideal column-test conditions.

The idea that extracolumn effects cause equal broadening for all bands is a common misconception. For isocratic separations, the later the bands come out, the broader they become. In other words, band broadening within the column is retention related, whereas extracolumn effects result mostly from the plumbing and the detector cell volume, which are constant for the set of one particular system.

Earlier peaks may broaden more than later ones if the value of k' for the early peak is less than 2 and particularly if it is less than 1. Now the problem lies with the k' value, and one should suspect that the wrong mobile phase (too strong as eluent) is being used. It is the k' value, not the retention time, that is the important factor. The k' value is independent of the column volume and flowrate, whereas retention is not.

5.3.9.4 Change in Retention Time. A gradual change in retention time of sample components can be caused by changes in either (1) the column or (2) the LC pumping system. Column changes that will affect retention time are as follows:

1. Coverage of active sites by extraneous sample material that is irreversibly adsorbed. An increase in operating pressure accompanies this problem. The remedy is to improve sample cleanup.

2. Gradual cleavage of the bonded phase from the silica support as a result of hydrolysis. The remedy is to increase the organic content of the mobile phase, lower the pH, decrease the ionic strength of the mobile phase, or switch to a neutral-bead polystyrene–divinylbenzene packing material.

3. Reaction of the column packing with a sample component is confined largely to primary alkylamine bonded phases when sample components contain aldehydes and reactive (unhindered) ketones. The remedy is to avoid this type of packing when working with carbonyl-containing compounds.

5.3.10 Safety

The three major areas for which proper safety precautions are imperative are (1) toxicity of liquids and fumes, (2) flammability of solvents, and (3) high fluid pressures.

5.3.10.1 Toxicity and Fumes. Toxicity of solvents and samples can be avoided by taking sensible safety precautions. All solvents should be stored in vented fireproof cabinets. A well-ventilated laboratory and the use of an exhaust hood will minimize exposure to these compounds. Careful handling of solvents, including the proper use of funnels and safety glasses, will protect the chromatographer.

Waste-collection vessels should be larger than the total volume of the solvent(s) to be pumped through the system; they should be kept covered to minimize the evaporation of waste solvent. Spills should be wiped up as quickly as possible; acids and bases should be neutralized prior to wiping.

5.3.10.2 Solvent Flammability. Another possible source of danger in HPLC exists from the use of highly flammable solvents. In this respect, leaks are of very great concern since ignition may occur without warning. Fire extinguishers should be kept in close proximity to the HPLC system. Smoking and open flames should be prohibited without exception in the laboratory, as should unnecessary operation of any electronic instrumentation or other potential sources of electrical discharges.

5.3.10.3 Mechanical Safety. When using highly flammable solvents with a pumping system that has a low-limit, shut-down feature, set that limit approximately 300 $lb \cdot in^{-2}$ (20.4 atm) below the minimum operating pressure necessary for the particular method. This shut-down feature automatically stops the pump when a leak occurs or when the system runs out of solvent, since the pressure drops below the limit setting as a result of fluid loss.

The compressibility of liquids is very small, so the dangers related to the high-pressure system are minimal. Generally a high-pressure system component will leak before the pressure involved causes the plumbing or components to fracture. However, it is good safety practice always to wear eye protection in the laboratory. A small fracture in tubing under high pressure can produce a stream of solvent that is capable of puncturing tissue. High pressure must not be used with glass columns since shattering of the column may easily occur.

Bibliography

Benson, J. R., "Modern Ion Chromatography," *Am. Lab.* **1985** (June): 30.

Dasgupta, P. K., "Ion Chromatography," *Anal. Chem.* **64**:775A (1992).

Dolan, J. W., and L. R. Snyder, *Troubleshooting LC Systems*, Humana, Clifton, NJ, 1989.

Fritz, J. S., "Ion Chromatography," *Anal. Chem.* **59**:335A (1987).

Gjerde, D. T., and J. S. Fritz, *Ion Chromatography*, 2d ed., Hüthig, Heidelberg, 1987.

Hancock, W. S., ed., *High Performance Liquid Chromatography in Biotechnology*, Wiley, New York, 1990.

Horvath, C., ed., *High-Performance Liquid Chromatography*, Academic, Orlando, FL, Vol. 1, 1980; Vol. 2, 1980; Vol. 3, 1983.

Jupille, T., "Column and Eluant Selection for Single-Column Ion Chromatography." *Am. Lab.* **1986** (May): 114.

Mulik, J. D., and E. Sawicki, *Ion Chromatographic Analysis of Environmental Pollutants*, Ann Arbor Science Publishers, Ann Arbor, MI, 1979.

Rubin, R. B., and S. S. Heberling, "Metal Determinations by Ion Chromatography," *Am. Lab.* **1987** (May):46.

Scott, R. P. W., ed., *Small Bore Liquid Chromatographic Columns: Their Properties and Uses*, Wiley, New York, 1984.

Scott, R. P. W., *Liquid Chromatography Column Theory*, Wiley, New York, 1992.

Simpson, C. F., ed., *Techniques in Liquid Chromatography*, Wiley-Heyden, New York, 1982.

Small, H., "Modern Inorganic Chromatography," *Anal. Chem.* **55**:235A (1983).

Small, H., *Ion Chromatography*, Plenum, New York, 1989.

Smith, F. C., Jr., and R. C. Chang, *The Practice of Ion Chromatography*, Wiley-Interscience, New York, 1983.

Snyder, L. R., and J. J. Kirkland, in *Introduction to Modern Liquid Chromatography*, 2d eds., K. Unger, and K. K. Unger, eds., *Packings and Stationary Phases in Chromatographic Techniques*, Dekker, New York, 1990.

Snyder, L. R., J. L. Glajch, and J. J. Kirkland, *Practical HPLC Method Development*, Wiley, New York, 1988.

Weiss, H., and E. L. Johnson, *Handbook of Ion Chromatography*, Dionex Corp., Sunnyvale, California, 1986.

Yau, W. W., J. J. Kirkland, and D. D. Bly, *Modern Size Exclusion Liquid Chromatography*, Wiley-Interscience, New York, 1979.

Yeung, E. S., ed., *Detectors for Liquid Chromatography*, Wiley-Interscience, New York, 1986.

5.4 PLANAR CHROMATOGRAPHY

Planar chromatography includes thin-layer and paper chromatography and electrophoresis. In *thin-layer chromatography* (TLC), a stationary phase is coated on an inert plate of glass, plastic, or metal. The samples are spotted or placed as streaks on the plate. Development of the chromatogram takes place as the mobile phase percolates through the stationary phase and the spot locations. The sample travels across the plate in the mobile phase, propelled by capillary action. Separation of components occurs through adsorption, partition, exclusion, or ion-exchange processes, or a combination of these. *Paper chromatography* employs special paper supports, but otherwise resembles TLC. *Electrophoresis* utilizes the electromigration of ions and molecules in an applied electric field as an auxiliary separation aid. Each charged species moves at a rate that is a function of its charge, size, and shape. In planar chromatography the position of the resultant bands or zones after development is observed or detected by appropriate methods. Because of its convenience and simplicity, sharpness of separations, high sensitivity, speed of separation, and ease of recovery of the sample components, planar chromatography finds many applications.

5.4.1 Thin-Layer Chromatography

Resolution of the sample in thin-layer chromatography is accomplished by passage of the mobile solvent mixture through the initial spot and beyond until the solvent front reaches within 2.5 cm of the opposite edge (top of plate in vertical development). Interaction of individual sample components with the stationary phase and the mobile phase, if each is properly selected, will cause the components to separate into individual spots. In TLC the variety of coating materials and solvent systems, as discussed in HPLC (Sec. 5.3), is also suitable for this technique.

The difference between HPLC and TLC is that HPLC operates with a closed system whereas TLC functions in an open thin-layer plate, the surface of which is exposed to the atmosphere. In TLC the separation is by development (the components remain in the chromatographic bed) and the detection is static (independent of time). HPLC is carried out in a steady-state mobile-phase environment, whereas in TLC the driving force is solely by capillary action; the latter may cause uneven flow velocity. TLC is a more economical separation technique than HPLC because multiple samples (including standards) may be introduced and run simultaneously. With a high-performance TLC plate, it is often sufficient to carry the development not more than 3 or 4 cm in order to achieve satisfactory separation, and this can be accomplished in 5 min, more or less, depending on the development solvent. Two-dimensional runs in TLC enable more complex samples to be analyzed.

A major advantage of TLC is the speed of analysis on a per sample basis. This is due to the short development distance (about 10 to 15 cm beyond the origin) and the resultant short development time. Spotting samples along with standards on the same plate allows them to be processed under identical conditions in contrast to sequential analysis on a column. By far the greatest number of chromatographic analyses are concerned with the measurement of only one or two components of a sample mixture. Therefore, development conditions are optimized for resolution of only the components of interest, while the remainder of the sample material is left at the origin or is moved away from the region of maximum resolution.

Modern high-performance TLC (HPTLC) rivals HPLC and GC in its ability to resolve complex mixtures and to provide analyte quantification. The choice of mobile-phase components is not restricted by concerns about deterioration of the coating material since layers are not reused or by compatibility with a detector; thus solvents that are highly absorbing in the ultraviolet region can be used. TLC provides for separations in the milligram to the picrogram range.

A complete view of a complex sample can be provided on a thin-layer chromatogram by use of a sequence of detection reagents. A variety of techniques can be used to optimize sensitivity and selectivity of detection. Often the entire plate can be examined to make sure that all of the sample has moved, whereas in HPLC one is never certain if every peak has been eluted at any given time. Separated substances that are tentatively identified by TLC can be eluted for further characterization by other microchemical techniques. Because there is less band broadening, compact zones are produced, leading to high detection sensitivity, resolution, and efficiency. Smaller samples are used so that a larger number of samples per plate can be applied.

Unlike GC or HPLC, the operations that make up the TLC system are not on-line. This lends considerable flexibility in the scheduling of the separate steps of sample application, development, and detection.

5.4.1.1 Retardation Factor. In systems, such as TLC, where the cross section and partition coefficient may not be constant along the length of the development path, the ratio of distances traveled by the solute and the mobile phase is called the retardation factor R_f:

$$R_f = \frac{\text{distance traveled by the solute}}{\text{distance traveled by the mobile phase}} \tag{5.41}$$

TLC can be used as a pilot technique for HPLC. The simple equation

$$k' = \frac{1}{R_f} - 1 \tag{5.42}$$

relates k', the capacity factor from a HPLC column, and the R_f value measured from the TLC plate. With this equation, isocratic experiments can sometimes be directly extrapolated from one technique to the other. R_f values of 0.1 to 0.5 correspond to the optimum 1 to 10 range for k' values in HPLC.

If the results of isocratic TLC are to be transferred to a gradient column HPLC, the gradient should be started about 20% weaker than the best mobile phase determined for TLC and run to a point about 20% stronger than the TLC solvent.

5.4.1.2 Solvent Selection. The selection of a solvent system to accomplish the required separation may involve a number of trials, but the choice of solvent is quite unrestricted by considerations of interference with detector response or of possible deterioration of the stationary phase. The solvent systems are usually a two-component mixture of water and a polar organic solvent miscible with water. Assuming that methanol was chosen with water as the strength-adjusting solvent, development is performed with different proportions of methanol and water to find the mixture that produces k' values for all compounds, or at least the ones of interest, within the optimum 1 to 10 range.

TABLE 5.25 Empirical Solvent Strength Parameters, S

Solvent	S
Hexane	0.0
1-Propanol	3.9
Tetrahydrofuran	4.2
Ethyl acetate	4.3
2-Propanol	4.3
Chloroform	4.4
1,4-Dioxane	4.8
Ethanol	5.2
Pyridine	5.3
Acetone	5.4
Acetic acid	6.2
Acetonitrile	6.2
N,N-Dimethylformamide	6.4
Methanol	6.6
Water	9.0

Source: A selection of solvents from L. R. Snyder, *J. Chromatogr.* **92**:223 (1974).

Once this solvent mixture is experimentally determined, one can obtain the total solvent strength (S') from the relationship:

$$S' = F_a S_a + F_b S_b + \cdots \tag{5.43}$$

where F is the volume fraction of the pure solvents (a, b, and so on) and S is their solvent strength parameter (Table 5.25).

Using similar equations, binary mixtures of acetonitrile–water and tetrahydrofuran–water, and even ternary and quaternary mixtures of solvents, are formulated so that they have the same overall strength (S' value) but different selectivities. Plates are then developed with these mixtures to determine the best overall mobile phase. Remember that the highest S values represent the strongest solute for adsorption (silica gel) TLC but the weakest for reversed-phase TLC.

For normal or reversed-phase TLC, benzene is a good solvent for separations on silica gel. If R_f values are too high, 5% or 10% of hexane is added to reduce polarity; similarly, 5% or 10% of methanol can be added to increase polarity. Another useful starting mobile phase for reversed-phase TLC is ethanol–water (80:20 v/v). Selectivity may be improved by substituting one of the solvents in Table 5.25 for ethanol and altering the ratio with water. To keep acidic and basic solutes nonionized so as to prevent tailing of zones, a small percentage of acetic acid or aqueous ammonia is added, respectively. To modify selectivity after locating the optimum ϵ^0, the most polar component is substituted in whole or part by one just above or just below in Table 5.19.

5.4.1.3 Efficiency. The number of theoretical plates in TLC is given by

$$N = 16\left(\frac{\text{distance of migration of spot center}}{\text{width of spot}}\right) \tag{5.44}$$

The number of effective theoretical plates in HPTLC is usually below 5000 for a 10-cm development; however, the normal development distance in HPTLC is 3 to 7 cm. For ordinary TLC, typical efficiencies are less than plates for separation distances of 10 to 15 cm.

5.4.1.4 TLC Plates. As a general rule, any of the stationary phases used in HPLC can be used in TLC provided that they are available in a uniformly fine particle size or can be bonded to the substrate. Conventional TLC plates are 20 × 20-cm, 5 × 20-cm, and 1 × 3-cm (microscope slide) sizes. Analytical plates are usually 0.100 to 0.250 mm in thickness. The absorbent particle size averages 20 μm with a distribution from 10 to 60 μm. Sample volume is 1 to 5 μL. The 20 × 20-cm plates can be purchased with 19 channels and/or a 3 cm distance of inert preadsorbent. Preparative TLC plates are 0.5 to 2.0 mm thick.

HPLTC plates are prepared with 5-μm particle sizes and held tightly within a narrow range. Plates are smaller (10 × 10 cm and 10 × 20 cm) and faster, and require smaller sample volumes (0.1 to 0.2 μL). The layer itself is usually somewhat thinner and the surface more uniform. The increased efficiency realized by the bed of smaller particles and tighter size control in HPTLC plates results in less band broadening and, hence, improved resolution and greater sensitivity of detection of the separated fraction. Although the small particle size reduces the velocity of mobile-phase flow, the length of the chromatographic bed required is markedly less than that encountered in conventional TLC.

5.4.1.5 ***Silica Gel.*** Silica gel is used more often than any other coating material. The silica-gel layer consists of an extremely dense packing of small particles of very uniform size (6 to 13 μm) with a smooth, homogeneous surface. The material is acid washed, water washed to neutrality, and dried to coalesce geminal hydroxyls into siloxane bonds. A common binder for silica-gel powders is 5% to 20% calcium sulfate hemihydrate, a refined form of gypsum (silica gel G). It is added to improve cohesion of the adsorbent particles and increase the adhesion of the adsorbent layer to the plate. Sometimes gypsum interferes with the separation of certain classes of compounds or their detection. In these situations a proprietary inorganic or unreactive organic binder can be used, which makes the layer hard and abrasion resistant. An organic binder makes the most durable silica-gel plate. Plates will withstand most solvent systems and any developing reagent without silica flaking or reacting with reagents. An inorganic binder is used when strong acid charring is used for visualization or when the presence of an organic binder interacts with the solvent system if special unreactive organic binders are not used.

Pore-size selection can be an important separation factor. Since the activity is a function of pore diameter, larger pore diameters provide a more nonpolar sorbent. The 15-nm diameter silica improves performance with larger complex molecules and offers a higher sample weight capacity. Silicas with a 6-nm pore diameter have standard polarity characteristics.

When preparing plates in the laboratory, the silica-gel is applied as an aqueous slurry approximately 1.5 to 2 parts water to 1 part powder. When properly purified, silica gel has little or no tendency to catalyze the reactivity of labile substances. The surface area of silica-gel adsorbent for TLC is typically 300 to 500 $m^2 \cdot g^{-1}$, and pore sizes range from 4.0 to 8.0 nm.

Silica gel is used for the resolution of acidic and neutral substances and mixtures with relatively low water solubility. Separation of organic acids is facilitated by the addition of a weak acid (usually acetic acid) to the mobile phase in order to give a pH value below the pK_a of the acid solutes. Adding specific material to the base silica adsorbent can aid in the resolution of certain analytes. The additive can be present in the slurry when the plate is prepared or it can be impregnated into the stationary phase of the plate. Examples of such additives are as follows:

1. Boric acid is used for the separation of sugar isomers since the borate ion forms complexes with the sugars to varying degrees.
2. Chelate-forming reagents are useful in the separation of inorganic cations and of phenol carboxylic acids.
3. Silver nitrate (argentation) is useful for compounds containing carbon–carbon double bonds.
4. Dilute (0.1M) sodium hydroxide increases basicity and improves the separation of organometallics, alkaloids, and some acidic classes.
5. Potassium oxalate is used for the separation of polyphosphoinositides and acidic magnesium acetate for phospholipids.
6. Ammonium sulfate provides a self-charring plate.

5.4.1.6 ***Cellulosic Sorbents.*** Cellulose is a highly polymerized polysaccharide that possesses a large number of free OH groups. TLC cellulose powders are either native fibrous or microcrystalline.

Plates coated with microcrystalline cellulose resembles paper chromatography in terms of separation properties, but the plates are generally more rugged, have shorter separation times, and give better resolution and sensitivity. Cellulose plates are suited for water-soluble (polar) materials and, in particular, carboxylic acids and carbohydrates. If preparing plates in the laboratory, use a 15% to 35% aqueous slurry of the powder that has been briefly homogenized. The coated plates should be air dried. As a rule, cellulose layers shrink to about one-half the coating thickness applied. Dry layers of about 125 μm are recommended. Activation prior to use is not recommended. Corrosive visualization agents must be avoided in cellulose TLC. Modified cellulose plates find special uses:

1. Acetylated cellulose can be used for reversed-phase chromatography. Caution: The degree of acetylation (10% or 30%) must be matched to the application.

2. Cellulose with diethylaminoethyl (DEAE) groups attached carries a positive charge on the amino ion-exchange group at neutral and acidic pH; it is used in the anion-exchange mode for separation of proteins and nucleic acids.

3. Polyethyleneimine (PEI) impregnated into the cellulose is used for the separation of nucleic acids and nucleic acid components. PEI plates must be refrigerated.

5.4.1.7 ***Alumina.*** Plates coated with alumina are most useful for the separation of neutral and acidic lipophilic substances. Alumina is an ideal media for many solvent cleanup applications. Commercial powders can be obtained as acid, basic, or neutral alumina and are available with or without a gypsum binder or fluorescent indicator. Test dyes are used to evaluate the activity of alumina layers according to the Brockmann activity (Table 5.26). Plates with activity I are used for the separation of polar samples in nonpolar systems. Plates with activity II and III contain 4% to 10% water, have medium activity, and are used as a replacement for organic polymeric ion exchangers when temperature becomes a problem.

Aromatic hydrocarbons are more strongly retained on alumina than silica gel. Alumina can catalyze certain labile substances. Chemical modification also provides some specialty products: Alumina P was developed for removal of pyrogens in solution; alumina C was prepared for analysis and removal of polychlorobenzenes (PCBs); alumina 200 is a large particle (>100 μm), chemically treated to remove metal ions, especially lead, from water; and alumina 5005 is a 50-μm spheroidal, macroporous alumina with high surface area for removal of color and dyes from water.

5.4.1.8 ***Polyamide Plates.*** The polyamide most frequently used is 2-poly(oxohexamethyleneimine) or ϵ-polycaprolactam. Separations on this support are achieved mainly by virtue of the reversible formation of hydrogen bonds of different strengths between solutes, eluents, and the peptide groups of the polyamide. These plates are well suited for separations of highly polar, water-soluble samples. The plates are not activated before use. Visualization is made possible by the inherent fluorescent properties of the support, and iodine can also be applied.

5.4.1.9 ***Chemically Bonded Reversed-Phase Plates.*** Reversed-phase TLC plates are available with C_2, C_8, C_{18}, and diphenyl bonded phases. A high-molecular weight, aliphatic cross-linked

TABLE 5.26 Brockmann Activity Scale

Dye	Brockmann activity*:	R_f I	II	III	IV
Azobenzene		0.59	0.74	0.85	0.95
4-Methoxyazobenzene		0.16	0.49	0.69	0.89
Sudan yellow		0.01	0.25	0.57	0.78
Sudan red		0.00	0.10	0.33	0.56
4-Aminoazobenzene		0.00	0.03	0.08	0.19

* Development is with CCl_4.

polymer is added to the formulation to form a durable coating. Since these coatings match the HPLC phases, TLC can be easily used for screening or methods development. Advantages of chemically bonded plates compared to classical impregnated plates include the absence of need to saturate mobile phases with the stationary phase; no contamination with stationary liquid or solutes recovered from the layer by elution; and uniform and reproducible R_f values. Unlike plates coated with silica gel or other adsorbents, reversed-phase plates do not normally require activation or other preparation prior to use. Reversed-phase plates do suffer from poor wettability with pure water or solvents with high water content, especially those containing shorter hydrocarbon chains. Reversed-phase TLC plates often contain different types of fluorescent indicators; the number following the letter "F" indicates the wavelength (in nm) for the activation peak (usually 254 or 366 nm).

Placed on a polarity scale, the chemically bonded phases can be arranged in decreasing polarity as follows:

$$\mathrm{Si} > \mathrm{NH_2} > \mathrm{CN} = \mathrm{diol} > \mathrm{C_2} > \mathrm{C_8} > \mathrm{C_{18}}$$

The CN plate can separate by reversed phase, normal adsorption phase, or ion pair, depending on the eluent system used.

Diphenyl reversed-phase plates are less hydrophobic than a C_{18} layer, are less load dependent, and generally run faster. Partition is based on hydrogen bonding from the aromatic rings. This plate is an effective media for biological separations within the 100 000-MW range.

Polar solvents for reversed-phase TLC typically consist of solvent mixtures such as water–methanol or water–acetonitrile. Substances migrate in a general order of decreasing polarity (the most polar solute moves the fastest), and mobile-phase strength increases with decreasing polarity (for example, acetonitrile is a stronger solvent than water).

For ion-pair chromatographic separations, lipophilic counterions are added to the mobile phase. The pH of the mobile phase is maintained at a value such that the solutes are ionized and can bond with the counterion.

Chiral compounds can be resolved with plates modified with Cu(II) and a proline derivative for ligand exchange separation.

5.4.1.10 Normal-Phase Bonded Plates. For normal-phase TLC, bonded phases available include the amino, cyano, and diol functional groups. The amino bonded phase displays weak-base ion-exchange properties and can be an alternative to PEI cellulose plates. The diol plates have some of the same characteristics of unmodified silica gel but differ in that the diol group resembles an alcohol as compared to a silanol.

5.4.1.11 Miscellaneous Sorbents. Swollen gels containing pores of carefully controlled dimensions are used for separations governed by size exclusion. Typical gels are Sephadex, Styragels, and polystyrene–divinylbenzene copolymers.

Cellulose-based ion exchangers, used mainly for separations of high-molecular-weight organic compounds, are made by chemical modification or impregnation of cellulose. See Sec. 2.3.1.3 for the various types of exchangers.

Resin-type exchangers are high-molecular-weight cross-linked polymers with attached functional groups that can exchange their bound ions with ions of the same charge in the mobile phase. The plates consist of resin exchangers bound to a flexible plastic sheet of polyethylene terephthalate. The layers are fixed to the plastic with silica gel plus low concentrations of organic binders.

5.4.1.12 Preparation of the Bed. Precoated plates are commercially available. Those on plastic or aluminum sheets can be cut easily with scissors. However, some workers prefer to prepare their own plates. A slurry of silica gel (or alumina) is prepared by adding 1.5 to 1 parts of deionized water to 1 part of adsorbent and mixing thoroughly. About 50 g of adsorbent are required to coat five 20×20-cm plates (or twenty 20×5-cm plates) of 0.25 mm thickness. Spreading is the most reproducible technique. The slurry of adsorbent and solvent is placed in some form of shallow trough with a dispensing slot. This trough is then used to disperse the slurry by pulling the trough of slurry over a series of plates laid out on a mounting board at a predetermined thickness above the plates.

After the plate is coated with adsorbent slurry, it is dried for 30 min in air and then heated to 110 to 120°C to give active (water-free) adsorbent. After activation, the plates must be stored in a special desiccator or storage cabinet with controlled humidity.

5.4.1.13 Sample Application. Since approximately 80% to 90% of all TLC analyses are done on reversed-phase plats, directions will be given for this method. Standards and samples should be dissolved in the weakest (most polar) possible solvent to minimize excessive band spreading. Because of stability and wettability limitation, sample solutions should be made up in totally or predominantly organic solvent rather than water. In most instances, methanol, which is quickly evaporated and wets the layer properly, is convenient for substances capable of hydrogen bonding. For less polar compounds, dichloromethane and acetone are convenient solvents for sample applications.

Standard TLC procedure involves these steps:

1. For manual spotting, a microcap micropipette of appropriate size or a capillary tube is used to apply samples and standards in concentrated spots on the plate, paper sheet, or paper strip. The spots should be no closer than 2.5 cm to the bottom edge (in the direction of development) and 2.5 cm to the edge that is parallel to the development. Spots are located 1 cm apart. If desired, a plate can be divided into individual columns by scoring the layer with a scriber.

2. The initial zones are dried completely before development using a heat gun (hair dryer) or infrared lamp. Care must be taken if the solute can decompose with heating, in which case a stream of room-temperature air or nitrogen can be used.

3. The plate (or sheet) is positioned in the developing tank. The solvent mixture should cover the bottom of the plate (but at least 1 cm below the sample spots); the top can lean against the side of the tank.

4. A sheet can be assembled in the shape of a cylinder with the adjoining edges held together with plastic clips. A simple arrangement for the development of strips involves placing the strip in a test tube closed with a cork.

5. The atmosphere of the tank need not necessarily be saturated with the solvent vapors. In adsorbent TLC solvent equilibration improves the reproducibility of separation but has an adverse effect on sample resolution.

6. When the solvent front has reached the desired position, usually 2 cm from the top of the plate, the plate (sheet or strip) is removed and dried in the same manner as were the initial spotted zones.

Sample volumes of 1 to 5 μL are used in ordinary TLC, whereas volumes of 100 nL are used in HPTLC when using fairly polar spotting solvents (200 nL for less polar solvents). A convenient means for manipulating such small volumes is a micropipette constructed from platinum–iridium capillary tubing fused into the end of a length of glass tubing.

In HPTLC spot diameters are less than 1 mm at the origin and about 2 to 3 mm diameter after development. The spot diameters in TLC will be 3 to 6 mm at the origin and increase to 6 to 15 mm after development. Small spots yield better separation and development. Multiple spotting of a sample, allowing the solvent to evaporate in between applications, keeps the initial spot small. A spotting template aids in reproducible placement of samples when not using plates with an inert preadsorbent area or concentrating zone.

The optimum sample size should not exceed 10 ng. The number of theoretical plates (efficiency) drops sharply with an increase in sample size up to 100 ng. A streaking pipette is used for applying larger amounts of sample as a streak parallel to the base of plate or paper.

In the sandwich technique a blank plate is placed over the spotted plate, the assembly clamped together, and placed in the developing chamber.

High-performance plates contain a 2-cm wide inert spotting strip (concentrating zone) composed of purified diatomaceous earth along the lower edge of the plate. Use of these plates allows TLC to be performed without the need for special spotting apparatus or laborious spotting techniques. A similar effect can be achieved after crude samples are spotted on conventional layers and then predeveloped for a distance of 1 to 2 cm with a polar solvent. After drying, the development is continued for a distance of 10 cm with a suitable but less polar solvent.

5.4.1.14 Development Methods. Development may be carried out in an ascending manner or with the plate in a horizontal position with any of the following techniques.

5.4.1.14.1 Continuous Development. A continuous flow of the mobile phase passes through the stationary phase. As the mobile phase reaches the opposite end of the plate, it is removed by wicks or simply allowed to drip off the edge where it is sucked up by an adsorbent pad. This technique is useful when the R_f values of the more rapidly moving spots are small but the spots show signs of separation. It allows further resolution among the components until the most rapidly moving spot reaches the opposite end of the plate. Mobile phases of low solvent strength are used to optimize selectivity. Continuous development occurs over a short distance (2 to 7 cm) with a constant, relatively high-velocity solvent flow. Since spots do not travel very far, they will be small and compact, thus enhancing sensitivity. Continuous development effectively lengthens the plate to improve the resolution of slowly moving solutes.

5.4.1.14.2 Stepwise Development. The sample is developed with a succession of solvents of different eluting strengths. This technique is useful for separating compounds in a mixture that contains a group of substances with low R_f values and other material with relatively large R_f values, that is, when the components have significantly different polarities. If initially the developer is the more polar solvent, compounds with the higher R_f values travel with the front. If development is stopped about midway and the plate is dried, development with a less polar second solvent serves to separate the compounds with high R_f values farther up the plate. The compounds originally separated in the lower half of the plate are not appreciably moved by the second solvent.

5.4.1.14.3 Two-Dimensional Development. The sample is spotted in the lower left-hand corner of the plate 2.5 cm from each edge. Standards are placed in a similar position in the lower right-hand corner and also in the upper left-hand corner of the plate. One edge of the plate is immersed in the first solvent and separation is conducted in one direction until the solvent front approaches the opposite edge of the plate but does not overrun the spot containing the standards. After thoroughly drying the plate to remove the first solvent, the other edge adjacent to the original spot is placed in the second solvent and the chromatogram is developed at right angles to the preceding development.

Two-dimensional development with two different solvent phases is recommended when additional resolving power is required for complex mixtures. This technique brings about more effective development of adjacent components that overlap after a one-dimensional separation using either solvent mixture alone. This method also provides separation of standards in both directions, so the migrations of the two reference mixtures form the coordinates of a graph that permits the unknowns to be located in two dimensions. Resolving power is nearly the square of that attained in one-dimensional TLC.

Whatman has a precoated plate with a strip of chemically bonded C_{18} reversed-phase silica gel alongside the major normal silica area for combination reversed-phase partition and adsorption two-dimensional TLC.

5.4.1.14.4 Multiple Development. Following a single development in the ascending mode, the plate(s) is removed from the chamber and dried in air for 5 to 10 min. The plate is then reinserted into the same solvent and redeveloped in the same direction. This procedure increases the resolution of components with R_f values less than 0.5. Repeated movement of the solvent front through the spots from the rear causes spot reconcentration and compression in the direction of the multiple developments, reducing the broadening that usually occurs during single-solvent development.

5.4.1.14.5 Radial (or Circular) Development. In the radial method the sample is applied to the center of the plate held in a closed development chamber. The solvent is fed at a constant, controlled rate through a small-bore pipette to the central point via a syringe controlled by a stepping motor. With radial development the component zones are continuously expanding as the mobile phase flows outward from the center. This compresses the ensuing zones into progressively narrower concentric rings. Furthermore, solvent feed is always faster at the trailing edge of the component zone, which tends to compress it. Separation is very fast, and resolution is high, especially in the low-R_f region. Samples can also be applied to the plate after the layer has been impregnated with mobile phase; in this case the TLC development more closely resembles HPLC. The relation between linear and radial migration is given by

$$R^2_{f,\mathrm{circ}} = R_{f,\mathrm{lin}} \tag{5.45}$$

5.4.2 Visualization and Qualitative Evaluation of Solutes

Detection in TLC is completely separate from the chromatographic operation and may be considered a static process. After chromatography and evaporation of the mobile phase, chromophoric substances can be located visually; colorless substances require other means. Any given chromatographic fraction can be examined for as long as necessary to extract the maximum amount of information. This freedom from time constraints is potentially the most important aspect of TLC because it permits utilization of a variety of techniques to enhance the sensitivity of detection and wavelength selection for optimum response of each component to be measured. In addition, in situ spectra can be obtained for component identification.

Spots that are neither naturally colored nor fluorescent can be visualized after reaction with a chromogenic or fluorogenic detection reagent or by fluorescence quenching. For fluorescence quenching, plates are coated with fluorescent materials, such as zinc silicate; the spot will obscure this fluorescence when the plate is irradiated. Chromogenic reagents (Table 5.27), selected to react with a particular functional group, can be applied by either dipping or aerosol spraying to reveal the spot location. After application of reagents, many detection reactions require uniform heating for a specified time at some controlled temperature or irradiation by ultraviolet light. Charring, after spraying with dilute sulfuric acid, or better, ammonium sulfate, is a general method that reveals organic material.

The spray should be very fine and applied with a horizontal motion, beginning in the upper left of the plate, proceeding to the right, down and across the plate to the left, in an alternating motion to the right and then left, and finally moving up and down the plate until the entire area has been covered twice.

After development, spots or zones can be removed from the plate by scraping. A miniature suction device collects scrapings directly in an extraction thimble held in a vacuum flask. Compounds are eluted from the scrapings and examined by an suitable means, often spectrophotometrically.

A permanent record of the developed chromatogram can be prepared by photography with a Polaroid camera or by spraying a plastic dispersion on the plate. After setting, the layer is peeled from the plate and stored as a flexible film.

R_f values of each substance range from 1.0 from zones migrating at the solvent front to 0.00 for a zone not leaving the point of application. Where the solvent front is not measurable, R_x values can be recorded:

$$R_{x,a} = \frac{\text{distance traveled by solute } a}{\text{distance traveled by standard substance } x} \tag{5.46}$$

R_x values are more reproducible than absolute R_f values.

The optimum method for obtaining tentative identification of a substance is to spot the sample and a series of reference compounds on the same chromatogram. A match in R_f values between a sample and standard is evidence for the identity of the sample. However, identification of a sample must be confirmed by chromatographing the sample and standards in several thin-layer systems. Even this provides corroborative evidence only if the systems are truly independent. Two-dimensional TLC on a two-phase silica gel and C_{18}-bonded silica-gel layer provides useful fingerprints of complex mixtures for confirmation and identity.

5.4.3 Quantitation

In the first method, quantitation is performed directly on the developed plate, sheet, or strip chromatograms by visual comparison, area measurement, or densitometry. The plate is mounted on

TABLE 5.27 Spray Reagents for Different Compound Groups

Compound class	Reagent (R)	Procedure	Results
Alcohols	$Ce(NH_4)(SO_4)_2$	6% R in 2*M* HNO_3.	Brown spots on yellow bkgd.
	2,2-Diphenyl-1-picryl-hydrazyl	15 mg R in 25 mL $CHCl_3$. Spray. Heat 5 to 10 min at 110°C.	Yellow spots on purple bkgd.
Aldehydes and ketones	2,4-Dinitrophenylhydrazine	Spray with 0.4 g R in 100 mL 2*M* HCl.	Yellow-red spots on pale orange bkgd.
	3,3′-Dimethoxybenzidene	1% solution R in HOAc; spray.	Yellow-brown spots
	Hydrazine sulfate	1% R in 1*M* HCl; spray. Observe by daylight and UV light. Heat to 100°C and observe by UV light.	
Alkaloids	Iodoplatinate	A: 5% $PtCl_4$ in H_2O; B: 10% aqueous KI. Mix 5 mL A with 45 mL B, dilute to 100 mL with H_2O.	Basic drugs yield blue or blue-violet spots that turn brown-yellow
Alkaloids and choline-containing compounds	Dragendorff-Munier	A: 1.7 g $BiONO_3$, 20 g tartaric acid, 80 mL H_2O. B: 16 g KI in 40 mL H_2O. Mix A and B 1:1, then mix 5 mL of this solution with 10 g tartaric acid in 50 mL H_2O.	Orange spots that are intensified by spraying with 0.025*M* H_2SO_4
Alkali and alkaline-earth metals	Violuric acid	1.5% aqueous solution R. After spraying, heat to 100°C for 20 min.	
Amides	Hydroxylamine-iron(III) nitrate	A: 1 g $NH_2OH \cdot HCl$ in 9 mL H_2O. B: 2 g NaOH in 8 mL H_2O. C: 4 g $Fe(NO_3)_3$ in 60 mL H_2O plus 40 mL HOAc. Mix 1 vol. A with 1 vol. B; spray and dry 10 min at 110°C; spray with 45 mL C and 6 mL 12*M* HCl.	Various colors on white bkgd.
Amines and amino acids	1,2,3-Indanetrione hydrate (ninhydrin)	Mix 95 parts 0.2% R in 1-butanol with 5 parts 10% HOAc.	Yellow, pink-red, or violet spots on white bkgd.
	Malonic acid	Spray 0.2 g R + 0.1 g salicylaldehyde in 100 mL absolute EtOH; heat at 120°C for 15 min. View by UV light.	Yellow spots
Amines, aromatic	$FeCl_3$	1:1 mix 0.1*M* R and 0.1*M* $K_3\{Fe(CN)_6]$ freshly prepared.	Blue spots
Antioxidants	Dichloroquinone-4-chlorimine	1 g R in 100 mL EtOH, spray; respray after 15 min with 2% borax in 4% ErOH.	Various colors on white bkgd.
Barbiturates	*s*-Diphenylcarbazone	0.1% R in 95% EtOH.	Purple spots
Bile acids	4-Methoxybenzaldehyde	0.5 mL R in 1 mL H_2SO_4 and 50 mL glacial HOAc. Spray. Heat 10 min at 125°C.	
Caffein	Chloramine-T	Spray with aqueous 10% R, then with 1*M* HCl. Heat at 98°C to remove Cl_2, expose to NH_3.	Rose color spot

(Continued)

TABLE 5.27 Spray Reagents for Different Compound Groups (*Continued*)

Compound class	Reagent (R)	Procedure	Results
Carbohydrates	4-Methoxybenzaklehyde	1 mL R and 1 mL H_2SO_4 in 18 mL EtOH. Spray and heat at 110°C.	Sugar phenylhydrazones: green-yellow spots in 3 min. Sugars: blue, green, violet spots in 10 min.
	4-Methoxyaniline plus phthalic acid	0.1*M* solution of 4-methoxyaniline and phthalic acid in EtOH.	Hexoses: green; pentoses: red-violet; methyl pentoses: yellow-green; uronic acids: brown
	4-Methoxyaniline HCl	3% R in 1-butanol; heat 2 to 10 min at 100°C.	Aldohexoses: green-brown; keto-hexoses: yellow; aldopentoses: green; uronic acids: red; 2-deoxyaldoses: gray-brown
	Thymol	1 g R in 10 mL 18*M* H_2SO_4 plus 190 mL EtOH. Spray. Heat 20 min at 120°C.	Dark pink to violet spots
Carboxylic acids	Bromocresol green	0.04 g R in 100 mL EtOH; add 0.05*M* NaOH until blue color just appears.	Acids: yellow spots on blue bkgd. Anions: blue spots on green bkgd.
Chlorinated hydrocarbons	Silver nitrate	0.1 g R in 1 mL H_2O + 10 mL 2-phenoxyethanol + 190 mL acetone; add 1 drop 30% H_2O_2; spray and expose 15 min to 254 nm light.	Gray spots on colorless bkgd.
Cholesterol and cholesteryl esters	Phosphotungstic acid	10% R in 90% EtOH. Heat 15 min at 100°C.	Red spots on white bkgd.
3,5-Dinitrobenzoate esters	1-Aminonaphthalene	1 g R in 100 mL EtOH.	Orange spots on white bkgd.
Ethanolamines	Benzoquinone	1 g R in 20 mL pyridine + 80 mL 1-butanol.	Red spots on pale bkgd.
Flavonoids	Uranyl(VI) acetate	1% aqueous solution; spray.	Brown spots
Glucosiduronates	1-(2-Pyridylazo)-2-naphthol	A: 0.4% R in EtOH. B: 0.8 g Co $(NO_3)_2$ in 100 mL H_2O. C: 2*M* NaOAc buffer, pH 4.6 (Fe free). D: Mix 4 mL C, dilute to 50 mL. Spray with A, dry, spray with D.	Violet on yellow bkgd.
Glycolipids	Diphenylamine	Dilute 20 mL 10% R in EtOH with 100 mL 12*M* HCl and 80 mL glacial HOAc. Spray. Cover with glass plate and heat 30–40 min at 110°C.	Blue-gray spots on light gray bkgd. for all lipids containing sugar
	Orcinol	0.1 g R, 40,7 mL 12*M* HCl, 1 mL 1% $FeCl_3$ solution, dilute to 50 mL with H_2O. Put chromatogram in HCl atmosphere to 1.5 h at 80°C. Spray. Heat again at 80°C.	Violet spots
	Trichloroacetic acid	Spray with 25% R in $CHCl_3$.Heat at 100°C for 2 min. UV light.	Yellow fluorescence
Halogen ions	Silver nitrate	0.05*M* $AgNO_3$. Spray, expose to UV light for 10 min, view immediately.	Halogen ions
Herbicides, triazine	Brilliant Green	0.5% R in acetone, spray, expose to Br_2 vapor.	Dark green spots

Heterocyclic nitrogen compounds	Malonic acid	Spray with 0.2 g R and 0.1 g salicicylaldehyde in 100 mL absolute EtOH. Heat 15 min at 120°C. View by UV light.	Yellow spots
Heterocyclic oxygen compounds	Aluminum chloride	1% R in EtOH; spray and view under 366-nm light.	Flavonoids produce yellow fluorescent spots
Hydrocarbons	Tetracyanoethylene	10% R in benzene; spray, heat at 100°C.	Aromatic hydrocarbons yield various colors
Hydroxamic acids	Iron(III) chloride	2 g R in 100 mL 0.5*M* HCl.	Red spots on colored bkgd.
Indoles	4-Dimethylaminobenzaldehyde	10% R in 12*M* HCl; mix 1 vol with 4 vol acetone; spray; color develops in 20 min.	Indoles: purple; hydroxyindoles: blue; aromatic amines and ureides: yellow; tyrosine: purple-red
	Prochazka reagent	Mix 10 mL 35% CH_2O, 10 mL 25% HCl, and 20 mL EtOH; spray, heat to 100°C, observe in daylight and UV light.	Yellow spots
Inorganic ions	Ammonium sulfide	Saturated aqueous H_2S made alkaline with aqueous ammonia.	Black: Ag, Co, Hg, Ni Brown: Au, Bi, Cu, Pb, Pd, Pt, Ti Yellow: As, Cd, Sn Yellow-orange: Sb
	Alizarin	Spray with saturated EtOH solution, then with 25% aqueous NH_3.	Al, Ba, Ca, Fe, Li, Mg, NH_4, Se, Th, Ti, Zn, Zr give colors
	Aurintricarboxylic acid, ammonium salt	Spray with 1% solution R in 1% aqueous NH_4OAc. Expose to NH_3 vapors.	Al, Cr, Li give spots
	Dithiooxamide	0.1% R in 1:1 EtOH/1-butanol. Dry at 100°C for 20 min.	Co, Cu, Ni
	Dithizone	0.1% R in $CHCl_3$. Note colors. Spray with 25% ammonia and not colors.	Heavy metal ions and organic atin salts
	Quercetin	2% R in 96% EtOH. Expose to NH_3 vapors. View by UV light.	Be, Cr, Cu, Fe, K, Li, Mn, Ni, Sb
	8-Hydroxyquinoline	10% R in ammoniacal EtOH.	Al, Co, Cr, Fe, Ga, Ni, Mn, Mo, Zn
Ketoses	Diphenylamine	2.3 g R in 100 mL aqueous 1-butanol. Air dry and they dry at 130°C for 20 min.	Blue spots
	Urea	5 g R in 20 mL 2*M* HCl plus 100 mL EtOH.	Blue spots
Lactams of α-guanidino acids	Picric acid	Spray with freshly pepared mixture (5:1) of 3% R in EtOH and 10% NaOH.	Orange spots
Lipids	2′,7′-Dichlorofluorescein	0.2% R in 96% EtOH; spray; observe in UV light.	Saturated and unsaturated polar lipids give green spots on purple bkgd; yellow-green fluorescence
	Crystal Violet	0.1% R in MeOH. Spray. Expose to bromine vapor.	Blue spots on yellow bkgd.
	Morin	0.05% R in MeOH. Dry 2 min at 100°C. View by UV light.	Yellow-green fluorescence or dark spots on fluorescent bkgd.
	Rhodamine B	0.05% R in EtOH. Spray. View by 254 nm light.	Purple spots on pink bkgd.

(*Continued*)

TABLE 5.27 Spray Reagents for Different Compound Groups (*Continued*)

Compound class	Reagent (R)	Procedure	Results
Nickel ions	Dimethylglyoxime	10% R in ammoniacal EtOH.	Red spot
Nitrate esters	Diphenylamine	1 g R in 100 mL EtOH; spray and expose to short-wave UV light.	Yellow-green spots on white bkgd.
Nitrosamines	Diphenylamine	A: 1.5% R in EtOH. B: 0.1% $PdCl_2$ in 0.2% saline solution. Spray with 5 parts A to 1 part B. Expose to 240 nm UV light.	Blue-violet spots
Olefins	Chlorosulfonic acid	Spray with R mixed with HOAc (1:2). Heat 1 min at 130°C.	
Oligosaccharides	Anthrone	Dissolves 300 mg R in 10 mL boiling HOAc, add 20 mL EtOH, 3 mL H_3PO_4, and 1 mL H_2O. Spray. Heat at 110°C for 5–6 min.	Yellow spots
	Thymol	5 g R in 20 mL 2*M* HCl plus 100 mL EtOH.	Blue spots
Organo-tin compounds	Pyrocatechol Violet	100 mg R in 100 mL EtOH. Expose chromatogram to UV light for 20 min. Spray with reagent.	Dark blue spots on gray-brown bkgd.
Penicillins	Iodine azide	Dissolve 3.5 g NaN_3 in 100 mL 0.1*M* I_2 (explosive when dry); spray.	Yellow spots
Peroxides, organic	Ammonium thiocyanate	A: 0.2 g NH_4SCN in 15 mL acetone. B: 4 g $FeSO_4$ in 100 mL H_2O. Add 10 mL B to A; spray.	Brownish red spots on pale bkgd.
	N,N-Dimethyl-*p*-phenylene-diamine dihydrochloride	1.5 g R in 128 mL MeOH, 25 mL H_2O, and 1 mL HOAc.	Reddish purple spots
Pesticides, chlorinated	2-Phenoxyethanol	0.1 g $AgNO_3$ in 1 mL H_2O with 10 mL E; dilute to 200 mL with acetone. Spray; dry 5 min, then 15 min at 75°C. Expose to UV light.	
	o-Toluidine	0.5% R in EtOH. Dry, expose to UV.	Green spots on white bkgd.
Pesticides, organic phosphorus	4-Methylumbelliferone	A: 0.5% I_2 in EtOH. B: 0.075 g R in 50 mL H_2O plus 50 mL EtOH. Make alkaline with 0.1*M* NH_4OH. Spray with A, record spots; spray with B, observe by UV light.	
Pesticides, organic sulfite	Malachite Green oxalate	A: 1 g KOH in 10 mL H_2O diluted to 100 mL with EtOH. B: 1 mL saturated acetone solution of R in 51 mL H_2O, 45 mL acetone, and 4 mL pH 7 buffer.	Spray A; heat at 150°C for 5 min, record spots, spray B, and observe by UV light
Pesticides, thiophosphate	*N*,2,6-Trichloro-*p*-benzoquinoneimine	Dissolve 0.3 g R in 100 mL spectro-grade cyclohexane. Prepare fresh weekly and store in brown bottle.	
Phenols	4-Aminoantipyrine	A: 3 g R in 100 mL EtOH; B: 8 g $K_3[Fe(CN)_6]$ in 100 mL H_2O. Spray with A, then B, expose to NH_3 vapor.	Red, orange, or pink spots on pale bkgd.
Phenols and aromatic amines that will couple	Pauly's reagent	0.5 g sulfanilic acid and 0.5 g KNO_3 in 100 mL 1*M* HCl. Spray; then spray with 1*M* NaOH.	Yellow-orange spots

Phosphorus-containing insecticides	4-(4-Nitrobenzyl)pyridine	A: 0.2 g R in 100 mL acetone; B: 10 g $(NH_4)_2CO_3$ in 50 mL H_2O and 50 mL acetone. Spray heavily with A, then lightly with B.	Blue spots on white bkgd.
Polynuclear aromatic hydrocarbons	Formaldehyde-sulfuric acid	2 mL 37% CH_2O in 100 mL 18*M* H_2SO_4.	Various colors on white bkgd.
Quinone and quinidine	Formic acid	Expose to vapors of R for 1 min; UV light.	Blue fluorescence
Sapogenins	Chlorosulfonic acid	Spray with R and HOAc (1:2). Heat 1 min at 130°C.	
Steroids	4-Toluenesulfonic acid	20% R in $CHCl_3$, spray; heat at 100°C for few minutes, observe under 360 nm UV.	Steroids, flanonoids, and catechins fluoresce
	2,3,5-Triphenyl-2*H*-tetrazolium chloride	Mix equal volumes of 4% R in MeOH and 1*M* NaOH in MeOH. Prepare fresh daily. Spray, heat 10 min at 110°C.	Red spots
Steroid glycosides	Trichloroacetic acid-chloramine T	A: 3% chloramine T in Hj_2O; B: 25% Cl_3CCOOH in 95% EtOH. Mix 10 mL A and 40 mL B; spray and heat 10 min at 110°C; observe under UV light.	Digitalis glycosides: blue spots
Sugars	Aniline phthalate	0.93 g R and 1.7 g *o*-phthalic acid in 100 mL 1-butanol satd with H_2O. Spray, heat to 105°C for 10 min.	Reducing sugars give various colors
Sugars, reducing	3,5-Dinitrosalicylic acid	0.5% R in 4% NaOH. Dry in air, then for 4–5 min at 100°C.	Brown spots on yellow bkgd.
Sulfonamides	*N*-(1-Naphthyl)ethylenediamine · 2HCl	Spray consecutively with 1*M* HCl, 5% $NaNO_2$, and 0.1% R in EtOH.	Red-purple spots
Terpenes	Diphenylpicrylhydrazyl	15 mg R in 25 mL $CHCl_3$. Spray, heat 10 min at 110°C.	Yellow spots on purple bkgd.
Terpenoids	Vanillin-sulfuric acid	1% vanillin in 18*M* H_2SO_4; spray, observe in daylight and UV light.	Various colored spots
Thiobarbituric acids and thioacids	Diethylamine	0.5 g $CuSO_4$ in 100 mL MeOH, add 3 mL R, shake and spray.	Green spots
Triphosphates (also hexoses)	1,3-Phenylenediamine-2HCl	Spray with 0.2*M* R in 76% EtOH. Heat 5 min at 110°C. Examine by UV light.	
Triterpenes and cardiac glycosides	Chlorosulfonic acid-acetic acid	1:2 solutions of acids; spray, heat at 130°C for 5 min; observe in UV light.	Violet-brown spots
Uranium(VI), UO_2^{2+}	1-(2-Pyridylazo)-2-naphthol	0.25% R in EtOH	
Vitamins	α,α^3-Dipyridyliron(III) chloride	A: 2% α,α^3-Dipyridyl in $CHCl_3$; B: 5% $FeCl_3$ in H_2O. Spray with A, then with B.	Detects vitamin E, phenols, and other compounds with reducing properties
	Iodine-starch	0.001-0.005% I_2 in 1% KI, with 0.4% starch solution; spray.	Ascorbic acid; white area on blue bkgd.
Vitamin B_6	*N*,2,6-Trichloro-*p*-benzoquinoneimine	0.1% R in EtOH. Expose to NH_3 vapor.	Blue spots

Note: These references list other visualization reagents that may be useful:
1. J. G. Kirchner, *Thin-Layer Chromatography*, Vol. XII in *Techniques of Organic Chemistry*, Wiley-Interscience, New York, 1967.
2. E. Stahl, *Thin-Layer Chromatography*, 2d ed., Academic, New York, 1969.

movable stage, usually motor-driven in the direction perpendicular to the slit length, and either manually operated or motor-driven in the orthogonal direction. In the second method, solutes are eluted from the sorbent before being examined further. The eluates from samples and standards can be analyzed by any convenient and appropriate procedure. Requirements for the various steps of TLC are more stringent when quantitation is to be carried out.

*5.4.3.1 **Absorption Densitometry.*** In absorption densitometry the spots on the TLC plate are scanned by a beam of monochromatic light formed into a slit image with the length of the slit selected according to the diameter of the largest spot. Because the response of reflectance–absorbance scans are nonlinear with concentration, calibration standards are included with each sample run. Consequently, all samples, both standards and unknowns, are subjected to exactly the same chromatographic conditions, and systematic errors remain very much at a minimum. Typical minimum detection levels for measurement of visible or ultraviolet absorption range from 100 pg to 100 ng per spot.

These are some tips on densitometric scanning of TLC plates.

1. *Single-beam scanners* often give excellent quantitative results. Baseline drift may be troublesome due to extraneous absorbed material in the thin layer that can move during chromatographic development or small irregularities in the plate surface.

2. *Double-beam scanners* have a reference beam scanning the intervening space between sample lanes. The difference signal eliminates the contribution of general plate background.

3. *Dual-wavelength scanners* have two monochromators that alternately furnish to the same sample lane a reference wavelength with minimal sample absorbance and a sample wavelength chosen for maximum absorbance by the analyte. This helps cancel out the general background and correct for plate irregularities.

*5.4.3.2 **Reflectance Mode.*** In the reflectance mode, the diffusely reflected (or scattered) light is detected with a photomultiplier. Those areas of the plate that are free of light-absorbing materials will yield a maximum signal, and the separated spots will cause a diminution of reflected light that is concentration dependent.

*5.4.3.3 **Fluorescence Measurement.*** When applicable, fluorescence measurement is preferred because of its greater selectivity and sensitivity and the high degree of linearity irrespective of the form of the zone. Often fluorimetry involves measuring a positive signal against a dark or zero background. Consequently, the signal can be highly amplified. A single-beam mode of operation of the fluorimeter is usually quite sufficient. Typical sensitivities for fluorescent compounds are in the low picogram range.

*5.4.3.4 **Standard Addition Method.*** The standard addition assay method compares each unknown sample with three standards. Four bands of the sample solution are applied first. After intermediate drying the first band is oversprayed with 10 μL of a standard solution containing known concentrations of the impurities. The second band is oversprayed with 4 μL, and the third with 2 μL. The fourth band is not oversprayed; it remains unchanged.

5.4.4 Paper Chromatography

In paper chromatography the substrate is a piece of porous paper with water adsorbed on it. The sample is placed on the paper as a spot or streak and then irrigated by the solvent system that percolates within the porous structure of the paper. Usually development of the chromatogram is stopped before the mobile phase reaches the farther edge of the paper, so the solute zones are distributed in space instead of time. The major limitations of paper chromatography are relatively long development times and less sharply defined zones as compared to thin-layer techniques. Thin-layer chromatography with powdered cellulose as stationary phase has displaced paper chromatography from many of its previous applications.

5.4.4.1 ***Solvent Systems.*** Chromatography on paper is essentially a liquid–liquid partition in which the paper serves as carrier for the solvent system. Aqueous systems are used for strongly polar or ionic solutes. Water is held stationary on the paper as a "water–cellulose complex" or puddles of water, organized and dense near the amorphous regions of the cellulose chains. The stationary phase is attained by exposing the suspended paper to an atmosphere saturated with water vapor in a closed chamber. If an aqueous buffer or salt solution is to be used as the stationary phase, the paper is drawn through the solution, allowed to dry, and then exposed to the atmosphere saturated with water vapor.

The mobile phase might be butanol for a neutral system, butanol–acetic acid–water (40:10:50) for an acidic system, or butanol – ammonia – water (75:8:17) for a basic system. The latter two systems are prepared by shaking all components in a separatory funnel; the less polar phase serves for development. Even water itself can serve as developer.

5.4.4.2 ***Equipment.*** The only essential piece of equipment is the developing chamber, often simply an enclosed container with an airtight lid. For one-dimensional paper chromatography, either ascending or descending development can be carried out in simple units. Descending development is more often used because it is faster and more suitable for long paper sheets, which give higher efficiencies.

Radial development is carried out by cutting a tab from a circular piece of paper, spotting the sample at the upper end of the table or streaked in a circle a short distance from the center of the paper, and then placing the tab in the solvent reservoir. Small chambers are made from two petri dishes, one inverted over the other.

5.4.4.3 ***Development.*** Each of the development techniques described for thin-layer chromatography is equally applicable for paper chromatography. One difference should be noted. The paper is equilibrated with both mobile and stationary phases for 1 to 3 h before development. The two phases are placed in separate reservoirs in the bottom of the chamber. The mobile phase should also be equilibrated with the stationary phase. Close temperature control is required for reproducible R_f values. With volatile solvents, careful equilibration is required to avoid band tailing.

5.4.4.4 ***Visualization and Evaluation of Paper Chromatograms.*** The discussion on the subject in TLC (Sec. 5.4.2, Table 5.27) is applicable for paper chromatography with the exception of fluorescent quenching using TLC plates impregnated with a phosphor. Special attachments for commercial transmission spectrophotometers provide a means for drawing the paper chromatogram across a window in front of a photodetector. The reflection mode can also be used. An assessment of the area under the photometric curve completes the measurement.

5.4.5 Zone Electrophoresis

Zone electrophoresis is a process for separating charged molecules in a supporting medium (bed) on the basis of their movement through a fluid under the influence of an applied electric field. Each charged species moves along the field gradient at a rate that is a function of its charge, size, and shape. The supporting medium (often a semisolid slab-gel) is soaked with a carrier electrolyte solution that maintains the requisite pH and provides sufficient conductivity to allow the passage of current (ions) necessary for the separation. Extra materials may be added to the electrolyte to adjust the separation's selectivity. The sample is spotted near the center of the bed. Upon application of a controlled dc source of potential to the ends of the supporting medium, each component of the sample begins to move according to its own mobility. Ideally, each sample component will eventually separate from its neighbors, forming a discrete zone. The chromatographic analog of zone electrophoresis is elution chromatography.

Electrophoresis is the premier method of separation and analysis of proteins and polynucleotides. The primary classification of protein components is based on the relative positions in a developed electrophoretic pattern.

5.4.5.1 ***Equipment.*** The basic procedure for performing gel electrophoresis is to prepare and pour the gel, apply the sample, run the separation, immerse the gel in a detection reagent, and photograph the gel for a permanent record or use a gel scanner. Electrodes are affixed at opposite ends of the gel bed and are separated from the bed by diffusion barriers, the function of which is to prevent diffusion and convection processes from carrying electrolytic decomposition products into the bed. The bed and electrodes are enclosed in an airtight chamber to prevent excessive evaporation of solvent.

Under the influence of the applied emf, usually between 5 to 10 $V \cdot cm^{-1}$, the charged species move with a velocity (ionic mobility) that depends upon the field strength. The power supply must develop between 50 to 150 V. With paper or cellulose acetate as supporting media the current drain is less than 1 mA per strip.

The temperature must be closely controlled because the mobility of molecules increases with temperature. Thermostating the bed minimizes the effect of electrical heating.

5.4.5.2 Supporting Media. Filter paper is now largely replaced by cellulose acetate strips or polyacrylamide gels, which provide superior resolution. The electrophoresis bed may be a trough or a vertical column of gel.

5.4.5.2.1 Cellulose Acetate Beds. Cellulose polyacetate forms strong and flexible membranes that possess a very uniform, foamlike structure. The pore size is closely controlled in the 1-μm range. Uniform porosity results in sharper boundaries and permits better resolution and shorter migration times as compared with filter paper. The inertness of the membrane virtually eliminates adsorptive effects and trailing boundaries. Running times are 20 to 90 min. After electrophoresis the membrane can be cleared to glasslike transparency for densitometry, thus permitting accurate quantitation. The developed strip is dipped into a mixture of acetic acid–ethanol or in Whitemor oil 120.

For a simple analysis of proteins in serum or other body fluids, zone electrophoresis on cellulose acetate membranes is sufficient. A variety of staining procedures is available.

5.4.5.2.2 Polyacrylamide Gels. Gels, such as polyacrylamide, are formed directly in the electrophoretic bed. By controlling the relative proportion of the cross-linking agent bisacrylamide, gels can be formed with cell-defined molecule-sieving properties. The sieving effect, unique to gels, provides extremely high resolving power. Techniques for gel sieving in polyacrylamide are the heart of the powerful and efficient methods developed for sequencing polynucleotides.

Polyacrylamide can also be used in slabs or rods and prepared so that the concentration of acrylamide increases in a continuous manner over the length of the gel. This is known as a *gradient gel.* In gradient electrophoresis the development occurs through a gradient of decreasing pore size, similar to separations by exclusion (gel-permeation) chromatography. Proteins are separated in the order of their molecular sizes. The shape of the gradient determines the resolution and is chosen to give a linear relationship between the migration distance of a globular protein and the logarithm of its molecular weight.

5.4.5.3 Gel Sieving. The most effective application of gel sieving to proteins involves the use of sodium dodecyl sulfate, which binds to a wide variety of proteins in a constant ratio, producing complexes with constant charge per unit mass. Protein aggregates are broken up (denatured or unfolded). Electrophoretic mobilities are a direct function of the protein subunit molecular weight. The rate of migration of a protein is compared to the rate of migration of standard proteins, and fairly accurate estimates of protein molecular weight are obtained.

5.4.5.4 Electrophoretic Mobilities. The positively charged species will move toward the cathode, and the negatively charged species will move toward the anode. The rate of movement of both species is determined by the product of the net charge of the species and the field strength X defined as

$$X = \frac{E}{s} \tag{5.47}$$

where E is the potential difference in volts between test probes inserted into the supporting medium and spaced s centimeters apart.

The electrophoretic mobility of an ion μ is given by

$$\mu = \frac{d}{t}\left(\frac{E}{s}\right) \tag{5.48}$$

where d is the linear distance (in centimeters) traveled by the migrating species relative to the bed in time t (in seconds) (that is, its velocity) in a field of unit potential gradient (in volts per centimeter). Electrophoretic migration is directly proportional to the field strength. Increasing the field strength will speed fractionation; it will also enhance the resolution and the decreased development time will minimize the opportunity for diffusion of the fractions. Permissible field strength is predicated by the removal of heat from the supporting medium by the cooling system.

The net charge carried by most species is pH-dependent. Buffer pH and ionic strength are important. The net mobility of a partially ionized solute in solution is given by

$$\mu' = \frac{\mu K_a}{[\mathrm{H}^+] + K_a} \tag{5.49}$$

where K_a is the ionization constant. The extent to which two incompletely ionized acids may be separated depends upon the difference in their apparent mobilities.

For cations, electrophoresis is generally run in the intermediate pH range using complexing agents such as lactic, tartaric, or citric acids to ensure solubility. Most procedures for separating anions or weakly acidic substances, such as carboxylic acids and phenols, require alkaline solutions to avoid any possible existence of two ionization states in equilibrium. Conversely, weakly ionized bases, such as amines and alkaloids, should be separated at low pH values.

The ionic strength of a buffer is adjusted to 0.05 to 0.1. This range minimizes the production of heat and the quantity of electrode products formed.

5.4.5.5 Electroosmosis. During the migration to the appropriate electrodes, hydrated ions carry with them their associated water molecules. This assemblage consists of water molecules held directly in the solvation sheath plus additional layers of water molecules attracted and held by the inner layers. Thus, it is possible for hundreds of solvent molecules to be dragged along by each migrating ion. Inasmuch as cations usually carry more water than anions, the net flow of liquid is almost always toward the cathode. This results in an apparent movement of neutral molecules toward the cathode. Because this movement is approximately directly proportional to the voltage gradient, it serves as a mobility correction that is added to anions and subtracted for cations. The value of this correction is determined by observing the migration of an uncharged species.

5.4.5.6 Detection of Separated Zones. Common approaches to detection involve the use of stains, autoradiogarphy of radiolabeled analytes, or immunoreaction with specially prepared antisera.

5.4.5.6.1 Staining. Staining procedures are quite lengthy and usually involve a fixative, such as trichloroacetic acid, which precipitates proteins and prevents their diffusion out of the gel during subsequent soaking with the staining reagent(s).

5.4.5.6.2 Autoradiography. In autoradiography the sample is run on a slab gel. Following development gel is covered with a thin sheet of plastic and placed on top of x-ray film. Regions of the gel where radioactivity is localized will create corresponding exposed regions on the film. For isotopes producing low-energy β particles, the gel is soaked first in a scintillation fluid, dried, and clamped on top of x-ray film between glass plates, and maintained at approximately −70°C for 1 day to 1 week.

5.4.5.6.3 Crossed Immunoelectrophoresis. In crossed immunoelectrophoresis a polyspecific antibody-containing gel is poured next to the developed electrophoresis gel. After this gel has solidified, the sample bands are electrophoretically migrated perpendicular to their original direction of migration and into the antibody-containing gel. Sample bands will continue to migrate through this gel until they have encountered sufficient antibody to result in precipitation of the antigen – antibody complex. Precipitated bands are then stained. Peak heights and areas are indicative of the quantity of protein in each band. The resolution of overlapping bands is remarkable.

5.4.6 Capillary Electrophoresis (CE)

CE separations are performed in capillaries, constructed from fused silica, with internal diameters between 50 and 100 μm with a practical lower limit of about 25 μm. High heat dissipation efficiencies in the narrow columns allow separations to be performed at high field strengths that vary from 100 to 500 $V \cdot cm^{-1}$. The distance from sample injection to the detector may be as short as 7 cm or as long as 1 m but most workers use capillaries with lengths from 15 to 40 cm. Run time is longer for the longer capillaries, and a higher voltage is required to achieve a given field strength. Resolution is maintained with increasing field strength owing to efficient heat dissipation. Analysis times are short and peak efficiencies range up to millions of theoretical plates. The detection limits, precision, interoperator reproducibility, linear dynamic range, and acquisition cost are all similar to HPLC. However, the solvent consumption is lower by 99+%, and the problems of both preparation and disposal of mobile phases are virtually eliminated with HPCE.

5.4.6.1 Instrumentation. The instrumentation consists of five main units: an injection system, a separation system in a temperature-regulated compartment, the capillary and buffer reservoirs, a detector, and a high-voltage power supply and controller, plus a data-processing system.

Instruments can blow out a capillary and fill it automatically with a different running buffer in seconds. This makes it easy to change the method.

Detection is generally accomplished by ultraviolet spectrophotometry, using modified HPLC detectors, although other detectors described for HPLC find use. Concentration detection limits are often one or two orders of magnitude poorer in HPCE than in HPLC. Solutions to the detection problem include the use of Z- or bubble-shaped cells and the use of focusing techniques.

5.4.6.2 Sample Injection. Sample injection is done electrokinetically (by electromigration) or hydrodynamically. In electrokinetic injection, the sample is introduced into the capillary by applying an electric field in a timed manner across the capillary while one end is immersed in the sample solution and the other in a buffer. For injection times between 1 and 20 s the field strengths are between 100 and 400 $V \cdot cm^{-1}$. The amount of material injected is a function of the electrophoretic mobility of each solute, the electrical conductivity of the sample buffer and the running buffer, and the electroosmotic flow.

In hydrodynamic injection, sample is introduced into the capillary by applying a pressure differential across the capillary while one end of the tube is immersed in the sample solution. The injection volume depends on the injection time, capillary dimensions, buffer viscosity, and pressure drop across the capillary. The injection volume should be equal to or less than 1% of the total separation length.

5.4.6.3 Columns. In free-solution capillary electrophoresis, separations are achieved as a result of the unequal rate of migration of different solutions under the influence of an externally applied electric field. There is no polymer network, superimposed pH gradient, or secondary phase.

In micellar electrokinetic chromatography the micelle is viewed as a pseudochromatographic phase for selective separation. Both hydrophobic and electrostatic selectivities are possible, as well as ligand exchange. Use of highly solubilizing detergents allows direct analysis of substances in complex matrices.

Gel-filled capillaries were introduced in 1990. In a sense, capillary gel electrophoresis is a hybrid of traditional slab-gel and free-solution capillary open-tube electrophoresis technology. Capillary diameter, capillary length, gel concentration, and field strength all affect separation efficiency and time. Sized-based CE separations use gels or polymer solutions. Several commercial sieving kits are available.

5.4.6.4 Summary. The reproducibility and quantitation achieved with CE are chief problems hampering its use as a routine method for real-world applications. The fused silica interior of the capillary tends to adsorb both small and large charged molecules and proteins. Permanently coating the silica with polyvinyl alcohol suppresses the interference and modifies the electroosmotic flow.

High-performance capillary electrophoresis (HPCE) solves many of the experimental problems of gels because gels are unnecessary with HPCE. The capillary walls provide mechanical support for the carrier electrolyte. Detection is on-line; the instrumental ouput resembles a chromatogram. The use of narrow 25- to 100-μm i.d. capillaries allows efficient dissipation of the heat produced by the passage of current. In turn this permits the use of high voltages to drive the separation that are faster than those in slab-gels. On the other hand, HPCE is a serial technique with one sample followed by another. The molecular-weight range of analytes separable by HPCE is enormous.

HPCE is growing rapidly and the preceding short introduction to the subject probably will be outdated when this *Handbook* is published. Further reading is suggested from these sources: M. Albin, P. D. Grossman, and S. E. Moring, *Anal. Chem.* **65**:489A (1993); B. L. Karger, "Capillary Electrophoresis: Current and Future Directions," *Am. Lab.* **1993** (October): 23; P. D. Grossman and J. C. Colburn, eds., *Capillary Electrophoresis: Theory and Practice*, Academic, San Diego, 1992; N. A. Guzman, *Capillary Electrophoresis Technology*, Dekker, New York, 1993; S. F. Y. Li, *Capillary Electrophoresis: Principles, Practice, and Applications*, Elsevier Science, New York, 1992; R. Weinberger, *Practical Capillary Electrophoresis*, Academic, Orlando, FL, 1993.

Bibliography

Braithwaite, A., and F. J. Smith, *Chromatographic Methods*, Chapman and Hall, London, 1985.

Fennimore, D. C., and C. M. Davis, "High Performance Thin-Layer Chromatography," *Anal. Chem.* **53**:252A (1981).

Fried, B., and J. Sherma, *Thin-Layer Chromatography: Techniques and Applications*, 3d ed., Dekker, New York, 1994.

Jorgenson, J. W., "Electrophoresis," *Anal. Chem.* **58**:743A (1986).

Jork, H., et al., *Thin-Layer Chromatography Reagents and Detection Methods*, VCH, Weinheim, Germany, 1990.

Kirchner, J. G., *Thin-Layer Chromatography*, 2d ed., Wiley, New York, 1978.

Poole, C. F., and S. K. Poole, "Instrumental Thin-Layer Chromatography," *Anal. Chem.* **66**:27A (1994).

Sherma, J., and B. Fried, eds., *Handbook of Thin-Layer Chromatography*, Dekker, New York, 1991.

Touchstone, J. C., *Practice of Thin-Layer Chromatography*, 3d ed., Wiley, New York, 1992.

SECTION 6

ELECTRONIC ABSORPTION AND LUMINESCENCE SPECTROSCOPY

6.1 SPECTROMETRIC AND PHOTOMETRIC INFORMATION

Energy can be transmitted by electromagnetic waves. They are characterized by their frequency ν, the number of waves passing a fixed point per second, and their wavelength λ, which is the distance between the peaks of any two consecutive waves. As the frequency increases, the wavelength decreases and, conversely, as the frequency decreases, the wavelength increases.

In any electromagnetic wave, wavelength and frequency are related to the energy of a photon, E, by Planck's constant h (6.62×10^{-34} J·s) and c (2.998×10^{10} cm·s^{-1}), the velocity of radiant energy in a vacuum:

$$E = h\nu = \frac{hc}{\lambda} \tag{6.1}$$

6.1.1 Types of Electromagnetic Radiation

The values of the wavelength and frequency differentiate one kind of radiation from another within the electromagnetic spectrum, the name given to the broad range of radiations that extend from cosmic rays with wavelengths as short as 10^{-9} nm all the way up to radio waves longer than 1000 km. The various regions in the electromagnetic spectrum are displayed in Fig. 6.1 along with the nature of the changes brought about by the interaction of matter and radiation.

Wavelengths are expressed in different units throughout the electromagnetic spectrum. This may lead to confusion and error when attempting to determine equivalent values, or when reading older literature that used values different from the International Union of Pure and Applied Chemistry (IUPAC) recommendations. Table 6.1 contains the symbols, SI units, and definitions for electromagnetic radiation. Table 6.2 shows how to interconvert wavelengths among different units.

6.1.2 Definitions and Symbols

6.1.2.1 Radiant Power. Radiant power (P) is the amount of energy transmitted in the form of electromagnetic radiation per unit time, expressed in watts. This is what a photodetector measures. It should not be confused with intensity (I), which is the radiant energy emitted within a time period per unit solid angle, measured in watts per steradian.

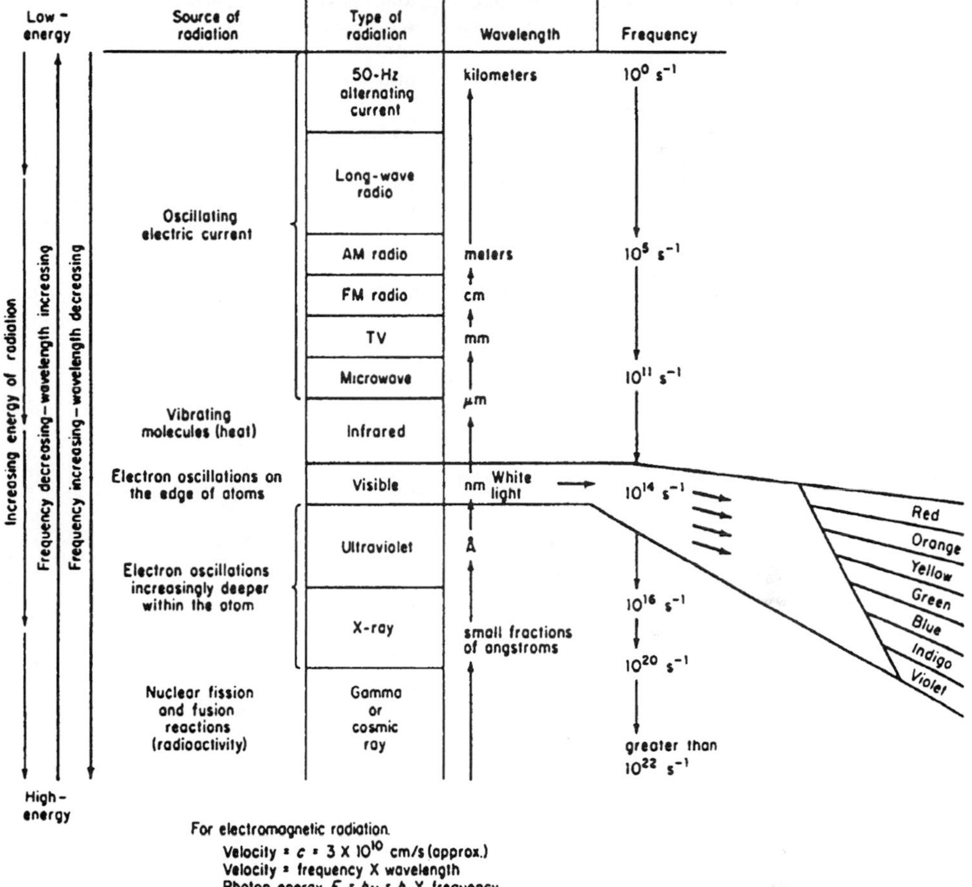

FIGURE 6.1 Electromagnetic radiation spectrum.

6.1.2.2 Transmittance. Transmittance (T) is the ratio of the radiant power (P) in a beam of radiation after it has passed through a sample to the power of the incident beam (P_0),

$$T = \frac{P}{P_0} \tag{6.2}$$

What is of more interest to the analyst is the internal transmittance of the cuvette contents, that is, the ratio of the radiant power of the unabsorbed (transmitted) radiation that emerges from the absorbing medium to that transmitted through a blank (usually a solvent blank)

$$T = \frac{P_{\text{sample}}}{P_{\text{blank}}} \tag{6.3}$$

when using the same or matched cuvettes. Percent transmittance is the transmittance multiplied by 100.

TABLE 6.1 Chemical Symbols and Definitions for Electromagnetic Radiation

Name	Symbol	SI unit	Definition
Absorbance	α		$\alpha = \Phi_{abs}/\Phi_0$
Absorption coefficient			
Linear (decaidic)	a, K	m^{-1}	$a = A/l$
Molar (decaidic)	ϵ	$m^2 \cdot mol^{-1}$	$\epsilon = B/l$
Angle of optical rotation	α	rad	
Circular frequency	ω	s^{-1}, $rad \cdot s^{-1}$	$\omega = 2\pi v$
Einstein transition probabilities			
Spontaneous emission	A_{nm}	s^{-1}	$dN_n/dt = -A_{nm}B_n$
Stimulated absorption	B_{mn}	$s \cdot kg^{-1}$	$dN_n/dt = \rho_v(v_{nm})B_{mn}N_m$
Stimulated emission	B_{nm}	$s \cdot kg^{-1}$	$dN_n/dt = \rho_v(v_{nm})B_{nm}N_n$
First radiation constant	c_1	$W \cdot m^2$	$c_1 = 2\pi hc_0^2$
Frequency	v	Hz	$v = c/\lambda$
Irradiance (radiant flux received)	E, (I)	$W \cdot m^{-2}$	$E = d\Phi/dA$
Molar refraction	R, R_m	$m^3 \cdot mol^{-1}$	$R = \frac{(n^2 - 1)}{(n^2 + 2)} V_m$
Path length (absorbing)	l	m	
Optical rotatory power	$[\alpha]_\lambda^\theta$	rad	$[\alpha]_\lambda^\theta = \alpha/\gamma l$
Planck constant	h	$J \cdot s$	
Planck constant/2π	$\hbar$	$J \cdot s$	$\hbar = h/2\pi$
Radiant intensity	I	$W \cdot sr^{-1}$	$I = d\Phi/d\Omega$
Radiant power, radiant energy per time	Φ, P	W	$P = dQ/dt$
Refractive index	n		$n = c_0/c$
Second radiation constant	c_2	$K \cdot m$	$c_2 = hc_0/k$
Speed of light			
In a medium	c	$m \cdot s^{-1}$	$c = c_0/n$
In vacuum	c_0	$m \cdot s^{-1}$	
Transmittance	T		$T = \Phi_{tr}/\Phi_0$
Wavelength	λ	m	
Wave number			
In a medium	σ	m^{-1}	$\sigma = 1/\lambda$
In vacuum	$\overline{v}$	m^{-1}	$\overline{v} = v/c_0 = 1/n\lambda$

6.1.2.3 Absorbance. Absorbance (A) is defined as

$$A = \log \frac{P_{blank}}{P_{sample}} = \log \frac{1}{T} \tag{6.4}$$

which is the logarithm (base 10) of the reciprocal of the transmittance. The conversion of transmittance to absorbance (and the reverse) is given in Table 6.3. Also useful are the data given in Table 6.4 for checking the photometric scale of a spectrophotometer.

TABLE 6.2 Wavelength Conversion

	Angstrom (Å)	Nanometer (nm)	Micrometer (μm)	Millimeter (mm)
1 Angstrom (Å)	1	10^{-1}	10^{-4}	10^{-7}
1 Nanometer (nm) [Millimicron (mμ)]*	10	1	10^{-3}	10^{-6}
1 Micrometer (μm) [Micron (μ)]*	10^4	10^3	1	10^{-3}
1 Millimeter (mm)	10^7	10^6	10^3	1
1 Centimeter (cm)	10^8	10^7	10^4	10
1 Meter (m)	10^{10}	10^9	10^6	10^3

* The old terms, millimicron and micron are the same as nanometer and micrometer, respectively.

TABLE 6.3 **Transmittance–Absorbance Conversion Table**

This table gives absorbance values to four significant figures corresponding to % transmittance values, which are given to three significant figures. The values of % transmittance are given in the left-hand column and in the top row. For example, 8.4% transmittance corresponds to an absorbance of 1.076.

Interpolation is facilitated and accuracy is maximized if the % transmittance is between 1 and 10, by multiplying its value by 10, finding the absorbance corresponding to the result, and adding 1. For example, to find the absorbance corresponding to 8.45% transmittance, note that 84.5% transmittance corresponds to an absorbance of 0.0731, so that 8.45% transmittance corresponds to an absorbance of 1.0731. For % transmittance values between 0.1 and 1, multiply by 100, find the absorbance corresponding to the result, and add 2.

Conversely, to find the % transmittance corresponding to an absorbance between 1 and 2, subtract 1 from the absorbance, find the % transmittance corresponding to the result, and divide by 10. For example, an absorbance of 1.219 can best be converted to % transmittance by noting that an absorbance of 0.219 would correspond to 60.4% transmittance; dividing this by 10 gives the desired value, 6.04% transmittance. For absorbance values between 2 and 3, subtract 2 from the absorbance, find the % transmittance corresponding to the result, and divide by 100.

% Transmittance	0.0	0.1	0.2	0.3	0.4	0.5	0.6	0.7	0.8	0.9
0	. . .	3.000	2.699	2.523	2.398	2.301	2.222	2.155	2.097	2.046
1	2.000	1.959	1.921	1.886	1.854	1.824	1.796	1.770	1.745	1.721
2	1.699	1.678	1.658	1.638	1.620	1.602	1.585	1.569	1.553	1.538
3	1.523	1.509	1.495	1.481	1.469	1.456	1.444	1.432	1.420	1.409
4	1.398	1.387	1.377	1.367	1.357	1.347	1.337	1.328	1.319	1.310
5	1.301	1.292	1.284	1.276	1.268	1.260	1.252	1.244	1.237	1.229
6	1.222	1.215	1.208	1.201	1.194	1.187	1.180	1.174	1.167	1.161
7	1.155	1.149	1.143	1.137	1.131	1.125	1.119	1.114	1.108	1.102
8	1.097	1.092	1.086	1.081	1.076	1.071	1.066	1.060	1.056	1.051
9	1.046	1.041	1.036	1.032	1.027	1.022	1.018	1.013	1.009	1.004
10	1.000	0.9957	0.9914	0.9872	0.9830	0.9788	0.9747	0.9706	0.9666	0.9626
11	0.9586	0.9547	0.9508	0.9469	0.9431	0.9393	0.9355	0.9318	0.9281	0.9245
12	0.9208	0.9172	0.9136	0.9101	0.9066	0.9031	0.8996	0.8962	0.8928	0.8894
13	0.8861	0.8827	0.8794	0.8761	0.8729	0.8697	0.8665	0.8633	0.8601	0.8570
14	0.8539	0.8508	0.8477	0.8447	0.8416	0.8386	0.8356	0.8327	0.8297	0.8268
15	0.8239	0.8210	0.8182	0.8153	0.8125	0.8097	0.8069	0.8041	0.8013	0.7986
16	0.7959	0.7932	0.7905	0.7878	0.7852	0.7825	0.7799	0.7773	0.7747	0.7721
17	0.7696	0.7670	0.7645	0.7620	0.7595	0.7570	0.7545	0.7520	0.7496	0.7471
18	0.7447	0.7423	0.7399	0.7375	0.7352	0.7328	0.7305	0.7282	0.7258	0.7235
19	0.7212	0.7190	0.7167	0.7144	0.7122	0.7100	0.7077	0.7055	0.7033	0.7011
20	0.6990	0.6968	0.6946	0.6925	0.6904	0.6882	0.6861	0.6840	0.6819	0.6799
21	0.6778	0.6757	0.6737	0.6716	0.6696	0.6676	0.6655	0.6635	0.6615	0.6596
22	0.6576	0.6556	0.6536	0.6517	0.6498	0.6478	0.6459	0.6440	0.6421	0.6402
23	0.6383	0.6364	0.6345	0.6326	0.6308	0.6289	0.6271	0.6253	0.6234	0.6216
24	0.6198	0.6180	0.6162	0.6144	0.6126	0.6108	0.6091	0.6073	0.6055	0.6038
25	0.6021	0.6003	0.5986	0.5969	0.5952	0.5935	0.5918	0.5901	0.5884	0.5867
26	0.5850	0.5834	0.5817	0.5800	0.5784	0.5766	0.5751	0.5735	0.5719	0.5702
27	0.5686	0.5670	0.5654	0.5638	0.5622	0.5607	0.5591	0.5575	0.5560	0.5544
28	0.5528	0.5513	0.5498	0.5482	0.5467	0.5452	0.5436	0.5421	0.5406	0.5391
29	0.5376	0.5361	0.5346	0.5331	0.5317	0.5302	0.5287	0.5272	0.5258	0.5243
30	0.5229	0.5214	0.5200	0.5186	0.5171	0.5157	0.5143	0.5129	0.5114	0.5100
31	0.5086	0.5072	0.5058	0.5045	0.5031	0.5017	0.5003	0.4989	0.4976	0.4962
32	0.4949	0.4935	0.4921	0.4908	0.4895	0.4881	0.4868	0.4855	0.4841	0.4828
33	0.4815	0.4802	0.4789	0.4776	0.4763	0.4750	0.4737	0.4724	0.4711	0.4698
34	0.4685	0.4672	0.4660	0.4647	0.4634	0.4622	0.4609	0.4597	0.4584	0.4572
35	0.4559	0.4547	0.4535	0.4522	0.4510	0.4498	0.4486	0.4473	0.4461	0.4449

(Continued)

TABLE 6.3 **Transmittance–Absorbance Conversion Table (*Continued*)**

% Transmittance	0.0	0.1	0.2	0.3	0.4	0.5	0.6	0.7	0.8	0.9
36	0.4437	0.4425	0.4413	0.4401	0.4389	0.4377	0.4365	0.4353	0.4342	0.4330
37	0.4318	0.4306	0.4295	0.4283	0.4271	0.4260	0.4248	0.4237	0.4225	0.4214
38	0.4202	0.4191	0.4179	0.4168	0.4157	0.4145	0.4134	0.4123	0.4112	0.4101
39	0.4089	0.4078	0.4067	0.4056	0.4045	0.4034	0.4023	0.4012	0.4001	0.3989
40	0.3979	0.3969	0.3958	0.3947	0.3936	0.3925	0.3915	0.3904	0.3893	0.3883
41	0.3872	0.3862	0.3851	0.3840	0.3830	0.3820	0.3809	0.3799	0.3788	0.3778
42	0.3768	0.3757	0.3747	0.3737	0.3726	0.3716	0.3706	0.3696	0.3686	0.3675
43	0.3665	0.3655	0.3645	0.3635	0.3625	0.3615	0.3605	0.3595	0.3585	0.3575
44	0.3565	0.3556	0.3546	0.3536	0.3526	0.3516	0.3507	0.3497	0.3487	0.3478
45	0.3468	0.3458	0.3449	0.3439	0.3429	0.3420	0.3410	0.3401	0.3391	0.3382
46	0.3372	0.3363	0.3354	0.3344	0.3335	0.3325	0.3316	0.3307	0.3298	0.3288
47	0.3279	0.3270	0.3261	0.3251	0.3242	0.3233	0.3224	0.3215	0.3206	0.3197
48	0.3188	0.3179	0.3170	0.3161	0.3152	0.3143	0.3134	0.3125	0.3116	0.3107
49	0.3098	0.3089	0.3080	0.3072	0.3063	0.3054	0.3045	0.3036	0.3028	0.3019
50	0.3010	0.3002	0.2993	0.2984	0.2976	0.2967	0.2958	0.2950	0.2941	0.2933
51	0.2924	0.2916	0.2907	0.2899	0.2890	0.2882	0.2874	0.2865	0.2857	0.2848
52	0.2840	0.2832	0.2823	0.2815	0.2807	0.2798	0.2790	0.2782	0.2774	0.2765
53	0.2757	0.2749	0.2741	0.2733	0.2725	0.2716	0.2708	0.2700	0.2692	0.2684
54	0.2676	0.2668	0.2660	0.2652	0.2644	0.2636	0.2628	0.2620	0.2612	0.2604
55	0.2596	0.2588	0.2581	0.2573	0.2565	0.2557	0.2549	0.2541	0.2534	0.2526
56	0.2518	0.2510	0.2503	0.2495	0.2487	0.2480	0.2472	0.2464	0.2457	0.2449
57	0.2441	0.2434	0.2426	0.2418	0.2411	0.2403	0.2396	0.2388	0.2381	0.2373
58	0.2366	0.2358	0.2351	0.2343	0.2336	0.2328	0.2321	0.2314	0.2306	0.2299
59	0.2291	0.2284	0.2277	0.2269	0.2262	0.2255	0.2248	0.2240	0.2233	0.2226
60	0.2218	0.2211	0.2204	0.2197	0.2190	0.2182	0.2175	0.2168	0.2161	0.2154
61	0.2147	0.2140	0.2132	0.2125	0.2118	0.2111	0.2104	0.2097	0.2090	0.2083
62	0.2076	0.2069	0.2062	0.2055	0.2048	0.2041	0.2034	0.2027	0.2020	0.2013
63	0.2007	0.2000	0.1993	0.1986	0.1979	0.1972	0.1965	0.1959	0.1952	0.1945
64	0.1938	0.1931	0.1925	0.1918	0.1911	0.1904	0.1898	0.1891	0.1884	0.1878
65	0.1871	0.1864	0.1858	0.1851	0.1844	0.1838	0.1831	0.1824	0.1818	0.1811
66	0.1805	0.1798	0.1791	0.1785	0.1778	0.1772	0.1765	0.1759	0.1752	0.1746
67	0.1739	0.1733	0.1726	0.1720	0.1713	0.1707	0.1701	0.1694	0.1688	0.1681
68	0.1675	0.1669	0.1662	0.1656	0.1649	0.1643	0.1637	0.1630	0.1624	0.1618
69	0.1612	0.1605	0.1599	0.1593	0.1586	0.1580	0.1574	0.1568	0.1561	0.1555
70	0.1549	0.1543	0.1537	0.1530	0.1524	0.1518	0.1512	0.1506	0.1500	0.1494
71	0.1487	0.1481	0.1475	0.1469	0.1463	0.1457	0.1451	0.1445	0.1439	0.1433
72	0.1427	0.1421	0.1415	0.1409	0.1403	0.1397	0.1391	0.1385	0.1379	0.1373
73	0.1367	0.1361	0.1355	0.1349	0.1343	0.1337	0.1331	0.1325	0.1319	0.1314
74	0.1308	0.1302	0.1296	0.1290	0.1284	0.1278	0.1273	0.1267	0.1261	0.1255
75	0.1249	0.1244	0.1238	0.1232	0.1226	0.1221	0.1215	0.1209	0.1203	0.1198
76	0.1192	0.1186	0.1180	0.1175	0.1169	0.1163	0.1158	0.1152	0.1146	0.1141
77	0.1135	0.1129	0.1124	0.1118	0.1113	0.1107	0.1101	0.1096	0.1090	0.1085
78	0.1079	0.1073	0.1068	0.1062	0.1057	0.1051	0.1046	0.1040	0.1035	0.1029
79	0.1024	0.1018	0.1013	0.1007	0.1002	0.0996	0.0991	0.0985	0.0980	0.0975
80	0.0969	0.0964	0.0958	0.0953	0.0947	0.0942	0.0937	0.0931	0.0926	0.0921
81	0.0915	0.0910	0.0904	0.0899	0.0894	0.0888	0.0883	0.0878	0.0872	0.0867
82	0.0862	0.0857	0.0851	0.0846	0.0841	0.0835	0.0830	0.0825	0.0820	0.0814
83	0.0809	0.0804	0.0799	0.0794	0.0788	0.0783	0.0778	0.0773	0.0768	0.0762
84	0.0757	0.0752	0.0747	0.0742	0.0737	0.0731	0.0726	0.0721	0.0716	0.0711
85	0.0706	0.0701	0.0696	0.0691	0.0685	0.0680	0.0675	0.0670	0.0665	0.0660

TABLE 6.3 Transmittance–Absorbance Conversion Table (*Continued*)

% Trans-mittance	0.0	0.1	0.2	0.3	0.4	0.5	0.6	0.7	0.8	0.9
86	0.0655	0.0650	0.0645	0.0640	0.0635	0.0630	0.0625	0.0620	0.0615	0.0610
87	0.0605	0.0600	0.0595	0.0590	0.0585	0.0580	0.0575	0.0570	0.0565	0.0560
88	0.0555	0.0550	0.0545	0.0540	0.0535	0.0531	0.0526	0.0521	0.0516	0.0511
89	0.0506	0.0501	0.0496	0.0491	0.0487	0.0482	0.0477	0.0472	0.0467	0.0462
90	0.0458	0.0453	0.0448	0.0443	0.0438	0.0434	0.0429	0.0424	0.0419	0.0414
91	0.0410	0.0405	0.0400	0.0395	0.0391	0.0386	0.0381	0.0376	0.0372	0.0367
92	0.0362	0.0357	0.0353	0.0348	0.0343	0.0339	0.0334	0.0329	0.0325	0.0320
93	0.0315	0.0311	0.0306	0.0301	0.0297	0.0292	0.0287	0.0283	0.0278	0.0273
94	0.0269	0.0264	0.0259	0.0255	0.0250	0.0246	0.0241	0.0237	0.0232	0.0227
95	0.0223	0.0218	0.0214	0.0209	0.0205	0.0200	0.0195	0.0191	0.0186	0.0182
96	0.0177	0.0173	0.0168	0.0164	0.0159	0.0155	0.0150	0.0146	0.0141	0.0137
97	0.0132	0.0128	0.0123	0.0119	0.0114	0.0110	0.0106	0.0101	0.0097	0.0092
98	0.0088	0.0083	0.0079	0.0074	0.0070	0.0066	0.0061	0.0057	0.0052	0.0048
99	0.0044	0.0039	0.0035	0.0031	0.0026	0.0022	0.0017	0.0013	0.0009	0.0004

Source: *From Meites,* Handbook of Analytical Chemistry, *1963, McGraw-Hill; by permission.*

TABLE 6.4 Standard Spectral Transmittance Values

The data given below refer to a solution containing 0.0400 $g \cdot L^{-1}$ *of* K_2CrO_4 *in 0.05*M *aqueous KOH in a cell having a thickness of 10.0 mm, and at a temperature of 25°C. See Haupt,* J. Res. Natl. Bur. Stds. ***48****:414 (1952). At 375 nm the temperature coefficients of the transmittance and absorbance are* $\Delta T/\Delta t = 0.0022$ *and 0.000 93* deg^{-1}*, respectively.*

Wavelength, nm	%*T*	*A*	Wavelength, nm	%*T*	*A*
220	35.8	0.446	340	48.3	0.316
225	60.1	0.221	345	37.3	0.428
230	67.4	0.171	350	27.6	0.559
235	61.6	0.210	355	19.9	0.701
240	50.7	0.295	360	14.8	0.830
245	40.2	0.396	365	11.6	0.935_5
250	31.9	0.496	370	10.3	0.987
255	26.8	0.572	375	10.2	0.991
260	23.3	0.653	380	11.7	0.932
265	20.2	0.696	385	15.0	0.824
270	18.0	0.745	390	20.2	0.695
275	17.5	0.757	395	29.4	0.532
280	19.4	0.712	400	40.1	0.396
285	25.7	0.590	410	63.2	0.199
290	37.3	0.428	420	75.1	0.124
295	53.3	0.273	430	82.4	0.084
300	70.9	0.149	440	88.2	0.054_5
305	83.4	0.079	450	92.7	0.033
310	89.5	0.048	460	96.0	0.018
315	90.0	0.046	470	98.0	0.009
320	86.4	0.063_5	480	99.1	0.004
325	80.4	0.095	490	99.7	0.001
330	71.0	0.149	500	100	0.000
335	60.0	0.222			

6.1.2.4 Specific Absorptivity. When absorbance is defined as

$$A = abC \tag{6.5}$$

the proportionality constant a is known as the absorptivity and is given in units $L \cdot g^{-1} \cdot cm^{-1}$. The optical path b is in centimeters and C is the molar concentration.

6.1.2.5 Molar Absorptivity. Molar absorptivity (ϵ), also known as the molar extinction coefficient, is the absorbance of a solution divided by the product of the optical path b in centimeters and the molar concentration C of the absorbing species:

$$A = \epsilon bC \tag{6.6}$$

where ϵ is in units of $L \cdot mol^{-1} \cdot cm^{-1}$.

6.1.3 Fundamental Laws of Spectrophotometry

6.1.3.1 Lambert–Beer Law. Lambert's law states that for parallel, monochromatic radiation that passes through an absorber of constant concentration, the radiant power decreases logarithmically as the path length b increases arithmetically.

Beer's law states that the transmittance of a stable solution is an exponential function of the concentration of the absorbing solute.

If both concentration and thickness are variable, the combined Lambert–Beer law (often known simply as Beer's law) becomes the form stated as Eq. (6.6). A plot of absorbance versus concentration should be a straight line passing through the origin whose slope is ϵb (or ab when C is $g \cdot L^{-1}$). Readout scales are often calibrated to read absorbance as well as transmittance.

6.1.3.2 Linearity and Beer's Law. Deviations from Beer's law fall into three categories: real, instrumental, and chemical. *Real deviations* arise from changes in the refractive index of the analytical system; these changes will be significant only in high-absorbance differential measurements.

6.1.3.2.1 Instrumental Deviations. Instrumental deviations arise primarily from the band pass (band width) of filters or monochromators. Beer's law assumed monochromatic radiation. Deviations from Beer's law are most serious for wide band passes and narrow absorption bands. What is important is the ratio between the spectral slit width and the band width of the absorption band.

6.1.3.2.2 Chemical Deviations. Chemical deviations are caused by shifts in the position of a chemical or physical equilibrium involving the absorbing species. Generally the pH and ionic strength of the system must be kept constant. For weakly ionized acids the pH should be adjusted to at least three units more (for the anion form) or three units less (for the acid form) than the pK_a value of the monoprotic acid. Alternatively, the wavelength corresponding to an *isosbestic* point can be used. An isosbestic point is a wavelength at which the molar absorptivity is the same for two materials that are interconvertible. When the absorbing species is a complex ion, the concentration of the free (excess) ligand must be constant and preferably in 100-fold excess.

6.1.3.3 Calibration Curves. Calibration curves, if known to be linear over the concentration range to be used, can be prepared in several ways:

1. Use a reagent blank to set zero absorbance and one standard (preferably near the concentration of the sample) to produce a finite absorbance.
2. Use a reagent blank to set zero absorbance and a known concentration factor (the slope of a previously determined calibration curve of absorbance versus concentration).
3. Use two standards whose concentrations bracket that of the sample(s).
4. Use one standard and a known concentration (slope) factor.

When establishing a linear calibration curve from a number of absorbance–concentration data points, a certain degree of data scatter usually occurs. It is necessary to perform a linear regression analysis on the standards to determine the calibration line that gives the best fit from the degree of correlation among the data. For this work a computer is invaluable. For most colorimetric analyses linear calibration curves are plotted for quantitation.

For *nonlinear calibration curves*, a stored program is needed. Values of absorbance–concentration data are stored. When point-to-point interpolation between entered standards is needed, the program generates a series of small straight lines that approximate the nonlinear curve.

6.1.4 Special Spectrophotometric Methodology

In the ordinary photometric method both ends of the transmittance scale are set precisely at 0% (with an occluder) and 100% T (with pure solvent in the beam). Under ordinary photometric methods, the maximum value of P_0 is limited by the length of the potentiometer slide-wire, since one end corresponds to zero radiant power and the other end to the full power at zero concentration, or P_0. This is an artificial requirement. The transmittance scale can also be calibrated by using two reference solutions that contain the absorbing species in different concentrations. The only condition is that one reference absorber must transmit more radiant energy than the sample to be measured and the other reference absorber must transmit less radiant energy than the sample. From these possible alterations of the ordinary method, three differential or scale-expansion techniques arise that can be used to increase precision. These techniques, summarized in Table 6.5 and displayed in Fig. 6.2, are applicable to direct methods, in which the absorbance increases with increasing concentration of the substance being determined.

For the modifications required in indirect methods see Table 6.6. In indirect methods the absorbance usually decreases with increasing concentration of the substance being determined.

6.1.4.1 Cell Corrections for the Ultimate Precision Method. The following procedure[1] is designed to minimize the errors that would otherwise result in the ultimate precision method from the fact that the optical path lengths in even "matched" cells may differ to an extent comparable with the attainable precision of the measurements. Cylindrical cells should never be used because the path length varies from one portion of the beam to another.

Prepare two reference solutions, one more concentrated and the other more dilute than the unknowns or standards to be used. Use these to make the 0% and 100% transmittance readings. Make both settings with the same cells throughout the measurements.

Prepare two standards having concentrations c_1 and c_2 between those of the previously prepared reference solutions, and such that the transmittance of c_2 at the wavelength of interest is roughly 0.30 transmittance unit (30% T) greater than that of c_1. Measure the percent transmittance of each solution in cells 1, 2, 3, etc.

For each cell, plot the transmittances of the two standard solutions obtained with that cell against the transmittances obtained in cell 1, which is henceforth to be considered as the reference cell. In an analysis subsequently carried out by measuring the transmittance of an unknown solution in cell 2, use the cell-correction plot to find, by vertical projection, the transmittance that would have been obtained in cell 1. If only a single unknown is to be analyzed, use cell 1.

6.1.5 Polycomponent Systems

Simultaneous spectrophotometric methods for the analysis of polycomponent systems are outlined in Table 6.7. Computer software programs can handle multicomponent mixtures and deviations from Beer's law. A fast algorithm for the resolution of overlapping spectral bands has been devised.[2]

---

[1] C. V. Banks, P. G. Grimes, and R. I. Bystroff, *Anal. Chim. Acta* **15**:367 (1956); C. M. Crawford, *Anal. Chem.* **31**:343 (1959).
[2] R. A. Caruana et al., *Anal. Chem.*, **58**:1162 (1986).

TABLE 6.5 Precision Photometric Measurements: Direct Methods

Requirements	Method: Transmittance-ratio method	Method: Trace-analysis method	Method: Ultimate-precision method
Transmittance range of unknown, % (vs. pure solvent)	0–50	50–100	0–100
0% T setting	Phototube in darkness.	Reference solution more concentrated than the most concentrated unknown or standard. A for this solution, measured against pure solvent, will vary depending on the sensitivity desired.	Reference solution more concentrated than the most concentrated unknown or standard.
100% T setting	Reference solution more dilute than the most dilute solution to be measured. For optimum precision in the 0–1% T range, use a reference solution for which A, measured against pure solvent, should preferably approach 2.0. Wider than usual slit widths will be necessary.	Pure solvent or blank.	Reference solution more dilute than the most dilute unknown or standard. The best accuracy is obtained when A for the unknown, measured against pure solvent, is about 0.43 (37% T), and when this is bracketed as closely as possible by the two reference solutions.
Calibration plot	Recommended, since Beer's law is not necessarily followed.	Recommended, since Beer's law is not usually followed.	Necessary, since Beer's law is not usually followed. If Beer's law is followed, the necessity is recalibration whenever the reference solution is changed may be avoided by plotting 10^{-abC} against instrument reading; see Ref. 1 for details.
Features	Increased precision, comparable to classical titrimetric and gravimetric methods for more concentrated solutions.	Increased precision for more dilute solutions than can be obtained with conventional photometric techniques. With very dilute solutions, however, relatively high errors are still to be expected.	At comparable concentrations, gives better precision than either the transmittance-ratio or the trace-analysis method. See text for cell corrections.
References	2, 3, 4	5	5

[1] C. V. Banks, J. L. Spooner, and J. W. O'Laughlin, *Anal. Chem.* **28**:1894 (1956).
[2] R. Bastian, *Anal. Chem.* **21**:972 (1949).
[3] C. F. Hiskey, *Anal. Chem.* **21**:1440 (1949).
[4] R. Bastian, R. Weberling, and F. Palilla, *Anal. Chem.* **22**:160 (1950).
[5] C. N. Reilley and C. M. Crawford, *Anal. Chem.* **27**:716 (1955).

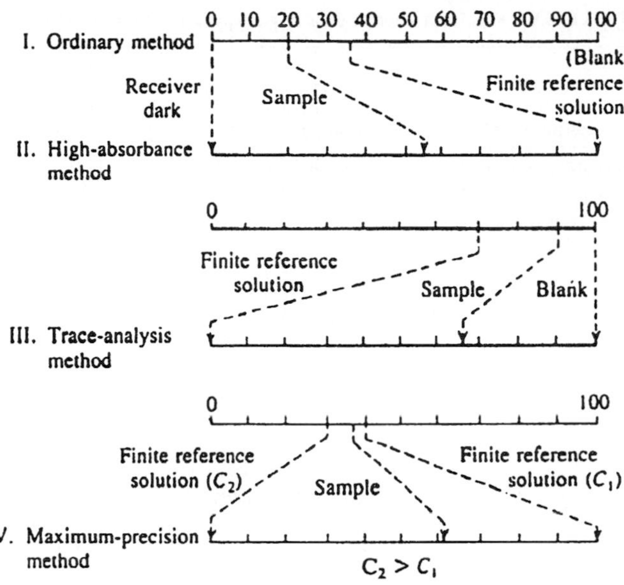

FIGURE 6.2 Differential (or expanded-scale) spectrophotometry.

6.1.6 Error and Its Minimization in Spectrophotometric Analysis

6.1.6.1 Photometric Accuracy. It is assumed herein that the error introduced into the final result by the chemical preparative treatment is negligible in comparison with the instrumental error.[3] The ultimate precision of a photometric measurement is determined by instrumental noise—that is, the statistical fluctuations of the signal that reaches the detector. Two extremes will be considered: (1) thermal-noise limited and (2) shot-noise limited. The former case arises in older instruments (with photovoltaic detectors, for example) with higher radiant powers. The latter is the more usual situation in modern UV-visible spectrophotometers with low radiant powers incident on the detector.

The conditions for maximum photometric accuracy in the conventional photometric method (case 1) can be deduced by equating the second derivative of the Beer's law expression

$$A = \log(P_0/P) = abC \tag{6.7}$$

to zero. If Beer's law is obeyed, then the relative error in concentration $\Delta C/C$ per unit photometric error ΔP is given by

$$\frac{\Delta C/C}{\Delta P} = -\frac{1}{2.303A(10^{-A})} \tag{6.8}$$

The left-hand side of this expression has a minimum value when $A = 0.434$ (36.8% T); this minimum value is

$$\left(\frac{\Delta C/C}{\Delta P}\right)_{\min} = -2.72 \tag{6.9}$$

Thus the relative error in concentration resulting from a 1% photometric error is 2.72% at 36.8% transmittance. The relative error between 20% and 60% T is not much larger than 2.7% per 1%

[3] C. F. Hiskey, *Anal. Chem.* **21**:1440 (1949).

TABLE 6.6 Precision Photometric Measurements: Indirect Methods

Requirements	Method		
	Expanded-scale method	Trace-analysis method	Transmittance-ratio indirect method
Transmittance range of unknown, %	0–50 (low concentration of unknown).	0–25 (traces of unknown).	10–50 (low concentrations of unknown).
0% T setting	Phototube in darkness.	Phototube in darkness.	Phototube in darkness.
100% T setting	Reference solution having a higher % T than any unknown or standard, prepared by bleaching the chromogenic reagent with a high concentration of the desired constituent than will be present in any unknown or standard.	Unknown or standard diluted with chromogenic reagent. A different slit width will be necessary for each standard or unknown.	As in trace-analysis method.
Unknown or standard % T reading		Pure chromogenic reagent.	Reference solution having a lower % T than any unknown or standard, prepared by bleaching the chromogenic reagent with a smaller concentration of the desired constituent than will be present in any standard or unknown.
Calibration plot	Necessary only if Beer's law is not obeyed in the conventional indirect method. % T increases with increasing concentration as in the conventional indirect method.	Recommended since Beer's law may not be followed. % T decreases with increasing concentration as in conventional direct methods.	Recommended since Beer's law may not be followed. % T decreases with increasing concentration as in conventional direct methods.
Features	Increased precision at low concentrations of unknown compared to the conventional indirect method. Avoids curvature in the region where the reaction between the desired constituent and the chromogenic reagent is incomplete (unfavorable equilibrium).	Increased precision for trace-analysis compared to conventional indirect methods, since readings are obtained with solutions having absorbances between 0 and 0.3. Time-consuming because different slit widths are necessary for each standard or unknown.	Increased precision as compared to the trace-analysis method at similar concentrations. Time-consuming because different slit widths are necessary for each standard or unknown.
References	1, 2, 3	3	3

[1] J. J. Lothe, *Anal. Chem.* **27**:1546 (1955).
[2] J. J. Lothe, *Anal. Chem.* **28**:949 (1956).
[3] C. N. Reilley and G. P. Hildebrand, *Anal. Chem.* **31**:1763 (1959).

TABLE 6.7 Simultaneous Spectrophotometric Methods for the Analysis of Polycomponent Systems

Type of measurement	Type of spectrum	Procedure
Absorbance measurement at n wavelengths for n-component system	No component absorbs at λ_{max} for any other component.	Prepare a calibration graph (or measure the absorptivity) for each component at its own λ_{max}. For a mixture, calculate the concentration of each component independently from the A at the appropriate wavelength.
	Only one component absorbs at one λ_{max}; at another λ_{max} there is absorption by two or more components.	For a two-component system, measure A at the λ_{max} where only one component absorbs, and calculate the concentration of that component from Beer's law and the previously measured absorptivity, or from a calibration graph. Find A of that component at the other λ_{max} from Beer's law or a calibration graph, and subtract this from the total A measured at the second λ_{max} to find A of the second component at its own λ_{max}. When more than two absorbing components are present, this effectively yields an $(n-1)$-component system.*
	Each component absorbs appreciably at λ_{max} for each other.	Measure the absorbance of solutions containing only one of the components at each of the n λ_{max} values for the n components, and for each component calculate the absorptivity at each of the n wavelengths. Adherence to Beer's law is assumed. Measure A of the mixture at each of the n wavelengths. Set up and solve the equations $A_{\lambda 1} = (a_x)_{\lambda 1} bC_x + (a_y)_{\lambda 1} bC_y + \cdots$ and $A_{\lambda 2} = (a_x)_{\lambda 2} bC_x + (a_y)_{\lambda 2} bC_y + \cdots$.
Compensation by reference solution		Compensate for the absorbance of one component by adding the same concentration of it to the reference cell.

* J. J. Lingane and J. W. Collat, *Anal. Chem.* **22**:166 (1950).

photometric error. Hence photometric measurements should be confined to the region of transmittances between about 20% and 60% (absorbances between about 0.2 and 0.7) for minimum error in the conventional photometric method (Fig. 6.3).

The minimum relative concentration error for most modern UV-visible photometers in which shot noise predominates (case 2) is determined in a similar manner. The noise (ΔP or ΔT) is proportional to the square root of the radiant power—that is, $\Delta T = k\sqrt{P}$. Replacement of ΔT by $k\sqrt{P}$ in Eq. (6.8) and proceeding as before yield

$$\frac{\Delta C}{C} = -0.434\,k\left(\sqrt{P}\log\frac{P_0}{P}\right)^{-1} \tag{6.10}$$

After differentiation with respect to P, the minimum for this function is when

$$\log\frac{P_0}{P} = 2(0.434) = 0.868 = A \qquad \text{or} \qquad T = 0.135 \tag{6.11}$$

For detectors in which shot noise predominates, the region of minimum error extends over a very wide range of absorbance values from 0.3 to 2.0($T = 0.5$ to 0.01). This conclusion is true for single- and double-beam spectrophotometers. Thus more concentrated solutions may be used, which, in

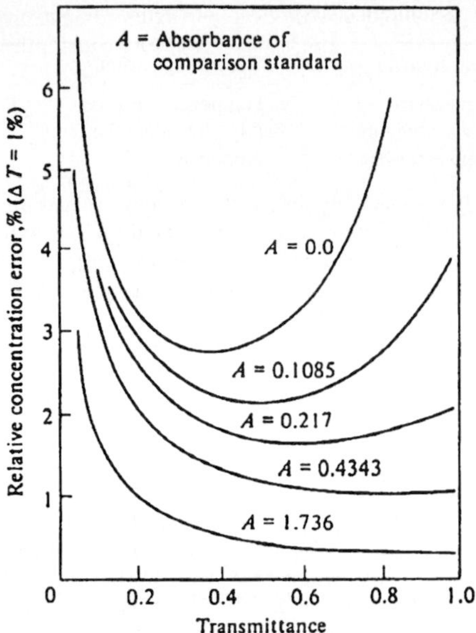

FIGURE 6.3 Plot of the relative concentration error.
(Curve $A = 0.0$ is for a thermal-noise-limited detector.)

turn, reduces errors in solution preparation and cuvette matching. Clean windows and scratch-free cuvette faces are less significant.

6.1.6.1.1 Transmittance-Ratio Method. In the *transmittance-ratio method* (also known as the *high-absorbance method*) the error function is given by

$$\frac{\Delta C}{C} = \frac{0.434\,\Delta T}{(TA)_{\text{sple}} + A_{\text{ref}}} \tag{6.12}$$

where the subscripts sple and ref refer to the sample and reference solutions, respectively. The position of minimum error gradually shifts to 100% T on the transmittance scale as the reference concentration increases; this amounts to comparing the reference and unknown at the same scale setting. To minimize volumetric errors, aliquots should be taken by weight. The precision attainable by the transmittance-ratio method may approach 0.01%.

6.1.6.1.2 Trace-Analysis Method. In the *trace-analysis method* the expression for the relative concentration error becomes

$$\frac{\Delta C}{C} = \frac{0.434(1 - T_{\text{ref}})}{TA}\Delta T \tag{6.13}$$

6.1.6.1.3 Ultimate Precision Method. In the *ultimate precision method* the relative concentration error is given by

$$\frac{\Delta C}{C} = \frac{0.434(T_{100\%\ \text{ref}} - T_{0\%\ \text{ref}})}{TA}\Delta T \tag{6.14}$$

In theory, the precision of this method can be increased by making the difference between the two reference solutions small. This assumes, however, the availability of an instrument with very high sensitivity and stability.

6.1.6.2 Wavelength Calibration. Instruments employed in the ultraviolet-visible regions of the spectrum can be calibrated with respect to wavelength with NIST traceable holmium oxide filters, which are commercially available.

6.1.7 Difference Spectroscopy

Two samples are used, one in the reference beam and the other in the sample beam of a double-beam spectrophotometer. The recordings are usually made with pen range expansion and are the difference in transmittance (or absorbance) of the two samples. Common features in the two spectra cancel. Usually the concentration of absorbing material in the two samples is identical but some solution parameter such as pH is different.

6.2 SPECTROPHOTOMETRIC METHODS OF ORGANIC ANALYSIS

Molecules with two or more isolated chromophores (absorbing groups) absorb light of nearly the same wavelength as does a molecule containing only a single chromophore of a particular type. The intensity of the absorption is proportional to the number of that type of chromophore present in the molecule. Representative chromophores are given in Table 6.8.

The solvent chosen must dissolve the sample, yet be relatively transparent in the spectral region of interest. In order to avoid poor resolution and difficulties in spectrum interpretation, a solvent should not be employed for measurements that are near the wavelength of or shorter than the wavelength of its ultraviolet cutoff, that is, the wavelength at which absorbance for the solvent alone approaches one absorbance unit. Ultraviolet cutoffs for solvents commonly used are given in Table 6.9.

Appreciable interaction among chromophores does not occur unless they are linked directly to each other, or forced into close proximity as a result of molecular stereochemical configuration. Interposition of a single methylene group, or *meta* orientation about an aromatic ring, is sufficient to insulate chromophores almost completely from each other. Certain combinations of functional groups afford chromophoric systems that give rise to characteristic absorption bands.

Sets of empirical rules, often referred to as Woodward's rules or the Woodward–Fieser rules, enable the absorption maxima of dienes (Table 6.10) and enones and dienones (Table 6.11) to be predicted. To the respective base values (absorption wavelength of parent compound) are added the increments for the structural features or substituent groups present. When necessary a solvent correction is also applied (Table 6.12).

Ring substitution on the benzene ring affords shifts to longer wavelengths (Table 6.13) and intensification of the spectrum. With electron-withdrawing substituents, practically no change in the maximum position is observed. The spectra of heteroaromatics are related to their isocyclic analogs, but only in the crudest way. As with benzene, the magnitude of substituent shifts can be estimated, but tautomeric possibilities may invalidate the empirical method.

When electronically complementary groups are situated *para* to each other in disubstituted benzenes, there is a more pronounced shift to a longer wavelength than would be expected from the additive effect due to the extension of the chromophores from the electron donating group through the ring to the electron-withdrawing group. When the *para* groups are not complementary, or when the groups are situated *ortho* or *meta* to each other, disubstituted benzenes show a more or less additive effect of the two substituents on the wavelength maximum. Calculation of the principal band of selected substituted benzenes is illustrated in Table 6.14.

6.3 SPECTROPHOTOMETRIC METHODS FOR DETERMINATION OF THE ELEMENTS

Selected methods for the determination of metals is given in Table 6.15 and for the determination of nonmetals in Table 6.16.

TABLE 6.8 Electronic Absorption Bands for Representative Chromophores

Chromophore	System	λ_{max}	ϵ_{max}
Acetylide	—C≡C—	175–180	6 000
Aldehyde	—CHO	210	strong
		280–300	11–18
Amine	$—NH_2$	195	2 800
Azido	>C=N—	190	5 000
Azo	—N=N—	285–400	3–25
Bromide	—Br	208	300
Carbonyl	>C=O	195	1 000
		270–285	18–30
Carboxyl	—COOH	200–210	50–70
Disulfide	—S—S—	194	5 500
		255	400
Ester	—COOR	205	50
Ether	—O—	185	1 000
Ethylene	—C=C—	190	8 000
Iodide	—I	260	400
Nitrate	$—ONO_2$	270 (shoulder)	12
Nitrile	—C≡N	160	. . .
Nitrite	—ONO	220–230	1 000–2 000
		300–400	10
Nitro	$—NO_2$	210	strong
Nitroso	—NO	302	100
Oxime	—NOH	190	5 000
Sulfone	$—SO_2—$	180	. . .
Sulfoxide	>S=O	210	1 500
Thiocarbonyl	>C=S	205	strong
Thioether	—S—	194	4 600
		215	1 600
Thiol	—SH	195	1 400
	$—(C=C)_2—$ (acyclic)	210–230	21 000
	$—(C=C)_3—$	260	35 000
	$—(C=C)_4—$	300	52 000
	$—(C=C)_5—$	330	118 000
	$—(C=C)_2—$ (alicyclic)	230–260	3 000–8 000
	C=C—C≡C	219	6 500
	C=C—C=N	220	23 000
	C=C—C=O	210–250	10 000–20 000
		300–350	weak
	$C=C—NO_2$	229	9 500
Benzene		184	46 700
		204	6 900
		255	170
Diphenyl		246	20 000
Naphthalene		222	112 000
		275	5 600
		312	175
Anthracene		252	199 000

TABLE 6.8 Electronic Absorption Bands for Representative Chromophores (*Continued*)

Chromophore	System	λ_{max}	ϵ_{max}
Anthracene		375	7 900
Phenanthrene		251	66 000
		292	14 000
Naphthacene		272	180 000
		473	12 500
Pentacene		310	300 000
		585	12 000
Pyridine		174	80 000
		195	6 000
		257	1 700
Quinoline		227	37 000
		270	3 600
		314	2 750
Isoquinoline		218	80 000
		266	4 000
		317	3 500

Source: J. A. Dean, ed., *Lange's Handbook of Chemistry*, 14th ed., McGraw-Hill, New York, 1992.

TABLE 6.9 Ultraviolet Cutoffs of Spectrograde Solvents

(Absorbance of 1.00 in a 10.0 mm cell vs. distilled water.)

Solvent	Wavelength, nm	Solvent	Wavelength, nm
Acetic acid	260	Hexadecane	200
Acetone	330	Hexane	210
Acetonitrile	190	Isobutyl alcohol	230
Benzene	280	Methanol	210
1-Butanol	210	2-Methoxyethanol	210
2-Butanol	260	Methylcyclohexane	210
Butyl acetate	254	Methylene chloride	235
Carbon disulfide	380	Methyl ethyl ketone	330
Carbon tetrachloride	265	Methyl isobutyl ketone	335
1-Chlorobutane	220	2-Methyl-1-propanol	230
Chloroform (stabilized with ethanol)	245	*N*-Methylpyrrolidone	285
		Nitromethane	380
Cyclohexane	210	Pentane	210
1,2-Dichloroethane	226	Pentyl acetate	212
Diethyl ether	218	1-Propanol	210
1,2-Dimethoxyethane	240	2-Propanol	210
N,N-Dimethylacetamide	268	Pyridine	330
N,N-Dimethylformamide	270	Tetrachloroethylene (stabilized with thymol)	290
Dimethylsulfoxide	265		
1,4-Dioxane	215	Tetrahydrofuran	220
Ethanol	210	Toluene	286
2-Ethoxyethanol	210	1,1,2-Trichloro-1,2,2-trifluoroethane	231
Ethyl acetate	255		
Ethylene chloride	228	2,2,4-Trimethylpentane	215
Glycerol	207	*o*-Xylene	290
Heptane	197	Water	191

Source: J. A. Dean, ed., *Lange's Handbook of Chemistry*, 14th ed., McGraw-Hill, New York, 1992.

TABLE 6.10 Absorption Wavelength of Dienes

Heteroannular and acyclic dienes usually display molar absorptivities in the 8 000 to 20 000 range, whereas homoannular dienes are in the 5 000 to 8 000 range.

Poor correlations are obtained for cross-conjugated polyene systems such as

$$-C{=}C\begin{matrix}\diagup C{=}C- \\ \diagdown C{=}C-\end{matrix}$$

The correlations presented here are sometimes referred to as Woodward's rules or the Woodward–Fieser rules.

Base value for heteroannular or open chain diene, nm	214
Base value for homoannular diene, nm	253
Increment (in nm) for	
Double bond extending conjugation	30
Alkyl substituent or ring residue	5
Exocyclic double bond	5
Polar groupings	
-*O*-acyl	0
-*O*-alkyl	6
-*S*-alkyl	30
-Cl, -Br	5
-*N*(alkyl)$_2$	60
Solvent correction (see Table 6.12)	——
Calculated wavelength =	total

Source: J. A. Dean, ed., *Lange's Handbook of Chemistry*, 14th ed., McGraw-Hill, New York, 1992.

6.4 PHOTOMETRIC TITRATIONS

The end point of a photometric titration is located by interpreting the changes in the concentration of an absorbing species, and hence in the absorbance of the solution, during the course of the titration. Usually the absorbance is plotted against the volume of titrant added to obtain the titration curve (Fig. 6.4). If the titrant is much more concentrated than the solution being titrated, the titration curve will consist of two straight lines (or, in some cases, a straight line and a curve) that intersect at the end point. Volume change is seldom negligible, and straight lines are obtained only if there is correction for dilution. This is done simply by multiplying the measured absorbance by the factor

$$A_{\text{corrected}} = \frac{V+v}{V} A_{\text{observed}} \tag{6.15}$$

where V is the initial volume and v is the volume for titrant added up to any point. Use of a microsyringe and a relatively concentrated titrant is desirable.

The absorbing species may be one of the substances taking part in or formed in the titration reaction (such cases are indicated by the appearance of "None" in the fifth column of Tables 6.17 and 6.18), it may be the product of the reaction between one of these substances and an added indicator, or it may be a free indicator (denoted by the appearance of "FI" in the seventh column of these tables. Advantages of photometric titrations, as compared to ordinary photometric measurements, are as follows:

1. Only a single absorber needs to be present among the reactant, titrant, or products. This extends photometric methods to a large number of nonabsorbing constituents. When self-indicating

TABLE 6.11 Absorption Wavelength of Enones and Dienones

$$O{=}C{-}\overset{\alpha}{C}{=}C\begin{smallmatrix}\beta\\ \beta\end{smallmatrix} \qquad O{=}C{-}\overset{\alpha}{C}{=}\overset{\beta}{C}{-}\overset{\gamma}{C}{=}C\begin{smallmatrix}\delta\\ \delta\end{smallmatrix}$$

Base values, nm	
Acyclic α,β-unsaturated ketones	215
Acyclic α,β-unsaturated aldehyde	210
Six-membered cyclic α,β-unsaturated ketones	215
Five-membered cyclic α,β-unsaturated ketones	214
α,β-Unsaturated carboxylic acids and esters	195
Increments (in nm) for	
Double bond extending conjugation	
Heteroannular	30
Homoannular	69
Alkyl group or ring residue	
α	10
β	12
γ, δ	18
Polar groups	
—OH	
α	35
β	30
γ	50
—O—CO—CH_3 and —O—CO—C_6H_5: α, β, γ, δ	6
—OCH_3	
α	35
β	30
γ	17
δ	31
—S—alkyl, β	85
—Cl	
α	15
β	12
—Br	
α	25
β	30
—N(alkyl)$_2$, β	95
Exocyclic double bond	5
Solvent correction (see Table 6.12)	—
Calculated wavelength =	total

Source: J. A. Dean, ed., *Lange's Handbook of Chemistry*, 14th ed., McGraw-Hill, New York, 1992.

systems are lacking, an indicator can be deliberately added, but in a relatively large amount to provide a sufficient linear segment on the titration curve beyond the equivalence point.

2. For reactions that are incomplete in the vicinity of the end point, extrapolation of the two linear segments of the titration curve establishes the intersection and end-point volume.
3. Precision of 0.5% or better is attainable because data are pooled in constructing the segments of the titration curve.

TABLE 6.12 Solvent Correction for Ultraviolet-Visible Spectroscopy

Solvent	Correction, nm
Chloroform	+1
Cyclohexane	
Diethyl ether	+11
1,4-Dioxane	+5
Ethanol	0
Hexane	+11
Methanol	0
Water	−8

Source: J. A. Dean, ed., *Lange's Handbook of Chemistry*, McGraw-Hill, New York, 1992.

Titration cells are readily designed for existing filter photometers of spectrophotometers. For titrations in the visible region, one needs a light source, a series of narrow-band-pass filters, a titration vessel (which can be an ordinary beaker firmly clamped in place), a photodetector, and a burette or microsyringe. Everything is housed in a light-tight compartment. By the use of Vycor beakers and a spectrophotometer, titrations may be conducted in the ultraviolet region. Two fiber-optic light pipes can be located facing each other across an area below the burette or syringe tip

TABLE 6.13 Primary Bands of Substituted Benzene and Heteroaromatics in Methanol

(Base value: 203.5 nm.)

Substituent	Wavelength shift, nm	Substituent	Wavelength shift, nm
$—CH_3$	3.0	—COOH	25.5
$—CH{=}CH_2$	44.5	$—COO^-$	20.5
—C≡CH	44	—CN	20.5
$—C_6H_5$	48	$—NH_2$	26.5
—F	0	$—NH_3^+$	−0.5
—Cl	6.0	$—N(CH_3)_2$	47.0
—Br	6.5	$—NH—CO—CH_3$	38.5
—I	3.5	$—NO_2$	57
—OH	7.0	—SH	32
$—O^-$	31.5	$—SO—C_6H_5$	28
$—OCH_3$	13.5	$—SO_2CH_3$	13
$—OC_6H_5$	51.5	$—SO_2NH_2$	14.0
—CHO	46.0	$—CH{=}CH—C_6H_5$	
$—CO—CH_3$	42.0	*cis*	79
$—CO—C_6H_5$	48	*trans*	92.0
		—CH=CH—COOH, *trans*	69.5

Heteroaromatic	Base value, nm	Heteroaromatic	Base value, nm
Furan	200	Pyridine	257
Pyrazine	257	Pyrimidine	ca. 235
Pyrazole	214	Pyrrole	209
Pyridazine	ca. 240	Thiophene	231

Source: J. A. Dean, ed., *Lange's Handbook of Chemistry*, 14th ed., McGraw-Hill, New York, 1992.

TABLE 6.14 Wavelength Calculation of the Principal Band of Substituted Benzene Derivatives in Ethanol

Base value of parent chromophore, nm	
C_6H_5COOH or C_6H_5COO—alkyl	230
C_6H_5—CO—alkyl (or aryl)	246
C_6H_5CHO	250
Increment (in nm) for each substituent on phenyl ring	
—Alkyl or ring residue	
o-, *m*-	3
p-	10
—OH and —O—alkyl	
o-, *m*-	7
p-	25
$—O^-$	
o-	11
m-	20
p-	78*
—Cl	
o-, *m*-	0
p-	10
—Br	
o-, *m*-	2
p-	15
$—NH_2$	
o-, *m*-	13
p-	58
$—NHCO—CH_3$	
o-, *m*-	20
p-	45
$—NHCH_3$	
p-	73
$—N(CH_3)_2$	
o-, *m*-	20
p-	85

* Value may be decreased markedly by steric hindrance to coplanarity.
Source: J. A. Dean, ed., *Lange's Handbook of Chemistry*, 14th ed., McGraw-Hill, New York, 1992.

in the titrating vessel. Provision for magnetic stirring or some type of overhead stirrer is desirable. Commercial photometric titrators are available. For discussions of the technique see Refs. 4 and 5.

Table 6.17 describes a number of selected methods of photometric titration of inorganic substances. Titrations of organic substances are listed in Table 6.18. The following special symbols and abbreviations are employed in these tables:

A = Automatic titration
Aliz C = 1,2-Dihydroxyanthraquinone-3-ylmethylamine-*N*,*N*-diacetic acid
Arsenazo = 2-(1,8-Dihydroxy-3,6-disulfo-1-naphthylazo)benzenearsonic acid
Calcon = Sodium 1-(2-hydroxy-1-naphthylazo)-2-naphthol-4-sulfonate
EDDHA = Ethylenediamine-*N*,*N*′-di(α-*o*-hydroxyphenylacetic acid)
F = Filter with transmission band centered around wavelength given

[4] R. F. Goddu and D. N. Hume, *Anal. Chem.* **26**:1740 (1954).
[5] J. B. Headridge, *Photometric Titrations*, Pergamon, New York, 1961.

TABLE 6.15 Spectrophotometric Methods for the Determination of Metals

(The letter R denotes the reagent.)

Element determined	Procedure	λ, nm	Range, μg · mL^{-1}	Interferences	References
Aluminum	(1) Ammonium aurintricarboxylate (Aluminon) (R): 0.03% R–H_2O, gelatin, pH 4.5 ± 0.5, heat at 100°C for 10 min, cool 25°C for 15 min. Mercaptoacetic acid complexes Fe(III); cupferron–$CHCl_3$ extraction removes Cu, Fe(III), Ti.	525	0.04–0.4	Turbidity: Ag, Bi, Pb, Sn. Colors: Be, Co, Cr(III,VI), Cu, Sc, Ti, Th, V(IV,V), Zr. Complex Al: citrate, fluoride, tartrate.	C. L. Luke, *Anal. Chem.* **24**:1122 (1952); J. A. Scherrer and W. D. Mogerman, *J. Res. Natl. Bur. Stds.* **21**:105 (1938).
	(2) Eriochrome cyanine R (R): 0.075% R–H_2O, pH 6 ± 1, stand 30 min. Cu masked by thiosulfate; Fe (III) masked by mercaptoacetic acid; Ti and V masked by H_2O_2.	535	0.04–0.4	Be, Cr(III), Cu, Fe(III), Mn, Ti, V, Zr.	U. T. Hill, *Anal. Chem.* **28**:1419 (1956); **38**:654 (1966).
	(3) Catechol Violet: can replace Eriochrome cyanine R without change in procedure.	585	0.04–0.4	Same as above.	A. Anton, *Anal. Chem.* **32**:725 (1960); A. D. Wilson and G. A. Sergeant, *Analyst* **88**:109 (1963); W. K. Dugan and A. L. Wilson, *ibid.* **99**:413 (1974).
	(4) 8-Hydroxyquinoline (R): 1% R–$CHCl_3$, pH 5.5 ± 1 or 9.5 ± 1.5.	395	0.1–3	Be, Cd, Ce(IV), Co, Cu, Fe(II,III), Hg(I,II), Ni, Pb, Sb, Sn(II,IV), Th, Ti, U, Zn.	C. H. R. Gentry and L. G. Sherrington, *Analyst* **71**:432 (1946); A. Claassen, L. Bastings, and J. Visser, *Anal. Chim. Acta* **10**:373 (1954).
	(5) Alizarinesulfonic acid (Alizarin Red S) (R): 0.04% R–H_2O, add 100 μg/mL Ca, heat at 60°C for 30 min, pH 4.5.	485	0.1–1.2	B, Be, Bi, Cr(III), Cu, Fe(III), Mo(VI), Sb(III), V(V), W(VI). Turbidities: Mn, Sn(IV), Ti.	A. P. Musakin, *Z. Anal. Chem.* **105**:351 (1936); C. A. Parker and A. P. Goddard, *Anal. Chim. Acta* **4**:517 (1950).
Antimony	(1) Iodide: 14% KI + 1% ascorbic acid–H_2O, 1M H_2SO_4.	330 or 425	0.4–4 3–25	Bi, Hg, Pb, Tl(I); HCl. Bi, Tl(I); HCl.	E. W. McChesney, *Ind. Eng. Chem., Anal. Ed.* **18**:146 (1946); A. Elkind, K. H. Gayer, and D. F. Boltz, *Anal. Chem.* **25**:1744 (1953).
	(2) Rhodamine B (R): 0.2% R HCl–H_2O, 6M HCl, 25°C, Ce(IV) sulfate oxidizes Sb(III) to Sb(V), $NH_2OH \cdot HCl$ destroys excess Ce(IV), extract Rhodamine B–$SbCl_6^-$ ion pair into benzene.	565	0.1–0.6	Au, Cr(VI), Fe(III), Ga, Hg(II), Pt, Tl, V, W. None if HBr–$HClO_4$ distillation precedes.	S. H. Webster and L. T. Fairhall, *J. Ind. Hyg. Toxicol.* **27**:184 (1945); R. W. Ramette and E. B. Sandell, *Anal. Chim. Acta* **13**:455 (1955); *Anal. Chem.* **40**:1146 (1968).
Arsenic	(1) Heteropoly blue: 0.5N H_2SO_4, 0.065% $(NH_4)_2MoO_4$ + $N_2H_4 \cdot H_2SO_4$, heat at 90°C for 10	840	0.3–3	Ba, Ge, Pb, Sb(III), Sn(II); P, Si. None if HBr–$HClO_4$ distillation precedes.	L. Duval, *Chim. Anal.* **51**:415 (1969); H. J. Morris and H. O. Calvery, *Ind. Eng. Chem., Anal. Ed.* **9**:447 (1937);

	min. Arsenic usually isolated as AsH_3, $AsCl_3$, or AsI_3 initially.				C. J. Rodden, *J. Res. Natl. Bur. Stds.* **24**:7 (1940); R. E. Stauffer, *Anal. Chem.* **55**:1205 (1983).
	(2) Silver diethyldithiocarbamate (R): Evolved AsH_3 absorbed in pyridine solution of reagent (0.5% w/v) or in $CHCl_3$ containing ethanolamine.	525; 540 if Sb present	0.2–15 μg	H_2S, Ge, Pt; elements hindering AsH_3 evolution.	H. Bode and K. Hachmann, *Z. Anal. Chem.* **229**:261 (1968); **241**:18 (1968); S. S. Sandhu and P. Nelson, *Anal. Chem.* **50**:322 (1978).
Barium	(1) Sulfonazo III (R): Use ethylene-glycol bis(β-aminoethylether)-*N,N,N′,N′*-tetraacetic acid (EGTA) as complexing agent, pH 3.	640	1.4–4	Sr forms useful colored complex. Presence of EGTA eliminates interference from Ca and most heavy metals.	P. J. Kemp and M. B. Williams, *Anal. Chem.* **45**:124 (1973).
	(2) *o*-Cresolphthalein (R): 0.1% R, aqueous ammonia, pH 11.3.	575	0.5–5	Separation of Ba from almost all other elements is required.	F. H. Pollard and J. V. Martin, *Analyst* **81**:348 (1956).
	(3) Chromate method: Precipitate $BaCrO_4$ in acetate buffer, dissolve in 4*M* HCl, compare color with acidified chromate solution.	350	10–800	Sr (unless double precipitation made), Pb; preconcentration necessary.	H. A. Fredoamo and B. J. Babler, *Ind. Eng. Chem., Anal. Ed.* **11**:487 (1939); J. Agterdenbos, *Z. Anal. Chem.* **159**:202 (1958).
Beryllium	(1) Acetylacetone (R): Be–R aqueous medium at pH 7, extracted by $CHCl_3$.	295	0.03–0.3	No interference: Ag, Ba, Ca, Ce, Cr, Cu, F, Mn, Mo, P, Pb, Sr, U, Zn.	J. A. Adam, E. Booth, and J. D. H. Strickland, *Anal. Chim. Acta* **6**:462 (1952).
	(2) 4-(*p*-Nitrophenylazo)orcinol (R): 0.03% R, 0.1*M* NaOH, EDTA, citrate–borate buffer.	515	1–8	Ca, Cu, Fe, Mg, Ni, Zn.	W. Stross and G. H. Osborn, *J. Soc. Chem. Ind.* **63**:249 (1944); L. C. Covington and M. J. Miles, *Anal. Chem.* **28**:1728 (1956); F. A. Vinci, *ibid.* **25**:1583 (1953).
Bismuth	(1) Dithizone (R): Extract aqueous phase, 0.03*M* KCN, 0.1% Na_2SO_3, pH 10.0, with 0.002% R in CCl_4 for 1 min. Measure *A* against CCl_4.	490	0.02–0.62	No interference: Ag, Cd, Co, Cu, Hg, Ni, Zn.	D. M. Hubbard, *Anal. Chem.* **20**:363 (1948); H. Onishi and N. Ishiwatari, *Talanta* **8**:753 (1961).
	(2) Iodide: 14% KI + 1% ascorbic acid in water, 1*M* ± 0.25*M* H_2SO_4. Color measured directly or extracted with 1-pentanol–ethyl acetate (several portions).	337 or 460	0.6–6	Ag, Hg(II), Pb, Sb, V(V), W(VI); insoluble iodides.	N. M. Lisicki and D. F. Boltz, *Anal. Chem.* **27**:1722 (1955); see also K. Hasebe and M. Taga, *Talanta* **29**:1135 (1982).
	(3) Thiourea (R): 1.2% R in aqueous medium, 1*M*–2*M* $HClO_4$.	322	0.6–6	Ag, Cu, Fe(III), Hg(II), Pb, Sb, W(VI).	N. M. Lisicki and D. F. Boltz, *Anal. Chem.* **27**:1722 (1955).
	(4) Diethyldithiocarbamate (R): 0.2% R in water, pH 7–10, extract with CCl_4.	400	5–30	Most metals do not react if EDTA and cyanide are used at pH 10 except Hg, Pb.	K. L. Cheng, R. H. Bray, and S. W. Melsted, *Anal. Chem.* **27**:24 (1955).
Cadmium	(1) Dithizone (R): 0.008% R in $CHCl_3$; tartrate buffer, 40% NaOH, 1% KCN, hydroxylamine · HCl, double extraction.	518	0.13	Hg, Tl.	B. E. Saltzman, *Anal. Chem.* **25**:493 (1953); H. Fischer and G. Leopoldi, *Mikrochim. Acta* **1**:30 (1937).

(Continued)

TABLE 6.15 Spectrophotometric Methods for the Determination of Metals (*Continued*)

Element determined	Procedure	λ, nm	Range, $\mu g \cdot mL^{-1}$	Interferences	References
Cadmium	(2) Cadion (*p*-nitrobenzenediazoaminoazobenzene) (R): 0.02% R, KOH, KNaTartrate, polyvinyl pyrrolidine, or Triton X-100.	560	0.05–0.5	Ag, Co, Cu, Fe(III), Hg, Mg, Ni, Sn(II), cyanide.	P. Chavanne and Cl. Geronimi, *Anal. Chim. Acta* **19**:377 (1958); H. Watanabe and H. Ohmori, *Talanta* **25**:959 (1979); **94**:39 (1969).
Calcium	(1a) Glyoxal bis(2-hydroxyanil) (R): In 1 : 1 water–methanol medium, 0.001*M* R, 0.04*M*–0.12*M* (but fixed) NaOH.	520	520	EDTA, fluoride, oxalate, or phosphate; Bi, Pb, Sb, Sn (removed as sulfides); Ba, Be, Mg, Sr, U.	F. Umland and K.-U. Meckenstock, *Z. Anal. Chem.* **176**:96 (1960); *Analyst* **94**:39 (1969).
	(1b) Ca–R complex extracted with 1 : 1 $CHCl_3$ and 1-hexanol, and back-extracted with 0.01*M* HCl.	520	1–8	EDTA, fluoride, oxalate, phosphate if Bi, Pb, Sb, Sn removed as sulfides.	K. T. Williams and J. R. Wilson, *Anal. Chem.* **33**:244 (1961); R Kiel, *Z. Anal. Chem.* **253**:15 (1971).
	(2) 8-Hydroxyquinoline (R): Extract Ca with 2% R in $CHCl_3$ at pH 11.3 ± 0.1 with 2% butylamine or 2-butoxyethanol present. Dry extract with anhydrous sodium sulfate, stabilize with 4% methanol.	380	0.1–15	Sr and Ba. Most metals require prior removal with diethyldithiocarbamate and $CHCl_3$ and cupferron and $CHCl_3$.	F. Umland and K.-U. Meckenstock, *Z. Anal. Chem.* **176**:96 (1960); C. L. Luke, *Anal. Chim. Acta* **32**:221 (1965).
	(3) Purpurate (Murexide) (R): 10 mL 0.015% R in 70% ethanol to sample, pH 11.2, shake 5 min, measure immediately.	506	0–1.2 for 10 mL R; 0–3 for 25 mL R	Ba, Cd, Co, Fe(II,III), Li, Mg, Ni, Sn, Sr, Zn, Zr, sulfate.	M. B. Williams and J. H. Moser, *Anal. Chem.* **25**:1414 (1953); F. H. Pollard and J. V. Martin, *Analyst* **81**:348 (1956); H. Gordon and G. Norwitz, *Talanta* **19**:7 (1972).
	(4) *o*-Cresolphthalein (R): ammoniacal buffer, pH 10.2, 0.03% R. Conditions must be carefully controlled; color not stable.	575	0.1–1.3	Ba, Mg, Sr.	F. H. Pollard and J. V. Martin, *Analyst* **81**:351 (1956).
	(5) Indirect oxalate: Precipitate CaC_2O_4 in usual way, filter, wash, dissolve in HCl. Add known amount of Ce(IV), then KI; measure tri-iodide.	400	3–30		J. Sendroy, Jr., *J. Biol. Chem.* **144**:243 (1942).
Cerium	(1) Peroxodisulfate (R): To 10 mL of sample, 0.5*M* H_2SO_4, 0.5 mg $AgNO_3$, 25 mg $K_2S_2O_8$, boil for 5 min, cool and dilute to 10 mL. For blank correction, add drop of H_2O_2 and measure.	320 or 350	3–20	Cr, Fe, Mn; V and U slightly; effect of Fe, U, Th, V less if measured at 350 nm.	A. I. Medalia and B. J. Byrne, *Anal. Chem.* **23**:453 (1951); L. A. Blatz, *Anal. Chem.* **33**:249 (1961).

	(2) 8-Hydroxyquinoline (R): Extract Ce from ammoniacal–citrate buffer, pH 9.9–10.5, with 3% R in $CHCl_3$.	505	2–20	Fe (unless converted to ferrocyanide), Mn, V; fluoride.	W. Westwood and A. Mayer, *Analyst* **73**:275 (1948).
	(3) 1,10-Phenanthroline (R): Ce(IV) oxidation with peroxodisulfate plus Ag^+ is reduced to Ce(III) by excess Fe(II)–R. Measure decrease in absorbance.	505	0.2–10	Cr, Mn, V, chloride, oxalate, citrate, tartrate, fluoride, and perchlorate.	F. Culkin and J. P. Riley, *Anal. Chim. Acta* **24**:167 (1961).
	(4) 2-Thenoyltrifluoroacetone (R): Oxidize with bromate to Ce(IV), 0.5*M* H_2SO_4, extract with 0.5*M* R in xylene.	440 or 450	0–0.01	V, Mn, Fe.	H. Onishi and C. V. Banks, *Anal. Chem.* **35**:1887 (1963).
Chromium	(1) Chromate: Oxidation with $K_2S_2O_8$ in NaOH, pH 10. Reduce any permanganate by heating with ethanol.	366	1–10	Ce(IV), U(VI).	S. Christow, *Z. Anal. Chem.* **125**:278 (1943).
	(2a) 1,5-Diphenylcarbohydrazide (R): 0.2*N* H_2SO_4 in 25 mL, 1 mL 0.25% R in acetone.	540	0.1–1	Cu, Fe(III), Hg, Mo(VI), V(V).	G. P. Rowland, *Ind. Eng. Chem., Anal. Ed.* **11**:442 (1939).
	(2b) Extract Cr(VI) from 1*M* HCl with 4-methyl-2-pentanone at 0–10°C, add R to organic phase, let stand 15 min.	540	0–1.7	None if organic phase is backwashed with 1*M* HCl.	J. A. Dean and M. L. Beverly, *Anal. Chem.* **30**:977 (1958); B. Morsches and G. Tölg, *Z. Anal. Chem.* **250**:81 (1970).
Cobalt	(1) Nitroso-R-salt (R): Evaporate to dryness to remove mineral acids, add 5 mL water, 1 mL 0.2*M* citric acid, and 1.2 mL phosphate–borate buffer (pH 8), add exactly 0.5 mL 0.2% R. Boil 1 min, add 1 mL HNO_3, boil 1 min, cool in dark, make to 10 mL.	420; 500–550 in presence of Fe	0.1–1	Fe(III), chloride, fluoride.	V. D. Anand et al., *Anal. Chem.* **33**:1933 (1961); G. Wünsch, *Talanta* **26**:177 (1979).
	(2a) Thiocyanate (R): 5% R and 50% acetone in final volume.	625		Bi, Cr, Cu, Fe(III), Ni (slight), U.	E. S. Tomula, *Z. Anal. Chem.* **83**:6 (1931); M. R. Verma and P. K. Gupta, *ibid.* **196**:187 (1963); **207**:18 (1965).
	(2b) 5 mL sample, 8 mL thiosulfate–phosphate, pH 3.5–4.0, 10 mL 50% R. Extract with 10 mL 1-pentanol–diethyl ether (3 : 1).	610	5–50	Fe(III), V.	R. S. Young and A. J. Hall, *Ind. Eng. Chem., Anal Ed.* **18**:264 (1946).
	(2c) 25 mL sample, pH 2–7, 5 mL 50% R, solid NH_4F until red Fe(III) thiocyanate fades and 200 mg excess, 1 mL 2% aqueous tetraphenylarsonium chloride.	620		Cu, Fe(III), Mo(V), V(V), and possibly Bi.	H. E. Affsprung et al., *Anal. Chem.* **23**:1680 (1951); L. P. Pepkowicz and J. L. Marley, *ibid.* **27**:1330 (1955).

(Continued)

TABLE 6.15 Spectrophotometric Methods for the Determination of Metals (*Continued*)

Element determined	Procedure	λ, nm	Range, $\mu g \cdot mL^{-1}$	Interferences	References
Cobalt	Extract with 10, 5, 5 mL portions $CHCl_3$ (adding 5 drops tetraphenylarsonium chloride each time).				
Copper	(1) Diethyldithiocarbamate (R): 20 mL sample, 20% citrate buffer, pH 8.5, 5 mL 0.1% R, 10 mL 5% EDTA, shake 1–2 min with 10 mL $CHCl_3$, extract twice more with 5-mL portions $CHCl_3$.	435	0.5–4	Ag, Au, Bi, Hg, Mo(VI), Os, Pd, Pt, Sb(III). Without EDTA also: Co, Fe, Mn, Ni. When Pb–R used: Ag, Bi, Hg, Tl(III).	V. Sedivec and V. Vasak, *Coll. Czech. Chem. Commun.* **15**:260 (1950); A. Claassen and L. Bastings, *Z. Anal. Chem.* **153**:30 (1956).
	(2a) 2,2′-Biquinoline (Cuproine) (R): 5 mL $NH_2OH \cdot HCl$, 5 mL H_2 tartrate, pH 5–6, extract with 10 mL 0.02% R in 1-pentanol, 1–2 min.	545	0.5–4	Cyanide, thiocyanate, oxalate.	(2a) R. J. Guest, *Anal. Chem.* **25**:1484 (1953); C. E. A. Shanahan and R. H. Jenkins, *Analyst* **86**:166 (1961).
	(2b) 2,9-Dimethyl-1,10-phenanthroline (2 mL 0.1% in ethanol) substituted for R.	457			(2b) A. R. Gahler, *Anal. Chem.* **26**:577 (1954).
Gallium	(1) Rhodamine B (R): Preliminary separation of Ga by Et_2O extraction from 6*M* HCl is recommended. 0.5% R in 1 : 1 HCl, 6.0*M* HCl, $TiCl_3$, extract with benzene [better: chlorobenzene–CCl_4 (3 : 1)].	562	0.1–1	Au, Fe, Sb, Tl.	F. Culkin and J. P. Riley, *Analyst* **83**:208 (1958); *Talanta* **2**:29 (1959).
	(2) 8-Hydroxyquinoline (R): Preliminary ether extraction as above. Back-extract into water, pH 8, 5 mL 10% NaCN, 20 mL 1% R in $CHCl_3$, shake 30 s.	400	0.2–2	Fluoride, citrate.	C. L. Luke and M. E. Campbell, *Anal. Chem.* **28**:1340 (1956).
Germanium	(1) Phenylfluorone (R): 0.03% R in ethanol, 0.2*M*–1.5*M* HCl, 0.5% gum arabic following isolation by extraction or distillation.	510	0.1–1	Ce(IV), Cr(VI), Ga, Mn(VII), Mo, Nb, Sb, Sn, Ta, Ti, W, Zr if Ge not isolated beforehand.	H. J. Cluley, *Analyst* **76**:523 (1951).
	(2) Heteropoly blue: acid solution, 10% $(NH_4)_2MoO_4$ aqueous medium, reduce with Fe(II) giving 12-molybdogermanic acid.	830	0.4–4	As, Ba, Bi, Fe(III), Pb, V; F^-, P, Si.	A. Halasz and E. Pungor, *Talanta* **18**:557, 569, 577 (1971); R. Jakubiec and D. F. Boltz, *Anal. Chem.* **41**:78 (1969).
Gold	(1) Rhodamine B (R): 0.04% R in water, 0.75*M* HCl, extract R-chloroaurate with diisopropyl ether or benzene. Usually follows coprecipitation with elemental Te produced by $SnCl_2$.	565	0.3–3	Sb(V) and Tl(III) interfere seriously.	E. B. Sandell, *Anal. Chem.* **20**:253 (1948); S. Natelson and J. L. Zuckerman, *Anal. Chem.* **23**:653 (1951).

	(2) Bromoaurate: To 10 mL sample and 5 mL conc. HBr. Extract twice with 15 mL diisopropyl ether, combine extracts, wash with 5 mL 4*M* HBr. Back-extract with one 20-mL and two 10-mL portions water. Add 1 mL conc. HBr and 0.5 mL H_3PO_4.	380	0.3–40	Chloride.	W. A. E. McBryde and J. H. Yoe, *Anal. Chem.* **20**:1094 (1948); A. Chow and F. E. Beamish, *Talanta* **10**:883 (1963).
Indium	(1) 8-Hydroxyquinoline (R): pH 3.2–4.5, 0.5% R in $CHCl_3$.	400	1–20	Al, Bi, Co (slightly), Cu, Ga, Fe(III), Mo(VI), Ni, Sn(II), Tl(III), V(V).	T. Moeller, *Ind. Eng. Chem., Anal. Ed.* **15**:270 (1943).
	(2) Extract In with butyl acetate from 1.5*M* KI–0.75*M* H_2SO_4, add 8-hydroxylquinoline to separated organic phase.	400	1–20	Tl(I), Sn(IV), and Fe(III) if not reduced with ascorbic acid prior to extraction.	S. G. Iyer et al., *Talanta* **23**:525 (1976).
	(3) Dithizone (R): 0.002% R in $CHCl_3$ from ammoniacal medium (pH 9) containing 0.04*M* KCN and 0.3*M* $H_2OH \cdot HCl$.	510	0.1–2	Bi, Pb, Tl unless prior removal by extraction.	V. T. Athavale et al., *Anal. Chim. Acta* **22**:56 (1960).
Iridium	(1) Tin(II) chloride–HBr: 5 mL sample <0.5*M* HCl, add 5 mL conc. HBr, heat at 95°C for 10 min, 5 mL 25% $SnCl_2 \cdot 2H_2O$ in conc. HBr for 2.0 min, cool quickly, measure immediately.	402	0.3–3	Pd, Pt, Rh severe; Au, Co, Cr(III), Cu, Fe(II), Ni, Sb(III),Ti.	S. S. Berman and W. A. E. McBryde, *Analyst* **81**:566 (1956); E. C. Cerceo and J. J. Markham, *Anal. Chem.* **38**:1426 (1966).
	(2) Leuco-crystal violet oxidized by Ir(IV).	590	0.3–3	Au, Fe severe; Pd, Pt, and Rh slight.	M. Ewen and E. B. T. Cook, *Natl. Inst. Metall. Rep. No. 1179* (1971).
Iron	(1) 1,10-Phenanthroline (R) or 2,2′ bipyridine: $NH_2OH \cdot HCl$, 0.01% R, pH 2–9. When citrate present, hydroquinone must be reducing agent.	510	0.5–5	Precipitates: Ag, Bi, Consume R: Cd, Hg, Zn. Colors: Co, Cu, Ni.	W. B. Fortune and M. G. Mellon, *Ind. Eng. Chem., Anal. Ed.* **10**:60 (1938); M. L. Moss and M. G. Mellon, *ibid.* **14**:862 (1942).
	(2a) Thiocyanate: 0.05*M* to 1*M* in HNO_3 or HCl, 0.3*M* KSCN (final volume), dilute to volume and measure at once. (2b) Complex can be extracted with 2-methyl-4-pentanone and measured.	480	1–10	Precipitates: Ag, Hg(I). Colors: Bi, Co, Cu, Ir, Mn, Mo, Os, Ru, Ti, U. Form SCN complexes: Cd, Hg(II), Sb(III), Zn, F^-, oxalate, $P_2O_7^{4-}$.	(2a) J. T. Woods and M. G. Mellon, *Ind. Eng. Chem., Anal. Ed.* **13**:551 (1941); T. C. J. Ovenston and C. A. Parker, *Anal. Chim. Acta* **3**:277 (1949). (2b) T. C. J. Ovenston and C. A. Parker, *Anal. Chim. Acta* **36**:122 (1966).
Lanthanoids	(1) Alizarinsulfonate (alizarin red S) (R): Acetate buffer—2 mL, pH 4.6, 2 mL 0.1% R, stand 5 min.	550	4–12	Other metals and other lanthanoids.	R. W. Reinhart, *Anal. Chem.* **26**:1820 (1956).
	(2) Arsenazo I (R): 2 mL 0.1% R, 5 mL triethanolamine buffer, pH 7.2 ± 0.1, 25 mL final volume.	570 or 580	0.1–4	Al, Co, Cr(III), Cu, Fe(III), Mn, Ni, Sc, Th, U(VI), Zr; F^-, phosphate.	J. S. Fritz, M. J. Richard, and W. J. Lane, *Anal. Chem.* **30**:1776 (1958).

(Continued)

TABLE 6.15 Spectrophotometric Methods for the Determination of Metals (*Continued*)

Element determined	Procedure	λ, nm	Range, $\mu g \cdot mL^{-1}$	Interferences	References
Lead	Dithizone (R): 0.006% R in $CHCl_3$, ammonium citrate plus KCN, pH 10.8, $NH_2OH \cdot HCl$, extract with $CHCl_3$.	510	0.4–3	Bi, Cu, In, Sn(II), Tl(I); much Ca (or Mg) and phosphate.	L. J. Snyder, *Anal. Chem.* **19**:684 (1947); O. B. Mathre and E. B. Sandell, *Talanta* **11**:295 (1964).
Lithium	Thoron (R): 20% KOH, 70% (v/v) acetone, 0.2% R.	482	0.1–1		P. F. Thompson, *Anal. Chem.* **28**:1527 (1956).
Magnesium	(1) Titan Yellow (R): In 50 mL water, 0.1 g sucrose, 5% $NH_2OH \cdot HCl$, 2.5 mL glycerol, 5 mL 0.2% R, 0.2*M*–0.3*M* NaOH. Let lake stand 1 h.	540	0.4–4	Al, Ba, Cd, Co, Cu, Fe(III), Mn, Sn, Ti, Zn, phosphate.	J. G. Hunter, *Analyst* **75**:91 (1950); E. J. Butler et al., *Anal. Chim. Acta* **30**:524 (1964); Analyst **86**:269 (1961); *ibid.* **92**:83 (1967).
	(2) 8-Hydroxyquinoline (R): Neutralize 40-mL sample to Congo Red with NH_3 and add 10 mL extra (pH 10), 5 mL 10% NaCN, 5 mL 1 : 1 butyl Cellosolve, 20 mL 3% R in $CHCl_3$, shake, filter through coarse paper.	400	1–5	Al, Ca, Cd, Fe, Ga, lanthanoids, Sn(IV), Zn, Zr.	C. L. Luke and M. E. Campbell, *Anal. Chem.* **26**:1778 (1954); F. Unland and W. Hoffmann, *Anal. Chim. Acta* **17**:234 (1957)
	(3) Eriochrome Black T (R): In ammoniacal buffer (pH 10.15), and 10 mL 0.1% R in methanol, mix, and dilute to mark. Measure immediately.	520	0.2–1.4	Al, Ca, Co, Cu, Sn, Zn, phosphate.	F. H. Pollard and J. V. Martin, *Analyst* **81**:348 (1956).
Manganese	(1) Permanganate method: 2*M* H_2SO_4, 5 mL H_3PO_4, 0.3 g KIO_3, boil—10 min (1 h for small amounts), cool, dilute to 100 mL.	522	1–25	Cr; Bi and Sn give turbidities; chloride; reducing substances.	H. H. Willard and L. H. Greathouse, *J. Am. Chem. Soc.* **39**:2366 (1917).
	(2) 0.75*M* HNO_3, 0.25*M* H_2SO_4, 0.3*M* H_3PO_4, 0.001*M* $AgNO_3$, 1 g $(NH_4)_2S_2O_8$, boil over open flame 2 min, stand 1 min, cool, dilute to 100 mL, measure in 15 min. Add 50 mg NaN_3 and obtain blank from colored ions.	522	1–25	Halides, As(III), Ba, Nb, Pb, Pt metals, Sn(II), Sb(III), Sr, Ta, Ti, Tl(I), Zr, H_2O_2, NO_2^-, tartrate, citrate, other reducing agents.	F. Nydahl, *Anal. Chim. Acta* **3**:144 (1949); G. Gottschalk, *Z. Anal. Chem.* **212**:303 (1965).
Mercury	(1) Dithizone (R) *Direct*: 0.5*M* H_2SO_4, 2 mL 6*M* HOAc, 5 mL $CHCl_3$, shake, discard $CHCl_3$ layer. Add 5.00 mL 0.001% R in $CHCl_3$, shake 1 min.	500 or 610 excess R	0.3–3	Ag, Au, Cu, Pd, Pt.	Analytical Methods Committee, *Analyst* **90**:515 (1965); K. L. Cheng, *Talanta* **9**:501 (1962).

	(2) Dithizone (R) *Reversion*: 0.25*N* H_2SO_4, 20 mL sample, 20 mL 0.0007% R in $CHCl_3$, 2 mL 6 *M* HOAc, shake 1 min. Measure extract; shake portion with equal volume reversion mixture (10 g KH phthalate and 30 g KI in 500 mL water), measure $CHCl_3$ layer.	610	0.3–3	Ag.	H. Irving, G. Andrew, And E. J. Risdon, *J. Chem. Soc.* **1949**, 541.
Molybdenum	(1) Thiocyanate: 10% KSCN, 1*M*–1.5*M* HCl, 0.05*M*–0.5*M* $SnCl_2$, extract with isopentyl alcohol.	465	0.1–5	Co, Cr(VI), Cu, Pb, Pt, Re, Rh, Ti(III), V, W(VI), F^-.	D. D. Perrin, *New Zealand J. Sci. Tech.* **27A**:396 (1946).
	(2) Dithiol (R): 0.15% R in 1% aqueous NaOH, 2*M* HCl or 1.5*M* H_2SO_4, let stand 2 h; extract with butyl acetate or CCl_4.	670	0.05–1	W unless masked with citrate, Fe(III) unless reduced with ascorbic acid, Cu masked with thiourea.	K. Tanaka and N. Takagi, *Jpn. Anal.* **19**:790 (1970); B. F. Quin and R. R. Brooks, *Anal. Chim. Acta* **74**:75 (1975).
Nickel	(1) Dimethylglyoxime (R): To sample add 10 mL 20% Na citrate, 2 mL 10% $NH_2OH \cdot HCl$, 2 mL 1% R in EtOH, make ammoniacal, dilute to 60 mL, shake 2 min with 20 mL $CHCl_3$. With $CHCl_3$ twice with 1 : 50 NH_3 for 1 min each, back-extract $CHCl_3$ with 15 mL 0.5*M* HCl. Add 2 mL 2.5% R, 1 mL 10*M* NaOH, 0.3 mL 10% $(NH_4)_2S_2O_8$, wait 10 min, make up to 25 mL.	460	0.2–5		A. Claassen and L. Bastings, *Rec. Trav. Chim.* **73**:783 (1954); W. Oelschläger, *Z. Anal. Chem.* **146**:339, 346 (1955).
	(2) Two $CHCl_3$ extracts of NiR in the presence of thiosulfate and tartrate (pH 6.5).	375	1–20	Au(III), Pd, Pt(II).	H.Christopherson and E. B. Sandell, *Anal. Chim. Acta* **10**:1 (1954).
	(3) Diethyldithiocarbamate (R): Separate Ni by dimethylglyoxime extraction as in procedure (1). Extract from ammoniacal citrate (pH 9.5) and 0.2% R with 10 mL isopentyl alcohol for 2 min.	325	1–20		O. R. Alexander, E. M. Godar, and N. J. Linde, *Ind. Eng. Chem., Anal. Ed.* **18**:206 (1946); see also A. M. Bond and G. G. Wallace, *Anal. Chem.* **55**:718 (1983); **56**:2085 (1984).
Niobium	(1) Thiocyanate: 20% KSCN, 4*M* HCl, $SnCl_2$, extraction within 5–10 min with diethyl ether or butyl acetate.	385	0.5–2.5	Cu, Mo, Pt, V, W; Ag, Hg, Te, Se precipitated; > 100-fold excess Fe, U, Ti; F^-, PO_4^{3-}, $C_2O_4^{2-}$.	D. N. Hume et al., *Anal. Chem.* **24**:1169 (1952); P. F. Sattler and I. E. Schreinlechner, *ibid.* **49**:80 (1977); *Talanta* **27**:537 (1980).
	(2) 4-(2-Pyridylazo)resorcinol (R): Add in order EDTA R, and 8% NH_4OAc buffer (pH 6); let stand 1 h.	550	0.7–3	Co and Ni (unless KCN added), Ta, U(VI), V(V); phosphate and F^-.	R. Belcher et al., *Talanta* **9**:943 (1962); M. Siroki and C. Djordjevic, *Anal. Chem.* **43**:1375 (1971).

(Continued)

TABLE 6.15 Spectrophotometric Methods for the Determination of Metals (*Continued*)

Element determined	Procedure	λ, nm	Range, $\mu g \cdot mL^{-1}$	Interferences	References
Niobium	(3) Sulfochlorophenol S (R): 1*M*–3*M* HCl, $HClO_4$, or HNO_3 (plus urea); tartrate, mercaptoacetic acid, EDTA, acetone, and R; heat 5 min at 70°, cool.	650	0.5–2.5	Mo.	I. P. Alimarin and S. B. Savvin, *Talanta* **13**:689 (1966); *Anal. Chim. Acta* **116**:185 (1980).
Osmium	(1) Thiourea (R): 10% R, 0.6*M* H_2SO_4; $SnCl_2$ if chloro- or bromoosmate present.	480	8–40	Pd, Ru.	R. D. Sauerbrunn and E. B. Sandell, *Anal. Chim. Acta* **9**:86 (1953).
	(2) Tetraphenylarsonium chloride (R): 2 mL 1% R to 0.2*M* HCl, dilute to 20 mL, shake 3–5 min with 10 mL $CHCl_3$.	375		> 0.5 NaCl, ClO_4^-, SO_3^{2-}.	R. Neeb, *Z. Anal. Chem.* **154**:23 (1957).
Palladium	(1) 4-Nitrosodiphenylamine (R): Acetate buffer (pH 4.8), 0.005% R in ethanol, let stand 5 min.	525	0.01–0.1	Other platinum metals, Ag, oxidants, CN^-, I^-.	L. G. Overholser and J. H. Yoe, *J. Am. Chem. Soc.* **63**:3224 (1941).
	(2) 4-Nitrosodiphenylamine (R): 0.005*M*–0.015*M* HCl, 1 mL 0.005% R, let stand 20 min, shake with 5.0 mL diethyl oxalate for 1 min.	525	0.01–0.1	Ag, Au, Se, W; Co, Cu, Fe, Ni interference removed by coprecipitation of Pd with Te.	E. E. R. Marhenke and E. B. Sandell, *Anal. Chim. Acta* **28**:259 (1963).
	(3) 2-Furildioxime (R): 0.1*M*–1*M* HCl, 1 mL 1% R in 30% ethanol, 10.0 mL $CHCl_3$, shake 0.5 min.	380	0.03–0.3	Cyanide.	O. Menis and T. C. Rains, *Anal. Chem.* **27**:1932 (1955).
Platinum	(1) Tin(II) chloride: In 50-mL flask add 5 mL 12*M* HCl and 10 mL 23% $SnCl_2 \cdot 2H_2O$. Measure at once if Rh is present.	403	0.5–20	Au, Ir, Os, Pd, Rh, Ru, Te.	S. S. Berman and E. C. Goodhue, *Can. J. Chem.* **37**:370 (1959).
	(2) 4-Nitrosodimethylaniline (R): 0.5% R in EtOH, pH 2–3, heat exactly 20 min at 100°C, cool immediately, dilute to 50 mL with EtOH.	525	0.1–1	Au, Co, Cr, Cu, Fe, Ni, other platinum metals.	J. H. Yoe and J. J. Kirkland, *Anal. Chem.* **26**:1335, 1340 (1954).
Potassium	(1) Dipicrylamine (R): 1% R plus 0.3% Li_2CO_3, neutral of slightly basic, wash precipitate with ice-cold water, dissolve in hot water, make to volume with dilute NaOH.	400	0.2–1	Ba, Cs, Hg, NH_4, Pb, Tl(I) and metals insoluble in basic medium.	P. R. Lewis, *Analyst* **80**:768 (1955).
	(2) 18-Crown-6 (R_1) and Bromocresol Green (R_2): For serum use 0.1 mL, add 4.9 mL 1.5% Cl_3CCOOH,	410			H. Sumiyoshi, K. Nakahara, and K. Ueno, *Talanta* **24**:763 (1977).

	centrifuge 5 min, transfer 2 mL of supernate, add 0.2 mL 11% LiOAc, 0.4 mL of acetate buffer (pH 3.9), 1.4 mL 0.16% R_2 in 20% EtOH, and 4 mL 0.3% R_1 in benzene, shake 5 min, centrifuge 1 min.				
Rhenium	(1) Thiocyanate: To sample 1.2*M* HCl, 20% KSCN, 1.5*M* Sn(II), let stand 5 min, extract with three 15-mL portions diethyl ether, wash combined ether layer with 10 mL 1 : 4 HCl, dilute ether to 50 mL.	432	1–4	Cr, Mo.	W. Geilmann and H. Bode, *Z. Anal. Chem.* **128**:489 (1948).
	(2) 2-Furildioxime (R): 0.35% R in acetone, 10% Sn(II), 1.2*M* HCl.	532	0.4–6	Cu, Mo, Pd.	V. W. Meloche et al., *Anal. Chem.* **29**:527(1957).
	(3) To 5–10 mL hot (80°C) ReO_4^- solution 3*M*–7*M* HCl, add 2 mL 1,4-diphenylthiosemicarbazide and heat at 80°C for 20 min. Extract with 25 mL $CHCl_3$.	510	1–4	Only alkali metals do not interfere.	*Z. Anal. Chem.* **151**:401 (1956); *Zh. Anal. Khim.* **29**:743 (1974).
Rhodium	(1) Tin(II) chloride (R): 1*M* R in 15 mL, heat at 95°C for 1 h, dilute to 50 mL with 2*M* HCl.	475	4–20	Au, Cr, Os, Pd, Pt, Ru.	A. D. Maynes and W. A. E. McBryde, *Analyst* **79**:230 (1954).
	(2) 2-Mercapto-4,5-dimethylthiazole (R): 0.3*M* HCl, 1 mL R per 100 μg Rh, boil 1 h, cool, dilute to 100 mL.	430	1–7	Ag, Bi, Cu(II), cyanide, iodide, sulfite.	D. E. Ryan, *Analyst* **75**:557 (1950); *Can. J. Chem.* **39**:2389 (1961).
Ruthenium	(1) Thiourea (R): 5 mL sample, 1 mL 10% R, 5 mL conc. HCl, 5 mL EtOH, heat at 85°C for 10 min, cool, make to 25 mL with 1 : 1 6*M* HCl and EtOH.	620	2–15	Co, Cr, Cu, Fe, Ni, Os, Pd.	G. H. Ayres and F. Young, *Anal. Chem.* **22**:1277 (1950); S. T. Payne, *Analyst* **85**:698 (1960).
	(2) 1,4-Diphenylthiosemicarbazide (R): 5.5*M*–6.5*M* HCl, Sn(II), heat at 100°C for 10 min, extract with 5 mL $CHCl_3$.	560	1–6	Re.	T. Hara and E. B. Sandell, *Anal. Chim. Acta* **25**:65 (1960).
	(3) 1,10-Phenanthroline (R): 20 mL sample, 15 mL 0.01*M* R, 5 mL 5% $NH_2OH \cdot HCl$, 10 mL 20% NaCl, pH 6.0, heat at 100°C for 2 h.	448	0.3–1.6	Distillation of RuO_4 must precede.	C. V. Banks and J. W. O'Laughlin, *Anal. Chem.* **29**:1412 (1957).
Scandium	(1) Xylenol Orange (R): 0.01 *M* $HClO_4$, 2 mL 0.05% R, 25 mL final volume.	553	0.3–1.7	Fe(III), Th, fluoride, phosphate.	S. S. Berman et al., *Anal. Chem.* **35**:1394 (1963).

(Continued)

TABLE 6.15 Spectrophotometric Methods for the Determination of Metals (*Continued*)

Element determined	Procedure	λ, nm	Range, $\mu g \cdot mL^{-1}$	Interferences	References
Scandium	(2) Arsenazo III (R): 2 mL 0.1% R in 25 mL, pH 2.	675	0.4–2	Fe(III), lanthanoids, Th, U(VI), Zr.	S. B. Savvin, *Talanta* **8**:673 (1961).
Silver	(1) Dithizone (R): 0.25*M* H_2SO_4, 2 mL 0.001% Cu–R reagent in CCl_4, shake 2 min.	540	0.2–2	Au, Hg, Pd.	V. Fano and L. Zanotti, *Anal. Chim. Acta* **72**:419 (1974).
	(2) 4-Dimethylaminobenzylidenerhodanine (R): 0.05*M* HNO_3, 0.01% R in EtOH.	460–495	0.004–0.04	Au(III), Cu(I), Hg, Pd(II), Pt (IV).	E. B. Sandell and J. J. Neumayer, *Anal. Chem.* **23**:1863 (1951).
Sodium	Indirect. Precipitation of triple acetate, dissolve in $(NH_4)_2CO_3$, add 3% H_2O_2.	520	8–80	Li.	E. A. Arnold and A. E. Pray, *Ind. Eng. Chem., Anal. Ed.* **15**:294 (1943).
Strontium	Murexide (R): In 25 mL, 15 mL ethylene glycol, 3 mL 0.05*M* NaOH, dilute to 22 mL, cool to 0°C, 3 mL 0.06% R.	510	1–8	Ca, most elements.	F. H. Pollard and J. V. Martin, *Analyst* **81**:348 (1956).
Tantalum	(1) Pyrogallol (R): 20% R, 4*M* HCl, 0.02*M* Sn(II), 0.018*M* $(NH_4)_2C_2O_4$.	325	4–40	Mo, Pt, Sb, U, W.	J. I. Dinnin, *Anal. Chem.* **25**:1803 (1953).
	(2) Phenylfluorone (R): 5 mL 10% EDTA, pH 4.5, acetate buffer, 10 mL 0.01% R, dilute to 50 mL.	530	0–2.4	Cr(VI), Ge, Nb, Sb(III), Te, Ti; none if double MIBK extraction from 1.3*M* HF–0.6*M* HCl.	C. D. Bingham et al., *Anal. Chem.* **41**:1144 (1969); C. L. Luke, *ibid.* **31**:904 (1959).
Thallium	(1a) Rhodamine B (R): 0.2% R, 2*M* HCl, Br_2 [or $Ce(SO_4)_2$ or $(NH_4)_2S_2O_8$], boil until bromine color just disappears, cool to 25°C, extract with 10 mL benzene (1 min shaking).	560	0.3–1.3	Au(III), Fe(III), Ga, Hg(II), Sb(V); Au and Hg interference reduced by washing diisopropyl either extract with $NH_2OH \cdot HCl$ before R extraction.	C. L. Luke, *Anal. Chem.* **31**:1680 (1950); M. Sager and G. Tölg, *Mikrochim. Acta* **1982(II)**:231; J. F. Woolley, *Analyst* **83**:477 (1958).
	(2a) $HTlBr_4$ is first extracted into diisopropyl ether, then shake ether layer with aqueous solution of R.				
	(3) Brilliant Green (R): 4*M* HCl, $Ce(SO_4)_2$, mix 1 min, reduce excess Ce(IV) with 1% $NH_2OH \cdot HCl$ until yellow color just disappears, 10 mL toluene, 1.0 mL 0.5% R in EtOH; repeat extraction.	640	0.3–1.3	Same as above.	A. G. Fogg et al., *Analyst* **98**:347 (1973); M. Ariel and D. Bach, *Analyst* **88**:30 (1963).
Thorium	(1) Arsenazo III (R): 8*M* HCl (6 mL conc. HCl), 10 mL 4% $H_2C_2O_4$, 1 mL 0.1% R, in 25 mL total volume.	665	0.3–1.3	Fe(III) and Zr masked in procedure given.	H. Onishi, *Jpn. Anal.* **12**:1153 (1963); H. Onishi and K. Sekine, *Talanta* **19**:473 (1972).

	(2) Thoron (R): 0.01% salt of R (final concentration) in water, pH 0.4–1.2 (HCl or $HClO_4$). Also determined with R in organic phase after tributyl phosphate extraction [P. J. Shirvington and T. M. Florence, *Anal. Chim. Acta* **27**:589 (1962)].	550	1–10	Serious; Ce(IV), Fe(III), Sb(III), U(IV), Zr, F^-, SO_4^{2-}, $C_2O_4^{2-}$, Sn(II,IV), phosphate, thiosulfate, sulfite; Ce(IV) removed with ascorbic acid, Mo(VI) and W(VI) removed with tartrate.	P. F. Thomason, M. A. Perry, and W. M. Byerly, *Anal. Chem.* **21**:1239 (1949); A. Mayer and G. Bradshaw, *Analyst* **77**:154 (1952).
Tin	(1a) Phenylfluorone (R): Extract as SnI_4 into benzene, back-extract with 0.25*M* H_2SO_4, KH phthalate–HCl buffer, gum arabic, 10 mL 1% R, dilute to 50 mL.	510	0.4–1.6	Fe(III), Ga, Mo, Sb(III), Ta, Ti, Zr, phosphate.	D. D. Gilbert and E. B. Sandell, *Microchem. J.* **4**:491 (1960).
	(1b) Use double diethylammonium diethyldithiocarbamate extraction with $CHCl_3$, the second with mercaptoacetic acid and KI–ascorbic acid present. Remove $CHCl_3$ layer, heat to H_2SO_4 fumes. Add H_2O_2, pH 5 acetate buffer, gum arabic, and R.	510	0.4–1.6	None if metals in milligram amounts.	C. L. Luke, *Anal. Chem.* **28**:1276 (1956).
	(2) Catchol Violet (R): pH 2.0–5.0, acetate buffer, 0.05% R.	552	0.2–1.2	Bi, Fe(III), Ga, Mo, Sb(III), Th, Ti, V, W, Zr, phosphate.	W. J. Ross and J. C. White, *Anal. Chem.* **33**:421 (1961); Analytical Methods Committee, *Analyst* **92**:320 (1967).
Titanium	(1) Hydrogen peroxide: 1.5*N*–3.5*N* H_2SO_4, 1.5*M* H_3PO_4, 1 mL 3% H_2O_2, dilute to 50 mL.	410	7–50	Fe(III), Mo, V.	A. Weissler, *Ind. Eng. Chem., Anal. Ed.* **17**:695 (1945).
	(2) Thiocyanate: 6*M* HCl (or H_2SO_4), > 20 mg/mL NH_4SCN, 5 mL 0.01*M* trioctylphosphine oxide in cyclohexane.	432	0.1–1.7	Mo, Nb, W, Zr; > 0.1*M* HNO_3.	J. P. Young and J. C. White, *Anal. Chem.* **31**:393 (1959).
	(3) Sulfosalicylic acid (R): pH 3–5, 5 mL 20% R in water, 5 mL 4% $HSCH_2COOH$, dilute to 25 mL.	445	0.7–5	Bi, Ce(IV), Co, Cr, Cu, Mo, Pd, Pt, Te, U, V(IV), W, citrate, tartrate, thiosulfate.	M. Zeigler and O. Glemser, *Z. Anal. Chem.* **139**:92 (1953).
Tungsten	(1) Thiocyanate: 2 mL 50% tartaric acid, 1.5 mL 50% NaSCN, dilute to 50 mL with 7% $SnCl_2 \cdot 2H_2O$ in conc. HCl. Stand 20 min.	420	1–17	As, Co, Cr, Cu, Ni, V, Se, Te.	W. Westwood and A. Mayer, *Analyst* **72**:464 (1947); A. G. Fogg et al., *Analyst* **95**:848 (1970).
	(2) Toluene-3,4-dithiol (R): 20% $SnCl_2 \cdot 2H_2O$ in 6*M* HCl, 0.5% R, extract with pentyl acetate.	640	0.5–8	Mo.	P. Y. Peng and E. B. Sandell, *Anal. Chim. Acta* **29**:325 (1963); K. Kawabuchi and R. Kuroda, *Talanta* **17**:67 (1970).

(Continued)

TABLE 6.15 Spectrophotometric Methods for the Determination of Metals (*Continued*)

Element determined	Procedure	λ, nm	Range, μg · mL^{-1}	Interferences	References
Uranium	(1) 4-(2-Pyridylazo)resorcinol (PAR) (R): Triethanolamine buffer, pH 8, mixture EDTA, NaF, and sulfosalicylic acid.	530	0.1–7	Cr(III), Fe(III), Si, V, Zr.	F. H. Pollard et al., *Anal. Chim. Acta* **20**:26 (1959).
	(2) Determined in organic phase with PAR after extraction with trioctylphosphine oxide.				E. N. Pollock, *Anal. Chim. Acta* **88**:399 (1977).
	(3) Arsenazo III (R): 4*M*–7*M* HCl, reduce U(VI) to U(IV) with Zn or Bi, 1 mL 0.1% R, final volume 25 mL.	665	0.1–3	Th, Zr.	E. Singer and M. Matucha, *Z. Anal. Chem.* **191**:248 (1962).
	(4) Dibenzoylmethane (R): Aqueous ethanol, pH 6.5–8.5, 1 mL 1% R, final volume 50 mL. Best to use preliminary extraction.	400	1–21	Most heavy metals, phosphate, citrate, oxalate, tartrate.	H. P. Holcomb and J. H. Yoe, *Anal. Chem.* **32**:612 (1960).
Vanadium	(1) *N*-Benzoyl-*N*-phenylhydroxylamine (R): 3*M*–4*M* HCl in 25 mL, 10 mL 0.1% R in $CHCl_3$, shake 1 min.	530	0.1–4	Mo(VI), Ti, U(VI), W(VI), Zr.	D. E. Ryan, *Analyst* **85**:569 (1960); H. Einaga and H. Ishii, *Jpn. Anal.* **17**:836 (1968).
	(2) 8-Hydroxyquinoline (R): pI_1 3.5–4.5, 0.5% R in $CHCl_3$.	550	0.2–7	Fe(III) and many other metals.	E. B. Sandell, *Ind. Eng. Chem., Anal. Ed.* **8**:336 (1936); N. A. Talvitie, *Anal. Chem.* **25**:604 (1953).
Zinc	(1) Dithizone (R): pH 4–5.5 with acetate buffer, 0.25 g $Na_2S_2O_3 \cdot 5H_2O$, extract with 10 mL 0.001% R in CCl_4 shaking 2 min.	540	0.1–0.7	Co, Ni.	D. W. Margerum and F. Santacana, *Anal. Chem.* **32**:1157 (1960).
	(2) Zincon (R): pH 8.5–9.5, borate buffer, 0.002*M* R.	620	0.3–2	Al, Be, Bi, Cd, Co, Cu, Fe(III), Mn, Mo(VI), Ni, Ti(IV).	R. M. Rush and J. H. Yoe, *Anal. Chem.* **26**:1345 (1954); G. Ackermann and J. Köthe, *Talanta* **26**:693 (1979).
Zirconium	(1) Alizarinsulfonate (R): 0.2*M* HCl, 2 mL 0.05% R, dilute to 100 mL, let stand 1 h.	525	0.2–3	Fe(III), Mo(VI), Sb, W; F^-, SO_4^{2-}, silicate.	D. E. Green, *Anal. Chem.* **20**:370 (1948); G. B. Wengert, *ibid.* **24**:1449 (1952).
	(2) Quercetin (R): 0.5*M* HCl, 5 mL EtOH, 3 mL 0.1% R in EtOH, dilute to 25 mL.	440	0.1–2.5	Cr(VI), Fe(III), Ga, Ge, Hf, Hg(I), Mo(VI), Nb, Pt metals, Sb(III), Sc, Sn(IV), Ta, Th, Ti, V, W(VI), F^-, PO_4^{3-}.	F. S. Grimaldi and C. E. White, *Anal. Chem.* **25**:1886 (1953).

TABLE 6.16 Spectrophotometric Methods for the Determination of Nonmetals

Element determined	Procedure	λ, nm	Range, $\mu g \cdot mL^{-1}$	Interferences	References
Boron	(1) Curcumin (R): Dilute sample to 50 mL with 0.035% R in EtOH plus 1.5% $H_2C_2O_4$ and 0.3*M* HCl.	555	0.01–0.1	Ge, fluoride, nitrate, nitrite, perchlorate.	G. S. Spicer and J. D. H. Strickland, *Anal. Chim. Acta* **18**:231 (1958); *Analyst* **91**:576 (1966).
	(2) Carminic acid (R): 0.1% R in 96% H_2SO_4, heat at 50°C for 20 min.	615	0.05–0.5	Ce(IV), Ge, Nb, Sc, Ti(IV), Zr, fluoride, NO_2^-, NO_3^-.	R. S. Brown, *Anal. Chim. Acta* **50**:157(1970): D. L. Callicoat et al., *Anal. Chem.* **31**:1434 (1959); *Mikrochim. Acta* **1**:445 (1980).
	(3) 1,1-Dianthrimide (R): 0.4% R in 96% H_2SO_4, heat at 80°C for 4.5 h.	635	0.05–0.5	>100 μg Te, >30 μg Ge, >200 μg Br^- and I^-, >0.9 μg F^-, >1 mg PO_4^{3-}.	L. Danielsson, *Talanta* **3**:138 (1959); H. K. L. Gupta and D. F. Boltz, *Anal. Lett.* **4**:161 (1971); *Water Res.* **5**:41 (1971).
	(4) 1,8-Dihydroxynaphthalene-4-sulfonic acid, pH 10.2, extract with 1,2-dichloroethane.	342		Al, Cu, Fe, Ti.	T. Korenaga et al., *Analyst* **105**:955 (1980).
Bromine	(1) Phenol red (R): Chloroamine-T to oxidize bromide, pH 5, 0.01% R.	590	0.1–1	NH_4^+, I_2, or I^-, reducing agents reacting with chloroamine-T.	E. Goldman and D. Byles, *J. Am. Water Works Assoc.* **51**:1051 (1959).
	(2) Methyl orange (R): Hypochlorite, pH 6, excess removed by formate; bleaching of R measured at pH 2.	522	0.02–0.15	Mn(III).	M. Taras, *Anal.Chem.* **19**:342 (1947).
	(3) Rosaniline (R): 0.006% R, 1.5*M*–3.5*M* H_2SO_4.	570	0.4–3		W. J. Turner, *Ind. Eng. Chem., Anal. Ed.* **14**:599 (1942).
Carbon (CO_2)	Phenolphthalein (R): Measure decrease in absorption of R.	515	0.0005–0.032%	Other acidic gases.	N. A. Spector and B. F. Dodge, *Anal. Chem.* **19**:55 (1947).
Carbon (CO)	Palladium(II) chloride–molybdophosphoric acid: 3 mL (0.2% $PdCl_2$, 0.022*M* HCl, 5% molybdophosphoric acid), 3 mL acetone, stopper, heat at 60°C for 60 min.	820	0.002–0.06%	Ethylene, acetylene, H_2S, H_2.	R. L. Beatty, *U.S. Bur. Mines Bull. No. 557* (1944).
Carbon (CN)	Pyridine-pyrazolone (R): To 50-mL flask, add 5 mL 1.4% phosphate buffer pH 6.8, 0.3 mL 1% chloramine-T, 15 mL R, stand 30 min.	620		Thiocyanate; metal hydrous oxides filtered off.	American Public Health Association. *Standard Methods for the Examination of Water and Wastewater*, 13th ed., Washington, D.C., 1971, pp. 397–406.

(Continued)

TABLE 6.16 Spectrophotometric Methods for the Determination of Nonmetals (*Continued*)

Element determined	Procedure	λ, nm	Range, $\mu g \cdot mL^{-1}$	Interferences	References
Carbon (SCN)	Thiocyanate: 5 mL neutralized sample, 1 mL 5% $Fe(NO_3)_3$ in 0.4*M* HNO_3, stand 5 min.	550		Ag, Bi, Cd, Cu, Hg, Mo, Ti, Zn; F^-, I^-, meta- and pyrophosphates, oxalate, phenol.	J. H. Karchmer, *The Analytical Chemistry of Sulfur and Its Compounds*, Wiley, New York, 1970, Part I.
Chlorine (Cl_2)	(1) *o*-Tolidine (R): 5 mL 0.1% R in 1.8*M* HCl per 100 mL, stand 9 min, pH 1.6.	438	0.01–1.0	Organic matter, Fe(III), Mn(VII or IV), nitrite, other oxidants.	H. C. Marks and R. R. Joiner, *Anal. Chem.* **20**:1197 (1948).
	(2) Methyl orange (R): Bleaching of 6×10^{-5}% R at pH 2.	505	0.02–0.6	Oxidized forms of Mn.	M. Taras, *Anal. Chem.* **19**:342 (1947); F. W. Sollo et al., *J. Am. Water Works Assoc.* **55**:1575 (1965).
	(3) König's reaction: Sulfanilic acid, pyridine, KCN, pH 8, heat at 60°C for 40 min.	395	0.02–2.0	Bromine, chloroamines, Mn(III).	I. Nusbaum and P. Skupeko, *Anal. Chem.* **23**:1881 (1951); P. K. Morris and H. A. Grant, *Analyst* **76**:492 (1951).
Chlorine (Cl^-)	(1) Indirect Hg(II) chloranilate (R): 100 mL containing 5 mL 1*M* HNO_3, 50 mL methyl Cellosolve, 0.2 g R; shake 15 min, centrifuge.	305	0.1–10	Interfering cations removed by Dowex 50-X8 in H form; bromide, fluoride, iodate, iodide, phosphate, SCN.	R. J. Bertolacini and J. E. Barney, *Anal. Chem.* **30**:202, 498 (1958).
	(2) $Hg(SCN)_2$: In 50-mL flask, add 10 mL sample, 5 mL 60% $HClO_4$, 1 mL $Hg(SCN)_2$, 2 mL $Fe(ClO_4)_3$.	460	1–10	Bromide, cyanide, iodide, nitrite, phosphate, sulfate, thiosulfate.	D. M. Zall et al., *Anal. Chem.* **28**:1665 (1956); *Anal. Chim. Acta* **123**:347 (1981).
Chlorine (ClO_3^-)	Sample plus 3 mL 0.5% benzidine in 0.06*M* HCl, 90 mL 5.8*M* HCl, dilute to 100 mL.	434	1–50	Other oxidants.	E. A. Burns, *Anal. Chem.* **32**:1800 (1960).
Chlorine (ClO_4^-)	Form ion-pair complex of perchloratobis(2,9-dimethyl-1,10-phenanthroline)copper(I) and extract into ethyl acetate, pH 3–5.	456	5–100 μg		W. J. Collinson and D. F. Boltz, *Anal. Chem.* **40**:1896 (1968).
Fluorine (F^-)	(1) Zr-Eriochrome Cyanine R (R): 50 mL sample, 5 mL 0.18% R in water, 5 mL $ZrOCl_2 \cdot 8H_2O$ solution (0.265 g plus 700 mL HCl per liter).	525	0–1.2	Al, phosphate, sulfate.	M. S. Frant and J. W. Ross, *Anal. Chem.* **40**:1169 (1968); S. Megregian, *ibid.* **26**:1161 (1954); *Analyst* **93**:643 (1968).

	(2) Zr-SPADNS* (R): 50.0 mL sample plus 10 mL mixed reagent [5 mL 0.2% SPADNS and 5 mL Zr solution described in method (1)].	570	0–1.4	Al, Fe(III), metaphosphate, phosphate, sulfate.	R. Belcher et al., *J. Chem. Soc.* **1959**:3577.
	(3) Alizarin fluorine blue: pH 4.6, 5 mL 1.2% succinate buffer, 10 mL 0.09% 1,2-dihydroanthraquinonylmethylamine-*N,N*-diacetic acid plus 0.8% $La(NO_3)_3$, 10 mL acetone, dilute to 50 mL. Let stand 30 min.	635	0.04–0.8	Preliminary separation advised. Most cations, acetate, carbonate, citrate, silicate, sulfate, sulfide, tartrate.	R. A. Kletsch and F. A. Richards, *Anal. Chem.* **31**:1435 (1970); E. J. Newman et al., *Analyst* **96**:384 (1971).
Iodine (I^-)	(1) Catalysis of Ce(IV)–As(III) reaction.	420	0.01–0.15	Ag, Hg(II), Os; Br^-, Cl^-, CN^-.	K. R. Meyer et al., *Am. J. Clin. Pathol.* **25**:1160 (1955); E. B. Sandell and I. M. Kolthoff, *Microchim. Acta* **1**:9 (1951).
	(2) Starch–I_2: Oxidize I^- to IO_3^- with $KMnO_4$, destroy excess with $NaNO_2$, add KI and 0.25% starch.	575	0.03–0.3		F. G. Houston, *Anal. Chem.* **22**:493 (1950); W. G. Gross et al., *ibid.* **20**:900 (1948).
	(3) Ultraviolet I_2: Oxidize I^- to IO_3^- with alkaline $KMnO_4$, add KI, extract I_2 into toluene, stand 10 min.	311	0.5–5		J. J. Custer and S. Natelson, *Anal. Chem.* **21**:1003, 1005 (1949).
Nitrogen (NH_3)	(1) Nessler: K_2HgI_4 reagent, EDTA, stand 10 min.	440	2–25	Fe, Mg, Mn(II); sulfide.	G. Moeller, *Z. Anal. Chem.* **245**:155 (1969). J. R. Polley, *Anal. Chem.* **26**:1523 (1954).
	(2) Indophenol: 10 mL sample, 5 mL 2% phenol (with 0.01% $Na[Fe(CN)_5NO]$), 5 mL NaOCl in 0.1% NaOH, warm to 37°C for 15 min.	635	0.05–1.0	Cu(II), NH_2OH, certain amino acids, urea.	M. W. Weatherburn, *Anal. Chem.* **39**:971 (1967); A. L. Chaney and E. P. Marbach, *Clin. Chem.* **8**:130 (1962).
	(3) Bispyrazolone method: To 50 mL sample, add 5 mL acetate buffer (pH 3.7), 2 mL 1% chloramine-T, stand 15.0 min at 15°C. Add 6 mL 0.2% bispyrazolone, mix, after 5 min add 10 mL 0.25% pyrazolone. Add 2 mL 0.5*M* HCl, extract 3 min with 10 mL CCl_4.	450	0.025–2.0	Ag, Co, Cu(II), Fe(II), Zn; cyanide, cyanate, thiocyanate.	L. Prochazkova, *Anal. Chem.* **36**:865 (1964); J. B. Lear and M. G. Mellon, *ibid.* **29**:293 (1957).

(Continued)

TABLE 6.16 Spectrophotometric Methods for the Determination of Nonmetals (*Continued*)

Element determined	Procedure	λ, nm	Range, μg · mL^{-1}	Interferences	References
Nitrogen (NO_2^-)	(1) Griess method: To 50 mL sample, add 1 mL each: 0.6% sulfanilic acid in 2.5*M* HCl, let stand 3–10 min, pH 1.4; 0.6% 1-aminonaphthalein in 0.12*M* HCl; 2*M* acetate buffer, pH 2.0–2.5. Stand 10–30 min.	520	0.01–0.15	Urea, aliphatic amines; ClO_3^-, ClO_4^-, Ce(IV), IO_4^-, MnO_4^-, $S_2O_8^{2-}$, W(VI); Fe(II), I^-, $SnCl_4^{2-}$, S^{2-}, $S_2O_3^{2-}$, SO_3^{2-}; Ag, Au, Bi, Fe(III), Pb, Pt, Sb(III), V.	B. F. Rider and M. G. Mellon, *Ind. Eng. Chem., Anal. Ed.* **18**:96 (1946); H. Barnes and A. R. Folkard, *Analyst* **76**:599 (1951).
	(2) Diazotized sulfanilic acid: Sample adjusted to pH 1.4, 1 mL 0.6% sulfanilic acid, dilute to 50 mL, stand 15 min.	270	0.05–1.0	Ag, Au, Bi, Cr(VI), Fe(III), Mn(VII), Mo(VI), V(V); chlorate, sulfite.	J. M. Pappenhagen, *Anal. Chem.* **25**:341 (1953).
Nitrogen (NO_3^-)	(1) 1,2,4-Trihydroxybenzenedisulfonic acid (R): Add 2 mL 12% R, sufficient NH_3 to develop maximum color, dilute to 100 mL.	410	0.1–2	Colored ions; chloride, nitrite, organic matter	E. M. Chamot and D. S. Pratt, *J. Am. Chem. Soc.* **32**:630 (1910).
	(2) Brucine (R): Add 0.2 mL 5% R in $CHCl_3$ to 10 mL sample, add 20 mL conc. H_2SO_4, stand 10 min, dilute to 50 mL.	410	0.2–7	Chlorine, nitrite	C. A. Noll, *Ind. Eng. Chem., Anal. Ed.* **3**:311 (1931); **17**:426 (1945).
	(3) Chromotropic acid (R): 2.5 mL sample, 1 drop 5% urea–4% sulfite solution, 2 mL 0.5% $Sb_2(SO_4)_3$ in conc. H_2SO_4, 1 mL R, stand 45 min.	410			P. W. West and T. P. Ramachandran, *Anal. Chim. Acta* **35**:317 (1966); J. J. Batten, *Anal. Chem.* **36**:939 (1964).
Oxygen (O_2)	Winkler method: Alkaline KI, Mn(II), acidification, liberated I_2 determined (a) starch, (b) extraction, (c) ultraviolet as triiodide.	(a) 580 (b) 450 (c) 352	(a) 0.005–0.5		(a) H. A. J. Pieters and W. J. Hanssen, *Anal. Chim. Acta* **2**:712 (1948). (b) L. Silverman and W. Bradshaw, *ibid.* **12**:526 (1955). (c) T. Ovenston and J. Watson, *Analyst* **79**:383 (1954).
Oxygen (O_3)	(1) Triiodide: KI solution plus acetate buffer, pH 3.8.	355	0.01–0.70	SO_2, H_2S; other oxidants.	S. Deutsch, *J. Air Pollut. Control Assoc.* **18**:78 (1968).
	(2) Iron(II)–SCN: 10 mL 0.05% Fe(II), draw in sample, add 2 mL 50% NH_4SCN, dilute to 25 mL.	480	>2	Other oxidants.	A. C. Egerton et al., *Anal. Chim. Acta* **10**:422 (1955); G. W. Todd, *Anal. Chem.* **27**:1490 (1955).

Element	Method	Wavelength, nm	Range, ppm	Interferences	References
Oxygen (H_2O_2)	(1) Peroxytitanic acid: Sample to 10-mL flask, add 1 mL 0.46% $TiOSO_4$ and 2.5 mL 9*M* H_2SO_4; heat at 60°C for 10 min.	407	1–30	Al, Cr(VI), Ni; borate, phosphate (> 0.005*M*); formaldehyde.	W. C. Wolfe, *Anal. Chem.* **34**:1328 (1962); C. B. Allsopp, *Analyst* **66**:371 (1941).
	(2) Oxidase: 1–5 mL sample, 1 mL 0.05% leuco crystal violet in 0.5% HCl, 0.5 mL (1 mg/mL) horseradish peroxidase, 5 mL (pH 4.5) 1*M* acetate buffer, dilute to 10.0 mL.	596	0.06–0.5		H. A. Mottola, B. E. Simpson, and G. Gorin, *Anal. Chem.* **42**:410 (1970).
Phosphorus (PO_4^{3-})	(1) Molybdophosphoric acid: (a) 0.04*M* Na_2MoO_4, 0.257*M* HNO_3. (b) Extract with 10 mL 20% 1-butanol in $CHCl_3$.	(a) 380 (b) 310	1–15	(a) Bi, Cu(II) (> 100 ppm), Fe(III), Ge, Ni (> 40 ppm), Sn(II); AsO_4^{3-}, F^- (> 25 ppm), SiO_3^{2-}, W(VI), V(V). (b) As, Ge, Si do not interfere.	(a) D. F. Boltz and M. G. Mellon, *Anal. Chem.* **29**:749 (1948). (b) C. Wadelin and M. G. Mellon, *ibid.* **25**:1668 (1953).
	(2) Heteropoly blue: 0.5*M* H_2SO_4, 0.12 g $Na_2MoO_4 \cdot 2H_2O$, 2 mL 0.15% $N_2H_4 \cdot H_2SO_4$, 50 mL total volume, 100°C bath for 10 min.	830	0.1–0.2	As(V), Ba, Bi, Fe(III) (< 200 ppm), Pb, Sb, Sn, W(VI); nitrate.	D. F. Boltz and M. G. Mellon, *Ind. Eng. Chem., Anal. Ed.* **19**:873 (1947); *Talanta* **27**:263 (1980).
	(3) Extraction heteropoly blue: 25 mL sample, 5 mL water, 5 mL $HClO_4$, 5 mL 1% molybdate, extract with 40 mL isobutyl alcohol. Shake organic phase with 25 mL 2.4% $SnCl_2 \cdot 2H_2O$, 2*M* in HCl.	725	0.2–1.5	As(V), As(III) (< 50 ppm), Au, Ge, Sn(II), Sn(IV) (< 25 ppm), V(V), W(VI); I^- (< 50 ppm), $S_2O_3^{2-}$, SCN^- (< 50 ppm).	P. Pakalns, *Anal. Chim. Acta* **40**:1 (1968); H. M. Theakston and W. R. Bandi, *Anal. Chem.* **38**:1764 (1966); C. H. Locke and D. F. Boltz, *ibid.* **30**:183 (1958).
	(4) Molybdovanadophosphoric acid: 0.5*M* HNO_3, 0.002*M* vanadate(V), 0.01*M* molybdate.	460	5–40	Ag, As (< 100 ppm), Bi, Ce(IV), Co, Cr(III,VI), Fe(II), Mn(VII), Sn(IV), Th; F^-, I^-, S^{2-}, SCN^-, $S_2O_3^{2-}$.	R. E. Kitson and M. G. Mellon, *Ind. Eng. Chem., Anal. Ed.* **16**:379 (1944); O. B. Michelson, *Anal. Chem.* **29**:60 (1957).
	(5) Extraction of method (4): 1*M* HCl, extract with 1-pentanol.	308	0.2–1.1		R. J. Jakubiec and D. F. Boltz, *Mikrochim. Acta* **1969**:181.
Selenium	(1) Sn(II): In 50 mL, 2*M*–3*M* HCl, 2 mL 10% $SnCl_2 \cdot 2H_2O$ in 1*M* HCl, 3 mL 4% gum arabic.	400	4–40		S. T. Volkov, *Zavod. Lab.* **5**:1429 (1939); V. S. Zemel, *ibid.* **5**:1433 (1939).

(Continued)

TABLE 6.16 Spectrophotometric Methods for the Determination of Nonmetals (*Continued*)

Element determined	Procedure	λ, nm	Range, $\mu g \cdot mL^{-1}$	Interferences	References
Selenium	(2) N_2H_4: To boiling solution (pH 8–10) add 2 mL 85% $N_2H_4 \cdot H_2O$, heat 2–5 min, cool, dilute to 50 mL.	260	2–18		R. W. Haisty, M. S. Thesis, University of Illinois, Urbana, 1955.
	(3) Iodine method: In 25.0 mL, 2*N* H^+. (a) Add 2 mL 5% CdI_2. (b) Add 2 mL 1% CdI_2–0.25% starch. Dilute both to volume, stand 5 min.	(a) 352 (b) 615	0.2–1		J. L. Lambert, P. Arthur, and T. E. Moore, *Anal. Chem.* **23**:1101 (1951).
	(4) 3,3′-Diaminobenzidine (R): 50 mL, pH 2–3 with formate buffer, 2 mL 0.5% R, stand 30–50 min, adjust to pH 6–7 with NH_3, extract with 10.0 mL toluene.	420	1–10	Cu(II) and Fe(III) unless masked with oxalate and fluoride, respectively. Cr(VI), V(V). Substances that reduce or complex Se(IV).	K. L. Cheng, *Anal. Chem.* **28**:1738 (1956); J. Hoste and J. Gillis, *Anal. Chim. Acta* **12**:58 (1955).
	(5) 2,2′-Dianthrimide (R): 5 mL 0.05% R, 12 mL conc. H_2SO_4, heat 5 h at 90°C.	605	56–2500	Specific method for Se(IV).	F. J. Langmyhr and I. Dahl, *Anal. Chim. Acta* **29**:377 (1963).
Silicon	(1) Molybdosilicic acid: To 50 mL sample, 2 mL 10% $(NH_4)_6Mo_7O_{24}$, stand 10 min, 1.5 mL 10% tartrate.	350	1–6	Ge, phosphate.	L. G. Hargis, *Anal. Chem.* **42**:1494, 1497 (1970).
	(2) Heteropoly blue: 90 mL sample, 1 mL 7.5% $(NH_4)_6Mo_7O_{24}$, stand 5 min, 4 mL 10% tartaric acid, 1 mL 0.0016% 1-amino-2-naphthol-4-sulfonic acid– 10% Na_2SO_3–10% $NaHSO_3$, pH 1.6, stand 20 min.	815	0.1–1.5	Ba, Bi, Pb, Sb cause turbidity.	D. F. Boltz and M. G. Mellon, *Ind. Eng. Chem., Anal. Ed.* **19**:873 (1947); L. A. Trudell and D. F. Boltz, *Anal. Chem.* **35**:2122 (1963).
Sulfur (H_2S)	(1) Methylene blue: 2% $Zn(OAc)_2$, 0.05% *p*-amino-*N,N*-dimethylaniline in 2.7*M* H_2SO_4, $FeCl_3$ catalyst followed by $(NH_4)HPO_4$.	745	0.005–0.5	Cu(II).	A. E. Sands et al., *U.S. Bur. Mines Rept. 4547,* Washington, D.C., 1949; R. Pomeroy, *Sew. Works J.* **13**:498 (1941).
	(2) Add *N,N*-Dimethyl-*p*-phenylene-diamine plus $Cr_2O_7^{2-}$ in acid solution, which gives methylene blue.	745	0.005–0.5		EPA approved method.

Sulfur (SO_2)	(1) Pararosaniline–SO_2–HCHO: SO_2 absorbed in 0.1*M* K_2HgCl_4, 2 mL 0.04% pararosaniline, 2 mL 0.2% HCHO, 25 mL total volume, stand 25 min.	560	0.2–1	NO_2 (removed by H_2NSO_2OH), O_3.	J. B. Pate et al., *Anal. Chem.* **34**: 1660 (1962); E. Lahman and K. E. Prescher, *Z. Anal. Chem.* **251**:300 (1970); R. V. Nauman, *ibid.* **32**:1307 (1960).
	(2) Fe(II)-1,10-phenanthroline: 10 mL 0.001*M* Fe(III), 10 mL 0.03*M* 1,10-phenanthroline, pH 5.5, 1 mL 1-octanol, 50°C, transfer to 100 mL, 2 mL NH_4HF_2.	510	(75 μL)		B. G. Stephens and F. Lindstrom, *Anal. Chem.* **36**:1308 (1964).
Sulfur (SO_4)	(1) Barium chloranilate (R): 10 mL 1% KH phthalate (pH 4), 45% EtOH, 100 mL total volume; 0.3 g R, centrifuge or filter.	320 or 520	0.5–10 10–300	Remove cations by cation exchange resin in H form.	R. J. Bertolacine and J. E. Barney, *Anal. Chem.* **29**:281 (1957); **30**:202, 498 (1958).
	(2) Benzidine sulfate precipitated; benzidine portion, 0.2*M* HCl, 0.1% $NaNO_2$, 0.5% $H_2NSO_2ONH_4$, 0.1% *N*-(1-naphthyl)ethylenediamine HCl.	550	0.3–3 as S	Chloride and phosphate.	B. Klein, *Ind. Eng. Chem., Anal. Ed.* **16**:536 (1944); T. V. Letnoff and J. G. Reinhold, *J. Biol. Chem.* **114**:147 (1936).
Tellurium	(1) Sn(II); see selenium, method (1)	400	2–20		
	(2) Iodotellurite: 2*M* KI, 0.15*M*–0.4*M* HCl, stand 20 min.	335	0.4–2		R. A. Johnson and J. P. Kwan, *Anal. Chem.* **23**:651 (1951).
	(3) HPH_2O_2: To boiling solution, 1–8 meq HCl, 5 mL 3*M* HPH_2O_2, digest 100°C for 15 min, 3 mL 4% gum arabic, dilute to 50 mL.	240–290	2–14		R. A. Johnson and B. R. Anderson, *Anal. Chem.* **27**:20 (1955); R. A. Johnson and J. P. Kwan, *ibid.* **25**:1017 (1953).

* SPADNS = 4,5-Dihydroxy-3-(2-hydroxy-5-sulfophenylazo)-2,7-naphthalenedisulfonic acid.

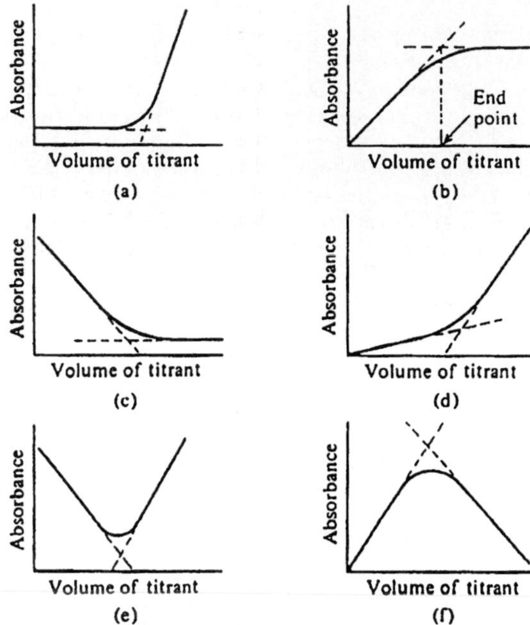

FIGURE 6.4 Possible shapes of photometric titration curves.

H = Heterometric titration

SPADNS = 2-(4-Sulfophenylazo)-1,8-dihydroxynaphthalene-3,6-disulfonic acid

Thoron = 2-(2-Hydroxy-3,6-disulfo-1-naphthylazo)benzenearsonic acid

Attention is especially called to the fact that a blank space in the column headed "Interferences" means merely that no information is being given; it does not mean that the titration in question is free from interferences.

6.5 TURBIDIMETRY AND NEPHELOMETRY

6.5.1 Principles

Turbidimetry and nephelometry involve the formation of a suspension and the measurement of the amount of radiation that passes through the sample in the forward direction (*turbidimetry*) or the amount of radiation scattered (*nephelometry*). Nephelometric measurements are made at an angle to the direction of the beam of radiation through the sample, usually 45° or 90°. These methods are applicable to colorless, opaque suspensions that do not show selective absorption, and so white light is generally used. There is no sharp division between these methods and absorptiometric methods involving true solutions, for example, procedures involving the formation of color "lakes."

Many elements can be determined turbidimetrically or, less often, nephelometrically. But as a rigid control of conditions is essential, the use of these techniques in practical analysis is restricted to the determination of elements that do not give good color reactions or that give precipitates that are difficult to separate from their mother liquors.

TABLE 6.17 Photometric Titrations of Inorganic Substances

Substance titrated	Scale	Solution composition	Reagent	Indicator	λ, nm	Absorbing species	Interferences
Ag	0.1–10 mg per 80 mL	Aq. NH_3 + excess $K_2Ni(CN)_4$, pH 11.3–11.8	EDTA (0.6–5 m*M*)	Murexide	435 F	Ni–murexide complex	Ba, Ca, Cu(II), Hg(II), Li, Mg, Pb, Sr, Tl(I), Zn
As(III)	0.5–50 mg per 75–100 mL	0.5*M* H_2SO_4 + OsO_4 catalyst	$Ce(SO_4)_2$ (0.0004*M*–0.1*M*)	None	320	$Ce(SO_4)_2$ excess	Other reducing agents
Ba	0.1–5 mg per 35–100 mL	NH_3 buffer, pH 10	EDTA (0.002*M*–0.1*M*) + Mg–EDTA complex	Eriochrome Black T	630–650	FI	Most other metals except alkali metals
Bi	0.2–0.7 mg per 100 mL	Chloroacetate buffer, pH 2	EDTA (0.001*M*–0.01*M*)	None	265	Bi–EDTA complex	Fe(III), Th
	0.5–10 mg per 100 mL	Chloroacetate buffer, pH 2	EDTA (0.01*M*)	Thiourea	400	Bi–thiourea complex	Fe(III)
Ca	0.1–0.6 mg per 20 mL	Diethylamine buffer, pH 12.3	EDTA (0.0004*M*)	Calcon	600 F	FI	
	0.02–0.4 mg per 6–100 mL	NH_3 buffer, pH 10	EDTA (1–2.5 m*M*) + Mg–EDTA complex	Eriochrome Black T	630–650	FI	
	5–200 μg per 1–10 mL	NaOH, pH 12.5–13.5	EDTA (0.5–10 m*M*)	Murexide	600–620	FI	
Cd	1–10 mg per 90 mL	NH_3 buffer, pH 10	EDTA (1–10 m*M*)	None	222–228	Cd–EDTA complex	Most other metal ions
Ce(III)	0.5–100 mg per 90 mL	10% $K_2P_2O_7$, pH 5.5–7.0	$KMnO_4$ (2–40 m*M*)	None	525	MnO_4^-	As(III), Cr(III), F^-, Hg(I), I_2, Sb(III), Tl(I), V(IV)
Co	0.2–0.5 mg per 20 mL	Na_3Citrate + HOAc	α-Nitroso-β-naphthol (0.01*M* in EtOH)	None	H	Co-α-nitroso-β-naphthol complex	Cu(II), Fe(III)
$Cr_2O_7^{2-}$	1.5–20 mg per 50 mL	0.75*M* H_2SO_4	0.05*M* $HAsO_2$	None	350	$Cr_2O_7^{2-}$	Other reducing agents
Cu(II)	0.6–6 mg per 25 mL	KI, OAc^- buffer, pH 4	$Na_2S_2O_3$ (0.025*M*)	None	390 F	I_3^-	
	0.5–4 mg per 10–200 mL	Citrate buffer, pH 7	0.1*M* Triethylene-tetramine SO_4^{2-}	None	575 F	Cu(II)–trien complex	
	10–60 mg per 90 mL	pH 2.6	0.1*M* EDTA	None	745	Cu(II)–EDTA complex	Bi, Co, Fe(III), Ni, Pb
F^-	5–50 μg per 35 mL	Chloroacetate buffer, pH 3.0	0.05 m*M* $Th(NO_3)_4$	Alizarin Red S	520	Th-Alizarin Red S lake	
	5–70 μg per 25 mL	pH 3.1	1 m*M* $Th(NO_3)_4$	SPADNS	580	Th–indicator lake	Bi(III), Cl^-, phosphate

(Continued)

TABLE 6.17 Photometric Titrations of Inorganic Substances (*Continued*)

Substance titrated	Scale	Solution composition	Reagent	Indicator	λ, nm	Absorbing species	Interferences
Fe(III)	10–60 mg per 90 mL	pH 2	0.1*M* EDTA	Salicylic acid	525	Fe(III)–salicylate complex	Bi, Co, Cu(II), Ni, Pb
	0.2–1 mg per 100 mL	Chloroacetate buffer, pH 3.5	0.01*M* EDDHA	None	470	Fe(III)–reagent complex	Ag, Al, Cu(II), Th
K	4–28 μg per 0.4 mL	At pH 4 add excess $NaB(C_6H_5)_4$, separate precipitate and back-titrate	Hexadecyltrimethylammonium bromide (2 m*M*)	None	570	Hexadecyltrimethyl ammonium tetraborate	
Mg	0.08–2 mg per 90 mL	NH_3 buffer, pH 10	EDTA (1–10 m*M*)	None	222–228	Mg–EDTA complex	Most other metal ions
	1 μg–2 mg per 100 mL	NH_3 buffer, pH 10	EDTA (1–10 m*M*)	Eriochrome Black T	630–660	FI	
Mo(V)	1–11 mg per 90 mL	OAc^- buffer, pH 4.2, excess EDTA	0.01*M* $ZnCl_2$	Aliz C	520 F	Zn–indicator complex	
Ni	10–60 mg per 90 mL	OAc^- buffer, pH 4.0	0.1*M* EDTA	None	1000	Ni–EDTA complex	
H_2O	0.5–40 μg per 3 mL	HOAc + H_2SO_4 catalyst	Acetic anhydride	None	250–260	Acetic anhydride	
Pb	0.2–2 mg per 100 mL	Chloroacetate buffer, pH 2	EDTA (1–10 m*M*)	None	240	Pb–EDTA complex	Co, Cu(II), Fe(III), Ni, Th
Rare earths	0.01–40 mg per 100 mL	Pyridine buffer, pH 6	EDTA (0.01–50 m*M*)	Arsenazo	570	Rare-earth–indicator complex	
	0.17–170 mg/200 mL	OAc^- buffer, pH 4.6; excess EDTA	2.5 m*M* $La(NO_3)_3$	Alizarin Red S	520 F	FI	
SO_4^{2-}	6–50 μg per 35 mL	H_2O–MeOH–isopentanol	1.25 m*M* $Ba(ClO_4)_2$	Thoron	520	Ba–thoron complex	
Sb(III)	3–30 mg per 80 mL	1*M*–1.5*M* H_2SO_4	Neutral 2–20 m*M* $KBrO_3$ + excess KBr	None	296–330	Br_3^-	Other substances that react with Br_2
Sc	4–11 mg per 75 mL	pH 3.0	0.03*M* EDTA	$Cu(NO_3)_2$	745	Cu(II)–EDTA complex	Bi, F^-, Hf, SO_4^{2-}, Zr
Sr	0.05–0.9 mg per 100 mL	NH_3 buffer, pH 10	2.5 m*M* EDTA + Mg–EDTA complex	Eriochrome Black T	630	FI	
H_6TeO_6	10–500 mg per 50 mL	Aqueous solution	0.02*M*–2*M* NH_3	None	250–280	$H_5TeO_6^-$	

Th	0.09–180 mg per 200 mL	pH 2.8	2.5 m*M* EDTA	Alizarin Red S	520 F	FI	
	2–60 μg per 35 mL	Aqueous EtOH, pH 3.0	0.2 m*M* EDTA	Quercetin	422	Th–quercetin complex	F^-, Fe(II,III), Mn, Mo(VI), PO_4^{3-}, SO_4^{2-}, Ti, OAc^-
V(V)	0.5 mg per 30 mL	0.3*M* HCl containing 1% H_2O_2	6 m*M* pyridine-2,6-dicarboxylic acid	None	436 F	V(V)–H_2O_2–reagent complex	Nb(V), Sn(IV), Ta
Zn	0.1–5 mg per 100 mL	NH_3 buffer, pH 9.5	2.5 m*M* EDTA	Eriochrome Black T	665	FI	Other metals titrated with EDTA at pH 9.5 unless masked
	30–50 μg per 50 mL	C_6H_6 saturated with NH_4OAc plus a little MeOH	Dithizone (2 m*M* in benzene)	None	640	FI	Co, FE(III), Ni, Pb
Zr	10–100 mg per 300 mL	Excess EDTA, pH 2.3	0.02*M* $FeCl_3$	K benzohydroxamate	520 F	Fe(III)–indicator complex	Al, Bi, Ce(IV), Sn(II,IV), Th, Ti(IV)

TABLE 6.18 Photometric Titrations of Organic Substances

Substance titrated	Scale	Solution composition	Reagent	Indicator	λ, nm	Absorbing species	Interferences
Al oxinate	10–200 mg per 100 mL	Glacial HOAc	0.1*M* $HClO_4$ in HOAc	None	450	Al oxinate	
Aniline	10 mg per 80 mL	MeOH–H_2O–12*M* HCl–40% aqueous KBr	Neutral 0.04*M* $KBrO_3$ containing excess KBr	None	350	Br_3^-	Other compounds that react with Br_2
4–Bromophenol	20–60 mg per 100 mL	Aqueous solution	0.1*M* NaOH	None	325	4-$BrC_6H_4O^-$	Other acidic compounds
Oleic acid	15–150 mg per 80 mL	MeOH–glacial HOAc–12*M* HCl–40% aqueous KBr	Neutral 0.04*M* $KBrO_3$ containing excess KBr	None	360	Br_3^-	Other compounds that react with Br_2
Phenol A	25 mg per 40 mL	> 90% acetone	Tributylmethylammonium OH (0.1*M* in 80% C_6H_6–20% 2-PrOH)	Azo violet	660 F	Basic form of azo violet	Other acidic compounds
4-Phenylphenol	20 mg per 100 mL	Butylamine	0.05*M* NaOH in absolute EtOH	None	372	4-$C_6H_5C_6H_4O^-$	Certain other phenols
Procaine HCl	420 mg per 150 mL	6*M* HCl	0.1*M* $NaNO_2$	None	385	HNO_2	
Propoxycaine HCl	1.5 mg per 150 mL	6*M* HCl	0.1*M* $NaNO_2$	None	385	Propoxycaine diazonium Cl^-	
Quinoline	130 mg per 100 mL	Glacial HOAc	0.1*M* $HClO_4$ in HOAc	None	350	Quinolinium ion	Other basic compounds
NaOAc	80 mg per 100 mL	Glacial HOAc	0.1*M* HCl in HOAc	2-Chloroaniline	312	FI	Other basic compounds

6.5.2 Sensitivity and Accuracy

Nephelometric methods are inherently more sensitive than turbidimetric (or colorimetric) ones, but are seldom used because the experimental conditions are more critical. The sensitivity is sometimes increased by the use of blue, instead of the customary white, light.

Because of the difference in the angle of measurement, turbidimetry is best suited for determining relatively high concentrations of suspended particles, whereas nephelometry is most suited for determining very low concentrations. The concentrations of solutions analyzed turbidimetrically usually lie between 0.05 and 0.5 mg per 100 mL. Under favorable conditions the range can be extended down to 0.02 mg per 100 mL and occasionally up to 2.0 mg per 100 mL. If a given suspension does not scatter strongly, that is, if the transmittance is greater than about 95% to 98%, turbidimetry should not be used. In this instance nephelometry would be much more sensitive since the small amount of scattered light would be measured against a black background.

For both techniques the accuracy is low because of the many variables involved and is usually about ±5%.

6.5.3 Calibration Curves

Calibration curves are usually empirical, since they depend on particle size as well as on concentration, and they should therefore be checked by absolute methods. In turbidimetry Beer's law is sometimes followed over only a limited range, but the addition of a protective colloid, such as gum arabic or gelatin, often stabilizes the suspension and extends the range.

The suspension that results from accurately weighing and dissolving 5 g of hydrazinium(2+) sulfate and 50 g of hexamethylenetetramine in 1 L of distilled water is defined as 4000 nephelometric turbidity units (NTU). After standing 48 h the insoluble polymer formazin, formed by the condensation reaction, develops a white turbidity. This turbidity can be prepared repeatedly with an accuracy of ±1%. The mixture can be diluted to prepare standards of any desired value.

6.5.4 Conditions

The suspensions should have these characteristics:

1. Low solubility.
2. High rate of formation.
3. Fair stability.
4. High opacity.
5. A refractive-index difference between the particle and its surrounding medium is needed if either reflection or scattering is to occur. It is sometimes advantageous to change solvents in order to increase the refractive-index differences.

Strict adherence to established conditions is essential if reproducible suspensions are to be obtained. The following factors affecting the primary and secondary particle size must be clearly defined:

1. Range of concentration of the substance being determined.
2. Concentration of reagent.
3. Rate and manner of addition of reagent.
4. pH of solution.
5. Temperature.
6. Rate and extent of agitation.
7. Addition of a protective colloid.
8. Effect of other salts in solution.

9. Time interval before measurement.
10. Wavelength of incident light (in turbidimetry). In turbidimetry it is important to choose a wavelength at which the sample solution does not absorb strongly.

6.5.5 Applications

Turbidimetry and nephelometry are used on gaseous, liquid, or even transparent solid samples. Whenever precipitates form that are difficult to filter, either due to small particle size or a gelatinous nature, they usually make ideal suspensions to be measured by light-scattering techniques, replacing gravimetric operations. A particularly valuable area of application is in air- and water-pollution studies, in which the two techniques are used to determine the clarity and to control the treatment of potable water, water-plant effluents, and other types of environmental waters. Similarly, light-scattering measurements are used to ascertain the concentration of smog, fog, smoke, and aerosols. Table 6.19 gives information on a number of the most useful procedures employing turbidimetric and nephelometric procedures.

Other applications include turbidimetric titrations, the measurements of the haziness of water, studies of the efficiency of filtration processes, and the examination of smokes and gases.

6.6 FLUORESCENCE ANALYSIS

The methods of fluorescence analysis outlined in this section are confined to the spectral region from 200 to 800 nm. Fluorescence methods are often more specific and more sensitive than colorimetric methods. Two types of spectra are determined in the development of a fluorometric procedure: the excitation spectrum and the emission spectrum. The excitation spectrum coincides with the absorption spectrum, and the excitation band of longest wavelength intersects the emission band. The excitation spectrum is found by measuring the intensity of the emission on exciting the substance over a wide range of radiant energy. The emission spectrum is obtained by measuring the intensity of the emission over a range of wavelengths while the substance is being irradiated with a monochromatic source. For organic compounds and metal chelates, both these measurements usually result in band spectra, and only the peak is reported in the tables cited in this section. The maximum excitation point is not necessarily chosen for analysis, since it may be too close to the emission for separation with filters in filter fluorimeters. An excitation should be chosen for which the compound under irradiation suffers the least decomposition. Much of the literature reports only the uncorrected excitation and emission maxima as given by a particular instrument. Data of this type are only approximate and require correction.[6] Most of the organic compounds and metal chelates that fluoresce above 400 nm can be excited with 350- to 360-nm radiation.

6.6.1 Photoluminescence Related to Concentration

Whether fluorescence (or phosphorescence), the luminescent power P_L is proportional to the number of molecules in excited states. In turn, this is proportional to the radiant power absorbed by the sample. Thus

$$P_L = \Phi_L (P_0 - P) \tag{6.16}$$

where Φ_L = luminescence efficiency or quantum yield of luminescence
P_0 = radiant power incident on the sample
P = radiant power emerging from the sample

[6] C. E. White, M. Ho, and E. Q. Weimer, *Anal. Chem.* **32**:438 (1960).

TABLE 6.19 Turbidimetric (T) and Nephelometric (N) Procedures for the Determination of the Elements

Element	Technique	Suspension	Reagent	Solution composition	Interferences	References
Ag	T or N	AgCl	NaCl	Dilute HNO_3		2
As	T	As	KPH_2O_2	HCl (1:1)	Se, Te	5
Au	T	Au	$SnCl_2$	HCl	Ag, Hg, Pd, Pt, Ru, Se, Te	1,2
Ba	T or N	$BaSO_4$	H_2SO_4 or Na_2SO_4 plus a protective colloid	HCl (1:200)	Pb	1,2
Ca	T	CaC_2O_4	$H_2C_2O_4$	HOAc	Mg, Na, SO_4^{2-}	1,2
	T	Ca oleate	Na oleate	Alkaline	Mg	
Cl^-	T or N	AgCl	$AgNO_3$	Dilute HNO_3 or H_2SO_4	Br^-, I^-	1,3,6
K	T	$K_2Na[Co(NO_2)_6]$	$Na_3[Co(NO_2)_6]$	HOAc	SO_4^{2-}	1,2
Li	T	Li stearate	Stearic acid	1-pentanol		1,2
Na	T or N	$NaM^{II}(UO_2)_3(OAc)_9$	Mg or Zn uranyl acetate	Water containing excess reagent	Li	2
SO_4^{2-}	T or N	$BaSO_4$	$BaCl_2$	HCl (1:200)	Pb	1,4
Se	T	Se	$NaPH_2O_2$ or $SnCl_2$	HCl (1:1)	Te	1,4,7
Sn	T	Arsonic compounds	Phenylarsonic or 4-hydroxy-3-nitrophenylarsonic acids	HCl + HNO_3 dilute or H_2SO_4 dilute	Fe	9
Te	T	Te	$NaPH_2O_2$	HCl (1:1)	Se	1,4,7,8
Zn	T or N	$K_2Zn_3[Fe(CN)_6]_2$	$K_4[Fe(CN)_6]$	0.4*M* HCl	Cu, Fe	1,2
	T or N	ZnS	H_2S or Na_2S	Neutral solution	Heavy metals	1,2,4

References

[1] Snell and Snell, *Colorimetric Methods of Analysis*, 3d ed., Van Nostrand, Princeton, NJ, Vol. I, 1949 and Vol. II, 1959.
[2] E. B. Sandell, *Colorimetric Determination of Traces of Metals*, 3d ed., Interscience, New York, 1959.
[3] D. F. Boltz, ed., *Colorimetric Determination of Nonmetals*, Interscience, New York, 1958.
[4] ASTM, *ASTM Methods for Chemical Analysis*, 1956.
[5] Steele and England, *Analyst* **82**:595 (1957).
[6] Challis and Jones, *Analyst* **81**:703 (1956).
[7] Challis, *Analyst* **67**:186 (1942).
[8] Crossley, *Analyst* **69**:209 (1944).
[9] Challis and Jones, *Anal. Chim. Acta* **21**:58 (1959).

Applying Beer's law to Eq. (6.16), one obtains

$$P_L = \Phi_L P_0 (1 - e^{-\epsilon bC}) \tag{6.17}$$

When expanded in a power series, this equation yields

$$P_L = \Phi_L P_0 \epsilon bC \left(1 - \frac{\epsilon bC}{2!} + \frac{(\epsilon bC)^2}{3!} - \cdots \right) \tag{6.18}$$

If ϵbC is 0.05 or less, only the first term in the series is significant and Eq. (6.18) can be written as

$$P_L = \Phi_L P_0 \epsilon bC \tag{6.19}$$

Thus, when the concentrations are very dilute and not over 2% of the incident radiation is absorbed, there is a linear relationship between luminescent power and concentration.

Of particular interest in Eq. (6.19) is the linear dependence of luminescence on the excitation power. This means sensitivity can be increased by working with high excitation powers. The b term is not the path length of the cell, but the solid volume of the beam defined by the excitation and emission slit widths together with the beam geometry. Therefore, silt widths are the critical factor and not the cell dimensions.

6.6.1.1 Problems with Photoluminescence

6.6.1.1.1 Self-Quenching. Self-quenching results when luminescing molecules collide and lose their excitation energy by radiationless transfer. Serious offenders are impurities, dissolved oxygen, and heavy atoms or paramagnetic species (aromatic substances are prime offenders). Always use "Spec-pure" solvents.

6.6.1.1.2 Absorption of Radiant Energy. Absorption either of the exciting or of the luminescent radiation reduces the luminescent signal. Remedies involve (a) diluting the sample, (b) viewing the luminescence near the front surface of the cell, and (c) using the method of standard additions for evaluating samples.

6.6.1.1.3 Self-Absorption. Attenuation of the exciting radiation as it passes through the cell can be caused by too concentrated an analyte. The remedy is to dilute the sample and note whether the luminescence increases or decreases. If the luminescence increases upon sample dilution, one is working on the high-concentration side of the luminescence maximum. This region should be avoided.

6.6.1.1.4 Excimer Formation. Formation of a complex between the excited-state molecule and another molecule in the ground state, called an excimer, causes a problem when it dissociates with the emission of luminescent radiation at longer wavelengths than the normal luminescence. Dilution helps lessen this effect.

6.6.2 Structural Factors Affecting Photoluminescence

Factors that affect photoluminescence are as follows:

1. Fluorescence is expected in molecules that are aromatic or contain multiple-conjugated double bonds with a high degree of resonance stability.
2. Fluorescence is also expected in polycyclic aromatic systems.
3. Substituents, such as $—NH_2$, —OH, —F, $—OCH_3$, $—NHCH_3$, and $—N(CH_3)_2$ groups, often enhance fluorescence.

4. On the other hand, these groups decrease or quench fluorescence completely: —Cl, —Br, —I, —$NHCOCH_3$, —NO_2, and —COOH.
5. Molecular rigidity enhances fluorescence. Substances fluoresce more brightly in a glassy state or viscous solution. Formation of chelates with metal ions also promotes fluorescence. However, the introduction of paramagnetic metal ions gives rise to phosphorescence but not fluorescence in metal complexes.
6. Changes in the system pH, if it affects the charge status of chromophore, may influence fluorescence.

6.6.3 Instrumentation for Fluorescence Measurement

The primary filter or excitation monochromator selects specific wavelengths of radiation from the source and directs them through the sample. The resultant luminescence, usually observed at 90° to the excitation radiation, is isolated by the secondary filter or fluorescence emission monochromator and directed to the photodetector. Front surface or small-angle viewing (37°) is used for high-absorbance samples or solids.

6.6.3.1 Radiation Sources. It is desirable to use a source as powerful as possible. High-pressure xenon arc lamps are used in nearly all commercial spectrofluorometers. The xenon lamp emits an intense and relatively stable continuum of radiation that extends from 300 to 1300 nm.

Low-pressure mercury vapor lamps are most frequently used in filter fluorometers. With a clear bulb of ultraviolet-transmitting material, individual mercury emission lines occur at 253.7, 296.5, 302.2, 312.6, 313.2 (doublet), 365.5 (triplet), 366.3, 404.7, 435.8, 546.1, 577.0, and 579.1 nm. Interference filters are used to select the desired mercury line. When the lamp bulb is coated with a phosphor, a more nearly continuous spectrum is emitted.

6.6.3.2 Fluorescence Measurements. Fluorescent measurements are usually made by reference to some arbitrarily chosen standard. The standard is placed in the instrument and the circuit is balanced with the reading scale at some chosen setting. Without readjusting any circuit components, the standard is replaced by known concentrations of the analyte and the fluorescence of each is recorded. Finally, the fluorescence of the solvent and cuvette alone is measured to establish the true zero-concentration readings, if the instrument is not equipped with a zero-adjust circuit. Fluorescence quantum yield values and secondary standards are given in Table 6.20.

With separate emission and excitation monochromators, the emission and excitation spectra can be ascertained. If no knowledge of the spectra is available, place a solution of the analyte into the cuvette. Select an emission wavelength that produces a fluorescent signal either visible to the eye or from the detector signal. While scanning through the spectrum with the excitation monochromator, plot the strength of the fluorescence signal. This gives an uncorrected excitation spectrum that should resemble the normal absorption spectrum obtained in the ultraviolet-visible region. Next, select a suitably strong excitation wavelength and scan the fluorescence spectrum with the emission monochromator.

6.6.4 Comparison of Luminescence and Ultraviolet-Visible Absorption Methods

If applicable, luminescence is usually the method of choice for quantitative analytical purposes, especially trace analysis. The significant advantages of luminescence over ultraviolet-visible absorption methods are enumerated below:

1. Fewer luminescing species exist than absorbing species in the ultraviolet-visible region.
2. Luminescence is more selective. A pair of wavelengths, excitation and emission, characterize the process instead of one.

TABLE 6.20 Fluorescence Quantum Yield Values

Compound	Solvent	Q_F value vs. Q_F standard
	Q_F standard	
9-Aminoacridine	Water	0.99
Anthracene	Ethanol	0.30
POPOP*	Toluene	0.85
Quinine sulfate dihydrate	$1N$ H_2SO_4	0.55
	Secondary standards	
Acridine orange hydrochloride	Ethanol	0.54 Quinine sulfate 0.58 Anthracene
1,8-ANS† (free acid)	Ethanol	0.38 Anthracene 0.39 POPOP
1,8-ANS (magnesium salt)	Ethanol	0.29 Anthracene 0.31 POPOP
Fluorescein	$0.1N$ NaOH	0.91 Quinine sulfate 0.94 POPOP
Fluorescein, ethyl ester	$0.1N$ NaOH	0.99 Quinine sulfate 0.99 POPOP
Rhodamine B	Ethanol	0.69 Quinine sulfate 0.70 Anthracene
2,6-TNS‡ (potassium salt)	Ethanol	0.48 Anthracene 0.51 POPOP

* POPOP denotes *p*-bis[2-(5-phenyloxazoyl)]benzene.
† ANS denotes anilino-8-naphthalene sulfonic acid.
‡ TNS denotes 2-*p*-toluidinylnaphthalene-6-sulfonate.
Source: J. A. Dean, ed., *Lange's Handbook of Chemistry*, 14th ed., McGraw-Hill, New York, 1992.

3. Luminescence is more sensitive. This is because (a) a luminescence signal is measured directly and against a very small background and (b) the signal is proportional to the intensity of the incident radiation. Greater sensitivity for weakly emitting compounds can be obtained by using more intense sources. Luminescence analyses can determine sub-part-per-billion concentrations of many substances. By contrast, in absorption spectrophotometry concentration is proportional to absorbance, which is the logarithm of the ratio between incident and transmitted radiant power. This corresponds to the measurement of a small difference between two large signals.

4. Luminescence lifetimes offer another factor for discrimination among compounds.

6.6.5 Applications

Fluorometric analysis is of greatest use in the determination of concentrations that are too small to be easily determined by spectrophotometry, colorimetry, or emission spectrography; determinations on the order of 0.1 μg in 10 mL are common. In Tables 6.21 and 6.22, the excitation and emission data are given as taken from the literature and are often incomplete and only approximate. The sensitivity of a given procedure is difficult to specify precisely, because it depends on such factors as the intensity of the exciting radiation, the sensitivity of the detector, and so on. The sensitivity values given in the following tables are interpreted from statements in the original articles and are not to be taken as absolute.

TABLE 6.21 Fluorometric Methods for the Determination of Inorganic Substances

Substance determined	Reagent	Conditions	Excitation, nm	Emission, nm	Sensitivity, $\mu g \cdot mL^{-1}$	Interferences	References
Al	Alizarin Garnet R	pH 4.6	470	590	0.007	Be, Co, Cr, Cu, F^-, Fe, NO_3^-, Ni, PO_4^{3-}, Th, Zr	*Anal. Chem.* **25**:960 (1953)
	Morin	pH 3.3	430	500	0.001	Fe, Th, U	*Anal. Chem.* **33**:1360 (1961); *Ind. Eng. Chem., Anal. Ed.* **12**:229 (1940)
	Pontachrome Blue Black R	pH 4.6	470, 580	630	0.001	Co, Cr, Cu, Fe, Ga, Ni, V	*Anal. Chem.* **18**:530 (1946); *Ind. Eng. Chem., Anal. Ed.* **18**:530 (1946)
	8-Hydroxyquinoline	$CHCl_3$ extraction; pH 5.7	365	520	0.1	Ga	*Anal. Chem.* **27**:961 (1955)
	Salicylidene-*o*-aminophenol	5.6	410	520	0.027	Cr(III), Sc, Th; F^-, citrate, $C_2O_4^{2-}$, tartrate	*Talanta* **13**:609 (1966)
B	Benzoin	pH 12.8 in EtOH	370	480	0.04	Be, Sb	*Analyst* **86**:62 (1961); **82**:606 (1957); *Anal. Chem.* **19**:802 (1947)
	Morin	Dilute $HCl–H_2C_2O_4$	365	460–540	1		*J. Chem. Soc.* **79**:231 (1958)
Be	1-Amino-4-hydroxy-anthraquinone	0.02*M* NaOH	530, 570	630	0.2	Li, Cr(VI)	*Ind. Eng. Chem., Anal. Ed.* **18**:179 (1946); **13**:809 (1941)
	Morin	0.05*M* NaOH	470	570	0.01	Ca, Cr(VI), Li, rare earths, Zn	*Anal. Chem.* **31**:598 (1959); **24**:1467 (1952)
Ca	Calcein	0.4*M* KOH	360	485	0.02	Ba, Sr	*Anal. Chem.* **35**:1238 (1963)
Ce	Ce(III) fluorescence	0.6*M*–2.9M $HClO_4$	260	355	0.1	NO_3^-	*Anal. Chem. Acta* **41**:404 (1968)
CN^-	Chloramine T + nicotinamide	1*M* KOH	365	Blue	0.3		*Anal. Chem.* **29**:879 (1957)
F^-	(Quenching) Al–Alizarin Garnet R	pH 4.6	470	590	0.001	Be, Co, Cr, Cu, Fe, Ni, PO_4^{3-}, Th, Zr	*Anal. Chem.* **25**:960 (1953)
Ga	8-Hydroxyquinoline	$CHCl_3$ extraction, pH 2.6	436	470–610	0.05 μg	Cu, Fe(III), Mo(VI), V(V); colored extractants	*Anal. Chem.* **27**:961 (1955)
	Rhodamine B	C_6H_6 extraction, 6*M* HCl	365	Orange-yellow	0.01 μg	Au, Fe, NO_3^-, Sb, Tl, W	*Anal. Chim. Acta* **13**:159 (1955); **24**:413 (1961)
Ge	Benzoin	Alkaline EtOH	365	Yellow-green	2	As(V), B, Be, Cr(VI), NO_2^-, silicate	*Nature* **175**:167 (1955)
Hf	Flavonol	0.1*M* H_2SO_4	365–400	460	0.1	Al, F^-, Fe, PO_4^{3-}, Zr	*Anal. Chem.* **23**:1149 (1951)
In	8-Hydroxyquinoline	$CHCl_3$ extraction, pH 5.1	365	535	0.04	Al, Be, Cu, Fe, Zr	*Z. Anal. Chem.* **138**:337 (1953)
Li	8-Hydroxyquinoline	Slightly alkaline EtOH	370	580	0.2	Mg	*Anal. Chem.* **23**:478 (1951)

(Continued)

TABLE 6.21 Fluorometric Methods for the Determination of Inorganic Substances (*Continued*)

Substance determined	Reagent	Conditions	Excitation, nm	Emission, nm	Sensitivity, $\mu g \cdot mL^{-1}$	Interferences	References
Li	Dibenzothiazolylmethane	2*M* KOH, 50% 1,4-dioxane	365	415		Insoluble hydroxides; Zn	*Anal. Chim. Acta* **37**:460 (1967)
Mg	8-Hydroxyquinoline–sulfonic acid	Aqueous	365	440	0.1	Ca unless masked with EGTA	*Clin. Chim. Acta* **136**:137 (1984)
	Bis(salicylidene-ethylene)diamine	Isobutylamine in dimethylformamide	355	577	0.0002	Be, In, Zn	*Anal. Chem.* **31**:2083 (1959)
Ru	5-Methyl-1,10-phenanthroline	Reduce to Ru(III), pH 6	465	416–436	1	Ag, Co, Cr(VI), Fe, Mn(VII), Pd	*Anal. Chem.* **32**:1426 (1960); *Zh. Anal. Khim.* **39:**1658 (1984)
S^{2-}	Pd complex of 5-sulfo-8-hydroxyquinoline	pH 9.2; add $MgCl_2$ and glycine	365	455	0.2	Cyanide	*Anal. Chem.* **30**:93 (1958)
Sc	Salicylaldehyde semicarbazone	pH 6.0	370	Greenish	0.05	Cupferron and tributyl phosphate extractions required	*Analyst* **91**:23 (1966); *Jpn. Anal.* **24**:321 (1975)
	Morin	pH 7; EtOAc extraction	365		0.1	Al, Be, F^-, Fe, Ga, In, phosphate	*Zh. Anal. Khim.* **22**:1812 (1967)
Se	2,3-Diaminonaphthalene extract	pH 2.0		Blue			*Analyst* **87**:558 (1962)
Sn(II)	7-Amino-3-nitronaphthalene-sulfonic acid	pH 10.6	365	495–500	0.1	Fe, Ti, U, V, dithionite	*Anal. Chim.Acta* **15**:246 (1956)
Sn(IV), organo	Morin	Hexane	415–420		0.001 (dialkyl) 0.1 (trialkyl)	Isolated by hexane or ethyl acetate extraction	*Anal. Chem.* **55**:1901 (1983)
Tb		0.5*M* HCl	230	545	3	Fe, nitrate, U	*Anal. Chem.* **26**:1134 (1954)
Th	Morin	0.01*M* HCl in 50% EtOH	420	520	0.02	Al, Ca, Fe, La, Zr	*J. Am. Chem. Soc.* **79**:5425 (1957); *Anal. Chem.* **29**:1426 (1957)
	1-Amino-4-hydroxy-anthraquinone	pH 2.3	550, 580	430			*Ind. Eng. Chem., Anal. Ed.* **13**:809 (1941)
Tl(I)		3.3*M* HCl, 0.8*M* KCl	250	580	0.01	Only Au, Bi, Pt(IV), Sb(V) after Et_2O extraction	*Talanta* **12**:517 (1965)
Tl(III)	Rhodamine B	2*M* HCl, C_6H_6 extraction	360		0.1	Au, Fe, Ga, Hg, Sb	*Anal. Chim. Acta* **9**:393 (1953)

U(VI)		Conc. H_3PO_4 or H_2SO_4	254	Yellow-green	0.1	Many	*Anal. Chem.* **19**:646 (1947)
	Na–K–Li–F fused button		365	560	0.001 μg	Ag, Au, Ca, Ce, Co, Cr, Mn, Ni, Pb, Pt, rare earths, Th	*Anal. Chem.* **25**:322 (1953); **28**:1651 (1956)
V(V)	Resorcinol	10*M* H_2SO_4	360	Red	2.5	Ce(IV)	*Z. Anal. Chem.* **161**:406 (1958)
W(VI)	Rhodamine B (quenching)	0.1*M* NaCl, pH 2.0	365	570–640	1	As, Au, Cr, F^-, Fe, Mo, PO_4^{3-}, Tl, V	*J. Chem. Soc. Jpn.* **77**:1259 (1956)
Y	8-Hydroxyquinoline	pH 9.5, $CHCl_3$ extraction			0.02	Ce, La	*J. Chem. Soc. Jpn.* **77**:1474 (1956)
Zn	8-Hydroxyquinoline	Acetate buffer, gum arabic	420	Green-yellow	1	Al, Fe, Mg, others	*Ind. Eng. Chem., Anal. Ed.* **16**:758 (1944)
Zr	Flavonol	0.1*M* H_2SO_4	400	465	0.1		*Anal. Chem.* **23**:1149 (1951)
	Morin	2*M* HCl, 80% EtOH	425	515	0.02	Al, Ga, Ge, Hf, Sb, Sc, Sn, Th, U; EDTA removes Zr fluorescence but not the interferences	*Anal. Chim. Acta* **16**:346 (1957); *Zh. Anal. Khim.* **28**:1331 (1973)

TABLE 6.22 Fluorescence Spectroscopy of Some Organic Compounds

Compound	Reagent	Conditions	λ_{ex}, nm	λ_{em}, nm	References
Acenaphthene		Pentane	291	341	
Acetoacetic acid	Resorcinol + HCl	12 h in dark; then pH 10	330	440	*Z. Physiol. Chem.* **286**:145 (1951)
Acetol	*o*-Aminobenzaldehyde in 0.2*M* NaOH	1 h at 100°C; then pH 6.6	365	440	*Anal. Chem.* **22**:902 (1950)
Acridine		F_3CCOOH or MeOH	358	475	*Ann. Chim.* (*Paris*) **5**:642 (1950)
Adenine		Water, pH 1	280	375	*Arch. Biochem. Biophys.* **68**:1 (1957)
Adenosine		Water, pH 1	285	395	*Ibid.*
Adenosine triphosphate		Water, pH 1	285	395	*Ibid.*
Adenylic acid		Water, pH 1	285	295	*Ibid.*
Adrenalin	$K_3[Fe(CN)_6]$, pH 6; then alkaline ascorbate		365	Yellow-green	*Anal. Chem.* **30**:1063 (1958)
Albumin	1-Anilinonaphthalene-9-sulfonic acid	pH 7	365		*Biochem. J.* **56**:xxxi (1954)
Alloxan	D-1-Ribitylamino-2-amino-4,5-dimethylbenzene in HOAc	pH 3.9	436	520	*Anal. Chem.* **22**:822 (1950)
4-Aminobenzoic acid		Water, pH 11	295	345	*Arch. Biochem. Biophys.* **68**:1 (1957)
2-Aminophenol	Benzoic acid	15 min at 100°C	365	Greenish yellow	*Ind. Eng. Chem., Anal. Ed.* **12**:403 (1940)
Aminopterin		Water, pH 7	280, 370	460	
1-Aminopyrene		F_3CCOOH	330, 342	415	
4-Aminosalicylic acid		Water, pH 11	300	405	*J. Pharm. Exptl. Therap.* **120**:20 (1957)
Amobarbital		Water, pH 14	265	410	*Ibid.*
Anilines		Water, pH 7–9	280, 291	344, 361	
Anthracene		Pentane	420	430	
Anthranilic acid		Water, pH 7	300	405	*Arch. Biochem. Biophys.* **68**:1 (1957)
Arginine	Ninhydrin	0.3*M* KOH	305, 390	495	*Nature* **183**:1053 (1959)
Aromatic polycyclic hydrocarbons		Petroleum ether	Various	Various	*Anal. Chem.* **32**:810, 1436 (1960); *J. Chem. Soc.* **1946**:1017; **1949**: 1683; **65**:1540 (1943)
Ascorbic acid	Na 1,2-Naphthoquinone-4-sulfonate	pH 7.5	365	Blue-white	*Z. Anal. Chem.* **139**:263 (1953)
Atabrine	Caffeine Na benzoate	0.05*M* H_2SO_4	365	540	*J. Biol. Chem.* **154**:597 (1944)
Azaindoles		Water, pH 10	290, 299	310, 347	
Benz[*c*]acridine		F_3CCOOH	295, 380	480	
Benz[*a*]anthracene		Pentane	284	382	
1,2-Benzanthracene			280, 340	390, 410	
Benzanthrone		F_3CCOOH	370, 420	550	
Benzil	3-(Diethylamino)phenol	EtOH solution	350	600	*Anal. Chem.* **23**:540 (1951)
Benzo[*b*]chrysene		Pentane	283	398	
11-*H*-Benzo[*a*]fluorene		Pentane	317	340	

Compound	Reagent	Conditions	Excitation	Emission	Reference
Benzoic acid		70% H_2SO_4	285	385	
3,4-Benzopyrene		Benzene solution	365	390, 480	*Anal. Chem.* **32**:819,1436 (1960)
Benzo[*e*]pyrene		Pentane	329	389	
Benzoquinoline		F_3CCOOH	280	425	
Benzoxanthane		Pentane	363	418	
3-Benzyl-4-methyl-7-hydroxycoumarin		EtOH solution	362	448	*J. Am. Chem. Soc.* **81**:1348 (1959)
Bromolysergic acid diethylamide		Water, pH 1	315	460	*J. Pharm. Exptl. Therap.* **120**:26 (1957)
Brucine		Water, pH 7	305	500	
Carbazole		*N,N*-Dimethylformamide	291	359	
Chlortetracycline			355	445	
Chrysene		Pentane	250, 300	260, 380	
Cinchonine		Water, pH 1	320	420	*J. Pharm. Exptl. Therap.* **120**:26 (1957)
Citric acid	Thionyl chloride, then NH_3, then 76% H_2SO_4	Ammoniacal solution	365	440	*Anal. Chem.* **21**:811 (1949)
Coproporphyrin		10% sulfuric acid	405	650	*Rec. Trav. Chim.* **74**:556 (1955)
Coumarin		Ethanol solution	280	352	*J. Am. Chem. Soc.* **81**:1348 (1959)
Creatine	Ninhydrin	0.3*M* KOH	305, 390	495	*Nature* **183**:1053 (1959)
Cysteine	Na 1,2-naphthoquinone-4-sulfonate	pH 7.5	365	Blue-white	*Z. Anal. Chem.* **139**:263 (1959)
Deoxyribonucleic acid	3,5-Diaminobenzoic acid in 0.6*M* $HClO_4$	30 min at 60°C	405	520	*J. Biol. Chem.* **233**:184, 483 (1958)
Deoxyribose	3,5-Diaminobenzoic acid in 0.6*M* $HClO_4$	30 min at 60°C	406	520	*Ibid.*
Dibenzo[*a,e*]anthracene		Pentane	280	381	
Dibenzo[*b,k*]chrysene		Pentane	308	428	
Dibenzo[*a,e*]pyrene		Pentane	370	401	
3,4,8,9-Dibenzopyrene			370, 335 390, 410	480, 510	
5,12-Dihydronaphthacene		Pentane	282	340	
3,4-Dihydroxyphenylacetic acid		pH 7	280	330	*Arch. Biochem. Biophys.* **68**:1 (1957)
3,4-Dihydroxyphenylalanine		0.005*M* H_2SO_4	285	325	*Ibid.*
3,4-Dihydroxyphenylethylamine		pH 1	285	325	*Ibid.*
3,4-Dihydroxyphenylserine		pH 1	280	320	*Ibid.*
4-Dimethylaminobenzaldehyde	Salicyloyl hydrazide	Ethanol solution, heat 1 h; then pH 10	390	470	*Ibid.*; *Anal. Chem.* **31**:296 (1959)
Dimethylguanidine	Ninhydrin	0.3*M* KOH	305, 390	495	*Nature* **183**:1053 (1959)

(Continued)

TABLE 6.22 **Fluorescence Spectroscopy of Some Organic Compounds (*Continued*)**

Compound	Reagent	Conditions	λ_{ex}, nm	λ_{em}, nm	References
1,4-Diphenylbutadiene		Pentane	328	370	
Epinephrine		Water, pH 7 or	295	335	
		$0.005M$ H_2SO_4	275	320	
Ethacridine		Water, pH 2	370, 425	515	
6-Ethoxy-1,2-dihydro-2,2,4-trimethylquinoline	Destroy fluorescence with 0.04% $KMnO_4$	Isooctane solution	435	480	*Anal. Chem.* **28**:376 (1956)
Fluoranthrene		Pentane	354	464	
Fluorene		Pentane	300	321	
Fluorescein		Water, pH 7–11	490	515	
Folic acid		Water, pH 7	365	450	*Arch. Biochem. Biophys.* **68**:1 (1957)
Gentisic acid		Water, pH 7	315	440	
Gibberellic acid		Cold 85% H_2SO_4; 1 h	405	465	
Griseofulvin		Water, pH 7	295, 335	450	*Nature* **184**:364 (1957)
Guanidinium compounds	Ninhydrin	$0.3M$ KOH	305, 390	495	*Nature* **183**:1053 (1959)
Guanine		Water, pH 1	285	365	*Arch. Biochem. Biophys.* **68**:1 (1957)
Guthion		Water, pH 11	250, 312	380	*J. Agr. Food Chem.* **6**:32 (1958)
Harmine		Water, pH 1	300, 365	400	*J. Pharm. Exptl. Therap.* **120**:26 (1957)
Hippuric acid		70% H_2SO_4	270	370	
Histamine	*o*-Phthalaldehyde		360	450	*Federation Proc.* **18**:444 (1950)
Homogentisic acid		Water, pH 7	290	340	*Arch. Biochem. Biophys.* **68**:1 (1957)
Homovanillic acid		Water, pH 7	270	315	*Ibid.*
Hydroxyamphetamine		pH 1	275	300	*J. Pharm. Exptl. Therap.* **120**:27 (1957)
3-Hydroxyanthranilic acid		Water, pH 7	340	430	*Ibid.*
o-Hydroxybenzaldehyde	Salicyloyl hydrazide	Ethanol solution; 1 h at 80°C; then pH 10	390	470	*Ibid.*
p-Hydroxybenzaldehyde	Salicyloyl hydrazide	Ethanol solution; 1 h at 80°C; then pH 10	390	470	*Anal. Chem.* **31**:296 (1959)
3-Hydroxybenzoic acid		pH 12	314	430	*Anal. Chem.* **30**:1361 (1958)
4-Hydroxycinnamic acid		pH 7	350	440	*Arch. Biochem. Biophys.* **68**:1 (1957)
7-Hydroxycoumarin		Ethanol solution	325	441	*J. Am. Chem. Soc.* **81**:1348 (1959)
5-Hydroxyindole		pH 2–10	295	330	*Arch. Biochem. Biophys.* **68**:1 (1957)
5-Hydroxyindoleacetic acid		pH 7	300	355	*Ibid.; Science* **125**:442 (1957)
3-Hydroxykynurenine (or 5-)		pH 11	365	460	*Arch. Biochem. Biophys.* **68**:1 (1957)
4-Hydroxymandelic acid		pH 7	300	380	*Ibid.*
4-Hydroxyphenylacetic acid		pH 7	280	310	*Ibid.*
4-Hydroxyphenylpyruvic acid		pH 7	290	345	*Ibid.*
4-Hydroxyphenylserine		pH 1	290	320	*Ibid.*
5-Hydroxytryptophan		pH 7	295	340	*Science* **125**:442 (1957)

Indole(s)		pH 7	280	355	*Ibid.; Arch. Biochem. Biophys.* **68**:1 (1957)
Indoleacetic acid		pH 7, water	285	345	*Science* **125**:442 (1957)
Isoniazid	CNBr	NaOH–ethanol	300	405	*Am. Rev. Respiratory Diseases* **81**:485 (1960)
Kynurenic acid		pH 7	325	405	*Arch. Biochem. Biophys.* **68**:1 (1957)
Lysergic acid diethylamide		Water, pH 7	325	465	*Science* **125**:364 (1959)
		Water, pH 9	315	440	
Malic acid	2-Naphthol	92% H_2SO_4, 30 min at 90°C	365	Green	*Anal. Chem.* **21**:1375 (1949)
4-Methyl-7-aminocarbostyril		Ethanol solution	365	407	*J. Am. Chem. Soc.* **81**:1348 (1959)
9-Methylanthracene		Pentane	382	410	
3-Methylcholanthrene		Pentane	297	392	
5-Methylcytosine	Br_2 water, then 2-aminobenzaldehyde in 0.15*M* NaOH	Phosphate buffer	365	420	*J. Biol. Chem.* **233**:483 (1958)
7-Methyldibenzopyrene		Pentane	460	467	
4-Methyl-7-diethylaminocoumarin		Ethanol solution	375	456	*J. Am. Chem. Soc.* **81**:1348 (1959)
Methylguanidine	Ninhydrin	0.3*M* KOH	305, 390	495	*Nature* **183**:1053 (1959)
4-Methyl-7-hydroxycarbostyril		Ethanol solution	338	412	*J. Am. Chem. Soc.* **81**:1348 (1959)
4-Methyl-7-hydroxycoumarin		Ethanol solution	325	442	*Ibid.*
2-Methylphenanthrene		Pentane	257	357	
3-Methylphenanthrene		Pentane	292	368	
1-Methylpyrene		Pentane	336	394	
4-Methylpyrene		Pentane	338	386	
Morphine		0.005*M* H_2SO_4	270–290	365	
Naphthacene			290, 310	480, 515	
Naphthaleneacetic acid		pH 11, water	230, 282	327	*J. Agr. Food Chem.* **6**:22 (1958)
1-Naphthol		0.1*M* NaOH in 20% ethanol	365	480	*Anal. Chem.* **30**:96 (1958)
2-Naphthol		0.1*M* NaOH in 20% ethanol	365	426	*Ibid.*
Nerve gases (sarin, soman, or tabun)	Indole + $NaBO_3$	Let stand 1 min	365	480	*Anal. Chem.* **29**:276 (1957)
Nicotinamide	CNBr at 80°C for 8 min, pH 7	1*M* NaOH	365	470	*Science* **114**:16 (1951); **116**:462 (1952)
2-Nitronaphthalene	60% oleum, then dilute and add Zn dust	pH 4.6	365	450	*Anal. Chem.* **23**:717 (1951)
2-Nitrophenol	Zn + HCl; then benzoic acid at 160°C	Benzene solution	365	Green-yellow	*Ind. Eng. Chem., Anal. Ed.* **12**:403 (1940)
Penicillin	2-Methoxy-6-chloro-9-(β-aminoethyl)aminoacridine	1*M* HCl	365	540	*J. Biol. Chem.* **164**:195 (1948)

(Continued)

TABLE 6.22 Fluorescence Spectroscopy of Some Organic Compounds (*Continued*)

Compound	Reagent	Conditions	λ_{ex}, nm	λ_{em}, nm	References
Pentobarbital		pH 13	265	449	*J. Pharm. Exptl. Therap.* **129**:26 (1957)
Pentothal		pH 13	315	530	*Ibid.*
Phenanthrene		Pentane	252	362	
Phenobarbital		pH 13	265	440	*J. Pharm. Exptl. Therap.* **120**:26 (1957)
Phenylalanine		Water	215, 260	282	*Biochem J.* **65**:476 (1957)
o-Phenylenepyrene		Pentane	360	506	
Phenylephrine			270	305	
Picene		Pentane	281	398	
Piperonylbutoxide		Methanol	248, 292	318	*J. Agr. Food Chem.* **6**:32 (1958)
Procaine		Water, pH 11	275	345	*J. Pharm. Exptl. Therap.* **120**:26 (1957)
Proteins		Water	280	313, 350	*Biochem. J.* **76**:381 (1960)
Pyrene		Pentane	330	382	
Pyridoxal		Water, pH 7	330	385	*Arch. Biochem. Biophys.* **68**:1 (1957)
Pyruvaldehyde	Chromotropic acid	H_2SO_4	436	540	*Anal. Chem.* **22**:899 (1950)
Pyruvic acid	Diphosphopyridine nucleotide, reduced + enzymic coupling		365	460	*Biochem. J.* **64**:56P (1956)
Quinacrine		Water, pH 11	285	500	*J. Pharm. Exptl. Therap.* **120**:26 (1957)
Quinidine		Water, pH 1	360	460	*J. Lab. Clin. Med.* **36**:478 (1950)
Quinine		Water, pH 1	250, 350	450	*J. Pharm. Exptl. Therap.* **120**:26 (1957)
Reserpine		Water, pH 1	300	375	*Ibid.*
Resorcinol		Water	265	315	
Riboflavin		Water, pH 7	270, 370	520	*Anal. Chem.* **27**:1178 (1955); **28**:1017 (1956)
Rutin		Water, pH 1	430	520	*Science* **105**:48 (1947)
Salicylic acid		Water, pH 11	310	435	*J. Pharm. Exptl. Therap.* **120**:26 (1957)
Salicyloyl hydrazide		pH 10	350	425	*Anal. Chem.* **31**:296 (1959)
Serotonin		3*M* HCl	295	550	*J. Pharm. Exptl. Therap.* **117**:82 (1956)
		Water	295	340	*Arch. Biochem. Biophys.* **68**:1 (1957)
Skatole		Water	290	370	*Science* **125**:442 (1957)
Streptomycin		Water, pH 13	366	445	
Succinic acid	Resorcinol	Conc. H_2SO_4, 1 h at 130ºC	470	Green	*Plant Physiol.* **23**:443 (1948)
p-Terphenyl		Pentane	284	338	

Thiamine	Alkaline $K_4[Fe(CN)_6]$	pH 7	365	450	*Anal. Chem.* **27**:1178 (1955); **29**:1017 (1956)
Thioglycolic acid	Na 1,2-naphthoquinone-4-sulfonate	pH 7.5	365	Blue-white	*Z. Anal. Chem.* **139**:263 (1953)
Thymidine and Thymine	Br_2 water, then *o*-aminobenzaldehyde in 0.15*M* NaOH	Phosphate buffer	365	420	*J. Biol. Chem.* **233**:483 (1958)
Thymol		pH 7	265	300	*J. Pharm. Exptl. Therap.* **120**:26 (1957)
Tocopherol		Hexane–ethanol	295	340	*Arch. Biochem. Biophys.* **84**:116 (1959)
Tribenzo[*a,e,i*]pyrene		Pentane	384	448	
Triphenylene		Pentane	288	357	
Tryptamine		Water, pH 7	290	360	*Arch. Biochem. Biophys.* **68**:1 (1957)
Tryptophan		Water, pH 11	285	365	*Ibid.*
Tyramine		Water, pH 1	275	310	*Ibid.*
Tyrosine		Water, pH 7 or 0.005*M* H_2SO_4	275	310	*Ibid.*; *Biochem. J.* **65**:476 (1957)
Uric acid		Water, pH 1	325	370	*Arch. Biochem. Biophys.* **68**:1 (1957)
Vitamin A		1-Butanol	340	490	*Ibid.*
Vitamin B_{12}		Water, pH 7	275	305	*Ibid.*
Warfarin		Methanol	290, 341	385	*J. Agr. Food Chem.* **6**:32 (1958)
Xanthine		Water, pH 1	315	435	*Arch. Biochem. Biophys.* **68**:1 (1957)
Xanthopterin		20% H_2SO_4	365	460, 520	*Analyst* **72**:383 (1947)
Xanthurenic acid		Water, pH 11	350	460	*Arch. Biochem. Biophys.* **68**:1 (1957)
2,6-Xylenol			275	305	
3,4-Xylenol			280	310	
Yohimbine		Water, pH 1	270	360	*J. Pharm. Exptl. Therap.* **120**:26 (1957)

Bibliography

Lakowicz, J. R., *Principles of Fluorescence Spectroscopy*, Plenum, New York, 1983.

Rendell, D., *Fluorescence and Phosphorescence*, Wiley, New York, 1987.

Schulman, S. G., ed., *Molecular Luminescence Spectroscopy*, Wiley, New York, 1985, Vol. 1; 1988, Vol. 2.

Wehry, E. L., ed., *Modern Fluorescence Spectroscopy*, Plenum, New York, 1981.

6.7 PHOSPHORIMETRY

Instrumentation for phosphorimetry is identical to that described for fluorescence measurements (Sec. 6.6) with the addition of a radiation interrupter and provision for immersion of the sample in a Dewar flask for liquid-nitrogen temperatures. Time resolution is considerably improved by the use of a microsecond-duration pulsed source in place of the older rotating can chopper or set of slotted disks with equally spaced ports. Phosphorescence decay times down to milliseconds can be observed and recorded with pulsed radiation.

The sample cuvette is a small Dewar flask made of fused silica and silvered, except in the region where the optical path traverses the Dewar. The solvent frequently used is a mixture of diethyl either, isopentane, and ethanol (often given the acronym EPA) in a volume ratio of 5:5:2. Another solvent mixture is isopentane and methylcyclohexane (1:4). When these mixtures are cooled to liquid-nitrogen temperatures, they give a clear transparent glass.

The phosphorescence spectroscopy of some organic compounds is given in Table 6.23. Some substances give phosphorescence at room temperatures when adsorbed on a solid.[7]

6.8 FLOW-INJECTION ANALYSIS

Flow-injection analysis (FIA) is based on the introduction of a precisely defined volume of sample (perhaps 30 μL) as a "plug" into a continuously flowing carrier or reagent stream (1 $\mathrm{mL \cdot min^{-1}}$) in a narrow-bore (0.5-mm) nonwetting tubing. These typical conditions produce laminar flow in which molecules of fluid flow in streamlines parallel to the walls of the tubing with a parabolic velocity gradient between the center streamline, which flows at twice the average linear velocity, and the wall, at which the velocity is zero.

Sample introduction is via a syringe or valve, the latter actuated by a microprocessor. The result is a sample plug bracketed by carrier. It is a nonsegmented stream in contrast to continuous-flow analysis. The absence of air segmentation leads to a higher sample throughput (30 to 300 $\mathrm{h^{-1}}$). There is no need to introduce and remove air bubbles, and an expensive high-quality pump is not necessary. FIA has been described as HPLC without the column and without the pressure.

The carrier stream is merged with a reagent stream to bring about a chemical reaction between sample and reagent. The total stream then flows through a detector. Experimental conditions are held constant for both standards and samples in terms of residence time, temperature, and dispersion. Precise pumping and injection lead to precise fluid flow, the fundamental feature of FIA that makes it a useful technique. The sample concentration is evaluated against appropriate standards treated identically. Sample volumes are in the range of 20 to 200 μL, but normally are 30 μL. The flow of reagents is in the range of 0.5 to 9 $\mathrm{mL \cdot min^{-1}}$. Typical FIA flowrates are in the range of 0.5 to 2.0 $\mathrm{mL \cdot min^{-1}}$; they can be easily varied by shifting pump tubes. A wide range of detectors has been described, among which are atomic absorption spectrometry, induction coupled plasmas,

[7] T. Vo-Dinh, *Room Temperature Phosphorimetry for Chemical Analysis*, Wiley, New York, 1984; R. J. Hurtubise, *Solid Surface Luminescence Analysis*, Dekker, New York, 1981; E. B. Asafu-Adajaye and S. Y. Yue, *Anal. Chem.* **58**:539 (1986); S. M. Ramasamy, Y. P. Senthilnathan, and R. J. Hurtubise, *ibid.* **58**:612 (1986).

TABLE 6.23 Phosphorescence Spectroscopy of Some Organic Compounds

(EPA, diethyl ether, isopentane, and ethanol (5:5:2) volume ratio)

Compound	Solvent	Lifetime, s	Excitation wavelength, nm	Emission wavelength, nm
Acenaphthene	Ethanol		300	515
3-Acetylpyridine	Ethanol	0.5	395	525
Adenine	Water–methanol (9:1)	2.9	278	406
Adenosine	Ethanol	0.8	280	422
p-Aminobenzoic acid	Ethanol		305	425
2-Aminofluorene	Ethanol	4.6	380	590
6-Amino-6-methylmercaptopurine	Water–methanol (9:1)	0.66	321	456
2-Amino-4-methylpyrimidine	Ethanol	2.1	302	438
2-Amino-5-nitrobenzothiazole	EPA	0.41	375	515
2-Amino-5-nitrobiphenyl	EPA	0.56	380	520
3-L-Aminotyrosine-2HCl	Ethanol	0.8	286	398
Anthracene	Ethanol		300	462
Aspirin	EPA	2.1	240	380
Atropine	Ethanol	1.4		410
8-Azaguanine	Ethanol	1.8	282	442
Benzaldehyde	Ethanol	3.4	254	433
1,2-Benzanthracene	Ethanol	2.2	310	510
Benzimidazole	Ethanol	2.3	280	406
Benzocaine	Ethanol	3.4	310	430
1,2-Benzofluorene	Ethanol		315	502
Benzoic acid	EPA	2.4	240	400
3,4-Benzopyrene	Ethanol		325	508
Benzyl alcohol	Ethanol		219	393
6-Benzylaminopurine	Water–methanol (9:1)	2.8	286	413
Biphenyl	Ethanol	1.0	270	385
6-Bromopurine	Water–methanol (9:1)	0.5	273	420
Brucine	Ethanol	0.9	305	435
Caffeine	Ethanol	2.0	285	440
Carbazole	Ethanol	7.8	341	436
2-Chloro-4-aminobenzoic acid	Ethanol	1.0	312	337
p-Chlorophenol	Ethanol	<0.2	290	505
o-Chlorophenoxyacetic acid	Ethanol	0.7	280	518
p-Chlorophenoxyacetic acid	Ethanol	<0.5	283	396
6-Chloropurine	Water–methanol (9:1)	0.64	273	419
Chlorpromazine · HCl	Ethanol	0.3	320	490
Chlorotetracycline	Ethanol	2.7	280	410
Cocaine · HCl	Ethanol	2.7	240	400
Codeine	Ethanol	0.3	270	505
Cytidine	Water–methanol (9:1)		290	420
Desoxypyridoxine · HCl	Ethanol	1.4	290	442
Diacetylsulfanilamide	Ethanol	1.3	280	405
2,6-Diaminopurine	Water–methanol (9:1)	2.7	288	410
2,6-Diaminopurine sulfate	Ethanol	1.7	294	424

(Continued)

TABLE 6.23 Phosphorescence Spectroscopy of Some Organic Compounds (*Continued*)

Compound	Solvent	Lifetime, s	Excitation wavelength, nm	Emission wavelength, nm
1,2,5,6-Dibenzanthracene	Ethanol	1.3	340	550
2,6-Dichloro-4-nitroaniline	EPA	0.5	368	525
2,4-Dichlorophenoxyacetic acid	Ethanol	<0.5	289	490
2,6-Diethyl-4-nitroaniline	EPA	0.66	388	525
3,4-Dihydroxymandelic acid	Ethanol	1.1	294	412
3,4-Dihydroxyphenylacetic acid	Ethanol	0.9	295	430
2,5-Dimethoxy-4-methylamphetamine	Water–methanol (9:1)	3.9	289	411
5,7-Dimethyl-1,2-benzacridine	Ethanol	0.6	310	555
N,N-Dimethyl-4-nitroaniline	EPA	0.54	398	525
N,N-Dimethyltryptamine	Water–methanol (9:1)	6.9	286	434
Dopamine	Ethanol	0.9	285	430
Ephedrine	Ethanol	3.6	225	390
Epinephrine	Ethanol	1.0	283	425
N-Ethylcarbazole	Ethanol	7.8	340	437
Ethyl 3-indoleacetate	Ethanol	3.3	290	440
Folic acid	Ethanol		367	425
Hippuric acid	EPA	4.9	311	450
Homovanillic acid	Ethanol	0.8	289	435
DL-5-Hydroxytryptophan	Ethanol	6.3	315	435
Indole-3-acetic acid	Ethanol	<0.5	290	438
3-Indoleacetonitrile	Ethanol	7.1	285	438
Indole-3-butanoic acid	Ethanol	0.6	284	510
Indolecarboxylic acid	Ethanol	5.5	290	429
Indole-2-propanoic acid	Ethanol	0.6	290	440
D-Lysergic acid	Water–methanol (9:1)	0.1	310	518
2-Methylcarbazole	Ethanol	8.1	333	442
N-Methylcarbazole	Ethanol	8.4	336	437
6-Methylmercaptopurine	Water–methanol (9:1)	0.6	291	420
N-Methyl-4-nitroaniline	EPA	0.5	390	522
6-Methylpurine	Water–methanol (9:1)	3.2	272	405
Morphine	Ethanol	0.3	285	500
Naphthacene	Ethanol		300	518
Naphthalene	EPA	1.8	310	475
1-Naphthaleneacetic acid	Ethanol	2.8	295	510
1-Naphthol	Ethanol	1.1	320	475
2-Naphthoxyacetic acid	Ethanol	2.6	328	497
2-Naphthylamine	Ethanol	2.3	270	303
Niacinamide	Ethanol		270	410
Nicotine	Ethanol	5.2	270	390
5-Nitroacenaphthene	EPA		380	540
4-Nitroaniline	EPA	0.6	380	510
9-Nitroanthracene	EPA		248	488
1-Nitroanthraquinone	EPA	0.3	250	490
4-Nitrobiphenyl	EPA		330	480
3-Nitro-*N*-ethylcarbazole	EPA	0.4	315	475
2-Nitrofluorene	EPA	0.4	340	517

TABLE 6.23 **Phosphorescence Spectroscopy of Some Organic Compounds** (*Continued*)

Compound	Solvent	Lifetime, s	Excitation wavelength, nm	Emission wavelength, nm
6-Nitroindole	EPA	0.4	372	520
1-Nitronaphthalene	EPA		340	520
2-Nitronaphthalene	EPA	0.4	260	500
4-Nitro-1-naphthylamine	EPA		400	578
4-Nitrophenol	Ethanol	<0.2	355	520
4-Nitrophenylhydrazine	EPA	0.5	390	520
4-Nitro-2-toluidine	EPA	0.5	375	520
Papaverine · HCl	Ethanol	1.5	260	480
Phenacetin	EPA			410
Phenanthrene	EPA	2.6	340	465
Phenobarbital	Ethanol	1.8	240	380
Phenylalanine	Ethanol		270	385
DL-2-Phenyllactic acid	Ethanol	5.4	262	383
Phthalylsulfathiazole	Ethanol	0.9	305	405
Procaine · HCl	Ethanol	3.5	310	430
Purine	Water–methanol (9:1)	2.2	272	405
Pyrene	Ethanol		330	515
Pyridine	Ethanol	1.4	310	440
Pyridine-3-sulfonic acid	Ethanol	1.2	272	408
Pyridoxine · HCl	Ethanol		290	425
Quercetin	Ethanol	2.1	345	480
Quinidine sulfate	Ethanol	1.3	340	500
Quinine · HCl	Ethanol	1.3	340	500
Salicylic acid	Ethanol	6.2	315	430
Strychnine phosphate	Ethanol	1.2	290	440
Sulfabenzamide	Ethanol	0.7	305	405
Sulfadiazine	Ethanol	0.7	275	410
Sulfanilamide	Ethanol	2.9	300	410
Sulfapyridine	Ethanol	1.4	310	440
Sulfathiazole	Ethanol	0.9	310	420
1,2,4,5-Tetramethylbenzene	EPA	4.5	275	390
2-Thiouracil	Ethanol	<0.5	310	430
2,4,5-Trichlorophenol	Ethanol	<0.2	305	485
2,4,5-Trichlorophenoxyacetic acid	Ethanol	1.1	295	475
Triphenylene	Ethanol	15	290	460
Tryptophan	Ethanol	1.5	295	440
Tyrosine	Ethanol	2.8	290	390
Vitamin K_1	Hexane	0.4	345	570
Vitamin K_3	Hexane	0.5	335	510
Vitamin K_5	Water–methanol (9:1)	1.3	310	535
Warfarin	Ethanol	0.8	305	460
Yohimbine · HCl	Ethanol	7.4	290	410

Source: J. A. Dean, ed., *Handbook of Organic Chemistry*, McGraw-Hill, New York, 1987, pp. 6–13 to 6–16.

ion-selective electrodes, fluorimetry, voltammetry, refractometry, and visible spectrophotometry; the latter the most widely used at this time. Detector volumes vary from 8 to 40 μL.

The distance between the manifold and detector should be kept as short as possible. Coil lengths are 0.1 to 2 m. Tubing diameter may be from 0.4 to 1.0 mm, although tubing of 0.5 mm i.d. is best. Larger-diameter tubing leads to increasing dispersion, whereas the smaller-diameter tubing may easily be blocked and high-pressure pumps are required for operation.

Dispersion, the intermixing of the solution, is readily controlled experimentally. In general, dispersion increases with tube length, tube diameter, detector volume, average flowrate, and the molecular diffusion coefficient. Dispersion decreases with sample volume injected, tight coiling, packed beds, and bead strings (that is, flow tortuosity).

FIA is best suited for analyses that require less than 30 s and the sequential addition of only one or two reagents. This technique requires extremely precise use of pumps, valves, and computerized control. FIA is excellent for automation of wet chemistry methods, such as the determination of ammonia, chloride, nitrate, and nitrite. The fast turnaround times are appreciated in environmental monitoring.

A recent development in FIA is sequential injection. Like FIA, the sample is still injected into a carrier, but it is not a continuous-flow system. A multiple-port valve connects the sample and any reagents to the detector.

Bibliography

Betteridge, D., "Flow Injection Analysis," *Anal. Chem.* **50**:832A (1978).

Karlberg, B. I., "Automation of Wet Chemical Procedures Using FIA," *Am. Lab.* **1983** (February):73.

Mottola, H. A., "Continuous Flow Analysis Revisited," *Anal. Chem.* **53**:1313A (1981).

Ruzicka, J., "Flow Injection Analysis," *Anal. Chem.* **55**:1041A (1983).

Ruzicka, J., and E. H. Hansen, *Flow Injection Analysis*, 2d ed., Wiley, New York, 1988.

Snyder, L., et al., "Automated Continuous Flow Analysis," *Anal. Chem.* **48**:942A (1976).

Stewart, K. K., "Flow Injection Analysis. New Tool for Old Assays," *Anal. Chem.* **55**:931A (1983).

Tyson, J., "Flow Injection Techniques for Analytical Atomic Spectrometry," *Spectroscopy* **7[3]**:14 (1992).

SECTION 7
INFRARED AND RAMAN SPECTROSCOPY

The infrared region of the electromagnetic spectrum includes radiation at wavelengths between 0.7 and 500 μm or, in wave numbers, between 14 000 and 20 cm^{-1}. The relationship between wave number and wavelength scales in the infrared region is given in Table 7.1. Molecules have specific frequencies that are directly associated with their rotational and vibrational motions. Infrared absorptions result from changes in the vibrational and rotational state of a molecular bond. Coupling with electromagnetic radiation occurs if the vibrating molecule produces an oscillating dipole moment that can interact with the electric field of the radiation. Homonuclear diatomic molecules such as hydrogen, oxygen, or nitrogen, which have a zero dipole moment for any bond length, fail to interact. These changes are subtly affected by interaction with neighboring atoms or groups, as are resonating structures, hydrogen bonds, and ring strain. This imposes a stamp of individuality on each molecule's infrared absorption spectrum as portions of the incident radiation are absorbed at specific wavelengths. The multiplicity of vibrations occurring simultaneously produces a highly complex

TABLE 7.1 Wave Number–Wavelength Conversion Table

(Wave number is reciprocal of wavelength. The wave number (in cm^{-1}) = 10 000/wavelength)(in μm). For example, 15.4 μm is equal to 649 cm^{-1}.)

Wavelength (μm)	0	0.1	0.2	0.3	0.4	0.5	0.6	0.7	0.8	0.9
									Wave number, cm^{-1}	
1.0	10 000	9091	8333	7692	7143	6667	6250	5882	5556	5263
2.0	5000	4762	4545	4348	4167	4000	3846	3704	3571	3448
3.0	3333	3226	3125	3030	2941	2857	2778	2703	2632	2564
4.0	2500	2439	2381	2326	2273	2222	2174	2128	2083	2041
5.0	2000	1961	1923	1887	1852	1818	1786	1754	1724	1695
6.0	1667	1639	1613	1587	1563	1538	1515	1493	1471	1449
7.0	1429	1408	1389	1370	1351	1333	1316	1299	1282	1266
8.0	1250	1235	1220	1205	1190	1176	1163	1149	1136	1124
9.0	1111	1099	1087	1075	1064	1053	1042	1031	1020	1010
10.0	1000	990	980	971	962	952	943	935	926	917
11.0	909	901	893	885	877	870	862	855	847	840
12.0	833	826	820	813	806	800	794	787	781	775
13.0	769	763	758	752	746	741	735	730	725	719
14.0	714	709	704	699	694	690	685	680	676	671
15.0	667	662	658	654	649	645	641	637	633	629
16.0	625	621	617	613	610	606	602	599	595	592
17.0	588	585	581	578	575	571	568	565	562	559
18.0	556	552	549	546	543	541	538	535	532	529
19.0	526	524	521	518	515	513	510	508	505	503
20.0	500	498	495	493	490	488	485	483	481	478
21.0	476	474	472	469	467	465	463	461	459	457
22.0	455	452	450	448	446	444	442	441	439	437
23.0	435	433	431	429	427	426	424	422	420	418
24.0	417	415	413	412	410	408	407	405	403	402
25.0	400	398	397	395	394	392	391	389	388	386
26.0	385	383	382	380	379	377	376	375	373	372
27.0	370	369	368	366	365	364	362	361	360	358
28.0	357	356	355	353	352	351	350	348	347	346
29.0	345	344	342	341	340	339	338	337	336	334
30.0	333	332	331	330	329	328	327	326	325	324
31.0	323	322	321	319	318	317	316	315	314	313
32.0	313	312	311	310	309	308	307	306	305	304
33.0	303	302	301	300	299	299	298	297	296	295
34.0	294	293	292	292	291	290	289	288	287	287
35.0	286	285	284	283	282	282	281	280	279	279
36.0	278	277	276	275	275	274	273	272	272	271
37.0	270	270	269	268	267	267	266	265	265	264
38.0	263	262	262	261	260	260	259	258	258	257
39.0	256	256	255	254	254	253	253	252	251	251
40.0	250									

Source: J. A. Dean, ed., *Lange's Handbook of Chemistry,* 14th ed., McGraw-Hill, New York, 1992.

absorption spectrum that is uniquely characteristic of the functional groups that make up the molecule and of the overall configuration of the molecule as well. It is therefore possible to identify substances from their infrared absorption spectrum.[1]

[1] N. B. Colthup, L. H. Daly, and S. E. Wiberley, *Introduction to Infrared and Raman Spectroscopy*, 3d. ed., Academic, New York, 1990.

For qualitative analysis, one of the best features of an infrared spectrum is that the absorption or the *lack of absorption* in specific frequency regions can be correlated with specific stretching and bending motions and, in some cases, with the relationship of these groups to the rest of the molecule. Thus, when interpreting the spectrum, it is possible to state that certain functional groups are present in the material and certain others are absent. The relationship of infrared spectra to molecular structure will be treated in Sec. 7.3.

7.1 THE NEAR-INFRARED REGION

The near-infrared region (NIR), which meets the visible region at about 12 500 cm^{-1} (800 nm) and extends to about 4000 cm^{-1} (2.50 μm), contains primarily overtones and combination bands of C—H, N—H, and O—H stretching frequencies, which are adequate for studying many organic compounds. The instruments used in NIR have fused quartz optics with either a quartz prism or a grating monochromator and photoconductor detectors. NIR uses a tungsten source that covers the range of 700 to about 2500 nm (4000 to 14 000 cm^{-1}). The techniques of near-infrared are closer to those of ultraviolet and visible spectrophotometry than to infrared in that long-path-length cells (0.1 to 10 cm) and dilute solutions are generally used. Quartz, glass, or Corex cells may be used up to 2.4 μm. Special grades of silica are readily available for use up to 3 μm. Because of the sharpness of the absorption bands in the near-infrared region, it is desirable to use high resolution for quantitative work, that is, spectral slit widths of the order of a few wave numbers.

7.1.1 Correlation of Near-Infrared Spectra with Molecular Structure

The absorptivity of near-infrared bands is from 10 to 1000 times less than that of mid-infrared bands. Thicker sample layers (0.5 to 10 mm) must be used to compensate. On the other hand, minor impurities in a sample are less troublesome.

Most analytical applications in the near-infrared region have been concerned with organic compounds and generally with quantitative functional-group analysis. This is because near-infrared spectra are mainly indicative of the hydrogen vibrations of the molecule, that is, C—H, S—H, O—H, N—H, etc., and few vibrations are dependent on the carbon or inorganic skeleton of the molecule. The data on the regions in which various functional groups absorb and the average intensities of the absorption bands have been collected in Fig. 7.1 and Table 7.2. Most of the data in Fig. 7.1 were obtained in carbon tetrachloride solution.[2]

Significant spectral features are as follows:

1. O—H stretching vibration near 7140 cm^{-1} (1.40 μm).
2. N—H stretching vibration near 6667 cm^{-1} (1.50 μm).
3. C—H stretching and deformation vibrations of alkyl groups at 4548 cm^{-1} (2.20 μm) and 3850 cm^{-1} (2.60 μm).
4. Absorption bands due to water at 2.76, 1.90, and 1.40 μm (3623, 5263, and 7143 cm^{-1}).
5. Aromatic amines: (a) Primary amines have bands near 1.97 and 1.49 μm (5076 and 6711 cm^{-1}). (b) Secondary amines show only the band at 1.49 μm. (c) Tertiary amines exhibit no appreciable absorption at either wavelength.

Rather than measure a property or a concentration singly, many more signals—a full spectrum—are taken, and a complex mathematical model (chemometrics) is used to calculate the parameters. Calibration involves taking spectra of many samples from various positions throughout the

[2] R. F. Goddu and D. A. Delker, *Anal. Chem.* **32**:140 (1960).

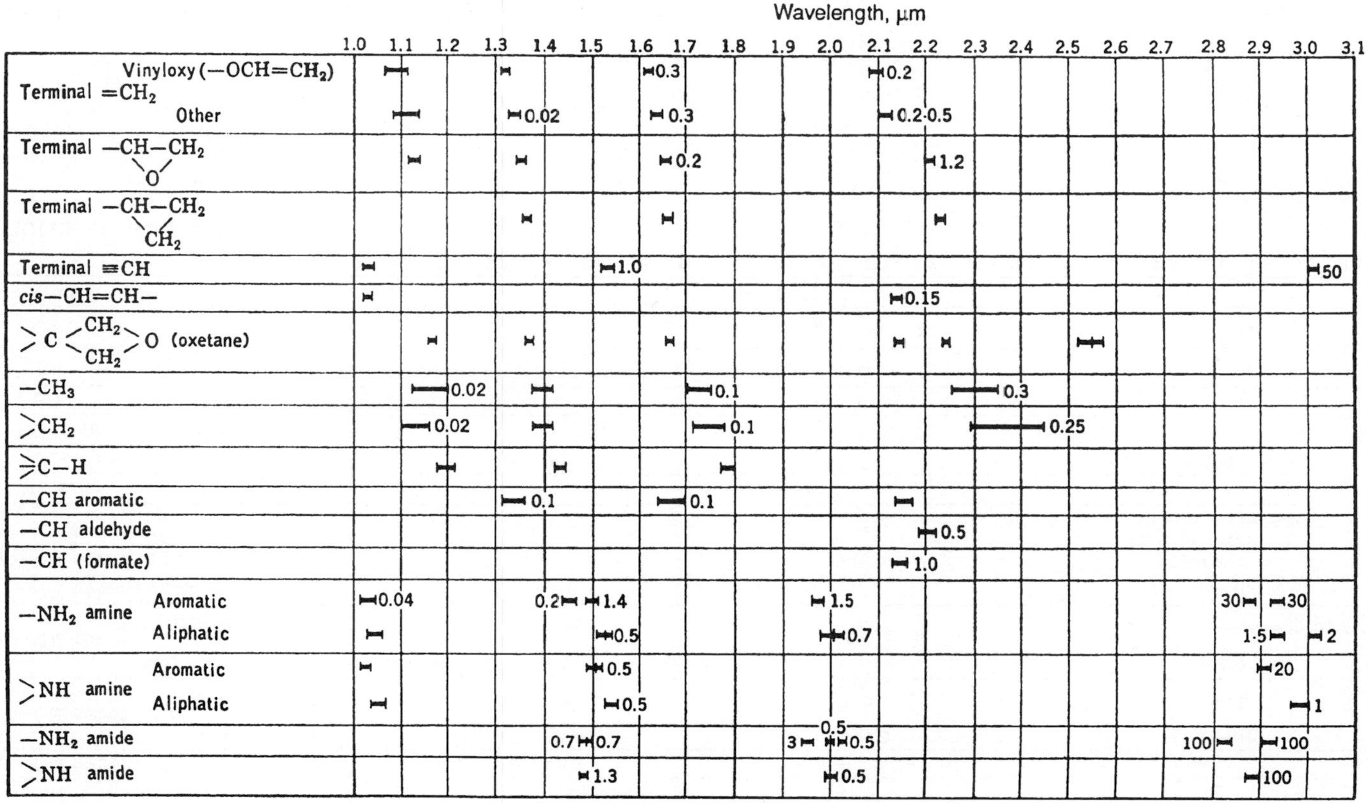

Wavelength, μm

1.0 1.1 1.2 1.3 1.4 1.5 1.6 1.7 1.8 1.9 2.0 2.1 2.2 2.3 2.4 2.5 2.6 2.7 2.8 2.9 3.0 3.1

Terminal $=CH_2$ Vinyloxy ($-OCH=CH_2$) 0.3 0.2

Other 0.02 0.3 0.2-0.5

Terminal $-CH-CH_2$ (O) 0.2 1.2

Terminal $-CH-CH_2$ (CH_2)

Terminal $\equiv CH$ 1.0 50

cis$-CH=CH-$ 0.15

$>C<^{CH_2}_{CH_2}>O$ (oxetane)

$-CH_3$ 0.02 0.1 0.3

$>CH_2$ 0.02 0.1 0.25

$\geqslant C-H$

$-CH$ aromatic 0.1 0.1

$-CH$ aldehyde 0.5

$-CH$ (formate) 1.0

$-NH_2$ amine Aromatic 0.04 0.2 1.4 1.5 30 30

Aliphatic 0.5 0.7 1.5 2

$>NH$ amine Aromatic 0.5 20

Aliphatic 0.5 1

$-NH_2$ amide 0.7 0.7 3 0.5 0.5 100 100

$>NH$ amide 1.3 0.5 100

FIGURE 7.1 Near-infrared spectra-structure correlations and average molar absorptivity data. The molar absorptivities are in units of liter · mol^{-1} · cm^{-1}. (*Courtesy of Analytical Chemistry.*)

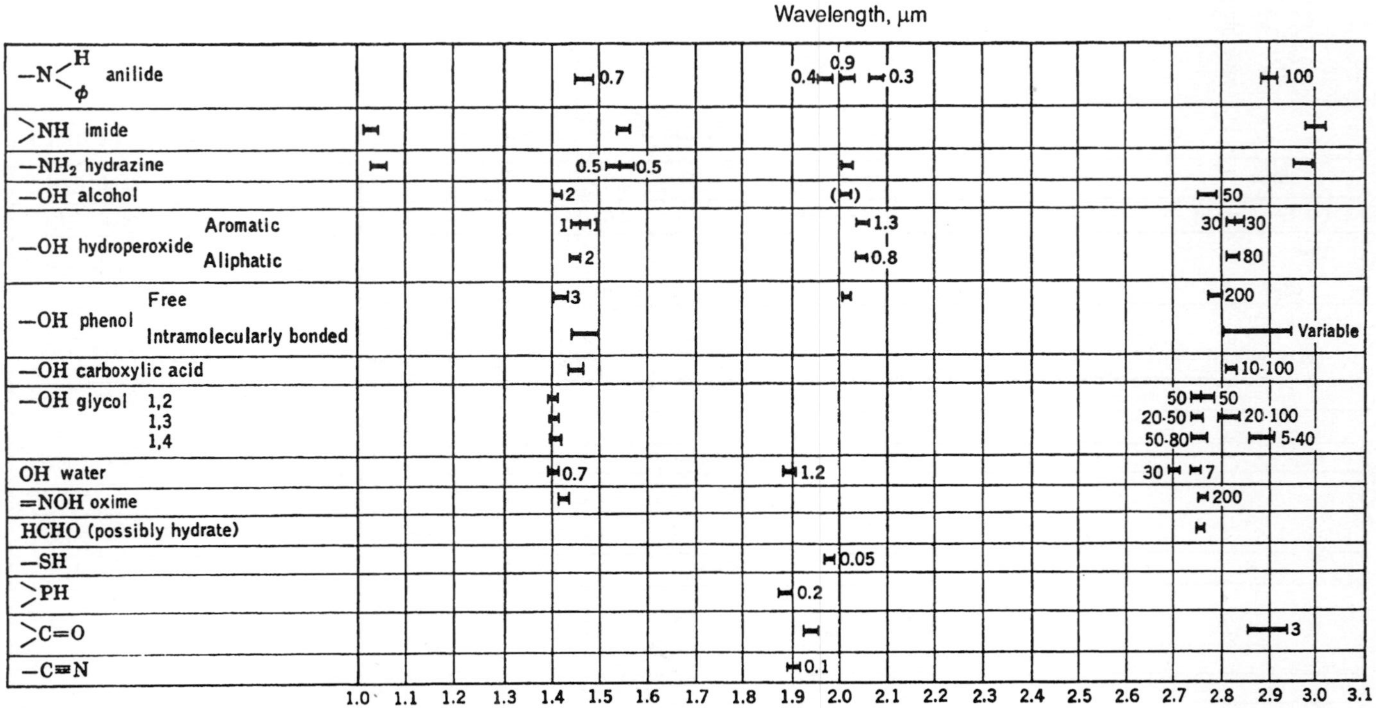

Wavelength, μm

Group	Bands (labels as printed, in order of increasing wavelength)
—N(H)(φ) anilide	0.7; 0.4, 0.9, 0.3; 100
>NH imide	(unlabeled bands)
—NH_2 hydrazine	0.5, 0.5
—OH alcohol	2; ( ); 50
—OH hydroperoxide, Aromatic	1, 1; 1.3; 30, 30
—OH hydroperoxide, Aliphatic	2; 0.8; 80
—OH phenol, Free	3; 200
—OH phenol, Intramolecularly bonded	Variable
—OH carboxylic acid	10-100
—OH glycol 1,2	50, 50
—OH glycol 1,3	20-50, 20-100
—OH glycol 1,4	50-80, 5-40
OH water	0.7; 1.2; 30, 7
=NOH oxime	200
HCHO (possibly hydrate)	(unlabeled band)
—SH	0.05
>PH	0.2
>C=O	3
—C≡N	0.1

1.0 1.1 1.2 1.3 1.4 1.5 1.6 1.7 1.8 1.9 2.0 2.1 2.2 2.3 2.4 2.5 2.6 2.7 2.8 2.9 3.0 3.1

FIGURE 7.1 (*Continued*)

TABLE 7.2 Absorption Frequencies in the Near Infrared

Values in parentheses are molar absorptivity.

Class	Band, cm^{-1}	Remarks
Acetylenes	9 800–9 430	
	6 580–6 400 (1.0)	Overtone of ≡CH stretching
Alcohols (nonhydrogen-bonded)	7 140–7 010 (2.0)	Overtone of OH stretching
Aldehydes		
Aliphatic	4 640–4 520 (0.5)	Combination of C=O and CH stretchings
Aromatic	ca. 8 000	
	ca. 4 525	
	ca. 4 445	
Formate	4 775–4 630 (1.0)	
Alkanes		
$—CH_3$	9 000 8 350 (0.02)	
	5 850–5 660 (0.1)	
	4 510–4 280 (0.3)	
$—CH_2—$	9 170–8 475 (0.02)	
	5 830–6 640 (0.1)	
	4 420–4 070 (0.25)	
$\geqslant$CH	8 550–8 130	All bands very weak
	7 000–6 800	
	5 650–5 560	
Cyclopropane	6 160–6 060	
	4 500–4 400	
Alkenes		
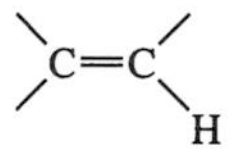 $>C=C<_{H}$	6 850–6 370 (1.0)	
$>C=CH_2$ and $—CH=CH_2$	7 580–7 300 (0.02)	
	6 140–5 980 (0.2)	
	4 760–4 700 (1.2)	
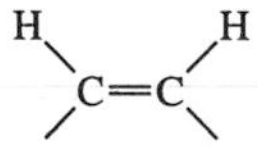 $\overset{H}{>}C=C\overset{H}{<}$	4 760–4 660 (0.15)	*Trans* isomers have no unique bands
$—O—CH=CH_2$	6 250–6 040 (0.3)	
$—CO—CH=CH_2$	7 580–7 410 (0.02)	
	6 190–5 990 (0.3)	
	4 820–4 750 (0.2–0.5)	
Amides		
Primary	7 400–6 540 (0.7)	Two bands; overtone of NH stretch
	5 160–5 060 (3.0)	Second overtone of C=O stretch; second overtone of NH deformation; combination of C=O and NH
	5 040–4 990 (0.5)	
	4 960–4 880 (0.5)	
Secondary	7 330–7 140 (0.5)	Overtone of NH stretch
	5 050–4 960 (0.4)	Combination of NH stretch and NH bending
Amines, aliphatic		
Primary	9 710–9 350	Second overtone of NH stretch
	6 670–6 450 (0.5)	Two bands; overtone of NH stretch
	5 075–4 900 (0.7)	Two bands; combination of NH stretch and NH bending
Secondary	9 800–9 350	Second overtone of NH stretch
	6 580–6 410 (0.5)	Overtone of NH stretch

(Continued)

TABLE 7.2 Absorption Frequencies in the Near Infrared (*Continued*)

Class	Band, cm^{-1}	Remarks
Amines, aromatic		
Primary	9 950–9 520 (0.4)	
	7 040–6 850 (0.2)	
	6 760–6 580 (1.4)	
	5 140–5 040 (1.5)	
Secondary	10 000–9 710	
	6 800–6 580 (0.5)	
Aryl-H	7 660–7 330 (0.1)	
	6 170–5 880 (0.1)	Overtone of CH stretch
Carbonyl	5 200–5 100	
Carboxylic acids	7 000–6 800	
Epoxide (terminal)	6 135–5 960 (0.2)	
	4 665–4 520 (1.2)	Cyclopropane bands in same region
Glycols	7 140–7 040	
Hydroperoxides		
Aliphatic	6 940–6 750 (2.0)	
	4 960–4 880 (0.8)	
Aromatic	7 040–6 760 (1.0)	Two bands
	4 950–4 850 (1.3)	
Imides	9 900–9 620	
	6 540–6 370	
Nitriles	5 350–5 200 (0.1)	
Oximes	7 140–7 050	
Phosphines	5 350–5 260 (0.2)	
Phenols		
Nonbonded	7 140–6 800 (3.0)	
	5 000–4 950	
Intramolecularly bonded	7 000–6 700	
Thiols	5 100–4 950 (0.05)	

measurement range and also measuring the parameters by standard reference methods. Each sample type needs a different model. The laboratory needs thousands of very similar samples to justify the labor and time involved.[3]

7.1.2 Solvents for the Near-Infrared Region

A wide variety of solvents can be used throughout most of the near-infrared region with the exception of the region from 2.7 to 3 μm. The spectral regions in which a few representative solvents are useful are shown in Fig. 7.2, together with the maximum desirable path lengths.[3] Since many of the near-infrared bands are greatly affected by the solvent because of intermolecular bonding, it is imperative that the same solvent be used for calibration or reference solution and for the analysis of unknowns.

[3] S. J. Swarin and C. A. Drumm, "Predicting Gasoline Properties Using Near-IR Spectroscopy," *Spectroscopy* **7**[7]:42 (1992).

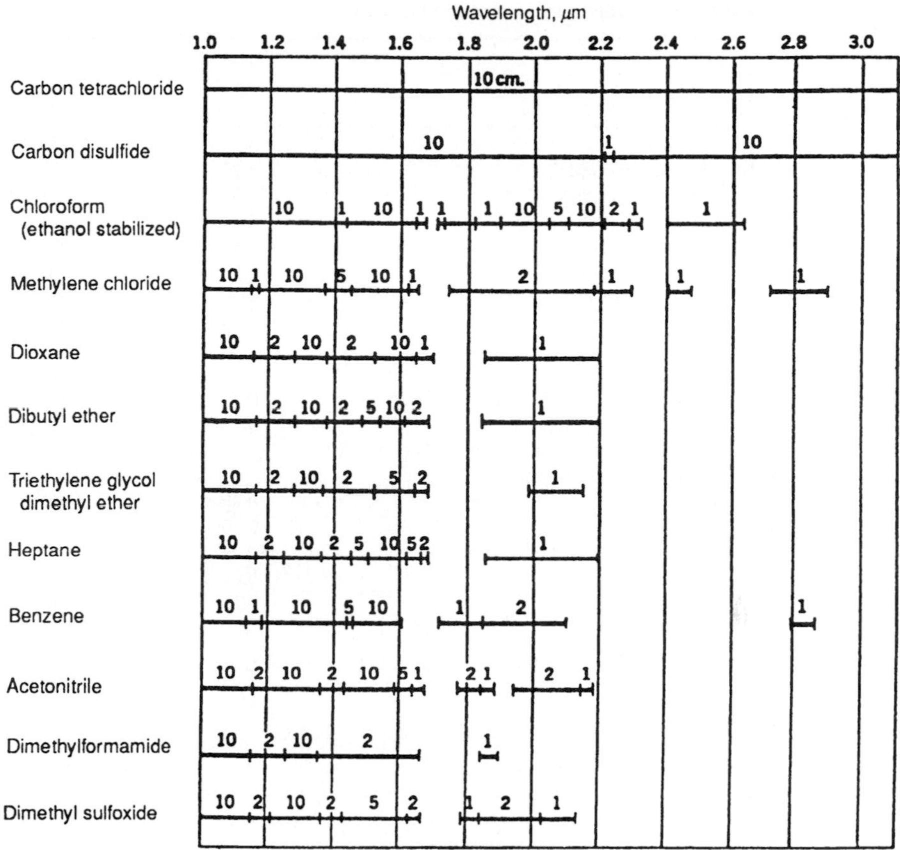

FIGURE 7.2 Solvents for near-infrared spectrophotometry. The solid lines indicate usable regions of the spectrum; the numbers above them indicate maximum desirable path lengths in centimeters. (*Courtesy of Analytical Chemistry.*)

7.2 THE MID-INFRARED REGION

The mid-infrared region is divided into the "group-frequency" region, 4000 to 1300 cm^{-1} (2.50 to 7.69 μm), and the fingerprint region, 1300 to 650 cm^{-1} (7.69 to 15.38 μm). In the group-frequency region the principal absorption bands are more or less dependent on only the functional group from which the absorption arises and not on the complete molecular structure. The fingerprint region involves motion of bonds linking a substituent group to the remainder of the molecule. These are single-bond stretching frequencies and bending vibrations of polyatomic systems.

7.2.1 Infrared-Transmitting Materials

The alkali halides are widely used; all are hygroscopic and should be stored in a desiccator when not in use. When working with wet or aqueous samples, windows of fused silica, calcium, or barium fluoride may be useful, though limited by their long-wavelength transmission. These transmission limitations can be overcome to some extent by using ZnS, ZnSe, or CdTe windows.

TABLE 7.3 Infrared-Transmitting Materials

Material	Wavelength range, μm	Wave number range, cm^{-1}	Refractive index at 2 μm
NaCl, rock salt	0.25–17	40 000–590	1.52
KBr, potassium bromide	0.25–25	40 000–400	1.53
KCl, potassium chloride	0.30–20	33 000–500	1.5
AgCl, silver chloride*	0.40–23	25 000–435	2.0
AgBr, silver bromide*	0.50–35	20 000–286	2.2
CaF_2, calcium fluoride (Irtran-3)	0.15–9	66 700–1 110	1.40
BaF_2, barium fluoride	0.20–11.5	50 000–870	1.46
MgO, magnesium oxide (Irtran-5)	0.39–9.4	25 600–1 060	1.71
CsBr, cesium bromide	1–37	10 000–270	1.67
CsI, cesium iodide	1–50	10 000–200	1.74
TlBr–TlI, thallium bromide–iodide (KRS-5)*	0.50–35	20 000–286	2.37
ZnS, zinc sulfide (Irtran-2)	0.57–14.7	17 500–680	2.26
ZnSe, zinc selenide* (vacuum deposited) (Irtran-4)	1–18	10 000–556	2.45
CdTe, cadmium telluride (Irtran-6)	2–28	5 000–360	2.67
Al_2O_3, sapphire*	0.20–6.5	50 000–1 538	1.76
SiO_2, fused quartz	0.16–3.7	62 500–2 700	
Ge, germanium*	0.50–16.7	20 000–600	4.0
Si, silicon*	0.20–6.2	50 000–1 613	3.5
Polyethylene	16–300	625–33	1.54

* Useful for internal reflection work.
Source: J. A. Dean, ed., *Handbook of Organic Chemistry*, McGraw-Hill, New York, 1987.

Although soft, silver chloride is useful for moist samples or aqueous solutions. Combined with its comparatively low cost and wide transmission range, silver bromide is attractive for use with aqueous solutions and wet samples. It is much less sensitive to visible light compared to silver chloride. Teflon has only C—C and C—F absorption bands. Properties of infrared-transmitting materials are compiled in Table 7.3.

In selecting a window material for an infrared cell, these factors must be considered:

1. The wavelength range over which the spectrum must be recorded and thus the transmission of the window material in this wavelength range.

2. The solubility of window material in the sample or solvent, and any reactivity of window with sample. Cell windows constructed from the alkali halides are easily fogged by exposure to moisture and require frequent repolishing.

3. The refractive index of the window material, particularly in internal reflectance methods. Materials of high refractive index produce strong, persistent interference fringes and suffer large reflectivity losses at air–crystal interfaces.

4. The mechanical characteristics of the window material. For example, silver chloride is soft, easily deformed, and also darkens upon exposure to visible light. Germanium is brittle as is zinc selenide (Irtran IV), which also releases H_2Se in acid solutions. There is no rugged window material for cuvettes or internal reflectance methods that is transparent and also inert over the entire infrared region.

7.2.2 Radiation Sources

7.2.2.1 ***Nernst Glower.*** A popular source is the Nernst glower, which is constructed from a fused mixture of zirconium, yttrium, and thorium oxides molded in the form of hollow rods 1 to 3 mm in diameter and 2 to 5 cm long. It is heated through platinum leads sealed in the ends of the cylinder and is fairly fragile. Because the glower has a negative temperature coefficient of electrical resistance, it must be preheated to be conductive. The circuit also requires a current-limiting device, otherwise burnout will occur. The glower may be operated between 900 and 1700°C and may be twice as intense as other sources mentioned here. It must be protected from drafts, but adequate ventilation is needed to remove surplus heat and evaporated oxides and binder.

7.2.2.2 ***Globar.*** The Globar is a silicon carbide rod 5 cm long and 5 mm in diameter with an operating temperature near 1300°C. One drawback is that the electric contacts of the Globar need water cooling to prevent arcing. It is a better choice than the Nernst glower below 5 μm and in the far-infrared region beyond 15 μm.

7.2.2.3 ***Incandescent Wire.*** An inexpensive, long-lived, and rugged source is a closely wound coil of Nichrome wire (a film of black oxide forms on the coil) around a ceramic core raised to its operating temperature (1000°C) by resistive heating. It requires no cooling and little maintenance. This source is recommended where reliability is essential, such as in-process, filter-type, and nondispersive spectrometers. The Nichrome coil emits less intense radiation than other infrared sources and the initial low energy is further diminished if gratings and mirrors are used. A rhodium wire heater sealed in a ceramic cylinder may be substituted for Nichrome; it is more intense and more costly.

Tungsten incandescent lamp sources are used primarily for near-infrared work. This source is reliable for up to 2000 h of continuous use and is quite inexpensive. The output is mostly between 780 and 2500 nm (12 800 and 4000 cm^{-1}).

7.2.2.4 ***Carbon Dioxide Lasers.*** Tunable CO_2 lasers produce radiation in the 1100 to 900 cm^{-1} (9 to 11 μm) range. The approximately 100 discrete lines in this region are extremely strong and pure, and occur where many materials have absorption bands. The power is amenable to the very long path lengths that are needed in environmental monitoring.

7.2.3 Infrared Spectrometers

Infrared instrumentation is divided into dispersive and Fourier-transform spectrometers. The dispersive instruments are similar to ultraviolet-visible spectrometers except that different sources and detectors are required for the infrared region.

7.2.3.1 ***Dispersive Spectrometers.*** Most dispersive spectrometers are double-beam instruments. Two equivalent beams of radiant energy from the source are passed alternately through the reference and sample paths. In the optical-null system, the detector responds only when the intensity of the two beams is unequal. An optical wedge or comb shutter coupled to the recording pen moves in or out of the reference beam to restore balance. The electrical beam-ratioing method is the other measuring technique. To cover the wide wavelength range, several gratings with different ruling densities and associated higher-order filters are necessary. Two gratings are mounted back to back; each is used in the first order. The gratings are changed at 2000 cm^{-1} (5.00 μm) in mid-infrared instruments. Undesired overlapping grating orders are eliminated with a fore prism or by suitable filters. The use of microprocessors has alleviated many of the tedious requirements necessary to obtain usable data. The operator selects a single recording parameter (scan time, slit setting, or pen response) and the microprocessor automatically optimizes these and other conditions.

7.2.3.2 ***Fourier-Transform Infrared (FT-IR) Spectrometer.*** The FT-IR spectrometer provides speed and sensitivity. A Michelson interferometer, a basic component, consists of two mirrors and a beam splitter. The beam splitter transmits half of all incident radiation from a source to a moving mirror and reflects half to a stationary mirror. Each component reflected by the two mirrors returns to the beam splitter, in which the amplitudes of the waves are combined either destructively or constructively to form an interferogram as seen by the detector. By means of algorithms the interferogram is Fourier-transformed into the frequency spectrum.

This technique has several distinct advantages over conventional dispersive techniques:

1. The FT-IR spectrometer scans the infrared spectrum in fractions of a second at moderate resolution, a resolution that is constant throughout its optical range. It is especially useful in situations that require fast, repetitive scanning (for example, in gas or high-performance liquid chromatography).

2. The spectrometer measures all wavelengths simultaneously. Scans are added. The signal is N times stronger and the noise is $N^{1/2}$ as great, so the signal-to-noise advantage is $N^{1/2}$.

3. An interferometer has no slits or grating; its energy throughput is high, and this means more energy at the detector where it is most needed.

7.2.4 Detectors

Below 1.2 μm, the detection methods are the same as those for ultraviolet-visible radiation. At longer wavelengths the detectors can be classified into two groups: (1) thermal detectors and (2) photon or quantum detectors.

7.2.4.1 Thermal Detectors. With thermal detectors the infrared radiation produces a heating effect that alters some physical property of the detector. The active element is blackened for maximum absorbance and thermally insulated from its substrate. When radiation ceases, the element radiates heat and returns to the temperature of the substrate within a finite time interval, usually milliseconds. This return to baseline follows a decay pattern that determines the response speed and leads to a limit to which the signal may be modulated (pulsed or chopped). This disadvantage is offset by their ability to work at room temperature and over a large range of wavelengths.

7.2.4.1.1 Thermocouple. A thermocouple is fabricated from two dissimilar metals such as bismuth and antimony. When incident radiation strikes the junction, a small voltage proportional to the temperature of the metal junction is produced. The junction surface is coated with gold oxide or bismuth black to enhance detection. Response time is about 30 ms. A *thermopile* is usually six thermocouples in series. Half the thermocouples receive radiation, while the other half are bonded to the substrate and serve as reference.

7.2.4.1.2 Thermistor. A thermistor is made up of sintered oxides of manganese, cobalt, and nickel. These have a high-temperature coefficient of resistance and function by changing resistance when heated. Two 10-μm flakes of the material are placed in the detector; one is blackened and active, while the other is shielded and acts as a reference or compensating detector against changes in ambient temperature. Connected in a bridge circuit, a steady bias voltage is maintained across the bridge. Response time is a few milliseconds.

7.2.4.1.3 Pyroelectric Detector. A pyroelectric detector depends on the rate of change of the detector temperature rather than on the temperature itself. Response time is much faster than that for the preceding types of detectors. It is the detector of choice for Fourier-transform spectrometers. But it also means that the pyroelectric detector responds only to changing radiation that is modulated (chopped or pulsed); it ignores steady background radiation.

7.2.4.1.4 Golay Pneumatic Detector. The Golay detector uses the expansion of xenon gas within an enclosed chamber to expand and deform a flexible blackened diaphragm that is silvered on the outside. The silvered surface reflects a light beam off its surface onto a photodiode. Distortions of the diaphragm diminish the light intensity striking the photodiode. Response time is about 20 ms. It is a superior detector for the far-infrared.

7.2.4.2 Photon Detectors. In a photon detector the incident photons interact with a semiconductor. The result produces electrons and holes—the internal photoelectric effect. An energetic photon strikes an electron in the detector, raising it from a nonconducting to a conducting state. These detectors require cryogenic cooling. Response times are less than 1 μs. When rapid Fourier-transform instruments are needed or sensitive measurements made, photon-type detectors are required.

7.2.4.2.1 Photoconductive Detector. In a photoconductive detector the presence of electrons in the conduction band lowers the resistance. This change is monitored through a bias current or voltage.

7.2.4.2.2 Photovoltaic Detectors. These detectors generate small voltages at a diffused *p–n* junction when exposed to radiation. A single crystal of InSb at liquid-nitrogen temperatures is only good to 5.5 μm. Lead tin telluride detectors cover the 5- to 13-μm region when cooled by liquid nitrogen, and when cooled by liquid helium have optimum performance in the 6.6- to 18-μm region. The most sensitive types are composed of mercury, cadmium, and tellurium. These latter detectors are used with a current-mode amplifier and have response speeds as high as 20 ns with comparable sensitivity to other members of this group.

7.2.5 Preparation of Samples

7.2.5.1 Liquids and Solutions. Pure liquids can be run directly as liquid samples, provided a cell of a suitable thickness is available. This represents a very thin layer, about 0.001 to 0.05 mm thick.

For solutions, concentrations of 10% and cell lengths of 0.1 mm are most practical. There are no nonabsorbing solvents in the infrared region. Transparent regions of selected solvents are shown in Fig. 7.3. When possible, the spectrum is obtained in a 10% solution in CCl_4 in a 0.1-mm cell in the region 4000 to 1333 cm^{-1} (2.50 to 7.50 μm) and in a 10% solution of CS_2 in the region 1333 to 650 cm^{-1} (7.50 to 15.38 μm). If the sample is insoluble in these solvents, chloroform, dichloromethane, acetonitrile, and acetone are useful solvents. For any solvent, a reference cell of the same path length as the sample cell is filled with pure solvent and placed in the reference beam.

Liquid-sample cells are very fragile and must be handled very carefully. Cells are usually filled by capillary action, and the solution or pure sample is introduced with a syringe. Each cell is labeled with its precise path length as measured by interference fringes. Permanent solution cells are constructed with two window pieces sealed and separated by thin gaskets of copper or lead that have been wetted with mercury. The whole assembly is clamped together and mounted in a holder. A demountable cell uses Teflon as gasket material and the cell is slipped into a mount and knurled nuts are turned by hand until tight.

The minicell consists of a threaded, two-piece plastic body and two silver chloride cell window disks with either a 0.025- or 0.100-mm circular depression. The windows fit into one portion of the cell; the second portion is then screwed in to form the seal (AgCl flows slightly under pressure). The windows can be (1) placed back-to-back for films or mulls, (2) arranged with one back to the circular depression, or (3) positioned with facing circular depressions.

7.2.5.2 Cell Path Length. In the interference fringe method for measuring the internal path length, the empty cell is placed in the spectrometer on the sample side and no cell in the reference beam. Cell windows must have a high polish. Operate the spectrometer as near as possible to the 100% line. Run sufficient spectra to produce 20 to 50 fringes. The cell internal thickness *b*, in centimeters, is calculated from

$$b = \frac{1}{2\eta}\left(\frac{n}{\overline{v}_1 - \overline{v}_2}\right) \tag{7.1}$$

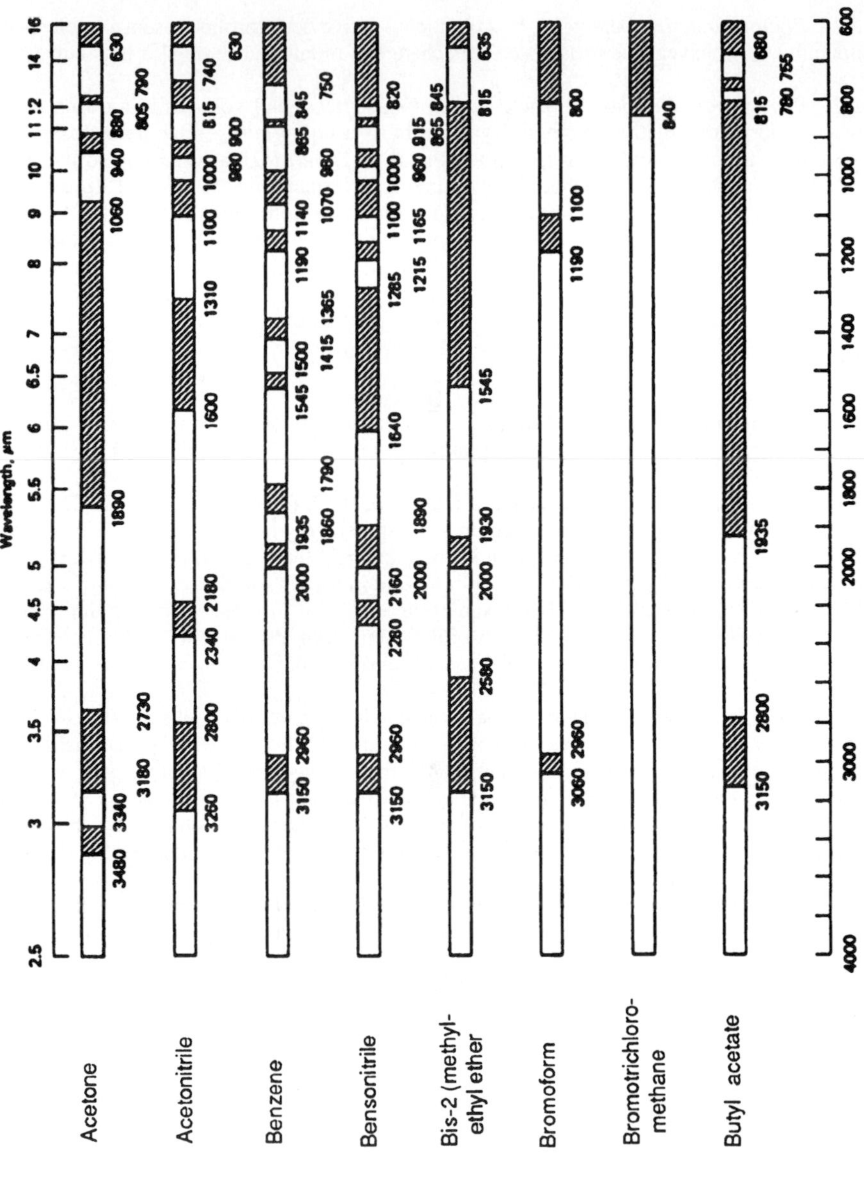

FIGURE 7.3 Infrared transmission characteristics of selected solvents. Transmission below 80% obtained with a 0.10-mm cell path, is shown as shaded area.

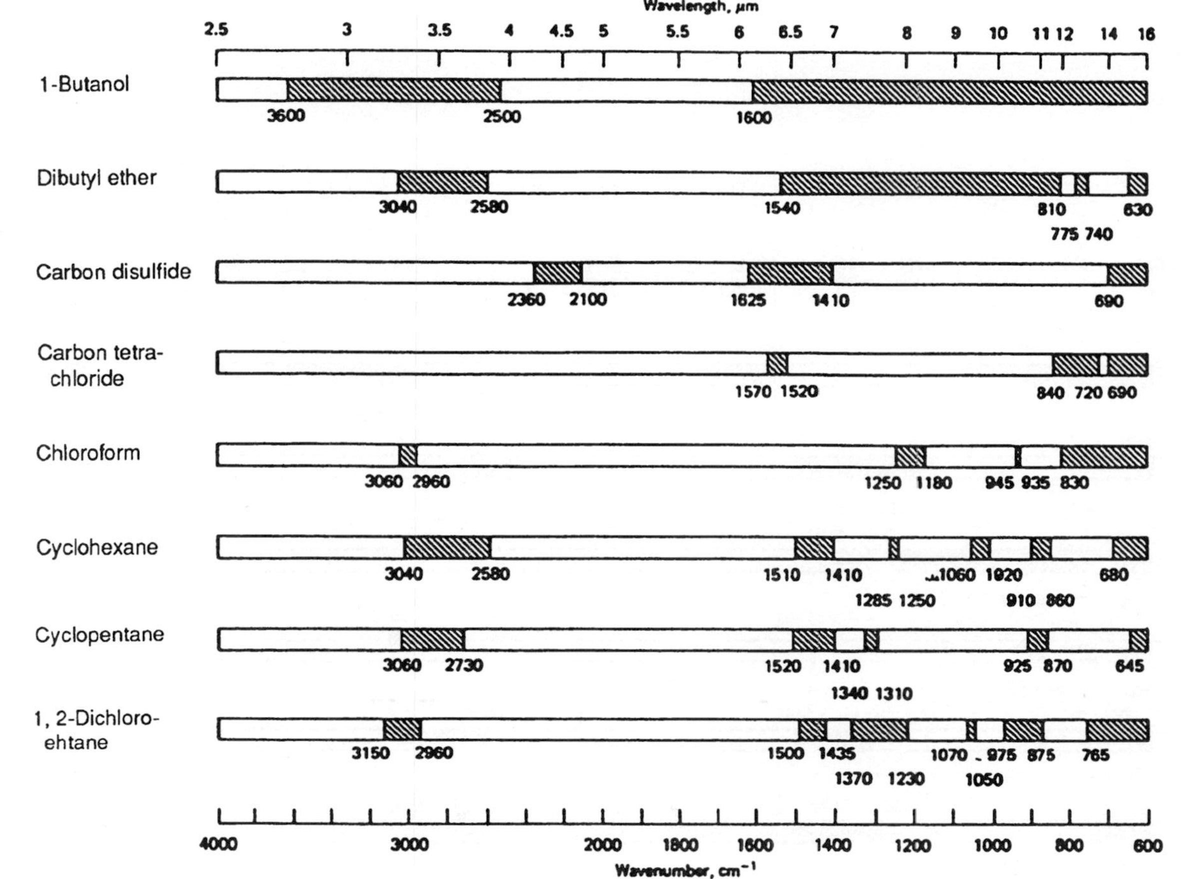

FIGURE 7.3 (*Continued*)

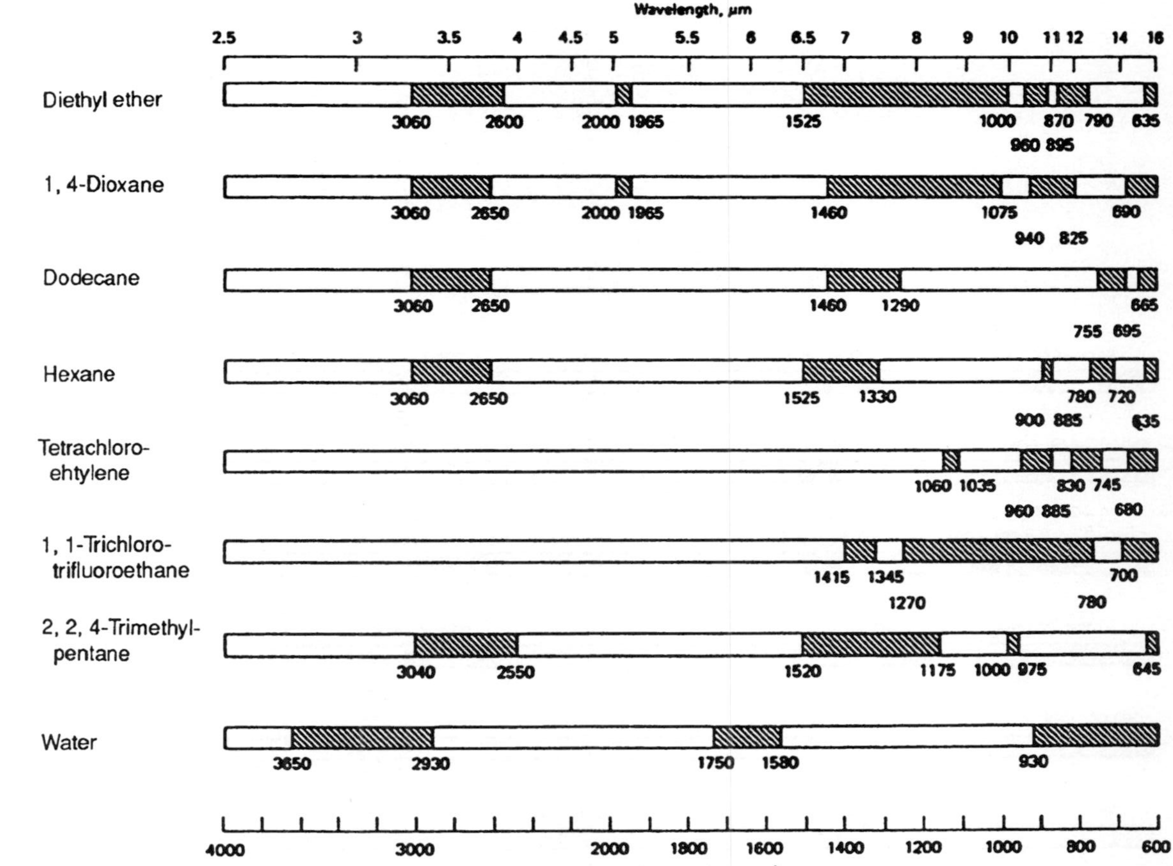

FIGURE 7.3 (*Continued*)

where n is the number of fringes (peaks or troughs) between two wave numbers, $\bar{\nu}_1$ and $\bar{\nu}_2$ (do not forget to count the first peak as 0), and η is the refractive index of the sample material. If measurements are made in wavelengths, the expression is

$$b = \frac{1}{2\eta}\left(\frac{n\lambda_1\lambda_2}{\lambda_2 - \lambda_1}\right) \tag{7.2}$$

where λ_2 is the starting wavelength and λ_1 is the final wavelength between which the fringes are counted. Film thickness can also be measured by the interference fringe method.

The *standard absorber method* may be used with a cell in any condition of polish, including cavity or minicells. Fill the cell with pure benzene. Use the 1960-cm^{-1} (5.10-μm) band of benzene for path lengths 0.1 mm or less for which benzene has an absorbance of 0.10 for every 0.01 mm of thickness. For longer path lengths, use the benzene band at 845 cm^{-1} (11.83 μm) for which the benzene absorbance is 0.24 for every 0.1 mm of thickness.

7.2.5.3 ***Film Technique.*** A large drop of the neat liquid is placed between two infrared-transmitting windows, which are then squeezed together and placed in a mount.

For polymers and noncrystalline solids, dissolve the sample in a volatile solvent, and pour the solution onto the window material. The solvent is evaporated by gentle heating to leave a thin film that then can be mounted in a holder. Sometimes polymers can be hot pressed onto window material or cut into a film of suitable thickness with a microtome.

7.2.5.4 ***Mull Technique.*** For qualitative analysis the mull technique is rapid and convenient, but quantitative data are difficult to obtain. The sample is finely ground in a clean mortar or a Wig-L-Bug. After grinding, the mulling agent (often mineral oil or Nujol, but may be perfluorokerosine or chlorofluorocarbon greases) is introduced in a small quantity just sufficient to convert the powder into the consistency of toothpaste. The cell is opened and a few drops of the pasty mull are placed on one plate, which is then covered with the other plate. The thickness of the cell is governed by squeezing the plates when the screws are tightened. Sample thickness should be adjusted so that the strongest bands display about 20% transmittance.

Always disassemble the cell by sliding the plates apart. *Do not attempt to pull the plates apart.* Be aware of changes in the sample that may occur during grinding.

7.2.5.5 ***Pellet Technique.*** Mix a few milligrams of finely ground sample with about 1 g of spectrophotometric grade KBr. Grinding and mixing are done in a vibrating ball mill or Wig-L-Bug. Place the mixture in an evacuable die at 60 000 to 100 000 lb·in^{-2} (4082 to 6805 atm). The pressure can be applied by using either a hydraulic press or a lever-screw press. Remove the pressed disk from the mold for insertion in the spectrometer.

KBr wafers can be formed, without evacuation, in a Mini-Press. Two highly polished bolts, turned against each other in a screw-mold housing with a wrench for about 1 min, produce a clear wafer. The housing with the wafer inside is inserted into the cell compartment of the spectrometer. Use 75 to 100 mg of powder. No moisture should be present, or water bands will obscure portions of the spectrum.

CsI or CsBr are used for measurements in the far-infrared region.

Caution Never apply pressure unless the powdered sample is in the mold. If no sample is present, the faces of the bolts or pistons will be scored.

7.2.6 Internal Reflectance

When a beam of radiation enters a plate surrounded by or immersed in a sample, it is reflected internally if the angle of incidence at the interface between sample and plate is greater than the critical angle (which is a function of refractive index). Although all the energy is reflected, the beam appears to penetrate slightly beyond the reflecting surface and then return. By varying the angle of incidence, the depth of penetration into the sample may be changed. At steep angles (near 30°) the depth of penetration is considerably greater by about an order of magnitude as compared with grazing angles (60°). This is significant in the study of surfaces, for example, a film or plastic in which chemical additives are suspected of migrating to the surface or where a surface coating has been exposed to weathering.

When a sample is placed in contact with the reflecting surface, the incident beam loses energy at those wavelengths for which the material absorbs due to an interaction with the penetrating beam. This attenuated radiation is an absorption spectrum that is similar to an infrared spectrum obtained in the normal transmission mode. Internal reflectance enables one to obtain a qualitative infrared absorption spectra from most solid materials or samples available only on a nontransparent support. This eliminates the need for grinding or dissolving or making a mull. Aqueous solutions are handled without compensating for very strong solvent absorption.

An internal reflectance attachment is inserted into the sampling space of an infrared spectrometer. One version has three standard positions of 30°, 45°, and 60°. Another version enables a range of angles to be selected by a scissor–jack assembly, linking the four mirrors and sample platforms in a pantograph system. Twenty-five internal reflections are standard for a 2-mm thick plate. A reflector plate with a relatively high index of refraction should be used. Thallium(I) bromide iodide (KRS-5) is satisfactory for most liquid and solid samples. The plate must not be brittle, as some pressure is required to bring some samples in contact with the plate. AgCl is suitable for aqueous samples.

In the single-pass plate, radiation enters through a bevel (effective aperture) at one end of the plate and, after propagation via multiple internal reflections down the length of the plate (1 to 10 cm in length), leaves through an exit bevel either parallel or perpendicular to the entrance bevel. In the double-pass technique, the radiation propagates down the length of the plate, is totally reflected at the opposite end by a surface perpendicular to the plate length, and returns to leave the plate at the entrance end. The end of the plate at which total reflection occurs can be dipped into liquids or powders and placed in closed systems.

7.3 CORRELATION OF INFRARED SPECTRA WITH MOLECULAR STRUCTURE IN THE MID-INFRARED REGION

Correlations pertinent to the near-infrared region are to be found is Sec. 7.1; likewise, those for the far-infrared region are discussed in Sec. 7.5.

Useful correlations in the mid-infrared region are shown in Fig. 7.4 and in Tables 7.4 to 7.19. It is best to divide this region into two parts—the group-frequency region (4000 to 1300 cm^{-1}; 2.50 to 7.69 μm) and the fingerprint region (1300 to 650 cm^{-1}; 7.69 to 15.38 μm). In searching infrared spectra, first note the absorption bands in the group-frequency region; these are more or less dependent on only the functional group that gives the absorption and not on the complete molecular structure, although structural influences do reveal themselves as shifts about the fundamental frequency. Proceed systematically through the group-frequency region. Hydrogen stretching frequencies with elements of mass 19 or less appear from 4000 to 2500 cm^{-1} (2.50 to 4.00 μm). Next observe the fingerprint region for here the major factors are single-bond stretching frequencies and skeletal frequencies of polyatomic systems that involve motions of bonds linking a substituent group to the remainder of the molecule. Collectively these absorption bands aid in identifying the material.

7.3.1 C—H Frequencies

The C—H stretching frequency of alkyl groups is less than 3000 cm^{-1} (Table 7.4), whereas for alkenes and aromatics it is greater than 3000 cm^{-1}. The CH_3 group gives rise to an asymmetric stretching mode at 2960 cm^{-1} and a symmetric mode at 2870 cm^{-1}. For —CH_2— these bands occur at 2930 and 2850 cm^{-1}.

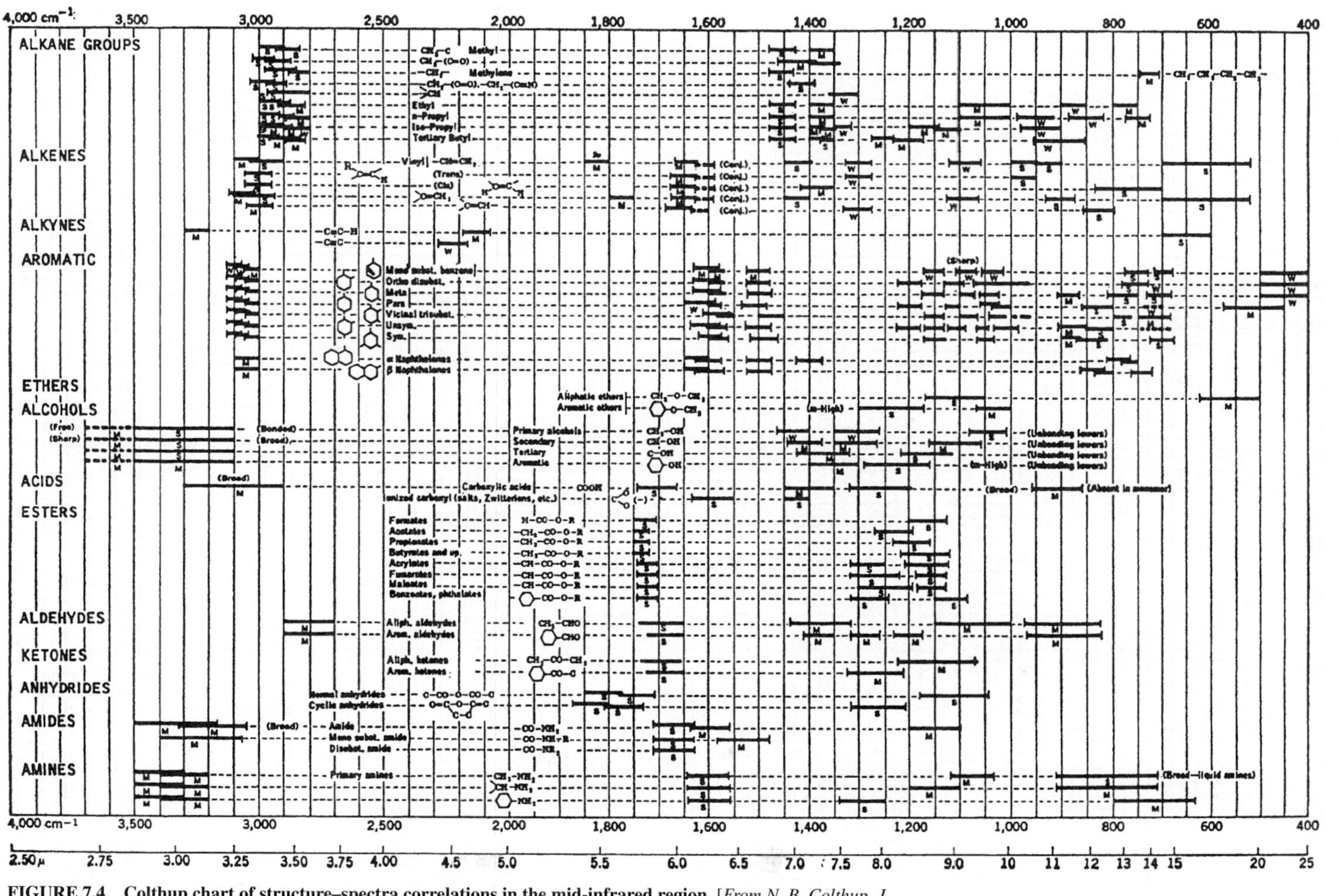

FIGURE 7.4 Colthup chart of structure–spectra correlations in the mid-infrared region. [*From N. B. Colthup, J. Opt. Soc. Am.* **40**:397 (1950). *Courtesy of N. B. Colthup and the editor of the Journal of the Optical Society of America.*]

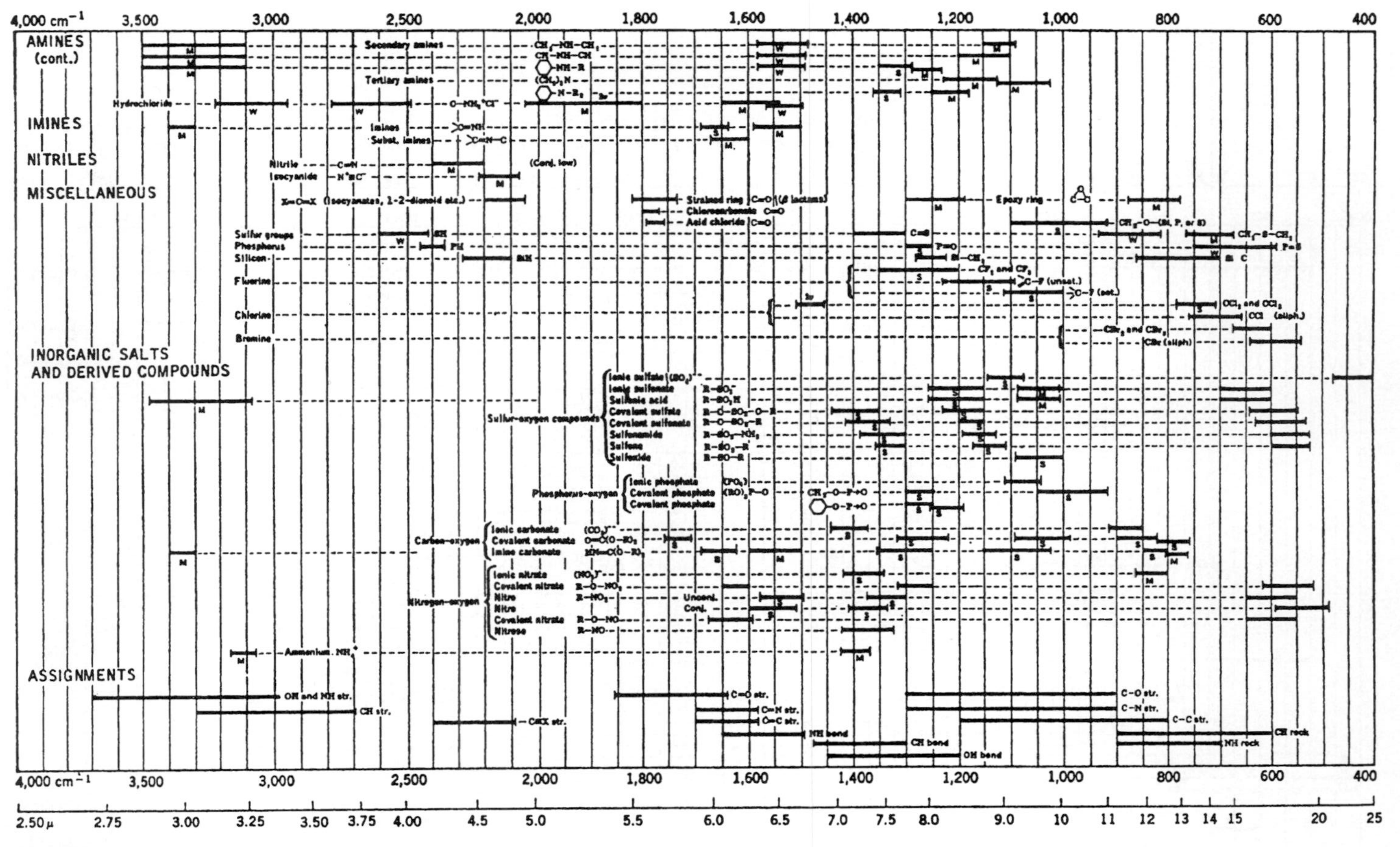

FIGURE 7.4 (*Continued*)

TABLE 7.4 Absorption Frequencies of Alkanes

Abbreviations used in the table

vw, very weak	*m-w moderate to weak*
w, weak	*ms, moderately strong*
w-m, weak to moderate	*s, strong*
m, moderate	*vs, very strong*

Vibrational group	Frequency, cm^{-1}	Remarks
Aliphatic–CH_3	2975–2950 (vs)	Asymmetric stretching
	2885–2865 (vs)	Symmetric stretch; doublet when double bond is adjacent
	1470–1440 (ms)	Asymmetric bending
	1380–1370 (m)	Symmetric bending of $—C(CH_3)$
	1385–1380 (m)	Symmetric bending of $—C(CH_3)_2$
	1375–1365 (m)	
	1395–1385 (m)	Symmetric bending of $—C(CH_3)_3$
	1375–1365 (ms)	
Aromatic–CH_3	2930–2920 (m)	
	2870–2860 (m)	
$—CH_2—$	2936–2915 (vs)	Asymmetric stretch
	2865–2840 (vs)	Symmetric stretch
	1475–1445 (ms)	Scissoring
	760–720 (m)	$—(CH_2)_n—$ in-phase rocking when $n \geq 4$; for $n = 3$, 2, or 1, the regions are 729–726, 743–734, and 785–770 cm^{-1}
Isopropyl group	1390–1380 (m)	Characteristic bending doublet about equal in intensity
	1372–1365 (m)	
	1175–1165 (w-m)	May shift to higher frequency if another branched carbon is adjacent
	1170–1140 (w-m)	
***gem*–Dimethyl group (in alkanes)**	1391–1381 (m)	Characteristic symmetric bending doublet; ratio is 4 to 5 for higher to lower frequency band
	1368–1366 (m)	
	1220–1206 (w)	
	1191–1185 (m)	
Tertiary butyl group	1400–1393 (m)	Unequal intensity doublet; very characteristic
	1374–1366 (s)	
	ca. 1245 (m-w)	
	ca. 1200 (m-w)	
	ca. 930 (m-w)	
Cyclopropane	3102 and 3082 (vs)	Antisymmetric stretching
	3038 and 3024 (vs)	Symmetric stretching
	1438 (m)	Scissoring
	1188 (w)	Ring breathing
	1028 (vs)	Wagging
	868 (vs)	Ring deformation
Cyclobutane	2974 and 2965 (vs)	Asymmetric stretching of CH_2
	2945 (s)	Symmetric stretching of CH_2
	1443 (m)	CH_2 scissoring
	1260 (s)	CH_2 wagging
	1224 (m)	CH_2 twisting
	901 (s)	Ring breathing
	626 (m)	CH_2 rocking
Cyclopentane	2965–2960 (s)	Asymmetric stretch of CH_2
	2880–2870 (s)	Symmetric stretch of CH_2
	1490–1430 (vs)	CH_2 scissoring

C≡C—H occurs around 3300 cm^{-1} (3.03 μm), and aromatic and unsaturated compounds around 3000 to 3100 cm^{-1} (3.33 to 3.23 μm).

For alkanes, bands at 1460 cm^{-1} (6.85 μm) and 1380 cm^{-1} (7.25 μm) are indicative of a terminal methyl group. If the latter band is split into a doublet at about 1397 and 1370 cm^{-1} (7.16 and 7.30 μm), geminal methyls are indicated. The latter band is shifted to lower frequencies if the methyl group is adjacent to =C=O (1360 to 1350 cm^{-1}), —S—(1325 cm^{-1}), and silicon (1250 cm^{-1}). A band at 1470 cm^{-1} indicates the presence of —CH_2—. Four or more methylene groups in a linear arrangement give rise to a weak band at about 720 cm^{-1}, and in the solid state, there will be a series of sharp bands around 1200 cm^{-1}, one band for every two methylene groups.

The substitution pattern of an aromatic ring can often be deduced from a series of weak bands in the region 2000 to 1670 cm^{-1} coupled with the position of strong bands between 900 and 650 cm^{-1}.

7.3.2 Alkenes

Double-bond frequencies fall in the region from 2000 to 1540 cm^{-1} (Table 7.5). An unsaturated C=C group introduces a band at 1650 cm^{-1}; it may be weak or nonexistent if symmetrically located in the molecule. Conjugation with a second C=C or C=O shifts the band 40 to 60 cm^{-1} to a lower frequency with a substantial increase in intensity.

Bands from the bending vibrations of hydrogen on a C=C bond are very valuable. A vinyl group gives rise to two bands at about 990 and 910 cm^{-1}. The =CH_2 band appears near 895 cm^{-1}. *Cis*- and *trans*-disubstituted olefins absorb in the region 685 to 730 cm^{-1} and 965 cm^{-1}, respectively. The single hydrogen in a trisubstituted olefin appears near 820 cm^{-1}.

7.3.3 Alkynes

Triple bonds, and little else, appear from 2500 to 2000 cm^{-1} (Table 7.6). The absorption band for —C≡C— is located around 2100 to 2140 cm^{-1} if terminal, but from 2260 to 2190 cm^{-1} if nonterminal. The intensity of nonterminal alkynes decreases as the symmetry of the molecule increases and may not appear. Conjugation with a carbonyl group increases the strength of the band markedly. The ethynyl hydrogen appears as a very sharp and intense band at 3300 cm^{-1}.

7.3.4 Alcohols and Phenols

The absorption due to the stretching of the O—H bond is most useful. In the unassociated state, it appears as a weak but sharp band at about 3600 cm^{-1}. Hydrogen bonding increases the band intensity and moves it to lower frequencies. If the hydrogen bonding is quite strong, the band becomes very broad.

The differentiation between the several types of alcohols is often possible on the basis of the C—O stretching bands. Saturated tertiary alcohols have a band in the region 1200 to 1125 cm^{-1}. For saturated secondary alcohols, the band lies between 1125 and 1085 cm^{-1}. Saturated primary alcohols show a band between 1085 and 1050 cm^{-1}. Table 7.7 contains the absorption frequencies of alcohols and phenols.

7.3.5 Amines

The absorption frequencies of primary, secondary, and tertiary amines are given in Table 7.8. Very useful are the N—H stretching frequencies at about 3500 and 3400 cm^{-1} for a primary amine (or amide), the N—H bending at 1610 cm^{-1}, and the bending of —NH_2 at about 830 cm^{-1} (which is broad for primary amines). A secondary amine exhibits a single band at about 3350 cm^{-1}.

TABLE 7.5 Absorption Frequencies of Alkenes >C=C<

Abbreviations used in the table

w, weak	*m-w, moderate to weak*
mw, moderately weak	*m-s, moderate to strong*
vw, very weak	*ms, moderately strong*
w-m, weak to moderate	*s, strong*
m, moderate	*vs, very strong*
	var, of variable strength

Vibrational group	Frequency, cm^{-1}	Remarks
$R_2C{=}CHR$	3040–3010 (w)	=CH stretching
	1680–1664 (w)	C=C stretching
	850–790 (m)	CH wag
$R_2C{=}CH_2$ vinylidene	3090–3075 (m)	$=CH_2$ asymmetric stretching
	3000–2980 (m)	$=CH_2$ symmetric stretching
	1810–1770 (w)	Overtone of CH_2 wag
	1660–1640 (ms)	C=C stretching
	1420–1400 (w)	CH_2 scissoring deformation
	900–885 (vs)	CH_2 wag
RHC=CHR *cis*-dialkyl	3020–2995 (m)	=CH stretching
	1662–1631 (m-w)	C=C stretching
	1429–1397 (m)	CH asymmetric rock
	1270–1250 (w)	CH symmetric rock
	730–650 (ms)	CH wag
RHC=CHR *trans*-dialkyl	3010–2995 (m)	=CH stretching
	1676–1665 (vw)	C=C stretching
	1325–1300 (vs)	CH symmetric rock
	1295 (mw)	CH asymmetric rock
	980–965 (vs)	CH wag
$RHC{=}CH_2$ vinyl	3095–3075 (m)	$=CH_2$ asymmetric stretching
	3000–2980 (m)	$=CH_2$ symmetric stretching
	3020–2995 (w)	=CH stretching
	1840–1805 (w)	Overtone of CH_2 wag
	1650–1638 (ms)	C=C stretching
	1420–1412 (mw)	CH_2 scissoring deformation
	1309–1288 (w)	CH rock
	995–985 (vs)	*trans* CH in-phase wag
	910–905 (vs)	CH_2 wag
	688–611 (w)	*cis* CH in-phase wag
$R_2C{=}CR_2$ nonconjugated, tetraalkyl	1680–1665 (vs)	May be weak if symmetrically substituted

Electronegative substituents on the C=C shift the =CH and $=CH_2$ stretching bands to higher wave numbers by approximately 30–60 cm^{-1}

C=C both carbons within a ring	CH=CH	CH=CR	CR=CR
Carbons in ring: 3	1656 (w-m)	1788	1900–1865
4	1566 (w-m)	1641	1675
5	1617–1611 (w-m)	1660–1640	1686–1671
6 or more	1655–1645 (w-m)	1686–1671	1685–1677
Exocyclic C=C—$(CH_2)_n$, $n = 2$	1780–1730 (m)	C=C stretching	
$= 3$	ca. 1680 (m)	C=C stretching	
$= 4$	1655–1650 (m)	C=C stretching	
$H_2C{=}CH{-}CH{=}CH_2$	1592 (s)	Antisymmetric stretching	

(Continued)

TABLE 7.5 Absorption Frequencies of Alkenes >C=C< (*Continued*)

Vibrational group	Frequency, cm^{-1}	Remarks
H_2C=CH—CR=CH_2, R: alkyl	1642–1632 (w)	Symmetric stretching; *tert*-butyl in *s-cis* conformation at 1611 cm^{-1} (s)
	1596–1590 (s)	Antisymmetric stretching; *tert*-butyl in *s-cis* conformation at 1645 cm^{-1} (w)
—C(=O)—CH=CH_2	1703–1674 (vs)	C=O stretching
	1648–1615 (var)	C=C stretching
	995–980 (s)	*trans* CH wag
	965–955 (m)	CH_2 wag
Conjugation with a phenyl ring	1637–1616 (m)	C=C stretching
	928–901	=CH_2 wag
2- and 6-ring substitution	1644–1623	C=C stretching
	947–918	=CH_2 wag
Conjugation with a nitrile group	1635–1609 (s)	C=C stretching

7.3.6 Carbonyl Group

The carbonyl group is often the strongest band in the spectrum. It will lie between 1825 and 1575 cm^{-1}; its exact position is dependent upon its immediate substituents (Table 7.9). Anhydrides usually show a double absorption band and, in addition, will exhibit a C—H stretching frequency of the CHO group at about 2720 cm^{-1}. The carboxyl group has bands at 2700, 1300, and 943 cm^{-1} that are associated with the carboxyl OH; these disappear when the carboxylate ion is formed. When a dimer exists, the band at 2700 cm^{-1} disappears.

The general trends of structural variation on the position of carbonyl stretching frequencies may be summarized:

1. The more electronegative the group X in the system R—CO—X—, the higher is the frequency.

2. α, β-Unsaturation causes a lowering of frequency of 15 to 40 cm^{-1}, except in amides, where little shift is observed and that usually to higher frequency.

3. Further conjugation has relatively little effect.

4. Ring strain in cyclic compounds causes a relatively large shift to higher frequency. This phenomenon provides a remarkably reliable test of ring size, distinguishing clearly between four-, five-, and larger-membered-ring ketones, lactones, and lactams. Six-ring and larger ketones, lactones, and lactams show the normal frequency found for open-chain compounds.

5. Hydrogen bonding to a carbonyl group causes a shift to lower frequency of 40 to 60 cm^{-1}. Acids, amides, enolized β-keto carbonyl systems, and *o*-hydroxyphenol and *o*-aminophenyl carbonyl compounds show this effect. All carbonyl compounds tend to give slightly lower values for the carbonyl stretching frequency in the solid state compared with the value for dilute solution.

6. Where more than one of the structural influences on a particular carbonyl group is operating, the net effect is usually close to additive.

7.3.7 Ethers

One important and quite strong band appears near 1100 cm^{-1}. Absorption frequencies of ethers and peroxides are given in Table 7.10.

TABLE 7.6 Absorption Frequencies of Triple Bonds

Abbreviations used in the table

w, weak	*w-m, weak to moderate*
m, moderate	*vs, very strong*
m-s, moderate to strong	*var, of variable strength*
s, strong	

Vibrational group	Frequency, cm^{-1}	Remarks
Alkynes		
Terminal, —C≡C—H	3320–3280 (s)	≡C—H stretching
	2140–2100 (w-m)*	C≡C stretching
	1375–1225 (w-m)	
	681–610 (vs)	≡C—H bending
	640–628 (s)	Alkyl substituted
	348–336 (w)	CH_3—$(CH_2)_n$—C≡CH, $n = 0 - 5$
Nonterminal, R_1—C≡C—R_2	2260–2150 (var)*	Symmetrical or nearly symmetrical substitution makes stretching frequency inactive
	540–465 (m)	The longer the chain, the lower the frequency
	1000–940 (w)	≡C—C stretching (monosubstituted) (disubstituted)
	1160–1105 (w)	
	2200–2170 (s)	≡C—C bending
	2245–2175 (s)	—C≡C—C≡C—, asymmetrical stretch symmetrical stretch
Nitriles —C≡N	2260–2200 (s)	C≡N stretch
	378 (s)	C—C≡N bending
Cyanamides N—C≡N	2225–2210 (s)	C≡N stretching; alkylation on nitrogen atom lowers frequency
Cyanates O—C≡N	2256–2245 (s)	C≡N stretching
Thiocyanates S—C≡N	2157–2155 (s)	C≡N stretching (aliphatic)
	2174–2160 (s)	(aromatic)
	513–453 (w)	S—C≡N bending
	416–405 (m)	
Selenocyanates Se—C≡N	ca. 2160 (s)	C≡N stretching
	365–360 (w)	Se—C≡N bending
Isocyanides —C—N≡C	2146–2134 (s)	N≡C stretching (aliphatic)
	2125–2109 (s)	(aromatic)
	2152 (s)	Aryl—CH_2—N≡C stretching
Nitrile *N*-oxides —C≡N→O	2305–2285 (s)	Aryl derivatives
	1395–1365 (s)	
Diazonium salts	2300–2230 (m-s)	

* Conjugation with olefinic or alkyne groups lowers the frequency and raises the intensity. Conjugation with carbonyl groups usually has little effect on the position of absorption.

7.3.8 Other Functional Groups

In addition to those functional groups already discussed, additional tables list the absorption frequencies of sulfur compounds (Table 7.11), aromatic and heteroaromatic bands (Table 7.12), the nitro group (Table 7.13), double bonds containing nitrogen atoms (Table 7.14), cumulated double bonds (Table 7.15), boron compounds (Table 7.16), phosphorus compounds (Table 7.17), silicon compounds (Table 7.18), and halogen compounds (Table 7.19).

TABLE 7.7 Absorption Frequencies of Alcohols and Phenols

Abbreviations used in the table

w, weak — *vs, very strong*
ms, moderately strong — *var, of variable strength*
s, strong,

Vibrational group	Frequency, cm^{-1}	Remarks
Primary aliphatic alcohols	3644–3630 (m)	Free OH stretch; only in very dilute, nonpolar solvents
	1430–1200 (w)	C—O—H bend
	1075–1000 (s)	Out-of-phase C—C—O stretch
	900–800 (m)	In-phase C—C—O stretch
Secondary aliphatic alcohols	3637–3620 (m)	Free OH stretch in nonpolar solvents
	1430–1200 (w)	C—O—H bend
	1130–1075 (s)	Out-of-phase C—C—O stretch
	900–800 (m)	In-phase C—C—O stretch; most ca. 820 cm^{-1}
Tertiary aliphatic alcohols	3625–3610 (m)	Free OH stretch in CCl_4
	1410–1310 (w)	C—O—H bend
	1210–1110 (s)	Out-of-phase C—C—O stretch
	800–750 (m)	In-phase C—C—O stretch
Tertiary bicyclic alcohols	3612–3606 (m)	Free OH stretch in CCl_4
Phenols	3612–3593 (m)	Free OH stretch in CCl_4
	1410–1310 (m)	C—O—H bend
	1260–1180 (s)	Out-of-phase C—C—O stretch
Hydrogen-bonded OH stretch		
Intermolecular		
Dimeric	3600–3450 (m)	Rather sharp band
Polymeric	3400–3200 (vs)	Broad
Intramolecular		
Single bridge	3600–3500 (m)	Sharp band
Chelation	3200–2500 (var)	Broad; occasionally weak; the lower the frequency, the stronger the intramolecular bond

7.3.9 Compound Identification[4–6]

The total structure of an unknown may not be readily identified from the infrared spectrum, but perhaps the type of class of compound can be deduced. Once the key functional groups have been established as present (or, equally important, as absent), the unknown spectrum is compared with spectra of known compounds. Several collections of spectra are available.[7,8]

[4] R. M. Silverstein, G. C. Bassler, and T. C. Morrill, *Spectrophotometric Identification of Organic Compounds*, 5th ed., Wiley, New York, 1991.

[5] D. H. Williams and J. Fleming, *Spectroscopic Methods in Organic Chemistry*, 4th ed., McGraw-Hill, New York, 1993.

[6] D. Lin-Vien, N. B. Colthup, W. G. Fateley, and J. G. Grasselli, *The Handbook of Infrared and Raman Characteristic Frequencies of Organic Molecules*, Academic, New York, 1991.

[7] C. J. Pouchert, ed., *The Aldrich Library of Infrared Spectra*, Aldrich, Milwaukee, WI. A series of dispersive and FT-IR spectra compilations.

[8] Sadtler Research Laboratories, *Catalog of Infrared Spectrograms*, Philadelphia, PA. A continuously updated series.

TABLE 7.8 Absorption Frequencies of Primary, Secondary, and Tertiary Amines

Abbreviations used in the table

w, weak	*ms, moderately strong*
mw, moderately weak	*s, strong*
m-w, moderate to weak	*s-m, strong to moderate*
m, moderate	*vs, very strong*
m-s, moderate to strong	*var, variable*

Vibrational group	Frequency, cm^{-1}	Remarks
Primary amines		
Aliphatic, $RCH_2—NH_2$	3398–3381 (mw)	Asymmetric NH_2 stretch; dilute solution
	3344–3324 (mw)	Symmetric NH_2 stretch; dilute solution
	1627–1590 (m)	NH_2 scissors
	1090–1040 (m-w)	C—N stretching
	854–778 (s)	Broad, liquid; NH_2 wag
	ca. 502 and 436	Skeletal CCN deformation
Aliphatic, R—CHR—NH_2	3370–3363 (mw)	Asymmetric NH_2 stretching; liquid
	3285–3280 (mw)	Symmetric NH_2 stretching; liquid
	1627–1590 (m)	NH_2 scissors
	1180–1170 (s-m)	C—N stretching and skeletal stretching modes
	1163–1153 (s-m)	
	1143–1130 (s-m)	
	1090–1080 (m-w)	
	1040–1000 (m-w)	
	791–784 (s)	Broad, liquid; NH_2 wag
	ca. 502 and 436	CCN deformation
Aliphatic, R—CR_2—NH_2	ca. 3350 (mw)	Asymmetric NH_2 stretch; dilute solution
	ca. 3280 (mw)	Symmetric NH_2 stretch; liquid
	1627–1590 (m)	NH_2 scissors
	1245–1235 (ms)	C—N stretch and skeletal stretching
	1218–1195 (ms)	
	1140–1110 (m-s)	
	1080–1060 (m-w)	
	1030–1000 (m-w)	
	897–848 (m)	NH_2 wag
	ca. 502 and 436	CCN deformation
Aromatic, Aryl—NH_2	3509–3460 (ms)	Asymmetric NH_2 stretch; dilute solution
	3416–3382 (ms)	Symmetric NH_2 stretch; dilute solution
	1638–1602 (s)	NH_2 scissors
	1330–1260 (s)	C—N and skeletal stretching
	700–600 (m)	NH_2 wag; liquid
	ca. 350	In plane CN bending
	ca. 200	Out-of-plane bending
Amides, C—C(=O)—NH_2	3475–3350	Asymmetric NH_2 stretch, Nujol
	3385–3180 (s)	Symmetric NH_2 stretch, Nujol
	1650–1620 (s)	NH_2 scissors, Nujol
	1430–1390	C—N and skeletal stretching
	720–600 (m)	NH_2 wag
	600–500	O=C—N skeletal deformation
Secondary amines		
Aliphatic, R—NHR	3320–3280 (w)	N—H stretching, liquid; 3360–3310 cm^{-1} in dilute solution
	1180–1130 (ms)	Asymmetric C—N—C stretching
	900–850 (var)	Symmetric C—N—C stretching
	415–377	C—N deformation

(Continued)

TABLE 7.8 Absorption Frequencies of Primary, Secondary, and Tertiary Amines (*Continued*)

Vibrational group	Frequency, cm^{-1}	Remarks
Aromatic, Aryl—NHR	ca. 3400 (m)	N—H stretch, liquid; ca. 3450 cm^{-1}, dilute solution
	ca. 1510 (ms)	C—N—C bending
	1350–1280 (var)	C—N stretching
Amide, —C(=O)—NHR	3320–3270 (m)	*trans* N—H stretch, liquid; 3480–3440 cm^{-1}, dilute solution
	ca. 3200 (m)	*cis* N—H stretch, liquid; 3180–3140 cm^{-1}, dilute solution
	1550–1510 (s)	C—N—H bend, *trans* amide II, dilute solution
	1490–1440 (m)	*cis* C—N—H bend
	1350–1310 (ms)	*cis* C—N stretch; for *trans* see amide II and amide III bands under N—H bend
	1300–1248 (m)	Amide III
	ca. 630	C—N deformation
Tertiary amines		
$—NR_2$	1250–1000 (s-m)	Asymmetric C—N stretch of $R—NR_2$
	833–740	Symmetric C—N stretch
	375–340	C—N—C deformation of $R—NR_2$
Amides	870–820	C—N stretch of $HC(=O)NR_2$
	ca. 650	O=C—N deformation of $HC(=O)NR_2$
	750–700	C—N stretch of $R—C(=O)NR_2$
	620–590	O=C—N deformation of $RC(=O)NR_2$
Amine salts		
Primary $—NH_3^+$	3200–2800 (s)	NH_3 stretchings
	ca. 2600 (s-m)	Combination bands
	2800–2100	Series of sharp bands, stronger intensity at higher frequencies
	2070–2000 (m-w)	Aryl amines at 2000–1960 cm^{-1}
	1625–1560	Asymmetric NH_3 deformations
	1550–1505	Symmetric NH_3 deformations
Secondary amine salts $—NH_2^+—$	3000–2700 (s)	Multiple stretching bands
	2700–2300	Combination bands
	1620–1560 (m)	NH_2^+ deformation band
Tertiary amine salts $>NH^+—$	3200 (s)	Free N—H moiety; associated at 3195 cm^{-1}
	2700–2300 (s)	Multibands
Quaternary amine salts R_4N^+		
Tetramethylammonium salts	ca. 3020	CH_3 band; also at 1485 and 1410 cm^{-1}
	ca. 950	Asymmetric $N—C_4$ stretchings; at 1060–1030 cm^{-1} for tetraethyl
	ca. 750	Symmetric $N—C_4$ stretchings; at 672–666 cm^{-1} for tetraethyl
Hydroxylamines	3251–3237	NH_2 symmetric stretch
	1595–1589	NH_2 scissoring
	906	O—N stretch of $HO—NH_2$
	851–840	C—O—N stretching
Hydrazine	3336–3190	NH_2 stretchings
	1628	NH_2 scissor
	882	NH_2 wag

TABLE 7.9 Absorption Frequencies of Carbonyl Bands

All bands quoted are strong

Group	Band, cm^{-1}	Remarks
Acid anhydrides —CO—O—CO—		
Saturated	1850–1800 1790–1740	Two bands usually separated by about 60 cm^{-1}. The higher-frequency band is more intense in acyclic anhydrides, and the lower-frequency band is more intense in cyclic anhydrides.
Aryl and α,β-unsaturated	1830–1780 1790–1710	
Saturated five-ring	1870–1820 1800–1750	
All classes	1300–1050	One or two strong bands due to CO stretching.
Acid chlorides —COCl		
Saturated	1815–1790	Acid fluorides higher, bromides and iodides lower.
Aryl and α,β-unsaturated	1790–1750	
Acid peroxide CO—O—O—CO—		
Saturated	1820–1810 1800–1780	
Aryl and α,β-unsaturated	1805–1780 1785–1755	
Esters and lactones —CO—O—		
Saturated	1750–1735	
Aryl and α,β-unsaturated	1730–1715	
Aryl and vinyl esters C═C—O—CO—alkyl	1800–1750	The C═C stretching band also shifts to higher frequency.
Esters with electronegative α substituents; e.g., >CCl—CO—O—	1770–1745	
α-Keto esters	1755–1740	
Six-ring and larger lactones	Similar values to the corresponding open-chain esters	
Five-ring lactone	1780–1760	
α,β-Unsaturated five-ring lactone	1770–1740	When α-CH is present, there are two bands, the relative intensity depending on the solvent.
β,γ-Unsaturated five-ring lactone, vinyl ester type	ca. 1800	
Four-ring lactone	ca. 1820	
β-Keto ester in H bonding enol form	ca. 1650	Keto from normal; chelate-type H bond causes shift to lower frequency than the normal ester. The C═C band is strong and is usually near 1630 cm^{-1}.
All classes	1300–1050	Usually two strong bands due to CO stretching.

(Continued)

TABLE 7.9 Absorption Frequencies of Carbonyl Bands (*Continued*)

Group	Band, cm^{-1}	Remarks
Aldehydes —CHO		
(See also Table 7.4 for C—H.) All values given below are lowered in liquid-film or solid-state spectra by about 10–20 cm^{-1}. Vapor-phase spectra have values raised about 20 cm^{-1}.		
Saturated	1740–1720	
Aryl	1715–1695	*o*-Hydroxy or amino groups shift this value to 1655–1625 cm^{-1} because of intramolecular H bonding.
α,β-Unsaturated	1705–1680	
$\alpha,\beta,\gamma,\delta$-Unsaturated	1680–1660	
β-Ketoaldehyde in enol form	1670–1645	Lowering caused by chelate-type H bonding.
Ketones $\gt C{=}O$		
All values given below are lowered in liquid-film or solid-state spectra by about 10–20 cm^{-1}. Vapor-phase spectra have values raised about 20 cm^{-1}.		
Saturated	1725–1705	
Aryl	1700–1680	
α,β-Unsaturated	1685–1665	
$\alpha,\beta,\alpha',\beta'$-Unsaturated and diaryl	1670–1660	
Cyclopropyl	1705–1685	
Six-ring ketones and larger	Similar values to the corresponding open-chain ketones	
Five-ring ketones	1750–1740	α,β Unsaturation, $\alpha,\beta,\alpha',\beta'$ unsaturation, etc., have a similar effect on these values as on those of open-chain ketones.
Four-ring ketones	ca. 1780	
α-Halo ketones	1745–1725	Affected by conformation; highest values are obtained when both halogens are in the same plane as the C=O.
α,α'-Dihalo ketones	1765–1745	
1,2-Diketones, *syn-trans*-open chains	1730–1710	Antisymmetrical stretching frequency of both C=O bands. The symmetrical stretching is inactive in the infrared but active in the Raman.
syn-cis-1,2-Diketones, six-ring	1760 and 1730	
syn-cis-1,2-Diketones, five ring	1775 and 1760	
o-Amino-aryl or *o*-hydroxyaryl ketones	1655–1635	Low because of intramolecular H bonding. Other substituents and steric hindrance affect the position of the band.
Quinones	1690–1660	C=C band is strong and is usually near 1600 cm^{-1}.
Extended quinones	1655–1635	
Tropone	1650	Near 1600 cm^{-1} when lowered by H bonding as in tropolones.

TABLE 7.9 **Absorption Frequencies of Carbonyl Bands** (*Continued*)

Group	Band, cm^{-1}	Remarks
Carboxylic acids —CO_2H		
All types	3000–2500	OH stretching; a characteristic group of small bands due to combination bands.
Saturated	1725–1700	The monomer is near 1760 cm^{-1}, but is rarely observed. Occasionally both bands, the free monomer, and the H-bonded dimer can be seen in solution spectra. Ether solvents give one band near 1730 cm^{-1}.
α,β-Unsaturated	1715–1690	
Aryl	1700–1680	
α-Halo-	1740–1720	
Carboxylate ions —CO_2^-		
Most types	1610–1550 1420–1300	Antisymmetrical and symmetrical stretching, respectively.
Amides —CO—N<		
(See also Table 7.8 for NH stretching and bending.)		
Primary —$CONH_2$		
In solution	ca. 1690	Amide I; C=O stretching.
Solid state	ca. 1650	
In solution	ca. 1600	Amide II: mostly NH bending.
Solid state	ca. 1640	
		Amide I is generally more intense than amide II. (In the solid state, amides I and II may overlap.)
Secondary —CONH—		
In solution	1700–1670	Amide I.
Solid state	1680–1630	
In solution	1550–1510	Amide II; found in open-chain amides only.
Solid state	1570–1515	Amide I is generally more intense than amide II.
Tertiary	1670–1630	Since H bonding is absent, solid and solution spectra are much the same.
Lactams		
Six-ring and larger rings	ca. 1670	
Five-ring	ca. 1700	Shifted to higher frequency when the N atom is in a bridged system.
Four-ring	ca. 1745	
R—CO—N—C=C		Shifted +15 cm^{-1} by the additional double bond.
C=C—CO—N		Shifted by up to +15 cm^{-1} by the additional double bond. This is an unusual effect by α,β unsaturation. It is said to be due to the inductive effect of the C=C on the well-conjugated CO—N system, the usual conjugation effect being less important in such a system.
Imides —CO—N—CO—		
Cyclic six-ring	ca. 1710 and ca. 1700	Shift of +15 cm^{-1} with α,β unsaturation.

(Continued)

TABLE 7.9 Absorption Frequencies of Carbonyl Bands (*Continued*)

Group	Band, cm^{-1}	Remarks
Imides —CO—N—CO—		
Cyclic five-ring	ca. 1770 and ca. 1700	
Ureas N—CO—N		
RNHCONHR	ca. 1660	
Six-ring	ca. 1640	
Five-ring	ca. 1720	
Urethanes R—O—CO—N	1740–1690	Also shows amide II band when nonsubstituted on N.
Thioesters and Acids RCO—S—R′		
RCOSH	ca. 1720	α,β-Unsaturated or aryl acid or ester shifted about $-25\ cm^{-1}$.
RCOS—alkyl	ca. 1690	
RCOS—aryl	ca. 1710	

TABLE 7.10 Absorption Frequencies of Ethers and Peroxides

Abbreviations used in the table

w, weak *s, strong*
m, moderate *vs, very strong*
var, variable

Vibrational group	Frequency, cm^{-1}	Remarks
Aliphatic ethers	1150–1060 (vs)	Asymmetric C—O—C stretch; α-branching reduces frequency, often multiple bands observed: diisopropyl ether, 1169, 1112, and 1076 cm^{-1}; *tert*-butyl, 1201, 1117, 1076 cm^{-1}
	890–820 (w)	Symmetric C—O—C stretch
Vinyl ethers	1225–1200 (vs)	Out-of-phase C—O—C stretch
	850–840 (w)	In-phase C—O—C stretch
Aromatic ethers		
Aryl-alkyl	1310–1210 (vs)	Aryl —O stretch
	1050–1010 (s)	O—CH_2 or O—CH_3 stretch
Diaryl	ca. 1240 (s)	Aryl —O stretch
Acetals		
Dialkoxymethanes	1140–1115 (s)	Antisymmetric C—O—C—O—C stretching
	1050–1040 (s)	
Ethylidene dialkyl ethers	1140–1130 (s)	Antisymmetric C—O—C—O—C stretching
	870–850 (s)	
Both dialkoxymethane and ethylidene dialkyl ethers	1115–1080 (s)	Symmetric C—O—C—O—C stretching
	870–800 (s)	
	660–600 (vs)	C—O—C—O—C deformation
	540–450 (s)	
	400–320 (s)	
Epoxides	3075–3030 (m)	Asymmetric CH_2 stretch
	3020–2990 (m)	Symmetric CH_2 stretch
	1280–1230 (m)	Ring symmetrical stretch
	950–815 (s)	Ring asymmetrical stretch
	880–750 (s)	Ring symmetrical deformation
Peroxides	900–800 (w)	O—O stretching; Raman much better

TABLE 7.11 Absorption Frequencies of Sulfur Compounds

Abbreviations used in the table

w, weak	*s, strong*
w-m, weak to moderate	*vs, very strong*
m, moderate	*vs-m, very strong to moderate*
ms, moderately strong	*var, of variable strength*
m-s, moderate to strong	

Vibrational group	Frequency, cm^{-1}	Remarks
Thiols		
—S—H	2600–2450 (w)	S—H stretch
Thiocarboxylic acids, —C(=O)—SH	2580–2540 (vs-m)	SH stretch; liquids
	726–626 (m)	Trans (higher) and gauche (lower) forms for R > 2 carbons; halogen substituents lower frequency
Dithioacids, —C(=S)—SH	2568–2552	Two S—H bands in solution
Sulfides		
R—S—R	750–690 (w-m)	
Aryl—S—R	ca. 722 (s)	
	ca. 698 (s)	
Aryl—S—aryl	ca. 701 (var)	
Cyclic—S—moieties	705–656 (var)	Often a doublet, lower frequency stronger
Disulfides	715–620 (var)	C—S stretching; S—S bands very weak
Thiocarbonyl		
C=S	1200–1050 (s)	Behaves similar to carbonyl band
Thioamides, N—C=S		
Primary and secondary	950–800 (ms)	—C=S stretch
	750–700 (m)	N—C=S deformation; 700–550 cm^{-1} for secondary
	500–400 (m)	
Tertiary	1563–1524 (vs)	C—N stretch
	1285–1210 (s)	
	1000–700 (var)	C—C, C=S, C—N—C stretch modes
	626–500 (m)	N—C=S deformation
	448–338 (m-w)	
S—C=S	ca. 580 (s)	
Sulfoxides S=O	1075–1030 (s)	S=O stretch; halogen bonded to sulfur increases frequency
	730–690 (var)	C—S stretch
	395–360 (var)	C—S=O bending
Thionyl halides	801 (vs)	S—F stretch
	721 (vs)	
	492 (vs)	S—Cl stretch
	455 (vs)	
Sulfones >SO_2	1335–1290 (vs)	Asymmetric stretch; halogen bonded to sulfur increases frequency
	1160–1120 (vs)	Symmetric stretch of SO_2
	586–505 (s)	Scissoring mode
	550–438 (s)	Wagging
	430–280	Rocking and twisting modes
Sulfonyl halides R—SO_2—X	1412–1365 (m-w)	F higher frequency than Cl; little difference between alkyl or aryl
	1197–1167 (vs)	See above
	ca. 373 (m)	S—Cl stretch

(Continued)

TABLE 7.11 Absorption Frequencies of Sulfur Compounds (*Continued*)

Vibrational group	Frequency, cm^{-1}	Remarks
Sulfuryl halides X—SO_2—X	1497–1414 (var)	Strength: F (w) and Cl (s)
	1263–1182 (var)	Strength: F (vs) and Cl (s)
Sulfonamides —SO_2—N	1380–1315 (vs)	
	1170–1140 (vs)	
	950–860 (m)	
	715–700 (w-m)	
Sulfonates —SO_2—O	1410–1335 (m)	Asymmetric stretch
	1200–1165 (vs)	Symmetric stretch
Thiosulfonates —SO_2—S	1335–1305 (s-m)	Asymmetric stretch
	1130–1125 (s)	Symmetric stretch
Sulfates O—SO_2—O	1415–1380 (s)	Electronegative substituents increases frequencies of stretch modes
	1200–1185 (s)	
Primary alkyl salts	1315–1220 (s)	Both bands strongly influenced by metal ion
	1140–1075 (m)	
Secondary alkyl salts	1270–1210 (vs)	Doublet; both bands strongly influenced by metal ion
	1075–1050 (s)	

TABLE 7.12 Absorption Frequencies of Aromatic and Heteroaromatic Bands

Abbreviations used in the table

w, weak	*m-s, moderate to strong*
mw, moderately weak	*ms, moderately strong*
w-m, weak to moderate	*s, strong*
m, moderate	*var, of variable strength*

Vibrational group	Frequency, cm^{-1}	Remarks
Benzene derivatives		
C—H stretch, substituted derivatives	3100–3000 (mw)	Phenyl group often has triplets
Carbon–carbon ring stretch		
Mono, di, and tri substituents	1620–1585 (m)	Often stronger than second band
	1590–1565 (m)	Enhanced by ring conjugation or halogen substitution
Mono, ortho, and meta substituted	1510–1470 (m)	
	1465–1430 (m)	
Para substituted	1524–1480 (m)	If different substituents, otherwise inactive; strong for electron donors
	1023–1003 (m)	
Mono substituted	1420–1400 (m)	
	1180–1170 (w)	
	1166–1146 (w)	
Ortho substituted	1170–1150 (m)	
	1150–1100 (m)	
	1055–1020 (m)	
Meta substituted	1180–1145 (w)	
	1140–1065 (mw)	
	1100–1060 (w)	
Para substituted	1128–1100 (w)	
	1023–1003 (m)	

TABLE 7.12 Absorption Frequencies of Aromatic and Heteroaromatic Bands (*Continued*)

Vibrational group	Frequency, cm^{-1}	Remarks
Adjacent hydrogen wag regions		
Mono (five adjacent hydrogens)	900–860 (w-m)	
	770–730 (s)	Diagnostic band
	720–680 (s)	Diagnostic band
	625–605 (w-m)	
	ca. 550 (w-m)	
1,2-Disubstitution (four adjacent hydrogens)	770–735 (s)	Diagnostic band
	550–495 (w-m)	
	470–415 (m-s)	
1,3-Disubstitution (three adjacent hydrogens)	810–750 (s)	Diagnostic band
	555–495 (w-m)	
	470–415 (m-s)	
1,4-Disubstitution (two adjacent hydrogens)	860–800 (s)	Diagnostic band
	650–615 (w-m)	
	460–415 (m-s)	490–460 cm^{-1} when substituents are electron-accepting groups
1,2,3-Trisubstitution (three adjacent hydrogens)	800–760 (s)	Diagnostic band
	720–685 (s)	
	570–535 (s)	
	ca. 485	
1,2,4-Trisubstitution (two adjacent hydrogens)	900–760 (m)	Diagnostic band; single H
	780–760 (s)	Diagnostic band; two adjacent H
1,3,5-Trisubstitution	950–925 (var)	
	865–810 (s)	
	730–680 (m-s)	
	535–495 (s)	
	470–450 (w-m)	
Pentasubstitution (one lone hydrogen)	900–860 (m-s)	Diagnostic band
	580–535 (s)	
Hexasubstitution	415–385 (m-s)	
Naphthalenes		
Alkyl-substituted	1520–1505	Doublet
	1400–1390	
Hydrogen wag		
1-Naphthalenes	805–775	Three adjacent H
	780–760	Four adjacent H
2-Naphthalenes	875–823	Lone H
	825–800	Two adjacent H
	760–735	Four adjacent H
More highly substituted naphthalenes	905–835	Lone H
	847–799	Two adjacent H
	820–730	Three adjacent H; often two bands
	800–726	Four adjacent H; often two bands
Anthracenes		
Alkyl-substituted	1640–1620	
	ca. 1550	Usually a band
	890–875 (ms)	Lone H on one or both 9- or 10-positions
	750–730 (s)	At least one four-adjacent-H group
Phenanthrenes		
Alkyl-substituted	ca.1600	Often a doublet
	1500	Distinguishes it from anthracenes
Substitution patterns	ca. 820	Two adjacent H on middle ring
	750–730 (s)	At least one four-adjacent-H group

(Continued)

TABLE 7.12 Absorption Frequencies of Aromatic and Heteroaromatic Bands (*Continued*)

Vibrational group	Frequency, cm^{-1}	Remarks
Pyridines		
2-Substituted	3100–3000	C—H stretch
	1620–1570 (s)	
	1580–1560 (s)	
	1480–1450 (s)	
	1440–1415 (s)	
	1050–1040 (m)	
	1000–985 (m)	
	780–740 (s)	
	740–720 (m)	
	630–615 (w)	
3-Substituted	3100–3000	C—H stretch
	1595–1570 (m)	
	1585–1560 (s)	
	1480–1465 (s)	
	1430–1410 (s)	
	1030–1010 (m)	
	820–770 (s)	
	730–690 (s)	
	630–615 (w)	
4-Substituted	3100–3000	
	1605–1565 (s)	
	1570–1555 (m)	C—H stretch
	1500–1480 (m)	
	1420–1410 (s)	
	1000–985 (m)	
	850–790 (s)	
	730–720 (m)	
Pyridine *N*-Oxides (also of pyrimidines and pyrazine)	1300–1200 (s)	N → O stretch
	880–845 (m)	
1,3,5-Triazine	3055 (s)	
	1550 (vs)	
	1410 (vs)	
	1172 (s)	
	730 (vs)	
	685 (vs)	
Pyrroles		
1-Substituted	3450–3208	Bonded N—H stretch
	3180–3090	C—H stretch
	1560–1540 (w)	Ring stretching bands
	1510–1490	
	1390–1380	
	1095–1080 (m)	
	1065–1055 (mw)	
2-Substituted	3450–3200	Bonded N—H stretch
	3180–3090	C—H stretch
	1570–1545	Ring stretching bands
	1475–1460	
	1420–1400	

TABLE 7.12 Absorption Frequencies of Aromatic and Heteroaromatic Bands (*Continued*)

Vibrational group	Frequency, cm^{-1}	Remarks
Pyrroles		
3-Substituted	3450–3200	Bonded N—H stretch
	3180–3090	C—H stretch
	1570–1560	Ring stretching bands
	1490–1480	
	1430–1420	
Furans, 2-substituted	3180–3090	C—H stretch
	1605–1590 (mw)	Ring stretching bands for unconjugated substituents
	1515–1490 (m)	
	1585–1560 (mw)	Ring stretching bands for conjugated C=C or C=O substituents
	1480–1460 (m)	
	1400–1370	
	1163–1136 (m)	
	1100–1072 (mw)	
	1030–1010 (m)	
	815–795 (m)	C—H wag
	ca. 755 (s)	C—H wag
Thiophenes		
2-Substituted	3120–3060	C—H stretch
	1535–1514	Ring stretching bands
	1454–1430	
	1361–1347	
	867–842 (m)	C—H wag; also 2,3-substitution
	740–690 (s)	C—H wag
3-Substituted	3120–3060	C—H stretch
	1542–1492	Ring stretching bands
	1410–1380	
	1376–1362	

TABLE 7.13 Absorption Frequencies of the Nitro Group

Abbreviations used in the table

w, weak	*m, moderate*
w-m, weak to moderate	*s, strong*
m-w, moderate to weak	*vs, very strong*
	var, of variable strength

Vibrational group	Frequency, cm^{-1}	Remarks
C-NO_2		
Nitroalkanes	1601–1531 (vs)	NO_2 asymmetric stretch
	1388–1297 (s)	NO_2 symmetric stretch
	894–847 (m-w)	C—N stretch for gauche form
	627–609 (w)	NO_2 scissors
	857–830	NO_2 wag
Aromatic	1555–1487 (vs)	NO_2 asymmetric stretch
	1357–1318 (s)	NO_2 symmetric stretch
	857–830 (m)	NO_2 scissors
Conjugation with a C═C group	1530–1510 (s)	Asymmetric NO_2 stretch
	1360–1355 (s)	Symmetric NO_2 stretch

The substitution of C═O groups, halogen atoms, and NO_2 groups on the α-carbon atom results in a frequency shift of +10, +17, and +29 cm^{-1}, respectively, of the asymmetric stretching mode, while the symmetric stretching frequencies are lowered by −8, −23, and −29 cm^{-1}, respectively.

Vibrational group	Frequency, cm^{-1}	Remarks
Nitroamines >N—NO_2	1300–1250 (vs)	Symmetric NO_2 stretch
	1634–1605 (s)	Asymmetric NO_2 stretch: alkyl
	1587–1575 (s)	aryl
Covalent nitrates —O—NO_2	1650–1620 (vs)	Asymmetric NO_2 stretch
	1285–1270 (vs)	Symmetric NO_2 stretch
	870–855 (vs)	O—N stretch
	760–755 (w-m)	NO_2 out-of-plane bending
	710–695 (w-m)	NO_2 in-plane bending

TABLE 7.14 Absorption Frequencies of Double Bonds Containing Nitrogen Atoms

Abbreviations used in the table

w, weak	*m-s, moderate to strong*
vw, very weak	*s, strong*
m-w, moderate to weak	*vs, very strong*
m, moderate	*var, of variable strength*

Vibrational group	Frequency, cm^{-1}	Remarks
Imines >C═N—		
Nonconjugated	1674–1638 (w)	C═N stretching
Conjugation with C═C bond	1635 and 1620 (m)	C═N stretching
Conjugation with phenyl group	1650–1600 (m)	C═N stretching
Conjugated cyclic systems	1660–1600 (var)	C═N stretching
Oximes		
Aliphatic	1670–1650 (m-w)	C═N stretching
Conjugation to C═O or aromatic ring	1650–1630	C═N stretching

TABLE 7.14 Absorption Frequencies of Double Bonds Containing Nitrogen Atoms (*Continued*)

Vibrational group	Frequency, cm^{-1}	Remarks	
Amidines N—C=N	1675–1580 (s)	C=N stretching; doublet for some	
Imidates RO(R′)C=NH			
	1655–1652	R,R′ = alkyl groups	
	1645–1630	R = alkyl; R′ = aryl	
	1676–1660	R′ = benzyl group	
Guanidines $(RHN)_2C{=}NH$	1592 (s)	R = alkyl group	
	ca. 1630 (s)	R = aryl group	
Imino carbonates $(RO)_2C{=}NH$	ca. 1658	R = alkyl group	
Nitrones >C=N(→O)—	1620–1540	C=N stretching	
	1280–1067	N→O stretching	
Azo compounds —N=N—			
Aliphatic			
trans	1576–1565 (vw)	—N=N— stretch; conjugated with C=O, ca. 1550 cm^{-1} (w)	
cis	1550–1540 (s)	—N=N— stretch	
Aromatic			
trans	1463–1380 (vs)	—N=N— stretch	
cis	ca. 1510 (vs)	—N=N— stretch	
Azoxy compounds R,R′N=N(→O)—	1530–1495 (m-s)	C=N stretch; *trans* R,R′ alkyl groups	
	1413 (vs)	C=N stretch; *trans* R,R′ phenyl groups	
	1468 (vs)	C=N stretch; *cis* R,R′ phenyl groups	
	1335–1250 (m-s)	N→O stretching	
Azothio compounds —N=N(→S)—	1465–1445 (w)	C=N stretching	
	1070–1058 (w)	N→S stretching	
Nitrosoamines >N—N=O			
Aliphatic	1460–1425 (m-s)	N=O stretch	
	1150–1030	N—N stretch	
Aromatic	1500–1450 (m-s)	N=O stretch	
	1025–925	N—N stretch	
***C*-Nitroso compounds** C—N=O			
Aliphatic	1621–1539 (s)	N=O stretch	
Aromatic	1523–1488 (s)	N=O stretch	
Nitrites —O—N=O			A_{trans}/A_{cis}
N=O stretch, *trans*			
Primary	1681–1669 (vs)	*cis*: 1625–1613 cm^{-1} (vs)	0.95–3.5
Secondary	1667–1664 (vs)	*cis*: 1618–1613 cm^{-1} (vs)	4.7–10
Tertiary	1655–1653 (vs)	*cis*: 1613–1610 cm^{-1} (vs)	35–50
N—O stretch, *trans*			
Primary	814–790 (s)		
Secondary	780–775 (s)		
Tertiary	764–751 (s)		
O—N=O bend, *trans*			
Primary	625–565	*cis*: 691–617 cm^{-1}	
Secondary	605–594	*cis*: 688–678 cm^{-1}	

TABLE 7.15 Absorption Frequencies of Cumulated Double Bonds

Abbreviations used in the table

w, weak	*s, strong*
vw, very weak	*m-s, moderate to strong*
m-w, moderate to weak	*ms, moderately strong*
m, moderate	*s-vs, strong to very strong*
	vs, very strong

Vibrational group	Frequency, cm^{-1}	Remarks
Carbon dioxide O=C=O	2349 (s)	Appears in many spectra as a result of inequalities in path length
Allenes >C=C=C<	1980–1915 (ms)	Out-of-phase stretch; two bands when terminal allene or when bonded to electron-attracting groups
	1096–1060 (vw)	In-phase stretching
Ketenes >C=C=O	2153–2085 (vs)	Out-of-phase stretching
	1420–1120 (s-m)	In-phase stretching
	675–668 (m-w)	C=C=O bending frequencies
	541–486 (ms)	
Ketene imines >C=C=N—		
$H_2C=C=N-CH_3$	2060 (s)	Out-of-phase stretching
	1233 (m)	In-phase stretching
$Y_2C=C=N-Y$ (Y: alkyl or aryl)	2050–2000 (s)	Out-of-phase stretching
Diazo compounds		
$H_2C=N=N$	2102 (vs)	Out-of-phase stretching
	1170 (s)	In-phase stretching
$R_1R_2C=N=N$	2049–2012 (vs)	Out-of-phase stretching (R = alkyl or hydrogen atom); Ar(C=O) and CF_3 substituents increase the frequency
	1389–1333 (s-vs)	In-phase stretching; see comments above
Isocyanates —N=C=O	2300–2250 (vs)	Out-of-phase stretching; not very sensitive to substitution
Isothiocyanates —N=C=S	2200–2000 (vs)	Out-of-phase stretching; broad and usually a doublet
	936–925 (m-s)	In-phase stretching of aromatic derivatives
Isoselenocyanates —N=C=Se	2182 (s)	Out-of-phase stretching: $CH_3-N=C=Se$
	2142–2100 (s)	Higher homologues
	983	In-phase stretching: $CH_3-N=C=Se$
	559–504	Higher homologues
Carbodiimides —N=C=N—	2155–2130 (s)	Out-of-phase stretching for alkyl and aryl compounds
	2260–2180 (s)	Out-of-phase stretch for silyl compounds

TABLE 7.16 Absorption Frequencies of Boron Compounds

Abbreviations used in the table

m, moderate	*s, strong*
m-s, moderate to strong	*vs, very strong*

Vibrational group	Frequency, cm^{-1}	Remarks
—BH_2	2640–2350	B—H stretch, usually a doublet; also B—H stretch for >B—H
	1205–1140	Deformation
	975–920	Wag
B···H···B bridge bonds	2220–1540	Series of bands; usually one band at 1900–1800 cm^{-1}
Diboranes, alkyl		
Terminal BH_2	2640–2571 (s)	Asymmetric stretch
	2532–2488 (s)	Symmetric stretch
	1170–1140 (m-s)	Deformation
	940–920 (m)	Wag
Single terminal BH	2565–2481 (s)	Stretch
	1180–1110 (s)	In-plane bend
	920–900 (m)	Out-of-plane bend
Borazines	2580–2450	B—H stretch
Borohydride salts and amine–borane coordination complexes	2400–2200 (s)	B—H stretch; two bands in BH_4^-
B—O	1380–1310 (vs)	B—O stretch; absent or very weak when a nitrogen atom is coordinated to boron (boron octet is complete)
B—OH	3300–3200	Broad, due to bonded O—H stretch
	ca. 1000 (m)	Usually broad; aromatic boronic acids not
	800–700 (m)	present in comparable anhydrides
B—N	1465–1330 (s)	Aliphatic groups adjacent to nitrogen have intensified deformation bands
	800–650	B←N dative bond in amine–borane complexes
B—Cl	1090–890 (s)	Plus other bands at lower frequencies
B—F		
—BF_2	1500–1410 (s)	
	1300–1200 (s)	
>BF	1360–1300 (s)	
BF_4^-	ca. 1030 (vs)	
BF_3 complexes	1260–1125 (s)	
	1030–800 (s)	

TABLE 7.17 Absorption Frequencies of Phosphorus Compounds

Abbreviations used in the table

w, weak	*s, strong*
m, moderate	*vs, very strong*
m-s, moderate to strong	*var, of variable strength*

Vibrational group	Frequency, cm^{-1}	Remarks
PH and PH_2	2505–2222 (s)	Sharp stretching bands
	1090–1080 (m-s)	Scissors bending or deformation
	840–810 (m-s)	Wag
P=O	1320–1140 (s)	Range 1415–1085 cm^{-1} for fluorine or OH substituents
P—OH	2700–2500 (m)	Broad
	2350–2100 (m)	Broad
	1700–1630	Occurs when there is one P—OH with one P=O in the molecule
	1040–910 (s)	
P—O—P	100–870 (s)	
	ca. 700 (w)	
P—O—C:		
Aliphatic	1050–970 (vs)	Asymmetric P—O—C stretch
	830–740 (s)	In methoxy and ethoxy compounds
Phenyl	1260–1160 (s)	
	994–914 (s)	In pentavalent phosphorus compounds
	875–855 (s)	In trivalent phosphorus compounds
P=S	835–713 (m)	
	675–568 (var)	
P—SH	2480–2440	Broad band, S—H stretching
	865–835	S—H bending
P —N—C		
Aliphatic C	1110–930	P—N bonds
	770–680	
Phenyl C	ca. 1290	Phenyl—N bond
	ca. 932	P—N bond
P=N	1320–1100 (s)	P=N stretch of cyclic compounds
	1385–1325 (s)	Compounds of type $(RO)_3P{=}N{-}C_6H_5$ and $(RO)_2RP{=}N{-}C_6H_5$
P—F	835–720	Phosphor–fluoridate salts
	890–805	For pentavalent phosphorus
P—Cl	587–435	One band for P—Cl and two bands for PCl_2 groups

TABLE 7.18 Absorption Frequencies of Silicon Compounds

Abbreviations used in the table

m, moderate — *s, strong* — *vs, very strong*

Vibrational group	Frequency, cm^{-1}	Remarks
—SiH_3	2153–2142 (s)	Monoalkyl R—SiH_3 stretchings
	2157–2152 (s)	Mono aromatic Ar—SiH_3 stretchings
	2190–2170 (s)	Alkyne substituent, —C≡C—SiH_3
	947–930 (s)	Asymmetric deformation
	930–910 (s)	Symmetric deformation
	720–680	Rocking
—SiH_2—	2150–2117 (s)	SiH_2 stretching; upper end for aryl substituents and when in a ring
	2200–2140	CH_3—SiH_2—halide
	950–928 (s)	Scissoring; halogenation 980–940 cm^{-1}
	900–843 (s)	Wagging; halogenation 955–875 cm^{-1}
	690–560 (s)	Twisting; halogenation 740–630 cm^{-1}
	540–480 (s)	Rocking; halogenation 520–460 cm^{-1}
Si—H group	2131–2094 (s)	SiH stretch; upper end for aryl substituents
	2215–2171 (s)	SiH stretch for $(CH_3)_2Cl$ and CH_3Cl_2 substituents
	2285–2235	SiH stretch for trihalide substituents
	2205–2190 (s)	SiH stretch for—$(OR)_3$ substituents
	842–800 (s)	SiH bending
Si—C groups		
Si—CH_3	1280–1255 (vs)	Sharp; CH_3 deformation
	860–760	Si—C stretching and CH_3 rocking: one methyl ca. 765 cm^{-1}, two methyls ca. 855 and 800 cm^{-1}, and three methyls ca. 840 and 765 cm^{-1}
Si—CH_2—R	1250–1200 (m)	Longer aliphatic chains at lower end
Si—aryl	1125–1100 (vs)	
Si—O—C	1110–1000 (s)	Asymmetric Si—O—C stretching
	850–800	Symmetric stretching
Si—O—phenyl	970–920	Stretching of Si—O bond
Si—O—Si	1130–1000 (s)	At least one band; asymmetric stretch
Si—halogen		
SiF_3	980–945 (s)	
	910–860 (m)	
SiF_2	945–915 (s)	
	910–870 (m)	
SiF	920–820	
$SiCl_3$	620–570 (s)	
	535–450 (m)	
$SiCl_2$	600–535 (s)	
	540–460 (m)	
SiCl	550–470	

TABLE 7.19 Absorption Frequencies of Halogen Compounds

Abbreviations used in the table

w, weak	*m-s, moderate to strong*
mw, moderately weak	*ms, moderately strong*
m-w, moderate to weak	*s, strong*
m, moderate	*vs, very strong*
	var, of variable strength

Vibrational group	Frequency, cm^{-1}	Remarks
	Fluorine compounds	
Mono C—F		
Aliphatic	1110–900 (vs)	
Aromatic	1270–1100 (m)	
—CF_2—	1280–1120 (vs)	Two bands
—CF_3		
Aliphatic	1350–1120 (vs)	
Aromatic	1330–1310 (m-s)	
HFH^- ion	1700–1400 (vs)	
	1260–1200 (vs)	
	Chlorine compounds	
C—Cl		
Primary alkanes	730–720 (s)	A carbon *trans* to chlorine
	685–680 (s)	A hydrogen *trans* to chlorine
	660–650 (s)	A second hydrogen *trans* to chlorine
Secondary alkanes	760–740 (m)	Two carbons *trans* to chlorine
	675–655 (m-s)	A carbon and a hydrogen *trans* to chlorine
	637–627 (mw)	Two hydrogens *trans* to chlorine
	615–605 (s)	Second hydrogen pair *trans* to chlorine
Tertiary alkanes	620–690 (ms)	CHH *trans* to chlorine
	590 (ms)	CHH *trans* to chlorine
	570–560 (vs)	HHH *trans* to chlorine
	540 (ms)	Second HHH *trans* to chlorine
Aryl		
1,2-Disubstitution	1060–1035 (m)	
1,3-Disubstitution	1000–1075 (m)	
1,4-Disubstitution	1100–1090 (m)	
—CCl_3	840–740	
	730–660	
=CCl	695	Anticonformer
	633	Gauche
Chloroformate	ca. 690 (s)	
	485–470 (s)	
=CCl_2	785	
	598	
HC=CCl	756	
XC≡CCl (X = CH_3, CHO, C≡CH, CN)	579–473	
Axial Cl	730–580 (s)	
Equatorial Cl	780–740 (s)	

TABLE 7.19 Absorption Frequencies of Halogen Compounds (*Continued*)

Vibrational group	Frequency, cm^{-1}	Remarks
	Bromine compounds	
C—Br		
Primary alkanes	650–635 (s)	A carbon *trans* to bromine
	565–555 (s)	A hydrogen *trans* to bromine
	625–615 (s)	Second hydrogen *trans* to bromine
Secondary alkanes	620–605 (s)	CH *trans* to bromine
	590–575 (m-w)	HH′ *trans* to bromine
	540–530 (s)	HH *trans* to bromine
Tertiary alkanes	590–580 (s)	CHH *trans* to bromine
	520–510 (var)	HHH *trans* to bromine
Aryl		
1,2-Disubstitution	1045–1025 (m)	
1,3- and 1,4-Disubstitution	1075–1065 (m)	
=CBr	594	Anticonformer
	546	Gauche
HC≡CBr	618	
XC≡CBr (X = CH_3, CHO, C≡C, CN)	474–395	
Axial	690–550 (s)	
Equatorial	750–685 (s)	
	Iodine compounds	
C—I		
Primary alkanes	600–590 (s)	One carbon *trans* to iodine
	590–580 (s)	Hydrogen *trans* to iodine
Secondary alkanes	595–585 (s)	Second hydrogen *trans* to iodine
	590–575 (s)	HH′ *trans* to iodine
	495–480 (s)	HH *trans* to iodine
Tertiary alkanes	595–585 (s)	CHH *trans* to iodine
	495–485 (s)	HHH *trans* to iodine
Aromatic	1060–1055 (m-s)	
	310–160 (s)	
	265–185	
—C≡CI	405–360	
Axial	ca. 640 (s)	
Equatorial	ca. 655 (s)	

7.4 QUANTITATIVE ANALYSIS

The baseline method for quantitative analysis involves the selection of an absorption band that is separated from the bands of other matrix components (Fig. 7.5). Draw a straight line tangent to the absorption band. From the illustration observe how the value of P_0 and P are obtained. The value of the absorbance, log (P_0/P), is then plotted against concentration for a series of standard solutions, and the unknown concentration is determined from this calibration curve. The use of such ratios eliminates many possible errors, such as changes in instrument sensitivity, source intensity, and adjustment of the optical system.

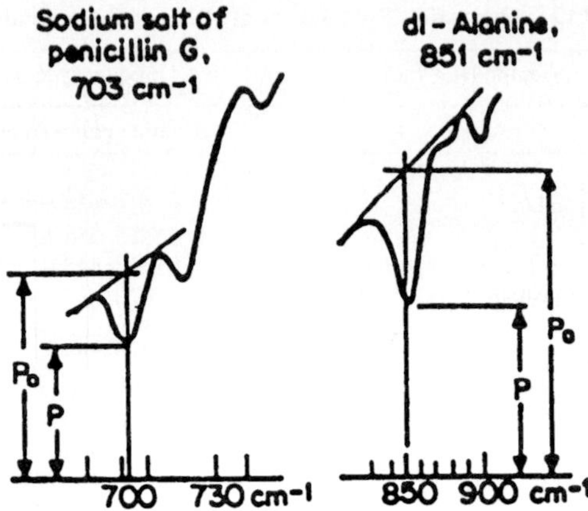

FIGURE 7.5 Baseline method for calculation of the transmittance ratio.

The KBr pellet technique, when combined with the internal standard method of evaluation, can be used. Potassium thiocyanate makes an excellent internal standard. After grinding and redrying, KSCN is reground with dry KBr to make a concentration of about 0.2% by weight of KSCN. A standard calibration curve is constructed by mixing known weights of the test substance (usually about 10% of the total weight) with a known weight of the KBr-KSCN mixture, then preparing the pellet or thin wafer. The ratio of the thiocyanate absorption at 2125 cm^{-1} to a chosen absorption band of the test substance is plotted against the concentration of the test substance.

7.5 THE FAR-INFRARED REGION

The far-infrared region comprises the portion of the electromagnetic spectrum between 15 and 35 μm (300 and 700 cm^{-1}). In this region infrared absorption is due to pure rotational and to vibrational–rotational transitions in gaseous molecules, to molecular vibrations in liquids and solids, and to lattice vibrations and molecular vibrations in crystals. The cesium bromide prism covers the entire range from 15 to 35 μm. The dispersion of the cesium bromide prism is not markedly inferior to that of the potassium bromide prism from 15 to 20 μm, and the convenience of being able to use a single prism for the whole 15- to 35-μm region is obvious.

7.5.1 Sources, Optical Materials, and Detectors

The transmission regions of most of the materials that are used as windows, cells, prisms, and filters for far-infrared instrumentation are given in Fig. 7.6. The black lines on the chart, which is plotted with a logarithmic abscissa scale, indicate the useful ranges of the optical materials. In general, the most suitable materials for use as windows, cells, and prisms from 15 to about 50 μm are KBr, CsBr, and CsI. These substances are moisture-sensitive and must be maintained in an area of low humidity.

The useful ranges of sources and detectors for the long-wavelength region are also included in Fig. 7.6. The problem of instrumentation for the far-infrared region is basically one of energy, and sources that provide a spectral distribution rich in long-wavelength radiation at a given temperature are the best choices. The Nernst glower has been successfully used on some double-beam instruments to

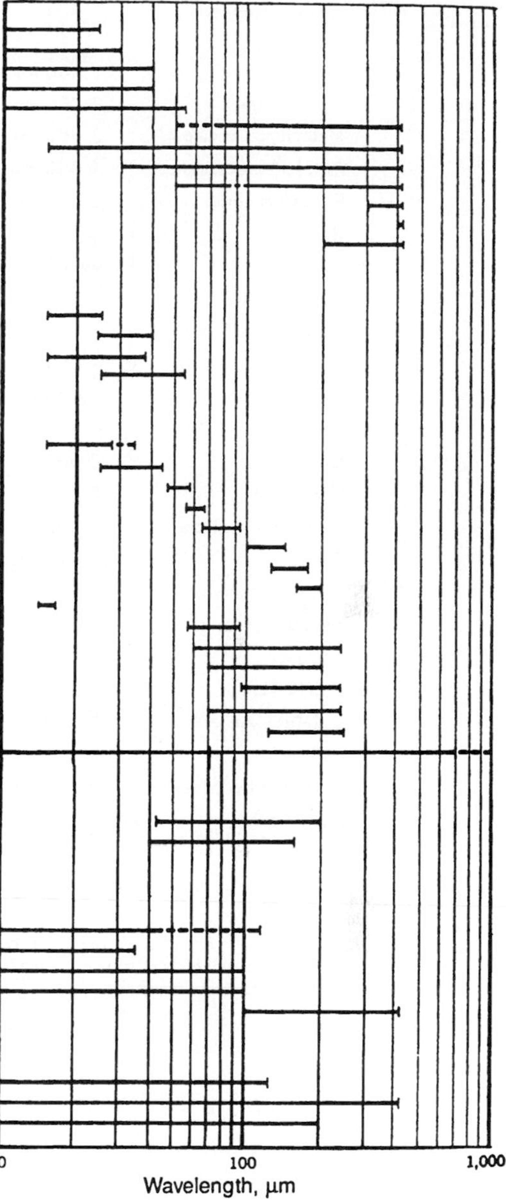

FIGURE 7.6 Transmission regions of selected optical materials, and useful ranges of sources and detectors in the far-infrared region.

about 30 μm, but the Globar more nearly approaches black-body radiators and appears to be a better choice for long-wavelength studies. A quartz mercury lamp serves as general source.

Thermal and pneumatic detectors are both used in the far-infrared region: the Golay cell more commonly on larger-grating instruments in which the mechanical slit widths are wide, and the thermocouple or thermopile on the small-grating and prism instruments.

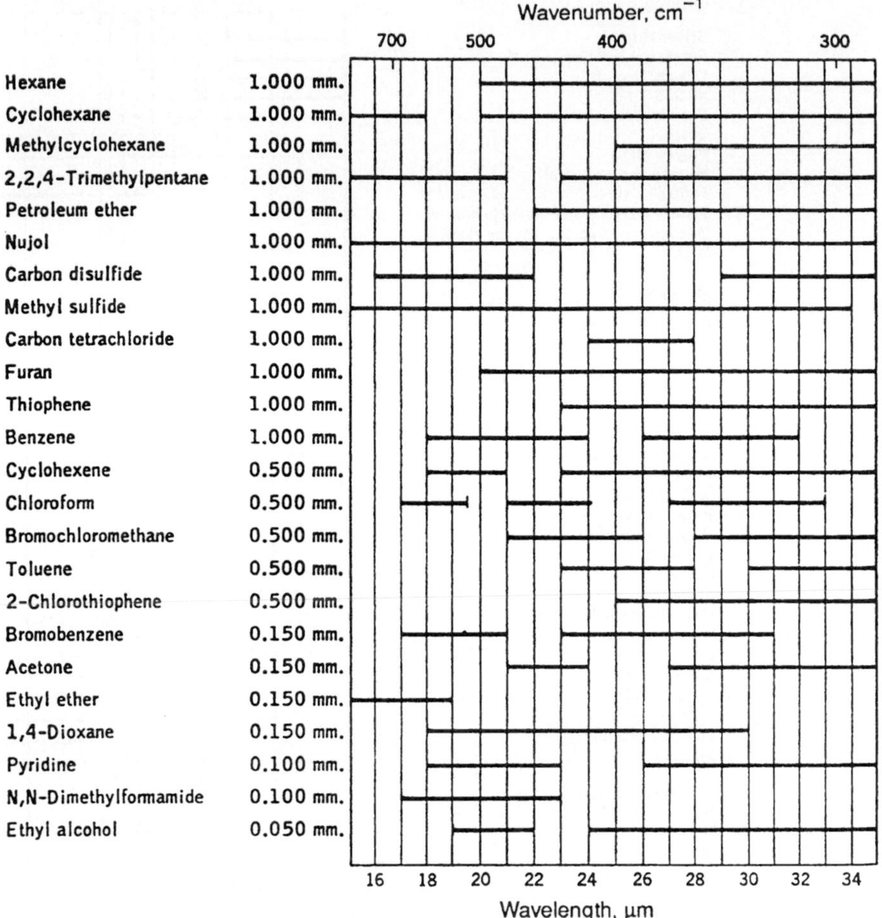

FIGURE 7.7 Solvents for the far-infrared region. The black lines indicate the regions of greatest utility. The transmission of these solvents in the indicated range is greater than 40% of the incident energy at the path length shown in millimeters.

7.5.2 Solvents and Sampling Techniques

Conventional infrared techniques are used, and in some instances simplified. Many common solvents can be used (Fig. 7.7). Nujol has no significant absorption bands in this region and serves as a mulling agent for solid materials. The spectra of greases, high polymers, and samples that are corrosive to cesium bromide plates can be obtained by using cells of polyethylene and KRS-5 to support the materials. Although suitable solvents can be found for most organic solids, cesium bromide pellets are easily prepared and can be used in qualitative and quantitative analyses of solids in this region. Transparent cesium and potassium bromide disks, prepared in pellet dies without a sample, can be used as demountable cells to support materials while obtaining their spectra.

Polymeric materials such as rubber, resins, and plastics are studied as pyrolyzates, films, or in solution. The spectra of liquid samples are obtained in cesium bromide or iodide and KRS-5 cells.

Since KRS-5 cells give interference patterns and reflect approximately 30% of the incident beam at each surface, cesium bromide or iodide cells are most desirable; these plates are soft and easily polished with paper towels. In an air-conditioned room, corrosion by atmospheric water vapor is not very serious, and with careful handling the cesium bromide cells are as easy to work with as potassium bromide or sodium chloride cells. Window materials for the far-infrared region include high-density polyethylene, silicon, and crystal quartz (cut with the optic axis parallel to the face of the window). Polyethylene has one weak, broad absorption band at approximately 70 cm^{-1}; its principal disadvantage is its lack of rigidity. High-resistivity silicon is rigid but its high index of refraction leads to large reflectivity losses.

Usually a cell with a longer path length is required for the far-infrared region. For the 15- to 35-μm region one of the most useful cells is 0.50 mm in path length. Highly polar materials require cells with path lengths of 0.05 mm or less, and the less polar materials require cells with path lengths of 2 mm or more.

A list of the more commonly used organic solvents is shown in Fig. 7.7. The black lines indicate the wavelength regions in which these solvents are most useful. The transmission of these solvents in the indicated range is greater than 40% of the incident energy at the path length shown.

7.5.3 Spectra–Structure Correlations

Many molecules have vibrational frequencies in the far-infrared region that are potentially useful in spectra–structure correlations for such substances as substituted benzenes, heterocyclics, and aliphatic and alicyclic hydrocarbons. Similarly, stretching and bending vibrations for heavy atoms in molecules, such as bromine, iodine, sulfur, silicon, and other organometallics, and inorganic radicals are often observed in the far-infrared region.

Spectra–structure correlation charts showing the probable positions of the characteristic absorption frequencies of aliphatic and aromatic compounds are shown in Figs. 7.8 and 7.9. Correlations for a number of inorganic ions are given in Fig. 7.10. Slight changes in structure produce considerable changes in the long-wavelength spectra, giving a more specific "fingerprint" of a molecule. Far-infrared spectra are useful in analytical and molecular structure studies in identifying and characterizing homologues and geometrical, optical (diastereoisomers), and rotational isomers. Shifts in absorption bands in this region give clues as to the nature of the molecule or, in many instances, the identity of the isomeric species present. The vibrational frequencies of the heavier molecules are more concentrated in the far-infrared region. As a consequence this region is uniquely important in analytical studies dealing with organometallic and heterocyclic systems as well as with compounds containing bromine, iodine, or sulfur.

Far-infrared spectra appear to be more sensitive to crystal structure, and molecules that differ by only a $—CH_2—$ group are readily distinguished by their spectra in the solid state. In most instances the solid molecules that exhibit this sensitivity to crystal-lattice vibrations possess entirely different spectra in solution. As a consequence, the physical state in which molecules are studied appears to be more critical in the far-infrared region. The long-chain fatty acids are a class of molecules that show identical spectra in solution, but exhibit spectral differences in their spectra when obtained in the solid state as Nujol mulls. In these molecules the spectra are greatly dependent on the unit cell of the crystal rather than on the molecular structure.

7.6 RAMAN SPECTROSCOPY

Raman spectroscopy is used to determine molecular structures and compositions of organic and inorganic materials. Materials in the solid and liquid are easily examined; even gas samples can be handled under special conditions. Normally the minimum sample requirements are on the order of tenths of a gram.

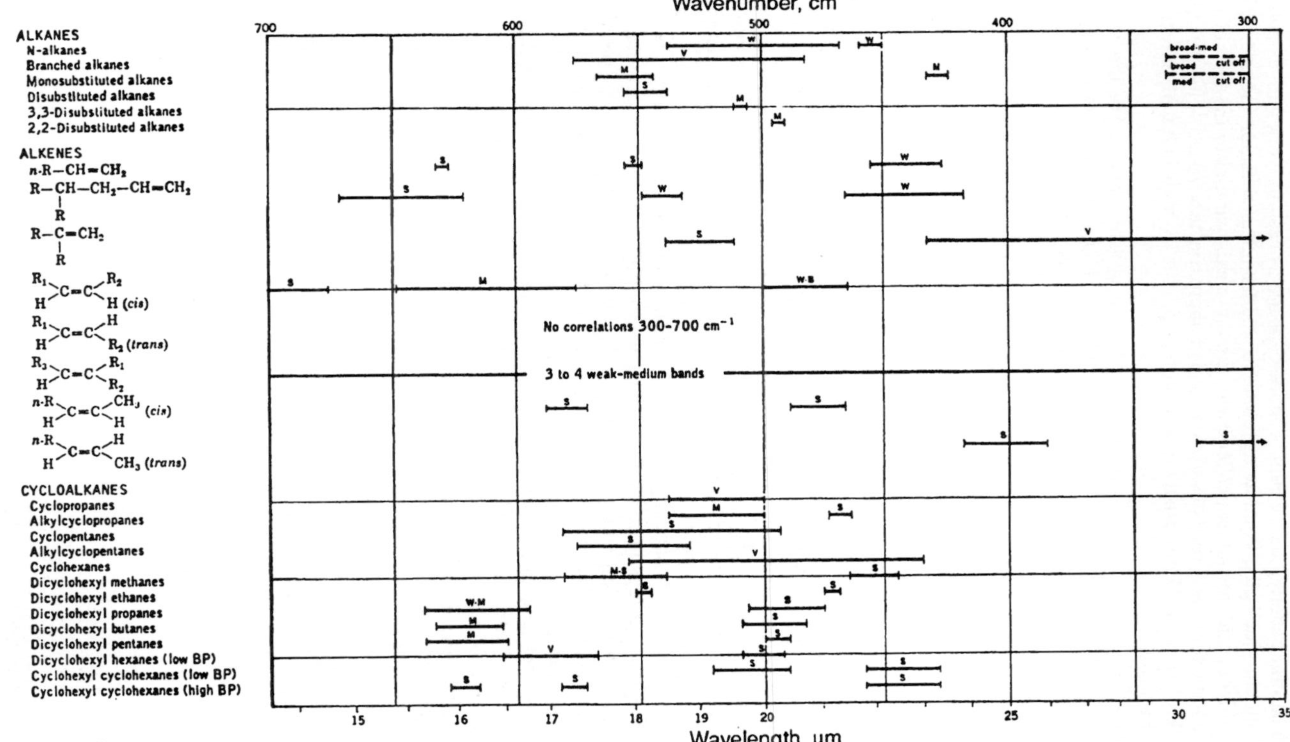

FIGURE 7.8 Spectra–structure correlation chart in the far-infrared region for alkanes, alkenes, cycloalkanes, and aromatic hydrocarbons. V = variable, W = weak, M = medium, S = strong, and M-S = medium to strong.

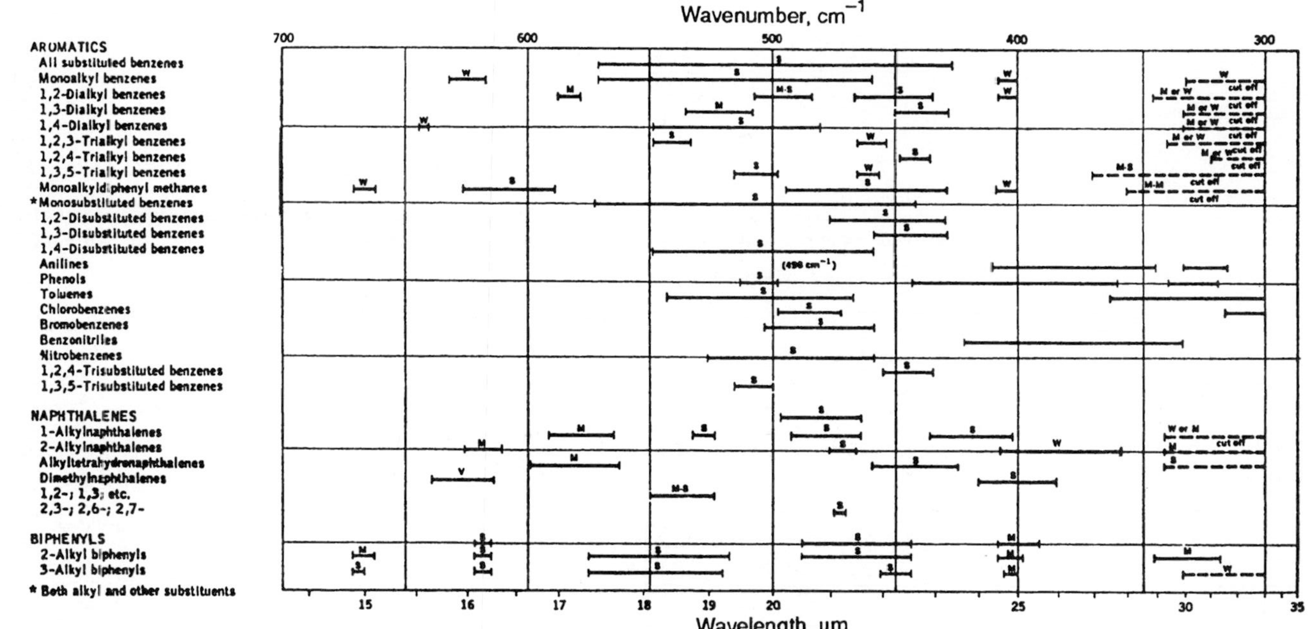

FIGURE 7.8 (*Continued*)

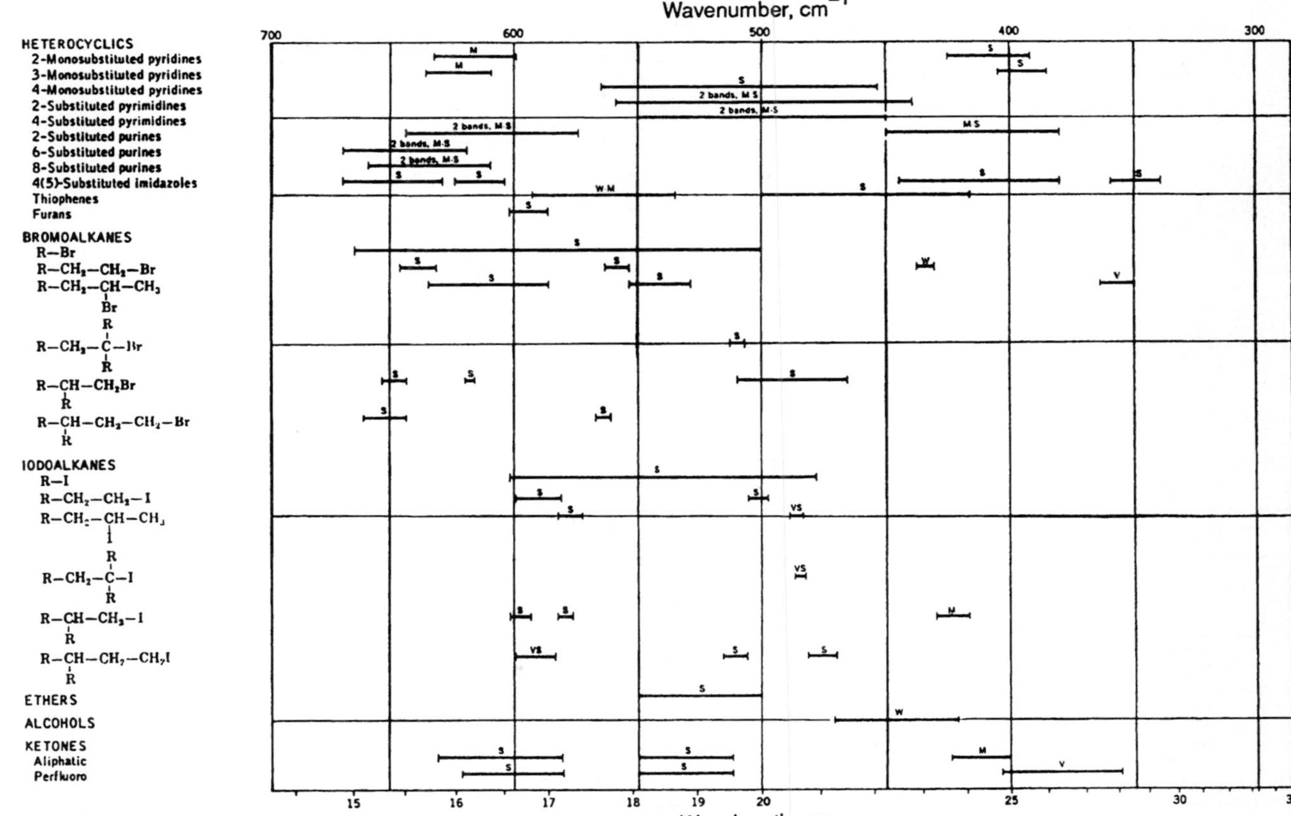

FIGURE 7.9 Spectra–structure correlation chart for the far-infrared for heterocyclic and organometallic compounds and aliphatic derivatives. (V = variable, W = weak, M = medium, S = strong, M-S = medium to strong.)

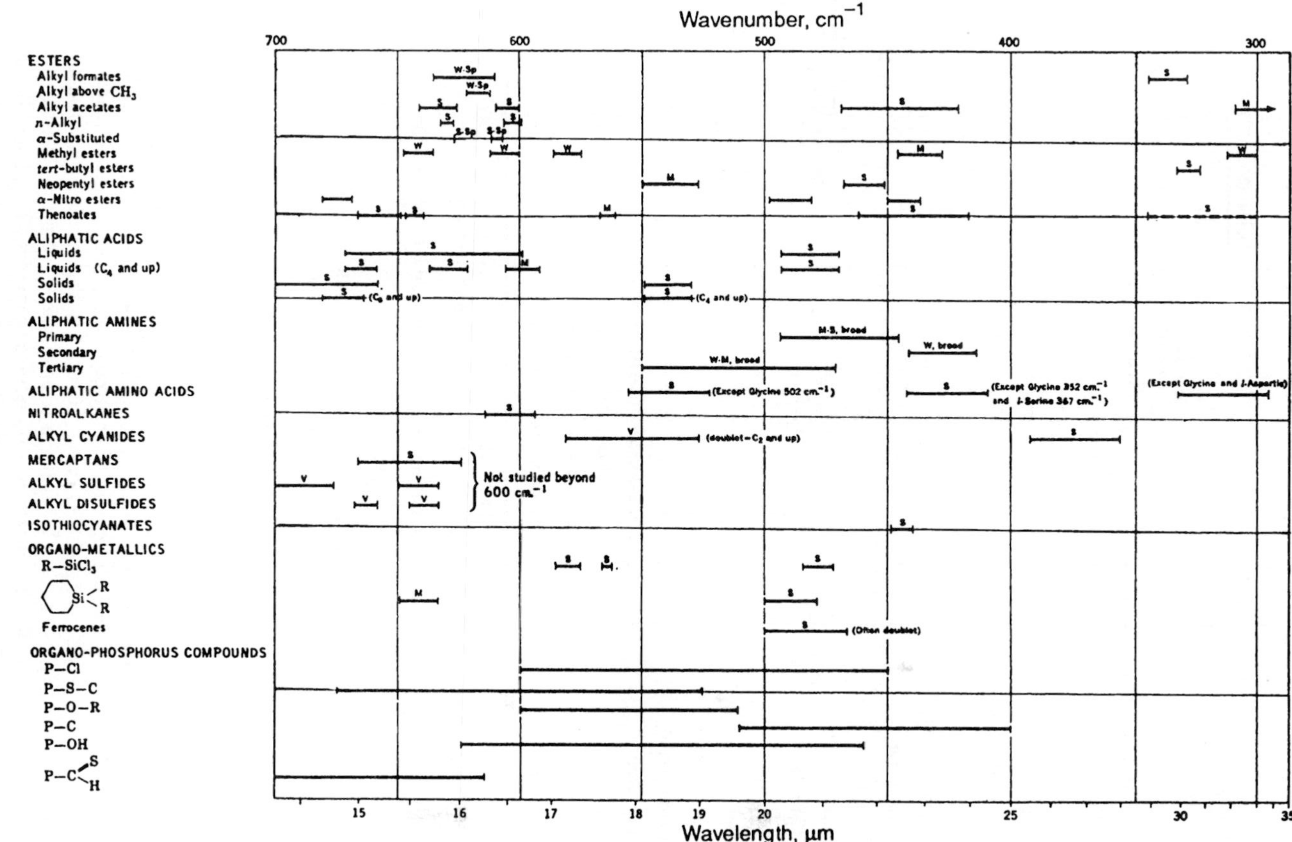

FIGURE 7.9 (Continued)

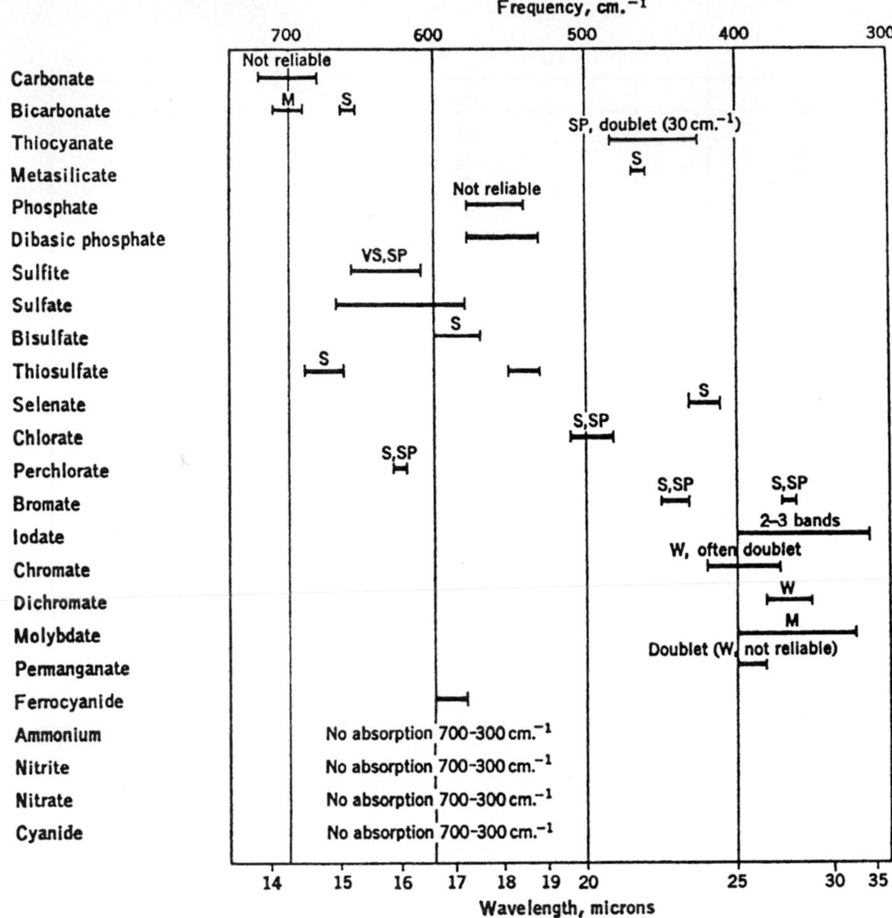

FIGURE 7.10 Spectra–structure correlation chart in far-infrared for inorganic ions.

Raman spectra embraces the entire vibrational spectrum with one instrument. It can be used to study materials in aqueous solutions. Sample preparation for Raman spectroscopy is generally much simpler than that for infrared.

7.6.1 Principles

In the basic Raman experiment, a sample is illuminated by a high-energy monochromatic light source (typically from a laser). Some of the incident photons collide with molecules in the sample and are scattered in all directions with unchanged energy; that is, most collisions are elastic with the frequency of the scattered light (v) being the same as that of the original light (v_0). The effect is known as *Rayleigh scattering.* However, a second type of scattering can also occur and is known as the *Raman effect.* The Raman effect arises when a beam of intense monochromatic radiation passes through a sample that contains molecules that can undergo a change in molecular polarizability as they vibrate. In other words, the electron cloud of the molecule must be more readily deformed in one extreme of the vibration than in the other extreme. By contrast, in the infrared region the vibration must cause a change in the permanent dipole moment of the molecule.

In Raman scattering, the molecule can either accept energy from the incident radiation being scattered, thus exciting the molecule into higher vibrational energy states (Stokes lines), or give up energy to the incident photons, causing the molecule to return to its ground vibrational state (anti-Stokes lines). The difference between the incident radiation and the Raman scattered radiations produces the vibrational spectrum of interest.

Rayleigh and Raman scattering are relatively inefficient processes. Approximately 10^{-3} of the intensity of the incident exciting frequency will appear as Rayleigh scattering, and only 10^{-6} as Raman scattering. As a result, very intense excitation sources are required. The incident radiation does not raise the molecule to any particular quantized level; rather, the molecule is considered to be in a virtual or quasiexcited state whose height above the initial energy level equals the energy of the exciting radiation. In fact, the wavelength of the incident radiation does not have to be one that is absorbed by the molecule. Through the induced oscillating dipole(s) that it stimulates, the radiation leads to the transfer of energy with the vibrational modes of the sample molecules. As the electromagnetic wave passes, the polarized molecule ceases to oscillate, and this quasiexcited state then returns to its original ground level by radiating energy in all directions except along the direction of the incident radiation. A vibrational quantum of energy usually remains with the scattering radiation so that there is a decrease in the frequency ($v_0 - v_v$) of the emitted radiation (*Stokes lines*). However, if the scattering molecule is already in an excited vibrational level of the ground electronic state, a vibrational quantum of energy may be abstracted from the scatterer, leaving the molecule in a lower vibrational level and thus increasing the frequency of the scattered radiation (*anti-Stokes lines*). The latter condition is less likely to prevail and, consequently, the anti-Stokes lines are less intense than the Stokes lines. For either case, the shift in frequency of the scattered Raman radiation is proportional to one of the vibrational energy levels involved in the transition. Thus the spectrum of the scattered radiation consists of a relatively strong component with frequency unshifted (Rayleigh scattering) and the Stokes and anti-Stokes lines. The Stokes (or anti-Stokes) lines in a Raman spectrum will have the general appearance of the corresponding infrared spectrum.

In the usual Raman method the excitation frequency of the Raman source is selected to lie below any singlet–singlet electronic transitions and above the most fundamental vibrational frequencies.

7.6.2 Instrumentation for Dispersive Raman Scattering

The laser Raman spectrometer consists of the laser excitation unit and the spectrometer unit using gratings with 1200 grooves · mm^{-1}. The laser beam enters from the rear of the spectrometer into the depolarization autorecording unit and, after passing through this unit, it illuminates the sample. The Raman scattering, collected at 90° to the exciting laser beam, is focused on the entrance slit of a grating double monochromator. Immediately ahead of the spectrometer is a polarization scrambler that overcomes grating bias caused by polarized radiation. When the polarization of the Raman spectrum is measured, a polarization analyzer is placed between the condenser lens and the polarization scrambler. The scattered radiation is detected by a photomultiplier tube that is placed in a thermoelectric cooler (–30°C) to lower the dark current and reduce noise. Often the signals from the detector are amplified and counted with a photon-counting system—the most effective means of recovering low-level Raman signals. Less expensive dc amplifiers are excellent for strong signals.

7.6.2.1 Laser Sources. The He–Ne laser line at 632.8 nm is favorably located in the spectrum for which the least amount of fluorescent problems appear in routine analyses. For some experiments the Ar–Kr laser is ideal; it has two intense argon lines at 488.0 and 514.5 nm and two major krypton lines at 568.2 and 647.1 nm.

7.6.2.2 Detectors. The choice of phototube depends on which laser line is used. Laser selection and detector choice are interwoven. Raman shifts of 3700 cm^{-1} require response to 827 nm for the He–Ne laser (excitation line at 632.8 nm) and slightly further for the Ar–Kr laser (Kr* excitation line at 647.1 nm). For these excitation lines, the extended red-sensitive multialkali-metal cathode and the gallium arsenide photocathode have the needed quantum efficiency in the far-red portion of the electromagnetic spectrum. Blue-sensitive photomultiplier tubes are near their peak sensitivity at 488.0 nm, one of the other emission lines of the Ar–Kr laser.

7.6.3 Instrumentation for Fourier-Transform Raman Spectroscopy

FT Raman spectroscopy uses the same basic instrumentation as does the FT infrared spectrometer (see Sec. 7.2.3.2) except for the light source and detector. The technique of FT Raman spectroscopy utilizes a near-infrared excitation source, a neodymium:yttrium aluminum garnet (Nd:YAG) laser, the primary emission of which is at 1.064 μm (9394 cm^{-1}). This laser offers several advantages over the shorter wavelength, visible lasers used in the dispersive Raman technique, the most important of these advantages being freedom from fluorescence and thermal decomposition. Many samples contain small amounts of materials that are highly fluorescent when excited with a visible source.

The longer wavelength used in FT Raman spectroscopy does result in a decrease in the Raman intensity relative to the visible excitation, such as with an argon ion laser, by a factor of 16. However, much of this disadvantage is overcome by the multiplex and throughput advantages of the Fourier-transform technique and by the utrasensitive near-infrared detectors employed. The nearly total elimination of the Rayleigh line at 9394 cm^{-1} is accomplished by optical filtering via a variety of techniques. This is necessary because the Rayleigh line is 1000 times stronger than any of the Raman scattered radiation. The availability of relatively low-cost, commercial optical fibers that function in the near-infrared region make the FT Raman technique a choice for on-line, process-monitoring applications.

Shorter-wavelength scattered radiation requires precise source alignment, which takes on a more critical nature in FT Raman spectroscopy.

7.6.4 Sample Handling

Raman spectroscopy can be performed on specimens in any state: liquid, solution, transparent or translucent solid, powder, pellet, or gas. Neat liquids are examined with a single pass of the laser beam either axially or transversely. Multiple passes offer considerable gain in Raman intensity and permit work with samples in the microliter range or even down to about 8 nL. Photo- or heat-labile materials are studied in spinning cells or, better, with FT Raman spectroscopy.

Water is a weak scatterer and therefore an excellent solvent for Raman work. Other solvents and their obscuration ranges are shown in Fig. 7.11.

Powders are tamped into an open-ended cavity for front-surface illumination or into a transparent glass capillary tube for transverse excitation. Sample illumination at 180° provides better signal-to-noise ratio whereas right-angle viewing improves the ratio of Raman to Rayleigh scattering.

For a translucent solid, the laser beam is focused into a cavity on the face of the sample, either a cast piece or a pellet formed by compression of the powder.

Gas samples are handled with powerful laser sources and efficient multiple passes or interlaser-cavity techniques. Gases are difficult to study because of their low scattering.

7.6.4.1 Sample Fluorescence. If fluorescence arises from impurities in the sample, one can clean up the sample by techniques such as chromatographic fractionation, recrystallization, or distillation.

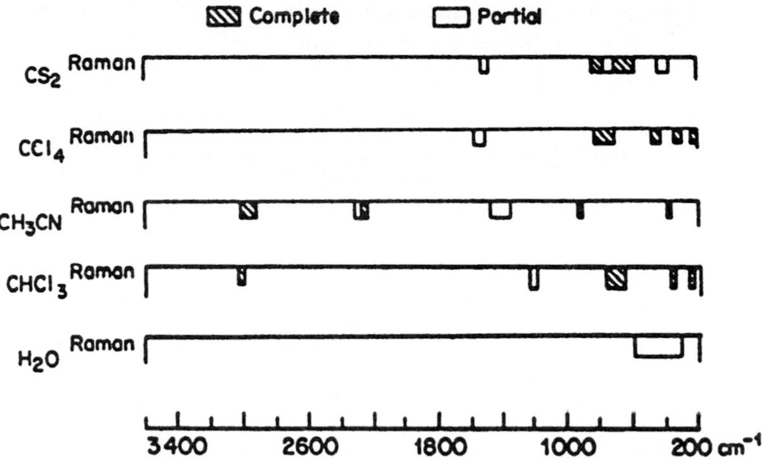

FIGURE 7.11 Obscuration ranges of the most useful solvents for Raman spectrometry.

However, if the fluorescence arises from the sample itself, one must select a different excitation line—a line that excites the Raman spectrum but not the fluorescent spectrum. The Nd:YAG laser used in FT Raman work will eliminate many fluorescent problems.

7.6.5 Diagnostic Structural Analysis

The Raman spectrum contains a number of distinct spectral features. Tables 7.20 to 7.32 contain detailed information on specific types of bonds. By comparison with infrared spectra, in Raman spectroscopy the most intense vibrations are those that originate in relatively nonpolar bonds with symmetrical charge distributions. Vibrations from —C=C—, —C≡C—, —C≡N, —C=S—, —C—S—, —S—S—, —N=N—, and —S—H bonds are readily observed. Raman spectroscopy has a distinct advantage in the detection of low-frequency vibrations. In most cases, information can be taken to within 20 to 50 cm^{-1} of the exciting line. This corresponds to the far-infrared region in which the important vibrations in metal bonding of inorganic and organometallic compounds reside.

Skeletal vibrations of finite chains and rings of saturated and unsaturated hydrocarbons are prominent in the region 800 to 1500 cm^{-1} and highly useful for cyclic and aromatic rings, steroids, and long chains of methylenes. All aromatic compounds have a strong band at 1600 ± 30 cm^{-1}. Monosubstituted compounds have an intense band at about 1000 cm^{-1}, a strong band at about 1025 cm^{-1}, and a weak band at 615 cm^{-1}. Meta- and 1,3,5-trisubstituted compounds have only one line at 1000 cm^{-1}. Ortho-substituted compounds have a line at 1037 cm^{-1}, and para-substituted compounds have a weak line at 640 cm^{-1}.

The band near 500 cm^{-1} is characteristic of the —S—S— linkage; the —C—S— group has a band near 650 cm^{-1}; and the intense band near 2500 cm^{-1} indicates the —S—H stretch.

Raman spectra are helpful whenever the infrared N—H and C—H stretching frequencies are obscured by intense O—H absorption. In Raman spectroscopy the O—H band is weak, whereas the N—H and C—H stretching frequencies exhibit moderate intensity.

The infrared spectrum provides evidence for the identification and location of substituent groups on the aromatic ring. For example, the highest intensity infrared band near 2280 cm^{-1} is

TABLE 7.20 Raman Frequencies of Alkanes

Abbreviations used in the table

m-w, moderate to weak	*s, strong*
m, moderate	*s-m, strong to moderate*
m-s, moderate to strong	*w, weak*
ms, moderately strong	*w-m, weak to moderately strong*
	vs, very strong

Vibrational group	Frequency, cm^{-1}	Remarks
$—CH_3$	2969–2952 (vs)	
	2884–2860 (vs)	
	1470–1440 (ms)	
	ca. 1205 (s)	In aryl compounds
	1150–1135	In unbranched alkyls
	1060–1056	In unbranched alkyls
	975–835 (s)	
$—CH_2—$	2949–2912 (vs)	
	2861–2843 (vs)	
	1473–1443 (m-s)	Intensity proportional to number of CH_2 groups
	1305–1295 (s)	
	1140–1040 (m)	
	900–800 (s-m)	Often multiple bands
	425–150	
$—CH(CH_3)_2$	1360–1350 (m-w)	
	1060–1040 (m)	
	950–900 (m)	
	830–800 (s)	If attached to aromatic ring, 740 cm^{-1}
	500–460 (s-m)	
	320–250 (s-m)	
$—C(CH_3)_3$	1265–1240 (m)	
	1220–1200 (m)	
	760–650 (s)	If attached to aromatic ring or alkene group, 760–720 cm^{-1}
Internal tertiary carbon	855–650 (s)	
	800–750 (s)	
	350–250	
Two adjacent tertiary carbon atoms	920–730	Often a band at 530–524 cm^{-1} indicates presence of adjacent tertiary and quaternary atoms
	770–725	
Internal quaternary carbon atom	ca. 1250	
	ca. 1200	
	750–650 (vs)	
	490–250	
Cyclopropane	3100–3080 (s)	
	3038–3024 (s)	
	1438 (m)	
	1188 (s)	Shifts to 1200 cm^{-1} for monoalkyl or 1,2-dialkyl substitution and to 1320 cm^{-1} for *gem*-1,1-dialkyl substitution
	868 (s)	
Cyclobutane	2965 (s)	
	2945 (m)	
	1443 (w)	

TABLE 7.20 Raman Frequencies of Alkanes (*Continued*)

Vibrational group	Frequency, cm^{-1}	Remarks
Cyclobutane	1224 (w)	
	1005 (vs)	Shifts to 933 cm^{-1} for monoalkyl, to 887 cm^{-1} for *cis*-1,3-dialkyl, and to 891 and 855 cm^{-1} for *trans*-1,3-dialkyl substitution
	926 (s)	
	901 (w)	
	626 (w)	
Cyclopentane	2965–2960 (s)	
	2880–2870 (s)	
	1490–1430 (m-s)	
	910–880 (vs)	Weakened by ring substitution
Cyclohexane	2933–2913 (vs)	
	2897–2850 (vs)	
	1460–1440 (m)	
	825–815 (vs)	Boat configuration
	810–795 (vs)	Chair configuration

TABLE 7.21 Raman Frequencies of Alkenes

Abbreviations used in the table

w, weak *s, strong*
m, moderate *vs, very strong*

Vibrational group	Frequency, cm^{-1}
Trialkyl, $R_2C{=}CHR$	3040–2995 (m)
	1680–1664 (vs)
	1360–1322 (w)
	522–488 (w)
1,1-Dialkyl, $R_2C{=}CH_2$	3095–3050 (w)
	2990–2983 (s)
	1660–1640 (vs)
	1420–1400 (m)
	900–885 (w)
cis-1,2-Dialkyl, RHC=CHR	3020–2995 (m)
	1662–1631 (vs)
	1270–1251 (s)
trans-1,2-Dialkyl, RHC=CHR	3010–2995 (m)
	1676–1665 (vs)
	1325–1290 (s)
Vinyl, $H_2C{=}CHR$	3090–3075 (m)
	3020–2995 (s)
	3000–2980 (w)
	1650–1640 (vs)
	1420–1410 (s)
	1309–1288 (s)
	995–985 (w)
	910–905 (w)
	688–611 (w)
Tetraalkyl, $R_2C{=}CR_2$	1680–1665 (vs)

TABLE 7.22 Raman Frequencies of Triple Bonds

Abbreviations used in the table

w, weak	*m, moderate*
m-w, moderate to weak	*s, strong*
	vs, very strong

Vibrational group	Frequency, cm^{-1}	Remarks
R—C≡C—H	3340–3290 (w) 2160–2100 (vs) 680–610 (m-w) 356–335 (m-w)	
R—C≡C—R	2300–2190 (vs)	Sometimes two bands
—C≡C—C≡C—	2264–2251 (vs)	
Nitriles, —C≡N	2260–2230 (vs)	Unsaturated alkyl substituents lower the frequency and enhance the intensity
	2234–2200 (vs)	Lowered ca. 30 cm^{-1} with aryl and conjugated aliphatics
	ca. 385 (vs) ca. 240 (s)	
HCN	2094 (vs)	
Diazonium salts	2300–2240 (s)	
Cyanamides, N—C≡N	2225–2210 (vs)	
Cyanates, O—C≡N	2256–2245 (s)	
Thiocyanates, S—C≡N	2157–2155 (s) 2174–2161 (s)	Aliphatic substituent Aromatic substituent
Isocyanides, C—N≡C	2146–2134 (s) 2124–2109 (s)	Aliphatic substituent Aromatic substituent

quite characteristic of the —N=C=O group. The Raman spectrum possesses a high-intensity band near 1510 cm^{-1} that is due to a symmetric stretching of the aromatic ring and the —N=C=O group. Although there are numerous coincidences between the two spectra, the differences in relative band intensities can be dramatic. For example, the 2280-cm^{-1} band is very intense in the infrared spectrum but very weak in the Raman spectrum.

7.6.5.1 Polarization Measurements. The depolarization ratio is defined as the ratio of the intensity of scattered light polarized perpendicular to the *xy* plane to that polarized parallel to the *xy* plane. The ratio of the radiant powers is measured by setting the analyzer prism in the path of the Raman scattered radiation at 0° and then at 90°. The depolarization ratio may vary from near 0 for highly symmetrical types of vibrations to a maximum of 0.75 for totally nonsymmetrical vibrations.

If the incident radiation is polarized in the *xy* plane (parallel illumination) and then in the *xz* plane (perpendicular illumination), the depolarization ratio may vary from 0 to a maximum of 0.86. The instrument operation should be checked with the known Raman bands. The 218-cm^{-1} band of

TABLE 7.23 Raman Frequencies of Cumulated Double Bonds

Abbreviations used in the table

w, weak	*s, strong*
ms, moderately strong	*vs, very strong*
	vs-w, very strong to weak

Vibrational group	Frequency, cm^{-1}	Remarks
Allenes, C=C=C	2000–1960 (s)	Aliphatic substituents
	1963–1953 (vs-w)	Halogen substituents
	1925 (s)	Aromatic substituents
	625–592	Haloalkenes
	548–485	Haloalkenes
	356 (s)	
Azides	2104 (ms)	
	1276 (vs)	
Diazo compounds	1367 (m)	
	2133–2087 (m)	R(C=O)—CH=N=N
Cyanamides	1460 (s)	Aliphatic substituents
	1150–1140 (vs)	Aromatic substituents
Isocyanates, NCO	1450–1400 (s)	
Isothiocyanates, NCS	2220–2100 (m)	Two bands
	1090–1035 (s)	
	690–650 (s)	
Ketenes, C=C=O	2060–2040 (vs)	
	ca. 1374 (s)	Alkyl derivatives
	1120 (s)	Aryl derivatives
Sulfinylamines, —N=S=O	1306–1214 (w)	
	1155–989 (s)	

carbon tetrachloride should have the maximum value, and the 459-cm^{-1} band should have a value that is essentially 0.

7.6.6 Quantitative Analysis

The radiant power of a Raman line is measured in terms of an arbitrarily chosen reference line, usually the line of carbon tetrachloride at 459 cm^{-1}, which is scanned before and after the spectral trace of the sample. Peak areas on the spectrum are coverted to *scattering coefficients* by dividing the area of the sample peak by the average of the areas of the dual traces of carbon tetrachloride. Both standards and samples must be recorded in cells of the same dimensions.

The scattering coefficient based on the area under a recorded peak is directly proportional to the volume fraction of the compound present. Although peak heights may be used for mixtures in which the components all have the same molecular type, peak areas compensate for band broadening.

TABLE 7.24 Raman Frequencies of Alcohols and Phenols

Abbreviations used in the table

w, weak	*m-s, moderate to strong*
w-m, weak to moderate	*s-m, strong to moderate*
m-w, moderate to weak	*vs-m, very strong to moderate*
m, moderate	*vs, very strong*

Vibrational group	Frequency, cm^{-1}
Free —OH	3650–3604 (w)
Intermolecularly bonded OH· · ·O	3400–3200 (w)
Primary alcohol	3644–3635 (w) 1430–1200 (m-w) 1075–1000 (m-s) 900–800 (vs-m) 460–430 (m-w)
Secondary alcohol	3637–3626 (w) 1430–1200 (m-w) 1150–1045 (m-s) ca. 820 (s-m) ca. 500 (m-w)
Tertiary alcohol	3625–3614 (w) 1410–1310 (m-w) 1210–1100 (m-s) ca. 1000 (w-m) 800–750 (vs) ca. 360 (w-m)
Aromatic alcohols	3612–3593 (w)

TABLE 7.25 Raman Frequencies of Amines and Amides

Abbreviations used in the table

vw, very weak	*s, strong*
w, weak	*s-m, strong to moderate*
mw, moderately weak	*vs-s, very strong to strong*
m-w, moderate to weak	*vs, very strong*
m, moderate	*var, of variable strength*

Vibrational group	Frequency, cm^{-1}	Remarks
Primary aliphatic amines		
No branching at α-carbon	3380–3361 (w) 3310–3289 (mw) 1090–1040 (s-m)	
Secondary branching at α-carbon	3370–3363 (w) 3285–3280 (mw) 1180–1130 (m) 1040–1000 (m)	Three bands

TABLE 7.25 Raman Frequencies of Amines and Amides (*Continued*)

Vibrational group	Frequency, cm^{-1}	Remarks
Primary aliphatic amines		
Tertiary branching at α-carbon	ca. 3350 (w)	
	ca. 3280 (mw)	
	1245–1235 (m-w)	
	1218–1195 (m-w)	
	1140–1110 (w)	
	1080–1060 (w)	
	1030–1000 (w)	
Aromatic primary amines	3500–3420 (vw)	
	1638–1602 (vw)	
	700–600 (vw)	
Secondary aliphatic amines	3320–3280 (w)	R—NHR
	ca. 3400 (w)	Ar—NHR
	3320–3270 (w)	*trans*—C(=O)—NHR
	1570–1515 (w)	Solid; amide II
	1300–1250 (vs-s)	Amide III
	1180–1130 (m)	R—NHR C—N stretchings
	900–850 (vs-s)	
	1350–1280 (var)	Aryl—NHR C—N stretching
Tertiary aliphatic amines	833–740 (s)	R—NR_2
	870–820 (s)	HC(=O)NR_2
	750–700 (s)	RC(=O)NR_2

TABLE 7.26 Raman Frequencies of Carbonyl Bands

Abbreviations used in the table

w, weak — *ms, moderately strong*
m-w, moderate to weak — *s, strong*
m, moderate — *vs, very strong*

Vibrational group	Frequency, cm^{-1}	Remarks
Acid anhydrides		
Saturated	1825–1815 (m)	
	1755–1745 (m)	
Noncyclic, conjugated	1780–1770 (m)	
	1725–1720 (m)	
Cyclic, unconjugated	1870–1845 (m)	
	1800–1775 (m)	
Cyclic, conjugated	1860–1850 (m)	
	1780–1760 (m)	
Acid fluorides	1840–1835	Alkyl
	1812–1800	Aryl
	ca. 832 (m)	
Acid chlorides		
Alkyl	1810–1795 (s)	
	731–565 (m-w)	
	450–420 (vs)	
Aryl	1785–1765	
	1750–1735	

(Continued)

TABLE 7.26 Raman Frequencies of Carbonyl Bands (*Continued*)

Vibrational group	Frequency, cm^{-1}	Remarks
Acid bromides	1812–1788	Alkyl
	1775–1754	Aryl
Acid iodides	ca. 1806	Alkyl
	ca. 1752	Aryl
Aldehydes	1740–1720 (w)	Aliphatic
	1710–1630 (vs)	Aromatic
	1666–1658 (s)	C≡C—CHO
Carboxylic acids		
Mono-	1800–1740	These α-substituents increase the frequency: F, Cl, Br, OH
Dimer	1720–1680 (w)	
Aromatic	1686–1625	
Amino acids	1743–1729	
Carboxylate ions	1650–1550 (w)	
	1440–1340 (vs)	
Esters		
Formates	1730–1715 (m)	
Acetates	1750–1735 (m)	
Saturated, C_3 or greater alkyl chain length	1740–1725 (m)	
Aryl and α,β-unsaturated	1740–1714 (m)	
Diesters		
Oxalates	1763–1761	
Phthalates	1738–1728	
Carbamates	1694–1688	
Thiols		
Unconjugated	1710–1680	
Conjugated	1700–1640	
Ketones		
Saturated	1725–1700 (ms)	
Singly conjugated	1700–1670 (m)	
Doubly conjugated	1680–1650 (m)	
Aryl	1700–1650 (m)	
Alicyclic		
$n = 4$	ca. 1782 (m)	
$n = 5$	ca. 1744 (m)	
$n \geq 6$	1725–1699 (m)	
Lactones		
5-membered ring	1850–1790	
6-membered ring	1750–1715	

TABLE 7.27 Raman Frequencies of Other Double Bonds

Abbreviations used in the table

w, weak	*s, strong*
s-m, strong to moderate	*s-vs, strong to very strong*
	vs, very strong

Vibrational group	Frequency, cm^{-1}	Remarks
Aldimines (azomethines)	1673–1639	
	1405–1400 (s)	
Aldoximes and ketoximes	1680–1617 (vs)	
	1335–1330 (w)	
Azines	1625–1608 (s)	Both alkyl and aryl
Azoxy, diphenyl	1468 (w)	*cis*
	1413 (s)	*trans*
Glyoximes	1650–1627 (s-vs)	
	1608–1500 (s-m)	
Hydrazones	1660–1610 (s-vs)	
Imidates (imido ethers)	1658–1648	
Semicarbazones and thiosemicarbazones	1665–1642 (vs)	Aliphatic derivatives; thiosemicarbazones fall in lower end of range
	1630–1610 (vs)	Aromatic derivatives
Azo compounds	1580–1555 (s)	
	1462–1380 (vs)	Conjugated to aromatic ring
	ca. 1140 (s)	In aryl compounds

TABLE 7.28 Raman Frequencies of Nitro Compounds

Abbreviations used in the table

w, weak	*m, moderate*
w-m, weak to moderate	*m-s, moderate to strong*
m-w, moderate to weak	*s, strong*
	vs, very strong

Vibrational group	Frequency, cm^{-1}	Remarks
Alkyl nitrites	1680–1620 (s)	
Alkyl nitrates	1660–1622 (w-m)	
	1285–1260 (vs)	
	710–690 (m)	
Nitroamines	1365–1290 (s-w)	
	1030–980 (vs)	
Nitroalkanes		
Primary	1560–1548 (m-w)	
	1395–1370 (vs)	Sensitive to substituents
	915–898 (m-s)	*trans*
	894–873 (m-s)	Gauche
	618–609 (m-w)	
	494–472 (w-m)	Broad; useful to distinguish from secondary nitroalkanes

(Continued)

TABLE 7.28 Raman Frequencies of Nitro Compounds (*Continued*)

Vibrational group	Frequency, cm^{-1}	Remarks
Secondary	1553–1547 (m-w)	
	1375–1360 (vs)	
	908–868 (m)	
	863–847 (s)	
	670–650 (w)	
	625–613 (m)	
	560–516 (s)	Sharp band
Tertiary	1553–1533 (m-w)	
	1355–1345 (vs)	
Nitroaryl compounds	1357–1318 (vs)	
	857–830 (m-w)	

TABLE 7.29 Raman Frequencies of Aromatic Compounds

Abbreviations used in the table

w, weak	*m-s, moderate to strong*
mw, moderately weak	*m-vs, moderate to very strong*
w-m, weak to moderate	*s, strong*
m, moderate	*vs, very strong*
ms, moderately strong	*var, of variable strength*

Vibrational group	Frequency, cm^{-1}	Remarks
	Substitution patterns of the benzene ring	
Monosubstituted	3100–3000 (s)	Also di- and trisubstituted
	1620–1585 (m)	Also di- and trisubstituted
	1590–1565 (m)	Also di- and trisubstituted
	1180–1170 (w)	
	1035–1015 (m)	
	1010–990 (vs)	Characteristic feature; found also with 1,3- and 1,3,5-substitutions
	630–605 (m-s)	
	420–390 (w)	
1,2-Disubstituted	3100–3000 (s)	Also mono- and trisubstituted
	1620–1585 (m)	Also mono- and trisubstituted
	1590–1565 (m)	Also mono- and trisubstituted
	1170–1150 (w)	
	1060–1020 (s)	
	760–715 (m-s)	
1,3-Disubstituted	3100–3000 (s)	
	1620–1585 (m)	
	1590–1565 (m)	
	1180–1145 (w)	
	1140–1065 (w)	
	1100–1060 (w)	
	1010–990 (vs)	
	750–640 (s)	

TABLE 7.29 Raman Frequencies of Aromatic Compounds (*Continued*)

Vibrational group	Frequency, cm^{-1}	Remarks
	Substitution patterns of the benzene ring	
1,4-Disubstituted	3100–3000 (s)	
	1620–1585 (m)	
	1590–1565 (m)	
	1180–1150 (m)	
	650–610 (m-s)	
1,2,3-Trisubstituted	3100–3000 (s)	
	1620–1585 (m)	
	1590–1565 (m)	
	1100–1050 (m)	
	670–500 (vs)	The lighter the mass of the substituent, the higher the frequency
	490–430 (w)	
1,2,4-Trisubstituted	3100–3000 (s)	
	1620–1585 (m)	
	1590–1565 (m)	
	750–650 (vs)	
	580–540 (var)	
	500–450 (var)	
1,3,5-Trisubstituted	3100–3000 (s)	
	1620–1585 (m)	
	1590–1565 (m)	
	1010–990 (vs)	
Isolated hydrogen	1379 (vs)	
	1290–1200 (s)	
	745–670 (m-vs)	
	580–480 (s)	
Completely substituted	1296 (s)	
	550 (vs)	
	450 (m)	
	381 (m)	
	Bands for monosubstituted pyridines	
2-Substituted	1620–1570 (ms)	
	1580–1560 (m)	
	1480–1450 (m)	
	1440–1415 (mw)	
	1050–1040 (ms)	
	1000–985 (vs)	Also in 2,4- and 2,4,6-substitution
	850–800 (ms)	
	630–615 (m)	
3-Substituted	1595–1570 (ms)	
	1585–1560 (m)	
	1480–1465 (m)	
	1430–1420 (mw)	
	1030–1010 (vs)	
	805–750 (m)	
	630–615 (m)	
4-Substituted	1605–1565 (ms)	
	1570–1555 (m)	
	1500–1480 (m)	

(Continued)

TABLE 7.29 Raman Frequencies of Aromatic Compounds (*Continued*)

Vibrational group	Frequency, cm^{-1}	Remarks
	Bands for monosubstituted pyridines	
4-Substituted	1420–1410 (mw)	
	1000–985 (vs)	Also in 2,4- and 2,4,6-substitution
	805–785 (ms)	
	670–660 (m)	
	Diazines and triazines	
Pyrimidine rings	1590–1555	
	1565–1520	
	1480–1400	
	1410–1375	
	1005–980 (s)	2-, 4-, 2,4-, or 2,4,6-substitution
	ca. 1050 (s)	Above band for 5-substitution
	645–625	Substitution insensitive for 2-substitution
	685–660	Substitution insensitive for 4-substitution
Pyrazines	1586–1559 (m)	For all substitution patterns
Monosubstituted pyrazines	1530–1517 (m)	
	1060–1050 (s)	
	1024–1003 (vs)	
	840–788 (s)	
	660–617 (w)	
2,3-Disubstituted pyrazines	1570–1525 (m)	
	1292–1252 (m)	
	1100–1081 (s)	
	758–686 (vs)	
2,5-Disubstituted pyrazines	1540–1520 (m)	
	865–838 (vs)	
	650–642 (m)	
2,6-Disubstituted pyrazines	1025–1021 (m)	
	ca. 708 (vs)	
Trisubstituted pyrazines	1540–1525 (m)	
	955–915 (m)	
	748–715 (m)	
	710–695 (m)	
Tetrasubstituted pyrazines	1550–1545 (m)	
	720–710 (s)	
1,3,5-Triazine	3042 (vs)	
	1555 (mw)	
	1410 (w)	
	1176 (w)	
	1132 (s)	
	992 (s)	
	676 (ms)	
	340 (m)	
	Five-membered ring heterocycles	
Pyrroles	1560–1540 (var)	
1-Substituted	1510–1490 (vs)	
	1390–1380 (s)	
2-Substituted	1570–1545 (var)	

TABLE 7.29 Raman Frequencies of Aromatic Compounds (*Continued*)

Vibrational group	Frequency, cm^{-1}	Remarks
	Five-membered ring heterocycles	
2-Substituted	1475–1460 (vs)	
	1420–1400 (s)	
3-Substituted	1570–1560 (var)	
	1490–1480 (vs)	
	1430–1420 (s)	
Furans, 2-substituted	1605–1560 (var)	
	1515–1460 (vs)	
	1400–1370 (s)	
	1230–1220 (m)	
	1160–1140 (m)	
	1080–1060 (m)	
	1020–992 (ms)	
Thiophenes		
2-Substituted	1535–1514 (var)	
	1454–1430 (vs)	
	1361–1347 (s)	
	867–842 (s)	
3-Substituted	1542–1492 (var)	
	1420–1380 (vs)	
	1376–1362 (s)	
	Fused-ring aromatics	
1-Naphthalene	1580–1560 (s)	
	1390–1370 (vs)	Characteristic for mono- and most disubstituted naphthalenes
	1080–1030 (w)	
	880–810 (w-m)	
	720–650 (m-s)	
	535–512 (w)	
2-Naphthalenes	1585–1570 (m)	
	1390–1380 (vs)	
	775–765 (s)	
	525–515 (ms)	
Anthracenes	1415–1385 (vs)	
	426–390 (vs)	

TABLE 7.30 Raman Frequencies of Sulfur Compounds

Abbreviations used in the table

w, weak	*s, strong*
mw, moderately weak	*s-m, strong to moderate*
w-m, weak to moderate	*vs, very strong*
m-w, moderate to weak	*vs-s, very strong to strong*
m, moderate	*vs-m, very strong to moderate*
ms, moderately strong	*m-s, moderate to strong*
	var, of variable strength

Vibrational group	Frequency, cm^{-1}	Remarks
C—S—H	2590–2560 (s)	Both aliphatic and aromatic
	910–830 (m-w)	
	465–430 (vs-m)	
Sulfides, R—S—R		
Acyclic	750–690 (m-s)	
	ca. 585 (vs)	
Cyclic C_n		
$n = 3$	688 (vs)	
$n = 4$	664 (vs)	
Disulfides —S—S—	750–630 (vs-s)	
	527–507 (vs-s)	Aryl at 540–520 cm^{-1} (s-m)
C(=O)—SH	2580–2540 (vs)	
	628–535 (s-m)	
C=S	1065–1050 (m)	
	735–690 (vs)	
S=O		
R_2SO	1070–1035 (w-m)	Aryl substituents at lower end
$(RO)_2SO$	1209–1198 (m-w)	One or two bands
	732 (s)	
	698 (m)	
$(R_2N)_2SO$	1108 (vs)	
	674 (s)	
	657 (s)	
SOF_2	1308 (vs)	
	801 (m)	
$SOBr_2$	1121 (w)	
	405 (s)	
	379 (m)	
Sulfones —SO_2—		
Aliphatic	1330–1295 (m-w)	
	1155–1135 (s)	
Aromatic	586–504 (var)	For aromatic also
	1334–1325 (w)	
	1160–1150 (w)	
Sulfonamides —SO_2—NH_2	ca. 1322 (m)	
	1163–1138 (vs)	Aryl group lowers strength to medium
Sulfonates —SO_2—O	1363–1338 (w-m)	Aryl substituents at higher end
	1192–1165 (vs)	Aryl substituents at higher end
	589–517 (w-m)	Aryl substituents at higher end
Thiosulfonates —SO_2—S	1334–1305 (m-s)	
	1128–1126 (s)	

TABLE 7.30 Raman Frequencies of Sulfur Compounds (*Continued*)

Vibrational group	Frequency, cm^{-1}	Remarks
Sulfonyl halides		
RSO_2F	1447 (w)	
	1412–1402 (mw)	
	1263 (vs)	
	1197–1167 (vs)	
RSO_2Cl	1414 (w)	
	1384–1361 (m-w)	
	1184–1169 (vs-s)	
Sulfates —O—SO_2—O—	1388–1372 (s)	
	1196–1188 (vs)	
Dithioesters —C(=S)—S—	1225–1203 (s)	
	910–857 (s)	
	585–572 (vs)	
	450 (w)	
Thioamides —C(=S)—NR_2		
Primary	950–800 (ms)	
	750–700 (vs)	
	500–400 (vs)	
Secondary	950–800 (ms)	
	700–550 (vs)	
	500–400 (s)	
Tertiary	1563–1524 (m)	
	1285–1210 (w)	
	1000–700 (var)	Multiple bands
	626–500 (vs)	
	448–338 (ms)	
Trithiocarbonates RSC(=S)SR		
Dialkyl acyclic	722 (m)	CH_3—S stretching
	517 (s)	S—C—S stretching modes
Cyclic 5-membered ring	1062 (vs)	C=S stretching
	882 (m)	C=S plus asymmetric S—C—S stretching modes
	832 (w)	C=S plus asymmetric S—C—S stretching modes
	674 (vs)	CH_2—S stretching
	503 (vs)	C=S plus asymmetric S—C—S stretching modes

TABLE 7.31 Raman Frequencies of Ethers

Abbreviations used in the table

mw, moderately weak | *m-s, moderate to strong*
w, weak | *s, strong*
m, moderate | *vs, very strong*
var, of variable strength

Vibrational group	Frequency, cm^{-1}	Remarks
Aliphatic acyclic R—O—R	1150–1060 (vs)	Higher frequencies with symmetric substitution
	890–830 (vs)	
	500–400 (vs)	
Aromatic	1310–1210 (w)	
	1050–1010 (w)	
Vinyl ethers	1225–1200 (w)	
	850–840 (s)	
Acetals	1145–1129 (var)	
R—O—CH_2—O—R	1115–1080 (m)	
	870–850 (m)	
	660–600 (m)	
	540–450 (m-s)	
	400–320 (m-s)	
Epoxy	3075–3030 (s)	
	3020–2990 (s)	
	1280–1240 (s)	
	980–815 (mw)	
	880–750 (m)	
Cyclic ethers $(CH_2)_n$		
$n = 3$	1040–1010 (s)	
$n = 4$	920–900 (s)	
$n = 5$	820–800 (s)	
Peroxides	900–800 (var)	

TABLE 7.32 Raman Frequencies of Halogen Compounds

Abbreviations used in the table

w, weak | *s, strong*
m, moderate | *ms, moderately strong*
m-s, moderate to strong | *vs, very strong*
var, of variable strength

Vibrational group	Frequency, cm^{-1}	Remarks
C—Cl		
Primary	730–720 (s)	
	690–680 (s)	
	660–650 (s)	
Secondary	760–740 (var)	
	675–655 (ms)	
	670 (s)	
	635–630 (ms)	
	615–605 (vs)	

TABLE 7.32 Raman Frequencies of Halogen Compounds (*Continued*)

Vibrational group	Frequency, cm^{-1}	Remarks
C—Cl		
Tertiary	620–590 (ms)	
	570–560 (vs)	
=C—Cl	ca. 805 (m)	*trans*
	ca. 758 (vs)	*cis*
$=CCl_2$	928 (w)	
	785	
	598	
	407 (vs)	
C—Br		
Primary	650–640 (vs)	
	625–615 (s)	
	565–560 (vs)	
Secondary	620–605 (m)	
	590–575 (m)	
	540–535 (s)	
Tertiary	590–580 (m)	
	520–510 (vs)	
$—CBr_2—$	597 (s)	
	486 (s)	
C—I	600–590 (vs)	
Primary	590–580 (s)	
	510–500 (vs)	
Secondary	590–575 (m)	Two bands, similar strength
	495–485 (s)	
Tertiary	580–570 (s)	
	495–485 (s)	

Bibliography

Baranksa, H., A. Labudzinska, and J. Terpinski, *Laser Raman Spectrometry, Analytical Applications*, Wiley, New York, 1988.

Colthup, N. B., L. H. Daly, and S. E. Wiberley, *Introduction to Infrared and Raman Spectroscopy*, 3d ed., Academic, New York, 1990.

Grasselli, J. G., and B. J. Bulkin, eds., *Analytical Raman Spectroscopy*, Wiley, New York, 1991.

Grasselli, J. G., M. K. Snavely, and B. J. Bulkin, *Chemical Applications of Raman Spectroscopy*, Wiley, New York, 1981.

Lin-Vien, D., N. B. Colthup, W. G. Fateley, and J. G. Grasselli, *The Handbook of Infrared and Raman Characteristic Frequencies of Organic Molecules*, Academic, New York, 1991.

Parker, F. S., *Applications of Infrared, Raman, and Resonance Raman Spectroscopy in Biochemistry*, Plenum, New York, 1983.

Spiro, T. G., *Biological Applications of Raman Spectroscopy*, Wiley, New York, 1987, Vols. 1 and 2.

Strommen, D. P., and K. Nakamoto, *Laboratory Raman Spectroscopy*, Wiley, New York, 1984.

SECTION 8
ATOMIC SPECTROSCOPY

8.1 INTRODUCTION

Atomic spectroscopy includes all analytical techniques that employ the emission and/or absorption of electromagnetic radiation by individual atoms. There are three kinds of emission spectra:

1. Continuous spectra, which are emitted by incandescent solids
2. Line spectra, which are characteristic of atoms that have been excited and are emitting their excess energy
3. Band spectra, which are emitted by excited molecules

The specific wavelength of the radiation (emitted or absorbed) identifies the element. The intensity of emitted (or absorbed) radiation at the specific wavelength is proportional to the amount of the element present.

Sections 8.2, 8.3, and 8.5 discuss methods that use combustion flames. A nonflame method, electrothermal atomic-absorption spectrometry, is discussed in Sec. 8.3.3. Three techniques can be used to observe the atomic vapor that is produced when a sample solution is nebulized and passed into a flame. These are atomic-absorption spectrometry (AAS), flame emission spectrometry (FES), and atomic fluorescence spectrometry (AFS). Of these, AAS and FES are the most widely used.

Atomic-emission spectroscopy and spectrometry use either electrical discharges or plasmas (Sec. 8.4) for excitation of emission spectra.

Flame spectrometric methods, qualitative and quantitative, can be applied to clinical materials (serum, plasma, and biological fluids), soils, plant materials, plant nutrients, and samples of inorganic and organic substances.

8.1.1 Instrumentation for Flame Spectrometric Methods

The basic instrumentation for flame atomic-emission spectrometric methods requires these items:

1. An atom source—a flame
2. A monochromator to isolate the specific wavelength of light to be used
3. A detector to measure the light transmitted
4. Electronics to treat the signal and a data logging device to display the results

For work in atomic absorption spectrometry, a primary light source must be added, either a hollow-cathode lamp or an electrodeless discharge lamp.

8.1.1.1 Atom Sources. The atom source must produce free atoms of analytical material from the sample. A high-energy acetylene–nitrous oxide flame (2800°C) is usually employed; its function is to rapidly desolvate and to efficiently dissociate the analyte in the sample in order to minimize interferences. This is particularly true if an oxide of the analyte has a high dissociation energy. A lower-temperature acetylene–air flame (2400°C) is sometimes desired when an element is too easily ionized in a hotter flame. The sample is introduced as an aerosol into the flame.

For FES and AAS a premixed, laminar-flow flame is employed that rests upon a slot burner. The flame burner head is aligned so that it intersects the light path of the spectrophotometer. Titanium burner heads provide maximum corrosion resistance when analyzing any type of sample. Before use gas supplies must be checked. Acetylene tanks must never be allowed to drop below 75 $lb \cdot in^{-1}$ (5.1 atm) to avoid contamination of the gas supply with acetone—the solvent for commercial acetylene.

8.1.1.2 Nebulization. In both FES and AAS, the sample solution is introduced as an aerosol (fine mist) into the flame. Pneumatic nebulization is the technique used in most flame spectrometric determinations. The sample solution is drawn through the inner annulus of two concentric annuli by the pressure differential generated by the high-velocity gas stream (usually the oxidant of the flame gases) passing across or concentric with the sample orifice. The nebulizer composition is stainless steel for most solutions; all plastic, corrosion resistant for very acidic samples; and platinum alloy with a tantalum venturi for corrosive samples.

A cloud of droplets of varying diameters is produced within an expansion chamber that precedes the burner. The larger droplets are removed from the sample stream upon collision with multivaned flow spoilers or the wall surfaces. Some nebulizer units break up the original droplets into smaller droplets by impact beads or a counter gas jet. The final aerosol, now a fine mist, is combined with the oxidizer–fuel mixture and carried into the burner. Only a small percentage (usually 2% to 3%) of the nebulized analyte solution reaches the flame. Unanticipated changes in viscosity and surface tension are avoided during aerosol generation by keeping the sample and standard matrices identical and by avoiding total acid or salt concentrations greater than about 0.5%.

Burner chambers are constructed of, or coated with, an inert, wettable plastic to allow for proper drainage of excess sample and to prevent burner "memory" of previous analyses.

Ultrasonic nebulizers are more efficient nebulization systems than the pneumatic nebulizers, but they possess a lower nebulization rate. By the use of these systems, it is possible to convert about 40% of small sample volumes to a mist of useful droplet size, but sensitivity is generally not increased because of the slowness of the nebulization. The stability and reproducibility of emission and absorption readings obtained with pneumatic nebulizers are better than with ultrasonic nebulizers.

8.1.1.3 Atomization. The flame atomization processes are complex. Everything leading to the production of free atoms must take place in the few milliseconds that correspond to the upward movement of the molecules and atoms through the flame gases. The atomization step must convert the analyte within the aerosol into free analyte atoms for AAS and flame atomic-emission spectrometry.

For AAS the following sequence of events occurs in rapid succession.

1. *Desolvation of the aerosol*: The water, or other solvent, is vaporized, leaving minute particles of dry salt.

2. *Vaporization of the resulting particles*: At the high temperature of the flame, the dry salt is vaporized (converted to gaseous molecules).

3. *Dissociation of gaseous molecules*: Part or all of the gaseous molecules are dissociated to give neutral atoms.

For flame atomic emission, two more events must occur.

4. *Excitation of atoms (and molecules)*: The neutral metal atoms are excited (and sometimes ionized) by the thermal energy of the flame. (Excitation occurs in AAS by thermal collisions as well as by absorption of radiation from the light source.)

5. *Emission*: From the excited electronic level(s) of the atom, a reversion takes place to the ground electronic state with the emission of light whose wavelength is characteristic of the element and whose intensity is proportional to the amount of analyte element present.

Unfortunately, some of the metal atoms unite with other radicals or atoms that are present in the flame gases or that are introduced into the flame along with the test element. For example, neutral metal atoms may unite with OH radicals or atomic oxygen present in the flame gases to form MOH, MH, and MO molecular species. These reactions weaken both the atomic absorption signal and the atomic-emission signal at the wavelength of the metal emission or absorption line(s). The MOH, MH, and MO molecules may be excited and, if so, will emit characteristic molecular band spectra that will clutter the general background. Some molecular band spectra are quite strong and have been used in flame emission as, for example, the band centered at 554 nm due to CaOH and band heads at 606 and 622 nm due to CaO molecules. If one desires to eliminate or minimize the interference from molecular bands, the use of hotter flames (acetylene–nitrous oxide) is recommended, along with the addition of an ionization buffer to minimize any ionization of metal atoms.

8.1.1.4 Spectrophotometers. The spectrophotometers are similar to those used in visible-ultraviolet instruments discussed in Sec. 5. The usual detectors are photomultiplier tubes or diode-array assemblies. A chopper will be added either to modulate the radiation from the sample in FES or to modulate the light source in AAS.

Care must be taken never to exceed the saturation limit of the photomultiplier tube by flooding it with light. This situation will arise when too much radiation emanates from the flame and is often caused by too high a concentration of sample components, even though these signals are eliminated by modulation in the final signal readout.

8.1.1.4.1 Scale Expansion. When the electronic noise in the detector–amplifier system can be reduced to a negligible level, scale expansion can be used. The zero point is displaced off-scale by applying a potential opposite to the signal arriving from the detector. Thus, the full scale of the readout device can be used as the upper end of a greatly expanded scale. Alternatively, the lower end of the instrument scale can be expanded by adjusting the amplification of the signal.

8.1.2 Flame Composition and Temperature

A premixed, laminar-flow flame with well-defined zones has lower background emission and less noise as a result of optical turbulence than a corresponding diffusion flame (total consumption atomizer–burner) where the zones are intermixed. Characteristics of common premixed flames are given in Table 8.1. The exact temperature of the flame depends on fuel–oxidant ratio and is generally highest for a stoichiometric mixture. Usually, no attempt is made to adjust flame temperatures subtly for a particular fuel–oxidant mixture other than the use of lean or fuel-rich flames.

Many of the interferences due to the formation of refractory oxides can be overcome or minimized by the use of the proper oxidant–fuel system, particularly the nitrous oxide–acetylene system. Fuel-rich conditions appear to be essential in the presence of refractory oxides. By using a mixture of 23% nitrous oxide in air with a slightly fuel-rich acetylene flame, the interference of 2000 $\mu g \cdot mL^{-1}$ of aluminum, silicon, or titanium in the analysis of 1 $\mu g \cdot mL^{-1}$ of magnesium was eliminated.

8.1.2.1 Burning Velocity. Flame propagation rate or burning velocity is important. If it exceeds approximately 40 $cm \cdot s^{-1}$, the flame likely will flash back into the mixing chamber and an explosion will result. That is why acetylene–oxygen flames must *never* be used with a premixed burner. When using acetylene–nitrous oxide flames, first ignite the flame with an acetylene–air mixture (turn on the air first). Then override the air with the nitrous oxide gas flow until the flame exhibits the characteristic reddish color in the interconal region. Reverse this sequence when extinguishing the flame; turn off the nitrous oxide while the air is flowing, then turn off the acetylene flow before turning off the air. Special burner heads (5 cm slot length, 0.5 mm width) and control units are required for igniting and extinguishing acetylene–nitrous oxide flames.

8.1.2.2 Flame Profile. The concentration of excited and unexcited atoms in a flame varies in different parts of the flame envelope. Although usually cooler, a fuel-rich flame (in which the ratio of fuel to oxidant exceeds that needed for stoichiometric combustion) provides the reducing atmosphere necessary for the production of a large free-atom population of those elements (alkaline-earth

TABLE 8.1 Characteristics of Some Common Flames

Flame	Flowrate ($L \cdot min^{-1}$) Oxidant	Fuel gas	Temperature* (°C)	Burning velocity† ($cm \cdot s^{-1}$)
Air–acetylene	8	1.2–2.2	2400	160–266 (160)
Nitrous oxide–acetylene	10	3.5–4.5	2800	260
Air–hydrogen	8	6	2045	324–440
Nitrous oxide–hydrogen	10	10	2690	390
Nitrous oxide–propane	10	4	2630	250
Air–propane	8	0.3–0.45	1925	43

* Stoichiometric mixture.
† Values in parentheses are probably the ones most applicable to laboratory burners.

TABLE 8.2 **Percent Ionization of Selected Elements in Flames***

Element	Ionization potential, eV	Propane–air, 1925°C	Acetylene–air, 2400°C	Acetylene–oxygen, 3140°C	Acetylene–nitrous oxide, 2800°C
Lithium	5.391	0.01	5	16	68
Sodium	5.139	1.1	15	26	82
Potassium	4.340	9	50	82	98
Rubidium	4.177	14	63	89	99
Cesium	3.894	29	82	96	100
Magnesium	7.646			0.01	6
Calcium	6.113		0.01	7.3	43
Strontium	5.694		0.01	17.2	84
Barium	5.211		1.9	42.3	88
Manganese	7.43				5

* Partial pressure of metal atoms in the flame is assumed to be 1×10^{-6} atm for acetylene–air and acetylene–oxygen flames, and approximately 10^{-8} atm for the acetylene–nitrous oxide flame.

elements, Al, B, Sb, and Ti) that have a tendency to form refractory oxides. The free-atom concentration of these metals is negligible unless analyzed in the reducing environment of a fuel-rich acetylene– nitrous oxide flame.

8.1.2.3 Observation Site. The region that is viewed within the flame is important. For example, the emission lines of boron (249.7 nm) and antimony (259.8 nm) are either absent or very weak in the outer mantle of a stoichiometric flame, but they appear in high concentrations in the reaction zone (blue cone) of a fuel-rich flame. For mixture strengths 0.40 and less (mixture strength refers to the actual ratio of oxygen to acetylene as compared to the ratio for a stoichiometric flame), the interconal gases will be bright green due to the high concentration of CH radicals that provide an excellent reducing atmosphere for refractory boron oxide. Some means for adjusting the burner height relative to the observation light path of the spectrometer will be needed to observe the interconal reaction zone.

8.1.2.4 Ionization Buffer. When flame or plasma temperatures are high enough to cause ionization of the analyte atoms (Table 8.2), an ionization buffer must be incorporated into the sample solution. The degree of ionization, α_i, is defined as

$$\alpha_i = \frac{[\mathrm{M}^+]}{[\mathrm{M}^+]+[\mathrm{M}]} \tag{8.1}$$

At equilibrium, when the ionization and recombination rates are balanced, the ionization constant K_i (in atm) is given by

$$K_i = \frac{[\mathrm{M}^+][e^-]}{[\mathrm{M}]} = \left(\frac{\alpha_i^2}{1-\alpha_i^2}\right) \rho_{\Sigma \mathrm{M}} \tag{8.2}$$

where $\rho_{\Sigma M}$ (in atm) is the total atom concentration of metal in all forms in the plasma. The ionization constant can be calculated from the *Saha equation*:

$$\log K_i = -5040 \frac{E_i}{T} + \frac{5}{2} \log T - 6.49 \log \frac{g_{\mathrm{M}} + g_{e^-}}{g_{\mathrm{M}}} \tag{8.3}$$

where E_i is the ionization potential of the metal in eV (see Table 8.5), T is the absolute temperature of the plasma (in kelvins), and the g terms are the statistical weights of the ionized atom, the electron, and the neutral atom. For the alkali metals the final term is zero; for the alkaline-earth metals, it is 0.6.

To suppress the ionization of a metal, another easily ionizable element (denoted a *deionizer* or *radiation buffer*) is added to the sample, but it must be an element which will not add any spectral line interference. Often easily ionized elements such as K, Cs, or Sr are added in approximately 100-fold excess (as compared to the analyte concentration) to suppress the ionization of easily ionized elements—those whose ionization potentials are less than about 8 eV. Actually the product $(K_i)\rho$ of the deionizer must exceed thc similar product for the test element 100-fold (for 1% residual ionization of the test element).

The use of ionization buffers is important when using the acetylene–nitrous oxide flame and desirable when handling samples that contain variable amounts of the alkali and alkaline-earth metals that are easily ionized and that will mutually act as ionization buffers upon each other.

8.1.2.5 Releasing and Shielding Agents. Releasing and shielding agents provide a chemical means for overcoming some vaporization interferences (Table 8.3). These agents may either combine with the interfering substance or deny the analyte to the interfering substance by mass action. For example, in calcium determinations a few hundred parts per million of lanthanum or strontium are often added routinely to solutions to minimize interference due to anions (such as aluminate,

TABLE 8.3 Releasing and Shielding Agents Used in Emission and Absorption

Analyte	Interferent	Reagent	Type	References
Mg, Ca	Al, Si, PO_4^{3-}, SO_4^{2-}	La	Releaser	*Anal. Chim. Acta* **19**:166 (1958); **22**:163 (1960); **29**:202 (1963). *Analyst* **85**:495 (1960); *Appl. Spectrosc.* **20**:214 (1966)
Mg, Ca, Ba	Al, B, Te, Se, NO_3^-, PO_4^{3-}, SO_4^{2-}	Sr	Releaser	*Appl. Spectrosc.* **20**:214 (1966); *Anal. Chem.* **32**:1475 (1960); *Anal. Chim. Acta* **36**:57 (1966); *Anal. Chem.* **32**:1475 (1960).
Ca	Al, Si, PO_4^{3-}, SO_4^{2-}	Mg	Releaser	*Anal. Chim. Acta* **41**:93 (1968)
Mg, Sr	Al, PO_4^{3-}	Ca	Releaser	*Anal. Chim. Acta* **41**:93 (1968); *Anal. Chem.* **26**:1060 (1954)
Na, K, Mg	Al, Fe	Ba	Releaser	*Chem. Abstr.* **59**:9325 (1935)
Sr	Al, PO_4^{3-}	Nd, Sm, Y, Pr	Releaser	*Anal. Chem.* **32**:1475 (1960)
Mg, Ca, Sr, Ba	Al, Fe, Th, rare earths, Si, B, Cr, Ti, PO_4^{3-}, SO_4^{2-}	Glycerol, $HClO_4$	Shielding	*Anal. Chem.* **34**:778 (1962); **33**:1722 (1961); *Talanta* **10**:367 (1963)
Na, Cr	Al	NH_4Cl	Shielding	*Anal. Chem.* **38**:1086 (1966)
Mo	Sr, Ca, Ba, PO_4^{3-}, SO_4^{2-}	NH_4Cl	Shielding	*Analyst* **93**:79 (1968)
Ca	PO_4^{3-}	Ethylene glycol	Shielding	*Talanta* **10**:367 (1963)
Ca	PO_4^{3-}	Mannitol	Shielding	*Talanta* **10**:367 (1963)
Ca, Sr	PO_4^{3-}	Dextrose, sucrose	Shielding	*Talanta* **10**:367 (1963); *J. Assoc. Off. Agr. Chemist* **37**:945 (1954)
Mg, Ca	Se, Te, B, Al, Si, NO_3^-, PO_4^{3-}, SO_4^{2-}	EDTA	Shielding	*Anal. Chim. Acta* **36**:57 (1966); *Anal. Chem.* **32**:1471 (1960)
Mg, Ca	Al	8-Hydroxyquinoline	Shielding	*Talanta* **14**:823, 933 (1967); *Analyst* **88**:259 (1963)

phosphate, silicate, sulfate, vanadate, and tungstate). The lanthanum or strontium preferentially binds the anion, thus releasing the calcium atoms.

Calcium and magnesium can be shielded by complexing the calcium and magnesium with EDTA. Once in the flame the EDTA is destroyed.

8.2 ATOMIC-EMISSION SPECTROMETRY WITH FLAMES

FES is a simple and convenient method for analyzing solutions of organic and inorganic materials for individual metals. FES is particularly suited to alkali and alkaline-earth metals, with 1 to 5 g of a dissolved sample being adequate for several analyses. Extremely low concentrations of metals can be determined (Table 8.4). The quantitative estimation for most metals, without prior concentration, is on the order of 0.1 to 2.0 $\mu g \cdot mL^{-1}$. Operating conditions for the determination of elements by FES are given in Table 8.9 (*q.v.*).

8.2.1 Principle

A solution of the sample is sprayed into a flame possessing the thermal energy required to excite the element to a level at which it will radiate its characteristic line-emission spectrum. For an atom or molecule in the ground electronic state to be excited to a higher electronic energy level, it must absorb energy from the flame via thermal collisions with the constituents of the partially burned flame gases. Upon their return to a lower or ground electronic state, the excited atoms and molecules emit radiation characteristic of the sample components.

Band spectra arise from electronic transitions involving molecules. For each electronic transition there will be a whole suite of vibrational levels involved. This causes the emitted radiation to be spread over a portion of the spectrum. Band emissions attributed to triatomic hydroxides (CaOH) at 554 nm and monoxides (AlO, strongest band at 484 nm) are frequently observed and occasionally employed in FES. The boron oxide system gives very sensitive bands at 518 and 546 nm. The dissociation energy of MO species is given in Table 8.5 along with the atomization efficiency.

The number of atoms in an excited state, or of molecules excited, is extremely small. The emitted radiation passes through a monochromator that isolates the specific wavelength for the desired analysis. A photodetector measures the radiant power of the selected radiation that is correlated with the concentration of analyte in the sample and in standards.

FES requires a monochromator capable of providing a band pass of 0.05 nm or less in the first order. The instrument should have sufficient resolution to minimize the flame background emission and to separate atomic-emission lines from nearby lines and molecular fine structure. A monochromator equipped with a laminar-flow burner for flame AAS serves equally well for FES. However, in FES it is often desirable to scan a portion of the spectrum.

8.2.2 Background Correction

Correction for background associated with flames and matrix components is a serious problem in FES and AAS methods. Background radiation from the flame itself arises from hydrogen molecules, OH radicals, and partially burned fuel and solvent molecules. These species plus continua from metals and metal oxides and hydroxides constitute the flame background.

8.2.2.1 Scanning Methods. A somewhat tedious correction method involves scanning the spectrum on both sides of the analyte emission line (or band). A baseline is drawn beneath the emission line from the background (extrapolated).

After a preliminary wavelength scan has been made, the line plus background signal is measured for a group of samples and standards at the wavelength of the emission line. The wavelength setting

TABLE 8.4 Detection Limits ($ng \cdot mL^{-1}$) in Atomic Spectroscopy

(The detection limits in the table correspond generally to the concentration of an element required to give a net signal equal to three times the standard deviation of the noise (background) in accordance with the International Union of Pure and Applied Chemistry (IUPAC) recommendations. Detection limits can be confusing when steady-state techniques such as flame atomic emission or absorption, and plasma atomic emission or fluorescence are compared with the electrothermal or furnace technique, which uses the entire sample and detects an absolute amount of the analyte element. To compare the several methods on the basis of concentration, the furnace detection limits assume a 20-μL sample. The detection limits in this table are expressed in ng/mL or ppb.

Data for the several flame methods assume an acetylene–nitrous oxide flame residing on a 5- to 10-cm slot burner. The sample is nebulized into a spray chamber placed immediately ahead of the burner. Detection limits are quite dependent on instrument and operating variables, particularly the detector, the fuel and oxidant gases, the slit width, and the method used for background correction and data smoothing.)

Element	Wavelength, nm	Flame emission	Flame atomic absorption	Electrothermal graphite furnace	Argon ICP
Aluminum	167.081				1.5
	309.28		30	0.25	11
	394.40	7.5	45		36
	396.15	3.6	30	0.05	20
Antimony	217.581		30		18
	231.13	70			30
	259.81	200		0.08	
Arsenic	188.985				12
	189.04		160		35
	193.76		120	0.5	50
	197.20		240		100
	234.90	250			
Barium	455.36	3			0.9
	493.41	4			1
	553.55	1.5	9	0.04	
Beryllium	234.86		1	0.025	0.4
	313.042		2	0.003	0.2
	313.11	100			1
Bismuth	223.061		18	0.35	12
	227.66			2	
Boron	182.59				8
	249.773		700	43	1.5
(as BO_2)	518.0	50			
(as BO_2)	547.6	50			
Bromine	154.07				50
Cadmium	214.44				1.0
	226.50				0.6
	228.802	6	1	0.01	1.5
	326.11	3	0.5	0.014	
Calcium	315.98				20
	393.366				0.03
	396.85				0.06
	422.67	1.5	1	0.03	
Carbon	193.09				44
	247.856				65
Cerium	413.38				30
	418.660				75
	569.92	150			

TABLE 8.4 **Detection Limits (ng · mL^{-1}) in Atomic Spectroscopy (*Continued*)**

Element	Wavelength, nm	Flame emission	Flame atomic absorption	Electrothermal graphite furnace	Argon ICP
Cesium	852.11	0.02	4	0.55	
	894.35	0.04	130		
Chlorine	134.72				50
Chromium	267.716				4
	283.58				20
	284.98				30
	357.87	6	6	0.075	
	359.35	7			
	360.53	13			
	425.44	3	6		66
	427.48	4			
	428.97	5			
Cobalt	228.616				5
	238.89				28
	240.73	5	5	0.21	7
	345.35	30			
Copper	324.754	1.5	3	0.3	2
	327.40	3	2	0.6	4
Dysprosium	353.170				0.3
	340.78				0.6
	404.60	30	50		
	421.17		30	2.3	
Erbium	337.271				0.7
	400.80	30	50	5	
	408.77		40		
Europium	381.967				0.3
	459.40	0.45	1.5	1.3	
Gadolinium	335.05				10
	342.247				2.5
	440.19	72	1000	8	
Gallium	294.36		100	0.23	30
	403.30	5	50		
	417.206	3	30	1	6.5
Germanium	209.43				50
	265.118	400	200	0.45	13
Gold	242.80		10	0.5	5
	267.595	500	8	0.22	5.5
Hafnium	264.141				4
	277.34				10
Holmium	345.600				0.5
	405.39	15	40	0.7	
Indium	303.94	100	40	6.8	
	325.609	22	8		18
	410.18	14	20		
	451.13	0.7	22		2

(Continued)

TABLE 8.4 Detection Limits ($ng \cdot mL^{-1}$) in Atomic Spectroscopy (*Continued*)

Element	Wavelength, nm	Flame emission	Flame atomic absorption	Electrothermal graphite furnace	Argon ICP
Iodine	178.276				60
	183.0			3	
Iridium	208.88	400	500	6.8	
	224.268				3.5
Iron	238.20				4
	248.33		6	0.06	
	259.940				1.5
	302.06	18	5		
	371.99	15	10		
	385.99	12	21		
Lanthanum	379.478				0.02
	408.67				2
	550.13	20	2000		
	579.13	5	2000	0.5	
Lead	217.00		10	0.28	
	220.353				14
	283.31	60	10	1	
	368.35	30			
	405.78	20			
Lithium	460.29	0.06	30		50
	610.36	0.001			
	670.784	0.003	2	0.2	0.6
Lutetium	261.542		300		0.05
	307.76				6
Magnesium	257.610				0.3
	279.553		2	0.03	0.1
	285.21	4.5	0.1	0.018	3.6
Manganese	257.610				0.3
	279.48	1	0.32	0.03	
	403.08	1.5	30		
Mercury	184.950				8.5
	253.65	150	200	7.5	50
Molybdenum	202.030				4
	313.26	220	20	0.35	
	390.30	75	50		
Neodymium	401.225				2
	492.45	150	1000		
Nickel	231.604				5.5
	232.00	8	4	0.24	10
	341.48	15	2		
	352.45	8	2		
Niobium	309.418				4
	334.9		2000		
	405.89	250	1000		
Nitrogen	174.272				50
Osmium	225.585				5
	263.71	2000	80		

TABLE 8.4 Detection Limits ($ng \cdot mL^{-1}$) in Atomic Spectroscopy (*Continued*)

Element	Wavelength, nm	Flame emission	Flame atomic absorption	Electrothermal graphite furnace	Argon ICP
Osmium	290.91		100		
Palladium	244.8	20	10	0.48	
	340.458	25	80		7
	363.47	50			60
Phosphorus	177.499				18
	213.62			110	50
(as HPO)	524.9	100			
Platinum	265.945	2000	100	0.35	20
	404.72	2.6			
	766.490	0.15	3	0.02	10
	769.90	0.3	2		
Praseodymium	417.939				0.8
	493.97	300			
Rhenium	227.525				11
	346.05	200	1000	10	
Rhodium	343.489	10	5	0.4	5
	369.24	20			
Rubidium	420.155				3
	780.02	0.0065	10	0.05	500
	794.76	0.013			
Ruthenium	267.876				5.5
	349.89	80	100	0.75	150
Samarium	442.434				7
	476.03	30	500		100
Scandium	357.24				1
	361.384				0.4
	391.18	21	50	6	120
	402.04	30			
Selenium	196.026		500	0.7	37
Silicon	251.611		300	0.75	5
Silver	328.065	2	0.9	0.035	3
	338.29	4			2
Sodium	330.23	125		0.7	15
	588.995	0.01	0.2	0.005	1
Strontium	407.771				0.02
	421.55				0.5
	460.73	0.1	2	0.1	
Sulfur	180.734		10		20
(as S_2)	394.0	1600			
Tantalum	268.517				9
	271.47		2000		
Tellurium	214.281	150	30	0.45	27
Terbium	350.917				5
	431.89	150	600		
Thallium	276.78		20	0.75	

(Continued)

TABLE 8.4 Detection Limits (ng · mL^{-1}) in Atomic Spectroscopy (*Continued*)

Element	Wavelength, nm	Flame emission	Flame atomic absorption	Electrothermal graphite furnace	Argon ICP
Thallium	351.924				16
	377.57	3		0.5	
	535.05	1.5			
Thorium	274.716				17
Thulium	346.220				1.5
	371.79	4	20		
	384.80				7
Tin	224.60		110	1	30
	235.48		100	0.5	
	242.949				15
	284.00	100	200		
	286.33		160	1.5	
Titanium	334.941				0.6
	364.27	210	100	2.5	
	365.35	180			
	399.86	150			
Tungsten	239.709				17
	400.87	450	1000		
Uranium	358.49	100	40 000	30	
	385.958				18
Vanadium	292.40				8
	309.311				2
	437.92	15			
Ytterbium	328.937				0.3
	369.42				2
	398.80	0.45	4	0.15	
Yttrium	360.07				3
	362.09	40	50	10	
	371.030				0.2
	410.24	30	200		
Zinc	202.55				4
	213.86	1000	1	0.0075	2
	231.856				0.9
Zirconium	339.198				1.5
	343.82				7
	360.12	1000	1000		

is moved to a suitable position on the background only, first on one side of the emission and then the other side. The average reading is subtracted from the line plus background reading. Computer-controlled equipment will make these measurements automatically.

8.2.2.2 ***Wavelength Modulation.*** The most common method uses an oscillating refractor plate which oscillates about the vertical axis.[1] The system is placed inside a monochromator after the entrance

[1] W. Snellman, T. C. Rains, K. W. Yee, H. D. Cook, and O. Menis, *Anal. Chem.* **42**:394 (1970).

TABLE 8.5 Ionization Potential and Atomization Efficiency of Elements and Dissociation Energy of MO Molecules

Element	Ionization potential, eV	Atomization efficiency	Dissociation energy of MO, eV
Aluminum	5.98	A-A,* < 10^{-5}; N-A,† 0.42	5.0
Antimony	8.64		3.2
Arsenic	9.8		4.9
Barium	5.21	A-A, 0.0011; N-A, 0.3	5.85
Beryllium	9.32		4.6
Bismuth	7.29		4.0
Boron	8.3		7.95
Cadmium	6.11	A-A, 0.50	3.8
Calcium	6.11	A-A, 0.14; N-A, 0.4	5.0
Cerium	5.6		8.3
Cesium	3.87		
Chromium	6.76	A-A, 0.064	4.2
Cobalt	7.86	A-A, 0.052	ca. 3.7
Copper	7.72	A-A, 0.98; N-A, 1.00	4.9
Dysprosium	6.2		
Erbium	6.08		
Europium	5.67		
Gadolinium	6.16		5.9
Gallium	6.0	A-A, 0.16	2.5
Germanium	7.88		6.7
Gold	9.22	A-A, 0.63; N-A, 0.71	
Hafnium	6.8		
Indium	5.78	A-A, 0.67	1.1
Iodine	10.45		
Iridium	9.3		
Iron	7.87	A-A, 0.66	4.0
Lanthanum	5.61		8.4
Lead	7.4	A-A, 0.44	4.1
Lithium	5.39	A-A, 0.2; N-A, 0.44	
Lutetium	6.15		5.3
Magnesium	7.64	A-A, 0.59	4.3
Manganese	7.43	A-A, 0.45; N-A, 0.76	4.0
Mercury	10.43		
Molybdenum	7.10		5.0
Neodymium	5.45		7.4
Nickel	7.63		4.3
Niobium	6.88		4.0
Osmium	8.73		
Palladium	8.33		
Phosphorus	10.49		
Platinum	9.0		
Potassium	4.34	A-A, 0.25	
Praseodymium	5.48		7.7
Rhenium	7.87		
Rhodium	7.45		
Rubidium	4.2	A-A, 0.16	
Ruthenium	7.34		
Samarium	5.6		6.1
Scandium	6.54		7.0
Selenium	9.75		3.5
Silicon	8.15		8.0

(Continued)

TABLE 8.5 Ionization Potential and Atomization Efficiency of Elements and Dissociation Energy of MO Molecules (*Continued*)

Element	Ionization potential, eV	Atomization efficiency	Dissociation energy of MO, eV
Silver	7.57	A-A, 0.66	1.4
Sodium	5.14	A-A, 1.00; N-A, 0.33	
Strontium	5.69	A-A, 0.13	4.85
Tantalum	7.88		
Tellurium	9.0		2.7
Terbium	5.98		
Thallium	6.11	A-A, 0.36	<3.9
Thorium	6.2		8.6
Thulium	6.2		
Tin	7.34	A-A, $< 10^{-4}$; N-A, 0.76	5.7
Titanium	6.82	N-A, 0.3	6.9
Tungsten	7.98		7.2
Uranium	6.1		7.7
Vanadium	6.74	N-A, 0.91	5.5
Ytterbium	6.22		
Yttrium	6.51		7.0
Zinc	9.39	A-A, 0.45; N-A, 0.91	4.0
Zirconium	6.84		7.8

* A-A denotes air–acetylene flame.
† N-A denotes nitrous oxide–acetylene flame.

slit or before the exit slit. The optical beam is refracted and will cause a small oscillating displacement (and therefore spectral distribution) of the light beam leaving the monochromator. Thus, the analyte line and background and adjacent background are measured alternately and their signals subtracted.

Other methods of achieving wavelength modulation involve using a vibrating or rotating mirror, oscillating the slits of the monochromator, or using a rotating refractor plate.

Using wavelength modulation, it is possible to measure the intensity of a weak line superimposed on an intense continuum background.

8.3 ATOMIC-ABSORPTION SPECTROMETRY

8.3.1 Introduction

Atomic absorption is the process that occurs when a ground-state atom absorbs energy in the form of electromagnetic radiation at a specific wavelength and is elevated to an excited state. The atomic absorption spectrum of an element consists of a series of resonance lines, all originating with the ground electronic state and terminating in various excited states. Usually the transition between the ground state and the first excited state is the line with the strongest absorptivity, and it is the line usually used.

Transitions between the ground state and excited states occur only when the incident radiation from a source is exactly equal to the frequency of a specific transition. Part of the energy of the incident radiation P_0 is absorbed. The transmitted radiation P is given by

$$P = P_0 e^{-(k_\nu b)} \tag{8.4}$$

where k_ν is the absorption coefficient of the analyte element and b is the horizontal path length of the radiation through the flame. Atomic absorption is determined by the difference in radiant power of the resonance line in the presence and absence of analyte atoms in the flame. The width of the line emitted by the light source must be smaller than the width of the absorption line of the analyte in the flame.

The amount of energy absorbed from a beam of radiation at the wavelength of a resonance line will increase as the number of atoms of the selected element in the light path increases. The relationship between the amount of light absorbed and the concentration of the analyte present in standards can be determined. Unknown concentrations in samples are determined by comparing the amount of radiation they absorb to the radiation absorbed by standards. Instrument readouts can be calibrated to display sample concentrations directly.

Detection limits are given for many elements in Table 8.4. Operating conditions are given in Table 8.9.

8.3.2 Light Sources

8.3.2.1 Hollow-Cathode Lamps. The basic components of a shielded-type hollow-cathode lamp are an anode, a cathode, and either a glass or quartz exit window, all sealed in a Pyrex cylinder filled with an inert gas (argon or neon) at a pressure of 4 to 10 torr. Neon gas provides a greater intensity of emitted element lines; argon is used only when a neon emission line lies in close proximity to a resonance line of the cathode element. An anode wire is positioned alongside the cylindrical cathode. The cathode is a hollow cylinder of the element whose spectrum is to be produced, or in some cases, an alloy or carefully selected mixture of metals that does not spectrally interfere. A protective shield (nonconductive) of mica around the outside of the cathode just behind the lip prevents spurious discharges around the outside of the cathode. Lamps are operated at currents below 30 mA.

Hollow-cathode lamps emit light by the following process. The fill gas is ionized when an electrical potential is applied between the anode and the cathode. The positively charged ions collide with the negatively charged cathode and dislodge individual metal atoms in a process known as *sputtering.* These gaseous metal atoms are excited through impact with fill-gas ions, and light of the specific wavelengths for that element is emitted when the atom decays from the excited atomic state back to a less excited state or the ground electronic state.

A liability of AAS is the need for a different lamp for each element to be analyzed. Lamps are available for 65 single elements (among the metals only lamps for Cs, Lu, Os, Th, U, and the radioactive elements, plus the noble gases, the halogens, C, N, O, and S, are unavailable). Twenty multielement lamps are available (Table 8.6). However, the narrow-emission linewidths of hollow-cathode lamps provide virtual specificity for each element. Nevertheless, spectral overlaps do exist (Table 8.7). Atomic-absorption and -emission profiles are not always as narrow as is usually expected. The spectral interference may be experienced at line separations 0.050 nm or less. Spectral overlap can only be

TABLE 8.6 Multielement Hollow-Cathode Lamps

Elements	Elements
Ag, Au	Cu, Fe, Ni
Ca, Mg	Ag, Cr, Cu, Ni
Ca, Mg*	Al, Cu, Fe, Ti*
Ca, Zn	Cu, Fe, Mn, Zn
K, Na	Ag, Cr, Cu, Fe, Ni
Pt, Ru	Co, Cr, Cu, Mn, Ni
Sn, Te	Co, Cu, Fe, Mn, Mo
Al,† Ca, Mg	Ag, Al, † Cr, Cu, Fe, Mg
Al, † Ca, Mg*	Co, Cr, Cu, Fe, Mn, Ni
Ca, Mg, Zn	Al, † Ca, Cu, Fe, Mg, Si, † Zn

* This lamp has a quartz window. Choose a quartz-window version when using secondary wavelengths below 245 nm.

† This element requires a nitrous oxide–acetylene flame.

Source: Perkin-Elmer Corporation.

TABLE 8.7 Spectral Overlap in Flame Atomic Absorption

Sensitivity is defined here as the concentration (in μg · mL^{-1}*) required to produce an absorbance of 0.0044 at the specified wavelength; exceptions are detection limits for those figures marked with a dagger. All sources were hollow-cathode lamps except those market EDL (electrodeless discharge lamps). Source lines marked with an asterisk are resonance lines.*

Source	Emission wavelength, nm	Analyte	Absorption wavelength, nm	Separation, nm	Sensitivity, μg · mL^{-1}
Aluminum	308.215	Vanadium	308.211	0.004	800
Antimony	217.023	Lead	216.999	0.024	250
Antimony	217.919	Copper	217.894	0.025	100
Antimony	231.147*	Nickel	231.095	0.052	35
Antimony	323.252	Lithium	323.261	0.009	200
Arsenic (EDL)	228.812	Cadmium	228.802	0.010	45
Copper	324.754*	Europium	324.753	0.001	75
Gallium	403.298*	Manganese	403.307	0.009	15
Germanium	422.657	Calcium	422.673	0.016	6
Iodine (EDL)	206.163	Bismuth	206.170	0.007	10
Iron	271.903*	Platinum	271.904	0.001	40
Iron	279.470	Manganese	279.482	0.012	0.04†
Iron	285.213	Magnesium	285.213	< 0.001	10.0†
Iron	287.417*	Gallium	287.424	0.007	250
Iron	324.728	Copper	324.754	0.026	0.8
Iron	327.445	Copper	327.396	0.049	1.1†
Iron	338.241	Silver	338.289	0.048	150
Iron	352.424	Nickel	352.454	0.030	0.1†
Iron	396.114	Aluminum	396.153	0.039	50
Iron	460.765	Strontium	460.733	0.032	20
Lead	241.173	Cobalt	241.162	0.011	15
Lead	247.638	Palladium	247.643	0.005	3.5
Manganese	403.307*	Gallium	403.298	0.009	25
Mercury	253.652*	Cobalt	253.649	0.003	100
Mercury	285.242	Magnesium	285.213	0.029	200
Mercury	359.348	Chromium	359.349	0.001	250
Neon (EDL)	359.352	Chromium	359.349	0.003	15
Silicon	250.690*	Vanadium	250.690	< 0.001	65
Zinc	213.856*	Iron	213.859	0.003	200

Source: J. D. Norris and T. S. West, *Anal. Chem.* **46**:1423 (1974), and sources cited therein.

overcome by removing either the interfering element or the desired analyte or, if possible, by using another absorption line for the analysis. Some of the spectral overlaps listed in Table 8.7, those which produced the better atomic-absorption sensitivities, can be utilized to achieve useful atomic-absorption signals when attempting to measure higher concentrations of an analyte.

8.3.2.2 Electrodeless Discharge Lamps. Electrodeless discharge lamps (EDLs) consist of an element or a salt of the element sealed in a quartz bulb that contains an inert-gas atmosphere. The bulb is contained in a ceramic cylinder on which an RF coil is wound. When an RF field of sufficient power is applied, the inert gas is ionized and the coupled energy vaporizes the element and excites the atoms inside the bulb, resulting in the emission of the characteristic spectrum. EDLs and mounts usually are interchangeable with hollow-cathode lamps.

EDLs are typically much brighter and, in some cases, provide better sensitivity than comparable hollow-cathode lamps, and they are preferred for certain volatile elements. They offer better precision and lower detection limits and are preferred for analyses that are noisy due to weak hollowcathode emission. Lamps are available for these elements: As, Bi, Cd, Cs, Ge, Hg, K, P, Pb, Rb, Sb, Se, Sn, Te, Ti, Tl, and Zn.

8.3.3 Electrothermal Atomizers

Electrothermal atomization is usually performed with a graphite furnace although vertical crucible furnaces and open filaments have been employed. Electrothermal (nonflame) atomizers offer several attractive features that complement flame AAS.

1. Only small amounts (10^{-8} to 10^{-11} g absolute) of analyte are required.
2. Solids can be analyzed directly, often without any pretreatment.
3. Small amounts of liquid samples, 5 to 100 μL, are needed.
4. Background noise is very low.
5. Sensitivity is increased because the production of free analyte atoms is more efficient than with a flame.

On the negative side, matrix effects are usually more severe, and precision, typically 5% to 10%, compares unfavorably with that of AAS and FES.

8.3.3.1 ***Principle.*** After insertion of the sample onto a platform mounted within the electrothermal atomizer, a heating sequence is initiated to take the sample through three steps: dry, ash or char, and atomize.

1. In the drying cycle, the sample is heated for 20 to 30 s at 110°C to evaporate any solvent or extremely volatile matrix components.

2. The ash or char step is performed at an intermediate temperature (often 500°C) to volatilize higher-boiling-point matrix components and to pyrolyze matrix materials that will crack and carbonize. The latter would be fats and oils. Loss of analyte may occur if the ashing temperature is too high or is maintained for too long. Recommended ash temperatures for selected elements are given in Table 8.8.

3. In the atomization step, maximum power is applied to raise the furnace temperature as quickly as possible to the selected atomization temperature or the maximum furnace temperature. The analyte residue is volatilized and dissociated into free atoms that will absorb light from the AAS source. The transient absorption signal must be measured rapidly.

8.3.3.2 ***Instrumentation.*** Three components make up an electrothermal atomizer: the workhead, the power unit, and the controls for the inert-gas supply. A heated graphite atomizer consists of a hollow graphite cylinder 28 mm long and 8 mm in diameter, the interior of which is coated with pyrolytic graphite. Electrodes at each end of the cylinder are connected to a power supply. A metal housing surrounding the furnace is water-cooled to allow the entire unit to be rapidly restored to ambient temperature after each run. Inert gas, usually argon, enters the cylinder at both ends and exits through the sample introduction port at the center of the cylinder. The gas flow removes the matrix components vaporized during the ashing step (to prevent subsequent vaporization during the atomization step and thereby a large background absorption signal) and prevents oxidation of the graphite cylinder during the heating cycles.

Liquid samples are introduced with a microsyringe through the small opening in the top of the cylinder and placed on a thin graphite plate or platform that is located within the cylinder. Care must be taken not to scratch the pyrolytic graphite surface. Solid samples can be introduced through the end of the cylinder on a microdish made of tungsten. The sample is heated by radiation from the walls of the cylinder, which enables the walls and vapor to reach a steady-state temperature before the sample is atomized.

TABLE 8.8 **Ash Temperatures with a Graphite Furnace**

Element	Recommended ash temperature, °C	Ash temperature with Pd modifier, °C
Antimony	800	1400
Arsenic	800*	1500
Bismuth	500	1100
Cadmium	300	550
Chromium	1300	1300
Cobalt	900	1200
Copper	900	1100
Gold	700	1100
Iron	800	1300
Lead	400	1000
Manganese	800	1200
Mercury	120	450
Nickel	900	1200
Selenium	700*	1100
Silver	500	950
Tellurium	500	1300
Tin	800	1300
Thallium	400	1500
Zinc	400	900

* Nickel modifier added.

A miniature version of the electrothermal analyzer is the carbon rod atomizer. A tube or cup unit is supported between two graphite electrodes, the outside ends of which are inserted in water-cooled terminal blocks. The tube is 9 mm long and 3 mm in diameter; sample capacity is 10 μL for smooth tubes. The tube can be replaced with a vertical cup for solid samples or samples that require preliminary chemical treatment that is performed directly in the cup. Detection limits are given for many elements in Table 8.4. Operating conditions are given in Table 8.9.

8.3.3.3 ***Matrix Modifier.*** Palladium(II) nitrate is a useful chemical modifier that stabilizes many elements to temperatures several hundred degrees higher than possible otherwise (Table 8.8). Palladium modifies the analyte element by converting it to a more thermally stable species. This lessens analyte volatility and permits a higher temperature for the ash step, which more efficiently removes bulk matrix constituents without loss of analyte element. Thus palladium reduces signal dependency on the sample composition and produces greater consistency in absorption analyte signals. The greatest temperature shifts are achieved for these elements: As, Bi, Ga, Ge, P, Pb, Sb, Se, Te, and Tl.

The palladium concentration is an important parameter, requiring careful investigation in methods development. It should be added as the nitrate (not as chloride). For relatively clean samples, lower levels of palladium (50 to 200 mg $\cdot$ L^{-1}) may be sufficient to shift analyte atomization away from the sample constituents causing background and chemical interferences. For more complex samples, higher concentrations of palladium (200 to 1000 mg $\cdot$ L^{-1}) may be necessary. If analyte signals produce late, broad, and irregular peaks, a higher atomization temperature may also be necessary. Hydrogen gas (5% hydrogen in 95% argon) is introduced during the ramp to the ash step from the dry step to assist in producing metallic palladium. Argon is then used for the ash and atomization steps. Magnesium nitrate (1000 mg $\cdot$ L^{-1}) has been combined with palladium(II) nitrate and hydrogen gas for the determination of a variety of elements.

8.3.3.4 ***Precautions.*** The proper temperature and timing parameters must be carefully selected for each step of the electrothermal process. The progress of each step should be observed by monitoring the absorption signal without background correction.

TABLE 8.9 Operating Conditions for the Determination of Elements by FES and AAS

Aluminum
Excitation conditions for FES are optimum with aqueous solutions when using a 4-mm red feather zone of nitrous oxide–acetylene flame. In AAS the fuel-rich, laminar nitrous oxide–acetylene flame is used with the red zone 20 mm in height. Sensitivity is roughly the same for the 396.15-nm line and for the unresolved 309.28-nm doublet. For the latter doublet when used in AAS, the band pass should not exceed 0.2 nm to minimize the intense flame background. The latter doublet provides the better signal-to-noise ratio. The detection limits are 100 times better in graphite-furnace (GFAAS) for the 396.15-nm line and 500 times better for the 309.28-nm doublet.
Unless suppressed with deionizers, the ionic lines of Ca at 393.3 and 396.8 nm interfere with the aluminum doublet in this region. In a nitrous oxide–acetylene flame, aluminum undergoes about 15% ionization, which can be suppressed if samples and standards contain 1 to 2 $mg \cdot mL^{-1}$ of an easily ionizable alkali salt.
Silicate has a slight depressive effect (about 4%) upon the aluminum emission in FES. In AAS the V 308.21-nm line interferes seriously, as does a 10-fold excess of Fe and chloride.
Antimony
For AAS a fuel-rich air–acetylene flame is slightly preferable to a nitrous oxide–acetylene flame. The best detection limit is achieved with the 217.58-nm line. Using the 259.81-nm line in FES, a light guide, and a mixture strength of 0.57, the optimum region of observation is 3 mm below the tip of the inner cone of the oxygen–acetylene flame. The GFAAS offers the best detection limit with the 259.81-nm line.
Copper exhibits a significant absorption at 217.58 nm in AAS. A spectral interference arises from the Pb line at 217.00 nm; the Sb 206.84- or 231.13-nm line should be used in the presence of lead. The acid concentrations must be matched in samples and standards. Chemical vaporization as SbH_3 overcomes many problems in the determination of trace amounts of antimony.
Arsenic
In FES the lines most frequently used are at 228.8 and 234.90 nm. A hydrocarbon species in the flame is required, supplied either by the fuel gas or by an organic solvent.
The most sensitive lines in AAS are at 189.04, 193.76, and 197.20 nm. To minimize absorption by the flame gases, a helium (entrained air)–hydrogen flame is preferable. Otherwise, a lean air–acetylene flame is used; the line at 193.76 nm is best. Serious interference arises from these elements: Al, Cr, Mg, Mn, Mo, Ni, Sn, Ti, and nitrate. The best approach is chemical vaporization as AsH_3 into a GFAAS as this procedure offers the maximum sensitivity and freedom from interferences.
Barium
In a nitrous oxide–acetylene flame the doublet ionic lines at 455.56 and 493.41 nm appear simultaneously with the resonance line at 553.55 nm. When using the 553.55-nm line in FES or AAS, it is necessary to use a radiation buffer to control the degree of ionization (approximately 90%). Although the chemical interferences from aluminum, phosphate, and silicate is lessened in the nitrous oxide–acetylene flame, the use of a lanthanum releasing agent is advised.
The determination of barium in lubricating oils is frequently performed by FES or AAS.
Beryllium
In AAS use the 234.86-nm line with a fuel-rich nitrous oxide–acetylene flame. The same line is used in GFAAS for which the sensitivity is significantly increased. Serious spectral interference arises from Co, Ni, Si, Sn, and V. A high concentration of aluminum suppresses the beryllium absorption slightly but this action is eliminated by fluoride ion and 8-hydroxyquinoline.
Bismuth
Because the 306.77-nm line coincides with the band head of the strongest OH band in the flame, the line at 223.06 nm is preferred for most AAS applications; however, this line cannot be used if copper is present in the hollow-cathode lamp as there are copper lines at 222.89, 223.01, and 223.09 nm. For work in GFAAS, the beryllium lines at 223.06, 227.66, or 306.77 nm are useful with the first and last mentioned the more sensitive lines.
Boron
The emission lines of boron are difficult to observe in a flame but do appear in fuel-rich acetylene flames and are strongest within the inner cone. For the BO_2 band heads at 492, 518, and 545 nm, adequate sensitivity is obtained only in nonaqueous solvents or in 50% methanolic solutions. When interferences occur, it usually is spectral in nature. General interference at the prominent band systems and the intervening minima is observed from Ba, Cr, Co, Fe, Mn, and Ni. Coincidences occur with magnesium and aluminum at the 492- and 518-nm bands. The calcium band systems cover the band at 546 nm and overlap the 518-nm band. Weak potassium emissions occur at the 505-nm minimum and at the 546-nm band unless a band pass 0.35 nm or less is used.

(Continued)

TABLE 8.9 Operating Conditions for the Determination of Elements by FES and AAS (*Continued*)

Boron

The principal applications of the AAS determination of boron have been to fertilizers, food, plants, and biological materials. Since the two resonance lines at 249.68 and 249.77 nm are separated by only 0.09 nm, a narrow band pass is necessary to isolate the desired 249.77-nm line. Usually boron is determined after separation using 20% 2-ethyl-1,3-hexanediol in a 4-methyl-2-pentanone solvent. The organic phase is nebulized into a fuel-rich nitrous oxide–acetylene flame.

Cadmium

The most sensitive emission line of cadmium lies at 228.80 nm. Because it lies in the far-ultraviolet region and is strongly self-absorbed, it finds more use in AAS than in FES. Another line at 326.11 nm is the line generally used in FES. In FES the air–hydrogen flame gives the highest absolute emission intensity at 326.11 nm and this flame has the lowest background emission. However, the nitrous oxide–acetylene flame provides reasonable sensitivity in both FES and AAS. In AAS cadmium hollow-cathode lamps must be operated at relatively low currents to prevent self-absorption. The ease of thermal atomization of cadmium renders it particularly suitable for GFAAS.

Spectral interferences arise from the As line at 228.81 nm in FES and AAS, also sodium at this wavelength. At 326.11 nm, the Cu line at 327.40 nm interferes when the band pass is 0.64 nm or larger. A Co line at 326.1 nm offers serious interference. In AAS only silicon exhibits any chemical interference.

Calcium

The atomic resonance line of Ca at 422.67 nm is free from discrete line or band interference arising from other elements except for the Ge line at 422.66 nm, but is subject to interference in FES from increased background or stray light at Na concentrations exceeding 100 μg/mL. For AAS measurements with decreased sensitivity, the 239.86-nm line may be used.

The most common chemical interferences are eliminated when the nitrous oxide–acetylene flame is used and La, Sr, or EDTA are added as releasing agent. With this flame KCl should be added to suppress calcium ionization.

Cesium

Either atomic emission line, 852.11 or 894.35 nm, is suitable for FES; only the former is useful for AAS. The Cs vapor discharge lamp should be operated at the lowest current consistent with good stability to minimize line broadening and reversal due to self-absorption. Since Cs is easily ionized, even in an air–acetylene flame, a large excess of potassium must be added to suppress Cs ionization. Interference in AAS occurs in the presence of high concentration of mineral acids.

Chromium

Although the 357.87-nm line is the most sensitive of the ultraviolet triplet, it lies in a region of cyanogen band emission. The AAS sensitivity in the air–acetylene flame is dependent upon the flame stoichiometry. Fuel-rich flames produce highest sensitivity, but interference is observed from Fe, Ni, and other elements although these interferences may be suppressed by the addition of NH_4Cl or NH_4HF_2. Although lower sensitivity is observed with the nitrous oxide–acetylene flame, fewer chemical interferences are encountered. The Cr 425.44-nm line is suitable for either FES or AAS.

With argon-filled hollow-cathode lamps, the Cr line at 357.87 nm may suffer interference from Ar lines at 357.66 and 358.23 nm. A neon-filled lamp is best, or the Cr 359.35-nm line should be used.

Cobalt

Cobalt is atomized in an air–acetylene flame, and is not subject to any significant interferences. The most sensitive line in AAS and FES is at 240.73 nm. For optimum sensitivity in AAS it is necessary to resolve the Co 240.73-nm line from nonabsorbing cobalt lines at 240.77 and 240.88 nm as they produce strong curvature in the calibration graph.

Copper

The resonance lines of copper at 324.75 and 327.40 nm are the most intense in FES and the most sensitive in AAS. In a nitrous oxide–acetylene flame the atomization of copper salts is complete. Spectral interferences in FES are possible from Cd 326.11-nm, Co 326.1-nm, and Ni 324.3-nm lines unless a band pass of 0.7 nm or less is used. The Ag 328.0-nm line will interfere with the Cu line at 327.40 nm if the band pass is greater than 0.3 nm.

There is no significant effect in FES or AAS from chemical- or physical-type interferences when using a nitrous oxide–acetylene flame.

Europium

The Eu line at 459.40 nm is suitable for FES and AAS. Little work has been reported on interferences.

TABLE 8.9 Operating Conditions for the Determination of Elements by FES and AAS (*Continued*)

Gallium

There are two usable emission lines, members of the same doublet: at 403.30 and 417.21 nm. A slightly reducing (3-mm red zone) nitrous oxide–acetylene flame gives the best detection limit with the maximum signal appearing near the tip of the inner cone.

In AAS a very lean nitrous oxide–acetylene flame is best. The line at 287.42 nm is most sensitive; however, the doublet at 294.4 nm and the line at 417.21 nm are also useful.

Germanium

The most satisfactory line for FES or AAS is at 265.12 nm. A slightly fuel-rich nitrous oxide–acetylene flame with a long-path burner is used for both methods to dissociate a stable monoxide (GeO). A variety of cations depress Ge absorbance but this reduction does not exceed 10% in the flame recommended.

Gold

The emission intensity of gold is rather weak. In AAS the resonance lines at 242.80 and 267.60 nm are the most sensitive. A serious depression in the AAS signal arises when the sulfuric acid concentration exceeds 0.25*M* and concentrations of hydrochloric and nitric acids exceed 0.5*M*. Since large amounts of lead and tellurium do not interfere, gold can be concentrated by coprecipitation with Te or PbS. There are no reported interferences unless very high concentrations (<10 mg/mL) of extraneous metals are present.

Indium

The emission lines at 410.18 and 451.13 nm are strongest, followed by lines at 325.61 and 303.94 nm. In emission a slightly reducing (2-mm red zone) nitrous oxide–acetylene flame is recommended. For AAS the line at 325.61 nm is most sensitive. Indium is about 30% ionized in a nitrous oxide–acetylene flame; therefore, potassium should be added as an ionization suppressor. Bismuth causes depression in AAS.

Iron

Iron is atomized (66%) satisfactorily in an air–acetylene flame and is free from interferences except by silica, the effect of which can be eliminated by the addition of $CaCl_2$. Differences between fuel-rich, stoichiometric, and lean flames are slight.In AAS the most sensitive line is at 248.33 nm but requires a monochromator with good resolution in order to prevent transmission of nearby weakly absorbing lines from the source. Slightly less sensitivity is achieved in a nitrous oxide–acetylene flame.

Iron is only slightly ionized in an air–acetylene flame. In a nitrous oxide–acetylene flame, iron ionization is substantial and an ionization buffer (usually potassium) should be added. Chemical interferences abound in an air–acetylene flame; some may be handled by the standard addition technique, others by use of releasing agents. Interference from Co, Cu, and Ni can be overcome by the addition of a 5% solution of 8-hydroxyquinoline. Chemical interferences can generally be avoided in FES with a nitrous oxide–acetylene flame.

Chemical interference problems are best overcome by extraction with 4-methyl-2-pentanone from 4*M* to 8*M* HCl; the iron can be stripped from the organic solvent with 0.001*M* HCl or the organic phase can be aspirated directly.

Lanthanum

Only FES offers a suitable method for lanthanum. The line at 579.13 nm offers the best sensitivity although the line at 550.13 nm may be useful. Extraction with acetylacetone in 4-methyl-2-pentanone circumvents most interferences.

Lead

The analytically significant line for FES is located at 405.78 nm, followed closely in sensitivity by lines at 368.35 and 283.31 nm. In AAS the lines most often used are those at 217.00 and 283.31 nm. An air–acetylene flame is adequate for the atomization of the metal, and there is little interference from other metals. Some of the common anions exhibit chemical interference; these effects can be overcome by the addition of EDTA.

For FES the Pb 405.78-nm line is subject to spectral interference from Mn 403.08- and 404.41-nm lines unless a very narrow spectral band pass is employed.

For AAS a neon-filled hollow-cathode lamp provides a much simpler background spectrum, whereas with an argon-filled lamp a spectral interference is possible with the Pb 217.00-nm line. The elements Al, Be, Th, and Zr produce a significant chemical interference, due most likely to formation of refractory compound in the flame. Anions that give precipitates with lead interfere. In most instances the addition of EDTA prevents the interference; only chloride, phosphate, and peroxodisulfate continue to diminish the absorption signal. A number of solvent extraction systems overcome these chemical interferences.

(*Continued*)

TABLE 8.9 Operating Conditions for the Determination of Elements by FES and AAS (*Continued*)

Lithium
Lithium shows great sensitivity in FES, a technique widely used in its determination. The most sensitive line is at 610.36 nm; lines at 670.78 and 460.29 nm are also very sensitive. An air–butane flame atomizes lithium satisfactorily, and the ionization is small and easily overcome by the addition of another easily ionized metal such as an alkali metal. The chief spectral interference is the SrOH band with a diffuse maximum at 671.0 nm. With narrow spectral bandwidths there are no significant interferences from line coincidences with Li at 670.78 nm. Spectral interference on the Li 460.29-nm line can arise from the strong Sr 460.73-nm and the weak Cr 460.08-nm lines. For AAS the preferred source for lithium is the hollow-cathode lamp. Both neon- and argon-filled lamps are available. With a neon filler gas, a glass filter should be employed to prevent second-order interference from the Ne 333.51-nm line when a low-dispersion grating monochromator is used.
Magnesium
The usual line used in FES and AAS lies at 285.21 nm. No spectral interferences have been reported. The air–acetylene flame is the most satisfactory. However, common chemical interference from aluminum and silicon is observed, and high concentrations of phosphate have a slight effect on the absorbance of magnesium. Lanthanum and strontium are effective as releasing agents. The use of the nitrous oxide–acetylene flame eliminates these interferences. The use of the ionic line at 279.55 nm for AAS is subject to interferences by alkali-metal ions and any other easily ionizable metal that depress the ionic population of magnesium.
Manganese
The line at 279.48 nm is used most often for AAS, while the line at 403.08 nm (part of a triplet) is usually used for FES although the line at 279.48 nm is almost as sensitive. Silicon suppresses the manganese signal unless $CaCl_2$ is added. The usual flame used in AAS is a fuel-rich air–acetylene mixture. For FES a fuel-lean nitrous oxide–acetylene flame is recommended.
Mercury
Mercury is determined by GFAAS or chemical vaporization using the line at 253.65 nm. It is unnecessary to employ a flame. Mercury compounds are reduced to elemental mercury with sodium borohydride, and mercury vapor is forced through a long-path absorption tube for measurement by AAS.
Molybdenum
For AAS the line at 313.26 nm is recommended for maximum sensitivity although the 390.30-nm line is useful. In the nitrous oxide–acetylene flame molybdenum emission is best measured at 390.30 nm. Vanadium and chromium give spectral interference at 390.30 nm. With a fuel-rich flame, chemical interferences are minimal. A fuel-rich air–acetylene flame or nitrous oxide–acetylene flame is required for the atomization of Mo for AAS. The absorption signal is quite dependent upon the height of observation in the flame. Interferences are numerous and include bromide, iodide, fluoride, sulfate, sulfite, nitrate, phosphate, selenium, tellurium, vanadium, silicon, arsenic lanthanum, calcium, strontium, barium, cobalt, nickel, palladium, copper, zinc, cadmium, mercury, tin, iron, chromium, lead, antimony, cerium, uranium, titanium, aluminum, zirconium, and beryllium. Addition of large amounts of ammonium chloride eliminates some of the chemical interferences. Solvent extraction methods are recommended for isolation of molybdenum from other troublesome metals and anions.
Nickel
In AAS the best sensitivity for nickel is obtained with a fuel-lean air–acetylene flame at 232.00 nm. Nickel is known for its nonlinear calibration curve unless the nonresonant nickel lines at 232.2 and 231.7 nm are removed by a high-resolution monochromator. The monochromator must also resolve the desired 230.00-nm line from the ionic nickel line at 231.99 nm that is emitted from the hollow-cathode lamp. The line at 342.48 nm does not suffer from these problems. A nitrous oxide–acetylene flame can be used for nickel, but with a loss of sensitivity. The addition of an ionization suppressor such as potassium is necessary. For FES, the lines at 232.00 and at 352.45 nm give the same sensitivity. Addition of potassium is necessary to suppress ionization. When a multielement lamp containing Ni and Fe is employed in AAS, interference at 232.00 nm may result when Ni is determined in a Fe matrix. In this case Ni lines at 352.45 or 341.48 nm should be used. Spectral interference from Co lines at 352.7 and 352.6 nm affects the Ni line at 352.45 nm unless a high-dispersion monochromator is used.

TABLE 8.9 Operating Conditions for the Determination of Elements by FES and AAS (*Continued*)

Palladium
For AAS the Pd line at 244.8 nm is favored, although a narrow band pass should be used to avoid transmission of the less sensitive Pd lines at 244.5 and 245.0 nm. Use of the fuel-lean air–acetylene flame is preferred. Few interference effects are observed and they can be suppressed by the addition of Cu or La. For FES any of the lines at 244.8, 340.46, or 363.47 nm are useful. There is interference from Ca, Fe, Mg, and Na. Sensitivity is improved and interferences eliminated by extracting Pd as the salicylaldoxime chelate into 4-methyl-2-pentanone and spraying into the flame.
Platinum
The platinum line at 265.95 nm is employed in AAS with nitrous oxide–acetylene flame. Interference from other noble metals can be suppressed by the addition of a high concentration of copper of a 1% lanthanum solution.
Potassium
The lines at 766.40 and 769.90 nm are used in AAS. The determination of potassium is relatively free from chemical interferences when the air–acetylene flame is employed. There is a suppression of the signal when a high concentration of mineral acids are present that can be compensated for by matching the samples and standards with respect to the acid concerned. In both AAS and FES an ion suppressor, such as cesium or sodium, should be present. For FES the doublet at 404.41 and 404.72 nm is useful although it is about one-tenth the sensitivity of the red doublet at 766.40 and 769.90 nm.
Rhodium
The line at 343.49 nm is used in FES and AAS. Spectral interference is not a problem with this line or the line at 369.24 nm. Although a lean air–acetylene flame gives the highest sensitivity, it also suffers from interference from the noble metals and mineral acids. These problems are overcome when using a nitrous oxide–acetylene flame except for Ir and Ru, which interfere seriously, but the interference can be suppressed using a 0.5% solution of zinc.
Rubidium
The most sensitive line is at 780.02 nm. Since rubidium is easily ionized, a low-temperature flame is best and the addition of KCl is required to suppress Rb ionization, but in so doing it is necessary to use a narrow bandwidth to avoid interference from the potassium doublet at 766.49 and 769.90 nm. Strong LaO bands between 740 and 780 nm and the strong CrO bands with band heads at 777.81 and 781.21 can affect Rb determination in FES. The alternative Rb doublet at 420.19 and 421.56 nm is unaffected by La or Cr, but Ca at 422.67 nm may interfere spectrally.
Scandium
A hot fuel-rich flame is required to permit detection of scandium absorption. Scandium ionization in the nitrous oxide–acetylene flame is controlled by the addition of excess potassium salts. In AAS fluoride and sulfate interfere. The line at 391.18 nm shows fair sensitivity in emission, less so in absorption.
Silver
Silver is a particularly easy metal to determine by AAS. It is atomized satisfactorily in an air–propane flame (although an air–acetylene flame is frequently used), shows high sensitivity, and is not subject to interferences except as noted later. In both FES and AAS, the two lines at 328.07 and 338.29 nm are the strongest. The line at 338.29 nm gives a more linear calibration curve for higher concentrations of silver. A band pass of 0.3 nm or less is necessary to isolate the silver line at 328.07 nm from the copper line at 327.40 nm. In AAS an appreciable reduction in signal occurs in the presence of Al and Th, and phosphoric and sulfuric acids depress the absorbance.
Sodium
The most sensitive sodium line is at 589.00 nm; it should be isolated from the 589.59-nm line using a 0.2- to 0.3-nm band pass to avoid curvature in the calibration curve. Argon is preferred to neon filler gas in hollow-cathode lamp sources as neon emits several lines in close proximity to the Na lines. Air–propane or air–acetylene flames are equally satisfactory, although less ionization occurs in the former flame. In any case, an ionization buffer composed of K or Cs salts is essential to avoid signal

(Continued)

TABLE 8.9 Operating Conditions for the Determination of Elements by FES and AAS (*Continued*)

Sodium
changes from other alkali metals in a sample. In the presence of high concentrations of Ca the emission and absorption produced by CaOH molecules in the flame may cause interferences and deterioration of the signal-to-noise ratio at the sodium lines.
Strontium
Fuel-rich conditions provide highest atomization efficiency and therefore good absorption and emission sensitivity when the air–acetylene flame is employed. The stoichiometric nitrous oxide–acetylene flame provides somewhat higher sensitivity. With any flame, alkali-metal addition must be used to suppress the substantial ionization in these flames. The strongest strontium line is at 460.73 nm for which no spectral interferences have been reported. When using the lower-temperature flames, it is advisable to add lanthanum to remove chemical interference effects from phosphate, silicon, and aluminum.
Sulfur
The high atomization efficiency obtained for sulfur in the nitrous oxide–acetylene flame allows AAS measurements at 180.73 nm with considerable freedom from chemical interferences. The S_2 emission bands are best excited with a nitrogen (entrained air)–hydrogen flame. Alternately, a fuel-rich air–hydrogen flame may be cooled by a Pyrex shield surrounding the flame. The strongest bands peak at 384.0 and 394.0 nm; bands of successively less intensity fall away on both sides and are approximately equally spaced (10 nm apart). These are the bands utilized in the gas chromatographic sulfur detector.
Tellurium
For AAS the line at 214.27 nm is most strongly absorbed. However, the neighboring and very intense nonabsorbing line at 214.7 nm will cause a reduction in sensitivity if the spectral band pass exceeds 0.7 nm. The 238.58-nm line is used in FES. Slight depressive effects were observed for Cd, Co, Cr, Cu, Mn, Ni, Pb, and Sb when present in 40-fold weight excess.
Thallium
Thallium has two sensitive lines in emission at 377.57 and 535.05 nm. The best sensitivity in FES is with a nitrous oxide–acetylene flame with a 4-mm red zone. The thallium atoms are present as the free atoms. However, the thallium is ionized about 40% in a nitrous oxide–acetylene flame; therefore, an ionization suppressor must be added to samples and standards. Many elements and anions appear to enhance the absorbance of the line at 276.78 nm, some undoubtedly through suppression of the thallium ionization. Enough interferences remain to warrant a preliminary separation of thallium, perhaps as the bromide complex into 4-methyl-2-pentanone or 2-octanone or as a chelate with diethyldithiocarbamate.
Thulium
Thulium requires a nitrous oxide–acetylene flame for adequate atomization. Suppression of ionization is required; it is accomplished with easily ionized metals such as the alkali metals. The line at 371.79 nm is the most sensitive in FES and AAS.
Tin
The nitrous oxide–acetylene flame is required to dissociate SnO effectively but, if used, potassium or cesium should be added as an ionization suppressant. With this flame the line at 235.48 nm offers the best sensitivity in AAS, followed by the lines at 286.33 and 284.00 nm. The 284.00-nm line is the best line for FES. Chemical interference problems with lower-temperature flames are often completely absent in the nitrous oxide–acetylene flame.
Titanium
Although the line at 364.27 nm has been most frequently recommended for absorption work, the 365.35-nm line is actually slightly more sensitive and exhibits equally intense hollow-cathode radiation. Both lines are free from flame background noise and spectral interferences. The 399.86-nm line provides the best sensitivity for flame emission in nitrous oxide–acetylene flame, while 398.98- and 365.35-nm lines are recommended as alternative lines. The optimum band pass is 0.08 nm. The nitrous oxide–acetylene flame is widely used for both FES and AAS. Slightly fuel-rich conditions are normally required for optimum AAS sensitivity; sensitivity is markedly affected by oxidant-to-fuel ratio. For the best FES sensitivity, a slightly less fuel-rich flame than that required for AAS should be used. Titanium is about 15% ionized but an equal amount of potassium is sufficient to prevent ionization of thallium. A number of elements interfere under certain conditions with the determination of titanium. Addition of aluminum chloride to samples and standards eliminates most of these interferences. In FES spectral interference occurs from the vanadium line at 399.87 nm although it can be excluded using a 0.033-nm band pass. Zirconium interferes at 398.98 nm, while chromium and cobalt interfere at 365.35 nm.

TABLE 8.9 **Operating Conditions for the Determination of Elements by FES and AAS (*Continued*)**

Uranium
In FES the line at 358.49 nm provides fair sensitivity. Ionization of uranium must be suppressed by adding potassium or cesium to samples and standards. A rich nitrous oxide–acetylene flame is required.
Vanadium
A fuel-rich nitrous oxide–acetylene flame is required for atomization of vanadium. In AAS the best sensitivity is obtained at 318.54 nm. Since vanadium is ionized about 10%, an excess of an ionization buffer is required to prevent enhancement of absorption when alkali metals are present. Enhancement of absorption is also observed in the presence of aluminum and titanium. In FES the best sensitivity is obtained with the triplet (318.34, 318.40, and 318.54 nm) and the line at 437.92 nm. A 0.033-nm band pass will resolve the first two triplet lines from the 318.54-nm line.
Ytterbium
The line at 398.90 nm provides the best sensitivity in FES and AAS. Ionization in the nitrous oxide–acetylene flame is suppressed by the addition of excess KCl to sample and standard solutions.
Yttrium
A fuel-rich nitrous oxide–acetylene flame is required. Since appreciable ionization occurs in this flame, potassium or cesium should be added to standard and sample solutions to suppress this effect. The line at 362.09 nm is more sensitive in AAS, whereas the 410.24-nm line is slightly better for FES and several times less sensitive for AAS. Some workers report interference from Al, Ce, and Si in AAS at 410.24 nm.
Zinc
The emission intensity of the zinc line at 213.86 nm is too weak for practical work. Consequently, AAS is the method of choice for zinc. Zinc in an air–acetylene flame is not subject to any significant interferences. However, molecular absorption and scattering are appreciable at 213.86 nm; a small correction may be needed when measuring traces of zinc in the presence of large concentrations of other material.

In the drying step the evaporation of solvent must be smooth and gentle. Foaming and splattering will result in loss of analyte. If observed, decrease the heating rate.

In the ashing step, most organic materials pyrolyze at around 350°C. If a residue of amorphous carbon (which resembles a spider web in appearance) is formed, a stream of air or oxygen may be introduced into the furnace to convert the carbon into carbon dioxide. Low-boiling-point elements, such as lead, may be partially lost if this step is prolonged unnecessarily or the final temperature is too high.

The atomization step must be rapid. The analyte should be converted into free atoms almost instantly so that the absorption peak is sharp.

Not every AAS spectrometer possesses a response time adequate to measure the transient absorption signal from an electrothermal analyzer. Lifetime of the free atoms in the optical path is 0.01 s or less.

The operational life of the graphite cylinder surface or tube is finite. A pyrolytic graphite coating prolongs the life and also prevents the sample from soaking into ordinary graphite. Formation of metal carbides by some elements is also prevented.

8.3.4 Chemical Vaporization

For elements (As, Ge, Sb, Se, Sn, and Te) that form volatile hydrides, the samples can be pretreated with sodium borohydride, dispensed in pellet form. The reducing agent is added to a 1*M* to 4*M* HCl solution. The gaseous hydride is released and injected into the atomizer in a stream of inert gas.

Either a hydrogen–air flame and conventional AAS instrument or a low-temperature tube-type quartz furnace is used to generate the gaseous free metal atoms.

Mercury compounds need only be heated to generate mercury vapor, although chemical reduction is an alternative.

8.3.5 Atomic-Absorption Spectrophotometer

The typical focal length is 0.33 to 0.50 m with a Czerny–Turner of Ebert configuration. Two gratings, mounted back to back on a turntable, cover the wavelengths of 190 to 440 nm and 400 to 900 nm, respectively. Wavelengths should be keyboard- or switch-selectable. In the keyboard mode, the instrument will be driven by a multitasking graphical interface featuring windows and pull-down menus. This allows the user to monitor the progress of any automated sequence including signal graphics, calibration graphics, method details, current solution results, and even the printed report simultaneously. Each module guides the operator through the pages needed to create and run methods and sequences. When used with a compatible sampling and dilution system, a number of rack locations permit up to 80 samples per rack (times the number of racks) to be loaded. With a diluter, over-range samples can be automatically diluted until they are within the calibration range.

The automatic burner adjuster allows the viewing height to be reset to the optimum viewing height for flame, furnace, or vapor. Once the correct height is determined, it is stored in the computer along with optimum gas flows and other method parameters.

8.3.6 Background Correction in Line-Source AAS

Background absorption occurs in AAS as molecular absorption and as light scattered by particulate matter (unevaporated droplets, but more likely unevaporated salt particles remaining after desolvation of the aerosol). The absorption is usually broad in nature as compared to the monochromator band pass and line-source emission widths. The analyte signal could be on a sloping or level background. Background correction is often required for elements with resonance lines in the far-ultraviolet region and is essential to achieve high accuracy in determining low levels of elements in complex matrices. Electrothermal atomization methods invariably require correction, and the background may change with time.

8.3.6.1 Use of a Continuum Source. This form of background correction uses a deuterium lamp for elements in the range of 180 to 350 nm and a tungsten–halide lamp for elements in the range of 350 to 800 nm. Both lamps are readily available. The continuum source is aligned in the optical path of the spectrometer so that light from the continuum source and light from the primary lamp source are transmitted alternately through the flame (or axis of the electrothermal analyzer tube). The continuum source has a broadband emission profile and is only affected by the background absorption, whereas the primary lamp source is affected by scatter and background absorption as well as by absorption resulting from the analyte in the flame. Subtraction of the two signals (electronically or manually) eliminates or at least lessens the effect of background absorption. This method does not correct for problems of scatter or spectral interferences. Highly structured molecular absorbance causes incorrect results. It is difficult to match the reference and sample beams exactly because different geometries and optical paths exist between the two beams. Most systems only correct up to 0.5 AU (absorbance unit).[2]

8.3.6.2 Smith–Hieftje Method. The Smith–Hieftje method[3] utilizes the phenomenon of self-reversal, which will occur when the light source is operated at high currents. In practice a hollow-cathode lamp line source is cycled at periods of low current for several milliseconds and then cycled at high current for several hundred microseconds. The time must be precisely controlled so

[2] S. R. Koirtyohann and E. E. Pickett, *Anal. Chem.* **37**:601 (1965).

[3] S. B. Smith and G. M. Hieftje, *Appl. Spectrosc.* **37**:419 (1983).

that the same number of photons reaches the detector during each phase of the cycle. At high currents the absorption signal caused by the analyte atoms is eliminated through self-reversal, but the background is not changed. The background signal can then be subtracted electronically to produce an accurate correction for structural background. Visible and ultraviolet regions are corrected. Optical alignment is simple and matching of two light sources is not required. Special hollow-cathode lamps are required to prevent gas cleanup and to avoid burnout at high currents. Not every hollow-cathode lamp (particularly for refractory elements) will produce a usable self-reversal effect. Sensitivity is reduced.

8.3.6.3 Zeeman–Effect Method. When under the influence of a magnetic field, the absorption line splits into three components for the normal Zeeman effect. The wavelength of one component coincides with the resonance emission line from the light source and the other components shift to shorter and longer wavelengths, respectively. When the light beam from the lamp is passed through a polarizer rotated so that the light is parallel to the central component, the emission line from the lamp is absorbed by any analyte atoms present in the flame, whereas the two wing components are unaffected. As the polarizer is rotated to the perpendicular position, there is no absorption by the atomic vapor of the sample. However, light scattering and broadband molecular absorption are measured in both polarizer configurations. Therefore, by using parallel emission lines as the sample beam and perpendicular emission lines as the reference beam, electronic subtraction of the two absorbances will produce the true absorbance of the analyte.[4]

The Zeeman effect can be produced in the light source [direct Zeeman atomic absorption (ZAA)] or in the atomic vapor (flame or furnace, inverse ZAA). The magnet placement around the light source in the direct ZAA imposes no restrictions on the atomizer and is compatible with any atomization technique, but it is generally inferior to inverse ZAA in baseline drift and dynamic range because it requires hollow-cathode lamps that are designed to operate in a magnetic field. Baseline drift may be greater than if the magnet is placed around the atomizer, as in the inverse ZAA. In the latter configuration, the magnetic system must be compatible with the atomizer and will be bulkier. Accuracy depends on background absorbance being unaffected by the magnetic field strength and polarization of the source. The polarizer contributes to the light loss. However, the same wavelength is used for the signal and the reference beam, which minimizes spectral interference.

Spectral interferences may be overcome if lines are 0.02 nm or more apart. Analytical calibration curves may be double-valued, that is, two widely different concentrations may give the same absorbance. Of the three correction systems discussed, it is the most costly.

8.4 OPTICAL-EMISSION SPECTROSCOPY WITH PLASMAS AND ELECTRICAL DISCHARGES

8.4.1 Introduction

Optical-emission spectroscopy (OES) with high-temperature atomization sources is a technique for simultaneously ascertaining qualitative information quickly and sensitively, and for determining the concentration of about 70 elements in inorganic and organic matrices (see Table 8.4). This capability is achieved with plasma and electrical discharges but with more spectral interferences and without the relative simplicity of atomic-absorption spectrometry. A minute amount of sample, often as little as a milligram, is vaporized and thermally excited to the point of atomic emission. For a quantitative determination, 20 mg of inorganic solid sample is usually adequate.

OES techniques can be applied to almost every type of sample. Areas of investigation include chemicals, minerals, soils, metals and alloys, plastics, agricultural products, foodstuffs, and water analysis. With preconcentration of the sample, most elements can be determined at the low parts-perbillion level.

[4] J. A. C. Broekaert, *Spectrochim. Acta* **37B:**65 (1982).

When a substance is excited by a plasma or electrical discharge (arc or spark), elements present emit light at wavelengths that are specific for each element (Table 8.10). The light emitted is dispersed by a grating or prism monochromator. The spectral lines produced are recorded either on a photographic plate or, in modern systems, by diode arrays or photomultiplier tubes linked directly to computer-driven data-processing systems.

8.4.2 Sampling Devices and Sources

A wide variety of sources are needed to handle the diverse sample matrices. For solid samples arc excitation is more sensitive, while spark sources are more stable. Plasma sources are the choice for solutions and gaseous samples. A nitrogen purge or a vacuum system is needed for determining B, C, P, and S.

8.4.2.1 Direct-Current Arcs. The dc arc consists of a high-current, low-voltage discharge between two solid electrodes, one of which may be the sample or an electrode supporting the sample while the other is the counterelectrode. In the United States the anode is generally the sample-containing electrode, whereas in Europe it is the cathode. High-purity graphite is a popular electrode material because it resists attack by strong acids or redox reagents and, being highly refractory, permits the volatilization of high-boiling-point sample components. Its emission spectrum contains few lines.

Several electrode configurations are used. Conducting samples are usually ground flat and used as one electrode with a pointed graphite counterelectrode. Powdered samples are mixed with graphite and pressed into a pellet or placed in a pedestal holder. Metal samples can be ground to a point and used with a graphite counterelectrode.

Because of its high sensitivity the dc arc is well suited for qualitative survey analyses of both trace and major elements or for semiquantitative analyses. An internal standard will minimize the effects of arc instability and the sample matrix. The element selected as the internal standard must have vaporization and excitation characteristics that closely match the element being determined. The tendency of the dc arc to wander around the sample surface seriously affects quantitative precision, unless a total burn is employed, but, on the other hand, it offers excellent qualitative surveys. To eliminate matrix problems, the sample may be fused with lithium tetraborate, then crushed to a powder before continuing with the analysis. Fractional volatilization separates the more volatile sample constituents from the more refractory elements. The more volatile elements can be detected by successive exposure of several spectra, above and below each other on the same photographic plate. The *carrier distillation* method is useful in the analysis of refractories. A low-boiling-point material is added to the sample to sweep (carry) the more volatile trace constituents up into the arc column while the base material remains unexcited. Popular carriers include copper hydroxy fluoride, silver fluoride, silver chloride, lithium fluoride, and gallium oxide.

8.4.2.2 Alternating-Current Arcs. The ac arc momentarily stops at the end of each half-cycle and then reignites during the next half-cycle when the applied voltage exceeds the voltage required for the dielectric breakdown of the gas between the electrodes. The sampling spot moves rapidly about on the electrode surface (120 times per second for a 60-Hz source). Consequently, reproducibility (precision) is superior to that of the dc arc; however, the sensitivity is less. The sustaining arc represents a good compromise between stability and sensitivity. The selective volatilization that is a problem with the dc arc does not occur with the ac arc.

The ac arc is used for the analysis of residuals in steels and other metal samples as well as for liquid samples that are evaporated on the flat ends of copper electrodes.

8.4.2.3 High-Voltage, Alternating-Current Spark. Although not as sensitive as the dc arc source, the high-voltage, ac spark provides the greatest precision and stability of all the electrical discharge sources. It also has a sampling spot that moves rapidly about on the electrode surface. High-energy presparking homogenizes metallurgical structures. Then the analytical spark discharge is used for the determination. Rather elaborate source units allow variations in capacitance and inductance values. Large values of inductance decrease the excitation energy and make the spark

TABLE 8.10 Sensitive Lines of the Elements

(In this table the sensitive lines of the elements are arranged in order of decreasing wavelengths. A Roman numeral II following an element designation indicates a line classified as being emitted by the singly ionized atom. In the column headed Sensitivity, the most sensitive line of the nonionized atom is indicated by U1, *and other lines by* U2, U3, *and so on, in order of decreasing sensitivity. For the singly ionized atom the corresponding designations are* V1, V2, V3, *and so on.)*

Wavelength, nm	Element	Sensitivity	Wavelength, nm	Element	Sensitivity
894.35	Cs	U2	492.45	Nd	U1
852.11	Cs	U1	488.91	Re	U4
819.48	Na	U4	487.25	Sr	U3
818.33	Na	U3	483.21	Sr	U2
811.53	Ar	U2	482.59	Ra	U1
794.76	Rb	U2	481.95	Cl II	V4
780.02	Rb	U1	481.67	Br II	V3
769.90	K	U2	481.05	Zn	U3
766.49	K	U1	481.01	Cl II	V3
750.04	Ar	U4	479.45	Cl II	V2
706.72	Ar	U3	478.55	Br II	V2
696.53	Ar	U3	476.03	Sm	U1
690.24	F	U3	470.09	Br II	V1
685.60	F	U2	467.12	Xe	U2
670.78	Li	U1	462.43	Xe	U3
656.28	H	U2	460.73	Sr	U1
649.69	Ba II	V4	460.29	Li	U4
624.99	La	U3	459.40	Eu	U1
614.17	Ba II	V3	459.32	Cs	U4
610.36	Li	U2	455.54	Cs	U3
593.06	La	U4	455.40	Ba II	V1
589.59	Na	U2	451.13	In	U1
589.00	Na	U1	450.10	Xe	U4
587.76	He	U3	445.48	Ca	U2
587.09	Kr	U2	442.43	Sm II	V4
579.13	La	U1	440.85	V	U4
569.92	Ce	U1	440.19	Gd	U1
567.96	N II	V2	439.00	V	U3
567.60	N II	V4	437.49	Y II	V4
566.66	N II	V3	437.92	V	U1
557.02	Kr	U3	435.84	Hg	U3
553.55	Ba	U1	431.89	Tb	U1
550.13	La	U2	430.36	Nd II	V2
546.55	Ag	U4	430.21	W	U1
546.07	Hg	U2	429.67	Sm	U1
545.52	La	U3	428.97	Cr	U3
535.84	Hg	U3	427.48	Cr	U2
535.05	Tl	U1	425.43	Cr	U1
521.82	Cu	U3	422.67	Ca	U1
520.91	Ag	U3	421.56	Rb	U4
520.84	Cr	U8	421.55	Sr II	V1
520.60	Cr	U7	421.17	Dy	U2
515.32	Cu	U4	420.19	Rb	U3
498.18	Ti	U1	418.68	Dy	U2
496.23	Sr	U2	418.66	Ce II	V1
493.97	Pr	U1	417.21	Ga	U1
493.41	Ba II	V2	414.31	Pr II	V2

(Continued)

TABLE 8.10 Sensitive Lines of the Elements (*Continued*)

Wavelength, nm	Element	Sensitivity	Wavelength, nm	Element	Sensitivity
414.29	Y	U4	385.99	Fe	U2
413.38	Ce II	V1	385.96	U II	V1
413.07	Ba II	V5	384.87	Tb II	V2
412.97	Eu II	V2	384.80	Tm II	V2
412.83	Y	U3	383.83	Mg	U2
412.38	Nb	U4	383.82	Mo	U2
412.32	La II	V4	382.23	Mg	U3
411.00	N	U2	382.94	Mg	U4
410.38	Ho	U1	381.97	Eu II	V1
410.24	Y	U1	379.94	Ru	U3
410.18	In	U2	379.63	Mo	U1
410.09	Nb	U3	379.48	La II	V2
409.99	N	U3	379.08	La II	V3
409.01	U II	V2	377.57	Tl	U3
408.77	Er	U1	377.43	Y II	V3
408.67	La II	V1	374.83	Fe	U4
407.97	Nb	U2	373.49	Fe	U2
407.77	Sr II	V2	372.80	Ru	U1
407.74	Y	U2	371.99	Fe	U1
407.74	La II	V2	371.79	Tm	U1
407.43	W	U2	371.03	Y II	V1
405.89	Nb	U1	369.42	Yb II	V2
405.78	Pb	U1	369.24	Rh	U2
405.39	Ho	U2	368.41	Gd	U2
404.72	K	U4	368.35	Pb	U2
404.66	Hg	U5	365.48	Hg	U4
404.60	Dy	U1	365.35	Ti	U2
404.41	K	U3	365.01	Hg	U3
403.45	Mn	U3	364.28	Sc II	V3
403.31	Mn	U2	364.27	Sn	U3
403.30	Ga	U2	363.47	Pd	U2
403.08	Mn	U1	363.07	Sc II	V2
402.37	Sc	U3	362.09	Y	U2
402.04	Sc	U3	361.38	Sc II	V1
401.91	Th II	V1	360.96	Pd	U2
401.23	Nd II	V1	360.12	Zr	U1
400.87	W	U1	360.07	Y II	V2
400.80	Er	U1	360.05	Cr	U6
399.86	Cr	U1	359.62	Ru	U3
399.86	Ti	U1	359.34	Cr	U5
398.80	Yb	U1	359.26	Sm II	V1
396.85	Ca II	V2	358.49	U	V1
396.15	Al	U1	357.87	Cr	U4
394.91	La II	V2	357.25	Zr II	V4
394.40	Al	U2	357.24	Sc II	V1
393.37	Ca II	V1	356.83	Sn II	V1
391.18	Sc	U1	355.31	Pd	U3
390.84	Pr II	V1	354.77	Zr	U3
390.75	Sc	U2	353.17	Dy II	V1
390.30	Mo	U1	352.98	Co	U3
389.18	Ba	V4	352.94	Tl	U4
388.86	He	U2	352.69	Co	U4
388.63	Fe	U5	352.45	Ni	U2
386.41	Mo	U2	351.96	Zr	U3

TABLE 8.10 Sensitive Lines of the Elements (*Continued*)

Wavelength, nm	Element	Sensitivity	Wavelength, nm	Element	Sensitivity
351.92	Tl	U2	323.26	Li	U3
351.69	Pd	U3	323.06	Er II	V2
351.36	Ir	U2	322.08	Ir	U1
350.92	Tb II	V1	318.54	V	U3
350.63	Co	U3	318.40	V	U2
350.23	Co	U2	317.93	Ca II	V3
349.89	Ru	U2	316.34	Nb II	V1
349.62	Zr II	V3	315.89	Ca II	V4
349.41	Er II	V1	313.26	Mo	U2
348.11	Pd	U5	313.13	Tm II	V1
347.40	Ni	U3	313.11	Be	U1
346.47	Re	U2	313.04	Be	U2
346.05	Re	U1	311.84	V II	V4
345.60	Ho II	V2	311.07	V II	V3
345.58	Co	U5	310.23	V II	V2
345.19	Re	U3	309.42	Nb II	V1
345.14	B II	V2	309.31	V II	V1
344.36	Co	U2	309.27	Al	U3
344.06	Fe	U2	308.22	Al	U4
343.82	Zr II	V2	307.76	Lu II	V2
343.67	Ru	U2	307.29	Hf	U1
343.49	Rh	U1	306.77	Bi	U3
342.83	Ru	U4	306.47	Pt	U1
342.12	Pd	U3	303.94	In	U4
341.48	Ni	U3	303.90	Ge	U2
341.23	Co	U4	303.41	Sn	U3
340.78	Dy II	V2	302.06	Fe	U3
340.51	Co	U2	300.91	Sn	U4
340.46	Pd	U2	294.91	Mn II	V4
339.90	Ho II	V1	294.44	W	U5
339.20	Zr II	V1	294.36	Ga	U3
338.29	Ag	U2	294.02	Ta	U3
337.28	Ti II	V3	293.30	Mn II	V4
336.12	Ti II	V2	292.98	Pt	U3
335.05	Gd II	V1	292.45	Nd	U2
334.94	Ti II	V1	292.40	V II	V1
334.50	Zn	U2	290.91	Os	U2
334.19	Ti	U4	289.80	Bi	U2
332.11	Be	U3	289.10	Mo II	V4
331.12	Ta	U3	288.16	Si	U1
330.03	Na	U6	287.42	Ga	U4
330.26	Zn	U3	287.15	Mo II	V3
330.23	Na	U5	286.33	Sn	U2
328.94	Yb II	V1	286.04	As	U2
328.23	Zn	U5	285.21	Mg	U1
328.07	Ag	U1	284.82	Mo II	V2
327.40	Cu	U2	284.00	Sn	U1
326.95	Ge	U3	283.73	Th II	V1
326.23	Sn	U3	283.58	Cr II	V2
326.11	Cd	U1	283.31	Pb	U3
325.61	In	U3	283.16	Si II	V1
324.75	Cu	U1	283.03	Pt	U3
324.27	Pd	U4	281.62	Al II	V2
323.45	Cr	V3	281.61	Mo II	V1

(*Continued*)

TABLE 8.10 Sensitive Lines of the Elements (*Continued*)

Wavelength, nm	Element	Sensitivity	Wavelength, nm	Element	Sensitivity
280.27	Mg II	V2	234.90	As	U4
280.20	Pb	U4	234.86	Be	U1
279.83	Mn	U3	232.00	Ni	U2
279.55	Mg II	V1	231.60	Ni II	V1
279.48	Mn	U3	231.15	Sb	U1
279.08	Mg II	V2	230.61	In II	V1
278.02	As	U1	228.81	As	U5
277.34	Hf II	V1	228.80	Cd	U2
276.78	Tl	U4	228.71	Ni II	V1
272.44	W	U4	228.62	Co II	V1
271.90	Fe	U5	228.23	Os II	V2
271.47	Ta	U1	227.66	Bi	U3
270.65	Sn	U4	227.02	Ni II	V2
267.72	Cr II	V1	226.50	Cd II	V2
267.60	Au	U2	226.45	Ni II	V3
266.92	Al II	V1	225.58	Os II	V1
265.95	Pt	U1	225.39	Ni II	V4
265.12	Ge	U1	224.70	Cu II	V3
265.05	Ba	U2	224.64	Ag II	V3
264.75	Ta	U2	224.60	Sn	U1
263.87	Hf II	V1	224.27	Ir II	V1
263.71	Os	U1	223.06	Bi	U1
260.57	Mn II	V3	220.35	Pb II	V1
259.94	Fe II	V1	219.87	Ge II	V2
259.81	Sb	U2	219.23	Cu II	V2
259.37	Mn	U2	217.58	Sb	U2
257.61	Mn II	V1	217.00	Pb II	V1
256.37	Mn II	V2	214.44	Cd II	V1
255.33	P	U3	214.42	Pt II	V1
255.24	Sc II	V3	214.27	Te	U1
253.65	Hg	U1	213.86	Zn	U1
253.57	P	U1	213.62	P	U1
252.85	Si	U2	213.60	Cu II	V1
252.29	Fe	U3	212.68	Ir II	V1
251.61	Si	U3	209.48	W II	V2
250.69	Si	U4	209.43	Ge II	V1
250.20	Zn II	V4	208.88	Ir	U1
249.77	B	U1	207.91	W II	V1
249.68	B	U2	207.48	Se	U4
248.33	Fe	U3	206.83	Sb	U1
247.86	C	U2	206.28	Se	U3
245.65	As	U4	206.19	Zn II	V2
243.78	Ag II	V2	203.99	Se	U1
242.80	Au	U1	203.84	Mo II	V3
241.05	Fe II	V4	202.55	Zn II	V1
240.73	Co	U1	202.03	Mo II	V2
240.49	Fe	V3	197.31	Re II	V1
240.27	Ru	V1	197.20	As	U3
240.06	Ta II	V1	196.03	Se	U2
239.56	Fe II	V2	194.23	Hg II	V1
238.89	Co II	V2	193.76	As	U1
238.58	Te	U2	193.09	C	U1
238.32	Te	U3	190.86	Tl II	V1

TABLE 8.10 **Sensitive Lines of the Elements (*Continued*)**

Wavelength, nm	Element	Sensitivity	Wavelength, nm	Element	Sensitivity
238.20	Fe II	V1	189.99	Sn II	V1
189.04	As	U2	178.38	I	U1
183.00	I	U2	178.28	P	U1
182.59	B II	V2	154.07	Br II	V4
180.73	S	U1	134.72	Cl II	V1

Source: J. A. Dean, ed., *Lange's Handbook of Chemistry*, 14th ed., McGraw-Hill, New York, 1992.

more arclike in its characteristics. The ac spark is the method of choice for analyses of ferrous metals in industrial operations.

Conducting samples are ground flat and used as one electrode in the point-to-plane method. Powdered samples are mixed with graphite powder and pressed into a pellet that is used as the plane electrode. Solutions are determined using a porous cup graphite electrode or a rotating disk electrode. The former consists of a porous-bottom graphite cup containing the sample solution. The counter-electrode beneath the cup discharges to the wet bottom of the porous cup.

8.4.2.4 Inductively Coupled Plasma Sources. The inductively coupled argon plasma (ICAP or ICP) torch derives its sustaining power from the interaction of a high-frequency magnetic field and ionized argon gas. The plasma is formed by a tangential stream of argon gas flowing between the outer two quartz tubes of the torch assembly. Radio-frequency power (27 to 30 MHz, 2 kW) is applied through the induction coil. This sets up an oscillating magnetic field. Power transfer between the induction coil (two-turn primary winding) and the plasma (one-turn secondary winding) is similar to a power transfer in a transformer. An eddy current of cations and electrons is formed as the charged particles are forced to flow in a closed annular path. The fast-moving ions and electrons collide with more argon atoms to produce further ionization and intense thermal energy as they meet resistance to their flow. A long, well-developed tail emerges from the flame-shaped plasma that forms near the top of the torch and above the induction coils. This tail is the spectroscopic source. The sample particles experience a gas temperature of about 7000 to 8000 K when they pass through the ICP, and by the time the sample decomposition products reach the analytical observation zone, they have had a residence time of about 2 ms at temperatures ranging downward from about 8000 to 5000 K. Both the residence time and temperatures experienced by the sample are approximately twice as large as those in a nitrous oxide–acetylene flame.

The argon gas stream is initially seeded with free electrons from a Tesla discharge coil. These seed electrons quickly interact with the magnetic field of the coil and gain sufficient energy to ionize argon atoms by collisional excitation. A second stream of argon provides a vortex flow of gas to cool the inside quartz walls of the torch. This flow also serves to center and stabilize the plasma. The sample is injected as an aerosol through the innermost concentric tube of the torch through the center of the plasma.[5] An ICAP source has these advantages:

1. The analytes are confined to a narrow region.

2. The plasma provides simultaneous excitation of many elements and, therefore, simultaneous analysis of several elements.

3. The analyst is not limited to analytical lines involving ground-state transitions but can select from first or even second ionization state lines. For the elements Ba, Be, Fe, Mg, Mn, Sr, Ti, and V, the ion lines provide the best detection limits.

4. The high temperature of the plasma ensures the complete breakdown of chemical compounds (even refractory compounds) and impedes the formation of other interfering compounds, thus virtually eliminating matrix effects.

[5] J. A. Koropchak, "Liquid Sample Introduction to ICP Spectrometries," *Spectroscopy* **8[8]**:20 (October 1993).

5. The ICAP torch provides a chemically inert atmosphere and an optically thin emission source.

6. Excitation and emission zones are spatially separated; this results in a low background. The optical window used for analysis lies just above the apex of the primary plasma and just under the base of the flamelike afterglow.

7. Low background, combined with a high signal-to-noise ratio of analyte emission, results in low detection limits, typically in the parts-per-billion range (Table 8.4).

8.4.2.5 Direct-Current Plasma Sources. In the three-electrode, direct-current plasma (DCP) source, the argon plasma jet is formed between two carbon anodes and a tungsten cathode in an inverted Y configuration. The sample excitation region and observation zone are centered in the crook of the Y where the temperature is approximately 6000 K. The plasma is initiated by moving the three electrodes into contact with argon-driven pistons and then withdrawing them. Samples are nebulized and the aerosol is introduced into the excitation region of the plasma. The plasma is stable in the presence of aerosols that contain large amounts of dissolved solids, organic solvents, and high concentrations of acids or bases.

As with the ICAP, the emission lines of atoms and ions are observed in a region separated from the main plasma. This leads to a high signal-to-background ratio and low detection limits. However, several limitations exist with the DCP source:

1. The electrodes are consumed and require reshaping after 2 h of continuous operation.
2. The excitation characteristics are greatly affected by high concentrations of easily ionized elements.
3. The DCP source cannot be incorporated into totally automated systems.
4. The relatively high consumption rates of argon pose a cost consideration.

8.4.3 Optical-Emission Spectrometers

High-resolution, high-luminosity monochromators are necessary to isolate a spectral line from its background without loss of radiant power. Both concave and plane diffraction gratings along with echelle gratings are used as dispersive elements. A fore prism as an order-sorter is an essential feature for use with an echelle grating.[6]

8.4.3.1 Direct-Reading Instruments. One type of a direct-reading (nonscanning) spectrometer has a holographic concave grating mounted in a Rowland circuit configuration with a focal length of 1 m. In this configuration the entrance slit, grating, and focal plane lie on the circumference of the Rowland circle (which has a radius of curvature half that of the grating). An array of slits with mirrors projects and focuses the selected spectral lines onto the cathodes of the photomultiplier tubes; alternatively, diode-array assemblies are used as detectors.

Direct readers are usually built for a specific analytical requirement by carefully positioning the exit slits (and accurate control of the slit width) and photomultiplier tubes to measure specific lines of elements of interest. Instruments generally measure from 8 to 24 elements in a single matrix. The array of receiver systems are often grouped into several bridges, one for each type of matrix, so that the instrument can be used to analyze several types of samples successively.

Integration times are from 25 to 40 s. The radiant power of the spectral line is converted by the photomultiplier tube into current that is used to charge a capacitor–resistor circuit. In integration with constant time, a voltage-to-frequency circuit converts the capacitor voltage into electrical pulses of equal height that are counted to give a number proportional to the radiant power. The output is in digital form and ready for computer processing.

---

[6] D. Noble, "ICP-AES From Fixed to Flexible," *Anal. Chem.* **66**:105A (1994).

The spectrophotometers are calibrated with a high- and low-concentration standard. The observed line intensities are related directly to concentrations. Typically, sample analysis, data output, and any associated data management functions are completed within 2 min after sample introduction.

8.4.3.2 Plane-Grating Spectrophotometers. Almost all scanning spectrometers use a plane grating in an Ebert mounting. In this mounting light of all wavelengths is brought to a focus on the detector without changing the detector-to-mirror distance. Wavelengths are easily changed by simply rotating the grating by a computer-controlled stepper motor. This permits the programmed selection of desired wavelengths and the sequential examination of spectral lines. The computer also checks the analytical parameters for each line analyzed (wavelength, slit width), makes the necessary background correction, performs required calculations, and prints the results.

If the monochromator is enclosed in a vacuum, lines of sulfur, phosphorus, and boron can be used. The spectral lines may be recorded on a photographic plate. The location and intensity of the lines produced by the sample are compared either visually or by means of a photoelectric densitometer with the lines produced by standards.

8.4.3.3 Fiber Optics. Fiber optics are used to transfer light from the analytical source to the individual spectrometer optics. This offers significant advantages:

1. The best wavelength for each element can be used, and the compromises of single optical systems are avoided.
2. Each optical system can be optimized for a specific wavelength range.
3. Simultaneous optics and their components (grating, detectors, slits) can be optimized for individual wavelengths.
4. The measurement of one element with the same spectral lines in different optical systems allows an additional control for important metallurgical or expensive alloying elements.
5. Interfering matrix effects and moving the location of optimum excitation in a plasma are avoided.

8.4.4 Comparison of AAS and ICP-OES Methods

From among the several methods discussed in this chapter, AAS and ICP-OES dominate analysis of solutions in atomic spectroscopy. Although ICP has become a major influence in solution analysis, AAS is still the backbone of elemental analysis. Both methods are actually complementary and not competitive. Each has strengths and weaknesses. Some applications are better suited to one technique for particular reasons—element type and detection limit. Gill[7] ably presents the strengths and weaknesses of both methods.

Considering AAS first, the atom source is limited to 3000 K for a nitrous oxide–acetylene flame. The residence time of an atom within the optical beam of the spectrophotometer, which determines the absorption, is extremely short, approximately 1 ms. For many elements, this is not a problem; compounds of the alkali metals, many of the heavy metals (lead, cadmium), and transition metals (manganese, nickel) are all atomized at high efficiency with flames. However, a number of refractory elements, such as molybdenum, tungsten, vanadium, and zirconium, cause problems in flame AAS. The flame temperature is insufficient to break down compounds of these elements and therefore sensitivity is poor. An ICP source atomizes and excites even the most refractory elements with high efficiency, so that detection limits for these elements can be well over an order of magnitude better than the corresponding values for flame AAS. For approximately half of the transition elements, the detection limits are within a factor of 2 for either flame AAS or ICP-OES. Of the other transition elements, ICP-OES produces superior detection limits. The lanthanides and phosphorus have significantly better

[7] R. J. Gill, "AAS or ICP-OES: Are They Competing Techniques?," *Am. Lab.* **1993** (November):24F.

detection limits for ICP, and sulfur can only be measured by ICP (and flame emission). With ICP several elements in a sample can be determined simultaneously without the need for repeated aspirations, adjustment of instrument parameters, and proper tracking of the samples.

When an AAS is fitted with a graphite-furnace electrothermal atomizer, the comparison changes dramatically. Although the graphite furnace is restricted to a maximum temperature similar to that of flame AAS, the atom residence time is markedly greater (up to 1000 ms), and this results in sensitivities up to 100 times better than flame AAS for many elements. Most elements, including refractory elements such as vanadium and molybdenum, are better analyzed by the graphite-furnace accessory. Nonetheless, ICP maintains an advantage when compared to graphite-furnace AAS for elements such as boron, phosphorus, sulfur, tungsten, and uranium.

When dealing with the rare earths, flame AAS detection limits are unacceptable for analysis. Similarly, graphite-furnace AAS shows poor sensitivity because of the very refractory, carbide-forming nature of the rare earths. ICP-OES is the obvious choice for solution-based analysis of these elements.

Graphite-furnace AAS is the primary method of analysis for body fluids as well as biological tissue samples. Owing to the viscous nature and high dissolved salts in body fluids, they are normally analyzed after a degree of dilution to reduce these two effects. ICP-OES cannot analyze these samples. Most published methods for biological samples involving ICP-OES is for tissue analysis where analyte levels may be concentrated during the sample dissolution stage of analysis.

Speed of analysis and the number of elements per sample are also important considerations. In general, simultaneous ICP-OES is approximately three times as fast as flame AAS when 10 elements per sample are being determined. On the other hand, a sequential ICP-OES system performs determinations at approximately the same rate as a flame AAS system, although if more than 10 elements per sample are required to be analyzed, sequential ICP-OES systems will prove faster. If only one or two elements are determined per sample flame, AAS is much faster then sequential ICP-OES. Graphite-furnace AAS will never compete on a speed basis because of the longer cycle time for this technique.

Spectral interferences in ICP are a serious problem when analyzing a wide variety of materials made of very different matrices. Flame AAS also has these problems but not as many spectral lines are excited in the lower-temperature flames. On the other hand, chemical interferences are few in ICP due to the high temperatures involved.

The final important consideration is the cost of the instrument. In terms of purchase cost and daily running costs, flame AAS remains the least expensive technique and ICP-OES the most expensive: four times for sequential ICP-OES, six times for simultaneous ICP-OES, and eight times for the combined techniques. Instrumentation for flame AAS is also much simpler. Applications of flame and furnace AAS and ICP are further discussed under environmental analysis of trace pollutants in Sec. 21.

8.5 ATOMIC FLUORESCENCE SPECTROMETRY

In atomic fluorescence spectrometry (AFS) the exciting source is placed at right angles to the flame and the optical axis of the spectrometer. Some of the incident radiation from the light source is absorbed by the free atoms of the analyte, which are formed by the thermal energy of the flame. The atoms are raised to the excited state that corresponds to the origin of the emission line of the source. Immediately after this absorption, energy is released as the excited atoms return to the ground state.

Instrumentation resembles that required for AAS except for placing the light source at right angles to the flame and optical axis of the spectrometer. In principle, the fluorescence is measured against a zero background of the flame; in practice, there may be a slight amount of scattered primary radiation from particulates in the flame gases. Use of modulated source and ac detection lessens this problem.

The intensity of the fluorescence is linearly proportional to the exciting radiation flux and thus the need for an intense source. The AFS technique is linear for an analytical concentration over several orders of magnitude and allows simultaneous multielement analyses with little or no spectral interferences. The marriage of the argon inductively coupled plasmas, as an exceedingly hot atomization system, and pulsed hollow-cathode lamps, as narrow-line excitation sources, is the heart of an instrument for AFS.

Bibliography

Barnes, R., ed., *Emission Spectroscopy*, Halsted, New York, 1976.

Boumans, P. W. J. W., ed., *Inductively Coupled Plasma Emission Spectroscopy*, Wiley, New York; Part 1: *Methodology, Instrumentation, and Performance*, 1987; Part 2, *Applications and Fundamentals*, 1987.

Dean, J. A., *Flame Photometry*, McGraw-Hill, New York, 1960.

Dean, J. A., and T. C. Rains, eds., *Flame Emission and Atomic Spectrometry: Theory*, Vol. 1, 1969; *Components and Techniques*, Vol. 2, 1971; *Elements and Matrices*, Vol. 3, 1975, Dekker, New York.

Ebdon, L., *An Introduction to Atomic Absorption Spectroscopy—A Self Teaching Approach*, Heydon, Philadelphia, 1982.

Keliher, P., et al., *Anal. Chem.* **58(5)**:334R (1986).

Lajunen, L. H. J., *Spectrochemical Analysis by Atomic Absorption and Emission*, Royal Society of Chemistry, Cambridge, England, 1992.

Mayer, B., *Guidelines to Planning Atomic Spectrometric Analysis*, Elsevier, New York, 1982.

Montaser, A., and D. Golightly, *Inductively Coupled Plasmas in Analytical Atomic Spectroscopy*, 2d ed., VCH, New York, 1992.

Ottaway, J., and A. Ure, *Practical Atomic Absorption Spectroscopy*, Pergamon, New York, 1983.

Patnaik, P. *Handbook of Environmental Analysis*, CRC Press, Boca Raton, Fla., 1997.

Sacks, R., "Emission Spectroscopy," Chap. 6, *Treatise on Analytical Chemistry*, 2d ed., P. J. Elving, E. J. Meehan, and I. M. Kolthoff, eds., Wiley, New York, 1981, Part I, Vol. 7.

Walsh, M., and M. Thompson, *A Handbook of Inductively Coupled Plasma Spectroscopy*, Methuen, New York, 1983.

SECTION 9
X-RAY METHODS

X-ray methods involve the excitation of an atom by the removal of an electron from an inner energy level, usually from the innermost K level or from one of the three L levels. Atoms can be excited either by direct bombardment of the sample with electrons (direct emission analysis, electron probe microanalysis, and Auger emission spectroscopy) or by irradiation of the sample with x rays of shorter wavelength than analyte elements (x-ray fluorescence analysis). Electron spectroscopy for chemical analysis (ESCA) measures the energy of the electrons ejected from inner electron levels when the sample is bombarded by a monochromatic x-ray beam. In x-ray absorption the intensity of an x-ray beam is diminished as it passes through material. X rays are also diffracted by the planes of a crystal, which provides a useful method for qualitative identification of crystalline phases.

9.1 PRODUCTION OF X RAYS AND X-RAY SPECTRA

When an atom is bombarded by sufficiently energetic electrons or x radiation, an electron may be ejected from one of the inner levels of the target atoms. The place of the ejected electron is promptly filled by an electron from an outer level whose place, in turn, is taken by an electron coming from still farther out. Each transition is accompanied by the release of an x-ray photon, the energy of which is characteristic of the element from which it originated. The measurement of the various photon energies produced by sample excitation provides a means of identifying its constituent elements. A count of the photons provides the quantification of each element.

Electron bombardment of the anode in an x-ray tube causes the emission of both a continuum of radiation and the characteristic emission lines of the anode material. When incident electrons (or x-ray photons) of sufficient energy impinge upon an atom, an electron from an inner shell may be photoejected from that atom. The energy required to initiate photoejection is called the absorption-edge energy. The absorption edge will be a sharp discontinuity in the plot of mass absorption coefficient versus wavelength.

There is successive ionization—first of electrons in the outermost levels of the sample or target (let us assume the M level), then of electrons in the L levels as the three L absorption edges are progressively exceeded, and finally in the K level as the K absorption edge is exceeded.

The energy required to lift a K electron out of the environment of the atom (to exceed the ionization limit) must exceed the energy of the K absorption edge. The relationship between the voltage applied across an x-ray tube (or the energy of incident x radiation, in volts) and the wavelength λ, in angstroms (Å), is given by the Duane–Hunt equation:

$$\lambda = \frac{hc}{eV} = \frac{12\,393}{V} \tag{9.1}$$

where V = x-ray tube voltage, V
e = charge on electron
h = Planck's constant
c = velocity of light

Following spectroscopic selection rules, electrons from outer shells (L and M) will undergo transitions to fill the K-shell vacancy. In so doing, they may emit an x-ray photon. (In a competing process, the energy released may be internally converted in the atom to cause the ejection of a secondary, or Auger, electron. Auger spectroscopy is discussed later.) The energy of the emitted radiation will be characteristic of the element and of the particular transition.

The wavelength of an absorption edge is always shorter than that of the corresponding emission lines. For example, in energy units, the $K\alpha_1$ line represents the difference: K edge minus L_{III} edge. By contrast, the K edge is the difference between the K energy level and the ionization limit.

The characteristic K or L emission lines (or absorption edges) of each element vary in a regular fashion from one element to another. The characteristic wavelengths decrease as the atomic number of the elements increases. Stated more exactly, the frequency of a given type of x-ray line increases approximately as the square of the atomic number of the element involved.

X-ray emission lines (Table 9.1) or absorption edges (Table 9.2) are quite simple because they consist of very few lines as compared to optical-emission or -absorption spectra observed in the visible-ultraviolet region. X-ray spectra are not dependent on the physical state of the sample nor on its chemical composition, except for the lightest elements, because the innermost electrons are not involved in chemical binding and are not significantly affected by the behavior of the valence electrons.

Use of energy-proportional detectors for x rays creates a need for energy values of K and L absorption edges (Table 9.3) and emission series (Table 9.4). These values were obtained by a conversion to keV of tabulated experimental wavelength values and smoothed by a fit to Moseley's law. Although values are listed to 1 eV, the chemical form may shift absorption edges and emission lines as much as 10 to 20 eV. Fine and Hendee[1] also give values for $K\beta_2$, $L\gamma_1$, and $L\beta_2$ lines.

The relative intensities of x-ray emission lines from targets vary for different elements. However, one can assume a ratio of $K\alpha_1 : K\alpha_2 = 2$ for the commonly used targets. The ratio of $K\alpha_2 : K\beta_1$ from these targets varies from 6 to 3.5. The intensities of $K\beta_2$ radiations amount to about 1% of that of the corresponding $K\alpha_1$ radiation. In practical applications these ratios have to be corrected for differential absorption in the window of the tube and air path, for the ratio of scattering factors and the differential absorption in the crystal, and for sensitivity characteristics of the detector. Generalizing, the intensities of radiations from the K and L series are as follows:

Emission line	$K\alpha_1$	$K\alpha_2$	$K\beta_1$	$K\beta_2$	$L\alpha_1$	$L\alpha_2$	$L\beta_1$	$L\beta_2$	$L\gamma_1$
Relative intensity	500	250	80 to 150	5	100	10	30	60	40

For angles at which the $K\alpha_1$, $K\alpha_2$ doublet is not resolved, a mean value $[K\bar{\alpha} = (2K\alpha_1 + K\alpha_2)/3]$ can be used.

9.2 INSTRUMENTATION

Analytically useful x radiation is generated by bombarding the sample with electrons in the range of 5 to 100 keV or with x-ray photons in a similar energy range. Instrumentation associated with x-ray emission methods, and specifically for a plane-crystal spectrometer as used in x-ray fluorescence, is shown in Fig. 9.1. The source of the x-ray photons may be an x-ray tube, a secondary target irradiated with photons from an x-ray tube, or emission from a radionuclide source. The x-ray photons are directed through a collimator onto a single analyzing crystal. The analyzing crystal acts as a diffraction grating. Scanning through the entire angular range of the goniometer permits radiation at a particular angular position to be correlated with wavelength through the Bragg condition, as will be discussed later. The detector is rotated at twice the angular rate of the analyzing crystal.

9.2.1 X-Ray Tubes

The x-ray tube is a high-vacuum, sealed-off unit. The target (anode), usually copper or molybdenum, is viewed from a very small glancing angle above the surface for diffraction work and at an angle of about 20° for fluorescence work. Dissipation of the generated heat is accomplished by water-cooling the target and occasionally by rotating it. The x-ray beam leaves the tube through a thin window of beryllium. For wavelengths of 6 to 70 Å, ultrathin films of aluminum or cast Parlodion are used as window material, and the equipment must be evacuated or flushed with helium.

[1] S. Fine and C. F. Hendee, *Nucleonics* **13**(3):36 (1955).

TABLE 9.1 Wavelengths of X-Ray Emission Spectra in Angstroms

Atomic no.	Element	$K\alpha_2$	$K\alpha_1$	$K\beta_1$	$L\alpha_1$	$L\beta_1$
3	Li		240			
4	Be		113			
5	B		67			
6	C		44			
7	N		31.60			
8	O		23.71			
9	F		18.31			
10	Ne		14.616	14.464		
11	Na		11.909	11.617	407.6	
12	Mg		9.889	9.558	251.0	
13	Al	8.3392	8.3367	7.981	169.8	
14	Si	7.1277	7.1253	6.7681	123	
15	P		6.1549	5.8038		
16	S	5.3747	5.3720	5.0317		
17	Cl	4.7305	4.7276	4.4031		
18	Ar	4.1946	4.1916	3.8848		
19	K	3.7446	3.7412	3.4538	42.7	
20	Ca	3.3616	3.3583	3.0896	36.32	35.95
21	Sc	3.0345	3.0311	2.7795	31.33	31.01
22	Ti	2.75207	2.7484	2.5138	27.39	27.02
23	V	2.5073	2.5035	2.2843	24.26	23.85
24	Cr	2.29351	2.28962	2.08480	21.67	21.28
25	Mn	2.1057	2.1018	1.9102	19.45	19.12
26	Fe	1.93991	1.93597	1.75653	17.567	17.255
27	Co	1.79278	1.78892	1.62075	15.968	15.667
28	Ni	1.66169	1.65784	1.50010	14.566	14.279
29	Cu	1.54433	1.54051	1.39217	13.330	13.053
30	Zn	1.4389	1.4351	1.2952	12.257	11.985
31	Ga	1.3439	1.3400	1.20784	11.290	11.023
32	Ge	1.2580	1.2540	1.1289	10.435	10.174
33	As	1.1798	1.1758	1.0573	9.671	9.414
34	Se	1.1088	1.1047	0.9921	8.990	8.736
35	Br	1.0438	1.0397	0.9327	8.375	8.125
36	Kr	0.9841	0.9801	0.8785	7.822	7.574
37	Rb	0.9296	0.9255	0.8286	7.3181	7.076
38	Sr	0.8794	0.8752	0.7829	6.8625	6.6237
39	Y	0.8330	0.8279	0.7407	6.4485	6.2117
40	Zr	0.7901	0.7859	0.7017	6.0702	5.8358
41	Nb	0.7504	0.7462	0.6657	5.7240	5.4921
42	Mo	0.713543	0.70926	0.632253	5.4063	5.1768
43	Tc	0.6793	0.6749	0.6014	5.1126	4.8782
44	Ru	0.6474	0.6430	0.5725	4.8455	4.6204
45	Rh	0.6176	0.6132	0.5456	4.5973	4.3739
46	Pd	0.5898	0.5854	0.5205	4.3676	4.1460
47	Ag	0.563775	0.559363	0.49701	4.1541	3.9344
48	Cd	0.5394	0.5350	0.4751	3.9563	3.7381
49	In	0.5165	0.5121	0.4545	3.7719	3.5552
50	Sn	0.4950	0.4906	0.4352	3.5999	3.3848
51	Sb	0.4748	0.4703	0.4171	3.4392	3.2256
52	Te	0.4558	0.4513	0.4000	3.2891	3.0767

TABLE 9.1 Wavelengths of X-Ray Emission Spectra in Angstroms (*Continued*)

Atomic no.	Element	$K\alpha_2$	$K\alpha_1$	$K\beta_1$	$L\alpha_1$	$L\beta_1$
53	I	0.4378	0.4333	0.3839	3.1485	2.9373
54	Xe	0.4204	0.4160	0.3685	3.016	2.807
55	Cs	0.4048	0.4003	0.3543	2.9016	2.8920
56	Ba	0.3896	0.3851	0.3408	2.7752	2.5674
57	La	0.3753	0.3707	0.3280	2.6651	2.4583
58	Ce	0.3617	0.3571	0.3158	2.5612	2.3558
59	Pr	0.3487	0.3441	0.3042	2.4627	2.2584
60	Nd	0.3565	0.3318	0.2933	2.3701	2.1666
61	Pm	0.3249	0.3207	0.2821	2.282	2.0796
62	Sm	0.3137	0.3190	0.2731	2.1994	1.9976
63	Eu	0.3133	0.2985	0.2636	2.1206	1.9202
64	Gd	0.2932	0.2884	0.2544	2.0460	1.8462
65	Tb	0.2834	0.2788	0.2460	1.9755	1.7763
66	Dy	0.2743	0.2696	0.2376	1.9088	1.7100
67	Ho	0.2655	0.2608	0.2302	1.8447	1.6488
68	Er	0.2572	0.2525	0.2226	1.7843	1.5873
69	Tm	0.2491	0.2444	0.2153	1.7263	1.5299
70	Yb	0.2415	0.2368	0.2088	1.6719	1.4756
71	Lu	0.2341	0.2293	0.2021	1.6194	1.4235
72	Hf	0.2270	0.2222	0.1955	1.5696	1.3740
73	Ta	0.2203	0.2155	0.1901	1.5219	1.3270
74	W	0.213813	0.208992	0.184363	1.4764	1.2818
75	Re	0.2076	0.2028	0.1789	1.4329	1.2385
76	Os	0.2016	0.1968	0.1736	1.3911	1.1972
77	Ir	0.1959	0.1910	0.1685	1.3513	1.1578
78	Pt	0.1904	0.1855	0.1637	1.3130	1.1198
79	Au	0.1851	0.1802	0.1590	1.2764	1.0836
80	Hg	0.1799	0.1750	0.1544	1.2411	1.0486
81	Tl	0.1750	0.1701	0.1501	1.2074	1.0152
82	Pb	0.1703	0.1654	0.1460	1.1750	0.9822
83	Bi	0.1657	0.1608	0.1419	1.1439	0.9520
84	Po	0.1608	0.1559	0.1382	1.1138	0.9222
85	At	0.1570	0.1521	0.1343	1.0850	0.8936
86	Rn	0.1529	0.1479	0.1307	1.0572	0.8659
87	Fr	0.1489	0.1440	0.1272	1.030	0.840
88	Ra	0.1450	0.1401	0.1237	1.0047	0.8137
89	Ac	0.1414	0.1364	0.1205	0.9799	0.7890
90	Th	0.1378	0.1328	0.1174	0.9560	0.7652
91	Pa	0.1344	0.1294	0.1143	0.9328	0.7422
92	U	0.1310	0.1259	0.1114	0.9105	0.7200
93	Np	0.1278	0.1226	0.1085	0.8893	0.6984
94	Pu	0.1246	0.1195	0.1058	0.8682	0.6777
95	Am	0.1215	0.1165	0.1031	0.8481	0.6576
96	Cm	0.1186	0.1135	0.1005	0.8287	0.6388
97	Bk	0.1157	0.1107	0.0980	0.8098	0.6203
98	Cf	0.1130	0.1079	0.0956	0.7917	0.6023
99	Es	0.1103	0.1052	0.0933	0.7740	0.5850
100	Fm	0.1077	0.1026	0.0910	0.7570	0.5682

Source: J. A. Dean, ed., *Lange's Handbook of Chemistry*, 14th ed., McGraw-Hill, New York, 1992.

TABLE 9.2 Wavelengths (in Angstroms) of Absorption Edges

Atomic no.	Element	K	L_I	L_{II}	L_{III}
3	Li	226.5			
4	Be	110.68			
5	B	66.289			
6	C	43.68			
7	N	30.99			
8	O	23.32			
9	F	17.913			
10	Ne	14.183			
11	Na	11.478			400
12	Mg	9.512	197.4		247.92
13	Al	7.951	142.5		170
14	Si	6.745	105.1		126.48
15	P	5.787	81.0		96.84
16	S	5.018	64.23		76.05
17	Cl	4.397	52.08	61.37	62.93
18	Ar	3.871	43.19	50.39	50.60
19	K	3.436	36.35	42.02	42.17
20	Ca	3.070	31.07	35.20	35.49
21	Sc	2.757	26.83	30.16	30.53
22	Ti	2.497	23.39	26.83	27.37
23	V	2.269	20.52	23.70	24.26
24	Cr	2.07012	16.7	17.9	20.7
25	Mn	1.896	16.27	18.90	19.40
26	Fe	1.74334	14.60	17.17	17.53
27	Co	1.60811	13.34	15.53	15.93
28	Ni	1.48802	12.27	14.13	14.58
29	Cu	1.38043	11.27	13.01	13.29
30	Zn	1.283	10.33	11.86	12.13
31	Ga	1.195	9.54	10.61	11.15
32	Ge	1.116	8.73	9.97	10.23
33	As	1.044	8.108	9.124	9.367
34	Se	0.9800	7.505	8.417	8.646
35	Br	0.9199	6.925	7.752	7.989
36	Kr	0.8655	6.456	7.165	7.395
37	Rb	0.8155	5.997	6.643	6.863
38	Sr	0.7697	5.582	6.172	6.387
39	Y	0.7276	5.233	5.756	5.962
40	Zr	0.6888	4.867	5.378	5.583
41	Nb	0.6529	4.581	5.025	5.223
42	Mo	0.61977	4.299	4.719	4.912
43	Tc	0.5888	4.064	4.427	4.629
44	Ru	0.5605	3.841	4.179	4.369
45	Rh	0.5338	3.626	3.942	4.130
46	Pd	0.5092	3.428	3.724	3.908
47	Ag	0.48582	3.254	3.514	3.698
48	Cd	0.4641	3.084	3.326	3.504
49	In	0.4439	2.926	3.147	3.324
50	Sn	0.4247	2.778	2.982	3.156
51	Sb	0.4066	2.639	2.830	3.000
52	Te	0.3897	2.510	2.687	2.855

TABLE 9.2 **Wavelengths (in Angstroms) of Absorption Edges (*Continued*)**

Atomic no.	Element	K	L_I	L_{II}	L_{III}
53	I	0.3738	2.390	2.553	2.719
54	Xe	0.3585	2.274	2.429	2.592
55	Cs	0.3447	2.167	2.314	2.474
56	Ba	0.3314	2.068	2.204	2.363
57	La	0.3184	1.973	2.103	2.258
58	Ce	0.3065	1.891	2.009	2.164
59	Pr	0.2952	1.811	1.924	2.077
60	Nd	0.2845	1.735	1.843	1.995
61	Pm	0.2743	1.668	1.766	1.918
62	Sm	0.2646	1.598	1.702	1.845
63	Eu	0.2555	1.536	1.626	1.775
64	Gd	0.2468	1.477	1.561	1.709
65	Tb	0.2384	1.421	1.501	1.649
66	Dy	0.2305	1.365	1.438	1.579
67	Ho	0.2229	1.319	1.390	1.535
68	Er	0.2157	1.269	1.339	1.483
69	Tm	0.2089	1.222	1.288	1.433
70	Yb	0.2022	1.181	1.243	1.386
71	Lu	0.1958	1.140	1.198	1.341
72	Hf	0.1898	1.099	1.154	1.297
73	Ta	0.1839	1.061	1.113	1.255
74	W	0.17837	1.025	1.074	1.215
75	Re	0.1731	0.9901	1.036	1.177
76	Os	0.1678	0.9557	1.001	1.140
77	Ir	0.1629	0.9243	0.9670	1.106
78	Pt	0.1582	0.8914	0.9348	1.072
79	Au	0.1534	0.8638	0.9028	1.040
80	Hg	0.1492	0.8353	0.8779	1.009
81	Tl	0.1447	0.8079	0.8436	0.9793
82	Pb	0.1408	0.7815	0.8155	0.9503
83	Bi	0.1371	0.7565	0.7891	0.9234
84	Po	0.1332	0.7322	0.7638	0.8970
85	At	0.1295	0.7092	0.7387	0.8720
86	Rn	0.1260	0.6868	0.7153	0.8479
87	Fr	0.1225	0.6654	0.6929	0.8248
88	Ra	0.1192	0.6446	0.6711	0.8027
89	Ac	0.1161	0.6248	0.6500	0.7813
90	Th	0.1129	0.6061	0.6301	0.7606
91	Pa	0.1101	0.5875	0.6106	0.7411
92	U	0.1068	0.5697	0.5919	0.7233
93	Np	0.1045	0.5531	0.5742	0.7042
94	Pu	0.1018	0.5366	0.5571	0.6867
95	Am	0.0992	0.5208	0.5404	0.6700
96	Cm	0.0967	0.5060	0.5246	0.6532
97	Bk	0.0943	0.4913	0.5093	0.6375
98	Cf	0.0920	0.4771	0.4945	0.6223
99	Es	0.0897	0.4636	0.4801	0.6076
100	Fm	0.0875	0.4506	0.4665	0.5935

Source: J. A. Dean, ed., *Lange's Handbook of Chemistry*, 14th ed., McGraw-Hill, New York, 1992.

TABLE 9.3 Critical X-Ray Absorption Energies in keV

Atomic no.	Element	K	L_I	L_{II}	L_{III}
1	H	0.0136			
2	He	0.0246			
3	Li	0.0547			
4	Be	0.112			
5	B	0.187			
6	C	0.284			
7	N	0.400			
8	O	0.532			
9	F	0.692			
10	Ne	0.874	0.048		0.022
11	Na	1.08	0.055		0.034
12	Mg	1.30	0.0628		0.0502
13	Al	1.559	0.0870		0.0720
14	Si	1.838	0.118		0.0977
15	P	2.142	0.153		0.128
16	S	2.469	0.193	0.163	0.162
17	Cl	2.822	0.238	0.202	0.201
18	Ar	3.200	0.287	0.246	0.244
19	K	3.606	0.341	0.295	0.292
20	Ca	4.038	0.399	0.350	0.346
21	Sc	4.496	0.462	0.411	0.407
22	Ti	4.966	0.530	0.462	0.456
23	V	5.467	0.604	0.523	0.515
24	Cr	5.988	0.679	0.584	0.574
25	Mn	6.542	0.762	0.656	0.644
26	Fe	7.113	0.849	0.722	0.709
27	Co	7.713	0.929	0.798	0.783
28	Ni	8.337	1.02	0.877	0.858
29	Cu	8.982	1.10	0.954	0.935
30	Zn	9.662	1.20	1.05	1.02
31	Ga	10.39	1.30	1.17	1.14
32	Ge	11.10	1.42	1.24	1.21
33	As	11.87	1.529	1.358	1.32
34	Se	12.65	1.66	1.472	1.431
35	Br	13.48	1.791	1.599	1.552
36	Kr	14.32	1.92	1.729	1.674
37	Rb	15.197	2.064	1.863	1.803
38	Sr	16.101	2.212	2.004	1.937
39	Y	17.053	2.387	2.171	2.096
40	Zr	17.998	2.533	2.308	2.224
41	Nb	18.986	2.700	2.467	2.372
42	Mo	20.003	2.869	2.630	2.525
43	Tc	21.050	3.045	2.796	2.680
44	Ru	22.117	3.227	2.968	2.839
45	Rh	23.210	3.404	3.139	2.995
46	Pd	24.356	3.614	3.338	3.181
47	Ag	25.535	3.828	3.547	3.375
48	Cd	26.712	4.019	3.731	3.541
49	In	27.929	4.226	3.929	3.732
50	Sn	29.182	4.445	4.139	3.911

TABLE 9.3 Critical X-Ray Absorption Energies in keV (*Continued*)

Atomic no.	Element	K	L_I	L_{II}	L_{III}
51	Sb	30.497	4.708	4.391	4.137
52	Te	31.817	4.953	4.621	4.347
53	I	33.164	5.187	4.855	4.559
54	Xe	34.551	5.448	5.103	4.783
55	Cs	35.974	5.706	5.360	5.014
56	Ba	37.432	5.995	5.629	5.250
57	La	38.923	6.264	5.902	5.490
58	Ce	40.43	6.556	6.169	5.728
59	Pr	41.99	6.837	6.446	5.968
60	Nd	43.57	7.134	6.728	6.215
61	Pm	45.19	7.431	7.022	6.462
62	Sm	46.85	7.742	7.316	6.720
63	Eu	48.51	8.059	7.624	6.984
64	Gd	50.23	8.383	7.942	7.251
65	Tb	52.00	8.713	8.258	7.520
66	Dy	53.77	9.053	8.587	7.795
67	Ho	55.61	9.395	8.918	8.074
68	Er	57.47	9.754	9.270	8.362
69	Tm	59.38	10.12	9.622	8.656
70	Yb	61.31	10.49	9.985	8.949
71	Lu	63.32	10.87	10.35	9.248
72	Hf	65.37	11.28	10.75	9.567
73	Ta	67.46	11.68	11.14	9.883
74	W	69.51	12.09	11.54	10.20
75	Re	71.67	12.52	11.96	10.53
76	Os	73.87	12.97	12.38	10.86
77	Ir	76.11	13.41	12.82	11.21
78	Pt	78.35	13.865	13.26	11.55
79	Au	80.67	14.351	13.731	11.92
80	Hg	83.08	14.838	14.205	12.278
81	Tl	85.52	15.344	14.695	12.65
82	Pb	87.95	15.861	15.200	13.03
83	Bi	90.54	16.386	15.709	13.42
84	Po	93.16	16.925	16.233	13.81
85	At	95.73	17.481	16.777	14.21
86	Rn	98.45	18.054	17.331	14.61
87	Fa	101.1	18.628	17.893	15.02
88	Ra	103.9	19.228	18.473	15.44
89	Ac	107.7	19.829	19.071	15.86
90	Th	109.8	20.452	19.673	16.278
91	Pa	112.4	21.096	20.295	16.720
92	U	115.0	21.757	20.944	17.163
93	Np	118.2	22.411	21.585	17.606
94	Pu	121.2	23.117	22.250	18.062
95	Am	124.3	23.795	22.935	18.524
96	Cm	127.2	24.502	23.629	18.992
97	Bk	131.3	25.231	24.344	19.466
98	Cf	133.6	26.010	25.070	19.954
99	Es	138.1	26.729	25.824	20.422
100	Fm	141.5	27.503	26.584	20.912

Source: J. A. Dean, ed., *Lange's Handbook of Chemistry*, 14th ed., McGraw-Hill, New York, 1992.

TABLE 9.4 X-Ray Emission Energies in keV

Atomic no.	Element	$K\beta_1$	$K\alpha_1$	$L\beta_1$	$L\alpha_1$
3	Li		0.052		
4	Be		0.110		
5	B		0.185		
6	C		0.282		
7	N		0.392		
8	O		0.523		
9	F		0.677		
10	Ne		0.851		
11	Na	1.067	1.041		
12	Mg	1.297	1.254		
13	Al	1.553	1.487		
14	Si	1.832	1.740		
15	P	2.136	2.015		
16	S	2.464	2.308		
17	Cl	2.815	2.622		
18	Ar	3.192	2.957		
19	K	3.589	3.313		
20	Ca	4.012	3.691	0.344	0.341
21	Sc	4.460	4.090	0.399	0.395
22	Ti	4.931	4.510	0.458	0.452
23	V	5.427	4.952	0.519	0.512
24	Cr	5.946	5.414	0.581	0.571
25	Mn	6.490	5.898	0.647	0.636
26	Fe	7.057	6.403	0.717	0.704
27	Co	7.649	6.930	0.790	0.775
28	Ni	8.264	7.477	0.866	0.849
29	Cu	8.904	8.047	0.948	0.928
30	Zn	9.571	8.638	1.032	1.009
31	Ga	10.263	9.251	1.122	1.096
32	Ge	10.981	9.885	1.216	1.186
33	As	11.725	10.543	1.317	1.282
34	Se	12.495	11.221	1.419	1.379
35	Br	13.290	11.923	1.526	1.480
36	Kr	14.112	12.649	1.638	1.587
37	Rb	14.960	13.394	1.752	1.694
38	Sr	15.834	14.164	1.872	1.806
39	Y	16.736	14.957	1.996	1.922
40	Zr	17.666	15.774	2.124	2.042
41	Nb	18.621	16.614	2.257	2.166
42	Mo	19.607	17.478	2.395	2.293
43	Tc	20.612	18.370	2.538	2.424
44	Ru	21.655	19.278	2.683	2.558
45	Rh	22.721	20.214	2.834	2.696
46	Pd	23.816	21.175	2.990	2.838
47	Ag	24.942	22.162	3.151	2.984
48	Cd	26.093	23.172	3.316	3.133
49	In	27.274	24.207	3.487	3.287
50	Sn	28.483	25.270	3.662	3.444
51	Sb	29.723	26.357	3.843	3.605
52	Te	30.993	27.471	4.029	3.769

TABLE 9.4 X-Ray Emission Energies in keV (*Continued*)

Atomic no.	Element	$K\beta_1$	$K\alpha_1$	$L\beta_1$	$L\alpha_1$
53	I	32.292	28.610	4.220	3.937
54	Xe	33.644	29.779	4.422	4.111
55	Cs	34.984	30.970	4.620	4.286
56	Ba	36.376	32.191	4.828	4.467
57	La	37.799	33.440	5.043	4.651
58	Ce	39.255	34.717	5.262	4.840
59	Pr	40.746	36.023	5.489	5.034
60	Nd	42.269	37.359	5.722	5.230
61	Pm	43.811	38.726	5.956	5.431
62	Sm	45.400	40.124	6.206	5.636
63	Eu	47.027	41.529	6.456	5.846
64	Gd	48.718	42.983	6.714	6.059
65	Tb	50.391	44.470	6.979	6.275
66	Dy	52.178	45.985	7.249	6.495
67	Ho	53.934	47.528	7.528	6.720
68	Er	55.690	49.099	7.810	6.948
69	Tm	57.487	50.730	8.103	7.181
70	Yb	59.352	52.360	8.401	7.414
71	Lu	61.282	54.063	8.708	7.654
72	Hf	63.209	55.757	9.021	7.898
73	Ta	65.210	57.524	9.341	8.145
74	W	67.233	59.310	9.670	8.396
75	Re	69.298	61.131	10.008	8.651
76	Os	71.404	62.991	10.354	8.910
77	Ir	73.549	64.886	10.706	9.173
78	Pt	75.736	66.820	11.069	9.441
79	Au	77.968	68.794	11.439	9.711
80	Hg	80.258	70.821	11.823	9.987
81	Tl	82.558	72.860	12.210	10.266
82	Pb	84.922	74.957	12.611	10.549
83	Bi	87.335	77.097	13.021	10.836
84	Po	89.809	79.296	13.441	11.128
85	At	92.319	81.525	13.873	11.424
86	Rn	94.877	83.800	14.316	11.724
87	Fr	97.483	86.119	14.770	12.029
88	Ra	100.136	88.485	15.233	12.338
89	Ac	102.846	90.894	15.712	12.650
90	Th	105.592	93.334	16.200	12.966
91	Pa	108.408	95.851	16.700	13.291
92	U	111.289	98.428	17.218	13.613
93	Np	114.181	101.005	17.740	13.945
94	Pu	117.146	103.653	18.278	14.279
95	Am	120.163	106.351	18.829	14.618
96	Cm	123.235	109.098	19.393	14.961
97	Bk	126.362	111.896	19.971	15.309
98	Cf	129.544	114.745	20.562	15.661
99	Es	132.781	117.646	21.166	16.018
100	Fm	136.075	120.598	21.785	16.379

Source: J. A. Dean, ed., *Lange's Handbook of Chemistry*, 14th ed., McGraw-Hill, New York, 1992.

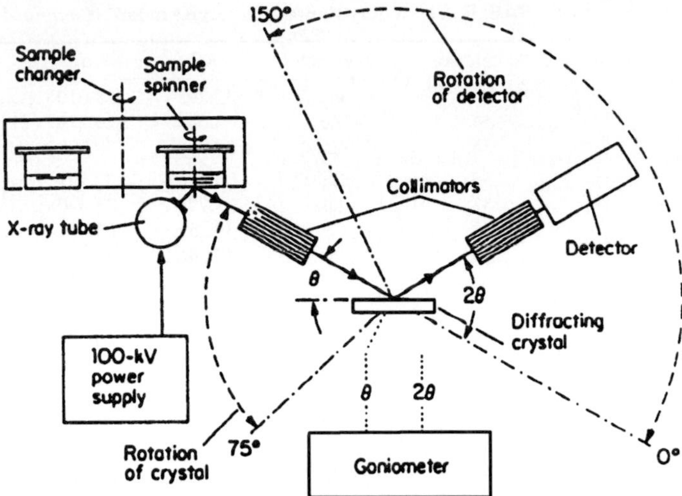

FIGURE 9.1 Geometry of a plane-crystal x-ray fluorescence spectrometer. *(Reprinted from G. F. Shugar and J. A. Dean, The Chemist's Ready Reference Handbook, McGraw-Hill, New York, 1990; courtesy of Philips Electronic Instruments.)*

9.2.2 Collimators and Filters

Radiation from an x-ray tube is collimated by a bundle of tubes, 0.5 mm or smaller in diameter and usually a few centimeters long. Collimators are placed between the sample and the analyzer crystal and between the analyzer crystal and the detector.

The *K* spectra of metals used as target material contain three strong lines, $K\alpha_1$, $K\alpha_2$, and $K\beta_1$. For x-ray diffraction, the $K\beta_1$ radiation can be reduced by using a thin foil filter, usually of the element of next lower atomic number to that of the target element. In general, an element with an absorption edge at a wavelength between two x-ray emission lines may be used as a filter to reduce the intensity of the line with the shorter wavelength; the longer-wavelength ($K\alpha$ in our example) lines are transmitted with a relatively small loss of intensity. Table 9.5, restricted to the *K* wavelengths of target elements in common use, lists the calculated thicknesses of β filters required to reduce the $K\beta_1$:$K\alpha_1$ integrated intensity ratio to 1:100. Filters are placed at the entrance slit of the monochromator (called a *goniometer* in x-ray work).

9.2.3 Analyzing Crystals

The analyzing crystal takes the place of a grating in a monochromator. Virtually monochromatic x radiation is obtained by reflecting x rays from crystal planes. The relationship between the wavelength of the x-ray beam λ, the angle of diffraction θ, and the distance between each set of atomic planes of the crystal lattice, *d*, is given by the Bragg condition:

$$m\lambda = 2d \sin \theta \tag{9.2}$$

The range of wavelengths usable with various analyzing crystals is governed by the *d* spacings of the crystal planes and by the geometric limits to which the goniometer can be rotated (see Fig. 9.1). The *d* value should be small enough to make the angle 2θ greater than 8°, even at the shortest wavelength used. A small *d* value is favorable for producing a large dispersion of the spectrum to give good separation of adjacent lines. On the other hand, a small *d* value imposes an upper limit to the range of wavelengths that can be analyzed. The goniometer is limited mechanically to about 150° for a 2θ value. A final requirement is the reflection efficiency.

TABLE 9.5 **β Filters for Common Target Elements**

Target element	$K\bar{\alpha}$, Å	Excitation voltage, keV	Absorber	$K\beta_1\ K\alpha_1 = {}^1/_{100}$ Thickness, mm	g/cm^2	% Loss $K\alpha_1$
Ag	0.560834	25.52	Pd	0.062	0.074	60
Mo	0.71069	20.00	Zr	0.081	0.053	57
Cu	1.54178	8.981	Ni	0.015	0.013	45
Ni	1.65912	8.331	Co	0.013	0.011	42
Co	1.79021	7.709	Fe	0.012	0.009	39
Fe	1.93728	7.111	Mn	0.011	0.008	38
			MnO_2	0.026	0.013	45
Cr	2.29092	5.989	V	0.011	0.007	37
			V_2O_5	0.036	0.012	48

Target element	$L\alpha_1$		Absorber	$L\beta_1\ L\alpha_1 = {}^1/_{100}$ Thickness, mm	g/cm^2	% Loss $L\alpha_1$
W	1.4763	10.200	Cu	0.035		77

Source: J. A. Dean, ed., *Lange's Handbook of Chemistry*, 14th ed., McGraw-Hill, New York, 1992.

Lithium fluoride is the optimum crystal for all wavelengths less than 3 Å. Pentaerythritol and potassium hydrogen phthalate are the crystals of choice for wavelengths from 3 to 20 Å. The long-wavelength analyzers are prepared by dipping an optical flat into the film of the metal fatty acid about 50 times to produce a layer of 180 molecules in thickness. Table 9.6 gives a list of crystals commonly used.

TABLE 9.6 **Analyzing Crystals for X-Ray Spectroscopy**

Crystal	Reflecting plane	2*d* Spacing, Å	Reflectivity
Quartz	$505\bar{2}$	1.624	Low
Aluminum	111	2.338	High
Topaz	303	2.712	Medium
Quartz	$20\bar{2}3$	2.750	Low
Lithium fluoride	220	2.848	High
Silicon	111	3.135	High
Quartz	112	3.636	Medium
Lithium fluoride	200	4.028	High
Sodium chloride	200	5.639	High
Calcium fluoride	111	6.32	High
Quartz	$10\bar{1}1$	6.686	High
Quartz	$10\bar{1}0$	8.50	Medium
Pentaerythritol (PET)	002	8.742	High
Ethylenediamine tartrate (EDT)	020	8.808	Medium
Ammonium dihydrogen phosphate (ADP)	110	10.648	Low
Gypsum	020	15.185	Medium
Mica	002	19.92	Low
Potassium hydrogen phthalate (KAP)	$10\bar{1}1$	26.4	Medium
Lead palmitate		45.6	
Strontium behenate		61.3	
Lead stearate		100.4	Medium

Source: J. A. Dean, ed., *Lange's Handbook of Chemistry*, 14th ed., McGraw-Hill, New York, 1992.

9.3 DETECTORS

9.3.1 Proportional Counters

A proportional counter consists of a central wire anode surrounded by a cylindrical cathode. The two electrodes are enclosed in a gastight envelope typically filled to a pressure of 80 torr of argon plus 20 torr of methane of ethanol (or 0.08 torr of chlorine). Radiation enters a thin window of mica, about 2 to 3 $mg \cdot cm^{-2}$ thick. For work at very long wavelengths, typical window materials are 1-μm aluminum screen dipped in Formvar (usable for sodium and magnesium x rays) and 1-μm cast Formvar or collodion films for oxygen, nitrogen, and boron x rays.

Each ionizing particle that enters the active volume of the counter collides with the filling gas to produce an ion pair (argon cation plus an electron). Under the influence of a potential gradient of 300 to 600 V applied across the electrodes, the initial electron soon acquires sufficient velocity to produce a new pair of ions upon collision with another atom of argon. This gives rise to an avalanche of electrons traveling toward the central anode. The output pulse is proportional to the number of primary pairs produced by the original ionizing particle.

The positive ions, if allowed to reach the cathode, would produce photons that would initiate a fresh discharge. Because the methane, ethanol, or halogen gas molecules have a lower ionization potential than argon, after a few collisions the ions moving toward the cathode consist of only these lower-energy particle that are unable to produce photons. When the organic filling-gas ions are discharged at the cathode, they decompose into various molecular fragments, and eventually the quenching gas is exhausted. Counter life is limited to about 10^{10} counts. Because chlorine atoms merely recombine, a halogen-quenched counter has a life in excess of 10^{13} counts.

The dead time, the time during which the counter will not respond to an entrant ionizing particle, is about 250 ns. Count rates up to 200 000 counts per second are possible; the upper limit is imposed by the associated electronic circuitry. About 30 eV is required for the production of an ion pair in this detector.

9.3.2 Scintillation Counters

Certain substances (called scintillators, phosphors, or fluors) will emit a pulse of visible light or near-ultraviolet radiation when they are subjected to x radiation. The light is observed by a photomultiplier tube (with light-amplification stages); the combination is called a *scintillation counter*. For x radiation a sodium iodide crystal doped with 1% thallium(I) is the best scintillator. When such radiation interacts with a NaI(Tl) crystal, iodine atoms are excited. Upon their return to the ground state, the reemitted ultraviolet radiation pulse is promptly absorbed by the thallium atom and, in turn, reemitted as fluorescent light at 410 nm (near the optimum wavelength response of a blue-sensitive photomultiplier tube). The crystal is sealed within an enclosure of aluminum foil that protects the crystal from atmospheric moisture and also serves as an internal reflector.

Dead time is 250 ns. Response is proportional to the energy of the x radiation; 500 eV is required to produce a photoelectron. The scintillation counter is usable throughout the important x-ray region, 0.3 to 2.4 Å, and possibly up to 4 Å.

9.3.3 Lithium-Drifted Semiconductor Detectors

The lithium-drifted germanium detector consists of a virtually windowless (Ge–Li) crystal, a vacuum cryostat maintained by cryosorption pumping, a liquid-nitrogen Dewar, and a preamplifier. Use of an electrically cooled detector (Peltier effect) eliminates the need for liquid nitrogen and allows the spectrometer to operate in any position. The solid-state Ge(Li) and Si(Li) detectors can be considered as a layered structure in which a lithium-diffused active region separates a *p*-type entry side from an *n*-type side. Detectors are fabricated by drifting lithium ions (a donor) into and through *p*-type germanium or silicon until only a layer of *p*-type material remains. All acceptors within the bulk material are compensated and this high-resistivity (or intrinsic) region becomes the radiation-sensitive region. Under reversed bias of approximately 800 to 1000 V, the active region serves as an insulator with an electric field gradient throughout its volume. When x radiation enters the intrinsic region, photoionization occurs with an electron–hole pair created for each 3.8 eV of photon energy for Si(Li) and 2.65 eV of photon energy for Ge(Li). The charge produced is rapidly collected under the influence of the bias voltage. The detector must be maintained at 77 K at all times (unless extremely pure germanium crystals, a recent development, are used). The charge collected each time an x-ray photon enters the detector is converted into a digital value representing the photon energy, which is interpreted as a memory address by a computer.

Silicon semiconductor detectors are preferred for x rays longer than 0.3 Å. For shorter wavelengths, Ge detectors are necessary since Si cannot absorb the x rays effectively because the available crystals are not deep enough. The response time is about 10 ns.

9.3.4 Comparison of X-Ray Detectors

For a given amount of energy absorbed, about 10 times as many electron–hole pairs are formed in solid-state Ge(Li) or Si(Li) detectors as the number of ion pairs in a gas proportional counter, and about 170 times as many electron–hole pairs as photoelectrons in NaI(Tl) scintillation counters. Since the relative resolution of a detector is proportional to the square root of the signal, the resolution of the Ge(Li) detector is about a factor of 13 better than the NaI(Tl) detector and 3 times better than a gas proportional counter.

9.4 PULSE-HEIGHT DISCRIMINATION

The measurement of pulse height provides the analyst with a tool for energy discrimination. The method is applicable whenever the amplitude of the pulse from an x-ray detector is proportional to the energy dissipation in the detector. All three types of detectors described meet this requirement.

To operate a pulse-height amplifier, the amplifier is adjusted to produce voltage output pluses of suitable magnitude. The pluses are then sorted into groups according to their pulse heights. The baseline discriminator is set to pass only those pulses above a certain amplitude. Finally, the second discriminator (variously called the window width, the channel width, or the acceptance slit) is adjusted to reject all pulses above the sum of the baseline discriminator and the acceptance slit. Only the pluses within the confines of these two discriminator settings pass on to the counting stages. A Si(Li) or Ge(Li) detector, because of its narrow pulse-amplitude discrimination, can discriminate (resolve) between elements one or two atomic numbers apart. The pulse-height analyzer serves, in effect, as a secondary monochromator. This process permits the accumulation of an emission spectrum and its subsequent display on a video screen.

An auxiliary electronic circuit, called the pulse pile-up rejection circuit, prevents charge collection from multiple photons that enter in rapid succession.

9.5 X-RAY ABSORPTION METHODS

9.5.1 Principle

Although complex experimental equipment is required, measurement in x-ray absorption methods is straightforward. Radiation traversing a layer of substance is diminished in intensity by a constant fraction per centimeter thickness x of material. The emergent radiant power P, in terms of incident radiant power P_0, is given by

$$P = P_0 \exp(-\mu x) \tag{9.3}$$

which defines the total linear absorption coefficient μ. Since the reduction of intensity is determined by the quantity of matter traversed by the primary beam, the absorber thickness is best expressed on a mass basis, in g · cm^2. The mass absorption coefficient μ/ρ, expressed in units cm^2·g, where ρ is the density of the material, is approximately independent of the physical state of the material and, to a good approximation, is additive with respect to the elements composing a substance.

Table 9.7 contains values of μ/ρ for the common target elements employed in x-ray work. A more extensive set of mass absorption coefficients for K, L, and M emission lines within the wavelength range from 0.7 to 12 Å is contained in Ref. 2.

In x-ray absorption work the specimen is irradiated with two or more monochromatic x rays while the incident (P_0) and transmitted (P) radiant energies are monitored. Because only one element has changed its mass absorption coefficient at an absorption edge, this relationship pertains:

$$2.3 \log \frac{P}{P_0} = (\mu'' - \mu')W\rho x \tag{9.4}$$

where the term in parentheses represents the difference in mass absorption coefficient at the edge discontinuity, W is the weight fraction of the element, and the product ρx is the mass thickness of the sample in grams per square centimeter. There is no matrix effect, which gives the absorption method an advantage over x-ray fluorescence analysis in some cases.

In contrast to absorption measurements in other portions of the electromagnetic spectrum, only a single attenuation measurement is made on each side of the edge because spectrometers that provide a continuously variable wavelength of x radiation are not commercially available. Instead, two x-ray emission lines are used for each edge. Primary excitation from an x-ray tube strikes a target. For example, a crystal of $SrCO_3$ would provide the Sr $K\alpha_1$ line and a crystal of RbCl would provide the Rb $K\alpha_1$ line. The Sr line lies on the short-wavelength side of the Br K edge and the Rb line on the long-wavelength side of the edge. After suitable standards have been run to calibrate the equipment, bromine can be determined in various materials such as dibromoethane in gasolines.

For the light elements whose valence electrons may be involved in x-ray absorption, the exact position of the absorption edge (fine structure) can give the oxidation state of the atom in question.

9.5.2 Radiography

The gross structure of various types of specimens may be examined by absorption techniques using relatively simple equipment. The microradiographic camera fits as an inset in the collimating system of commercial x-ray equipment. Photographic film with an extremely fine grain makes magnification up to 200 times possible. Sample thicknesses vary from a few hundredths to a few tenths of a millimeter; only a few seconds of exposure are necessary.

9.6 X-RAY FLUORESCENCE METHOD

9.6.1 Principle

When a sample is bombarded by a beam of x radiation that contains wavelengths shorter than the absorption edge of the spectral lines desired, characteristic secondary fluorescent x-ray spectra are emitted. The atom is in a highly excited state after such photoelectric absorption. An electron jumps from a higher to the lower shell to fill that vacancy emitting an x-ray photon. These x-rays

[2] K. F. J. Heinrich, in T. D. McKinley, K. F. J. Heinrich, and D. B. Wittry, eds., *The Electron Microprobe*, Wiley, New York, 1966, pp. 351–377.

TABLE 9.7 Mass Absorption Coefficients for $K\alpha_1$ Lines and W $L\alpha_1$ Line

Emitter Wavelength, Å / Absorber	Ag $K\alpha_1$ 0.559	Mo $K\alpha_1$ 0.709	Cu $K\alpha_1$ 1.541	Ni $K\alpha_1$ 1.658	Co $K\alpha_1$ 1.789	Fe $K\alpha_1$ 1.936	Cr $K\alpha_1$ 2.290	W $L\alpha_1$ 1.476
1 H	0.37	0.38	0.43	0.4	0.4	0.5	0.5	0.4
2 He	0.16	0.18	0.37	0.4	0.4	0.5	0.7	0.3
3 Li	0.18	0.22	0.50	0.6	0.7	0.9	1.5	0.4
4 Be	0.22	0.30	1.2	1.5	1.9	2.3	3.7	1.1
5 B	0.30	0.45	2.5	3.1	3.9	4.9	7.9	2.2
6 C	0.42	0.50	4.6	5.7	7.1	8.8	14.2	4.1
7 N	0.60	0.83	7.5	9.3	11.5	14.4	23.1	6.7
8 O	0.80	1.45	12.9	15.8	19.5	24.5	39.4	11.4
9 F	1.00	1.9	16.5	20.3	25.2	31.4	50.3	14.6
10 Ne	1.41	2.6	22.8	27.9	34.6	43.1	69.0	20.1
11 Na	1.75	3.5	30.3	37.2	45.9	57.2	91.4	26.8
12 Mg	2.27	4.6	39.5	48.4	59.8	74.6	119.1	34.9
13 Al	2.74	5.8	49.6	60.7	75.0	93.4	149.0	43.9
14 Si	3.44	7.3	61.4	75.2	92.8	115.5	183.8	54.4
15 P	4.20	8.8	74.7	91.4	112.9	140.5	223.6	66.2
16 S	5.15	10.6	89.2	109.2	134.7	167.4	266.1	79.1
17 Cl	5.86	12.4	104.8	128.2	158.1	196.6	312.4	92.8
18 Ar	6.40	14.5	121.4	148.5	183.0	227.3	360.7	107.6
19 K	8.0	16.7	139.8	171	211	262	415	124
20 Ca	9.7	18.9	158.6	194	239	296	469	141
21 Sc	10.5	21.8	180.5	221	272	337	534	160
22 Ti	11.8	25.3	203	247	304	378	597	180
23 V	13.3	27.7	228	278	342	424	77	202
24 Cr	15.7	31.0	254	311	382	474	88	226
25 Mn	17.4	34.5	282	344	423	63.5	101	250
26 Fe	19.9	38.1	311	380	57.6	71.4	113	276
27 Co	21.8	42.1	341	52.8	64.9	80.6	127	303
28 Ni	25.0	46.4	48.3	58.9	72.5	90.0	142	333
29 Cu	26.4	50.7	53.7	65.5	80.6	100.0	158	47.6
30 Zn	28.2	55.4	59.5	72.7	89.4	110.9	175	52.8
31 Ga	30.8	60.1	65.9	80.5	99.0	122.8	194	58.5
32 Ge	33.5	65.2	72.3	88.2	108.6	134.7	213	64.1
33 As	36.5	70.5	79.1	96.6	118.9	147	233	70.2
34 Se	38.5	76.0	86.1	105.1	129.4	161	254	76.4
35 Br	42.3	82.5	93.9	114.7	141.2	175	277	83.4
36 Kr	45.0	88.3	101.9	124.5	153.2	190	300	90.5
37 Rb	48	95	84	103	127	158	252	98
38 Sr	52	102	90	110	137	170	271	106
39 Y	56	109	97	119	147	183	292	114
40 Zr	61	17	104	128	158	197	314	122
41 Nb	66	18	112	138	170	212	338	132
42 Mo	71	19	119	146	180	225	358	140
43 Tc	76	20	128	157	194	241	384	150
44 Ru	12	22	137	168	207	258	410	160
45 Rh	13	23	146	179	221	275	438	171
46 Pd	14	24	155	190	235	292	466	182
47 Ag	15	26	165	202	249	310	493	193

(Continued)

TABLE 9.7 Mass Absorption Coefficients for $K\alpha_1$ Lines and W $L\alpha_1$ Line (*Continued*)

Emitter Wavelength, Å / Absorber	Ag $K\alpha_1$ 0.559	Mo $K\alpha_1$ 0.709	Cu $K\alpha_1$ 1.541	Ni $K\alpha_1$ 1.658	Co $K\alpha_1$ 1.789	Fe $K\alpha_1$ 1.936	Cr $K\alpha_1$ 2.290	W $L\alpha_1$ 1.476
48 Cd	15	28	174	213	263	327	520	204
49 In	16	30	185	227	280	347	553	217
50 Sn	17	32	195	239	295	367	583	229
51 Sb	19	34	206	252	310	386	612	241
52 Te	19	36	216	265	326	405	644	253
53 I	21	37	230	281	346	431	684	269
54 Xe	22	39	239	293	361	448	710	280
55 Cs	24	42	332	404	495	612	822	295
56 Ba	25	44	349	425	522	645	622	311
57 La	26	46	365	444	545	673	647	325
58 Ce	28	48	383	466	571	603	216	341
59 Pr	29	51	401	487	597	453	229	356
60 Nd	31	54	420	510	534	473	241	373
61 Pm	32	56	440	535		164	254	392
62 Sm	33	59	456	473	417	173	268	406
63 Eu	35	61	L_I 405	354	148	182	282	423
64 Gd	36	64	424	370	156	191	296	
65 Tb	38	67	L_{II} 316	135	164	201	311	393 L_I
66 Dy	39	70	329	141	172	211	327	293 L_{II}
67 Ho	41	72	L_{III} 123	148	181	222	343	304
68 Er	43	75	129	156	189	233	360	316
69 Tm	45	79	135	163	199	244	377	120 L_{III}
70 Yb	46	82	141	171	208	256	395	126
71 Lu	48	84	148	179	218	267	414	132
72 Hf	51	88	155	187	228	280	433	138
73 Ta	52	91	162	196	238	293	453	144
74 W	55	95	169	204	249	306	473	151
75 Re	57	98	176	213	260	319	494	157
76 Os	59	102	184	223	271	333	515	164
77 Ir	61	106	192	232	283	347	538	171
78 Pt	64	109	200	242	295	362	560	179
79 Au	67	113	209	252	307	377	584	186
80 Hg	69	117	218	263	321	394	609	194
81 Tl	72	121	227	275	334	411	635	203
82 Pb	74	125	236	286	348	428	662	211
83 Bi	78	129	247	298	363	446	690	220
84 Po		131	258	311	380	466	721	230
85 At			269	325	397	487	753	240
86 Rn	85		281	340	414	509	787	251
87 Fr		89	294	356	433	532	823	262
88 Ra	91		307	372	453	556	861	274
89 Ac			322	389	474	582	900	287
90 Th	97		337	408	497	610	944	301
91 Pa			353	427	520	639	988	315
92 U	104		372	450	548	673	898	332
93 Np			392	474	578	709	945	350
94 Pu		54	418	505	615	755	835	373

Source: J. A. Dean, ed., *Lange's Handbook of Chemistry*, 14th ed., McGraw-Hill, New York, 1992.

are characteristic of an element as the energy or wavelength is different for each element. Besides x rays, electron bombardment is used in the scanning electron microscope and the electron microprobe. Certain radionuclides are x-ray emitters and can be used as excitation sources, particularly in portable equipment.

The x-ray fluorescence method rivals the accuracy of wet chemical techniques in the analysis of major constituents. However, for trace analyses, it is difficult to detect an element present in less than one part in 10 000. In absolute terms the limit is about 10 ng. The method is attractive for elements that lack reliable wet chemical methods and for the analysis of nonmetallic specimens. X-ray fluorescence is one of the few techniques that offers the possibility of quantitative determinations with about 5% to 10% relative error without the use of a suite of standards. X-ray fluorescence method allows direct analysis of an element in solid substances.

9.6.2 Instrumentation

The general arrangement for exciting, dispersing, and detecting fluorescent radiation with a plane-crystal spectrometer is shown in Fig. 9.1. The sample is irradiated with an unfiltered beam of primary x rays. A portion of the fluorescence x radiation is collimated by the entrance slit of the goniometer and directed onto the surface of the analyzing crystal. The radiations reflected according to the Bragg condition [Eq. (9.2)] pass through the exit collimator to the detector.

For elements of atomic number less than 21, a vacuum of 0.1 torr is needed or the system must be flushed with helium. Below magnesium (atomic number 12) the transmission becomes seriously attenuated.

9.6.3 Sample Handling

Samples are best handled as liquids if they can be conveniently dissolved. Sample depth should be at least 5 mm so that the sample will appear infinitely thick to the primary x-ray beam. If possible, solvents that do not contain heavy atoms should be used. Water and nitric acid are superior to hydrochloric or sulfuric acid. Powders are best converted into a solid solution by fusion with lithium borate (both light elements). Powders can also be pressed into a wafer but should be heavily diluted with a material that has a low absorption, such as powdered starch, lithium carbonate, lampblack, or gum arabic to avoid matrix effects (as done by a borax fusion).

9.6.4 Matrix Effects

Matrix effects arise from the interaction of elements in the sample to affect the x-ray emission intensity in a nonlinear manner. They are often negligible when thin samples, which may be collected on a filter, mesh, or membrane, are used for analyses. The most practical way to correct for matrix effects is to use the internal standard technique. Even so, this technique is only valid if the matrix elements affect the reference line and analytical line in exactly the same way. The following list presents the potential problems:

1. If a disturbing element has an absorption edge between the comparison lines, preferential absorption of the line on the short-wavelength side of the edge will occur.
2. If fluorescence from a matrix line lies between the absorption edges of the analytical and reference elements, selective enhancement results for the element whose absorption edge lies at the longer wavelength.
3. Line intensity can be enhanced if a matrix element absorbs primary radiation and then, by fluorescence, emits radiation that in turn is absorbed by a sample element and causes that latter to fluoresce more strongly.

9.7 ENERGY-DISPERSIVE X-RAY SPECTROMETRY

Energy-dispersive x-ray spectrometry differs from x-ray fluorescence in that x rays emitted from elements in the sample are observed by a solid-state detector and pulse-height analyzer without using a crystal analyzer and collimators with their attendant energy losses. Equipment is very compact. Computer-based multichannel analyzers are then used to acquire a spectrum of counts versus energy and to perform data analysis. As this information is being collated, it can be simultaneously displayed as a spectrum on a video screen. Computer-generated emission lines can be superimposed on the spectrum displayed. Often elements can be identified by the line positions alone. When overlap of the emission lines from one or more elements may be possible, one should use the relative intensities of the lines as well as their positions to confirm the elemental identification. Detection limits in bulk material are typically a few parts per million.

Energy-dispersive x-ray spectrometry offers the advantage of being a simultaneous multielement method for samples with elements separated by one or two atomic numbers. Quantitative determinations can range from the use of simple intensity–concentration standard working curves to computer programs that convert intensity to concentration. Emission lines that have no spectral overlap can be measured from a spectrum using software to perform integration of the net peak intensity above the background for the peak of interest. Cases in which peak overlap is of concern require spectrum-fitting techniques from a library of reference spectra. Care must be taken to maintain the accuracy of the energy calibration of the instrumentation.

Primary x-ray tubes are operated at low power and can be air-cooled. The tube anode material is usually Rh, Ag, or Mo. A filter wheel, typically fitted with six filters, is located between the tube and the sample. The x-ray tube can also be focused on secondary targets to produce specific fluorescent x-ray lines.

Portable instruments are designed around x rays emitted from a radionuclide source. A variety of isotopes is needed to provide radiation over the energy range needed.

Unfortunately, the resolution of an energy-dispersion instrument is as much as 50 times less than the wavelength-dispersion spectrometer using a crystal analyzer. For precise quantitative measurements, wavelength dispersion followed by energy-dispersive detectors and pulse-height analyzers must be used to get sufficient resolution.

9.8 ELECTRON-PROBE MICROANALYSIS

When materials are bombarded by a high-energy electron beam, as in an electron microscope, x-ray fluorescence radiation is produced. By incorporating an x-ray fluorescence spectrometer directly into the instrument, it is possible to obtain the same sort of elemental data as obtained by normal x-ray fluorescence directly on the area viewed or scanned by the electron beam. It is possible to obtain qualitative and quantitative elemental data from a volume of 0.1 μm^3 for elements of atomic number 5 (boron) and higher. When the electron beam is scanned across the sample, a point-by-point spatial distribution of elements across the surface (or area) of the sample is produced. Excitation is restricted to thin surface layers because the electron beam penetrates to a depth of only 1 or 2 μm into the specimen.

The electron optical system consists of an electron gun followed by two electromagnetic focusing lenses to form the electron-beam probe. The specimen is mounted as the target inside the high-vacuum column of the instrument. If not conductive, the surface of the specimen must be treated with a coating to make its surface conductive in order to avoid the problems of specimen charging with electron bombardment.

9.9 ELECTRON SPECTROSCOPY FOR CHEMICAL APPLICATIONS (ESCA)

Electron spectroscopy for chemical applications, or x-ray photoelectron spectroscopy (XPS), is a nondestructive spectroscopic tool for studying the surfaces of solids. Any solid material can be studied and all elements (except hydrogen) can be detected by this technique, usually at 0.1 atomic percent abundance.

9.9.1 Principle

When a specimen is exposed to a flux of x-ray photons of known energy, all electrons whose binding energies E_b are less than the energy of the exciting x rays are ejected. The kinetic energies E_{kin} of these photoelectrons are then measured by an energy analyzer in a high-resolution electron spectrometer. For a free molecule or atom, conservation of energy requires that

$$E_b = h\nu - E_{\text{kin}} - \phi \tag{9.5}$$

where $h\nu$ is the energy of the exciting radiation and ϕ is the spectrometer work function, a constant for a given analyzer. The binding energy is indicative of a specific element and a particular structural feature of electron distribution.

9.9.2 Chemical Shifts

The binding energies of core electrons are affected by the valence electrons and therefore by the chemical environment of the atom. Chemical shifts, as they are called, are observed for every element except hydrogen and helium. Applicability to carbon, nitrogen, and oxygen makes ESCA an important structural tool for organic materials. In general, the ESCA chemical shifts lie in the range 0 to 1500 eV; peak widths vary from 1 to 3 eV. To make assignments, one must refer to a catalog of element reference spectra or to correlation charts.

9.9.3 ESCA Instrumentation

Instrumentation for ESCA involves (1) a source of soft x rays, usually Mg $K\alpha_{1,2}$ and Al $K\alpha_{1,2}$, (2) a device that collects the emitted electrons, counts them, and carefully measures their kinetic energy (an energy analyzer), and (3) a vacuum system capable of providing an operating pressure of about 5×10^{-6} torr. Samples must be low-pressure solids or liquids condensed onto a cryogenic probe. The two alternative sources are needed to distinguish ESCA peaks from Auger peaks. When a different x-ray source is used, the ESCA peaks shift in kinetic energy but the kinetic energies of Auger peaks remain constant and appear at the same energy position in the spectrum.

One type of ESCA spectrometer places the x-ray anode, a spherically bent crystal disperser, and the sample on a Rowland circle. In the energy analyzer an electrostatic field sorts the electrons by their kinetic energies and focuses them at the detector. A continuous channel electron multiplier counts the electrons at each step as the electrostatic field is increased in a series of small steps. A plot of the counting rate as a function of the focusing field gives the spectrum.

9.9.4 Uses

The intensity of a photoelectron line is proportional not only to the photoelectric cross section of a particular element but also to the number of atoms of that particular element present in the sample. Analyses of mixtures are often accurate to ±2%.

Because of the unique ability of ESCA to study surfaces, the technique is used to study heterogeneous catalysts, polymers and polymer adhesion problems, and materials such as metals, alloys, and semiconductors. The probe depth is 1 to 2 nm.

9.10 AUGER EMISSION SPECTROSCOPY (AES)

Once an atom is ionized, it must relax by emitting either an x-ray photon or an electron. The latter is the nonradiative Auger process that nature chooses in most instances, and, increasingly so, for elements of atomic number less than 30. A combination unit involves an electron gun centered with an energy analyzer unit. The electron gun is focused on the sample; the detector is at the opposite end of the energy analyzer.

A *KLL* Auger transition involves these processes: A *K* electron undergoes the initial ionization. An *L*-level electron moves in to fill the *K*-level vacancy and, at the same time, gives up the energy of that transition to another *L*-level electron. The latter becomes the ejected Auger electron. The energy of the ejected electron is a function only of the atomic energy levels involved in the Auger transition and is thus characteristic of the atom from which it came. Most elements have more than one intense Auger peak; *LMM* and *MNN* are other transitions. A recording of the spectrum of energies of Auger electrons released from any surface is compared with the known spectra of pure elements. Typically the sampling depth is about 2 nm.

Shifts from one element to the next are about 25 eV; peak positions can be measured to an accuracy of ± 1 eV. Spectra of all the elements lie between 50 and 1000 eV. Both qualitative and quantitative analyses of the elements in the immediate surface atomic layers are possible with Auger emission spectroscopy. When combined with a controlled removal of surface layers by ion sputtering, AES provides the means to solve some very important problems involving surfaces.

9.11 X-RAY DIFFRACTION

9.11.1 Principle

When a beam of monochromatic x radiation is directed at a crystalline material, one observes reflection or diffraction of the x rays at various angles with respect to the primary beam. The relationship between the wavelength of the x radiation, the angle of diffraction, and the distance between each set of atomic planes of the crystal lattice is given by the Bragg condition [Eq. (9.2)]. From the Bragg condition one can calculate the interplanar distances of the crystalline material. The interplanar spacings depend solely on the geometry of the crystal's unit cell while the intensities of the diffracted x rays depend on the type of atoms in the crystal and the location of the atoms in the fundamental repetitive unit, the unit cell.

From the rearranged Bragg equation

$$\theta = \sin^{-1}\left(\frac{\lambda}{2d}\right) \tag{9.6}$$

the two factors within the parentheses control the choice of x radiation. Because the parenthetical term cannot exceed unity, the use of long-wavelength radiation limits the number of reflections that are observed, whereas short-wavelength radiation tends to crowd individual reflections very close together. Furthermore, radiation just shorter in wavelength than the absorption edge of an element in the specimen should be avoided because the resulting fluorescent radiation increases the background. For these reasons, a multiwindow x-ray tube with anodes of Ag, Mo, W, and Cu is used.

Diffractometer alignment procedures require the use of a well-prepared polycrystalline specimen. Two standard samples found to be suitable are silicon and α-quartz (including Novaculite). The 2θ values of several of the most intense reflections for these materials are listed in Table 9.8. To convert to *d* for average $K\bar{\alpha}$ values or to *d* for $K\alpha_2$, multiply the tabulated *d* value (Table 9.8) for $K\alpha_1$ by the factor given below:

Element	$K\bar{\alpha}$ (average)	$K\alpha_2$
W	1.007 69	1.023 07
Ag	1.002 63	1.007 89
Mo	1.002 02	1.006 04
Cu	1.000 82	1.002 48
Ni	1.000 77	1.002 32
Co	1.000 77	1.002 16
Fe	1.000 67	1.002 04
Cr	1.000 57	1.001 70

TABLE 9.8 Interplanar Spacings for $K\alpha_1$ Radiation, d versus 2θ

					α-Quartz (including novaculite)					
hkl	100	101	110	102	200	112	202	211	203	301
d, Å	4.260	3.343	2.458	2.282	2.128	1.817	1.672	1.541	1.375	1.372
W $K\alpha_1$: 2θ	2.81	3.58	4.87	5.25	5.63	6.59	7.17	7.78	8.72	8.74
Ag $K\alpha_1$: 2θ	7.53	9.60	13.07	14.08	15.10	17.71	19.26	20.91	23.47	23.52
Mo $K\alpha_1$: 2θ	9.55	12.18	16.59	17.88	19.19	22.51	24.49	26.61	29.89	29.96
Cu $K\alpha_1$: 2θ	20.83	26.64	36.52	39.45	42.44	50.16	54.86	59.98	68.14	68.31
Ni $K\alpha_1$: 2θ	22.44	28.71	39.42	42.60	45.85	54.28	59.44	65.08	74.15	74.34
Co $K\alpha_1$: 2θ	24.24	31.04	42.68	46.15	49.71	58.98	64.68	70.96	81.16	81.38
Fe $K\alpha_1$: 2θ	26.27	33.66	46.38	50.20	54.11	64.38	70.75	77.83	89.50	89.74
Cr $K\alpha_1$: 2θ	31.18	40.05	55.52	60.22	65.09	78.11	86.42	95.96	112.73	113.11

					Silicon					
hkl	111	220	311	400	331	422	511.333	440	531	620
d, Å	3.1353	1.91997	1.63736	1.357630	1.24584	1.1085	1.0451	0.0959986	0.917922	0.858637
W $K\alpha_1$: 2θ	3.82	6.24	7.32	8.83	9.62	10.82	11.48	12.50	13.07	13.98
Ag $K\alpha_1$: 2θ	10.24	16.75	19.67	23.78	25.95	29.23	31.04	33.88	35.48	38.02
Mo $K\alpha_1$: 2θ	12.99	21.29	25.02	30.28	33.08	37.32	39.67	43.36	45.45	48.79
Cu $K\alpha_1$: 2θ	28.44	47.30	56.12	69.13	76.38	88.03	94.96	106.71	114.10	127.55
Ni $K\alpha_1$: 2θ	30.66	51.16	60.83	75.26	83.42	96.80	104.96	119.42	129.12	149.76
Co $K\alpha_1$: 2θ	33.15	55.53	66.22	82.42	91.77	107.59	117.71	137.42	154.04	
Fe $K\alpha_1$: 2θ	35.97	60.55	72.48	90.96	101.97	121.67	135.70			
Cr $K\alpha_1$: 2θ	42.83	73.21	88.72	114.97	133.53					

Source: *Tables of Interplanar Spacings d vs. Diffraction Angle 2θ for Selected Targets*, Picker Nuclear, White Plains, NY, 1966.

The applications of x-ray diffraction can be conveniently considered under three main headings: powder diffraction, polymer characterization, and single-crystal structure studies. In addition there are many specialized uses.

9.11.2 Powder Diffraction

In the powder method, the sample is a large collection of very small crystals, randomly oriented. A polycrystalline aggregate is formed into a cylinder whose diameter is smaller than the diameter of the incident x-ray beam. The diffraction pattern in a series of nonuniformly spaced cones (as intercepted on the photographic film) whose spacings are determined by the prominent planes of the crystallites.

Metal samples may be machined to a cylindrical configuration, plastic materials can often be extruded through suitable dies, and all other samples are best ground to a fine powder (200 to 300 mesh) and shaped into thin rods after mixing with collodion binder or simply tamped into a uniform glass capillary. Liquids must be converted into crystalline derivatives.

The x-ray pattern of a pure crystalline substance can be considered as a "fingerprint" with each crystalline material having, within limits, a unique diffraction pattern. The ASTM has published the powder diffraction patterns of some 50 000 compounds. An identification of an unknown compound is made by comparing the interplanar spacings of the powder pattern of a sample to the compounds in the ASTM file. The d value for the most intense line of the unknown is looked up first in the file. In a single compound, the d values of the next two most intense lines are matched against the file values. After a suitable match for one component is obtained, all the lines of the identified component are omitted from further consideration. The intensities of the remaining lines are rescaled by setting the strongest intensity equal to 100 and repeating the entire procedure. If x-ray fluorescence data are also added, the comparison is even more definitive. A systematic search, either manual or by computer, usually leads to an identification within an hour. Mixtures of up to nine compounds can often be completely identified. The minimum limit of detection is about 1% to 2% of a single phase.

In addition to identifying the compounds in a powder, the diffraction pattern can also be used to determine the degree of crystalline disorder, crystalline size, texture, and other parameters associated with the state of the crystalline materials. Differentiation among various oxides such as MnO, Mn_2O_3, MnO_2, and Mn_3O_4, or between materials present in such mixtures as NaCl + KBr or KCl + NaBr, is easily accomplished by x-ray diffraction. Identification of various hydrates, such as $Na_2CO_3 \cdot H_2O$ and $Na_2CO_3 \cdot 10H_2O$, is another application.

9.11.3 Polymer Characterization

The following information can be obtained from wide-angle and small-angle x-ray studies of polymers:

1. Degree of crystallinity
2. Crystallite size
3. Degree and type of preferred orientation
4. Polymorphism
5. Microdiffraction patterns
6. Information concerning the macrolattice of the crystallites

Fibers and partially oriented samples show spotty diffraction patterns rather than uniform cones; the more oriented the specimen, the spottier the pattern. The degree of crystallinity and crystallite size can be measured in powders, films, and fibers. This determination usually takes a few hours. The degree of orientation in uniaxially oriented materials, such as fibers, can also be determined in a few hours. If the orientation is other than uniaxial, the type of orientation and an approximate determination of the degree of orientation require similar times.

Polymorphism, the phenomenon in which a chemical compound can exist in more than one crystalline form, is often exhibited by polymers. From a study of the diffraction pattern, the presence of polymorphism can be ascertained and the approximate percentages of the polymorphs present determined. Such a study usually requires 1 to 2 days.

Micro x-ray diffraction can be used to identify inclusions in polymers and other materials. It can also be used to study the relationship of the skin of fibers to their interiors in regard to crystallinity, crystallite size, and orientation. Areas as small as 25 μm^2 can be studied. One limitation is that the specimen must be transparent to the x-ray beam. For most materials, a thickness up to 1 mm can be tolerated. Studies require 1 to 2 days.

Small-angle x-ray scattering is used to obtain information about the macrolattice, which can be defined as a periodic array of matter in space that is greater than 5.0 nm in size. In many polymers the chains fold back on themselves, forming crystallites. This leads to crystalline and amorphous domains ranging in size from less than 10 nm to almost 100 nm. Small-angle scattering studies are used to determine the size, distribution, and orientation of these domains.

9.11.4 Single-Crystal Structure Determination

Three-dimensional molecular structures can be determined using single-crystal x-ray diffraction procedures. The structures of molecules of considerable complexity (up to about 100 atoms) can be completely elucidated. In addition to the chemical structure, a crystal determination reveals the configuration and conformation of the molecule in the solid state.

The crystal, less than 1 mm in size, is affixed to a thin glass capillary (with shellac) that in turn is fastened to a brass pin that is mounted in the x-ray diffraction unit. In the rotating-crystal method, monochromatic x radiation is incident on the crystal that is rotated about one of its axes. The reflected beams lie as spots on the surface of cones that are coaxial with the rotation axis.

Bibliography

Bertin, E. P., *Principles and Practice of X-Ray Spectrometric Analysis*, 2d ed., Plenum, New York, 1978.

Jenkins, R., R. W. Gould, and D. Gedcke, *Quantitative X-Ray Spectrometry*, Dekker, New York, 1981.

Tertian, R., and F. Claisse, *Principles of Quantitative X-Ray Fluorescence Analysis*, Heyden, London, 1982.

Thompson, M., M. D. Baker, A. Christie, and J. F. Tyson, "Auger Emission Spectroscopy," in Vol. 74 of *Chemical Analysis*, Wiley, New York, 1985.

SECTION 10
MASS SPECTROMETRY

Mass spectrometry is the analytical technique that provides the most structural information for the least amount of analyte material. It provides qualitative and quantitative information about the atomic and molecular composition of inorganic and organic materials and their chemical structures. As an analytical technique it possesses distinct advantages:

1. Increased sensitivity over most other analytical techniques because the analyzer, as a mass-charge filter, reduces background interference.
2. Excellent specificity from characteristic fragmentation patterns to identify unknowns or confirm the presence of suspected compounds.
3. Information about molecular weight.
4. Information about the isotopic abundance of elements.

Mass spectrometry often fails to distinguish between optical and geometrical isomers and the positions of substituent in *o*-, *m*- and *p*- positions in an aromatic ring. Also, its scope is limited in identifying hydrocarbons that produce similar fragmented ions.

Application of mass spectrometry for the analysis of pesticides and herbicides is discussed in Sec. 20 and in the environmental analysis of trace organic pollutants is highlighted in Sec. 21.

10.1 INSTRUMENT DESIGN

Functionally, all mass spectrometers have these components (Fig. 10.1): (1) inlet sample system, (2) ion source, (3) ion acceleration system, (4) mass (ion) analyzer, (5) ion-collection system, usually an electron multiplier detector, (6) data-handling system, and (7) vacuum system connected to components (1) through (5). To provide a collision-free path for ions once they are formed, the pressure in the spectrometer must be less than 10^{-6} torr.

10.1.1 Inlet Sample Systems

Gas samples are transferred from a vessel of known volume (3 mL), where the pressure is measured, into a reservoir (3 to 5 L). Volatile liquids are drawn through a sintered disk into the low-pressure reservoir in which they are vaporized instantly. Oftentimes a nonvolatile compound can be converted into a derivative that has sufficient vapor pressure.

The gaseous sample enters the source through a pinhole in a piece of gold foil. For analytical work, molecular flow (where the mean free path of gas molecules is greater than the tube diameter) is usually preferred. However, in isotope-ratio studies viscous flow (where the mean free path is

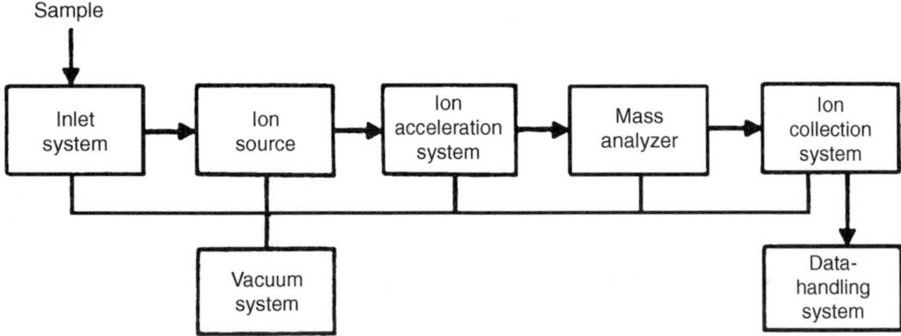

FIGURE 10.1 Components of a mass spectrometer. (*From Shugar and Dean, The Chemist's Ready Reference Handbook, McGraw-Hill, New York, 1990.*)

smaller than the tube diameter) is employed to avoid any tendency for various components to flow differently from the others.

10.2 IONIZATION METHODS IN MASS SPECTROMETRY

Ionization methods in mass spectrometry are divided into gas-phase ionization techniques and methods that form ions from the condensed phase inside the ion source. All ion sources are required to produce ions without mass discrimination from the sample and to accelerate them into the mass analyzer. The usual source design has an ion withdrawal and focusing system. The ions formed are removed electrostatically from the chamber. Located behind the ions is the repeller, which has the same charge as the ions to be withdrawn. A strong electrostatic field between the first and second accelerating slits of 400 to 4000 V, which is opposite in charge to the ions, accelerates the ions to their final velocities.

10.2.1 Electron Ionization

The electron ionization source is a commonly used ionization method. The ionizing electrons from the cathode of an electron gun located perpendicular to the incoming gas stream collide with the sample molecules to produce a molecular ion. A source operating at 70 V, the conventional operating potential, also has sufficient energy to cause the characteristic fragmentation of sample molecules.

Some compounds do not produce a molecular ion in an electron ionization source. This is a disadvantage of this source.

A mass spectrometer is calibrated in the electron ionization mode. Perfluoroalkanes are often used as markers because they provide a peak at intervals of masses corresponding to CF_2 groups.

10.2.2 Chemical Ionization[1]

Chemical ionization results from ion–molecule chemical interactions that involve a small amount of sample with an exceedingly large amount of a reagent gas. The source must be tightly enclosed with an inside pressure of 0.5 to 4.0 torr. The pressure outside the source is kept about 4 orders of magnitude less than the inside by a differential pumping system.

Often the primary reason for using this technique is to determine the molecular weight of a compound. For this purpose a low-energy reactant, such as *tert*-$C_4H_9^+$ (from isobutane) is frequently used. In the first step the reagent gas is ionized by electron ionization in the source. Subsequent reactions between the primary ion and additional reagent gas produce a stabilized reagent gas plasma. When a reagent ion encounters a sample molecule (MH), several products may be formed:

MH_2^+ by proton transfer

M^+ by hydride abstraction

MH^+ by charge transfer

Practically all the spectral information will be clustered around the molecular ion, or one mass unit larger or smaller, with little or no fragmentation. This type of ionization is desirable when an analysis of a mixture of compounds is needed and the list of possible components is limited. The general absence of carbon–carbon cleavage reactions for the chemical ionization spectra means that they provide little skeletal information.

[1] B. Munson, "Chemical Ionization Mass Spectrometry," *Anal. Chem.* **49**:772A (1977).

Negative chemical ionization[2] can be conducted with hydroxide and halide ions. For these studies the charges on the repeller and accelerating slits in the ion source are reversed with the repeller having a negative charge.

10.2.3 Other Ionization Methods

The less frequently used ionization methods receive only brief mention here. For more details consult the references cited.

Field ionization[3] and *field desorption*[4] are techniques used for studying surface phenomena, such as adsorbed species and trapped samples, and the results of chemical reactions on surfaces; they are also suitable for handling large lipophilic polar molecules.

Fast atom bombardment[5] and *plasma (californium-252) desorption*[6] techniques deal rather effectively with polar substances (usually of higher molecular weight) and salts. Samples may be bulk solids, liquid solutions, thin films, or monolayers.

In *thermal ionization* the sample is put on a filament substrate (a metal ribbon), which is heated in the mass spectrometer source until the sample evaporates (ca. 2000°C). Filament-loading procedures tend to be element-specific. Both positive and negative ions are produced, and thermal ionization usually results in the formation of long-lived, stable ion beams. Thermal ionization is appropriate for inorganic compounds that have ionization potentials in the range from 3 to 6 eV. On the other hand, the technique is inefficient for organic compounds because their ionization potentials usually range from 7 to 16 eV.

Laser desorption methods[7–9] produce a microplasma that consists of neutral fragments together with elementary molecular and fragment ions. Suitable mass spectrometers are limited to time-of-flight and Fourier-transform spectrometers.

The recent development of *electrospray ionization*[10] has extended the range of masses amenable to study by mass spectrometery to above several hundred kilodaltons, and commercial instruments are available.

10.3 MASS ANALYZERS

The function of the mass analyzer is to separate the ions produced in the ion source according to their different mass–charge ratios. The analyzer section is continuously pumped to a very low vacuum so that ions may be passed through it without colliding with the gas molecules. The energies and velocities v of the ions moving into the mass analyzer are determined by the accelerating voltage V from the ion source slits and the charge z on the ions of mass m:

$$\frac{1}{2}m_1v_1^2 = \frac{1}{2}m_2v_2^2 = \frac{1}{2}m_3v_3^2 = \cdots = zV \tag{10.1}$$

[2] R. C. Dougherty, "Negative Chemical Ionization Mass Spectrometry," *Anal. Chem.* **53**:625A (1981).
[3] M. Anbar and W. H. Aberth, "Field Ionization Mass Spectrometry," *Anal. Chem.* **46**:59A (1974).
[4] W. D. Reynold, "Field Desorption Mass Spectrometry," *Anal. Chem.* **51**:283A (1979).
[5] M. Barber et al., "Fast Atom Bombardment Mass Spectrometry," *Anal. Chem.* **54**:645A (1982).
[6] R. D. MacFarlane, "Californium-252 Plasma Desorption Mass Spectrometry," *Anal. Chem.* **55**:1247A (1983).
[7] R. J. Cotter, "Lasers and Mass Spectrometry," *Anal. Chem.* **56**:485A (1984).
[8] E. R. Denoyer et al., "Laser Microprobe Mass Spectrometry: Basic Principles and Performance Characteristics," *Anal. Chem.* **54**:26A (1982).
[9] D. M. Hercules et al., "Laser Microprobe Mass Spectrometry: Applications to Structural Analysis," *Anal. Chem.* **54**:280A (1980).
[10] C. M. Whitehouse et al., *Anal. Chem.* **57**:675 (1985).

10.3.1 Magnetic-Deflection Mass Analyzer

In a single-focusing magnetic-sector mass analyzer, the ion source, the collector slit, and the apex of the sector shape (usually 60°) are colinear. Upon entering the magnetic field, the ions are classified and segregated into beams, each with a different m/z ratio.

$$\frac{m}{z} = \frac{H^2 r^2}{2V} \tag{10.2}$$

where H is the strength of the magnetic field and r is the radius of the circular path followed by the ions. Since the radius and the magnetic field strength are fixed for the particular sector instrument, only ions with the proper m/z ratio will pass through the analyzer tube without striking the walls, where they are neutralized and pumped out of the system as neutral gas molecules. Focusing is accomplished by changing either the electrostatic accelerating voltage or the magnetic field strength; often the former is allowed to diminish while the spectrum is scanned. Each m/z ion from light to heavy is successively swept past the detector slit at a known rate. The detector current is amplified and displayed on a strip-chart recorder. Since the ion paths are separated from one another, the recorder signal will fall to the baseline and then rise as each mass strikes the detector. The height of the peaks on the chart will be proportional to the number of ions of the corresponding mass–charge ratio.

A magnetic-sector mass analyzer has a mass range of 2500 Da at 4-kV ion accelerating voltage. Mass resolution is continuously variable up to 25 000 (10% valley definition). Metastable peaks that aid in structural elucidation are also recorded.

10.3.2 Double-Focusing Sector Spectrometers

Because single-focusing mass analyzers are not velocity focusing for ions of a given mass, their resolving power is limited. In double-focusing mass spectrometers an electrostatic deflection field is incorporated between the ion source and the magnetic analyzer. Resolving power lies in the range of 100 000. Additional focusing is achieved with quadrupole lenses placed before the electrostatic field and between the electrostatic and magnetic fields.

10.3.3 Quadrupole Mass Analyzer

In the quadrupole mass analyzer, ions from the ion source are injected into the quadrupole array, shown in Fig. 10.2. Opposite pairs of electrodes are electrically connected; one pair at $+U_{dc}$ volts and the other pair at $-U_{dc}$ volts. An rf oscillator supplies a signal to each pair of rods, but the signal to the second pair is retarded by 180°. When the ratio U_{dc}/V_{rf} is controlled, the quadrupole field can be set to pass ions of only one m/z ratio down the entire length of the quadrupole array. When the dc and rf amplitudes are changed simultaneously, ions of various mass–charge ratios will pass successively through the array to the detector and an entire mass spectrum can be produced.

Registration of negative ions, as from a chemical ionization source, is possible with two electron multipliers, one for positive and one for negative ions.

Scan rates can reach 780 Da · s^{-1} before resolution is significantly affected. The quadrupole mass analyzer is ideal for coupling with a gas chromatograph. Practical m/z limits are 4000 Da.

10.3.4 Time-of-Flight Spectrometer

In the time-of-flight (TOF) mass spectrometer, the ions leave the source as discrete ion packets by pulsing the voltage on the accelerating slits at the exit of the ion source. Upon leaving the accelerating slits, the ions enter into the field-free region (drift path) of the flight tube, 30 to 100 cm long,

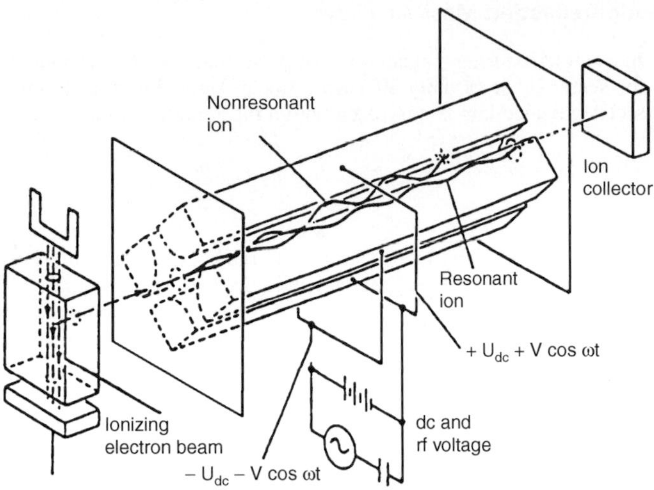

FIGURE 10.2 Quadrupole mass analyzer. (*From Shugar and Dean, 1990.*)

with whatever velocity they have acquired [Eq. (15.1)]. Because their velocities are inversely proportional to the square roots of their masses, the lighter ions travel down the flight tube faster than the heavier ions. The original ion packet becomes separated into "wafers" of ions according to their mass–charge ratio. The wafers are collected sequentially at the detector.

A spectrum can be recorded every 10 s. This makes the TOF mass spectrometer suitable for kinetic studies and for coupling with a gas chromatograph to examine effluent peaks.

10.3.5 Ion-Trap Mass Spectrometer

A quadrupole ion-trap consists of three electrodes; two end-cap electrodes normally are held at ground potential and between them a ring electrode to which an rf potential, often in the megahertz range, is applied to generate a quadrupole electric field. These components can be held in the palm of the hand. Ionization in ion traps is commonly achieved by electron ionization, which occurs within the trap. Chemical ionization uses the variable time scale of the ion trap first to generate reagent ions via electron impact and then allows these reagent ions to react with the vaporized analyte molecules. Both ionization methods are limited to gaseous samples.

Desorption ionization methods enable mass spectrometry application to fragile nonvolatile compounds, which can be implemented by forming ions in an external source by fast ion bombardment or secondary ion mass spectrometry, and then injecting them into the trap. Although trapped ions can be mass-analyzed by several methods, a mass-selective instability scan is used most commonly. In this procedure, a change in operating voltages is used to cause trapped ions of a particular m/z ratio to adopt unstable trajectories. By scanning the amplitude of the rf voltage applied to the ring electrode, ions of successively increasing m/z are made to adopt unstable trajectories and to exit the ion trap, where they can be detected by using an externally mounted electron multiplier. Other methods for mass analysis have been described.[11]

[11] R. G. Cooks et al., *Chem. Eng. News* **1991**(March 25):26.

10.3.6 Additional Mass Analyzers

Space precludes more than mention of the more sophisticated mass analyzers, such as the Fourier-transform (ion-trap) mass spectrometer,[12,13] tandem mass spectrometers,[14] triple quadrupole mass spectrometer,[15] and inductively coupled plasma–mass spectrometer.[16] Triple quadrupole instruments are now routinely used in protein structure determinations, pesticide residue analysis, and drug metabolism studies.

10.3.7 Resolving Power

The most important parameter of a mass analyzer is its resolving power. Using the *10% valley definition*, two adjacent peaks (whose mass differences are Δm) are said to be separated when the valley between them is 10% or less of the peak height (and the peak heights are approximately equal). For this condition, $\Delta m/m$ equals the peak width at a height that is 5% of the individual peak height.

A resolution of 1 part in 800 adequately distinguishes between m/z values 800 and 801 so long as the peak intensity ratio is not greater than 10 to 1. However, if one wanted to distinguish between the parent peaks of 2,2-naphthylbenzothiophene (260.0922) and 1,2-dimethyl-4-benzoylnaphthalene (260.1201), the required resolving power is

$$\frac{m}{\Delta m} = \frac{260}{260.1201 - 260.0922} = 9319 \tag{10.3}$$

10.4 DETECTORS

After leaving a mass analyzer, the resolved ion beams sequentially strike some type of detector. The electron multiplier, either single or multichannel, is most commonly used.

10.4.1 Electron Multiplier

In the electron multiplier the ion beam strikes a conversion dynode, which converts the ion beam to an electron beam. A discrete dynode multiplier has 15 to 18 individual dynodes arranged in a venetian blind configuration and coated with a material that has high secondary-electron-emission properties. A magnetic field forces the secondary electrons to follow circular paths, causing them to strike successive dynodes.

A microchannel plate is a solid-state electron multiplier composed of a hexagonal close-packed array of millions of independent, continuous, single-channel electron multipliers all fused together in a rigid parallel array. With channel densities on the order of 10^6 per cm^2, these devices are one of the highest pixel density sensors known. Pore diameters range from 10 to 25 μm. The inside of each pore, or channel, is coated with a secondary-electron-emissive material; thus each channel constitutes an independent electron multiplier. The onset of ion feedback within the channel can be staved off by curving each channel in the plate but at the cost of considerable spatial distortion.

[12] M. V. Buchanan and R. L. Hettich, "Fourier Transform Mass Spectrometry of High-Mass Biomolecules," *Anal. Chem.* **65**:245A (1993).

[13] A. G. Marshall and L. Schweikhard, *Int. J. Mass Spectrom. Ion Proc.* **118/119**:37 (1992).

[14] J. V. Johnson and R. A. Yost, "Tandem Mass Spectrometry for Trace Analysis," *Anal. Chem.* **57**:758A (1985).

[15] R. A. Yost and C. G. Enke, "Triple Quadrupole Mass Spectrometry for Direct Mixture Analysis and Structure Elucidation," *Anal. Chem.* **51**:1251A (1979).

[16] R. S. Houk, "Mass Spectrometry of Inductively Coupled Plasma," *Anal. Chem.* **58**:97A (1986).

10.4.2 Faraday Cup Collector

The Faraday cup collector consists of a cup with suitable suppressor electrodes, to suppress secondary-ion emission, and guard electrodes. It is placed in the focal plane of the mass spectrometer.

10.5 CORRELATION OF MASS SPECTRA WITH MOLECULAR STRUCTURE

10.5.1 Molecular Identification

In the identification of a compound, the most important information is the molecular weight. The mass spectrometer is able to provide this information, often to four decimal places. One assumes that no ions heavier than the molecular ion form when using electron-impact ionization. The chemical ionization spectrum will often show a cluster around the nominal molecular weight.

Several relationships aid in deducing the empirical formula of the parent ion (and also molecular fragments). From the empirical formula hypothetical molecular structures can be proposed using the entries in the formula indices of Beilstein (*Beilsteins Handbuch der Organischen Chemie*) and *Chemical Abstracts*.

10.5.2 Natural Isotopic Abundances

The relative abundances of natural isotopes produce peaks one or more mass units larger than the parent ion (Table 10.1*a*). For a compound $C_wH_xN_yO_z$, there is a formula that allows one to calculate the percentage of the heavy isotope contributions from a monoisotopic peak P_M to the P_{M+1} peak:

$$100\,\frac{P_{M+1}}{P_M} = 0.015x + 1.11w + 0.37y + 0.037z \tag{10.4}$$

TABLE 10.1 Isotopic Abundances and Masses of Selected Elements

(*a*) Abundances of some polyisotopic elements, %

Element	Abundance	Element	Abundance	Element	Abundance
^{1}H	99.985	^{16}O	99.76	^{33}S	0.76
^{2}H	0.015	^{17}O	0.037	^{34}S	4.22
^{12}C	98.892	^{18}O	0.204	^{35}Cl	75.53
^{13}C	1.108	^{28}Si	92.18	^{37}Cl	24.47
^{14}N	99.63	^{29}Si	4.71	^{79}Br	50.52
^{15}N	0.37	^{30}Si	3.12	^{81}Br	49.48

(*b*) Selected isotope masses

Element	Mass	Element	Mass
^{1}H	1.0078	^{31}P	30.9738
^{12}C	12.0000	^{32}S	31.9721
^{14}N	14.0031	^{35}Cl	34.9689
^{16}O	15.9949	^{56}Fe	55.9349
^{19}F	18.9984	^{79}Br	78.9184
^{28}Si	27.9769	^{127}I	126.9047

Source: J. A. Dean, ed., *Lange's Handbook of Chemistry*, 14th ed., McGraw-Hill, New York, 1992.

Tables of abundance factors have been calculated for all combinations of C, H, N, and O up to mass 500.[17]

Compounds that contain chlorine, bromine, sulfur, or silicon are usually apparent from prominent peaks at masses 2, 4, 6, and so on, units larger the nominal mass of the parent or fragment ion. For example, when one chlorine atom is present, the $P + 2$ mass peak will be about one-third the intensity of the parent peak. When one bromine atom is present, the $P + 2$ mass peak will be about the same intensity as the parent peak. The abundance of heavy isotopes is treated in terms of the binomial expansion $(a + b)^m$, where a is the relative abundance of the light isotope, b is the relative abundance of the heavy isotope, and m is the number of atoms of the particular element present in the molecule. If two bromine atoms are present, the binomial expansion is

$$(a+b)^2 = a^2 + 2ab + b^2 \tag{10.5}$$

Now substituting the percent abundance of each isotope (^{79}Br and ^{81}Br) into the expansion:

$$(0.505)^2 + 2(0.505)(0.495) + (0.495)^2$$

gives

$$0.255 + 0.500 + 0.250$$

which are the proportions of P:($P + 2$):($P + 4$), a triplet that is slightly distorted from a 1:2:1 pattern. When two elements with heavy isotopes are present, the binomial expansion

$$(a+b)^m(c+d)^n$$

is used.

Sulfur-34 enhances the $P + 2$ peak by 4.22%; silicon-29 enhances the $P + 1$ peak by 4.71% and the $P + 2$ peak by 3.12%.

10.5.3 Exact Mass Differences

If the exact mass of the parent or fragment ions is ascertained with a high-resolution mass spectrometer, this relationship is often useful for combinations of C, H, N, and O (Table 10.1*b*):

$$\frac{\text{Exact mass difference from nearest integral mass} + 0.0051z - 0.0031y}{0.0078} = \text{number of hydrogens} \tag{10.6}$$

One substitutes integral numbers (guesses) for z (oxygen) and y (nitrogen) until the divisor becomes an integral multiple of the numerator within 0.0002 mass unit.

For example, if the exact mass is 177.0426 for a compound containing only C, H, O, and N (note the odd mass which indicates an odd number of nitrogen atoms), thus

$$\frac{0.0426 + 0.0051z - 0.0031y}{0.0078} = 7 \text{ hydrogen atoms}$$

[17] J. H. Beynon and A. E. Williams, *Mass and Abundance Tables for Use in Mass Spectrometry*, Elsevier, Amsterdam, 1963.

when $z = 3$ and $y = 1$. The empirical formula is $C_9H_7NO_3$ since

$$\frac{177 - 7(1) - 1(14) - 3(16)}{12} = 9 \text{ carbon atoms}$$

10.5.4 Number of Rings and Double Bonds

The total number of rings and double bonds can be determined from the empirical formula ($C_wH_xI_zN_y$) by the relationship

$$\frac{1}{2(2w - x + y + 2)}$$

when covalent bonds comprise the molecular structure. Remember the total number for a benzene ring is 4 (one ring and three double bonds); for a triple bond it is 2.

10.5.5 General Rules

1. If the nominal molecular weight of a compound containing only C, H, O, and N is even, so is the number of hydrogen atoms it contains.

2. If the nominal molecular weight is divisible by 4, the number of hydrogen atoms is also divisible by 4.

3. When the nominal molecular weight of a compound containing only C, H, O, and N is odd, the number of nitrogen atoms must be odd.

10.5.6 Metastable Peaks

A further means of ion characterization is achieved by monitoring specific fragmentations of a chosen parent ion. This approach involves monitoring of metastable peaks that correspond to fragmentation that occurs in the first field-free region of a double-focusing mass spectrometer (also of a 60° sector instrument). The field-free region is between the exit of the ion source and the entrance to the mass analyzer. Signal detection is dictated by the mass-to-charge ratios of both parent and daughter ions. Metastable peaks m^* appear as a weak, diffuse (often humped-shaped) peak, usually at a nonintegral mass. The one-step decomposition process takes the general form

$$\text{Original ion} \rightarrow \text{daughter ion} + \text{neutral fragment} \qquad (10.7)$$

The relationship between the original ion and daughter ion is given by

$$m^* = \frac{(\text{mass of daugher ion})^2}{\text{mass of original ion}} \qquad (10.8)$$

For example, a metastable peak appeared at 147.9 mass units in a mass spectrum with prominent peaks at 65, 91, 92, 107, 108, 155, 172, and 200 mass units. After trying all possible combinations in the above expression, the fit is given by

$$147.9 = \frac{(172)^2}{200}$$

which provides this information:

$$200^+ \rightarrow 172^+ + 28$$

The probable neutral fragment lost is either $CH_2{=}CH_2$ or CO.

10.6 MASS SPECTRA AND STRUCTURE

The mass spectrum is a fingerprint for each compound because no two molecules are fragmented and ionized in exactly the same manner on electron-impact ionization. When mass spectra are reported, the data are normalized by assigning the most intense peak (denoted as base peak) a value of 100. Other peaks are reported as percentages of the base peak.

A very good general survey for interpreting mass spectral data is given by Silverstein et al.[18]

10.6.1 Initial Steps in Elucidation of a Mass Spectrum

1. Tabulate the prominent ion peaks, starting with the highest mass.
2. usually only one bond is cleaved. In succeeding fragmentations a new bond is formed for each additional bond that is broken.
3. When fragmentation is accompanied by the formation of a new bond as well as by the breaking of an existing bond, a rearrangement process is involved. These will be even mass peaks when only C, H, and O are involved. The migrating atom is almost exclusively hydrogen; six-membered cyclic transition states are most important.
4. Tabulate the probable groups that (a) give rise to the prominent charged ion peaks and (b) list the neutral fragments.

10.6.2 General Rules for Fragmentation Patterns

1. Bond cleavage is more probable at branched carbon atoms: tertiary > secondary > primary. The positive charge tends to remain with the branched carbon.
2. Double bonds favor cleavage in a beta position to the carbon (but see rule 6).
3. A strong parent peak often indicates a ring.
4. Saturated ring systems lose side chains at the alpha position carbon. Upon fragmentation, two ring atoms are usually lost.
5. A heteroatom induces cleavage at the bond in the beta position to it.
6. Compounds that contain a carbonyl group tend to break at this group; the positive charge remains with the carbonyl portion.
7. For linear alkanes, the initial fragment lost is an ethyl group (never a methyl group), followed by propyl, butyl, and so on. An intense peak at mass 43 suggests a chain longer than butane.
8. The presence of Cl, Br, S, and Si, can be deduced from the unusual isotopic abundance patterns of these elements. These elements can be traced through the positively charged

[18] R. M. Silverstein, G. C. Bassler, and T. C. Morrill, *Spectrophotometric Identification of Organic Compounds*, 5th ed., Wiley, New York, 1991.

fragments until the pattern disappears or changes due to the loss of one of these atoms to a neutral fragment.

9. When unusual mass differences occur between some fragment ions, the presence of F (mass difference 19), I (mass difference 127), or P (mass difference 31) should be suspected.

10.6.3 Characteristic Low-Mass Fragment Ions

Mass 30 = Primary amines
Masses 31, 45, 59 = Alcohol or ether
Masses 19 and 31 = Alcohol
Mass 66 = Monobasic carboxylic acid
Masses 77 and 91 = Benzene ring

10.6.4 Characteristic Low-Mass Neutral Fragments from the Molecular Ion

Mass 18 (H_2O) = From alcohols, aldehydes, ketones
Mass 19 (F) and 20 (HF) = Fluorides
Mass 27 (HCN) = Aromatic nitriles or nitrogen heterocycles
Mass 29 = Indicates either CHO or C_2H_5
Mass 30 = Indicates either CH_2O or NO
Mass 33 (HS) and 34 (H_2S) = Thiols
Mass 42 = CH_2CO via rearrangement from a methyl ketone or an aromatic acetate or an aryl -$NHCOCH_3$ group
Mass 43 = C_3H_7 or CH_3CO
Mass 45 = COOH or OC_2H_5

Table 10.2 is condensed, with permission, from the Catalog of Mass Spectral Data of the American Petroleum Institute Research Project 44. These, and other tables, should be consulted for further and more detailed information.

Included in the table are all compounds for which information was available through the C_7 compounds. The mass number for the five most important peaks for each compound are listed, followed in each case by the relative intensity in parentheses. The intensities in all cases are normalized to the *n*-butane 43 peak taken as 100. Another method for expressing relative intensities is to assign the base peak a value of 100 and express the relative intensities of the other peaks as a ratio to the base peak. Taking ethyl nitrate as an example, the tabulated values would be

Ethyl nitrate 91(0.01)(*P*) 46(*100*) 29(44.2) 30(30.5) 76(24.2)

The compounds are arranged in the table according to their molecular formulas. Each formula is arranged alphabetically, except that C is first if carbon occurs in the molecules, followed by H if it occurs. The formulas as then arranged alphabetically and according to increasing number of atoms of each kind, all C_4 compounds being listed before any C_5 compounds, and so on.

TABLE 10.2 Mass Spectra of Some Selected Compounds

Molecular formula	Name	Mass numbers (and intensities) of:				
		Parent peak	Base peak	Three next most intense peaks		
B_2H_6	Diborane	28(0.13)	26(54)	27(52)	24(48)	25(30)
$B_3H_6N_3$	Triborine triamine	81(21)	80(58)	79(37)	53(29)	52(22)
B_5H_9	Pentaborane	64(15)	59(30)	60(30)	62(24)	61(21)
$CBrClF_2$	Difluorochlorobromomethane	164(0.23)	85(86)	87(27)	129(17)	131(16)
CBr_2F_2	Difluorodibromomethane	208(1.7)	129(70)	131(68)	79(18)	31(18)
CCl_2F_2	Difluorodichloromethane	120(0.07)	85(33)	87(11)	50(3.9)	101(2.8)
CCl_3F	Fluorotrichloromethane	136(0.04)	101(54)	103(35)	66(7.0)	35(5.8)
CCl_4	Tetrachloromethane	152(0.0)	117(39)	119(37)	35(16)	47(16)
CF_3I	Trifluoroiodomethane	196(51)	196(51)	127(49)	69(40)	177(16)
CF_4	Tetrafluoromethane	88(0.0)	69(57)	50(6.8)	19(3.9)	31(2.8)
$CHBrClF$	Fluorochlorobromomethane	148(5.5)	67(120)	69(38)	31(13)	111(11)
$CHBrF_2$	Difluorobromomethane	130(13)	51(83)	31(18)	132(13)	79(13)
$CHCl_3$	Trichloromethane	118(1.3)	83(69)	85(44)	47(24)	35(13)
CHF_3	Trifluoromethane	70(0.25)	69(20)	51(18)	31(9.9)	50(2.9)
CHN	Hydrogen cyanide	27(92)	27(92)	26(15)	12(3.8)	28(1.6)
CH_2ClF	Fluorochloromethane	68(48)	68(48)	33(25)	70(15)	49(11)
CH_2Cl_2	Dichloromethane	84(41)	49(71)	86(26)	51(21)	47(13)
CH_2F_2	Difluoromethane	52(2.7)	33(26)	51(25)	31(7.3)	32(2.9)
CH_2O	Methanal (formaldehyde)	30(19)	29(21)	28(6.6)	14(0.94)	13(0.92)
CH_2O_2	Methanoic acid (formic)	46(72)	29(118)	45(56)	28(20)	17(20)
CH_3Cl	Chloromethane	50(66)	50(66)	15(54)	52(21)	49(6.6)
CH_3F	Monofluoromethane	34(29)	15(31)	33(28)	14(5.3)	31(3.2)
CH_3I	Iodomethane	142(78)	142(78)	127(29)	141(11)	15(10)
CH_3NO_2	Nitromethane	61(35)	30(65)	15(34)	46(23)	29(5.3)
CH_4	Methane	16(67)	16(67)	15(58)	14(11)	13(5.5)
CH_4O	Methanol	32(26)	31(38)	29(25)	28(2.4)	18(0.7)
CH_4S	Methanethiol	48(49)	47(65)	45(40)	46(9.5)	15(8.9)
CH_5N	Aminomethane (methylamine)	31(30)	30(53)	28(47)	29(8.7)	27(8.6)
CO	Carbon monoxide	28(78)	28(78)	12(3.7)	16(1.3)	29(0.9)
COS	Carbonyl sulfide	60(83)	60(83)	32(48)	28(6.9)	12(5.0)
CO_2	Carbon dioxide	44(76)	44(76)	28(5.0)	16(4.7)	12(1.9)
CS_2	Carbon disulfide	76(184)	76(184)	32(40)	44(33)	78(16)
C_2F_4	Tetrafluoroethene	100(20)	31(47)	81(34)	50(14)	12(3.6)
C_2F_6	Hexafluoroethane	138(0.14)	69(95)	119(39)	31(17)	50(9.6)
C_2F_6Hg	Hexafluorodimethylmercury	340(0.83)	69(111)	202(26)	271(22)	200(21)
C_2H_2	Ethyne	26(102)	26(102)	25(20)	24(5.7)	13(5.7)
C_2H_2ClN	Chloroethanenitrile	75(51)	75(51)	48(46)	40(23)	77(16)
$C_2H_2Cl_2$	*cis*-1,2-Dichloroethene	96(53)	61(72)	98(34)	63(23)	26(22)
$C_2H_2Cl_2$	*trans*-1,2,Dichloroethene	96(49)	61(73)	98(32)	26(25)	63(23)
$C_2H_2Cl_4$	1,1,2,2-Tetrachloroethane	166(5.9)	83(95)	85(60)	95(11)	87(9.7)
$C_2H_2F_2$	1,1-Difluoroethene	64(32)	64(32)	45(21)	31(16)	33(13)
$C_2H_3Cl_3$	1,1,1-Trichloroethane	132(0.0)	97(37)	99(24)	61(19)	117(7.1)
$C_2H_3Cl_3$	1,1,2-Trichloroethane	132(3.9)	97(43)	83(41)	99(27)	85(26)
$C_2H_3F_3$	1,1,1-Trifluoroethane	84(0.94)	69(81)	65(31)	15(13)	45(10)
C_2H_3N	Ethanenitrile	41(89)	41(89)	40(46)	39(17)	38(10)
C_2H_4	Ethene (ethylene)	28(66)	28(66)	27(43)	26(41)	25(7.8)
C_2H_4BrCl	1-Chloro-2-bromoethane	142(7.9)	63(93)	27(82)	65(30)	26(24)
$C_2H_4Br_2$	1,2-Dibromoethane	186(1.6)	27(93)	107(72)	109(67)	26(23)
$C_2H_4Cl_2$	1,1-Dichloroethane	98(5.7)	63(89)	27(64)	65(28)	26(21)
$C_2H_4Cl_2$	1,2-Dichloroethane	98(1.7)	62(12)	27(11)	49(4.9)	64(3.9)
$C_2H_4N_2$	Diazoethane	56(16)	28(27)	27(25)	26(21)	41(5.2)

(Continued)

TABLE 10.2 Mass Spectra of Some Selected Compounds (*Continued*)

Molecular formula	Name	Mass numbers (and intensities) of: Parent peak	Base peak	Three next most intense peaks		
C_2H_4O	Ethanal (acetaldehyde)	44(30)	29(66)	43(18)	42(6.1)	26(6.1)
C_2H_4O	Ethylene oxide	44(30)	29(46)	15(30)	14(12)	43(7.1)
$C_2H_4O_2$	Ethanoic acid (acetic)	60(19)	43(37)	45(33)	15(21)	14(8.0)
$C_2H_4O_2$	Methyl formate	60(27)	31(96)	29(60)	32(33)	28(6.8)
C_2H_5Br	Bromoethane	108(35)	29(54)	27(48)	110(33)	26(16)
C_2H_5Cl	Chloroethane	64(36)	64(36)	28(32)	29(30)	27(27)
C_2H_5F	Fluoroethane	48(2.4)	47(24)	27(8.9)	33(8.2)	26(3.0)
C_2H_5N	Ethylenimine	43(31)	42(56)	28(44)	15(20)	41(11)
$C_2H_5NO_2$	Nitroethane	75(0.0)	29(85)	27(74)	30(19)	26(11)
$C_2H_5NO_3$	Ethyl nitrate	91(0.01)	46(95)	29(42)	30(29)	76(23)
C_2H_6	Ethane	30(26)	28(99)	27(33)	26(23)	29(21)
C_2H_6O	Ethanol	46(9.7)	31(63)	45(22)	29(14)	27(14)
C_2H_6O	Dimethyl ether	46(32)	45(71)	29(56)	15(41)	14(8.9)
$C_2H_6O_2$	Dimethyl peroxide	62(28)	29(47)	31(45)	15(16)	30(12)
C_2H_6S	2-Thiapropane	62(56)	47(69)	45(42)	46(29)	35(24)
C_2H_6S	Ethanethiol	62(44)	62(44)	29(43)	47(36)	27(35)
$C_2H_6S_2$	2,3-Dithiabutane	94(95)	94(95)	45(59)	79(56)	46(34)
$C_2H_6S_3$	2,3,4-Trithiapentane	126(54)	126(54)	45(32)	79(27)	47(19)
C_2H_7N	Aminoethane (ethylamine)	45(18)	30(96)	28(28)	44(19)	27(13)
C_2H_7N	*N*-Methylaminomethane	45(36)	44(71)	28(48)	15(14)	42(13)
$C_2H_8N_2$	1,2-Diaminoethane	60(2.7)	30(111)	18(14)	42(6.9)	43(5.9)
C_3F_6	Hexafluoropropene	150(16)	31(56)	69(44)	131(41)	100(20)
C_3F_8	Octafluoropropane	188(0.0)	69(171)	31(49)	169(42)	50(16)
C_3H_3N	Propenenitrile	53(55)	26(55)	52(41)	51(18)	27(10)
C_3H_4	Propadiene	40(72)	40(72)	39(69)	38(29)	37(23)
C_3H_4	Propyne (methylacetylene)	40(79)	40(79)	39(73)	38(29)	37(22)
C_3H_4ClN	3-Chloropropanenitrile	89(12)	49(68)	54(54)	51(29)	26(20)
C_3H_4O	Propenal (acrolein)	56(16)	27(25)	26(15)	28(13)	55(11)
C_3H_5Cl	1-Chloro-1-propene	76(30)	41(70)	39(43)	40(10)	78(9.6)
C_3H_5ClO	3-Chloro-1,2-epoxypropane	92(0.19)	57(55)	27(53)	29(40)	31(21)
$C_3H_5ClO_2$	Methyl chloroacetate	109(0.23)	59(56)	49(44)	15(43)	29(37)
$C_3H_5Cl_3$	1,2,3-Trichloropropane	146(0.71)	75(61)	110(22)	77(19)	61(18)
C_3H_5N	Propanenitrile	55(8.3)	28(83)	54(51)	26(17)	27(15)
C_3H_6	Cyclopropane	42(64)	42(64)	41(58)	39(44)	27(23)
C_3H_6	Propene	42(39)	41(58)	39(41)	27(22)	40(17)
$C_3H_6Cl_2$	1,1-Dichloropropane	112(0.0)	63(27)	41(25)	77(22)	62(19)
$C_3H_6Cl_2$	1,2-Dichloropropane	112(2.6)	63(51)	62(36)	27(29)	41(25)
C_3H_6O	1-Propen-3-ol (allyl alc.)	58(12)	57(43)	29(34)	31(26)	27(19)
C_3H_6O	Propanal	58(25)	29(66)	28(46)	27(38)	26(14)
C_3H_6O	Propanone (acetone)	58(24)	43(85)	15(26)	27(5.9)	42(5.9)
C_3H_6O	1,2-Epoxypropane	58(19)	28(44)	29(30)	27(28)	26(18)
$C_3H_6O_2$	1,3-Dioxolane	74(3.1)	73(52)	43(36)	44(30)	29(30)
$C_3H_6O_2$	Propanoic acid	74(27)	28(34)	29(28)	27(21)	45(19)
$C_3H_6O_2$	Ethyl formate	74(5.8)	31(82)	28(60)	29(54)	27(36)
$C_3H_6O_2$	Methyl acetate	74(22)	43(148)	29(16)	42(15)	59(8.4)
$C_3H_6O_3$	Methyl carbonate	90(3.3)	15(93)	45(54)	29(43)	31(34)
C_3H_7Br	1-Bromopropane	122(14)	43(94)	27(55)	41(47)	39(22)
C_3H_7Br	2-Bromopropane	122(11)	43(100)	27(50)	41(47)	39(24)
C_3H_7Cl	1-Chloropropane	78(3.6)	42(60)	29(27)	27(22)	41(14)
C_3H_7Cl	2-Chloropropane	78(14)	43(58)	27(20)	63(15)	41(13)
C_3H_7F	2-Fluoropropane	62(1.0)	47(84)	46(24)	61(12)	27(7.6)

TABLE 10.2 Mass Spectra of Some Selected Compounds (*Continued*)

Molecular formula	Name	Mass numbers (and intensities) of: Parent peak	Base peak	Three next most intense peaks		
C_3H_7N	2-Methylethylenimine	57(22)	28(76)	56(34)	30(24)	29(19)
C_3H_7N	*N*-Methylethylenimine	57(31)	42(94)	15(46)	28(25)	27(17)
C_3H_7NO	*N,N*-Dimethylformamide	73(54)	44(63)	42(29)	28(25)	15(24)
$C_3H_7NO_2$	1-Nitropropane	89(0.0)	43(68)	27(67)	41(58)	39(24)
$C_3H_7NO_2$	2-Nitropropane	89(0.0)	43(75)	41(55)	27(53)	39(23)
C_3H_8	Propane	44(25)	29(85)	28(50)	27(33)	43(19)
C_3H_8O	1-Propanol	60(7.2)	31(115)	27(18)	29(17)	59(10)
C_3H_8O	2-Propanol	60(0.45)	45(112)	43(19)	27(18)	29(11)
C_3H_8O	Methyl ethyl ether	60(24)	45(94)	29(46)	15(23)	27(19)
$C_3H_8O_2$	Dimethoxymethane	76(1.6)	45(117)	29(51)	75(51)	15(48)
$C_3H_8O_2$	2-Methoxy-1-ethanol	76(7.3)	45(122)	29(44)	15(38)	31(32)
C_3H_8S	2-Thiabutane	76(47)	61(73)	48(40)	47(30)	27(27)
C_3H_8S	1-Propanethiol	76(30)	47(43)	43(34)	27(34)	41(32)
C_3H_8S	2-Propanethiol	76(41)	43(65)	41(44)	27(41)	61(26)
C_3H_9N	1-Aminopropane	59(1.5)	30(20)	28(2.5)	27(1.3)	41(1.0)
C_3H_9N	Trimethylamine	59(37)	58(95)	42(44)	15(32)	30(17)
$C_3H_{12}B_3N_3$	*B,B′,B″*-Trimethylborazole	123(30)	108(102)	107(77)	67(38)	66(34)
C_4F_6	Hexafluorocyclobutene	162(21)	93(80)	31(51)	143(15)	74(6.9)
C_4F_6	Hexafluoro-1,3-butadiene	162(27)	93(90)	31(45)	74(10)	112(10)
C_4F_6	Hexafluoro-2-butyne	162(18)	93(47)	143(38)	31(25)	69(20)
C_4F_8	Octafluorocyclobutane	200(0.12)	100(97)	131(84)	31(53)	69(24)
C_4F_8	Octafluoromethylpropene	200(14)	69(74)	181(54)	31(44)	93(22)
C_4F_8	Octafluoro-1-butene	200(11)	131(122)	31(86)	69(44)	93(16)
C_4F_{10}	Decafluorobutane	238(0.0)	69(178)	119(33)	31(22)	100(15)
$C_4HF_7O_2$	Heptafluorobutanoic acid	214(0.0)	45(26)	69(24)	119(17)	100(14)
C_4H_2	1,3-Butadiyne	50(133)	50(133)	49(57)	48(14)	25(12)
C_4H_4	1-Buten-3-yne	52(55)	52(55)	51(28)	50(23)	49(7.2)
C_4H_4O	Furan	68(36)	39(58)	38(9.7)	29(9.3)	40(6.7)
C_4H_4S	Thiophene	84(93)	84(93)	58(56)	45(49)	39(24)
$C_4H_4S_2$	2-Thiophenethiol	116(68)	116(68)	71(64)	45(31)	39(11)
C_4H_5N	3-Butenenitrile	67(27)	41(80)	39(36)	27(30)	40(20)
C_4H_5N	Pyrrole	67(67)	67(67)	39(46)	41(42)	40(36)
C_4H_6	1,2-Butadiene	54(65)	54(65)	27(35)	53(29)	39(28)
C_4H_6	1,3-Butadiene	54(46)	39(53)	27(36)	53(31)	28(24)
C_4H_6	1-Butyne	54(64)	54(64)	39(49)	53(27)	27(26)
C_4H_6	2-Butyne	54(93)	54(93)	27(42)	53(41)	39(24)
$C_4H_6Cl_2O_2$	Ethyl dichloroacetate	156(0.12)	29(192)	27(58)	83(23)	28(19)
$C_4H_6O_2$	2,3-Butanedione	86(13)	43(118)	15(40)	14(12)	42(8.6)
$C_4H_6O_2$	Methyl 2-propenoate	86(2.0)	55(98)	27(66)	15(27)	26(22)
$C_4H_7BrO_2$	2-Bromoethyl acetate	166(0.03)	43(158)	27(35)	106(31)	108(30)
C_4H_7Cl	2-Chloro-2-butene	90(27)	55(68)	27(21)	39(21)	29(18)
$C_4H_7ClO_2$	2-Chloroethyl acetate	122(0.0)	43(162)	73(43)	15(36)	27(29)
$C_4H_7ClO_2$	Ethyl chloroacetate	122(0.96)	29(130)	27(41)	77(37)	49(29)
C_4H_7N	2-Methylpropanenitrile	69(1.7)	42(79)	68(38)	28(26)	54(19)
C_4H_7N	*n*-Butanenitrile	69(0.15)	41(112)	29(70)	27(38)	28(11)
C_4H_8	Cyclobutane	56(41)	28(65)	41(58)	27(27)	26(15)
C_4H_8	2-Methylpropene	56(36)	41(85)	39(37)	28(18)	27(17)
C_4H_8	1-Butene	56(32)	41(87)	39(30)	27(26)	28(26)
C_4H_8	*cis*-2-Butene	56(36)	41(76)	39(27)	27(25)	28(24)
C_4H_8	*trans*-2-Butene	56(37)	41(80)	27(27)	39(26)	28(26)
$C_4H_8Cl_2$	1,2-Dichlorobutane	126(0.30)	41(39)	77(35)	27(20)	76(16)

(Continued)

TABLE 10.2 Mass Spectra of Some Selected Compounds (*Continued*)

Molecular formula	Name	Parent peak	Base peak	Three next most intense peaks		
		Mass numbers (and intensities) of:				
$C_4H_8Cl_2$	1,4-Dichlorobutane	126(0.03)	55(87)	41(29)	27(24)	90(23)
$C_4H_8Cl_2$	*dl*-2,3-Dichlorobutane	126(0.95)	63(63)	62(58)	27(57)	55(29)
$C_4H_8Cl_2$	*meso*-2,3-Dichlorobutane	126(0.95)	63(64)	27(57)	62(54)	55(31)
$C_4H_8N_2$	Acetaldazine	84(23)	42(92)	15(47)	28(46)	69(38)
C_4H_8O	Butanal	72(19)	27(41)	29(38)	44(34)	43(32)
C_4H_8O	2-Butanone	72(17)	43(97)	29(24)	27(15)	57(6.0)
C_4H_8O	Ethyl ethenyl ether	72(27)	44(64)	43(56)	29(49)	27(43)
C_4H_8O	*cis*-2,3-Epoxybutane	72(3.6)	43(67)	44(39)	27(35)	29(33)
C_4H_8O	*trans*-2,3-Epoxybutane	72(3.5)	43(69)	44(35)	29(32)	27(31)
C_4H_8O	Tetrahydrofuran	72(22)	42(76)	41(39)	27(25)	71(20)
$C_4H_8O_2$	2-Methyl-1,3-dioxacyclopentane	88(0.33)	73(67)	43(48)	45(44)	29(34)
$C_4H_8O_2$	1,4-Dioxane	88(42)	28(138)	29(51)	58(33)	31(24)
$C_4H_8O_2$	2-Methylpropanoic acid	88(8.1)	43(77)	41(33)	27(26)	73(19)
$C_4H_8O_2$	*n*-Butanoic acid	88(1.0)	60(40)	73(12)	27(9.6)	41(9.1)
$C_4H_8O_2$	*n*-Propyl formate	88(0.41)	31(123)	42(89)	29(38)	27(36)
$C_4H_8O_2$	Ethyl acetate	88(7.1)	43(181)	29(46)	45(24)	27(24)
$C_4H_8O_2$	Methyl propanoate	88(23)	29(110)	57(83)	27(40)	59(27)
C_4H_8S	3-Methylthiacyclobutane	88(42)	46(101)	45(31)	39(24)	47(21)
C_4H_8S	Thiacyclopentane	88(44)	60(82)	45(29)	46(29)	47(22)
C_4H_9Br	1-Bromobutane	136(7.0)	57(86)	41(63)	29(50)	27(46)
C_4H_9Br	2-Bromobutane	136(0.72)	57(108)	41(65)	29(61)	27(36)
C_4H_9N	Pyrrolidine	71(24)	43(102)	28(38)	70(33)	42(20)
$C_4H_9NO_2$	*n*-Butyl nitrite	103(0.0)	27(55)	43(54)	41(50)	30(47)
C_4H_{10}	2-Methylpropane	58(3.2)	43(117)	41(45)	42(39)	27(33)
C_4H_{10}	*n*-Butane	58(12)	43(100)	29(44)	27(37)	28(33)
$C_4H_{10}Hg$	Diethylmercury	260(12)	29(188)	27(54)	28(21)	231(15)
$C_4H_{10}O$	2-Methyl-1-propanol	74(7.5)	43(84)	31(56)	42(48)	41(47)
$C_4H_{10}O$	2-Methyl-2-propanol	74(0.0)	59(92)	31(31)	41(19)	43(14)
$C_4H_{10}O$	1-Butanol	74(0.37)	31(52)	56(44)	41(31)	43(30)
$C_4H_{10}O$	2-Butanol	74(0.30)	45(116)	31(23)	59(22)	27(20)
$C_4H_{10}O$	Diethyl ether	74(22)	31(73)	59(34)	29(29)	45(28)
$C_4H_{10}O$	Methyl isopropyl ether	74(8.3)	59(126)	29(42)	43(37)	15(32)
$C_4H_{10}O_2$	1,1-Dimethoxyethane	90(0.06)	59(93)	29(52)	15(37)	31(37)
$C_4H_{10}O_2$	1,2-Dimethoxyethane	90(12)	45(177)	29(53)	15(50)	60(16)
$C_4H_{10}O_2$	2-Ethoxyethanol	90(0.49)	31(112)	29(57)	59(56)	27(31)
$C_4H_{10}O_2$	Diethyl peroxide	90(20)	29(116)	15(42)	45(34)	62(30)
$C_4H_{10}S$	3-Methyl-2-thiabutane	90(41)	41(49)	75(47)	43(41)	48(38)
$C_4H_{10}S$	2-Thiapentane	90(58)	61(126)	48(50)	41(43)	27(43)
$C_4H_{10}S$	3-Thiapentane	90(41)	75(59)	47(51)	27(39)	61(33)
$C_4H_{10}S$	2-Methyl-1-propanethiol	90(35)	41(60)	43(46)	56(34)	47(29)
$C_4H_{10}S$	2-Methyl-2-propanethiol	90(34)	41(68)	57(61)	29(44)	39(21)
$C_4H_{10}S$	1-Butanethiol	90(40)	56(74)	41(65)	27(42)	47(31)
$C_4H_{10}S$	2-Butanethiol	90(34)	41(56)	57(50)	61(46)	29(46)
$C_4H_{10}S_2$	2,3-Dithiahexane	122(37)	80(53)	43(36)	41(27)	27(25)
$C_4H_{10}S_2$	3,4-Dithiahexane	122(73)	29(82)	66(81)	27(57)	94(53)
$C_4H_{10}SO_3$	Ethyl sulfite	138(3.3)	29(131)	31(59)	45(42)	27(39)
$C_4H_{11}N$	*N*-Ethylaminoethane	73(17)	58(83)	30(81)	28(30)	27(24)
$C_4H_{11}N$	1-Amino-2-methylpropane	73(1.0)	30(22)	28(2.0)	41(1.2)	27(1.1)
$C_4H_{11}N$	2-Amino-2-methylpropane	73(0.25)	58(127)	41(26)	42(20)	15(18)
$C_4H_{11}N$	1-Aminobutane	73(12)	30(200)	28(23)	27(16)	18(12)
$C_4H_{11}N$	2-Aminobutane	73(1.2)	44(170)	18(25)	41(18)	58(18)

TABLE 10.2 Mass Spectra of Some Selected Compounds (*Continued*)

Molecular formula	Name	Parent peak	Base peak	Three next most intense peaks		
$C_4H_{12}Pb$	Tetramethyllead	268(0.14)	253(69)	223(59)	208(46)	251(36)
C_5F_{10}	Decafluorocyclopentane	250(0.62)	131(173)	100(41)	31(40)	69(28)
C_5F_{12}	Dodecafluoro-2-methylbutane	288(0.0)	69(277)	119(45)	131(23)	31(18)
C_5F_{12}	Dodecafluoropentane	288(0.08)	69(259)	119(76)	169(25)	31(24)
C_5HF_9	Nonafluorocyclopentane	232(0.07)	131(61)	113(49)	69(34)	31(19)
C_5H_5N	Pyridine	79(135)	79(135)	52(95)	51(48)	50(35)
C_5H_6	Cyclopentadiene	66(95)	66(95)	65(40)	39(35)	40(30)
C_5H_6	*trans*-2-Penten-4-yne	66(77)	66(77)	39(54)	65(38)	40(35)
$C_5H_6N_2$	2-Methylpyrazine	94(81)	94(81)	67(48)	26(33)	39(30)
$C_5H_6O_2$	Furfuryl alcohol	98(3.4)	98(3.4)	41(3.3)	39(3.3)	42(2.6)
C_5H_6S	2-Methylthiophene	98(68)	97(125)	45(26)	39(17)	53(11)
C_5H_6S	3-Methylthiophene	98(74)	97(138)	45(35)	39(14)	27(11)
C_5H_8	Methylenecyclobutane	68(38)	40(67)	67(48)	39(47)	53(21)
C_5H_8	Spiropentane	68(8.9)	67(58)	40(56)	39(52)	53(23)
C_5H_8	Cyclopentene	68(41)	67(99)	39(36)	53(23)	41(19)
C_5H_8	3-Methyl-1,2-butadiene	68(53)	68(53)	53(40)	39(28)	41(26)
C_5H_8	2-Methyl-1,3-butadiene	68(40)	67(48)	53(41)	39(34)	27(23)
C_5H_8	1,2-Pentadiene	68(39)	68(39)	53(38)	39(37)	27(31)
C_5H_8	*cis*-1,3-Pentadiene	68(40)	67(53)	39(43)	53(38)	41(25)
C_5H_8	*trans*-1,3-Pentadiene	68(41)	67(52)	39(43)	53(39)	41(26)
C_5H_8	1,4-Pentadiene	68(40)	39(47)	67(35)	53(33)	41(30)
C_5H_8	2,3-Pentadiene	68(62)	68(62)	53(42)	39(36)	41(31)
C_5H_8	3-Methyl-1-butyne	68(8.5)	53(74)	67(45)	27(35)	39(21)
C_5H_8	1-Pentyne	68(8.7)	67(50)	40(44)	39(42)	27(34)
C_5H_8	2-Pentyne	68(67)	68(67)	53(61)	39(32)	27(27)
$C_5H_8N_2$	3,5-Dimethylpyrazole	96(47)	96(47)	95(37)	39(16)	54(12)
$C_5H_8O_2$	2,4-Pentanedione	100(22)	43(120)	85(33)	15(23)	27(11)
$C_5H_8O_2$	2-Propenyl acetate	100(0.16)	43(177)	41(30)	39(29)	15(28)
$C_5H_8O_2$	Methyl methacrylate	100(26)	41(78)	69(52)	39(31)	15(16)
$C_5H_9ClO_2$	Ethyl 3-chloropropanoate	136(0.70)	27(65)	29(62)	91(42)	63(37)
C_5H_{10}	*cis*-1,2-Dimethylcyclopropane	70(39)	55(77)	42(35)	39(32)	41(32)
C_5H_{10}	*trans*-1,2-Dimethylcyclopropane	70(42)	55(79)	42(34)	41(33)	39(30)
C_5H_{10}	Ethylcyclopropane	70(26)	42(93)	55(47)	41(39)	39(35)
C_5H_{10}	Cyclopentane	70(44)	42(148)	55(43)	41(43)	39(31)
C_5H_{10}	2-Methyl-1-butene	70(30)	55(97)	42(36)	39(34)	41(28)
C_5H_{10}	3-Methyl-1-butene	70(26)	55(102)	27(31)	42(28)	29(27)
C_5H_{10}	2-Methyl-2-butene	70(31)	55(88)	41(31)	39(28)	42(27)
C_5H_{10}	1-Pentene	70(27)	42(89)	55(53)	41(39)	39(31)
C_5H_{10}	*cis*-2-Pentene	70(30)	55(89)	42(41)	39(30)	29(26)
C_5H_{10}	*trans*-2-Pentene	70(31)	55(93)	42(41)	39(30)	41(28)
$C_5H_{10}O$	3-Methyl-1-butanal	86(3.0)	41(30)	43(26)	58(20)	29(20)
$C_5H_{10}O$	2-Pentanone	86(16)	43(106)	29(23)	27(23)	57(20)
$C_5H_{10}O$	3-Pentanone	86(15)	57(87)	29(87)	27(32)	28(9.4)
$C_5H_{10}O$	Ethyl-2-propenyl ether	86(6.2)	41(52)	29(48)	58(44)	57(42)
$C_5H_{10}O$	Ethenyl isopropyl ether	86(21)	43(87)	44(69)	41(46)	27(45)
$C_5H_{10}O$	2-Methyltetrahydrofuran	86(8.9)	71(57)	43(55)	41(40)	27(27)
$C_5H_{10}O_2$	Tetrahydrofurfuryl alcohol	102(0.02)	71(8.9)	43(6.8)	41(4.8)	27(3.8)
$C_5H_{10}O_2$	2-Methoxyethyl ethenyl ether	102(3.0)	29(69)	45(58)	15(48)	58(45)
$C_5H_{10}O_2$	2,2-Dimethylpropanoic acid	102(2.0)	57(83)	41(38)	29(27)	39(12)
$C_5H_{10}O_2$	2-Methylbutanoic acid	102(0.32)	74(54)	57(34)	29(33)	41(28)
$C_5H_{10}O_2$	*n*-Butyl formate	102(0.27)	56(80)	41(48)	31(47)	29(42)

Mass numbers (and intensities) of: Parent peak, Base peak, Three next most intense peaks.

(*Continued*)

TABLE 10.2 Mass Spectra of Some Selected Compounds (*Continued*)

Molecular formula	Name	Mass numbers (and intensities) of: Parent peak	Base peak	Three next most intense peaks		
$C_5H_{10}O_2$	Isobutyl formate	102(0.27)	43(58)	56(48)	41(46)	31(38)
$C_5H_{10}O_2$	*sec*-Butyl formate	102(0.17)	45(99)	29(49)	27(32)	41(31)
$C_5H_{10}O_2$	*n*-Propyl acetate	102(0.07)	43(176)	61(34)	31(31)	27(26)
$C_5H_{10}O_2$	Isopropyl acetate	102(0.17)	43(155)	45(50)	27(22)	61(18)
$C_5H_{10}O_2$	Ethyl propanoate	102(10)	29(151)	57(97)	27(52)	28(24)
$C_5H_{10}O_2$	Methyl 2-methylpropanoate	102(8.9)	43(69)	71(23)	41(19)	59(17)
$C_5H_{10}O_2$	Methyl butanoate	102(1.0)	43(53)	74(37)	71(29)	27(23)
$C_5H_{10}O_3$	Ethyl carbonate	118(0.30)	29(114)	45(80)	31(60)	27(46)
$C_5H_{10}S$	2-Methylthiacyclopentane	102(37)	87(88)	41(30)	45(29)	59(18)
$C_5H_{10}S$	3-Methylthiacyclopentane	102(40)	60(45)	41(31)	45(25)	74(23)
$C_5H_{10}S$	Thiacyclohexane	102(43)	87(44)	68(33)	61(32)	41(28)
$C_5H_{10}S$	Cyclopentanethiol	102(19)	41(48)	69(47)	39(26)	67(18)
$C_5H_{11}N$	Piperidine	85(22)	84(43)	57(22)	56(22)	44(17)
$C_5H_{11}NO$	*N*-Methylmorpholine	101(4.4)	43(18)	42(8.6)	15(3.4)	71(2.9)
$C_5H_{11}NO_2$	3-Methylbutyl nitrite	117(0.0)	29(75)	41(68)	57(43)	30(42)
C_5H_{12}	2,2-Dimethylpropane	72(0.01)	57(126)	41(52)	29(49)	27(20)
C_5H_{12}	2-Methylbutane	72(4.7)	43(74)	42(64)	41(49)	57(40)
C_5H_{12}	*n*-Pentane	72(10)	43(114)	42(66)	41(45)	27(39)
$C_5H_{12}O$	2-Methyl-1-butanol	88(0.18)	57(57)	29(55)	41(53)	56(50)
$C_5H_{12}O$	3-Methyl-1-butanol	88(0.02)	55(47)	42(42)	43(39)	41(38)
$C_5H_{12}O$	2-Methyl-2-butanol	88(0.0)	59(43)	55(37)	45(25)	73(22)
$C_5H_{12}O$	1-Pentanol	88(0.0)	42(41)	55(30)	41(25)	70(23)
$C_5H_{12}O$	Methyl *n*-butyl ether	88(3.1)	45(211)	56(36)	29(36)	27(28)
$C_5H_{12}O$	Methyl isobutyl ether	88(12)	45(186)	41(30)	29(30)	15(27)
$C_5H_{12}O$	Methyl *sec*-butyl ether	88(2.0)	59(142)	29(50)	27(27)	41(25)
$C_5H_{12}O$	Methyl *tert*-butyl ether	88(0.02)	73(119)	41(33)	43(32)	57(32)
$C_5H_{12}O$	Ethyl isopropyl ether	88(2.6)	45(143)	43(46)	73(40)	27(24)
$C_5H_{12}O_2$	Diethoxymethane	104(2.1)	31(104)	59(99)	29(62)	103(39)
$C_5H_{12}O_2$	1,1-Dimethoxypropane	104(0.05)	75(84)	73(62)	29(43)	45(37)
$C_5H_{12}S$	3,3-Dimethyl-2-thiabutane	104(30)	57(83)	41(62)	29(42)	39(16)
$C_5H_{12}S$	4-Methyl-2-thiapentane	104(37)	41(46)	56(38)	27(29)	39(23)
$C_5H_{12}S$	2-Methyl-3-thiapentane	104(82)	89(119)	62(79)	43(63)	61(58)
$C_5H_{12}S$	2-Thiahexane	104(38)	61(77)	56(50)	41(39)	27(33)
$C_5H_{12}S$	3-Thiahexane	104(30)	75(72)	27(53)	47(50)	62(33)
$C_5H_{12}S$	2,2-Dimethyl-1-propanethiol	104(31)	57(100)	41(55)	55(48)	29(42)
$C_5H_{12}S$	2-Methyl-1-butanethiol	104(28)	41(65)	29(44)	57(40)	70(40)
$C_5H_{12}S$	2-Methyl-2-butanethiol	104(18)	43(88)	71(54)	41(46)	55(34)
$C_5H_{12}S$	3-Methyl-2-butanethiol	104(23)	61(73)	43(55)	27(33)	55(28)
$C_5H_{12}S$	1-Pentanethiol	104(35)	42(91)	55(44)	41(39)	70(39)
$C_5H_{12}S$	2-Pentanethiol	104(28)	43(72)	61(52)	27(39)	55(38)
$C_5H_{12}S$	3-Pentanethiol	104(23)	43(56)	41(48)	75(29)	47(23)
$C_5H_{12}S_2$	4,4-Dimethyl-2,3-dithiapentane	136(12)	57(74)	41(38)	29(36)	80(13)
$C_5H_{12}S_2$	2-Methyl-3,4-dithiahexane	136(20)	94(49)	27(46)	43(39)	66(37)
$C_5H_{14}Pb$	Trimethylethyllead	282(0.64)	223(61)	253(52)	208(51)	221(33)
C_6F_6	Hexafluorobenzene	186(95)	186(95)	117(59)	31(58)	93(23)
C_6F_{12}	Dodecafluorocyclohexane	300(0.96)	131(138)	69(97)	100(40)	31(30)
C_6F_{14}	Tetradecafluoro-2-methylpentane	338(0.0)	69(317)	131(41)	119(36)	169(29)
C_6F_{14}	Tetradecafluorohexane	338(0.13)	69(268)	119(74)	169(51)	131(37)
C_6H_5Br	Bromobenzene	156(75)	77(98)	158(74)	51(41)	50(36)
C_6H_5Cl	Chlorobenzene	112(102)	112(102)	77(49)	114(33)	51(17)
$C_6H_5NO_2$	Nitrobenzene	123(39)	77(93)	51(55)	50(23)	30(15)

TABLE 10.2 Mass Spectra of Some Selected Compounds (*Continued*)

Molecular formula	Name	Mass numbers (and intensities) of: Parent peak	Base peak	Three next most intense peaks		
C_6H_6	Benzene	78(113)	78(113)	52(22)	77(20)	51(18)
C_6H_6	1,5-Hexadiyne	78(58)	39(65)	52(38)	51(32)	50(26)
C_6H_6	2,4-Hexadiyne	78(108)	78(108)	51(55)	52(38)	50(31)
C_6H_6S	Benzenethiol	110(68)	110(68)	66(26)	109(17)	51(15)
C_6H_7N	Aminobenzene (aniline)	93(19)	93(19)	66(6.5)	65(3.6)	39(3.5)
C_6H_7N	2-Methylpyridine	93(86)	93(86)	66(36)	39(28)	51(16)
C_6H_7NO	1-Methyl-2-pyridone	109(71)	109(71)	81(49)	39(34)	80(29)
C_6H_8	Methylcyclopentadiene	80(53)	79(87)	77(29)	39(19)	51(11)
C_6H_8	1,3-Cyclohexadiene	80(53)	79(92)	77(35)	39(21)	27(18)
C_6H_8O	2,5-Dimethylfuran	96(57)	43(65)	95(48)	53(37)	81(24)
C_6H_8S	2,3-Dimethylthiophene	112(44)	97(53)	111(44)	45(16)	27(9.4)
C_6H_8S	2,4-Dimethylthiophene	112(27)	111(36)	97(18)	45(9.4)	39(7.0)
C_6H_8S	2,5-Dimethylthiophene	112(67)	111(95)	97(59)	59(23)	45(19)
C_6H_8S	2-Ethylthiophene	112(27)	97(68)	45(16)	39(8.9)	27(5.4)
C_6H_8S	3-Ethylthiophene	112(54)	97(147)	45(38)	39(20)	27(12)
C_6H_9N	2,5-Dimethylpyrrole	95(73)	94(127)	26(52)	80(22)	42(19)
C_6H_{10}	Isopropenylcyclopropane	82(20)	67(92)	41(47)	39(46)	27(22)
C_6H_{10}	1-Methylcyclopentene	82(26)	67(98)	39(21)	81(16)	41(16)
C_6H_{10}	Cyclohexene	82(33)	67(83)	54(64)	41(31)	39(30)
C_6H_{10}	2,3-Dimethyl-1,3-butadiene	82(41)	67(60)	39(55)	41(44)	54(22)
C_6H_{10}	2-Methyl-1,3-pentadiene	82(23)	67(48)	39(30)	41(26)	27(13)
C_6H_{10}	1,5-Hexadiene	82(1.3)	41(98)	67(80)	39(60)	54(52)
C_6H_{10}	3,3-Dimethyl-1-butyne	82(0.57)	67(101)	41(57)	39(31)	27(11)
C_6H_{10}	4-Methyl-1-pentyne	82(2.3)	67(82)	41(74)	43(64)	39(55)
C_6H_{10}	1-Hexyne	82(1.0)	67(131)	41(88)	27(85)	43(67)
C_6H_{10}	2-Hexyne	82(56)	67(58)	53(50)	27(39)	41(36)
C_6H_{10}	3-Hexyne	82(55)	67(59)	41(55)	39(37)	53(20)
$C_6H_{10}O$	Cyclohexanone	98(32)	55(102)	42(86)	41(35)	27(34)
$C_6H_{10}O$	4-Methyl-3-penten-2-one	98(40)	55(82)	83(82)	43(64)	29(38)
$C_6H_{10}O_2$	2,5-Hexanedione	114(4.0)	43(148)	15(25)	99(22)	14(14)
$C_6H_{10}O_3$	Propanoic anhydride	130(0.0)	57(190)	29(119)	27(62)	28(26)
$C_6H_{10}O_3$	Ethyl acetoacetate	130(8.3)	43(150)	29(52)	27(32)	15(27)
$C_6H_{11}N$	4-Methylpentanenitrile	97(0.13)	55(98)	41(51)	43(45)	27(39)
$C_6H_{11}N$	Hexanenitrile	97(0.54)	41(73)	54(49)	27(43)	55(40)
C_6H_{12}	1,1,2-Trimethylcyclopropane	84(38)	41(132)	69(81)	39(34)	27(24)
C_6H_{12}	1-Methyl-1-ethylcyclopropane	84(25)	41(78)	55(58)	69(53)	27(33)
C_6H_{12}	Isopropylcyclopropane	84(2.0)	56(114)	41(84)	39(30)	43(28)
C_6H_{12}	Ethylcyclobutane	84(3.8)	56(138)	41(89)	27(35)	55(34)
C_6H_{12}	Methylcyclopentane	84(18)	56(116)	41(74)	69(37)	42(33)
C_6H_{12}	Cyclohexane	84(58)	56(75)	41(44)	55(25)	42(21)
C_6H_{12}	2,3-Dimethyl-1-butene	84(27)	41(117)	69(96)	39(36)	27(24)
C_6H_{12}	3,3-Dimethyl-1-butene	84(23)	41(112)	69(107)	39(28)	27(26)
C_6H_{12}	2-Ethyl-1-butene	84(30)	41(74)	69(66)	55(56)	27(38)
C_6H_{12}	2,3-Dimethyl-2-butene	84(32)	41(108)	69(88)	39(35)	27(20)
C_6H_{12}	2-Methyl-1-pentene	84(29)	56(91)	41(73)	55(39)	39(36)
C_6H_{12}	3-Methyl-1-pentene	84(25)	55(85)	41(67)	69(60)	27(43)
C_6H_{12}	4-Methyl-1-pentene	84(12)	43(110)	41(80)	56(47)	27(37)
C_6H_{12}	2-Methyl-2-pentene	84(36)	41(120)	69(111)	39(35)	27(28)
C_6H_{12}	3-Methyl-*cis*-2-pentene	84(37)	41(104)	69(82)	55(46)	27(36)
C_6H_{12}	3-Methyl-*trans*-2-pentene	84(38)	41(102)	69(81)	55(47)	27(35)
C_6H_{12}	4-Methyl-*cis*-2-pentene	84(35)	41(122)	69(114)	39(35)	27(26)

(Continued)

TABLE 10.2 Mass Spectra of Some Selected Compounds (*Continued*)

Molecular formula	Name	Parent peak	Base peak	Three next most intense peaks		
C_6H_{12}	4-Methyl-*trans*-2-pentene	84(34)	41(123)	69(112)	39(34)	27(26)
C_6H_{12}	1-Hexene	84(20)	41(70)	56(60)	42(52)	27(48)
C_6H_{12}	*cis*-2-Hexene	84(27)	55(91)	42(51)	41(45)	27(45)
C_6H_{12}	*trans*-2-Hexene	84(32)	55(112)	42(54)	41(46)	27(41)
C_6H_{12}	*cis*-3-Hexene	84(28)	55(81)	41(62)	42(54)	27(32)
C_6H_{12}	*trans*-3-Hexene	84(32)	55(89)	41(72)	42(62)	27(35)
$C_6H_{12}N_2$	Acetone azine (ketazine)	112(31)	56(99)	15(31)	97(31)	39(26)
$C_6H_{12}O$	Cyclopentylmethanol	100(0.02)	41(35)	68(32)	69(31)	67(24)
$C_6H_{12}O$	4-Methyl-2-pentanone	100(12)	43(115)	58(37)	41(22)	57(22)
$C_6H_{12}O$	Ethenyl *n*-butyl ether	100(5.7)	29(80)	41(59)	56(45)	57(35)
$C_6H_{12}O$	Ethenyl isobutyl ether	100(5.8)	29(73)	41(65)	57(58)	56(40)
$C_6H_{12}O_2$	4-Hydroxy-4-methyl-2-pentanone	116(0.0)	43(149)	15(45)	58(32)	27(14)
$C_6H_{12}O_2$	*n*-Butyl acetate	116(0.03)	43(172)	56(58)	41(30)	27(27)
$C_6H_{12}O_2$	*n*-Propyl propanoate	116(0.03)	57(147)	29(84)	27(57)	75(47)
$C_6H_{12}O_2$	Isopropyl propanoate	116(0.26)	57(116)	43(88)	29(54)	27(46)
$C_6H_{12}O_2$	Methyl 2,2-dimethylpropanoate	116(3.2)	57(85)	41(32)	29(24)	56(21)
$C_6H_{12}O_2$	Ethyl butanoate	116(2.2)	43(50)	71(45)	29(43)	27(31)
$C_6H_{12}O_3$	2,4,6-Trimethyl-1,3,5-trioxacyclo-hexane	132(0.12)	45(196)	43(107)	29(35)	89(23)
$C_6H_{12}S$	1-Cyclopentyl-1-thiaethane	116(31)	68(72)	41(64)	39(37)	67(37)
$C_6H_{12}S$	*cis*-2,5-Dimethylthiacyclopentane	116(32)	101(85)	59(34)	41(26)	74(24)
$C_6H_{12}S$	*trans*-2,5-Dimethylthiacyclopentane	116(32)	101(85)	59(34)	74(25)	41(25)
$C_6H_{12}S$	2-Methylthiacyclohexane	116(42)	101(81)	41(37)	27(32)	67(30)
$C_6H_{12}S$	3-Methylthiacyclohexane	116(41)	101(55)	41(47)	39(33)	45(28)
$C_6H_{12}S$	4-Methylthiacyclohexane	116(46)	116(46)	101(44)	41(40)	27(39)
$C_6H_{12}S$	Thiacycloheptane	116(60)	87(75)	41(66)	67(48)	47(46)
$C_6H_{12}S$	1-Methylcyclopentanethiol	116(20)	83(76)	55(58)	41(39)	67(33)
$C_6H_{12}S$	*cis*-2-Methylcyclopentanethiol	116(32)	55(55)	83(54)	60(48)	41(47)
$C_6H_{12}S$	*trans*-2-Methylcyclopentanethiol	116(28)	67(48)	55(46)	41(42)	83(40)
$C_6H_{12}S$	Cyclohexanethiol	116(21)	55(56)	41(45)	67(35)	83(32)
$C_6H_{13}N$	Cyclohexylamine	99(8.9)	56(92)	43(25)	28(13)	30(13)
$C_6H_{13}N$	3-Methylpiperidine	99(23)	44(49)	30(34)	28(27)	57(26)
$C_6H_{13}NO$	*N*-Ethylmorpholine	115(2.0)	42(9.8)	57(7.0)	100(5.2)	28(4.3)
C_6H_{14}	2,2-Dimethylbutane	86(0.04)	43(85)	57(82)	71(61)	41(51)
C_6H_{14}	2,3-Dimethylbutane	86(5.3)	43(157)	42(136)	41(49)	27(40)
C_6H_{14}	2-Methylpentane	86(4.4)	43(147)	42(78)	41(47)	27(40)
C_6H_{14}	3-Methylpentane	86(3.2)	57(105)	56(80)	41(67)	29(64)
C_6H_{14}	*n*-Hexane	86(12)	57(87)	43(71)	41(64)	29(55)
$C_6H_{14}N_2$	*cis*-2,5-Dimethylpiperazine	114(0.38)	58(10)	28(7.7)	30(4.7)	44(4.2)
$C_6H_{14}O$	2-Ethyl-1-butanol	102(0.0)	43(114)	70(40)	29(39)	27(38)
$C_6H_{14}O$	2-Methyl-1-pentanol	102(0.0)	43(110)	41(40)	29(34)	27(33)
$C_6H_{14}O$	3-Methyl-1-pentanol	102(0.0)	56(26)	41(20)	29(19)	55(18)
$C_6H_{14}O$	4-Methyl-2-pentanol	102(0.08)	45(111)	43(34)	41(17)	27(14)
$C_6H_{14}O$	1-Hexanol	102(0.0)	56(63)	43(52)	41(37)	55(36)
$C_6H_{14}O$	Ethyl *n*-butyl ether	102(3.8)	59(108)	31(87)	29(61)	27(42)
$C_6H_{14}O$	Ethyl *sec*-butyl ether	102(1.5)	45(150)	73(76)	29(51)	27(39)
$C_6H_{14}O$	Ethyl isobutyl ether	102(8.7)	59(124)	31(95)	29(53)	27(38)
$C_6H_{14}O$	Diisopropyl ether	102(1.4)	45(125)	43(66)	87(23)	27(19)
$C_6H_{14}O_2$	1,1-Diethoxyethane	118(0.0)	45(132)	73(69)	29(36)	27(27)
$C_6H_{14}O_2$	1,2-Diethoxyethane	118(1.2)	31(124)	59(88)	29(72)	45(53)
$C_6H_{14}O_3$	*bis*-(2-Methoxyethyl) ether	134(0.0)	59(140)	29(74)	58(57)	15(56)

Mass numbers (and intensities) of: Parent peak, Base peak, Three next most intense peaks.

TABLE 10.2 Mass Spectra of Some Selected Compounds (*Continued*)

Molecular formula	Name	Mass numbers (and intensities) of: Parent peak	Base peak	Three next most intense peaks		
$C_6H_{14}S$	2,2-Dimethyl-3-thiapentane	118(33)	57(147)	41(70)	29(54)	27(40)
$C_6H_{14}S$	2,4-Dimethyl-3-thiapentane	118(33)	43(94)	61(85)	41(48)	103(44)
$C_6H_{14}S$	2-Methyl-3-thiahexane	118(206)	43(540)	41(317)	42(301)	27(287)
$C_6H_{14}S$	4-Methyl-3-thiahexane	118(195)	89(585)	29(343)	27(296)	41(279)
$C_6H_{14}S$	5-Methyl-3-thiahexane	118(171)	75(520)	41(230)	47(224)	56(217)
$C_6H_{14}S$	3-Thiaheptane	118(35)	75(55)	29(33)	27(33)	62(28)
$C_6H_{14}S$	4-Thiaheptane	118(47)	43(86)	89(74)	41(57)	27(55)
$C_6H_{14}S$	2-Methyl-1-pentanethiol	118(19)	43(96)	41(51)	56(32)	27(31)
$C_6H_{14}S$	4-Methyl-1-pentanethiol	118(30)	56(142)	41(57)	43(57)	27(32)
$C_6H_{14}S$	4-Methyl-2-pentanethiol	118(6.3)	43(68)	69(61)	41(56)	84(42)
$C_6H_{14}S$	2-Methyl-3-pentanethiol	118(20)	41(64)	43(63)	75(50)	27(28)
$C_6H_{14}S$	1-Hexanethiol	118(16)	56(66)	41(41)	27(40)	43(38)
$C_6H_{14}S_2$	2,5-Dimethyl-3,4-dithiahexane	150(31)	43(152)	108(41)	41(36)	27(30)
$C_6H_{14}S_2$	5-Methyl-3,4-dithiaheptane	150(14)	29(86)	94(66)	66(57)	27(41)
$C_6H_{14}S_2$	6-Methyl-3,4-dithiaheptane	150(4.9)	29(42)	66(40)	122(30)	94(29)
$C_6H_{14}S_2$	4,5-Dithiaoctane	150(44)	43(167)	27(65)	41(64)	108(35)
$C_6H_{15}N$	Triethylamine	101(21)	86(134)	30(46)	27(36)	58(35)
$C_6H_{15}N$	Di-*n*-propylamine	101(7.1)	30(89)	72(70)	44(36)	43(28)
$C_6H_{15}N$	Diisopropylamine	101(5.0)	44(171)	86(52)	58(24)	42(22)
$C_6H_{16}Pb$	Dimethyldiethyllead	296(0.98)	267(89)	223(83)	208(79)	221(44)
C_7F_{14}	Tetradecafluoromethylcyclohexane	350(0.0)	69(244)	131(107)	181(48)	100(38)
C_7F_{16}	Hexadecafluoroheptane	388(0.0)	69(330)	119(89)	169(68)	131(44)
C_7H_5N	Benzonitrile	103(246)	103(246)	76(80)	50(42)	51(24)
C_7H_7Br	1-Methyl-2-bromobenzene	170(48)	91(97)	172(46)	39(21)	63(20)
C_7H_7Br	1-Methyl-4-bromobenzene	170(46)	91(97)	172(45)	39(20)	65(19)
C_7H_7Cl	1-Methyl-2-chlorobenzene	126(44)	91(121)	63(20)	39(19)	89(18)
C_7H_7Cl	1-Methyl-3-chlorobenzene	126(51)	91(120)	63(19)	39(18)	128(16)
C_7H_7Cl	1-Methyl-4-chlorobenzene	126(44)	91(120)	125(19)	63(18)	39(17)
C_7H_7F	1-Methyl-3-fluorobenzene	110(79)	109(129)	83(17)	57(12)	39(12)
C_7H_7F	1-Methyl-4-fluorobenzene	110(73)	109(122)	83(16)	57(12)	39(9.3)
C_7H_8	Methylbenzene (toluene)	92(82)	91(108)	39(20)	65(14)	51(10)
C_7H_8S	1-Phenyl-1-thiaethane	124(76)	124(76)	109(34)	78(25)	91(19)
C_7H_9N	2,4-Dimethylpyridine	107(76)	107(76)	106(29)	79(16)	92(13)
$C_7H_{10}S$	2,3,4-Trimethylthiophene	126(50)	111(81)	125(47)	45(22)	39(18)
C_7H_{12}	Ethenylcyclopentane	96(13)	67(118)	39(44)	68(38)	54(35)
C_7H_{12}	Ethylidenecyclopentane	96(40)	67(180)	39(44)	41(30)	27(30)
C_7H_{12}	Bicyclo[2.2.1]heptane	96(12)	67(64)	68(50)	81(44)	54(30)
C_7H_{12}	3-Ethylcyclopentene	96(29)	67(193)	39(36)	41(35)	27(26)
C_7H_{12}	1-Methylcyclohexene	96(32)	81(83)	68(38)	67(37)	39(33)
C_7H_{12}	4-Methylcyclohexene	96(28)	81(84)	54(50)	39(44)	55(34)
C_7H_{12}	4-Methyl-2-hexyne	96(13)	81(71)	67(52)	41(48)	39(35)
C_7H_{12}	5-Methyl-2-hexyne	96(42)	43(49)	81(43)	27(39)	39(38)
C_7H_{12}	1-Heptyne	96(0.44)	41(75)	81(70)	29(65)	27(47)
C_7H_{14}	1,1,2,2-Tetramethylcyclopropane	98(21)	55(92)	83(90)	41(69)	39(41)
C_7H_{14}	*cis*-1,2-Dimethylcyclopentane	98(19)	56(85)	70(77)	41(65)	55(65)
C_7H_{14}	*trans*-1,2-Dimethylcyclopentane	98(25)	56(93)	41(63)	55(61)	70(54)
C_7H_{14}	*cis*-1,3-Dimethylcyclopentane	98(12)	56(81)	70(78)	41(64)	55(59)
C_7H_{14}	*trans*-1,3-Dimethylcyclopentane	98(13)	56(81)	70(68)	41(63)	55(58)
C_7H_{14}	1,1-Dimethylcyclopentane	98(6.7)	56(81)	55(63)	69(56)	41(55)
C_7H_{14}	Ethylcyclopentane	98(14)	69(83)	41(78)	68(60)	55(46)
C_7H_{14}	Methylcyclohexane	98(41)	83(94)	55(78)	41(55)	42(34)

(*Continued*)

TABLE 10.2 Mass Spectra of Some Selected Compounds (*Continued*)

Molecular formula	Name	Parent peak	Base peak	Three next most intense peaks		
		Mass numbers (and intensities) of:				
C_7H_{14}	Cycloheptane	98(37)	41(57)	55(54)	56(50)	42(49)
C_7H_{14}	2,3,3-Trimethyl-1-butene	98(20)	83(101)	55(83)	41(61)	39(33)
C_7H_{14}	3-Methyl-2-ethyl-1-butene	98(22)	41(71)	69(71)	55(62)	27(38)
C_7H_{14}	2,3-Dimethyl-1-pentene	98(13)	41(92)	69(86)	55(40)	39(35)
C_7H_{14}	2,4-Dimethyl-1-pentene	98(9.1)	56(117)	43(68)	41(61)	39(39)
C_7H_{14}	3,3-Dimethyl-1-pentene	98(9.4)	69(104)	41(85)	55(42)	27(36)
C_7H_{14}	3,4-Dimethyl-1-pentene	98(0.61)	56(75)	55(62)	43(55)	41(54)
C_7H_{14}	4,4-Dimethyl-1-pentene	98(2.6)	57(161)	41(86)	29(52)	55(49)
C_7H_{14}	3-Ethyl-1-pentene	98(19)	41(116)	69(91)	27(43)	39(37)
C_7H_{14}	2,3-Dimethyl-2-pentene	98(31)	83(80)	55(75)	41(63)	39(34)
C_7H_{14}	2,4-Dimethyl-2-pentene	98(26)	83(97)	55(71)	41(52)	39(34)
C_7H_{14}	3,4-Dimethyl-*cis*-2-pentene	98(30)	83(87)	55(82)	41(52)	27(32)
C_7H_{14}	3,4-Dimethyl-*trans*-2-pentene	98(31)	83(89)	55(83)	41(52)	27(34)
C_7H_{14}	4,4-Dimethyl-*cis*-2-pentene	98(27)	83(96)	55(92)	41(62)	39(35)
C_7H_{14}	4,4-Dimethyl-*trans*-2-pentene	98(28)	83(105)	55(89)	41(58)	39(31)
C_7H_{14}	3-Ethyl-2-pentene	98(33)	41(86)	69(80)	55(74)	27(33)
C_7H_{14}	2-Methyl-1-hexene	98(4.6)	56(105)	41(54)	27(30)	39(27)
C_7H_{14}	3-Methyl-1-hexene	98(7.7)	55(76)	41(60)	69(57)	56(48)
C_7H_{14}	4-Methyl-1-hexene	98(4.9)	41(98)	57(94)	56(80)	29(70)
C_7H_{14}	5-Methyl-1-hexene	98(1.6)	56(91)	41(75)	55(47)	27(42)
C_7H_{14}	2-Methyl-2-hexene	98(28)	69(113)	41(99)	27(36)	39(33)
C_7H_{14}	3-Methyl-*cis*-2-hexene	98(30)	41(95)	69(90)	55(42)	27(36)
C_7H_{14}	4-Methyl-*trans*-2-hexene	98(23)	69(118)	41(106)	55(40)	39(35)
C_7H_{14}	5-Methyl-2-hexene	98(13)	56(90)	55(74)	43(71)	41(57)
C_7H_{14}	2-Methyl-*trans*-3-hexene	98(24)	69(86)	41(74)	55(62)	56(37)
C_7H_{14}	3-Methyl-*cis*-3-hexene	98(28)	69(98)	41(82)	39(33)	27(33)
C_7H_{14}	3-Methyl-*trans*-3-hexene	98(28)	69(97)	41(86)	55(63)	39(35)
C_7H_{14}	1-Heptene	98(15)	41(91)	56(79)	29(64)	55(54)
C_7H_{14}	*trans*-2-Heptene	98(27)	55(64)	56(59)	41(50)	27(35)
C_7H_{14}	*trans*-3-Heptene	98(27)	41(98)	56(65)	69(55)	55(47)
$C_7H_{14}O$	2,4-Dimethyl-3-pentanone	114(13)	43(226)	71(62)	27(49)	41(42)
$C_7H_{14}O_2$	*n*-Butyl propanoate	130(0.03)	57(152)	29(98)	56(54)	27(52)
$C_7H_{14}O_2$	Isobutyl propanoate	130(0.07)	57(187)	29(87)	56(52)	27(47)
$C_7H_{14}O_2$	*n*-Propyl *n*-butanoate	130(0.05)	43(96)	71(90)	27(54)	89(48)
$C_7H_{14}O_3$	*n*-Propyl carbonate	146(0.02)	43(171)	27(61)	63(55)	41(49)
$C_7H_{14}S$	*cis*-2-Methylcyclohexanethiol	130(28)	55(138)	97(70)	81(44)	41(44)
$C_7H_{15}N$	2,6-Dimethylpiperidine	113(5.3)	98(73)	44(43)	42(34)	28(26)
C_7H_{16}	2,2,3-Trimethylbutane	100(0.03)	57(110)	43(84)	56(67)	41(64)
C_7H_{16}	2,2-Dimethylpentane	100(0.06)	57(130)	43(95)	41(59)	56(52)
C_7H_{16}	2,3-Dimethylpentane	100(2.1)	43(94)	56(93)	57(67)	41(64)
C_7H_{16}	2,4-Dimethylpentane	100(1.6)	43(139)	57(93)	41(59)	56(50)
C_7H_{16}	3,3-Dimethylpentane	100(0.03)	43(166)	71(103)	27(38)	41(36)
C_7H_{16}	3-Ethylpentane	100(3.1)	43(175)	70(77)	71(77)	29(45)
C_7H_{16}	2-Methylhexane	100(5.9)	43(154)	42(59)	41(57)	85(49)
C_7H_{16}	3-Methylhexane	100(4.0)	43(110)	57(52)	71(52)	41(50)
C_7H_{16}	*n*-Heptane	100(17)	43(126)	41(65)	57(60)	29(58)
$C_7H_{16}O$	2-Heptanol	116(0.01)	45(131)	43(29)	27(25)	29(23)
$C_7H_{16}O$	3-Heptanol	116(0.01)	59(61)	69(41)	41(29)	31(25)
$C_7H_{16}O$	4-Heptanol	116(0.02)	55(102)	73(72)	43(45)	27(32)
$C_7H_{16}O$	*n*-Propyl *n*-butyl ether	116(3.7)	43(120)	57(102)	41(51)	29(49)
$C_7H_{16}O_2$	Di-*n*-propoxymethane	132(0.58)	43(194)	73(114)	27(45)	41(34)

TABLE 10.2 **Mass Spectra of Some Selected Compounds (*Continued*)**

Molecular formula	Name	Mass numbers (and intensities) of: Parent peak	Base peak	Three next most intense peaks		
$C_7H_{16}O_2$	Diisopropoxymethane	132(0.16)	43(133)	45(84)	73(71)	27(28)
$C_7H_{16}O_2$	1,1-Diethoxypropane	132(0.0)	59(138)	47(88)	87(84)	29(74)
$C_7H_{16}S$	2,2,4-Trimethyl-3-thiapentane	132(30)	57(149)	41(74)	29(35)	43(32)
$C_7H_{16}S$	2,4-Dimethyl-3-thiahexane	132(30)	61(94)	103(60)	41(51)	43(46)
$C_7H_{16}S$	2-Thiaoctane	132(34)	61(73)	56(53)	27(46)	41(44)
$C_7H_{16}S$	1-Heptanethiol	132(14)	41(48)	27(40)	56(39)	70(38)
$C_7H_{18}Pb$	Methyltriethyllead	310(0.84)	281(86)	208(76)	223(66)	237(60)
$C_7H_{18}Pb$	*n*-Butyltrimethyllead	310(0.14)	253(76)	223(75)	208(68)	295(52)
$C_7H_{18}Pb$	*sec*-Butyltrimethyllead	310(1.8)	253(94)	223(85)	208(74)	251(45)
$C_7H_{18}Pb$	*tert*-Butyltrimethyllead	310(0.09)	252(95)	223(82)	208(65)	250(46)
C_8H_{10}	1,2-Dimethylbenzene	106(52)	91(91)	105(22)	39(15)	51(14)
C_8H_{10}	1,3-Dimethylbenzene	106(58)	91(93)	105(26)	39(17)	51(14)
C_8H_{10}	1,4-Dimethylbenzene	106(52)	91(85)	105(25)	51(13)	39(13)
C_8H_{10}	Ethylbenzene	106(45)	91(146)	51(19)	39(14)	65(12)
F_3N	Nitrogen trifluoride	71(10)	52(33)	33(13)	14(3.0)	19(2.7)
HCl	Hydrogen chloride	36(54)	36(54)	38(17)	35(9.2)	37(2.9)
H_2S	Hydrogen sulfide	34(75)	34(75)	32(33)	33(32)	1(4.1)
H_3N	Ammonia	17(32)	17(32)	16(26)	15(2.4)	14(0.7)
H_3P	Phosphine	34(59)	34(59)	33(20)	31(19)	32(7.5)
H_4N_2	Hydrazine	32(48)	32(48)	31(23)	29(19)	30(15)
NO	Nitric oxide	30(76)	30(76)	14(5.7)	15(1.8)	16(1.1)
NO_2	Nitrogen dioxide	46(6.6)	30(18)	16(4.0)	14(1.7)	47(0.02)
N_2	Nitrogen	28(65)	28(65)	14(3.3)	29(0.47)	…
N_2O	Nitrous oxide	44(60)	44(60)	30(19)	14(7.8)	28(6.5)
O_2	Oxygen	32(54)	32(54)	16(2.7)	28(1.7)	34(0.22)
O_2S	Sulfur dioxide	64(47)	64(47)	48(23)	32(4.9)	16(2.4)

Source: L. Meites, ed., Handbook of Analytical Chemistry, McGraw-Hill, New York, 1963.

Nearly all these spectra have been recorded using 70-V electrons to bombard the sample molecules.

10.7 SECONDARY-ION MASS SPECTROMETRY[19,20]

Secondary-ion mass spectrometry (SIMS) is used for the analysis of surface layers and their composition to a depth of 1 to 3 nm. A focused ion beam strikes the sample surface and releases secondary ions, which are detected by a mass spectrometer. Typical instrumentation might involve a plasma-discharge source coupled with a quadrupole mass analyzer. The plasma discharge also serves as a sputtering device to remove successive layers of sample for profiling the material.

The SIMS technique affords qualitative identification of all surface elements and permits identification of isotopes and the structural elucidation of molecular compounds present on a surface. Detection sensitivity is in parts per million. SIMS is also useful for analyzing nonvolatile and thermally labile molecules, including polymers and large biomolecules.

[19] K. F. J. Heinrich and D. E. Newbury, eds., *Secondary Ion Mass Spectrometry*, NBS Spec. Publ. No. 427, U.S. Government Printing Office, Washington, D.C., 1975.

[20] R. J. Day, S. E. Unger, and R. G. Cooks, "Molecular Secondary Ion Mass Spectrometry," *Anal. Chem.* **52**:557A (1980).

10.8 ISOTOPE-DILUTION MASS SPECTROMETRY (IDMS)

Stable isotopes can be used to "tag" compounds and thus serve as tracers to determine the ultimate fate of the compound in chemical or biological systems and also as an analytical method. A number of stable isotopes in sufficiently concentrated form are available for studying organic and inorganic systems: H, B, C, N, O, S, and Cl. In principle, IDMS is applicable to all 60 elements that have more than one available stable isotope.[21] These isotopes complement the relatively larger number of radioactive isotopes. The isotope-dilution method (Sec. 11.2.3) can be employed equally well with stable isotopes.[22] IDMS is based on the addition of a known amount of enriched isotope (called the spike) to a sample. After equilibration of the spike isotope with the natural element in the sample, MS is used to measure the altered isotopic ratio(s). It is only necessary to know the ratio of isotopes present in the added sample of the substance, the ratio present in the final sample isolated from the mixture, and the weight of the added sample.

The measured ratio (R_m) of isotope A to isotope B can be calculated as follows:

$$R_m = \frac{A_x C_x W_x + A_s C_s W_s}{B_x C_x W_x + B_s C_s W_s} \tag{10.9}$$

where A_x and B_x are the atom fractions of isotopes A and B in the sample, A_s and B_s are the atom fractions of isotopes A and B in the spike, C_x and C_s are the concentrations of the element in the sample and spike, respectively, and W_x and W_s are the weights of the sample and spike, respectively. The concentration of the element in the sample can then be calculated:

$$C_x = \left(\frac{C_s W_s}{W_x}\right)\left(\frac{A_s - R_m B_s}{R_m B_x - A_x}\right) \tag{10.10}$$

Because IDMS requires equilibration of the spike isotope and the natural isotope(s), the sample must be dissolved. If the sample does not completely dissolve, if the spike or sample isotopes are selectively lost before equilibration, or if contamination occurs in the dissolution process, the measured isotopic ratios will not reflect the accurate ratio of added spike atoms to sample atoms for that element.

Thermal ionization is the ionization method of choice for precise and accurate IDMS. Precision and accuracy are typically 0.1% or better. Other useful types of mass spectrometry include electron ionization with thermal probes, spark source, secondary ions, resonance ionization, and field desorption.

The isotope-ratio mass spectrometer, a less expensive adaptation of the usual mass spectrometer, is available for work in this field. In the modified instrument the ion currents from two ion beams—for example, the ion beams from $^{32}SO_2$ and $^{34}SO_2$, are collected simultaneously by means of a double exit slit and are amplified simultaneously by two separate amplifiers. The larger of the two amplified currents is then attenuated by the operator until it exactly balances the smaller signal from the other amplifier. The ratio of the two signals is determined from the attenuation required. This is a null method and practically eliminates the effect of other variables in the system.

10.9 QUANTITATIVE ANALYSIS OF MIXTURES

Sensitivity and specificity are the major advantages of mass spectrometry as a quantitative analytical technique. An ion incorporating the intact molecule (molecular ion peak) is most characteristic. Production of molecular ions (or at least high-mass fragment ions) is favored by the use of low-energy electron-impact ionization or by the use of chemical ionization. By judicious choice of reagent gases,

[21] "Relative Abundance of Naturally Occurring Isotopes," in J. A. Dean, ed., *Lange's Handbook of Chemistry*, 14th ed., McGraw-Hill, New York, 1992, pp. 4.53 to 4.56.

[22] J. D. Fassett and P. J. Paulsen, "Isotope Dilution Mass Spectrometry for Accurate Elemental Analysis," *Anal. Chem.* **61**:643A (1989).

the latter ionization method provides the opportunity for selective ionization of certain components of complex mixtures. Detection of analytes with high electron affinities through negative chemical ionization provides another useful technique.

The flow of each kind of molecule through the leak in the inlet system is molecular; that is, the rate of flow is proportional to the partial pressure of the species behind the leak and independent of the presence of other kinds of molecules. Consequently, the intensities of the various ion beams from the source are also proportional to the partial pressures of the substance behind the leak. If two or more species yield ion beams having the same *m/z* ratio, the beam intensities, which are usually measured in arbitrary units and conventionally called peak heights, are additive. Thus, for a mixture of x number of substances at a total pressure of P_0 in the reservoir behind the leak, x peaks are selected for measurement. Spectra are recorded on pure samples of each component. From inspection of the individual mass spectra, analysis peaks are selected on the basis of both intensity and freedom from interference. It possible, monocomponent peaks (perhaps molecular-ion peaks) are selected. Computation is simplified if the components of the mixture give at least one peak whose intensity is entirely due to the presence of one component.

From the mass spectrum of each pure compound, the sensitivity is obtained by dividing thc peak height of each significant peak by the pressure of the pure compound in the sample reservoir of the mass spectrometer. From the simplified case in mixtures when the intensity of one peak is entirely due to the presence of one component, the height of the monocomponent peak is measured and divided by the appropriate sensitivity factor to give its partial pressure. Then division by the total pressure in the sample reservoir at the time of analysis yields the mole fraction of the particular component.

If the mixture has no monocomponent peaks, simultaneous linear equations are then set up from the coefficients (percent of base peak) at each analysis peak. For example, the significant portion of the mass spectral data is given in Table 10.3 for individual C_1 to C_3 alcohols. Using mass peaks at 32, 39, 46, and 59, four equations are written:

$$68.03x_1 + 1.14x_2 + 2.25x_3 = M_{32} \tag{10.11}$$

$$4.00x_3 + 5.52x_4 = M_{39} \tag{10.12}$$

$$16.23x_2 = M_{46} \tag{10.13}$$

$$9.61x_3 + 3.58x_4 = M_{59} \tag{10.14}$$

TABLE 10.3 Mass Spectral Data (Relative Intensities) for the C_1 to C_3 Alcohols

	Percent of base peak (italic)				
m/z	Methyl	Ethyl	*n*-Propyl	Isopropyl	Unknown
15	35.48	9.44	3.77	10.70	
27	—	21.62	15.20	15.50	
29	58.80	21.24	14.14	9.49	
31	*100*	*100*	*100*	5.75	
32	68.03(*P*)	1.14	2.25	—	600
39	—	—	4.00	5.52	3000
43		7.45	3.18	16.76	
45		37.33	4.39	*100*	
46		16.23(*P*)	—	—	
59			9.61	3.58	1100
60			6.36(*P*)	0.44(*P*)	2300
Sensitivity, divisions per 10^{-3} torr	8.76	17.98	26.51	23.47	

Next substitute the values for the mixture peaks of the unknown into the above equations.

$$16.23x_2 = M_{46} = 1100, \qquad x_2 = 67.78$$

$$4.00x_3 + 5.52x_4 = M_{39} = 3000, \qquad x_3 = 50.50$$

$$9.61x_3 + 3.58x_4 = M_{56} = 2300, \qquad x_4 = 506.9$$

$$68.03x_1 + 1.14(67.78) + 2.25(50.50) = M_{32} = 600; \qquad x_1 = 6.01$$

Division by the appropriate sensitivity factor (Table 10.1) yields the partial pressure of each component ($\times 10^{-3}$ torr):

Methanol $6.01 \div 8.76 = 0.686$

Ethanol $67.78 \div 17.98 = 3.770$

1-Propanol $50.50 \div 26.51 = 1.905$

2-Propanol $506.90 \div 23.47 = 21.60$

Finally, each partial pressure is divided by the total pressure (27.96×10^{-3} torr), yielding the fractional mole content which, when multiplied by 100, gives the percent in the sample:

Methanol $0.686 \div 27.96 = 0.0245$ (or 2.45%)

Ethanol $3.770 \div 27.96 = 0.1348$ (or 13.48%)

1-Propanol $1.905 \div 27.96 = 0.0681$ (or 6.81%)

2-Propanol $21.60 \div 27.96 = 0.7725$ (or 77.25%)

The sum of the partial pressures should equal the total sample pressure. A discrepancy would indicate an unsuspected component or a change in operating sensitivity. It must be emphasized that sensitivities are not precisely reproducible from instrument to instrument or from time to time, and that calibrations with pure substances must be run on the same instrument under the same conditions and at as nearly as possible the same time as the mixture analysis.

10.10 HYPHENATED GC-MS AND LC-MS TECHNIQUES

The combination of chromatographic and mass spectrometric techniques makes available the benefits and advantages of both analytical fields. The chromatograph does the separating and the mass spectrometer does the identifying and quantitation. Benefits include near-universal analyte response, low detection limits, and high information content provided from the mass spectra of organic compounds. Since infrared is nondestructive, it is possible to combine the three instruments into a GC-FTIR-MS.[23]

10.10.1 GC-MS

There are three requirements for a GC-MS interface: (1) The volume of gas from the gas chromatograph must be reduced to that compatible with the inlet of the mass spectrometer and, furthermore, this pressure reduction should be accomplished with reducing the analyte concentration. (2) The spectra of analytes should be obtained "on the fly," often on the order of milliseconds. (3) A final

[23] C. L. Wilkins, *Science* **222**:291 (1983).

requirement is a data system capable of handling the volume of data generated by a fast-scanning mass spectrometer.

The direct type of interface with open tubular columns involves extending the end of the open tubular column from the gas chromatography directly into the ion source of the mass spectrometer. The GC flows are low enough, and the vacuum pumping high enough that the vacuum required by the MS can be maintained without any other interfaces. Two disadvantages prevail: all the column effluent is deposited in the ion source of the MS, causing it to become contaminated rather quickly, and the GC column cannot be changed without shutting down the MS because there is no way to isolate one from the other.

In the open split interface, the space between the GC column and the MS inlet is maintained at about atmospheric pressure by the use of a second source of gas and a separate vacuum. The amount of purge gas can be controlled to enable the column to be disconnected without shutting down the MS and undesirable sample components can be removed before they enter the MS.

Elemental speciation methods in which GC, LC, and SFC are coupled with ICP mass spectrometry have been reported.[24] The glow discharge source can replace the ICP unit.

10.10.2 LC-MS[25,26]

The problem encountered when interfacing LC with MS is the mismatch between the mass flows involved in conventional HPLC which are two or three orders of magnitude larger than can be accommodated by conventional MS vacuum systems. Another problem is the difficulty of vaporizing involatile and thermally labile molecules without degrading them excessively.

10.10.2.1 Thermospray Method.[27,28] The HPLC effluent is fed into a microfurnace maintained at up to 400°C that protrudes into a region of reduced pressure (approximately 1 torr). The heat creates a supersonic, expanding aerosol jet that contains a mist of fine droplets of solvent vapor and sample molecules. The droplets vaporize downstream and the excess vapor is pumped away. Ions of the sample molecules are formed in the spray either by direct desorption or by chemical ionization when used with polar mobile phases that contain buffers such as ammonium acetate. A conventional electron beam is used to provide gas-phase reagent ions for the chemical ionization of solute molecules.

10.10.2.2 Aerosol-Generation Interface.[29] The high-pressure effluent from the HPLC column passes through a small-diameter orifice to form a fine liquid jet that breaks up under natural forces to form uniform drops that are dispersed with a gas stream introduced at a right angle to the liquid flow. The solvent evaporates in a desolvation chamber. A two-stage aerosol-beam separator, which consists of two nozzle and skimmer devices, reduces the column pressure to the pressure in the ion source.

10.10.2.3 Electrospray Interface.[30] A voltage of 2 to 3 kV is applied to the metal capillary tip, which is typically 0.2 mm o.d. and 0.1 mm i.d. and located 1 to 3 cm from a large planar counter electrode. This counterelectrode has an orifice leading to the MS sampling system. The very high electric field imposed causes an enrichment of positive electrolyte ions at the meniscus of the solution at the capillary tip. At a sufficiently high field, the cone is not stable and a liquid filament

[24] N. P. Vela, K. K. Olson, and J. A. Caruso, *Anal. Chem.* **65**:585A (1993).

[25] T. R. Covey et al., *Anal. Chem.* **58**:1451A (1986).

[26] A. L. Yergey, C. G. Edmonds, I. A. S. Lewis, and M. L. Vestal, *Liquid Chromatography/Mass Spectrometry*, Plenum, New York, 1990.

[27] C. R. Blakeley and M. L. Vestal, *Anal. Chem.* **55**:750 (1983).

[28] L. Yang, G. J. Ferguson, and M. L. Vestal, *Anal. Chem.* **56**:2632 (1984).

[29] R. C. Willoughby and R. C. Browner, *Anal. Chem.* **56**:2626 (1984).

[30] P. Kebarie and L. Tang, *Anal. Chem.* **65**:972A (1993).

with a diameter of a few micrometers, whose surface is enriched on positive ions, is emitted from the cone tip. At some distance downstream the filament becomes unstable and forms a fine mist of positively charged droplets. The charged droplets shrink by solvent evaporation and repeated droplet disintegrations, leading to very small, highly charged droplets capable of producing gas-phase ions which enter the MS.

When coupling capillary electrophoresis to MS, electrical connection is established at the gold-plated CE capillary terminus. Various designs feature sheath flow (coaxial) interface, a liquid-junction interface, and a sheathless interface.[31]

10.10.2.4 Mechanical Transport Interface. In the moving-belt interface, chromatographic effluent is deposited or sprayed onto a continuous moving belt that is woven from ultrafine quartz fiber. The belt passes below an infrared heater that evaporates most of the mobile phase before the belt reaches the entrance slit to the first of two successive vacuum locks. The pressure is reduced to 0.1 torr before the belt moves into a chamber where a flash vaporizer vaporizes the sample into the ion source. The belt then exits the chamber, passes over a final heater that cleans the belt by evaporation of any residue, and then moves out through the locks to recycle. In some versions the belt passes through a wash bath after leaving the vacuum locks. The moving-belt interface can be used with magnetic sector or quadrupole instruments and in either the electron or chemical ionization mode.

Bibliography

Adams, F., R. Gijbels, and R. Van Grieken, eds., *Inorganic Mass Spectrometry*, Wiley, New York, 1988.

Harrison, W., *Chemical Ionization Mass Spectrometry*, CRC Press, Boca Raton, Fla., 1992.

McLafferty, F. W., and D. B. Stauffer, *Registry of Mass Spectral Data*, 2d ed.,Wiley, New York, 1988.

McLafferty, F. W., and D. B. Stauffer, *The Important Peak Index of the Registry of Mass Spectral Data*, Wiley, New York, 1991.

NIST/EPA/NIH Mass Spectral Database, NIST Standard Reference Data, Gaithersburg, Maryland, 1992. Continuously upgradable data compatible with PCs.

Watson, J. T., *Introduction to Mass Spectrometry*, Raven, New York, 1985.

[31] R. D. Smith et at., *Anal. Chem.* **65**:574A (1993).

SECTION 11
RADIOCHEMICAL METHODS

11.1 INTRODUCTION

Radioactivity is the spontaneous disintegration of an atom that is accompanied by emission of radiation. There are many radioactive elements that are isotopes (having the same atomic number but different atomic mass) of nonradioactive elements. An atom of a radioactive isotope has the same number of orbital electrons as an atom of its nonradioactive counterpart and, in general, will behave chemically and biologically like the nonradioactive species. Therefore experimental and diagnostic as well as analytical procedures can utilize atoms of radioactive isotopes as tracers. The difference between the radioactive and the nonradioactive atoms of identical elements is the number of neutrons in the nucleus, the number of protons and electrons being the same for all. (Some elements have more than two isotopes.)

11.1.1 Modes of Radioactive Decay

All radioactive nuclides exhibit the phenomenon of radioactive decay with emission of one or more of several types of radiation. The modes of decay with associated radiations are summarized in Table 11.1. In many cases, the product nucleus is left in an excited state after a decay [α, β, electron capture (EC)]. Stability is achieved by emission of gamma radiation from the nucleus. Sometimes this process takes place very quickly; otherwise gamma radiation may be delayed, perhaps by many days. The latter situation is known as an *isomeric transition*, and the nucleus is said to be in a *metastable* state, denoted by *m* after the mass number.

Alpha, gamma, and x radiations have discrete energies; by suitable methods one can detect these energies as *photopeaks*. Beta particles, either a very energetic negatively charged electron (negatron) or a positively charged electron (positron), have a continuous distribution of kinetic energies up to a maximum energy E_{max}, the value given in tables of nuclides. For example, the energy of the beta particles emitted in the decay of ^{32}P lie in a continuum from 0 to 1.71 MeV; the average energy is 0.69 MeV.

All positron decays are accompanied by *annihilation radiation* resulting from destruction of the positrons. The energy of this radiation is 0.511 MeV, and two such photons arise from the destruction of each positron. Every electron-capture and internal-conversion process is followed immediately by emission of x rays.

A positron-emitting atom may decay by capturing one of its own orbital *K* electrons (*K* capture or internal conversion). The excess energy is emitted as γ radiation. The daughter element (one atomic number less than its parent) has a vacant *K* orbital; x radiation characteristic of the daughter is emitted when *L*- and *M*-level electrons fall into the *K* level.

Radioactive decay of a nucleus often results in the formation of a daughter nucleus in an excited unstable state. As the daughter nucleus changes from the excited state to a state of lower energy, energy is released, usually as electromagnetic radiation in the form of γ radiation. The change is termed an *isomeric transition* (IT) because the nucleus decays with no change in atomic mass or number.

A selection of radionuclides and their characteristics is given in Table 11.2. Radionuclides of elements occurring naturally in biological organisms, such as ^{3}H, ^{14}C, ^{32}P, ^{35}S, and ^{131}I, are used for biological and medical research as tracers or in labeled compounds wherein one or more radioactive atoms have replaced stable atoms of the same element in the compound.

TABLE 11.1 Modes of Radioactive Decay

Mode	Reaction	Radiation
Alpha (α)	$^{210}Po \rightarrow {}^{206}Pb + {}^{4}He$	α
Negatron (beta, β^-)	$^{32}P \rightarrow {}^{32}S + \beta^-$	β^- (beta)
Positron (β^+)	$^{22}Na \rightarrow {}^{22}Ne + \beta^+$	β^+
Electron capture (EC)	$^{188}Pt \rightarrow {}^{188}Ir$ + x rays	Ir x rays
Gamma (γ)	$^{110m}Ag \rightarrow {}^{110}Ag + \gamma$	γ (electromagnetic)

TABLE 11.2 **Table of Nuclides**

Explanation of column headings

Nuclide. *Each nuclide is identified by element name and the mass number A, equal to the sum of the numbers of protons Z and neutrons* N *in the nucleus. The* m *following the mass number (for example,* ^{69m}Zn*) indicates a metastable isotope. An asterisk preceding the mass number indicates that the radionuclide occurs in nature. The digits in parentheses following a numerical value represent the standard deviation of that value in terms of the final listed digits.*

Half-life. *The following abbreviations for time units are employed: y = years, d = days, h = hours, m = minutes, and s = seconds.*

Natural abundance. *The natural abundances listed are on an "atom percent" basis for the stable nuclides present in naturally occurring elements in the earth's crust.*

Thermal neutron absorption cross section. *Simply designated* cross section, *it represents the ease with which a given nuclide can absorb a thermal neutron (energy less than or equal to 0.025 eV) and become a different nuclide. The cross section is given here in units of barns (1 b =* 10^{-24} *cm*^{2}*). If the mode of reaction is other than (n,g), it is so indicated.*

Major radiations. *In the last column are the principal modes of disintegration and energies of the radiations in million electronvolts (MeV) in parentheses. Symbols used to represent the various modes of decay are*

α, *alpha particle emission*	K, *electron capture*
β^-, *beta particle (negatron)*	IT, *isomeric transition*
β^+, *positron*	x, x *rays of indicated element (e.g., O-x, oxygen* x *rays)*
γ, *gamma radiation*	

For β^- and β^+, *values of* E_{max} *are listed. Radiation types and energies of minor importance are omitted unless useful for identification purposes. For detailed decay schemes the literature should be consulted.*

Element	*A*	Half-life	Natural abundance, %	Cross section, b	Radiation, MeV
Hydrogen	1		99.985(1)	0.332	
	2		0.015(1)	0.0005	
	3	12.33(6) y			β^- (0.0186)
Beryllium	7	53.29(7) d			*K*, γ(0.478)
	9		100	0.009	
	10	1.51(6) × 10^6 y			β^- (0.555)
Boron	10		19.9(2)	3837(*n,a*)	
Carbon	11	20.385(20) m			β^+ (0.961)
	14	5730(40) y			β^- (0.156)
Nitrogen	13	9.965(4) m			β^+ (1.190); γ(0.511)
	14		99.63(2)	1.81(*n,p*)	
	16	7.13(2) s			β^- (10.40, 4.27); γ(6.13)
Oxygen	19	26.91(8) s			β^- (4.60); γ(0.197, 1.37)
Fluorine	18	109.77(5) m			β^+ (0.635); *K*, O-x
	20	11.00(2) s			β^- (5.41), γ(1.63)
Neon	22		8.82	0.04	
	23	37.24(12) s			β^- (0.438); γ(0.439)
Sodium	22	2.6088(14) y			β^+ (0.545, 1.83); *K* Ne-x,
			γ(1.275)		
	23		100	0.53	
	24	14.9590(12) h			β^- (1.39); γ(2.75, 1.37)
Magnesium	25		10.00(1)	0.3	
	27	9.46(1) m			β^- (1.75), γ(0.84, 1.01)
	28	20.91(3) h			β^- (0.46); γ(1.34, 0.94, 0.40, 0.031)

(Continued)

TABLE 11.2 Table of Nuclides (*Continued*)

Element	*A*	Half-life	Natural abundance, %	Cross section, b	Radiation, MeV
Aluminum	27		100	0.235	
	28	2.2414(12) m			β^- (2.85); γ (1.780)
Silicon	30		3.10(1)	0.11	
	31	153.3(3) m			β^- (1.48); γ (1.27)
	32	172(4) y			β^- (0.213)
Phosphorus	31		100	0.19	
	32	14.262(14) d			β^- (1.71)
	33	25.34(12) d			β^- (0.25)
Sulfur	34		4.21(8)	0.27	
	35	87.51(12) d			β^- (0.167)
	36	3.08×10^5 y		100	β^- (0.714)
	37	5.05(2) m			β^- (4.7, 1.6); γ(3.09)
	38	170.3(7) m			β^- (1.0, 3.0); γ(1.94)
Chlorine	35		75.53(5)	44	
	36	$3.01(2) \times 10^5$ y			β^- (0.71); *K*, S-x
	37		24.23(5)	0.4	
	38	37.24(5) m			β^- (4.81); γ(2.17, 1.60)
	39	55.6(2) m			β^- (1.91, 2.18, 3.45); γ(1.27, 0.25, 1.52)
Argon	37	35.04(4) d			*K*, Cl-x
	41	1.822(2) h			β^- (1.20, 2.49); γ(1.29)
Potassium	*40	$1.28(1) \times 10^9$ y	0.0117(1)	70	β^- (1.34); *K*, Ar-x; γ(1.46)
	41		6.730(3)	1.2	
	42	12.360(3) h			β^- (3.52, 1.97); γ(1.46)
	43	22.3(1) h			β^- (0.83, 1.22, 1.82); γ(0.618, 0.373, 0.39, 0.59)
Calcium	44		2.086(5)	0.7	
	45	163.8(8) d			β^- (0.255)
	47	4.536(2) d			β^- (1.98, 0.67); γ(1.30)
	49	8.71(2) m			β^- (1.95); γ(3.10, 4.1)
Scandium	46	83.81(1) d			β^- (0.357); γ(1.12, 0.889); Ti-x
	46*m*	18.75(4) s			γ(0.142)
Titanium	45	3.08(1) h			β^+ (1.044); *K*, Sc-x
	50		5.4(1)	0.14	
	51	5.76(1) m			β^- (2.14); γ(0.320, 0.928)
Vanadium	48	15.973(3) d			β^+ (0.696), γ(0.983, 1.312)
	52	3.75(1) m			β^- (2.47); γ(1.434)
Chromium	51	27.702(4) d			*K*, V-x; γ(0.32)
	52		83.7(1)	0.8	
Manganese	52	5.591(3) d			β^+ (0.575), γ(0.744, 0.935, 1.434)
	54	312.1(1) d			γ(0.835)
	55		100	13.3	
	56	2.5785(2) h			β^- (2.84), γ(0.847, 1.81, 2 2.11)

TABLE 11.2 Table of Nuclides (*Continued*)

Element	*A*	Half-life	Natural abundance, %	Cross section, b	Radiation, MeV
Iron	55	2.73(3) y	2.1(1)		*K*, Mn-x
	56		91.72(15)		
	59	44.496(7) d			β^- (0.273, 0.475), γ(1.10, 1.29)
Cobalt	57	271.80(5) d			*K*, Fe-x; γ(0.136, 0.122)
	58	70.82(3) d			*K*, β^+ (0.474), Fe-x; γ (0.811)
	59		100	19	
	60	5.2714(5) y			β^- (0.318), γ(1.173, 1.332)
	60*m*	10.47(4) m			β^- (1.55); γ (0.059)
Nickel	63	100(2) y			β^- (0.067)
	64		0.926(1)	1.5	
	65	2.520(1) h			β^- (2.14, 0.65, 1.02), γ(1.48, 0.367, 1.12)
Copper	63		69.17(2)	4.5	
	64	12.701(2) h			β^- (0.571); β^+ (0.657); Ni-x; γ(0.511)
	66	5.10(1) m			β^- (2.63), γ (1.039)
Zinc	64		48.6(3)	0.46	
	65	243.9(1) d			*K*, β^+ (0.325), Cu-x; γ(1.12)
	68		18.8(4)	1.0	
	69*m*	13.76(2) h			IT, Zn-x, γ (0.439)
	69	55.4(9) m			β^- (0.90)
Gallium	67	3.261(1) d			*K*, Zn-x; γ (0.093, 0.184, 0.30)
	69		60.108(6)	1.9	
	70	21.14(3) m			β^- (1.65); γ (0.173)
	71		39.892(6)	5.0	
	72	14.10(2) h			β^- (3.17); γ(0.835, 0.63, 0.894)
Germanium	74		35.94(2)	0.3	
	75	82.78(4) m			β^- (1.19), γ (0.265, 0.199)
	75*m*	47.7(5) s			γ(0.139)
	77	11.30(1) h			β^- (2.2), γ (0.21, 0.268, 0.368, 0.417, 0.568, 0.632, 0.73)
Arsenic	75		100	4.5	
	76	26.32(7) h			β^- (2.97), γ (0.559, 0.657)
	78	90.7(2) m			β^- (4.1), γ (0.614, 0.70, 0.83, 1.31)
Selenium	74		0.89(2)	30	
	75	119.779(4) d			*K*, γ (0.265, 0.136, 0.280, 0.121, 0.401); As-x
	77*m*	17.36(5) s			γ (0.161)
	80		49.6(3)	0.5	
	81	18.5(1) m			β^- (1.58); γ(0.28)
	81*m*	57.28(5) m			γ (0.108)

(Continued)

TABLE 11.2 Table of Nuclides (*Continued*)

Element	A	Half-life	Natural abundance, %	Cross section, b	Radiation, MeV
Bromine	79		50.69(5)	8.5	
	80	17.68(2) m			β^- (1.997); K, β^+ (0.85), Se-x; γ (0.616, 0.667)
	80*m*	4.42(1) h			IT, Br-x; γ (0.037)
	81		49.31(5)	3	
	82	95.30(2) h			β^- (0.444); γ (0.554, 0.619, 0.698, 0.777, 0.818, 1.04, 1.32, 1.48)
Krypton	81*m*	13(1) s			IT, Kr-x; γ (0.19)
	84		57.3(3)	0.10	
	85	10.756(18) y			β^- (0.67); γ (0.514)
	85*m*	4.480(8) h			β^- (0.84); γ (0.151, 0.305)
Rubidium	85		72.17(1)	0.9	
	86	18.63(2) d			β^- (1.78, 0.71); γ (1.08)
	86*m*	61.0(2) s			γ (0.56)
	87		27.83(1)	0.12	
	88	17.8(1) m			β^- (5.080); γ (1.863, 0.898)
Strontium	85	64.84(2) d			K, Rb-x; γ (0.514)
	85*m*	67.63(4) s			γ(0.150, 0.231)
	87*m*	2.804(3) h			IT, γ (0.388)
	90	29.1(3) y			β^- (0.546)
Yttrium	89		100	1/3	
	89*m*	16.06(4) s			γ (0.909)
	90*m*	3.19(1) h			γ (0.202, 0.482)
Zirconium	95	64.02(4) d			β^- (0.89, 0.396); γ (0.724, 0.756)
	97	16.90(5) h			β^- (1.91); γ (0.743)
Niobium	93		100	1	
	94*m*	6.26(1) m			γ (0.871)
	95	34.97(3) d			β^- (0.160); γ (0.765)
Molybdenum	98		24.13(6)	0.51	
	99	2.7477(4) d			β^- (1.214); Tc-x; γ(0.181, 0.740, 0.780)
	101	14.6(1) m			β^- (2.23); γ (0.191, 0.51, 0.59, 0.70, 0.89, 1.02, 1.18, 1.39, 1.56, 2.08)
Technetium	95	20.0(1) h			K, Mo-x; γ (0.766, 0.84)
	96	4.28(6) d			K, Mo-x; γ (0.778, 0.813, 0.850, 1.12)
	97	$2.6(4) \times 10^6$ y			K, Mo-x
	99	$2.111(12) \times 10^5$ y			β^- (0.292)
	99*m*	6.01(1) h			IT, Tc-x; γ (0.141)
Ruthenium	96		5.54(2)	0.2	
	97	2.9(1) d			γ (0.216, 0.324)
	102		31.6(2)	1.4	
	103	39.26(1) d			β^- (0.12, 0.22); γ (0.497)
	105	4.44(2) h			β^- (1.187); γ (0.469, 0.724)

TABLE 11.2 Table of Nuclides (*Continued*)

Element	*A*	Half-life	Natural abundance, %	Cross section, b	Radiation, MeV
Rhodium	103		100	144	
	103*m*	56.12(1) m			IT, Rh-x; γ(0.129)
	104	42.3(4) s			β^- (2.44); γ(0.56)
	104*m*	4.34(5) m			γ (0.051)
	105	36.35(6) h			β^- (0.568, 0.249); γ (0.306, 0.319)
	106	29.80(8) s			β^- (3.53); γ (0.512, 0.622)
Palladium	106		27.33(3)	0.29	
	107*m*	21.3(3) s			γ(0.321)
	108		26.46(9)	12	
	109	13.7(1) h			β^- (1.028); Ag-x; γ (0.088, 0.311, 0.636)
	109*m*	4.69(1) m			γ(0.188)
Silver	107		51.839(5)	35	
	108	2.37(1) m			β^- (1.64); β^+ (0.90); γ (0.434, 0.511, 0.614, 0.632)
	109		48.161(5)	89	
	109*m*	39.6(2)			γ(0.088)
	110	24.6(2) s			β^- (2.89); γ(0.658)
	110*m*	249.76(4) d			β^-(0.087, 0.53); IT, γ (0.658)
Cadmium	111*m*	48.6 m			γ(0.159, 0.247)
	113		12.22(8)	20 000	
	115	53.46(10) h			β^- (1.11, 0.58); In-x; γ(0.336, 0.528)
	115*m*	44.6(3) d			β^- (1.62); γ(0.934, 1.29, 0.485)
	117	2.49(4) h			β^- (2.33); γ(0.273, 0.314, 0.434, 1.303, 1.577)
Indium	111	2.8049(1) d			*K*, Cd-x; γ(0.172, 0.247)
	113*m*	1.6582(6) h			IT, In-x; γ(0.393)
	114	71.9 s			β^- (1.99, 0.67); *K*, β^+ (0.40), Cd-x, γ(1.30)
	*115	$4.41(25) \times 10^{14}$ y			β^- (0.495)
	116*m*	54.41(3) m			β^- (1.00); γ(0.417, 0.819, 1.09, 1.293, 1.509, 2.111)
Tin	124		5.79(5)	0.1	
	125	9.64(3) d			β^-(2.35); γ(1.067, 0.811)
	125*m*		9.52(5) m		γ(0.325)
Antimony	121		57.36(15)	6	
	122	2.70(1) d			β^- (1.414, 1.980, 0.723); γ(0.564); *K*, Sn-x
	123		42.64(15)	3.3	
	124	60.20(3) d			β^- (2.301); γ(0.603, 1.69, 0.722)
Tellurium	127	9.35(7) h			β^- (0.70); I-x; γ(0.360, 0.418)

(Continued)

TABLE 11.2 Table of Nuclides (*Continued*)

Element	*A*	Half-life	Natural abundance, %	Cross section, b	Radiation, MeV
Tellurium	129	69.6(2) m			β^- (1.453, 0.989, 0.69); I-x; γ (0.460, 0.487)
	131	25.0 m			β^- (2.14, 1.69, 1.35); I-x; γ (0.150)
	131*m*	30(2) h			β^- (2.46); IT, Te-x, I-x; γ (0.150, 0.774, 0.794)
Iodine	123	13.2(1) h			*K*, Te-x; γ (0.159)
	125	60.14(11) d			*K*, Te-x; γ (0.035)
	127		100	6.4	
	128	24.99(2) m			β^- (2.12); γ (0.441)
	129	$1.57(4) \times 10^7$ y			β^- (0.150); Xe-x; γ (0.038)
	131	8.04(1) d			β^- (0.806, 0.606); Xe-x, γ (0.364)
Xenon	132		26.9(5)	<5	
	133	5.243(1) d			β^- (0.346); Cs-x; γ (0.081)
	134		10.4(2)	5	
	136		8.9(1)	0.15	
	137	3.82(1) m			β^- (4.1); γ (0.455)
Cesium	133		100	28	
	134	2.062(5) y			β^- (0.658); γ (0.605, 0.796, 0.57)
	134*m*	2.91(1) h			IT, β^- (0.55), Cs-x; γ (0.127)
	137	30.1(2) y			β^- (0.514, 1.18); Ba-x, γ (0.662)
Barium	135		6.59(2)	5	
	135*m*	28.7(2) h			IT, Ba-x; γ (0.268)
	137		11.23(5)	4	
	137*m*	2.552(1) m			IT, Ba-x; γ (0.662)
	138		71.70(9)	0.4	
	139	83.06(28) m			β^- (2.38); La-x; γ (0.166, 1.421)
	140	12.752(3) d			β^- (1.02, 0.83); La-x; γ (0.537)
Lanthanum	139		99.9098(2)	8.9	
	140	1.6781(3) d			β^- (2.164, 1.680, 1.365); γ (0.487, 1.596)
Cerium	138		0.25(1)	1	
	139*m*	54.8(10)			γ (0.746)
	140		88.43(10)	0.6	
	141	32.501(5) d			β^- (0.582); Pr-x; γ (0.145)
	142		11.13(10)	1	
	143	33.10(5) h			β^- (1.40); Pr-x; γ (0.293)
	144	284.893(8) d			β^- (0.316), Pr-x, γ (0.080, 0.134)
Praseodymium	141		100	12	
	142	19.12(4) h			β^- (2.164); γ (1.576)
	143	13.5(2) d			β^- (0.932)

TABLE 11.2 Table of Nuclides (*Continued*)

Element	*A*	Half-life	Natural abundance, %	Cross section, b	Radiation, MeV
Neodymium	146		17.19(8)	2	
	147	10.98(1) d			β^- (0.810, 0.369); γ (0.090, 0.531)
	148		5.76(3)	4	
	149	1.72(1) h			β^- (1.5); γ (0.114, 0.211, 0.27)
	150		5.46(3)	1.5	
	151	12.44(7) m			β^- (2.0); γ (0.110, 0.174, 0.256)
Promethium	144	363(14) d			*K*, Nd-x; γ (0.477, 0.618, 0.696)
	146	5.53(5) y			*K*, β^- (0.795); Nd-x; γ (0.453, 0.75)
	147	2.6234(2) y			β^- (0.224); γ (0.122)
	148*m*	41.29(11) d			β^- (0.69); γ (0.550, 0.630, 0.727)
	149	53.08(5) h			β^- (1.064, 0.784); γ (0.286)
	150	2.68(2) h			β^- (3.260); γ (0.344, 0.831, 0.88, 1.165, 1.33)
Samarium	151	90(8) y			β^- (0.076); γ (0.022)
	152		26.7(2)	210	
	153	46.27(1) h			β^- (0.81); γ (0.103)
	154		22.7(5)	5	
	155	22.3(2)			β^- (1.53); γ (0.104)
Europium	151		47.8(5)	5900	
	152	13.54(1) y			*K*, β^- (1.492); β^+ (0.727); Gd-x, Sm-x; γ (0.122, 0.344, 0.779, 0.964, 1.086, 1.112, 1.408)
	152*m*	9.274(9) h			β^- (1.89); γ (0.122, 0.84, 0.96)
	153		52.2(5)	320	
	154	8.592(8) y			β^- (0.843); γ (0.123, 1.005, 1.274)
Gadolinium	158		24.84(12)	3.4	
	159	18.56(8) h			β^- (0.95); Tb-x, γ (0.363)
	160		21.86(4)	0.8	
	161	3.66 m			β^- (1.61); γ (0.102, 0.315, 0.361)
Terbium	159		100	46	
	160	72.3(2) d			β^- (1.76, 0.87); γ (0.299, 0.879, 0.966)
Dysprosium	164		28.2(2)	2 000	
	165	2.334(6)			β^- (1.305); γ (0.095, 0.362)
	165*m*	1.257(6) m			β^- (1.04, 0.89); γ (0.108, 0.362, 0.514)

(Continued)

TABLE 11.2 Table of Nuclides (*Continued*)

Element	*A*	Half-life	Natural abundance, %	Cross section, b	Radiation, MeV
Holmium	165		100	64	
	166	26.80(2) h			β^- (1.85, 1.78); Er-x; γ (0.081)
Erbium	167*m*	2.269(3) s			γ (0.208)
	170		14.9(1)	9	
	171	7.52(3) h			β^- (1.49, 1.06); Tm-x; γ (0.112, 0.296, 0.308)
Thullium	169		100	125	
	170	128.6(3) d			β^- (0.968, 0.884); γ (0.084)
Ytterbium	168		0.13	11 000	
	169	32.022(8) d			γ (0.177, 0.198)
	174		31.8(4)	46	
	175	4.19(1) d			β^- (0.466); Lu-x; γ (0.114, 0.283, 0.396)
	176		12.7(1)	7	
	177	1.91 h			β^- (1.40); Lu-x; γ (0.151)
Lutetium	175		97.41(2)	18	
	176*m*	3.635(3) h			β^- (1.31); Hf-x; γ (0.0884)
	177	6.71(1) d			β^- (0.497); γ (0.113, 0.208)
Hafnium	179		13.629(5)	65	
	179*m*	18.67(3) s			γ (0.217)
	180		35.1000(6)	10	
	180*m*	5.5(1) h			IT, Hf-x; γ (0.058, 0.215, 0.333, 0.444)
	181	42.39(6) d			β^- (0.408); Ta-x; γ (0.133, 0.346, 0.482)
Tantalum	181		99.988(2)	21	
	182	114.43(3) d			β^- (1.713, 1.470); γ (0.068, 1.121, 1.189, 1.221)
	182*m*	15.84(10) m			γ (0.147, 0.172, 0.184)
Tungsten	185	75.1(3) d			β^- (0.433); γ (0.125)
	186		28.6(2)	40	
	187	23.72(6) h			β^- (1.312, 0.622; Re-x; γ (0.480)
Rhenium	185		37.40(2)	110	
	186	3.777(4) d			β^- (1.07, 0.933); *K*, W-x, Os-x; γ (0.137, 0.632, 0.768)
	187		62.60(2)	70	
	188	16.98(2) h			β^- (2.12, 1.96); Os-x; γ (0.155)
	188*m*	18.6(1) m			γ (0.092, 0.106)
Osmium	190		26.4(4)	8.6	
	190*m*	9.9(1) m			IT, Os-x; γ (0.187, 0.361, 0.502, 0.616)
	191	15.4(1) d			β^- (0.143); Os-x; γ (0.129)
	192		41.0(3)	1.6	
	193	30.5(4) h			β^- (1.13); Ir-x; γ (0.073, 0.139, 0.460)

TABLE 11.2 **Table of Nuclides (*Continued*)**

Element	*A*	Half-life	Natural abundance, %	Cross section, b	Radiation, MeV
Iridium	191		37.3(5)	750	
	192	73.831(8) d			β^- (0.672); γ (0.296, 0.308, 0.316, 0.468)
	193		62.7(5)	110	
	194	19.15(3) h			β^- (2.24); γ (0.328)
Platinum	195		33.8(3)	27	
	195*m*	4.02(1) d			IT, Pt-x; γ (0.0311, 0.0991, 0.130)
	196		25.3(5)	0.9	
	197*m*	95.41(18) m			IT, β^- (0.737); Pt-x, γ (0.279, 0.346)
	197	18.3(3) h			β^- (0.719, 0.642, 0.451); Au-x, γ (0.077, 0.191)
	198		7.2(2)	4	
	199	30.8(4) m			β^- (1.69); γ (0.475, 0.540)
	199*m*	13.6(4) s			γ (0.393)
Gold	197		100	98.7	
	197*m*	7.73(6) s			IT, Au-x; γ (0.279)
	198	2.6935(4) d			β^- (1.371); γ (0.412, 0.676)
	199	3.139(7) d			β^- (0.296, 0.250, 0.462); Hg-x, γ (0.158, 0.208)
Mercury	196		0.15(1)	880	
	197	2.672(2) d			*K*, Au-x; γ (0.077, 0.191)
	197*m*	23.8(1) h			IT, *K*, Hg-x; γ (0.134)
	199		16.87(10)	2000	
	199*m*	42.6(2) m			γ (0.158, 0.375)
	202		29.86(20)	4	
	203	46.61(2) d			β^- (0.214); γ (0.279)
	205	5.2(1) m			β^- (1.7); γ (0.205)
Thallium	201	3.046(1) d			*K*, Hg-x; γ (0.135, 0.167)
	202	12.23(2) d			γ (0.439)
	203		29.524(9)	11	
	204	3.78(2) y			β^- (0.763); *K*, Hg-x
	205		70.476(9)	0.11	
	206	4.20(1) m			β^- (1.53)
Lead	204*m*	67.2(3) m			IT, Pb-x, γ (0.375, 0.899)
	207*m*	0.805(10) s			γ (0.570, 1.064)
	209	3.253(14) h			β^- (0.645)
	210	22.3(2) y			α (3.72); γ (0.0465)
	212	10.64(1) h			β^- (0.569, 0.331); Bi-x; γ (0.239)
	214	26.8(9) m			β^- (0.59, 0.65, 1.03); γ (0.352)
Bismuth	209		100	0.019	
	210	5.013(5) d			β^- (1.16); α (4.69, 4.65)
	212	60.55(4) m			β^- (2.25); γ (0.727), TI-x; γ (6.05, 6.09)
	214	19.9(4) m			β^- (3.26, 1.88, 1.51); γ γ (0.609); α (5.512, 5.448)

(Continued)

TABLE 11.2 Table of Nuclides (*Continued*)

Element	*A*	Half-life	Natural abundance, %	Cross section, b	Radiation, MeV
Polonium	208	2.898(2) y			α (5.11); *K*, Bi-x; γ (0.292, 0.571, 0.603, 0.862)
	209	102(5) y			α (4.88), I, Bi-x; γ (0.26)
	210	138.376(2) d			α (5.30); γ (0.803)
	212	298(3) ns			α (8.78)
	214	0.164(2) ms			α (7.69); γ (0.799)
	216	145(2) ms			α (6.78)
	218	3.10(1) m			α (6.00)
Astatine	207	1.80(4) h			*K*, α (5.76); γ (0.588, 0.814)
	208	1.63(3) h			*K*, α (5.641); Po-x, γ (0.177, 0.660, 0.685)
	209	5.41(5) h			*K*, α (5.65), Po-x, γ (0.545, 0.782, 0.790)
	210	8.1(4) h			*K*, α (5.52, 5.44, 5.36), Po-x, γ (0.245, 1.181, 1.483)
	211	7.214(7) h			*K*, α (5.87), Po-x, γ (0.67)
Radon	220	55.6(1) s			α (6.29); γ (0.550)
	222	2.8235(3) d			α (5.49); γ (0.510)
Radium	*224	3.66(4) d			α (5.68, 5.45); Rn-x; γ (0.241)
	*226	1600(7) y			α (4.78, 4.60); Rn-x; γ (0.187)
	*228	5.75(3) y			γ (0.135)
Actinium	*227	21.773(3) y			β^- (0.046), α (4.95); γ (0.086, 0.100, 0.160)
	*228	6.15(3) h			β^- (2.18, 1.85); Th-x; γ (0.339, 0.911, 0.969)
Thorium	228	1.913(1) y			α (5.43, 5.34); Ra-x; γ (0.084, 0.132, 0.167, 0.214)
	*230	$7.54(3) \times 10^4$ y			α (4.68, 4.62), Ra-x; γ (0.068)
	*232	$1.405(6) \times 10^{10}$ y			α (4.01, 3.95), γ (0.059)
	233	22.3(1) m			β^- (1.23), γ (0.029, 0.087, 0.171, 0.453)
	*234	24.10(3) d			β^- (0.199, 0.104); Pa-x; γ (0.063, 0.093)
Protactinium	*231	$3.276(11) \times 10^4$ y			α (5.06, 4.95, 4.73); γ (0.027, 0.284, 0.300, 0.303)
	233	26.967(2) d			β^- (0.257, 0.568); γ (0.312)
	234*m*	1.17(3) m			β^- (2.29); IT, U-x; γ (0.765, 1.00)
Uranium	233	$1.592(2) \times 10^5$ y			α (4.82, 4.78); Th-x; γ (0.042, 0.055, 0.097, 0.119, 0.146, 0.164, 0.22, 0.291, 0.32)
	*234	$2.45(2) \times 10^5$ y			α (4.77, 4.72); Th-x; γ (0.053, 0.121)
	*235	$7.038(5) \times 10^8$ y			α (4.40, 4.37, 4.22); Th-x; γ (0.186)
	236	$2.3415(14) \times 10^7$ y			α (4.49, 4.44); γ (0.049)
	*238	$4.468(3) \times 10^9$ y			α (4.20, 4.15); γ (0.050)
	239	23.50(5) m			β^- (1.21, 1.29); Np-x, γ (0.044, 0.075)

TABLE 11.2 Table of Nuclides (*Continued*)

Element	A	Half-life	Natural abundance, %	Cross section, b	Radiation, MeV
Neptunium	236	$1.15(12) \times 10^5$ y			$K, \beta^-; \gamma$ (0.160)
	237	2.141×10^6 y			α (4.79, 4.77); Pa-x; γ (0.029, 0.086)
Plutonium	236	2.87(1) y			α (5.77, 5.66); U-x
	238	87.74(4) y			α (5.50, 5.46); U-x; γ (0.0435, 0.998, 0.153)
	239	$2.412(3) \times 10^4$ y			α (5.16, 5.14, 5.11); U-x; γ (0.052, 0.129)
	241	14.35(10) y			α (4.90, 4.85); U-x; γ(0.149)
	242	$3.733(12) \times 10^5$ y			α (4.90, 4.86); γ(0.045, 0.103)
Americium	241	432.7(6) y			α (5.49, 5.44); Np-x; γ (0.060)
	243	7380(40) y			α (5.28, 5.23); Np-x; γ (0.075)
Curium	242	162.79(9) d			α (6.12, 6.07); Pu-x; γ (0.561, 0.605)
	244	18.10(2) y			α (5.81, 5.77); γ (0.043)
Berkelium	249	320(6) d			α (5.42); β^- (0.125); γ (0.327)
Californium	252	2.645(8) y			α (6.12, 6.08); Cm-x; γ (0.043)
Einsteinium	253	20.47(3) d			α (6.64); Bk-x; γ (0.387, 0.389, 0.429)
	254	275.7(5) d			α (6.44); Bd-x; γ (0.034, 0.036, 0.043)
Fermium	257	100.5(2) d			α (6.52); Cf-x; γ (0.115, 0.241)

Source: C. M. Lederer and V. S. Shirley, eds., *Table of Isotopes*, 7th ed., Wiley-Interscience, New York, 1978; V. S. Shirley, ed., *Table of Radioactive Isotopes*, 9th ed., Wiley-Interscience, New York, 1986; National Nuclear Data Center, *Nuclear Wallet Cards*, Brookhaven National Laboratory, Upton, New York, 1990.

11.2 UNITS AND CHARACTERISTICS OF RADIOACTIVITY

Activity is expressed in terms of the *Curie* (Ci) where 1 Ci is 3.700×10^{10} disintegrations per second (dps). *Specific activity* is the term used to describe the rate of radioactive decay of a substance the energies of which are measured in millions of electronvolts, or megaelectron volts (MeV). It is expressed by disintegration per second per unit mass or volume, or in units such as microcurie or millicurie per milliliter, per gram, or per millimole. The latter is preferable for labeled compounds.

11.2.1 Radioactive Decay

The decay of a radionuclide follows the first-order rate law:

$$\frac{dN}{dt} = -\lambda N \tag{11.1}$$

where N is the number of radionuclide atoms that remain at time t and λ is the characteristic decay constant. The activity A, the quantity usually observed or computed, is related to N by the equation

$$A = \lambda N \tag{11.2}$$

After integration, the rate equation from which the decay of radioactivity may be calculated is

$$A_t = A_0 e^{-\lambda t} \tag{11.3}$$

where A_0 is the activity at some initial time and A_t is the activity after elapsed time t. Table 11.3 gives the fraction A_t/A_0 remaining, or $\exp(-\lambda t)$, as a function of half-life.

11.2.2 Half-Life

Each radionuclide has a characteristic *half-life* which is the time period in which a certain initial mass of a radioactive substance decays to half its mass. Radioactive substances decay (disintegrate) at a statistical rate that cannot be altered by any chemical or physical treatment of the radionuclide. In order to measure this rate, one can determine how long it takes for one-half of the radionuclide to decay. This is called the half-life of the radionuclide. It follows first order kinetics and can be expressed as

$$t_{1/2} = \frac{1}{\lambda} \ln \frac{A}{A/2} = \frac{0.693}{\lambda} \tag{11.4}$$

One-half of the radionuclide will decay in that time; then one-half of what is left will decay in that same time interval; and so on. After 10 half-lives only 0.1% of the radionuclide remains. An accurate knowledge of the characteristic decay constant is essential when working with short-lived radionuclides to correct for the decay that occurs while the experiment is in progress.

Example 11.1 The activity of a manganese 56 ($t_{1/2}$ = 2.576 h) sample at the beginning of an experiment was 3.68×10^6 counts per second. What would be the activity 2 h later at the conclusion of the experiment?

$$A_t = 3.68 \times 10^{10} \exp\left[\frac{-0.693(2.0 \text{ h})}{2.576 \text{ h}}\right]$$
$$= 2.153 \times 10^6 \text{ counts per second}$$

It is often necessary to correct for decay during the course of an experiment. Table 11.3 is useful for this purpose; it shows the fractions of a nuclide remaining, $\exp(-\lambda t)$, after various numbers of its half-lives have elapsed.

Some radionuclides decay to products that are themselves radioactive. Such a product is called a *daughter* nuclide, and the original species is called the *parent*. The daughter activity, in terms of disintegration rate, is given by

$$A_2 = \frac{\lambda_2 A_1^0}{\lambda_2 - \lambda_1} [\exp(-\lambda_1 t) - \exp(-\lambda_2 t)] + A_2^0 \exp(-\lambda_2 t) \tag{11.5}$$

where A_2 = daughter activity
λ_1, λ_2 = parent and daughter decay constants
A_1^0, A_2^0 = initial activities of parent and daughter

If the parent is much longer-lived than the daughter, the equation simplifies to

$$A_2 = A_1 [1 - \exp(-\lambda_2 t)] + A_2^0 \exp(-\lambda_2 t) \tag{11.6}$$

TABLE 11.3 Decay of a Radionuclide

F = *fraction remaining* = $e^{-\lambda t}$. *Note that* $F_{a+b} = F_a F_b$. *Thus, after 14 half-lives, the fraction remaining* (F_{14}) *is equal to* $F_{10}F_4 = 0.000\,98 \times 0.0625 = 0.000\,061$.

Half-lives	*F*	Half-lives	*F*	Half-lives	*F*	Half-lives	*F*
0	1.000	0.90	0.536	1.80	0.287	3.75	0.0743
0.02	0.986	0.92	0.529	1.82	0.283	3.80	0.0718
0.04	0.973	0.94	0.521	1.84	0.279	3.85	0.0694
0.06	0.959	0.96	0.514	1.86	0.275	3.90	0.0670
0.08	0.946	0.98	0.507	1.88	0.272	3.95	0.0647
0.10	0.933	1.00	0.500	1.90	0.268	4.00	0.0625
0.12	0.920	1.02	0.493	1.92	0.264	4.05	0.0604
0.14	0.908	1.04	0.486	1.94	0.261	4.10	0.0583
0.16	0.895	1.06	0.480	1.96	0.257	4.15	0.0563
0.18	0.883	1.08	0.473	1.98	0.253	4.20	0.0544
0.20	0.871	1.10	0.467	2.00	0.250	4.25	0.0525
0.22	0.859	1.12	0.460	2.05	0.241	4.30	0.0507
0.24	0.847	1.14	0.454	2.10	0.233	4.35	0.0490
0.26	0.835	1.16	0.447	2.15	0.225	4.40	0.0474
0.28	0.824	1.18	0.441	2.20	0.218	4.45	0.0458
0.30	0.812	1.20	0.435	2.25	0.210	4.50	0.0442
0.32	0.801	1.22	0.429	2.30	0.203	4.55	0.0427
0.34	0.790	1.24	0.423	2.35	0.196	4.60	0.0412
0.36	0.779	1.26	0.418	2.40	0.189	4.65	0.0398
0.38	0.768	1.28	0.412	2.45	0.183	4.70	0.0385
0.40	0.758	1.30	0.406	2.50	0.177	4.75	0.0372
0.42	0.747	1.32	0.401	2.55	0.171	4.80	0.0359
0.44	0.737	1.34	0.395	2.60	0.165	4.85	0.0347
0.46	0.727	1.36	0.390	2.65	0.159	4.90	0.0335
0.48	0.717	1.38	0.384	2.70	0.154	4.95	0.0324
0.50	0.707	1.40	0.379	2.75	0.149	5.00	0.0313
0.52	0.697	1.42	0.374	2.80	0.144	5.10	0.0292
0.54	0.688	1.44	0.369	2.85	0.139	5.20	0.0272
0.56	0.678	1.46	0.363	2.90	0.134	5.30	0.0254
0.58	0.669	1.48	0.358	2.95	0.129	5.40	0.0237
0.60	0.660	1.50	0.354	3.00	0.125	5.50	0.0221
0.62	0.651	1.52	0.349	3.05	0.121	5.60	0.0206
0.64	0.642	1.54	0.344	3.10	0.117	5.70	0.0192
0.66	0.633	1.56	0.339	3.15	0.113	5.80	0.0179
0.68	0.624	1.58	0.334	3.20	0.109	5.90	0.0167
0.70	0.616	1.60	0.330	3.25	0.105	6.00	0.0156
0.72	0.607	1.62	0.325	3.30	0.102	6.10	0.0146
0.74	0.599	1.64	0.321	3.35	0.0981	6.20	0.0136
0.76	0.590	1.66	0.316	3.40	0.0947	6.30	0.0127
0.78	0.582	1.68	0.312	3.45	0.0915	6.40	0.0118
0.80	0.574	1.70	0.308	3.50	0.0884		
0.82	0.566	1.72	0.304	3.55	0.0854	6.50	0.0110
0.84	0.559	1.74	0.299	3.60	0.0825	6.60	0.0103
0.86	0.551	1.76	0.295	3.65	0.0797	6.70	0.00962
0.88	0.543	1.78	0.291	3.70	0.0769	6.80	0.00897

(Continued)

TABLE 11.3 Decay of a Radionuclide (*Continued*)

Half-lives	F	Half-lives	F	Half-lives	F	Half-lives	F
6.90	0.00837	7.70	0.00481	8.50	0.00276	9.30	0.00159
		7.80	0.00449	8.60	0.00258	9.40	0.00148
7.00	0.00781	7.90	0.00419	8.70	0.00240		
7.10	0.00729			8.80	0.00224	9.50	0.00138
7.20	0.00680	8.00	0.00390	8.90	0.00209	9.60	0.00129
7.30	0.00635	8.10	0.00364			9.70	0.00120
7.40	0.00592	8.20	0.00340	9.00	0.00195	9.80	0.00112
7.50	0.00552	8.30	0.00317	9.10	0.00182	9.90	0.00105
7.60	0.00515	8.40	0.00296	9.20	0.00170	10.00	0.00098

Source: L. Meites, ed., *Handbook of Analytical Chemistry*, McGraw-Hill, New York, 1963.

The term in brackets is called the *saturation factor*. When a nuclide is produced at a constant rate, either from a long-lived parent or by a nuclear reaction, the saturation factor is the ratio of the activity produced at a given time t to that which would be produced in a very long time. Table 11.4 presents values of the factor at times from 0 to 10 half-lives of the daughter nuclide.

11.3 MEASUREMENT OF RADIOACTIVITY

The random nature of nuclear disintegrations requires that a large number of individual disintegrations be observed to obtain a counting rate or total sample count with a desired statistical significance. Several problems must be considered in measurement of radioactivity.

1. The radionuclide emission may be absorbed in the air path or in the walls of the detector.
2. The dead time (unresponsive period between radioactive events) may lead to uncounted events.
3. The energy of the ionizing particle may be insufficient to produce an ion in the sensitive volume of the detector.

11.4 DETECTORS

The detectors described in Sec. 10.3 are also suitable for measurement of radioactivity. When resolution is the highest priority, measurement should be made with a solid-state detector [such as lithium-drifted Ge(Li) or Si(Li) or ultrapure germanium] coupled to a multichannel analyzer system. If sensitivity is paramount, a proportional counter in conjunction with a scintillation crystal or liquid scintillation system is the choice.

11.4.1 Scintillation Counters

A good match should exist between the emission spectrum of the scintillator and the response curve of the photocathode of the multiplier phototube. The decay time is 250 ns for a sodium iodide crystal doped with 1% thallium(I) iodide, a crystal useful for counting beta particles but particularly useful for counting gamma radiation. When radiation interacts with a NaI(Tl) crystal, the transmitted energy excites the iodine atoms. Upon their return to the ground electronic state, this energy is reemitted in the form of a light pulse in the ultraviolet region, which is promptly absorbed by the thallium atom and reemitted as fluorescent light at 410 nm. The well-type

TABLE 11.4 Growth of a Radionuclide Produced at a Constant Rate

$F = 1 - e^{-\lambda t}$ = *fraction of saturation value (saturation factor).*

Half-lives	*F*	Half-lives	*F*	Half-lives	*F*	Half-lives	*F*
0.01	0.0069	0.74	0.401	1.46	0.636	2.45	0.817
0.02	0.0138	0.76	0.410	1.48	0.641	2.50	0.823
0.04	0.0273	0.78	0.418	1.50	0.646	2.55	0.829
0.06	0.0407	0.80	0.426	1.52	0.651	2.60	0.835
0.08	0.0539	0.82	0.434	1.54	0.656	2.65	0.841
0.10	0.0670	0.84	0.441	1.56	0.661	2.70	0.846
0.12	0.0798	0.86	0.449	1.58	0.666	2.75	0.851
0.14	0.0925	0.88	0.457	1.60	0.670	2.80	0.856
0.16	0.105	0.90	0.464	1.62	0.675	2.85	0.861
0.18	0.117	0.92	0.471	1.64	0.679	2.90	0.866
0.20	0.129	0.94	0.479	1.66	0.684	2.95	0.871
0.22	0.141	0.96	0.486	1.68	0.688	3.00	0.875
0.24	0.153	0.98	0.493	1.70	0.692	3.10	0.883
0.26	0.165	1.00	0.500	1.72	0.696	3.20	0.891
0.28	0.176	1.02	0.507	1.74	0.701	3.30	0.898
0.30	0.188	1.04	0.514	1.76	0.705	3.40	0.905
0.32	0.199	1.06	0.520	1.78	0.709	3.50	0.912
0.34	0.210	1.08	0.527	1.80	0.713	3.60	0.918
0.36	0.221	1.10	0.533	1.82	0.717	3.70	0.923
0.38	0.232	1.12	0.540	1.84	0.721	3.80	0.928
0.40	0.242	1.14	0.546	1.86	0.725	3.90	0.933
0.42	0.253	1.16	0.553	1.88	0.728	4.00	0.938
0.44	0.263	1.18	0.559	1.90	0.732	4.25	0.947
0.46	0.273	1.20	0.565	1.92	0.736	4.50	0.956
0.48	0.283	1.22	0.571	1.94	0.739	4.75	0.963
0.50	0.293	1.24	0.577	1.96	0.743	5.00	0.969
0.52	0.303	1.26	0.582	1.98	0.747	5.25	0.974
0.54	0.312	1.28	0.588	2.00	0.750	5.50	0.978
0.56	0.322	1.30	0.594	2.05	0.759	5.75	0.981
0.58	0.331	1.32	0.599	2.10	0.767	6.00	0.984
0.60	0.340	1.34	0.605	2.15	0.775	6.50	0.989
0.62	0.349	1.36	0.610	2.20	0.782	7.00	0.992
0.64	0.358	1.38	0.616	2.25	0.790	7.50	0.994
0.66	0.367	1.40	0.621	2.30	0.797	8.00	0.996
0.68	0.376	1.42	0.626	2.35	0.804	9.00	0.998
0.70	0.384	1.44	0.631	2.40	0.811	10.00	0.999
0.72	0.393						

Source: L. Meites, ed., *Handbook of Analytical Chemistry*, McGraw-Hill, New York, 1963.

arrangement of the scintillation crystal increases the counting efficiency to approximately 100% by surrounding the sample with the detector crystal. This arrangement is best for counting gamma radiation.

Two-inch crystal detectors are typically used for lower-energy gamma rays, such as cobalt-57, chromium-51, iodine-125 and iodine-131, and 3-in crystals for higher-energy gamma rays such as cobalt-60, cesium-137, iron-59, and sodium-22.

11.4.2 Proportional Counters

A sample of radioactive material can be placed inside the active volume of a flow proportional counter, thus avoiding losses due to window absorption. The chamber is purged with a rapid flow of counter gas (P-10 gas, a mixture of 10% methane in argon) and the flow is maintained during counting of samples. Counter life is virtually unlimited. Such a counter is particularly suited for distinguishing and counting low-energy alpha and beta particles.

11.4.3 Radioactivity Flow Detectors

Radioactivity emissions from the flowing sample are detected by photomultiplier tubes on opposite sides of the flowing stream. The resultant signals are summed, checked for coincidence, and sorted by either a pulse-height analyzer or a multichannel analyzer. The sample counts in a flowing system are dependent on the flowrate and on the volume of the counting cell. Sensitivity is maximized by increasing the flow-cell volume and decreasing the flowrate.

The overall sensitivity is determined by the flow-cell type and the residence time. (1) The liquid cell is most commonly used because it gives the highest efficiency (and therefore sensitivity), but this type of cell requires mixing the sample with scintillation solution and thus precludes sample recovery. Efficiency is counting tritium is greater than 50% and greater than 90% for carbon-14. (2) The solid cell is packed with inorganic scintillator crystals. The efficiency is roughly 25% for carbon-14 when the packing is plastic (200 μm) or glass (80 μm); efficiency increases to greater than 80% for carbon-14 when the packing is CaF_2(Eu) (100 μm) but that for tritium drops to about 8%. Solid cells are easily contaminated but are often used in quality-control analysis or when it is known that the radiolabeled compound will not bind to the scintillator crystals. (3) The high-energy isotope cell is normally used to quantitate very high-energy beta emitters (^{32}P) and low-energy gamma emitters (^{125}I).

Radioactivity flow detectors provide fast, efficient, and cost-effective analysis of single samples or for monitoring ongoing processes. However, these detectors are not designed for a high throughput of different samples; they lack the flexibility and high sample capacity to ever replace conventional liquid scintillation analysis (see Sec. 11.6).

11.5 STATISTICS OF RADIOACTIVITY MEASUREMENTS

Because of the randomness of radioactive decay, the true average count of any sample is not known. However, the relative error of a particular number of counts can be estimated. We need to know the counting rate N_s/t_s or the total counts N_s where t_s is the counting time. There is always some background activity (background counts N_b and counting time for background only t_b) that the detector registers along with the sample activity.

11.5.1 Fractional Error

The fractional error F_y of the sample, after correction for background ($N_s - N_b$), is given by

$$F_y = \frac{K}{N_s - N_b}\sqrt{\frac{N_s}{t_s} + \frac{N_b}{t_b}} \tag{11.7}$$

where K is the number of standard deviations (Table 11.5) for a particular error or confidence limit.

TABLE 11.5 Table of Constants of Relative Error

Error (confidence limit)	Probability of occurrence, %	K
Probable	50.0	0.675
Standard deviation (1σ)	31.7	1.000
0.95	5.00	1.960
2σ	4.55	2.000
0.99	1.00	2.576
3σ	0.27	3.000

11.5.2 Distribution of Counting Time

The optimum distribution of counting time between background and sample is given by

$$\frac{t_b}{t_s} = \sqrt{m\left(\frac{N_b}{N_s}\right)} \tag{11.8}$$

where m is the number of experimental samples measured in that batch for a single radionuclide, not including the background measurements.

11.5.3 Preset Time or Preset Count

The counting time required for a sample to achieve a predetermined precision is given by the expression

$$t_s = \frac{K^2}{F_y^2}\left[\frac{N_s + N_b/c}{(N_b - N_s)^2}\right] \tag{11.9}$$

where $c = mN_b/N_s$. For a preset count, the counting time [Eq. (11.9)] is multiplied by the counting rate N_s/t_s.

11.6 LIQUID-SCINTILLATION COUNTING

Liquid scintillation counting has as its primary application the counting of weak beta emitters, such as tritium and carbon-14. The energy transfer from these relatively low-energy beta particles is maximized when the radiolabeled sample is in close proximity to the scintillator, as in a homogeneous solution. However, many other radioactive isotopes can also be measured, sometimes with higher counting efficiency and greater ease than by other techniques. Higher-energy beta emitters can be readily counted as well as weak x-ray, alpha, and gamma emitters. Some 60 isotopes of over 40 elements have been measured by liquid-scintillation counting; these are summarized in Table 11.6.

Commonly, the use of intermediate half-life isotopes is preferred. Isotopes with long half-lives have low specific activity and require larger amounts of the isotope to give an adequate counting rate. Short half-life isotopes require prompt shipment and delivery from the source; those having half-lives greater than 12 h can usually be obtained and employed without difficulty.

TABLE 11.6 Isotopes Measured by Liquid-Scintillation Counting

Abbreviations include s, seconds; m, minutes; h, hours; d, days; y, years; EC, electron capture; and CR indicates Ĉerenkov radiation counting is feasible.

Isotope	Emission	Energy, MeV	Half-life	Remarks, counting conditions
		Isotopes of nonmetals		
Carbon-14	β	0.156	5730(4) y	
Chlorine-36	β	0.71	$3.01(2) \times 10^5$ y	CR
Hydrogen-3	β	0.0186	12.3 y	
Iodine-125	γ	0.035	60.14(11) d	
Iodine-129	β, γ	0.15, 0.040	$1.57(4) \times 10^7$ y	
Iodine-131	β, γ	0.607, 0.364	8.04(1) d	CR
Krypton-85	β, γ	0.67, 0.517	10.56(18) y	CR, gas soluble in toluene
Phosphorus-32	β	1.71	14.262(14) d	CR
Phosphorus-33	β	0.25	25.34(12) d	
Sulfur-35	β	0.167	87.51(12) d	$BaSO_4$ in gel or H_2SO_4 on glass fibers
		Isotopes of metals		
Barium-140	β, γ	1.01, 0.54	12.752(3) d	
Cadmium-109	EC		462.0(6) d	See silver-109*m*
Calcium-45	β	0.255	163.8(8) d	$CaCl_2$ in dibutyl phosphate
Calcium-47	β, γ	1.98, 1.30	4.536(2) d	CR
Cerium-144	β, γ	0.316, 0.134	284.893(8) d	See also praseodymium-144
Cesium-137	β, γ	0.514, 0.662	30.1(2) y	CR
Chromium-51	EC, γ	0.32	27.72(4) d	
Cobalt-57	EC, γ	0.136	271.80(5) d	
Cobalt-60	β, γ	0.318, 1.173	5.2714(5) y	
Copper-66	β, γ	2.63, 1.039	5.10(1) m	CR
Curium-242	α	6.12	162.79(9) d	
Gold-198	β, γ	1.371, 0.412	2.6935(4) d	
Indium-113*m*	γ	0.393	1.6582(6) h	
Iron-55	EC, x ray	0.0065	2.73(3) y	Iron(III) phosphate complex in gel
Iron-59	β, γ	0.475, 1.29	44.496(7) d	
Lead-210	β, γ	0.017, 0.0465	22.3(2) y	Aqueous concentrate, plastic vials
Magnesium-27	β, γ	1.75, 0.84	9.46(1) m	CR
Magnesium-28	β, γ	0.46, 1.34	20.91(3) h	
Manganese-54	EC, γ	0.835	312.1(1) d	
Manganese-56	β, γ	2.84, 0.847	2.5785(2) h	CR
Mercury-203	β, γ	0.214, 0.279	46.61(2) d	
Nickel-63	β	0.067	100(2) y	Tetrapyridinenickel dithiocyanate
Niobium-95	β, γ	0.160, 0.765	34.97(3) d	
Plutonium-236	α	5.75	2.87(1) y	Iron(III) phosphate complex in gel
Plutonium-238	α	5.50	87.74(4) y	
Plutonium-239	α	5.16	$2.412(3) \times 10^4$ y	
Plutonium-240	α	5.16	6563(7) y	
Plutonium-241	β, γ	0.02, 4.91	14.35(10) y	
Potassium-40	β, γ	1.34, 1.46	$1.277(8) \times 10^9$ y	CR
Potassium-42	β, γ	3.52, 1.46	12.360(3) h	CR
Praseodymium-144	β, γ	2.996, 2.186	17.28(5) m	CR

TABLE 11.6 Isotopes Measured by Liquid-Scintillation Counting (*Continued*)

Isotope	Emission	Energy, MeV	Half-life	Remarks, counting conditions
		Isotopes of metals		
Promethium-147	β	0.224	2.6234(2) y	Bis(2-ethylhexyl) phosphate complex
Promethium-149	β	1.07	53.08(5) h	
Rhodium-106	β, γ	3.53, 0.622	29.80(8) s	CR
Rubidium-86	β	1.78	18.63(2)	CR, 2-ethylhexanoic acid complex
Rubidium-87	β	0.275	$4.75(4) \times 10^{10}$ y	
Ruthenium-106	β	0.0392	373.59(15) d	See also rhodium-106
Samarium-151	β, γ	0.76, 0.022	90(8) y	
Silver-108	β	1.64	2.37(1) m	
Silver-109*m*	γ	0.088	39.6(2) s	
Sodium-22	β^+, γ	0.545, 1.275	2.6088(14) y	
Sodium-24	β, γ	1.39, 2.75	14.9590(12) h	CR
Strontium-85	EC, γ	0.514	64.84(2) d	2-Ethylhexanoic acid complex
Strontium-89	β	1.463	50.53(7) d	CR
Strontium-90	β	0.546	29.1(3) y	See also yttrium-90
Tin-113	EC, γ	0.391	115.09(4) d	See also indium-113*m*
Thorium-232	α	4.01	$1.405(6) \times 10^{10}$ y	Alkyl phosphate complex
Uranium-233	α	4.82	$1.592(2) \times 10^{5}$ y	Alkyl phosphate complex
Uranium-236	α	4.50	$2.341(1) \times 10^{7}$ y	
Yttrium-90	β	2.288	58.51(6) d	CR
Yttrium-91	β	1.545	58.51(6) d	
Zinc-65	β^+, γ	0.325, 1.12	243.9(1) d	
Zirconium-95	β, γ	0.89, 0.756	64.02(4) d	Alkyl phosphate complex; see also niobium-95

In the first step of liquid-scintillation counting, the radionuclide emits a beta particle (low-energy electron) that transfers its energy (upon collision with) to an aromatic organic solvent molecule. The π-electron cloud of these solvent molecules accepts this energy and becomes excited to a higher energy level. Next the energy from the activated solvent molecule is transferred to another organic molecule called a fluor or scintillator (Table 11.7). This activated scintillator releases its energy when it returns to the ground state and emits photons of light. Pulses of light are thus produced whose intensity, determined by the number of photons, is directly proportional to the energy of the beta particle. The light, whose wavelength lies in the responsive range of a photomultiplier tube, is detected and amplified. The photomultiplier tube counts the pulses of light and measures the intensity of each pulse. Thus the scintillator must be capable of absorbing light at the wavelength emitted by the bulk organic solvent molecules, and reemitting it at a longer wavelength that matches the spectral sensitivity of the blue-sensitive photomultiplier tubes.

11.6.1 Liquid-Scintillation Solvents

For liquid-scintillation counting (LSC), samples are dissolved in a "liquid cocktail" that contains a suitable solvent, a primary scintillator and perhaps a secondary scintillator, and additives to improve water miscibility and to permit counting at low temperatures. Modern LSC cocktails must be designed to handle multipurpose applications, most of which are of an aqueous nature. These cocktails must have an efficient energy-transfer system that will convert sample radioactivity into measurable light and, for aqueous samples, must have a surfactant (emulsifier) system that will

TABLE 11.7 LSC Scintillator Guide

Chemical name	Acronym	Scintillator function	Fluorescence emission maximum, nm*
1,5-Bis(5″-*tert*-butyl-2-benzoxazolyl)-thiophene	BBOT	Primary and secondary	425–435
2-(4-Biphenylyl-5-(*p*-*tert*-butylphenyl)-1,3,4-oxadiazole	Butyl-PBD	Primary	360–365
2-(4-Biphenylyl)-5-phenyl-1,3,4-oxadiazole	PBD	Primary	360–370
2,5-Diphenyloxazole	PPO	Primary	360–365
p-Terphenyl	*p*-TP	Primary	338–346
p-Bis(*o*-methylstyryl)benzene	Bis-MSB	Secondary	420–430
2,2′-*p*-Phenylene-bis(4-methyl-5-phenyl-oxazole)	Dimethyl POPOP	Secondary	425–430
2-(1-Naphthyl)-5-phenyloxazole	α-NPO	Secondary	395–405
2-(4-Biphenylyl)-6-phenylbenzoxazole	PBBO	Secondary	390–400
2,2′-*p*-Phenylenebis(5-phenyloxazole)	POPOP	Secondary	410–420
Naphthalene		Secondary	335–340 (in 1,4-dioxane)

* Range for wavelength maximum for fluorescence in toluene.

allow the incorporation of a variety of aqueous sample types. Lipophilic samples do not require these surfactants.

Although traditional cocktails were based on toluene or xylene, changing restrictions or transportation and concern for the environment has led to the availability of new, safer solvents. Cocktails based on linear alkylbenzenes (LAB) and phenylxylylethane (PXE) have emerged. They show no permeation through the walls of polyethylene counting vials, whereas in 10 weeks pseudocumene loses about 9%, xylene about 16%, and toluene about 29%. The characteristics essential to a solvent for use within LSC include high flash point, low vapor pressure, biodegradability, low toxicity and irritancy, odorlessness, high counting efficiency, high quench resistance, no permeation through plastics, low photo- and chemiluminescence, and low viscosity. The characteristics of scintillation solvents in use are summarized in Table 11.8. High counting efficiency and good quench resistance are vitally important criteria for LSC solvents. Pseudocumene has a higher initial counting efficiency but an overall lower quench resistance than LAB. PXE is classified as nontoxic and imposes no acute health hazard to humans. LAB shows similar nontoxic properties, whereas the toxicity of pseudocumene and other classical solvents used in LSC cocktails is well known. Pseudocumene has a highly aromatic penetrating odor, which, if inhaled over a relatively short period of time, can lead to headaches and narcosis; it is also a skin irritant and is irritating to the eyes.

TABLE 11.8 Characteristics of Scintillation Solvents

Solvent	Boiling point, °C	Flash Point, °C	Vapor pressure, mmHg at 25°C	Viscosity, $mN \cdot s \cdot m^{-2}$
Toluene	110.6	4	28	0.62 (15°C)
m-Xylene	139.1	25	8	0.62 (20°C)
1,2,4-Trimethylbenzene (pseudocumene)	169.4	48	2	0.73 (30°C)
LAB (linear alkylbenzenes)	300	149	<1	11.4
PXE (phenylxylylethane)	305	149	<1	16.2

11.6.2 Sample Vials

Sample vials should have a low background count from naturally occurring potassium-40, for example, less than 10 counts per minute when determining tritium. High-density polyethylene vials are economical and are used for counting samples. Borosilicate glass vials are best for sample storage because they are impervious to toluene, xylene, and 1,4-dioxane.

11.6.3 Color and Chemical Quenching

Color quenching is caused by the presence of color in the sample when it absorbs a portion of the fluorescent photons emitted by the scintillator. Color quenching is common in colored specimens such as blood or plant tissues. It is best eliminated prior to counting. The sample can be digested using a mixture of hydrogen peroxide and perchloric acid or completely combusted to water (for tritium), to CO_2 and absorbed, perhaps as a carbonate, which is soluble in an organic liquid-scintillation solution (for carbon-14), or to a phosphate (for phosphorus-32) (see Sec. 1.7.3). Sample oxidation provides a homogeneous solution with the color removed from blood, green plants, and urine. The complete separation of tritium and carbon-14 can be achieved from other sample materials.

Chemical quenching reduces the amount of light produced. Some of the energy released by the emission of the beta particle during decay of the radionuclide is absorbed as it is transferred from the beta particle to the solvent, or from the solvent to the scintillator. The method of standard additions is helpful.

11.6.4 Instrumentation

The conversion of nuclear energy to light (Fig. 11.1) is the basis of liquid-scintillation counting. The first step is the conversion of the photons to electrical pulses by a pair of photomultiplier tubes (PMTs). Pulses from the PMTs pass through a preamplifier and an amplifier; these increase the amplitude of the pulses and shape them for easy quantitation. A coincidence circuit

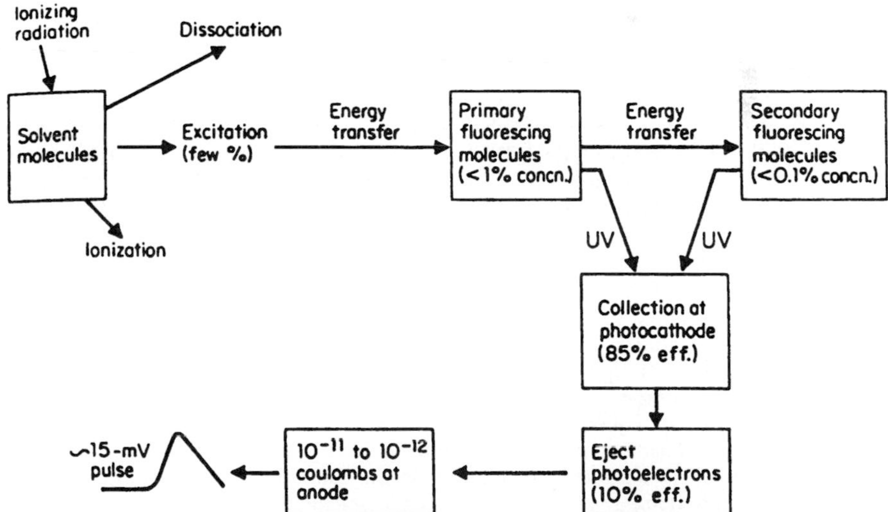

FIGURE 11.1 Block diagram of the energy transfer in liquid-scintillation counting. (*From Shugar and Dean, The Chemist's Ready Reference Handbook, McGraw-Hill, New York, 1990.*)

of the two photomultiplier tubes eliminates spurious pulses (background) produced by phototube thermal noise, cosmic radiation, and other sources. Coincident pulses registered by both PMTs within 20 ns (unlikely to be background) are sorted by their intensity and counted by either pulse-height analysis (the older method) or by the multichannel analyzer. In the latter, all pulses from the coincidence circuit are analyzed on the basis of their number and energy. The complete energy spectrum is digitized and stored for each sample and used to determine the absolute activity, that is, the actual number of particles produced by nuclear decay (disintegrations per minute).

Three-dimensional analysis allows the liquid-scintillator analyzer to discriminate background counts from true radioactivity present in the sample. Thereafter pulses of each pulse analyzed are counted. True beta-particle pulses exhibit a sharp (ca. 2-ns wide) peak. The same prompt pulse arises from background radiation, but, in addition, from 1 to 15 after pulses that may occur up to 5 μs later. By assessing the number of after pulses through a three-dimensional plot (pulse index discrimination), background pulses can be discriminated from true beta-particle pulses and excluded from the count. This is especially valuable for low-level counting.

Unwanted luminescence produces artificially high results. A special delayed coincidence circuit determines the number of "accidental" coincidences caused by the single photons generated by luminescence, which are then stripped from the sample spectra to correct for luminescence. This technique is valuable in analyses that use tissue solubilizers or analyses in which chemicals are present that cause significant luminescence.

11.6.5 Counting Čerenkov Radiation

Čerenkov radiation occurs only for higher-energy charged particles that have very high velocities. When the velocity of the charged particle exceeds the velocity of light in a medium, energy is lost by the charged particle as it slows down to the speed of light. Its energy is transferred to molecules of the medium. The energy lost is emitted as electromagnetic radiation in the violet–ultraviolet region. These photons of light are then converted to electronic pulses and counted, as for conventional liquid-scintillation analysis.

The threshold energy for the emission of Čerenkov radiation (E_{min}) increases as the refractive index of the medium increases. It can be calculated as

$$\text{threshold (in MeV)} = 0.511\left[\frac{1}{(1-1/\eta^2)^{1/2}}\right] - 1 \tag{11.10}$$

when η is the refractive index of the solvent. Water, with a refractive index of 1.3325, gives a threshold of 0.263 MeV. Thresholds for common reagents (HCl, HNO_3, H_2SO_4, and NaCl) vary slightly with concentration, and are approximately as follows: 0.5*M*, 258 keV; 1.0*M*, 256 keV; 2.0*M*, 250 keV; 4.0*M*, 240 keV; and 6.0*M*, 230 keV. For example, the beta particles emitted in the decay of ^{32}P is in a continuum (from 0 to 1710 keV), and the average energy of 690 keV is fairly high. A large fraction of the beta particles emitted in this decay lie above the threshold value for solutions of common reagents. Hence, ^{32}P and radionuclides with similar energies can be expected to be counted with good efficiency. Conversely, low-energy beta emitters that have E_{max} between the threshold and about 600 keV cannot be expected to be counted by this technique. Although conventional liquid-scintillation analysis can count ^{32}P at 98% to 10% efficiency in a toluene cocktail, the dissolved sample cannot be used for further tests. However, in Čerenkov counting, larger volumes of aqueous solution can be used to offset the lower counting efficiency. Wavelength shifters (such as β-naphthol, 4-methylumbelliferone, and Triton X-100), if used in an external compartment, increase counting efficiency and function as in a liquid-scintillation medium without introducing problems of chemical quenching.

Čerenkov counting is not subject to chemical quenching because no internal scintillator is used. Color quenching may be a problem if materials that absorb in the violet–ultraviolet region are present, but can often be overcome by bleaching, as mentioned in Sec. 11.6.3. Thus this technique offers the advantage of simple sample preparation. Aqueous solutions can be counted directly; this eliminates the necessity for disposing of an expensive scintillation cocktail and, also, the associated waste-disposal problem.

11.7 APPLICATIONS OF RADIONUCLIDES AS TRACERS

Radioactive tracers are used to follow the behavior of atoms or groups of atoms in any analytical scheme or other chemical reaction, in an industrial system, or in a biologic process. A radioactive tracer unequivocally labels the particular atoms to be traced. The specificity and sensitivity of detection of radionuclides is extremely good; less than 10^{-18} g can often be detected. However, a tagging radionuclide must not be exchangeable with similar atoms in other compounds.

A chemical separation from inactive, and occasionally active, contaminants generally precedes the activity measurement. Ordinary analytical techniques and reactions form the basis for most radiochemical separations, which might involve carrying a microconstituent by coprecipitation with a macroconstituent (called a carrier), solvent extraction, volatilization, adsorption, ion exchange, electrodeposition, and chromatography.

The active material must be spread in a uniform layer over a definite area unless 4-π geometry (such as a well-type scintillation counter) is employed. Electrodeposition onto a flat surface gives excellent deposits for many metals; a thin film of plastic sprayed with gold can serve as the active electrode. The sample may be spread as a slurry in a solvent that is later evaporated; deposits that are not coherent can be stabilized with a binder such as collodion.

11.8 ISOTOPE DILUTION ANALYSIS

Isotope dilution analysis is useful if no quantitative procedure is known, such as in radiocarbon dating of archaeological specimens and in the analysis of complex biochemical mixtures. The procedure is as follows:

1. To the sample add a known weight W_1 of the same compound that is in the sample; the added compound is tagged with the radioactive element.
2. Determine the specific activity A_1 of the tagged compound separately.
3. Isolate from the mixture a small amount of pure compound that is sufficient for an accurate determination.
4. Measure the specific activity A of the isolated material.

The amount W of inactive material present is given by

$$W = W_1\left(\frac{A_1}{A} - 1\right) \tag{11.11}$$

11.9 ACTIVATION ANALYSIS

11.9.1 Principle

Activation analysis is an extremely sensitive technique of elemental analysis based upon selectively inducing radioactivity in some of the atoms of the elements comprising the sample and then selectively measuring the radiations emitted by the radionuclides. After bombardment with suitable nuclear particles, the induced radionuclides are identified or quantitatively measured.

Neutron activation analysis is the most common and widely used form of the activation analysis methods. It involves bombarding the sample with neutrons, usually thermal (slow) neutrons generated in a nuclear reaction, but in some cases fast neutrons generated by an accelerator. In neutron activation analysis the sample is kept in a neutron flux for a length of time that is sufficient to produce radionuclide product(s) in amounts that can be measured with the desired statistical precision. At the termination of the irradiation, the activity A_0 of a particular nuclide will be equal to the difference between the rate of formation and the rate of decay (term in brackets):

$$A_0 = \Phi \sigma N \left[1 - \exp\left(-\frac{0.693t}{t_{1/2}} \right) \right] \tag{11.12}$$

where Φ = number of bombarding particles, or flux, in $cm^{-2} \cdot s^{-1}$
σ = reaction cross section, cm^2 per target atom (10^{-24} cm^2 per nucleus, also denoted barns, or b)
N = number of target nuclei available
$t_{1/2}$ = half-life of the radionuclide
t = duration of irradiation period

The induced activity will reach 98% of the maximum value for an irradiation period equal to six half-lives (Table 11.4). In fact, at $t/t_{1/2}$ ratios of 1, 2, 3, 4, 5, and 6, the corresponding values of the term within the brackets of Eq. (11.12) are 0.5, 0.75, 0.87, 0.94, 0.97, and 0.98, respectively.

Example 11.2 What is the activity for a 10.0-mg sample (from a painting) of pigment containing 1.50% chromium after a 24.0-h irradiation in a flux of 1×10^{14} neutrons $cm^{-2} \cdot s^{-1}$? Chromium-51 has a half-life of 27.8 days; chromium-50, the target isotope, has a natural abundance of 4.31% and a thermal neutron cross section of 17 b (17×10^{-24} $cm^2 \cdot nuclei^{-1}$).

$$\begin{aligned} A_0 &= \frac{(0.0150)(0.0100\ g)(0.0431)(6.02 \times 10^{23}\ \text{nuclei} \cdot \text{mol}^{-1})}{52.00\ g \cdot \text{mol}^{-1}} \\ &\quad \times (1 \times 10^{14}\ \text{neutrons} \cdot \text{cm}^{-1} \cdot \text{s}^{-1})(17 \times 10^{-24}\ \text{cm}^2 \cdot \text{nuclei}^{-1}) \\ &\quad \times \left\{ 1 - \exp\left[\frac{-0.693(1.00\ \text{day})}{27.8\ \text{days}} \right] \right\} \\ &= 3.13 \times 10^6\ \text{disintegrations per second} \end{aligned}$$

While such calculations may be useful in estimating the feasibility of neutron activation analysis, the technique is rarely used in an absolute manner. Samples and standards are irradiated and measured under identical conditions. Samples and standards are sealed into polyethylene or quartz containers that are further encapsulated prior to bombardment. The doubly encapsulated samples and standards are brought to the reactor core either by pneumatic transfer tubes of by manual placement.

Longer-lived nuclides, such as chromium-52, are enhanced by the use of a longer irradiation period, followed, if necessary, by an appreciable delay for the decay of interfering short-lived activities, before counting the chromium-52. The reverse is true for short-lived activities.

Common sources of neutrons for thermal neutron activation analysis include the following:

1. Low-level sources use either ^{88}Y or ^{124}Sb embedded in ^9Be and provide a neutron yield of approximately 1×10^5 $n \cdot s^{-1} \cdot Ci^{-1}$.
2. Intermediate-level sources use either ^{239}Pu or ^{241}Am with ^9Be and provide 10^7 or 10^6 $n \cdot s^{-1} \cdot Ci^{-1}$, respectively.
3. Nuclear reactors of various types (TRIGA, General Atomics, Inc., and SLOWPOKE, Atomic Energy of Canada, Ltd.) provide a neutron flux of 10^{12} to 10^{15} $n \cdot cm^{-2} \cdot s^{-1}$.

The weight of a given element required to produce a preassigned activity is readily calculated.

$$W = \frac{AM}{0.602\ \sigma \Phi[1-\exp(-\lambda t)]} \tag{11.13}$$

where W = weight of target element, g
M = atomic weight
σ = neutron activation cross section, b
Φ = thermal neutron, flux $n \cdot cm^{-2} \cdot s^{-1}$

If the activity A is the lowest quantity that can be readily measured, W becomes the *limit of measurement*. Note that the limit is directly proportional to M, a function of nuclide half-life, and inversely proportional to the capture cross section and the neutron flux. Thus the highest flux and cross section and a short (but not inconveniently short) half-life lead to the lowest limit or greatest sensitivity. Table 11.9 lists some approximate neutron activation analysis sensitivities, expressed as the micrograms of the element that must be present in the sample to be detected and determined when using a neutron flux of $4 \times 10^{12}\ n \cdot cm^{-2} \cdot s^{-1}$. The sensitivity values involve the arbitrary assumptions that the limit of counting detection is 1000 photopeak counts per minute for induced activities of half-life <1 m, 100 counts/min for those with half-lives of 1 to 60 m, and 10 counts/min for those with half-lives >1 h. The counting was done with a 3 × 3-in solid NaI(Tl) detector. The microgram limits based upon beta-particle detection assume that the limit of counting detection is 100 counts/min for induced activities with half-lives of 1 h, and 10 counts/min for those with half-lives >1 h. They also assume a 50% chemical recovery in the radiochemical separations, a 40% counting efficiency, and a decay period prior to counting of one half-life or 1 min, whichever is longer. These values should serve as a guide for the analyst in regard to the feasibility of neutron activation analysis.

Sensitivities are a function of the neutron flux and will be improved when using a higher flux. The sensitivities enumerated in Table 11.9 are values achievable in an interference-free matrix.

11.9.2 Fast-Neutron Activation Analysis

Fast neutrons, which encompass a range of energies from 0.5 MeV and higher, are responsible for threshold reactions, such as (n,p), (n,α), and $(n,2n)$ reactions. For a few elements, particularly the light elements, activation by fast neutrons is more sensitive than that with thermal neutrons. These elements are N, O, F, Al, Si, P, Cr, Mn, Fe, Cu, Y, Mo, Nb, and Pb.

A fast-neutron flux is present in the core of a nuclear reactor, but this source is seldom available for general laboratory use. The diversity of reactions produced by 14-MeV neutrons often dictates the use of a Cockroft–Walton-type accelerator. Pulsed generators provide very high, fast-neutron fluxes, well over 10^{16} neutrons $cm^{-2}\ s^{-1}$, of about 30 ms duration. The amount of a particular activity A_p produced in a pulse, relative to the amount of activity produced in the usual steady-state operation of the same reactor to saturation activity A_s is approximately $A_p/A_s = 70/t_{1/2}$.

Table 11.10 lists elements that are detectable with 14-MeV neutrons. Detectability is based on a 100-count photopeak, corresponding to a statistical precision of about 10%.

11.9.3 Instrumental (Nondestructive) Method

The instrumental or nondestructive approach is preferred whenever applicable. A set of samples and standards are irradiated under exactly the same flux conditions. After the irradiation step the samples are counted. Beta particles are counted as described under liquid-scintillation analysis. Because gamma rays are emitted at characteristic energies, the sample is transferred to a multichannel gamma-ray spectrometer and counted for a suitable period of time. An illustrative gamma-ray spectrum (or pulse-height spectrum) is shown in Fig. 11.2. The shaded area indicated under the photopeak at 0.847 MeV would be used for quantitative work. The two less intense photopeaks at 1.811

TABLE 11.9 Estimated Sensitivities by Neutron Activation Analysis

See text for irradiation and counting conditions.

Radionuclide measured	β particle	γ radiation		
	Sensitivity, μg	Photopeak, MeV	Sensitivity, μg	Interferences
Aluminum-28	0.1	1.780	0.004 9	^{28}Si, ^{31}P, ^{26}Mg
Antimony-122	0.005	0.564	0.002	^{124}Sb
Antimony-124		0.603	0.068	^{122}Sn, ^{124}Te, ^{124}Te
Argon-37		Cl x rays	0.2	^{40}Ca
Argon-41		1.29	0.000 41	^{41}K, ^{44}Ca
Arsenic-76	0.001	0.559	0.001 5	^{76}Se, ^{79}Br, ^{74}Ge
Barium-137*m*		0.662	0.11	
Barium-139	0.05	0.163	0.003 0	^{139}La, ^{142}Ce
Bismuth-210	0.5		No γ	
Bromine-80	0.005	0.616	0.000 70	^{82}Br, ^{80}Kr
Bromine-82	0.005	0.554 + 0.619	0.004 8	^{82}Kr, ^{85}Rb, ^{80}Se
Cadmium-111*m*		0.247	0.003 9	
Cadmium-115	0.05	0.336	0.038	^{115}In, ^{118}Sn
Calcium-49	1.0	3.10	1.5	
Cerium-139*m*	No β	0.74	0.65	
Cerium-141	1	0.145	0.33	^{141}Pr, ^{144}Nd, ^{139}La
Cerium-143	0.1	0.293	0.097	^{146}Nd
Cesium-134	0.5	0.605	0.090	^{134}Ba
Cesium-134*m*		0.127	0.000 49	
Chlorine-38	0.01	1.60	0.031	^{38}Ar, ^{41}K
Chromium-51	No β	0.32	0.45	^{54}Fe
Cobalt-60	0.5	1.173	0.27	^{60}Ni, ^{63}Cu
Cobalt-63*m*	0.005	0.059	0.003 2	
Copper-64	0.001	0.51	0.001 0	^{64}Zn, ^{62}Ni
Copper-66	0.01	1.039	0.020	
Dysprosium-165	0.000 001	0.095	0.000 018	^{165}Ho, ^{168}Er
Dysprosium-165*m*		0.108	0.000 058	^{165}Dy
Erbium-167*m*	No β	0.208	0.008 6	
Erbium-169	0.1			
Erbium-171	0.001	0.308	0.001	^{174}Yb
Europium-152*m*	0.000 005	0.963	0.000 033	^{152}Gd
Fluorine-20	. . .	1.63	0.55	^{23}Na, ^{20}Ne
Gadolinium-159	0.01	0.363	0.006 0	^{159}Tb, ^{162}Dy
Gadolinium-161		0.361	0.1	^{164}Dy
Gallium-70		0.173	0.009	^{70}Ge, ^{68}Zn
Gallium-72	0.005	0.835	0.001 1	^{72}Ge, ^{75}As, ^{70}Zn
Germanium-75	0.005	0.265	0.029	^{75}As, ^{78}Se
Germanium-75*m*		0.139	0.11	
Germanium-77	0.5	0.215	0.4	^{80}Se
Gold-198	0.000 5	0.412	0.000 070	^{198}Hg, ^{196}Pt
Hafnium-179*m*	No β	0.215	0.000 46	
Hafnium-180*m*		0.444	0.009 2	
Hafnium-181		0.482	0.045	^{181}Ta, ^{184}W
Holmium-166	0.000 1	0.081	0.000 67	^{166}Er, ^{169}Tm, ^{164}Dy
Indium-144		1.30	0.1	^{114}Sn
Indium-116*m*	0.000 05	1.293	0.000 028	^{116}Sn
Iodine-128	0.005	0.441	0.000 082	^{128}Xe
Iridium-192		0.316	0.001 0	^{192}Pt, ^{190}Os
Iridium-194		0.328	0.000 24	^{192}Ir

TABLE 11.9 Estimated Sensitivities by Neutron Activation Analysis (*Continued*)

	β particle	γ radiation		
Radionuclide measured	Sensitivity, μg	Photopeak, MeV	Sensitivity, μg	Interferences
Iron-59	50	1.10	100	^{59}Co, ^{62}Ni
Krypton-85		0.514	0.1	^{85}Rb, ^{88}Sr
Krypton-87		0.403	1	^{87}Sr
Lanthanum-140	0.001	1.596	0.003 6	^{140}Ce, ^{139}Ba
Lead-207*m*	No β	0.570	ca. 200	
Lead-209	10	No γ		
Lutetium-176*m*	0.000 05	0.088 4	0.000 19	
Lutetium-177	0.005	0.208	0.003 0	^{176}Yb, ^{177}Hf, ^{180}Ta, ^{180}Hf
Magnesium-27	0.5	0.84	0.14	^{27}Al, ^{30}Si
Manganese-56	0.000 05	0.847	0.000 036	^{56}Fe, ^{59}Co, ^{54}Cr
Mercury-197	No β	0.191	0.79	
Mercury-197*m*	No β	0.134	0.083	
Mercury-199*m*	No β	0.158	0.11	^{197}Hg
		0.375	0.77	
Mercury-203		0.279	0.25	^{203}Tl, ^{206}Pb
Mercury-205		0.205	0.41	^{197m}Hg, ^{199m}Hg, ^{205}Tl, ^{208}Pb, ^{203}Hg
Molybdenum-99	0.5	0.181	0.047	^{102}Ru
Molybdenum-101	0.1	0.191	0.088	
Neodymium-147	0.1	0.090	0.1	^{150}Sn
Neodymium-149		0.211	0.006 3	^{152}Sn
Neodymium-151		0.110	0.002 8	^{149}Nd
Neon-23		0.439	1	^{23}Na, ^{26}Mg
Nickel-65	0.05	1.48	0.180	^{65}Cu, ^{68}Zn, ^{63}Cu
Niobium-94	0.005	0.871	0.29	^{94}Mo
Niobium-94*m*	No β	0.871	1	
Osmium-190*m*	No β	0.616	2.1	
Osmium-191		0.129	0.003	^{191}Ir, ^{194}Pt
Osmium-193	0.05		No γ	
Palladium-107*m*	No β	0.21	0.15	^{109m}Pd
Palladium-109	0.000 5	0.088	0.18	^{109}Ag, ^{112}Cd
Palladium-109*m*	. . .	0.188	0.016	
Phosphorus-32	0.05		No γ	^{32}S, ^{35}Cl
Platinum-195*m*	No β	0.099	0.4	^{198}Hg
Platinum-197		0.077	0.1	^{197}Au, ^{200}Hg
Platinum-199	0.05	0.318	0.086	
Platinum-199*m*	No β	0.393	11	
Potassium-42	0.05	1.46	0.094	^{42}Ca, ^{45}Sc, ^{40}Ar
Praseodymium-142	0.000 5	1.576	0.022	^{142}Nd
Rhenium-186	0.001	0.137	0.021	^{186}Os
Rhenium-188	0.000 5	0.155	0.000 60	^{186}Re, ^{188}Os, ^{191}Ir, ^{186}W
Rhodium-104	0.001	0.56	0.01	^{104}Pd, ^{107}Ag
Rhodium-104*m*	No β	0.051	0.001 0	
Rubidium-86	0.05	1.08	2.0	^{86}Sr, ^{89}Y, ^{89}Kr
Rubidium-86*m*	No β	0.56	0.011	
Rubidium-88		1.863	0.24	^{88}Sr, ^{86}Kr
Ruthenium-97	No β	0.216	0.24	
Ruthenium-103	0.5	0.497	0.46	^{103}Rd, ^{106}Pd

(*Continued*)

TABLE 11.9 Estimated Sensitivities by Neutron Activation Analysis (*Continued*)

	β particle	γ radiation		
Radionuclide measured	Sensitivity, μg	Photopeak, MeV	Sensitivity, μg	Interferences
Ruthenium-105	0.01	0.724	0.011	^{108}Pd
Samarium-153	0.000 5	0.103	0.000 30	^{153}Eu, ^{156}Gd
Samarium-155		0.104	0.000 68	^{153}Sm, ^{158}Gd
Scandium-46	0.01	1.12	0.011	^{46}Ti
Scandium-46*m*	No β	0.142	0.001 2	
Selenium-75	No β	0.265	0.41	^{78}Kr
Selenium-77*m*	No β	0.161	0.013	
Selenium-81		0.28	0.23	^{75}Se, ^{81}Br, ^{84}Kr
Selenium-81*m*	No β	0.104	0.021	^{75}Se, ^{86}Kr
Silicon-31	0.05	1.27	30	^{31}P, ^{34}S
Silver-108	0.005	0.511	0.003 4	^{108}Cd
Silver-110	. . .	0.658	0.004 3	^{110}Cd, ^{113}In, ^{108}Pd
Silver-110*m*		0.658	0.52	
Sodium-24	0.005	1.37	0.002 9	^{24}Mg, ^{27}Al
Strontium-85	No β	0.514	50	
Strontium-85*m*		0.92	0.054	
Strontium-87*m*		0.388	0.001 3	^{90}Zr
Sulfur-35	10		No γ	
Sulfur-37	5	3.09	70	^{37}Cl, ^{40}Ar
Tantalum-182	0.05	1.122–1.222	0.048	^{182}W, ^{185}Re, ^{180}Hf
Tantalum-182*m*	No β	0.147–0.184	0.041	
Tellurium-127		0.360	0.1	^{127}I, ^{130}Xe
Tellurium-131	0.05	0.150	0.018	^{134}Xe
Terbium-160	0.05	0.299	0.028	^{160}Dy
Thallium-204			No γ	
Thorium-233	0.05	0.087	0.05	^{236}U
Thulium-170	0.01	0.084	0.2	^{170}Yb *K* x ray
Tin-125*m*	0.5	0.325	0.5	
Titanium-51	0.5	0.320	0.040	^{51}V, ^{54}Cr
Tungsten-187	0.001	0.480	0.001 5	^{187}Re, ^{190}Os
Vanadium-52	0.005	1.434	0.000 38	^{52}Cr, ^{55}Mn
Xenon-133		0.081	0.003	^{133}Ce, ^{136}Ba
Xenon-135		0.250	0.9	^{138}Ba
Xenon-137		0.455	0.01	
Ytterbium-169		0.177 + 0.198	0.049	
Ytterbium-175	0.001	0.396	0.011	^{175}La, ^{178}Hf
Ytterbium-177		0.151	0.005 1	
Yttrium-90*m*	No β	0.202	0.24	
Zinc-65		1.12	5.4	^{63}Cu
Zinc-69*m*	0.1	0.439	0.083	
Zirconium-95		0.722 + 0.754 + 0.768 (^{95}Nb)	83	^{98}Mo
Zirconium-97 + niobium-97	1	0.750 0.666	1	

and 2.11 MeV aid in qualitative identification of manganese. The x rays of iodine arise from the NaI(TII) crystal detector used to obtain the ^{56}Mn spectrum. Observation of the decay rates by sequential measurements of the spectrum over a period of time will supply the half-life of the nuclide and confirm the identification.

TABLE 11.10 Elements Detectable with 14-MeV Neutrons

Less than 0.1 mg	0.1–0.3 mg	0.3–1.0 mg
Aluminum	Chromium	Bromine
Antimony	Fluorine	Cobalt
Arsenic	Gallium	Hafnium
Barium	Mercury	Indium
Cerium	Oxygen	Iron
Copper	Scandium	Iodine
Germanium	Selenium	Magnesium
Palladium	Sodium	Manganese
Phosphorus	Tantalum	Molybdenum
Rubidium	Tungsten	Nitrogen
Silicon		Potassium
Strontium		Tellurium
Vanadium		Titanium
Zirconium		Zinc

To take advantage of the energy resolution of modern detectors, it is necessary to use thousands of energy channels. Modern systems achieve multichannel analysis by digitizing the detected pulse height. An analog-to-digital converter produces the digital output. A core memory array stores the multichannel data and provides the desired description of the gamma-ray energy spectrum. Each channel tabulates counts at a given energy level, in 0.5-keV segments. The energy level increases sequentially from zero to perhaps 2.000 MeV.

For the analysis of mixtures, spectrum stipping may be performed. The spectrum of one element may dominate and obliterate one or more other spectra. Other elements may generate spectra that are dominant and easily identifiable. To remove these, the recorded spectra from pure samples (stored in memory) of the elements may be subtracted from the composite spectrum. As these recognized elements are removed, other spectra become evident and also may be subtracted until a single spectrum remains. This procedure accurately determines each element contained in the sample.

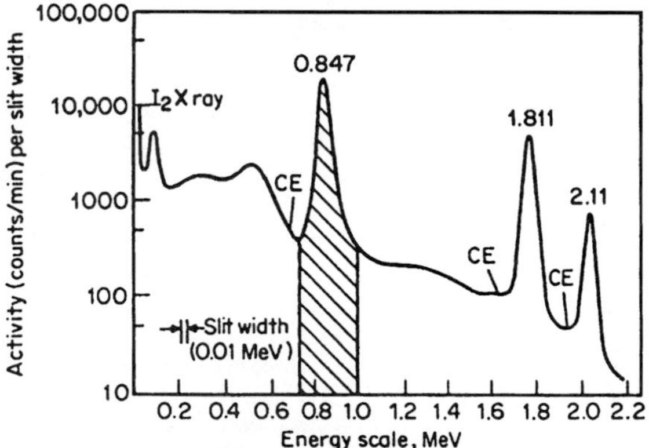

FIGURE 11.2 Gamma spectrum of manganese-56. The area indicated under the photopeak at 0.847 MeV would be used in quantitative work. CE indicates the Compton edges for each photopeak. (*From Shugar and Dean, 1990.*)

11.9.4 Radiochemical Separations

If interferences from other induced activities cannot be removed by either a suitable decay period or use of filters (for beta-particle activities), a postirradiation chemical separation will be needed. The sample (and standard) is dissolved and chemically equilibrated with an accurately known amount of carrier (usually about 10.0 mg) for each of the elements of interest. Then the analyte is separated and purified by any suitable separation procedure and finally counted. The amount of carrier element, if added as a holdback carrier of the analyte radionuclide, is recovered and measured quantitatively. Results are normalized to 100% recovery. Complete recovery is unnecessary; however, the higher the recovery, the better the counting statistics.

Many separation schemes use solvent-extraction and ion-exchange procedures. A good separation of a single radionuclide would enable a NaI(Tl) detector, or even liquid-scintillation counter, with its high detection efficiency to be used. If the sample is not radiochemically pure, a detector with good energy resolution [Ge(Li) or high-purity Ge] will discriminate against residual interferences.

11.10 RADIATION SAFETY

11.10.1 Radiation Characteristics of Radionuclides

Alpha radiation consists of heavy, positively charged particles (helium nuclei or alpha particles) that travel only 5 to 7 cm in air and have very little penetrating effect. A piece of paper, a rubber glove, or the dead outer layer of the human skin stops them. Their energies are very high. As a result, the ionizing power of an alpha particle is high.

A 0.5-MeV beta particle has a range in air of 1 m and produces 60 ion pairs per centimeter of its path. The most energetic can penetrate about 8.5 mm of skin, but almost all are stopped by 3.2 mm of aluminum or 13 mm of Lucite.

Gamma and X radiation are made up of electromagnetic waves of higher frequency and much greater penetrating ability than those in the ultraviolet spectrum. They cannot be entirely stopped, but they can be attenuated (i.e., their intensity can be reduced) as they pass through lead or concrete barriers. Their ionizing power is less than that of alpha or beta radiation. Substances emitting gamma radiation are dangerous and must be handled with due regard to all safety precautions.

11.10.2 Labeling of Radioactive Substances

All radioactive substances are supplied in approved containers that are designed to specific standards and that must pass extreme testing for safety to fire and shock. There are type A and B packages used for small and large quantities. Only a specified quantity of the radionuclide can be shipped in a particular package. Labeling is standardized; in the case of an emergency the hazard can be quickly assessed.

The following labeling procedures have been adopted in the United States:

Category I	White label	The dose rate at the surface of the package is $0.5\ mSv \cdot h^{-1}$ or less. (The abbreviation mSv stands for millisievert.)
Category II	Yellow label	The dose rate at the surface of the package does not exceed $10\ mSv \cdot h^{-1}$ and the dose rate 1 m from the center of the package does not exceed $0.5\ mSv \cdot h^{-1}$.
Category III	Red label	The dose rate does not exceed $200\ mSv \cdot h^{-1}$ at the surface of the package, and the dose rate 1 m from the center of the package does not exceed $10\ mSv \cdot h^{-1}$.

11.10.3 Handling of Radioactive Substances

Small quantities of various radioactive isotopes are available without license because such quantities are considered to be harmless under general laboratory conditions. All radioactive substances must be handled so that any unnecessary exposure or excessive exposure is prevented. Usually the greatest exposure occurs during the transfer of the radionuclide from its container to the experimental equipment. The type of shielding for personnel depends on the type of radioactive emission. Alpha emitters do not require any shielding. Beta emitters can be shielded by sheets of plastic such as transparent methacrylate. Gamma emitters require lead or steel shielding.

11.10.4 Storage of Radioactive Substances

The storage requirements of radioactive materials depends upon the type of emitter and the quantity involved. Normally, laboratory quantities of radionuclides may not require specialized storage areas but the containers should be positioned away from any counting areas so as to prevent stray radiation from affecting and increasing the counter background.

Disposal regulations for radioactive wastes can be obtained from the U.S. Nuclear Regulatory Commission.

Bibliography

Barkouskie, M. A., "Liquid Scintillation Counting," *Am. Lab.* **8:**101 (May 1976).

Bernstein, K., "Neutron Activation Analysis," *Am. Lab.* **12:**151 (September 1980).

De Soete, D., R. Gijbels, and J. Hoste, *Neutron Activation Analysis*, Vol. 34 in *Chemical Analysis*, P. J. Elving and I. M. Kolthoff, eds., Wiley-Interscience, New York, 1972.

Ehmann, W. D., and D. E. Vance, *Radiochemistry and Nuclear Methods of Analysis*, Vol. 116 in *Chemical Analysis*, J. D. Winefordner and I. M. Kolthoff, eds., Wiley-Interscience, New York, 1991.

Finston, H. L., "Radioactive and Isotopic Methods of Analysis," in *Treatise on Analytical Chemistry*, I. M. Kolthoff and P. J. Elving, eds., Wiley-Interscience, New York, 1971, Part I, Vol. 9.

Friedlander, G., J. W. Kennedy, E. S. Macias, and J. M. Miller, *Nuclear and Radiochemistry*, 3d ed., Wiley, New York, 1981.

Leclerc, J.-C., and A. Cornu, *Neutron Activation Analysis Tables*, Heyden, London.

Katz, S. A., "Neutron Activation Analysis," *Am. Lab.* **17:**16 (June 1985).

Kessler, M. J., "Recent Advances in Detectors for Radioisotopes," Part 1, *Am. Lab.* **20**:86 (June 1988); Part 2, *Am. Lab.* **20**:76 (August 1988).

Kolthoff, I. M., and P. J. Elving, eds., "Nuclear Activation and Radioisotopic Methods of Analysis," in *Treatise on Analytical Chemistry*, 2d ed., Wiley, New York, 1986, Part I, Vol. 14, Sec. K, Chap. 1–8.

Yule, H. P., *Anal. Chem.* **37**:128 (1965).

SECTION 12

NUCLEAR MAGNETIC RESONANCE SPECTROSCOPY AND ELECTRON SPIN RESONANCE

12.1 INTRODUCTION

Nuclear magnetic resonance (NMR) spectroscopy is a powerful method for elucidating the structure of a molecule. The spectra obtained answer many questions such as the following (referring to specific nuclei):[1]

[1] J. A. Dean, "Use of Nuclear Magnetic Resonance in Determining Molecular Structure of Urinary Constituents of Low Molecular Weight," *Clin. Chem.* **14**:326 (1968).

1. What are you?
2. Where are you located in the molecule?
3. How many of you are there?
4. What and where are your neighbors?
5. How are you related to your neighbors?

The result is often the delineation of complete sequences of groups or arrangements of atoms in the molecule. The sample is not destroyed in the process.

NMR can also be used for a particular facet of a structure, such as chain length or moles of ethoxylation, and in the study of polymer motion by relaxation measurements. Kinetic studies of reactions at temperatures in the range from −150 to 200°C are another application. Useful information can also be obtained from complex mixtures such as, for example, the total aldehydic content of a perfume, the phenyl–methyl ratio in polysiloxanes, the monoester–diester ratios in emulsifiers, and the alcohol–water ratios in colognes and aftershaves. In solvent systems an understanding of the basic types of solvents present can be had by scanning the product in the NMR region. Of course, the most important applications of NMR spectroscopy are in organic syntheses to determine the class of compounds to which the molecule belongs and to provide more detailed information about its structure.

Integration of areas under the absorption peaks and the peak of the internal standard enables quantitative analysis to be performed.

12.2 BASIC PRINCIPLES

The nuclei of certain isotopes have an intrinsic spinning motion around their axes that generates a magnetic moment along the axis of spin. The simultaneous application of a strong external magnetic field H_0 and the energy from a second and weaker radio-frequency source (H_1) (applied perpendicular to H_0) to the nuclei results in the rotation of the macroscopic nuclear magnetization away from its equilibrium position parallel to the applied magnetic field. The transitions between energy states of the nuclear spin are shown in Fig. 12.1. Absorption occurs when these nuclei undergo transition from one alignment in the applied field to an apposite one. The energy needed to excite these transitions can be measured. The resonance frequency v that causes the transitions between energy levels is given by

$$\Delta E = hv = \frac{\mu H_0}{I} \tag{12.1}$$

where μ = magnetic moment of the nucleus
I = spin quantum number, h/π
h = Planck's constant

Nuclei with $I = \frac{1}{2}$ give the best resolved spectra because their electric quadrupole moment is zero. These nuclei include ^{1}H, ^{13}C, ^{19}F, ^{29}Si, and ^{31}P.

As indicated by Eq. (12.1), the frequency of the resonance condition varies with the value of the applied magnetic field. For example, in a magnetic field of 14 092 G, protons require a frequency of 60 MHz to realign their spins. In a field of 23 490 G, the frequency rises to 100 MHz. Higher values of field strength lead to a stronger absorption signal (roughly in proportion to the square of the magnetic field strength). Nuclear properties of selected elements are given in Table 12.1.

12.2.1 Relaxation Processes

Energy absorbed and stored in the higher-energy state must be dissipated and the nuclei returned to the lower-energy state. Otherwise, the rf system equalizes the populations of the energy states, the spin system becomes saturated, and the absorption signal disappears.

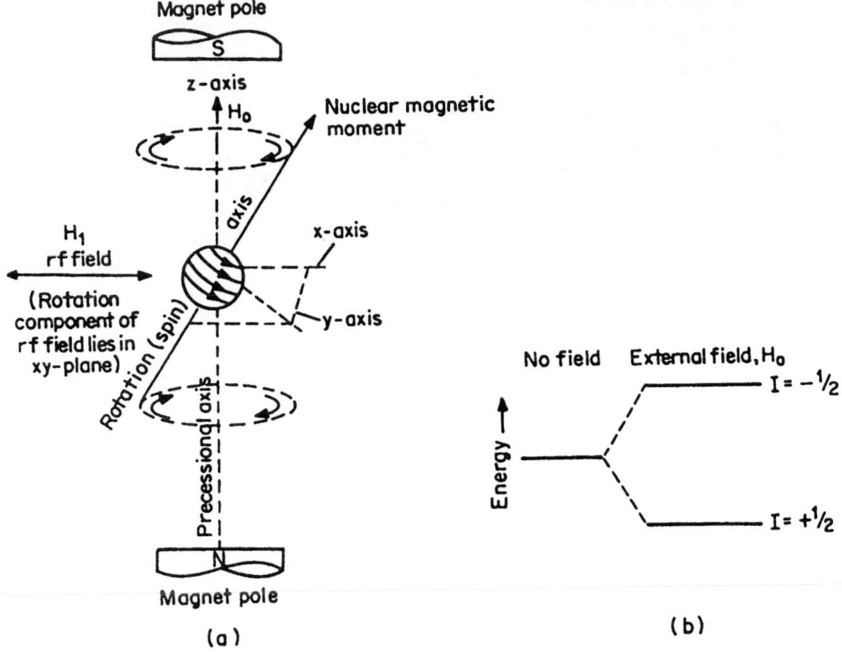

FIGURE 12.1 **(*a*) Spinning nucleus in a magnetic field. (*b*) Energy-level diagram for a nucleus with a spin of 1/2.** *(From Shugar and Dean, The Chemist's Ready Reference Handbook, McGraw-Hill, New York 1990.)*

TABLE 12.1 Nuclear Properties of the Elements

In the following table the magnetic moment μ is in multiples of the nuclear magneton $\mu_N(eh/4\pi Mc)$ with diamagnetic correction. The spin I is in multiples of $h/2\pi$, and the electric quadrupole moment Q is in multiples of 10^{-28} m². Nuclei with spin 1/2 have no quadrupole moment. Sensitivity *is for equal numbers of nuclei at constant field.* NMR frequency *at any magnetic field is the entry for columns 5 multiplied by the value of the magnetic field in kilogauss. For example, in a magnetic field of 23.490 kG, protons will process at 4.2576 × 23.490 kG = 100.0 MHz. Radionuclides are denoted with an asterisk.*

The data were extracted from A. H. Wapstra and G. Audi, "The 1983 Atomic Mass Evaluation," Nucl. Phys. ***A432***:*1–54 (1985), and M. Lederer and V. S. Shirley, Table of Isotopes, 7th ed., Wiley Interscience, New York, 1978.*

Nuclide	Natural abundance, %	Spin I	Sensitivity at constant field relative to ^{1}H	NMR frequency for a 1-kG field, MHz	Magnetic moment μ/μ_N, J · T^{-1}	Electric quadrupole moment Q, 10^{-28} m^2
^{1}n		1/2	0.322	2.916 5	−1.913 0	
^{1}H	99.985(1)	1/2	1.000	4.257 6	2.792 846	
^{2}H	0.015(1)	1	0.009 64	0.653 6	0.857 438	0.002 875
^{3}H*		1/2	1.21	4.541 4	2.978 960	
^{3}He	1.38×10^{-4}	1/2	0.443	3.243 4	−2.127 624	
^{6}Li	7.5(2)	1	0.008 51	0.626 5	0.822 047	0.000 645
^{7}Li	92.5(2)	3/2	0.294	1.654 7	−3.256 424	−0.036 6
^{9}Be	100	3/2	0.013 9	0.598 3	−1.177 9	0.053
^{10}B	19.9(2)	3	0.019 9	0.457 5	1.800 65	0.084 73
^{11}B	80.1(2)	3/2	0.165	1.366 0	2.688 637	0.040 65
^{13}C	1.10(3)	1/2	0.015 9	1.070 5	0.702 411	

TABLE 12.1 Nuclear Properties of the Elements (*Continued*)

Nuclide	Natural abundance, %	Spin I	Sensitivity at constant field relative to ^{1}H	NMR frequency for a 1-kG field, MHz	Magnetic moment μ/μ_N, $J \cdot T^{-1}$	Electric quadrupole moment Q, 10^{-28} m^2
^{14}N	99.634(9)	1	0.001 01	0.307 6	0.403 761	0.015 6
^{15}N	0.366(9)	1/2	0.001 04	0.431 5	−0.283 189	
^{17}O	0.003 8(3)	5/2	0.029 1	0.577 2	−1.893 80	−0.025 78
^{19}F	100	1/2	0.834	4.005 5	2.628 867	
^{21}Ne	0.27(1)	−3/2	0.027 2	0.336 11	−0.661 966	0.103 0
^{22}Na*		3	0.018 1	0.443 4	1.745	
^{23}Na	100	3/2	0.092 7	1.126 2	2.217 520	0.102
^{25}Mg	10.00(1)	5/2	0.026 8	0.260 6	−0.855 46	0.22
^{27}Al	100	5/2	0.207	1.109 4	3.641 504	0.140
^{29}Si	4.67(1)	1/2	0.078 5	0.846 0	−0.555 29	
^{31}P	100	1/2	0.066 4	1.723 5	1.131 60	
^{33}S	0.75(1)	3/2	0.022 6	0.326 6	0.643 821	−0.064
^{35}S*		3/2	0.008 50	0.508	1.00	0.045
^{35}Cl	75.77(5)	3/2	0.004 71	0.417 2	0.821 874	−0.082 49
^{36}Cl*		2	0.012 1	0.489 3	1.283 8	−0.016 8
^{37}Cl	24.23(5)	3/2	0.002 72	0.347 2	0.684 123	−0.064 93
^{39}K	93.258(3)	3/2	0.000 508	0.198 7	0.391 466	0.049
^{40}K*	0.011 9(1)	4	0.005 21	0.247 0	−1.298 099	−0.067
^{41}K	6.730(3)	3/2	0.000 083 9	0.109 2	0.214 870	0.060
^{43}Ca	0.135(3)	7/2	0.063 9	0.286 5	−1.317 27	
^{45}Sc	100	7/2	0.301	1.034 3	4.756 483	−0.22
^{47}Ti	7.3(1)	5/2	0.002 10	0.240 0	−0.788 48	0.29
^{49}Ti	5.5(1)	7/2	0.003 76	0.240 1	−1.104 17	0.24
^{50}V	0.250(2)	6	0.055 3	0.424 5	3.347 45	0.07
^{51}V	99.75(2)	7/2	0.383	1.119 3	5.151 4	−0.052
^{53}Cr	9.50(1)	3/2	0.000 10	0.240 6	−0.473 45	0.022
^{55}Mn	100	5/2	0.178	1.053 3	3.453 2	0.40
^{59}Co	100	7/2	0.281	1.010 3	4.627	0.404
^{61}Ni	1.13(1)	3/2	0.003 50	0.380 48	−0.750 02	0.162
^{63}Cu	69.17(2)	3/2	0.093 8	1.128 5	2.223 3	−0.209
^{65}Cu	30.83(2)	3/2	0.116	1.209 0	2.381 7	−0.195
^{67}Zn	4.1(1)	5/2	0.002 86	0.263 5	0.875 479	0.150
^{69}Ga	60.1(2)	3/2	0.069 3	1.021 8	2.016 59	0.168 8
^{71}Ga	39.9(2)	3/2	0.142	1.298 4	2.562 27	0.106
^{73}Ge	7.8(2)	9/2	0.001 40	0.148 5	−0.879 467	−0.173
^{75}As	100	3/2	0.025 1	0.729 2	1.439 47	0.29
^{77}Se	7.6(2)	1/2	0.006 97	0.813 1	0.535 06	
^{79}Br	50.69(5)	3/2	0.078 6	1.066 7	2.106 399	0.293
^{81}Br	49.31(5)	3/2	0.098 4	1.149 8	2.270 560	0.27
^{83}Kr	11.5(1)	9/2	0.001 89	0.016 4	−0.970 669	0.270
^{85}Rb	72.16(1)	5/2	0.010 5	0.411 1	1.353 03	0.274
^{87}Rb	27.83(1)	3/2	0.177	1.393 2	2.751 24	0.132
^{87}Sr	7.00(1)	9/2	0.002 69	0.184 5	−1.092 83	0.16
^{89}Y	100	1/2	0.000 117	0.208 6	−0.137 415	
^{91}Zr	11.22(2)	5/2	0.009 4	0.40	−1.303 62	
^{93}Nb	100	9/2	0.482	1.040 7	6.170 5	−0.37
^{95}Mo	15.92(4)	5/2	0.003 22	0.277 4	−0.914 2	−0.019
^{97}Mo	9.55(2)	5/2	0.003 42	0.283 3	−0.933 5	−0.102
^{99}Ru	12.7(1)	5/2	0.000 195	0.144 3	−0.641 3	0.077

(Continued)

TABLE 12.1 Nuclear Properties of the Elements (*Continued*)

Nuclide	Natural abundance, %	Spin I	Sensitivity at constant field relative to ^1H	NMR frequency for a 1-kG field, MHz	Magnetic moment μ/μ_N, $J \cdot T^{-1}$	Electric quadrupole moment Q, 10^{-28} m^2
^{101}Ru	17.0(1)	5/2	0.001 4	0.210 4	−0.718 9	0.44
^{103}Rh	100	1/2	0.000 031 2	0.134 0	−0.088 40	
^{105}Pd	22.33(8)	5/2	0.000 779	0.174	−0.642	0.80
^{107}Ag	51.839(5)	1/2	0.000 066 9	0.172 2	−0.113 570	
^{109}Ag	48.161(5)	1/2	0.000 101	0.198 1	−0.130 691	
^{111}Cd	12.80(6)	1/2	0.009 54	0.902 8	−0.594 886	
^{113}Cd	12.22(6)	1/2	0.010 9	0.944 4	−0.622 300	
^{113}In	4.3(2)	9/2	0.345	0.931 0	5.528 9	0.846
^{115}In	95.7(2)	9/2	0.347	0.932 9	5.540 8	0.861
^{115}Sn	0.36(1)	1/2	0.035 0	1.322	−0.918 84	
^{117}Sn	7.68(7)	1/2	0.045 3	1.577	−1.001 05	
^{119}Sn	8.58(4)	1/2	0.051 8	1.587	−1.047 29	
^{121}Sb	57.3(9)	5/2	0.160	1.019	3.363 4	−0.20
^{123}Sb	42.7(9)	7/2	0.045 7	1.551 8	2.549 8	−0.26
^{123}Te	0.908(3)	1/2	0.018	1.115 7	−0.736 79	
^{125}Te	7.14(1)	1/2	0.031 6	1.345	−0.888 28	
^{127}I	100	5/2	0.093 5	0.851 9	2.813 28	−0.789
^{129}Xe	26.4(6)	1/2	0.021 2	1.178	−0.777 977	
^{131}Xe	21.2(4)	3/2	0.002 77	0.349 0	0.691 861	−0.12
^{133}Cs	100	7/2	0.047 4	0.558 5	2.582 024	−0.3
^{135}Ba	6.59(2)	3/2	0.004 99	0.425	0.837 943	0.18
^{137}Ba	11.23(4)	3/2	0.006 97	0.476	0.937 365	0.28
^{139}La	99.91(1)	7/2	0.059 2	0.601 4	2.783 2	0.22
^{141}Pr	100	5/2	0.234	1.13	4.136	−0.058 9
^{143}Nd	12.18(5)	7/2	0.002 81	0.22	−1.065	−0.484
^{145}Nd	8.30(5)	7/2	0.000 670	0.14	−0.656	−0.253
^{147}Sm	15.0(2)	7/2	0.000 88	0.147	−0.814 9	−0.18
^{149}Sm	13.8(1)	7/2	0.000 47	0.119	−0.671 8	0.053
^{151}Eu	47.81(5)	5/2	0.168	1.0	3.471 8	1.16
^{153}Eu	52.2(5)	5/2	0.014 5	0.46	1.533 1	2.94
^{155}Gd	14.80(5)	3/2	0.000 279	0.162 59	−0.259 1	1.59
^{157}Gd	15.65(3)	3/2	0.000 544	0.203 26	−0.339 9	2.03
^{159}Tb	100	3/2	0.058 3	0.965 45	2.014	1.18
^{161}Dy	18.9(1)	5/2	0.000 417	0.140 26	−0.480 6	2.44
^{163}Dy	24.9(2)	5/2	0.001 12	0.195 12	0.672 6	2.57
^{165}Ho	100	7/2	0.181	0.873 27	4.173	2.74
^{167}Er	22.95(13)	7/2	0.000 507	0.123 05	−0.566 5	2.827
^{169}Tm	100	1/2	0.000 566	0.353 12	−0.231 6	
^{171}Yb	14.3(2)	1/2	0.004 19	0.69	0.493 67	
^{173}Yb	16.1(2)	5/2	0.001 19	0.198	−0.679 89	2.8
^{175}Lu	97.41(2)	7/2	0.049 4	0.57	2.232 7	5.69
^{177}Hf	18.606(3)	7/2	0.000 638	0.132 85	0.793 6	4.50
^{179}Hf	13.629(5)	9/2	0.000 216	0.079 61	−0.640 9	5.1
^{181}Ta	99.988(2)	7/2	0.026 0	0.46	2.371	3.9
^{183}W	14.3(1)	1/2	0.000 069 8	0.175	0.117 785	
^{185}Re	37.40(2)	5/2	0.133	0.958 6	3.187 1	2.36
^{187}Re	62.60(2)	5/2	0.137	0.986 4	3.219 7	2.24
^{189}Os	16.1(3)	3/2	0.002 24	0.330 7	0.659 933	0.91
^{191}Ir	37.3(5)	3/2	0.000 035	0.081	0.146 2	0.78

TABLE 12.1 **Nuclear Properties of the Elements (*Continued*)**

Nuclide	Natural abundance, %	Spin I	Sensitivity at constant field relative to ^{1}H	NMR frequency for a 1-kG field, MHz	Magnetic moment μ/μ_N, J · T^{-1}	Electric quadrupole moment Q, 10^{-28} m^2
^{193}Ir	62.7(5)	3/2	0.000 042	0.086	0.159 2	0.70
^{195}Pt	33.9(5)	1/2	0.009 94	0.915 3	0.609 50	
^{197}Au	100	3/2	0.000 021 4	0.069 1	0.148 159	0.547
^{199}Hg	16.8(11)	1/2	0.005 72	0.761 2	0.505 885	
^{201}Hg	13.23(11)	3/2	0.001 90	0.308	−0.560 225	0.455
^{203}Tl	29.524(9)	1/2	0.187	2.433	1.622 257	
^{205}Tl	70.476(9)	1/2	0.192	2.457	1.638 214	
^{207}Pb	22.1(1)	1/2	0.009 13	0.889 9	0.582 19	
^{209}Bi	100	9/2	0.137	0.684 2	4.110 6	−0.46
^{235}U	0.720(1)	7/2	0.000 121	0.076 23	−0.35	4.55

* Radioactive nuclide.
Source: J. A. Dean, ed., *Lange's Handbook of Chemistry*, 14th ed., McGraw-Hill, New York, 1992.

The *spin–lattice relaxation* involves interaction of the spin with the fluctuating magnetic fields produced by the random motions of neighboring nuclei (called the *lattice*). Nuclei lose their excess energy as thermal energy to the lattice. The relaxation process is first order and decreases exponentially with time. T_1 is a rate constant called the spin–lattice relaxation time. It depends on the average distance of the nucleus from magnetic neighboring nuclei and on the types of molecular motion that the functional group is undergoing. Internal rotations and segmental motions in polymers are detected in this way. In typical organic liquids and dilute solutions the time is in the range 0.01 to 100 s.

The *spin–spin relaxation* process involves a second time constant T_2. Nuclei exchange spins with neighboring nuclei through the interaction of their magnetic moments, and they thereby lose phase coherence. T_2 can be calculated from the dispersion mode signal as a function of time.

Automated T_1 and T_2 measurements are done with computer software. The time constants aid in understanding physical properties of polymers. Dissolved oxygen must be removed because it is paramagnetic and competes with the dipole–dipole mechanism in inducing relaxation.

12.3 INSTRUMENTATION

NMR instrumentation involves these basic units:

1. A magnet to separate the nuclear spin energy states.

2. (a) One rf channel for field or frequency stabilization, which produces stability for long-term operation, (b) one rf channel to furnish irradiating energy to the sample, and (c) a third channel that may be added for decoupling nuclei.

3. A sample probe that houses the sample and also coils for coupling the sample with the rf transmitter and the phase-sensitive detector. It is inserted between the pole faces of the magnet.

4. A detector to collect and process the NMR signals.

5. A sweep generator for sweeping the rf field through the resonance frequencies of the sample. Alternatively, the magnetic field may be swept and the rf field held constant.

6. A recorder to display the spectrum.

12.3.1 Sample Handling

The sample for solution-state NMR is contained in cylindrical, thin-walled precision-bore glass tubes (purchased from supply houses) that have outer diameters of 5 and 10 mm and that require from 0.5 to 2 mL of solution, respectively. The tube is filled until the length–diameter ratio is about 5. An air-bearing turbine rotates the sample tube at a rate of 20 to 40 revolutions per second. The ideal sample for solution-state NMR is a fairly concentrated solution (>3 m*M* for proton NMR and >0.1*M* for carbon NMR) of the material of interest dissolved in a volume that barely exceeds the dimensions of the NMR coil. The sample is generally dissolved in a deuterated solvent (Table 12.2), such as chloroform-*d*, benzene-d_6, or D_2O. The same solvents are used to dissolve samples when the ^{13}C spectrum is desired. The deuterium nuclei provides the signal that is used to lock the ratio of H_0/v of the NMR spectrometer over long periods of time. Methanol-d_4 is used for work at very low temperatures, and mixed solvents (such as 1,2,4-trichlorobenzene plus 1,4-dioxane-d_8 or benzene-d_6 as

TABLE 12.2 Properties of Deuterated and Other Solvents for NMR

The abbreviations mp, bp, and Fp denote melting point, boiling point, and flash point, respectively.

Compound	mp, °C	bp, °C	Fp, °C
Acetic acid-d_1	17	118	40
Acetic acid-d_4	17	118	40
Acetone-d_6	–94	55.5	–17
Acetonitrile-d_3	–45	80.7	5
Benzene-d_6	5	79.1	–11
Bromobenzene-d_6	–31	156	65
Carbon tetrachloride	–23	77	None
Chloroform-d_1	–64	60.9	None
Cyclohexane-d_{12}	6	78	–18
Deuterium oxide	3.8	20.0	None
1,2-Dichloroethane-d_4	–40	103	None
Diethyl-d_{10} ether	–116	35	–40
N,N-Dimethylformamide-d_7	–61	153	57
Dimethyl-d_6 sulfoxide	18	189	95
1,4-Dioxane-d_8	12	101	12
Ethanol-d_1	–114.5	78.8	12
Ethanol-d_6	–130	79	12
Hexachloro-1,3-butadiene	–19	210 to 220	None
Hexamethyldisiloxane	–67	99 to 100	1
Hexamethylphosphoramide-d_{18}	7	233	105
Methanol-d_1	–98	65.5	11
Methanol-d_4	–98	65.4	11
Methylene-d_2 dichloride	–95	40	None
Methyl-d_3 iodide	–67	42	None
Nitrobenzene-d_5	6	211	87
Nitromethane-d_3	–29	100	35
Phosphoric acid-d_3, 85% in D_2O	42	d 213	None
2-Propanol-d_8	–86	83	12
Pyridine-d_5	–42	114.4	20
1,1,2,2-Tetrachloroethane-d_2	–44	146.5	None
Tetrahydrofuran-d_8	–109	66	–17
Tetramethylsilane	–27	26.3	–27
Toluene-d_8	–95	110	10
Trifluoroacetic acid-d_1	–15	75	None
2,2,2-Trifluoroethanol-d_3	–44	103	29

the lock material) are used for polymers. Deuterated solvents are best stored in a dry box under a nitrogen atmosphere and opened and transferred only in the dry box.

The deuterium replaces the hydrogen nuclei that would generate a background solvent signal that would overwhelm the proton signals from the sample. Because the isotopic abundance of ^{13}C is only 1.3%, the sample size is larger than that required for proton spectra—between a few milligrams and a few hundred milligrams.

12.3.2 The Magnet

Less-expensive instruments may use permanent magnets. These are simple and inexpensive to operate, but they do require extensive shielding and must be thermostatted to ±0.001°C.

Electromagnets require elaborate power supplies and cooling systems. They offer the opportunity to use different field strengths to disentangle chemical shifts from multiplet structures and to study different nuclei. The upper limit of magnet strength is 23.49 kG.

For higher field strengths, one must turn to cryogenic superconducting solenoids. Instruments with fields of 93.9 and 117.4 kG permit rf fields of 400 and 500 MHz, respectively. Superconducting magnets allow the spectrum to be spread out and thereby distinguish chemical shifts that might nearly be coincident at low magnetic fields. This facilitates the interpretation of spectra from complex organic compounds.

12.3.3 Pulsed (Fourier-Transform) NMR

In Fourier-transform NMR all the spectral lines are excited simultaneously. A strong pulse of energy (H_1) is applied to the sample for a very short time (1 to 100 μs). In the interval between pulses, the free precession of the magnetic moments of the nuclei gradually decay. This free induction decay requires several spin–lattice time constants ($3T_1$ to $5T_1$). The decay signal following each repetitive pulse (about 0.82 s for 8192 data points) is digitized by a fast analog-to-digital converter, and the successive digitized transient signals are coherently added in the computer until an adequate signal-to-noise ratio is obtained. Using the Cooley–Tukey algorithm, the computer then performs a fast Fourier transformation to the frequency domain and plots a normal spectral presentation of the NMR spectrum in 10 to 20 s.

Pulsed Fourier-transform NMR makes possible the study of less-sensitive nuclei, such as ^{13}C, ^{14}N, ^{15}N, ^{17}O, ^{31}P, and even unstable, radioactive nuclei with spins. Proper selection of pulse sequences under computer control allows measurement of the various T_1 values.

12.3.4 Magic-Angle Spinning and Cross Polarization

Line broadening from polycrystalline materials can largely be removed by spinning the sample very rapidly (ca. 5 to 15 kHz) around an axis that forms an angle of 54.7° (the magic angle) with respect to the magnetic field H_0 direction. The resolution observed for a solid is increased by averaging the chemical shift anisotropies to their isotropic values. The averaging that occurs is similar to tumbling in liquids.

Solids are machined into rotors or packed as powders (sometimes with pressure) into hollow rotors. The rotor is dropped into the probe and spun at greater than 2 kHz (120 000 r/min). There is no single standard reference material for solid-state NMR; see Table 12.19 which lists some secondary standards. A pressed wafer of the reference material is put in the sample rotor with the solid sample.[2]

For ^{13}C studies, cross polarization transfers magnetization from one nuclear species to another. This technique overcomes the problems of long spin–lattice relaxation times, particularly when handling solids. Details are given by Miknis and colleagues.[3]

[2] W. L. Earl and D. L. VanderHart, *J. Magn. Reson.* **48**:35 (1982), discuss the problems and methods.

[3] F. P. Miknis, V. J. Bartuska, and G. E. Maciel, "Crosspolarization ^{13}C NMR with Magic-Angle Spinning," *Am. Lab.* **11** (November):19 (1979).

12.3.5 Two-Dimensional NMR Spectroscopy

In two-dimensional (2D) NMR spectroscopy the nuclear magnetizations are allowed to precess during an initial time period. Next various pulse sequences are applied, and a *free induction decay* is recorded and then Fourier-transformed. The sequence is pulse, evolution, pulse, detection. Two-dimensional Fourier transformation of the two independent time domains produces a spectrum that can be displayed along two orthogonal frequency dimensions. Literally hundreds of different 2D methods have been devised, each aimed at correlating resonances at different frequencies on the basis of some interaction between the nuclei responsible for the resonances. Structures of complex organic molecules that previously had required the tedious application of double-resonance methods can now be determined systematically and efficiently by 2D methods.

12.4 SPECTRA AND MOLECULAR STRUCTURE

For most purposes, NMR spectra are described in terms of *chemical shifts* and *coupling constants*. These are discussed in considerable detail for proton spectra but the general discussion applies equally well to boron-11, nitrogen-15, fluorine-19, silicon-29, and phosphorus-31 spectra. The spin–lattice (T_1) and spin–spin (T_2) relaxation times have already been discussed.

12.4.1 Chemical Shifts of Protons

On the basis of nuclear properties alone, all protons would precess at exactly the same frequency in a fixed magnetic field and would be indistinguishable from each other. Fortunately, in different chemical environments, shifts in resonant frequency arise from partial shielding of the nuclei from the applied magnetic field by the electron cloud around it. The density of this cloud varies with the number and nature of the neighboring atoms. The specific location of the shifted resonant frequencies indicates the chemical nature of the nuclei and provides some information concerning their neighbors. The resonance position of proton spectra is always stated with respect to the resonance of the protons in a reference compound. This overcomes the problem of NMR spectrometers with different field strengths. The magnitude of the chemical shift is expressed in parts per million.

$$\delta = \frac{H_r - H_s}{H_r} \times 10^6 \qquad (12.2)$$

where H_r and H_s are the positions of the absorption lines for the reference and sample, respectively, expressed in magnetic units.

For ^{1}H, ^{13}C, and ^{29}Si spectra the reference material is usually tetramethylsilane, $(CH_3)_4Si$, abbreviated TMS. A very small amount is added internally when organic solvents are used. 4,4-Dimethyl-4-silapentane sodium carboxylate or similar compounds are used for aqueous solutions. The position of TMS is assigned as exactly 0.0 on the delta (δ) scale. Table 12.3 lists some reference materials of proton NMR measurements in solution.

Tables 12.4 and 12.5 list chemical shifts for protons. For compounds not specifically included in these tables, Table 12.6 provides information for the estimation of proton chemical shifts of methylene and methane groups, and Table 12.7 does the same for protons attached to a double bond.

To avoid confusion from spurious peaks in the NMR spectrum when using various deuterated solvents, Table 12.8 lists the solvent positions of residual protons in incompletely deuterated solvents.

TABLE 12.3 Proton Chemical Shifts of Reference Compounds

Relative to tetramethylsilane, $\delta_H = 0.000$.

Compounds	Boiling point or melting point, °C	δ, ppm
4,4-Dimethyl-4-silapentane sodium carboxylate	mp > 300	0.000
4,4-Dimethyl-4-silapentane sodium sulfonate	mp = 200	0.015
Hexamethyldisilane	bp = 112	0.037
Hexamethyldisiloxane	bp = 100	0.055
Hexamethyldisilazane	bp = 125	0.042
1,1,3,3,5,5-Hexakis(trideuteromethyl)-1,3,5-trisilacyclohexane	bp = 208	–0.327
Octamethylcyclotetrasiloxane	bp = 175 mp = 16.8	0.085
Tetrakis(trimethylsilyl)methane	mp = 307	0.236
Tetramethylsilane	bp = 26.3	0.000

TABLE 12.4 Proton Chemical Shifts

Values are given on the officially approved δ scale; τ = 10.00 – δ. The abbreviations R and Ar denote alkyl aryl groups, respectively.

Substituent group	Methyl protons	Methylene protons	Methine proton
HC—C—CH_2	0.95	1.20	1.55
HC—C—NR_2	1.05	1.45	1.70
HC—C—C=C	1.00	1.35	1.70
HC—C—C=O	1.05	1.55	1.95
HC—C—NRAr	1.10	1.50	1.80
HC—C—H(C=O)R	1.10	1.50	1.90
HC—C—(C=O)NR_2	1.10	1.50	1.80
HC—C—(C=O)Ar	1.15	1.55	1.90
HC—C—(C=O)OR	1.15	1.70	1.90
HC—C—Ar	1.15	1.55	1.80
HC—C—OH	1.20	1.50	1.75
HC—C—OR	1.20	1.50	1.75
HC—C—C≡CR	1.20	1.50	1.80
HC—C—C≡N	1.25	1.65	2.00
HC—C—SR	1.25	1.60	1.90
HC—C—OAr	1.30	1.55	2.00
HC—C—O(C=O)R	1.30	1.60	1.80
HC—C—SH	1.30	1.60	1.65
HC—C—(S=O)R and HC—C—SO_2R	1.35	1.70	
HC—C—NR_3^+	1.40	1.75	2.05
HC—C—O—N=O	1.40		
HC—C—O(C=O)CF_3	1.40	1.65	
HC—C—Cl	1.55	1.80	1.95
HC—C—F	1.55	1.85	2.15
HC—C—NO_2	1.60	2.05	2.50

(Continued)

TABLE 12.4 Proton Chemical Shifts (*Continued*)

Substituent group	Methyl protons	Methylene protons	Methine proton
HC—C—O(C=O)Ar	1.65	1.75	1.85
HC—C—I	1.75	1.80	2.10
HC—C—Br	1.80	1.85	1.90
HC—CH_2	0.90	1.30	1.50
HC—C=C	1.60	2.05	
HC—C≡C	1.70	2.20	2.80
HC—(C=O)OR	2.00	2.25	2.50
HC—(C=O)NR_2	2.00	2.25	2.40
HC—SR	2.05	2.55	3.00
HC—O—O	2.10	2.30	2.55
HC—(C=O)R	2.10	2.35	2.65
HC—C≡N	2.15	2.45	2.90
HC—I	2.15	3.15	4.25
HC—CHO	2.20	2.40	
HC—Ar	2.25	2.45	2.85
HC—NR_2	2.25	2.40	2.80
HC—SSR	2.35	2.70	
HC—(C=O)Ar	2.40	2.70	3.40
HC—SAr	2.40		
HC—NRAr	2.60	3.10	3.60
HC—SO_2R and HC—(SO)R	2.60	3.05	
HC—Br	2.70	3.40	4.10
HC—NR_3^+	2.95	3.10	3.60
HC—NH(C=O)R	2.95	3.35	3.85
HC—SO_3R	2.95		
HC—Cl	3.05	3.45	4.05
HC—OH and HC—OR	3.20	3.40	3.60
HC—PAr_3	3.20	3.40	
HC—NH_2	3.50	3.75	4.05
HC—O(C=O)R	3.65	4.10	4.95
HC—OAr	3.80	4.00	4.60
HC—O(C=O)Ar	3.80	4.20	5.05
HC—O(C=O)CF_1	3.95	4.30	
HC—F	4.25	4.50	4.80
HC—NO_2	4.30	4.35	4.60
Cyclopropane		0.20	0.40
Cyclobutane		2.45	
Cyclopentane		1.65	
Cyclohexane		1.50	1.80
Cycloheptane		1.25	

Substituent group	Proton shift	Substituent group	Proton shift
HC≡CH	2.35	HO—C=O	10–12
HC≡CAr	2.90	HO—SO_2	11–12
HC≡C—C=C	2.75	HO—Ar	4.5–6.5
HAr	7.20	HO—R	0.5–4.5
HCO—O	8.1	HS—Ar	2.8–3.6
HCO—R	9.4–10.0	HS—R	1–2
HCO—Ar	9.7–10.5	HN—Ar	3–6
HO—N=C (oxime)	9–12	HN—R	0.5–5

TABLE 12.4 Proton Chemical Shifts (*Continued*)

Saturated heterocyclic ring systems

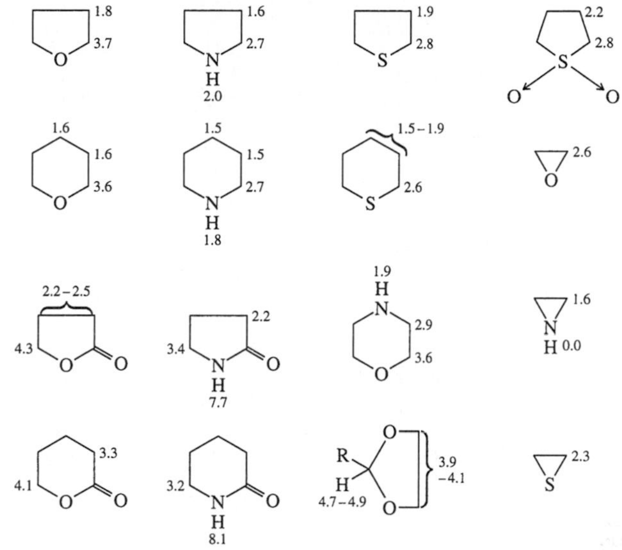

1.8 3.7 O
1.6 2.7 N H 2.0
1.9 2.8 S
2.2 2.8 S O O

1.6 1.6 3.6 O
1.5 1.5 2.7 N H 1.8
1.5–1.9 2.6 S
2.6 O

2.2–2.5 4.3 O O
2.2 3.4 N O H 7.7
1.9 H N 2.9 3.6 O
1.6 N H 0.0

3.3 4.1 O O
3.2 N O H 8.1
O R H 4.7–4.9 O 3.9 –4.1
2.3 S

Unsaturated cyclic systems

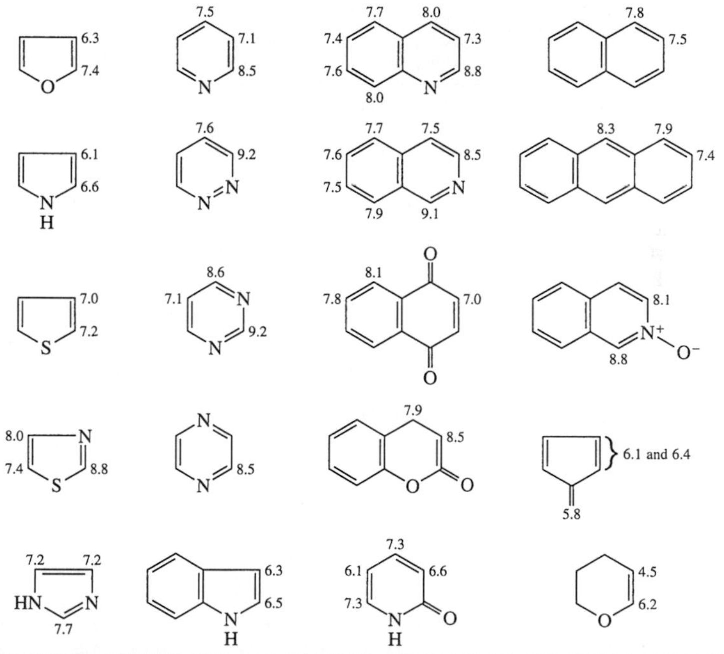

6.3 7.4 O
7.5 7.1 8.5 N
7.7 8.0 7.4 7.3 7.6 8.8 N 8.0
7.8 7.5

6.1 6.6 N H
7.6 9.2 N N
7.7 7.5 7.6 8.5 7.5 N 7.9 9.1
8.3 7.9 7.4

7.0 7.2 S
8.6 7.1 N 9.2 N
O 8.1 7.8 7.0 O
8.1 N+ O− 8.8

8.0 N 7.4 8.8 S
N 8.5 N
7.9 8.5 O O
6.1 and 6.4 5.8

7.2 7.2 HN N 7.7
6.3 6.5 N H
7.3 6.1 6.6 7.3 N O H
4.5 6.2 O

Source: J. A. Dean, ed., *Lange's Handbook of Chemistry*, 14th ed., McGraw-Hill, New York, 1992.

TABLE 12.5 Chemical Shifts in Monosubstituted Benzene

$$\delta = 7.27 + \Delta_i$$

Substituent	Δ_{ortho}	Δ_{meta}	Δ_{para}
NO_2	0.94	0.18	0.39
CHO	0.58	0.20	0.26
COOH	0.80	0.16	0.25
$COOCH_3$	0.71	0.08	0.20
COCl	0.82	0.21	0.35
CCl_3	0.8	0.2	0.2
$COCH_3$	0.62	0.10	0.25
CN	0.26	0.18	0.30
$CONH_2$	0.65	0.20	0.22
$\overset{+}{N}H_3$	0.4	0.2	0.2
CH_2X*	0.0–0.1	0.0–0.1	0.0–0.1
CH_3	–0.16	–0.09	–0.17
CH_2CH_3	–0.15	–0.06	–0.18
$CH(CH_3)_2$	–0.14	–0.09	–0.18
$C(CH_3)_2$	–0.09	0.05	–0.23
F	–0.30	–0.02	–0.23
Cl	0.01	–0.06	–0.08
Br	0.19	–0.12	–0.05
I	0.39	–0.25	–0.02
NH_2	–0.76	–0.25	–0.63
OCH_3	–0.46	–0.10	–0.41
OH	–0.49	–0.13	–0.2
OCOR	–0.2	0.1	–0.2
$NHCH_3$	–0.8	–0.3	–0.6
$N(CH_3)_2$	–0.60	–0.10	–0.62

* X = Cl, alkyl, OH, or NH_2.
Source: J. A. Dean, ed., *Lange's Handbook of Chemistry*, 14th ed., McGraw-Hill, New York, 1992.

TABLE 12.6 Estimation of Chemical Shift for Protons of —CH_2— and Methine Groups

$\delta_{CH_2} = 0.23 + C_1 + C_2$ $\qquad$ $\delta_{CH} = 0.23 + C_1 + C_2 + C_3$

X*	C	X*	C	X*	C
—CH_3	0.5	—SR	1.6	—OR	2.4
—CF_3	1.1	—C≡C—Ar	1.7	—Cl	2.5
>C=C<	1.3	—CN	1.7	—OH	2.6
—C≡C—R	1.4	—CO—R	1.7	—N=C=S	2.9
—COOR	1.5	—I	1.8	—OCOR	3.1
—NR_2	1.6	—Ph	1.8	—OPh	3.2
—$CONR_2$	1.6	—Br	2.3		

* R, alkyl group; Ar, aryl group; Ph, phenyl group.
Source: J. A. Dean, ed., *Lange's Handbook of Chemistry*, 14th ed., McGraw-Hill, New York, 1992.

TABLE 12.7 Estimation of Chemical Shift for Proton Attached to a Double Bond

Positive Z values indicate a downfield shift, and an arrow indicates the point of attachment of the substituent group to the double bond.

$$\delta_{\mathrm{C=C\diagdown H}} = 5.25 - Z_{gem} + Z_{cis} + Z_{trans}$$

R_{cis}(H)C=C(R_{trans})(T_{gem})

R	Z_{gem}, ppm	Z_{cis}, ppm	Z_{trans}, ppm
→H	0	0	0
→alkyl	0.45	−0.22	−0.28
→alkyl—ring (5- or 6-member)	0.69	−0.25	−0.28
→CH_2O—	0.64	−0.01	−0.02
→CH_2S—	0.71	−0.13	−0.22
→CH_2X (X: F,Cl,Br)	0.70	0.11	−0.04
→CH_2N<	0.58	−0.10	−0.08
>C=C (isolated)	1.00	−0.09	−0.23
>C=C (conjugated)	1.24	0.02	−0.05
→C≡N	0.27	0.75	0.55
→C≡C—	0.47	0.38	0.12
>C=O (isolated)	1.10	1.12	0.87
>C=O (conjugated)	1.06	0.91	0.74
→COOH (isolated)	0.97	1.41	0.71
→COOH (conjugated)	0.80	0.98	0.32
→COOR (isolated)	0.80	1.18	0.55
→COOR (conjugated)	0.78	1.01	0.46
→C(H)=O	1.02	0.95	1.17
→C(N<)=O	1.37	0.98	0.46
→C(Cl)=O	1.11	1.46	1.01
→OR (R: aliphatic)	1.22	−1.07	−1.21
→OR (R: conjugated)	1.21	−0.60	−1.00
→OCOR	2.11	−0.35	−0.64
→CH_2—C(|)=O; →CH_2—C≡N	0.69	−0.08	−0.06
→CH_2—aromatic ring	1.05	−0.29	−0.32
→F	1.54	−0.40	−1.02
→Cl	1.08	0.18	0.13
→Br	1.07	0.45	0.55
→I	1.14	0.81	0.88
→N—R (R: aliphatic)	0.80	−1.26	−1.21

(Continued)

TABLE 12.7 Estimation of Chemical Shift for Proton Attached to a Double Bond (*Continued*)

R	Z_{gem}, ppm	Z_{cis}, ppm	Z_{trans}, ppm
→N—R (R: conjugated)	1.17	−0.53	−0.99
→N—C=O	2.08	−0.57	−0.72
→aromatic	1.38	0.36	−0.07
→CF_3	0.66	0.61	0.32
→aromatic (*o*-substituted)	1.65	0.19	0.09
→SR	1.11	−0.29	−0.13
→SO_2	1.55	1.16	0.93

Source: J. A. Dean, ed., *Lange's Handbook of Chemistry*, 14th ed., McGraw-Hill, New York, 1992.

TABLE 12.8 Solvent Positions of Residual Protons in Incompletely Deuterated Solvents

Relative to tetramethylsilane.

Solvent	Group	δ, ppm
Acetic-d_3 acid-d_1	Methyl	2.05
	Hydroxyl	11.5*
Acetone-d_6	Methyl	2.057
Acetonitrile-d_3	Methyl	1.95
Benzene-d_6	Methine	6.78
tert-Butanol-d_1 $(CH_3)_3COD$	Methyl	1.28
Chloroform-d_1	Methine	7.25
Cyclohexane-d_{12}	Methylene	1.40
Deuterium oxide	Hydroxyl	4.7*
N,N-Dimethyl-d_6-formamide-d_1	Methyl	2.75; 2.95
	Formyl	8.05
Dimethyl-d_6 sulfoxide	Methyl	2.51
	Absorbed water	3.3*
1,4-Dioxane-d_8	Methylene	3.55
Hexamethyl-d_{18}-phosphoramide	Methyl	2.60
Methanol-d_4	Methyl	3.35
	Hydroxyl	4.8*
Dichloromethane-d_2	Methylene	5.35
Pyridine-d_5	C-2 Methine	8.5
	C-3 Methine	7.0
	C-4 Methine	7.35
Toluene-d_8	Methyl	2.3
	Methine	7.2
Trifluoroacetic acid-d_1	Hydroxyl	11.3*

* These values may vary greatly, depending upon the solute and its concentration.

Source: J. A. Dean, ed., *Lange's Handbook of Chemistry*, 14th ed., McGraw-Hill, New York, 1992.

The location of proton resonances for C—H bonds may be summarized as follows:

1. When only aliphatic groups are substituents, the range is $\delta = 0.9$ to 1.5.
2. Protons in CH_2 and CH groups appear slightly further downfield (in that order) from protons in CH_3 groups.
3. An adjacent, unsaturated bond (whether alkene or aryl) shifts the resonance position of CH_3 to $\delta = 1.6$ to 2.7.
4. An adjacent nitrogen atom causes the proton resonance to lie in the range 2.1 to 2.3 for alkyl—N—CH groups, 2.8 to 3.0 for aryl—N—CH groups, and 2.8 to 3.1 for amide groups.
5. An adjacent oxygen atom markedly shifts proton signals downfield to $\delta = 3.2$ to 3.4 for aliphatic entities and to $\delta = 3.6$ to 3.9 for aryl—O—CH situations.
6. Aryl protons are found at $\delta = 7.2$.
7. Aldehyde protons in aliphatic compounds lie in the range $\delta = 9.4$ to 10.0 and those for aryls at $\delta = 9.7$ to 10.5.

12.4.2 Spin–Spin Coupling of Protons

The real power of NMR derives from its ability to define complete sequences of groups or arrangements of atoms in the molecule. The absorption band multiplicities (splitting patterns) give the spatial positions of the nuclei. These splitting patterns arise through reciprocal magnetic interaction between spinning nuclei in a molecular system facilitated by the strongly magnetic binding electrons of the molecule in the intervening bonds. This coupling, called *spin–spin coupling*, causes mutual splitting of the otherwise sharp resonance lines into multiplets. The strength of the spin–spin coupling, denoted by J, is given by the spacings between the individual lines of the multiplets, which will be the same in both coupled multiplets—an aid in structure elucidation. Proton–proton spin coupling constants are listed in Table 12.9.

Proton–proton couplings in aliphatic organic compounds are usually transmitted through only two or three bonds. Longer-range coupling occurs between protons separated by unsaturated systems. Couplings also depend on geometry. Adjacent axial–axial protons of cyclohexane are strongly coupled, whereas axial–equatorial and equatorial–equatorial protons are coupled only moderately. Coupling constants of *trans* and *cis* protons on olefinic double bonds have a ratio of approximately 2, which is useful in assigning structures of geometric isomers.

The number of lines in multiplets from protons is $n + 1$ lines, where n is the number of nuclei producing the splitting. Thus, one neighboring proton splits the observed resonance of a proton on an adjacent group into a doublet, the intensities of which are in the ratio 1:1, two produce a triplet (1:2:1), three a quartet (1:3:3:1), four a quintet (1:4:6:4:1), and so on. The peaks within a multiplet are proportional to the coefficients of the binomial expansion. The relative intensity of each of the multiplets, integrated over the whole multiplet, is proportional to the number of nuclei in the group—useful for quantitative work and for structural assignments.

As the external magnetic field strength increases, the multiplets move farther apart but the spacing of the peaks within each multiplet remains the same. This is one distinct advantage in the use of high-field strengths obtainable with 300-, 400-, and 500-MHz NMR spectrometers. The ratio J/δ is crucial and determines the appearance of the spectrum. If J/δ is 0.05 or less, the spectrum consists of well-separated multiplets with the theoretical intensities of the peaks. If J/δ is 0.15 or greater, the intensities are no longer binomial and the spectrum deviates noticeably from the ideal appearance. If the chemical shift difference vanishes, the multiplet collapses to a singlet. For example, a H_3C—O—CH_2—CH_2—O—CH_3 structure would show only a single peak and not a pair of triplets; the chemical shift for each methylene group is identical.

Other nuclei with spins of $^1/_2$ interact with protons (and each other) and cause observable spin–spin couplings. Only significant numbers of fluorine and phosphorus atoms occur naturally; the presence of

TABLE 12.9 Proton–Proton Spin Coupling Constants

Structure		J, Hz
>C(H)(H)		12–15
>CH—CH< (free rotation)		6–8
>CH—OH (no exchange)		5
>CH—NH		4–8
>CH—SH		6–8
>CH—C(H)=O		1–3
—N=C(H)(H)		8–16
H_t(H_c)C=C(H_g)H	gem	0–3
	cis	6–14
	trans	11–18
H_c(H_t)C=C(H_g)CH<	cis	0.5–3
	trans	0.5–3
	gem	4–10
>C=CH—CH=C<		10–13
=CH—C(H)=O		6
—CH_2—C≡C—CH<		0–3
>CH—C≡CH		0–3
HC=CH (ring)	3–member	0–2
	4–member	2–4
	5–member	5–7
	6–member	6–9
	7–member	10–13
[three-membered ring with O]	cis	4–5
	trans	3
	gem	5–6
[three-membered ring with S]	cis	0
	trans	7
	gem	6

Structure		J, Hz
[three-membered ring with N—H]	cis	2
	trans	6
	gem	4
[five-membered ring with O; positions 2–5]	2–3	1.8
	3–4	3.5
	2–4	0–1
	2–5	1–2
[five-membered ring with S; positions 2–5]	2–3	5–6
	3–4	3.5–5.0
	2–4	1.5
	2–5	3.4
[five-membered ring with N—H; positions 1–5]	1–2	2–3
	1–3	2–3
	2–3	2–3
	3–4	3–4
	2–4	1–2
	2–5	1–3
[chair six-membered ring with axial and equatorial H]	a–a	8–10
	a–e	2–3
	e–e	2–3
Cyclopentane	cis	4–6
	trans	4–6
Cyclobutane	cis	8
	trans	8
Cyclopropane	cis	9–11
	trans	6–8
	gem	4–6
[benzene ring with H]	o	6–10
	m	1–3
	p	0–1
[naphthalene; positions 1, 2, 3]	1–2	8–9
	2–3	6
[six-membered ring with N; positions 2–6]	2–3	5–6
	3–4	7–9
	2–4	1–2
	3–5	1–2
	2–5	0–1
	2–6	0–1

Source: J. A. Dean, ed., *Lange's Handbook of Chemistry*, 14th ed., McGraw-Hill, New York, 1992.

TABLE 12.10 Hydrogen-1 to Fluorine-19 Spin Coupling Constants

Structure	J_{HF}, Hz
$H_{(2)}C—CFH_{(1)}$	
$H_{(1)}—F$	48
$H_{(2)}—F$	26
$H_{(2)}C—CH_{(1)}F_2$	
$H_{(1)}—F$	57
$H_{(2)}—F$	21
$HC—CF_3$	13
$H_{(2)}F_2(2)C—CF_{2(1)}H_{(1)}$	
$H_{(1)}—F_{(1)}$	52
$H_{(1)}—F_{(2)}$	5
$F_{3(2)}C—CF_{(1)}H_{2(1)}$	
$H_{(1)}—F_{(1)}$	46
$H_{(1)}—F_{(2)}$	8
$F_3C—CH_2X$ (X = Cl, Br, OH)	8.5–9
$CH_3—CF_2X$ (X = Cl, Br)	15–16
$HFC{=}CCl_2$	81
$H_2C{=}CF_2$	
cis	1
trans	34
$HFC{=}CFH$	
cis	4
trans	20
gem	74
$H_2C{=}CFH$	
cis	20
trans	52
gem	85
$F_2C{=}CHCl$	
cis	<3
trans	13
$H(CH_3)C{=}CFH$	
cis	20
trans	40
gem	85–90
$FC{\equiv}CH$	21
$H—C{=}C—C—F$	0–6
Fluorobenzene	
ortho	6–10
meta	4–8
para	0–2.5
4-Fluorotoluene	
ortho	7
meta	9
2,2-Difluorocyclohexane	
H_a-F_a	34
H_e-F_e	12
H_a-F_e	5–8
Fluorocyclohexane	
gem	49
H_a-F	44
H_e-F	<3
$H_2C{=}CFCl$	
cis	8
trans	40
$F_2C{=}CFH$	
cis	<3
trans	12
gem	72

one of these elements may be deduced from an otherwise unexplained coupling effect in a proton spectrum. Hydrogen to fluorine-19 spin coupling constants are given in Table 12.10. Phosphorus-31 to hydrogen-1 spin coupling constants are in Table 12.37.

12.4.3 Chemical Shifts of Carbon-13

In a general way, the chemical shifts of carbon-13 atoms (Tables 12.11 through 12.15) follow the same shielding and deshielding characteristics that are found in proton NMR spectra; the major exception is that the ring current effects observed for aromatic protons are not observed for aromatic carbons. The hybridization of a carbon determines to a great extent the range within which its ^{13}C NMR signal is found. The ^{13}C resonances of sp^3 carbon atoms absorb at highest field (–20 to 100 ppm), followed by sp carbon atoms (70 to 130 ppm), while sp^2 hybridized centers are shifted farthest (120 to 240 ppm) in organic compounds.

Electron-withdrawing substituents, heteroatoms, and alkyl groups that are adjacent to a carbon shifts the ^{13}C signal downfield. The $^{13}C—X$ signal (C–α, and to a lesser extent, C–β) shifts to a lower field with increasing electronegativity of the constituent X. However, upfield shifts are observed for C–γ. The influence of alkyl groups is additive in alkanes and can be approximated by the addition of the ^{13}C shift increments of all neighboring alkyl substituents (Tables 12.16 and 12.17). Increasing the crowding of alkyl or electronegative substituents at a carbon atom causes a successive downfield shift of its ^{13}C resonance; whereas crowding due to iodine causes upfield shifts.

Unshared electron pairs at a carbon atom cause downfield shifts of its ^{13}C resonance. Electron deficiency at a carbon atom causes drastic deshielding. Examples are the *sp*^{2} carbons (typical of carbenium ions and for carbene metal complexes) and for the *sp* carbons of metal carbonyls.

Configuration of substituents will also influence ^{13}C shifts. Carbons bearing equatorial substituents are more strongly deshielded than those bearing axial substitutents; the shift difference is about 5 ppm. Lower ^{13}C shift differences are observed for the olefinic carbons of *cis* and *trans* isomers in branched and substituent alkenes. The *trans* configuration often causes a stronger deshielding of carbon atoms that are in an alpha position to the double bond.

Downfield shifts occur when going from nonpolar solvents to those susceptible to hydrogen bonding. To avoid confusion when using deuterated solvents, the carbon-13 chemical shifts of deuterated solvents are given in Table 12.18. Secondary references for solid-state ^{13}C NMR are summarized in Table 12.19.

12.4.4 ^{13}C Coupling Constants

One-bond carbon–hydrogen spin coupling constants are given in Table 12.20, and two-bond carbon–hydrogen spin coupling constants are given in Table 12.21. Table 12.22 gives carbon–carbon spin coupling constants. Spin coupling constants for carbon–fluorine are given in Table 12.23. Carbon-13 spin coupling constants with various other nuclei are given in Table 12.24.

TABLE 12.11 Carbon-13 Chemical Shifts

Values given in ppm on the δ scale, relative to tetramethylsilane.

Substituent group	Primary carbon	Secondary carbon	Tertiary carbon	Quaternary carbon
Alkanes				
C—C	5–30	25–45	23–58	28–50
C—O	45–60	42–71	62–78	73–86
C—N	13–45	44–58	50–70	60–75
C—S	10–30	22–42	55–67	53–62
C—halide (I to Cl)	3–25	3–40	34–58	35–75

Substituent group	δ, ppm	Substituent group	δ, ppm
Cyclopropane	−5–5	Sulfoxides, sulfones	35–55
Cycloalkane C_4–C_{10}	5–25	Alcohols R—OH	45–87
Mercaptanes	5–70	Ethers R—O—R	57–87
Amines		Nitro R—NO_2	60–78
R_2N—C	20–70		
Aryl—N	128–138		

TABLE 12.11 Carbon-13 Chemical Shifts (*Continued*)

Substituent group	δ, ppm	Substituent group	δ, ppm
Alkynes		Amides	154–178
HC≡CR	63–73	Oximes	155–165
RC≡CR	72–95	Esters	
Acetals, ketals	88–112	Saturated	158–165
Thiocyanates R—SCN	96–118	α,β-Unsaturated	165–176
Alkenes		Isocyanides R—NC	162–175
H_2C=	100–122	Carboxylic acids	
R_2C=	110–150	Nonconjugated	162–165
Heteroaromatics		Conjugated	165–184
C=N	100–152	Salts (anion)	175–195
C_α	142–160	Ketones	
Cyanates R—OCN	105–120	α-Halo	160–200
Isocyanates R—NCO	115–135	Nonconjugated	192–202
Isothiocyanates R—NCS	115–142	α,β-Unsaturated	202–220
Nitriles, cyanides	117–124	Imides	165–180
Aromatics		Acyl chlorides	
Aryl-C	125–145	R—CO—Cl	165–183
Aryl-P	119–128	Thioureas	165–185
Aryl-N	128–138	Aldehydes	
Aryl-O	133–152	α-Halo	170–190
Azomethines	145–162	Nonconjugated	182–192
Carbonates	159–162	Conjugated	192–208
Ureas	150–170	Thioketones R—CS—R	190–202
Anhydrides	150–175	Carbonyl $M(CO)_n$	190–218
		Allenes =C=	197–205

Saturated heterocyclic ring systems

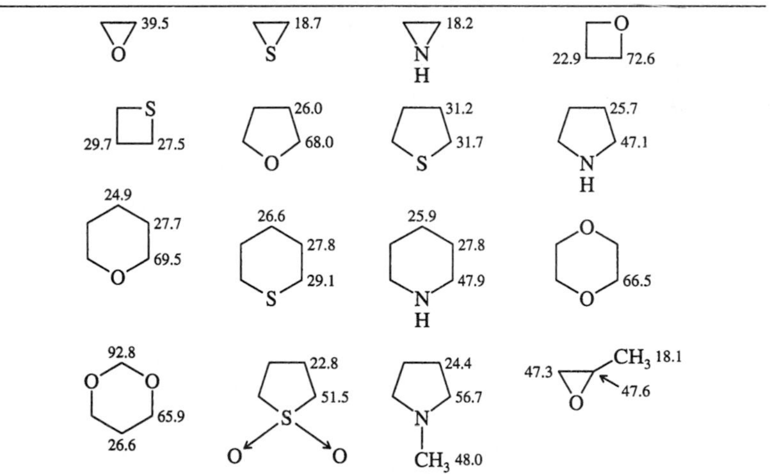

(Continued)

TABLE 12.11 Carbon-13 Chemical Shifts (*Continued*)

Unsaturated cyclic systems

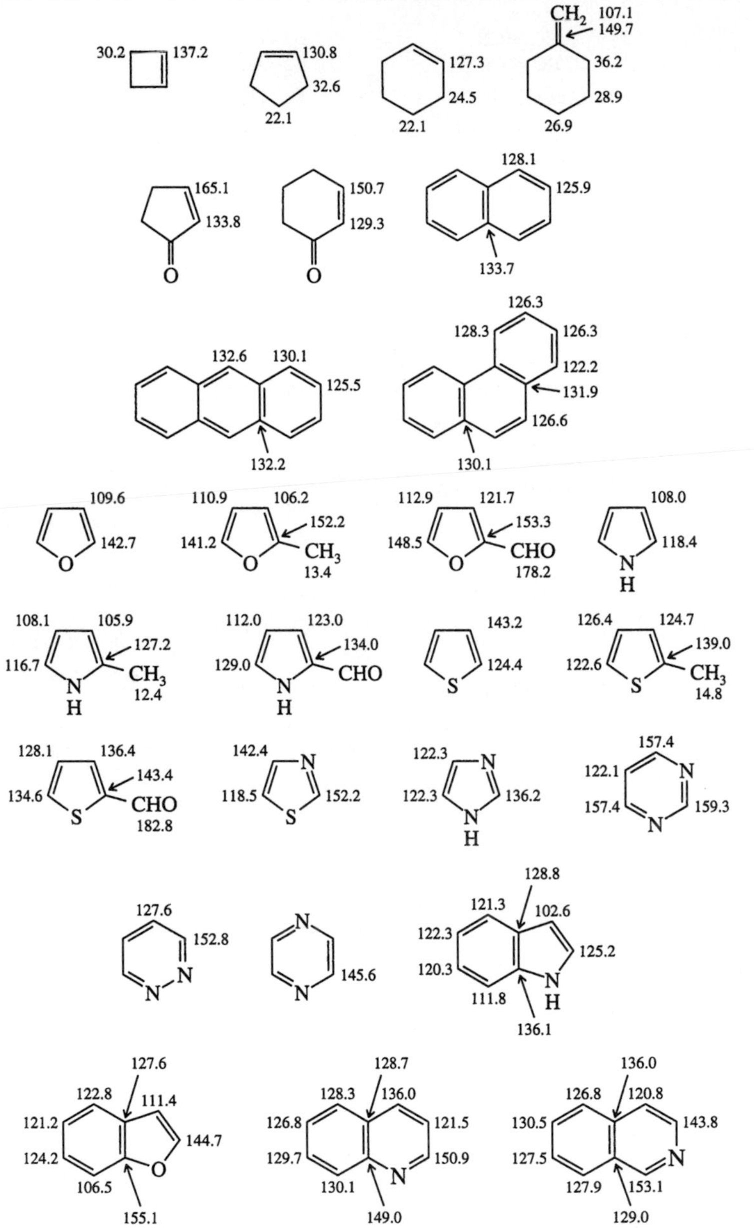

30.2 137.2 130.8 32.6 22.1 127.3 24.5 22.1 CH_2 107.1 149.7 36.2 28.9 26.9

165.1 133.8 O 150.7 129.3 O 128.1 125.9 133.7

132.6 130.1 125.5 132.2 126.3 128.3 126.3 122.2 131.9 126.6 130.1

109.6 142.7 O 110.9 106.2 152.2 141.2 O CH_3 13.4 112.9 121.7 153.3 148.5 O CHO 178.2 108.0 118.4 N H

108.1 105.9 127.2 116.7 N H CH_3 12.4 112.0 123.0 134.0 129.0 N H CHO 143.2 124.4 S 126.4 124.7 139.0 122.6 S CH_3 14.8

128.1 136.4 143.4 134.6 S CHO 182.8 142.4 N 118.5 S 152.2 122.3 N 122.3 136.2 N H 157.4 122.1 N 157.4 N 159.3

127.6 152.8 N N N N 145.6 128.8 121.3 102.6 122.3 125.2 120.3 N H 111.8 136.1

127.6 122.8 111.4 121.2 144.7 124.2 O 106.5 155.1 128.7 128.3 136.0 126.8 121.5 129.7 150.9 N 130.1 149.0 136.0 126.8 120.8 130.5 143.8 127.5 N 127.9 153.1 129.0

TABLE 12.11 **Carbon-13 Chemical Shifts (*Continued*)**

Saturated alicyclic ring systems

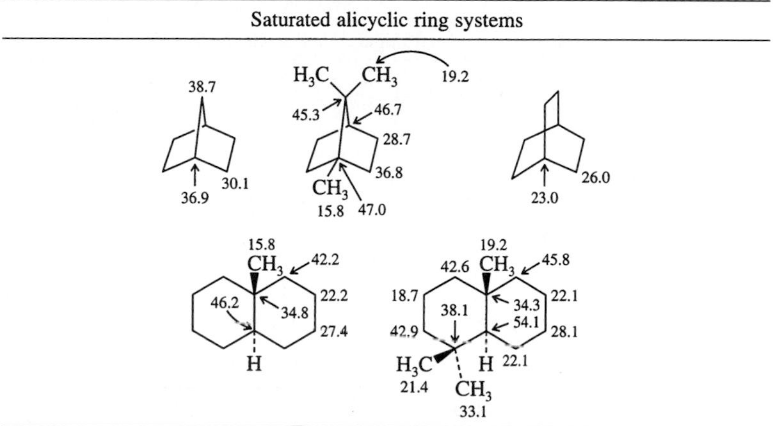

Source: J. A. Dean, ed., *Lange's Handbook of Chemistry*, 14th ed., McGraw-Hill, New York, 1992.

TABLE 12.12 **Effect of Substituent Groups on Alkyl Chemical Shifts**

These increments are added to the shift value of the appropriate carbon atom as calculated from Table 12.16.

Straight: Y—CH_2—CH_2—CH_3 (α, β)

Branched: —CH_2—CH_2—CH(Y)—CH_2—CH_2— (γ, β, α, β, γ)

	α-carbon		β-carbon		
Substituent group Y*	Straight	Branched	Straight	Branched	γ-carbon
—CO—OH	20.9	16	2.5	2	−2.2
—COO⁻ (anion)	24.4	20	4.1	3	−1.6
—CO—OR	20.5	17	2.5	2	−2
—CO—Cl	33	28		2	
—CO—NH_2	22	2.5			−0.5
—CHO	31		0		−2
—CO—R	30	24	1	1	−2
—OH	48.3	40.8	10.2	7.7	−5.8
—OR	58	51	8	5	−4
—O—CO—NH_2	51		8		
—O—CO—R	51	45	6	5	−3
—C—CO—Ar	53				
—F	68	63	9	6	−4
—Cl	31.2	32	10.5	10	−4.6
—Br	20.0	25	10.6	10	−3.1
—I	−8	4	11.3	12	−1.0
—NH_2	29.3	24	11.3	10	−4.6
—NH_3^+	26	24	8	6	−5
—NHR	36.9	31	8.3	6	−3.5
—NR_2	42		6		−3
—NR_3^+	31		5		−7
—NO_2	63	57	4	4	
—CN	4	1	3	3	−3

(*Continued*)

TABLE 12.12 Effect of Substituent Groups on Alkyl Chemical Shifts (*Continued*)

	α-carbon		β-carbon		
Substituent group Y*	Straight	Branched	Straight	Branched	γ-carbon
—SH	11	11	12	11	−6
—SR	20		7		−3
$—CH{=}CH_2$	20		6		−0.5
$—C_6H_5$	23	17	9	7	−2
—C≡CH	4.5		5.5		−3.5

*R, alkyl group; Ar, aryl group.
Source: J. A. Dean, ed., *Lange's Handbook of Chemistry*, 14th ed., McGraw-Hill, New York, 1992.

TABLE 12.13 Carbon-13 Chemical Shifts in Substituted Benzenes

$\delta_C = 128.5 + \Delta$

Substituent group	Δ_{C-1}	Δ_{ortho}	Δ_{meta}	Δ_{para}
$—CH_3$	9.3	0.8	−0.1	−2.9
$—CH_2CH_3$	15.6	−0.4	0	−2.6
$—CH(CH_3)_2$	20.2	−2.5	0.1	−2.4
$—C(CH_3)_3$	22.4	−3.1	−0.1	−2.9
$—CH_2O—CO—CH_3$	7.7	0	0	0
$—C_6H_5$	13.1	−1.1	0.4	−1.2
$—CH{=}CH_2$	9.5	−2.0	0.2	−0.5
—C≡CH	−6.1	3.8	0.4	−0.2
$—CH_2OH$	12.3	−1.4	−1.4	−1.4
—CO—OH	2.1	1.5	0	5.1
$—COO^-$ (anion)	8	1	0	3
$—CO—OCH_3$	2.1	1.1	0.1	4.5
$—CO—CH_3$	9.1	0.1	0	4.2
—CHO	8.6	1.3	0.6	5.5
—CO—Cl	4.6	2.4	1	6.2
$—CO—CF_3$	−5.6	1.8	0.7	6.7
$—CO—C_6H_5$	9.4	1.7	−0.2	3.6
—CN	−15.4	3.6	0.6	3.9
—OH	26.9	−12.7	1.4	−7.3
$—OCH_3$	31.4	−14.0	1.0	−7.7
$—OC_6H_5$	29.2	−9.4	1.6	−5.1
$—O—CO—CH_3$	23.0	−6.4	1.3	−2.3
$—NH_2$	18.0	−13.3	0.9	−9.8
$—N(CH_3)_2$	22.4	−15.7	0.8	−11.5
$—N(C_6H_5)_2$	19	−4	1	−6
$—NHC_6H_5$	14.6	−10.7	0.7	−7.7
$—NH—CO—CH_3$	11.1	−9.9	0.2	−5.6
$—NO_2$	20.0	−4.8	0.9	5.8
—F	34.8	−12.9	1.4	−4.5
—Cl	6.2	0.4	1.3	−1.9
—Br	−5.5	3.4	1.7	−1.6
—I	−32.2	9.9	2.6	−1.4
$—CF_3$	−9.0	−2.2	0.3	3.2
—NCO	5.7	−3.6	1.2	−2.8
—SH	2.3	1.1	1.1	−3.1
$—SCH_3$	10.2	−1.8	0.4	−3.6
$—SO_2—NH_2$	15.3	−2.9	0.4	3.3
$—Si(CH_3)_3$	13.4	4.4	−1.1	−1.1

Source: J. A. Dean, ed., *Lange's Handbook of Chemistry*, 14th ed., McGraw-Hill, New York, 1992.

TABLE 12.14 Carbon-13 Chemical Shifts of the Carbonyl Group

$$X-\overset{\overset{\displaystyle O}{\|}}{C}-Y$$

X	Y	δ_C	X	Y	δ_C
H—	—CH_3	199.7	CH_3—	—CH=CH_2	196.9
H—	—CCl_3	175.3	CH_3—	—C_6H_5	197.6
H—	—NH_2	165.5	CH_3—	—CH_2—CO—CH_3	201.9 (keto)
H—	—$N(CH_3)_2$	162.4			191.4 (enol)
H—	2-Furyl	153.3	CH_3—	—CH_2CHO	167.7
H—	2-Pyrrolyl	134.0	CH_3—	—C_6H_5—CH_3	196 (*m*, *p*)
H—	2-Thienyl	143.3			199 (*o*)
$(CH_3)_2CH$—	—OH	184.8	CH_3—	—2,6-$(CH_3)_2C_6H_5$	206
C_6H_5—	—OH	172.6	CH_3—	—OH	178
CF_3—	—OH	163.0	CH_3—	—O^- (anion)	181.5
CCl_3—	—OH	168.0	CH_3—	—OCH_3	170.7
$CH_3CH(NH_2)$—	—OH	176.5	CH_3—	—O—CH=CH_2	167.7
CF_3—	—OCH_2CH_3	158.1	CH_3—	—O—$CH(CH_3)_2$	170.3
H_2N—	—OCH_2CH_3	157.8	CH_3—	—O—CO—CH_3	167.3
2-Furyl	—OCH_3	159.1	CH_3—	—NH_2	172.7
$(CH_3)_2N$—	—C_6H_5	170.8	CH_3—	—$NHCH_3$	172
CH_2=$CHCH_2O$—CO—	—OCH_2CH=CH_2	157.6	CH_3—	—$N(CH_3)_2$	169.5
CH_3CH_2—	—CH_2CH_3	211.4	CH_3—	—Cl	169.6
CH_3—CH_2—	—O—CO—CH_2CH_3	170.3	CH_3—	—Br	165.6
CH_3—	—CH_3	205.8	CH_3—	—I	158.9
CH_3—	—CH_2CH_3	207			

$(CH_2)_n$ C=O (cyclic ketones)

n	δ_C
3	207.9
4	218.2
5	211.3
6	211.4
7	216.0

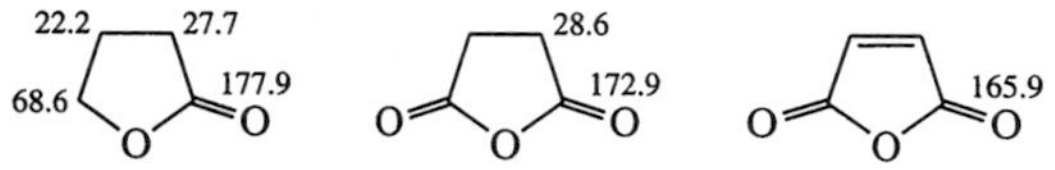

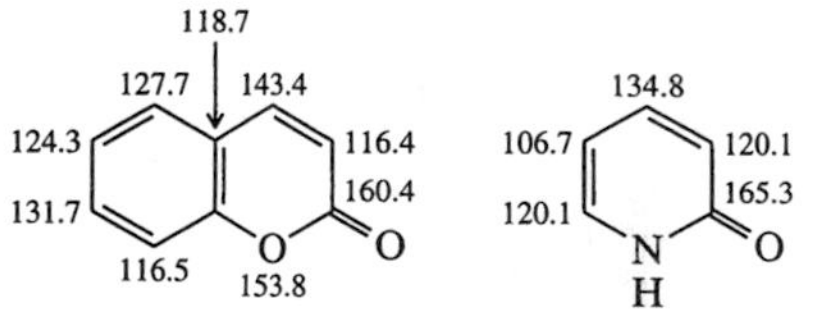

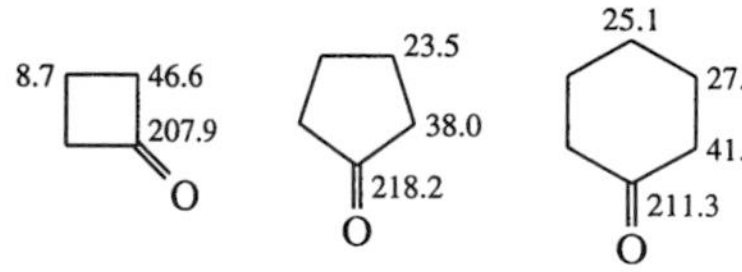

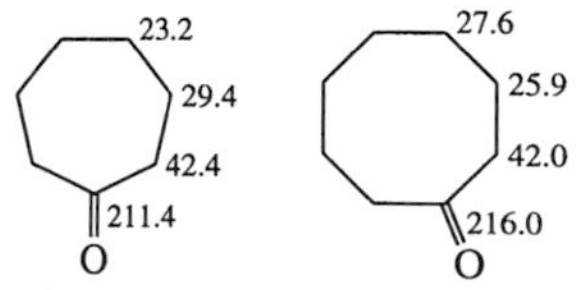

Source: J. A. Dean, ed., *Lange's Handbook of Chemistry*, 14th ed., McGraw-Hill, New York, 1992.

TABLE 12.15 Carbon-13 Chemical Shifts in Substituted Pyridines*

$$\delta_C(k) = C_k + \Delta_i$$

Substituent group	$C_2 = C_6 = 149.6$ ppm $\Delta_{C\text{-}2}$ or $\Delta_{C\text{-}6}$	Δ_{23}	Δ_{24}	Δ_{25}	Δ_{26}
—CH_3	9.1	−1.0	−0.1	−3.4	−0.1
—CH_2CH_3	14.0	−2.1	0.1	−3.1	0.2
—CO—CH_3	4.3	−2.8	0.7	3.0	−0.2
—CHO	3.5	−2.6	1.3	4.1	0.7
—OH	14.9	−17.2	0.4	−3.1	−6.8
—OCH_3	15.3	−13.1	2.1	−7.5	−2.2
—NH_2	11.3	−14.7	2.3	10.6	−0.9
—NO_2	8.0	−5.1	5.5	6.6	0.4
—CN	−15.8	5.0	−1.7	3.6	1.9
—F	14.4	−14.7	5.1	−2.7	−1.7
—Cl	2.3	0.7	3.3	−1.2	0.6
—Br	−6.7	4.8	3.3	−0.5	1.4

Substituent group	Δ_{32}	$C_3 = C_5 = 124.2$ ppm $\Delta_{C\text{-}3}$ or $\Delta_{C\text{-}5}$	Δ_{34}	Δ_{35}	Δ_{36}
—CH_3	1.3	9.0	0.2	−0.8	−2.3
—CH_2CH_3	0.3	15.0	−1.5	−0.3	−1.8
—CO—CH_3	0.5	−0.3	−3.7	−2.7	4.2
—CHO	2.4	7.9	0	0.6	5.4
—OH	−10.7	31.4	−12.2	1.3	−8.6
—NH_2	−11.9	21.5	−14.2	0.9	−10.8
—CN	3.6	−13.7	4.4	0.6	4.2
—Cl	−0.3	8.2	−0.2	0.7	−1.4
—Br	2.1	−2.6	2.9	1.2	−0.9
—I	7.1	−28.4	9.1	2.4	0.3

Substituent group	$\Delta_{42} = \Delta_{46}$	$\Delta_{43} = \Delta_{45}$	$C_4 = 136.2$ ppm $\Delta_{C\text{-}4}$
—CH_3	0.5	0.8	10.8
—CH_2CH_3	0	−0.3	15.9
—CH=CH_2	0.3	−2.9	8.6
—CO—CH_3	1.6	−2.6	6.8
—CHO	1.7	−0.6	5.5
—NH_2	0.9	−13.8	19.6
—CN	2.1	2.2	−15.7
—Br	3.0	3.4	−3.0

* May be used for disubstituted, polyheterocyclic, and polynuclear systems if deviations due to steric and mesomeric effects are allowed for.

Source: J. A. Dean, ed., *Lange's Handbook of Chemistry*, 14th ed., McGraw-Hill, New York, 1992.

12.4.5 Boron-11

Table 12.25 gives the boron-11 chemical shifts; data are limited.

12.4.6 Nitrogen-15

Extensive coverage of nitrogen-15 chemical shifts is provided in Tables 12.26 and 12.27. These values are relative to NH_3 (liquid at 23°C), the reference standard generally employed in the older literature.

TABLE 12.16 Estimation of Chemical Shifts of Alkane Carbons

Relative to tetramethylsilane.
Positive terms indicate a downfield shift.

$$\delta_C = -2.6 + 9.1n_\alpha + 9.4n_\beta - 2.5n_\gamma + 0.3n_\delta + 0.1n_\epsilon$$
(plus any correction factors)

where n_α is the number of carbons bonded directly to the ith carbon atom and n_β, n_γ, n_δ, and n_ϵ are the number of carbon atoms two, three, four, and five bonds removed. The constant is the chemical shift for methane.

Chain branching*	Correction factor	Chain branching*	Correction factor
1°(3°)	−1.1	4°(1°)	−1.5
1°(4°)	−3.4	2°(4°)	−7.2
2°(3°)	−2.5	3°(3°)	−9.5
3°(2°)	−3.7	4°(2°)	−8.4

* 1° signifies a CH_3— group; 2°, a —CH_2— group; 3°, a >CH— group; and 4°, a >C< group. 1° (3°) signifies a methyl group bound to a >CH— group, and so on.

Examples: For 3-methylpentane, CH_3—CH_2—$CH(CH_3)$—CH_2—CH_3,

$\delta_{C-2} = -2.6 + 9.1(2) + 9.4(2) - 2.5 - 1(1)[2°(3°)] = 29.4$
$\delta_{C-3} = -2.6 + 9.1(3) + 9.4(2) + (2)[3°(2°)] = 36.2$

Source: J. A. Dean, ed., *Lange's Handbook of Chemistry*, 14th ed., McGraw-Hill, New York, 1992.

TABLE 12.17 Estimation of Chemical Shifts of Carbon Attached to a Double Bond

The olefinic carbon chemical shift is calculated from the equation

$$\delta_C = 123.3 + 10.6n_\alpha + 7.2n_\beta - 7.9n_{\alpha'} - 1.8n_{\beta'}$$
(plus any steric correction terms)

where n is the number of carbon atoms at the particular position, namely,

$$\begin{matrix}\beta & \alpha & \alpha' & \beta' \\ \text{C—} & \text{C=} & \text{C—} & \text{C}\end{matrix}$$

Substituents on both sides of the double bond are considered separately. Additional vinyl carbons are treated as if they were alkyl carbons. The method is applicable to alicyclic alkenes; in small rings carbons are counted twice, i.e., from both sides of the double bond where applicable. The constant in the equation is the chemical shift for ethylene. The effect of other substituent groups is tabulated below.

Substituent group	β	α	α'	β'
—OR	2	29	−39	−1
—OH	6			−1
—O—CO—CH_3	−3	18	−27	4
—CO—CH_3		15	6	
—CO—OH		5.2	9.1	
—CO—OR		6	7	

(Continued)

TABLE 12.17 Estimation of Chemical Shifts of Carbon Attached to a Double Bond (*Continued*)

Substituent group	β	α	α'	β'
—CN		−15.4	14.3	
—F		24.9	−34.3	
—Cl	−1	3.3	−5.4	2
—Br	0	−7.2	−0.7	2
—I		−37.4	7.7	
—C_6H_5		12	−11	

Substituent pair		Steric correction term
α, α'	*trans*	0
α, α'	*cis*	−1.1
α, α	*gem*	−4.8
$\alpha'\alpha'$		+2.5
β, β		+2.3

Source: J. A. Dean, ed., *Lange's Handbook of Chemistry*, 14th ed., McGraw-Hill, New York, 1992.

TABLE 12.18 Carbon-13 Chemical Shifts of Deuterated Solvents

Relative to tetramethylsilane.

Solvent	Group	δ, ppm
Acetic-d_3 acid-d_1	Methyl	20.0
	Carbonyl	205.8
Acetone-d_6	Methyl	28.1
	Carbonyl	178.4
Acetonitrile-d_3	Methyl	1.3
	Carbonyl	117.7
Benzene-d_6		128.5
Carbon disulfide		193
Carbon tetrachloride		97
Chloroform-d_1		77
Cyclohexane-d_{12}		25.2
Dimethyl sulfoxide-d_6		39.5
1,4-Dioxane-d_6		67
Formic-d_1 acid-d_1	Carbonyl	165.5
Methanol-d_4		47–49
Methylene chloride-d_2		53.8
Nitromethane-d_3		57.3
Pyridine-d_5	C_3, C_5	123.5
	C_4	135.5
	C_2, C_6	149.9
		13.6

Source: J. A. Dean, ed., *Lange's Handbook of Chemistry*, 14th ed., McGraw-Hill, New York, 1992.

TABLE 12.19 ^{13}C **Chemical Shift References for Solid-State NMR**

Compound	Chemical Shifts, ppm			
Adamantane	29.5	38.6		
4,4-Dimethyl-4-silapentane sodium sulfonate	0.4	18.5	20.4	57.4
Dodecamethylcyclohexasilane	−7.7			
Hexamethylbenzene	17.4	132.2		
Poly(dimethylsilane)	−6.0			
Poly(dimethylsiloxane)	1.5			
Polyethylene	33.6			
Poly(oxymethylene)	89.1			

Source: W. L. Earl and D. L. VanderHart, *J. Magn. Resonance* **48**:35 (1982).

TABLE 12.20 One-Bond Carbon–Hydrogen Spin Coupling Constants

Structure	J_{CH}, Hz
$H{-}CH_3$	125.0
$H{-}CH_2CH_3$	124.9
$CH_3{-}\underline{CH_2}{-}CH_3$	119.2
$H{-}C(CH_3)_2$	114.2
$H{-}CH_2CH_2OH$	126.9
$H{-}CH_2CH{=}CH_2$	122.4
$H{-}CH_2C_6H_5$	129.4
$H{-}CH_2C{\equiv}CH$	132.0
$H{-}CH_2CN$	136.1
$H{-}CH(CN)_2$	145.2
$H{-}CH_2{-}$halogen	149–152
$H{-}CHF_2$	184.5
$H{-}CHCl_2$	178.0
$H{-}CH_2NH_2$	133.0
$H{-}CH_2NH_3^+$	145.0
$H{-}CH_2OH$ (or $H{-}CH_2OR$)	140–141
$H{-}CH(OR)_2$	161–162
$H{-}C(OR)_3$	186
$H{-}C(OH)R_2$	143
$H{-}CH_2NO_2$	146.0
$H{-}CH(NO_2)_2$	169.4
$H{-}CH_2COOH$	130.0
$H{-}CH(COOH)_2$	132.0
$H{-}CH{=}CH_2$	156.2
$H{-}C(CH_3){=}C(CH_3)_2$	148.4
$H{-}CH{=}C(\textit{tert-}C_4H_9)_2$	152
$H{-}C(\textit{tert-}C_4H_9){=}C(\textit{tert-}C_4H_9)_2$	143
Methylenecycloalkane C_4–C_7	153–155
$H{-}CH{=}C{=}CH_2$	168
$H{-}C(C_6H_5){=}CH(C_6H_5)$	
cis	155
trans	151
Cyclopropene	220

Structure		J_{CH}, Hz
$(H_t)(H_c)C{=}C(H_g)(F)$	*gem*	200
	cis	159
	trans	162
$(H_t)(H_c)C{=}C(H_g)(Cl)$	*gem*	195
	cis	163
	trans	161
$(H_t)(H_c)C{=}C(H_g)(CHO)$	*gem*	162
	cis	157
	trans	162
$(H_t)(H_c)C{=}C(H_g)(CN)$	*gem*	177
	cis	163
	trans	165
$(H)(CH_3)C{=}N(OH)$	*cis*	163
	trans	177
$H{-}CH{=}O$; $CH_3{-}\underline{CH}{=}O$		172
$H_2N{-}CH{=}O$		188.3
$(CH_3)_2N{-}\underline{CH}{=}O$		191
$H{-}COOH$		222
$H{-}COO^-$ (anion)		195
$H{-}CO{-}OCH_3$		226
$H{-}CO{-}F$		267
$CH_3CH_2{-}O{-}\underline{CHO}$		225.6
$Cl_3{-}CHO$		207
$H{-}C{\equiv}CH$		249
$H{-}C{\equiv}CCH_3$		248
$H{-}C{\equiv}CC_6H_5$		251
$H{-}C{\equiv}CCH_2OH$		241

(Continued)

TABLE 12.20 One-Bond Carbon–Hydrogen Spin Coupling Constants (*Continued*)

Structure		J_{CH}, Hz	Structure		J_{CH}, Hz
H—CN		269	2,4,6-Trimethylpyridine		158
Cyclopropane		161	(ring: N, H)	2,5	183
Cyclobutane		136		3,4	170
Cyclopentane		131			
Cyclohexane		123			
Tetrahydrofuran	2,5	149	(ring: O)	2,5	201
	3,4	133		3,4	175
1,4-Dioxane		145			
Benzene		159	(ring: S)	2,5	185
Fluorobenzene	2,6	155		3,4	167
	3,5	163			
	4	161	(ring: N, N, H)	3,5	190
Bromobenzene	2,6	171		4	178
	3,5	164			
	4	161			
Benzonitrile	2,6	173	(ring: N, N, H)	2	208
	3,6	166		4	199
	4	163			
Nitrobenzene	2,6	171	(ring: N, N, N, H)		205
	3,5	167			
	4	163			
Mesitylene		154			
(ring: N)	2,6	170	(ring: N—N, N, N, H)		216
	3,5	163			
	4	152			

Source: J. A. Dean, ed., *Lange's Handbook of Chemistry*, 14th ed., McGraw-Hill, New York, 1992.

TABLE 12.21 Two-Bond Carbon–Hydrogen Spin Coupling Constants

Structure	$^2J_{CH}$, Hz	Structure		$^2J_{CH}$, Hz
$\underline{C}H_3—CH_2—H$	−4.5	$(CH_2)_n$ (ring) $C=CH_2$	$n = 4$	4.2
$\underline{C}Cl_3—CH_2—H$	5.9		$n = 5$	5.2
$ClCH_2—CH_2Cl$	−3.4		$n = 6$	5.5
$Cl_2CH—CHCl_2$	1.2	$\underline{H}(Cl)C=\underline{C}(H)Cl$	*cis*	16.0
$\underline{C}H_3—C\underline{H}O$	26.7		*trans*	0.8
$CH_2=CH_2$	−2.4			
$(C\underline{H}_3)_2\underline{C}=O$	5.5			
$CH_2=\underline{C}H—C\underline{H}=O$	26.9	$H\underline{C}≡C\underline{H}$		49.3
$(C_2H_5)_2\underline{C}H—C\underline{H}O$	26.9	$C_6H_5O—\underline{C}≡C\underline{H}$		61.0
$H_2NCH=C\underline{H}—\underline{C}HO$	6.0	$HC≡\underline{C}—C\underline{H}O$		33.2
$H_2NCH—\underline{C}H—C\underline{H}O$	20.0	$Cl\underline{C}H_2—C\underline{H}O$		32.5
C_6H_6	1.0	$Cl_2\underline{C}H—C\underline{H}O$		35.3
		$Cl_3\underline{C}—C\underline{H}O$		46.3
		$C_6H_5—C≡\underline{C}≡C\underline{H}_3$		10.8

Source: J. A. Dean, ed., *Lange's Handbook of Chemistry*, 14th ed., McGraw-Hill, New York, 1992.

TABLE 12.22 Carbon–Carbon Spin Coupling Constants

Structure*	J_{CC}, Hz
$H_3C—CH_3$	35
$H_3C—CHR_2$	37
$H_3C—CH_2Ar$	34
$H_3C—CH_2CN$	33
$H_3C—CH_2—CH_2OH$	
C-1, C-2	38
C-2, C-3	34
$H_3C—CH_2NH_2$	37
$C—C{=}O$	38–40
$\underline{C}—\underline{C}—C{=}O$	36
$\underline{C}—\underline{C}—Ar$	43
$C—CO—O^-$ (anion)	52
$C—CO—N$	52
$C—CO—OH$	57
$C—CO—OR$	59
$C—CN$	52–57
$C—C{\equiv}C$ $^2J_{CC} = 11.8$	67
$H_2C{=}CH_2$	68
$>\underline{C}{=}\underline{C}—CO—OH$	70–71
$>\underline{C}{=}\underline{C}—CN$	71
$>\underline{C}{=}\underline{C}—Ar$	67–70
C_6H_6	57
$C_6H_5NO_2$	
1–2	55
2–3, 3–4	56
$^3J_{2-5}$	7.6

Structure	J_{CC}, Hz
C_6H_5I	
1–2	60
2–3	53
3–4	58
$^3J_{2-5}$	8.6
$C_6H_5—OCH_3$	
2–3	58
3–4	56
$C_6H_5NH_2$	
1–2	61
2–3	58
3–4	57
$^3J_{2-5}$	7.9
$C_6H_5CH_3$	44
Pyridine	
2–3	54
3–4	56
$^3J_{2-5}$	14
Furan	69
Pyrrole	69
Thiophene	64
$H_2\underline{C}{=}\underline{C}{=}C(CH_3)_2$	100
$—C{\equiv}C—$	170–176

Structure	$^2J_{CC}$, Hz
$\underline{C}H_3—CO—\underline{C}H_3$	16
$\underline{C}H_3—C{\equiv}\underline{C}H$	11.8
$\underline{C}H_3CH_2—\underline{C}N$	33

* R, alkyl group; Ar, aryl group.

Source: J. A. Dean, ed., *Lange's Handbook of Chemistry*, 14th ed., McGraw-Hill, New York, 1992.

Preference is now learning toward nitromethane (neat). Nitrogen-15 chemical shifts for standards are listed in Table 12.28.

Nitrogen-15 spin coupling constants with protons are given in Table 12.29, with carbon-13 in Table 12.30, and those with fluorine-19 in Table 12.31.

12.4.7 Fluorine-19

Fluorine-19 is the only isotope of fluorine, which is fortunate because the NMR signal response for a given number of fluorine nuclei is about 20% weaker than the signal for the same number of hydrogen nuclei. Chemical shifts for fluorine-19 (Table 12.32) are relative to trifluoroacetic acid as reference standard. The total range of chemical shifts is over 1000 ppm. The chemical shifts are quite dependent upon the nature of the other geminal atoms and upon the nature of the atoms attached to the adjacent carbon atoms. The least-shielded ^{19}F nuclei are those attached directly to nitrogen and sulfur atoms. Accumulation of halogen atoms on the same carbon also results in marked deshielding.

TABLE 12.23 Carbon–Fluorine Spin Coupling Constants

Structure*	J_{CF}, Hz	Structure*	J_{CF}, Hz
$F(H)C(H)H$	−158	$F_2C{=}CH_2$	−287
$F(H)C(F)H$	+235	$F_2C{=}O$	−308
$F(F)C(F)H$	+274	$F(R)C{=}O$	−353
$F(F)C(F)F$	+259	$F(H)C{=}O$	−369
$F(F)C(F)CH_3$	+271	$F(H)C(F)CH_2OH$	−241
$F(H)C(H)Ar$	+165	$F(F)C(F)CH_2OH$	−278
$F{-}CH_2CH_2{-}$ or $F{-}CR_3$	+167	$F(F)C(F)OCF_3$	−265
p-$F{-}C_6H_4{-}OR$	+237		
p-$F{-}C_6H_4{-}R$	+241		
p-$F{-}C_6H_4{-}CF_3$	+252		
p-$F{-}C_6H_4{-}CO{-}CH_3$	+253	$F(F)C(F){-}CO{-}CH_3$	−289
p-$F{-}C_6H_4{-}NO_2$	+257		
$F{-}C_6H_5$	252		
$^2J_{CF} = 21.0$			
$^3J_{CF} = 7.7$			
$^4J_{CF} = 3.4$			

* Ar, aryl group; R, alkyl group.

Source: J. A. Dean, ed., *Lange's Handbook of Chemistry*, 14th ed., McGraw-Hill, New York, 1992.

Reference standards for fluorine-19 are listed in Table 12.33. Trifluoroacetic acid can only be used as an external standard because of its chemical reactivity. Bulk susceptibility corrections must be applied when correlating chemical shift measurements with different standards. Trichlorofluoromethane would be an alternative standard to trichlorofluoromethane were it not for the signal broadening caused by the three chlorine atoms.

Spin coupling constants for fluorine to fluorine are given in Table 12.34, those with carbon-13 are given in Table 12.23, those with hydrogen are given in Table 12.10, and those with nitrogen-15 are given in Table 12.31. An extensive review of fluorine coupling constants is provided by Emsley and Phillips.[4]

[4] J. W. Emsley and L. Phillips, *Prog. Nucl. Magn. Reson. Spectrosc.* **7** (1971).

TABLE 12.24 Carbon-13 Spin Coupling Constants with Various Nuclei

Nuclei	Structure	^1J, Hz	^2J, Hz	^3J, Hz	^4J, Hz
^2H	$CDCl_3$	32			
	CD_3—CO—CD_3	20			
	$(CD_3)_2SO$	22			
	C_6D_6	26			
^7Li	CH_3Li	15			
^{11}B	$(C_6H_5)_4B^-$	49		3	
^{14}N	$(CH_3)_4N^+$	10			
	CH_3NC	8			
^{29}Si	$(CH_3)_4Si$	52			
^{31}P	$(CH_3)_3P$	14			
	$(C_4H_9)_3P$	11	12	5	
	$(C_6H_5)_3P$	12	20	7	0
	$(CH_3)_4P^+$	56			
	$(C_4H_9)_4P^+$	48	4	15	
	$(C_6H_5)_4P^+$	88	11	13	3
	$R(RO)_2P{=}O$	142	5–7		
	$(C_4H_9O)_3P{=}O$		6	7	
^{77}Se	$(CH_3)_2Se$	62			
	$(CH_3)_3Se^+$	50			
^{113}Cd	$(CH_3)_2Cd$	513, 537			
^{119}Sn	$(CH_3)_4Sn$	340			
	$(CH_3)_3SnC_6H_5$	474	37	47	11
^{125}Te	$(CH_3)_2Te$	162			
^{199}Hg	$(CH_3)_2Hg$	687			
	$(C_6H_5)_2Hg$	1186	88	102	18
^{207}Pb	$(CH_3)_2Pb$	250			
	$(C_6H_5)_4Pb$	481	68	81	20

Source: J. A. Dean, ed., *Lange's Handbook of Chemistry*, 14th ed., McGraw-Hill, New York, 1992.

TABLE 12.25 Boron-11 Chemical Shifts

Values given in ppm on the δ scale, relative to $B(OCH_3)_3$.

Structure	δ, ppm	Structure	δ, ppm
R_3B	−67 to −68	$B(BF_4)$	19 to 20
Ar_3B	−43	HB(NH—BH)(NH—BH)NH (ring)	−12
BF_3	24		
BCl_3	−12		
BBr_3	−6	$H_2B(\mu$-H$)(\mu$-$NR_2)BH_2$	37
BI_3	41		
$B(OH)_3$	36		
$B(OR)_3$	0 to 1		
$B(NR_2)_3$	−13	$H_2B(\mu$-$NR_2)_2BH_2$	15
$C_6H_5BCl_2$	−36		
$C_6H_5B(OH)_2$	−14		
$C_6H_5B(OR)_2$	−10	$(CH_3)_2N$—$B(CH_3)_2$	62
$M(BH_4)$	55 to 61		

(*Continued*)

TABLE 12.25 Boron-11 Chemical Shifts (*Continued*)

Structure	δ, ppm	Structure	δ, ppm	
Addition complexes		**Boranes**		
$R_2O \cdot BH_3$	18 to 19	B_2H_6	1	
$R_3N \cdot BH_3$	25	B_4H_{10}		
$R_2NH \cdot BH_3$	33	(BH_2)	25	
		(BH)	60	
N · BH_3	31		Base	Apex
R_2O(or ROH) · BF_3	17 to 19	B_5H_9	31	70
R_2O(or ROH) · BCl_3	−7 to −8	B_5H_{11}	−16	50
R_2O(or ROH) · BBr_3	23 to 24	$B_{10}H_{14}$	7	54
R_2O(or ROH) · BI_3	74 to 82			
N · BBr_3	24			

Source: J. A. Dean, ed., *Lange's Handbook of Chemistry*, 14th ed., McGraw-Hill, New York, 1992.

TABLE 12.26 Nitrogen-15 Chemical Shifts

Values given in ppm on the δ scale, relative to NH_3 liquid.

Substituent group	δ, ppm	Substituent group	δ, ppm
Aliphatic amines		Amides	
Primary	3 to 59	R = secondary	104 to 148
Secondary	7 to 81	R = tertiary	96 to 133
Tertiary	14 to 44	HCO—NH—Aryl	138 to 141
Cyclo, primary	29 to 44	RCO—NHR or $RCO—NR_2$	103 to 130
Aryl amines	40 to 100	RCO—NH—Aryl	131 to 136
Aryl hydrazines	40 to 100	Aryl—CO—H—Aryl	ca. 126
Piperidines, decahydroquinolines	30 to 82	Guanidines	
Amine cations		Amino	30 to 60
Primary	19 to 59	Imino	166 to 207
Secondary	40 to 74	Thioureas	85 to 111
Tertiary	30 to 67	Thioamides	135 to 154
Quaternary	43 to 70	Cyanamides	
Enamines, tertiary type		R_2N—	−12 to −38
Alkyl	29 to 82	—CN	175 to 200
Cycloalkyl	55 to 104	Carbodiimides	95 to 120
Aminophosphines	59 to 100	Isocyanates	
Amine *N*-oxides	95 to 122	Alkyl, primary	14 to 32
Ureas		Alkyl, secondary and tertiary	54 to 57
Aliphatic	63 to 84	Aryl	ca. 46
Aryl	105 to 108	Isothiocyanates	90 to 107
Sulfonamides	79 to 164	Azides	52 to 80
Amides			108 to 122
HCO—NHR			240 to 260
R = primary	100 to 115		

TABLE 12.26 Nitrogen-15 Chemical Shifts (*Continued*)

Substituent group	δ, ppm	Substituent group	δ, ppm
Lactams	113 to 122	Nitrones	270 to 285
Hydrazones		Imides	170 to 178
Amino	141 to 167	Imimes	310 to 359
Imino	319 to 327	Oximes	340 to 380
Cyanates	155 to 182	Nitramines	
Nitrile *N*-oxides, fulminates	195 to 225	Amine	252 to 280
Isonitriles		$—NO_2$	328 to 355
Alkyl, primary	162 to 178	Nitrates	310 to 353
Alkyl, secondary	191 to 199	*gem*-Polynitroalkanes	310 to 353
Aryl	ca. 180	Nitro	
Nitriles		Aryl	350 to 382
Alkyl	235 to 246	Alkyl	372 to 410
Aryl	258 to 268	Hetero, unsaturated	354 to 367
Thiocyanates	265 to 280	Azoxy	330 to 356
Diazonium		Azo	504 to 570
Internal	222 to 230	Nitrosamines	222 to 250
Terminal	315 to 322		525 to 550
Diazo		Nitrites	555 to 582
Internal	226 to 303	Thionitrites	720 to 790
Terminal	315 to 440	Nitroso	
Nitrilium ions	123 to 150	Aliphatic amines, NO	535 to 560
Azinium ions	185 to 220	Aryl	804 to 913
Azine *N*-oxides	230 to 300		
Saturated cyclic systems			
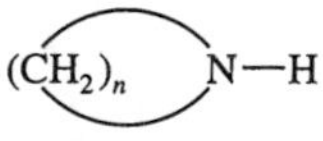 $(CH_2)_n$ N—H		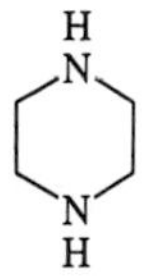 H N N H	35.5
$n = 2$	−8.5		
$n = 3$	25.3		
$n = 4$	36.7		
$n = 5$	37.7	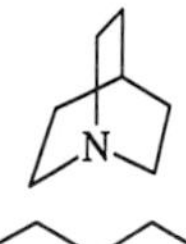 N	7.5 (in C_6H_6) 18.0 (in H_2O)
O N H	32.1	N H	
		cis	42.4
		trans	52.9

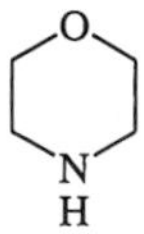

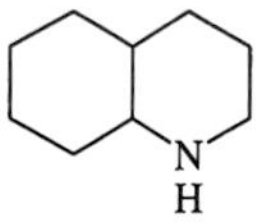

(*Continued*)

TABLE 12.26 Nitrogen-15 Chemical Shifts (*Continued*)

Unsaturated cyclic systems

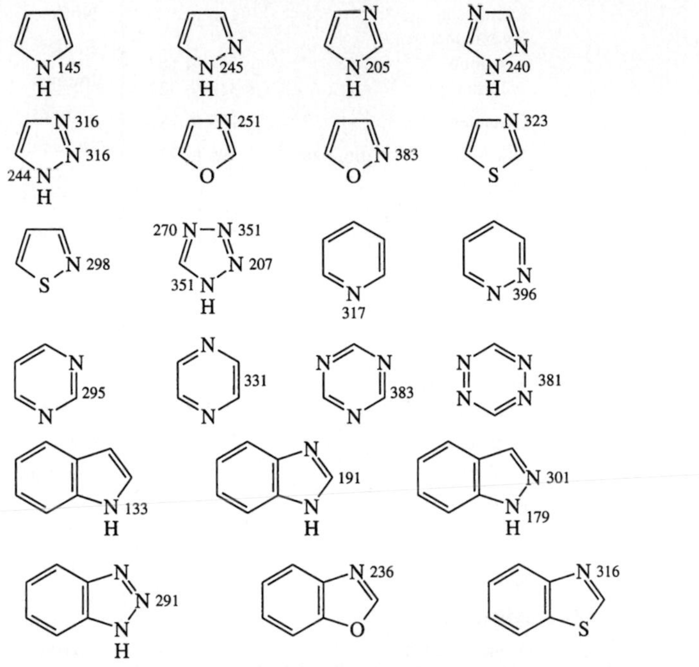

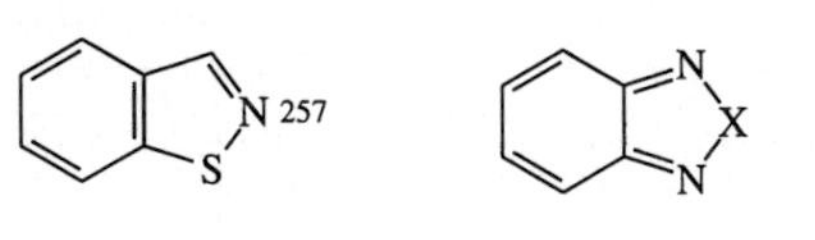

N 145
H

N
N 245
H

N
N 205
H

N
N
N 240
H

N 316
N 316
244 N
H

N 251
O

N 383
O

N 323
S

N 298
S

270 N—N 351
N 207
351 N
H

N
317

N
N 396

N
295
N

N
331
N

N N
383
N

N N
381
N N

N 133
H

N
191
N
H

N 301
N 179
H

N
N 291
N
H

N 236
O

N 316
S

N 257
S

N
X
N

X	δ, ppm
O	517
S	331
Se	373

N
316

N 316

N 293
N
281

N
330
N

N
N 412

N
361
N

N
283

N
H 114

N
N
308

N
N 280
N
H
N
261

N 301
N 399
N—N
351

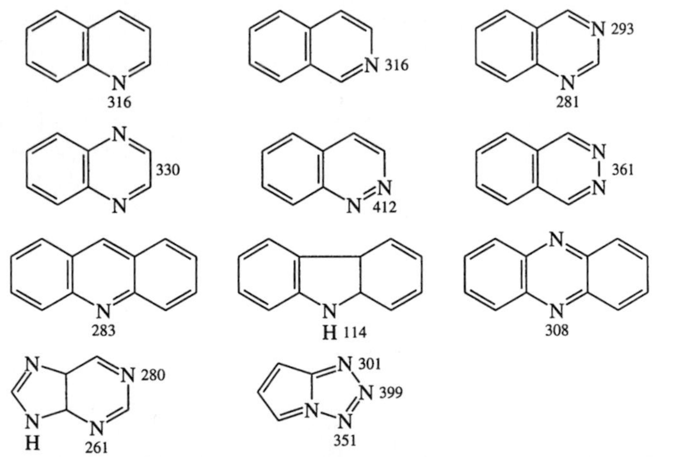

Source: J. A. Dean, ed., *Lange's Handbook of Chemistry*, 14th ed., McGraw-Hill, New York, 1992.

TABLE 12.27 Nitrogen-15 Chemical Shifts in Monosubstituted Pyridine

$\delta = 317.3 + \Delta_i$

Substituent	Δ_{C-2}	Δ_{C-3}	Δ_{C-4}
$—CH_3$	−0.4	0.3	−8.0
$—CH_2CH_3$	−1.8		−6.6
$—CH(CH_3)_2$	−5.1		−5.9
$—C(CH_3)_3$	−2.5		−5.8
—CN	−0.9	−0.8	10.6
—CHO	10	11	29
$—CO—CH_3$	−9	15	11
$—CO—OCH_2CH_3$	11.8		−5
$—OCH_3$	−49	0	−23
—OH	−126	−2	−118
$—NO_2$	−23	1	22
$—NH_2$	−45	10	−46
—F	−42	−18	
—Cl	−4	4	−6
—Br	2	8	7

Source: J. A. Dean, ed., *Lange's Handbook of Chemistry*, 14th ed., McGraw-Hill, New York, 1992.

TABLE 12.28 Nitrogen-15 Chemical Shifts for Standards

Values given in ppm, relative to NH_3 liquid at 23°C.

Substance	δ, ppm	Conditions
Nitromethane (neat)	380.2	For organic solvents and acidic aqueous solutions
Potassium (or sodium) nitrate (saturated aqueous solution)	376.5	For neutral and basic aqueous solutions
$C(NO_2)_4$	331	For nitro compounds
$(CH_3)_2$—CHO (neat)	103.8	For organic solvents and aqueous solutions
$(C_2H_5)_4N^+Cl^-$	64.4	Saturated aqueous solution
$(CH_3)_4N^+Cl^-$	43.5	Saturated aqueous solution
NH_4Cl	27.3	Saturated aqueous solution
NH_4NO_3	20.7	Saturated aqueous solution
NH_3	0.0	Liquid, 25°C
	−15.9	Vapor, 5 atm

Source: J. A. Dean, ed., *Lange's Handbook of Chemistry*, 14th ed., McGraw-Hill, New York, 1992.

TABLE 12.29 Nitrogen-15 to Hydrogen-1 Spin Coupling Constants

Structure	J_{NH}, Hz	Structure	J_{NH}, Hz
R—NH_2 and R_2NH	61–76	R(C=O)CNHR	88–92
Aryl—NH_2	78	H*(C=O)C—NH_2 N-H*	16
p-CH_3O—aryl—NH_2	79	HC≡NH^+	134
p-O_2N—aryl—NH_2	90–93	CH_3(or C_6H_5)C≡NH^+	136
Amine salts (alkyl and aryl)	73–76	>P—NH_2	82–90
Aryl—NHOH	79	$(R_3Si)_2NH$	67
Aryl—$NHCH_3$	78	CF_3—S—NH_2	81
Aryl—$NHCH_2F$	90	$(CF_3—S)_2NH$	99
Aryl—$NHNH_2$	90	Aryl—SO_2—NH_2	81
p-O_2N—aryl—$NHNH_2$	99	Aryl—SO_2—NHR	86
Quinoline N-H_2	11	Quinolinium ion	
		N-H_1	96
		N-H_2	2
Pyrole		Imidazole	
N-H_1	97	N-H_2	10
N-$H_{2,5}$	5	N-$H_{4,5}$	3
N-$H_{3,4}$	4		
Pyridine		Pyridinium ion	
N-$H_{2,6}$	11	N-$H_{2,6}$	3
N-$H_{3,5}$	2	N-$H_{3,5}$	4
N-H_4	<1	N-H_4	1

TABLE 12.30 Nitrogen-15 to Carbon-13 Spin Coupling Constants

Structure	*J*, Hz	Structure	*J*, Hz
Alkyl amines	4–4.5	Alkyl—NO_2	11
Cyclic alkyl amines	2–2.5	R—CN	18
Alkyl amines protonated	4–5	CH_3—$\overset{+}{N}$≡$\overline{C}$	
Aryl amines	10–14	H_3C—N	10
Aryl amines protonated	9	—N≡C	9
CH_3CO—NH_2	14–15	Diaryl azoxy	
H_2N—CO—NH_2	20	*anti*	18
Aryl—NO_2	15	*syn*	13

Source: J. A. Dean, ed., *Lange's Handbook of Chemistry*, 14th ed., McGraw-Hill, New York 1992.

12.4.8 Silicon-29

Chemical shifts for silicon-29 are given in Table 12.35. The reference standard is tetramethylsilane.

12.4.9 Phosphorus-31

Phosphorus-31 chemical shifts are given in Table 12.36, and phosphorus-31 spin coupling constants with hydrogen-1, boron-11, and carbon-13 are given in Table 12.37.

TABLE 12.31 Nitrogen-15 to Fluorine-19 Spin Coupling Constants

Structure	J, Hz	Structure	J, Hz
NF_3	155	Pyridine	
F_4N_2	164	2-F	52
FNO_2	158	3-F	4
F_3NO	190	2,6-di-F	37
$F_3C—O—NF_2$	164–176	Pyridinium ion	
$FCO—NF_2$	221	2-F	23
$(NF_4)^+SbF_6^-$	323	3-F	3
$(NF_4)^+AsF_6^-$	328	Quinoline, 8-F	3
$(N_2F)^+AsF_6^-$	459	Aniline	
$F_3C—NO_2$	215	2-F	0
		3-F	0
F—N=N—F ($^2J = 10$)	190	4-F	1.5
		Anilinium ion	
		2-F	1.4
		3-F	0.2
F—N=N—F (both F on same side) ($^2J = 52$)	203	4-F	0

Source: J. A. Dean, ed., *Lange's Handbook of Chemistry*, 14th ed., McGraw-Hill, New York, 1992.

TABLE 12.32 Fluorine-19 Chemical Shifts

Values given in ppm on the δ scale, relative to CCl_3F.

Substituent group	δ, ppm	Substituent group	δ, ppm
$—SO_2—F$	−67 to −42 (aryl)(alkyl)	$HO—CO—CF_3$	77
$—CO—F$	−29 to −20	$—CHF—C\underline{F}_3$	81
$>N—CO—F$	−5	$—CF_2—C\underline{F}_3$	78 to 88
$Aryl—CF_2Cl$	49	$—CS—F$	81
$—CF_2I$	56	$CF_3—C—N<$	84 to 96
$—CF_2Br$	63	$—CO—CF_2—CF_3$	83
$R—CF_2Cl$	61 to 71	$—CF_2—$	86 to 126
$>C—CF_3$ and $aryl—CF_3$	56 to 73	$—CF_2Br$	91
$—CS—CF_3$	70	$—C—CF_2—S—$	91 to 98
$>CF—CF_3$	71 to 73	$—CF=$	180 to 192
$—S—CF_3$	41	$—C\underline{F}_2—CF_3$	111
$—S—CF_2—S—$	39	$—CO—CF_2—$	116 to 131
$>P—CF_3$	46 to 66	$—C(halide)—CF_2—$	119 to 128
$>N—CF_3$	40 to 58	$—C\underline{F}_2—CF_3$	121 to 125
$>N—CF_2—C$	85 to 127	$—CF_2—CF_2—$	121 to 129
$—O—CF_2—R$	70 to 91	$—CF_2—CH_2—$	122 to 133
$—O—CF_2—C\underline{F}_3$	70 to 91	$—C\underline{F}_2—CHF_2$	128 to 132
$—CH_2—CF_3$	76 to 77	$—CF_2H$	136 to 143

(*Continued*)

TABLE 12.32 Fluorine-19 Chemical Shifts (*Continued*)

Substituent group	δ, ppm	Substituent group	δ, ppm
Cyclopropane F_2	151 to 156	F-1	126
Cyclobutane F_2	147	F-2	155
Cyclopentene F_2	96 to 133	F-3	162
Cyclopentene F	159	$ClFC{=}CH{-}CF_3$	61
Cyclohexane-F	210 (axial) to 240 (equatorial)	Cycloalkenes	
Perfluorocycloalkane	131 to 138	$=CF{-}C\underline{F}_2{-}C(CF_3 \text{ or } H){-}$	101 to 113
$>C\underline{F}{-}CF_3$	163 to 198	$-CF_2{-}C\underline{F}_2{-}C(CF_3 \text{ or } CH_3){=}$	110 to 114
$>C\underline{F}(CF_3)_2$	180 to 191	$-CF_2{-}C\underline{F}_2{-}CH{=}$	113 to 116
$-CFH-$	198 to 231	$-CF_2{-}C\underline{F}_2{-}CF{=}$	119 to 122
$-CFH_2$	235 to 244	Aryl—F	113
$F_2C{=}CF_2$	133	$C_{10}H_7{-}F$	
F_c, F_t, F_g on $C{=}C$ with $CF_2{-}CF_2H$		F-1	127
cis	108	F-2	114
trans	92	$C_6H_5{-}C_6H_4{-}F$	
gem	192	F-2	117
$H_2C{=}CF_1{-}CF_2{=}CF_3H$ (H, H, F_2, F_3, H, F_1 on $C{=}C{-}C{=}C$)		F-3	113
		F-4	109
		C_6F_6	163

Source: J. A. Dean, ed., *Lange's Handbook of Chemistry*, 14th ed., McGraw-Hill, New York, 1992.

TABLE 12.33 Fluorine-19 Chemical Shifts for Standards

Substance	Formula	δ, ppm
Trichlorofluoromethane	$CFCl_3$	0.0
α,α,α-Trifluorotoluene	$C_6H_5CF_3$	63.8
Trifluoroacetic acid	CF_3COOH	76.5
Carbon tetrafluoride	CF_4	76.7
Fluorobenzene	C_6H_5F	113.1
Perfluorocyclobutane	C_4F_8	138.0

Source: J. A. Dean, ed., *Lange's Handbook of Chemistry*, 14th ed., McGraw-Hill, New York, 1992.

TABLE 12.34 Fluorine-19 to Fluorine-19 Spin Coupling Constants

Structure	J_{FF}, Hz
F_2C cycloalkane	
gem	212–260
Unsaturated compounds $>C=C<$	
gem	30–90
trans	115–130
cis	9–58
Aromatic compounds, monocyclic	
ortho	18–22
meta	0–7
para	12–15
Alkanes	
$C\underline{F}Cl_2—C\underline{F}_2—CFCl_2$	6
$C\underline{F}Cl_2—C\underline{F}_2—CCl_3$	5
$C\underline{F}_2Cl—C\underline{F}_2—CF_2Cl$	1
$C\underline{F}_3—C\underline{F}_2—CF_2Cl$ (or $—C\underline{F}_3$)	<1
$CF_3—C\underline{F}_2—C\underline{F}_2Cl$	2
$C\underline{F}_3—CF_2—C\underline{F}_2Cl$	9
$C\underline{F}_3—CF_2—C\underline{F}_3$	7

Source: J. A. Dean, ed., Lange's *Handbook of Chemistry*, 14th ed., McGraw-Hill, New York, 1992.

TABLE 12.35 Silicon-29 Chemical Shifts

Values given in ppm on the δ scale relative to tetramethysilane.

Substituent group X in $(CH_3)_{4-n}SiX_n$	*n*			
	1	2	3	4
—F	35	9	−52	−109
—Cl	30	32	13	−19
—Br	26	20	−18	−94
—I	9	−34	−18	−346
—H	−19	−42	−65	−93
$—C_2H_5$	2	5	7	8
$—C_6H_5$	−5	−9	−12	
$—CH=CH_2$	−7	−14	−21	−23
—O alkyl	14–17	−3 to −6	−41 to −45	−79 to −83
—O aryl	17	−6	−54	−101
—O—CO—alkyl	22	4	−43	−75
$—N(CH_3)_2$	6	−2	−18	−28

(*Continued*)

TABLE 12.35 Silicon-29 Chemical Shifts (*Continued*)

Structure	δ, ppm
Hydrides	
$H_3Si—$	−39 to −60
$—H_2Si—$	−5 to −37
$HSi\lessgtr$	−2 to −39
Silicates	
Orthosilicate anions	−69 to −72
Silicon in end position	−77 to −81
Silicon in middle	−85 to −89
Branching silicons	−93 to −97
Cross-linked silicons	−107 to −120
Methyl siloxanes	
$(CH_3)_2Si—O—$ (end position)	6 to 8
$(CH_3)_2Si(O—)_2$ (middle)	−18 to −23
$CH_3Si(H)(O—)_2$ (middle)	−35 to −36
$CH_3Si(O—)_2—O—$ (branching)	−65 to −66
$—O—Si(O—)_2—O—$ (cross-linked)	−105 to −110
Polysilanes	
$F_3Si—SiF_3$	−74
$Cl_3Si—SiCl_3$	−8
$(CH_3O)_3Si—Si(OCH_3)_3$	−53
$(CH_3)_3Si—Si(CH_3)_3$	−20
$(CH_3)_2\underline{Si}[Si(CH_3)_3]_2$	−48
$H\underline{Si}[Si(CH_3)_3]_3$	−117
$\underline{Si}[Si(CH_3)_3]_4$	−135

Source: J. A. Dean, ed., *Lange's Handbook of Chemistry,* 14th ed., McGraw-Hill, New York, 1992.

TABLE 12.36 Phosphorus-31 Chemical Shifts

Values given in ppm on the δ scale, relative to 85% H_3PO_4.

Structure	Identical atoms attached directly to phosphorus	Nonidentically substituted phosphorus		
		R═CH_3	R═C_2H_5	R═C_6H_5
P_4	–488			
PR_3		–62	–20	–7
$P(C_3H_7)_3$, $P(C_4H_9)_3$	–33			
PHR_2		–99	–56	–41
PH_2R		–164	–128	–122
PH_3	–241			
PF_3	97			
PRF_2			168	207
PCl_3	220			
$PRCl_2$		192	196	162
PR_2Cl		94	119	81
PBr_3	227			
$PRBr_2$		184	194	152
PR_2Br		91	116	71
PI_3	178			
$P(CN)_3$	–136			
$P(SiR_3)_3$		–251		
$P(OR)_3$		141	139	127
$P(OR)_2Cl$		169	165	157
$P(OR)Cl_2$		114	177	173
$P(SR)_3$		125	115	132

TABLE 12.36 Phosphorus-31 Chemical Shifts (*Continued*)

Structure	Identical atoms attached directly to phosphorus	Nonidentically substituted phosphorus		
		R=CH_3	R=C_2H_5	R=C_6H_5
$P(SR)_2Cl$		188	186	183
$P(SR)Cl_2$		206	211	204
$P(SR)_2Br$				184
$P(SR)Br_2$		204		
$P(NR_2)_3$		123	118	
$P(NR_2)Cl_2$		166	162	151
$PR(NR_2)_2$		86	100	100
$PR_2(NR_2)$		39	62	
$H_2P—PH_2$(–80°C)	–204			
$F_2P—PF_2$	226			
$Cl_2P—PCl_2$	155			
$I_2P—PI_2$	170			
$PH_2^-K^+$	–255			
$P(CF_3)_3$	–3			
P_4O_6	113			

Structure	Identical atoms attached directly to phosphorus	Nonidentically substituted phosphorus		
		X=F	X=Cl	X=Br
$P(NCO)_3$	97			
$P(NCO)_2X$		128	128	127
$P(NCO)X_2$		131	166	
$P(NCS)_3$	86			
$P(NCS)_2X$			114	112
$P(NCS)X_2$			155	153

Structure	Identical atoms attached directly to phosphorus	Nonidentically substituted phosphorus		
		R=CH_3	R=C_2H_5	R=C_6H_5
O=PR_3		36	48	25
O=PHR_3		63		23
O=PF_3	–36			
O=PRF_2		27	29	11
O=PR_2F		–66		
O=PCl_3	2–5			
O=$PRCl_2$		44	53	34
O=PR_2Cl		63–68	77	43
O=$PRBr_2$		8		
O=PR_2Br		–51		
O=$P(OR)_3$		1	–1	–18
O=$P(OR)_2Cl$		6	3	–6
O=$P(OR)Cl_2$		3–6	6	2
O=$PH(OR)_2$		19	15	
O=$PH_2(OR)$		19	15	
O=$PR_2(OC_2H_5)$		50	52	31
O=$PR(OC_2H_5)_2$		32	33	17

(*Continued*)

TABLE 12.36 Phosphorus-31 Chemical Shifts (*Continued*)

Structurc	Identical atoms attached directly to phosphorus	Nonidentically substituted phosphorus		
		$R{=}CH_3$	$R{=}C_2H_5$	$R{=}C_6H_5$
$O{=}P(NR_2)_3$		23	24	2
$O{=}PR_2(NR_2)$		44		26
$O{=}P(OR)_2NH_2$		15	12	3
$O{=}P(OR)_2(NCS)$			−19	−29
$O{=}P(SR)_2$		66	61	55
$O{=}PBr_3$	−103			
$O{=}P(NCO)_3$	−41			
$O{=}P(NCS)_3$	−62			
$O{=}P(NH_2)_3$	22			
$S{=}PR_3$		59	55	43
$S{=}PCl_3$	29			
$S{=}PRCl_2$		80	94	75
$S{=}PR_2Cl$		87	109	80
$S{=}PBr_3$	−112			
$S{=}PRBr_2$		21	42	20
$S{=}PR_2Br$		64	98	
$S{=}P(OR)_3$		73	68	53
$S{=}P(OR)Cl_2$		59	56	54
$S{=}P(OR)_2Cl$		73	68	59
$S{=}PH(OR)_2$		74	69	59
$S{=}P(SR)_3$		98	92	92
$S{=}P(NH_2)_3$	60			
$S{=}P(NR_2)_3$		82	78	
$Se{=}P(OR)_3$		78	71	58
$Se{=}P(SR)_3$		82	76	
$P(OR)_5$			−71	−86
PRF_4		−30	−30	−42
PR_2F_3		9	6	

Structure	δ	Structure	δ
PF_5	−35	Phosphonium cations	
		Alkyl	43 −32
		Aryl	35 −18
$PF_6^-H^+$	−144	$(O_3P{-}PO_3)^{4-}$	9
PBr_5	−101	Polyphosphates	
		End group	ca. −6
		Middle group	ca. −18
		Branch group	ca. −30
$P(OC_2H_5)_5$	−71	Adenosine-3′-triphosphate	P_α −11
			P_β −23
			P_γ −7
PO_4^{3-}	6	Adenosine-3′-diphosphate	P_α −11
			P_β −7
$O{=}P[OSi(CH_3)_3]_3$	−33	$P(CH_3)_4OCH_3$	−89
$H_4P_2O_7$	−11		
Phosphonates	24 to 2		

TABLE 12.37 Phosphorus-31 Spin Coupling Constants

Substituent group	J_{PH}, Hz
PH_3	181
$—PH_2—$	134
PH_2CH_3	210
$PH_2C_6H_5$	200
ortho (ring)	7
para (ring)	2
$P(CH_3)_3$	3
$P(CH_3)Cl_2$	16–17
$P(C_2H_5)_3$	14
$—P—CH_2—$	14
$(Halogen)_2P—CH$	16–20
P—NH	10–28
P—N—CH	8–25
P—C—CH	0–4
$P—C_6H_5$	
ortho	7–10
meta	2–4
$H_2P—PH_2$	
PH	186
PPH	12
$O{=}PHR_2$	210–500
O—PH(S)R	490–540
O_2PHR	500–575
$O{=}PH_2(OCH_3)$	575
P—O—CH	13
$O{=}PH_2(OC_2H_5)$	567
O_2PHR	500–575
$O_2PH(N)$	560–630
$O_2PH(S\text{ or }Se)$	630–655
O_3PH	630–760

Substituent group	J_{PH}, Hz
$S(\text{or }Se){=}P^{v}—H$	490–650
$S(\text{or }Se){=}PHR_2$	420–454
$O{=}P^{V}—CH_3$	7–15
$O{=}P^{V}—CH—Aryl$	
(or C=O)	15–30
$O{=}P^{V}—CH{=}C$	15–30
$(Halogen)_2$	
P—N—CH	9–18
$S{=}P^{V}—CH$	11–15
$P—CH_3^+$	12–17
$P—H^+$	490–600

Substituent group	J_{PB}, Hz
$H_3B—P—N$	80

Substituent group	J_{PC}, Hz
$P(C_6H_5)_3$	
ipso	12
ortho	20
meta	7
para	0.3
$O{=}P(\text{cyclo }C_4H_8)$	
C_6H_5	
$P—C_1$	67
$P—C_2$	8
+	
$(C_6H_5)_3—PCH_3$	
ipso	89
ortho	11
meta	13
para	4
$P—C(CH_3)$	57

12.5 QUANTITATIVE ANALYSIS

The area under an absorption band (or multiplet) is proportional to the number of nuclei responsible for the absorption. A device for integrating the absorption signal is a standard item on commercial NMR spectrometers. Thus, we have available a nuclei counter to supplement the mass spectrometer. The integrator allows the particular nuclei to be assigned within each resonance group, if the total is known. If the total is not known, quite often a particular singlet or multiplet can be tentatively assigned the number of nuclei, say, protons, from which the proportional numbers in the remaining groups can be found. If the numbers are rational in each multiplet, the assignments can be accepted. Accuracy is typically within ±2%.

For quantitative analysis, a known amount of a reference compound can be included with the sample. Ideally, the signal of the reference compound should be a strong singlet lying in a region of the NMR spectrum unoccupied by sample peaks. Quantitative analysis requires that at least one resonance band from each component in a mixture be free from overlap by other absorption signals. The amount of the unknown is calculated by

$$W_{unk} = W_{std} \frac{N_{std} M_{unk} A_{unk}}{N_{unk} M_{std} A_{std}} \qquad (12.3)$$

where A's = two peak areas
N's = numbers of protons in groups giving rise to absorption peaks
M's = molecular weights of unknown and standard
W's = weights of unknown and internal standard

Whenever the empirical formula is known, the total height divided by the number of protons yields the increment of height per proton. If the assignment of a particular absorption band has been deduced, the increment per height per proton is calculated from the height for the assigned group divided by the number of protons in the particular group. The same procedure would be used for other nuclei.

12.6 ELUCIDATION OF NMR SPECTRA

The application of NMR to structure analysis is based primarily on the empirical correlation of structure with observed chemical shifts and coupling constants. Extensive NMR spectra and surveys have been published.[5–8] A unique advantage of NMR is that spectra can often be interpreted by using reference data from structurally related compounds. Internal simulation programs allow the NMR spectrum of an unknown to be compared against a postulated structure.

Example 12.1 Let us perform the structural elucidation of the compound with the proton NMR spectrum shown in Fig. 12.2.

1. The singlet at $\delta = 2.1$ has to be an isolated methyl group; its downfield position indicates that it is adjacent to a carbonyl group.

2. The triplet–quartet pattern are coupled with each other; the triplet is a methyl group split by two adjacent protons, which in turn are split into a quartet by the three methyl protons.

3. The pair of doublets in the aryl region imply strong coupling between protons *ortho* to each other, that is, a *para*-substituted aryl ring. Furthermore, different atoms must be connected directly to the ring. After consulting Table 12.5, we see that nitrogen could cause the downfield shift to $\delta = 7.4$ and that oxygen could cause the upfield shift to $\delta = 6.85$. Unperturbed benzene protons appear at $\delta = 7.2$.

4. The "hump" at $\delta = 7.9$ can now be assigned to a proton connected to a nitrogen atom; the smearing of the proton signal is caused by the electric quadruple moment of nitrogen.

Now our pieces are

CH_3—CH_2— CH_3—C(=O)— —NH— —N—C_6H_4—O (*para-*)

The methylene group is far downfield; an adjacent oxygen would place it around $\delta = 3.6$. Something in the beta position is causing it to be further downfield; a likely candidate is the phenyl ring's double bonds. Putting the pieces together, we have

CH_3—CH_2—O—C_6H_4—NH—C(=O)—CH_3

which is phenacetin, often used in headache preparations.

[5] *The Aldrich Library of NMR Spectra*, 7th ed., Aldrich Chemical Co., Milwaukee, Wis., 1983, Vols. 1 and 2.
[6] Sadtler Research Laboratories, *Nuclear Magnetic Resonance Spectra*, Philadelphia, Pa., a continuously updated subscription service.
[7] R. M. Silverstein, G. C. Bassler, and T. C. Morrill, *Spectrometric Identification of Organic Compounds*, 5th ed., Wiley, New York, 1991.
[8] Asahi Research Center Co., Ltd., Japan, *Handbook of Proton-NMR Spectra and Data*, Academic, Orlando, Fla., 1985.

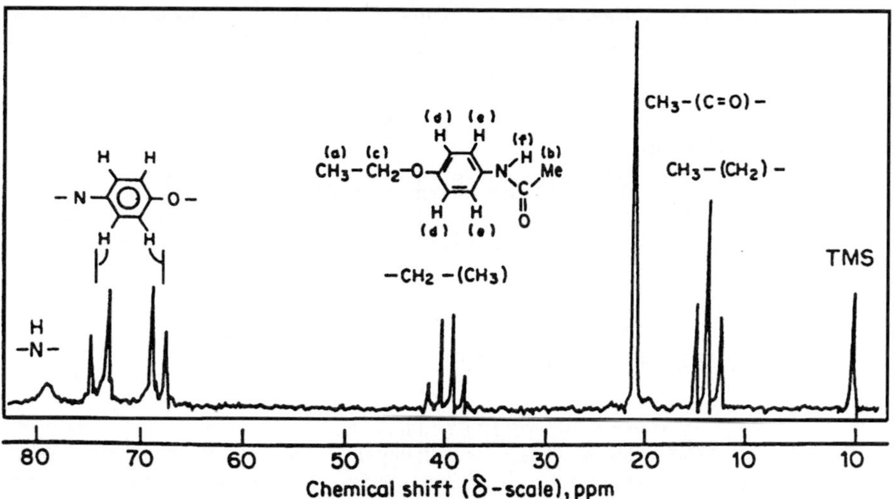

FIGURE 12.2 NMR spectrum of phenacetin, illustrating the origin of resonance positions and multiplet patterns. (*From Shugar and Dean, 1990.*)

12.6.1 Other Techniques Useful with Complex Spectra

12.6.1.1 Shaking with Deuterium Oxide. Brief vigorous shaking with a few drops of D_2O results in a complete exchange of deuterium for the labile protons attached to O, N, and S and the collapse of their absorption signal and any spin–spin coupling associated with the absorption signal.

12.6.1.2 Spin Decoupling. Spin decoupling can simplify spectra and aid in identification of spin–spin coupling partners. In homonuclear decoupling, the sample is simultaneously irradiated with a second, relatively strong rf field that is perpendicular to H_0 and that is at resonance with the nuclei to be decoupled; these nuclei collapse into single lines (or to the multiplicity from the normal coupling with a third nucleus). Irradiation of the methyl group of crotonaldehyde ($CH_3CH{=}CHCHO$) simplifies the complex pattern of the vinyl protons from a quarter of doublets into simple doublets. If the aldehyde proton were to be irradiated, the small doublets (due to the weak coupling with the aldehyde proton) on the other multiplet patterns would disappear.

Heteronuclear decoupling involves irradiating the sample at the NMR frequency characteristic of the heteroatom. The coupling between proton and fluorine spins, or the perturbations caused by nitrogen in the spectrum of dimethylformamide, could be disentangled with heteronuclear decoupling.

12.6.1.3 Shift Reagents.[9] Shift reagents are usually fluorinated β-diketones containing a lanthanide. They act as Lewis acids to complex preferentially with oxygen, nitrogen, or other Lewis base sites. Shift reagents cause large shifts of groups adjacent to these centers, thereby improving the J/δ ratio. Increments of shift reagent are added until sufficient resolution is attained. Europium and ytterbium chelates induce downfield shifts (away from the TMS reference position), and those based on praesodymium induce upfield shifts.

Chiral shift reagents provide the method of choice for determining the optical purity of organic materials if chiral impurities are present. The NMR method is preferable to polarimetric methods if the sample possibly contains other optically active impurities.

[9] J. Reuben, in *Progress in Nuclear Magnetic Resonance Spectroscopy*, J. W. Emsley, J. Feeney, and L. H. Sutcliffe, eds., Pergamon, Oxford, 1973, Vol. 9, p. 1.

12.6.1.4 Solvent Influence. Through screening of particular protons, solvents will cause shifts. For example, the addition of 50% benzene to a sample of 4-butyrolactone dissolved in CCl_4 causes the ring protons to appear at a higher field as well-separated multiplets.

12.6.2 Troubleshooting

In comparison to other spectroscopic methods, NMR is highly dependent on the performance of the spectrometer, the sample, the sample tube, and solute conditions. To obtain the appropriate information from spectra and to understand the basic instrumental problems, the discussion by Tchapla et al.[10] is informative.

12.7 ELECTRON SPIN RESONANCE

In electron spin resonance (ESR), radiation of microwave frequency induces transitions between magnetic energy levels of electrons with unpaired spins. The magnetic energy splitting is created by a static magnetic field. Unpaired electrons, relatively unusual in occurrence, are present in odd molecules, free radicals, triplet electronic state, and transition-metal and rare-earth ions.

12.7.1 Electron Behavior

Imposition of an external magnetic field H_0 establishes two electron energy levels. The difference in energy between the two levels is given by

$$\Delta E = \mu_e H_0 / M_s = h\nu \tag{12.4}$$

where μ_e is the electron magnetic moment and M_s is the angular momentum quantum number, which can have values of $+\,^1\!/_2$ or $-^1\!/_2$. Substituting the values in Eq. (12.4), and rearranging,

$$\nu = \frac{\mu_e H_0}{hM_s} = \frac{(9.285 \times 10^{-24}\ \mathrm{J \cdot T^{-1}})H_0}{(6.626 \times 10^{-34}\,\mathrm{J \cdot s})(\frac{1}{2})} = 2.803 \times 10^{10}\,\mathrm{s^{-1} \cdot T^{-1}}(H_0) \tag{12.5}$$

The magnetic field strength unit is named the tesla (T); the conversion to gauss is 10 000 G≡1 T.

12.7.2 ESR Spectrometer

The principal components of an ESR spectrometer are (1) a source of microwave radiation of constant frequency and variable amplitude; (2) a means of applying the microwave power to the sample—the microwave bridge; (3) a homogeneous and steady magnetic field; (4) an ac field superimposed on the steady field so as to sweep continuously through the resonance absorption of the sample; (5) a detector to measure the microwave power absorbed from the microwave field; and (6) a graphic *x-y* recorder.

The sample, contained in a cylindrical quartz tube, is held in a cavity between the poles of the magnet. Tubing of 3 to 5 mm internal diameter with a sample volume of 0.15 to 0.5 mL can be used for samples that do not possess a high dielectric constant. For samples with a high dielectric constant, flat cells with a thickness of about 0.25 mm and sample volume of 0.05 mL, are often used. For aqueous solutions, $10^{-7}M$ represents a reasonable lower limit. For structure determinations and quantitative analysis, the concentration should be about $10^{-6}M$. Working at higher magnetic field strengths gives a higher sensitivity. Rotatable cavities are used for studying anisotropic effects in solid samples.

---

[10] A. Tchapla, G. Emptoz, A. Aspect, and A. Nahon, "Troubleshooting in Proton NMR," *Am. Lab*. **17** (November):38 (1985).

12.7.3 Hyperfine Interaction in ESR Spectra

In a homogeneous magnetic field an unpaired electron in an assemblage of other atoms, themselves possessing nuclear spins, can have a number of energy states. Since the radical electron is usually delocalized over the entire molecule or at least a large part of it, the unpaired electron comes into contact and interacts with many nuclei. Nuclei possessing a magnetic moment may interact and cause a further splitting of the electron resonance line. The energies of a coupled level are given by

$$E = g\mu_B H_0 M_s + a_i h M_I \tag{12.6}$$

where a_i is called the *hyperfine coupling constant* and M_I is the *spin quantum number* of the coupling nucleus. On a spectrum the hyperfine coupling constant is the distance between associated peaks of a submultiplet, measured in gauss. A single nucleus of spin $I = 1/2$ will cause a splitting into two lines of equal intensity. Common nuclei with spin $1/2$ are ^1H, ^{13}C, ^{15}N, ^{19}F, and ^{31}P. Interaction with a single deuterium or nitrogen nucleus (^2H or ^{14}N, $I = 1$) will cause a splitting into three lines of equal intensity.

Hydrogen atoms that are attached to carbon atoms adjacent to the unpaired spin undergo interaction with the unpaired spin. This interaction is at a maximum when the carbon–hydrogen bond and orbital are coplanar. In general, hyperfine splitting by hydrogen atoms is important only at the 1 or 2 position in an alkyl radical. Hyperfine splitting by β-hydrogens in acrylic radicals can be barely detected in most cases. In rigid molecules possessing a highly bridged structure, hydrogen atoms in a beta position to the radical site undergo a strong interaction. If a double bond or a system of double bonds is followed by a single bond, as for example in methyl-substituted aromatic radicals, the double-bond character may be partially transferred to the single bond. In the ESR spectrum of 2-methyl-1,4-benzosemiquinone radical ion, the spin density is the same at the methyl protons as at the ring protons, and the spectrum consists of seven lines with an intensity ratio of 1:6:15:20:15:6:1, corresponding to the interaction with six equivalent protons. Sometimes the radical electron density penetrates even two C—C bonds, but with greatly diminished spin density. Other geometries can give rise to a long-range hyperfine splitting by hydrogen atoms. Any arrangement that places the back side of a carbon atom in close proximity to an orbital containing unpaired spin density will be expected to lead to interaction. A number of hyperfine splitting constants are gathered together in Table 12.38.

If several magnetic nuclei are present, the situation is somewhat more complicated because the electron experiences an interaction with each nucleus, and the spectrum is the result of a superposition of the hyperfine splitting for each nucleus. Several general types will be considered. When two equivalent nuclei are involved, the number of possible fields is reduced. Generally, $2nI + 1$ lines result from n equivalent nuclei; the relative intensity of these lines follow the coefficients of the binomial expansion. Two equivalent protons split each of the original electronic energy levels into three hyperfine levels in an intensity ratio 1:2:1.

If the unpaired electron couples with nonequivalent protons, each proton will have its own coupling constant. In general, n nonequivalent protons will produce a spectrum with 2^n hyperfine lines.

12.7.4 The *g* Factor

The g-factor is a dimensionless constant and equal to 2.002 319 for the unbound electron. The exact value of the g factor reflects the chemical environment, particularly when heteroatoms are involved, because the orbital angular momentum of the electron can have an effect on the value of the transition $\Delta M_s = \pm 1$. In many organic free radicals, the g value of the odd electron is close to that of a free electron because the electron available for spin resonance is usually near or at the periphery of the species with which it is associated. However, In mental ions g values are often greatly different from the free-electron value.

Hyperfine splitting is independent of the microwave frequency employed. This enables a nuclear hyperfine reaction to be distinguished from the effects of differences in g factors. If, for instance, the separation of two peaks is observed to vary with the microwave frequency, they correspond to two transitions with different g factors, not to interaction with a nuclear spin. To measure the g factor for free radicals, determine the field separation between the center of the unknown spectrum and that of a reference substance whose g value is accurately known. A dual-sample cavity simplifies the measurement. Two signals will be observed simultaneously with a field separation of ΔH. The g factor for the unknown is given by

$$g = g_{\mathrm{std}}\left(1 - \frac{\Delta H}{H}\right) \tag{12.7}$$

where H is the resonant frequency.

TABLE 12.38 Spin–Spin Coupling (Hyperfine Splitting Constants)

Values of coupling constant a_i given in gauss

Involves protons unless otherwise indicated.

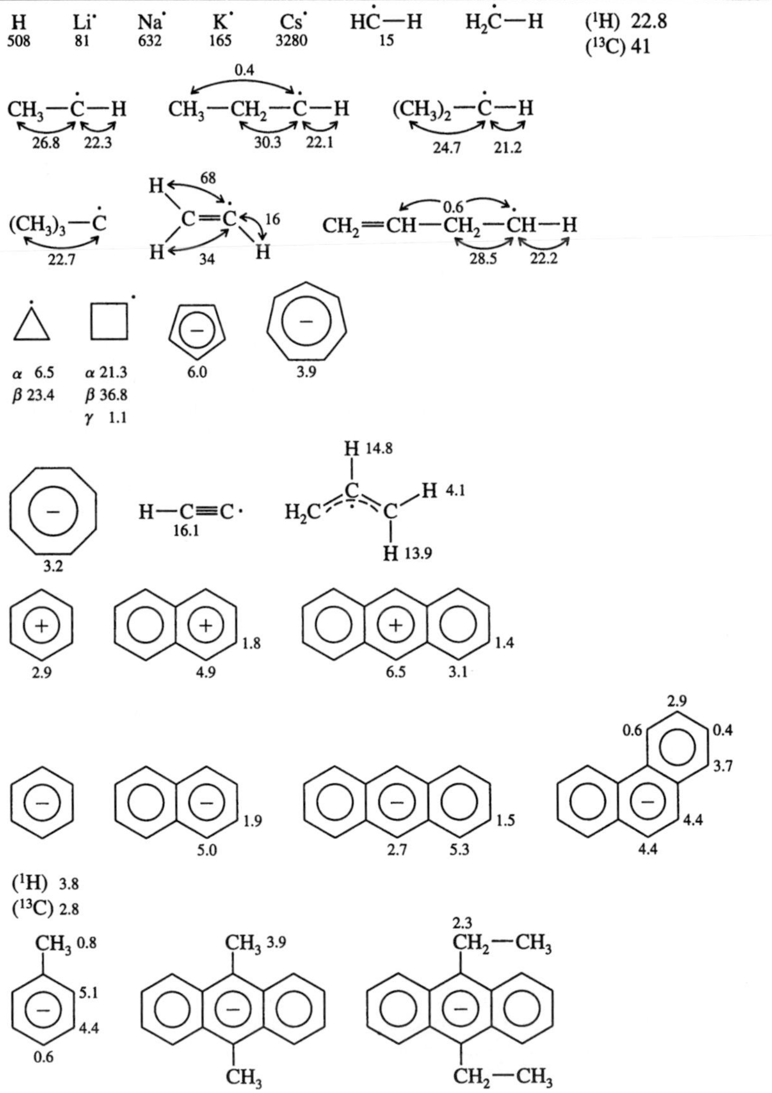

H 508; Li· 81; Na· 632; K· 165; Cs· 3280; $H\dot{C}{-}H$ 15; $H_2\dot{C}{-}H$ (^{1}H) 22.8, (^{13}C) 41

$CH_3{-}\dot{C}{-}H$: 26.8, 22.3

$CH_3{-}CH_2{-}\dot{C}{-}H$: 0.4, 30.3, 22.1

$(CH_3)_2{-}\dot{C}{-}H$: 24.7, 21.2

$(CH_3)_3{-}\dot{C}$: 22.7

$H_2C{=}\dot{C}H$ (vinyl): 68, 34, 16

$CH_2{=}CH{-}CH_2{-}\dot{C}H{-}H$: 0.6, 28.5, 22.2

Cyclopropyl: α 6.5, β 23.4

Cyclobutyl: α 21.3, β 36.8, γ 1.1

Cyclopentadienyl (−): 6.0

Cycloheptatrienyl (−): 3.9

Cyclooctatetraene (−): 3.2

$H{-}C{\equiv}C\cdot$: 16.1

Allyl ($H_2C{\cdots}C{\cdots}C$): H 14.8, H 4.1, H 13.9

Benzene (+): 2.9

Naphthalene (+): 4.9, 1.8

Anthracene (+): 6.5, 3.1, 1.4

Benzene (−): (^{1}H) 3.8, (^{13}C) 2.8

Naphthalene (−): 5.0, 1.9

Anthracene (−): 2.7, 5.3, 1.5

Phenanthrene (−): 2.9, 0.6, 0.4, 3.7, 4.4, 4.4

Toluene (−): CH_3 0.8, 5.1, 4.4, 0.6

9,10-Dimethylanthracene (−): CH_3 3.9

9,10-Diethylanthracene (−): $CH_2{-}CH_3$ 2.3

TABLE 12.38 Spin–Spin Coupling (Hyperfine Splitting Constants) (*Continued*)

$^{14}N—O^{\cdot}$ (^{14}N) 14·3

$\cdot C^{14}N$ (^{14}N) 174

(*Continued*)

TABLE 12.38 Spin–Spin Coupling (Hyperfine Splitting Constants) (*Continued*)

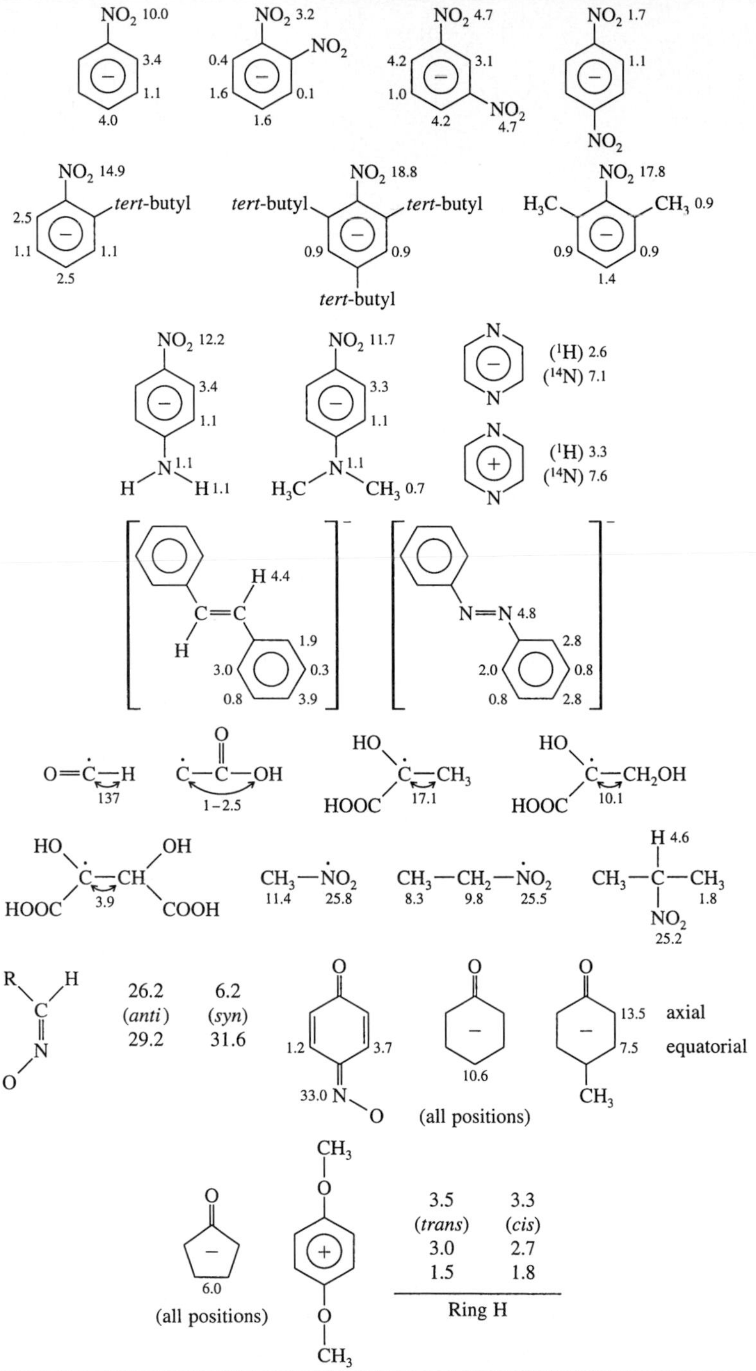

NO_2 10.0 3.4 1.1 4.0
NO_2 3.2 NO_2 0.4 1.6 0.1 1.6
NO_2 4.7 4.2 3.1 1.0 4.2 NO_2 4.7
NO_2 1.7 1.1 NO_2

NO_2 14.9 *tert*-butyl 2.5 1.1 1.1 2.5
NO_2 18.8 *tert*-butyl *tert*-butyl 0.9 0.9 *tert*-butyl
NO_2 17.8 H_3C CH_3 0.9 0.9 0.9 1.4

NO_2 12.2 3.4 1.1 N 1.1 H H 1.1
NO_2 11.7 3.3 1.1 N 1.1 H_3C CH_3 0.7
(^{1}H) 2.6 (^{14}N) 7.1
(^{1}H) 3.3 (^{14}N) 7.6

H 4.4 C=C H 1.9 3.0 0.3 0.8 3.9
N=N 4.8 2.8 2.0 0.8 0.8 2.8

O=C–H 137
C–C(=O)–OH 1–2.5
HO HOOC C–CH_3 17.1
HO HOOC C–CH_2OH 10.1

HO HOOC C–CH(OH) COOH 3.9
CH_3–NO_2 11.4 25.8
CH_3–CH_2–NO_2 8.3 9.8 25.5
H 4.6 CH_3–C–CH_3 1.8 NO_2 25.2

R H C=N O
26.2 (*anti*) 29.2
6.2 (*syn*) 31.6
O 1.2 3.7 33.0 N O
O 10.6 (all positions)
O 13.5 axial 7.5 equatorial CH_3

O 6.0 (all positions)
CH_3 O O CH_3

3.5	3.3
(*trans*)	(*cis*)
3.0	2.7
1.5	1.8
Ring H	

Source: J. A. Dean, ed., *Handbook of Organic Chemistry*, 14th ed., McGraw-Hill, New York, 1987.

12.7.5 Linewidths in ESR Spectra

ESR spectroscopy presents a time-averaged view of the geometry of a paramagnetic species. As the temperature is lowered the rate of conformational interconversion decreases and certain lines in the spectrum become broader. The lines that broaden correspond to transitions for which conformational interconversion results in a change in spin state. Eventually with further lowering of temperature the spectrum approaches that of a blocked or frozen conformation (when the conformation lifetime is more than 10^{-6} s).

12.7.6 Interpretation of ESR Spectra

An ESR spectrum is assigned by perceiving the magnitudes of the coupling constants and correctly counting the lines. The field sweep must be known. The measured spectrum may not contain all the lines expected because the g factor and coupling values can be such that two lines come very close to one another and are not resolved. When many equivalent nuclei interact, relative peak heights become very large and it is difficult to see the smallest peaks. These smaller peaks are important because the outer portions of a spectrum are invariably the simplest, and the correct interpretation of these outer lines often provides the key to unraveling the more complex central parts. With dividers and beginning at the left, ascertain the coupling constant from the leftmost line and the first larger (more intense) line to its right. Check to see if this separation is found between any other pairs of lines. If so, from the center of each multiplet the individual members of the other multiplet are located. Remember that coupling an electron spin with a single proton produces a doublet (with 1:1 intensity ratio), coupling with two equivalent protons produces a triplet (with intensity ratio 1:2:1), and coupling with three equivalent protons produces a quartet (with intensity ratio 1:3 :3:1). Another way to analyze extremely complicated spectra is to introduce approximate coupling constants into a computer program and to compare the computed spectra with the experimental ones.

12.7.7 Spin Labeling

Chemical compounds that do not contain an unpaired electron can be studied if they are chemically bonded to a stable free radical, or spin label. The organic free radicals normally employed in spin-labeling experiments are protected nitroxide free radicals (nitroxide radicals with no α-hydrogens). The ESR spectrum that results provides detailed information concerning the molecular environment of the label. Specific sites on a molecule can be tagged by selecting the appropriate spin label and reaction conditions.

When the nitroxide radical is present in low concentration in a nonviscous solvent its ESR spectrum consists of three equally spaced lines. These are further characterized by the hyperfine splitting constant, the g factor, and the linewidth. The ESR spectrum is sensitive to changes in the polarity of the label environment, the molecular motion of the label, and the orientation of the label with respect to the applied magnetic field.

12.7.8 Electron Nuclear Double Resonance (ENDOR)

Electron nuclear double resonance is a method for improving the effective resolution of an ESR spectrum. The sample is irradiated simultaneously with a microwave frequency suitable for electron resonance and a radio frequency suitable for nuclear resonance. The radio frequency is swept while one point of the ESR spectrum is observed under conditions of microwave saturation. The ENDOR display is ESR signal height as a function of the swept nuclear radio frequency. The ENDOR technique is useful when a large variety of nuclear energy levels broadens the normal electron resonance line.

12.7.9 Electron Double Resonance (ELDOR)

In electron double resonance the sample is irradiated simultaneously with two microwave frequencies. One of these is used to observe an ESR signal at some point of the spectrum, while the other is swept through other parts of the spectrum to display the ESR signal height as a function of the difference of the two microwave frequencies. ELDOR is used to separate overlapping multiradical spectra and to study relaxation phenomena such as chemical and spin exchange.

12.7.10 Quantitative Analysis

The total area enclosed by either the absorption or dispersion signal is proportional to the number of unpaired electron spins in the sample. Comparison is made with a standard containing a known number of unpaired electrons and having the same line shape as the unknown. A solid frequency used is 1,1-diphenyl-2'-picrylhydrazyl (DPPH) or solutions of peroxylamine disulfonate. DPPH contains 1.53×10^{21} unpaired spins per gram; substandards are prepared by dilution with carbon black. Secondary standards include charred dextrose or synthetic ruby attached to the cavity.

Bibliography for NMR

Allerhand, A., and S. R. Maple, "Ultra-High Resolution NMR," *Anal. Chem.* **59**:441A (1987).

Axenrod, T., and G. A. Webb, eds., *Nuclear Magnetic Resonance Spectroscopy of Nuclei Other Than Protons*, Wiley, New York, 1974.

Becker, C. D., *High Resolution NMR*, Academic, New York, 1980.

Beritmaier, E., and W. Voelter, *Carbon-13 NMR Spectroscopy*, 3d ed., VCH, Weinheim, Germany, 1990.

Bovey, F. A., *Nuclear Magnetic Resonance Spectroscopy*, 2d ed., Academic, San Diego, Calif., 1988.

Derome, A. E., *Modern NMR Techniques for Chemical Research*, Pergamon, New York, 1987.

Emsley, J. W., and L. Phillips, *Fluorine Chemical Shifts*, Pergamon, Elmsford, N. Y., 1971.

Gorenstein, D. G., ed., *Phosphorus-31 NMR: Principles and Applications*, Academic, Orlando, Fla., 1984.

Harris, R. K., *Nuclear Magnetic Resonance Spectroscopy*, Wiley, New York, 1986.

Levy, G. C., and R. L. Lichter, *Nitrogen-15 Nuclear Magnetic Resonance Spectroscopy*, Wiley, New York, 1979.

Levy, G. C., R. Lichter, and G. Nelson, *Carbon-13 Nuclear Magnetic Resonance Spectroscopy*, 2d ed., Wiley, New York, 1980.

Macomber, R. S., "A Primer on Fourier Transform NMR," *J. Chem. Educ.* **62**:213 (1985).

North, H., *NMR Spectroscopy of Boron Compounds*, Springer, Berlin, 1978.

Sanders, J. K. M., and B. K. Hunter, *Modern NMR Spectroscopy: A Guide for Chemists*, Oxford University Press, New York, 1986.

Williams, D. A. R., and D. J. Mowthorpe, *Nuclear Magnetic Resonance Spectroscopy*, Wiley, New York, 1986.

Williams, E. A., "Silicon-29 NMR," *Ann. Rep. NMR Spectrosc.* **15**:235 (1983).

Bibliography for ESR

Berliner, L. J., ed., *Spin Labeling Theory and Applications*, Academic, New York, 1976.

Bovkin, D. W., ed., *^{17}O NMR Spectroscopy in Organic Chemistry*, CRC, Boca Raton, Fla., 1990.

Dorio, M. M., and J. H. Freed, eds., *Multiple Electron Resonance Spectroscopy*, Plenum, New York, 1979.

Poole, C. S., Jr., *Electron Spin Resonance: A Comprehensive Treatise on Experimental Techniques*, 2d ed., Wiley, New York, 1983.

Ranby, B., and J. F. Rabek, *ESR Spectroscopy in Polymer Research*, Springer, New York, 1977.

SECTION 13
MAGNETIC SUSCEPTIBILITY

The magnetic susceptibility and the magnetic moment are often used to describe the magnetic behaviors of substances. Some important related concepts, formulas, and tables of magnetic properties useful to measure magnetic susceptibility are presented in this section. For theoretical and experimental details the reader should refer to the Bibliography.

13.1 BASIC CONCEPTS AND DEFINITIONS

13.1.1 Magnetic Flux and Magnetic Flux Density

The principal measure of the strength of a magnetic field is the magnetic flux density (or magnetic induction) vector $\boldsymbol{B}$ per unit area taken perpendicularly to the direction of the magnetic flux. The SI unit is the tesla T, where one tesla is one weber per square meter (or one newton per ampere per meter).

The magnetic flux Φ is the magnitude of a magnetic field, as given by the product of the magnetic flux density and area involved. It is also the total number of lines of force emanating from the (north) pole face of a magnet.

$$\Phi = \int \boldsymbol{B}\, dA \tag{13.1}$$

13.1.2 Magnetic Dipole Moment

A magnetic dipole is a macroscopic or microscopic magnetic system in which the north and south poles of a magnet, equal and opposite in character, are separated by a short but definite distance. A magnetic dipole tends to orient itself parallel to an applied magnetic field in the same way that an electric dipole behaves in an electric field.

In the presence of a magnetic field, magnetic dipoles within a material experience a turning effect and become partially oriented; the orientation is proportional to their magnetic moment m whose

SI unit is A · m^2 or J · T^{-1}. The magnetic dipole moment per unit volume is referred to as the magnetization M. The unit of the magnitude of the magnetization vector $\boldsymbol{M}$ is A · m^{-1}. The magnetic moment refers to the turning effect produced when a magnetic dipole is placed in a magnetic field, which is proportional to the magnetic dipole moment.

It is convenient to introduce another vector quantity, the magnetic field strength $\boldsymbol{H}$ (SI units, A · m^{-1}), which is defined by

$$\boldsymbol{H} \equiv \frac{\boldsymbol{B}}{\mu_0} - \boldsymbol{M} \tag{13.2}$$

where μ_0 is the permeability of vacuum. The permeability of vacuum has the value $4\pi \times 10^{-7}$ N · A^{-2} exactly. The physical significance of $\boldsymbol{H}$ is better seen by writing Eq. (13.3) as

$$\boldsymbol{B} = \mu_0(\boldsymbol{H} + \boldsymbol{M}) \tag{13.3}$$

Now we realize that the magnetic flux density $\boldsymbol{B}$ is determined by $\mu_0\boldsymbol{H}$ arising from the electric currents producing the field and $\mu_0\boldsymbol{M}$ arising from the magnetic moments induced by the magnetizing field.

The fundamental unit of magnetic moment is the Bohr magneton μ_B, which is equal to $eh/4\pi mc$, where e is the charge and m the mass of the electron, h is Planck's constant, and c is the velocity of light. Introducing the values of these quantities gives $\mu_B = 9.274 \times 10^{-24}$ J · T^{-1} (joules per tesla).

13.1.3 Magnetic Susceptibility

For isotropic substances the magnetic susceptibility χ is defined by

$$\chi = \frac{M}{H} \tag{13.4}$$

where M and H are the magnitudes of the magnetization and magnetic field vectors. Thus the magnetic susceptibility is a dimensionless quantity that can be expressed in units of per unit mass or volume. For an isotopic body the susceptibility is the same in all directions. However, for anisotropic crystals the susceptibilities along the three principal magnetic axes are different, and measurements on powdered samples give the average of the three values.

13.1.4 Atomic and Molar Susceptibilities

The atomic susceptibility χ_A and the molar susceptibility χ_M are simply defined as the susceptibility per gram-atom and per gram-mole, respectively. Hence,

$$\chi_A = \chi \times \text{atomic weight}$$

$$\chi_M = \chi \times \text{molecular weight}$$

The ionic susceptibility is similarly defined as the susceptibility per gram-ion.

If ρ is the density of a material, then the susceptibility for 1 cm^3 is equal to the volume susceptibility κ, which is the quantity measured with a Gouy balance. Hence, the susceptibility per gram of the material, called the mass or specific susceptibility χ, as measured by the Faraday method, is given by

$$\chi = \kappa/\rho \tag{13.5}$$

The magnetic susceptibility observed in bulk matter represents the contributions from both the electrons and nuclei within the system (atoms, ions, molecules, and free radicals). However, the contributions of the nuclei to the susceptibility are negligible in comparison with those of the electrons, and therefore the term *magnetic susceptibility* is usually taken to represent only the electronic property. However, some authors prefer the term *electronic susceptibility* to distinguish it from *nuclear susceptibility*.

13.2 TYPES OF MAGNETIC BEHAVIORS

If a substance has no permanent magnetic dipole, but has one induced in it by an external field, this induced magnetic field will oppose the applied field. This effect is known as *diamagnetism* and is a universal property that is shown by most inorganic compounds. It is most perceptible when all electrons are paired, that is, when they have no permanent spin moment. For a *diamagnetic* substance χ is negative, small, independent of the magnetic field intensity, and independent of temperature.

Molecules with a permanent magnetic dipole will behave like small bar magnets; they will align themselves with an applied field, thus reinforcing it. This effect is known as *paramagnetism.* Salts and certain complexes of transition elements, "odd" electron molecules like NO_2, O_2, and free radicals such as triphenylmethyl exhibit this effect, an effect sufficiently large to mask the underlying diamagnetism. For a *paramagnetic* substance χ is positive, small, independent of the magnetic field intensity, and decreases with increasing temperature.

If the permanent magnetic dipoles in a substance are so close together as to interact and support each other, the result is a group or cooperative effect known as *ferromagnetism.* For a *ferromagnetic* substance χ is positive, large, dependent on the magnetic field and temperature, and dependent on previous history. Beyond a certain temperature (the Curie point), magnetism drops and the material shows paramagnetic behavior.

For an *antiferromagnetic* substance χ is small and positive, is dependent on previous history, and has a complex temperature dependence. Up to a critical temperature, magnetization increases, then decreases past the transition temperature (known as the Néel point) as the material becomes diamagnetic.

13.3 ADDITIVITY RELATIONSHIPS

13.3.1 Mixtures and Solutions

The mass (or specific) susceptibility χ of a mixture or solution of n components having susceptibilities $\chi_1, \chi_2, \ldots, \chi_n$ and weight fractions $P_1, P_2, \ldots, P_n$ is given by Wiedemann's additivity law:

$$\chi = P_1\chi_1 + P_2\chi_2 + \cdots + P_n\chi_n \tag{13.6}$$

This is obeyed quite closely by mechanical mixtures and solutions of diamagnetic substances, in which generally little or no interaction takes place either between molecules or ions of the components, or between these and the solvent. However, caution must be exercised in deducing the susceptibility of a solute from that of its solution.

13.3.2 Additivity of Atomic Susceptibilities

According to Pascal, the molecular susceptibility χ_M of a compound can be expressed by

$$\chi_M = \sum_i (N_i\chi_i + \lambda) \tag{13.7}$$

where N_i is the number of atoms of the ith element in each molecule of the compound, and χ_i is the atomic susceptibility of that element (values of χ_i are listed in Table 13.1), while λ is a constitutive correction constant depending on the nature of the chemical bonds between the atoms. Values of λ for various important types of bonds are listed in Tables 13.2 and 13.3; those for a few ligands are listed in Table 13.4.

TABLE 13.1 **Atomic Susceptibility Constants**

Atom	χ, $\times 10^{-6}$	Atom	χ, $\times 10^{-6}$
Ag	−31.0	N: Monoamides	−1.54
Al	−13.0	Diamides and imides	−2.11
As(III)	−20.9	Na	−9.2
As(V)	−43.0	O: Alcohol or ether	−4.61
B	−7.00	Aldehyde or ketone	+1.73
Bi	−192.0	Carboxylic=O in esters and acids	−3.36
Br	−30.6		
C	−6.00	3 O atoms in acid anhydrides	−11.23
Ca	−15.9		
Cl	−20.1	P	−26.30
F	−6.3	Pb(II)	−46.0
H	−2.93	S	−15.0
Hg(II)	−33.0	Sb(III)	−74.0
I	−44.6	Se	−23.0
K	−18.5	Si	−20.0
Li	−4.2	Sn(IV)	−30.0
Mg	−10.0	Te	−37.3
N: Open chain	−5.57	Tl(I)	−40.0
Closed chain (ring)	−4.61	Zn	−13.5

Source: L. Meites, ed., *Handbook of Analytical Chemistry*, McGraw-Hill, New York, 1963.

TABLE 13.2 **Constitutive Correction Constants**

Groups containing only carbon atoms are listed first, then groups containing other atoms in the order of the symbols of these other atoms. For constitutive correction constants for cyclic systems see Table 13.3.

Group	λ, $\times 10^{-6}$	Group	λ, $\times 10^{-6}$
C=C, ethylenic linkage	+5.5	—C—I, monoiodo derivative	+4.1
C=C—C=C, diethylenic linkage	+10.6		
CH_2=CH—CH_2—, allyl group	+4.5		
C≡C, acetylenic linkage	+0.8		
Ar—C≡C—	+2.30		
Ar—C≡C—Ar	+3.85	>C=NR	+8.2
C in one aromatic ring (*e.g.*, benzene)	−0.24		
C in two aromatic rings (*e.g.*, naphthalene)	−3.1	RC≡N	+0.8
		RN≡C	±0.00
C in three aromatic rings (*e.g.*, pyrene)	−4.0	>C=N—N=C<, azines	+10.2
—C—Br, monobromo derivative	+4.1		
Br—C—C—Br, dibromo derivative	+6.24	RC≡C—C(=O)R′ or RC≡C—C(=O)OR′	+0.8
—C—Cl, monochloro derivative	+3.1		
Cl—C—C—Cl, dichloro derivative	+4.3	C bound to other C atoms with 3 bonds and in α, γ, δ, or ϵ position with respect to a carbonyl group	−1.3
>CCl_2	+1.44	C bound to other C atoms with 4 bonds and in α, γ, δ, or ϵ position with respect to a carbonyl group	−1.54
—$CHCl_2$	+6.43	C bound to other C atoms with 3 or 4 bonds and in β position with respect to a carbonyl group	−0.5
		—N=N—, azo group	+1.8
		—N=O	+1.7

Source: L. Meites, ed., *Handbook of Analytical Chemistry*, McGraw-Hill, New York, 1963.

TABLE 13.3 Constitutive Correction Constants for Cyclic Systems

Most of the values listed below are taken from Pacault, Rev. Sci. ***84****:169 (1946);* ***86****:38 (1948).*

Structure	$\lambda, \times 10^{-6}$	Structure	$\lambda, \times 10^{-6}$
Benzene	−1.41	Piperidine	+3.0
Cyclobutane	+7.2	Pyramidon	±0.0
Cyclohexadiene	+10.56	Pyrazine	+9.0
Cyclohexane	+3.0	Pyrazole	+8.0
Cyclohexene	+6.9	Pyridine	+0.5
Cyclopentane	±0.0	Pyrimidine	+6.5
Cyclopropane	+7.2	α- or γ-Pyrone	−1.4
Dicyclohexyl (C_6H_{11}—C_6H_{11})	+7.8	Pyrrole	−3.5
Dioxane	+5.5	Pyrrolidine	±0.0
Furan	−2.5	Tetrahydrofuran	±0.0
Imidazole	+8.0	Thiazole	−3.0
Isoxazole	+1.0	Thiophene	−7.0
Morpholine	+5.5	Triazine	−1.4
Piperazine	+7.0	Urazol	±0.0

Source: L. Meites, ed., *Handbook of Analytical Chemistry*, McGraw-Hill, New York, 1963.

TABLE 13.4 Diamagnetic Corrections for Various Ligands

Ligand	$\chi, \times 10^{-6}$
Dipyridyl	−105
Phenanthroline	−128
o-Phenylene-*bis*-dimethylarsine	−194
Water	−13

Source: L. Meites, ed., *Handbook of Analytical Chemistry*, McGraw-Hill, New York, 1963.

For salts, it is assumed that $\chi_M = \chi_{cation} + \chi_{anion}$. The susceptibility constants for many inorganic ions are given in Table 13.5.

13.4 MAGNETIC SUSCEPTIBILITY MEASUREMENTS

Of the various methods for measuring magnetic susceptibility of material, the Faraday and Gouy methods are the most widely used and known. Each method offers some distinct advantages. Both methods use a sensitive balance to detect the susceptibility change (as a weight change) when a sample is suspended within a magnetic field for which the product of H and dH/dx is constant and expressed by

$$F = m\chi\left(\frac{H\,dH}{dx}\right) \tag{13.8}$$

System sensitivity of commercial instruments is quoted at 0.5×10^{-6} per gram (remember susceptibility is dimensionless).

TABLE 13.5 Diamagnetic Susceptibilities of Inorganic Ions

The ionic susceptibilities listed below are reproduced, by permission of the publisher, from Selwood, Magnetochemistry, *Interscience, New York, 1956, p. 78, and are based on compilations by Klemm,* Z. anorg. u. allgem. Chem. ***244****:377 (1940);* ***246****:347 (1941). The values in parentheses are from Jagannadham,* Proc. Rajasthan. Acad. Sci. India ***1****:6 (1950). Asterisks are used to denote values for the underlying diamagnetism of paramagnetic ions.*

Ion	χ, ×10^{-6}	Ion	χ, ×10^{-6}	Ion	χ, ×10^{-6}	Ion	χ, ×10^{-6}
Ag^{+}	–24	*Eu^{2+}	–22	*Nd^{3+}	–20	SeO_3^{2-}	–44
*Ag^{2+}	–24?	*Eu^{3+}	–20	*Ni^{2+}	–12	SeO_4^{2-}	–51
Al^{3+}	– 2	F^{-}	–11	O^{2-}	–12	Si^{4+}	– 1
As^{3+}	– 9?	*Fe^{2+}	–13	OH^{-}	–12	SiO_3^{2-}	–36
As^{5+}	– 6	*Fe^{3+}	–10	*Os^{2+}	–44	*Sm^{2+}	–23
AsO_3^{3-}	–51	Ga^{3+}	– 8	*Os^{3+}	–36	*Sm^{3+}	–20
AsO_4^{3-}	–60	*Gd^{3+}	–20	*Os^{4+}	–29	Sn^{2+}	–20
Au^{+}	–40?	Ge^{4+}	– 7	*Os^{5+}	–18	Sn^{4+}	–16
Au^{3+}	–32	H^{+}	± 0	Os^{6+}	–11	Sr^{2+}	–15
B^{3+}	– 0.2	Hf^{4+}	–16	P^{3+}	– 4	Ta^{5+}	–14
BF_4^{-}	–39	Hg^{2+}	–37	P^{5+}	– 1	*Tb^{3+}	–19
BO_3^{3-}	–35	*Ho^{3+}	–19	PO_3^{-}	–30	*Tb^{4+}	–17
Ba^{2+}	–32	I^{-}	–52	PO_3^{3-}	–42	Te^{2-}	–70
Be^{2+}	– 0.4	I^{5+}	–12	Pb^{2+}	–28	Te^{4+}	–14
Bi^{3+}	–25?	I^{7+}	–10	Pb^{4+}	–26	Te^{6+}	–12
Bi^{5+}	–23	IO_3^{-}	–50	*Pd^{2+}	–25	TeO_3^{2-}	–63
Br^{-}	–36	IO_4^{-}	–54	*Pd^{4+}	–18	TeO_4^{2-}	–55
Br^{5+}	– 6	In^{3+}	–19	*Pm^{3+}	(–27)	Th^{4+}	–23
BrO_3^{-}	–40	*Ir^{+}	–50	*Pr^{3+}	–20	Ti^{2+}	(–22)
C^{4+}	– 0.1	*Ir^{2+}	–42	*Pr^{4+}	–17	Ti^{3+}	– 9
CN^{-}	–18	*Ir^{3+}	–35	*Pt^{2+}	–40	Ti^{4+}	– 5
CNO^{-}	–21	*Ir^{4+}	–29	*Pt^{3+}	–33	Tl^{+}	–34
CNS^{-}	–35	*Ir^{5+}	–20	*Pt^{4+}	–28	Tl^{3+}	–31
CO_3^{2-}	–34	K^{+}	–13	Rb^{+}	–20	*Tm^{3+}	–18
Ca^{2+}	– 8	La^{3+}	–20	*Re^{3+}	–36	*U^{3+}	–46
Cd^{2+}	–22	Li^{+}	– 0.6	*Re^{4+}	–28	*U^{4+}	–35
*Ce^{3+}	–20	Lu^{3+}	–17	*Re^{6+}	–16	*U^{5+}	–26
Ce^{4+}	–17	Mg^{2+}	– 3	Re^{7+}	–12	U^{6+}	–19
Cl	–26	Mn^{2+}	–14	*Rh^{3+}	–22	*V^{2+}	–15
Cl^{5+}	– 2	Mn^{3+}	–10	*Rh^{4+}	–18	*V^{3+}	–10
ClO_3^{-}	–32	*Mn^{4+}	– 8	*Ru^{3+}	–23	*V^{4+}	– 7
ClO_4^{-}	–34	*Mn^{6+}	– 4	*Ru^{4+}	–18	V^{5+}	– 4
*Co^{2+}	–12	*Mn^{7+}	– 3	S^{2-}	–38?	*W^{2+}	–41
*Co^{3+}	–10	*Mo^{2+}	–31	S^{4+}	– 3	*W^{3+}	–36
*Cr^{2+}	–15	*Mo^{3+}	–23	S^{6+}	– 1	*W^{4+}	–23
*Cr^{3+}	–11	*Mo^{4+}	–17	SO_3^{2-}	–38	*W^{5+}	–19
*Cr^{4+}	– 8	*Mo^{5+}	–12	SO_4^{2-}	–40	W^{6+}	–13
*Cr^{5+}	– 5	Mo^{6+}	– 7	$S_2O_8^{2-}$	–78	Y^{3+}	–12
Cr^{6+}	– 3	N^{5+}	– 0.1	Sb^{3+}	–17?	Yb^{2+}	–20
Cs^{+}	–31	NH_4^{+}	–11.5	Sb^{5+}	–14	*Yb^{3+}	–18
Cu^{+}	–12	NO_2^{-}	–10	Sc^{3+}	– 6	Zn^{2+}	–10
*Cu^{2+}	–11	NO_3^{-}	–20	Se^{2+}	–48?	Zr^{4+}	–10
*Dy^{3+}	–19	Na^{+}	– 5	Se^{4+}	– 8		
*Er^{3+}	–18	Nb^{5+}	– 9	Se^{6+}	– 5		

Source: L. Meites, ed., *Handbook of Analytical Chemistry*, McGraw-Hill, New York, 1963.

13.4.1 Faraday Method

The range of field strengths is the same as for the Gouy method, but the field must be nonuniform with a constant field gradient. The Faraday method is applicable to diamagnetic, paramagnetic, and ferromagnetic materials. Small samples are used, typically 5 to 30 mg in a quartz pan, up to 100 mg in plastic capsules; microtechniques are also available. This is advantageous if only small amounts of very expensive materials, unstable chemicals, and reaction intermediates are available. Samples may be powders or solids; 0.1 mL liquids in plastic capsules. Temperatures are possible over a wide range from those of liquid helium or nitrogen, up to +360°C. Temperature control of small samples is considerably easier than that for large samples, for temperatures both above and below ambient. This is important when studying susceptibility versus temperature for Curie and Curie–Weiss laws. In measurements mass susceptibility is determined directly and is most useful for the chemist. The accuracy of the Faraday method is ±0.1%. The flux density at the center of a 1-in gap between the pole caps is approximately 5700 G; in the constant $H\ dH/dx$ zone is approximately 4200 G.

13.4.2 Gouy Method

In the Gouy method a uniform magnetic field is employed; it is easily available with electromagnets and even with permanent magnets. The method is applicable only to the study of diamagnetic and paramagnetic materials. The Gouy method can only detect the presence of ferromagnetic materials since not all of the sample is in the same magnetic field. Powders, solids, pure liquids, and solutions can be handled; the latter without the need for special containers. The Gouy method can be adapted to the measurement of a gas surrounding a known sample. The convenient sample size is 0.5 g of solid or powder, and 5 mL of liquid; with a special apparatus, a few milligrams or even micrograms can be handled. Samples must be uniformly packed into a tube, an experimental difficulty not encountered in the Faraday method. The Gouy method yields volume susceptibility and hence the density must be known to convert it to mass susceptibility. Powder densities are difficult to measure. Temperature range is the same as for the Faraday method. Accuracy is generally ±1% but may be improved to ±0.1%.

Bibliography

Farkas, ed., *Adsorption and Collective Paramagnetism*, Academic, New York, 1962.

Michaelis, "Determination of Magnetic Susceptibility," in A. Weissberger, ed., *Technique of Organic Chemistry*, 2d ed., Interscience, New York, 1949, Vol. 1.

Mulay, L. N., "Analytical Applications of Magnetic Susceptibility," in I. M. Kolthoff and P. J. Elving, eds., *Treatise on Analytical Chemistry*, Wiley, New York, 1963, Part I, Vol. 4.

Mulay, L. N., and E. A. Boudreaux, eds., *Theory and Applications of Molecular Diagmagnetism and Paramagnetism*, Wiley-Interscience, New York, 1976.

SECTION 14
ELECTROANALYTICAL METHODS

14.1 ELECTRODE POTENTIALS

14.1.1 Introduction

Many chemical reactions can be classified as oxidation–reduction reactions (redox reactions) and can be considered as the resultant of two reactions, one oxidation and the other reduction. An element is said to have undergone oxidation if it loses electrons or if its oxidation state has increased, that is, it has attained a more positive charge. An element is said to have undergone reduction if it gains electrons or if its oxidation state has been reduced, that is, it has attained a more negative charge. Atoms of elements in their elemental state have a zero charge.

Table 14.1 reflects the relative affinity for electrons between an element and its ion or between two intermediate oxidation states of an element. The values given in Table 14.1 are reduction potentials, where E^0 is the single electrode potential when each substance involved in the oxidation–reduction reaction is at unit "activity." If a particular substance is more easily oxidized than hydrogen, its E^0 is assigned a negative value. If a substance is not oxidized as easily as hydrogen, its E^0 is positive in sign.

14.1.2 Standard and Formal Potentials

The potentials listed in Table 14.1 are given in volts *versus* the normal hydrogen electrode. Standard potentials are denoted by the absence of any entry in the column headed "Solution composition." All other values are formal potentials, *F*. The values refer to a temperature of 25°C. Many additional formal potentials may be found in Tables 14.19 and 14.20.

TABLE 14.1 Potentials of the Elements and Their Compounds at 25°C

Standard potentials are tabulated except when a solution composition is stated; the latter are formal potentials and the concentrations are in mol/liter.

Half-reaction	Standard or formal potential, V	Solution composition
Actinium		
$Ac^{3+} + 3e^- = Ac$	−2.13	
Aluminum		
$Al^{3+} + 3e^- = Al$	−1.676	
$AlF_6^{3-} + 3e^- = Al + 6F^-$	−2.07	
$Al(OH)_4^- + 3e^- = Al + 4OH^-$	−2.310	
Americium		
$AmO_2^{2+} + 4H^+ + 2e^- = Am^{4+} + 2H_2O$	1.20	
$AmO_2^{2+} + e^- = AmO_2^+$	1.59	
$AmO_2^+ + 4H^+ + e^- = Am^{4+} + 2H_2O$	0.82	
$AmO_2^+ + 4H^+ + 2e^- = Am^{3+} + 2H_2O$	1.72	
$Am^{4+} + e^- = Am^{3+}$	2.62	
$Am^{4+} + 4e^- = Am$	−0.90	
$Am^{3+} + 3e^- = Am$	−2.07	
Antimony		
$Sb(OH)_6^- + 2e^- = SbO_2^- + 2OH^- + 2H_2O$	−0.465	1 NaOH
$SbO_2^- + 2H_2O + 3e^- = Sb + 4OH^-$	0.639	1 NaOH
$Sb + 3H_2O + 3e^- = SbH_3 + 3OH^-$	−1.338	1 NaOH
$Sb_2O_5 + 6H^+ + 4e^- = 2SbO^+ + 3H_2O$	0.605	
$Sb_2O_5 + 4H^+ + 4e^- = Sb_2O_3 + 2H_2O$	0.699	
$Sb_2O_5 + 2H^+ + 2e^- = Sb_2O_4 + H_2O$	1.055	
$Sb_2O_4 + 2H^+ + 2e^- = Sb_2O_3 + H_2O$	0.342	
$SbO^+ + 2H^+ + 3e^- = Sb + H_2O$	0.204	
$Sb + 3H^+ + 3e^- = SbH_3$	−0.510	
Arsenic		
$H_3AsO_4 + 2H^+ + 2e^- = HAsO_2 + 2H_2O$	0.560	
$HAsO_2 + 3H^+ + 3e^- = As + 2H_2O$	0.240	
$As + 3H^+ + 3e^- = AsH_3$	−0.225	
$AsO_4^{3-} + 2H^+ + 2e^- = AsO_2^- + 2OH^-$	−0.67	
$AsO_2^- + 2H_2O + 3e^- = As + 4OH^-$	−0.68	
$As + 3H_2O + 3e^- = AsH_3 + 3OH^-$	−1.37	
Astatine		
$HAtO_3 + 4H^+ + 4e^- = HAtO + 2H_2O$	ca. 1.4	
$2HAtO + 2H^+ + 2e^- = At_2 + 2H_2O$	ca. 0.7	
$At_2 + 2e^- = 2At^-$	0.20	
Barium		
$BaO_2 + 4H^+ + 2e^- = Ba^{2+} + 2H_2O$	2.365	
$Ba^{2+} + 2e^- = Ba$	−2.92	
Berkelium		
$Bk^{4+} + 4e^- = Bk$	−1.05	
$Bk^{4+} + e^- = Bk^{3+}$	1.67	
$Bk^{3+} + 3e^- = Bk$	−2.01	
Beryllium		
$Be^{2+} + 2e^- = Be$	−1.99	

TABLE 14.1 Potentials of the Elements and Their Compounds at 25°C (*Continued*)

Half-reaction	Standard or formal potential, V	Solution composition
Bismuth		
Bi_2O_4 (bismuthate) + $4H^+ + 2e^- = 2BiO^+ + 2H_2O$	1.59	
$Bi^{3+} + 3e^- = Bi$	0.317	
$Bi + 3H^+ + 3e^- = BiH_3$	−0.97	
$BiCl_4^- + 3e^- = Bi + 4Cl^-$	0.199	
$BiBr_4^- + 3e^- = Bi + 4Br^-$	0.168	
$BiOCl + 2H^+ + 3e^- = Bi + H_2O + Cl^-$	0.170	
Boron		
$B(OH)_3 + 3H^+ + 3e^- = B + 3H_2O$	−0.890	
$BO_2^- + 6H_2O + 8e^- = BH_4^- + 8OH^-$	−1.241	
$B(OH)_4^- + 3e^- = B + 4OH^-$	−1.811	
Bromine		
$BrO_4^- + 2H^+ + 2e^- = BrO_3^- + H_2O$	1.853	
$BrO_3^- + 6H^+ + 6e^- = Br^- + 3H_2O$	1.478	
$BrO_3^- + 5H^+ + 4e^- = HBrO + 2H_2O$	1.444	
$2BrO_3^- + 12H^+ + 10e^- = Br_2 + 6H_2O$	1.5	
$2HBrO + 2H^+ + 2e^- = Br_2 + 2H_2O$	1.604	
$HBrO + H^+ + 2e^- = Br^- + H_2O$	1.341	
$BrO^- + H_2O + 2e^- = Br^- + 2OH^-$	0.76	1 NaOH
$Br_3^- + 2e^- = 3Br^-$	1.050	
$Br_2(aq) + 2e^- = 2Br^-$	1.087	
Cadmium		
$Cd^{2+} + 2e^- = Cd$	−0.403	
$Cd^{2+} + Hg + 2e^- = Cd(Hg)$	−0.352	
$CdCl_4^{2-} + 2e^- = Cd + 4Cl^-$	−0.453	
$Cd(CN)_4^{2-} + 2e^- = Cd + 4CN^-$	−0.943	
$Cd(NH_3)_4^{2+} + 2e^- = Cd + 4NH_3$	−0.622	
$Cd(OH)_4^{2-} + 2e^- = Cd + 4OH^-$	−0.670	
Calcium		
$CaO_2 + 4H^+ + 2e^- = Ca^{2+} + 2H_2O$	2.224	
$Ca^{2+} + 2e^- = Ca$	−2.84	
$Ca + 2H^+ + 2e^- = CaH_2$	0.776	
Californium		
$Cf^{3+} + 3e^- = Cf$	−1.93	
$Cf^{3+} + e^- = Cf^{2+}$	−1.6	
$Cf^{2+} + 2e^- = Cf$	−2.1	
Carbon		
$CO_2 + 2H^+ + 2e^- = CO + H_2O$	−0.106	
$CO_2 + 2H^+ + 2e^- = HCOOH$	−0.20	
$2CO_2 + 2H^+ + 2e^- = H_2C_2O_4$	−0.481	
$C_2O_4^{2-} + 2H^+ + 2e^- = 2HCOO^-$	0.145	
$HCOOH + 2H^+ + 2e^- = HCHO + H_2O$	0.034	
$C_2N_2 + 2H^+ + 2e^- = 2HCN$	0.373	
$HCHO + 2H^+ + 2e^- = CH_3OH$	0.2323	
$CNO^- + H_2O + 2e^- = CN^- + 2OH^-$	−0.97	

(*Continued*)

TABLE 14.1 Potentials of the Elements and Their Compounds at 25°C (*Continued*)

Half-reaction	Standard or formal potential, V	Solution composition
Cerium		
Ce(IV) + e^- = Ce(III)	1.70	1 $HClO_4$
	1.61	1 HNO_3
	1.44	0.5 H_2SO_4
	1.28	1 HCl
$Ce^{3+} + 3e^- = Ce$	−2.34	
Cesium		
$Cs^+ + e^- = Cs$	−2.923	
Cesium		
$Cs^+ + Hg + e^- = Cs(Hg)$	−1.78	
Chlorine		
$ClO_4^- + 2H^+ + 2e^- = ClO_3^- + H_2O$	1.201	
$2ClO_4^- + 16H^+ + 14e^- = Cl_2 + 8H_2O$	1.392	
$ClO_4^- + 8H^+ + 8e^- = Cl^- + 4H_2O$	1.388	
$ClO_3^- + 2H^+ + e^- = ClO_2(g) + H_2O$	1.175	
$ClO_3^- + 3H^+ + 2e^- = HClO_2 + H_2O$	1.181	
$2ClO_3^- + 12H^+ + 10e^- = Cl_2 + 6H_2O$	1.468	
$ClO_3^- + 6H^+ + 6e^- = Cl^- + 3H_2O$	1.45	
$ClO_2(g) + H^+ + e^- = HClO_2$	1.188	
$HClO_2 + 2H^+ + 2e^- = HClO + H_2O$	1.64	
$HClO_2 + 3H^+ + 4e^- = Cl^- + 2H_2O$	1.584	
$2HClO_2 + 6H^+ + 6e^- = Cl_2(g) + 4H_2O$	1.659	
$2ClO^- + 2H_2O + 2e^- = Cl_2(g) + 4OH^-$	0.421	1 NaOH
$ClO^- + H_2O + 2e^- = Cl^- + 2OH^-$	0.890	1 NaOH
$Cl_3^- + 2e^- = 3Cl^-$	1.415	
$Cl_2(aq) + 2e^- = 2Cl^-$	1.396	
Chromium		
$Cr_2O_7^{2-} + 14H^+ + 6e^- = 2Cr^{3+} + 7H_2O$	1.36	
	1.15	0.1 H_2SO_4
	1.03	1 $HClO_4$
$CrO_4^{2-} + 4H_2O + 3e^- = Cr(OH)_4^- + 4OH^-$	−0.13	1 NaOH
$Cr^{3+} + e^- = Cr^{2+}$	−0.424	
$Cr^{3+} + 3e^- = Cr$	−0.74	
$Cr^{2+} + 2e^- = Cr$	0.90	
Cobalt		
$CoO_2 + 4H^+ + e^- = Co^{3+} + 2H_2O$	1.416	
$Co(H_2O)_6^{3+} + e^- = Co(H_2O)_6^{2+}$	1.92	
$Co(NH_3)_6^{3+} + e^- = Co(NH_3)_6^{2+}$	0.058	7 NH_3
$Co(OH)_3 + e^- = Co(OH)_2 + OH^-$	0.17	
$Co(en)_3^{3+} + e^- = Co(en)_3^{2+}$ [en = ethylenediamine]	−0.2	0.1 en
$Co(CN)_6^{3-} + e^- = Co(CN)_5^{2-} + CN^-$	−0.8	0.8 KOH
$Co^{2+} + 2e^- = Co$	−0.277	
$Co(NH_3)_6^{2+} + 2e^- = Co + 6NH_3$	−0.422	
$[Co(CO)_4]_2 + 2e^- = 2Co(CO)_4^-$	−0.40	
Copper		
$Cu^{2+} + 2e^- = Cu$	0.340	
$Cu^{2+} + e^- = Cu^+$	0.159	

TABLE 14.1 Potentials of the Elements and Their Compounds at 25°C (*Continued*)

Half-reaction	Standard or formal potential, V	Solution composition
Copper		
$Cu^+ + e^- = Cu$	0.520	
$Cu^{2+} + Cl^- + e^- = CuCl$	0.559	
$Cu^{2+} + 2Br^- + e^- = CuBr_2^-$	0.52	1 KBr
$Cu^{2+} + I^- + e^- + CuI$	0.86	
$Cu^{2+} + 2CN^- + e^- = Cu(CN)_2^-$	1.12	
$Cu(NH_3)_4^{2+} + e^- = Cu(NH_3)_2^+ + 2NH_3$	0.10	1 NH_3
$Cu(en)_2^{2+} + e^- = Cu(en)^+ + en$	–0.35	
$Cu(CN)_2^- + e^- = Cu + 2CN^-$	–0.44	
$CuCl_3^{2-} + e^- = Cu + 3Cl^-$	0.178	1 HCl
$Cu(NH_3)_2^+ + e^- = Cu + 2NH_3$	–0.100	
Curium		
$Cm^{4+} + e^- = Cm^{3+}$	3.2	1 $HClO_4$
$Cm^{3+} + 3e^- = Cm$	–2.06	
Dysprosium		
$Dy^{3+} + 3e^- = Dy$	–2.29	
$Dy^{3+} + e^- = Dy^{2+}$	–2.5	
$Dy^{2+} + 2e^- = Dy$	–2.2	
Einsteinium		
$Es^{3+} + 3e^- = Es$	–2.0	
$Es^{3+} + e^- = Es^{2+}$	–1.5	
$Es^{2+} + 2e^- = Es$	–2.2	
Erbium		
$Er^{3+} + 3e^- = Er$	–2.32	
Europium		
$Eu^{3+} + 3e^- = Eu$	–1.99	
$Eu^{3+} + e^- = Eu^{2+}$	–0.35	
$Eu^{2+} + 2e^- = Eu$	–2.80	
Fermium		
$Fm^{3+} + 3e^- = Fm$	–1.96	
$Fm^{3+} + e^- = Fm^{2+}$	–1.15	
$Fm^{2+} + 2e^- = Fm$	–2.37	
Fluorine		
$F_2 + 2H^+ + 2e^- = 2HF$	3.053	
$F_2 + H^+ + 2e^- = HF_2^-$	2.979	
$F_2 + 2e^- = 2F^-$	2.87	
$OF_2 + 3H^+ + 4e^- = HF_2^- + H_2O$	2.209	
Francium		
$Fr^+ + e^- = Fr$	ca. –2.9	
Gadolinium		
$Gd^{3+} + 3e^- = Gd$	–2.28	
Gallium		
$Ga^{3+} + 3e^- = Ga$	–0.529	
$Ga^{3+} + e^- = Ga^{2+}$	–0.65	
$Ga^{2+} + 2e^- = Ga$	–0.45	
Germanium		
$GeO_2(tetr) + 2H^+ + 2e^- = GeO(yellow) + H_2O$	–0.255	
$GeO_2(tetr) + 4H^+ + 2e^- = Ge^{2+} + 2H_2O$	–0.210	

(Continued)

TABLE 14.1 Potentials of the Elements and Their Compounds at 25°C (*Continued*)

Half-reaction	Standard or formal potential, V	Solution composition
Germanium		
$GeO_2(hex) + 4H^+ + 2e^- = Ge^{2+} + 2H_2O$	−0.132	
$H_2GeO_3 + 4H^+ + 4e^- = Ge + 3H_2O$	0.012	
$Ge^{4+} + 2e^- = Ge^{2+}$	0.0	
$Ge^{2+} + 2e^- = Ge$	0.247	
$GeO + 2H^+ + 2e^- = Ge + H_2O$	−0.255	
$Ge + 4H^+ + 4e^- = GeH_4$	−0.29	
Gold		
$Au^{3+} + 3e^- = Au$	1.52	
$Au^{3+} + 2e^- = Au^+$	1.36	
$Au^+ + e^- = Au$	1.83	
$AuCl_4^- + 2e^- = AuCl_2^- + 2Cl^-$	0.926	
$AuBr_4^- + 2e^- = AuBr_2^- + 2Br^-$	0.802	
$Au(SCN)_4^- + 2e^- = Au(SCN)_2^- + 2SCN^-$	0.623	
$AuBr_4^- + 3e^- = Au + 4Br^-$	0.854	
$AuCl_4^- + 3e^- = Au + 4Cl^-$	1.002	
$Au(SCN)_4^- + 3e^- = Au + 4SCN^-$	0.662	
$Au(OH)_3 + 3H^+ + 3e^- = Au + 3H_2O$	1.45	
$AuBr_2^- + e^- = Au + 2Br^-$	0.960	
$AuCl_2^- + e^- = Au + 2Cl^-$	1.15	
$AuI_2^- + e^- = Au + 2I^-$	0.576	
$Au(CN)_2^- + e^- = Au + 2CN^-$	−0.596	
$Au(SCN)_2 + e^- = Au + 2SCN^-$	0.69	
Hafnium		
$Hf^{4+} + 4e^- = Hf$	−1.70	
$HfO_2 + 4H^+ + 4e^- = Hf + 2H_2O$	−1.57	
Holmium		
$Ho^{3+} + 3e^- = Ho$	−2.23	
Hydrogen		
$2H^+ + 2e^- = H_2$	0.0000	
$2D^+ + 2e^- = D_2$	0.029	
$2H_2O + 2e^- = H_2 + 2OH^-$	−0.828	
Indium		
$In^{3+} + 3e^- = In$	−0.338	
$In^{3+} + 2e^- = In^+$	−0.444	
$In^+ + e^- = In$	−0.126	
Iodine		
$H_5IO_6 + H^+ + 2e^- = IO_3^- + 3H_2O$	1.603	
$IO_3^- + 5H^+ + 4e^- = HIO + 2H_2O$	1.14	
$HIO_3 + 5H^+ + 2Cl^- + 4e^- = ICl_2^- + 3H_2O$	1.214	
$2IO_3^- + 12H^+ + 10e^- = I_2(c) + 6H_2O$	1.195	
$IO_3^- + 3H_2O + 6e^- = I^- + 6OH^-$	0.257	
$2IBr_2^- + 2e^- = I_2Br^- + 3Br^-$	0.821	
$2IBr_2^- + 2e^- = I_2(c) + 4Br^-$	0.874	
$2IBr + 2e^- = I_2Br^- + Br^-$	0.973	
$2IBr + 2e^- = I_2 + 2Br^-$	1.02	
$2ICl + 2e^- = I_2(c) + 2Cl^-$	1.20	
$2ICl_2^- + 2e^- = I_2(c) + 4Cl^-$	1.07	
$2ICN + 2H^+ + 2e^- = I_2(c) + 2HCN$	0.695	

TABLE 14.1 Potentials of the Elements and Their Compounds at 25°C (*Continued*)

Half-reaction	Standard or formal potential, V	Solution composition
Iodine		
$2ICN + 2H^+ + 2e^- = I_2(aq) + 2HCN$	0.609	
$2HIO + 2H^+ + 2e^- = I_2 + 2H_2O$	1.45	
$HIO + H^+ + 2e^- = I^- + H_2O$	0.985	
$I_3^- + 2e^- = 3I^-$	0.536	
$I_2(aq) + 2e^- = 2I^-$	0.621	
$I_2(c) + 2e^- = 2I^-$	0.5355	
Iridium		
$IrBr_6^{2-} + e^- = IrBr_6^{3-}$	0.805	
$IrCl_6^{2-} + e^- = IrCl_6^{3-}$	0.867	
$IrI_6^{2-} + e^- = IrI_6^{3-}$	0.49	
$IrO_2 + 4H^+ + e^- = Ir^{3+} + 2H_2O$	0.223	
$IrO_2 + 4H^+ + 4e^- = Ir + 2H_2O$	0.935	1 H_2SO_4
$Ir^{3+} + 3e^- = Ir$	1.156	
$IrCl_6^{2-} + 4e^- = Ir + 6Cl^-$	0.835	
$IrCl_6^{3-} + 3e^- = Ir + 6Cl^-$	0.77	
Iron		
$FeO_4^{2-} + 8H^+ + 3e^- = Fe^{3+} + 4H_2O$	2.2	
$FeO_4^{2-} + 2H_2O + 3e^- = FeO_2^- + 4OH^-$	0.55	10 NaOH
$Fe^{3+} + e^- = Fe^{2+}$	0.771	
	0.70	1 HCl
	0.67	0.5 H_2SO_4
	0.44	0.3 H_3PO_4
$Fe(CN)_6^{3-} + e^- = Fe(CN)_6^{4-}$	0.361	
	0.71	1 HCl
$Fe(EDTA)^- + e^- = Fe(EDTA)^{2-}$	0.12	0.1 EDTA, pH 4–6
$Fe(OH)_4^- + e^- = Fe(OH)_4^{2-}$	−0.73	1 NaOH
$Fe^{2+} + 2e^- = Fe$	−0.44	
$[Fe(CO)_4]_3 + 6e^- = 3Fe(CO)_4^{2-}$	−0.70	
Lanthanum		
$La^{3+} + 3e^- = La$	−2.38	
Lawrencium		
$Lr^{3+} + 3e^- = Lr$	−2.0	
Lead		
$Pb^{4+} + 2e^- = Pb^{2+}$	1.65	
$PbO_2(\alpha) + SO_4^{2-} + 4H^+ + 2e^- = PbSO_4 + 2H_2O$	1.690	
$PbO_2 + 4H^+ + 2e^- = Pb^{2+} + 2H_2O$	1.46	
$PbO_2 + 2H^+ + 2e^- = PbO + H_2O$	0.28	
$PbO_3^{2-} + 2H_2O + 2e^- = HPbO_2^- + 3OH^-$	0.3	2 NaOH
$Pb^{2+} + 2e^- = Pb$	−0.126	
$HPbO_2^- + H_2O + 2e^- = Pb + 3OH^-$	−0.54	
$PbHPO_4 + 2e^- = Pb + HPO_4^{2-}$	−0.465	
$PbSO_4 + 2e^- = Pb + SO_4^{2-}$	−0.356	
$PbF_2 + 2e^- = Pb + 2F^-$	−0.344	
$PbCl_2 + 2e^- = Pb + 2Cl^-$	−0.268	
$PbBr_2 + 2e^- = Pb + 2Br^-$	−0.280	
$PbI_2 + 2e^- = Pb + 2I^-$	−0.365	
$Pb + 2H^+ + 2e^- = PbH_2$	−1.507	

(Continued)

TABLE 14.1 Potentials of the Elements and Their Compounds at 25°C (*Continued*)

Half-reaction	Standard or formal potential, V	Solution composition
Lithium		
$Li^+ + e^- = Li$	−3.040	
$Li^+ + Hg + e^- = Li(Hg)$	−2.00	
Lutetium		
$Lu^{3+} + 3e^- = Lu$	−2.30	
Magnesium		
$Mg^{2+} + 2e^- = Mg$	−2.356	
$Mg(OH)_2 + 2e^- = Mg + 2OH^-$	−2.687	
Manganese		
$MnO_4^- + e^- = MnO_4^{2-}$	0.56	
$MnO_4^- + 4H^+ + 3e^- = MnO_2(\beta) + 2H_2O$	1.70	
$MnO_4^- + 2H_2O + 3e^- = MnO_2 + 4OH^-$	0.60	
$MnO_4^- + 8H^+ + 5e^- = Mn^{2+} + 4H_2O$	1.51	
$MnO_4^{2-} + e^- = MnO_4^{3-}$	0.27	
$MnO_4^{2-} + 2H_2O + 2e^- = MnO_2 + 4OH^-$	0.62	
$MnO_4^{3-} + 2H_2O + e^- = MnO_2 + 4OH^-$	0.96	
$MnO_2 + 4H^+ + e^- = Mn^{3+} + 2H_2O$	0.95	
$MnO_2(\beta) + 4H^+ + 2e^- = Mn^{2+} + 2H_2O$	1.23	
$Mn^{3+} + e^- = Mn^{2+}$	1.5	
$Mn(H_2P_2O_7)_3^{3-} + 2H^+ + e^- = Mn(H_2P_2O_7)_2^{2-} + H_4P_2O_7$	1.15	0.4 $H_2P_2O_7^{2-}$
$Mn(CN)_6^{3-} + e^- = Mn(CN)_6^{4-}$	−0.24	1.5 NaCN
$Mn^{2+} + 2e^- = Mn$	−1.17	
Mendelevium		
$Md^{3+} + 3e^- = Md$	−1.7	
$Md^{3+} + e^- = Md^{2+}$	−0.15	
$Md^{2+} + 2e^- = Md$	−2.4	
Mercury		
$2Hg^{2+} + 2e^- = Hg_2^{2+}$	0.911	
$2HgCl_2 + 2e^- = Hg_2Cl_2 + 2Cl^-$	0.63	
$Hg^{2+} + 2e^- = Hg(1q)$	0.8535	
$HgO(c,red) + 2H^+ + 2e^- = Hg + H_2O$	0.926	
$Hg_2^{2+} + 2e^- = 2Hg$	0.7960	
$Hg_2F_2 + 2e^- = 2Hg + 2F^-$	0.656	
$Hg_2Cl_2 + 2e^- = 2Hg + 2Cl^-$	0.2682	
$Hg_2Br_2 + 2e^- = 2Hg + 2Br^-$	0.1392	
$Hg_2I_2 + 2e^- = 2Hg + 2I^-$	−0.0405	
$Hg_2SO_4 + 2e^- = 2Hg + SO_4^{2-}$	0.614	
Molybdenum		
$MoO_4^{2-} + 4H_2O + 6e^- = Mo + 8OH^-$	−0.913	
$H_2MoO_4 + 6H^+ + 6e^- = Mo + 4H_2O$	0.114	
$H_2MoO_4 + 2H^+ + 2e^- = MoO_2 + 2H_2O$	0.646	
$MoO_2 + 4H^+ + 4e^- = Mo + 2H_2O$	−0.152	
$H_2MoO_4 + 6H^+ + 3e^- = Mo^{3+} + 4H_2O$	0.428	
$Mo(CN)_8^{3-} + e^- = Mo(CN)_8^{4-}$	0.725	
$Mo^{3+} + 3e^- = Mo$	−0.2	
Neodymium		
$Nd^{3+} + 3e^- = Nd$	−2.32	
$Nd^{3+} + e^- = Nd^{2+}$	−2.6	
$Nd^{2+} + 2e^- = Nd$	−2.2	

TABLE 14.1 Potentials of the Elements and Their Compounds at 25°C (*Continued*)

Half-reaction	Standard or formal potential, V	Solution composition
Neptunium		
$NpO_3^+ + 2H^+ + e^- = NpO_2^{2+} + H_2O$	2.04	
$NpO_2^{2+} + e^- = NpO_2^+$	1.34	
$NpO_2^{2+} + 4H^+ + 2e^- = Np^{4+} + 2H_2O$	0.95	
$Np^{4+} + e^- = Np^{3+}$	0.18	
$Np^{4+} + 4e^- = Np$	−1.30	
$Np^{3+} + 3e^- = Np$	−1.79	
Nickel		
$NiO_4^{2-} + 4H^+ + 2e^- = NiO_2 + 2H_2O$	1.8	
$NiO_2 + 4H^+ + 2e^- = Ni^{2+} + 2H_2O$	1.593	
$NiO_2 + 2H_2O + 2e^- = Ni(OH)_2 + 2OH^-$	0.490	
$Ni(CN)_4^{2-} + e^- = Ni(CN)_3^{2-} + CN^-$	−0.401	
$Ni^{2+} + 2e^- = Ni$	−0.257	
$Ni(OH)_2 + 2e^- = Ni + 2OH^-$	−0.72	
$Ni(NH_3)_6^{2+} + 2e^- = Ni + 6NH_3$	−0.49	
Niobium		
$Nb_2O_5 + 10H^+ + 4e^- = 2Nb^{3+} + 5H_2O$	−0.1	
$Nb_2O_5 + 10H^+ + 10e^- = 2Nb + 5H_2O$	−0.65	
$Nb^{3+} + 3e^- = Nb$	−1.1	
Nitrogen		
$2NO_3^- + 4H^+ + 2e^- = N_2O_4 + 2H_2O$	0.803	
$NO_3^- + 3H^+ + 2e^- = HNO_2 + H_2O$	0.94	
$N_2O_4 + 2H^+ + 2e^- = 2HNO_2$	1.07	
$HNO_2 + H^+ + e^- = NO + H_2O$	0.996	
$2HNO_2 + 4H^+ + 4e^- = N_2O(g) + 3H_2O$	1.297	
$2HNO_2 + 4H^+ + 4e^- = H_2N_2O_2 + 2H_2O$	0.86	
$2NO + 2H^+ + 2e^- = H_2N_2O_2$	0.71	
$2NO + 2H^+ + 2e^- = N_2O + H_2O$	1.59	
$H_2N_2O_2 + 6H^+ + 4e^- = 2HONH_3^+$	0.496	
$N_2O + 2H^+ + 2e^- = N_2 + H_2O$	1.77	
$N_2O + 6H^+ + H_2O + 4e^- = 2HONH_3^+$	−0.05	
$N_2 + 2H_2O + 4H^+ + 2e^- = 2HONH_3^+$	−1.87	
$N_2 + 5H^+ + 4e^- = N_2H_5^+$	−0.23	
$HONH_3^+ + 2H^+ + 2e^- = NH_4^+ + H_2O$	1.35	
$2HONH_3^+ + H^+ + 2e^- = N_2H_5^+ + 2H_2O$	1.41	
$N_2H_5^+ + 3H^+ + 2e^- = 2NH_4^+$	1.275	
$3N_2 + 2H^+ + 2e^- = 2HN_3$	−3.40	
Nobelium		
$No^{3+} + 3e^- = No$	−1.2	
$No^{3+} + e^- = No^{2+}$	1.4	
$No^{2+} + 2e^- = No$	−2.5	
Osmium		
$OsO_4(aq) + 4H^+ + 4e^- = OsO_2 \cdot 2H_2O$	0.964	
$OsO_4(c, yellow) + 8H^+ + 8e^- = Os + 4H_2O$	0.85	
$OsO_2 + 4H^+ + 4e^- = Os + 2H_2O$	0.687	
$OsCl_6^{2-} + e^- = OsCl_6^{3-}$	0.45	
$OsBr_6^{2-} + e^- = OsBr_6^{3-}$	0.35	
Oxygen		
$O_3 + 2H^+ + 2e^- = O_2 + H_2O$	2.075	

(*Continued*)

TABLE 14.1 Potentials of the Elements and Their Compounds at 25°C (*Continued*)

Half-reaction	Standard or formal potential, V	Solution composition
Oxygen		
$O_3 + H_2O + 2e^- = O_2 + 2OH$	1.240	1 NaOH
$O_2 + 4H^+ + 4e^- = 2H_2O$	1.229	
$O_2 + 2H^+ + 2e^- = H_2O_2$	0.695	
$O_2 + H_2O + 2e^- = HO_2^- + OH^-$	−0.076	
$H_2O_2 + 2H^+ + 2e^- = 2H_2O$	1.763	
$HO_2^- + H_2O + 2e^- = 3OH^-$	0.867	1 NaOH
$O_2 + 2H_2O + 4e^- = 4OH^-$	0.401	
Palladium		
$PdO_3 + 2H^+ + 2e^- = PdO_2 + H_2O$	2.030	
$PdCl_6^{2-} + 2e^- = PdCl_4^{2-} + 2Cl^-$	1.470	
$PdBr_6^{2-} + 2e^- = PdBr_4^{2-} + 2Br^-$	0.99	
$PdI_6^{2-} + 2e^- = PdI_4^{2-} + 2I^-$	0.48	
$Pd^{2+} + 2e^- = Pd$	0.915	
$PdCl_4^{2-} + 2e^- = Pd + 4Cl^-$	0.62	1 HCl
$PdBr_4^{2-} + 2e^- = Pd + 4Br^-$	0.49	
$Pd(NH_3)_4^{2+} + 2e^- = Pd + 4NH_3$	0.0	1 NH_3
$Pd(CN)_4^{2-} + 2e^- = Pd + 4CN^-$	−1.35	1 KCN
Phosphorus		
$H_3PO_4 + 2H^+ + 2e^- = H_3PO_3 + H_2O$	−0.276	
$2H_3PO_4 + 2H^+ + 2e^- = H_4P_2O_6 + 2H_2O$	−0.933	
$H_4P_2O_6 + 2H^+ + 2e^- = 2H_3PO_3$	0.380	
$H_3PO_3 + 2H^+ + 2e^- = HPH_2O_2 + H_2O$	−0.499	
$HPH_2O_2 + H^+ + e^- = P + 2H_2O$	−0.365	
$H_3PO_3 + 3H^+ + 3e^- = P + 3H_2O$	−0.502	
$2P(white) + 4H^+ + 4e^- = P_2H_4$	−0.100	
$P_2H_4 + 2H^+ + 2e^- = 2PH_3$	−0.006	
$P(white) + 3H^+ + 3e^- = PH_3$	−0.063	
Platinum		
$PtO_3 + 2H^+ + 2e^- = PtO_2 + H_2O$	2.0	
$PtO_2 + 2H^+ + 2e^- = PtO + H_2O$	1.045	
$PtCl_6^{2-} + 2e^- = PtCl_4^{2-} + 2Cl^-$	0.726	
$PtBr_6^{2-} + 2e^- = PtBr_4^{2-} + 2Br^-$	0.613	1 KBr
$PtI_6^{2-} + 2e^- = PtI_4^{2-} + 2I^-$	0.321	1 KI
$Pt^{2+} + 2e^- = Pt$	1.188	
$PtCl_4^{2-} + 2e^- = Pt + 4Cl^-$	0.758	
$PtBr_4^{2-} + 2e^- = Pt + 4Br^-$	0.698	
Plutonium		
$PuO_2^{2+} + e^- = PuO_2^+$	1.02	
$PuO_2^{2+} + 4H^+ + 2e^- = Pu^{4+} + 2H_2O$	1.04	
$Pu^{4+} + e^- = Pu^{3+}$	1.01	
	0.80	1 H_3PO_4
	0.50	1 HF
$Pu^{4+} + 4e^- = Pu$	−1.25	
$Pu^{3+} + 3e^- = Pu$	−2.00	
Polonium		
$PoO_2 + 4H^+ + 2e^- = Po^{2+} + 2H_2O$	1.1	
$Po^{4+} + 4e^- = Po$	0.73	
$Po^{2+} + 2e^- = Po$	0.37	
$Po + 2H^+ + 2e^- = H_2Po$	ca. −1.0	

TABLE 14.1 Potentials of the Elements and Their Compounds at 25°C (*Continued*)

Half-reaction	Standard or formal potential, V	Solution composition
Potassium		
$K^+ + e^- = K$	−2.924	
$K^+ + Hg + e^- = K(Hg)$	ca. −1.9	
Praseodymium		
$Pr^{4+} + e^- = Pr^{3+}$	3.2	
$Pr^{3+} + e^- = Pr$	−2.35	
Promethium		
$Pm^{3+} + 3e^- = Pm$	−2.42	
Protoactinium		
$PaOOH^{2+} + 3H^+ + e^- = Pa^{4+} + 2H_2O$	−0.10	
$PaOOH^{2+} + 3H^+ + 5e^- = Pa + 2H_2O$	−1.19	
$Pa^{4+} + 4e^- = Pa$	−1.46	
Radium		
$Ra^{2+} + 2e^- = Ra$	−2.916	
Rhenium		
$ReO_4^- + 2H^+ + e^- = ReO_3 + H_2O$	0.768	
$ReO_4^- + 4H^+ + 3e^- = ReO_2 + 2H_2O$	0.51	
$ReO_4^- + 2H_2O + 3e^- = ReO_2 + 4OH^-$	−0.594	
$ReO_4^- + 6Cl^- + 8H^+ + 3e^- = ReCl_6^{2-} + 4H_2O$	0.12	
$2ReO_4^- + 10H^+ + 8e^- = Re_2O_3 + 5H_2O$	−0.808	
$ReO_3 + 2H^+ + 2e^- = ReO_2 + H_2O$	0.63	
$ReO_2 + 4H^+ + 4e^- = Re + 2H_2O$	0.22	
$ReCl_6^{2-} + 4e^- = Re + 6Cl^-$	0.51	
$Re + e^- = Re^-$	−0.10	
Rhodium		
$RhO_2 + 4H^+ + e^- = Rh^{3+} + 2H_2O$	1.881	
$Rh^{3+} + 3e^- = Rh$	0.76	
$RhCl_6^{3-} + 3e^- = Rh + 6Cl^-$	0.5	
Rubidium		
$Rb^+ + e^- = Rb$	−2.924	
$Rb^+ + Hg + e^- = Rb(Hg)$	−1.81	
Ruthenium		
$RuO_4 + e^- = RuO_4^-$	0.89	
$RuO_4 + 4H^+ + 4e^- = RuO_2 + 2H_2O$	1.4	
$RuO_4 + 8H^+ + 8e^- = Ru + 4H_2O$	1.04	
$RuO_4^- + e^- = RuO_4^{2-}$	0.593	
$RuO_4^{2-} + 4H^+ + 2e^- = RuO_2 + 2H_2O$	2.0	
$RuO_2 + 4H^+ + 4e^- = Ru + 2H_2O$	0.68	
$Ru(H_2O)_6^{3+} + e^- = Ru(H_2O)_6^{2+}$	0.249	
$Ru(NH_3)_6^{3+} + e^- = Ru(NH_3)_6^{2+}$	0.10	
$Ru(CN)_6^{3-} + e^- = Ru(CN)_6^{4-}$	0.86	
$Ru^{3+} + e^- = Ru^{2+}$	0.249	
Samarium		
$Sm^{3+} + 3e^- = Sm$	−2.30	
$Sm^{3+} + e^- = Sm^{2+}$	−1.55	
$Sm^{2+} + 2e^- = Sm$	−2.67	
Scandium		
$Sc^{3+} + 3e^- = Sc$	−2.03	

(*Continued*)

TABLE 14.1 Potentials of the Elements and Their Compounds at 25°C (*Continued*)

Half-reaction	Standard or formal potential, V	Solution composition
Selenium		
$SeO_4^{2-} + 4H^+ + 2e^- = H_2SeO_3 + H_2O$	1.151	
$H_2SeO_3 + 4H^+ + 4e^- = Se + 3H_2O$	0.74	
$Se(c) + 2H^+ + 2e^- = H_2Se(aq)$	–0.115	
$Se + H^+ + 2e^- = HSe^-$	–0.227	
$Se + 2e^- = Se^{2-}$	–0.670	1 NaOH
Silicon		
$SiO_2(quartz) + 4H^+ + 4e^- = Si + 2H_2O$	–0.909	
$SiO_2 + 2H^+ + 2e^- = SiO + H_2O$	–0.967	
$SiO_2 + 8H^+ + 8e^- = SiH_4 + 2H_2O$	–0.516	
$SiF_6^{2-} + 4e^- = Si + 6F^-$	–1.37	
$SiO + 2H^+ + 2e^- = Si + H_2O$	–0.808	
$Si + 4H^+ + 4e^- = SiH_4(g)$	–0.143	
Silver		
$AgO^+ + 2H^+ + e^- = Ag^{2+} + H_2O$	1.360	
$Ag_2O_3 + 2H^+ + 2e^- = 2AgO + H_2O$	1.569	
$Ag_2O_3 + H_2O + 2e^- = 2AgO + 2OH^-$	0.739	1 NaOH
$Ag_2O_3 + 6H^+ + 4e^- = 2Ag^+ + 3H_2O$	1.670	
$Ag^{2+} + e^- = Ag^+$	1.980	
$AgO + 2H^+ + e^- = Ag^+ + H_2O$	1.772	
$Ag^+ + e^- = Ag$	0.7991	
$Ag_2SO_4 + 2e^- = 2Ag + SO_4^{2-}$	0.653	
$Ag_2C_2O_4 + 2e^- = 2Ag + C_2O_4^{2-}$	0.47	
$Ag_2CrO_4 + 2e^- = 2Ag + CrO_4^{2-}$	0.447	
$Ag(NH_3)_2^+ + e^- = Ag + 2NH_3$	0.373	
$AgCl + e^- = Ag + Cl^-$	0.2223	
$AgBr + e^- = Ag + Br^-$	0.071	
$AgCN + e^- = Ag + CN^-$	–0.017	
$AgI + e^- = Ag + I^-$	–0.152	
$Ag(CN)_2 + e^- = Ag + 2CN^-$	–0.31	
$AgSCN + e^- = Ag + SCN^-$	0.09	
$Ag_2S + 2e^- = 2Ag + S^{2-}$	–0.71	
Sodium		
$Na^+ + e^- = Na$	–2.713	
$Na^+ + Hg + e^- = Na(Hg)$	–1.84	
Strontium		
$SrO_2 + 4H^+ + 2e^- = Sr^{2+} + 2H_2O$	2.33	
$Sr^{2+} + 2e^- = Sr$	–2.89	
Sulfur		
$S_2O_8^{2-} + 2e^- = 2SO_4^{2-}$	1.96	
$S_2O_8^{2-} + 2H^+ + 2e^- = 2HSO_4^-$	2.08	
$2SO_4^{2-} + 4H^+ + 2e^- = S_2O_6^{2-} + 2H_2O$	–0.25	
$SO_4^{2-} + 4H^+ + 2e^- = SO_2(aq) + 2H_2O$	0.158	
$SO_4^{2-} + H_2O + 2e^- = SO_3^{2-} + 2OH^-$	–0.936	
$S_2O_6^{2-} + 4H^+ + 2e^- = 2H_2SO_3$	0.569	
$S_2O_6^{2-} + 2e^- = 2SO_3^{2-}$	0.037	
$2HSO_3^- + 2H^+ + 2e^- = S_2O_4^{2-} + 2H_2O$	0.099	
$2SO_3^{2-} + 2H_2O + 2e^- = S_2O_4^{2-} + 4OH^-$	–1.13	

TABLE 14.1 Potentials of the Elements and Their Compounds at 25°C (*Continued*)

Half-reaction	Standard or formal potential, V	Solution composition
Sulfur		
$4H_2SO_3 + 4H^+ + 6e^- = S_4O_6^{2-} + 6H_2O$	0.507	
$4HSO_3^- + 8H^+ + 6e^- = S_4O_6^{2-} + 6H_2O$	0.577	
$2SO_2(aq) + 2H^+ + 4e^- = S_2O_3^{2-} + H_2O$	0.400	
$2SO_3^{2-} + 3H_2O + 4e^- = S_2O_3^{2-} + 6OH^-$	–0.576	1 NaOH
$SO_3^{2-} + 3H_2O + 4e^- = S + 6OH^-$	–0.59	1 NaOH
$S_4O_6^{2-} + 2e^- = 2S_2O_3^{2-}$	0.080	
$S_2O_3^{2-} + 6H^+ + 4e^- = 2S + 3H_2O$	0.5	
$SF_4(g) + 4e^- = S + 4F^-$	0.97	
$S_2Cl_2(g) + 2e^- = 2S + 2Cl^-$	1.19	
$S + H^+ + 2e^- = HS^-$	0.287	
$S + 2H^+ + 2e^- = H_2S(aq)$	0.144	
$S + 2H^+ + 2e^- = H_2S(g)$	0.174	
$S + 2e^- = S^{2-}$	–0.407	
Tantalum		
$Ta_2O_5 + 10H^+ + 10e^- = 2Ta + 5H_2O$	–0.81	
$TaF_7^{2-} + 5e^- = Ta + 7F^-$	–0.45	
Technetium		
$TcO_4^- + 4H^+ + 3e^- = TcO_2 + 2H_2O$	0.738	
$TcO_4^- + 2H^+ + e^- = TcO_3 + H_2O$	0.700	
$TcO_4^- + e^- = TcO_4^{2-}$	0.569	
$TcO_4^- + 8H^+ + 7e^- = Tc + 4H_2O$	0.472	
$TcO_4^{2-} + 4H^+ + 2e^- = TcO_2 + 2H_2O$	1.39	
$TcO_2 + 4H^+ + 4e^- = Tc + 2H_2O$	0.272	
$Tc + e^- = Tc^-$	ca. –0.5	
Tellurium		
$H_2TeO_4 + 6H^+ + 2e^- = Te^{4+} + 4H_2O$	0.929	
$H_2TeO_4 + 2H^+ + 2e^- = TeO_2(c) + 2H_2O$	1.02	
$TeO_4^{2-} + 2H^+ + 2e^- = TeO_3^{2-} + H_2O$	0.897	
$TeOOH^+ + 3H^+ + 4e^- = Te + 2H_2O$	0.559	
$H_2TeO_3 + 4H^+ + 4e^- = Te + 3H_2O$	0.589	
$TeO_3^{2-} + 6H^+ + 4e^- = Te + 3H_2O$	0.827	
$TeO_3^{2-} + 3H_2O + 4e^- = Te + 6OH^-$	–0.415	
$TeO_2(c) + 4H^+ + 4e^- = Te + 2H_2O$	0.521	
$Te + 2H^+ + 2e^- = H_2Te(aq)$	–0.740	
$Te + H^+ + 2e^- = HTe^-$	–0.817	
$Te^{2-} + 2H^+ + 2e^- = 2HTe^-$	–0.794	
Terbium		
$Tb^{3+} + 3e^- = Tb$	–2.31	
Thallium		
$Tl^{3+} + 2e^- = Tl^+$	1.25	1 $HClO_4$
	0.77	1 HCl
$Tl^{3+} + 3e^- = Tl$	0.72	
$Tl^+ + e^- = Tl$	–0.336	
$TlCl + e^- = Tl + Cl^-$	–0.557	
$TlBr + e^- = Tl + Br^-$	–0.658	
$TlI + e^- = Tl + I^-$	–0.752	

(Continued)

TABLE 14.1 Potentials of the Elements and Their Compounds at 25°C (*Continued*)

Half-reaction	Standard or formal potential, V	Solution composition
Thorium		
$Th^{4+} + 4e^- = Th$	−1.83	
Thullium		
$Tm^{3+} + 3e^- = Tm$	−2.32	
Tin		
$Sn^{4+} + 2e^- = Sn^{2+}$	0.154	
$SnCl_6^{2-} + 2e^- = SnCl_4^{2-} + 2Cl^-$	0.14	
$SnO_3^{2-} + 6H^+ + 2e^- = Sn^{2+} + 3H_2O$	0.849	
$SnF_6^{2-} + 4e^- = Sn + 6F^-$	−0.200	
$Sn^{2+} + 2e^- = Sn$	−0.1375	
$SnCl_4^{2-} + 2e^- = Sn + 4Cl^-$	−0.19	1 HCl
$HSnO_2^- + H_2O + 2e^- = Sn + 3OH^-$	−0.91	
$Sn + 4H^+ + 4e^- = SnH_4$	−1.07	
Titanium		
$TiO^{2+} + 2H^+ + e^- = Ti^{3+} + H_2O$	−0.10	
$TiO^{2+} + 2H^+ + 4e^- = Ti + H_2O$	−0.86	
$Ti^{3+} + e^- = Ti^{2+}$	−0.37	
$Ti^{3+} + 3e^- = Ti$	−1.21	
$Ti^{2+} + 2e^- = Ti$	−1.63	
Tungsten		
$2WO_3 + 2H^+ + 2e^- = W_2O_5 + H_2O$	−0.029	
$WO_3 + 6H^+ + 6e^- = W + 3H_2O$	−0.090	
$WO_4^{2-} + 4H_2O + 6e^- = W + 8OH^-$	−1.074	
$WO_4^{2-} + 2H_2O + 2e^- = WO_2 + 4OH^-$	−1.259	
$W_2O_5 + 2H^+ + 2e^- = 2WO_2 + H_2O$	−0.031	
$W(CN)_8^{3-} + e^- = W(CN)_8^{4-}$	0.457	
$WO_2 + 4H^+ + 4e^- = W + 2H_2O$	−0.119	
$WO_2 + 2H_2O + 4e^- = W + 4OH^-$	−0.982	
Uranium		
$UO_2^{2+} + e^- = UO_2^+$	0.16	
$UO_2^{2+} + 4H^+ + 2e^- = U^{4+} + 2H_2O$	0.27	
$UO_2^+ + 4H^+ + e^- = U^{4+} + 2H_2O$	0.38	
$U^{4+} + e^- = U^{3+}$	−0.52	
$U^{4+} + 4e^- = U$	−1.38	
$U^{3+} + 3e^- = U$	−1.66	
Vanadium		
$VO_2^+ + 2H^+ + e^- = VO^{2+} + H_2O$	1.000	
$VO_2^+ + 4H^+ + 2e^- = V^{3+} + 2H_2O$	0.668	
$VO_2^+ + 4H^+ + 3e^- = V^{2+} + 2H_2O$	0.361	
$VO_2^+ + 4H^+ + 5e^- = V + 4H_2O$	−0.236	
$VO^{2+} + 2H^+ + e^- = V^{3+} + H_2O$	0.337	
$V^{3+} + e^- = V^{2+}$	−0.255	
$V^{2+} + 2e^- = V^{2+}$	−1.13	
Xenon		
$H_4XeO_6 + 2H^+ + 2e^- = XeO_3 + 3H_2O$	2.42	
$HXeO_6^{3-} + 2H_2O + e^- = HXeO_4 + 4OH^-$	0.9	
$XeO_3 + 6H^+ + 2F^- + 4e^- = XeF_2 + 3H_2O$	1.6	
$XeO_3 + 6H^+ + 6e^- = Xe(g) + 3H_2O$	2.10	

TABLE 14.1 Potentials of the Elements and Their Compounds at 25°C (*Continued*)

Half-reaction	Standard or formal potential, V	Solution composition
Xenon		
$XeF_2 + e^- = XeF + F^-$	0.9	
$XeF_2 + 2H^+ + 2e^- = Xe(g) + 2HF$	2.64	
$XeF + e^- = Xe(g) + F^-$	3.4	
Ytterbium		
$Yb^{3+} + e^- = Yb^{2+}$	−1.05	
$Yb^{2+} + 2e^- = Yb$	−2.8	
$Yb^{3+} + 3e^- = Yb$	−2.22	
Yttrium		
$Y^{3+} + 3e^- = Y$	−2.37	
Zinc		
$Zn^{2+} + 2e^- = Zn$	−0.7626	
$Zn(NH_3)_4^{2+} + 2e^- = Zn + 4NH_3$	−1.04	
$Zn(CN)_4^{2-} + 2e^- = Zn + 4CN^-$	−1.34	
$Zn(tartrate)_4^{6-} + 2e^- = Zn + 4(tartrate)^{2-}$	−1.15	
$Zn(OH)_4^{2-} + 2e^- = Zn + 4OH^-$	−1.285	
Zirconium		
$Zr^{4+} + 4e^- = Zr$	−1.55	
$ZrO_2 + 4H^+ + 4e^- = Zr + 2H_2O$	−1.45	

Source: A. J. Bard, R. Parsons, and J. Jordan (eds.), *Standard Potentials in Aqueous Solution* (prepared under the auspices of the International Union of Pure and Applied Chemistry), Dekker, New York, 1985; G. Charlot et al., *Selected Constants: Oxidation-Reduction Potentials of Inorganic Substances in Aqueous Solution*, Butterworths, London, 1971.

14.1.2.1 Deriving the Potential of a Third Half-Reaction from Values Given for Two Others. The following equations are useful in calculating the potentials of half-reactions not listed in Table 14.1. Given

$$C + n_{C,B}e = B, \qquad E^0 = E^0_{C,B} \tag{14.1}$$

$$B + n_{B,A}e = A, \qquad E^0 = E^0_{B,A} \tag{14.2}$$

then

$$C + (n_{C,B} + n_{B,A})e = A, \qquad E^0 = E^0_{C,A} = \frac{n_{B,A}E^0_{B,A} + n_{C,B}E^0_{C,B}}{n_{B,A} + n_{C,B}} \tag{14.3}$$

Example 14.1.

$$N_2 + 2H_2O + 4H^+ + 2e^- = 2NH_3OH^+, \qquad E^0 = -1.87\ \text{V}$$

$$2NH_3OH^+ + H^+ + 2e^- = N_2H_5^+ + 2H_2O, \qquad E^0 = +1.41\ \text{V}$$

$$N_2 + 5H^+ + 4e^- = N_2H_5^+, \qquad E^0 = \frac{(2)(+1.41) + (2)(-1.87)}{2+2} = -0.23\ \text{V}$$

14.1.2.2 Deriving the Potential of a Half-Reaction Involving a Solid Salt. Given

$$M^{n+} + ne = M, \qquad E^0 = E^0_{M^{n+},M} \tag{14.4}$$

$$M_pX_q \rightleftharpoons pM^{n+} + qX^{(pn/q)-}, \qquad K = K_{sp} \text{ for } M_pX_q \tag{14.5}$$

then

$$M_pX_q + pne = pM + qX^{(pn/e)-} \tag{14.6}$$

$$E^0 = E^0_{M^{n+},M} + \frac{0.05915}{pn} \log K_{sp} \tag{14.7}$$

Example 14.2.

$$Ag^+ + e^- \longrightarrow Ag, \qquad E^0 = 0.7991 \text{ V}$$

$$AgBrO_3 \rightleftharpoons Ag^+ + BrO_3^-, \qquad K_{sp} = 5.3 \times 10^{-5}$$

$$AgBrO_3 + e^- \longrightarrow Ag + BrO_3^-, \qquad E^0 = 0.7991 + 0.05915 \log 5.3 \times 10^{-5}$$

$$= 0.546 \text{ V for } AgBrO_3, \text{ Ag system}$$

14.1.2.3 Effect of Complexation on Electrode Potentials. Reagents that react with one or both participants of an electrode process can significantly affect the electrode potential. Two cases will be examined. In the presence of significant concentrations of triethylenetetramine (trien), the copper–copper(II) system will be converted essentially to a copper–copper(II) trien system. Derive the potential of a half-reaction involving a single complex ion:
Given

$$M^{n+} + ne^- \longrightarrow M, \qquad E^0 = E^0_{M^{n+1},M} \tag{14.8}$$

$$M^{n+} + pX^{q-} \rightleftharpoons MX_p^{(n-pq)}, \qquad K_f \tag{14.9}$$

then

$$MX_p^{(n-pq)} + ne^- \longrightarrow M + pX^{q-} \tag{14.10}$$

$$E^0 = E^0_{M^{n+},M} - (0.05915/n) \log K_f \tag{14.11}$$

Example 14.3.

$$Cu^{2+} + 2e^- \longrightarrow Cu, \quad E^0 = 0.340 \text{ V}$$

$$Cu^{2+} + \text{trien} \rightleftharpoons Cu(\text{trien})^{2+}, \quad K_f = 2.51 \times 10^{20}$$

$$Cu(\text{trien})^{2+} + 2e^- \longrightarrow Cu + \text{trien}$$

$$E^0 = 0.340 - (0.05915/2) \log 2.51 \times 10^{20}$$

$$= -0.26 \text{ V}$$

For a redox couple involving two oxidation states in solution, such as the aquo-cobalt species (not explicitly shown in the following expressions),

$$Co^{3+} + e^- \longrightarrow Co^{2+}, \qquad E^0 = 1.84 \text{ V} \tag{14.12}$$

In the presence of aqueous ammonia, both the cobalt(II) hexammine and the cobalt(III) hexammine species predominate. The respective formation constants are

$$\frac{[Co(NH_3)_6^{2+}]}{[Co^{2+}][NH_3]^6} = K_{f,1} = 10^5 \tag{14.13}$$

$$\frac{[Co(NH_3)_6^{3+}]}{[Co^{3+}][NH_3]^6} = K_{f,2} = 10^{34} \tag{14.14}$$

Substitution of these values into the Nernst equation for the cobalt system gives

$$E = E^0 + 0.059\ 15\ \log \frac{K_{f,1}}{K_{f,2}} + 0.059\ 15\ \log \frac{[Co(NH_3)_6^{3+}]}{[Co(NH_3)_6^{2+}]} \tag{14.15}$$

$$E^0 = 1.84 + 0.059\ 15\ \log \frac{10^5}{10^{34}} = 0.12\ \text{V} \tag{14.16}$$

Here the shift in potential is a function of the ratio of the formation constants for each electroactive species. Generally, the higher oxidation state will form the more stable complex and, if it does, the shift in electrode potential will be in the negative direction. In the example above, the shift is significant, being 1.72 V.

14.1.3 Standard Reference Electrodes

A reference electrode is an oxidation–reduction half-cell of known and constant potential at a particular temperature. It consists of three parts:

1. An internal element.
2. A filling solution that also constitutes the salt-bridge electrolyte.
3. An area in the tip of the electrode that permits a slow, controlled flow of filling solution to escape the electrode and maintain electrical conductance with the remainder of the electrochemical cell.

14.1.3.1 Internal Elements. The choice of a reference electrode is between the calomel or mercury–mercury(I) chloride half-cell and the silver–silver chloride half-cell as the internal element. Both electrodes are anion reversible.

The *silver–silver chloride* electrode is often a silver wire coated with silver chloride that is immersed in a filling solution of potassium chloride of known concentration, usually 1.0*M*, and saturated with silver chloride. In the cartridge form the metal is in contact with a paste of the salt moistened with electrolyte, and all is enclosed in an inner glass tube.

One equilibrium involved is

$$Ag^+ + e^- \rightleftharpoons Ag \tag{14.17}$$

The expression for the electrode potential is

$$E = 0.799 - 0.059\ 16\ \log\ [Ag^+] \tag{14.18}$$

In addition, there is a chemical equilibrium,

$$AgCl(s) \rightleftharpoons Ag^+ + Cl^-, \quad K_{sp} = 1.8 \times 10^{-10} \tag{14.19}$$

Combining Eqs. (14.17) and (14.18) gives the Nernst expression for the chloride ion:

$$E = 0.799 + 0.059\ 16\ \log K_{sp} - 0.059\ 16\ \log\ [Cl^-] \tag{14.20}$$

This simplifies to

$$E = 0.2222 - 0.059\,16 \log [\mathrm{Cl^-}] \tag{14.21}$$

Thus when the electrode system is immersed in a filling solution that contains a constant (and known) amount of the chloride ion, the electrode potential is constant (and known). Table 14.2 lists potentials of common reference electrodes (in volts) in aqueous solutions as a function of temperature. Similar compilations for water–organic solvent mixtures and for nonaqueous media are given in Tables 14.3 and 14.4, respectively.

Calomel electrodes have a construction that is similar to the cartridge type of silver–silver chloride electrodes. They comprise a platinum wire in contact with mercury and a paste of mercury(I) chloride, mercury, and potassium chloride, moistened with the filling solution, which is enclosed in an inner glass tube and makes contact with a filling solution of potassium chloride (usually 0.1*M*, 1.0*M*, or saturated, 4.2*M*) through a porous plug.

14.1.3.2. Comparison of Reference Electrodes. Calomel electrodes are easy to prepare and maintain. They must never be used at temperatures higher than 80°C. The saturated calomel electrode (SCE), although easiest to prepare, reaches equilibrium more slowly following temperature changes.

Silver–silver chloride electrodes can be used up to the boiling point of water (and higher under special conditions). The temperature coefficient of the electrode potential is less than that for calomel electrodes. These electrodes are also stabler over long periods of time.

14.1.4 Liquid-Junction Potentials

Electrical contact between the sample and the reference electrode is established by a slow but continuous leak of filling solution through a constricted orifice. The liquid junction must be clean, must be free flowing, and must make good contact with the sample. Liquid junctions come in several physical forms: (1) a sleeve or tapered ground-glass junction, (2) a porous ceramic or quartz fiber junction, and (3) a double-chamber salt bridge.

1. *Sleeve junctions* have an electrolyte flow of about 0.1 $\mathrm{mL \cdot h^{-1}}$, a self-cleaning facility. This junction is useful when working with slurries, emulsions, suspensions, pastes, gels, and nonaqueous solvent systems.

2. *Porous ceramic* or *fiber junctions* have a very small flowrate of filling solution (about 8 $\mu\mathrm{L \cdot h^{-1}}$). This type of junction should be used when contamination of the sample by the filling solution must be avoided.

3. *Double-junction* salt bridges overcome all problems with leakage of electrolyte into the sample or compatibility of filling and sample solution. The leakage from the internal filling solution is retained in the outer (chamber) salt bridge that contains an innocuous electrolyte that can be flushed frequently. This type of junction must be used if the test solution contains ions that would precipitate or complex with silver or mercury(II) ions found in the filling solution or contains strong reducing agents that might reduce silver ions to silver metal at the junction.

For a liquid junction of the type $M_aX_b(c_1) \parallel M_aX_b$, the liquid-junction potential E_i is given by

$$E_i = \frac{RT}{abF}\,[1-(a+b)t_-]\ln\frac{c_1}{c_2} \tag{14.22}$$

where t_- is the transference number (fraction of current carried by the anion) of the anion X. For a 1:1 electrolyte at 25°C, this may be written

$$E_i = 0.05915(1-2t_-)\log\frac{c_1}{c_2} \tag{14.23}$$

TABLE 14.2 Potentials of Reference Electrodes in Volts as a Function of Temperature

Liquid-junction potential included.

Temp., °C	0.1*M* KCl calomel*	1.0*M* KCl calomel*	3.5*M* KCl calomel*	Satd. KCl calomel*	1.0*M* KCl Ag–AgCl†	1.0*M* KBr Ag–AgBr‡	1.0*M* KI Ag–AgI§
0	0.3367	0.2883		0.25918	0.23655	0.08128	–0.14637
5					0.23413	0.07961	–0.14719
10	0.3362	0.2868	0.2556	0.25387	0.23142	0.07773	–0.14822
15	0.3361			0.2511	0.22857	0.07572	–0.14942
20	0.3358	0.2844	0.2520	0.24775	0.22557	0.07349	–0.15081
25	0.3356	0.2830	0.2501	0.24453	0.22234	0.07106	–0.15244
30	0.3354	0.2815	0.2481	0.24118	0.21904	0.06856	–0.15405
35	0.3351			0.2376	0.21565	0.06585	–0.15590
38	0.3350		0.2448	0.2355			
40	0.3345	0.2782	0.2439	0.23449	0.21208	0.06310	–0.15788
45					0.20835	0.06012	–0.15998
50	0.3315	0.2745		0.22737	0.20449	0.05704	–0.16219
55					0.20056		
60	0.3248	0.2702		0.2235	0.19649		
70					0.18782		
80				0.2083	0.1787		
90					0.1695	0.0251	

* R. G. Bates et al., *J. Research Natl. Bur. Stand.* (U.S.) **45**:418 (1950).
† R. G. Bates and V. E. Bower, *J. Res. Natl. Bur. Stand.* (U.S.) **53**:283 (1954).
‡ M. B. Hetzer, R. A. Robinson and R. G. Bates, *J. Phys. Chem.* **66**:1423 (1962).
§ H. B. Hetzer, R. A. Robinson and R. G. Bates, *J. Phys. Chem.* **68**:1929 (1964).

Temp., °C	125	150	175	200	225	250	275
1.0*M* KCl Ag–AgCl*	0.1330	0.1032	0.0708	0.0348	–0.0051	–0.054	–0.090
1.0*M* KBr Ag–AgBr†	–0.0048	–0.0312	–0.0612	–0.0951			

* R. S. Greeley et al., *J. Phys. Chem.* **64**:652 (1960).
† Towns et al., *J. Phys. Chem.* **64**:1861 (1960).

Several additional potentials of reference electrodes at 25°C	
Ag–AgCl, satd. KCl	0.198
Ag–AgCl, 0.1*M* KCl	0.288
Hg–HgO, 1.0*M* NaOH	0.140
Hg–HgO, 0.1*M* NaOH	0.165
Hg–Hg_2SO_4, satd. K_2SO_4 (22°C)	0.658
Hg–Hg_2SO_4, satd. KCl	0.655

Source: J. A. Dean, ed., *Lange's Handbook of Chemistry*, 14th ed., McGraw-Hill, New York, 1992.

The liquid-junction potential across a boundary between two different electrolytes containing equal concentrations of a common ion depends on the structure of the boundary; it may be approximated by the relationship, for 1 : 1 electrolytes,

$$E_i = \frac{RT}{F_y} \ln \frac{\Lambda_2}{\Lambda_1} \tag{14.24}$$

where the subscripts 1 and 2 represent the equivalent conductance (Λ) of the solutions on the left- and right-hand sides of the boundary, respectively.

TABLE 14.3 Potentials of Reference Electrodes (in Volts) at 25°C for Water–Organic Solvent Mixtures

Electrolyte solution of 1M HCl.

Solvent, wt %	Methanol, Ag–AgCl	Ethanol, Ag–AgCl	2-Propanol, Ag–AgCl	Acetone, Ag–AgCl	Dioxane, Ag–AgCl	Ethylene glycol, Ag–AgCl	Methanol, calomel	Dioxane, camolel
5			0.2180	0.2190		0.2190		
10	0.2153	0.2146	0.2138	0.2156		0.2160		
20	0.2090	0.2075	0.2063	0.2079	0.2031	0.2101	0.255	0.2501
30		0.2003				0.2036		
40	0.1968	0.1945		0.1859		0.1972	0.243	
45					0.1635			0.2104
50		0.1859		0.158				
60	0.1818	0.173				0.1807		
70		0.158			0.0659		0.216	0.1126
80	0.1492	0.136						
82					–0.0614			–0.0014
90	0.1135	0.096		–0.034				
94.2	0.0841							
98		0.0215						
99							0.103	
100	–0.0099	–0.0081		–0.53				

Source: J. A. Dean, ed., *Lange's Handbook of Chemistry*, 14th ed., McGraw-Hill, New York, 1992.

Table 14.5 gives values of E_j at boundaries between 3.5F or 0.1F potassium chloride and various other electrolyte solutions. They refer to a temperature of 25°C and give at least approximate ideas of the liquid-junction potentials that may be encountered in using some common reference electrodes and salt bridges. The first column gives the formula and concentration of the electrolyte on one side of the boundary. The second column gives values of E_j in millivolts for the boundary electrolyte || KCl (3.5F); the third column gives values of E_j in milllvolts for the boundary electrolyte || KCl (0.1F). A positive value of E_j signifies that the boundary has the polarity – || +.

14.1.5 Maintenance of Reference Electrodes

Reference electrodes should be stored separately in a dilute KCl solution of approximately 0.1M. To prevent back-flow and possible contamination of the electrolyte, always maintain the level of the filling solution in the reference electrode above the level of both soaking and sample solutions and above the internal elements.

14.1.5.1 Blocked or Clogged Liquid Junctions. A blocked or clogged liquid junction can be detected in several ways:

1. Apply air pressure to the filling hole and observe the dried junction—a bead of electrolyte should form readily.

2. Measure the junction resistance with an ohmmeter—it should read less than 20 000 Ω.

3. Place the glass–reference electrode assembly into 250 mL of deionized water (with stirring) and note the reading. Add 50 mg of ultrapure solid KCl and note the reading. The reading should change less than 0.08 pH.

The junction of the reference electrode can be unblocked as follows;

1. Gel-filled electrodes need only to be immersed in warm water for several minutes or overnight in 0.1M KCl.

TABLE 14.4 Reference Electrodes in Nonaqueous Media

This table lists some electrodes that have been used or proposed as primary or working standards in various nonaqueous media and gives approximate information concerning their potentials.

Solvent	Electrode	Potential			Reference
		Volts	vs.	Temp., °C	
Acetic acid	Ag–AgCl(s), KCl(s)	+0.23	aq SCE	22 ± 0.5	1*a*
	Ag–$AgNO_3$(s)	+0.87	aq SCE	22 ± 0.5	1*a*
	Hg–Hg_2Cl_2(s), KCl(s)	+0.27	aq SCE	22 ± 0.5	1*b*
	Hg–Hg_2SO_4(s), K_2SO_4(s)	+0.69	aq SCE	22 ± 0.5	1*b*
Acetone	Hg–Hg_2Cl_2(s), LiCl(s)				2
Acetonitrile	Ag–$AgNO_3$ (0.01*F*)	$+0.30_0$	aq SCE	25	3, 4
	Ag–AgCl (0.015*F*), Me_3EtNCl (0.118*F*)	$-0.638 - 6 \times 10^{-4}(t - 25)$	Preceding electrode		5
	Hg–Bu_4NI (0.1*F*)*	ca. −0.89	Preceding electrode	25	5
Ammonia	Cd–$CdCl_2$(s)	−0.93	Hg–HgI_2(s) in NH_3	−36.5	6
	Hg–$HgCl_2$(s)	−0.068 ± 0.004	Hg–HgI_2(s) in NH_3	−36.5	6
	Hg–HgI_2(s)			−36.5	6
Ethylenediamine	Hg–Hg_2Cl_2(s), LiCl(s)				7
Formic acid	Hg–Hg_2Cl_2(s), KCl(s)				8
	Pt–quinhydrone (0.05*F*), Na formate (0.25*F*)	$+0.538_4 \pm 0.000_5$	Preceding electrode	25	8
2,4-Lutidine	Hg–Hg_2Cl_2(s), KCl(s)	+0.33	aq SCE	22 ± 0.5	1*b*
	Hg–Hg_2SO_4(s), K_2SO_4(s)	+0.29	aq SCE	22 ± 0.5	1*b*
2,6-Lutidine	Hg–Hg_2Cl_2(s), KCl(s)	+0.45	aq SCE	22 ± 0.5	1*b*
	Hg–Hg_2SO_4(s), K_2SO_4(s)	+0.36	aq SCE	22 ± 0.5	1*b*
2-Picoline	Hg–Hg_2Cl_2(s), KCl(s)	+0.42	aq SCE	22 ± 0.5	1*b*
	Hg–Hg_2SO_4(s), K_2SO_4(s)	+0.39	aq SCE	22 ± 0.5	1*b*
Pyridine	Hg–Hg_2SO_4(s), K_2SO_4(s)	+0.34	aq SCE	22 ± 0.5	1*b*
Quinoline	Ag–AgCl(s), KCl(s)	+0.17	aq SCE	22 ± 0.5	1*a*

* Readily polarizable, as are Hg–Hg_2Cl_2 and Hg–Hg_2SO_4 electrodes in this medium.[5]

References

1. Tutundzic and Putanov, *Glasnik Khem. Drushtva Beograd* **21**:(*a*) 19, (*b*) 257 (1956).
2. Arthur and Lyons, *Anal. Chem.* **24**:1422 (1952).
3. Pleskov, *Zhur. Fiz. Khim.* **22**:351 (1948).
4. Larson, Iwamoto, and Adams, *Anal. Chim. Acta* **25**:371 (1961).
5. Popov and Geske, *J. Am. Chem. Soc.* **79**:2074 (1957).
6. Watt and Sowards, *J. Electrochem. Soc.* **102**:545 (1955).
7. Gran and Althin, *Acta Chem. Scand.* **4**:967 (1950).
8. Pinfold and Sebba, *J. Am. Chem. Soc.* **78**:2095 (1956).

Source: L. Meites, ed., *Handbook of Analytical Chemistry*, McGraw-Hill, New York, 1963.

2. Sleeve-type junctions can be unblocked by simply loosening the sleeve to drain the electrolyte, rinsing the cavity with distilled water, and then refilling with fresh KCl solution.

3. Porous ceramic or cracked bead junctions should be immersed in warm distilled water to clear the junction. Apply air pressure to the fill hole to reestablish the electrolyte flow.

4. For porous plugs, try boiling the junction in dilute KCl for 5 to 10 min or soak overnight in 0.1*M* KCl. If this fails, carefully sand or file the porous plug.

TABLE 14.5 Liquid-Junction Potentials

Electrolyte	E_j, in mV, at boundary electrolyte ‖ KCl(c), when c = 3.5F	0.1F
HCl		
0.01F	1.4	9.3
0.1F	3.1	26.8
1.0F	16.6	56.2
H_2SO_4		
0.05F	4.0	25.0
0.5F	14.0	53.0
KCl		
0.01F	1.0	0.4
0.1F	0.6	0.0
1.0F	0.2	
KOH		
0.1F	−1.7	−15.4
1.0F	−8.6	−34.2
LiCl		
0.1F		−8.9
NH_4Cl		
0.1F		2.2
NaCl		
0.1F	−0.2	−6.4
1F	−1.9	−11.2
NaOH		
0.1F	−2.1	−18.9
1F	−10.5	−45.0

14.1.5.2 Contamination on Tip and Surface. Contamination on the surface and tip can be removed as follows:

1. For protein layers, wash with pepsin or 0.1M HCl.
2. For inorganic deposits, wash with EDTA or dilute acids.
3. For grease films, wash with acetone, methanol, or diethyl ether.

14.1.5.3 Faulty Internal Elements or Contaminated Filling Solutions. Faulty internal elements, contaminated filling solution, or a too-great junction potential can be checked by plugging the suspect electrode into the reference jack of your pH meter and plugging a reference electrode known to be working properly into the pH-indicating electrode jack (pin–jack adapter needed). Immerse both electrodes in a saturated KCl solution. The observed pH meter reading should be 0 ± 5 mV for similar electrodes, or about +44 ±5 mV if a calomel reference is tested against a silver–silver chloride type.

14.1.6 Outline of Electroanalytical Methods

Each basic electrical measurement of current (i), resistance (R), and potential (V) has been used alone or in combination for analytical purposes.

14.1.6.1 Steady-State Methods. Steady-state or static methods entail measurements under equilibrium conditions existing throughout the bulk of the solution. Time is eliminated as a variable, and equilibrium is assured by vigorously stirring the solution. Each of the following steady-state methods will be discussed in succeeding sections.

Ion-selective potentiometry: The electrode potential is measured at zero current.

Null-point potentiometry: Analyte additions or subtractions are done to null any potential difference at zero current.

Potentiometric titrations: The electrode potential is measured as a function of volume of titrant; current is zero.

Amperometric titrations: Current is measured as a function of the volume of titrant; the potential is kept constant.

14.1.6.2 Transient Methods. Our analytical will can be imposed on many electrochemical systems by subjecting them to impressed current or voltage signals and observing qualitatively and quantitatively how they respond. This is done by applying a variety of voltage–time patterns to a microelectrode in a sample solution and analyzing the resultant currents. These transient methods constitute the field of *voltammetry* and its subset *polarography*. The concentration of the electroanalyte is the variable.

14.1.6.3 Controlled Potential Methods. The weight of a separated phase or the integrated current, that is, the current multiplied by time or coulombs, is a measure of the total amount of material converted to another electrochemical form. The potential at the working electrode is strictly controlled. In *controlled-potential electroanalysis* the weight of the separated phase is measured. In *controlled-potential coulometry* the total number of coulombs flowing through the system is measured.

14.1.6.4 Charge Transport by Migration. There is a group of methods, the basis of which depends upon charge transport by migration. Electron-transfer reactions are unimportant. These methods are as follows.

Conductance measurements: The conductance ($1/R$) is measured as a function of concentration.

Conductometric titrations: Conductance versus the volume of titrant is measured.

14.2 MEASUREMENT OF pH

14.2.1 Operational Definition of pH

It is universally agreed that the definition of pH difference is an operational one. The pH difference is ascertained from the emf of an electrochemical cell of the following design:

Electrode reversible to hydrogen ions	Unknown (x) or standard (s) buffer solution	Salt bridge	Reference electrode

The two bridge solutions may be any molality of potassium chloride greater than 3.5m. To a good approximation, the pH-responsive electrodes in both cells may be replaced by other hydrogen-ion-responsive electrodes. In most measurements, a single glass electrode–reference electrode probe assembly is transferred between the cells. E_x is the emf of the probe assembly when immersed in the solution of unknown pH_x, and E_s is the emf of the probe when immersed in a standard reference material whose value is pH_s. pH_x is given by the relationship

$$pH_x = pH_s + \frac{E_x - E_s}{2.3026\ RT/F} \tag{14.25}$$

where R = the gas constant
T = temperature, K
F = the Faraday constant

The pH_x of the unknown is calculated from that of an accepted standard (pH_s) and the measured difference in the emf (E) of the electrode combination when the standard solution is removed from the cell and replaced by the unknown. The double vertical line marks a liquid junction. Electrodes as fabricated exhibit variations in the reproducibility of the reference electrode, in the liquid-junction potential, and, with glass electrodes, in the asymmetry potential. These differences are all eliminated in the standardizing procedure with standard reference pH buffers.

14.2.2 Standard Reference Solutions

The recommended values of pH_s are given in Tables 14.6 and 14.7. The operator of pH equipment should be aware of the following limitations in the use of pH reference buffer standards:

1. The uncertainty in pH_s is estimated as 0.005 pH unit (0 to 60°C) and 0.008 pH unit (60 to 95°C).

2. The operational definition of pH is valid for only dilute solutions and for the pH range 2 to 12. For ionic strengths greater than 0.1 the reproducibility of the liquid-junction potential is seriously impaired and errors as large as several tenths of a pH unit can result.

3. To use the pH probe assembly in the very acid or very basic regions requires two secondary standards (potassium tetraoxalate and calcium hydroxide solutions), which are included in Table 14.6

4. The temperature must be known to ±2°C for an accuracy of ±0.01 pH unit. Not only does the proportionality factor between the cell emf and pH vary with temperature, but dissociation equilibria and junction potentials also have significant temperature coefficients. Solutions should be measured at their operating temperatures and not cooled to room temperature.

Values for a 0.1m solution of HCl are given to extend the pH scale up to 275°C (see Fn. 1):

t, °C	25	60	90	125	150	175	200	255–275
pH	1.10	1.11	1.12	1.13	1.14	1.15	1.16	1.2

Uncertainties in the values are ±0.03 pH unit from 25 to 90°C, ±0.05 pH unit from 125 to 200°C, and ±0.1 pH unit from 225 to 275°C.

The buffer values for the NIST (National Institute for Science and Technology, formerly National Bureau of Standards) reference pH buffer solutions are given below:

Buffer solution	KH tartrate	0.05M KH_2 citrate	0.05M KH phthalate	0.025M KH_2PO_4, 0.25M Na_2HPO_4	0.0087M KH_2PO_4, 0.0302M Na_2HPO_4	0.01M $Na_2B_4O_7$	0.025M $NaHCO_3$, 0.025M Na_2CO_3
Buffer value β	0.027	0.034	0.016	0.029	0.016	0.020	0.029

For the secondary pH reference standards in Table 14.6, the buffer value is 0.070 for potassium tetraoxalate and 0.09 for calcium hydroxide.

[1] R. S. Greeley, *Anal. Chem.* **32**:1717 (1960).

TABLE 14.6 NIST (U.S.) Reference pH Buffer Solutions

Temperature, °C	Secondary standard 0.05*M* K tetraoxalate	KH tartrate (saturated at 25°C)	0.05*M* KH_2 citrate	0.05*M* KH phthalate	0.025*M* KH_2PO_4, 0.025*M* Na_2HPO_4	0.0087*M* KH_2PO_4, 0.0302*M* Na_2HPO_4	0.01*M* $Na_2B_4O_7$	0.025*M* $NaHCO_3$, 0.025*M* Na_2CO_3	Secondary standard $Ca(OH)_2$ (saturated at 25°C)
0	1.666		3.860	4.003	6.984	7.534	9.464	10.317	13.423
5	1.668		3.840	3.999	6.951	7.500	9.395	10.245	13.207
10	1.670		3.820	3.998	6.923	7.472	9.332	10.179	13.003
15	1.672		3.802	3.999	6.900	7.448	9.276	10.118	12.810
20	1.675		3.788	4.002	6.881	7.429	9.225	10.062	12.627
25	1.679	3.557	3.776	4.008	6.865	7.413	9.180	10.012	12.454
30	1.683	3.552	3.766	4.015	6.853	7.400	9.139	9.966	12.289
35	1.688	3.549	3.759	4.024	6.844	7.389	9.102	9.925	12.133
38	1.691	3.548		4.030	6.840	7.384	9.081		12.043
40	1.694	3.547	3.753	4.035	6.838	7.380	9.068	9.889	11.984
45	1.700	3.547		4.047	6.834	7.373	9.038		11.841
50	1.707	3.549	3.749	4.060	6.833	7.367	9.011	9.828	11.705
55	1.715	3.554		4.075	6.834		8.985		11.574
60	1.723	3.560		4.091	6.836		8.962		11.449
70	1.743	3.580		4.126	6.845		8.921		
80	1.766	3.609		4.164	6.859		8.885		
90	1.792	3.650		4.205	6.877		8.850		
95	1.806	3.674		4.227	6.886		8.833		
Dilution value $\Delta pH_{1/2}$	+0.186	+0.049	0.024	+0.052	+0.080	+0.070	+0.01	0.079	−0.28

Source: R. G. Bates, *J. Res. Natl. Bur. Stand. (U.S.)* **66A**:179 (1962); B. R. Staples and R. G. Bates, *ibid.* **73A**:37 (1969).

TABLE 14.7 Composition and pH Values of Buffer Solutions

Values based on the conventional activity pH scale as defined by the National Bureau of Standards (U.S.) and pertain to a temperature of 25°C. Buffer value is denoted by column headed β.

25 mL 0.2*M* KCl + *x* mL 0.2*M* HCl, diluted to 100 mL

pH	x	β
1.00	67.0	0.31
1.20	42.5	0.34
1.40	26.6	0.19
1.60	16.2	0.077
1.80	10.2	0.049
2.00	6.5	0.030
2.20	3.9	0.022

50 mL 0.1*M* KH phthalate + *x* mL 0.1*M* HCl, diluted to 100 mL

pH	x	β
2.20	49.5	
2.40	42.2	0.036
2.60	35.4	0.033
2.80	28.9	0.032
3.00	22.3	0.030
3.20	15.7	0.026
3.40	10.4	0.023
3.60	6.3	0.018
3.80	2.9	0.015

50 mL 0.1*M* KH phthalate + *x* mL 0.1*M* NaOH, diluted to 100 mL

pH	x	β
4.20	3.0	0.017
4.40	6.6	0.020
4.60	11.1	0.025
4.80	16.5	0.029
5.00	22.6	0.031
5.20	28.8	0.030
5.40	34.1	0.025
5.60	38.8	0.020
5.80	42.3	0.015

50 mL 0.1*M* KH_2PO_4 + *x* mL 0.1*M* NaOH, diluted to 100 mL

pH	x	β
5.80	3.6	
6.00	5.6	0.010
6.20	8.1	0.015
6.40	11.6	0.021
6.60	16.4	0.027
6.80	22.4	0.033
7.00	29.1	0.031
7.20	34.7	0.025
7.40	39.1	0.020
7.60	42.4	0.013
7.80	44.5	0.009
8.00	46.1	

50 mL 0.1*M* Tris (hydroxymethyl)aminomethane + *x* mL of 0.1*M* HCl, diluted to 100 mL, $\Delta pH/\Delta t \simeq -0.028$, $I = 0.001\,x$

pH	x	β
7.00	46.6	
7.20	44.7	0.012
7.40	42.0	0.015
7.60	38.5	0.018
7.80	34.5	0.023
8.00	29.2	0.029
8.20	22.9	0.031
8.40	17.2	0.026
8.60	12.4	0.022
8.80	8.5	0.016
9.00	5.7	

50 mL of a mixture 0.1*M* with respect to both KCl and H_3BO_3 + *x* mL 0.1*M* NaOH, diluted to 100 mL

pH	x	β
8.00	3.9	
8.20	6.0	0.011
8.40	8.6	0.015
8.60	11.8	0.018
8.80	15.8	0.022
9.00	20.8	0.027
9.20	26.4	0.029
9.40	32.1	0.027
9.60	36.9	0.022
9.80	40.6	0.016
10.00	43.7	0.014
10.20	46.2	

50 mL 0.025*M* borax + *x* mL 0.1*M* HCl, diluted to 100 mL, $\Delta pH/\Delta t \simeq -0.008$, $I = 0.025$

pH	x	β
8.00	20.5	
8.20	19.7	0.010
8.40	16.6	0.012
8.60	13.5	0.018
8.80	9.4	0.023
9.00	4.6	0.026
9.10	2.0	

50 mL 0.025 borax + *x* mL 0.1*M* NaOH, diluted to 100 mL, $\Delta pH/\Delta t \simeq -0.008$, $I = 0.001(25 + x)$

pH	x	β
9.20	0.9	
9.40	3.6	0.026
9.60	11.1	0.022
9.80	15.0	0.018
10.00	18.3	0.014
10.20	20.5	0.009
10.40	22.1	0.007
10.60	23.3	0.005

50 mL 0.05*M* $NaHCO_3$ + *x* mL 0.1*M* NaOH, diluted to 100 mL, $\Delta pH/\Delta t \simeq -0.009$, $I = 0.001(25 + 2x)$

pH	x	β
9.60	5.0	
9.80	6.2	0.014
10.00	10.7	0.016
10.20	13.8	0.015
10.40	16.5	0.013
10.60	19.1	0.012
10.80	21.2	0.009
11.00	22.7	

TABLE 14.7 Composition and pH Values of Buffer Solutions (*Continued*)

50 mL 0.05*M* Na_2HPO_4 + *x* mL 0.1*M* NaOH, diluted to 100 mL, $\Delta pH/\Delta t \simeq -0.025$, $I = 0.001(77 + 2x)$			25 mL 0.2*M* KCl + *x* mL 0.2*M* NaOH, diluted to 100 mL $\Delta pH/\Delta t \simeq -0.033$, $I = 0.001(50 + 2x)$		
pH	*x*	β	pH	*x*	β
11.00	4.1	0.009	12.00	6.0	0.028
11.20	6.3	0.012	12.20	10.2	0.048
11.40	9.1	0.017	12.40	16.2	0.076
11.60	13.5	0.026	12.60	25.6	0.12
11.80	19.4	0.034	12.80	41.2	0.21
11.90	23.0	0.037	13.00	66.0	0.30

Source: Bower and Bates, *J. Res. Natl. Bur. Stand.* (*U.S.*) **55**:197 (1955); Bates and Bower, *Anal. Chem.* **28**:1322 (1956).

14.2.2.1 Preparation of Standard Buffer Solutions. To prepare the standard pH buffer solutions recommended by NIST, the indicated weights of the pure materials in Table 14.8 should be dissolved in water of specific conductivity not greater than 5 micromhos. The tartrate, phthalate, and phosphates can be dried for 2 h at 110°C before use. Potassium tetraoxalate and calcium hydroxide need not be dried. Fresh-looking crystals of borax should be used. Before use, excess solid potassium, hydrogen tartrate and calcium hydroxide must be removed. Buffer solutions pH 6 or above should be stored in plastic containers and should be protected from carbon dioxide with soda-lime traps. The solutions should be replaced within 2 to 3 weeks or sooner if formation of mold is noticed. A crystal of thymol may be added as a preservative.

14.2.2.2 Standards for pH Measurement of Blood and Biological Materials. Blood is a well-buffered medium. In addition to the NIST phosphate standard of 0.025*M* ($pH_s = 6.480$ at 38°C),

TABLE 14.8 Compositions of Standard pH Buffer Solutions [National Institute for Science and Technology (U.S.)]

Air weight of material per liter of buffer solutions.

Standard	Weight, g
$KH_3(C_2O_4)_2 \cdot 2H_2O$, 0.05*M*	12.61
Potassium hydrogen tartrate, about 0.034*M*	Saturated at 25°C
Potassium hydrogen phthalate, 0.05*M*	10.12
Phosphate	
KH_2PO_4, 0.025*M*	3.39
Na_2HPO_4, 0.025*M*	3.53
Phosphate	
KH_2PO_4, 0.008 665*M*	1.179
Na_2HPO_4, 0.030 32*M*	4.30
$Na_2B_4O_7 \cdot 10H_2O$, 0.01*M*	3.80
Carbonate	
$NaHCO_3$, 0.025*M*	2.10
Na_2CO_3, 0.025*M*	2.65
$Ca(OH)_2$, about 0.0203*M*	Saturated at 25°C

Source: J. A. Dean, ed., *Lange's Handbook of Chemistry*, 14th ed., McGraw-Hill, New York, 1992.

another solution containing the same salts, but in the molal ratio 1 : 4, has an ionic strength of 0.13. It is prepared by dissolving 1.360 g of KH_2PO_4 and 5.677 g of Na_2HPO_4 (air weights) in carbon dioxide–free water to make 1 L solution. The pH_s is 7.416 ± 0.004 at 37.5 and 38°C.

The compositions and pH_s values of tris(hydroxymethyl)aminomethane, covering the pH range 7.0 to 8.9, are listed in Table 14.7.

The phosphate–succinate system gives the values of pH_s shown below:

Molality KH_2PO_4 = Molality $Na_2HC_6H_5O_7$	pH_s	$\Delta(pH_s/\Delta t)$
0.005	6.251	−0.000 86 deg^{-1}
0.010	6.197	−0.000 71
0.015	6.162	
0.020	6.131	
0.025	6.109	−0.000 4

14.2.2.3 Standard Reference Solutions for Alcohol–Water Systems. Standard buffer solutions for use in methanol–water and ethanol–water systems are given in Tables 14.9*a* and 14.9*b*.

14.2.2.4 Buffer Solutions Other Than Standards for Control Purposes. The range of the buffering effect of a single weak acid group is approximately one pH unit on either side of the pK_a. The ranges of some useful buffer systems are collected in Table 14.10. After all the components have been brought together, the pH of the resulting solution should be determined at the temperature to be employed with reference to standard reference solutions. Buffer components should be compatible with other components in the system under study; this particularly significant for buffers employed in biological studies. Check tables of formation constants (Tables 2.1 and 2.2) to ascertain whether any metal-binding character exists.

When there are two or more acid groups per molecule, or a mixture is composed of several overlapping acids, the useful range is larger. Universal buffer solutions consist of a mixture of acids groups that overlap such that successive pK_a values differ by 2 pH units or less. The Prideaux–Ward mixture comprises phosphate, phenylacetate, and borate plus HCl and covers the range from 2 to 12 pH units. The McIlvaine buffer is a mixture of citric acid and Na_2HPO_4 that covers the range from pH 2.2 to 8.0. The Britton–Robinson system consists of acetic acid, phosphoric acid, and boric acid plus NaOH and covers the range from pH 4.0 to 11.5. A mixture composed of Na_2CO_3, NaH_2PO_4, citric acid, and 2-amino-2-methyl-1,3-propanediol covers the range from pH 2.2 to 11.0.

TABLE 14.9*a* Standard Reference Values pH_s^* for the Measurement of Acidity in 50 wt % Methanol–Water

OAc = acetate; Suc = succinate. The asterisk in pH_s^* denotes that this is not an aqueous system entirely; it seems not to be universally used today.

Temperature, °C	0.02*m* HOAc, 0.02*m* NaOAc, 0.02*m* NaCl	0.02*m* NaHSuc, 0.02*m* NaCl	0.02*m* KH_2PO_4, 0.02*m* Na_2HPO_4, 0.02*m* NaCl
10	5.560	5.806	7.937
15	5.549	5.786	7.916
20	5.543	5.770	7.898
25	5.540	5.757	7.884
30	5.540	5.748	7.872
35	5.543	5.743	7.863
40	5.550	5.741	7.858

Reference: R. G. Bates, *Anal. Chem.* **40**(6):35A (1968).

TABLE 14.9*b* pH* Values for Buffer Solutions in Alcohol–Water Solvents at 25°C

Liquid-junction potential not included. Suc = succinate; Sal = salicylate. The asterisk in pH denotes that this is not an aqueous system entirely.*

Solvent composition (wt % alcohol)	0.01*M* $H_2C_2O_4$, 0.01*M* $NH_4HC_2O_4$	0.01*M* H_2Suc, 0.01*M* LiHSuc	0.01*M* HSal, 0.01*M* NaSal
	Methanol–water solvents		
0	2.15	4.12	
10	2.19	4.30	
20	2.25	4.48	
30	2.30	4.67	
40	2.38	4.87	
50	2.47	5.07	
60	2.58	5.30	
70	2.76	5.57	
80	3.13	6.01	
90	3.73	6.73	
92	3.90	6.92	
94	4.10	7.13	
96	4.39	7.43	
98	4.84	7.89	
99	5.20	8.23	
100	5.79	8.75	7.53
	Ethanol–water solvents		
0	2.15	4.12	
30	2.32	4.70	
50	2.51	5.07	
71.9	2.98	5.71	
100			8.32

Source: J. A. Dean, ed., *Lange's Handbook of Chemistry*, 14th ed., McGraw-Hill, New York, 1992.

General directions for the preparation of buffer solutions of varying pH but fixed ionic strength are given by Bates.[2] Preparation of McIlvaine buffered solutions at ionic strengths of 0.5 and 1.0 and Britton–Robinson solutions of constant ionic strength have been described by Elving et al.[3] and Frugoni,[4] respectively. Zwitterionic biological buffers are included in Table 14.10.

14.2.3 Glass-Indicating Electrodes

Typical pH-sensitive glass membranes are either sodium–calcium silicate or lithium silicates with lanthanum and barium ions added to retard silicate hydrolysis and lessen sodium ion mobility. All glass electrodes must be conditioned for a time by soaking in water or in a dilute buffer solution. Upon immersion of the membrane in water, the surface layer becomes involved in an ion-exchange process between the hydrogen ions in the solution and the sodium or lithium ions of the membrane.

The glass pH electrode has an internal reference electrode immersed in a chloride salt buffer solution that is sealed within the tip or bulb of the electrode. Often a phosphate buffer at pH 7 is used.

[2] R. G. Bates, *Determination of pH, Theory and Practice*, Wiley, New York, 1964, pp. 121–122.

[3] P. J. Elving, J. M. Mackowitz, and I. Rosenthal, *Anal. Chem.* **28**:1179 (1956).

[4] Frugoni, *Gazz. Chim. Ital.* **87**:1403 (1957).

TABLE 14.10 pH Values of Buffer Solutions for Control Purposes

Materials	pK_a 20°C	Useful pH range	Solubility, g/100 mL (0°C)
p-Toluenesulfonic acid and NaOH	<0	1.1–3.3	67
Glycine and HCl	2.31(+1)	1.0–3.7	25
Citrate and HCl	3.13 (pK_1) 4.76 (pK_2)	1.3–4.7	59
Sodium formate and HCl	3.75	2.8–4.6	81
Succinic acid and borax	4.21 (pK_1) 5.63 (pK_2)	3.0–5.8	6.3 (borax) 7.7 (succinic)
Sodium acetate and acetic acid	4.76	3.7–5.6	75
Sodium succinate and succinic acid	5.64 (pK_2)	4.8–6.3	7.7
2-(*N*-Morpholino)ethanesulfonic acid and NaOH	6.15	5.8–6.5	12.7
2,2-Bis(hydroxymethyl)-2,2′,2″-nitrilotriethanol and HCl	6.46(+1)	5.8–7.2	
Potassium dihydrogen phosphate and borax	7.20 (pK_2)	5.8–9.2	22.6 (KH_2PO_4) 6.3 (borax)
N-(2-Acetamido)iminodiacetic acid and NaOH	6.62	6.2–7.2	1.7
N-(2-Acetamido)-2-aminoethanesulfonic acid and NaOH	6.88	6.4–7.4	6.6
Potassium dihydrogen phosphate and disodium hydrogen phosphate	7.20 (pK_2)	6.1–7.5	22.6 (KH_2PO_4) 25.0 (Na_2HPO_4)
N-2-Hydroxyethylpiperazine-*N*′-2-ethanesulfonic acid and NaOH	7.50	7.0–8.0	59.6
Triethanolamine and HCl	7.76(+1)	6.9–8.5	Miscible
5,5-Diethylbarbiturate (veronal) and HCl	8.02	7.0–8.5	0.7
Tris(hydroxymethyl)aminomethane and HCl	8.08(+1)	7.0–8.5	
N,*N*-Bis(2-hydroxyethyl)-2-aminoethanesulfonic acid and NaOH	7.15	6.6–7.6	68.2
3-(*N*-Morpholino)propanesulfonic acid and NaOH	7.20	6.5–7.9	6.5
N-2-Hydroxyethylpiperazine-*N*′-3-propanesulfonic acid and NaOH	8.0	7.6–8.8	39.9
N-Tris(hydroxymethyl)methylglycine and HCl	8.15	7.6–8.8	0.8
N,*N*′-Bis(2-hydroxyethyl)glycine and HCl	8.35	7.8–8.8	1.1
Borax and HCl	9.24	7.6–8.9	6.3
Ammonia (aqueous) and ammonium chloride	9.25	8.3–9.2	26.0 (NH_4Cl)
Ethanolamine and HCl	9.50(+1)	8.6–10.4	Miscible
2-(Cyclohexylamino)ethanesulfonic acid and NaOH	9.55	9.0–10.1	23.6
Glycine and NaOH	9.70(0)	8.2–10.1	25
Borax and NaOH	9.24	9.4–11.1	6.3
Sodium carbonate and sodium hydrogen carbonate	10.33 (pK_2)	9.2–11.0	8.0 ($NaHCO_3$) 25.0 (Na_2CO_3)
3-(Cyclohexylamino)propanesulfonic acid and NaOH	10.40	9.7–11.1	10.4
Disodium hydrogen phosphate and NaOH	11.90 (pK_3)	11.0–12.0	25

TABLE 14.10 pH Values of Buffer Solutions for Control Purposes (*Continued*)

x mL of 0.2*M* sodium acetate (27.199 g $NaOAc \cdot 3H_2O$ per liter) plus *y* mL of 0.2*M* acetic acid

pH	NaOAc, mL	Acetic acid, mL
3.60	7.5	92.5
3.80	12.0	88.0
4.00	18.0	82.0
4.20	26.5	73.5
4.40	37.0	63.0
4.60	49.0	51.0
4.80	60.0	40.0
5.00	70.5	29.5
5.20	79.0	21.0
5.40	85.5	14.5
5.60	90.5	9.5

x mL of 0.1*M* KH_2PO_4 (13.617 $g \cdot L^{-1}$) plus *y* mL of 0.05*M* borax solution (19.404 g $Na_2B_4O_7 \cdot 10H_2O$ per liter)

pH	KH_2PO_4, mL	Borax, mL	pH	KH_2PO_4, mL	Borax, mL
5.80	92.1	7.9	7.60	51.7	48.3
6.00	87.7	12.3	7.80	49.2	50.8
6.20	83.0	17.0	8.00	46.5	53.5
6.40	77.8	22.2	8.20	43.0	57.0
6.60	72.2	27.8	8.40	38.7	61.3
6.80	66.7	33.3	8.60	34.0	66.0
7.00	62.3	37.7	8.80	27.6	72.4
7.20	58.1	41.9	9.00	17.5	82.5
7.40	55.0	45.0	9.20	5.0	95.0

x mL of veronal (20.6 g Na diethylbarbiturate per liter) plus *y* mL of 0.1*M* HCl

pH	Veronal, mL	HCl, mL
7.00	53.6	46.4
7.20	55.4	44.6
7.40	58.1	41.9
7.60	61.5	38.5
7.80	66.2	33.8
8.00	71.6	28.4
8.20	76.9	23.1
8.40	82.3	17.7
8.60	87.1	12.9
8.80	90.8	9.2
9.00	93.6	6.4

x mL of 0.2*M* aqueous NH_3 solution plus *y* mL of 0.2*M* NH_4Cl (10.699 $g \cdot L^{-1}$)

pH	Aq NH_3, mL	NH_4Cl, mL
8.00	5.5	94.5
8.20	8.5	91.5
8.40	12.5	87.5
8.60	18.5	81.5
8.80	26.0	74.0
9.00	36.0	64.0
9.25	50.0	50.0
9.40	58.5	41.5
9.60	69.0	31.0
9.80	78.0	22.0
10.00	85.0	15.0

x mL of 0.1*M* citrate (21.0 g citric acid monohydrate + 200 mL 1*M* NaOH per liter) plus *y* mL of 0.1*M* NaOH

pH	Citrate, mL	NaOH, mL
5.10	90.0	10.0
5.30	80.0	20.0
5.50	71.0	29.0
5.70	67.0	33.0
5.90	62.0	38.0

x mL of 0.2*M* NaOH added to 100 mL of stock solution (0.04*M* acetic acid, 0.04*M* H_3PO_4, and 0.04*M* boric acid)

pH	NaOH, mL	pH	NaOH, mL	pH	NaOH, mL	pH	NaOH, mL
1.81	0.0	4.10	25.0	6.80	50.0	9.62	75.0
1.89	2.5	4.35	27.5	7.00	52.5	9.91	77.5
1.98	5.0	4.56	30.0	7.24	55.0	10.38	80.0
2.09	7.5	4.78	32.5	7.54	57.5	10.88	82.5
2.21	10.0	5.02	35.0	7.96	60.0	11.20	85.0
2.36	12.5	5.33	37.5	8.36	62.5	11.40	87.5
2.56	15.0	5.72	40.0	8.69	65.0	11.58	90.0
2.87	17.5	6.09	42.5	8.95	67.5	11.70	92.5
3.29	20.0	6.37	45.0	9.15	70.0	11.82	95.0
3.78	22.5	6.59	47.5	9.37	72.5	11.92	97.5

(Continued)

TABLE 14.10 pH Values of Buffer Solutions for Control Purposes (*Continued*)

x mL of 0.1*M* HCl plus *y* mL of 0.1*M* glycine (7.505 g glycine + 5.85 g NaCl per liter)			*x* mL of 0.1*M* HCl plus *y* mL of 0.1*M* citrate (21.008 g citric acid monohydrate + 200 mL 1*M* NaOH per liter)			*x* mL of 0.05*M* succinic acid (5.90 g · L^{-1}) plus *y* mL of borax solution (19.404 g $Na_2B_4O_7 \cdot 10H_2O$ per liter)		
pH	HCl, mL	Glycine, mL	pH	HCl, mL	Citrate, mL	pH	Succinic acid, mL	Borax, mL
1.20	84.0	16.0	3.50	52.8	47.2	3.60	90.5	9.5
1.40	71.0	29.0	3.60	51.3	48.7	3.80	86.3	13.7
1.60	61.8	38.2	3.80	48.6	51.4	4.00	82.2	17.8
1.80	55.2	44.8	4.00	43.8	56.2	4.20	77.8	22.2
2.00	49.1	50.9	4.20	38.6	61.4	4.40	73.8	26.2
2.20	42.7	57.3	4.40	34.6	65.4	4.60	70.0	30.0
2.40	36.5	63.5	4.60	24.3	75.7	4.80	66.5	33.5
2.60	30.3	69.7	4.80	11.0	89.0	5.00	63.2	36.8
2.80	24.0	76.0				5.20	60.5	39.5
3.00	17.8	82.2				5.40	57.9	42.1
3.30	10.8	89.2				5.60	55.7	44.3
3.60	6.0	94.0				5.80	54.0	46.0

x mL of 0.2*M* $Na_2HPO_4 \cdot 2H_2O$ (35.599 g · L^{-1}) plus *y* mL of 0.1*M* citric acid (19.213 g · L^{-1})

pH	Na_2HPO_4, mL	Citric acid, mL	pH	Na_2HPO_4, mL	Citric acid, mL	pH	Na_2HPO_4, mL	Citric acid, mL
2.20	2.00	98.00	4.20	41.40	58.60	6.20	66.10	33.90
2.40	6.20	93.80	4.40	44.10	55.90	6.40	69.25	30.75
2.60	10.90	89.10	4.60	46.75	53.25	6.60	72.75	27.25
2.80	15.85	84.15	4.80	49.30	50.70	6.80	77.25	22.75
3.00	20.55	79.45	5.00	51.50	48.50	7.00	82.35	17.65
3.20	24.70	75.30	5.20	53.60	46.40	7.20	86.95	13.05
3.40	28.50	71.50	5.40	55.75	44.25	7.40	90.85	9.15
3.60	32.20	67.80	5.60	58.00	42.00	7.60	93.65	6.35
3.80	35.50	64.50	5.80	60.45	39.55	7.80	95.75	4.25
4.00	38.55	61.45	6.00	63.15	36.85	8.00	97.25	2.75

An external reference electrode completes the probe assembly. The combination electrode contains both the pH-sensing and reference electrodes in a single probe body, as shown in Fig. 14.1.

Glass electrodes are available in a variety of configurations. For medical studies, special electrodes are available that can be inserted into blood vessels, the stomach, and other parts of the body. Syringe and capillary electrodes require only one or two drops of solution, whereas others penetrate soft solids (such as leather) or pastes. The normal-size electrode operates with a solution volume of 1 to 5 mL. Only the pH-sensitive membrane needs to be completely immersed; the remainder of the glass electrode is made of rugged, inert glass.

The chemical composition of the glass membrane may dictate its usage. Universal glass electrodes have a resistance of about 100 MΩ at 25°C; they can be used down to 0°C. The rugged membrane withstands abuse and rough handling typical of many industrial applications. Glass membranes may suffer chemical attack by the test solution, so chemical durability is important. For longer membrane life at extreme pH values, a full-range, high-pH glass electrode should be used. This is the membrane that should be used in solutions with high alkali-metal content and high pH values; however, its lower temperature limit is 10°C.

The glass electrode is not influenced by oxidizing or reducing agents. It is not poisoned by heavy metals, and it functions well in numerous partly aqueous media. However, the glass membrane tends to adsorb materials from the test solution.

*14.2.3.1 **Precautions When Using Glass Electrodes.***

1. When not in use, keep the pH electrode assembly immersed in a buffer solution. For long-term storage, carefully dry the electrode and place it in a protective container.
2. Thoroughly wash the electrode with distilled water after each measurement. Carefully blot dry.
3. Rinse the electrode with several portions of the test solution before making the final measurement.
4. Vigorously stir poorly buffered solutions during measurement; otherwise, the stagnant layer of solution at the glass–solution interface tends toward the composition of the particular membrane.
5. Wipe suspensions and colloidal material from the membrane surface with a soft tissue. Avoid scratching the delicate membrane.
6. Never use glass electrodes in acid fluoride solutions because the membrane will be attacked chemically.

*14.2.3.2 **Troubleshooting Tips.*** When poor or suspect readings are encountered and the reference electrode has been eliminated as a problem, (1) inspect the electrode for visible cracks or scratches on the membrane surface. If either exists, the electrode must be discarded. (2) Perform a standard calibration for one buffer, then transfer the probe assembly to a second buffer. A reading accurate to within ±0.05 pH unit from the standardization value should be obtained within 30 s. If this condition is not met, the electrode should be rejuvenated.

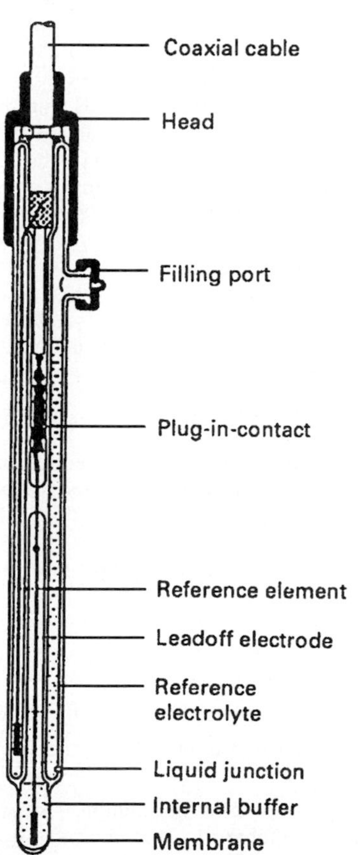

FIGURE 14.1 Combination of a pH-responsive glass and a reference electrode. (*From Shugar and Dean, The Chemist's Ready Reference Handbook, McGraw-Hill, 1990.*)

Rejuvenation of the sensing membrane may succeed after trying the following procedures:

1. Immerse the membrane into 0.1*M* HCl for about 15 s. Rinse with distilled water, then immerse into 0.1*M* KOH for 15 s, and rinse with distilled water. Cycle the membrane through these solutions several times.

2. Immerse the membrane into 20% ammonium hydrogen fluoride for exactly 3 min. Thoroughly rinse the electrode with distilled water, dip into concentrated hydrochloric acid, and rinse with distilled water. Soak the electrode in pH 4 buffer solution for 24 h. If proper electrode response has not been restored, discard the electrode.

14.2.4 The Antimony Electrode

The antimony–antimony(III) oxide electrode consists of a stick of metallic antimony cemented into a hard rubber sleeve. An invisible surface coating of antimony(III) oxide seems always to be present. The solutions in which the antimony electrode is used should preferably contain dissolved air.

The standard potential of the electrode is not very reproducible and the slope of the emf–pH curve is not constant over a very wide

TABLE 14.11 Values of $2.3026RT/F$ at Several Temperatures, in Millivolts

t, °C	Value	t, °C	Value	t, °C	Value	t, °C	Value
0	54.197	25	59.157	50	64.118	80	70.070
5	55.189	30	60.149	55	65.110	85	71.062
10	56.181	35	61.141	60	66.102	90	72.054
15	57.173	38	61.737	65	67.094	95	73.046
18	57.767	40	62.133	70	68.086	100	74.038
20	58.165	45	63.126	75	69.078		

Source: Report of the National Academy of Sciences: National Research Council Committee of Fundamental Constants, 1963.

pH range; the slope usually appears to be less than theoretical. Calibration with several different standard buffer solutions is essential. A properly calibrated antimony electrode can be used from pH 1 to about 10. It is unsafe to use this electrode in solutions of unknown composition because of many chemical interferences. Oxidizing and reducing agents normally cause difficulty, and the electrode exhibits a marked sensitivity to complexing agents, notably the anions of hydroxyl acids. The presence in the test solution of metals more noble than antimony causes large deviations.

14.2.5 pH Meter

Electronic pH meters are simply voltmeters with scale divisions in pH units that are equivalent to the values of $2.3026RT/F$ (in mV) per pH unit. Values of this function at several temperatures are given in Table 14.11. There is no compensation incorporated in the meter for the changes in pH of the test solution as a function of temperature. An electrometric pH measurement system (pH meter) consists of (1) a pH-responsive electrode, (2) a reference electrode, and (3) a potential-measuring device—some form of a high-impedance electronic voltmeter for glass-electrode combinations and this or a potentiometer arrangement for other pH-responsive electrodes.

To achieve a reproducibility of ±0.005 pH unit, the pH meter must be reproducible to at least 0.2 mV. The current drawn from the probe assembly should be 10^{-12} A, or less.

The schematic circuit diagram of a pH meter, based on an operational amplifier, is shown in Fig. 14.2. The relationship for the emf of a pH probe assembly is

$$E = k - KT\,(\text{pH}) \tag{14.26}$$

This is the equation of a straight line with slope $-KT$ and a zero intercept of k on the E axis. The glass electrode with its internal reference electrode and filling solution (usually pH 7.00) must have an isopotential point at 0 V.

14.2.5.1 Operating Procedure. Proceed as follows in making pH measurements:

1. Turn on the pH meter. Different manufacturers designate the neutral position of the selector switch with various markings, such as *Standby*, *Bal*, and so on.
2. Raise the pH probe assembly from the storage solution.
3. Rinse the electrodes thoroughly with distilled water and then with the standard reference solution.
4. Immerse the pH probe assembly in the standard reference buffer that is nearest the expected pH value of the sample.
5. Adjust the temperature control to the temperature of the solution. This amounts to adjusting the KT factor in Eq. (14.26) (and enumerated in Table 14.11) by means of the emf–pH slope control about the isopotential point.
6. Rotate the selector switch to pH.
7. Bring the meter reading into juxtaposition with the pH_s value by adjusting the intercept (variously labeled *Zero, Standardization, Calibration*, or *Asymmetry*) control.

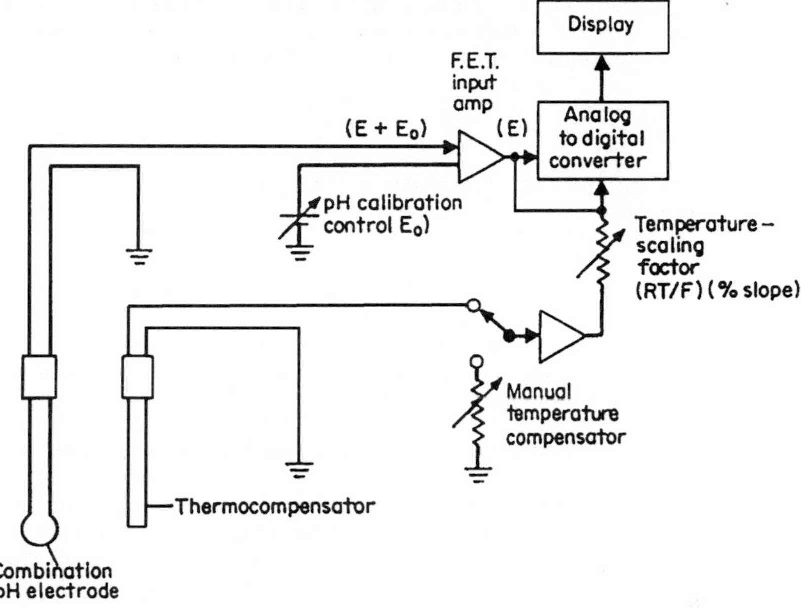

FIGURE 14.2 **Schematic circuit diagram of a pH meter.** (*From Shugar and Dean, 1990.*)

8. Remove the pH probe assembly, and rinse with distilled water and then with portions of the test solution. Blot the electrodes between rinses with adsorbent, lint-free tissue.

9. Lower the pH probe assembly into the test solution. With the selector switch still at pH, read the pH of the test solution.

After the standardization with the standard reference buffer, it is always wise to check the functionality of the probe system by measuring the pH of a second reference buffer; the two reference buffers should bracket the expected pH of the sample. Faulty glass electrodes or plugged external reference electrodes may seem to operate correctly upon initial standardization, but they will not give the proper pH reading for the second pH reference buffer.

Microcomputer-based pH meters measure the output of the electrode assembly and subject it to appropriate algorithms based on Eq. (14.26). A prescribed operating protocol must be carefully followed for calibration and standardization.

14.3 ION-SELECTIVE ELECTRODES

An ion-selective electrode (ISE) is a potentiometric probe whose output potential, when measured against a suitable reference electrode, is logarithmically proportional to the activity of the selected ion in the test solution. Ion activities are the thermodynamically effective free-ion concentrations. In dilute solutions, ion activity approaches the ion concentration. Activity measurements are valuable because the activities of ions determine rates of reactions and chemical equilibria.

ISEs have several valuable features that make them superior to other methods of analysis. Measurements are simple, often rapid, nondestructive, inexpensive, and applicable to a wide range of concentrations. Many analyses can be performed directly without the need for time-consuming sample preparation, such as centrifugation and filtration. Solution color or turbidity does not affect results. ISEs are unresponsive to oxidation–reduction couples in the test solution. Portability enables them to be used in field work. However, ISEs have some major disadvantages. The classical ion-selective membranes cannot be used at high temperature or pressure and must be used in an upright position. Certain anionic pollutants and dissolved gases in aqueous environmental samples may be measured conveniently by ISE methods (see Sec. 21).

14.3.1 Membrane Sensors

There are three types of membrane sensors: glass, solid state, and solid matrix (liquid ion exchange). Gas-sensing and biocatalytic electrodes are merely special designs that incorporate one of the three types into the system.

14.3.1.1 Glass Membrane Electrodes. The glass-electrode construction is the same as shown in Fig. 14.1 for pH-responsive electrodes. Substitution of aluminum for part of the silicon in the alkali-metal–silicate glasses produces cation-selective glass membranes. Glasses of the composition 11% Na_2O:18% Al_2O_3:71% SiO_2 are highly sodium (and silver) -selective with respect to other alkali-metal ions. Glasses that have a composition about 27% Na_2O:5% Al_2O_3:68% SiO_2 show a general univalent cation response ($H^+ > K^+ > Na^+ > NH_4^+ > Li^+ \cdots \gg Ca^{2+}$). This type also displays considerable response to thallium(I), copper(I), and alkyl quaternary ammonium ions.

14.3.1.2 Solid-State Sensors. The electrode body is composed of a chemically resistant epoxy formulation. Bonded to the electrode body is the sensing membrane, a solid-state ionic conductor. Completing the electrode is an internal reference electrode, usually silver–silver chloride, and an internal filling solution. A cross-sectional view is shown in Fig. 14.3. In newer versions of the solid-state sensors, the sensing membrane is composed of pure, nonporous material, unlike the old-style, pressed granular powder membrane. The membrane has a homogeneous, mirrorlike surface of low microporosity that keeps sample retention to a minimum. The lower limits of useful response are imposed by the solubility of the sensor's membrane material in the sample solution.

The active membrane of the fluoride electrode has a single crystal of LaF_3, doped with europium(II), and an internal solution that is 0.1*M* in NaF and in NaCl. The fluoride ion activity controls the potential of the inner surface of the LaF_3 membrane, and the chloride ion activity fixes the potential of the internal Ag–AgCl wire reference electrode. When in contact with the sample at 25°C, the external surface of the membrane responds to the fluoride ion activity in the sample:

$$E = \text{const} - 0.0592 \log [F^-] \tag{14.27}$$

The fluoride-responsive electrode follows a logarithmic response when fluoride concentrations are as low as $10^{-5}M$ and a useful response to at least $10^{-6}M$ fluoride ion.

A group of ISEs based on the Ag_2S membrane are available. By itself, such a membrane can be used to detect silver ions or to measure sulfide ion levels. The dynamic range of the electrode extends from saturated solutions down to silver and sulfide levels on the order of $10^{-8}M$. For sulfide measurements a special buffer must be mixed with the samples to raise the pH of the sample solution, to free sulfide bound to hydrogen, to fix the total ionic strength, and to retard oxidation of the sulfide ion.

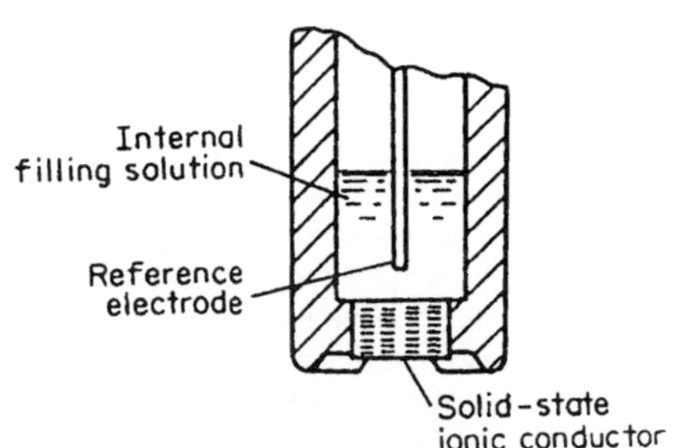

FIGURE 14.3 Cross-sectional view of solid-state sensor. (*Courtesy of Orion Research, Inc.*) (*From Shugar and Dean, 1990.*)

If the silver sulfide (or HgS) is altered by dispersing within it another metal sulfide, the corresponding metal-selective electrode is obtained. Two solid-phase equilibria must now be established—the solubility-product equilibria of silver sulfide and the metal sulfide, which must be larger than that of the silver sulfide. Useful copper-, cadmium-, and lead-selective electrodes have been prepared from CuS, CdS, and PbS, respectively, dispersed in a silver sulfide matrix.

Other more soluble silver salts, dispersed within the silver sulfide matrix, make up the anion-selective electrodes of chloride, bromide, iodide, and thiocyanate, respectively. The halide electrodes exhibit useful ranges (in pI or -log [ion] units) up to 5, 6, and 7, respectively, for chloride, bromide, and iodide ions.

The cyanide-selective electrode is an anomaly. After an AgI (or AgI–Ag_2S) ISE is soaked in a cyanide solution, sufficient chemical attack occurs to form pockets of $Ag(CN)_2^-$. Now when dipped into a test solution, the electrode is responsive to cyanide ions.

14.3.1.3 Solid- (or Liquid-) Matrix Electrodes. In the solid-matrix electrode, the sensing membrane is an ion exchanger permanently embedded in a plastic material [typically poly(vinyl chloride)] that is sealed to the inert electrode body. These electrodes never need rebuilding or solution replenishing. Such polymer membrane electrodes can be fabricated in miniature or flow-through forms. Typically, 1 to 5 wt % of ionophore or ion-binding sites within the membrane phase is required to achieve optimal potentiometric ion response. In the older liquid-matrix electrodes, the ion-selective membranes consisted of wet organic solution phases that contained appropriate ion-exchanger or ionophore molecules that was supported by some hydrophobic porous material, often porous poly(tetrafluoroethylene). The constrution is shown in Fig. 14.4.

The calcium-selective electrode uses the calcium salt of bis(2-ethylhexyl)phosphoric acid or bis(decyl)phosphoric acid as the ion exchanger; the aqueous internal filling solution consists of a fixed concentration of calcium and chloride ions. Uranium(VI) is also responsive.

The perchlorate-sensing membrane uses the tris(substituted 1,10-phenanthroline)iron(II) perchlorate as the ion-exchange site group. For the nitrate and tetrafluoroborate electrodes, a tris(substituted 1,10-phenanthroline)nickel(II) nitrate or tetrafluoroborate site group, respectively, is used. 2-Nitro-*p*-cymene is the solvent. Nitrate electrodes are also made in which the exchanger is the large tetraheptylammonium ion.

The chloride-ion-sensing membrane employs a long-chain amine, such as dimethyldioctadecylammonium chloride or dimethyldistearylammonium chloride as the site group.

A potassium-selective electrode incorporates a neutral carrier, the valinomycin molecule. This molecule is a doughnut-shaped complex with an electron-rich pocket in the center into which potassium ions are selectively bound through ion-dipole interaction to replace the hydration shell around potassium. The potassium electrode exhibits 1000:1 selectivity for K^+ in preference to Na^+ and no pH dependence.

Other examples of neutral-carrier-based ISEs include a nonactin–monactin liquid phase for NH_4^+ and a mixture of nonylphenoxypoly(ethyleneoxyl)ethanol and barium tetraphenylborate for Ba^{2+}.

14.3.1.4 Solid-State Gas Sensors. Gas-sensing electrodes are available for the measurement of NH_3, CO_2, NO_2, HCN, SO_2, and H_2S. This type of sensor exposes a gas-permeable membrane to the solution to be analyzed. The gaseous analyte present in the solution partitions into this membrane, diffuses

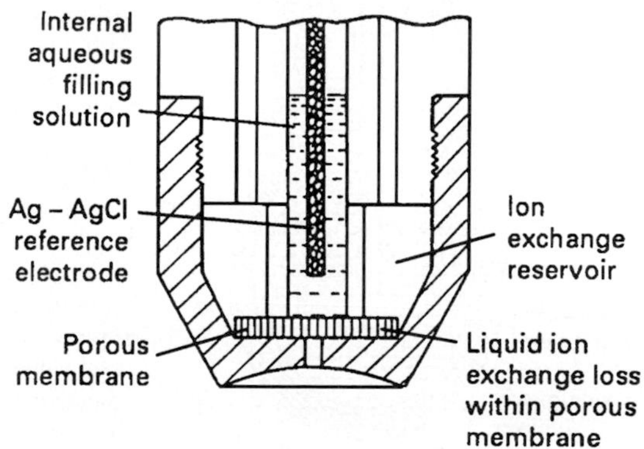

FIGURE 14.4 Construction of a liquid ion-exchange electrode. *(Courtesy of Orion Research, Inc.) (From Shugar and Dean, 1990.)*

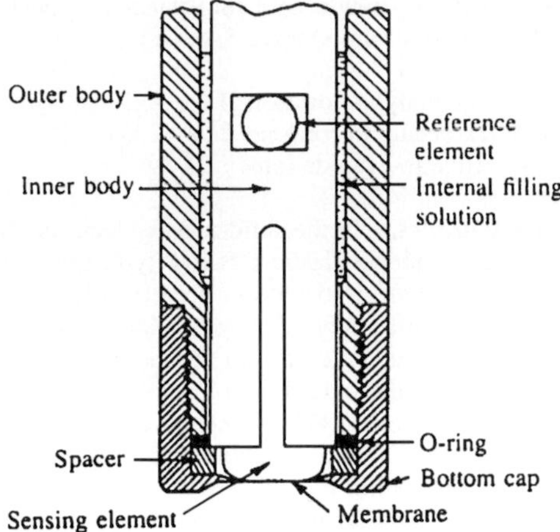

FIGURE 14.5 Construction of a gas-sensing electrode. (*Courtesy of Orion Research, Inc.*) (*From Shugar and Dean, 1990.*)

through it, and then partitions into an electrolyte solution on the other side of the membrane. The effect of this gas on the ionic content of this electrolyte solution is then monitored with an appropriate ISE. The response and recovery time is relatively long; often 30 s to 5 min. The construction of a solid-state gas sensor is shown in Fig. 14.5.

In the ammonia-selective electrode, dissolved ammonia from a sample diffuses through a fluorocarbon membrane into the internal filling solution that consists of $0.01M$ NH_4Cl in which the ammonia reacts with water to form hydroxide and ammonium ions until a reversible equilibrium is established between the ammonia level of the sample and the internal solution. Either the ammonium-ion concentration can be measured with an ammonium-ion-selective glass electrode or the pH of the system can be monitored with a pH-responsive glass electrode. Standards and samples are adjusted to a fixed pH or to a pH greater than 11. Sensitivity extends from $10^{-6}M$ to $1M$ ammonia.

The carbon dioxide sensor possesses a microporous Teflon membrane that separates the sample from an internal electrolyte of $0.01M$ $NaHCO_3$. An internal pH glass electrode monitors the HCO_3^-–CO_2 equilibrium as it is affected by the CO_2 dissolved in the sample.

In a similar manner the SO_2 content of a sample in an HSO_3^- buffer can be followed through the ratio of SO_2 to $0.01M$ HSO_3^- (internal electrolyte) with use of a pH-responsive electrode.

The NO_2 content of samples buffered with citrate can be followed through the equilibrium:

$$2NO_2 + H_2O \rightleftharpoons NO_3^- + NO_2^- + 2H^+ \tag{14.28}$$

An internal pH glass electrode or a nitrate-ion-sensing electrode completes the assembly with $0.02M$ $NaNO_2$ serving as the internal electrolyte.

In a similar manner the concentration of H_2S can be followed using an internal citrate buffer and a sulfide ion sensing electrode.

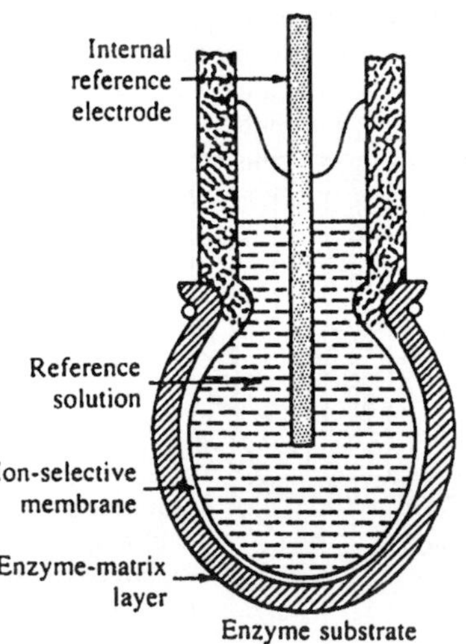

FIGURE 14.6 Biocatalytic membrane electrode schematic.

The concentration of HCN can be followed through the equilibrium:

$$Ag(CN)_2^- \rightleftharpoons Ag^+ + 2CN^- \tag{14.29}$$

with use of a silver-ion-sensing electrode. The sample is adjusted to pH < 7 and Pb^{2+} is added to remove interference from H_2S. The internal electrolyte is $KAg(CN)_2$.

14.3.1.5 Biocatalytic Membrane Electrodes. The membranes in biocatalytic electrodes are multilayered composites that contain one or more biocatalysts (often enzymes) in a gel layer that coats a conventional ion-selective electrode (Fig. 14.6). Enzyme electrodes are fabricated by immobilizing or covalently binding an enzyme to an ion- or gas-selective electrode. The electrode will respond after the substrate diffuses to the immobilized enzyme and reacts. This reaction results in the release of a specific by-product that can be detected directly by the ISE. Numerous enzyme–substrate combinations are possible. One example will suffice. A urea-responsive electrode has the enzyme urease fixed in a layer of acrylamide gel held in place around the pH glass electrode bulb by a nylon mesh or a thin cellophane film. The urease acts specifically upon urea in the sample solution to yield ammonium ions (and hydroxide ions) that diffuse through the gel layer and are sensed by the internal pH electrode (or ammonium-ion-selective electrode). The major limitations of these systems is the stability of the enzyme in this new environment and the response time.

14.3.1.6 Coated-Wire Electrodes (CWEs). Coated-wire electrodes consist of a thin polymeric film containing the electroactive species, coated directly on a platinum substrate. The design of CWEs eliminates the need for an internal reference electrode. The exact mechanism of the CWE behavior continues to elude explanation, the membrane coating appears to play a pivotal role in maintaining an internal reference potential. The CWEs are inexpensive to produce and can readily be miniaturized. Their response is similar to that of classical ISEs with regard to sensitivity and range of concentration. The disadvantage is the limited lifetime and durability of the membrane coating.

14.3.1.7 Field-Effect Transistors (ISFETs). With ISFETs the ion-selective membrane is directly coated onto the gate of an FET. The reference electrode is defined by the potential of the encapsulated silicon substrate with respect to the analyte solution. The main advantage of ISFETs is that it is possible to prepare small multisensory systems with multiple gates that can sense several ions simultaneously. However, major compatibility problems between the electrochemical element and the electronic componet of the ISFETs have led to a persistent and unpredictable drift for these systems.

14.3.2 Ion Activity Evaluation Methods

For an electrode responsive to the activity of ion X, a_X, the electrode response is

$$E_X = \text{const} + S \log a_X \tag{14.30}$$

where E_X is the potential reading on a pION meter and S is the slope of the calibration plot (approximately 60 mV for a monovalent ion and 30 mV for a divalent ion). Calibration procedures must use solutions of known activity or concentration, depending on which parameter is required. At less than $10^{-3}M$ to $10^{-4}M$ the two quantities are practically indistinguishable.

14.3.2.1 Direct Calibration Plot. In this method several measurements of E are made in solutions of known activity (or concentration). These solutions are prepared by serial dilution of a concentrated standard. The recommended ionic strength buffer is added to each standard and to each unknown. One buffer for fluoride-ion determination consists of 0.25M acetic acid, 0.75M sodium acetate, 1M sodium chloride, and 1mM sodium citrate [for masking aluminum(III) and iron(III), which interfere by complexing with the fluoride ion]. Appropriate ionic strength buffers (and perhaps combined with pH buffer) is added for other selective-ion determinations. The calibration plot of electrode potential versus activity will be linear over several orders of magnitude; the concentration plot will deviate from linearity above $10^{-4}M$ to $10^{-3}M$. The potentials of unknown sample solutions can then be measured, and their activities (or concentrations) can be read directly from the calibration plot. The lower limit of detection reflects the experimental difficulty of preparing extremely dilute solutions without extensive adsorption on and desorption from the surfaces of containment vessels and the electrodes. Solubility equilibria and ion-pair association equilibria also affect the lower limit of electrode utility.

14.3.2.2 Method of Additions. In the method of additions (standard addition or sample addition), the approximate concentration (or activity) of the sample, C_u, is first estimated from the observed electrode potential (rough calibration plot):

$$E_1 = \text{const} + S \log C_u \tag{14.31}$$

A known volume of the standard solution, C_s (which should be about 10 times the estimated sample activity or concentration) is added. After correction for dilution of the original sample by the addition, and the corresponding correction of the added standard:

$$E_2 = \text{const} + S \log (C'_u + C'_s) \tag{14.32}$$

Combining equations, one gets

$$\Delta E = E_2 - E_1 = S \log \frac{C'_u + C'_s}{C'_u} \tag{14.33}$$

Volume corrections are

$$C'_u = C_u \frac{V_u}{V_u + V_s} \quad \text{and} \quad C'_s = C_s \frac{V_s}{V_u + V_s} \tag{14.34}$$

The concentration of the unknown can be explicitly determined by solving the equation

$$C_u = C_s \left(\frac{V_s}{V_u + V_s} \right) \left[10^{\Delta E/S} \left(\frac{V_u}{V_u + V_s} \right) \right]^{-1} \tag{14.35}$$

Sample addition is the inverse of the standard addition method. It is useful for samples that are small, highly concentrated, or dirty. It cannot be used when the unknown species is complexed. With this technique the sample is added to a known volume or a standard solution. The unknown concentration is determined by solving this equation:

$$C_u = C_s \left[10^{\Delta E/S} \left(\frac{V_u + V_s}{V_u} \right) \left(\frac{V_s}{V_u} \right) \right] \tag{14.36}$$

Known addition and subtraction methods are particularly suitable for samples with a high unknown total ionic strength. If the species being measured is especially unstable, known subtraction is preferred over known addition.

14.3.3 Selectivity and Interference

As their name implies, ion-selective electrodes are selective rather than specific for the particular ion. Thus a potassium-selective electrode responds not only to the activity of potassium ions in solution but also to some fraction of the sodium ions present. In a first approximation, the effect of foreign cations on the electrode potential may be fitted by an extended Nikolsky equation:

$$E = \text{const} + \frac{RT}{z_i F} \ln \left(a_i + \Sigma K_{ij} a_j^{(z_i/z_j)} \right) \tag{14.37}$$

where a_i = activity of the primary ion with charge of z_i
a_j = activity of interfering ion with charge of z_j
K_{ij} = selectivity ratio characteristic of a given membrane

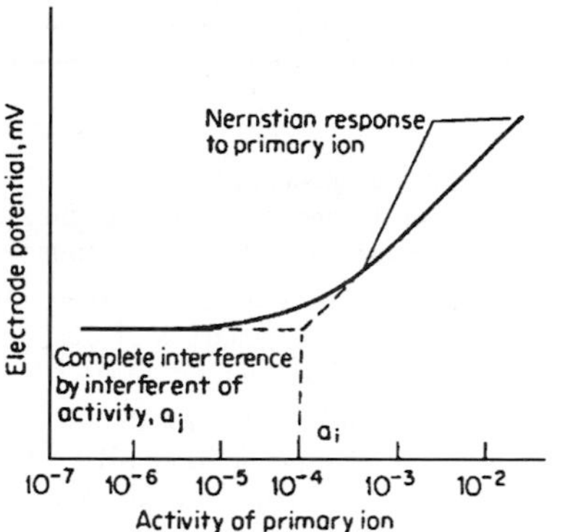

FIGURE 14.7 A method for calculating selectivity ratios. (*From Shugar and Dean, 1990.*)

The value of the constant in the equation depends on the experimental conditions (that is, the choice of external reference electrode, electrolyte concentration, liquid-junction potentials).

The simplest way to determine the selectivity ratio is shown in Fig. 14.7 Potential measurements are made in a series of solutions, with the activity of the ion to be measured varying in the presence of a constant background level of the interference. The region over which the electrode response is linear with respect to the test ion is clearly demarcated; this is the slope factor (S), here equal to RT/z_iF. The intercept of the extension of the response line (slope) with the horizontal line in the region of high interference defines a particular intercept activity for the primary ion, a_i. The particular activity is related to the selectivity ratio by the second term in the large parentheses of Eq. (14.37):

$$a_i = K_{ij} a_i^{(z_i/z_j)} \tag{14.38}$$

where a_j is the constant background level of the interference. Knowing the activity of a_i and a_j, one can solve for the selectivity ratio. The selectivity (the inverse of the selectivity ratio) is quoted for various ion-selective electrodes in Table 14.12. From the foregoing discussion, it should be obvious that selectivity ratios are not constant but vary with the concentration of both the primary and interference ion and, therefore, are often stated at a particular concentration.

Example 14.4 Valinomycin membranes show excellent selectivity for potassium—about 3800 times that of sodium and 18 000 times that of hydrogen ions. The selectivity ratio ($1/3800 = 2.632 \times 10^{-4}$) indicates that a $0.1M$ sodium solution will give the same response as a $2.6 \times 10^{-5}M$ potassium solution; that is, a 100% error in the apparent potassium concentration. To lower the error due to the presence of sodium ions to 2%, the sodium-ion concentration should not exceed $0.02 \times 0.1M$, or $0.002M$ for this level of potassium ion concentration. If the potassium concentration were tenfold greater, the sodium-ion concentration could also be tenfold greater and the error in potassium readings would be the same.

Continuing the example, but for hydrogen-ion interference, the hydrogen-ion selectivity ratio is $1/18\,000 = 5.56 \times 10^{-5}$. Thus, the pH of the sample should not exceed 6.26 if the potassium readings in the $1M$ range are to be accurate within 1%.

In solid-state sensors, the interference arises from surface reactions that can convert one of the components of the solid membrane to a second insoluble compound. As a result, the membrane loses sensitivity to the test ion. For example, chloride ion can interfere with bromide-ion measurements if the following reaction takes place:

$$Cl^- + AgBr(s) \longrightarrow AgCl(s) + Br^- \tag{14.39}$$

It will occur if the ratio of chloride-ion activity to bromide-ion activity exceeds the ratio of the solubility product of silver chloride to the solubility product of silver bromide, or

$$\frac{1}{K_{ij}} = \frac{a_{Cl}}{a_{Br}} = \frac{1.8 \times 10^{-10}}{5.0 \times 10^{-13}} = 360 \tag{14.40}$$

An expected response also fails to occur when the concentration of a primary ion approximates the solubility of the membrane material. This is the basis for the statement made earlier that the bromide-ion-selective electrode exhibited a useful response down to pBr = 6 (or, more precisely, $0.5\sqrt{K_{sp}} = 6.15$).

TABLE 14.12 Ion-Selective Electrode Characteristics

Electrode	Type	Measuring range, M	Preferred pH range	Interferences
Ammonia	Gas	1–(5 × 10^{-7})	11–13	Volatile amines or those soluble in the polymer of the membrane, such as mono-, di-, and trialkylamines, piperidine, and cyclohexylamine
Barium	Liquid	>5μg/L	2–7	Selectivity coefficients: H^+, 0.06; NH_4^+, 0.004; Li^+, 0.0005; Na^+, 0.003; K^+, 0.02; Rb^+, 0.01; Cs^+, 0.003; Mg^{2+}, 8 × 10^{-6}; Ca^{2+}, 0.0002; Sr^{2+}, 0.03
Bromide	Solid	1–(5 × 10^{-6})	2–12	$[S^{2-}] \le 10^7 M$; $[I^-] < 2 \times 10^{-4}[Br^-]$; $[CN^-] < 8 \times 10^{-5}$ $[Br^-]$; $[OH^-] < 3 \times 10^4$ $[Br^-]$; $[Cl^-] < 400$ $[Br^-]$
Cadmium	Solid	1–10^{-7}	3–7	$[Ag^+]$, $[Hg^{2+}]$, $[Cu^{2+}] \le 10^{-7}M$; high levels of Pb^{2+}, Fe^{3+} Selectivity coefficients: Fe^{2+}, 214; Tl^+, 126; Mn^{2+}, 2.5; Al, 0.8; Ni, 0.03; Cu^{2+}, 0.02; Zn, 4 × 10^{-4}; Ca, 4 × 10^{-4}; Mg, 1.7 × 10^{-4}
Calcium	Neutral PVC membrane	1–(5 × 10^{-6})	6–8	Selectivity coefficients: Zn, 10^{-4}; Fe^{2+}, 0.005; Pb^{2+}, 10; Mg^{2+}, 10^{-4}; (Na, K, NH_4^+), 0.003; Cu^{2+}, 0.0025; Li^+, 0.081; Ba^{2+}, 0.010; Sr^{2+}, 0.017; (I^-, ClO_4^-), < $10^{-3}M$; H^+, 0.0062
Carbon dioxide	Gas	(3 × 10^{-2})–10^{-5}	<pH 4	Volatile organic acids
Chloride	Solid	1–(5 × 10^{-6})	0–14	$[S^{2-}] \le 10^{-7}M$; $[OH^-] < 80$ $[Cl^-]$; $[Br^-] < 300$ $[Cl^-]$; $[S_2O_3^{2-}] <$ 100 $[Cl^-]$; $[NH_3] < 8$ $[Cl^-]$; traces of I^- and CN^-
	Liquid	10–10^{-5}	2–11	Selectivity coefficients: I^-, 17; NO_3^-, 4.2; Br^-, 1.6; ClO_4^-, 32; OH^-, 1.0; HCO_3^-, 0.19; SO_4^{2-}, 0.14; F^-, 0.10; OAc^-, 0.32
Copper(II)	Solid	1–(5 × 10^{-7})	3–7	$[S^{2-}]$, $[Ag^+]$, $[Hg^{2+}] \le 10^{-7}M$; high levels of Cl^-, Br^-, Cd^{2+}, Fe^{3+}; reagents that reduce Cu^{2+}
Cyanide	Solid	10^{-2}–(5 × 10^{-7})	11–13	$[S^{2-}] \le 10^{-7}M$ (add Pb^{2+}); $[I^-] < 0.1$ $[CN^-]$; $[Br^-] <$ 5000 $[CN^-]$; $[Cl^-] < 10^6$ $[CN^-]$
Fluoride	Solid	1–10^{-6}		Maximum level: $[OH^-] < 0.1$ $[F^-]$
Iodide	Solid	1–(2 × 10^{-7})	3–12	$[S^{2-}] < 10^{-7}M$; $[Cl^-] < 10^6$ $[I^-]$; $[Br^-] < 5000$ $[I^-]$; $[CN^-] <$ 0.4 $[I^-]$
Lead	Solid	1–10^{-7}	4–7	$[Ag^+]$, $[Cu^{2+}]$, $[Hg^{2+}] \le 10^{-7}$; $[Cd^{2+}]$, $[Fe^{3+}] > [Pb^{2+}]$
Lithium	Liquid	0.7–70 ng/L		Selectivity coefficients: H^+, 1; NH_4^+, 0.05; Na, 0.05; K, 0.007; Rb, 0.004; Cs, 0.003; Mg, 0.0002; Ca, 0.0006
Nitrate	Liquid	1–(5 × 10^{-6})	3–10	Selectivity coefficients: I^-, 20; Br^-, 0.1; Cl^-, 0.004; NO_2^-, 0.04; SO_4^{2-}, 0.000 03; CO_3^{2-}, 0.0002; ClO_4^-, 1000; F^-, 0.000 06; ClO_3^-, 2; CN^-, 0.02; HCO_3^-, 0.02; OAc^-, 0.006; $H_2HPO_4^-$, 0.003; HPO_4^{2-}, 0.000 08
Nitrite	Gas	0.02–(5 × 10^{-8})	Citrate buffer	Volatile organic acids; SO_2 must be destroyed; Cr(VI), CO_2 interferes
Perchlorate	Liquid	0.1–10^{-5}	3–10	Selectivity coefficients: I^-, 0.012; NO_3^-, 0.0015; Br^-, 0.000 56; F^-, 0.000 25; Cl^-, 0.000 22; OH^-, 1.0; OAc^-, 0.000 51; HCO_3^-, 0.000 35; SO_4^{2-}, 0.000 16
Potassium	Liquid	1–10^{-5}	3–10	Selectivity coefficients: Cs^+, 1.0; NH_4^+, 0.03; H^+, 0.01; Ag^+, 0.001; Na^+, 0.0002; Li^+, 0.0001; Mg, 0.001; Cu^{2+}, 0.002
Silver/sulfide	Solid	1–10^{-7} Ag^+ or S^{2-}	2–9 (Ag^+) 13–14 (S^{2-})	$[Hg^{2+}] \le 10^{-7}M$ as silver sensor None as sulfide sensor if ascorbic acid is added to remove O_2

TABLE 14.12 Ion-Selective Electrode Characteristics (*Continued*)

Electrode	Type	Measuring range, M	Preferred pH range	Interferences
Sodium	Glass	$1–10^{-6}$	9–10	Selectivity coefficients: Li^+, 0.002; K^+, 0.001; Rb^+, NH_4^+, 0.000 03; H^+, 100; Cs^+, 0.0015; Tl^+, 0.0002; Ag^+, 350; $(C_2H_5)_4N^+$, 0.005
	Neutral ion carrier			Selectivity coefficients: H^+, 0.5; Li^+, 0.04; K^+, 0.5; NH_4^+, 0.2; Cs^+, 0.002; Ca^{2+}, 0.002; Mg^{2+}, 0.0008
Sulfide	Ag_2S	$\geq 10^{-14}$	13	$[CN^-] < 500\ [S^{2-}]$
Sulfur dioxide	Gas	$10^{-2}–10^{-6}$	0–2	Volatile acids; Cl_2, NO_2 must be destroyed with N_2H_4
Tetrafluoroborate	Liquid	$0.1–10^{-5}$	3–10	Selectivity coefficients: NO_3^-, 0.1; Br^-, 0.04; OAc^-, 0.004; HCO_3^-, 0.004; Cl^-, 0.001; SO_4^{2-}, 0.001; OH^-, 0.001; I^-, 20; F^-, 0.001
Thiocyanate	Solid	$1–(5 \times 10^{-6})$	2–12	$[OH^-] < 100\ [SCN^-]$; $[Br^-] < 0.003\ [SCN^-]$; $[Cl^-] < 20\ [SCN^-]$; $[NH_3] < 0.13\ [SCN^-]$; $[S_2O_3^{2-}] < 0.01\ [SCN^-]$; $[CN^-] < 0.007\ [SCN^-]$; $[I^-]$, $[S^{2-}] \leq 10^{-7}M$; Bi and transitional-metal ions form weak SCN^- complexes; Au(I) and Hg(II) must be absent

14.4 POTENTIOMETRIC TITRATIONS

Potentiometric titrations involve following the changes in the cell emf brought about by the addition of a titrant of an accurately known concentration to the test solution. The method is applicable to any titrimetric reaction for which an indicator electrode is available to follow the activity (concentration) of at least one of the substances involved. Rapid changes in cell emf signal the equivalence point of the titration. A general advantage of potentiometric titrations is that no visual error can be made in the detection of the end point, and a high degree of precision can be reached. Especially in the titrations of colored or turbid systems, or of those for which there are no suitable visual indicators, the potentiometric methods have great importance. Precision is limited by the sharpness of the potential change at the end point and by the accuracy with which the volume of titrant can be delivered. Generally, solutions more dilute than $10^{-3}M$ do not give satisfactory end points. This is a limitation of potentiometric titrations.

Requirements for reference electrodes are greatly relaxed. Accuracy is increased because measured potentials are used to detect rapid changes in activity that occur at the equivalence point. This rate of emf change is usually considerably greater than the response slope, which limits precision in direct potentiometry. Furthermore, it is the change in emf versus titration volume rather than the absolute value of emf that is of interest. Thus, the influence of liquid-junction potentials and activity coefficients is minimized.

The equivalence point may be calculated or it can be located by inspection from the inflection point of the titration curve. It is the point that corresponds to the maximum rate of change of cell emf per unit volume of titrant added (usually 0.05 or 0.1 mL).

14.4.1 Potentiometric Precipitation Titrations

Extensive work has been done on potentiometric methods of detecting the end points of precipitation titrations because relatively few chemical indicators for such titrations are available and because most of those are not suitable for the titration of mixtures. Potentiometric precipitation titrations are carried out with a silver, mercury, or platinum indicator electrode, or an ion-selective electrode (Sec. 14.3), together with an appropriate reference electrode. The end point is typically located by finding the point of maximum potential change for a given increment of titrant (the inflection point of the titration curve). Although this method is often theoretically incorrect, as in the case of unsymmetrical precipitates, the error involved is usually slight. Graphical methods are also extensively used, especially in the titration of mixtures. Titration to a calculated or predetermined end-point potential is also of use, but it must be remembered that changes in ionic strength, in liquid-junction potentials, or in the surface of the indicator electrode may render this method unreliable unless care is taken to compensate for their effects.

The accuracy of the titration increases with increasing concentration of the species being determined and also with increasing insolubility of the precipitate. If we take 60 mV as a minimum ΔE for which an accurate titration is possible, and the case in which the ratio of cation to anion is 1, then the solubility product for the precipitate when titrating solutions of a monovalent ion should be no greater than

2.5×10^{-5} when titrating $10^{-2}M$ solutions

2.5×10^{-9} when titrating $10^{-4}M$ solutions

2.5×10^{-13} when titrating $10^{-6}M$ solutions

Table 14.13 outlines procedures for the execution of a number of potentiometric precipitation titrations, selected on the basis of their reliability and usefulness. The reader should also consult the material in Sec. 3.4.

14.4.2 Potentiometric Oxidation–Reduction Titrations

As one proceeds down in Table 14.14, the oxidizing agents decrease in strength and the reducing agents increase in strength. In general, if two half-reactions are represented by the following equations,

$$(\text{Oxidizing agent})_1 + ne^- = (\text{reducing agent})_1 \tag{14.41}$$

$$(\text{Oxidizing agent})_2 + ne^- = (\text{reducing agent})_2 \tag{14.42}$$

and the second equation occurs lower in the table than first equation, then a reaction may occur between $(\text{oxidant})_1$ and $(\text{reductant})_2$, whereas no reaction is possible between $(\text{oxidant})_2$ and $(\text{reductant})_1$. If two or more reactions between two substances are possible, usually the reaction involving half-reactions that are farthest apart in Table 14.14 will occur first. The reader is also referred to the discussion in Sec. 3.5.1. No predictions concerning the rate of the reaction are possible.

Potentiometric redox titrations are carried out using some "noble"-metal electrode, the potential of which is measured versus a suitable reference electrode. The noble-metal indicator electrode theoretically serves simply to transfer electrons from dissolved species in solution to the external circuit. Usually platinum or gold is used as an indicator electrode. It should always be remembered that both platinum and gold can become oxidized by strong oxidants. This oxidation sometimes obscures end points or at least causes steady potentials to be obtained rather slowly. Certain difficulties are also observed with platinum electrodes in strongly reducing solutions such as chromium(II) solutions. The reduction of hydrogen ion is catalyzed at platinum surfaces, causing the observed potentials to be less reducing than they should be. Mercury electrodes do not have the latter disadvantage.

The general technique of redox titrations is essentially the same as that for other types of titrations. One difference arises from the fact that the indicator electrode responds to the ratio of the dominant oxidation–reduction couple and not to a single ion. Usually the end point is located as the point of maximum potential change for a specific volume increment. Sometimes potential changes are so great that accurate results may be obtained by approaching the end point dropwise and establishing the drop that causes a large potential change. The end point may also be established by other methods such as titrating to a calculated or, better, a predetermined potential. Successive titrations can be handled more easily than would be the case with visual end points.

Table 14.15 gives information about a few of the most useful or interesting potentiometric redox titrations. Most of the procedures outlined in Table 3.24 are adaptable to potentiometric titrations.

TABLE 14.13 Potentiometric Precipitation Titrations

Analyte	Titrant	Indicator electrode	Precipitate formed formed, and pK_{sp}	Supporting electrolyte, procedure
Ag^+	KI, KBr, KCl	Ag wire, Ag_2S ISE	AgI, 16.08; AgBr, 12.3; AgCl, 9.75	0.01*M* HNO_3. Bi, Cd, Cu, Fe, Pb, and Zn may be masked with EDTA at pH 4–5.5.
	Na diethyldithiocarbamate (NaDDC)	Ag wire, A_2S ISE	Ag diethyldithiocarbamate	Solution 50% ethanol.
Al^{3+}	NaF; $La(NO_3)_3$	F^- ISE	AlF_3	Add excess NaF, back-titrate excess with $La(NO_3)_3$.
Amines	Ag^+ (of excess $NaB(C_6H_5)_4$	Ag metal, Ag_2S ISE	$AgB(C_6H_5)_4$	To sample solution add excess $NaB(C_6H_5)_4$, filter, wash precipitate with 50% aqueous acetone, and titrate excess reagent in filtrate and washings.
AsO_4^{3-}	$AgNO_3$ $La(NO_3)_3$; NaF	Ag wire, Ag_2S ISE F^- ISE	Ag_3AsO_4, 22.0 $LaAsO_4$	Maintain pH at 9–11 (with NaOH) during titration. Back-titrate excess $La(NO_3)_3$ with NaF.
Ba^{2+}	K_2CrO_4	Pt	$BaCrO_4$, 9.93	Solvent: 30% aqueous EtOH. Titrate at 70°C. End-point break is small unless air is excluded.
	Na_2SO_4, then $Pb(ClO_4)_2$	Pb^{2+} ISE	$BaSO_4$, 9.96	Add excess Na_2SO_4, then $Pb(ClO_4)_2$; back-titrate excess Pb^{2+}.
Br^-	≥0.001*M* $AgNO_3$	Ag wire, Br^- or Ag_2S ISE	AgBr, 12.30	0.01*M* HNO_3.
	$Hg_2(NO_3)_2$	Hg pool or Hg-plated Au wire	Hg_2Br_2, 22.24	Reagent solution must be acid to prevent hydrolysis.
$C_2O_4^{2-}$	$Hg_2(NO_3)_2$	Hg (as above)	$Hg_2C_2O_4$, 12.7	May be applied to determination of Ca, Cd, Pb, or Sr by adding excess oxalate and back-titration.
Ca^{2+}	$C_2O_4^{2-}$	Ag–Ag_2Ox, CaOx, Ca^{2+}	CaC_2O_4, 8.4	Third class electrode.
Cd^{2+}	Same as for Ca^{2+}		CdC_2O_4, 7.04	Add excess oxalate and back-titrate.
Cl^-	Ag^+	Ag wire, Ag^+ or Cl^- ISE	AgCl, 9.75	0.01*M* HNO_3; in mixtures with Br^- and I^-, also add 5% (w/v) $Ba(NO_3)_2$ to decrease adsorption.
ClO_4^-	0.05*M* $(C_6H_5)_4As^+$, Cl^-	Perchlorate ISE		End-point break is 150–200 mV. By operating at 2°C useful limit is extended to 0.1 g/L of perchlorate. Iodide, periodate, and permanganate must be absent.
CN^-	$AgNO_3$	Ag wire, CN ISE	AgCN, 15.92	pH ≥ 9. The complexometric $Ag(CN)_2^-$ end point is sharper.
Co(II)	Dithiooxamide	CuS ISE	Co dithiooxamide	Ammoniacal solution; pH 8.1–8.4.
CrO_4^{2-}	$AgNO_3$	Ag wire, Ag_2S ISE	Ag_2CrO_4, 11.95	Solution 1 : 1 with methanol.
Cs^+	$NaB(C_6H_5)_4$	Cs PVC ISE (crown ethers)		Sodium and divalent ions have little effect; potassium, rubidium, and ammonium ions must be absent.
F^-	≥0.01*M* $La(NO_3)_3$	F^- ISE	LaF_3, 16.2	≤0.01*M* titrant; dilute solution 1 : 1 with methanol.

(Continued)

TABLE 14.13 Potentiometric Precipitation Titrations (*Continued*)

Analyte	Titrant	Indicator electrode	Precipitate formed formed, and pK_{sp}	Supporting electrolyte, procedure
I^-	Ag^+	Ag wire, Ag^+ or I^- ISE	AgI, 16.08	0.01*M* HNO_3.
	$HgCl_2$	Hg	HgI_2	Br^- and Cl^- do not interfere if less concentrated than I^-.
Hg_2^{2+}	KCl, KBr, or KI	Hg	Hg_2Cl_2, 17.88; Hg_2Br_2, 22.24; Hg_2I_2, 28.35	
Hg^{2+}	≥0.001*M* KI	AgI ISE	HgI_2	Direct or add excess Ag^+ and back-titrate with I^-.
	Thioacetamide	Ag_2S ISE		0.033*M* EDTA, 0.66*M* NaOH, 0.4% gelatin.
K^+	$NaB(C_6H_5)_4$; Ag^+	K ISE	$KB(C_6H_5)_4$, 7.65	Add excess reagent; wash precipitate; titrate filtrate and washings with Ag^+.
La^{3+}	NaF	F ISE	LaF_3	
Li^+	NaF	F ISE	LiF, 2.42	Solvent 1 : 1 methanol.
Mercaptan	≥0.001*M* $AgNO_3$	Ag_2S ISE		
MoO_4^{2-}	$Pb(NO_3)_2$	Pb ISE	$PbMoO_4$, 13.0	
NO_3^-	$[(C_6H_5)_2\ Tl]_2SO_4$	Nitrate ISE		Medium: 0.1*M* K_2SO_4 plus 0.05*M* H_2SO_4.
Ni^{2+}	Na diethyldithiocarbamate	CuS or Ag_2S ISE		Medium: 50% ethanol, pH 4–6.
PO_4^{3-}	Ag^+	Ag wire, Ag^+ ISE	Ag_3PO_4, 15.84	
	$La(NO_3)_3$; KF	F^- ISE	$LaPO_4$, 22.43	Add excess La^{3+}; back-titrate with F^-.
Pb^{2+}	Na_2MoO_4	Pb ISE	$PbMoO_4$, 13.00	Medium: 50% methanol.
Rb^+	$NaB(C_6H_5)_4$	K ISE	$RbB(C_6H_5)_4$	Interference from Cs, K, and ammonium ions.
ReO_4^-	$Pb(NO_3)_2$	Pb ISE	$PbReO_4$	
Rare earths	NaF	F ISE	(Rare earth)F_3	
S^{2-}	$AgNO_3$ or $Pb(NO_3)_2$	Ag_2S ISE or Pb ISE	Ag_2S, 49.2; PbS, 27.9	
SCN^-	$AgNO_3$	Ag wire, Ag^+ ISE	AgSCN, 12.00	0.01*M* HNO_3 containing 5% $Ba(NO_3)_2$.
SO_4^{2-}	$BaCl_2$	Ba^{2+} PVC ISE	$BaSO_4$, 9.96	$BaCl_2$ excess; back-titrate with EDTA
	$Pb(ClO_4)_2$	Pb ISE	$PbSO_4$, 7.79	Medium: 50% 2-propanol, acetone, or methanol.
Sr^{2+}				See method for Ca^{2+}.
Th(IV)	NaF; $La(NO_3)_3$	F^- ISE	ThF_4, 7.23	Add excess NaF; back-titrate with La^{3+}.
Tungstate	$Pb(NO_3)_2$	Pb ISE	$PbWO_4$, 6.36	Interference from molybdate and perrhenate.
Zn^{2+}	$K_4[Fe(CN)_6]$	Pt	$K_2Zn_3[Fe(CN)_6]_2$	pH 2–3; add 3–4 drops 1% $K_3[Fe(CN)_6]$. Many interferences eliminated by EDTA.

TABLE 14.14 Potentials of Selected Half-Reactions at 25°C

A summary of oxidation–reduction half-reactions arranged in order of decreasing oxidation strength and useful for selecting reagent systems.

Half-reaction	E^0, volts
$F_2(g) + 2H^+ + 2e^- = 2HF$	3.053
$O_3 + H_2O + 2e^- = O_2 + 2OH^-$	1.246
$O_3 + 2H^+ + 2e^- = O_2 + H_2O$	2.075
$Ag^{2+} + e^- = Ag^+$	1.980
$S_2O_8^{2-} + 2e^- = 2SO_4^{2-}$	1.96
$HN_3 + 3H^+ + 2e^- = NH_4^+ + N_2$	1.96
$H_2O_2 + 2H^+ + 2e^- = 2H_2O$	1.763
$Ce^{4+} + e^- = Ce^{3+}$	1.72
$MnO_4^- + 4H^+ + 3e^- = MnO_2(c) + 2H_2O$	1.70
$2HClO + 2H^+ + 2e^- = Cl_2 + 2H_2O$	1.630
$2HBrO + 2H^+ + 2e^- = Br_2 + 2H_2O$	1.604
$H_5IO_6 + H^+ + 2e^- = IO_3^- + 3H_2O$	1.603
$NiO_2 + 4H^+ + 2e^- = Ni^{2+} + 2H_2O$	1.593
Bi_2O_4(bismuthate) $+ 4H^+ + 2e^- = 2BiO^+ + 2H_2O$	1.59
$MnO_4^- + 8H^+ + 5e^- = Mn^{2+} + 4H_2O$	1.51
$2BrO_3^- + 12H^+ + 10e^- = Br_2 + 6H_2O$	1.478
$PbO_2 + 4H^+ + 2e^- = Pb^{2+} + 2H_2O$	1.468
$Cr_2O_7^{2-} + 14H^+ + 6e^- = 2Cr^{3+} + 7H_2O$	1.36
$Cl_2 + 2e^- = 2Cl^-$	1.3583
$2HNO_2 + 4H^+ + 4e^- = N_2O + 3H_2O$	1.297
$N_2H_5^+ + 3H^+ + 2e^- = 2NH_4^+$	1.275
$MnO_2 + 4H^+ + 2e^- = Mn^{2+} + 2H_2O$	1.23
$O_2 + 4H^+ + 4e^- = 2H_2O$	1.229
$ClO_4^- + 2H^+ + 2e^- = ClO_3^- + H_2O$	1.201
$2IO_3^- + 12H^+ + 10e^- = I_2 + 3H_2O$	1.195
$N_2O_4 + 2H^+ + 2e^- = 2HNO_3$	1.07
$2ICl_2^- + 2e^- = 4Cl^- + I_2$	1.07
$Br_2(1q) + 2e^- = 2Br^-$	1.065
$N_2O_4 + 4H^+ + 4e^- = 2NO + 2H_2O$	1.039
$HNO_2 + H^+ + e^- = NO + H_2O$	0.996
$NO_3^- + 4H^+ + 3e^- = NO + 2H_2O$	0.957
$NO_3^- + 3H^+ + 2e^- = HNO_2 + H_2O$	0.94
$2Hg^{2+} + 2e^- = Hg_2^{2+}$	0.911
$Cu^{2+} + I^- + e^- = CuI$	0.861
$OsO_4(c) + 8H^+ + 8e^- = Os + 4H_2O$	0.84
$Ag^+ + e^- = Ag$	0.7991
$Hg_2^{2+} + 2e^- = 2Hg$	0.7960
$Fe^{3+} + e^- = Fe^{2+}$	0.771
$H_2SeO_3 + 4H^+ + 4e^- = Se + 3H_2O$	0.739
$HN_3 + 11H^+ + 8e^- = 2NH_4^+$	0.695
$O_2 + 2H^+ + 2e^- = H_2O_2$	0.695
$Ag_2SO_4 + 2e^- = 2Ag + SO_4^{2-}$	0.654
$Cu^{2+} + Br^- + e^- = CuBr(c)$	0.654
$Au(SCN)_4^- + 3e^- = Au + 4SCN^-$	0.636
$2HgCl_2 + 2e^- = Hg_2Cl_2(c) + 2Cl^-$	0.63
$Sb_2O_5 + 6H^+ + 4e^- = 2SbO^+ + 3H_2O$	0.605
$H_3AsO_4 + 2H^+ + 2e^- = HAsO_2 + 2H_2O$	0.560
$TeOOH^+ + 3H^+ + 4e^- = Te + 2H_2O$	0.559
$Cu^{2+} + Cl^- + e^- = CuCl(c)$	0.559
$I_2^- + 2e^- = 3I^-$	0.536

(Continued)

TABLE 14.14 Potentials of Selected Half-Reactions at 25°C (*Continued*)

Half-reaction	E^0, volts
$I_2 + 2e^- = 2I^-$	0.536
$Cu^+ + e^- = Cu$	0.53
$4H_2SO_3 + 4H^+ + 6e^- = S_4O_6^{2-} + 6H_2O$	0.507
$Ag_2CrO_4 + 2e^- = 2Ag + CrO_4^{2-}$	0.449
$2H_2SO_3 + 2H^+ + 4e^- = S_2O_3^{2-} + 3H_2O$	0.400
$UO_2^+ + 4H^+ + e^- = U^{4+} + 2H_2O$	0.38
$Fe(CN)_6^{3-} + e^- = Fe(CN)_6^{4-}$	0.361
$Cu^{2+} + 2e^- = Cu$	0.340
$VO^{2+} + 2H^+ + e^- = V^{3+} + H_2O$	0.337
$BiO^+ + 2H^+ + 3e^- = Bi + H_2O$	0.32
$UO_2^{2+} + 4H^+ + 2e^- = U^{4+} + 2H_2O$	0.27
$Hg_2Cl_2(c) + 2e^- = 2Hg + 2Cl^-$	0.2676
$AgCl + e^- = Ag + Cl^-$	0.2223
$SbO^+ + 2H^+ + 3e^- = Sb + H_2O$	0.212
$CuCl_3^{2-} + e^- = Cu + 3Cl^-$	0.178
$SO_4^{2-} + 4H^+ + 2e^- = H_2SO_3 + H_2O$	0.158
$Sn^{4+} + 2e^- = Sn^{2+}$	0.15
$S + 2H^+ + 2e^- = H_2S$	0.144
$Hg_2Br_2(c) + 2e^- = 2Hg + 2Br^-$	0.1392
$CuCl + e^- = Cu + Cl^-$	0.121
$TiO^{2+} + 2H^+ + e^- = Ti^{3+} + H_2O$	0.100
$S_4O_6^{2-} + 2e^- = 2S_2O_3^{2-}$	0.08
$AgBr + e^- = Ag + Br^-$	0.0711
$HCOOH + 2H^+ + 2e^- = HCHO + H_2O$	0.056
$CuBr + e^- = Cu + Br^-$	0.033
$2H^+ + 2e^- = H_2$	0.0000
$Hg_2I_2 + 2e^- = 2Hg + 2I^-$	−0.0405
$Pb^{2+} + 2e^- = Pb$	−0.125
$Sn^{2+} + 2e^- = Sn$	−0.136
$AgI + e^- = Ag + I^-$	−0.1522
$N_2 + 5H^+ + 4e^- = N_2H_5^+$	−0.225
$V^{3+} + e^- = V^{2+}$	−0.255
$Ni^{2+} + 2e^- = Ni$	−0.257
$Co^{2+} + 2e^- = Co$	−0.277
$Ag(CN)_2^- + e^- = Ag + 2CN^-$	−0.31
$PbSO_4 + 2e^- = Pb + SO_4^{2-}$	−0.3505
$Cd^{2+} + 2e^- = Cd$	−0.4025
$Cr^{3+} + e^- = Cr^{2+}$	−0.424
$Fe^{2+} + 2e^- = Fe$	−0.44
$H_3PO_3 + 2H^+ + 2e^- = HPH_2O_2 + H_2O$	−0.499
$2CO_2 + 2H^+ + 2e^- = H_2C_2O_4$	−0.49
$U^{4+} + e^- = U^{3+}$	−0.52
$Zn^{2+} + 2e^- = Zn$	−0.7626
$Mn^{2+} + 2e^- = Mn$	−1.18
$Al^{3+} + 3e^- = Al$	−1.67
$Mg^{2+} + 2e^- = Mg$	−2.356
$Na^+ + e^- = Na$	−2.714
$K^+ + e^- = K$	−2.925
$Li^+ + e^- = Li$	−3.045
$3N_2 + 2H^+ + 2e^- = 2HN_3$	−3.10

Source: J. A. Dean, ed., *Lange's Handbook of Chemistry*, 14th ed., McGraw-Hill, New York, 1992.

TABLE 14.15 Potentiometric Redox Titrations

Substance determined	Reagent	End-point potential, V vs SCE	Titration conditions and remarks
As(III)	Ce(IV)	ΔE max.	$4M$ HCl; ICl catalyst.
	$KMnO_4$	ΔE max.	Na_2CO_3 medium.
	$KBrO_3$	ΔE max.	Medium: 10% HCl.
Au(III)	Ascorbic acid	ΔE max.	pH 1.6–3; 50°C. Cu(II), Fe(III), Hg(II), and $0.1M$ Cl^- do not interfere.
Br^-	$KMnO_4$	ΔE max.	10 mL sample + 5 mL 10% KCN + 10 mL conc. H_2SO_4 + 100 mL water. Equiv. wt. is Br/2.
Oxalate	$KMnO_4$	ΔE max.	$0.2M$–$1M$ H_2SO_4; 70°C.
	Ce(IV)	ΔE max.	Add 20 mL conc. HCl + 10 mL $0.005M$ ICl per 70 mL sample.
Ce(III)	$K_3[Fe(CN)_6]$ or $KMnO_4$	0.04	Add sample to enough $4M$ K_2CO_3 to ensure $1.5M$ K_2CO_3 at end point. Titrate in absence of air.
Ce(IV)	Fe(II)		Medium: $1M$ H_2SO_4.
	$H_2C_2O_4$		Add 20 mL conc. HCl + 10 mL $0.005M$ ICl per 70 mL sample.
Cr(VI)	As(III)	ΔE max.	Medium: 20% H_2SO_4.
	Ti(III)	ΔE max.	Medium: 10% H_2SO_4; exclude air.
Fe(II)	$KMnO_4$	1.09	Medium: $0.2M$ H_2SO_4; H_3PO_4 may be added to improve break.
	Ce(IV)	0.95	Medium: $1M$ H_2SO_4. End point depends on acid used.
$Fe(CN)_6^{3-}$	Ce(IV)		Medium: $>1M$ HCl or H_2SO_4.
I^-	$KMnO_4$	1.0	Medium: $0.1M$–$0.25M$ H_2SO_4. Br^- and Cl^- may be present.
Mn(II)	$KMnO_4$	0.53 (pH 6.0)	Add acid solution of sample to 250 mL fresh saturated $Na_4P_2O_7$ solution, adjust pH to 6–7, and titrate. Equiv. wt. is Mn/4. Titration is specific except for vanadium; vanadium does not interfere if pH is 3–3.5 although end-point break is smaller.
	Fe(II)	1.3; 0.7	To 50 mL sample containing $2M$–$5M$ HNO_3, add small portions of AgO until solution is black; dilute to 150 mL with $1M$ H_2SO_4 and titrate. First break = Ag(II); second break = Mn (as MnO_2). Equiv. wt. is Mn/2.
MnO_4^-	Fe(II)	1.09	Medium: $0.2M$ H_2SO_4; add excess standard Fe(II) and back-titrate with standard $KMnO_4$.
NO_2^-	$KMnO_4$		Add nitrite sample slowly to 10% excess $KMnO_4$ + $0.75M$ H_2SO_4; then add excess KI and titrate with $KMnO_4$.
H_2O_2	Ce(IV)		Medium: $0.5M$–$3M$ HOAc or HCl.
Sb(III)	$KBrO_3$		Medium: $3M$ (5%) HCl.
Sb(V)	$KBrO_3$	0.3–0.5; ΔE max.	To sample in 5% HCl add excess Ti(III), then immediately three drops 3% $CuSO_4$ solution, stir, and titrate with $KBrO_3$. First break = excess Ti(III); second break = Sb. No interference from As(V).
Sulfite	$KMnO_4$		Add excess of $KMnO_4$ to alkaline sulfite; add H_2SO_4 to give $0.5M$ and excess KI, then titrate excess with $KMnO_4$.
Sn(II)	Iron(III)		Medium: 10% HCl, 75°C.
Vanadium	$KMnO_4$	0.05	V(II) $\rightarrow$ V(III).
		0.44	V(III) $\rightarrow$ V(IV).
		1.1	V(IV) $\rightarrow$ V(V) done at 70°C.

14.4.2.1 Potentiometric Titrations at Constant Current. A number of redox couples normally used in titrimetry, such as thiosulfate–tetrathionate and dichromate (VI)-chromium(III), are irreversible and thus establish steady potentials at a platinum electrode rather slowly if the ordinary zero-current method of measurement is used. Steady potentials are obtained more rapidly and larger potential breaks are obtained by the use of polarized indicator electrodes. Interpretations of the titration curves obtained are best discussed on the basis of current–potential curves (see Sec. 14.5).

Potentiometric titrations at constant current may be divided into two classes, depending on whether one or two indicator electrodes are used. In the former case the potential of a single polarized platinum indicator electrode is measured versus a reference electrode. The indicator electrode may be polarized either anodically or cathodically, depending on the situation. Alternatively, two platinum electrodes may be used, in which case one will act as an anode and the other as a cathode. The potential between these two platinum electrodes is measured.

If thiosulfate were titrated with iodine, some idea of the various possible titration curves could be established by knowing that the iodide–iodine couple is reversible while the thiosulfate–tetrathionate couple is irreversible. The three possible curves are obtained with (1) two similar polarized platinum electrodes, (2) one platinum electrode polarized anodically, and (3) one platinum electrode polarized cathodically.

In case (1), before the equivalence point, the cathode would reduce oxygen (air) or water, while the anode would oxidize iodide ion. Thus a fairly large difference in potential between the two platinum electrodes would exist. After the equivalence point, excess iodine would be present and would be reduced at the cathode. The difference in the potential between the two electrodes would therefore decrease rapidly around the end point.

In case (2), the single electrode would oxidize iodide ion both before and after the equivalence point. Only a small potential change would be noticed when excess iodine was added. The results in this case would be much worse than in standard zero-current potentiometry.

If the single electrode is polarized cathodically [case (3)], it will first reduce oxygen or water, but after the equivalence point will reduce iodine and thereby undergo a substantial potential shift.

Despite the advantages of using one polarized electrode for cases in which electrode potentials are established slowly, caution is in order because of the possibility of error due to the fact that the point of greatest potential change does not correspond to the equivalence point. This error is proportional to the size of the current, and thus small currents ($<10\ \mu A$) are normally used. In fact, it is recommended that as small a current be used as suffices to produce a reasonably sharp potential break. In addition, the composition of the solution may be changed by the passage of electric current, but this is negligible under normal conditions. Methods involving two polarized electrodes suffer from the same disadvantages unless both the couples involved are reversible. In this case the generation of an oxidant at the anode is just balanced by the generation of a reductant at the cathode.

The apparatus used for potentiometric titrations at constant current consists of some potential-measuring device, such as a pH meter, and a circuit to pass a small constant current between two small platinum-wire electrodes. For two-electrode potentiometry the potential is measured between the two platinum electrodes, but if the potential of only one electrode is to be measured, the pH meter is connected between the appropriate platinum electrode and some reference electrode such as SCE, constructed in such a manner that current can be passed through it as in voltammetry. The constant-current circuit should consist of several high-voltage batteries (900 to 1200 V) in series with several hundred megohms of resistance such that a very steady current on the order of $1\ \mu A$ is obtained.

14.4.3 Potentiometric Complexometric Titrations

The material in Sec. 3.6, particularly Sec. 3.6.4, is pertinent to this section. The magnitude of the end-point break will be larger if more concentrated solutions are titrated and if strong complexing agents with high β values are used. Best results can be obtained with reagents such as EDTA and tetraethylenepentamine, which form only one complex with the metal ion.

An important type of complexometric titration involves the titration of a solution containing two metal ions, both of which form a complex with a particular reagent. If sufficient difference exists between the formation constants of the two metals, successive end points will be obtained. This type of titration is very useful as it allows an ion-selective electrode sensitive to the metal forming the more stable complex to be used to titrate any other metal that forms a less stable complex with the reagent, even though the ion-selective metal may not be directly responsive to the second metal. In practice, the sample containing the unresponsive metal ion is spiked with an indicator concentration of the metal ligand complex ($10^{-5}M$ to $10^{-4}M$), which need not be accurately measured.

The titrations mentioned employ a mercury indicator electrode in the presence of approximately $10^{-6}M$ Hg(II)–EDTA complex. Reilley and coworkers[5,6] showed how to use an electrode of known reversibility to measure activities of ions for which no electrode of the first kind exists. They used a small mercury electrode (or gold amalgam wire) in contact with a solution containing metal ions to be titrated with a chelon Y, such as EDTA. A small quantity of mercury(II) chelonate, HgY^{2-}, saturated the solution and established the half-cell:

$$\mathrm{Hg} \mid \mathrm{HgY}^{2-}, \mathrm{MY}^{(n-4)+}, \mathrm{M}^{n+}$$

where the electrode potential is given by

$$E = E^0 + \frac{0.059\ 15}{2} \log \frac{[\mathrm{M}^{n+}]\,[\mathrm{HgY}^{2-}]}{[\mathrm{MY}^{(n-4)+}]} \qquad (14.43)$$

Because a fixed amount of HgY^{2-} is present, the potential is dependent upon the ratio $[M^{4+}]/[MY^{(n-4)+}]$. The species HgY^{2-} must be considerably more stable than $MY^{(n-4)+}$. Halides, sulfide, cyanide, and other substances that react extensively with mercury(I) or (II) must be absent.

Ion-selective electrodes are also very useful as indicator electrodes in complexometric titrations. Almost any metal for which there is an ion-selective electrode can be determined by titration with a chelating agent, but not at such low concentration as can be achieved by direct potentiometry. Moreover, complexometric titrations can be used to determine metals for which no electrodes exist. For most purposes, copper(II) ions and copper-selective electrodes are used. Metals that form stronger complexes than copper(II) are determined by back-titration. A known concentration of chelating agent is added in excess to the sample and the unreacted portion is determined by titration with copper(II). Metals that form weaker complexes than the copper(II) ion can be determined in the presence of a known concentration of that ion by titration with a chelating agent. The copper(II) ion is added to the sample either as (1) a simple salt or (2) a solution containing the Cu(II) chelate. In the first case, two steps will be observed in the titration curve; in the second case, only one step occurs in the titration curve.

The information tabulated in the following is useful to assess whether one metal ion will interfere with the determination of another. In general, the metal ions that can be titrated at pH 4 to 5.5, or at pH 8 to 10 will not interfere with titrations at pH 2, and alkaline-earth ions will not interfere with titrations at pH 4 to 5.5. For specific procedural information consult Table 14.16.

[5] C. N. Reilley and R. W. Schmid, *Anal. Chem.* **30**:947 (1958).

[6] C. N. Reilley, R. W. Schmid, and D. W. Lamson, *Anal. Chem.* **30**:953 (1958).

pH	Buffer systems	Procedure	Metals titrated
2	HNO_3 or $ClCH_2COOH$	Direct titration with EDTA	Bi, Hg(II), Th
4–5.5	HOAc or hexamethylenetetramine	Direct titration with EDTA	Bi, Cd, Cu, Hg(II), La, Mn(II), Pb, R.E.(III)*, Sc, V(IV), Y, Zn
		Addition of excess EDTA, back-titration with standard Cd, Cu, Hg(II), Pb, or Zn solution	Al, Bi, Cd, Cr(III), Cu, Fe(III), Ga, Hf, Hg(II), In, La, Mn(II), Ni, Pb, R.E. (III), Sc, Th, V(IV), Y, Zn, Zr

* R.E. denotes rare-earth elements.

pH	Buffer systems	Procedure	Metals titrated
8–10	NH_3, ethanolamine, or triethanolamine	Direct titration with EDTA	Ca, Cd, Co, Cu, In, Mg, Ni, Pb, Zn
		Addition of excess EDTA, back-titration with standard Ca, Cd, Cu, Mg, or Zn solution	Bi, Ca, Cd, Co, Cr(III), Cu, Hg(II), La, Mg, Ni, Pb, R.E.(III), Sc, TI(III), Y, Zn
10	NH_3 or ethanolamine	Direct titration with EDTA	Ba, Sr, metals titrated at pH 8–10
		Addition of excess EDTA, back-titration with standard Ca, Cd, Cu, Mg, or Zn solution	Ba, Sr, metals titrated at pH 8–10

14.4.4 Potentiometric Acid-Base Titrations

Extensive discussion of acid–base titrations, both in aqueous and in nonaqueous media, will be found in Secs. 3.2 and 3.3. The general technique is the same as in pH measurements (Sec. 14.2.3). Usually a glass pH-indicator electrode is used in conjunction with an SCE. An antimony-burette electrode finds use as the reference electrode when nonaqueous systems involve a strongly basic solvent. The end point is located as the point of maximum potential change for a specific volume increment of titrant.

14.5 VOLTAMMETRIC METHODS[7]

Voltammetric techniques allow the determination of those components in a solution that can be electrochemically oxidized or reduced. In these methods a potential is applied to the sample via a conductive electrode, called the working electrode, which is immersed in the sample solution. The potential, which serves as the driving force, is scanned over a region of interest. If at a particular potential a component of the solution is oxidized or reduced, then a current will flow at the working electrode. The potential at which this occurs identifies the component, and the amount of current produced is proportional to the concentration of that component in the solution. Voltammetric methods rank among the most sensitive analytical techniques available. They are routinely used for the determination of electroactive inorganic elements and organic substances (Tables 14.17 and 14.18) in nanogram and even picogram amounts; often the analysis takes place in seconds. Many electroinactive functional groups can be converted into electroactive species after performing suitable chemistry (Table 14.19).

14.5.1. Instrumentation

14.5.1.1 The Mercury-Drop Microelectrode. The mercury working electrode has found widespread use due to the high overpotential for hydrogen-ion reduction (1.2 V) and the reproducibility of the metal surface. The original mercury electrode was the dropping-mercury electrode (dme) in which the growth of the drop and the drop lifetime are controlled by gravity. A stream of mercury is forced through a glass capillary (0.05 to 0.08 mm i.d.) under the pressure of an elevated reservoir of mercury connected to the capillary by flexible tubing (Fig. 14.8). Mercury issues from the capillary at the rate of one drop every 2 to 5 s. The constant renewal of the drop surface (as it expands) eliminates poisoning effects. Mercury forms amalgams with many metals and thereby lowers their reduction potentials. The diffusion current assumes a steady value immediately and is reproducible.

[7] J. B. Flato, "The Renaissance in Polarographic and Voltammetric Analysis," *Anal. Chem.* **44**:75A (1972).

TABLE 14.16 Potentiometric Complexometric Titrations

*A mercury indicator electrode implies the presence of 10^{-6}M Hg(II) EDTA (1 drop of a 10^{-3}M solution for each 50 mL volume at the end point. EGTA = 2,2′-ethylenedioxybis[ethyliminodi(acetic acid)]; trien = triethylenetetramine; tetren = tetraethylenepentamine; CDTA = trans-cyclohexane-1,2-diamine-*N,N,N′,N′*-tetraacetic acid; TEPA = tetraethylenepentamine.*

Analyte	Titrant	Electrode	Procedure and remarks
Al(III)	EDTA	Hg	Acidify 10 mL 0.1*M* Al(III) solution to pH 1 to 2, boil 1 min, add 15 mL 0.1*M* EDTA while hot, cool, add acetate buffer and Hg(II)–EDTA, and back-titrate with Zn(II) solution.
	KF	F ISE	Medium: 100 mL standard aluminum solution or blank at pH 4 (acetate buffer). Add six (or more) increments of KF solution and record steady-state emf reading for at least five concentrations of standard solution. Select a target emf; titrate sample to target emf value. Optimum range: 2–100 ppm aluminum.
Ba(II)	EDTA	Hg	To 15–25 mL ≤ 0.05*M* Ba(II) add 10–25 mL 0.5*M* NH_3 buffer (pH 9.5–10) and the Hg(II)–EDTA. Deaerate with N_2 (exclude O_2) and titrate with 0.005*M*–0.05*M* EDTA (slowly near end point).
	CDTA	Cu ISE	Add a known amount of $Cu(NO_3)_2$, then titrate with CDTA. First end-point break = Cu(II); second end-point break = Ba(II).
Bi(III)	EDTA	Hg	Add NH_3 or HNO_3 to pH 1.5 and titrate with EDTA. Use NH_4NO_3 salt bridge to prevent BiOCl precipitation.
Ca(II)	EGTA	Cu ISE	Add known amounts of $Cu(NO_3)_2$ or CuEGTA and titrate with EGTA.
	EDTA	Hg	To 15–25 mL ≤ 0.05*M* Ca(II) add 10–25 mL 0.5*M* NH_3 buffer (or 10 mL 0.5*M* triethanolamine buffer, pH 8.5) and titrate with 0.005–0.05*M* standard EDTA.
	EGTA	Hg	To sample add enough triethanolamine buffer to consume EGTA liberated during titration, add most of the EGTA, wait 2 min, and titrate slowly to inflection point.
Cd(II)	EDTA	Cd ISE	Add 2 mL 2*M* acetate buffer per 100 mL sample and titrate with EDTA.
	EDTA	Hg	To 15–25 mL ≤ 0.05*M* Cd(II) add 10–25 mL 0.5*M* acetate buffer (pH 4.6) and titrate with standard EDTA.
	Trien or tetren	Hg	To sample with 0.002 mmol Cd(II) in 60–100 mL, add 2 mL 1*M* NH_3 buffer (pH 10) and 1 drop Hg(II)–chelonate, and titrate with 0.01*M* reagent.
	Trien or tetren	Hg	In presence of Cu: After titration of Cu(II) at pH 4.8 add 3 mL 1*M* NH_3 buffer (pH 10) and titrate with 0.01*M* reagent.
	Trien or tetren	Hg	In presence of Al, Ca, La, and Mg: Dilute sample containing 0.2 mmol Cd(II) to 60–100 mL, add 1 mL 1*M* NH_3 buffer (pH 10), 1 g Na_2tartrate · $2H_2O$, and titrate with 0.01*M* reagent.
Ce(III)	EDTA	Hg	To 10 mL 0.005*M* Ce(III) solution, add 15 mL 0.005*M* EDTA, then 15*M* NH_3 to pH 9.5–10, and back-titrate with standard Zn(II) solution.
Co(II)	EDTA	Cu ISE	Add known amounts of $Cu(NO_3)_2$ (or CuEDTA) and titrate with EDTA; second end point = Co(II) [or the only end point is for Co(II)].
	EDTA	Hg	To 15–25 mL ≤ 0.05*M* Co(II) add 10–25 mL 0.5*M* NH_3 buffer and titrate with 0.005*M*–0.05*M* EDTA.
Cr(III)	EDTA	Hg	Heat 5 mL 0.02*M* Cr(III) solution, 10 mL 0.02*M* EDTA, and 10 mL 0.2*M* acetate buffer (pH 3.5) to boiling for 10 min, cool, adjust pH to 4.8, and back-titrate with Zn(II).
Cu(II)	EDTA	Cu ISE	Titrate directly with EDTA.
Cu(II)	EDTA	Hg	In the presence of Cd or Ni: Dilute sample containing 0.2 mmol Cu(II) to 60–100 mL, add 1 mL 1*M* acetate buffer (pH 4.8), and titrate with EDTA.

(Continued)

TABLE 14.16 Potentiometric Complexometric Titrations (*Continued*)

Analyte	Titrant	Electrode	Procedure and remarks
	EDTA	Hg	In the presence of Al, Ca, La, and Mg: Proceed as above but use 2 mL acetate buffer.
Ga(III)	EDTA	Hg	To 5 mL 0.02*M* Ga(III) solution, add 10 mL 0.02*M* EDTA and 10 mL 0.2*M* acetate buffer (pH 4.6), heat to boiling 1 min, chill, and back-titrate with Zn(II) solution.
Hf(IV)	EDTA	Hg	To 5 mL 0.05*M* Hf(IV) solution add excess 0.05*M* EDTA, adjust pH to 4 with acetate buffer, and back-titrate with standard Cu(II) solution.
Hg(II)	EDTA	Hg	Proceed as for cadmium with EDTA.
	trien or tetren	Hg	Dilute acidic sample containing 0.2 mmol Hg(II) to 60–100 mL, add 1 mL 1*M* triethanolamine buffer (pH 8), and titrate with reagent.
In(III)	EDTA	Hg	To 10 mL 0.01*M* In(III) solution add 25 mL 0.2*M* acetate buffer (pH 4.6), 10 mL 0.02*M* EDTA, and back-titrate with standard Zn(II) solution.
La(III)	EDTA	Hg	Proceed as for cerium(III).
Mg(II)	EDTA	Hg	Proceed as for calcium using triethanolamine buffer and titrate with EDTA.
	EDTA	Cu ISE	Add known amount of $Cu(NO_3)_2$ and titrate with EDTA to second end point.
Mn(II)	EDTA	Cu ISE	Add known amount of $Cu(NO_3)_2$ and titrate with EDTA to second end point.
	EDTA	Hg	Proceed as for cadmium using EDTA as titrant.
Nd(III)	EDTA	Hg	Proceed as for cerium(III).
Ni(II)	TEPA	Cu ISE	Add $Cu(NO_3)_2$ and titrate with EDTA; use the second end point.
	EDTA	Hg	Proceed as for calcium using EDTA.
	Trien or tetren	Hg	Proceed as for cadmium (first method using trien or tetren).
Pb(II)	EDTA	Hg	Proceed as for cadmium with EDTA.
Pr(III)	EDTA	Hg	Proceed as for Ce(III).
Sc(III)	EDTA	Hg	Proceed as for Ce(III).
Sm(III)	EDTA	Hg	Proceed as for Ce(III).
Sr(II)	EDTA	Hg	Proceed as for calcium using the ammoniacal medium.
	EDTA	Cu ISE	Add a small amount of $Cu(NO_3)_2$ and titrate with EDTA; second end point is SR(II).
Th(IV)	EDTA	Hg	Proceed as for hafnium.
Tl(III)	EDTA	Hg	Proceed as for indium but adjust pH to 4.0; perform titration rapidly.
V(IV)	EDTA	Pt	Proceed as for cadmium using EDTA but adjust pH to 3.9 and use Pt indicator electrode.
Y(III)	EDTA	Hg	Proceed as for cerium(III).
Zn(II)	TEPA	Cu ISE	Add a small amount of $Cu(NO_3)_2$ and titrate with EDTA; second end point is Zn(II).
	EDTA	Hg	Proceed as for cadmium using EDTA as titrant.
Zr(IV)	EDTA	Hg	Proceed as for hafnium.

The dme is useful over the range +0.3 to −2.7 V versus SCE in aqueous solutions with use of a tetrabutylammonium salt as the supporting electrolyte. At potentials more positive than 0.3 V, mercury is oxidized, which results in an anodic wave.

TABLE 14.17 Electroreducible Organic Functional Groups

Acetylenes	Hydroquinones
Acyl sulfides	Hydroxylamines
Aldehydes	Imines
Aromatic carboxylic acids	Ketones
Azomethines	Nitrates
Azoxy compounds	Nitriles
Conjugated alkenes	Nitro compounds
Conjugated aromatics	Nitroso compounds
Conjugated carboxylic acids	Organometallics
Conjugated halides	Oximes
Conjugated ketones	Peroxides
Diazo compounds	Quinones
Dienes: conjugated double bonds	Sulfones
Disulfides	Sulfonium salts
Heterocycles	Thiocyanates

TABLE 14.18 Electrooxidizable Organic Functional Groups

Alcohols	Ethers
Aliphatic halides	Heterocyclic amines
Amines	Heterocyclics
Aromatic amines	Nitroaromatics
Aromatic halides	Olefins
Aromatics	Organometallics
Carboxylic acids	Phenols

TABLE 14.19 Conversion of Organic Functional Groups into Electroactive Species

Functional group	Reagent	Active voltammetric group
Alcohols	Chromic acid	Aldehyde
Carbonyl	Girard T and D	Azomethine
	Semicarbazide	Carbazide
	Hydroxylamine	Hydroxylamine
Carboxyl	Transform to thiouronium salts	—SH (anodic wave)
1,2-Diols	Periodic acid	Aldehyde
Phenyl	Nitration	$—NO_2$
Primary amine	Piperonal	Azomethine
	CS_2	Dithiocarbonate (anodic wave)
	$Cu_3(PO_4)_3$ suspension	Copper(II) amine
Secondary amine	HNO_2	Nitrosoamine

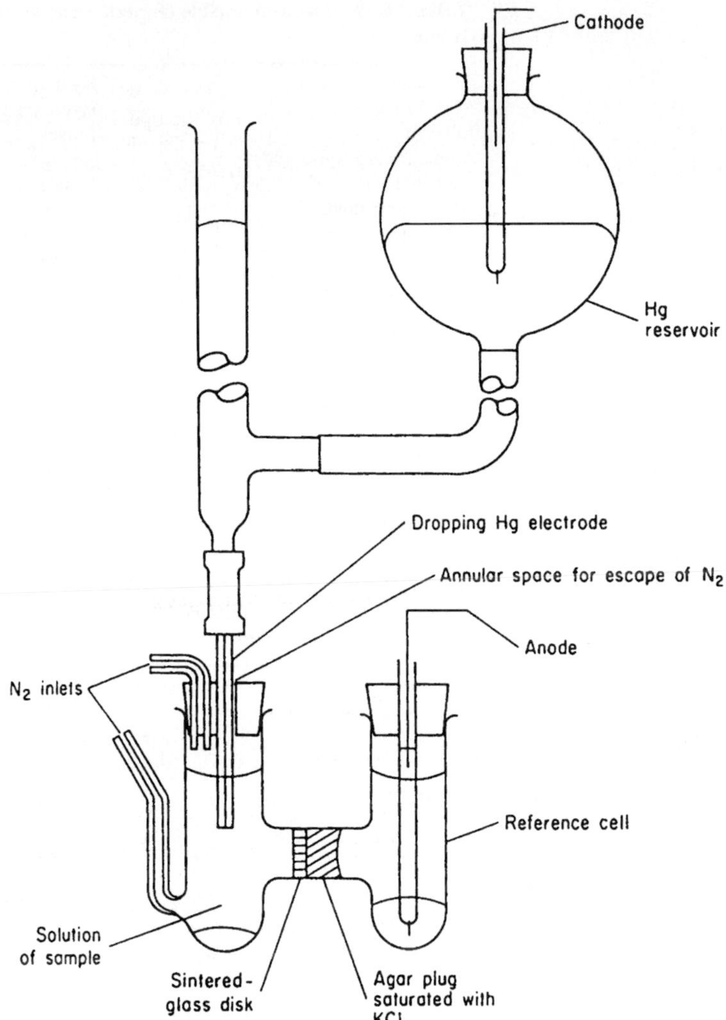

FIGURE 14.8 A dropping-mercury electrode and voltammetric cell. (*From Shugar and Dean, 1990.*)

In contrast to the dme, the static-mercury-drop electrode (smde) uses a valve to obtain a static mercury drop on the capillary tip, which is removed by a drop knocker. The hanging-mercury-drop electrode (hmde) is a stationary electrode used extensively for quantitative analysis. In the controlled-growth mercury-drop electrode (cgme), the drop growth is controlled by a fast response valve. The opening of the valve is controlled by a computer-generated pulse sequence, which leads to a stepped increase in the drop size. The drop size can be varied by changing the number of pulses and/or the pulse width, so a wide range of different drop sizes is available.

To achieve drop synchronization, essential for reproducible measurements in many voltammetric methods, a drop of mercury is detached from the capillary at regular time intervals. One type moves the capillary away from the drop at a fixed time interval; others deliver a sharp knock to the capillary to dislodge the drop. At the same moment a trigger signal is sent to the time base. These intervals are less than the normal drop life. Each drop is permitted to grow until its area changes the least, usually after the first 1.5 to 2.0 s of drop life.

14.5.1.2 Solid Electrodes. Solid electrodes may be fabricated from platinum, gold, or glassy carbon. At periodic intervals, as selected by the operator, the stationary electrode tip is rapidly raised and lowered to renew the diffusion layer on the electrode tip. Glassy carbon electrodes find use to monitor electro-oxidation reactions, particularly on-line as in liquid-column chromatography.

14.5.1.3 Rotating-Disk Electrodes. A simple form of a rotating-disk electrode consists of a platinum or gold wire or glassy carbon that is sealed in an insulating material with the sealed end ground smooth and perpendicular to the rod axis. In the laboratory rotating-disk electrodes are used primarily for ultratrace analysis and electrode kinetics research. Rotating-disk electrodes provide superior control for anodic stripping voltammetry and for amperometric titrations.

14.5.1.4 Thin-Film Mercury Electrode. This type of electrode is prepared by depositing a thin film of mercury on a glassy carbon electrode. The thin film limits the use of this type of electrode to analyte concentrations less than $10^{-7}M$. These electrodes are used only when maximum sensitivity is required as for stripping voltammetry and as amperometric detectors for liquid chromatography.

14.5.1.5 Dearation of Solutions. Oxygen must be removed from analyte solutions by bubbling pure nitrogen through the solution via a glass dispersion frit for 1 to 2 min. The nitrogen gas should be presaturated with solvent vapor to prevent any change in the sample concentration. During experiments the bubbling is stopped; however, a slight positive pressure of nitrogen should be maintained over the solution.

14.5.1.6 Three-Electrode Potentiostat. Potentiostatic control of the working-electrode potential accompanied by the measurement of the current at that electrode is the basis of voltammetric methods. The schematic diagram of a three-electrode potentiostat is shown in Fig. 14.9. It consists of a working (indicator) electrode, a reference electrode positioned as close as possible to the working electrode, and an auxiliary (counter-) electrode to complete the electrochemical cell. If the working electrode is the cathode, then the auxiliary electrode will be the anode. The function of the potentiostat is to observe the potential of the reference-electrode–working-electrode pair through a circuit that draws essentially no current. Since the potential of the reference electrode is constant, the circuit really senses the potential of the working electrode without the need to correct for any *iR* drop across the cell between the anode and cathode. This makes it possible to use nonaqueous solvents of high resistance and quite dilute aqueous electrolytes. Distortion of wave shapes is less pronounced, if not entirely eliminated.

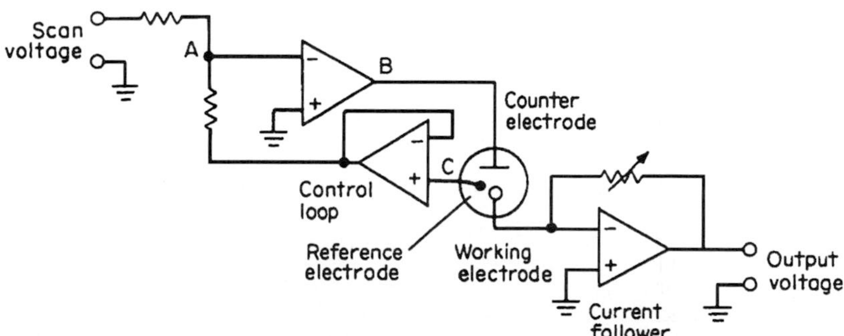

FIGURE 14.9 Schematic diagram of a three-electrode potentiostat. (*From Shugar and Dean, 1990.*)

14.5.2 Linear Potential Sweep (DC) Voltammetry

In classical polarography, a linear dc potential ramp is applied between two electrodes, one small and easily polarized working electrode (the dme) and the other relatively large and relatively constant in potential (often a pool of mercury). Before proceeding further, some terms require definition.

14.5.2.1 Diffusion Current. An electroactive species undergoing an electron exchange at a microelectrode may be brought to the electrode surface by three modes:

1. *Migration* of charged ions or dipolar molecules capable of undergoing directed movement in an electrical field
2. *Convection* due to thermal currents and by density gradients; convection occurs whenever the solution is stirred into the path of the electrode (hydrodynamic transport)
3. *Diffusion* under the influence of a concentration gradient

The effect of electrical migration can be eliminated by adding some salt ("supporting electrolyte") in a concentration at least 100-fold greater than the analyte. Usually a potassium or quaternary ammonium salt is added; their discharge potential is more negative than most cations. In the presence of a supporting electrolyte, the resistance of the solution and thus the potential gradient (iR drop) through it are made desirably small.

Convection effects can be minimized by using quiescent (unstirred) solutions during current measurements.

Under these conditions, the maximum diffusion current (or limiting current) is the current flowing as a result of an oxidation–reduction process at the electrode surface. The limiting current is controlled entirely by the rate at which the reactive substance can diffuse through the solution to the working electrode. The diffusion current is given by the Ilkovic equation

$$i_d = 708nCD^{1/2}m^{2/3}t^{1/6} \tag{14.44}$$

where i_d = diffusion current, μA
D = diffusion coefficient of analyte, $cm^2 \cdot s^{-1}$
C = concentration of analyte, mM
m = mass of mercury flowing through capillary tip, $mg \cdot s^{-1}$
t = drop life, s

In classical polarography, measurements are made just at the end of the drop's life when the diffusion current is greatest. Reproducibility is improved by using an electromechanical drop dislodger.

The Ilkovic equation permits the calculation of C for an unknown solution if m and t for the capillary are known and if a value of I, the diffusion current constant, is available:

$$I = 708nD^{1/2} \tag{14.45}$$

However, this requires that the experimental conditions employed in the measurement of I be exactly reproduced in the subsequent analysis and, in addition, it introduces the possibility of an error on the order of ±3% to 5% because I is not truly independent of variations in m and t. Methods of analysis that obviate this source of uncertainty are as follows:

1. The *standard addition* method, in which the calculation is based on the equation

$$C_u = \frac{i_1 v C_s}{i_2 v + (i_2 - i_1)V} \tag{14.46}$$

where i_1 is the diffusion current of the wave obtained with V mL of unknown solution and i_2 is the diffusion current obtained after addition of v mL of a known solution whose concentration (in the desired units) is C_s.

2. The *pilot-ion* method, in which a known amount of a substance not present in the sample is added to give an additional wave. If the heights of both the wave of the unknown and the wave of the pilot ion are proportional to the respective concentrations, then

$$C_u = k\frac{(i_d)_u}{(i_d)_p}C_p \tag{14.47}$$

where C_p is the concentration of the pilot ion in the desired units. The calibration factor k must be obtained by measurements on known solutions of the two substances, but is almost always independent of variations in capillary characteristics or other experimental conditions.

14.5.2.2 Current-Potential Curves. The potential of an oxidation–reduction system under conditions of diffusion control is given by

$$E = E^0 - \frac{0.059\ 16}{n}\log\frac{i}{i_d - i} + \frac{0.059\ 16}{n}\log\left(\frac{D_{\text{red}}}{D_{\text{ox}}}\right)^{1/2} \tag{14.48}$$

A current–potential curve (Fig. 14.10) shows the damped sawtooth pattern that arises from the growth and detachment of each mercury drop. As the potential (the dc potential ramp) is slowly increased in the negative direction, the voltammogram (polarogram) exhibits a sigmoidal curve for each electroactive species in the solution. By definition, the *half-wave potential* is the point at which the current is one-half the limiting diffusion current (Fig. 14.11). At this point the first logarithmic

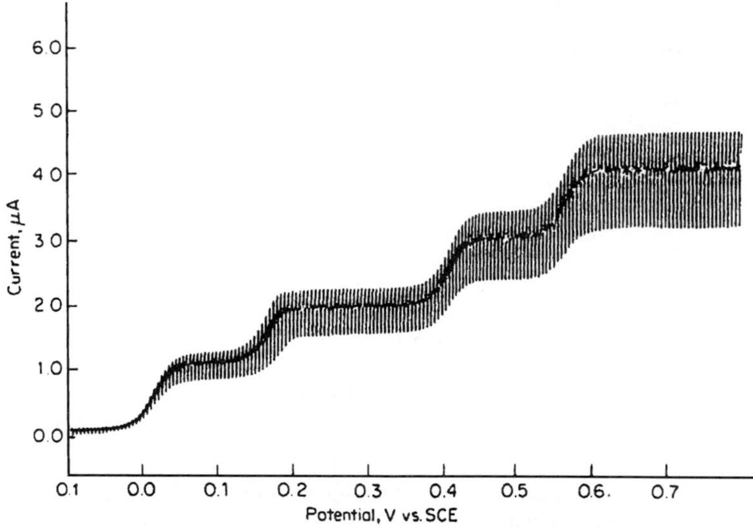

FIGURE 14.10 A typical voltammogram obtained with a dropping-mercury electrode. *(From Shugar and Dean, 1990.)*

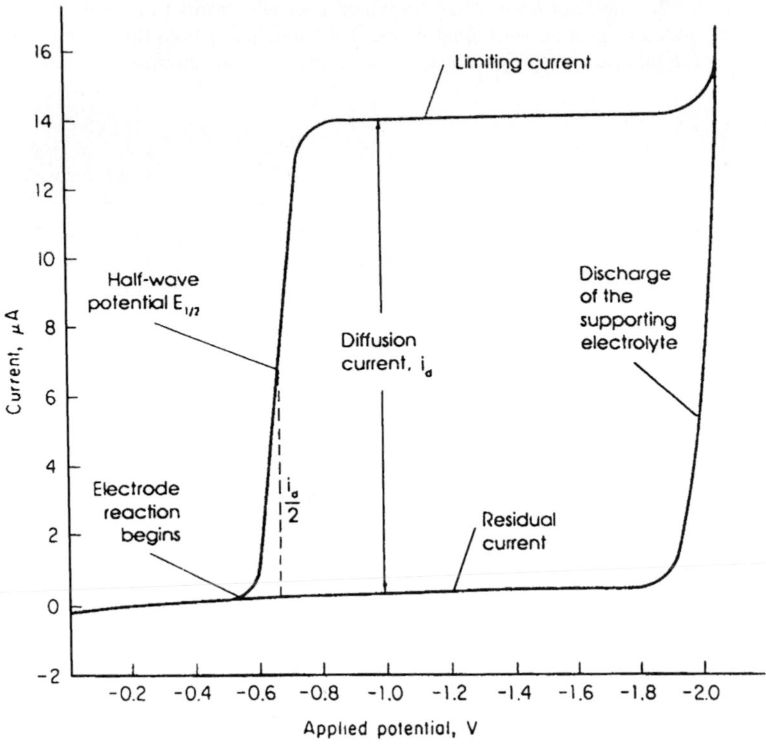

FIGURE 14.11 Current-voltage curve. (*From Shugar and Dean, 1990.*)

term in Eq. (14.48) becomes zero, and

$$E_{1/2} = E^0 + \frac{0.059\ 16}{n} \log \left(\frac{D_{\text{red}}}{D_{\text{ox}}} \right)^{1/2} \qquad (14.49)$$

Many pertinent organic and inorganic half-wave potentials are listed in Tables 14.20 and 14.21.

In order for two polarographic waves of equal height to be sufficiently well resolved for the ordinary purposes of quantitative polarographic analysis, the difference between their half-wave potentials should be no less than about three times the sum of their $E_{3/4} - E_{1/4}$ values. A greater separation is required if adjacent wave heights are considerably disparate.

To obtain the true diffusion (faradaic) current, a correction must be made for the charging or capacitive current, often called the *residual current*. Detection limits are about $10^{-5}M$. The charging current arises from the current that flows to charge each fresh mercury drop as it grows. The correction may be made in two ways:

1. The residual current of the supporting electrolyte alone (see Fig. 14.11) is evaluated in a separate voltammogram.

2. Extrapolate the residual current portion of the voltammogram immediately preceding the rising part of the current–potential curve. Take the difference between this extrapolated line and the

TABLE 14.20 Half-Wave Potentials of Inorganic Materials

All values are in volts vs. the saturated calomel electrode.

Element	$E_{1/2}$, volts	Solvent system
Aluminum		
3+	−0.5	0.2*M* acetate, pH 4.5–4.7, plus 0.07% azo dye Pontochrome Violet SW; reduction wave of complexed dye is 0.2 V more negative than that of the free dye
Antimony		
3+ to 0	−0.15	1*M* HCl
	−0.31(1)	1*M* HNO_3 (or 0.5*M* H_2SO_4)
	−0.8	0.5*M* tartrate, pH 4.5
	−1.0; −1.2	0.5*M* tartrate, pH 9 (waves not distinct)
	−1.26	1*M* NaOH; also anodic wave (3+ to 5+) at −0.45
	−1.32	0.5*M* tartrate plus 0.1*M* NaOH
5+	0.0; −0.257	6*M* HCl; first wave (5+ to 3+) starts at the oxidation potential of Hg; second wave is 3+ to 0
5+ to 0	−0.35	1*M* HCl plus 4*M* KBr
Arsenic		
3+ to 5+	−0.26	0.5*M* KOH (anodic wave); only suitable wave
3+	−0.8; −1.0	0.1*M* HCl; ill-defined waves
	−0.7; −1.0	0.5*M* H_2SO_4 (or 1*M* HNO_3)
Barium		
2+ to 0	−1.94	0.1*M* $(C_2H_5)_4NI$
Bismuth		
3+ to 0	−0.025(15)	1*M* HNO_3 (or 0.5*M* H_2SO_4)
	−0.09	1*M* HCl
	−0.29	0.5*M* tartrate, pH 4.5
	−0.7	0.5*M* tartrate (pH 9); wave not well developed
	−1.0	0.5*M* tartrate plus 0.1*M* NaOH; poor wave
Bromine		
5+ to 1−	−1.75	0.1*M* alkali chlorides (or 0.1*M* NaOH)
	0.13	0.05*M* H_2SO_4
0 to 1−	0.0	Wave (anodic) starts at zero; Hg_2Br_2 forms
Br^-	0.1	Oxidation of Hg to form mercury(I) bromide
Cadmium		
2+ to 0	−0.60	0.1*M* KCl, or 0.5*M* H_2SO_4, or 1*M* HNO_3
	−0.64	0.5*M* tartrate at pH 4.5 or 9
	−0.81	1*M* NH_4Cl plus 1*M* NH_3
Calcium		
2+ to 0	−2.22	0.1*M* $(C_2H_5)_4NCl$
	−2.13	0.1*M* $(C_2H_5)_4NCl$ in 80% ethanol
Cerium		
3+ to 0	−1.97	0.02*M* alkali sulfate
Cesium		
1+ to 0	−2.05	0.1*M* $(C_2H_5)_4NOH$ in 50% ethanol
Chlorine		
Cl^-	0.25	Oxidation of Hg to form Hg_2Cl_2
Chromium		
6+ to 3+	−0.85	CrO_4^{2-} to CrO_2^- in 0.1 to 1*M* NaOH
3+ to 0	−0.35; −1.70	1*M* NH_4Cl-NH_3 buffer (pH 8–9); 3+ to 2+ to 0

(Continued)

TABLE 14.20 Half-Wave Potentials of Inorganic Materials (*Continued*)

Element	$E_{1/2}$, volts	Solvent system
Chromium		
3+ to 2+	−0.95	0.1*M* pyridine–0.1*M* pyridinium chloride
2+ to 0	−1.54	1*M* KCl
2+ to 3+	−0.40	1*M* KCl (anodic wave)
Cobalt		
3+ to 0	−0.5; −1.3	1*M* NH_4Cl plus 1*M* NH_3; 3+ to 2+ to 0
2+ to 0	−1.07	0.1*M* pyridine plus pyridinium chloride
	−1.03	Neutral 1*M* potassium thiocyanate
	−1.4	$Co(H_2O)_6^{2+}$ in noncomplexing systems
3+ to 2+	0.0	1*M* sodium oxalate in acetate buffer (pH 5); diffusion current measured between 0 and −0.1 V
Copper		
2+ to 0	0.04	0.1*M* KNO_3, 0.1*M* NH_4ClO_4, or 1*M* Na_2SO_4
	−0.085	0.1*M* $Na_4P_2O_7$ plus 0.2*M* Na acetate, pH 4.5
	−0.09	0.5*M* Na tartrate, pH 4.5
	−0.20	0.1*M* potassium oxalate, pH 5.7 to 10
	−0.22	0.5*M* potassium citrate, pH 7.5
	−0.4	0.5*M* Na tartrate plus 0.1*M* NaOH (pH 12)
	−0.568	0.1*M* KNO_3 plus 1*M* ethylenediamine
2+	0.04; −0.22	1*M* KCl; consecutive waves: 2+ to 1+ to 0
	−0.02; −0.39	0.1*M* KSCN; consecutive waves: 2+ to 1+ to 0
	0.05; −0.25	0.1*M* pyridine plus 0.1*M* pyridinium chloride; consecutive waves: 2+ to 1+ to 0
	−0.24; −0.50	1*M* NH_4Cl plus 1*M* NH_3; consecutive waves
Gallium		
3+ to 0	−1.1	Not more than 0.001*M* HCl or wave masked by hydrogen wave which immediately follows
Germanium		
2+ to 0	−0.45	6*M* HCl; prior reduction with HPH_2O_2 to 2+
Gold		
3+ to 1+	0	1*M* KCN; wave starts at 0 V
1+ to 0	−1.4	$Au(CN)_2^-$ wave best for analytical purposes
Indium		
3+ to 0	−0.60	1*M* KCl
		In Na acetate, pH 3.9 to 4.2
Iodine		
IO_4^-	0.36	First wave at pH 0 (shifts to −0.08 at pH 12); second wave corresponds to iodate reduction
IO_3^-	−0.075	0.2*M* KNO_3 (shifts −0.13 V/pH unit increase)
	−0.305	0.1*M* hydrogen phthalate, pH 3.2
	−0.500	0.1*M* acetate plus 0.1*M* KCl, pH 4.9
	−0.650	0.1*M* citrate, pH 5.95
	−1.050	0.2*M* phosphate, pH 7.10
	−1.20	0.05*M* borax + 0.1*M* KCl, pH 9.2; or NaOH plus 0.1*M* KCl, pH 13.0
0 to 1−	0.0	Wave starts from zero in acid media; Hg_2I_2 formed
1−	−0.1	Oxidation of Hg to form Hg_2I_2
Iron		
3+	−0.44; −1.52	1*M* $(NH_4)_2CO_3$; two waves: 3+ to 2+ to 0

TABLE 14.20 Half-Wave Potentials of Inorganic Materials (*Continued*)

Element	$E_{1/2}$, volts	Solvent system
Iron		
	−0.17; −1.50	0.5*M* Na tartrate, pH 5.8; two waves: 3+ to 2+ to 0
	−0.9; −1.5	0.1 to 5*M* KOH plus 8% mannitol; 3+ to 2+ to 0
3+ to 2+	−0.13	0.1*M* EDTA plus 2*M* Na acetate, pH 6–7
	−0.27	0.2*M* Na oxalate, pH 7.9 or less
	−0.28	0.5*M* Na citrate, pH 6.5
	−1.46(2)	1*M* NH_4ClO_4
	−1.36	0.1*M* KHF_2, pH 4 or less
2+ to 3+	−0.28	0.5*M* Na citrate, pH 6.5
	−0.27	0.2*M* Na oxalate, pH 7.9 or less
	−0.17	0.5*M* Na tartrate, pH 5.8
	−1.36	0.1*M* KHF_2, pH 4 or less
Lead		
2+ to 0	−0.405	1*M* HNO_3
	−0.435	1*M* KCl (or HCl)
	−0.49(1)	0.5*M* Na tartrate, pH 4.5 or 9
	−0.72	1*M* KCN
	−0.75	1*M* KOH or 0.5*M* Na tartrate plus 0.1*M* NaOH
Lithium		
1+ to 0	−2.31	0.1*M* $(C_2H_5)_4NOH$ in 50% ethanol
Magnesium		
2+ to 0	−2.2	0.1*M* $(C_2H_5)_4NCl$ (poorly defined wave)
Manganese		
2+ to 0	−1.65	1*M* NH_4Cl plus 1*M* NH_3
	−1.55	1*M* KCNS
	−1.33	1.5*M* KCN
Molybdenum		
6+	−0.26; −0.63	0.3*M* HCl, two waves: 6+ to 5+ to 3+
Nickel		
2+ to 0	−0.70	1*M* KSCN
	−0.78	1*M* KCl plus 0.5*M* pyridine
	−1.09	1*M* NH_4Cl plus 1*M* NH_3
	−1.1	$Ni(H_2O)_6^{2+}$ in NH_4ClO_4 or KNO_3
	−1.36	$Ni(CN)_4^{2-}$ in 1*M* KCN (alkaline media)
Niobium		
5+ to 3+	−0.80(4)	1*M* HNO_3
Nitrogen		
Nitrate	−1.45	0.017*M* $LaCl_3$ (reduced to hydroxylamine)
HNO_2	−0.77	0.1*M* HCl
C_2N_2	−1.2; −1.55	0.1*M* Na acetate, two waves
Oxamic acid	−1.55	0.1*M* Na acetate
Cyanide	−0.45	0.1*M* NaOH; anodic wave starts at −0.45
Thiocyanate	0.18	Anodic wave; neutral or weakly alkaline medium
Osmium		
OsO_4	0.0; −0.41; −1.16	Sat'd $Ca(OH)_2$; three waves: first starts at 0; second wave is OsO_4^{2-} to Os(V); and third wave is Os(V) to Os(III)
Oxygen		
O_2	−0.05; −0.9	Buffer solutions of pH 1 to 10; two waves: O_2 to H_2O_2 and H_2O_2 to H_2O; second wave extends from −0.5 to −1.3

(Continued)

TABLE 14.20 Half-Wave Potentials of Inorganic Materials (*Continued*)

Element	$E_{1/2}$, volts	Solvent system
Oxygen		
H_2O_2	–0.9	Very extended wave (see above); sharper in presence of Aerosol OT
Palladium		
2+ to 0	–0.31	1*M* pyridine plus 1*M* KCl
	–0.64	0.1*M* ethylenediamine plus 1*M* KCl
	–0.72	1*M* NH_4Cl plus 1*M* NH_3
Potassium		
1+ to 0	–2.10	0.1*M* $(C_2H_5)_4NOH$ in 50% ethanol
Rhenium		
7+ to 4+	–0.44	2*M* HCl or (better) 4*M* $HClO_4$
4+ to 3+	–0.51	$ReCl_6^{2-}$ ion in 1*M* HCl
Rhodium		
3+ to 2+	–0.41	1*M* pyridine plus 1*M* KCl
Rubidium		
1+ to 0	–1.99	0.1*M* $(C_2H_5)_4NOH$ in 50% ethanol
Scandium		
3+ to 0	–1.80	0.1*M* LiCl, KCl, or $BaCl_2$
Selenium		
4+ to 2–	–1.44	1*M* NH_4Cl plus NH_3, pH 8.0
	–1.54	Same system adjusted to pH 9.5
2–	–0.49	Anodic wave at pH 0 due to HgSe
	–0.94	Anodic wave at pH 12 (0.01*M* NaOH)
Silver		
1+ to 0		Wave starts at oxidation potential of Hg
1+ to 0	–0.3	0.0014*M* $KAg(CN)_2$ without excess cyanide
Sodium		
1+ to 0	–2.07	0.1*M* $(C_2H_5)_4NOH$ in 50% ethanol
Strontium		
2+ to 0	–2.11	0.1*M* $(C_2H_5)_4NI$, water or 80% ethanol
Sulfur		
SO_2	–0.38	1*M* HNO_3 (or other strong acid); 4+ to 2+
$S_2O_4^{2-}$	–0.43	0.5*M* $(NH_4)_2HPO_4$ plus 1*M* NH_3 (anodic wave)
$S_2O_3^{2-}$	–0.15	1*M* strong acid; anodic mercury wave
0 to 2–	–0.50	90% methanol, 9.5% pyridine, 0.5% HCl (pH 6)
HS^-	–0.76	0.1*M* NaOH (anodic mercury wave)
Tellurium		
4+ to 0	–0.4	Citrate buffer, pH 1.6 (second of two waves)
	–0.63	Ammoniacal buffer, pH 9.4
4+ to 2–	–1.22	0.1*M* NaOH
2– to 0	–0.72	1*M* HCl (true anodic reversible wave)
	–0.08	1*M* NaOH (same as above; intermediate values at pH 1 to 13)
Thallium		
3+ to 0	–0.48	1*M* KCl, KNO_3, K_2SO_4, KOH, or NH_3
Tin		
4+ to 2+	–0.25; –0.52	4*M* NH_4Cl + 1*M* HCl; two waves: 4+ to 2+ to 0
2+ to 0	–0.59	0.5*M* tartrate, pH 4.3
	–1.22	1*M* NaOH (stannite ion to tin)

TABLE 14.20 Half-Wave Potentials of Inorganic Materials (*Continued*)

Element	$E_{1/2}$, volts	Solvent system
Tin		
2+ to 4+	−0.28	0.5*M* Na tartrate, pH 4.3 (anodic wave)
	−0.73	1*M* NaOH (stannite ion to stannate ion)
Titanium		
4+ to 3+	−0.173	0.1*M* $K_2C_2O_4$ plus 1*M* H_2SO_4
	−1.22	0.4*M* tartrate, pH 6.5
Tungsten		
6+	0.0; −0.64	6*M* HCl; two waves: first wave starts at zero and is W(VI) to W(V), and the second wave is W(V) to W(III)
Uranium		
6+	−0.180; −0.92	UO_2^{2+} to UO_2^{+}, then U^{3+} in 0.02*M* HCl
Vanadium		
5+ to 4+ to 2+	−0.97; −1.26	1*M* NH_4Cl plus 1*M* NH_3 and 0.08*M* Na_2SO_3
4+ to 2+	−0.98	0.05*M* H_2SO_4
3+ to 2+	−0.55	0.5*M* H_2SO_4
4+ to 5+	−0.32	1*M* NH_4Cl, 1*M* NH_3, and 0.08*M* Na_2SO_3
4+ to 5+	0.76	0.05*M* H_2SO_4; anodic wave starting from zero
2+ to 3+	−0.55	0.5*M* H_2SO_4; anodic wave
Zinc		
2+ to 0	−0.995	0.1*M* KCl
	−1.01	0.1*M* KSCN
	−1.15	0.5*M* tartrate, pH 9
	−1.23	0.5*M* tartrate, pH 4.5
	−1.33	1*M* NH_4Cl plus 1*M* NH_3
	−1.53	1*M* NaOH

Source: J. A. Dean, ed., *Lange's Handbook of Chemistry*, 14th ed., McGraw-Hill, New York, 1992.

current–potential plateau as the faradaic (limiting) current. This method is questionable at low concentrations (less than $10^{-4}M$).

Maxima may arise from convection around the growing mercury drop. Sometimes they are recognizable peaks, but other types of maxima simply enhance the plateau current without visibly distorting the wave shape. As a precaution, surfactants, such as gelatin or Triton X-100, are routinely added in small quantities (0.005% to 0.01%) to all test solutions.

14.5.2.3 Current-Sampled (Tast) Voltammetry. In current-sampled voltammetry the current is measured (sampled) at a fixed time (and for a short period of time) near the end of the drop life. This sampled-current measurement is held and recorded until the same point in time is reached on the succeeding drop when the measurement is updated to the new measured value of current. A voltammogram is free of drop growth oscillations. The voltammogram resembles a series of steps the width of which is the drop time. To control drop life and current-sampling time, a synchronized time circuit and drop knocker are required. Detection limits are near $10^{-6}M$. The improvement in detection limits over those of classical polarography is largely a result of freedom from the current fluctuations caused by drop growth and fall.

TABLE 14.21 Half-Wave Potentials (vs. Saturated Calomel Electrode) of Organic Compounds at 25°C

The solvent systems in this table are listed below:

: *A. acetonitrile and a perchlorate salt such as $LiClO_4$ or a tetraalkyl ammonium salt*
: *B. acetic acid and an alkali acetate, often plus a tetraalkyl ammonium iodide*
: *C. 0.05*M *to 0.175*M *tetraalkyl ammonium halide and 75% 1,4-dioxane*
: *D. buffer plus 50% ethanol (EtOH)*

Abbreviations used in the table

Bu, butyl	*Me, methyl*
Et, ethyl	*MeOH, methanol*
EtOH, ethanol	*PrOH, propanol*
M, *molar*	

Compound	Solvent system	$E_{1/2}$
	Unsaturated aliphatic hydrocarbons	
Acrylonitrile	C but 30% EtOH	−1.94
Allene	C	−2.29
1,3-Butadiene	A	−2.03
	C	−2.59
1,3-Butadiyne	C	−1.89
1-Buten-2-yne	C	−2.40
1,4-Cyclohexadiene	A	−1.6
Cyclohexene	A	−1.89
1,3,5,7-Cyclooctatetraene	B	−1.42
	C	−1.51
Diethyl fumarate	B, pH 4.0	−0.84
Diethyl maleate	B, pH 4.0	−0.95
2,3-Dimethyl-1,3-butadiene	A	−1.83
Dimethylfulvene	C	−1.89
Diphenylacetylene	C	−2.20
1,1-Diphenylethylene	B	−1.52
	C	−2.19
Ethyl methacrylate	0.1*N* LiCl + 25% EtOH	−1.9
2-Methyl-1,3-butadiene	A	−1.84
2-Methyl-1-butene	A	−1.97
1-Piperidino-4-cyano-4-phenyl-1,3-butadiene	$LiClO_4$ in dimethylformamide	−0.16
trans-Stilbene	B	−1.51
Tetrakis(dimethylamino)ethylene	A	−0.75
	Aromatic hydrocarbons	
Acenaphthene	A	−0.95
	B	−1.36
	C	−2.58
Anthracene	A	−0.84
	B	−1.20
	C	−1.94
Azulene	A	−0.71
	C	−1.66, −2.26, −2.56
1,2-Benzanthracene	C	−2.03, −2.54

TABLE 14.21 Half-Wave Potentials (vs. Saturated Calomel Electrode) of Organic Compounds at 25°C (*Continued*)

Compound	Solvent system	$E_{1/2}$
	Aromatic hydrocarbons	
2,3-Benzanthracene	A	−0.54, −1.20
Benzene	A	−2.08
1,2-Benzo[α]pyrene	A	−0.76
Biphenyl	A	−1.48
	B	−1.91
	C	−2.70
Chrysene	A	−1.22
1,2,5,6-Dibenzanthracene	A	−1.00, −1.26
1,2-Dihydronaphthalene	C	−2.57
9,10-Dimethylanthracene	A	−0.65
2,3-Dimethylnaphthalene	A	−1.08, −1.34
9,10-Diphenylanthracene	A	−0.92
Fluorene	A	−1.25
	B	−1.65
	C	−2.65
Hexamethylbenzene	A	−1.16
	B	−1.52
Indan	A	−1.59, −2.02
Indene	A	−1.23
	C	−2.81
1-Methylnaphthalene	A	−1.24
	B	−1.53
	C	−2.46
2-Methylnaphthalene	A	−1.22
	B	−1.55
	C	−2.46
Naphthalene	A	−1.34
	B	−1.72
Pentamethylbenzene	A	−1.28
	B	−1.62
Phenanthrene	A	−1.23
	B	−1.68
	C	−2.46, −2.71
Phenylacetylene	C	−2.37
Pyrene	A	−1.06, −1.24
trans-Stilbene	B	−1.51
	C	−2.26
Styrene	C	−2.35
1,2,3,5-Tetramethylbenzene	A	−1.50, −1.99
1,2,4,5-Tetramethylbenzene	A	−1.29
Tetraphenylethylene	C	−2.05
1,4,5,8-Tetraphenylnaphthalene	A	−1.39
Toluene	A	−1.98

(*Continued*)

TABLE 14.21 Half-Wave Potentials (vs. Saturated Calomel Electrode) of Organic Compounds at 25°C (*Continued*)

Compound	Solvent system	$E_{1/2}$
	Aromatic hydrocarbons	
1,2,3-Trimethylbenzene	A	−1.58
1,2,4-Trimethylbenzene	A	−1.41
1,3,5-Trimethylbenzene	A B	−1.50 −1.90
Triphenylene	A	−1.46, −1.55
Triphenylmethane	C	−1.01, −1.68, −1.96
o-Xylene	A	−1.58, −2.04
m-Xylene	A	−1.58
p-Xylene	A	−1.56
	Aldehydes	
Acetaldehyde	B, pH 6.8–13	−1.89
Benzaldehyde	McIlvaine buffer, pH 2.2	−0.96, −1.32
Bromoacetaldehyde	pH 8.5 pH 9.8	−0.40 −1.58, −1.82
Chloroacetaldehyde	Ammonia buffer, pH 8.4	−1.06, −1.66
Cinnamaldehyde	Buffer + EtOH, pH 6.0	−0.9, −1.5, −1.7
Crotonaldehyde	B, pH 1.3–2.0 Ammonia buffer, pH 8.0	−0.92 −1.30
Dichloroacetaldehyde	Ammonia buffer, pH 8.4	−1.03, −1.67
3,7-Dimethyl-2,6-octadienal	0.1*M* Et_4NI	−1.56, −2.22
Formaldehyde	0.05*M* KOH + 0.1*M* KCl, pH 12.7	−1.59
2-Furaldehyde	pH 1–8 pH 10	−0.86 −1.43
Glucose	Phosphate buffer, pH 7	−1.55
Glyceraldehyde	Britton–Robinson buffer, pH 5.0 Britton–Robinson buffer, pH 8.0	−1.47 −1.55
Glycolaldehyde	0.1*M* KOH, pH 13	−1.70
Glyoxal	B, pH 3.4	−1.41
4-Hydroxybenzaldehyde	Britton–Robinson buffer, pH 1.8 Britton–Robinson buffer, pH 6.8	−1.16 −1.45
4-Hydroxy-2-methoxybenzaldehyde	McIlvaine buffer, pH 2.2 McIlvaine buffer, pH 5.0 McIlvaine buffer, pH 8.0	−1.05 −1.16, −1.36 −1.47
o-Methoxybenzaldehyde	Britton–Robinson buffer, pH 1.8 Britton–Robinson buffer, pH 6.8	−1.02 −1.49
p-Methoxybenzaldehyde	Britton–Robinson buffer, pH 1.8 Britton–Robinson buffer, pH 6.8	−1.17 −1.48
Methyl glyoxal	A, pH 4.5	−0.83
m-Nitrobenzaldehyde	Buffer + 10% EtOH, pH 2.0	−0.28, −1.20
Phthalaldehyde	Buffer, pH 3.1 Buffer, pH 7.3	−0.64, −1.07 −0.89, −1.29
2-Propenal (acrolein)	pH 4.5 pH 9.0	−1.36 −1.1

TABLE 14.21 Half-Wave Potentials (vs. Saturated Calomel Electrode) of Organic Compounds at 25°C (*Continued*)

Compound	Solvent system	$E_{1/2}$
	Aldehydes	
Propionaldehyde	0.1*M* LiOH, pH 13	−1.93
Pyrrole-2-carbaldehyde	0.1*M* HCl + 50% EtOH	−1.25
Salicylaldehyde	McIlvaine buffer, pH 2.2	−0.99, −1.23
	McIlvaine buffer, pH 5.0	−1.20, −1.30
	McIlvaine buffer, pH 8.0	−1.32
Trichloroacetaldehyde	Ammonia buffer, pH 8.4	−1.35, −1.66
	0.1*M* KCl + 50% EtOH	−1.55
	Ketones	
Acetone	B, pH 9.3	−1.52
	C	−2.46
Acetophenone	D + McIlvaine buffer, pH 4.9	−1.33
	D + McIlvaine buffer, pH 7.2	−1.58
	D + McIlvaine buffer, pH 1.3	−1.08
7*H*-Benz[*de*]anthracen-7-one	0.1*N* H_2SO_4 + 75% MeOH	−0.96
Benzil	D + McIlvaine buffer, pH 1.3	−0.27
	D + McIlvaine buffer, pH 4.9	−0.50
Benzoin	D + McIlvaine buffer, pH 1.3	−0.90
	D + McIlvaine buffer, pH 8.6	−1.49
Benzophenone	D + McIlvaine buffer, pH 1.3	−0.94
	D + McIlvaine buffer, pH 8.6	−1.36
Benzoylacetone	Buffer, pH 2.6	−1.60
	Buffer, pH 5.3 and pH 7.6	−1.68
	Buffer, pH 9.7	−1.72
Bromoacetone	0.1*M* LiCl	−0.29
2,3-Butanedione	0.1*M* HCl	−0.84
3-Buten-2-one	0.1*M* KCl	−1.42
Butyrophenone	0.1*M* NH_4Cl + 50% EtOH	−1.55
D-Carvone	0.1*M* Et_4NI + 80% EtOH	−1.71
txChloroacetone	0.1*M* LiCl	−1.18
Coumarin	McIlvaine buffer, pH 2.0	−0.95
	McIlvaine buffer, pH 5.0	−1.11, −1.44
Cyclohexanone	C	−2.45
cis-Dibenzoylethylene	D, pH 1	−0.30
	D, pH 11	−0.62, −1.65
trans-Dibenzoylethylene	D, pH 1	−0.12
	D, pH 11	−0.57, −1.52
Dibenzoylmethane	D, pH 1.3	−0.59
	D, pH 11.3	−1.30, −1.62
9,10-Dihydro-9-oxoanthracene	D, pH 2.0	−0.93
1,5-Diphenyl-1,5-pentanedione	A	−2.10
1,5-Diphenylthiocarbazone	D, pH 7.0	−0.6

(*Continued*)

TABLE 14.21 Half-Wave Potentials (vs. Saturated Calomel Electrode) of Organic Compounds at 25°C (*Continued*)

Compound	Solvent system	$E_{1/2}$
	Ketones	
Flavanone	Acetate buffer + Me_4NOH + 50% 2-PrOH, pH 6.1	−1.30
	Acetate buffer + Me_4NOH + 50% 2-PrOH, pH 9.6	−1.51
Fluorescein	Acetate buffer, pH 2.0	−0.50
	Phthalate buffer, pH 5.0	−0.65
	Borate buffer, pH 10.1	−1.18, −1.44
Fructose	0.02*M* LiCl	−1.76
Girard derivatives of aliphatic ketones	pH 8.2	−1.52
o-Hydroxyacetophenone	D, pH 5	−1.36
p-Hydroxyacetophenone	D, pH 5	−1.46
1,2,3-Indantrione (ninhydrin)	Britton–Robinson buffer, pH 2.5	−0.67, −0.83
	Britton–Robinson buffer, pH 4.5	−0.73, −1.01
	Britton–Robinson buffer, pH 6.8	−0.10, −0.90, −1.20
	Britton–Robinson buffer, pH 9.2	−1.35
α-Ionone	C	−1.59, −2.08
Isatin	Phosphate buffer + citrate buffer, pH 2.9	−0.3, −0.5
	Phosphate buffer + citrate buffer, pH 4.3	−0.3, −0.5, −0.8
	Phosphate buffer + citrate buffer, pH 5.4	−0.8
4-Methyl-3,5-heptadien-2-one	A	−0.64
4-Methyl-2,6-heptanedione	A	−1.28
4-Methyl-3-penten-2-one	D + McIlvaine buffer, pH 1.3	−1.01
	D + McIlvaine buffer, pH 11.3	−1.60
4-Phenyl-3-buten-2-one	D, pH 1.3	−0.72
	D, pH 8.6	−1.27
Phthalide	0.1*M* Bu_4NI + 50% dioxane	−0.20
Phthalimide	pH 4.2	−1.1, −1.5
	pH 9.7	−1.2, −1.4
Pulegone	C	−1.74
Quinalizarin	Phosphate buffer + 1% EtOH, pH 8.0	−0.56
Testosterone	D + Britton–Robinson buffer, pH 2.6	−1.20
	D + Britton–Robinson buffer, pH 5.8	−1.40
	D + Britton–Robinson buffer, pH 8.8	−1.53, −1.79
	Quinones	
Anthraquinone	Acetate buffer + 40% dioxane, pH 5.6	−0.51
	Phosphate buffer + 40% dioxane, pH 7.9	−0.71
o-Benzoquinone	Britton–Robinson buffer, pH 7.0	+0.20
	Britton–Robinson buffer, pH 9.0	+0.08
2,3-Dimethylnaphthoquinone	D, pH 5.4	−0.22
1,2-Naphthoquinone	Phosphate buffer, pH 5.0	−0.03
	Phosphate buffer, pH 7.0	−0.13
1,4-Naphthoquinone	Britton–Robinson buffer, pH 7.0	−0.07
	Britton–Robinson buffer, pH 9.0	−0.19

TABLE 14.21 Half-Wave Potentials (vs. Saturated Calomel Electrode) of Organic Compounds at 25°C (*Continued*)

Compound	Solvent system	$E_{1/2}$
	Acids	
Acetic acid	A	−2.3
Acrylic acid	pH 5.6	−0.85
Adenosine-5′-phosphoric acid	$HClO_4$ + $KClO_4$, pH 2.2	−1.13
4-Aminobenzenesulfonic acid	0.05*M* Me_4NI	−1.58
3-Aminobenzoic acid	pH 5.6	−0.67
Anthranilic acid	pH 5.6	−0.67
Ascorbic acid	Britton–Robinson buffer, pH 3.4	+0.17
	Britton–Robinson buffer, pH 7.0	−0.06
Barbituric acid	Borate buffer, pH 9.3	−0.04
Benzoic acid	A	−2.1
Benzoylformic acid	Britton–Robinson buffer, pH 2.2	−0.48
	Britton–Robinson buffer, pH 5.5	−0.85, −1.26
	Britton–Robinson buffer, pH 7.2	−0.98, −1.25
	Britton–Robinson buffer, pH 9.2	−1.25
Bromoacetic acid	pH 1.1	−0.54
2-Bromopropionic acid	pH 2.0	−0.39
Crotonic acid	C	−1.94
Dibromoacetic acid	pH 1.1	−0.03, −0.59
Dichloroacetic acid	pH 8.2	−1.57
5,5-Diethylbarbituric acid	Borate buffer, pH 9.3	0.00
Flavanol	D, pH 5.6	−1.25
	D, pH 7.7	−1.40
Folic acid	Britton–Robinson buffer, pH 4.6	−0.73
Formic acid	0.1*M* KCl	−1.66
Fumaric acid	HCl + KCl, pH 2.6	−0.83
	Acetate buffer, pH 4.0	−0.93
	Acetate buffer, pH 5.9	−1.20
2,4-Hexadienedioic acid	Acetate buffer, pH 4.5	−0.97
Iodoacetic acid	pH 1	−0.16
Maleic acid	Britton–Robinson buffer, pH 2.0	−0.70
	Britton–Robinson buffer, pH 4.0	−0.97
	Britton–Robinson buffer, pH 6.0	−1.11, −1.30
	Britton–Robinson buffer, pH 10.0	−1.51
Mercaptoacetic acid	B, pH 6.8	−0.38
Methacrylic acid	D + 0.1*M* LiCl	−1.69
Nitrobenzoic acids	Buffer + 10% EtOH, pH 2.0	−0.2, −0.7
Oxalic acid	B, pH 5.4–6.1	−1.80
2-Oxo-1,5-pentanedioic acid	HCl + KCl, pH 1.8	−0.59
2-Oxo-1,5-pentanedioic acid	Ammonia buffer, pH 8.2	−1.30
2-Oxopropionic acid	Britton–Robinson buffer, pH 5.6	−1.17
	Britton–Robinson buffer, pH 6.8	−1.22, −1.53
	Britton–Robinson buffer, pH 9.7	−1.51

(Continued)

TABLE 14.21 Half-Wave Potentials (vs. Saturated Calomel Electrode) of Organic Compounds at 25°C (*Continued*)

Compound	Solvent system	$E_{1/2}$
	Acids	
Phenolphthalein	Phthalate buffer, pH 2.5	–0.67
	Phthalate buffer, pH 4.7	–0.80
	D, pH 9.6	–0.98, –1.35
Picric acid	pH 4.2	–0.34
	pH 11.7	–0.36, –0.56, –0.96
1,2,3-Propenetricarboxylic acid	pH 7.0	–2.1
Trichloroacetic acid	Ammonia buffer, pH 8.2	–0.84, –1.57
	Phosphate buffer, pH 10.4	–0.9, –1.6
3,4,5-Trihydroxybenzoic acid	Phosphate buffer, pH 2.9	+0.50
	Phosphate buffer, pH 8.8	+0.1
p-Aminophenol	Britton–Robinson buffer, pH 6.3	+0.14
	Britton–Robinson buffer, pH 8.6	–0.04
	Britton–Robinson buffer, pH 12.0	–0.16
o-Chlorophenol	pH 5.6	–0.63
m-Chlorophenol	pH 5.6	–0.73
p-Chlorophenol	pH 5.6	–0.65
o-Cresol	pH 5.6	–0.56
m-Cresol	pH 5.6	–0.61
p-Cresol	pH 5.6	–0.54
1,2-Dihydroxybenzene	pH 5.6	–0.35
1,3-Dihydroxybenzene	pH 5.6	–0.61
1,4-Dihydroxybenzene	pH 5.6	–0.23
o-Methoxyphenol	pH 5.6	–0.46
m-Methoxyphenol	pH 5.6	–0.62
p-Methoxyphenol	pH 5.6	–0.41
1-Naphthol	A	–0.74
2-Naphthol	A	–0.82
1,2,3-Trihydroxybenzene	Britton–Robinson buffer, pH 3.1	+0.35
	Britton–Robinson buffer, pH 6.5	+0.10
	Britton–Robinson buffer, pH 9.5	–0.10
	Halogen compounds	
Bromobenzene	A	–1.98
	C	–2.32
1-Bromobutane	C	–2.27
Bromoethane	C	–2.08
Bromomethane	C	–1.63
1-Bromonaphthalene (also 2-bromonaphthalene)	A	–1.55, –1.60
3-Bromo-1-propene	C	–1.29
p-Bromotoluene	A	–1.72
Carbon tetrachloride	C	–0.78, –1.71
Chlorobenzene	A	–2.07

TABLE 14.21 Half-Wave Potentials (vs. Saturated Calomel Electrode) of Organic Compounds at 25°C (*Continued*)

Compound	Solvent system	$E_{1/2}$
	Halogen compounds	
Chloroform	C	−1.63
Chloromethane	C	−2.23
3-Chloro-1-propene	C	−1.91
α-Chlorotoluene	C	−1.81
p-Chlorotoluene	A	−1.76
N-Chloro-*p*-toluenesulfonamide	0.5*M* K_2SO_4	−0.13
9,10-Dibromoanthracene	A	−1.15, −1.47
p-Dibromobenzene	C	−2.10
1,2-Dibromobutane	D + 1% Na_2SO_3	−1.45
Dibromoethane	C	−1.48
meso-2,3-Dibromosuccinic acid	Acetate buffer, pH 4.0	−0.23, −0.89
Dichlorobenzenes	C	−2.5
Dichloromethane	C	−1.60
Diiodomethane	C	−1.12, −1.53
Hexabromobenzene	C	−0.8, −1.5
Hexachlorobenzene	C	−1.4, −1.7
Iodobenzene	A	−1.72
Iodoethane	C	−1.67
Iodomethane	A	−2.12
	C	−1.63
Tetrabromomethane	C	−0.3, −0.75, −1.49
Tetraidomethane	C	−0.45, −1.05, −1.46
Tribromomethane	C	−0.64, −1.47
α,α,α-Trichlorotoluene	C	−0.68, −1.65 −2.00
	Nitro and nitroso compounds	
1,2-Dinitrobenzene	Phthalate buffer, pH 2.5	−0.12, −0.32, −1.26
	Borate buffer, pH 9.2	−0.38, −0.74
1,3-Dinitrobenzene	Phthalate buffer, pH 2.5	−0.17, −0.29
	Borate buffer, pH 9.2	−0.46, −0.68
1,4-Dinitrobenzene	Phthalate buffer, pH 2.5	−0.12, −0.33
	Borate buffer, pH 9.2	−0.35, −0.80
Methyl nitrobenzoates	Buffer + 10% EtOH, pH 2.0	−0.20 to −0.25
		−0.68 to −0.74
p-Nitroacetophenone	Britton–Robinson buffer, pH 2.2	−0.16, −0.61, −1.09
	Britton–Robinson buffer, pH 10.0	−0.51, −1.40, −1.73
o-Nitroaniline	0.03*M* LiCl + 0.02*M* benzoic acid in EtOH	−0.88
m-Nitroaniline	Britton–Robinson buffer, pH 4.3	−0.3, −0.8
	Britton–Robinson buffer, pH 7.2	−0.5
	Britton–Robinson buffer, pH 9.2	−0.7
p-Nitroaniline	pH 2.0	−0.36
	Acetate buffer, pH 4.6	−0.5

(*Continued*)

TABLE 14.21 Half-Wave Potentials (vs. Saturated Calomel Electrode) of Organic Compounds at 25°C (*Continued*)

Compound	Solvent system	$E_{1/2}$
	Nitro and nitroso compounds	
o-Nitroanisole	Buffer + 10% EtOH, pH 2.0	–0.29, –0.58
p-Nitroanisole	Buffer + 10% EtOH, pH 2.0	–0.35, –0.64
1-Nitroanthraquinone	Britton–Robinson buffer, pH 7.0	–0.16
Nitrobenzene	HCl + KCl + 8% EtOH, pH 0.5	–0.16, –0.76
	Phthalate buffer, pH 2.5	–0.30
	Borate buffer, pH 9.2	–0.70
Nitrocresols	Britton–Robinson buffer, pH 2.2	–0.2 to –0.3
	Britton–Robinson buffer, pH 4.5	–0.4 to –0.5
	Britton–Robinson buffer, pH 8.0	–0.6
Nitroethane	Britton–Robinson buffer + 30% MeOH, pH 1.8	–0.7
	Britton–Robinson buffer + 30% MeOH, pH 4.6	–0.8
2-Nitrohydroquinone	Phosphate buffer + citrate buffer, pH 2.1	–0.2
	Phosphate buffer + citrate buffer, pH 5.2	–0.4
	Phosphate buffer + citrate buffer, pH 8.0	–0.5
Nitromethane	Britton–Robinson buffer + 30% MeOH, pH 1.8	–0.8
	Britton–Robinson buffer + 30% MeOH, pH 4.6	–0.85
o-Nitrophenol	Britton–Robinson buffer + 10% EtOH, pH 2.0	–0.23
	Britton–Robinson buffer + 10% EtOH, pH 4.0	–0.4
	Britton–Robinson buffer + 10% EtOH, pH 8.0	–0.65
	Britton–Robinson buffer + 10% EtOH, pH 10.0	–0.80
m-Nitrophenol	Britton–Robinson buffer + 10% EtOH, pH 2.0	–0.37
	Britton–Robinson buffer + 10% EtOH, pH 4.0	–0.40
	Britton–Robinson buffer + 10% EtOH, pH 8.0	–0.64
	Britton–Robinson buffer + 10% EtOH, pH 10.0	–0.76
p-Nitrophenol	Britton–Robinson buffer + 10% EtOH, pH 2.0	–0.35
	Britton–Robinson buffer + 10% EtOH, pH 4.0	–0.50
	Britton–Robinson buffer + 10% EtOH, pH 8.0	–0.82
1-Nitropropane	Britton–Robinson buffer + 30% MeOH, pH 1.8	–0.73
	Britton–Robinson buffer + 30% MeOH, pH 8.6	–0.88
	Britton–Robinson buffer + 30% MeOH, pH 8.0	–0.95
2-Nitropropane	McIlvaine buffer, pH 2.1	–0.53
	McIlvaine buffer, pH 5.1	–0.81
Nitrosobenzene	McIlvaine buffer, pH 6.0	–0.03
	McIlvaine buffer, pH 8.0	–0.14
1-Nitroso-2-naphthol	D + buffer, pH 4.0	+0.02
	D + buffer, pH 7.0	–0.20
	D + buffer, pH 9.0	–0.31
N-Nitrosophenylhydroxylamine	pH 2.0	–0.84
o-Nitrotoluene	Phthalate buffer, pH 2.5	–0.35, –0.66
	Phthalate buffer, pH 7.4	–0.60, –1.06
m-Nitrotoluene (also *p*-nitrotoluene)	Phthalate buffer, pH 2.5	–0.30, –0.53
	Phthalate buffer, pH 7.4	–0.58, –1.06
Tetranitromethane	pH 12.0	–0.41
1,3,5-Trinitrobenzene	Phthalate buffer, pH 4.1	–0.20, –0.29, –0.34
	Borate buffer, pH 9.2	–0.34, –0.48, –0.65

TABLE 14.21 Half-Wave Potentials (vs. Saturated Calomel Electrode) of Organic Compounds at 25°C (*Continued*)

Compound	Solvent system	$E_{1/2}$
	Heterocyclic compounds containing nitrogen	
Acridine	D, pH 8.3	–0.80, –1.45
Cinchonine	B, pH 3	–0.90
2-Furanmethanol	Britton–Robinson buffer, pH 2.0	–0.96
	Britton–Robinson buffer, pH 5.8	–1.38, –1.70
2-Hydroxyphenazine	Britton–Robinson buffer, pH 4.0	–0.24
8-Hydroxyquinoline	B, pH 5.0	–1.12
	Phosphate buffer, pH 8.0	–1.18, –1.71
3-Methylpyridine	D + 0.1*M* LiCl	–1.76
4-Methylpyridine	D + 0.1*M* LiCl	–1.87
Phenazine	Phosphate buffer + citrate buffer, pH 7.0	–0.36
Pyridine	Phosphate buffer + citrate buffer, pH 7.0	–1.75
Pyridine-2-carboxylic acid	B, pH 4.1	–1.10
	B, pH 9.3	–1.48, –1.94
Pyridine-3-carboxylic acid	0.1*M* HCl	–1.08
Pyridine-4-carboxylic acid	Britton–Robinson buffer, pH 6.1	–1.14
	pH 9.0	–1.39, –1.68
Pyrimidine	Citrate buffer, pH 3.6	–0.92, –1.24
	Ammonia buffer, pH 9.2	–1.54
Quinoline-8-carboxylic acid	pH 9	–1.11
Quinoxaline	Phosphate buffer + citrate buffer, pH 7.0	–0.66, –1.52
	Azo, hydrazine, hydroxylamine, and oxime compounds	
Azobenzene	D, pH 4.0	–0.20
	D, pH 7.0	–0.50
Azoxybenzene	Buffer + 20% EtOH, pH 6.3	–0.30
Benzoin 1-oxime	Buffer, pH 2.0	–0.88
	Buffer, pH 5.6	–1.08
	Buffer, pH 8.2	–1.67
Benzoylhydrazine	0.13*M* NaOH, pH 13.0	–0.30
Dimethylglyoxime	Ammonia buffer, pH 9.6	–1.63
Hydrazine	Britton–Robinson buffer, pH 9.3	–0.09
Hydroxylamine	Britton–Robinson buffer, pH 4.6	–1.42
	Britton–Robinson buffer, pH 9.2	–1.65
Oxamide	Acetate buffer	–1.55
Phenylhydrazine	McIlvaine buffer, pH 2	+0.19
	0.13*M* NaOH, pH 13.0	–0.36
Phenylhydroxylamine	McIlvaine buffer + 10% EtOH, pH 2	–0.68
	McIlvaine buffer + 10% EtOH, pH 4–10	–0.33 0.061 pH
Salicylaldoxime	Phosphate buffer, pH 5.4	–1.02
Thiosemicarbazide	Borate buffer, pH 9.3	–0.26
Thiourea	0.1*M* sulfuric acid	+0.02

(Continued)

TABLE 14.21 Half-Wave Potentials (vs. Saturated Calomel Electrode) of Organic Compounds at 25°C (*Continued*)

Compound	Solvent system	$E_{1/2}$
	Indicators and dyestuffs	
Brilliant Green	HCl + KCl, pH 2.0	−0.2, −0.5
Indigo carmine	pH 2.5	−0.24
Indigo disulfonate	pH 7.0	−0.37
Malachite Green G	HCl + KCl, pH 2.0	−0.2, −0.5
Metanil yellow	Phosphate buffer + 1% EtOH, pH 7.0	−0.51
Methylene blue	Britton–Robinson buffer, pH 4.9	−0.15
	Britton–Robinson buffer, pH 9.2	−0.30
Methylene green	Phosphate buffer + 1% EtOH, pH 7.0	−0.12
Methyl orange	Phosphate buffer + 1% EtOH, pH 7.0	−0.51
Morin	D, pH 7.6	−1.7
Neutral red	Britton–Robinson buffer, pH 2.0	−0.21
	Britton–Robinson buffer, pH 7.0	−0.57
	Peroxide	
Ethyl peroxide	0.02*M* HCl	−0.2

Source: J. A. Dean, ed., *Lange's Handbook of Chemistry*, 14th ed., McGraw-Hill, New York, 1992.

14.5.3 Potential Step Methods[8]

Potential step methods are based on the measurement of current as a function of time after applying a potential to the working electrode. These methods seek to optimize the ratio of faradaic to charging current by applying a sudden change (pulse) in applied potential and sampling the faradaic current just before the drop is detached but after the capacitative current has largely decayed. The potential step methods discriminate against the charging current by delaying the current measurement until close to the end of the pulse.

14.5.3.1 Normal Pulse Voltammetry. In normal pulse voltammetry a series of square-wave voltage pulses of successively increasing magnitude is superimposed upon a constant dc voltage signal (Fig. 14.12). Near the end of each pulse (perhaps 50 ms in duration) and before the drop is dislodged, the current is sampled for perhaps 17 ms. The sampled current is presented to a recorder as a constant signal until the current sample taken in the next drop lifetime replaces it. A staircase plot traces the potential–current curve. The time delay between pulses must be long enough to restore all concentration gradients at a dropping-mercury electrode or a solid microelectrode to their original state before the next potential pulse is applied. Each potential pulse is made a few millivolts higher for each drop (or for each pulse when a solid electrode is used).

14.5.3.2 Differential Pulse Voltammetry. In differential pulse voltammetry a series of potential pulses of fixed but small amplitude (10 to 100 mV) is superimposed on a constant dc voltage ramp (Fig. 14.13) near the end of the drop life and after the drop has attained the bulk of its growth. The current is sampled immediately before applying the potential pulse (perhaps 17 ms) and again (for 17 ms) just before the drop is dislodged. Subtraction (instrumentally) of the first current sampled from the second provides a stepped peak-shaped derivative voltammogram.

Differential pulse voltammetry discriminates effectively against the capacitive component of current signal. Detection limits are improved by the return of the signal to the baseline after each peak.

[8] S. A. Borman, "New Electroanalytical Pulse Techniques," *Anal. Chem.* **54**:698A (1982).

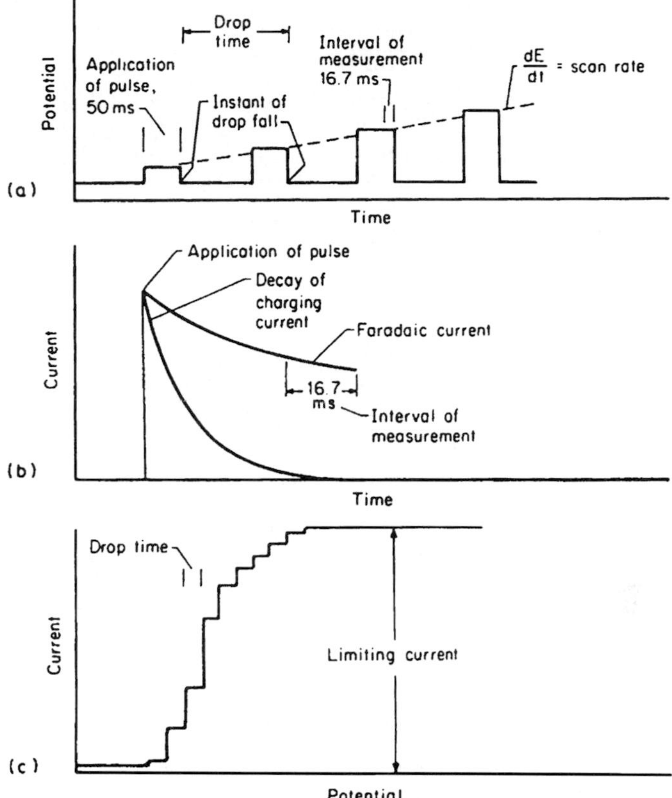

FIGURE 14.12 Normal pulse voltammetry. (*a*) Schematic showing the pulse application, drop time, and interval of current measurement; (*b*) the variation of current during the pulse; (*c*) stepped voltammogram. (*From Shugar and Dean, 1990.*)

14.5.3.3 Square-Wave Voltammetry.[9] Square-wave voltammetry is performed on a single mercury drop; a new drop is dispensed for each analysis cycle. The mercury drop is maintained at a constant surface area throughout the analysis, which eliminates the problem of baseline drift due to continuous mercury flow. The potential waveform employed is that obtained by superimposing a symmetrical square-wave potential pulse E_{sw} on a staircase waveform. The forward pulse (point 1 in Fig. 14.14) of the square wave is coincident with the staircase step. The current is sampled twice during each square-wave cycle, once at the end of the forward pulse and again at the end of the reverse pulse (point 2). Plotting the difference between the current observed at point 1 (forward current) and that at point 2 (reverse current) yields current–voltage peaks with height proportional to the concentration of the species in solution. Frequencies of 1 to 120 square-wave cycles per second permit extremely rapid scan rates that may be as high as 1200 $mV \cdot s^{-1}$. Because the amplitude of the square-wave modulation is so large, the reverse pulse causes reoxidation of the product produced on the forward pulse back to the original state with the resulting anodic current. Concentration levels of parts per billion are easily detectable.

[9] J. G. Osteryoung and R. A. Osteryoung, "Square Wave Voltammetry," *Anal. Chem.* **57**:101A (1985).

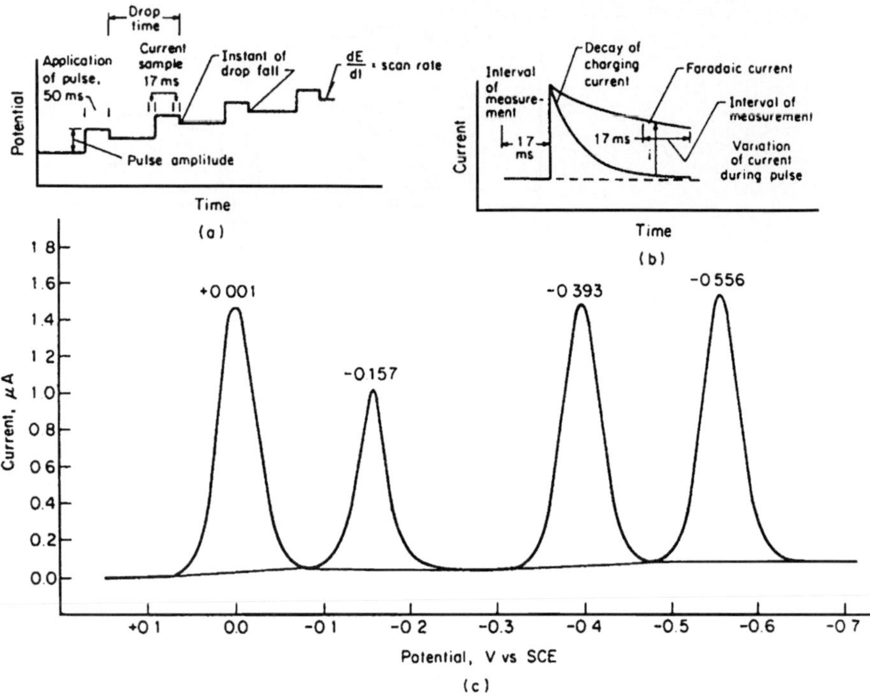

FIGURE 14.13 Differential pulse voltammetry. (*a*) A linearly increasing scan voltage on which a 35-mV pulse is superimposed during the last 50 ms of the drop life. (*b*) Variation of current during the pulse application. (*c*) The net signal observed between the two intervals of current measurement, for four electroactive substances. Individual steps not shown. (*From Shugar and Dean, 1990.*)

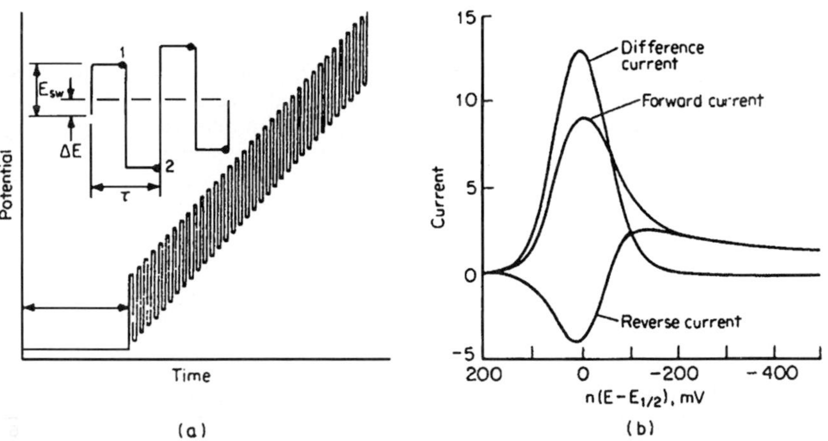

FIGURE 14.14 Square-wave voltammetry. (*a*) Excitation signal. (*b*) Forward, reverse, and difference current. (*From Shugar and Dean, 1990.*)

14.5.4 Cyclic Potential Sweep Voltammetry[10]

Cyclic voltammetry consists of cycling the potential of a stationary electrode immersed in a quiescent solution and measuring the resulting current (Fig. 14.15). An isosceles-triangular or staircase potential waveform is used. Scans can be initiated either in the cathodic or anodic direction. In the example, the initial potential (+ 0.150 V versus Ag–AgCl) is chosen to avoid any electrolysis of electroactive species in the sample when the cycling is initiated. Then the potential is scanned in the negative direction at 20 mV · s^{-1} until the desired negative potential is reached. For the first electroactive species present, a cathodic current commences to flow at about +0.07 V. The cathodic current increases rapidly until the surface concentration of oxidant at the electrode surface approaches zero, which is signaled by the current peaking at –0.029 V. The current then decays as the solution surrounding the electrode is depleted of oxidant. Additional rises occur as additional electroactive species undergo reaction or the system reaches the discharge potential of the supporting electrolyte. At the selected point (–0.800 V), and about 45 s after initiation of the forward scan, the potential is switched to scan in the positive direction. Briefly, reduction of oxidant continues to occur until the potential becomes sufficiently positive to bring about oxidation of the reductant that had been accumulating adjacent to the electrode surface. An anodic current begins to flow and increases rapidly until the surface concentration of the accumulated reductant approaches zero, at which point the anodic current peaks. The anodic current then decays as the solution surrounding the electrode is depleted of reductant formed during the forward scan. Similarly, the anodic peaks grow and decay for all other reversible electrode reactions.

Although peak currents can be measured and are linearly related to the analyte concentration, it is sometimes difficult to establish the correct baseline. Cyclic voltammetry finds its greatest use in the study of electrochemical reversibility, kinetics, and transient intermediates. The number of electrons transferred in the electrode reaction can be calculated if the diffusion coefficient is known (or the estimation of the diffusion coefficient if n is known). Cathodic and anodic peak potentials are separated by $57/n$ mV for a reversible electrode reaction (provided the switching potential is more negative than $100/n$ mV of the reduction peak). Also, for a reversible wave, the peak potential is independent of the scan rate (usually 20 to 100 mV · s^{-1}) and the peak current is proportional to the square root of the

[10] W. R. Heineman and P. T. Kissinger, "Cyclic Voltammetry: Electrochemical Equivalent of Spectroscopy," *Am. Lab.* **14** (June):27 (1981).

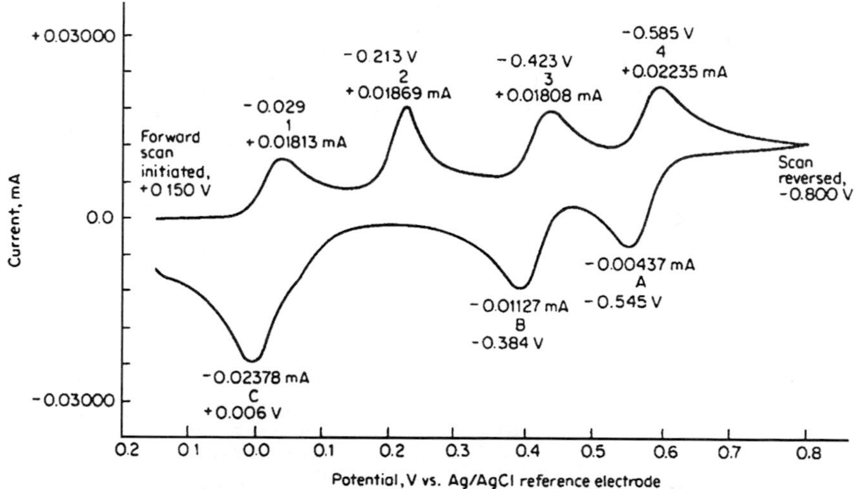

FIGURE 14.15 Cyclic voltammogram of four electroactive species. (*From Shugar and Dean, 1990.*)

scan rate. Quasireversibility is characterized by a separation of cathodic and anodic peak potentials larger than $57/n$ mV and irreversibility by the disappearance of a reverse peak (as shown by the disappearance of the anodic wave for the species whose cathodic peak occurred at –0.213 V).

14.5.5 Stripping Voltammetry[11]

Stripping voltammetry is the most sensitive electrochemical technique currently available. The technique is applicable to analytes that oxidize or reduce reversibly at a solid (thin-film mercury) electrode or that form an insoluble species with the electrode material that can subsequently be removed electrochemically.

Basically, stripping voltammetry is a two-step operation. During the first step the ion or ions of interest are electrolytically deposited on the working electrode by controlled potential electrolysis (see Sec. 14.6). After a quiescent period, a reverse potential scan, or stripping step, is applied in which the deposited analyte(s) is removed from the electrode. The preconcentration or electrodeposition step provides the means for substantially improving the detection limit for the stripping step, often by a factor of up to 1 million.

14.5.5.1 Anodic Stripping Voltammetry. Anodic stripping voltammetry (ASV) is used to determine the concentration of trace amounts of metal ions that can be preconcentrated at an electrode by reduction to the metallic state. Very electropositive metal ions, such as mercury(II), gold(III), silver, and platinum(IV), are deposited on glassy carbon electrodes. Other metal ions are deposited on a thin-film mercury electrode or a mercury-drop electrode.

In the plating step, a deposition potential is chosen that is more negative than the half-wave potential of the most electronegative metal to be determined; in Fig. 14.16 the plating step was done at –0.900 V versus SCE. Potential–current voltammograms are helpful in selecting the deposition potential that should lie on the diffusion current plateau of the most electronegative metal system. The deposition step is seldom carried to completion; often the plating time is 60 s. Usually only a fraction of the metal ions need to be deposited. Thus during the plating step, the temperature and rate of stirring must be kept as constant and reproducible as possible, and the deposition time must be strictly controlled so that the same fraction of metal ion is removed during each experiment with samples and standards.

[11] W. M. Peterson and R. V. Wong, "Fundamentals of Stripping Voltammetry," *Am. Lab.* **13**(November):116 (1981).

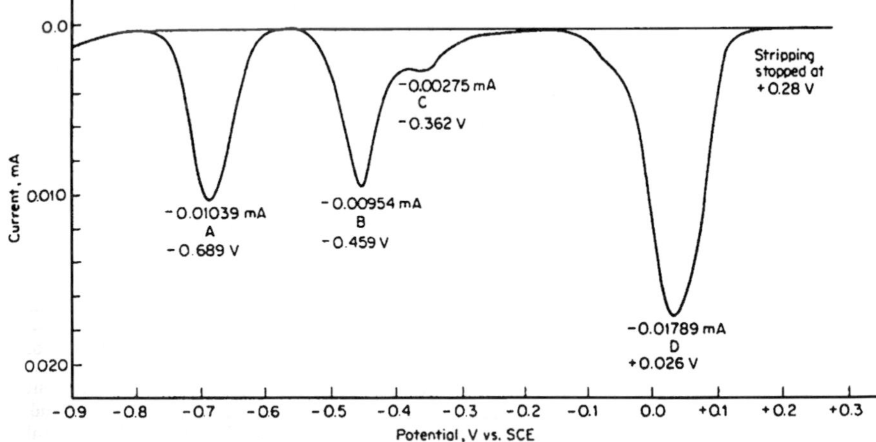

FIGURE 14.16 Anodic stripping voltammogram using differential pulse voltammetry during the stripping cycle. (*From Shugar and Dean, 1990.*)

Interference from dissolved oxygen is eliminated by purging the system with purified nitrogen for 2 to 10 min and maintaining a blanket of nitrogen gas over the solution during the entire operation.

Following the deposition step, stirring is halted for a period of 30 to 60 s to allow convection currents to decrease to a negligible level and to allow time for any amalgam to stabilize. The deposition potential is still applied to the working electrode. Then, in the stripping step, the potential is scanned in the positive direction (from –0.9 to +0.28 V in Fig. 14.16) using usually differential pulse voltammetry. The potentials of the stripping peaks identify the respective metals and the area under each current peak is proportional to the concentration of the metal species.

Standards and a blank are carried through identical plating and stripping steps. The method of standard addition is often used for evaluation. If desirable, the supporting electrolyte may be changed after the deposition step to one better suited for the stripping process. Prior to each experiment, the supporting electrolyte must be conditioned (or purified) by applying a potential of 0.0 V versus SCE for 1 to 2 min to clean the electrode by removing metal contaminants.

Elements that can be determined by anodic stripping voltammetry include Ag, Au, Bi, Cd, Cu, Ga, Ge, Hg, In, Pb, Sb, Sn, Tl, and Zn.

14.5.5.2 Cathodic Stripping Voltammetry. Cathodic stripping voltammetry follows the same steps as outlined for anodic stripping voltammetry: (1) preconcentration by applying a controlled oxidation potential to an electrode, (2) a quiescent period, and (3) subsequent stripping via a negative potential scan using differential pulse polarography. Cathodic stripping voltammetry is used to determine anions that form insoluble mercury(I) salts at the mercury surface or insoluble silver salts on a silver electrode. A partial list of species that can be determined include arsenate, chloride, bromide, iodide, chromate, sulfate, selenide, sulfide, mercaptans, thiocyanate, and thio compounds.

14.5.6 Amperometric Titrations

In amperometric titrations the potential is maintained at some constant value, usually on a diffusion current plateau, and the current is measured and plotted against the volume of titrant. Either the indicator (working) electrode is rotated at a constant rate or the solution is stirred at a constant velocity past a stationary solid electrode. The reference electrode should be large in area so that when appreciable currents pass through the electrochemical cell its potential will remain constant.

Only one of the reactants need be electroactive at the working potential. To avoid corrections for dilution, the titrant should be tenfold more concentrated than the solution being titrated. The shape of the titration curve can be predicted from hydrodynamic voltammograms of the analyte obtained at various stages of the titration (Fig. 14.17). Only three or four experimental points need to be accumulated to establish each branch of the titration curve (Fig. 14.18). Since

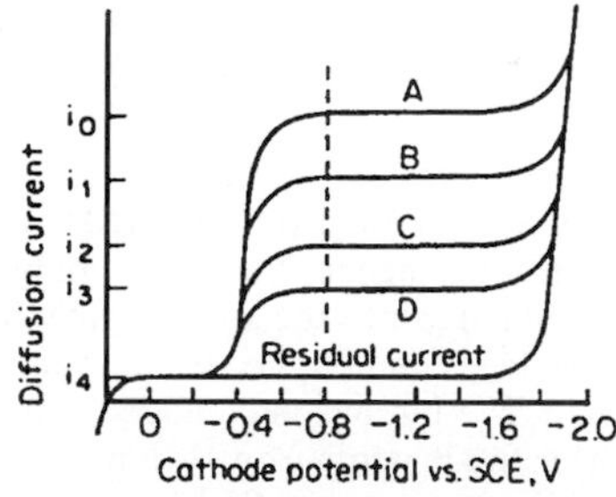

FIGURE 14.17 Successive current-voltage curves of lead ion made after increments of sulfate ion were added. (*From Shugar and Dean, 1990.*)

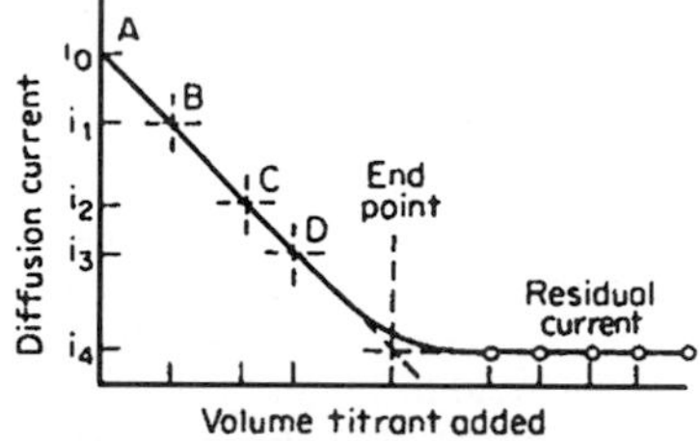

FIGURE 14.18 Amperometric titration curve for the reaction of lead ions with sulfate ions. See Fig. 14.17 for the corresponding current-voltage curves. Performed at –0.8 V versus SCE. (*From Shugar and Dean, 1990.*)

amperometric titration curves are linear on either side of the equivalence point, the equivalence point is found by extrapolation of data points away from the vicinity of the equivalence point. Data points selected between 0% and 50% and between 150% and 200% of the end-point volume will lie in regions in which the common ion effect suppresses dissociation of complexes and solubility of precipitates.

Several shapes of titration curves are possible. An L-shaped curve is obtained when only the analyte is electroactive; after the end point the only signal will be the residual current. If both analyte and titrant are electroactive, the current will decrease up to the end point and then increase again as unreacted titrant is added—a V-shaped curve is obtained. If the analyte is not electroactive but the titrant is, a horizontal line is obtained that rises after the end point is attained. When possible, the latter shape is preferred because the titrant can be added rather rapidly until the diffusion current starts to increase; then three or four data points are taken and the titration is done.

Automation is easy when the titration curve has the "reversed" L shape. The titrator can be programmed to shut off when a specified current level is reached. Titrant is run into a blank supporting electrolyte until the specified current level is reached, the sample is added, and the titrant flow is continued until the specified current is attained once again. Selected methods of amperometric titrations are given in Table 14.22.

14.5.6.1 Amperometric Titrations with Two Polarizable Electrodes. Amperometric titrations with two polarizable electrodes (usually two small similar platinum electrodes) was known at its inception as the "dead-stop-end-point" technique. The areas of the two electrodes and the constant applied voltage are selected so that the indicator current will be in the microampere range. This is done to ensure that the reactions at the indicator electrodes do not alter the composition of the solution (a problem if an irreversible couple is involved). Generally the area of the electrode is between 0.1 and 3 cm^2. The applied voltage seldom exceeds 500 mV, but may be as small as 10 mV. A magnetic stirrer is usually used to provide efficient and relatively constant stirring. The latter point is important since the value of the indicator current depends to some extent on the rate of stirring. The constant polarizing voltage can conveniently be supplied by a potential divider composed of a variable potentiometer (1000 ohms) powered by a 1.5-V dry cell. Because the indicator current is generally sensitively dependent on the potential applied across the electrodes, it is important that the microammeter used have as small a resistance as possible so that the iR drop across it will be insignificant. For certain applications, mercury-plated gold electrodes are used rather than platinum.

These titration curves may be explained by considering the reactions occurring at each electrode during the titration. When iodine in a solution containing iodide is titrated with thiosulfate, the platinum anode will oxidize iodide ion and the cathode will reduce iodine. This is possible since the iodide–iodine couple is reversible. As more and more iodine is removed from the solution by reaction with thiosulfate, the current will decrease, since there is less iodine available for reduction at the cathode. Finally, at the point at which the iodine is completely removed from the solution, little or no current can flow. The small value of the applied potential and the irreversibility of the tetrathionate–thiosulfate couple make it impossible for anything except iodine to undergo reduction at the cathode.

If both couples involved in the titration are reversible, the shape of the titration curve is quite different. Suppose that an acidic solution of iron(II) ion is to be titrated with a standard solution of cerium(IV) ion. Before the addition of reagent has begun, the anode could oxidize iron(II) ion but the cathode can reduce nothing but hydrogen ion. Since the applied potential is much smaller than the potential difference between the iron and cerium couples, little or no reaction will take place and no current will flow. As soon as cerium(IV) ion has been added, iron(III) ion will be produced. The iron(III) ion can be reduced at the cathode while iron(II) ion is oxidized at the anode because the iron(III)–iron(II) couple is reversible. As more iron(III) ion is produced, the current will rise until the concentration of iron(III) ion equals that of the iron(II) ion. If the titration is continued further, the current will decrease since the amount of iron(II) ion available for reaction at the anode becomes smaller. Finally, at the equivalence point, the current will decrease to almost zero since there will no longer be iron(II) ion for the anode to oxidize. The oxidation of cerium(III) ion is prevented by the small value of the applied potential. After the equivalence point, the current will immediately rise owing to the presence of both members of a new reversible couple, the cerium(IV)–cerium(III) couple. Beyond the equivalence point the cathode will reduce cerium(IV) ion rather than iron(III) ion and the anode will be able to oxidize cerium(III) ion. Thus the current will rise from nearly zero to a maximum at the midpoint of the titration, decrease to zero at the equivalence point, and finally rise again.

The main advantage of amperometric titrations with two polarized electrodes (Table 14.23) is that the apparatus and electrode system are extremely simple and easy to operate. This may be particularly important in work with nonaqueous solutions since reference electrodes suitable for such work are difficult to prepare.

14.5.7 Voltammetric Methods Compared

A very effective way to compare some of the voltammetric techniques discussed is to examine the voltammograms obtained using the same solution. The solution selected contained 0.0004*M* each of copper(II), lead(II), and cadmium plus 0.000 28*M* bismuth(III), all in an acetate–tartrate buffer.

A classical dc polarogram of this solution is shown in Fig. 14.10. The polarogram shows individual waves, in order of increasing negative half-wave potentials, for copper, bismuth, lead, and cadmium. This technique has the poorest detection limits (about $10^{-5}M$) of any of the voltammetric methods. The culprit is the capacitive current the compensation of which is difficult at low concentrations.

A current-sampled (Tast) voltammogram (not shown) provides the same information that is obtained from the dc polarogram. However, the Tast voltammogram is substantially free of the fluctuating pattern (sawtooth waves) of the classical dc polarogram. Consequently, detection limits are near $10^{-6}M$.

A normal pulse voltammogram (Fig. 14.12) overcomes the problem with the capacitative current. The diffusion-current plateau is clearly delineated and detection limits are similar to those obtained in Tast polarography.

A differential pulse voltammogram is shown in Fig. 14.13. While the current scale here is about one-third that for the two preceding voltammograms, the relative peak heights and separation from the residual (background) current are greater. The current is not sampled until the capacitive current has decayed to a negligible value. Also, with a stepped dc ramp the charging current value attributable to the dc potential scan is eliminated. Reversibly reducible ions can be detected at concentrations down to $10^{-8}M$. Because the current signal is peak-shaped and returns to the baseline, the selectivity is improved and the signal can be amplified without driving subsequent peaks off-scale.

Square-wave voltammetry shortens the observation time to about 6 s compared with perhaps 3 min with differential pulse voltammetry. Peak heights are also somewhat larger in square-wave voltammetry. Its rapid response makes it a useful detector in chromatography.

The merits of cyclic voltammetry are largely confined to qualitative and diagnostic experiments. Cyclic voltammetry is capable of rapidly generating a new oxidation state (or reduction state) during the forward scan and then probing its fate on the reverse scan (Fig. 14.15). Systems that exhibit a wide range of rate constants can be studied, and transient species with half-lives of milliseconds are readily detected.

The major advantage of stripping voltammetry is the preconcentration (by factors of 100 or more) of the analyte(s) on or in the small volume of a microelectrode before a voltammetric analysis. Combined with differential pulse voltammetry in the stripping step (Fig. 14.16), solutions as dilute as $10^{-11}M$ can be analyzed.

TABLE 14.22 Selected Methods of Amperometric Titration

Substance determined	Reagent	Titration medium	Indicator electrode, V	Electrodes
Al(III)	0.6*M* NaF	50% aqueous EtOH; NaCl + 1 m*M* $FeCl_3$ (indicator)	0.00	Dme; SCE
Amides, aliphatic and aromatic	0.5*M* $Ca(ClO)_2$	20% aqueous 1,4-dioxane, 1*M* HCl	+0.40	Rotating Pt; SCE
m-Aminophenol		See *p*-Aminosalicylic acid		
p-Aminosalicylic acid	0.02*M* $KBrO_3$	0.05*M* KBr + 0.08*M* HOAc + 1*M* HCl	+0.20	Rotating Pt; SCE
As(III)	0.04*M* $Ca(ClO)_2$	0.001*M*–0.24*M* KBr + 0.17*M* $NaHCO_3$	+0.20	Rotating Pt; SCE
	0.002*M* $KBrO_3$	0.05*M* KBr + 1*M* HCl	+0.20	Vibrating Pt; SCE
As(V)		See phosphate		
$B(C_6H_5)_4$ (1−)	0.03*M*–0.1*M* $AgNO_3$	50%–100% aqueous acetonitrile	−0.10	Dme; Hg pool
Ba(II)	0.01*M* Li_2SO_4	Water, 0.05*M* Et_4NOH in EtOH	−2.00	Dme; SCE
Bi(III)	0.02*M* EDTA	HNO_3, pH 1–2	−0.18	Dme; SCE
Br^-	0.05*M* $AgNO_3$	50% aqueous acetone, 0.8*M* HNO_3	+0.05	Rotating Pt; SCE
Ca(II)	0.05*M* EDTA	0.2*M* KOH + 0.08*M* KCl + 2 m*M* K_2ZnO_2 (indicator)	−1.70	Dme; SCE
	0.01*M* EDTA	0.1*M* ammoniacal buffer, pH 11	+1.2	PbO_2; SCE
Cd(II)	EDTA	0.01*M*–0.1*M* KNO_3	−0.70	Hg pool; SCE
	0.1*M* (1,2-cyclohexanedinitrilo)tetraacetic acid	2-Propanol 2*M* with ethanolamine, 0.01*M* $LiNO_3$	−0.63	Dme; SCE
Ce(III)	0.01*M* $K_3Fe(CN)_6^{3-}$	3.6*M* K_2CO_3	0.0	Rotating Pt; SCE
Cl^-	0.001*M*–0.5*M* $AgNO_3$	50%–75% acetone, 0.8*M* HNO_3	+0.5	Rotating Pt; SCE
ClO_4^-	0.09*M* $[(C_6H_5)_4Sb]_2SO_4$	Acid or neutral medium	−0.88	Dme; SCE
Co(II)	0.01*M*–0.1*M* 1-nitroso-2-naphthol in 50%–60% HOAc	0.2*M* NaOAc + 0.1*M* HOAc	−0.60	Dme; SCE
Cr(VI)	0.001*M* $Fe(NH_4)_2SO_4$	0.5*M* H_2SO_4	−1.0	Pyrolytic graphite; SCE
Cu(II)	0.02*M* cupferron	OAc^-, citrate, phthalate, or tartrate buffer, pH 5–6	−0.50	Dme; SCE
	EDTA	0.1*M* KNO_3	−0.18	Hg pool; SCE
	0.02*M* α-benzoinoxime	0.05*M* NH_3 + 0.1*M* NH_4Cl	−0.80 or −1.70	Dme; SCE
		See also Co(II)		
Cysteine	0.005*M*–0.05*M* Cu(II)	0.05*M* NH_3 + 0.15*M* Na_2SO_3	−0.40	Rotating Pt; Hg–HgI
	0.001*M* $AgNO_3$	0.1*M* NH_3 + 0.2*M* NH_4NO_3	−0.30	Rotating Pt; Hg–HgI
Cystine	Cystine can be determined by either of the above methods after a preliminary reduction to cysteine.			
EDTA	Cu(II)	0.2*M* acetate buffer	−0.45	Rotating Pt; SCE

Fe(III)	0.03*M* cupferron	0.05*M* citrate or tartrate, pH 3–4	–0.50	Dme; SCE
	1-Nitroso-2-naphthol; see Co(II)			
Hg(II)	0.003*M* bis(2-hydroxyethyl)dithiocarbamate	0.05*M* borax, 0.1*M* KCl, pH 9.0	–0.20	Rotating Pt; SCE
I^-	0.002*M* KIO_3	1*M* H_2SO_4	+0.65	Rotating Pt; SCE
	0.05*M* $AgNO_3$	0.8*M* HNO_3	+0.05	Rotating Pt; SCE
I_2	0.005*M* $Na_2S_2O_3$	0.02*M* H_2SO_4	+0.20	Rotating Pt; SCE
IO_3^-		Add excess KI and determine as I_2		
K^+	See tetraphenylborate ion for determination after precipitation of $KB(C_6H_5)$			
Mg(II)	0.001*M*–0.01*M* EDTA	0.1*M* ammoniacal buffer, pH 11	+1.2	PbO_2; SCE
Mo(VI)	0.01*M* $Pb(NO_3)_2$	0.1*M* acetate buffer plus 0.1*M* KCl	–0.80	Dme; SCE
NH_3	0.01*M* $Ca(ClO)_2$ (or NaOBr)	0.0013*M* KBr + 0.15*M* $NaHCO_3$	+0.20	Rotating Pt; SCE
N_2H_4	0.002*M* $KBrO_3$	0.05*M* KBr + 2*M* HCl	0.0	Rotating Pt; SCE
NO_3^-	0.11*M* $(C_2H_5)_2Tl(III)\ SO_4$	0.1*M* K_2SO_4 + 0.05*M* H_2SO_4, pH 1.4–1.6	–1.00	Dme; SCE
Ni(II)	0.1*M* dimethylglyoxime in 96% EtOH	0.1*M* NH_4Cl + 0.5*M* HCl	–1.85	Dme; SCE
PO_4^{3-}	0.1*M* $UO_2(OAc)_2$ in 0.1*M* HOAc	0.1*M* KCl in 20% EtOH	–0.7	Dme; Hg pool
Pb(II)	0.05*M* $K_2Cr_2O_7$	0.01*M* KNO_3	–1.00	Dme; SCE
	0.1*M* (1,2-dyclohexanedinitrilo)tetraacetic acid	2-Propanol with 2*M* ethanolamine, 0.01*M* $LiNO_3$	–0.39	Dme; SCE
Pd(II)	0.6*M* 1,2,3-benzotriazole	0.4*M* acetate buffer	–0.50	Dme; SCE
		See also Co(II)		
	0.01*M* $Ca(ClO)_2$	0.5*M* KCl + 0.05*M* NaN_3; pH 2–3	+0.5	Rotating Pt; SCE
SCN^-	0.005*M*–0.05*M* $Ca(ClO)_2$	0.2*M* KBr + 0.16*M* $NaHCO_3$	+0.20	Rotating Pt; SCE
SO_4^{2-}	0.01*M*–0.1*M* $Pb(NO_3)_2$	20%–30% aqueous ethanol	–1.20	Dme; SCE
Sb(III)	0.002*M* $KBrO_3$	0.05*M* KBr + 1*M* HCl	+0.20	Vibrating Pt; SCE
Styrene	0.002*M* $KBrO_3$ containig 0.1*M* KBr	75% aqueous methanol; 0.6*M* HCl + 0.08*M* KBr	+0.20	Rotating Pt; SCE
Th(IV)	0.04*M* *m*-nitrophenylarsonic acid	0.01*M* HCl	–0.30	Dme; SCE
α-Tocopherol	0.01*M* $AuCl_3$	75% aqueous ethanol; 0.1*M* benzoate buffer + 0.1*M* NaCl in 75% EtOH	–0.07	Dme; SCE
U(IV)	0.1*M* $Ce(HSO_4)_4$	1*M* H_2SO_4	+1.00	Graphite; SCE
	0.02*M*–0.2*M* $FeNH_4(SO_4)_2$	0.2*M* H_2SO_4	–0.05	Rotating Pt; SCE
U(VI)	0.04*M* *m*-nitrophenylarsonic acid	0.025*M* NaOAc + 0.05*M* HOAc + 0.3*M* NaCl	–0.55	Dme; SCE
Zn(II)	Potassium tetracyano-mono-(1,10-phenanthroline)-ferrate (II)	0.1*M* NH_4Cl	–1.00	Rotating Pt; SCE
	0.1*M* (1,2-cyclohexanedinitrilo)tetraacetic acid	2-Propanol with 2*M* ethanolamine, 0.01 $LiNO_3$	–1.20	Dme; SCE
Zr(IV)	0.1*M* cupferron	2*M* H_2SO_4	–0.84	Dme; SCE
	0.04*M* *m*-nitrophenylarsonic acid	20% aqueous ethanol; 1.8*M* HCl	–0.50	Dme; SCE

TABLE 14.23 Amperometric Titrations with Two Polarizable Electrodes

Method number	Substance determined	Reagent	Applied potential, mV	Procedure and remarks	References
1	As(III)	Br_2	200	Generated from 0.1*M* H_2SO_4 + 0.2*M* KBr	*J. Am. Chem. Soc.* **70**:1047 (1948)
		Ag(II)		See method 15	
	$C_2O_4^{2-}$	Ag(II)		See method 15	
2	Ca(II)	EDTA	100	0.1*M* NH_3 buffer; Hg-plated Au electrodes	*Anal. Chem.* **31**:993 (1959)
	Ce(III)	Ag(II)		See method 15	
3	Ce(IV)	Fe(II)	100	3*M* H_2SO_4; V(V) determined with 50 mV; see also method 13	*Anal. Chim. Acta* **17**:329 (1957)
	$Cr_2O_7^{2-}$			See methods 8 and 12	
	Cu(II)	EDTA		See method 2	
	Fe(III)			See methods 10 and 13	
4	I^-	Ag(I)	230	Add sample to 25 mL buffer: 0.32*M* OAc^- + 0.027*M* HOAc; Ag anode (or Ag–Pt) cleaned in CN^-	*Anal. Chem.* **26**:627 (1954)
5		Br_2	130	Reagent generated from 2*M* HCl + 0.1*M* KBr	*Anal. Chem.* **21**:1457 (1949)
6		Ce(IV)	100	Reagent generated from 1*M* H_2SO_4 + 0.025*M* Ce(III)	*Anal. Chim. Acta* **16**:165 (1957)
7	I_2	Sn(II)	150	Reagent generated from 4*M* NaBr, 0.008*M* KI, 0.2*M* HCl, 0.2*M* $SnCl_4$; method applicable to iodometric titrations	*Anal. Chim. Acta* **20**:463 (1959)
8	IO_3^-	Cu(I)	50	Reagent generated from 3.2*M* HCl + 0.02*M* $CuSO_4$	*Anal. Chem.* **28**:1871 (1956)
9	NH_3	BrO^-	150	Borate buffer + 0.1*M* NaBr, pH 8.5; $2NH_3 + 3BrO^- \rightarrow 3Br^- + 3\ H_2O$	*Anal. Chem.* **28**:440 (1956)
10	$S_2O_3^{2-}$	I_2	135	Reagent generated from 0.1*M* HCl + 0.1*M* KBr	*Anal. Chem.* **26**:373 (1954)
11	Sb(III)	Br_2	200	See method 1	*J. Am. Chem. Soc.* **71**:2717 (1949)
12	Tl(I)	Br_2	200	See method 5	*Anal. Chem.* **24**:1195 (1952)
13	U(VI)	Ti(III)	250	Reagent generated from 7*M* H_2SO_4 + 0.6*M* $TiOSO_4$ + 10 mg $Fe(NH_4)_2(SO_4)_2$ (catalyst)	*Anal. Chim. Acta* **18**:240 (1958)
14	Unsaturated hydrocarbons	Br_2	200	Reagent generated from 30 mL glacial HOAc, 13 mL MeOH, 7 mL 1*M* KBr, and 0.1 g $Hg(OAc)_2$; titrate to predetermined constant current	*Anal. Chem.* **28**:1553 (1956)
15	V(IV)	Ag(II)	200	Reagent generated from 5*M* HNO_3 + 0.1*M* $AgNO_3$ at 0°C; Au generator anode; preoxidized Pt indicator electrodes	*Anal. Chim. Acta* **18**:245 (1958)
16	V(V)	Ti(III)	350	See method 13; see also methods 3 and 8	
17	Zn(II)	$Fe(CN)_6^{4-}$	100	To sample add slight excess EDTA, adjust to pH 2.1, add 10 drops 10% $K_3Fe(CN)_6$ solution and titrate with standard reagent	*Talanta* **2**:124 (1959)

Bibliography

Bard, A. J., and L. R. Faulkner, *Electrochemical Methods*, Wiley-Interscience, New York, 1980.

Bond, A. M., *Modern Polarographical Methods in Analytical Chemistry*, Dekker, New York, 1980.

Kissinger, P. T., and W. R. Heineman, eds., *Laboratory Techniques in Electroanalytical Chemistry*, Dekker, New York, 1984.

Koryta, J., and J. Dvorak, *Principles of Electrochemistry*, Wiley, New York, 1987.

14.6 ELECTROGRAVIMETRY

When electrodes are immersed in an electrolyte and the applied potential exceeds the cell emf, an electrochemical reaction is initiated. Oxidation takes place at the anode, and reduction occurs at the cathode. The cathode will react most completely with the strongest oxidizing agent in the solution. In a solution of ionic salts of various metals, such as silver, copper, and lead, the silver ion will be reduced first. It is the strongest oxidizing agent, and all the silver will be converted to the metal form. Copper is next in line (refer to Table 14.14), and the copper ions will be converted into metallic copper. Finally, the lead, which is the least powerful oxidant of these three, will be reduced. However, to plate out selectively each metal quantitatively before the next would commence to plate, the potential of the cathode must be controlled during the electrolysis.

14.6.1 Instrumentation

In addition to the platinum gauze anode and cathode, the equipment needed for controlled potential electrolytic separations includes an auxiliary reference electrode which is positioned as close as possible to the working electrode (either cathode or anode). The auxiliary reference-electrode–working-electrode couple is connected to a vacuum-tube voltmeter (for manual control) or a potentiostat (Sec 14.5.1.6). In manual control the potential of the working electrode is achieved by adjusting the voltage applied to the electrolytic cell (dc output voltage) by manual adjustment of the autotransformer (Variac) at the input of commercial electroanalyzers. A variety of electronic circuits are available for decreasing or increasing automatically the applied emf in order to maintain a constant working-electrode potential.

While deposition is in progress, the solution is vigorously stirred with a magnetic stirrer or the anode is rotated. When deposition is complete, the stirring is discontinued but the electrical circuit remains connected while the electrodes are slowly raised from the solution and washed thoroughly with a stream of distilled water or other appropriate wash solution.

14.6.2 Separations with Controlled Potential Electrolysis

An approximate value of the limiting potential of the working electrode can be calculated from the Nernst equation, but lack of knowledge concerning the overpotential term(s) for a system severely limits its usefulness. A more reliable method utilizes the information obtained from current–potential curves (See. 14.5) at solid electrodes.

Example 14.5 What range of cathode potentials in needed to deposit silver quantitatively (that is, lower the silver concentration to at least $10^{-6}M$) from a solution $0.0100M$ in silver nitrate?

Plating of silver will commence when the cathode potential is

$$E = 0.790 - 0.0592 \log \frac{1}{10^{-2}} = 0.672 \text{ V}$$

The removal of silver ions would be considered complete at

$$E = 0.790 - 0.0592 \log \frac{1}{10^{-6}} = 0.435 \text{ V}$$

Thus, the deposition would begin at 0.672 V and would be completed at 0.435 V, a difference of 0.237 V. When a saturated calomel electrode (0.245 V versus the standard hydrogen electrode or SHE) is used as the auxiliary reference electrode, the range would be from 0.427 to 0.190 V.

Example 14.6 Would the silver removal be complete before the copper would commence to plate from a solution 0.0100*M* in copper(II) ions?

Copper(II) ions would commence to deposit at

$$E = 0.337 - \frac{0.0592}{2} \log \frac{1}{10^{-2}} = 0.278 \text{ V} \qquad \text{(or 0.033 V versus SCE)}$$

The removal of silver would be complete at 0.190 V (versus SCE), whereas copper would start to deposit at 0.033 V versus SCE. Controlling the cathode potential somewhere in the interval between 0.2 and 0.05 V versus SCE should be satisfactory.

The current through the system steadily decreases as the deposition proceeds. However, the maximum permissible current is used at all times so the electrolysis proceeds at the maximum rate.

The evolution of hydrogen gas accompanies many electroseparations. Fortunately, the overpotential of hydrogen-gas evolution on some electrode materials is significant and delays the discharge of hydrogen (Table 14.24). Any overpotential term is added to the Nernst expression and takes the sign of the electrode.

Example 14.7 Consider the quantitative removal of cadmium from a solution 0.0100*M* in cadmium ions. To what pH must the solution be adjusted in order to prevent interference from the evolution of hydrogen gas?

In a 0.1*M* nitric acid solution, hydrogen would commence to evolve at a platinum electrode when

$$E = 0.000 - \frac{0.0592}{1} \log \frac{1}{10^{-1}} = -0.059 \text{ V}$$

However, cadmium would not commence plating until

$$E = -0.403 - \frac{0.0592}{2} \log \frac{1}{10^{-2}} = -0.444 \text{ V}$$

The separation is impossible at pH 1 and with use of platinum electrodes at which the hydrogen overpotential is only about 0.09 V. However, two alternations in the procedure will enable cadmium to be deposited. Adding acetate ions and adjusting the pH of the solution to pH 5 will shift the electrode potential at which hydrogen evolves to a more negative value.

$$E = 0.000 - \frac{0.0592}{1} \log \frac{1}{10^{-5}} = -0.296 \text{ V}$$

Second, use an electrode that has been precoated with copper because the hydrogen overpotential is about 0.4 V on copper. Now hydrogen will not interfere until the cathode potential reaches about –0.7 V. At this potential,

$$-0.7 = -0.403 - \frac{0.0592}{2} \log \frac{1}{[Cd^{2+}]}$$

the residual cadmium concentration remaining in solution is theoretically $\leq 1.4 \times 10^{-10} M$.

TABLE 14.24 Overpotentials for Common Electrode Reactions at 25°C

The overpotential is defined as the difference between the actual potential of an electrode at a given current density and the reversible electrode potential for the reaction. The overpotential required for the evolution of O_2 from dilute solutions of $HClO_4$, HNO_3, H_3PO_4 or H_2SO_4 onto smooth platinum electrodes is approximately 0.5 V.

	Current density, A/cm²					
	0.001	0.01	0.1	0.5	1.0	5.0
Electrode	Overpotential, volts					
	Liberation of H_2 from 1*M* H_2SO_4					
Ag	0.097	0.13	0.3		0.48	0.69
Al	0.3	0.83	1.00		1.29	
Au	0.017		0.1		0.24	0.33
Bi	0.39	0.4			0.78	0.98
Cd		1.13	1.22		1.25	
Co		0.2				
Cr		0.4				
Cu			0.35		0.48	0.55
Fe		0.56	0.82		1.29	
Graphite	0.002		0.32		0.60	0.73
Hg	0.8	0.93	1.03		1.07	
Ir	0.0026	0.2				
Ni	0.14	0.3			0.56	0.71
Pb	0.40	0.4			0.52	1.06
Pd	0	0.04				
Pt (smooth)	0.0000	0.16	0.29		0.68	
Pt (platinized)	0.0000	0.030	0.041		0.048	0.051
Sb		0.4				
Sn		0.5	1.2			
Ta		0.39	0.4			
Zn	0.48	0.75	1.06		1.23	
	Liberation of O_2 from 1*M* KOH					
Ag	0.58	0.73	0.98		1.13	
Au	0.67	0.96	1.24		1.63	
Cu	0.42	0.58	0.66		0.79	
Graphite	0.53	0.90	1.09		1.24	
Ni	0.35	0.52	0.73		0.85	
Pt (smooth)	0.72	0.85	1.28		1.49	
Pt (platinized)	0.40	0.52	0.64		0.77	
	Liberation of Cl_2 from saturated NaCl solution					
Graphite			0.25	0.42	0.53	
Platinized Pt	0.006		0.026	0.05		
Smooth Pt	0.008	0.03	0.054	0.161	0.236	
	Liberation of Br_2 from saturated NaBr solution					
Graphite		0.002	0.027	0.16	0.33	
Platinized Pt		0.002	0.012	0.069	0.21	
Smooth Pt		0.002	0.006*	0.26	0.38†	

(Continued)

TABLE 14.24 Overpotentials for Common Electrode Reactions at 25°C (*Continued*)

	Current density, A/cm^2					
	0.001	0.01	0.1	0.5	1.0	5.0
Electrode	Overpotential, volts					
	Liberation of I_2 from saturated NaI solution					
Graphite	0.002	0.014	0.097			
Platinized Pt		0.006	0.032		0.196	
Smooth Pt		0.003	0.03	0.12	0.22	

* At 0.23 A/cm^2.
† At 0.72 A/cm^2.
Source: J. A. Dean, ed., *Lange's Handbook of Chemistry*, 14th ed., McGraw-Hill, New York, 1992.

Actually, as soon as a layer of cadmium metal completely covers the electrode surface, the overpotential value of hydrogen gas on a cadmium surface (ca. 1.2 V) should be used to calculate the value at which hydrogen would begin to evolve.

From the foregoing examples, several facts emerge:

1. The formal potentials of two oxidation–reduction systems must be $0.178/n$ mV or greater in difference to be able to deposit the metal completely from the more positive oxidation–reduction system before the metal from the next system would commence to plate.

2. A pH change of one unit alters the potential of hydrogen evolution by 0.059 V.

3. The hydrogen overpotential on certain metal surfaces, particularly on mercury, cadmium, and zinc, can be very advantageous.

The addition of a masking agent to form a complex with a metal ion usually shifts the deposition potential in a more negative direction. Tin(II) and lead(II) have almost identical deposition potentials in a chloride medium, but oxidation of tin to the tetravalent state and then complexation of tin(IV) with tartrate ions at pH 5 enables lead to be deposited in the presence of tin. After the solution is acidified to destroy the tartrate complex of tin(IV), the tin can be deposited.

14.6.2.1 Factors Governing the Current. During controlled potential electrolysis the current decreases logarithmically with time. The decadic change in concentration of reactant is independent of the initial concentration of reactant. The concentration C (or cell current i) at any time t is a function of the initial concentration C_0 (or initial current i_0):

$$C_t = C_0 e^{-kt} \quad \text{or} \quad 2.3 \log \frac{C_0}{C_t} = kt \tag{14.50}$$

The constant k is given by $DA/V\delta$, where A is the area of the working electrode (in cm^2), D is the diffusion coefficient of the electroactive species (in cm$^2 \cdot$ min^{-1}), V is the volume of the solution (in cm^3), and δ is the thickness of the diffusion layer (in cm). From the terms comprising the constant k, several factors should be considered in order to optimize an electrodeposition:

1. Use only a volume of electrolyte that is needed to cover the electrodes; a tall conical beaker is recommended for this purpose.
2. Keep the area of the working electrode as large as possible; here a gauze electrode is superior to a perforated metal electrode.

3. Stir the solution because it increases the mass transfer of analyte up to the electrode–solution interface and thereby decreases the thickness of the diffusion layer.

4. An elevated temperature, perhaps about 60°C, usually increases the rate of mass transfer.

Electrolysis is usually discontinued when the current has diminished to 10 or 20 mA. A logarithmic plot of current versus time will enable the value of k to be calculated from the slope of the curve; then the time required to diminish the initial concentration to $10^{-6}M$ can be estimated. Table 14.25 enumerates the treatment of deposits after electrodeposition.

14.6.2.2 Electrolyte Composition. When two metals have similar discharge potentials, the addition of a masking agent will be beneficial if the discharge potential of one of the ions becomes more negative. For example, by adjusting an alkaline solution of copper and bismuth to $1M$ in cyanide ions (plus sufficient tartrate to keep bismuth in solution), copper and bismuth can be separated easily. The discharge potential of bismuth is hardly affected; however, the discharge potential of the tricyanocuprate(I) ion is shifted to −1.05 V.

The temperature dependence of homologous metal complexes sometimes provides a means for separating discharge potentials. In an ammonical solution at 20°C the discharge potentials of nickel and zinc are too close together, being −0.90 and −1.14 V, respectively. They differ markedly at 90°C, now being −0.60 and −1.05 V. In general, the less dissociated the complex at room temperature, the greater the change in the degree of dissociation as the temperature rises.

Adherence to the electrode is the most important physical characteristic of a deposit. Generally a smooth deposit and adherence are congruent. Flaky, spongy, or powdery deposits adhere only loosely to an electrode. Simultaneous evolution of a gas is often detrimental. The discharge of hydrogen gas frequently causes the film of solution in the vicinity of the cathode to become alkaline with the consequent formation of hydrous oxides or basic salts.

The chemical nature of the anions in solution often has an important influence on the physical form of the deposited metal. A pure, bright, and adherent deposit of copper is obtained by electrolyzing a nitric acid solution of copper(II) ions. By contrast, a coarse, treelike deposit of silver is obtained under similar conditions. If a suitable deposit of silver is to be obtained, the electrolysis must be carried out from a solution of $Ag(CN)_2^-$. The best deposits of iron are obtained from an oxalate complex, and those of nickel from an ammoniacal solution.

Electrogravimetric methods for the elements are outlined in Table 14.26.

14.6.2.3 Potential Buffers. Suitable oxidation–reduction systems that are preferentially reduced at the cathode (or oxidized at the anode) may be employed to limit and maintain a constant potential at the working electrode. For example, the uranium(III)–(IV) system will prevent the cathode potential from exceeding −0.5 V.

Nitrate ions have long been employed in the constant-current deposition of copper and lead dioxide to prevent the formation of metallic lead at the cathode. Nitrate ions are less easily reduced than copper ion, thus permitting the quantitative removal of copper.

The successful deposition of copper from chlorocuprate(I) ions and the prevention of the competing oxidation of these ions to copper(II) ions at the anode are due to the redox buffering action of hydrazine (or hydroxylamine). The oxidation of hydrazine ($E^0 = 0.17$ V) takes place in preference to chlorocuprate(I) ions ($E^0 = 0.51$ V); as long as hydrazine is present in excess, the anodic oxidation of chloride and copper(I) ions will be prevented.

14.6.3 The Mercury Cathode

A widely used type of electroreduction consists of a pool of mercury (35 to 50 mL) to which a constant potential is applied. The anode is a platinum wire formed into a flat spiral. Agitation is accomplished with a mechanical stirrer the impeller blades of which are only partially immersed in the mercury. The supporting electrolyte is usually a $0.1M$ to $0.5M$ solution of sulfuric acid or perchloric acid.

TABLE 14.25 Treatment of Deposits in Electrogravimetric Analysis

At the conclusion of the electrolysis, the stirrer is stopped and the electrolysis beaker is lowered away from the electrodes without breaking the electric circuit. The electrodes are washed with a stream of water from a wash bottle as they emerge from the solution. Prompt removal of the electrolyte is necessary to prevent redissolution of the deposit. (Significant losses on washing deposits of Cd, Co, Pb, Sn, and Zn cannot be prevented.) The deposit is finally washed with a volatile water-miscible solvent; then dried. Overdrying is to be avoided, as the fresh deposits are usually susceptible to air oxidation at elevated temperatures. After weighing, the deposit can usually be removed from the electrode with nitric acid.

	Treatment of deposit		
Deposit	Washing	Drying	Remarks
Ag, or Ag with known Hg	H_2O, then EtOH	100°C, 5 min	
As as Cu_3As_2 with known Cu	H_2O, then EtOH	100°C, 5 min	If unknown Cu is in sample, determine total weight of deposit, dissolve, and determine Cu separately.
Au	H_2O, then EtOH	100°C, 5 min	Remove from cathode with dilute KCN solution containing H_2O_2.
Bi	H_2O, then EtOH	110°C, 3 min	
Br as AgBr on Ag	H_2O	250°C, 1 h	Remove from Ag anode with NH_3.
Cd on Pt or on Cu–Pt	H_2O (loss), then EtOH	100°C, 5 min	Cd deposited on Cu–Pt except for cyanide electrolyte. If trace Cd only, precoat electrode with Cu, then Cd, before determination.
Cl as AgCl on Ag	H_2O	250°C, 1 h	Remove from Ag anode with NH_3.
Co	H_2O (loss), then EtOH	90°C, 5 min	New positive error (1 mg with 50 mg Co) despite washing loss. Remove from cathode with cold HNO_3 containing H_2O_2.
Cu	H_2O, then EtOH	90°C, 3 min	
Fe	H_2O, then EtOH	100°C, 3 min	
Ga with known Cu	H_2O, then EtOH	110°C, 10 min	
Hg on Au or on Ag–Pt	H_2O, then EtOH	20°C (air dry)	
In with known Cu	H_2O, then EtOH	110°C, 5 min	If In deposited in absence of Cu, dry deposit 5 min at 80°C.
Mn as hydrated Mn oxide	H_2O	Ignite at 1200°C	Ignition gives Mn_3O_4. $3[Mn]/[Mn_3O_4] = 0.7203$.
Ni	H_2O, then EtOH	80°C, 5 min	Remove from cathode with boiling HNO_3.
Pb on Cu–Pt	H_2O (loss), then EtOH	100°C, 5 min	1.5 mg washing loss on 160 cm^2 cathode.
Pb as PbO_2	H_2O	120°C, 30 min	$[Pb]/[PbO_2] = 0.8662$.
Pd	H_2O, then EtOH	100°C, 5 min	Remove from cathode with hot concentrated KCl solution containing H_2O_2.
Pt on Cu–Pt	H_2O, then EtOH	100°C, 5 min	Use thick Cu coating to protect cathode.
Rh on Pt or on Cu–Pt	H_2O		Deposit contaminated with oxygen. Reduce in H_2 at 450°C before weighing. Remove Rh from cathode by fusion with $K_2S_2O_8$.
Sb on Pt or on Cu–Pt	H_2O, then EtOH	100°C, 5 min	Remove from cathode with HCl.
Se as CuSe with known Cu	H_2O, then EtOH	100°C, 3 min	
Sn on Cu–Pt	H_2O (loss), then EtOH	100°C, 5 min	Remove from cathode with HCl.
Te	O_2-free H_2O, then EtOH	90°C, 3 min	
Te as CuTe with known Cu	H_2O, then EtOH	100°C, 3 min	
Tl on Hg–Pt	H_2O, then acetone		Dry in stream of CO_2 at room temperature.

TABLE 14.25 Treatment of Deposits in Electrogravimetric Analysis (*Continued*)

Deposit	Treatment of deposit		Remarks
	Washing	Drying	
Tl as Tl_2O_3	H_2O, then EtOH	160°C, 20 min	2 [Tl]/[Tl_2O_3] = 0.8949. Some Cu codeposits. Determine total weight of deposit, dissolve in 3 mL freshly boiling HNO_3, add 5 mL H_2SO_4, dilute to 200 mL and determine Cu.
U as U oxide	Dilute NH_4OAc	Ignite at 1200°C	3[U]/[U_3O_8] = 0.8480.
Zn on Cu–Pt	H_2O (loss), then EtOH	90°C, 5 min	

This technique finds extensive use for the removal of elements, particularly major constituents of certain samples that interfere in various methods for the minor constituents. Selectivity is poor unless an attempt is made to control the potential of the working electrode with a three-electrode potentiostat. Iron in steel and cast-iron samples is frequently removed in this way; an alternative method is the use of solvent extraction. No attempt is made to determine any of the metals deposited.

These elements are quantitatively deposited in the mercury: Ag, Au, Bi, Cd, Co, Cr, Cu, Fe, Ga, Ge, Hg, In, Ir, Mo, Ni, Pd, Po, Pt, Re, Rh, Sn, Tl, and Zn. In addition these elements are quantitatively separated from the electrolyte but not quantitatively deposited in the mercury: As, Os, Pb, Se, and tellurium. Antimony, manganese, and ruthenium are incompletely separated from the electrolyte.

14.6.4 Internal Electrolysis

Internal electrolysis employs an attackable anode that is connected directly to the cathode. In reality the arrangement is nothing but a short-circuited galvanic cell. The electrolysis proceeds spontaneously without the application of an external voltage, and the choice of an attackable anode limits the cathode potential without elaborate instrumentation or the operator's attention. The driving force is the difference between the potential of the system plating at the cathode and the dissolution of the anode. The cell resistance is a critical factor in determining the rate of metal deposition.

The selection of an anode is made with a knowledge of the reversible potentials of the various metal-ion–metal oxidation–reduction systems. A typical application is the removal of small amounts of copper and bismuth from pig lead. Because the reduction potential of lead is sufficiently far apart from the reduction potentials of the copper and bismuth systems, the anodes can be constructed from helices of pure lead wire. Dual anodes are used to provide a larger electrode area. These are inserted within a porous membrane (Alundum shell) in order to isolate them from the sample and forestall any direct plating on the lead itself. A short platinum gauze electrode is placed between the anode compartments. A stirrer is placed within the platinum gauze electrode (or a magnetic stirring bar is used). The entire assembly is placed within a 150-mL beaker. The electrolysis is begun by short-circuiting the cathode to the anode.

To keep the ohmic resistance small (and thereby increase the current flowing in the system), the anode solution must have a high concentration of electrolyte. In addition it must contain a higher concentration of the ions formed from the dissolution of the anode (in this example lead ions) than does the catholyte containing the dissolved sample. For an anolyte solution that is $1M$ in Pb^{2+} ion, the cathode potential cannot exceed –0.12 V. The only metal ions that will deposit are those with electrode potentials more positive than this value.

TABLE 14.26 Electrogravimetric Methods for the Elements

The abbreviation C.p. refers to controlled-potential electrolysis. Electrode potentials are referred to the saturated calomel electrode (SCE). Table 14.25 should be consulted for directions regarding the treatment of the deposit at the conclusion of the electrolysis.

Preparation of solution	Conditions of electrolysis: Cathode	Anode	Current, time, temperature	Codeposit or interference	Separation from
Ag					
Dissolve sample of $AgNO_3$ in 100 mL H_2O plus 2 mL HNO_3. If >0.1 g Ag, add $Hg(NO_3)_2$ solution (≈0.1 wt of Ag).	Pt	Pt	2 A, 20 min, 95°C; Ag codeposits with known Hg.	Cu, Hg (As, Au, Bi, Cd, Pb, Pt, Sb, Se, Sn, Te)	Zn
0.1–0.5 g Ag in Ag–Cu alloys. Dissolve sample in 20 mL 1:1 HNO_3, dilute to 100 mL, add 6 g $NaNO_3$.	Pt	Pt	2 A·dm^{-2}, 1 h.	As, Hg, Se, Te	Bi, Cd, Co, Cr, 0.5 g Cu, Fe, Mn, Ni, Pb, Zn
0.5 g Ag as $AgNO_3$. Dissolve in 200 mL H_2O, add 1*M* KOH solution to complete precipitation of Ag, redissolve precipitate by addition of 10% KCN solution.	Pt	Pt	2 A, 20 min.	Au, Bi, Cd, Co, Cu, Hg, Ni, Zn	Fe
0.06–0.6 g Ag in Ag solder. Dissolve sample in 6 mL 1:1 HNO_3, boil, cool, dilute to 150 mL, add 8 mL NH_3.	Pt, −0.24 V	Pt	C.p., 25 min; O_2 bubbled through solution to keep Cu oxidized and prevent colloidal Ag.	Hg	Cd, Cu, Zn
HOAc–OAc^- buffer	Pt, +0.1 V	Pt	C.p.		Cu and baser metals
As					
25 mg As as As_2O_3. Dissolve in 10 mL 1*M* NaOH; dilute to 200 mL, add 20 mL HCl plus 8 g $NH_2OH \cdot HCl$ plus 0.5 g Cu as standard $CuSO_4$ solution.	Pt	Pt	2 A · dm^{-2}, 1.5 h. As codeposits as Cu_3As_2 with known Cu.	Cu, Ag, Bi, Cd, Hg, Pb, Pt, Sb, Se, Sn, Te	As(V)
50 mg As as As_2O_3. Dissolve in 15 mL HCl, dilute to 150 mL, add 1 g $N_2H_4 \cdot 2HCl$ plus 0.25 g Cu as standard $CuSO_4$ solution.	Pt, −0.40 V	Pt	C.p., 20 min, 50°C. As codeposits as Cu_3As_2 with known Cu.	Cu, Ag, Be, Hg, Sb	As(V), Cd, Sn, Zn
Au					
Dissolve 0.1 g Au in aqua regia and evaporate to dryness twice with least excess HCl; dilute to 100 mL, add 5 mL HCl plus 1 g $NH_2OH \cdot HCl$.	Pt	Pt	2–3 A, 20 min.	Bi, Cd, Cu, Hg, Pb, Pt, Sb, Sn, NO_3^-	
Dissolve 30–60 mg Au in 5 mL HCl plus 5 mL HNO_3 with warming, add 20 mL water, boil, cool, dilute to 100 mL, neutralize with 4*M* NaOH plus 15 mL in excess.	Pt	Pt	2 A · dm^{-2}, 45 min.	Heavy metals, CN^-, Fe, Ga, In, Zn	

Dissolve 0.1 g Au in aqua regia, evaporated to dryness twice with least excess HCl, dilute to 100 mL, add 1 to 2 g KCN.	Pt	Pt	0.3–0.5 A, 30 min, 30–50°C.	Cu, Pd, Pt, Ag, Bi, Cd, Co, Hg, Ni, Zn	Fe
50 mg Au in Au–Ni, Au–Ag, Cd, Cu, Zn alloys. Dissolve sample in aqua regia, evaporate twice to dryness with least excess HCl, dilute to 120 mL, filter off any AgCl, add 2 mL HCl plus 2 g NH_4OAc.	Pt	Pt	0.7 V applied to cell (0.1 A initially), 100 min, 60°C; pass N_2 or CO_2 through solution to displace Cl_2 formed.	Nitrate	Ag, Cd, 0.06 g Cu, Pd, 0.06 g Pt, Zn
Bi					
Dissolve 0.1 g Bi in 5 mL HNO_3, add 10 mL $HClO_4$, fume, dilute to 200 mL, add 5 mL saturated $N_2H_4 \cdot H_2SO_4$ solution.	Pt	Pt	1 A, 1 h.	Ag, As, Cd, Cu, Hg, Pb, Sb, Sn	
Dissolve 0.25 g Bi in 3 mL HNO_3 or 5 mL HCl, dilute to 100 mL, add 3 g NaOH plus 2 g NaHtartrate plus 1 g $NH_2OH \cdot HCl$ (plus 3 g KCN if Cu present).	Pt, −0.75 V then −0.90 V	Pt	C.p., 75°C.	Ag, Cu, Pb	0.25 g Cu, Sb
Dissolve 0.1 to 0.4 g Bi in 10 mL 1:2 HNO_3 (or 5 mL HCl plus minimum HNO_3 if Sb or Sn present), dilute to 200 mL, add 1 g urea plus 12 g Na_2tartrate plus 2 g $N_2H_4 \cdot 2HCl$ plus 1 g succinic acid; adjust to pH 5.9.	Pt, −0.40 V	Pt	C.p.	Ag, Cu, Hg (can be separated from Bi at −0.30 V)	0.1 g Al, Cd, 0.04 g Fe, Mg, Mn, 0.1 g Pb, Sb, 0.2 g Sn, Zn
0.1–0.3 g Bi in Bi–Pb, Sn alloys. Dissolve 0.4 g sample in 1–2 mL HNO_3, add 10 mL HCl, boil, add 5 mL more HCl, dilute to 100 mL, add 5 g oxalic acid plus 0.5 g $N_2H_4 \cdot 2HCl$.	Pt, −0.15 V then up to −0.30 V	Pt	C.p., 85°C.		Pb, Sn and baser metals
Br					
30–300 mg Br as NaBr in 100 mL containing 2.7 g NaOAc, adjust to pH 4.7 with HOAc.	Pt	Ag, −0.22 V	C.p., AgBr deposited on anode.	Cl, I	
Cd					
Dissolve 0.25 g Cd in 10 mL 2.5*M* H_2SO_4 or 5 mL $HClO_4$, dilute to 200 mL, add 10 mL 0.1% gelatin.	Cu/Pt	Pt	3 A, 40 min.	Cu, Ag, Au, Bi, Hg, Pb, Pt, Sb, Sn	0.5 g $NiSO_4$, 0.5 g $ZnSO_4$
0.1–0.4 g Cd in metal alloys. Dissolve sample in 3 mL H_2O plus 5 mL $HClO_4$	Pt, −0.40 for 15 min,	Pt	C.p.	Ag, Bi, Cu, Pb (can be separated from	Zn can be deposited at −1.45 V.

(Continued)

TABLE 14.26 Electrogravimetric Methods for the Elements (*Continued*)

Preparation of solution	Conditions of electrolysis: Cathode	Anode	Current, time, temperature	Codeposit or interference	Separation from
Cd					
plus 5 mL HNO_3, evaporate almost to dryness, take up in 75 mL H_2O, add 5 g Na_2tartrate plus 1.5 g $N_2H_4 \cdot 2HCl$ plus 20 mL 0.1% gelatin. Adjust to pH 4.5–5.0 with NH_3, dilute to 175 mL, add 5–10 mg standard Cu solution.	then −1.00 to −1.15 V			Cd at −0.60 V)	
HOAc–OAc^- buffer, pH 4	Pt, −0.80 V	Pt	C.p.		Zn
Cl					
30–150 mg Cl in 100 mL containing 2.7 g NaOAc, adjust to pH 4.7 with HOAc.	Pt	Ag, −0.25 V	C.p., AgCl deposited on anode.	Br, I	
Co					
Dissolve $CoSO_4$ in 100 mL H_2O containing 15 mL NH_3 plus 3 g NH_4Cl plus 0.1 g $NH_2OH \cdot HCl$.	Pt	Pt	2–5 $A \cdot dm^{-2}$, 45 min.	Cu, Ni, Zn, Pd, Tl	
Cu					
Cu in brass, bronze, and Cu–Sn alloys. Dissolve 1 g sample in 20 mL 1:1 HNO_3 without heating, add 10 mL H_3PO_4, boil 2 min, dilute to 200 mL, add 2 drops 0.1*M* HCl.	Pt	Pt	2 A, 1.5 h.	Ag, Bi	Sb, Sn, Zn
Dissolve 0.3 g Cu in 12 mL HNO_3, boil, dilute to 150 mL, add 10 mL NH_3 plus 2 g N_2H_4.	Pt, −0.73 V	Pt	C.p., purge O_2 by bubbling N_2 through solution.	Ag, Hg (both separated from Cu at −0.24 with O_2 present).	Cd, Zn
Dissolve 0.1–0.4 g Cu in 10 mL 1:2 HNO_3 (if Sb or Sn present, dissolve in 5 mL HCl plus minimum HNO_3), dilute to 200 mL, add 1 g urea plus 12 g Na_2 tartrate plus 2 g $N_2H_4 \cdot 2HCl$ plus 1 g succinic acid; adjust pH to 5.9.	Pt, −0.30 V	Pt	C.p.	Ag, Hg	0.1 g Al, 0.4 g Bi, Cd, 0.04 g Fe, Mg, Mn, 0.1 g Pb, 0.2 g Sn, Zn
Cu in brass, bronze, white metals. Dissolve in 5 mL HNO_3 plus 5 mL HCl, add 10 mL H_3PO_4, evaporate to remove HCl and HNO_3, dilute to 120 mL.	Pt, −0.35 V	Pt	C.p.	Ag, Bi	Cd, Cr, Fe, Mn, Ni, Pb, Sb, Sn, Zn
3–30 mg Cu in Fe. Dissolve sample in 1.5 mL H_2SO_4, dilute to 150 mL, add 0.2 g $N_2H_4 \cdot H_2SO_4$.	Pt, −0.40 V	Pt	Divided cell to prevent cyclic $Fe^{2+} \rightleftarrows Fe^{3+}$. Analyte: 100 mL containing 10 g Na_2SO_4 and 1 mL H_2SO_4.		10 g Fe

Fe					
0.05 g Fe. Dissolve in 100 mL H_2O containing 1 mL H_2SO_4, add 5 g ammonium oxalate.	Pt	Pt	6 A, 30 min, deposition quantitative at pH 4–9.	Co, Cu, Mn, Ni	
Ga					
0.2 g Ga. Dissolve in 100 mL H_2O, add 10 mL NaOH plus 5 g $(NH_4)_2SO_4$ plus 0.04 g standard Cu.	Pt	Pt	4–5 A, 70°C. Ga codeposits with known Cu.	Cu, Co, Fe, In, Ni, Pd, Tl, Zn	
Hg					
0.3 g Hg dissolved in 10 mL HNO_3 or H_2SO_4. Add 90 mL H_2O.	Au	Pt	1 A, 45 min.	Ag, Au, Bi, Cd, Cu, Pb, Pt, Sb, Sn	
In					
0.02 g In in salt; dissolve in 100 mL water, add 0.04 g Cu (standard solution). Make alkaline with NH_3, then add HCOOH to discharge blue color and redissolve precipitate.	Pt	Pt	4–5 A, In codeposits with known Cu.	Co, Cu, Fe, Ni, Pd, Tl, Zn	
Mn					
0.2 Mn in 100 mL water containing 5 mL 77% HCOOH plus 1 g NaOOCH.	Pt	Pt dish	1.4 A · dm^{-2}, 1.5 h unstirred. Hydrated Mn oxide deposits on anode.	Pb, Tl	
Ni					
Fume sample (0.1–0.3 g Ni) with 8 mL H_2SO_4, cool, dilute, neutralize with NH_3, cool, add 35 mL NH_3, dilute to 200 mL.	Cu/Pt	Pt	0.5 A, 3 h. If Co is present, add 2 g $NaHSO_3$ to ensure its quantitative codeposition.	Ag, As, Co, Cu, Mo, Zn, nitrate	
To dissolved sample (0.15 g Ni) in 100 mL, add 5 g Na_2SO_4 plus 30 mL NH_3 plus 3 g Na_2SO_3. Dilute to 200 mL.	Cu/Pt, −0.95 V	Pt	C.p., 90°C.	Cu, Co, Fe, In, Pd, Tl	0.15 g Zn
Ammoniacal tartrate plus Na_2SO_3.	Pt, −1.10 V	Pt	C.p.		Al, Fe, Zn
Pb					
0.1–0.3 g Pb in metallurgical material. Dissolve in 3 mL H_2O plus 5 mL HNO_3 plus 5 mL $HClO_4$. Evaporate almost to dryness, add 75 mL H_2O, 5 g Na_2tartrate plus 1.5 g $N_2H_4 \cdot 2HCl$ plus 20 mL 0.1% gelatin. Adjust pH to 4.5–5.0 with NH_3.	Cu/Pt, −0.60 V	Pt	C.p. Use cathode precoated with Cu.	Ag, Bi, Cu, Hg	Al, Cd, Fe, Mg, Mn, Ni, Sn, Zn
60 mg Pb as $Pb(NO_3)_2$ in 100 mL containing 5 to 9 mL HNO_3.	Pt	Pt	2.4 A, 1.5 h, 80°C. PbO_2 deposited on anode.	>25 mg Fe, Mn, Tl	<25 mg Fe, 0.75 g Zn

(Continued)

TABLE 14.26 Electrogravimetric Methods for the Elements (*Continued*)

Preparation of solution	Conditions of electrolysis: Cathode	Anode	Current, time, temperature	Codeposit or interference	Separation from
Pb					
8–60 mg Pb as $Pb(NO_3)_2$ in 100 mL containing 5 to 10 mL HNO_3 plus 0.08 g Cu (as $CuSO_4$ solution).	Pt	Pt	2 A · dm^{-2}, 1 h, 15°C. PbO_2 deposited on anode.	Cl^-, Tl	60 mg Mn
Pb–Sn aloys. Dissolve sample in 30 mL HNO_3 plus 5 mL HF, dilute to 125 mL, add $K_2Cr_2O_7$ solution until yellow color is apparent.	Pt	Pt	2 A, 20 min, PbO_2 deposited on anode.	Sn	Cu, Sb
Pd					
0.06 g Pd as $PdCl_2$ in 100 mL containing 2 mL HCl plus 2 g NH_4OAc.	Pt	Pt	1.25 V applied to cell (0.02 A initially), 3 h, 60°C.		
Pt					
0.05 g Pt as H_2PtCl_6 in 100 mL containing 2 mL HCl plus 2 g NH_4OAc.	Pt	Pt	1.4 V applied to cell (0.02 A initially), 3 h, 60°C.		
Rh					
0.06 g Rh as Na_2RhCl_6. Add 2.5 mL 1:10 H_2SO_4 and dilute to 100 mL with boiling water.	Ag/Pt dish	Pt	8 A, 15 min.	Ag, Au, Bi, Cd, Cu, Hg, Pb, Pt	Zn
15–30 mg Rh as chloride salt. Evaporate sample to dryness once with 2 mL HNO_3, twice with 2 mL HCl. Add 15 mL H_2O plus 5 mL 1:10 HC1; bubble in Cl_2 for 5 min. Heat to 50°C for 5 min, cool, add 18.7 g NH_4Cl, dilute to 50 mL, bubble in Cl_2 for 5 min. Add 1 g $NH_2OH \cdot HCl$ and 1 g $(NH_4)_2SO_4$, dilute to 100 mL.	Pt, –0.25 V then –0.40 V	Pt	C.p. Rh deposit contaminated with O.		60 mg Ir as chloride salt
Sb					
0.4 g Sb. Dissolve in 20 mL HCl, dilute to 200 mL, add 2 g $NH_2OH \cdot HCl$.	Cu/Pt, –0.40 V	Pt	C.p., 50°C	Bi, Cu	0.5 g Sn
1:1 H_2SO_4 plus $N_2H_4 \cdot H_2SO_4$.	Pt, –0.07 V then –0.22 V	Pt			Sn
Se					
0.05–0.2 g Se. Dissolve in 15 mL 1:1 HNO_3, add 0.5 g Cu (known), boil, cool, dilute to 190 mL, add 1 drop 0.1*M* HCl.	Pt	Pt	2 A · dm^{-2}, 1.5 h. CuSe deposits with Cu from Se(IV) solution.	Cu, Te(IV), Ag, As, Au, Bi, Cd, Hg, Pb, Pt, Sb, Sn, Tl	Se(VI), Zn

Sn					
Dissolve sample (0.1–0.3 g as Sn) in 15 mL HNO_3 plus 10 mL H_2SO_4 plus 5 mL $HClO_4$, fume, dilute to 200 mL, add 5 mL HCl plus 5 g $NH_2OH \cdot HCl$.	Cu/Pt	Pt	$2\ A \cdot dm^{-2}$.	Ag, As, Au, Bi, Cd, Cu, Hg, Pb, Pt, Sb, Tl	Zn
Dissolve 0.1 to 0.3 g [as Sn(II) or (IV)] in 12 mL HCl, dilute to 200 mL, add 2 g $N_2H_4 \cdot H_2SO_4$.	Cu/Pt, –0.60 to –0.75 V	Pt	C.p., 45°C.	>2 g NH_4 salts, Bi, Cu, Sb	
Sn					
Dissolve sample (0.2 g Sn) in 10 mL HCl with minimum HNO_3, dilute to 200 mL, add 12 g Na_2 tartrate plus 2 g $N_2H_4 \cdot 2HCl$ plus 1 g succinic acid. Adjust to pH 5.9 with NaOH. Add 10 mg (known) Cu. Deposit Cu at –0.40 V for 15 min, then add 20 mL HCl.	Pt, –0.60 V	Pt	C.p.	Bi, Cu, Pb (see separation from Sn at pH 4.5 to 5.0), Ag, Hg, Sb	Cd, Zn
Te					
Dissolve 0.5 g Cu (known) in 15 mL 1:1 HNO_3, add Te (5 to 15 mg), boil, cool, dilute to 100 mL, add 1 drop 0.1*M* HCl.	Pt	Pt	$3.5\ A \cdot dm^{-2}$, 1 h. CuTe codeposits with Cu from Te(IV) solution.	Cu, Se(IV), Ag, As, Au, Bi, Cd, Hg, Pb, Pt, Sb, Sn, Tl	Te(VI), Zn
Tl					
0.02–0.2 g Tl as Tl(I). Dissolve in 5 mL HNO_3, dilute to 175 mL, add 1 g benzoic acid.	Hg/Pt	Pt	5 A, 15 min, 45°C.	Ag, As, Au, Bi, Cd, Hg, Pb, Pt, Sb, Sn	Zn
0.002–0.2 g Tl(I or III). Dissolve in 20 mL HNO_3, add 0.2 g $Cu(NO_3)_2$ in 100 mL H_2O (cathode depolarizer). Make alkaline with NH_3, add 3 mL excess for 0.002 g and 10 mL excess for 0.2 g.	Pt	Pt	2 A, 1h. Tl_2O_3 deposits on anode. Some Cu is codeposited with Tl_2O_3.	As, Bi, Cu, Mn, Pb, 3 mg Sb, Cl^-, PO_4^{3-}	
U					
0.2 g U as U(VI). Dissolve in 100 mL H_2O containing 4 g NH_4OAc. Adjust pH to 6–7.	Pt dish	Pt	$0.2\ A \cdot dm^{-2}$, 90°C. Electrolyze to negative $K_4[Fe(CN)_6]$ test (2 h).	Mo	
Zn					
Dissolve 0.2 g Zn in 10 mL 3*M* H_2SO_4, dilute to 200 mL, add 1.5 g citric acid, adjust to pH 4–5 with NaOH.	Cu–Pt	Pt	$1\ A \cdot dm^{-2}$, 2 h.	Ag, As, Bi, Cd, Co, Cu, Fe, Hg, Mn, Ni, Sb, NO_3^-	Tl

(Continued)

TABLE 14.26 Electrogravimetric Methods for the Elements (*Continued*)

Preparation of solution	Conditions of electrolysis			Codeposit or Interference	Separation from
	Cathode	Anode	Current, time, temperature		
Zn					
0.4 g Zn as $Zn(SO_4)$. Dissolve in 50 mL H_2O, add NH_3 or NaOH to incipient precipitation, then 1 g KCN plus 20 mL NH_3, dilute to 125 mL.	Cu–Pt	Pt	3 A, 30 min.	Ag, Au, Bi, Cd, Co, Cu, Hg, Ni	Fe
0.15 g Zn as $ZnSO_4$. Dissolve in 50 mL H_2O, add 1 g KNa tartrate, then KOH to just redissolve precipitate.	Cu–Pt	Pt	0.3 A, 45 min.	Bi, Cu, Fe, Pb, Sn	0.15 g Ni as $NiSO_4$, Co, Sb
Dissolve 0.1–0.35 g Zn in 10 mL HCl plus 5 mL H_2O, add 10 mL 12*M* NaOH, cool, add 14 g NaCl, dilute to 500 mL. Add 14.3 g Na_2 tartrate plus 1 g succinic acid plus 2 g urea, plus 3 g $N_2H_4 \cdot 2HCl$; adjust to pH 9 with NH_3.	Cu–Pt, C.p.	Pt	−1.40 V until current becomes constant (60 to 100 mA, 40 min), then −1.50 V for 20 min.	Bi, Cu, Fe, Pb, Sn	Al, Mn, Co, Ni, Sb

TABLE 14.27 Determinations by Internal Electrolysis

Element	Cathode solution	Anode system	Separation from
Ag	HNO_3	Cu metal plus $Cu(NO_3)_2$	Bi, Cu, Pb
Ag	Ammoniacal solution plus Na_2SO_3	Cu metal plus $Cu(NH_3)_4^{2+}$, ammoniacal solution	Cu
Ag	H_2SO_4 plus $N_2H_4 \cdot H_2SO_4$	Cu metal plus $CuSO_4$	Cu, Fe, Ni, Zn
Bi	H_2SO_4	Pt metal in V(II)–V(III)	
Bi	HNO_3	Zn metal plus $ZnCl_2$	
Cd	HOAc–NaOAc buffer plus NH_4Cl plus $N_2H_4 \cdot HCl$	Zn metal plus $ZnCl_2$	Zn
Co	HOAc–NaOAc buffer	Mg metal plus NH_4Cl and HCl	
Cu	H_2SO_4 plus $N_2H_4 \cdot H_2SO_4$	Fe metal plus $FeSO_4$	Fe
Cu	H_2SO_4	Zn metal plus $ZnCl_2$	Fe, Zn
Cu	H_2SO_4 plus HNO_3	Mg metal plus $MgCl_2$ and HCl	
Cu + Bi	HNO_3 plus tartrate buffer plus NH_2OH	Pb metal plus $Pb(NO_3)_2$	Pb, Sb
Cu + Bi	HNO_3 plus HF plus tartate buffer	Pb metal plus $Pb(NO_3)_2$	As, Cd, Fe, Pb, Sb, Sn, Te
Hg	HNO_3	Zn metal plus $Zn(NO_3)_2$	Zn
Hg	H_2SO_4	Cu metal plus $CuSo_4$	Cu, Zn
In	Tartrate buffer, pH 3.6	Zn metal plus $ZnCl_2$	Zn
Ni	Acetate buffer	Mg metal plus $MgSO_4$ and $(NH_4)_2SO_4$ or HCl and NH_4Cl	
Pb	HCl plus 0.1% gelatin, or acetate buffer	Zn metal plus $Zn(NO_3)_2$	Zn
Sb	HCl	Zn metal plus $ZnCl_2$	
Sb	H_2SO_4 plus tartrate buffer	Mg metal plus $MgSO_4$ or $(NH_4)_2SO_4$	
Sn	HCl	Zn metal plus $ZnCl_2$ or Mg metal plus $MgCl_2$	
Zn	HOAc–acetate buffer	Mg metal plus NH_4Cl and HCl	

The anode need not always be constructed of the material that constitutes the matrix of the sample. For selective reduction of several trace constituents in zinc, for example, four separate samples would be dissolved for the separation of traces of silver, copper, lead, and cadmium. In the first, an attackable anode of copper would permit the complete removal of silver. Similarly, a lead anode would make it possible to remove silver plus copper; a cadmium electrode would remove silver, copper, and lead; a zinc anode would remove all four elements.

The amount of deposit is generally limited to quantities not exceeding 25 mg. With larger quantities the deposit is apt to be spongy and there is danger that some of the metal ions may diffuse to the anode. Little attention is required during an electrolysis except to flush the anolyte compartment once or twice. Average running time is 30 min per sample.

Table 14.27 states the experimental conditions for the internal electrolysis of several metals.

14.7 ELECTROGRAPHY

14.7.1 Principle

Electrography is used mainly to identify anions and cations capable of being released from specimen surfaces by controlled electrolysis into paper or other porous media and to map the distribution of these ions.

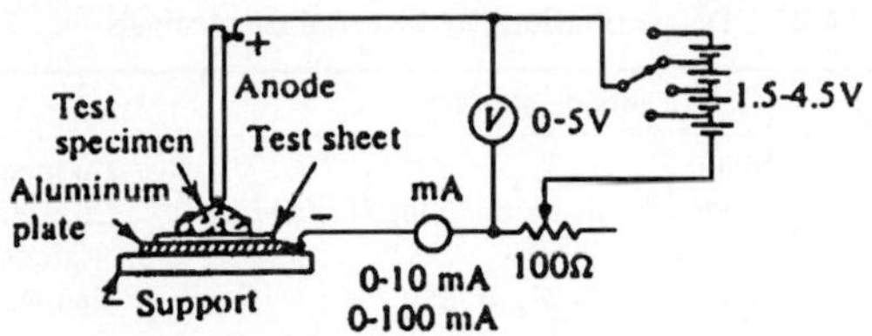

FIGURE 14.19 Schematic arrangement of equipment and electrical circuit for electrographic analysis.

Basically, the electrographic arrangement is simple. It consists of two metal surfaces between which is sandwiched a layer of absorbent paper (or for better rendition, gelatin-coated paper) moistened with electrolyte. The working electrode may be a flat square electrode for flat surfaces, a long narrow electrode for use on metal ribbons, or sponge rubber covered with aluminum foil for uneven sample surfaces. Pressure is applied to the surface to ensure intimate contact. A general laboratory circuit is shown in Fig. 14.19. If, as is most common, the specimen is the anode, its ions move into the paper where they react, either with the ions of the electrolyte or with an added reagent, to produce an identifiable color product. When anions in thin conducting films, such as chloride or sulfide, are to be identified, the specimen is made cathodic.

The current source may be a 12-V storage battery or a 7.5-V C battery. The flat plate in most cases may be aluminum or stainless steel. The second electrode connection may be a flexible cord or cords terminating in any of a variety of clips and probes suitable for making contact on either flat or irregularly shaped specimens. In general, 50 μg of most metals will produce brilliantly colored products if the reaction is confined to an area of 1 cm^2. These conditions require a current of 15 mA and an exposure time of 10 s.

The electrographic method can be applied for the inspection of lacquer coating and of plated metals for pinholes and cracks in their surface. It can be used for many alloy identifications, such as the differentiation of lead-containing brass from ordinary brass, nickel in steel, and the distribution of metal constituents within an alloy. In the biological field the method is applicable to the localization of constituents that are normally present within the tissue in an ionic state. Portable field kits have found extensive use in inspection and sorting work in the stockroom and in mineralogical field work.

Analogous to the anodic oxidation transfer is the cathodic reduction of certain anions of tarnish or corrosion films on metals. These are often tied up as basic insoluble salts and are not detectable in simple contact printing.

14.7.2 Procedure

A general outline for producing an electrographic print is as follows:

1. Select the printing condition.
   a. *Standard pad.* Use a hardened filter paper similar to Whatman 50 or S&S 576, and a thick, soft backing paper such as Eastman Kodak blotting paper. Gelatin paper or imbibition paper similar to Eastman Kodak transfer paper is sometimes used instead of filter paper.
   b. *Reagent papers.* Prepare by immersing hardened filter paper in the first specified solution rapidly and uniformly. Follow by drying. Then immerse in the second solution uniformly, wash well with running water, and dry. If gelatin paper is used, immerse and wash for longer periods of time. The preparation of fixed reagent papers along with the color of reaction products for 12 metals is described in Table 14.28; analogous reagent papers for three anions and printing conditions are described in Table 14.29.
2. Immerse the printing medium in electrolyte. Table 14.30 lists some electrolytes and their uses.

TABLE 14.28 Preparation of Fixed Reagent Papers for Metals

Prints are to be prepared by using anodic specimens and current densities of 25 mA·cm^{-2} for 10–30 s. The following abbreviations are used for the colors of the papers and of the reaction products:

Bk	*black*	*Dk*	*dark*	*Lt*	*light*	*Wh*	*white*
Bl	*blue*	*Fl*	*fluorescence*	*Or*	*orange*	*Wk*	*weak*
Bn	*brown*	*Gn*	*green*	*Rd*	*red*	*Yl*	*yellow*
Br	*bright*	*Gy*	*gray*	*Vi*	*violet*		

Electrolyte abbreviations are A, 0.5M Na_2CO_3 + 0.5M NaN_3, 3 : 1 (v/v); or 0.5M Na_2CO_3 + 0.5M NaCl, 3 : 1 (v/v) where Co or Fe predominate; and a, 0.25M ammonium citrate.

Reagent paper	Preparation	Electrolyte	Color of reaction product with Ag	Al	Bi	Cd	Co	Cu	Fe	Mn	Ni	Pb	Sn	Zn
Antimony sulfide (orange)	2% Na sulfantimoniate; then dilute HCl	A	Bk		Bn			Bk		*		Bn		
α-Benzoinoxime	1% solution in EtOH	A†	*					Gn		*				
Cd diethyldithiocarbamate	1% solution in acetone	a	*					Lt Bn		*				
Cadmium sulfide (yellow)	0.25*M* $Cd(OAc)_2$, then 0.25*M* Na_2S	A	Bk		Bn			Bk		*		Bn		
Cinchonine	1% solution in EtOH	KI‡	Yl		Or		Or	Bn		*	Or	Yl		
Dimethylglyoxime	1% solution in EtOH	A§	*							*	Rd			
Morin¶ (yellow-white)	0.5% solution in EtOH		*	Br Fl	Wk Fl	Wk Fl	Bn	Lt Yl	Bn	*		Wk Fl	Wk Fl	Br Fl
Salicylic acid	2% solution in EtOH	A		Fl			Gn	Gn Yl	Vi					Fl
$Zn_2[Fe(CN)_6]$	0.25*M* $Zn(OAc)_2$, then 0.25*M* $K_4Fe(CN)_6$	A						Rd Br	Bl	Bn	Lt Gn			
Zinc sulfide	0.25*M* $Zn(OAc)_2$, then 0.25*M* Na_2S	A	Bk		Bn	Yl		Bk		*		Bn		
Zinc xanthate	0.1*M* $Zn(OAc)_2$, then 0.1*M* K xanthate	A	*				Gy Gn	Yl	Bn	Vi	Or			

* Dark spot due to reduction of Ag or Mo.
† Wash print with dilute HOAc.
‡ Bleach print with H_2SO_3.
§ Wash print with ammoniacal ammonium citrate solution.
¶ Fluorescent reagent; examine under ultraviolet light.

TABLE 14.29 Preparation of Fixed Reagent Papers and Their Reaction with Anions

				Printing conditions (cathodic reduction)				
Anion	Reagent paper	Preparations	Paper color	Pad	Electrolyte	Current density, $mA \cdot cm^{-2}$	Time, s	Print color
Cl^-	Ag_2CrO_4	0.1*M* $AgNO_3$, then 0.1*M* Na_2CrO_4	Red	Standard plus veiling	0.5*M* $Mg(OAc)_2$	5–10	10–60	Red is bleached
S^{2-}	$PbCO_3$	0.1*M* $Pb(OAc)_2$, then 0.1*M* Na_2CO_3	White	Standard	0.5*M* Na_2CO_3	10–25	10–60	Black to brown
SO_4^{2-}	Ba rhodizonate	0.5*M* Na rhodizonate, then 0.25*M* $Ba(OAc)_2$	Orange-red	Standard plus veiling	0.5*M* NH_4OAc	10–20	60–120	Orange-red is bleached

TABLE 14.30 Composition of Electrolytes for Electrography

Electrolyte	Composition	Principal use
A	0.5*M* Na_2CO_3 plus 0.5*M* $NaNO_3$, 3 : 1 v/v	Fixing electrolyte for latent prints
	0.5*M* Na_2CO_3 plus 0.5*M* NaCl, 3 : 1 v/v	If iron or cobalt predominates
B	0.5*M* Na_2CO_3 plus 0.5*M* Na_2SO_4, 2 : 1 v/v	Fixing electrolyte to remove interference of lead by masking
C	0.5*M* $Ba(OAc)_2$	Solution of iron(II) alloys, fixation of chromate ion
D	0.5*M* NaK tartrate	Solution of iron(II) alloys where reactions of iron are to be masked
E	0.5*M* $NaNO_3$	For tin

3. Drain excess electrolyte from the printing medium, blot, and place between the test specimen and the other electrode. Apply pressure to ensure adequate contact.
4. Connect the circuit and apply the voltage needed for the length of time specified in the tables.
5. Release the pressure, wash in running water, and inspect the print. If a colored product is not obtained directly by the migration of ions from the specimen surface into the printing medium, develop the print with other reagents as directed in the tables. Direct electrographic metal prints and confirming tests for seven common metals are described in Table 14.31. In Table 14.32 the schematic examination of a single print of a pure metal surface for nine elements is outlined.

Table 14.33 summarizes the electrographic tests for metals in aluminum, copper, and iron alloys. The table also contains the printing conditions recommended in these electrographic tests. The development procedures are described in detail in Table 14.34. Finally, the electrographic reactions of minerals is summarized in Table 14.35.

A brief discussion of electrography may be found in Ref. 12. A thorough treatment of electrography by these same authors is given in Ref. 13.

[12] H. W. Hermance and H. V. Wadlow, in *Standard Methods of Chemical Analysis*, 6th ed., F. J. Welchor, ed., Van Nostrand, New York, 1966, Vol. 3, Pt. A, pp. 500–520.

[13] H. W. Hermance and H. V. Wadlow, in *Physical Methods in Chemical Analysis*, G. Berl, ed., Academic, New York, 1951, Vol. II, pp. 155–228.

TABLE 14.31 Direct Electrographic Metal Prints and Confirming Tests

		Color of confirming test		
		Fuming with		
Metal	Color with electrolyte A* and standard pad	NH_3	HCl	Heat and light
Ag	Colorless			Brown to black
Co	Dirty brown	Brown	Light blue	HCl blue deepens
Cr	Yellow (CrO_4^{2-})	Yellow	Yellow	Yellow
Cu	Greenish blue	Deep blue	Green-yellow	Green-blue
Fe	Brown	Brown	Orange-yellow	Brown
Mo	Deep blue-violet	Gray	Gray	Gray
Ni	Light green	Light violet	Green	Light green

* See Table 14.30.

TABLE 14.32 Schematic Examination of a Single Print of a Pure Metal Surface for Nine Common Elements

Prepare a print using the standard pad, electrolyte A, and a current density of 25 mA · cm^{-2} for 10 s. Hold the print over concentrated aqueous NH_3 until it is thoroughly permeated. Classify by color. If group II or III metals are present, cut the print into at least three parts; use two of these for the group treatments and the other for confirmatory tests. In the following outline the symbol of each element is set in boldface type at the point where the element may be positively identified.

Group I (strong colored)	Group II (weakly colored)	Group III (colorless)
Clear yellow: **Cr** (as CrO_4^{2-}); confirm with 1,5-diphenylcarbohydrazide.	Blue-green: Cu, Ni.	Cd, Pb, Sn, Zn.
Light brown: **Fe** [as $Fe(OH)_3$]; confirm with $K_4[Fe(CN)_6]$.	Gray after exposure to light: **Ag**.	

Group II	Group III
Immerse in warm photographic developer (e.g., Eastman D-76), wash well with 1% Na_2CO_3, then H_2O.	Immerse in solution containing 1 g Na_2S plus 2 g NH_4OAc in 100 mL H_2O, wash in suction apparatus with 1% HOAc until all sulfide is removed.
Gray to black: **Ag**; confirm with $Cr_2O_7^{2-}$, giving red Ag_2CrO_4.	Yellow: **Cd**; colorless: Zn; brown: Sn; brown-black: Pb.
Blot and immerse print in 1% K ethylxanthate for 60 s, wash thoroughly and note color.	If print is colorless, spot with 5% $Pb(NO_3)_2$ or 2% $AgNO_3$. Black: **Zn**.
Bright yellow: **Cu**, confirm with α-benzoinoxime, giving green color.	If print is suspected to contain Pb or Sn, immerse in solution containing 5 g NaOH plus 5 g Na_2S plus 1 g S in 25 mL H_2O, then wash in suction apparatus with several small portions of this reagent, and then wash with H_2O.
Orange: **Ni**, confirm with dimethylglyoxime (red color).	Brown color disappears: **Sn**.
If results are negative, proceed with group III.	Brown to black color remains: **Pb**, confirm with I^- or CrO_4^{2-}.

14.8 COULOMETRIC METHODS

The major advantage of coulometric methods accrues from the fact that standardized reagent solutions are unnecessary. Titration with electrons eliminates the laborious preparation of primary standard materials. The analyst needs only to obtain a good-quality electrical supply and timer plus small platinum electrodes (occasionally an attackable silver electrode). It is essential that every electron passing through the generating circuit be effective in producing the desired reagent. There must be no side reaction, nothing to reduce the current efficiency from 100%. With the exception of the very different means employed for adding the reagent and for measuring the quantity of reagent used, the coulometric titration method differs but little in either theory or practice from conventional titration methods. The measurement of current and time, and hence coulombs, can be made with extremely high accuracy and precision. The method is applicable in the range from milligram quantities down to microgram quantities. Sensitivity is usually limited only by problems of end-point detection. The minimum amount of material that can be titrated successfully is determined not by difficulties in regulating and measuring small electrolysis current but rather by difficulties encountered in connection with the location of the equivalence point and by side reactions resulting from impurities in the reagents or in the sample. The end-point detection system need not be different from that used in volumetric titrimetry. It may be photometric, potentiometric, or amperometric, to name the most widely used approaches.

TABLE 14.33 Electrographic Tests for Metals in Aluminum, Copper, and Iron Alloys

This table summarizes the printing conditions recommended in electrographic tests for some common constituents in various alloys. The first column lists the alloys in alphabetical order, the second lists the elements to be detected. The third column lists the elements that interfere in the procedure, but that are not normally present. The "electrolyte" column refers to Table 14.30. The last column gives the code number used in Table 14.34 to identify the recommended procedure for the development and interpretation of the print obtained.

			Printing conditions				
Alloy	Metal sought	Interferences	Pad	Elec-trolyte	Current density, $mA \cdot cm^{-2}$	Time, s	Development procedure number
Al-based	Cu		a	D	20	60	5
	Fe		std	A	20 to 30	60–120	6
	Mn		std	A	15	60	7
	Zn		std	A	15 to 20	60	11
Cu-based Al bronzes	Al	Be, Bi, Fe, Pb, Sn	std	A	20 to 25	60	1
	Fe		std	A	20 to 30	60–120	6
Be alloys	Be	Al, Bi, Pb, Sn	std	A	20 to 25	60	3
Brasses and bronzes	Fe		std	A	20 to 30	60–120	6
	Pb		std	A	15 to 20	60	9
	Sn		b	E	15 to 20	60	10
	Zn	Bi	std	A	15 to 20	60	11
Constantin, Ni brass, Ni coinage	Al	Be, Bi, Fe, Pb, Sn	std	A	20 to 25	30–60	2
	Mn	Same	std	B	20 to 25	60	7
	Ni	Same	c	A	20 to 25	60	8
	Zn	Same	std	A	15 to 20	30	11
Mn bronzes, manganin	Mn	Pb	std	B	20 to 25	60	7
	Ni	Pb	c	A	20 to 25	30–60	8
Fe-based steels	Cr		std	C	15	30–60	4
	Cu		a	D	20	60–120	5
	Ni		d	C	15	60	4A

[a] Cd diethyldithiocarbamate paper.
[b] Phosphormolybdate paper plus veiling.
[c] Dimethylglyoxime paper.
[d] Dimethylglyoxime paper plus veiling.

The error due to impurities in the supporting electrolyte can be eliminated by pretitration, and in most cases errors due to impurities in the reagents used in preparing the sample for titration can be eliminated by blank titrations.

The preparation, storage, and standardization of standard solutions are eliminated. Reagents difficult to use or unstable reagents present no problems. They are produced in situ and immediately consumed.

Coulometric titration is best suited for small samples. For larger samples, either the time must be increased to inconvenient lengths, or the electrolysis current must be increased to the point at which side reactions are difficult to prevent, so that the necessary 100% current efficiency is lost. At the other extreme, the smallness of samples is limited by one's ability to control and measure very small currents (less than 100 μA) and to measure short time intervals. The usual difficulties in handling very small samples also apply.

TABLE 14.34 Electrographic Tests for Metals in Aluminum, Copper, and Iron Alloys: Development Procedures

The tests described in this table are applicable to the prints obtained by the procedures outlined in Table 14.33. The first column of this table gives the code numbers that appeared in the last column of Table 14.33; the second column gives the symbol of the element tested for; the third column summarizes the procedure for developing the print; and the fourth column describes the appearance of a positive test and also contains various supplementary remarks.

Procedure number	Metal sought	Procedure	Interpretation and remarks
1	Al	Immerse in 1*M* KCN contg. 1% aq NH_3, wash with H_2O, using suction apparatus, suck dry. With the suction continued, add dropwise to the print on the pad a satd. soln. of either alizarin or morin in 50% EtOH contg. 5% aq NH_3. Wash with 50% EtOH and dil. aq NH_3.	Al is indicated by bright red color with alizarin or brilliant yellow-green fluorescence under UV light with morin
2	Al	Wash print in suction apparatus with 0.5*M* NH_4Cl–0.5*M* KCN soln. contg. 10% aq NH_3, wash with H_2O, suck dry, and treat with alizarin or morin as in preceding method.	See preceding method.
3	Be	Immerse in 1*M* KCN until blue-green color (Cu) has disappeared. Wash with H_2O in suction apparatus. Add a satd. soln. of alizarin in 50% EtOH dropwise on the suction pad, sucking through the paper each time. Wash with dil. aq NH_3 until the background color is decreased sufficiently to permit recognition of Be lake.	Be is indicated by the lavender color of its alizarin lake, which persists on washing with aq NH_3.
4	Cr	Wash the print with H_2O in suction apparatus to remove reddish ferric acetate.	Cr is indicated by the presence of yellow $BaCrO_4$.
4*A*	Cr, Ni	Cr and Ni can be detected simultaneously by using two papers in the sandwich. The one in contact with the specimen removes CrO_4^{2-} as yellow $BaCrO_4$ (see preceding method), but allows Fe and Ni to pass through to the second paper. This is impregnated with dimethylglyoxime, which ppts. Ni as the red complex. Prepare the printing pad as follows: Immerse the dimethylglyoxime paper and the backing paper in the electrolyte, blot fairly dry, and place on the cathode plate. Immerse the top printing paper in the electrolyte, blot, place on the pad, and print immediately. This prevents excessive soln. and diffusion of dimethylglyoxime into the top sheet.	
		Separate the printing sheets and wash each with H_2O in suction apparatus.	Cr is indicated by a yellow color on the top print, as in preceding method. Ni is indicated by a red color on the second print.
5	Cu	Wash on suction apparatus alternately with H_2O and 0.25*M* KNaTart soln. until color is no longer lost.	0.1–0.3% Cu can be detected by appearance of a light brown color. 0.05% Cu can be detected by first etching the metal in 10% HCl 15–30 min, washing, and blotting dry without rubbing to conc. alloyed Cu on the surface.

TABLE 14.34 Electrographic Tests for Metals in Aluminum, Copper, and Iron Alloys: Development Procedures (*Continued*)

Procedure number	Metal sought	Procedure	Interpretation and remarks
6	Fe	Wash on suction apparatus with 5% aq NH_3–0.5*M* KNO_3 until the blue color (Cu) has entirely disappeared, then wash with H_2O. Add dropwise 0.2*M* $K_4[Fe(CN)_6]$ contg. 5% HOAc while continuing suction; wash with H_2O.	Fe is indicated by the blue color of prussian blue. A pinkish color indicates incomplete removal of Cu; this may usually be discounted if the Fe content ≥1%, especially if a red filter is used to view the print. In doubtful cases print again and wash more thoroughly.
7	Mn	Immerse in 1*M* KCN contg. 1% aq NH_3, wash with 1% aq NH_3 on suction apparatus, and add 5% H_2O_2 dropwise to print on suction pad. Repeat H_2O_2 treatment several times, sucking dry each time, wash with H_2O, suck dry, and immerse in 1% aq benzidine acetate soln.	Mn is indicated by a green-blue color (benzidine oxdn. product) changing slowly to yellow-brown. Pb is masked by SO_4^{2-}.
8	Ni	Wash dimethylglyoxime print with 0.2*M* NH_4Cl contg. 10% aq NH_3 to remove Cu and Mn.	Ni is indicated by the red color of its dimethylglyoxime complex.
9	Pb	Immerse 2–3 min in 0.2*M* $K_2Cr_2O_7$ + 5% HOAc with agitation. Wash with 5% HOAc on suction pad until all $Cr_2O_7^-$ color is gone from area outside print.	Pb is indicated by the yellow color of $PbCrO_4$, not removed by washing with HOAc.
10	Sn	Use a top veiling sheet of S. & S. No. 576 paper to prevent contact of the phosphomolybdate paper with the surface of the metal. (All metals above Ag in the electromotive series will red. this paper.) Thus only Sn^{2+} can effect the redn. The paper is light-sensitive and should be freshly prepared and preserved away from light and metals. Prepare paper by impregnation with a soln. contg. 5 g NH_4 phosphomolybdate + a little NH_3 in 100 mL H_2O, dry, immerse in 5% HNO_3 in subdued light, wash thoroughly with H_2O, dry in darkness, and preserve under compression away from light.	
		Immerse print in 2% KOH until the yellow background color is bleached; wash with H_2O.	Sn is indicated by the blue color resulting from redn. of phosphomolybdate by Sn^{2+}. Excess rgt. is removed by washing with dil. alk., which gives a light-stable print.
11	Sn, Zn	Immerse in fresh 1*M* KCN–1*M* Na_2S soln. 2–3 min with agitation. Using suction apparatus, wash with 0.25*M* KCN–0.25*M* NH_4Cl–5% aq NH_3, then with 5% Na_2S_x (to remove Sn), then H_2O, then 2% HOAc. Suck dry, blot, and immerse in 0.5*M* $Pb(OAc)_2$ (or use spotting or streaking technique if a residual color is obtained).	If Bi, Pb, etc., are absent, and if washings are properly made, the print should bleach to white. Small amounts of Pb give an off-white to yellowish brown color. Sn is still detectable if spotting is used so that the increase in color can be noticed. Zn is indicated by the brown spot of PbS or increased color developing on the bleached original print area.

Source: L. Meites, ed., *Handbook of Analytical Chemistry,* McGraw-Hill, New York, 1963.

TABLE 14.35 Electrographic Reactions of Minerals

This table summarizes electrographic methods for the detection of various important constituents in a number of common minerals. The symbol "CR" (cathodic reduction) in the third column signifies that the print is made with the specimen as the cathode; otherwise the specimen is understood to be the anode. The abbreviations used to denote the various colors in the last column are defined in the introduction to Table 14.28.

Conducting mineral	Element sought	Printing conditions: Electrolyte	Voltage, V	Time, s	Reagent for development	Print color
Bismuth (native)	Bi	HCl (1:20)	4	30	KI-cinchonine	Or
Bornite	Cu	aq NH_3	4	15	Rubeanic acid	Dk Gn
Breithauptite	As	HCl (1:1)	4	30	$SnCl_2$ + HCl	Bn
	Sb	H_2Tart + H_3PO_4	4	30	Methyltrioxyfluorone	Rd
Chalcopyrite	Fe	HCl (1:20)	4–8	30	Chromotropic acid	Gn
	S	5% NaOH (CR)	4	30	$SbCl_3$ + HCl	Or
Chalcosine	Cu	dil. aq NH_3	4	5	Rubeanic acid	Dk Gn
Chloanthite	Ni	aq NH_3	8	15	Dimethylglyoxime	Rd
Cobaltite	As	aq NH_3 + H_2O_2 (5:1)	8	30	$AgNO_3$	Bn
	Co	aq NH_3	4–8	30	α-Nitroso-β-naphthol	Bn
	S				NaN_3 + I_2	
Copper (gray)	Ag	NHO_3 (1:4)	8–12	60–180	Redg. agent	Bk
	As	HCl (1:1)	8–12	60	$SnCl_2$ + HCl	Bn
	Cu	aq NH_3	4–8	15	Rubeanic acid	Dk Gn
	S	5% NaOH (CR)	8–12	30	$SbCl_3$ + HCl	Or
	Sb	10% H_2Tart + H_3PO_4	8–12	60	Methyltrioxyfluorone	Rd
Covellite	S	5% NaOH (CR)	4	15	$SbCl_2$ + HCl	Or
Danaite	As	aq NH_3 + H_2O_2 (5:1)	4–8	30	$AgNO_3$	Bn
Danaite	Co	aq NH_3	8–12	60	Rubeanic acid	Yl-Bn
	S				NaN_3 + I_2	
Galena	Pb	HOAc	4	30	KI + $SnCl_2$	Yl-Or
	S	5% NaOH (CR)	4	15	$SbCl_3$ + HCl	Or
Gersdorffite	Ni	aq NH_3	4	15	Dimethylglyoxime	Rd
Ilmenite	Fe	HCl or HNO_3 (1:10)	12–16	60	$K_4[Fe(CN)_6]$	Bl
	Ti	25% H_2SO_4 + H_3PO_4	16	180	Chromotropic acid	Rd-Bn
Linnaeite	Co	5% KCN	8	30	(Direct print)	Yl-Or
	Cu	aq NH_3	8	30	α-Benzoinoxime	Gn
	Fe	HCl (1:20)	8	30	Chromotropic acid	Gn
	Ni	aq NH_3	8	15	Dimethylglyoxime	Rd
	S	5% NaOH (CR)	8	15	$SbCl_3$ + HCl	Or
Lollingite	As	HCl (1:1)	4	30	$SnCl_2$ + HCl	Bn
	Fe	HCl or HNO_3 (1:20)	4	30	$K_4[Fe(CN)_6]$	Bl
Magnetite	Fe	HCl or HNO_3 (1:10)	4–8	30	$K_4[Fe(CN)_6]$	Bl
Marcasite	Fe	HCl or HNO_3 (1:20)	4	20	$K_4[Fe(CN)_6]$	Bl
	S	5% NaOH (CR)	4	15	$SbCl_3$ + HCl	Or
Millerite	Ni	aq NH_3	4	10	Dimethylglyoxime	Rd
	S	5% NaOH (CR)	4	15	$SbCl_3$ + HCl	Or
Mispickel	Fe	HCl or HNO_3 (1:20)	4–8	30	$K_4[Fe(CN)_6]$	Bl
Nickeline	Ni	aq NH_3	4	10	Dimethylglyoxime	Rd
Pentlandite	Fe	HCl or HNO_3 (1:20)	4	30	$K_4[Fe(CN)_6]$	Bl
	Ni	aq NH_3	4	15	Dimethylglyoxime	Rd
	S	5% NaOH (CR)	4	15	$SbCl_3$ + HCl	Or

TABLE 14.35 Electrographic Reactions of Minerals (*Continued*)

Conducting mineral	Element sought	Printing conditions: Electrolyte	Voltage, V	Time, s	Reagent for development	Print color
Pyrites			See Marcasite above			
Pyrrhotine			See Marcasite above			
Rammelsbergite	As	aq $NH_3 + H_2O_2$ (5 : 1)	4	30	$AgNO_3$	Bn
Safflorite	Fe	HCl or HNO_3 (1 : 20)	8	60	$K_4[Fe(CN)_6]$	Bl
Smaltite	Co	dil. aq NH_3	8	30	α-Nitroso-β-naphthol	Bn
Ullmannite	As	HCl (1 : 1)	4–8	30	$SnCl_2$ + HCl	Bn
	S	5% NaOH (CR)	8	30	$SbCl_3$ + HCl	Or
	Sb	$H_2Tart + HNO_3$	8	60	Methyltrioxyfluorone	Rd

14.8.1 Controlled-Potential Coulometry

14.8.1.1 General Principles. Controlled-potential coulometry employs the same equipment as used in controlled-potential electrolysis with the addition of a coulometer. A mercury pool is often used for reduction processes; oxidations can be performed at a cylindrical platinum electrode. The potential of the working electrode is controlled within 1 to 5 mV of the control value by a potentiostat. Current-potential diagrams must be determined for analyte system and for any possible interfering system. The necessary data can be obtained in two ways:

1. Set the potentiostat to one cathode-reference potential after another in sequence, allowing only enough time at each setting for the current indicator to balance.

2. Perform the coulometric analysis in the usual manner. Periodically adjust the potential to a value that stops the current flow. Note the net charge transferred up to each adjustment point; a plot of number of coulombs versus the potential provides a coulogram.

A controlled-potential coulometric electrolysis is like a first-order reaction, with the concentration and the current decaying exponentially with time during the electrolysis [Eq. (14.50)] and eventually attaining the residual current of the supporting electrolyte. The concentration limits vary from about 2 meq down to about 0.05 meq (set by the magnitude of the residual current).

Advantages of controlled-potential coulometry are these: (1) No indicator-electrode system is necessary. (2) It proceeds virtually unattended with automatic instruments. (3) Optimum conditions for successive reactions are easily obtained.

As in voltammetry, it is necessary to prereduce the supporting electrolyte in a coulometric reduction. To do this step, add the sample and deaerate the system before the reduction is started. Standards should be run under the same conditions that will be used with the samples. In order to achieve 100% current efficiency, the generating current must not exceed the diffusion current (even with stirring). As the desired constituent is removed from solution by reaction at the electrode surface, the diffusion current decreases from a relatively large value at the start of the determination to essentially zero at the equivalence point. Thus the potential of the working electrode must be controlled at some value on the diffusion current plateau in order to achieve the necessary 100% current efficiency. In many ways coulometric titrations at controlled potentials are very similar to electrogravimetric determinations at controlled potentials, the major difference being that the desired constituent is estimated from the number of coulombs in the former method and from the weight of the deposit in the latter method. The coulometric method is more versatile, since it can be applied to determinations in which the electrolysis product is a gas, a species in solution, or an amalgam, none of which can be weighed readily (see Table 14.36).

TABLE 14.36 Controlled-Potential Coulometric Methods

Abbreviations: PES, preelectrolyze sample solution at potential stated; PESE, preelectrolyze supporting electrolyte alone at potential stated, add sample and electrolyze at potential stated (if different).

Substance determined	Supporting electrolyte; working electrode potential; electrodes; notes	Interferences	No interference	References
Ag(I)	2.4*M* $HClO_4$; +0.15 V; Pt–SCE–Pt			*Anal. Chem.* **30**:487 (1958)
Au(III)	1*M* HCl; PES +0.75 V, then +0.45 V, Pt–SCE ‖ Pt; deposited Au may be anodically stripped at +1.20 V	Br^-, CN^-, Fe(III), >1*M* HNO_3, Pu(III), Ru	Ag, OAc^-, F^-, Hg(II), PO_4^{3-}, Pb(II), Pd(II), SO_4^{2-}, strong oxidants	*J. Electroanal. Chem.* **3**:112 (1962)
	In slags and Cu–Au alloys, potentiometrically			*Anal. Chim. Acta* **63**:129 (1973)
Bi(III)	0.4*M* Na_2tart, 0.1*M* NaHtart, 0.2*M* NaCl; PESE −0.7 V, PES −0.24 V, then −0.35 V; Hg–SCE–Ag		Cu(II)	*J. Am. Chem. Soc.* **67**:1916 (1945)
Br^-	0.1*M* NaOAc, 0.1*M* HOAc; +0.16 ± 0.05 V; Pt,Ag–SCE–Pt		I^- at pH 5 in presence of 5% $Ba(NO_3)_2$	*Anal. Chem.* **25**:274 (1953)
Cd(II)	0.1*M* KCl; PES −1.0 V, scan −1.0 to −0.3 V; Pt,Hg or Ag,Hg–SCE; anodic stripping, 0.02 V per second			*Anal. Chem.* **25**:1393 (1953)
Cl^-	Acetate buffer, pH 5; +0.25 V; Pt,Ag–SCE–Pt		See Br^-	*Anal. Chem.* **25**:274 (1953)
Co(II)	1*M* pyridine, 0.3*M* HCl, 0.2*M* $N_2H_4 \cdot H_2SO_4$, pH 7; PESE, PES −0.95 V, −1.20; Hg–SCE–Ag		Ni(II)	*Anal. Chim. Acta* **13**:281 (1955)
	1*M* NH_3, 1*M* NH_4Cl; PES −1.10 V, −1.45 V; Hg–SCE ‖ Pt	Zn	Cu, Ni	*Anal. Chem.* **28**:404 (1956)
Cr(VI)	0.5*M* H_2SO_4 containing NaI; +0.1 V; Pt–SCE ‖ Pt			*Anal. Chem.* **30**:487 (1958)
Cu(II)	0.1*M* NH_3, 0.1*M* NH_4Cl; PESE, −0.75 V; Hg–SCE ‖ Pt			*Anal. Chem.* **27**:1116 (1955)
	0.1*M* HCl (or $HClO_4$) or 0.5*M* citrate buffer, pH 5.5; PESE −0.1 V (or −0.5 V); Hg–SCE ‖ Pt			*Anal. Chem.* **27**:1116 (1955)
	0.5*M* H_2SO_4; PESE, −0.3 to −0.5 V; Hg–SCE ‖ Pt (or C)			*Anal. Chem.* **31**:492 (1959)
	2.4*M* $HClO_4$; −0.2 V; Pt–SCE ‖ Pt			*Anal. Chem.* **30**:487 (1958)
	In slags and Cu–Au alloys, potentiometrically			*Anal. Chim. Acta* **63**:129 (1973)
Eu(III)	0.1*M* HCl (or $HClO_4$); PES −0.9 V, −0.1 V; Hg–Ag,AgCl ‖ Pt	SO_4^{2-}, >0.016*M* HNO_3	Al, Ca, Ce, traces Fe, Cd, La, Si, Y, Yb	*Anal. Chem.* **31**:1095 (1959)
Fe(II)	1*M* H_2SO_4; +0.9 to +1.0 V; Pt–SCE ‖ Pt			*Anal. Chem.* **24**:986 (1952)
	In slags and Cu–Au alloys, potentiometrically			*Anal. Chim. Acta* **63**:129 (1973)
Fe(III)	0.2*M* $HClO_4$, 0.008*M* Ce(III); scan +0.8 to +0.2 V; Au–SCE	Ag, F^-, Hg, PO_4^{3-}, Pd, Pt, Pu, Ru	Au, Cl^-, Cu, NO_3^-, Np, Pb, SO_4^{2-}	*Anal Chem.* **31**:1095 (1959)
$Fe(CN)_6^{3-}$	0.025*M* acetate buffer, pH 5; +0.22 V; Pt–SCE ‖ Pt,Ag	Cl^-		*Z. Anal. Chem.* **169**:102 (1959)

$Fe(CN)_6^{4-}$	0.5*M* acetate buffer, pH 5; PESE, +0.32 V; Ag–SCE ‖ Pt	Anions giving insoluble Ag salts		*Z. Anal. Chem.* **179**:342 (1961)
In(III)	1*M* $HClO_4$, 1*M* NaI (12 : 1 ratio I^-:In); –0.615 V for In(III) to In(Hg)	In $HClO_4$, Ag(I) reduced 0.0 V and Cd(II) at –0.63		*Anal. Chem.* **43**:607 (1971)
Ir(IV)	0.2*M* HCl; +0.25 V to Ir(III); Pt cathode		Rh(III)	*Anal. Chem.* **43**:602 (1971)
Mn(II)	0.25*M* $Na_4P_4O_7$ at pH 2; to Mn(III) at +1.10 V	As(III), Ce(III), Sb(III), Tl(I)	Cr(III), Cl^-	*Anal. Chem.* **41**:758 (1969)
Mo(VI)	0.3*M* HCl, NaOAc, pH 1.5–2.0; –0.4; Hg–SCE–Ag		Cr(III)	*Chem. Abst.* **52**:12668b (1958)
Ni(II)	1*M* pyridine, 0.3*M* HCl, 0.2*M* $N_2H_4 \cdot H_2SO_4$, pH 7; PESE –0.95 V; Hg–SCE–Ag; Co may be subsequently determined in same solution		20–200 mg Co	*Anal. Chim. Acta* **13**:281 (1955)
Os(VI)	>3*M* HCl; –0.3 V; Hg–SCE ‖ Pt			
	0.1*M*–10*M* NaOH; –0.35 V [to Os(IV)], then –1.0 V [to Os(II)]; Hg–SCE ‖ Pt			*J. Am Chem. Soc.* **79**:4631 (1957)
Pb(II)	0.5*M* KCl; –0.50 V; Hg–SCE–Ag		Cd	*J. Am. Chem. Soc.* **67**:1916 (1945)
	0.1*M* KCl or KNO_3; PES –1.0 V; scan –0.7 to –0.2 V; Pt,Hg–SCE	Tl	Cd	*Anal. Chim. Acta* **11**:574 (1954)
Pu(IV)	1*M* H_3Cit, 0.1*M* $Al_2(SO_4)_3$, KOH, pH 4.5; PES [to Pu(III)], –0.07 V; Hg–SCE ‖ Pt		Cu, little Fe, Hg, Pb, Pd, U(VI)	U.S. Atomic Energy Commission Report No. HW-58491 (1958)
	Oxidation at 0.73 V vs. AgCl of Pu(III); reduction at 0.33 V Pu(IV) to Pu(III); mixed H_2SO_4 and HNO_3; Au electrode			*Talanta* **19**:1321 (1972)
	0.03*M* H_2SO_4, 0.03*M* bathophenanthroline; reduction Pu(IV) to (III) and Fe(III) to (II) at +0.30 V; stand 30 min, then oxidize Pu to (IV) at +0.66 V		Fe	*Anal. Chem.* **43**:603 (1971)
Rh(III)	0.2*M* HCl; –0.20 V to Rh(0); Pt cathode		Rh	*Anal. Chem.* **43**:602 (1971)
Sb	Sb(V) to (III) at –0.21 V and (III) to (0), Sb(Hg) at –0.35 V; 0.4*M* tartaric acid, 6*M* HCl		As, Fe, Ni, Pb, Sn, U	*Anal. Chem.* **34**:499 (1962)
SCN^-	0.2*M* KNO_3; +0.38 V (H_2O), +0.28 V (MeOH); Ag–SCE–Pt			*Chem. Abst.* **51**:16201a (1957)
Se^{2-}	1*M* NH_4Cl, NH_3, pH 8; –0.4 V; Hg–SCE ‖ Pt			*J. Am. Chem. Soc* **70**:4115 (1948)
Se(IV)	1*M* NH_4Cl, NH_3, pH 8; –1.65 V; Hg–SCE ‖ Pt			*J. Am. Chem. Soc* **71**:196 (1949)
Sn(IV)	3*M* NaBr, 0.3*M* HCl; PESE –0.8 V, PES –0.4 V [to Sn(II)], –0.7 V; Hg–SCE ‖ Pt			*Anal. Chim. Acta* **22**:577 (1960)
Te^{2-}	1*M* NaOH; –0.6 V (to Te metal); Hg–SCE ‖ Pt			*J. Am. Chem. Soc.* **70**:4115 (1948)

(Continued)

TABLE 14.36 Controlled-Potential Coulometric Methods (*Continued*)

Substance determined	Supporting electrolyte; working electrode potential; electrodes; notes	Interferences	No interference	References
Te(IV)	0.5*M* NH_4Cl, NH_3, pH 9.4; –0.9 V; Hg–SCE ‖ Pt			*J. Am. Chem. Soc.* **71**:196 (1949)
	0.5*M* H_3Cit, pH 1.6; –0.65 V; Hg–SCE ‖ Pt			*J. Am. Chem. Soc.* **71**:196 (1949)
Ti(IV)	9*M* H_2SO_4; Ti(IV) to (III) at –0.20 V; reoxidation at +0.22 V allows both oxidation states to be determined	As(III), Bi(II), Cu(II), Mo(VI), Se(VI), Te(IV)	Fe, Nb, V, W, Zr	*Anal. Chem.* **43**:747 (1971)
Tl(I)	1*M* H_2SO_4; PESE +1.38 V, then +1.34 V; Pt–SCE–Pt			*Anal. Chem.* **28**:1101 (1956)
U(IV)	1*M* $HClO_4$ (or HNO_3 or H_2SO_4 plus 3*M* H_3PO_4), sulfamic acid; +1.4 V; Pt–Ag,AgCl ‖ Pt	As(III), Br^-, CN^-, Ce(III), Cl^-, Cu(I), Fe(II), Hg(I), I^-, Mn(II), Mo(III), Ru, Ti(III), V(IV)	Co(II), mg amounts Cr(III), Th	*Anal. Chem.* **33**:1016 (1961)
	Oxidation at 0.73 V vs. AgCl; Au electrode; H_2SO_4 and HNO_3			*Talanta* **19**:1321 (1972)
U(VI)	1*M* H_3Cit, 0.1*M* $Al_2(SO_4)_3$, KOH, pH 4.5; PES –0.2, then –0.6 V; Hg–Ag,AgCl ‖ Pt		Al, Ce, Cu, Fe, HCl, Hg(II), HNO_3	*Anal. Chem.* **31**:10 (1959)
	0.5*M* H_2SO_4; PES + 0.175 V, –0.2 V; Hg–Ag,AgCl ‖ Pt		Cr, Cu, Mo, Sb	*Anal. Chem.* **31**:492 (1959)
V(V)	Reduction to V(IV); media effects studied	Interferences studied		*Analyst (London)* **98**:553 (1973)
Yb(III)	0.1*M* Et_4NBr in MeOH containing known amount Eu(III), –1.20 V; Hg–Ag,AgBr ‖ Ag. Determine Yb by difference (reaction induced by Eu)			*Anal. Chem.* **32**:1417 (1960)
Zn(II)	2*M* NH_3, 1*M* $(NH_4)_3Cit$; PES –1.1 V, change Hg, then –1.45 V; –0.5 to –1.0 V; Hg–SCE ‖ Pt	Co		*Anal. Chim. Acta* **2**:456 (1959)

14.8.1.2 Application Examples. In a mixture of uranium and chromium, both metals can be prereduced at –0.15 V to uranium(III) and chromium(II). Now, if electrolysis is carried out at –0.55 V, only uranium(III) is oxidized to uranium(IV), not with 100% current efficiency but completely. Chromium(II) is then determined by oxidation to chromium(III) at –0.15 V.

Mixtures of two reversible oxidation–reduction states are handled as follows for vanadium(IV)–vanadium(V). At 0.7 V versus SCE, vanadium(IV) is oxidized to vanadium(V); the reduction of vanadium(V) will occur quantitatively at 0.3 V versus SCE. In a mixture of vanadium(IV) and vanadium(V), control the anode potential at 0.75 V and measure the number of coulombs involved in the oxidation of vanadium(IV). Reverse the working-electrode potential, control the cathode potential at 0.3 V versus SCE, and measure the coulombs required for the reduction of original and generated vanadium(V). The difference in number of coulombs between the cathodic reduction and the anodic oxidation gives the original vanadium(V) concentration.

Table 14.36 summarizes the conditions that have been used and the results that have been obtained in determining various inorganic species by controlled-potential coulometric analysis.

14.8.2 Constant-Current Coulometry

Chemical reagents are generated within the supporting electrolyte in constant-current coulometry (often called coulometric titrations). A constant current is maintained through the electrochemical cell throughout the reaction period. The quantity of unknown present is given by the number of coulombs (measuring the product of current and time) of electricity used. The use of chemical intermediates not only facilitates the determination of many substances that can be determined by primary processes, but makes the coulometric method applicable to many determinations that cannot be carried out at all by primary electrode processes. The problem is to find electrode reactions that proceed with 100% current efficiency and suitable end-point detection systems.

14.8.2.1 Primary Coulometric Titrations. Only electrodes of silver metal, mercury, or mercury amalgams, or electrodes coated with silver-silver halide, are suitable sources of the electrogenerated species, For example, the silver ions generated at a silver anode will react with mercaptans dissolved in a mixture of aqueous methanol and benzene to which aqueous ammonia and ammonium nitrate are added to buffer the solution and supply the supporting electrolyte. The end point is determined amperometrically; excess silver ions will generate a signal at a platinum indicator electrode. Before the mercaptan sample is added, free silver ion is generated to a predetermined amperometric (current) signal. The sample is added and the generation continued until the same amperometric signal is attained again. Chloride ion in biological samples is determined in a similar manner. Combustion in an oxygen flask precedes the titration step for nonionic halides in organic compounds.

14.8.2.2 Secondary Coulometric Titrations. Secondary coulometric titrations are the most frequently used coulometric technique. These conditions must be met:

1. An active intermediate ion must be generated from an oxidation–reduction buffer (titrant precursor) added in excess to the supporting electrolyte.
2. The intermediate must be generated with 100% efficiency.
3. The intermediate must react rapidly and stoichiometrically with the substance being determined.
4. The standard potential of the titrant precursor must lie between the potential "window" of the unknown redox system and the potential at which the supporting electrolyte or another sample constituent undergoes a direct electrode reaction.
5. An end-point detection system must be available to indicate when the coulometric generation should be terminated.

An example will aid the discussion. Consider the coulometric titration of iron(II) to iron(III). The direct coulometric method will not succeed with 100% efficiency unless the potential is carefully controlled. When a finite current is forced to flow through the electrochemical cell, the current transported by the iron(II) ions soon falls below that demanded by the imposed current flow. However, if excess cerium(III) ions are added as the titrant precursor, they will begin and continue to transport the current. At the anode the cerium(III) ions will be oxidized to cerium(IV) ions, which will immediately react with unoxidized iron(II) ions. The reaction is stoichiometric and cerium(III) is reformed. The total coulombs ultimately required will be the sum needed for the direct oxidation of iron(II) and for the indirect oxidation via cerium(IV). Because there is an inexhaustible supply of cerium(III), the anode potential is stabilized at a value less positive than the oxidation of water, which would destroy the coulometric efficiency required. The end point is signaled either potentiometrically with a platinum reference-electrode pair or spectrophotometrically at the wavelength where the first excess of unused cerium(IV) absorbs strongly.

14.8.2.3 Instrumentation. The instrumentation required consists of an operational amplifier to force a constant current through the generator cell and some means to measure the electrogeneration time. A manual circuit can be assembled easily. The current source can be a heavy-duty dry cell (6 V) or several B batteries connected in series with an adjustable rheostat (to control the current level), a precision resistor with a potentiometer connected across its terminals (to measure the current), and the generator electrodes. If electrolytic products generated at the counterelectrode interfere with the reactions at the working electrode, the counterelectrode must be isolated from the remainder of the electrochemical cell by a porous glass frit or another type of salt bridge. The solution in the electrochemical cell is stirred throughout the titration.

14.8.2.4 Applications. Secondary coulometric methods enable uncommon, but useful, titrants such as bromine, chlorine, chromium(II), copper(I), silver(II), titanium(III), and uranium(III and V) to be generated in situ. Ordinarily these solutions would be difficult or impossible to prepare and store as standard solutions. Even electrolytic generation of hydroxyl ion offers the advantages of preparing very small amounts for the determination of very dilute acid solutions, such as would result from adsorption of acidic gases, and in a carbonate-free condition. Coulometric reagents and their precursors are listed in Table 14.37. In an aqueous solution the strongest oxidant is silver(II) ($E^0 \approx 2$ V) and the strongest reductant is uranium(III) ($E^0 = -0.63$ V). The electrogeneration of these titrants requires that the kinetics of their reactions with water and hydrogen ion, respectively, be slow.

Internally generated halogens, particularly bromine, have widespread applications, especially in organic analysis. Bromine can be easily generated with 100% current efficiency from a solution consisting of dilute acid and 0.2*M* sodium bromide. Sodium and lithium bromides are quite soluble in various organic solvents in which bromination can be conducted. Bromine is generated at the platinum anode, while hydrogen is generated at a platinum cathode that is isolated in a fritted glass tube. Iodine can similarly be generated from iodide solutions ranging from strongly acid to approximately pH 8 with 100% current efficiency. Strict control of reagent concentration is required for the generation of chlorine.

The coulometric Karl Fischer titration allows the determination of microgram amounts of water in organic liquids and of moisture in gases. Iodine is generated from an iodide salt in anhydrous methanol plus amine solvents that contain sulfur dioxide.

Azo dyestuffs can be titrated with titanium(III) generated externally and delivered to the hot dye solution via a capillary delivery tube. External generation guarantees that an optimum set of generation and reaction conditions prevail for each step. A double-arm electrolytic cell with separate anode and cathode delivery tubes (one to the sample and other to waste) is used for external generation.

The coulometric titrations listed in Tables 14.38*a* and 14.38*b* do not include all examples available in the literature. If several reagents are applicable, only the most common and convenient reagents are listed. Not all organic compounds have been listed; however, many determinations can be selected by finding a compound with the same functional group. Table 14.37 should be consulted for details of reagent generation. The titration conditions are given in detail if they differ from those specified for generation of the reagent. The precursor is often not listed unless its concentration varies from the usual generation conditions. Detailed titration conditions are not given for those cases in which consultation of the original reference is desirable. Inert atmospheres should be used for all strongly reducing systems and will be specified only for titrations in which their use might not be readily apparent.

TABLE 14.37 Coulometric Reagents and Their Precursors

Unless otherwise indicated, precursor concentrations are 0.05M–0.1M and generator electrode (platinum except as otherwise indicated) areas or 2–5 cm² are employed with generating currents up to about 50 mA. Variations from these conditions may usually be calculated by estimating the ratio of current density to precursor concentrations as 0.5 $mA \cdot cm^{-2} \cdot mmol^{-1} \cdot L^{-1}$. In some cases, 100% current efficiency can be obtained only by using current densities smaller than this criterion would indicate; such cases are indicated by the appearance of a current density ($mA \cdot cm^{-2}$) in the second column (represented by the symbol i/A). Nitrogen atmospheres must be used in work with reducing agents as strongly reducing as Fe(II).

For a general reference, see J. J. Lingane, Electroanalytical Chemistry, *2d ed., Interscience, New York, 1958.*

Reagent	Precursor; solution composition; conditions	References
Ag(I)	Ag anode; 0.5*M* $HClO_4$ or 0.2*M* NH_3 + 0.05*M* NH_4NO_3	*Anal. Chem.* **26**:622 (1954)
Ag(II)	0.1*M* $AgNO_3$; 5*M* HNO_3; Au anode, 0°C, *i/A* = 2 to 25 $mA \cdot cm^{-2}$	*Anal. Chim. Acta* **18**:245 (1958)
Br_2	0.2*M* Br^-; H_2SO_4, $HClO_4$, or HCl (pH < 5)	*J. Am. Chem. Soc.* **70**:1047 (1948)
Br^-	Ion-exchange membrane	*Anal. Chem.* **32**:1240 (1960)
BrO^-	1*M* Br^-; $Na_2B_4O_7$ buffer, pH 8–8.5 (avoid NH_3 in reagents)	*Anal. Chem.* **28**:440 (1956)
Ca(II)	Ion-exchange membrane	*Anal. Chem.* **32**:1240 (1960)
Ce(IV)	Saturated $Ce_2(SO_4)_3$; >3*M* H_2SO_4; *i/A* = 1 – 10	*Anal. Chim. Acta* **16**:165 (1957)
Cl_2	0.1*M*–2*M* Cl^-; HCl, H_2SO_4, or $HClO_4$ (pH < 1)	*Anal. Chem.* **22**:889 (1950)
Cl^-	Ion-exchange membrane	*Anal. Chem.* **32**:1240 (1960)
Cu	Saturated CuI; 1*M* NaI plus HOAc or KH phthalate buffer, pH 3.5	*Anal. Chem.* **28**:1510 (1956)
Cu(I)	Cu(II); 1*M*–3*M* HCl; $CuCl_3^{2-}$ is reductant	*J. Am. Chem. Soc.* **71**:2340 (1949)
Fe(II)	Fe(III); 1*M* H_2SO_4 plus 0.1*M* H_3PO_4; for potentiometric end point [Fe (III)/unknown] <10^3	*Anal. Chem.* **24**:1057 (1952)
$Fe(CN)_6^{3-}$	$Fe(CN)_6^{4-}$; no pH limits	*Anal. Chim. Acta* **13**:184 (1955)
$Fe(CN)_6^{4-}$	$Fe(CN)_6^{3-}$ (solution unstable); no pH limits	*Anal. Chim. Acta* **11**:475 (1954)
$Fe(EDTA)^{2-}$	$Fe(EDTA)^-$; NaOAc to pH > 2	*Anal. Chem.* **28**:520 (1956)
H^+	H_2O; 1*M* Na_2SO_4	*Anal. Chem.* **23**:941 (1951); **27**:1475 (1955)
	H_2O; 0.05*M* $LiClO_4$ in acetonitrile plus hydroquinone (0.1 g in 100 mL)	*Anal. Chem.* **28**:916 (1956)
	HOAc; 0.1*M* $NaClO_4$ in acetic anhydride–HOAc (6:1); Hg anode	*Anal. Chim. Acta* **21**:468 (1959)
$Hg_2(II)$	Hg or Hg-plated Au anode; 0.5*M* $NaClO_4$ plus 0.02*M* $HClO_4$	*Anal. Chem.* **28**:797, 799 (1956)
Hg(II)	Hg or Hg-plated Au anode; phosphate buffer, pH 9–12; internal generation only	*Anal. Chem.* **30**:65, 1064 (1958)
$HSCH_2COO^-$	$Hg(SCH_2COO)_2$;HOAc–NaOAc or NH_3–NH_4OAc buffer (pH 5–10); Hg cathode, N_2 atmosphere	*Anal. Chem.* **32**: 524 (1960)
H_2EDTA^{2-}	Ion-exchange membrane	*Anal. Chem.* **32**:1240 (1960)
$H(EDTA)^{3-}$	0.02*M* $HgNH_3(EDTA)^{2-}$; 0.05*M* NH_4NO_3 plus NH_3 to pH 8.5; Hg cathode, N_2 atmosphere, *i/A* <1.5	*Anal. Chem.* **28**:443 (1956)
I^-	Ion-exchange membrane	*Anal. Chem.* **32**:1240 (1960)
I_3^-	I^-; strongly acid (remove O_2) to pH 8.5	*Anal. Chem.* **22**:332 (1950)
Karl Fischer reagent		*Anal. Chem.* **31**:215 (1959)
Mn(III)	>0.2*M* Mn(II); >1.8*M* H_2SO_4; N_2 atmosphere, *i/A* = 1 to 4	*Anal. Chim. Acta* **21**:536 (1959); **12**:382, 390 (1955)
OH^-	H_2O; 1*M* Na_2SO_4	*Anal. Chem.* **23**:938, 941 (1951)
	Internal generation: H_2O; 0.05*M* KBr; Ag anode, *i/A* < 5	*Anal. Chim. Acta* **11**:283 (1954)

(Continued)

TABLE 14.37 Coulometric Reagents and Their Precursors (*Continued*)

Reagent	Precursor; solution composition; conditions	References
$S_2O_4^{2-}$	0.01*M* HSO_3^-; phthalic acid buffer, pH 3–5; Hg cathode, *i/A* = 1	*Talanta* **1**:110 (1958)
Sn(II)	0.2*M* $SnCl_4$; 3*M*–4*M* NaBr plus 0.2*M* HCl; do not clean Pt cathode, *i/A* = 10 to 85	*Anal. Chim. Acta* **20**:463 (1959)
Ti(III)	0.6*M* $TiOSO_4$; 6*M*–8*M* H_2SO_4 or >7*M* HCl; *i/A* < 3; remove O_2	*Anal. Chim. Acta* **15**:465 (1956); *Anal. Chem.* **27**:741 (1955)
U(V)	UO_2Cl_2 or $UO_2(ClO_4)_2$; HCl to pH 1.5–2.5; *i/A* < 2.5 at U(VI) = 0.1*M;* >0.03*M* nitrate interferes	*Anal. Chem.* **28**:1876 (1956)
U(IV)	UO_2SO_4; 0.25*M* H_2SO_4; *i/A* < 2 at U(VI) = 0.1*M*	*Anal. Chem.* **27**:1750 (1955)
V(IV)	$NaVO_3$; 0.5*M*–3*M* H_2SO_4	*Anal. Chem.* **31**:1460 (1959)

14.8.2.5 Corrosion or Tarnish Films. The thickness of corrosion or tarnish films can be measured coulometrically. The specimen is made the cathode and the film is reduced with a constant known current to the metal. By following the cathode potential, the end point is taken as the point of inflection of the voltage–time curve. Anodic dissolution is used to determine the successive coatings on a metal surface. For example, the thickness of a tin undercoating and a copper–tin surface layer on iron can be measured because the two coatings exhibit individual step potentials. From the known current *i*, expressed in milliamperes, and the elapsed time *t*, in seconds, the film thickness *d*, in nanometers, can be calculated from the known film area *A*, in cm^2, and the film density ρ, according to the equation

$$d = \frac{10^4 Mit}{AnF\rho} \qquad (14.51)$$

where *M* is the molecular weight of the tarnish film and *F* is the faraday.

14.9 CONDUCTANCE METHODS

One of the oldest and in many ways simplest of the electrochemical methodologies is the measurement of electrolytic conductance. Practical applications are of three types: direct analysis, stream monitoring, and titration.

14.9.1 Electrolytic Conductivity

Solutions of electrolytes conduct an electric current because the ions migrate under the influence of a potential gradient applied to two electrodes immersed in the solution. The positive ions (cations) are attracted to the negative electrode (cathode) while the negative ions (anions) are attracted to the positive electrode (anode). The flow of current depends upon the magnitude of the applied potential and the resistance of the solution between the electrodes, as expressed by Ohm's law.

The reciprocal of the resistance 1/*R* is called the *conductance S* and expressed in siemens (or reciprocal ohms, or mhos, older terms but not SI nomenclature). The conductance is directly proportional to the cross-sectional area *A* of the electrodes and inversely proportional to the distance *d* between them:

$$\frac{1}{R} = S = \kappa \frac{A}{d} \qquad (14.52)$$

TABLE 14.38*a* Procedures for Coulometric Titrations: Inorganic Substances

No.	Substance determined	Reagents; titration conditions; end point; notes	References
1	Ag(I)	Br^-, Cl^-, or I^-; ion-exchange membrane	*Anal. Chem.* **32**:1240 (1960)
2	Al(III)	OH^-; 48% EtOH, 0.05*M* Na_2SO_4, 0.006*M* oxine; $E_{glass\text{-}SCE}$	*Anal. Chim. Acta* **19**:272 (1958)
3	As(III)	Br_2; 0.1*M*–1*M* H_2SO_4; $E_{Pt\text{-}SCE}$, $i_{Pt\text{-}Pt}$ (0.2 V)	*J. Am. Chem. Soc.* **70**:1047 (1948)
4		Ce(IV); H_2SO_4, OsO_4 catalyst; $A_{320,360,375}$	*Anal. Chem.* **28**:515 (1956)
5		I_2; HCO_3^-, pH 8; $E_{Pt\text{-}SCE}$, $i_{Pt\text{-}Pt}$ (0.15 V), A_{342}	*Anal. Chem.* **22**:332 (1950)
6	Au(III)	Cu(I); 1*M*–2*M* HCl, 0.04*M* $CuSO_4$; $E_{Au\text{-}SCE}$	*Anal. Chim. Acta* **19**:394 (1958)
6a		$Hg(EDTA)^{2-}$; measure Hg(II) displaced	*Talanta* **26**:445 (1979)
7		$HSCH_2COO^-$; NH_3–NH_4OAc buffer, pH 7.5; *E* Hg–Hg,Hg_2SO_4, $i_{Hg\text{-}Hg}$ (0.15V)	*Anal. Chem.* **32**:524 (1960)
8		Sn(II); 4*M* NaBr, 0.3*M* HCl; $E_{Au\text{-}SCE}$, $i_{Au\text{-}Au}$ (0.15 V)	
9	$Br_2(BrO_3^-)$	Cu(I); 1*M* NaBr, 0.02*M* $CuSO_4$, 0.3*M* $HClO_4$; $i_{Pt\text{-}Pt}$ (0.15 V)	
10		Sn(II); NaBr, HCl; $E_{Pt\text{-}SCE}$ (end point at +0.38 V), $i_{Pt\text{-}Pt}$ (0.15 V); pretitrate Br_2	
11	Br^-	Ag(I); 0.5*M* $HClO_4$, use 75% acetone for Br^- <0.5 mg per 50 mL; $E_{Ag\text{-}SCE}$, eosin; see also no. 60	
12		Hg(I); $HClO_4$, $NaClO_4$, 80% MeOH; $E_{Hg\text{-}SCE}$	
13	Ca(II)	$H(EDTA)^{3-}$; NH_4NO_3, NH_3, pH 8.5; $E_{Hg\text{-}SCE}$; N_2 atm.	
14	Ce(III)	Ag(II); 0.1*M* HNO_3, 0°C; $i_{Pt\text{-}Pt}$ (0.075 V), $E_{Pt\text{-}SCE}$ = +1.43 V	*Anal. Chim. Acta* **18**:245 (1958)
15	Ce(IV)	Fe(II); H_2SO_4, H_3PO_4; $E_{Pt\text{-}SCE}$, *E* Pt–Hg,Hg_2SO_4	
16		Sn(II): NaBr, HCl, 0.008*M* KI; $i_{Pt\text{-}Pt}$ (0.15 V), $E_{Pt\text{-}SCE}$ (end point at + 0.28 V)	
17		Ti(III); H_2SO_4; $E_{Pt\text{-}SCE}$, $i_{Pt\text{-}Pt}$ (0.15 V); Cd cathode in isolated compartment	
17a		U(IV); see no. 23a	
18	CN^-	Hg(II); 0.5*M* Na_3PO_4, pH 9; $E_{Hg\ on\ Au\text{-}SCE}$; pretitrate KCN to –0.05 V before adding sample; N_2 atm.	*Anal. Chem.* **30**:65 (1958)
18a	CO	Continuous monitoring in steel works stack gases	*Analyst* **103**:1185 (1978)
19	CO_2	OH^-; 0.04*M* $BaCl_2$, 0.005*M* H_2O_2, 0.5% EtOH; $E_{Pt\text{-}SCE}$	*Anal. Chim. Acta* **17**:247 (1957)
19a	CO_3^{2-}	Ba(II); dissolve in boiling H_3PO_4, scrub to remove H_2S; titrate CO_2 with generated $Ba(OH)_2$	*Z. Anal. Chem.* **285**:369 (1977)
20	Cl^-	Ag(I); $HClO_4$, 80% EtOH or acetone; $E_{Ag\text{-}SCE}$, dichlorofluorescein	
21		Hg(I); see no. 12	
22	Cr(VI)	Cu(I); 1.3*M* HCl, 0.02*M* $CuSO_4$; $i_{Pt\text{-}Pt}$ (0.2 V); pretitrate dichromate	*J. Am. Chem. Soc.* **71**:2340 (1949)
23		Fe(II); see no. 15	
23a		U(IV); 0.1*M*–1*M* UO_2SO_4, 0.1*M*–0.5*M* H_2SO_4; $E_{Pt\text{-}SCE}$	*Anal. Chem.* **27**:1750 (1955)
24	Cu(I)	Br_2; 1*M* HCl; $i_{Pt\text{-}Pt}$ (0.15–0.2 V), $E_{Pt\text{-}SCE}$	
25		Cl_2; 1*M* HCl; $i_{Pt\ Pt}$ (0.2 V), $E_{Pt\text{-}SCE}$	
26	Cu(II)	Sn(II); HCl, NaBr; $E_{Pt\text{-}SCE}$	*Anal. Chim. Acta* **21**:227 (1959)
27		$H(EDTA)^{3-}$; see no. 13	
28		$HSCH_2COO^-$; HOAc–NaOAc buffer, pH 5; *E* Hg–Hg,Hg_2SO_4	

(Continued)

TABLE 14.38a Procedures for Coulometric Titrations: Inorganic Substances (*Continued*)

No.	Substance determined	Reagents; titration conditions; end point; notes	References
29	F^-	H^+; 90 mL acetic anhydride, 0.1*M* $NaClO_4$, add sample in 1–2 mL HOAc; *E* glass–Hg,$Hg_2(OAc)_2$,HOAc,$NaClO_4$; allow 2-min intervals near end point	*Anal. Chem.* **33**:132 (1961)
30	Fe(II)	Ce(IV); H_2SO_4; $E_{Pt\text{-}SCE}$	*Anal. Chem.* **23**:945 (1951)
31		Mn(III); 6.3*M* H_2SO_4; *E* Pt–Hg, Hg_2SO_4 (+0.4 V), 1,10-phenanthroline-iron(II)	*Anal. Chim. Acta* **21**:536 (1959); **12**:390 (1955)
31a		Cl_2; 2*M*–3*M* HCl, 0.05*M* $CuSO_4$; $E_{Pt\text{-}SCE}$ (+0.65 V)	*Anal. Chem.* **33**:1318 (1961)
32	Fe(III)	Sn(II); see no. 16	
33		Ti(III); see no. 17	
33a		U(V); 0.02*M* UO_2SO_4, HCl, pH 1.5–2.5; *i/A* = 0.5; N_2 atmosphere; $i_{Pt\text{-}Pt}$ (0.3 V)	*Anal. Chem.* **28**:1876 (1956)
34		$Fe(EDTA)^{2-}$; NaOAc to pH 2.5; $E_{Pt\text{-}AgCl}$; pretitrate Fe(III), N_2 atmosphere	*Anal. Chem.* **28**:520 (1956)
35	$Fe(CN)_6^{3-}$	$HSCH_2COO^-$; NH_3–NH_4OAc buffer, pH 7.5; *E* Hg–Hg,Hg_2SO_4	*Anal. Chem.* **27**:1275 (1955)
36	$Fe(CN)_6^{4-}$	Ce(IV); 0.5*M* H_2SO_4; *E* Pt-Pb(Hg), $PbSO_4$ (+1.21 V); remove O_2 with N_2 before adding sample	
37		Br_2; 2*M* HCl; $i_{Pt\text{-}Pt}$ (0.2 V)	
38	H^+	OH^-; $E_{glass\text{-}SCE}$	
39	Hg(II)	$HSCH_2COO^-$; see no. 7	
40	H_2O	Karl Fischer reagent; see Reference	*Anal. Chem.* **31**:215 (1959)
41		P_2O_5; see Reference	*Anal. Chem.* **31**:2043 (1959)
42	H_2O_2	ClO^-	*Anal. Chem.* **38**:1400 (1966)
42a		I_2	*Anal. Chem.* **37**:1418 (1965)
43	$I_2(IO_3^-)$	Sn(II); see no. 16	
44	I^-	Ag(I); $HClO_4$; $E_{Ag\text{-}SCE}$, eosin; pretitrate to +0.20 V, stop at +0.12 V to allow for drift at end point	*Anal. Chem.* **26**:622, 626 (1954)
45		Hg(I); $HClO_4$, $NaClO_4$; $E_{Hg\text{-}SCE}$	
46		Br_2; 2*M* HCl; $i_{Pt\text{-}Pt}$ (0.13 V), $E_{Pt\text{-}SCE}$	*Anal. Chem.* **21**:1457 (1949)
47		Ce(IV); 1*M* H_2SO_4, 0.025*M* $Ce_2(SO_4)_3$; $i_{Pt\text{-}Pt}$ (0.10 V), $E_{Pt\text{-}SCE}$	*Anal. Chim. Acta* **16**:165 (1957)
48	KH phthalate	H^+; 0.1*M* $NaClO_4$ in acetic anhydride–HOAc (6:1); *E* glass-Hg,$Hg_2(OAc)_2$; pretitrate basic impurities	*Anal. Chim. Acta* **21**:468 (1959)
49	Mn(VII)	Fe(II); 0.25*M* H_2SO_4, H_3PO_4; $E_{Pt\text{-}SCE}$, *E* Pt–Pg(Hg), $PbSO_4$ (+1.14 V); CO_2 atm.	*Anal. Chem.* **23**:1662 (1951); **24**:205 (1952)
50		V(IV); 0.5*M* H_2SO_4, 40°C; $i_{Pt\text{-}Pt}$ (0.25 V); allow 2-min intervals near end point	*Anal. Chem.* **31**:1460 (1959)
51	NH_3	BrO^-; $Na_2B_4O_7$ buffer, pH 8.0–8.5; see Reference	*Anal. Chem.* **28**:440 (1956)
52	$N_2H_4 \cdot H_2SO_4$	Br_2; 0.3*M* HCl, 0.1*M* KBr; $i_{Pt\text{-}Pt}$ (0.2 V)	*Anal. Chem.* **32**:1545 (1960)
53	NaOAc	H^+; see no. 48	
54	Pb(II)	$H(EDTA)^{3-}$; see no. 13	
55	PO_4^{3-}	OH^-; convert to H_3PO_4 by ion exchange, see Reference	*Anal. Chem.* **27**:122 (1955)
56	Pt(IV)	Sn(II)-Br_2; 0.3*M* HCl, 4*M* NaBr; $E_{Pt\text{-}SCE}$, A_{400}; generate 10% excess Sn(II), wait 5 min, generate Br_2	*Anal. Chem.* **32**:623 (1960)
57	S^{2-}	I_2; see no. 5; for very small amounts, see Reference	*Chem. Abst.* **48**:7494a (1954)
58		Br_2; 4*M* H_2SO_4, 0.1*M* KBr; $E_{Pt\text{-}SCE}$	*Ind. Eng. Chem.* **46**:1422 (1954)

TABLE 14.38*a* Procedures for Coulometric Titrations: Inorganic Substances (*Continued*)

No.	Substance determined	Reagents; titration conditions; end point; notes	References
59		Hg(II); 0.1*M* NaOH, 80°C; $E_{Hg\text{-}SCE}$; N_2 atmosphere; pretitrate S^{2-} to $E = -0.5$ V; stir vigorously	
60	SCN^-	Ag(I); 0.5*M* $HClO_4$, 90% acetone; $E_{Pt\text{-}SCE}$; consecutive determination of Br^- and SCN^-	
61	$S_2O_3^{2-}$	I_2; pH 1–8; $i_{Pt\text{-}Pt}$ (0.135 V), starch; remove O_2 with N_2	*Anal. Chem.* **26**:373 (1954)
62	SO_2	Br_2; absorb SO_2 in ferricyanide (1 g/L), add to 4 volumes 2*M* KBr; $i_{Pt\text{-}Pt}$ (0.2 V); see no. 58	*Anal. Chim. Acta* **20**:344 (1959)
63	Sb(III)	Br_2; 2*M* HCl or 0.01*M* H_2SO_4, 0.1*M* KBr; $i_{Pt\text{-}Pt}$ (0.02 V); absorbed in NaOH	*Anal. Chem.* **34**:138 (1962)
64		I_2; 0.025*M* K_2tartrate, 0.1*M* phosphate buffer, pH 8; $i_{Pt\text{-}Pt}$, $E_{Pt\text{-}SCE}$, starch	*Anal. Chim. Acta* **16**:271 (1957)
65	Se(IV)	I_2; 0.1*M* HCl, add excess standard thiosulfate, then 0.1*M* KI, generate I_2; $i_{Pt\text{-}Pt}$ (0.2 V)	*Anal. Chem.* **27**:818 (1955)
66	Ti(III)	Ce(IV); 0.5*M* H_2SO_4; *E* Pt-Pb(Hg),$PbSO_4$ (+1.25 V); pretitrate, see Reference for mixtures of Ti and Fe	*Anal. Chem.* **27**:1596 (1955)
67		Ti(III); H_2SO_4, HBF_4, add excess standard Fe(III), generate Ti(III); leuco methylene blue, A_{665}	*Anal. Chem.* **28**:1884 (1956)
68	Tl(I)	Br_2, 1*M* $HClO_4$, 0.13*M* NaBr; $i_{Pt\text{-}Pt}$ (0.2 V)	*Anal. Chem.* **24**:1195 (1952)
69		Cl_2; 1*M* HCl; $i_{Pt\text{-}Pt}$ (0.3 V)	
70		Ferricyanide; 2*M* NaOH; $E_{Pt\text{-}SCE}$, $i_{Pt\text{-}Pt}$ (−0.15 to −0.30 V)	*Anal. Chim. Acta* **13**:183 (1955)
71	U(IV)	Ce(IV); H_2SO_4 + 0.01*M* $FeNH_4(SO_4)_2$; see no. 30; CO_2 atm., Cd reductor for preparation of U(IV) from U(VI)	*Anal. Chem.* **25**:482 (1953)
72	U(VI)	Ti(III); 8*M* H_2SO_4, 0.003*M* $FeNH_4(SO_4)_2$; $i_{Pt\text{-}Pt}$ (0.25 V); N_2 atm., pretitrate the Fe(III)	*Anal. Chim. Acta* **18**:249 (1958)
73	V(IV)	Ti(III); see no. 17	
74		Ag(II); HNO_3; *E* Pt–Hg,Hg,Hg_2SO_4; preoxidize Pt indicator electrode, see Reference; $i_{Pt\text{-}Pt}$ (0.200 V)	*Anal. Chim. Acta* **18**:245 (1958)
75	V(V)	Cu(I); 2.6*M* HCl, 0.04*M* $CuSO_4$; remove O_2 with CO_2; $i_{Pt\text{-}Pt}$ (0.2 V)	*J. Am. Chem. Soc.* **71**:2340 (1949)
76		Fe(II); H_2SO_4, H_3PO_4; *E* Pt–Hg,Hg_2SO_4 (+ 0.22 V)	*Anal. Chem.* **23**:1665 (1951)
77		Sn(II); 0.25*M* HCl, 4*M* NaBr, 0.004*M* $VOSO_4$; $E_{Au\text{-}SCE}$	*Anal. Chim. Acta* **20**:581 (1959)
78		Ti(III); 8*M* H_2SO_4; $i_{Pt\text{-}Pt}$ (0.35 V), $E_{Pt\text{-}SCE}$; if U(VI) is present, add Fe(III) after the U is titrated	*Anal. Chim. Acta* **18**:249 (1958)
79	Zn(II)	Ferrocyanide; ferricyanide, H_2SO_4 or HCl, pH 1–3; $E_{Pt\text{-}SCE}$; pretitrate Zn(II) to $E = +0.58$ V	*Anal. Chim. Acta* **11**:475 (1954)
80		$H(EDTA)^{3-}$; see no. 13	

where κ is the *specific conductance* expressed in $S \cdot cm^{-1}$ (or $\Omega^{-1} \cdot cm^{-1}$). For a given electrolytic cell with fixed electrodes, the ratio *d/A* is a constant, called the *cell constant*, θ. From Equation (14.52), it follows that

$$\kappa = S\Theta \tag{14.53}$$

The *d/A* ratio is determined by measuring the resistance of a standard solution of known specific conductance (Table 14.39), and the cell constant is then computed.

TABLE 14.38*b* Procedures for Coulometric Titrations: Organic Substances

No.	Substance determined	Reagents; titration conditions; end point; notes	References
81	Acetic acid	OH^-; 0.001*M* LiCl, 70% 2-propanol; $E_{glass\text{-}SCE}$; for titration of acetate see no. 53	*Anal. Chem.* **23**:1019 (1951)
82	Acetic acid, mercapto-	Hg(II); $Na_2B_4O_7$ buffer, pH 9.2; $E_{Hg\text{-}SCE}$, pretitrate to –0.200 V; remove O_2 with N_2	*Anal. Chim. Acta* **18**:596 (1958)
83	Acetophenone	H_2; 95% ETOH, 10% Pd on C; constant pressure	*Anal. Chem.* **30**:295 (1958)
84	Aniline	Br_2–Cu(I); 1*M* HCl, 0.1*M* NaBr, 0.02*M* $CuSO_4$; $i_{Pt\text{-}Pt}$ (0.2 V); generate excess Br_2, wait 1 min, then titrate to reference indicator current with Cu(I)	*Anal. Chem.* **24**:499 (1952)
85	Ascorbic acid	I_2; 0.1*M* KI, 0.6*M* NaCl, 0.007*M* HCl; $i_{Pt\text{-}Pt}$ (0.15 V); saturate solution with CO_2	*Chem. Anal. (Warsaw)* **2**:453 (1957); *Chem. Anal.* **52**:5209b (1958)
86	Benzene	H_2; HOAc, PtO_2; constant pressure	*Anal. Chem.* **30**:295 (1958)
87	Benzylamine	H^+; 0.05*M* $LiClO_4$ in acetonitrile; $E_{glass\text{-}SCE}$	*Anal. Chem.* **28**:130, 916 (1956)
88	1,3-Butadiene, 2-methyl-; 2-butene, 2-methyl (or 1-phenyl)	Br_2; HOAc, MeOH, HCl, $HgCl_2$, KBr; A_{360}	*Anal. Chem.* **29**:475 (1957)
89	1-(or 2-)butene, 2,3-dimethyl	Br_2; HOAc, MeOH, $Hg(OAc)_2$, KBr; $i_{Pt\text{-}Pt}$ (0.2 V); pretitrate to arbitrary current	*Anal. Chem.* **28**:1553 (1956)
90	Butanethiol (and other thiols)	Ag(I); C_6H_6, EtOH, NH_3, NH_4NO_3; $i_{Au\text{-}Pt}$ (0.25 V)	*Anal. Chem.* **26**:1607 (1954)
91		Ag(I); 0.5*M* KNO_3, 70% EtOH, 0.1% gelatin; $i_{Pt\text{-}SCE}$, rotating Pt electrode	*Anal. Chim. Acta* **17**:247 (1957)
92		Br_2, see no. 58	
93		Hg(II), see no. 82	
94	Butanoic acid	OH^-; see no. 81	
95	Butyrophenone	H_2; see no. 83	
96	Cinnamic acid	H_2, see no. 83	
97	*o*-Cresol (*p*-cresol)	Br_2; pH 1 to 5; $I_{Pt\text{-}Pt}$ (0.3 V)	*Chem. Listy* **52**:595 (1955)
98	Cyclohexene, 4-vinyl	Br_2, see nos. 88, 89	
99	Cyclohexylamine (secondary amines)	Hg(II); 0.5*M* $NaClO_4$, 25% CS_2, 75% acetone; $E_{Hg\text{-}SCE}$; pretitrate to –0.200 V; remove O_2 with N_2, react primary amines with salicylaldehyde before titration	*Anal. Chim. Acta* **18**:596 (1958)
100	Cysteine	Hg(II); see no. 82	
101	Dialkylamines	Hg(II); see no. 99	
102	Disulfides	Br_2; see no. 58	
103	Diisobutene	Br_2; see no. 89	
104	1,3-Diphenylguanidine	H^+; see no. 87	

TABLE 14.38*b* **Procedures for Coulometric Titrations: Organic Substances (*Continued*)**

No.	Substance determined	Reagents; titration conditions; end point; notes	References
105	1-Dodecene	Br_2; see no. 89	
106	13-Docosenoic acid	Cl_2; 0.2*M*–1.2*M* HCl, 80%–90% HOAc; $i_{Pt\text{-}Pt}$ (0.36 V)	*Chem. Listy* **47**:1166 (1953)
107	EDTA	Ca(II); ion-exchange membrane	
108	Fatty acids, long chain, 1 double bond	Cl_2; see no. 106	
109	Fumaric acid	H_2; see no. 83	
110	1-Hexene; 1,5-hexadiene	Br_2; see no. 88; also for substituted 1-heptenes	
111	Hydrazine and hydrazides	Br_2; HOAc, MeOH, $Hg(OAc)_2$, KBr; $i_{Pt\text{-}Pt}$ (0.2 V)	*Anal. Chem.* **32**:1545 (1960)
112	Hydroquinone	Br_2 or Cl_2; 0.1*M* HBr or HCl; $i_{Pt\text{-}Pt}$ (0.3 V)	*Chem. Listy* **52**:595 (1958)
113		Br_2–Sn(II); 0.2*M* HCl, 3*M* NaBr, 0.2*M* $SnCl_2$; $i_{Pt\text{-}Pt}$ (0.15 V); generate excess Br_2 and back-titrate with Sn(II)	*Anal. Chim. Acta* **20**:463 (1959)
114		Ce(IV); 2*M* H_2SO_4, 1*M* H_3PO_4; $E_{Pt\text{-}Pb(Hg)}$ (+1.25 V)	*Anal. Chem.* **25**:1564 (1953)
114a		Cu(II); MeOH	*Talanta* **27**:989 (1980)
115	Indigo carmine	$S_2O_4^{2-}$; HOAc buffer, pH 4; A_{610}	*Talanta* **1**:110 (1958)
116	Isonicotinic acid hydrazide	Br_2; 0.3*M* HCl; $i_{Pt\text{-}Pt}$ (0.3 V), $E_{Pt\text{-}SCE}$; see no. 146	
117	Methylene blue	$S_2O_4^{2-}$; see no. 115	
118	Morpholine	Hg(II); see no. 99	
119	Naphthalene	H_2; see no. 86	
120	1-Naphthyl-amine	H_2; 0.05*M* $LiClO_4$, 0.01*M* hydroquinone, acetonitrile; $E_{glass\text{-}Ag,AgCl}$	*Anal. Chem.* **28**:916 (1956)
121	2-Naphthyl-amine	Ce(IV)–Fe(II); >3*M* H_2SO_4, 0.1*M* H_3PO_4; $E_{Pt\text{-}SCE}$	
122	1-Octene	Br_2; see no. 89	
122a	Olefins	Br_2; 20–1000 ppm vapor, 82 mL HOAc, 15 mL H_2O, 3 mL ethylene glycol, 2 g KBr	*Anal. Chem.* **34**:418 (1962)
123	Oleic acid	Br_2; 0.5*M* HBr, 85% HOAc; $i_{Pt\text{-}Pt}$ (0.3 V); styrene does not interfere	*Chem. Listy* **52**:1899 (1958)
124		Cl_2; 1.2*M* HCl, 80% HOAc; $i_{Pt\text{-}Pt}$ (0.35 V); styrene titrates	
125	Oxalic acid	Ce(IV)–Fe(II); see no. 121	
126		Ag(II): HNO_3; $i_{Pt\text{-}Pt}$ (0.075 V); preoxidize indicator electrodes; insert after visual end point	*Anal. Chim. Acta* **18**:245 (1958)
127		Mn(III); H_2SO_4; *E* Pt–Hg,Hg_2SO_4, 1,10-phenanthroline iron(II)	*Anal. Chim. Acta* **12**:382, 390 (1955)
128		OH^-; see no. 81	
129	1-(or 2-)Pentene	Br_2; see no. 89; includes 2,3,3- (or 2,4,4,-)trimethyl substituents	
130	2-Pentene, 4-methyl (*cis* and *trans*)	Br_2; see no. 88	

(Continued)

TABLE 14.38*b* Procedures for Coulometric Titrations: Organic Substances (*Continued*)

No.	Substance determined	Reagents; titration conditions; end point; notes	References
131	*o*-Phenylene-diamine	H^+; see no. 120	
132	Phenol	Br_2; pH 0.5; $i_{Pt\text{-}Pt}$ (0.2–0.3 V); see no. 146	
133	Phenol, *p*-amino-	Ce(IV); see no. 114	
134		H^+; see no. 120; *p*-aminophenol also titrates	
135	Phenol, 4-nitro-	H_2; see no. 83	
136	Phenol, 4-methylamino-	Ce(IV); see no. 114	
137	Propanoic acid	OH^-; see no. 81	
138	Pyridine	H^+; see no. 87	
139	Pyrocatechol	Br_2; pH 4–5; $i_{Pt\text{-}Pt}$ (0.3 V)	
140	8-Quinolinol	Br_2; 0.2*M* NaBr, 0.001*M*–0.0001*M* HCl; $i_{Pt\text{-}Pt}$ (0.25 V); 0.4–2 mg oxine per titration	*Anal. Chem.* **22**:1565 (1950)
141	Quinone	Sn(II)-Br_2; 0.2*M* HCl, 3*M* KBr, 0.2*M* $SnCl_4$; $i_{Pt\text{-}Pt}$ (0.15 V); generate excess Sn(II), back-titrate with Br_2	*Anal. Chim. Acta* **20**:463 (1959)
142	Resorcinol	Br_2; see no. 139	
143	Salicylic acid	Br_2-Cu(I); 0.3*M* HCl, 0.05*M* $CuSO_4$, 0.1*M* KBr; $i_{Pt\text{-}Pt}$ (0.25 V); generate excess Br_2, wait 2 min, then generate Cu(I); see no. 84	
144	Styrene	Cl_2; see no. 124	
145	Thioethers; thiophenes	Br_2; see no. 58	
146	Thiodiglycol	Br_2, 50% HOAc; $i_{Pt\text{-}Pt}$ (0.3 V), $E_{Pt\text{-}SCE}$	*Anal. Chem.* **19**:197 (1947)
146a	Thiols	Cu(II); MeOH	*Talanta* **27**:989 (1980)
147	*o*-(*p*-)Toluidine	H^+; see no. 120	
148	Triethylamine	H^+; see no. 87	
149	Urea, thio-	Hg(II); 0.03*M* H_2SO_4; 0.1*M* K_2SO_4; $i_{Hg\text{-}Hg}$ (0.01–0.03 V); remove O_2 with N_2	*Z. Anal. Chem.* **161**:348 (1958)
150		Ag(I); add excess saturated AgBr in concentrated NH_3; warm to 70°C until no odor of NH_3; add $HClO_4$, titrate Br^-; $E_{Ag\text{-}SCE}$; protect from light	*Bull. Chem. Soc. Jpn.* **26**:394 (1953)

14.9.2 Instrumentation

14.9.2.1 Conductance Cells. A conductivity cell with large electrodes very close together has a low cell constant. If the distance is increased and/or the area of the electrodes is decreased, the cell constant increases. For example, a cell with a constant of 10 cm^{-1} would have electrodes perhaps 0.5 cm^2 in area and spaced 5 cm apart. This arrangement would be suitable for measuring the conductance of 0.005% to 2.0% sulfuric acid solutions the specific conductance of which ranges from about 0.000 44 to 0.176 $S \cdot cm^{-1}$. The resistance readings would range from 22 700 Ω for the dilute acid to 57 Ω for the 2% acid solution.

TABLE 14.39 Standard Solutions for Calibrating Conductivity Vessels

The values of conductivity κ are corrected for the conductivity of the water used. The cell constant θ of a conductivity cell can be obtained from the equation

$$\theta = \frac{\kappa R R_{solv}}{R_{solv} - R}$$

where R is the resistance measured when the cell is filled with a solution of the composition stated in the table below, and R_{solv} is the resistance when the cell is filled with solvent at the same temperature.

Grams KCl per kilogram solution (in vacuo)	Conductivity in $\text{ohm}^{-1}\cdot\text{cm}^{-1}$ at 0°C	18°C	25°C
71.135 2	$0.065\ 14_4$	$0.097\ 79_0$	$0.111\ 28_7$
7.419 13	$0.007\ 134_4$	$0.011\ 161_2$	$0.012\ 849_7$
0.745 263*	$0.000\ 773\ 2_6$	$0.001\ 219\ 9_2$	$0.001\ 408\ 0_8$

* Virtually 0.0100*M*.

Data from Jones and Bradshaw, *J. Am. Chem. Soc.* **55**:1780 (1933). The original data have been converted from (int. ohm)$^{-1}$ cm^{-1}.

Source: J. A. Dean, ed., *Lange's Handbook of Chemistry,* 14th ed., McGraw-Hill, New York, 1992.

The dip-type cell is simplest to use whenever the liquid to be tested is in an open container. Whatever the configuration of the cell, the test solution must completely cover the electrodes. Pipette cells permit measurements with as little as 0.01 mL of solution. A pair of individual square platinum electrodes on glass wands is useful in conductometric titrations.

14.9.2.2 Conductivity Meters. In the classical mode of conductance measurements, resistance measurements are made using some variation of a Wheatstone bridge. To balance the capacitive effects in the conductance cell, the bridge circuit must also contain a variable capacitance (8 to 200 pF) in parallel with the balancing resistor. A built-in generator provides bridge current at frequencies of 100, 1000, and 3000 Hz. A lower frequency is preferred when the measured resistance is high and a higher frequency when the measured resistance is low. Use of an alternating current eliminates the effects of faradaic processes; that is, the deposition potential is not exceeded. The cell constant of the conductivity cell should be selected to maintain the measured resistance between 100 Ω and 1.1 MΩ. Any smooth metal surface can serve as an electrode at an operating frequency of 3000 Hz. Stainless steel electrodes are frequently used for industrial on-line applications.

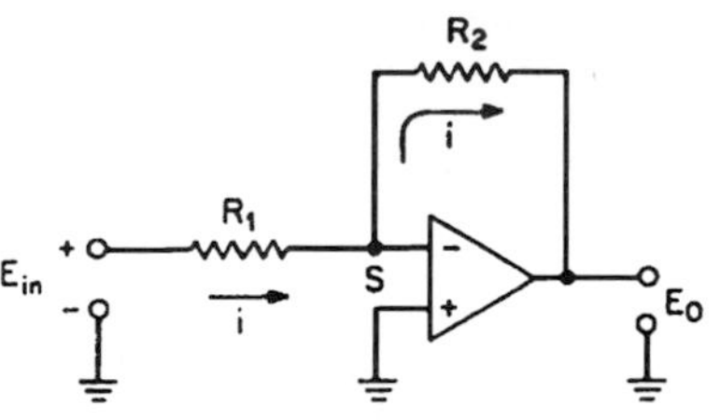

FIGURE 14.20 Operational amplifier used for resistance and conductance measurements. (*From Shugar and Dean, 1990.*)

The Wheatstone bridge can be replaced by operational amplifier circuitry, as shown in Fig. 14.20. For conductance measurements, the cell is connected in place of R_1; the result is a current $i = E/R_1$, which is proportional to the conductance of the cell. For resistance measurements, a fixed resistance R_1 is used in the input current-generating circuit and the cell is connected in place of R_2. The synchronous detector automatically rejects the current component resulting from capacitative coupling.

Differential conductivity measurements permit the measurement of differences in conductivity even when the total conductance is high. Also, a small discrete conductance change can be measured in the presence of a steadily changing conductance, as in gradient elution chromatography. A second cell monitoring the gradient can provide a flat baseline—except when a discrete change occurs due to elution of the sample. If the two cells are held at the same temperature, then even if that temperature changes, very precise measurement of differences in conductance can be made.

A second example of difference conductance measurements involves the monitoring of a flowing stream to which a reagent or contaminant is added. If conductance changes are measured by two cells—one upstream and one downstream from the point at which the additive enters the streams—then only differences in conditions caused by the additive will appear at the meter output, and changes in temperature and concentration of the stream will not affect the measurement.

14.9.3 Direct Analysis

Direct analysis involves the measurement of concentration of various electrolytes as a function of conductivity. Table 14.40 lists the conductivity of common industrial solutions at 25°C. The use of such data is restricted to solutions of a single solute at a specified temperature. When plotted from the tabulated data, the curves are markedly nonlinear at higher concentrations at which ionization may be incomplete and activity coefficients cannot be neglected. Maxima and minima are observed with many materials. Conductance increases with temperature at the rate of 0.024 per degree Celsius (at 25°C) for salts such as NaCl.

The direct method of analysis is valuable for many industrial purposes such as pickling bath concentration, rinsing after plating and cleaning operations, caustic degreasing baths, fruit peeling baths, concentration of process liquids, strengths of concentrated acids and oleum, and waters of various types and origins. A widespread application of a conductance monitor is in water-purification systems. Conductance liquid chromatography detectors are very important in monitoring solvent composition in gradient elution involving change of ionic strength or buffer characteristics. The restriction to a single solute can be removed if the system is well defined and the calibration made with similar mixtures.

14.9.4 Conductometric Titrations

In conductometric titrations the variation of the electrical conductivity of a solution during the course of a titration is followed. This technique is of wide applicability and good precision. Since the property measured bears a direct (not logarithmic) relation to the concentration, the titration curves (volume-corrected) consist of intersecting straight lines.

A conductometric titration is devised so that the ionic species to be determined can be replaced by another ionic species of significantly different conductance. The end point is obtained by the intersection of two straight lines that are drawn through a suitable number of points (usually four for each linear branch) obtained by measurement of the conductivity after each addition of titrant. The titrant should be at least 10 times as concentrated as the analyte in order to keep the volume change small. If necessary a correction may be applied. All conductance readings are multiplied by the ratio

$$\frac{V+v}{V}$$

where V is the initial volume and v is the volume of titrant added up to the particular conductance reading.

The major applications are to acid–base titrimetry. Titrant concentrations can be as low as 0.0001M. Under optimum conditions the end point can be located with a relative error of approximately 0.5%. Typical acid–base titrations will be considered. Limiting equivalent ionic conductances in aqueous solutions at 25°C are given in Table 14.41.

14.9.4.1 Strong Acid Titrated With Strong Base. Consider the titration of a 0.001M solution of HCl with 0.1M NaOH. During the formation of water, the highly conducting hydronium ion ($\lambda_+ = 350$)

TABLE 14.40 Conductivity of Common Industrial Solutions at 25°C

Values listed are in units of $\mu S \cdot cm^{-1}$ until the italicized value, which begins units of $S \cdot cm^{-1}$. 1% by weight is equivalent to 10 000 ppm, etc. Underlined values indicate that conductivity passes through a maximum between the two listed concentrations.

	Conductivity at 25°C*											
% by weight	Acetic acid	CO_2	HCl	HF	HNO_3	H_3PO_4	H_2SO_4	NH_3	NaCl	NaOH	Sea salt	SO_2
0.0001	4.2	1.2	11.7		6.8		8.8	6.6	2.2	6.2	2.2	
0.0003	7.4	1.9	35.0		20		26.1	14	6.5	18.4	6.5	
0.001	15.5	3.9	116		67		85.6	27	21.4	61.1	21.3	
0.003	30.6	6.8	340	290	199		251	49	64	182	64	
0.01	63	12	1140	630	657	342	805	84	210	603	208	
0.03	114	20	3390	1490	1950	890	2180	150	617	1780	612	
0.1	209	39	*0.0111*	2420	6380	2250	6350	275	1990	5820	1930	3600
0.3	368	55	0.0322	5100	*0.0189*	4820	*0.0158*	465	5690	*0.0169*	5550	7900
1	640		0.103	*0.0117*	0.0600	*0.0105*	0.0485	810	*0.0176*	0.0532	*0.0170*	*0.0170*
3	1120		0.283	0.0347	0.172	0.0230	0.141	<u>1110</u>	0.0486	0.144	0.0462	0.0327
10	<u>1730</u>		<u>0.709</u>	0.118	0.498	0.0607	0.432	1120	0.140	<u>0.358</u>		0.0610
30	1620		0.732	0.390	0.861	0.182	0.822	210		0.292		

* Nitric acid possesses a maximum at 30.7%, and minimum at 97.6%.
Sulfuric acid possess maxima at 31.4%, 92.0%, and 102.9% (oleum); and minima at 86.0% and 100.0%.

TABLE 14.41 Limiting Equivalent Ionic Conductances in Aqueous Solutions

In mho · cm² · equiv⁻¹.

Ion	Temperature, °C		
	0	18	25
	Inorganic cations		
Ag^+	33	54.5	61.9
Al^{3+}	29		61
Ba^{2+}	33.6	54.3	63.9
Be^{2+}			45
Ca^{2+}	30.8	51	59.5
Cd^{2+}	28	45.1	54
Ce^{3+}			70
Co^{2+}	28	45	53
$Co(NH_3)_6^{3+}$			100
$Co(en)_3^{3+}$			74.7
Cr^{3+}			67
Cs^+	44	68	77.3
Cu^{2+}	28	45.3	56.6
D^+ (deuterium)		213.7	
Dy^{3+}			65.7
Er^{3+}			66.0
Eu^{3+}			67.9
Fe^{2+}	28	45.3	53.5
Fe^{3+}			69
Gd^{3+}			67.4
H^+	224.1	315.8	350.1
Hg_2^{2+}			68.7
Hg^{2+}			63.6
Ho^{3+}			66.3
K^+	40.3	64.6	73.50
La^{3+}	35.0	59.2	69.6
Li^+	19.1	33.4	38.69
Mg^{2+}	28.5	46	53.06
Mn^{2+}	27	44.5	53.5
NH_4^+	40.3	64	73.7
$N_2H_5^+$ (hydrazinium 1+)			59
Na^+	25.85	43.5	50.11
Nd^{3+}			69.6
Ni^{2+}	28	45	50
Pb^{2+}	37.5	60.5	71
Pr^{3+}			69.6
Ra^{2+}	33	56.6	66.8
Rb^+	43.5	67.5	77.8
Sc^{3+}			64.7
Sm^{3+}			65.8
Sr^{2+}	31	51	59.46
Tl^+	43.3	66	74.9
Tm^{3+}			65.5
UO_2^{2+}			32
Y^{3+}			62
Yb^{3+}			65.2
Zn^{2+}	28	45.0	53.5

TABLE 14.41 Limiting Equivalent Ionic Conductances in Aqueous Solutions (*Continued*)

Ion	Temperature, °C		
	0	18	25
	Inorganic anions		
$Au(CN)_2^-$			50
$Au(CN)_4^-$			36
$B(C_6H_5)_4^-$			21
Br^-	43.1	67.6	78.4
Br_3^-			43
BrO_3^-	31.0	49.0	55.8
Cl^-	41.4	65.5	76.35
ClO_2^-			52
ClO_3^-	36	55.0	64.6
ClO_4^-	37.3	59.1	67.9
CN^-			78
CO_3^{2-}	36	60.5	72
$Co(CN)_6^{3-}$			98.9
CrO_4^{2-}	42	72	85
F^-		46.6	55.4
$Fe(CN)_6^{4-}$			111
$Fe(CN)_6^{3-}$			101
$H_2AsO_4^-$			34
HCO_3^-			44.5
HF_2^-			54.4
HPO_4^{2-}			57
$H_2PO_4^-$		28	36
HS^-	40	57	65
HSO_3^-	27		50
HSO_4^-			50
$H_2SbO_4^-$			31
I^-	42.0	66.5	76.9
IO_3^-	21.0	33.9	41.0
IO_4^-		49	54.5
MnO_4^-	36	53	62.8
MoO_4^{2-}			74.5
N_3^-			69.5
$Ni(CN)_4^{2-}$			54.5
NO_2^-	44	59	72
NO_3^-	40.2	61.7	71.42
$NH_2SO_3^-$ (sulfamate)			48.6
OCN^- (cyanate)		54.8	64.6
OH^-	117.8	175.8	199.2
PF_6^-			56.9
PO_3F^{2-}			63.3
PO_4^{3-}			69.0
$P_2O_7^{4-}$			81.4
$P_3O_9^{3-}$			83.6
$P_3O_{10}^{5-}$			109
ReO_4^-		46.5	54.9
SCN^- (thiocyanate)	41.7	56.6	66.5
$SeCN^-$			64.7
SeO_4^{2-}		65	75.7
SO_3^{2-}			79.9

(Continued)

TABLE 14.41 Limiting Equivalent Ionic Conductances in Aqueous Solutions (*Continued*)

Ion	Temperature, °C		
	0	18	25
	Inorganic anions		
SO_4^{2-}	41	68.3	80.0
$S_2O_3^{2-}$			85.0
$S_2O_4^{2-}$	34		66.5
$S_2O_6^{2-}$			93
$S_2O_8^{2-}$			86
WO_4^{2-}	35	59	69.4
	Organic cations		
Decylpyridinium			29.5
Diethylammonium			42.0
Dimethylammonium			51.5
Dipropylammonium			30.1
Dodecylammonium			23.8
Ethylammonium			47.2
Ethyltrimethylammonium			40.5
Isobutylammonium			38.0
Methylammonium			58.3
Piperidinium			37.2
Propylammonium			40.8
Tetrabutylammonium			19.1
Tetraethylammonium			33.0
Tetramethylammonium			45.3
Tetrapropylammonium			23.5
Trimethylammonium			34.3
Triethylsulfonium			36.1
Trimethylammonium			46.6
Trimethylsulfonium			51.4
Tripropylammonium			26.1
	Organic anions		
Acetate	20	34	41
Benzoate			32.4
Bromobenzoate			30
Butanoate			32.6
Chloroacetate			39.7
Chlorobenzoate			33
Citrate(3–)			70.2
Cyanoacetate			41.8
Cyclohexanecarboxylate			28.7
Cyclopropane-1, 1-dicarboxylate			53.4
Decylsulfonate			26
Dichloroacetate			38.3
Diethylbarbituate(2–)			26.3
Dihydrogen citrate			30
3,5-Dinitrobenzoate			28.3
Dodecylsulfonate			24
Ethylsulfonate			39.6
Fluorobenzoate			33
Formate		47	54.6

TABLE 14.41 Limiting Equivalent Ionic Conductances in Aqueous Solutions (*Continued*)

Ion	Temperature, °C		
	0	18	25
	Organic anions		
Hydrogen oxalate(1−)			40.2
Lactate			38.8
Methylsulfonate			48.8
Octylsulfonate			29
Phenylacetate			30.6
Propanoate			35.8
Propylsulfonate			37.1
Salicylate			36
Succinate(2−)			58.8
Tartrate(2−)		55	64
Trichloroacetate			36.6

Source: J. A. Dean, ed., *Lange's Handbook of Chemistry,* 14th ed., McGraw-Hill, New York, 1992.

is replaced by a less highly conducting sodium ion ($\lambda_+ = 50$). λ is the limiting equivalent ionic conduction, λ_+ or λ_-, for cations or anions, respectively. The conductivity falls linearly, as shown in curve 1 of Fig. 14.21, reaching a minimum if the solution consists of only NaCl. Unused NaOH and previously formed NaCl constitute the conductance of the rising branch of the titration curve. The conductance of the solution at any point on the descending branch of the titration curve is given by the expression

$$\frac{1}{R} = \frac{1}{1000\Theta}(C_{\mathrm{H}}\lambda_{\mathrm{H}} + C_{\mathrm{Na}}\lambda_{\mathrm{Na}} + C_{\mathrm{Cl}}\lambda_{\mathrm{Cl}}) \tag{14.54}$$

This equation can be expressed in terms of the initial concentration of HCl, C_i, and the fraction of the acid titrated, f:

$$C_{\mathrm{H}} = C_i(1-f), \qquad C_{\mathrm{Na}} = C_i f, \qquad C_{\mathrm{Cl}} = C_i \tag{14.55}$$

Substituting these values into Eq. (14.53),

$$\frac{1}{R} = \frac{C_i}{1000\Theta}[\lambda_{\mathrm{H}} + \lambda_{\mathrm{Cl}} + f(\lambda_{\mathrm{Na}} - \lambda_{\mathrm{H}})] \tag{14.56}$$

The term within the parentheses in Eq. (14.56) expresses the steepness of the drop in conductivity up to the end point.

14.9.4.2 Incompletely Dissociated Acids (or Bases). Titration of incompletely dissociated acids (or bases) is somewhat more difficult. Initially the solution has a low conductivity, which is due to

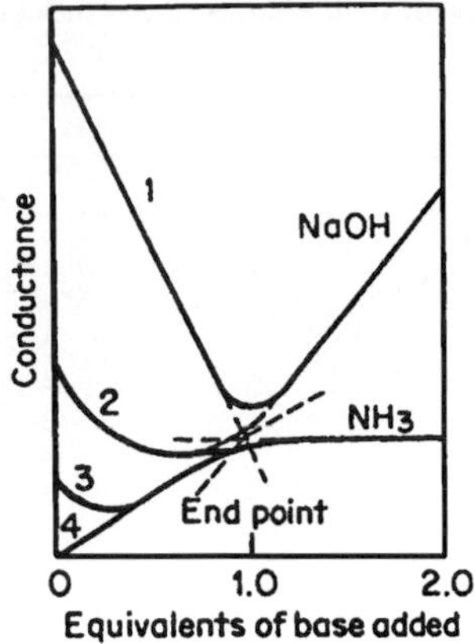

FIGURE 14.21 Acids of different strengths titrated with either sodium hydroxide or aqueous ammonia. The numbered curves are (1) HCl, (2) an acid of $pK_a = 3$, (3) an acid of $pK_a = 5$, and (4) acids whose $pK_a = 7$ or greater. *(From Shugar and Dean, 1990.)*

the few ionized ions present. As neutralization proceeds, the common ion formed (anion for a weak acid) represses the dissociation so that an initial fall in conductivity may occur. Consequently, the shape of the initial portion of these titration curves will vary with the strength of the weak acid (or base) and its concentration, as indicated in Figs. 14.21 and 14.22. Pronounced hydrolysis in the vicinity of the end point makes it necessary to select the experimental points for the construction of the two branches considerably removed from the end point.

Sometimes no linear region is obtained preceding the end point. A clever stratagem overcomes the problem. For example, in the titration of a weak acid, sufficient aqueous ammonia is added to neutralize about 80% of the weak acid. Then the titration is carried out with standard NaOH. After the remaining weak acid has been neutralized, the titration involves the reaction of the ammonium ion (formed during the neutralization of the weak acid) with NaOH to form ammonia, water, and sodium ions. The conductance falls owing to the replacement of the ammonium ion ($\lambda_+ = 73$) by the sodium ion ($\lambda_+ = 50$). When the replacement is complete, the conductivity increases abruptly due to the unused NaOH.

14.9.4.3 Mixtures of Different Acid Strength. A mixture of a strong and a weak acid can be determined in one titration, as illustrated in Fig. 14.23 for the titration of the first and the second protons in oxalic acid. The curve would be similar for the titration of a mixture of HCl and acetic acid (or other carboxylic acid). A more distinct second end point is obtained when the titrant is aqueous ammonia (or propyl amine). Weak acids can be titrated only if the product of the ionization constant and the acid concentration exceeds 10^{-11}.

The conductometric titration technique is also useful in the titration of the conjugate base of a weakly ionized acid, and vice versa. This extends the technique to organic salts such as acetates, benzoates, nicotinates, and so on. One caveat is that the ionization constant of the displaced acid or base divided by the original salt concentration must not exceed 5×10^{-3}. Conductometric titration methods are given in Table 14.42.

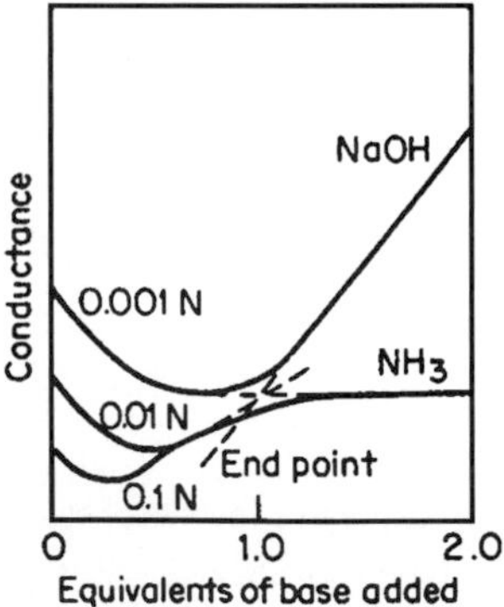

FIGURE 14.22 Effect of concentration upon the shape of the titration curve for an acid whose $pK_a = 4.8$. *(From Shugar and Dean, 1990.)*

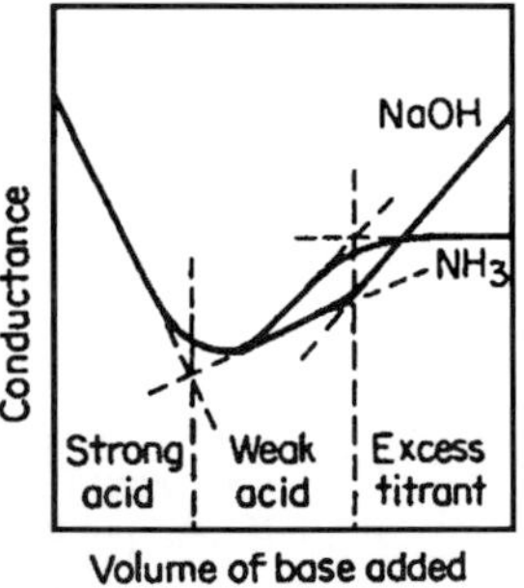

FIGURE 14.23 Titration of a mixture of a strong acid and a weak acid; the specific example involves the two protons of oxalic acid. *(From Shugar and Dean, 1990.)*

TABLE 14.42 Conductometric Titration Methods

Substance titrated	Concentration range	Reagent	Remarks
Acids ($pK_a \leq 2$)	0.1*M*–0.001*M*	NaOH	CO_2 absent.
Acids (pK_a 2 to 6)	0.1*M*–10^{-4}*M*	Aqueous NH_3	CO_2 absent.
Acids ($pK_a > 7$)	0.1*M*–10^{-4}*M*	LiOH	CO_2 absent. Titrate water-insoluble acids in 75% EtOH.
Ag(I)	0.1*M*–10^{-4}*M*	NaCl or LiCl	Neutral or dilute acid solution.
	0.1*M*–0.001*M*	Na_2CrO_4 or $Li_2C_2O_4$	Neutral solution.
Al(III)	10^{-2}*M*–10^{-3}*M*	NaOH	Titrate in H_2SO_4. Three breaks appear in titration curve. Analysis based on reagent consumed between first (neutralization of excess H^+) and third breaks (AlO_2^- formed).
As(III)	10^{-4}*M*–10^{-5}*M*	I_2 in EtOH	Titrate in dilute $NaHCO_3$ solution.
Ba(II)	0.1*M*–10^{-3}*M*	Li_2SO_4 or Li_2CrO_4	30% EtOH. Wait several minutes for stable readings after each addition of reagent.
Bases, strong	0.1*M*–10^{-3}*M*	HCl	CO_2 absent.
Bases, weak	0.1*M*–10^{-4}*M*	HOAc	CO_2 absent.
Bases, very weak	10^{-3}*M*	HCl	CO_2 absent. Titrate water-insoluble bases in 75% EtOH.
Bases, salts of weak acids	10^{-2}*M*–10^{-5}*M*	HCl or Cl_3CCOOH	CO_2 absent.
Be(II)	10^{-2}*M*–10^{-4}*M*	NaOH	Cl^- or NO_3^- solution, CO_2 absent. Two breaks appear: first is for neutralization of excess H^+.
Br^-	0.1*M*–10^{-5}*M*	$AgNO_3$	Neutral or slight acid solution. Add up to 90% EtOH for titration of very dilute solution.
ClO_4^-	0.05*M*	$(C_6H_5)_4AsCl$	pH 7; 30-s pause between increments of titrant.
CN^-	0.1*M*–10^{-3}*M*	$AgNO_3$	Neutral solution. Two breaks in titration curve; second [formation of $Ag(CN)_2^-$] gives more precise results.
	0.01*M*	$Hg(ClO_4)_2$	Neutral solution. Two breaks in titration curve corresponding to $Hg(CN)_4^{2-}$ and $Hg(CN)_2$. Using first break, CN^- may be determined in presence of Cl^-.
CNO^-	0.1*M*–0.01*M*	$AgNO_3$	Neutral solution.
CO_3^{2-}	10^{-2}*M*–10^{-4}*M*	HCl	In OH^- solution, three breaks occur: neutralization of OH^-, CO_3^{2-}, and HCO_3^-. Analysis based on third break most precise.
Cd(II)	0.1*M*	$Li_4[Fe(CN)_6]$	Pb(II) does not interfere.
	0.1*M*	EDTA	Acetate buffer, pH 5.
Cl^-	0.1*M*–10^{-5}*M*	$AgNO_3$	Add up to 90% EtOH for titration of very dilute solutions.
Co(II)	0.01*M*	EDTA	Acetate buffer, pH 5.
	0.01*M*	$Li_3[Fe(CN)_6]$	Dilute acid solution.
F^-	0.1*M*–10^{-3}*M*	$AlCl_3$	Titrate in 30%–50% EtOH containing 50-fold excess for NaCl. Precipitate is Na_3AlCl_6.
Fe(II)	0.01*M*–10^{-3}*M*	$K_2Cr_2O_7$	Titrate in 0.1*M* H_2SO_4.
$Fe(CN)_6^{3-}$	0.1*M*–0.01*M*	$AgNO_3$	Neutral solution.

(Continued)

TABLE 14.42 Conductometric Titration Methods (*Continued*)

Substance titrated	Concentration range	Reagent	Remarks
$Fe(CN)_6^{4-}$	0.1*M*–0.01*M*	$Pb(NO_3)_2$	Neutral or dilute acid solution.
H^+		See Acids above	
I^-	0.1*M*–10^{-4}*M*	$AgNO_3$	Titration in presence of 2% NH_3 prevents interference from Cl^- and small amounts of Br^-.
IO_3^-	0.01*M*–10^{-4}*M*	HCl	Neutral solution containing small excess of KI and $Na_2S_2O_3$.
K^+	5×10^{-3}*M*	$NaB(C_6H_5)_4$	pH 5–10.
Mg(II)	0.1*M*–0.01*M*	NaOH	
	0.1*M*–0.01*M*	EDTA	NH_3 buffer, pH 10.
NO_3^-	0.05*M*	Nitron in HOAc	Dilute acid solution.
Ni(II)	0.01*M*	Dimethylglyoxime in EtOH	NH_3 buffer. Preferable to add excess DMG and back-titrate with standard Ni(II).
OH^-		See Bases above	
Pb(II)	0.1*M*–10^{-4}*M*	$Li_2C_2O_4$ or Na_2CrO_4	Neutral solution.
	0.01*M*	EDTA	Acetate buffer, pH 5.
Peroxide	0.1*M*–10^{-3}*M*	$KMnO_4$	0.05*M* H_2SO_4 solution.
PO_4^{3-}	0.1*M*–10^{-3}*M*	$BiOCl_4$	0.3*M* $HClO_4$ solution. As absent. Small amounts of most metals do not interfere.
SCN^-	0.1*M*–10^{-5}*M*	$AgNO_3$	Neutral or slightly acid solution. Add up to 90% EtOH for titration of very dilute solutions.
	0.1*M*–10^{-3}*M*	$Hg(ClO_4)_2$	Neutral solution. Second break [formation of $Hg(SCN)_2$] gives more precise results.
SO_4^{2-}	0.1*M*–10^{-4}*M*	$Ba(OAc)_2$ in 1% HOAc	20% EtOH. Large amounts of nitrate interfere.
$S_2O_3^{2-}$	0.1*M*–0.01*M*	$Pb(NO_3)_2$	Neutral solution.
Se(IV)	0.01*M*–10^{-2}*M* as selenite	$AgNO_3$ or $Pb(NO_3)_2$	Neutral solution.
Se(VI)	0.01*M*–10^{-2}*M* as selenate	$BaCl_2$ or $Pb(NO_3)_2$	50% EtOH solution.
Sr(II)	0.01*M*	EDTA	NH_3 buffer, pH 10.
Tl(I)	5×10^{-3}*M*	Na_2CrO_4	Neutral solution.
	0.02*M*	$NaB(C_6H_5)_4$	Neutral solution.
	0.1*M*–10^{-2}*M*	KSCN	Neutral solution.
U(IV)	10^{-2}*M*–10^{-3}*M*	$KMnO_4$	0.2*M* to 0.5*M* H_2SO_4 solution.
V(III)	10^{-2}*M*–10^{-3}*M*	$KMnO_4$	Dilute H_2SO_4 solution.
V(V)	0.01*M* VO_4^{3-}	$AgNO_3$	Neutral solution.
V(V)	0.01*M*–0.005*M* VO_4^{3-}	$Co(NH_3)_6Cl_3$	Neutral solution. Precipitate has composition $[Co(NH_3)_6][V_2O_7]_3$.
	0.01*M* VO_3^-	$Co(NH_3)_6Cl_3$	Neutral solution. Precipitate has composition $[Co(NH_3)_6][VO_3]_3$.
Zn(II)	10^{-2}*M*–10^{-3}*M*	NaOH	Solution should contain H^+ equivalent to Zn(II) present. Two breaks in titration curve: first for H^+, second for precipitation of $Zn(OH)_2$.

SECTION 15
THERMAL ANALYSIS

15.1 INTRODUCTION

Thermal analysis includes a group of techniques in which specific physical properties of a material are measured as a function of temperature. The techniques include the measurement of temperatures at which changes may occur, the measurement of the energy absorbed (*endothermic* transition) or evolved (*exothermic* transition) during a phase transition or a chemical reaction, and the assessment of physical changes resulting from changes in temperature.

Various environments (vacuum, inert, or controlled gas composition) and heating rates from 0.1 to 500°C · min^{-1} are available for temperatures ranging from −190 to 1400°C. The analysis of gas(es) released by the specimen as a function of temperature is possible when thermal analysis equipment is coupled with Fourier-transform infrared detection or with a mass spectrometer.

The applications of thermal analysis are many and varied. For environmental measurements, these parameters can be measured: vapor pressure, thermal stability, flammability, softening temperatures, and boiling points. Compositional analysis offers phase diagrams, free versus bound water, solvent retention, additive analysis, mineral characterization, and polymer system analysis. In the important area of product reliability, thermal methods provide heat-capacity data, liquid-crystal transitions, solid fat index, purity, polymer cures, polymer quality control, glass transitions, Curie point, and fiber properties. Information on stability can be obtained from modulus changes, creep studies, expansion coefficients, and antioxidant evaluation. Dynamic properties of materials are found from viscoelastic measurements, impact resistance, cure characteristics, elastic modulus, loss modulus, and shear modulus. Lastly, chemical reactions can be followed through heats of transition, reaction kinetics, catalyst evaluation, metal–gas reactions, and crystallization phenomena.

15.2 DIFFERENTIAL SCANNING CALORIMETRY AND DIFFERENTIAL THERMAL ANALYSIS

Differential scanning calorimetry (DSC) and quantitative differential thermal analysis (DTA) measure the rate and degree of heat change as a function of time or temperature. In addition to these direct energy measurements, the precise temperature of the sample material at any point during the experiment is also monitored. Since DSC can measure both temperatures and heats of transitions or reactions, it has replaced DTA as the primary thermal analysis technique, except in certain high-temperature applications.

These methods are used to investigate the thermal properties of inorganic and organic materials. The procedure involves recording a temperature (DTA) or power (DSC) difference between a sample and a reference container as both are carried through a temperature program. DTA detects the physical and chemical changes that are accompanied by a gain or loss of heat in a substance as the temperature is altered. DSC provides quantitative information about any heat changes, including the rate of heat transfer.

When a thermal transition occurs in the sample, thermal energy is added either to the sample or the reference holders in order to maintain both the sample and reference at the sample temperature. Because the energy transferred is exactly equivalent in magnitude to the energy absorbed or evolved in the transition, the balancing energy yields a direct calorimetric measurement of the transition energy at the temperature of the transition.

Elevated pressure or reduced pressure is often beneficial to DSC experiments for such applications as the study of pressure-sensitive reactions, accelerated oxidation of lubricants or polymers, evaluation of catalysts, and resolution of overlapping transitions.

15.2.1 Instrumentation for DSC

In a cell designed for quantitative DSC measurements, two separate sealed pans, one containing the material of interest and the other containing an appropriate reference, are heated (or cooled) uniformly. The enthalpic difference between the two is monitored either at any one temperature (isothermal) or the temperature can be raised (or lowered) linearly. If maximum calorimetric accuracy is desired, the sample and reference thermocouples must be removed from direct contact with the materials.

The gradient temperature experiments can be run slowly (0.1°C · min^{-1}) or rapidly (up to 300°C · min^{-1}). The electronic circuitry detects any change in heat flow in the sample versus the reference cell. This event in turn is sent as an analog signal to an output device such as a strip-chart recorder, digital integrator, or a computer. Information may be obtained with samples as small as 0.1 mg.

However, quantitative studies usually require at least 1 mg of sample. Expanding the upper temperature limit to 1500°C allows for DSC analyses to be done on materials such as ceramics, metal alloy systems, silicates, and high-temperature composites.

The DSC cell uses a constantan disk as the primary means of transferring heat to the sample and reference holders and also as one element of the temperature-sensing thermoelectric junction. Samples in powder, sheet, film, fiber, crystal, or liquid form are placed in disposable aluminum sample pans of high thermal conductivity and weighed on a microbalance. The sample is placed in one sample holder and an empty sample holder serves as reference. Sample sizes range from 0.1 to 100 mg.

The differential heat flow to the sample and reference through the disk is monitored by the chromel–constantan thermocouples formed by the junction of the constantan disk and the chromel wafer covering the underside of each platform. Chromel and alumel wires connected to the underside of the wafers form a chromel–alumel thermocouple, which is used to monitor the sample temperature. Thin-layer, large-area sample distribution minimizes thermal gradients and maximizes temperature accuracy and resolution.

Generally, holders are covered with domed, aluminum sample-holder covers. The covers are essentially radiation shields that enhance baseline linearity and reproducibility from run to run. An airtight sample-holder enclosure isolates the holders from external thermal disturbances and allows vacuum or controlled atmosphere operation. A see-through window permits observation of physical changes in unencapsulated samples during a scan. Side ports allow addition of catalysts and seeding of supercooled melts.

Three different DSC cells are available: standard (as described), dual sample, and pressure capability. The dual-sample cell provides an immediate twofold increase in sample throughput, while the pressure DSC cell provides capability from vacuum to 7 MPa (1000 $lb \cdot in^{-2}$).

15.2.2 Instrumentation for DTA

In DTA a thermocouple is inserted into the center of the material in each sample holder. The sample is in one holder and reference material is placed in the other sample holder. The sample blocks are then heated. The difference in temperature between sample and reference thermocouples is continuously measured. Furnace temperature is measured by an independent thermocouple. Any transition that the sample undergoes results in liberation or absorption of energy by the sample with a corresponding deviation of its temperature from that of the reference. A plot of the differential temperature versus the programmed temperature indicates the transition temperature(s) and whether the transition is exothermic or endothermic.

For high-temperature studies (1200 to 1600°C), the sample holders are fabricated from platinum-iridium.

15.2.3 Applications of DSC and DTA

Applications include melting-point and glass transition measurements, curing studies, oxidative stability testing, and phase-transition studies.

15.2.3.1 Glass Transition Temperature. The DSC trace of a polymeric material will show a step in the recorded signal by the polymer undergoing a second-order heat-capacity change at the glass transition temperature (T_g). As the temperature continues to increase, viscosity drops, and the polymer becomes mobile enough to realign into a crystalline structure. The heat released from the so-called "cold crystallization" produces an exothermic peak on the DSC curve. The area under the peak is proportional to the heat of crystallization. At higher temperatures the crystalline material melts, and an endothermic peak is recorded. The glass transition determines the useful lower temperature range of an elastomer, which is normally used well above this temperature to ensure the desired softness. In thermoset materials the glass transition temperature is related to the degree of cure, with the glass transition temperature steadily increasing as the degree of cure increases. Additionally, the glass transition temperature is often used to determine the storage temperature of uncured or partially cured material,

TABLE 15.1 Glass Transition Temperatures (T_g) and Crystalline Melting Points of Selected Polymers

Values in parentheses are approximate or uncertain.

Polymer	T_g, °C	mp, °C
Polybutadiene	−86	(−20)
Polyisobutylene	−73	(44)
Poly(ethylene vinyl acetate) copolymer	−20 to 20*	40 to 100*
Polyethylene, low density	(−100)	120
Polybutene		130
Polyethylene, high density	(−70)	135
Poly(oxymethylene) copolymer		164 to 168
Polypropylene	(−30)	165
Poly(vinyl chloride) soft	−40 to 10†	
Poly(vinylidene chloride)	−17	
Poly(oxymethylene) homopolymer		175 to 180
Poly(cinylidene fluoride)		178
Polyamide 11		186
Poly(vinyl acetate)	30	
Poly(vinyl chloride)	85	(190)
Poly(butylenes terephthalate)	65	220
Polyamide 6	(40)	220
Polyamide 6,10	(46)	226
Poly(vinyl alcohol)	85	
Polystyrene	90 to 100	
Poly(methyl methacrylate)	105	
Epoxy resin	50 to 150‡	
Poly(phenylene oxide)		230
Polycarbonate	155	(235)
Polyamide 6,6	(50)	255
Poly(ethylene terephthlate)	(69)	256
Poly(ethylene tetrafluoroethylene) copolymer		270
Poly(fluoroethylene propylene)		280
Poly(phenylene sulfide)	80	280
Polyacrylonitrile	100	(320)
Polytetrafluoroethylene	(−20)	327

* Depending on ethylene content.
† Depending on plasticizer content.
‡ Depending on formula.

since curing does not begin until the material is heated above its glass transition temperature. Table 15.1 contains glass transition temperatures and crystalline melting points of selected polymers.

15.2.3.2 Crystallization and Fusion. The exothermal crystallization enthalpy as well as the endothermal heat of fusion (at −25°C) can be evaluated by integrating curve areas. In a semicrystalline polymer the crystallinity can be calculated from the measured heat of fusion, which is a characteristic property and widely used for quality control.

15.2.3.3 Curing Reactions. During a curing reaction, energy is released from the sample as cross-linking occurs and a large exothermic peak follows the glass transition.

15.2.3.4 Curie Point. An interesting application of DSC is the determination of the Curie point temperatures of ferromagnetic materials. The Curie point is discussed in more detail in Sec. 15.3. Selected Curie point standards are alumel, 163°C; nickel, 354°C; perkalloy, 596°C, and iron, 780°C.

15.2.3.5 Oxidative Stability. DSC is a useful tool for generating data on oxidative stability of fats and oils. In either the isothermal or the temperature-programmed mode, the onset of a deviation from the baseline can be related to the oxidative induction period. In practice, the system would be brought to the test temperature and held isothermally in a nitrogen atmosphere. Oxygen is then introduced and the time necessary to give the first indication of oxidation (exotherm) is noted. DSC can differentiate quickly between stable and unstable systems and determine which antioxidant system would best preserve the specimen.

15.2.3.6 Chemical Reactions. Depending upon the polymer and its intended use, various types of chemical reactions may be studied. Exotherms produced by polymerization, curing, oxidative degradation, and other reactions can be studied. Such information is useful in processing, in shelf-life estimates, and in hazards evaluation for energetic polymers.

In many cases it is possible to derive kinetics information from the DSC curves. One of the simplest and most broadly applicable methods is based upon the change in the DSC peak maximum temperature with heating rate. Samples are temperature-programmed at several rates (β) and the corresponding exothermic peak temperatures (T) are plotted as

$$\frac{d \log \beta}{d(1/T)} \approx 1.912 E/R \tag{15.1}$$

where E is the Arrhenius activation energy in joules per mole. The reaction frequency factor (Z) is calculated from the same curves as

$$Z = \frac{\beta E e^{E/RT}}{RT^2} \tag{15.2}$$

The polymer reaction rate (k) at any temperature is available from the Arrhenius equation

$$k = Z e^{-E/RT} \tag{15.3}$$

The equation for Z is a first-order form; however, the error incurred for other reaction orders is small and can usually be neglected.

15.3 THERMOGRAVIMETRIC ANALYSIS[1]

Thermogravimetric analysis (TGA) monitors the change in mass of a substance as a function of temperature or time as the sample is subjected to a controlled temperature program. TGA has often been used to rank polymeric materials in order of their thermal stability by comparing losses of weight versus temperature. Many important TGA procedures involve isothermal monitoring of the mass loss from a material. Thus, when one or more isothermal holds are involved in the total heating program, the mass loss is often monitored versus time. In addition, plotting heating rate versus the temperature for a specified degree of conversion yields kinetic data for the decomposition process. The weight-loss method for determining kinetic values avoids difficulties that occur for DSC when exothermic and endothermic events coincide and interfere with the analysis.

A second use of TGA is the determination of the loss rate of moisture, diluent, and unreacted monomer that must be removed from the polymer. Polymers can also be pyrolyzed in the TGA equipment to enable determination of carbon black fillers or residual inorganic material. Another important use of TGA is in helping in the interpretation of DSC and DTA thermograms. For example, early endothermic activity in a programmed DSC curve might represent low-melting-point

[1] C. M. Earnest, "Modern Thermogravimetry," *Anal. Chem.* **56**:1471A (1984).

polymer, or it may be due to volatilization of low-molecular-weight material. A TGA run would resolve the question.

15.3.1 TGA Instrumentation

To perform TGA, the equipment must be capable of both simultaneous heating and weighing. Instrumentation includes the following:

1. A sensitive recording microbalance capable of detecting weight changes as small as 0.1 μg.
2. A furnace and an appropriate enclosure for heating the sample specimen. For most applications the temperature range will be from ambient to 1000°C. A small low-mass designed furnace permits rapid linear heating and cooling rates as fast as 200°C per minute.
3. A temperature programmer, heat-control circuitry, and associated electronics. Linear heating rates from 5 to 10°C $\cdot$ min^{-1} are typical, although much higher rates are available. Many important TGA procedures involve isothermal monitoring of the mass loss from a material.
4. A pneumatic system for dynamic purging of the furnace and sample chamber.
5. A data acquisition system.

Additional features might be a purge-gas switching capability for those applications in which the purge gas is changed during the experiment, and software to generate a first-derivative differential thermogravimetric curve from the TG data after storage.

Thermal curves obtained by TGA offer what is probably the most accurate ordinate scale of weight or weight percent of all thermal analysis techniques. The temperature axis is often less well defined. First, the thermocouple is generally close but not in contact with the sample; also the dynamic gaseous atmosphere surrounding the TG sample may influence the sample temperature.

15.3.2 Applications of TGA

Thermogravimetry provides the laboratory chemist with a number of important testing applications. The most important applications include compositional analysis and decomposition profiles of multicomponent systems studied at varying temperatures and atmospheric conditions, parameters that can be tailored and switched at any point during the experiment. Other important applications include the rapid proximate analysis of coal, quantitative separation of the main sample components in multicomponent mixtures, the determination of the volatile and moisture components in a sample material, kinetic studies, accelerated aging tests, and oxidation–reduction reactions.

Curie point measurements by TGA provide an accurate method for the calibration of the temperature axis of the thermogram since the Curie point temperature of many materials is well known and characterized. The ferromagnetic material is suspended in a magnetic field that is oriented such that a vertical component of magnetic force acts on the sample. This magnetic force acts as an equivalent magnetic mass on the TGA microbalance beam to indicate an apparent sample weight. When the sample is heated through its Curie point, the magnetic mass will be lost and the microbalance will indicate an apparent weight loss.

Moisture determination is an important application of TGA. In many industries even small amounts of moisture have serious consequences. When the sample is rapidly heated to 105°C and held at this temperature, any moisture present in the sample is lost. Moisture levels at 0.5% and often below can be determined.

Determination of the temperature of oxidation of a sample is another TGA application. For example, if magnesium powder is heated from 300 to 900°C in an oxidizing (air) atmosphere, at approximately 682°C a sharp increase in sample weight is noted that corresponds to the rapid oxidation of the material.

15.4 THERMOMECHANICAL ANALYSIS

Thermomechanical analysis (TMA) measures either the dimension or deformation of a substance under nonoscillatory load as a function of temperature or time. It is used to measure changes in the physical properties of sample materials, such as compression, coefficients of expansion, softening temperatures, viscosity/flow, stress relaxation/viscoelasticity, composite delamination, glass transition temperature, heat deflection temperature measurements, and creep, fiber, or film shrinkage. Useful information is provided with regard to behavior of the material at either elevated or reduced temperatures while under an external load, which can be varied. Very little chemical information can be determined by TMA.

15.4.1 TMA Instrumentation

The analyzer measures the change in dimension as a function of temperature and stress. One of several selectable probes makes contact with a sample. The probe touches the upper surface of the sample and applies a well-defined stress. As the sample expands, contracts, or softens, the position of the probe will change. This position is monitored by a linear variable differential transformer (LVDT) that provides a signal proportional to the probe displacement. Several probe configurations are available for TMA.

In the simplest arrangement, with a sharp probe placed under load on the sample, glass transition temperatures as well as softening and flow temperatures can be extrapolated. The T_g determination is generally more sensitive than that by DSC.

A flat probe on the sample with little or no load allows determination of a linear coefficient of expansion. In the expansion, penetration, and hemispherical configurations, the outer member is the quartz platform and is fixed. The inner members are movable and connected to the LVDT core.

Another probe, in which the sample expands in a glass bulb filled with microbeads, serves as a dilatometer to determine cubical expansion versus temperature.

The tension and fiber probes put the sample in tension by pulling instead of pushing. The dilatometer probe is designed to measure volume changes.

15.4.2 Applications of TMA

In the penetration and expansion modes, the sample rests on a quartz stage surrounded by the furnace. Under no load, expansion with temperature is observed. At the glass transition temperature, there is a discontinuous change in expansion coefficient, as evidenced by the elbow in the TMA scan. The thermal coefficient of linear expansion is calculated directly from the slope of the resulting curve. In the penetration mode the penetration of the probe tip into the sample is observed under a fixed weight placed in the weight tray. Sample sizes may range from a 0.1-mil coating to a 0.5-in thick solid; sensitivities down to a few microinches are observable. In the penetration mode, TMA is a sensitive tool for the characterization and quality control of thin films and coatings. For the measurement of samples in tension, the sample holder consists of stationary and movable hooks constructed of fused silica. This permits extension measurements on films and fibers; these measurements are related to the tensile modulus of a sample.

For a swelling measurement, the sample is placed on the bottom of a small cup and is covered by a small disk of aluminum oxide. In an isothermal experiment, the cup is filled with solvent at time zero. The swelling of the sample increases its thickness, which is measured by the probe.

15.5 DYNAMIC MECHANICAL ANALYSIS

Dynamic mechanical analysis (DMA) is the measurement of the mechanical properties of materials as they are deformed under periodic stress. Of special importance is the modulus of materials as well as their damping characteristics under oscillatory load as a function of temperature. It involves measuring the resonant frequency and mechanical damping of a material forced to flex at a selected amplitude. These parameters define the inherent stiffness of a material and its tendency to convert mechanical energy into heat when stressed.

Two parallel, balanced sample-support arms are free to oscillate around flexure pivots. A specimen material of known dimensions is clamped between the two arms. The force modulator is an electromagnet used in place of the weights required in the static TMA mode. The magnet receives its current impulses through the circuit with an integrated function generator and power amplifier; this oscillates the sample–arm–pivot system. The frequency and amplitude of this oscillation are detected by a linear variable differential transformer positioned at the opposite end of the active arm. The frequency of oscillation is directly related to the modulus of the sample, while the energy needed to maintain constant amplitude oscillation is a measure of damping within the sample.

DMA allows the user to observe elastic and inelastic deformations of materials, providing information relative to changes in moduli of shear and elasticity, as well as changes in loss modulus. Microprocessor data-reduction techniques provide graphical and tabular outputs of these properties as functions of time or temperature. It also makes possible the detection of glass transitions generally before such are detectable by other means.

15.6 ENTHALPIMETRIC ANALYSIS[2,3]

Enthalpimetric analysis, which includes the titrimetric and calorimetric modes, utilizes the temperature change in a system while a titrant is gradually added or measures the thermal energy released during a controlled reaction of the specimen. The main types of enthalpimetric analysis are thermometric enthalpy titrations and direct injection enthalpimetry. In practical terms they can be differentiated by the way the reactant is introduced to the adiabatic cell.

15.6.1 Thermometric Enthalpimetric Titrations

Thermometric enthalpimetric titrations (TET) are characterized by the continuous addition of the titrant to the sample under effectively adiabatic conditions. The total amount of heat evolved (if the reaction is exothermic) or absorbed (if the reaction is endothermic) is monitored using the unbalance potential of a Wheatstone bridge circuit, incorporating a temperature-sensitive semiconductor (thermistor) as one arm of the bridge. Simple styrofoam-insulated reaction cells will maintain pseudo-adiabatic conditions for the short period of a titration.

The heat capacity of the system will remain essentially constant if the change in volume of the solution is minimized and if the titrant and titrate are initially at the same temperature (usually room temperature). The TET ethalpogram (a thermometric titration curve), shown in Fig. 15.1, illustrates an exothermic titration reaction. The base line AB represents the temperature–time blank, recorded prior to the start of the actual titration. B corresponds to the beginning of the addition of the titrant, C is the end point, and CD is the excess reagent line. In order to minimize variations in heat capacity during titrations, it is customary to use a titrant that is 50 to 100 times more concentrated than the specimen being titrated. Thus the volume of the titrate solutions is maintained virtually constant, but the titrant is diluted appreciably. Correction for the latter is conveniently made by linear back-extrapolation CB'. Under these conditions, the extrapolated ordinate height, BB', represents a measure of the change of temperature due to the titration reaction.

---

[2] A. J. Hogarth and J. D. Stutts, "Thermometric Titrimetry: Principles and Instrumentation," *Am. Lab.* **13** (January):18 (1981).

[3] J. Jordan et al., "Enthalpimetric Analysis," *Anal. Chem.* **48**:427A (1976).

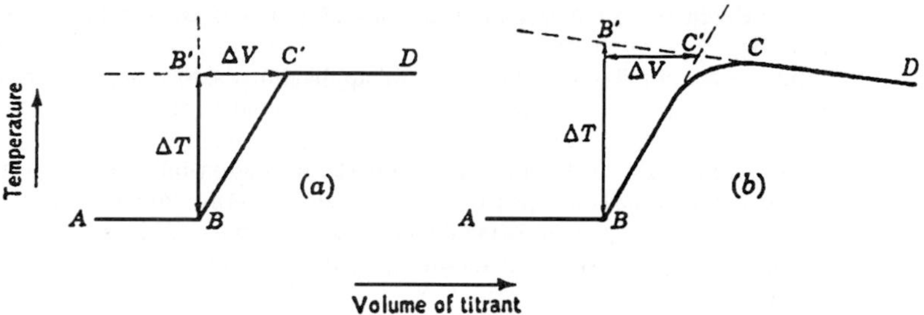

FIGURE 15.1 Typical thermometric titration curves for an exothermic process: (*a*) idealized curve, and (*b*) actual curve, illustrating extrapolation correction for curvature due to incompleteness of reaction. *AB*—temperature–time blank, slope due to heat leakage into or out of titration cell; *B*—start of titration; *BC*—titration branch; *C′*—end point; *CD*—excess reagent line, slope due to heat leakage and temperature difference between reagent and solution being titrated; ΔT—corrected temperature change; ΔV—end-point volume. (*From L. Meites, ed., Handbook of Analytical Chemistry, McGraw-Hill, New York, 1963.*)

In contrast to most analytical procedures that depend on a property related solely to equilibrium constants (i.e., free energy methods), TET depend on the heat of the reaction as a whole, viz.,

$$\Delta H = \Delta F + T\,\Delta S \tag{15.4}$$

Consequently, thermometric titration curves may yield a well-defined end point when all free energy methods fail if the entropy term in Eq. (15.4) is favorable. This is indeed the case in many alkalimetric titrations of weak acids. The corresponding enthalpograms for the titration of hydrochloric acid and boric acid are strikingly similar, because the heats of neutralization are comparable: –56.5 and −42.7 J/mol, respectively, for HCl and boric acid. For boric acid ($pK_a = 9.24$) the direct free energy (potentiometric titration) method for boric acid fails to provide a sharp end point.

15.6.1.1 Applications. Since heat of reaction is the most general property of chemical processes, thermometric titrations have a very wide range of applicability in quantitative analysis. Nonaqueous systems are well suited for this method. Thermometric titrations are very useful in titrating acetic anhydride in acetic acid–sulfuric acid acetylating baths, water in concentrated acids by titration with fuming acids, and free anhydrides in fuming acids. Precipitation and ion-combination reactions such as the halides with silver and cations with ethylene–diaminetetraacetate are other possibilities. Even halide titrations in fused salts have been done.

15.6.2 Direct-Injection Enthalpimetry

Direct-injection enthalpimetry (DIE) involves the (virtually) instantaneous injection of a single "shot" of reactant into the solution under investigation, the reagent being in stoichiometric excess. The corresponding heat evolved (or absorbed) is directly proportional to the number of moles of analyte reacted. Unstandardized reagents can be used, provided they are added in sufficient excess to make the reaction at least 99% complete. Circuitry is identical to that employed with TET. The amount of heat evolved or absorbed is calculated from Joule heating calibration experiments. Precise volume measurements are not a prerequisite of DIE.

The calorimeter is calibrated by passing a constant current i through a calibration heater resistor (which is immersed in the solution) and a standard resistor in the external circuit. The heat dissipated within the calorimeter may then be calculated from the expression:

$$Q(\text{J}) = i^2 R_H t = \frac{V_S V_H t}{R_S} \tag{15.5}$$

where V_S and V_H are potential drops measured across the standard resistor R_S and the calibration heater R_H.

DIE is a method of virtually universal applicability. Any rapid process that involves a heat of reaction that is equal or greater than 4.18 kJ per mole is amenable for use, whether exothermic or endothermic.

Continuous-flow enthalpimetry is utilized for the on-line analysis of industrial process streams. The technique consists of passing two reactant solutions at a constant rate through a mixing chamber and continuously monitoring the heat output of the product stream. The reagent must be in stoichiometric excess. The direct thermodynamic characterization of biological molecular interactions has been summarized.[4]

15.7 THERMOMETRY

The international temperature scale, known as ITS-90, was adopted in September 1989. Neither the definition of thermodynamic temperature nor the definition of the kelvin or the Celsius temperature scales has changed; it is the way in which we are to realize these definitions that has changed. The changes concern the recommended thermometers to be used in different regions of the temperature scale and the list of secondary standard fixed points. The changes in temperature determined using ITS-90 from the previous IPTS-68 are always less than 0.4 K, and almost always less than 0.2 K, over the range 0 to 1300 K.

The ultimate definition of thermodynamic temperature is in terms of pV (pressure × volume) in a gas thermometer extrapolated to low pressure. The kelvin (K), the unit thermodynamic

[4] E. Treire, O. L. Mayorga, and M. Straume, "Isothermal Titration Calorimetry," *Anal. Chem.* **62**:950A (1990).

TABLE 15.2 Fixed Points in the ITS-90

Fixed points	T, K	t, °C
Triple point of hydrogen	13.8033	−259.3467
Boiling point of hydrogen at 33 321.3 Pa	17.035	−256.115
Boiling point of hydrogen at 101 292 Pa	20.27	−252.88
Triple point of neon	24.5561	−248.5939
Triple point of oxygen	54.3584	−218.7916
Triple point of argon	83.8058	−189.3442
Triple point of mercury	234.3156	−38.8344
Triple point of water	273.16	0.01
Melting point of gallium	302.9146	29.7646
Freezing point of indium	429.7458	156.5985
Freezing point of tin	505.078	231.928
Freezing point of zinc	692.677	419.527
Freezing point of aluminum	933.473	660.323
Freezing point of silver	1234.93	961.78
Freezing point of gold	1337.33	1064.18
Freezing point of copper	1357.77	1084.62
Secondary reference points to extend the scale (IPTS-68)		
Freezing point of platinum	2042	1769
Freezing point of rhodium	2236	1963
Freezing point of iridium	2720	2447
Melting point of tungsten	3660	3387

Source: J. A. Dean, ed., *Lange's Handbook of Chemistry*, 14th ed., McGraw-Hill, New York, 1992.

temperature, is defined by specifying the temperature of one fixed point on the scale—the triple point of water which is defined to be 273.16 K. The Celsius temperature scale (°C) is defined by the equation

$$°C = K - 273.15 \tag{15.6}$$

where the freezing point of water at 1 atm is 273.15 K.

The fixed points in the ITS-90 are given in Table 15.2. Platinum resistance thermometers are recommended for use between 14 and 1235 K (the freezing point of silver), calibrated against the fixed points. Below 14 K either the vapor pressure of helium or a constant-volume gas thermometer is to be used. Above 1235 K radiometry is to be used in conjunction with the Planck radiation law,

$$L_\lambda = c_1 \lambda^{-5} (e^{c_2/\lambda T} - 1)^{-1} \tag{15.7}$$

where L_λ is the spectral radiance at wavelength λ. The first radiation constant c_1 is $3.741\ 83 \times 10^{-16}$ $W \cdot m^2$ and the second radiation constant c_2 has a value of 0.014 388 m · K.

15.8 THERMOCOUPLES

The thermocouple reference data in Tables 15.3 to 15.11 give the thermoelectric voltage in millivolts with the reference junction at 0°C. Note that the temperature for a given entry is obtained by adding the corresponding temperature in the top row to that in the left-hand column, regardless of whether the latter is positive or negative.

The noble metal thermocouples, types B, R, and S, are all platinum or platinum–rhodium thermocouples and hence share many of the same characteristics. Metallic vapor diffusion at high temperatures can readily change the platinum wire calibration; hence platinum wires should only be used inside a nonmetallic sheath such as high-purity alumina.

Type B thermocouples (Table 15.4) offers distinct advantages of improved stability, increased mechanical strength, and higher possible operating temperatures. They have the unique advantage that the reference junction potential is almost immaterial, as long as it is between 0 and 40°C. Type B is virtually useless below 50°C because it exhibits a double-value ambiguity from 0 to 42°C.

Type E thermoelements (Table 15.5) are very useful down to about liquid-hydrogen temperatures and may even be used down to liquid-helium temperatures. They are the most useful of the commercially standardized thermocouple combinations for subzero temperature measurements because of their high Seebeck coefficient (58 μV/°C), low thermal conductivity, and corrosion resistance. They also have the largest Seebeck coefficient (voltage response per degree Celsius) above 0°C of any of the standardized thermocouples, which makes them useful for detecting small temperature changes. They are recommended for use in the temperature range from –250 to 871°C in oxidizing or inert atmospheres. They should not be used in sulfurous, reducing, or alternately reducing and oxidizing atmospheres unless suitably protected with tubes. They should not be used in vacuum at high temperatures for extended periods of time.

Type J thermocouples (Table 15.6) are one of the most common types of industrial thermocouples because of the relatively high Seebeck coefficient and low cost. They are recommended for use in the temperature range from 0 to 760°C (but never above 760°C due to an abrupt magnetic transformation that can cause decalibration even when returned to lower temperatures). Use is permitted in vacuum and in oxidizing, reducing, or inert atmospheres, with the exception of sulfurous atmospheres above 500°C. For extended use above 500°C, heavy-gauge wires are recommended. They are not recommended for subzero temperatures. These thermocouples are subject to poor conformance characteristics because of impurities in the iron.

TABLE 15.3 Thermoelectric Values in Millivolts at Fixed Points for Various Thermocouples

Abbreviations used in the table

fp, freezing point — *bp, boiling point*
nbp, normal boiling point — *tp, triple point*

Fixed point	°C	Type B	Type E	Type J	Type K	Type N	Type R	Type S	Type T
Helium nbp	−268.934		−9.8331		−6.4569	−4.345			−6.2563
Hydrogen tp	−259.347*		−9.7927		−6.4393	−4.334			−6.2292
Hydrogen nbp	−252.88*		−9.7447		−6.4167	−4.321			−6.1977
Neon tp	−248.594*		−9.7046		−6.3966	−4.271			−6.1714
Neon nbp	−246.048		−9.6776		−6.3827	−4.300			−6.1536
Oxygen tp	−218.792*		−9.2499		−6.1446	−4.153			−5.8730
Nitrogen tp	−210.001		−9.0629	−8.0957	−6.0346	−4.083			−5.7533
Nitrogen nbp	−195.802		−8.7168	−7.7963	−5.8257	−3.947			−5.5356
Oxygen nbp	−182.962		−8.3608	−7.4807	−5.6051	−3.802			−5.3147
Carbon dioxide sp	−78.474		−4.2275	−3.7187	−2.8696	−1.939			−2.7407
Mercury tp	−38.834*		−2.1930	−1.4849		−0.985	−0.1830	−0.1895	−1.4349
Ice point	0.000	−0.000	0.000	0.000	0.000	0.000	0.000	0.000	0.000
Diphenyl ether tp	26.87	−0.0024	1.6091	1.3739	1.076	0.698	0.1517	0.1537	1.0679
Water bp	100.00	0.0332	6.3171	5.2677	4.0953	2.774	0.6472	0.6453	4.2773
Benzoic acid tp	122.37	0.0561	7.8468	6.4886	5.0160	3.446	0.8186	0.8129	5.3414
Indium fp	156.598*	0.1019	10.260	8.3743	6.0404	4.508	1.0956	1.0818	7.0364
Tin fp	231.928*	0.2474	15.809	12.552	9.4201	6.980	1.7561	1.7146	11.013
Bismuth fp	271.442	0.3477	18.821	14.743	11.029	8.336	2.1250	2.0640	13.219
Cadmium fp	321.108	0.4971	22.684	17.493	13.085	10.092	2.6072	2.5167	16.095
Lead fp	327.502	0.5182	23.186	17.846	13.351	10.322	2.6706	2.5759	16.473
Mercury bp	356.66	0.6197	25.489	19.456	14.571		2.9630	2.8483	18.218
Zinc fp	419.527*	0.8678	30.513	22.926	17.223		3.6113	3.4479	
Cu–Al eutectic fp	548.23	1.4951	40.901	30.109	22.696		5.0009	4.7140	
Antimony fp	630.74	1.9784	47.561	34.911	26.207		5.9331	5.5521	
Aluminum fp	660.37	2.1668	49.941	36.693	27.461		6.2759	5.8591	
Silver fp	961.93*	4.4908	73.495	55.669	39.779		10.003	9.1482	
Gold fp	1064.43*	5.4336		61.716	43.755		11.364	10.334	
Copper fp	1084.5	5.6263		62.880	44.520		11.635	10.570	
Nickel fp	1455	9.5766					16.811	15.034	
Cobalt fp	1494	10.025					17.360	15.504	
Palladium fp	1554	10.721					18.212	16.224	
Platinum fp	1772	13.262					21.103	18.694	

* Defining fixed points of the International Temperature Scale of 1990 (ITS-90). Except for the triple points, the assigned values of temperature are for equilibrium states at a pressure of one standard atmosphere (101 325 Pa).

Source: J. A. Dean, ed., *Lange's Handbook of Chemistry*, 14th ed., McGraw-Hill, New York, 1992.

TABLE 15.4 Type B Thermocouples: Platinum–30% Rhodium Alloy vs. Platinum –6% Rhodium Alloy

Thermoelectric voltage in millivolts; reference junction at 0°C.

°C	0	10	20	30	40	50	60	70	80	90
0	0.00	−0.0019	−0.0026	−0.0021	−0.0005	0.0023	0.0062	0.0112	0.0174	0.0248
100	0.0332	0.0427	0.0534	0.0652	0.0780	0.0920	0.1071	0.1232	0.1405	0.1588
200	0.1782	0.1987	0.2202	0.2428	0.2665	0.2912	0.3170	0.3438	0.3717	0.4006
300	0.4305	0.4615	0.4935	0.5266	0.5607	0.5958	0.6319	0.6690	0.7071	0.7462
400	0.7864	0.8275	0.8696	0.9127	0.9567	1.0018	1.0478	1.0948	1.1427	1.1916
500	1.2415	1.2923	1.3440	1.3967	1.4503	1.5048	1.5603	1.6166	1.6739	1.7321
600	1.7912	1.8512	1.9120	1.9738	2.0365	2.1000	2.1644	2.2296	2.2957	2.3627
700	2.4305	2.4991	2.5686	2.6390	2.7101	2.7821	2.8548	2.9284	3.0028	3.0780
800	3.1540	3.2308	3.3084	3.3867	3.4658	3.5457	3.6264	3.7078	3.7899	3.8729
900	3.9565	4.0409	4.1260	4.2119	4.2984	4.3857	4.4737	4.5624	4.6518	4.7419
1000	4.8326	4.9241	5.0162	5.1090	5.2025	5.2966	5.3914	5.4868	5.5829	5.6796
1100	5.7769	5.8749	5.9734	6.0726	6.1724	6.2728	6.3737	6.4753	6.5774	6.6801
1200	6.7833	6.8871	6.9914	7.0963	7.2017	7.3076	7.4140	7.5210	7.6284	7.7363
1300	7.8446	7.9634	8.0627	8.1724	8.2826	8.3932	8.5041	8.6155	8.7273	8.8394
1400	8.9519	9.0648	9.1780	9.2915	9.4053	9.5194	9.6338	9.7485	9.8634	9.9786
1500	10.0940	10.2097	10.3255	10.4415	10.5577	10.6740	10.7905	10.9071	11.0237	11.1405
1600	11.2574	11.3743	11.4913	11.6082	11.7252	11.8422	11.9591	12.0761	12.1929	12.3100
1700	12.4263	12.5429	12.6594	12.7757	12.8918	13.0078	13.1236	13.2391	13.3545	13.4696
1800	13.5845	13.6991	13.8135							

Source: J. A. Dean, ed., *Lange's Handbook of Chemistry*, 14th ed., McGraw-Hill, New York, 1992.

The type K thermocouple (Table 15.7) is more resistant to oxidation at elevated temperatures than the type E, J, or T thermocouple and consequently finds wide application at temperatures above 500°C. It is recommended for continuous use at temperatures within the range −250 to 1260°C in inert or oxidizing atmospheres. It should not be used in sulfurous or reducing atmospheres, or in vacuum at high temperatures for extended times.

The type N thermocouple (Table 15.8) is similar to type K but it has been designed to minimize some of the instabilities in the conventional chromel–alumel combination. Changes in the alloy

TABLE 15.5 Type E Thermocouples: Nickel–Chromium Alloy vs. Copper–Nickel Alloy

Thermoelectric voltage in millivolts; reference junction at 0°C.

°C	0	10	20	30	40	50	60	70	80	90
−200	−8.824	−9.063	−9.274	−9.455	−9.604	−9.719	−9.797	−9.835		
−100	−5.237	−5.680	−6.107	−6.516	−6.907	−7.279	−7.631	−7.963	−8.273	−8.561
−0	0.000	−0.581	−1.151	−1.709	−2.254	−2.787	−3.306	−3.811	−4.301	−4.777
0	0.000	0.591	1.192	1.801	2.419	3.047	3.683	4.394	4.983	5.646
100	6.317	6.996	7.683	8.377	9.078	9.787	10.501	11.222	11.949	12.681
200	13.419	14.161	14.909	15.661	16.417	17.178	17.942	18.710	19.481	20.256
300	21.033	21.814	22.597	23.383	24.171	24.961	25.754	26.549	27.345	28.143
400	28.943	29.744	30.546	31.350	32.155	32.960	33.767	34.574	35.382	36.190
500	36.999	37.808	38.617	39.426	40.236	41.045	41.853	42.662	43.470	44.278
600	45.085	45.891	46.697	47.502	48.306	49.109	49.911	50.713	51.513	52.312
700	53.110	53.907	54.703	55.498	56.291	57.083	57.873	58.663	59.451	60.237
800	61.022	61.806	62.588	63.368	64.147	64.924	65.700	66.473	67.245	68.015
900	68.783	69.549	70.313	71.075	71.835	72.593	73.350	74.104	74.857	75.608
1000	76.358									

Source: J. A. Dean, ed., *Lange's Handbook of Chemistry*, 14th ed., McGraw-Hill, New York, 1992.

TABLE 15.6 Type J Thermocouples: Iron vs. Copper–Nickel Alloy

Thermoelectric voltage in millivolts; reference junction at 0°C.

°C	0	10	20	30	40	50	60	70	80	90
−200	−7.890	−8.096								
−100	−4.632	−5.036	−5.426	−5.801	−6.159	−6.499	−6.821	−7.122	−7.402	−7.659
−0	0.000	−0.501	−0.995	−1.481	−1.960	−2.431	−2.892	−3.344	−3.785	−4.215
0	0.000	0.507	1.019	1.536	2.058	2.585	3.115	3.649	4.186	4.725
100	5.268	5.812	6.359	6.907	7.457	8.008	8.560	9.113	9.667	10.222
200	10.777	11.332	11.887	12.442	12.998	13.553	14.108	14.663	15.217	15.771
300	16.325	16.879	17.432	17.984	18.537	19.089	19.640	20.192	20.743	21.295
400	21.846	22.397	22.949	23.501	24.054	24.607	25.161	25.716	26.272	26.829
500	27.388	27.949	28.511	29.075	29.642	30.210	30.782	31.356	31.933	32.513
600	33.096	33.683	34.273	34.867	35.464	36.066	36.671	37.280	37.893	38.510
700	39.130	39.754	40.482	41.013	41.647	42.283	42.922			

Source: J. A. Dean, ed., *Lange's Handbook of Chemistry*, 14th ed., McGraw-Hill, New York, 1992.

content have improved the order–disorder transformations occurring at 500°C, and a higher silicon content of the positive element improves the oxidation resistance at elevated temperatures.

The type R thermocouple (Table 15.9) was developed primarily to match a previous platinum–10% rhodium British wire, which was later found to have 0.34% iron impurity in the rhodium. Comments on type S also apply to type R thermocouples.

The type S thermocouple (Table 15.10) is so stable that it remains the standard for determining temperatures between the antimony point (630.74°C) and the gold point (1064.43°C). The other fixed point used is that of silver. The type S thermocouple can be used from −50°C continuously up

TABLE 15.7 Type K Thermocouples: Nickel–Chromium Alloy vs. Nickel–Aluminum Alloy

Thermoelectric voltage in millivolts; reference junction at 0°C.

°C	0	10	20	30	40	50	60	70	80	90
−200	−5.891	−6.035	−6.158	−6.262	−6.344	−6.404	−6.441	−6.458		
−100	−3.553	−3.852	−4.138	−4.410	−4.669	−4.912	−5.141	−5.354	−5.550	−5.730
−0	0.000	−0.392	−0.777	−1.156	−1.517	−1.889	−2.243	−2.586	−2.920	−3.242
0	0.000	0.397	0.798	1.203	1.611	2.022	2.436	2.850	3.266	3.681
100	4.095	4.508	4.919	5.327	5.733	6.137	6.539	6.939	7.338	7.737
200	8.137	8.537	8.938	9.341	9.745	10.151	10.560	10.969	11.381	11.793
300	12.207	12.623	13.039	13.456	13.874	14.292	14.712	15.132	15.552	15.974
400	16.395	16.818	17.241	17.664	18.088	18.513	18.839	19.363	19.788	20.214
500	20.640	21.066	21.493	21.919	22.346	22.772	23.198	23.624	24.050	24.476
600	24.902	25.327	25.751	26.176	26.599	27.022	27.445	27.867	28.288	28.709
700	29.128	29.547	29.965	30.383	30.799	31.214	31.629	32.042	32.455	32.866
800	33.277	33.686	34.095	34.502	34.909	35.314	35.718	36.121	36.524	36.925
900	37.325	37.724	38.122	38.519	38.915	39.310	39.703	40.096	40.488	40.879
1000	41.269	41.657	42.045	42.432	42.817	43.202	43.585	43.968	44.349	44.729
1100	45.108	45.486	45.863	46.238	46.612	46.985	47.356	47.726	48.095	48.462
1200	48.828	49.192	49.555	49.916	50.276	50.633	50.990	51.344	51.697	52.049
1300	52.398	52.747	53.093	53.439	53.782	54.125	54.466	54.807		

Source: J. A. Dean, ed., *Lange's Handbook of Chemistry*, 14th ed., McGraw-Hill, New York, 1992.

TABLE 15.8 Type N Thermocouples: Nickel–14.2% Chromium–1.4% Silicon Alloy vs. Nickel–4.4% Silicon–0.1% Magnesium Alloy

Thermoelectric voltage in millivolts; reference junction at 0°C.

°C	0	10	20	30	40	50	60	70	80	90
−200	−3.990	−4.083	−4.162	−4.227	−4.277	−4.313	−4.336	−4.345		
−100	−2.407	−2.612	−2.807	−2.994	−3.170	−3.336	−3.491	−3.634	−3.766	−3.884
−0	0.000	−0.260	−0.518	−0.772	−1.023	−1.268	−1.509	−1.744	−1.972	−2.193
0	0.000	0.261	0.525	0.793	1.064	1.339	1.619	1.902	2.188	2.479
100	2.774	3.072	3.374	3.679	3.988	4.301	4.617	4.936	5.258	5.584
200	5.912	6.243	6.577	6.914	7.254	7.596	7.940	8.287	8.636	8.987
300	9.340	9.695	10.053	10.412	10.772	11.135	11.499	11.865	12.233	12.602
400	12.972	13.344	13.717	14.091	14.467	14.844	15.222	15.601	15.981	16.362
500	16.744	17.127	17.511	17.896	18.282	18.668	19.055	19.443	19.831	20.220
600	20.609	20.999	21.390	21.781	22.172	22.564	22.956	23.348	23.740	24.133
700	24.526	24.919	25.312	25.705	26.098	26.491	26.885	27.278	27.671	28.063
800	28.456	28.849	29.241	29.633	30.025	30.417	30.808	31.199	31.590	31.980
900	32.370	32.760	33.149	33.538	33.926	34.315	34.702	35.089	35.476	35.862
1000	36.248	36.633	37.018	37.402	37.786	38.169	38.552	38.934	39.315	39.696
1100	40.076	40.456	40.835	41.213	41.590	41.966	42.342	42.717	43.091	43.464
1200	43.836	44.207	44.577	44.947	45.315	45.682	46.048	46.413	46.777	47.140
1300	47.502									

Source: J. A. Dean, ed., *Lange's Handbook of Chemistry*, 14th ed., McGraw-Hill, New York, 1992.

TABLE 15.9 Type R Thermocouples: Platinum–13% Rhodium Alloy vs. Platinum

Thermoelectric voltage in millivolts; reference junction at 0°C.

°C	0	10	20	30	40	50	60	70	80	90
(Below zero)		−0.0515	−0.100	−0.1455	−0.1877	−0.2264				
0	0.0000	0.0543	0.1112	0.1706	0.2324	0.2965	0.3627	0.4310	0.5012	0.5733
100	0.6472	0.7228	0.8000	0.8788	0.9591	1.0407	1.1237	1.2080	1.2936	1.3803
200	1.4681	1.5571	1.6471	1.7381	1.8300	1.9229	2.0167	2.1113	2.2068	2.3030
300	2.4000	2.4978	2.5963	2.6954	2.7953	2.8957	2.9968	3.0985	3.2009	3.3037
400	3.4072	3.5112	3.6157	3.7208	3.8264	3.9325	4.0391	4.1463	4.2539	4.3620
500	4.4706	4.5796	4.6892	4.7992	4.9097	5.0206	5.1320	5.2439	5.3562	5.4690
600	5.5823	5.6960	5.8101	5.9246	6.0398	6.1554	6.2716	6.3883	6.5054	6.6230
700	6.7412	6.8598	6.9789	7.0984	7.2185	7.3390	7.4600	7.5815	7.7035	7.8259
800	7.9488	8.0722	8.1960	8.3203	8.4451	8.5703	8.6960	8.8222	8.9488	9.0758
900	9.2034	9.3313	9.4597	9.5886	9.7179	9.8477	9.9779	10.1086	10.2397	10.3712
1000	10.5032	10.6356	10.7684	10.9017	11.0354	11.1695	11.3041	11.4391	11.5745	11.7102
1100	11.8463	11.9827	12.1194	12.2565	12.3939	12.5315	12.6695	12.8077	12.9462	13.0849
1200	13.2239	13.3631	13.5025	13.6421	13.7818	13.9218	14.0619	14.2022	14.3426	14.4832
1300	14.6239	14.7647	14.9056	15.0465	15.1876	15.3287	15.4699	15.6110	15.7522	15.8935
1400	16.0347	16.1759	16.3172	16.4583	16.5995	16.7405	16.8816	17.0225	17.1634	17.3041
1500	17.4447	17.5852	17.7256	17.8659	18.0059	18.1458	18.2855	18.4251	18.5644	18.7035
1600	18.8424	18.9810	19.1194	19.2575	19.3953	19.5329	19.6702	19.8071	19.9437	20.0797
1700	20.2151	20.3497	20.4834	20.6161	20.7475	20.8777	21.0064			

Source: J. A. Dean, ed., *Lange's Handbook of Chemistry*, 14th ed., McGraw-Hill, New York, 1992.

TABLE 15.10 Type S Thermocouples: Platinum–10% Rhodium Alloy vs. Platinum

Thermoelectric voltage in millivolts; reference junction at 0°C.

°C	0	10	20	30	40	50	60	70	80	90
(Below zero)		−0.0527	−0.1028	−0.1501	−0.1944	−0.2357				
0	0.0000	0.0552	0.1128	0.1727	0.2347	0.2986	0.3646	0.4323	0.5017	0.5728
100	0.6453	0.7194	0.7948	0.8714	0.9495	1.0287	1.1089	1.1902	1.2726	1.3558
200	1.4400	1.5250	1.6109	1.6975	1.7849	1.8729	1.9617	2.0510	2.1410	2.2316
300	2.3227	2.4143	2.5065	2.5991	2.6922	2.7858	2.8798	2.9742	3.0690	3.1642
400	3.2597	3.3557	3.4519	3.5485	3.6455	3.7427	3.8403	3.9382	4.0364	4.1348
500	4.2336	4.3327	4.4320	4.5316	4.6316	4.7318	4.8323	4.9331	5.0342	5.1356
600	5.2373	5.3394	5.4417	5.5445	5.6477	5.7513	5.8553	5.9595	6.0641	6.1690
700	6.2743	6.3799	6.4858	6.5920	6.6986	6.8055	6.9127	7.0202	7.1281	7.2363
800	7.3449	7.4537	7.5629	7.6724	7.7823	7.8925	8.0030	8.1138	8.2250	8.3365
900	8.4483	8.5605	8.6730	8.7858	8.8989	9.0124	9.1262	9.2403	9.3548	9.4696
1000	9.5847	9.7002	9.8159	9.9320	10.0485	10.1652	10.2823	10.3997	10.5174	10.6354
1100	10.7536	10.8720	10.9907	11.1095	11.2286	11.3479	11.4674	11.5871	11.7069	11.8269
1200	11.9471	12.0674	12.1878	12.3084	12.4290	12.5498	12.6707	12.7917	12.9127	13.0338
1300	13.1550	13.2762	13.3975	13.5188	13.6401	13.7614	13.8828	14.0041	14.1254	14.2467
1400	14.3680	14.4892	14.6103	14.7314	14.8524	14.9734	15.9042	15.2150	15.3356	15.4561
1500	15.5765	15.6967	15.8168	15.9368	16.0566	16.1762	16.2956	16.4148	16.5338	16.6526
1600	16.7712	16.8895	17.0076	17.1255	17.2431	17.3604	17.4474	17.5942	17.7105	17.8264
1700	17.9417	18.0562	18.1698	18.2823	18.3937	18.5038	18.6124			

Source: J. A. Dean, ed., *Lange's Handbook of Chemistry*, 14th ed., McGraw-Hill, New York, 1992.

to about 1400°C, and intermittently at temperatures up to the freezing point of platinum (1769°C). The thermocouple is most reliable when used in a clean oxidizing atmosphere, but may also be used in inert gaseous atmospheres or in a vacuum for short periods of time. It should not be used in reducing atmospheres nor in those containing metallic vapor (such as lead or zinc), nonmetallic vapors (such as arsenic, phosphorus, or sulfur), or easily reduced oxides, unless suitably protected with nonmetallic protecting tubes.

The type T thermocouple (Table 15.11) is popular for the temperature region below 0°C (but see under type E). It can be used in vacuum or in oxidizing, reducing, or inert atmospheres.

TABLE 15.11 Type T Thermocouples: Copper vs. Copper–Nickel Alloy

Thermoelectric voltage in millivolts; reference junction at 0°C.

°C	0	10	20	30	40	50	60	70	80	90
−200	−5.603	−5.753	−5.889	−6.007	−6.105	−6.181	−6.232	−6.258		
−100	−3.378	−3.656	−3.923	−4.177	−4.419	−4.648	−4.865	−5.069	−5.261	−5.439
−0	0.000	−0.383	−0.757	−1.121	−1.475	−1.819	−2.152	−2.475	−2.788	−3.089
0	0.000	0.391	0.789	1.196	1.611	2.035	2.467	2.908	3.357	3.813
100	4.277	4.749	5.227	5.712	6.204	6.702	7.207	7.718	8.235	8.757
200	9.286	9.820	10.360	10.905	11.456	12.011	12.572	13.137	13.707	14.281
300	14.860	15.443	16.030	16.621	17.217	17.816	18.420	19.027	19.638	20.252
400	20.869									

Source: J. A. Dean, ed., *Lange's Handbook of Chemistry*, 14th ed., McGraw-Hill, New York, 1992.

Bibliography

Brennen, W. P., R. B. Cassel, and M. P. DiVito, "Materials and Process Characterization by Thermal Analysis," *Am. Lab.*, **20** (January):32 (1988).

Burros, B. C., "Thermal Analysis and Food Chemistry," *Am. Lab.* **18** (January):20 (1986).

Daly, K. F., "Applications of Thermal Analysis to Pharmaceutical Compounds and Related Materials," *Am. Lab.* **7** (January):57 (1975).

DiVito, M. P., W. P. Brennan, and R. L. Fyans, "Thermal Analysis: Trends in Industrial Applications," *Am. Lab.* **18** (January):82 (1986).

Gibbons, J. J., "Applications of Thermal Analysis Methods to Polymers and Rubber Additives," *Am. Lab.* **19** (January):33 (1987).

Gill, P. S., "Thermal Analysis Developments in Instrumental and Applications," *Am. Lab.* **16** (January):39 (1984).

Kolthoff, I. M., P. J. Elving, and C. Murphy, eds., *Treatise on Analytical Chemistry*, 2d ed., Part I, Vol. 12, *Thermal Methods*, Wiley, New York, 1983.

Marcozzi, C., and K. Reed, "Simultaneous TGA-DTA, Theory and Applications," *Am. Lab.* **25** (January):33 (1993).

Staub, F., "Application of TA to Elastomers," *Am. Lab.* **18** (January):56 (1986).

Svehla, G., ed., *Wilson and Wilson's Comprehensive Analytical Chemistry*, Vol. XII, *Thermal Analysis*, Elsevier, New York, 1984.

Welton, R. E., et al., "Third Generation DMTA Instrumentation," *Am. Lab.* **25** (January):15 (1993).

Wendlandt, W., *Thermal Methods of Analysis*, 3d ed., Wiley, New York, 1986.

Widmann, G., "Thermal Analysis of Plastics," *Am. Lab.* **19** (January):98 (1987).

Wunderlich, B., *Thermal Analysis*, Academic, New York, 1990.

SECTION 16
OPTICAL ACTIVITY AND ROTATORY DISPERSION

16.1 INTRODUCTION

Optical activity (the property of rotating the plane of polarization of light) is characteristic of many organic compounds and of a few inorganic complexes. The determination of the angle of optical rotation is a classical technique of quantitative analysis. More recently, the technique has been extended to structural and stereochemical compounds, and in these the rotatory dispersion (the variation of specific rotation of plane-polarized light with the wavelength of the light) has been found most revealing.

16.2 CHIRALITY AND OPTICAL ACTIVITY

A compound is chiral (the term *dissymmetric* was formerly used) if it is not superimposable on its mirror image. A chiral compound does not have a plane of symmetry. Each chiral compound possesses one (or more) of three types of chiral elements, namely, a chiral center, a chiral axis, or a chiral plane.

16.2.1 Chiral Center

The chiral center, which is the chiral element most commonly met, is exemplified by an asymmetric carbon with a tetrahedral arrangement of ligands about the carbon. The ligands comprise four different atoms or groups. One "ligand" may be a lone pair of electrons; another, a phantom atom of atomic number zero. This situation is encountered in sulfoxides or with a nitrogen atom. Lactic acid is an example of a molecule with an asymmetric (chiral) carbon.

16.2.2 Enantiomers

Two nonsuperimposable structures that are mirror images of each other are known as *enantiomers*. Enantiomers are related to each other in the same way that a right hand is related to a left hand. Except for the direction in which they rotate the plane of polarized light, enantiomers are identical in all physical properties. Enantiomers have identical chemical properties except in their reactivity toward optically active reagents.

Enantiomers rotate the plane of polarized light in opposite directions but with equal magnitude. If the light is rotated in a clockwise direction, the sample is said to be dextrorotatory (to the right) and is designated as (+). When a sample rotates the plane of polarized light in a counterclockwise direction, it is said to levorotatory (to the left) and is designated as (−). Use of the former designations *d* and *l* (or D and L) is discouraged.

16.2.3 Optically Inactive Chiral Compounds

Although chirality is a necessary prerequisite for optical activity, chiral compounds are not necessarily optically active. With an equal mixture of two enantiomers, no net optical rotation is observed. Such a mixture of enantiomers is said to be *racemic* and is designated as (±) and not as *dl* or DL.

The number of stereoisomers increases rapidly with an increase in the number of chiral centers in a molecule. A molecule possessing two chiral atoms should have four optical isomers, that is, four structures consisting of two pairs of enantiomers. However, if a compound has two chiral centers but both centers have the same four substituents attached, the total number of isomers is three rather than four. One isomer of such a compound is not chiral because it is identical with its mirror image; it has an internal mirror plane. This is an example of a *diastereromer*. The achiral structure is denoted as a *meso* compound. Diastereomers have different physical and chemical properties from the optically active enantiomers.

16.3 SPECIFIC ROTATION

Optical rotation is caused by individual molecules of the optically active compound. The amount of rotation depends upon how many molecules the light beam encounters in passing through the sample tube. When allowances are made for the length of the tube that contains the sample and the sample concentration, it is found that the amount of rotation, as well as its direction, is a characteristic of each individual optically active compound.

Specific rotation is the number of degrees of rotation observed if a 1-dm tube is used and the compound being examined is present to the extent of 1 g per 100 mL. For a pure liquid, its density replaces the solution concentration.

$$\text{Specific rotation} = [\alpha] = \frac{\text{observed rotation (degrees)}}{\text{length (dm)} \times (\text{g}/100\ \text{mL})} \tag{16.1}$$

The temperature of the measurement is indicated by a superscript and the wavelength of the light employed by a subscript written after the bracket; for example, $[\alpha]^{20}_{589}$ implies that the measurement was made at 20°C using 589-nm radiation. Values are always understood to apply to a wavelength of 589 nm (sodium D line) except as otherwise indicated by the appearance of a different wavelength as a superscript.

Rotatory power is also expressed in terms of molar rotation $[\phi]$ (formerly $[M]$):

$$[\phi] = \frac{[\alpha] \times \text{molecular weight}}{100} \tag{16.2}$$

The effect of solvent is often small, but it may become extremely large if it affects the state of ionization or conformation.

The effect of temperature on rotation is usually small, but it, too, may become larger whenever it affects the structure of the compound. This is particularly true in those cases in which two forms are in an equilibrium that shifts with the temperature. In carbohydrates, such equilibria give rise to mutarotation. Whenever a sample of the α or β form is dissolved, the rotation changes until the equilibrium is established. For carbohydrates (see Table 16.2), the initial values of rotation are listed with arrows pointing to the final values. Where no temperature is given, the data cited were obtained at room temperature.

The optical rotation of pure organic liquids is given in Table 16.1. Table 16.2 contains the optical rotation of carbohydrates and related compounds; Table 16.3 has similar information for natural amino acids and related compounds. Optical rotations of selected organic compounds are in Table 16.4.

16.4 OPTICAL ROTATORY DISPERSION

An optical rotatory dispersion curve (ORD) is related to the change in refractive index of a compound as a function of wavelength. The observed rotation is

$$\alpha = \frac{\pi}{\lambda}(\eta_l - \eta_r) \tag{16.3}$$

where λ = wavelength
η_l = refractive index for left-circularly polarized light
η_r = refractive index for right-circularly polarized light

An optically inactive substance will retard the speeds of the two circularly polarized components to the same extent, and no net rotation can be observed. However, the speed of the two circularly polarized components are retarded by an optically active substance to a different extent, and this results in the rotation of the plane of polarization. Since changes in refractive index are greatest in the vicinity of absorption bands, ORD spectra are most pronounced in these wavelength regions.

For each ORD spectrum there is an anomalous curve within the spectral region of the optically active absorption band that is called the *Cotton effect*. The tailings outside the absorption band are called the *plain curve*. Each Cotton effect consists of two extremes, a geometric maximum called a *peak* and a geometric minimum called a *trough*. A Cotton effect is further characterized by its *amplitude*, defined as the vertical distance between peak and trough, its *breadth*, which is the horizontal distance between the peak and trough, and its *sign*. A positive Cotton effect curve has its peak at the longer wavelength region, while a negative Cotton effect curve is defined as having its trough appear at the longer wavelength.

16.5 CIRCULAR DICHROISM

Circular dichroism (CD) is concerned with the intensities of the right- and left-circularly polarized components of the linearly polarized light beam. CD is related directly to the molecular absorption

TABLE 16.1 Optical Rotation of Pure Organic Liquids

Compound	Molecular weight	[α]	Temperature, °C	Wavelength, nm
(+)-2-Bromobutane	137.03	+10.8	25	
(−)-2-Bromobutane	137.03	−12.2	25	
(+)-2-Bromooctane	193.13	+27.5	17	
(−)-2-Bromooctane	193.13	−27.5	17	
2-Bromopropanoic acid	152.98	−27	20	D*
		+70	20	400
(+)-2-Butanol	74.12	+13.52	27	D
(−)-2-Butanol	74.12	−13.51	25	D
sec-(+)-Butyl acetate	116.16	+25.4	20	
(+)-Carvone	150.21	+61.2	20	D
(−)-Carvone	150.21	−62.46	20	D
(−)-Diethyl hydroxysuccinate	190.19	−10.2	20	
(+)-Diethyl tartrate	206.19	+7.5	20	
(−)-Ethyl lactate	118.13	−10	14	D
(+)-Fenchone	152.23	+66.9	20	D
(−)-Fenchone	152.23	−51.2	16	D
		−77.2	18	546
		−154	18	436
(−)-2-Heptanol	116.20	−10.48	17	D
(+)-2-Hydroxy-2-butanone	88.10	+4.5	20	671
		+8.0	20	546
		+17.6	20	436
		+57.3	20	340
(+)-12-Hydroxyoleic acid	298.45	+6.67	22	D
(+)-Limonene	136.23	+85	22	671
		+123.8	19.3	D
		+136	22	546
		+233	22	436
(−)-Limonene	136.23	−103.1	19.3	D
(+)-Linalool	154.24	+19.3	20	D
(−)-Linalool	154.24	−20.1	20	D
(+)-*N*-Methylbenzylamine	121.18	+40.3	22	D
(+)-3-Methylbutanal	86.13	+23.6	20	
(+)-3-Methylbutanoic acid	102.13	+18.0	17	
1-Methyl-2-propylpiperidine	141.25	+81	24	D
Nicotine	162.23	−169	20	D
		−203	20	546
Nicotine hydrochloride 10% in H_2O	198.71	+104	20	D
(+)-2-Octanol	130.22	+7.5	20	671
		+9.89	20	D
		+26.2	20	382
(+)-α-Pinene hydrochloride	172.71	+51.14	20	D
(−)-α-Pinene hydrochloride	172.71	−51.28	20	D
(+)-β-Pinene	136.23	+28.59		
(−)-β-Pinene	136.23	−22.4		D
(+)-Propylenediamine	74.13	+29.7	25	
(−)-Propylenediamine	74.13	−28	25	
2-Propylpiperidine (coniine)	127.22	−14.2	23	D
Thiolactic acid	106.14	−45.5	15	
Thujol	154.2	+68	20	
(−)-α-Thujone	152.23	−19.2	20	D
(+)-β-Thujone	152.23	+72.5	15	D

* Sodium D line, 589 nm.

TABLE 16.2 Optical Rotation of Carbohydrates and Related Compounds

The value of [α] was measured at 589 nm (sodium D line).

Compound	Molecular weight	[α]	Solvent	Concentration, g/100 mL	Temperature, °C
α-Acetobromoglucose	411.21	+199.3	$CHCl_3$	3	19
		+230	Benzene	9	15
(+)-β-Allopyranose (β-allose)	180.16	+0.58 (4 min) → +3.26 (10 min) → +14.41 (20 h, max)	Water	6	20
(+)-Altropyranose (β-altrose)	180.16	+32.6	Water	7.6	20
(−)-β-Arabinose	150.13	+173 (6 min) → +105.1 (22.5 h)	Water	3	12
(+)-Arabitol	152.15	+7.7	Satd. aq. $Na_2B_4O_7$	9.26	20
Cellobiose	342.30	+14.2 → +34.6 (15 h)	Water	8	20
Chondrosine	355.31	+39	Water		20
2-Deoxy-(+)-glucose	164.16	+38.5 → +45.5 (35 min)	Water	0.52	17.5
6-Deoxy-(+)-glucose (isorhamnose)	164.16	+73 (6 min) → +30 (3 h, final)	Water	8.3	20
2-Deoxy-(+)-ribose	134.13	−56.2 (final value)	Water	1	22
Fructose	180.16	−132 → −92 (rapid)	Water	2	20
Fructose-6-dihydrogen phosphate	260.14	+2.5	Water	3	21
		+1.2	Water	0.9	21
(+)-Fucose	164.16	+127.0 (7 min) → +89.4 (31 min) → +76.0 (146 min, final value)	Water	10	19
(−)-Fucose	164.16	−124.1 (10 min) → −108.0 (20 min) → −75.6 (24 h)	Water	9	20
(+)-α-Galactose	180.16	+150.7 → +80.2	Water	5	20
(+)-β-Galactose	180.16	+52.8 → +80.2	Water	4	20
(+)-α-Galacturonic acid	194.14	+98.0 → +50.9	Water	10	20
(+)-β-Galacturonic acid	194.14	+27 → +55.6	Water		20
α-Gentiobiose	342.30	+16 (3 min) → +8.3 (3.5 h)		4	22
β-Gentiobiose	342.30	−5.9 (6 min) → +9.6 (6 h)	Water	3	22
Gluconic acid	196.16	−6.7	Water	1	20
Glucosamine	179.17	+100 → +47.5 (30 min)	Water	1	20
α-(+)-Glucose	180.16	+112.2 → +52.7	Water	10	20
β-(+)-Glucose	180.16	+18.7 → +52.7	Water	10	20
α-Glucose-l-phosphate	260.14	+120	Water	1	25
α-Glucose-1-phosphate, potassium salt		+78	Water	4	20
Glucose-6-phosphate, potassium salt	336.33	+21.2	Water	1.3	24
(+)-Glucuronic acid	194.14	+11.7 → +36.3 (2 h)	Water	6	24
(+)-Glucuronic acid γ-lactone	176.12	+19.8	Water	5.19	25
(+)-Glyceraldehyde	90.08	+8.7	Water	2	25
(−)-Glyceraldehyde	90.08	−8.7	Water	2	25
Glycogen (high-mol.-wt. polymer)		+196 to +197	Water		25

(Continued)

TABLE 16.2 Optical Rotation of Carbohydrates and Related Compounds (*Continued*)

Compound	Molecular weight	[α]	Solvent	Concentration, g/100 mL	Temperature, °C
(+)-Gulose	180.16	−20.4	Water		20
(−)-Gulose	180.16	+21.3	Water	4.58	
(+)-Idose	180.16	+15.8	Water	2.3	13
(−)-Idose	180.16	−17.4	Water	3.6	20
α-Lactose monohydrate	360.32	+92.6 → +83.5 (10 min) → +69 (50 min) → +52.3 (22 h)	Water	4.5	20
β-Lactose	342.30	+34 (2 min) → +39 (6 min) → +46 (1 h) → +52.3 (22 h)	Water	4	25
α-(+)-Lyxose	150.13	+5.5 → −14.0	Water	0.82	20
Maltose monohydrate	360.33	+111.7 → +130.4	Water	4	20
α-(+)-Mannose	180.16	+29.3 → +14.2	Water	4	20
β-(+)-Mannose	180.16	−17.0 → +14.2	Water	4	20
β-(+)-Mannose phenylhydrazone	270.3	+26.3 → +33.8	Pyridine		20
Melibiose dihydrate	378.33	+111.7 → +129.5	Water	4	20
N-Methyl-α-(−)-glucosamine HCl	228.68	−103 → −88	Water	0.6	25
Octaacetyl-β-cellobiose	678.61	−14.7	$CHCl_3$	5	20
(+)-Psicose	180.16	+4.7	Water	4.3	25
Raffinose pentahydrate	594.54	+105.2	Water	4	20
α-(−)-Rhamnose	164.16	−7.7 → +8.9	Water	4	20
β-(−)-Rhamnose	164.16	+31.5 (1 min) → +8.9	Water	4	20
(+)-Ribose	150.13	−25	Water	4	24
(+)-Tetrahydroxyhexanedioic acid	210.14	+6.86 → +20.60	Water		19
Sorbitol	182.17	−2.0	Water		20
(−)-Sorbose	180.16	−42.7	Water	5	30
Sucrose	342.30	+66.47 to +66.49	Water	26	20
		+78.3 (546), +128.4 (436), +309.5 (302), +541.2 (284)	Water	26	20
(+)-Tagatose	180.16	−2.3	Water	2.19	20
(+)-Threose	120.10	−12.3 (20 min, final)	Water	4	20
(−)-Threose	120.10	+13.10 (final)	Water	4.5	20
Trehalose dihydrate	378.34	+178.3	Water	7	20
Turanose	342.30	+27.3 → +75.8	Water	4	20
Xylose	150.13	+92 → +18.6 (16 h)	Water	10	20

TABLE 16.3 Optical Rotations of Natural Amino Acids and Related Compounds

Compound	Molecular weight	$[\alpha]$	Solvent	Concentration, g/100 mL	Temperature, °C
(−)-Alanine	89.09	+2.8	Water	6	25
		+10.3 (660), +14.3 (589), +23.3 (500), +39.3 (440)	3*M* HCl	4.4	22
α-(−)-Aminobutyric acid	103.12	+8.40	Water	4	16
α-(−)-Aminobutyric acid · HCl	139.60	+12.90	Water	3.64	19
β-(+)-Aminobutyric acid	103.12	+35.20	Water		20
Anserine	140.26	+12.3	Water	5	30
(−)-Arginine	174.20	+12.5	Water	3.5	20
		+26.9	6.0*N* HCl	1.65	20
Arginine · HCl	210.68	+12.0	Water	4	20
(−)-Asparagine · HCl	168.60	−5.42	Water	1.3	20
		+20.0	1*M* HCl	(1 mol)	20
(+)-Aspartic acid	133.10	−23.0	6*N* HCl	2.3	27
(−)-Aspartic acid	133.10	+25.0	6*N* HCl	1.97	20
Cysteine	121.16	+9.8	Water	1.3	30
(+)-Cystine	140.30	+223	1.0*N* HCl	1	20
(−)-Cystine	140.30	−223.4	1.0*N* HCl	1	20
(−)-2,3-Diaminopropionic acid · HCl	140.59	+25.3	1.0*M* HCl	5	20
(+)-2,3-Diaminopropionic acid · HCl	140.59	−25.0	1.0*M* HCl	5	20
(−)-3,5-Diiodotyrosine	432.97	+2.89	5 g of 4% HCl	0.246	20
		+2.27	5 g of 25% NH_3	0.227	20
(+)-3-(3,4-Dihydroxyphenyl)alanine (DOPA)	197.19	+13.0	1*N* HCl	5.27	11
(−)-3-(3,4-Dihydroxyphenyl)alanine	197.19	−13.1	1*N* HCl	5.12	13
(−)-Glutamic acid	147.13	+31.4	6*N* HCl	1	22.4
(+)-Glutamic acid	147.13	−30.5	6*N* HCl	1	20
Glutamine	146.15	+6.1	Water	3.6	23
Glutathione	307.33	−21	Water	2.74	27
Histidine	155.16	−39.74	Water	1.13	20
		+47.6	Hydrochloride in H_2O	2	20
(−)-Homoserine	119.12	−8.8	Water	5	26
		+18.3	2*N* HCl	5	26
Hydroxyglutamic acid	163.14	+17.6	6*N* HCl	2	20
		+1.2	Water	2	20
4-Hydroxy-(−)-proline	131.13	−76.5	Water	1	
cis form		−58.1	Water	5.2	18
(−)-Isoleucine	131.17	+11.29	Water	3	20
		+40.61	6.1*N* HCl	4.6	20

(Continued)

TABLE 16.3 Optical Rotations of Natural Amino Acids and Related Compounds (*Continued*)

Compound	Molecular weight	[α]	Solvent	Concentration, g/100 mL	Temperature, °C
(–)-Isovaline	117.15	+11.13	Water	5	25
(+)-Isovaline	117.15	–11.28	Water	5	25
(–)-Leucine	131.17	–10.8	Water	2.2	25
(–)-Lysine	146.19	+14.6	Water	6.5	20
		+25.9	6.0*N* HCl	2	23
(–)-Methionine	149.21	–8.2	Water	1	25
		+19.3	1*M* HCl	3.4	28
N-Methyl-(–)-tryptophan (Abrine)	218.25	+44	0.5*M* HCl	2.8	21
(–)-Norleucine	131.17	+23.1	5*N* HCl	4.25	20
		+6.26	Water	0.7	20
(–)-Norvaline	117.15	+23.0	20% HCl	10	20
(+)-Norvaline	117.15	–24.2	20% HCl	10	20
Ornithine	132.16	+11.5	Water	6.5	25
Ornithine · 2HCl		+16.6	Water	5.3	23
(–)-Phenylalanine	165.19	–35.1	Water	1.94	20
(+)-Phenylalanine	165.19	+35.0	Water	2.04	20
		+7.1	18% HCl	3.8	20
(–)-Proline	115.13	–85.0	Water	1	23.4
		–52.6	0.50*N* HCl	0.58	20
(–)-Serine	105.09	–6.83	Water	10	20
		+14.45	5.6*N* HCl	0.5	25
(–)-Threonine	119.12	–28.3	Water	1.1	26
(–)-Thyroxine	776.93	–4.4	0.13*N* NaOH in 70% EtOH	3%	20
(+)-Thyroxine	776.93	+2.97 at 546 nm	6 g 0.5*N* NaOH and 14 g EtOH	0.74	21
(–)-Tryptophan	204.22	–31.5	Water	1	23
		+2.4	0.5*N* HCl	1	20
(–)-Tyrosine	181.19	–10.6	1.0*N* HCl	4	22
		–13.2	3*N* NaOH	4	18
(–)-Valine	117.15	+13.9	Water	0.9	26
		+22.9	20 HCl	0.8	23

TABLE 16.4 Optical Rotations of Selected Organic Compounds

Compound	Molecular weight	[α]	Solvent	Concentration, g/100 mL	Temperature, °C
Abietic acid	302.44	−106	Absolute alcohol	1	24
(+)-*cis,trans*-Abscisic acid	264.31	+411.10	EtOH	1	20
		+426.5	0.005*N* H_2SO_4 in MeOH	1	20
(−)-*cis,trans*-Abscisic acid	264.31	−426.2	0.005*N* H_2SO_4	1	20
N^2-Acetyl-(−)-glutamine	188.18	−12.5	Water	2.9	20
(−)-Acetylcarnitine	203.24	−19.52	Water	6	20
(+)-*N*-Acetylmethionine	191.26	+20.3	Water	4	25
(+)-*N*-Acetylpenicillamine	191.25	+18	50% EtOH	1	25
Aconitine	645.72	+17.3	Chloroform		
Aconitine nitrate	708.8	−35	Water	2	20
Adenosine	267.24	−58.2	Water	0.658	9
(−)-Adenosine 3′-monophosphate hydrate (3′-Adenylic acid)	365.24	−41.6	Water	1	22
(−)-Adenosine 5′-monophosphate hydrate (5′-Adenylic acid, AMP)	365.24	−46.6	2% NaOH	2	22
Adenosine triphosphate	507.21	−26.7	Water	3.095	22
Adrenosterone	300.40	+300	$CHCl_3$	1	20
Agaricic acid	416.56	−9.8	1*N* NaOH	1	25
(−)-Alanyl-(−)-alanine	160.17	−35.9	6*N* HCl	2	22
Alborixin	885.20	−7	Acetone	4	20
Aldosterone	360.45	+150	$CHCl_3$	0.1	25
Alfadolone acetate	390.52	+97	$CHCl_3$	1.02	26
Alfaxalone	332.49	+113.4	$CHCl_3$	1.2	26
Algestone	346.45	+95	$CHCl_3$	0.81	23
Allotetrahydrocortisone	364.47	+94	1,4-Dioxane	1.45	25
Alstonine	348.39	+141	Water	0.4	25
N-Allyl-3-hydroxymorphinan	283.40	−88.9	Methanol	3	20
N-Allylnormorphine	311.39	−155.3	Methanol	3	25
(+)-*N*-(4-Aminobenzoyl)glutamic acid	266.25	+16.7	0.1*N* HCl	2	20
(+)-4-Amino-3-hydroxybutyric acid	119.12	+18.3	Water	2	20
(+)-Aminopentamide	296.40	+98.9	Methanol	1	23
(−)-Aminopentamide	296.40	−101.9	Methanol	1	23
(*R*)-(+)-1-Amino-2-(methoxy-methyl) pyrrolidine	130.19	+79	(neat)		18
(*R*)-(−)-2-Amino-3-methyl-butanol	103.17	−16	Ethanol	10	20
(+)-*threo*-2-Amino-1-(4-nitro-phenyl)-1,3-propanediol	212.21	−30	6*N* HCl	1	25
(+)-6-Aminopenicillanic acid (6-APA)	216.26	+276.3	0.1*N* HCl	1.2	22
Amoxicillin	365.42	+246	Water	0.1	20

(Continued)

TABLE 16.4 Optical Rotations of Selected Organic Compounds (*Continued*)

Compound	Molecular weight	[α]	Solvent	Concentration, g/100 mL	Temperature, °C
Amphomycin	1290.46	+7.5	Water, pH 6	1	25
Amphotericin B.	924.11	+333	Acidic dimethylformamide (DMF)	1	24
		−33.6	0.1*N* HCl in MeOH	1	24
Ampicillin	349.42	+281	Water	1	21
Amygdalin	457.42	−42	Water	1	20
α-Amyrin	426.70	+91.6	Benzene	1.3	17
β-Amyrin	426.70	+99.8	Benzene	1.3	19
Anabasine	162.24	−83.1	Water	1	20
Anagestone (as acetate)	330.49	+24	$CHCl_3$	1	20
(−)-Anatabine	160.21	−177.8	Water	1	17
Androstenediol	290.43	−55.5	2-Propanol	0.4	18
Anisomycin	265.30	−30	Methanol	1	23
Anthramycin	315.32	+930	Dimethylformamide	1	25
Aplasmomycin	798.72	+225	Chloroform	1.24	22
Artimisinin	282.35	+66.3	Chloroform	1.64	17
(−)-Ascorbic acid	176.12	+48	Methanol	1	23
Azaserine	173.13	−0.5	Water, pH 8.6	8.46	27.5
(−)-Benzoin	212.22	−118	Acetone	1.2	12
(+)-Benzoin	212.22	+120.5	Acetone	1.2	12
Biotin	244.31	+91	0.1*N* NaOH	1	21
(+)-Borneol	154.24	+37.7	Ethanol	5	20
3-Bromo-(+)-camphor	231.14	+122.7	Benzene, 100 g	14.5	20
Brucine	394.47	−127	Chloroform	1	
Bufotalin	444.55	+5.4	Chloroform	0.5	20
(+)-Butaclamol	361.54	+218.5 (as HCl salt)	Methanol	1	25
Calcitriol	416.65	+48	Methanol	1	25
(+)-Camphene	136.23	+103.5	Diethyl ether	9.67	17
(−)-Camphene	136.23	−119.11	Benzene	2.33	21
Camphor	152.23	+41 to +43	U.S.P. alcohol	10	25
Camphoric acid	200.23	+47 to +48	Ethanol	2.5	20
Camphor-β-sulfonic acid	232.31	+21.5	Water	4.3	20
Carbomycin A	842.00	−58.6	Chloroform	1	25
Carbomycin B	826.00	−35	Chloroform	1	25
3-Carene	136.24	+15	(neat)		20
(−)-Carnitine	161.20	−23.9	Water	0.86	30
(*R*)-(−)-Carvone	150.22	−58	(neat)		20
(*S*)-(+)-Carvone	150.22	+58	(neat)		20
(+)-Catechin hydrate	290.28	+16.0	Water	1	21

Cephaeline	466.60	−43.4	Chloroform	2	20
(+)-Chalcose	162.18	+120 (2 min) → +97 (10 min) → +76 (3 h)	Water	1.5	24
Chelidonine	353.36	+117	Chloroform	3	20
α-Chloralose	309.54	+19	98% Ethanol	5	22
Chloramphenicol	323.14	+18.6	Ethanol	4.86	25
3-Chloro-1-butene	90.55	−2.52 D (−) form	(neat)		20
		+5.87 L (+) form	(neat)		25
3-Chloro-*d*-camphor	186.68	+96.1 *endo* form	Ethanol	5	20
		+35 *exo* form	Ethanol	5	
Chlorogenic acid	354.30	−35 (hemihydrate)	Water	2.8	26
Chlorophyll a	893.5	−262	Acetone		20
Chlorophyll b	907.5	−267	Acetone + methanol		20
Chlortetracycline	478.88	−275.0	Methanol		23
		−240 (hydrochloride)	Water		23
Cholesterol	386.64	−31.5	Diethyl ether	2	20
Cholic acid	408.56	+37	Ethanol	0.6	20
Cinchonidine	294.38	−109.2	Ethanol		20
Cinchonine	294.38	+229	Ethanol	0.6	20
Citramalic acid	148.11	+23.6 (+) form			
		−23.4 (−) form	Water	3	22
Citronellal	154.24	+11.50	Ethanol		25
β-Citronellol	156.26	+5.22 (+) form			
		−4.76 (−) form	Ethanol		20
Cocaine	303.35	−16	Chloroform	4	20
Codeine	299.36	−136	Ethanol	2	15
		−112	Chloroform	2	15
Codeine *N*-oxide	315.36	−97.1 (monohydrate)	Water	2.1	18
Colchicine	399.43	−121	Chloroform	0.9	17
		−429	Water	1.72	17
Coniferin	342.35	−68	Water	0.5	20
Cortisone	360.46	+209	Ethanol	1.2	25
		+269 (546 nm)	Benzene	0.125	25
Corydaline	369.44	+311	Ethanol	0.8	20
Cuprein	310.38	−176	Methanol	1.8	17
Cymarose	162.18	+54.7 (24 h)	Water	3.2	20
Cytidine	243.22	+31	Water	0.7	25
2′-Cytidylic acid	323.19	+20.7	Water	1	20
3′-Cytidylic acid	323.19	+49.4	Water	1	20
Dehydroascorbic acid	174.11	+56	Water	1	20
Demecolcine	371.42	−129.0	Chloroform	1	20
Deoxycholic acid	392.56	+55	Ethanol		20
Deoxycorticosterone acetate	372.49	+168 to +175	1,4-Dioxane		20
Deoxydihydrostreptomycin	567.62	−102.5	Water	1	20

(Continued)

TABLE 16.4 Optical Rotations of Selected Organic Compounds (*Continued*)

Compound	Molecular weight	[α]	Solvent	Concentration, g/100 mL	Temperature, °C
Dextromoramide	392.52	+25.5	Benzene	5	20
Diacetylmorphine	369.40	−166	Methanol	1.49	25
Dideoxyadenosine	235.25	−25.2	Water	1.01	25
Dideoxycytidine	211.22	+81	Water	0.635	25
Digitalose	178.18	+109 (15 min; 546 nm)	Water	1.7	27
Digitogenin	448.62	−81	Chloroform	1.4	19
Digitonin	1229.30	−54	Methanol	2.8	20
Digitoxigenin	374.50	+19	Methanol	1.36	25
Digitoxin	764.92	+4.8	1,4-Dioxane	1.2	20
Dihydrocodeine	301.37	−72 to −75 (acid tartrate)	Water	1	25
Dihydroergotamine	583.67	−79 (546 nm)	Pyridine	0.5	20
Dihydromorphine	287.35	−112 (as HCl)	Water	1.6	25
4,5-Dihydroorotic acid	158.11	+33.23 (−) form	1% $NaHCO_3$	1.992	25.3
		−31.54 (+) form	1% $NaHCO_3$	2.01	25.3
Doxifluridine	246.20	+18.4	Water	0.419	25
Emetine	480.63	−50	Chloroform	2	20
Enterobactin	669.57	+7.40	Ethanol		25
Ephedrine	165.23	+62 (+) form HCl	Water	0.8	20
		−33 to −35.5 (−) form HCl	Water	5	25
Epinephrine	183.20	−53.5	0.5*M* HCl	5	25
Ergonovine	325.39	+90	Water	1	20
Ergotamine	581.65	−160	Chloroform	1	20
Erythromycin	733.92	−78	Ethanol	2	25
(+)-Erythrose	120.11	+1 to −14.5 (3 d)	Water	1	20
Esculin	340.28	−78.4 (sesquihydrate)	50% 1,4-Dioxane	2.5	18
Estrone	270.36	+152	Chloroform	1	22
Folic acid	441.40	+23	0.1*M* NaOH	0.5	25
Fumagillin	458.53	−26.6	Ethanol	1	25
Gibberellic acid	346.37	+86	Water	2.12	19
Giractide	2218.75	−51.4 ± 1.9	0.1*N* Acetic acid	0.472	23.5
(+)-Glaucine	355.42	+115	Ethanol	3	20
Glucamine	181.19	−7.95	Water	10	15
Gluconolactone	178.14	+61.7	Water	1	20
Glucosamine	179.17	+100 → +47.5 (30 min)	Water	1	20
Glutathione	307.33	−18.9	Water	4.653	25
Glutethimide	217.26	+176 (+) form	Methanol		20
		−181 (−) form	Methanol		20
Glycocholic acid	465.61	+30.8	Ethanol	7.5	23
Grandisol	154.25	+18.5	Hexane	1	21.5
Griseofulvin	352.77	+370	Chloroform	Saturated	17

Guanosine	283.24	−60.5	0.1*N* NaOH	3	20
3′-Guanylic acid	363.23	−65	5% NaOH	2	20
		−8	Water	2	20
Guaran		+53	1*N* NaOH		25
Hesperidin	610.55	−76	Pyridine	2	20
Heterophylline (Aricine)	382.44	−91	Chloroform	1.4	20
Hydrastine	383.39	−50	Ethanol	0.3	20
3-Hydroxybutanoic acid	104.10	+24.3	Water	2.226	10
25-Hydroxycholesterol	402.67	−39.0	Chloroform	1.05	25
β-Hydroxyglutamic acid	163.14	+17.6	6*N* HCl	2	20
17*α*-Hydroxyprogesterone	330.45	+105.6	Chloroform	1.042	17
5-Hydroxytryptophan	220.22	−32.5 (−) form	Water	1	20
		+16.0 (−) form	4*N* HCl	1	20
Hyoscyamine	289.36	−21.0	Ethanol	1	20
Inosine	268.23	−49.2	Water	0.9	20
5′-Inosinic acid	348.22	−18.5	2.5% HCl	3 g Ba salt	20
Inulin	ca. 5000	−40	Water	2	20
Iopamidol	777.09	−2.01	Water	10	20
Iopanoic acid	570.93	−5.2 (−) form	Ethanol	2	20
		5.1 (+) form	Ethanol	2	20
(+)-*cis*-*α*-Irone	206.32	+109	Dichloromethane		20
(+)-*trans*-*α*-Irone	206.32	+420	Dichloromethane		20
(+)-*β*-Irone	206.32	+59	Dichloromethane		20
(+)-*cis*-*γ*-Irone	206.32	+2	Dichloromethane		20
Isatropic acid	296.31	+9.94	Ethanol	6.7	20
Isepamicin	569.61	+110.9	Water (disulfate hydrate)	1	20
Isoascorbic acid	176.12	−16.6	Water	1	16.5
Isobutol	666.77	+3	Water	5%	20
Isoflupredone	378.45	+108	Ethanol	0.611	23
(−)-Isoglutamine	146.15	+20.5	Water	6.1	21
Isoladol	273.32	−150 (−) form	Ethanol	0.952	20
Isolysergic acid	268.34	+281 (dihydrate)	Pyridine	1	20
Isomethadone	309.43	+20.8 (+) form	(neat)		25
		−20 (−) form	Ethanol	1.5	25
cis-Isopilosine	286.32	+83.9	Ethanol	1	20
Isothebaine	311.37	+285	Ethanol	1	18
Isradipine	371.39	+6.7 *S*(+) form	Ethanol	1.5	20
Kainic acid	213.23	−14.8	Water	1.01	24
Kallidin	1188.44	−57	1*N* Acetic acid	1	21
Kanamycin A	484.51	+146	0.1*N* H_2SO_4	1	24
11-Ketoprogesterone	328.46	+270	Chloroform		25
(−)-Kynurenine	208.21	−29 (hydrate)	Water	0.4	20
(+)-Lactic acid	90.08	−2.6 (546 nm)	Water	8	21.5

(Continued)

TABLE 16.4 Optical Rotations of Selected Organic Compounds (*Continued*)

Compound	Molecular weight	[α]	Solvent	Concentration, g/100 mL	Temperature, °C
(−)-Lactic acid	90.08	+2.6 (546 nm)	Water	2.5	21–22
Lactobionic acid	358.30	+53.0 → +22.6 (6 h)	Water	8	20
Lactulose	342.30	−51.4 (24 h)	Water, pH 4.8	4	
Lanosterol	426.70	+62.0	Chloroform	1	20
Lanthionine	208.24	+9.4 (−) form	2.4*N* NaOH	1.4	25
Lasalocid A.	590.80	−7.55	Methanol	1	25
Leuprolide	1209.42	−31.7	1% Acetic acid	1	25
Levodopa	197.19	−13.1	1*N* HCl	5.12	13
Limonene	136.23	+123.8 (+) form	(neat)		19.5
		−101.3 (−) form	(neat)		19.5
Linalool	154.24	−20.1 (−) form			20
		+19.3 (+) form			20
Lincomycin	406.56	+137 (HCl, hemihydrate)	Water	1	25
Liothyronine	651.01	+21.5	33% 1*N* HCl + 67% ethanol	4.75	29.5
Lisuride	338.46	+31.3	Pyridine	0.60	20
Lithocholic acid	376.56	+33.7	Ethanol	1.5	20
Lobeline	337.47	−43 (−) form	Ethanol	1	15
Lovastatin	404.55	+323	Acetonitrile	0.5	25
Lupeol	426.70	+27.2	Chloroform	4.8	20
Lysergamide	267.32	15 (546 nm)	Pyridine	0.5	20
Lysergic acid	268.32	+40	Pyridine	0.5	20
Malic acid	134.09	−2.3 (−) form	Water	8.5	19
Mandelonitrile	133.14	43.75 (+) form	Benzene	5	25
Medrogestone	340.51	+79	Chloroform	1	23
Melittin	2846.54	−89.52	Water	0.409	21
Menthol	156.26	−50	10% Ethanolic solution	5.5	18
(−)-Menthone	154.24	−24.8	Ethanol	5.5	20
Methadone hydrochloride	345.90	−145 (−) form	Water	2.5	20
		−169 (−) form	Ethanol	2.1	20
Methotrimeprazine	328.46	−17	Chloroform	5	20
(−)-Methyldopa	211.21	−4.0 (sesquihydrate)	0.1*N* HCl	1	20
Methylergonovine	339.42	−45	Pyridine	0.4	20
N-Methylglucamine	195.22	−23	Water	1	20
N-Methyl-α-(−)-glucosamine	193.20	−62	Methanol	1	25
α-Methylglucoside	194.18	+158.9	Water	10	20
Methylprednisolone	374.46	+83	1,4-Dioxane	1	20
17-Methyltestosterone	302.44	+69 to +75	1,4-Dioxane	1	25
Mildiomycin	514.45	+100	Water	0.5	20
Morphine	285.33	−132 (monohydrate)	Methanol	1	20
Morphine hydrochloride	321.81	−113.5 (trihydrate)	Water	2.2 (anhyd.)	25

Muscone	238.40	−13(−) form			17
Nandrolone	274.39	+55	Chloroform	0.93	23
Narcotoline	399.39	−189	Chloroform (20-cm tube)	0.4	20
Neomycin	322.4	+112.8	Water	1	20
Neopine	299.36	−28 (hydrobromide)	Chloroform	7.5	23
Netilmicin	475.60	164	Water	3	26
Nicotine	162.23	+104 (hydrochloride)	Water	10	20
Nilvadipine	385.38	+222.42 (+) form	Methanol	1	20
		−219.62 (−) form	Methanol	1	20
Nimodipine	418.45	+7.9 (+) form	1,4-Dioxane	0.439	20
		−7.93 (−) form	1,4-Dioxane	0.374	20
Norepinephrine	169.18	−37.3 (−) form	Water +1 equiv. HCl	5	25
Norethindrone	298.41	−31.7	Chloroform	1	20
Norgestrel	312.44	−42.5 (−) form	Chloroform	1	25
Nornicotine	148.20	−89	Water	100	22
Norpseudoephedrine	151.20	+37.9 (+) form	Methanol	3	20
Novobiocin	612.65	−63.0	Ethanol	1	24
Nystatin		−10	Glacial acetic acid	1	25
Ondansetron	293.37	−14 3*S* form	Methanol	0.19	25
		+16 3*R* form	Methanol	0.34	24
Oxycondone	315.36	−125 (hydrochloride)	Water	2.5	20
Oxytetracycline	460.44	−196.6 (dihydrate)	0.1*N* HCl	1	25
Pantolactone	130.14	−50.7 (−) form	Water	2.05	25
Pantothenic acid	219.23	+37.5	Water		25
Parasorbic acid	112.12	+210	Ethanol	2	19
Penicillamine	149.21	−63 (hydrochloride)	1*N* NaOH	1	25
Penicillin G benzathine	909.11	+206 (hydrated)	Formamide	0.105	25
Penicillin G benzhydrylamine	517.63	206	Water	1	20
Penicillin G hydrabamine	1265.79	+115.3	Chloroform	10	25
Penicillin G potassium	372.47	+285 to +310	Water	0.7	22
Penicillin N	359.40	+187 (barium salt)	Water	0.6	20
Penicillin V	350.38	+223 (potassium salt)	Water	0.2	25
2-Pentenylpenicillin sodium	334.37	+316	Water	0.88	15
Pentostatin	268.27	+76.4	Water	1	25
Perillaldehyde	150.21	+127	Carbon tetrachloride	13.1	20
α-Phellandrene	136.23	−217 (−) form	(neat)		20
		+86.4 (+) form	(neat)		16
β-Phellandrene	136.23	+65.2 (+) form	(neat)		20
		−51.9 (−) form	(neat)		20
Pheneturide	206.24	+54.0 (+) form	Ethanol	1	17
Phloridzin	436.40	−52 (dihydrate)	Ethanol	3.2	25
Physostigmine	275.34	−76	Chloroform	1.3	17
Picrotoxin	602.57	−29.3	Ethanol	2.31	16

(Continued)

TABLE 16.4 Optical Rotations of Selected Organic Compounds (*Continued*)

Compound	Molecular weight	[α]	Solvent	Concentration, g/100 mL	Temperature, °C
Pilocarpine	208.25	+106	Water	2	18
		+91 (hydrochloride)	Water	2	18
β-Pinene	136.23	+28.59			
Pivampicillin	463.55	+196	Water	1	20
Prednisolone	360.44	+102	1,4-Dioxane	1	25
		+116 (21-acetate)			
Prednisone	358.44	+186 (21-acetate)	1,4-Dioxane	1	25
α-(+)-Propoxyphene	339.48	+59.8 (hydrochloride)	Water	0.6	25
Protoveratrine A	794.0	−40.5	Pyridine	1	25
Protoveratrine B	810.0	−37	Pyridine	1	25
Pseudococaine	303.35	+42	Chloroform	5	20
Pulegone	152.23	−22.5	(neat)		23
Pyrethrin I	328.4	−14	Isooctane	1	20
Pyrethrin II	372.4	+14.7	Isooctane–diethyl ether	1	19
(−)-Pyroglutamic acid	129.11	−11.9	Water	2	20
Quillaic acid	486.67	+56.1	Pyridine	2.9	20
Quinacillin	416.42	+183.5	Water	1	23
Quinidine	324.41	+230	Chloroform	1.8	15
Quinine	324.41	−169	Ethanol	2	15
Quinine sulfate	746.93	−220 (dihydrate)	0.5*N* HCl	5	15
Quinovose	164.16	+75 (5 min) → +30 (3 h, final)	Water	8.3	20
Racemethorphan	271.41	+27.6 (hydrobromide)	Water	1.5	20
Rescinnamine	634.71	−27	Chloroform	1	24
Reserpine	608.70	−118	Chloroform	1	23
Riboflavine	376.36	−112 to −122	0.02*M* NaOH in ethanol	0.4	25
Ricinoleic acid	298.45	+7.15	Acetone	5	26
Rutin	610.51	+13.82	Ethanol	1	23
Rutinose	326.30	+3.2 → +0.8	Water	4	20
Salicin	286.27	−45.6	Abs. ethanol	0.6	20
Salinomycin	751.02	−37 (sodium salt)	Ethanol	1	25
α-Santonin	246.29	−170 to −175 (−) form	Ethanol	2	25
		+165.9 (+) form	Ethanol	1.92	20
(−)-Sarcolysine	305.20	−31.5	Methanol	0.67	22
Scopolamine	303.35	−28	Water	2.7	22
Serpentine	348.39	+292	Methanol	0.27	25
Shikimic acid	174.15	−183.8	Water	4.03	18
β-Sitosterol	414.69	−37	Chloroform	2	25
Sparteine	234.37	−16.4	Abs. ethanol	10	23
Streptomycin	581.58	−84 (trihydrochloride)	Water	1	25
Streptomycin B	743.75	−47 (trihydrochloride monohydrate)	Water	1.35	25
Streptothricin F	363.4	−51.3 (hydrochloride)	Water	1.4	25

Strophanthidin	404.49	+43.1	Methanol	2.8	25
Strychnine	334.40	−139.3	Chloromethane	1	18
Sucralose	397.64	+68.2	Ethanol	1	
Sucrose octaacetate	678.58	+58.5	Abs. ethanol	2.56	25.4
Tartaric acid	150.09	−12 (−) form, +12 (+) form	Water	20	20
Taxol	853.92	−49	Methanol	1	20
Terpenylic acid	172.19	+56.3 (+) form, −56.5 (−) form			25
α-Terpineol	154.24	+92.45 (+) form, −100 (−) form	Ethanol	20	20
Testosterone	288.41	+100	Ethanol	4	24
Thebaine	311.37	−219	Ethanol	2	15
Thioctic acid	206.32	+104 (+) form, −113 (−) form	Benzene	0.88	23
5-Thio-(+)-glucose	196.22	+188	Water	1.56	20
Thioguanosine	299.31	−64	0.1*M* NaOH	1.3	22
Thymidine	242.23	+30.6	Water	1.029	25
Thyroxine	776.93	+2.97 (+) form	6 g 0.5*M* NaOH and 14 g ethanol	0.74	21
		−4.4 (−) form	0.13*M* NaOH in 70% EtOH	3%	20
β-Tocopherol	416.66	+6.37			20
γ-Tocopherol	416.66	−2.4	Ethanol		20
δ-Tocopherol	416.66	+3.4 (546 nm)	Ethanol	16	25
Tubocurarine chloride	681.66	+190	Water	0.5	22
Uridine	144.20	+4	Water	2	20

coefficients of a compound. The difference between ϵ_l and ϵ_r, the absorption coefficients for left- and right-circularly polarized light, respectively, is a measure of the intensity of circular dichroism. If the medium is optically inactive, the decrease of their intensities is equal. However, an optically active medium absorbs the two components to a different extent, resulting in an absorption difference. The CD band can be positive or negative. The shape of the CD band is the same as the absorption band and is observable only within the spectral region in which absorption occurs. The maximum or minimum of the CD band corresponds to the maximum of the absorption band.

16.6 RELATION OF ORD AND CD

ORD and CD are both manifestations of the same underlying phenomenon arising from the interaction of polarized light with asymmetric structural elements of a molecule. CD is a measure of the differential absorption of left- and right-handed circularly polarized light. ORD appears as a rotation of the direction of vibration of linearly polarized light. The point of inflection of the Cotton effect is located at the same wavelength as the maximum of the CD band, and the sign of the Cotton effect is the same as that of the corresponding CD. On a purely theoretical basis both methods will provide the same amount of structural information. However, practical instrumentation does impose some different limitations, so that each has its advantages and each may complement the other.

Sometimes resolution of the ORD spectra is difficult because of overlapping of the tailings of each of the Cotton effects. In contrast, relatively weak transitions can be identified by CD. The ready resolution of CD bands make it superior to ORD, particularly in the structural studies of complex molecules and biomolecules that may possess many asymmetric elements. On the other hand, some optically active absorption bands may be located in a spectral region inaccessible to the instruments. Under such circumstances, while no CD measurements can be made, the continuing tailing of the Cotton effect of an ORD spectrum that usually extends into the instrumentally accessible spectral region can be turned to advantage in providing structural information originating from the inaccessible Cotton effect.

16.7 MEASUREMENT OF OPTICAL ROTATION

16.7.1 Polarimeters

When ordinary white light, which is vibrating in all possible planes, is passed through a Nicol prism (the polarizer), two polarized beams of light are generated. One of these beams passes through the prism, while the other beam is reflected and does not interfere with the plane-polarized beam, the one that is coincident with the propagation axis of light.

If the beam of plane-polarized light is also passed through a second Nicol prism (the analyzer), it can be transmitted only if the second Nicol prism has its axis oriented so that it is parallel to the plane-polarized light. If its axis is perpendicular to that of the plane-polarized light, the light will not pass through.

The sample cell is placed between the two Nicol prisms. If an optically active substance is in the sample tube, the light is deflected. The analyzer prism is rotated to permit maximum passage of light and is then said to be aligned. The angle of rotation (in degrees) is measured.

The standard wavelength is that of the green mercury line at 546.1 nm, although the sodium doublet has been widely employed, especially in the older measurements. Additional wavelengths can be obtained from the mercury lamp and suitable filters: 365, 405, 436, and 633 nm. The standard temperature is 20°C.

Although the standard length of sample tubes is 100 ± 0.03 mm, lengths of 50 and 200 mm are available. Sample volumes will range from 0.1 up to 12.0 mL, depending both on length and internal diameter of the sample tubes.

16.7.2 Spectropolarimeters

The basic requirements for ORD measurements are (1) a means of producing an intense linearly polarized beam of monochromatic light at various wavelengths, and (2) a detection system with its associated electronic circuitry.

16.7.2.1 Light Source. A high-pressure, 450-W, xenon arc lamp provides high-intensity, continuous emission throughout the range 185 to 600 nm. UV-transmitting quartz is used to construct the lamp envelope. The entire optical train must be purged with dry nitrogen (about 1 $L \cdot min^{-1}$) to remove corrosive ozone from the system and to minimize the absorption due to atmospheric oxygen, water vapor, and carbon dioxide when working at wavelengths less than 200 nm.

16.7.2.2 Monochromator. A double monochromator is used in order to reduce stray light, which will affect the apparent rotation detected by the analyzer since some of this stray light may be polarized. The type of linear polarizer most suitable for ORD and CD measurements in the ultraviolet and visible region is the *birefringent polarizer*. It produces light within a highly defined plane of polarization and possesses high transmittance at a wide range of wavelengths. This type of polarizer is made from double refracting uniaxial crystals such as calcite or quartz. Most spectropolarimeters employ Rochon-type polarizers, which are made from two prisms cut from a uniaxial crystal with their optic axes perpendicular to one another and one of them parallel to the direction of the incident light. Specially cut crystal quartz prisms permit the prism to be used simultaneously as a monochromator (Jasco J710–J720 series).

When a polarized light beam passes through an optically active sample, its plane of vibration is rotated by an angle that can be measured by the use of an analyzer. In visual polarimeters, constant light intensity reaching the eye is used to determine the closeness to crossed position of the polarizer and the analyzer. In recording spectropolarimeters the polarizer can be oscillated through a few degrees about a mean angle, which results in extinction when the mean angle is equal to zero. Another technique uses a *Faraday cell* to rotate the plane of the polarized beam so as to compensate for rotation due to an optically active sample.

16.7.2.3 Detector. The detector is a photomultiplier tube with a working range of 185 to 800 nm (an S-20 surface, end-on configuration).

16.7.2.4 Calibration. ORD instruments can be calibrated or checked with standard sucrose or dextrose solution [NIST, standard reference materials (SRMs) 17d and 41c] at three wavelengths: 546, 589, and 633 nm.

Bibliography

Charney, E., *The Molecular Basis of Optical Activity*, Wiley, New York, 1979.

Crabbe, P., *ORD and CD in Chemistry and Biochemistry: An Introduction*, Academic, New York, 1972.

Djerassi, C., *Optical Rotatory Dispersion*, McGraw-Hill, New York, 1960.

Wong, K-P., "Optical Rotatory Dispersion and Circular Dichroism," *J. Chem. Educ*. **51**:A573 (1974); **52**:A9 (1975); **52**:A83 (1975).

SECTION 17
REFRACTOMETRY

17.1 INTRODUCTION

When light passes from one medium into another, its velocity is changed. The ratio of the velocity of light in a vacuum to that in a substance is known as the *index of refraction* or *refractive index* of that substance. The index of refraction varies with the wavelength of light employed, with temperature, and with pressure (for gases).

A variety of instruments, called refractometers, permits the measurement of indices of refraction of gases, liquids, and solids. Refractometry is the term applied to the group of optical methods for the analysis of either relatively pure substances or complex mixtures, based on refractive-index measurements. It, usually, is applied to identify pure substances.

Some substances, called isotropic materials, transmit light with equal velocity in all directions and have only one index of refraction. Gases, liquids, glasses, and most solids of the isometric system belong to the isotropic group of materials. Other solids, which do not transmit light with equal velocity in all directions, are called anisotropic materials.

17.1.1 Refractive Index

The refractive index of a liquid is the ratio of the velocity of light in a vacuum to the velocity of light in the liquid. The angle of refraction varies with the wavelength of the light used. Usually the yellow sodium doublet lines are used; they have a weighted mean of 589.26 nm and are symbolized by D. A typical refractive index (η) would be expressed as

$$\eta_{\mathrm{D}}^{5} = 1.4567$$

where the superscript indicates the temperature and the subscript indicates the wavelength of the light source. The refractive indices for many thousands of compounds will be found in Ref. 1.

When only a single refractive index is available, approximate values over a small temperature range may be calculated using a mean value of 0.000 45 per degree for $d\eta/dt$, remembering that η decreases with an increase in temperature. If a transition point lies within the temperature range, extrapolation is not reliable.

The refractive index of moist air can be calculated from the expression

$$(\eta - 1) \times 10^6 = \frac{103.49}{T} p_1 + \frac{177.4}{T} p_2 + \frac{86.26}{T}\left(1 + \frac{5748}{T}\right) p_3 \tag{17.1}$$

[1] J. A. Dean, ed., *Lange's Handbook of Chemistry*, 14th ed., McGraw-Hill, New York, 1992.

where p_1 is the partial pressure of dry air (in mmHg), p_2 is the partial pressure of carbon dioxide (in mmHg), p_3 is the partial pressure of water vapor (in mmHg), and T is the temperature (in kelvins).

17.1.1.1 Specific Refraction. The specific refraction r_D is independent of the temperature and pressure and may be calculated by the Lorentz and Lorenz equation:

$$r_D = \frac{\eta_D^2 - 1}{\rho(\eta_D^2 + 2)} \tag{17.2}$$

where ρ is the density at the same temperature as the refractive index.

The empirical Eykman equation

$$\left(\frac{\eta_D^2 - 1}{\eta_D + 0.4}\right)\frac{1}{\rho} = \text{const} \tag{17.3}$$

offers a more accurate means for checking the accuracy of experimental densities and refractive indices, and for calculating one from the other.

The molar refraction is equal to the specific refraction multiplied by the molecular weight. It is a more or less additive property of the groups or elements comprising the compound. A set of atomic refractions is given in Table 17.1; an extensive discussion is given in Ref. 2.

[2] N. Bauer, K. Fajans, and S. Lewin, in A. Weissberger, ed., *Physical Methods of Organic Chemistry*, 3d ed., Wiley-Interscience, New York, 1960, Vol. 1, Pt. II, Chap. 28.

TABLE 17.1 Atomic and Group Refractions

Group	Mr_D	Group	Mr_D
H	1.100	N (primary aliphatic amine)	2.322
C	2.418	N (*sec*-aliphatic amine)	2.499
Double bond (C═C)	1.733	N (*tert*-aliphatic amine)	2.840
Triple bond (C≡C)	2.398	N (primary aromatic amine)	3.21
Phenyl (C_6H_5)	25.463	N (*sec*-aromatic amine)	3.59
Naphthyl ($C_{10}H_7$)	43.00	N (*tert*-aromatic amine)	4.36
O (carbonyl)(C═O)	2.211	N (primary amide)	2.65
O (hydroxyl)(O—H)	1.525	N (*sec*-amide)	2.27
O (ether, ester)(C—O—)	1.643	N (*tert*-amide)	2.71
F (one fluoride)	0.95	N (imidine)	3.776
F (polyfluorides)	1.1	N (oximido)	3.901
Cl	5.967	N (carbimido)	4.10
Br	8.865	N (hydrazone)	3.46
I	13.900	N (hydroxylamine)	2.48
S (thiocarbonyl)(C—O)	7.97	N (hydrazine)	2.47
S (thio)(S—H)	7.69	N (aliphatic cyanide)(C≡N)	3.05
S (dithia)(—S—S—)	8.11	N (aromatic cyanide)	3.79
Se (alkyl selenides)	11.17	N (aliphatic oxime)	3.93
Three-membered ring	0.71	NO (nitroso)	5.91
Four-membered ring	0.48	NO (nitrosoamine)	4.37
		NO_3 (alkyl nitrate)	7.59
		NO_2 (alkyl nitrite)	7.44
		NO_2 (aliphatic nitro)	6.72
		NO_2 (aromatic nitro)	7.30
		NO_2 (nitramine)	7.51

17.1.2 Refractometers

Two types of refractometers are available, the *differential* and the *critical angle.* In the differential refractometer, a light beam is transmitted through a partitioned cell that refracts the beam at an angle that depends on the difference in refractive index between the sample liquid in one part and a standard liquid in the other. In the critical-angle refractometer the light incident on the surface of the solution changes sharply from reflected to transmitted light at a critical angle. Refractometers differ in the ranges they cover, the accuracy obtainable, the type of light source employed, and the presence or absence of color-compensating prisms. In general, when white light is used for illumination, the color-compensating prisms are provided.

17.1.2.1 Abbé Refractometer. The Abbé refractometer, an example of a critical-angle refractometer, compares the angles at which light from a point source passes through the test liquid and into a prism the refractive index of which is known. A drop of the sample is placed between the upper and lower prisms and, following the directions supplied by the manufacturer, the refractive index of the sample is read from the dial.

The critical angle is the angle from the perpendicular at which the beam changes from light transmitted into the liquid to light totally reflected at the liquid surface. At angles smaller than the critical angle the light is transmitted into the liquid. The critical angle depends not only on the solution composition but also on the prism material.

The significant feature of a critical angle refractometer is that it measures the refractive index at the surface of a solution. Since surface reflection requires no penetration of the light beam into the solution, this type of refractometer may be used for highly opaque samples and various murky solutions and suspensions, as well as transparent samples. The range of the Abbé refractometer is normally 1.3000 to 1.7000, the maximum precision attainable being 0.0001. This refractometer reads the refractive index directly and requires only a drop of sample.

Process versions of the critical-angle refractometer make it uniquely suitable for many types of binary mixtures. Applications, with an indication of sensitivities attainable, are shown in Table 17.2.

TABLE 17.2 Refractometer Sensitivity for Solutions and Organic Mixtures

System	Minimum full-scale span, wt. %
Water	
In acetic acid	0–0.40
In ethanol	0–1.07
In methanol	0–0.79
Ethanol	
In water	0–0.59
In benzene	0–0.26
Methanol in water	0–1.78
Ethylene glycol in water	0–0.42
Propylene glycol in water	0–0.37
Glycerol in water	0–0.33
Acetone in water	0–0.58
Benzene	
In ethanol	0–0.31
In cyclohexane	0–0.75
Cyclohexane in benzene	0–0.40
Trichlorfluoromethane in dichlorodifluoromethane	0–0.36
Sodium chloride in water	0–0.41
Ammonium sulfate in water	0–0.28

17.1.2.2 Differential Refractometers. Differential refractometers are intended primarily for the analysis of liquid mixtures. They are applicable to any mixture whose refractive index is a single-valued function of the composition; as such they are uniquely applicable as detectors in high-performance liquid chromatography in which they monitor the difference in refractive index between the mobile phase (reference) and the column effluent. The sensitivity of differential refractometers is 0.000 001 refractive-index units. Liquid samples must be clear and clean. Different refractometers operate on one of two principles, as discussed in the following sections.

17.1.2.2.1 Deflection Type. The *deflection type* measures the deflection of a beam of monochromatic light by a double prism. The sample (or column eluent) is placed in (or flows through) half of the prism; the reference liquid (or pure mobile phase) fills the other half. The reference and sample compartments are separated by a diagonal glass divider. If the refractive index of the sample differs from that of the reference, or the eluent from a chromatographic column differs from the pure mobile phase, the beam from the sample compartment is slightly deflected. The cell volume is 15 to 25 μL.

Deflection refractive-index detectors have the advantage of a wide range of linearity. One cell covers the entire refractive-index range.

17.1.2.2.2 Reflection Type. In the optical path of the *reflection-type* refractometer, two collimated beams from the light source (with masks and lens) illuminate the reference (mobile phase only) and sample (eluent) cells. The cells' volume (3 μL) is a depression formed with a Teflon gasket that is clamped between the prism and a reflecting backplate (finely ground to diffuse the light). The diffuse reflected light passes through the flowing liquid film and is imaged onto dual photodetectors. Since the percentage of reflected light at the glass–liquid interface changes as the refractive index of the liquid changes, a signal arises when a solute emerges. The detector is adjusted to zero with mobile phase in both cells.

This type of refractometer has a relatively limited range. Two different prisms must be used to cover the usual refractive-index range.

SECTION 18
ELEMENTAL ANALYSIS OF ORGANIC COMPOUNDS

18.1 MICRODETERMINATION OF CARBON, HYDROGEN, AND NITROGEN

Since all organic compounds contain carbon and hydrogen, and a large number of them also additionally contain nitrogen, it can be seen that the ability to measure these elements accurately is of extreme importance for characterization and identification of such organic compounds. The microcombustion technique is the principal means for determining carbon, hydrogen, and nitrogen. Automated elemental analyzers that are commercially available offer multisample and unattended operation. The combustion operation is completely automated and is followed by an on-line measurement of the components in the combustion gases. Computerization permits extensive data reduction, calculation, reporting, and storage capabilities. The technique involves several steps.

1. In the purge mode, the weighed sample is dropped into the loading head, which is then sealed and all the interfering gases are purged from the combustion path.

2. In the burn mode, the sample is moved onto a ceramic crucible and into the furnace for combustion in a flowing stream of pure oxygen at 900°C. The sample boat can subsequently be removed for weighing any residue. Alternatively, the sample can be mixed with cobalt(III) oxide [or a mixture of manganese dioxide and tungsten(VI) oxide] to provide the oxygen and heated to the same combustion temperature. Carbon dioxide, water vapor, nitrogen and some oxides of nitrogen, and oxides of sulfur are possible products of combustion of an organic compound. The burn time is 10 to 12 min.

3. The removal of interfering elements is effected.

4. The measurement of the carbon dioxide, nitrogen, and water vapor formed is performed.

18.1.1 Removal of Interfering Substances

The interfering substances encountered in the determination of carbon and hydrogen are sulfur, the halogens, and nitrogen.

Hot copper at 550 to 670°C reduces the nitrogen oxides to nitrogen and removes residual oxygen. In carbon–hydrogen analyzers, oxides of nitrogen are removed with manganese dioxide.

Copper oxide converts any carbon monoxide to carbon dioxide.

A magnesium oxide layer in the middle of the furnace removes fluorine.

A silver-wool plug at the exit removes chlorine, iodine, and bromine, and also any sulfur or phosphorus compounds that result from the combustion of the sample.

Calcium oxide removes oxides of sulfur in a secondary combustion zone so that water vapor cannot combine to form sulfurous or sulfuric acid.

18.1.2 Measurement of Combustion Gases

A variety of techniques have been used to separate and measure the components in the combustion gases. These include gas chromatography (Sec. 5.2), thermal conductivity (Sec. 5.2.5.1), infrared spectrometry (Sec. 7), and coulometry (Sec. 14.8).

18.1.2.1 Gas-Chromatographic Method. In one approach excess oxygen is removed from the combustion gases and the nitrogen oxides are reduced to nitrogen with copper. Helium, used as carrier gas, sweeps the carbon dioxide, water, and nitrogen onto a chromatographic column for separation. Signals from the three chromatographic peaks are integrated to ascertain the quantities present in the sample. In this method relatively small samples must be used so that the combustion products represent a "slug" injection on the chromatographic column.

In another approach the gases pass through a charge of calcium carbide in which water vapor is converted to acetylene. A nitrogen cold trap freezes the sample gases and isolates them in a loop of tubing. A valve seals off the combustion train, which is then ready for another sample. The chromatographic separation is begun by removing the cold trap and heating the loop. Another stream of dry helium gas carries the gases from the trap into the chromatographic column where the three gases (nitrogen, carbon dioxide, and acetylene) are completely separated. The chromatographic separation requires about 10 min.

18.1.2.2 Thermal Conductivity Detection. Three pairs of thermal conductivity detectors are used in a differential manner. Specific absorbents are placed between the detectors. The helium carrier gas fills a mixing volume to a specified pressure. The mixed gas is passed through the sample side of detector 1. Water is then removed by a magnesium perchlorate trap from the gases, which are then passed through the reference side of the same detector. Similarly carbon dioxide is determined by passing the effluent from the first detector into the sample side of detector 2, removing carbon dioxide from the gas with a soda–asbestos trap, and passing the stripped gas through the reference side of detector 2. Nitrogen is determined by detector 3, which compares the effluent gas from the second detector after removal of carbon dioxide (and water in the earlier measurement) with pure helium.

18.1.2.3 Infrared Detection Methods. Dispersive (Sec. 7.2.3.1) and nondispersive infrared detectors are available for water and carbon dioxide. One type of nondispersive spectrometer uses filters to isolate the wavelength desired. A simple filter infrared analyzer is designed around a multisegment circular interference filter drive system.

18.1.2.4 Coulometric Detection. Coulometric detectors for carbon dioxide provide 100% efficiency and an absolute digital readout in terms of micrograms of carbon. Hydrogen can be determined by trapping the water from the combustion step on calcium chloride. The water is desorbed by heating and passed through a proprietary material that quantitatively converts water to carbon dioxide for measurement by coulometry.

18.2 TOTAL CARBON AND TOTAL ORGANIC CARBON ANALYZERS

The total carbon (TC) and total organic carbon (TOC) analyzers are designed for natural and wastewater samples and seawater. Periodically (often every 2.5 min for TC and 5 min for TOC) an aspirated sample is injected into a high-temperature (900°C) reaction chamber through which flows a nitrogen carrier gas with a constant level of oxygen. All oxidizable components are combusted to their stable oxides and all inorganic and organic carbon in the aqueous sample are converted to carbon dioxide. The carbon dioxide generated is transferred by the carrier gas through a scrubber to remove corrosive impurities and interferences. Then the carrier gas flows through a nondispersive infrared analyzer.

To differentiate between total carbon and total organic carbon, a separate sample is drawn and mixed with acid at a constant sample-to-acid ratio. Inorganic carbon in the sample is converted to carbon dioxide and is removed by nitrogen sparging. Then the sample no longer containing any inorganic carbon is treated as described above for total carbon to determine the total organic carbon.

18.3 KJELDAHL DETERMINATION OF NITROGEN

In the Kjeldahl method the sample is digested with sulfuric acid and a catalyst. Organic material is destroyed and the nitrogen is converted to ammonium hydrogen sulfate. The heating is continued until the solution becomes colorless or light yellow. Selenium, copper, and mercury, and salts of each, have been used as the catalyst. Potassium sulfate added to the catalyst raises the temperature and thereby speeds the decomposition. A blank should be carried through all the steps of the analysis.

After the digestion is complete, allow the flask to cool. Cautiously dilute with distilled water and cool to room temperature. Arrange a distillation apparatus with a water-jacketed condenser the adapter tip of which extends just below the surface of the solution in the receiver. Carefully pour a concentrated solution of NaOH down the side of the digestion flask so that little mixing occurs with the solution in the flask. Add several pieces of granulated zinc and a piece of pH test paper. Immediately connect the flask to a spray trap and the condenser. Swirl the solution until mixed; the test paper should indicate an alkaline value. Bring the solution to a boil and distill at a steady rate until only one-third of the original solution remains.

When the reaction mixture is made alkaline, ammonia is liberated and removed by steam distillation. (1) In the classical method, the distillate is collected in a known excess of standard HCl solution. The unused HCl is titrated with a standard solution of NaOH, using as indicator methyl red or bromocresol green. These indicators change color at the pH that corresponds to a solution of ammonium ions. (2) In the boric acid method, the distillate is collected in an excess of boric acid (crystals). The borate ion formed is titrated with a standard solution of HCl, using methyl red or bromocresol green as the indicator.

18.3.1 Special Situations

Compounds containing N—O or N—N linkages must be pretreated or subjected to reducing conditions prior to the Kjeldahl digestion. The N—O linkages are reduced with zinc or iron in acid. There is no general technique for the N—N linkages.

Samples containing very high concentrations of halide can in some instances cause trouble because of the formation of oxyacids known to oxidize ammonia to nitrogen.

For nitrate-containing compounds, salicylic acid is added to form nitrosalicylic acid which is reduced with thiosulfate. Then the digestion can proceed as described.

18.3.2 Automated Kjeldahl Method

The Technicon AutoAnalyzer utilizes a procedure based on the Kjeldahl digestion. The sample is digested using a mixture containing selenium dioxide, sulfuric acid, and perchloric acid. The digest is then made up to a given volume and placed in the autoanalyzer. The ammonium hydrogen sulfate in the digest is automatically sampled and treated with sodium hydroxide. The ammonia liberated is mixed with a phenol–hypochlorite reagent to produce a blue color, which is then measured with a filter photometer.

18.4 DETERMINATION OF SULFUR

The determination of sulfur follows the same basic steps outlined earlier for the determination of carbon and hydrogen. A solid sample up to 1 g is weighed in a combustion boat on the integral electronic balance. Coal or coke samples are then covered with a layer of vanadium pentaoxide powder. The vanadium pentaoxide acts as a flux to moderate sample combustion. The boat with sample is inserted through the open port combustion tube of the resistance furnace where it is correctly positioned under the oxygen inlet by a mechanical stop. Oxidative combustion converts the organic material into carbon dioxide, water, sulfur dioxide, and sulfur trioxide. Raising the temperature to 1350°C ensures the production of sulfur dioxide with no sulfur trioxide. Moisture and dust are removed by appropriate traps.

Liquids, such as petroleum samples, are loaded dropwise onto a bed of vanadium pentoxide contained in a crucible with cover. The use of a crucible cover serves to retard the combustion of volatile samples.

18.4.1 Measuring the Sulfur in the Combustion Products

Various methods exist for determining sulfur after oxidation. A simple, straightforward method measures the sulfur dioxide gas by a selective, solid-state, infrared detector.

In the amperometric titration method, the sulfur dioxide is pumped from the furnace to a reaction vessel in the analyzer via a heated manifold. The sulfur dioxide is bubbled through a specially formulated diluent and determined directly through an iodometric titration. An electrical current is preset for a platinum electrode in the diluent. As the diluent absorbs sulfur dioxide, the current decreases. The current decrease triggers an integral burette, which automatically releases a precisely monitored quantity of titrant to restore the current in the diluent to the preset level. A sensitivity of 0.005% sulfur can be achieved for a nominal 100-mg sample. The specificity of the method eliminates any interference from chlorine, organic nitrogen, phosphorus, and lead antiknock compounds.

18.4.2 Tube Combustion (Manual)

In the manual method the sample is burned with the aid of a vanadium(V) oxide [or tungsten(VI) oxide] catalyst and pure oxygen in an alundum tube maintained at 1000°C. The combustion products pass successively through magnesium perchlorate, 8-hydroxyquinoline, and free copper (heated at 840°C), which remove water, the halogens, and oxygen. The residual gases are absorbed in neutral hydrogen peroxide, which converts all the sulfur oxides to sulfuric acid, which is determined by titration with standard base.

18.4.3 Schöninger Combustion

In the Schöninger combustion technique, the sulfur is converted by oxidation to sulfur dioxide and sulfur trioxide and subsequently oxidized to sulfuric acid with hydrogen peroxide. The methodology is described in Sec. 1.7.3.1. This method is useful for nonvolatile compounds only.

18.5 DETERMINATION OF HALOGENS

Here the term *halogen* refers only to chlorine, bromine, and iodine.

18.5.1 Decomposition of the Organic Material

The sample is decomposed in an oxygen atmosphere at 700°C in the presence of a platinum catalyst. The evolved gases are absorbed in a sodium carbonate solution with hydrazine present.

18.5.2 Measurement of the Halides by Amperometric Titration

Iodide, bromide, and chloride can be successively titrated in mixtures with silver nitrate, using a rotating microelectrode (see Sec. 14.5.6, Amperometric Titrations). In a 0.1*M* to 0.3*M* solution of ammonia only silver iodide precipitates. The indicator electrode is held at –0.2 V versus SCE. During the titration of iodide, the current remains constant at zero, or nearly so, until the iodide ions are consumed, and then it rises. After three or four points have been recorded past the end point, the solution is acidified to make it 0.8*M* in nitric acid. Immediately the silver ions added in excess and now released from the silver ammine complex combine with the bromide ions and precipitate as silver bromide, and the current drops to zero. The indicator electrode is held at +0.2 V versus SCE. A second rise in the current indicates the end point of the bromide titration. A chloride end point can be obtained by adding sufficient gelatin to make the solution 0.1% in gelatin. Gelatin suppresses the current due to silver chloride, and the titration is continued until the current again rises after the chloride end point.

18.5.3 Separation of the Halides by Ion Exchange

The anion-exchange separation of the halides is carried out on a column of Dowex 1-X10 in the nitrate form by elution with sodium nitrate solutions. For example, a Dowex 1-X10 column, 3.7 $cm^2 \times 7.4$ cm, is eluted at 1.0 $mL \cdot min^{-1}$ with 0.5*M* sodium nitrate for chloride ion, which elutes within 50 mL.

The eluant is then changed to 2.0M sodium nitrate. Bromide elutes in the next 50 mL. Finally, iodide elutes as a peak extending over the volume from 75 to 275 mL of the stronger nitrate solution.

The individually separated halides may be determined by potentiometric titration with silver nitrate.

18.6 OXYGEN DETERMINATION

For oxygen determination a quartz pyrolysis tube that contains platinized carbon is employed. This is followed by a tube that contains copper(II) oxide. The operating temperature is 900°C and a helium atmosphere is used. Any oxygen in the sample forms carbon monoxide, which is converted in the copper(II) oxide tube to carbon dioxide. The carbon dioxide is measured by GC, IR, or coulometric techniques as previously described for the carbon determination.

18.7 DETERMINATION OF OTHER NONMETALS

18.7.1 Antimony

For organic materials that contain antimony, wet digestion in a Kjeldahl flask with H_2SO_4 and H_2O_2 is often employed. If the compound contains chlorine, the digestion is started with (1:2) HNO_3 and then 18M H_2SO_4 and H_2O_2 are added until the solution becomes clear. The excess H_2O_2 is decomposed by heating; then antimony(V) is reduced to antimony(III) with Na_2SO_3 or $N_2H_4 \cdot 2HCl$. After decomposition of the excess sulfite ions, antimony(III) is titrated with 0.01N I_2 containing Na_2CO_3.

When using the Rhodamine B colorimetric procedure, destruction of organic matter and conversion of antimony to the quinquevalent state is best effected with nitric, sulfuric, and perchloric acids. If iron is present, $SbCl_5$ is extracted with diisopropyl ether from 1.5M HCl solution. The organic phase is shaken with a 1M HCl solution of Rhodamine B and the red-violet color is measured in the organic layer at 545 to 555 nm.[1]

18.7.2 Arsenic

Dry-ashing is not suitable for decomposing organic arsenic (or antimony) compounds or organic substances that contain these metals because of the danger of losses of volatile arsenic or antimony compounds. An exception is dry-ashing at 600°C for materials high in lipids after addition of MgO and $Mg(NO_3)_2$. A strong oxidizing attack with a mixture of nitric and sulfuric acids or of nitric, sulfuric, and perchloric acids is usually employed for most organic samples. Oils, fats, and tobacco are hard to decompose completely by wet oxidation. In these instances, the oxygen flask method is used in which the sample is combusted in quartz wool. The arsenic oxides formed are absorbed in alkaline solution.

The most rapid method for the determination of arsenic in the acid digest consists of the distillation of arsenic(V) after the addition of bromide, followed by the direct application of the molybdenum blue method. Other methods for the isolation of arsenic in the digest consist of the distillation as $AsCl_3$, evolution as AsH_3, or extraction with diethylammonium diethyldithiocarbamate.

Colorimetric methods complete the determination.[2] In the heteropoly molybdenum blue procedure, isolated arsenic is oxidized to the pentavalent state, ammonium molybdate is added to the acid solution, and then hydrazine is added to reduce the heteropoly acid to molybdenum blue.[3] The silver

[1] R. W. Ramette and E. B. Sandell, *Anal. Chim. Acta* **13**:455 (1955).

[2] A. D. Wilson and D. T. Lewis, *Analyst* **100**:54 (1975); Z. Stefanac, *Mikrochim. Acta* **1962**:1108,1115; B. Griepink and W. Krijgsman, *ibid.* **1973**:574.

[3] R. E. Stauffer, *Anal. Chem.* **55**:1205 (1983).

diethyldithiocarbamate method involves evolution of AsH_3 and its absorption in a pyridine solution of the reagent (0.5% w/v). The red reaction product is colloidal silver; the absorbance is measured at 540 nm.[4]

Evolved AsH_3 can be collected and the gas released into an atomic absorption spectrometer. The complexometric titration method follows the precipitation of arsenate ions with $AgNO_3$. The precipitate is dissolved in a reagent solution containing tetracyanonickelate(II) complex, the nickel ions being liberated, and titrated with EDTA in the presence of murexide indicator.

18.7.3 Bismuth

The destruction of organic matter is effected by digestion with nitric and sulfuric acid, often together with perchloric acid. Organobismuth compounds can also be decomposed by closed-flask combustion using a silica spiral, with HCl as absorbent. After combustion by the latter method, H_2SO_4 is added and the mixture is heated to incipient dryness. Treat the residue with H_2SO_4, ascorbic acid, and KI; Bi is determined by measuring the absorbance at 465 nm. This method is suitable when the bismuth content of the sample is greater than 5 to 10 $\mu g \cdot mL^{-1}$. For bismuth content less than 10 μg, the dithizone method is preferable.[5]

18.7.4 Boron

Oxygen flask combustion is suitable for the mineralization of most organic boron compounds.[6] Volatile compounds are mixed with Na_2CO_3 and glucose and placed in a methylcellulose capsule. Low results for boron are reported for fluorine-containing organic boron compounds, but the Wickbold combustion is satisfactory.[7] Wet digestion in a Kjeldahl flask equipped with a reflux condenser is practical if the boron content is low, and thus larger amounts have to be used for analysis. For very reactive compounds, oxidation is done in a Carius bomb with fuming nitric acid or trifluoroperoxoacetic acid to give boric acid. Nonvolatile boron compounds can be fused with Na_2CO_3 in a platinum crucible.

After the dissolution step, the boric acid formed is titrated with $0.01N$ to $0.1N$ NaOH in the presence of excess mannitol to an end point of pH 8.6 with a glass–calomel electrode pair; for small amounts of boric acid, use an ultramicro burette. For spectrophotometric methods, see Table 6.16.

18.7.5 Fluoride

For the determination of fluoride after closed flask combustion, the absorbent solution is titrated with $0.01M$ $Th(NO_3)_4$ at pH 3.0 with sodium alizarinsulfonate as indicator.

18.7.6 Phosphorus

For the decomposition of organophosphorus compounds, the principally used methods are oxygen flask combustion, with water as absorbent, or digestion with HNO_3. In the oxygen flask method, the paper containing the compound is placed in a silica holder since platinum is attacked. The absorbing solution is $0.4N$ H_2SO_4 containing peroxodisulfate as the oxidant or a mixture of NaOH with Br_2. Phosphate ions are determined spectrophotometrically using the molybdophosphate method or, better, the molybdovanadophosphoric acid method measured at 430 nm.[8] Phosphorus can also be determined by induction coupled plasma–optical emission spectroscopy (ICP-OES).

---

[4] V. Vasak and V. Sedivec, *Collect. Czech. Chem. Commun.* **18**:64 (1953).
[5] D. M. Hubbard, *Anal. Chem.* **20**:363 (1948); J. C. Gage, *Analyst* **83**:672 (1958).
[6] S. K. Yasuda and N. R. Rogers, *Microchem. J.* **4**:155 (1960).
[7] B. Schreiber and R. W. Frei, *Mikrochim. Acta* **1**:219 (1975).
[8] J. P. Dixon, *Modern Methods in Organic Microanalysis*, Van Nostrand, London, 1968, pp. 160–162.

18.7.7 Silicon

Organosilicon compounds can be digested with HNO_3–H_2SO_4 and programmed heating: dissolution at 200°C, wet-ashing at 300°C, dry-ashing at 600°C, and calcination at 950°C. The resulting SiO_2 is determined gravimetrically. Another method involves digesting nonvolatile silicon compounds in a platinum dish by heating with a mixture of H_2SO_4 and peroxodisulfate. The dehydrated silica is filtered, ignited, and weighed. Phosphorus and titanium, if present, can be determined in the filtrate.

Two methods are available for the colorimetric determination of silica. In strongly acidic HCl solutions, a yellow complex is formed with ammonium molybdate, a reliable but not very sensitive method. Much more sensitive is the formation of molybdenum blue in a less acidic solution. The first method suffers no interference from fluoride, which does affect the formation of the molybdenum blue color.

A review details problems in the analysis of silanes and siloxanes.[9]

18.7.8 Selenium and Tellurium

Organic selenium compounds can be digested with an oxidizing acid or a mixture of acids. After digestion, the selenium can be precipitated as the metal by hydrazine in strong HCl solution (Sec. 2.5.2). In the oxygen flask combustion, bromine water is used for absorption. The finish method can be the iodometric determination of selenate ions or spectrophotometrically with 3,3′-diaminobenzidine (see Table 6.16). The results of four different methods suggested for the determination of selenium have been compared.[10]

Tellurium can be determined by using methods similar to those for selenium.

18.8 DETERMINATION OF TRACE METALS IN ORGANIC MATERIALS

Metal ions of some organic compounds may react directly in solution with certain reagents, but most of them must be digested before the metals are determined by inorganic microanalytical methods. The pure metal is obtained after pyrolytic decomposition of gold, platinum, and silver organic compounds. The list extends to nickel and cobalt if, after pyrolytic decomposition, the residue is heated in a stream of hydrogen to reduce their oxides. Heating Al, Cr, Cu, Fe, Mg, Sn, and Zn compounds in air gives metal oxides with stoichiometric compositions.

Organic metal compounds can be digested with oxidizing acids or a mixture of acids, fused with oxidizing reagents, or combusted in the oxygen flask method. In addition to closed-flask combustion methods already described, microwave digestion is gaining acceptance because of operational ease and the saving of time.

18.9 METHODS FOR MULTIELEMENT TRACE ANALYSES

Anodic stripping voltammetry is applicable to traces of Cd, Co, Cu, Ni, Pb, and Zn in biological and plant materials after prior sample decomposition or digestion. The major advantage of stripping voltammetry in the preconcentration by factors of 100 or more of the analyte(s) on or in the small volume of a microelectrode before a voltammetric analysis. Combined with differential pulse voltammetry in the stripping step, solutions as dilute as $10^{-11}M$ can be analyzed.

Atomic absorption spectrometry is frequently used as the method of finish for metals; see Table 8.8. Elements analyzed in biological, feedstuff, food, marine, petroleum, pharmaceutical,

[9] J. C. Smith, *Analyst* **85**:465 (1960).
[10] Z. Stefanac, M. Tomaskovic, and I. Bregovec. *Microchem. J.* **16**:226 (1971).

plant, and polymer samples are listed below. Atomic emission (ICP-OES) extends the elements applicable to include those in parentheses.

1. Biological: Ca, Cu, Fe, K, Mg, Na, Zn (Cd, Co, Mn, P)
2. Fodder: Ca, Cu, Fe, K, Mg, Mn, Na, Zn (Al, Co, Cd, P, Pb, Sr)
3. Food: Al, Ca, Cd, Cr, Cu, Fe, K, Na, Ni, Sn, Zn (As, Mg, Mo, P, Se, Sr)
4. Marine: Ag, Au, Cd, Co, Cu, Fe, Hg, Mn, Ni, Pb, Zn
5. Pharmaceutical: Cd, Co, Cr, Cu, Mn, Ni, Pb, Zn
6. Plant: Al, Co, Cu, Fe, Mn, Ni, Sb, V, Zn (As, Cd, Ce, La, Mo, Pb, Sn, Tl)
7. Wine: As, Cd, Cr, Hg, Pb, Se (Ag, Ba, Ca, Cu, K, Li, Mg, Na, Ru, Sb, Sr)

Ion-selective electrodes find use for these elements when biological materials are being analyzed: Ca, Cl, Fe, K, Li, Mg, Na; see Table 14.12. ISEs are adaptable to monitoring specific metals during processing industrial liquids. Neutron activation is one of the preferred methods if a large number of elements are required to be determined in a sample of organic material, particularly for biological and marine samples, and if the applicable elements are desired.

Spectrophotometric methods in the visible and ultraviolet portions of the spectrum are often used when analyzing for trace levels of many elements; see Table 6.15.

X-ray fluorescence has been used on biological materials for the elements Br, Ca, Cr, Cu, Fe, K, Mn, Ni, Pb, Ti, Zn; and on plant materials for Br, Ca, Cl, Cu, Fe, K, Ni, Pb, Rb, Sr, Ti, Zn.

Bibliography

R. Boch, *A Handbook of Decomposition Methods in Analytical Chemistry*, International Textbook Company, London, 1979.

E. C. Dunlap and C. R. Grinnard, Chap. 2, "Decomposition and Dissolution of Samples: Organic," in I. M. Kolthoff and P. J. Elving, eds., *Treatise on Analytical Chemistry*, Part I, Vol. 5, 2d ed., Wiley-Interscience, New York, 1978.

L. Mázor, *Methods of Organic Analysis*, Vol. XV in G. Svehla, ed., Wilson and Wilson's *Comprehensive Analytical Chemistry*, Elsevier, Oxford, 1983.

SECTION 19

DETERMINATION OF FUNCTIONAL GROUPS IN ORGANIC COMPOUNDS

19.1 UNSATURATION

As a class, unsaturated compounds represent the majority of organic structures, particularly if structures containing other functional groups are included. The analytical methods outlined in Tables 19.1 and 19.2 and elsewhere in this section are generally useful for compounds containing carbon-to-carbon unsaturated hydrocarbon compounds. They may or may not apply to structures containing other functional groups. In some cases, either chemical or instrumental methods will be applicable; the analyst may then choose a method compatible with the circumstances.

TABLE 19.1 Characteristic Infrared and Raman Bands for Measurement of Unsaturation

Type of compound	Nature of vibration	Type of spectrum	Wavelength, μm	Wave number, cm^{-1}	Remarks
Acetylene gas	≡C—H sym stretch	R (not IR)	2.964	3374	
Monosubstituted acetylenes	≡C—H stretch	IR, R	3.012–3.049	3320–3280	Strong IR, weak R
Acetylene gas	≡C—H asym stretch	IR (not R)	3.042	3287	
Ethylene gas	=C—H stretch	R (not IR)	3.218	3108	
		IR (not R)	3.220	3106	
Aromatic compounds	=C—H stretch	IR	3.226–3.333	3100–3000	Moderately weak
C_6H_6 gas	=C—H stretch	IR	3.227	3099	
RCH=CH_2	=C—H stretch	IR, R	3.231–3.252	3095–3075	Moderate strength
R_1R_2C=CH_2	=C—H stretch	IR, R	3.247–3.252	3090–3075	Moderate IR, weak R
Mono-, di-, and tri-substituted benzenes	=C—H stretch	R	3.257–3.284	3070–3045	Strong
C_6H_6 liquid	=C—H stretch	R	3.266–3.282	3062–3047	Strong
R_1CH=CHR_2, *cis*	=C—H stretch	IR, R	3.311–3.339	3020–2995	Moderate strength
trans	=C—H stretch	IR, R	3.322–3.339	3010–2995	Moderate strength
Ethylene gas	=C—H stretch	R (not IR)	3.312	3019	
Dialkyl acetylenes	—C≡C—stretch	R	4.340	2304	Second band at 2227 cm^{-1}
Disubstituted acetylenes	—C≡C—stretch	IR	4.425–4.651	2260–2150	Variable strength
Dialkyl acetylenes	—C≡C—stretch	R	4.490	2227	Second band at 2304 cm^{-1}
Cyclic acetylenes	—C≡C—stretch	IR	4.490–4.541	2227–2202	
Monoalkyl acetylenes	—C≡C—stretch	IR	4.673–4.762	2140–2100	Weak to moderate strength
		R	4.630–4.760	2125–2118	Very strong
Benzene derivatives	Overtone and combination bands	IR	5.000–6.061	2000–1650	Weak but characteristic
Allene and derivatives	—C=C=C—asym stretch	IR, R	5.050–5.222	1980–1915	Strong IR and R
Acetylene gas	—C≡C—stretch	R	5.066	1974	Very strong
RCH=CH_2	=C—H bend overtone	IR	5.435–5.540	1840–1805	Weak
R_1R_2C=CH_2	=C—H bend overtone	IR	5.525–5.650	1810–1770	Weak
R_1CH=CHR_2, *trans*	—C=C—stretch	IR, R	5.967–6.006	1676–1665	Var IR; vs R
R_1R_2C=CHR_3	—C=C—stretch	IR, R	5.952–6.010	1680–1664	Weak IR; vs R
R_1R_2C=CR_3R_4	—C=C—stretch	IR, R	5.952–6.006	1680–1665	Very strong; not IR for symmetrically substituted compounds
R_1CH=CHR_2, *cis*	—C=C—stretch	IR, R	6.017–6.131	1662–1631	Cyclohexene, 1646 cm^{-1}; cyclopentene, 1611 cm^{-1}; cyclobutene, 1571 cm^{-1}
R_1R_2C=CH_2	—C=C—stretch	IR, R	6.053–6.068	1662–1648	Very strong R
RCH=CH_2	—C=C—stretch	IR, R	6.061–6.105	1650–1638	IR moderate; vs R
Aromatic hydrocarbons	Ring vibration	IR, R	6.211–6.289	1610–1590	Strong in R
		IR	6.667–6.757	1500–1480	Weak or absent in R

(Continued)

TABLE 19.1 Characteristic Infrared and Raman Bands for Measurement of Unsaturation (*Continued*)

Type of compound	Nature of vibration	Type of spectrum	Wavelength, μm	Wave number, cm^{-1}	Remarks
Alkylbenzene derivatives	$—CH_3$ asym bending	IR	6.803–7.042	1470–1420	
Substituted ethylenes	—C=CH in-plane C—H bend	IR, R	6.897–8.333	1450–1200	Strong R, weak IR
$RCH{=}CH_2$	—C=CH in-plane C—H bend	IR, R	7.042–7.062	1420–1416	
$R_1R_2C{=}CH_2$	—C=CH in-plane C—H bend	IR	7.042–7.082	1420–1412	Moderately weak
$R_1CH{=}CHR_2$	—C=CH in-plane C—H bend	IR	7.092–7.143	1420–1400	
Monosubstituted acetylenes	—C≡C—H bend overtone	IR	7.937	1260	
Allene and alkyl derivatives	—C=C=C—sym stretch	IR, R	9.346	1070	Strong IR and R
Monosubstituted benzene; also 1,3-		R	9.90–10.10	1010–990	Very strong
$RCH{=}CH_2$	—C=CH out-of-plane C—H bend	IR	10.05–10.44	995–985	Very strong; also at 910 cm^{-1}
$R_1CH{=}CHR_2$, *trans*	—C=CH out-of-plane C—H bend	IR	10.20–10.36	980–965	
$RCH{=}CH_2$	—C=CH out-of-plane C—H bend	IR	10.99–11.05	910–905	Very strong; also at 990 cm^{-1}
$R_1R_2C{=}CH_2$	—C=CH out-of-plane C—H bend	IR	11.24	890	
1,2,4-Trisubstituted benzene derivatives	C—H bend	IR	11.11–11.49	900–870	Moderate strength; also at 780–760 cm^{-1}
Pentasubstituted benzene derivatives	C—H bend	IR	10.04–11.63	980–860	Moderately strong
1,2,4,5-Tetrasubstituted benzene derivatives	C—H bend	IR	11.76–11.90	850–840	
1,3,5-Trisubstituted benzene derivatives	C—H bend	IR	11.56–12.35	865–810	Strong
$R_1R_2C{=}CHR_3$	—C=CH out-of-plane C—H bend	IR	11.76–12.66	850–790	Moderate strength
1,2,4-Trisubstituted benzene derivatives	C—H bend	IR	12.12–12.42	825–805	Also at 900–870 cm^{-1}
1,2,3,4-Tetrasubstituted benzene derivatives	C—H bend	IR	12.35–12.50	810–800	
1,3-Disubstituted benzene derivatives	C—H bend	IR (not R)	12.50–12.99	800–770	Also at 710–690 cm^{-1}
Disubstituted ethylene group in six-membered ring	C—H bend	IR	12.50–15.38	800–650	
1,2,3-Trisubstituted benzene derivatives	C—H bend	IR (not R)	12.82–13.16	780–760	Also at 745–705 cm^{-1}
1,2-Disubstituted benzene derivatives	C—H bend	IR (not R)	13.16–13.51	760–740	
Monosubstituted benzene derivatives	C—H bend	IR (not R)	13.39–13.57	747–737	Also at 701–671 cm^{-1}
1,2,3-Trisubstituted benzene derivatives	C—H bend	IR (not R)	13.42–14.18	745–705	Also at 780–760 cm^{-1}

1,2,4-Trisubstituted benzene derivatives	C—H bend	R (not IR)	13.48–13.97	742–716	
1,2-Disubstituted benzene derivatives	C—H bend	R (not IR)	13.59–14.06	736–711	
$R_1CH{=}CHR_2$, *cis*	—C=CH out-of-plane C—H bend	IR	13.79–14.81	725–675	
1,3-Disubstituted benzene derivatives	C—H bend	R (not IR)	13.90–14.06	719–711	
		IR (not R)	14.08–14.49	710–690	Also at 800–770 cm^{-1}
Monosubstituted benzene derivatives	C—H bend	IR (not R)	14.27–14.91	701–671	Also at 747–737 cm^{-1}
1,3,5-Trisubstituted benzene derivatives	C—H bend	IR (not R)	13.70–14.71	730–680	Also at 850–830 cm^{-1}
Benzene	C—H bend	IR (not R)	14.90	671	
1,2,3-Trisubstituted benzene derivatives	C—H bend	R (not IR)	14.93–20.00	670–500	The lighter the mass of the substituent, the higher the frequency
Monosubstituted acetylenes	—C≡C—H bend	IR, R	15.62–15.92	640–628	Strong; overtone, 1260 cm^{-1}
Monosubstituted benzene derivatives	C—H bend	R (not IR)	15.87–16.52	630–605	Moderately strong
$R_1CH{=}CHR_2$, *cis*		IR, R	17.24	580	Also in R at 413 and 297 cm^{-1}
1,3,5-Trisubstituted benzene derivatives	C—H bend	R (not IR)	17.54–18.05	570–554	
$RCH{=}CH_2$		IR	18.18	550	
$R_1CH{=}CHR_2$, *trans*		IR, R	20.41	490	Also in R at 210 cm^{-1}
$RCH{=}CH_2$		R	22.99	435	
$R_1R_2C{=}CH_2$		R	23.04	434	Also at 394–361 cm^{-1}

TABLE 19.2 Measurement of Unsaturation by Spectral Methods: Ultraviolet Absorption Bands

Type of compound	Structure	λ_{max}, nm	$\log_{10} \epsilon_{max}$	Remarks
Alkyl monoolefins (as iodine complexes in i-C_8H_{18})				
(*a*) Terminal	$RHC{=}CH_2$	275	12*	
(*b*) Terminal, substituted	$RR_1C{=}CH_2$	290–295	25*	
(*c*) Internal	*cis*-$RCH{=}CHR$	295–300	19*	
	trans-$RCH{=}CHR$	295–300	11*	
(*d*) Internal, substituted	$RR_1C{=}CHR$	317	27*	
(*e*) Tetrasubstituted ethylenes	$RR_1C{=}CR_2R_3$	337	23*	
Cyclic monoolefins (as iodine complexes in i-C_8H_{18})				
(*a*) Cyclopentene		300	21*	
(*b*) Cyclohexene		295	29*	
Acyclic dienes				
(*a*) Only acyclic substituents		217–228	4.23–4.43	One or more alkyl substituents
(*b*) One cyclic substituent		235.5–236.5	3.7–4.02	
(*c*) Two cyclic substituents		245.0–248.0	4.21–4.54	Cyclic substituents are cyclohexyl rings, which may be substituted
Semicyclic dienes		230.0–242.0	3.11–4.3	In hexane or EtOH
Monocyclic dienes				
(*a*) Cyclopentadiene		238.5	3.53	
(*b*) Cyclohexa-1,3-diene		256.5	3.90	In cyclohexane

(*c*) Cyclodienes-1,3 (C_7–C_{12})		219.5–248	3.4–3.78	In hexane, *i*-octane, or EtOH
Bicyclic dienes				
Two double bonds in different rings		236.0	4.25	In EtOH
Polycyclic dienes				
(*a*) Double bonds in same ring†		260.0–282.0	3.72–4.13	In EtOH
(*b*) Double bonds in different rings†		235.0–248.0	3.97–4.36	In EtOH
Arenes				
(*a*) Benzene	[benzene ring]	255.0	2.33	
(*b*) Monosubstituted benzenes	[benzene ring]–R	260–261	2.34–2.48	R = Me, Et, *n*-Pr, *n*-Bu
(*c*) Polysubstituted benzenes	[benzene ring]–$(R)_n$	266–272	2.48–2.91	R = CH_3; $n = 2–6$
Polycyclic arenes				
(*a*) Linear series (includes naphthalene, anthracene, naphthacene, pentacene)		314–580	2.50–5.20	
(*b*) Angular series (includes phenanthrene, chrysene, pyrene, dibenzanthracene, benzpyrene, methylcholanthrene)		210–403	2.3–5.1	
Polyphenyl compounds				
(*a*) Diphenyl alkanes	ϕ_2CHR	262	2.58–2.69	In EtOH R = H or CH_3; ϕ =
Polyphenyl compounds				
(*b*) Triphenyl alkanes	ϕ_3CR	262	2.94	In EtOH; R = H only
(*c*) Tetraphenylmethane	ϕ_4C	262	3.31	In EtOH
Phenyl ethylenes				
Styrene	[benzene ring]–CH=CH_2	282	2.7	Also band at 244 nm ($\log_{10} \epsilon_{max} = 4.10$)
Phenyl acetylenes				
Phenylacetylene	[benzene ring]–C≡CH	278	2.8	Also band at 236 nm ($\log_{10} \epsilon_{max} = 4.20$)
Azulenes		238–738	2.01–4.78	Azulene and Me derivatives in EtOH
Conjugated diynes	R—C≡C—C≡C—R	218.5–254	2.2–2.6	In EtOH; R = H, Me, Et, or *n*-Bu only

* Apparent molecular absorptivity (*not* its logarithm) in liters per mole of olefins per gram per liter of I_2 for a 1-cm path length.
† Steroid and triterpenoid dienes.
Source: L. Meites, ed., *Handbook of Analytical Chemistry*, McGraw-Hill, New York, 1963.

19.1.1 Characteristic Infrared and Raman Wavelengths

Infrared and Raman absorption bands are useful for the identification of compounds containing carbon–carbon double or triple bonds. Table 19.1 lists the characteristic wavelengths associated with stretching or deformation vibrations in various unsaturated structures. Before applying these correlations to analytical problems, it will usually be desirable to consult sources of more detailed spectral data. Pertinent information is contained in Sec. 7 on infrared absorption, Raman frequencies, and spectra–structure correlation charts.

The bands listed in Table 19.1 are arranged in order of increasing wavelength (in the fourth column) or decreasing frequency (in the fifth column). The first column gives the type of compound responsible for the band, and sometimes also its physical state. The second column gives the nature of the vibration; *sym* is symmetrical, while *asym* is asymmetrical. The third column gives the type of spectrum in which the band is observed: *R* is Raman and *IR* is infrared. The entry *R* (*not IR*) in this column means that the band in question is observed in the Raman spectrum but not in the infrared spectrum.

The characteristic infrared bands in the region from 995 to 685 cm^{-1} can be used for quantitative determinations. As the neighboring groups have less influence on Raman bands than on infrared bands, Raman spectroscopy is useful for the determination of substituted olefins at 1640 and 1680 cm^{-1}.

19.1.2 Ultraviolet Absorption Bands

Such chromophoric groups as carbon–carbon double and triple bonds absorb ultraviolet radiation. This absorption is weak for structures containing only one ethylenic linkage, except in the far-ultraviolet region below 200 nm. Structures containing two or more isolated double bonds behave like simple olefins. On the other hand, conjugation leads to high-intensity absorption in the region from 200 to 1000 nm. (See also Sec. 6.2.1.)

The first column of Table 19.2 lists the type of compound; the second column shows the typical structure. The third column gives the position (or, for classes of compounds, the range of positions) of the peak of the absorption band. The fourth column gives values of $\log_{10} \epsilon_{max}$, where ϵ is the molar absorbance (see Sec. 6.1). Data have been included, for the sake of convenience, for the iodine complexes of a number of simple olefins that do not absorb in the readily accessible portion of the spectrum above about 200 nm. For these compounds the values given in the fourth column are of the apparent molecular absorptivity (*not* its logarithm), in liters per mole of olefins per gram per liter of I_2 for a 1-cm path length. These values are denoted by asterisks.

19.1.3 Detection and Determination of Unsaturation (—C═C—; —C≡C—)

For the analysis of unsaturated compounds there are a number of chemical methods described in a review.[1] Of these methods, three methods most commonly used will be discussed in some detail. For the investigation of structural problems or the identification of more complicated bond systems, instrumental methods will be needed. These tables in earlier sections should be examined: the infrared absorption frequencies of alkenes (Table 7.5), triple bonds (Table 7.6), and aromatic and heteroaromatic compounds (Table 7.12); the Raman frequencies of alkenes (Table 7.21), alkynes (Table 7.22), and aromatic compounds (Table 7.29); and the ultraviolet absorption wavelengths of dienes (Table 6.10), enones and dienones (Table 6.11), and benzene and its derivatives (Tables 6.13 and 6.14).

19.1.3.1 Analytical Hydrogenation. Analytical hydrogenation is the most general chemical method for the determination of olefinic unsaturation. The method depends upon the measurement of the amount of hydrogen consumed by the sample when the reaction is complete. Errors due to substitution do not occur. The rate at which hydrogen reacts with different olefins varies considerably. In general, monosubstituted olefins react rapidly, followed in sequence by conjugated olefins and di-, tri-, and tetrasubstituted

[1] K. Muller, *Z. Anal. Chem.* **181**:126 (1961).

compounds. Aromatic double bonds are the most difficult to hydrogenate. By the proper choice of catalyst and operating parameters, selective hydrogenation can be achieved, thereby avoiding interference due to aromatic compounds. Platinum dioxide catalyst generally causes hydrogenations to proceed more rapidly than Raney nickel. The time required per analysis is rather high because of the long reaction time generally needed. The hydrogenated sample can be recovered if necessary.

In general, the weighed sample in a capsule is isolated from the catalyst and solvent in the constant-temperature hydrogenation cell. After flushing, the apparatus is filled with hydrogen at the desired pressure. When the take-up of hydrogen by the catalyst ceases, the volume or pressure is noted; then the sample capsule is broken. The amount of hydrogen consumed by the sample is measured volumetrically or manometrically when the reaction is complete. For accurate results, the volume of the apparatus is needed to correct hydrogen-volume readings if the temperature at the end of an analysis is different from the temperature at the beginning. The reactivity of hydrogen with many compounds depends on operating conditions, so that the time of analysis may vary from one-half hour to several hours. Compounds containing carbonyl, epoxy, sulfur, nitrogen, or other reducible groups may interfere. Under the conditions employed, platinum dioxide will saturate the double bonds in a benzene ring whereas Raney nickel will not.

19.1.3.2 Bromination. Bromination procedures are usually too slow and not quantitative unless a catalyst [mercury(II) sulfate] is present. A 25-mL aliquot of a water-soluble sample is used (0.002 equivalent in unsaturation). Carbon tetrachloride is the solvent for hydrocarbon-soluble samples. When volatile, the sample is weighed in a sealed glass ampoule, the ampoule placed under the solvent in a volumetric flask, and the ampoule crushed. Although bromination is the most useful halogenation reaction, knowledge of the types of compounds in the sample is required for interpretation of the results.

In the analysis procedure[2] a 10% to 15% excess of 0.1000*N* bromate–bromide solution (about 25 mL) is introduced into the reaction flask. The flask is evacuated with a three-way stopcock, and 5 mL of 6*N* H_2SO_4 is added via a funnel at the other stopcock position. Allow 2 to 3 min to elapse for the bromine to be liberated. Next, 10 to 20 mL of 0.2*N* $HgSO_4$ is added, followed by 25 mL of the sample solution and rinsing the funnel with solvent. If the solvent is CCl_4, 20 mL of glacial HOAc is also added. The flask is wrapped in a black cloth and shaken for 7 min (or longer for some samples). Then 15 mL of 2*N* NaCl and 15 mL of 20% KI are added and the flask is shaken for 30 s. The vacuum is broken, and the free iodine is titrated with 0.05*N* $Na_2S_2O_3$, starch indicator being used. A blank is run under the same conditions. The accuracy generally varies from 0.5% to 10%, depending on the reactivity of the compound determined and its concentration in the sample.

An accurate method for trace unsaturation is the coulometric titration of the sample with generated bromine to an amperometric end point. It is rapid, only 5 min per determination.

Bromine number is the number of grams of bromine reacting with 100 g of sample.

19.1.3.3 Iodine Number. There are instances when the bromination procedure cannot be used because the bromine not only adds on the unsaturated linkage but also substitutes some of the hydrogen atoms. This situation can be averted, or at least minimized, if iodine monobromide is used as the brominating agent. This procedure is used mostly on unsaturated hydrocarbons, fatty acids, esters, vinyl esters, and some unsaturated alcohols.

In the Wijs method, excess iodine monochloride reagent is added along with $Hg(OAc)_2$ catalyst. The reaction is allowed to stand, then KI is added, and the liberated iodine is titrated with $Na_2S_2O_3$. It is widely used for fats and oils, but does not determine conjugated unsaturation.[3]

Iodine number (or *value*) is the number of milligrams of iodine reacting with 1 g of sample under specified conditions.

19.1.3.4 Miscellaneous Methods. For the gas-chromatographic separation and determination of olefin mixtures silicone oils and squalane can be used as nonpolar stationary phases in capillary columns. Ethylene glycol–silver nitrate, Carbowax 1500, and trimethyl phosphate are suitable as polar stationary phases.

---

[2] H. J. Lucas and D. Pressman, *Ind. Eng. Chem., Anal. Ed.* **10**:140 (1938).

[3] F. A. Norris and R, J. Buswell, *Ind. Chem. Eng., Anal. Ed.* **15**:258 (1943).

TABLE 19.3 Recommended Conditions for Visual Nonaqueous Titrations of Carboxylic Acids

(References are presented below Table 19.4)

Abbreviations

b	*blue*	*rv*	*red-violet*
c	*colorless*	*v*	*violet*
m	*magenta*	*y*	*yellow*

Acid	Solvent	Titrant	Indicator	Color change	Reference
Acetic	Acrylonitrile	NaOH/MeOH	Bromothymol blue/MeOH	y-b	14
	Dioxane	NaOMe/benzene–MeOH	Thymol blue/1,4-dioxane	y-b	16
	Acetone, acetonitrile, benzene–EtOH	$(Butyl)_4NOH$/benzene–MeOH	Thymol blue/2-propanol	y-b	1
Phenyl-	Acetone	$(Butyl)_4NOH$/benzene–MeOH	Thymolphthalein/MeOH	c-b	12
Acrylic	Acrylonitrile	NaOH/MeOH	Bromothymol blue/MeOH	y-b	14
β-Phenyl-	Chloroform		Phenolphthalein/EtOH	c-rv	5
Benzoic	Acetone	KOMe/benzene–MeOH	*p*-Hydroxyazobenzene/MeOH	c-y	12
		$(Butyl)_4NOH$/benzene–MeOH	Thymolphthalein/MeOH	c-b	5
	Acetonitrile	KOMe/benzene–MeOH	*p*-Hydroxyazobenzene/MeOH	c-y	2*
	Benzene	Diphenylguanidine/benzene	Bromophthalein magenta E/benzene	y-m	15
	1,4-Dioxane or EtOH	NaOMe/benzene–MeOH	Thymol blue/1,4-dioxane	y-b	15
	Acetone, acetonitrile, pyridine, or benzene–EtOH	$(Butyl)_4NOH$/benzene–MeOH	Thymol blue/2-propanol	y-b	1
p-Amino-	Benzene–MeOH	NaOMe/benzene–MeOH	Thymol blue/MeOH	y-b	6
m-Hydroxy-	Pyridine	$(Butyl)_4NOH$/benzene–MeOH	Azo violet/benzene	y-v	1
o-Hydroxy-	Acetonitrile or benzene–MeOH	NaOMe/benzene–MeOH	Thymol blue/MeOH	y-b	5
	Acetone, acetonitrile pyridine, or benzene–EtOH	$(Butyl)_4NOH$/benzene–MeOH	Thymol blue/MeOH	y-b	1
p-Hydroxy-	Pyridine	$(Butyl)_4NOH$/benzene–MeOH	Azo violet	y-v	1
m-Nitro-	Benzene–MeOH	NaOMe/benzene–MeOH	Thymol blue/MeOH	y-b	6
p-Nitro-	Acetone	$(Butyl)_4NOH$/benzene–MeOH	Thymolphthalein/MeOH	c-b	12
trans-Butenedioic	Pyridine	$(Butyl)_4NOH$/benzene–MeOH	Azo violet/benzene	y-v	1
Citric	Acetonitrile	$(Butyl)_4NOH$/benzene–MeOH	Thymol blue/MeOH	y-b	1
Hexanedioic	Chloroform	NaOEt/EtOH	Phenolphthalein/EtOH	c-rv	4

Lactic	Chloroform	NaOEt/EtOH	Phenolphthalein/EtOH	c-rv	4
Malonic	1-Butylamine	NaOMe/benzene–MeOH	Thymol blue/MeOH	y-b	6
Mandelic	Acetone or acetonitrile	KOMe/benzene–MeOH	*p*-Hydroxyazobenzene/benzene	c-y	5
Oleic	Chloroform	NaOEt/EtOH	Phenolphthalein/EtOH	c-rv	4
Palmitic	Chloroform	NaOEt/EtOH	Phenolphthalein/EtOH	c-rv	4
trans-3-Phenyl-propenoic	$CHCl_3$ or acetonitrile	NaOEt/EtOH	Bromothymol blue/MeOH	y-b	4
m-Phthalic	Ethylene glycol–EtOH	KOH/MeOH	*m*-Cresol purple/EtOH	y-v	3
o-Phthalic	Pyridine	$(Butyl)_4NOH$/benzene–MeOH	Azo violet/benzene	y-v	1
p-Phthalic	Toluene–*N,N*-dimethylformamide		Bromothymol blue in solvent	y-b	10
Propanoic	Chloroform	NaOEt/EtOH	Phenolphthalein/EtOH	c-rv	4
3-Pyridine carboxylic	Acetone, acetonitrile, pyridine, or benzene–EtOH	$(Butyl)_4NOH$/benzene–MeOH	Thymol blue/2-propanol	y-b	1
Stearic	Chloroform	NaOEt/EtOH	Phenolphthalein/EtOH	c-rv	4
	Benzene–MeOH	NaOMe/benzene–MeOH	Thymol blue/MeOH	y-b	6, 17
Succinic	Pyridine	$(Butyl)_4NOH$/benzene–MeOH	Azo violet/benzene	y-v	1
Tartaric	Chloroform–EtOH	NaOEt/EtOH	Phenolphthalein/EtOH	c-rv	4

* Includes information on titrations of 40 substituted aromatic acids.

Chromatography columns prepared from sulfonic acid resins in which sulfonic acid protons have been partially replaced with silver ions have been found to be effective in the separation of unsaturated fatty acids and glycerides.[4] A reversible olefin trap consisting of a silver-containing microporous copolymer of divinylbenzene and styrene will separate olefins from saturated hydrocarbons. Subsequent analysis is done by gas chromatography.[5] A procedure for the separation and double-bond position determination of unsaturated alcohols, aldehydes, and carboxylic acid involves derivatization using dimethyl disulfide and iodine followed by gas chromatography and mass spectral detection.[6]

The problems and conditions required for the carbon-13 NMR analysis of hexane–hexene mixtures have been discussed.[7]

Gasolines have been run on a mass spectrometer, after which the olefins were removed from the sample with benzenesulfonyl chloride (which reacts quantitatively with olefins to form a high-boiling-point addition product) and the spectrum was obtained on the residue.[8]

19.2 CARBOXYL GROUPS

A wide variety of organic compounds show acid properties and might be classed as acids if salt formation and the neutralization of alkalies were the sole criteria. The functional group that most consistently displays all the phenomena of acidity is the carboxyl or —COOH group, and it is to substances containing this characteristic group that this subsection is confined.

The degree of dissociation of a carboxyl compound and the pH of its solution depend largely upon the nature of R in R—COOH. The dissociation of simple aliphatic acids is minimal even at high dilutions, and decreases with increasing length of R. The acidity is increased by the presence of a triple bond. Salt formation is characteristic of most acids, with the sodium salts being more soluble in water than the corresponding acids. The hydroxyl group of the carboxyl can be replaced by a halogen atom, giving an acid halide, and acid anhydrides can be formed by the loss of a molecule of water from two molecules of acid.

pK_a values of a very large number of organic materials in water are available in Table 8.8 of Ref. 9. Organic acids may be titrated if their pK_a are not greater than 5 to 6, at a concentration of 0.01*N* with 0.1*N* NaOH solution in the presence of phenolphthalein (or thymolphthalein when $pK_a > 5$) with a sharp end point. If the pK_a is >6, the titration is best performed in a nonaqueous solvent as discussed in Sec. 4.3. Carboxylic acids with aqueous pK_a values of 6 to 9 can usually be titrated in the solvent *N,N*-dimethylformamide using as titrant potassium methoxide in benzene–methanol (10 : 1) and thymol blue (or azo violet) as indicator. Very weak acids have been titrated in *tert*-butanol with potassium *tert*-butoxide as titrant.[10] Recommended conditions for visual nonaqueous titrations of organic carboxylic acids are given in Table 19.3. Table 19.4 summarizes the conditions that have been recommended for the nonaqueous titration of these acids when employing a potentiometric end point.

Aliphatic monobasic acids are found in plant and animal products, both free and as glycerides (fats). In this group of straight-chain acids, those having an odd number of carbon atoms have melting points lower than their neighbors with even numbers of carbon atoms, and similar differences are observed in their surface tensions and molecular volumes.

19.2.1 Spectral Methods

It is easy to recognize the presence of carboxylic acids in the infrared spectrum. A characteristic group of small bands (due to combination bands) appear between 3000 and 2500 cm^{-1}. For quantitative purposes, the absorption of the monomeric carboxyl group at 1760 cm^{-1} (not often observed) and the

[4] R. O. Adiof and E. A. Emken, *J. Am. Oil Chem. Soc.* **57**:276 (1980).
[5] E. G. Boeren et al., *J. Chromatogr.* **349**:377 (1985).
[6] B. A. Leonhardt and E. D. DeVilbiss, *J. Chromatogr.* **322**:484 (1985).
[7] D. A. Forsyth et al., *Anal. Chem.* **54**:1896 (1982).
[8] L. Mikkelsen, R. L. Hopkins, and D. Y. Yee, *Anal. Chem.* **30**:317 (1958).
[9] J. A. Dean, ed., *Lange's Handbook of Chemistry,* 14th ed., McGraw-Hill, New York, 1992.
[10] J. S. Fritz and L. W. Marple, *Anal. Chem.* **34**:921 (1962).

TABLE 19.4 Recommended Conditions for Potentiometric Nonaqueous Titrations of Carboxylic Acids

SCE denotes saturated (aqueous) calomel electrode and SMCE saturated methanolic calomel electrode.

Acid	Solvent	Titrant	Electrode system	Reference
Acetic	Methyl ethyl ketone	$(Butyl)_4NOH$/benzene–MeOH	Glass–SMCE	9
Phenyl-	Acetone	$(Butyl)_4NOH$/benzene–2-propanol	Pt(10% Rh)–graphite	16
Acrylic	Pyridine	$(Butyl)_4NOH$/benzene–MeOH	Glass–SMCE	17
Benzoic	Acetone or pyridine	$(Butyl)_4NOH$/benzene–MeOH	Glass–SMCE	7
p-Amino-	1,2-Ethylenediamine	$NaOC_2H_4NH_2$/1,2-ethylenediamine	H_2–SCE or H_2–Sb	13
m-, *o*-, or *p*-Hydroxy-	Pyridine	$(Butyl)_4NOH$/benzene–MeOH	Glass–SMCE	1
p-Nitro-	Acetone	$(Butyl)_4NOH$/benzene–MeOH	Glass–SMCE	7
cis-Butenedioic	Pyridine	$(Butyl)_4NOH$/benzene–MeOH	Glass–SMCE	17
trans-Butenedioic	Pyridine	$(Butyl)_4NOH$/benzene–MeOH	Glass–SMCE	17
Citric	Pyridine	$(Butyl)_4NOH$/benzene–MeOH	Glass–SMCE	1
Crotonic	Pyridine	$(Butyl)_4NOH$/benzene–MeOH	Glass–SMCE	17
Dodecanoic	EtOH	KOH/water	Glass–SCE	8
Formic	Pyridine	$(Butyl)_4NOH$/benzene–MeOH	Glass–SMCE	17
Heptanedioic	Pyridine	$(Butyl)_4NOH$/benzene–MeOH	Glass–SMCE	17
Hexanedioic	Pyridine	$(Butyl)_4NOH$/benzene–MeOH	Glass–SMCE	17
Hydroxybutanedioic	Pyridine	$(Butyl)_4NOH$/benzene–MeOH	Glass–SMCE	1, 17
Lactic	Pyridine	$(Butyl)_4NOH$/benzene–MeOH	Glass–SMCE	17
Nonanedioic	Pyridine	$(Butyl)_4NOH$/benzene–MeOH	Glass–SMCE	17
Oleic	*N,N*-dimethylformamide	KOH/water	Pt–SCE	11
Oxalic	Pyridine	$(Butyl)_4NOH$/benzene–MeOH	Glass–SMCE	17
Palmitic	EtOH	KOH/water	Glass–SCE	8
Pentanedioic	Pyridine	$(Butyl)_4NOH$/benzene–MeOH	Glass–SMCE	17
Phthalic (*m*-, *o*-, *p*-)	Pyridine	$(Butyl)_4NOH$/benzene–MeOH	Glass–SMCE	17
Methyl-	*N,N*-Dimethylformamide	KOH/water	Pt–SCE	11
Propanedioic	Pyridine	$(Butyl)_4NOH$/benzene–MeOH	Glass–SMCE	17
Stearic	EtOH	KOH/water	Glass–SCE	16
Succinic	Pyridine	$(Butyl)_4NOH$/benzene–MeOH	Glass–SMCE	1, 17
Tartaric	Pyridine	$(Butyl)_4NOH$/benzene–MeOH	Glass–SMCE	17
Tetradecanoic	EtOH	KOH/water	Glass–SCE	8

References for Tables 19.3 and 19.4

1. Cundiff, R. H., and P. C. Markunas, *Anal. Chem.* **28**:792 (1956).
2. Davis and Hetzer, *J. Res. Natl. Bur. Stand.* (*U.S.*) **60**:569 (1958).
3. Esposito, G. G., and M. H. Swann, *Anal. Chem.* **32**:49 (1960).
4. Folin and Flanders, *J. Am. Chem. Soc.* **34**:774 (1912).
5. Fritz, J. S., and R. T. Keen, *Anal. Chem.* **25**:179 (1953).
6. Fritz, J. S., and N. M. Lisicki, *Anal. Chem.* **23**:589 (1951).
7. Fritz, J. S., and S. S. Yamamura, *Anal. Chem.* **29**:1079 (1957).
8. Grunbaum, B. W., F. I. Schaffer, and P. L. Kirk, *Anal. Chem.* **25**:480 (1953).
9. Harlow, G. A., C. M. Noble, and G. E. A. Wyld, *Anal. Chem.* **28**:787 (1956).
10. Hensley, A. L., *Anal. Chem.* **32**:542 (1960).
11. Kirrmann and Daune-Dubois, *Compt. rend.* **236**:1361 (1953).
12. Malmstadt, H. V., and D. A. Vassallo, *Anal. Chem.* **31**:862 (1959).
13. Moss, M. L., J. H. Elliot, and R. T. Hall, *Anal. Chem.* **20**:784 (1948).
14. Owens, Jr., M. L., and R. L. Maute, *Anal. Chem.* **17**:1177 (1955).
15. Patchernik, A., and S. Erhlich-Rogozinski, *Anal. Chem.* **31**:985 (1959).
16. Radell, J., and E. T. Donahue, *Anal. Chem.* **26**:590 (1954).
17. Streuli, C. A., and R. R. Miron, *Anal. Chem.* **30**:1978 (1958).

dimeric carboxyl group between 1725 and 1700 cm^{-1} (but often at 1710 cm^{-1}) are useful. More detailed infrared frequencies are given in Table 7.9; those in the Raman spectrum are given in Table 7.26.

The colorimetric determination of carboxylic acids using a carbodiimide, a hydroxylammonium salt, and iron(III) ions can be carried out in organic solvents containing as much as 20% water at pH 3 to 6.[11]

19.2.2 Separation of Mixtures of Acids (or Their Esters)

Procedures for the chromatographic determination of carboxylic acids involve derivatization to volatile compounds followed by the use of liquid or gas chromatography. Alkanoic, alkanedioic, and alkenedioic acids and tricarboxylic acids can be analyzed as their *tert*-butyldimethylsilyl derivatives[12]; aromatic dicarboxylic acids and higher fatty acids as their acetonyl esters[13]; and alkanoic, alkanedioic, and aromatic di- and polycarboxylic acids and hydroxyl acids as their butyl esters.[14] Selected chromatographic procedures for the separation of organic acids should be perused in these tables in Sec. 5: gas chromatography, Table 5.14; liquid–liquid chromatography, Table 5.20; ion chromatography, Tables 5.21 and 5.22; and planar chromatography, Table 5.27. Of the methods commonly used to analyze mixtures of monomeric and oligomeric fatty acids such as GC, HPLC, and TLC, the TLC method was found to be the best one.[15]

Detection sensitivity is improved by the use of fluorescent derivatives; detection limits have been as low as 20 fmol.

Supercritical fluid chromatography is useful for the determination of fatty acids without derivatization in the presence of their esters; CO_2 was the mobile phase.[16]

19.2.3 Determination of Salts of Carboxylic Acids

A 1 : 1 mixture by volume of ethylene or propylene glycol with 2-propanol is used as the solvent medium. The titrant is perchloric acid (or hydrochloric acid) dissolved in the same solvent mixture.[17] Indicators or a pH meter is used to indicate the end point. The indicator is an alcoholic solution of methyl red (or methyl orange); the end point is a pink color.

Glacial acetic acid is another solvent that has been used for the titration of carboxylic acid salts.[18] Very good indicator end points are obtained, but potentiometric titration can also be used with glass and calomel electrodes. This solvent system usually gives sharper titration curves than the glycol–2-propanol medium, but the latter medium has the advantage of the flexibility of solvent for dissolving many different types of samples.

19.3 METHODS FOR THE DETERMINATION OF ALCOHOLS AND PHENOLS (—OH GROUP)

The determination of the hydroxyl group is the most complicated task in organic chemical analysis. The behavior of the hydroxyl group varies, depending on the rest of the molecule to which it is attached. Hydroxyl groups on primary and secondary carbon atoms, as well as phenolic and enolic hydroxyl groups, can be determined by esterification in pyridine solution with an acid anhydride, usually acetic anhydride, added in known excess. After completing the reaction, the excess can be hydrolyzed with water and the resulting acid is titrated with alkali.

[11] Y. Kasai, T. Tanimura, and T. Tamura, *Z. Anal. Chem.* **47**:34 (1975); M. Pesez and J. Bartos, *Talanta* **21**:1306 (1974).
[12] T. P. Mawwhinney et al., *J. Chromatogr.* **361**:117 (1986).
[13] D. V. McCalley.et al., *Chromatographia* **20**:664 (1985).
[14] I. Molnar-Perl et al., *Chromatographia* **20**:421 (1985).
[15] I. Zeman, M. Ranny, and L. Winterova, *J. Chromatogr.* **354**:283 (1986).
[16] A. Nomura et al., *Anal. Chem.* **61**:2076 (1988).
[17] S. Palit, *Ind. Eng. Chem., Anal. Ed.* **18**:246 (1946).
[18] J. S. Fritz, *Anal. Chem.* **22**:1028 (1950).

TABLE 19.5 Scope and Limitations of Procedures Employed for the Determination of Hydroxyl Groups

Procedure	Scope	Interferences
1. Acetylation (Ac_2O)	Alcohols, essential oils, fats, glycols, hydroxy acids, phenols, sugars, waxes	Acetylated compounds not hydrolyzed with water, low-molecular-weight aldehydes, primary and secondary amines, sulfhydryl groups
2. Bromination ($KBrO_3 + KBr$)	Phenols (replacement of H by Br at *o*- and *p*-positions)	Aliphatic hydrazines, unsaturated compounds
3. Coupling	Phenolic compounds	Amines, active methylene groups
4. Esterification ($HOAc + BF_3$)	Aliphatic and alicyclic alcohols, aromatic alcohols with OH in aliphatic side chain, hydroxy acids	Acetals, aldehydes, amines, ketals, ketones
5. Formylation (HOAc–HCOOH)	Citronellol, essential oils, linalool, easily dehydrated terpene alcohols, terpineol	
6. Grignard reaction (CH_3HgI)	Tertiary OH groups	Other compounds or groups containing active H atoms
7. Infrared and Raman	General	Functional groups that absorb in same area of infrared or Raman spectrum
8. $LiAlH_4$ reduction	Alcohols, phenols	Other compounds or groups containing active H atoms
9. Oxidation (HIO_4)	OH groups on adjacent C atoms	OH and CO, OH and NH_2, or CO groups on adjacent C atoms
10. Phthalation (phthalic anhydride)	Alcohols, essential oils	

A more powerful reagent is acetyl chloride in pyridine solution.[19] 3,5-Dinitrobenzoyl chloride reacts faster.[20] The excess reagent is hydrolyzed with water, and HCL is titrated in benzene–methanol (7 : 1) solution with 0.200*N* $(C_4H_9)_4NOH$ using the color change of the titrant from yellow to red or potentiometric end-point detection.

Phthalic anhydride is the most selective of all the acylating reagents; aldehydes, ketones, and aromatic hydroxyl compounds do not interfere. The reaction with 3-nitrophthalic anhydride in the presence of triethylamine catalyst in *N,N*-dimethylformamide solution is complete in 10 min at room temperature.

Boron trifluoride is a very strong catalyst in acylation reactions with acetic acid; the water formed during the reaction is determined by the Karl Fischer titration. This method determines the total hydroxyl group content of tertiary aliphatic and alicyclic alcohols including polyhydroxy alcohols. Only the phenolic hydroxyl group does not react.

Enolic and some aromatic hydroxyl groups are acidic enough to be titrated with KOH, particularly in nonaqueous solvents. The scope and limitations of procedures employed for the determination of hydroxyl groups are set forth in Table 19.5. Table 19.6 gives the conditions frequently employed in volumetric methods for the determination of hydroxyl groups. Finally, Table 19.7 provides a summary of the methods that have been found to be most useful in dealing with various important classes of compounds containing hydroxyl groups. When several methods are available for dealing with any one class of compounds, these are listed in the same order as they appear in Table 19.6.

The infrared (Table 7.7) or Raman (Table 7.24) methods are also applicable for the determination of the hydroxyl group. Primary alcohols have intense absorption bands at 3640 to 3636 cm^{-1},

[19] G. A. Olah and M. B. Comisarow, *J. Am. Chem. Soc.* **88**:4442 (1966).
[20] W. T. Robinson, *Anal. Chem.* **33**:1030 (1961).

TABLE 19.6 Conditions Frequently Employed in Volumetric Methods for Determination of Hydroxyl Groups

	Conditions				
		Reagent			
Procedure	Sample	Concentration	Amount	Time	Temperature
1. Acetylation	Depends on OH content and compound analyzed	Acetic anhydride + pyridine 1 : 3 (v/v)	5 mL; maintain at least 100% excess acetic anhydride	1 h	Steam bath
2. Bromination	Equivalent of 0.5 g phenol	$KBrO_3$ (0.017*M*) + excess KBr	50 mL	15 min	Room
4. Esterification	1–2 mL in 1,4-dioxane	100 g BF_3 + 1–2 mL water diluted to 1 L with HOAc	20 mL	2 h	67°C
8. $LiAlH_4$	8–15 mmol OH	0.25*M* $LiAlH_4$ in tetrahydrofuran	20 mL	15–30 min	Room
9. Oxidation	Depends on OH content				
a. OH on two adjacent C atoms		$H_5IO_6 + H_2O$ + HOAc, 5.4 : 100 : 1900 (w/v/v)	50 mL	30 min	Room
b. OH on three adjacent C atoms		60 g $NaIO_4$ + 120 mL 0.05*M* H_2SO_4 diluted to 1 L with water	50 mL	30 min	Room
10. Phthalation	Depends on OH content	Solid phthalic anhydride	2 g (100% molar excess needed)	2 h	100°C

TABLE 19.7 Methods for the Determination of Hydroxyl Groups

Compound determined	Reagent	Procedure	Reference
OH groups	CH_3COCl + pyridine	Acetylate, hydrolyze, and titrate excess reagent	28
	Acetic anhydride + pyridine	Micro technique	30
		Measure infrared absorption at 5000–3125 cm^{-1}	27
		Gas chromatography	9, 29
OH groups (in presence of α-epoxide)	$LiAlH_4$	Volumetric estimation of H_2 formed	20
Alcohols	HOAc + BF_3	Esterify and titrate water formed	6
	$LiAlH_4$	Titrate excess $LiAlH_4$ with 1-propanol	15
	Phthalic anhydride in pyridine	Esterify and titrate excess anhydride	11
Alcohols (in amine mixtures)	Acetic anhydride + pyridine (1 : 3, v/v)	Acetylate and determine sponification value of esters formed	12
Alcohols (primary and secondary)	Acetic anhydride + pyridine (1 : 3, v/v)	Acetylate, hydrolyze, and titrate excess reagent	6
	Acetic anhydride + ethyl acetate + $HClO_4$	Acetylate, hydrolyze, and titrate excess reagent	13
Alcohols (tertiary)	HOAc + BF_3	Esterify and titrate water formed	6
	Grignard reagent	Measure volume of CH_4 evolved	14
Carbohydrates	Acetic anhydride + pyridine	Acetylate, hydrolyze, and titrate excess reagent	23
Cellulose derivatives	Acetic anhydride + pyridine (1 : 19, v/v)	Acetylate and titrate excess reagent	26
Cellulose esters (total and primary OH)	Phenyl isocyanide	Ultraviolet absorption of carbanilate at 280 nm	25
Essential oils (alcohols)	Acetic anhydride + NaOAc	Acetylate and determine saponification value	17
Essential oils (primary alcohols)	Phthalic anhydride + benzene	Esterify and titrate excess anhydride	16
Essential oils (easily dehydrated alcohols)	HCOOH + HOAc	Esterify and determine saponification value	18
Glycerol	Acetic anhydride + pyridine	Acetylate, hydrolyze, and titrate excess HOAc	1
	Acetic anhydride	Acetylate and titrate excess reagent	4
	$NaIO_4 + H_2SO_4$	Oxidize and titrate HCOOH formed	1
	Schiff reagent	Colorimetric measurement	3
Glycols	Acetic anhydride + pyridine (1 : 3, v/v)	Acetylate, hydrolyze, and titrate excess reagent	1
	0.025M H_5IO_6	Oxidize and titrate excess reagent iodometrically	7
	0.1M $NaIO_4$	Oxidize and tirate acid formed with standard alkali	10
Glycols (mixtures)	0.025M $NaIO_4$	Oxidize and apply combination of acidimetric and iodometric titrations	7, 8
Hydroxy acids	Acetic anhydride + pyridine (1 : 3, v/v)	Acetylate, hydrolyze, and titrate HOAc	1
		Semimicroprocedure similar to above	5
	HOAc + BF_3	Esterify and titrate water formed	6

(Continued)

TABLE 19.7 Methods for the Determination of Hydroxyl Groups (*Continued*)

Compound determined	Reagent	Procedure	Reference
Monoglycerides	H_5IO_6 + HOAc (1 : 370, w/v)	Oxidize and determine excess H_5IO_6 iodometrically	1
Natural fats	Acetic anhydride + pyridine (1 : 9, v/v)	Acetylate, hydrolyze, and titrate excess reagent	1
	Acetic anhydride	Acetylate and determine saponification values of sample before and after acetylation	1
	Acetic anhydride	Acetylate in sealed tube and titrate excess reagent	2
Phenols	Acetic anhydride + pyridine (1 : 3, v/v)	Acetylate, hydrolyze, and titrate excess reagent	5
	$KBrO_3$ + KBr	Brominate and titrate excess Br_2 iodometrically	19
	Diazo compounds	Coupling	22
	$LiAlH_4$	Titrate excess $LiAlH_4$ with 1-propanol	21
Phenols (*o*-substituted)	$LiAlH_4$	Measure volume of H_2 evolved	20
Polyesters	F_3CCOOH	Measure water formed by Karl Fischer method	31
Sterols	Acetic anhydride + pyridine	Acetylate, hydrolyze, and titrate excess reagent	24
	Gas chromatography–liquid chromatography	(Also for sterol esters and waxes in oils and fats)	32

1. *Official and Tentative Methods*, 2d ed., American Oil Chemist's Society, Chicago, 1946.
2. K. Helrich and W. Rieman, *Ind. Eng. Chem., Anal. Ed.* **19**:691 (1947).
3. Basset, *Ind. Eng. Chem.* **2**:389 (1910).
4. Mehlenbacker, in J. Mitchell et al., eds., *Organic Analysis*, Interscience, New York, 1953, Vol. 1, p. 18.
5. C. L. Ogg, W. L. Porter, and C. O. Willets, *Ind. Eng. Chem., Anal. Ed.* **17**:394 (1945).
6. W. M. D. Bryant, J. Mitchell, Jr., and D. M. Smith, *J. Am. Chem. Soc.* **62**:1 (1940).
7. N. Allen, H. Y. Charbonnier, and R. M. Coleman, *Ind. Eng. Chem., Anal. Ed.* **12**:384 (1940).
8. Pohle and Mehlenbacker, *J. Am. Oil Chem. Soc.* **24**:155 (1947).
9. L. Ginsburg, *Anal. Chem.* **31**:1822 (1959).
10. S. Dal Nogare and A. N. Oemler, *Anal. Chem.* **24**:902 (1952).
11. P. J. Elving and B. Warshowsky, *Ind. Eng. Chem., Anal. Ed.* **19**:1006 (1947).
12. S. Siggia and I. R. Kervenski, *Anal. Chem.* **23**:117 (1951).
13. Fritz, J. S., and G. H. Schenk, *Anal. Chem.* **31**:1808 (1959).
14. W. Fuchs, N. H. Ishler, and A. G. Sandhoff, *Ind. Eng. Chem., Anal. Ed.* **12**:507 (1940).
15. C. J. Linter, R. H. Schleif, and T. Higuche, *Anal. Chem.* **22**:534 (1950).
16. Gunter, *The Essential Oils*, Van Nostrand, Princeton, N.J., 1948, Vol. 1, p. 275.
17. Ibid., p. 272.
18. Ibid., p. 276.
19. Furman, ed., *Scott's Standard Methods of Chemical Analysis,* 5th ed., Van Nostrand, Princeton, N.J., 1939, Vol. II, p. 2253.
20. G. A. Stenmark and F. T. Weiss, *Anal. Chem.* **28**:1784 (1956).
21. T. Higuche, in J. Mitchell et al., eds., *Organic Analysis*, Interscience, New York, 1954, Vol. 2, p. 123.
22. Arndt, in J. Mitchell et al., eds., *Organic Analysis*, Interscience, New York, 1953, Vol. 1, pp. 197–239.
23. B. E. Christensen and R. A. Clarke, *Ind. Eng. Chem., Anal. Ed.* **17**:265 (1945).
24. M. Freed and A. M. Wynne, *Ind. Eng. Chem., Anal. Ed.* **8**:278 (1936).
25. C. J. Malm et al., *Anal. Chem.* **26**:188 (1954).
26. C. J. Malm, L. B. Ganung, and R. F. Williams, Jr., *Ind. Eng. Chem., Anal. Ed.* **14**:935 (1942).
27. C. L. Hilton, *Anal. Chem.* **31**:1610 (1959).
28. Smith and Bryant, *J. Am. Chem. Soc.* **57**:61 (1935).
29. H. S. Knight, *Anal. Chem.* **30**:2030 (1958).
30. J. W. Petersen, K. W. Hedberg, and B. T. Christensen, *Ind. Eng. Chem., Anal. Ed.* **15**:225 (1943).
31. C. A. Lucchesi, B. Bernstein, and P. Ronald, *Anal. Chem.* **47**:173 (1975).
32. K. Grob, M. Lanfranchi, and C. Mariani, *J. Chromatogr.* **471**:397 (1989).

secondary alcohols at 3636 to 3630 cm^{-1}, and tertiary alcohols at 3623 to 3620 cm^{-1}. The carboxylic group shows intense absorption in these regions. The association of hydroxyl groups in polymers with tetrahydrofuran, which absorbs strongly at 3450 cm^{-1}, forms the basis of a hydroxyl determination in polymers.[21] Chemical, infrared, and NMR spectral methods for the determination of the hydroxyl group have been discussed.[22]

Hydroxyl groups on adjacent carbon atoms (glycolic hydroxyl) can be readily determined by oxidation with periodic acid. A known excess of the reagent is used, and after the oxidation the iodine liberated is titrated with thiosulfate. In an alternative finish, the unreacted periodate is reduced to iodide, which is then titrated with silver ion using an iodide-selective electrode. The time required for the determination of hydroxyl groups in various polymer polyols can be cut to 15 min by the addition of imidazole or *N*-methylimidazole as catalyst. The time required for the periodate reaction can be followed with a perchlorate-selective ion electrode and related to the concentration of glycol.[23]

Spectrophotometric methods are available for the determination of small amounts of alcoholic hydroxyl groups. Secondary alcohols, in the presence of primary alcohols, can be determined by oxidizing the secondary alcohols to ketones with $K_2Cr_2O_7$ and determining the ketones with 2,4-dinitrophenylhydrazine.[24] Primary alcohols can be determined with vanadium 8-hydroxyquinoline.[25] For tertiary alcohols, the alcohol can be transformed into alkyl iodide with HI, and the absorption of the iodide is measured.[26] Phenols react with nitrous acid to form nitrosophenol, which, on treatment with alcoholic ammonia solution, form an intense color owing to the formation of a quinone. Coupling reactions of phenols with diazonium compounds display an absorption band at 270 to 280 nm, which in alkaline solution shifts to 295 to 300 nm, with an increase in absorbance. For the determination of very small amounts of phenols, such as those in surface waters, 4-aminoantipyrine is added in alkaline solution and in the presence of $K_3[Fe(CN)_6]$ with the formation of colored compounds.[27] The hydroxyl groups of poly(ethylene glycols) are silanized with dimethylaminosilanes, and the derivatives determined photometrically with a sensitivity 1000-fold more sensitive but with the same precision as the acetylation method.[28]

(*R*)-(+)-2-Phenylselenopropionic acid is a useful reagent for determining the enantiomeric composition of secondary chiral alcohols.[29] For gas chromatography, both carboxylic acids and alcohols (or hydroxy acids) can be derivatized simultaneously with the use of heptafluorobutyric anhydride, pyridine, and ethanol.[30] Reference gas-chromatographic data have been recorded for the analysis of alcohols and phenols as trimethylsilyl ethers.[31]

Both the ^{19}F NMR of trifluoroacetates and the ^{13}C NMR of CH_2 and CH groups serve to distinguish primary from secondary alcohols and ethylene oxide from propylene oxide units in the chains of various polymer polyols. Adduct formation between alcohols and hexafluoroacetone allows the determination of hydroxyl groups by the use of ^{19}F NMR.[32] Mixtures of different alcohols or mixtures of alcohols and water can be determined simultaneously.

19.4 METHODS FOR THE DETERMINATION OF ALDEHYDES AND KETONES

Compounds containing the carbonyl group, such as aldehydes and ketones, can be determined selectively because of their reducing and complex-forming properties. Acetals (the condensation products of alcohols and aldehydes) are grouped with the aldehydes because they can be determined with the same type of reactions. Methods that serve for the determination of both aldehydes and ketones are

[21] C. S. Y. Kim et al., *Anal. Chem.* **54**:232 (1982).
[22] S. Siggia, J. G. Hanna, and T. R. Stengle, in S. Patal, ed., *Chemistry of the Hydroxyl Group*, Pt. 1, Wiley, London, 1971.
[23] C. H. Efstathiou and T. P. Hadjiioannou, *Anal. Chem.* **47**:864 (1975).
[24] F. E. Critchfield and J. A. Hutchinson, *Anal. Chem.* **32**:862 (1960).
[25] R. Amos, *Anal. Chim. Acta* **40**:401 (1968).
[26] M. W. Scoggins and J. W. Miller, *Anal. Chem.* **38**:612 (1966).
[27] M. B. Ettinger, R. J. Ruchhoft, and R. J. Liska, *Anal. Chem.* **23**:1783 (1951); F. W. Ochynsky, *Analyst* **85**:278 (1960).
[28] D. F. Fritz et al., *Anal. Chem.* **51**:7 (1979).
[29] P. Michelsen and G. Odham, *J. Chromatogr.* **331**:295 (1985).
[30] J. B. Brooks, C. C. Alley, and J. A. Liddle, *Anal. Chem.* **46**:1930 (1930).
[31] M. Mattsson and G. Petersson, *J. Chromatogr. Sci.* **15**:546 (1977).
[32] F. F.-L. Ho, *Anal. Chem.* **45**:603 (1973); *ibid.* **46**:496 (1974).

TABLE 19.8 Methods for the Determination of Aldehydes and Ketones

For references, see list below Table 19.10.

Type of compound determined	Technique	Procedure and reference	Interferences
General	Chromatographic	After derivatization with 2,4-dinitrophenylhydrazine, use HPLC.	
	Chromatographic	Convert volatile carbonyl compounds to their *o*-benzyloximes, then separate by GC.	
	Gasometric (Grignard reagent) (1–10 meq)	Reaction with CH_3MgI and decomposition of excess reagent with water. Measure volume CH_4 evolved [1,7].	Compounds containing active H, acid halides, alkyl halides, esters, isonitriles, nitriles
	Gravimetric (1–5 meq)	2,4-Dinitrophenylhydrazine forms precipitate, filter and weigh [1].	Hydrocarbons
	HPLC–mass spectrometry	A moving belt interface introduces 2,4-dinitrophenylhydrazones of aldehydes and ketones into a mass spectrometer after separation by HPLC. CH_4 chemical ionization mode; aldehydes were detected by negative ion mass spectrometry.	
	Potentiometric (0–0.25 meq)	Reaction with excess H_2NON, titration of excess reagent with standard HCl [1].	Acids, bases, peroxides
	(1–10 meq)	Reaction with excess $H_2NOH \cdot HCl$, titration of liberated HCl with standard NaOH solution [1].	
	(1–10 meq)	After reaction with excess $H_2NOH \cdot HCl$, titrate liberated water with Karl Fischer reagent.	Peroxides
	Spectrophotometric (0–500 μg/mL)	After reaction with excess 2,4-dinitrophenylhydrazine, make alkaline, measure absorption at 480 nm [1].	Acids, other colored compounds
Aliphatic	Volumetric (1–5 meq)	After reaction with excess standard peroxotrifluoroacetic acid in 1,2-dichloroethane, add excess KI and titrate liberated I_2 with standard $Na_2S_2O_3$ [9].	Aromatic aldehydes and ketones, other reducing agents
α,β-Unsaturated	Spectrophotometric (0–500 μg/mL)	Reaction with *m*-phenylenediamine, then measure absorbance of colored Schiff's base [1].	HCHO, glyoxal, isobutyraldehyde
β-Dicarbonyl	Volumetric (1–5 meq)	After reaction with standard $Cu(OAc)_2$, remove Cu(II) complex by filtration and extraction with $CHCl_3$, and determine excess Cu(II) iodometrically [1].	Other complexing agents, excess base
Acetaldehydes, methyl ketones	Spectrophotometric (0–400 μg/mL)	After reaction with alkaline I_2 (NaOI), measure absorbance of CHI_3 at 347 nm [1].	EtOH, peroxides

listed in Table 19.8; methods for the determination of aldehydes are listed in Table 19.9; and methods for the determination of ketones are listed in Table 19.10.

19.5 METHODS FOR THE DETERMINATION OF ESTERS

The esters of carboxylic acids are relatively stable compounds, and therefore no direct reactions are available for their determination. Esters are most commonly determined by saponification if the corresponding carboxylic acid and alcohol are formed. If a known excess of alkali is added, the excess can be titrated with standard acid to the phenolphthalein end point or performed potentiometrically with the glass–calomel electrode pair.

For easily saponifiable esters, the solvent can be C_1 to C_4 alcohols and water or hydrocarbon–alcohol mixtures. The system is refluxed for 30 min using 0.05*M* to 1.0*M* KOH, then cooled, and the unused base is titrated with 0.05*M* to 1.0*M* H^+.[33–36]

For difficultly saponifiable esters, the solvents are alcohols higher than C_5, glycols, glycol ethers, polyglycols, polyglycol ethers, and other high-boiling-point solvents. Small amounts of water are needed. Reaction conditions are the same as described for easily saponifiable esters.

When the alkaline hydrolysis is done in aqueous dimethylsulfoxide solution, many esters react even at room temperature while others require heating for 5 min on a water bath.[37] Using the alkaline hydrolysis method, the "ester number" or "saponification number" can be determined. These data are used in industry to characterize fats and waxes. The ester number is the amount of potassium hydroxide (in mg) that is necessary for the saponification of the esters found in 1 g of the sample.

Gas chromatography is an excellent method for the separation and determination of esters. Methyl esters of the C_{12} to C_{18} fatty acids have been separated on Chromosorb R and Celite 545 as the stationary phase, which are treated with dimethylchlorosilane and then moistened with poly(vinyl acetate).[38]

A colorimetric method involves reaction of the ester with hydroxylamine to form the corresponding oxime, after which Fe(III) ion is added to form a red chelate.[39,40]

In the infrared spectrum most esters will show a C—O stretching vibration at 1100 to 1250 cm^{-1}. Normal saturated esters have a C=O band at 1750 to 1735 cm^{-1} whereas in unsaturated and aryl ester the same band appears at 1730 to 1717 cm^{-1}.

The details of the procedure that is best suited to any particular analytical problem depend so much on the nature of the sample and on the identity of the ester in question that the original literature must be consulted before making a selection.

19.6 METHODS FOR THE DETERMINATION OF OTHER OXYGEN-BASED FUNCTIONAL GROUPS

19.6.1 Ethers

Aliphatic ethers are boiled with concentrated HI. The alkyl iodide formed is volatile (relative to the boiling point of HI) when methoxy, ethoxy, and isopropoxy groups are the reactants. Higher alkyloxy entities are difficult or impossible to remove by distillation.

---

[33] D. T. Englis and J. E. Reinschreiber, *Anal. Chem.* **21**:602 (1949).
[34] W. Reiman, *Ind. Eng. Chem., Anal. Ed.* **15**:325 (1943).
[35] S. Siggia and J. G. Hanna, *Quantitative Organic Analysis Via Functional Groups*, 4th ed., Wiley, New York, 1977.
[36] D. M. Smith, J. Mitchell, Jr., and A. M. Billmeyer, *Anal. Chem.* **24**:1847 (1952).
[37] J. A. Vinson, J. S. Fritz, and G. A. Kingsbury, *Talanta* **13**:1673 (1966).
[38] I. Hornstein and P. F. Crowe, *Anal. Chem.* **33**:310 (1961); see also F. H. M. Nestler and D. F. Zinkel, *ibid.* **39**:1118 (1967).
[39] Hestrin, *J. Biol. Chem.* **180**:249 (1949).
[40] U. T. Hill, *Ind. Eng. Chem., Anal. Ed.* **18**:317 (1946); *Anal. Chem.* **19**:932 (1947).

TABLE 19.9 Methods for the Determination of Aldehydes

For references, see list below Table 19.10.

Type of aldehyde determined	Technique and range	Procedure and reference	Interferences
General	Fluorescence	1,2-Diaminonaphthalene reacts with aldehydes to give a fluorescent product.	
	Gravimetric (0.1–10 meq)	React with 3,5-dimethylcyclohexane-1,3-dione in solution buffered at pH 4.6, filter, dry and weigh precipitate [1].	Acids, bases
	Polarographic (0.001*M*–0.1*M*)	Reduction at dme in aqueous 0.1*M* $(CH_3)_4NOH$ solution. $E_{1/2}$ = ca. –2.1 V vs. SCE [1,6].	Other reducible compounds
	Potentiometric titration (1–5 meq)	Reaction with 1-dodecylamine in ethylene glycol–2-propanol medium followed by titration of excess amine with standard salicylic acid solution [14].	Strong acids, acyl anhydrides, acyl halides
	Spectrophotometric	Details for colorimetric, fluorimetric, and phosphorimetric analysis of over 74 aldehydes given.	
		Reaction product of an aldehyde, diethylamine and chloranil is measured at 640 to 660 nm.	Aromatic aldehydes, ketones, and formaldehyde do not interfere
	Volumetric (1–5 meq)	After reaction with excess Ag_2O, add excess standard alkali and back-titrate with standard acid [1].	Acids, esters, peroxides
	(20–40 meq)	After reaction in neutral solution with Na_2SO_3, titrate with standard acid [1].	Cyclohexanone, some other ketones if >10%, some organic acids, peroxides
Acetals	Spectrophotometric	After hydrolysis with dilute HCl, the corresponding aldehyde is reacted with diethylamine and chloranil and color measured at 640 to 660 nm.	
Aliphatic	Spectrophotometric (0.01–30 μg/mL)	After reaction with 2-hydrazinobenzothiazole, develop color with $K_3[Fe(CN)_6]$ plus KOH, measure absorbance [18].	Acrolein, nitromethane
	Spectrophotometric	Measure absorbance at 290–293 nm for 0.001*M*–1*M*, and at 180–200 nm for $10^{-6}M$–$10^{-3}M$ [1,16,17].	Other absorbing compounds; some ketones, aromatics, and unsaturates
Lower aliphatic	Spectrophotometric (0–500 μg/mL)	Basic fuchsin (red) plus aqueous SO_2 gives colorless leuco Schiff's base which is coupled with aldehydes to give red dye [1–3].	Higher aldehydes, some ketones, bases, oxidizing reagents
	Spectrophotometric	Color formed from the reaction of 4-amino-3-hydrazino-5-mercapto-1,2,4-triazole with simple aldehydes shows an absorption maximum at 520 to 550 nm.	
	Volumetric (0.1–0.2 meq)	Reaction with Ag_2O in column or flask; titration of Ag salts in eluate or unreacted Ag in flask with standard KSCN [+ Fe(III) indicator] [1,8].	Acids, esters
Unsaturated	Volumetric (1–5 meq)	After reaction with dodecanethiol at room temperature, titrate excess mercaptan iodometrically [1].	Other unsaturated compounds, peroxides

α,β-Unsaturated	Spectrophotometric ($10^{-7}M$–$10^{-4}M$)	Measure absorbance at 215, 265, 315, or 320 nm, depending on extent of conjugation [1,16,17].	Other absorbing compounds
	Spectrophotometric	Reaction with anthrone gives benzanthrone derivatives, detectable by ultraviolet spectroscopy but better detected by fluorescence spectroscopy.	No interference from saturated aldehydes
	Spectrophotometric–flame atomic absorption	Tollen's reagent method is used; the Ag is filtered, dissolved in HNO_3 and measured at 328 nm by flame AAS.	Ketones
Aromatic	Fluorometric ($10^{-8}M$–$10^{-7}M$)	Reaction with 4,5-dimethyoxy-1,2-diaminobenzene in dilute acid. Fluorescence is developed by addition of alkali and stabilized by the addition of 2-mercaptoethanol.	
	HPLC–fluorescence	Form fluorescent derivatives with 1,2-diamino-4,5-ethylene-dioxybenzene and separate by reversed-phase HPLC.	Selective for aromatic aldehydes
	Thermometric titration	After reaction with 2,4-dinitrophenylhydrazine or *N,N*-dimethylhydrazine, titrate excess thermometrically with standard *o*-methoxybenzaldehyde in (1 : 1 : 23) sulfuric acid : water : isobutyl alcohol medium.	
Specific compounds	Infrared ($10^{-3}M$–$0.2M$)	Measure the C═O stretching band in the region 1834–1653 cm^{-1} [4,5].	Carboxylic acids, esters, ketones
	Mass spectrometry (>0.2 meq)	Obtain mass spectrum and calculate concentration(s) from heights of characteristic peaks [10,18].	
Formaldehyde	Colorimetric	Modified pararosaniline method for formaldehyde in air. NIOSH*-recommended procedure uses chromotropic acid and H_2SO_4.	Unaffected by phenol
	Monitoring	Two monitoring methods based on 2,4-dinitrophenylhydrazine and *N*-benzylethanolamine were evaluated.	
	Polarographic (10–100 μg/mL)	Obtain polarogram in aqueous $0.1M$ LiOH–$0.01M$ LiCl; $E_{1/2} = -1.6$ V vs. SCE.	Acetaldehyde and acrolein above limit determined by formaldehyde concentration
	Spectrophotometric (0–100 μg/mL)	After reaction with chromotropic acid (1,8-dihydroxy-3,6-naphthalenedisulfonic acid) plus $18M$ H_2SO_4 at 100°C for 30 min, dilute and measure absorbance at 670 nm.	Alcohols, formals, some ketones
	Volumetric (0–25 mg)	After reaction with excess KCN, determine excess by titration with standard $AgNO_3$; or destroy CN^- with Br_2, add excess KI, and titrate liberated I_2 with standard $Na_2S_2O_3$ [1].	>5% acetaldehyde, >50% acetone
	Volumetric (0.1–5 meq)	After reaction with excess standard I_2 in alkaline solution, make acid and titrate excess I_2 with standard $Na_2S_2O_3$ [1].	Acetaldehyde, acrolein, EtOH, furfural, methyl ketones, 2-oxopropionic acid
	General methods	Review of methods for air and forest products.	

* NIOSH = National Institute for Occupational Safety and Health.

TABLE 19.10 Methods for the Determination of Ketones

Type of ketone	Technique	Procedures and reference	Remarks
General	Polarographic ($10^{-4}M$–$10^{-2}M$)	Obtain polarogram of N_2H_4 adduct in 0.1M $N_2H_4 \cdot H_2SO_4$ plus 0.05M H_2SO_4 at $E_{1/2} = -1.1$ and -1.4 V vs. SCE [11].	For aliphatic ketones; aldehydes and other reducible compounds interfere
	Polarographic ($10^{-5}M$–$10^{-2}M$)	After reaction with betaine hydrazide HCl, obtain polarogram in buffered alkaline chloride solution; $E_{1/2} = -1.4$ to -1.5 vs. internal Hg pool [11].	Aldehydes, other reducible compounds interfere
	Infrared ($10^{-3}M - 0.2M$)	Measure absorbance at C═O stretching band at 1835–1653 cm^{-1}	Carboxylic acids, aldehydes, esters interfere
	Ultraviolet ($10^{-3}M$–$1M$)	Measure absorbance at 279 to 289 nm for aliphatic and >289 nm for aromatic ketones [13,22].	Aromatic compounds interfere
	Volumetric (1–10 meq)	Oxidize aldehydes selectively with Ag_2O; react remaining ketones with excess H_2NOH, and back-titrate excess with standard HCl solution [13].	Strong bases interfere
Aliphatic	Spectrophotometric	Dissociation of dimer of dye cation from Brilliant Green occurs in presence of aliphatic ketones resulting in an increase in absorbance; detected eluants in reversed-phase HPLC [27].	
Diketones, aliphatic	Ultraviolet	For α-diketones ($10^{-3}M$–$1M$), measure absorbance at ca. 286 nm; for β-diketones ($10^{-5}M$–$10^{-2}M$), at ca. 270 nm [13,22].	Some aromatics interfere with β-diketones
Specific compounds	Mass spectrometry (>0.2 meq)	Obtain mass spectrum and use heights of characteristic peaks and calibration from pure compounds [24].	
Acetone	Spectrophotometric (0.02M–0.2M)	Determine purple color after reaction with furfural plus H_2SO_4.	Aliphatic aldehydes

Volumetric (0.1–2 meq)	After reaction with standard I_2 in alkaline solution, titrate excess I_2 with standard $Na_2S_2O_3$ [13].	Acetaldehyde, acrolein, EtOH, furfural, methyl ketones, 2-oxopropionic acid

References for Tables 19.8 to 19.10

1. L. S. Bark and P. Bate, *Analyst* **96**:881 (1971); L. S. Bark and P. Prachuabpaibul, *Anal. Chem. Acta* **84**:207 (1976).
2. W. F. Chao et al., *Anal. Chim. Acta* **215**:259 (1988); S. Hara et al., *ibid.* **215**:267 (1988).
3. J. Chrastil and R. M. Reinhardt, *Anal. Chem.* **58**:2848 (1986).
4. D. L. DuVal, M. Rogers, and J. S. Fritz, *Anal. Chem.* **57**:1583 (1985).
5. K. Fung and D. Grosjean, *Anal. Chem.* **53**:168 (1981).
6. J. A. Gilpin and F. W. McLafferty, *Anal. Chem.* **29**:990 (1957).
7. L. Gollob and J. D. Willons, *For. Prod. J.* **30**:27 (1980).
8. M. F. Hawthorne, *Anal. Chem.* **28**:540 (1956).
9. N. W. Jacobsen and R. G. Dickinson, *Anal. Chem.* **46**:298 (1974).
10. S. E. Know and S. S. Q. Hee, *Ind. Hyg. Assoc. J.* **45**:325 (1984).
11. I. M. Kolthoff and J. J. Lingane, *Polarography*, 2d ed., Interscience, New York, 1952.
12. B. Miller and N. D. Danielson, *Anal. Chem.* **60**:622 (1988).
13. J. Mitchell et al., eds., *Organic Analysis*, Interscience, New York, 1953, Vol. 1.
14. M. Nakamura et al., *Anal. Chim. Acta* **134**:39 (1982).
15. S. I. Obtemperanskaya and E. K. R. Mohamed, *Z. Anal. Khim.* **35**:1982 (1980); *C.A.* **93**:230330g (1980).
16. Y. Ohkura and K. Zaitsu, *Talanta* **21**:554 (1974).
17. P. J. Oles and S. Siggia, *Anal. Chem.* **46**:911 (1974).
18. D. G. Ollett, A. B. Attygalle, and E. D. Morgan, *J. Chromatogr.* **367**:207 (1986).
19. K. L. Olson and S. J. Swarin, *J. Chromatogr.* **333**:337 (1985).
20. E. Priha and I. Ahonen, *Anal. Chem.* **58**:1195 (1986).
21. E. L. Saier and R. H. Hughes, *Anal. Chem.* **30**:513 (1958).
22. E. Sawicki and T. R. Hauser, *Anal. Chem.* **32**:1434 (1960).
23. E. Sawicki and C. R. Sawicki, *The Analysis of Organic Materials: Aldehydes Photometric Analysis*, Academic, New York, 1975, Vols. 2 and 3.
24. A. G. Sharkey, J. L. Shultz, and R. A. Friedel, *Anal. Chem.* **28**:934 (1956).
25. H. Siegal and F. T. Weiss, *Anal. Chem.* **26**:917 (1954).
26. S. Siggia and E. Segal, *Anal. Chem.* **25**:640, 830 (1953).
27. A. Trujilo, S. W. Kang, and H. Freiser, *Anal. Chim. Acta* **182**:71 (1986).
28. M. T. M. Zaki, *Anal. Lett.* **18**:1697 (1985).

More recent procedures, especially for microamounts, use a volumetric version to complete the determination. In one procedure the alkyl iodide vapor is conducted into a glacial acetic acid solution of bromine with carbon dioxide or other inert gas. Before entering this solution, the vapor is washed to remove HI and I_2. The bromine oxidizes the iodide portion of the alkyl iodide to iodate ions. On adding excess iodide ions, a sixfold amount of iodine atoms are liberated, making the procedure very sensitive. The liberated iodine is titrated with standard $Na_2S_2O_3$ solution. In another procedure the distilled alkyl iodide is absorbed by pyridine and titrated with tetrabutylammonium hydroxide solution.[41] In a third variation the alkyl iodides are absorbed in a known amount of benzene, which is then reacted with aniline; the aniline iodide formed is titrated with standard sodium methoxide solution.[42] This latter method is suitable for the determination of the C_4 to C_{20} alkyl iodides.

Rather than distillation, the alkyl iodides produced from the higher alkoxyl groups can be extracted with cyclohexane and their ultraviolet absorption bands measured.

All ethers that contain an α-hydrogen atom can be oxidized with bromine water; the excess of bromine is determined by iodometric titration.[43]

Polyglycol ethers are determined in a manner similar to the alkoxyl groups.[44] A two-phase titration procedure for the determination of poly(oxyethylene) nonionic surfactants involves the extraction of the surfactant as a sodium tetraphenylborate complex, which is then (1) titrated with tetradecyldimethylbenzylammonium chloride,[45] (2) determined in a two-phase system by titration with sodium tetrakis(4-fluorophenyl)borate,[46] or (3) determined by ultraviolet spectroscopy after conversion into a red complex by reaction with $Fe(SCN)_3$. Nonionic poly(oxyethylene) can be oxidized by V_2O_5 in H_2SO_4 solution and the excess V_2O_5 determined potentiometrically with Fe(II) ion.[47] Ultra-trace levels of these nonionic surfactants have been determined in the presence of cationic surfactants in water by extraction with excess potassium picrate in dichloromethane; the potassium complex of the poly(oxyethylene) complex is concentrated in the dichloromethane layer.[48]

The oxygen ether content of coals and humic substances has been determined by carbon-13 NMR.[49] If the alkyl iodides formed by hydrolysis with HI are absorbed in a 2-methylpyridine solution, 2,6-dimethylpyridine iodide is formed. For a spectrophotometric determination, this is reacted with an alkaline solution that contains 2,7-dihydroxynaphthol, $K_3[Fe(CN)_6]$, and KCN [211].[50]

Gas chromatography easily separates the alkyl iodides formed by the reaction with HI because of their volatility. The column is tricresyl phosphate on Chromosorb R at 90°C; the eluant is passed through a thermal conductivity detector.

Vinyl ethers are hydrolyzed with H_2SO_4 at room temperature for 15 to 30 min. The acetaldehyde formed is treated with $NaHSO_3$ and the method finished volumetrically.[51] The acetaldehyde can also be treated with excess hydroxylamine hydrochloride and the excess determined volumetrically.

Infrared-absorption characteristics of various ethers will be found in Table 7.10. The Raman frequencies are given in Table 7.31.

19.6.2 The Epoxy Group (Oxiranes)

A detailed description of the analysis of the epoxy group can be found in a monograph.[52] On treatment with a nucleophilic reagent, such as hydrohalic acids, the oxirane ring of 1,2-epoxy compounds is opened and the corresponding chlorohydrin is formed. The methods consist of the addition of a

[41] S. Ehrlich-Rogozinski and A. Patchornik, *Anal. Chem.* **36**:849 (1964).
[42] J. Schole, *Z. Anal. Chem.* **193**:321 (1963).
[43] N. C. Deno and N. H. Potter, *J. Am. Chem. Soc.* **89**:3350 (1967).
[44] P. W. Morgan, *Ind. Eng. Chem., Anal. Ed.* **18**:500 (1946); S. Siggia et al., *Anal. Chem.* **30**:115 (1958).
[45] M. Tsubouchi and Y. Tanaka, *Talanta* **31**:633 (1984).
[46] M. Tsubouchi, N. Yamasaki, and K. Yanagisawa, *Anal. Chem.* **57**:763 (1985).
[47] C. Dauphin et al., *Anal. Chim. Acta* **149**:313 (1983).
[48] L. Favretto, B. Stancher, and F. Tunis, *Int. J. Environ. Anal. Chem.* **14**:201 (1983).
[49] T. Yoshida et al., *Fuel* **63**:282 (1984); E. Bayer, *Angew. Chem.* **96**:151 (1984).
[50] R. F. Makeus, L. R. Rothringer, and A. R. Donia, *Anal. Chem.* **31**:1265 (1959).
[51] S. Siggia, *Ind. Eng. Chem., Anal. Ed.* **19**:1025 (1947); S. Siggia, *Quantitative Organic Analysis Via Functional Groups*, 3d ed., Wiley, New York, 1963 pp. 98–101.
[52] B. Dobinson, W. Hoffmann, and B. P. Stark, *The Determination of Epoxide Groups*, Pergamon, London, 1969.

known excess of the hydrohalic acid and the determination of the excess by titration with standard base. The use of various organic solvents makes possible the analysis of oxirane compounds insoluble in water. Table 19.11 gives methods for the determination of oxiranes.

The reaction of oxiranes with sodium sulfide followed by interaction of the product with taurine and *o*-phthalic anhydride produces an intense fluorescence on excitation with a detection limit of 0.1 mmol/100 μL.[53]

A method based on proton NMR, and not subject to chemical interfering reactions, has been applied to epoxy resins.[54]

For water-soluble and reactive epoxy compounds, HCl saturated with $MgCl_2$ is the reagent. The reaction time is 15 to 30 min at room temperature. On completion the excess HCl is titrated with standard NaOH to a methyl orange end point.[55] An alcoholic $MgCl_2$ solvent system with the same reagents can also be used; in this case the indicator is bromocresol green.

In glacial acetic acid HBr reacts readily with the less reactive epoxy compounds. By adding HBr in known excess, the appropriate reaction time (15 to 60 min) can be selected. The volumetric finish involves titrating with a standard solution of sodium acetate until the color of the crystal violet indicator changes to blue-green. Chlorobenzene has been suggested as a superior solvent to glacial acetic acid. Epoxy compounds that form aldehydes on reaction with acids (such as styrene oxide) and those in which the epoxy ring contains a tertiary carbon atom cannot be determined by this method.[56]

Several other volumetric methods are based on the same general reaction scheme already outlined. The sample is added to an excess of HCl in 1,4-dioxane, allowed to stand at room temperature for 15 min, and the excess HCl is titrated with 0.1*M* NaOH in methanol to the cresol red end point. With pyridine as the solvent, the reaction time is 20 min at reflux; the back titration with NaOH is to the phenolphthalein end point.[57] In yet another method, tetraethylammonium bromide and sample in chloroform, acetone, or benzene solution are titrated with standard perchloric acid dissolved in glacial acetic acid to the crystal violet end point or by potentiometry.[58]

The epoxide group can be cleaved by sulfuric acid to give glycols, which are in turn cleaved with excess H_5IO_6.[59] A newer version, applicable to nanomole quantities of a variety of epoxides, 50% glyme is used as solvent but the excess H_5IO_6 is allowed to react with CdI_2 to generate free iodine, which is then determined photometrically as its colored complex with starch.[60]

19.6.3 Peroxides

The chemical methods for the determination of peroxides are outlined in Table 19.12. Peroxides will oxidize iodide ion to free iodine, arsenite ion to arsenate, and titanium(III) to titanium(IV). The liberation of iodine from potassium iodide is a rapid method, but it can be used only in a few organic solvents and cannot generally be used in the presence of unsaturated compounds. The arsenious oxide method suffers no interference from unsaturated compounds. Titanium(III) chloride will determine almost any peroxide; however, the reagent requires special handling and must be isolated from oxygen.

Acyl peroxides in the presence of peresters, other peroxides, and hydrogen peroxide can be determined by reaction with hydroxylamine at pH 7 followed by formation of the colored iron(III) complex.[61] By carrying out the hydroxylamine reaction at pH 14, both peroxoesters and diacyl peroxides can be determined.

---

[53] A. Sana and S. Takitani, *Anal. Chem.* **57**:1687 (1985).

[54] B. Davis, *Anal. Chem.* **49**:832 (1977).

[55] A. Elek, *Ind. Eng. Chem., Anal. Ed.* **11**:174 (1939); Furter, *Helv. Chem. Acta* **21**:873, 1144 (1938); W. Deckert, *Z. Anal. Chem.* **82**:297 (1930).

[56] A. J. Durbetaki, *Anal. Chem.* **28**:2000 (1956).

[57] G. King, *Nature* **164**:706 (1949); S. Siggia, *Quantitative Organic Analysis Via Functional Groups*, 3d ed., Wiley, New York, 1963.

[58] W. Selig, *Mikrochim. Acta* **1**:112 (1980); R. R. Jay, *Anal. Chem.* **36**:667 (1964).

[59] F. E. Critchfield and J. B. Johnson, *Anal. Chem.* **29**:797 (1957).

[60] H. E. Mishmash and C. E. Meloan, *Anal. Chem.* **44**:835 (1972).

[61] N. A. Kozhikhova et al., *Zh. Anal. Khim.* **34**:1217 (1979).

TABLE 19.11 Methods for the Determination of Epoxides (Oxiranes)

Solvent system	Reagent	Reaction time, min	Temperature	Titrant	Indicator	Reference
Aqueous	0.1*M* HCl in saturated $MgCl_2$	15–30	Room	0.1*M* NaOH	Methyl orange	4, 9
Aqueous	H_2SO_4 hydrolysis to glycol which is cleaved by H_5IO_6					1
Alcoholic $MgCl_2$	0.5*M* HCl + $MgCl_2$ in EtOH	30	Room	0.5*M* NaOH	Bromocresol green	
Cellusolve	0.2*M* HCl in Cellosolve	240	65°C	0.1*M* NaOH	Bromothymol blue	
Chloroform (or acetone or benzene)	$(C_2H_5)_4NBr$			$HClO_4$ in glacial acetic acid	Crystal violet or potentiometry	7, 8
Diethyl ether	0.2*M* HCl in diethyl ether	180	Room	0.1*M* NaOH	Phenolphthalein	9
1,4-Dioxane	0.2*M* HCl in 1,4-dioxane	15	Room	0.1*M* NaOH in MeOH	Cresol red	5
Glacial HOAc or chlorobenzene	0.1*M* HBr	15–60	(Styrene oxide and epoxy ring with a tertiary carbon atom in ring cannot be determined)	0.1*M* NaOAc in HOAc	Crystal violet	2
50% Glyme	H_2SO_4, then excess H_5IO_6, CdI_2 added			Free I_2 plus starch	Photometrically	6
Pyridine	0.2*M* HCl in pyridine	20	Reflux	0.1*M* NaOH in MeOH	Phenolphthalein	5, 9
Pyridine–$CHCl_3$	0.2*M* HCl in pyridine–$CHCl_3$	30 to 120	Reflux	0.1*M* NaOH in MeOH	Phenolphthalein	

1. Critchfield, F. E., and J. B. Johnson, *Anal. Chem.* **29**:797 (1957).
2. Durbetaki, A. J., *Anal. Chem.* **28**:2000 (1956).
3. Jay, R. R., *Anal. Chem.* **36**:667 (1964).
4. Kerchov, F. W., *Z. Anal. Chem.* **108**:249 (1937).
5. King, G., *Nature* **164**:706 (1949).
6. Mishmash, H. E., and C. E. Meloan, *Anal. Chem.* **44**:835 (1972).
7. Selig, W., *Mikrochim. Acta* **1**:112 (1980).
8. Siggia, S. and J. G. Hanna, *Quantitative Organic Analysis Via Functional Groups*, 4th ed., Wiley, New York, 1977.
9. Dobinson, B., W. Hofmann, and B. P. Stark, *The Determination of Epoxide Groups*, Pergamon, London, 1969.

TABLE 19.12 Chemical Methods for the Determination of Peroxides

Reducing agent	Reduction time, min	Temperature	Solvent	Final reagent employed	Reference
		Volumetric			
NaI	5–20	Room	Acetic anhydride	$Na_2S_2O_3$	8
NaI	15	Reflux	2-Propanol	$Na_2S_2O_3$	15
KI	15–60	Room	*tert*-BuOH + CCl_4	$Na_2S_2O_3$	5
$Fe(NH_4)_2(SO_4)_2$	15		HOAc	$K_2Cr_2O_7$	12
$FeSO_4$			Acetone–water (1 : 1)	$TiCl_3$	13
As_2O_3				Excess I_2 added and back-titrated with $Na_2S_2O_3$	11
		Colorimetric			
Fe(II) ions	0–5	Room to incipient boiling	Absolute MeOH	SCN^-	14, 16
Fe(II) ions	15	Room	Benzene–MeOH	1,10-Phenanthroline	7
N,N-Dimethyl-*p*-phenylene-diamine sulfate	5	Room	MeOH	None needed	4
		Miscellaneous techniques for various compounds			
Hydroxylamine, pH 7 (Acyl peroxides)				Forms colored iron (III) complex	6, 9
Hydroxylamine, pH 14 (Peresters and diacyl peroxides)				Forms colored iron (III) complex	6
Triethylamine (Benzoyl peroxide)				Chemiluminescence measured	2
N,N-Dimethyl-*p*-phenylene-diamine (Peroxides derived from cyclohexanone)				None needed; color measured	10
p-Tolyl methyl sulfide (Peracids)				Gas chromatography of sulfoxide formed or unused sulfide	3
p-Phenetidine (Peracids)			KH_2PO_4	Color measured	1

1. Blazheevskii, N. E., and V. K. Zinchuk, *Chem. Abstr.* **105**:17620m (1986).
2. Bowyer, J. R., and S. R. Spurlin, *Anal. Chim. Acta* **192**:289 (1987).
3. DiFuria, F., et al., *Analyst* **109**:985 (1984).
4. Dugan, P. R., *Anal. Chem.* **33**:1630 (1961).
5. Hartman, L., and M. D. L. White, *Anal. Chem.* **24**:527 (1952).
6. Kozhikhova, N. A., et al., *Zh. Anal. Khim.* **34**:1217 (1979).
7. Laitinen, H. A., and J. S. Nelson, *Ind. Eng. Chem., Anal. Ed.* **18**:422 (1946).
8. Nozaki, K., *Ind. Eng. Chem., Anal. Ed.* **18**:583 (1946).
9. Robey, R. F., and H. K. Wiese, *Ind. Eng. Chem., Anal. Ed.* **17**:425 (1945).
10. Sevast'yanova, E. M., and Z. S. Smirnova, *Chem. Abstr.* **106**:143701x (1987).
11. Siggia, S., *Ind. Eng. Chem., Anal. Ed.* **19**:872 (1947).
12. Tanner and Brown, *J. Inst. Petrol.* **32**:341 (1946).
13. Wagner, C. D., R. H. Smith, and E. D. Peters, *Ind. Eng. Chem., Anal. Ed.* **19**:982 (1947).
14. Wagner, C. D., H. L. Clever, and E. D. Peters, *Ind. Eng. Chem., Anal. Ed.* **19**:980 (1947).
15. Wagner, C. D., R. H. Smith, and E. D. Peters, *Ind. Eng. Chem., Anal. Ed.* **19**:976 (1947).
16. Young, C. A., R. R. Vogt, and J. A. Nieuwland, *Ind. Eng. Chem., Anal. Ed.* **8**:198 (1936).

Peracids in the presence of a large excess of hydrogen peroxide react selectively with *p*-tolyl methyl sulfide to produce *p*-tolyl methyl sulfoxide, which is determined by gas chromatography of the sulfoxide or the unreacted sulfide.[62]

The interaction of peroxides with amines forms the basis of peroxide analysis in several methods. Benzoyl peroxide can be determined by measuring the chemiluminescence arising from reaction with triethylamin.[63] Peroxides derived from cyclohexanone were determined colorimetrically by reaction of the peroxide with *N,N*-dimethyl-*p*-phenylenediamine.[64] Organic peracids in the presence of hydrogen peroxide were determined by measuring the absorbance of the reaction of the peroxide with *p*-phenetidine in KH_2PO_4.[65]

19.6.4 Quinones

Quinones are moderately strong oxidizing agents ($E_{1/2} = 0.7$ V for quinone itself). About 100 mg of the sample is dissolved in 20 mL of ethanol. An ethanolic solution of 2.5% KI and 2.5*M* in HCl is added. The iodine formed is titrated with 0.1*N* $Na_2S_2O_3$. Good results can also be obtained with stronger oxidants as titrants. Systems include cerium(IV), vanadium(V), dichromate(VI), and hexacyanoferrate(III) (with Zn). A critical summary of known procedures has been published.[66]

19.7 METHODS FOR THE DETERMINATION OF FUNCTIONAL GROUPS CONTAINING NITROGEN AND OXYGEN

19.7.1 Determination of Nitrates, Nitro, and Nitroso Compounds

Using 3 to 6 mg of an aromatic nitro compound, the reduction is carried out at room temperature with 0.04*N* titanium(III) chloride (or sulfate) in a medium buffered with sodium citrate. Use titanium sulfate if dealing with easily chlorinated compounds. By performing the reaction in 12*M* HCl, nitroso compounds (except *N*-nitrosoamines) can be determined selectively in the presence of nitro compounds.[67] An alternative procedure involves adding excess titanium(III) and then back-titrating the excess with iron(III) ions using thiocyanate as indicator.

Primary nitroalkanes can be determined by reacting with nitrous acid. The formed nitro acids are then titrated with NaOH solution.[68]

Polarographic half-wave potentials for many nitro and nitroso compounds are given in Table 14.21. *N*-Nitrosoamines can be detected down to $3 \times 10^{-8}M$ using differential pulse voltammetry.[69] Half-wave potentials become more negative as the organic content of the solvent is increased. A coulometric method using controlled-potential electroreduction will determine nitro and nitroso compounds in MeOH–LiCl solution.[70] Aromatic nitroso compounds dissolved in ethanolic HCl can be titrated with $SnCl_2$ in glycerol to an amperometric, potentiometric, or visual end point without interference by nitro compounds.[71] The nitro group can be reduced with Fe(II) in a direct titrimetric procedure using alkaline sorbitol media with detection of the end point by potentiometric or amperometric means.[72]

---

[62] F. DiFuria et al., *Analyst* **109**:985 (1984).
[63] J. R. Bowyer and S. R. Spurlin, *Anal. Chim. Acta* **192**:289 (1987).
[64] E. M. Sevast'yanova and Z. S. Smirnova, *Chem. Abstr.* **106**:143701x (1987).
[65] N. E. Blazheevskii and V. K. Zinchuk, *Chem. Abstr.* **105**:17620m (1986).
[66] U. A. Th. Brinkman and H. A. M. Snelders, *Talanta* **11**:47 (1964).
[67] T. S. Ma and J. V. Early, *Mikrochim. Acta* **1959**:129.
[68] C. A. Reynolds and D. C. Underwood, *Anal. Chem.* **40**:1983 (1968).
[69] R. Samuelsson, *Anal. Chim. Acta* **108**:213 (1979).
[70] J. M. Kruse, *Anal. Chem.* **31**:1854 (1959).
[71] E. Ruzicka, M. Paleskova, and J. A. Jilek, *Collect. Czech, Chem. Commun.* **45**:1677 (1980).
[72] B. Velikov, J. Dolezal, and J. Zyka, *Anal. Chim. Acta* **94**:149 (1977).

19.8 DETERMINATION OF AMINES AND AMINE SALTS

Amines with ionization constants equal to or greater than approximately 10^{-9} can be satisfactorily titrated in water or alcohol–water mixtures using either an indicator or a potentiometric procedure. Since aqueous acidimetric procedures are accurate and precise and usually require no special equipment, many are still used for the quality control of commercially available materials. However, these methods are much more limited in scope than are nonaqueous techniques. For example, a great number of amines that are water-insoluble or that are too weakly basic to be titrated in aqueous solution react as strong bases toward perchloric acid in acetic acid medium. In addition, mixtures of two or more amines can frequently be analyzed by differentiating titrations in water-free solvents. Tertiary amines can be determined in the presence of primary and secondary amines after acetylation, and secondary amines can be determined in the presence of primary amines by first treating the mixture with an aldehyde.

Acid–base titrations in nonaqueous media are discussed in Sec. 4.3. The reader is referred to this material for the various organic solvent systems, the preparation and standardization of titrants, and procedures for selected titrations in nonaqueous media (Table 4.17).

The leveling effect of glacial acetic acid for all types of bases has made this solvent an exceedingly useful reagent in nonaqueous titrimetry. However, this property is also a disadvantage in that it prevents the individual titration of two amines of widely different basic strengths. Differentiating titrations of certain amine mixtures are possible by the proper choice of solvents. Chloroform and acetonitrile have been suggested for this purpose, with perchloric acid in 1,4-dioxane as the titrant. This technique permits the analysis of binary mixtures of some aromatic and aliphatic amines as well as the determination of several aromatic amine mixtures. The latter also can be titrated with perchloric acid in acetic acid, since aromatic amines are usually too weak to be leveled by acetic acid. The following procedures can be used for these determinations.

Procedure A: Dissolve 0.6 to 1.0 meq of sample (total amines) in 20 mL of acetonitrile, and titrate potentiometrically with 0.1*M* $HClO_4$ in 1,4-dioxane. Perform a blank determination on 20 mL of each lot of solvent.

Procedure B: Use the same sample size and solvent volume as in procedure A. Add six drops of eosin Y indicator and titrate with 0.1*M* $HClO_4$ to a pale-yellow end point. Add two drops of methyl violet and 20 mL of HOAc and continue the titration until a blue-green end point is reached. Determine a blank by the method given in procedure A.

Procedure C: Follow procedure A, but use 0.1*M* $HClO_4$ in HOAc as titrant. Determine a blank on 20 mL of acetonitrile, using the same titrant. Subtract the blank from the first end point only.

Table 19.13 lists a number of amine mixtures that have been successfully determined by these procedures. The titration curves of 55 amines reported by Hall[73] are helpful in predicting whether other amine mixtures can be successfully titrated.

19.8.1 Determination of Primary Amines

Primary amines react with salicylaldehyde to form Schiff bases, which are weaker bases than the parent compound. Salicylaldehyde does not react with tertiary amines and generally does not affect the basicity of secondary amines. This permits the determination of primary amines in the presence of other amines and also provides a procedure for the analysis of secondary plus tertiary amines in the presence of primary amines.

Procedure: Pipette 25 mL of $CHCl_3$ and 5 mL of salicylaldehyde into a glass-stoppered flask. Add 4 to 6 drops of bromocresol green indicator and weight into the flask an amount of amine mixture containing not more than 12.5 meq of primary and 12.5 meq of secondary plus tertiary amines. If the solution becomes turbid, add enough 1,4-dioxane to effect solution. After 15 min, titrate with 0.5*M* $HClO_4$ in 1,4-dioxane just to the disappearance of the green color. Record the volume of titrant

[73] N. F. Hall, *J. Am. Chem. Soc.* **52**:5115 (1930).

TABLE 19.13 Typical Amine Mixtures That Can Be Analyzed by Differentiating Titrations in Nonaqueous Solvents

Components of mixtures	pK_b of amine in H_2O	Procedure
Dibutylamine	2.81	A
Pyridine	8.85	
2-Phenylethylamine	4.17	B
Aniline	9.42	
Pyridine	8.85	C
Caffeine	13.39	
Aniline	9.42	C
o-Chloroaniline	11.32	
Aniline	9.42	C
Sulfathiazole	11.64	

used. Add 75 mL of 1,4-dioxane, 8 to 10 drops of congo red indicator, and titrate with the $HClO_4$ solution to the appearance of a pure green color. The volume of titrant required for the first end point is a measure of the secondary plus tertiary amine content of the sample. The volume required for the second end point is a measure of the amount of primary amine.[74]

The method cannot be used for the analysis of mixtures of aromatic primary and secondary amines or of mixtures of primary aliphatic and primary aromatic amines. Morpholine, secondary alcohol amines, and certain polyethylene amines (such as diethylenetriamine and triethylenetetramine but not ethylenediamine) interfere with the determination. Ammonia does not react quantitatively and, if present, must be separated before analysis. In general, the method cannot be used for the determination of compounds containing both primary and secondary or tertiary amine groups in the same molecule.

Primary aromatic amines can be diazotized with a known excess of sodium nitrite solution, and the excess back-titrated with *p*-nitroaniline solution. The method has been modified by using 4,4-sulfanilic dianiline as the reagent, the indicator being diphenylamine. On back-titration with 0.1*N* $NaNO_2$ solution the color changes from red to yellow sharply at the end point.[75]

19.8.2 Determination of Secondary Amines

Primary and secondary amines reach with carbon disulfide to form dithiocarbamic acids. These acids can be titrated quantitatively with a standard base without interference from ammonia and tertiary amines.

Imines, formed by the reaction of primary amines with an aldehyde, do not react with carbon disulfide. Secondary aliphatio amines do not react with 2-ethylhexaldehyde and can be converted to dithiocarbamic acids in the presence of imines.[76]

Procedure: Pipette 10 mL of 2-ethylhexaldehyde into a flask and add 50 mL of 2-propanol. Weigh an amount of sample containing no more than 13 meq of secondary amine. After 5 min at room temperature, add additional solvent and cool the contents of the system to –10°C in a suitable bath (but not dry ice–acetone). Remove the flask from the cooling bath; add 5 mL of CS_2 with a pipette and 5 to 6 drops of phenolphthalein indicator. Titrate immediately with 0.5*M* NaOH to the first definite pink color that persists for 1 min. During the titration keep the flask in a bath of crushed ice and MeOH.

Aromatic amines and highly branched aliphatic amines cannot be determined by this method. 2-Ethylhexaldehyde does not react quantitatively with primary alcohol amines, aromatic amines, highly

[74] F. E. Critchfield and J. B. Johnson, *Anal. Chem.* **29**:957 (1957); C. D. Wagner, R. H. Brown, and E. D. Peters, *J. Am. Chem. Soc.* **69**:2611 (1947).

[75] E. Szekely, A. Brande, and M. Flitman, *Talanta* **19**:1429 (1972).

[76] F. E. Critchfield and J. B. Johnson, *Anal. Chem.* **29**:957 (1957).

branched primary amines such as *tert*-butylamine and isopropylamine, and polyamines. These amines interfere in the determination, and so do materials that are acidic or basic under the titration conditions used.

19.8.3 Determination of Tertiary Amines

The direct determination of tertiary amines involves acetylation of primary and secondary amines and NH_3 with acetic anhydride, and then potentiometric titration of the unreacted tertiary amine with $HClO_4$ in HOAc. The method is generally applicable to all aliphatic amines except certain sterically hindered secondary amines,[77] and to aromatic amine mixtures.[78]

19.8.4 Determination of Primary Plus Secondary Amines

Primary and secondary amines may be determined by acetylation and measurement of the excess acetic anhydride or by nonaqueous titration of the total amine content before and after acetylation. Ammonia, if present, interferes seriously in both procedures and should be removed before analysis.

A specific method has been developed for the direct determination of the total primary and secondary amine content in the presence of NH_3 and tertiary amines. An excess of CS_2 is reacted with the primary or secondary amine in 2-propanol, alone or mixed with pyridine. The dithiocarbamic acid formed is titrated with a standard base using phenolphthalein as indicator.[79] This procedure can be combined with a total base and tertiary amine determination to obtain, indirectly, the NH_3 present in an amine–NH_3 mixture.

Interference with the foregoing procedure include compounds acidic or basic to phenolphthalein. However, suitable corrections for these materials can be made before addition of CS_2. Aromatic amines, *tert*-butylamine, and diisopropylamine do not react quantitatively with the reagent and interfere in the analysis.

19.8.5 Determination of Secondary Plus Tertiary Amines

The reaction of salicylaldehyde with an aliphatic primary amine is not only useful for the determination of primary amines, but was also proposed primarily for the determination of secondary plus tertiary amines. The azomethane formed in the reaction is a much weaker base than the secondary amine, and the total secondary and tertiary amine content of the reaction mixture may be titrated potentiometrically in a nonaqueous solvent. Neither NH_3 nor H_2O, if present alone in the amine mixture, interferes in the titration. However, if both NH_3 and H_2O are present in large amounts, NH_3 must be removed.

Procedure: Into a tall-form beaker containing 5 mL salicylaldehyde and 80 mL MeOH, weigh about 3.5 meq of sample. Cover the beaker, mix the contents thoroughly, and allow to stand at room temperature for 30 min. Titrate the mixture potentiometrically to the first end point with 0.5*M* HCl in 2-propanol.[80]

19.8.6 Determination of Primary, Secondary, and Tertiary Amines in Mixtures

The tertiary amine is determined by the direct addition of acetic anhydride to a weighed sample, solution of the reaction mixture in 1 : 1 ethylene glycol–2-propanol, and finally titration with HCl dissolved in the same solvent. The primary amine in the mixture is determined by a total base titration before and after the addition of salicylaldehyde. The difference between the two titrations is a measure of the primary amine content. The secondary amine is determined from the titration value obtained after the addition of salicylaldehyde. This value is the tertiary plus secondary amine content. The amount of secondary amine is computed by subtracting the tertiary amine content from the tertiary plus secondary amine value.[81]

---

[77] C. D. Wagner, R. H. Brown, and E. D. Peters, *J. Am. Chem. Soc.* **69**:2609 (1947).
[78] S. Siggia, J. G. Hanna, and I. R. Kervenski, *Anal. Chem.* **22**:1295 (1950).
[79] F. E. Critchfield and J. B. Johnson, *Anal. Chem.* **28**:430 (1956).
[80] C. D. Wagner, R. H. Brown, and E. D. Peters, *J. Am. Chem. Soc.* **69**:2611 (1947).
[81] S. Siggia, J. G. Hanna, and I. R. Kervenski, *Anal. Chem.* **22**:1295 (1950).

N,N-Di-(2-hydroxyethyl)aniline cannot be determined by the tertiary amine procedure. Diphenylamine and triphenylamine also cannot be determined because these compounds are too weakly basic to be titrated in this nonaqueous system.

19.9 DETERMINATION OF AMINE AND QUATERNARY AMMONIUM SALTS

The method described in Sec. 19.8.1 for the determination of total amines can also be used to determine many amine salts. Salts of organic bases and most acids (other than halogen and sulfonic) can be titrated directly with $HClO_4$ provided the salt is soluble in the selected solvent. Halogen acid salts of organic bases can be titrated in HOAc after adding $Hg(OAc)_2$ to the sample–solvent mixture due to the formation of slightly dissociated mercury halide salt and an equivalent amount of free acetate ion. Amine sulfates and nitrates can be titrated without the addition of $Hg(OAc)_2$; the sulfate end point is reached when the sulfate is neutralized to hydrogen sulfate ion.[82]

19.10 DETERMINATION OF AMINO ACIDS

A number of amino acids can be titrated in a differentiating solvent in which the basicity of the amino group is enhanced while the acidity of the carboxyl group is decreased. Amino acids can be titrated in glacial acetic acid solution with $HClO_4$ in the presence of crystal violet indicator. Often it is better to dissolve the amino acid in a known excess of $HClO_4$ and to back-titrate the excess with NaOAc dissolved in glacial acetic acid.

If the amino group is masked with formaldehyde, the carboxyl group can be titrated with NaOH using phenolphthalein as indicator.

An almost specific reagent for amino acids is ninhydrin, which reacts with all compounds that contain a free amino group with the formation of a colorless hydrindantin. At pH 3 to 4, an excess of ninhydrin reacts with hydrindantin and ammonia to form a blue product. The colored product is a useful spot test when developing thin-layer chromatographic (TLC) plates and can be used for the spectrophotometric determination of amino acids.

19.11 METHODS FOR THE DETERMINATION OF COMPOUNDS CONTAINING OTHER NITROGEN-BASED FUNCTIONAL GROUPS

19.11.1 Determination of Amides

Amides have been determined by alkaline and acid hydrolysis, by hydroxamic acid formation and colorimetric estimation, by reaction with 3,5-dinitrobenzoyl chloride and titration of the excess benzoyl chloride, by several reductometric procedures, and by nonspecific total nitrogen procedures.

Acetamide and a few other amides can be titrated as bases in nonaqueous solvents.[83] However, it remained for Wimer[84] to demonstrate the usefulness of these observations and to develop a general method for the titration of amides and acetylated and formylated amines.

19.11.1.1 Determination of Amides by Potentiometric Titration in Acetic Anhydride. *Procedure:* Weigh 6 to 9 mmol of sample into a 100-mL volumetric flask; dissolve in and dilute to volume with acetic anhydride. Transfer a 10-mL aliquot to a tall-form beaker, add 100 mL acetic anhydride and

[82] C. W. Pifer and E. G. Wollish, *Anal. Chem.* **24**:300 (1952).

[83] N. F. Hall and T. H. Werner, *J. Am. Chem. Soc.* **50**:2367 (1928); J. S. Fritz and M. O. Fulda, *Anal. Chem.* **25**:1837 (1953); A. F. Gremillion, *ibid.* **27**:133 (1955).

[84] C. D. Wimer, *Anal. Chem.* **30**:77 (1958); see also T. Higuchi et al., *ibid.* **34**:400 (1962).

titrate potentiometrically with 0.1*M* $HClO_4$ in acetic anhydride using a glass–calomel electrode pair. Replace the aqueous solution in the sleeve-type calomel with 0.1*M* $LiClO_4$ in acetic anhydride.

N-Phenyl- or α-phenyl-substituted amides, trifluoromethylformamide, cyanamide, and certain tertiary amides are too weakly basic to be titrated by this procedure. Unsaturated amides in which the double bond is conjugated with the carbonyl group appear to react with the anhydride but do not yield stoichiometric results. Diamides of dibasic acids, except malonamide and tetrasubstituted phthalamides, are too insoluble in acetic anhydride to be determined by this method.

19.11.1.2 Determination of Unsubstituted Amides by the Dinitrobenzoyl Chloride–Pyridine Method. The method of Mitchell and Ashby[85] is based upon the reaction of 3,5-dinitrobenzoyl chloride with amides in pyridine followed by an acidimetric determination of the excess benzoyl chloride.

Procedure: Transfer about 10 meq of accurately weighed sample to a 250-mL glass-stoppered flask containing 15 mL of 3,5-dinitrobenzoyl chloride and 5 mL of pyridine. Place the flask together with a blank in a water bath maintained at 60°C for 30 min. (For amides of dibasic acids use a reaction time of 1 h at 70°C.) Remove the flasks and cool in an ice-water bath. Add 2 mL of methanol to each flask, wait 5 min, and add 25 mL more. If the solutions are highly colored, add ethyl-bis(2,4-dinitrophenyl) acetate indicator and titrate both sample and blank with 0.5*M* sodium methoxide. Phenolphthalein may be used as indicator if the solutions are only lightly colored. The net increase in acidity of the sample over the blank, after correction for free acid and water present in the sample, is equivalent to the amide content.

Generally, secondary and tertiary amides do not interfere. Neither do amines and alcohols, beyond consuming some of the reagent. However, free acid and water do interfere. If present, these materials should be determined in the original sample and corrections applied to the results of the amide determination.

19.11.2 Methods for the Determination of Azo Compounds

19.11.2.1 Reduction with Titanium(III) Salts. For most azo compounds, four equivalents of titanium(III) are required. However, some chlorinated azobenzenes have been found to require only two equivalents of titanium(III). Furthermore, there is no one procedure that is completely satisfactory for all azo compounds. However, the following one can be applied to most samples.

Procedure: Deaerate the reaction flask by adding several small pieces of dry ice or by passing CO_2 into it for 5 min. Continue to pass the inert gas through the flask, and transfer to it a weight sample or an aliquot of a solution of the sample that will require about 2.5 meq of titrant. Dissolve the sample in 25 mL ethanol or glacial acetic acid. Add 25 mL 1 : 1 HCl and 50 mL 0.2*M* titanium(III) salt, and reflux for 5 to 10 min. With the CO_2 still bubbling through the solution, cool the flask and contents to about room temperature, and titrate with 0.15*M* iron(III) ammonium sulfate solution until the purple color is very faint, add 10 mL 10% NH_4SCN solution, and continue the titration until the pink color persists for 1 min. Run a blank.

Nitro, nitroso, and many other compounds are also reduced by titanium(III) under the same conditions. Oxygen is a troublesome interferant and every precaution must be taken to keep it excluded.

19.11.2.2 By Reduction with Copper. A simple indirect gravimetric method for the determination of nitro, nitroso, and azo compounds is based upon the loss in weight of copper during reduction of the compound to the amine.[86]

Procedure: Add 20 mL of ethanol and 20 mL 3*M* H_2SO_4 to the reaction flask and deaerate the mixture by adding small pieces of dry ice. Weigh a sample of the azo compound that will cause 0.5 to 1.0 g Cu to dissolve during the reaction, and transfer it to the reaction flask. Add 6 to 15 g Cu and reflux the mixture for 2.5 h while passing a slow stream of CO_2 through the condenser. Rinse rapidly 5 to 6 times with deaerated water, then 3 times with deaerated acetone. Dry the Cu under vacuum and weight. Run a blank.

---

[85] J. Mitchell, Jr., and Ashby, *J. Am. Chem. Soc.* **67**:161 (1945).

[86] Juvet, Twickler, and Afremow, *Anal. Chim. Acta* **22**:87 (1960).

The quantitative reduction of nitro and nitroso compounds also occurs. Halides interfere, as do oxidizing agents such as nitrate ion and iron(III). Aldehydes, ketones, carboxylic acids, phenols, alcohols, amines, and nitriles do not interfere.

19.11.3 Methods for the Determination of Hydrazines

Procedures for the determination of hydrazine and its derivatives are based on their behavior as either weak bases or reducing agents. The available methods involve titration with acids or oxidation of the hydrazine group to nitrogen. Among the oxidants used are I_2, Br_2, ferricyanide, permanganate, bromate, iodate, cerium(IV), dichromate, periodate, vanadate, and chloramine T. In thc presencc of other bases an acidimetric procedure will not indicate the true hydrazine content. However, in combination with an oxidimetric method a measure can be obtained of both total base and hydrazine.

Either an aqueous or a nonaqueous acidimetric procedure may be used for the titration of these bases. Water-insoluble hydrazines are best determined by titration with $HClO_4$ in HOAc.

In the presence of ≥4*M* HCl, hydrazine and monosubstituted hydrazines can be titrated directly with KIO_3 to the ICl equivalence point. In the presence of H_2SO_4, the reduction of iodate proceeds to I_2. Both of these procedures have been used for the determination of monosubstituted hydrazines and their hydrazones and acyl derivatives.[87]

Other derivatives of hydrazine, such as hydrazones, guanidines, triazoles, and tetrazoles, also react with iodate ion and interfere in the titration. Polysubstituted hydrazine undergo reproducible oxidations that vary from a two- to six-electron change per hydrazine group.

19.11.4 Determination of Primary Hydrazides

Primary hydrazides are sufficiently basic to be titrated with $HClO_4$ in HOAc. The procedures described for the determination of total primary, secondary, and tertiary amines is applicable to primary hydrazides.

19.11.5 Determination of Oxazolines

Oxazolines can be titrated potentiometrically in glacial acetic acid with $HClO_4$ if they behave as a strong base in HOAc.[88]

19.11.6 Determination of Isocyanates, Isothiocyanates, and Isocyanides

Both aliphatic and aromatic isocyanates and isothiocyanates react quantitatively with amines (e.g., 0.3*M* butylamine in 1,4-dioxane) in about 45 min at room temperature to form a substituted carbamide (urea) or thiocarbamide (thiourea). Since the ureas are neutral compounds, the excess of added amine can be determined acidimetrically with 0.1*N* H_2SO_4 using methyl red as indicator. This provides the basis for the method developed by Siggia and Hanna,[89] which was adapted to a micro and semimicro scale by Karten and Ma.[90] Since water and alcohols also react with isothiocyanates and isocyanates, a nonhydroxylic solvent, 1,4-dioxane, is used as the reaction medium.

Isocyanides (R—N═C═) are reacted with a known excess of HSCN to form a substituted triazine dithione. The excess of HSCN is titrated with triethylamine dissolved in ethyl acetate using a methanolic solution of methylene blue–neutral red mixed indicator.[91]

---

[87] W. R. McBride, R. A. Henry, and S. Skolnik, *Anal. Chem.* **25**:1042 (1953).
[88] P. C. Markunas and J. A. Riddick, *Anal. Chem.* **23**:337 (1951).
[89] S. Siggia and J. G. Hanna, *Ind. Eng. Chem., Anal. Ed.* **20**:1084 (1948).
[90] B. S. Karten and T. S. Ma, *Microchem. J.* **3**:507 (1959).
[91] A. S. Arora, E. Hinrichs, and I. Ugi, *Z. Anal. Chem.* **269**:9 (1974).

19.11.7 Determination of *vic*-Dioximes

In the presence of $Hg(OAc)_2$, I_2 dehydrogenates *vic*-dioximes to furoxanes. The basis of this reaction forms a simple titrimetric method for the determination of aliphatic and alicyclic *vic*-dioximes.[92] The reaction proceeds rapidly in chloroform medium, and back-titration of the excess I_2 yields the amount of dioxime originally present. Materials easily oxidized by iodine, easily halogenated compounds, primary amines, ketoximes, aldoximes, and monoximes of *vic*-diketones interfere in the titration.

19.11.8 Determination of Hydroxylamine

Hydroxylamine can be titrated with 0.1*M* HCl using methyl orange or bromophenol blue as indicator. Its salts can be titrated with 0.1*M* NaOH in the presence of phenolphthalein. However, a more accurate procedure uses the reduction of iron(III) ammonium sulfate by hydroxylamine in HCl medium; the iron(II) formed is titrated with $KMnO_4$.

19.12 DETERMINATION OF NITRILES

The determination of nitriles is possible by hydrolysis with a known excess of water in glacial acetic acid solution in the presence of BF_3 catalyst. The remaining water is titrated with Karl Fischer solution.[93]

Some α,β-unsaturated nitriles react with mercaptans; an excess is added and the excess is titrated by an iodometric or argentimetric method.[94]

19.13 METHODS FOR THE DETERMINATION OF PHOSPHORUS-BASED FUNCTIONS

This subsection includes tables dealing with both the determination of phosphorus-based functional groups in organic molecules and the determination of phosphorus-based inorganic groups. Table 19.14 summarizes methods for the determination of phosphorus in the elemental form and in its inorganic compounds and ions and also includes some methods for the separation of these. Table 19.15 summarizes methods for the determination of organic phosphorus-based functional groups. References are given at the end of Table 19.15. Some additional methods are briefly discussed below.

The determination of dibutylphosphoric acid in the presence of monobutylphosphoric acid and tributyl phosphate depends on its separation from the mixture by TLC. The separated analyte is mixed with a Th–Thoron complex and its concentration measured by the extent to which it reduces the color the complex.[95]

The analysis of diethyl dithiophosphate in aqueous solution is based on its conversion to the lead salt followed by an extraction of the salt into chloroform at pH 4 to 5. The concentration of the diethyl dithiophosphate is proportional to the UV absorbance of the $CHCl_3$ extract.[96] A technique for the determination of phosphonates in water involving supercritical fluid extraction with CO_2 and supercritical fluid chromatography has been described.[97]

---

[92] Banks and Richard, *Talanta* **2**:235 (1959).
[93] D. H. Whitehurst and J. B. Johnson, *Anal. Chem.* **30**:1332 (1958).
[94] W. D. Beesing et al., *Anal. Chem.* **21**:1073 (1949).
[95] S. C. Tripathi, *Analyst* **111**:239 (1986).
[96] H. Socio, P. Garrigues, and M. Ewald, *Analusis* **14**:344 (1986).
[97] J. Hedrick and L. T. Taylor, *Anal. Chem.* **61**:1986 (1989).

TABLE 19.14 Methods for the Determination of Inorganic Phosphorus Groups

For references, see list below Table 19.15.

Substance determined	Technique	Procedure	Reference
Hypophosphate, diphosphate(IV) $[(HO)_2OP]_2$	Volumetric	Oxidation with cerium(IV) in hot acid solution and determination of resulting orthophosphate (see below).	15
Phosphinate (hypophosphite) HPH_2O_2	Volumetric	Oxidation to P(III) with cerium(IV) or $KBrO_3$; or oxidation to P(V) (see phosphite below).	15
	Gravimetric	See phosphite below.	
	Colorimetric	Formation of blue color with molybdate iron.	18
	Paper chromatography	See phosphite below.	
Orthophosphate, PO_4^{3-}	Volumetric	Precipitation with $(NH_4)_2MoO_4$ from HNO_3 solution; dissolve precipitate in excess standard NaOH, and titrate excess alkali with standard acid.	See Table 4.4
	Gravimetric	Precipitate as $MgNH_4PO_4$, ignite, and weigh as $Mg_2P_2O_7$.	
	Colorimetric	As reduced molybdate, molybdovanadate [9], or acetone molybdate [6] after precipitation of ammonium phosphomolybdate.	9 6
	Ion exchange	Adsorption on Dowex 1-X8 resin, elution with KCl solution.	10
	Paper chromatography	Ascending technique, S&S No. 589. Solvent: 25 mL 20% (w/v) aqueous Cl_3CCOOH, 5 mL H_2O, 0.25 mL concentrated aqueous NH_3 and 70 mL [5] or 2-propanol [14,23]. Spray solution: 1% (w/v) $(NH_4)_2MoO_4$ in 0.12*M* HCl–0.60*M* $HClO_4$. Develop blue bands under ultraviolet light or by spraying with 0.1% (w/v) $SnCl_2$ in 0.1*M* HCl.	
	Infrared	In KBr pellet, PO_4^{3-} shows no band that can be measured in presence of pyro-, trimeta-, and tri-polyphosphate.	7
Peroxophosphates: mono- and di-	Volumetric	Treatment with excess KI in acid solution and titration of liberated I_2, or titration with $KMnO_4$ or Ce(IV).	
	Ion exchange	See orthophosphate above.	
	Paper chromatography	See orthophosphate above.	
	Ultraviolet	In aqueous solution P—O—O—P absorbs at 230 nm; P—O—O—H absorbs at 230 and also at 290 nm.	
Phosphine, PH_3	Colorimetric	Reaction with 0.5% 2-mercaptobenzimidazole in pyridine.	20
Phosphite, HPO_3^{2-}	Volumetric	Titration of phosphate with $KMnO_4$, I_2, Ce(IV), or $KBrO_3$.	3, 11
	Paper chromatography	Ascending technique. S&S No. 589 paper. Solvent: 70 mL acetone +25 mL H_2O + 5 mL concentrated aqueous NH_3. Spray solution: see orthophosphate above.	
Phosphorus, red		Oxidation with $HNO_3 + HClO_4$ and determine as orthophosphate.	
Phosphorus, white or yellow		Dissolve in benzene, treat with $Cu(NO_3)_2$, oxidize with $HNO_3 + HClO_4$, and determine as orthophosphate.	13
	Ultraviolet	Extract from aqueous emulsion with benzene and measure absorbance at 290 nm against benzene reference.	

Phosphorus chloride oxide, PClO		After reaction with large excess of H_2O (giving HCl + H_3PO_4), determine as orthophosphate.	
	Infrared	Measure P—Cl or P═O bands; Table 7.17.	
Phosphorus pentachloride, PCl_5	Volumetric	1 mol PCl_5 plus solid KI gives 1 mol I_2,	
	Infrared	P—Cl band; see Table 7.17.	
Phosphorus trichloride, PCl_3	Volumetric	After reaction with large excess of H_2O (giving HCl + $HPO(OH)_2$), determine as phosphite.	
	Infrared	P—Cl band; see Table 7.17.	
P–S compounds, P_4S_3, P_4S_7, P_4S_{10}	Volumetric	Dissolve in hot NaOH + H_2O_2, acidify, boiling with Br_2 (giving H_3PO_4 + H_2SO_4), and determine products.	
	Paper chromatography	Dissolve in CS_2. Ascending technique using S&S No. 589 paper. Solvent: CS_2; spray solution: 10% aqueous $AgNO_3$.	
	Infrared	KBr pellet; see Table 7.17 for bands 714–625 cm^{-1}.	
Polyphosphates	Paper chromatography	See orthophosphate above.	19
Pyrophosphate, $P_2O_7^{4-}$	Volumetric	Adjust pH to 3.8, add excess $ZnSO_4$, and titrate liberated H^+ with standard alkali to pH 3.8.	1
	Colorimetric	Indirect by bleaching of Fe(II) 1,10-phenanthroline color.	8
	Ion exchange	See orthophosphate above.	
	Paper chromatography	See orthophosphate above.	
	Infrared	KBr pellet, 1031 and 735 cm^{-1}.	
Tetrametaphosphate	Ion exchange	See orthophosphate above.	
	Paper chromatography	See orthophosphate above.	
Trimetaphosphate	Separation	Precipitation with Ba(II); no other inorganic phosphate gives a Ba salt soluble above pH 9.0.	
	Ion exchange	See orthophosphate above.	
	Paper chromatography	See orthophosphate above.	
	Infrared	KBr pellet, 772 cm^{-1}.	
Tripolyphosphate	Volumetric	Like volumetric method for pyrophosphate.	2
	Colorimetric	Measure absorbance at 455 nm of cobalt ammine solution before and after precipitation of tripolyphosphate at pH 3.6.	21
	Ion exchange	See orthophosphate above.	
	Paper chromatography	See orthophosphate above.	
	Infrared	KBr pellet. Phase I: 752 and 707 cm^{-1}; phase II: 1015, 735, and 662 cm^{-1}.	

TABLE 19.15 Methods for the Determination of Organic Phosphorus Groups

Substance determined	Technique	Procedure	Reference
Phosphates $ROP(O)(OH)_2$ $(RO)_2P(O)OH$ $(RO)_3P{=}O$	Volumetric	$(RO)_2P(O)OH + NaOH \rightarrow (RO)_2POONa + H_2O$; 1 equivalence point in acetone-H_2O (9 : 1, v/v). $ROP(O)(OH)_2 + 2NaOH \rightarrow ROP(O)(ONa)_2 + 2\ H_2O$; 2 equivalence points in acetone–H_2O (9 : 1, v/v). If R = Aryl, or if R contains >6 C atoms, 2nd equivalence point can be detected by adding 15 mL 10% (w/v) $BaCl_2$ at pH 9.0 to release HCl.	
	Gravimetric	$ROP(O)(OH)_2$ gives Ba Salt insoluble above pH 7.0.	
	Paper chromatography	See phosphonic compounds below.	
	Infrared	See Table 7.17.	
	Ultraviolet	P—O—Ar absorbs at 260–280 nm.	
Phosphines R_3P, Ar_3P, R_2PH, Ar_2PH, RPH_2, $ArPH_2$	Volumetric	Titrate in HOAc, using 1% methyl violet in HOAc as indicator. 1 mol R_3P or Ar_3P reacts with 2 mol $HClO_4$; 1 mol R_2PH or Ar_2PH reacts with 1 mol $HClO_4$. To 0.3 g sample is 50 mL EtOH, add 15 mL CCl_4, mix, let stand 10 min, add 10 mL 1*M* NaOH, acidify, and titrated liberated Cl^- with standard $AgNO_3$. $R_2PH + 3\ NaOH + 2\ CCl_4 \rightarrow R_2P(O)ONa + H_2O + 3\ CHCl_3 + NaCl$ [Ar_2PH behaves similarly]. $RPH_2 + 3\ CCl_4 + 5\ NaOH \rightarrow RP(O)(ONa)_2 + 2\ H_2O + 3\ CHCl_3 + 3\ NaCl$ [$ArPH_2$ behaves similarly].	
	Infrared	P—H and Ar—P bands; see Table 7.17.	
	Ultraviolet	Ar—P, 260 to 270 nm; Ar_2P and Ar_3P, 220 to 240 nm.	
Phosphine oxides, $R_2P{=}O$, $As_2P{=}O$	Infrared	P=O and Ar—P bands.	
	Ultraviolet	Ar_3P, 220–240 nm.	
Phosphinous (P(III)) compounds, R_2POH, R_2POR (R may be Ar)	Volumetric	Oxidize to P(V) by shaking 15 min with excess $KBrO_3$–KBr in HOAc–HCl (25 : 10) volution; determine excess $KBrO_3$ with KI and $Na_2S_2O_3$.	
	Gravimetric	Add excess $HgCl_2$ to hot acid solution of sample, dry and weigh precipitated Hg_2Cl_2.	
	Infrared	P—O—C and P—O—Ar bands; see Table 7.17.	
	Ultraviolet	P—O—Ar at 260 to 280 nm; Ar_2P at 220–240 nm.	
Phosphites $(RO)_3P$, $(RO)_2PH({=}O)$, ROPH(O)OH	Volumetric, gravimetric, colorimetric, infrared, ultraviolet	See phosphonic compounds and phosphonous compounds.	
Phosphonic compounds ROPH(O)OH, RP(O)(OR)(OH), $RP(O)(OR)_2$, R may be Ar	Volumetric	$RP(O)(OR)OH + NaOH \rightarrow RP(O)(OR)ONa + H_2O$; 1 equivalence point in acetone–H_2O (9 : 1, v/v). $RP(O)(OH)_2 + 2\ NaOH \rightarrow RP(O)(ONa)_2 + 2\ H_2O$; 2 equivalence points under same conditions.	
	Paper chromatography	Ascending technique, S&S No. 589 paper. Solvent: 75 mL acetone +25 mL H_2O + 5 mL concentrated NH_3. For spray solution, see orthophosphate, Table 19.14.	
	Infrared and ultraviolet	See phosphine compounds above.	

Phosphonous compounds RPH(O)OH, RPH(O)OR, $RP(OR)_2$, R may be Ar	Volumetric	$RPH(O)OH + NaOH \rightarrow RPH(O)ONa + H_2O$; one equivalence point in acetone–H_2O (9:1, v/v).	5
		$RPH(O)OR + NaOH \rightarrow RPH(O)ONa + H_2O$. Reaction occurs in EtOH containing excess NaOH; let stand 15 min for R, 30 min for Ar.	
		$RPH(O)OR + 2\ NaOH + CCl_4 \rightarrow RP(O)(OR)ONa + H_2O + CHCl_3 + NaCl$; see phosphines above for procedure.	
		$RP(OR)_2 + H_2O \rightarrow RPH(O)OR + ROH$. Reaction occurs in 95% EtOH containing 5 mL 0.1*M* HCl; let stand 30 min to R, 60 min for Ar.	
	Gravimetric	See phosphinous compound below.	
	Colorimetric	—PH(O) group forms color with trinitrobenzene in alkaline medium.	17
	Infrared	See Table 7.17	
	Ultraviolet	Ar—P, Ar—O—P, 260–280 nm.	
Phosphorus–halogen compounds	Volumetric	Reaction with peroxide in alkaline solution giving peroxophosphate; determine excess peroxide.	16
		Hydrolysis with aqueous NaOH in acetone solution giving NaX; determine product.	
	Infrared	See P—Cl compounds, Table 19.14.	
Phosphorus–nitrogen compounds		See P—N compounds, Table 19.14.	
Phosphorus–sulfur compounds			
Thio acids	Volumetric	$P^v—SH + Ag^+ \rightarrow P^v—SAg + H^+$. Titrate 0.3 g sample in 100 mL acetone +5 mL HOAc with 0.05*M* $AgNO_3$ using glass–Ag electrode system; equivalence point is at 0.0 mV.	
		$P^v—SH + I_2 \rightarrow P^v—S—S—P^v + 2\ H^+ + 2\ I^-$. Shake 15 min with excess of standard KIO_3–KI in 5% aqueous HCl, back titrate with standard $Na_2S_2O_3$.	
	Infrared	S—H band at 2500 cm^{-1}.	
	Ultraviolet	220–240 nm.	
Thiol groups, P^v—SR	Volumetric	Shake 15 min with excess standard $KBrO_3$–KBr in 5% aqueous HCl giving RSO_2; determine excess bromate with KI and $Na_2S_2O_3$.	
Phosphorus-sulfur compounds			
Thiono groups, P^v=S	Volumetric	Shake 15 min with excess standard $KBrO_3$–KBr in 5% HCl giving sulfate ion; determine excess bromate with KI and $Na_2S_2O_3$.	
	Gravimetric	Reflux 15 to 20 min with 55% HNO_3 and determine sulfate formed by precipitation with $BaCl_2$ and weighing of $BaSO_4$.	
	Infrared	P=S, 714 to 625 cm^{-1}.	

(Continued)

TABLE 19.15 Methods for the Determination of Organic Phosphorus Groups (*Continued*)

Substance determined	Technique	Procedure	Reference
Pyrophosphates	Volumetric	Reaction with peroxide in alkaline solution giving peroxophosphate; determine excess peroxide.	16
		Hydrolyze to orthophosphate by refluxing with pyridine + water (9:1), and determine phosphate formed.	
	Paper chromatography	See phosphonic compounds above.	
	Infrared	See Table 7.17.	

References for Tables 19.14 and 19.15

1. R. N. Bell, *Ind. Eng. Chem., Anal. Ed.* **19**:97 (1947).
2. R. N. Bell, A. R. Wreath, and W. T. Culess, *Anal. Chem.* **24**:1997 (1952).
3. D. N. Bernhart, *Anal. Chem.* **26**:1798 (1954).
4. D. N. Bernhart and W. B. Chess, *Anal. Chem.* **31**:1026 (1959).
5. D. N. Bernhart and K. H. Rattenbury, *Anal. Chem.* **26**:1765 (1956).
6. D. N. Bernhart and A. R. Wreath, *Anal. Chem.* **27**:440 (1955).
7. D. E. C. Corbridge and E. J. Lowe, *Anal. Chem.* **27**:1383 (1955).
8. W. B. Chess and D. N. Bernhart, *Anal. Chem.* **30**:111 (1958).
9. A. Gee and V. R. Dietz, *Anal. Chem.* **25**:1320 (1953).
10. J. A. Grande and J. Beukenkamp, *Anal. Chem.* **28**:1495 (1956).
11. R. T. Jones and E. H. Swift, *Anal. Chem.* **25**:1272 (1953).
12. E. Karl-Kroupa, *Anal. Chem.* **28**:1091 (1956).
13. R. A. Keeler, C. J. Anderson, and D. Satriana, *Anal. Chem.* **26**:933 (1954).
14. I. M. Kolthoff, *Rec. trav. chim.* **46**:350 (1927).
15. T. Moeller and G. H. Quinty, *Anal. Chem.* **24**:1354 (1952).
16. S. Sass et al., *Anal. Chem.* **32**:285 (1960).
17. S. Sass and J. Cassidy, *Anal. Chem.* **28**:1968 (1956).
18. A. P. Scancillo, *Anal. Chem.* **26**:411 (1954).
19. M. J. Smith, *Anal. Chem.* **31**:1023 (1959).
20. Vasak, *Chem. Listy* **50**:1116 (1956).
21. H. J. Weiser, Jr., *Anal. Chem.* **28**:477 (1956).

An HPLC method has been developed for the analysis of inositol phosphates using a strong anion-exchange column, which avoids the hydrolysis of the esters on the column.[98] The conditions for the ion chromatographic separation of poly(methylenephosphonic acids) have been investigated.[99]

19.14 METHODS FOR THE DETERMINATION OF SULFUR-BASED FUNCTIONAL GROUPS

A monograph in three volumes discusses the analysis of sulfur compounds.[100] The infrared-absorption frequencies of sulfur compounds are given in Table 7.11 and the Raman frequencies in Table 7.30.

19.14.1 Determination of Mercaptans (Thiols) and Hydrogen Sulfide

The determination of both H_2S (Table 19.16) and thiols (the mercapto group) (Table 19.17) is often necessary in technical analysis. Both are easily oxidized. Both react with certain metal ions [silver, copper(II), and mercury(II)] with the formation of insoluble metal mercaptides or sulfides. An old and simple method is the oxidation with iodine as titrant added in known excess (because the reaction is slow in dilute solution), and the excess of iodine is then titrated with sodium thiosulfate.

[98] R. A. Minear et al., *Analyst* **113**:645 (1988).

[99] G. Tschaebunin, P. Fischer, and G. Schwedt, *Z. Anal. Chem.* **333**:117 (1989).

[100] J. H. Karchmer, *Analytical Chemistry of Sulfur and Its Compounds*, Interscience, New York, 1969; M. R. F. Ashworth, *The Determination of Sulfur-Containing Groups*, Academic, New York, 1972.

TABLE 19.16 Methods for the Determination of Hydrogen Sulfide

Technique	Procedure	Range	Interferences
Potentiometric titration	Dissolve sample in a mixture of C_6H_6, MeOH, NaOAc, and aqueous NH_3. Titrate with alcoholic $AgNO_3$ using Ag_2S ISE and SCE.* Dissolve sample in EtOH containing NaOAc (or, if water soluble, in aqueous NaOH + NH_3), and titrate with $AgNO_3$, using Ag_2S ISE and glass electrodes. Break at ca. –0.6 V is H_2S end point; 1 mol H_2S is equivalent to 2 mol of Ag^+.†	Micro	Free H_2S, RSH, CN^-
Volumetric	Absorb in acidic $CdSO_4$ solution, dissolve CdS precipitate in excess standard I_2 solution and back-titrate with $Na_2S_2O_3$.*	Micro	
Spectrophotometric	Absorb and precipitate H_2S as ZnS; dissolve and react this with *p*-aminodimethylaniline + $FeCl_3$ giving methylene blue; measure at 670 nm and compare with standards.‡	3–500 μg/mL	

* J. H. Karchmer, *Anal. Chem.* **30**:80 (1958).
† M. W. Tamele, L. B. Ryland, and R. N. McCoy, *Anal. Chem.* **32**:1007 (1960).
‡ J. K. Fogo and M. Popowsky, *Anal. Chem.* **21**:732 (1949).

TABLE 19.17 Methods for the Determination of RSH Groups

Technique	Procedure and reference	Range	Interferences
Potentiometric titration	If free S is absent, RSH and H_2S can be determined simultaneously by methods given in Table 19.16. Break at ca. –0.3 V is RSH end point; 1 mol of RSH is equivalent to 1 mol Ag^+ [1,2].	Micro	See Table 19.16
	If free S is present, remove H_2S with acidic $CdSO_4$ solution, then dissolve sample in a mixture of C_6H_6, MeOH, HOAc, and NaOAc, and titrate as above. Use total titer even though two breaks may appear [1].		See Table 19.16
	Titrate with 0.01*N* $Hg(ClO_4)_2$ solution using a bromide ISE or an ISE with a HgS matrix [10].		
Conductometric titration	Thiols in water and DMF solution are titrated with $HgCl_2$ [15].		
Atomic absorption flame	The thiol is precipitated as silver salt with alcoholic $AgNO_3$. After isolation of the precipitate, it is converted to Ag^+ and analyzed by FAAS [16].		
Amperometric titration	Dissolve 5 to 50 mL sample in 100 mL acetone +5 mL supporting electrolyte (NH_4NO_3 + aqueous NH_3), and titrate with aqueous $AgNO_3$, using rotating Pt and calomel electrodes [3].	0–400 μg/mL	H_2S, CN^-, I^-
	The RSH content of biological liquids is titrated with 0.002*M* $AgNO_3$ dissolved in tris(hydroxymethyl)methylamine [6].		
Volumetric	Add excess $AgNO_3$ and back-titrate with standard NH_4SCN, using iron(III) alum indicator.	Micro	H_2S, Cl^-
	Titrate with $AgNO_3$ in ammoniacal alcoholic medium, using NH_4 dithizonate indicator [4].	0.01–1 mmol	Sulfide
	Titrate with $Hg(ClO_4)_2$ [7] directly, or titrate the HNO_3 liberated from a known excess of $Hg(NO_3)_2$ [8].		
	Dissolve sample in toluene and titrate with 0.05*N* *p*-tolylmercury chloride in presence of EtOH and KOH, using thiofluorescein as indicator. *o*-Hydroxymercurybenzoic acid is an alternate titrant [9].		
	Tritrate with CrO_2O_2 by direct enthalpimetry [11].	>3 μg/mL	
Spectrophotometric	Add sample to CCl_4 solution of Ag dithiozonate and measure green color at 615 nm [4].	0.002–0.08 mmol	Sulfide
	Based on color developed when an acetic acid solution of the thiol is treated with *p*-aminodimethylaniline in presence of $FeCl_3$ and $K_3Fe(CN)_6$ [13].		

HPLC	Low-molecular-weight thiols determined by HPLC after derivatization with 7-chloro-4-nitrobenz-2,1,3-oxadiazole [12].
Mass spectrometric	Extract RSH compound (C_1 to C_7) from sample with K isobutyrate solution, acidify, and reextract into isooctane. Analzye isooctane solution by low ionizing voltage (or CI technique) [5].
	Negative ion MS used in hydrocarbon solvents for a direct determination [14].

1. J. H. Karchmer, *Anal. Chem.* **30**:80 (1958).
2. M. W. Tamele, L. B. Ryland, and R. N. McCoy, *Anal. Chem.* **32**:1007 (1960).
3. M. D. Grimes et al., *Anal. Chem.* **27**:152 (1955).
4. R. K. Kunkel, J. E. Buckley, and G. Gorin, *Anal. Chem.* **13**:1098 (1959).
5. W. P. Hoogendonk and F. W. Porsche, *Anal. Chem.* **32**:941 (1960).
6. S. K. Bhattachaya, *Nature* **183**:1327 (1959).
7. D. C. Gregg, P. E. Bonffard, and R. Barton, *Anal. Chem.* **32**:269 (1961).
8. B. Saville, *Analyst* **86**:29 (1961).
9. M. Wronsky, *Z. Anal. Chem.* **206**:352 (1964).
10. W. Selig, *Mikrochim. Acta* **1973**:453.
11. M. Wronsky and A. S. Abbas, *Analyst* **111**:1073 (1986).
12. Y. Nishikawa and K. Kuwata, *Anal. Chem.* **57**:1864 (1985).
13. V. B. Dorogova and V. A. Khomutova, *Chem. Abstr.* **99**:10196b (1983).
14. H. Knof, R. L. Arge, and G. Albers, *Anal. Chem.* **48**:2130 (1976).
15. L. M. Doane and J. T. Stock, *Anal. Chem.* **50**:1891 (1978).
16. J. S. Marhevka and S. Siggia, *Anal. Chem.* **51**:1259 (1979).

An extensive review of methods for the thiol group is available.[101] An automated potentiometric titration system has been developed for the determination of thiols and sulfides in petroleum systems.[102] Thiols in the presence of sulfides and disulfides in lubricating oils have been determined by a potentiometric titration with silver ammoniate solution.[103]

19.14.2 Determination of Thioethers (R—S—R) and Disulfides (R—S—S—R)

Thioethers easily form sulfoxides, and less easily sulfones, upon oxidation with bromine. The sample is oxidized in acetic acid solution, containing hydrochloric acid, with a known excess of 0.02*N* $KBrO_3$–KBr reagent, and, after a time for completion of the reaction, the excess of bromine is back-titrated by iodimetry. Table 19.18 gives methods for thioethers.

Disulfides can be reduced to mercaptans with zinc amalgam or sodium borohydride (tetrahydridoborate); the mercaptan formed is determined by argentimetry or iodimetry. Other methods include reducing disulfides in methanolic solution with triphenylphosphine,[104] and a mercurimetric titration after reduction with butyllithium.[105]

19.14.3 Determination of Thioketones

No reliable specific methods for the determination of thioketones are known. The methods listed below are useful for the qualitative detection of these compounds.

Procedure and reference	*Interferences*
Add $NaN_3 + I_2$. Disappearance of free I_2 indicates presence of monomeric thiones.[106]	Divalent sulfur compounds
Treat sample with $H_2O_2 + N_2H_4$. Thiones give sulfate ion on oxidation by peroxide and hydrazones (with H_2S evolution) on reaction with hydrazine; apply tests for these products.[107]	

19.14.4 Determination of Sulfoxides and Sulfones

A review contains 147 references describing methods for the determination of sulfoxides and sulfones.[108]

Since they are weak bases, sulfoxides can be titrated in acetic anhydride with $HClO_4$ dissolved in glacial acetic acid. Sulfoxides can also be reduced with strong reducing agents, usually with a period of standing in contact with an excess of the reducing agent [titanium(II) or tin(II)], and the excess reductant ascertained by titration. Methods for the determination of sulfoxides are given in Table 19.19.

Only very strong oxidizing agents will transform sulfones into sulfates. Sulfones can be reduced to the sulfide with metallic reductants and some can be reduced with lithium aluminum hydride. Procedures for sulfones are given in Table 19.19.

---

[101] M. R. F. Ashworth, *Determination of Sulfur-Containing Groups*, Vol. 2: *Analytical Methods for Thiol Groups*, Academic, London, 1976.
[102] S. W. Batson, G. J. Moody, and J. D. R. Thomas, *Analyst* **111**:3 (1986).
[103] N. G. Tyshchenko et al., *Chem. Abstr.* **92**:44273v (1980).
[104] R. E. Humprey and J. M. Hawkins, *Anal. Chem.* **36**:1812 (1964).
[105] S. Veibel and M. Wronsky, *Anal. Chem.* **38**:910 (1966).
[106] F. Feigl and Docorse, *Chem. Abstr.* **38**:2585 (1944).
[107] Kambara, Okita, and Tajima, *Chem. Abstr.* **46**:1795i (1952).
[108] M. R. F. Ashworth, in S. Patal et al. eds., *Chemistry of Sulfones and Sulfoxides*, Wiley, Chichester, UK, 1988.

TABLE 19.18 Methods for the Determination of R—S—R Groups

Technique	Procedure and reference	Range	Interferences
Spectrophotometric (for aliphatic and cyclic sulfides)	Add isooctane–I_2 solution to sample; measure absorbance at 308 nm [1,2].	Micro	Slight interference from aromatic sulfides
	Prepare sample–I_2 blend and measure absorbance at 310 nm against CCl_4 reference [3].	Macro	See above
Chemical	Remove RSH and H_2S with 0.05*M* $AgNO_3$ and elemental S with Hg. Determine total S by lamp (O_2 combustion, Sec. 18.4). Extract with solid $HgNO_3$, and again determine total sulfur. Difference approximates aliphatic and cyclic sulfides [4].	Micro and macro	
	Remove RSH and H_2S with 0.05*M* $AgNO_3$, free S with Hg, and aliphatic sulfides with $HgNO_3$. Determine total S by lamp (see Sec. 18.4). Extract with solid $Hg(NO_3)_2$, and again determine total S. Difference approximates aromatic sulfides [4].	Micro or macro	Thiophenes
	Extract with aqueous $Hg(OAc)_2$ solution (21.2 g per 100 mL). Suitable for separation of alkyl and cyclic sulfides and isolation of each from petroleum fractions [5].	Macro	Thiols and olefins
Volumetric	Oxidize with saturated aqueous Br_2 giving sulfoxides and sulfones; titrate acid formed with standard NaOH solution using bromocresol purple indicator [6].	Macro	Thiols and olefins
	Direct titration with standard (0.1*N*) $KBrO_3$–Br^- solution. $R_2S + Br_2 \rightarrow R_2SBr_2$; $R_2SBr_2 + H_2O \rightarrow R_2SO + 2\ HBr$ [7].	Macro	Olefins, thiols, and disulfides
Electrokinetic	Alkyl phenyl sulfides and alkyl benzyl sulfides are separated by an electrokinetic method in a micellar solution (aqueous–MeOH solution containing sodium dodecyl sulfate) [8].		

1. S. R. Hastings, *Anal. Chem.* **25**:420 (1953).
2. S. R. Hastings and B. H. Johnson, *Anal. Chem.* **27**:564 (1955).
3. H. V. Drushel and J. F. Miller, *Anal. Chem.* **27**:495 (1955).
4. J. S. Ball, *U.S. Bur. Mines Rept. Invest.* **3591** (1941).
5. Birch and McAllan, *J. Inst. Petrol.* **37**:443 (1951).
6. J. H. Sampey, K. H. Slagle, and E. E. Reid, *J. Am. Chem. Soc.* **52**:2401 (1932).
7. S. Siggia and B. L. Edsberg, *Anal. Chem.* **20**:938 (1948).
8. K. Otsuka, S. Terabe, and T. Ando, *Chem. Abstr.* **105**:12648r (1986).

TABLE 19.19 Methods for the Determination of Sulfoxides and Sulfones

Technique	Procedure and reference	Range	Remarks
	Sulfoxides		
Potentiometric titration	Dissolve about 1 mmol of sulfoxide in 75 mL of acetic anhydride and titrate with freshly standardized $HClO_4$ in 1,4-dioxane, using glass and calomel electrodes [1,2].		Sulfides, sulfones, and diethyl sulfite do not interefere.
Volumetric	To 20 mL of air-free aqueous or alcoholic solution of samples under CO_2, add 10 mL of $SnCl_2$ solution (27 g/L $SnCl_2$ in 0.9*M* HCl, standardized against I_2 solution) plus 10 mL hot 12*M* HCl. Boil 15 min, add 25 mL H_2O + 5 mL 12*M* HCl and boil 30 min. Add 20 mL 12*M* HCl + 10 mL H_2O + 5 to 10 drops 1% K indigotrisulfonate solution. Cool and titrate in artificial light with 0.1*M* $FeNH_4(SO_4)_2$ in 0.5 *M* H_2SO_4 very slowly as end point is approached [4].	Mean error < 1%	At end point, solution is a reddish-blue color, which does not deepen further on addition of excess iron(III) solution. A blank is essential.
	In a N_2 atmosphere, treat sample with 15 mL 0.1*M* $TiCl_3$ solution. Let stand 1 h at 80°C. Oxidize excess Ti(III) with boiling Fe(III) solution, let stand 30 s, cool rapidly, add 10 mL 2.5*M* H_3PO_4–3*M* H_2SO_4, extract sulfides twice with BuOH and 3 times with 15 mL portions of CCl_4. Add 55 mL water and titrate Fe(II) with 0.05*N* $K_2Cr_2O_7$ using diphenhylaminesulfonic acid indicator [3,6].	0.7 to 1.0 mequiv	Other oxidizing agents interfere.
	Add KI and acetyl chloride to an acetic acid solution of the sulfoxide. After 2 to 5 min, dilute the solution with HCl and titrate the iodine formed with $Na_2S_2O_3$ solution [5].		
	Treat sample with excess KI in a trifluoroacetic acid–acetone medium to produce I_2 which is then titrated with $Na_2S_2O_3$ [7].		
	In a N_2 atmosphere, treat a solution of the sample in glacial acetic acid with excess standard $TiCl_3$ solution + 1.5 mL 3*M* NH_4SCN. Let stand 1 h in water bath at 80°C, and titrate while hot with standard 0.05*M* iron(III) alum solution [3].	Macro range	Other oxidizing agents interfere.
	Sulfones		
Colorimetric	In blood or urine: Mix 1 mL of oxalated blood or suitable dilution of urine with 6 mL H_2O, 5 mL 2*M* HCl, and 4 mL 12% Cl_3CCOOH, and filter. To 10 mL of filtrate add 3 drops fresh 0.3% $NaNO_2$ solution; 3 min later add 3 drops 1.5% NH_4 sulfamate solution; 2 min later add 3 drops 0.1% *N*-(1-naphthyl) ethylenediamine hydrochloride, let stand 20 min in the dark, compare with standards and correct for a blank [8].		

1. D. D. Wimer, *Anal. Chem.* **30**:77, 2026 (1958).
2. C. A. Streuli, *Anal. Chem.* **30**:997 (1958).
3. E. Barnard and K. R. Hargrove, *Anal. Chim. Acta* **5**:536 (1951).
4. E. Glynn, *Analyst* **72**:248 (1947).
5. S. Allenmark, *Acta Chem. Scand.* **20**:910 (1968).
6. R. R. Legault and K. Groves, *Anal. Chem.* **29**:1495 (1957).
7. W. Ciesielski, *Talanta* **35**:969 (1988).
8. Simpson, *Intern. J. Leprosy* **17**:208 (1949).

TABLE 19.20 Determination of Sulfinic and Sulfonic Acids

Technique	Procedure and reference	Range	Remarks
	Sulfinic acids		
Volumetric	Make sample (ca. 0.06*M* sulfinic acid) slightly alkaline with NaOH. Remove SO_3^{2-}, if present, by digestion with excess $BaCl_2$ solution, filtration, and dilution to known volume. Use this to titrate a mixture of 50 mL 0.03*M* NaOCl (standardized at 2 h intervals vs. standard 0.05*M* $NaAsO_2$), 50 mL 10% Na_2CO_3 and 100 mL H_2O at 15°C using starch–iodide paper as external indicator [1].	Macro (ca. 2.5 g)	Sulfonates do not interfere.
	Titrate with standard NaOH, using cresol red, thymol blue, or phenolphthalein as indicator. In the presence of sulfonic acids, titrate potentiometrically in acetic acid medium.		Suitable for assay of pure compounds, but sulfonic and many other acids interfere.
	Titrate directly to sulfonic acid with either NaOCl [1] or $KMnO_4$ solution [2].		
Gasometric	Exchange the hydrogen of the sulfinic acid group with iodine, and the iodine subsequently replaced with potassium by treatment with KOH to give HIO. In presence of H_2O_2 the HIO is reduced and one molecule of O_2 is liberated and gas evolved is measured [3].		Although an old method, it still finds use.
	Sulfonic acids		
Volumetric	Titrate with standard NaOH in aqueous solution using phenolphthalein as indicator. Less strong acids are titrated in nonaqueous medium (ethanol, ethylene glycol, or pyridine) using potentiometric end-point detection [4].		
Gas chromatography	Sulfonic acids and sulfonates can be transformed with thionyl chloride into sulfonyl chloride and then determined by gas chromatography [6,7].		
	Sulfonic acids are converted to sulfonyl chlorides with PCl_5 followed by reduction to thiols with $LiAlH_4$; the thiols were determined by GC using a 3% OV-1 column [8].		
General	Evaporate sample with H_2SO_4, then determine the sulfate (usually alkyl sulfate) formed [5].		Alkali sulfates must be absent.

1. I. Ackerman, *Ind. Eng. Chem., Anal. Ed.* **18**:243 (1946).
2. P. Allen, *J. Org. Chem.* **7**:23 (1942).
3. S. Krishna and Q. B. Das, *J. Indian Chem. Soc.* **4**:367 (1927).
4. V. Z. Deal and G. E. A Wild, *Anal. Chem.* **27**:47 (1955).
5. S. Siggia and J. G. Hanna, *Quantitative Organic Analysis Via Functional Groups*, 4th ed., Wiley, New York, 1977.
6. J. J. Kirkland, *Anal. Chem.* **32**:1388 (1960).
7. A. Amer, E. G. Alley, and C. U. Pittman, Jr., *J. Chromatogr.* **362**:413 (1986).
8. H. Oka and T. Kojima, *Chem. Abstr.* **91**:203867q (1979).

TABLE 19.21 Methods for the Determination of Specific Substances

Functional group determined	Procedure and reference	Range	Remarks
Diethyldithiocarbamate	*Colorimetric*. (To determine CS_2 in air.) Absorb CS_2 in a solution prepared by adding 1 mL $(C_2H_5)_2NH$ + 20 mL triethanolamine + 50 mg $Cu(OAc)_2$ to 1 L 85% to 100% EtOH (or methyl Cellusolve) and letting stand until clear. Measure absorbance of golden-yellow color at 420 nm [1,2].	0–50 μg CS_2	Remove H_2S. Dimethyl sulfide, thiophene, and mercaptans should not interfere. Thiocarbonyls react similarly.
Dixanthogen	*Volumetric*. To 10 mL of solution of sample in acetone, add 5 mL 10% (w/v) aqueous NH_4NO_3, 10 mL 2% aqueous KCN, and 10 mL acetone. Heat 25 min at 40 to 50°C. Extract twice with benzene, add 10 mL 1:4 H_3PO_4 to aqueous solution and heat 30 min at 100°C. Cool, add Br_2 dropwise to permanent yellow color. Decolorize by careful addition of $FeSO_4$, add 3 g KI and a little $NaHCO_3$, let stand 10 min in the dark, and titrate liberated I_2 with $Na_2S_2O_3$ [3].		
Ethyl xanthate	*Ultraviolet*. Measure absorbance at 301 nm in 50% EtOH containing 0.05*M* KOH [4].		
Saccharin	*Spectrophotometric*. Dissolve sample in dilute aqueous HOAc and extract a measured aliquot into diethyl ether. Evaporate to dryness, dissolve residue in 1 mL 10% NaOH, dilute to 100 mL with water. Take an aliquot containing 50–100 μg saccharin, evaporate to dryness at 100°C, and fuse residue 2 h at 140°C with 2 mL phenolsulfonic acid. Cool, dissolve in water, dilute to 50 mL, add 10 mL 10% NaOH, dilute to 100 mL, and measure the phenol red color at 480 nm [5].	5–30 mg	Caffeine and vanillin interfere. Separate caffeine by extraction from first NaOH solution with $CHCl_3$; separate vanillin by extraction from first residue with light petroleum.
Sulfamate	*Gravimetric*. Add dilute HNO_3 or $HClO_4$ + excess $NaNO_2$ (giving sulfate ion) and determine as $BaSO_4$ [6].		If sulfate ion is present originally, run second aliquot but omitting $NaNO_2$.
Sulfamic acids	*Potentiometric titration*. Under N_2 atmosphere, titration a solution of sample in pyridine with standard $(butyl)_4$ NOH using glass–calomel electrodes [7].		No interference from H_2SO_4.
	Volumetric. To 100 mL sample in an iodine flask, add 20 mL 1:1 H_2SO_4, then excess standard 0.01*M* $NaNO_2$, stopper, and let stand 30 min. Add excess 0.02*M* Ce(IV) sulfate and back titrate with standard $Fe(NH_4)_2(SO_4)_2$ using 1,10-phenanthroline indicator [8].		HNO_3 does not interfere.
Sulfenyl chlorides	See under sulfonyl chlorides.		
Sulfinyl chlorides	See under sulfonyl chlorides.		

Sulfonyl chlorides	*Titrimetric*. Titrate an acetone–water (2 : 3 v/v) solution, 0.02–0.1*M* in RSO_2Cl with standard Na_2S (a) to first persistent yellow color, (b) to a two-electrode amperometric and point at 0.15 V, or (c) to a constant-current potentiometric end point using Pt electrodes [9,10].	Sulfenyl and sulfinyl chlorides interfere and may be similarly determined.
	Volumetric. Add a small excess of pyridine, aniline, or 8-hydroxyquinoline to sample, cool, add a few drops of water, let stand a few minutes, dilute with 30 to 50 mL water, make acid with HNO_3, add excess standard $AgNO_3$, and back titrate with standard KSCN, using Fe(III) as indicator [11].	Reaction mixture gives low results if warmed. Interference from compounds with easily replaceable halogens, and may be determined similarly.
Thio acids	*Volumetric*. Titrate with standard NaOH, using phenolphthalein indicator.	Other acidic compounds interfere. Use intermediate salt bridge if done potentiometrically.
Thioamide, —C(=S)NH_2	*Volumetric*. To an alkaline solution of sample, add excess I_2, let stand, acidify, and back titrate excess I_2 with standard $NA_2S_2O_3$ [12].	Interference from thiocyanates, thiols, and other reducible compounds.
	Potentiometric. In aqueous solution the sample is titrated with $AgNO_3$ using a silver ion selective electrode [23].	
Thiol esters	*Volumetric*. Dissolve 30 mg in hot water, add excess standard ICl, let stand 15 min. Extract I_2 into $CHCl_3$, was with water and titrated the combined aqueous solutions with thiosulfate after adding excess KI [13].	
Thiolsulfinate	*Colorimetric*. To 1 mL 2-propanol containing 0.05–5 μmol sample, add 1 mL 0.05*M* *N*-ethylmalemide in 2-propanol, then add 1 mL 0.25*M* KOH in 2-propanol. Let stand 10–20 min, and measure absorbance at 515 nm [14].	
	Volumetric. Dissolve 20 to 80 mg sample in 15 mL HOAc, dearate with CO_2, add 2 mL air-free saturated aqueous KI, stir 2 min, and titrate with thiosulfate using starch indicator (continue to exclude air) [15].	
Thiolsulfonates	*Volumetric*. Dissolve 1 mmol sample in 5 mL neutral EtOH, add 2 mL 20% (w/v) thiophenol in EtOH, swirl, and titrate RSO_2H with 0.1*N* NaOH using bromophenol blue indicator [15].	
Thione esters	*Volumetric*. Let stand 12 h at room temperature in the dark with excess alcoholic $AgNO_3$ solution. Filter, wash, and dissolve precipitated Ag_2S by warming with 6*M* HNO_3. Determine silver content by titration with KSCN using Fe(III) as indicator [16].	Applicable in presence of thiol esters, which react only when heated.
Thiosemicarbazones	*Volumetric*. The sample in aqueous acetonitrile is treated with excess $Cu(NO_3)_2$; protons released are titrated with a base [24].	

(Continued)

TABLE 19.21 Methods for the Determination of Specific Substances (*Continued*)

Functional group determined	Procedure and reference	Range	Remarks
Thionyl	*Volumetric.* Reflux with KOH solution (giving sulfite ion), acidify with H_3PO_4, distill SO_2 into excess standard I_2, and back-titrate with thio-sulfate [17].		
Thiourea	*Colorimetric.* To a neutral solution of sample, add excess $NaHCO_3$ solution (pH 9 to 10), $Na_2[Fe(NO)(CN)_5]$, NH_2OH, and Br_2. Compare resulting color to standards [18].		
	Volumetric. To 35 mL neutral solution of sample, add 1 g powdered KBr, then 20 mL 12*M* HCl, warm to 50°C, add 1 mL 0.1% $AuCl_3$ solution, and titrate with 0.0167*M* $KBrO_3$ potentiometrically [19].	1 to 3 meq	
Xanthate	*Potentiometric.* Titrate an alkaline solution of the sample with standard $AgNO_3$ using an Ag indicator electrode [20].	Applicable to $10^{-5}M$ – $10^{-4}M$	
	Volumetric. To a neutral solution of sample, add a small excess of standard 0.1*N* $AgNO_3$, then immediately add Fe(III) indicator and back titrate with KSCN [21].		
	Volumetric. Add 1*M* HOAc until just acid to phenolphthalein, then starch indicator and titrated with 0.01–0.2*N* I_2 solution. Cooling to 0°C before titration improves the end point [22].	0.05 to 3 mmol	

1. Viles, *J. Ind. Hyg. Toxicol.* **22**:188 (1940).
2. F. F. Morehead, *Ind. Eng. Chem., Anal. Ed.* **12**:373 (1940).
3. Shankaranarayana and S. Patal, *Analyst* **86**:98 (1961).
4. Maurice and Mulder, *Mikrochim. Acta* **1957**:661.
5. Whittle, *Analyst* **69**:45 (1944).
6. Barnard and Hargrave, *Anal. Chim. Acta* **5**:536 (1951).
7. Cundiff and Markunas, *Anal. Chim. Acta* **20**:506 (1959).
8. Whitman, *Anal. Chem.* **29**:1684 (1957).
9. Ashworth, Walisch, and Kronz-Dienhart, *Anal. Chim. Acta* **20**:96 (1959).
10. Walish, Hertel, and Ashworth, *Chim. Anal.* **43**:234 (1961).
11. Drahowzal and Klamann, *Monatsh.* **82**:470 (1951).
12. Wojahn and Wempe, *Arch. Pharm.* **285**:375 (1952).
13. Carson and Wong, *Nature* **183**:1673 (1959).
14. Rapaport, *Chem. Abstr.* **53**:15856h (1959).
15. Barnard and Cole, *Anal. Chim. Acta* **20**:540 (1959).
16. Karjala and McElvain, *J. Am. Chem. Soc.* **55**:2966 (1933).
17. Hennert and Merlin, *Chem. Anal.* **39**:429 (1957).
18. Grote, *J. Biol. Chem.* **93**:25 (1931).
19. Szbelledy and Madia. *Z. Anal. Chem.* **114**:253 (1938).
20. Plaksin et al., *Zav. Lab.* **22**:28 (1956).
21. Makins, *J. Am. Chem. Soc.* **57**:405 (1935).
22. Matuszak, *Ind. Eng. Chem., Anal. Ed.* **4**:98 (1932).
23. H. Sikorska-Tomicka and G. Strutynsaka, *Chem. Anal.* (*Warsaw*) **30**:547 (1985).
24. M. Argueso et al., *Microchem. J.* **31**:74 (1985).

19.14.5 Determination of Sulfinic and Sulfonic Acids

The sulfinic acid group can be easily oxidized to the sulfonic acid group. Since sulfonic acids are strong acids, they can be determined by an acidimetric titration. Less strong sulfonic acids can be titrated in nonaqueous medium. Methods for sulfinic and sulfonic acids are given in Table 19.20.

19.14.6 Determination of Miscellaneous Sulfur-Based Functional Groups

Table 19.21 is a collection of methods for sulfur-based function groups that have not been covered in earlier tables of this subsection. Compounds or functional groups are listed alphabetically.

In a review, the instrumental methods available for the determination of linear alkylbenzenesulfonates in sewage, sludges, soils, and sediments have been discussed.[109] Of all the methods used for the determination of α-olefins sulfonates, reversed-phase HPLC was found to be the most valuable since it provided more qualitative and quantitative information about the analyte.[110] A critical review of the methods used in the analysis and separation of sulfonates has been published.[111]

---

[109] H. DeHenau, E. Mathijs, and W. D. Hopping, *Int. J. Environ. Anal. Chem.* **26**:279 (1986).
[110] V. Castro, J. P. Canseller, and J. L. Boyer., *Commun. Jorn. Com. Esp. Deterg.* **16**:373 (1985).
[111] M.-S. Kuo and H. A. Mottola, *CRC Crit. Rev. Anal. Chem.* **9**:297 (1980).

SECTION 20
ANALYSIS OF PESTICIDES AND HERBICIDES

20.1 INTRODUCTION

The terms *pesticides* and *herbicides* refer to a wide class of substances used for pest and weed control, respectively, and include several thousand formulations. Because of their wide-scale uses in household pest control, as well as in garden, crop, and other agricultural applications, trace residues of many pesticides are susceptible to persist in the environment and may be found in many wastewaters, groundwaters, soils, and sediments and also in crops, fruits, and other food products.

Although these substances vary widely in their chemical structures and properties, many common pesticides and herbicides fall under the following categories:

1. Organochlorine pesticides
2. Organophosphorus pesticides
3. Carbamate pesticides
4. Triazine herbicides
5. Chlorophenoxy acid herbicides
6. Urea-type herbicides

Except for the organochlorine pesticides, compounds of each of these classes may have some similar structural features. Their general structures and methods of analyses are discussed below.

The most common instrumental methods for trace analyses of pesticides and herbicides involve gas chromatography (GC), mass spectrometry (MS), and high-performance liquid chromatography (HPLC). Other techniques, such as enzyme immunoassay, are also known for a limited number of compounds, but such methods are susceptible to interference from other pesticides of similar structures. The GC, GC/MS, and HPLC methods are also susceptible to interference. Therefore, it is always recommended that once a pesticide is detected in a sample extract, its presence should be confirmed on an alternative column (GC or HPLC) or by an alternative technique, such as, derivatization or by mass spectrometry.

Sample extraction is probably the most important step in pesticide analysis. Extraction not only transfers the analytes into a suitable organic solvent for chromatographic determination but also increases their concentrations by several order of magnitude and thereby lowers their detection levels accordingly. Also, extraction processes separate interfering substances from analytes. Various extraction methods are highlighted below.

20.2 SAMPLE EXTRACTIONS AND CLEANUP

Sample extractions constitute the most critical steps in the analysis of pesticides and herbicides. Such steps involve extractions of the analytes from their bulk matrices into an appropriate solvent and then removal of potentially interfering substances from the solvent extracts by various cleanup procedures and concentrating the extracts to small volumes—usually 1–2 mL—before analysis. Often the extraction and cleanup steps can be achieved simultaneously, in part or fully, by many techniques. For example, supercritical fluid extraction (SFE) can extract and separate the analytes in a single step by appropriately controlling the pressure, modifier, or flow. Such techniques, however, are not commonly applied in pesticides analyses. The common and traditional approach is to carry out extractions and cleanup separately in multiple stages. While the method of extraction usually depends on the sample matrix, the cleanup procedure may depend on the nature of the interfering substances that may be present in the extract. For example, for the GC analysis of certain organochlorine pesticides, such as aldrin or lindane, using an electron-capture detector, interfering sulfur that may mask these peaks in the chromatogram are removed from the extract by treatment with copper powder.

20.2.1 Aqueous Samples

Trace residues of pesticides that have very low solubilities in water may be extracted by (1) liquid–liquid extraction (LLE) or (2) solid-phase extraction (SPE). Various extraction techniques are highlighted in Table 20.1.

20.2.1.1 Liquid–Liquid Extraction. In liquid–liquid extraction, a 1-L aliquot of the aqueous sample is repeatedly extracted with three 50- to 60-mL portion of methylene chloride in a 2-L separatory

TABLE 20.1 Pesticides/Herbicides Extraction Processes

Extraction process	Technique	Matrix and methods
Liquid–liquid extraction (LLE)	Separatory funnel	All aqueous samples; usually 1-L volume, manually shaken with a solvent heavier than water; extract concentrated to 1–2 mL
	Continuous	All aqueous samples; 1-L or larger volume extracted continuously by intimate mixing with a suitable immiscible solvent heavier than water; no manual shaking but longer extraction time; extract concentrated to 1–2 mL
	Microextraction	All aqueous samples; much smaller volume, usually 25–35 mL in a 40-mL vial; shaken with 2 mL hexane or other solvent lighter than water; a few microliters of top layer injected directly onto GC
Solid-phase extraction (SPE)	Cartridge or column adsorption	Aqueous samples, biological fluids, or fruits and vegetables (extracted with water-miscible solvents) filtered through SPE cartridge or column to retain either pesticides or interfering substances; pesticides retained on cartridge eluted with an appropriate extracting solvent; selection of solvent based on polarity and K_{ow} of analyte molecules; if interference removed on cartridge, sample passed through is subjected to further extraction
	Disk-type device	Aqueous samples or food extracts passed through disk-type device (i.e., glass fiber filter with C-18 or C-8 sorbent) under pressure or suction; disk removed from device and shaken with an appropriate solvent; hydrophobic disks (C-18, C-8) require activation by passing a small volume of a water-miscible solvent
	SPE microextraction	Water, juices, and biological fluids; a fiber coated with extractant or a droplet of extractant suspended in sample, transferred to a GC for analysis
Supercritical fluid extraction (SFE)	Static or dynamic mode (or both)	Soils, sediments, solid wastes, fruits, vegetables, oil, meats and animal tissues; a supercritical fluid such as CO_2 or N_2O is pumped (under supercritical temperature and pressure conditions) through sample matrix in an extraction vessel to transfer the analytes from the matrix into a collection device, usually either an organic solvent or a SPE device
Soxhlet extraction		Soil, sediment, and animal tissues; pesticides extracted by an appropriate solvent by intimate mixing of solvent and sample in several cycles of batch processes
Passive sampling device (PSD)		Air, water, and soil; the sampling device is left in the environmental medium to sorb analyte, which is then extracted separately by a variety of techniques
Microwave-assisted solvent extraction (MASE)		Solid samples; analytes removed from sample by heating with a solvent under microwave energy.

funnel. All nonpolar trace residues of pesticides partition from the aqueous phase into methylene chloride at the bottom immiscible phase. The methylene chloride extract is then concentrated down to a much smaller volume, usually 1 mL, by slowly and carefully evaporating the solution on a water bath. Many concentrators, such as Kuderna-Danish type, are commercially available. Thus, a 1-L sample is extracted and concentrated down to a volume of 1 mL, thereby lowering the method detection limit of the pesticides by almost 1000 times.

Continuous liquid–liquid extraction units are commercially available that eliminate the need for manual extraction using a separatory funnel. Such units are designed to extract a large number of samples and in larger volumes.

It may be noted that the selection of the solvent for extraction is important. Methylene chloride or other chlorinated solvents cannot be used if the GC analysis involves the use of an electron-capture detector. Chlorinated solvents, however, may readily be employed if the pesticides are to be detected by mass spectrometry. Also, solvent selection may depend on other factors, such as the density, volatility, and of course the solubility of the pesticides in the solvent. Solvents that are nonpolar or have low polarity are effective for extracting analytes that are nonpolar or have low polarity. Also, when a separatory funnel is used in LLE, the solvent should have a density greater than that of the water, so that the water-immiscible, denser solvent can be drained out conveniently from the separatory funnel and separated from the top aqueous phase for repeated extraction.

Many hydrocarbon solvents have proven to be effective in the extraction of organochlorine pesticides such as aldrin, lindane, and DDT. Such solvents include *n*-hexane, isooctane, and toluene. The former is a solvent of choice in LLE microextraction for potable water. In such extractions, a much smaller volume of aqueous sample, i.e., 35 mL, is placed in a 40-mL vial with a Teflon cap. To this sample, 1 or 2 mL of *n*-hexane is added. The mixture is shaken vigorously for 1 min and the solution is allowed to stand. Hexane-soluble pesticide residues partition into the lighter solvent layer on the top, from which a few microliters of the extract are carefully withdrawn for injection onto the GC. Although such extraction is simple and faster than that using a separatory funnel or a continuous extractor, the extent of sample concentration is much lower than with the latter types of extraction.

The octanol–water partition coefficient K_{ow} may serve as a guidance for selecting a solvent, its volume, and the number of repetitive extractions. For example, if the K_{ow} of the pesticide falls in the range 10^4–10^6, any of the solvents mentioned above may provide a good extraction. If the K_{ow} is ~10^3 or less, certain additional steps may be necessary to achieve the same degree of extraction as that with a $K_{ow} > 10^4$. In such a case the degree of extraction may be increased by (1) increasing the volume of the solvent or (2) by performing multiple extractions. Alternatively, the solubility of the pesticide in the water may be decreased by (1) adding a salt, such as sodium chloride, to "salt out" the analyte, (2) choosing a slightly more polar solvent, or (3) adjusting the pH to ~3 for acidic analytes and to ~10 for basic analytes.

20.2.1.2 Solid-Phase Extraction. Both solid-phase extraction (SPE) and solid-phase microextraction (SPME) are widely applied for extracting many types of pesticides from aqueous matrix. SPE devices offer certain advantages over LLE methods:

1. The speed of extraction is faster than with LLE.
2. The technique is more versatile and can be applied to a wide array of pesticides and herbicides of varying molecular structures and polarities.
3. Interference substances are also separated from the analytes during extraction. Thus, the SPE technique may be applied for sample cleanup.
4. SPE or SPME reduces or often eliminates the use of much larger volumes of organic solvents that may be toxic, environmentally polluting, or expensive.
5. The SPE device may be carried to the field for on-site analysis.

Solid-phase extraction is carried over SPE cartridges, columns, or disks, which are commercially available. The sample is passed through the cartridge or column. While the analytes are retained on the cartridge or the column, the interfering substances are allowed to pass through. The analytes are subsequently eluted from the cartridge or column bed using a suitable solvent or solvent mixture.

Another approach is to retain the interfering substances on the cartridge, allowing the pesticides to pass though for subsequent analysis.

Selection of cartridge materials and elution solvents depends on the polarity of the analyte. For nonpolar and low-polar analytes, hydrophobic octadecyl ($-C_{18}$), octyl ($-C_8$), and cyclohexyl ($-C_6H_{11}$) cartridges are commonly used, in which the analytes are retained on the adsorbent bed and latter eluted out for analysis. Alternatively, such cartridges may be employed for the removal of nonpolar or low-polar interfering substances from moderately polar pesticides. Other packing materials that are used in solid-phase extraction and cleanup include phenyl, cyanoprohyl, diol, aminopropyl, carboxymethyl, and sulfonylpropyl groups. Silica, alumina, and florisil cartridges are known to be effective for many polar pesticides or interfering substances. All types of cartridges and columns can be readily activated, loaded, washed, and eluted.

Empore disks were introduced in the 1990s for solid-phase extraction of pesticides and other pollutants from aqueous matrix. These disks constitute reverse-phase bonded HPLC silica particles held tightly into a porous Teflon or glass fiber disk. Disks are commercially available with octadecyl, octyl, and other bonded phases on silica and plastic bead particles. The disks are placed in a filtration apparatus and the sample is passed through. Vacuum is then applied for complete drying of the disk. The disk is then shaken in a solvent to elute out the pesticides adsorbed over it. Hydrophobic disks containing C_{18}- or C_8-bound silica particles require activation. This is achieved by passing a small volume of methanol through the disk. Disks containing hydrophilic particles do not require activation.

The SPE disk technique has both advantages and disadvantages relative to the cartridge and column devices. The major advantages include faster flow rates of samples without breakthrough of pesticides, rapid elution from the disk, and the ease of storage and transportation of samples. A disadvantage, however, is possible plugging of the disk when extracting dirty samples.

Many solvents and combinations in varying proportions have accomplished excellent elution of pesticides from SPE cartridges, columns, and disks. Such solvents include acetonitrile, methanol, hexane, diethyl ether, petroleum ether, and methylene chloride.

20.2.2 Fruits and Vegetables

Fruits and vegetables consist largely of water. The pesticide residues in such plant extracts are therefore extracted with a water-miscible solvent, such as acetone or acetonitrile. A convenient method of extraction involves the use of a SPE cartridge. The method is suitable for pesticides having moderate degree of polarity and K_{ow} values ranging between 100 and 1000. The water content of the extract is carefully adjusted by adding an appropriate volume of extracting solvent. Hydrophobic $-C_{18}$, $-C_8$, or -phenyl-bonded SPE cartridges are used for the purpose. Aqueous solutions of plant extracts containing oils, waxes, lipids, sugars, and other interfering substances are filtered through the cartridge. While such interferences are retained on the hydrophobic cartridge, the pesticide extract solution passes through. Furthermore, retention of pesticides of low polarity on the SPE cartridge or column may be avoided or minimized by adjusting the pH and the water content of the extracting solvent mixture. The extract solution is then analyzed for pesticides by HPLC methods. Polar interferences in the plant extract may be removed in a similar manner by using a polar SPE cartridge. The packing materials in such cartridges include aminopropyl, cyanopropyl, quaternary methylamine, and diol groups.

20.3 ORGANOCHLORINE PESTICIDES

20.3.1 Methods of Analysis

Some chlorinated pesticides with their alternative names and chemical formulas are listed in Table 20.2. Pesticides referred to under this broad class do not have any common specific structural features. They all contain chlorine atoms in their molecules and therefore may be determined by GC using a

TABLE 20.2 Formulas and Synonyms of Some Organochlorine Pesticides

Pesticide	Molecular formula	Some other names
Alachlor	$C_{14}H_{20}ClNO_2$	Metachlor, Lasso, Alanox
Aldrin	$C_{12}H_8Cl_6$	Aldrex, Octalene, Seedrin
α-BHC	$C_6H_6Cl_6$	α-Benzene hexachloride, α-hexachlorocyclohexane, α-hexachloran
β-BHC	$C_6H_6Cl_6$	β-Benzene hexachloride, β-hexachlorocyclohexane,
γ-BHC	$C_6H_6Cl_6$	γ-Benzene hexachloride, γ-hexachlorocyclohexane
δ-BHC	$C_6H_6Cl_6$	δ-Hexachlorocyclohexane, δ-benzene hexachloride
Captafol	$C_{10}H_9Cl_4NO_2S$	Difolatan, Folcid, Mycodifol
α-Chlordane	$C_{10}H_6Cl_8$	*cis*-Chlordan, α-Chlordan
γ-Chlordane	$C_{10}H_6Cl_8$	*trans*-Chlordan, *trans*-γ-Chlordane, β-Chlordane
Chlorobenzilate	$C_{16}H_{14}Cl_2O_3$	Acar, Benzilan, Folbex
Chloropropylate	$C_{17}H_{16}Cl_2O_3$	Chlormite, Rospin, Acaralate
Chlorothalonil	$C_8Cl_4N_2$	Daconil, Sweep, Nopcocide
DBCP	$C_3H_5Br_2Cl$	Nemagon, Fuzmazone, Nemabrom
4,4′-DDD	$C_{14}H_{10}Cl_4$	*p,p′*-DDD, Rothane, Dilene
4,4′-DDE	$C_{14}H_8Cl_4$	*p,p′*- DDE, DDT dehydrochloride
4,4′-DDT	$C_{14}H_9Cl_5$	*p,p′*-DDT, Chlorphenotane, Neocid
Dieldrin	$C_{12}H_8Cl_6O$	Dieldrite, Termitox, Aldrin epoxide
Endosulfan I	$C_9H_6Cl_6O_3S$	α-Endosulfan, α-Thiodan
Endosulfan II	$C_9H_6Cl_6O_3S$	β-Endosulfan, β-Thiodan
Endosulfan sulfate	$C_9H_6Cl_6O_4S$	Thiodan sulfate
Endrin	$C_{12}H_8Cl_6O$	Endrex, Mendrin, Hexadrin
Endrin aldehyde	$C_{12}H_8Cl_6O$	Endrex aldehyde
Endrin ketone	$C_{12}H_8Cl_6O$	δ-Ketoendrin
Heptachlor	$C_{10}H_5Cl_7$	Rhodiachlor, Hetamul, Drinox
Heptachlor epoxide	$C_{10}H_5Cl_7O$	Epoxyheptachlor, HCE
Hexachlorobenzene	C_6Cl_6	Sanocide, Amatin, HCB
Hexachlorocyclopentadiene	C_5Cl_6	Graphlox, perchloro-1,3-cyclopentadiene
Isodrin	$C_{12}H_8Cl_6$	Latka 711, SD 3418
Kepone	$C_{10}Cl_{10}O$	Chlordecone, Merex
Methoxychlor	$C_{16}H_{15}Cl_3O_2$	Metox, Marlate, Methoxcide
Mirex	$C_{10}Cl_{12}$	Dechlorane, Paramex
Nitrofen	$C_{12}H_7Cl_2NO_3$	Nitrochlor, TOK, NIP
Permathrin	$C_{21}H_{20}Cl_2O_3$	Elimite, Nix
Perthane	$C_{18}H_{20}Cl_2$	Ethylan
Propachlor	$C_{11}H_{14}ClNO$	Bexton, Ramrod, Satecid
Strobane	A mixture of terpene isomers	Terpene polychlorinates, Dichloricide aerosol
Toxaphene	$C_{10}H_{10}Cl_8$	Camphechlor, Alltex, Octachlorocamphene
Trifluralin	$C_{13}H_{16}F_3N_3O_4$	Nitran, Trifloran, Trefanocide

halogen-specific detector. Many environmental samples, after extraction into a suitable nonhalogen solvent such as hexane or isooctane, and after the solvent extract is concentrated down to a small volume, are analyzed by GC-ECD (electron-capture detector). These detectors are most sensitive to organochlorine pesticides. After the compound is identified by GC, its presence in the sample may be confirmed by mass spectrometry using GC/MS or LC/MS. HPLC and enzyme immunoassay are the other methods of detecting these pesticides. Some of these methods are briefly discussed below:

GC/ECD analyses may be performed using either packed or capillary columns. A fused-silica, open-tubular capillary column provides improved resolution, increased sensitivity, and better selectivity than a packed column. Both types of columns are commercially available. A 30-m-length capillary column having an inside diameter (ID) of 0.53 or 0.32 mm and a film thickness ranging between 0.83 and 1.5 μm should exhibit high resolution in separating pesticides in mixtures. Such columns under different trade names include DB-5, DB-17, Rxt-5, SPB-608, and DB-608. Helium and nitrogen may be used as carrier and makeup gases, respectively. Alternatively a 1.8-m × 4-mm-ID glass column packed with 1.5% SP-2250, 1.95% SP-2401, or 3% OV-1 on Supelcoport or equivalent may be used as a packed column. The chromatographic conditions for such packed column separation of pesticides are outlined below to serve as a rough guideline. Other conditions may be employed as appropriate.

TABLE 20.3 **Characteristic Mass Ions of Some Chlorinated Pesticides**

Pesticides	Primary ion	Secondary mass ions
Alachlor	45	160, 188, 146, 237, 118, 77, 132
Aldrin	66	263, 79, 91, 101, 265, 65, 39, 293
α-BHC	181	183, 219, 111, 217, 51, 221, 185, 109
β-BHC	109	183, 181, 111, 51, 219, 85, 217, 73
γ-BHC	181	183, 219, 111, 109, 217, 51, 221, 185
δ-BHC	181	183, 219, 111, 109, 217, 51, 218, 221
Captafol	79	77, 80, 78, 39, 27, 107, 51
α-Chlordane	39	65, 75, 27, 49, 373, 375, 51, 109
γ-Chlordane	373	375, 109, 272, 119, 377, 75, 117, 235
Chlorobenzilate	251	139, 253, 111, 141, 75, 29, 252
Chloropropylate	139	251, 253, 111, 141, 43, 252, 75
Chlorothalonil	266	264, 268, 267, 270, 109, 124, 231
DBCP	57	157, 75, 155, 28, 39, 27, 77, 41
4,4′-DDD	235	237, 165, 236, 239, 199, 178, 238
4,4′-DDE	246	318, 248, 316, 320, 176, 105, 123
4,4′-DDT	235	237, 165, 236, 212, 199, 28, 282
Dieldrin	79	81, 263, 277, 279, 77, 345, 380
Endosulfan I	195	241, 197, 243, 207, 277, 75, 69, 339
Endosulfan II	39	48, 41, 49, 63, 50, 85, 195
Endosulphan sulfate	272	274, 229, 387, 239, 227, 422
Endrin	81	39, 83, 67, 263, 53, 113, 281, 345
Endrin aldehyde	67	345, 250, 347, 66, 173, 252, 243
Endrin ketone	67	27, 39, 66, 87, 79, 55, 147
Heptachlor	100	65, 272, 39, 102, 274, 135, 237, 337
Heptachlor epoxide	353	355, 351, 81, 357, 263, 237, 388
Hexachlorobenzene	284	286, 282, 288, 142, 249, 107, 71
Hexachlorocyclopentadiene	237	239, 235, 95, 130, 60, 272, 119
Isodrin	193	195, 263, 66, 261, 197, 265, 147, 91
Kepone	272	274, 270, 237, 276, 239, 143, 355
Methoxychlor	227	229, 152, 169, 212, 115, 274, 344
Mirex	272	274, 270, 237, 239, 235, 276, 143, 119
Nitrofen	283	285, 202, 50, 63, 75, 139, 162
Permathrin	183	163, 165, 77, 91, 184, 127, 51
Perthane	223	224, 167, 179, 236, 193, 306
Propachlor	120	77, 176, 93, 43, 51, 169, 196, 211
Toxaphene	197	125, 159, 75, 99, 195, 100, 83, 233, 343
Trifluralin	306	264, 43, 335, 290, 248, 206, 151

Carrier gas: 5% methane/95% argon

Flow rate: 60 mL/min

Temperature:
Oven 200°C (isothermal)
Injector 250°C, ECD 320°C

Many target analytes may co-elute even on a 30-m capillary column. For example, perthane and endrin or methoxychlor and dicofol co-elute on a DB-5 column. Different co-eluting pesticides that may co-elute similarly on other columns are also known. Dual-column analysis using two different columns is therefore preferred over single-column analysis. Also, narrow-bore capillary columns of film thickness 0.32 μm or less give better resolution of pesticides than a wide-bore column having a film thickness of 0.53 μm. The latter, however, has a greater loading capacity and is therefore recommended for dirtier environmental samples.

20.3.2 Mass Spectrometric Analysis

Although the mass-selective detector (MSD) is not as sensitive in detecting pesticides at the level that ECD does, such compounds may be identified by mass spectrometry at relatively higher concentration levels. The solvent extract of the sample must be appropriately concentrated. The pesticides may be quantitated from the area response of their primary mass ions. Table 20.3 gives the primary and secondary mass ions of some selected organochlorine pesticides. For many compounds, seven to eight secondary ions are presented in the table, to help distinguish between isomers or different compounds that are structurally similar. The secondary ions shown in Table 20.3 are usually listed in order of decreasing abundance. For some compounds the molecular ions, even with low abundance, are included in the table to aid in identifying a compound. The characteristic ions for these and other pesticides in this section are obtained under electron impact ionization mode.

20.4 ORGANOPHOSPHORUS PESTICIDES

20.4.1 Structural Features

Organophosphorus pesticides are esters of phosphoric acid with structures containing the $(RO)_2P(=O$ or $S)—O($or $S)$ unit, where R is an alkyl or aryl group. Some common pesticides of this class and their alternative names are listed in Table 20.4.

20.4.2 Extraction and Analysis

Aqueous samples are extracted with methylene chloride at neutral pH by liquid–liquid extraction using a separatory funnel or a continuous liquid–liquid extractor. Oils, fats, and nonaqueous liquid wastes may be appropriately diluted in a suitable solvent and analyzed. Solid environmental samples, such as soils, sediments, hazardous wastes, or food and agricultural products, may be extracted by Soxhlet extraction using 1:1 methylene chloride/acetone or other solvent mixtures. Such solvent mixtures should be a combination of miscible nonpolar and polar organic solvents that can extract the analytes that are of nonpolar and polar-type phosphate esters from solid matrices. Fruits and vegetables may be extracted by solid-phase extraction (SPE). A detailed procedure is given in Sec. 20.5. Organic phosphates in biological fluids may be extracted into an appropriate solvent by liquid–liquid extraction or waste dilution methods, depending on to what degree the sample is miscible in the solvent. If the sample is expected to contain elemental sulfur, the latter may be removed by treating the sample

TABLE 20.4 Common Organophosphorus Pesticides and Their Formulas and Synonyms

Pesticide	Molecular formula	Alternative Names
Abate	$C_{16}H_{20}O_6P_2S_3$	Difenphos, Temefos, Bithion
Acephate	$C_4H_{10}NO_3PS$	Acetamidophos, Orthene
Akton*	$C_{12}H_{14}Cl_3O_3PS$	Axiom
Aspon	$C_{12}H_{28}O_5P_2S_2$	ASP 51, Propyl thiopyrophosphate
Azinphos methyl	$C_{10}H_{12}N_3O_3PS_2$	Gusathion, Guthion, Carfene
Azinphos ethyl	$C_{12}H_{16}N_3O_3PS_2$	Gusathion ethyl, Ethyl guthion, Bionex
Bolstar sulfone	$C_{12}H_{19}O_4PS_3$	—
Bomyl	C H O P	GC 3707
Carbofenthion	$C_{11}H_{16}ClO_2PS_3$	Trithion, Hexathion, Acarithion
Chlorfenvinphos	$C_{12}H_{14}Cl_3O_4P$	Birlane, Enolofos, Sapecron
Chlorofos	$C_4H_8Cl_3O_4P$	Metrifonate, Anthon, Trichlorfon
Coumaphos	$C_{14}H_{16}ClO_5PS$	Muscatox, Asuntol, Baymix
Crotoxyphos	$C_{14}H_{19}O_6P$	Ciodrin, Decrotox
Demeton-O	$C_8H_{19}O_3PS_2$	Mercaptophos, Thiodemcton
Demeton-S	$C_8H_{19}O_3PS_2$	Isosystox, Thioldemeton
Diazinon	$C_{12}H_{21}N_2O_3PS$	Basudin, Dimpylate, Neocidol
Dicapthon	$C_8H_9ClNO_5PS$	Isochlorthion, Chlorthion, Dicaptan
Dichlofenthion	$C_{10}H_{13}Cl_2O_3PS$	Nemacide, Mobilawn, Bromex
Dichlorvos	$C_4H_7Cl_2O_4P$	Chlorvinphos, Cyanophos, Atgard
Dicrotophos	$C_8H_{16}NO_5P$	Carbicron, Ektafos, Bidrin
Dimethoate	$C_5H_{12}NO_3PS_2$	Fosfotox, Cygon, Phosphamid
Dioxathion	$C_{12}H_{26}O_6P_2S_4$	Navadel, Ruphos, Delnatex
Disulfoton	$C_8H_{19}O_2PS_3$	Dithiodemeton, Glebofos, Dithiosystox
Dursban	$C_9H_{11}Cl_3NO_3PS$	Chlorpyrifos, Lorsban, Eradex
EPN	$C_{14}H_{14}NO_4PS$	Santox, EPN 300
Ethion	$C_9H_{22}O_4P_2S_4$	Rodocide, Phosphotox E, Ethanox
Ethephon	$C_2H_6ClO_3P$	Camposan, Ethrel
Famophos	$C_{10}H_{16}NO_5PS_2$	Famphur, Warbex, Cyflee
Fenamiphos	$C_{13}H_{22}NO_3PS$	Nemacur, Phenamiphos
Fensulfothion	$C_{11}H_{17}O_4PS_2$	Terracur P, Desanit, Hexazir
Fenthion	$C_{10}H_{15}O_3PS_2$	Baycid, Mercaptophos, Baytex
Folithion	$C_9H_{12}NO_5PS$	Fenitrothion, Metathion, Nitrophos
Fonofos	$C_{10}H_{15}OPS_2$	Dyphonate, Difonate
Isazophos	$C_9H_{17}ClN_3O_3PS$	Miral
Isofenphos	$C_{15}H_{24}NO_4PS$	Oftanol, Amaze
Leptophos	$C_{13}H_{10}BrCl_2O_2PS$	Phosvel, Abar, MBCP
Malathion	$C_{10}H_{19}O_6PS_2$	Fosfothion, Carbofos, Malafos, Cythion
Merphos	$C_{12}H_{27}PS_3$	Folex, Tributyl trithiophosphite
Methamidophos	$C_2H_8NO_2PS$	ENT, Tamaron, Monitor
Methidathion	$C_6H_{11}N_2O_4PS_3$	Supracide, Medathion, Ultracide
Mevinphos	$C_7H_{13}O_6P$	Phosdrin, Duraphos, Mevinex
Monocrotophos	$C_7H_{14}NO_5P$	Azodrin, Bilobran
Montrel	$C_{12}H_{19}ClNO_3P$	Crufomate, Amidophos, Ruelene
Naled	$C_4H_7Br_2Cl_2O_4P$	Dibrom, Bromex, Dibromfos
Oxydemeton-methyl	$C_6H_{15}O_4PS_2$	Metasystox R, Metaisosystox sulfoxide
Oxydemetonmethyl sulfone	$C_6H_{15}O_5PS_2$	Metasystox R sulfone
Parathion-ethyl	$C_{10}H_{14}NO_5PS$	Parathion, Thiophos, Foliodol
Parathion-methyl	$C_8H_{10}NO_5PS$	Metaphos, Azofos, Metron, Nitrox
Paraoxon	$C_{10}H_{14}NO_6P$	Phosphacol, Diethyl paraoxon
Phenamiphos	$C_{13}H_{22}NO_3PS$	Nemacur
Phorate	$C_7H_{17}O_2PS_3$	Thimate, Aastar, Granutox
Phosalone	$C_{12}H_{15}ClNO_4PS$	Rubitox, Zolone, Benzophosphate
Phosfolan	$C_7H_{14}NO_3PS_2$	Cyolane, Cylan

(Continued)

TABLE 20.4 Common Organophosphorus Pesticides and Their Formulas and Synonyms (*Continued*)

Pesticide	Molecular formula	Alternate Names
Phosphamidon	$C_{10}H_{19}ClNO_5P$	Famfos, Dimecron, Apamidon
Phosmet	$C_{11}H_{12}NO_4PS$	Imidan, Decemthion, Safidon
Profenofos	$C_{11}H_{15}BrClO_3PS$	Polycron, Selecron
Prophos	$C_8H_{19}O_2PS_2$	Ethoprophos, Mocap, Ethoprop
Sulfotepp	$C_8H_{20}O_5P_2S_2$	Dithiophos, Thiotepp, Pirofos
TEPP	$C_8H_{20}O_7P_2$	Fosvex, Nifos, Hexamite
Terbufos	$C_9H_{21}O_2PS_3$	Counter, ST-100
Tetrachlorvinphos	$C_{10}H_9Cl_4O_4P$	Stirofos, Rabon, Gardona
Tokuthion	$C_{11}H_{15}Cl_2O_2PS_2$	Protothiophos
Trichloronate	$C_{10}H_{12}Cl_3O_2PS$	Agritox, Phytosol, Phenophosphon
Zinophos	$C_8H_{13}N_2O_3PS$	Cyanophos, Menafos, Thionazine

extract with tetrabutylammonim sulfite solution. Many interfering substances may be removed from sample extracts by Florisil column cleanup procedure. However, such cleanup steps may also remove certain organic phosphates and therefore should not be applied for analyzing such substances. The extracts may be analyzed directly, without any cleanup, if a flame photometric detector (FPD) or a mass spectrometer is used.

The organic phosphates are best measured by a gas chromatograph with a flame photometric or a nitrogen–phosphorus detector (NPD). Both these detectors are operated in phosphorus mode to minimize interference from materials that do not contain phosphorus or sulfur, especially when using an FPD, or that do not contain phosphorus when using an NPD.

The compounds may be separated on a 30-m-long capillary column with 0.32-mm ID. Shorter columns with 0.25-mm or 0.53-mm ID may also be used, depending on the resolution desired. Several such columns are commercially available that contain various chemically bonded polysiloxane mixtures in varying proportions. They include phenyl methyl polysiloxane, trifluoropropyl polysiloxane, methyl polysiloxane, and phenyl polysiloxane under different trade names, such as DB-1, DB-5, DB-608, DB-210, SPB-1, SPB-5, SPB-608, RTx-1, RTx-5, RTx-35, and RTx-1701. Although a 15-m × 0.53-mm fused-silica capillary column gives adequate resolution of most organic phosphates, a few compounds, such as ethyl azinphos, crotoxyphos, dicrotophos, famphur, fonofos, leptophos, terbufos, and zinofohs, are not well separated from compounds that have closely related structures in phosphate mixtures using such columns. A 30-m or longer column provides better resolution of these and certain other phosphates.

Once a compound is detected from its retention time, its presence is confirmed on an alternate column or using another detector. The most common approach is to use two fused-silica open-tubular columns of different polarities connected to an injection tee, with each connected to a detector. On the other hand, if two detectros are to be used, FPD and NPD are the ideal choice. FPD is more sensitive and selective to phosphorus- or sulfur-containing compounds. Both the FPD and NPD are flame ionization-type detectors. The performance of the FPD may be optimized by selecting the proper optical filter and by adjusting the flows of air and hydrogen to the flame. Although FPD is more sensitive than NPD, an advantage of the latter is that it can also measure triazine herbicides and other nitrogen-containing substances in sample extracts. As mentioned earlier, presence of elemental sulfur can interfere with FPD analysis and therefore must be removed by treating the sample extract with tetrabutylammonium sulfite solution.

Many organophosphorus pesticides, such as dichlorvos, trichloronate, ronnel, coumaphos, chlorpyrifos, stirophos, and naled, contain halogen atoms in their molecules. Such compounds, marked with asterisks in Table 20.4, may also be determined by a halogen-specific detector, such as electrolytic conductivity or microcoulometric detector. Also, many organic phosphates may be determined by ECD, although the latter may not be as specific as FPD or NPD.

Among other analytical techniques, HPLC and mass spectrometry are as common as GC-FPD and GC-NPD. HPLC measurement usually requires a mobile phase of acetonitrile:water solvent mixture (e.g., 60:40 mixture) at a flow rate such as 1 mL/min at ambient temperature using a UV detector at 214 nm. Pesticides in fruits and vegetables are commonly extracted by SPE or SPME and separated on columns such as Discovery RP-AmideC_{16} and analyzed by HPLC-UV techniques.

Although the MSD is the least sensitive of all detectors, unknown compounds can be identified from their mass spectra. The characteristic mass ions (both the primary and the secondary mass ions) of some selected organic phosphates are given in Table 20.5.

20.5 CARBAMATE PESTICIDES

20.5.1 Extraction and Analysis: General Outline

Carbamate pesticides/herbicides are nitrogen-containing substances having the following structural feature:

$$\begin{matrix} R_1 \\ \quad\diagdown \\ \quad\diagup \\ R_2 \end{matrix} N-\overset{\overset{\displaystyle O}{\|}}{C}-O-\text{(leaving group)}$$

where R_1 and R_2 are alkyl or aryl groups or hydrogen. These compounds can be detected by GC-NPD in nitrogen mode, GC-FID, GC/MS, LC/MS, and HPLC. Carbamates are extracted from sample matrices by various extraction procedures discussed earlier. Most common methods of extraction involve the SPE and SPME techniques, which may be applied to all aqueous samples including wastewaters, drinking waters, biological fluids, and water-soluble plant extracts.

Among instrumental techniques, HPLC using postcolumn derivatization and UV detection are the most common methods. Numerous methods are reported in the literature, a few of which are outlined below.

20.5.2 HPLC Postcolumn Derivatization and Fluorometric Detection

Carbamates in sample extracts are separated on a C-18 reverse-phase HPLC column (e.g., 25-cm × 4.6-mm stainless steal packed with 5-μm Beckman Ultrasphere, 15-cm × 2.9-mm stainless steel packed with 4-μm Nova Pac C-18, 25-cm × 4.6-mm stainless steel packed with 5-μm Supelco LC-1, or equivalent). The separated carbamates are then hydrolyzed with 0.05*N* NaOH to convert them to their methyl esters. The esters are then reacted with *o*-phthalaldehyde and 2-mercaptoethanol at 40°C (at a flow rate of 0.7 mL/min and a residence time of 35 s) to give highly fluorescent derivatives which are detected by a fluorescent detector. The excitation wavelength for flourometric detection is 330 nm (cutoff filter). The solvents A and B are reagent-grade water acidified with phosphoric acid and 1:1 methanol/acetonitrile mixture, respectively, and the flow rate is 1 mL/min.

The derivatization reagent *o*-phthalaldehyde is prepared by adding a 10-mL aliquot of 1% *o*-phthalaldehyde in methanol to 10 mL of acetonitrile containing 100 mL of 2-mercaptoethanol. The mixture is then diluted to 1 L with 0.05*N* sodium borate solution.

Carbamates can also be determined by HPLC without derivatization. A number of LC columns are commercially available. They include Discovery C18, 15 m × 4.6-mm ID, 5-μm particles (Supelco, Inc.), Luna 5μ CN 150 × 4.6 mm, or Luna 3μ C18, and such columns provide excellent separation of carbamate, triazine, urea, and other types of pesticides and herbicides using a mobile phase of acetonitrile/water under varying gradient programs. The compounds may be detected by a UV detector at 214 nm.

20.5.3 GC/MS Analysis

Although UV detection is commonly employed for analyzing carbamates, other detectors are equally versatile. Mass spectrometry is the single most confirmatory test. Either a GC or an LC is interfaced

TABLE 20.5 Characteristic Mass Ions of Some Organophosphorus Pesticides

Compound	Primary ion	Secondary mass ions
Azinphos-ethyl	132	77, 160, 105, 104, 65, 29, 76
Azinphos-methyl	160	77, 132, 104, 105, 93, 51, 50
Bolstar sulfone	188	43, 113, 312, 63, 141, 354
Carbofenthion	157	45, 97, 121, 159, 199, 342
Chlorfenvinphos	267	323, 269, 325, 81, 109, 295, 170
Chlorofos	109	185, 79, 47, 145, 187, 220
Coumaphos	362	109, 226, 97, 210, 364, 334, 228
Crotoxyphos	127	105, 193, 166, 104, 77, 79, 67
Demeton-O	88	60, 29, 115, 171, 143, 45, 258
Demeton-S	88	89, 60, 171, 97, 115, 28, 29, 258
Diazinon	179	137, 152, 199, 304, 29, 135, 276
Dicapthon	262	125, 79, 47, 109, 93, 63, 216
Dichlofenthion	279	97, 223, 251, 88, 162, 281, 109
Dichlorvos	109	185, 79, 47, 145, 187, 220
Dicrotophos	127	67, 72, 193, 109, 44, 237
Dimethoate	87	93, 125, 79, 58, 47, 229, 172
Dioxathion	97	125, 270, 73, 45, 65, 197, 153
Disulfoton	88	29, 89, 60, 97, 142, 125, 153, 274
Dursban	97	197, 199, 29, 314, 316, 258, 125
EPN	157	169, 63, 141, 185, 77, 29, 323
Ethion	231	153, 97, 125, 121, 65, 384, 93
Ethephon	82	81, 109, 27, 28, 65, 47, 91, 145
Fampohos	218	93, 125, 28, 44, 109, 63, 282
Fenamiphos	303	154, 217, 44, 288, 260, 80, 122
Fensulfothion	293	308, 141, 97, 125, 153, 109, 265
Fenthion	278	125, 109, 153, 169, 93, 79, 279, 280
Fonofos	109	137, 246, 110, 81, 28, 63, 174
Leptophos	171	377, 28, 375, 77, 155, 60, 379
Malathion	173	127, 93, 158, 99, 29, 143, 79, 256
Merphos	57	209, 41, 298, 153, 88, 56, 55, 242
Methamidophos	15	94, 47, 95, 141, 45, 64, 30
Methidathion	85	145, 93, 58, 47, 125, 63, 302
Mevinphos	127	192, 109, 67, 43, 164, 193, 79
Monocrotophos	127	67, 97, 58, 192, 109, 43, 193
Montrel	256	108, 276, 182, 169, 278, 41, 291
Naled	15	109, 145, 79, 47, 18, 29, 31
Oxydemetonmethyl sulfone	169	109, 125, 168, 79, 110, 142
Parathion-ethyl	97	291, 109, 137, 139, 125, 29, 186, 65
Parathion-methyl	109	125, 263, 79, 42, 61, 92, 28
Paraoxon	109	29, 149, 81, 275, 139, 65, 99, 75
Phenamiphos	303	154, 288, 44, 80, 260, 43, 304
Phorate	75	121, 260, 97, 28, 47, 65, 93
Phosalone	182	121, 97, 184, 154, 65, 111, 367
Phosfolan	92	140, 196, 60, 168, 81, 29, 255, 227
Phosphamidon	127	72, 264, 138, 109, 28, 67, 193
Phosmet	160	61, 76, 77, 133, 104, 161, 50
Profenofos	139	97, 208, 206, 125, 339, 337, 63
Prophos	158	43, 97, 139, 126, 93, 41, 200, 242
Sulfotepp	322	202, 97, 266, 65, 174, 121, 238, 294
TEPP	99	155, 43, 127, 81, 109, 82, 111
Terbufos	57	231, 97, 29, 153, 41, 103, 233
Tetrachlorvinphos	329	109, 331, 333, 79, 240, 93
Tokuthion	43	27, 29, 113, 267, 63, 309, 162, 155
Trichloronate	109	297, 269, 299, 28, 93, 271, 137
Zinophos	107	96, 106, 97, 143, 248, 140, 79, 68

TABLE 20.6 **Characteristic Mass Ions of Some Common Carbamates**

Carbamates	Synonyms	Primary ion	Secondary ions
Aldicarb	Temik	41	58, 86, 76, 144
Aldicarb sulfone	Aldoxycarb, Standak	86	143, 85, 41
Aldicarb sulfoxide	—	86	41, 28, 58, 143
Banol	Carbanolate	156	121, 91, 158, 141, 65
Barban	Barbamate, Carbyne	222	51, 87, 143, 153, 224, 257
Bendiocarb	Ficam	151	126, 166, 51, 223
Benomyl	Benlate	191	159, 40, 105, 42, 132
Carbaryl	Sevin, Arylam	144	115, 116, 28, 89, 201
Carbofuran	Furadan	164	149, 122, 57, 221
Chlorpropham	Chloro IPC	43	127, 213, 171, 129, 154, 41
Matacil	Aminocarb	151	136, 77, 120, 77, 208
Methiocarb	Mesurol	168	153, 225, 109
Methomyl	Lannate	58	105, 32, 42, 88, 91, 45
Oxamyl	—	72	44, 32, 30, 162, 58, 115, 88, 145
Pirimicarb	Pirimor	166	72, 238, 138
Promecarb	Carbamult	135	150, 91, 58, 41, 39
Propham	IPC	93	43, 179, 137, 120, 65
Propoxur	Baygon, Aprocarb	110	152, 81, 27, 43, 39
Swep	—	219	221, 174, 59, 89, 187, 176
Terbucarb	Terbutol	205	57, 220, 58, 206, 41, 105

to a mass spectrometer. The pesticides are identified from their characteristic mass ions and retention times. The characteristic mass ions of some selected carbamates under electron-impact ionization are listed in Table 20.6.

20.6 UREA-TYPE HERBICIDES

20.6.1 Methods of Analysis

Urea-type herbicides, like carbamates, are all nitrogen-containing organic compounds. The structural features are also similar to those of carbamates, except that the terminal oxygen atom of carbamate is replaced by a nitrogen atom in urea-type substances. Thus, the common structural feature in all urea herbicides is

$$(R_1)(R_2)N{-}C({=}O){-}N\langle \text{ (leaving group)}$$

Trace residues of these herbicides in environmental samples, biological fluids, and food products may be determined by HPLC using a UV detector, GC using a NPD in nitrogen mode, or by mass spectrometry. Among these methods, HPLC-UV and LC/MS are the most common techniques. Urea-type herbicides are polar compounds. They may be effectively extracted by solid-phase extraction. Traditional LLE is also applicable for extracting aqueous samples. SPE tubes containing ENVI-Carb carbon-based packing or charcoal/celite packing have been reported to provide superior and more uniform recovery of such polar analytes over the octyl (C8)- or octadecyl (C18)-bonded silica packing. Alternatively, these herbicides may be extracted by solid-phase microextraction using a SPME fiber coated with 100-μm film of polydimethylsiloxane.

TABLE 20.7 Characteristic Mass Ions of Some Common Urea-Type Herbicides

Name	Synonyms	Primary ion	Secondary ions
Chlorbromuron	Maloran	61	46, 294, 206, 124
Chloroxuron	Tenoran	72	245, 290, 44, 182
Diuron	Karmex, Herbatox	72	232, 234, 44, 73, 187, 124
Fenuron	Dybar, Beet-kleen	72	164, 44, 119, 77, 65, 91
Fluometuron	Cotoran, Lanex	72	232, 44, 42, 28, 145, 187
Linuron	Sinuron, Cephalon	61	46, 248, 250, 160, 62, 162
Metobromuron	Patoran	61	46, 91, 258, 260, 172, 32
Monuron	Telvar	72	198, 38, 153, 200, 61
Monuron TCA	Urox	72	82, 81, 198, 127, 98, 153, 270
Neburon	Kloben	57	44, 114, 41, 58, 125, 274
Siduron	Tupersan	93	55, 56, 41, 39, 77, 232
Tebuthiuron	Tebulan, Perflan	156	171, 74, 41, 88, 57, 157

Two typical extraction procedures are given below as examples for fruits and vegetables and water samples, respectively. Other extraction methods referred to in this handbook are equally efficient.

For fruits and vegetables, 50 g of chopped sample are mixed with 100 mL of acetonitrile and homogenized in a mixer for 5 min. To this solution, 10 g of sodium chloride are added and the resulting solution is homogenized for another 5 min. About 13 mL of acetonitrile solution is transferred from the top layer to a 15-mL graduated centrifuge tube. To the latter is then added about 3 g of anhydrous sodium sulfate to bring it up to the 15-mL mark. The solution is then shaken to remove water from the acetonitrile. It is then centrifuged for 5 min. A 10-mL aliquot of this solution is transferred to a clean 15-mL tube. The solution is evaporated on a water bath at 35°C under nitrogen flow to a final volume of 0.5 mL. The latter is then transferred onto an SPE tube containing 500 mg (6-mL tube) ENVI-Carb packing. The urea and carbamate herbicides are then eluted with 20-mL of 3 : 1 acetonitriletoluene mixture. The extract is concentrated down to a volume of 2 mL in a rotary evaporator and then solvent-exchanged to acetone by adding 10 mL acetone successively. The solution is then analyzed by a suitable technique.

As mentioned earlier, HPLC-UV and LC/MS are commonly employed in analyzing urea herbicides in sample extracts. Compound can be separated on many types of LC columns, and several such columns are commercially available. They include C-18 Novapk (Waters Corp.), Pinnacle II C8 (Resteck), Discovery C18 (Supelco, Inc.), Luna 3μ Phenyl-Hexyl, and many other equivalent columns. The mobile phase usually consists of acetonitrile/water under varying gradient conditions. The analysis may be carried out at ambient temperature using a UV detector at wavelengths ranging between 220 and 240 nm.

Identification of compounds by mass spectrometry is the most confirmative test. Characteristic mass ions of some common urea-type herbicides are presented in Table 20.7.

20.7 TRIAZINE HERBICIDES

Triazine herbicides are compounds containing a triazine ring as follows:

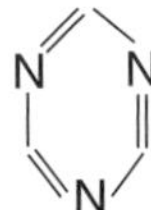

These heterocyclic nitrogen compounds usually exhibit low toxicity. They may be determined by GC-NPD in nitrogen mode, by HPLC, or by GC/MS. Although the latter is not as sensitive as the

TABLE 20.8 Names, Synonyms, and Charcteristic Mass Ions of Triazine Herbicides

Compound	Synonyms(s)	Primary ion	Secondary ions
Ametryne	Ametrex, Gesapax	227	212, 58, 44, 170, 98, 185, 71
Anilazine	Dyrene, Triasyn	239	241, 178, 143, 62, 75, 274, 276
Atraton	Primatol, Gesatamin	196	58, 211, 169, 44, 69, 91, 154, 141
Atrazine	Gesaprim, Zeazine	58	43, 44, 200, 68, 71, 69, 215, 173
Cyanazine	Bladex, Cyanazone	68	44, 225, 43, 173, 198, 240, 96
Dipropetryne	Sancap, Cotofer	43	58, 68, 255, 85, 113, 152, 184, 240
Metribuzine	Sencor, Lexone	198	41, 57, 74, 61, 144, 103, 199
Procyazine	Cycle	41	28, 39, 68, 81, 237, 210, 170
Prometon	Primatol, Gesafram	210	168, 225, 58, 43, 183, 141, 69, 98
Prometryne	Gesagard, Caparol	241	58, 184, 226, 43, 68, 199, 69
Propazine	Gesamil, Milogard	58	214, 229, 43, 172, 187, 68, 69, 104, 152
Simazine	Tafazine, Herbazin	201	44, 186, 173, 68, 28, 96, 138
Simetryne	Gybon	213	170, 155, 68, 198, 43, 96
Terbuthylazine	Gardoprim	43	214, 173, 229, 58, 216, 68, 100
Terbutryne	Igran, Perbane	226	185, 241, 170, 43, 83, 106

GC and HPLC methods, it is the most confirmatory technique. Certain triazines may also be determined by colorimetric methods. The common trade names, synonyms, and the characeristic mass ions of some triazine herbicides for identification by mass spectrometry are listed in Table 20.8.

20.8 CHLOROPHENOXY ACID HERBICIDES

20.8.1 General Discussion

Chlorophenoxy acid herbicides are used in various forms, such as acids, salts, or esters. The acid forms of these herbicides are strong organic acids that can readily react with basic compounds and may be lost during analysis. Because of this, becausc of their occurrence in various forms, and because of potential interference from other chlorinated organic acids, phenolic compounds, and phthalate esters, the most crucial steps in the analysis of chlorophenoxy acid herbicides are sample extraction, cleanup of the solvent extracts, and derivatization of the chlorophenoxy acids. Complete analysis of these herbicides in all forms (as acids, salts, and esters) involves several tedious steps, including hydrolysis and esterification and repeated extractions with diethyl ether in between. The extraction and analytical steps are briefly outlined below.

20.8.2 Extraction, Derivatization, and Analysis

The aqueous samples are acidified with HCl to a pH below 2 to convert any salt or ester forms of herbicides present in the samples into their acids. The acidified sample is then extracted with diethyl ether (peroxide free and stabilized with butylated hydroxytoluene) in a separatory funnel. Chlorophenoxy acids, being soluble in ether, partition into ether. The aqueous phase, containing water-soluble organic interfering substances, is discarded. The ether extract containing the herbicides in acid form is heated carefully with an aqueous solution of potassium hydroxide. This converts the herbicide acids into their water-soluble potassium salts.

The product mixture again is repeatedly extracted with ether. While most organic contaminants dissolve in ether, the potassium salts of herbicide acids pass into the aqueous phase. The ether extract, which may contain many types of organic contaminants that are soluble in ether, is discarded. The aqueous solution is then acidified to a pH below 2 to convert the potassium salts of the herbicides back

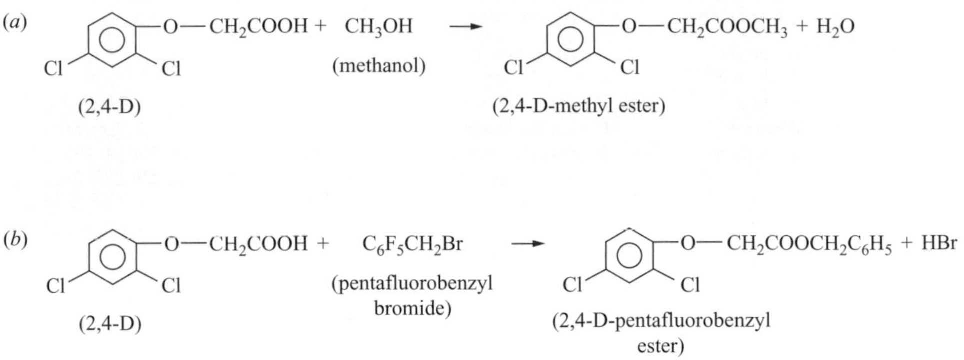

(a) Cl, Cl-substituted ring —O—CH_2COOH + CH_3OH (methanol) → Cl, Cl-substituted ring —O—CH_2COOCH_3 + H_2O

(2,4-D) (2,4-D-methyl ester)

(b) Cl, Cl-substituted ring —O—CH_2COOH + $C_6F_5CH_2Br$ (pentafluorobenzyl bromide) → Cl, Cl-substituted ring —O—$CH_2COOCH_2C_6H_5$ + HBr

(2,4-D) (2,4-D-pentafluorobenzyl ester)

FIGURE 20.1 Esterification of 2,4-D: (*a*) methylation; (*b*) pentafluorobenzylation.

again into their acids. The acidified solution is then repeatedly extracted with diethyl ether. The acid herbicides partition into ether. The aqueous phase is now discarded. The ether extract is concentrated down to a small volume by carefully evaporating the solvent.

After removing the interfering substances in the above steps and concentrating the ether extract of the acids to a small volume, the chlorophenoxy acids are converted into their methyl esters. Esterification is carried out either by treatment with boron trifluoride in methanol solution or with diazomethane. In the former, BF_3 serves as a catalyst for methylation. The esterification is carried out by heating the mixture for a few minutes in a water bath. The excess methanol and BF_3, both water-soluble, are removed from the organic solvent by treatment with water. The methyl esters of chlorophenoxy acids are solvent-exchanged to toluene or hexane. An aliquot of the organic solvent from the top immiscible layer is then injected onto a GC equipped with an ECD for analysis of the methyl ester derivatives.

If diazomethane is used for esterification it may be generated in situ from Diazald (*N*-methyl-*N*-nitroso-*p*-toluenesulfonamide) or other equivalent kits that are commercially available. Alternatively, carbitol (diethylene glycol monoethyl ether) may be employed to produce alcohol-free diazomethane. The sample extracts should be dry before methylation, or the recoveries will be poor. Also, the diethyl ether used for extraction should be stabilized with BHT (butylated hydroxytoluene) and not ethanol. The latter can form ethyl esters of herbicides, lowering the yield of methyl esters.

Alternatively, the chlorophenoxy herbicides in their acid forms may be esterified to their pentafluorobenzyl derivatives instead of methyl esters. Derivatization can be carried out with 2,3,4,5,6-pentafluorobenzyl bromide. The reactions are shown in Fig. 20.1 for methylation and perfluorobenzylation with 2,4-D as an example.

The ester derivatives are determined by GC-ECD or GC/MS. The latter is a confirmatory test, although it is much less sensitive than the GC-ECD. The characteristic mass ions of chlorophenoxy acids and their esters for their identification by GC/MS under electron-impact ionization is listed in Table 20.9.

The methyl esters also can be determined by GC-FID. Fused-silica capillary columns containing a stationary phase made of phenyl silicone, methyl silicone, and cyanopropyl phenyl silicone in varying compositions provide excellent resolution. Such columns, commercially available, include, DB-5, DB-608, DB-1701, SPB-5, SPB-608, SPB-608, SPB-1701, HP-608, BP-608, AT-1701, Rtx-5, and equivalent. The herbicides are determined as acids, stoichiometrically calculated from the concentration of their esters as follows:

$$\text{Concentration of acid} = \frac{\text{concentration of ester} \times \text{molecular wt. of acid}}{\text{molecular wt. of ester}}$$

TABLE 20.9 Primary and Secondary Mass Ions of Herbicide Acids and Esters

Chlorophenoxy acids and esters	Primary ion	Secondary ions
2,4-D	162	164, 220, 222, 161, 133, 63, 175
2,4-D methyl ester	199	175, 234, 111, 147, 133, 161, 201
2,4-D butoxyethyl ester	57	29, 56, 85, 220, 145, 320
2,4-DB	57	29, 41, 43, 185, 111, 220, 276
2,4-DB butyl ester	87	41, 43, 29, 143, 69, 57, 162, 231
Dicamba	173	175, 220, 222, 174, 203, 191, 97
Dicamba methyl ester	203	205, 188, 97, 234, 190, 160
Dichlorprop	28	162, 164, 234, 45, 98, 189
MCPA	141	200, 77, 125, 143, 51, 155, 202
MCPA methyl ester	141	214, 77, 155, 125, 89, 143, 45
MCPA isooctyl ester	57	43, 41, 29, 200, 125, 155, 312
MCPB	142	107, 43, 45, 87, 144, 77, 51
MCPP	142	214, 77, 107, 45, 141, 169, 216
MCPP methyl ester	169	142, 107, 77, 228, 141, 59, 89
2,4,5-T	196	198, 254, 256, 200, 209, 167, 97
2,4,5-T butyl ester	57	41, 40, 43, 109, 196, 312
2,4,5-T isopropyl ester	209	211, 196, 198, 74, 109, 145, 296, 298
2,4,5-T methyl ester	233	235, 45, 268, 270, 73, 209, 145
2,4,5-TB	87	196, 198, 45, 43, 40, 284, 282
2,4,5-TP (Silvex)	196	198, 97, 270, 268, 169, 167, 45
Silvex methyl ester	196	198, 59, 225, 223, 55, 87, 200, 284
Silvex isooctyl ester	57	43, 71, 55, 41, 196, 198, 69, 223

Source: NIST Mass Spectral Library.

Herbicides in their acid form may also be determined by HPLC techniques. Aqueous samples are extracted by SPE methods using ENVI-Carb or equivalent sorbent. Polymeric-coated silica-based HPLC specialty columns are commercially available. A photodiode array detector is interfaced to HPLC for determination of such compounds. No esterification or derivatization is required for analysis when the compounds are present in the samples in their acid form. To determine herbicides in all forms, the sample may be acidified, upon which the esters and salts of chlorophenoxy acids are converted to their acids and then analyzed by HPLC as described above.

Typical SPE/HPLC analytical conditions for measuring herbicide acids in water are given as an example in Table 20.10.

TABLE 20.10 SPE/HPLC Analysis of Herbicides

Extraction:	SPE, ENVI-Carb, 6 mL, 250 mg	
Column:	Polymeric-coated silica-based PAH specialty column, 20 cm × 3-μm ID, 5-μm particles	
Mobile phase:	Gradient, A = water/0.05%H_3PO_4, B = acetonitrile	
Flow rate:	0.5 mL/min	
Gradient program:	Time (min)	% B
	2.5	40
	5.0	60
	13.0	60
	13.5	40
Temperature:	50°C	
Detector:	Photodiode array at 210 and 225 nm; peak width 0.320 s (0.053-min sampling interval)	

Source: *Supelco Catalog for Chromatography Products for Analysis and Purification,* 2003.

Bibliography

American Public Health Association, American Water Works Association and Water Environment Federation, *Standard Methods for the Examination of Water and Wastewater,* 20th ed., American Public Health Association, Washington, DC, 1998.

Fong, W. G., H. N., Moye, J. N., Seiber, and J. P. Toth, *Pesticide Residues in Foods*, Wiley, New York, 1999.

Milne, G. W. A., ed., *CRC Handbook of Pesticides*, CRC Press, Boca Raton, FL, 1995.

Patnaik, P., *Handbook of Environmental Analysis*, CRC Press, Boca Raton, FL, 1997.

U.S. Environmental Protection Agency, *Test Methods for Evaluating Solid Waste, SW-846*, National Technical Information Service, Washington, DC, 1997.

U.S. Food and Drug Administration, *Pesticide Analytical Manual*, National Technical Information Service, Washington, DC, 1975.

SECTION 21

ANALYSIS OF TRACE POLLUTANTS IN THE ENVIRONMENT

21.1 INTRODUCTION

The analysis of trace pollutants in environmental matrices, such as wastewaters, groundwaters and surface waters, soils, sediments and solid wastes, and the ambient and atmospheric air, has grown into a major, fully pledged discipline of analytical science in recent years. This may be attributed to growing concern about pollution of the environment and numerous legislations and regulatory requirements that have been adopted to abate this problem. The first step in understanding the problem and subsequently controlling it rests on the identification and accurate measurements of toxic pollutants and their degradation products at trace levels. This involves measuring various organic substances, metal ions, inorganic anions, radionuclides, microorganisms, and certain physicochemical properties of environmental waters and solid wastes by various physical, chemical, and instrumental methods. A full discussion of the subject is beyond the scope of this text.

21.2 SAMPLING

If water from a faucet is to be collected, allow the water to run into the sink for a few minutes before collecting the sample. If water is to be sampled from a river or a lake, collect the samples from an appropriate depth from the middle of the river or the lake or select the locations where the levels of pollutants need to be monitored. If the sample is taken in a single lot for analysis, it is called a *grab* sample. However, it may often be necessary to collect a more representative sample than a grab sample. In such cases, several samples are collected at different but nearby locations or are taken from the same locations at different intervals. Such samples are homogenized and mixed thoroughly. They are termed *composite* samples.

Samples, especially soils, sediments, and solid wastes, are often composited for analysis. For such analysis, any stone, rock, grass, or any foreign material such as a piece of wood that may be found in the soil or the sediment but does not represent the sample matrix, should be removed from the sample before homogenizing.

21.2.1 Sample Containers, Preservatives, and Holding Time

Samples are collected in wide-mouthed glass or plastic bottles with screw caps. Substances in the matrix that are susceptible to photochemical degradation must be collected in amber glass bottles or vials. Usually, glass containers are suitable for all types of sampling; in certain cases, however, such as trace analysis of boron metal, borosilicate glass containers must not be used. Plastic containers are suitable for collecting samples, for most inorganic analytes but not for measuring total organic carbon (TOC) or any individual organic analytes.

Many substances are susceptible to decomposition at ambient conditions because of oxidation, volatilization, or microbial degradation. Such degradation can significantly affect the accuracy of measurement at trace concentrations. To prevent such loss of certain types of analytes from the samples, preservatives may be added to the samples or to the sample container bottles before sampling, to retain the integrity of the analytes. For example, samples for metal analysis are acidified to prevent any loss of metals due to possible ion exchange. This acidification enhances the stability of the samples and their holding times. Even with preservatives added, samples tend to degrade over time and therefore must be tested for the respective analytes within their holding times. All samples must be placed in a refrigerator at a temperature of 4°C or below to minimize any volatilization, if tests cannot be performed immediately. The samples must be brought back to room temperature before analysis. The nature of the sample containers, the preservatives to be added, and the holding times of samples for various tests are summarized in Table 21.1.

21.3 SAMPLE EXTRACTION AND DIGESTION

Most analyses of trace pollutants require extracting the pollutants into an appropriate solvent and concentration of the solutions into smaller volumes before analysis. This is essential for practically all soil, sediment, and solid-waste analysis and for many types of water testing, especially those pertaining to organic analytes. Also, reduction of the volume of the solvent extracts by careful evaporation can lower the detection limits of trace pollutants by severalfold. Some common extraction methods are briefly outlined below.

21.3.1 Solid Samples

An accurately weighed quantity of a solid sample, such as soil, sediment, or waste, is mixed with a solvent and subject to sonication or Soxhlett extraction. Alternatively, the sample may be shaken with solvent on a mechanical shaker or stirred using a magnetic stirrer. If the analytes are soluble in water, the latter may be used as the solvent for extraction. For the extraction of organic pollutants, acetone, methanol, methylene chloride, and hexane are some of the common solvents that are used.

21.3.2 Nonaqueous Liquid Samples

Many nonaqueous liquid samples, such as waste oil, may simply be diluted with a solvent in which the liquid is soluble. Dilution is used in determining certain types of organic pollutants such as polychlorinated biphenyls (PCBs). Since fuel oils and other petroleum products are essentially hydrocarbon in nature, a hydrocarbon solvent such as *n*-hexane, isooctane, or toluene may be used to dilute liquid wastes before their analysis.

21.3.3 Aqueous Samples

Trace organic pollutants are extracted from aqueous samples into appropriate organic solvents by several methods. These techniques fall primarily into two categories, liquid–liquid extraction (LLE)

TABLE 21.1 Sample Containers, Preservatives, and Holding Times

Analyte	Container	Preservation	Maximum holding time
		Inorganics and microbial tests	
Acidity	P, G	Cool, 4°C	14 d
Alkalinity	P, G	Cool, 4°C	14 d
Bacterias, coliform (total and fecal)	P, G	Cool, 4°C; add 0.008% $Na_2S_2O_3$ if residual chlorine is present	6 h
Biochemical oxygen demand	P, G	Cool, 4°C	48 h
Bromide	P, G	None required	28 d
Chloride	P, G	None required	28 d
Chlorine, residual	P, G	None required	Analyze immediately
Chemical oxygen demand	P, G	Cool, 4°C, H_2SO_4 to pH < 2	28 d
Color	P, G	Cool, 4°C	48 h
Cyanide	P, G	Cool, 4°C, pH > 12, 0.6 g ascorbic acid	14 d
Fluoride	P	None required	28 d
Hardness	P, G	pH < 2 with HNO_3 or H_2SO_4	6 mo
Iodine	P, G	None required	Analyze immediately
Kjeldahl nitrogen	P, G	Cool, 4°C, pH < 2 with H_2SO_4	28 d
Metals (except chromium-VI, boron and mercury)	P, G	HNO_3 to pH < 2	6 mo
Chromium-VI	P, G	Cool, 4°C	24 h
Mercury	P, G	HNO_3 to pH < 2	28 d
Boron	P	HNO_3 to pH < 2	28 d
Nitrate	P, G	Cool, 4°C, H_2SO_4 to pH < 2	28 d
Nitrite	P, G	Cool, 4°C	48 h
Odor	G	None required	Analyze immediately
Oil and grease	G	Cool, 4°C, H_2SO_4 or HCl to pH < 2	28 d
Oxygen, dissolved	G (BOD bottle)	None required	Analyze immediately
pH	P, G	None required	Analyze immediately
Phenolics	G	Cool 4°C, H_2SO_4 to pH < 2	28 d
Phosphorus			
Elemental	G	Cool, 4°C	48 h
Orthophosphate	P, G	Cool, 4°C	48 h
Total	P, G	Cool, 4°C, H_2SO_4 to pH < 2	28 d
Residue			
Total	P, G	Cool, 4°C	7 d
Filterable	P, G	Cool, 4°C	7 d
Nonfilterable (TSS)	P, G	Cool, 4°C	7 d
Settleable	P, G	Cool, 4°C	48 h
Volatile	P, G	Cool, 4°C	7 d
Silica	P	Cool, 4°C	28 d
Specific conductance	P, G	Cool, 4°C	28 d
Sulfate	P, G	Cool, 4°C	28 d

TABLE 21.1 Sample Containers, Preservatives, and Holding Times (*Continued*)

Analyte	Container	Preservation	Maximum holding time
		Inorganics and microbial tests	
Sulfide	P, G	Cool, 4°C, zinc acetate plus NaOH to pH > 9	7 d
Sulfite	P, G	None required	Analyze immediately
Surfactants	P, G	Cool, 4°C	48 h
Taste	G	Cool, 4°C	24 h
Temperature	P, G	None required	Analyze
Total organic carbon	G	Cool, 4°C, HCl or H_2SO_4 to pH < 2	28 d
Total organic halogen	G (amber bottles)	Cool, 4°C, store in dark, HNO_3 to pH < 2, add Na_2SO_3 if residual chlorine present	14 d
Turbidity	P, G	Cool, 4°C	48 h
		Organics tests	
Purgeable halocarbons	G (Teflon-lined septum)	Cool, 4°C, no headspace (add 0.008% $Na_2S_2O_3$ if residual chlorine is present)	14 d
Purgeable aromatics	G (Teflon-lined septum)	Cool, 4°C, no headspace (add 0.008% $Na_2S_2O_3$ if residual chlorine is present), HCl to pH < 2	14 d
Pesticides, chlorinated	G (Teflon-lined cap)	Cool, 4°C, pH 5–9	7 d until extraction; 40 d after extraction
PCBs	G (Teflon-lined cap)	Cool, 4°C	7 d until extraction; 40 d after extraction
Phthalate esters	G (Teflon-lined cap)	Cool, 4°C	7 d until extraction; 40 d after extraction
Nitroaromatics	G (Teflon-lined cap)	Cool, 4°C (add 0.008% $Na_2S_2O_3$, if residual chlorine present), store in dark	7 d until extraction 40 d after extraction
Nitrosamines	G (Teflon-lined cap)	Cool, 4°C (add 0.008% $Na_2S_2O_3$, if residual chlorine present), store in dark	7 d until extraction; 40 d after extraction
Polynuclear aromatic hydrocarbons	G (Teflon-lined cap)	Cool, 4°C (add 0.008% $Na_2S_2O_3$, if residual chlorine present), store in dark	7 d until extraction; 40 d after extraction
Haloethers	G (Teflon-lined cap)	Cool, 4°C (add 0.008% $Na_2S_2O_3$ if residual chlorine present)	7 d until extraction; 40 d after extraction

(Continued)

TABLE 21.1 Sample Containers, Preservatives, and Holding Times (*Continued*)

Analyte	Container	Preservation	Maximum holding time
		Organics tests	
Phenols	G (Teflon-lined cap)	Cool, 4°C (add 0.008% $Na_2S_2O_3$ if residual chlorine present)	7 d until extraction; 40 d after extraction
Dioxins and dibenzofurans	G (Teflon-lined cap)	Cool, 4°C (add 0.008% $Na_2S_2O_3$ if residual chlorine present)	7 d until extraction; 40 d after extraction

Note: P, polyethylene; G, glass. If there is no residual chlorine in the sample, the addition of $Na_2S_2O_3$ may be omitted.
Source: P. Patnaik, *Handbook of Environmental Analysis*, CRC Press, Boca Raton, Fla., 1997.

and solid-phase extraction (SPE). Extraction units are commercially available for continuous liquid–liquid extraction. In the LLE method, extraction may be carried out manually using a separatory funnel. In this procedure, an aqueous sample, 1 L in volume, is put into a 2-L separatory funnel. The sample is then repeatedly shaken with three 50–60-mL aliquots of methylene chloride. The solvent extracts are then combined and the solution is evaporated slowly in a water bath into a small volume, usually 1–2 mL. The solvent extract is then passed through a thin bed of anhydrous Na_2SO_4 to remove any water present in it. Thus a 1-L sample is concentrated down to a volume of 1 mL, thereby lowering the detection limits for trace pollutants by a factor of 1000.

Alternatively, LLE may be carried out using a continuous liquid–liquid extractor. Such apparatus are commercially available. Although the process is automatic, proceeds continuously, and does not involve manual shaking of the samples with the extraction solvent in a separatory funnel, the process takes much longer and requires a relatively large volume of solvent.

In recent years, solid-phase extraction has been slowly replacing the LLE methods. The advantages of solid-phase extraction are low cost, faster speed, and ease of operation. In SPE methods, sample aliquots, usually in much smaller volumes than the amounts employed in the LLE methods, are passed through small cartridges packed with adsorbent materials (such as C-18 or other hydrocarbons, often functionalized). The packing material is designed to trap the pollutants or interfering substances in the samples. Analytes retained on the cartridges are eluted out with an appropriate solvent or a solvent mixture.

In contrast to separatory funnel extraction, microextraction processes are very rapid and may be applied to extract organic analytes in potable and nonpotable waters. In such extractions about 30 mL of sample are treated with 1–2 mL of hexane in a 40-mL vial. The vial is capped and shaken for 2–3 min. The solution is then allowed to stand until the immiscible hexane layer on top separates out from the denser water layer at the bottom. A few microliters of solution from the top hexane layer are then carefully withdrawn and injected onto a gas chromatograph (GC). Although microextraction in environmental analysis usually involves separation of anlytes into a lighter, water-immiscible solvent such as hexane, heavier, water-immiscible solvents, such as methylene chloride, have been found to be effective as well. In using the latter, a small aliquot of the solvent extract is carefully withdrawn from the bottom layer and allowed to pass through anhydrous Na_2SO_4. Water removal from the solvent extract is essential if methylene chloride is used in such a microextraction.

21.4 CLEANUP OF SAMPLE EXTRACTS

Sample extracts often contain interfering substances, which may be concentrated along with the analytes. Such substances may co-elute or produce extraneous peaks in the chromatograms of the analytes, affecting the resolution efficiency of the column. Also, they may cause loss of detector sensitivity and shorten the life of the column. Solvent extracts obtained from aqueous or nonaqueous samples may be purified by one or more of the following techniques.

21.4.1 Acid–Base Partitioning

Acid–base partitioning is used to separate acidic pollutants from basic pollutants or vice versa. Some U.S. Environmental Protection Agency (EPA) methods, such as Methods 625, 8250, and 8270, have adopted acid–base partitioning techniques for the extraction of organic pollutants from aqueous matrices. The method applies to separate acidic pollutants from basic and neutral pollutants during sample extraction. The technique may also be applied for cleanup purposes based on the acidic or basic properties of the interfering substances. In such cleanup procedures, the solvent extract is shaken with water that is highly basic. The acidic analytes, such as phenols, partition from the organic phase into the basic aqueous phase. The organic phase now contains the basic and neutral analytes. The aqueous solution is then acidified with HCl to a pH below 2 and shaken with ethylene chloride. The acidic analytes from the aqueous phase partition into ethylene chloride, separating out from the acidified water. Thus, we now have two separate ethylene chloride solutions, one containing basic analytes and the other acidic compounds only.

21.4.2 Alumina Column Cleanup

Alumina column cleanup is based on separating analytes from interfering substances in the solvent extracts by virtue of differences in polarity. The column is packed with highly porous granular aluminum oxide. The latter is available in three pH ranges—acidic, neutral, and basic. All three grades of alumina are used in column cleanup. The acidic form of alumina, which has a pH of 4–5, is usually used to separate strong acids and acid pigments. Basic alumina, pH 9–10, is applied to separate basic compounds, such as alkaline solutions, amines, and alkaloids. The neutral form of alumina is less active than the basic form and may be used to separate aldehydes, ketones, and esters. Some applications of various grades of alumina are summarized in Table 21.2. Alumina of various activity grades can be selectively prepared by adding water to dehydrated alumina. The latter is obtained by heating alumina above 400°C until no more water is lost.

TABLE 21.2 Separation of Organic Substances on Alumina Columns

Alumina grades	Organic compounds
Basic alumina (pH 9–10)	Alkaloids, amines, pyridine nitrosamines, steroids, and basic pigments
Neutral alumina (pH 6–8)	Aldehydes, ketones, alcohols, alkyl esters, phthalates, lactones, and petroleum waste
Acidic alumina (pH 4–5)	Carboxylic acids, chlorophenoxy acids, phenolics, and acid pigments

Alumina is covered under anhydrous Na_2SO_4 in the column before loading the extract. Analytes are eluted with a suitable solvent or solvent mixtures, leaving behind interfering substances adsorbed onto the column.

21.4.3 Silica Gel Cleanup

Silica gel is a weakly acidic form of amorphous silica. For cleanup purposes it may be prepared by treating sodium sulfate with H_2SO_4. An activated form of silica gel may be made by heating the latter at 150°C for several hours. Incorporating 10–20% water may produce a deactivated form of silica gel. The latter is used to separate alkaloids, steroids, terpenoids, dyes, plasticizers, sugars, esters, and lipids. Activated silica gel may be used to separate hydrocarbons.

21.4.4 Florisil Column Cleanup

Florisil, a form of magnesium silicate, is used to remove interfering substances from sample extracts in the analysis of many types of pesticides, phthalates, haloethers, nitrosamines, and chlorinated hydrocarbons. Like silica gel, florisil exhibits weak acid properties. In addition to cleanup of sample extracts, florisil is also used to separate esters, ketones, alkaloids, steroids, glycerides, and certain types of carbohydrates.

21.4.5 Gel Permeation Cleanup

Gel permeation cleanup (GPC) is applicable for separating macromolecules such as proteins, lipids, polymers, and steroids from sample extracts. The gel is porous and hydrophobic, having a pore size greater than the size of the molecules to be separated. The solvent extract is loaded onto the GPC column and the analytes are eluted out using a suitable solvent.

21.4.6 Permanganate–Sulfuric Acid Cleanup

Permanganate–sulfuric acid cleanup is applied to separate oxidizable interfering substances in the sample extract from analytes such as polychlorinated biphenyls, which are highly stable under strong oxidizing conditions. The extract is heated with a strong oxidant, such as a mixture of $KMnO_4$ and H_2SO_4, for a short time, during which interfering substances in the extract are destroyed.

21.4.7 Sulfur Cleanup

Sulfur occurs in many industrial wastewaters, sludges, sediments, and marine algae. Its presence can mask the regions of interest. Also, its mass spectra, corresponding to the mass ions S_2, S_4, and S_6, can interfere in the determination of compounds that produce similar primary or secondary characteristic mass ions. In addition, the solubility of sulfur in most organic solvents is similar to that in some of the organochlorine and organophosphorus pesticides. It therefore cannot be removed by florisil cleanup.

Sulfur present in the organic solvent extract of a sample can be effectively removed by shaking the extract with a small quantity of either copper (powder or turning) or mercury, which settles to the bottom and leaves a clean extract solution on the top that can be injected onto a GC column. Alternatively, sulfur may be separated from the extract by vigorously shaking the extract with tetrabutyl ammonium–sodium sulfite reagent.

Some common cleanup methods in trace organic analysis are summarized in Table 21.3.

TABLE 21.3 **Some Common Cleanup Methods in Organic Analysis**

Technique	Analyte groups
Acid–base partitioning	Amines, imines, amides, nitrosamines, phenols, carboxylic acids, chlorophenoxy acid herbicides
GPC	Polynuclear aromatics, nitroaromatics, nitrosamines, phenols, phthalate esters, chlorinated hydrocarbons, PCBs, organophosphorus pesticides, organochlorine pesticides, cyclic ketones
Alumina	Amines, alkaloids, nitrosamine aldehydes, ketones, esters, alcohols, phenols, carboxylic acids, chlorophenoxy acid herbicides, polynuclear aromatics
Silica gel	Polynuclear aromatics, PCBs, certain chlorinated pesticides, derivatized phenolic compounds
Florisil	Organochlorine pesticides, organophosphorus pesticides, PCBs, phthalates, nitrosamines, nitroaromatics, haloethers, chlorinated hydrocarbons
Sulfur cleanup	PCBs
Permanganate–sulfuric acid	PCBs and compounds that are chemically stable under strong oxidizing conditions

21.5 DETERMINATION OF TRACE ORGANIC POLLUTANTS BY INSTRUMENTAL TECHNIQUES

Trace organic pollutants extracted from sample matrices into an appropriate solvent as outlined above are then analyzed by one or more of the following instrumental techniques. Because of the concentration of the solvent extract to a much smaller final volume and the high sensitivity of the instrument, a detection level below 1 ng/L (ppt) may be obtained for many types of pollutants. The detection levels of analytes depend on the instrument and the detector used. Gas chromatography and high-performance liquid chromatography (HPLC) are most commonly employed in environmental analysis. While the most sensitive GC detectors for determining halogenated organics are the electron-capture detector (ECD) and the Hall electrolytic conductivity detector (HECD), the photoionization detector (PID) is sensitive to substances containing double bonds, such as olefins and aromatics, the nitrogen–phosphorus detector (NPD) is sensitive to nitrogen- or phosphorus-containing organics in N or P mode, respectively, and the flame photometric detector (FPD) can accurately determine trace organics containing sulfur or phosphorus. If an analyte is suspected to be present in the sample extract based on its retention time on the GC column, determined from injecting its standard solution, its presence must be confirmed either on an alternative GC column or preferably by mass spectrometry (MS). In laboratories that are not equipped with a mass spectrometer, a dual-column, dual-injector GC should be adequate to carry out these organic analyses. In all such analyses, an aliquot of the sample extract is simultaneously injected onto the second (or confirmatory) GC column.

Mass spectrometry is by far the best confirmatory test for the presence of a substance in a sample. A mass spectrometer is interfaced to a GC or a liquid chromatography (LC) column to identify products after their separation on the GC or LC column. The presence of a compound is determined from its characteristic mass spectra, as well as from its retention time. The characteristic mass ions of various types of common pollutants are listed in Tables 21.4 and 21.5. The mass ions of pesticide and herbicide classes of compounds are not presented in these tables; they are listed separately in Sec. 20.

Table 21.4 lists the primary and secondary mass ions of volatile organic compounds (VOCs) that may be purged from an aqueous sample with an inert gas for identification. Table 21.5 lists the characteristic mass ions of nonvolatile or semivolatile substances.

TABLE 21.4 Characteristic Mass Ions for Identifying Purgeable Organics

Compounds	Primary ion (*m/z*)	Secondary ions (*m/z*)
Alcohols		
Allyl alcohol	57	58, 39
1-Butanol	56	41
2-Butanol	74	43
Isobutyl alcohol	43	41, 42, 74
Ethanol	31	45, 27, 46
Propargyl alcohol	55	39, 38, 53
Aldehydes and ketones		
Acetone	58	43
Acrolein	56	55, 58
Methyl ethyl ketone	72	43
Methyl isobutyl ketone	100	43, 58, 85
2-Hexanone	43	58, 57, 100
Aromatics		
Benzene	78	—
n-Butylbenzene	91	92, 134
sec-Butylbenzene	105	134
tert-Butylbenzene	119	91, 134
Ethylbenzene	91	106
Isopropylbenzene	105	120
n-Propylbenzene	91	120
Styrene	104	78
Toluene	92	91
o-Xylene	106	91
m-Xylene	106	91
p-Xylene	106	91
p-Isopropyltoluene	119	134, 91
1,2,4-Trimethylbenzene	105	120
1,3,5-Trimethylbenzene	105	120
Esters		
Ethyl acetate	88	43, 45, 61
Ethyl methacrylate	69	41, 99, 86, 114
Methyl acrylate	55	85
Methyl methacrylate	69	41, 100, 39
Vinyl acetate	43	86
Nitriles		
Acetonitrile	40	40, 39
Acrylonitrile	53	52, 51
Malononitrile	66	39, 65, 38
Methacrylonitrile	41	67, 39, 52, 66
Propionitrile	54	52, 55, 40
Halogenated organics		
Allyl chloride	76	41, 39, 78
Benzyl chloride	91	126, 65, 128
Bromoacetone	136	43, 138, 93, 95
Bromobenzene	156	77, 158
Bromochloromethane	128	49, 130
Bromodichloromethane	83	85, 127
Bromoform	173	175, 254

TABLE 21.4 Characteristic Mass Ions for Identifying Purgeable Organics (*Continued*)

Compounds	Primary ion (*m/z*)	Secondary ions (*m/z*)
Halogenated organics		
Bromomethane	94	96
Chlorobenzene	112	77, 114
1-Chlorobutane	56	49
Chlorodibromomethane	129	208, 206
Chloroethane	64	66
2-Chloroethyl vinyl ether	63	65, 106
2-Chloroethanol	49	44, 43, 51, 80
Chloroprene	53	88, 90, 51
2-Chlorotoluene	91	126
1,2-Dibromo-3-chloropropane	75	155, 157
Dibromochloromethane	129	127
1,2-Dibromoethane	107	109, 188
Dibromomethane	93	95, 174
1,2-Dichlorobenzene	146	111, 148
1,3-Dichlorobenzene	146	111, 148
1,4-Dichlorobenzene	146	111, 148
cis-1,4-Dichloro-2-butene	75	53, 77, 124, 89
trans-1,4-Dichloro-2-butene	53	88, 75
Dichlorodifluoromethane	85	87
1,1-Dichloroethane	63	65, 83
1,2-Dichloroethane	62	98
1,1-Dichloroethene	96	61, 63
cis-1,2-Dichloroethene	96	61, 98
trans-1,2-Dichloroethene	96	61, 98
1,2-Dichloropropane	63	112
1,3-Dichloropropane	76	78
2,2-Dichloropropane	77	97
1,3-Dichloro-2-propanol	79	43, 81, 49
1,1-Dichloropropene	75	110, 77
cis-1,3-Dichloropropene	75	77, 39
trans-1,3-Dichloropropene	75	77, 39
Epichlorohydrin	57	49, 62, 51
Hexachlorobutadiene	225	223, 227
Hexachloroethane	201	166, 199, 203
Methylene chloride	84	86, 49
Methyl iodide	142	127, 141
Pentachloroethane	167	130, 132
1,2,3-Trichlorobenzene	180	182, 145
1,2,4-Trichlorobenzene	180	182, 145
1,1,1,2-Tetrachloroethane	131	133, 119
1,1,2,2-Tetrachloroethane	83	131, 85
Tetrachloroethene	164	129, 131, 166
1,1,1-Trichloroethane	97	99, 61
1,1,2-Trichloroethane	83	97, 85
Trichloroethene	95	97, 130, 132
Trichlorofluoromethane	151	101, 153
1,2,3-Trichloropropane	75	77
Vinyl chloride	62	64
Miscellaneous compounds		
Carbon disulfide	76	78

(*Continued*)

TABLE 21.4 Characteristic Mass Ions for Identifying Purgeable Organics (*Continued*)

Compounds	Primary ion (*m*/*z*)	Secondary ions (*m*/*z*)
Miscellaneous compounds		
Nitrobenzene	123	51, 77
Ethylene oxide	44	43, 42
1,4-Dioxane	88	58, 43, 57
n-Propylamine	59	41, 39
Pyridine	79	52
bis(2-Chloroethyl) sulfide	109	111, 158, 160
2-Picoline	93	66, 92, 78
β-Propiolactone	42	43, 44
Diethyl ether	74	45, 59
Methyl-*tert*-butyl ether	73	57
Internal standards/surrogates		
Benzene-d_6	84	83
Bromobenzene-d_5	82	162
Bromochloromethane-d_2	51	131
1,4-Difluorobenzene	114	—
Chlorobenzene-d_5	117	—
1,4-Dichlorobenzene-d_4	152	115, 150
4-Bromofluorobenzene	95	174, 176
Dibromofluoromethane	113	—
Toluene-d_8	98	—
Pentafluorobenzene	168	—
Fluorobenzene	96	77

Another technique to confirm the presence of trace organic pollutants is GC interfaced with Fourier-transform infrared spectrometry (FT-IR). This method may serve as a useful complement to GC/MS analysis. The analytes in the sample extracts are separated by capillary GC and the target analytes are detected and quantified by FT-IR. The compound classes are determined from infrared group absorption frequencies. The possible presence of a particular compound class may be inferred from the presence of an infrared band in the appropriate group frequency region. In this technique, a temperature-programmable gas chromatograph equipped with a capillary column is interfaced to an FT-IR spectrometer. The infrared spectrum of the analyte is visually compared with the search library spectrum of the most promising online library search hits. The five most intense, sharp, and well-resolved IR bands of the analytes are compared with the corresponding bands in the library spectrum. The FT-IR stretching frequencies of the compound groups in the gas phase may be 0–30 cm^{-1} higher in frequency than those of the condensed phase. Additionally, as a faster confirmation of the presence of a compound in the sample extract, the relative retention time of the analyte can be compared with an authentic standard of the same compound.

Quantitation may be done by two methods: (1) integrated absorbance technique or (2) maximum absorbance IR band technique. In both these methods, the concentrations of compounds in the samples are determined from standard calibration curves. The standard calibration curves are constructed by plotting concentrations versus integrated infrared absorbance or maximum infrared band intensity. The working range of analyte concentration should fall within the calibration curve, which should span at least one order of magnitude. A medium-band mercury–cadmium–tellurium (MCT) detector that can reach 650 cm^{-1} should be suitable for the purpose.

TABLE 21.5 Characteristic Mass Ions for Identifying Some Semivolatile Organics

Compound	Primary ion (*m*/*z*)	Secondary ions (*m*/*z*)
Acenaphthene	154	153, 152
Acenaphthylene	152	151, 153
Acetopheonone	105	71, 51, 120
2-Acetylaminofluorene	181	180, 223, 152
1-Acetyl-2-thiourea	118	43, 42, 76
2-Aminoanthraquinone	223	167, 195
Aminoazobenzene	197	92, 120, 65, 77
4-Aminobiphenyl	169	168, 170, 115
3-Amino-9-ethylcarbazole	195	210, 181, 127
Aniline	93	66, 65
o-Anisidine	108	80, 123, 52
Anthracene	178	176, 179
Benzidine	184	92, 185
Benzoic acid	122	105, 77
Benz(a)anthracene	228	229, 226
Benzo(b)fluoranthene	252	253, 125
Benzo(k)fluoranthene	252	253, 125
Benzo(g,h,i)perylene	276	138, 277
Benzo(a)pyrene	252	253, 125
p-Benzoquinone	54	108, 82, 80, 52
Benzyl alcohol	108	79, 77
bis(2-Chloroethoxy)methane	93	95, 123
bis(2-Chloroethyl)ether	93	63, 95
bis(2-Chloroisopropyl)ether	45	77, 121
bis(2-Ethylhexyl)phthalate	149	167, 279
4-Bromophenyl phenyl ether	51	77, 248, 250, 50, 63
Bromoxynil	277	279, 88, 275, 168
Butyl benzyl phthalate	149	91, 206
Carbofuran	164	149, 131, 122
4-Chloroaniline	127	129, 65, 92
1-Chloronaphthalene	162	127, 164
2-Chloronaphthalene	162	127, 164
2-Chlorophenol	128	64, 130
4-Chloro-1,2-phenylenediamine	142	80, 114, 144
4-Chlorophenyl phenyl ether	204	206, 141
Chrysene	228	226, 229
p-Cresidine	122	94, 137, 77, 93
2-Cyclohexyl-4,6-dinitrophenol	231	185, 41, 193, 266
2,4-Diaminotoluene	121	122, 94, 77, 104
Dibenz(a,j)acridine	279	280, 277, 250
Dibenz(a,h)anthracene	278	139, 279
Dibenzofuran	168	139
Dibenzo(a,e)pyrene	302	151, 150, 300
Di-*n*-butyl phthalate	149	150, 104
Dichlorobenzene (all isomers)	146	148, 111
3,3′-Dichlorobenzidine	252	254, 126
Dichlorophenol (2,4- and 2,6-)	162	164, 98
Diethyl phthalate	149	177, 150
Diethyl stilbestrol	268	145, 107, 239, 121, 159
Diethyl sulfate	139	45, 59, 99, 111, 125
Dihydrosaffrole	135	64, 77

(Continued)

TABLE 21.5 Characteristic Mass Ions for Identifying Some Semivolatile Organics (*Continued*)

Compound	Primary ion (*m*/*z*)	Secondary ions (*m*/*z*)
3,3′-Dimethoxybenzidine	244	201, 229
Dimethylaminoazobenzene	225	120, 77, 105, 148, 62
7,12-Dimethylbenz(a)anthracene	256	241, 239, 120
3,3′-Dimethylbenzidine	212	106, 196, 180
2,4-Dimethylphenol	122	107, 121
Dimethyl phthalate	163	194, 164
1,2-Dinitrobenzene	168	50, 63, 74
1,3-Dinitrobenzene	168	76, 50, 75, 92, 122
1,4-Dinitrobenzene	168	75, 50, 76, 92, 122
4,6-Dinitro-2-methylphenol	198	51, 105
2,4-Dinitrophenol	184	63, 154
2,6-Dinitrophenol	162	164, 126, 98, 63
Dinitrotoluene (all isomers)	165	63, 89
Diphenylamine	169	168, 167
5,5-Diphenylhydantoin	79	77, 80, 107
1,2-Diphenylhydrazine	77	105, 182
Di-*n*-octyl phthalate	149	167, 43
Ethyl carbamate	62	44, 45, 74
Ethyl methanesulfonate	79	109, 97, 45, 65
Fluoranthene	202	101, 203
Fluorene	166	165, 167
Hexachlorobenzene	284	142, 249
Hexachlorobutadiene	225	223, 227
Hexachlorocyclopentadiene	237	235, 272
Hexachloroethane	117	201, 199
Hexachlorophene	196	198, 209, 211, 406, 408
Hexachloropropene	213	211, 215, 117, 106, 141
Hexamethylphosphoramide	135	44, 179, 92, 42
Hydroquinone	110	81, 53, 55
Indeno(1,2,3-cd)pyrene	276	138, 227
Isodrin	193	66, 195, 263, 265, 147
Isophorone	82	95, 138
Isosafrole	162	131, 104, 77, 51
Maleic anhydride	54	98, 53, 44
3-Methylcholanthrene	268	252, 253, 126, 134, 113
4,4′-Methylenebis(2-chloroaniline)	254	134, 253, 210, 118
Methyl methanesulfonate	80	79, 65, 95
2-Methylnaphthalene	142	141
Methylphenol (all isomers)	107	108, 77, 79, 90
Naphthalene	128	129, 127
1,4-Naphthoquinone	158	104, 102, 76, 50, 130
1-Naphthylamine	143	115, 89, 63
2-Naphthylamine	143	115, 116
Nicotine	84	133, 161, 162
5-Nitroacenaphthene	199	152, 169, 141, 115
2-Nitroaniline	65	92, 138
3-Nitroaniline	138	108, 92
5-Nitro-*o*-anisidine	168	79, 52, 138, 153, 77
Nitrobenzene	77	123, 65
4-Nitrobiphenyl	199	152, 141, 169, 151
Nitrophenol (all isomers)	139	109, 65

TABLE 21.5 Characteristic Mass Ions for Identifying Some Semivolatile Organics (*Continued*)

Compound	Primary ion (*m/z*)	Secondary ions (*m/z*)
5-Nitro-*o*-toluidine	152	77, 79, 106, 94
Nitroquinoline-1-oxide	174	101, 128, 75, 116
N-Nitrosodi-*n*-butylamine	84	57, 41, 116, 158
N-Nitrosodiethylamine	102	42, 57, 44, 56
N-Nitrosodimethylamine	42	74, 44
N-Nitrosomethylethylamine	88	42, 43, 56
N-Nitrosodiphenylamine	169	168, 167
N-Nitrosodi-*n*-propylamine	70	42, 101, 130
N-Nitrosomorpholine	56	116, 86, 30, 42
N-Nitrosopiperidine	114	42, 55, 56, 41
N-Nitrosopyrrolidine	100	41, 42, 68, 69
Octamethyl pyrophosphoramide	135	44, 199, 286, 153, 243
4,4′-Oxydianiline	200	108, 171, 80, 65
Pentachlorobenzene	250	252, 108, 248, 215, 254
Pentachloronitrobenzene	237	142, 214, 249, 295, 265
Pentachlorophenol	266	264, 268
Phenacetin	108	180, 179, 109, 137
Phenanthrene	178	176, 179
Phenobarbital	204	117, 232, 146, 161
Phenol	94	65, 66
1,4-Phenylenediamine	108	80, 53, 54, 52
Phthalic anhydride	104	76, 50, 148
2-Picoline	93	66, 92
Piperonyl sulfoxide	162	135, 105, 77
Pronamide	173	175, 145, 109, 147
Propylthiouracil	170	142, 114, 83
Pyrene	202	200, 203
Pyridine	79	52, 51
Resorcinol	110	81, 82, 53, 69
Safrole	162	104, 77, 103, 135
Strychnine	334	335, 333
1,2,4,5-Tetrachlorobenzene	216	214, 179, 108, 143, 218
2,3,4,6-Tetrachlorophenol	232	131, 230, 166, 234, 168
Tetraethyl pyrophosphate	99	155, 127, 81, 109
Thiophenol	110	66, 109, 84
Toluene diisocyanate	174	145, 173, 146, 132, 91
o-Toluidine	106	107, 77, 51, 79
1,2,4-Trichlorobenzene	180	182, 145
2,4,6-Trichlorophenol	196	198, 200
2,4,5-Trimethylaniline	120	135
Trimethyl phosphate	110	79, 95, 109, 140
1,3,5-Trinitrobenzene	75	74, 213, 120, 91, 63
Internal standards and surrogates		
1,4-Dichlorobenzene-d_4 (IS)	152	150, 115
Naphthalene-d_8 (IS)	136	68
Acenaphthene-d_{10} (IS)	164	162, 160
Phenanthrene-d_{10} (IS)	188	94, 80
Chrysene-d_{12} (IS)	240	120, 236
Perylene-d_{12} (IS)	264	260, 265
2-Fluorophenol (surr)	112	64

(Continued)

TABLE 21.5 Characteristic Mass Ions for Identifying Some Semivolatile Organics (*Continued*)

Compound	Primary ion (*m/z*)	Secondary ions (*m/z*)
Internal standards and surrogates		
2-Fluorobiphenyl (surr)	172	171
Nitrobenzene-d_5 (surr)	82	128, 54
Phenol-d_6 (surr)	99	42, 71
Terphenyl-d_{14} (surr)	244	122, 212
2,4,6-Tribromophenol (surr)	330	332, 141

The gas-phase group frequencies of various types of compounds containing different functional groups are presented in Table 21.6. The most intense IR peak for some common base/neutral- and acid-extractable pollutants for quantitation are tabulated in Table 21.7.

21.6 ANALYSIS OF INORGANIC ANIONS

The anions in aqueous and solid samples can be measured by several techniques; the most common methods involve colorimetry, capillary ion electrophoresis, ion chromatography, ion-selective electrode methods, and titrimetry. Some of these techniques are discussed briefly in the following sections and highlighted in Table 21.8.

21.6.1 Colorimetric Methods

Colorimetric methods are widely employed in environmental wet analysis. Aqueous samples and the aqueous extracts of soils, sediments, and hazardous wastes are treated with various reagents to form colored complexes with the anions. The absorbance or the transmittance of the solutions after adding the color-forming reagents are then measured by a spectrophotometer or a filter photometer. Within a narrow range of concentrations, the absorbance or the transmittance of the colored complexes formed in the solutions should be proportional to the concentrations of specific anions in the solutions, within the limits of Beers' law. The concentrations of analytes in samples are determined directly from standard calibration curves constructed by plotting the absorbance or transmittance of calibration standard solutions against their concentrations.

The colorimetric methods, however, have certain limitations, the major ones being that they cannot be applied to dirty samples and that the presence of other substances in the samples may interfere in the tests. For example, the presence of chloride ion at a concentration greater than 200 mg/L in the sample may interfere in the analysis of iodide ion when using a colorless indicator, such as 4,4′,4″-methylidynetris (*N,N*-dimethylaniline also known as leucocrystal violet). Dilution of samples, or pretreatments such as distillation, may reduce the effects of interfering substances in samples. With these limitations, colorimetric procedures nevertheless offer certain advantages over other methods because of low cost, simplicity, and low detection limits.

Table 21.9 outlines briefly one or two colorimetric methods each for some common anions found in environmental matrices.

21.6.2 Titrimetry

Certain inorganic anions in aqueous samples may conveniently be determined by titration. Titrimetric procedures often are short and simple, although the detection limits are not as low as

TABLE 21.6 Gas-Phase Group Frequencies of Various Classes of Compounds

Functional group	Compound type	Frequency range ν, cm^{-1}
Acid	Aliphatic	3574–3580
		1770–1782
	Aromatic	3574–3586
		1757–1774
Alcohol	Primary aliphatic	3630–3680
		1206–1270
		1026–1094
	Secondary aliphatic	3604–3665
		1231–1270
	Tertiary aliphatic	3640–3670
		1213–1245
Aldehyde	Aliphatic	1742–1744
		2802–2877
		2698–2712
	Aromatic	1703–1749
		2820–2866
		2720–2760
Alkane		2930–2970
		2851–2884
		1450–1475
		1355–1389
Alkyne	Aliphatic	3323–3329
Amide	Substituted acetamide	1710–1724
Amine	Aliphatic	760–785
	Primary aromatic	3480–3532
	Secondary aromatic	3387–3480
Benzene	Monosubstituted	675–698
		735–790
		831–893
		1470–1510
		1582–1630
		1707–1737
Ester	Unsubstituted aliphatic	1748–1761
	Aromatic	1703–1759
	Monosubstituted acetate	1753–1788
Ether	Alkyl, aryl	1215–1275
	Alkyl, benzyl	1103–1117
	Dialkyl	1084–1130
	Diaryl	1238–1250
	Alkyl, vinyl	1204–1207
		1128–1142
Ketone	Aliphatic (acyclic)	1726–1732
	α,β-Unsaturated	1638–1699
	Aromatic	1701–1722
Nitrile	Aliphatic	2240–2265
	Aromatic	2234–2245

(Continued)

TABLE 21.6 Gas-Phase Group Frequencies of Various Classes of Compounds (*Continued*)

Functional group	Compound type	Frequency range ν, cm^{-1}
Nitro	Aliphatic	1566–1594
		1548–1589
		1377–1408
		1327–1381
	Aromatic	1535–1566
		1335–1358
Phenol	1,2-Disubstituted	3582–3595
		1255–1274
	1,3-Disubstituted	3643–3655
		1256–1315
		1157–1198
	1,4-Disubstituted	3645–3657
		1233–1269
		1171–1190

those attained by colorimetric and electrode methods. Because of the relatively higher detection limits for analytes, applications of titrimetric techniques in environmental sample analyses are very limited. Also, the presence of certain substances in the samples may interfere, and pretreatment of samples may be required. Titrimetric methods for the analysis of some common anions in environmental samples are highlighted in Table 21.10.

21.6.3 Capillary Ion Electrophoresis

Many common inorganic anions can be determined in a single analysis by capillary ion electrophoresis. The method is rapid and similar to ion chromatography, giving a "fingerprint" of anions present in the sample matrix. Additionally, it can provide information on organic acids that may not be available from isocratic ion chromatography. The presence of cations and neutral organics does not interfere in the analysis of anions. No specific sample preparation step is necessary, although aqueous samples or their aqueous extracts may be diluted with reagent water before analysis. Total suspended solids in the sample may be removed by simple filtration, while oil and grease may be removed using solid-phase extraction cartridges. Unlike many colorimetric and titrimetric methods, capillary ion electrophoresis can detect halide ions or nitrate and nitrite ions in the presence of each other in the sample matrix. The anions are detected by indirect UV detection, as they displace the UV-absorbing electrolyte anion (chromate) charge for charge, in a silica capillary, thus decreasing the UV absorbance of the analyte anion compared to the background electrolyte. The anions are identified from their migration time on an electropherogram of the sample in a manner similar to retention time in chromatography. The analyte anions are quantified from their peak areas relative to standards. This capillary ion electrophoresis technique using indirect UV detection should be able to measure inorganic anions and organic acid anions at a minimum detectable concentration of about 0.1 mg/L.

21.6.4 Ion Chromatography

Ion chromatography, like capillary ion electrophoresis, is a single-instrument technique that may be applied to determine several anions sequentially in a single analysis. The method distinguishes anions in different oxidation states, such as SO_4^{2-} and SO_3^{2-} or NO_3^- and NO_2^-, as well as halides,

TABLE 21.7 Most Intense IR Peak for Compound Identification and Quantitation

Compound	v_{max}, cm^{-1}
Base/neutral-extractables	
Acenaphthene	799
Acenaphthylene	799
Anthracene	874
Benzo(a)anthracene	745
Benzo(a)pyrene	756
bis(2-Chloroethyl)ether	1115
bis(2-Chloroethoxy)methane	1084
bis(2-Chloroisopropyl)ether	1088
Butyl benzyl phthalate	1748
1-Bromophenyl phenyl ether	1238
2-Chloronaphthalene	851
2-Chloroaniline	1543
4-Chlorophenyl phenyl ether	1242
Chrysene	757
Di-*n*-butyl phthalate	1748
Dibenzofuran	1192
Diethyl phthalate	1748
Dimethyl phthalate	1751
Di-*n*-octyl phthalate	1748
Di-*n*-propyl phthalate	1748
1,2-Dichlorobenzene	1458
1,3-Dichlorobenzene	779
1,4-Dichlorobenzene	1474
2,4-Dinitrotoluene	1547
2,6-Dinitrotoluene	1551
bis(2-Ethylhexyl) phthalate	1748
Fluoranthene	773
Fluorene	737
Hexachlorobenzene	1346
Hexachlorocyclopentadiene	814
Hexachloroethane	783
1,3-Hexachlorobutadiene	853
Isophorone	1690
2-Methyl naphthalene	3069
Naphthalene	779
Nitrobenzene	1539
N-Nitrosodimethylamine	1483
N-Nitrosodi-*n*-propylamine	1485
N-Nitrosodiphenylamine	1501
2-Nitroaniline	1564
3-Nitroaniline	1583
4-Nitroaniline	1362
Phenanthrene	729
Pyrene	820
1,2,4-Trichlorobenzene	750
Acid-extractables	
Benzoic acid	1751
2-Chlorophenol	1485

(Continued)

TABLE 21.7 Most Intense IR Peak for Compound Identification and Quantitation (*Continued*)

Compound	ν_{max}, cm^{-1}
Acid-extractables	
4-Chlorophenol	1500
4-Chloro-3-methylphenol	1177
2-Methylphenol	748
4-Methylphenol	1177
2,4-Dichlorophenol	1481
2,4-Dinitrophenol	1346
4,6-Dinitro-2-methylphenol	1346
2-Nitrophenol	1335
4-Nitrophenol	1350
Pentachlorophenol	1381
Phenol	1184
2,4,6-Trichlorophenol	1470
2,4,5-Trichlorophenol	1458

which often interfere with each other in certain colorimetric and titrimetric analyses. Oxyhalides and many carboxylate anions may also be determined by ion chromatography. In environmental analysis the technique is usually applied to determine common anions in potable and nonpotable waters, air impinger solutions, and air particulate extracts, and in aqueous extracts of soils, sediments, and solid wastes. Some common eluants used in ion chromatography for the analysis of anions include solutions of sodium hydroxide, sodium carbonate–bicarbonate, and sodium tetraborate–boric acid.

Ion-exchange columns are commercially available from many suppliers. The stationery phases usually are composed of microporous polymeric resins that may contain styrene or ethyl vinyl benzene cross-linked with divinylbenzene, a chemically stable and porous resin core onto which a layer of ion-exchange coating is bonded to produce a reactive surface. A guard column is used to protect the ion-exchange column from contamination by organic and particulate matter in the samples.

21.7 DETERMINATION OF GENERAL AND AGGREGATE PROPERTIES OF SAMPLES

Determination of certain general or aggregate properties of samples is often far more important for understanding and evaluating the properties of samples than measuring any individual analytes in the samples. This may also be essential for regulatory requirements. For example, the measurement of pH, acidity, alkalinity, hardness, conductivity, chemical and biochemical oxygen demand, and total dissolved solids in a sample aliquot may provide very useful information about the nature and composition of a sample and may predict its harmful effects on the environment and the ecosystem. Similarly, the characteristics of a hazardous waste may be determined by testing its corrosivity, ignitability, and toxicity. Testing of such common general and aggregate properties of aqueous and nonaqueous samples is briefly outlined in this section.

21.7.1 pH

The pH of a sample measures its hydrogen-ion concentration and indicates whether a sample is acidic, neutral, or basic. The pH may be measured accurately using a pH meter. Meters are commercially available from a number of suppliers. The pH meter must be calibrated before making pH measurements or calibrated daily using three standard buffer solutions, usually at pH 4.00, 7.00, and

TABLE 21.8 **Measurement of Inorganic Anions at Trace Concentrations by Various Analytical Techniques**

Anion	Analytical techniques	Comments
Bromide, Br^-	Simple colorimetric	Many substances interfere
	Flow injection, colorimetric	Use a FIA bromide manifold
	Ion chromatography	Rapid; presence of other halides does not interfere
Chloride, Cl^-	Titrimetry, silver nitrate (argentometric titration)	Many substances interfere; detection limits not as low as other methods can provide
	Titrimetry, mercuric nitrate	Many substances interfere; the endpoint in the titration is sharper and easier to detect than in argentometric titration
	Chloride electrode	Iodide, bromide, chromate, and dichromate interfere; colored and turbid samples may also be analyzed
	Colorimetric, automated ferricyanide	An automated method; very little chemical interference
	Colorimetric, flow injection	Automated colorimetric method; large number of samples may be analyzed
	Ion chromatography	Rapid, accurate, and interference from other substances in the sample minimal
Fluoride, F^-	Colorimetric, zirconium dye	Presence of several substances may interfere; preliminary distillation of sample may be required
	Ion-selective electrode	Sample distillation may be required to minimize interference effect
	Colorimetric, automated complexone	An automated method; interference normally removed by distillation; has a lower detection limit than the electrode method
	Ion chromatography	Use weaker eluents to separate fluoride from interfering peaks
	Capillary ion electrophoresis	Rapid; a detection limit of 0.1 mg/L can be achieved; formate may interfere
Cyanide, CN^-	Titrimetry, silver nitrate	Sample should be distilled into an alkaline solution for the removal of interfering substances; detection limit is higher than for colorimetric and electrode methods
	Colorimetry, chloramine–pyridine–barbituric acid	Sample distillation is recommended to remove interfering substances; the method is tedius; the formation of red-blue color upon addition of pyridine–barbituric acid after treating the sample with chloramine-T is sharp; avoid inhalation of toxic cyanogen chloride
	Cyanide-selective electrode	Sample distillation eliminates interference; the method measures cyanide in the concentration range 0.05–10 mg/L
Cyanate, CNO^-	Cyanate hydrolysis	Cyanate hydrolyzes to ammonia when heated with acid; ammonia produced may be analyzed by colorimetry, titrimetry, or ammonia-selective electrode; oxidizing substances in the sample may oxidize cyanate to CO_2 and N_2; this interference effect may be reduced by treating the sample with sodium thiosulfate
Iodide, I^-	Voltammetry	Most suitable and sensitive method; measures iodide concentration in the range 0.10–10 μg/L; presence of iodate, iodine, and organic iodine does not affect the test; sulfide may interfere but which can

(Continued)

TABLE 21.8 Measurement of Inorganic Anions at Trace Concentrations by Various Analytical Techniques (*Continued*)

Anion	Analytical techniques	Comments
Iodide, I^-		be removed as H_2S by acidification and purging and then readjusting the pH back to 8 before analysis; the voltammetry analyzer system consists of a potentiostat, static mercury drop electrode, a saturated calomel electrode as the reference electrode, a stirrer, and a plotter
	Leuco crystal violet, colorimetry	This colorimetry method can measure iodide concentration in water in the range 50–5000 μg/L; chloride at concentrations over 200 mg/L may interfere
	Catalytic reduction method, colorimetry	Iodide ion catalyzes the reduction of ceric ions by arsenious acid; after a specific time interval the reaction is stopped by addition of ferrous ammonium sulfate; the resulting ferric ions, proportional to the remaining ceric ions, form a colored complex with potassium thiocyanate; intensity of color proportional to iodide concentration; plot a calibration curve for quantitation; iodate, elemental iodine, and hypoiodite ion interfere; this procedure measures iodide concentrations below 100 μg/L; sensitivity of this test falls between that of leuco crystal violet and voltammetric methods
	Ion chromatography	Iodide ion may be measured simultaneously with other anions in the sample; other halide ions do not interfere
Iodate, IO_3^-	Polarography	Iodate is reduced to iodide under mild basic conditions; measured by a polarographic analyzer system consisting of a static mercury drop electrode, a saturated calomel electrode, a potentiostat, and a plotter; the method is highly sensitive and can measure iodate at 5 μg/L in the presence of iodide and other iodine species; dissolved oxygen and zinc interfere and must be removed before analysis
Nitrate, NO_3^-	Colorimetry, cadmium reduction, and hydrazine reduction	Nitrite ion, NO_2^-, interferes; samples must be measured separately for nitrite ion before carrying out any cadmium or hydrazine reduction
	Ultraviolet spectrophotometric method	Usually applied for screening samples to monitor their nitrate levels; UV absorption is measured at two different wavelengths, 220 and 275 nm; dissolved organic matter in samples also may absorb UV light at 220 nm
	Nitrate ion-selective electrode method	Nitrate electrode can measure NO_3^- concentrations in samples in a wide range between 0.5 and 5000 mg/L; samples must be buffered over the pH range 3–9
	Capillary ion electrophoresis with indirect UV detection	Rapid and probably the most confirmatory technique; the interference effects from other ions are easy to overcome by changing the conditions to alter the migration time of co-eluting peaks; the minimum detection limit, however, is about

TABLE 21.8 **Measurement of Inorganic Anions at Trace Concentrations by Various Analytical Techniques (*Continued*)**

Anion	Analytical techniques	Comments
		0.1 mg/L, about 10 times higher than that of the cadmium-reduction colorimetric method
	Ion chromatography	The test is rapid and readily distinguishes nitrate from nitrite in the sample; the limit of detection is about 0.10 mg NO_3^-/L
Nitrite, NO_2^-	Colorimetric, diazotization coupling	The method is very sensitive; the reddish-purple color of the azo dye produced can be detected at 543 nm at a lowest concentration of 10 μg/L
	Capillary ion electrophoresis	Not as sensitive as the colorimetric method; the analysis, however, is fast and free from interference
	Ion chromatography	Rapid; interference effects are minimal; the detection level is not as low as what may be achieved by colorimetric techniques
Orthophosphate, PO_4^{3-}	Colorimetric, vanadomolybdo-phosphoric acid	This test is not as sensitive as the other two colorimetric tests; yellow color of the product may be easily masked in dirty or colored samples; if the sample is dirty, shake it with activated carbon and then filter the carbon to remove color before analysis; the detection limit is in the range of 0.5 mg PO_4^{3-}/L
	Colorimetric, stannous chloride	The molybdophosphoric acid obtained from treating phosphate with ammonium molybdate may be reduced by stannous chloride to give an intensely colored molybdenum blue, which may be extracted into benzene–isobutanol solvent; the test is more sensitive than the above method; a detection level of 10 μg/L may be achieved
	Colorimetric, ascorbic acid	Instead of stannous chloride, ascorbic acid may be used to produce molybdenum blue; somewhat less sensitive than the stannous chloride reduction test
	Capillary ion electrophoresis with indirect UV detection	Although the minimum detection limit of phosphate obtained by this technique is well above those from colorimetric tests, the analysis is faster, simpler, and almost free from interference
	Ion chromatography	The test is rapid, and the presence of interfering substances does not affect the test; the detection limit, however, is higher than that obtained with the molybdenum blue colorimetric test
Sulfate, SO_4^{2-}	Gravimetry	Sulfate is precipitated in HCl solution as $BaSO_4$ by addition of $BaCl_2$; the gravimetric method is susceptible to many errors; also, the analytical procedure is very tedious
	Turbidimetric method	Sulfate is precipitated as $BaSO_4$ of uniform size in acetic acid medium and the light absorbance of the suspension is measured; the lower detection limit is in the range 2–5 mg/L
	Colorimetry, methylthymol blue	The method is less sensitive than the turbidimetric procedure; the use of continuous-flow automated analytical equipment is ideal to analyze a large number of samples

(Continued)

TABLE 21.8 **Measurement of Inorganic Anions at Trace Concentrations by Various Analytical Techniques (*Continued*)**

Anion	Analytical techniques	Comments
Sulfate, SO_4^{2-}	Capillary ion electrophoresis with indirect UV detection	A detection level in the range of 2 mg/L may be achieved, which is lower than with the turbidimetric method; the analysis time is very short, less than 5 min
	Ion chromatography	Analysis time is short; other anions do not interfere; detection limit may be lower than with the turbidimetric method
Sulfide, S^{2-}	Titration, iodometric	The method is accurate for measuring S^{2-} at concentrations above 1 mg/L; oxidizing substances in the sample may interfere
	Colorimetric, methylene blue and gas dialysis, automated methylene blue	The methylene blue method is based on the reaction of sulfide in the presence of $FeCl_3$ with dimethyl-*p*-phenylenediamine to produce methylene blue; this test can measure sulfide at concentrations ranging from 0.1 to 20.0 mg/L; also, continuous flow automated analytical equipment based on the above reaction may be used; a gas dialysis technique is used to separate sulfide from the sample matrix and most interfering substances; the automated methylene blue method is suitable for analyzing a large number of samples and can measure sulfide at concentrations in the range 0.002–0.10 mg/L
	Ion-selective electrode	The electrode method can measure sulfide in the sample in a much wider range of concentrations, between 0.05 and 100 mg/L
Sulfite SO_3^{2-}	Colorimetry, phenanthroline	This method is applied to measure sulfite at low concentrations; the minimum detectable concentration is 0.01 mg/L
	Titration, iodometric	This method can analyze sulfite only at much higher concentrations, the detection limit may be in the range of 2 mg/L; also, the presence of other oxidizable substances in the sample can give erroneous results
Thiocyanate, SCN^-	Colorimetry, ferric nitrate	The test measures thiocyanate in concentrations between 0.1 and 2.0 mg/L; several substances such as reducing agents or hexavalent chromium interfere

10.00, to cover more or less the entire pH range. Solid samples may be crushed and treated with an equal mass of water, shaken, and the pH of the supernatant aqueous phase measured by a pH meter. A more accurate method of determining the pH of a solid sample may involve its extraction with an aqueous solution of calcium chloride instead of reagent water, where calcium ions exchange with the hydrogen ions bound to the pores and release the hydrogen ions from the solid matrix into the solution. The pH of the supernatant solution is then measured.

21.7.2 Acidity

Acidity of water measures its quantitative capacity to react with a strong base to a designated pH, that is, any pH of interest. Acids influence the rates of chemical reactions, biological processes, and

TABLE 21.9 Determination of Inorganic Anions by Colorimetric Methods

Anion	Method	A brief outline of the method
Bromide, Br^-	Phenol red	Sample treated with a dilute solution of chloramine-T in the presence of phenol red at pH 4.5–4.7 turns the solution reddish to violet, depending on the concentration of bromide ion; presence of free chlorine, higher concentrations of Cl^- and HCO_3^- can interfere; measured at 590 nm
Chloride, Cl^-	Ferricyanide (automated)	Sample treated with mercuric thiocyanate; chloride liberates thiocyanate ion, which in the presence of ferric ion forms red ferric thiocyanate; absorbance or transmittance measured at 480 nm
Fluoride, F^-	Zirconium-dye lake	Fluoride reacts with a zirconium-dye lake, a mixture of sodium 2-(parasulfophenylazo)-1,8-dihydroxy-3,6-naphthalene disulfonate and zirconyl chloride octahydrate solutions in concentrated HCl; fluoride dissociates the dye lake, forming the colorless complex anion, ZrF_6^{2-}; color becomes lighter with increase in the amount of F^- in the sample; measured at 570 nm; samples may be distilled before analyses for removal of interfering substances
Iodide, I^-	Leuco crystal violet	Iodide is selectively oxidized to iodine by potassium peroxymonosulfate, $KHSO_5$, to produce iodine; the latter reacts instantaneously with the colorless indicator reagent containing 4,4′,4″-methylidynetris (*N*,*N*-dimethylaniline), also known as leucocrystal violet, to produce a highly colored crystal violet dye; absorbance or transmittance measured at 592 nm within the pH range 3.5–4.0 in the presence of a buffer (citric acid–ammonium hydroxide–ammonium dihydrogen phosphate)
Nitrate, NO_3^-	Cadmium reduction	Nitrate is reduced to nitrite, NO_2^-, in the presence of cadmium; the NO_2^- produced is diazotized with sulfanilamide and coupled with *N*-(1-naphthyl)-ethylenediamine dihydrochloride to form a highly colored reddish-purple azo dye that may be measured at 543 nm; certain metals, such as iron or copper at high concentrations, may lower the reduction efficiency of cadmium; EDTA may be added to samples to eliminate such interference; the test measures the concentration of both nitrate and nitrite in the sample; measure the concentration of nitrite in the sample on another sample aliquot without cadmium reduction to determine the amount of nitrate
	Hydrazine reduction	The method is similar to cadmium reduction; however, instead of cadmium granules, hydrazine hydrate is employed to reduce nitrate to nitrite
Nitrite, NO_2^-	Diazotization and coupling	The procedure is the same as above except that no reducing agent, such as cadmium or hydrazine hydrate, is added to the sample
Orthophosphate, PO_4^{3-}	Ammonium molybdate	Ammonium molybdate reacts with phosphate under acid conditions to form a heteropolyacid, molybdophosphoric acid, which in the presence of vanadium gives yellow vanadomolybdophosphoric acid; the color is measured at 400–490 nm (usually at 470 nm); certain anions at high concentrations may interfere in the test
	Molybdenum blue	Molybdophosphoric acid as formed above may be reduced by stannous chloride or ascorbic acid to produce an intensely colored molybdenum blue; the method is more sensitive than the one above; the absorbance or transmittance of the solution may be measured at 650 nm; alternatively, the molybdenum blue in

(Continued)

TABLE 21.9 Determination of Inorganic Anions by Colorimetric Methods (*Continued*)

Anion	Method	A brief outline of the method
Orthophosphate, PO_4^{3-}		aqueous solution is extracted with benzene–isobutanol and measured at 625 nm; for ascorbic acid reduction, use potassium antimonyl tartrate along with ammonium molybdate in acid medium to produce phosphomolybdic acid; arsenate present in the sample may interfere
Sulfate, SO_4^{2-}	Barium chloride (turbidimetry)	In acetic acid medium, barium chloride precipitates barium sulfate crystals of uniform size; white suspension of $BaSO_4$ is measured at 420 nm; silica in excess (above 500 mg/L) may interfere
Sulfide, S^{2-}	Methylene blue	Sample is treated with ferric chloride solution, followed by diammonium hydrogen phosphate and methylene blue reagent; an intense blue color develops; absorbance or transmittance is measured at 664 nm
Sulfite, SO_3^{2-}	Phenanthroline	Sample is acidified and purged with nitrogen gas; SO_2 gas liberated is trapped in an absorbing solution containing ferric ion and 1,10-phenanthroline; SO_2 reduces Fe^{3+} to Fe^{2+}, which complexes with 1,10-phenanthroline to produce the orange complex tris(1,10-phenanthroline)iron(II); the intensity of the color, proportional to SO_3^{2-} concentration, is measured at 510 nm
Thiocyanate, SCN^-	Ferric nitrate	In acid medium, thiocyanate reacts with ferric ion, forming an intense red color; the concentration of SCN^- in the sample is determined from the intensity of color measured at 664 nm
Cyanide, CN^-	Chloramine T-pyridine–barbituric acid	Reaction of cyanide in alkaline medium with chloramine-T reagent produces cyanogen chloride, which forms a red-blue color upon addition of pyridine–barbituric acid reagent; the intensity of color of the solution is measured between 575 and 582 nm; interference may be removed by distilling an acidified sample and purging the liberated hydrogen cyanide with nitrogen into an aqueous solution of caustic soda; calibration standards are prepared in NaOH solution
Cyanate, CNO^-	Acid hydrolysis	Cyanate hydrolyzes to ammonia (ammonium) when heated at a low pH; ammonia produced may be measured by various methods

contribute to corrosiveness. The acidity of water is determined by titration against a standard solution of an alkali and depends on the endpoint pH or indicator used. Traditionally, standard acidity is measured by titration to endpoint of pH 3.7 or 8.3. The former is known as *methyl orange acidity*, and the latter as *phenolphthalein acidity* or the *total acidity*. While the former may be determined using a color indicator, such as methyl orange or bromophenol blue, the standard endpoint of titration for acidity may be measured by using an indicator such as phenolphthalein or metacresol purple.

Acidity of a sample with respect to any pH of interest may be accurately determined from a titration curve, constructed by plotting the sample pH against the volume of titrant, added successively in small increments and recording the inflection point.

21.7.3 Alkalinity

Alkalinity measures the acid-neutralizing capacity of water. Alkalinity, like acidity, is a measure of an aggregate property of water. It is attributed to the presence of hydroxide, carbonate, and bicarbonate ions in the sample. Weak bases, such as phosphates, silicates, and borates, may also contribute to alkalinity. Alkalinity is measured by titrating a measured volume of a sample aliquot against a standard acid solution to a designated pH endpoint, usually 8.3 (phenolphthalein alkalinity) or 4.5

TABLE 21.10 Determination of Common Inorganic Anions by Titrimetric Methods

Anion	Method	A brief outline of the method
Chloride, Cl^-	Argentometric	Chloride reacts with silver nitrate to quantitatively precipitate silver chloride; in neutral or slightly alkaline solution, potassium chromate can indicate the endpoint, forming red silver chromate after all silver chloride is precipitated; bromide, iodide, and cyanide interfere; interfering effects of sulfide, sulfite, and thiosulfate ions can be removed by treatment with hydrogen peroxide; orthophosphate and iron may interfere at high concentrations, above 25 and 10 mg/L, respectively
Cyanide, CN^-	Silver nitrate	Hydrogen cyanide distilled from an acidified sample into caustic soda solution is titrated with a standard solution of $AgNO_3$, forming the soluble cyanide complex $Ag(CN)_2^-$; at the endpoint of titration, after all CN^- ions complex, the excess Ag^+ is detected by the silver-sensitive detector, *p*-dimethylaminobenzalrhodamine present in the solution; the color changes from yellow to salmon
Dichromate, $Cr_2O_7^{2-}$	Redox titration	In acid medium, orange $Cr_2O_7^{2-}$ ion is reduced to green Cr^{3+} ion; diphenylamine sulfonic acid is used as an indicator; at the endpoint the green color of the reduced form of the indicator changes to red-violet in its oxidized form
Sulfide, S^{2-}	Iodometric titration	Sample is acidified and a measured excess amount of iodine solution is added; iodine oxidizes sulfide to sulfur: $I_2 + S^{2-} \xrightarrow{acid} S + 2I^-$; the excess unreacted iodine is back-titrated with a standard solution of sodium thiosulfate or phenylarsine oxide using starch indicator; at the endpoint the blue color of the solution disappears; presence of other oxidizable substances, such as sulfite, SO_3^{2-}, and thiosulfate, $S_2O_3^{2-}$, may interfere; sulfide may be separated from the sample by precipitation with zinc acetate to give zinc sulfide; the acidified solution of the latter may be subjected to iodometric titration as described above

(total alkalinity). A color indicator such as phenolphthalein or metacresol purple may be used for the titration to pH 8.3, and bromocresol green or a mixed bromocresol green–methyl red indicator for titration to the pH endpoint 4.5. The acid titrant used for such titration is usually a standard solution of 0.02*N* or 0.1*N* sulfuric or hydrochloric acid. Alternatively, a pH meter may be used instead of a color indicator. When a pH meter is used, the standard solution of the acid titrant is added in measured amounts in successive small increments to the sample to attain the designated endpoint. Alkalinity of water is expressed as mg $CaCO_3$ per liter of sample.

If both the total and the phenolphthalein alkalinity of the water are known, then the alkalinity attributed to hydroxide, carbonate, and bicarbonate may be calculated from the relationship shown in Table 21.11.

TABLE 21.11 Alkalinity Relationships

Result from titration	Hydroxide alkalinity (as $CaCO_3$)	Carbonate alkalinity (as $CaCO_3$)	Bicarbonate alkalinity (as $CaCO_3$)
$P = T$	T	0	0
$P \geq \frac{1}{2}T$	2P – T	2(T – P)	0
$P = \frac{1}{2}T$	0	2P	0
$P \leq \frac{1}{2}T$	0	2P	T – 2P
$P = 0$	0	0	T

P = phenolphthalein alkalinity; T = total alkalinity.

21.7.4 Conductivity

Conductivity of water measures its ability to conduct electric current. It is another important aggregate property of water and depends on the presence of ions and their concentrations, their oxidation states, and the mobility and the temperature of the water. The purity of water is assessed from its conductivity. Reagent water of high quality, typically prepared by distillation, deionization, or reverse osmosis treatment of feedwater, should have a conductivity below 0.1 micromhos/cm (or resistivity above 10 megohm-cm) at 25°C, while the conductivity of potable water should be below 2 micromhos/cm. Conductivity is the reciprocal of resistivity and expressed in the unit micromhos per centimeter (μmhos/cm) or mhos per centimeter (mho/cm). In SI units, conductivity is reported as millisiemens per meter (mS/m) or microsiemens per centimeter (μS/cm). The conversion of these units is as follows:

$$1 \text{ mS/m} = 10\ \mu\text{mhos/cm}$$

$$1\ \mu\text{S/m} = 1\ \mu\text{mhos/cm}$$

Conductivity of aqueous samples in the laboratory or in the field may be measured directly using a conductivity meter. Conductivity meters equipped with temperature sensors that can measure and display a direct readout of conductivity are commercially available. Before taking a measurement, the instrument must be calibrated with conductivity standard solutions. Such standard solutions are also commercially available or may be prepared in the laboratory. A 0.0100*M* potassium chloride solution (made by dissolving 745.6 mg of anhydrous KCl in reagent water to 1 L at 25°C and stored in CO_2-free atmosphere) may be used as a standard reference solution that has a conductivity of 1412 micromhos/cm at 25°C.

21.7.5 Solids

The presence of solids in water, including total dissolved and suspended materials, may adversely affect the water quality. High concentrations of solids may affect the taste of drinking water or produce ill effects or make the water unsuitable for bathing or industrial applications. The total dissolved solids in a potable water is recommended not to exceed 500 mg/L.

The total solids in water may be determined by simple gravimetry. A measured volume of well-mixed water is evaporated in a weighed dish and dried in an oven at 103–105°C to constant weight. The difference in weights between that of the empty dish and the weight taken after evaporating all water should be equal to the weight of the total solids in the sample aliquot.

The total suspended solids in the water may similarly be determined by gravimetry at 103–105°C. The sample is well mixed and then filtered through a weighed standard glass-fiber filter. The residue collected on the filter is dried in an oven at 103–105°C to constant weight. The increase in weight in the filter is due to the total suspended solids in the sample.

The total dissolved solids in water are also determined by a similar procedure. A measured volume of well-mixed sample is filtered through a standard glass-fiber filter. While the suspended particles remain on the filter, the filtrate containing dissolved solids is evaporated to dryness in a weighed dish and dried to a constant weight at 180°C. The total dissolved solids in the sample is equal to the increase in the weight of the dish. It may be noted that dissolved solids in water are usually determined at 180°C and not at 103–105°C.

21.7.6 Hardness

Hardness measures the capacity of water to precipitate soap and is attributed primarily to the presence of calcium and magnesium ions in the water and to a small extent to the presence of other polyvalent metal ions. In general, the total hardness is defined as the sum of the concentrations of calcium and magnesium with both expressed as mg calcium carbonate per liter solution.

Hardness can be measured by titrating a measured volume of an aliquot of the sample against a standard solution of ethylenediaminetetraacetic acid (EDTA) or its sodium salt using an indicator dye, such as Eriochrome Black T or Calmagite. The pH of the sample is adjusted and maintained at 10.0 with a solution of $NH_4Cl–NH_4OH$ buffer. Upon addition of the indicator, the color of the solution turns wine red in the presence of calcium and magnesium ions in the sample. The solution is then titrated with a standard solution of EDTA. At the endpoint the color of the solution turns from wine red to blue. Usually a small amount of neutral magnesium salt of EDTA is added to the buffer before titration to yield a satisfactory endpoint.

The hardness of water can also be calculated from the concentrations of Ca^{2+} and Mg^{2+} in the solution using the following relationship. Both metals can be analyzed by atomic absorption or atomic emission spectrophotometry.

$$\text{Hardness, as mg equivalent } CaCO_3/L = 2.497[\text{Ca (mg/L)}] + 4.118[\text{Mg (mg/L)}]$$

21.7.7 Chemical Oxygen Demand

Chemical oxygen demands (COD) is a measure of the oxygen equivalent of organic matter in the sample that is susceptible to oxidation by a strong oxidizing agent. Certain inorganic oxidizable substances that may be present in the sample, such as nitrite or sulfite ion, may readily undergo oxidation under such conditions and thus may interfere in the test. The COD determination is usually carried out using a boiling mixture of potassium dichromate and H_2SO_4 that can oxidize practically all organics under refluxing condition. Other strong oxidizing agents, such as a $KMnO_4–H_2SO_4$ mixture are also effective.

A measured volume of an aliquot of a sample is refluxed with a known excess of $K_2Cr_2O_7$ in a strong acid solution in the presence of Ag_2SO_4 and $HgSO_4$. During oxidation the dichromate ion is reduced to Cr^{3+}, that is, Cr^{6+} is reduced to Cr^{3+}. The amount of Cr^{6+} consumed in the reaction is determined either by titration or by colorimetry. The Cr^{6+} is measured by titration against a standard solution of ferrous ammonium sulfate using ferroin indicator. At the endpoint of titration the color of the solution changes from greenish blue to reddish brown.

The colorimetric analysis is faster if the sample is digested with $K_2Cr_2O_7–H_2SO_4$ in an ampule or a vial under closed reflux conditions. The concentration of dichromate is measured by a spectrophotometer at 600 nm by comparing the absorbance of the solution against a standard calibration curve. COD vials are commercially available for measuring the COD of a sample in low, medium, and high ranges.

Potassium hydrogen phthalate (KHP), the theoretical COD value of which is 1.175 mg for 1.00 mg KHP, is used as a reference standard in COD analysis.

21.7.8 Biochemical Oxygen Demand

Biochemical oxygen demand (BOD) is an empirical test that measures the amount of oxygen required for microbial oxidation of organic compounds in water. Usually the BOD is measured for a 5-day incubation period, during which the amount of oxygen utilized by the microorganism to oxidize organics and the oxidizable inorganics in the sample are determined. Incubation is carried out in the dark at a temperature of 20 ± 1°C. BOD oxidation reactions follow first-order kinetics, and therefore the BOD for any incubation period may be calculated from the following relationship if the rate constant k, for the reaction is known:

$$B_t = U(1 - 10^{-kt})$$

where B_t = BOD for t-day incubation period
U = ultimate BOD, which should be more or less equal to the COD
K = rate constant for the reaction

The BOD values of wastewaters are used to calculate the waste loading capacity in the design of wastewater treatment plants.

Different volumes of sample aliquots are diluted with "seeded" dilution water to 300 mL in BOD incubation bottles filled to their full capacity, without leaving any headspace. The bottles are tightly closed and placed in an air incubator or water bath at 20 ± 1°C in the dark for 5 days. At the end of the incubation period the concentration of dissolved oxygen (DO) in the water is measured. The concentration of the DO in the dilution water before incubation is also measured. Thus, the amount of oxygen consumed by microbes in the oxidation of organics in the sample is determined from the difference between these two DO values.

The DO in the sample may be measured either by using a DO probe (electrode method) or by titration (Winkler titration). The electrode method is simple and faster and gives a direct readout of the concentration of dissolved oxygen. Oxygen-sensitive membrane electrodes are commercially available. Alternatively, the DO may be measured by Winkler iodometric titration. In this technique, divalent manganese ($MnSO_4$) is added to a measured volume of the sample. This is followed by addition of caustic soda solution, resulting in the precipitation of divalent $Mn(OH)_2$. The dissolved oxygen in the sample rapidly oxidizes an equivalent amount of dispersed Mn^{2+} precipitate to higher-valent Mn^{4+}. In the presence of iodide ion and in acid medium, Mn^{4+} is reduced back to Mn^{2+} with liberation of iodine. The liberated iodine is titrated against a standard solution of sodium thiosulfate or phenyl arsine oxide using starch indicator to a colorless endpoint.

21.7.9 Organic Carbon

Total organic carbon (TOC) is a measure of total organic content in the sample. Measurement of TOC, like that of COD or BOD, provides vital information on waste loading and operation of water and waste treatment plants.

All analytical methods are based on breaking down organically bound carbon in organic molecules and converting the carbon to carbon dioxide utilizing high temperatures, catalysts, oxidizing agents, or UV radiation. The CO_2 may be analyzed for quantitative measurement. Alternatively, the CO_2 may be analyzed by coulometric titration or by a CO_2–selective electrode.

The inorganic carbon, such as CO_3^{2-} and HCO_3^-, in the sample may be removed by treating the sample with an acid and sparging out the CO_2 produced before measuring TOC. Treating the sample with persulfate under heating or in the presence of UV radiation may also convert the organic carbon to CO_2.

Many types of TOC analyzers are commercially available.

21.7.10 Oil and Grease

Oil and grease in aqueous samples may be measured by two different methods: (1) partition gravimetry and (2) partition infrared. In the partition gravimetry method, a measured volume of sample is shaken with trichlorotrifluoroethane or, preferably, with a mixture of *n*-hexane and methyl-*tert*-butyl ether (80 : 20). The solvent(s) from the extract is distilled out in a water bath, the flask is cooled in a desiccator, and the residue is weighed.

Trichlorotrifluoroethane is used as an extraction solvent in the partition infrared method. The absorbance of the carbon–hydrogen bond in the infrared is used to measure oil and grease. An infrared spectrophotometer with a cell of 1-cm path length may be used to measure the absorbance at 3200–2700 cm^{-1} with solvent in the reference beam.

21.7.11 Kjeldahl Nitrogen

The total Kjeldahl nitrogen (TKN) is an indirect measure of most types of organic nitrogen in the water. It is the sum of organic nitrogen and the ammonia nitrogen in the sample, so the difference between TKN and ammonia nitrogen measured separately in two sample aliquots should give an estimation of organic nitrogen in the sample. TKN, however, does not measure certain nitroorganics, including oxime, semicarbazone, hydrazone, azide, nitrile, and azo, nitro, and nitroso derivatives. On the other hand, the method readily measures the amino nitrogen of many organic materials along with free ammonia. The method involves digesting on a heating device a measured volume of sample in a Kjeldahl flask in the presence of potassium sulfate, sulfuric acid, and cupric sulfate as catalyst. This converts amino organics and any free ammonia in the sample to ammonium ion. A base is then added. The ammonia is then distilled from the alkaline medium and absorbed in boric acid solution. Ammonia is then measured in the boric acid solution by titration with a standard mineral acid, or by colorimetric methods or by using an ammonia-selective electrode (*see* ammonia).

21.8 DISSOLVED GASES IN WATER

Water may contain a variety of gaseous substances at trace but significant concentrations, which may affect aquatic life or may produce adverse health effects upon human consumption. These gases may include the toxic hydrogen sulfide, chlorine, and ammonia, or the common atmospheric gases such as oxygen, nitrogen, argon, and carbon dioxide, concentrations of which above or below the normal levels may affect the water quality for drinking purposes or for industrial processes. Various analytical methods for the determination of some of the dissolved gases in water are highlighted in Table 21.12.

21.9 METALS

Metals in general can be analyzed by various techniques, including colorimetric methods, atomic absorption and atomic emission spectrometry, X-ray diffraction and fluorescence methods, neutron activation analysis, ion-selective electrode methods, ion chromatography, eletrophoresis, redox titration, and gravimetry. Although these techniques all have their own merits and limitations, atomic spectroscopy is the most widely used analytical tool for industrial analysis of metals and is currently applied exclusively in environmental analysis for the determination of metals in aqueous and nonaqueous matrices. Atomic absorption and emission spectroscopy are discussed fully in Sec. 8. Their applications in environmental analysis are briefly highlighted here.

Both atomic absorption spectrometry (AAS) and atomic emission spectrometry (AES) provide rapid, multielement determination of metals at low ppm and ppb levels as found in ground-, surface-, and wastewaters and in soils, sediments, and hazardous wastes. Some of these metals are toxic and are regulated by the U.S. Environmental Protection Agency. Some of them, known as priority pollutant metals and classified under various EPA programs include aluminum, arsenic, antimony, barium, beryllium, cadmium, chromium, cobalt, copper, iron, lead, manganese, molybdenum, nickel, silver, selenium, thallium, vanadium, and zinc. Analysis of these metals by atomic spectroscopic methods is discussed below in brief.

21.9.1 Flame and Furnace Atomic Absorption Spectrophotometry

The AAS methods are of two types; (1) flame and (2) furnace, differing only in atomization process. In flame AA spectrometry the heat source is a flame, air–acetylene or air–nitrous oxide. The sample is aspirated into the flame and atomized. The metal atoms absorb energy at their own characteristic wavelengths from a light beam directed through the flame. The energy absorbed is proportional to the concentration of the metal in the sample. In the furnace mode of AAS the heat source is a graphite furnace heated electrically, producing a temperature much higher than that obtained from flame. Thus, better sensitivity and a much lower detection limit for metals are obtained in furnace mode.

TABLE 21.12 Determination of Dissolved Gases in Water

Dissolved gas	Analytical methods
Oxygen (DO)	(1) Winkler or iodometric titration: Divalent manganese solution is added to the sample, followed by a strong alkali in a glass-stoppered bottle. The divalent manganous hydroxide, obtained as a precipitate and dispersed in the sample mixture, is rapidly oxidized by the dissolved oxygen in the sample, giving an equivalent amount of manganese in higher valency states. In the presence of iodide and in acid medium, the higher-valent manganese reverts to Mn^{2+} with the liberation of iodine equivalent to the original dissolved oxygen in the sample. Iodine is titrated with a standard solution of sodium thiosulfate or phenyl arsine oxide using starch indicator. At the endpoint of titration the blue color of the solution decolorizes. Alternatively, the liberated iodine can be measured directly by an absorption spectrophotometer. The iodometric method has been modified to minimize the effect of interfering substances. In the presence of Fe^{3+} and NO_2^- ions, the sample should be treated with sodium azide and sodium hydroxide solution after adding Mn^{2+} solution and potassium fluoride before acidification. Interference from Fe^{2+} may be removed by addition of a small amount of $KMnO_4$–H_2SO_4 and then potassium oxalate solution to remove the permanganate color completely before analysis as above. Interference from suspended solids may be removed by alum flocculation modification, in which a solution of aluminum potassium sulfate and ammonium hydroxide is added, the sample is allowed to settle, and the clear supernate is siphonned off for analysis as before. (2) Membrane electrode method: Unlike Winkler titration, membrane electrodes provide a simple, rapid, and excellent method for measuring dissolved oxygen in situ in all kinds of polluted waters, including highly colored waters and strong wastes. These oxygen-sensitive membrane electrodes are used to measure DO both in the field and in the laboratory. Portable DO meters equipped with direct display of the concentrations of DO in water are commercially available.
Ozone	Ozone (or residual ozone after water treatment) in water may be measured by the indigo colorimetric method. In acidic solution, ozone rapidly decolorizes the color of indigo (potassium indigo trisulfonate) solution. The decrease in absorbance is inversely proportional to increasing ozone concentration and is measured at 600 nm. A minimum detectable concentration of 10–20 μg ozone per liter may be measured by this method.
Chlorine	Chlorine (total or free chlorine) may be found in water in the form of molecular chlorine (Cl_2), hypochlorous acid (HClO), and hypochlorite ion (ClO^-), and the relative proportion of these are pH- and temperature-dependent. HClO, the hydrolyzed form of Cl_2 and the ClO^-, however, predominate at the pH of most waters. (1) DPD ferrous titration: The sample is titrated with a standard solution of ferrous ammonium sulfate (FAS) using *N,N*-diethyl-*p*-phenylenediamine (DPD) as indicator. In the absence of iodide ion, free chlorine reacts with DPD, producing a red color. A phosphate buffer solution is prepared from Na_2HPO_4 (24 g), KH_2PO_4 (46 g), and 0.8 g of EDTA disodium salt dissolved in 1 L of water. To 5 mL of this phosphate buffer solution, 5 mL of DPD indicator solution are added. This is followed by addition of 100 mL of sample (or a suitable volume of diluted sample). The mixture is titrated with a standard solution of FAS (1.106 g/L of distilled water containing 1 mL of 1 + 3 H_2SO_4) until the red color disappears. 1 mL FAS standard solution = 1.00 mg Cl as Cl_2/L. (2) DPD colorimetric test: This method is very similar to the above DPD titration, except that the mixture is not titrated against a standard solution of FAS. To a small volume of phosphate buffer and DPD reagent mixture as prepared in the above test, and in a test tube, add 10 mL of sample and mix well. The absorbance is read at 515 nm. The concentration of chlorine is determined from the chlorine standard calibration curve. Chlorine standard solutions are prepared in the range 0.05–4 mg/L and standardized by iodometric titration with a standard solution of $Na_2S_2O_3$ using starch indicator. The DPD colorimetric test can measure chlorine as Cl_2 at approximately a minimum detection level of 10 μg/L. (3) Syringaldazine colorimetric test: The free chlorine reacts with syringaldazine (3,5-dimethoxy-4-hydroxybenzaldazine) in 2-propanol to produce a colored product. The intensity of color is measured at 530 nm. The optimum color in the solution is obtained at a pH between 6.5 and 6.8. This method measures free chlorine over the range 0.1–10 mg/L. (4) Amperometric titration: Free chlorine in the sample may be determined by titrating the sample at pH 6.5–7.5 against a standard solution of phenyl arsine oxide, observing current changes on a microammeter. Titrant is added progressively in smaller increments until all needle movement ceases, that is, until there is no needle response. Sluggish needle movements signal the approach of the endpoint.

TABLE 21.12 Determination of Dissolved Gases in Water (*Continued*)

Dissolved gas	Analytical methods
Chlorine dioxide (ClO_2)	DPD colorimetric or titrimetric test: Chlorine dioxide may be distinguished from chlorine, chlorite, or hypochlorite by DPD colorimetric or titrimetric test. To 100 mL of sample add 2 mL of glycine solution (10 g glycine/100 mL distilled water). In another container, mix 5 mL of phosphate buffer solution and 5 mL of DPD indicator solution (see Chlorine). Add sample–glycine mixture to buffer–indicator mixture. A red color forms immediately. The red solution is titrated with a standard solution of FAS until the color disappears. Alternatively, the absorbance of the red solution is measured at 515 nm and ClO_2 content in the sample is determined from a standard calibration curve. The standard solutions of ClO_2 are prepared from dissolving $NaClO_2$ in water and adding 5*N* H_2SO_4; the ClO_2 gas is passed through a saturated $NaClO_2$ scrubber solution and then collected in distilled water. The FAS titration method measures ClO_2 as chlorine (Cl). The results, therefore, have to be multiplied by 1.9.
Cyanogen chloride (CNCl)	Pyridine–barbituric acid colorimetric test: Cyanogen chloride is unstable and hydrolyzes to cyanate (CNO^-) at a pH of 12 or more. The analysis should therefore be done immediately after the collection of sample. To 20 mL of sample in a 50-mL volumetric flask, add 1 mL of phosphate buffer (138 g $NaH_2PO_4 \cdot H_2O$ in 1 L of water), stopper, and mix by inversion one time. Allow the solution to stand for a minute. This is followed by addition of 5 mL pyridine–barbituric acid reagent. Stopper and mix the solution one time by inversion. Allow the color to develop over 3 min. Dilute the solution with reagent-grade water to 50 mL. Mix thoroughly and allow to stand for another 5 min. Measure absorbance of the red-blue solution at 578 nm in a 1-cm cell using distilled water as reference. Determine the concentration of CNCl in the sample from a cyanide calibration standard (CN^- standard solution + NaOH dilution solution + chloramine – T reagent + phosphate buffer; mix and then add pyridine–barbituric acid reagent; dilute to the same volume with water and let stand for color development). [See Cyanide, under Colorimetric Tests.]
Ammonia (NH_3)	Ammonia in water may be measured by several methods, including colorimetry, titration, and electrode techniques, and can be done by both manual and automated methods. A preliminary distillation of sample is required for titrimetric analysis. However, in other methods the distillation step may be omitted. It is recommended that the sample be distilled if it is dirty or interfering substances are present. If the test cannot be performed within 24 h after collection, then acidity the sample to a pH below 2 and store in a refrigerator at 4°C. In such a case, bring the sample to room temperature and neutralize before analysis. (1) Ammonia-selective electrode method: Ammonia-selective electrodes are commercially available to measure ammonia in water. A chloride-ion-selective electrode serves as the reference electrode. A pH meter having an expanded millivolt scale or a specific ion meter is used for potentiometric measurements. Dissolved ammonia and NH_4^+ ion are measured by this method in the range 0.03–1500 mg NH_3 – N/L. Amines interfere in the test. (2) Phenate colorimetric test: The sample is treated with phenol solution followed by addition of sodium nitroprusside solution and an oxidizing solution (a mixture of sodium hypochlorite, trisodium citrate, and sodium hydroxide in deionized water). The solutions are thoroughly mixed and allowed to stand. An intense blue color of indophenol developes. The intensity of color is measured at 640 nm. The concentration of ammonia is determined from a calibration standard curve. (3) Titration: A suitable volume of sample, accurately measured, is distilled in a distillation flask. Ammonia distilled out is absorbed and collected in a solution of boric acid containing the mixed color indicators methyl red and methylene blue. The distillate is titrated against a standard solution of 0.02*N* H_2SO_4 to a pale lavender color endpoint. The lower detection limit of NH_3 – N in the sample depends on the volume of the sample distilled and the normality of the standard titrant.
Carbon dioxide (CO_2)	Titrimetric method: A measured volume of sample is titrated with a 0.01*N* standard solution of sodium bicarbonate ($NaHCO_3$) containing phenolphthalein indicator to an endpoint where the solution turns pink. Alternatively, add a few drops of phenolphthalein indicator solution (in ethanol) to the sample and titrate against a standard solution of sodium hydroxide to the pink endpoint. The sample may also be titrated potentiometrically to a pH of 8.3. The titration should be carried out immediately after the sample is collected. Free CO_2 in water may also be evaluated by a nomographic method if the pH and the temperature of the sample, as well as the bicarbonate alkalinity and the total dissolved solids content of the sample, are known.

TABLE 21.13 Recommended Wavelength, Flame Type, and Technique for Flame Atomic Absorption Analysis

Element	Wavelength (nm)	Flame	Technique
Aluminum	309.3	N_2O–acetylene	DA, CE
Antimony	217.6	Air–acetylene	DA
Arsenic	193.7	Air–hydrogen	H
Barium	553.6	N_2O–acetylene	DA
Beryllium	234.9	N_2O–acetylene	DA, CE
Bismuth	223.1	Air–acetylene	DA
Cadmium	228.8	Air–acetylene	DA, CE
Chromium	357.9	Air–acetylene	DA, CE
Cobalt	240.7	Air–acetylene	DA, CE
Copper	324.7	Air–acetylene	DA, CE
Iron	248.3	Air–acetylene	DA, CE
Lithium	670.8	Air–acetylene	DA
Lead	283.3, 217.0	Air–acetylene	DA, CE
Magnesium	285.2	Air–acetylene	DA
Manganese	279.5	Air–acetylene	DA, CE
Molybdenum	313.3	N_2O–acetylene	DA
Nickel	232.0	Air–acetylene	DA, CE
Silver	328.1	Air–acetylene	DA, CE
Selenium	196.0	Air–hydrogen	H
Silicon	251.6	N_2O–acetylene	DA
Strontium	460.7	Air–acetylene	DA
Tin	224.6	Air–acetylene	DA
Titanium	365.3	N_2O–acetylene	DA
Vanadium	318.4	N_2O–acetylene	DA
Zinc	213.9	Air–acetylene	DA, CE

Note: DA, direct aspiration; CE, chelation extraction; H, hydride generation.
Source: P. Patnaik, *Handbook of Environmental Analysis,* CRC Press, Boca Raton, FL, 1997.

Another advantage of the graphite furnace technique over the conventional flame method is that the former requires a smaller volume of sample. On the other hand, a disadvantage of the furnace technique is that because of its high sensitively, interference from other substances present in the sample is often manifested. Such interference can be removed or reduced by adding a matrix modifier to the sample or by correcting for background absorbance. Analysis in flame mode is simple, and if the metals are present in relatively high concentrations—above 1 mg/L—the method gives satisfactory results. The wavelengths recommended for the metals of environmental interest, the flame type, and the techniques for sample introduction are presented in Table 21.13.

21.9.2 Inductively Coupled Plasma Atomic Emission Spectroscopy (ICP-AES)

ICP-AES offers certain advantages over furnace AAS. Although the latter has lower detection limits than the former, ICP-AES provides simultaneous or sequential multielement analysis in a simple analysis. Thus, several metals can be determined rapidly in a sample in a simple run. Also, a single point calibration suffices in IEC-AES analysis and, unlike furnace AA, chemical interference is very low. The principle of the ICP method is discussed in Sec. 8, Atomic Spectroscopy. The recommended wavelengths and the instrument detection levels for some selected metals are highlighted in Table 21.14.

TABLE 21.14 **Recommended Wavelength and Instrument Detection Level for ICP-AES**

Element	Wavelength recommended (nm)	Alternative wavelength (nm)	Approximate detection limit (μg/L)
Aluminum	308.22	237.32	50
Antimony	206.83	217.58	30
Arsenic	193.70	189.04	50
Barium	455.40	493.41	2
Beryllium	313.40	234.86	0.5
Boron	249.77	249.68	5
Cadmium	226.50	214.44	5
Calcium	317.93	315.89	10
Chromium	267.72	206.15	10
Cobalt	228.62	230.79	10
Copper	324.75	219.96	5
Iron	259.94	238.20	10
Lead	220.35	217.00	50
Lithium	670.78	—	5
Magnesium	279.08	279.55	30
Manganese	257.61	294.92	2
Molybdenum	202.03	203.84	10
Nickel	231.60	221.65	15
Potassium	766.49	769.90	100
Selenium	196.03	203.99	75
Silica (SiO_2)	212.41	251.61	20
Silver	328.07	338.29	10
Sodium	589.00	589.59	25
Strontium	407.77	421.55	0.5
Thallium	190.86	377.57	50
Vanadium	292.40	—	10
Zinc	213.86	206.20	2

Source: P. Patnaik, *Handbook of Environmental Analysis,* Boca Raton, FL, 1997.

21.9.3 Inductively Coupled Plasma Mass Spectrometry (ICP-MS)

ICP-MS provides a method for multielement determination of dissolved metals at ultratrace levels. ICP-MS combines inductively coupled plasma with a quadrupole mass spectrometer. ICP of high energy generates charged ions from the atoms of the elements present in the sample. The ions generated are directed onto a mass spectrometer, separated, and measured according to their mass-to-charge ratio. The method is highly sensitive, and the detection limits for some metals may be 100 times lower than that obtained by graphite furnace AA technique. Nonmetals and isotopes can also be measured by ICP-MS. However, ICP-MS has not yet found much application in routine environmental analysis of metals because of the high cost of the instrument and also because furnace AA and the ICP-AES can effectively achieve the detection levels for metals required for regulatory purpose.

21.9.4 Sample Digestion

To measure metals by the instrumental methods discussed above, a nonaqueous sample must be brought into aqueous phase. The metals must be solubilized in water. Even for aqueous samples, such as wastewaters or aqueous sludges that may contain high suspended solids, such sample digestion for extraction of metals is essential. Additionally, by reducing the volume of the final sample extract to a

TABLE 21.15 Acid Combinations for Sample Preparation

Acid combination	Suggested use
HNO_3–HCl	Sb, Sn, Ru, and readily oxidizable organic matter
HNO_3–H_2SO_4	Ti and readily oxidizable organic matter
HNO_3 $HClO_3$	For organic materials that are difficult to oxidize
HNO_3–HF	Siliceous meterials

Source: P. Patnaik, *Handbook of Environmental Analysis,* CRC Press, Boca Raton, FL, 1997.

small volume, the detection levels of metals present in the samples may be lowered accordingly. Normally, a sample, aqueous or nonaqueous, is digested with a strong acid, usually nitric acid or its combination with another acid or a strong oxidizing agent, to solubilize metals and bring the solution into aqueous phase. Various acid combinations for sample preparation are suggested in Table 21.15.

21.9.5 Chelation Extraction Method

Many metals, such as Cd, Cr, Co, Pb, Ni, Mn, and Ag, at low concentrations may be extracted with a chelating agent, such as ammonium pyrrolidine dithiocarbamate (APDC). The metal chelate formed is then extracted with a suitable solvent, such as methyl isobutyl ketone (MIBK). The MIBK extract is then aspirated directly into the air–acetylene flame. Other chelating agents, such as 8-hydroxyquinoline, are often used. If an emulsion forms at the interface between the MIBK and water, use anhydrous Na_2SO_4 to break up the emulsion.

21.9.6 Hydride Generation Method

Metals such as arsenic and selenium are converted into their hydrides in an HCl medium by treatment with sodium borohydride. The hydrides formed are purged into the atomizer with argon or nitrogen for conversion into gas-phase atoms. When the sample is digested with nitric acid these metals are oxidized to their higher valence states, As(V) and Se(VI). The digested sample should therefore be heated with concentrated HCl (and sodium iodide for As determination) to reduce these metals into their lower oxidation states, As(III) and Se(IV), for conversion into their hydrides. The calibration standards for these metals should also be converted into their hydrides in the same manner.

21.9.7 Cold Vapor Method for Measuring Mercury

Mercury is analyzed by cold vapor AA technique. Nitric acid digestion of sample converts mercury into its nitrate. When the solution is treated with stannous chloride, mercury is reduced to its elemental form and volatilizes to vapor. Under aeration the vapor is carried into an absorption cell. The absorbance is measured at 253.7 nm. The calibration standards are subjected to similar oxidation, reduction, and vaporization. Interference from sulfide and chloride is removed prior to reduction by treatment with $KMnO_4$. Free chlorine formed from chloride is removed by treating with hydroxylamine sulfate and sweeping the sample gently with air.

Bibliography

American Public Health Association, American Water Works Association, and Water Environment Federation, *Standard Methods for the Examination of Water and Wastewater*, 20th ed., American Public Health Association, Washington, DC, 1998.

American Society for Testing and Materials, *Annual Book of ASTM Standards, Volume 11.01, Water and Environmental Technology*, American Society for Testing & Materials, Philadelphia, PA, 1989.

Burrell, D. C., *Atomic Spectrometric Analysis of Heavy-Metal Pollutants in Water*, Ann Arbor Science Publishers, Ann Arbor, MI, 1975.

Patnaik, P., *Handbook of Environmental Analysis*, CRC Press, Boca Raton, FL, 1997.

U.S. Environmental Protection Agency, *Methods for Organic Chemical Analysis of Municipal and Industrial Wastewater*, Environmental Monitoring and Support Laboratory, Cincinnati, OH, 1982.

U.S. Environmental Protection Agency, *Methods for Chemical Analysis of Water and Wastes*, Environmental Monitoring and Support Laboratory, Cincinnati, OH, 1983.

U.S. Environmental Protection Agency, *40 CFR Part 136, Federal Register, 49, No. 209*, 1984.

U.S. Environmental Protection Agency, *Methods for the Determination of Metals in Environmental Samples—Supplement I,* EPA 600/R-94-111, 1994.

U.S. Environmental Protection Agency, *Test Methods for Evaluating Solid Waste: Physical/Chemical Methods, SW-846,* National Technical Information Service, Washington, DC, 1997.

SECTION 22
AIR ANALYSIS

22.1 INTRODUCTION

Analysis of ambient air is critical to identify pollutants present in the air and to determine their concentrations. These analyses involve several steps that begin with extensive preplanning depending on the objective—the foremost being whether the air is indoor and confined or outdoor or atmospheric. Indoor air analysis is usually carried out to ascertain occupational safety in a workplace, to maintain industrial hygiene, or to address possible health-related problems with respect to a specific case history. The monitoring of atmospheric air, in contrast, is usually done year around, especially in urban areas, to measure the extent of air pollution. A comprehensive sampling plan, therefore, is the starting point for air analysis.

The air sampling plan involves determination of proper sites for sampling, and the time, duration, and number of samples to be collected. For outdoor air, weather conditions, topography, humidity, and altitude are the critical factors in selecting sampling sites. A detailed discussion of sampling plans is beyond the scope of this text. Readers interested in further information may refer to the *Annual Book of ASTM Standards* and Patnaik's *Handbook of Environmental Analysis* as listed in the Bibliography. Various techniques employed in air sampling, especially pertaining to indoor workplaces and the methods of analysis, are briefly highlighted in this section. The National Institute of Occupational Safety and Health (NIOSH) has developed and validated a number of methods for measuring organic and inorganic pollutants in indoor air. Some of these methods are cited below. Also, the U.S. Environmental Protection Agency (EPA) has developed a series of sampling and analytical methods for a small number of pollutants belonging to certain classes of organics. They too are discussed briefly below. Also presented below are some general guidelines that may be followed to analyze pollutants for which no method is known.

22.2 AIR SAMPLING AND ANALYSIS

The term *air sampling* refers to collection of air or trapping of the air for analysis. Usually, it refers to trapping pollutants in the air by various techniques, identifying them, and measuring their concentrations in the air. Direct air sampling methods include collection of air from the sites in Tedlar bags, canisters, or any appropriate container following repeated evacuation of the containers to flush out the existing air, or collecting the air in a canister under pressure. Also, air may be liquified under low temperature and high pressure and collected in a canister and brought to the laboratory for analysis. In this case, the existing air in the container is first flushed out repeatedly

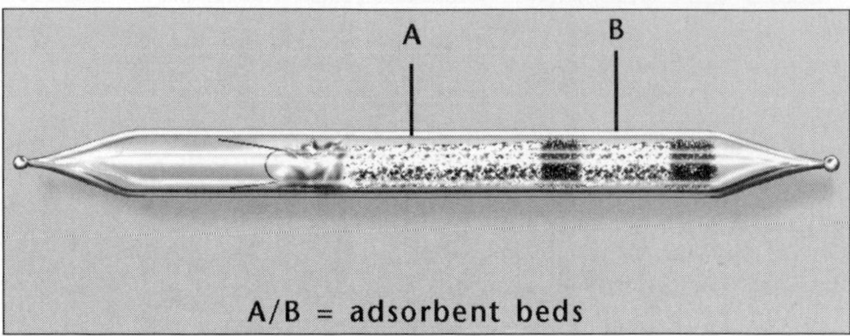

FIGURE 22.1 Solvent desorption tube. (*Supelco Catalog 2003, Supelco Inc., Bellefonte, Pa.*)

at the sampling site and then the sampled air in the container is placed in a liquid argon or oxygen bath. Such direct air sampling methods may only be applicable to determine organic pollutants present in the air or to measure pollutants that are gaseous at ambient temperature.

The most common sampling technique, however, involves passing a measured volume of air either through a tube packed with adsorbent materials or through a filter cassette or through an impinger solution, depending on the nature of the analyte. Among the adsorbent materials, activated charcoal is most commonly used to trap many types of organic pollutants in air. Other adsorbents include carbon molecular sieves, Tenax (2,6-diphenylene oxide), many types of porous polymers under various trade names, and silica gel. This last is used to trap polar organic molecules such as lower aliphatic alcohols, aldehydes, and ketones. The adsorbent may alternatively constitute a derivatizing substance to convert the pollutants to derivatives for easier determination. For example, formaldehyde may be converted to 2-benzyloxazolidine by passing air through a solid sorbent tube containing 2-(benzylamino)ethanol on Chromosorb 102 or XAD-2 . Glass beads sometimes are packed in cryogenically cooled traps to increase surface area, providing inert and thermally stable adsorption sites. Often they are used as a filter at the inlet end of multibed adsorbent tube to condense large molecules from air.

The adsorbent tubes usually have two sections, the front and the back sections, separated by a thin porous pad. The back portion is meant to trap any material that escapes from the front or when the latter is fully saturated with analyte molecules. A diagram of a typical sampling tube is shown in Fig. 22.1. The open ends of the adsorbent tube are packed with glass wood to prevent any mechanical loss of adsorbent during air sampling. A sampling pump, usually a small vacuum pump, is attached to the back end of the adsorbent tube. Air is sucked in by turning on the pump. Air is allowed to flow at the desired rate through the tube from the front to the back section, passing through the adsorbent packing. The pump is calibrated before use to regulate the flow rate of the air and to measure the total volume of air sampled. At the end of sampling, the sampling tube is disconnected from the pump and both ends of the tube are capped for storage or shipment.

When air is sampled to measure particulate matter, metal dusts, or other solid particles, a membrane filter of appropriate pore size, such as 0.45 μm, is placed on a cassette and the latter is attached to a sampling pump. The solid particles are retained on the filter. The total volume of air sampled is calculated from the flow rate and the total time of sampling.

Water-soluble pollutants present in air may be collected as aqueous solutions by bubbling the air through a measured volume of water in an impinger. Acid or alkaline vapors or gases may be absorbed in the solution by this technique rather than being adsorbed over a solid adsorbent. Also, organic solvents may be used instead of water in such impingers to collect organic pollutants that are readily soluble in such solvents. The vapor pressure, toxicity, and all safety factors, however, must be taken into consideration before using such organic solvents to absorb soluble pollutants from air. Solid adsorbents are usually preferred over organic solvents to trap pollutants from air.

The general sampling techniques discussed above are summarized in Table 22.1.

22.2.1 Desorption of Pollutants from Sorbent Tubes

Pollutants trapped over the adsorbents may be desorbed either by using an appropriate solvent or by thermal desorption. Solvent desorption methods are more common in air analysis, especially for

TABLE 22.1 Air Sampling Techniques

Sampling technique	Types of pollutants	Outline of the method
Direct sampling:		
At ambient temperature and pressure	Many common gases, e.g., CH_4, CO_2, N_2O	Air is collected from sampling sites in Tedlar bags or glass flasks after repeated evacuation and flushing of the containers with air from the site; the air is injected directly onto a GC column for separation of pollutants and their determination by thermal conductivity detection (TCD) or mass spectrometry
Under cooling	Organic vapors and gases	Air is collected in a canister and placed in a liquid argon or oxygen bath; organic vapors and many gaseous substances condense and collect in the canister; the pollutants are transported from the canisters and interfaced onto a GC injector port with heating under helium or nitrogen flow
Under pressure	Organic vapors and gases, and many other common gases	Canisters are pressurized with air from the sampling site using an air pump after flushing out its inside; the containers are brought to the laboratory and interfaced to a GC; the air in the canisters is transferred to the GC port under cryogenic cooling or by other techniques.
Adsorbent tubes:		
Activated charcoal	Most organic pollutants	Air is passed through the front and back sections of an adsorbent tube; adsorbed organics are desorbed out from activated carbon by desorption with a solvent such as CS_2 or by heating under vacuum; desorbed organics are measured by GC, HPLC, or GC/MS
Tenax	Many organic substances	Sampling procedure is the same as above; adsorbed compounds are desorbed using a suitable solvent and analyzed
Silica gel	Acid vapors or acidic gases, such as HCl; polar organic molecules	Desorption by a weak base, such as, Na_2CO_3 or $NaHCO_3$; the solution is analyzed by an appropriate instrumental method
Adsorbent coated with derivatizing agents	Many organics that can readily be derivatized, such as aldehydes and ketones	Sampling procedures are the same as above; the organic derivatives in the sorbent tube are desorbed into an appropriate solvent using an ultrasonic bath; the solution is analyzed by GC, GC/MS, HPLC, or other instrumental technique
Adsorbing solvents or solution in an impinger two-necked flask or other suitable apparatus	Water-soluble inorganic pollutants or organic substances that are soluble in water or nonvolatile organic solvents	Using an air sampling pump, a measured volume of air is bubbled slowly at a constant rate through an appropriate solvent or solution; the pollutants in the air dissolve in the solvents (or solutions) and the solutions may be analyzed by various wet methods or instrumental techniques

(Continued)

TABLE 22.1 Air Sampling Techniques (*Continued*)

Sampling technique	Types of pollutants	Outline of the method
Filter cassettes	Particulate matter, dust particles, metal powder	A measured volume of air is passed through a membrane filter of appropriate pore size placed on a cassette; the particulate matter and dust collected on the filter is weighed to determine its concentration in air; the powder may be digested in nitric acid alone or in combination with another acid or oxidizing agent; the solution is diluted and analyzed for metals by atomic absorption (AA) or inductively coupled plasma (ICP) spectrophotometry

measuring pollutants in the workplace. The methods for analyzing a number of organic pollutants in the air developed by NIOSH using gas chromatography/flame ionization detection (GC/FID) or gas chromatography/mass spectrometry (GC/MS) are based on solvent desorption. An important criterion in solvent desorption is to know the *desorption efficiency* of the solvent for the pollutant. That is, one must know how much of the pollutants retained on the adsorbent material will desorb out into that solvent when transferred into a small measured volume of that solvent. Also, the conditions under which the desorption attained reaches maximum must be known. Such conditions may include contact time of adsorbent material with the solvent, swirling rate, the volume of solvent, and the temperature required to achieve best desorption. The desorption efficiency of a solvent for a compound may be determined by "spiking" a known amount of the compound to the adsorbent, placing the latter in contact with the solvent, and analyzing the resulting solution by GC to measure percent spike recovery.

Thermal desorption is an alternative to solvent desorption. The adsorbent tube is placed in a thermal desorption device connected to the GC injector port. The tube is electrically heated and the adsorbed molecules are desorbed under vacuum or with an inert carrier gas to be transported onto the injector port of a temperature-programmed GC for analysis.

22.2.2 Analysis

Organic pollutants are mostly analyzed by GC, GC/MS, or high-performance liquid chromatography (HPLC) techniques. GC methods commonly employ flame ionization detectors. Halogenated substances are detected by electron-capture detection or Hall electrolytic conductivity detection. GC detectors that may measure nitrogen-, phosphorus-, or sulfur-containing organics other than FID include the nitrogen–phosphorus detector and the flame photometric detector. Thermal conductivity detectors are employed to analyze common gases such as methane or nitrogen oxides. Compounds containing double bonds such as olefins or aromatics may conveniently be determined with a photoionization detector (PID). HPLC detectors in common use are UV and fluorescence types. Many U.S. EPA methods are based on GC/MS determination. Although its sensivity is lower than with most GC or HPLC detectors, an advantage is that it identifies unknown compounds from their mass spectra. Various portable instruments, mostly consisting of infrared detectors, are commercially available to measure in-situ volatile organic substances, solvent vapors, and organic and inorganic gases.

Metal dusts and fumes are usually determined by atomic absorption and emission spectrometry. Particulate matter in the air is collected on filter cassettes and their weight determined by gravimetry. Fibrous materials and dusts may also be analyzed by electron microscopy and X-ray methods. Other general analytical techniques for many types of pollutants include colorimetry, ion-specific electrodes, and titrimetry. Badges or tubes are commercially available to routinely monitor specific pollutants in the workplace from color change in the material.

Various air sampling techniques are outlined in Table 22.1. Air analysis of a number of pollutants by NIOSH and EPA methods are summarized in Table 22.2 and Table 22.3, respectively.

TABLE 22.2 **Analysis of Common Pollutants by NIOSH Methods**

Specific compound or class of pollutants	Examples	NIOSH method(s)	Outline of the method
Acetonitrile (cyanomethane, methyl cyanide)	—	1606	1–25 L of air at a flow rate of 0.01–0.2 L/min is passed through coconut shell charcoal (400 mg/200 mg); analyte is desorbed with 2 mL of methylene chloride–methanol mixture (85:15) in an ultrasonic bath for 45 min; the solution is analyzed by GC/FID using a fused-silica capillary column, cross-bonded polyethylene glycol (PEG), or equivalent
Acids, mineral	Hydrochloric, hydrobromic, sulfuric, nitric, phosphoric, hydrofluoric acids	7903	Between 3 and 100 L of air at a flow rate of 0.2–0.5 L/min are passed through a solid sorbent tube containing washed silica gel (400 mg/200 mg) with a glass-fiber filter plug; the analytes are desorbed from the solid sorbent with 10 mL of 0.0017*M* $NaHCO_3$/0.0018*M* Na_2CO_3 solution; the acid anions are determined by ion chromatography using an appropriate anion separator column and $NaHCO_3/Na_2CO_3$ eluent at the above concentrations
Alcohols	Ethanol, 2-propanol *n*-butanol	1400, 1401	Adsorbed over coconut charcoal; desorbed with a 1% solution of another alcohol in CS_2; analyze by GC/FID (volume of air sampled is usually lower for lower alcohols)
Aldehydes	Formaldehyde, acetaldehyde, acrolein, furfural	2539, 2538	Adsorbed on 10% 2-(hydroxy methyl) piperidine on XAD-2; derivatized to oxazolidine derivative; desorbed into toluene under ultrasonication; analyzed by GC/FID or GC/MS
Alkaline dust	Caustic soda, caustic potash, lye	7401	Between 70 and 1000 L of air at a flow rate of 1–4 L/min are passed through a 1-μm PTFE membrane filter; after sampling, the filter placed in 5 mL of 0.01*N* HCl for 15 min under nitrogen with stirring; the alkalinity of the solution is measured by acid–base titration using a pH electrode
Amines, aliphatic	Dimethylamine, triethylamine	2010	Between 3 and 30 L of air at a flow rate of 0.01–1.0 L/min are passed through a solid sorbent tube containing silica gel (150 mg/75 mg); analytes are desorbed with 1 mL of dilute H_2SO_4 aqueous methanol during 3 h of ultrasonication; analytes are analyzed by GC/FID using a column containing 4% Carbowax 250M/0.8% KOH on Carbosieve B (60/80 mesh)
Amines, aromatic	Aniline, 2-amino toluene, *N*,*N*-dimethyl aniline	2002	Air is passed through a silica gel sorbent tube (150 mg/75 mg); analytes are desorbed into 1 mL of 95% ethanol in an ultrasonic bath over 1 h; the solution is analyzed by GC/FID using a packed column such as Chromosorb 103 or equivalent

(Continued)

TABLE 22.2 Analysis of Common Pollutants by NIOSH Methods (*Continued*)

Specific compound or class of pollutants	Examples	NIOSH method(s)	Outline of the method
Amines, aromatic	Aniline, aminotoluene	2017	5–50 L of air at a flow rate of 0.2 L/min are passed through an assembly of a glass-fiber filter and a sorbent tube packed with H_2SO_4-treated silica gel (520 mg/260 mg); the analytes are desorbed into 2 mL of ethanol; the solution is analyzed by GC/FID using an appropriate column
Asbestos and other fibers	Actinolite, amosite, crocidolite, tremolite, fibrous glass	7400, 7402	Airborne fibers are collected on a 0.45–1.2-μm cellulose ester membrane placed on a filter cassette; fibers are counted by phase-contrast light microscopy to differentiate asbestos and nonasbestos fibers; air flow should be between 0.5 and 16 L/min, and the total volume of air to be sampled may be adjusted to give 100–1300 fiber/mm^2
Benzoyl peroxide	—	5009	Pass 40–400 L of air at a flow rate of 1–3 L/min through a 0.8-μm cellulose ester membrane filter; analyte is desorbed with 10 mL of ethyl ether and analyzed by HPLC/UV at 254 nm using a pressure column and a 70 : 30 methanol/water mobile phase
Biphenyl	—	2530	Analyte is adsorbed on Tenax in a solid sorbent tube; desorbed with CCl_4; analyzed by GC/FID
Boron carbide	—	7506	Particles are collected on a filter consisting of a10-mm Higgins-Dewell or nylon-type cyclone and a 5-μm PVC membrane; ashed in RF plasma asher; suspended in 2-propanol and measured by X-ray powder diffraction; between 100 and 1000 L of air at a flow rate of about 2 L/min may be sampled
1,3-Butadiene	—	1024	Between 3 and 25 L of air at a flow rate of 0.01–0.5 L/min are passed through a solid sorbent tube containing coconut charcoal; desorbed into methylene chloride over 30 min standing; analyzed by GC/FID using a fused-silica capillary column
Bromine and chlorine	—	6011	Air is passed through an assembly of a prefilter and a filter consisting of PTFE, 0.5 μm, and a silver membrane, 0.45 μm and 25 mm; halogens are converted to halide ions; extracted with 3 mL of 6 mM $Na_2S_2O_3$ solution; analyzed by ion chromatography using a conductivity detector
Bromoxynil and its octanoate	Dibromocyanophenol	5010	Analytes are collected on a 2-μm PTFE membrane filter; extracted with 3 mL of acetonitrile; analyzed by reverse-phase HPLC using a UV detector at 254 nm; the compounds may also be determined by GC

TABLE 22.2 Analysis of Common Pollutants by NIOSH Methods (*Continued*)

Specific compound or class of pollutants	Examples	NIOSH method(s)	Outline of the method
Carbaryl (Sevin)	—	5006	Between 20 and 400 L of air at a flow rate of 1–3 L/min are passed through a glass fiber filter; analyte is extracted with a solution made up of 2 mL of 0.1*M* methanolic KOH, 17 mL of glacial acetic acid, and 1 mL of *p*-nitrobenzenediazonium tetrafluoroborate; the latter converting the analyte to a complex; absorbance of the solution measured by a spectrophotometer at 475 nm; calibration standard made from carbaryl in methylene chloride
Carbon dioxide	—	6603	Air is collected in a gas sampling bag at a flow rate of 0.02–0.1 L/min, the bag being filled to 80% or less capacity; the air is injected onto a portable GC equipped with a TCD and Porapak QS or equivalent column
Carbon disulfide	—	1600	Between 2 and 25 L of air at a flow rate of 0.01–0.2 L/min are passed through an assembly of a solid sorbent tube containing coconut shell charcoal (100 mg/50 mg) and a drying tube containing sodium sulfate (270 mg); analyte is desorbed with 1 mL toluene over 30 min standing; the solution is analyzed by GC/FPD in sulfur mode using GasChrom Q or equivalent GC column
Cyanuric acid	—	5030	A volume of 10–1000 L of air at a flow rate of 1–3 L/min is passed through a 5-μm PVC membrane filter; analyte is extracted with 3 mL of 0.005*M* Na_2HPO_4 at neutral pH for 10 min in an ultrasonic bath; the solution is analyzed by HPLC/UV at 225 nm using u-Bondapak C_{18} or equivalent column
1,3-Cyclopentadiene	—	2523	Between 1 and 5 L of air at a flow rate of 0.01–0.05 L/min are passed though a solid sorbent tube packed with maleic anhydride on Chromosorb 104 (100 mg/50 mg); maleic anhydride forms an adduct with 1,3-cyclopentadiene; the contents of the sorbent tube after sampling are placed in 10 mL of ethyl acetate and allowed to stand for 15 min; the adduct dissolves in the solvent; the solution is analyzed by GC-FID using 5% OV-17 on Chromosorb WHP or equivalent column
Diesel particulate (elemental carbon)	—	5040	An appropriate volume of air (about 140 L) at a flow rate of 2–4 L/min is passed through a 37-mm quartz-fiber filter; diesel particulate is determined with a thermal optical analyzer using evolved gas analysis (EGA technique); a cyclone should be used if

(Continued)

TABLE 22.2 Analysis of Common Pollutants by NIOSH Methods (*Continued*)

Specific compound or class of pollutants	Examples	NIOSH method(s)	Outline of the method
Diesel particulate (elemental carbon)			heavy loadings of carbonate are anticipated; to minimize collection of coal dust (as in coal mines), an impactor with a submicrometer cut point must be used
1,1-Dimethylhydrazine	—	3515	An air volume of 2–100 L at a flow rate of 0.2–1 L/min is bubbled though 0.1*M* HCl; the solution is treated with phosphomolybdic acid, heated at 95°C for 60 min, and then cooled; analyte forms a complex; absorbance is measured at 730 nm with a spectrophotometer
Dimethyl sulfate	—	2524	An air volume of 0.25–12 L is passed through Porapak P (100 mg/50 mg); analyte is desorbed with 1 mL of diethyl ether over 30 min contact; the solution is analyzed by GC using an electrolytic conductivity detector in sulfur mode
Dioxane	—	1602	A volume of 0.5–15 L of air at a flow rate of 0.01–0.2 L/min is passed through coconut shell charcoal; analyte is desorbed into 1 mL of CS_2 over 30 min contact; the solution analyzed by GC/FID
Dyes, aminobiphenyl	Benzidine, *o*-tolidine, *o*-dianisidine	5013, 5509	Air is passed through a 5-μm PTFE membrane filter or a glass-fiber filter; analytes are desorbed with water under ultrasonic condition and reduced to free amine with sodium hydrosulfite; the solution is analyzed by HPLC using a UV detector at 280 nm and a C-18 column (method 5509 may be applied to measure benzidine only; a 13-mm glass-fiber filter is used for sampling; the analyte is desorbed with triethylamine in methanol and measured by HPLC/UV at 254 nm)
Epichlorohydrin	—	1010	Between 2 and 30 L of air are passed over coconut shell charcoal (100 mg/50 mg) at a flow rate of 0.01–0.2 L/min; the analyte is desorbed into 1 mL of CS_2 over 30 min contact of the charcoal with the solvent; the solution is analyzed by GC/FID
Esters	*n*-Propyl acetate, *n*-butyl acetate, isoamyl acetate, ethyl acrylate	1450 (method 1454 describes a similar procedure for isopropyl acetate; method 1457, for ethyl acetate; and method 1458, for methyl acetate)	Between 1 and 10 L of air is sampled over coconut shell charcoal (100 mg/50 mg); esters are desorbed into 1 mL of CS_2 by placing the adsorbent in CS_2 for 30 min; the solution is analyzed by GC/FID using an appropriate column

TABLE 22.2 **Analysis of Common Pollutants by NIOSH Methods (*Continued*)**

Specific compound or class of pollutants	Examples	NIOSH method(s)	Outline of the method
Ethers	Diethyl ether	1610	An air volume of 0.25–3 L is sampled over coconut shell charcoal (100 mg/50 mg) at a flow rate of 0.01–0.2 L/min; analyte is desorbed with 1 mL of ethyl acetate over 30 min contact time; the solution analyzed by GC/FID
	Diisopropyl ether	1618	Similar to above method; recommended volume and flow rate of air are 0.1–3 L and 0.01–0.05 L/min, respectively; analyte is desorbed into 1 mL of CS_2 by placing charcoal in the solvent for 30 min; the solution is analyzed by GC/FID
Ethylene dibromide	—	1008	Between 0.1 and 25 L of air are sampled over coconut shell charcoal (100 mg/50 mg) at a flow rate of 0.02–0.2 L/min; analyte is desorbed with 10 mL of benzene–methanol (99:1) over 1 h standing; the solution analyzed by GC/ECD (^{63}Ni)
Ethylenediamine	—	2540	Between 1 and 20 L of air are passed through a solid sorbent tube containing XAD-2 coated with 1-naphthylisothio-cyanate (80 mg/40 mg) at a flow rate of 0.01–0.1 L/min; analyte is converted to its naphthylisothiourea derivative; it is desorbed with 2 mL of dimethylfor-mamide under 30 min ultrasonication; the solution is analyzed by HPLC/UV
Ethylenimine	—	3514	1–48 L of air are bubbled through a solution of Follin's reagent at a flow rate of 0.2 L/min; analyte is derivatized to 4-(1-aziridinyl)-1, 2-naphthoquinone; it is desorbed with 4 mL of chloroform; the derivative is analyzed by HPLC/UV at 254 nm
Ethylene oxide	—	1614	1–24 L of air are sampled at a flow rate of 0.05–0.15 L/min; analyte is adsorbed over HBr-coated petroleum charcoal (100 mg/ 50 mg), converted to 2-bromoethanol, desorbed with dimethylformamide, and analyzed by GC/ECD
		3702	Ambient air, as is, or collected in a bag, is analyzed directly by a portable GC equipped with a PID
Ethylene thiourea	—	5011	Between 200 and 800 L of air at a flow rate of 1–3 L/min is passed through a 5-μm PVC or cellulose ester membrane filter; analyte is extracted with distilled water at 60°C and complexed with pentacyanoaminoferrate; absorbance of the solution is measured at 590 nm by a spectrophotometer

(*Continued*)

TABLE 22.2 Analysis of Common Pollutants by NIOSH Methods (*Continued*)

Specific compound or class of pollutants	Examples	NIOSH method(s)	Outline of the method
Glycols	1,2-Ethanediol, 1,2-propanediol, 1,3-butanediol	5523	Between 5 and 60 L of air are passed through an XAD-7 OVS tube (13-mm glass-fiber filter and 200/100 mg XAD-7); analytes are desorbed into 2 mL of methanol under 30 min ultrasonication; the solution is analyzed by GC/FID using a fused-silica capillary column
Glycol ethers	Cellosolve, methyl cellosolve, butyl cellosolve	1403	Air is passed through a solid sorbent tube containing coconut shell charcoal (100 mg/50 mg); analytes are desorbed into 1 mL of 5% methanol in methylene chloride; the solution is analyzed by GC/FID
Herbicides, chlorophenoxy acid	2,4-D (2,4-dichlorophenoxy) acetic acid, 2,4,5-T (2,4,5-trichlorophenoxy) acetic acid	5001	15–200 L of air at a flow rate 1–3 L/min is passed through a glass-fiber filter (binderless); after sampling, the filter is placed in 15 mL of methanol and allowed to stand for 30 min; the analytes are dissolved in methanol; the solution is analyzed for the chlorophenoxy acid anions by HPLC/UV at 284 nm using a Zipax SAX HPLC column or equivalent and $NaClO_4$–$Na_2B_4O_7$ eluent (0.001*M* mix); this method measures chlorophenoxy acids and their salts, but not their esters
Hydrocarbons, aliphatics, and aromatics (volatile, boiling between 36 and 126°C)	*n*-Pentane, *n*-hexane, cyclohexane, benzene, toluene	1500	Air is passed through a solid sorbent tube containing coconut shell charcoal (100 mg/50 mg); hydrocarbons trapped over the charcoal are desorbed with 1mL of CS_2 over 30 min standing; the solution is analyzed by GC/FID
Hydrocarbons, aromatic	Benzene, toluene, xylene, cumene, styrene, naphthalene	1501	Sampling and analytical procedure similar to method 1500 above; the flow rate and the volume of air sampled may vary; the method involves adsorption of analytes over charcoal, desorption with CS_2, and analysis by GC/FID
Hydrocarbons, polynuclear aromatic (PAH)	Anthracene, phenanthrene, chrysene, benzopyrene, fluoranthene, acenaphthene	5506	Between 200 and 1000 L of air at a flow rate of 2 L/min are passed through a 37-mm, 2-μm PTFE filter and a sorbent tube containing washed XAD-2 (100 mg/50 mg); the analytes are extracted with 5 mL of acetonitrile in an ultrasonic bath for 30–60 min; the solution is analyzed by HPLC using a fluorescence detector at 340 nm (excitation) or 425 nm (emission), or by UV detection at 254 nm; a reversed-phase C_{15} (5-μm) column is suitable for HPLC analysis (the method is applicable to analyze pollutants that can be extracted with acetonitrile only)

TABLE 22.2 **Analysis of Common Pollutants by NIOSH Methods (*Continued*)**

Specific compound or class of pollutants	Examples	NIOSH method(s)	Outline of the method
		5515	Uses the same sampling technique as method 5506 but GC/FID measurement; a fused-silica capillary column such as DB-5 or equivalent is suitable for analysis
		5800	Uses the same sampling technique and a HPLC flow injection method to measure the total PAH at two different sets of fluorescent wavelengths
Hydrocarbons, halogenated	Chloroform, bromoform, carbon tetrachloride, ethylene, dichloride, 1,1,1-trichloroethane, benzyl chloride	1003	Air is passed through a solid sorbent tube containing coconut shell charcoal (100 mg/50 mg) at a flow rate of 0.01–0.2 L/min; analytes are desorbed with CS_2 over 30 min contact; the solution is analyzed by GC/FID; there are several other methods for one or more such compounds
Hydrogen cyanide	—	6010	Between 2 and 90 L of air at a flow rate of 0.05–0.2 L/min are passed through a solid sorbent tube containing soda lime (600 mg/200 mg); the sample is desorbed into water over a period of 60 min contact; the solution is treated with succinimide and pyridine–barbituric acid; color is measured by a spectrophotometer at 580 nm
		7904 (for aerosol cyanides and HCN gas)	Between 10 and 180 L of air at a flow rate of 0.5–1 L/min is passed through a 0.8-μm PVC membrane and 15 mL of 0.1*N* KOH solution; the combined extract and solution are analyzed for CN by a cyanide ion-specific electrode (the method cannot distinguish HCN from aerosol cyanide; also, there is interference from several other ions in cyanide measurement)
Hydrogen sulfide	—	6013	Between 1 and 40 L of air are passed through an assembly of a 0.5-μm Zefluor filter and a solid sorbent tube packed with coconut shell charcoal (400 mg/200 mg) at a flow rate of about 0.2 L/min; H_2S trapped from the air is desorbed from the sampler assembly into a solution containing 2 mL of 0.2*M* NH_4OH and 5 mL of 30% H_2O_2; H_2S oxidizes to sulfate (SO_4^{2-}) ion; the latter is measured by ion chromatography using a conductivity detector, Ion-Pac AS4A separator or equivalent column, and 40m*M* NaOH eluent; SO_2 is a positive interference this above test
Hydroquinone	—	5004	Between 30 and 180 L of air at a flow rate of 1–4 L/min are passed through a 0.8-μm cellulose ester membrane; analyte is extracted from the filter with 10 mL of 1% acetic acid; the solution is analyzed by HPLC-UV at 290 nm; a C_{18} μ-Bondapak or equivalent column may be used

(Continued)

TABLE 22.2 Analysis of Common Pollutants by NIOSH Methods (*Continued*)

Specific compound or class of pollutants	Examples	NIOSH method(s)	Outline of the method
Isocyanates	2,4- and 2,6-Toluene diisocyanate	5522	Between 15 and 360 L of air are passed through 20 mL of tryptamine solution in dimethyl sulfoxide at a flow rate of 1–2 L/min; isocyanates are converted to their tryptamine derivatives; the solution is analyzed by HPLC using a fluorescence or an electrochemical detector and a μ-Bondapak C_{18} or equivalent column and acetonitrile–sodium acetate buffer mobile phase
	Monomeric isocyanates (most of the above isocyanates may also be analyzed by this method)	5521	Air is passed through an impinger solution containing 1-(2-methoxyphenyl)-piperazine in toluene; isocyanates are converted to their urea derivatives; the solution is analyzed by HPLC using a UV or electrochemical detector (at 242 nm)
Isophorone	—	2508	Between 2 and 25 L of air are passed through a solid sorbent tube packed with petroleum-based charcoal (100 mg/50 mg) at a flow rate between 0.01 and 1 L/min; isophorone is extracted into 1 mL of CS_2 over 30 min contact time of the absorbent with the solvent; the CS_2 solution is analyzed by GC/FID using an appropriate column
Ketones	Acetone, cyclohexanone, methyl isobutyl ketone, camphor, mesityl oxide, ethyl butyl ketone	1300, 1301	An appropriate volume of air (depending on the analyte to be measured) at flow rates varying between 0.01 and 0.2 L/min is sampled over coconut shell charcoal (100 mg/50 mg); ketones are desorbed into 1 mL of CS_2 (or CS_2 containing a small amount of methanol to desorb out higher ketones); the solution is analyzed by GC/FID
Lead	—	7700 (a qualitative spot test)	Between 10 and 240 L of air at a flow rate of 2 L/min are passed through a 0.8-μm cellulose ester membrane; lead dusts collected on the membrane filter are tested qualitatively by chemical spot test using a rhodizonate test kit; lead forms an orange to pink-red complex under acid conditions; some other metals, e.g., Hg^+, Tl^+, Sn^{2+}, Cd^{2+}, and Ba^{2+}, also form colored compounds, but only lead–rhodizonate complex gives a characteristic red color
		7701	20–1500 L of air at a flow rate of 1–4 L/min are passed through a 37-mm, 0.8-μm mixed cellulose ester membrane; lead dusts are extracted with 10 mL of 10% HNO_3 under ultrasonication; the solution is diluted to 50 mL with 2% HNO_3; lead is analyzed by a field-portable anodic stripping voltammeter

TABLE 22.2 **Analysis of Common Pollutants by NIOSH Methods (*Continued*)**

Specific compound or class of pollutants	Examples	NIOSH method(s)	Outline of the method
			using mercury film on glassy carbon as working electrode and a Ag/AgCl or calomel reference electrode; the method is quantitative; thallium may interfere in the test
		7702 (field-portable XRF instrument)	570–1900 L of air at a flow rate of 1–4 L/min are passed through a 37-mm, 0.8-μm mixed cellulose ester membrane filter; lead is analyzed by a field-portable X-ray fluorescence instrument with a cadmium-109 source; bromine may interfere in the test, giving a higher XRF reading; the method is quantitative, nondestructive, and applicable to all elemental forms of lead
		7082 (Flame AA)	200–1500 L of air at a flow rate of 1–4 L/min are passed through a 0.8–μm cellulose ester membrane filter; the metal dust collected on the filter is digested with a mixture of 6 mL conc. HNO_3 and 1 mL 30% H_2O_2; the solution diluted to 10 mL with 10% HNO_3; lead is determined by flame AA at 283.3 nm
		7105 (Furnace AA)	The procedure is similar to the above method; after sample digestion with HNO_3–H_2O_2 at 140°C and dilution and cooling, lead is measured by furnace AA at 283.3 nm
		7300 (ICP-AES)	The procedure is similar to methods 7082 and 7105; the sample is extracted with an HNO_3–$HClO_3$ mixture (4:1) at 130°C or, alternatively, by microwave heating; lead is analyzed by ICP/atomic emission spectroscopy at 220.4 nm
Lead tetraalkyls	Tetramethyl lead, tetraethyl lead	2534 (tetramethyl lead) 2533 (tetraethyl lead)	An appropriate volume of air between 15 and 200 L is passed through a solid sorbent tube packed with XAD-2 resin; after sampling, the adsorbent resin is transferred into 1–2 mL of pentane and allowed to stand for 30 min; lead tetraalkyls in the solution are analyzed by GC/PID using a Carbowax 20M on Chromosorb WHP column or equivalent
Maleic anhydride	—	3512	An air volume of 40–500 L is bubbled through 15 mL of distilled water at a flow rate of 0.2–1.5 L/min; the solution is analyzed by HPLC/UV at 254 nm using a C-18 Bondapak or equivalent column; mobile phase, 0.5% dicyclohexylamine/0.5% formic acid/25% methanol/74% water, 1.7 mL/min

(Continued)

TABLE 22.2 Analysis of Common Pollutants by NIOSH Methods (*Continued*)

Specific compound or class of pollutants	Examples	NIOSH method(s)	Outline of the method
Mercaptans	Methyl-, ethyl-, and *n*-butyl mercaptans	2542	Between 10 and 150 L of air are passed through a 37-mm glass-fiber filter impregnated with mercuric acetate at a flow rate of 0.1–0.2 L/min; analytes are extracted into 25 mL of a solution consisting of 20 mL HCl (25% v/v) and 5 mL of 1,2-dichloroethane by shaking for 2 min; analysis is by GC in FPD mode using a narrow-bore fused-silica capillary column
		2525 (for *n*-butyl mercaptan)	Between 1 and 4 L of air are passed over Chromosorb-104 (150 mg/75 mg) in a solid sorbent tube at a flow rate of 0.01–0.05 L/min; analyte is desorbed into 1 mL of acetone over 15 min contact with the solvent; the solution is analyzed by GC/FPD in sulfur mode using a Chromosorb-104 column
Mercury	—	6009	Between 2 and 100 L of air at a flow rate of 0.15–0.25 L/min are passed through a Hopcalite sorbent tube; mercury is desorbed from the sorbent material into a mixture of conc. HNO_3/HCl; the solution is diluted to 50 mL; mercury is determined by AA spectrophotometry using cold vapor technique
Metals	Aluminum, arsenic, calcium, copper, iron, lead, nickel, magnesium, manganese, sodium, silver, titanium, vanadium, zinc, and many other metals	7300	An appropriate volume of air depending on the metal to be determined is passed through a 0.8-μm cellulose ester membrane filter at a flow rate of 1–4 L/min; the metal deposited on the filter is extracted with 5 mL of HNO_3–$HClO_3$ (4 : 1) mixture under heating; the extract solution is cooled, diluted, and analyzed by ICP/AES technique; metals may alternatively be measured by flame or furnace AA; furnace AA gives greater sensitivity over ICP/AES for many metals; the above method may also be applied to determine phosphorus, a nonmetal
Methanol	—	2000	An air volume of 1–5 L at a flow of 0.02–0.2 L/min is passed through silica gel (100 mg/50 mg) in a solid sorbent tube; methanol adsorbed onto the silica gel is desorbed with 1 mL of water–isopropanol mixture (95 : 5); 1 μL of solution is injected into GC a capillary column containing 35% diphenyl–65% dimethyl polysiloxane or equivalent and detected by FID

TABLE 22.2 Analysis of Common Pollutants by NIOSH Methods (*Continued*)

Specific compound or class of pollutants	Examples	NIOSH method(s)	Outline of the method
Methylal (dimethoxymethane)	—	1611	Between 1 and 3 L of air at a flow of 0.01–0.2 L/min are passed through coconut shell charcoal (100 mg/50 mg); analyte is desorbed into 1 mL of hexane over a contact time of 30 min; the solution is analyzed by GC/FID
Naphthylamines	α-Naphthyl amine, β-naphthylamine	5518	Between 30 and 100 L of air at a flow rate of 0.2–0.8 L/min are passed through a glass-fiber filter and a solid sorbent tube containing silica gel (100 mg/50 mg); analytes are desorbed into a 0.5 mL solution of acetic acid (0.05% v/v) in 2-propanol; the solution is analyzed by GC/FID using an appropriate column
Nicotine	—	2544	Between 60 and 400 L of air are passed through XAD-2 (100 mg/50 mg) at a flow rate 1 L/min; the adsorbent is then transferred into 1 mL of ethyl acetate; the solution is allowed to stand for 30 min and then analyzed by GC/NPD
		2551	The procedure is similar to the above method except that XAD-40 (80 mg/40 mg) is used as adsorbent; also, the method is applicable to measure air at a lower flow rate, between 0.1 and 1 L/min, and a wider volume range, 0.5–600 L; the analyte is desorbed into 1 mL of ethyl acetate containing 0.01% triethylamine; the solution is analyzed by GC/NPD
Nitroaromatics	2-, 3-, and 4-nitrotoluenes, nitrobenzene, 4-chloronitrobenzene	2005	An appropriate volume of air, depending on the types of nitroaromatics (usually between 10 and 150 L for most compounds, but 1 to 30 L for nitrotoluenes) is passed through a solid sorbent tube packed with silica gel (150 mg/75 mg) at a flow rate of 0.01–0.02 L/min; analytes are desorbed out from the silica gel into 1 mL of methanol in an ultrasonic bath over 30 min; the methanol solution is analyzed for nitroaromatics by GC/FID; a fused-silica capillary column is suitable for the purpose
Nitrogen oxides	Nitrous oxide (N_2O)	6600	Ambient air or bag samples may be measured for N_2O by a field-readout, long-path-length, portable infrared spectrophotometer at 4.48 μm; samples are usually stable for 2 h at ambient temperature
	Nitric oxide (NO) and nitrogen dioxide (NO_2)	6014	Between 1.5 and 6 L of air are passed at a rate of 0.025 L/min (for NO) or 0.025–0.20 L/min (for NO_2) through an assembly of three sorbent tubes connected

(Continued)

TABLE 22.2 Analysis of Common Pollutants by NIOSH Methods (*Continued*)

Specific compound or class of pollutants	Examples	NIOSH method(s)	Outline of the method
Nitrogen oxides			in series, the first and the third tubes packed with triethanolamine-coated molecular sieve (400 mg) and the second tube packed with 800 mg of chromate (oxidizer) to convert NO to NO_2; contents of the tubes are extracted separately with 10 mL each of an absorbing solution containing 15 g of triethanolamine and 0.5 mL of *n*-butanol in 1 L of water; the extract is treated with 1 mL of 0.02% H_2O_2, 10 mL of sulfanilamide solution (2% strength in ~4% H_3PO_4 solution), and 1.5 mL of *N*-(1-naphthyl) ethylenediamine dihydrochloride solution (0.1% w/v); NO_2 is converted to NO_2^- ion; the solution is allowed to stand for 10 min to complete color development; absorbance is measured at 540 nm with a spectrophotometer; concentrations of NO_x are determined from calibration standard solutions of nitrite ion ($NaNO_2$)
	Nitrogen dioxide (NO_2)	6700	The procedure is similar to method 6014 above except that a Palmes tube with three triethanolamine-treated screens is used without any oxidizer to sample air
Nitroglycerine	—	2507	Between 3 and 100 L of air at a flow rate of 0.2–1 L/min are passed through GC-grade Tenax (100 mg/50 mg); analyte is desorbed into 2 mL of ethanol over 30 min contact time; the solution is analyzed by GC/ECD using an appropriate detector
Nitrosamines	*N*-nitrosodimethylamine, *N*-nitrosodiethylamine	2522	Between 15 and 1000 L of air at a flow rate of 0.2–2 L/min are passed through the commercially available Thermosorb/N sorbent tube attached to a sampling pump; analytes are desorbed into 2 mL of methylene chloride/methanol mixture (3 : 1) over a sorbent-solvent contact time of 30 min; the solution is analyzed for nitrosamines by GC equipped with a thermal energy analyzer (TEA); 10% Carbowax 20M + 2% KOH on Chromosorb W-AW or equivalent column may be used
Nuisance dusts (particulates not regulated, must not contain asbestos and quartz >1%)	—	0500	Between 7 and 133 L of air at a flow rate of 1–2 L/min are passed through a 37-mm, 5-μm PVC filter; airborne particulate material collected on the filter is measured by gravimetry using an analytical balance; organic and volatile matter may be removed by dry ashing

TABLE 22.2 Analysis of Common Pollutants by NIOSH Methods (*Continued*)

Specific compound or class of pollutants	Examples	NIOSH method(s)	Outline of the method
Organoarsenic acids	Methanearsonic acid, cacodylic acid, atoxylic acid	5022	Between 50 and 1000 L of air are passed through a 1-μm PTFE filter at a flow rate of 1–3 L/min; analytes are extracted from the filter with 25 mL of borate carbonate buffer; the solution is analyzed for respective organoarsenic anions by ion chromatography; alternatively, the solution can be analyzed by hydride atomic absorption spectroscopy at 193.7 nm
Pesticides, organochlorine	Endrin, chlordane, aldrin, methoxychlor	5519 (endrin) 5510 (chlordane)	Air is passed through an assembly of a 0.8-μm cellulose ester membrane and a solid sorbent tube packed with Chromosorb 102 (100 mg/50 mg) at a flow rate of 0.5–1 L/min; analytes are desorbed out from the adsorbent into 10 mL of toluene over 30 min contact time; the toluene solution of pesticide(s) is analyzed by GC using a Ni-63 ECD and an appropriate packed or capillary column (the same techniques may be applied to measure similar chlorinated pesticides in the air)
Pesticides, organophosphorus	Malathion, ronnel, mevinphos	5600	An air volume of 12–480 L at a flow of 0.2–1 L/min is passed through a 13-mm quartz filter and a solid sorbent tube packed with XAD-2 (270 mg/140 mg); pesticides are desorbed with 2 mL of tolune–acetone mixture (90 : 10); the solution is analyzed by GC/FPD in P mode using a fused-silica capillary column
Petroleum naphtha (mineral spirits, Stoddard solvent)	—	1550	Between 1 and 20 L of air at a flow rate of 0.01–0.2 L/min are passed through a solid sorbent tube containing coconut shell charcoal (100 mg/50 mg); analytes are desorbed with 1 mL of CS_2 over a solvent; contact time of 30 min; the naphtha hydrocarbons in the solution are measured by GC/FID using a fused-silica capillary column
Phthalates	Dibutylphthalate; *bis*(2-ethylhexyl) phthalate	5020	Between 6 and 200 L of air are passed through a 0.8-μm cellulose ester membrane at a flow rate of 1–3 L/min; the analytes on the filter are desorbed into 2 mL of CS_2 in an ultrasonic bath over 30 min; the solution is analyzed by GC/FID
Phosphine	—	6002	1–16 L of air at a flow rate of 0.01–0.2 L/min are passed through a sorbent tube packed with $Hg(CN)_2$-coated silica gel (300 mg/150 mg); analyte is extracted

(Continued)

TABLE 22.2 Analysis of Common Pollutants by NIOSH Methods (*Continued*)

Specific compound or class of pollutants	Examples	NIOSH method(s)	Outline of the method
Phosphine	—		with 10 mL of hot acidic permanganate solution at about 70°C; the adsorbance of the solution is measured at 625 nm with a spectrometer; analyte is quantified from standard solutions of KH_2PO_4; PCl_3 and PCl_5 vapors and certain organic phosphorus compounds may interfere
		OSHA Method 180	Employs KOH-impregnated carbon medium for air sampling
Phosphorus trichloride	—	6402	10–100 L of air at a flow rate of 0.05–0.2 L/min are bubbled through 15 mL of water in a bubbler; the aqueous solution of analyte is treated with 3 mL of Br_2 water, 5 mL of sodium molybdate into molybdenum blue; the absorbance of the solution is measured at 830 nm by a spectrophotometer; analyte is quantified from calibration standard solutions of KH_2PO_4; phosphorus (V) compounds do not interfere in the test
Polychlorinated biphenyls (PCBs)	Aroclor-1242, Aroclor-1254	5503	1–50 L of air at a flow rate of 0.05–0.2 L/min are passed through an assembly of a 13-mm glass-fiber filter and a solid sorbent tube packed with Florisil (100 mg/50 mg); PCBs are desorbed from the filter and front section with 5 mL of hexane and from the back section with 2 mL of hexane; the solution is analyzed by GC/ECD (Ni-63); chlorinated pesticides such as DDT and DDE and sulfur-containing compounds in petroleum products may interfere in the test
Propylene oxide	—	1612	0.5–5 L of air at a flow rate of 0.01–0.2 L/min are passed through a solid sorbent tube packed with coconut shell charcoal (100 mg/50 mg); propylene oxide is desorbed into 1 mL of CS_2 over 30 min contact time of adsorbent with CS_2; the solution is analyzed by GC/FID using a fused-silica capillary column, such as DB-5 or equivalent
Pyridine	—	1613	20–150 L of air at a flow rate of 0.01–1 L/min are passed through a solid sorbent tube packed with coconut shell charcoal (100 mg/50 mg); pyridine is desorbed into 1 mL of methylene chloride over 30 min contact of charcoal with the solvent; the solution is analyzed by GC/FID using a packed column, 5% Carbowax 20M on acid-washed DMCS Chromosorb W

TABLE 22.2 Analysis of Common Pollutants by NIOSH Methods (*Continued*)

Specific compound or class of pollutants	Examples	NIOSH method(s)	Outline of the method
Refrigerants	Trifluorobromomethane, chlorodifluoromethane, dichlorodifluoromethane	1017, 1018	An appropriate volume of air (less than 4 L) is passed at a flow rate of 0.01–0.05 L/min through two coconut shell charcoal tubes in series (400 mg/200 mg and 100 mg/50 mg); analytes are desorbed with methylene chloride and analyzed by GC/FID
Sulfur dioxide	—	6004	4–200 L of air are passed through a filter assembly of a 0.8-μm cellulose ester membrane and a Na_2CO_3-treated filter at a flow rate of 0.5–1.5 L/min; analyte is extracted with 10 mL of 0.00175*M* $NaHCO_3$/0.002*M* Na_2CO_3 solution mixture; the solution is analyzed for sulfite and sulfate ions by ion chromatography using a conductivity detector; SO_3, if present in the air, interferes in the test
		5308	Alternatively, the air may be bubbled through 0.3*N* H_2O_2; the solution is titrated with NaOH or barium perchlorate
		P&CAM 160	Air is bubbled through tetrachloromercurate solution; color developed due to SO_2 reaction with reagent is measured by visible spectrophotometry
		P&CAM 204	SO_2 is trapped over a 5A molecular sieve, thermally desorped, and analyzed by mass spectrometry
Sulfur hexafluoride	—	6602	Air is collected in a Tedlar or other gas sampling bag at a flow rate of 0.02–0.1 L/min until the bag is filled to <80% of capacity; the air is analyzed using portable GC equipped with an ECD
		5244	An alternative method; the bag sample is analyzed by GC/TCD; this method is less sensitive than Method 6602; lower detection limit, 500 ppm
Sulfuryl fluoride	—	6012	1–10 L of air at a flow rate of 0.05–0.1 L/min are passed through an adsorbent bed of coconut shell charcoal (800 mg/200 mg); sulfuryl fluoride is extracted with 20 mL of 0.04*M* NaOH solution under sonication for 60 min; the solution is analyzed for fluoride ion by ion chromatography using a conductivity detector; other fluoride compounds may interfere in the test
		5245	The test involves collecting air in gas bag samples and analyzing by GC/FPD
Stibine	—	6008	4–5 L of air at a flow rate of 0.01–0.2 L/min are passed through a solid sorbent tube packed with $HgCl_2$-coated silica gel (1000 mg/500 mg); stibine converted into

(Continued)

TABLE 22.2 Analysis of Common Pollutants by NIOSH Methods (*Continued*)

Specific compound or class of pollutants	Examples	NIOSH method(s)	Outline of the method
Stibine			antimony(III) chloride is extracted into 15 mL of conc. HCl by placing the contents of the sorbent tube in HCl for 30 min; the absorbance of the solution is measured at 552 nm by a spectrophotometer; antimony is measured quantitatively from standard calibration solutions of Sb(III) in conc. HCl; antimony in the extract solution may also be determined by AA spectrophotometry or ICP emission spectrometry
Terpenes	Limonene, α-pinene, β-pinene, 3-carene	1552	Between 2 and 30 L of air are passed through a solid sorbent tube packed with coconut shell charcoal (100 mg/50 mg) at a flow rate of 0.01–0.2 L/min; analytes are desorbed into 1 mL of CS_2 by allowing the charcoal to stand in the solvent for 30 min; the solution is analyzed by GC/FID using an appropriate column
Tetrahydrofuran	—	1609	1–9 L of air at a flow rate of 0.01–0.2 L/min are passed through a sorbent tube packed with coconut shell charcoal (100 mg/50 mg); adsorbent is transferred into 0.5 mL of CS_2 and allowed to stand for 30 min; tetrahydrofuran desorbs into CS_2; the solution is analyzed by GC/FID using a packed Porapak Q or equivalent column or a fused-silica capillary coated column
Tetramethyl thiourea	—	3505	50–250 L of air at a flow rate 0.2–1 L/min are bubbled through water in an impinger; the aqueous solution after sampling is treated with pentacyanoaminoferrate reagent; tetramethyl thiourea forms a colored complex with the reagent; absorbance of the solution is measured at 590 nm with a spectrophotometer; analyte is quantified from calibration standard solutions of prepared complex
Tetranitromethane	—	3513	20–250 L of air at a flow rate of 0.5–1 L/min are bubbled through 15 mL of ethyl acetate in an impinger; the solution is analyzed for tetranitromethane by GC/NPD in N mode or by GC/FID; NPD is more sensitive than the FID; a DB-1 or equivalent capillary column may be used
Toluenesulfonic acid (tosic acid)	—	5043	10–1000 L of air at a flow rate of 1–3 L/min are passed through a 13-mm glass-fiber

TABLE 22.2 Analysis of Common Pollutants by NIOSH Methods (*Continued*)

Specific compound or class of pollutants	Examples	NIOSH method(s)	Outline of the method
			filter; analyte is extracted with 2 mL of 2% isopropanol in an ultrasonic bath for 10 min; the solution is analyzed by HPLC/UV at 222 nm using a μ-Bondapak C_{18} or equivalent column
Thiram (a thiocarbamate pesticide)	—	5005	10–400 L of air at a flow rate of 1–4 L/min are passed through a 1-μm-PTFE membrane filter; analyte is extracted from the filter cassette with 10 to 20 mL of acetonitrile over 30 min contact time; the solution is analyzed by HPLC/UV at 254 nm using a C-18 μ-Bondapak or equivalent column
Vinyl chloride	—	1007	0.7–5 L of air at a flow rate of 0.05 L/min are passed through two tandem tubes, each containing 150 mg of activated coconut charcoal; the latter is transferred into 1 mL of CS_2 and allowed to stand for 30 min; analyte is desorbed into CS_2 and the solution is analyzed by GC/FID
Volatile organic compounds	Aromatics, e.g., benzene, toluene, xylene; aliphatic hydrocarbons, e.g., *n*-pentane, *n*-hexane, *n*-octane; ketones, e.g., acetone, methyl ethyl ketone; aldehydes, e.g., hexanol, benzaldehyde; alcohols, e.g., methanol, ethanol, isopropanol, and butanol; phenolics, e.g., phenol and cresols; esters, e.g., ethyl and butyl acetate; chlorinated hydrocarbons, e.g., methylene chloride, 1,1,1-trichloroethane, tetrachloroethylene, dichlorobenzene, and Freon-113; and miscellaneous volatile compounds	2549	1–6 L of air at a flow rate of 0.01–0.05 L/min are passed through multi-bed sorbent tubes containing graphitized carbons and carbon molecular sieve; volatile organics are thermally desorbed out from the sorbent tubes at 300°C for 10 min; compounds are determined by GC/MS, identified from their mass spectra and retention times; a fused-silica capillary column such as DB-1 or equivalent may be used to separate the compounds
Warfarin	—	5002	200–1000 L of air at a flow rate of 1–4 L/min are passed through a 1-μm-PTFE membrane filter; analyte is extracted with 5 mL of methanol; the solution is analyzed for warfarin by HPLC/UV at 280 nm using a C-18 reverse-phase column

TABLE 22.3 U.S. EPA Methods for Air Analysis

Method	Types of pollutants	Sampling and analytical techniques
TO-1	Volatile and nonpolar organics in the b.p. range 80 to 200°C	Tenax adsorption and GC/MS analysis
TO-2	Highly volatile nonpolar organics in the b.p. range −15 to +120°C	Carbon molecular sieve adsorption and GC/MS analysis
TO-3	Volatile nonpolar organics in the b.p. range −10 to +200°C	Cryogenic trapping and GC-FID or ECD analysis
TO-4	Chlorinated pesticides and PCBs	Requires polyurethane foam, high-volume of air; GC-ECD analysis
TO-5	Aldehydes and ketones	Derivatization with 2,4-dinitrophenylhydrazine in impinger solution; HPLC-UV determination
TO-6	Phosgene	Air bubbled through aniline solution; forms carbanilide; HPLC-UV analysis
TO-7	Nitrosamines, e.g., *N*-nitrosodimethylamine	Thermosorb/N absorption; desorbed into methylene chloride; GC/MS analysis
TO-8	Phenol and cresol	Absorbed into caustic soda solution in an impinger; HPLC-UV analysis
TO-9	Polychlorinated dibenzo-*p*-dioxin	High-volume polyurethane foam sampling; high-resolution GC/high-resolution MS (HRGC/HRMS) analysis
TO-10	Chlorinated pesticides	Low-volume polyurethane foam sampling; GC-ECD analysis
TO-11	Formaldehyde	Trapped over dinitrophenylhydrazine-coated cartridge; DNPH derivative eluted with acetonitrile; HPLC-UV analysis
TO-12	Many organic compounds, but not methane	Cryogenic preconcentration and direct flame ionization detection
TO-13	Polynuclear aromatics	Absorbed over polyurethane foam/XAD-2; desorbed with a suitable organic solvent; HPLC-UV or GC-FID detection
TO-14	Organics of various types, including chlorinated and aromatic compounds	SUMMA Passivated Canister sampling; GC analysis using FID, ECD, NPD, and PID, or GC/MS analysis

Source: P. Patnaik, *Handbook of Environmental Analysis*, CRC Press, Boca Raton, FL, 1997.

Bibliography

American Society for Testing and Materials, "Atmospheric Analysis," in *Annual Book of ASTM Standards,* Vol. 11.03, American Society for Testing and Materials, Philadelphia, PA, 1993.

National Institute of Occupational Safety and Health, *NIOSH Manual of Analytical Methods,* National Institute of Occupational Health and Safety, Cincinnati, OH, 1994.

Patnaik, P., *Handbook of Environmental Analysis,* CRC Press, Boca Raton, FL, 1997.

U.S. Environmental Protection Agency, *Compendium of Methods for the Determination of Toxic Organic Compounds in Ambient Air,* National Technical Information Service, U.S. Department of Commerce, Springfield, 1988.

SECTION 23
MINERAL ANALYSIS

23.1 IDENTIFICATION AND CHARACTERIZATION OF MINERALS

23.1.1 Infrared Spectroscopy

Infrared spectroscopy provides a rapid, simple, and convenient nondestructive means of characterizating and identifying minerals. It can readily indicate the presence of specific atomic groupings within the crystal structure. It is probably the best technique for detecting the presence of water in a mineral and for indicating the form in which the water is present. It is also a powerful means of detecting carbonate and of providing clues to the nature of the silicate anion in the mineral structure. Infrared studies are able to detect noncrystalline phases, and some minerals are more easily identified from their IR spectra than from their x-ray diffraction patterns.

23.1.1.1 Identification of Minerals. Infrared spectroscopy is largely complementary to x-ray diffraction as a tool for the identification of mineral species. The method has the great advantage of x-ray diffraction, however, of being able to detect and characterize noncrystalline compounds, since these absorb as strongly as crystalline material. Moreover, the infrared spectrum points more directly to the general nature of an unknown substance, such information being obtainable from the presence or absence of characteristic absorption bands, as are listed in Tables 23.1 and 23.2. Because the positions of the absorption bands in a spectrum are sensitive to the mass, charge, and bonding characteristics of the constituent ions, infrared spectroscopy can often place a mineral species within the range of compositions over which it exists. In mixtures it may be found that some components are more readily recognized in the X-ray diffraction pattern; others may show up more clearly in the infrared spectrum.

Of course, the empirical identification of a mineral from its infrared spectrum requires access to collections of infrared spectra prepared from well-characterized specimens. Lists of the frequencies of absorption maxima are seldom adequate for identification purposes, since band contours and relative intensities are often important distinguishing features. Sadtler's spectra include 1500 inorganic compounds, the data ranging down to 200 cm^{-1}; while Nyquist and Kagel[1] give 900 spectra covering 3800 to 45 cm^{-1}.

[1] R. A. Nyquist and R. O. Kagel, *Infrared Spectra of Inorganic Compounds*, Academic, New York, 1971.

TABLE 23.1 Vibrational Frequencies (in cm^{-1}) of Molecular Entities in Minerals

Group	Stretching	Bending
XOH	3750–2000	1300–400
H_2O	3660–2800	1650–1590
NH_4^+	3330–2800	1500–1390
CO_3^{2-}; HCO_3^-	1650–1300	890–700
$(BO_3)_n$	1460–1200	800–600
$(BO_4)_n$	1100–850	800–600
$(SiO_4)_n$	1250–900	<500
$(SiO_6)_n$	950–600	
SO_4^{2-}	1200–1100	700–600
$(PO_4)_n$	1200–900	600–500
VO_4^{3-}	915–730	< 500
CrO_4^{2-}	870–700	< 500
WO_2^{4-}; MoO_4^{2-}	850–740	< 500
$(MoO_6)_n$	1000–750	< 500
$(WO_6)_n$	900–700	< 500
AsO_4^{3-}	850–730	

TABLE 23.2 Absorption Frequencies (in cm^{-1}) for Metal–Oxygen Stretching for Tetrahedral and Octahedral Coordinations

	XO_4		XO_6	
Metal ion (X)	Isolated	Condensed	Isolated	Condensed
Al	650–800	700–900	400–450	500–650
Be		700–950		
Fe^{2+}				ca. 400
Fe^{3+}	550–650	550–750	300–450	400–550
Li	400–500	400–600		< 300
Mg	500–550			350–480
Ti^{4+}	650–800			600–650
Zn	400–500	400–650		ca. 400

The most distinctive of all infrared absorption bands are those arising from water of crystallization and from structural hydroxyl groups. The range of stretching frequencies of water in crystals overlaps that of structural hydroxyl, but the bending vibration of the water molecule near 1630 cm^{-1} is a very distinctive feature that allows the ready recognition of water molecules in a mineral in the presence of structural hydroxyl groups. A sharp band at 3532 cm^{-1} can be assigned to a weakly hydrogen-bonded free hydroxyl group and two broad bands at 2847 and 2450 cm^{-1} are typical of strongly hydrogen-bonded acidic hydroxyl groups.

23.1.1.2 Characterization of Solid-State Reactions. Infrared spectroscopy is particularly appropriate to the study of reactions involving hydroxyl groups and can give information that would be difficult to obtain by any other method. The technique is not limited to the study of hydroxyl reactions.

Infrared studies can also be coupled with differential thermal analysis and thermogravimetric analysis, allowing weight losses to be correlated with the loss of water molecules, structural hydroxyl, or carbon dioxide from carbonate groups. Infrared spectroscopy complements x-ray diffraction methods in following the crystallization of glasses and the subsequent ordering processes that occur.

23.1.1.3 Techniques and Instrumentation. It is essential to examine the region 4000 to 200 cm^{-1} in order to characterize inorganic compounds adequately. In most practical applications, minerals must be examined in powder form, but the higher refractive indices of most minerals (as compared to organic compounds) require that they be ground more finely than organic compounds to avoid excessive light scattering. Excessive particle size results in spectra of low relief with broad, distorted absorption bands.

Mineral hardness makes the necessary fine particle size more difficult to achieve. It is necessary to grind the sample to a size less than the wavelength of the incident radiation. In general this involves grinding to a particle size below 2 μm. Grinding should be done under an inert liquid, such as absolute alcohol, to avoid loss of components that could occur during grinding since local temperatures may reach 500°C. Agate and mullite mortars are preferred. Note that some substances, notably quartz, carbonates, etc., absorb very strongly and as little as 1% to 2% of these as impurities can present difficulties in interpreting the resulting spectra.

Samples for IR viewing are either pressed into a transparent disk with a large excess of an alkali halide or mixed into a mull with suitable organic oil and held between alkali halide disks. In both cases the surrounding material reduces scattering of the infrared beam by virtue of the close match between its refractive index and that of the powder. Potassium bromide is most frequently used in the disk method and is transparent down to 250 cm^{-1}, but CsI is necessary to reach 150 cm^{-1}.

Typically, about 1 mg of sample, previously crushed by hand in an agate mortar slurried with absolute alcohol, is added to about 350 mg of KBr and mixed in a vibratory grinder, of the Wig-L-Bug type, fitted with a steel or agate vial and beads. A mixing time of 10 min normally suffices to reduce the sample to a particle size of less than 2 μm. The mixture is then transferred to a vacuum die and subjected to a pressure of about 130 000 psi (8840 atm) to give a clear die.

If problems are expected to arise in using the disk technique, the disk spectra should be confirmed by using paraffin oil mulls or disks pressed from polyethylene powder. Nujol is commonly used for mulls. Fluorocarbon oil is used to examine regions that are obscured by Nujol.

23.1.2 Thermal Analysis

Thermal analysis methods are extensively used in mineralogical studies, particularly in clay mineralogy. Weight changes will occur only in certain reactions, such as those involving decomposition and oxidation, but energy changes occur in all reactions since finite amounts of energy are required to drive off a volatile, to institute a polymorphic change, and to trigger off a solid-state reaction. Thermoanalytical results should always be considered in conjunction with information from x-ray diffraction and infrared-absorption spectroscopy.

23.1.2.1 Differential Scanning Calorimetry and Differential Thermal Analysis. Various difficulties may arise when mixtures are involved. Peaks may overlap, but more serious is the occurrence of solid-state reactions between the components that can give rise to completely new peaks, not belonging to any of the individual components. Another difficulty arises from the fact that the nature of a saturating cation can alter the appearance of the DTA curves for some minerals. Sometimes curves given by fine-grained minerals differ from those obtained from massive crystals of the same composition.

Despite such difficulties it is often possible to attribute clay and other minerals to the correct group on the basis of the general appearance of their DTA curves. DTA is often valuable in carrying out quantitative, or semiquantitative, assessments of the relative abundance of components once the peaks associated with the various minerals present have been identified. DTA has definite advantages over TGA since it enables phase changes and solid-state reactions to be detected as well as decomposition or oxidation reactions.

23.1.2.2 Thermogravimetric Analysis. The information given by a TGA curve is quantitative. Differential TGA (DTGA) curves reveal minor changes in slope (reaction rate) that might be missed on the TGA curve, and they show accurately the procedural initial and final temperatures in addition to the temperature of the maximum rate of reaction. Comparison of DTA and DTGA curves immediately shows which reactions are associated with weight change and which are not.

*23.1.2.3 **Evolved Gas Analysis.*** Evolved gas analysis (EGA) can be extremely useful in mineral studies since it immediately reveals whether any specific DTA or DTGA peak is associated with the evolution of water, carbon dioxide, or other molecular species.

23.1.3 X-Ray Diffraction

X-ray diffraction techniques provide fast, convenient, and simple methods not only for identifying crystalline phases at room temperatures and pressures, but also for characterizing materials at high and low temperatures and at high pressure and as an aid to studying solid-state reactions. The basic sample for powder x-ray diffraction is a small quantity of the material, generally 1 to 2 mg but up to 100 mg for a diffractometer sample, which has been crushed to $< 53\ \mu m$, but which has not been overground until it contains an appreciable amount of extremely fine material. The latter gives rise to problems of line broadening in estimating line positions. A cylinder of the powder, of radius about 0.1 to 0.5 mm, is prepared.

The most commonly used powder camera is based on the Starumanis-Ievins configuration. The camera is made in two sizes, the smaller having a diameter of 57.3 mm and the larger of 114.6 mm. The smaller camera is useful when results are required quickly without a high degree of resolution, often the case in routine work, whereas the larger instrument is better suited for accurate work and for situations where good resolution of the diffraction lines are important. Two factors control the choice of x radiation: the use of long-wavelength radiation limits the number of reflections that are observed and, conversely, when the unit cell is very large, short-wavelength radiation tends to crowd individual reflections very close together. The choice of radiation is also affected by the absorption characteristics of a sample. Radiation that has a wavelength just shorter than the absorption edge of an element contained in the sample should be avoided because then the element absorbs the radiation strongly and reemits the absorbed energy as fluorescent radiation in all directions, thus increasing the background. The scan over the range 70° through 5° of 2θ is commonly used for minerals.

Specimen preparation that is useful for more heavily absorbing samples involves diluting the sample with a gum, such as gum acacia, in the ratio 4 samples to 1 gum, moistening the mixture, and rolling a ball of the resulting thick paste first between the fingers and then between two glass slides to form a thin, straight cylinder about 0.3 mm in diameter. Alternatively, a paste made from the sample plus fish glue or collodion is extruded from a hypodermic needle. For many purposes, it suffices to coat the specimen onto a thin glass fiber using vacuum grease or shellac as the adhering medium. Borosilicate glass tubes are particularly useful for substances that might be affected by additives and are imperative for substances that are unstable in air. In any case, the cylinder of powdered sample must be smaller than the diameter of the incident x-ray beam.

The most important ability is that of identifying crystalline phases rapidly and relatively simply, Its drawbacks lie in its inability to detect glassy phases and in the difficulty sometimes encountered with complex patterns. X-ray diffraction methods normally cannot give any direct indication of the elemental analysis of the sample. One should especially feed in chemical analytical data, to give the stoichiometry of the system, and optical microscopic finding, to give some indication of the presence of glassy phases or of the probable number of separate crystalline phases in the mixture.

23.1.4 Microanalysis with Electrons and X Rays

There are four main kinds of information that the scanning electron microprobe (SEM) is able to supply concerning minerals. First, there is the study of surface topography. Second, the bulk microstructure may be investigated in the form of chosen sections through it. Third, quantitative studies may be made although with less accuracy than with the electron probe microanalyzer (EPMA). Finally, it may be possible to obtain crystallographic information from single-crystal specimens by varying the incident-beam angle at a given spot.

23.1.4.1 Electron Probe Microanalyzer. The EPMA is designed primarily to provide quantitative chemical analyses, image formation being of secondary concern. Primary x rays penetrate about 1 μm in the sample, and so the minimum volume that can be analyzed is of the order of a cubic micrometer. The characteristic x rays, generated by an electron beam, are primary x rays that are superimposed on a continuous background from which they must be selected before measurement. X-ray intensity is not strictly proportional to concentration. Correction factors must be applied to the observed intensities to account for factors as differences in the atomic numbers of the elements determined, for the matrix absorption effect, and for fluorescence. Apart from the light elements ($Z = 10$ and below), the relative sensitivity of EPMA is in the range 10 to 100 ppm, while the absolute detection sensitivity is about 10^{-14} g. For mineral specimens, an overall accuracy of 3% to 5% can be expected.

The virtue of EPMA lies in its ability to perform analysis on a very fine scale; that is, if the microstructure is composed of very small particles or grains, then the EPMA is often the only available technique.

23.1.4.2 Electron Microscope Microanalyzer. The electron microscope microanalyzer (EMMA) offers the opportunity of detecting very small particles, less than 1 μm in size, together with larger ones that give rise to poor contrast in either the electron backscatter or the x-ray scanning modes in EPMA. The usefulness of EMMA would appeal to electron microscopists, who now have the facility of obtaining some form of compositional information, which is at worst semiquantitative, in addition to their major microstructural information.

23.1.4.3 Ion Probe Microanalyzer. Ion probe microanalysis uses ion beams in place of the electrons used in EPMA and EMMA, but the information that these beams can supply is very similar. A primary-ion beam is directed at the specimen to produce secondary-ion emission, or ion sputtering. These secondary ions are then analyzed by means of a mass spectrometer. The specimen surface is undergoing continual erosion at a rate that can be controlled to between a few angstroms and some hundreds of angstroms per second. The area of sample examined can vary in the range from 1 to 250 μm, with an image resolution of 1 to 5 μm and a depth resolution about 10 nm. One advantage of the ion probe lies in its ability to analyze low-atomic-number elements, including hydrogen; its good depth resolution; its ability to produce depth profiles; and the measurement of isotopic ratios. Its disadvantages are poorer area resolution and direct image-producing qualities.

23.1.4.4 Auger Electron Spectroscopy (AES). If the derivative of the $N(E)$ versus energy curve is plotted, the detection of the Auger peaks becomes much easier. Since the escape depth of Auger electrons is roughly 1 to 4 atomic layers, the technique is very much one of surface analysis, although concentration profiles into the bulk may be obtained by ion sputtering of the surface layers with argon. A relative sensitivity from 0.3% of a monolayer for carbon or sulfur up to 5% of a monolayer for sodium and aluminum can be obtained. The electron energy resolution is sufficiently good in some cases to distinguish between oxidation states of the same element. Applications include the detection of surface segregants, the investigation of surface reactions, and the analysis of internal surfaces produced by *in situ* fracture or cleavage.

23.1.4.5 Electron Spectroscopy for Chemical Analysis (ESCA). The technique can be used for analysis. More importantly it can be used to distinguish between different binding or oxidation states. The photoelectrons originate from a depth of up to 20 atomic layers, about 5 nm below the specimen surface. Thus the analogous information is presented for a depth level between that of AES and that of the ion microprobe. The area resolution is similar to that of AES.

23.2 METHODS FOR THE ANALYSIS OF MINERALS

Tables 23.3 and 23.4 present outlines of methods for the determination of the most important and most frequently determined constituents of approximately 270 minerals. Table 23.3 lists the common names and formulas of the minerals. A very few frequently encountered alternative names are included as cross-references. For various reasons, the formulas will often correspond only roughly to the actual result of an analysis. Table 23.3 also provides each mineral with a code number that is used to identify it in Table 23.4. In addition, it lists the constituents of each mineral for which methods of analysis are included in Table 23.4. Although analyses for other constituents may occasionally be required for various special purposes, those included here are the ones most often performed in actual practice.

The choice of a procedure for the determination of any particular element depends considerably on the environment in which that element occurs. Different minerals must be decomposed in different ways, and necessitate the execution of different separations. For these reasons, 13 or more or less different methods for the determination of Al are included in Table 23.4. The procedure that is best suited to the determination of A1 in analcite may be identified by locating the code number of analcite (8) in the second column of Table 23.4. The remaining columns of this table will then provide information concerning the decomposition of the mineral, the nature and some details of the separation that must be carried out, and the final isolation and determination of the element sought.

To the "so-called" classical methods, those involving gravimetry, colorimetry or spectrophotometry, titrimetry, and arc or spark emission methods, must now be added those involving ICP-OES, flame atomic absorption spectrometry (FAAS), graphite-furnace AAS, x-ray fluorescence spectroscopy, and flame emission spectroscopy. Furthermore, a different arsenal of methods will usually be needed for trace methods. Arc OES has been a tool for routine geoanalysis for more than six decades. The greatest advantage of arc OES is the simplicity of sample preparation, which involves only grinding and mixing with graphite and possibly a buffer. Spark OES is most extensively used where very rapid analyses are required. Sample preparation for XRFS is relatively easy, involving direct briquetting of ground samples or various fusion methods. Generally elements present at high concentration can be easier and better determined by XRFS. Gravimetry, titrimetry, and colorimetric methods provide a corpus of well-tried methods of maximum accuracy.

The sample dissolution procedure must completely dissolve the sample; typical troublemakers are Cr, Ti, Zr, and W. Nebulizer problems arising from the fluxes used often lead to clogging problems with any nebulizer system from the heavy salt loads. The analysis of silica-rich materials is especially troublesome. Solution techniques such as colorimetry and FAAS generally require a sample-to-flux ($LiBO_2$) ratio of 1 : 5 or even 1 : 7. It is possible to fuse silicate rocks and slags with a 1 : 1 ratio and follow with ICP-OES but not with FAAS. The hot plasma with ICP-OES will decompose the dissolved macrocomplexes. Polymerization of silicate and aluminate proceeds in sample solutions; this leads to low results if colorimetric analyses are delayed more than a few hours. Preconcentration by means of ion-exchange columns or removal of interferences by solvent extraction are useful in trace analysis.

$HF–HClO_4$ dissolution is frequently used for low-level trace-element determinations. The mixture should not be evaporated to dryness or near-dryness with too much heat to avoid reformation of refractory oxides. Use of a water bath or an infrared heater is best. Only a closed digestion system, such as HF–HCl with boric acid, avoids loss of silica (as SiF_4).

Sediments rich in a carbonate phase are easily dissolved in cold to moderately warm dilute acid. Carbonates are separated from the rest of the sample. Warm HCl and hydroxyammonium chloride or citric acid preferentially dissolve metal hydroxide phases.

TABLE 23.3 Minerals and Their Formulas

For a description of the contents of this table and of the manner in which it is to be used in conjunction with Table 23.4 see the text. Both the minerals and the symbols of the substances for which methods of determination will be found in Table 23.4 are listed in alphabetical order below.

Mineral number	Mineral	Formula	Analyze for
1	Aegirite	$NaFeSi_2O_6$	Ca, Fe, K, Mg, Na, Si
2	Albite	$NaAlSi_3O_2$	Al, K, Na, Si
3	Allanite	$(Ca,Ce,La,Na)_2(Al,Fe,Be,Mn,Mg)_3(SiO_4)_3(OH)$	Al, Be, Ca, Ce, Fe, La, Mg, Mn, Na, R.E., Si
4	Almandine	$Fe_3Al_2Si_3O_{12}$	Al, Fe, Si
5	Alunite	$KAl_3(SO_4)_2(OH)_6$	Al, H_2O, K, S
6	Amblygonite	$Li(AlF)PO_4$	Al, F, Li, P
7	Amethyst	See mineral 210	
8	Analcite	$NaAlSi_2O_6 \cdot H_2O$	Al, Na, Si
9	Anatase	See mineral 214	
10	Andalusite	Al_2SiO_5	Al, Ca, Mg, Na, Si
11	Anglesite	$PbSO_4$	Pb, S
12	Anhydrite	$CaSO_4$	Ca, S
13	Anorthite	$CaAl_2Si_2O_8$	Al, Ca, K, Na, Si
14	Anorthoclase	$(Na,K)AlSi_3O_8$	Al, Ca, K, Na, Si
15	Antlerite	$Cu_3(SO_4)(OH)_4$	Cu, S
16	Apatite	$(CaF)Ca_4(PO_4)_3$	Ca, F, P
16*a*	Apatite	$(CaCl)Ca_4(PO_4)_3$	Ca, Cl, P
17	Apophyllite	$KCa_4FSi_4O_{10} \cdot 8H_2O$	Ca, F, H_2O, K, Na, Si
18	Aragonite	$CaCO_3$	CO_2, Ca, Mg
19	Argentite	Ag_2S	Ag, S
20	Arsenopyrite	FeAsS	As, Fe, S
21	Atacamite	$Cu_2Cl(OH)_3$	Cl, Cu, H_2O
22	Augite	$Ca(Mg,Fe,Al)(Si,Al)_2O_6$	Al, Ca, Fe, Mg, Si
23	Autunite	$Ca(UO_2)_2P_2O_8 \cdot 10\text{–}12H_2O$	Ca, H_2O, P, U
24	Axinite	$H(Ca,Mn,Fe)_3Al_2B(SiO_4)_4$	Al, B, Ca, Fe, H_2O, Mn, Si
25	Azurite	$CuCO_3 \cdot Cu(OH)_2$	CO_2, Cu, H_2O
26	Baddeleyite	ZrO_2	Hf, Si, Ti, Zr
27	Barite	$BaSO_4$	Ba, Ca, S, Sr
28	Barysilite	$Pb_3Si_2O_7$	Fe, Pb, Si
29	Barytocalcite	$BaCO_3 \cdot CaCO_3$	Ba, CO_2, Ca
30	Bastnaesite	$(Ce,La,Dy)(CO_3)F$	CO_2, F, R.E.
31	Bauxite	$Al_2O_3 \cdot 2H_2O$	Al, Fe, H_2O, Si, Ti
32	Beryl	$Be_3Al_2Si_6O_{18}$	Al, Be, Si
33	Beryllonite	$NaBePO_4$	Be, Na, P
34	Betafite	$(U,Ca)(Nb,Ta,Ti)_3O_9 \cdot nH_2O$	Ca, H_2O, Nb, Ta, Ti, U
35	Biotite	$(H,K)_2(Mg,Fe)_2Al_2Si_3O_{12}$	Al, Fe, H_2O, K, Si
36	Bismuthinite	Bi_2S_3	Bi, S
37	Bismutite	$Bi_2O_3 \cdot CO_2 \cdot H_2O$	Bi, CO_2, H_2O
37*a*	Boehmite	See mineral 82	
38	Boracite	$Mg_7Cl_2B_{16}O_{20}$	B, Cl, Mg
39	Borax	$Na_2B_4O_7 \cdot 10H_2O$	B, H_2O, Na
40	Bornite	Cu_5FeS_4	Cu, Fe, S
41	Boulangerite	$Pb_5Sb_4S_{11}$	Ag, Pb, S, Sb
42	Bournonite	$PbCuSbS_3$	As, Cu, Pb, S, Sb
43	Braunite	Mn_7SiO_{12}	Mn, Si
44	Brochantite	$CuSO_4 \cdot 3Cu(OH)_2$	Cu, Fe, S
45	Bromlite	See mineral 29	

(*Continued*)

TABLE 23.3 Minerals and Their Formulas (*Continued*)

Mineral number	Mineral	Formula	Analyze for
46	Bromyrite	AgBr	Ag, Br
47	Brookite	TiO_2	Ti
48	Brucite	$Mg(OH)_2$	H_2O, Mg
49	Calamine	$H_2Zn_2SiO_5$	Si, Zn
50	Calaverite	$AuTe_2$	Ag, Au, Te
51	Calcite	See mineral 18	
52	Calomel	HgCl	Cl, Hg
53	Carnallite	$KMgCl_3 \cdot 6H_2O$	Cl, K, Mg
53*a*	Carnotite	$K_2(UO_3)_2(VO_4)_2 \cdot 3H_2O$	U, V
54	Cassiterite	SnO_2	Sn
55	Celestite	$SrSO_4$	S, Sr
56	Cerargyrite	AgCl	Ag, Cl
57	Cerite	$(Ca,Fe)(CeO)(OH)_3Ce_2(SiO_3)_3$	Ce, H_2O, La, R.E., Si
58	Cerussite	$PbCO_3$	CO_3, H_2O, Pb
59	Cervantite	$Sb_2O_3 \cdot Sb_2O_5$	Sb
60	Chabazite	$(Ca, Na,K)_7Al_{12}(Al,Si)_2Si_{24}O_{50} \cdot 40H_2O$	Al, Ca, H_2O, K, Na, Si
61	Chalcanthite	$CuSO_4 \cdot 5H_2O$	Cu, H_2O, S
62	Chalcocite	Cu_2S	Cu, S
63	Chalcophyllite	$Cu_{18}Al_2(AsO_4)_3(SO_4)_3(SO_4)_3(OH)_{27} \cdot 33H_2O$	Al, As, Cu, H_2O, S
64	Chalcopyrite	$CuFeS_2$	Au, Cu, Fe, S
65	Chloropal	$H_6Fe_2Si_3O_{12} \cdot 2H_2O$	Fe, H_2O, Si
66	Chondrodite	$Mg_5(SiO_4)_2(F,OH)_2$	F, H_2O, Mg, Si
67	Chromite	$(Fe,Mg)(Cr,Al,Fe)_2O_4$	Al, Cr, Fe, Mg, Si
68	Chrysocolla	$CuSiO_3 \cdot 2H_2O$	Cu, H_2O, Si
69	Chrysolite	$(Mg,Fe)_2SiO_4$	Fe, Mg, Si
70	Cinnabar	HgS	Hg, S
71	Clinochlore	$H_5(Mg,Fe)_5Al_2Si_3O_{18}$	Al, Fe, H_2O, Mg, Si
72	Coal	C	C, S
73	Cobaltite	CoAsS	As, Co, Ni, S
74	Colemanite	$Ca_2B_6O_{11} \cdot 5H_2O$	B, Ca, H_2O
75	Columbite	$(Fe,Mn)(Nb,Ta)_2O_6$	Fe, Mn, Nb, Sn, Ta, Ti
76	Copper (native)	Cu	Cu
77	Corundum	Al_2O_3	Al, Fe, Si
78	Cryolite	Na_3AlF_6	Al, F, Na
79	Cuprite	Cu_2O	Cu
80	Datolite	$HCaBSiO_5$	B, Ca, Si
81	Descloizite	$(Zn,Cu)Pb(VO_4)OH$	Cu, H_2O, Pb, V, Zn
82	Diaspore	AlO(OH)	Al, H_2O, Si
83	Diopside	$CaMgSi_2O_6$	Ca, Mg, Si
84	Dioptase	H_2CuSiO_4	Cu, H_2O, Si
85	Dufrenite	$Fe_5(PO_4)_3(OH)_5 \cdot 2H_2O$	Fe, H_2O, P
86	Dolomite	$(Ca,Mg)CO_3$	CO_2, Ca, Mg
87	Dufrenoysite	$Pb_2As_2S_5$	As, Pb, S
88	Durdenite	$Fe_2(TeO_3)_3 \cdot 4H_2O$	Fe, H_2O, Te
89	Dyscrasite	Ag_3Sb	Ag, Sb
90	Enargite	$3Cu_2S \cdot As_2S_5$	As, Cu, S
91	Enstatite	$MgSiO_3$	Mg, Si
92	Eosphorite	$(Mn,Fe)Al(PO_4)(OH)_2 \cdot H_2O$	Al, Fe, H_2O, Mn, P
93	Epidote	$Ca_2(Al,Fe)_3(SiO_4)_3(OH)$	Al, Ca, Fe, Si
94	Epsomite	$MgSO_4 \cdot 7H_2O$	H_2O, Mg, S
95	Erythrite	$Co_3As_2O_8 \cdot 8H_2O$	As, Co, H_2O, Ni
96	Eucairite	$Cu_2Se \cdot Ag_2Se$	Ag, Cu, Se

TABLE 23.3 Minerals and Their Formulas (*Continued*)

Mineral number	Mineral	Formula	Analyze for
97	Euclase	$HBeAlSiO_5$	Al, Be, Si
98	Euxenite	$(Y,Ca,Ce,U,Th)(Nb,Ta,Ti)_2O_6$	Ca, Ce, Nb, R.E., Ta, Th, Ti, U, Y
99	Ferberite	$FeWO_4$	Ca, Fe, Mn, Mo, Sn, W
100	Fergusonite	$(Y,Er,Ce,Fe)(Nb,Ta,Ti)O_4$	Ce, Er, Fe, Nb, R.E., Ta, Ti, Y
101	Fluorite	CaF_2	CO_2, Ca, F, Si
102	Franklinite	$(Fe,Zn,Mn)O \cdot (Fe,Mn)_2O_3$	Fe, Mn, Zn
103	Gadolinite	$Be_2FeY_2Si_2O_{10}$	Be, Fe, R.E., Si, Y
104	Galena	PbS	Cu, Fe, Ge, In, Pb, S, Zn
105	Garnet	$(Ca,Mg,Fe,Mn)_3(Al,Fe,Cr)_2(SiO_4)_3$	Al, Ca, Cr, Fe, FeO, Mg, Mn, Si
106	Genthite	$2NiO \cdot 2MgO \cdot 3SiO_2 \cdot 6H_2O$	H_2O, Mg, Ni, Si
106*a*	Germanite	(Cu,Ge)(S,As)	Ge
107	Geradorffite	NiAsS	As, Co, Ni, S
108	Gibbsite	$Al(OH)_3$	Al, Fe, H_2O, Si
109	Gilsonite	Natural asphalt	C (also ash and volatile matter)
110	Glauberite	$Na_2Ca(SO_4)_2$	Ca, Na, S
111	Glaucodot	(Co,Fe)AsS	As, Co, Fe, Ni, S
112	Glaucophane	$NaAl(SiO_3)_2 \cdot (Fe,Mg)SiO_3$	Al, Fe, Mg, Na, Si
113	Gold (native)	Au (usually in quartz)	Au
114	Graphite	C	C (also ash and volatile matter)
115	Greenockite	CdS	Cd, Zn
116	Gummite	U oxides + H_2O	U
117	Gypsum	$CaSO_4 \cdot 2H_2O$	Ca, H_2O, S
118	Halite	NaCl	Cl, Na
119	Halloysite	$H_4Al_2Si_2O_9$	Al, H_2O, Si
120	Hanksite	$Na_{22}K(SO_4)_9(CO_3)_2Cl$	CO_2, K, Na, S
121	Hausmannite	$MnMn_2O_4$	Mn, MnO_2
122	Hedenbergite	$CaFeSi_2O_6$	Ca, Fe, Si
123	Hematite	Fe_2O_3	Fe, P, S, Si
124	Hemimorphite	$Zn_4Si_2O_7(OH)_2 \cdot H_2O$	H_2O, Si, Zn
125	Heulandite	$(Ca,Na,K)_6Al_{10}(Al,Si)Si_{29}O_{30} \cdot 25H_2O$	Al, Ca, H_2O, K, Na, Si
126	Hornblende	$(Ca,Na)(Mg,Fe)_4(Al,Fe,Ti)_3Si_6O_{22}(O,OH)_2$	Al, Ca, Fe, H_2O, Mg, Na, Si, Ti
127	Hubnerite	$MnWO_4$	Ca, Fe, Mn, Mo, Sn, W
128	Hyalophane	$(K_2,Ba)Al_2Si_4O_{12}$	Al, Ba, K, Si
129	Hydrocerussite	$2PbCO_3 \cdot Pb(OH)_2$	See Cerussite, mineral 58
130	Hyperathene	$(Fe,Mg)SiO_3$	Fe, Mg, Si
131	Ilmenite	$FeTiO_3$	Cr, Fe, Si, Ti, V
132	Iodyrite	AgI	Ag, I
133	Jadeite	$NaAl(SiO_3)_2$	Al, Na, Si
134	Jamesonite	$Pb_2Sb_2S_5$	Pb, S, Sb
135	Jarosite	$KFe_3(SO_4)_2(OH)_4$	Fe, H_2O, K, S
136	Kainite	$MgSO_4 \cdot KCl \cdot 3H_2O$	Cl, K, Mg, S
137	Kaolin	$Al_2Si_2O_5(OH)_4$	Al, H_2O, Si
138	Kernite	$Na_2B_4O_7 \cdot 4H_2O$	B, H_2O, Na
139	Kieserite	$MgSO_4 \cdot H_2O$	H_2O, Mg, S
140	Krennerite	Au_8Te_{16}	Ag, Au, Te
141	Kyanite	See mineral 10	
142	Labradorite	Intermediate between albite and anorthite	Al, Ca, K, Na, Si

(*Continued*)

TABLE 23.3 Minerals and Their Formulas (*Continued*)

Mineral number	Mineral	Formula	Analyze for
143	Lazulite	$(Mg,Fe)Al_2(PO_4)_2(OH)_2$	Al, Fe, H_2O, Mg, P
144	Lazurite	$Na_{4-5}Al_3Si_3O_{12}S$	Al, Na, S, Si
145	Leadhillite	$Pb_4(SO_4)(CO_3)_2(OH)_2$	CO_2, H_2O, Pb, S
146	Lepidolite	$KLi[Al(OH,F)_2]Al(SiO_3)_3$	Al, Cs, F, H_2O, K, Li, Rb, Si
147	Leucite	$KA1(SiO_3)_2$	Al, K, Si
148	Libethenite	$Cu_2(OH)PO_4$	Cu, H_2O, P
149	Limonite	$FeO(OH) \cdot nH_2O$	Fe, H_2O, P, Si
150	Linnaeite	Co_3S_4	Co, Cu, Fe, Ni, Si
151	Lithiophilite	$Li(Fe,Mn)PO_4$	Cs, Fe, Li, Mn, Na, P, Rb
152	Magnesite	$MgCO_3$	CO_2, Ca, Mg
153	Magnetite	Fe_3O_4	Fe, FeO, S, Ti
154	Malachite	$CuCO_3 \cdot Cu(OH)_2$	CO_2, Cu, H_2O
155	Manganite	$MnO(OH)$	H_2O, Mn, MnO_2
156	Marcasite	FeS_2	Fe, S
157	Margarite	$H_2CaAl_4Si_2O_{12}$	Al, Ca, H_2O, Si
158	Marialite	$Na_4Al_3Si_3O_{24}(Cl,CO_3,SO_4)$	Al, CO_2, Ca, Cl, Na, S, Si
158*a*	Marmatite	Mineral 229 containing FeS	Fe, Pb, S, Zn
159	Meionite	$Ca_4Al_5Si_6O_{24}(Cl,CO_3,SO_4)$	Al, CO_2, Ca, Cl, Na, S, Si
160	Meneghinite	$Pb_4Sb_2S_7$	Pb, S, Sb
161	Miargyrite	$AgSbS_2$	Ag, S, Sb
162	Microcline	$KAlSi_3O_8$	Al, K, Na, Si
163	Microlite	$(Na,Ca)_2Ta_2O_6(O,OH,F)$	Ca, F, H_2O, Na, Nb, Sn, Ta, Ti
164	Millerite	NiS	As, Co, Cu, Ni, S
165	Mimetite	$Pb_6(AsO_4,PO_4)_3Cl$	As, Cl, P, Pb
166	Mirabilite	$Na_2SO_4 \cdot 10H_2O$	H_2O, Na, S
167	Molybdenite	MoS_2	Cu, Mo, Re, S
168	Monazite	$(Ce,La,Y,Th)(PO_4)$	Ce, La, R.E., P, Th, Zr
169	Montmorillonite	$H_2Al_2Si_4O_{12}$	Al, Ca, Fe, H_2O, K, Na, Si
170	Mottramite	$(Cu,Zn)Pb(VO_4)OH$	Cu, H_2O, Pb, V, Zn
171	Muscovite	$KAl_3Si_3O_{10}(OH)_2$	Al, H_2O, K, Si
172	Natrolite	$Na_2Al_2Si_3O_{10} \cdot 2H_2O$	Al, Ca, H_2O, K, Na, Si
173	Nepheline	$(Na,K)(Al,Si)_2O_4$	Al, K, Na, Si
174	Niccolite	NiAs	As, Ni, S, Sb
175	Niter	KNO_3	K, N, Na
176	Oligoclase	Intermediate between albite and anorthite	Al, Ca, K, Na, Si
177	Olivenite	$Cu_2(AsO_4)(OH)$	As, Cu, H_2O
178	Olivine	$(Mg,Fe)_2SiO_4$	Mg, Fe, Si
179	Opal	$SiO_2 \cdot nH_2O$	H_2O, Si
180	Orpiment	As_2S_2	As, S
181	Orthite	See mineral 3	
182	Orthoclase	$KAlSi_3O_8$	Al, Ca, K, Mg, Na, Si
182*a*	Osmiridium	(Os,Ir)	Ir, Os, Pd, Pt, Rh, Ru
183	Palladium (native)	Pd	Ir, Pd, Pt
184	Paragonite	$H_2NaAl_3Si_3O_{12}$	Al, H_2O, K, Na, Si
185	Parisite	$(Ce,La)_2Ca(CO_3)_3F_2$	CO_2, Ca, Ce, F, La, R.E.
186	Pectolite	$Ca_2NaSi_3O_3(OH)$	Ca, H_2O, Na, Si
187	Periclase	MgO	Fe, H_2O, Mg
188	Perovskite	$CaTiO_3$	Ca, Ce, Fe, La, R.E., Ti
189	Petalite	$LiAl(Si_2O_5)_2$	Al, Li, Na, Si
190	Phenacite	Be_2SiO_4	Al, Be, Si

TABLE 23.3 Minerals and Their Formulas (*Continued*)

Mineral number	Mineral	Formula	Analyze for
191	Philipsite	$(K_2,Ca)Al_2Si_4O_{12} \cdot 4^1/_2\ H_2O$	Al, Ca, H_2O, K, Na, Si
192	Phlogopite	$KMg_3AlSi_3O_{10}(OH)_2$	Al, F, H_2O, K, Mg, Na, Si
193	Phosgenite	$Pb_2(CO_3)Cl_2$	CO_2, Cl, Pb
194	Phosphuranylite	$Ca(UO_2)PO_4 \cdot nH_2O$	Ca, H_2O, P, U
195	Pitchblende	See mineral 260	
196	Platinum (native)	Pt	Ir, Os, Pd, Pt
197	Polianite	See mineral 209	
198	Pollucite	$H_2Cs_4Al_4(SiO_3)_9$	Al, Cs, H_2O, K, Rb, Si
199	Polybasite	$(Ag,Cu)_{16}Sb_2S_{11}$	Ag, Cu, S, Sb
200	Polycrase	$(Y,Ca,Ce,U,Th)(Ti,Nb,Ta)_2O_6$	Ca, Ce, Nb, R.E., Ta, Th, Ti, U, Y
201	Polyhalite	$K_2Ca_2Mg(SO_4)_4 \cdot 2H_2O$	Ca, H_2O, K, Mg, S
202	Powellite	$CaMoO_4$	Ca, Mo, W
203	Prehnite	$Ca_2Al_2Si_3O_{10}(OH)_2$	Al, Ca, H_2O, Si
204	Proustite	Ag_3AsS_3	Ag, As, S, Sb
205	Pailomelane	$BaMnMn_3O_{16}(OH)_4$	Ba, Mn, MnO_2, P, Si
206	Pyrargyrite	Ag_3SbS_3	Ag, S, Sb
207	Pyrite	FeS_2	Au, Fe, S, Tl
208	Pyrochlore	$NaCaNb_2O_6F$	Ca, Ce, F, La, Na, Nb, R.E., Sn, Ta, Ti
209	Pyrolusite	MnO_2	Ba, Mn, MnO_2, P, Si
210	Quartz	SiO_2	Al, Au, Fe, Si
211	Realgar	AsS	As, S
212	Rhodochrosite	$MnCO_3$	CO_2, Ca, Fe, Mg, Mn, Zn
213	Rhodonite	$MnSiO_3$	Ca, Fe, Mg, Mn, Si, Zn
214	Rutile	TiO_2	Cr, Fe, Si, Ti, V, Zr
215	Samarskite	$(Y,Er,Ce,U,Ca,Fe,Pb,Th)(Nb,Ta,Ti,Sn)_2O_6$	Ca, Ce, Er, Fe, Nb, Pb, R.E., Sn, Ta, Th, Ti, U, Y
216	Scheelite	$CaWO_4$	As, Ca, Fe, Mn, Mo, S, Sn, W
217	Scorodite	$Fe(AsO_4) \cdot 2H_2O$	As, Fe, H_2O
218	Scorsalite	$(Fe,Mg)Al_2(PO_4)_2(OH)_2$	Al, Fe, H_2O, Mg, P
219	Selenium (native)	Se	Se, Te
219*a*	Selenite	See mineral 117	
220	Siderite	$FeCO_3$	CO_2, Fe
221	Silliminite	See mineral 10	
222	Silver (native)	Ag	Ag
223	Simpsonite	$Al_2Ta_2O_3$	Al, Nb, Sn, Ta, Ti
224	Skutterudite	$(Co,Ni)As_3$	As, Co, Fe, Ni, S
225	Smaltite	See mineral 224	
226	Smithsonite	$ZnCO_3$	CO_2, Ca, Fe, Mg, Mn, Zn
227	Soda niter	$NaNO_3$	K, N, Na
228	Sodalite	$Na_4Al_3Si_3O_{12}Cl$	Al, Ca, Cl, K, Na, Si
229	Sphalerite	ZnS	Cd, Cu, Fe, Ga, Ge, In, Mn, Pb, S, Tl, Zn
230	Spinel	$MgAl_2O_4$	Al, Fe, Mg, Si
231	Spodumene	$LiAlSi_2O_6$	Al, Fe, K, Li, Na, Si
232	Stannite	Cu_2FeSnS_4	Cu, Fe, S, Sn, Zn
233	Staurolite	$(Fe,Al)_4Si_2O_{10}(OH)_2$	Al, Fe, H_2O, Mg, Si
234	Stephanite	Ag_5SbS_3	Ag, S, Sb
235	Stibnite	Sb_2S_3	As, Pb, S, Sb
236	Stilbite	$(Ca,Na)_3Al_5(Al,Si)Si_{14}O_{40} \cdot 15H_2O$	Al, Ca, H_2O, Na, Si

(*Continued*)

TABLE 23.3 Minerals and Their Formulas (*Continued*)

Mineral number	Mineral	Formula	Analyze for
237	Strontianite	$SrCO_3$	Ba, CO_2, Ca, Sr
238	Sulfur	S	S, Se
239	Sylvanite	$(Au,Ag)Te_2$	Ag, Au, Pb, Tc
240	Sylvite	KCl	Cl, K, Na
241	Talc	$Mg_3Si_4O_{10}(OH)_2$	Al, Fe, H_2O, Mg, Si
242	Tantalite	See mineral 75	
243	Tapiolite	$FeTa_2O_6$	Fe, Mn, Nb, Sn, Ta, Ti
244	Tellurium (native)	Te	Se, Te
245	Tennantite	$(Cu,Fe)_{12}As_4S_{13}$	As, Cu, Fe, S
246	Tetradymite	Bi_2Te_2S	Bi, S, Se, Te
247	Tetrahedrite	$(Cu,Fe)_{12}Sb_4S_{13}$	Cu, Fe, S, Sb
248	Thenardite	Na_2SO_4	Na, S
249	Thorite	$ThSiO_4$	Ce, La, Pb, R.E., Si, Th, U
250	Titanite	$CaTiSiO_5$	Ca, Fe, Si, Ti
251	Topaz	$Al_2SiO_4(F,OH)_2$	Al, F, H_2O, K, Na, Si
252	Torbernite	$Cu(UO_2)(PO_4)_2 \cdot 8\text{–}12H_2O$	As, Cu, H_2O, P, U
253	Tourmaline	$(Na,Ca)(Fe,Mg)_3B_3Al_3(Al_3Si_6O_{27})(OH)_4$	Al, B, Ca, Fe, Li, Mg, Na, Si
254	Tremolite	$Ca_2Mg_5Si_8O_{22}(OH)_2$	Ca, H_2O, Mg, Si
255	Triphylite	$LiFePO_4$	Fe, Li, Mn, P
256	Trona	$Na_2CO_3 \cdot NaHCO_2$	CO_2, H_2O, Na
257	Turquoise	$CuAl_6(PO_4)_4(OH)_8 \cdot 4H_2O$	Al, Cu, H_2O, P
258	Tyuyammite	$Ca(UO_2)_2(VO_4)_2 \cdot nH_2O$	Ca, H_2O, U, V
259	Ulexite	$NaCaB_5O_9 \cdot 8H_2O$	B, Ca, H_2O, Na
260	Uraninite	UO_2	He, Pb, U
261	Uranophane	$CaU_2Si_2O_{11} \cdot 7H_2O$	Ca, H_2O, Si, U
262	Vanadinite	$Pb_5(VO_4)_3Cl$	Cl, Pb, V
263	Variscite	$AlPO_4 \cdot 2H_2O$	Al, P
264	Vivianite	$Fe_3(PO_4)_2 \cdot 8H_2O$	Fe, H_2O, P
265	Wavellite	$Al_3(OH)_3(PO_4)_2 \cdot 5H_2O$	Al, F, H_2O, P
266	Willemite	Zn_2SiO_4	Fe, Mn, Si, Zn
267	Witherite	$BaCO_3$	Ba, CO_2, S, Sr
268	Wolframite	$(Fe,Mn)(WO_4)$	Ca, Fe, Mn, Mo, S, Sn, W
269	Wollastonite	$CaSiO_3$	Ca, H_2O, Mg, Si
270	Wulfenite	$PbMoO_4$	Ca, Mo, Pb
271	Xenotime	YPO_4	Er, La, R.E., P, Th, U, Y, Zr
272	Yttrotantalite	$(Fe,Y,U,Ca)(Nb,Ta,Zr,Sn)O_4$	Ca, Fe, Nb, R.E., Sn, Ta, U, Y
273	Zinnwaldite	$H_2K_4Li_4Fe_3Al_5F_8Si_{14}O_{42}$	Al, Fe, H_2O, Li, Si
274	Zircon	$ZrSiO_4$	Fe, Hf, Si, Ti, Zr
275	Zoisite	$Ca_2Al_3(SiO_4)_3OH$	Al, Ca, H_2O, Mg, Si

Source: L. Meites, ed., *Handbook of Analytical Chemistry*, McGraw-Hill, New York, 1963.

TABLE 23.4 Procedures for the Analysis of Minerals

For a description of the contents of this table and of the way in which code numbers are used to identify the minerals to which the procedures given below are applicable, see the text. The elements and substances being determined are listed in the first column in the alphabetical order of their symbols; the code numbers are arranged in order in each entry in the second column. For the sake of completeness, a few procedures (e.g., for the determination of the rare gases, for the determination of iodine in brine and sea water, etc.) have been included in this table although they are not cross-indexed in Table 23.3.

Element determined	Mineral numbers	Decomposition with	Separations required from	Type of separation	Procedure
A	Air; minerals contg. K			Fractional distn. of liq. air	Mass spectroscopy.
Ag	19, 41, 46, 50, 56, 89, 96, 132, 140, 161, 199, 204, 206, 222, 234, 239	Fusion with flux of Na_2CO_3, $Na_2B_2O_7$, PbO + redg. agt.	All elements	Scorification and cupellation; see Au	Scorification: heat Pb button resulting from fusion to ox. part of the Pb to PbO. Place in a cupel of bone ash and heat until all Pb is absbd., weigh resulting Ag + Au bead, and "part" with HNO_2 to dissolve Ag. Difference in wt. = Ag.
Al	2, 4, 8, 10, 13, 14, 22, 24, 35, 60, 71, 93, 105, 112, 125, 126, 133, 137, 142, 144, 146, 147, 157, 158, 159, 162, 169, 171, 172, 173, 176, 182, 184, 189, 191, 192, 198, 203, 228, 230, 231, 233, 236, 241, 251, 253, 273, 275	Na_2CO_3 fusion	Si	Dehydration in HCl soln.	Leach cooled CO_3^- melt with dil. HCl, evap. soln. to dryness, and filt. off dehydrated SiO_2.
			Na, K, Li, Ca, Mg	NH_3 pptn. of Al	
			Fe	Pptn. of Fe with cupferron	Ppt. Al, Fe, and Ti with aq NH_3 (re-ppt. if large amounts of Ca are present), fume ppt. with HNO_3 + $HClO_4$, dil. with H_2O, remove Fe and Ti by pptn. with cupferron, ppt. Al from filtrate with aq NH_3, ignite, and weigh as Al_2O_3.
	128	Same	Ca, Fe, K, Li, Mg, Na, Si	Same	See preceding method.
			Ba	Pptn. as $BaSO_4$	
	32, 97, 190	Same	Ca, Fe, K, Li, Mg, Na, Si	Same	As above, but ppt. Al with 8-hydroxyquinoline instead of NH_3 and weigh as $Al(C_9H_6ON)_3$.
			Be	8-Hydroxyquinoline pptn. of Al	
	6, 92, 143, 218, 263, 265	Same	Ca, Fe, K, Li, Mg, Na, Si	Same	Remove SiO_2 as above, ppt. PO_4^{3-} with excess NH_4 molybdate, then ppt. Al, Fe, and Ti, and continue as above.
			PO_4^{3-}	Pptn. of PO_4^{3-} with NH_4 molybdate	
	3	Same	Ca, Fe, K, Li, Mg, Na, Si	Same	As above, but dissolve Al-Fe-Ti ppt. in acid and ppt. R.E. with excess H_2Ox. Filt., destroy

(*Continued*)

TABLE 23.4 Procedures for the Analysis of Minerals (*Continued*)

Element determined	Mineral numbers	Decomposition with	Separations required from	Type of separation	Procedure
Al			R.E.	Ox pptn.	H_2Ox with $HNO_3 + HClO_4$, and continue with cupferron pptn. as above.
	63, 257	Hot concd. H_2SO_4	Cu, As	H_2S pptn.	Remove H_2S group elements, then ppt. PO_4^{3-} with NH_4 molybdate.
	78	Hot concd. H_2SO_4	Na	NH_3 pptn.	Expel HF, then ppt. Al with aq NH_3, ignite, and weigh as Al_2O_3.
	5	Dil. HCl	SO_4^-, K	NH_3 pptn.	Ignite ppt. and weigh as Al_2O_3.
	31, 82, 108, 119	$HNO_3 + HCl + H_2SO_4$	Si	Evapn. and dehydration	Evap. soln. to fumes, dil., and filt. off SiO_2. Ppt. R_2O_3 group with aq NH_3, ignite, and correct for $Fe_2O_3 + TiO_2 + P_2O_5$. Or det. Fe_2O_3, P_2O_5, SiO_2, and TiO_2, and calc. Al_2O_3 by difference.
	67	Hot $HClO_4$	Si	Evapn. and dehydration	Ppt. with aq NH_3 and weigh as Al_2O_3.
			Cr	NH_3 pptn. of Al or Hg-cathode electrolysis	
			Mg	NH_3 pptn. of Al	
			Fe	Cupferron pptn. of Fe or Hg-cathode electrolysis	
	77	$KHSO_4$ fusion	Si	Heating with H_2SO_4, dilution and filtn.	Same.
			Fe, Ti	Cupferron pptn.	
	223	$KHSO_4$ fusion	Ta, Nb, Fe, Sn	Cupferron pptn.	Same.
	210	HF decomposition	Fe	Cupferron pptn.	Same.
As	20, 42, 63, 73, 87, 90, 95, 107, 111, 164, 165, 174, 177, 180, 204, 211, 216, 217, 224, 235, 245, 252	Hot $H_2SO_4 + KHSO_4$	All elements	Distn. of $AsCl_3$	Dist. $AsCl_3$ at 100°C. from 9*M* HCl contg. a redg. agt. ($FeSO_4$, $N_2H_4 \cdot 2HCl$, or KBr). Neut. distillate with aq NH_3, acidify with HCl, add $NaHCO_3$, and titr. with std. I_2 soln. Or (if < 50 mg As is present) ppt. As_2S_3 from distillate with H_2S, filt., and weigh.
Au	50, 64, 113, 140, 207, 210, 239	Fusion with PbO, SiO_2 and/or Na_2CO_3 + redg. agt.	Pb, Cu, Ag	Fire-assay meths.	Scorification (oxdn.) of Pb button, cupellation (absbn. of PbO by bone ash), and parting (treatment with HNO_3 to remove Ag); weigh Au residue.
B	38, 39, 74, 138, 259	H_2O and/or dil. HCl	R_2O_3 elements	Pptn. with excess $BaCO_3$ or $CaCO_3$	Neut. sl. acidic soln. to pH 5.5 with dil. NaOH, add excess mannitol, and titr. boric acid with 0.1*M* NaOH to pH 7.5.

	24, 80, 253	Na_2CO_3 fusion	SiO_2, R_2O_3	Acidify leached and filtd. melt with HCl; ppt. R_2O_3 + SiO_2 with $BaCO_3$ or $CaCO_3$	Same.
Ba	29, 205, 209, 237, 267	HCl	Mn Ca	Pptn. as $BaSO_4$ Pptn. as $BaCrO_4$ or $BaCl_2$	Ppt. $BaSO_4$ by addn. of H_2SO_4 to dil. HCl soln., filt., and ignite.
	27, 128	Na_2CO_3 fusion	Si, SO_4^-	Filtn. of Na_2CO_3 melt	Same.
Be	3, 32, 33, 97, 103, 190	Na_2CO_3 fusion	Si PO_4^{3-} Y Al, Fe, Mn, Ti, Y	Dehydration in HCl soln. and filtn. NH_4 molybdate pptn. of P Ox pptn. of Y Oxine pptn. in buffered soln. (pH 5.5)	Ppt. Be from filtrate from oxine pptn. with aq NH_3 and ignite to BeO. Or, if Al, Fe, Mn, Ti, and Y are masked with EDTA, Be may be detd. grav. as phosphate.
Bi	36, 37, 246	HNO_3 + $HClO_4$	Fe Cu, SO_4	H_2S pptn. of Bi in 2*M* HCl soln. $(NH_4)_2CO_3$ pptn. of Bi	Titr. in dil. $HClO_4$ soln. with EDTA, using SCN^- or another indicator. Or ppt. BiOCl from dil. HNO_3 soln., filt., dry, and weigh as BiOCl.
Br	46	Na_2CO_3 fusion	Cl	Distn. of Br_2	Transfer alk. soln. to a distn. flask, acidify with HOAc, add $KMnO_4$, and dist. Br_2 formed into a NaOH soln. Finally titr. with std. $AgNO_3$.
C	72, 109, 114	Combustion in O_2		Formn. of CO_2	Absb. CO_2 in ascarite; meas. gain in wt.
CO_2	18, 25, 29, 30, 37, 58, 86, 101, 120, 145, 152, 154, 158, 159, 185, 193, 212, 220, 226, 237, 256, 267	Dil. HCl	None	Evolution of CO_2	Absb. evolved CO_2 in ascarite; meas. gain in wt.
Ca	1, 3, 10, 13, 14, 17, 22, 24, 60, 80, 83, 93, 105, 122, 125, 126, 142, 157, 158, 159, 169, 172, 176, 182, 185, 186, 191, 203, 212, 213, 228, 236, 250, 253, 254, 261, 269, 275	Na_2CO_3 fusion	Si R_2O_3 group, Mn and alkalies Mg	Dehydration in HCl soln. NH_3 pptn. in presence of Br_2 Pptn. of CaOx	Ignite CaOx to CaO and weigh, or dissolve it in acid and titr. with $KMnO_4$. Or titr. filtrate from R_2O_3 ppt. with std. EDTA at pH 12.
	12, 18, 86, 110, 117, 152, 201, 226	Dil. HCl	R_2O_3 elements SO_4^-, Mg	NH_3 pptn. Ox pptn. of Ca	Ppt. CaOx from sl. ammoniacal soln., filt., wash with H_2O, and either dissolve in acid and titr. with $KMnO_4$ or ignite and weigh as CaO.

(*Continued*)

TABLE 23.4 Procedures for the Analysis of Minerals (*Continued*)

Element determined	Mineral numbers	Decomposition with	Separations required from	Type of separation	Procedure
Ca	16, 16*a*	Dil. HNO_3	PO_4^{3-}	Cation exchange in dil. acid soln. to elute PO_4^{3-}, or anion exchange to elute Ca	Vol. or grav. as oxalate.
	74, 259	Dil. HCl	R_2O_3, if present B	NH_3 pptn. Double pptn. of CaOx	Vol. or grav. as oxalate.
	23, 194, 258	Concd. HNO_3 + H_2SO_4	U, PO_4^{3-}, V	Pptn. of $CaSO_4$ from 80% EtOH soln.	Filt. off $CaSO_4$, dissolve it in dil. HCl, and ppt. CaOx. Finish vol. or grav.
	34, 98, 163, 188, 200, 208, 215, 272	$KHSO_4$ fusion in quartz	Ta, Ti, Nb, Fe, Zr, Sn U, R.E., PO_4^{3-}	Pptn. of cupferrates in acid medium Pptn. of $CaSO_4$	Same.
	27, 29, 237	Dil. HCl	Ba, Sr	Cation exchange, or pptn. of Ba and Sr as chlorides or chromates	Vol. or grav. as oxalate.
	99, 127, 216, 268	Concd. HCl	W R_2O_3 + Mn	Filtn. of WO_3 NH_3 pptn. in presence of Br_2	Vol. or grav. as oxalate or by flame photometry.
	101	Hot concd. H_2SO_4	R_2O_3	NH_3 pptn.	As oxalate.
	202	Concd. H_2SO_4	Mo	Pptn. of $CaSO_4$	Convert $CaSO_4$ to CaOx as above, and finish vol. or grav.
	270	Concd. H_2SO_4	Mo, Pb	Anion exchange in 5*M* HCl soln. to retain Mo, Pb	As oxalate.
Cd	115, 229	HNO_3 + H_2SO_4	R_2O_3, if present Pb Cu Zn, Fe, Mn, Ca, Mg	NH_3 pptn. Pptn. of $PbSO_4$ Pptn. with met. Fe Retention on cation-exchange column in I^- medium	In I^- eluate contg. only Cd, det. Cd by titrn. with EDTA, electrograv., or spectrophotometrically with dithizone. Polarographic i_d of Cd in HCl soln. can be measd. without any sepns.
Ce (see also R.E.)	3, 57, 185, 249	Na_2CO_3 fusion	SiO_2 Ca, Mg	Dehydration in acidic soln. NH_3 pptn.	Ignite oxalate ppt. contg. Ce and other R.E. (→ oxides), weigh, dissolve in HCl, convert Cl^- to SO_4^-, ox. Ce with $NaBiO_3$, filt., add measd. excess of std. $FeSO_4$ soln., and back-titr. with std. $KMnO_4$.

	98, 100, 188, 200, 208, 215	Na_2O_2 fusion in Ni crucible	Ni, Ca, Mg	NH_3 pptn.	
			Si	Dehydration in acidic soln.	
			Ta, Nb, Ti, Fe	Cupferron pptn., or pptn. of Ce as fluoride	Same.
	168	Hot concd. H_2SO_4	PO_4^{3-}, R_2O_3	Pptn. as Ox	Same.
Cl	16*a*, 21, 38, 52, 56, 158, 159, 165, 193, 228, 262	Na_2CO_3 fusion	Insol. carbonates	Fusion leached in water and filtd.	Acidify CO_3^- soln. with dil. HNO_3 and titr. Cl^- with std. $AgNO_3$. Or ppt., filt. off, and weigh AgCl.
	53, 118, 136, 240	H_2O	None	None	Titr. with std. $AgNO_3$.
Co	73, 95, 107, 111, 150, 164, 224	$HNO_3 + H_2SO_4$	As	Expulsion with HBr	Co may be detd. electrograv. by deposition from ammoniacal SO_4^- medium or by titrn. with Feic or EDTA. Small amts. may be detd. photometrically with nitrose-R salt or SCN^-.
			Ni	Anion-exchange elution with 9*M* HCl	
			Fe	Elution of Co with 4*M* HCl	
			Cu	Pptn. of CuS with H_2S	
Cr	67	Na_2O_2 fusion in Fe crucible	None	None	Leach melt in H_2O, b. to destroy H_2O_2, acidify with $H_2SO_4 + HNO_3$, and b. with $AgNO_3$ + $KMnO_4 + (NH_4)_2S_2O_8$ to ox. Cr. Ppt. Ag^+ with HCl, add a measd. excess of $FeSO_4$, and back-titr. with std. $KMnO_4$.
	105, 131, 214	$Na_2CO_3 + KNO_3$ fusion	Ti, Fe	Filt. leached melt	Acidify alk. soln. with dil. H_2SO_4 and det. Cr spectrophotometrically with diphenyl-carbazide.
Cs	146, 198, 151	$HF + HClO_4$	Al, Fe, Ca, Mg, Na, Li	Ext. sol. perchlorates with BuOH-EtOAc mixt.	Weigh combined Cs, K, and Rb perchlorates. Chem. sepn. of Cs from K and Rb is difficult. Detn. may be completed by flame photometry or (better) x-ray fluorescence. See Rb below.
I	Brine and sea water		Na, Cl, Br	Distn. of I_2 from H_2SO_4 soln. contg. $NaNO_2$	Add dil. H_2SO_4, then $NaNO_2$, and dist. I_2 into $NaOH–H_2O_2$ soln. B. to destroy excess H_2O_2, acidify with dil. H_2SO_4, red. IO_3^- to I^- with SO_2, and ppt. and weigh as AgI.
	132	Na_2CO_3 fusion	Br, Cl	Distn. of I_2 as above	Same.
In	104, 158*a*, 229	$HNO_3 + H_2SO_4$	Pb	Pptn. of $PbSO_4$	The trace amts. of In occurring in minerals are best detd. spectrographically or polarographically.
			Zn	NH_3 pptn. of In	
			Al	NaOH pptn. of In	
			Fe, Ga	Cupferron pptn. of Fe and Ga	
			H_2S group	Pptn. of H_2S group in acidic soln., $pH < 1$	

(*Continued*)

TABLE 23.4 Procedures for the Analysis of Minerals (*Continued*)

Element determined	Mineral numbers	Decomposition with	Separations required from	Type of separation	Procedure
Ir	182*a*, 183, 196	Fusion with Zn or Pb, followed by Na_2O_2 fusion	Os, Ru	Expulsion of OsO_4 and RuO_4 from sl. acidic (H_2SO_4) soln. contg. oxdg. agt.	For final detn., Ir may be pptd. with H_2S from boiling 9*M* HCl, or with satd. NH_4Cl [$\rightarrow (NH_4)_2IrCl_5$], or by BrO_3^- hydrolysis. Ppt. is always ignited in H_2 atm. ($\rightarrow$ met. Ir). In mixts. with other Pt metals, Ir will remain in an aqua regia–insol. residue.
			Pt	"BrO_3^- hydrolysis" of Ir	
			Rh	Pptn. of Rh with $TiCl_3$	
			Pd	Pptn. of Pd with dimethylglyoxime	
K	1, 2, 35, 128, 146, 147, 162, 169, 171, 172, 173, 182, 184, 192, 198, 228, 231, 251	HF + H_2SO_4	None		Flame-photometric comparison of H_2SO_4 soln. with stds. contg. known amts. of K and impurities.
	13, 14, 17, 60, 125, 142, 176, 191	HF + $HClO_4$	None		Flame-photometric comparison of $HClO_4$ soln. with stds. Because of possible radiation interference from Ca, PO_4^{-3} should be added to mask it, or stds. should contain equal concns. of Ca.
	5, 53, 120, 135, 136, 175, 201, 227, 240	HCl	All elements	Pptn. of K as $KB(C_6H_5)_4$	Add excess $NaB(C_4H_5)_4$ to a sl. acidic soln. [$\rightarrow KB(C_6H_5)_4$ ppt.], and either filt. and weigh the ppt., titr. its $B(C_6H_5)_4^-$ content, or titr. the excess rgt. in the filtrate. [Titr. with $AgNO_3$, using eosin indicator, or with (n-$C_{16}H_{33}$)–Me_3NBr, using bromophenol blue indicator.] Cs, NH_4^+, Rb, and Ti^+ interfere. K can also be detd. grav. as K_3PtCl_4 or $KClO_4$.
La	3, 57, 249, 185, 271 See also R.E. below	Na_2CO_2 fusion	SiO_2 Ca, Mg R_2O_3	Dehydration in acidic soln. NH_3 pptn. Pptn. of La as Ox	Ignite Ox ppt. contg. La + R.E. and weigh mixed oxides. Det. La by emission or x-ray spectroscopy.
	98, 100, 188, 208 See also R.E. below	Na_2O_2 fusion in Ni crucible	Ni, Ca, Mg Si Ta, Nb, Ti, Fe	NH_3 pptn. Dehydration in acidic soln. Cupferron pptn., or pptn. of La as fluoride	Same.
	168	Hot concd. H_2SO_4	PO_4^{-2}, R_2O_3	Pptn. as Ox	Same.

Li	6, 146, 151, 189, 231, 253, 255, 273	HF + H_2SO_4	None		Flame-photometric comparison of H_2SO_4 soln. with stds. contg. known concns. of Li and impurities.
Lu	See R.E. below				
Mg	1, 10, 22, 69, 71, 83, 91, 105, 106, 112, 126, 130, 178, 182, 192, 230, 233, 241, 253, 254, 269, 275	Na_2CO_3	SiO_2 R_2O_2 group Ca	Dehydration and filtn. of SiO_2 NH_3 sepn. Ox pptn. of Ca	After removing SiO_2, R_2O_3, and Ca, ppt. Mg from ammoniacal filtrate with $(NH_4)_2HPO_4$, ignite pptd. $MgNH_4PO_4$, and weigh as $Mg_2P_2O_7$. Or titr. Mg in ammoniacal soln. with std. EDTA.
	38	Dil. HCl	B R_2O_3 group	Evapn. with MeOH NH_3 sepn.	Grav. as $Mg_2P_2O_7$ or vol. with EDTA. Correct for Ca present.
	48, 53, 94, 136, 139, 152, 187	Dil. HCl	R_2O_3 group	NH_3 sepn.	Titr. with std. EDTA as above.
	18, 86, 201, 226	Dil. HCl	R_2O_3 group CaO	NH_3 sepn. Ox pptn. of Ca	Grav. as $Mg_2P_2O_7$; see above.
	66	HF + H_2SO_4	R_2O_3	NH_5 pptn.	After evapn. to near dryness and removal of R_2O_3 group, titr. Mg in ammoniacal filtrate with std. EDTA.
	67	$HClO_4$	Cr, Fe Al	Hg-cathode electrolysis Pptn. of Al with 8-hydroxyquinoline at pH 5.5	Ppt. Mg, in filtrate from Al pptn., as 8-hydroxyquinolate, filt. on a Gooch crucible, and weigh as $Mg(C_9H_6ON)_2$.
	143, 218	HCl + H_2SO_4	PO_4^{3-} R_2O_3 group	Pptn. of PO_4^{3-} with NH_4 molybdate NH_3 pptn.	Grav. as $Mg_2P_2O_7$; see above. Correct for Ca present.
	212, 213	HCl	SiO_2 Mn, R_2O_3 Ca	Dehydration in acidic soln. Pptn. with NH_3 + Br_2 Ox pptn.	Grav. as $Mg_2P_2O_7$; see above.
Mn	24, 105, 151	Na_2CO_3 fusion	SiO_2	Dehydration in H_2SO_4 medium	If Mn < 2%, ox. to MnO_4^- with KIO_4 or with $AgNO_3$ + $(NH_4)_2S_2O_8$ in mixed HNO_2–H_3PO_4–H_2SO_4 medium, and titr. with std. Na arsenite. If Mn > 2%, neut. H_2SO_4 soln. with $NaHCO_3$, add excess ZnO, and titr. with std. $KMnO_4$.
	75, 243	$KHSO_4$ fusion	Ta, Nb	Filtn. of hydrolyzed earth acids in H_2SO_4 medium	Same.
	99, 127, 216, 268	HCl + HNO_3	W	Evapn. to fumes of H_2SO_4 and filtn.	Same.
	92, 102, 226, 229, 255	HCl + H_2SO_4			Same.
	266	HCl + H_2SO_4	SiO_2	Filtn.	Same.

(*Continued*)

TABLE 23.4 Procedures for the Analysis of Minerals (*Continued*)

Element determined	Mineral numbers	Decomposition with	Separations required from	Type of separation	Procedure
Mn	43, 213	HCl	SiO_2	Filtn.	Evap. to SO_2 fumes, neut. an aliquot to pH 6.5 with NaOH, then sat. with $Na_4P_2O_7 \cdot 10H_2O$. Titr. potentiometrically with std. $KMnO_4$ [→ Mn(III)], using a Pt indicator electrode. Or ox. Mn to MnO_4^- with NaB_1O_3 in HNO_3 soln., filt., and titr. with std. $FeSO_4$.
	121, 155, 205, 209, 212	HCl + H_2SO_4 + few drops of HF			Same.
MnO_2	121, 155, 205, 209	H_2SO_4 + H_2Ox (measd. excess)			When decompn. (→ Mn^{2+} + CO_2) is complete, titr. excess H_2Ox with std. $KMnO_4$.
Mo	167, 202, 270	HNO_3 + H_2SO_4	Pb, Ca Fe	Filtn. of insol. sulfates NH_3 pptn.	Evap. the filtrate from the NH_3 sepn. to small vol., acidify with H_2SO_4, pass through a Jones reductor [→ Mo(III)] into a soln. contg. Fe(III) and H_3PO_4, and titr. with std. $KMnO_4$. Or ppt. and weigh as $PbMoO_4$.
	99, 127, 216, 268	Na_2CO_3 fusion	Insol. matter	Filtn.	To an aliquot of the Na_2CO_3 soln. add H_2Tart and acidify with H_2SO_4; add $FeCl_3$, then KSCN, then $SnCl_2$; ext. Mo(V)–SCN^- complex into BuOAc and meas. A_{520}.
N	175, 227	H_2SO_4 + salicylic a. + $Na_2S_2O_3$ + Zn + HgO		Distn.	Fume strongly with H_2SO_4, transfer to a distn. flask, make alk. with NaOH, dist. NH_3 into a measd. excess of std. acid, and back-titr. with std. alk.
Na	1, 2, 8, 13, 14, 17, 60, 78, 112, 120, 125, 126, 133, 142, 144, 151, 158, 159, 162, 169, 172, 173, 176, 182, 184, 186, 189, 191, 192, 228, 231, 236, 251, 253	HF + H_2SO_4, or HF + $HClO_4$	None		Flame-photometric comparison of H_2SO_4 or $HClO_4$ soln. with stds. Because of possible radiation interference from Ca, PO_4^{3-} should be added to mask it, or stds. should contain equal concns. of Ca.
	10, 33, 163, 208	Fusion with a mixt. of $CaCO_3$ and NH_4Cl	Insol. matter Ca	Filtn. Pptn. with $(NH_4)_2CO_3$	After removal of Ca with $(NH_4)_2CO_3$, evap. soln. to dryness, expel NH_4^+ salts by heating, dissolve residue in dil. HCl, and compare with stds. by flame photometry. Or ppt. and weigh $NaM^{II}(UO_2)_3(OAc)_9 \cdot aq$ [M^{II} = Zn or Mg].
	110	Dil. HCl	Ca	Pptn. with $(NH_4)_2CO_3$	Same.
	39, 138, 256, 259	Dil. HCl	Ca B Ca	Pptn. with $(NH_4)_2CO_3$ Evapn. with MeOH Pptn. with $(NH_4)_2CO_3$	After removing Ca, evap. soln. to dryness, heat to expel NH_4^+ salts, and weigh as NaCl. Correct for K (detd. by flame photometry).

	118, 166, 227, 248	Dil. H_2SO_4	Insol. matter	Filtn.	Evap. the soln. of the samp. to dryness and ignite in a tared Pt dish to decomp. $NaHSO_4$. Weigh as Na_2SO_4 and correct for K (detd. by flame photometry).
	175, 240	HCl	None	None	By flame photometry as above.
Nb	34, 75, 98, 100, 163, 200, 208, 215, 223, 243, 272	Fusion with $NaHSO_4$ or Na_2O_2	Fe, Mn, Mo, Ti, W, Zr	Anion exchange in 5 + 4 + 11 mixt. of HCl–HF–H_2O	Add boric a. to eluate (to mask F^-), ppt. Nb with cupferron, and ignite to Nb_2O_8.
			Ta	Elution with 1*M* HF–3*M* NH_4Cl	
Nd	See R.E. below				
Ne	Air			Fractional distn. of liq. air	Mass spectroscopy.
Ni	106, 107, 164, 174	$HNO_3 + H_2SO_4$	As	Volatilization with HBr	Add H_2 Tart or H_3Cit, make sl. alk. with aq NH_3, ppt. with dimethylglyoxime, filt., and weigh.
			Insol. matter	Filtn.	
	73, 95, 111, 150, 224	$HNO_3 + H_2SO_4$	As	Volatilization with HBr	Grav. as Ni dimethylglyoxime; see preceding method.
			Insol. matter	Filtn.	
			Co	9*M* HCl soln. passed through anion-exchange column (Co retained)	
O	In native metals				By vacuum or inert-gas fusion.
Os	182*a*, 196	Heating with Zn in graphite crucible; fusion of residue with Na_2O_2	All elements	See final detn.	Leach cooled Na_2O_2 melt in H_2O, add NaOCl, acidify with H_2SO_4, add a soln. contg. NaBr + $NaClO_3$, and dist. out Os and Ru together as tetroxides. RuO_4 is absbd. in dil. HCl (→ $RuCl_3$); OsO_4 passes unabsbd. through this but is subsequently absbd. in HCl contg. SO_2, or in NaOH soln. Evap. the distillate several times to dryness with intermittent addn. of HCl. Take up in dil. acid, adjust pH to 7 with $NaHCO_3$, filt. off the hydrated oxide, and ignite it in H_2 (→ met. Os). Final detn. may also be made colorimetrically with thiourea.
P	16, 16a	HNO_3	Ca	Pptn. of P as $MgNH_4PO_4$	Ppt. $MgNH_4PO_4$ from a sl. ammoniacal Cit soln. with $MgCl_2$. Re-ppt., ignite, and weigh as $Mg_2P_2O_7$.

(*Continued*)

TABLE 23.4 Procedures for the Analysis of Minerals (*Continued*)

Element determined	Mineral numbers	Decomposition with	Separations required from	Type of separation	Procedure
P	23, 85, 92, 143, 148, 151, 165, 194, 218, 252, 255, 257, 263, 264, 265	HCl	H_2S group	H_2S sepn.	After removal of H_2S group and reoxdn. of the soln., ppt. with NH_4 molybdate from dil. HNO_3 soln. Filt. off phoephomolybdate ppt., dissolve in dil. NH_3, and ppt. with $MgCl_2$. Finally weigh as $Mg_2P_2O_7$.
	6, 33, 168, 271	Heating with H_2SO_4	Al, Be, R.E.	Pptn. as phosphomolybdate	Grav. as $Mg_2P_2O_7$.
	123, 149, 205, 209	HCl	Insol. matter	Filtn.	Ppt. P as phosphomolybdate (see above), filt., wash free from acid with KNO_3 soln., dissolve in exccess std. acid, and back-titr. with std. NaOH. Or det. colorimetrically as the heteropoly Mo blue complex.
Pb	11, 41, 42, 87, 134, 160, 165, 229, 235	$HNO_3 + H_2SO_4$	Sb, As	Expulsion of Sb and As by evapn. with HBr	After expelling As and Sb, dissolve sol. sulfates in H_2O, filt. off $PbSO_4$, dissolve it in NH_4OAc, and ppt. and weigh as $PbCrO_4$.
	58, 81, 104, 145, 158*a*, 170, 193, 262, 270	$HNO_3 + H_2SO_4$	Cu, Fe, Zn, V, Mo	Evapn. to SO_3 fumes and filtn. of insol. $PbSO_4$	Dissolve the $PbSO_4$ in NH_4OAc soln. and ppt. and weigh $PbCrO_4$.
	28	$HF + H_2SO_4$	Mn, Fe, Zn	Filtn. of $PbSO_4$	Same.
	239	$HNO_3 + H_2SO_4$	Te	With HBr	Small amts. of Pb are best detd. polarographically, photometrically (with dithizone), or spectrographically.
			Ag	As AgCl	
			Au	Redn. to met. Au	
	260	$HNO_3 + HSO_4$	U	Collection of Pb with Fe^{3+} and $(NH_4)_2CO_3$	Same.
Pd	183, 196	Aqua regia	Pt	Pptn. of Pt with NH_4Cl	Ppt. Pt with excess NH_4Cl from dil. HCl soln. Ppt. Pd from filtrate with dimethylglyoxime, filt., and weigh. Pd may also be sepd. from Pt by "BrO_2^- hydrolysis."
	182*a*	See Os above	Pt, Ir	Pptn. of Pt and Ir with NH_4Cl	Ppt. Ir, Pd, Pt, and Rh with H_2S from soln. remaining after Os distn.; dissolve sulfides in aqua regia, ppt. Ir and Pt from dil. HCl soln. with excess NH_4Cl, and ppt. Pd from filtrate with dimethylglyoxime as in preceding meth.
			Rh	Dimethylglyoxime pptn. of Pd	
Pr	See R.E. below				
Pt	182*a*	See Os above	Ir	BrO_3^- hydrolysis	Recover Pt in filtrate from hydrolysis (Pt is pptd. only if pH > 7) with NaOH + EtOH, or with H_2S. Ignite in H_2 and weigh as met. Pt. Final detn. may also be made colorimetrically with $SnCl_2$.

	183, 196	Aqua regia	Pd	Pptn. of Pt with NH_4Cl	Ignite the pptd. $(NH_4)_2PtCl_6$ and weigh as metallic Pt.
R.E.	3, 57, 103, 188, 249	Na_2CO_8 fusion	SiO_2 Na, Ca, Mg R_2O_3	Dehydration and filtn. Pptn. of R.E. with NH_3 Pptn. of R.E. with H_2Ox	Dissolve the NH_3 ppt. in acid, ppt. R.E. with excess H_2Ox, filt., and ignite the Ox ppt. to oxide.
	30, 168, 185, 271	Heating with H_2SO_4	Insol. matter PO_4^{3-}	Filtn. Ox pptn. of R.E.	Heat samp. strongly with H_2SO_4, transfer soln. into ice H_2O, filt., and ppt. R.E. from an aliquot with excess H_2Ox.
	98, 100, 200, 208, 215, 272	$KHSO_4$ fusion	Salts, Ca, Mg Ta, Nb, Ti, Fe, U, Zr	NH_3 pptn. of R.E. Treatment with HF	Leach $KHSO_4$ melt in acid, remove sol. salts by pptn. with aq NH_3, ignite ppt., and treat it repeatedly with HF. Filt. off insol. fluorides, convert to sol. salts by fuming with H_2SO_4 or $HClO_4$, and finally ppt. with H_2Ox.
Ra	Minerals contg. U	Acids	All elements	Various, finally co-pptn. with Ba	Radiochemically.
Rb	146, 151, 198	HF + H_2SO_4	None		Flame-photometric comparison of H_2SO_4 soln. with stds. contg. known concns. of Rb and interfering elements. Larger amounts of Rb can be detd. grav. by pptn. of $RbClO_4$, $RbB(C_6H_5)_4$, or Rb_2PtCl_4. Sepn. of Rb from Cs and K is difficult, and mixts. of these elements are best analyzed by x-ray fluorescence.
Re	167	Na_2O_2 fusion	All elements	Distn.	Leach melt in H_2O, acidify with $H_2SO_4 + H_3PO_4$, and dist. Re out as bromide. Det. Re colorimetrically in distillate by a SCN^- meth. similar to that outlined above for Mo.
Rh	182*a*	See Pd above.	All elements	See Os, Ru, Ir, Pt, Pd.	In filtrate from Pd detn., recover Rh with H_2S, ignite ppt. in H_2, and weigh as met. Rh.
Rn	Air				By mass spectroscopy or radiochemically.
Ru	182*a*	See Os above.			Evap. the HCl fraction contg. the Ru, filt., and ppt. the Ru with $NaHCO_3$ at pH 7.0. Filt., ignite the hydrated oxide in H_2, and weigh as met. Ru. Ru may also be detd. colorimetrically with thiourea.
S	5, 12, 94, 110, 117, 120, 136, 139, 166, 201	Dil. HCl	Insol. matter	Filtn.	To the clear HCl soln. of the samp. add excess $BaCl_2$ soln., filt., and weigh as $BaSO_4$.

(*Continued*)

TABLE 23.4 Procedures for the Analysis of Minerals (*Continued*)

Element determined	Mineral numbers	Decomposition with	Separations required from	Type of separation	Procedure
S	15, 44, 61, 63, 135	Dil. HCl	Cu, As	Treatment with met. Fe	After removing As, Cu, and other elements with met. Fe, ppt. SO_4^- with $BaCl_2$ and finish as above.
	11, 145	HCl	Pb	Pptn. of Pb with $(NH_4)_2CO_3$	After removing Pb with $(NH_4)_2CO_3$, b. filtrate until neut., add a little HCl, ppt. with $BaCl_2$, and finish as above.
	27, 55, 158, 159	Fusion with Na_2CO_3	Ba, Sr	Filtn. of Na_2CO_3 soln.	Leach Na_2CO_3 melt with H_2O, filt., acidify filtrate with HCl, evap. to dryness, filt. off SiO_2, take up in dil. HCl, and finish with $BaCl_2$ as above.
			SiO_2	Dehydration in HCl soln.	
	20, 36, 40, 41, 42, 62, 64, 70, 73, 87, 90, 104, 107, 111, 134, 160, 161, 164, 174, 180, 199, 204, 211, 224, 232, 234, 235, 245, 246, 247	Br_2 + HCl + HNO_3	Cu, As, Sb, Ag, Bi, Pb, Hg, Te	Treatment with met. Fe in dil. HCl soln.	Decomp. samp. as indicated, evap. several times on a steam bath with intermittent addn. of HCl, treat with met. Fe, filt., and finish with $BaCl_2$ as above.
	19, 206	HNO_3 + HCl	Ag, Sb	Treatment with met. Fe	As preceding meth., but treat with HNO_3 before adding HCl.
	156, 158*a*, 167, 207, 229	Br_2 + HCl + HNO_3	Fe	NH_3 sepn.	Make the very sl. acidic (HCl) soln. ammoniacal, filt., re-ppt. the $Fe(OH)_3$, b. to expel excess NH_3, acidify with HCl, and finish with $BaCl_3$ as above.
	98, 200, 215	$KHSO_4$ fusion	Salts, Ca, Mg	NH_3 pptn.	Same.
			Ta, Nb, Ti, Fe, U, Zr	Treatment with HF (see R.E.)	
			F	Pptn. as Ox (see R.E.)	
	249	Na_2CO_3 fusion	SiO_2	Dehydration in HCl soln. and filtn.	Dissolve the NH_3 ppt. in dil. HCl, ppt. Th (+R.E.) with H_2Ox, and ignite to ThO_2.
				NH_3 pptn.	
Ti	47, 131, 153, 214	$KHSO_4$ fusion	None		Leach melt in dil. H_2SO_4, pass soln. through a Jones reductor (→ Ti(III)], and titr. with std. $Fe(NH_4)(SO_4)_2$ soln.
	34, 75, 98, 100,163, 200, 208, 215, 223, 243	$KHSO_4$ fusion	None		Leach melt in a soln. contg. 1*M* H_2SO_4, 2% (w/v) succinic acid, and 0.5% H_2O_2, and meas. $A_{400-420}$.

	26, 274	KOH fusion in Ni crucible	Salts and Ni SiO_2	NH_3 pptn. Filtn. after dehydration	Leach melt in H_2O, acidify with HCl, ppt. ZrO_2, TiO_2, etc., with NH_3, and filt. Fume ppt. with $HNO_3 + H_2SO_4$, filt. off SiO_2, and det. Ti colorimetrically with H_2O_2.
	31	$HNO_3 + HCl + H_2SO_4$	SiO_2	Dehydration in H_2SO_4 soln. and filtn.	Det. Ti colorimetrically in filtrate from SiO_2 detn.
	126	Na_2CO_3 fusion	SiO_2	Dehydration in HCl soln.	Same.
	188, 250	Na_2CO_3 fusion	Ca	NH_3 pptn.	Leach melt in H_2O, acidify with HCl, ppt. Ti with NH_3, and ignite ppt. Fuse in $KHSO_4$, leach in dil. H_2SO_4, pass through a Jones reductor, and titr. with std. Fe(III) soln.
Tl	207, 229				Spectrographically.
Tm	See R.E. above.				
U	23, 53*a*, 116, 194, 252, 258, 260, 261	$HNO_3 + HCl + H_2SO_4$	H_2S group Fe, Mo	H_2S sepn. Cupferron–$CHCl_3$ sepn.	Ppt. H_2S group elements from H_2SO_4 soln., filt., b. to expel H_2S, and ox. soln. with $KMnO_4$. Ppt. Fe, Mo, etc., with cupferron and ext. them into $CHCl_3$. Evap. aq phase to SO_4 fumes, destroy org. matter with HNO_3, and evap. several times with intermittent addn. of H_2O to remove N oxides. Dil., pass through a Jones reductor, aerate the red. soln. to ox. U(III) to U(IV), and titr. with std. $K_2Cr_2O_7$, using diphenylamine sulfonate as indicator.
	34, 98, 200, 215, 249, 271, 272	Fusion with $KHSO_4$ + NaF	Ta, Nb, Ti, R.E., Th	$(NH_4)_2CO_3$ sepn.	Heat melt with H_2SO_4 to expel HF quant. In subsequent $(NH_4)_2CO_3$ sepns., U remains in filtrate, which is evapd., acidified with H_2SO_4, red. in a Jones reductor, and titrd. as in preceding meth.
V	53*a*, 81, 170, 258, 262	$HNO_3 + H_2SO_4$	None		To the H_2SO_4 soln. add $FeSO_4$ [$\rightarrow$ V(IV)], then a sl. excess of $KMnO_4$ [$\rightarrow$ V(V)], then $NaNO_2$ (to red. excess MnO_4^-), then urea (to destroy HNO_2), and titr. with std. $FeSO_4$ soln.
	131, 214	$Na_2CO_3 + KNO_3$ fusion	Ti, Fe Cr	Filtn. Cupferron pptn. of V	Filt. the CO_3^- soln., acidify the filtrate, and ppt. V with cupferron. Ignite ppt., fuse it with $KHSO_4$, leach in dil. H_2SO_4, and det. V colorimetrically with H_2O_2.

(*Continued*)

TABLE 23.4 Procedures for the Analysis of Minerals (*Continued*)

Element determined	Mineral numbers	Decomposition with	Separations required from	Type of separation	Procedure
W	99, 127, 216, 268	$HCl + HNO_3$(10:1)	Ca, Fe, Mn, Mo Insol. matter	Filtn. of WO_3 Extn. of WO_3 with aq. NH_3	Decomp. samp. by heating with 100 mL HCl, add 10 mL HNO_3, and evap. to 10 mL Dil., let stand, filt. off WO_3, dissolve it in aq NH_3, and filt. B. the soln. to expel NH_3, ppt. W with cinchonine, ignite, and weigh as WO_2.
	202	HCl	Mo	Anion exchange	In a 50:10 HCl–HF soln., Mo is retained on anion-exchange resin. In the eluate, det. W either grav. with cinchonine as above, or photometrically with hydroquinone or dithiol.
Xe	Air			Fractional distn. of liq. air	Mass spectroscopy.
Y	See R.E. above				
Yb	See R.E. above				
Zn	102, 158*a*, 226, 229	$HNO_3 + HCl$	Fe, Mn, Al	Pptn. with $NH_3 + (NH_4)_2S_2O_8$	Evap. the soln. of samp. to dryness several times with intermittent addn. of HCl, add aq $NH_3 + (NH_4)_2S_2O_8$ to ppt. Al, Fe, and Mn, filt., and b. to expel NH_3. Remove Cu by treatment with met. Pb and titr. Zn with std. K_4Feoc soln.; or titr. Zn with std. EDTA after addn. of CN^- (to mask Cu) and HCHO (to demask Zn).
	49, 124, 266	$HF + H_2SO_4$	Fe, Mn, Al	NH_3 pptn.	Evap. the H_2SO_4 soln. of samp. to dryness, dissolve residue in dil. HCl, and continue with $NH_3 + (NH_4)_2S_2O_8$, etc., as in preceding method.
	104	$HNO_3 + H_2SO_4$	Pb Fe, Mn	Filtn. of $PbSO_4$ NH_3 pptn.	Evap. filtrate from the $PbSO_4$ to dryness and continue as in preceding meth.
	115	$HNO_3 + H_2SO_4$	Cd	Anion exchange	Pass a dil. H_2SO_4-HI soln. of samp. through an anion-exchange column. If present, remove Fe, etc., from the eluate with aq NH_3 after evapn. to dryness. Det. Zn in eluate by titrn. with EDTA.
	81, 170, 212, 213	$HNO_3 + H_2SO_4$	Pb, insol. matter Cu, V, Mn, Fe	Filtn. Anion exchange	Evap. filtrate from the $PbSO_4$ or insol. matter to dryness, dissolve the salts in 1*M* HCl, pass soln. through an anion-exchange column to remove the elements indicated, elute Zn with 3*M* HNO_3, and det. it as above.
	232	Same	Same + Sn	Same Expulsion with HBr	Same.

Zr	26, 274	KOH fusion in Ni crucible	Salts and Ni SiO_2 Fe	NH_3 pptn. Dehydration and filtn. $(NH_4)_2S$ pptn. in presence of H_2Tart	Leach melt in H_2O, acidify with HCl, and ppt. Zr, etc., with aq NH_3. Filt., evap. ppt. to fumes with $HNO_3 + H_2SO_4$, dil., filt., add H_2Tart, make soln. ammoniacal, and ppt. with H_2S. Filt., acidify the filtrate, and ppt. Zr + Ti with cupferron. Ignite and weigh the mixed oxides; correct for TiO_2 (detd. colorimetrically with H_2O_2).
	168, 271	Heating with hot concd. H_2SO_4	R.E.	Filtn. of insol. $ZrSiO_4$ (see R.E.)	The zircon is not attacked by the H_2SO_4 treatment. After filtn., fuse residue with KOH as in preceding meth. and continue as described there.
	214	KOH fusion (see treatment for minerals 26 and 274)	SiO_2	Filtn.	Filt. off SiO_2. To the H_2SO_4 soln. add excess H_2O_2, then 5 g $(NH_4)_2HPO_4$. Filt., ignite, and weigh as ZrP_2O_7. Or ppt. Zr with bromomandelic acid and ignite to ZrO_2.

Source: L. Meites, ed., *Handbook of Analytical Chemistry*, McGraw-Hill, New York, 1963.

SECTION 24

METHODS FOR DETERMINATION OF WATER IN GASES, LIQUIDS AND SOLIDS

There is no single technique applicable to measure water in solid, liquid, and gaseous substances although the Karl Fischer titrimetric procedure is widely used and the method to which all other methods are often compared. The analytical chemist must review need with respect to required precision and accuracy, water concentration, and skills available, as well as equipment on hand. Often speed is the most important criterion, particularly in production facilities. An excellent treatise on methods for the determination of water is the three-part monograph by Mitchell and Smith.[1] There is also a much abbreviated treatment by Mitchell.[2]

Selected methods and techniques are outlined in Tables 24.1 through 24.3. Some comments on various methods are now given.

The Karl Fischer method is perhaps the most widely used procedure for the determination of water. Although this method works well in many cases, the commercial reagents are rather costly, the visual titration end point is difficult to discern, and there are numerous interferences, including oxidizing agents, unsaturated compounds, and thio compounds. Liang[3] discusses automatic on-line monitoring by flow-injection sampling.

Infrared spectrometry is broadly useful for determining water in the gas, liquid, or solid phase. Several absorption bands can be used; the most useful are located in the near-infrared region at about 1.9 μm and in the fundamental region at about 2.7 and 6 μm. The reviews by Kaye[4] and Wheeler[5] and the report by Keyworth[6] provide useful information on the near-infrared region.

Procedures based on colorimetry usually employ $CoCl_2$ (blue when anhydrous) or $CoBr_2$ (green when anhydrous) that change to red for the fully hydrated salts. Cobalt chloride in ethanol gives an absorption maximum at 671 nm. Anhydrous ethanol can be used to extract water from the solid sample. Other substances that have found specific uses as colorimetric reagents have been methylene blue for traces of water in jet fuels, halides, ketones, and hydrocarbons; cobalt bromide–impregnated strips for testing halogenated refrigerants, gasoline, and oils; fuchsine for estimating water in granulated sugar and refinery pastes; and chloranilic acid for organic solvents except those containing amino-nitrogen.

[1] John Mitchell, Jr., and D. M. Smith, *Aquametry*, 2d ed., Wiley, New York, 1977–1980, Parts I–III.

[2] J. Mitchell, Jr., in F. J. Welcher, ed., *Standard Methods of Chemical Analysis*, 6th ed., Van Nostrand, New York, 1966, Vol. 3, Part B, Chap. 64.

[3] Y. Y. Liang, *Anal. Chem.* **62**:2504 (1990).

[4] W. Kaye, *Spectrochim. Acta* **6**:257 (1954); **7**:181 (1955).

[5] O. H. Wheeler, *Chem. Rev.* **59**:629 (1959).

[6] D. A. Keyworth, *Talanta* **8**:461 (1961).

TABLE 24.1 Methods for the Determination of Water in Gases

Method and technique	Procedure and references	Range	Interferences
Karl Fischer (volumetric)	Condensation; alcohol, acid, or *tert*-amine extraction; titration to electrometric or visual end point [1]. $H_2O \equiv I_2$.	ppm to several percent	RCHO, RSH
Succinyl chloride (volumetric)	Pass gas through molten reagent at 60°C, absorb released HCl in water, and titrate with standard Na borate [3]. $H_2O \equiv 2HCl$.	0.1% to few percent	ROH, RNH_2, R_2NH, acids
Magnesium nitride (volumetric)	Pass through reagent at 100°C, absorb NH_3 in standard H_2SO_4 [2]. $NH_3 \equiv 3H_2O$.	0.5% to few percent	Other volatile bases or acids
Cobalt(II) chloride (colorimetric)	Alcoholic extraction; measure absorbance at 671 nm [4].	0.5% to several percent	Colored substances
Calcium carbide (colorimetric)	Pass gas through CaC_2 at 180 to 200°C; pass C_2H_2 released through ammoniacal Cu_2SO_4 and determine as red-colored CuC_2 [2,16].	0.1%–1%	ROH
Absorption (gravimetric)	Pass through tared tube containing P_2O_5 or other suitable desiccant. Increase in weight is proportional to H_2O [2].	ppm	Other substances absorbed by desiccant
Infrared spectrophotometry	Measure absorbance at suitable wavelength [2,5]. Absorbance of moisture in air at 27.97 μm using a water vapor laser [13].	ppm	ROH, RNH_2 Other components of air have negligible effect
Vacuum ultraviolet	Measure absorbance at 127 nm [6].	Low ppm	Substances absorbing 105–150 nm, such as CH_4, H_2S
Mass spectrometry	Measure $m/e = 18$ [2].	0.5%–5%	Other compounds contributing to $m/e = 18$
Gas chromatography	Separation through packed column of Carbowax 20M or Porapak Q at 150°C [2,7,12].	0.5%–5%	Other compounds with same retention time
	Reaction of H_2O with CaC_2; separation of resulting C_2H_2 by gas–liquid (OV-11 or DC 710) or gas–solid (silica gel at 80°C) chromatography [2].	ppm to few percent	Other compounds with same retention time
	Separation through packed column of Porapak N [17].	2–100 ppm	
Thermal conductivity	Measure heat transfer [8].	Several volume percent	Hydrocarbons
Mass spectrometry (indirect)	Pass gas sample through a cold trap containing CaC_2 on which water is condensed; heat trap to 90°C and determine C_2H_2 produced [14].	≥1 μg	
Calcium hydride (thermometric)	Measure temperature before and after passage of gas through CaH_2 [2].	ppm	ROH, RCOCH3, RCHO, NH_3
Magnesium nitride (conductometric)	Absorbed the released NH_3 in H_3BO_3 and measure electrical conductivity [9].	High ppm	Other volatile bases or acids

Electrolysis	Pass through electrolyte cell containing P_2O_5 and measure current [10].	1 ppm to 0.1%	ROH, RCHO, NH_3, CH_3COCH_3, HF
Hygrometry	Measure relative humidity with psychrometer or hygrometer [2,11].	Several volume percent	
	Supersaturation hygrometer utilizing a thermo-optical system that senses growth of salt particles optically and controls their growth by heating the substrate with an infrared source. Output of heater is a measure of ambient relative humidity [15].	Area of 100% relative humidity	
Dew-point measurement	Pass over cooled polished metal surface; measure temperature at which dew forms, as observed photometrically or visually [2,18].	1–1000 ppm	Other compounds that condense
Vapor pressure (manometric)	Measure pressure before and after removal of H_2O [2].	Low percent	Other substances that condense

1. J. Mitchell, Jr., and D. M. Smith, *Aquametry*, 2d ed., Wiley-Interscience, New York, 1977–1980, three volumes.
2. J. Mitchell, Jr., in I. M. Kolthoff and P. J. Elving, eds, *Treatise in Analytical Chemistry*, Part II, Vol. 1, Interscience, New York, 1961.
3. C. B. Belcher, Thompson, and T. S. West, *Anal. Chim. Acta,* **19**:148 (1958).
4. Singliar and Zubák, *Chem. prumysl* **6**:426 (1956).
5. Curcio and Petty, *J. Opt. Soc. Am.* **41**:302 (1951).
6. Garton, Webb, and Wildy, *J. Sci. Instrum.* **34**:496 (1957).
7. S. Dal Nogare and Safranski, in J. Mitchell et al., eds., *Organic Analysis*, Interscience, New York, 1960, Vol. 4.
8. R. H. Cherry, *Anal. Chem.* **20**:958 (1948).
9. Peck, Zedek, and Wittova, *Chem. prumysl* **5**:219 (1955).
10. F. A. Keidel, *Anal. Chem.* **31**:2043 (1959).
11. Weaver, Hughes, and Diniak, *J. Res. Natl. Bur. Stand.* (*U.S.*) **60**:489 (1958).
12. V. M. Sakharov, G. S. Beskova, and A. I. Butusova, *Zh. Anal. Khim.* **31**:250 (1976) (English, p. 214).
13. P. B. Lund and L. Kinnunen, *J. Phys. E. Sci. Instrum.* **9**:528 (1976).
14. G. L. Carlson and W. R. Morgan, *Appl. Spectrosc.* **31**:48 (1977).
15. H. Gerber, *Res. Dev.* **28**:17 (Nov. 1977).
16. W. Boller, *Chemiker-Ztg.* **50**:537 (1983).
17. F. F. Andrawes, *Anal. Chem.* **55**:1869 (1983).
18. E. Flaherty, C. Herold, and D. Murray, *Anal. Chem.* **58**:1903 (1986).

TABLE 24.2 Methods for the Determination of Water in Liquids

Method and technique	Procedure and references	Range	Interferences
Karl Fischer (volumetric)	Titration in inert solvent (e.g., MeOH or pyridine) to electrometric or visual end point [1,2].	ppm to 100%	RCHO, RSH, NH_2OH, $(RCOO)_2$, quinone, ascorbic acid
Karl Fischer (flow injection)	Water in organic solvents determined by flow injection and measurement at 546 nm [21].	0.001%–0.1%	Ketals
Acetyl chloride (volumetric)	Reaction with reagent at room temperature, treatment of excess reagent with MeOH, and titration of sample and blank with standard base [1,2].	0.02% to several percent	HCOOH, RCHO, strong R_3N, high concentrations of ROH, RNH_2 or R_2NH
Magnesium nitride (volumetric)	After reaction with reagent in Kjeldahl assembly, remove NH_3 by steam distillation and determine acidimetrically [2].	0.05% to several percent	Steam-distillable bases or acids; high concentrations of MeOH
Acetic anhydride (volumetric)	Hydrolyze with strong acid and catalyst. Titrate sample and blank with standard NaOMe [3].	0.1% to several percent	Same as acetyl chloride method
	After hydrolysis, add known excess aniline, and titrate sample and blank with standard $HClO_4$ in HOAc [4].	0.1% to several percent	Same as acetyl chloride method
(conductometric)	Titrate sample with acetic anhydride in HOAc, using strong acid as catalyst [2].	0.01% to few percent	ROH, RNH_2, R_2NH
(spectrophotometric)	Hydrolyze at 110°C and determine excess acetic anhydride by measuring absorbance at 252 nm [2,5].	0.01% to few percent	ROH, RNH_2, R_2NH
Infrared spectrophotometry	Measure absorbance in near-infrared region, 14 286 to 5000 cm^{-1} [2].	0.1% to several percent	ROH, RNH_2
	Measure absorbance in fundamental region at or near 3590 cm^{-1}. A drying technique was established employing vacuum distillation onto 4A molecular sieves [14].	ppm to few percent	ROH, RNH_2
Nuclear magnetic resonance spectroscopy	Measure proton resonance. Chemical shift varies with water concentration and H bonding [2].	Several percent	
Paper chromatography	Measure intensity of blue color from water absorption in paper impregnated with $FeSO_4$ and K ferricyanide [6].	0.1% to few percent	

Gas chromatography	Column packed with Porapak Q and kept at 150°C.	0.03%–2%	
	Reaction with 2,2-dimethoxypropane and a solid acid catalyst (Nafion resin) for 5 min, followed by capillary column GC for acetone formed [16].	As low as 0.001% or 13.4 ppm	
	Instantaneous reaction with a 0.5 n*M* ortho ester (triethyl orthoformate) and catalyst (methanesulfonic acid) followed by capillary column GC for ethanol formed [17].	As low as 0.001% or 13.4 ppm	
HPLC	Columns using microparticulate normal phase, reversed phase, and ion exchange with NaCl in eluant and using a conductivity detector [22].	2.5 ppm up to 50%	
Size exclusion and adsorption chromatography	Water and alcohols in gasoline blends separated on HPLC ion exclusion columns (Ultrastyragel 100 and 500 Å) with toluene as mobile phase [20].	0.04–0.2 μL	
Voltammetric sensor	Thin-film perfluorosulfonate ionomer sensors overcoated with cellulose triacetate, polyvinyl alcohol–H_3PO_4 composite films operated in a pulsed voltammetric mode. Water in sample equilibrates with sensing film between pulses and is then electrolyzed by the pulse [17].		
Voltammetry	Molten salt ($AlCl_3$–*N*-butylpyridinium chloride) and water generates HCl which is electrochemically reduced at a rotating Pt disk electrode [19].	Linear up to 50 m*M*	
Acetic anhydride (thermometric)	Measure temperature rise on mixing sample with acetic anhydride and $HClO_4$ catalyst [7].	0.5% to few percent	ROH, RNH_2, R_2NH
Distillation	Use a preliminary distillation step with a low-efficiency column, heating up to 135–150°C, then cooling the two-phase distillate at –10 to –20°C. The water will solidify and reject the dissolved hydrocarbons from the crystal matrix, which can then be decanted [12].		
Dielectric constant	Measure dielectric constant directly in solution using high-frequency technique [8]. A microwave resonance method for water in oil emulsions uses the difference in dielectric properties between water and oil [15].	1% to high percent	Other compounds having high dielectric constant
Conductivity	Measure current flowing through electrodes at fixed potential [9].	0.3% to several percent	Other conducting substances
Conductometric titration	Titrant is LiH in dimethylsulfoxide for water in organic solvents [11].		
Calcium carbide (manometric)	Measure pressure rise when sample and CaC_2 react in closed vessel [2].	Several percent	ROH
Calcium hydride (gasometric)	Measure volume H_2 evolved when sample and CaH_2 react in gas-volumetric apparatus [9].	Several percent	ROH, RCHO, NH_3
Luminescence lifetime	Water in dimethylformamide and dimethylsulfoxide determined by luminescence lifetime measurements of Eu(III) [18].	0.05–5 mol %	

(Continued)

TABLE 24.2 Methods for the Determination of Water in Liquids (*Continued*)

Method and technique	Procedure and references	Range	Interferences
Turbidity	Measure turbidity or cloud point on titration with water-immiscible liquid (e.g., xylene or mineral oil) [2].	0.3% to few percent	Other compounds only slightly soluble in sample solution
Density or specific gravity	Measure specific gravity directly in known system where water is only variable [2].	Variable	Unknown constituents
Refractometry	Measure refractive index of known system where water is only variable [2].	Variable	Unknown constituents
Radiochemical (β-ray absorption)	β-Rays passed through sample and measured with appropriate counter [10].	0.1%–1%	ROH

1. J. Mitchell, Jr., and D. M. Smith, *Aquametry*, 2d ed., Wiley-Interscience, New York, 1977–1980, in three parts.
2. J. Mitchell, Jr., in I. M. Kolthoff and P. J. Elving, eds., *Treatise on Analytical Chemistry*, Interscience, New York, 1961, Part II, Vol. 1.
3. Toennies and Elliott, *J. Am. Chem. Soc.*, **57**:2136 (1935); **59**:902 (1967).
4. Das, *J. Indian Chem. Soc.* **34**:247 (1957).
5. S. Bruckenstein, *Anal. Chem.* **28**:1920 (1956).
6. Stringer, *Nature*, **167**:1071 (1951).
7. L. H. Greathouse, H. J. Janssen, and C. H. Haydel, *Anal. Chem.* **28**:356 (1956).
8. Oehme, *Angew. Chem.* **68**:457 (1956).
9. Perryman, *Analyst* **70**:45 (1945).
10. Friedman, Zisman, and Sullivan, U.S. Patent No. 2,487,797 (1949).
11. C. Yoshimura, K. Miyamoto, and K. Tamura, *Bunseki Kagaku* **27**:310 (1978); *Chem. Abstr.* **89**: 16126 (1978).
12. T. H. Gouw, *Anal. Chem.* **49**:1887 (1977).
13. J. Kovarik, *Chem. Abstr.* **86**:56052 (1977).
14. A. Barbetta and W. Edgell, *Appl. Spectrosc.* **32**:93 (1978).
15. D. A. Doherty, *Anal. Chem.* **49**:690 (1977).
16. K. D. Dix, P. A. Sakkinen, and J. S. Fritz, *Anal. Chem.* **61**:1325 (1989).
17. H. Huang and P. K. Dasgupta, *Anal. Chem.* **64**:2406 (1992).
18. S. Lis and G. R. Choppin, *Anal. Chem.* **63**:2542 (1991).
19. S. Sakami and R. A. Osteryoung, *Anal. Chem.* **55**:1970 (1983).
20. M. Zinbo, *Anal. Chem.* **56**:244 (1984).
21. I. Norden-Andersson and A. Edergren, *Anal. Chem.* **57**:2571 (1985).
22. T. S. Stevens and K. M. Chritz, *Anal. Chem.* **59**:1716 (1971).

TABLE 24.3 Methods for the Determination of Water in Solids

Method and technique	Procedure and references	Range	Interferences
Karl Fischer (volumetric)	Extraction with, or solution in, an inert liquid (MeOH, pyridine, $CHCl_3$, or 1,4-dioxane) [1,2].	0.01% to high percent	MOH, MO, MCO_3, Cu(II), $FeCl_3$, NH_2OH, borates
Oven drying (gravimetric)	Heat sample to constant weight at ≥100°C at 1 atm or ≤100°C at <1 atm [2].	0.1% to high percent	Other volatiles, thermally unstable materials, incompletely dehydrated hydrates
Thermogravimetry	Measure weight loss as function of time and temperature [2].	Few to several percent	See above
Thermogravimetry and differential thermal analysis	At ambient to 250°C results for grains were same whether made in static or flowing air or N_2 at atmospheric or reduced pressure [10].		
Desiccation (gravimetric)	Dry sample at room temperature at ≤1 atm; absorb water with P_2O_5 or $Mg(ClO_4)_2$ and determine loss in weight of sample or gain in weight of $Mg(ClO_4)_2$ [2].	0.1% to high percent	Other volatile materials
Microwave drying (gravimetric)	Constant weight loss is obtained in 5 min [9].		
Freeze drying (gravimetric)	Sublimation of water from sample at <0°C in vacuo [3].	Few to several percent	
Calcium carbide (gravimetric)	Add known weight of CaC_2 to sample in tared vessel; heat mixture to ca. 100°C, cool, weigh. Net loss in weight is equivalent to water [4].	Several percent	Other volatile substances
Infrared spectrophotometry	Measure absorbance of iodine samples at 3665 cm^{-1} and subtract baseline value at 4000 cm^{-1} [11].	0.1% to several percent	ROH, RNH_2
	Reflectance measurements in near-infrared region used for moisture in malt [12], powdered pharmaceutical products [13], and in clay powders and flakes [14]. Measure reflectance at 1.93 μm for water and at 1.7 μm for background.	<1%	
	The integrated absorption of the scissoring vibrations of water at 1640 cm^{-1} varied linearly with water content of polymer membrane irrespective of the nature of the cation [15].		
	Peak height absorbance of band at 5236 μm used for poly(methylmethacrylate) studies [16].		

(Continued)

TABLE 24.3 Methods for the Determination of Water in Solids (*Continued*)

Method and technique	Procedure and references	Range	Interferences
NMR spectroscopy	Measure proton signal from low-resolution instrument [6].	Few to high percent	Other proton-containing liquids
Fluorescent x radiation	Measure the extent of attenuation of fluorescent x radiation excited from a sheet of iron or copper on which paper is placed [17].	Up to 500 g/m^2	
Distillation	Separate water by azeotropic distillation at ≤1 atm [2].	0.1% to several percent	Other volatile water-miscible substances
Dielectric constant	Measure dielectric constant of sample or extract with 1,4-dioxane [2].	Few to several percent	Other substances having high dielectric constants
Calcium carbide (manometric)	Treat sample with CaC_2 in pressure flask; measure increase in pressure.	Several percent	Other substances reacting with CaC_2
Vapor pressure (manometric)	Evacuate closed system containing sample, then heat sample and collect water vapor in manometer [2].	0.1% to few percent	Other volatile substances
Displacement	To known weight of wet sample, add known volume of water in graduated vessel; measure total volume and correct for volumes of dry solid and water added.	High percent	Other liquids
Radiochemical methods	Pass γ radiation from source (e.g., Cs-137) through sample; detect by ionization chamber [2].	High percent	
	Irradiate a powdered sample in a neutron flux of about 4.8×10^7 n $\cdot$ cm^{-2} $\cdot$ s^{-1}; resulting H γ radiation is measured at 2.232 MeV and that from the TiO_2 internal standard at 1.381 MeV [18].		
	Sample of ion exchange resin is mixed with tritium-labeled water of known activity and extracted with dry 1,4-dioxane in a Soxhlet apparatus. Residual ^{3}HHO activity measured following addition of weight amount of inactive H_2O. Activity of solution is measured after reducing ^{3}HHO to ^{3}HH with Zn amalgam and counting the gas [19].	≥0.1 pg or <0.5% water	
	Hydrogen is more capable than other elements of slowing and scattering fast neutrons (Po–Be source); used to determine moisture in soils [21].	Several percent	Other substances containing hydrogen

Conductivity (relative)	300–500-mg glass sample is heated with 500–800 mg of CuO at 1250°C in a tube furnace; released water is carried by N_2 stream over Ag wool and into cell containing PCl_5 to form $POCl_3$ (retained in cold trap) and HCl which is absorbed in 14 m*M* HCl. Relative conductivity measured [20].	20–500 μg

1. J. Mitchell, Jr., and D. M. Smith, *Aquametry*, 2d ed., Wiley-Interscience, New York, 1977–1980, in three parts.
2. J. Mitchell, Jr., in I. M. Kolthoff and P. J. Elving, eds., *Treatise on Analytical Chemistry*, Interscience, New York, 1961, Part II, Vol. 1.
3. B. Makower and E. Nielsen, *Anal. Chem.* **20**:856 (1948).
4. Williams, McComb, and Washauer, *Food Ind.* **22**:458 (1950).
5. L. Shapiro and W. W. Brannock, *Anal. Chem.* **27**:560 (1955).
6. Shaw and Elsken, *J. Assoc. Offic. Anal. Chem.* **36**:1070 (1953).
7. Martin and Mounfield, U.S. Patent No. 2,874,564 (1959).
8. Serger, *Chemiker-Ztg.* **78**:681 (1954).
9. S. Yamaguchi, M. Kubo, and K. Konishi, *J. Am. Oil Chem. Soc.* **54**:539 (1977).
10. M. B. Neher, R. W. Pheil, and C. A. Watson, *Cereal Chem.* **50**:617 (1973).
11. R. N. P. Farrow and A. G. Hill, *Analyst* **102**:480 (1977).
12. M. Moll, R. Flayeux, and J-M. Leheude, *Bios* (*France*) **7**:3 (1976).
13. H. Seager, J. Burt, and H. Fisher, *J. Pharm. Pharmacol.* **1976**:28, Suppl. 62P.
14. G. Vondracek, *Ber. Dt. Keram. Ges.* **52**:185 (1975).
15. L. Levy, H. D. Hurwitz, and A. Jenard, *Anal. Chim. Acta* **88**:377 (1977).
16. A. S. Gilbert, R. A. Pethrick, and D. W. Phillips, *J. Appl. Polymer Sci.* **21**:319 (1977).
17. S. Aksela, *Paperi Puu* **57[4a]**:183, 187 (1975); *Chem. Abstr.* **31[1]**:1C55 (1976).
18. M. Heurtebise and J. A. Lubkowitz, *Anal. Chem.* **48**:2143 (1976).
19. E. Blasius and R. Schmitt, *Z. Anal. Chem.* **241**:4 (1968).
20. H. Malissa, E. Pell, and H. Puxbaum, *Z. Anal. Chem.* **278**:353 (1976).
21. Gardner and Kirkham, *Soil Sci.* **73**:391 (1952).

APPENDIX
SYMBOLS, ABBREVIATIONS, AND DEFINITIONS

TABLE A.1 SI Prefixes

Submultiple	Prefix	Symbol	Multiple	Prefix	Symbol
10^{-1}	deci	d	10	deka	da
10^{-2}	centi	c	10^{2}	hecto	h
10^{-3}	milli	m	10^{3}	kilo	k
10^{-6}	micro	μ	10^{6}	mega	M
10^{-9}	nano	n	10^{9}	giga	G
10^{-12}	pico	p	10^{12}	tera	T
10^{-15}	femto	f	10^{15}	peta	P
10^{-18}	atto	a	10^{18}	exa	E

Numerical (multiplying) prefixes

Number	Prefix	Number	Prefix	Number	Prefix
0.5	hemi	19	nonadeca	39	nonatriaconta
1	mono	20	icosa	40	tetraconta
1.5	sesqui	21	henicosa	41	hentetraconta
2	di (bis)*	22	docosa	42	dotetraconta
3	tri (tris)*	23	tricosa	43	tritetraconta
4	tetra (tetrakis)*	24	tetracosa	44	tetratetraconta
5	penta	25	pentacosa	45	pentatetraconta
6	hexa	26	hexacosa	46	hexatetraconta
7	hepta	27	heptacosa	47	heptatetraconta
8	octa	28	octacosa	48	octatetraconta
9	nona	29	nonacosa	49	nonatetraconta
10	deca	30	triaconta	50	pentaconta
11	undeca	31	hentriaconta	60	hexaconta
12	dodeca	32	dotriaconta	70	heptaconta
13	trideca	33	tritriaconta	80	octaconta
14	tetradeca	34	tetratriaconta	90	nonaconta
15	pentadeca	35	pentatriaconta	100	hecta
16	hexadeca	36	hexatriaconta	110	decahecta
17	heptadeca	37	heptatriaconta	120	icosahecta
18	octadeca	38	octatriaconta	130	triacontahecta

* In the case of complex entities such as organic ligands (particularly if they are substituted) the multiplying prefixes bis-, tris-, tetrakis-, pentakis-, . . . are used, i.e., -kis is added starting from tetra-. The modified entity is often placed within parentheses to avoid ambiguity.

TABLE A.2 Greek Alphabet

Capital	Lower-case	Name	Capital	Lower-case	Name
A	α	Alpha	N	ν	Nu
B	β	Beta	Ξ	ξ	Xi
Γ	γ	Gamma	O	o	Omicron
Δ	δ	Delta	Π	π	Pi
E	ϵ	Epsilon	P	ρ	Rho
Z	ζ	Zeta	Σ	σ	Sigma
H	η	Eta	T	τ	Tau
Θ	θ	Theta	Y	υ	Upsilon
I	ι	Iota	Φ	ϕ	Phi
K	κ	Kappa	X	χ	Chi
Λ	λ	Lambda	Ψ	ψ	Psi
M	μ	Mu	Ω	ω	Omega

TABLE A.3 Abbreviations and Standard Letter Symbols

Abampere	abamp
Absolute	abs
Absolute activity	λ
Absorbance (decaidic)	A
Absorbance (napierian)	B
Absorptance	α
Absorption coefficient, linear decaidic	α, K
Absorption coefficient, linear napierian	α
Absorption coefficient, molar decaidic	μ, ϵ
Absorption coefficient, molar napierian	κ
Absorption index	k
Acceleration	a
Acceleration due to gravity	g, g_n
Acetyl	Ac
Acoustic absorption factor	α_a
Acoustic dissipation factor	δ
Acoustic reflection factor	ρ
Acoustic transmission factor	τ
Activation energy	E_a
Activity (referenced to Raoult's law)	a
Activity (referenced to Henry's law):	
Concentration basis	a_c
Molality basis	a_m
Activity (referenced to Henry's law):	
Mole fraction basis	a_x
Activity (radioactive)	A
Activity coefficient (referenced to Raoult's law)	f
Activity coefficient (referenced to Henry's law):	
Concentration basis	γ_c
Molality basis	γ_m
Mole fraction basis	γ_x
Adjusted retention time	t'_R
Adjusted retention volume	V'_R
Admittance	Y
Affinity of reaction	A
Alcohol	alc
Alfvén number	Al
Alkaline	alk
Alpha particle	α
Alternating current	ac
Amorphous	am
Amount concentration	c
Amount of substance	n
Ampere	A
Amplification factor	μ
Angle of optical rotation	α
Angstrom	Å
Angular dispersion	$d\theta/d\lambda$
Angular momentum	π
Angular momentum terms	j, J, l, L, N
Angular velocity	ω
Anhydrous	anhyd
Approximate (circa)	ca.
Aqueous solution	aq
Aqueous solution at infinite dilution	aq, ∞
Area, unit of area	a
Area	A, S
Area per molecule	a, σ
Astronomical unit	AU
Asymmetry parameter	κ
Atmosphere, unit of pressure	atm
Atomic mass	m_a
Atomic mass constant	m_u
Atomic mass unit	amu
Atomic number	Z
Atomic percent	at. %
Atomic weight	at. wt.
Average	av
Average linear gas velocity	μ
Avogadro constant	L, N_A
Axial angular momentum	λ, Λ, Ω
Axial spin angular momentum	σ, Σ
Bandwidth (10%) of a spectral filter	$\Delta\lambda_{0.1}$
Band variance	σ^2
Bar, unit of pressure	bar
Barn, unit of area	b
Barrel	bbl
Base of natural logarithms	e
Becquerel	Bq
Bed volume	V_g
Beta particle	β^g
Bloch function	$u_k(r)$
Body-centered cubic	bcc
Bohr	b
Bohr magneton	μ_B
Bohr radius	a_0
Boiling point	bp
Boltzmann constant	k, k_B
Bragg angle	θ
Breadth	b
British thermal unit	Btu
Bulk modulus	K
Bulk strain	θ
Burgers vector	$\boldsymbol{b}$
Butyl	Bu
Calorie, unit of energy	cal
Calorie, international steam table	cal_{IT}
Candela	cd
Capacitance	C

(*Continued*)

TABLE A.3 Abbreviations and Standard Letter Symbols (*Continued*)

Capacity, volume	Q_V
Capacity, weight	Q_w
Cartesian space coordinates	x, y, z
Celsius temperature	t, θ
Centimeter-gram-second system	cgs
Centrifugal distortion constants	
A reduction	Δ, δ
S reduction	D, d
Charge density of electrons	ρ
Charge number of electrochemical reaction	n
Charge number of an ion	z
Chemically pure	CP
Chemical potential	μ
Chemical shift	δ
Circa (approximate)	ca.
Circular frequency	ω
Circular wave vector	
For particles	$\boldsymbol{k}$
For phonons	$\boldsymbol{q}$
Circumference divided by the diameter	π
Citrate	Cit
Coefficient of heat transfer	h
Collision cross section	σ
Collision diameter	d
Collision frequency	Z
Collision frequency factor	z
Collision number	Z
Column volume	V_{col}
Compare (confer)	cf.
Complex refractive index	$\hat{n}$
Component of angular momentum	k, K, m, M
Compressibility	
Isentropic	κ_s
Isothermal	κ_T
Compression factor	Z
Compression modulus	K
Compton wavelength of electron	λ_c
Compton wavelength of neutron	$\lambda_{c,n}$
Compton wavelength of proton	$\lambda_{c,p}$
Concentration (amount of substance)	c
Concentration (mass)	γ
Concentration at peak maximum	C_{max}
Concentration of solute in mobile phase	C_M
Concentration of solute in stationary phase	C_S
Condensed phase (solid or liquid)	cd
Conductance	G
Conductivity	γ, κ
Conductivity cell constant	K_{cell}
Conductivity tensor	σ_{ik}
Contact angle	θ
Coordinate, position vector	r
Coulomb	C
Counts per minute	cpm, c/m
Coupling constant, direct dipolar	D_{AB}
Critical density	d_c
Critical temperature	t_c
Cross section	σ
Crystalline	cr, cryst
Cubic	cub
Cubic expansion coefficient	α, α_v, γ
Curie	Ci
Cycles per second	Hz
Curie temperature	T_c
Dalton (atomic mass unit)	Da
Day	d
Debye, unit of electric dipole	D
Debye circular frequency	ω_D
Debye circular wavenumber	q_D
Debye-Walker factor	D, B
Decay constant (radioactive)	λ
Decibel	dB
Decompose	dec
Degeneracy, statistical weight	d, g, β
Degree of dissociation	α
Degrees Baume	°Be
Degrees Celsius	°C
Degrees Fahrenheit	°F
Density (mass)	ρ, γ
Density, critical	d_c
Density, relative	d
Density of liquid phase	ρ_L
Density of states	N_E, ρ
Density of vibrational modes (spectral)	N_ω
Detect, determine(d)	det(d)
Determination	detn
Deuteron	d
Diamagnetic shielding factor	$1 + \sigma$
Diameter	d
Dielectric polarization	$\boldsymbol{P}$
Differential thermal analysis	DTA
Diffusion coefficient	D
Diffusion coefficient, liquid film	D_f
Diffusion coefficient, mobile phase	D_M
Diffusion coefficient, stationary phase	D_S
Diffusion current	i_d
Diffusion length	L
Diffusion rate constant, mass transfer coefficient	k_d

TABLE A.3 Abbreviations and Standard Letter Symbols (*Continued*)

Term	Symbol
Dilute	dil
Dirac delta function	δ
Direct current	dc
Direct dipolar coupling constant	D_{AB}
Disintegration energy	Q
Disintegrations per minute	dpm
Displacement vector of an ion	$\boldsymbol{u}$
Dissociation energy	D, E_d
From ground state	D_0
From the potential minimum	D_e
Distribution ratio	D
Donor ionization energy	E_d
Dropping-mercury electrode	dme
Dyne, unit of force	dyn
Einstein transition probabilities	A, B
Spontaneous emission	A_{nm}
Stimulated absorption	B_{mn}
Stimulated emission	B_{nm}
Electric charge	Q
Electric current	I
Electric current density	j, J
Electric dipole moment of a molecule	$\boldsymbol{p}$, μ
Electric displacement	$\boldsymbol{D}$
Electric field gradient	q
Electric field strength	E
Electric flux	Ψ
Electric mobility	u, μ
Electric polarizability of a molecule	α
Electric potential	V, ϕ
Electric potential difference	U, ΔV
Electric susceptibility	χ_e
Electric conductivity	σ
Electrical conductance	G
Electrical resistance	R
Electrochemical transfer coefficient	α
Electrokinetic potential	ζ
Electromagnetic unit	emu
Electromotive force	E, emf
Electron	e^-, e
Electron affinity	E_{ea}
Electron magnetic moment	μ_e
Electron paramagnetic resonance	EPR
Electron radius	r_e
Electron rest mass	m_e
Electron spin resonance	ESR
Electronvolt	eV
Electrostatic unit	esu
Elementary charge	e
Elution volume, exclusion chromatography	V_e
Emittance	ϵ
By blackbody	M_{bb}
Energy	E
Energy density	w, ρ
Energy per electron hole pair of ion pair in detector	ϵ
Enthalpy	H
Entropy	S
Entropy unit	e.u.
Equilibrium constant	K, K^0
Concentration basis	K_c
Molality basis	K_m
Pressure basis	K_p
Equilibrium position vector of an ion	R_o
Equivalent weight	equiv wt
Erg, unit of energy	erg
Especially	esp.
et alii (and others)	et al.
et cetera (and so forth)	etc.
Ethyl	Et
Ethylenediamine	en
Ethylenediamine-*N,N,N′,N′*-tetraacetic acid	EDTA
Euler number	*Eu*
Exempli gratia (for example)	e.g.
Expansion coefficient	α
Exponential	exp
Extent of reaction	ξ
Fano factor	F
Farad	F
Faraday constant	F
Fermi, unit of length	f
Fermi energy	E_F
Film tension	Σ_f
Film thickness	h, t
Fine structure constant	α
Finite change	Δ
First radiation constant	c_1
Flowrate	q
Flowrate, column chromatography	F_c
Fluid phase (gas or liquid)	fl
Fluidity	ϕ
Fluorescent efficiency	Φ_{F}
Fluorescent power	P_{F}
Flux	F, J
Focal length	f
Foot	ft
For example (exempli gratia)	e.g.
Force	F
Force constant (vibrational levels)	k

(*Continued*)

TABLE A.3 Abbreviations and Standard Letter Symbols (*Continued*)

Formal concentration	F
Fourier number	Fo
Franklin, unit of electric charge	Fr
Freezing point	fp
Frequency	f, ν
Friction coefficient	f, μ
Froude number	Fr
Fugacity	f
Fugacity coefficient	ϕ
Gallon	gal
Galvani potential difference	$\Delta\phi$
Gamma, unit of mass	γ
Gamma radiation	γ
Gap energy (solid state)	E_g
Gas (physical state)	g
Gas constant	R
Gauss	G
g factor	g
Gibbs energy	G
Grade	grad
Grain, unit of mass	gr
Gram	g
Grand partition function	Ξ
Grashof number	Gr
Gravimetric	grav
Gravitational constant	G
Gray	Gy
Grüneisen parameter	γ, Γ
Half-life	$t_{1/2}$
Half-wave potential	$E_{1/2}$
Hall coefficient	A_H, R_H
Hamilton function	H
Harmonic vibration wave number	ω
Hartmann number	Ha
Hartree energy	E_h
Heat	q, Q
Heat capacity	C
At constant pressure	C_p
At constant volume	C_v
Heat flowrate	ϕ
Heat flux	J
Hectare, unit of area	ha
Height	h
Helion	h
Helmholtz energy	A
Henry	H
Hertz	Hz
Hexagonal	hex
Horsepower	hp
Hour	h
Hygroscopic	hygr
Hyperfine coupling constant	a, A
Hyperfine coupling tensor	$\boldsymbol{T}$
ibidem (in the same place)	ibid.
id est (that is)	i.e.
Ignition	ign
Impedance	Z
Inch	in
Indices of a family of crystallographic planes	hkl
Indirect spin–spin coupling constant	J_{AB}
Inductance	L
Inertial defect	Δ
Infinitesimal change	δ
Infrared	ir, IR
Inner column volume	V_i
Inner electric potential	ϕ
Inner electrode potential	ϕ
Inorganic	inorg
Inside diameter	i.d.
Insoluble	insol
Interatomic distances	
Equilibrium distance	r_e
Ground-state distance	r_0
Substitution structure distance	r_s
Zero-point average distance	r_z
Internal energy	U
Interstitial (outer) volume	V_o
In the place cited (loco citato)	loc. cit.
In the same place	ibid.
In the work cited	op. cit.
Ionic conductivity	λ, Λ
Ionic strength	I
Concentration basis	I_c
Molality basis	I_m
Ionization energy	E_i
Irradiance	E
Joule	J
Joule–Thomson coefficient	μ, μ_{JT}
Kelvin	K
Kilocalorie	kcal
Kilogram	kg
Kilogram-force	kgf
Kilowatt-hour	kWh
Kinematic viscosity	ν, ϕ
Kinetic energy	K, T, E_k
Knudsen number	Kn
Kovats retention indices	RI
Lagrange function	L
Lambda, unit of volume	λ

TABLE A.3 Abbreviations and Standard Letter Symbols (*Continued*)

Landé *g* factor	g, g_e	Mass transfer coefficient	k_d
Larmor circular frequency	ω_L	Matrix volume	V_g
Larmor frequency	ν_L	Maximum	max
Lattice plane spacing	d	Maxwell, unit of magnetic flux	Mx
Lattice vector	$\boldsymbol{R}$, $\boldsymbol{R}_0$	Mean ionic activity	$a_\pm$
Lattice vectors	$\boldsymbol{a}, \boldsymbol{b}, \boldsymbol{c}$	Mean ionic activity coefficient	$\gamma_\pm$
Length	l, L	Mean ionic mobility	$W_\pm$
Length of arc	s	Melting point	mp
Lewis number	Le	Metallic	met
Light year	l.y.	Metastable	m
Limit (mathematics)	lim	Metastable peaks	$m*$
Linear expansion coefficient	α_1	Meter	m
Linear reciprocal dispersion	D^{-1}, $d\lambda/dx$	Methyl	Me
Linear strain	e, ϵ	Micrometer	μm
Liquid	l, lq	Micron	μ
Liquid crystal	lc	Mile	mi
Liter	L, l	Miller indices	h, l, k
loco citato (in the place cited)	loc. cit.	Milliequivalent	meq
Logarithm, common	log	Millimeters of mercury, unit of pressure	mmHg
Logarithm, base *e*	ln	Millimole	mM
Longitudinal relaxation time	T_1	Minimum	min
Lorenz coefficient	L	Minute	m, min
Loss angle	δ	Mixture	mixt
Lumen	lm	Mobility	μ
Luminous intensity	I	Mobility ratio	b
Lux	lx	Modulus of elasticity	E
Mach number	Ma	Molal	m
Madelung constant	α	Molality	b
Magnetic dipole moment of a molecule	$\boldsymbol{m}$, μ	Molar	M, M
Magnetic field strength	H	Molar (decadic) absorption coefficient	ϵ
Magnetic flux	Φ	Molar ionic conductivity	λ, Λ
Magnetic flux density	B	Molar magnetic susceptibility	χ_{m}
Magnetic moment of protons in water	μ_p/μ_B	Molar mass	M
Magnetic quantum number	M_j	Molar quantity *X*	X_{m}
Magnetic Reynolds number	Rm	Molar refraction	R, R_{m}
Magnetic susceptibility	κ, χ	Molar volume	V_{m}
Magnetic vector potential	A	Mole	mol
Magnetizability	ξ	Mole fraction, condensed phase	x
Magnetization	M	Gaseous mixtures	y
Magnetogyric ratio	γ	Mole percent	mol %
Mass	m	Molecular weight	mol wt
Mass absorption coefficient	μ/ρ, μ_m	Moment of inertia	I, J
Mass concentration	γ, ρ	Momentum	p
Mass density	ρ	Monoclinic	mn
Mass flowrate	q_m	Monomeric form	mon
Mass fraction	w	Muon, negative	μ^-
Massieu function	J	Muon, positive	μ^+
Mass number	A	Mutual inductance	M, L
Mass of atom	m, m_a	Napierian absorbance	B

(*Continued*)

TABLE A.3 Abbreviations and Standard Letter Symbols (*Continued*)

Name	Symbol
Napierian base	e
Napierian molar absorption coefficient	κ
Néel temperature	T_N
Net retention volume	V_N
Neutrino	ν_e
Neutron	n
Neutron magnetic moment	μ_N
Neutron number	N
Neutron rest mass	m_n
Newton	N
Normal concentration	N
Normal stress	σ
Nuclear magnetic resonance	NMR
Nuclear magneton	μ_N
Nuclear spin angular momentum	I
Nucleon number	A
Number concentration	C
Number density	n
Number of entities	N
Numerical aperture	NA
Nusselt number	*Nu*
Obstruction factor	γ
Oersted, unit of magnetic field	Oe
Ohm	Ω
opere citato (in the work cited)	op. cit.
Optical speed	*f*/number
Orbital angular momentum	
Quantum number	$L = 0, 1, 2, 3, \ldots$
Series symbol	$S, P, D, F, \ldots$
Orbital angular momentum (molecules)	
Quantum number	$\Lambda = 0, 1, 2, \ldots$
Symbol	$\Sigma, \Pi, \Delta, \ldots$
Orbital angular momenta of individual electrons	$l = 0, 1, 2, 3, \ldots$
	$s, p, d, f, \ldots$
Order of Bragg reflection	n
Order of reaction	n
Order of reflection	n
Order parameters (solid state)	
Long range	s
Short range	σ
Organic	org
Orthorhombic	o-rh
Osmotic coefficient	ϕ
Molality basis	ϕ_m
Mole fraction basis	ϕ_x
Osmotic pressure (ideal dilute solution)	Π
Ounce	oz
Outer diameter	o.d.
Outer electric potential	ψ
Overall order of reaction	n
Overpotential	η
Oxalate	Ox
Oxidant	ox
Packing uniformity factor	λ
Page(s)	p. (pp.)
Parsec, unit of length	pc
Partial molar quantity	X
Partial order of reaction	n_B
Particle diameter	d_p
Particle position vector	
Electron	$\boldsymbol{r}$
Ion position	$\boldsymbol{R}_j$
Partition coefficient	K
Partition function	q, Q, z, Z, Ω
Partition ratio	k'
Parts per billion, volume	ng/mL
Parts per billion, weight	ng/g
Parts per million, volume	μg/mL
Parts per million, weight	μg/g
Pascal	Pa
Path length (absorbing)	l
Peak asymmetry factor	*AF*
Peak resolution	*Rs*
Péclet number	*Pe*
Peltier coefficient	Π
Percent	%
Period of time	T
Permeability	μ
Permeability of vacuum	μ_0
Permittivity	ϵ
Permittivity of vacuum	ϵ_0
pH, expressed in activity	p*a*H
Expressed in molarity	pH
Phenyl	Ph, ϕ
Phosphorescent efficiency	Φ_P
Phosphorescent power	P_P
Photochemical yield	ϕ
Photoluminescence power	P
Photon	γ
Pion	π
Planck constant	h
Planck constant/2π	$\hbar$
Planck function	Y
Plane angle	$\alpha, \beta, \gamma, \theta, \phi$
Plate height	H
Plate number, effective	N_{eff}
Poise	P
Polymeric form	pol

TABLE A.3 Abbreviations and Standard Letter Symbols (*Continued*)

Porosity, column	ϵ
Positron	β^+
Potential energy	V, Φ, E_p
Pound	lb
Pounds per square inch	psi
Powder	pwd
Power	p
Poynting vector	**S**
Prandtl number	*Pr*
Pressure (partial)	p
Pressure (total)	p, P
Pressure coefficient	β
Pressure, column inlet	p_i
Pressure, column outlet	p_o
Pressure, critical	p_c
Pressure drop	ΔP
Pressure-gradient correction	j
Principal moments of inertia	I_A; I_B; I_C
Principal quantum number	n
Probability	P
Probability density	P
Product sign	Π
Propyl	Pr
Proton	p
Proton magnetic resonance	pmr
Proton magnetogyric ratio	γ_p
Proton number	Z
Proton rest mass	m_p
Pyridine	py
Quadrupole interaction energy tensor	χ
Quadrupole moment of a molecule	Q, Θ
Quantity of electricity, electric charge	Q
Quantum of energy	$h\nu$
Quantum of yield	ϕ
Rad, unit of radiation dose	rad
Radian	rad
Radiant energy	Q, W
Radiant energy density	ρ, w
Radiant energy flux	dQ/dt
Radiant exitance	M
Radiant flux received	E
Radiant intensity	I
Radiant intensity at time t after termination of excitation	$I(t)$
Radiant power	Φ
Radiant power incident on sample	P_0
Radio frequency	*rf*
Radius	r
Rate of concentration change	r
Rate constant	k
Rate of reaction	v
Ratio of heat capacities	γ
Reactance	X
Reciprocal lattice	$\boldsymbol{a}*$, $\boldsymbol{b}*$, $\boldsymbol{c}*$
Reciprocal lattice vector (circular)	G
Vectors for	$\boldsymbol{a}*$, $\boldsymbol{b}*$, $\boldsymbol{c}*$
Reciprocal radius of ionic atmosphere	κ
Reciprocal temperature parameter, $1/kT$	β
Reciprocal thickness of double layer	κ
Reduced column length	λ
Reduced mass	μ
Reduced plate height	h
Reduced velocity	v
Reductant	red
Reference	ref
Reflectance	ρ
Reflection plane	σ
Refractive index	n
Relative permeability	μ_r
Relative permittivity (dielectric constant)	ϵ_r
Relative pressure coefficient	α_p
Relative retention ratio	α
Relaxation time	τ
Rem, unit of dose equivalent	rem
Residual resistivity (solid state)	ρ_R
Resistivity tensor	ρ
Retardation factor	R_f
Retarded van der Waals constant	B, β
Retention time	t_R
Retention volume	V_R
Revolutions per minute	rpm, r/min
Reynolds number	*Re*
Rhombic	rh
Rhombohedral	rh-hed
Roentgen	R
Root-mean-square	rms
Rotational constants	
In frequency	A, B, C
In wave number	$\tilde{A}$, $\tilde{B}$, $\tilde{C}$
Rotational term (spectroscopy)	F
Rotation-reflection	S_n
Rydberg, unit of energy	Ry
Rydberg constant	R, R_∞
Saturated	satd
Saturated calomel electrode	SCE
Schmidt number	*Sc*
Second	s
Second radiation constant	c_2

(*Continued*)

TABLE A.3 Abbreviations and Standard Letter Symbols (*Continued*)

Term	Symbol
Second virial coefficient	B
Sedimentation coefficient	s
Selectivity coefficient	k
Self-inductance	L
Separation factor	α
Shear modulus	G
Shear strain	γ
Shear stress	τ
Sherwood number	Sh
Shielding constant (NMR)	σ
Short-range order parameter	σ
Siemens	S
Sievert	Sv
Signal-to-noise ratio	S/N
Slightly	sl
Solid	c, s
Solid angle	ω, Ω
Solid angle over which luminescence is measured (F, fluorescence; P, phosphorescence; DF, delayed fluorescence)	$\Omega_{F(P,DF)}$
Solid angle over which radiation is absorbed in cell	Ω_A
Solubility	s
Soluble	sol
Solution	soln, sln
Solvent	solv
Sound energy flux	P, P_a
Spacing between crystal diffracting planes	d
Species adsorbed on a substance	ads
Specific gravity	sp gr
Specific retention volume	V_g^0
Specific surface area	s
Specific volume	v, υ
Spectral bandwidth of emission monochromator	$\Delta\lambda_{em}$
Spectral bandwidth of excitation monochromator	$\Delta\lambda_{ex}$
Spectral bandwidth of monochromator	$\Delta\lambda_m$
Spectral radiant energy	Q_λ, $dQ/d\lambda$
Spectral radiant energy density	
In terms of frequency	ρ_ν, w_ν
In terms of wavelength	ρ_λ, w_λ
In terms of wave number	$\rho_{\tilde{\nu}}$, $w_{\tilde{\nu}}$
Spectral radiant energy flux	$d\phi/d\lambda$
Spectroscopic splitting factor	g
Speed	u, w
Speed of light	
In a medium	c
In vacuum	c_0
Spherical polar coordinates	r, θ, ϕ
Spin angular momentum	s, S
Spin–lattice relaxation time	T_1
Spin–orbit coupling constant	A
Spin–spin coupling constant	J_{AB}
Spin–spin (or transverse) relaxation time	T_2
Spin wave functions	α, β
Square	sq
Standard	std
Standard enthalpy of activation	$H\ddagger$
Standard enthalpy of formation	ΔHf^0
Standard entropy	S^0
Standard entropy of activation	$\Delta S\ddagger$
Standard Gibbs energy of activation	$\Delta G\ddagger$
Standard Gibbs energy of formation	ΔGf^0
Standard heat capacity	C_p
Standard hydrogen electrode	SHE
Standard partial molar enthalpy	H^0
Standard partial molar entropy	S^0
Standard potential of electrochemical cell reaction	E^0
Standard reaction enthalpy	$\Delta_r H^0$
Standard reaction entropy	$\Delta_r S^0$
Standard reaction Gibbs energy	$\Delta_r G^0$
Standard temperature and pressure	STP
Stanton number	St
Statistical weight	W, β, ω
Statistical weight of atomic states	g
Stefan–Boltzmann constant	σ
Steradian	sr
Stoichiometric number	ν
Stokes	St
Summation sign	Σ
Surface charge density	σ
Surface concentration	Γ
Surface coverage	θ
Surface density	ρ_A, ρ_S
Surface electric potential	χ
Surface pressure	π
Surface tension	γ, σ
Susceptance	B
Svedberg, unit of time	Sv
Symmetrical	sym
Symmetry coordinate	S
Symmetry number	s, σ
Tartrate	Tart
Temperature	θ, Θ
Temperature, thermodynamic	T
Temperature at boiling point	T_b

TABLE A.3 Abbreviations and Standard Letter Symbols (*Continued*)

Term value spectroscopy	T
Tesla	T
Tetragonal	tetr
Thermal conductance	G
Thermal conductivity	λ, k
Thermal diffusivity	a
Thermal resistance	R
Thermoelectric force	E
Thickness of diffusion layer	δ
Thickness of layer	t
Thickness (effective) of stationary phase	d_f
Thickness of surface layer	τ
Thickness of various layers	δ
Thomson coefficient	μ, τ
Thomson cross section	σ_e
Time	t
Time interval, characteristic	T, τ
Tonne	t, ton
Torr (mm of mercury)	Torr
Torque	$\boldsymbol{T}$
Total bed volume	V_{tot}
Total term (spectroscopy)	T
Transconductance	g_m
Transfer coefficient	α
Transit time of nonretained solute	t_M, t_0
Transition	tr
Transition dipole moment of a molecule	$\boldsymbol{M}$, $\boldsymbol{R}$
Transition frequency	ν
Transition wavenumber	$\tilde{\nu}$
Translation (circular)	b_1, b_2, b_3
Translation vectors for crystal lattice	$\boldsymbol{a}_1$, $\boldsymbol{a}_2$, $\boldsymbol{a}_3$ $\boldsymbol{a}$, $\boldsymbol{b}$, $\boldsymbol{c}$
Transmission factor	τ
Transmittance	T, τ
Transport number	t
Transverse relaxation time	T_2
Triclinic	tric
Trigonal	trig
Triton (tritium nucleus)	t
Ultrahigh frequency	uhf
Ultraviolet	uv, UV
Unified atomic mass unit	u
United States Pharmacopoeia	USP
Vacuum	vac
van der Waals constant	λ
Vapor pressure	p, vp
Velocity	u, w
Versus	vs.
Vibrational anharmonicity constant	χ
Vibrational coordinates	
Internal coordinates	R_i, r_I, θ_j, etc.
Normal coordinates, dimensionless	q_r
Mass adjusted	Q_r
Vibrational force constants	
Diatomic	f
Polyatomic, dimensionless normal coordinates	$\phi_{rst\ldots}$, $k_{rst\ldots}$
Vibrational force constants	
Internal coordinates	f_{ij}
Symmetry coordinates	F_{ij}
Vibrational quantum number	υ
Vibrational term	G
Viscosity	η, μ
Vitreous substance	vit
Volt	V
Volt ampere-reactive	var
Volta potential difference	$\Delta\psi$
Volume	V, υ
Volume flowrate	q_υ
Volume fraction	ϕ
Volume in space phase	Ω
Volume liquid phase in column	V_L
Volume mobile phase in column	V_M
Volume of activation	Δ‡V
Volume percent	vol %
Volume per volume	v/v
Volume strain	θ
Watt	W
Wave function	ϕ, ψ, Ψ
Wavelength	λ
Wave number (in a medium)	σ
Wave number in vacuum	$\tilde{\nu}$
Weber	Wb
Weber number	*We*
Weight	W
Weight of liquid phase	w_L
Weight percent	wt %
Weight per volume	w/v
Wien displacement constant	b
Work	w, W
Work function	Φ
x unit	X
Yard	yd
Young's modulus	E
Zeeman splitting constant	μ_B/hc
Zone width at baseline	W_b
Zone width at one-half peak height	$W_{1/2}$

TABLE A.4 Symbols and Definitions in Chromatography

Name	Symbol	Definition
Adjusted retention time	t'_R	$t'_R = t_R - t_M$
Adjusted retention volume	V'_R	$V'_R = V_R - V_M$
Average linear gas velocity	μ	$\mu = L/t_M$
Band variance	σ^2	
Bed volume	V_g	
Capacity, volume	Q_v	
Capacity, weight	Q_W	
Column length	L	
Column temperature	θ	
Column volume	V_{col}	$V_{col} = \pi D d_c^2/4$
Concentration at peak maximum	C_{max}	
Concentration of solute in mobile phase	C_M	
Concentration of solute in stationary phase	C_s	
Density of liquid phase	ρ_L	
Diffusion coefficient, liquid film	D_f	
Diffusion coefficient, mobile phase	D_M	
Diffusion coefficient, stationary phase	D_S	
Distribution ratio	D_c	$= [A^+]_s/[A^+]_M$
		$= \dfrac{\text{amount of A per cm}^3\text{ stationary phase}}{\text{amount of A per cm}^3\text{ of mobile phase}}$
	D_g	$= \dfrac{\text{amount A per gram dry stationary phase}}{\text{amount A per cm}^3\text{ of mobile phase}}$
	D_v	$= \dfrac{\text{amount A, stationary phase per cm}^3\text{ bed volume}}{\text{amount A per cm}^3\text{ of mobile phase}}$
	D_S	$= \dfrac{\text{amount of A per m}^2\text{ of surface}}{\text{amount of A per cm}^3\text{ of mobile phase}}$
Elution volume, exclusion chromatography	V_e	
Flowrate, column	F_c	$F_c = (\pi d_c^2/4)(\epsilon_{tot})(L/t_M)$
Gas–liquid volume ratio	β	
Inner column volume	V_i	
Interstitial (outer) volume	V_o	
Kovats retention indices	RI	
Matrix volume	V_g	
Net retention volume	V_N	$V_N = jV'_R$
Obstruction factor	γ	
Packing uniformity factor	λ	
Particle diameter	d_p	$d_p = L/Nh$
Partition coefficient	K	$K = C_S/C_M = (V_R - V_M)/V_S$
Partition ratio	k'	$k' = C_S V_S/C_M V_M = K(V_S/V_M)$
Peak asymmetry factor	AF	Ratio of peak half-widths at 10% peak height
Peak resolution	Rs	$Rs = (t_{R,2} - t_{R,1})/0.5(W_2 + W_1)$
Plate height	H	$H = L/N_{eff}$
Plate number	N_{eff}	$N_{eff} = L/H = 16(t'_R/W_b)^2 = 5.54(t'_R/W_{1/2})^2$
Porosity, column	ϵ	

TABLE A.4 Symbols and Definitions in Chromatography (*Continued*)

Name	Symbol	Definition
Pressure, column inlet	p_i	
Pressure, column outlet	p_o	
Pressure drop	ΔP	
Pressure-gradient correction	j	$j = \frac{3[(p_i/p_o)^2 - 1]}{2[(p_i/p_o)^3 - 1]}$
Recovery factor	R_n	$R_n = 1 - (rD_c + 1)^{-n}$; $r = V_{org}/V_{sq}$
Reduced column length	λ	$\lambda = L/d_p$
Reduced plate height	h	$h = H/d_p$
Reduced velocity	v	$v = \mu d_p/D_M = Kd_p/t_M D_M$
Relative retention ratio	α	$\alpha = (k'_2/k'_1)$
Retardation factor*	R_f	$R_f = d_{solute}/d_{mobile\ phase}$
Retention time	t_R	$t_R = t_M(1 + k') = L/\mu$
Retention volume	V_R	$V_R = t_R F_c$
Selectivity coefficient†	$k_{A,B}$	$k_{A,B} = [A^+]_r[B^+]/[B^+]_r[A^+]$
Separation factor	$\alpha_{A/B}$	$\alpha_{A/B} = (D_c)_A/(D_c)_B$
Specific retention volume	V_g^0	$V_g^0 = 273R/(p^0 M w_L)$
Thickness (effective) of stationary phase	d_f	
Total bed volume	V_{tot}	
Transit time of nonretained solute	t_M, t_0	
Vapor pressure	p	
Volume liquid phase in column	V_L	
Volume mobile phase in column	V_M	
Weight of liquid phase	w_L	
Zone width at baseline	W_b	$W_b = 4\sigma$
Zone width at 1/2 peak height	$W_{1/2}$	

* The distance d corresponds to the movement of solute and mobile phase from the starting (sample spotting) line.

† Subscript r represents an ion-exchange resin phase. Two immiscible liquid phases might be represented similarly using subscripts 1 and 2.

Source: J. A. Dean, ed., *Lange's Handbook of Chemistry*, 14th ed., McGraw-Hill, New York, 1992.

TABLE A.5 Symbols and Definitions in Electromagnetic Radiation

Name	Symbol	SI unit	Definitions
Absorbance	α		$\alpha = \Phi_{ab\sigma}/\Phi_0$
Absorbance (decaidic)	A		$A = -\log(1 - \alpha_i)$
Absorbance (napierian)	B		$B = -\ln(1 - \alpha_i)$
Absorption coefficient			
Linear (decaidic)	α, K	m^{-1}	$\alpha = A/l$
Linear (napierian)	α	m^{-1}	$\alpha = B/l$
Molar (decaidic)	ϵ	$m^2 \cdot mol^{-1}$	$\epsilon = a/d = A/cl$
Molar (napierian)	κ	$m^2 \cdot mol^{-1}$	$\kappa = \alpha/c = B/cl$
Absorption index	k		$k = \alpha/4\pi\tilde{\nu}$
Angle of optical rotation	α	rad	
Circular frequency	ω	s^{-1}, $rad \cdot s^{-1}$	$\omega = 2\pi\nu$
Complex refractive index	$\hat{n}$		$\hat{n} = \eta + ik$
Concentration, amount of substance	c	$mol \cdot m^3$	

(*Continued*)

TABLE A.5 Symbols and Definitions in Electromagnetic Radiation (*Continued*)

Name	Symbol	SI unit	Definitions
Concentration, mass	γ	$kg \cdot m^3$	
Einstein transition probabilities			
Spontaneous emission	A_{nm}	s^{-1}	$dN_n/dt = -A_{nm}N_n$
Stimulated absorption	B_{mn}	$s \cdot kg^{-1}$	$dN_n/dt = \rho_{\tilde{\nu}}(\tilde{\nu}_{nm})B_{mn}N_m$
Stimulated emission	B_{nm}	$s \cdot kg^{-1}$	$dN_n/dt = \rho_{\tilde{\nu}}(\tilde{\nu}_{nm})B_{nm}N_m$
Emittance	ϵ		$\epsilon = M/M_{bb}$
By blackbody	M_{bb}		
First radiation constant	c_1	$W \cdot m^2$	$c_1 = 2\pi hc_0^2$
Frequency	ν	Hz	$\nu = c/\lambda$
Irradiance (radiant flux received)	E, (I)	$W \cdot m^{-2}$	$E = d\Phi/dA$
Molar refraction	R, R_m	$m^3 \cdot mol^{-1}$	$R = \dfrac{(n^2-1)}{(n^2+2)}V_m$
Path length (absorbing)	l	m	
Optical rotatory power	$[\alpha]_\lambda^\theta$	rad	$[\alpha]_\lambda^\theta = \alpha/\gamma l$
Planck constant	h	$J \cdot s$	
Planck constant/2π	$\hbar$	$J \cdot s$	$\hbar = h/2\pi$
Radiant energy	Q, W	J	
Radiant energy density	ρ, w	$J \cdot m^{-3}$	$\rho = Q/V$
Radiant exitance, emitted radiant flux	M	$W \cdot m^{-2}$	$M = d\Phi/dA_{source}$
Radiant intensity	I	$W \cdot sr^{-1}$	$I = d\Phi/d\Omega$
Radiant power, radiant energy per time	Φ, P	W	$\Phi = dQ/dt$
Refractive index	n		$n = c_0/c$
Reflectance	ρ		$\rho = \Phi_{refl}/\Phi_0$
Second radiation constant	c_2	$K \cdot m$	$c_2 = hc_0/k$
Spectral radiant energy density			
In terms of frequency	ρ_ν, w_ν	$J \cdot m^{-3} \cdot Hz^{-1}$	$\rho_\nu = dp/d\nu$
In terms of wavelength	ρ_λ, w_λ	$J \cdot m^{-4}$	$\rho_\lambda = dp/d\lambda$
In terms of wave number	$\rho_{\tilde{\nu}}$, $w_{\tilde{\nu}}$	$J \cdot m^{-2}$	$\rho_{\tilde{\nu}} = dp/d\tilde{\nu}$
Speed of light			
In a medium	c	$m \cdot s^{-1}$	$c = c_0/n$
In vacuum	c_0	$m \cdot s^{-1}$	
Stefan–Boltzmann constant	σ	$W \cdot m^{-2} \cdot K^{-4}$	$M_{bb} = \sigma T^4$
Transmittance	τ, T		$\tau = \Phi_{tr}/\Phi_0$
Wavelength	λ	m	
Wave number			
In a medium	σ	m^{-1}	$\sigma = 1/\lambda$
In vacuum	$\tilde{\nu}$	m^{-1}	$\tilde{\nu} = \nu/c_0 = 1/n\lambda$

Source: J. A. Dean, ed., *Lange's Handbook of Chemistry*, 14th ed., McGraw-Hill, New York, 1992.

TABLE A.6 Symbols and Definitions in Electrochemistry

Name	Symbol	SI unit	Definition
Charge density (surface)	σ	$C \cdot n^{-2}$	$\sigma = Q/A$
Charge number of an ion	z		$z_B = Q_B/e$
Charge number of electrochemical cell reaction	n (z)		
Conductivity (specific conductance)	κ	$S \cdot m^{-1}$	$\kappa = j/E$
Conductivity cell constant	K_{cell}	m^{-1}	$K_{cell} = \kappa R$
Current density (electric)	j	$A \cdot m^{-2}$	$j = I/A$

TABLE A.6 Symbols and Definitions in Electrochemistry (*Continued*)

Name	Symbol	SI unit	Definition
Diffusion rate constant, mass transfer coefficient	k_d	$m \cdot s^{-1}$	$k_{d,B} = \lvert v_B \rvert I_{1,B}/nFcA$
Electric current	I	A	$I = dQ/dt$
Electric mobility	μ	$m^2 \cdot V^{-1} \cdot s^{-1}$	$\mu_B = v_B/E$
Electric potential difference (of a galvanic cell)	$\Delta V, E, U$	V	$\Delta V = V_R - V_L$
Electrochemical potential	$\tilde{\mu}$	$J \cdot mol^{-1}$	$\tilde{\mu}_B^\alpha = (\partial G/\partial n_B^\alpha)$
Electrode reaction rate constant	k	(varies)	$k_{ox} = I_a \Big/ \left(nFA \prod_i c_i^{n_i} \right)$
Electrokinetic potential (zeta potential)	ζ	V	
Elementary charge (proton charge)	e	C	
emf, electromotive force	E	V	$E = \lim_{I \to 0} \Delta V$
emf of the cell	E	V	$E = E^0 - (RT/nF) \Sigma v_i \ln \alpha_i$
Faraday constant	F	$C \cdot mol^{-1}$	$F = eL$
Galvani potential difference	$\Delta\phi$	V	$\Delta_\alpha^\beta \phi = \phi^\beta - \phi^\alpha$
Inner electrode potential	ϕ	V	$\nabla\phi = -E$
Ionic conductivity	λ	$S \cdot m^2 \cdot mol^{-1}$	$\lambda_B = \lvert z_B \rvert F u_B$
Ionic strength	I_c, I	$mol \cdot m^{-3}$	$I_c = 1/2 \, \Sigma \, c_i z_c^2$
Mean ionic activity	$a_\pm$		$a_\pm = m_\pm \gamma_\pm / m^0$
Mean ionic activity coefficient	$\gamma_\pm$		$\gamma_\pm^{(\gamma_+ + \gamma_-)} = (\gamma_\pm^{v_+})(\gamma_-^{v_-})$
Mean ionic mobility	$m_\pm$	$mol \cdot kg^{-1}$	$m_\pm^{(v_+ + v_-)} = (m_+^{v_+})(m_-^{v_-})$
Molar conductivity (of an electrolyte)	Λ	$S \cdot m^{-2} \cdot mol^{-1}$	$\Lambda_B = \kappa c_B$
pH	pH		$pH \simeq -\log \left[\frac{c(H^+)}{mol \cdot dm^{-3}} \right]$
Outer electrode potential	ψ	V	$\psi = Q/4\pi\epsilon_0 r$
Overpotential	η	V	$\eta = E_I - E_{I-0} - IR_u$
Reciprocal radius of ionic atmosphere	κ	m^{-1}	$\kappa = (2F^2 I/\epsilon RT)^{1/2}$
Standard emf, standard potential of electrochemical cell reaction	E^0	V	$E^0 = -\Delta_r G^0/nF = (RT/nF)(\ln K)$
Surface electric potential	χ	V	$\chi = \phi - \psi$
Thickness diffusion layer	δ	m	$\delta_B = D_B/k_{d,B}$
Transfer coefficient	α		$\alpha_c = \frac{-\lvert v \rvert RT}{nF} \frac{\partial \ln \lvert I_c \rvert}{\partial E}$
Transport number	t		$t_B = j_B/\Sigma j_i$
Volta potential difference	$\Delta\psi$	V	$\Delta_\alpha^\beta = \psi^\beta - \beta^\alpha$

Source: J. A. Dean, ed., *Lange's Handbook of Chemistry*, 14th ed., McGraw-Hill, New York, 1992.

TABLE A.7 Pressure Conversion Chart

1 bar = 10^5 pascal.

psi	Inches H_2O at 4°C	Inches Hg at 0°C	mmH_2O at 4°C	mmHg at 0°C	atm	pascals $(N \cdot m^{-2})$
0.01	0.2768	0.0204	7.031	0.517	0.0007	68.95
0.02	0.5536	0.0407	14.06	1.034	0.0014	137.90
0.03	0.8304	0.0611	21.09	1.551	0.0020	206.8
0.04	1.107	0.0814	28.12	2.068	0.0027	275.8
0.05	1.384	0.1018	35.15	2.586	0.0034	344.7
0.06	1.661	0.1222	42.18	3.103	0.0041	413.7
0.07	1.938	0.1425	49.22	3.620	0.0048	482.6
0.08	2.214	0.1629	56.25	4.137	0.0054	551.6
0.09	2.491	0.1832	63.28	4.654	0.0061	620.5
0.10	2.768	0.2036	70.31	5.171	0.0068	689.5
0.20	5.536	0.4072	140.6	10.34	0.0136	1 379.9
0.30	8.304	0.6108	210.9	15.51	0.0204	2 068.5
0.40	11.07	0.8144	281.2	20.68	0.0272	2 758
0.50	13.84	1.018	351.5	25.86	0.0340	3 447
0.60	16.61	1.222	421.8	31.03	0.0408	4 137
0.70	19.38	1.425	492.2	36.20	0.0476	4 826
0.80	22.14	1.629	562.5	41.37	0.0544	5 516
0.90	24.91	1.832	632.8	46.54	0.0612	6 205
1.00	27.68	2.036	703.1	51.71	0.0689	6 895
2.00	55.36	4.072	1 072	103.4	0.1361	13 790
3.00	83.04	6.108	2 109	155.1	0.2041	20 684
4.00	110.7	8.144	2 812	206.8	0.2722	27 579
5.00	138.4	10.18	3 515	258.6	0.3402	34 474
6.00	166.1	12.22	4 218	310.3	0.4083	41 369
7.00	193.8	14.25	4 922	362.0	0 4763	48 263
8.00	221.4	16.29	5 625	413.7	0.5444	55 158
9.00	249.1	18.32	6 328	465.4	0.6124	62 053
10.0	276.8	20.36	7 031	517.1	0.6805	68 948
14.7	406.9	29.93	10 332	760.0	1.000	101 325
15.0	415.2	30.54	10 550	775.7	1.021	103 421
20.0	553.6	40.72	14 060	1 034	1.361	137 895
25.0	692.0	50.90	17 580	1 293	1.701	172 369
30.0	830.4	61.08	21 090	1 551	2.041	206 843
40.0	1 107	81.44	28 120	2 068	2.722	275 790
50.0	1 384	101.8	35 150	2 586	3.402	344 738
60.0	1 661	122.2	42 180	3 103	4.083	413 685
70.0	1 938	142.5	49 220	3 620	4.763	482 633
80.0	2 214	162.9	56 250	4 137	5.444	551 581
90.0	2 491	183.2	63 280	4 654	6.124	620 528
100.0	2 768	203.6	70 307	5 171	6.805	689 476
150.0	4 152	305.4		7 757	10.21	1 034 214
200.0	5 536	407.2		10 343	13.61	1 378 951
250.0	6 920	509.0			17.01	1 723 689
300.0	8 304	610.8			20.41	2 068 427
400.0					27.22	2 757 903
500.0					34.02	3 447 379

INDEX

ABOUT THE AUTHOR

Pradyot Patnaik, Ph.D., is the Director of the Analytical and Environmental Chemistry Laboratory of the Interstate Environmental Commission, Staten Island, New York, and a Research Investigator for the Center for Environmental Science of the City University of New York at the College of Staten Island. A respected author in chemistry, he also wrote McGraw-Hill's *Handbook of Inorganic Chemicals; A Comprehensive Guide to the Hazardous Properties of Chemical Substances; Handbook of Environmental Analysis*; and numerous journal articles. He is also an Adjunct Professor at the New Jersey Institute of Technology and at the Community College of Philadelphia.